Phylogeny of the Living World–Bacteria

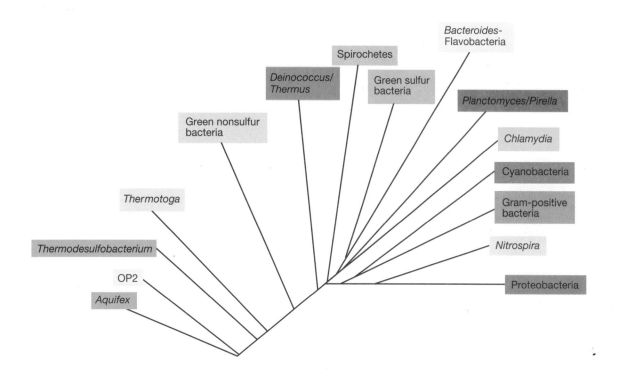

PHYLOGENETIC TREE OF BACTERIA. This tree is derived from 16S ribosomal RNA sequences. Fifteen major groups of Bacteria can be defined as indicated. See Sections 12.4–12.8 for further information on ribosomal RNA-based phylogenies. *Data for the tree obtained from the Ribosomal Database Project* http://www.cme.msu.edu/RDP/

BROCK

Biology of Microorganisms

MICHAEL T. MADIGAN dedicates this book to all of his microbiology professors who revealed to him, each in their own way, the fascinating world of microorganisms. These include in particular Joseph Harris and Robert Simpson at UW-Stevens Point, Thomas Brock and Jerry Ensign at UW-Madison, and Howard Gest at Indiana University. A very special acknowledgement goes to Tom Brock, who besides guiding me through my graduate studies and introducing me to the world of bacterial diversity, taught me the value of hard work and instilled in me a love of writing. And I will never forget Tom's willingness nearly 20 years ago to give a young, untested assistant professor a chance to contribute to a book that has turned out to be the experience of a lifetime.

JOHN M. MARTINKO dedicates this book to the inspiring teachers and mentors he had throughout his education. To Peter Baker, Jerry Senturia, and Richard Dickerson at Cleveland State University, you shared your enthusiasm as well as your knowledge. To Willard Schmidt at Case Western Reserve University, you taught me about practical science. To Richard Bankert and George Mayers at Roswell Park Memorial Institute and SUNY at Buffalo, you showed me how research should be done. To Stanley Nathenson at Albert Einstein College of Medicine, you showed me the single-minded dedication and focus necessary to be successful.

JACK PARKER dedicates this edition to the wonderful teachers he has had. There have been too many to list them all, but in particular there was Wesley Stieg at North Central College, Ed Umbarger at Purdue University, Frederck C. Neidhardt at Purdue and at the University of Michigan, and James D. Friesen at York University. Thank you all for the time you spent, and for your love of learning.

About the Authors

MICHAEL T. MADIGAN (in photo with Willie and Plum) received a bachelor's degree in biology and education from Wisconsin State University at Stevens Point in 1971 and M.S. and Ph.D. degrees in 1974 and 1976, respectively, from the University of Wisconsin, Madison, Department of Bacteriology. His graduate work involved study of the biology of hot spring photosynthetic bacteria under the direction of Thomas D. Brock. Following three years of postdoctoral training in the Department of Microbiology, Indiana University, where he worked on photosynthetic bacteria with Howard Gest, he moved to Southern Illinois University at Carbondale, where he is now Professor of Microbiology. He has been a coauthor of *Biology of Microorganisms* since the fourth edition (1984) and teaches courses in introductory microbiology and bacterial diversity. In 1988 he was selected as the outstanding teacher in the College of Science, and in 1993 its outstanding researcher. His research has dealt almost exclusively with anoxygenic phototrophic bacteria, especially those species that inhabit extreme environments. He has published nearly 85 research papers, has coedited a major treatise on photosynthetic bacteria, and is Chief Editor for North America of the journal *Archives of Microbiology*. His nonscientific interests include reading, hiking, tree planting, and caring for his dogs and horses. He lives aside a quiet lake about five miles from the SIU campus with his wife, Nancy, two dogs, Willie and Plum, and King and Feenkönig (horses).

JOHN M. MARTINKO attended The Cleveland State University and majored in biology with a chemistry minor. As an undergraduate student he participated in a cooperative education program, gaining research experience in several microbiology and immunology laboratories. He then worked for two years at Case Western Reserve University as a laboratory manager, continuing his cooperative education research on the structure, serology, and epidemiology of *Streptococcus pyogenes*. He next went to the State University of New York at Buffalo where he did research on antibody specificity and idiotypes for his M.A. and Ph.D. (1978) in Microbiology. As a postdoctoral fellow, he worked at Albert Einstein College of Medicine in New York on the structure of major histocompatibility complex proteins. Since 1981, he has been in the Department of Microbiology at Southern Illinois University at Carbondale where he is currently the Chair and Associate Professor. His research interests include the effects of growth hormone on the immune response and the immunological identification of soybean brown stem rot disease. His teaching interests include undergraduate and graduate courses in immunology and a team-taught general microbiology course, where he is responsible for immunology, host defense, and infectious diseases. He lives with his wife Judy, a junior high school science teacher, and their daughters, Martha and Helen, in Carbondale where he has been active in coaching his daughters' soccer and softball teams. He tries to find time to play soccer and golf.

JACK PARKER received his bachelor's degree in biology and also received his doctoral degree in a biology program (Ph.D., Purdue University, 1973). However, his research project dealt with bacterial physiology and he completed his Ph.D. research while in the microbiology department at the University of Michigan. Following this he spent four years studying bacterial genetics at York University in Toronto, Ontario. He has taught courses in bacterial genetics, general genetics, human genetics, molecular biology, and molecular genetics, and has participated in courses in introductory microbiology, medical microbiology, and virology primarily at Southern Illinois University at Carbondale, where he is now a Professor in the Department of Microbiology and Dean of the College of Science. His research has been in the broad area of molecular genetics and gene expression and has been focused most specifically on studies of how cells control the accuracy of protein synthesis. He is the author of approximately 50 research papers. His home is on the edge of the Shawnee National Forest in deep southern Illinois where he lives with his wife, Beth, and three children, Justine, D'Arcy, and Grant.

Ninth Edition

BROCK
Biology of Microorganisms

Michael T. Madigan
John M. Martinko
Jack Parker

Southern Illinois University Carbondale

Prentice Hall
Upper Saddle River, NJ 07458

Library of Congress Cataloging-in-Publication Data

MADIGAN, MICHAEL T.,
    Brock biologyof microorganisms / Michael T. Madigan,
John M. Martinko, Jack Parker. — 9th ed.

    ISBN 0-13-081922-0
    1. Microbiology.   I. Martinko, John M.   II. Parker,
Jack.   III. Title.
QR41.2, B77   2000   579—DC21   99-30064

Editions of Biology of Microorganisms
First Edition, 1970, Thomas D. Brock
Second Edition, 1974, Thomas D. Brock
Third Edition, 1979, Thomas D. Brock
Fourth Edition, 1984, Thomas D. Brock, David W. Smith, and
    Michael T. Madigan
Fifth Edition, 1988, Thomas D. Brock and Michael T.
    Madigan
Sixth Edition, 1991, Thomas D. Brock and Michael T.
    Madigan
Seventh Edition, 1994, Thomas D. Brock, Michael T.
    Madigan, John M. Martinko, and Jack Parker
Eighth Edition, 1997, Michael T. Madigan, John M.
    Martinko, and Jack Parker
Ninth Edition, 2000, Michael T. Madigan, John M. Martinko,
    and Jack Parker

Editor in Chief: Paul F. Corey
Editorial Director: Tim Bozik
Assistant Vice President of Production & Manufacturing:
    David W. Riccardi
Executive Managing Editor: Kathleen Schiaparelli
Assistant Managing Editor: Lisa Kinne
Production Editor: Debra A. Wechsler
Creative Director: Paula Maylahn
Associate Creative Director: Amy Rosen
Art Director: Heather Scott
Assistant to Art Director: John Christiana
Art Manager: Gus Vibal
Art Editor: Karen Branson
Interior Design: Anne Flanagan

Illustrators: Imagineering
Cover Designer: John Christiana
Cover Art Researcher: Karen Sanatar
Cover Art: Paintings by Henriëtte Wilhelmina Beijerinck;
    photographed by Lesley A. Robertson for the Kluyver
    Laboratory Museum, Technical University of Delft, The
    Netherlands
Marketing Manager: Jennifer Welchans
Manufacturing Manager: Trudy Pisciotti
Editor in Chief, Development: Ray Mullaney
Associate Editor in Chief, Development: Carol Trueheart
Copy Editor: Jane Loftus
Editorial Assistant: Damian Hill

Printed in the United States of America
10 9 8 7 6 5 4 3

ISBN 0-13-081922-0

Prentice-Hall International (UK) Limited, *London*
Prentice-Hall of Australia Pty. Limited, *Sydney*
Prentice-Hall Canada Inc., *Toronto*
Prentice-Hall Hispanoamericana, S.A., *Mexico*
Prentice-Hall of India Private Limited, *New Delhi*
Prentice-Hall of Japan, Inc., *Tokyo*
Prentice-Hall (*Singapore*) Pte Ltd
Editora Prentice-Hall do Brasil, Ltda., *Rio de Janeiro*

Brief Contents

Preface xvi

CHAPTER 1   Microorganisms and Microbiology   1

CHAPTER 2   Macromolecules   29

CHAPTER 3   Cell Biology   48

CHAPTER 4   Nutrition and Metabolism   102

CHAPTER 5   Microbial Growth   135

CHAPTER 6   Principles of Microbial Molecular Biology   163

CHAPTER 7   Regulation of Gene Expression   212

CHAPTER 8   Viruses   236

CHAPTER 9   Microbial Genetics   289

CHAPTER 10   Genetic Engineering and Biotechnology   343

CHAPTER 11   Industrial Microbiology/Biocatalysis   384

CHAPTER 12   Microbial Evolution and Systematics   422

CHAPTER 13   Prokaryotic Diversity: Bacteria   453

CHAPTER 14   Prokaryotic Diversity: The Archaea   545

CHAPTER 15   Metabolic Diversity   573

CHAPTER 16   Microbial Ecology   642

CHAPTER 17   Eukaryotic Microorganisms   720

CHAPTER 18   Microbial Growth Control   740

CHAPTER 19   Host–Parasite Relationships   773

CHAPTER 20   Concepts of Immunology   801

CHAPTER 21   Clinical and Diagnostic Microbiology and Immunology   854

CHAPTER 22   Epidemiology and Public Health Microbiology   891

CHAPTER 23   Person-to-Person Microbial Diseases   922

CHAPTER 24   Animal-Transmitted, Vectorborne, and Common-Source Microbial Diseases   956

APPENDIX 1   Energy Calculations in Microbial Bioenergetics   A-1

APPENDIX 2   Bergey's Manual of Systematic Bacteriology, Second Edition   A-5

Glossary   G-1

Index   I-1

Overview

Brock Biology of Microorganisms, Ninth Edition

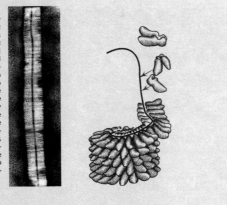

Tobacco Mosaic Virus (TMV), whose structure is shown here, was the first virus discovered. Work began on it in 1886 and the Russian microbiologist Ivanowsky showed that it passed through filters that could retain all known bacteria. The Dutch microbiologist Martinus Beijerinck proposed in 1899 that the tobacco mosaic agent was a new type of microorganism that we now call a virus. TMV was also the first virus purified and crystallized (by Wendell Stanley in 1935), and later the first large biological structure whose subunits could be shown to spontaneously reassemble. Interestingly, TMV is also a virus whose genetic material is RNA and gave early evidence of the incredible variety of strategies that viruses use to replicate themselves.

CHAPTER 8 Viruses

8.1	General Properties of Viruses 237	8.12	Temperate Bacteriophages: Lysogeny and Lambda 259
8.2	Nature of the Virion 238	8.13	A Transposable Phage: Bacteriophage Mu 265
8.3	The Virus Host 242	8.14	Overview of Animal Viruses 267
8.4	Quantification of Viruses 242	8.15	Positive-Strand RNA Animal Viruses 270
8.5	General Features of Virus Reproduction 244	8.16	Negative-Strand RNA Viruses 272
8.6	Steps in Virus Multiplication 246	8.17	Double-Stranded RNA Viruses: Reoviruses 275
8.7	Overview of Bacterial Viruses 250	8.18	Replication of DNA Viruses of Animals 275
8.8	RNA Bacteriophages 251	8.19	Herpesviruses 277
8.9	Single-Stranded Icosahedral DNA Bacteriophages 252	8.20	Pox Viruses 279
8.10	Single-Stranded Filamentous DNA Bacteriophages 254	8.21	Adenoviruses 280
8.11	Double-Stranded DNA Bacteriophages: Lytic Viruses 255	8.22	Retroviruses 281
		8.23	Viroids and Prions 285

The CHAPTER OUTLINE provides an overview of the chapter's main concepts.

SECTION NUMBERS keyed to page numbers provide easy reference points

The WORKING GLOSSARY provides definitions of important terms within each chapter.

WORKING GLOSSARY

Anticodon a sequence of three bases in a tRNA molecule that base-pairs with a codon during protein synthesis

Aminoacyl-tRNA synthetases a group of enzymes each one of which catalyzes the attachment of the correct amino acid to a tRNA

Antiparallel in reference to double-stranded DNA, one strand runs $5' \rightarrow 3'$ and other $3' \rightarrow 5'$

Chromosome a genetic element, usually circular in prokaryotes and linear in eukaryotes, carrying genes essential to cellular function

Codon a sequence of three bases in mRNA that encodes an amino acid

DNA polymerase an enzyme that synthesizes a new strand of DNA in the $5' \rightarrow 3'$ direction using an antiparallel DNA strand as a template

Exon the coding DNA sequences in a split gene (contrast with intron)

Gene a segment of DNA specifying a protein (via mRNA), a tRNA, or an rRNA

Genome the total complement of genes contained in a cell or virus

Hybridization formation of a duplex nucleic acid molecule with strands derived from different sources by complementary base pairing

Intron the intervening noncoding DNA sequences in a split gene (contrast with exon)

Messenger RNA (mRNA) an RNA molecule that contains the genetic information necessary to encode a particular protein

Molecular chaperones a group of proteins that help other proteins fold or refold from a partially denatured state

Operon a cluster of genes whose expression is controlled by a single operator

Primary transcript an unprocessed RNA molecule that is the direct product of transcription

Primer a molecule (usually a polynucleotide) to which DNA polymerase can attach the first deoxyribonucleotide during DNA replication

Promoter a site on DNA to which RNA polymerase can bind and begin transcription

Replication synthesis of DNA using DNA as a template

Restriction enzyme an enzyme that recognizes and makes double-stranded breaks at specific DNA sequences

Ribosomal RNA (rRNA) types of RNA found in the ribosome; some participate actively in the process of protein synthesis

Ribosome a cytoplasmic particle composed of ribosomal RNA and protein, which is a central part of the protein-synthesizing machinery of the cell

Ribozyme an RNA molecule that can catalyze chemical reactions

RNA polymerase an enzyme that synthesizes RNA in the $5' \rightarrow 3'$ direction using an antiparallel DNA strand as a template

RNA processing the conversion of a precursor RNA to its mature form

Semiconservative replication DNA synthesis yielding new double helices, each consisting of one parental and one progeny strand

Transcription the synthesis of RNA using a DNA template

Transfer RNA (tRNA) an adaptor molecule used in translation that has specificity for both a particular amino acid and for one or more codons

Translation the synthesis of protein using the genetic information in messenger RNA as a template

W
e now begin a study of the flow of information in microorganisms that will extend over the next five chapters. As we noted in Chapter 1, two hallmarks of life are *energy transformation* and *information flow*. In Chapter 4 we dealt with the problem of energy transformation: *metabolism*. Now we deal with the problem of information flow: *genetics*. **Genetics** is the discipline that deals with the mechanisms by which traits are passed from one organism to another and how they are expressed.

Since biological information flow is the basis of cellular function, genetics is a major research tool in attempts to understand the molecular mechanisms by which cells function. The study of genetics at the molecular level is also central to an understanding of the variability of organisms and the evolution of species.

Understanding how information is transmitted through biological systems also has tremendous practical applications. For instance, genetics provides us with techniques to specifically modify organisms for

TABLES have been redesigned to make key information even more accessible to students.

CONCEPT CHECKS summarize each section and provide quiz questions, so students can evaluate their understanding as they progress through the chapter.

TABLE 9.1   Kinds of mutants

Description	Nature of change	Detection of mutant
Auxotroph	Loss of enzyme in biosynthetic pathway	Inability to grow on medium lacking the nutrient
Cold-sensitive	Alteration of an essential protein so it is inactivated at low temperature	Inability to grow at a low temperature (for example, 20°C) that normally supports growth
Drug-resistant	Alteration of permeability to drug or drug target or detoxification of drug	Growth on medium containing a growth-inhibitory concentration of the drug
Noncapsulated	Loss or modification of surface capsule	Small, rough colonies instead of larger, smooth colonies
Nonmotile	Loss of flagella; nonfunctional flagella	Compact colonies instead of flat, spreading colonies
Pigmentless	Loss of enzyme in biosynthetic pathway leading to loss of one or more pigments	Presence of different color or lack of color
Rough colony	Loss or change in lipopolysaccharide outer layer	Granular, irregular colonies instead of smooth, glistening colonies
Sugar fermentation	Loss of enzyme in degradative pathway	Lack of color change on agar containing sugar and a pH indicator
Temperature-sensitive	Alteration of an essential protein so it is more heat-sensitive	Inability to grow at a temperature normally supporting growth (for example, 40°C) but still growing at a lower temperature (for example, 30°C)
Virus-resistant	Loss of virus receptor	Growth in presence of large amounts of virus

Some of the most common kinds of mutants and the means by which they are detected are listed in Table 9.1.

✓ 9.1 Concept Check

Mutation, a heritable change in DNA, can lead to a change in phenotype. Selectable mutations are those that give the mutant a growth advantage under certain environmental conditions and are especially useful in genetics research.

✓ Distinguish between *mutation* and *mutant*.
✓ Distinguish between *screening* and *selection*.

9.2

Molecular Basis of Mutation

As previously mentioned, mutations arise in cells because of changes in the *base sequence* of an organism's genetic material. In many cases, mutations lead to phenotypic changes in the organism; these changes are mostly harmful or neutral although beneficial changes do occur occasionally.

Mutation can be either spontaneous or induced. **Spontaneous mutations** can occur as a result of the action of natural radiation (cosmic rays, and so on) which alters the structure of bases in the DNA. Spontaneous mutations can also occur during replication, as a result of errors in the pairing of bases, leading to changes in the replicated DNA. In fact, such errors occur at a frequency of about 10^{-7}–10^{-11} per base pair during a single round of replication. (A typical gene has about 1000 base pairs, therefore, the frequency of these errors in a

normal, fully grown culture of organisms having approximately 10^8 cells/ml, there are probably a number of different mutants in each milliliter of culture.

Mutations involving one (or a very few) base pairs are sometimes referred to as **point mutations.** Point mutations can result in *base-pair substitutions* in the DNA or in the insertion or deletion of a base pair (called *microinsertions* and *microdeletions*). As is the case with all mutations, the phenotypic change that comes about because of a point mutation depends on exactly where the mutation took place in the gene, what the nucleotide change was, and what product the gene normally encodes.

Base-Pair Substitutions

If a point mutation occurs within the coding region of a gene that encodes a protein, any change in the phenotype of the cell is almost certainly the result of a change in the amino acid *sequence* of the protein being produced. Figure 9.3 shows a number of base-pair substitutions that can occur in a short region of DNA within a gene that encodes a protein. The error in the DNA is transcribed into mRNA, and this erroneous mRNA in turn is used as a template and translated into protein. (Because only one strand of the DNA is used as template for the mRNA, an AT base pair does not have the same meaning as a TA base pair.) The triplet code that directs the insertion of an amino acid via a transfer RNA will thus be incorrect. What are the consequences of base-pair substitutions?

In interpreting the results of mutation, we must first recall that the genetic code is degenerate (⇔ Section

because the genes encode proteins that can diffuse through the cytoplasm of the cell. (One cannot complement mutations in regulatory sites such as promoters because these function at the DNA level.) If each homologous DNA molecule contributes a different required gene, then the cell will have all the enzymes it requires to synthesize tryptophan. Notice that complementation *does not* involve recombination. To do the test, the mutations must be in *trans*. (If one molecule has both mutations, the other is wild type and should be sufficient itself to confer the wild-type phenotype unless one of the mutations is *dominant*. Therefore, having the mutations in cis serves as a control.)

This type of complementation test, called a *cis–trans test*, is used to define whether two mutations are in the same genetic (functional) unit. The genetic unit defined by the cis–trans test is sometimes called a **cistron** (a term essentially equivalent to a gene). As noted, two mutations in the *same* cistron *cannot* complement each other, and so when complementation is found to exist, this implies that the two mutations lie in *different* cistrons (that is, different genes). The term *cistron* is now rarely used except when describing whether an mRNA has the genetic information from one gene (monocistronic mRNA) or from more than one gene (polycistronic mRNA) (⇔ Section 6.8).

✓ 9.5 Concept Check

Homologous recombination arises when closely related sequences from two genetically distinct elements are combined together in the same element. Recombination is an important evolutionary process, and cells have specific mechanisms for ensuring that recombination takes place. In eukaryotes, genetic recombination occurs as a consequence of the sexual cycle. Mechanisms of recombination also occur in prokaryotes but involve DNA transfer during the processes of transformation, transduction, and conjugation.

✓ What protein, found in all prokaryotes, facilitates the pairing required for homologous recombination?
✓ Complementation tests do not involve recombination. Explain.

9.6

Genetic Transformation

As we have noted, genetic transformation is a process by which free DNA is incorporated into a recipient cell and brings about genetic change. The discovery of genetic transformation in bacteria was one of the outstanding events in biology, as it led to experiments demonstrating that DNA is the genetic material (see the box, Origins of Bacterial Genetics). This discovery became the keystone of molecular biology and modern genetics.

A number of prokaryotes have been found to be naturally transformable, including certain species of

both gram-negative and gram-positive Bacteria and some species of Archaea. However, even within transformable genera, only certain strains or species are transformable. Since the DNA of prokaryotes is present in the cell as a large single molecule, when the cell is gently lysed, the DNA pours out (Figure 9.13). Because of its extreme length (1700 μm in *Bacillus subtilis*), the DNA molecule breaks easily; even after gentle extraction the *B. subtilis* chromosome of 4.2 megabase pairs is converted to fragments of about 15 kilobase pairs. Because the DNA that corresponds to an average gene is about 1000 nucleotides, each of the fragments of purified DNA has about 15 genes. A single cell usually incorporates only one or a few DNA fragments so only a small proportion of the genes of one cell can be transferred to another by a single transformation event.

Competence

A cell that is able to take up a molecule of DNA and be transformed is said to be **competent.** Only certain strains are competent; the ability seems to be an inherited property of the organism. Competence in most naturally transformable bacteria is regulated, and special proteins play a role in the uptake and processing of DNA. These competence-specific proteins may include a membrane-associated DNA binding protein, a cell

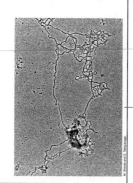

FIGURE 9.13 The prokaryotic chromosome, as shown in the electron microscope. The circular chromosome is from the hyperthermophile *Sulfolobus*, a member of the Archaea (⇔ Section 14.8). See also Figure 9.18.

CONCEPT LINKS alert students to material that builds on previous concepts and provides a useful cross-referencing system for the entire book.

Outstanding MICROGRAPHS are included throughout.

FEATURE BOXES provide additional, relevant information. Some take an historical perspective, some focus on techniques and applications, and others explore a text topic in greater depth.

A FOCUS ON . . .   Selenocysteine: The Twenty-first Amino Acid

The genetic code has codons for 20 amino acids that are assembled into proteins during translation. However, many proteins contain other amino acids. In fact, there are well over 100 different amino acids found in at least a few proteins. Until recently, it was thought that these "extra" amino acids were made by modifying one of the standard amino acids *after* it was incorporated into protein, a process called *posttranslational modification*. However, it is now clear that one of these extra amino acids is put into protein by the translational machinery itself. This one exception is *selenocysteine*.

Selenocysteine has the same structure as cysteine, but it has a selenium atom rather than a sulfur atom. It was known for some time that a few proteins contain this unusual amino acid. For example, *Escherichia coli* makes two different formate dehydrogenase enzymes and both contain a single selenocysteine residue. When the gene encoding one of these enzymes was sequenced, it was found that the codon that corresponded to the selenocysteine was a UGA. UGA is normally an efficient stop codon in *E. coli*, but it has now been demonstrated that it can be translated directly as selenocysteine in certain mRNA molecules, not only in *E. coli* but also in other prokaryotes and in eukaryotes, including humans. Therefore, selenocysteine is the twenty-first amino acid known to be encoded by the genetic code.

How can a codon sometimes be a stop codon and sometimes a sense codon *in the same chromo-*

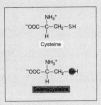

some? The answer apparently lies in the *context* of the codon, the sequence of the bases surrounding the UGA codon and in their secondary structure. In certain contexts, the translational machinery interprets UGA as "selenocysteine." In all other contexts, UGA means "stop translation." Selenocysteine has its own tRNA (as do all the standard amino acids) and also has a special protein factor that brings only this tRNA to the ribosome.

Selenocysteine is even more readily oxidized than cysteine. Therefore, enzymes that contain this amino acid must be protected from oxygen. It has been proposed that UGA might once have been a normal sense codon, calling only for selenocysteine, but that the increase in oxygen in our environment following the evolution of photosynthesis (⟷ Chapter 12) selected for proteins that contain cysteine (whose codons are UGU and UGC). This allowed the coding assignment of UGA to be altered except in a few special cases. ■

Other Genetic Codes

When the genetic code was cracked during the 1960s, all the prokaryotes and eukaryotes examined were found to use the same code. When the mRNA for mammalian hemoglobin was given to the *Escherichia coli* protein-synthesizing machinery (such as ribosomes and tRNA), mammalian hemoglobin was synthesized. Therefore, the genetic code appeared to be a universal code in that the exact same code was used by all living systems. Once techniques became available to sequence DNA easily, genes from many organisms began to be sequenced. However, comparison of these DNA sequences to the amino acid sequence of the proteins they encode has led to a few unexpected surprises. One of these, the discovery of introns in eukaryotic genes, was very surprising but did not change our ideas about the genetic code. However, it has also been discovered that

are slight variations of the "universal" genetic code (see the box, Selenocysteine: The Twenty-first Amino Acid).

The original findings of these alternative codes were in the genomes of mitochondria. So far as is known, only the mitochondria of plants use the universal code without change. The other organelles in plants, the chloroplasts, also use this standard code. The mitochondria of all other eukaryotes use codes with one or a few slight differences. A few of these variations are shown in Table 6.7. Note that there is *not* simply a mitochondrial code, although there are a few common themes, such as the general use of UGA as a tryptophan codon. It is also clear that these alternate codes are very closely related to the universal code and are almost certainly derived from it evolutionarily. A few genomes of cells are now known to use one or more of these slightly different codes, and a few examples are also given in Table 6.7.

BOLDFACED TERMS are defined in the Glossary.

The ART in the ninth edition has been thoroughly revisd, but still maintains the use of consistent color coding.

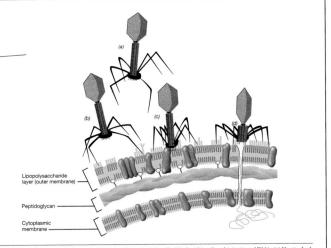

FIGURE 8.10 Attachment of T4 bacteriophage particle to the cell wall of *Escherichia coli* and injection of DNA: (a) Unattached particle. (b) Attachment to the wall by the long tail fibers interacting with core polysaccharide. (c) Contact of cell wall by the tail pins. (d) Contraction of the tail sheath and injection of the DNA. For a detailed description of the gram-negative cell wall see Section 3.8.

✓ 8.6a Concept Check

The attachment of a virion to a host cell is a highly specific process, involving the interaction of receptors on the host with proteins on the surface of the virus particle. Only after attachment has occurred can the virus or its nucleic acid penetrate the host cell.

Virus Restriction and Modification by the Host

We have already seen that one form of host resistance to virus arises when there is no receptor site on the cell surface to which the virus can attach. Another and more specific kind of host resistance occurs in prokaryotes and involves destruction of the double-stranded DNA genome of a virus after it has been injected. This destruction is brought about by host enzymes that cleave the viral DNA at one or several places, thus preventing its replication. This phenomenon is called *restriction* and is part of a general host mechanism to prevent the invasion of foreign nucleic acid. We have already discussed *restriction enzymes* and their action in some detail (⟷ Section 6.5) and noted that their cellular role is in defending against foreign DNA. Restriction enzymes are highly specific, attacking only certain sequences (generally four or six base pairs). The host protects its own DNA from the action of restriction enzymes by *modifying* its DNA at the sites where the restriction enzymes act. Modification of host DNA is brought about by methylation of purine or pyrimidine bases (in such a way that their base-pairing properties are not altered).

Some viruses can overcome host restriction mechanisms by modifications of their nucleic acids so that they are no longer subject to enzymatic attack. Two kinds of chemical modifications of viral DNA have been recognized, glucosylation and methylation. For instance, the T-even bacteriophages (T2, T4, and T6) have their DNA glucosylated to varying degrees, and the glucosylation prevents or greatly reduces endonuclease attack. Many other viral nucleic acids have been found to be modified by methylation, but glucosylation has been

REVIEW QUESTIONS

1. If an enzyme can be effectively inhibited by feedback inhibition, why would cells also have mechanisms to regulate its synthesis?

2. Compare the processing of proinsulin and the product of the *gyrA* gene of *Mycobacterium leprae* (see box, Protein Processing). In both cases an "interior" peptide is removed. However, only the latter case is considered protein splicing. Explain.

3. Describe why a protein that binds to a specific sequence of double-stranded DNA is unlikely to bind to the same sequence if the DNA is single-stranded.

4. Describe the regulation of two different operons, one having an effector that is an *inducer* and the other having an effector that is a *corepressor*.

5. The maltose regulon is inducible and is regulated by an activator protein. The lactose operon is inducible but is regulated by a repressor protein. Explain how induction can be brought about by either positive control (activator protein) or negative control (repressor protein).

6. In most cases operators are very close to the promoters they control, while activator binding sites can be some distance away. Explain why this should be so.

7. Describe how transcriptional attenuation works. What is actually being "attenuated"? Why hasn't the type of attenuation that controls several different amino acid biosynthetic pathways in *Escherichia coli* also been found in eukaryotes?

8. Describe the mechanism by which catabolic activator protein (CAP), the regulatory protein for catabolite repression, functions using the lactose operon as an example. For this operon the CAP protein is not a repressor. Describe the regulatory region of a gene for which the CAP protein *is* a repressor. (*Hint*: Think about your answer to Question 6.)

9. What are the two components that give the name to signal transduction regulation in prokaryotes? What is the function of each of the components?

10. One of the members of a two-component system is typically located in the cell membrane. What reason can you think of why this might be so?

11. Many genes are under multiple control systems. In the lactose operon, there is lactose-specific regulation and regulation by a global control system. Describe how each of the controls on the lactose operon actually functions. Why do you think both systems are necessary?

APPLICATION QUESTIONS

1. The amino acids isoleucine and valine share a common pathway for most steps in their biosynthesis. In *Escherichia coli* the first common step can be subject to feedback inhibition by valine but not by isoleucine. In most strains, though, the addition of valine does not cause isoleucine deprivation. However, in other strains it does (that is, adding valine causes isoleucine starvation and the cells stop growing). What explanation can you give for the difference between the normal "valine-resistant" strains and those whose growth is sensitive to valine?

2. What would happen to regulation from a promoter under negative control if the region where the regulatory protein binds were deleted? What if the promoter were under positive control?

3. Promoters from *Escherichia coli* under positive control are not close matches to the DNA consensus sequence for *E. coli* (⊃⊃ Section 6.7). Why?

4. Interestingly, the attenuation control of some of the pyrimidine biosynthetic pathway genes in *Escherichia coli* actually involves coupled transcription and translation. Can you describe a mechanism whereby the cell could somehow make use of translation to help it measure the level of pyrimidine nucleotides?

REVIEW QUESTIONS challenge the student's mastery of chapter concepts.

APPLICATION QUESTIONS allow students to test their analytical and problem-solving skills.

APPENDICES provide useful tutorial and reference information.

APPENDIX 1   Energy Calculations in Microbial Bioenergetics

The information in Appendix 1 is intended to help students calculate changes in free energy accompanying chemical reactions carried out by microorganisms. It begins with definitions of the terms required to make such calculations and proceeds to show how knowledge of redox state, atomic and charge balance, and other factors are necessary to calculate free-energy problems successfully.

Definitions

1. ΔG^0 = standard free-energy change of the reaction at 1 atm pressure and 1 M concentrations; ΔG = free-energy change under the conditions specified; $\Delta G^{0\prime}$ = free-energy change under standard conditions at pH 7.

2. Calculation of ΔG^0 for a chemical reaction from the free energy of formation, G_f^0, of products and reactants:

$$\Delta G^0 = \Sigma \, \Delta G_f^0 \, \text{(products)} - \Sigma \, \Delta G_f^0 \, \text{(reactants)}$$

That is, sum the ΔG_f^0 of products, sum the ΔG_f^0 of reactants, and subtract the latter from the former.

3. For energy-yielding reactions involving H^+, converting from standard conditions (pH 0) to biochemical conditions (pH 7):

$$\Delta G^{0\prime} = \Delta G^0 + m \, \Delta G_f'(H^+)$$

where m is the net number of protons in the reaction (m is negative when more protons are consumed than formed) and $\Delta G_f'(H^+)$ is the free energy of formation of a proton at pH 7 = -39.83 kJ at 25°C.

4. Effect of concentrations on ΔG: with soluble substrates, the concentration ratios of products formed to exogenous substrates used are generally equal to or greater than 10^{-2} at the beginning of growth and equal to or less than 10^{-2} at the end of growth. From the relation between ΔG and the equilibrium constant (see item 8), it can be calculated that ΔG for the free-energy yield in practical situations differs from the free-energy yield under standard conditions by at most 11.7 kJ, a rather small amount, and so for a first approximation, standard free-energy yields can be used in most situations. However, with H_2 as a product, H_2-consuming bacteria present may keep the concentration of H_2 so low that the free-energy yield is significantly affected. Thus, in the fermentation of ethanol to acetate and H_2 ($C_2H_5OH + H_2O \rightarrow C_2H_3O_2^- + 2\,H_2 + H^+$), the $\Delta G^{0\prime}$ at 1 atm H_2 is +9.68 kJ, but at 10^{-4} atm H_2 it is -36.03 kJ. With H_2-consuming bacteria present, therefore, the ethanol fermentation becomes useful. (See also item 9.)

5. Reduction potentials: by convention, electrode equations are written in the direction, oxidant + $ne^- \rightarrow$ reductant (that is, as reductions), where n is the number of electrons transferred. The standard potential (E_0) of the hydrogen electrode, $2\,H^+ + 2\,e^- \rightarrow H_2$ is set by definition at 0.0 V at 1.0 atm pressure of H_2 gas and 1.0 M H^+, at 25°C. E_0' is the standard reduction potential at pH 7. See also Table A1.2.

6. Relation of free energy to reduction potential:

$$\Delta G^{0\prime} = -nF \, \Delta E_0'$$

where n is the number of electrons transferred, F is the Faraday constant (96.48 kJ/V), and $\Delta E_0'$ is the E_0' of the electron-accepting couple minus the E_0' of the electron-donating couple.

7. Equilibrium constant, K. For the generalized reaction $aA + bB \rightleftharpoons cC + dD$,

$$K = \frac{[C]^c \, [D]^d}{[A]^a [B]^b}$$

where A, B, C, and D represent reactants and products; a, b, c, and d represent number of molecules of each; and brackets indicate concentrations. This is true only when the chemical system is in equilibrium.

8. Relation of equilibrium constant, K, to free-energy change. At constant temperature and pressure,

$$\Delta G = \Delta G^0 + RT \ln K$$

where R is a constant (8.29 J/mol/°K) and T is the absolute temperature (in °K).

9. Two substances can react in a redox reaction even if the standard potentials are unfavorable, provided that the concentrations are appropriate.

Assume that normally the reduced form of A would donate electrons to the oxidized form of B. However, if the concentration of the reduced form of A were low and the concentration of the reduced form of B were high, it would be possible for the reduced form of B to donate electrons to the oxidized form of A. Thus, the reaction would proceed in the direction opposite that predicted from standard potentials. A practical example of this is the utilization of H^+ as an electron acceptor to produce H_2. Normally, H_2 production in fermentative bacteria is not extensive because H^+ is a poor electron acceptor; the E_0' of the $2\,H^+/H_2$ pair is -0.41 V. However, if the concentration of H_2 is kept low by continually removing it (a process done by methanogenic prokaryotes, which use $H_2 + CO_2$ to produce methane, CH_4, or by many other anaerobes capable of consuming H_2 anaerobically), the potential will be more positive and then H^+ will serve as a suitable electron acceptor.

Contents

Preface xvi

CHAPTER 1
Microorganisms and Microbiology   1

1.1 Microbiology and You   2
1.2 Microorganisms as Cells   4
1.3 Elements of Cell and Viral Structure   6
1.4 Evolutionary Relationships Among Living Organisms   8
1.5 Microbial Populations, Communities, and Ecosystems   11
1.6 Laboratory Culture of Microorganisms   12
1.7 The Impact of Microorganisms on Human Affairs   15
1.8 The Historical Roots of Microbiology: van Leeuwenhoek, Pasteur, and Koch   17
1.9 Transition into the Twentieth Century   24

CHAPTER 2
Macromolecules   29

2.1 Strong and Weak Bonds   30
2.2 An Overview of Macromolecules and Water as the Solvent of Life   33
2.3 Polysaccharides   35
2.4 Lipids   36
2.5 Nucleic Acids   37
2.6 Amino Acids   40
2.7 Proteins: The Peptide Bond, Primary and Secondary Structure   43
2.8 Proteins: Higher Order Structure and Denaturation   44

CHAPTER 3
Cell Biology   48

3.1 Light Microscopy   50
3.2 Three-Dimensional Imaging: Interference Contrast, Atomic Force, and Confocal Scanning Laser Microscopy   54
3.3 Electron Microscopy   55
3.4 Overview of Cell Structure and the Significance of Smallness   57
3.5 Cytoplasmic Membrane: Structure   60
3.6 Cytoplasmic Membrane: Function   63
3.7 The Cell Wall of Prokaryotes: Peptidoglycan and Related Molecules   69

3.8 The Outer Membrane of Gram-Negative Bacteria   73
3.9 Cell Wall Synthesis and Cell Division   76
3.10 Arrangement of DNA in Prokaryotes   77
3.11 Flagella and Motility   79
3.12 Bacterial Behavior: Chemotaxis, Phototaxis, and Other Taxes   83
3.13 Bacterial Cell Surface Structures and Cell Inclusions   85
3.14 Gas Vesicles   89
3.15 Endospores   91
3.16 The Nucleus and Organelles of Eukaryotic Microorganisms   95
3.17 Comparison of the Prokaryotic and Eukaryotic Cell   99

CHAPTER 4
Nutrition and Metabolism   102

4.1 An Overview of Metabolism   103
4.2 Microbial Nutrition   104
4.3 Culture Media   106
4.4 Energetics   108
4.5 Catalysis and Enzymes   110
4.6 Oxidation–Reduction   112
4.7 Electron Carriers   114
4.8 High Energy Compounds and Energy Conservation   116
4.9 Fermentation: The Embden–Meyerhof Pathway (Glycolysis)   118
4.10 Respiration and Electron Transport   121
4.11 Energy Conservation from Electron Transport   123
4.12 Carbon Flow: The Citric Acid Cycle   126
4.13 The Balance Sheet of Aerobic Respiration and Energy Storage   128
4.14 An Overview of Alternate Modes of Energy Generation   129
4.15 Biosynthesis of Monomers   130

CHAPTER 5
Microbial Growth   135

5.1 Overview of Cell Growth   136
5.2 Population Growth   137
5.3 Growth Cycle of Populations   139
5.4 Measurement of Growth   141

5.5 Continuous Culture: The Chemostat 145
5.6 Effect of Environment on Growth 146
5.7 Effect of Temperature on Growth 147
5.8 Microbial Growth at Cold Temperatures 148
5.9 Microbial Growth at High Temperatures 151
5.10 Microbial Growth at Low or High pH 154
5.11 Osmotic Effects on Microbial Growth 156
5.12 Oxygen as a Factor in Microbial Growth 158

CHAPTER 6

Principles of Microbial Molecular Biology 163
6.1 Macromolecules and Genetic Information 165
6.2 DNA Structure: The Double Helix 167
6.3 DNA Structure: Supercoiling 177
6.4 Genetic Elements 179
6.5 Restriction and Modification of DNA 182
6.6 DNA Replication 184
6.7 Transcription: The Basic Process in Bacteria 191
6.8 Transcription: Other Patterns and Inhibition 195
6.9 RNA Processing and Ribozymes 197
6.10 The Genetic Code 199
6.11 Transfer RNA 201
6.12 Translation: The Process of Protein Synthesis 204
6.13 Translation: Alternatives and Errors 208

CHAPTER 7

Regulation of Gene Expression 212
7.1 Regulation of Enzyme Activity 213
7.2 Regulation of Transcription: Induction and Repression 217
7.3 Regulation of Transcription: Positive Control 220
7.4 DNA Binding Proteins 221
7.5 Attenuation 224
7.6 Global Control 226
7.7 Signal Transduction and Two-Component Regulatory Systems 230
7.8 Contrasts in Gene Expression between Prokaryotes and Eukaryotes 233

CHAPTER 8

Viruses 236
8.1 General Properties of Viruses 237
8.2 Nature of the Virion 238
8.3 The Virus Host 242
8.4 Quantification of Viruses 242
8.5 General Features of Virus Reproduction 244

8.6 Steps in Virus Multiplication 246
8.7 Overview of Bacterial Viruses 250
8.8 RNA Bacteriophages 251
8.9 Single-Stranded Icosahedral DNA Bacteriophages 252
8.10 Single-Stranded Filamentous DNA Bacteriophages 254
8.11 Double-Stranded DNA Bacteriophages: Lytic Viruses 255
8.12 Temperate Bacteriophages: Lysogeny and Lambda 259
8.13 A Transposable Phage: Bacteriophage Mu 265
8.14 Overview of Animal Viruses 267
8.15 Positive-Strand RNA Animal Viruses 270
8.16 Negative-Strand RNA Viruses 272
8.17 Double-Stranded RNA Viruses: Reoviruses 275
8.18 Replication of DNA Viruses of Animals 275
8.19 Herpesviruses 277
8.20 Pox Viruses 279
8.21 Adenoviruses 280
8.22 Retroviruses 281
8.23 Viroids and Prions 285

CHAPTER 9

Microbial Genetics 289
9.1 Mutations and Mutants 291
9.2 Molecular Basis of Mutation 294
9.3 Mutagens 297
9.4 Mutagenesis and Carcinogenesis: The Ames Test 300
9.5 Genetic Recombination 302
9.6 Genetic Transformation 306
9.7 Transduction 311
9.8 Plasmids 314
9.9 Conjugation and Chromosome Mobilization 319
9.10 Transposons and Insertion Sequences 324
9.11 The *Escherichia coli* Chromosome 329
9.12 Comparative Prokaryotic Genomics 333
9.13 Genetics in Eukaryotic Microorganisms 336
9.14 Yeast Genetics 337

CHAPTER 10

Genetic Engineering and Biotechnology 343
10.1 Molecular Cloning 345
10.2 Plasmids as Cloning Vectors 345
10.3 Bacteriophage Lambda as a Cloning Vector 347
10.4 Other Vectors 349
10.5 Hosts for Cloning Vectors 352

10.6 Finding the Right Clone 353
10.7 Expression Vectors 356
10.8 Synthetic DNA 359
10.9 Amplifying DNA: The Polymerase Chain Reaction 360
10.10 Cloning and Expression of Mammalian Genes in Bacteria 362
10.11 *In Vitro* and Site-Directed Mutagenesis 366
10.12 Practical Applications of Genetic Engineering 368
10.13 Production of Mammalian Products and Vaccines by Genetically Engineered Microorganisms 369
10.14 Genetic Engineering in Plant Agriculture 374
10.15 Genetic Engineering in Animal and Human Genetics 378
10.16 Genetic Engineering as a Microbial Research Tool 380
10.17 Summary of Principles at the Basis of Genetic Engineering 381

CHAPTER 11
Industrial Microbiology/ Biocatalysis 384

11.1 Industrial Microorganisms and Products 385
11.2 Growth and Product Formation in Biocatalyses 387
11.3 Characteristics of Large-Scale Fermentations 389
11.4 Fermentation Scale-Up 391
11.5 Antibiotics: Isolation and Characterization 392
11.6 Industrial Production of Penicillins and Tetracyclines 396
11.7 Vitamins and Amino Acids 399
11.8 Microbial Bioconversion 401
11.9 Enzymes 402
11.10 Vinegar 405
11.11 Citric Acid and Other Organic Compounds 406
11.12 Yeast as an Agent of Fermentation and as Food 407
11.13 Alcohol and Alcoholic Beverages 409
11.14 Mushrooms as a Food Source 413
11.15 Sewage and Wastewater Microbiology 416

CHAPTER 12
Microbial Evolution and Systematics 422

12.1 Evolution of Earth and Earliest Life Forms 424
12.2 Primitive Organisms: Molecular Coding and Energy Generation 427
12.3 Eukaryotes and Organelles 430
12.4 Evolutionary Chronometers 432
12.5 Ribosomal RNA Sequences and Cellular Evolution 434
12.6 Signature Sequences, Phylogenetic Probes, and Molecular Microbial Ecology 436
12.7 Microbial Phylogeny as Revealed by Ribosomal RNA Sequencing 439
12.8 Characteristics of the Primary Domains 442
12.9 Conventional and Molecular Taxonomy 444
12.10 The Species Concept, Nomenclature, and *Bergey's Manual* 449

CHAPTER 13:
Prokaryotic Diversity: Bacteria 453

KINGDOM I: PROTEOBACTERIA
13.1 Purple Phototrophic Bacteria 455
13.2 The Nitrifying Bacteria 461
13.3 Sulfur- and Iron-Oxidizing Bacteria 462
13.4 Hydrogen-Oxidizing Bacteria 466
13.5 Methanotrophs and Methylotrophs 467
13.6 *Pseudomonas* and the Pseudomonads 470
13.7 Acetic Acid Bacteria 474
13.8 Free-Living Aerobic Nitrogen-Fixing Bacteria 475
13.9 *Neisseria, Chromobacterium,* and Relatives 476
13.10 Enteric Bacteria 477
13.11 *Vibrio* and *Photobacterium* 482
13.12 Rickettsias 484
13.13 Spirilla 485
13.14 Sheated Proteobacteria: *Sphaerotilus* and *Leptothrix* 489
13.15 Budding and Prosthecate/Stalked Bacteria 491
13.16 Gliding Myxobacteria 495
13.17 Sulfate- and Sulfur-Reducing Proteobacteria 498

KINGDOM II: GRAM-POSITIVE BACTERIA
13.18 Nonsporulating, Low GC, Gram-Positive Bacteria 502
13.19 Endospore-Forming, Low GC, Gram-Positive Bacteria 507
13.20 Cell Wall–Less, Low GC, Gram-Positive Bacteria: The Mycoplasmas 513
13.21 High GC, Gram-Positive Bacteria: Coryneform and Propionic Acid Bacteria 515
13.22 High GC, Gram-Positive Bacteria: *Mycobacterium* 517
13.23 Filamentous, High GC, Gram-Positive Bacteria: The Actinomycetes 519

KINGDOM III: CYANOBACTERIA, PROCHLOROPHYTES, AND CHLOROPLASTS
13.24 Cyanobacteria 524
13.25 Prochlorophytes and Chloroplasts 527

KINGDOM IV: CHLAMYDIA
13.26   The Chlamydia   529

KINGDOM V: *PLANCTOMYCES/PIRELLA*
13.27   *Planctomyces:* A Phylogenetically Unique
            Stalked Bacterium   532

KINGDOM VI: *BACTEROIDES/FLAVOBACTERIA*
13.28   *Bacteroides, Flavobacterium,* and *Cytophaga*   532

KINGDOM VII: GREEN SULFUR BACTERIA
13.29   *Chlorobium* and Other Green Sulfur Bacteria   533

KINGDOM VIII: TIGHTLY COILED BACTERIA:
THE SPIROCHETES
13.30   Spirochetes   537

KINGDOM IX: DEINOCOCCI
13.31   *Deinococcus/Thermus*   540

KINGDOM X: THE GREEN NONSULFUR BACTERIA
13.32   *Chloroflexus* and *Heliothrix*   542

KINGDOMS XI, XII, AND XIII: HYPERTHERMOPHILES
13.33   *Thermotoga* and *Thermodesulfobacterium*   543
13.34   *Aquifex* and Relatives   544

CHAPTER 14

Prokaryotic Diversity: The Archaea   545

14.1   Phylogenetic Overview of the Archaea   546

KINGDOM EURYARCHAEOTA
14.2   Extremely Halophilic Archaea   548
14.3   Methane-Producing Archaea: Methanogens   553
14.4   Thermoplasmatales: *Thermoplasma* and
           Picrophilus   556
14.5   Hyperthermophilic Euryarchaeota:
           Thermococcales and *Methanopyrus*   558
14.6   Hyperthermophilic Euryarchaeota: The
           Archaeoglobales   560

KINGDOM CRENARCHAEOTA
14.7   Habitats and Energy Metabolism of
           Crenarchaeotes   561
14.8   Hyperthermophiles from Terrestrial Volcanic
           Habitats: Sulfolobales and
           Thermoproteales   564
14.9   Hyperthermophiles from Submarine Volcanic
           Habitats: Desulfurococcales   566

EVOLUTION AND LIFE AT HIGH TEMPERATURES
14.10   Heat Stability of Biomolecules   568
14.11   Hyperthermophilic Archaea and Microbial
            Evolution   571
14.12   Korarchaeota, Hydrogen, and Microbial Life on
            Other Planets   571

CHAPTER 15

Metabolic Diversity   573

15.1   Energy Conservation and Carbon Metabolism
           574
15.2   Photosynthesis   575
15.3   The Role of Chlorophyll and
           Bacteriochlorophyll in Photosynthesis   576
15.4   Carotenoids and Phycobilins   580
15.5   Anoxygenic Photosynthesis   582
15.6   Oxygenic Photosynthesis   587
15.7   Autotrophic CO_2 Fixation: The Calvin Cycle   590
15.8   Autotrophic CO_2 Fixation: Reverse Citric Acid
           Cycle and the Hydroxypropionate Cycle   592
15.9   Chemolithotrophy: Energy from the Oxidation
           of Inorganic Electron Donors   592
15.10   Hydrogen Oxidation   594
15.11   Oxidation of Reduced Sulfur Compounds   595
15.12   Iron Oxidation   598
15.13   Nitrification   601
15.14   Methanotrophy and Methylotrophy   603
15.15   Anaerobic Respiration   605
15.16   Nitrate Reduction and the Denitrification
            Process   606
15.17   Sulfate Reduction   608
15.18   Acetogenesis   611
15.19   Methanogenesis   613
15.20   Ferric Iron, Manganese, Chlorate, and Organic
            Electron Acceptors   617
15.21   Fermentations: Energetic and Redox
            Considerations   620
15.22   Fermentative Diversity   622
15.23   Syntrophy   624
15.24   Hexose, Pentose, and Polysaccharide
            Utilization   626
15.25   Organic Acid Metabolism   629
15.26   Lipids as Microbial Nutrients   630
15.27   Molecular Oxygen (O_2) as a Reactant in
            Biochemical Processes   631
15.28   Hydrocarbon Transformations   632
15.29   Nitrogen Fixation   634

CHAPTER 16

Microbial Ecology   642

16.1   Populations, Guilds, and Communities   643
16.2   Microorganisms in Nature   645
16.3   Methods in Microbial Ecology   648
16.4   Enrichment and Isolation Methods   648
16.5   Viability and Quantification Using Staining
           Techniques   653
16.6   Genetic Stains, Community Analysis, and
           Optical Tweezers   655
16.7   Microbial Activity Measurements:
           Radioisotopes and Microelectrodes   658

16.8 Microbial Activity Measurements: Stable Isotopes 660

16.9 Terrestrial Environments 662

16.10 Aquatic Habitats 665

16.11 Deep-Sea Microbiology 667

16.12 Hydrothermal Vents 670

16.13 The Carbon Cycle 676

16.14 Ecology of Syntrophy and Methanogenesis 677

16.15 The Rumen Microbial Ecosystem 681

16.16 The Nitrogen Cycle 685

16.17 The Sulfur Cycle 686

16.18 The Iron Cycle 689

16.19 Microbial Leaching of Ores 691

16.20 Mercury and Heavy Metal Transformations 694

16.21 Petroleum Biodegradation 696

16.22 Biodegradation of Xenobiotics 698

16.23 Plant–Microorganism Interactions: Lichens and Mycorrhizae 704

16.24 *Agrobacterium* and Crown Gall Disease 706

16.25 Root Nodule Bacteria and Symbiosis with Legumes 709

CHAPTER 17

Eukaryotic Microorganisms 720

17.1 Eukaryotic Cell Structure 721

17.2 Phylogenetic Overview of Eukarya 724

17.3 Protozoa 725

17.4 Fungi 729

17.5 Slime Molds 733

17.6 Algae 735

CHAPTER 18

Microbial Growth Control 740

18.1 Heat Sterilization 742

18.2 Radiation Sterilization 745

18.3 Filter Sterilization 747

18.4 Chemical Growth Control 749

18.5 Antiseptics, Disinfectants, and Sterilants 751

18.6 Synthetic Antibacterial Chemotherapeutic Agents 753

18.7 Antibiotics 758

18.8 β -Lactam Antibiotics: Penicillins and Cephalosporins 759

18.9 Antibiotics from Prokaryotes 760

18.10 Viral Control 762

18.11 Fungal Control 764

18.12 Antimicrobial Drug Resistance 765

18.13 The Search for New Antimicrobial Drugs 770

CHAPTER 19

Host–Parasite Relationships 773

19.1 Microbial Interactions with Higher Organisms 775

19.2 Normal Flora of the Skin 776

19.3 Normal Flora of the Oral Cavity 777

19.4 Normal Flora of the Gastrointestinal Tract 780

19.5 Normal Flora of Other Body Regions 782

19.6 Entry of the Pathogen into the Host 784

19.7 Colonization and Growth 786

19.8 Exotoxins 788

19.9 Enterotoxins 791

19.10 Endotoxins 793

19.11 Virulence 794

19.12 Nonspecific Host Defenses 796

19.13 Inflammation and Fever 799

CHAPTER 20

Concepts of Immunology 801

20.1 Cells and Organs of the Immune System 804

20.2 Immunogens and Antigens 806

20.3 Nonspecific Immunity: Phagocytes and Phagocytosis 808

20.4 Specific Immunity: Lymphocytes 811

20.5 Immunoglobulins (Antibodies) 812

20.6 T Cell Receptors 823

20.7 Histocompatibility Proteins 824

20.8 Cytokines 826

20.9 Cell-Mediated Immunity 829

20.10 Clonal Selection and Immune Tolerance 830

20.11 Mechanism of Immunoglobulin Formation 832

20.12 The Complement System 835

20.13 Polyclonal and Monoclonal Antibodies 837

20.14 Antigen–Antibody Reactions 840

20.15 Immune Diseases 843

20.16 Immunity to Infectious Diseases 847

20.17 Alternative Immunization Strategies 851

CHAPTER 21

Clinical and Diagnostic Microbiology and Immunology 854

21.1 Isolation of Pathogens from Clinical Specimens 855

21.2 Growth-Dependent Identification Methods 861

21.3 Testing Cultures for Antimicrobial Drug Sensitivity 865

21.4 Immunodiagnostics 867

21.5 Agglutination 870

21.6 Immunoelectron Microscopy 871

21.7 Fluorescent Antibodies   871
21.8 Enzyme-Linked Immunosorbent Assay and Radioimmunoassay   875
21.9 Immunoblot Procedures   879
21.10 Nucleic Acid Probes in Clinical Diagnostics   882
21.11 Diagnostic Virology   887
21.12 Safety in the Clinical Laboratory   888

CHAPTER 22

Epidemiology and Public Health Microbiology   891

22.1 The Science of Epidemiology   892
22.2 Epidemiological Terminology   893
22.3 Disease Reservoirs   895
22.4 Epidemiology of AIDS: An Example of How Epidemiological Research Is Done   898
22.5 Infectious Disease Transmission   900
22.6 The Host Community   902
22.7 Hospital-Acquired (Nosocomial) Infections   906
22.8 Public Health Measures for the Control of Disease   908
22.9 Global Health Considerations   911
22.10 Emerging and Reemerging Infectious Diseases   913

CHAPTER 23

Person-to-Person Microbial Diseases   922

23.1 Airborne Transmission of Pathogens   923
23.2 Respiratory Pathogens: Bacterial   925
23.3 *Mycobacterium* and Tuberculosis   931
23.4 Respiratory Pathogens: Viral   934
23.5 Why Are Respiratory Infections So Common?   940
23.6 Sexually Transmitted Diseases   940
23.7 Acquired Immunodeficiency Syndrome   946

CHAPTER 24

Animal-Transmitted, Vectorborne, and Common-Source Microbial Diseases   956

ANIMAL-TRANSMITTED AND VECTORBORNE DISEASES
24.1 Animal-Transmitted Diseases: Rabies   957
24.2 Animal-Transmitted Diseases: Hantavirus Pulmonary Syndrome   959
24.3 Insect- and Tick-Transmitted Diseases: Rickettsias   960
24.4 Tick-Transmitted Diseases: Lyme Disease   963
24.5 Insect-Transmitted Diseases: Malaria   966
24.6 Insect-Transmitted Diseases: Plague   969

COMMON-SOURCE DISEASES
24.7 Soil Microorganisms: The Pathogenic Fungi   971
24.8 Public Health and Water Quality   974
24.9 Waterborne Diseases   976
24.10 Microbial Growth in Food   980
24.11 Foodborne Diseases: Food Poisoning   983
24.12 Foodborne Diseases: Food Infection   986

APPENDIX 1

Energy Calculations in Microbial Bioenergetics   A-1

APPENDIX 2

Bergey's Manual of Systematic Bacteriology, Second Edition   A-5

Glossary   G-1

Index   I-1

Preface

As we enter the new millennium the rapid pace of basic research and the enormous possibilities for applications have thrust the field of microbiology into the forefront of the biological sciences. Through the years the science of microbiology has kept its roots firmly planted in fundamental principles. But now with the explosion of new molecular methods for both field and laboratory studies microbiologists can ask questions and do experiments that they could only dream of years ago. Indeed, as one notable microbiologist put it recently, the science of microbiology is "on a roll." It is during this exciting time that we present to students and instructors of microbiology a snapshot of microbiology today in the form of the ninth edition of *Brock Biology of Microorganisms (BBOM)*.

What's New in Organization?

In the 30 years since the first edition of this book, originally titled *Biology of Microorganisms*, was published, many things have changed in the field of microbiology. The new edition of *BBOM* has also seen many changes—some minor, some not so minor—from that of the previous edition. In terms of chapter organization, the ninth edition follows along the lines of the eighth with two exceptions: (1) The chapter on "Microbial Evolution and Systematics" has been moved forward to Chapter 12 to set the stage for later chapters on microbial diversity, metabolic diversity, and ecology. With this move, every chapter in the book that deals with organisms and their activities in nature (including medical microbiology) can be seen, as it should be, within a phylogenetic context. And (2), the chapter entitled "Microbial Growth Control" has been moved back to Chapter 18 in this edition to immediately precede the medical/immunology block of chapters where it fits better as an introduction to this material.

Because medical microbiology is such an important area of our science, the disease chapter has been restructured and split into two chapters. Chapter 23 is now entitled "Person-to-Person Microbial Diseases" and exclusively covers microbial diseases transmitted in this fashion. Here one will find up-to-the-minute coverage of airborne and sexually transmitted pathogens and the diseases they cause. In Chapter 24, entitled "Animal-Transmitted, Vectorborne, and Common-Source Microbial Diseases," one finds coverage of important diseases transmitted by insects or animals such as malaria, Lyme, hantavirus and rabies, and diseases caused by a common vehicle such as foodborne and waterborne diseases. Those instructors who structure their introductory courses around a theme of medical microbiology will now find more than enough coverage of diseases and the disease process in the ninth edition.

Finally, it should also be mentioned that the new organization in *BBOM 9/e* conforms very closely to the recommendations of the Education Division of the American Society for Microbiology (ASM) for teaching introductory microbiology courses. This includes the content outline of the new ASM telecourse in microbiology, whose 12-part series produced by Oregon Public Broadcasting and funded by the Annenberg Corporation closely mirrors the organization of material in the ninth edition.

What's New in Content?

What *isn't* new? Every chapter in this book has seen revision, in some cases, very extensive revision. The authors' goal in producing this new edition was to maintain the breadth and depth of coverage that users have come to expect from this book while at the same time keeping things within bounds, both in terms of what students should be expected to master in one semester, and the overall length of an "introductory" textbook. As for previous editions of this book, the ninth has tried to strike a balance between concepts and details such that the beginning student is not overwhelmed while the more experienced student can still use the book as a general resource.

A very important point for instructors to note is the numbered head system used in this book. As for all eight previous editions, *BBOM 9/e* is constructed in a modular fashion, using numbered heads to group off major topics within each chapter. Instructors should take advantage of this system by assigning to their students as much or as little (depending on the depth and breadth desired) of the material of a given chapter as necessary. Instructors know that most textbooks can not (and should not) be covered in a single semester course, and *BBOM 9/e* is no exception. While some chapters of this book should be covered in their entirety in the introductory course, others need only be sampled. The numbered head system makes sampling convenient for instructors and at the same time partitions concepts into more digestible portions for students.

Highlights of the revision include: (1) a more streamlined chapter on cell chemistry (now entitled "Macromolecules") to put the focus on the structure of the cells' major molecules; (2) a major new section on prokaryotic genomics (Chapter 9) to reflect the great excitement and advances occurring in this area today; (3) a major reorganization of the material on prokaryotes (Chapters 13 and 14) along phylogenetic lines to better integrate the evolutionary material in Chapter 12 with the properties of the organisms themselves; (4) extensive coverage of the exciting areas of phylogenetic probes and other fluorescent probes and their use in clinical medicine and microbial ecology (Chapters 16 and 21); (5) an expanded treatment of eukaryotic microorganisms and eukaryotic cell structure (Chapter 17); (6) a heavily reorganized immunology chapter which, following the pattern of the successful "nucleic acids" box in Chapter 6, separates out the molecular details of immunology into a box ("Molecular Biology of the Immune Response") for those instructors that wish to cover the field at the molecular level, while retaining the basic concepts of immunology in the text for those who wish to take a more practical approach to teaching this important area; and (7) an exciting description of the recent emergence of hantavirus pulmonary syndrome (in the *new* Chapter 24).

In summary, long-time users of this book will find a number of old friends between the covers and will meet a number of new ones as they go along. New users of this book will likely find it to be a refreshing, modern approach to teaching the science of microbiology. Either way, the book itself, coupled with the helpful pedagogical and instructional aids that make up the complete package (see below), should make the ninth edition of *BBOM* the most "teachable" for the instructor and the most useful to the student, of them all.

Pedagogical Aids

Previous users of this text will quickly notice the entirely new art program in *BBOM 9/e*. Every piece of art has been re-done, with special care to introduce more three dimensionality where pedagogically useful, and with a better eye for color usage. As before, color is a learning tool in *BBOM 9/e* and instructors will find that students quickly grasp the significance of color for reinforcing important concepts.

High quality photos and photomicrographs have been a mainstay of *Biology of Microorganisms* since the first edition appeared in 1970. *BBOM 9/e* is no exception and contains over 75 *new* color and black and white photos, most of which, like in previous editions, were supplied by top researchers in the field. Indeed, the

combination of a more pleasing art style and superb photos should be well received by today's "visual" generation of students.

As has become a tradition with this book, *BBOM 9/e* incorporates several study aids into the body of each chapter. *Concept links* (signaled by the blue "chain link" icon) are the ties between what has just been read and related material found elsewhere in the book; *Concept Checks* (found at the end of each numbered head) are a brief overview of the previous material followed by several short questions designed to ensure understanding of critical concepts; and the *Working Glossary* (that begins each chapter) is a dictionary of the essential terms to be encountered therein. With regular use of these study aids students will (1) know where to go within the book to read more about a particular topic; (2) instantly know whether they have mastered a given concept; and (3) have a fingertip reference to the key terminology. In addition to these study aids built into each chapter, *BBOM 9/e* contains an extensive list of end of chapter study questions designed to both recall important concepts and apply them to solving problems.

This new edition of *Brock Biology of Microorganisms* comes complete with a companion Website (www.prenhall.com/brock). On this site you will find supplementary materials for each chapter, an advanced readings list, and a new learning aid introduced with the ninth edition, the "Testing Center." Instructors are well aware that students nowadays like to prepare for upcoming examinations by taking "practice exams." In the *BBOM 9/e* Testing Center students can do this by taking a "Virtual Exam" for each unit of material, testing their preparedness with thousands of on-line questions designed to help them succeed. The questions themselves have been taken from authentic examinations given in introductory microbiology courses taught at universities throughout the United States that use *BBOM* as their textbook resource. The instant feedback available on these exams will allow students to assess just how ready they are for exam day in their own classes.

Supplements

A number of supplements for instructor use accompany the ninth edition. A set of 275 full color transparencies, far more than is available with any other textbook of microbiology, accompany every adoption. Although classroom instruction is going more and more toward computer-generated presentation, the transparency is still the visual aid workhorse in many courses in microbiology; in recognition of this, the publishers are pleased to include 25 more in this edition than its predecessor. But for those instructors whose classrooms are wired for the use of laptop computers, we offer a CD-ROM containing virtually all of the art, photos, and tables from *BBOM 9/e*. This CD, driven by a new version of the Prentice Hall "Presentation Manager" software system, gives instructors instant access to almost any visual aid in the book and allows them to be organized to fit an instructor's particular needs.

Students were very receptive to the free Website that accompanied the eighth edition of *BBOM*. Based on their comments and other reviews, we have revised the Website to be even more relevant and helpful. Along with the "Testing Center" mentioned previously, the quiz section of the Website allows students to take self-graded quizzes to test their knowledge, but the number of questions has been expanded and now contains much of the content formerly in the Study Guide. Finally a new feature called "Advances" encourages students to look ahead to the future in microbiology and to research current events using Weblinks we provide.

Between the book itself, the instructors' supplements, and the enhanced Website, we feel we offer an authoritative yet readable account of microbiology as we know it in the twenty-first century and have made available the means to teach it in a first-class way; indeed, *BBOM 9/e* is a winning package for both students and instructors alike.

Acknowledgments

This revision is not a product of just the authors. Many college and university instructors of microbiology gave valuable input through reviews of an earlier draft or particular chapters (or sections within a chapter) in the draft, while others made special efforts to provide color photographs taken directly from their laboratory research. We are extremely grateful for their efforts and list them below. Errors and omissions in the text are, of course, the responsibility of the authors, and we would greatly appreciate receiving comments, suggestions, and corrections.

Charles Abella, University of Girona, Spain
Laurie Achenbach, Southern Illinois University
John H. Andrews, University of Wisconsin
Jeanette A. Baker, Archer Daniels Midland Co.
Linda Barnett, University of East Anglia
Carl E. Bauer, Indiana University
Dennis A. Bazylinski, Iowa State University
Mary Bateson, Montana State University
Carl A. Batt, Cornell University
Sharisa Beek, Southern Illinois University
John A. Breznak, Michigan State University
Cheryl Broadie, Southern Illinois University
Ron Browning, Southern Illinois University
Clare Bunce, Cold Spring Harbor Laboratory
Craig Cary, University of Delaware
Richard W. Castenholz, University of Oregon
Feng Chen, University of Georgia

Thomas Christianson, Southern Illinois University
David Clark, Southern Illinois University
Rita Colwell, National Science Foundation
Morris Cooper, Southern Illinois University
Stephen Cooper, University of Michigan
Christian Jeanthon, Centre National de la Recherche Scientifique, Roscoff, France
Leanne Constantine, Affymetrix GeneChip®
Roy Curtiss III, Washington University
Philip R. Cunningham, Wayne State University
Shiladitya DasSarma, University of Massachusetts
Edward Delong, Monterey Bay Aquarium Research Institute
Ravin Donald, Northern Arizona University
Nicole Eis, University of Regensburg, Germany
Imre Friedmann, Florida State University
Brian J. Ford, Rothay House, Cambridgeshire, England
George M. Garrity, Michigan State University
William C. Ghiorse, Cornell University
Kodzo Gbewonyo, Merck and Co., Inc.
John Haddock, Southern Illinois University
Mary Hall, Southern Illinois University
Gary Heil, University of Iowa
Hans Hippe, Deutsche Sammlung von Mikroorganismen und Zellkulturen, Braunschweig, Germany
Robert E. Hodson, University of Georgia
David A. Hopwood, John Innes Centre, Norwich, England
Johannes Imhoff, Universität Kiel, Germany
Edward E. Ishiguro, University of Victoria
Timothy C. Johnston, Murray State University
Brian Jones, Genencor International B. V., Leiden, The Netherlands
Jason A. Kahana, Harvard Medical School
Suzanne V. Kelly, Scottsdale Community College
Susan Koval, University of Western Ontario
Robert G. Kranz, Washington University
Brian Lanoil, Oregon State University
Martine Legrand, Dunod Editeur, Paris
Le Ma, Harvard University
John C. Makemson, Florida International University
Renee D. Mastrocco, Rockefeller University Archives
Michael McGlananan, Florida International University
Diane McGovern, Roche Molecular Systems
Ortwin Meyer, Universität Bayreuth, Germany
Cindy Morris, INRA, Station de Pathologie Vegétale, Montfavet, France
Edward Moticka, Southern Illinois University
Miklós Müller, Rockefeller University
Dianne K. Newman, Princeton University
Aharon Oren, Hebrew University, Jerusalem

Jörg Overmann, University of Oldenburg, Germany
Norman R. Pace, University of Colorado
Cathy Patton, Molecular Probes, Inc., Eugene, OR
Norbert Pfennig, University of Konstanz, Germany
Hans Paerl, University of North Carolina
Suzette Pereira, Ohio State University
Sue Pollicott, Finnfeeds International, England
Kirsten Price, Harvard University
Reinhard Rachel, University of Regensburg, Germany
D. J. Read, University of Sheffield, England
John N. Reeve, Ohio State University
Hans Reichenbach, Gesellschaft für Biotechnologie, Braunschweig, Germany
Gary P. Roberts, University of Wisconsin
Lesley Robertson, Technical University of Delft, The Netherlands
Bernhard Schink, University of Konstanz, Germany
James Shapiro, University of Chicago
Helen Shio, Rockefeller University
Pamela A. Silver, Harvard Medical School
Jolynn F. Smith, Southern Illinois University
Jiri Snaidr, University of München, Germany
Barton Spear, Spear Studio, Erie, PA
Stefan Spring, Universität München, Germany
Gary Stacy, University of Tennessee
David A. Stahl, Northwestern University
Karl O. Stetter, University of Regensburg, Germany
Philip S. Stewart, Montana State University
Karthikeyan Subramanian, University of Saskatchewan
Hideto Takami, Japan Marine Science and Technology Center, Kanagawa
Nancy Trun, National Institutes of Health
David M. Ward, Montana State University

Kounosuke Watabe, Southern Illinois University
Fritz Widdel, Max Planck Institute of Marine Microbiology, Germany
Carl R. Woese, University of Illinois
John Vercillo, Southern Illinois University
Stephen H. Zinder, Cornell University

In addition to those listed above, the photo credits aside each photo in this book lists those who provided the photograph. Special thanks go to Lesley Robertson and the Kluyver Laboratory Museum, Technical University of Delft, The Netherlands, for sharing with us the wonderful illustrative materials about the scientific life of Martinus Beijerinck (see for example, the front and back cover).

Finally, the authors are grateful to all the people at Prentice Hall who have contributed in significant ways to this edition, including in particular Linda Schreiber and David K. Brake (editorial), and Debra Wechsler (production). It was through Linda's strong efforts that the art program was revamped, and the authors are very grateful for her foresight in this regard. Debra deserves high praise for her professionalism in all aspects of this book and was simply a pleasure to work with—the appearance of the final product owes much to Debra's efforts. We also wish to acknowledge the contributions of Jane Loftus (Clackamas, OR) to copyediting, Robie Grant (Hadley, MA), who composed the index, and Toni Huppert, Southern Illinois University, for her expert word processing skills.

Michael T. Madigan (madigan@micro.siu.edu):
Author of Chapters 1–5 and 11–17, and
BBOM 9/e coordinator
John M. Martinko (martinko@micro.siu.edu)
Author of Chapters 18–24
Jack Parker (parker@cos.siu.edu)
Author of Chapters 6–10

The science of microbiology grew out of the key contributions of a handful of "giants" including van Leeuwenhoek, Pasteur, Koch, Beijerinck, and Winogradsky. These early microbiologists were blessed with a keen sense of observation that allowed them to discover, often using rather simple equipment, the major principles of microbiology and many of the major groups of microorganisms. For example, the Russian microbiologist Sergei Winogradsky studied photosynthetic sulfur bacteria in the late 1880s, an era before modern photomicroscopy. Instead of capturing his images on film or digitally on computer disk, Winogradsky saved his images on paper with figures drawn by hand as shown here. However, these drawings were remarkably accurate as later photomicrographs of the same organisms have shown.

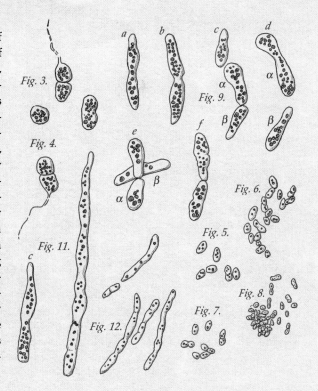

CHAPTER **1** # Microorganisms and Microbiology

1.1   Microbiology and You   2
1.2   Microorganisms as Cells   4
1.3   Elements of Cell and Viral Structure   6
1.4   Evolutionary Relationships Among Living Organisms   8
1.5   Microbial Populations, Communities, and Ecosystems   11
1.6   Laboratory Culture of Microorganisms   12
1.7   The Impact of Microorganisms on Human Affairs   15
1.8   The Historical Roots of Microbiology: van Leeuwenhoek, Pasteur, and Koch   17
1.9   Transition into the Twentieth Century   24

Archaea a group of phylogenetically related prokaryotes distinct from Bacteria

Aseptic technique methods for maintaining sterile culture media and other sterile objects free from microbial contamination during manipulations

Bacteria a group of phylogenetically related prokaryotes distinct from Archaea

Cell the fundamental unit of living matter

Chromosome a genetic element carrying genes essential to cell function

Cytoplasm the fluid portion of a cell, bounded by the cell membrane but excluding the nucleus (if present)

DNA deoxyribonucleic acid, the hereditary material of cells and some viruses

Ecology the study of organisms in their natural environments

Enrichment culture a method for isolating microorganisms from nature using specific culture media and incubation conditions

Entropy a measure of the degree of disorder in a system; entropy always increases in a closed system

Enzymes protein catalysts that function to speed up chemical reactions

Eukarya all eukaryotic organisms

Eukaryote a cell possessing a membrane-enclosed nucleus and usually other organelles

Evolution change in a line of descent over time leading to the production of new species or varieties within a species

Genetics heredity and variation of living organisms

Metabolism all biochemical reactions in a cell

Microorganism a microscopic organism consisting of a single cell or cell cluster, including the viruses

Pathogen a disease-causing microorganism

Plasmid a small genetic element that exists separately from the chromosome

Prokaryote a cell lacking a nucleus and other organelles

Pure culture a culture containing a single kind of microorganism

RNA ribonucleic acid, involved in protein synthesis as messenger RNA, transfer RNA, and ribosomal RNA

Spontaneous generation the hypothesis that living organisms can originate from nonliving matter

Sterile absence of all living organisms and viruses

Microbiology is the study of microorganisms, a large and diverse group of microscopic organisms that exist as single cells or cell clusters; it also includes viruses, which are microscopic but not cellular. Microbial cells are thus distinct from the cells of animals and plants, which are unable to live alone in nature and can exist only as parts of multicellular organisms (Figure 1.1). In contrast to macroorganisms, microorganisms are generally able to carry out their life processes of growth, energy generation, and reproduction independently of other cells, either of the same kind or of a different kind.

1.1

Microbiology and You

Microbiology is about living cells and how they work. It is about microorganisms, especially about bacteria, a large group of cells of enormous basic and practical significance. It is about microbial diversity and evolution, about how different kinds of microorganisms arose and why. It is about what microorganisms do in the world at large, in human society, in our own bodies, and in the bodies of animals and plants. What does this all have to do with you? Let's look at who studies microbiology and why you should know some microbiology.

Who Studies Microbiology and Why?

Microbiologists study microbiology for two major reasons:

1. As a *basic*, biological science, microbiology provides some of the most accessible research tools for probing the nature of life processes. Our most sophisticated understanding of the chemical and physical basis of life has arisen from studies of microorganisms. This is in large part because microbial cells share many biochemical properties with cells of multicellular organisms; indeed, *all* cells have much in common. And this, coupled with the fact that microbial cells can grow to extremely high densities in laboratory culture and are readily amenable to biochemical and genetic study, makes them excellent models for understanding cell function in higher organisms.

2. As an *applied* biological science, microbiology deals with many important practical problems in medicine, agriculture, and industry. Some of the most important diseases of humans, other animals, and plants are caused by microorganisms. Microorganisms also play major roles in soil fertility and domestic animal production. In addition, many large-scale industrial processes, such as the production of antibiotics, are microbially based. And the combination of genetic engineering and industrial microbiology has even spawned a whole new discipline, *biotechnology*.

FIGURE 1.1 Living organisms are composed of cells. (a) Plants and (b) animals are composed of many cells; they are thus *multicellular*. A single plant or animal cell cannot have an independent existence; each of its cells is dependent on the other. (c, d) Microorganisms are free-living cells. A single microbial cell can have an independent existence. Shown are photomicrographs of photosynthetic microorganisms called (c) *purple bacteria*, and (d) *cyanobacteria*. Cyanobacteria were the first O_2-evolving phototrophs on Earth and were responsible for oxygenating the atmosphere.

In this book, both the basic and applied aspects of microbiology are covered. We discuss the experimental basis of microbiology, the general principles of cell structure and function, the classification and diversity of microorganisms, biochemical processes in cells, the genetic basis of microbial growth and evolution, and the ecological activities of microorganisms in nature. From an applied viewpoint we discuss microbial diseases, the nature of the immune response, the roles of microorganisms in food and agriculture, and industrial and biotechnological processes employing microorganisms.

Why Should *You* Study Microbiology?

Not everyone that studies microbiology aspires to be a microbiologist. But everyone should know some "microbial facts," and you will learn these along with their supporting concepts as you continue through this book. For example, you will come to understand the central

role microorganisms play in your life as well as the whole web of life on Earth; indeed, without microorganisms, all higher life forms on Earth would cease to exist. Humans, animals, and plants are intimately tied to microbial activities for the recycling of key nutrients and for degrading organic matter; even the oxygen we breathe is the result of microbial activity (see Figure 1.1d). You will also learn that although individual microorganisms are very small, collectively their metabolic power is very great, and that the sum total of their protoplasm constitutes the greatest source of biomass on Earth.

If this is not enough, consider the following. Your study of microbiology will teach you that contrary to popular belief, only a very limited number of microorganisms are pathogens, that is, capable of causing disease. But of course some of these pathogens are so powerful that their activities can lead to the death of plants and animals. You will learn that microorganisms do not exist as individuals in nature, but instead live with other organisms in microbial *communities,* and that their small size and rapid growth, coupled with their ready ability to exchange genes, allow them to adapt rapidly to changing environmental conditions. You will also learn how microorganisms were the first life forms on Earth and that all living organisms, including humans, share an evolutionary link to the microbial world. Finally, you will learn that microorganisms show the greatest genetic and physiological diversity of all living organisms; indeed, microorganisms, especially the bacteria, are Earth's greatest chemists.

Simply put, you should know some microbiology because microorganisms are a key part of your life. No other group of organisms approaches the importance of microorganisms in supporting and maintaining life on Earth. Microorganisms not only prepared the planet for higher life forms (see Figure 1.1), they are the very reason that higher life forms can exist today. If we as humans are to survive on planet Earth, we clearly need to know everything we can about microbial life.

We begin our journey in microbiology now with a consideration of the cellular nature of microorganisms.

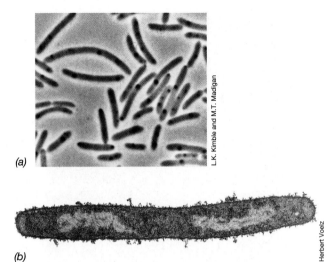

(a)

(b)

FIGURE 1.2   Cells. (a) Photomicrograph of rod-shaped bacterial cells as seen in the light microscope; a single cell is about 1 μm in diameter. (b) Longitudinal section through a bacterial cell as viewed with an electron microscope. The two lighter areas represent the nucleoid, regions in the cell containing aggregated DNA.

1.2

Microorganisms as Cells

The **cell** is the fundamental unit of all living matter. A single cell is an entity, isolated from other cells by a cell membrane (and perhaps a cell wall) and containing within it a variety of chemicals and subcellular structures (Figure 1.2). The **cell membrane** is the barrier that separates the inside of the cell from the outside. Inside the cell membrane are the various structures and chemicals that make it possible for the cell to function. Key structures are the **nucleus** or **nucleoid,** where the *genetic information,* DNA, needed to make more cells is stored, and the cytoplasm, where the *machinery* for cell growth and function is present.

All cells contain certain types of chemical components: **proteins, nucleic acids, lipids,** and **polysaccharides.** Collectively, these are called *macromolecules.* It is the precise chemical nature and arrangement of macromolecules in a cell of one organism that makes it distinct from those of another. Although each kind of cell has a definite structure and size, a cell is a dynamic unit, constantly undergoing change and replacing its parts. Even when it is not growing, a cell may be taking materials from its environment and working them into its own fabric. At the same time, it discards waste products into its environment. A cell is thus an *open system,* forever changing yet generally remaining the same.

Where did the first cells come from? In some way the first cell must have come from a noncell, something before

the cell, a procellular structure. Although formation of the first cell over 3.5 billion years ago was an improbable event that may have taken several hundred million years to occur, once the first cell arose, a series of highly probable events followed, including growth and division to form populations of cells from which evolution could begin to select for improvements and diversification. Then, through billions of years of evolutionary change, the tremendous diversity of cell types that exist today arose. And, because the cells of all living organisms today are constructed from the four classes of macromolecules mentioned earlier and share many other things in common, it is hypothesized that all cells have descended from a common ancestor, the *universal ancestor* of all life.

Characteristics of Living Systems

What are the essential characteristics of life? What differentiates cells from inanimate objects? Our concept of what is "alive" is constrained by what we can observe on Earth today or can deduce from the fossil record. But from our knowledge of biology thus far, we can identify several characteristics shared by most living systems, and these are summarized in Figure 1.3.

All living organisms show some form of **metabolism.** That is, cells take up chemical substances from their environment and transform them, conserving some of the energy of these substances in a form the cell can use, and then eliminate waste products. All cells show **reproduction,** that is, they are capable of directing the series of

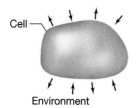

1. Metabolism
Uptake of chemicals from the environment, their transformation within the cell, and elimination of wastes into the environment. The cell is thus an *open* system.

2. Reproduction (growth)
Chemicals from the environment are turned into new cells under the direction of preexisting cells.

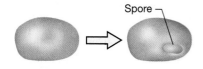

3. Differentiation
Formation of a new cell structure such as a spore, usually as part of a cellular *life cycle*.

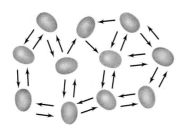

4. Communication
Cells *communicate* or *interact* primarily by means of chemicals that are released or taken up.

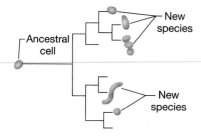

5. Evolution
Cells *evolve* to display new biological properties. Phylogenetic trees show the evolutionary relationships between cells.

FIGURE 1.3   The hallmarks of cellular life.

biochemical events that results in their own synthesis. As a result of metabolic processes a cell grows and divides to form two cells. Many cells undergo **differentiation,** a process by which new substances or structures are formed. Cell differentiation is often part of a cellular life cycle in which cells form special structures involved in reproduction, dispersal, or survival. Cells respond to chemical signals in their environment, including those produced by other cells. Cells can thus undergo **communication** and even assess their own numbers in the surrounding environment by way of small diffusible molecules that pass freely into and out of cells. Finally, unlike nonliving structures, cells can *evolve.* Through the process of **evolution** cells can permanently change their characteristics and transmit these new properties to their offspring.

Although other hallmarks of living systems could probably be included in this list, these five properties: metabolism, reproduction, differentiation, chemical signaling, and evolution, are the major hallmarks of a cell (Figure 1.3).

Cells as Machines and as Coding Devices

Cells can be looked at in two ways. On one hand, cells can be considered *chemical machines* that carry out chemical transformations within the confines of a cellular structure. The catalysts of this chemical machine are **enzymes,** proteins capable of greatly accelerating the rate of specific chemical reactions. On the other hand, cells can be considered *coding devices,* analogous to computers, which serve as storehouses and processors of genetic information (DNA) that is eventually passed on to offspring during reproduction (Figure 1.4). Replication of the stored genetic information and its processing will be considered in Chapter 6 where we cover the core cellular functions of *DNA replication, transcription,* and *translation* in detail.

In reality, cells are both chemical machines *and* coding devices, and the link between these two attributes of a cell is *growth.* Under proper conditions, a viable cell will grow larger and then divide to form two cells (Figure 1.4). In the orderly process that results in cell division, the amount of all the constituents of the cell must *double.* This requires the chemical machinery of the cell function to supply energy and precursors for biosynthesis of macromolecules. But when a cell divides, each of the two resulting cells must contain all of the genetic information necessary for the formation of more cells, and so during the growth process there must also be a replication of DNA (Figure 1.4). The machine and coding functions of the cell must thus be highly coordinated in order for the cell to reproduce itself faithfully. We will see later that this is indeed the case and that in addition to *coordination,* the various machine and coding functions of the cell are subject to *regulation,* such that the cell stays optimally attuned to its environment.

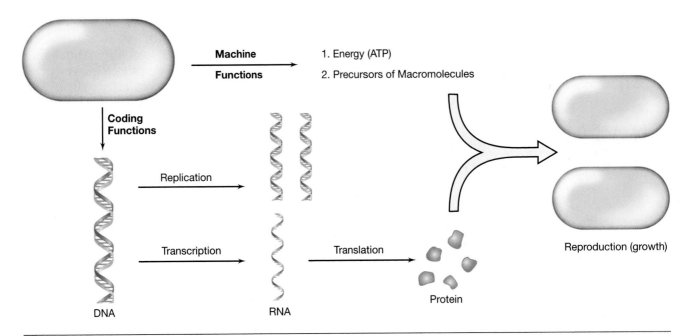

FIGURE 1.4   The machine and coding functions of the cell. In order for a cell to reproduce itself there must be an adequate supply of energy and precursors for the synthesis of new macromolecules, the genetic instructions must be replicated such that upon division each cell receives a copy, and genes must be expressed (the processes of transcription and translation) to form the proper amounts of necessary proteins and other macromolecules that will make up the new cell.

✓ 1.2 Concept Check

The cell has a barrier, the cytoplasmic membrane, that separates the cytoplasm from the environment. Other cell features include the nucleus or nucleoid, cell wall, and cytoplasm. Certain key features such as metabolism and reproduction are associated with the living state, and cells can be thought of conceptually as both chemical machines and as biological coding devices.

- ✓ List the five features associated with living organisms. Why is each feature important to the survival of a cell?
- ✓ Compare the machine and coding functions of a microbial cell. Why is neither of value to a cell without the other?
- ✓ Name the four classes of cellular macromolecules.

1.3

Elements of Cell and Viral Structure

What is the structure of a cell? All cells have a barrier separating inside from outside that is called the **cytoplasmic (cell) membrane** (Figure 1.5). It is through the cell membrane that all nutrients and other substances of vital importance to the cell pass in, and it is through this same membrane that waste materials and other cell products pass out. Within a cell, and bounded by the cytoplasmic membrane, is a complicated mixture of substances and structures called the **cytoplasm.** These materials and structures, bathed in water, carry out the functions of the cell. The major components of the cytoplasm, other than water, include macromolecules, ribosomes, small organic molecules (mainly precursors

of macromolecules), and various inorganic ions. **Ribosomes,** the cell's protein-synthesizing factories, are key cytoplasmic components and interact with several proteins and RNA in the process of protein synthesis.

The cell gains structural strength from the **cell wall.** The cell wall is located outside the membrane (Figure 1.5) and is a much stronger layer than the cell membrane. Plant cells and most microorganisms have cell walls, generally quite rigid cell walls. Animal cells, however, do not have walls; these cells have developed other means of support and protection.

Prokaryotic and Eukaryotic Cells

Upon careful study of the internal structure of cells, two basic types have been recognized: **prokaryote** and **eukaryote** (Figure 1.5). A major feature of eukaryote cells, absent from prokaryotic cells, is the presence in the cytoplasm of membrane-enclosed structures called *organelles.* These include the membrane-bound nucleus, of course, but also structures such as the mitochondria and chloroplasts (the latter in photosynthetic cells only) (Figure 1.5). Eukaryotic cells also have an interior cytoskeleton, a series of scaffolding structures that physically support the cell and help organize and move its internal components. By contrast, prokaryotic cells usually have a quite simple internal structure, lacking membrane-bound organelles (Figure 1.5). Prokaryotes are represented by the **Bacteria** and the **Archaea** while eukaryotes include the **algae, fungi,** and **protozoa.** In addition, vir-

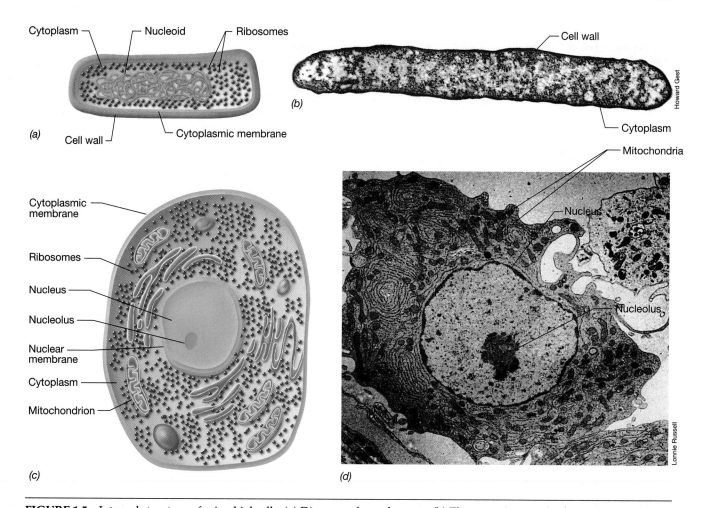

FIGURE 1.5 Internal structure of microbial cells. (a) Diagram of a prokaryote. (b) Electron micrograph of a prokaryote. The cell is about 1 μm in diameter and the light-colored areas in the cell are DNA (the nucleoid). (c) Diagram of a eukaryote. (d) Electron micrograph of a eukaryote (animal cell). The cell is about 25 μm in diameter.

tually all multicellular life forms (for example, the plants and animals) are constructed of eukaryotic cells. We consider eukaryotic cells in more detail in Chapter 17.

Microorganisms in general are very small. A rod-shaped prokaryote is typically about 1–5 micrometers (μm) long and about 1 μm wide and thus is completely invisible to the naked eye. To illustrate how small a bacterium is, consider that 500 bacteria 1 μm long could be placed end to end across the period at the end of this sentence. Eukaryotes are typically much larger than prokaryotes but the range can vary dramatically. We revisit the subject of cell size in more detail early in Chapter 3.

Arrangement of DNA in Prokaryotic and Eukaryotic Cells and the Genetic Potential of the Cell

In prokaryotic cells DNA is present in a large double-stranded molecule called the **bacterial chromosome** and aggregates to form a visible mass in the electron microscope called the *nucleoid* (Figure 1.5a and b). We will see in later chapters that this DNA is circular in most prokaryotes and that most prokaryotes contain only a *single* chromosome. Moreover, prokaryotes contain only a single copy of each gene (there are a few exceptions) on their chromosome and are therefore genetically *haploid*. Most prokaryotes also contain small amounts of extrachromosomal DNA, again usually arranged in circular fashion, called **plasmids.** Plasmids generally contain genes that confer special properties on a cell as opposed to essential ("housekeeping") genes, which are needed for basic survival and are located on the chromosome.

In eukaryotes, DNA is present within the nucleus in linear molecules, arranged to form several **chromosomes** (the baker's yeast *Saccharomyces cerevisiae*, for example, contains 16 chromosomes). The latter contain more than just DNA; they include special proteins that assist in folding and packing the DNA and other proteins that are required for gene expression. A key genetic difference between prokaryotes and eukaryotes is that eukaryotes typically contain *two copies* of each gene (again there are some exceptions) and are thus genetically *diploid*. During cell division in eukaryotic cells the nucleus divides following a doubling of chromosome number in the well-known process of **mitosis** (Figure 1.6). The normal diploid state in eukaryotic cells is halved in the process of **meiosis** to form haploid gametes for sexual reproduction. Fusion of two gametes during zygote formation then restores the diploid state. We discuss these processes in more detail in Chapter 9.

How many genes and proteins does a cell have? *Escherichia coli*, a fairly typical bacterial cell, has a chromosome of about 4.6 million base pairs of DNA (4600 kilobases) (∞ Section 9.11). Because the *E. coli* chromosome has been completely sequenced, we know that it contains almost 4300 potential coding sequences, or *genes*. Eukaryotic cells generally have many more genes than this; a human cell, for example, contains over 1000 times the DNA of *E. coli* and over 10 times as many genes. A single cell of *E. coli* contains about 1900 different kinds of proteins and a total of about 2.4 million individual protein molecules. However, some proteins in the cell are present in very high copy number, others at moderate copy number, and some in only very low copy number. Thus, the cell has mechanisms for controlling the *expression* (transcription and translation, see Figure 1.4) of its genes so that not all genes are expressed to the same extent or at the same time; we focus on these mechanisms of gene expression in Chapter 7.

Viruses

Viruses are not cells. Viruses lack many of the attributes of cells, the most important of which is that they are not dynamic open systems. A single virus particle is a static structure, quite stable, and unable to change or replace its parts (Figure 1.7). Only when it is associated with a cell does a virus acquire a key attribute of a living system, reproduction (see Figure 1.3). However, unlike cells viruses have no metabolic abilities of their own. And although they possess genetic information (*either* DNA or RNA but not both), viruses lack ribosomes and therefore depend on the cell's biosynthetic machinery for protein synthesis.

Viruses are known to infect virtually all cells, including microbial cells. Many viruses cause disease in the organisms they infect, but virus infection does not always lead to disease. We will see in Chapter 8 that besides causing disease, viruses can have other profound effects on cells, including genetic alterations that actually improve the capabilities of the cell to carry out some function or survive under some particular set of conditions. Viruses are much smaller than cells, even much smaller than prokaryotic cells, and a relative feel for the sizes of cells and viruses can be obtained from examination of Figure 1.7.

✓ 1.3 Concept Check

The architecture of a cell can be considered either prokaryotic or eukaryotic, but despite significant structural differences, all cells have many things in common. Viruses, by contrast, are not cells, but show certain features of cellular organisms. Chromosomes are the structures containing the genes necessary for survival.

- ✓ What are the major structural differences between prokaryotic and eukaryotic cells?
- ✓ What function do *ribosomes* play in the cell?
- ✓ How does the bacterial chromosome differ from chromosomes in eukaryotic cells?
- ✓ Do viruses contain nucleic acid? If so, what kinds?

1.4

Evolutionary Relationships Among Living Organisms

Although cells can be differentiated structurally as being prokaryotic or eukaryotic, is structure a good predictor of evolutionary (phylogenetic) relationship? Yes and no. As we will see, prokaryotic and eukaryotic cells are phylogenetically distinct, but not all prokaryotic cells are closely related. Phylogenetic relation-

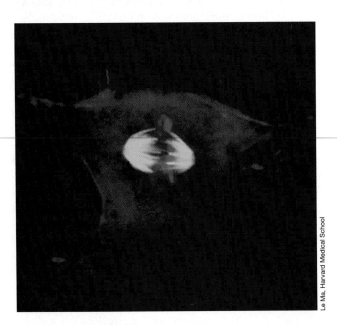

FIGURE 1.6 Mitosis in stained kangaroo rat cells. The cell was photographed while in the *metaphase* stage of mitotic division. The green color stains a protein called *tubulin*, important in pulling chromosomes apart. The blue color is from a DNA-binding dye and shows the chromosomes. Although an integral part of the cell cycle of eukaryotic cells, mitosis does not occur in prokaryotic cells.

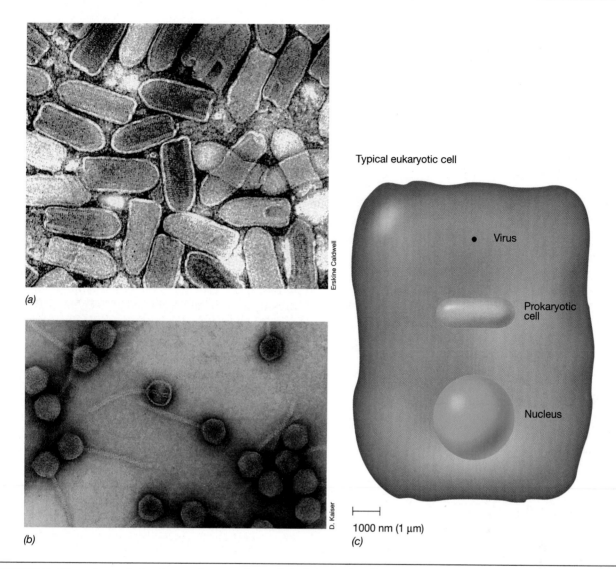

FIGURE 1.7   Virus size and structure and comparison of viral size with cell size. (a) Rhabdovirus particles. A single virus particle is about 65 nm in diameter. (b) Bacterial virus (bacteriophage) lambda. The head of each particle is about 65 nm in diameter. (c) The size of the viruses shown in (a) and (b) in comparison to a bacterial and eukaryotic cell.

ships between microorganisms can be determined using comparative gene-sequencing methods. A particularly useful gene to sequence in this connection has been the gene encoding 16S or 18S (S refers to size) ribosomal RNA—structural RNA of the ribosome, the key cell structure involved in protein synthesis. Ribosomal RNA gene phylogenies have revealed the major evolutionary relationship between microbial groups and can be used to generate phylogenetic trees that summarize the evolutionary connections among species of microorganisms (Figure 1.8b).

From comparative ribosomal RNA gene sequences three phylogenetically distinct cellular lineages have been identified, two of which are prokaryotic in cell structure and one eukaryotic. The groups are called the **Bacteria,** the **Archaea,** and the **Eukarya** (Figure 1.8a). All three groups are thought to have diverged from a common ancestral organism, the "universal ancestor" mentioned previously, early in the history of life on Earth. Because the cells of higher animals and plants are all eukaryotic, it follows that eukaryotic microorganisms were the ancestors of multicellular organisms, while representatives of the Bacteria and Archaea never evolved past the microbial stage (Figures 1.8 and 1.9a). The organelles of eukaryotic cells, the mitochondria and chloroplast, became part of the eukaryotic cell eons ago in a process called *endosymbiosis* (◠◡ Sections 3.16 and 12.3).

With molecular tools like ribosomal RNA sequencing, microbiologists can begin to understand the natural evolutionary relationships among microorganisms and between microorganisms and higher organisms; we discuss these ideas in more detail in Chapter 12.

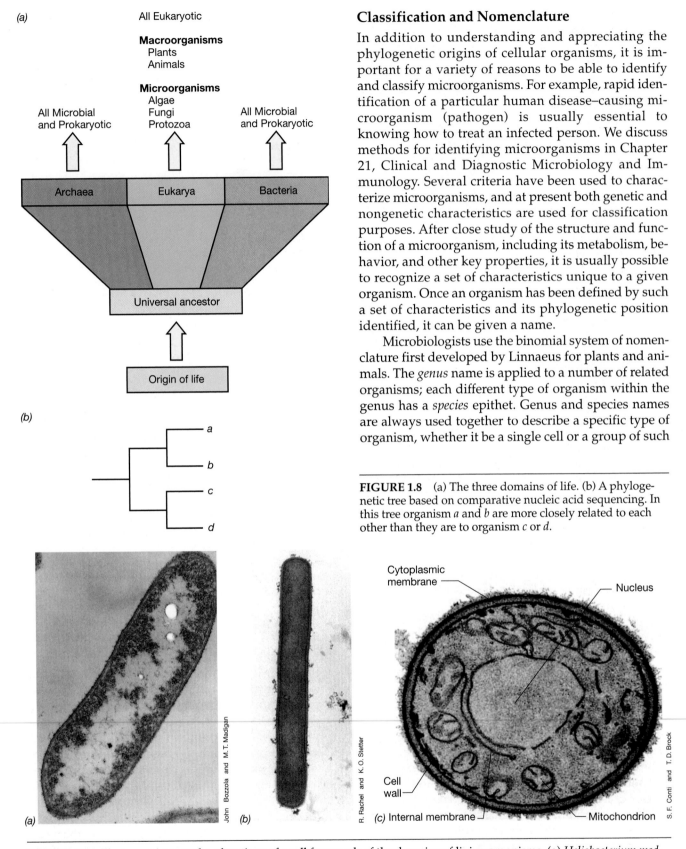

(a)

All Eukaryotic

Macroorganisms
Plants
Animals

Microorganisms
Algae
Fungi
Protozoa

All Microbial
and Prokaryotic

All Microbial
and Prokaryotic

| Archaea | Eukarya | Bacteria |

Universal ancestor

Origin of life

(b)

a

b

c

d

Classification and Nomenclature

In addition to understanding and appreciating the phylogenetic origins of cellular organisms, it is important for a variety of reasons to be able to identify and classify microorganisms. For example, rapid identification of a particular human disease–causing microorganism (pathogen) is usually essential to knowing how to treat an infected person. We discuss methods for identifying microorganisms in Chapter 21, Clinical and Diagnostic Microbiology and Immunology. Several criteria have been used to characterize microorganisms, and at present both genetic and nongenetic characteristics are used for classification purposes. After close study of the structure and function of a microorganism, including its metabolism, behavior, and other key properties, it is usually possible to recognize a set of characteristics unique to a given organism. Once an organism has been defined by such a set of characteristics and its phylogenetic position identified, it can be given a name.

Microbiologists use the binomial system of nomenclature first developed by Linnaeus for plants and animals. The *genus* name is applied to a number of related organisms; each different type of organism within the genus has a *species* epithet. Genus and species names are always used together to describe a specific type of organism, whether it be a single cell or a group of such

FIGURE 1.8   (a) The three domains of life. (b) A phylogenetic tree based on comparative nucleic acid sequencing. In this tree organism *a* and *b* are more closely related to each other than they are to organism *c* or *d*.

Cytoplasmic membrane

Nucleus

Cell wall

(a)

(b)

John Bozzola and M. T. Madigan

R. Rachel and K. O. Stetter

(c) Internal membrane

Mitochondrion

S. F. Conti and T. D. Brock

FIGURE 1.9   Electron micrographs of sections of a cell from each of the domains of living organisms. (a) *Heliobacterium modesticaldum* (Bacteria); the cell measures 1 × 3 μm. (b) *Methanopyrus kandleri* (Archaea); the cell measures 0.5 × 4 μm. (c) *Saccharomyces cerevisiae* (Eukarya); the cell measures about 8 μm in diameter.

cells. (In writing, the genus and species names are either underlined or printed in italics. For example, the bacterium *Escherichia coli*, or *E. coli* for short, has a genus name, *Escherichia*, and a species epithet, *coli*.)

Microbial Diversity

An understanding of microbial diversity requires an appreciation of the evolutionary roots of cells. Because evolution has shaped all life on Earth, the structural and functional diversity we see in cells represents successful evolutionary events that have conferred survival value (fitness) on the microorganisms extant today. Microbial diversity can be seen in terms of variations in cell size and shape (morphology), metabolic strategies, motility, cell division, developmental biology, adaptation to environmental extremes, and many other structural and functional aspects of the cell. We describe the major groups of Bacteria, Archaea, and Eukarya in Chapters 13, 14, and 17, respectively, and consider there many of the aspects that make microbial life so amazingly diverse. Here we give only a quick overview of things to come.

Several evolutionary branches occur within the **Bacteria,** including all known disease-causing (pathogenic) prokaryotes and most of the bacteria commonly found in soil, water, animal digestive tracts, and many other environments. Some of these organisms contain pigments that allow them to use light as an energy source in a process called *phototrophy,* others rely on organic chemicals as energy sources, and some can even use inorganic chemicals as fuel to drive cellular processes. Oxic environments (those containing O_2), as well as various anoxic habitats, are inhabited by various species of Bacteria.

With the prokaryotes called **Archaea,** in contrast, we see a much different picture. Most Archaea are anaerobes, cells incapable of living in air. Many also thrive under unusual growth conditions, inhabiting what humans would consider extreme environments: hot springs (to temperatures *above* the boiling point of water), freezing waters like Antarctic sea ice, extremely salty bodies of water, and highly acidic or alkaline soils and water. Indeed, certain species of Archaea currently define the limits of biological tolerance to physiochemical extremes. Certain Archaea also show unusual biochemical features, such as the methanogens, which are prokaryotes that produce methane (natural gas) as an integral part of their energy metabolism.

The microbial **Eukarya** include the algae, fungi, and protozoa. *Algae* contain chlorophyll, a pigment that serves as a light-gathering molecule, making it possible for them to carry out photosynthesis. Algae are common in aquatic habitats and can also be found in soil. *Fungi*—molds, yeasts, and mushrooms—lack chlorophyll and obtain their energy from organic compounds in soil and water. Fungi play a major role in the breakdown of dead organic mat-

ter in these and other environments. *Protozoa* are colorless, motile Eukarya that obtain food by ingesting other organisms or organic particles. Protozoa lack the cell walls of algae and fungi and in this respect resemble animal cells. Many protozoa are free-living microorganisms, but several cause disease in humans and other animals.

✓ 1.4 Concept Check

Living organisms can be divided into prokaryotes and eukaryotes based on cell structure, but from an evolutionary viewpoint, living organisms form *three* major groups: Bacteria, Archaea, and Eukarya. Structural and functional diversity among microorganisms is substantial.

- ✓ Which of the cells shown in Figure 1.9 are prokaryotic? Which are eukaryotic? What does cell structure say about evolutionary relationships?
- ✓ *Bacillus subtilis* is a common soil bacterium. What genus does *B. subtilis* belong to?

1.5

Microbial Populations, Communities, and Ecosystems

Up to now, we have been discussing cells as if they lived in isolation in their environments. Nothing could be further from the truth. Cells live in nature in association with other cells in assemblages that we call **populations.** Such populations are composed of groups of related cells, generally derived by successive cell divisions from a single parent cell. The location in an environment where a population lives is called the **habitat.** In nature, populations of cells rarely live alone. Rather, they live in association with other populations of cells in assemblages called **microbial communities** (Figure 1.10).

The Effect of Organisms on Each Other and on Their Habitats

Populations in microbial communities interact in various ways, and these interactions may be either harmful or beneficial. In many cases the populations interact and cooperate in their feeding efforts, with the waste products of the metabolic activities of some cells serving as the nutrients for others. Organisms in a habitat also interact with their physical and chemical environment. Habitats differ markedly in their characteristics, and a habitat that is favorable for the growth of one organism may actually be harmful for another organism. Thus, the makeup of the microbial community in a given habitat is determined to a great extent by the physical and chemical characteristics of that environment. Collectively, we speak of the living organisms together with the physical and chemical constituents of their environment as an **ecosystem.**

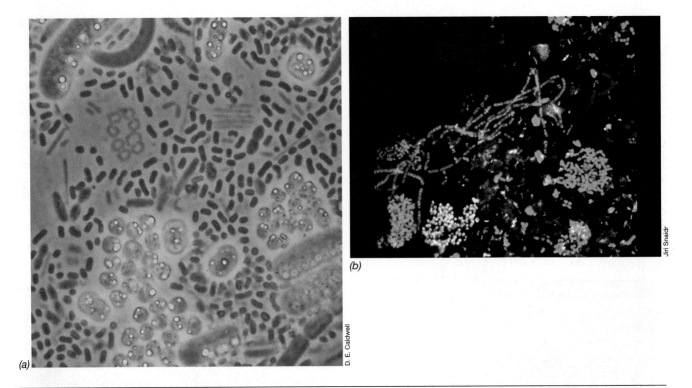

FIGURE 1.10   Examples of microbial communities. (a) Photomicrograph of a bacterial community that developed in the depths of a small lake (Wintergreen Lake, Michigan), showing cells of various sizes. (b) A bacterial community in a sewage sludge sample. The sample was stained with a series of dyes, each of which stained a different bacterial group (∞ Section 16.6 and Figure 16.11*b* for details of how the staining was done).

The properties of an ecosystem are often controlled to a significant extent by microbial activities. Organisms carrying out metabolic processes remove nutrients from the environment and use them to build new cells. At the same time, organisms excrete waste products of their metabolism into the environment. Therefore, over time, a microbial ecosystem can gradually change, both chemically and physically, through living processes. The presence or absence of oxygen is a good example of this. We will see later that molecular oxygen, O_2, is a vital nutrient for some microorganisms while being a poison to others. However, the oxygen-consuming activities of one group of organisms can make an oxic habitat anoxic and suitable for growth of organisms that were previously kept in check.

Since single microbial cells are too small to be seen with the unaided eye, our knowledge of microorganisms in nature begins with studies using the microscope. Examination of natural material such as soil or water always reveals microbial cells. Although such tiny cells may seem inconsequential, single cells are capable of multiplying rapidly and producing a massive number of progeny that may have a major effect on the habitat. Indeed, some of the most important and impressive attributes of microorganisms are their rapid growth and high population numbers. Thus, although microorganisms may seem to be minor components in nature, they are extremely important parts of virtually every ecosystem. In later chapters, after we have learned some of the details of microbial structure and function, genetics, evolution, and diversity, we will return to a discussion of the ways in which microorganisms affect animals, plants, and the whole global ecosystem.

✓ 1.5 Concept Check

Microorganisms exist in nature in populations that interact with other populations in microbial communities. The activities of microbial communities can greatly affect the chemical and physical properties of their habitats.

✓ What is a *microbial habitat?*
✓ How do microorganisms change the chemical and physical properties of their habitats?

1.6

Laboratory Culture of Microorganisms

Although we can get an idea of what microorganisms look like from a microscopic study of a natural habitat, we can best study their characteristics by obtaining them in a pure culture. A **pure culture** is a culture consisting of only one kind of microorganism. In order to obtain a pure culture, we must be able to grow the organism in the laboratory. This requires that we provide it with the proper nutrients and environmental conditions so that it can grow. It is also essential that we keep other organisms from entering the culture. Such unwanted organisms, called **contaminants,** are ubiquitous, and microbiological technique revolves around the avoidance of contaminants. Once we have isolated a pure culture, we can then proceed to study its biochemistry, physiology, genetics, and other characteristics.

Culture Media

What are the conditions required for microbial growth? Microorganisms are cultured in water to which appropriate nutrients have been added. The aqueous solution containing such necessary nutrients is called a **culture medium.** The nutrients present in the culture medium provide the microbial cell with the ingredients required

for the cell to produce more cells like itself. Besides a source of energy, which can be an organic or an inorganic chemical, or light, a culture medium must have a source of carbon, nitrogen, and several other necessary nutrients to be described in detail in Chapter 4. Culture media can be prepared for use in either a liquid state or a gel (semisolid) state. A liquid culture medium is converted to the semisolid state by the addition of a gelling agent, usually agar. Culture media containing agar are dispensed in flat, covered dishes called **Petri dishes,** where microbial cells can grow and form visible masses called *colonies* (Figure 1.11). The history of the development of solid culture media and its impact on microbiology are discussed later in the box, Solid Media and Development of the Petri Plate.

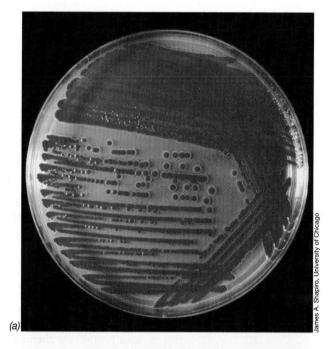
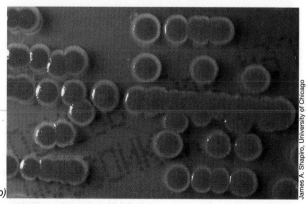

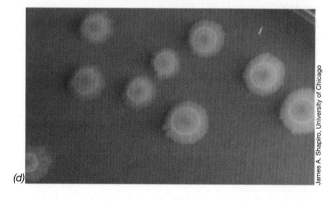

FIGURE 1.11 Examples of bacterial colonies. Colonies are visible masses of cells formed from the subsequent division of one or a few cells. The size, shape, texture, and color of a bacterial colony is a function of the organism that made them. Depending on its size and the arrangement of cells within the colony, a colony can have highly variable cell numbers. But colonies containing over one *billion* individual cells are not at all unusual. (a) *Serratia marcesens,* grown on MacConkey agar. (b) Close-up of colonies shown in (a). (c) *Pseudomonas aeruginosa,* grown on Trypticase-Soy agar. (d) *Shigella flexneri,* grown on MacConkey agar.

Aseptic Technique

Before actually proceeding to the culture of microorganisms, we must first consider how to exclude contaminants. Microorganisms are everywhere. Because of their small size, they are easily dispersed in the air and on surfaces. Therefore, we must **sterilize** the culture medium soon after its preparation to eliminate microorganisms already contaminating it; this is usually done by heat. However, it is equally important to take precautions during the subsequent *handling* of a sterile culture medium to exclude from it all but the desired organisms. Thus, other materials that come into contact with a sterile culture medium must themselves be sterile.

The technique used in the prevention of contamination during manipulations of cultures and sterile culture media is called **aseptic technique.** Its mastery is required for success in the microbiology laboratory, and it is one of the first methods learned by the novice microbiologist. Airborne contaminants are the most common problem because the air always contains dust particles that generally have a community of microorganisms on them. When containers are opened, they must be handled in such a way that contaminant-laden air does not enter (Figures 1.12 and 1.13). Aseptic transfer of a culture from one tube of medium to another is usually accomplished with an inoculating loop or needle that has been sterilized by incineration in a flame (Figure 1.12). Cultures in which growth has taken place can also be transferred to the surface of agar plates (Figure 1.13), where colonies develop from the growth and division of single cells. Picking and restreaking from an isolated colony is a major method of obtaining pure cultures from microbial communities containing many different organisms.

✓ 1.6 Concept Check

Microorganims can be grown in the laboratory in culture media containing the nutrients they require. Successful cultivation of pure cultures of microorganisms can be done only if aseptic technique is practiced.

✓ What is meant by the word *sterile?* What would happen if freshly prepared culture media were not sterilized?
✓ Why is aseptic technique necessary for successful cultivation of pure cultures in the laboratory?

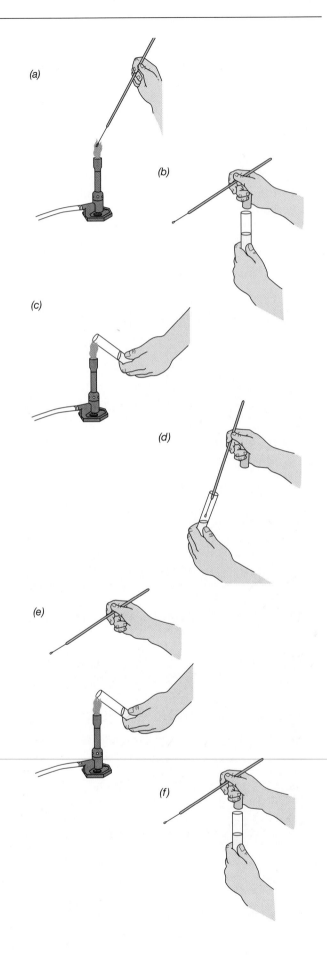

FIGURE 1.12 Aseptic transfer. (a) Loop is heated until red-hot and cooled in air briefly. (b) Tube is uncapped. (c) Tip of tube is run through the flame. (d) Sample is removed on sterile loop. (e) After removing sample on loop, the tube is reflamed and the sample transferred to a sterile medium. (f) The tube is recapped. Loop is reheated before being taken out of service.

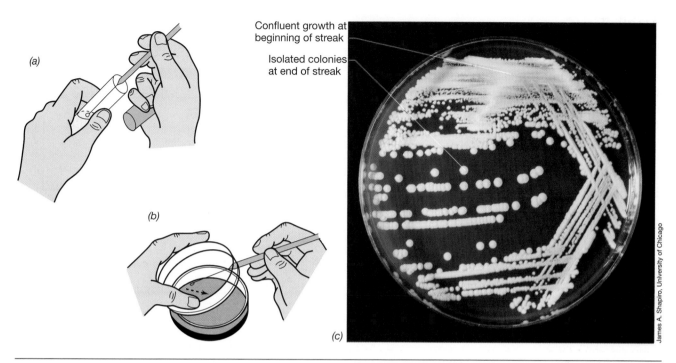

FIGURE 1.13 Method of making a streak plate to obtain pure cultures. (a) Loop is sterilized, and then a loopful of inoculum is removed from tube. (b) Streak is made over a sterile agar plate, spreading out the organisms. Following the initial streak, subsequent streaks are made at angles to it, the loop being resterilized between streaks. (c) Appearance of the streaked plate after incubation. Colonies of the bacterium *Micrococcus luteus* grown on blood agar plates. It is from such well-isolated colonies that pure cultures can usually be obtained.

1.7

The Impact of Microorganisms on Human Affairs

One goal of the microbiologist is to understand how microorganisms work and, through this understanding, to devise ways in which benefits may be increased and damages curtailed. Microbiologists have been eminently successful in achieving these goals, and microbiology has played a major role in the advancement of human health and welfare. An overview of the impact of microorganisms on human affairs is shown in Figure 1.14.

Microorganisms as Disease Agents

One measure of the microbiologist's success is shown by the statistics in Figure 1.15, which compare the present causes of death in the United States to those at the beginning of the twentieth century. At the beginning of the twentieth century, the major causes of death were infectious diseases; currently, such diseases are of only minor importance. Control of infectious disease has come as a result of our comprehensive understanding of disease processes as well as improved sanitary practices and the discovery and use of antimicrobial agents. As we will see later in this chapter, microbiology as a science had its beginnings in these studies of disease.

However, although we now live in a world where many pathogenic microorganisms are under control, for the individual dying slowly of acquired immune deficiency syndrome (AIDS), the cancer patient whose immune system has been devastated as a result of treatment with an anticancer drug, or the individual in-

Agriculture	Food
N$_2$ fixation	Food preservation
Nutrient cycling	Fermented foods
Animal husbandry	Food additives

Disease
Identifying new diseases
Treatment and cure
Disease prevention

Energy/Environment	Biotechnology
Biofuels (methane and ethanol)	Genetically modified organisms
Bioremediation	Production of pharmaceuticals
Microbial mining	Gene therapy for certain diseases

FIGURE 1.14 The impact of microorganisms on human affairs. Although many people think of microorganisms in the context of infectious diseases, few microorganisms actually cause disease. Microorganisms affect many aspects of our lives in addition to playing a role as disease agents.

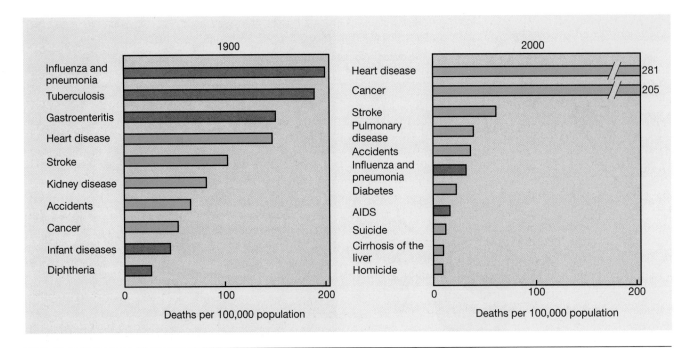

FIGURE 1.15 Death rates for the 10 leading causes of death in the United States: 1900 and 2000. Infectious diseases were the leading causes of death in 1900, whereas today they are much less important. Microbial diseases are shown in red, nonmicrobial diseases in green. Data from the United States National Center for Health Statistics.

fected with a multiple-drug–resistant pathogen, microorganisms can still be a major threat to survival. Further, microbial diseases still constitute the major causes of death in many of the developing countries of the world. Although eradication of smallpox from the world has been a stunning triumph for medical science, millions still die yearly from such pervasive illnesses as malaria, tuberculosis, cholera, African sleeping sickness, and severe diarrheal diseases.

Thus, microorganisms are still serious threats to human existence. But on the other hand we must emphasize that most microorganisms are *not* harmful to humans. In fact, most microorganisms cause no harm at all and instead are actually *beneficial,* carrying out processes that are of immense value to human society. Even in the health care industry, microorganisms play beneficial roles. For instance, the pharmaceutical industry is a multibillion dollar industry built in part on the large-scale production of antibiotics by microorganisms. A number of other major pharmaceutical products are also derived, at least in part, from the activities of microorganisms.

Microorganisms and Agriculture

Our whole system of *agriculture* depends in many important ways on microbial activities. A number of major crops are members of a plant group called the **legumes,** which live in close association with special bacteria that form structures called **nodules** on their roots. In these

root nodules, atmospheric nitrogen (N_2) is converted to fixed nitrogen compounds that the plants can use for growth. In this way, the activities of the root nodule bacteria reduce the need for costly plant fertilizer. Also of major agricultural importance are the microorganisms that are essential for the digestive process in ruminant animals such as cattle and sheep. These important farm animals have a special digestive organ called the **rumen** in which microorganisms carry out the digestive process. Without these microorganisms, cattle and sheep could not digest their food and thus could not thrive on substances like grass and hay. Microorganisms also play key roles in the cycling of important nutrients in plant nutrition, particularly carbon, nitrogen, and sulfur. Microbial activities in soil and water convert these elements to forms that are readily accessible to plants. In addition to benefits to agriculture, microorganisms also have harmful effects. Animal and plant diseases due to microorganisms have major economic impact.

Microorganisms and the Food Industry

Once food crops and animals are produced, they must be delivered in wholesome form to consumers. Microorganisms play important roles in the *food industry.* We note first that because of food spoilage vast amounts of money are wasted every year. The canning, frozen-food, and dried-food industries exist to prepare foods in such ways that they will not undergo microbial spoilage.

However, not all microorganisms have harmful effects on foods. Dairy products manufactured, at least in part, via microbial activity include cheese, yogurt, and buttermilk, all products of major economic value. Sauerkraut, pickles, and some sausages also owe their existence to microbial activity. Baked goods are made using yeast. Even more pervasive in our society are alcoholic beverages, also based on the activities of yeast. Many mushrooms (which are fungi) are also edible, and the production of some of the tastiest varieties is a big business in the food industry. Many of these topics are covered in Chapter 11 of this book.

Microorganisms, Energy, and the Environment

Developed countries are energy-driven, and here also microorganisms play major roles. Much natural gas (methane) is a product of bacterial action, arising from the activities of methanogenic bacteria. A few other mineral and energy products are also the result of microbial activity, but of even greater interest is the negative impact of microorganisms on the petroleum industry. Crude oil is subject to vigorous microbial attack, and drilling, recovery, and storage of crude oil all have to be done under conditions that minimize microbial damage.

In the future, microorganisms may provide major alternative energy sources. Phototrophic microorganisms can harvest light energy for the production of **biomass,** energy stored in living organisms. Microbial biomass and existing waste materials such as domestic refuse, surplus grain, and animal wastes, can be converted to "biofuels," such as methane and ethanol, by the degradative activities of other microorganisms.

Microorganisms can also be used to help clean up pollution created by human activities, a process called *bioremediation*. Various organisms have now been isolated from nature that consume spilled oil, solvents, and other environmentally toxic pollutants, either directly at the site of the spill or later on after the toxic materials have pervaded soils or entered the groundwater. The great diversity of microorganisms on Earth contains vast genetic resources for solutions to cleaning up the environment, and much research in this area is taking place at present.

Microorganisms and the Future

Biotechnology entails the use of microorganisms in large-scale industrial processes, usually using genetically modified microorganisms capable of synthesizing specific products of high commercial value.

Biotechnology is highly dependent on **genetic engineering,** the discipline that concerns the artificial manipulation of genes and their products. Genes from human or other sources can be broken into pieces and modified in various ways, using microorganisms and

their enzymes as precise and sophisticated molecular tools. It is even possible to make completely artificial genes using genetic engineering techniques. Once the desired gene has been selected or created, it can be inserted into a microorganism where it can be expressed to make the desired gene products. For instance, human insulin, a hormone found in abnormally low amounts in people with the disease diabetes, can be produced microbiologically from a human insulin gene engineered into a microorganism. We discuss genetic engineering and biotechnology in detail in Chapter 10.

The overwhelming influence of microorganisms in human society is clear. We have many reasons to be aware of microorganisms and their activities (Figure 1.14). As the eminent French scientist Louis Pasteur, one of the founders of microbiology, expressed it: "The role of the infinitely small in nature is infinitely large." Therefore, before we begin a detailed study of microbiology, let us consider briefly the contributions that Pasteur and others have made to the foundation of the science of microbiology.

✓ 1.7 Concept Check

Microorganisms can be both beneficial and harmful to humans. Although we tend to emphasize the harmful microorganisms (infectious disease agents), many more microorganisms are beneficial than harmful.

✓ In what ways are microorganisms important in the food and agricultural industries?
✓ What is biotechnology and how might it improve the lives of humans?

1.8

The Historical Roots of Microbiology: van Leeuwenhoek, Pasteur, and Koch

Although the existence of creatures too small to be seen with the eye had long been suspected, their discovery was linked to the invention of the microscope. Robert Hooke described the fruiting structures of molds (eukaryotic cells) in 1664 (Figure 1.16), but the first person to see microorganisms in any detail was the Dutch amateur microscope builder Antoni van Leeuwenhoek, who in 1684, used simple microscopes of his own construction (Figure 1.17a). Leeuwenhoek's microscopes were extremely crude by today's standards, but by careful manipulation and focusing he was able to see organisms as small as prokaryotes. He reported his observations in a series of letters to the Royal Society of London, which published them in 1684 in English translation. Drawings of some of Leeuwenhoek's "wee animalcules," as he referred to them, are shown in Figure 1.17b. His observations were confirmed by other workers, but progress in under-

(a)

(b)

FIGURE 1.16   (a) The microscope used by Robert Hooke. (b) A drawing by Robert Hooke, which represents one of the first microscopic descriptions of microorganisms: a blue mold growing on the surface of leather; the round structures contain spores of the mold.

standing the nature and importance of these tiny organisms came only slowly. Only in the nineteenth century did improved microscopes become available and widely distributed. During most of its history, the science of microbiology has traditionally taken the greatest steps forward when better microscopes have been developed, for these have enabled scientists to penetrate ever deeper into the mysteries of the cell. By contrast, in recent years, despite the development of ever more powerful microscopes including confocal laser and atomic force microscopes (∞ Section 3.2), the biggest leaps in microbiology have come from advances in molecular biology and our ability to culture new organisms in the laboratory.

Microbiology as a science did not develop until the latter part of the nineteenth century. This long delay occurred because, in addition to microscopy, certain basic techniques for the study of microorganisms needed to be devised. In the nineteenth century, investigation of two perplexing questions led to the development of these techniques and laid the foundation of microbiological science: (1) Does spontaneous generation occur? (2) What is the nature of contagious disease? To answer these pressing questions, we jump ahead here almost 200 years from the days of van Leeuwenhoek to the mid- to late 1800s, when two giants in the early days of microbiology, Louis Pasteur and Robert Koch, made their major contributions. From their keen experimental work, the science of microbiology was firmly established as a distinct and growing field.

Pasteur and the Downfall of Spontaneous Generation

The basic idea of spontaneous generation can easily be understood. If food is allowed to stand for some time, it putrefies. When the putrefied material is examined microscopically, it is found to be teeming with bacteria. Where do these bacteria come from, since they are not seen in fresh food? Some people said they developed from seeds or germs that had entered the food from the air, whereas others said that they arose spontaneously from nonliving materials.

The most powerful opponent of spontaneous generation was the French chemist Louis Pasteur (1822–1895), whose work on this problem was the most exacting and convincing. Pasteur first showed that structures were present in air that resembled closely the microorganisms seen in putrefying materials. He did this by passing air through guncotton filters, the fibers of which stopped solid particles. After the guncotton was dissolved in a mixture of alcohol and ether, the particles that it had trapped fell to the bottom of the liquid and were examined on a microscope slide. Pasteur found that in ordinary air there constantly exists a variety of microbial cells and that they could not

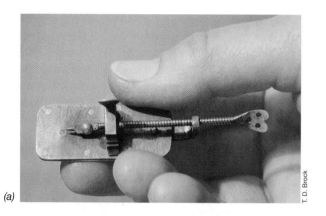

(a)

T. D. Brock

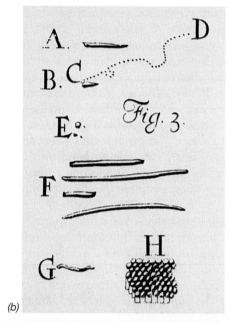

(b)

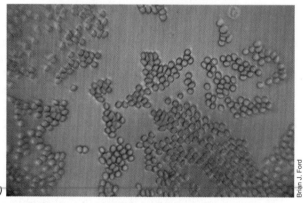

(c)

Brian J. Ford

FIGURE 1.17 (a) Photograph of a replica of Leeuwenhoek's microscope. The lens is mounted in the brass plate adjacent to the tip of the adjustable focusing screw. (b) Leeuwenhoek's drawings of bacteria, published in 1684. Even from these crude drawings we can recognize several morphological types of common bacteria. A, C, F, and G, rod-shaped; E, spherical or coccus-shaped; H, cocci packets. (c) Photomicrograph of a human blood smear taken through a van Leeuwenhoek microscope. Red blood cells are clearly apparent.

be distinguished from the organisms found in much larger numbers in putrefying materials. Pasteur concluded that the organisms found in putrefying materials originated from microorganisms present in the air. He postulated that these cells are constantly being deposited on all objects. If this conclusion was correct, Pasteur deduced that food treated in such a way as to destroy all living organisms contaminating it, should not putrefy.

Pasteur used heat to eliminate contaminants since it had already been established that heat effectively kills living organisms. In fact, other workers had shown that when a nutrient solution was sealed in a glass flask and heated to boiling, it never supported microbial growth. The proponents of spontaneous generation criticized such experiments by declaring that fresh air was necessary for spontaneous generation and that the air itself inside the sealed flask was affected in some way by heating so that it could no longer support spontaneous generation. Pasteur skirted this objection simply and brilliantly by constructing a swan-necked flask, now called a *Pasteur flask* (Figure 1.18). In such a flask nutrient solutions could be heated to boiling; after the flask was cooled, air could reenter but the bends in the neck prevented particulate matter containing bacteria or other microorganisms from getting into the main body of the flask. Material sterilized in such a flask did not putrefy, and no microorganisms ever appeared as long as the neck of the flask did not contact the sterile liquid. However, if the flask was tipped to allow the sterile liquid to contact the neck of the flask, putrefaction occurred and the liquid soon teemed with microorganisms. This simple experiment effectively settled the controversy surrounding the theory of spontaneous generation.

Killing all the bacteria or other microorganisms in or on objects is a process we now call **sterilization,** and the procedures that Pasteur and others used were eventually refined and carried over into microbiological research (see Section 1.6). Disproving the theory of spontaneous generation thus led to the development of effective sterilization procedures, without which microbiology as a science could not have developed. Food science also owes a debt to Pasteur, as his principles are applied in the canning and preservation of many foods.

Pasteur went on to many other triumphs in microbiology and medicine. Chief among these was his development of vaccines for the diseases anthrax, fowl cholera, and rabies during a very productive period from 1880 to 1890. These medical and veterinary breakthroughs were not only highly significant in their own right, but helped solidify the concept of the germ theory of disease whose principles were being developed at this time by a contemporary of Pasteur, Robert Koch. We review this important period in the history of microbiology now.

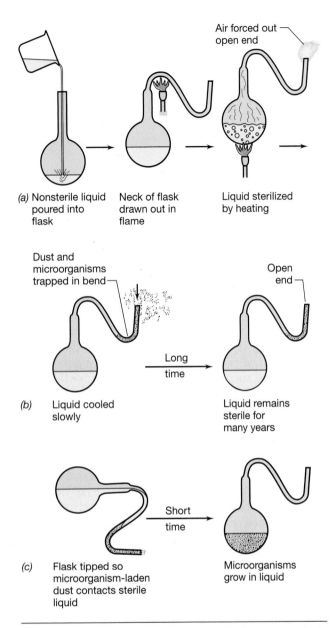

FIGURE 1.18 Pasteur's experiment with the swan-necked flask. (a) Sterilizing the contents of the flask. (b) If the flask remained upright, no microbial growth occurred. (c) If microorganisms trapped in the neck reached the sterile liquid, microbial growth ensued.

Koch and the Germ Theory of Disease

Proof that microorganisms could cause disease provided the greatest impetus for the development of the science of microbiology. Indeed, even in the sixteenth century it was thought that something could be transmitted from a diseased person to a well person that induced the disease. Many diseases seemed to spread through populations and were called *contagious*; the unknown agent that did the spreading was called the *contagion*. After the discovery of microorganisms, it was widely held that these organisms were responsible for contagious diseases, but proof was lacking. Discoveries by Ignaz Semmelweis and Joseph Lister provided indirect evidence for the importance of microorganisms in causing human diseases, but it was not until the work of Robert Koch (1843–1910), a physician by training, that the *germ theory of disease* was clearly conceptualized and given experimental support.

In his early work, published in 1876, Koch studied *anthrax*, a disease of cattle that sometimes also occurs in humans. Anthrax is caused by a spore-forming bacterium now called *Bacillus anthracis,* and the blood of an animal infected with anthrax teems with cells of this large bacterium. Koch established by careful microscopy that the bacteria were always present in the blood of an animal that was succumbing to the disease. However, mere association of the bacterium with the disease did not prove that it actually *caused* the disease; it might instead be a *result* of the disease. Therefore, Koch demonstrated that it was possible to take a small amount of blood from a diseased mouse, which he used as an experimental animal, and inject it into a second mouse, which subsequently became diseased and died. He could then take blood from this second animal, inject it into another, and again obtain the characteristic disease symptoms. By repeating this process as often as 20 times, successively transferring small amounts of blood containing bacteria from one animal to another, he proved that the bacteria did indeed cause anthrax: The twentieth mouse died just as rapidly as the first, and in each case Koch could demonstrate by microscopy that the blood of the dying animal contained large numbers of the spore-forming bacterium.

However, Koch carried this experiment further. He found that the bacteria could also be cultivated in nutrient fluids outside the animal body and that even after many transfers in culture the bacteria could still cause the disease when reinoculated into an animal. Bacteria from a diseased animal and bacteria in culture both induced the same disease symptoms upon injection. On the basis of these and other experiments Koch formulated the following criteria, now called **Koch's postulates,** for proving that a specific type of microorganism causes a specific disease:

1. The organism should be constantly present in animals suffering from the disease and should not be present in healthy individuals.

2. The organism must be cultivated in a pure culture away from the animal body.

3. Such a culture, when inoculated into susceptible animals, should initiate the characteristic disease symptoms.

4. The organism should be reisolated from these experimental animals and cultured again in the labo-

ratory, after which it should still be the same as the original organism.

Koch's postulates are summarized in Figure 1.19. Koch's postulates not only supplied a means of demonstrating that specific organisms cause specific diseases but also provided a tremendous spur for the development of the science of microbiology by stressing the importance of laboratory culture. Using Koch's postulates as a guide, investigators following Koch revealed the causes of many important diseases of humans and other animals. These discoveries led to the development of

successful treatments for the prevention and cure of many infectious diseases, thereby greatly improving the scientific basis of clinical medicine.

Koch and Pure Cultures

As we have noted, in order to successfully study the activities of a microorganism, such as a microorganism that causes a disease, one must be sure that it alone is present in a culture. That is, the culture must be *pure* (see Section 1.6). With objects as small as microorganisms, ascertaining purity is not easy, for even a very tiny sam-

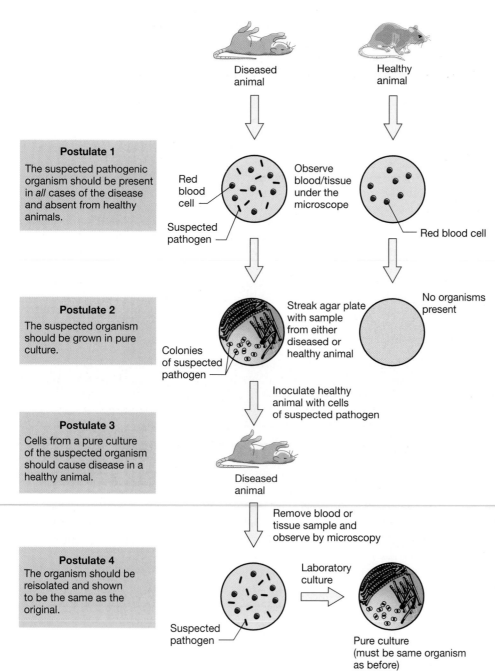

FIGURE 1.19 Koch's postulates for proving that a specific microorganism causes a specific disease. Note how it is essential that following isolation of a pure culture of the suspected pathogen, a laboratory culture of the organism should both initiate the disease *and* be recovered from the diseased animal. However, a caveat to keep in mind is that one must select the correct medium or media (and incubation conditions) for growth of the pathogen, otherwise it will be missed.

LEARNING FROM THE PAST   Solid Media, the Petri Plate, and Pure Cultures

Robert Koch was the first to grow bacteria on solid culture media. Koch initially employed gelatin as a solidifying agent for the various nutrient fluids he used to culture pathogenic bacteria and developed a method for preparing horizontal slabs of solid media which were kept free of contamination by covering them with a bell jar or glass box (see Figure 1.20*c*).

Nutrient gelatin was a marvelous culture medium for the isolation and study of various bacteria, but it had several drawbacks, the most important being that gelatin does not remain solid at body temperature (37°C), the optimum temperature for growth of most human pathogens. Thus, a more versatile solidifying agent was needed, and this turned out to be agar.

Agar is a polysaccharide derived from red algae. It was used widely in the nineteenth century, especially in tropical countries, as a gelling agent. The first use of agar as a solidifying agent for bacteriological culture media was made by Walter Hesse, an associate of Koch. The actual suggestion that agar be used instead of gelatin was made by Hesse's wife, Fannie. Fannie Hesse had used agar in the preparation of fruit jellies, and when it was tried as a solidifying agent in nutrient media, it was immediately found to be superior to gelatin. Hesse wrote to Koch about this discovery, and Koch quickly adapted agar to his own studies, including his classic studies on the isolation of the bacterium *Mycobacterium tuberculosis*, the cause of the disease tuberculosis (see text and Figure 1.20).

In 1887 Richard Petri published a brief paper describing a modification of Koch's flat plate technique. Petri's enhancement, which turned out to be amazingly useful, was the development of the double-sided dishes that bear his name. The advantage of Petri dishes was apparent; they could be easily stacked and sterilized separately from the medium, and, following the addition of molten medium to the smaller of the two double dishes, the larger dish could be used as a cover to prevent contamination. Colonies that formed on the surface of the agar in the Petri dish remained fully exposed to air and could easily be manipulated for further study. The original idea of Petri has not been improved on to this day, as the Petri dish, made either of reusable glass and sterilized by dry heat or of disposable plastic and sterilized by ethylene oxide (a gaseous sterilant), is the mainstay of the microbiology laboratory.

Finally, it should also be noted that Koch was keenly aware of the implications his pure culture methods had for the study of microbial systematics. Koch observed that different colonial forms (differing in color, colony morphology, size, and the like) developed on solid media exposed to a contaminated object, and that these colonial forms bred true and could be distinguished from one another by their colony characteristics. Cells from different colonies also differed microscopically and often in their temperature or nutrient requirements as well. Koch realized that these differences among microorganisms met all the requirements that taxonomists had established for the classification of larger organisms, such as plant and animal species. In Koch's own words: "*All bacteria which maintain the characteristics which differentiate one from another when they are cultured on the same medium and under the same conditions, should be designated as species, varieties, forms, or other suitable designation.*" Thus, Koch's discovery of solid culture media and his emphasis on pure culture microbiology reached far beyond the realm of medical bacteriology; his discoveries supplied critically needed tools for development of the field of bacterial taxonomy, genetics, and several related disciplines. Indeed, the entire field of microbiology owes a huge debt of gratitude to Koch and his associates for the intuition they displayed in grasping the significance of pure cultures and developing some of the most basic methods in microbiology. ■

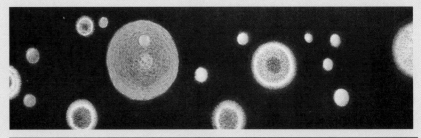

A hand-colored photograph of colonies formed on agar taken by Walter Hesse, an associate of Robert Koch. The colonies include those of fungi (molds) and bacteria and were obtained during studies Hesse initiated on the microbiological content of air in Berlin, Germany, in 1882. From Hesse, W. 1884. "Ueber quantitative Bestimmung der in der luft enthaltenen Mikroorganismen," in Struck (ed.), *Mittheilungen aus dem Kaiserlichen Gesundheitsamte*. August Hirschwald.

ple of blood or animal fluid may contain several kinds of organisms that may all grow together in culture. Koch realized the importance of pure cultures. He developed several ingenious methods of obtaining them, of which the most useful is that involving the isolation of single *colonies* on solid media (see the box). Koch observed that when a solid nutrient surface, such as a potato slice, was exposed to air and then incubated, bacterial colonies developed, each having a characteristic shape and color. He inferred that each colony had arisen from a single bacterial cell that had fallen on the surface, found suitable nutrients, and begun to multiply. Because the solid surface prevented the bacteria from moving around, all the offspring of the initial cell had remained together, and when a large enough number of organisms was present, the mass of cells became visible to the naked eye. He assumed that colonies with different shapes and colors were derived from different kinds of microorganisms. When the cells of a single colony were spread out on a fresh surface, many colonies developed, each with the same shape and color as the original.

Koch realized that this discovery provided a simple way of obtaining pure cultures: He found that if mixed cultures were spread on solid nutrient surfaces, the individual cells were so far apart that the colonies they produced did not mingle. Many organisms could not grow on potato slices, and so he devised semisolid media in which gelatin and later *agar* (see the box) was used as a solidifying agent for various nutrient solutions. Agar is the major agent used for solidifying culture media today.

It is important to realize that Koch's postulates have relevance beyond simply identifying organisms that cause specific diseases. The essential point here is that from the study of pure cultures, one can show that *specific organisms have specific effects*. This principle that different organisms have unique biological activities was important in establishing microbiology as an independent biological science (see the box). Because of these major contributions of Koch, by the beginning of the twentieth century the disciplines of bacteriology and microbiology were on a firm footing.

Koch and Tuberculosis

The greatest accomplishment of Robert Koch's work in medical bacteriology was with tuberculosis. At the time Koch began this work (1881), one-seventh of all reported deaths of humans were caused by tuberculosis. Even at this time there was strong evidence that tuberculosis was a contagious disease, but the suspected causal organism had never been seen, either in diseased tissues or in culture. Koch's aim from the beginning of his work on tuberculosis was to demonstrate the causal agent of tuberculosis, and to this end he employed all the meth-

ods he had so carefully developed previously: microscopy, staining of tissues, pure culture isolation, and animal inoculation.

As is now well-known, *Mycobacterium tuberculosis,* the "tubercle bacillus," is very difficult to stain because of large amounts of lipid present on its outer surface. But Koch devised a staining procedure for *M. tuberculosis* in tissue samples using alkaline methylene blue in conjunction with a second stain (bismarck brown) that stained only the tissue (Koch's method was the forerunner of the Ziehl–Nielsen stain used today for staining acid-fast bacteria like *M. tuberculosis,* (Section 13.22). Using his newly developed staining method, Koch observed bright-blue, rod-shaped cells of *M. tuberculosis* in tuberculous tissues, the latter staining a light brown (Figure 1.20). However, from his previous work on anthrax, Koch realized that simply identifying an organism associated with tuberculosis was not enough; he must *culture* the organism in order to prove that it was the specific cause of tuberculosis.

Producing cultures of *Mycobacterium tuberculosis* was not easy, but eventually Koch was successful in obtaining colonies of this organism on coagulated blood serum. Later he used agar, which had just been introduced as a solidifying agent (see the box). Under the best of conditions, *M. tuberculosis* grows slowly in culture, but Koch's persistence and patience eventually led to pure cultures of this organism from a variety of human and animal sources. From here it was relatively easy to obtain definitive proof that the organism he had isolated was the true cause of the disease tuberculosis. Guinea pigs can be readily infected with *M. tuberculosis* and eventually succumb to systemic tuberculosis. Koch showed that diseased guinea pigs contained masses of *M. tuberculosis* cells in their tissues and that pure cultures obtained from such animals transmitted the disease to uninfected animals. Thus, Koch successfully satisfied all four criteria of his famous postulates (Figure 1.19), and the cause of tuberculosis was understood.

Koch announced his discovery of the tubercle bacillus in a famous lecture given in Berlin in 1882. The news of Koch's discovery spread quickly. And for his work on tuberculosis, including not only discovery of the causative agent but also the development of methods for its specific staining (Figure 1.20) and preparation of a substance called *tuberculin,* useful in the diagnosis of tuberculosis, Koch received the 1905 Nobel Prize for Physiology or Medicine just 5 years before his death.

Koch had many other triumphs in microbiology besides discovering the causes of anthrax and tuberculosis. These included the discovery and isolation of the causative agent of cholera, *Vibrio cholerae,* the discovery of the importance of water filtration in the control of cholera, the development of the concept of disease "carriers," and the publication of the first photomicrographs

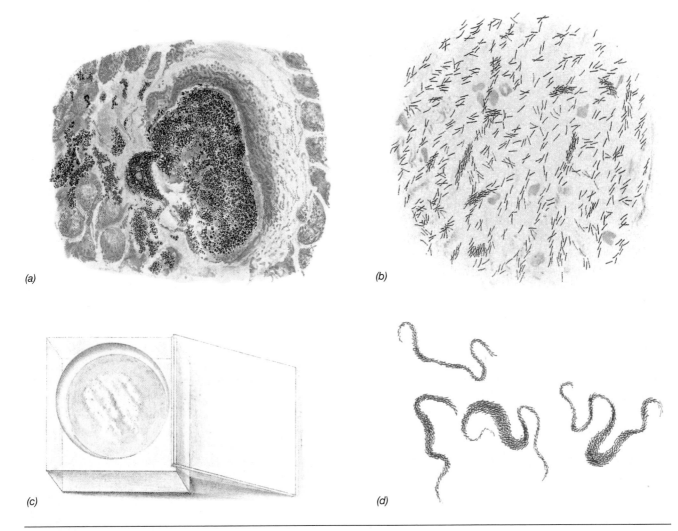

FIGURE 1.20 Robert Koch's drawings of cells of *Mycobacterium tuberculosis* in tissues and in laboratory culture. (a) Section through a tubercle from lung tissue. Cells of *M. tuberculosis* stain blue, whereas the lung tissue stains brown. (b) Cells of *M. tuberculosis* in a sputum sample of a tuberculous patient. (c,d) Growth of *M. tuberculosis* in pure culture: (c) Growth on a glass plate of coagulated blood serum inside a glass box (with lid open). (d) A colony of *M. tuberculosis* cells taken from the plate in (c) and observed microscopically at 700×; cells appear as long "cordlike" forms (compare with Figure 13.71*b*). Original drawings appeared in Koch, R. 1884. "Die Aetiologie der Tuberkulose." *Mittheilungen aus dem Kaiserlichen Gesundheitsamte* 2:1–88.

of bacteria. Clearly, Robert Koch's contributions to the development of modern microbiology were monumental.

1.9

Transition into the Twentieth Century

As the science of microbiology developed further in the late 1800s and entered the twentieth century, new specialties arose leading to the present era of "molecular microbiology," where the powerful tools of molecular biology can be applied to the study of virtually any microbial system. Two giants spanned the early part of this transition to the twentieth century, the Dutch microbiologist, Martinus Beijerinck, and the Russian,

Sergei Winogradsky. Both were interested in bacteria that inhabit soil and water (rather than medically relevant bacteria) and both can be considered the founders of the field of *general microbiology*.

Beijerinck and Winogradsky

Martinus Beijerinck (1851–1931) was a professor at the Delft Polytechnic School in his later years (recall that van Leeuwenhoek was also from Delft), but was originally trained in botany and began his career in microbiology studying the microbiology of plants. Perhaps the greatest contribution of Beijerinck to the field of microbiology was his clear formulation of the concept of the **enrichment culture.** Instead of isolating microorganisms from nature by exposing nonselective culture media to some environment and allowing chance to dictate what grew, Beijerinck proposed a different approach. He proposed *selecting* specific microorganisms from a natural sample through the use of specific culture media and incubation conditions that favored growth of one type of organism while constraining the growth of others. Using his enrichment culture (or "selective culture," as he liked to call it) technique, Beijerinck isolated the first pure cultures of many soil and aquatic microorganisms including aerobic nitrogen-fixing bacteria (Figure 1.21), sulfate-reducing and sulfur-oxidizing bacteria, nitrogen-fixing root nodule bacteria, *Lactobacillus* species, green algae, and many other microorganisms. And from his studies of tobacco mosaic disease, Beijerinck showed that the infectious agent (a virus) was not bacterial but somehow became incorporated into the cells of the host plant and required the living plant to reproduce; in essence, Beijerinck described the basic tenets of virology.

Sergei Winogradsky (1856–1953) had interests similar to Beijerinck's and was also successful in isolating several key bacteria for the first time. Winogradsky was interested in soil bacteria, particularly those involved in the cycling of nitrogen and sulfur compounds (Figure 1.22). In this connection, Winogradsky isolated pure cultures of nitrifying bacteria, showing clearly that the process of nitrification (the oxidation of ammonia to nitrate) was the result of bacterial action, and studied the oxidation of hydrogen sulfide by sulfur-oxidizing bacteria directly in their natural habitats.

Besides demonstrating that bacteria can be geochemical agents, Winogradsky is also remembered for his keen insight into the *metabolic significance* of these processes. For example, from his studies of the sulfur-oxidizing bacteria, Winogradsky proposed the concept of *chemolithotrophy*, the oxidation of *inorganic* compounds coupled to the release of energy (⚙ Sections 4.14 and 15.9). And from studies of the nitrifying bacteria Winogradsky concluded that these organisms ob-

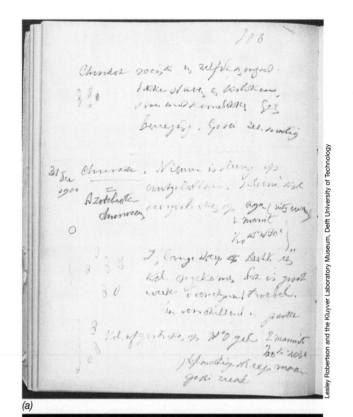
(a)

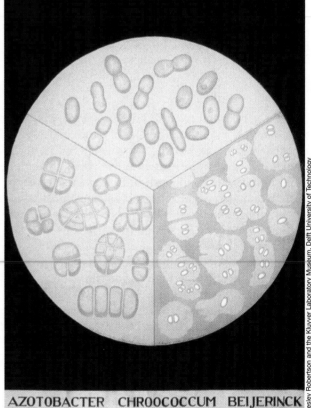
(b)

AZOTOBACTER CHROOCOCCUM BEIJERINCK

FIGURE 1.21 Martinus Beijerinck and *Azotobacter*. (a) A page from the laboratory notebook of M. Beijerinck dated December 31, 1900, that describes his observations on the aerobic nitrogen-fixing bacterium *Azotobacter chroococcum*. It is on this page that Beijerinck uses this bacterial name for the first time. Compare Beijerinck's drawings of pairs of *A. chroococcum* cells with a photomicrograph of cells of *Azobacter* shown in Figure 13.20a. (b) A painting by M. Beijerinck's sister, Henriëtte Beijerinck, showing cells of *Azotobacter chroococcum*. Beijerinck used such paintings to illustrate his lectures, because this was long before the days of the overhead, slide, and computer projectors used in lectures and seminars today.

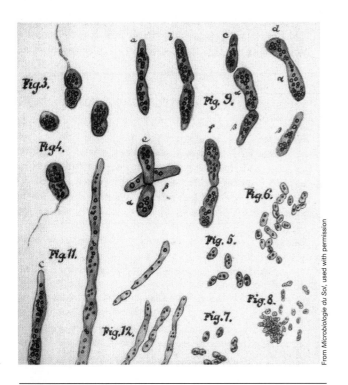

From *Microbiologie du Sol*, used with permission

FIGURE 1.22 Hand-colored drawings of cells of purple sulfur phototrophic bacteria included in the monograph *Microbiologie du Sol*, by Sergei Winogradsky. The original drawings were made by S. Winogradsky about 1887 and then copied and hand-colored by his wife Hélène for publication in the monograph. These drawings included cells of the genus *Chromatium*, such as *C. okenii* (Figs. 3 and 4) and *C. vinosum* (Figs. 5–8). These species are still recognized today. Note the prominent flagella on cells of *C. okenii*. Compare Figs. 3 and 4 with a photomicrograph of living cells of *C. okenii* shown in Figure 15.17a of this book.

tained their carbon from CO_2 in air, that is, that they were *autotrophs*. Although neither of these concepts was readily accepted at the time, we now know that bacterial chemolithotrophy and autotrophy are extremely important processes on Earth and can even support the growth of higher organisms (∞ Section 16.12). Using an enrichment method, Winogradsky also isolated the first nitrogen-fixing bacterium (the anaerobe *Clostridium pasteurianum*) and by so doing developed the concept of bacterial N_2 fixation. Winogradsky lived to be almost 100, publishing many scientific papers, along with a major monograph, *Microbiologie du Sol (Soil Microbiology)*; the latter, a true milestone in microbiology, contained his original drawings of many of the organisms he had isolated or otherwise studied in enrichment culture or natural material during his career (Figure 1.22).

Table 1.1 summarizes some of the important discoveries in the field of microbiology from van Leeuwenhoek to the present.

Development of the Major Subdisciplines of Microbiology

In more recent years the field of microbiology has developed rapidly in two separate directions—applied microbiology and basic microbiology. On the applied side, the practical advances initially made by Koch led to extensive developments in *medical microbiology* and *immunology* in the early part of the twentieth century, with the discovery of many new bacterial pathogens (∞ Discoverers of the Main Bacterial Pathogens, Chapter 23) and the working out of the principles by which these pathogens infect the body and are in turn resisted by the body's defenses. Other early practical advances, bolstered by the discoveries of Beijerinck and Winogradsky, were in the field of *agricultural microbiology*, which led to an understanding of microbial processes in the soil that are beneficial or harmful to plant growth. Later in the twentieth century, such studies on soil microbiology led to the discovery of important uses of microorganisms, such as in the synthesis of antibiotics and commodity chemicals. This led, especially after World War II, to the field of *industrial microbiology*.

Soil microbiology also provided an important foundation for studies on microbial processes in water bodies such as lakes, rivers, and oceans, studies classified under the field of *aquatic microbiology*. One branch of aquatic microbiology deals with the development of processes for providing safe water for humans. The handling of human wastes, especially domestic sewage, has required the development of microbial processes for sewage treatment. To provide safe drinking water, procedures for eliminating harmful bacteria from water supplies were developed and are now in widespread practice worldwide. As interest in the biodiversity and activities of microorganisms in their natural environments grew, the field of *microbial ecology* emerged as a major discipline in microbiology.

In addition to the *applied* aspects of microbiology that have provided such important advances for human society, extensive developments have occurred in our understanding of the *basic* principles of microbial function. In the early part of the twentieth century, the most important developments in basic microbiology involved the discovery of new kinds of bacteria and their proper classification (*bacterial taxonomy*). Bacterial classification required a study of the nutrients that bacteria consume and the products that they make, studies comprising part of the field of *bacterial physiology*. One part of physiology that became of major importance as the twentieth century progressed involved the study of the physical and chemical structure of bacteria, studies included in the field of *bacterial cytology* (the word *cytology* refers to the study of the cell). Another major development from physiology was the study of bacterial en-

TABLE 1.1	Three hundred years of microbiology: Some key papers in microbiology, 1684–1999[a]	

Year	Investigator(s)	Discovery
1684	Antoni van Leeuwenhoek	Discovery of bacteria
1798	Edward Jenner	Smallpox vaccination
1857	Louis Pasteur	Microbiology of the lactic acid fermentation
1860	Louis Pasteur	Role of yeast in alcoholic fermentation
1864	Louis Pasteur	Settled spontaneous generation controversy
1867	Robert Lister	Antiseptic principles in surgery
1881	Robert Koch	Methods for study of bacteria in pure culture
1882	Robert Koch	Discovery of cause of tuberculosis
1884	Robert Koch	Koch's postulates
1884	Christian Gram	Gram-staining method
1889	Sergei Winogradsky	Concept of chemolithotrophy
1889	Martinus Beijerinck	Concept of a virus
1890	Sergei Winogradsky	Autotrophic growth of chemolithotrophs
1901	Martinus Beijerinck	Enrichment culture method
1901	Karl Landsteiner	Human blood groups
1908	Paul Ehrlich	Chemotherapeutic agents
1928	Frederick Griffith	Discovery of pneumococcus transformation
1929	Alexander Fleming	Discovery of penicillin
1944	Oswald Avery, Colin Macleod, Maclyn McCarty	Explanation of Griffith's work—DNA is genetic material
1944	Selman Waksman and Albert Schatz	Discovery of streptomycin
1946	Edward Tatum and Joshua Lederberg	Bacterial conjugation
1951	Barbara McClintock	Discovery of transposable elements
1953	James Watson, Francis Crick, Rosalind Franklin	Structure of DNA
1959	Arthur Pardee, Francois Jacob, Jacques Monod	Gene regulation by a repressor protein
1959	Rodney Porter	Immunoglobulin structure
1959	F. Macfarlane Burnet	Clonal selection theory
1960	Francois Jacob, David Perrin, Carmon Sanchez, Jacques Monod	Concept of an operon
1960	Rosalyn Yalow and Solomon Bernson	Development of radioimmunoassay (RIA)
1966	Marshall Nirenberg and H. Gobind Khorana	Discovery of the genetic code
1967	Thomas Brock	Discovery of bacteria growing in boiling hot springs
1969	Howard Temin, David Baltimore, Renato Dulbecco	Discovery of retroviruses/reverse transcriptase
1969	Thomas Brock and Hudson Freeze	Isolation of *Thermus aquaticus*, source of *Taq* DNA polymerase
1970	Hamilton Smith	Specificity of action of restriction enzymes
1975	Georges Kohler, Cesar Milstein	Monoclonal antibodies
1976	Susumu Tonegawa	Rearrangement of immunoglobulin genes
1977	Carl Woese and George Fox	Discovery of the Archaea
1977	Fred Sanger, Steven Niklen, Alan Coulson	Methods for sequencing DNA
1981	Stanley Prusiner	Characterization of prions
1982	Karl Stetter	Isolation of first prokaryote with temperature optimum $> 100°C$
1983	Luc Montagnier	Discovery of HIV, the cause of AIDS
1988	Kary Mullis	Discovery of the polymerase chain reaction (PCR)
1995	Craig Venter and Hamilton Smith	Complete sequence of a bacterial genome
1999	The Institute for Genomic Research (TIGR), and others	Over 100 microbial genomes sequenced or in progress

a Major reference sources here include Brock, T. D. (1961), *Milestones in Microbiology*, Prentice Hall, Englewood Cliffs, NJ; Brock, T. D. (1990). *The Emergence of Bacterial Genetics*, Cold Spring Harbor Press, Cold Spring Harbor, NY. *Year* refers to the year in which the discovery was published.

zymes and the chemical reactions that they carry out, a field called *bacterial biochemistry.*

Another very important area of basic research involved the study of heredity and the variation that bacteria undergo during their growth and development, studies that fall under the discipline of *bacterial genetics.* Although some ideas of bacterial variation were known early in the twentieth century, it was not until the discovery of genetic exchange in bacteria in about 1950 that bacterial genetics really became a major field of study. Bacterial genetics, biochemistry, and physiology developed mainly during the 1950s, leading by the

early 1960s to an advanced understanding of DNA, RNA, and protein synthesis. The field of *molecular biology* arose to a great extent from these bacterial studies.

Another important development in the twentieth century involved the study of viruses. Although Beijerinck discovered the first virus over 100 years ago, it was not until the middle of the twentieth century that the true nature of viruses was understood. Much of this work involved the study of viruses that infect bacteria, called *bacteriophages.* An important development was the realization that virus infection was analogous to genetic transfer, and the relationships between viruses and other genetic elements was worked out primarily from research on bacteriophages.

By the 1970s, our knowledge of the basic processes of bacterial physiology, biochemistry, and genetics had advanced to such a great extent that it was possible to manipulate the genetic material of cells experimentally, using bacteria a tools. It also became possible to introduce genetic material (DNA) from foreign sources into bacteria and control its replication and characteristics. This led to development of the field of *biotechnology.* Although biotechnology originally arose from basic studies, its use in promoting human welfare required application of the principles of physiology and industrial microbiology, a good example of how basic and applied research advance together. Also at about this same time, nucleic acid sequencing was worked out and used as a tool to discern phylogenetic relationships among prokaryotes, which led to revolutionary new concepts in the field of biological classification and to the first true understanding of the evolutionary history of microorganisms. Now entire genomes can be rapidly sequenced and the age of genomic analysis is clearly upon us (∞ Sections 9.11

and 9.12). The new millennium will clearly be an exciting one for microbiology and microbiologists!

✓ 1.8 and 1.9 Concept Check

Louis Pasteur's work on spontaneous generation led to the development of methods for control of the growth of microorganisms. Robert Koch developed criteria for the study of infectious microorganisms, and developed the first methods for the growth of pure cultures and microorganisms. Beijerinck and Winogradsky studied bacteria in soil and water and developed the enrichment culture technique. In the twentieth century, basic and applied microbiology have worked hand in hand to yield a number of important practical advances and a revolution in molecular biology.

- ✓ How did Pasteur's famous experiment defeat the theory of spontaneous generation?
- ✓ How can Koch's postulates prove cause and effect in a disease?
- ✓ Who was the first person to use solid culture media in microbiology? What advantages do solid media offer for the culture of microorganisms?
- ✓ What is the enrichment culture technique and why was it a useful new method in microbiology?

REVIEW QUESTIONS

1. List five key properties associated with the living state. Which of these are properties of *all* cells? Which are properties of only *some types* of cells?

2. Cells can be thought of as both machines and coding devices. Explain how these two attributes of a cell differ.

3. How many genes does a typical prokaryotic cell have? Do eukaryotes have a smaller or a larger number of genes than prokaryotes?

4. Compare and contrast the prokaryotic and the eukaryotic cell. List the properties that are common to both types. List the properties that are different in these two cell types.

5. Are viruses cells? List three ways in which viruses differ from cells.

6. What is needed for translation to occur in a cell? What is the product of the translational process?

7. In what ways are Bacteria and Archaea similar? In what ways are they different? How can one distinguish a cell of the Bacteria from a cell of the Archaea?

8. What is an ecosystem? Do microorganisms live in pure cultures in an ecosystem? What effects can microorganisms have on their ecosystem?

9. What is a pure culture and how can one be obtained? Why was knowledge of how to obtain a pure culture important for development of the science of microbiology?

10. How would you convince a friend that microorganisms are much more than just agents of disease?

11. Explain the principle behind the use of the Pasteur flask in studies on spontaneous generation.

12. Explain why the invention of solid culture media was of great importance to the development of microbiology as a science.

13. Describe a major contribution to microbiology of the early microbiologist Martinus Beijerinck.

14. Using Table 1.1 as a guide, contrast the focus of microbiological research before and after World War II.

APPLICATION QUESTIONS

1. Observe the organisms shown in Figure 1.1. Describe how the cells in the organisms shown in parts (a) and (b) differ from the organisms in parts (c) and (d). List as many differences as you can.

2. Pasteur's experiments on spontaneous generation were of enormous importance for the advance of microbiology, having an impact on the methodology of microbiology, ideas on the origin of life, and the preservation of food,

to name just a few. Explain briefly how the impact of his experiments was felt on each of the topics listed.

3. Describe the various lines of proof Robert Koch used to definitively associate the bacterium *Mycobacterium tuberculosis* with the disease tuberculosis. How would his proof have been flawed if any of the tools he developed for studying bacterial diseases had not been available for his study of tuberculosis?

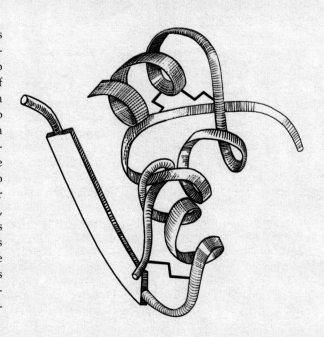

Cells are constructed of macromolecules whose chemistry and physical arrangement form larger structures unique to each species. Study of the structure of specific macromolecules isolated from cells often yields important clues to their function in the intact organism ("form follows function" is an old saying in biology). This is expecially true of proteins. The sequence of amino acids in a protein molecule allows for only certain types of folding to occur, and structural biologists depict this folding as helical coils or flat sheets as shown here. Elucidation of the structure of many different microbial proteins has revealed the great unity as well as diversity that exists in these macromolecules throughout the microbial world.

CHAPTER 2 # Macromolecules

2.1   Strong and Weak Bonds   30
2.2   An Overview of Macromolecules and Water as the Solvent of Life   33
2.3   Polysaccharides   35
2.4   Lipids   36
2.5   Nucleic Acids   37
2.6   Amino Acids   40
2.7   Proteins: The Peptide Bond, Primary and Secondary Structures   43
2.8   Proteins: Higher Order Structure and Denaturation   44

Covalent bond a chemical bond in which electrons are shared between two atoms

Denaturation destructon of the folding properties of a protein leading (usually) to loss of biological activity

Glycosidic bond a type of covalent bond that links sugar units together in a polysaccharide

Hydrogen bond a weak chemical bond between a hydrogen atom and a second, more electronegative element, usually an oxygen or nitrogen atom

Lipid glycerol bonded to fatty acids and other groups such as phosphate by an ester or ether linkage

Macromolecule polymer of covalently linked monomeric units

Molecule two or more atoms chemically bonded to one another

Nonpolar possessing hydrophobic (water-repelling) characteristics and not easily dissolved in water

Nucleotide a monomer of a nucleic acid containing a nitrogen base (adenine, guanine, cytosine, thymine, or uracil), a molecule of phosphate, and a sugar, either ribose (in RNA) or deoxyribose (in DNA)

Peptide bond a type of covalent bond joining amino acids in a polypeptide

Phosphodiester bond a type of covalent bond linking nucleotides together in a polynucleotide

Polar possessing hydrophilic characteristics and generally water-soluble

Polymer a chemical compound formed by polymerization and consisting of repeating units called monomers

Polynucleotide a polymer of nucleotides bonded to one another by phosphodiester bonds

Polypeptide a polymer of amino acids bonded to one another by peptide bonds

Polysaccharide a polymer of sugar units bonded to one another by glycosidic bonds

Primary structure in an informational macromolecule, such as a polypeptide, the precise sequence of monomeric units

Protein a polypeptide or group of polypeptides that form a molecule of specific biological function

Quaternary structure in proteins, the number and arrangement of individual polypeptides in the final protein molecule

Secondary structure the initial pattern of folding of a polypeptide or a polynucleotide, usually dictated by opportunities for hydrogen bonding

Stereoisomer one form of a molecule that is the mirror image of another form of the same molecule

Tertiary structure the final folded structure of a polypeptide that has previously attained secondary structure

---

To understand microbiology and how cells work, one must have some understanding of the molecules present and chemical processes that take place within cells. Molecules, especially *macromolecules,* are the "guts" of the cell and are the subject of this chapter. It is assumed that the reader has some background in elementary chemistry, especially regarding the chemical nature of atoms and atomic bonding. In this chapter we will expand on this background with a primer on relevant biochemical bonds followed by a detailed discussion of the structure and function of the four classes of macromolecules: polysaccharides, lipids, nucleic acids, and proteins.

2.1

Strong and Weak Bonds

The major chemical elements of life include hydrogen, oxygen, carbon, nitrogen, phosphorous, and sulfur; these can bond in various ways to form the molecules of life. What is a **molecule?** A molecule can be defined as two or more atoms chemically bonded to one another. Thus, two oxygen (O) atoms can combine to form a molecule of oxygen (O_2). The chemical elements of life are able to form strong bonds in which electrons are shared more or less equally between atoms; these are called **covalent bonds.** To envision a covalent bond, consider the formation of a molecule of water from its constituent elements, O and H:

$$\overset{\circ\circ}{\underset{\circ\circ}{O}}\circ + 2H\bullet \longrightarrow H\overset{\circ\circ}{\underset{\circ\circ}{O}}H$$

Oxygen contains six electrons in its outermost shell while hydrogen has but a single electron. When they combine to form H_2O they do so by way of covalent bonds that maintain the two atoms in tight association and a stable state. In similar fashion, depending on the elements, of course, double and triple covalent bonds can form, and the strength of these bonds increases dramatically with their number (Figure 2.1).

Hydrogen Bonds and Other Weak Associations

Besides covalent bonds, a variety of much weaker bonds also plays an important role in biological molecules. Foremost among these are **hydrogen bonds.** Hydrogen bonds (Figure 2.2) form between hydrogen atoms and more electronegative elements, like oxygen or nitrogen. An individual hydrogen bond is very weak. However, when many hydrogen bonds are formed within and between molecules, the stability of the molecules may be greatly increased.

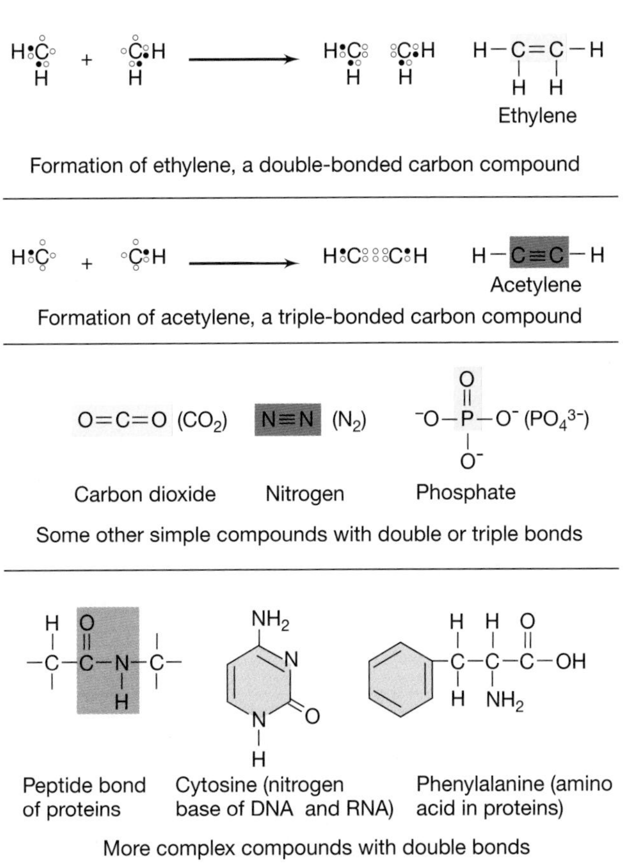

H:C° + °C:H ⟶ H:C° °C:H   H—C=C—H
 | | | |
 H   H   H H
                                   Ethylene

Formation of ethylene, a double-bonded carbon compound

H:C° + °C:H ⟶ H:C°°°C:H   H—C≡C—H
                                Acetylene

Formation of acetylene, a triple-bonded carbon compound

O=C=O (CO$_2$)   N≡N (N$_2$) $^-$O—P—O$^-$ (PO$_4$$^{3-}$)

Carbon dioxide   Nitrogen   Phosphate

Some other simple compounds with double or triple bonds

Peptide bond of proteins   Cytosine (nitrogen base of DNA and RNA)   Phenylalanine (amino acid in proteins)

More complex compounds with double bonds

FIGURE 2.1   Covalent bonding of some biologically important molecules containing double or triple bonds. Bond strength is measured in units of kilojoules (kJ), which is the amount of heat needed to break a bond. The bond strengths of some of the above molecules are: C=C, 616 kJ; C≡C, 805 kJ; N≡N, 955 kJ; C=O, 704 kJ.

Water molecules readily undergo hydrogen bonding (Figure 2.2a). Since an oxygen atom is rather electronegative (electron withdrawing), while a hydrogen atom is not, the covalent bond between oxygen and hydrogen is one in which the shared electrons in the outer shells actually orbit slightly nearer the oxygen nucleus than the hydrogen nucleus. Because electrons carry a negative charge, this creates a slight charge separation, oxygen negative and hydrogen positive (Figure 2.2a). As water molecules orient themselves in solution, the slight positive charge on the hydrogen atom can bridge negative charges on two oxygen atoms; this bridge is the hydrogen bond. Hydrogen bonds also form between atoms in macromolecules (Figure 2.2b and c). When these weak electrical forces accumulate in a large molecule like a protein, they improve the stability of the molecule and can also affect its overall structure. We will see later that hydrogen bonds play major roles in the biological properties of macromolecules, especially in the higher order struc-

(a) Hydrogen bonding between water molecules

(b) Hydrogen bonds between amino acids in protein chain. The R groups represent the unique side chain of each amino acid.

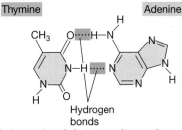

(c) Hydrogen bonds between nitrogen bases in DNA

FIGURE 2.2   Hydrogen bonding of (a) water, (b) proteins, and (c) nucleic acids.

ture of proteins (Figure 2.2b) and nucleic acids (Figure 2.2c) (see Sections 2.5 and 2.7).

Biomolecules form other types of weak interactions. For instance, van der Waals forces are weak attractive forces that occur between atoms when they obtain an interatomic distance of 3–4 angstroms (Å). These forces occur because of momentary charge asymmetries around atoms due to electron movement. However, if atoms become closer than 3–4 Å, overlap

TABLE 2.1	Functional groups of biochemical importance	
Chemical species	**Structure**	**Biological importance**
Carboxylic acid	$\overset{O}{\overset{\|}{-C-OH}}$	Organic, amino, and fatty acids; lipids, proteins
Aldehyde	$\overset{O}{\overset{\|}{-C-H}}$	Functional group of reducing sugars such as glucose; polysaccharides
Alcohol	$\overset{H}{\underset{H}{\overset{\|}{-C-OH}}}$	Lipids; carbohydrates
Keto	$\overset{O}{\overset{\|}{-C-}}$	Pyruvate, citric acid cycle intermediates
Ester	$\overset{H}{\underset{H}{\overset{\|}{-C}}}-O-\overset{O}{\overset{\|}{C-}}$	Lipids of Bacteria and Eukarya, amino acid attachment to tRNAs
Phosphate ester	$\overset{O^-}{\underset{O}{\overset{\|}{^-O-P}}}-O-\overset{\|}{\underset{\|}{C-}}$	Nucleic acids, DNA and RNA
Thioester	$R_1-\overset{O}{\overset{\|}{C}}\sim S-R_2$	Energy metabolism, biosynthesis of fatty acids
Ether	$\overset{H}{\underset{H}{\overset{\|}{-C}}}-O-\overset{H}{\underset{H}{\overset{\|}{C-}}}$	Lipids of Archaea; sphingolipids
Acid anhydride	$R-\overset{O}{\overset{\|}{C}}\sim O-\overset{O^-}{\underset{O}{\overset{\|}{P}}}=O$	Energy metabolism, for example, acetyl phosphate
Phosphoanhydride	$^-O-\overset{O^-}{\underset{O}{\overset{\|}{P}}}\sim O-\overset{O^-}{\underset{O}{\overset{\|}{P}}}-O^-$	Energy metabolism, for example, ATP

of the atom's electron shells causes repulsion rather than attraction. The action of van der Waals forces often play a significant role in the binding of substrates to enzymes (f Section 4.5) and in protein-nucleic acid interactions.

Hydrophobic interactions are also important in biomolecules. Hydrophobic interactions occur because nonpolar (water-repelling) molecules tend to cluster in an aqueous environment. Because of this, nonpolar portions of a macromolecule will associate if possible. Hydrophobic interactions can play a major role in the folding patterns of proteins, and, along with van der Waals interactions, play important roles in the binding of substrates to enzymes. In addition, hydrophobic interactions often control how different subunits in a multisubunit protein associate with one another to form the biologically active molecule.

Bonding Patterns in Biomolecules

The element **carbon** is a major component of all macromolecules. Carbon is able to combine not only with itself, but with many other elements as well, to yield large structures of considerable diversity and complexity. In different organic (carbon-containing) compounds, a variety of chemical combinations reappear with high frequency, and an awareness of and appreciation for these combinations will make later discussion of macromolecular structure, cell physiology, and biosynthesis easier to track. Table 2.1 lists several of these bonding patterns and the molecules or macromolecules they appear in most often. Each of these *functional groups,* as they are called, has unique chemical properties that may be important in their biological role within the cell.

✓ 2.1 Concept Check

Covalent bonds are strong bonds that hold together elements in macromolecules. Weak bonds, such as hydrogen bonds, van der Waals forces, and hydrophobic interactions, also affect macromolecular structure, but through more subtle atomic interactions. A variety of functional groups containing carbon atoms are common in important biomolecules.

- ✓ Why are *covalent* bonds stronger than *hydrogen* bonds?
- ✓ How can a hydrogen bond play a role in macromolecular structure?

2.2

An Overview of Macromolecules and Water as the Solvent of Life

If you were to biochemically dissect a prokaryotic cell like the common intestinal bacterium *Escherichia coli,* what would you find? You would find water as the major constituent, but after removing the water you would find large amounts of *macromolecules,* smaller amounts of *monomers* (the precursors of macromolecules), and a variety of inorganic ions (Table 2.2). In fact, 96% of the dry weight of a cell consists of macromolecules, and of these, proteins are by far the most abundant class (Table 2.2). **Proteins** are polymers of monomers called *amino acids.* Our *Escherichia coli* cell is capable of producing several thousand different proteins, but not all are produced at the same time or in the same amount; we will see in Chapter 7 that protein synthesis is such an energetically demanding process for the cell that considerable control is necessary of both the *types* and *amounts* of each protein the call can make. Proteins are found throughout the cell, in both structural as well as catalytic (enzymatic) roles (Figure 2.3*a*).

Nucleic acids are polymers of *nucleotides* and are found in the cell in two forms, **RNA** and **DNA**. Next to protein, ribonucleic acid (RNA) is the next most abundant macromolecule in an actively growing prokaryotic cell (Table 2.2 and Figure 2.3*b*). This is because there are thousands of ribosomes (the machines that make new proteins) in each cell, and ribosomes are composed of RNA and protein. In addition, smaller amounts of RNA are present in the form of messenger and transfer RNAs, other key players in protein synthesis. In contrast to RNA, DNA makes up a relatively insignificant (by weight) fraction of the bacterial cell (Table 2.2). However, although quantitatively rather minor, DNA is of course critical to cell function as the repository of genetic information needed to make a new cell.

TABLE 2.2	Chemical composition of a prokaryotic cell[a]		
Molecule	**Percent of dry weight[b]**	**Molecules per cell**	**Different kinds**
Total macromolecules	96	24,610,000	~2500
Protein	55	2,350,000	~1850
Polysaccharide	5	4,300	2[c]
Lipid	9.1	22,000,000	4[d]
Lipopolysaccharide	3.4	1,430,000	1
DNA	3.1	2.1	1
RNA	20.5	255,500	~660
Total monomers	3.0		~350
Amino acids and precursors	0.5		~100
Sugars and precursors	2		~50
Nucleotides and precursors	0.5		~200
Inorganic ions	1		18
Total	100%		

a Data from Neidhardt, F. C., et al. (eds.), 1996. *Escherichia coli* and *Salmonella typhimurium—Cellular and Molecular Biology,* 2nd edition. American Society for Microbiology, Washington, DC.

b Dry weight of an actively growing cell of *E. coli* $\cong$ 2.8×10^{-13} g; total weight (70% water) = 9.5×10^{-13} g.

c Assuming peptidoglycan and glycogen to be the major polysaccharides present.

d There are several classes of phospholipids, each of which exists in many kinds because of variability in fatty acid composition between species and because of different growth conditions.

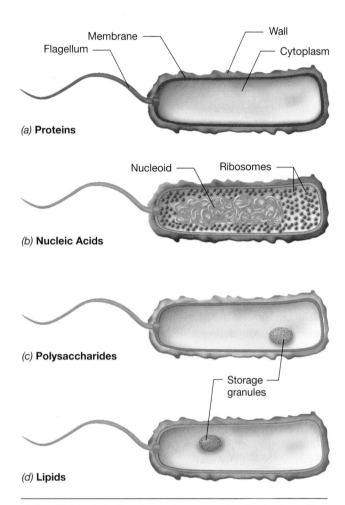

(a) **Proteins**

(b) **Nucleic Acids**

(c) **Polysaccharides**

(d) **Lipids**

FIGURE 2.3 The locations of macromolecules in the cell. (a) *Proteins* (brown) are found throughout the cell as both parts of cell structures and as enzymes. (b) *Nucleic acids.* DNA (green) is found in the nucleoid of prokaryotic cells and in the nucleus of eukaryotic cells. RNA (orange) is found in the cytoplasm (mRNA, tRNA) and in ribosomes (rRNA). (c) *Polysaccharides* (yellow) are located in the cell wall and occasionally in internal storage granules. (d) *Lipids* (blue) are found in the cytoplasmic membrane, the cell wall, and in storage granules. Note the color-coding scheme used here; these same colors will be used to depict these macromolecules throughout this book. For DNA, see also the legend to Figure 2.11.

Lipids are an important component of the bacterial cell and play crucial roles in membrane structure as well as in serving as storage depots for excess carbon (Figure 2.3c). Unlike other macromolecules, lipids are not simply polymers of monomers, but contain various, primarily hydrophobic components like *fatty acids,* as will be discussed later in this chapter. **Polysaccharides** are polymers of *sugars* and are present primarily in the cell wall; however, as for lipids, polysaccharides like glycogen (see later in this chapter) can be major forms of carbon and energy storage in the cell (Figure 2.3d).

Water as a Biological Solvent

Macromolecules and all other molecules in cells are bathed in water. Water has several important chemical features that make it an ideal biological solvent; indeed, water is a necessary prerequisite for life as we know it. The key chemical properties of water that make it such a good solvent are its *polarity* and *cohesiveness.*

The polar properties of water are important because many biologically important molecules are themselves polar and thus readily dissolve in water. As we will see in Chapter 3, dissolved substances are continually being passed into and out of the cell through transport activities of the cytoplasmic membrane (∞ Section 3.6). The polar properties of water also promote hydrogen bonding. Water forms three-dimensional networks, both with itself (Figure 2.2a) and within macromolecules, and by so doing, helps to spatially position atoms within biomolecules for potential interactions. The high polarity of water is also beneficial to the cell because it tends to force *nonpolar* substances to aggregate and remain together. Membranes, for example, contain nonpolar macromolecules such as lipids, and these tend to aggregate and function as a barrier to the free flow of polar molecules in and out of the cell.

In addition to hydrogen bonding, the polar nature of water makes it highly cohesive, meaning that water molecules tend to have a high affinity for one another and form chemically ordered arrangements in which hydrogen bonds (Figure 2.2) are constantly forming, breaking, and re-forming. The cohesive nature of water is responsible for many of its biologically important properties, such as *high surface tension* and *high specific heat.* Also, the fact that water expands on freezing to yield a less dense solid form (ice) has a profound effect on life in temperate and polar aquatic environments. In a lake, for example, ice on the surface insulates the water beneath the ice and prevents its freezing, thus allowing aquatic organisms to survive despite the overlying ice.

Life originated in water, and anywhere on Earth where liquid water exists, microorganisms are likely to be found (and we will see some spectacular examples of this in ∞ Section 5.9 and Chapter 14). We now consider the structure of the major macromolecules of life (Table 2.2).

✓ 2.2 Concept Check

Proteins, polymers of amino acids, are the most abundant class of macromolecule in the cell. Other macromolecules include the nucleic acids (DNA and RNA), lipids, polysaccharides, and lipopolysaccharides. Water is a particularly good solvent for living organisms because of its polarity and cohesiveness.

✓ Why does RNA make up such a large proportion of an actively growing cell?

✓ Why does the high polarity of water make it useful as a biological solvent?

Polysaccharides

Carbohydrates (sugars) are organic compounds containing carbon, hydrogen, and oxygen in a ratio of 1:2:1. The structural formula for glucose, the most abundant of all sugars, is $C_6H_{12}O_6$ (Figure 2.4). The most biologically relevant carbohydrates are those containing 4, 5, 6, and 7 carbon atoms (designated as C_4, C_5, C_6, and C_7). C_5 sugars (pentoses) are of special significance because of their role as structural backbones of nucleic acids. Likewise, C_6 sugars (hexoses) are the monomeric constituents of cell wall polymers and energy reserves. Figure 2.4 shows the structural formulas of a few common sugars. Derivatives of simple carbohydrates can be formed by replacing one or more of the hydroxyl groups

by other chemical species. For example, the important bacterial cell wall polymer **peptidoglycan** (👓 Section 3.7) contains the glucose derivatives *N*-acetylglucosamine and *N*-acetylmuramic acid (Figure 2.5). Besides sugar derivatives, sugars having the same *structural* formula can still differ in their *stereoisomeric* properties (see Section 2.6). Hence, a large number of different sugars are potentially available to the cell for the construction of polysaccharides.

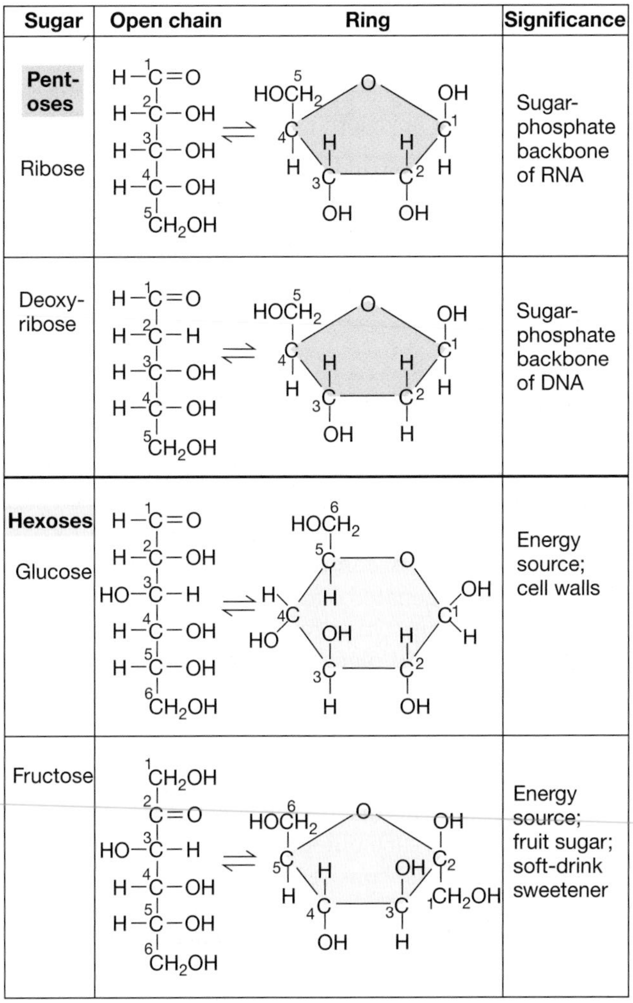

The Glycosidic Bond

Polysaccharides are high-molecular-weight carbohydrates containing many (sometimes hundreds or even thousands) monomeric units connected to one another by covalent bonds referred to as **glycosidic bonds** (Figure 2.6). If two sugar units (monosaccharides) are joined by a glycosidic linkage, the resulting molecule is called a *disaccharide*. The addition of one more monosaccharide yields a *trisaccharide*, and several more an *oligosaccharide*; an extremely long chain of monosaccharides in glycosidic linkage is called a **polysaccharide.**

The glycosidic bond can exist in two different orientations, referred to as alpha (α) and beta (β) (Figure 2.6*a*). Polysaccharides with a repeating structure composed of glucose units linked between carbons 1 and 4 in the *alpha* orientation (for example, glycogen and starch, Figure 2.6*b*) function as important carbon and energy reserves in bacteria, plants, and animals. Alternatively, glucose units

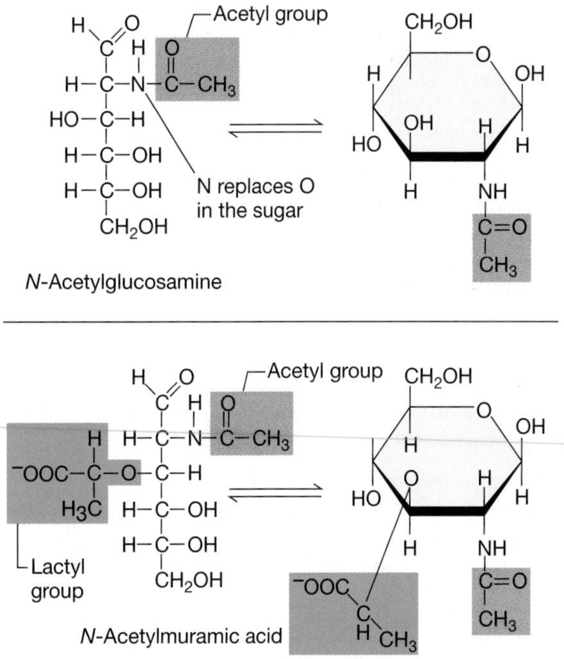

FIGURE 2.4 Structural formulas of a few common sugars. The formulas can be represented in two alternate ways, open chain and ring. The open chain is easier to visualize, but the ring form is the commonly used structure. Note the numbering system on the ring.

FIGURE 2.5 Sugar derivatives found in the cell walls of most Bacteria, in the polysaccharide called *peptidoglycan*. Note that the parent structure is *glucose* in both cases.

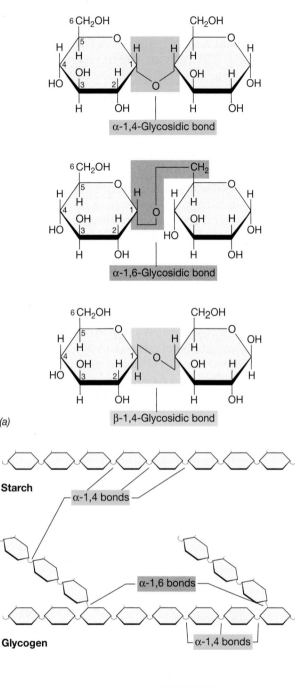

(a)

Starch

α-1,4 bonds

Glycogen

α-1,6 bonds

α-1,4 bonds

β-1,4 bonds

Cellulose

(b)

FIGURE 2.6   Polysaccharides (a) Structure of different glycosidic bonds. Note that both the *linkage* and the *geometry* of the linkage can vary about the glycosidic bond. (b) General structures of some common polysaccharides. Note color coding to (a).

joined by *beta*-1,4 linkages are present in cellulose (Figure 2.6*b*), a stiff plant and algal cell wall component functionally unrelated to glycogen or starch. Thus, even though starch and cellulose are both composed solely of glucose units, their functional properties are entirely different because of the different configurations, α and β, of their glycosidic bonds. Polysaccharides can also combine with other classes of macromolecules, such as protein and lipid, to form complex polysaccharides like **glycoproteins** and **glycolipids.** These compounds play important roles in cell membranes as cell surface receptor molecules. The compounds reside on the external surfaces of the membrane where they are in contact with the environment. Glycolipids also constitute a major portion of the cell wall of gram-negative bacteria and as such impart a number of unique surface properties to these organisms (∞ Section 3.8).

✓ 2.3 Concept Check

Sugars (carbohydrates) combine into long polymers called polysaccharides. Polysaccharides can also contain other molecules such as protein or lipid, forming complex polysaccharides.

✓ How can glycogen and cellulose differ so much in their physical properties when they both consist of 100% glucose?

2.4

Lipids

Fatty acids are the main constituents of **lipids** of Bacteria and Eukarya. Fatty acids have interesting chemical properties because they contain both highly hydrophobic (water-repelling) and highly hydrophilic (water-soluble) regions. Palmitate*, for example (Figure 2.7), is a 16-carbon fatty acid composed of a chain of 15 saturated (fully hydrogenated) carbon atoms and a single carboxylic acid group. Other common fatty acids are stearic (C_{18} saturated) and oleic (C_{18} monounsaturated) acids (Figure 2.7).

Triglycerides and Complex Lipids

Lipids are not true macromolecules as the monomers are not linked to each other by covalent bonds. Instead, simple lipids (fats) consist of fatty acids bonded to the C_3 alcohol *glycerol* (Figure 2.7). Simple lipids are also referred to as **triglycerides** because three fatty acids are linked to the glycerol molecule.

---

*Fatty acids can exist in both protonated (RCOOH) and unprotonated (RCOO⁻) forms, depending on pH. Because at pH 7 fatty acids are generally unprotonated, this is indicated by adding the suffix *-ate* to the root term for the fatty acid. Thus, palmitic acid is $C_{15}H_{31}COOH$, and palmitate is $C_{15}H_{31}COO^-$.

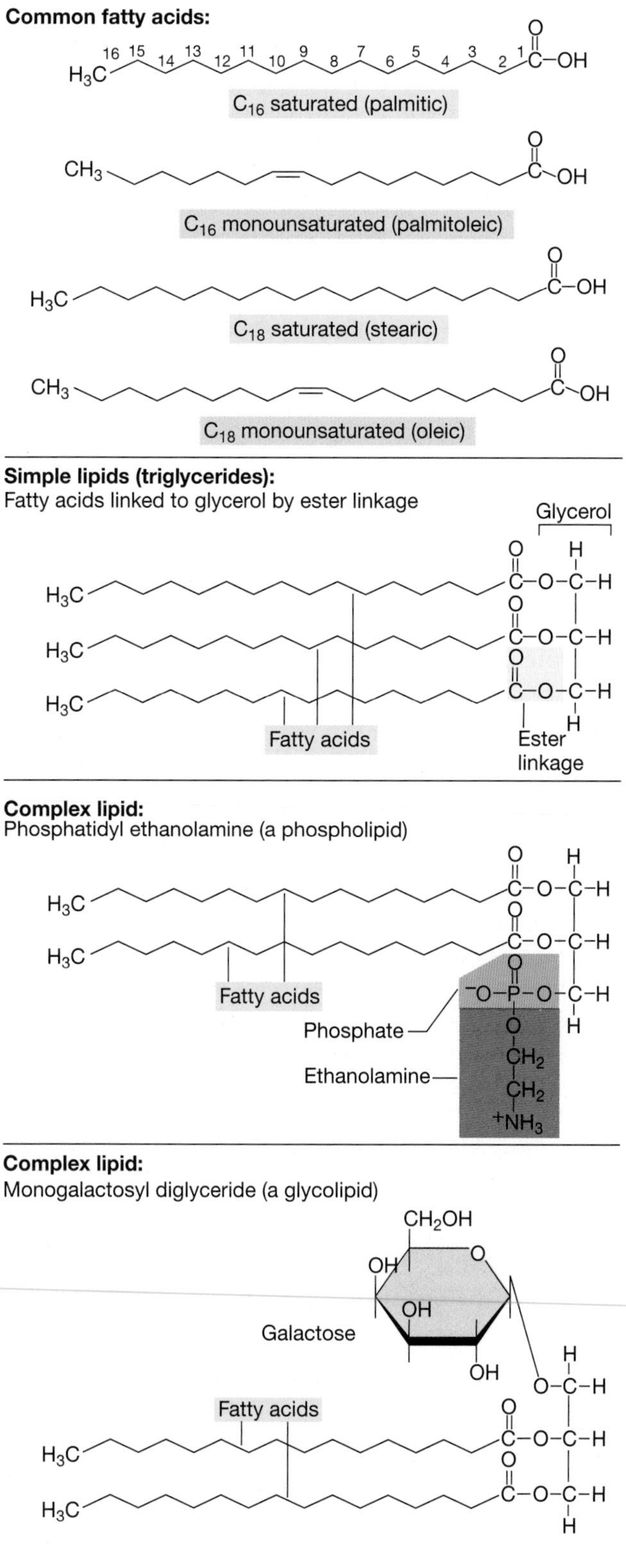

Common fatty acids:

C$_{16}$ saturated (palmitic)

C$_{16}$ monounsaturated (palmitoleic)

C$_{18}$ saturated (stearic)

C$_{18}$ monounsaturated (oleic)

Simple lipids (triglycerides):
Fatty acids linked to glycerol by ester linkage

Glycerol

Fatty acids

Ester linkage

Complex lipid:
Phosphatidyl ethanolamine (a phospholipid)

Fatty acids

Phosphate

Ethanolamine

Complex lipid:
Monogalactosyl diglyceride (a glycolipid)

Galactose

Fatty acids

FIGURE 2.7 Fatty acids, simple lipids (fats), and complex lipids. Simple lipids are formed by a dehydration reaction between fatty acids and glycerol to yield the ester linkage.

Complex lipids are simple lipids that contain additional elements such as phosphorous, nitrogen, or sulfur, or small hydrophilic carbon compounds such as sugars, ethanolamine, serine, or choline (Figure 2.7). Lipids containing a phosphate group, called **phospholipids,** are a very important class of complex lipids, as they play a major structural role in the cytoplasmic membrane (∞ Section 3.5).

The chemical properties of lipids make them ideal structural components of membranes. Because they are *amphipathic,* that is, show properties of hydrophobicity and hydrophilicity, lipids aggregate in membranes with the hydrophilic portions toward the external or internal (cytoplasmic) environment, while maintaining their hydrophobic portions away from the aqueous milieu (∞ Section 3.5). Such structures are ideal permeability barriers because of the inability of water-soluble substances to flow through the hydrophobic fatty acid portion of the lipids. Indeed, the major function of the cytoplasmic membrane is to serve as a barrier to the diffusion of substances into or out of the cell.

✓ 2.4 Concept Check

Lipids contain both hydrophobic and hydrophilic units; their chemical properties make them ideal structural components for cell membranes.

✓ What part of a fatty acid molecule is hydrophobic? Hydrophilic?
✓ How does a *phospholipid* differ from a *triglyceride*?
✓ Draw the chemical structure of butyrate, a C$_4$ fully saturated fatty acid.

2.5

Nucleic Acids

The nucleic acids, deoxyribonucleic acid (DNA), and ribonucleic acid, (RNA), are macromolecules of monomers called **nucleotides.** DNA and RNA are thus both **polynucleotides.** DNA carries the genetic blueprint for the cell, and RNA acts as an intermediary molecule to convert the blueprint into defined amino acid sequences in proteins. Each nucleotide is composed of three separate units: a five-carbon sugar, either ribose (in RNA) or deoxyribose (in DNA), a nitrogen base, and a molecule of phosphate, PO_4^{3-}. Figure 2.8 shows the general structure of nucleotides of DNA and RNA.

Nucleotides

The nitrogen bases of nucleic acids belong to either of two chemical classes. *Purine* bases, **adenine** and **guanine,** contain two fused heterocyclic rings, whereas *pyrimidine* bases, **thymine, cytosine,** and **uracil,** contain a single six-

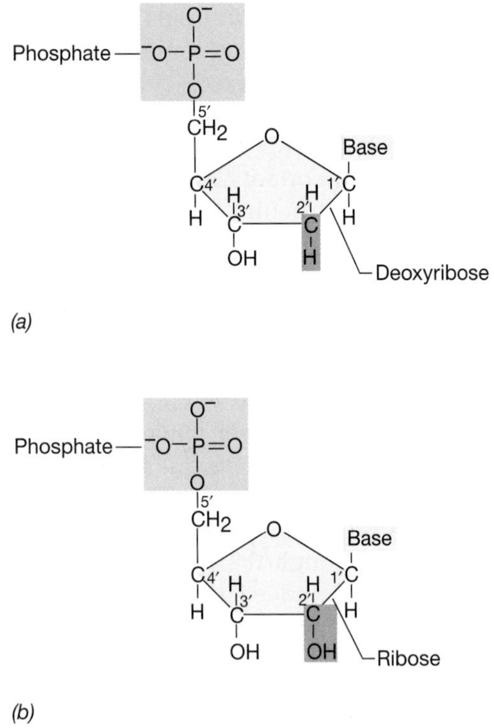

(a)

(b)

FIGURE 2.8   Nucleotides. (a) Of DNA. (b) Of RNA. Note the numbering system employed and the chemical differences on carbon atom 2′ between deoxyribose and ribose. The numbers on the pentose sugars contain a prime (′) after them because the ring structure in the nitrogen base is also numbered (1, 2, 3, etc.) (see Figure 2.9). Both deoxyribonucleotides (in DNA) and ribonucleotides (in RNA) contain a 5′-phosphate.

membered heterocyclic ring (Figure 2.9). Guanine, adenine, and cytosine are found in both DNA and RNA; thymine is present (with minor exceptions) only in DNA,

and uracil is present only in RNA. In a nucleotide, a base is attached to a pentose sugar by a glycosidic linkage between carbon atom 1 of the sugar and a nitrogen atom of the base, either the nitrogen atom labeled atom 1 (pyrimidine base) or nitrogen atom 9 (purine base). Without a phosphate, a base bonded to its sugar is referred to as a **nucleoside.** Nucleotides are thus nucleosides containing one or more phosphates (Figure 2.10).

Nucleotides play other roles in the cell besides their major role as constituents of nucleic acids. Nucleotides, especially adenosine triphosphate (ATP) (Figure 2.10), function as sources of chemical energy, releasing sufficient energy during the hydrolytic cleavage of a phosphate bond to drive energy-requiring reactions in the cell (∞ Section 4.8). Other nucleotides or nucleotide derivatives function in oxidation–reduction reactions in the cell (∞ Section 4.6) as carriers of sugars in the biosynthesis of polysaccharides (∞ Section 4.15) and as regulatory molecules inhibiting or stimulating the activities of certain enzymes or metabolic events. However, here we are discussing the role of a nucleotide as a building block of nucleic acid, the major *informational* function of nucleotides.

Nucleic Acids

Nucleic acids are polymers in which nucleotides are covalently bonded to one another in a defined sequence, forming structures called **polynucleotides.** The backbone of the nucleic acid is a polymer in which sugar and phosphate molecules alternate (Figure 2.11). Hence, when we refer to a specific *sequence* of nucleotides in a nucleic acid, we are really referring to the variable portions of the nucleotide—the nitrogen bases; the sugar–phosphate backbone is always the same.

In chemical terms nucleic acids are composed of nucleotides covalently attached to one another via phosphate from carbon 3 [referred to as the 3′ (3 prime)

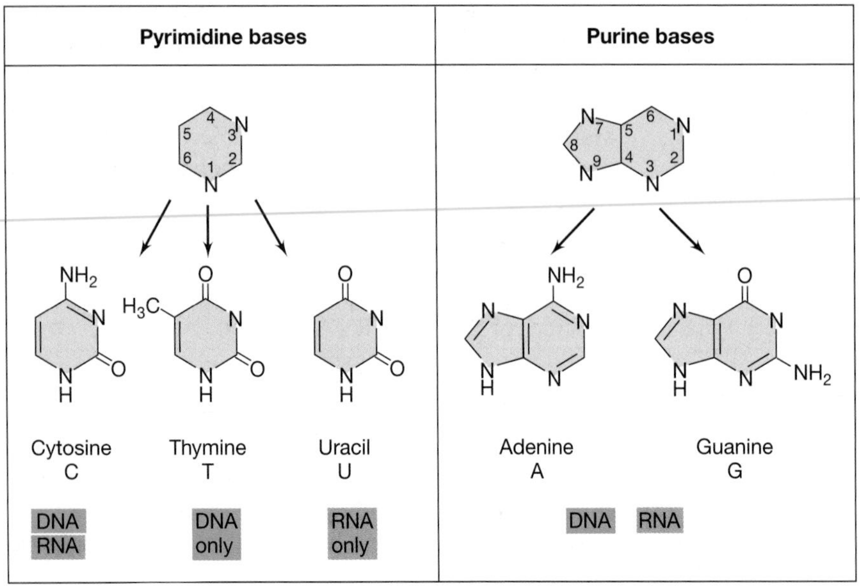

FIGURE 2.9   Structure of the bases of DNA and RNA. The letters C, T, U, A, and G are used to designate the individual bases. Note the numbering system of the rings. In attaching the base to the 1′ carbon of the sugar phosphate shown in Figure 2.8, pyrimidine bases are bonded through N-1 of the ring and purine bases through N-3 of the ring.

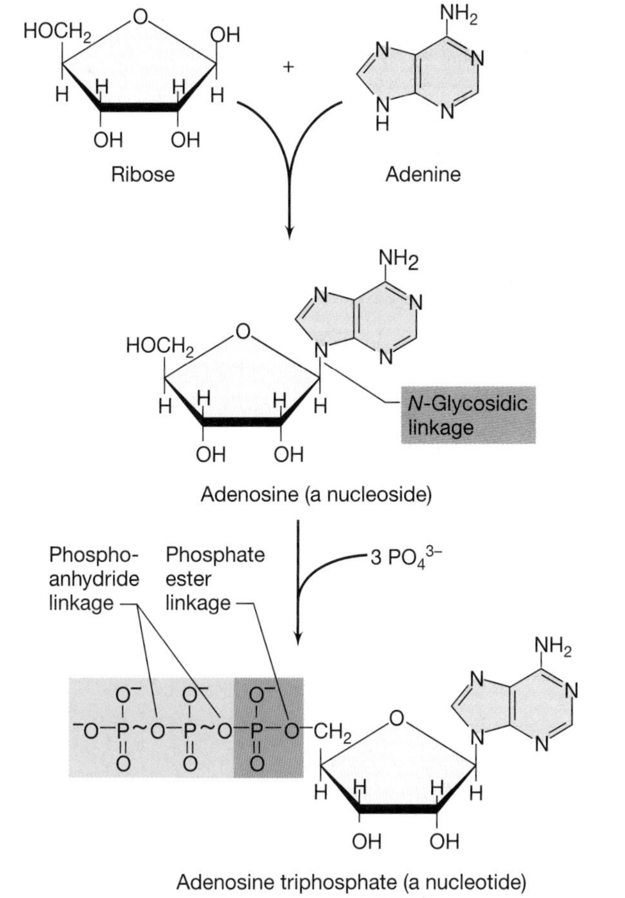

FIGURE 2.10 Components of the important nucleotide triphosphate, adenosine triphosphate. The energy of hydrolysis of a phosphoanhydride bond (shown as squiggles) is greater than that of a phosphate ester and will have significance in Chapter 4 (co Section 4.8).

carbon] of one sugar to carbon 5 (5′) of the adjacent sugar (Figure 2.11a). The phosphate linkage is chemically a **phosphodiester** since a single phosphate is connected by ester linkage to two separate sugars.

The precise sequence of nucleotides in a DNA or RNA molecule is referred to as its *primary structure*. As we have discussed, the sequence of bases in a DNA or RNA molecule is *informational*, representing the genetic information necessary to reproduce an identical copy of the organism. We will see later that the replication of DNA and the production of RNA are highly complex processes (co Chapter 6) and that a virtually error-free mechanism is necessary to ensure the faithful transfer of genetic traits from one generation to another.

DNA

In cells, DNA is present in double-stranded form. Each cellular chromosome contains two strands of DNA, each strand containing several million nucleotides linked by phosphodiester bonds. The strands them-

selves associate with one another by hydrogen bonds that form between the nucleotides of one strand and the nucleotides of the other. When positioned adjacent to one another, purine and pyrimidine bases can undergo hydrogen bonding (see Figure 2.2c). The most stable hydrogen bonding configuration occurs when guanine (G) forms hydrogen bonds with cytosine (C), and adenine (A) forms hydrogen bonds with thymine (T) (see Figure 2.2c). Specific base pairing, A with T and G with C, means that the two strands of DNA are *complementary* in base sequence; wherever a G is found in one strand, a C is found in the other, and wherever a T is present in one strand, its complementary strand has an A (Figure 2.11b). The molar amounts of guanine and cytosine are therefore *identical* in double-stranded DNA from any source; likewise, the molar amounts of adenine and thymine are the same. It should also be noted that although DNA is generally double-stranded, some viruses contain only single-stranded DNA (co Section 8.1).

RNA

With the exception of certain viruses that contain double-stranded RNA, all ribonucleic acids are *single-stranded molecules*. However, RNA molecules can fold back upon themselves in regions where complementary base pairing can occur to form a variety of highly folded structures. The pattern of folding observed in RNA is referred to as its *secondary structure* (Figure 2.11c).

RNA plays three crucial roles in the cell. **Messenger RNA** (mRNA) contains the genetic information of DNA in a single-stranded molecule *complementary* in base sequence to a portion of the base sequence of DNA. **Transfer RNA** (tRNA) molecules are the "adaptor" molecules in protein synthesis. The tRNA molecule effectively adapts the genetic information from the language of nucleotides to the language of amino acids, the building blocks of proteins. **Ribosomal RNA** (rRNA) molecules, of which several distinct types are known, are important structural and catalytic components of the ribosome, the protein-synthesizing system of the cell. These various RNA molecules are discussed in detail in Chapter 6.

✓ 2.5 Concept Check

The informational content of a nucleic acid is determined by the sequence of nitrogen bases along the polynucleotide chain. Both RNA and DNA are informational macromolecules. RNA can fold into various configurations to obtain secondary structure.

✓ What is a *nucleotide*?
✓ How does a nucleo*side* differ from a nucleo*tide*?
✓ Distinguish between the *primary* and *secondary* structure of RNA.

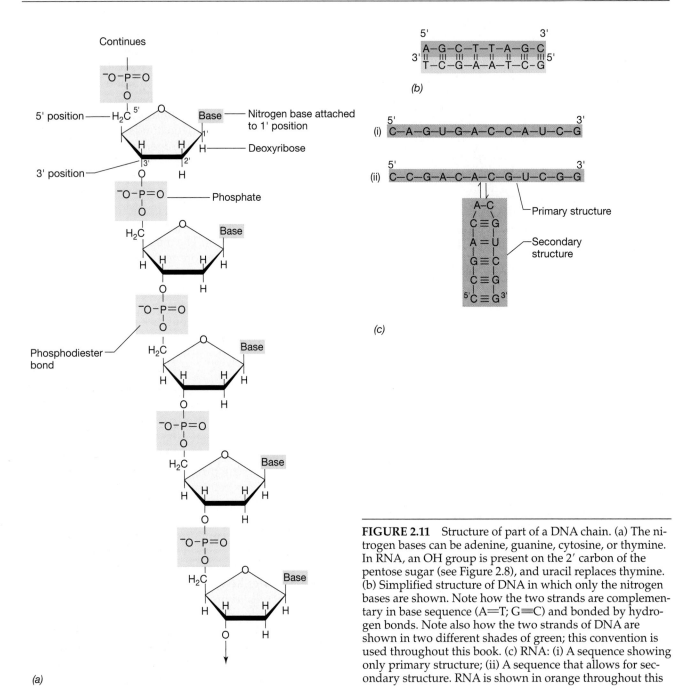

(a)

(b)

(c)

FIGURE 2.11 Structure of part of a DNA chain. (a) The nitrogen bases can be adenine, guanine, cytosine, or thymine. In RNA, an OH group is present on the 2′ carbon of the pentose sugar (see Figure 2.8), and uracil replaces thymine. (b) Simplified structure of DNA in which only the nitrogen bases are shown. Note how the two strands are complementary in base sequence (A=T; G≡C) and bonded by hydrogen bonds. Note also how the two strands of DNA are shown in two different shades of green; this convention is used throughout this book. (c) RNA: (i) A sequence showing only primary structure; (ii) A sequence that allows for secondary structure. RNA is shown in orange throughout this

2.6

Amino Acids

Amino acids are the monomeric units of proteins. Most amino acids consist of only carbon, hydrogen, oxygen, and nitrogen, but 2 of the 21 common amino acids found in cells also contain sulfur and 1 contains selenium. All amino acids contain two important functional groups, a *carboxylic acid* group (—COOH) and an *amino* group (—NH₂) (Figure 2.12). These groups are functionally important because covalent bonds, between the carbon of the carboxyl group of one amino acid and the nitrogen of the

amino group of a second amino acid (with elimination of a molecule of water), form the **peptide bond,** a type of covalent bond characteristic of proteins (see Figure 2.13).

All amino acids conform to the general structure shown in Figure 2.12. Amino acids differ in the nature of the side group (abbreviated R in Figure 2.12) attached to the α-carbon. The α-carbon is the carbon atom *immediately adjacent* to the carboxylic acid group. The side chains on the alpha carbon vary considerably, from as simple as a hydrogen atom in the amino acid glycine, to aromatic ringed structures in amino acids such as phenylalanine (Figure 2.12). The chemical properties of an amino acid are to a major degree governed by the nature of the side

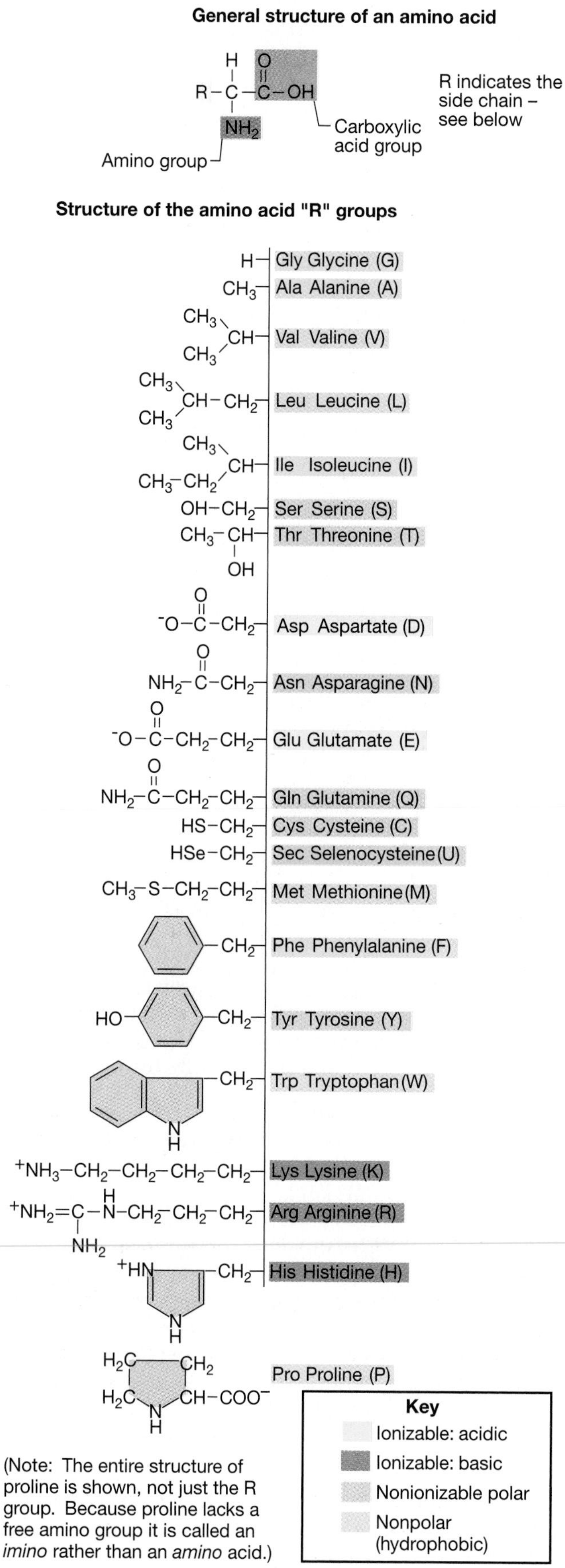

General structure of an amino acid

R indicates the side chain – see below

Amino group — NH₂

Carboxylic acid group

Structure of the amino acid "R" groups

H— Gly Glycine (G)
CH₃— Ala Alanine (A)
Val Valine (V)
Leu Leucine (L)
Ile Isoleucine (I)
OH–CH₂— Ser Serine (S)
Thr Threonine (T)
Asp Aspartate (D)
Asn Asparagine (N)
Glu Glutamate (E)
Gln Glutamine (Q)
HS–CH₂— Cys Cysteine (C)
HSe–CH₂— Sec Selenocysteine (U)
CH_3–S–CH₂–CH₂— Met Methionine (M)
Phe Phenylalanine (F)
Tyr Tyrosine (Y)
Trp Tryptophan (W)
Lys Lysine (K)
Arg Arginine (R)
His Histidine (H)
Pro Proline (P)

Key
Ionizable: acidic
Ionizable: basic
Nonionizable polar
Nonpolar (hydrophobic)

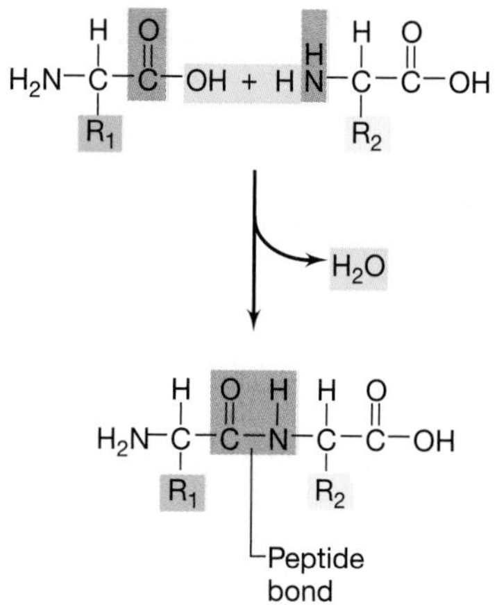

(Note: The entire structure of proline is shown, not just the R group. Because proline lacks a free amino group it is called an *imino* rather than an *amino* acid.)

FIGURE 2.13 Peptide bond formation. R_1 and R_2 refer to the variable portion (side chain) of the amino acid (see Figure 2.12).

chain, and thus amino acids that show similar chemical properties can be grouped into amino acid "families" as shown in Figure 2.12. For example, the side chain may itself contain a carboxylic acid group, such as in aspartic acid or glutamic acid, rendering the amino acid acidic. Alternatively, several amino acids contain nonpolar hydrophobic side chains and are grouped together as nonpolar amino acids. The amino acid cysteine contains a sulfhydryl group (— SH), which is frequently important in connecting one chain of amino acids to another by *disulfide linkage* (R—S—S—R). In summary, the large number of chemically distinct amino acids that exist makes it possible for cells to produce an enormous number of proteins with different amino acid sequences and with widely different biochemical properties.

Stereoisomerism

Two molecules may have the same molecular formula but exist in different structural forms. These related but not identical molecules are referred to as **isomers** (Figure 2.14) (see the box, Optical Isomers and Life). Isomers are important in biology, especially in the chemistry of sugars and amino acids. Many isomers of common sugars are found as constituents of the cell walls of Bacteria and Archaea (∞ Section 3.7). Some sugar isomers are known that contain the same molecular and structural formulas except that one is a "mirror image" of the other, just as the left hand is a mirror image of the right. Two identical sugars that are mirror images of one another are called **stereoisomers** or **enantiomers** and have been given the designations

FIGURE 2.12 Structure of the 21 common amino acids. The three-letter codes for the amino acids are to the left of the names, and the one-letter codes are in parentheses to the right of the names.

A FOCUS ON . . .   Optical Isomers and Life

Louis Pasteur was trained as a chemist, and his early research work was on the chemistry of stereoisomers, specifically, on the optical properties of crystals of tartaric acid (see the accompanying figure), published in 1848. Because stereoisomers are mirror images (that is, they show handedness), they are also *chiral,* meaning that they are optically active: crystals or solutions of one stereoisomer rotate polarized light in one direction, whereas the other stereoisomer rotates light in the opposite direction. Pasteur also noticed that racemic mixtures of tartaric acid were often contaminated with fungi, but that as the organisms grew on the tartaric acid, they consumed only one of the two stereoisomeric forms. When this occurred, an optically inactive solution of tartaric acid became optically active! From these studies Pasteur recognized the significance of asymmetry in living systems, and his interest in this problem led him from chemistry into biology, a field that would occupy him for the rest of his life.

For Pasteur, the fact that living organisms discriminated between stereoisomers was profoundly significant. He saw living processes as inherently asymmetric, whereas nonliving chemical processes were not asymmetric. Pasteur's vision of the asymmetry of life has been well confirmed over the nearly 150 years since he did his work. In fact, when Pasteur began to study spontaneous generation (∞ Section 1.8), he used his knowledge of the asymmetry of living organisms to bolster his confidence that spontaneous generation could not exist, a concept that in many ways brought forth the field of microbiology as an experimental science. In addition, Pasteur's early conclusions that fermentations such as the lactic

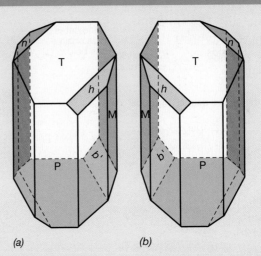

(a)   (b)

Pasteur's drawings of tartaric acid ($C_4H_6O_6$) crystals, used to illustrate his famous paper on optical activity. (a) Left-handed crystal (L form). (b) Right-handed crystal (D form). Note that the two crystals are mirror images. The letters on the faces of the crystals were Pasteur's way of labeling the mirror image faces of the two crystals. Color has been added here to show the mirror image faces more clearly.

acid and alcoholic fermentations are due to the activities of living organisms were based in part on his discovery that chemical reactions in living organisms, in contrast to abiotic systems, yield highly asymmetric products.

Why is it that the amino acids of proteins in living organisms are of only the L-configuration, whereas when the same amino acids are made chemically, they are racemic, that is, they consist of mixtures of D and L forms? We know that this is because chemical reactions in living organisms are carried out by asymmetric catalysts, **enzymes.** Asymmetry begets asymmetry, and because enzymes are asymmetric, they introduce asymmetry into the molecules they produce. (It should be noted that D-amino acids are found in a few biological molecules, most notably bacterial cell walls and a few antibiotics, but are absent from cellular proteins.)

The preceding statement begs the question of how the *first*

asymmetry arose in the living world. Was it by chance or by design? Is there somewhere in the universe a planet with living organisms whose proteins are comprised of only D-amino acids? Possibly. There is no biochemical reason such proteins would not be just as functional as ones made from L-amino acids, and in fact, experiments in which enzymes have been totally synthesized from D-amino acids have shown them to be mirror images of their L-form counterparts and to show the opposite chiral specificity in their substrate requirements and products formed.

Although the origin of the L-form preference for amino acids in life on Earth may never be understood, recent radiotelescopic observations of stars have shown that L-form amino acids may indeed predominate throughout the universe. If true, their dominance in life on Earth may just have been a matter of chance and not such a conundrum after all. ■

D and L (Figure 2.14*b*). D Sugars predominate in biological systems.

Like sugars, amino acids can exist as D or L stereoisomers. However, in the case of proteins, life has evolved to use the L form rather than the D (Figure 2.14*c*) (see the box). Nevertheless, D-amino acids are found occasionally in nature, most commonly in the cell wall polymer peptidoglycan (⚬ Section 3.7) and in certain peptide antibiotics (⚬ Section 18.6). Prokaryotes are equipped to handle the conversions of D-amino acids to the L form or L-sugars to the D form by way of enzymes that specifically catalyze this transformation. Cells contain enzymes called **racemases** whose function is to convert the unusual form (L sugar or D-amino acid) to the readily metabolizable form (D sugar or L-amino acid).

✓ 2.6 Concept Check

Twenty-one common amino acids are found in cells and can bond to each other via the *peptide bond*. Stereoisomeric (mirror-image) forms of sugars and amino acids exist, but only one stereoisomer of each is found in cell polysaccharides and proteins, respectively.

✓ Why can it be said that all amino acids are structurally similar yet different at the same time?
✓ Draw the structure of a dipeptide containing the amino acids alanine and tyrosine. Outline the peptide bond.
✓ What stereoisomeric form of sugars and amino acids are commonly found in living organisms? Why doesn't the amino acid *glycine* have different enantiomers?

2.7

Proteins: The Peptide Bond, Primary and Secondary Structure

Proteins play key roles in cell function. Two major classes of proteins are *catalytic* proteins (enzymes) and *structural* proteins. Enzymes serve as catalysts for the wide variety of chemical reactions that occur in cells (⚬ Chapters 4 and 15). Structural proteins are those that become integral parts of the structures of cells in membranes, walls, and cytoplasmic components. In essence, a cell is what it is because of the kinds of proteins it contains. Therefore, an understanding of protein structure is essential for an understanding of cell function.

Primary Structure

Proteins are polymers of variable length containing defined sequences of amino acids covalently bonded by peptide bonds (Figure 2.13). Two amino acids bonded together constitute a *dipeptide*, three amino acids a *tripeptide*, and so on. Many amino acids covalently linked via peptide bonds constitute a **polypeptide,** and proteins consist of one or more polypeptides. The number of amino acids in a protein varies from one protein to another. Proteins with as few as 15 and as many as 10,000 amino acids are known. Since proteins may vary in their composition, sequence, and number of amino acids, it is easy to see that enormous variation in protein structure (and thus function) is possible.

Proteins are folded molecules and show complex arrangements of structure. The linear array of amino acids is referred to as the **primary structure** of the polypeptide; primary structure thus gives a complete description of all covalent bonds present in the molecule. In many ways the primary structure of a polypeptide can be considered the most important, because a

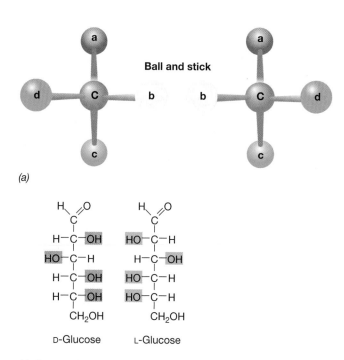

FIGURE 2.14 Stereoisomers. (a) Ball-and-stick model showing mirror images (stereoisomers). (b) Stereoisomers of glucose. (c) Stereoisomers of the amino acid alanine.

given primary structure allows only certain types of higher order structure to occur.

Secondary Structure

The juxtaposition of α-carbon R groups dictated by the primary structure forces the polypeptide to twist and fold in a specific way. This process leads to formation of the **secondary structure** of the protein (Figure 2.15). Hydrogen bonds, the weak noncovalent linkages discussed earlier (see Section 2.1), play important roles in the type of secondary structure that a protein attains. A typical secondary structure for many polypeptides is the α-*helix*. To envision a protein helix, imagine a linear polypeptide wound around a cylinder (Figure 2.15a). Under these conditions oxygen and nitrogen atoms from different amino acids become positioned close enough together in the twisted structure to allow hydrogen bonding to occur. This opportunity for H bonding (and the inherent stability associated with it) helps direct many polypeptides to take on an α-helix secondary structure (Figure 2.15a).

Many polypeptides conform to a different type of secondary structure referred to as the β-*sheet*. In the β-sheet, the chain of amino acids in the polypeptide folds back and forth upon itself instead of forming a helix; this type of folding exposes hydrogen atoms that can un-

dergo extensive hydrogen bonding (Figure 2.15b). Some polypeptides contain both regions of α-helix and regions of β-sheet secondary structure, the type of folding being determined by the available opportunities for hydrogen bonding and hydrophobic interactions (recall that these will ultimately be dictated by the primary structure—the amino acid sequence—of the polypeptide). Since β-sheet secondary structure generally yields a rather rigid structure, whereas α-helical secondary structures are usually more flexible, the secondary structure of a given polypeptide to some degree dictates a functional role for the protein in the cell. Many polypeptides fold into two or more segments, each displaying α-helix or β-sheet secondary structure (see Figure 2.16). These segments, referred to as *domains*, are regions of the polypeptide that have specific functions in the final protein molecule.

2.8

Proteins: Higher Order Structure and Denaturation

Once a polypeptide has achieved a given secondary structure it can fold back upon itself to form an even more stable molecule. This folding leads to formation

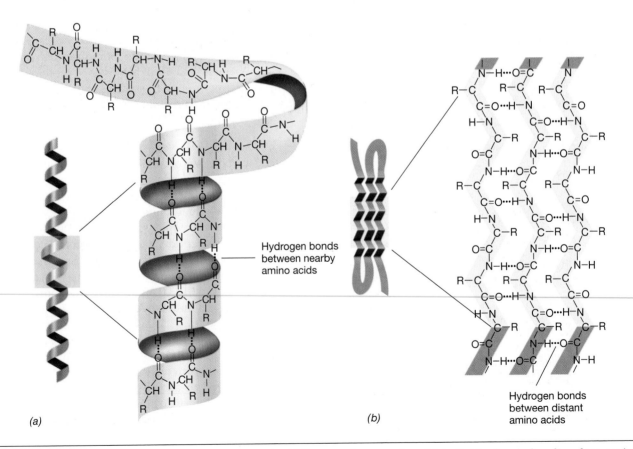

(a)   *(b)*

Hydrogen bonds between nearby amino acids

Hydrogen bonds between distant amino acids

FIGURE 2.15   Secondary structure of polypeptides. (a) α-Helix secondary structure. Note that hydrogen bonding does not involve the R groups but instead occurs between atoms in the peptide bonds. (b) β-Sheet secondary structure.

of the **tertiary structure** of the protein. Like secondary structure, tertiary structure of a protein is ultimately determined by primary structure, but tertiary structure is also governed to some extent by the secondary structure of the molecule. As a result of the formation of secondary structure, the side chain of each amino acid in the polypeptide is positioned in a specific way. If additional hydrogen bonds, covalent bonds, hydrophobic interactions, or other atomic interactions are able to form, the polypeptide will fold to accommodate them and attain a unique three-dimensional shape (Figure 2.16).

Frequently a polypeptide folds in such a way that adjacent sulfhydryl ($—SH$) groups of cysteine residues are exposed (Figure 2.16*a*). These free $—SH$ groups can join covalently to form a disulfide ($—S—S—$) bridge between the two amino acids. If the two cysteine residues are located in different polypeptides in a protein, the disulfide bond physically links the two molecules (Figure 2.16*a*). In addition, a single polypeptide can spontaneously fold and bond to itself covalently if two cysteine residues form a disulfide linkage within the molecule. The tertiary folding of the polypeptide ultimately forms exposed regions or grooves in the molecule (Figures 2.16 and 2.17), which may be of importance in binding other molecules (for example, the binding of a substrate to an enzyme) (∞ Section 4.5).

Quaternary Structure

If a protein consists of two or more polypeptides, and many proteins do, the arrangement of the subunits in space to form the final protein molecule is referred to as the **quaternary structure** of the protein (Figure 2.17). It

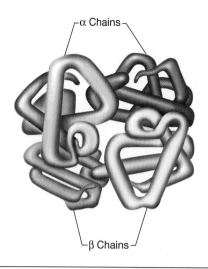

FIGURE 2.17 Quaternary structure of hemoglobin. There are two *kinds* of polypeptide in hemoglobin, α chains (shown in blue and red) and β chains (shown in orange and yellow), but a total of four polypeptides in the final protein molecule. Separate colors are used to distinguish the four distinct chains.

should be remembered that in proteins showing quaternary structure, each subunit of the final protein itself contains primary, secondary, and tertiary structure. Some proteins contain identical subunits, while others contain nonidentical subunits. The subunits of multisubunit proteins are held together either by noncovalent interactions (hydrogen bonding, van der Waals forces, or hydrophobic interactions) or by covalent linkages, generally intersubunit disulfide bonds.

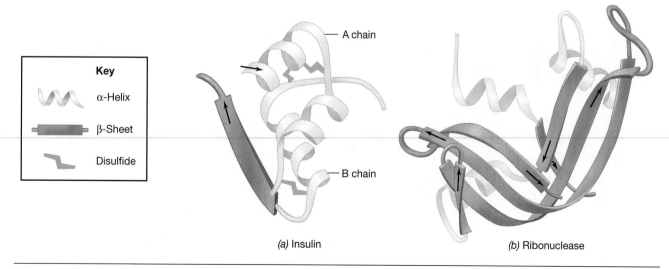

FIGURE 2.16 Tertiary structure of polypeptides showing where regions of α-helix or β-sheet secondary structure might be located. (a) Insulin, a protein containing two polypeptide chains; note how the B chain contains both α-helix and β-sheet secondary structure and how disulfide linkages ($—S—S—$) may help in dictating folding patterns (tertiary structure). (b) Ribonuclease, a large protein with several regions of α-helix and β-sheet.

Denaturation of Proteins

When proteins are exposed to extremes of heat or pH, or to certain chemicals or metals that affect their folding properties, they are said to undergo **denaturation** (Figure 2.18). In general, the biological properties of a protein are lost when it is denatured. When proteins are denatured, peptide bonds are unaffected. However, denaturation causes the polypeptide chain to unfold, destroying the higher order structure of the molecule. The denatured polypeptide retains its primary structure because it is held together by covalent peptide bonds. Depending on the severity of the denaturing conditions, refolding of the polypeptide may occur after removal of the denaturant (Figure 2.18). However, the fact that denaturation is usually associated with loss of biological activity of the protein clearly shows that biological activity is not inherent in the primary structure of a protein but instead is a result of the unique folding of the molecule as ultimately directed by primary structure. Folding of a polypeptide therefore accomplishes two things: (1) the polypeptide obtains a unique shape that is compatible with a *specific* biological function, and (2) the folding process converts the molecule to its most chemically stable form.

Macromolecules and Cells

We started this chapter with a consideration of macromolecules as parts of a cell (Table 2.2). We then examined the structure of each class of macromolecule with an eye toward understanding how structure relates to function. We now proceed in Chapter 3, Cell Biology, to a detailed examination of the major structures that make up a microbial cell—the structures like cell membranes and cell walls that actually make up cells. But the student should keep in mind while traveling through Chapter 3 that the large and often complex structures encountered are really only composites of one or more macromolecules arranged in different ways to yield the various components of the cell, each of which has a specific function. Thus, it may be helpful from time to time to return to Chapter 2 and review the structure of the macromolecules of life. To a great extent, the biochemical makeup of these important molecules dictates their function, and thus a real understanding of cell biology can only come with a firm grasp of the structure of macromolecules.

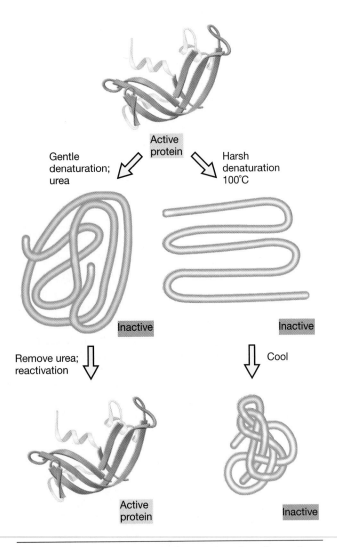

✓ 2.7 and 2.8 Concept Check

The primary structure of a protein is determined by its amino acid sequence, but it is the folding (higher order structure) of the polypeptide that determines how the protein functions in the cell.

- ✓ Define the terms *primary, secondary,* and *tertiary* with respect to protein structure.
- ✓ How does a *polypeptide* differ from a *protein*?
- ✓ What secondary structural features tend to make β-sheet proteins more rigid than α-helices?
- ✓ What would be the quaternary structure of a *homodimeric* protein? A *heterodimeric* protein?
- ✓ Describe the structural and biological effects of the denaturation of a protein.

FIGURE 2.18   Denaturation of a protein using ribonuclease (whose structure was discussed in Figure 2.16*b*) as an example. Note how harsh denaturation generally yields a permanently destroyed molecule from the standpoint of biological function because of improper folding.

REVIEW QUESTIONS

1. Which elements are the major elements found in living organisms? Why are oxygen and hydrogen particularly abundant in living organisms?

2. Define the word *molecule*. How many atoms are in a molecule of hydrogen gas? In a molecule of glucose?

3. Refer to the structure of the nitrogen base *cytosine* shown in Figure 2.1. Draw this structure and then label the positions of all single bond and double bonds in the cytosine molecule.

4. Compare and contrast the words *monomer* and *polymer*. Give three examples of biologically important polymers and list the monomers of which they are composed.

5. List the components that would make up a simple lipid. How does a triglyceride differ from a complex lipid?

6. Examine the structures of the triglyceride and of phosphatidyl ethanolamine shown in Figure 2.7. How might the substitution of phosphate and ethanolamine for a fatty acid alter the chemical properties of the lipid?

7. RNA and DNA are similar types of macromolecules but show distinct differences as well. List three ways in which RNA differs chemically or physically from DNA. What is the cellular function of DNA and RNA?

8. Why are *amino acids* so named? Write a general structure for an amino acid. What is the importance of the R group to final protein structure? Why does the amino acid *cysteine* have special significance for protein structure?

9. Chemically, what type of reaction between two amino acids leads to formation of the peptide bond? (You may wish to review Figure 2.13 before answering.)

10. Draw the peptide bond. Now redraw this structure in atomic form (see Figure 2.1) showing the arrangement of electrons within the atoms of the peptide bond.

APPLICATION QUESTIONS

1. Observe the following nucleotide sequences of RNA: (a) GUCAAAGAC, (b) ACGAUAACC. Can either of these RNA molecules have secondary structure? If so, draw the potential secondary structure(s).

2. A few soluble (cytoplasmic) proteins contain a high content of hydrophobic amino acids. How would you predict these proteins would fold as to their tertiary structure and why?

3. Cells of the genus *Halobacterium,* an archaeon that lives in very salty environments, contain over 5 molar (M) potassium (K^+). Because of this high K^+ content, many cytoplasmic proteins of *Halobacterium* cells are enriched in two specific amino acids that are present in much higher proportions in *Halobacterium* proteins than in functionally similar proteins from *Escherichia coli* (which has only very low levels of K^+ in its cytoplasm.) Which amino acids are enriched in *Halobacterium* proteins and why? *Hint:* Which amino acids could best neutralize the positive charges due to K^+?

4. It is often the case that proteins that show α-helix secondary structure are more flexible than proteins showing β-sheet secondary structure. Discuss why this could be the case.

5. When an egg is placed in a beaker of boiling water, changes in the egg occur almost immediately. Describe what happens and why the contents of a boiled egg look so different from the contents of a fresh egg.

6. In light of your answer to the preceding question, explain how it can be that certain prokaryotes, called *hyperthermophiles,* thrive (and indeed grow optimally) in boiling hot springs. How must the proteins of hyperthermophiles differ from proteins in the egg?

Membranes are critical cell structures. Membranes function first and foremost as the cell's permeability barrier—the gatekeeper, so to speak, of substances that enter and leave the cell—but also as an anchor for many proteins and as the structural basis for ion gradient-mediated energy conservation reactions. Cell membranes have been studied from many organisms and for many different reasons, but one unifying principle emerges from every study: The general architecture of all cell membranes is basically the same. Membranes consist of proteins and lipids with hydrophilic external surfaces and a hydrophobic internal matrix. These form a *lipid bilayer* (as shown here), which is an ideal structural solution to the most basic of problems all cells must solve: concentrating and maintaining dissolved substances in their cytoplasm.

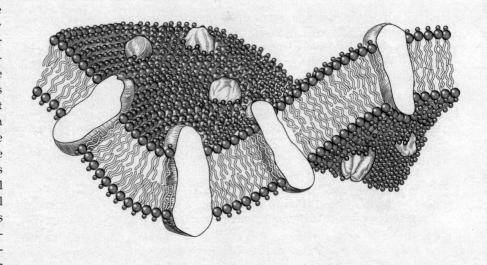

CHAPTER 3 # Cell Biology

3.1   Light Microscopy   50
3.2   Three-Dimensional Imaging: Interference Contrast, Atomic Force, and Confocal Scanning Laser Microscopy   54
3.3   Electron Microscopy   55
3.4   Overview of Cell Structure and the Significance of Smallness   57
3.5   Cytoplasmic Membrane: Structure   60
3.6   Cytoplasmic Membrane: Function   63
3.7   The Cell Wall of Prokaryotes: Peptidoglycan and Related Molecules   69
3.8   The Outer Membrane of Gram-Negative Bacteria   73
3.9   Cell Wall Synthesis and Cell Division   76
3.10 Arrangement of DNA in Prokaryotes   77
3.11 Flagella and Motility   79
3.12 Bacterial Behavior: Chemotaxis, Phototaxis, and Other Taxes   83
3.13 Bacterial Cell Surface Structures and Cell Inclusions   85
3.14 Gas Vesicles   89
3.15 Endospores   91
3.16 The Nucleus and Organelles of Eukaryotic Microorganisms   95
3.17 Comparison of the Prokaryotic and Eukaryotic Cell   99

ABC (Antigen-Binding Cassette) transporter a membrane transport system consisting of three proteins, one of which hydrolyzes ATP as an energy source to drive the transport event, one of which binds the substrate on the outside of the cell, and one of which functions as the transport channel through the membrane

Chemotaxis movement of an organism toward (*positive*) or away from (*negative*) a chemical gradient

Chloroplast the chlorophyll-containing photosynthetic organelle of eukaryotic photosynthetic organisms

Chromosome a DNA molecule, usually circular in prokaryotes and linear in eukaryotes, carrying genes essential to cellular function

Cytoplasmic membrane the permeability barrier of the cell, separating the cytoplasm from the environment

Endospore a highly heat-resistant, thick-walled, differentiated cell produced by certain gram-positive Bacteria

Eukaryote a cell containing a membrane-enclosed nucleus and usually other organelles

Flagellum a long, thin cellular appendage capable of rotation in prokaryotic cells and responsible for swimming motility

Gas vesicles gas-filled cytoplasmic structures bounded by protein and conferring buoyancy on cells

Gram-negative a prokaryotic cell whose cell wall contains relatively little peptidoglycan but contains an outer membrane composed of lipopolysaccharide, lipoprotein, and other complex macromolecules

Gram-positive a prokaryotic cell whose cell wall consists chiefly of peptidoglycan and lacks the outer membrane of gram-negative cells

Group translocation an energy-dependent transport process in which the substance transported is chemically modified during the transport process

Lipopolysaccharide (LPS) lipid in combination with polysaccharide and protein forming the major portion of the cell wall in gram-negative Bacteria

Magnetosomes particles of magnetite (Fe_3O_4) organized into nonunit membrane-enclosed structures in the cytoplasm of magnetotactic Bacteria

Mitochondrion (mitochondria) an organelle found in most eukaryotic cells in which respiration and energy generation occurs

Nucleoid an aggregated state of the circular chromosome of prokaryotic cells

Nucleus a membraned-enclosed structure in cells of Eukarya that contains the genetic material, arranged in chromosomes

Organelle a unit membrane-enclosed structure found in the cytoplasm of eukaryotic cells

Peptidoglycan a polysaccharide composed of alternating repeats of acetylglucosamine and acetylmuramic acid with the latter in adjacent layers cross-linked by short peptides

Periplasm a gellike region between the outer surface of the cytoplasmic membrane and the inner surface of the lipopolysaccharide layer of gram-negative Bacteria

Peritrichous in reference to flagellation pattern; flagella located in many places around the surface of the cell

Phototaxis movement of an organism toward light

Poly-β-hydroxybutyrate (PHB) a common storage material of prokaryotic cells consisting of a polymer of β-hydroxybutyrate or another β-alkanoic acid

Prokaryote a cell that lacks a membrane-enclosed nucleus and that usually has a single circular DNA molecule as its chromosome

Protoplast an osmotically protected cell whose cell wall has been removed

Ribosome small particles composed of RNAs and proteins that function in protein synthesis

---

In this chapter we present the principles of structure and function relationships in microbial cells. We emphasize the biology of the prokaryotic cell, but we compare and contrast prokaryotes with eukaryotes and discuss in detail a few eukaryotic cell structures.

Cells, like houses, are built by connecting simple building blocks in various ways to create more complex structures. We discussed the chemical nature of cellular building blocks in Chapter 2 and emphasized how these simple molecules can be polymerized to form macromolecules. Cells are basically well-defined assemblages of macromolecules. Despite great diversity in the chemical composition of the macromolecules found in different cells, from a structural perspective, all cells solve biological problems in common ways. For example, all cells contain a lipid bilayer or structurally related layer as the structural foundation of the cytoplasmic membrane. Although the bilayer may vary in its exact chemical composition from species to species, all cytoplasmic membranes have the same basic structure and all function as permeability barriers. Also, ribosomes, the structures on which proteins are synthesized in all organisms, differ somewhat in chemical detail from species to species but not in their general structure or cellular function. Thus, *unity* exists in many structure and function relationships in all cells.

Because cells are microscopic, we begin this chapter with a discussion of microscopes and microscopy. The microscope is a major tool of the microbiologist. Historically it was the microscope that first revealed the secrets of cell structure, and even today it remains a powerful tool in cell biology studies.

3.1

Light Microscopy

Microscopic examination of microorganisms makes use of either the **light microscope** or the **electron microscope.** For most routine work, the light microscope is used, whereas for special research purposes, especially in studies on internal cell structure, the electron microscope is used in addition to the light microscope. All microscopes employ the principle that specific lenses magnify the image of a cell such that details of its structure are more apparent. In addition to magnification, however, is *resolution*, the ability to distinguish two adjacent points as separate. Although magnification can be increased virtually without limit, resolution cannot; resolution is dictated by the physical properties of light. It is thus resolution and not magnification that ultimately defines the limits of what we are able to see with a microscope. We begin our discussion with the light microscope, for which the limits of resolution are about 0.2 μm [200 nanometers (nm)], and then proceed to describe the electron microscope, for which resolution is improved over that of the light microscope by about 1000-fold.

The Compound Light Microscope

The **light microscope** has been of crucial importance for the development of microbiology as a science and remains a basic tool of routine microbiological research. Several types of light microscopes are commonly used in microbiology; *bright-field*, *phase contrast*, *dark-field*, and *fluorescence*. The **bright-field microscope** is most commonly used in elementary biology and microbiology courses and consists of two series of lenses (objective lens and ocular lens), which function together to resolve the image (Figure 3.1). With this microscope, specimens are made visible because of the differences in contrast that exist between them and the surrounding medium. Contrast differences arise because cells absorb or scatter light in varying degrees. Many bacterial cells are difficult to see well with the bright-field microscope because of their lack of contrast with the surrounding medium. Pigmented organisms are an exception, however, because the color of the organism adds contrast, thus improving visualization of the cells (Figure 3.2).

Magnification and Resolution

The total magnification of a compound microscope is the *product* of the magnification of its objective and ocular lenses (Figure 3.1*b*). Magnifications of about 1500× are near the upper limit obtainable with a compound light microscope. This limit is set because of a property of a lens called **resolution.** Resolving power is a function of the wavelength of light used and an innate property of the objective

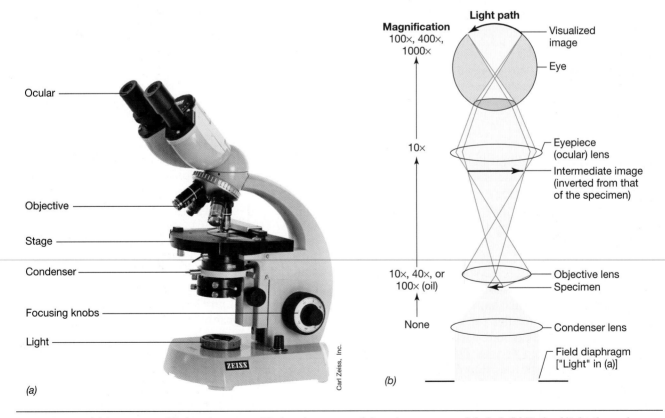

(a)

Carl Zeiss, Inc.

(b)

FIGURE 3.1   (a)A compound light microscope. Various key parts of the microscope are labeled. (b) Path of light through a compound light microscope. Besides 10×, eyepieces (oculars) are available in 15–30×.

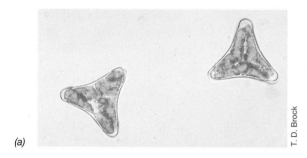

(a)

T. D. Brock

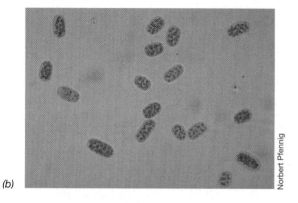

(b)

Norbert Pfennig

FIGURE 3.2 Photomicrographs of pigmented microorganisms by bright-field microscopy. (a) A green alga (eukaryote). (b) A purple phototrophic bacterium (prokaryote). The algal cells are about 15 μm in diameter, and the bacterial cells are about 5 μm in diameter.

lens known as its *numerical aperture* (a measure of light-gathering ability). In general, there is a correspondence between the magnification of a lens and its numerical aperture: Lenses with higher magnification usually have higher numerical apertures. The diameter of the smallest resolvable object is equal to 0.5λ/numerical aperture, where λ is the wavelength of light used. Based on this formula, resolution is greatest when blue light is used to illuminate a specimen and the objective that is used has a very high numerical aperture.

The highest resolution possible in a compound light microscope is about 0.2 μm. This means that two objects closer together than 0.2 μm are not resolvable as distinct and separate. Most microscopes used in microbiology have oculars that magnify 10–15× and objectives of 10–100× (Figure 3.1*b*); at 1000×, objects 0.2 μm in diameter can just be resolved. With the 100× objective, and with certain other objectives of very high numerical aperture, a high-grade optical oil is used between the specimen and the objective. Lenses on which oil is used are called *oil-immersion lenses*. Immersion oil increases the light-gathering ability of a lens by allowing rays emerging from the specimen at higher angles (and that would otherwise be lost to the objective lens) to be collected and viewed.

Staining: Increasing Contrast for Bright-Field Microscopy

As previously mentioned, one of the limitations to bright-field microscopy is insufficient contrast. Dyes can be used to stain cells and increase their contrast so that they can be more easily seen in the bright-field microscope. Dyes are organic compounds, and each class of dye has an affinity for specific cellular materials. Many dyes commonly used in microbiology are positively charged (cationic) and combine strongly with negatively charged cellular constituents such as nucleic acids and acidic polysaccharides. Examples of cationic dyes include *methylene blue, crystal violet*, and *safranin*. Because cell surfaces are generally negatively charged, these dyes combine with structures on the surfaces of cells and hence are excellent general-purpose stains.

The simplest staining procedures are done with dried preparations (Figure 3.3). A slide containing a dried suspension of microorganisms is flooded for a

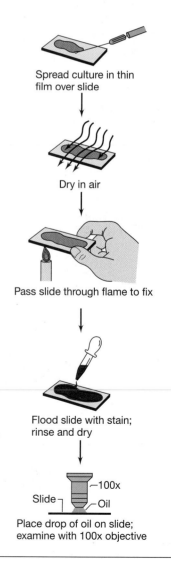

Spread culture in thin film over slide

Dry in air

Pass slide through flame to fix

Flood slide with stain; rinse and dry

Slide — 100x — Oil

Place drop of oil on slide; examine with 100x objective

FIGURE 3.3 Staining cells for microscopic observation.

minute or two with a dilute solution of a dye, rinsed several times in water, and blotted dry. It is usual to observe dried stained preparations of bacteria with a high-power (oil-immersion) lens (Figure 3.3).

Differential stains are so named because they are used in procedures that do not stain all kinds of cells equally. An important differential staining procedure widely used in bacteriology is the *Gram stain* (Figure 3.4*a*). On the basis of their reaction to the Gram stain, bacteria can be divided into two major groups: **gram-positive** and **gram-negative.** After Gram staining, gram-positive bacteria appear purple and gram-negative bacteria appear red (Figure 3.4*b*). This difference in reaction to the Gram stain arises because of differences in the cell wall structure of gram-positive and gram-negative cells (as discussed later in this chapter). The Gram stain is one of the most useful staining procedures in the bacteriological laboratory; it is almost essential in identifying an unknown bacterium to determine first whether it is gram-positive or gram-negative. If a fluorescent microscope is available (see later discussion of fluorescence microscopy), the Gram stain can be reduced to a one-step procedure where gram-positive and gram-negative cells fluoresce different colors (Figure 3.4*c*).

Phase Contrast, Dark-Field, and Fluorescence Microscopy

The **phase contrast microscope** was developed to improve contrast differences between cells and the surrounding medium, making it possible to see cells without staining them (Figure 3.5). Phase contrast microscopy is based on the principle that cells differ in refractive index from their surroundings and hence bend some of the light rays that pass through them. Light passing through a specimen of refractive index different from that of the surrounding medium is retarded. This effect is amplified by a special ring in the objective lens of a phase contrast microscope, leading to the formation of a dark image on a light background (Figure 3.5*b*). The phase contrast microscope is widely employed in research applications because it can be used to observe wet-mount (living) preparations. Staining, on the other hand, although a widely used procedure

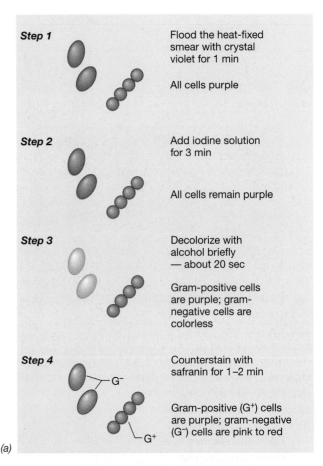

Step 1 — Flood the heat-fixed smear with crystal violet for 1 min

All cells purple

Step 2 — Add iodine solution for 3 min

All cells remain purple

Step 3 — Decolorize with alcohol briefly — about 20 sec

Gram-positive cells are purple; gram-negative cells are colorless

Step 4 — Counterstain with safranin for 1–2 min

G⁻

G⁺

Gram-positive (G⁺) cells are purple; gram-negative (G⁻) cells are pink to red

(a)

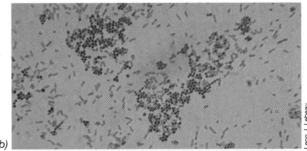

(b)

Leon J. Lebeau

(c)

Molecular Probes, Inc., Eugene, Oregon

FIGURE 3.4   The Gram stain. (a) Steps in the Gram stain procedure. (b) Photomicrograph of Gram-stained Bacteria that are gram-positive (blue-purple) and gram-negative (pink-red). The species are *Staphylococcus aureus* and *Escherichia coli*, respectively. (c) Photomicrograph of cells of *Pseudomonas aeruginosa* (gram-negatives, green) and *Bacillus cereus* (gram-positives, orange) stained with the one step fluorescent staining method **LIVE *Bac* Light**™. This method allows for differentiating gram-positive from gram-negative cells in a single staining step.

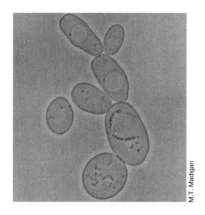
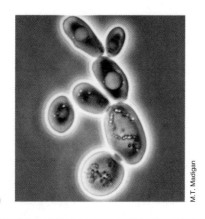
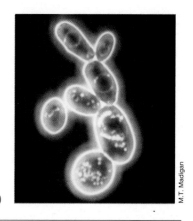

(a) M.T. Madigan   *(b)* M.T. Madigan   *(c)* M.T. Madigan

FIGURE 3.5 Photomicrographs of the same field of cells of the baker's yeast *Saccharomyces cerevisiae,* taken by different types of light microscopy. (a) Bright-field. (b) Phase contrast. (c) Dark-field.

in light microscopy, generally kills cells and can distort their features.

The **dark-field microscope** is a light microscope in which the lighting system has been modified to reach the specimen from the sides only. The only light reaching the lens is light scattered by the specimen, and thus the specimen appears light on a dark background (Figure 3.5*c*). Resolution by dark-field microscopy is quite high, and objects can frequently be resolved by dark-field that are not resolvable in bright-field or phase contrast microscopes. Dark-field microscopy is also an excellent way to observe the motility of microorganisms, as bundles of flagella are often resolvable with this technique (see Figure 3.47*a*).

The **fluorescence microscope** is used to visualize specimens that *fluoresce,* that is, emit light of one color when light of another color shines upon them. Fluorescence occurs either because of the presence within cells of naturally fluorescent substances such as chlorophyll or other fluorescing components (*autofluorescence*) (see Figure 3.6*a*) or because the cells have been treated with a fluorescent dye (Figures 3.4*c* and 3.6*b*). Fluorescence microscopy is widely used in clinical diagnostic microbiology and also in microbial ecology (⟳ Chapters 17 and 21).

✓ 3.1 Concept Check

Microscopes are essential for microbiological studies. Various types of light microscopes exist, including bright-field, phase contrast, and fluorescence microscopes. In bright-field microscopy, stains are necessary to increase contrast.

✓ Define the term *resolution.*
✓ What is the upper limit of magnification for a light microscope?
✓ What light microscopic techniques can sometimes improve resolution?
✓ What color would a gram-negative bacterium be after Gram staining by the conventional method?

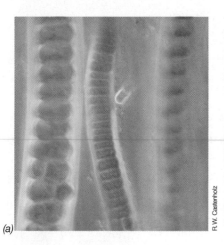
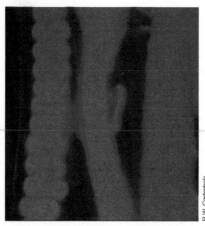
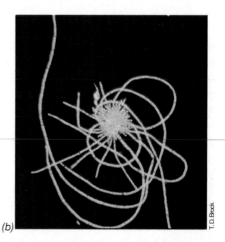

(a) R.W. Castenholz   *(b)* R.W. Castenholz   T.D. Brock

FIGURE 3.6 Photomicrographs of various microorganisms as visualized by fluorescence microscopy. (a) Cyanobacteria. Left, cells observed by bright-field microscopy. Right, same cells observed by fluorescence after shining light of 546 nm on them. The red color is due to autofluorescence of chlorophyll and other pigments. (b) Cells of the filamentous bacterium *Leucothrix mucor* stained with the fluorescent dye, acridine orange, which fluoresces green.

3.2

Three-Dimensional Imaging: Interference Contrast, Atomic Force, and Confocal Scanning Laser Microscopy

One of the drawbacks to the forms of light microscopy just considered is that the images obtained are essentially two dimensional. How can this limitation be overcome? We will see in the next section that the scanning electron microscope offers one solution to this problem, but other forms of microscopy do as well, and we discuss these options here.

Differential Interference Contrast Microscopy

Differential interference contrast (DIC) is a form of light microscopy that employs a polarizer to produce plane-polarized light. The polarized light then passes through a prism that generates two distinct beams, and these are what traverse the specimen and enter the objective lens. Here the two beams are recombined into one, and because of slight differences in refractive index of the substances each beam passed through, the combined beams are not totally in phase but instead create an interference effect. This effect intensifies subtle differences in cell structure, and thus, by DIC microscopy, things like the nucleus of eukaryotic cells (Figure 3.7*a*), spores, vacuoles, granules, and the like attain a pseudo–three-dimensional appearance. DIC microscopy is particularly useful for observing *unstained* cells because of its ability to generate images that reveal internal cell structures that are less apparent (or even invisible) by bright-field techniques (compare Figures 3.5*a* and *b* with Figure 3.7*a*).

Atomic Force Microscopy

Another form of microscopy useful for three-dimensional imaging of biological structures is the **atomic force microscope (AFM)**. In atomic force microscopy, a tiny stylus is positioned extremely close to the specimen such that weak repulsive atomic forces are established between the probe and the specimen. As the specimen is scanned in both the horizontal and vertical directions the stylus rides up and down the hills and valleys, constantly recording its interactions with the surface. This pattern is monitored by a series of detectors that feed the digital information into a computer that generates an image (Figure 3.7*b*). Although the images obtained from an atomic force microscope appear similar to those from the scanning electron microscope (compare Figure 3.7*b* with Figure 3.10*b*), the AFM has the big advantage that specimen preparation is similar to that for light microscopy (that is, no fixatives or coat-

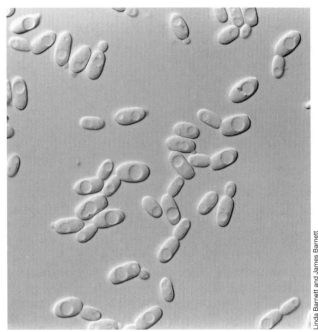

(a)

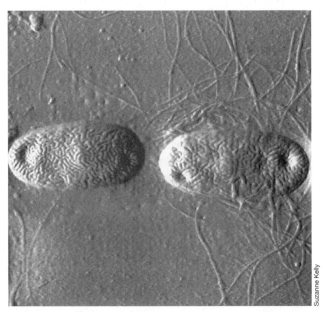

(b)

FIGURE 3.7   Three-dimensional imaging of cells using (a) interference contrast microscopy and (b) atomic force microscopy. The yeast cells in (a) are about 8 μm in diameter. Note how the nucleus is clearly visible here (compare Figure 3.7*a* with Figure 3.5*a*). The bacterial cells in (b) are about 2.2 μm in length, and the micrograph was taken from a natural biofilm that developed on the surface of a glass slide immersed for 24 h in a dog's water bowl. The slide was air dried before viewing with an atomic force microscope.

ings are required). The AFM also allows living and hydrated specimens to be viewed, something that is generally not possible with electron microscopes.

Confocal Scanning Laser Microscopy

Confocal scanning laser microscopy (CSLM) is a computerized microscope that couples a laser light source to a light microscope; this technique allows for the generation of three-dimensional digital images of microorganisms and other biological specimens (Figure 3.8). In CSLM a laser beam is bounced off a mirror that directs the beam through a scanning device and then a pinhole that precisely adjusts the plane of focus of the beam to a given vertical layer within a specimen. By precisely illuminating only a single plane of the specimen, illumination intensity drops off rapidly above and below the plane of focus, and because of this, stray light from other planes of focus are minimized. Thus, in a relatively thick specimen such as a microbial biofilm, for example (Figure 3.8), not only are cells on the surface of the biofilm apparent, as would be the case with conventional light microscopy, but cells in the various layers can also be observed by adjusting the plane of focus of the laser beam.

Cells in CSLM preparations are frequently stained with fluorescent dyes to make them more visible (Figure 3.8). Alternatively, false color images can be generated by adjusting the microscope in such a way as to make different layers take on different colors. The laser confocal microscope is equipped with computer software to assemble digital images for subsequent image processing. Thus, images obtained from different layers can be stored and then digitally overlaid to reconstruct a three-dimensional image of the entire specimen (Figure 3.8). CSLM has found widespread use in microbial ecology, especially for identifying phylogenetically distinct populations of cells present in a microbial

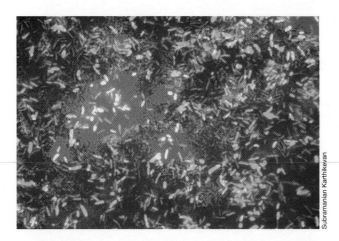

FIGURE 3.8 Confocal scanning laser microscopy. Confocal image of a mixed microbial biofilm community cultivated in the laboratory. The green rod-shaped cells are *Pseudomonas aeruginosa* experimentally introduced into the biofilm. Other cells that are different colors are present at different depths in the biofilm.

habitat (see for example, Figure 16.11*b*), but is useful anywhere thick specimens need to be examined for their microbial content with depth.

✓ 3.2 Concept Check

Interference contrast (DIC) and confocal scanning (CSLM) are forms of light microscopy that allow for greater three-dimensional imaging than other forms of light microscopy, and confocal microscopy allows imaging through thick specimens. The atomic force microscope yields a detailed three-dimensional image of live preparations.

✓ What structure in eukaryotic cells is more easily imaged using DIC microscopy?
✓ How is CSLM able to view different layers in a thick preparation?

3.3

Electron Microscopy

Electron microscopes are widely used for studying the detailed structure of cells. To study the internal structure of cells, a **transmission electron microscope (TEM)** is essential. In the TEM, electrons are used instead of light rays and electromagnets function as lenses, the whole system operating in a high vacuum (Figure 3.9). The resolving power of the electron microscope is much greater than that of the light microscope, and thus the electron microscope enables one to see many structures of even molecular size, such as proteins and nucleic acids (see Figure 3.43*b*). However, electron beams do not penetrate very well, and if one is interested in seeing internal cell structure, even a single cell is too thick to be viewed directly. Consequently, special techniques of *thin sectioning* are needed to prepare specimens for the electron microscope. A single bacterial cell, for instance, is cut into many very thin slices, which are then examined individually with the electron microscope (see Figure 3.10*a*). To obtain sufficient contrast, the preparations are treated with a special electron microscope stain such as osmic acid, permanganate, uranium or lanthanum salts, or lead. Because these substances are composed of atoms of high atomic weight, they scatter electrons well and thus improve contrast (see Figure 3.10*a*).

Scanning Electron Microscopy

If only the *external* features of an organism need be observed, thin sections are not necessary, and intact cells or cell components can be observed directly by TEM with a technique called *negative staining* (see, for exam-

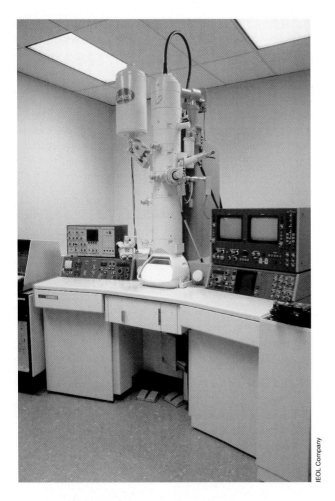

JEOL Company

ple, Figure 3.46). Alternatively, one can use the **scanning electron microscope (SEM)** (Figures 3.9 and 3.10*b*). With this tool, the specimen is coated with a thin film of a heavy metal such as gold. An electron beam from the SEM is then directed down on the specimen and scans back and forth across it. Electrons scattered by the metal are collected, and they activate a viewing screen to produce an image (Figure 3.10*b*). In the SEM, even fairly large specimens can be observed, and the depth of field is extremely good. A wide range of magnifications can be obtained with the SEM, from as low as 15× up to about 100,000×, but only the *surface* of an object can be visualized. All electron microscopes are fitted with cameras to allow a photograph, called an *electron micrograph*, to be taken.

✓ 3.3 Concept Check

Electron microscopes have far greater resolving power than do light microscopes, the limits of resolution being about 0.2 nm. Two major types of electron microscopy are performed, transmission electron microscopy, for observing internal cell structure down to the molecular level, and scanning electron microscopy, useful for three-dimensional imaging.

FIGURE 3.9   An electron microscope. This instrument encompasses both transmission and scanning electron microscope functions.

Membrane ⌐   Wall ⌐   DNA ⌐

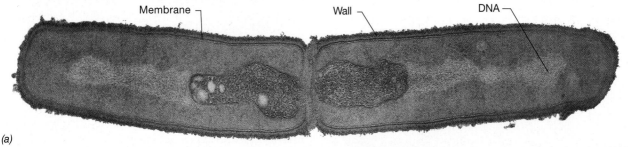

(a)

Stanley C. Holt

FIGURE 3.10   Electron micrographs of bacterial cells taken with (a) transmission and (b) scanning electron microscopes. (a) Thin section of a typical gram-positive bacterium, *Bacillus subtilis*. The cell has just divided, and two membrane-containing structures are attached to the cross-wall. Note the light region in the middle (DNA or the *nucleoid*). The cell is about 0.8 μm in diameter. (b) Cells of the phototrophic bacterium *Rhodovibrio sodomensis*. A single cell is about 0.75 μm wide.

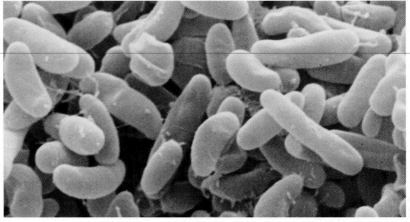

(b)

F. R. Turner

✓ Keeping in mind that chemical fixatives are necessary and that electron microscopes must operate at high vacuum, what major *disadvantage* do electron microscopes have compared with light microscopes?

✓ What type of electron microscope would you use to observe the bacterial nucleoid?

✓ What is an *electron micrograph?*

3.4

Overview of Cell Structure and the Significance of Smallness

Careful study with light and electron microscopes has revealed the detailed structure of cells. In Chapter 1 we learned that cellular structure can be either prokaryotic or eukaryotic; we pick up from this introduction here.

The Prokaryotic Cell

A typical prokaryotic cell of either the Bacteria or the Archaea generally has the following major structures: cell wall, cytoplasmic membrane, ribosomes, inclusions, and nucleoid (Figure 3.10a). What are these structures and what are their functions?

The **cytoplasmic membrane** is the critical permeability barrier, separating the inside from the outside of the cell. The **cell wall** is a rigid structure outside the cytoplasmic membrane, which provides support and protection from osmotic lysis. **Ribosomes** are small particles composed of protein and ribonucleic acid (RNA), which are visible in the transmission electron microscope. A single prokaryotic cell may have as many as 10,000 ribosomes. Ribosomes are part of the translation apparatus, and synthesis of cell proteins takes place on these structures. (We will discuss ribosomes, proteins, and RNA in detail in Chapter 6.) Prokaryotes occasionally contain **inclusions** consisting of storage material made up of compounds of carbon, nitrogen, sulfur, or phosphorus. Such inclusions can be formed when these nutrients are in excess in the environment and function as repositories of these nutrients when limitations occur.

The nuclear region of the prokaryotic cell differs quite significantly from that of the eukaryotic cell. Prokaryotic cells do not possess a true nucleus, the function of the nucleus being carried out by a single molecule of deoxyribonucleic acid (DNA). DNA is present in a more-or-less free state within the prokaryotic cell but is often seen in electron micrographs (Figure 3.10a) in an aggregated form referred to as the **nucleoid.** In analogy to the eukaryote, the DNA molecule of the prokaryote is called a **chromosome.**

Many, but not all, bacteria are able to move. Movement of a prokaryotic cell is usually by means of a structure called a **flagellum** (plural, **flagella**). Each flagellum consists of a single, coiled tube of protein. The *rotation* of flagella propels the cell through liquids. Bacterial flagella can be seen with the light microscope if special staining techniques are used (see Figure 3.45) and are readily visible with the electron microscope (see Figure 3.46).

Morphology of Prokaryotes

The *shape* of a cell is referred to as its *morphology*. Several distinct shapes of bacteria can be recognized and have been given different names. Schematic examples of some of these bacterial shapes along with phase photomicrographs are shown in Figure 3.11. A bacterium that is spherical or ovoid in morphology is called a **coccus** (plural, **cocci**). A bacterium with a cylindrical shape is called a **rod.** Some rods are curved, frequently forming spiral-shaped patterns and are then called **spirilla.**

In many prokaryotes, the cells remain together in groups or clusters after division, and the arrangements in these groups are often characteristic of different organisms. For instance, cocci or rods may occur in long chains. Some cocci form thin sheets of cells, whereas others occur in three-dimensional cubes or irregular cube-like clusters. Several groups of bacteria are immediately recognizable by their unusual shapes. Examples include **spirochetes,** which are tightly coiled bacteria, **appendaged bacteria,** which possess extensions of their cells as long tubes or stalks, and **filamentous bacteria,** which form long, thin cells or chains of cells (Figure 3.11). It should be noted that the morphologies of prokaryotic cells shown in Figure 3.11 are *representative* ones; many variations of each of these basic morphological types are possible.

The Eukaryotic Cell

Eukaryotic cells are larger and more complex in structure than prokaryotic cells, and a key difference is that eukaryotes contain *true nuclei*. The **nucleus** is a special membrane-enclosed structure within which DNA is located (Figure 3.12). The DNA in the nucleus is organized into **chromosomes,** structures that remain essentially invisible except at the time of cell division. Before cell division occurs, the chromosomes are duplicated and then condense, become thicker, and undergo division as the nucleus divides. The process of nuclear division in eukaryotes is called **mitosis** (see Figures 1.6 and 6.24), which is a complex but highly organized process. Two identical daughter cells result from the division of one parent cell; each daughter cell receives a nucleus with an identical set of chromosomes.

Eukaryotic cells also contain distinct structures called **organelles,** within which important cellular functions occur (Figure 3.12). Organelles are absent from prokaryotes, although major physiological processes that take place in organelles, such as respiration and photosynthesis, may still occur in prokaryotic cells. One kind of organelle found in most eukaryotes is the **mitochondrion** (plural, **mitochondria**) (Figure 3.12; see also Figure 3.70). Mitochondria are the organelles within which the energy-generating functions of the eukaryotic cell occur. The energy produced in mitochondria is then used throughout the cell.

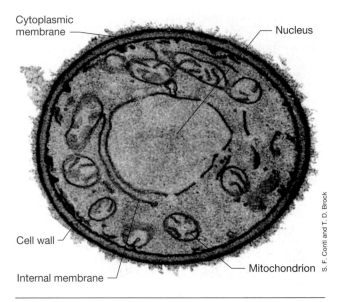

FIGURE 3.12   The yeast cell, a typical eukaryotic microorganism. Transmission electron micrograph of a thin section through a yeast cell, showing various internal structures. A single cell is about 8 μm in diameter.

Algae are eukaryotic microorganisms that carry out the process of *photosynthesis*. In these organisms, as well as in green plants, an additional type of organelle is found: the **chloroplast.** The chloroplast is green and is the site where chlorophyll is localized and where the light-gathering functions involved in photosynthesis occur (see Figures 3.2*a* and 3.71).

The Size of Microbial Cells and the Significance of Being Small

Prokaryotes vary in size from cells as small as 0.1–0.2 μm in diameter to those more than 50 μm in diameter; a few very large prokaryotes, such as the surgeonfish symbiont *Epulopiscium fishelsoni* (Figure 3.13), are up to 50 μm in diameter and can be more than 0.5 millimeters (mm) in length (Figures 3.13 and 3.14). However, the dimensions of an average rod-shaped prokaryote, the bacterium *Escherichia coli,* for example, are about 1 × 3 μm (Figure 3.14). Typical eukaryotic cells may be 2 μm to more than 200 μm in diameter. Thus, prokary-

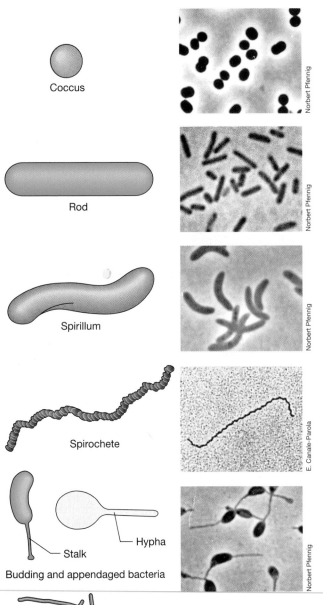

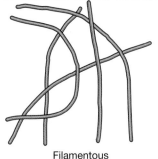

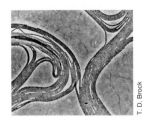

FIGURE 3.11   Representative cell shapes (morphology) in prokaryotes. Next to each drawing is a phase photomicrograph showing an example of that morphology. Organisms are coccus, *Thiocapsa roseopersicina* (diameter of a single cell = 1.5 μm); rod, *Desulfuromonas acetoxidans* (diameter = 1 μm); spirillum, *Rhodospirillum rubrum* (diameter = 1 μm); spirochete, *Spirochaeta stenostrepta* (diameter = 0.25 μm); budding and appendaged, *Rhodomicrobium vannielii* (diameter = 1.2 μm); filamentous, *Chloroflexus aurantiacus* (diameter = 0.8 μm).

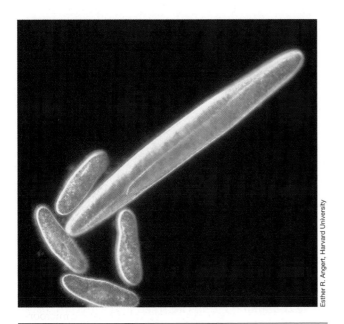

FIGURE 3.13 Photomicrograph of a giant prokaryote, the surgeonfish symbiont *Epulopiscium fishelsoni*. The rod-shaped *E. fishelsoni* cell in this field is about 600 μm (0.6 mm) long and is shown with four cells of the protozoan (eukaryote) *Paramecium*, each of which measures about 150 μm in length. *E. fishelsoni* is phylogenetically related to *Clostridium* species.

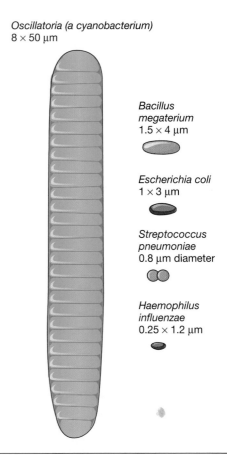

Oscillatoria (a cyanobacterium)
8 × 50 μm

Bacillus megaterium
1.5 × 4 μm

Escherichia coli
1 × 3 μm

Streptococcus pneumoniae
0.8 μm diameter

Haemophilus influenzae
0.25 × 1.2 μm

FIGURE 3.14 Comparison of sizes of a variety of prokaryotes. Most known prokaryotes have cell diameters in the range of 0.5–2 μm.

otes are very small cells compared to eukaryotes, and the small size of prokaryotes affects a number of their biological properties. For example, the rate at which nutrients and waste products pass into and out of a cell, a factor that can greatly affect cellular metabolic rates and growth rates, is in general *inversely* proportional to cell size. This is because transport rates are to some degree a function of the amount of *membrane surface area* available, and relative to cell volume, small cells have more surface available than do large cells. This point can be seen most readily in the case of a sphere, in which the *volume* is a function of the cube of the radius ($V = \frac{4}{3}\pi r^3$), whereas the *surface area* is a function of the square of the radius ($SA = 4\pi r^2$). The surface-to-volume ratio of a sphere can thus be expressed as $3/r$ (Figure 3.15). A cell with a smaller r value therefore has a *higher* ratio of surface area to volume than a larger cell and thus can have a more efficient exchange of nutrients with its surroundings than a large cell. This advantage of the small cell typically allows for more rapid growth rates and larger populations of prokaryotic cells than eukaryotic cells in most microbial habitats. This in turn affects an organism's ecology in that high numbers of rapidly metabolizing cells can cause major physiochemical changes in an ecosystem over a relatively short period of time. We will pick this theme up again when we consider microorganisms in their natural habitats in Chapters 16 and 19.

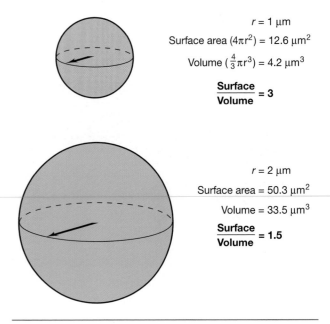

r = 1 μm
Surface area $(4\pi r^2)$ = 12.6 μm²
Volume $(\frac{4}{3}\pi r^3)$ = 4.2 μm³
$\dfrac{\text{Surface}}{\text{Volume}} = 3$

r = 2 μm
Surface area = 50.3 μm²
Volume = 33.5 μm³
$\dfrac{\text{Surface}}{\text{Volume}} = 1.5$

FIGURE 3.15 Surface area and volume relationships in cells. As a cell *increases* in size, its surface area-to-volume ratio *decreases*.

✓ 3.4 Concept Check

Although prokaryotes and eukaryotes are distinguished by nuclear structure, other important differences exist between these two cell types. Prokaryotes are smaller in size than eukaryotes, and eukaryotes contain a membrane-enclosed nucleus and organelles within which many important functions are carried out. The small size of prokaryotic cells affects their physiology, growth rate, and ecology

- ✓ List three morphological types of prokaryotes.
- ✓ What is a *flagellum* and what does it do?
- ✓ What physical property of cells increases as cells become smaller?

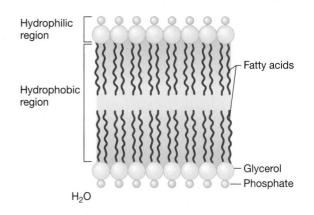

FIGURE 3.16 Fundamental structure of a phospholipid bilayer. The cytoplasmic membrane is about 8 nm (80 Å) wide.

3.5

Cytoplasmic Membrane: Structure

The **cytoplasmic membrane** is a thin structure that completely surrounds the cell. Only about 8 nm thick, this vital structure is the critical barrier separating the inside of the cell (the cytoplasm) from its environment. If the membrane is broken, the integrity of the cell is destroyed, the internal contents leak into the environment, and the cell dies. The cytoplasmic membrane is also a *highly selective barrier,* enabling a cell to concentrate specific metabolites and excrete waste materials.

Chemical Composition of Membranes

The general structure of most biological membranes is a **phospholipid bilayer** (Figure 3.16). As discussed in Section 2.4, phospholipids contain both highly hydrophobic (fatty acid) and relatively hydrophilic (glycerol) moieties and can exist in many different chemical forms as a result of variation in the nature of the fatty acids or phosphate-containing groups attached to the glycerol backbone. As phospholipids aggregate in an aqueous solution, they tend to form bilayer structures spontaneously—the fatty acids point inward toward each other in a hydrophobic environment, and the hydrophilic portions remain exposed to the aqueous external environment (Figure 3.16); the bilayer character of membranes probably represents the most stable arrangement of lipid molecules in an aqueous environment.

Thin sections of the cytoplasmic membrane can be seen with the electron microscope; a representative example is seen in Figure 3.17a. By careful high resolution electron microscopy, the cytoplasmic membrane appears as two light-colored lines separated by a darker area (Figure 3.17a). This **unit membrane,** as it is called, consists of a phospholipid (∞ Figure 2.7) bilayer and proteins embedded within it (Figure 3.18). The major

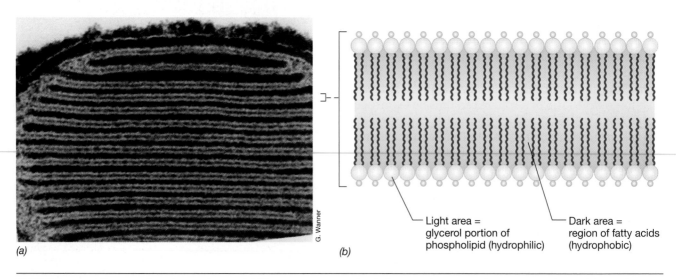

FIGURE 3.17 The cytoplasmic membrane. (a) Electron micrograph of photosynthetic membrane stacks derived from the cytoplasmic membrane in the phototrophic bacterium *Ectothiorhodospira halochloris.* Note the distinct lipid bilayers (unit membranes). Each bilayer is about 8 nm thick. (b) Enlarged schematic view of a single unit membrane shown in (a).

proteins of the cell membrane generally have very hydrophobic external surfaces in the regions of the protein that span the membrane and have surfaces exposed on both the inside and the outside of the cell (Figure 3.18). The overall structure of the cytoplasmic membrane is stabilized by hydrogen bonds and hydrophobic interactions. In addition, cations such as Mg^{2+} and Ca^{2+} also help stabilize the membrane by combining ionically with negative charges of the phospholipids.

Other Features of the Cytoplasmic Membrane

The outer surface of the cytoplasmic membrane faces the environment and in certain bacteria makes contact with a variety of proteins that serve to bind substrates or process large molecules for transport into the cell (periplasmic proteins) (see discussion in Sections 3.6 and 3.8). The inner side of the cytoplasmic membrane faces the cytoplasm and interacts with proteins involved in energy-yielding reactions and other important cellular functions. Some proteins, such as those in the periplasm (see Section 3.8) and some cytoplasmic proteins, may associate quite firmly with the surface of the membrane and actually function as if they were membrane-bound proteins. Although not themselves integral membrane proteins, such proteins usually interact

directly with integral membrane proteins in various cellular processes. Some of these *peripheral membrane proteins*, as they are called, are lipoproteins and contain a lipid tail on the amino terminus of the protein, which anchors the protein into the membrane.

Although appearing somewhat rigid when viewed diagrammatically (Figure 3.18), the cytoplasmic membrane is actually quite fluid; phospholipid and protein molecules have significant freedom to move about the membrane surface. Measurements of membrane viscosities indicate that membranes have a viscosity approximating that of a light grade oil. Thus, membranes can be thought of as *fluid mosaics* in which various globular proteins oriented in a specific manner span a highly mobile, yet ordered, phospholipid bilayer. This arrangement confers a number of important functional properties on membranes, and we discuss these properties in the next section.

Membrane Strengthening Agents: Sterols and Hopanoids

One major difference in chemical composition of membranes between eukaryotic and prokaryotic cells is that the eukaryotes have **sterols** in their membranes (Figure 3.19a,b). Sterols are absent from the membranes of virtually

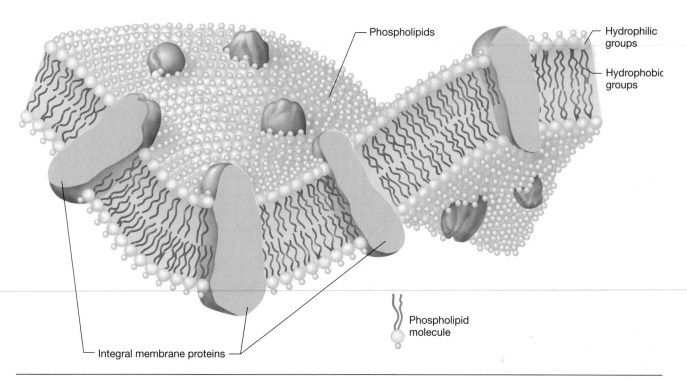

Phospholipids

Hydrophilic groups

Hydrophobic groups

Integral membrane proteins

Phospholipid molecule

FIGURE 3.18   Diagram of the structure of the cytoplasmic membrane. The matrix of the unit membrane is composed of phospholipids, with the hydrophobic groups directed inward and the hydrophilic groups toward the outside, where they associate with water. Embedded in the matrix are proteins that have considerable hydrophobic character in the region that traverses the fatty acid bilayer. Hydrophilic proteins and other charged substances, such as metal ions, may be attached to the hydrophilic surfaces. Although there are some chemical differences, the overall structure of the cytoplasmic membrane shown is similar in both prokaryotes and eukaryotes (but see an exception to the bilayer design in Figure 3.21).

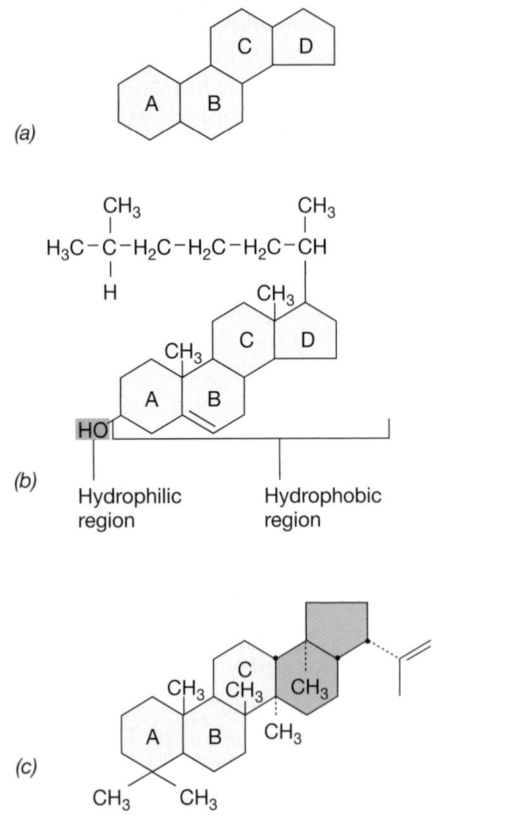

FIGURE 3.19 Sterols and hopanoids. (a) The general structure of a sterol. All sterols contain the same four rings, labeled A, B, C, and D. (b) The structure of cholesterol. (c) The structure of the hopanoid diploptene. Note the structural resemblance to cholesterol in rings A through C. Sterols are found in the membranes of eukaryotes and hopanoids in the membranes of some prokaryotes.

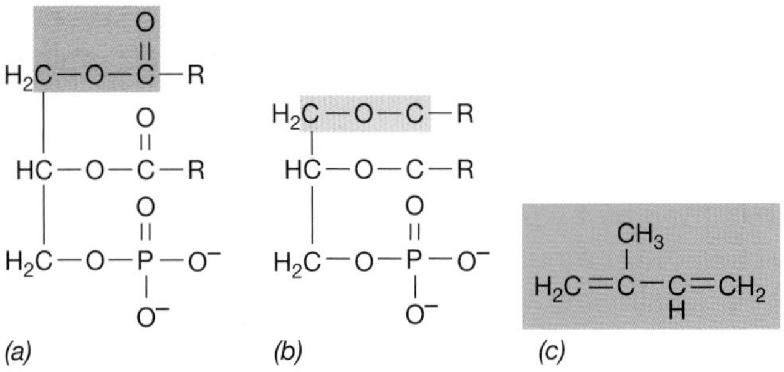

FIGURE 3.20 Chemical bonds in lipids. (a) The *ester* linkage as found in the lipids of Bacteria and Eukarya. (b) The *ether* linkage of lipids from Archaea. (c) Isoprene, the parent structure of the hydrophobic side chains (R) of archaeal lipids. By contrast, in lipids of Bacteria and Eukarya, R are fatty acids.

all prokaryotes (methanotrophic bacteria are a major exception, ⌘ Sections 13.5 and 15.14). Depending on the cell type, sterols can make up from 5 to 25% of the total lipids of eukaryotic membranes. Sterols are rigid, planar molecules, whereas fatty acids are flexible. The association of sterols with the membrane serves to stabilize its structure and make it less flexible. Membrane rigidity may be necessary in eukaryotes because many of them lack a rigid cell wall. Moreover, eukaryotes are much larger than prokaryotes and thus must endure greater physical stresses on the membrane, necessitating a more rigid membrane structure in order to keep the cell stable and functional. Molecules similar to sterols, called *hopanoids*, are present in several bacteria and may play a role similar to that of sterols in eukaryotic cells. One widely distributed hopanoid is the C_{30} hopanoid *diploptene* (Figure 3.19c).

Archaeal Membranes

The lipids in Archaea are chemically unique. In contrast to the lipids in Bacteria and Eukarya in which *ester* link-ages bond the fatty acids to the glycerol molecule (Figure 3.20a) (⌘ Section 2.4), lipids from Archaea have *ether* linkages between glycerol and their hydrophobic side chains. In addition, archaeal lipids lack fatty acids and instead have side chains composed of repeating units of the hydrocarbon molecule *isoprene* (Figure 3.20c). However, the overall structure of archaeal lipid membranes, forming inner and outer hydrophilic surfaces with a hydrophobic interior, is maintained.

Glycerol *diethers* and glycerol *tetraethers* (Figure 3.21a,b) are the major classes of lipids present in Archaea. Note that in the tetraether molecule the phytanyl side chains from each glycerol molecule are covalently bonded together (Figure 3.21b). Employed within a membrane structure this yields a lipid *monolayer* instead of a lipid bilayer (Figure 3.21d). Lipid monolayers are quite resistant to peeling apart, and it is thus not surprising that this membrane structure is widespread among hyperthermophilic Archaea, prokaryotes that grow at very high temperatures (⌘ Sections 5.9 and 14.4–14.12). We will encounter many other features that set Bacteria and Archaea apart, but the chemistry of membrane lipids is a major defining feature of each phylogenetic group.

✓ 3.5 Concept Check

The cytoplasmic membrane is a highly selective permeability barrier constructed of lipid and protein that forms a bilayer with hydrophilic exteriors and a hydrophobic interior. Other molecules such as sterols and hopanoids may strengthen the membrane. Unlike Bacteria and Eukarya, Archaea contain ether-linked lipids, and some species have membranes of monolayer instead of bilayer construction.

✓ Draw the basic structure of a lipid bilayer.
✓ Why are compounds like sterols and hopanoids good at stabilizing the cytoplasmic membrane?
✓ Contrast the linkage between glycerol and the hydrophobic portion of lipids in Bacteria and Archaea.

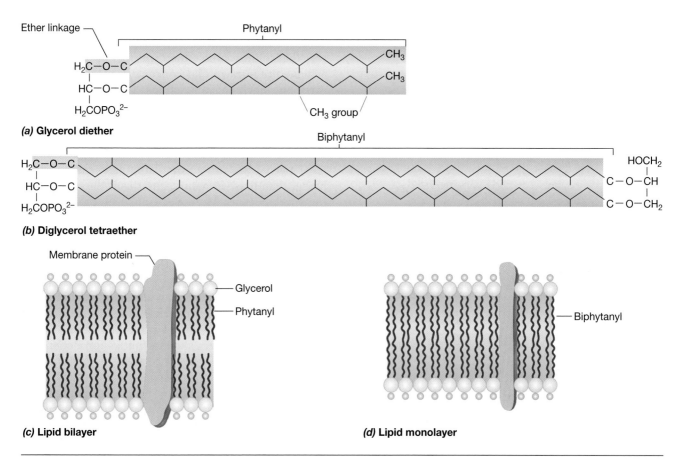

FIGURE 3.21   Major lipids of Archaea and the structure of archaeal membranes. (a) Glycerol diethers. (b) Diglycerol tetraethers. Note that in both cases, the hydrocarbon is attached to the glycerol by *ether* linkages. Hydrocarbon in (a) phytanyl (C_{20}) and (b) dibiphytanyl (C_{40}) (c, d) Membrane structure in Archaea. (c) Lipid bilayer. (d) Lipid monolayer.

3.6

Cytoplasmic Membrane: Function

The cytoplasmic membrane is more than just a barrier separating the inside from the outside of the cell. The membrane plays several critical roles in cell function. First and foremost, the membrane functions as a *permeability barrier*, preventing the passive leakage of cytoplasmic constituents into or out of the cell (Figure 3.22). In addition, the membrane is the site of many proteins, some of which are enzymes and many of which are involved in one way or another in the transport of substances into and out of the cell.

We will learn in Chapter 4 that the cytoplasmic membrane is also a device for energy conservation in the cell. The membrane can exist in an energetically "charged" form in which a separation of protons (H^+) from hydroxyl ions (OH^-) occurs across its surface (Figure 3.22). This charge separation is a form of metabolic energy, analogous to the potential energy present in a charged battery. The energized state of the membrane, referred to as a *proton-motive force* (PMF), is responsible for driving many energy-requiring functions in the cell including some forms of transport, motility, and the biosynthesis of the cell's energy currency, ATP (Figure 3.22).

The Cytoplasmic Membrane as a Permeability Barrier

The interior of the cell (the cytoplasm) consists of an aqueous solution of salts, sugars, amino acids, vitamins, coenzymes, and a wide variety of other soluble materials. The hydrophobic nature of the cytoplasmic membrane makes it a tight barrier; although some small hydrophobic molecules may pass through the membrane by diffusion, hydrophilic and charged molecules do not readily pass but instead must be specifically transported. Even a substance as small as a hydrogen ion (H^+) does not diffuse across the cytoplasmic membrane. One molecule that does penetrate the membrane is water itself, which is sufficiently small and uncharged to pass between phospholipid molecules. The relative permeability of a few biologically important substances

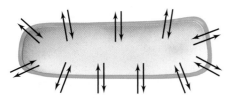

Permeability Barrier — Prevents leakage and functions as a gateway for transport of nutrients into and out of the cell

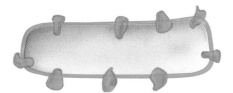

Protein Anchor — Site of many proteins involved in transport, bioenergetics, and chemotaxis

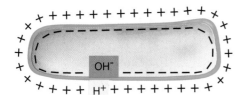

Energy Conservation — Site of generation and use of the proton motive force

FIGURE 3.22 The major functions of the cytoplasmic membrane.

is shown in Table 3.1. As can be seen, most substances do not passively enter the cell, and thus transport processes are critical to cellular function.

The Necessity for Transport Proteins

Transport proteins do more than just ferry things across the membrane; they are able to accumulate solutes inside the cell *against* the concentration gradient. The ne-

cessity for carrier-mediated transport in microorganisms is easy to understand. If diffusion were the only way that solutes could enter the cell, cells would never achieve the intracellular concentrations necessary to carry out biochemical reactions. This is because both the rate of uptake and the intracellular level of diffusable solutes are proportional to their external concentration (Figure 3.23). However, the concentration of nutrients in nature is often very low. Thus, cells must have mechanisms for accumulating nutrients to levels higher than those in nature, and this is the function of transport systems. Moreover, unlike simple diffusion, carrier-mediated transport shows a *saturation effect*; if the concentration of substrate is high enough to saturate the carrier, which is usually the case even at very low substrate concentration, the rate of uptake becomes maximal (Figure 3.23).

One characteristic of carrier-mediated transport processes is the highly specific nature of the transport event. Many carrier proteins react only with a single molecule while others show affinities for a chemical class of molecules. For instance, there are carriers that transport a variety of related sugars or amino acids. This economy in uptake reduces the need for separate transport proteins for every single amino acid or every single sugar the cell needs to transport. In addition, the synthesis of transport proteins is *regulated* by the cell such that the specific complement of transporters present in the membrane is a function of both the nutrients present and their concentration. The latter is a factor because oftentimes transport of a particular nutrient occurs via one type of transporter when the nutrient is present at high concentration and by a different transporter when present at low concentration.

TABLE 3.1	Comparative permeability of membranes to various molecules
Substance	**Rate of permeability**[a]
Water	100
Glycerol	0.1
Tryptophan	0.001
Glucose	0.001
Chloride ion (Cl^-)	0.000001
Potassium ion (K^+)	0.0000001
Sodium ion (Na^+)	0.00000001

a Relative scale—permeability with respect to permeability of water, given as 100.

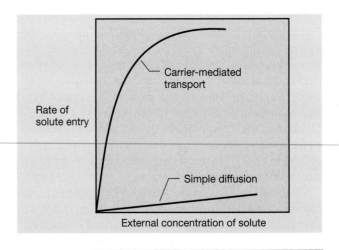

FIGURE 3.23 Relationship between uptake rate and external concentration in diffusion and transport. Note that in the carrier-mediated process the uptake rate shows saturation at relatively low external concentrations.

Structure and Function of Membrane Transport Proteins

There are at least three classes of membrane-transporting systems, those involving only a membrane-spanning component, those involving a periplasmic-binding component plus a membrane-spanning component, and those, like the phosphotransferase system, that involve a series of proteins that cooperate to mediate the transport event (Figure 3.24). All of these transport systems require energy, either in the form of the proton-motive force, ATP, or some other high-energy compound.

The membrane-spanning proteins of all bacterial transport systems show significant similarities in both their primary and secondary structure, undoubtedly a testament to their common evolutionary roots. Structurally, these transporters form 12 alpha helices that wind back and forth through the membrane to form a channel through which the transported substance is carried into the cell (Figure 3.25a). The actual transport event involves a conformational change in the protein following binding of its specific substrate and this event shuttles the compound across the membrane.

What *types* of transport events can occur? These are summarized in Figure 3.25b. *Uniporters* are proteins that simply transport a molecule in a unidirectional fashion across the membrane. *Symporters* are proteins that transport a substance *along with* another substance, frequently a proton (H^+). *Antiporters* are proteins that, as their name implies, transport a substance across the membrane in one direction while at the same time transporting a second substance in the *opposite* direction (Figure 3.25b).

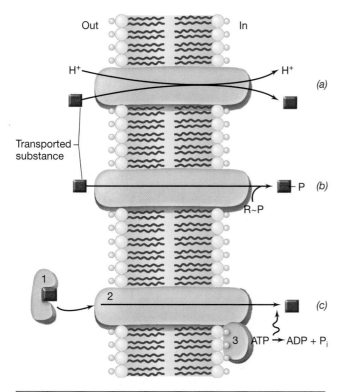

FIGURE 3.24 The three classes of membrane-transporting systems. (a) Simple transporter (symporter, see also Figure 3.25). The transport event is driven by the energy in the proton motive force. (b) Group translocation, such as the phosphotransferase system. In the phosphotransferase system of *Escherichia coli*, the necessary energy comes from phosphoenolpyruvate (see also Figure 3.27) (c) The ABC system. A periplasmic-binding protein 1 binds the substance to be transported, the membrane-spanning protein 2 transports the substance, and the energy necessary to drive the transport event comes from the hydrolysis of ATP by protein 3 (see also Figure 3.28). Note how simple transporters and the ABC system transport substances *without* chemically modifying them, while group translocation results in the chemical modification (phosphorylation) of the transported substance.

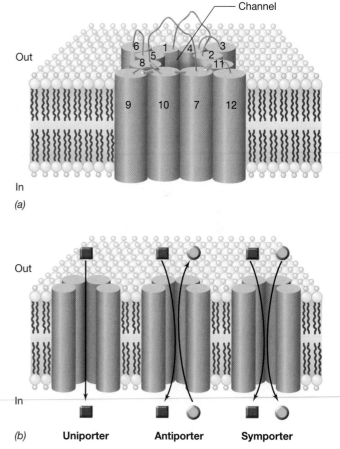

FIGURE 3.25 Structure of membrane-spanning transporters and types of transport events. (a) In prokaryotes, membrane-spanning transport proteins typically contain 12 alpha helices (shown as connected barrels here) that form a channel through the membrane. (b) The three transport events that can occur; the molecule transported into the cell is shown in red. For antiporters and symporters the cotransported molecule is shown in yellow.

LacY Permease: A Simple Transporter

The bacterium *Escherichia coli* can grow on the disaccharide lactose. Lactose is taken up by cells of *E. coli* by a symporter called the LacY permease (encoded by the *lacY* gene). LacY is a typical symporter, taking up one molecule of lactose along with one proton. This is shown in Figure 3.26 where the activity of LacY is compared with other simple transporters, including uniporters and antiporters. Note that as each lactose molecule is transported by the LacY permease, the energy in the proton motive force is slowly diminished by the influx of protons into the cell. However, the proton motive force is being constantly reestablished in the cell through energy-yielding reactions that we will describe in later chapters (☞ Chapters 4 and 15). The final result of a simple transporter such as the LacY permease is the accumulation of a solute, in this case lactose, to high concentrations where its metabolism yields energy for the cell (☞ Section 4.9).

Group Translocation

Group translocation is a transport process in which the transported substance is *chemically altered* during passage across the membrane. The best-studied cases of group translocation involve transport of the sugars glucose, mannose, and fructose, which are *phosphorylated* during transport by the **phosphotransferase system.**

The phosphotransferase system in the bacterium *Escherichia coli* is composed of 24 proteins, at least 4 of which are necessary to transport a given sugar. Before the sugar is transported into the cell, the proteins in the

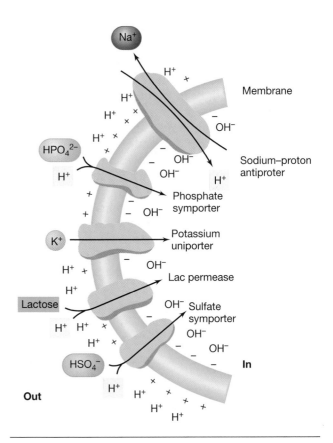

FIGURE 3.26 Function of the Lac permease (a symporter) of *Escherichia coli*, and several other well-characterized simple transporters. Although for simplicity the membrane-spanning proteins are drawn here in globular form, note that their structure is actually as depicted in Figure 3.25*a*. To review the action of transport proteins, see Figure 3.25*b*.

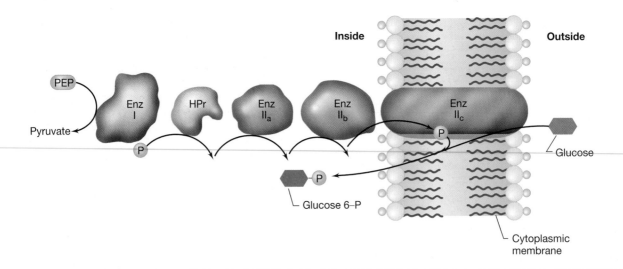

FIGURE 3.27 Mechanism of the phosphotransferase system of *Escherichia coli*. For glucose uptake, the system consists of five proteins: Enzyme (Enz) I; Enzymes II$_a$, II$_b$, and II$_c$; and HPr. Sequential phosphate transfer occurs from phosphoenolpyruvate (PEP) through the proteins shown to Enzyme II$_c$. The latter actually transports (and phosphorylates) the sugar.

phosphotransferase system are themselves alternately phosphorylated and dephosphorylated in a cascading fashion until the membrane-spanning protein, called Enzyme II$_c$, receives the phosphate group and phosphorylates the sugar in the actual transport event (Figure 3.27). A small protein called HPr, the enzyme that phosphorylates it (Enzyme I), and Enzyme II$_a$ are cytoplasmic proteins, while Enzyme II$_b$ and II$_c$ are membrane proteins (Figure 3.27). HPr and Enzyme I are nonspecific components of the phosphotransferase system and participate in the uptake of various sugars, while specific Enzymes II exist for each individual sugar.

The high-energy phosphate bond that supplies the energy for the phosphotransferase system comes from a key metabolic intermediate called *phosphoenolpyruvate.* However, it should be noted that although energy in the form of one high energy phosphate bond is consumed in the process of transporting the glucose molecule (Figure 3.27), the phosphorylation of glucose to glucose-6-P is the first step in its intracellular metabolism anyway (Section 4.9). Thus, in the case of glucose phosphorylation by the phosphotransferase system, the uptake of glucose is energetically neutral.

Periplasmic Binding Protein-Dependent Transport: The "ABC" System

We will learn a bit later in this chapter (see Section 3.8) that gram-negative bacteria contain a space between the cytoplasmic membrane and a lipid-rich outer membrane layer, called the *periplasm* (see Figure 3.37). The periplasm contains various proteins, many of which function in transport; the latter are called *periplasmic-binding proteins.* Transport systems that employ periplasmic-binding proteins also have membrane-spanning components that actually mediate the transport event and a third component that supplies the necessary energy, obtained by hydrolysis of ATP. These types of transporters have been called **ABC transport systems,** the *ABC* standing for *ATP-binding cassette.* The mechanism of action of an ABC transporter is shown in Figure 3.28.

More than 200 different ABC transport systems have been identified in prokaryotes, and analyses of the structures of functionally related components from the different systems show that they are clearly a *family* of related proteins. One of the interesting properties of ABC-type transport systems is the extremely high substrate affinity of the periplasmic-binding proteins. These proteins are mobile within the periplasm and are able to bind their substrates even when they are present at extremely low concentration; for example, substrate concentrations of 1 micromolar (10^{-6} M) are easily bound and transported. Once trapped by the binding protein, the complex interacts with its respective membrane-spanning component, and the actual transport event occurs

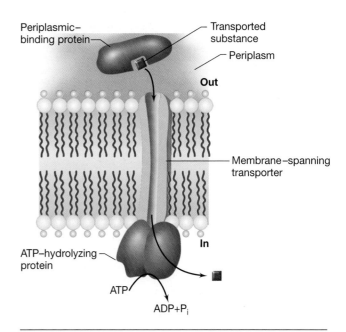

FIGURE 3.28 Mechanism of an Antigen-Binding Cassette (ABC-type) transporter. The periplasmic binding protein has high affinity for substrate, the membrane-spanning protein is the transport channel, and the cytoplasmic ATP-hydrolyzing protein supplies the energy for the transport event. In *Escherichia coli,* the maltose (a disaccharide sugar) transport system is an example of an ABC system.

driven by the energy of ATP (Figure 3.28). Interestingly, even though gram-positive bacteria lack a periplasm, binding protein-dependent transport systems are also present in many of these organisms as well. In gram-positives, specific binding proteins are not mobile but instead are anchored to the cytoplasmic membrane. However, as in gram-negative bacteria, once these binding proteins bind their substrate, they interact with a membrane-spanning component where, at the expense of ATP, transport across the membrane occurs.

✓ 3.6 Concept Check

Transport of nutrients into the cell is a major physiological function of the cytoplasmic membrane. Transport requires energy from either the proton motive force or from ATP. Three types of transporters are known, simple transporters, phosphotransferase-type transporters, and ABC systems, which contain three interacting components.

- ✓ Discuss two reasons why a cell cannot depend on diffusion as a means of getting nutrients into the cell.
- ✓ Contrast the energy requirements of simple transporters, the phosphotransferase system, and the ABC transport system.
- ✓ Contrast the three types of transport systems in terms of any chemical alterations that occur in the transported molecule.
- ✓ Which transport system is best suited for the transport of nutrients present in the environment in extremely low amounts, and why?

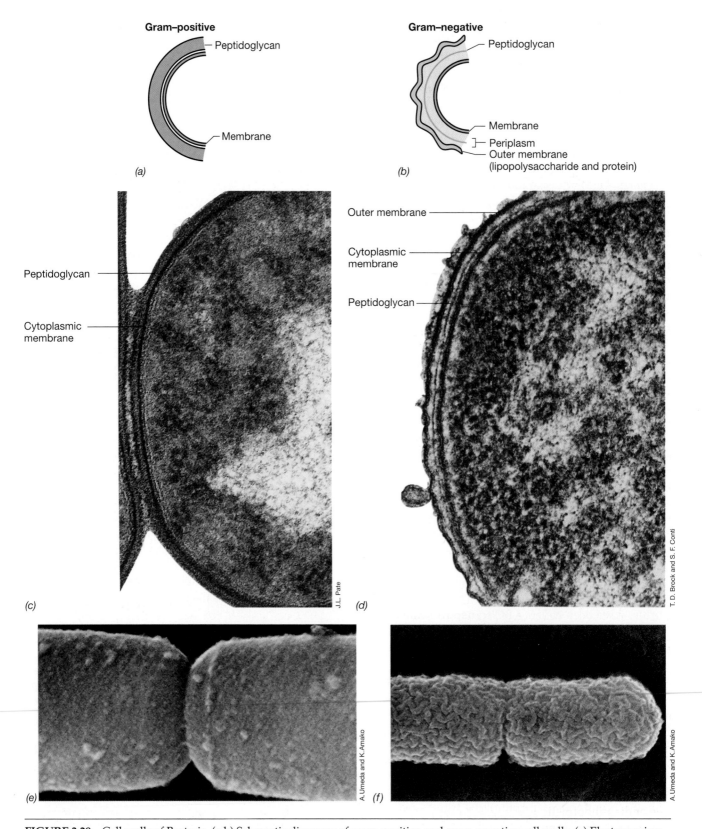

FIGURE 3.29   Cell walls of Bacteria. (a,b) Schematic diagrams of gram-positive and gram-negative cell walls. (c) Electron micrograph showing the cell wall of a gram-positive bacterium, *Arthrobacter crystallopoietes*. (d) Gram-negative bacterium, *Leucothrix mucor*. (e,f) Scanning electron micrographs of gram-positive (*Bacillus subtilis*) and gram-negative (*Escherichia coli*) Bacteria. Note the surface texture in the cells shown in (e) and (f). A single cell of *B. subtilis* or *E. coli* is about 1 μm in diameter.

3.7

The Cell Wall of Prokaryotes: Peptidoglycan and Related Molecules

Because of the concentration of dissolved solutes inside the bacterial cell, a considerable turgor pressure develops, estimated at 2 atmospheres (atm) in a bacterium like *Escherichia coli;* this is roughly the same as the pressure in an automobile tire. To withstand these pressures, bacteria contain **cell walls,** which also function to give shape and rigidity to the cell. The prokaryotic cell wall is difficult to see well with the light microscope but can be readily seen in thin sections of cells with the electron microscope.

Bacteria can be divided into two major groups, called **gram-positive** and **gram-negative.** The original distinction between gram-positive and gram-negative was based on a special staining procedure, the *Gram stain* (see Section 3.1), but differences in cell wall structure are at the base of these differences in the Gram-staining reaction. Gram-positive and gram-negative cells differ markedly in the appearance of their cell walls, as is shown in Figure 3.29. The gram-negative cell wall is a multilayered structure and quite complex, whereas the gram-positive cell wall consists of primarily a single type of molecule and is often much thicker. Close examination of Figure 3.29 shows that there is also a significant textural difference between the surfaces of gram-positive and gram-negative Bacteria, as revealed by the scanning electron microscope.

The focus of this section is on the polysaccharide component of the cell walls of prokaryotes, both Bacteria and Archaea. These include, in particular, peptidoglycan, but also a variety of related and unrelated polysaccharides found in Archaea. In Section 3.8 we describe the special wall components found in gram-negative Bacteria.

Peptidoglycan

In the cell walls of Bacteria there is one rigid layer that is primarily responsible for the strength of the wall. In gram-negative Bacteria, additional layers are present outside this rigid layer. The rigid layer of both gram-negative and gram-positive Bacteria is very similar in chemical composition. Called **peptidoglycan** (or **murein**), this layer is a thin sheet composed of two sugar derivatives, *N-acetylglucosamine* and *N-acetylmuramic acid,* and a small group of amino acids consisting of L-alanine, D-alanine, D-glutamic acid, and either lysine or diaminopimelic acid (DAP) (Figure 3.30). These constituents are connected to form a repeating structure, the *glycan tetrapeptide* (Figure 3.31).

The basic structure of peptidoglycan is a thin sheet in which the glycan chains formed by the sugars are

FIGURE 3.30 a) Diaminopimelic acid. (b) Lysine. The only difference in the two molecules is highlighted in green.

connected by *peptide cross-links* formed by the amino acids. The glycosidic bonds connecting the sugars in the glycan chains are very strong, but these chains alone cannot provide rigidity in all directions. The full strength of the peptidoglycan structure is realized only when these chains are cross-linked by amino aids. This cross-linking occurs to characteristically different extents in different Bacteria, with greater rigidity coming from more complete cross-linking. In gram-negative Bacteria, cross-linkage usually occurs by direct peptide linkage of the amino group of diaminopimelic acid to the carboxyl group of the terminal D-alanine (Figure 3.32a). In gram-positive Bacteria, cross-linkage is usually

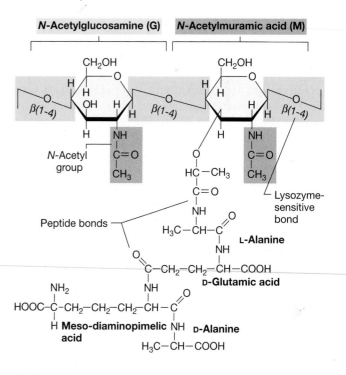

FIGURE 3.31 Structure of one of the repeating units of the peptidoglycan cell wall structure, the glycan tetrapeptide. The structure given is that found in *Escherichia coli* and most other gram-negative Bacteria. In some Bacteria, other amino acids are found.

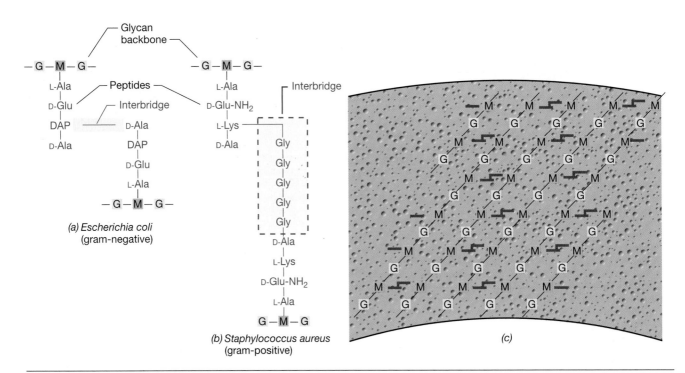

FIGURE 3.32   Manner in which the peptide and glycan units are connected in formation of the peptidoglycan sheet. (a) Direct interbridge in gram-negative Bacteria. (b) Glycine interbridge in *Staphylococcus aureus* (gram-positive). (c) Overall structure of peptidoglycan. The diagram depicts several ribbons of peptidoglycan cross-linked to one another. To visualize an entire single layer of peptidoglycan, imagine these cross-linked ribbons extending around a cylinder or sphere representing the cell as shown. G, *N*-acetylglucosamine; M, *N*-acetylmuramic acid; bold lines in (c) indicate peptide cross-links.

by a peptide interbridge, the kinds and numbers of cross-linking amino acids varying from organism to organism. In *Staphylococcus aureus*, a well-studied gram-positive organism, each interbridge peptide consists of five molecules of the amino acid glycine connected by peptide bonds (Figure 3.32*b*). The overall structure of a peptidoglycan molecule is shown in Figure 3.32*c*.

In gram-positive Bacteria, as much as 90% of the cell wall consists of peptidoglycan, although another kind of constituent, teichoic acid (see discussion later in this section) is usually present in small amounts. And, although some bacteria have only a single layer of peptidoglycan surrounding the cell, many Bacteria, especially gram-positive Bacteria, have several (up to about 25) peptidoglycan layers. In gram-negative Bacteria only about 10% of the wall is peptidoglycan, the majority of the wall consisting of the outer membrane as discussed in Section 3.8. However, the shape of both gram-positive and gram-negative cells is thought to be determined by the lengths of the peptidoglycan chains and by the manner and extent of cross-linking of the chains.

Diversity in Peptidoglycan

Peptidoglycan is present only in Bacteria; the sugar *N*-acetylmuramic acid and the amino acid diaminopimelic acid are never found in the cell walls of Archaea or Eukarya. However, not all Bacteria have DAP in their peptidoglycan. This amino acid is present in all gram-negative Bacteria and in some gram-positive species, but most gram-positive cocci have lysine instead of DAP, and a few other gram-positive Bacteria have other amino acids. Another unusual feature of the bacterial cell wall is the presence of two amino acids that have the D configuration, D-alanine and D-glutamic acid. As we saw in Chapter 2, in proteins amino acids are always of the L stereoisomer.

Several generalizations regarding peptidoglycan structure can be made. The glycan portion is uniform, with only the sugars *N*-acetylglucosamine and *N*-acetylmuramic acid being present, and these sugars are always connected in β-1,4 linkage. The tetrapeptide of the repeating unit shows major variation in only one amino acid, the lysine–diaminopimelic acid alternation. However, the D-glutamic acid at position 2 can be hydroxylated in some organisms, whereas substitutions occur in amino acids at positions 1 and 3 in a few others.

More than 100 different peptidoglycan types are known, and the greatest variation among them occurs in the interbridge. Any of the amino acids present in the tetrapeptide can also occur in the interbridge, but in addition, a number of other amino acids can be found there, such as glycine, threonine, serine, and aspartic acid. However, certain amino acids are never found in the interbridge: branched-chain amino acids, aromatic amino acids, sulfur-containing amino acids, and histidine, arginine, and proline. Thus, it can be stated that although the precise chemistry of peptidoglycan can vary, the structural makeup of peptidoglycan is the same in all forms of the molecule: Glucosamine and muramic acid form the backbone, and the muramic acid molecules are cross-linked with amino acids.

Teichoic Acids and a Summary of the Gram-Positive Wall

Gram-positive Bacteria frequently have acidic polysaccharides called **teichoic acids** (from the Greek word *teichos*, meaning "wall") attached to their cell wall. The term *teichoic acids* includes all wall, membrane, or capsular polymers containing glycerophosphate or ribitol phosphate residues. These polyalcohols are connected by phosphate esters and usually have other sugars and D-alanine attached (Figure 3.33a). Because they are negatively charged, teichoic acids are partially responsible for the negative charge of the cell surface as a whole and may function to effect passage of ions through the cell wall. Certain glycerol-containing acids are bound to membrane lipids of gram-positive Bacteria; because these teichoic acids are intimately associated with lipid, they have been called *lipoteichoic acids.*

Figure 3.33b summarizes the structure of the cell wall of gram-positive Bacteria and shows how teichoic acids and lipoteichoic acids are arranged in the overall wall structure.

Protoplast Formation

Peptidoglycan, the signature molecule of Bacteria, can be destroyed by certain agents. One such agent is the enzyme **lysozyme,** a protein that breaks the 1,4-glycosidic bonds between N-acetylglucosamine and N-acetylmuramic acid in peptidoglycan (Figure 3.31), thereby weakening the wall. Water then enters the cell, and the cell swells and eventually bursts, a process called **lysis** (Figure 3.34a). Lysozyme is found in animal secretions including tears, saliva, and other body fluids and presumably functions as a major line of defense against infection by Bacteria.

If the proper concentration of a solute that does not penetrate the cell, such as sucrose, is added to the medium, the solute concentration outside the cell balances that inside. Under these conditions, lysozyme still digests peptidoglycan, but lysis does not occur, and instead a protoplast is formed (Figure 3.34b). If such sucrose-stabilized protoplasts are placed in water, lysis occurs immediately. The word *spheroplast* is often used as a synonym for protoplast, although the two words have slightly different meanings: Protoplasts are generally free of residual cell wall material, whereas spheroplasts usually contain pieces of wall material attached to the otherwise membrane-enclosed structure.

Although most prokaryotes cannot survive without their cell walls, a group of Bacteria are able to do so; these are the mycoplasmas, a group that cause certain infectious diseases (∞ Section 13.20). Mycoplasmas are essentially free-living protoplasts and are able to survive

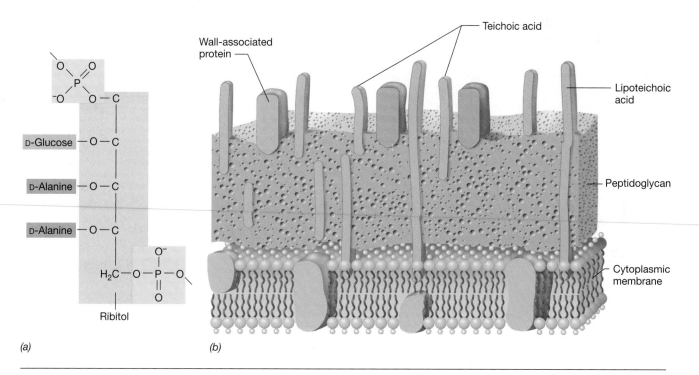

(a)   (b)

FIGURE 3.33   Teichoic acids and the overall structure of the gram-positive cell wall. (a) Structure of the ribitol teichoic acid of *Bacillus subtilis.* The teichoic acid is a polymer of the repeating ribitol units shown here. (b) Summary diagram of the gram-positive cell wall.

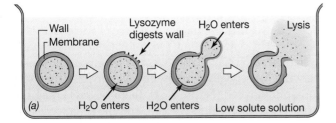

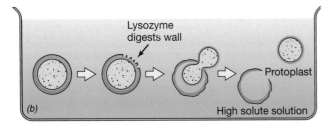

FIGURE 3.34   Protoplasts. (a) In dilute solution break-down of the wall releases the protoplast, but it immediately lyses because the cytoplasmic membrane is very weak. (b) In a solution containing a high concentration of a solute such as sucrose, water does not enter the protoplast and it remains stable.

without cell walls either because they have unusually tough membranes or because they live in osmotically protected habitats, such as the animal body. Certain mycoplasmas have sterols in their cell membranes, which lends strength and rigidity to this structure.

Pseudopeptidoglycan and Other Cell Walls of Archaea

Certain Archaea contain cell walls constructed of a polysaccharide very similar to that of peptidoglycan. This material is called *pseudopeptidoglycan* (Figure 3.35*b*). The backbone of pseudopeptidoglycan is composed of alternating repeats of *N*-acetylglucosamine and *N*-acetyltalosaminuronic acid (the latter replaces the *N*-acetylmuramic acid of peptidoglycan) (compare Figures 3.31 and 3.35*b*). The backbone of pseudopeptidoglycan also varies from peptidoglycan in that the glycosidic bonds are 1,3 instead of 1,4 found in true peptidoglycan (Figures 3.31 and 3.35*b*).

Cell walls of other Archaea lack both peptidoglycan and pseudopeptidoglycan and consist of polysaccharide, glycoprotein, or protein. *Methanosarcina* species, also

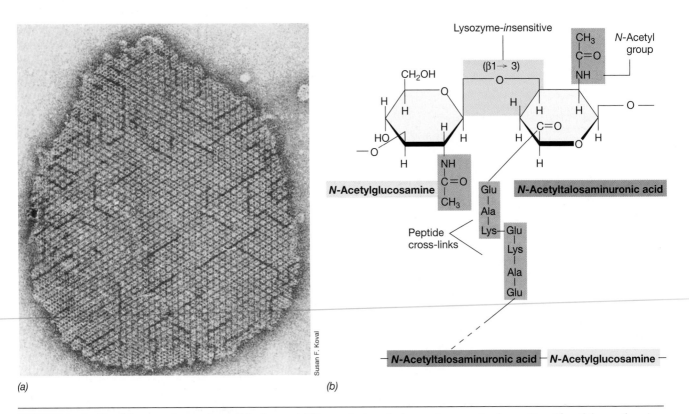

FIGURE 3.35   The S-layer and pseudopeptidoglycan. (a) Transmission electron micrograph of a portion of an S-layer showing the paracrystalline nature of this cell wall layer. Shown is the S-layer from the prokaryote *Aquaspirillum serpens* (a member of the Bacteria); this S-layer displays hexagonal symmetry as do many of the S-layers found in Archaea. (b) Structure of pseudopeptidoglycan, the cell wall polymer of *Methanobacterium* species. Note the resemblance to the structure of peptidoglycan shown in Figure 3.31, especially the peptide cross-links, in this case between *N*-acetyltalosaminuronic acid residues instead of muramic acid residues.

methanogenic Archaea, contain thick polysaccharide walls composed of glucose, glucuronic acid, galactosamine, and acetate. Extremely halophilic (salt-loving) Archaea such as *Halococcus* produce walls similar to that of *Methanosarcina* but that also contain an abundance of sulfate (SO_4^{2-}) residues. However, the most common wall type among Archaea is the paracrystalline surface layer (S-layer) (see Section 3.13 for more discussion) consisting of protein or glycoprotein, generally of hexagonal symmetry (Figure 3.35*a*). S-layers have been found among species of all groups of Archaea, the extreme halophiles, the methanogens, and the hyperthermophiles.

We thus see in species of Archaea a great variety of cell wall types, varying from molecules that closely resemble peptidoglycan to cell walls totally lacking a polysaccharide component. But with one exception, *Thermoplasma* (∞ Section 14.4), all Archaea contain a cell wall of some sort, and as in Bacteria, the archaeal cell wall functions to prevent osmotic lysis and to define cell shape. In addition, because they lack peptidoglycan in their cell walls, all Archaea are naturally resistant to the action of lysozyme (see earlier) and penicillin, agents that destroy this molecule.

✓ 3.7 Concept Check

The cell walls of Bacteria contain a polysaccharide called peptidoglycan. This material consists of strands of alternating repeats of *N*-acetylglucosamine and *N*-acetylmuramic acid, with the latter cross-linked between strands by short peptides. Archaea lack peptidoglycan but contain walls made of other polysaccharides or of protein. The enzyme lysozyme destroys peptidoglycan, leading to cell lysis.

✓ List the monomeric components of peptidoglycan.
✓ Why is peptidoglycan such a strong macromolecule?
✓ How does pseudopeptidoglycan resemble peptidoglycan? How do the two molecules differ?
✓ How is a protoplast generated?

3.8

The Outer Membrane of Gram-Negative Bacteria

Besides peptidoglycan, gram-negative Bacteria contain an additional wall layer made of **lipopolysaccharide.** This layer is effectively a second lipid bilayer, but it is not constructed solely of phospholipid, as is the cytoplasmic membrane; instead it contains polysaccharide and protein. The lipid and polysaccharide are intimately linked in the outer layer to form specific lipopolysaccharide structures. Because of the presence of lipopolysaccharide, this outer layer is frequently called the **lipopolysaccharide layer,** or simply **LPS.** Another term in widespread use for this structure is the **outer membrane.**

Chemistry of LPS

Although complex, the chemistry of the LPS of several bacteria is now understood. As seen in Figure 3.36, the polysaccharide consists of two portions, the core polysaccharide and the *O*-polysaccharide. In *Salmonella,* where it has been best studied, the **core polysaccharide** consists of ketodeoxyoctonate (KDO), seven-carbon sugars (heptoses), glucose, galactose, and *N*-acetylglucosamine. Connected to the core is the *O-polysaccharide,* which usually contains galactose, glucose, rhamnose, and mannose (all six-carbon sugars), as well as one or more unusual dideoxy sugars such as abequose, colitose, paratose, or tyvelose. These sugars are connected in four- or five-membered sequences, which often are branched. When the sequences are repeated, the long *O*-polysaccharide is formed.

The relationship of the *O*-polysaccharide to the rest of the LPS is shown in Figure 3.37. The lipid portion of the lipopolysaccharide, referred to as **lipid A** (Figure 3.36) is not a glycerol lipid, but instead the fatty

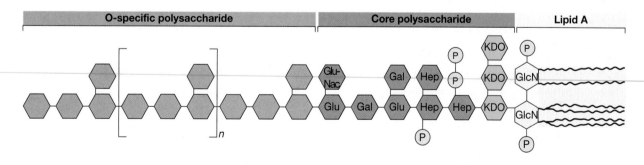

FIGURE 3.36   Structure of the lipopolysaccharide of gram-negative Bacteria. The precise chemistry of lipid A and the polysaccharide components varies among species of gram-negative Bacteria, but the sequence of major components (lipid A–KDO–core–O-specific) is generally uniform. The O-specific polysaccharide varies among species. KDO, ketodeoxyoctonate; Hep, heptose; Glu, glucose; Gal, galactose; GluNac, *N*-acetylglucosamine; GlcN, glucosamine; P, phosphate. The lipid A portion of LPS can be toxic to animals and comprises the *endotoxin complex* (∞ Section 19.10). Compare this figure with Figures 3.37 and 3.38 and note the color coding of different portions of the LPS.

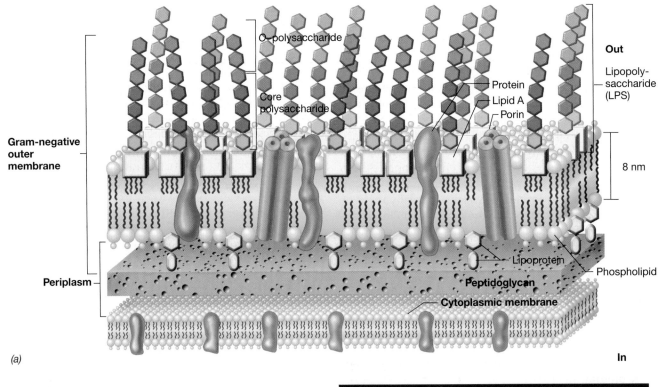

FIGURE 3.37 The gram-negative cell wall. (a) Arrangement of lipopolysaccharide, lipid A, phospholipid, porins, and lipoprotein in the outer membrane. See Figure 3.36 for details of the structure of LPS. (b) Molecular model of porin proteins. Note the three pores present, one formed from each of the proteins forming a porin molecule. The view is perpendicular to the plane of the membrane. Model based on X-ray diffraction studies of *Rhodobacter blasticus* porin.

acids are connected by ester amine linkage to a disaccharide composed of *N*-acetylglucosamine phosphate (Figure 3.36). The disaccharide is attached to the core *O*-polysaccharides through KDO (Figure 3.36). Fatty acids commonly found in lipid A include caproic, lauric, myristic, palmitic, and stearic acids. In the outer membrane, the LPS associates with various proteins to form the *outer* half of the unit membrane structure. A **lipoprotein** complex is found on the *inner* side of the outer membrane of a number of gram-negative Bacteria (Figure 3.37a). This lipoprotein is a small protein that functions as an anchor between the outer membrane

and peptidoglycan. In the *outer* leaf of the outer membrane, LPS replaces phospholipids; the latter are found predominantly in the inner leaf (Figure 3.37a).

Endotoxin

One important biological property of the outer membrane layer of many gram-negative Bacteria is that it is frequently *toxic* to animals. Gram-negative Bacteria that are pathogenic for humans and other mammals include members of the genera *Salmonella*, *Shigella*, and *Escherichia*, among others. The toxic property of the outer membrane layer of these bacteria is responsible for some of the symptoms of infection that they bring about. The toxic properties are associated with part of the lipopolysaccharide layer, in particular, lipid A, of these organisms. The term **endotoxin** is used to refer to this toxic component of LPS, as we will discuss in Section 19.10. However, interestingly, LPS from several nonpathogenic bacteria can also show endotoxin activity. Thus, the organism itself need not be pathogenic to contain toxic cell wall components.

Porins and the Periplasm

Unlike the cytoplasmic membrane, the outer membrane of gram-negative Bacteria is relatively permeable to small molecules even though it is basically a lipid bilayer. This is because proteins called **porins** are present in the outer membrane of gram-negative Bacteria and function as channels for the entrance and exit of hydrophilic low-molecular-weight substances (Figure 3.37). Several porins have now been identified, and both specific and nonspecific classes are known. *Nonspecific porins* form water-filled channels through which small substances of any type can pass. By contrast, some porins are highly specific because they contain a specific binding site for one or more substances.

Structural studies have shown that most porins are proteins containing *three* identical subunits. Porins are transmembrane proteins (Figure 3.37*a*) and associate to form small membrane holes about 1 nm in diameter (Figure 3.37*b*). Through the action of porins, the outer membrane is relatively permeable to small molecules. However, the outer membrane is *not* permeable to enzymes or other large molecules. In fact, one of the major functions of the outer membrane may be to keep certain enzymes, which are present outside the cytoplasmic membrane, from diffusing away from the cell. These enzymes are present in a region called the **periplasm** (see Figures 3.37 and 3.38). In *Escherichia coli* this space

between the outer surface of the cytoplasmic membrane and the inner surface of the LPS-containing outer membrane occupies a distance of about 12–15 nm and is gel-like in consistency, presumably because of the abundance of periplasmic proteins found there (Figure 3.38). The periplasm of gram-negative Bacteria generally contains several proteins including *hydrolytic enzymes*, which function in the initial degradation of food molecules, *binding proteins*, which begin the process of transporting substrates (see Section 3.6), and *chemoreceptors*, which are proteins involved in the chemotaxis response (see Section 3.12 and ∞ Section 7.7).

Relationship of Cell Wall Structure to the Gram Stain

Are the structural differences between the cell walls of gram-positive and gram-negative Bacteria responsible in any way for the Gram stain reaction? In the Gram stain (see Section 3.1), an insoluble crystal violet–iodine complex is formed inside the cell, and this complex is extracted by alcohol from gram-*negative* but not from gram-*positive* Bacteria. Gram-positive Bacteria, which have very thick cell walls consisting of several layers of peptidoglycan, become dehydrated by the alcohol. This causes the pores in the walls to close, preventing the insoluble crystal violet–iodine complex from escaping. In gram-negative Bacteria, alcohol readily penetrates the lipid-rich outer layer, and the thin peptidoglycan layer also does not prevent solvent passage, thus, the crystal violet–iodine complex is easily removed.

Now that we have a picture of the basic structure of the cell walls of Bacteria and Archaea, we consider how the wall is synthesized, focusing on the peptidoglycan layer from which most of the information on this subject is available. Because final synthesis of this crucial cellular molecule occurs *outside* the cytoplasmic membrane, careful control and coordination of peptidoglycan synthesis is necessary for cellular integrity to remain. We see how this happens now.

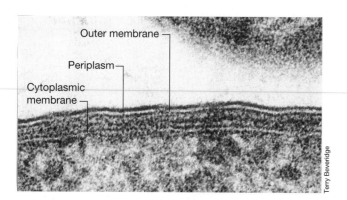

FIGURE 3.38 High magnification thin section of the cell envelope of *Escherichia coli* showing the periplasmic gel bounded by the outer membrane and the cytoplasmic membrane. The large, dark particles in the cytoplasm are ribosomes.

Outer membrane
Periplasm
Cytoplasmic membrane

Terry Beveridge

✓ 3.8 Concept Check

In addition to peptidoglycan, gram-negative Bacteria contain an outer membrane consisting of lipopolysaccharide, protein, and lipoprotein. Proteins called porins allow for permeability across the outer membrane, and a space called the periplasm is present, which contains various proteins involved in important cellular functions.

✓ What components constitute the LPS layer of gram-negative Bacteria?
✓ What is the function of porins and where are they located in a gram-negative cell wall?
✓ Why does alcohol readily decolorize gram-negative bacteria?

3.9

Cell Wall Synthesis and Cell Division

When a cell enlarges during the division process, new cell wall synthesis must take place, and this new wall material must be added to the preexisting wall without loss of structural integrity. This process occurs as shown in Figure 3.39a. Small openings in the macromolecular structure of the wall are created by enzymes called **autolysins,** similar in function to lysozyme, which are produced within the cell. New wall material is then added across the openings (Figure 3.39). The junction between new and old peptidoglycan forms a ridge on the cell surface of gram-positive Bacteria (Figure 3.39b), analogous to a scar. Thus, it is essential that new peptidoglycan be spliced onto preexisting peptidoglycan *before* severing bonds within the latter to ensure that cell turgor pressure does not burst the cell wall at a splice point. If this does not take place, a process of spontaneous lysis called **autolysis** can occur.

Biosynthesis of Peptidoglycan

The peptidoglycan layer can be thought of as a stress-bearing fabric, much like a sheet of rubber. Synthesis of new peptidoglycan during cell growth involves controlled cutting by autolysins of bonds connecting small areas of preexisting peptidoglycan, with the simulta-

neous insertion of new pieces of peptidoglycan, much as a patch is inserted into a piece of woven fabric. As this process continues during cell division, cell volume increases until a cross-wall (septum) forms and the cell divides into two cells (Figure 3.39).

Two carrier molecules participate in peptidoglycan synthesis, uridine diphosphate and a lipid carrier. The lipid carrier, called **bactoprenol,** is a C_{55} isoprenoid alcohol that is connected via phosphodiester linkage to N-acetylmuramic acid to which a pentapeptide is attached (Figure 3.40). The second amino sugar of peptidoglycan, *N*-acetylglucosamine, is then added followed by addition of the pentaglycine bridge.

The assembly of polymers such as peptidoglycan *outside* the cytoplasmic membrane presents special problems of transport and control. Bactoprenol is involved in transport of peptidoglycan building blocks across the membrane, where the disaccharide pentapeptide is then inserted into a growing point of the cell wall (Figure 3.41). The function of bactoprenol is to render sugar intermediates sufficiently hydrophobic so that they will pass through the hydrophobic cytoplasmic membrane. The lipid carrier inserts the disaccharide pentapeptide complex into the glycan backbone and then moves back inside the cell to pick up another peptidoglycan precursor unit (Figure 3.41).

Transpeptidation: The Penicillin Target

The final step in cell wall synthesis is formation of the peptide cross-links between adjacent glycan chains. The formation of peptide cross-links involves peptide bond formation, called **transpeptidation,** which is also noteworthy because it is the reaction inhibited by the antibiotic *penicillin.* This cross-linking reaction involves peptide formation with one of several different amino acids, depending on the organism involved. In gram-negative Bacteria, such as *Escherichia coli,* the cross-linking is between diaminopimelic acid (DAP) on one peptide and D-alanine on an adjacent peptide (Figure 3.41b). Initially, there are *two* D-alanine groups at the

Coccus

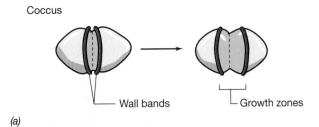

Wall bands   Growth zones

(a)

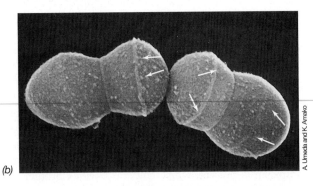

(b)

FIGURE 3.39 Cell wall synthesis in gram-positive Bacteria. (a) Localization of new cell wall synthesis during cell division. In cocci, new cell wall synthesis (shown in green) is localized at only one point. (b) Scanning electron micrograph of cells of *Streptococcus hemolyticus* showing wall bands (arrows). A single cell is about 1 μm in diameter.

$$H_3C-\underset{CH_3}{\overset{}{C}}=CHCH_2(CH_2\underset{CH_3}{\overset{}{C}}=CHCH_2)_9CH_2\underset{CH_3}{\overset{}{C}}=CHCH_2$$

$$O=P-O^-$$

$$O=P-O^-$$

[N-Acetylmuramic acid] —O

FIGURE 3.40 Bactoprenol (undecaprenolphosphate), the lipid carrier of the cell wall peptidoglycan building blocks.

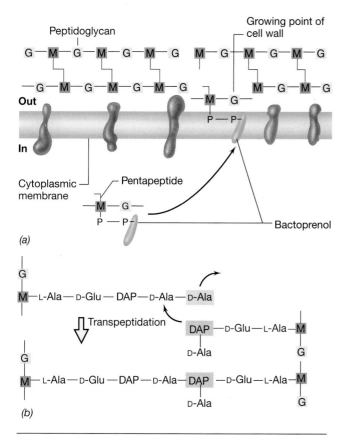

(a)

(b)

FIGURE 3.41 Peptidoglycan synthesis. (a) Transport of peptidoglycan precursors across the cytoplasmic membrane to the growing point of the cell wall. (b) The transpeptidation reaction that leads to the final cross-linking of two peptidoglycan chains. Penicillin inhibits this reaction.

end of the peptidoglycan precursor, but one D-alanine group is split off during the transpeptidation reaction (Figure 3.41). This reaction occurs outside the cytoplasmic membrane, where energy is not available, and the transpeptidation reaction functions to drive the reaction forward.

Inhibition of transpeptidation by penicillin thus leads to the formation of a weakened peptidoglycan. Further damage to the cell, resulting in lysis and death, occurs as autolysins continue to function, but because new peptidoglycan cross-links cannot occur, the cell wall becomes progressively weaker and osmotic lysis takes place. However, because peptidoglycan *synthesis* must be occurring for penicillin to act, penicillin-induced lysis occurs only in *growing* cells. In nongrowing cells, the action of autolysins does not occur, and so breakdown of the cell wall peptidoglycan is prevented. It is fascinating to consider that one of the key developments in human medicine, the discovery of penicillin, is linked to a specific biochemical reaction involved in cell wall synthesis, transpeptidation.

✓ 3.9 Concept Check

New cell wall is synthesized during bacterial growth by inserting new glycan units into preexisting wall material. A long-chain alcohol called bactoprenol facilitates transport of new glycan units through the cytoplasmic membrane to become part of the growing cell wall.

✓ What are *autolysins* and why are they necessary?
✓ What is the function of bactoprenol?
✓ What is *transpeptidation* and why is it important?

3.10

Arrangement of DNA in Prokaryotes

Now that we have developed an understanding of the structure and function of the cytoplasmic membrane and cell wall, two key structures in the cell, we proceed to a consideration of the arrangement of DNA in prokaryotes. We briefly mentioned in Chapter 1 that unlike the chromosomes of eukaryotic cells, the bacterial chromosome is usually a single circular molecule and that small circles, called *plasmids*, are also often present. Here we consider the details of how chromosomal and plasmid DNA is packaged within the prokaryotic cell.

Supercoiling and Chromosome Structure

The chromosome of prokaryotes is typically a covalently closed *circular* molecule, although some Bacteria are known that have linear chromosomal DNA (∞ Section 6.4). The total amount of DNA in the chromosome of a bacterium such as *Escherichia coli* is about 4600 kilobase pairs. Not surprisingly, this is considerably less than that of eukaryotic cells, but it is greater than that of viruses or organelles (Figure 3.42). Although prokaryotic DNA is not confined to a nucleus as it is in eukaryotes, it does tend to aggregate as a distinct structure within the cell and is visible when observed with the electron microscope (see Figure 3.10*a*). The term **nucleoid** is used to describe aggregated DNA in the prokaryotic cell, and under special staining conditions the nucleoid can actually be observed in cells examined with just the light microscope (Figure 3.43*a*). Ribosomes are absent in the region of the cell where the bacterial nucleoid is situated, probably because nucleoid DNA exists in a gellike form that tends to exclude particulate matter.

If gently lysed, DNA can be released from prokaryotic cells (Figure 3.43*b*), and the extensive folding and twisting necessary to store the DNA in the cell becomes readily apparent. The amount of twisting and folding can be appreciated when it is considered that the 4.6 *million* base pairs in the genome of *Escherichia coli*, if

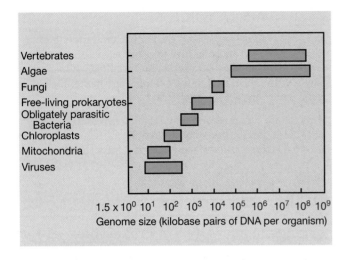

FIGURE 3.42 Range of genome sizes in various groups of organisms and the organelles of Eukarya.

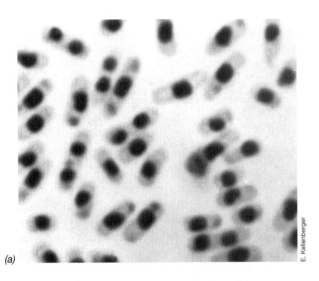

(a)

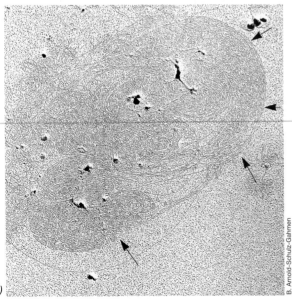

(b)

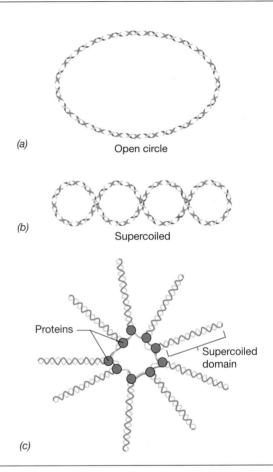

(a)

Open circle

(b)

Supercoiled

Proteins

Supercoiled domain

(c)

FIGURE 3.44 The bacterial chromosome and supercoiling. (a) Open circular form of the bacterial chromosome. (b) Supercoiled form. (c) In actuality, the double-stranded DNA in the bacterial chromosome is arranged not in one supercoil but in several *supercoiled domains,* as shown here. In *Escherichia coli* over 50 supercoiled domains are thought to exist, each of which is stabilized by binding to specific proteins.

opened and linearized, would be about 1 *mm* in length, yet the *E. coli* cell is only about 2–3 μm long! To package this much DNA into the cell requires that the DNA be **supercoiled** (Figure 3.44; ⟳ Section 6.3). Supercoiled DNA takes on a considerably more compact shape than its freely circularized counterpart. However, the bacterial chromosome is not a simple supercoil as shown in Figure 3.44b. There are over 50 **supercoiled domains** in the *E. coli* chromosome, and these domains are stabilized by association with structural proteins

FIGURE 3.43 The *Escherichia coli* nucleoid. (a) Light photomicrograph of cells of *E. coli* treated in such a way as to make the nucleoid visible. (b) Electron micrograph of an isolated nucleoid released from a cell of *E. coli*. The cell was gently lysed to allow the highly compacted nucleoid to emerge intact. Arrows point to edge of DNA strands.

(Figure 3.44c). This structure allows the very long DNA molecule to be folded and twisted to fit into the cell. In some of the Archaea the chromosomal DNA is extensively complexed with proteins that bear a strong resemblance to the histone proteins of eukaryotic cells (∞ Section 14.10).

Chromosomal Copy Number

Organisms like prokaryotes that reproduce asexually are typically **haploid** in genetic complement; that is, the cell's minimum genome contains a single copy of each chromosome. Because most prokaryotes seem to have only a single chromosome (∞ Section 6.4), this means that all the cell's genetic information is present on this single chromosome. Typically this means one copy of each gene.

However, rapidly growing prokaryotic cells usually contain multiple copies or partially completed copies (generally two to four) of the bacterial chromosome, and only when cell growth has ceased does the chromosome number approach one per cell. The reason for this is that rapidly growing cells can actually divide faster than the DNA replication machinery can make new copies of the bacterial chromosome. Thus, to ensure that a complete copy of the bacterial chromosome is ready for each daughter cell at cell division, new rounds of DNA synthesis are initiated before the old round is completed. This leads to multiple or partial copies of the chromosome in rapidly growing cells.

✓ 3.10 Concept Check

The DNA of the typical prokaryotic cell exists in a very long, single, circular molecule, called the prokaryotic chromosome, which is present in the cell in a highly aggregated state called the nucleoid. The nucleoid is not surrounded by a membrane and is present free in the cytoplasm in a highly supercoiled form.

✓ What is a prokaryotic chromosome and how many chromosomes does a prokaryote have?
✓ What is a nucleoid?

3.11

Flagella and Motility

Many prokaryotes are motile, and this ability to move independently is usually due to a special structure, the **flagellum** (plural, **flagella**) (Figure 3.45). Certain bacterial cells can move along solid surfaces by *gliding* (∞ Section 13.16), and certain aquatic microorganisms can regulate their position in a water column by gas-filled structures called gas vesicles (see Sections 3.14 and 13.24). However, the majority of motile prokaryotes

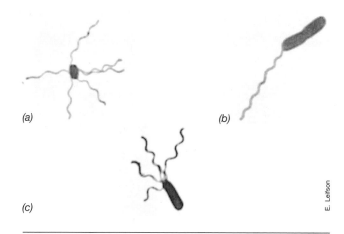

FIGURE 3.45 Light photomicrographs of prokaryotes containing different flagellar arrangements. Cells are stained with Leifson flagella stain. (a) Peritrichous. (b) Polar. (c) Lophotrichous.

move by means of flagella. Motility allows the cell to reach different regions of its environment. In the struggle for survival, movement to a new location may mean the difference between survival and death of the cell. But, as in any physical process, cell movement is closely tied to an energy expenditure, and the movement of flagella is no exception. We begin now with a detailed consideration of flagellar motility in prokaryotes.

Bacterial Flagella

Bacterial flagella are long, thin appendages free at one end and attached to the cell at the other end. They are so thin (about 20 nm) that a single flagellum can never be seen directly with the light microscope but only after staining with special flagella stains that increase their diameter (Figure 3.45). Flagella are readily seen with the electron microscope (Figure 3.46).

Flagella are arranged differently on different bacteria. In **polar flagellation** the flagella are attached at one or both ends of the cell (Figures 3.45b and 3.46a). Occasionally a tuft (group) of flagella may arise at one end of the cell, an arrangement called **lophotrichous** (*lopho* means "tuft"; *trichous* means "hair") (Figure 3.45c). Tufts of flagella of this type can be seen in living cells by dark-field microscopy (see Section 3.1 and Figure 3.47a), where the flagella appear light and are attached to light-colored cells against a dark background. In extremely large prokaryotes, tufts of flagella can also be observed by phase contrast microscopy (Figure 3.47b). In **peritrichous flagellation** the flagella are inserted at many places around the cell surface (*peri* means "around"). The type of flagellation, polar or peritrichous, is often used as a characteristic in the classification of bacteria (see Figures 3.45–3.47).

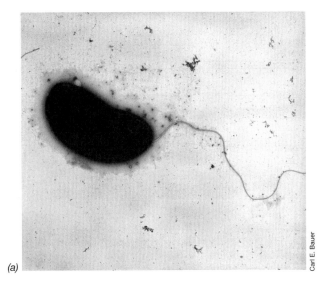

(a)

Carl E. Bauer

FIGURE 3.46 Bacteria flagella as observed by negative staining in the transmission electron microscope. (a) Polar flagella. (b) Peritrichous flagella. Both micrographs are of cells of the phototrophic bacterium *Rhodospirillum centenum.* Cells of *R. centenum* are normally polarly flagellated but under certain growth conditions form peritrichously flagellated "swarmer" cells. See also Figure 3.52*b.*

Flagellar Structure

Flagella are not straight but helically shaped; when flattened, they show a constant distance between two adjacent curves, called the *wavelength,* and this wavelength is constant for a given organism (Figures 3.45–3.47). The filament of bacterial flagella is composed of subunits of a protein called **flagellin.** The shape and wavelength of the flagellum are in part determined by the structure of the flagellin protein and also to some extent by the direction of rotation of the filament. The basic flagellar structure to be described here varies little among species of Bacteria, however in Archaea several different flagellins are known and the flagellar structure is probably quite different from that of Bacteria. In the latter domain flagellin is highly conserved, suggesting that flagellar motility has deep evolutionary roots within this prokaryotic group.

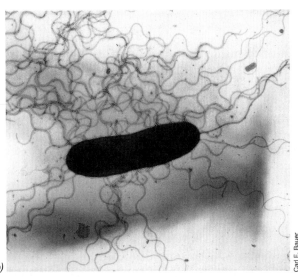

(b)

Carl E. Bauer

FIGURE 3.47 Bacterial flagella as observed in living cells. (a) Dark-field photomicrograph of a group of large rod-shaped bacteria with flagellar tufts at each pole. A single cell is about 2 μm wide. Dark-field microscopy uses horizontal illumination to yield reflected light (see Section 3.1 and Figure 3.5). (b) Phase contrast photomicrograph of the large phototrophic purple bacterium *Rhodospirillum photometricum.* A single cell measures about 3 × 30 μm. Note the tuft of lophotrichous flagella that emanates from one of the poles.

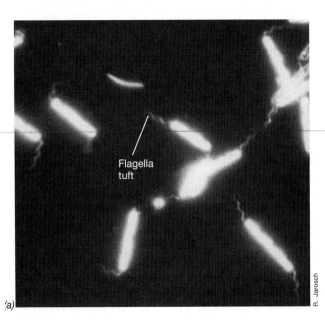

Flagella
tuft

(a)

R. Jarosch

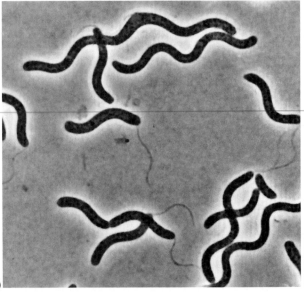

(b)

Norbert Pfennig

The base of the flagellum is different in structure from that of the filament (Figure 3.48). There is a wider region at the base of the flagellum called the *hook.* The hook consists of a single type of protein and functions to connect the filament to the motor portion of the flagellum. This motor, called the *basal body,* is anchored in the cytoplasmic membrane and cell wall. The basal body consists of a small central rod that passes through a system of rings. In gram-negative Bacteria, an outer ring is anchored in the lipopolysaccharide layer and another in the peptidoglycan layer of the cell wall, and an inner ring is located within the cytoplasmic membrane (Figure 3.48). In gram-positive Bacteria, which lack the outer lipopolysaccharide layer, only the inner pair of rings is present. Surrounding the inner ring and an-

chored in the cytoplasmic membrane are a pair of proteins called *Mot* (Figure 3.48). These proteins actually drive the flagellar motor causing a torque that rotates the filament. A final set of proteins, called the *Fli* proteins (Figure 3.48) function as the motor switch, reversing rotation of the flagella in response to intracellular signals.

Several genes are required for flagellar synthesis and subsequent motility. In *Escherichia coli* and *Salmonella typhimurium,* where studies have been most extensive, over 40 genes, called *fla, fli,* and *flg,* are necessary for motility. These genes have several functions, including encoding structural proteins of the flagellar apparatus, export of flagellar components through the membrane to the outside of the cell, and

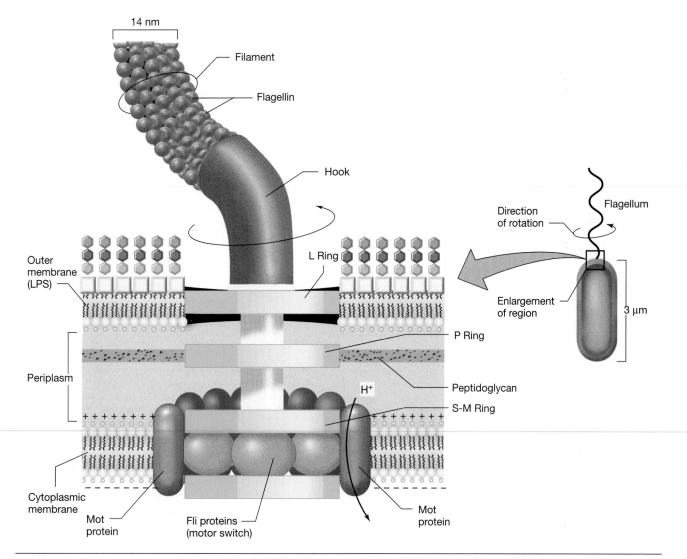

FIGURE 3.48   Structure of the prokaryotic flagellum and attachment to the cell wall and membrane in a gram-negative bacterium. The L ring is embedded in the LPS, and the P ring in peptidoglycan. The S-M ring is embedded in the cytoplasmic membrane. The Mot proteins function as the flagellar motor, whereas the Fli proteins function as the motor switch. The flagellar motor rotates the filament to propel the cell through the medium.

regulation of the many biochemical events surrounding the synthesis of new flagella. Control of flagella synthesis is tightly regulated in the cell both by metabolic factors and by signals emerging from the cell division cycle.

The individual flagellum grows not from the base, as does an animal hair, but from the tip. Flagellin molecules formed in the cell pass up through the hollow core of the flagellum and add on at the terminal end. Interestingly, the synthesis of a flagellum from its flagellin protein molecules occurs by a process called *self-assembly;* all the information for the final structure of the flagellum resides in the protein subunits themselves. Growth of the flagellum occurs more-or-less continuously, although the rate of growth slows as the filament elongates and approaches its final length. Broken flagella can continue to rotate while new flagellin is passed through the filament to repair the broken off piece.

Flagellar Movement

How is motion imparted to the flagellum? Each individual flagellum is actually a semirigid structure that does not flex but, as mentioned previously, moves by rotation, in the manner of a propeller. Visual evidence of this can be obtained by observing the behavior of motile cells tethered by their flagella to microscope slides. Such cells rotate around the point of attachment at rates of revolution consistent with those inferred for flagellar movement in free-swimming cells.

The rotary motion of the flagellum is imparted from the basal body, which functions as a motor. The energy required for rotation of the flagellum comes from the proton motive force (see Sections 3.6 and 4.11). Proton movement across the membrane through the Mot complex (Figure 3.48) drives rotation of the flagellum, and calculations have shown that about 1000 protons must be translocated per single rotation of the flagellum.

Flagella do not rotate at a constant speed but instead can increase or decrease their rotational speed in relation to the strength of the proton motive force. Flagella rotation can move bacteria through liquid media at speeds of up to 60 cell lengths/second (sec). Although this is only about 0.00017 kilometer/hour (km/hr), when comparing this speed with that of higher organisms in terms of the number of lengths moved per second, it is extremely fast. The fastest animal, the cheetah, moves at a maximum rate of about 110 km/hr, but this represents only about 25 body lengths/sec. Thus, when size is accounted for, prokaryotic cells swimming at 50–60 lengths/sec are actually moving much faster than larger organisms.

The motions of polarly and lophotrichously flagellated organisms are different from those of peritrichously flagellated organisms. Peritrichously flagellated organisms generally move in a straight line in a slow, stately fashion. Polarly flagellated organisms, on the other hand, move more rapidly, spinning around and dashing from place to place. The different behavior of flagella on polar and peritrichous organisms is illustrated in Figure 3.49.

✓ 3.11 Concept Check

Motility in microorganisms is usually associated with flagella. In prokaryotes the flagellum is a complex structure made of several proteins, most of which are anchored in the cell wall and membrane. The flagellum filament, which is made of a single kind of protein, rotates at the expense of the proton motive force, which drives the flagellar motor.

✓ How does *polar flagellation* differ from *peritrichous flagellation?*
✓ What is *flagellin* and where is it found?
✓ How does a bacterial flagellum move a cell forward?

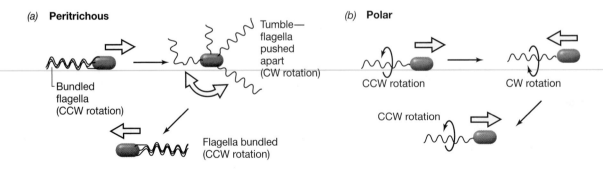

FIGURE 3.49 Manner of movement in polarly and peritrichously flagellated prokaryotes. (a) Peritrichous: Forward motion is imparted by all flagella rotating counterclockwise (CCW) in a bundle. Clockwise (CW) rotation causes the cell to tumble, and then a return to counterclockwise rotation leads the cell off in a new direction. (b) Polar: Cells change direction by reversing flagellar rotation (thus pulling instead of pushing the cell) and then return to pushing. The large yellow arrows show the direction the cell is traveling.

3.12

Bacterial Behavior: Chemotaxis, Phototaxis, and Other Taxes

Although not all prokaryotes are motile, many are, and it is reasonable to assume that motility confers a selective advantage on cells under certain environmental conditions. Prokaryotes often encounter *gradients* of physical and chemical agents in nature, and the motility machinery in the cell has evolved to respond in a positive or negative way to these gradients by directing movement of the cell either toward or away from the signal molecule, respectively. Such directed movements are called *taxes,* and a variety of such responses occur in microorganisms. **Chemotaxis,** a response to chemicals, and **phototaxis,** a response to light, are two well-known taxes, and we focus on these here.

Chemotaxis

To understand chemotaxis we can focus on the behavior of a single bacterial cell faced with a chemical gradient of an attractant (Figure 3.50). Unlike larger organisms, prokaryotes are too small to sense a gradient along their body length. They must instead, while moving, compare the chemical or physical state of their environment with that sensed a few seconds before. In other words, bacteria respond to the *temporal* (rather than *spatial*) *gradient* of signal molecules as they swim along.

In the absence of a gradient, cells move in a random fashion that includes **runs,** where the cell is swimming forward in a smooth fashion, and **tumbles,** when the cell stops and jiggles about (Figure 3.50*a*). Following a tum-

ble, the direction of the next run is random (Figure 3.50*a*). Thus, by means of runs and tumbles the organism moves about randomly in its environment but does not really go anywhere. However, if a gradient of a chemical attractant is present, these random movements become biased. As the organism senses higher concentrations of the attractant (through periodic sampling of the concentration of the chemical in its environment), runs become longer and tumbles less frequent. The net result of this behavior is that the organism moves up the concentration gradient of the attractant (Figure 3.50*b*). If the organism is sensing a repellant, the same general mechanism applies, although in this case it is the *decrease* in concentration of the repellant (rather than the *increase* in concentration of an attractant) that promotes runs. Forward movement in a run occurs when the flagellar motor is rotating counterclockwise. When the flagella rotate clockwise, the bundle pushes apart, forward motion ceases, and the cells tumble (Figure 3.49).

Measuring Chemotaxis

How do bacteria use temporal changes in chemical concentrations to control flagellar rotation? This is a complex story and involves several regulatory events at the genetic and biochemical levels. We therefore reserve detailed discussion of the mechanism of chemotaxis for Chapter 7 (∞ Section 7.7) where this topic can be better understood in a biochemical-genetic context. However, suffice it for now to say that the molecular mechanism of chemotaxis involves sensory proteins in the membrane called *chemoreceptors* that sense the chemical gradient with time and interact with cytoplasmic proteins to affect flagellar motor direction. Thus, because the direction of flagellar rotation governs whether the cell runs or tumbles, chemo-

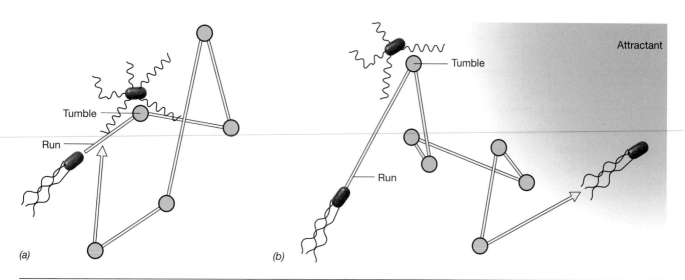

FIGURE 3.50   Chemotaxis. (a) In the absence of a chemical attractant the cell swims randomly in runs, changing direction during tumbles. (b) In the presence of an attractant runs become biased, and the cell moves up the gradient of the attractant.

taxis can be thought of as a chemically driven *sensory response system* affecting flagellar function.

Bacterial chemotaxis can be most easily demonstrated by immersing a small glass capillary containing an attractant in a suspension of motile bacteria that does not contain the attractant. From the tip of the capillary, a gradient is set up into the surrounding medium, with the concentration of chemical gradually decreasing with distance from the tip (Figure 3.51). If the capillary contains an attractant, the bacteria will move toward the capillary, forming a swarm around the open tip (Figure 3.51); subsequently many of the motile bacteria will move into the capillary. Of course some bacteria will move into the capillary even if it contains a solution of the same composition as the medium because of random movements. But if an attractant is present, the concentration of bacteria within the capillary can be many times higher than the external concentration. On the other hand, if the capillary contains a repellant, the concentration of bacteria within the capillary will be considerably less than the concentration outside (Figure 3.51). Using this simple method, it is possible to rapidly screen chemicals for their ability to act as attractants or repellants for a given bacterium.

Phototaxis

Many phototrophic (photosynthetic) microorganisms move toward light, a process called *phototaxis*. The advantage of phototaxis is that it allows a phototrophic organism to orient itself for the most efficient photosynthesis. This can be shown if a light spectrum is spread across a microscope slide on which there are motile phototrophic bacteria; the bacteria accumulate at the wavelengths at which their photosynthetic pigments absorb (Figure 3.52) (∞ Sections 15.3 and 15.4 for a discussion of photosynthetic pigments).

Two different taxes are observed in phototrophic prokaryotes. One, called *scotophobotaxis,* can be observed only microscopically and occurs when a phototrophic bacterium happens to swim outside the illuminated field of view of the microscope into darkness. This event signals the cell to tumble, reverse direction, and once again swim in a run, thus reentering the light. The fact that phototrophic cells show scotophobic behavior and accumulate in regions of the spectrum in which their pigments absorb (Figure 3.52*a*) strongly suggests that the scotophobotactic response is somehow triggered by changes in energy generation (ATP synthesis or the state of the proton motive force).

However, in addition to scotophobotaxis, phototrophic microorganisms can carry out true phototaxis, a directed movement up a light gradient toward an increasing intensity of light. This can be thought of as analogous to positive chemotaxis except that in this case the attractant is not a chemical but instead is light. In some species, such as the highly motile phototrophic organism *Rhodospirillum centenum,* (see Figure 3.46), *entire colonies* of cells show phototaxis and move in unison toward the light (Figure 3.52*b*). Although the molecular details of phototaxis are not yet known, there is

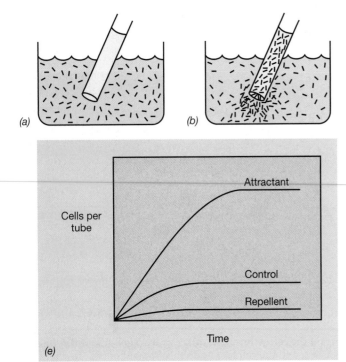

(a) (b) (c) (d)

(e)

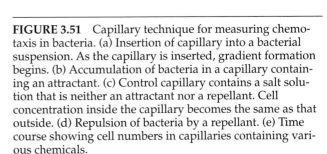

FIGURE 3.51 Capillary technique for measuring chemotaxis in bacteria. (a) Insertion of capillary into a bacterial suspension. As the capillary is inserted, gradient formation begins. (b) Accumulation of bacteria in a capillary containing an attractant. (c) Control capillary contains a salt solution that is neither an attractant nor a repellant. Cell concentration inside the capillary becomes the same as that outside. (d) Repulsion of bacteria by a repellant. (e) Time course showing cell numbers in capillaries containing various chemicals.

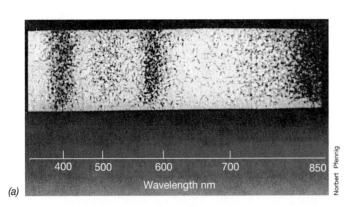

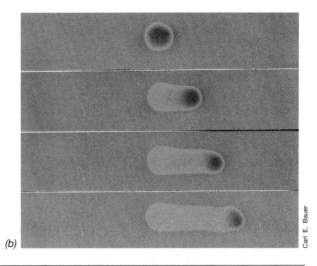

FIGURE 3.52   Phototaxis. (a) Scotophobic accumulation of the phototrophic bacterium *Thiospirillum jenense* at light wavelengths at which its pigments absorb. A light spectrum was displayed on a microscope slide containing a dense suspension of the bacteria; after a period of time, the bacteria had accumulated selectively and the photomicrograph was taken. The wavelengths at which accumulations occur are those at which bacteriochlorophyll *a* absorbs (compare with Figure 15.3*b*). (b) Phototaxis of an entire colony of the purple phototrophic bacterium *Rhodospirillum centenum* over a 2-hr time course (time 0 at top). These strongly phototactic cells move in unison toward the light source on the right. See Figure 3.46 for electron micrographs of *R. centenum* cells.

good evidence that several parts of the regulatory system that govern chemotaxis are also involved in phototaxis. These include in particular cytoplasmic proteins (Che proteins) that control the direction of rotation of the flagella (∞ Section 7.7). This evidence has emerged from the study of mutants of phototrophic bacteria defective in phototaxis; such mutants frequently have defective chemotaxis systems as well. A *photoreceptor,* analogous to a chemoreceptor but able to sense a gradient of *light* instead of chemicals, is responsible for orchestrating the phototaxis response. It is hypothesized that the photoreceptor can in some way interact with the proteins that affect flagella rotation to maintain the cell in a run if it is swimming toward an increasing intensity of light.

Other Taxes

Other bacterial taxes, such as movement toward or away from oxygen (*aerotaxis*) or toward or away from conditions of high ionic strength (*osmotaxis*), are also beginning to be understood in molecular terms now that some of the general principles of sensory response systems have been elucidated, primarily from work on chemotaxis. In most of these cases a common mechanism applies: Cells periodically sample their environment and process this information through a signal transduction pathway (∞ Section 7.7) that leads to control of the direction of flagellar rotation. These can be considered simple behavioral responses, and a rationale for elucidating the molecular mechanisms of bacterial

taxes is to gain a better understanding of similar responses, such as nerve transmission, in higher organisms. Thus, from a behavioral point of view, motile prokaryotes are well attuned to the chemical and physical state of their environment and as such can move toward or away from various stimuli presumably as a means of remaining competitively successful.

✓ 3.12 Concept Check

Motile bacteria can respond to chemical and physical gradients in their environment. In the processes of chemotaxis and phototaxis, random movement of a prokaryotic cell can be biased either toward or away from a chemical (or toward light in a phototrophic microorganism) by controlling the degree to which runs or tumbles occur. The latter are controlled by the direction of rotation of the flagellum, which in turn is controlled by a network of sensory and response proteins.

- ✓ Define the word *chemotaxis.*
- ✓ What causes a *run* versus a *tumble?*
- ✓ How does *scotophobotaxis* differ from *phototaxis?*

3.13

Bacterial Cell Surface Structures and Cell Inclusions

Prokaryotes can produce a variety of structures that are attached to or in some way protrude from the cell surface, and also several different types of internal cell structures. We survey these structures here, but it should be understood that not all bacteria will contain all these structures, and many bacteria may contain none of them. Thus, these are *optional* structures produced by some kinds of prokaryotes but not others.

Fimbriae and Pili

Fimbriae and pili are structurally similar to flagella but are not involved in motility. **Fimbriae** are considerably shorter than flagella and are more numerous (Figure 3.53) but, like flagella, consist of protein. Not all organisms have fimbriae, and the ability to produce them is an inherited trait. The functions of fimbriae are not known for certain in all cases, but there is good evidence that they enable organisms to stick to surfaces including animal tissues in the case of some pathogenic bacteria or to form pellicles or biofilms (⚬⚬ Section 16.2) on surfaces.

Pili are similar structurally to fimbriae but are generally longer, and only one or a few pili are present on the surface. Pili can be seen under the electron microscope because they serve as specific receptors for certain types of virus particles, and when coated with virus they can be easily seen (Figure 3.54). There is strong evidence that pili are involved in the process of conjugation in prokaryotes, as will be discussed in Section 9.9. Pili are also involved in attachment to human tissues by some pathogenic bacteria.

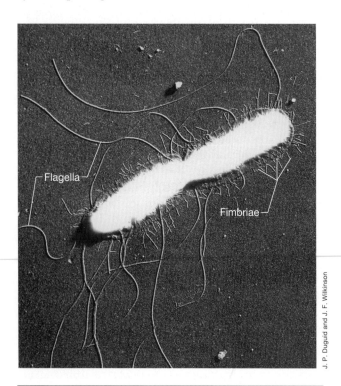

FIGURE 3.53 Electron micrograph of a dividing cell of *Salmonella typhi*, showing flagella and fimbriae. A single cell is about 0.9 μm in diameter.

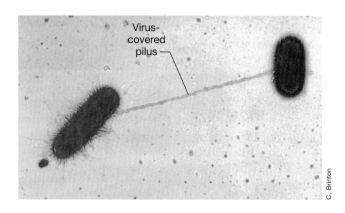

FIGURE 3.54 The presence of pili on an *Escherichia coli* cell is revealed by the use of viruses that specifically adhere to the pilus. The cell is about 0.8 μm in diameter.

Paracrystalline Surface Layers (S-Layers)

Many prokaryotes contain a cell surface layer composed of a two-dimensional array of protein. These layers are called *S-layers*. S-layers have been detected in representatives of virtually every phylogenetic grouping of Bacteria and are nearly universal among Archaea. In some species of Archaea the S-layer is also the cell wall (see Section 3.7). S-layers have a crystalline appearance and show various symmetries, such as hexagonal, tetragonal, or trimeric, depending upon the number and structure of the protein or glycoprotein subunits of which they are composed (see Figure 3.35*a* to view an electron micrograph of an S-layer). By virtue of their wide phylogenetic distribution, S-layers can obviously associate with a variety of cell wall structures, including the LPS of gram-negative Bacteria (see Section 3.8), the peptidoglycan layers of gram-positive Bacteria, and even directly with the cell membrane in certain Archaea (see Section 3.7).

The major function of S-layers is unknown. However, as the interface between the cell and its environment it is likely that in cells that produce them the S-layer at least functions as an external permeability barrier, allowing the passage of low-molecular-weight substances while excluding large molecules. Evidence also exists that in pathogenic (disease-causing) bacteria that contain S-layers, this structure may confer protection on the bacterium against certain host defense mechanisms.

Capsules and Slime Layers: The Glycocalyx

Many prokaryotic organisms secrete on their surfaces slimy or gummy materials (Figure 3.55). A variety of these structures consist of polysaccharide, and a few consist of protein. The terms **capsule** and **slime layer** are frequently used to describe polysaccharide layers;

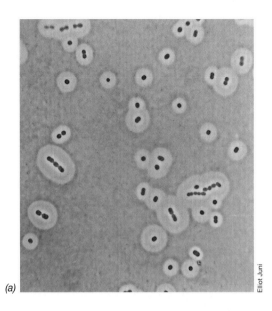

(a)

Elliot Juni

(b)

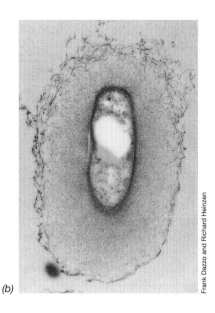

Frank Dazzo and Richard Heinzen

FIGURE 3.55 Bacterial capsules. (a) Demonstration of the presence of a capsule in an *Acinetobacter* species by negative staining with india ink observed by phase contrast microscopy. The india ink does not penetrate the capsule, and so it is revealed in outline as a light structure on a dark background. (b) Electron micrograph of a thin section of a *Rhizobium trifolii* cell stained with ruthenium red to reveal the capsule. The diameter of the cell proper (not including the capsule) is about 0.7 μm.

the more general term **glycocalyx** is also used. Glycocalyx is defined as the polysaccharide-containing material lying outside the cell. The composition of these layers varies in different organisms but can contain glycoproteins and different polysaccharides, including polyalcohols and amino sugars. These layers may be thick or thin, rigid or flexible, depending on their chemical nature in a specific organism. The rigid layers are organized in a tight matrix that excludes particles, such as india ink; this form is referred to as a *capsule*. If the glycocalyx is more easily deformed, it will not exclude particles and is more difficult to see; this arrangement is referred to as a *slime layer*.

Glycocalyx layers have several functions in bacteria. Outer polysaccharide layers play an important role in the *attachment* of certain pathogenic microorganisms to their hosts. As we will see(⌬ Section 19.6), pathogenic microorganisms that enter the animal body by specific routes usually do so because of binding reactions that occur between outer cell surface components (such as the glycocalyx) and specific host tissues. The glycocalyx plays other roles as well. There is some evidence that encapsulated bacteria are more difficult for phagocytic cells of the immune system (⌬ Section 20.3) to recognize and subsequently destroy. In addition, because outer polysaccharide layers probably bind a significant amount of water, there is reason to believe a glycocalyx layer plays some role in resistance to desiccation.

Carbon Storage Polymers

Granules or other inclusions are often seen within cells. Their nature differs in different organisms, but they almost always function in the storage of energy or as a reservoir of structural building blocks. Inclusions can often be seen directly with the phase contrast microscope,

but their contrast can usually be increased by using dyes. Inclusions often show up very well with the electron microscope (Figure 3.56). Most cellular inclusions are bounded by a thin *nonunit* membrane consisting of lipid separating the inclusion from the cytoplasm proper.

In prokaryotic organisms, one of the most common inclusion bodies consists of **poly-β-hydroxybutyric acid (PHB)**, a lipidlike compound that is formed from β-hydroxybutyric acid units (Figure 3.56a). The monomers of this acid are connected by ester linkages, forming the long PHB polymer, and these polymers aggregate into granules. The length of the monomer in the polymer can vary considerably, from as short as C_4 to as long as C_{18} in certain organisms. Thus, the collective term *poly-β-hydroxyalkanoate* (PHA) is used to describe this whole class of carbon/energy storage polymers. A wide variety of prokaryotes, including representatives of both the Bacteria and the Archaea, produce PHAs; however, Eukarya do not naturally produce PHAs.

Another storage product formed by prokaryotes is **glycogen,** which is a starchlike polymer of glucose subunits (we discussed the chemistry of glycogen in Section 2.3). Glycogen granules are usually smaller than PHB granules and can be seen only with the electron microscope, but the presence of glycogen in a cell can be detected in the light microscope because the cell appears a red-brown color when treated with dilute iodine because of a glycogen–iodine reaction. Like PHB and other PHAs, glycogen is a storage depot for carbon and energy.

Other Storage Materials and Inclusions

Many microorganisms accumulate large reserves of inorganic phosphate in the form of granules of **polyphos-**

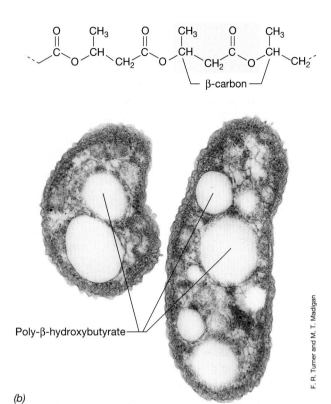

(b)

F. R. Turner and M. T. Madigan

Poly-β-hydroxybutyrate—

FIGURE 3.56 Poly-β-hydroxybutyrate (PHB). (a) Chemical structure of PHB, a common poly-β-hydroxyalkanoate. A monomeric unit is shaded. Other alkanoate polymers are made by substituting longer-chain hydrocarbons for the −CH$_3$ group on the β carbon. (b) Electron micrograph of a thin section of cells of the phototrophic bacterium *Rhodovibrio sodomensis* containing granules of PHB.

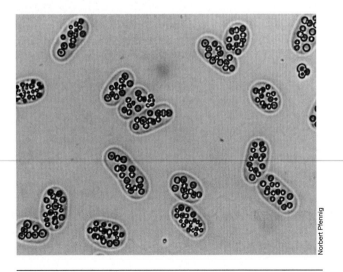

Norbert Pfennig

FIGURE 3.57 Bright-field photomicrograph of cells of the purple sulfur bacterium *Chromatium buderi*. Note the sulfur globules inside the cell. A single cell measures about 4 × 7 μm.

phate. These granules are stained by many basic dyes; one of these dyes, toluidine blue, becomes reddish violet in color when combined with polyphosphate. This phenomenon is called *metachromasy* (color change), and granules that stain in this manner are often called *metachromatic granules.*

A variety of prokaryotes are capable of oxidizing reduced sulfur compounds such as hydrogen sulfide and thiosulfate. These oxidations are linked to either reactions of energy metabolism (∞ Sections 15.9 and 15.11) or biosynthesis (∞ Section 15.7), but in both instances **elemental sulfur** frequently accumulates inside the cell in large, readily visible granules (Figure 3.57). The granules of elemental sulfur remain as long as the source of reduced sulfur is still present. However, as the reduced sulfur source becomes limiting, the sulfur in the granules is oxidized, usually to sulfate, and the granules slowly disappear as this reaction proceeds.

Magnetosomes are intracellular crystal particles of the iron mineral magnetite, Fe$_3$O$_4$ (Figure 3.58). Magnetosomes impart a permanent magnetic dipole to a cell, allowing it to respond to a magnetic field. Bacteria that produce magnetosomes (Figure 3.58a) exhibit *magnetotaxis*, the process of orienting and migrating along geomagnetic field lines (∞ Section 13.13). Although the combining form *-taxis* is used in the word *magnetotaxis*, there is no evidence that magnetotactic bacteria employ the sensory systems of chemotactic or phototactic bacteria (see Sections 3.12 and ∞ 7.7). Instead, the alignment of magnetosomes in the cell simply imparts magnetic properties to it, which then orient the cell in a particular direction in its environment. Thus, a better term to describe these organisms would be *magnetic bacteria.*

Magnetosomes are surrounded by a membrane containing phospholipids, proteins, and glycoproteins (Figure 3.58b). Magnetosome membrane proteins probably play a role in precipitating Fe^{3+} (brought into the cell in soluble form by chelating agents) as Fe$_3$O$_4$ in the developing magnetosome. The morphology of magnetosomes appears to be species-specific, varying in shape from square to rectangular to spike-shaped in certain bacteria.

Magnetosomes have been described from a variety of different primarily aquatic bacteria (Figure 3.58a) and have even been found in some algae (eukaryotes). Algal magnetosomes presumably function to make the algal cells magnetic, as in bacteria. Measurements of the magnetic moment of magnetotactic algal cells indicates that the magnetic force within these cells is much greater than that of magnetotactic bacteria, which is consistent with the larger size of the algal cells.

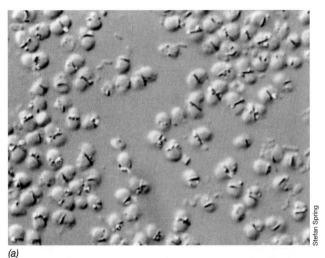

(a)

Stefan Spring

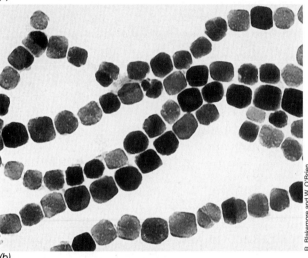

(b)

R. Blakemore and W. O'Brien

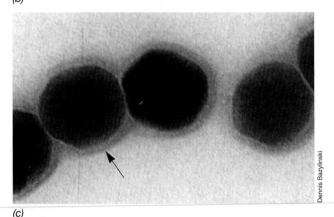

(c)

Dennis Bazylinski

FIGURE 3.58 Magnetotactic bacteria and magnetosomes. (a) Interference contrast micrograph of coccoid magnetotactic bacteria. Note magnetosomes. A single cell is about 2.2 μm in diameter. (b) Magnetosomes isolated from the magnetotactic bacterium *Magnetospirillum magnetotacticum*. Each particle is about 50 nm in length (⚬⚬ Figure 13.32). (c) High magnification electron micrograph of magnetosomes from a magnetic coccus-shaped bacterium. The arrow points to the membrane surrounding each magnetosome. A single magnetosome is about 90 nm wide.

✓ 3.13 Concept Check

Prokaryotic cells often contain various structures that either emerge from or are present inside the cell. These include fimbriae and pili, S-layers, capsules or slime layers, carbon and other storage polymers, and magnetosomes. However, unlike the organelles of eukaryotic cells, these structures are not bounded by a unit membrane.

- ✓ How do fimbriae differ from pili, both structurally and functionally?
- ✓ What is *poly-β-hydroxybutyrate* and what is its function in the cell? Under what conditions is this material made?
- ✓ What are *magnetosomes,* what are they made of, and what property do they confer on cells that contain them?

3.14

Gas Vesicles

A number of prokaryotic organisms that live a floating existence in lakes and the sea produce **gas vesicles,** which confer buoyancy on the cells. Gas vesicles are a means of motility, allowing cells to float up and down in a water column in response to environmental factors. The most dramatic instances of flotation due to gas vesicles are seen in cyanobacteria that form massive accumulations (blooms) in lakes (Figure 3.59). Gas-vesiculate cells rise to the surface of the lake and are blown by winds into dense masses. Gas vesicles are also present in certain purple and green phototrophic bacteria (⚬⚬

T. D. Brock

FIGURE 3.59 Flotation of gas vesiculate cyanobacteria from a bloom on a nutrient-rich lake, Lake Mendota, Madison, Wisconsin.

Sections 13.1 and 13.29) and in some nonphototrophic bacteria that live in lakes and ponds. Some species of Archaea also contain gas vesicles.

Structure of Gas Vesicles

Gas vesicles are spindle-shaped structures made of protein, hollow but rigid, that are of variable lengths and diameters. Gas vesicles in different organisms vary in length from about 300 to over 1000 nm and in width from 45 to 120 nm, but the vesicles of any given organism are more or less of constant size. They are present in the cytoplasm and may number from a few to hundreds per cell. The gas vesicle membrane is composed only of protein, is about 2 nm thick, and is impermeable to water and solutes but permeable to most gases; thus, gas vesicles exist as gas-filled structures surrounded by the constituents of the cytoplasm (Figure 3.60).

The rigidity of the gas vesicle membrane is essential for the structure to resist the pressures exerted on it from without; it is probably for this reason that it is composed of a protein able to form a rigid membrane rather than of lipid, which would form a fluid, highly mobile membrane. However, even the gas vesicle membrane can-

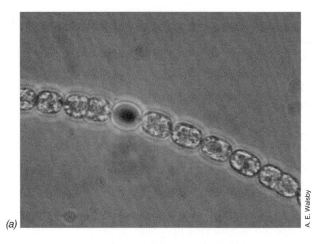

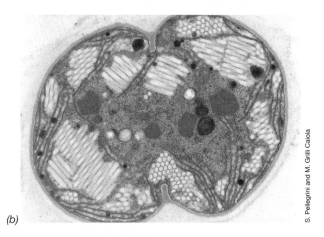

FIGURE 3.61 Gas vesicles of the cyanobacteria *Anabaena* and *Microcystis*. (a) *Anabaena flos-aquae*. The cell in the center (a heterocyst) lacks gas vesicles. In the other cells, the vesicles group together as phase-bright objects that scatter light. (b) Transmission electron micrograph of the cyanobacterium *Microcystis*. Gas vesicles are arranged in bundles, here observable in both longitudinal and cross section.

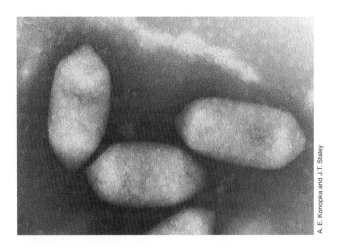

FIGURE 3.60 Transmission electron micrographs of gas vesicles purified from the bacterium *Ancyclobacter aquaticus* and examined in negatively stained preparations. A single gas vesicle is about 100 nm in diameter.

not resist high hydrostatic pressure and can be collapsed, leading to a loss of buoyancy. Once collapsed, gas vesicles cannot be reinflated. The presence of gas vesicles can be determined by either light or electron microscopy (Figure 3.61), but their identity is never certain unless they disappear when the cells are subjected to high hydrostatic pressure.

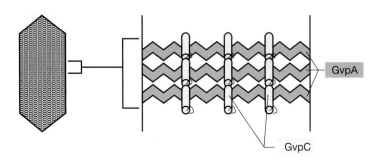

FIGURE 3.62 Model of how the two proteins that make up the gas vesicle, GvpA and GvpC, interact to form a watertight but gas-permeable structure. GvpA makes up the rib and is a rigid β-sheet. GvpC is the cross-linker and is of an α-helix structure.

Molecular Structure of Gas Vesicles

Gas vesicles contain only two different types of protein (Figure 3.62). The major gas vesicle protein, called GvpA, is a small, highly hydrophobic protein. GvpA is the shell protein and makes up 97% of the total protein of the gas vesicle. The second protein, called GvpC, is a larger protein but is present in much smaller amounts; the function of GvpC protein is to strengthen the shell of the gas vesicle (Figure 3.62). Gas vesicles are constructed of several copies of the GvpA protein aligned as parallel "ribs" forming a watertight surface. GvpA protein folds as a β-sheet and thus gives considerable rigidity to the overall vesicle structure (Figure 3.62). The ribs of GvpA protein are strengthened by GvpC protein, which acts as a cross-linker, binding several GvpA ribs together like a clamp (Figure 3.62). The final shape of the gas vesicle, which can vary in different organisms from long and thin to short and fat (compare Figures 3.60 and 3.61), is a function of how the GvpA and GvpC proteins are arranged to form the intact vesicle.

Because the gas vesicle membrane is freely permeable to gases, the composition of the gas inside a gas vesicle is the same as that of the gas in which the organism is suspended, and gas is present in the vesicle at about 1 atm pressure. And, because the gas vesicle attains a density of about 5–20% of that of the cell proper, intact gas vesicles decrease the density of the cell, thereby increasing its buoyancy. Aquatic phototrophic organisms in particular benefit from this motility strategy because it allows them to adjust their position rapidly in a water column to regions where the light intensity for photosynthesis is optimal.

✓ 3.14 Concept Check

Gas vesicles are small gas-filled structures made of protein that function to confer buoyancy on cells. Gas vesicles contain two different proteins arranged to form a gas permeable, but watertight structure.

✓ How are the two proteins that make up the gas vesicle, GvpA and GvpC, arranged to form such a water impermeable structure?

3.15

Endospores

Certain species of Bacteria produce special structures called **endospores** within their cells (Figure 3.63). Endospores are differentiated cells that are very resistant to heat and cannot be destroyed easily, even by harsh chemicals. Endospore-forming bacteria are found most commonly in the soil, and virtually any sample of soil has some endospores present. The genera *Bacillus* and *Clostridium* are the best studied of endospore-forming bacteria.

The discovery of bacterial endospores was of immense importance to microbiology because the knowledge of such remarkably heat-resistant forms was essential for the development of adequate methods of sterilization, not only of culture media but also of foods and other perishable products. Although many organisms other than bacteria form spores, the bacterial endospore is unique in its degree of heat resistance. Endospores are also resistant to other harmful agents such as drying, radiation, acids, and chemical disinfectants, and can remain dormant for extremely long periods of time (see the box, How Long Can an Endospore Survive?).

Endospore Structure

Endospores (so called because the spore is formed *within* the cell) are readily seen under the light microscope as strongly refractile bodies (see Figure 3.63).

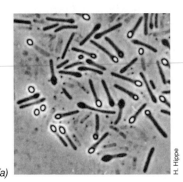 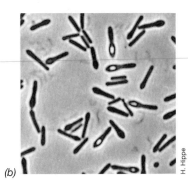 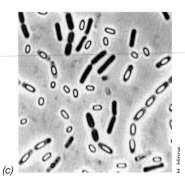

(a)   (b)   (c)

FIGURE 3.63 The bacterial endospore. Phase contrast photomicrographs illustrating several types of endospore morphologies and intracellular locations. (a) Terminal. (b) Subterminal. (c) Central.

A FOCUS ON . . .   How Long Can an Endospore Survive?

In this chapter we have discussed the dormancy and resistance properties of bacterial endospores and have pointed out that endospores can survive for long periods in a dormant state. But how long is long?

Published evidence for endospore longevity has shown that endospores can remain viable (that is, capable of germination into vegetative cells) for at least several decades and probably for much longer than that. A suspension of spores of the bacterium *Clostridium aceticum* (see photograph) prepared in 1947 was placed in growth medium in 1981, 34 years later, and in less than 12 hr growth commenced, leading to a robust culture. *Clostridium aceticum* was originally isolated by the Dutchman K. T. Wieringa in 1940 but was thought to have been lost until this vial of *C. aceticum* spores was found in a storage room at the University of California at Berkeley and revived.[a]

Other more extreme examples of endospore longevity have been documented. Bacteria of the genus *Thermoactinomyces* are thermophilic endospore formers that are widespread in nature in soil, plant litter, and fermenting plant material. Microbiological examination of a Roman archaeological site in the United Kingdom that was dated to over 2000 years ago yielded significant numbers of viable *Thermoactinomyces* spores in various pieces of debris. Additionally, *Thermoactinomyces* spores were recovered in fractions of sediment cores from a Minnesota lake known to be over 7000 years old. Although contamination is always a possibility in such studies, samples in both of these cases were processed in such a way as to virtually rule out contamination with "recent" spores.[b]

What factors could limit the age of an endospore? Cosmic radiation has been considered a major factor because it can introduce mutations in DNA.[b] It has been hypothesized

Photograph of a test tube containing spores of the bacterium *Clostridium aceticum* prepared on May 7, 1947. After remaining dormant for over 30 years, the spores were suspended in a culture medium after which growth occurred within 12 hr.

that over periods of thousands of years, the cumulative effects of cosmic radiation could introduce so many mutations into the genome of an organism that even highly radiation-resistant structures such as endospores would succumb to the genetic damage. However, extrapolations from actual experimental assessments of the effect of natural radiation on endospores suggest that if suspensions of endospores were partially shielded from cosmic radiation, for example, by being embedded in layers of organic matter, they could retain viability over periods as great as *several hundred thousand years* and perhaps even longer. Amazing, but is this the upper limit?

In 1995 a group of scientists reported the revival of bacterial spores they claimed were 25–40 million years old.[c] The spores were allegedly preserved in the gut of an extinct bee trapped in amber of known geological age. The presence of endospore-forming bacteria in these bees was previously sus-

pected from electron microscopic studies of the insect gut which showed endospore-like structures and because *Bacillus*-like DNA was recovered from the insect. DNA stored under the proper conditions is known to be quite resistant to decay, but could actual *viable cells* have survived this long? Although this was originally considered highly unlikely, the results of growth experiments showed otherwise. Samples of bee tissue incubated in a sterile culture medium quickly yielded endospore-forming bacteria. Rigorous precautions were taken to demonstrate that the endospore-forming bacterium revived from the amber-encased bee was not a modern-day contaminant. Moreover, comparisons of the nucleotide sequence in a specific gene obtained from DNA from the insect's gut and from DNA prepared from the revived bacterium showed that the bacterium isolated was the likely source of the DNA in the insect gut.

If this claim of almost unbelievable endospore longevity is supported by repetition of the results in independent laboratories (and such confirmation is crucial for verifying such a highly controversial finding), then endospores stored under the proper conditions can remain viable indefinitely. This is a remarkable testimony to the endospore, a structure that undoubtedly evolved to help cells remain viable for relatively short periods but that turned out to be such a well-designed structure that dormancy for hundreds of thousands, if not millions, of years may be possible. ■

[a]Braun, M., F. Mayer, and G. Gottschalk. 1981. *Clostridium aceticum* (Wieringa), a microorganism producing acetic acid from molecular hydrogen and carbon dioxide. *Arch. Microbiol.* 128:288–293.

[b]Gest, H., and J. Mandelstam. 1987. Longevity of microorganisms in natural environments. *Microbiol. Sci.* 4:69–71.

[c]Cano, R. J., and M. K. Borucki. 1995. Revival and identification of bacterial spores in 25- to 40-million-year-old Dominican amber. *Science* 268: 1060–1064.

Gerhard Gottschalk

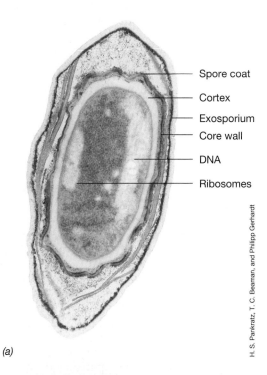

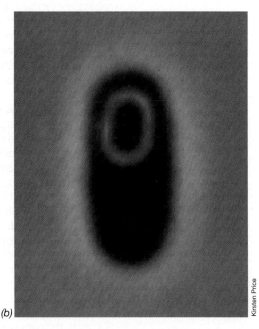

FIGURE 3.64 The bacterial endospore. (a) Transmission electron micrograph of a mature endospore from *Bacillus megaterium*. (b) Fluorescent photomicrograph of a cell of *Bacillus subtilis* undergoing sporulation. The green area is due to a dye that specifically stains a sporulation protein in the spore coat.

Spores are very impermeable to dyes, so occasionally they are seen as unstained regions within cells that have been stained with basic dyes such as methylene blue. To stain spores specifically, special spore-staining procedures must be used. The structure of the spore as seen with the electron microscope is vastly

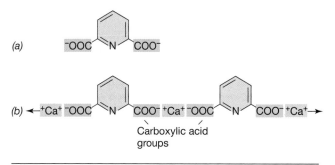

FIGURE 3.65 Dipicolinic acid (DPA). (a) Structure of DPA. (b) How Ca^{2+} cross-links DPA molecules to form a complex.

different from that of the vegetative cell (Figure 3.64). The structure of the spore is much more complex than that of the vegetative cell in that it has many layers. The outermost layer is the **exosporium,** a thin, delicate covering made of protein. Within this are the **spore coats,** composed of layers of spore-specific proteins (Figure 3.64*b*). Below the spore coat is the **cortex,** which consists of loosely cross-linked peptidoglycan, and inside the cortex is the **core** or **spore protoplast,** which contains the usual cell wall (core wall), cytoplasmic membrane, cytoplasm, nucleoid, and so on. Thus the spore differs structurally from the vegetative cell primarily in the kinds of structures found outside the core wall.

One chemical substance that is characteristic of endospores but not present in vegetative cells is **dipicolinic acid** (Figure 3.65). This substance has been found in all endospores examined and is located in the core. Spores are also high in calcium ions, most of which are combined with dipicolinic acid. The calcium–dipicolinic acid complex of the core represents about 10% of the dry weight of the endospore.

Properties of the Endospore Core

The core of a mature endospore differs greatly from the vegetative cell from which it was formed. Besides having an abundant calcium dipicolinate (Figure 3.65) content, the core is in a partially dehydrated state. The core of a mature endospore contains only 10–30% of the water content of the vegetative cell, and thus the consistency of the core cytoplasm is that of a gel. Dehydration of the core greatly increases the heat resistance of the endospore but has also been shown to confer resistance to chemicals, such as hydrogen peroxide (H_2O_2), and causes enzymes remaining in the core to become inactive.

In addition to the low water content of the spore, the pH of the core cytoplasm is about one unit lower than that of the vegetative cell and contains high levels of core-specific proteins called *small acid-soluble spore proteins* (SASPs). These are made during the sporulation process and have at least two functions. SASPs bind tightly to DNA in the core and protect it from potential damage from ultraviolet radiation, dessication, and dry heat. However, in addition, SASPs function as a carbon and energy source for the outgrowth of a new vegetative cell from the endospore, a process called *germination* (discussed later in this section).

Endospore Formation

During endospore formation, a vegetative cell is converted to a nongrowing, heat-resistant structure—the endospore (Figure 3.66). As previously described and as summarized in Table 3.2, the differences between the endospore and the vegetative cell are profound. Sporulation involves a very complex series of events in *cellular differentiation*. Bacterial sporulation does not occur when cells are dividing exponentially but only when growth ceases owing to the exhaustion of an essential nutrient. Thus, cells of *Bacillus*, a typical endospore-forming bacterium, cease vegetative growth and begin sporulation when a key nutrient such as the carbon or nitrogen source becomes limiting.

Many genetically directed changes in the cell underlie the conversion from vegetative growth to sporulation. The structural changes occurring in sporulating cells of *Bacillus* are shown in Figure 3.67, and the process can be divided into several stages. In *Bacillus subtilis*, where detailed studies of sporulation have been done, the entire sporulation process takes about 8 hr. Genetic studies of mutants of *Bacillus*, each blocked at one of the various stages of sporulation shown in Figure 3.67, have

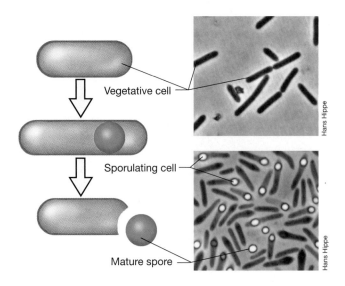

FIGURE 3.66 Formation of the endospore. Phase contrast photomicrographs are of cells of *Clostridium pascui*.

shown that as many as 200 genes are involved in the sporulation process. Sporulation requires that the synthesis of some proteins involved in vegetative cell functions cease and that specific spore proteins be made (see Figure 3.64*b*). This is accomplished by activation of a variety of spore-specific genes including *spo*, *ssp* (which encodes SASPs), and many other genes in response to an environmental trigger to sporulate. The proteins encoded by these genes catalyze the series of events leading from a moist, metabolizing vegetative cell to a dry, metabolically inert but extremely resistant endospore (Table 3.2 and Figure 3.67).

TABLE 3.2	Differences between endospores and vegetative cells	
Characteristic	**Vegetative cell**	**Endospore**
Structure	Typical gram-positive cell; a few gram-negative cells	Thick spore cortex Spore coat Exosporium
Microscopic appearance	Nonrefractile	Refractile
Calcium content	Low	High
Dipicolinic acid	Absent	Present
Enzymatic activity	High	Low
Metabolism (O_2 uptake)	High	Low or absent
Macromolecular synthesis	Present	Absent
mRNA	Present	Low or absent
Heat resistance	Low	High
Radiation resistance	Low	High
Resistance to chemicals (for example, H_2O_2) and acids	Low	High
Stainability by dyes	Stainable	Stainable only with special methods
Action of lysozyme	Sensitive	Resistant
Water content	High, 80–90%	Low, 10–25% in core
Small acid-soluble proteins (product of *ssp* genes)	Absent	Present
Cytoplasmic pH	About pH 7	About pH 5.5–6.0 (in core)

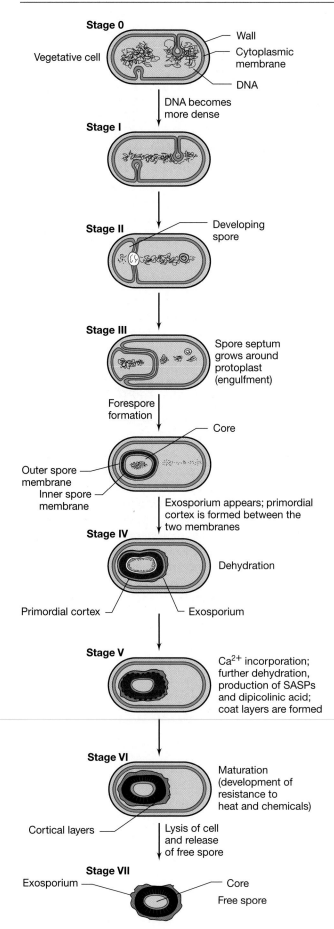

Stage 0

Vegetative cell — Wall, Cytoplasmic membrane, DNA

DNA becomes more dense

Stage I

Stage II — Developing spore

Stage III — Spore septum grows around protoplast (engulfment)

Forespore formation

Core

Outer spore membrane, Inner spore membrane

Exosporium appears; primordial cortex is formed between the two membranes

Stage IV — Dehydration

Primordial cortex — Exosporium

Stage V — Ca^{2+} incorporation; further dehydration, production of SASPs and dipicolinic acid; coat layers are formed

Stage VI — Maturation (development of resistance to heat and chemicals)

Cortical layers — Lysis of cell and release of free spore

Stage VII — Exosporium, Core, Free spore

FIGURE 3.67 Stages in endospore formation. The stages listed (0 through VII) are defined from both genetic studies and microscopic analyses.

Germination

An endospore is able to remain dormant for many years (see the box), but it can convert back to a vegetative cell relatively rapidly. This process involves three steps: activation, germination, and outgrowth (Figure 3.68).

Activation is most easily accomplished by heating freshly formed endospores for several minutes at a sublethal but elevated temperature. Activated spores are then conditioned to germinate when placed in the presence of specific nutrients. *Germination*, usually a rapid process (on the order of several minutes), involves loss of microscopic refractility of the spore, increased ability to be stained by dyes, and loss of resistance to heat and chemicals. Loss from the spores of calcium dipicolinate and cortex components occurs during this stage, and the SASPs are degraded. The next stage, *outgrowth*, involves visible swelling due to water uptake and synthesis of new RNA, proteins, and DNA. The cell emerges from the broken spore coat and eventually begins to divide (Figure 3.68). The cell then remains in vegetative growth until environmental signals that trigger sporulation are once again sensed.

✓ 3.15 Concept Check

The endospore is a highly resistant differentiated bacterial cell produced by certain types of primarily gram-positive Bacteria. Spore formation leads to a nearly dehydrated spore coat that contains essential macromolecules and a variety of substances such as calcium dipicolinate and small acid-soluble proteins, absent from vegetative cells. Spores can remain dormant indefinitely but germinate quickly when the appropriate trigger is applied.

✓ What is *dipicolinic acid* and where is it found?
✓ What are *SASPs* and what is their function?
✓ What happens when an endospore germinates?

3.16

The Nucleus and Organelles of Eukaryotic Microorganisms

We learned earlier that cells are either prokaryotic (Archaea and Bacteria) or eukaryotic (Eukarya) in their general structure (∞ Sections 1.3–1.5). Cells with a membrane-enclosed nucleus are eukaryotes, and the nucleus contains the eukaryotic cell's DNA.

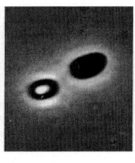
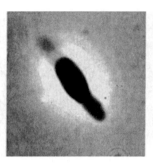

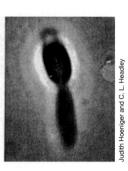

Judith Hoeniger and C. L. Headley

FIGURE 3.68 Endospore germination: Conversion of the endospore to a vegetative cell; photomicrographs showing the sequence of events starting from a highly refractile mature spore.

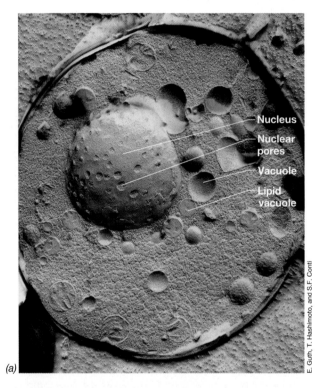

(a)

E. Guth, T. Hashimoto, and S.F. Conti

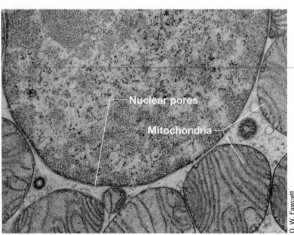

(b)

D. W. Fawcett

DNA within the nucleus is present in the form of chromosomes and the intricate packaging of DNA necessary to form these structures will be described in Chapter 6 (🔗 Section 6.3). In many eukaryotic cells the nucleus is a large organelle many micrometers in diameter, easily visible with the light microscope even without staining (see Figure 3.7a). In smaller eukaryotes, however, special staining procedures often are required to see the nucleus.

Nuclear Structure

The nuclear membrane consists of a pair of unit membranes separated by a space of variable thickness. The inner membrane is usually a simple sac, but the outer membrane is in many places continuous with the cytoplasmic membrane. The dual-membrane arrangement does, however, facilitate functional specificity because the inner and outer membranes specialize in interactions with the nucleoplasm and cytoplasm, respectively. The nuclear membrane contains many pores (Figure 3.69), which are formed from holes in both unit membranes at places where the inner and outer membranes are joined. The pores are composed of a complex of several proteins whose function is to import and export substances into and out of the nucleus. Some of these proteins have been called *importins,* to indicate their transport function. Like transport across the cytoplas-

FIGURE 3.69 The nucleus and nuclear pores. (a) Electron micrograph of a yeast cell by the freeze-etch technique, showing a surface view of the nucleus. The cell is about 8 μm wide. (b) Thin section of mouse adipose tissue showing a portion of the nucleus and several mitochondria. The nucleus is about 2 μm wide. Note the pores in the nuclear membrane in both (a) and (b).

mic membrane (see Section 3.6), nuclear transport events require energy and this comes from the hydrolysis of guanosine triphosphate (GTP).

A structure often seen within the nucleus is the *nucleolus*, an area rich in RNA that is the site of ribosomal RNA synthesis. Ribosomal proteins synthesized in the cytoplasm are transported into the nucleolus and combined with ribosomal RNA to form the small and large subunits of the eukaryotic ribosome. These are then exported to the cytoplasm where they combine to form the intact ribosome and function in protein synthesis.

Mitochondria

In eukaryotic cells the processes of respiration and oxidative phosphorylation (a mechanism of ATP formation) (∞ Sections 4.10 and 4.11) are localized in membrane-enclosed organelles called **mitochondria** (singular, **mitochondrion**). Mitochondria are of prokaryotic size and can be rod-shaped or nearly spherical (Figure 3.70, see also Figure 3.12b). A typical animal cell can contain 1000 mitochondria, but the number per cell depends somewhat on the cell type and size; a yeast cell may have as few as two mitochondria per cell, while some eukaryotic cells actually lack mitochondria altogether (∞ Sections 12.3, 17.1, and 17.2). The mitochondrial membrane, which lacks sterols, is much less rigid than the eukaryotic cell's cytoplasmic membrane. Mitochondria thus show a considerable plasticity, which makes their shape as seen in electron micrographs highly variable (Figure 3.70b,c).

The mitochondrial membrane is constructed in a manner similar to other unit membranes: a bilayer of phospholipid with embedded proteins. However, unlike the cytoplasmic membrane (see Section 3.5), the outer mitochondrial membrane is rather permeable, with channels present that allow passage of ions and small organic molecules. It is for this reason that ATP, produced within the mitochondrion, can move to the cytoplasm, where it is used in energy-requiring reactions. In addition to the outer membrane, mitochondria possess a system of folded inner membranes called *cristae*. These inner membranes, formed by invagination of the outer membrane, are the site of enzymes involved in respiration and ATP production and of specific transport proteins that regulate the passage of metabolites into and out of the *matrix* of the mitochondrion (Figure 3.70a). The matrix contains a number of enzymes involved in the oxidation of organic compounds, in particular, enzymes of the citric acid cycle (∞ Section 4.12). The mitochondrion can thus be viewed as the energy storehouse of the eukaryotic cell.

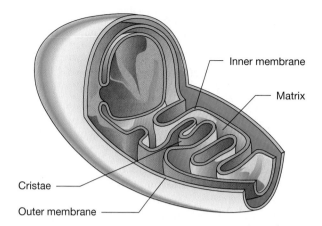

Inner membrane

Matrix

Cristae

Outer membrane

(a)

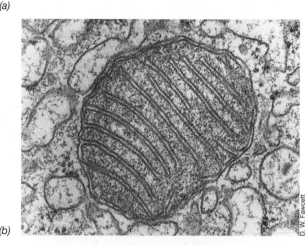

(b)

D. W. Fawcett

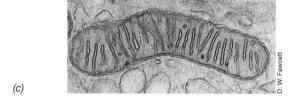

(c)

D. W. Fawcett

FIGURE 3.70 Structure of the mitochondrion. (a) Diagram showing the overall structure of the mitochondrion. Note inner and outer membranes. (b,c) Transmission electron micrographs of mitochondria from rat tissue, showing the variability in morphology of typical mitochondria.

In Chapter 17, where we discuss eukaryotic microorganisms in more detail, we will describe the **hydrogenosome,** an organelle functionally related to the mitochondrion but which is present only in certain eukaryotic microorganisms that inhabit anoxic (O_2-free) environments. The hydrogenosome carries out the oxidation of pyruvate to acetyl-CoA, CO_2 and H_2 (∞ Figure 17.2b) and, as we will see, originated from endosymbiotic events, just as the mitochondrion and chloroplast have (see later in this section and Section 17.2).

Chloroplasts

Chloroplasts are chlorophyll-containing organelles found in all eukaryotic organisms able to carry out photosynthesis. Chloroplasts of many algae are relatively large and hence are readily visible with the light microscope (Figure 3.71). The size, shape, and number of chloroplasts per algal cell vary markedly but, unlike mitochondria, they are generally much larger than bacteria.

Like mitochondria, chloroplasts have a very permeable outer membrane, a much less permeable inner membrane, and an intermembrane space. The inner membrane surrounds the lumen of the chloroplast, called the *stroma*, but is not folded into cristae like the inner membrane of the mitochondrion. Instead, chlorophyll and all other components needed for photosynthesis are located in a series of flattened membrane discs called **thylakoids** (Figure 3.72). The thylakoid mem-

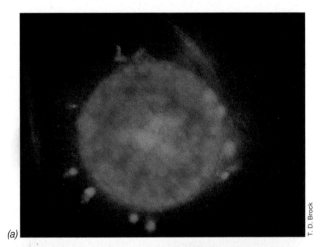

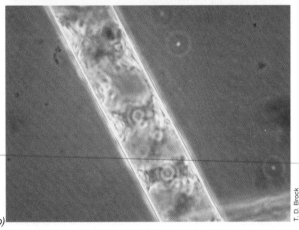

FIGURE 3.71   Photomicrographs of algal cells showing chloroplasts. (a) Fluorescence photomicrograph of the diatom *Stephanodiscus*. The chlorophyll in the chloroplasts absorbs light and fluoresces red. (b) Phase contrast photomicrograph of *Spirogyra* showing the characteristic spiral-shaped chloroplasts.

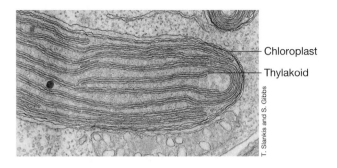

FIGURE 3.72   Transmission electron micrograph showing a chloroplast of the golden brown alga *Ochromonas danica*. Note the thylakoids.

brane is highly impermeable to ions and other metabolites because its function is to establish the proton motive force necessary for ATP synthesis (⟳ Sections 4.11 and 15.6). In green algae and green plants, thylakoids are usually associated in stacks of discrete structural units called *grana* (⟳ Figure 15.5).

The chloroplast stroma contains large amounts of the enzyme *ribulose bisphosphate carboxylase,* called *RubisCO* for short. This enzyme is the key enzyme of the *Calvin cycle,* the series of reactions by which most photosynthetic organisms convert CO_2 to organic form (⟳ Section 15.7). RubisCO makes up over 50% of the total chloroplast protein and catalyzes the formation of phosphoglyceric acid, a key compound in the biosynthesis of glucose (⟳ Sections 4.9 and 4.15). The permeability of the outer chloroplast membrane allows glucose and ATP produced during photosynthesis to diffuse into the cytoplasm where they can be used to build new cell material.

Relationships of Organelles to Prokaryotes

On the basis of their relative autonomy and morphological resemblance to bacteria, it was suggested long ago that mitochondria and chloroplasts were descendants of ancient prokaryotic organisms. This theory of *endosymbiosis* (*endo* means "within") says that eukaryotes arose from the engulfment of a prokaryotic cell by a larger cell (⟳ Sections 12.3, 17.1 and 17.2). Several pieces of evidence support this hypothesis:

1. *Mitochondria and chloroplasts contain DNA.* Although most of their functions are encoded by nuclear DNA, a few organellar components are encoded within the organellar genome, most notably ribosomal RNA, transfer RNAs, and certain proteins of the respiratory chain. Moreover, mitochondrial and chloroplast DNA exists in a *covalently closed circular form,* as it does in prokaryotes (see Section 3.10). Mi-

tochondrial DNA can be seen in cells by using special staining methods (Figure 3.73).

2. *Mitochondria and chloroplasts contain their own ribosomes.* Ribosomes, the cell's structures for protein synthesis (Section 6.12), exist in either a large form [80 Svedberg (S) units] typical of the cytoplasm of eukaryotic cells or a smaller form (70S), unique to prokaryotes. Mitochondrial and chloroplast ribosomes are 70S in size, the same as those of prokaryotes.

3. *Antibiotic specificity.* Several antibiotics (streptomycin is one example) kill or inhibit Bacteria by specifically interfering with 70S ribosome function; these same antibiotics also inhibit protein synthesis in mitochondria and chloroplasts.

4. *Molecular phylogeny.* Phylogenetic studies using comparative ribosomal RNA sequencing methods (Sections 12.4–12.8) have convincingly shown that the chloroplast and mitochondrion are evolutionarily related to Bacteria. These studies clearly point to the modern eukaryotic cell as having arisen from an association of two organisms, presumably by endosymbiosis (Section 12.3).

From the above summary of their relationships to prokaryotes, it is now considered certain that the eukaryotic mitochondrion and chloroplast arose from the endosymbiotic uptake of free-living prokaryotes. Over evolutionary time, more and more genetic functions of the endosymbionts became transferred to the host nucleus to eventually yield highly dependent structures, that is, modern organelles, whose main functions were either respiration or photosynthesis. In return for

archiving and directing most of the genetic events of their endosymbionts, host cells obtained permanent intracellular partners specializing in the production of energy. That this arrangement was an evolutionary success can be attested to by the fact that with rare exception, all eukaryotic cells contain mitochondria, and if photosynthetic, also chloroplasts.

✓ 3.16 Concept Check

The nucleus of eukaryotic cells contains DNA in linear form. The nucleus contains pores that allow for the transport of substances into or out from the cytoplasm. Two key organelles of eukaryotes are the chloroplast, involved in photosynthesis, and the mitochondrion, involved in respiration. It is likely that these organelles were originally Bacteria that established permanent residence inside another cell (endosymbiosis).

✓ Summarize the molecular evidence that supports the relationship of organelles to Bacteria.

✓ Why does streptomycin but not penicillin affect organelles?

3.17

Comparison of the Prokaryotic and Eukaryotic Cell

As we end this chapter it might be useful to draw some comparisons between the prokaryotic and eukaryotic cell (Table 3.3). It should be clear by now that there are profound differences in the internal structure of these two cell types. But in addition to structural distinctions, one should keep in mind the major *evolutionary* differences between prokaryotes and eukaryotes: Two major lineages of prokaryotes and a single eukaryotic lineage exist (Section 1.4).

Table 3.3 compares the prokaryotic and eukaryotic cell in several ways and emphasizes the great structural differences between prokaryotes and eukaryotes. However, we should recall here at the end of our journey through cell biology that all cells, regardless of their structure or evolutionary history, are composed of common types of molecules—proteins, nucleic acids, polysaccharides, and lipids—and that all use many of the same kinds of metabolic machinery. Chemical differences in the building blocks and variations in the assembly of macromolecules to form different types of cells has lead to the diversity in living organisms we see today.

We now turn from the structure of cells to the important aspects of nutrition and energy metabolism. How does the cell obtain the energy that it needs to create structurally complex macromolecules from simpler chemical constituents? We consider the nature of energy and energy conservation in the next chapter.

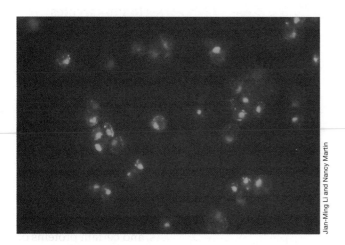

Jian-Ming Li and Nancy Martin

FIGURE 3.73 Cells of the yeast *Saccharomyces cerevisiae* stained to show mitochondrial DNA. Each mitochondrion has two to four circular chromosomes that stain blue with the fluorescent dye used.

TABLE 3.3	Comparison of the prokaryotic and eukaryotic cell	
Properties	**Prokaryote**	**Eukaryote**
Phylogenetic groups	Bacteria, Archaea	Eukarya: Algae, fungi, protozoa, plants, animals
Size	Generally small, usually <2 μm in diameter	Usually larger, 2 to >100 μm in diameter
Nuclear structure and function:		
Nuclear membrane	Absent	Present
Nucleolus	Absent	Present
DNA	Single molecule generally covalently closed and circular, not complexed with histones (other DNA in plasmids)	Linear, present in several chromosomes, usually complexed with histones
Division	No mitosis	Mitosis; mitotic apparatus with microtubular spindle
Sexual reproduction	Fragmentary process, unidirectional; no meiosis; usually only portions of genetic complement reassorted	Regular process; meiosis; reassortment of whole chromosome complement
Introns in genes	Rare	Common
Cytoplasmic structure and organization:		
Cytoplasmic membrane	Usually lacks sterols; hopanoids may be present	Sterols usually present; hopanoids absent
Internal membranes	Relatively simple; limited to specific groups	Complex; endoplasmic reticulum; Golgi apparatus
Ribosomes	70S in size	80S, except for ribosomes of mitochondria and chloroplasts, which are 70S
Membranous organelles	Absent	Several present
Respiratory system	Part of cytoplasmic membrane; mitochondria absent	In mitochondria; hydrogenosomes in certain anaerobic species
Photosynthetic pigments	In internal membranes or chlorosomes; chloroplasts absent	In chloroplasts
Cell walls	Present (in most), composed of peptidoglycan (Bacteria), other polysaccharides, protein, glycoprotein (Archaea)	Present in plants, algae, fungi, usually polysaccharide; absent in animals, most protozoa
Endospores	Present (in some), very heat-resistant	Absent
Gas vesicles	Present (in some)	Absent
Magnetosomes	Present in some species	Rarely present
Forms of motility:		
Flagellar movement	Flagella in Bacteria composed of a single type of protein arranged in a fiber and anchored into the cell wall and membrane; in Archaea, several flagellin proteins may be present; flagella rotate	Flagella or cilia; do not rotate
Nonflagellar movement	Gliding motility; gas vesicle–mediated	Cytoplasmic streaming and ameboid movement; gliding motility
Cytoskeleton containing microtubules	Absent	Present; microtubules are present in flagella, cilia, basal bodies, mitotic spindle apparatus, centrioles

REVIEW QUESTIONS

1. What is the function of staining in light microscopy? Why are cationic dyes used for general staining purposes?

2. What is the advantage of a *differential interference contrast* microscope over a bright-field microscope? A *phase-contrast* microscope over a bright-field microscope?

3. What is the major advantage of electron microscopes over light microscopes? What type of electron microscope would be used to view the three-dimensional features of a cell?

4. What are the major morphologies of prokaryotes? Draw cells for each morphology you list.

5. Describe in a single sentence the manner in which a unit membrane is formed from phospholipid molecules.

6. Explain in a single sentence why ionized molecules do not readily pass through the membrane barrier of a cell. How do such molecules get through the cytoplasmic membrane?

7. Describe a major chemical difference between membranes of Bacteria and of Archaea.

8. Cells of *Escherichia coli* take up lactose via the LacY permease system, glucose via the phosphotransferase system, and maltose via an ABC-type transporter. For each of these sugars describe: (1) the components of their transport system, and (2) the source of energy that drives the transport event.

9. Why is the bacterial cell wall rigid layer called *peptidoglycan?* What are the chemical reasons for the rigidity that is conferred on the cell wall by the peptidoglycan structure?

10. Since a single peptidoglycan molecule is very thin, explain in chemical terms how the very *thick* peptidoglycan-containing cell wall of gram-positive Bacteria is formed.

11. List several functions for the outer membrane in gram-negative Bacteria. What is the chemical composition of the outer membrane?

12. Both lysozyme and penicillin bring about bacterial cell lysis but by different mechanisms. Describe the mechanism by which each of these agents causes cell lysis.

13. Write a clear explanation (two or three sentences) of why sucrose is able to stabilize bacterial cells from lysis by lysozyme.

14. Describe the structure and function of a bacterial flagellum. What is the energy source for the flagellum?

15. In a few sentences, write an explanation for how a motile bacterium is able to sense the direction of an attractant and move toward it.

16. What types of cytoplasmic inclusions are formed by prokaryotes? How does an inclusion of poly-β-hydroxybutyric acid (PHB) differ from a magnetosome in composition and metabolic role?

17. What is the function of gas vesicles? How are these structures made such that they can remain gas tight?

18. In a few sentences, indicate how the bacterial endospore differs from the vegetative cell in structure, chemical composition, and ability to resist extreme environmental conditions.

19. The discovery of the bacterial endospore was of great practical importance. Why?

20. How does the eukaryotic nucleus differ from the prokaryotic nucleoid? In what ways are these two structures similar?

21. Set up a table following the format of Table 3.3, with the second and third columns blank, and then fill in the blanks. As you do so, think back to the figures in this chapter that illustrate the properties being considered.

APPLICATION QUESTIONS

1. Calculate the size of the smallest resolvable object if 600-nm light is used to observe a specimen with a 100× oil-immersion lens having a numerical aperture of 1.32? How could resolution be improved using this same lens?

2. Calculate the surface-to-volume ratio of a spherical cell 15 μm in diameter and a cell 2 μm in diameter. What are the consequences of these differences in surface-to-volume ratio for cell function?

3. Imagine a planet where life evolved in a totally non-aqueous environment. Cells on this planet contain highly hydrophobic cytoplasm and live in water-free environments. Predict and draw the structure of the type of cytoplasmic membranes organisms on this planet would have and discuss why such a membrane would be best suited to these organisms.

4. From what you know about the nature of the bacterial cell wall and membrane, explain why a rod-shaped bacterial cell becomes a spherical structure when its wall is removed under conditions such that cell lysis cannot occur.

5. Assume you are given two cultures, one of a species of gram-negative Bacteria and one of a species of Archaea. Other than by sequencing ribosomal RNA, discuss at least five different ways you could tell which culture is which.

In this chapter we consider the "machine" functions of a cell, the major metabolic reactions that conserve chemical energy needed to make a new cell. A key component of energy conservation is ATP synthase, or "ATPase" for short, shown in the diagram here. ATPases are nature's smallest machines, functioning to interconvert ATP and the proton motive force. These membrane-integrated protein complexes function as tiny motors, using rotational energy from the central core of the complex to create torque that ultimately drives ATP synthesis. The evolution of this remarkable energy converter likely occurred quite early in the history of life, as the proteins that make up ATPases and their interactions to form the final structure are highly conserved across living systems.

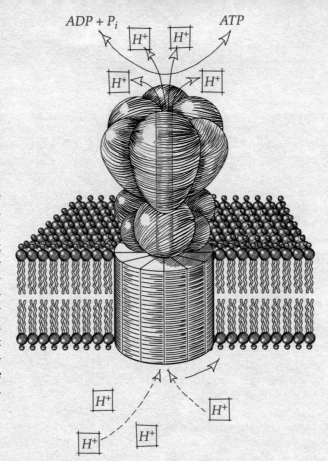

CHAPTER 4 Nutrition and Metabolism

4.1 An Overview of Metabolism   103
4.2 Microbial Nutrition   104
4.3 Culture Media   106
4.4 Energetics   108
4.5 Catalysis and Enzymes   110
4.6 Oxidation–Reduction   112
4.7 Electron Carriers   114
4.8 High Energy Compounds and Energy Conservation   116
4.9 Fermentation: The Embden–Meyerhof Pathway (Glycolysis)   118
4.10 Respiration and Electron Transport   121
4.11 Energy Conservation from Electron Transport   123
4.12 Carbon Flow: The Citric Acid Cycle   126
4.13 The Balance Sheet of Aerobic Respiration and Energy Storage   128
4.14 An Overview of Alternate Modes of Energy Generation   129
4.15 Biosynthesis of Monomers   130

Activation energy   the energy required to bring substrates to the reactive state

Aerobe   a microorganism able to use O_2 in respiration

Anabolism   the sum total of all biosynthetic reactions in the cell

Autotroph   an organism capable of biosynthesizing all cell material from CO_2 as the sole carbon source

Catabolism   biochemical reactions leading to the production of usable energy (usually ATP) by the cell

Chemolithotroph   an organism that uses inorganic chemicals as energy sources (electron donors)

Chemoorganotroph   an organism that uses organic chemicals as energy sources (electron donors)

Citric acid cycle   a cyclical series of reactions resulting in the conversion of acetate to two CO_2

Coenzyme   a small nonprotein molecule that participates in a catalytic reaction as part of an enzyme

Complex medium   a culture medium composed of digests of chemically undefined substances such as yeast and meat extracts

Culture medium   an aqueous solution of various nutrients suitable for the growth of microorganisms

Defined medium   a culture medium whose precise chemical composition is known

Electron acceptor   a substance that can accept electrons from some other substance, thereby becoming reduced in the process

Electron donor   a substance that can donate electrons to some electron acceptor, thereby becoming oxidized in the process

Enzyme   a protein that has the ability to speed up (catalyze) a specific chemical reaction

Fermentation   anaerobic catabolism in which an organic compound serves as both an electron donor and an electron acceptor and in which ATP is produced by substrate-level phosphorylation

Free energy   (G) energy available to do work

Oxidative phosphorylation   the production of ATP at the expense of a proton motive force formed by electron transport

Phototroph   an organism capable of using light as an energy source

Proton motive force   an energized state of the membrane resulting from the separation of charge and the elements of water (H^+ versus OH^-) across the membrane

Reduction potential   the inherent tendency (measured in volts) of a compound to donate electrons

Respiration   the process in which a compound is oxidized with O_2 or an O_2 substitute functioning as the terminal electron acceptor, usually accompanied by ATP production by oxidative phosphorylation

Siderophores   iron chelators that can bind iron present at very low concentrations

Substrate-level phosphorylation   production of ATP by the direct transfer of a high energy phosphate molecule from a phosphorylated organic compound to ADP

A key feature of living systems is their ability to direct chemical reactions and organize molecules into specific structures. The ultimate expression of this organization is self-replication (growth). The term **metabolism** is used to refer to all the chemical processes taking place within a cell. In this chapter we focus on metabolism and in the next chapter consider the actual process of cell growth.

Microbial cells are built of chemical substances of a wide variety of types, and when a cell grows, all these chemical constituents increase in amount. The basic constituents of a cell come from outside the cell—from the environment—and these substances are transformed by the cell into the characteristic constituents of which the cell is composed. This process by which a cell is built up from the simple nutrients obtained from its environment is called **anabolism.** Because anabolism results in the biochemical synthesis of new cell material, it is also called **biosynthesis.**

Biosynthesis is an *energy-requiring process,* and each cell must thus have a means of obtaining energy. Cells also need energy for other functions, such as transport (∞ Section 3.6) and motility (∞ Section 3.11). Although a number of organisms, including some microorganisms, obtain their energy from light, most microorganisms obtain energy from the oxidation of chemical compounds. Chemicals used as energy sources are broken down into simpler forms, and as this breakdown occurs, energy is released that can be conserved by the cell. The process by which chemicals are broken down and energy released is called **catabolism.** Catabolic reactions are the focus of this chapter.

4.1

An Overview of Metabolism

A simplified view of cell metabolism is shown in Figure 4.1, which depicts how catabolic degradative reactions supply energy needed for cell functions and how anabolic reactions bring about the synthesis of cell components from nutrients. Note that in anabolism, nutrients from the environment or those generated from catabolic reactions are converted to cell *components,* whereas in catabolism, energy sources from the environment are converted to *waste products* (Figure 4.1).

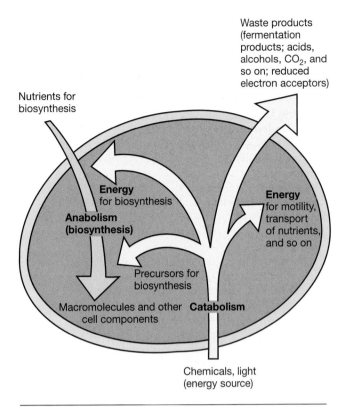

FIGURE 4.1 A simplified view of cell metabolism. Note the coupling between catabolic and anabolic processes.

Energy Classes of Microorganisms

It is conventional to group microorganisms in metabolic classes depending on the sources of *energy* they use. All the terms used to describe these classes employ the combining form *troph,* derived from a Greek word meaning "to feed." Thus, organisms that use *light* as an energy source are called **phototrophs** (*photo* is from the Greek word for "light"), and organisms that use *chemicals* as energy sources are called **chemotrophs**. Most of the organisms we deal with in microbiology use *organic* compounds as energy sources and thus are types of chemotrophs called **chemoorganotrophs**. Organisms able to use *inorganic* chemicals as energy sources are called **chemolithotrophs.**

A knowledge of cell metabolism is essential for understanding the biochemistry of microbial growth. Also, a knowledge of metabolism aids in developing laboratory procedures for culturing microorganisms and in developing suitable procedures for preventing the growth of unwanted microorganisms. Because many of the important practical consequences of microbial growth, such as infectious disease or the production of useful products, are linked to microbial metabolism, a knowledge of microbial nutrition and metabolism is also of great use in medical and industrial microbiology. We begin with an overview of nutrition before considering metabolism.

✓ 4.1 Concept Check

Metabolism involves two classes of chemical transformations, building up (biosynthetic) processes, called anabolism, and breaking down processes, called catabolism. Two kinds of energy sources can be used by cells—light and chemicals.

✓ What class of metabolic reactions in the cell are energy-*yielding* reactions?
✓ Why are *chemoorganotrophs* so named?

4.2

Microbial Nutrition

Recall from Chapter 2 that cells consist mainly of macromolecules and water and that macromolecules consist of smaller units called monomers (⬭⬮ Section 2.2). Thus, microbial nutrition is really all about supplying cells with the chemical tools they need to make monomers. These chemical tools are called **nutrients.** Different organisms need different sets of nutrients and often need these nutrients in one or another specific forms. And not all nutrients are required in the same amounts; some nutrients, called *macronutrients* (Table 4.1), are required in large amounts, while others, called *micronutrients,* are required in lesser, sometimes even trace, amounts. We begin with a consideration of the major macronutrients *carbon* and *nitrogen*.

Carbon and Nitrogen

Many prokaryotes require an organic compound of some sort as their source of **carbon.** Nutritional studies have shown that bacteria can assimilate various organic carbon compounds and use them to make new cell material. Amino acids, fatty acids, organic acids, sugars, nitrogen bases, aromatic compounds, and countless other organic compounds have been shown to be used by one bacterium or another. Some prokaryotes are *autotrophs,* able to build all of their organic structures from carbon dioxide (CO_2) with energy obtained from either light or inorganic chemicals. On a dry weight basis, a typical cell is about 50% carbon and carbon is the major element in all classes of macromolecules.

After carbon, the next most abundant element in the cell is **nitrogen.** A typical bacterial cell is about 12% nitrogen (by dry weight), and nitrogen is a major element in proteins, nucleic acids, and several other constituents in the cell. Nitrogen can be found in nature in both organic and inorganic forms (Table 4.1). However, the bulk of available nitrogen in nature is in *inorganic* form, either as ammonia (NH_3), nitrate (NO_3^-), or N_2. Most bacteria are capable of using ammonia as the sole nitrogen source, and many can also use nitrate. However, nitrogen gas (N_2) can be a nitrogen source for cer-

TABLE 4.1	Macronutrients in nature and in culture media

Element	Usual form of nutrient found in the environment	Chemical form supplied in culture media
Carbon (C)	CO_2, organic compounds	Glucose, malate, acetate, pyruvate, hundreds of other compounds, or complex mixtures (yeast extract, peptone, and so on)
Hydrogen (H)	H_2O, organic compounds	H_2O, organic compounds
Oxygen (O)	H_2O, O_2, organic compounds	H_2O, O_2, organic compounds
Nitrogen (N)	NH_3, NO_3^-, N_2, organic nitrogen compounds	*Inorganic:* NH_4Cl, $(NH_4)_2SO_4$, KNO_3, N_2 *Organic:* Amino acids, nitrogen bases of nucleotides, many other N-containing organic compounds
Phosphorus (P)	PO_4^{3-}	KH_2PO_4, Na_2HPO_4
Sulfur (S)	H_2S, SO_4^{2-}, organic S compounds, metal sulfides (FeS, CuS, ZnS, NiS, and so on)	Na_2SO_4, $Na_2S_2O_3$, Na_2S, cysteine, or other organic sulfur compounds
Potassium (K)	K^+ in solution or as various K salts	KCl, KH_2PO_4
Magnesium (Mg)	Mg^{2+} in solution or as various Mg salts	$MgCl_2$, $MgSO_4$
Sodium (Na)	Na^+ in solution or as NaCl or other Na salts	NaCl
Calcium (Ca)	Ca^{2+} in solution or as $CaSO_4$ or other Ca salts	$CaCl_2$
Iron (Fe)	Fe^{2+} or Fe^{3+} in solution or as FeS, $Fe(OH)_3$, or many other Fe salts	$FeCl_3$, $FeSO_4$, various chelated iron solutions (Fe^{3+} EDTA, Fe^{3+} citrate, and so on)

tain bacteria, the *nitrogen-fixing bacteria,* and we discuss the properties of these organisms in detail later (Sections 13.8, 15.29, and 16.16).

Other Macronutrients: P, S, K, Mg, Ca, Na, Fe

Phosphorus occurs in nature in the form of organic and inorganic phosphates and is required by the cell primarily for synthesis of nucleic acids and phospholipids. **Sulfur** is required because of its structural role in the amino acids cysteine and methionine (Section 2.6) and because it is present in a number of vitamins, such as thiamine, biotin, and lipoic acid, as well as in coenzyme A. Sulfur undergoes a number of chemical transformations in nature carried out exclusively by microorganisms (Section 16.17) and is available to organisms in a variety of forms. Most cell sulfur originates from inorganic sources, either sulfate (SO_4^{2-}) or sulfide (HS^-) (Table 4.1).

Potassium is required by all organisms. A variety of enzymes, including some of those involved in protein synthesis, specifically require potassium. **Magnesium** functions to stabilize ribosomes, cell membranes, and nucleic acids and is also required for the activity of many enzymes. **Calcium** (which is not an essential nutrient for the growth of many microorganisms) helps stabilize the bacterial cell wall and plays a key role in the heat stability of endospores (Section 3.15). **Sodium** is required by some but not all organisms, and its need often reflects the habitat of the organism. For example, seawater has a high sodium content and marine mi-

croorganisms generally require sodium for growth, whereas closely related freshwater species are usually able to grow in the absence of sodium.

Iron plays a major role in cellular respiration, being a key component of the cytochromes and iron–sulfur proteins involved in electron transport (see Section 4.10 and Table 4.2). However, because most inorganic iron compounds are highly insoluble, many organisms produce specific iron-binding agents called **siderophores,** which solubilize iron and transport it into the cell. One major group of siderophores consists of derivatives of hydroxamic acid, which chelate ferric (Fe^{3+}) iron very strongly (Figure 4.2*a*). Once the iron–hydroxamate complex has passed into the cell, the iron is released and the hydroxamate can exit the cell and be used again for iron transport. Enteric bacteria such as *Escherichia coli* and *Salmonella typhimurium* produce structurally complex phenolic siderophores called **enterobactins** (Figure 4.2*b*). These siderophores are derivatives of the aromatic compound catechol and have an extremely high binding affinity for iron.

Micronutrients (Trace Elements)

Although required in just tiny amounts, micronutrients are nevertheless just as critical to cell function as are macronutrients. Micronutrients are metals, many of which play a structural role in various enzymes, the cells' catalysts. Table 4.2 summarizes the major micronutrients of living systems and gives examples of enzymes in which each plays a role.

TABLE 4.2	Micronutrients (trace elements) needed by living organisms[a]
Element	**Cellular function**
Chromium (Cr)	Required by mammals for glucose metabolism; no known microbial requirement
Cobalt (Co)	Vitamin B_{12}; transcarboxylase (propionic acid bacteria)
Copper (Cu)	Certain proteins, notably those involved in respiration, for example, cytochrome *c* oxidase; or in photosynthesis, for example, plastocyanin; some superoxide dismutases
Manganese (Mn)	Activator of many enzymes; present in certain superoxide dismutases and in the water-splitting enzyme of photosystem II in oxygenic phototrophs
Molybdenum (Mo)	Present in various flavin-containing enzymes; also in molybdenum nitrogenase, nitrate reductase, sulfite oxidase, DMSO-TMAO reductases, some formate dehydrogenases, oxotransferases
Nickel (Ni)	Most hydrogenases; coenzyme F_{430} of methanogens; carbon monoxide dehydrogenase; urease
Selenium (Se)	Formate dehydrogenase; some hydrogenases; the amino acid selenocysteine
Tungsten (W)	Some formate dehydrogenases; oxotransferases of hyperthermophiles (for example, aldehyde:ferredoxin oxidoreductase of *Pyrococcus furiosus*)
Vanadium (V)	Vanadium nitrogenase; bromoperoxidase
Zinc (Zn)	Present in the enzymes carbonic anhydrase, alcohol dehydrogenase, RNA and DNA polymerases, and many DNA-binding proteins
Iron (Fe)[b]	Cytochromes, catalases, peroxidases, iron–sulfur proteins (for example, ferredoxin), oxygenases, all nitrogenases

a Not every micronutrient listed is required by all cells; some metals listed are found in enzymes present in only specific microorganisms.

b Needed in greater amounts than other metals—not generally considered a trace element.

Because the requirement for trace elements is so small, for the laboratory culture of microorganisms it is frequently unnecessary to add trace elements to the culture medium. However, if a culture medium contains highly purified chemicals dissolved in high purity distilled water, a trace element deficiency can occur. In such cases a small amount of a solution of trace metals (Table 4.2) is added to the medium to make available the necessary metals.

Growth Factors

Growth factors are *organic* compounds that, like micronutrients, are required in very small amounts and only by some cells. Growth factors include vitamins, amino acids, purines, and pyrimidines. Although most microorganisms are able to synthesize all of these compounds, some microorganisms require one or more of them preformed from the environment.

Vitamins are the most commonly needed growth factors. Most vitamins function as parts of coenzymes (see, for instance, Figures 4.8, 4.13, and 4.21), and these are summarized in Table 4.3. Many microorganisms are able to synthesize all the components of their coenzymes, but some are unable to do so and must be provided with certain parts of these coenzymes in the form of vitamins. Lactic acid bacteria, which include the genera *Streptococcus*, *Lactobacillus*, *Leuconostoc*, and others

(⌘ Section 13.18), are renowned for their complex vitamin requirements, which are even more extensive than those of humans (see Table 4.4)! The vitamins most commonly required by microorganisms are thiamine (vitamin B_1), biotin, pyridoxine (vitamin B_6), and cobalamin (vitamin B_{12}).

✓ 4.2 Concept Check

The hundreds of chemical compounds present inside a living cell are formed from starting materials called nutrients. Elements required in fairly large amounts are called macronutrients, while metals and organic compounds needed in very small amounts are called micronutrients and growth factors, respectively.

✓ What two classes of macromolecules contain the bulk of the *nitrogen* in a cell?
✓ Why is an element like Co^{2+} considered a *micro*nutrient whereas an element like C is considered a *macro*nutrient?
✓ What roles do iron play in cellular metabolism?

4.3

Culture Media

We can summarize our discussion of microbial nutrition by examining the chemical composition of several culture media (Table 4.4). **Culture media** are the nutrient

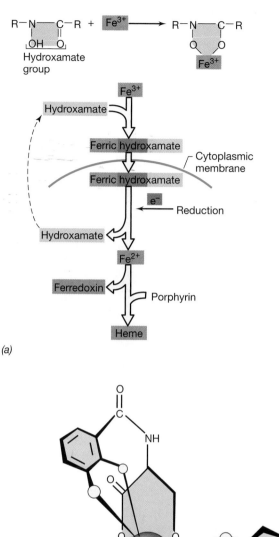

FIGURE 4.2 Iron-chelating agents produced by microorganisms. (a) Hydroxamate. Iron is bound as Fe^{3+} and released inside the cell as Fe^{2+}. (b) Ferric enterobactin of *Escherichia coli*. The oxygen atoms of each catechol molecule are shown in yellow.

TABLE 4.3	Vitamins and their functions
Vitamin	**Function**
p-Aminobenzoic acid	Precursor of folic acid
Folic acid	One-carbon metabolism; methyl group transfer
Biotin	Fatty acid biosynthesis; β-decarboxylations; some CO_2 fixation reactions
Cobalamin (B_{12})	Reduction of and transfer of single carbon fragments; synthesis of deoxyribose
Lipoic acid	Transfer of acyl groups in decarboxylation of pyruvate and α-ketoglutarate
Nicotinic acid (niacin)	Precursor of NAD^+; electron transfer in oxidation–reduction reactions
Pantothenic acid	Precursor of coenzyme A; activation of acetyl and other acyl derivatives
Riboflavin	Precursor of FMN, FAD in flavoproteins involved in electron transport
Thiamine (B_1)	α-Decarboxylations; transketolase
Vitamins B_6 (pyridoxal-pyridoxamine group)	Amino acid and keto acid transformations
Vitamin K group; quinones	Electron transport; synthesis of sphingolipids
Hydroxamates	Iron-binding compounds; solubilization of iron and transport into cell

organic chemicals to distilled water. Therefore, the *exact chemical composition* of a defined medium is known. In many cases, however, knowledge of the exact composition of a medium is not critical. In these instances complex media may suffice or for various reasons may even be advantageous. Complex media employ digests of casein (milk protein), beef, soybeans, yeast cells, or any of a number of other highly nutritious (yet chemically undefined) substances. Such digests are available commercially in powdered form and can be weighed out rapidly and dissolved in distilled water to give a medium. However, a major concession in using a complex medium is loss of control of its precise nutrient composition.

Nutritional Requirements and Biosynthetic Capacity

Table 4.4 shows three recipes for culture media, two defined and one complex. The complex medium is easiest to prepare and supports good growth of either organism shown in the table, the enteric bacterium *Escherichia coli* or the lactic acid bacterium *Leuconostoc mesenteroides*, an extremely fastidious (nutritionally demanding) bacterium. The simple defined medium supports excellent growth of *E. coli* but not of *L. mesenteroides*; growth of the latter organism in defined medium requires the addition of several organic nutrients and growth factors not

solutions used to grow microorganisms in the laboratory. Two broad classes of culture media are used in microbiology: **chemically defined** and **undefined (complex).** Chemically defined media are prepared by adding precise amounts of highly purified inorganic or

TABLE 4.4	**Examples of culture media for microorganisms with simple and demanding nutritional requirements[a]**	
Defined culture medium for *Escherichia coli*	**Defined culture medium for** *Leuconostoc mesenteroides*	**Complex culture medium for** *either E. coli or L. mesenteroides*
K_2HPO_4 7 g KH_2PO_4 2 g $(NH_4)_2SO_4$ 1 g $MgSO_4$ 0.1 g $CaCl_2$ 0.02 g Glucose 4–10 g Trace elements (Fe, Co, Mn, Zn, Cu, Ni, Mo) 2–10 μg each Distilled water 1000 ml pH 7	K_2HPO_4 0.6 g KH_2PO_4 0.6 g NH_4Cl 3 g $MgSO_4$ 0.1 g Glucose 25 g Sodium acetate 20 g Amino acids (alanine, arginine, asparagine, aspartate, cysteine, glutamate, glutamine, glycine, histidine, isoleucine, leucine, lysine, methionine, phenylalanine, proline, serine, threonine, tryptophan, tyrosine, valine) 100–200 μg of each Purines and pyrimidines (adenine, guanine, uracil, xanthine) 10 mg of each Vitamins (biotin, folate, nicotinic acid, pyridoxal, pyridoxamine, pyridoxine, riboflavin, thiamine, pantothenate, *p*-aminobenoic acid) 0.01–1 mg of each Trace elements (see first column) 2–10 μg each Distilled water 1000 ml pH 7	Glucose 15 g Yeast extract 5 g Peptone 5 g KH_2PO_4 2 g Distilled water 1000 ml pH 7
(a)		(b)

a The photos are tubes of (a) the defined medium described, and (b) the complex medium described. Note how the complex medium is colored from the various organic extracts and digests that it contains. *Photos courtesy of Cheryl L. Broadie.*

needed by *E. coli* (Table 4.4). With this in mind, which organism, *E. coli* or *L. mesenteroides,* has a greater *biosynthetic* capacity? Obviously, *E. coli,* since its ability to grow on a simple defined culture medium means that it has the ability to synthesize *all* its organic cellular constituents from a single carbon compound, in this case glucose (Table 4.4). By contrast, *L. mesenteroides* has multiple growth factor requirements indicative of limited biosynthetic capacity. The complex nutritional needs of *L. mesenteroides* can be satisfied by either preparing a defined medium as shown in Table 4.4 (although in this case it could take quite a bit of time to do so) or by using a complex medium (Table 4.4), which by contrast can usually be quickly prepared.

It is important to understand when examining recipes for culture media such as those shown in Table 4.4 that *different microorganisms can have vastly different nutritional requirements.* Thus, for successful culture of a given microorganism it is necessary to understand its nutritional requirements and then supply it with its essential nutrients in the proper form and proportions in a culture medium. If care is taken in preparing culture media, it is usually quite easy to culture microorganisms in the laboratory.

With this foundation in nutritional principles, we turn our attention to the cell's use of nutrients to drive energy-yielding reactions—catabolism.

✓ 4.3 Concept Check

Culture media supply the nutritional needs of microorganisms and can be either chemically defined or undefined (complex).

✓ Why is the routine culture of *Leuconostoc mesenteroides* easier in a complex medium than in a chemically defined medium?

✓ In which medium, simple, defined, or complex (shown in Table 4.4), do you think *Escherichia coli* would grow faster? Why?

4.4

Energetics

Energy is defined as the ability to do work. In this chapter, we discuss how living organisms conserve chemical energy. **Chemical energy** is the energy released when organic or inorganic compounds are oxidized. In biology the most commonly used energy units are the kilo-calorie (kcal) and the kilojoule (kJ). A kilocalorie is defined as the quantity of heat energy necessary to raise the temperature of 1 kilogram of water 1°C. One kilocalorie is equivalent to 4.184 kJ. Because the kJ is widely used in microbial energetics, we will use this convention throughout this book.

Free Energy

Chemical reactions are accompanied by changes in energy. Although in any chemical reaction some energy is lost as heat, in microbiology we are interested in **free energy** (abbreviated *G*), which is defined as the energy released *that is available to do useful work*. The change in free energy during a reaction is expressed as $\Delta G^{0\prime}$, where the symbol Δ should be read "change in." The superscripts "0" and "′" mean that the free-energy value was obtained under "standard" conditions: pH 7, 25°C, all reactants and products initially at 1 *M* concentration.

If in the reaction:

$$A + B \rightarrow C + D$$

the $\Delta G^{0\prime}$ is *negative*, the reaction will proceed with the *release* of free energy, energy that the cell can conserve in the form of ATP. Such energy-yielding reactions are called **exergonic.** However, if $\Delta G^{0\prime}$ is *positive*, the reaction *requires* energy in order to proceed; such reactions are called **endergonic.** Thus, from the standpoint of the microbial cell, exergonic reactions yield energy while endergonic reactions require energy.

Free energy calculations using standard conditions are estimations, but usually fairly accurate estimations, of the free energy changes that would actually occur when a reaction takes place in nature, this despite the fact that nutrient concentrations of 1 *M* rarely if ever occur in nature. Although for our purposes here, calculations of $\Delta G^{0\prime}$ are reasonable estimates, we will see later that factors such as the actual concentration of products and reactants and the pH can occasionally dramatically alter the bioenergetics of reactions in microbiologically important ways (∞ Sections 15.23 and 16.14). These issues are also discussed in more detail in Appendix 1.

Free Energy of Formation and Calculating $\Delta G^{0\prime}$

In addition to speaking of the free energy *yield* of reactions, it is also necessary to talk about the free energy of individual substances. This is the *free energy of formation*, the energy yielded or energy required for the *formation* of a given molecule from its constituent elements. By convention, the free energy of formation (G^0_f) of the elements (for instance, C, H_2, N_2) is zero. If the formation of a *compound* from elements proceeds exergonically, then the free energy of formation of the compound is negative (energy is released), whereas if the reaction is

endergonic (energy is required), then the free energy of formation of the compound is positive. A few examples of free energies of formation are given in Table 4.5. For most compounds G^0_f is *negative*, reflecting the fact that compounds tend to form spontaneously from elements. The positive G^0_f for nitrous oxide ($+104.2$ kJ/mol) tells us that this molecule does not form spontaneously but rather decomposes to nitrogen and oxygen. The free energies of formation of a variety of compounds of microbiological interest are given in Appendix 1.

Using free energies of formation, it is possible to calculate the *change* in free energy occurring in a given reaction. For a simple reaction such as $A + B \rightarrow C + D$, $\Delta G^{0\prime}$ is calculated by subtracting the *sum* of the free energies of formation of the reactants (in this case A and B) from that of the products (C and D). Thus,

$$\Delta G^{0\prime} \text{ of } A + B \rightarrow C + D$$
$$= G^0_f[C + D] - G^0_f[A + B]$$

The saying "products minus reactants" summarizes the necessary steps for calculating changes in free energy during chemical reactions. However, it is necessary to balance the reaction chemically before free-energy calculations can be made. Appendix 1 details the steps involved in calculating free energies for any hypothetical reaction.

✓ 4.4 Concept Check

The chemical reactions of the cell are accompanied by changes in energy, expressed in kJ. A chemical reaction can occur with the release of free energy (exergonic), or with the consumption of free energy (endergonic).

✓ What is free energy?
✓ In general, are *catabolic* reactions exergonic or endergonic?
✓ Using the data in Table 4.5, calculate $\Delta G^{0\prime}$ for the reaction $CH_4 + \frac{1}{2}O_2 \rightarrow CH_3OH$.

TABLE 4.5	Free energy of formation for a few compounds of biological interest
Compound	**Free energy of formation**[a]
Water (H_2O)	-237.2
Carbon dioxide (CO_2)	-394.4
Hydrogen gas (H_2)	0
Oxygen gas (O_2)	0
Ammonium (NH_4^+)	-79.4
Nitrous oxide (N_2O)	$+104.2$
Acetate ($C_2O_2O_3^-$)	-369.4
Glucose ($C_6H_{12}O_6$)	-917.3
Methane (CH_4)	-50.8
Methanol (CH_3OH)	-175.4

a The free-energy values (G^0_f) are in *kJ/mol*.

4.5

Catalysis and Enzymes

A free-energy calculation tells us only whether energy is released or required in a given reaction; it tells us nothing about the *rate* of the reaction. Consider the formation of water from gaseous oxygen and hydrogen. The energetics of this reaction is quite favorable: $H_2 + \frac{1}{2}O_2 \rightarrow H_2O$, $\Delta G^{0'} = -237$ kJ. However, if we were to simply mix O_2 and H_2 together, no measurable formation of water would occur for many years. This is because the rearrangement of oxygen and hydrogen atoms to form water requires that the chemical bonds of the reactants be broken first. The breaking of bonds requires energy, and this energy is referred to as **activation energy.** Activation energy is the amount of energy required to bring all molecules in a chemical reaction to the reactive state. For a reaction that proceeds with a net release of free energy (that is, an exergonic reaction), the situation is as diagrammed in Figure 4.3.

Enzymes

The idea of activation energy leads us to the concept of catalysis. A **catalyst** is a substance that serves to *lower* the activation energy of a reaction. A catalyst serves to *increase* the rate of reaction even though it itself is not changed. It is important to note that catalysts do not affect the energetics or the equilibrium of a reaction; catalysts affect only the *speed* at which reactions proceed.

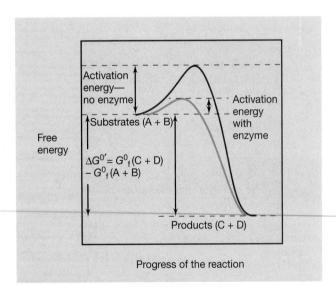

FIGURE 4.3 Progress of a hypothetical exergonic reaction: $A + B \rightarrow C + D$ and the concept of activation energy. Chemical reactions may not proceed spontaneously even though energy would be released, because the reactants must first be activated. Once activation has occurred, the reaction then proceeds spontaneously. Catalysts such as enzymes lower the required activation energy.

Most reactions in living organisms would not occur at appreciable rates without catalysis. The catalysts of biological reactions are proteins called **enzymes.** Enzymes are highly specific in the reactions that they catalyze. That is, each enzyme catalyzes only a *single type* of chemical reaction, or in the case of certain enzymes, a class of closely related reactions. This specificity is related to the precise three-dimensional structure of the enzyme molecule. In an enzyme-catalyzed reaction, the enzyme temporarily combines with the reactant, which is termed a **substrate** (S) of the enzyme, forming an **enzyme–substrate complex.** Then, as the reaction proceeds, the **product** (P) is released and the enzyme (E) is returned to its original state:

$$E + S \rightleftharpoons E-S \rightleftharpoons E + P$$

The enzyme is generally much larger than the substrate(s), and the combination of enzyme and substrate(s) usually depends on weak bonds, such as hydrogen bonds, van der Waals forces, and hydrophobic interactions (Section 2.1) to join the enzyme to the substrate. The small portion of the enzyme to which substrates bind is referred to as the **active site** of the enzyme.

Enzyme Catalysis

The catalytic power of enzymes is impressive. Enzymes typically increase the rate of chemical reactions from 10^8 to 10^{20} times the rate that would occur spontaneously. To catalyze a specific reaction, an enzyme must do two things: (1) bind the correct substrate, and (2) position the substrate relative to the catalytically active groups at the enzyme's active site. Binding of substrate to enzyme produces the enzyme–substrate complex (Figure 4.4). This serves to align reactive groups and places strain on specific bonds in the substrate(s). The result of enzyme–substrate complex formation is a reduction in the activation energy required to make the reaction proceed (Figure 4.3) with the conversion of substrate(s) to product(s). These steps are summarized diagrammatically in Figure 4.4 for the glycolytic enzyme *fructose bisphosphate aldolase* (see Section 4.9).

Note that the reaction depicted in Figure 4.3 is exergonic because the free energy of formation of the substrate is *greater* than that of the product; that is, product formation proceeds with the release of energy. Enzymes can also catalyze endergonic reactions, converting energy-poor substrates to energy-rich products. In this case, not only must an activation energy barrier be overcome, but sufficient free energy must also be put *into* the system to raise the energy level of the substrates to that of the products. Although theoretically all enzymes are reversible in their action, in practice, enzymes catalyzing highly exergonic or highly endergonic reactions are essentially unidirectional. If a particularly exergonic

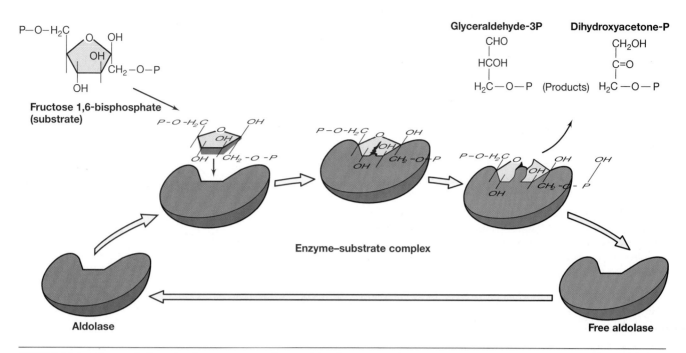

Glyceraldehyde-3P

CHO
|
HCOH
|
H$_2$C—O—P

Dihydroxyacetone-P

CH$_2$OH
|
C=O
|
H$_2$C—O—P

(Products)

Fructose 1,6-bisphosphate
(substrate)

Enzyme–substrate complex

Aldolase

Free aldolase

FIGURE 4.4 The catalytic cycle of an enzyme as depicted for the enzyme fructose bisphosphate aldolase. This enzyme catalyzes the following reaction: fructose 1,6-bisphosphate ⇌ glyceraldehyde 3-phosphate + dihydroxyacetone phosphate in glycolysis (see Figure 4.12). Following binding of fructose 1,6-bisphosphate in the formation of the enzyme–substrate complex, the conformation of the enzyme is altered, placing strain on certain bonds of the substrate, which break and yield the two products.

reaction needs to be reversed during cellular metabolism, a distinctly different enzyme is frequently involved in the reaction.

Structure of Enzymes

As we have discussed, enzymes are proteins, polymers of amino acids (∞ Sections 2.6–2.8). Each enzyme has a specific three-dimensional shape. The linear array of amino acids (primary structure) folds and twists into a specific configuration to achieve secondary and tertiary structure. A specifically folded protein thus assumes specific binding and physical properties. The precise conformation of an enzyme may be seen more easily in a computer-generated space-filling model (Figure 4.5). In this example of the peptidoglycan-cleaving enzyme *lysozyme* (∞ Section 3.7), the large cleft is the site where the substrate binds (the active site).

Many enzymes contain small nonprotein molecules that participate in the catalytic function but are not considered substrates in the usual sense. These small enzyme-associated molecules are divided into two categories on the basis of the nature of their association with the enzyme, *prosthetic groups* and *coenzymes*. **Prosthetic groups** are bound very tightly to their enzymes, usually permanently. The heme group present in cytochromes is an example of a prosthetic group, and cytochromes will be described in detail later in this chapter. **Coenzymes** are bound rather loosely to enzymes, and a single coenzyme molecule may associate with a num-

ber of different enzymes at different times during growth. Coenzymes serve as intermediate carriers of small molecules from one enzyme to another. Most coenzymes are derivatives of vitamins (see Table 4.3).

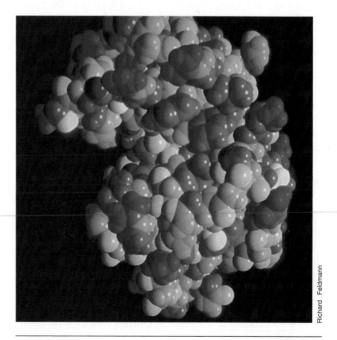

FIGURE 4.5 Computer-generated space-filling model of the enzyme lysozyme. The substrate (peptidoglycan) binding site (active site) is in the large cleft on the left side of the model (∞ Section 3.7).

Enzymes are named either for the substrate they bind or for the chemical reaction they catalyze, by addition of the combining form *-ase*. Thus cellul*ase* is an enzyme that attacks cellulose, glucose oxid*ase* is an enzyme that catalyzes the oxidation of glucose, and ribonucle*ase* is an enzyme that decomposes ribonucleic acid. A more formal nomenclature system employing a specific numbering system is used to classify enzymes more precisely.

✓ 4.5 Concept Check

The reactants in a chemical reaction must first be activated before the reaction can take place, and this requires a catalyst. Enzymes are catalytic proteins that are highly specific in the reactions they catalyze, and this specificity resides in the folding pattern of the polypeptide(s) in the protein.

- ✓ What is the function of a *catalyst*?
- ✓ What *class* of macromolecules are enzymes?
- ✓ Where on an enzyme does its substrate bind?

4.6

Oxidation–Reduction

The conservation of energy from chemical reactions in living organisms involves **oxidation–reduction** (also called **redox**) reactions. Chemically, an oxidation is defined as the *removal* of an electron or electrons from a substance. A reduction is defined as the *addition* of an electron (or electrons) to a substance. In biochemistry, oxidations and reductions frequently involve the transfer of not just electrons but whole hydrogen atoms. A hydrogen atom (H) consists of an electron plus a proton. When the electron is removed, the hydrogen atom becomes a *proton* (or hydrogen ion, H^+). We will on occasion need to distinguish between oxidation–reduction reactions involving electrons only or hydrogen atoms only, but reserve this distinction for the appropriate time (see Sections 4.10 and 4.11).

Electron Donors and Acceptors

Oxidation–reduction reactions involve electrons being donated by an electron donor and being accepted by an electron acceptor. For example, hydrogen gas, H_2, can release electrons and hydrogen ions (protons) and become oxidized:

$$H_2 \rightarrow 2\,e^- + 2\,H^+$$

However, electrons cannot exist alone in solution; they must be part of atoms or molecules. The equation as drawn thus gives us chemical information but does not itself represent a real reaction. The above reaction is only a *half reaction*, a term that implies the need for a second half reaction. This is because for any *oxidation* to occur, a subsequent *reduction* must also occur. For example, the oxi-

dation of H_2 could be coupled to the reduction of many different substances including O_2 in a second reaction:

$$\tfrac{1}{2}\,O_2 + 2\,e^- + 2\,H^+ \rightarrow H_2O$$

This half reaction, which is a reduction, when coupled to the oxidation of H_2 above, yields the following overall balanced reaction:

$$H_2 + \tfrac{1}{2}\,O_2 \rightarrow H_2O$$

In reactions of this type, we will refer to the substance *oxidized*, in this case H_2, as the **electron donor**, and the substance *reduced*, in this case O_2, as the **electron acceptor** (Figure 4.6). The key to understanding biological oxidations and reductions is to keep straight the proper half reactions—there must always be one reaction involving an electron *donor* and another reaction involving an electron *acceptor*.

Reduction Potentials

Substances vary in their tendency to become oxidized or to become reduced. This tendency is expressed as the **reduction potential** (E_0') of the substance. This potential is measured electrically in reference to a standard substance, H_2. By convention, reduction potentials are expressed for half reactions written as *reductions*. Thus, oxidized form + $e^- \rightarrow$ reduced form. If protons are involved in the reaction, as is often the case, then the reduction potential is to some extent influenced by the hydrogen ion concentration (pH). By convention in biology, reduction potentials are given for neutrality (pH 7) because the cytoplasm of most cells is neutral or nearly so. Using these conventions, at pH 7 the reduction potential (E_0') of

$$\tfrac{1}{2}\,O_2 + 2\,H^+ + 2\,e^- \rightarrow H_2O$$

is +0.816 volts (V), and that of

$$2\,H^+ + 2\,e^- \rightarrow H_2$$

is −0.421 V (see Figure 4.7).

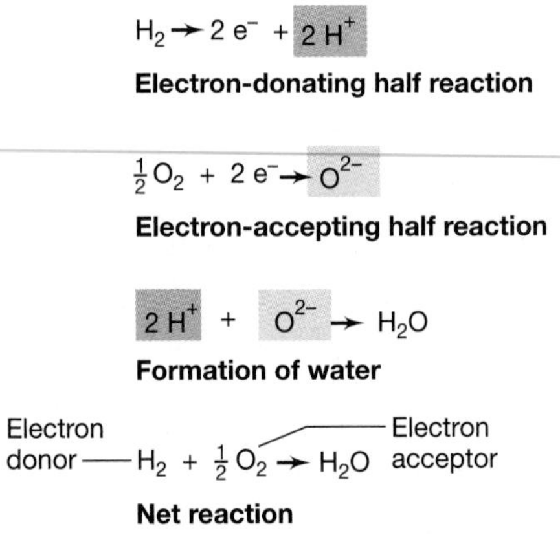

$$H_2 \rightarrow 2\,e^- + \boxed{2\,H^+}$$

Electron-donating half reaction

$$\tfrac{1}{2}O_2 + 2\,e^- \rightarrow \boxed{O^{2-}}$$

Electron-accepting half reaction

$$\boxed{2\,H^+} + \boxed{O^{2-}} \rightarrow H_2O$$

Formation of water

Electron donor —— $H_2 + \tfrac{1}{2}O_2 \rightarrow H_2O$ Electron acceptor

Net reaction

FIGURE 4.6 Example of an oxidation–reduction reaction: The formation of H_2O from H_2 and O_2.

Oxidation–Reduction Couples and Complete Redox Reactions

Most molecules can be either electron donors or electron acceptors under different circumstances, depending on what other substances they react with. The same atom on each side of the arrow in the half reactions can be thought of as representing a redox couple, such as $2 H^+/H_2$ or $\frac{1}{2} O_2/H_2O$. When writing a redox couple, the *oxidized* form is always placed on the left.

In constructing complete oxidation–reduction reactions from their constituent half reactions, it is simplest to remember that the reduced substance of a redox couple whose reduction potential is more negative *donates* electrons to the oxidized substance of a redox couple whose potential is more positive. Thus, in the couple $2 H^+/H_2$, which has a potential of -0.42 V, H_2 has a great tendency to *donate* electrons. On the other hand, in the couple $\frac{1}{2} O_2/H_2O$, which has a potential of $+0.82$

V, H_2O has a very slight tendency to donate electrons, but O_2 has a great tendency to *accept* electrons. It follows then that in a reaction of H_2 and O_2, H_2 will be the electron *donor* and become oxidized, and O_2 will be the electron *acceptor* and become reduced (Figure 4.6). Even though by chemical convention both half reactions are written as reductions, in an actual redox reaction one of the two half reactions must be written as an oxidation and therefore proceeds in the reverse direction. Thus, note that in the reaction shown in Figure 4.6, the oxidation of H_2 to $2 H^+ + 2 e^-$ is reversed from the formal half reaction, written as a reduction.

The Electron Tower

A convenient way of viewing electron transfer in biological systems is to imagine a vertical tower (Figure 4.7). The tower represents the range of reduction potentials

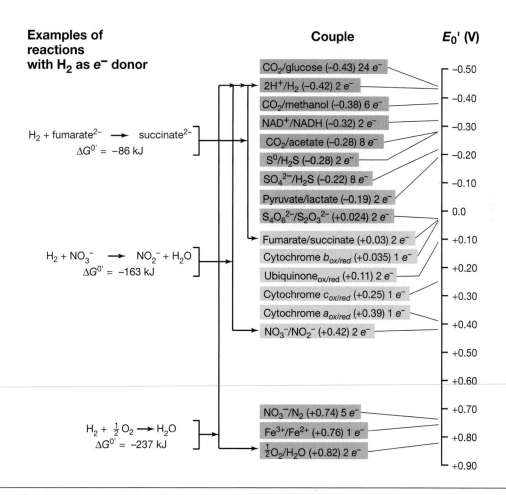

FIGURE 4.7 The electron tower. Redox couples are arranged from the strongest reductants (negative reduction potential) at the top to the strongest oxidants (positive reduction potentials) at the bottom. As electrons are donated from the top of the tower, they can be "caught" by acceptors at various levels. The farther the electrons fall before they are caught, the greater the difference in reduction potential between electron donor and electron acceptor and the more energy is released. As an example of this, on the left is shown the differences in energy released when a single electron donor, H_2, reacts with any of three different electron acceptors, fumarate, nitrate, and oxygen. The E_0' of the Fe^{3+}/Fe^{2+} couple is highly pH dependent. At acid pH, the E_0' is $+0.76$V but at neutral pH, the E_0' is substantially less positive, about $+0.2$V.

for redox couples from the most negative at the top to the most positive at the bottom. The reduced substance in the redox pair at the top of the tower has the greatest amount of potential energy, whereas the oxidized substance in the couple at the bottom of the tower has the greatest tendency to accept electrons.

As electrons from the electron donor at the top of the tower fall, they can be "caught" by acceptors at various levels. The difference in potential between two substances is expressed as $\Delta E_0'$. The farther the electrons drop from a donor before they are caught by an acceptor, the greater the amount of energy released; that is, $\Delta E_0'$ *is proportional to* $\Delta G^{0'}$ (Figure 4.7). O_2, at the bottom of the tower, is the most favorable electron acceptor used by organisms. In the middle of the tower, redox couples can act as either electron donors or acceptors. For instance, the $2\,H^+/H_2$ couple has a reduction potential of -0.42 V. The fumarate–succinate couple has a potential of $+0.02$ V. Hence, the oxidation of hydrogen (the electron donor) can be coupled to the reduction of fumarate (the electron acceptor):

$$H_2 + \text{fumarate}^{2-} \rightarrow \text{succinate}^{2-}$$

On the other hand, the oxidation of succinate to fumarate can be coupled to the reduction of NO_3^- or $\frac{1}{2}\,O_2$:

$$\text{Succinate}^{2-} + NO_3^- \rightarrow \text{fumarate}^{2-} + NO_2^- + H_2O$$

$$\text{Succinate}^{2-} + \tfrac{1}{2}\,O_2 \rightarrow \text{fumarate}^{2-} + H_2O$$

Hence, under conditions where oxygen is absent (called *anoxic*) in the presence of H_2, fumarate can be an electron acceptor (producing succinate), and under other conditions (for example, anoxic in the presence of NO_3^-, or aerobic) succinate can be an electron donor (producing fumarate). Indeed, all the transformations involving fumarate and succinate described here are actually carried out by various microorganisms under certain nutritional and environmental conditions.

In catabolism the electron donor is often referred to as an **energy source.** Many potential electron donors exist in nature (∞ Chapters 15 and 16), but for now it is essential to understand that it is not the electron donor *per se*, that contains energy, but it is the *chemical reaction* in which the electron donor gets oxidized, that actually releases energy. As discussed in the context of the electron tower, the amount of energy released in a redox reaction depends on the nature of *both* the electron donor and the electron acceptor: The greater the difference between reduction potentials of the two half reactions, the more energy there will be released when they react (Figure 4.7) (see also Appendix 1).

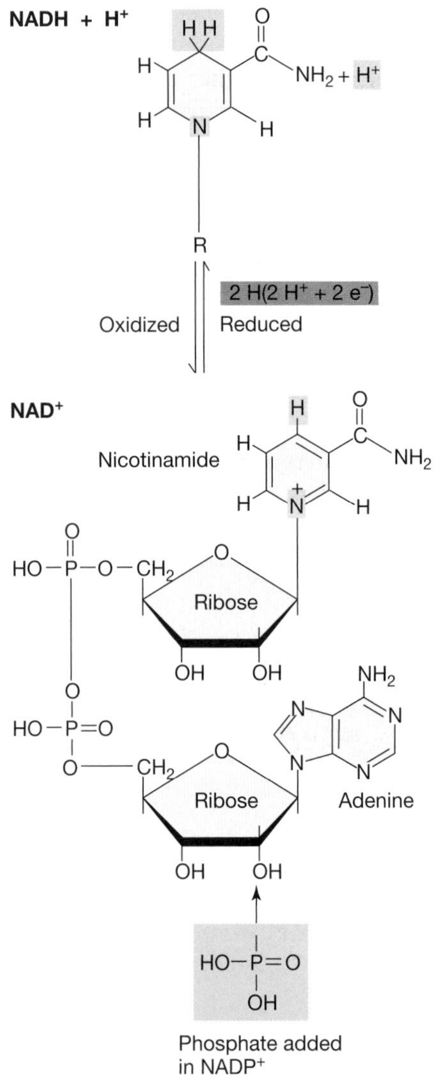

✓ 4.6 Concept Check

Oxidation–reduction reactions, which are involved in the energy-yielding reactions of cells, involve the transfer of electrons from one substance to another. The tendency of a

FIGURE 4.8   Structure of the oxidation–reduction coenzyme nicotinamide adenine dinucleotide (NAD^+). In $NADP^+$, a phosphate group is present, as indicated. Both NAD^+ and $NADP^+$ undergo oxidation–reduction as shown.

compound to accept or release electrons is expressed quantitatively by its reduction potential.

✓ In the reaction $H_2 + \frac{1}{2}\,O_2 \rightarrow$ what is the electron *donor* and what is the electron *acceptor?*

✓ What is the E_0' of the $2\,H^+/H_2$ couple?

✓ Why is NO_3^- a better electron acceptor than fumarate?

4.7

Electron Carriers

In the cell, the transfer of electrons in an oxidation–reduction reaction from donor to acceptor involves one or more intermediates referred to as **carriers.** When such carriers are used, we refer to the initial donor as the **pri-**

mary electron donor and to the final acceptor as the **terminal electron acceptor.** The net energy change of the complete reaction sequence is determined by the *difference* in reduction potentials between the primary donor and the terminal acceptor.

Electron carriers can be divided into two general classes: those freely diffusible and those firmly attached to enzymes in the cytoplasmic membrane. The fixed carriers function in membrane-associated electron transport reactions and are discussed in Section 4.10. Freely diffusible carriers include the coenzymes nicotinamide-adenine dinucleotide (NAD^+) and NAD-phosphate ($NADP^+$) (Figure 4.8). NAD^+ and $NADP^+$ are *hydrogen atom* carriers and always transfer two hydrogen atoms to the next carrier in the chain. Such hydrogen atom transfer is referred to as a dehydrogenation.†

†Strictly speaking NAD^+ or $NADP^+$ carries two electrons and one proton, the second H^+ being released to solution. Therefore, $NAD^+ + 2 e^- + 2 H^+$ actually yields $NADH + H^+$. However, for simplicity, we write $NADH + H^+$ as NADH.

The reduction potential of the NAD^+/NADH (or $NADP^+$/NADPH) couple is -0.32 V, which places it fairly high on the electron tower; that is, NADH (or NADPH) is a good electron *donor.* However, although the NAD^+ and $NADP^+$ couples have the same reduction potentials, they generally function in different capacities in the cell. NAD^+/NADH is directly involved in energy-generating (catabolic) reactions, whereas $NADP^+$/NADPH is involved primarily in biosynthetic (anabolic) reactions.

Coenzymes increase the diversity of redox reactions by making it possible for chemically dissimilar molecules to interact as initial electron donor and ultimate electron acceptor, with the coenzyme acting as intermediary. As we have discussed, most biological reactions are catalyzed by specific enzymes that can react with only one or a very limited range of substrates. Oxidation–reduction reactions may be considered to proceed in three stages: removal of electrons from the primary donor, transfer of electrons through one or a series of electron carriers, and addition of electrons to the terminal acceptor. Each step in the reaction is catalyzed by a different enzyme, each of which binds to its substrate and to its specific coenzyme. Figure 4.9 is a schematic

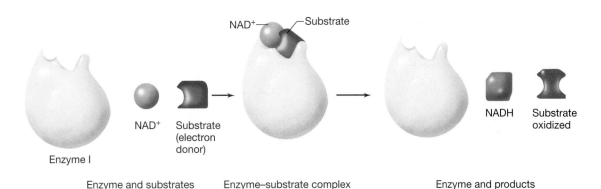

Reaction 1. Enzyme I reacts with substrate (electron donor) and oxidized form of coenzyme, NAD^+.

Reaction 2. Enzyme II reacts with substrate (electron acceptor) and reduced form of coenzyme, NADH.

FIGURE 4.9 Schematic example of an oxidation–reduction reaction involving the oxidized and reduced forms of the coenzyme nicotinamide-adenine dinucleotide, NAD^+ and NADH.

diagram showing the functioning of the coenzyme NAD^+ in a two-part reaction. Note that after a coenzyme has performed its chemical function in one reaction, it can diffuse through the cytoplasm until it collides with another enzyme that requires the coenzyme in that form. Following conversion of the coenzyme back to its original form, the whole process can be repeated (Figure 4.9).

✓ 4.7 Concept Check

The transfer of electrons from donor to acceptor in a cell often involves the participation of one or more electron carriers. Some electron carriers are membrane-bound, whereas others are freely diffusible, transferring electrons from one place to another in the cell.

✓ What is the difference between an *electron* and a *hydrogen* atom?
✓ Is NADH a better electron donor than H_2? Why or why not?

4.8

High Energy Compounds and Energy Conservation

Energy released as a result of oxidation–reduction reactions must be conserved for cell functions. In living organisms, chemical energy released in redox reactions is usually conserved in the form of **high energy phosphate bonds;** these compounds then function as the energy source to drive energy-requiring reactions in the cell.

In phosphorylated compounds, phosphate groups are attached via oxygen atoms by *ester* or *anhydride* bonds, as illustrated in Figure 4.10. However, not all phosphate bonds are high energy bonds. As a means of expressing the energy of phosphate bonds, the free energy released when the phosphate is hydrolyzed can be given. As seen in Figure 4.10, the $\Delta G^{0\prime}$ of hydrolysis of the phosphate bond in glucose 6-phosphate is only -13.8 kJ/mol, whereas the $\Delta G^{0\prime}$ of hydrolysis of the phosphate bond in phosphoenolpyruvate is -51.6 kJ/mol, almost four times that of glucose 6-phosphate. Thus, phosphoenolpyruvate, a phospho-anhydride, is considered a *high energy compound* and glucose 6-phosphate, a phosphate ester, is not.

Adenosine Triphosphate (ATP)

The most important high energy phosphate compound in living organisms is adenosine triphosphate (ATP). ATP consists of the ribonucleoside adenosine, to which three phosphate molecules are bonded in series (Figure 4.10). ATP serves as the prime energy carrier in living organisms, being generated during exergonic reactions and being used to drive endergonic reactions. From the

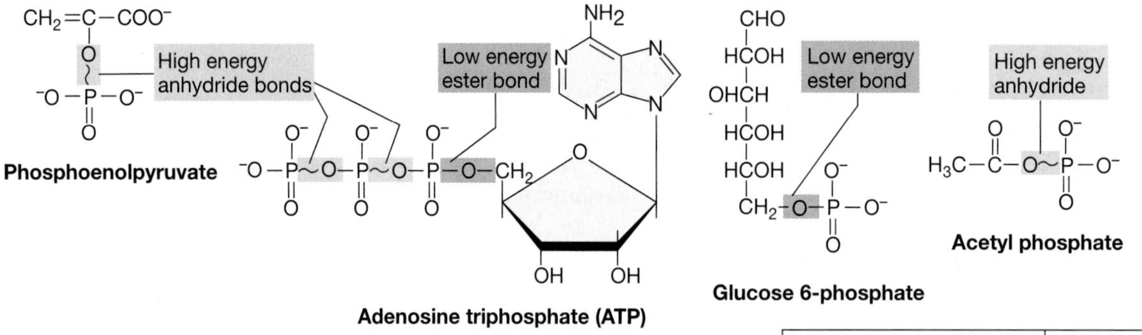

Compound	$G^{0\prime}$kJ/mol
High energy	
Phosphoenolpyruvate	−51.6
1,3-Bisphosphoglycerate	−52.0
Acetyl phosphate	−44.8
ATP	−31.8
ADP	−31.8
Low energy	
AMP	−14.2
Glucose 6-phosphate	−13.8

FIGURE 4.10 High energy phosphate bonds. The table shows the free energy of hydrolysis of some of the key phosphate esters and anhydrides, indicating that some of the phosphate ester bonds are of higher energy than others. Structures of four of the compounds are given to indicate the position of low energy and high energy bonds. ATP contains three phosphates, but only two of them are high energy (shown in blue). ADP contains two phosphates of which only one is high energy. AMP does not contain a high energy phosphate bond.

structure of ATP (Figure 4.10) it can be seen that two of the phosphate bonds of ATP are phosphoanhydrides and have high free energies of hydrolysis.

It should be emphasized that although we express the energy of high energy phosphate bonds in terms of the free energy of hydrolysis, in actuality it is undesirable for these bonds to hydrolyze in cells in the absence of a second reaction that can use the energy released because the free energy of hydrolysis would then be lost to the cell as heat. The free energy of high energy phosphate bonds is generally used to drive biosynthetic reactions and other aspects of cell function through carefully regulated processes in which the energy released from ATP hydrolysis is coupled to energy-requiring reactions.

Coenzyme A

In addition to high energy phosphate compounds, certain other high energy compounds are produced in the cell and can conserve the energy released in exergonic reactions. These include derivatives of coenzyme A (for example, acetyl-CoA; see structure in Figure 4.21). Coenzyme A derivatives contain *sulfo*anhydride (thioester) instead of *phospho*anhydride (Figure 4.10) bonds and yield sufficient free energy on hydrolysis to drive the synthesis of a high energy phosphate bond (👁️ Table 2.1). For example, in the reaction: acetyl-S-CoA + H_2O + ADP + P → acetate$^-$ + HS-CoA + ATP + H^+, the energy released from the hydrolysis of coenzyme A is conserved in the synthesis of ATP. Coenzyme A derivatives (acetyl-CoA is just one of many) are especially important to the energetics of anaerobic microorganisms, in particular those whose energy metabolism involves fermentation (see Sections 4.19, 15.21, and 15.22); we will thus return to the importance of these compounds in Chapter 15.

Fermentation and Respiration

The pathways for the oxidation of organic compounds and conservation of energy in ATP can be divided into two major groups: (1) **fermentation,** in which the redox process occurs in the *absence* of any added terminal electron acceptors; and (2) **respiration,** in which molecular oxygen or some other oxidant serves as the terminal electron acceptor. The oxidation in a fermentation is coupled to the subsequent reduction of an organic compound generated from the initial substrate; thus, no externally supplied electron acceptor is required.

ATP is produced in fermentations by a process called **substrate-level phosphorylation.** In substrate-level phosphorylation, ATP is synthesized during specific enzymatic steps in the catabolism of the organic compound (Figure 4.11*a*). This is in contrast to **oxida-**

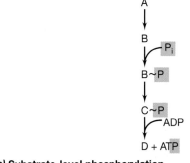

(a) **Substrate-level phosphorylation**

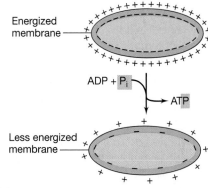

(b) **Oxidative phosphorylation**

FIGURE 4.11 Energy conservation in fermentation and respiration. (a) In fermentation, ATP synthesis occurs as a result of *substrate-level phosphorylation;* a phosphate group gets added to some intermediate in the biochemical pathway and eventually gets transferred to ADP to form ATP. (b) In respiration, the cytoplasmic membrane, energized by the proton motive force, dissipates some of that energy in the formation of ATP from ADP and inorganic phosphate (P_i) in the process called *oxidative phosphorylation.* The coupling of the proton motive force to ATP synthesis occurs by way of a membrane protein complex called ATP synthase (ATPase) (see Section 4.11 and Figure 4.19).

tive (or **electron transport**) **phosphorylation** (discussed later), where ATP is produced via the proton motive force (Figure 4.11*b*). We now contrast these modes of ATP synthesis by considering the details of a fermentation (see Section 4.9) with those of respiration (see Sections 4.10–4.12).

✓ 4.8 Concept Check

The energy released in oxidation–reduction reactions is conserved in the formation of certain compounds that contain high energy phosphate or sulfur bonds. One of the most common high energy phosphate compounds is ATP, which serves as a prime energy carrier in the cell.

✓ Why are ATP and ADP considered high energy phosphate compounds whereas AMP is not?

✓ How does ATP synthesis differ in fermentation versus respiration?

4.9

Fermentation: The Embden–Meyerhof Pathway (Glycolysis)

A fermentation is an internally balanced oxidation–reduction reaction in which some atoms of the energy source (electron donor) become more reduced whereas others become more oxidized, and energy is produced by substrate-level phosphorylation. A common biochemical pathway for the fermentation of glucose is **glycolysis,** also named the **Embden–Meyerhof pathway** for its major discoverers. Glycolysis can be divided into three major stages, each involving a series of individually catalyzed enzymatic reactions (Figure 4.12).

Stage I of glycolysis is a series of preparatory rearrangements, reactions that do not involve oxidation–reduction and do not release energy but that lead to the production from glucose of two molecules of the key intermediate, *glyceraldehyde 3-phosphate*. In Stage II, oxidation–reduction occurs, energy is conserved in the form of ATP, and two molecules of pyruvate are formed. In Stage III, a second oxidation–reduction reaction occurs and *fermentation products* (for example, ethanol and CO_2, or lactic acid) are formed (Figure 4.12).

Stages I and II: Preparatory and Redox Reactions

In Stage I, glucose is phosphorylated by ATP yielding glucose 6-phosphate; this is then converted to an isomeric form, fructose 6-phosphate, and a second phosphorylation leads to the production of *fructose 1,6-bisphosphate,* which is a key intermediate product of glycolysis. The enzyme **aldolase** splits fructose 1,6-bisphosphate into two three-carbon molecules, glyceraldehyde 3-phosphate and its isomer, dihydroxyacetone phosphate (see also Figure 4.4).[‡] Note that thus far, there have been no oxidation–reduction reactions and that all the reactions, including the consumption of ATP, proceed without electron transfers.

The first redox reaction of glycolysis occurs in Stage II during the conversion of glyceraldehyde 3-phosphate to 1,3-bisphosphoglyceric acid. In this reaction (which occurs twice, once for each molecule of glyceraldehyde 3-phosphate), an enzyme whose coenzyme is NAD^+, accepts two hydrogen atoms and NAD^+ is converted to NADH; the enzyme catalyzing this reaction is called **glyceraldehyde-3-phosphate dehydrogenase.** Simul-

taneously, each glyceraldehyde-3-P molecule is phosphorylated by the addition of a molecule of inorganic phosphate. This reaction, in which inorganic phosphate is converted to organic form, sets the stage for energy conservation by substrate-level phosphorylation; ATP formation is possible because each of the phosphates on a molecule of 1,3-bisphosphoglyceric acid is a high energy phosphate (see Figure 4.10). The synthesis of ATP occurs when each molecule of 1,3-bisphosphoglyceric acid is converted to 3-phosphoglyceric acid and later on in the pathway, when each molecule of phosphoenolpyruvate is converted to pyruvate (Figure 4.12).

In glycolysis, *two* ATP molecules are consumed in the two phosphorylations of glucose, and *four* ATP molecules are synthesized (two from each 1,3-bisphosphoglyceric acid converted to pyruvate). Thus, the *net gain* to the organism is two molecules of ATP per molecule of glucose fermented.

Stage III: Production of Fermentation Products

During the formation of two molecules of 1,3-bisphosphoglyceric acid, two molecules of NAD^+ are reduced to NADH (see Figure 4.12). However, a cell contains only a small amount of NAD^+, and if all of it were converted to NADH, the oxidation of glucose would stop; the continued oxidation of glyceraldehyde 3-phosphate can proceed only if there is a molecule of NAD^+ present to accept released electrons. This "roadblock" is overcome in fermentation by the oxidation of NADH back to NAD^+ through reactions involving the reduction of pyruvate to any of a variety of **fermentation products.** In the case of yeast, pyruvate is reduced to ethanol with the release of CO_2. In lactic acid bacteria, pyruvate is reduced to lactate (see lower part of Figure 4.12). Many routes of pyruvate reduction in various fermentative prokaryotes are known (∞ Chapter 15), but the net result is the same; NADH must be returned to the oxidized form, NAD^+, for the energy-yielding reactions of fermentation to continue. As a diffusible coenzyme, NADH can move away from glyceraldehyde-3-phosphate dehydrogenase, attach to an enzyme that reduces pyruvate to lactic acid (lactate dehydrogenase), and diffuse away once again following the oxidation of NADH to NAD^+ to repeat the cycle all over again (see Figure 4.9 for details of this mechanism).

In any energy-yielding process, oxidation must balance reduction and there must be an electron acceptor for each electron removed. In this case, the *reduction* of NAD^+ at one enzymatic step in glycolysis is balanced with its *oxidation* at another. The final product(s) must also be in oxidation–reduction balance with the starting substrate, glucose. Hence, the products discussed here, ethanol plus CO_2 or lactate plus protons, are in electrical and atomic balance with the starting glucose.

[‡]There is an enzyme that catalyzes the interconversion of dihydroxyacetone phosphate and glyceraldehyde 3-phosphate. For simplicity, we consider here only glyceraldehyde 3-phosphate since it is the compound that is further metabolized.

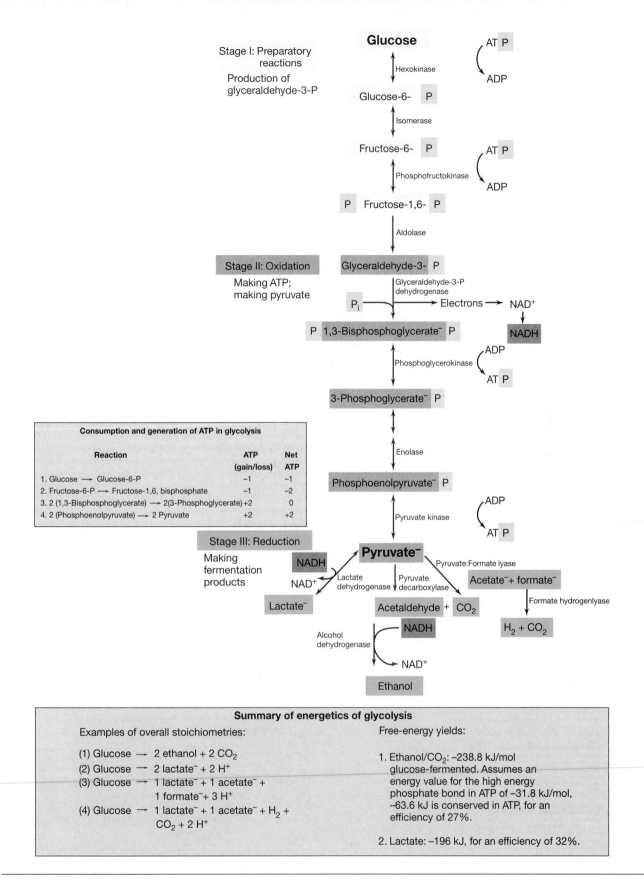

FIGURE 4.12 Embden–Meyerhof pathway (glycolysis), the sequence of enzymatic reactions in the conversion of glucose to pyruvate and then to fermentation products (enzymes are shown in small type). Note that when fructose bisphosphate is split, *two* molecules of glyceraldehyde 3-P are formed, although for simplicity, only one molecule is shown in the figure. Also note that pyruvate is the central "hub" of glycolysis—all fermentation products are made from pyruvate and just a few common examples are given.

TECHNIQUES & APPLICATIONS...   The Products of Yeast Fermentation

The aerobic and anaerobic processes of energy generation may seem dull and prosaic, but they are at the basis of one of the most striking discoveries of the human race, fermented foods and beverages (see photo).

In the production of breads and most alcoholic beverages, it is the yeast *Saccharomyces cerevisiae* that is exploited to produce ethanol plus CO_2. Found in various sugar-rich environments such as fruit juices and nectar, yeasts can carry out the two opposing modes of metabolism discussed in this chapter, fermentation and respiration. When oxygen is present, yeasts grow efficiently on the sugar substrate, making yeast cells and CO_2. However, when O_2 is absent, yeasts switch to an anaerobic metabolism, resulting in a reduced cell yield but significant amounts of alcohol. Every home wine maker or brewer is an amateur microbiologist, perhaps without even realizing it (◌◌ Home Brew, Chapter 11). When grapes are squeezed to make juice, small numbers of yeast cells present on the grapes in the vineyard are transferred to the must. During the first several days of the wine-making process, these yeast cells grow by respiration and consume O_2, making the juice anoxic. As soon as the oxygen is depleted, fermentation can begin, and the process of alcohol formation takes over. This switch from aerobic to anaerobic metabolism is crucial, and special care must be taken to make sure air

Major products in which fermentation by the yeast *Saccharomyces cerevisiae* is critical.

is kept out of the fermenting vessel.

Wine is only one of many products made with yeast. Others include beer and distilled spirits such as brandy, whisky, vodka, and gin (see photo). In distilled spirits, the ethanol, produced in relatively low amounts (10–15% by volume) by the yeast, is concentrated by distilling to make a beverage containing 40–60% alcohol. Even alcohol for motor fuel is made with yeast in parts of the world where sugar is plentiful but petroleum is in short supply (such as Brazil). Yeast also serves as the leavening agent in bread, although

here it is not the *alcohol* that is important but CO_2, the other product of the alcohol fermentation. We discuss yeast and yeast products in some detail in Chapter 11.

We can thus appreciate how the yeast cell, forced to carry out a fermentative lifestyle because the oxygen it needs for respiration is absent, has impacted the lives of humans. Besides being "waste products" of the glycolytic pathway in yeast, ethanol and CO_2 are the key ingredients in the products of the alcoholic beverage and baking industries, respectively. ■

Glucose Fermentation:
Net and Practical Results

The ultimate result of glycolysis is the consumption of glucose, the net synthesis of two ATPs, and the production of fermentation products. For the organism the crucial product is ATP, which is used in a wide variety of energy-requiring reactions, and fermentation products are merely waste products. However, the latter substances are hardly considered waste products by the distiller, the brewer, the cheese-maker, or the baker (see

the box, The Products of Yeast Fermentation). Thus, fermentation is more than just an energy-yielding process. It is a means of producing natural products useful to humans. We discuss industrial production of fermentation products in more detail in Chapter 11.

✓ 4.9 Concept Check

Glycolysis is a major pathway for fermentation and is widespread in living organisms. The end result of glycolysis is the release of a small amount of energy that is conserved as ATP and used for various cell functions, and the production of fermentation products. Common fermentation products of glycolysis include ethanol, lactic acid, and a variety of other acids, alcohols, and gaseous substances, depending upon the organism.

✓ Which reaction(s) in glycolysis involve oxidations and reductions?

✓ What is the role of NAD^+ in glycolysis?

✓ Why are fermentation products made during glycolysis?

4.10

Respiration and Electron Transport

We have just discussed the fermentation of glucose, a process that occurs in the absence of external electron acceptors. A relatively small amount of energy is released in fermentation and only a few ATP molecules synthesized. This small energy release may be understood in terms of the formal principles of oxidation–reduction reactions. Fermentations yield little energy for two reasons: (1) the carbon atoms in the starting compound are only partially oxidized [that is, highly reduced and still energy-rich fermentation products are excreted from the cell, (see Figure 4.12)], and (2) the difference in reduction potentials between the primary electron donor and terminal electron acceptor (that is, the vertical distance on the electron tower) is small. However, if O_2 or some other terminal acceptor is present, all the substrate molecules can be oxidized completely to CO_2 and a far higher yield of ATP is theoretically possible. The process by which a compound is oxidized using O_2 as external electron acceptor is called **aerobic respiration.**

Our discussion of respiration deals with both the carbon and electron transformations: (1) the biochemical pathways involved in the transformation of organic carbon to CO_2 and (2) the way electrons are transferred from the organic compound to the terminal electron acceptor, driving ATP synthesis at the expense of the proton motive force (Figure 4.11b). We begin with a discussion of electron flow.

Electron Transport

Electron transport systems are composed of *membrane-associated* electron carriers. These systems have two basic functions: (1) to accept electrons from an electron donor and transfer them to an electron acceptor, and (2) to conserve some of the energy released during electron transfer for synthesis of ATP.

Several types of oxidation–reduction enzymes are involved in electron transport: (1) NADH dehydrogenases, which transfer hydrogen atoms from NADH; (2) riboflavin-containing electron carriers, generally called flavoproteins [which contain flavin mononucleotide (FMN) or flavin-adenine dinucleotide (FAD)]; (3) iron–sulfur proteins; and (4) cytochromes, which are proteins containing an iron–porphyrin ring called *heme*. In addition, one class of *nonprotein* electron carriers is known, the lipid-soluble quinones. Quinones can diffuse freely through the membrane, generally transferring electrons from iron–sulfur proteins to cytochromes. We now consider each of these classes of electron transport components in slightly more detail.

NADH dehydrogenases are proteins bound to the inside surface of the cell membrane. They accept hydrogen atoms from NADH (Figure 4.8), generated in various cellular reactions, and pass the hydrogen atoms to flavoproteins.

Flavoproteins are proteins containing a derivative of riboflavin (Figure 4.13); the flavin portion, which is bound to a protein, is a prosthetic group that is alternately reduced as it accepts hydrogen atoms and oxidized when electrons are passed on. Note that flavoproteins *accept* hydrogen atoms and *donate* electrons; we will consider what happens to the two protons later. Two flavins are commonly observed in cells, flavin mononucleotide and flavin-adenine dinucleotide, in which FMN is bonded to ribose and adenine through a second phosphate. Riboflavin, also called vitamin B_2, is a required growth factor for some organisms (see Section 4.2).

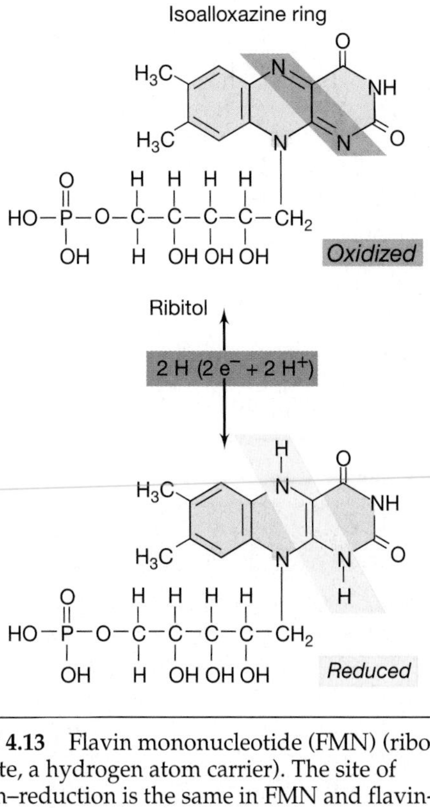

FIGURE 4.13 Flavin mononucleotide (FMN) (riboflavin phosphate, a hydrogen atom carrier). The site of oxidation–reduction is the same in FMN and flavin-adenine dinucleotide (FAD).

The **cytochromes** are proteins with iron-containing porphyrin ring (heme) prosthetic groups attached to them (Figure 4.14). They undergo oxidation and reduction through loss or gain of a *single electron* by the iron atom at the center of the cytochrome:

$$\text{Cytochrome–Fe}^{2+} \rightleftharpoons \text{cytochrome–Fe}^{3+} + e^-$$

Several classes of cytochromes are known, differing in their reduction potentials. One cytochrome can transfer electrons to another that has a more positive reduction potential and can itself accept electrons from a cytochrome or quinone molecule with a less positive reduction potential. The different cytochromes are designated by letters, such as cytochrome *a*, cytochrome *b*, cytochrome *c*. The cytochromes of one organism may

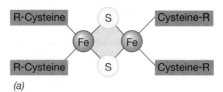

(a)

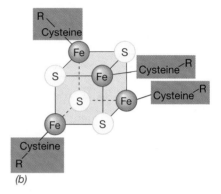

(b)

FIGURE 4.15 Arrangement of the iron–sulfur centers of nonheme iron–sulfur proteins. (a) Fe_2S_2 center. (b) Fe_4S_4 center. The cysteine linkages are from the protein portion of the molecule. Iron–sulfur proteins typically carry electrons only.

differ slightly from those of another, and so there are designations such as cytochrome a_1, cytochrome a_2, cytochrome aa_3, and so on. Occasionally, cytochromes form tight complexes with other cytochromes or with iron–sulfur proteins. An example of such a complex is the cytochrome bc_1 complex, which contains two different *b*-type cytochromes and one *c*-type cytochrome. It plays an important role in energy metabolism (see Section 4.11).

In addition to the cytochromes, where iron is bound to heme, several **nonheme iron–sulfur proteins** are associated with electron transport chains. Various arrangements of sulfur and iron have been found in different nonheme iron–sulfur proteins, but Fe_2S_2 and Fe_4S_4 clusters are the most common (Figure 4.15). The iron atoms are bonded to free sulfur and to the protein via sulfur atoms from cysteine residues (Figure 4.15). *Ferredoxin*, a common iron–sulfur protein in biological systems, has

(a)   (b)

Pyrrole

Porphyrin
(a tetrapyrrole)

Heme (a porphyrin)

(c)

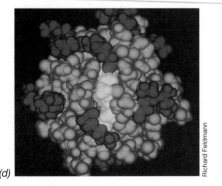

(d)

FIGURE 4.14 Cytochrome and its structure. (a) Structure of the pyrrole ring. (b) Four pyrrole rings are condensed, leading to formation of the porphyrin ring. Various metals can be incorporated into the porphyrin ring system. (c) In some cytochromes, like cytochrome *c*, the porphyrin ring is covalently linked via disulfide bridges to cysteine molecules in the protein. Note the presence of iron in the center of the ring. (d) Computer-generated model of cytochrome *c*. The protein completely surrounds the porphyrin ring (light color) in the center. Cytochromes carry electrons only, not entire hydrogen atoms; the redox site is the iron atom, which can alternate between the Fe^{2+} and Fe^{3+} oxidation state.

an Fe_2S_2 configuration. The reduction potentials of iron–sulfur proteins vary over a wide range depending on the number of iron atoms and sulfur atoms present and how the iron centers are attached to protein. Thus different iron–sulfur proteins can function at different points in the electron transport process. Like cytochromes, iron–sulfur proteins carry *electrons only*, not hydrogen atoms.

The **quinones** (Figure 4.16) are highly hydrophobic molecules involved in electron transport. Some quinones found in bacteria are related to vitamin K, a growth factor for higher animals. Like flavoproteins, quinones serve as *hydrogen atom* acceptors and *electron* donors.

✓ 4.10 Concept Check

Electron transport systems consist of a series of membrane-associated electron carriers that function in an integrated fashion to carry electrons from the electron donor to an electron acceptor such as oxygen. The carriers are bound to membrane proteins (except for quinones) and are arranged in the membrane in the order of their reduction potentials, from most electronegative to most electropositive.

✓ How is iron present in cytochromes compared with a protein like ferredoxin?
✓ In what major way do quinones differ from other electron carriers in the membrane?

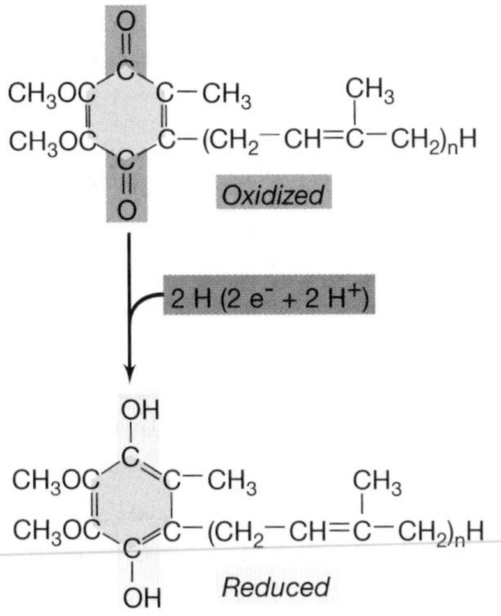

FIGURE 4.16 Structure of oxidized and reduced forms of coenzyme Q, a quinone. The five-carbon unit in the side chain (an isoprenoid) occurs in a number of multiples. In bacteria, the most common number is $n = 6$; in eukaryotes, $n = 10$. Note that oxidized quinone requires two hydrogen atoms (2 H) to become fully reduced. An intermediate form, the semiquinone (one H more reduced than oxidized quinone), is formed during the reduction of a quinone (see Figure 4.18b).

4.11

Energy Conservation from Electron Transport

The overall process of electron transport in the electron transport chain is shown in Figure 4.17. During electron transport, ATP is produced by the process of oxidative phosphorylation. The production of ATP is linked directly to the establishment of a **proton motive force** across the membrane, electron transport reactions serving to establish this energized state of the membrane. We now consider the details of this process.

The Proton Motive Force: Chemiosmosis

To understand the manner in which electron transport is linked to ATP synthesis, we must first discuss the manner in which the electron transport system is oriented in the cell membrane. The overall structure of the membrane was outlined in Section 3.5 (∞ Figure 3.18). It was shown there that proteins are embedded in the lipid bilayer of membranes and that the orientation of proteins in the membrane is such that most have access to both the outside and the inside of the cell (that is, they are transmembrane proteins).

The electron transport carriers discussed earlier are oriented in the membrane in such a way that a *separation* of protons from electrons occurs across the membrane during the transport process. Hydrogen atoms, removed from hydrogen carriers such as NADH, are separated into electrons and protons, the electrons being transported through the chain by specific carriers and the protons being extruded outside the cell into the environment (in gram-negative prokaryotes protons are extruded to the periplasm), resulting in a slight acidification of the external milieu (Figure 4.18). At the end of the electron transport chain, the electrons are passed to the final electron acceptor (in the case of aerobic respiration, this is O_2) and reduce it.

When O_2 is reduced to H_2O, it requires H^+ from the cytoplasm to complete the reaction, and these protons originate from the dissociation of water into H^+ and OH^-. The use of H^+ in the reduction of O_2 to H_2O and the extrusion of H^+ cause a net accumulation of OH^- on the *inside* of the membrane. Despite their small size, because they are charged, neither H^+ nor OH^- freely passes through the membrane, and so equilibrium cannot be spontaneously restored. Thus, although electron transport to O_2 can be thought of as producing water, what is actually produced are the *elements* of water, H^+ and OH^-, which accumulate on opposite sides of the membrane. The net result is the generation of a *pH gradient* and an *electrochemical* potential across the membrane, with the *inside* of the cytoplasm electrically

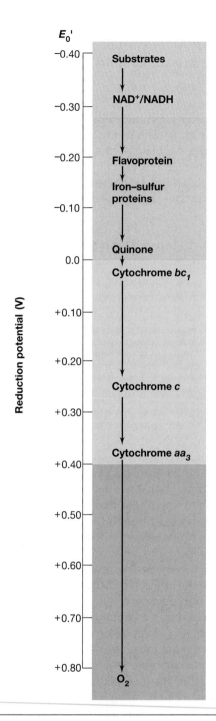

E_0'

- −0.40 — **Substrates**
- −0.30 — **NAD⁺/NADH** — $NAD^+/NADH$
- −0.20 — **Flavoprotein**
- **Iron–sulfur proteins**
- −0.10
- 0.0 — **Quinone**
- **Cytochrome bc_1**
- +0.10
- +0.20
- +0.30 — **Cytochrome c**
- +0.40 — **Cytochrome aa_3**
- +0.50
- +0.60
- +0.70
- +0.80 — **O_2**

Reduction potential (V)

FIGURE 4.17 One example of an electron transport system, leading to the transfer of electrons from substrate to O_2. This particular sequence is typical of the electron transport chain of the mitochondrion and some Bacteria (for example, *Paracoccus denitrificans*). The chain in *Escherichia coli* lacks cytochromes c and aa_3, and instead electrons go directly from cytochrome b to cytochrome o or d, which is the terminal oxidase (∞ Figure 15.39*a,b*). By breaking up the complete oxidation into a series of discrete steps, energy conservation is possible through proton motive force formation leading to ATP synthesis. Compare color-coding here with those in Figure 4.7.

negative and alkaline, and the *outside* of the membrane electrically positive and acidic. This pH gradient and electrochemical potential cause the membrane to be energized (much like a battery), and some of this electrical energy can be conserved by the cell.

In the same way that the energized state of a battery is expressed as its electromotive force (in volts), the energized state of a membrane is expressed as the **proton motive force** (also in volts). The energized state of the membrane induced as a result of electron transport processes can be used directly to do useful work such as ion transport (∞ Section 3.6) or flagellar rotation (∞ Section 3.11), or it can be used to drive the formation of high energy phosphate bonds in ATP, as will be described later. The idea of a proton gradient driving ATP synthesis was first proposed as the *chemiosmotic theory* in 1961 by the English scientist Peter Mitchell; Mitchell later received the Nobel Prize for this important contribution.

Generation of the Proton Motive Force

The key steps in proton motive force formation involve the activities of the flavin enzymes, quinones, and the cytochrome bc_1 complex (Figure 4.18). The series of oxidation–reduction reactions occurring during electron transport may be analyzed by examining each pair of carriers sequentially (Figure 4.18*a*). Following the donation of two hydrogen atoms from NADH to FAD, two H^+ are extruded when FADH donates two electrons (only) to an iron–sulfur protein. Two protons are taken up from the dissociation of water in the cytoplasm when the nonheme iron protein reduces coenzyme Q. Coenzyme Q passes electrons one at a time to the cytochrome bc_1 complex. The cytochrome bc_1 complex contains several proteins and is present in the electron transport chain of most organisms able to respire. It also plays a role in photosynthetic electron flow (∞ Sections 15.5 and 15.6). The major function of the cytochrome bc_1 complex is to transfer electrons from quinones to cytochrome c linked to the translocation of protons across the membrane. The cytochrome bc_1 complex is oriented in the membrane in such a way that protons are discharged to the environment when electrons are transferred to an acceptor, resulting in the accumulation of OH^- in the cytoplasm and protons on the outer surface of the membrane (Figure 4.18). How does this process actually occur?

Reduced coenzyme Q (QH_2) donates one electron to the bc_1 complex, extruding a proton and converting QH_2 to $QH\cdot$, the semiquinone form of coenzyme Q. The semiquinone can be reduced to QH_2 by one of the b-type cytochromes in the bc_1 complex, along with the uptake of a proton. For every two $QH\cdot$ molecules that enter the complex, one is reduced to QH_2 while the other is oxidized to Q. This "Q cycle" (Figure 4.18*b*), increases the number of protons extruded across the membrane at the Q-bc_1 site. Electrons travel from the bc_1 complex to cytochrome c and cytochrome a, the latter of which serves in conjunction with the terminal oxidase (Figure 4.18b). Finally, the reduction of $\frac{1}{2}$ O_2 to H_2O occurs and the electron transport reactions are completed (Figure 4.18).

The electron transport scheme shown in Figure 4.18 is just one of many different carrier sequences observed in different organisms. However, the important feature of all of them is the generation of a *proton gradient*, acidic outside and alkaline inside. The gradient results in a proton motive force that actually drives the synthesis of ATP.

The Proton Motive Force and ATP Formation

How does the proton motive force drive ATP synthesis? Interestingly, there is a strong parallel here between the mechanism of ATP synthesis and the mechanism of the motor that drives rotation of the bacterial flagellum (∞ Section 3.11). The catalyst for conversion of the proton motive force into ATP is a large membrane protein complex called *ATP synthase*, or *ATPase* for short. The ATPase contains two major parts, a multi-subunit headpiece called F_1, located on the inside surface of the membrane, and a proton-conducting channel, called F_0, that spans the membrane (Figure 4.19). The F_1/F_0 complex catalyzes a reversible reaction between ATP and ADP + P_i (inorganic phosphate) as shown in Figure 4.19.

The F_1/F_0 ATPase is the smallest known biological motor. The flow of protons through F_0 generates a torque that is transmitted to F_1 by the γ subunit of this complex. This torque is caused by the rotation of γ within the F_1 complex; in essence, γ is a rotating shaft that mediates the energy conversion between the proton influx (via F_0) and ATP synthesis at the surface of the F_1 complex (Figure 4.19). Thus, in analogy to the flagellar motor (∞ Figure 3.48), γ is the rotor and F_1 is the stator of the ATPase motor. But, unlike the flagellum, the rotational energy of the ATPase motor is not used for propulsion, but instead to synthesize ATP (Figure 4.19). ATPase-catalyzed ATP synthesis is referred to as **oxidative phosphorylation,** and measurements of the stoichiometry between protons and ATP show that four protons are consumed per ATP produced.

The tiny F_1/F_0 molecular motor can be reversed: The hydrolysis of ATP supplies torque for γ to rotate in the opposite direction, causing protons to be pumped from inside to outside the cell through F_0, thereby creating, instead of dissipating, a proton motive force. This reversibility explains why obligately fermentative organisms unable to carry out oxidative phosphorylation still contain an ATP synthase. Since many reactions in the cell such as motility and transport require the ener-

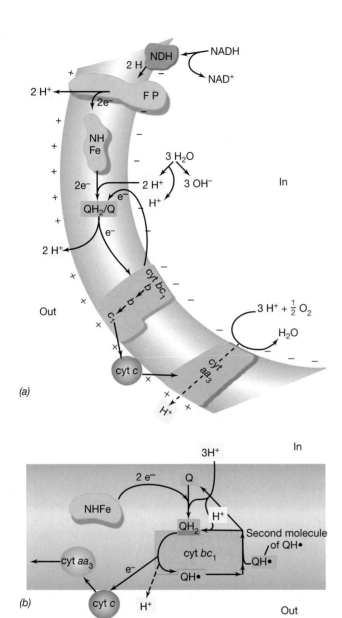

FIGURE 4.18   Generation of a proton motive force. (a) Orientation of electron carriers in the bacterial membrane, showing the manner in which electron transport can lead to charge separation and generation of a proton gradient. The + and − charges originate from the protons (H^+) or hydroxyl ions (OH^-) accumulating on opposite sides of the membrane. NDH, NADH dehydrogenase; FP, flavoprotein; NHFe, nonheme iron–sulfur protein; Q, coenzyme Q; cyt, cytochromes. IN and OUT refer to the cytoplasm and the environment (the periplasm of gram-negative bacteria), respectively. (b) The Q cycle. Electrons shuttle between QH_2 and the cytochrome bc_1 complex. For each molecule of QH_2 oxidized to QH· (semiquinone), one electron is passed to cyt c and one proton is passed to the outside. The bc_1 complex can also disproportionate two QH· to QH_2 plus Q, with the uptake of a proton from the cytoplasm. This mechanism increases the number of protons pumped at the Q-bc_1 site of the electron transport chain.

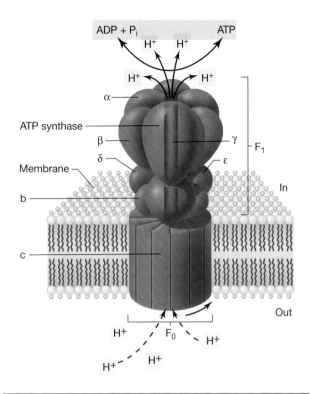

FIGURE 4.19 Structure and function of the membrane-associated ATP synthase (ATPase), which functions as a proton channel between the cytoplasm and the cell exterior (environment). F_1 consists of five polypeptides, α (three copies), β (three copies), γ, δ, and ε. This is the catalytic protein complex responsible for the interconversion of ATP and ADP + P_i. F_0 is integrated in the membrane and consists of three polypeptides, a, b (two copies), and c (four copies). It is responsible for channeling protons across the membrane. As protons enter, the dissipation of the proton motive force drives ATP synthesis from ADP + P_i. The reverse reaction, ATP → ADP + P_i, drives the extrusion of protons to the cell exterior. Thus, ATP synthase is reversible in its action.

gy of a proton motive force, the ATP synthase in organisms like the lactic acid bacteria, that do not respire, functions unidirectionally to generate a proton motive force to drive these necessary cell functions.

The events of electron transport can be studied with certain chemicals that affect oxidative phosphorylation. Two such classes of chemicals are known: *inhibitors* and *uncouplers*. Inhibitors block electron flow and thus establishment of the proton motive force. Examples include carbon monoxide (CO) and cyanide (CN^-), both of which bind tightly to certain cytochromes and prevent their functioning. In contrast, uncouplers prevent ATP synthesis *without* affecting electron transport. These lipid soluble substances, such as dinitrophenol and dicumarol, make membranes leaky, thereby destroying the proton motive force and its ability to drive ATP synthesis.

✓ 4.11 Concept Check

When electrons are transported through a membrane-integrated electron transport system, protons are extruded to the outside of the membrane forming the proton motive force. Key electron carriers include flavins, quinones, the cytochrome bc_1 complex, and other cytochromes, depending on the organism. The cell uses the proton motive force via rotating ATPases to make ATP.

✓ How do electron transport reactions generate the proton motive force?

✓ What structure in the cell converts the proton motive force to ATP? How does it operate?

4.12

Carbon Flow: The Citric Acid Cycle

We now consider the metabolic aspects of carbon flow in respiration. The early steps in the respiration of glucose involve the same biochemical steps as those of glycolysis (see Figure 4.12). As we noted, a key intermediate in glycolysis is pyruvate. Whereas in fermentation pyruvate is converted to fermentation products, in respiration pyruvate is oxidized fully to CO_2. One major pathway by which pyruvate is completely oxidized to CO_2 is called the **citric acid cycle** (CAC) as outlined in Figure 4.20.

Pyruvate is first decarboxylated, leading to the production of one molecule of NADH and an acetyl molecule coupled to coenzyme A (acetyl-CoA) (Figure 4.21). The acetyl group of acetyl-CoA combines with the four-carbon compound oxalacetate, leading to the formation of citric acid, a six-carbon organic acid, the energy of the high energy acetyl-CoA bond being used to drive this synthesis (Figure 4.21). Dehydration, decarboxylation, and oxidation reactions follow, and two additional CO_2 molecules are released. Ultimately, oxalacetate is regenerated and can function again as an acetyl acceptor, thus completing the cycle.

CO_2 Release and Fuel for Electron Transport

For each pyruvate molecule oxidized through the cycle, three CO_2 molecules are released, one during the formation of acetyl-CoA, one by the decarboxylation of isocitrate, and one by the decarboxylation of α-ketoglutarate (Figure 4.20). As in fermentation, the electrons released during the oxidation of intermediates in the CAC are transferred to enzymes containing the coenzyme NAD^+ or FAD. However, respiration differs from fermentation in the manner in which NADH and FADH are oxidized. In respiration, the electrons from NADH, instead of being used to reduce an intermediate such as pyruvate, are transferred to oxygen or other

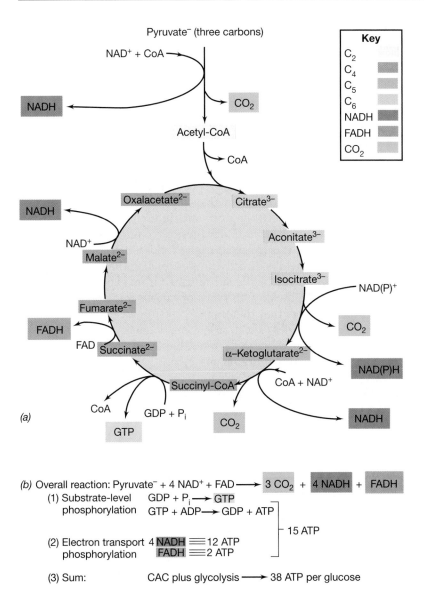

(a)

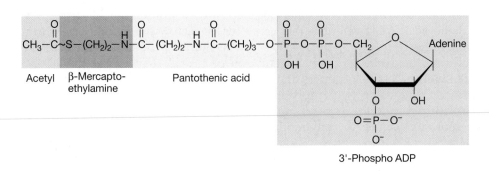

(b) Overall reaction: Pyruvate⁻ + 4 NAD⁺ + FAD ⟶ 3 CO₂ + 4 NADH + FADH

(1) Substrate-level phosphorylation
$$GDP + P_i \longrightarrow GTP$$
$$GTP + ADP \longrightarrow GDP + ATP$$

(2) Electron transport phosphorylation
$$4 \; NADH \equiv 12 \; ATP$$
$$FADH \equiv 2 \; ATP$$

15 ATP

(3) Sum: CAC plus glycolysis ⟶ 38 ATP per glucose

FIGURE 4.20 The citric acid cycle (CAC). The three-carbon compound pyruvate is oxidized to CO_2, with the electrons used to make NADH and FADH. The reoxidation of NADH and FADH in the electron transport chain leads to the generation of a proton motive force and thus to the synthesis of ATP. The CAC begins when the two-carbon compound acetyl-CoA condenses with the four-carbon compound ox-alacetate to form the six-carbon compound cit-rate. Through a series of oxidations and transformations, this six-carbon compound is ultimately converted back to the four-carbon compound oxalacetate, which then begins an-other cycle with addition of the next molecule of acetyl-CoA. (b) The overall balance sheet of fuel (NADH/FADH) for the electron transport chain and CO_2 generated in the CAC.

FIGURE 4.21 Structure of acetyl-CoA. The coenzyme A portion consists of the β-mercaptoethy-lamine–pantothenic acid–ADP complex. Note that the C—S bond between the acetyl portion and the β-mercaptoethylamine portion is a high energy thioester bond.

terminal electron acceptors through the action of the *electron transport system* described in Section 4.10. Thus, unlike the situation in fermentation, the presence of an electron acceptor in respiration allows for the complete oxidation of glucose to CO_2 with a much greater yield of energy.

Biosynthesis and the Citric Acid Cycle

Besides playing a key role in catabolic reactions, the citric acid cycle is important to the cell for biosynthet-ic reasons as well. This is because the cycle is com-posed of a number of key intermediates that can be drawn off for biosynthetic purposes when needed. Particularly important in this regard are the intermediates α-ketoglutarate and oxalacetate, which are the precursors of a number of amino acids (see Section 4.15), and succinyl-CoA, needed to form the porphyrin ring of the cytochromes, chlorophyll, and several other tetrapyrrole compounds. Oxalacetate is also important because it can be converted to phosphoenolpyruvate, a precursor of glucose (see Figure 4.28). In addition to these, acetyl-CoA provides the starting material for fatty acid biosynthesis (see Section 4.25). Thus, the citric acid cycle plays two major roles in the cell: *bioenergetic* and *biosynthetic*. Much the same can be said about the glycolytic pathway, as intermediates from this pathway are drawn off for various biosynthetic needs as well.

✓ 4.12 Concept Check

Respiration involves the complete oxidation of an organic compound with much greater energy release than during fermentation. The citric acid cycle plays a major role in the respiration of organic compounds.

- ✓ How many molecules of CO_2 and pairs of hydrogen atoms are released *per acetate consumed* in the citric acid cycle?
- ✓ What two major roles do the citric acid cycle and glycolysis have in common?

4.13

The Balance Sheet of Aerobic Respiration and Energy Storage

The net result of reactions of the citric acid cycle is the complete oxidation of pyruvic acid to three molecules of CO_2 with the production of four molecules of NADH and one molecule of FADH. As we have seen, the NADH and FADH molecules can be reoxidized through the electron transport system; the yield of ATP is up to 3 ATP molecules per molecule of NADH and 2 ATPs per molecule of FADH. In addition, the oxidation of α-ketoglutarate to succinate involves a substrate-level phosphorylation, producing guanosine triphosphate (GTP), which can be converted to ATP (Figure 4.20). Thus, a total of 15 ATP molecules can be synthesized for each turn of the cycle. Since the oxidation of glucose yields 2 molecules of pyruvic acid, a total of 30 molecules of ATP can be synthesized in the citric acid cycle. Also, when oxygen is available, the 2 NADH molecules produced during glycolysis can be reoxidized by the electron transport system, yielding 6 more molecules of ATP. Finally, 2 molecules of ATP are produced by substrate-level phosphorylation during the conversion of glucose to pyruvic acid. Thus, aerobes can form up to *38* ATP molecules from 1 glu-

cose molecule in contrast to the *2* molecules of ATP produced fermentatively.

ATP and Cell Yield

The amount of ATP produced by an organism has a direct effect on cell yield. The yield of cells obtained in a culture, that is, the biomass produced, can vary greatly with the composition of the growth medium and the environmental conditions under which growth occurs. For example, if one measures the grams of yeast cells obtained from growth on glucose aerobically versus fermentatively, the dramatic difference in the amount of ATP produced under these two conditions (as just calculated) is reflected in the yield of cells obtained. That cell yield is directly proportional to the amount of ATP produced has been confirmed from experimental studies on the growth yields of various microorganisms and implies that the energy costs for assembly of macromolecules (which are the major energy costs for a growing microbial cell) (∞ Chapter 6) are much the same for all microorganisms.

Energy Storage

ATP is present in relatively low concentration in the cell, about 2 millimolar (mM) in an actively growing cell, and thus ATP plays only a *catalytic* role during growth of the cell; ATP is continuously being broken down to drive biosynthetic reactions and resynthesized at the expense of catabolic reactions. For energy *storage*, most microorganisms produce insoluble polymers that can later be oxidized for the production of ATP. Examples include the polyglucose polymers starch and glycogen, the lipid polymer poly-β-hydroxybutyrate and other polyhydroxyalkanoates, and elemental sulfur, stored by many sulfur chemolithotrophs. These polymers are deposited within the cell as large granules that can often be seen with the light or electron microscope (∞ Section 3.13). In the absence of an external energy source, the cell may oxidize these polymers and thus be able to make new cell material or simply supply itself with maintenance energy, even when nutrients are temporarily unavailable in the environment.

Polymer formation is important to the cell for two reasons. First, potential energy is stored in a stable form, and second, insoluble polymers have little effect on the internal osmotic pressure of cells. If the same number of units were present as monomers in the cell, the high solute concentration would increase cellular osmotic pressure, resulting in an influx of water and possible lysis. Thus, storage polymers make possible the storage of energy in a readily accessible form that does not interfere with other cellular processes.

✓ 4.13 Concept Check

Although only two high energy phosphate bonds are typically conserved during glycolysis, as many as 38 ATPs can be formed during aerobic respiration. Storage polymers are common in prokaryotes and function as carbon and/or energy reserves.

✓ Considering just the reactions of glycolysis, why can 8 ATPs be made under aerobic conditions but only 2 ATPs anaerobically by fermentation?

✓ Of what advantage to the cell are storage polymers?

4.14

An Overview of Alternate Modes of Energy Generation

This chapter has thus far only dealt with energy generation by either fermentation or respiration of organic compounds, but microorganisms have many other possibilities for obtaining energy. These alternate modes of energy metabolism form the theme of Chapter 15 but are summarized here to provide an overview of catabolic processes.

Anaerobic Respiration

One alternate mode of energy generation is a variation on respiration in which electron acceptors *other than* oxygen are used. Because of the analogy to aerobic respiration, these processes are called **anaerobic respiration.** Electron acceptors used in anaerobic respiration include nitrate (NO_3^-), ferric iron (Fe^{3+}), sulfate (SO_4^{2-}), carbonate (CO_3^{2-}), and even certain organic compounds. Because of their positions on the electron tower (none of these acceptors have as electropositive an E_0' as the O_2/H_2O couple) (see Figure 4.7), less energy is released when these electron acceptors are used instead of oxygen. However, the use of these alternate electron acceptors permits microorganisms to respire in environments where oxygen is absent. The contrasts between aerobic and anaerobic respiration are presented in Figure 4.22*a* and *b*.

Chemolithotrophy

A second mode of energy generation involves the use of *inorganic* rather than organic chemicals. Organisms able to use inorganic chemicals as electron donors are a type of chemotroph called **chemolithotrophs.** Examples of inorganic electron donors include hydrogen sulfide (H_2S), hydrogen gas (H_2), ferrous iron (Fe^{2+}), and ammonia (NH_3). Chemolithotrophic metabolism usually involves aerobic respiratory processes such as those described in this chapter but using an inorganic energy source rather than an organic one (Figure 4.22). Chemolithotrophs have electron transport components

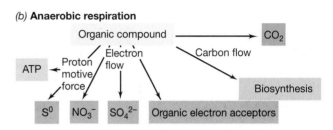

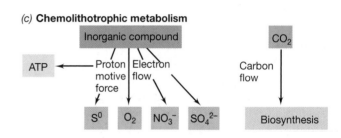

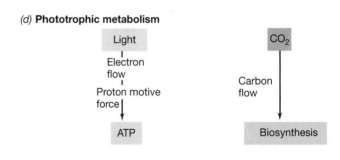

FIGURE 4.22 Energetics and carbon flow in (a) aerobic respiration, (b) anaerobic respiration, (c) chemolithotrophic metabolism, and (d) phototrophic metabolism. Note the importance of electron transport leading to proton motive force formation in each case.

similar to those of chemoorganotrophs and form a proton motive force, which drives ATP synthesis. However, one important distinction between chemolithotrophs and chemoorganotrophs is in their sources of *carbon* for biosynthesis. Chemoorganotrophs can generally use compounds such as glucose as carbon sources as well as energy sources, but chemolithotrophs cannot use their inorganic electron donors as sources of carbon.

Most chemolithotrophs use carbon dioxide as a carbon source and are, hence, **autotrophs.**

Phototrophy

A large number of microorganisms, as well as the green plants, are *phototrophic,* using light as an energy source in the process of photosynthesis. We call such organisms **phototrophs** (literally, *light-eating*). The mechanisms by which light is used as an energy source are unique and complex, but the underlying result is the generation of a proton motive force that can be used in the synthesis of ATP. Most phototrophs use energy conserved in ATP for the assimilation of carbon dioxide as the carbon source for biosynthesis. Such phototrophs are also, hence, autotrophs. However, as we will see in Chapter 15, photosynthesis in microorganisms has some special features and complications. There are, for instance, two types of photosynthesis in microorganisms, one form similar to that of higher plants, and a unique type of photosynthesis found only in certain prokaryotes.

Importance of the Proton Motive Force to Alternate Bioenergetic Strategies

One overall conclusion to be drawn at this point is that in terms of energy metabolism, microorganisms show an amazing diversity of bioenergetic strategies. Thousands of organic compounds, many reduced inorganic compounds, and light can be used by one or another microorganism as an energy source. However, with the exception of most fermentations, where substrate-level phosphorylations prevail, metabolic diversity in respiration and photosynthesis revolves around a common theme; *generation of a proton motive force.* Thus, regardless of whether electrons come from the oxidation of organic or inorganic chemicals or from phototrophic processes, they all traverse a membrane-associated electron transport chain and in so doing generate a proton motive force (Figure 4.22); energy conservation in all cases occurs through function of the ATPase (Figure 4.19). In Chapter 15 we will examine some of the details of the different bioenergetic strategies that result in generation of the proton motive force.

✓ 4.14 Concept Check

Electron acceptors other than oxygen can function as terminal electron acceptors for energy generation. Because oxygen is absent, this is called anaerobic respiration. Chemolithotrophs use inorganic compounds as electron donors, while phototrophs use light to form a proton motive force. The proton motive force is involved in all forms of respiration and photosynthesis.

- ✓ In terms of their electron donor(s), how do chemoorganotrophs differ from chemolithotrophs?
- ✓ What is the carbon source for autotrophic organisms?

4.15

Biosynthesis of Monomers

We have just discussed **catabolism,** the processes by which microorganisms obtain energy from organic compounds. We now briefly consider the other major reactions series that occurs in the cell, **anabolism,** the processes by which microorganisms build up the vast array of chemical substances of which they are composed.

Energy for anabolism is provided by ATP or the proton motive force, different forms of chemical energy (Figure 4.23). Energy conserved during catabolic reactions as ATP or the proton motive force is consumed in both the formation of monomers and during the polymerization of monomers to form macromolecules. Our coverage of biosynthetic pathways here is intended only to be an overview of monomer synthesis; the biosynthesis of macromolecules is considered in Chapter 6.

Monomers of Polysaccharides: Sugars

Polysaccharides are key constituents of the cell walls of many organisms and in Bacteria, the peptidoglycan cell wall has a polysaccharide backbone. In addition, cells often store carbon and energy in the form of the polysaccharides *glycogen* and *starch* (⚭ Sections 2.3 and 3.13). The monomeric units of these polysaccharides are six-carbon sugars called *hexoses,* in particular, glucose or glucose derivatives. In addition to hexoses, five-carbon sugars called *pentoses* are common in the cell. Most

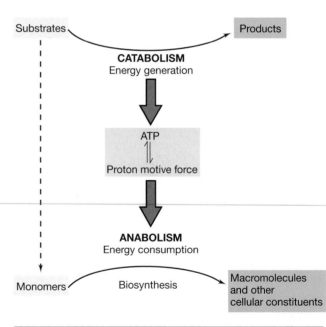

FIGURE 4.23 Scheme of anabolism and catabolism showing the key role of ATP and the proton motive force in integrating the processes. Monomers can come preformed as nutrients from the environment or from catabolic pathways like glycolysis and the citric acid cycle.

notably, these include ribose and deoxyribose, present in the backbone of RNA and DNA, respectively.

In prokaryotes, polysaccharides are synthesized from either *uridine diphosphoglucose* (UDPG, Figure 4.24*a*) or *adenosine diphosphoglucose* (ADPG), both of which are *activated* forms of glucose. ADPG is the precursor for the biosynthesis of glycogen while UDPG is the precursor of various glucose derivatives needed for the biosynthesis of other polysaccharides in the cell like peptidoglycan or the lipopolysaccharide of the gram-negative outer membrane (⌘ Sections 3.7 and 3.8).

When a cell is growing on a hexose like glucose, obtaining glucose for polysaccharide synthesis is obviously not a problem. But when the cell is growing on other carbon compounds, glucose must be synthesized. This process, called *gluconeogenesis*, uses as a starting material the compound *phosphoenolpyruvate*, one of the key intermediates in glycolysis (see Figure 4.12). An overview of hexose metabolism is shown in Figure 4.24.

Pentoses are formed by the removal of one carbon atom from a hexose, commonly as CO_2 (Figure 4.24*c*). The pentoses needed for nucleic acid synthesis, ribose and deoxyribose, are formed as shown in Figure 4.24*c*. The important enzyme *ribonucleotide reductase* converts ribose into deoxyribose by reduction of the 2′ carbon on the ring. Interestingly, this reaction occurs *after*, not before, synthesis of nucleotides, such that ADP, GDP, CTP, and TTP are biosynthesized first, and later some of each are reduced to their deoxy forms for use as precursors of DNA.

Monomers of Proteins: Amino Acids

Organisms that cannot obtain some or all of their amino acids preformed from the environment must synthesize them from other sources. Amino acids can be grouped into structurally related *families* that share biosynthetic features (Figure 4.25). The carbon skeletons for amino acids come almost exclusively from intermediates of glycolysis or the citric acid cycle (Figure 4.25).

The *amino group* of amino acids is often derived from some inorganic nitrogen source in the environment, such as ammonia (NH_3). Ammonia is typically incorporated in the formation of the amino acids *glutamate* or *glutamine* by the enzymes *glutamate dehydrogenase* or *glutamine synthetase*, respectively (Figure 4.26*a*, *b*). Once ammonia is incorporated, it can be transferred to form other needed nitrogenous compounds. For example, glutamate can donate its amino group to oxalacetate in a *transaminase* reaction to yield α-ketoglutarate and aspartate (Figure 4.26*c*). Alternatively, glutamine can react with α-ketoglutarate to form two molecules of glutamate in an *aminotransferase* reaction (Figure 4.26*d*). The end result of these types of reactions is the shuttling of ammonia into various key carbon skeletons from which further biosynthetic reactions can occur to form all the 20 amino acids needed to make proteins.

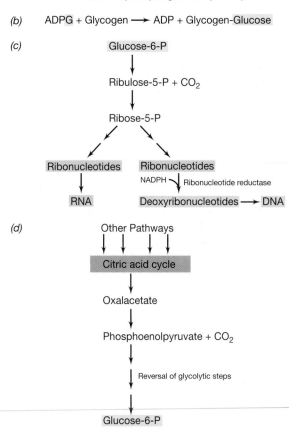

FIGURE 4.24 Sugar metabolism. (a) Polysaccharides are synthesized from activated forms of hexoses such as UDPG, whose structure is shown here. (b) Glycogen is biosynthesized from adenosine-phosphoglucose (ADPG) by the sequential addition of glucose. (c) Pentoses for nucleic acid synthesis are formed by decarboxylation of hexoses like glucose-6-phosphate. Note how the precursors of DNA are produced from the precursors of RNA by the enzyme ribonucleotide reductase. (d) Gluconeogenesis. When glucose is needed it can be biosynthesized from other carbon compounds, generally by the reversal of steps in glycolysis.

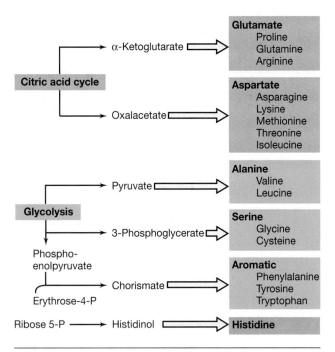

FIGURE 4.25 Amino acid families. Note how the carbon skeletons for most amino acids are derived from either the citric acid cycle or from glycolysis. Synthesis of the various amino acids in a family frequently requires many separate enzymatically catalyzed steps starting from the parent amino acid (shown in bold).

Monomers of Nucleic Acids: Nucleotides

The biosynthesis of the purines and pyrimidines is quite complex. Purines are constructed, literally atom by atom, from several carbon and nitrogen sources, including CO_2 (Figure 4.27a). The first key purine, *inosinic acid* (Figure 4.27b), is the precursor of all of the other

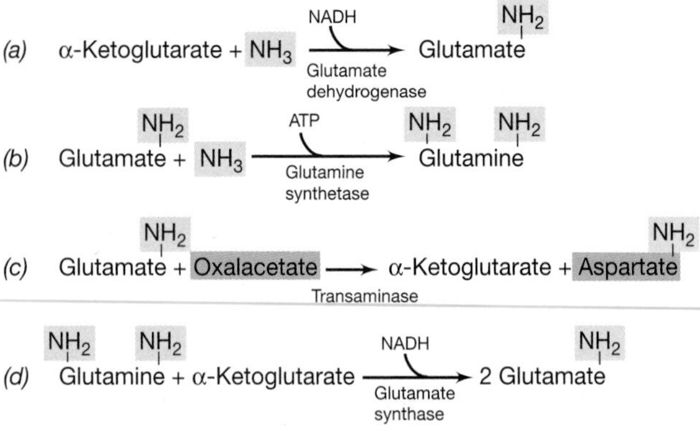

(a)   α-Ketoglutarate + NH_3 $\xrightarrow[\text{Glutamate dehydrogenase}]{\text{NADH}}$ Glutamate–NH_2

(b)   Glutamate–NH_2 + NH_3 $\xrightarrow[\text{Glutamine synthetase}]{\text{ATP}}$ Glutamine (NH_2, NH_2)

(c)   Glutamate–NH_2 + Oxalacetate $\xrightarrow{\text{Transaminase}}$ α-Ketoglutarate + Aspartate–NH_2

(d)   Glutamine (NH_2, NH_2) + α-Ketoglutarate $\xrightarrow[\text{Glutamate synthase}]{\text{NADH}}$ 2 Glutamate–NH_2

FIGURE 4.26 Ammonia incorporation in bacteria. Two major pathways for NH_3 assimilation in bacteria are those catalyzed by the enzymes (a) glutamate dehydrogenase, and (b) glutamine synthetase. (c) Transaminase reactions simply transfer an amino group from an amino acid to an organic acid. (d) In the reaction catalyzed by the enzyme glutamate synthase, two glutamates are formed from one glutamine and one α-ketoglutarate.

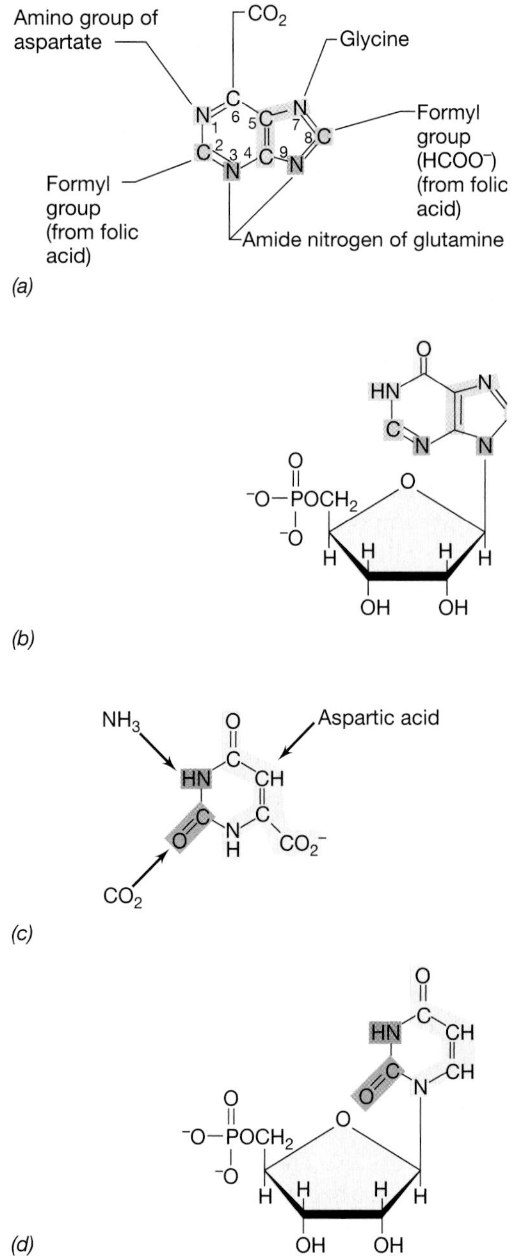

FIGURE 4.27 Biosynthesis of purines and pyrimidines. (a) The precursors of the purine skeleton. (b) Inosinic acid, the precursor of all purine nucleotides. (c) The precursors of the pyrimidine skeleton, orotic acid. (d) Uridylate, the precursor of all pyrimidine nucleotides. Uridylate is formed from orotate following a decarboxylation and the addition of ribose-5-phosphate.

purine nucleotides; once these are synthesized (in their triphosphate forms), they are ready to be incorporated into DNA or RNA (∞ Section 6.6). Like the purine ring, the pyrimidine ring is also constructed from several sources (Figure 4.27c) and the first key pyrimidine is the compound *uridylate,* from which all other pyrimidines are derived (Figure 4.27d).

Monomers of Lipids: Fatty Acids

Fatty acids are required for the biosynthesis of lipids in Bacteria and Eukarya (∞Sections 2.4 and 3.5). Fatty acids are biosynthesized two carbon atoms at a time with the help of a small protein called *acyl carrier protein* (ACP). ACP holds the growing fatty acid as it is being synthesized and releases it once it has reached the required length, such as the very common C_{16} fatty acid palmitate (Figure 4.28). Interestingly, although the fatty acid chain is constructed *two* carbons at a time, the two carbons are donated from a *three*-carbon compound called *malonate*; as each malonate is donated, one molecule of CO_2 is released (Figure 4.28).

The fatty acid composition of cells vary from species to species and can also vary somewhat within a species due to temperature (growth at low temperatures favors shorter-length fatty acids), but C_{14}–C_{18} are the most common fatty acids found in Bacteria. In addition to saturated, even-carbon-number fatty acids, fatty acids can also be unsaturated, contain branches, or have an odd number of carbon atoms. *Unsaturated* fatty acids contain one or more double bonds in the long hydrophobic portion of the molecule. The number and position of these double bonds is often species or group specific.

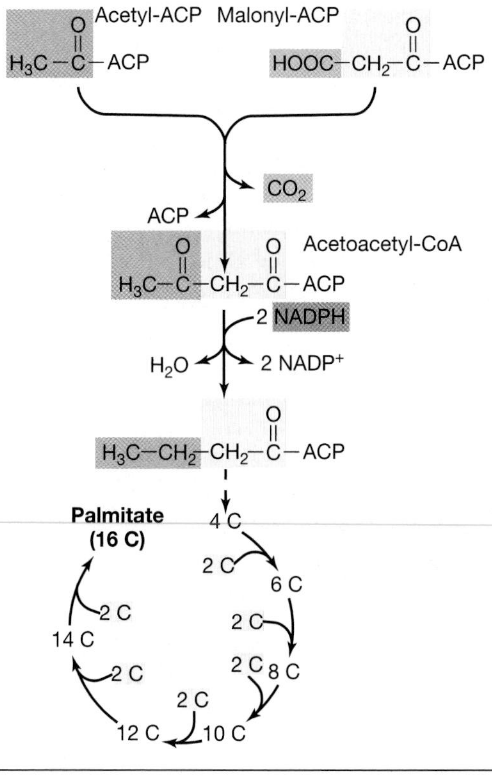

Branched chain and *odd-carbon-number* fatty acids are made using an initiating molecule that contains a branched chain fatty acid or a propionyl (C_3) group, respectively.

The final assembly of lipids in Bacteria and Eukarya involves the addition of fatty acids to a glycerol molecule. For simple triglycerides, all three glycerol carbons are esterified with fatty acids, but for complex lipids, one of the carbons of glycerol contains a molecule like phosphate, ethanolamine, a sugar, or some other polar substance (∞ Figure 2.7). In Archaea, lipids contain phytanyl side chains instead of fatty acids (∞ Section 3.5) but, as in Bacteria or Eukarya, the third carbon of the glycerol backbone usually contains a polar group of some sort.

Biosynthesis and Growth

We have now discussed the basic principles of biosynthesis of the monomers needed for the synthesis of macromolecules, the bulk of a cell's substance (∞ Table 2.2). However, before we consider how the cell directs synthesis of its macromolecules, in particular its informational macromolecules, we examine the phenomenon of cell growth. Growth (multiplication) is the final result of all the catabolic and anabolic reactions we have considered in this chapter. Once we have mastered the principles of microbial growth and understand the environmental factors that control it, we will be ready to study the details of "information flow" and genetics and appreciate how the syntheses of nucleic acids and proteins are coordinated in the prokaryotic cell.

✓ 4.15 Concept Check

The biosynthesis of monomers from either nutrients in the environment or from catabolic intermediates prepares the cell for the final step in biosynthesis, production of macromolecules. A variety of enzyme systems and biosynthetic pathways exist for the biosynthesis of sugars, amino acids, nucleotides, and fatty acids.

✓ What activated forms of glucose are involved in polysaccharide synthesis or in the synthesis of other hexoses?

✓ Where do most of the carbon skeletons for amino acid biosynthesis come from?

✓ Why can it be said that of all the enzymes whose activities are shown in Figure 4.26, only transaminase has no energy requirement?

✓ Explain why in fatty acid synthesis, fatty acids get built *two* carbon atoms at a time while the immediate donor for these carbons contains *three* carbon atoms.

FIGURE 4.28 The biosynthesis of fatty acids; shown is the biosynthesis of the C_{16} fatty acid, *palmitate*. The condensation of acetyl-ACP and malonyl-ACP forms acetoacetyl-CoA. Each successive addition of an acetyl unit comes from malonyl-CoA.

REVIEW QUESTIONS

1. Define the terms *chemoorganotroph, chemolithotroph, photoautotroph,* and *photoheterotroph.*

2. Why are *carbon* and *nitrogen* macronutrients while cobalt is a micronutrient?

3. What are siderophores and why are they necessary?

4. Why would the following medium *not* be considered a chemically defined medium: glucose, 5 grams (g); NH_4Cl, 1 g; KH_2PO_4, 1 g; $MgSO_4$, 0.3 g; yeast extract, 5 g; distilled water, 1 liter.

5. Describe how would you calculate $\Delta G^{0\prime}$ for the reaction: glucose $+ 6 O_2 \rightarrow 6 CO_2 + 6 H_2O$. If you were told that this reaction is highly *exergonic*, what would be the sign (negative or positive) of the $\Delta G^{0\prime}$ you would expect for this reaction?

6. Distinguish between $\Delta G^{0\prime}$ and G^0_f.

7. Why are enzymes needed by the cell?

8. Describe a rationale for why an enzyme from the bacterium *Escherichia coli* loses its catalytic ability after being boiled.

9. Describe the difference between a *coenzyme* and a *prosthetic group.*

10. The following is a series of coupled electron donors and electron acceptors. Using just the data given in Figure 4.7, order this series from most energy-yielding to least energy-yielding. H_2/Fe^{3+}, H_2S/O_2, methanol/NO_3^- (producing NO_2^-), H_2/O_2, Fe^{2+}/O_2, NO_2^-/Fe^{3+}, H_2S/NO_3^-.

11. What is an electron carrier? Give three examples of electron carriers and indicate their oxidized and reduced forms.

12. Is the reaction glucose 6-phosphate + ADP → glucose + ATP exergonic or endergonic (refer to Figure 4.10 to help answer this question)?

13. Where in glycolysis is NADH *produced?* Where is NADH *consumed?*

14. Iron plays an important role within the cell in energy-generating processes. Give three examples in which iron plays a role as an electron carrier. How is iron provided as a nutrient in culture media?

15. What is meant by the term *proton motive force* and why is this concept so important in biology?

16. The chemicals dinitrophenol and cyanide both act as cellular poisons but in quite different ways. Compare and contrast the modes of action of these two chemicals.

17. Give in simplified form the equation for the oxidation–reduction reaction that takes place in the cytochrome molecule. Can you suggest a role for the portions of the cytochrome molecule that are *not* involved in oxidation–reduction?

18. Knowing the function of the electron transport chain, can you imagine an organism that could live if it completely lacked the components (for example, cytochromes) needed for an electron transport chain? (*Hint:* Focus your answer on the mechanism of ATPase.)

19. Work through the energy balance sheets for fermentation and respiration and account for all sites of ATP synthesis. Organisms can obtain nearly 20 times more ATP when growing aerobically on glucose than anaerobically. Write one sentence that accounts for this difference.

20. Why can it be said that the citric acid cycle plays *two* major roles in the cell?

21. What are the similarities and differences in aerobic respiration in *Escherichia coli*, a chemoorganotroph, and *Thiobacillus thioparus*, a sulfur chemolithotroph?

22. Figure 4.25 indicates that there are only a few intermediate compounds that serve as the starting points for amino acid biosynthesis. For each family of amino acids, identify the starting compound and from what major pathway it originates.

23. Describe the process by which a fatty acid such as palmitate (C_{16} saturated) is synthesized in a cell.

APPLICATION QUESTIONS

1. Design a defined culture medium for an organism that can grow aerobically on acetate as a carbon and energy source. Make sure all the nutrient needs of the organism are accounted for and in the correct relative proportions.

2. *Desulfovibrio* can grow anaerobically with H_2 as electron donor and SO_4^{2-} as electron acceptor (which is reduced to H_2S). Based on this information and the data in Table A1.2 (Appendix 1), indicate which of the following components *could not* exist in the electron transport chain of this organism and why: cytochrome *c*, ubiquinone, cytochrome c_3, cytochrome aa_3, ferredoxin.

3. Again, using the data in Table A1.2, predict the sequence of electron carriers in an organism growing aerobically and producing the following electron carriers: ubiquinone, cytochrome aa_3, cytochrome *b*, NADH, cytochrome *c*, FAD.

4. Explain the following observation: cells of *Escherichia coli* fermenting glucose grow faster when NO_3^- is supplied to the culture (NO_2^- is produced), and then grow even faster (and stop producing NO_2^-) when the culture is highly aerated.

5. Applying what you know about glycolysis, the citric acid cycle, and gluconeogenesis, explain why a mutant of *Escherichia coli* containing a defective malate dehydrogenase (that is, cannot carry out the reaction malate^{2-} + NAD$^+$ → oxaloacetate^{2-} + NADH + H$^+$) (see Figure 4.20) is able to grow on glucose but not on acetate, whereas nonmutated (wild-type) *E. coli* can be grown on either substrate.

Microbial cells grow in an exponential fashion. During this process cell numbers double in a regular and orderly manner, often yielding huge populations in a short period of time. In a closed growth system, like a tube or a flask, a typical unicellular organism shows several phases of growth, as depicted in the classical bacterial "growth curve," shown here. The mathematics of microbial growth, the principles of which you will learn in this chapter, apply only to the exponential phase. But before growth begins cells need to acclimate to a new environment (this is the lag phase). And after a period of exponential growth cells either run out of food or become inhibited by their own wastes (triggering onset of the stationary phase). A firm grasp of the principles of microbial growth are important for an understanding of several basic and applied problems in microbiology.

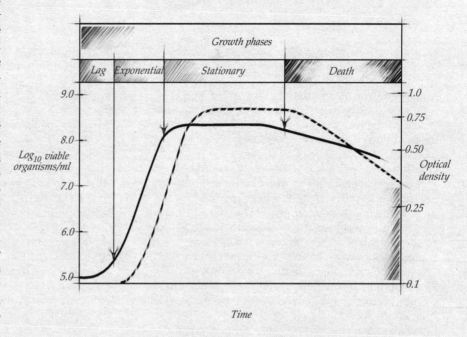

CHAPTER 5 Microbial Growth

5.1   Overview of Cell Growth   136
5.2   Population Growth   137
5.3   Growth Cycle of Populations   139
5.4   Measurement of Growth   141
5.5   Continuous Culture: The Chemostat   145
5.6   Effect of Environment on Growth   146
5.7   Effect of Temperature on Growth   147
5.8   Microbial Growth at Cold Temperatures   148
5.9   Microbial Growth at High Temperatures   151
5.10   Microbial Growth at Low or High pH   154
5.11   Osmotic Effects on Microbial Growth   156
5.12   Oxygen as a Factor in Microbial Growth   158

WORKING GLOSSARY

Acidophile   an organism that grows best at low pH

Aerobe   an organism that can use O_2 in respiration; some require O_2 for growth

Alkaliphile   an organism that grows best at high pH

Anaerobe   an organism that cannot use O_2 in respiration and whose growth may be inhibited by O_2

Batch culture   a closed-system microbial culture of fixed volume

Chemostat   a device that allows for the continuous culture of microorganisms in which both growth rate and cell number can be controlled independently

Compatible solute   a molecule that is accumulated in the cytoplasm for adjustment of water activity but that does not inhibit biochemical processes

Exponential growth   growth of a microorganism where the cell number doubles within a fixed time period

Extremophile   an organism that grows optimally under one or more chemical or physical extremes, such as high or low temperature or pH

Facultative   with respect to O_2, an organism that can grow in either its presence or absence

Generation time   the time required for a population of microbial cells to double

Growth   an increase in cell number

Halophile   a microorganism that requires NaCl for growth

Hyperthermophile   a microorganism that has a growth temperature optimum of 80°C or greater

Lag phase   a period preceding the exponential growth phase when cells may be metabolizing but are not yet growing

Lysis   loss of cellular integrity with release of cytoplasmic constituents

Mesophile   an organism that grows best at temperatures between 20 and 45°C

Microaerophile   an aerobic organism that can grow only when oxygen tensions are reduced from that in air

Psychrophile   an organism with a growth temperature optimum of 15°C or lower and a maximum growth temperature below 20°C

Psychrotolerant   an organism capable of growth at low temperatures but whose growth temperature optimum is above 20°C

Stationary phase   the period immediately following exponential growth when the growth rate of the population falls to zero

Thermophile   an organism whose growth temperature optimum lies between 45 and 80°C

Viable   capable of reproducing

Xerophile   an organism that is able to live, or that lives best, in very dry environments

---

We have thus far discussed cellular macromolecules (Chapter 2), cell biology (Chapter 3), and the general principles of microbial nutrition and metabolism (Chapter 4). Before we begin our study of the biosynthesis of macromolecules and the molecular genetics of microorganisms (Chapter 6), we consider here microbial growth.

In microbiology, the word **growth** is defined as *an increase in the number of cells.* Growth is an essential component of microbial function, as any given cell has a finite life span in nature and the species is maintained only as a result of continued growth of the population. However, in addition to understanding the basic science of microbial growth, many practical situations call for the *control* of microbial growth. Knowledge of how microbial populations can rapidly expand is quite useful in designing methods to control microbial growth; we will study these methods in Chapter 18.

In the present chapter we first discuss the general principles of microbial growth and describe the growth cycle of microbial populations. After consideration of means to measure growth and a special tool, the chemostat, for manipulating growth of a microbial population, we focus on the major environmental factors influencing microbial growth: temperature, pH, water potential, and oxygen. We will then be ready to take our knowledge of microbial growth and place it in the context of the fundamental molecular biological processes that occur as a cell prepares to divide and become two cells.

5.1

Overview of Cell Growth

The bacterial cell is a synthetic machine that is able to duplicate itself. The synthetic processes of bacterial cell growth involve as many as 2000 chemical reactions of a wide variety of types. Some of these reactions involve energy transformations. Other reactions involve biosynthesis of small molecules—the building blocks of macromolecules—as well as the various cofactors and coenzymes needed for enzymatic reactions. However, the main reactions of cell synthesis are *polymerization reactions,* the processes by which polymers (macromolecules) are made from monomers. The major reactions of macromolecular synthesis will be discussed in Chapter 6: DNA synthesis, RNA synthesis, and protein synthesis. Once polymers are made, the stage is set for the final events of cell growth: assembly of macromolecules and formation of cellular structures such as the cell wall, cytoplasmic membrane, flagella, ribosomes, inclusion bodies, enzyme complexes, and so on.

Binary Fission

In most prokaryotes, growth of an individual cell continues until the cell divides into two new cells, a process called *binary fission* (*binary* to express the fact that *two* cells have arisen from one cell). In a growing culture of a rod-shaped bacterium such as *Escherichia coli*, for example, cells are observed to elongate to approximately twice the length of the smallest cell and then form a partition that eventually separates the cell into two daughter cells (Figure 5.1). This partition is referred to as a *septum* and is a result of the inward growth of the cytoplasmic membrane and cell wall from opposing directions until the two daughter cells are pinched off (Figure 5.1). During the growth cycle all cellular constituents increase in number such that each daughter cell receives a complete chromosome and sufficient copies of all other macromolecules, monomers, and inorganic ions to exist as an independent cell. Partitioning of the replicated DNA molecule between the two daughter cells depends on the DNA remaining attached to membranes during division, with septum formation leading to sep-

aration of chromosome copies, one going to each daughter cell (Figure 5.1).

The time required for a complete growth cycle in bacteria is highly variable and is dependent on a number of factors, both nutritional and genetic. Under the best nutritional conditions the bacterium *Escherichia coli* can complete the cycle in about 20 min; a few bacteria can grow even faster than this, but many grow much slower. The control of cell division is a complex process and appears to be intimately tied to chromosomal replication events.

✓ 5.1 Concept Check

Microbial growth involves an increase in the *number* of cells rather than in the size of individual cells. Growth of most microorganisms occurs by binary fission. Cell division and chromosome replication are usually coordinately regulated.

✓ Define *microbial growth*.
✓ Why is it essential that the bacterial chromosome be replicated before binary fission occurs?

5.2

Population Growth

As we have mentioned, *growth* is defined as an increase in the *number* of microbial cells in a population, which can also be measured as an increase in microbial *mass*. **Growth rate** is the change in cell number or cell mass *per unit time*. During this cell division cycle, all the structural components of the cell double. The interval for the formation of two cells from one is called a **generation,** and the time required for this to occur is called the **generation time.** The generation time is thus the time required for the cell population to double. Because of this, the generation time is also sometimes called the *doubling time*. Note that during a single generation, both the cell number and cell mass double. Generation times vary widely among organisms. Many bacteria have generation times of 1–3 hr, but a few very rapidly growing organisms are known that divide in as little as 10 min, and others have generation times of several hours or even days.

Exponential Growth

A growth experiment beginning with a single cell having a doubling time of 30 min is presented in Figure 5.2. This pattern of population increase, where the number of cells *doubles* during each unit time period, is referred to as **exponential growth.** When the cell number from such an experiment is graphed on arithmetic coordinates as a function of elapsed time, one obtains a curve with a constantly increasing slope (Figure 5.2*b*). How-

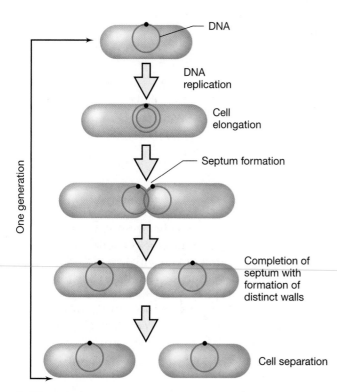

FIGURE 5.1 The process of binary fission in a rod-shaped prokaryote. For simplicity, the nucleoid is depicted as a single circle in green.

(DNA — DNA replication — Cell elongation — Septum formation — Completion of septum with formation of distinct walls — Cell separation)

One generation

Time (hr)	Total number of cells
0	1
0.5	2
1	4
1.5	8
2	16
2.5	32
3	64
3.5	128
4	256
4.5	512
5	1,024
5.5	2,048
6	4,096
.	.
.	.
.	.
10	1,048,576

(a)

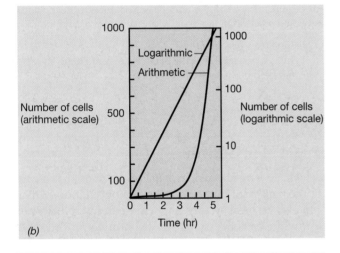

(b)

FIGURE 5.2   The rate of growth of a microbial culture. (a) Data for a population that doubles every 30 min. (b) Data plotted on an arithmetic (left ordinate) and a logarithmic (right ordinate) scale.

ever, deriving growth rate information from such curves is difficult. The number of cells on a logarithmic ($\log_{10}$) scale is presented in Figure 5.2*b* in a graph in which cell number is plotted logarithmically and time is plotted arithmetically (a *semilogarithmic* graph), resulting in a straight line. This straight-line function is an immediate indicator that the cells are growing exponentially. Semilogarithmic graphs are also convenient and simple to use for estimating generation times from a set of results. The doubling time may be read directly from the graph (Figure 5.3; see also Figure 5.8).

One of the characteristics of exponential growth is that the *rate* of increase in cell number is slow initially but increases at an ever faster rate. This results, in the later stages, in an explosive increase in cell numbers. For example, in the experiment in Figure 5.2, the *rate* of cell production in the first 30 min of growth is one cell per 30 min. However, between 4 and 4.5 hr of growth, the rate of cell production is considerably faster at 256 cells per 30 min

(Figure 5.2). A practical implication of exponential growth is that when a nonsterile product such as milk is allowed to stand under conditions such that microbial growth can occur, a few hours during the early stages of exponential growth are not detrimental, whereas standing *for the same length of time* during the later stages is disastrous.

Calculating Generation Times

The increase in cell number that occurs in an exponentially growing bacterial culture is a geometric progression of the number 2. As two cells double (to become four cells), we can express this as $2^1 \rightarrow 2^2$. As four cells become eight, we express this as $2^2 \rightarrow 2^3$, and so on (Figure 5.2). Because of this geometric progression, there is a direct relationship between the number of cells present in a culture initially and the number present after a period of exponential growth:

$$N = N_0 2^n$$

where N = final cell number, N_0 = initial cell number, and n = *number of generations* that have occurred during the period of exponential growth. The generation time g of the cell population is calculated as t/n, where t is simply the hours or minutes of exponential growth. Thus, from a knowledge of the initial and final cell numbers in an exponentially growing cell population, it is possible to calculate n, and from n and knowledge of t, the generation time g.

To express the equation $N = N_0 2^n$ in terms of n, the following transformations are necessary:

$$N = N_0 2^n$$

$$\log N = \log N_0 + n \log 2$$

$$\log N - \log N_0 = n \log 2$$

$$n = \frac{\log N - \log N_0}{\log 2} = \frac{\log N - \log N_0}{0.301}$$

$$= 3.3 (\log N - \log N_0)$$

With n now expressed in terms of readily measurable quantities, N and N_0, generation times can be calculated. As an example of how to perform a calculation, we use actual data from the lower graph in Figure 5.3. The generation time of 2 hr, which in this case was determined directly from the graph, can also be derived from the facts that $N = 10^8$, $N_0 = 5 \times 10^7$, and $t = 2$. Thus,

$$n = 3.3[\log 10^8 - \log (5 \times 10^7)] = 3.3 (8 - 7.69)$$
$$= 3.3(0.301) = 1$$

Thus, the generation time $t/n = 2/1 = 2$ hr. The generation time g can also be calculated from the slope of the line obtained in the semilogarithmic plot of exponential growth, as slope = $0.301/g$.

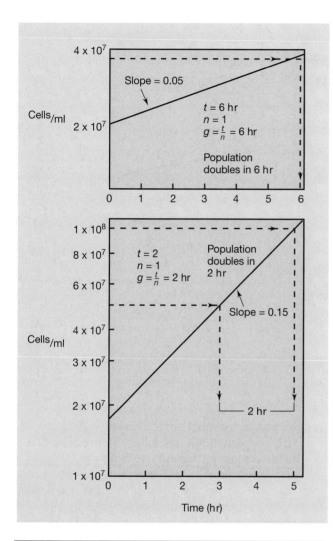

FIGURE 5.3 Method of estimating the generation times (g) of exponentially growing populations with generation times of 6 and 2 hr, respectively, from data plotted on semilogarithmic graphs. The slope of each line is $0.301/g$. All numbers are expressed in scientific notation; that is, 10,000,000 is 1×10^7, 60,000,000 is 6×10^7, and so on.

Another index of growth rate is the *growth rate constant*, abbreviated k. The growth rate constant can be calculated as $k = \frac{\ln 2}{g} = \frac{0.693}{g}$ and has units of reciprocal hours (h^{-1}). While the term g is a measure of the *time* it takes for a population to double in cell number, k is a measure of the *number of generations* that occur per unit time in an exponentially growing culture.

Armed with knowledge of n and t, one can calculate g and k for different microorganisms growing under different culture conditions. This is often useful for optimizing culture conditions for a particular organism and also for testing the positive or negative effect of some treatment on the bacterial culture.

✓ **5.2 Concept Check**

Microbial populations show a characteristic type of growth pattern called exponential growth, which is best seen by plotting the number of cells at various time periods on a semilogarithmic graph. From knowledge of the initial and final cell numbers and the time of exponential growth, the generation time of the cell population can be calculated directly.

✓ Distinguish between the terms *growth rate* and *generation time*.
✓ Why does exponential growth lead to large cell populations in so short a period of time?
✓ What is a *semilogarithmic* plot?

5.3

Growth Cycle of Populations

The data presented in Figure 5.2 reflect only part of the growth cycle of a microbial population, the part called *exponential growth*. In an enclosed vessel, referred to as a *batch culture,* a typical *growth curve* for a population of cells is obtained as illustrated in Figure 5.4. This growth curve can be divided into several distinct phases called the **lag phase, exponential phase, stationary phase,** and **death phase.**

Lag Phase

When a microbial population is inoculated into a fresh medium, growth usually does not begin immediately but only after a period of time called the *lag phase,* which may be brief or extended depending on the history of the culture and growth conditions. If an exponentially growing culture is inoculated into the same medium under the same conditions of growth, a lag is not seen and exponential growth begins immediately. However, if the inoculum is taken from an old (stationary phase, which is discussed later) culture and inoculated into the same medium, a lag usually occurs even if all the cells in the inoculum are *viable,* that is, able to reproduce. This is because the cells are usually depleted of various essential constituents and time is required for their resynthesis. A lag also ensues when the inoculum consists of cells that have been damaged (but not killed) by treatment with heat, radiation, or toxic chemicals because of the time required for the cells to repair the damage.

A lag is also observed when a population is transferred from a rich culture medium to a poorer one. This happens because for growth to occur in a particular culture medium the cells must have a complete complement of enzymes for synthesis of the essential metabolites not present in that medium. On transfer to a new medium, time is required for synthesis of the new enzymes.

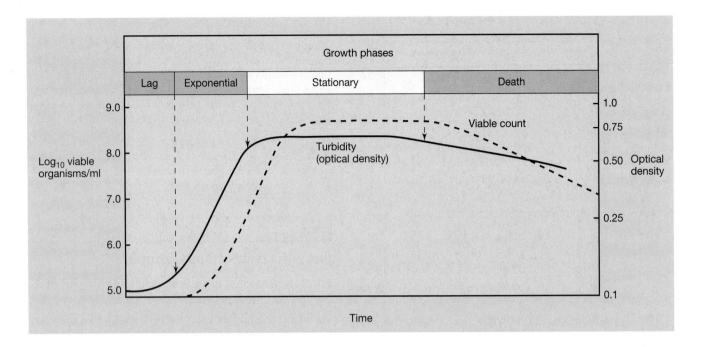

FIGURE 5.4   Typical growth curve for a bacterial population. See Section 5.4 for a description of the counting methods employed.

Exponential Phase

The *exponential phase* of growth has already been discussed. As noted, it is a consequence of the fact that each cell divides to form two cells, each of which also divides to form two more cells, and so on. Cells in exponential growth are usually in their healthiest state, and thus cells in "midexponential" phase are often desirable for studies of enzymes or other cell components.

Most unicellular microorganisms grow exponentially, but rates of exponential growth vary greatly. The rate of exponential growth is influenced by environmental conditions (temperature, composition of the culture medium) as well as by genetic characteristics of the organism itself. In general, prokaryotes grow faster than eukaryotic microorganisms, and small eukaryotes grow faster than large ones.

Stationary Phase

In a batch culture (see Section 5.5), exponential growth cannot occur indefinitely. One can calculate that a single bacterium with a generation time of 20 min would, if it continued to grow exponentially for 48 hr, produce a population that weighed about 4000 times the weight of Earth! This is particularly impressive because a single bacterial cell weighs only about one-trillionth (10^{-12}) of a gram (∞ Table 2.2). Obviously, something must happen to limit growth of the population long before this time. What generally happens is that either (1) an essential nutrient of the culture medium is used up or (2) some waste product of the organism builds up in the medium to an inhibito-

ry level and exponential growth ceases, or both. At this point the population has reached the **stationary phase.**

In the stationary phase there is no net increase or decrease in cell number. However, although no growth usually occurs in the stationary phase, many cell functions may continue, including energy metabolism and some biosynthetic processes. In some organisms, slow growth may even occur during the stationary phase; some cells in the population grow, whereas others die, the two processes balancing out so that no net increase or decrease in cell number occurs (this is a phenomenon called *cryptic growth*).

Studies with *Escherichia coli* have identified several genes that are necessary for the survival of cells that have entered the stationary phase. Some of these genes, called *sur* (for *survival*) genes, have as yet unknown functions, but mutations in *sur* genes lead to rapid cell death as cells enter the stationary phase. Because in nature it is likely that many bacterial cells are in a nongrowing or very slow-growing state (∞ Chapter 16), it is not surprising that several genes have evolved to deal with conditions in which the cells in a population have reached the stationary phase or for other reasons are growing at extremely low rates.

Death Phase

If incubation continues after a population reaches the stationary phase, the cells may remain alive and continue to metabolize, but they may also die. If the latter occurs, the population is said to be in the *death phase*. In some cases death is accompanied by actual cell **lysis.** Note in Figure 5.4 how the death phase of the growth cycle is also an exponential function; however, in most cases, the rate of cell death is much slower than that of exponential growth.

Before we conclude this section it should be emphasized that the phases of the bacterial growth curve shown in Figure 5.4 are reflections of the events in a *population* of cells, not in individual cells. The terms *lag phase, exponential phase, stationary phase,* and *death phase* do not apply to individual cells but only to *populations* of cells.

Now let us turn our attention to methods for determining cell numbers in microbial cultures.

✓ 5.3 Concept Check

Microorganisms typically show a characteristic growth pattern when inoculated into a fresh culture medium. There is an initial lag phase, and then growth commences in an exponential fashion. As essential nutrients are depleted, or toxic products build up, growth ceases and the population enters the stationary phase. If incubation continues, cells may begin to die and the population is said to be in the death phase.

✓ To what phase of the growth curve do the mathematics of growth apply?
✓ When does a lag phase usually *not* occur?
✓ Why do cells enter the stationary phase?

5.4

Measurement of Growth

Population growth is measured by following changes in the number of cells or weight of cell mass. There are several methods for counting cell numbers or estimating cell mass, suited to different organisms or different problems.

Total Cell Count

The number of cells in a population can be measured by counting a sample under the microscope, a method called the **direct microscopic count.** Two kinds of direct microscopic counts are done, either on samples dried on slides or on samples in liquid. With liquid samples, special *counting chambers* must be used. In such a counting chamber, a grid is marked on the surface of the glass slide, with squares of known small area (Figure 5.5). Over each square on the grid is a volume of known size, very small but precisely measured. The number of cells per unit area of grid can be counted under the microscope, giving a measure of the number of cells per small chamber volume. Converting this value to the number of cells per milliliter of suspension is easily done by multiplying by a conversion factor based on the volume of the chamber sample (Figure 5.5).

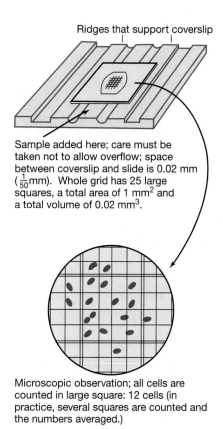

Ridges that support coverslip

Sample added here; care must be taken not to allow overflow; space between coverslip and slide is 0.02 mm ($\frac{1}{50}$mm). Whole grid has 25 large squares, a total area of 1 mm^2 and a total volume of 0.02 mm^3.

Microscopic observation; all cells are counted in large square: 12 cells (in practice, several squares are counted and the numbers averaged.)

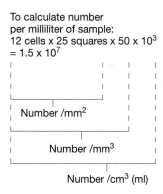

To calculate number per milliliter of sample:
12 cells x 25 squares x 50 x 10^3
= 1.5 x 10^7

Number /mm^2

Number /mm^3

Number /cm^3 (ml)

FIGURE 5.5   Direct microscopic counting procedure using the Petroff–Hausser counting chamber.

Direct microscopic counting is a quick way of estimating microbial cell number. However, it has certain limitations: (1) Dead cells are not distinguished from living cells. (2) Small cells are difficult to see under the microscope, and some cells are probably missed. (3) Precision is difficult to achieve. (4) A phase contrast microscope is required when the sample is not stained. (5) The method is not usually suitable for cell suspensions of low density. With bacteria, if a cell suspension has less than 10^6 cells/milliliter (ml), few if any bacteria will be seen in the microscope field. However, dilute suspensions may be counted if a sample is first concentrated and then resuspended in a small volume.

Viable Count

In the method just described, both living and dead cells are counted. In many cases we are interested in counting only live cells, and for this purpose *viable* cell counting methods have been developed. A viable cell is defined as one that is able to divide and form offspring, and the usual way to perform a viable count is to determine the number of cells in the sample capable of forming *colonies* on a suitable agar medium. For this reason, the viable count is often called the **plate count,** or **colony count.** The assumption made in this type of counting procedure is that *each viable cell can yield one colony.*

There are two ways of performing a plate count: the spread plate method and the pour plate method (Figure 5.6). With the **spread plate method,** a volume of an appropriately diluted culture usually no greater than 0.1 ml is spread over the surface of an agar plate using a sterile glass spreader. The plate is then incubated until the colonies appear, and the number of colonies is counted. It is important that the surface of the plate be dry so that the liquid that is spread soaks in. Volumes greater than 0.1 ml are rarely used because the excess liquid does not soak in and may cause the colonies to coalesce as they form, making them difficult to count. In the **pour plate method** (Figure 5.6), a known volume (usually 0.1–1.0 ml) of culture is pipetted into a sterile Petri plate; melted agar medium is then added and mixed well by gently swirling the plate on the table top. Because the sample is mixed with the molten agar medium, a larger volume can be used than with the spread plate; however, with the pour plate method the organism to be counted must be able to briefly withstand the temperature of melted agar, 45°C.

Dilutions

With both the spread plate and pour plate methods, it is important that the number of colonies developing on the plates not be too large because on crowded plates some cells may not form colonies and some colonies may fuse, leading to erroneous measurements. It is also essential that the number of colonies not be too small, or the statistical significance of the calculated count will be low. The usual practice, which is the most valid statistically, is to count colonies only on plates that have between 30 and 300 colonies.

To obtain the appropriate colony number, the sample to be counted must almost always be *diluted.* Since one rarely knows the approximate viable count ahead of time, it is usually necessary to make more than one dilution. Several 10-fold dilutions of the sample are commonly used (Figure 5.7). To make a 10-fold (10^{-1}) dilution, one can mix 0.5 ml of sample with 4.5 ml of diluent, or 1.0 ml sample with 9.0 ml diluent. If a 100-fold (10^{-2}) dilution is needed, 0.05 ml can be mixed with 4.95 ml diluent, or 0.1 ml with 9.9 ml diluent. Alternatively, a 10^{-2} dilution can be made by making two successive 10-fold dilutions. In most cases, such *serial dilutions* are needed to reach the final dilution desired. Thus, if a 10^{-6} ($1/10^6$) dilution is needed, it can be achieved by making three successive 10^{-2} ($1/10^2$) dilutions or six successive 10^{-1} dilutions (Figure 5.7).

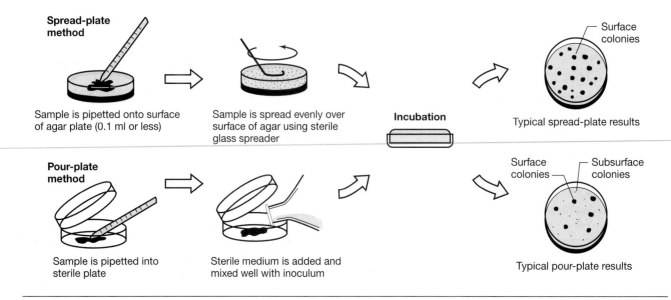

FIGURE 5.6   Two methods of performing a viable count (plate count). In either case the sample must usually be diluted before plating.

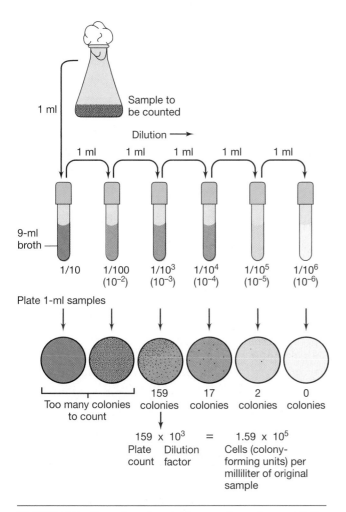

1 ml

Sample to
be counted

Dilution ⟶

1 ml   1 ml   1 ml   1 ml   1 ml

9-ml
broth

1/10   1/100   1/10³   1/10⁴   1/10⁵   1/10⁶
         (10^{-2})   (10^{-3})   (10^{-4})   (10^{-5})   (10^{-6})

Plate 1-ml samples

Too many colonies   159   17   2   0
to count   colonies   colonies   colonies   colonies

$$159 \times 10^3 = 1.59 \times 10^5$$
Plate   Dilution   Cells (colony-
count   factor   forming units) per
                         milliliter of original
                         sample

FIGURE 5.7   Procedure for viable counting using serial dilutions of the sample. The sterile liquid used for making dilutions can simply be water, but a balanced salt solution or growth medium may yield a higher recovery. The dilution factor is the reciprocal of the dilution.

Sources of Error in Plate Counting

The number of colonies obtained in a viable count depends not only on the inoculum size but also on the suitability of the culture medium and the incubation conditions used; it also depends on the length of incubation. The cells deposited on the plate will not all develop into colonies at the same rate, and if a short incubation time is used, less than the maximum number of colonies will be obtained. Furthermore, the size of colonies often varies. If some tiny colonies develop, they may be missed during the counting. It is usual to determine the incubation conditions (medium, temperature, time) that will give the maximum number of colonies of a given organism and then use these conditions throughout. Viable counts can be subject to large error, and if accurate counts are desired, great care must be taken and replicate plates of key dilutions must be

prepared. Note that two or more cells in a clump form only a single colony, and so a viable count may be erroneously low. To more clearly state the result, viable counts are often expressed as the number of *colony-forming units* obtained rather than as the number of *viable cells* (since a colony-forming unit may contain one or more cells).

Despite the difficulties associated with viable counting, the procedure gives the best information on the number of viable cells and so is widely used. In food, dairy, medical, and aquatic microbiology, viable counts are employed routinely. The method has the virtue of high sensitivity: Samples containing very few cells can be counted, thus permitting sensitive detection of microbial contamination of products or materials. Moreover, the use of highly selective culture media and growth conditions (∞ Section 21.2) in viable counting procedures allows for the counting of only particular cell types in a mixed population of microorganisms.

Turbidimetric Measurements of Cell Number

A more rapid and quite useful method of obtaining an estimate of cell number is by use of *turbidity measurements*. A cell suspension looks cloudy (turbid) to the eye because cells scatter light passing through the suspension. The more cells present, the more light scattered and hence the more turbid the suspension. Turbidity can be measured with a *photometer* or a *spectrophotometer*, devices that pass light through a cell suspension and detect the amount of unscattered light that emerges (Figure 5.8). The major difference between these two instruments is that a photometer employs a simple filter (usually red, green, or blue) to generate incident light of relatively broad wavelength, whereas a spectrophotometer employs a prism or diffraction grating to generate incident light in a narrow band of wavelengths to impinge on the sample (Figure 5.8a). However, both devices measure only *unscattered* light, and readings are recorded in photometer units (for example, "Klett units" for the Klett-Summerson photometer) (see Figure 5.8b) or optical density (OD) units for a spectrophotometer.

For unicellular organisms, photometer units or OD are proportional (within certain limits) to cell number; therefore, turbidity readings can be used as a substitute for direct counting methods. However, before using turbidity as an estimate of cell number, a *standard curve* must first be prepared for each organism to be studied, relating some direct measurement of cell number (microscopic or viable count) or mass (dry weight) to the *indirect* measurement obtained from turbidity (Figure 5.8c). Such a curve can contain data for both cell number and cell mass, allowing for an estimate of both parameters from a single turbidity reading (Figure 5.8c).

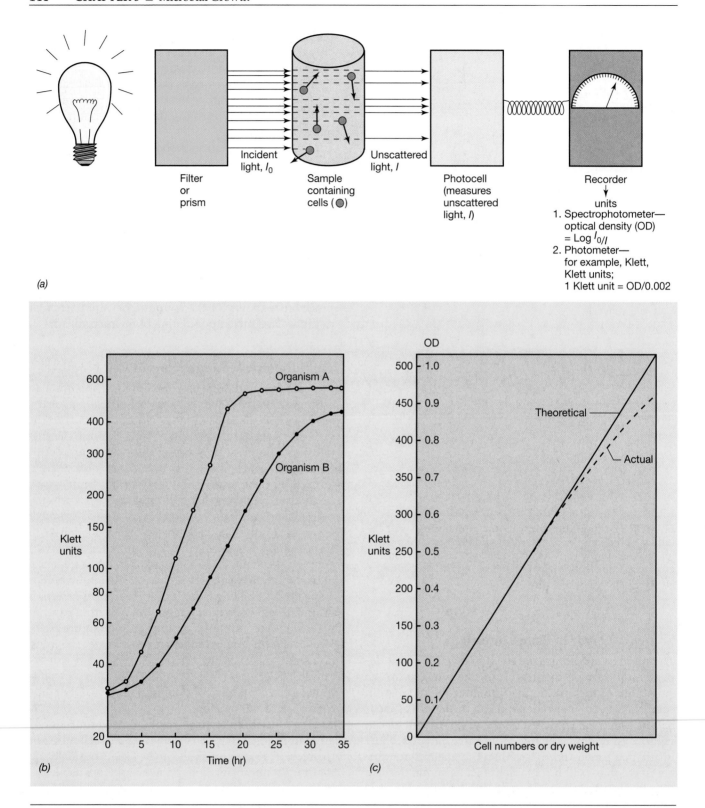

FIGURE 5.8 Turbidity measurements of microbial growth. (a) Measurements of turbidity are made in a spectrophotometer or photometer. The photocell measures incident light unscattered by cells in suspension and gives readings in optical density or photometer units. (b) Typical growth curve data obtained in Klett units for two organisms growing at different growth rates. For practice, calculate the generation time (g) of the two cultures using the formula $n = 3.3(\log N - \log N_0)$ where N and N_0 are two different Klett values taken between a time interval t. Which organism is growing faster, A or B? (c) Relationship between cell number or dry weight and turbidity readings. Note that the one-to-one correspondence between these relationships breaks down at high turbidities.

At high concentrations of cells, light scattered away from the detecting unit by one cell can be rescattered back by another (and thus appear to the photocell as if it had never been scattered), and when this occurs, the one-to-one correspondence between cell number and turbidity loses linearity (Figure 5.8c). Nevertheless, within limits turbidity measurements can be reasonably accurate, and they have the virtue of being quick and easy to perform. In addition, turbidity measurements can usually be made without destroying or significantly disturbing the sample. For these reasons turbidity measurements are widely employed to follow the growth rate of microbial cultures; the same sample can be checked repeatedly, and the measurements plotted on a semilogarithmic plot versus time (Figure 5.8b) and used to calculate the generation time of the growing culture.

✓ 5.4 Concept Check

Growth is measured by the change in number of cells with time. Cell counts done microscopically measure the total number of cells in a population, whereas viable cell counts (plate counts) measure only the living population. Turbidity measurements are an indirect but very useful measure of cell numbers.

✓ Why is a *viable count* more sensitive than a *microscopic count*?
✓ What is the major assumption made in relating plate count results to cell number?
✓ Describe how you would dilute a bacterial culture by 10^{-7}.
✓ Of what use is a spectrophotometer in the study of microbial growth?

5.5

Continuous Culture: The Chemostat

Our discussion of population growth thus far has been confined to *batch cultures*, growth occurring in a fixed volume of a culture medium that is continually being altered by the actions of the growing organisms until it is no longer suitable for growth. In the early stages of exponential growth in batch cultures, conditions may remain relatively constant, but in later stages when cell numbers become quite large, drastic changes in the chemical composition of the culture medium usually occur. However, for many studies, it is desirable to keep cultures in constant environments for long periods, and this is done by employing *continuous cultures*. A continuous culture is essentially a flow system of constant volume to which medium is added continuously and from which continuous removal of any overflow can occur. Once such a system is in equilibrium, cell number and nutrient status remain *constant*, and the system is said to be in **steady state.**

The Chemostat

The most common type of continuous culture device used is a **chemostat** (Figure 5.9), which permits control of both the population density and the growth rate of the culture. Two elements are used in the control of a chemostat—the *dilution rate* and the *concentration of a limiting nutrient,* such as a carbon or nitrogen source. In a batch culture nutrient concentration can affect both the growth rate and the growth yield of a microorganism (Figure 5.10). At very low concentrations of a given nutrient, the growth rate is reduced, probably because the nutrient cannot be transported into the cell fast enough to satisfy metabolic demand, whereas at moderate or higher levels of nutrient, growth *rates* may not be affected while cell *yield* continues to increase (Figure 5.10). In contrast to a batch culture, in a chemostat, growth rate and growth yield can be controlled *independently of each other,* the former by adjusting the dilution rate and the latter by varying the concentration of a nutrient present in a limiting amount.

Effects of varying dilution rate and concentration of the growth-limiting nutrient are given in Figure 5.11. As seen, there are rather wide limits over which the dilution rate controls growth rate, although at both very low and very high dilution rates, the steady state breaks down. At high dilution rates, the organism cannot grow fast enough to keep up with its dilution, and the culture is washed out

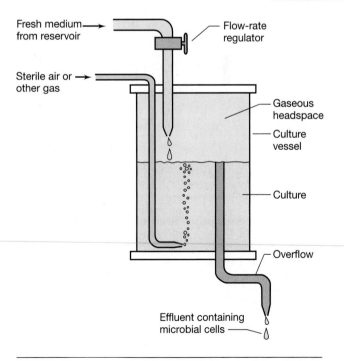

FIGURE 5.9 Schematic for a continuous culture device (chemostat). In such a device, the population density is controlled by the concentration of limiting nutrient in the reservoir, and the growth rate is controlled by the flow rate (see Figure 5.11). Both parameters can be set by the experimenter.

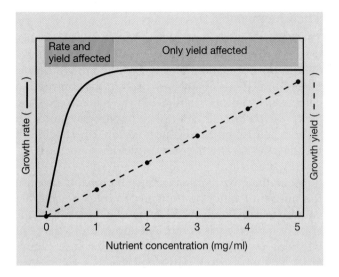

FIGURE 5.10 Relationship among nutrient concentration, growth rate (solid line), and growth yield (dashed line) in a batch culture (closed system). At low nutrient concentrations both growth rate and growth yield are affected.

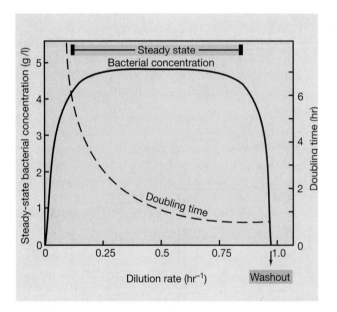

FIGURE 5.11 Steady-state relationships in the chemostat. The dilution rate is determined from the flow rate and the volume of the culture vessel. Thus, with a vessel of 1000 ml and a flow rate through the vessel of 500 ml/hr, the dilution rate would be 0.5 hr^{-1}. Note that at high dilution rates, growth cannot balance dilution, and the population washes out. Note also that although the population density remains constant during steady state, the growth rate (doubling time) can vary over a wide range. Thus, the experimenter can obtain populations with widely varying growth rates without affecting population density.

of the chemostat. At the other extreme, at very low dilution rates, a large fraction of the cells may die from starvation because the limiting nutrient is not being added fast enough to permit maintenance of cell metabolism.

The *cell density* (cells/ml) in the chemostat is controlled by the level of the limiting nutrient, just as cell yield was controlled in a batch culture (Figure 5.10). If the concentration of this nutrient in the incoming medium is raised, with the dilution rate remaining constant, the cell density will increase. Thus, by adjusting dilution rate and nutrient level, the experimenter can obtain at will a variety of population densities growing at a variety of growth rates.

Experimental Uses of the Chemostat

One of the major advantages of a chemostat is that this device allows the experimenter to control growth rate and population density *independently* of each other. As was shown in Figure 5.11, even over rather wide ranges, any desired growth rate can be obtained in the chemostat by simply varying the dilution rate. Similarly, the population density may be determined by varying the concentration of a single nutrient (the growth-limiting nutrient) in the medium reservoir. Independent control of these two crucial growth parameters is impossible with batch cultures because the batch culture is a *closed system* from the standpoint of nutrient resources and waste product removal, and thus growth conditions are constantly changing with time.

A practical advantage to the chemostat is that a population may be maintained in the exponential growth phase for long periods, for days and even weeks. Because exponential phase cells are usually most desirable for physiological experiments, the experimenter using the chemostat can have such cells available at any time. Thus, experiments can be planned in detail and then performed whenever most convenient. Moreover, repetition of experiments can be done with the knowledge that the cell population will be as close to the same each time as possible.

✓ 5.5 Concept Check

Continuous culture devices (chemostats) are a means of maintaining cell populations in exponential growth for long periods. In a chemostat, the rate at which the culture is diluted governs the growth rate and the population size is governed by the concentration of the growth-limiting nutrient entering the vessel.

✓ How do microorganisms in a *chemostat* differ from microorganisms in a *batch culture*?
✓ How is growth rate controlled in a chemostat?

5.6

Effect of Environment on Growth

Up to now we have described growth of microorganisms under essentially ideal laboratory conditions. However, the activities of microorganisms are greatly affected by the chemical and physical conditions of their environments. Understanding environmental influences helps us to explain the distribution of microorganisms in nature and makes it possible for us to devise methods for controlling microbial activities and destroying undesirable organisms. Not all organisms respond equally to a given environmental factor. In fact, an environmental condition may be harmful to one organism and actually beneficial to another. Organisms can tolerate some adverse conditions under which they cannot grow, and hence we must distinguish between the effects of environmental conditions on the *viability* of an organism and on growth, differentiation, and reproduction.

Regardless of whether organisms are interacting with other organisms in natural communities or with each other in pure culture in the laboratory, the environment can significantly affect their ability to carry out metabolic reactions and grow. Many environmental factors could be considered in this connection, however, four main factors have been identified that clearly play major roles in controlling microbial growth: temperature, pH, water availability, and oxygen. We consider each of these factors in detail here.

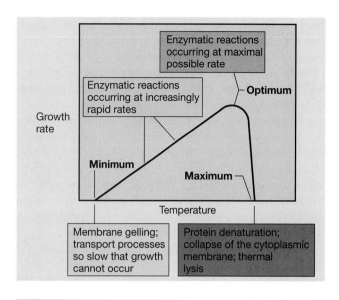

FIGURE 5.12   Effect of temperature on growth rate and the molecular consequences for the cell.

5.7

Effect of Temperature on Growth

Temperature is one of the most, if not *the* most, important environmental factors affecting growth and survival of microorganisms. At either too cold or too hot a temperature, microorganisms will not be able to grow. But what absolute values these minimum and maximum temperatures take vary greatly among different microorganisms and are usually a reflection of the temperature range and average temperature of their habitats. We examine the effect of temperature on microbial growth here.

Cardinal Temperatures

Temperature can affect living organisms in either of two opposing ways. As the temperature rises, chemical and enzymatic reactions in the cell proceed at more rapid rates, and growth becomes faster. However, *above* a certain temperature, particular proteins may be irreversibly damaged. Thus, as the temperature is increased within a given range, growth and metabolic function increase up to a point where inactivation reactions set in. Above this point, cell functions fall sharply to zero. Thus, we find that for every organism there is a **minimum temperature** below which growth no longer occurs, an **optimum temperature** at which growth is most rapid, and a **maximum temperature** above which growth is not possible (Figure 5.12). The optimum temperature is always nearer the *maximum* than the minimum. These

three temperatures, called the **cardinal temperatures,** are generally characteristic of each type of organism but are not completely fixed, as they can be modified slightly by other factors of the environment—in particular, the composition of the growth medium.

The maximum growth temperature of a given organism most likely reflects the inactivation of one or more key proteins in the cell. However, the factors controlling an organism's *minimum* growth temperature are not as clear. As mentioned earlier (∞ Section 3.6), the cytoplasmic membrane must be in a fluid state for proper functioning. Perhaps the minimum temperature of an organism results from "freezing" of the cytoplasmic membrane so it no longer functions properly in nutrient transport or proton gradient formation. This explanation is supported by experiments in which the minimum temperature for an organism is altered to some extent by adjustments in membrane lipid composition (see Section 5.8). It is also observed that the cardinal temperatures of different microorganisms differ widely; some organisms have temperature optima as low as 4°C, and some higher than 100°C. The temperature range throughout which growth occurs is even wider than this, from below freezing to greater than boiling (the archaeon *Pyrolobus fumarii* has a temperature maximum of 113°C!). However, no single organism can grow over this whole temperature range, and the usual range for a given organism is about 30°, although some have a broader temperature range than others.

Temperature Classes of Organisms

Although there is a continuum of organisms, from those with very low temperature optima to those with high temperature optima, it is possible to broadly distinguish *four groups* of microorganisms in relation to their temperature optima: **psychrophiles,** with low temperature optima, **mesophiles,** with midrange temperature optima, **thermophiles,** with high temperature optima, and **hyperthermophiles,** with very high temperature optima (Figure 5.13). Mesophiles are found in warm-blooded animals and in terrestrial and aquatic environments in temperate and tropical latitudes. Psychrophiles and thermophiles are found in unusually cold and unusually hot environments, respectively. Hyperthermophiles are found in extremely hot habitats such as hot springs, geysers, and deep-sea hydrothermal vents (see Sections 5.9 and 16.12).

In *Escherichia coli,* a typical mesophile, a detailed study of growth as a function of temperature has precisely defined the cardinal temperatures. The optimum temperature of *E. coli* in a rich complex medium is 39°C, the maximum is 48°C, and the minimum is 8°C. These values are subject to slight strain differences, and in general, the maximum and minimum temperatures supporting growth of an organism are higher and lower, respectively, when tested in complex rather than defined media.

✓ 5.7 Concept Check

Temperature is a major environmental factor controlling microbial growth. Various microorganisms differ greatly in their temperature requirements for growth.

✓ What are the approximate cardinal temperatures for *Escherichia coli*? To what temperature class does it belong?
✓ How does a *hyperthermophile* differ from a *psychrophile*?
✓ *Escherichia coli* can grow at a higher temperature in complex than in defined medium. Why?

5.8

Microbial Growth at Cold Temperatures

Because humans live and work on the surface of Earth where temperatures are generally moderate, it is natural to consider very hot and very cold environments as being "extreme." And they are extreme for human habitation because humans would die quickly if immersed in boiling or freezing water. However, the natural habitats of many microorganisms can be either extremely hot or extremely cold, and the organisms that live there, referred to as *extremophiles,* have evolved to grow optimally under these conditions. We consider organisms here that grow at cold temperatures.

Cold Environments

Much of Earth's surface experiences fairly low temperatures. The oceans, which make up over half of Earth's surface, have an average temperature of 5°C, and the depths of the open oceans have constant temperatures of about 1–3°C. Vast land areas of the Arctic and Antarctic are permanently frozen or are unfrozen for only a few weeks in summer (Figure 5.14a). These cold environments are rarely sterile, and some microorganisms can be found alive and growing at any low temperature at which liquid water still exists. Even in many frozen materials there are usually microscopic pockets of liquid water present where microorganisms can grow. It is important to distinguish between environments that are cold *throughout* the year and those that are cold *only* in winter. The latter, characteristic of continental temperate climates, may have summer temperatures as high as 40°C and winter temperatures of −20°C or colder.

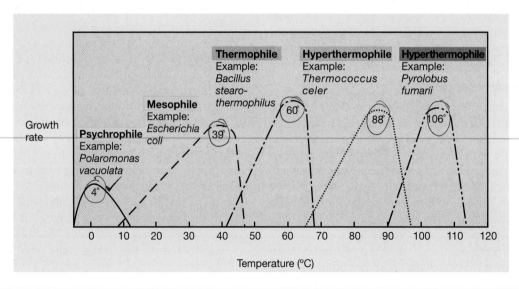

FIGURE 5.13   Relation of temperature to growth rates of a typical psychrophile, a typical mesophile, a typical thermophile, and two different hyperthermophiles. The temperature optima of the example organisms are shown on the graph.

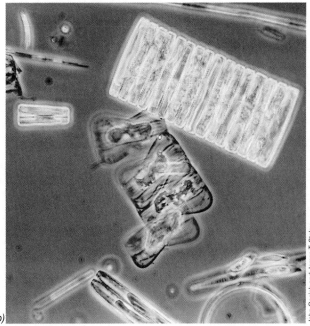

(a)

(b)

John Gosink and James T. Staley

John Gosink and James T. Staley

FIGURE 5.14   Microorganisms from Antarctic sea ice. (a) A core of permanently frozen seawater from McMurdo Sound, Antarctica. Note the dense coloration due to pigmented microorganisms, and the boot included to show scale. (b) Phase contrast micrograph of phototrophic microorganisms from the core shown in (a). Most organisms are either diatoms or green algae (both eukaryotic microorganisms).

Such highly variable environments are much less favorable for cold-adapted organisms than are the constantly cold environments found in polar regions, at high altitudes, and in the depths of the oceans.

Psychrophilic and Psychrotolerant Microorganisms

As noted earlier, organisms with low temperature optima are called **psychrophiles.** A psychrophile can be defined as an organism with an optimal temperature for growth of 15°C or lower, a maximum growth temperature below 20°C, and a minimal temperature for growth at 0°C or lower. Organisms that grow at 0°C but have optima of 20–40°C are called **psychrotolerant.**

Psychrophiles are found in environments that are constantly cold, and they may be rapidly killed even by brief warming to room temperature. For this reason, their laboratory study requires that great care be taken to ensure that they never warm up during sampling, transport to the laboratory, plating, pipetting, or other manipulations. Some of the best-studied psychrophiles have been algae that grow in dense masses within and under the ice in polar regions (Figure 5.14b). Psychrophilic algae are also often seen on the surfaces of snowfields and glaciers in such large numbers that they impart a distinctive red or green coloration to the surface (Figure 5.15a). The most common

snow alga is *Chlamydomonas nivalis;* its brilliant red spores are responsible for the red color (Figure 5.15b). This green alga grows within the snow as a green-pigmented vegetative cell and then sporulates; as the snow dissipates by melting, erosion, and vaporization, the spores become concentrated on the surface. Snow algae are most commonly seen on melting permanent snowfields in midsummer to late summer and are especially common in sunny, dry areas, probably because in more rainy areas they are washed away from the snowfields.

Psychrotolerant microorganisms are much more widely distributed than psychrophiles and can be isolated from soils and water in temperate climates as well as from meat, milk and other dairy products, cider, vegetables, and fruit stored under refrigeration (4°C). As noted, psychrotolerant microorganisms grow best at a temperature between 20 and 40°C. Because temperate environments warm up in summer, it is understandable that they cannot support the heat-sensitive psychrophiles, the warming essentially providing a selective force favoring psychrotolerant species and excluding psychrophilic forms. It should be emphasized that although psychrotolerant microorganisms do grow at 0°C, they do not grow very well, and one must often wait several weeks before visible growth is seen in culture media. Various genera of Bacteria, fungi, algae, and protozoa have members that are psychrotolerant.

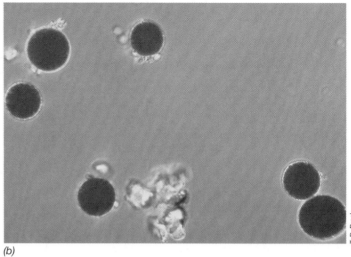

(b)

T. D. Brock

Katherine M. Brock

(a)

FIGURE 5.15   Snow algae. (a) Snow bank in the Sierra Nevada, California, with red coloration caused by the presence of snow algae. Pink snow such as this is common on summer snow banks at high altitudes throughout the world. (b) Photomicrograph of red-pigmented spores of the snow alga *Chlamydomonas nivalis*.

Molecular Adaptations to Psychrophily

Psychrophiles produce enzymes that function optimally in the cold and that are often denatured or otherwise inactivated at even very moderate temperatures. The molecular basis for this is not entirely understood, but it has been observed that, on average, cold-active enzymes have greater amounts of α-helix and lesser amounts of β-sheet secondary structure (⊙⊙ Section 2.7) than enzymes that are inactive in the cold. Because the β-sheet tends to form a more rigid structure, the greater α-helix content of cold-active enzymes allows these proteins greater flexibility in the cold. Cold-active enzymes also tend to have greater polar and lesser hydrophobic amino acid contents than their counterparts from mesophiles and thermophiles, and this may also assist in keeping the protein flexible (and thus enzymatically active) at cold temperatures.

Another feature of psychrophiles is that compared to mesophiles, active transport occurs well at low temperature, an indication that the cytoplasmic membranes of psychrophiles are constructed in such a way that low temperatures do not inhibit membrane phenomena. Studies on the composition of cytoplasmic membranes from psychrophiles have shown them to contain a higher content of *unsaturated* fatty acids (⊙⊙ Section 4.15), which help to maintain a semifluid state of the membrane at low temperatures (membranes composed of predominantly saturated fatty acids would become waxy and nonfunctional at low temperatures). The lipids of some psychrophilic bacteria also contain *polyunsaturated* fatty acids and long-chain hydrocarbons with multiple double bonds. In the latter connection, a hydrocarbon with nine double bonds ($C_{31:9}$) has been identified from the lipids of several Antarctic bacteria.

Freezing

Despite the ability of some organisms to grow at low temperatures, there is a lower limit below which reproduction is impossible. Pure water freezes at 0°C and seawater at −2.5°C, but freezing is not continuous and microscopic pockets of water continue to exist at much lower temperatures. Although freezing prevents microbial growth, it does not always cause microbial death. In addition, the medium in which the cells are suspended considerably affects sensitivity to freezing.

Water-miscible liquids such as glycerol and dimethylsulfoxide (DMSO), when added at about 10% (final concentration) to the suspending medium, penetrate the cells and protect by reducing the severity of dehydration effects and preventing ice crystal formation. In fact, the addition of such agents, called *cryoprotectants*, is a common way of *preserving* microbial cultures at very low temperatures (usually −70 to −196°C).

✓ 5.8 Concept Check

Organisms with cold temperature optima are called *psychrophiles* and the most extreme examples inhabit permanently cold environments. Psychrophiles have evolved biomolecules that function best at cold temperatures but that are unusually sensitive to warm temperatures.

- ✓ How do *psychrotolerant* organisms differ from *psychrophilic* organisms?
- ✓ What molecular adaptations to the cytoplasmic membrane are seen in psychrophiles and why are they necessary?

5.9

Microbial Growth at High Temperatures

Microbial life flourishes in high-temperature environments, up to and including boiling water. And above about 65°C, only *prokaryotic* life forms survive, but even here, a huge diversity of both Bacteria and Archaea exist. We consider some of these hot environments and their microbial life here.

Thermal Environments

Recall that organisms whose growth temperature optimum is above 45°C are called **thermophiles,** and those whose optimum is above 80°C are called **hyperthermophiles** (Figure 5.13). Temperatures as high as these are found in nature only in certain restricted areas. For example, soils subject to full sunlight are often heated to temperatures above 50°C at midday, and some soils may become warmed to even 70°C, although a few centimeters under the surface the temperature is much lower. Fermenting materials such as compost piles and silage usually reach temperatures of 60–65°C. However, the most extensive and extreme high temperature environments found in nature are in association with volcanic phenomena.

Many hot springs have temperatures near or at boiling, and steam vents (fumaroles) may reach 150–500°C. Hydrothermal vents in the bottom of the ocean have temperatures of 350°C or greater (⌀ Section 16.12). Hot springs occur throughout the world, but are especially concentrated in the western United States, New Zealand, Iceland, Japan, Italy, Indonesia, Central America, and central Africa. However, the area with the largest single concentration of hot springs in the world is Yellowstone National Park, Wyoming. Although some hot springs vary in temperature, others are very constant, not varying more than 1–2°C over many years. In addition, these springs have variable chemical compositions and pH values, generally containing sufficient levels of nutrients to support populations of both chemoorganotrophs and chemolithotrophs.

Hyperthermophiles in Hot Springs

Many hot springs are at the boiling point for the altitude (92–93°C at Yellowstone, 99–100°C at locations where the springs are close to sea level). In boiling hot springs (Figure 5.16), a variety of hyperthermophiles

(a)

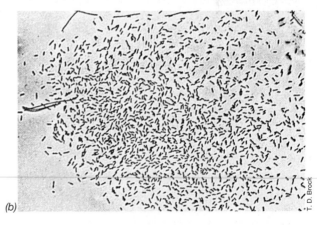
(b)

FIGURE 5.16 Growth of hyperthermophiles in boiling water. (a) A small boiling spring in Yellowstone National Park, Boulder Spring. This spring is superheated, having a temperature 1–2°C above the boiling point. The mineral deposits around the spring consists mainly of silica and sulfur. (b) Photomicrograph of a microcolony of prokaryotes that developed on a microscope slide immersed in a boiling spring such as that shown in (a).

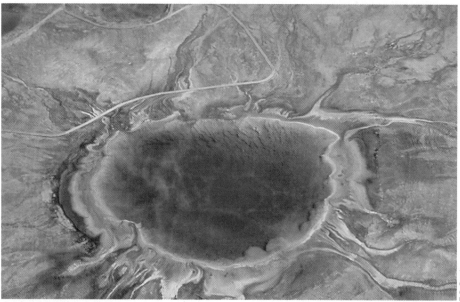

(a)

(b)

FIGURE 5.17   Growth of thermophilic cyanobacteria in hot springs in Yellowstone National Park. (a) Aerial photograph of a very large boiling spring, Grand Prismatic Spring. The orange color in the outflow channels is due to the rich carotenoid pigments of bacteria and cyanobacteria. (b) Characteristic V-shaped pattern (arrows) formed by cyanobacteria at the upper temperature for photosynthetic life, 70–74°C, in the thermal gradient formed from a boiling hot spring. The pattern develops because the water cools more rapidly at the edges than in the center of the channel. The spring flows from the back of the picture toward the foreground. Organisms can be isolated along the thermal gradient that show various temperature optima (see Figure 5.13), but in general, isolates from higher temperatures show higher growth temperature optima than isolates obtained further from the source.

are typically present. The growth of such organisms can be studied by immersing microscope slides into the spring and retrieving them after a few days. Microscopic examination of the slides reveals colonies of prokaryotes (Figure 5.16b) that have developed from single bacterial cells that attached to and grew on the glass surface.

Ecological studies of organisms living in boiling springs have shown that growth rates are fairly rapid, and doubling times of as short as 1 hr have been recorded. Cultures of many of these prokaryotes have been obtained and a variety of morphological and physiological types exist (∞ Chapter 14). Phylogenetic studies using ribosomal RNA sequencing (∞ Section 12.5) have shown great evolutionary diversity among these hyperthermophiles including various Bacteria and, especially, Archaea. Some of these hyperthermophiles show growth-temperature optima greater than 100°C and thus are grown in the laboratory in pressurized vessels to reach temperatures above the boiling point.

Thermophiles

Many thermophiles (optima < 80°C) are present in hot springs as well as other thermal environments. In hot springs, as boiling water overflows the edges of the spring and flows away from the source, it gradually cools, setting up a *thermal gradient*. Along this gradient, various microorganisms grow (Figure 5.17), with different species growing in the different temperature ranges. By studying the species distribution along such thermal gradients and by examining hot springs and other thermal habitats at different temperatures around the world, it is possible to determine the upper temperature limits for each kind of organism (Table 5.1). From this information we conclude that (1) prokaryotic organisms in general are able to grow at temperatures higher than those at which eukaryotes can grow: (2) the most thermophilic of all prokaryotes are certain species of Archaea; and (3) nonphototrophic organisms are able to grow at higher temperatures than can phototrophic forms.

Thermophilic prokaryotes have also been found in artificial thermal environments. The hot water heater, domestic or industrial, has a temperature of 55–80°C and is therefore a favorable habitat for the growth of thermophilic prokaryotes. Organisms resembling *Thermus aquaticus,* a common hot spring thermophile, have been isolated from hot water heaters. Electric power plants, hot industrial process water, and other artificial thermal sources also provide sites where thermophiles can grow and such organisms can be easily isolated using the appropriate enrichment techniques.

Molecular Adaptations to Thermophily

How can thermophiles and hyperthermophiles thrive at high temperatures? First, their enzymes and other proteins are much more stable to heat than are those of mesophiles, and their macromolecules actually function *optimally* at high temperatures. How is heat sta-

TABLE 5.1 Presently known upper temperature limits for growth of living organisms

Group	Upper temperature limits (°C)
Animals	
Fish and other aquatic vertebrates[a]	38
Insects	45–50
Ostracods (crustaceans)	49–50
Plants	
Vascular plants	45
Mosses	50
Eukaryotic microorganisms	
Protozoa	56
Algae	55–60
Fungi	60–62
Prokaryotes	
Bacteria	
Cyanobacteria	70–74
Anoxygenic phototrophs	70–73
Chemoorganotrophic/chemolithotrophic Bacteria	95
Archaea	
Chemoorganotrophic/chemolithotrophic Archaea	113

a See a possible exception in Section 16.12 and Figure 16.34.

bility achieved? Amazingly, studies of several thermostable enzymes have shown that they often differ very little in amino acid sequence from an enzyme that catalyzes the same reaction in a mesophile. It appears that a critical amino acid substitution in one or a few locations in the enzyme allows it to fold in a way that is consistent with heat stability. Heat stability of proteins in hyperthermophiles is also improved as a result of the increased number of *salt bridges* (ionic bonds between the positive and negative charges of various amino acids) present and the densely packed highly hydrophobic interiors of the proteins, which naturally resist unfolding in the aqueous milieu.

In addition to enzymes and other proteins in the cell, the protein-synthesizing machinery itself (that is, ribosomes and other constituents) of thermophiles and hyperthermophiles, as well as such structures as the cytoplasmic membrane, need to be heat-stable. Concerning membranes, we mentioned earlier that psychrophiles have membrane lipids rich in *unsaturated* fatty acids, thus making the membranes fluid and functional at low temperatures. Conversely, thermophiles have lipids rich in *saturated* fatty acids, thus allowing the membranes to remain stable and functional at high temperatures. Saturated fatty acids form a much stronger hydrophobic environment than do unsaturated fatty acids, which helps account for the membrane stability. Hyperthermophiles, most of which are Archaea, do not contain fatty acids at all in their membranes but instead have C_{40} hydrocarbons composed of repeating units of the five-carbon compound phytane (∞ Figure 3.20*c*) bonded by ether linkage to glycerol phosphate. In addition, the overall structure of these membranes forms a *lipid monolayer* (∞ Figure 3.21*d*), and this structure is undoubtedly much more heat resilient than the lipid bilayer of Bacteria and Eukarya. We discussed the details of the unique membrane architecture of hyperthermophilic Archaea in Section 3.5 and will consider other aspects of heat stability, including that of DNA, in hyperthermophiles in Section 14.10.

Why are eukaryotes absent from environments with temperatures above about 60°C (Table 5.1)? This most likely involves the stability of organellar membranes, which must remain fairly porous to permit passage of large molecules like ATP and RNA; porous membranes such as these would be more temperature labile than the typical lipid bilayers of prokaryotes (or lipid monolayers of some hyperthermophiles) (∞ Section 3.5). Thus, above about 60°C, the organelles of eukaryotes probably cannot survive and the only life forms observed are prokaryotes.

Biotechnological Aspects of Thermophily

Thermophilic and hyperthermophilic microorganisms are interesting for more than just basic biological reasons. These organisms offer some major advantages for industrial and biotechnological processes, many of which run more rapidly and efficiently at high temperatures. Enzymes from thermophiles and hyperthermophiles are capable of catalyzing biochemical reactions at high temperatures (∞ Section 11.9 and Figure 11.16) and are generally more stable than enzymes from mesophiles, thus prolonging the shelf life of enzyme preparations. A practical example of a heat-stable enzyme of great applied importance is the DNA polymerase isolated from the thermophile *Thermus aquaticus. Taq polymerase*, as this enzyme is known, has been used to automate the repetitive steps involved in the *polymerase chain reaction (PCR)* technique, a method of amplifying specific DNA sequences (∞ Section 10.9). Because *Taq* polymerase does not denature at the high temperatures needed to melt DNA in the PCR method, it is possible to perform several repetitive melting and polymerization steps without having to add fresh DNA polymerase, as would be the case if polymerases from mesophilic microorganisms were used. Several other uses of heat-stable enzymes and other thermostable cell products are known or are being developed for industrial applications.

✓ 5.9 Concept Check

Organisms with growth temperature optima between 45 and 80°C are called *thermophiles*, while those with optima >80°C are called *hyperthermophiles*. These organisms inhabit hot environments up to and including boiling hot springs, as well as undersea hydrothermal vents that can have temperatures in excess of 100°C. Thermophiles and hyperthermophiles produce heat stable macromolecules.

✓ What is the structure of membranes of hyperthermophilic Archaea and why might this structure be useful for growth at high temperature?

✓ What is *Taq polymerase* and why has it been of use to biotechnology?

✓ What is the current upper temperature limit for growth of a prokaryote? Is the organism that grows at this temperature a member of the Bacteria or the Archaea?

5.10

Microbial Growth at Low or High pH

Acidity or alkalinity of a solution is expressed by its **pH** on a scale on which neutrality is pH 7 (Figure 5.18). Those pH values that are less than 7 are said to be *acidic*, and those greater than 7 are *alkaline* (or *basic*). It is important to remember that pH is a *logarithmic function; a*

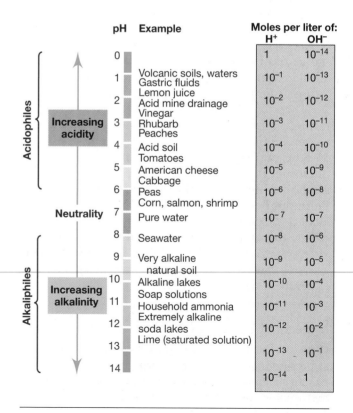

FIGURE 5.18 The pH scale. Note that although some microorganisms can live at very low or very high pH, the cell's internal pH remains near neutrality.

change of 1 pH unit represents a *10-fold* change in hydrogen ion concentration. Thus, vinegar (pH near 2) and household ammonia (pH near 11) differ in hydrogen ion concentration by a billionfold.

pH and Microbial Growth

Each organism has a pH range within which growth is possible and usually has a well-defined pH optimum. Most natural environments have pH values between 5 and 9, and organisms with optima in this range are most common. Only a few species can grow at pH values of less than 2 or greater than 10. Organisms that live at low pH are types of extremophiles called **acidophiles.** Fungi as a group tend to be more acid-tolerant than bacteria. Many fungi grow optimally at pH 5 or below, and a few grow well at pH values as low as 2. Several bacteria are also acidophilic. In fact, some of these bacteria are *obligate* acidophiles, unable to grow at all at neutral pH. Obligately acidophilic bacteria include several species of *Thiobacillus* (∞ Section 13.3) and several genera of Archaea, including *Sulfolobus* and *Thermoplasma* (∞ Sections 14.4 and 14.8).

Thiobacillus species, such as *T. ferrooxidans,* and *Sulfolobus,* exhibit an interesting property related to their acidophilic nature: They oxidize sulfide minerals and produce sulfuric acid. We discuss the role of these organisms in mining processes in Section 16.19. Probably the most critical factor for obligate acidophily is the cytoplasmic membrane. When the pH is raised to neutrality, the cytoplasmic membrane of obligately acidophilic bacteria actually dissolves and the cells lyse, suggesting that high concentrations of hydrogen ions are required for membrane stability. The most acidophilic prokaryote known, *Picrophilus oshimae,* is also a thermophile (growth temperature optimum, 60°C) and grows optimally at pH 0.7; above pH 4 cells of *P. oshimae* lyse rapidly!

A few extremophiles have very high pH optima for growth, sometimes as high as pH 10–11, and are known as **alkaliphiles.** Alkaliphilic microorganisms are usually found in highly basic habitats such as soda lakes and high carbonate soils. However, most alkaliphilic prokaryotes studied have been aerobic nonmarine bacteria, and many are *Bacillus* species. Some extremely alkaliphilic bacteria are also halophilic (salt-loving), and most of these are Archaea (∞ Section 14.2). Some alkaliphiles have found industrial uses because they produce hydrolytic enzymes, such as proteases and lipases, which function well at alkaline pH and are used as supplements for household detergents (∞ Section 11.9).

Alkaliphiles are also interesting because of the bioenergetics problems they face living at such high pH. From studies of an alkaliphilic *Bacillus* species it has been shown that a Na^+ gradient (instead of the usual proton motive force) supplies the energy for transport and motility but that a proton motive force can indeed be established and is responsible for driving respiratory ATP synthesis. Exactly how this occurs is an interesting problem in alkaliphile research today.

Finally, concerning pH and microbial growth, it should be emphasized that despite the pH requirements of a particular organism for a specific pH for growth, the optimal growth pH represents the pH of the *extracellular* environment only; the *intracellular* pH must remain near neutrality in order to prevent destruction of acid- or alkali-labile macromolecules in the cell. In extreme acidophiles or extreme alkaliphiles the intracellular pH may vary by several pH units from neutrality, but for the majority of microorganisms, whose pH optimum for growth is between pH 6 and 8 (referred to as **neutrophiles**), the cytoplasm remains neutral or very nearly so (Figure 5.18).

In the previously mentioned acidophile *P. oshimae,* the internal pH has been measured at pH 4.6, and in extreme alkaliphiles an intracellular pH of as high as 9.5 has been measured. If these are not the lower and upper limits of cytoplasmic pH, respectively, they must be very close to the limits, since macromolecular stability would almost certainly be compromised above or below these pH values.

Buffers

In a batch culture the pH can change during growth as the result of metabolic reactions that consume or produce acidic or basic substances. Thus, chemicals called *buffers* are frequently added to microbial culture media to keep the pH relatively constant. Such pH buffers generally work over only a narrow pH range; hence different buffers must be used to buffer at different pH values. For near-neutral pH ranges (pH 6–7.5), phosphate, usually supplied as KH_2PO_4, is an excellent buffer. Many other buffers for use in microbial growth media or for the assay of enzymes extracted from microbial cells are available, and the best buffering system for one organism or enzyme may be considerably different from that of another. Thus, the optimal buffer for use in a particular situation must usually be determined empirically, although for assaying enzymes *in vitro,* a certain buffer that works well in an assay for the enzyme from one organism will usually work well for assaying this same enzyme from other organisms.

✓ 5.10 Concept Check

The acidity or alkalinity of an environment can greatly affect microbial growth. Some organisms have evolved to grow best at low or high pH, but most organisms grow best between pH 6 and 8. The internal pH of a cell must stay relatively close to neutrality even though the external pH is highly acidic or basic.

✓ What is the increase in concentration of protons (H^+) when going from pH 7 to pH 4?
✓ What are *buffers* and why are they needed?

5.11

Osmotic Effects on Microbial Growth

Water is the solvent of life. All organisms require water, and water availability is an important factor affecting the growth of microorganisms in nature. Water availability not only depends on the water content of an environment, that is, how moist or dry a solid microbial habitat may be, but is also a function of the concentration of solutes such as salts, sugars, or other substances that are dissolved in water. This is because dissolved substances have an affinity for water, which makes the water associated with solutes unavailable to organisms.

Water Activity, Osmosis, and Halophiles

Water availability is generally expressed in physical terms such as **water activity.** Water activity, abbreviated a_w, is a ratio of the vapor pressure of the air in equilibrium with a substance or solution to the vapor pressure of pure water. Thus values of a_w vary between 0 and 1 and some representative values are given in Table 5.2. Water activities in agricultural soils generally range between 0.90 and 1.00.

Water diffuses from a region of high water concentration (low solute concentration) to a region of lower water concentration (higher solute concentration) in the process of *osmosis*. In most cases, the cytoplasm of a cell has a higher solute concentration than the environment, so water tends to diffuse into the cell, and the cell is said to be in *positive water balance*. However, when a cell is in an environment of low water activity, there is a tendency for water to flow out of the cell.

In nature, osmotic effects are of interest mainly in habitats with high concentrations of salts. Seawater contains about 3% sodium chloride (NaCl) plus small amounts of many other minerals and elements. Microorganisms found in the sea usually have a specific requirement for the sodium ion in addition to growing optimally at the water activity of seawater (Figure 5.19). Such organisms are called

halophiles. The growth of halophiles requires at least some NaCl, but the optimum varies with the organism; thus the terms *mild halophile* and *moderate halophile* are used to describe halophiles with low (1–6%) and moderate (6–15%) NaCl requirements, respectively (Figure 5.19).

Most microorganisms are unable to cope with environments of very low water activity and either die or become dehydrated and dormant under such conditions. **Halotolerant** organisms can tolerate some reduction in the a_w of their environment but generally grow best in the absence of the added solute (Figure 5.19). By contrast, some organisms thrive at very low water activity, and these organisms are of interest not only from the standpoint of their adaptation to life under these conditions but also from an applied standpoint, such as that of the food industry, where solutes such as salt and

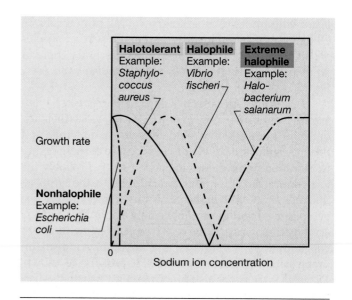

FIGURE 5.19   Effect of sodium ion concentration on growth of microorganisms of different salt tolerances. The optimum NaCl concentration for marine microorganisms such as *V. fischeri* is about 3%; for extreme halophiles, it is between 15 and 30%, depending on the organism.

TABLE 5.2	Water activity of several substances	

Water activity, a_w	Material	Examples of organisms growing at stated water activity
1.000	Pure water	*Caulobacter, Spirillum*
0.995	Human blood	*Streptococcus, Escherichia*
0.980	Seawater	*Pseudomonas, Vibrio*
0.950	Bread	Most gram-positive rods
0.900	Maple syrup, ham	Gram-positive cocci such as *Staphylococcus*
0.850	Salami	*Saccharomyces rouxii* (yeast)
0.800	Fruit cake, jams	*Saccharomyces bailii, Penicillium* (fungus)
0.750	Salt lakes, salted fish	*Halobacterium, Halococcus*
0.700	Cereals, candy, dried fruit	*Xeromyces bisporus* and other xerophilic fungi

sucrose are commonly used as food additives to inhibit microbial growth. Organisms capable of growth in very salty environments are called **extreme halophiles** (Figure 5.19); extreme halophiles generally require 15–30% NaCl, depending on the species, for optimum growth (⌘ Section 14.2). Organisms able to live in environments high in sugar are called /osmophiles,/ and those able to grow in very dry environments (made dry by lack of water) are called \xerophiles.\ Examples of these various organisms are given in Table 5.2.

Compatible Solutes

How do organisms grow under conditions of low water activity? When an organism grows in a medium with a low water activity, it can obtain water from its environment only by increasing its *internal* solute concentration. An increase in internal solute concentration can be accomplished by either pumping inorganic ions into the cell from the environment or synthesizing or concentrating an organic solute. Organisms are known that employ either of these mechanisms, and several examples are given in Table 5.3.

The solute used inside the cell for adjustment of cytoplasmic water activity must be noninhibitory to biochemical processes within the cell; such compounds are called **compatible solutes.** Several different compatible solutes are known in microorganisms (Table 5.3 and Figure 5.20). These substances are all highly water-soluble

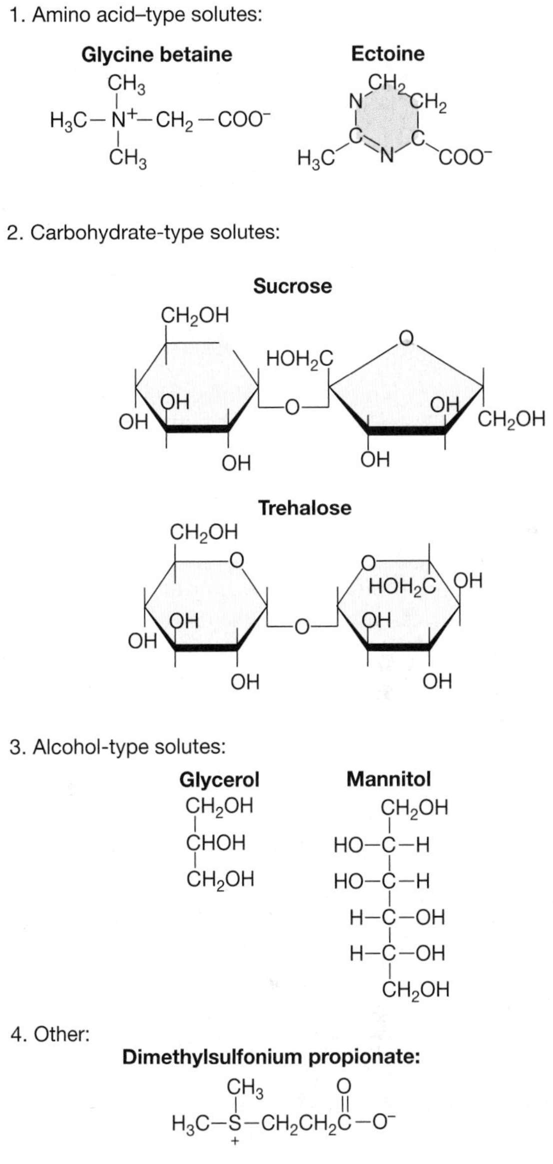

FIGURE 5.20 Structures of some common compatible solutes in microorganisms. The structures of glutamate and proline, other common solutes, were shown in Figure 2.12. The formal name of ectoine is 1,4,5,6-tetrahydro-2-methyl-4-pyrimidine carboxylate. Note that all the compounds shown here are very water-soluble.

1. Amino acid–type solutes:

Glycine betaine　　　　**Ectoine**

2. Carbohydrate-type solutes:

Sucrose

Trehalose

3. Alcohol-type solutes:

Glycerol　　**Mannitol**

4. Other:

Dimethylsulfonium propionate:

TABLE 5.3	Compatible solutes of microorganisms	
Organism	**Major solute(s) accumulated**	**Minimum a_w for growth**
Bacteria, nonphototrophic	Glycine betaine, proline (mainly gram-positive), glutamate (mainly gram-negative)	0.97–0.90
Freshwater cyanobacteria	Sucrose, trehalose	0.98
Marine cyanobacteria	α-Glucosylglycerol	0.92
Marine algae	Mannitol, various glycosides, proline, dimethylsulfonium propionate	0.92
Salt lake cyanobacteria	Glycine betaine	0.90–0.75
Halophilic anoxygenic phototrophic Bacteria (*Ectothiorhodospira/Halorhodospira* and *Rhodospirillum* species)	Glycine betaine, ectoine, trehalose	0.90–0.75
Extremely halophilic Archaea (for example, *Halobacterium*) and some Bacteria (for example, *Haloanaerobium*)	KCl	0.75
Dunaliella (halophilic green alga)	Glycerol	0.75
Xerophilic yeasts	Glycerol	0.83–0.62
Xerophilic filamentous fungi	Glycerol	0.72–0.61

sugars or sugar alcohols, other alcohols, amino acids or their derivatives, or in the case of extremely halophilic Archaea and a very few extremely halophilic Bacteria, potassium ions (K^+) (Table 5.3). Compatible solutes are either synthesized by the microorganisms directly or in some cases (such as glycine betaine or K^+) accumulated from the environment. The concentration of compatible solutes in the cell is a function of the level of external solutes, and in each organism the maximal amount of compatible solute(s) made or that can be accumulated is a genetically directed characteristic; this results in different organisms tolerating different ranges of water potential (Tables 5.2 and 5.3). Thus, nonhalotolerant, halotolerant, halophilic, and extremely halophilic microorganisms are to a major extent defined by their genetic capacity to produce or accumulate compatible solutes.

Gram-positive cocci of the genus *Staphylococcus* are notoriously halotolerant (in fact, a common isolation procedure for them is to use media containing 7.5% NaCl), and these organisms use the amino acid *proline* as a compatible solute. *Glycine betaine* is a derivative of the amino acid glycine in which the protons on the amino group are replaced by three methyl groups; this leaves a permanent positive change on the N atom, which increases its solubility. Glycine betaine is widely distributed as a compatible solute, especially among halophilic Bacteria and cyanobacteria (Table 5.3). Some extremely halophilic bacteria produce a novel compatible solute called *ectoine*, which is a derivative of the cyclic amino acid proline (Figure 5.20). A variety of glycosides are produced by marine algae, but with rare exception they accumulate only in low amounts because the cells are not very halophilic. Xerophilic yeasts and halophilic green algae produce mainly *glycerol* as a compatible solute. Other examples of compatible solutes are listed in Table 5.3, and structures are shown in Figure 5.20.

✓ 5.11 Concept Check

Water can become limiting to an organism when the dissolved solute concentration in its environment increases. To counteract this situation organisms produce or accumulate intracellular compatible solutes that function to maintain the cell in positive water balance. Some microorganisms have evolved to grow best at reduced water potential, and some even require high levels of salts in order to grow.

✓ What is the a_w of pure water?
✓ What is a *compatible solute* and why is it needed?

5.12

Oxygen as a Factor in Microbial Growth

Microorganisms vary in their need for, or tolerance of, oxygen. In fact, microorganisms can be divided into several groups depending on the effect of oxygen, as outlined in Table 5.4. **Aerobes** are species capable of growth at full oxygen tensions (air is 21% O_2), and many can even tolerate elevated concentrations of oxygen (hyperbaric oxygen). **Microaerophiles,** by contrast, are aerobes that can use O_2 only when it is present at levels reduced from that in air, usually because of their limited capacity to respire or because they contain some oxygen-sensitive molecule such as an oxygen-labile enzyme. Many aerobes are **facultative,** meaning that, under the appropriate nutrient and culture conditions, they can grow under *either* aerobic or anaerobic conditions.

Organisms that lack a respiratory system cannot use oxygen as a terminal electron acceptor. Such organisms are called **anaerobes,** but there are two kinds of anaerobes: **aerotolerant anaerobes,** which can tolerate oxygen and grow in its presence even though they cannot use it, and **obligate** (or **strict**) **anaerobes,** which

TABLE 5.4	Oxygen relationships of microorganisms			
Group	**Relationship to O_2**	**Type of metabolism**	**Example**	**Habitat**[a]
Aerobes				
Obligate	Required	Aerobic respiration	*Micrococcus luteus*	Skin, dust
Facultative	Not required, but growth better with O_2	Aerobic, anaerobic respiration, fermentation	*Escherichia coli*	Mammalian large intestine
Microaerophilic	Required but at levels lower than atmospheric	Aerobic respiration	*Spirillum volutans*	Lake water
Anaerobes				
Aerotolerant	Not required, and growth no better when O_2 present	Fermentation	*Streptococcus pyogenes*	Upper respiratory tract
Obligate	Harmful or lethal	Fermentation or anaerobic respiration	*Methanobacterium formicicum*	Sewage sludge digestors, anoxic lake sediments

a Listed are typical habitats of the example organism.

are killed by oxygen (Table 5.4). The reason obligate anaerobes are killed by oxygen is probably because they are unable to detoxify some of the products of oxygen metabolism (see later discussion). When oxygen is reduced, several toxic products such as hydrogen peroxide (H_2O_2), superoxide (O_2^-), and hydroxyl radical ($OH\cdot$) are formed. Many obligate anaerobes are rich in flavin enzymes (∞ Section 4.10), which can react spontaneously with O_2 to yield these toxic products. By contrast, aerobes have enzymes that decompose toxic oxygen products, whereas anaerobes seem to lack all or some of these enzymes.

Microbial Culture and the Effects of Oxygen

For the growth of many aerobes, it is necessary to provide extensive aeration. This is because O_2 is only poorly soluble in water and the O_2 used up by the organisms during growth is not replaced fast enough by diffusion from the air. Forced aeration of cultures is therefore frequently desirable and can be achieved either by vigorously shaking the flask or tube on a shaker or by bubbling sterilized air into the medium through a fine glass tube or porous glass disc. Usually aerobes grow much better with forced aeration than when O_2 is provided by simple diffusion.

For anaerobic culture, the problem is to *exclude*, not provide, oxygen. And because oxygen is present in the air, special methods are needed to culture anaerobic microorganisms. Obligate anaerobes vary in their sensitivity to oxygen, and a number of procedures are available for reducing the O_2 content of cultures—some simple and suitable mainly for less sensitive organisms, others more complex but necessary for growth of strict anaerobes.

Bottles or tubes filled completely to the top with culture medium and provided with tightly fitting stoppers provide anoxic conditions for organisms not too sensitive to small amounts of oxygen. It is also possible to add a chemical called a *reducing agent* that reacts with oxygen and reduces it to H_2O. A good example is *thioglycolate*, which is added to a medium called *thioglycolate broth*, commonly used to test an organism's requirements for O_2 (Figure 5.21). After thioglycolate reacts with oxygen throughout the tube, oxygen can penetrate only near the top of the tube where the medium contacts air. Obligate aerobes grow only at the top of such tubes. Facultative organisms grow throughout the tube but best near the top. Microaerophiles grow near the top but not right at the top (Figure 5.21). Anaerobes grow only near the bottom of the tube, where oxygen cannot penetrate (Figure 5.21). A redox indicator dye called *resazurin* is added to the medium because the dye changes color in the presence of oxygen and thereby indicates the degree of penetration of oxygen into the medium (Figure 5.21).

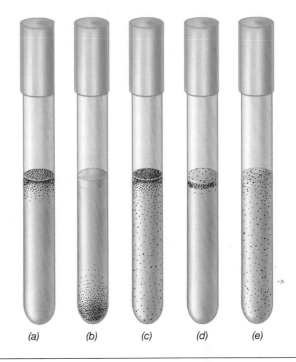

FIGURE 5.21 Aerobic, anaerobic, facultative, microaerophilic, and aerotolerant anaerobe growth, as revealed by the position of microbial colonies (depicted here as black dots) within tubes of a culture medium. A small amount of agar has been added to keep the liquid from becoming disturbed and the redox dye, resazurin, which is pink when oxidized and colorless when reduced, is added as a redox indicator. (a) Oxygen penetrates only a short distance into the tube, so obligate aerobes grow only at the surface. (b) Anaerobes, being sensitive to oxygen, grow only away from the surface. (c) Facultative aerobes are able to grow in either the presence or the absence of oxygen and thus grow throughout the tube. (d) Microaerophiles grow away from the most oxic zone. (e) Aerotolerant anaerobes grow throughout the tube.

To remove all traces of O_2 for the culture of anaerobes, it is possible to place an O_2-consuming gas in a jar holding the tubes or plates. One of the simplest devices for this is an *anaerobic jar*, a heavy-walled jar with a gastight seal within which tubes, plates, or other containers to be incubated are placed. The air in the jar is replaced with a mixture of H_2 and CO_2, and in the presence of a chemical catalyst the traces of O_2 left in the vessel or culture medium are consumed, thus leading to anoxic conditions (∞ Section 21.1 and Figure 21.6).

For strict anaerobes, such as methanogenic bacteria, it is necessary not only to carefully remove all traces of O_2 but also to carry out all manipulations of cultures in an anoxic atmosphere, as these organisms can be killed by even a brief exposure to O_2. In these cases, a culture medium is first boiled to render it oxygen-free, and then a reducing agent such as H_2S is added and the mixture is sealed under an oxygen-free gas. All manipulations are carried out under a jet of oxygen-free hy-

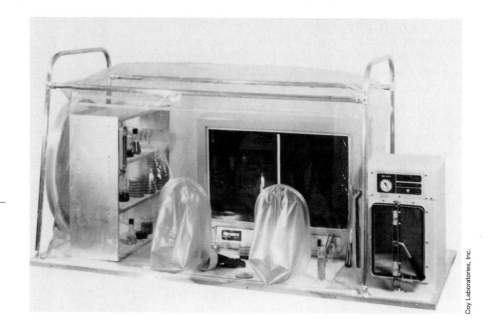

FIGURE 5.22 Anoxic glove bag for manipulating and incubating cultures under anoxic conditions. The airlock on the right, which can be evacuated and filled with oxygen-free gas, serves as a port for adding and removing materials to and from the glove bag.

drogen or nitrogen gas that is directed into the culture vessel when it is open, thus driving out any O_2 that might enter. For extensive research on anaerobes, special boxes fitted with gloves, called *anoxic glove boxes (bags)*, permit work with open cultures in completely anoxic atmospheres (Figure 5.22).

Toxic Forms of Oxygen

Oxygen is a powerful oxidant and an excellent electron acceptor for respiration (∞ Section 4.10). Oxygen in its normal ground state is referred to as **triplet oxygen.** However, one major form of toxic oxygen is called **singlet oxygen,** a higher energy form of oxygen in which outer shell electrons surrounding the nucleus become highly reactive and are able to carry out a variety of spontaneous and undesirable oxidations within the cell. Singlet oxygen is produced both photochemically and biochemically, the latter through the action of various peroxidase enzymes. Organisms that frequently encounter singlet oxygen, such as airborne bacteria and phototrophic microorganisms, often contain pigments called **carotenoids,** which function to convert singlet oxygen to nontoxic forms (∞ Section 15.4).

Other highly toxic forms of oxygen include **superoxide anion (O_2^-), hydrogen peroxide (H_2O_2),** and **hydroxyl radical (OH·),** all of which are produced as inadvertent by-products during the reduction of O_2 to H_2O in respiration (Figure 5.23). Flavoproteins, quinones, thiols, and iron–sulfur proteins (∞ Section 4.10) can also carry out the reduction of O_2 to O_2^-. Superoxide is highly reactive and can oxidize virtually any organic compound in the cell, including macromolecules. Peroxides such as H_2O_2 can damage cell components but are generally not as toxic to the cell as superoxide or hydroxyl radical. The latter is the most reactive of all toxic

$$O_2 + e^- \longrightarrow O_2^- \quad \textbf{Superoxide}$$
$$O_2^- + e^- + 2\,H^+ \longrightarrow H_2O_2 \quad \textbf{Hydrogen peroxide}$$
$$H_2O_2 + e^- + H^+ \longrightarrow H_2O + OH· \quad \textbf{Hydroxyl radical}$$
$$OH· + e^- + H^+ \longrightarrow H_2O \quad \textbf{Water}$$

$$\boxed{\textbf{Overall: } O_2 + 4\,e^- + 4\,H^+ \longrightarrow 2\,H_2O}$$

FIGURE 5.23 Four-electron reduction of O_2 to water by stepwise addition of electrons. All the intermediates formed are reactive and toxic to cells.

oxygen species and can instantly oxidize any organic substance in the cell. However, the hydroxyl radical is only a transient species in most cells because the major source of OH· is ionizing radiation, to which most cells are not commonly exposed. Small amounts of OH· can also be produced from H_2O_2 (Figure 5.23), but when peroxides do not accumulate in the cell (because of the action of catalase, which is discussed later), this source of hydroxyl radical is virtually eliminated. We will see later that various toxic oxygen species can be produced by certain immune cells in the animal body and used to kill microbial invaders (∞ Section 20.3).

Enzymes That Destroy Toxic Oxygen

With such an array of toxic oxygen derivatives, it is perhaps not surprising that organisms have evolved enzymes that destroy toxic oxygen products (Figure 5.24). The most common enzyme in this category is **catalase,** which attacks hydrogen peroxide; the activity of catalase is illustrated in Figures 5.24*a* and 5.25. Another enzyme that destroys hydrogen peroxide is **peroxidase** (Figure 5.24*b*), which differs from catalase in requiring a reductant, usually NADH, producing H_2O_2 as a prod-

(a) Catalase:
$$H_2O_2 + H_2O_2 \rightarrow 2\,H_2O + O_2$$

(b) Peroxidase:
$$H_2O_2 + NADH + H^+ \rightarrow 2\,H_2O + NAD^+$$

(c) Superoxide dismutase:
$$O_2^- + O_2^- + 2\,H^+ \rightarrow H_2O_2 + O_2$$

(d) Superoxide dismutase/catalase in combination:
$$4\,O_2^- + 4\,H^+ \rightarrow 2\,H_2O + 3\,O_2$$

FIGURE 5.24 Enzymes that destroy toxic oxygen species. (a) Catalases and (b) peroxidases are porphyrin-containing proteins, although some flavoproteins may consume toxic oxygen species as well. (c) Superoxide dismutases are metal-containing proteins, the metals being copper and zinc, manganese, or iron. (d) Combined reaction of superoxide dismutase and catalase.

T. D. Brock

FIGURE 5.25 Method for testing a microbial culture for the presence of catalase. A heavy loopful of cells from an agar culture was mixed on a slide with a drop of 30% hydrogen peroxide. The immediate appearance of bubbles is indicative of the presence of catalase. The bubbles are O_2 produced by the reaction $H_2O_2 + H_2O_2 \rightarrow 2\,H_2O + O_2$.

uct. Superoxide is destroyed by the enzyme **superoxide dismutase** (Figure 5.24*c*), which combines two molecules of superoxide to form one molecule of hydrogen peroxide and one molecule of oxygen. Superoxide dismutase and catalase working together can thus bring about the conversion of superoxide back to oxygen (Figure 5.24).

Aerobes and facultative aerobes generally contain both superoxide dismutase and catalase, although a few obligate aerobes lack catalase. Superoxide dismutase is indispensable to aerobic cells, and the low levels (or complete absence) of this enzyme in obligate anaerobes is likely the major reason why oxygen is toxic to them. Some aerotolerant anaerobes, such as lactic acid bacteria, also lack superoxide dismutase, but they are some-

how able to use protein-free Mn^{2+} complexes to carry out the dismutation of O_2^- to H_2O_2 and O_2. Such a reaction may have functioned as a primitive form of superoxide dismutase in ancient organisms. This is supported by the fact that all known SODs contain a metal cofactor, usually Mn^{2+}, but also Fe^{2+} or Cu^{2+} plus Zn^{2+}, at the enzyme's active site.

Anaerobic Microorganisms

Considering our discussion of toxic forms of oxygen and the fact that anaerobes frequently lack the means to defend against them, the student may get the impression that anaerobic organisms are quite rare. However, nothing could be further from the truth. Anoxic environments abound on Earth and include muds and other sediments of lakes, rivers, and oceans; bogs and marshes; waterlogged soils; canned foods; intestinal tracts of animals; the oral cavity of animals, especially around the teeth; certain sewage treatment systems; deep underground areas such as oil pockets; and some underground waters (see Chapter 16 for descriptions of some of these habitats). In most of these habitats, anoxic conditions and the accompanying low reduction potentials are due to the activities of organisms, mainly bacteria, that consume oxygen during respiration and produce highly reducing substances such as H_2 and H_2S. If no replacement oxygen is available, the habitat becomes anoxic.

So far as is known, obligate anaerobiosis occurs in three groups of microorganisms: a wide variety of prokaryotes, a few fungi, and a few protozoa. One of the best-known groups of obligately anaerobic Bacteria belongs to the genus *Clostridium*, a group of gram-positive spore-forming rods. Clostridia are widespread in soil, lake sediments, and intestinal tracts and are often responsible for spoilage of canned foods (∞ Sections 13.19 and 24.10). Other obligately anaerobic bacteria are found among the methanogens and many other archaeans, the sulfate-reducing and homoacetogenic bacteria (∞ Chapters 13–15), and many of the bacteria that inhabit the animal gut (∞ Section 19.4). Among obligate anaerobes, however, the sensitivity to oxygen varies greatly; some organisms are able to tolerate traces of oxygen, whereas others are not.

Growth and Molecular Biology

You have now completed your study of microbial growth and the major environmental factors that control it, and we are ready to examine the molecular biological processes that occur during microbial growth: DNA replication, transcription, and translation.

These fundamental molecular processes must be coordinated in such a way that at the time of cell division each daughter cell receives an equal share of newly synthesized cell materials and contains at least one copy of every essential protein and nucleic acid needed to exist as an independent entity. These molecular processes occur in similar ways in all organisms. Thus, a psychrophilic organism inhabiting Antarctic sea ice (Figure 5.14) and a hyperthermophile growing in boiling water (Figure 5.16) employ similar mechanisms in processing genetic information, but the actual genes in the two organisms are very different; these genetic differences are at the heart of microbial diversity, and will be considered in Chapters 13, 14, and 17. However a detailed understanding of the coding functions of a cell and how genetics influences cellular evolution is necessary to bring the diversity of microbial life we have seen in this chapter into clear focus, and we embark on this journey now.

✓ 5.12 Concept Check

Aerobes require oxygen to live, whereas anaerobes do not and may even be killed by O_2. Several toxic forms of oxygen can be formed in the cell, but enzymes are present that can neutralize most of them. Special methods may be necessary to grow strictly aerobic or anaerobic bacteria.

✓ What is a *facultative aerobe*?
✓ What is the chemical structure of superoxide anion?
✓ How does a reducing agent work?
✓ How does superoxide dismutase protect a cell?

REVIEW QUESTIONS

1. What is the difference between the growth rate constant (k) of an organism and its generation time (g)?

2. Why is it useful to plot growth data from a growing microbial culture?

3. Describe the growth cycle of a population of bacterial cells from the time this population is first inoculated into fresh medium. How can the growth pattern differ when it is measured by total count or by viable count?

4. Describe one direct and one indirect method by which microbial growth can be measured. Make sure that the methods you choose agree with your definition.

5. Describe briefly the process by which a single cell develops into a visible colony on an agar plate. With this explanation as a background, describe the principle behind the viable count method.

6. How can a chemostat regulate growth rate and cell numbers independently?

7. Examine the graph describing the relationship between growth rate and temperature (Figure 5.12). Give an explanation, in biochemical terms, of why the optimum temperature for an organism is usually closer to its maximum than its minimum.

8. Would you expect to find a psychrophilic microorganism alive in a hot spring? Why? It is frequently possible to isolate thermophilic microorganisms from cold-water environments. Give an explanation of how this can be.

9. Concerning the pH of the environment and of the cell, in what ways are acidophiles and alkaliphiles different? In what ways are they similar?

10. Write an explanation in molecular terms for how a cell of a halophile is able to make water molecules flow *into* itself.

11. List three *chemical classes* of compatible solutes produced by various microorganisms. List at least two things they all have in common.

12. Contrast an aerotolerant and an obligate anaerobe in terms of sensitivity and ability to grow in the presence of oxygen (O_2). How does an aerotolerant anaerobe differ from a microaerophile?

13. Compare and contrast the enzymes *catalase* and *superoxide dismutase* from the following points of view: substrates, oxygen products, organisms containing them, role in oxygen tolerance of the cell.

APPLICATION QUESTIONS

1. Starting with four bacterial cells per milliliter in a rich nutrient medium, with a 1-hr lag phase and a 20-min generation time, how many cells will there be in 1 l of this culture after 1 hr? After 2 hr? After 2 hr if one of the initial four cells was dead?

2. Calculate g and k in a growth experiment in which a medium was inoculated with 5×10^6 cells/ml of *Escherichia coli* cells and, following a 1-hr lag, grew exponentially for 5 hr, after which the population was 5.4×10^9 cells/ml.

3. Return to Chapter 2 and locate a figure that best describes what happens to a cell of a mesophile like *Escherichia coli* when placed in a culture medium at 80°C. Contrast this with a figure from Chapter 5 that best describes what would happen if cells of *Pyrolobus fumarii* were placed under the same conditions. Describe why neither organism would grow.

4. From what you know concerning growth at reduced water potential and the phenomenon of compatible solutes, describe what would happen if you took a cell of the extreme halophile *Halobacterium salinarum* from its growth medium (containing 25% NaCl) and placed it in distilled water. Also, predict what a cell of *Escherichia coli*, a typical nonhalophile, would do if you did the reverse experiment (distilled water to 25% NaCl).

By the early 1950s it had become clear that DNA was the genetic material in cells and that DNA was in some way decoded to give rise to proteins, the cell's enzymatic machinery. One important step in the understanding of this process was the discovery that RNA played several important roles in protein synthesis. By the end of the 1950s it was clear that RNA was transcribed from DNA and that some RNA was a short-lived message that was translated to yield proteins. It was further learned that translation itself occurred on ribosomes. By the early 1960s it had proved possible to isolate polyribosomes: a series of ribosomes simultaneously translating a single messenger RNA, as shown here. Although the overall flow of genetic information had been deciphered by the end of the 1960s, many astonishing details, such as the fact that some RNA has catalytic activity, remained to be discovered.

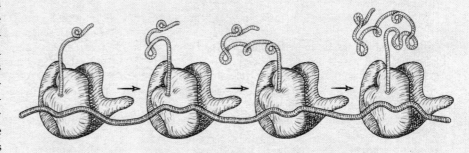

CHAPTER 6

Principles of Microbial Molecular Biology

6.1   Macromolecules and Genetic Information   165
6.2   DNA Structure: The Double Helix   167
6.3   DNA Structure: Supercoiling   177
6.4   Genetic Elements   179
6.5   Restriction and Modification of DNA   182
6.6   DNA Replication   184
6.7   Transcription: The Basic Process in Bacteria   191
6.8   Transcription: Other Patterns and Inhibition   195
6.9   RNA Processing and Ribozymes   197
6.10 The Genetic Code   199
6.11 Transfer RNA   201
6.12 Translation: The Process of Protein Synthesis   204
6.13 Translation: Alternatives and Errors   208

Working Glossary

Anticodon a sequence of three bases in a tRNA molecule that base-pairs with a codon during protein synthesis

Aminoacyl-tRNA synthetases a group of enzymes each one of which catalyzes the attachment of the correct amino acid to a tRNA

Antiparallel in reference to double-stranded DNA, one strand runs $5' \rightarrow 3'$ and other $3' \rightarrow 5'$

Chromosome a genetic element, usually circular in prokaryotes and linear in eukaryotes, carrying genes essential to cellular function

Codon a sequence of three bases in mRNA that encodes an amino acid

DNA polymerase an enzyme that synthesizes a new strand of DNA in the $5' \rightarrow 3'$ direction using an antiparallel DNA strand as a template

Exon the coding DNA sequences in a split gene (contrast with intron)

Gene a segment of DNA specifying a protein (via mRNA), a tRNA, or an rRNA

Genome the total complement of genes contained in a cell or virus

Hybridization formation of a duplex nucleic acid molecule with strands derived from different sources by complementary base pairing

Intron the intervening noncoding DNA sequences in a split gene (contrast with exon)

Messenger RNA (mRNA) an RNA molecule that contains the genetic information necessary to encode a particular protein

Molecular chaperones a group of proteins that help other proteins fold or refold from a partially denatured state

Operon a cluster of genes whose expression is controlled by a single operator

Primary transcript an unprocessed RNA molecule that is the direct product of transcription

Primer a molecule (usually a polynucleotide) to which DNA polymerase can attach the first deoxyribonucleotide during DNA replication

Promoter a site on DNA to which RNA polymerase can bind and begin transcription

Replication synthesis of DNA using DNA as a template

Restriction enzyme an enzyme that recognizes and makes double-stranded breaks at specific DNA sequences

Ribosomal RNA (rRNA) types of RNA found in the ribosome; some participate actively in the process of protein synthesis

Ribosome a cytoplasmic particle composed of ribosomal RNA and protein, which is a central part of the protein-synthesizing machinery of the cell

Ribozyme an RNA molecule that can catalyze chemical reactions

RNA polymerase an enzyme that synthesizes RNA in the $5' \rightarrow 3'$ direction using an antiparallel DNA strand as a template

RNA processing the conversion of a precursor RNA to its mature form

Semiconservative replication DNA synthesis yielding new double helices, each consisting of one parental and one progeny strand

Transcription the synthesis of RNA using a DNA template

Transfer RNA (tRNA) an adaptor molecule used in translation that has specificity for both a particular amino acid and for one or more codons

Translation the synthesis of protein using the genetic information in messenger RNA as a template

We now begin a study of the flow of information in microorganisms that will extend over the next five chapters. As we noted in Chapter 1, two hallmarks of life are *energy transformation* and *information flow*. In Chapter 4 we dealt with the problem of energy transformation: *metabolism*. Now we deal with the problem of information flow: *genetics*. **Genetics** is the discipline that deals with the mechanisms by which traits are passed from one organism to another and how they are expressed.

In the present chapter we discuss how genetic information is organized and expressed in cells, and in the next chapter we deal with how this expression is regulated. Subsequent chapters concern the genetic information of some noncellular forms, the transfer of genetic information, and how genetic information can be manipulated.

Since biological information flow is the basis of cellular function, genetics is a major research tool in attempts to understand the molecular mechanisms by which cells function. The study of genetics at the molecular level is also central to an understanding of the variability of organisms and the evolution of species.

Understanding how information is transmitted through biological systems also has tremendous practical applications. For instance, genetics provides us with techniques to specifically modify organisms for our own purposes. Many important recent advances in agriculture and industry are dependent on this technology. Medicine is also becoming increasingly reliant on molecular biology and genetics for diagnosis and treatment of disease. An understanding of the processes of biological information flow in a cell not only allows us to understand how normal organisms function, but also how these functions are altered by disease. Armed with such knowledge, scientists are developing not only methods for controlling the important infectious diseases but also genetic-based cures for metabolic disorders once thought incurable. We will have much to say about the application of genetics (including the genetics of microorganisms) to human affairs in subsequent chapters.

6.1

Macromolecules and Genetic Information

All the processes that take place in the cell involve molecules. Many of the molecules involved in the steps of genetic information flow are very large; they are called **macromolecules.** However, long before we knew what these molecules were, or even what steps in the flow might be, it was clear that there was a functional unit of genetic information. This unit has come to be called the **gene.**

What Is a Gene and What Is Its Function?

An oversimplified definition of a gene is an entity that specifies the structure of a single polypeptide or *protein* chain. We discussed the chemistry of proteins in Chapter 2 and noted that they consist of one or more polypeptide chains. A polypeptide is composed of a series of amino acids connected in peptide linkage. There are usually 20 different amino acids present in proteins, and a single protein molecule typically has several hundred amino acid residues (∞ Sections 2.6–2.8). The gene is the element of information that specifies the *sequence* of amino acids of the protein. Genes are stored information, whereas proteins are the cell's functional entities.

In all cells the genes themselves are composed of *deoxyribonucleic acid* (DNA). The information in the gene is present as the sequences of bases in the DNA. Like protein, DNA is a macromolecule. Interestingly, the information stored in the DNA specifies the sequence of a protein only through the intermediary of another macromolecule, *ribonucleic acid* (RNA). RNA can serve either as a true informational intermediate (a messenger) or in some cases as a more active part of the cell's machinery. Because all three of these molecules, DNA, RNA, and protein, contain biological information in their sequences, they are often called **informational macromolecules** (∞ Section 2.2).

In DNA and RNA the information is encoded in the *base sequence* of the purine and pyrimidine bases of the polynucleotide chain. When we discuss the information content of a nucleic acid, we thus speak of the *coding* properties of this material. The amino acid sequence of the polypeptide is *coded* by the sequence of purine and pyrimidine bases within the nucleic acid, with *three* bases encoding a single amino acid. We will discuss this coding function in detail in this chapter.

The Steps in Information Flow

When a cell divides and forms two cells, all types of informational macromolecules are synthesized. The molecular processes underlying genetic information flow can be divided into three stages, which are described briefly here (Figure 6.1).

1. *Replication.* The DNA molecule, which serves as the cell's genetic material, is a **double helix** of two long chains (∞ Section 2.5). During replication, DNA, containing the master genetic blueprint, duplicates. The products of DNA replication are two double helices, the two strands thus becoming four strands.

2. *Transcription.* DNA does not function directly in protein synthesis but through an RNA intermediate. The transfer of the information to RNA is called **transcription,** and the RNA molecule carrying the information to encode a protein is called **messenger RNA** (mRNA). In most cases, at any particular location on the chromosome, only one strand of the DNA is transcribed, and the information of this strand is then contained in the mRNA. Some regions of DNA that are transcribed do not encode proteins but rather contain information for other types of RNA, such as **transfer RNA** (tRNA) and **ribosomal RNA** (rRNA). Therefore, we must expand our definition of a gene to include regions of

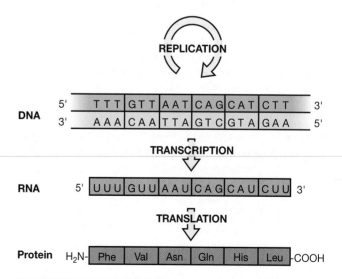

FIGURE 6.1 Synthesis of the three types of informational macromolecules. Note that in any particular region, only one of the two strands of the DNA double helix is transcribed.

DNA that encode one of these types of RNA. As we shall see, these other types of RNA molecules also have important functions in the cell.

3. *Translation.* The specific sequence of amino acids in each protein is determined by a specific sequence of bases in the mRNA (which was transcribed from the DNA). There is a linear correspondence between the base sequence of a gene and the amino acid sequence of a polypeptide (Figure 6.1). It takes *three* bases on the mRNA to encode a single amino acid, and each triplet of bases is called a **codon.** This **genetic code** is actually translated into protein by means of the protein-synthesizing system. This system consists of **ribosomes** (which are themselves made up of proteins and rRNA), transfer RNA, and a number of enzymes. The ribosomes are the structures to which messenger RNA attaches and on which protein synthesis takes place. Transfer RNA is the link between codon and amino acid. There are one or more separate tRNA molecules corresponding to each amino acid, and the tRNA has a triplet of three bases, the **anticodon,** which is *complementary* to the codon of the messenger RNA. An enzyme brings about the attachment of the correct amino acid to the correct tRNA.

In the processes of replication, transcription, and translation, the information in the sequence of the nucleic acid may specify either the sequence of another nucleic acid or of a protein. However, the transfer of sequence information from nucleic acid to protein is unidirectional; the sequence of a protein **does not** specify the sequence of a nucleic acid. This one-way transfer of genetic information from nucleic acid to protein is sometimes referred to as the *central dogma of molecular biology* because it holds for all life forms on our planet.

The three transfer steps shown in Figure 6.1 are those used in all cells. In Chapter 7 we will learn that information transfer in the reproduction of some viruses can involve two other types of transfer. One is *RNA replication*, where RNA is used as a template for RNA synthesis. The other is *reverse transcription*, where the use of sequence information in RNA specifies a sequence in DNA. Note that in both cases the information transfer is from nucleic acid to nucleic acid, and therefore neither violates the central dogma.

Prokaryotic and Eukaryotic Genetics

The three information transfer steps, replication, transcription, and translation, are used in all cells. However, as we shall see, there are some differences in the mechanisms of these processes in prokaryotes and in eukaryotes. In part this is due to differences in the organization of the genetic information and to the fact that eukaryotes have a nucleus.

We emphasized in Chapter 3 the basic differences in the organization of DNA in prokaryotes and eukaryotes. In summary, the typical bacterial genome consists

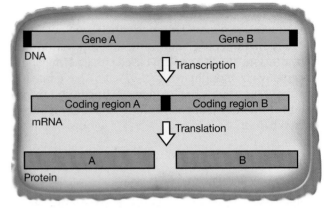

(a) PROKARYOTE

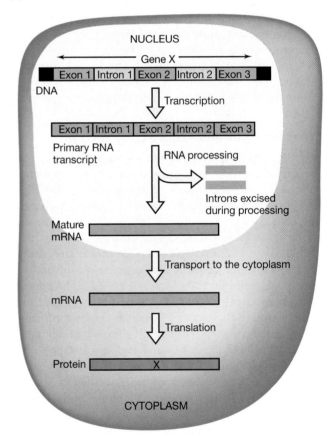

(b) EUKARYOTE

FIGURE 6.2 Contrast of information transfer in prokaryotes and eukaryotes. (a) Prokaryote. A single mRNA often contains more than one coding region (such mRNAs are called *polycistronic*). (b) Eukaryote. Noncoding regions (*introns*) are removed from the primary RNA transcript before translation. The mRNAs of eukaryotes are almost always *monocistronic*.

of a single, covalently closed *circular* molecule of DNA distributed in the cytoplasm of the cell, and the eukaryotic genome consists of several *linear* pieces of DNA present in individual chromosomes in the cell nucleus.

In prokaryotes there is no membrane separating the chromosome from the cytoplasm (∞ Section 3.16), and therefore the production of messenger RNA from the DNA template can be closely linked to translation of the messenger RNA by the ribosomes. In eukaryotes the chromosomes are located inside the nucleus and the ribosomes in the cytoplasm, so transcription and translation are spatially separated. Transcription occurs inside the nucleus, and the RNA molecules must then be transported into the cytoplasm. In all types of cells the definition of a gene is the same: a segment of DNA specifying a protein (via mRNA), a tRNA, or an rRNA. However, in eukaryotes protein-encoding genes are frequently split into two or more coding regions, with noncoding regions separating coding regions. The coding sequences are called **exons,** and the intervening noncoding regions, **introns.** Both intron and exon regions are transcribed into the **primary transcript,** or **pre-mRNA,** and the functional mRNA is subsequently formed by enzymatic removal of noncoding regions. A summary contrasting genetic phenomena in prokaryotes and eukaryotes is given in Figure 6.2.

✓ 6.1 Concept Check

The three key processes of macromolecular synthesis are DNA replication; transcription, the synthesis of RNA from a DNA template; and translation, the synthesis of proteins using messenger RNA as template. Although the basic processes are the same in both prokaryotes and eukaryotes, the organization of the genetic information is more complex in eukaryotes because many genes have distinct coding regions (exons) and noncoding regions (introns).

✓ What three informational macromolecules are involved in genetic information flow?
✓ In all cells there are three processes involved in genetic information flow. What are they?

6.2

DNA Structure: The Double Helix

We dealt with the general structure of nucleic acids in Chapter 2. In the next few sections of this chapter we shall discuss various subjects related to DNA, the nucleic acid that is the genetic material of all cells: (1) details of DNA structure necessary for an understanding of molecular genetics, (2) the types of genetic elements containing DNA that are found in cells, (3) methods for studying DNA experimentally, and (4) the mechanism of DNA replication. With this information as a basis, we will then be able to turn to a discussion of how the information of DNA is transcribed into RNA.

As we have noted, only four different nucleic acid bases are found in DNA: adenine (A), guanine (G), cytosine (C), and thymine (T). The genetic information for all cellular processes is stored in DNA in the *sequence* of bases along the polynucleotide chain. As already shown

in Figure 2.11, the backbone of the DNA chain consists of alternating units of phosphate and the sugar *deoxyribose;* connected to each sugar is one of the nucleic acid *bases.* Note especially the numbering system for the positions of sugar and base; the phosphate connecting two sugars spans from the 3'-carbon of one sugar to the 5'-carbon of the adjacent sugar. This numbering system is frequently used in discussing DNA replication and should be kept in mind. The phosphate linkage in DNA is a phospho*diester* because a single phosphate is connected by ester linkage to two separate sugars. At one end of the DNA molecule the sugar has a phosphate on the 5'-hydroxyl, whereas at the other end the sugar has a free hydroxyl at the 3'-position.

The biochemistry of DNA replication is shown in Figure 6.3. As seen, the precursor of the new unit added is a deoxyribonucleoside *tri*phosphate. Replication of DNA proceeds by insertion of a new nucleoside triphos-

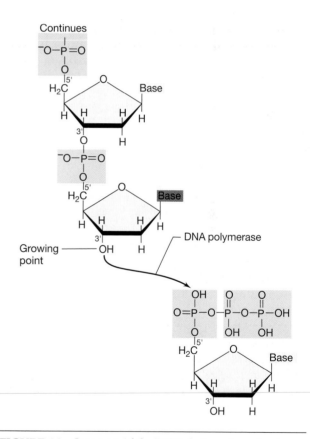

FIGURE 6.3 Structure of the DNA chain and mechanism of growth by addition from a deoxyribonucleoside triphosphate at the 3'-end of the chain. Growth always proceeds from the 5'-phosphate to the 3'-hydroxyl end. The enzyme DNA polymerase catalyzes the addition reaction. The four deoxyribonucleotides that serve as precursors are deoxythymidine triphosphate (dTTP), deoxyadenosine triphosphate (dATP), deoxyguanosine triphosphate (dGTP), and deoxycytidine triphosphate (dCTP). The two terminal phosphates of the triphosphate are split off as pyrophosphate (PP$_i$). Thus, two high energy phosphate bonds are consumed on the addition of a single nucleotide.

phate at the free 3'- (hydroxyl) end, with the subsequent loss of two phosphates (generating a deoxyribonucleoside *mono*phosphate); thus, DNA synthesis *always* proceeds toward the 3'-end of the molecule (5' → 3'). As we will see, this requirement that DNA synthesis always proceeds 5' → 3' has important consequences in the replication of double-stranded DNA for both cells and viruses.

DNA as a Double Helix

As we shall discuss in Chapter 8, the chromosomes of some viruses are single-stranded. However, in all **cellular organism chromosomes,** DNA does not exist as a single-stranded polynucleotide but as two polynucleotide strands that are not identical in base sequence but instead are **complementary.** The complementarity of DNA arises because of the specific pairing of the purine and pyrimidine bases: Adenine always pairs with thymine, and guanine always pairs with cytosine (Figure 6.4). The two strands in the resulting **double-stranded** molecule are arranged in an *antiparallel* fashion (see Figure 6.5). This means the two strands are in a "head-to-toe" arrangement. In Figure 6.5 the strand on the left is arranged 5' to 3' top to bottom, whereas the other strand is 5' to 3' bottom to top. However, the two strands are not arranged side by side in the DNA molecule but instead are wrapped around each other in a helix, the **double helix** (Figure 6.6). In this double helix, DNA has

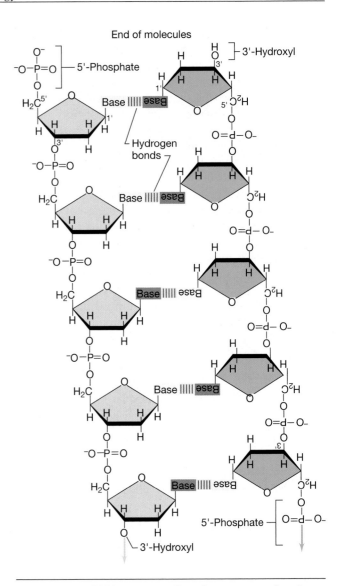

FIGURE 6.5   DNA structure. Complementary and antiparallel nature of DNA. Note that one chain ends in a 5'-phosphate group, whereas the other ends in a 3' hydroxyl. The red bases represent pyrimidines, and the yellow bases represent purines.

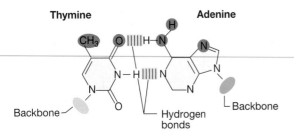

FIGURE 6.4   Specific pairing between adenine (A) and thymine (T) and between guanine (G) and cytosine (C) via hydrogen bonds. These two base pairs are the base pairs typically found in double-stranded DNA. Atoms that are found in the major groove of the double helix and that interact with proteins are highlighted in red. The deoxyribose phosphate backbones of the two strands of DNA are also indicated.

two distinct grooves, the *major groove* and the *minor groove.* There are many important proteins that interact specifically with DNA (as we shall see in Chapter 7). In general, these proteins interact predominantly with the major groove, where there is a considerable amount of space. Because of the regularity of the double helix, some atoms of the bases are always exposed in the major groove (and some in the minor groove). Atoms in the major groove that are known to be important in interactions with proteins are shown in Figure 6.4.

The size of a DNA molecule can be expressed in terms of its *molecular weight,* but because a single nucleotide has a molecular weight of about 330, and because DNA molecules are many nucleotides long, the

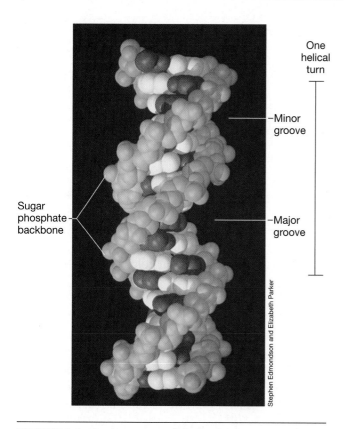

One
helical
turn

–Minor
groove

Sugar
phosphate–
backbone

–Major
groove

Stephen Edmondson and Elizabeth Parker

FIGURE 6.6   A computer model of a short segment of DNA showing the overall arrangement of the double helix. One of the sugar–phosphate backbones is shown in blue and the other in green. The pyrimidine bases are shown in red and the purines in yellow. Note the locations of the major and minor grooves. The model was produced using software from the Computer Graphics Laboratory, University of California at San Francisco.

molecular weight mounts up rapidly. (The nucleic acid in even small viruses, for instance, may have a molecular weight in the millions, the DNA in cells in the billions.) A more convenient way of expressing the sizes of DNA molecules is in terms of the *number of thousands* of nucleotide bases, or base pairs, per molecule. Thus, a DNA molecule with 1000 bases contains 1 *kilobase* of DNA. If the DNA is a double helix, then one speaks of *kilobase pairs.* Thus, a double helix 5000 base pairs in length would have a length of 5 kilobase pairs. The bacterium *Escherichia coli* has about 4600 kilobase pairs of DNA in its chromosome. With the increasing amount of genomic sequence information becoming available, it is often useful to speak of *millions* of base pairs, or *megabase pairs.* The genome of *E. coli* is 4.6 megabase pairs. Each base pair takes up 0.34 nanometers (nm) in length along the helix, and each turn of the helix contains 10 base pairs. Therefore, 1 kilobase of DNA has 100 turns of the helix and is 0.34 μm long. Calculations such as this can be very interesting, as we shall soon see.

Other Important Features of DNA Structure

In regions of the chromosome that encode proteins, the sequence of the DNA is dictated in large measure by the amino acid sequence of the encoded proteins and the nature of the genetic code (see Section 6.10). However, there are frequently base sequences in DNA that are present not because of their coding properties but because they influence the *secondary structure* of DNA, or the way in which DNA interacts with proteins.

Long DNA molecules are quite flexible, but stretches of DNA less than 100 base pairs are much more rigid. Some short segments of DNA can be bent by proteins that interact with them. However, certain sequences themselves result in bends in the DNA. The sequences of this **bent DNA** often involve several runs of five or six adenines (in the same strand), each separated by four or five bases. In addition, some sequences contribute to the ability of DNA to bend when certain proteins interact with the DNA. DNA bending seems to be commonly involved in the regulation of gene expression, as we shall discuss in Chapter 7.

Short, repeated sequences are often found in DNA molecules. Many proteins have been found that interact with regions of DNA containing repeated sequences (∞ Chapter 7) but that are repeated in inverse orientation. This type of repeat is called an **inverted repeat.** Inverted repeats give the DNA sequence a twofold symmetry. As shown in Figure 6.7, nearby inverted repeats could theoretically lead to the formation of **stem-loop** (cruciform) structures in DNA. (Note that the stem of a stem-loop is a short double helix with normal base pairing and antiparallel strands.)

By examining Figure 6.7 it can be seen that either strand from this region could form a stem-loop. Therefore, some inverted repeats found in DNA that is transcribed may be important because of the stem-loops found in the RNA. Such *secondary structure* formed by base pairing within a single strand of nucleic acid is very important in both transfer RNA (see Section 6.11) and ribosomal RNA (∞ Section 12.5 and Figure 12.7).

The *ends* of linear DNA molecules can also have interesting sequences that lead to changes in structure. Some DNA molecules have single-stranded regions at each end that are complementary. This leads to the possibility that the two ends can find each other and associate by complementary base pairing, as illustrated in Figure 6.8 for the formation of a circle. The genome of the bacterial virus lambda circularizes in this fashion (∞ Section 8.12). DNA with single-strand complementary sequences at the ends is said to have "sticky ends." Some linear DNA molecules have **hairpin** structures at each end. A hairpin is like a stem-loop but with almost no loop. Hairpins can be formed from a single-stranded region at the end of a molecule that contains an inverted repeat, as illustrated in Figure 6.9. We shall discuss other

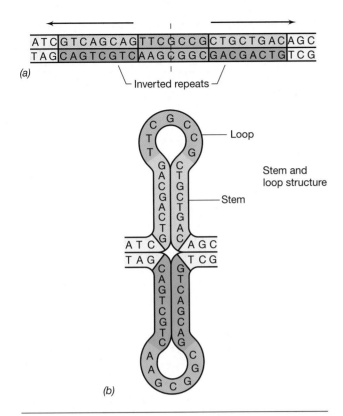

(a)

Inverted repeats

Loop

Stem and loop structure

Stem

(b)

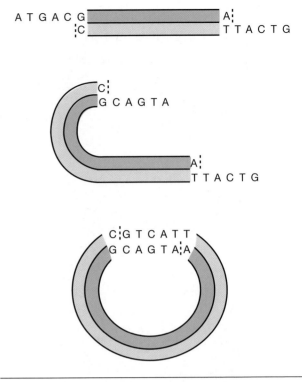

FIGURE 6.7 Inverted repeats and the formation of a stem-loop structure. (a) Nearby inverted repeats in DNA. The arrows indicate the symmetry around the imaginary axis (dashed line). (b) Formation of stem-loop structures (cruciform structures) by pairing of complementary bases on the *same* strand.

FIGURE 6.8 Linear DNA with complementary single-stranded ends ("sticky ends") can cyclize by base pairing of the complementary ends.

important sequences found at the ends of the linear DNA in eukaryotic chromosomes in Section 6.6.

Because of the enormous length of the DNA in a cell, it can almost never be handled experimentally as a complete unit. The mere manipulation of DNA in the test tube leads to its fragmentation into molecules of smaller size. Since this involves breaking covalent bonds, this was a very unexpected finding, and for many years this breakage obscured the fact that DNA molecules inside cells were immensely long. The need to study the size and shape of DNA is evident. Some of the tools for studying DNA are presented in the box, Working with Nucleic Acids: The Tools. As described, a useful technique for studying the sizes of DNA molecules is electrophoresis. As noted, the nucleic acid molecules migrate through the pores of the gel at rates depending on their molecular weight or molecular shape. Small or compact molecules migrate more rapidly than large or loose molecules. In one figure (see the box), a number of DNA fragments have been separated out in the gel.

The Effect of Temperature on DNA Structure

Although the hydrogen bonds between the base pairs are individually very weak (◌ Section 2.1), there are many such bonds holding together the strands of a typ-

ical double-stranded DNA molecule. There may be millions or even hundreds of millions of such bonds, depending on the number of base pairs in the molecule. Remember that each adenine–thymine base pair has two such bonds, and each guanine–cytosine base pair has three. This makes GC pairs stronger than AT pairs.

When isolated from cells and kept at temperatures near room temperature and at physiological salt concentrations, DNA remains in a double-stranded form. However, if the temperature is raised, the hydrogen bonds will break, but the covalent bonds holding a chain together will not, and so the DNA strands will separate. This process is generally called *melting* and, as shown in the box, can be measured experimentally. DNA with high

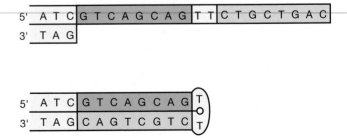

FIGURE 6.9 A hairpin structure at one end of a linear DNA molecule. If the linear DNA had been completely double-stranded, the sequences shown in green would have been inverted repeats.

TECHNIQUES & APPLICATIONS . . .   Working with Nucleic Acids: The Tools

Our knowledge of molecular biology and genetics has depended on the development of adequate research tools. Advances in knowledge of how nucleic acids work have generally been tied to the development of new methods. We discuss here some of these methods.

1. Extraction and purification of DNA The first requirement is a sample of DNA free of other cellular chemicals. The steps in the purification of DNA by one standard method are shown in Fig. 1. The

for detecting the presence of DNA in a solution. One of the most widely used is by its absorption of ultraviolet radiation. DNA strongly absorbs ultraviolet radiation at a wavelength of 260 nm. The absorption is due to the purine and pyrimidine bases. As seen in Fig. 2, double-stranded DNA absorbs less strongly than single-stranded DNA. This is because the interaction between the bases on the opposite strands of the double-stranded DNA (hydrogen bonding) reduces the ultraviolet absorbance.

termined by centrifugation at very high speed in a *gradient* of cesium chloride (CsCl). The DNA solution is added to a solution of CsCl and centrifuged at high speed for several hours until equilibrium is reached. The CsCl forms a density gradient from the top to bottom of the tube, and DNA molecules form bands at appropriate densities. At equilibrium, the DNA molecules become positioned in the gradient at positions corresponding to their densities. If ethidium bromide has been added, observation of the cen-

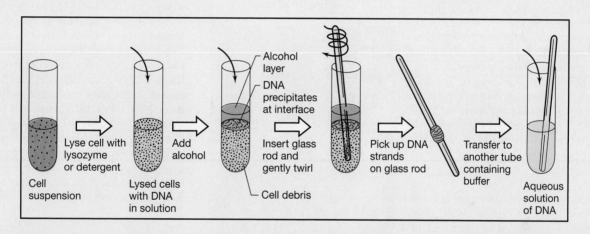

FIG. 1 Isolation of DNA

aqueous solution in the final step is treated with RNAse to remove RNA. Proteins are then removed by use of denaturing solvents (usually phenol). By repeating the purification steps a number of times, a solution can be obtained that is virtually free of any components other than DNA.

Note that the solution of DNA obtained never consists of native DNA molecules of the length found in the cell. The purification process causes the DNA to be broken into fragments of various (random) lengths. However, if the DNA has been handled gently during purification, the fragments will be tens of thousands of base pairs in length.

2. Detecting the presence of DNA There are several methods

When nucleic acids are treated with dyes that are fluorescent and are able to combine firmly with the nucleic acid chain, the nucleic acid is rendered fluorescent. The dye *ethidium bromide* is widely used to render DNA fluorescent because it combines tightly within the DNA molecule. Ethidium bromide interacts with double-stranded DNA. If the DNA is then observed with an ultraviolet source, it will fluoresce.

3. Density-gradient centrifugation of DNA DNA molecules vary in density, depending on their exact chemical composition. DNA molecules with a higher content of guanine plus cytosine (GC) are denser than molecules with low GC. The density of DNA can be de-

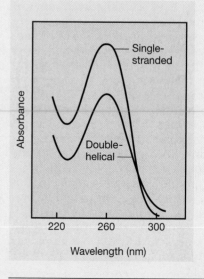

FIG. 2 Ultraviolet absorbance of DNA

trifuge tube with ultraviolet radiation after the centrifugation reveals bands of DNA by fluorescence. This method is called the *buoyant density method* and permits both determination of density and separation of molecules of differing density. DNA is also often purified by column chromatography.

4. Gel electrophoresis One of the most widespread methods of studying nucleic acids is gel electrophoresis. Introduction of electrophoresis methods has revolutionized research on molecular genetics. *Electrophoresis* is the procedure by which charged molecules are allowed to migrate in an electric field, the rate of migration being determined by the size of the molecules and their electric charge. In gel electrophoresis, the nucleic acid is suspended in a gel, usually made of polyacrylamide or agarose. The gel is a complex network of fibrils, and the *pore size* of the gel can be controlled by the way in which the gel is prepared. The nucleic acid molecules migrate through the pores of the gel at rates dependent on their molecular weight and molecular shape. Small molecules or compact molecules migrate more rapidly than large or loose molecules. After a defined period of time of migration (usually a few hours), the locations of the DNA molecules in the gel are assessed by making the DNA molecules fluorescent and observing the gel with ultraviolet radiation.

Shown in Fig. 3 is a photograph of an electrophoresis apparatus. The horizontal frame, made of lucite plastic, holds the gel. The gel is submerged in buffer that makes an electrical connection to the power supply (shown in background). The gel is stained after electrophoresis with ethidium bromide and observed by use of ultraviolet radiation (Fig. 4). In each lane, a mixture of DNA fragments was applied.

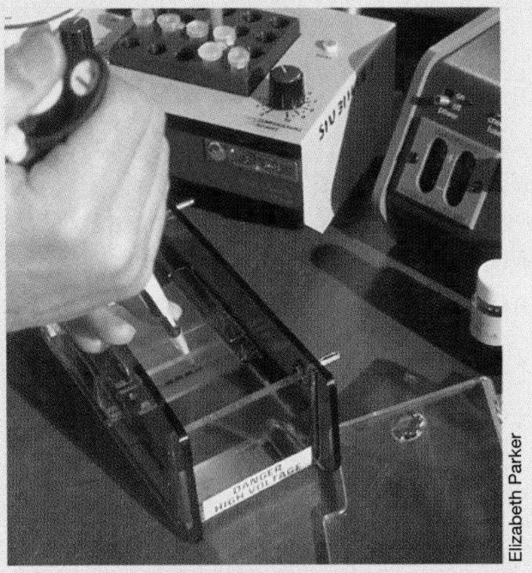

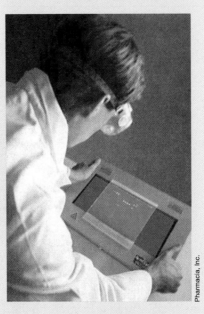

FIG. 3   Loading an agarose gel

FIG. 4   Using fluorescence to locate nucleic acid bands on a gel

Typically, the mixture of fragments is generated by digesting the DNA with one or more **restriction endonucleases.** These enzymes recognize and cut within specific short sequences of DNA. Therefore, the mixture of fragments obtained after digestion of a particular sample of DNA is nonrandom and reproducible. Separation of DNA fragments by these

relatively simple means has proven extremely useful (see Fig. 5). However, large molecules (greater than 40,000 base pairs) are not separated from one another. New electrophoretic techniques have been developed that allow separation of large fragments. One of these methods, called *pulse field gel electrophoresis* or **PFGE,** involves sending short pulses of electricity to an array of electrodes surrounding the agarose gel. PFGE and related techniques are very valuable for analyzing DNA molecules the size of those found in a small eukaryotic chromosome.

5. Labeling nucleic acids Radioactivity is widely used in nucleic acid research because radioactivity can be detected in extremely tiny amounts. Radioactive nucleic acids can be detected either directly with a scintillation counter or indirectly via their effect on photographic film (autoradiography). Autoradiography of radioactive nucleic acids is one of the most widely used techniques in molecular genetics because it can be used

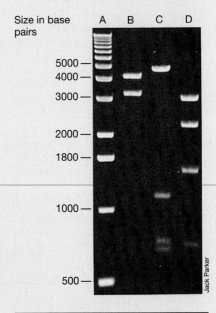

FIG. 5   A typical gel

to detect nucleic acid molecules after gel electrophoresis.

If radioactive phosphate is added to a culture while the cells are synthesizing nucleic acid, the newly synthesized nucleic acid becomes radioactively labeled.

$$^{32}PO_4 \rightarrow {}^{32}P\text{-labeled nucleotides}$$
$$\rightarrow {}^{32}P\text{-labeled nucleic acid}$$

Alternatively, end-labeling of purified DNA that contains a free hydroxyl group at the 5'-position can be done in a test tube, using radioactive ATP labeled in the third phosphate. (Typically, one first removes the phosphate commonly found at the 5'-end of a strand using an enzyme such as bacterial alkaline phosphotase.) The enzyme polynucleotide kinase specifically removes the third phosphate from ATP and attaches it to the free hydroxyl group at the 5'-end of the molecule.

$$^{32}P\text{-P-P-adenosine} + HO\text{-}$$
$$\text{deoxyribose-DNA} \rightarrow$$
$$^{32}P\text{-O-deoxyribose-DNA} + ADP$$

End labeling is an extremely useful technique as it permits labeling of preformed molecules. By tracing the radioactivity through subsequent chemical steps, the end of the molecule can be followed.

New methods of labeling nucleic acids have been developed that make use of nonradioactive chemicals that can be incorporated into DNA and can be detected by a variety of reagents that give either a colored product or even emit light (which can be detected with X-ray film). Some of these methods have nearly the sensitivity of a radioactive label but are more convenient to use and do not generate radioactive waste.

6. Denaturing nucleic acids
The strands of a double helix can be separated by heating, a process generally called *melting* (Fig. 6). As shown in Fig. 6, double-stranded

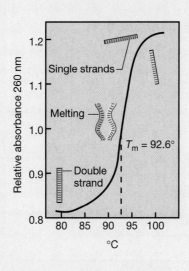

FIG. 6   Thermal denaturation of DNA

molecules show lower ultraviolet absorbance than single-stranded molecules. Therefore, if the ultraviolet absorbance of a nucleic acid solution is measured while it is being heated, the increase in absorbance when the double-stranded molecules are converted to single-stranded molecules will show the temperature at which strand separation occurs. Fig. 6 shows the change in absorbance at 260 nm when a solution containing double-stranded DNA is gradually heated. The midpoint of the transition, called T_m, is a function of the GC content of the DNA. If the heated DNA is allowed to cool slowly, the double-stranded native DNA may reform.

Strands can also be separated at room temperature by changing the ionic conditions of a solution of DNA. Because temperature is not involved, the process is usually referred to as *denaturation*. However, the end result is the same. And just as with strands separated by heat, if ionic conditions are slowly returned to normal, then the double helix can reform.

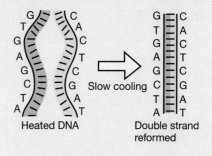

FIG. 7   Reassociation of denatured DNA

7. Nucleic acid hybridization
Hybridization is the artificial construction of a double-stranded nucleic acid by complementary base pairing of two single-stranded nucleic acids. When a DNA solution that has been heated (see earlier discussion) is allowed to cool slowly, many of the complementary strands reassociate and the original double-stranded complex reforms (Fig. 7). Hybrids can also be formed if the base sequences of the two strands are complementary. Thus, nucleic acid hybridization permits the formation of artificial double-stranded hybrids of DNA, RNA, or DNA:RNA. Nucleic acid hybridization provides a powerful tool for studying the genetic relatedness between nucleic acids. It also permits the detection of pieces of nucleic acid that are complementary to a single-stranded molecule of known sequence. Such a single-stranded molecule of known sequence is called a **probe.** For instance, a radioactive nucleic acid probe can be used to *locate*, in an unknown mixture, a nucleic acid sequence complementary to the probe (Fig. 8). Detection of nucleic acid hybridization is usually done with membrane filters constructed of nitrocellulose. Single-stranded DNA is first bound to the filter, and then the probe is added. Probe that does not base-pair with the DNA on the filter is then washed off. If necessary, hy-

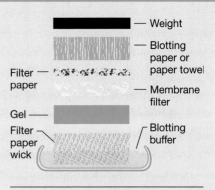

FIG. 9   Procedure for a Southern blot

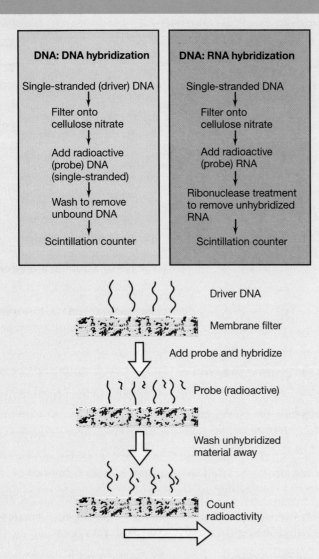

FIG. 8   Nucleic acid hybridization

bridization conditions can be manipulated to favor the formation of DNA:DNA or DNA:RNA hybrids. Nucleic acid hybridization is a powerful tool in genetics.

Hybridization can also be done after gel electrophoresis. The nucleic acid molecules are transferred by blotting from the gel to a sheet of membrane filter material, and the probe is then added to the filter. The procedure when DNA is in the gel and RNA or DNA is the probe is often called a *Southern blot procedure*, named for the scientist

E. M. Southern, who first developed it (Fig. 9). When RNA is in the gel and DNA or RNA is the probe, the procedure is called a *Northern blot*. A Western blot (sometimes called an immunoblot) involves protein–antibody binding rather than nucleic acids; ⌀ Section 21.9. Fig. 10 shows the use of a nucleic acid probe to search for complementary sequences in a mixture. The DNA fragments have been spread out by gel electrophoresis and then transferred to the membrane filter. The RNA

probe, which is radioactively labeled, is allowed to reanneal to the DNA on the filter and its position is determined by autoradiography.

8. Determining the sequence of DNA Although the base sequences of both DNA and RNA can be determined, it turns out for chemical reasons that it is easier to sequence DNA. Automated machines are now available for determining the sequences of DNA molecules. Appropriate treatments are used to generate DNA fragments that end at the four bases and that are radioactive. Then the fragments are subjected to electrophoresis so molecules with one nucleotide difference in length are separated on the gel. This electrophoresis procedure involves *four* separate lanes, one for fragments ending at each of the four bases of the DNA, adenine, guanine, cytosine, and thymine. The positions of these fragments are located by autoradiography, and from a knowledge of which base is represented by each lane, the sequence of the DNA can be read off.

Two different procedures have been developed to accomplish such determinations, called the *Maxam–Gilbert* and the *Sanger dideoxy* procedures. In the Maxam–Gilbert procedure chemicals are used that break the DNA preferentially at each of the four nucleotide bases under conditions in which only one break per chain is

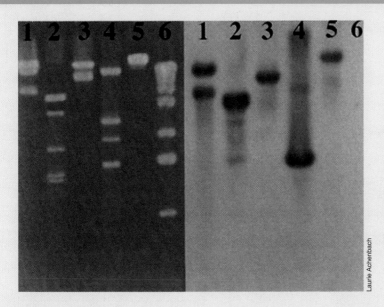

FIG. 10 (Left panel) Agarose gel electrophoresis of DNA molecules. Purified molecules of DNA from several different plasmids were treated with restriction enzymes and then subjected to electrophoresis. (Right panel) Southern blot of the DNA gel shown to the left. After blotting, hybridization with a radioactively labeled probe was carried out. The positions of the bands have been detected by X-ray autoradiography. Note that only some of the DNA fragments have sequences complementary to the labeled probe. Lane 6 contained DNA used as a size marker and none of the bands hybridized to the probe.

Laurie Achenbach

made. (Thus, four separate test tubes are prepared.)

In the Sanger dideoxy procedure the sequence is actually determined by making a *copy* of the single-stranded DNA, using the enzyme *DNA polymerase*. This enzyme uses deoxyribonucleoside triphosphates as substrates and adds them to a *primer*. In the incubation mixtures (four separate test tubes) are small amounts of each of the dideoxy analogs of the deoxyribonucleoside triphosphates

(see Fig. 11). Because the dideoxy sugar lacks the 3'-hydroxyl, continued lengthening of the chain cannot occur. The dideoxy analog thus acts as a *specific chain-termination reagent*. Fragments of variable length are obtained, depending on the incubation conditions. The nucleic acid fragments formed are radioactive from using either a radioactive primer or a radioactive deoxynucleoside triphosphate in the reactions. Electrophoresis of these fragments is then carried out,

and the positions of the radioactive bands are determined by autoradiography. By aligning the four dideoxynucleotide lanes and noting the vertical position of each fragment relative to its neighbor, the sequence of the DNA copy can be read directly from the gel (Fig. 12).

A major advantage of the Sanger method is that it can be used to sequence RNA as well as DNA. To sequence RNA, a single-stranded DNA copy is made (using the RNA

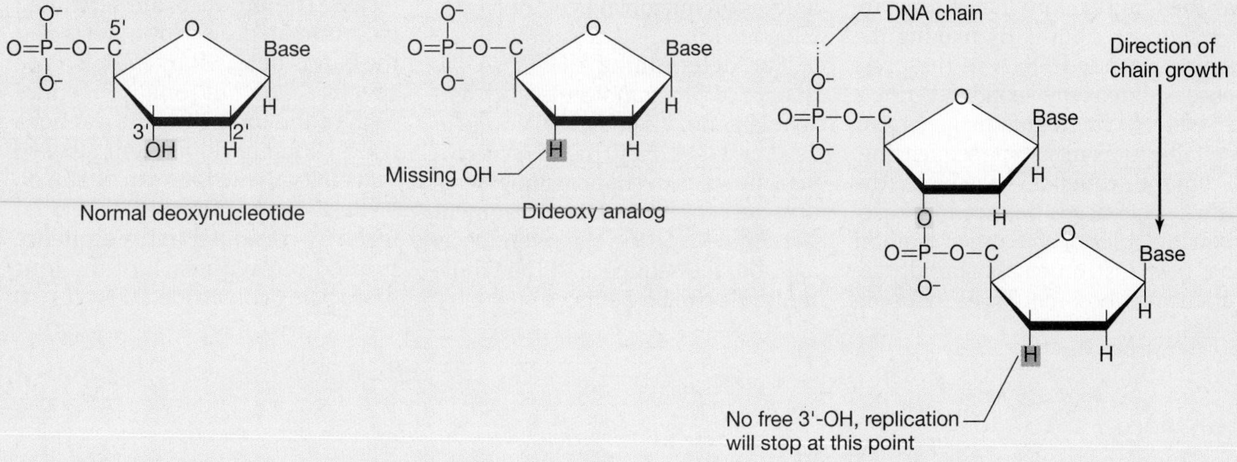

FIG. 11 Dideoxynucleotides and Sanger sequencing

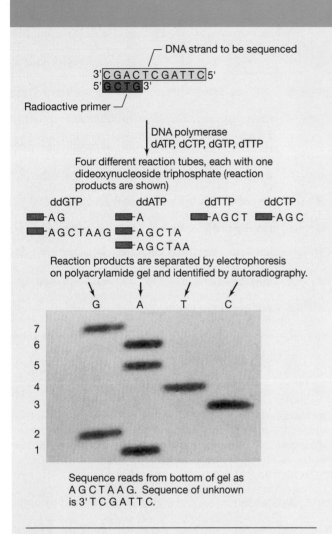

FIG. 12 DNA sequencing using the Sanger method

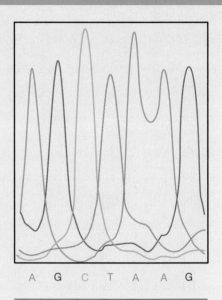

FIG. 13 Results of sequencing the same DNA as shown in Fig. 12, but using an automated sequencer and fluorescent labels.

as the template) by the enzyme reverse transcriptase. By making the single-stranded DNA in the presence of dideoxynucleotides, various-sized DNA fragments are generated suitable for Sanger-type sequencing. From the sequence of the DNA, the RNA sequence is deduced by base-pairing rules. The Sanger method has been instrumental in rapidly sequencing ribosomal RNAs for use in studies on microbial evolution (∞ Chapter 15).

For determining the DNA sequence of a long molecule, such as a whole gene, it is necessary to proceed in stages. First, the DNA is broken into small overlapping fragments and the sequence of each fragment determined. Using the overlaps as a guide, the sequence of the whole molecule can be deduced.

The demands of projects that involve sequencing entire genomes (∞ Sections 9.12 and 10.15) have led to the development of automated DNA sequencing systems. With such systems the sequencing reactions are still based on the dideoxy methodology, but fluorescent dye–labeled primers (or bases) are used so that bands can be easily detected. The products are separated by automated electrophoresis and the bands detected by fluorescence spectroscopy. In one procedure each of the four different reactions use a different fluorescent label so that all four reactions can be run on a single lane. The results are analyzed by computer and a sequence printed out with each of the four bases being color coded (Fig. 13). ■

numbers of GC pairs melts at a higher temperature than a similar-sized molecule with more AT pairs. (The fact that DNA with even high amounts of GC melts below 100°C is why the structure of DNA in the chromosomes of organisms that live at extremely high temperatures is of such interest.) If the heated DNA is allowed to cool slowly, the double-stranded native DNA can reform. Interestingly, such a process can be used not simply to reform native DNA but also to form *hybrid* molecules whose two strands come from separate sources.

Hybridization of Nucleic Acids

Hybridization is the artificial construction of a double-stranded nucleic acid by complementary base pairing of two single-stranded nucleic acids. The procedure for constructing nucleic acid hybrids is shown in the box. Since adenine pairs with uracil as well as with thymine, both DNA:DNA and DNA:RNA hybrids can be made. There must be a high degree of complementarity between two single-stranded nucleic acid molecules if they are to form a stable hybrid. In the most common use of hybridization, one of the molecules is used as a radioactive *probe* to detect a specific nucleic acid sequence, and formation of hybrids is detected by observing the formation of double-stranded molecules containing radioactivity. For example, short DNA sequences are often used to detect or quantify mRNA molecules. Detection of DNA:RNA hybridization is usually done with membrane filters made of nitrocellulose or nylon (see the box).

✓ 6.2 Concept Check

DNA is generally arranged as a double-stranded molecule that assumes a helical configuration. The two strands in the double helix are antiparallel. The strands of a double-helical DNA molecule can be separated experimentally by heat in a process called melting. Two complementary single strands can hybridize to form a stable double-stranded molecule. RNA can also hybridize with single-stranded DNA.

- ✓ Explain what antiparallel means in regard to the structure of double-stranded DNA.
- ✓ Why can some nucleic acids hybridize?

6.3

DNA Structure: Supercoiling

Large DNA molecules, representing thousands or millions of base pairs, are often represented in figures very simply as rods or, as in the case of the bacterial chromosome, uniform circles. It would be simplest to imagine that double-stranded DNA molecules have exactly the "correct" number of turns that one would predict by knowing the number of base pairs. Such a DNA mol-

ecule is said to be *relaxed*. However, consideration of the length of a simple, relaxed double helix and the size of microbial cells and viruses indicate that there must be some higher order structure, because such a molecule could not be packed into a cell. For instance, if we calculate the length of DNA in the *Escherichia coli* chromosome, we will find it to be more than 1 mm, about 500 times longer than the *E. coli* cell itself! Such packaging problems also occur in viral DNA and the DNA of eukaryotic cells. How is it possible to pack so much DNA into such a little space? The solution: *supercoiling.*

Supercoiled DNA

Supercoiling is a state in which double-stranded DNA molecules are further twisted. Figure 6.10 shows a diagram of how this could happen in a circular DNA duplex. Supercoiling puts the DNA molecule under torsion. (Take a rubber band and twist it about itself. This twisting generates a tightly coiled structure that is under considerable torsion. This torsion is held, however, only if the circular structure is maintained. Cut the twisted rubber band and see what happens!) DNA can be supercoiled in either a *positive* or *negative* direction. **Negative supercoiling** occurs when the DNA is twisted about its axis in the *opposite* direction from that of the right-handed double helix. It is in this form that supercoiled DNA is predominantly found in nature.

Although there are proteins attached to the DNA of prokaryotes, one can almost consider the chromosomes of prokaryotes to be "naked" DNA. This is not the case for the chromosomes of eukaryotes. Eukaryotic chromosomes are complex structures in which large amounts of protein are bound to the DNA in a very regular fashion. As has been mentioned (see Section 6.1) each eukaryotic chromosome contains a linear double-stranded DNA molecule. This linear DNA mol-

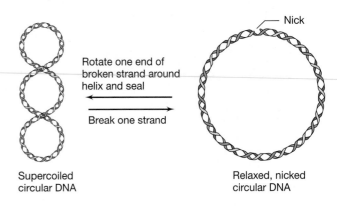

FIGURE 6.10 Supercoiled circular DNA and relaxed, nicked circular DNA interconversions. A nick is a break in a phosphodiester bond of one strand.

ecule is wound around proteins called **histones** in a very regular way to form structures called **nucleosomes** (Figure 6.11). In eukaryotic chromosomes, the formation of the nucleosome introduces negative supercoils. Histones are positively charged proteins that can neutralize the negative charge of DNA (resulting from the phosphate groups). The nucleosomes are spaced along the double helix at very regular intervals, but can aggregate to form a fibrous material called *chromatin*. Chromatin itself can be further contracted by folding and looping to eventually form a very compact structure. It is these compact structures which are most easily visible during cell division (see Figure 6.24).

Topoisomerases

Supercoiling of DNA in prokaryotes is typically brought about in a much different manner. In Bacteria and in Archaea, there is a special enzyme called **DNA gyrase,** which introduces negative supercoils. The process can be thought to occur in several stages. First, the circular DNA molecule is twisted, then a break occurs where the two chains come together, and then the broken double helix is resealed on the opposite side of the intact strand (Figure 6.12). DNA gy-

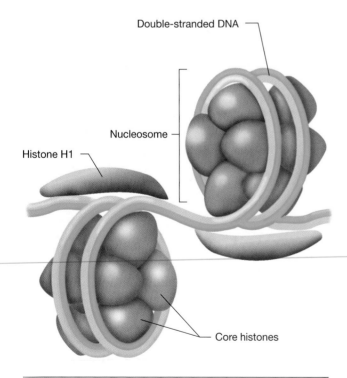

Double-stranded DNA

Nucleosome

Histone H1

Core histones

FIGURE 6.11 Packaging of DNA around a core of histone proteins to form a nucleosome. Nucleosomes are arranged along the DNA strand somewhat like beads on a string. This arrangement is typical of DNA in eukaryotic cells.

rase is a *topoisomerase,* specifically a topoisomerase II. Note the derivation of the name *topoisomerase. Topology* is the branch of mathematics that deals with the properties of geometric figures that are unaltered when the figures are twisted or contorted. We are dealing here with the topology of DNA, and a topoisomerase is an enzyme that affects this topology. It is interesting that some of the antibiotics that act on Bacteria, such as the quinolones (e.g., *nalidixic acid*), fluoroquinolones (e.g., *ciprofloxacin*) and *novobiocin,* inhibit the action of DNA gyrase. Novobiocin is also effective against several species of Archaea, where it also seems to inhibit DNA gyrase.

There is another enzyme that is able to *remove* supercoiling in DNA. This enzyme, called *topoisomerase I,* introduces a single-strand break in the DNA and causes the rotation of one single strand of the double helix around the other. As was shown in Figure 6.10, a break in the backbone (a **nick**) of either strand allows the DNA to return to the relaxed state. This is true whether the supercoiling is positive or negative. Such enzymes are found in both prokaryotes and eukaryotes. Linear DNA, as in eukaryotic chromosomes, is prevented from returning to the relaxed state by the proteins bound to it. To prevent the entire bacterial chromosome from becoming relaxed every time a nick is made, the chromosome contains approximately 50 *supercoiled domains.* A nick in the DNA in one of these domains does not relax the DNA in the others. It is unclear what holds the DNA in these domains, but it is likely to involve proteins.

Through the action of these topoisomerases, the DNA molecule can be alternately supercoiled and relaxed. Because supercoiling is necessary for packing the DNA into the confines of a cell and relaxing is necessary so DNA can be replicated, these two complementary processes clearly play an important role in the behavior of DNA in the cell. In most prokaryotes, the actual level of negative supercoiling is the result of a balance between the activity of DNA gyrase and topoisomerase I. In addition, however, supercoiling is known to affect gene expression. Certain genes are more actively transcribed when DNA is supercoiled, whereas transcription of other genes is inhibited by excessive supercoiling.

A few prokaryotes contain an enzyme called *reverse gyrase.* This topoisomerase is capable of introducing *positive* supercoils in DNA. The organisms that contain this enzyme are among those that grow at extremely high temperatures (∞ Section 5.9). Some DNA in these organisms seems to be "relaxed," that is, with neither positive nor negative supercoils. This may be a matter of a balance between reverse gyrase and the activity of histonelike proteins. Interestingly, one organism, the hyperthermophilic archaeon *Methanothermus fervidus,* has

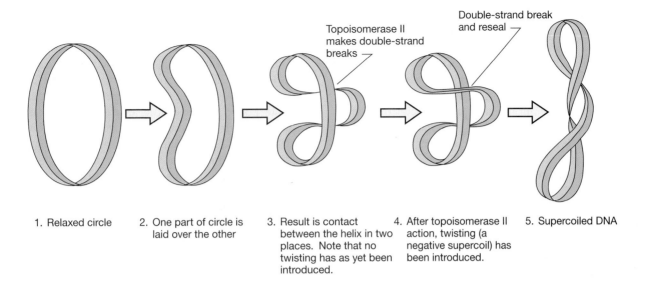

1. Relaxed circle
2. One part of circle is laid over the other
3. Result is contact between the helix in two places. Note that no twisting has as yet been introduced.
4. After topoisomerase II action, twisting (a negative supercoil) has been introduced.
5. Supercoiled DNA

FIGURE 6.12   Introduction of supercoiling in a circular DNA by activity of topoisomerase II, which makes double-strand breaks.

nucleosome-like structures in which the DNA seems to be positively supercoiled(⟳ Section 14.10).

Positive supercoiling, brought about by reverse gyrase or other means, is a kind of "overwinding" and might play an important role in protecting the DNA from being denatured. However, in order to be of use, the information in DNA must be accessible to the cell's machinery. Therefore, in all cells, the structure of DNA is likely to be quite dynamic.

✓ 6.3 Concept Check

The very long DNA molecule is able to be packaged into the cell because it is supercoiled. In most prokaryotes this supercoiling is brought about by enzymes called topoisomerases. Eukaryotic chromosomes contain very regular structures called nucleosomes.

✓ Why is supercoiling important?
✓ What function do topoisomerases serve inside cells?

6.4

Genetic Elements

As we have seen, the genetic material of all cells is double-stranded DNA (⟳ Section 2.5). Before proceeding to describe how cells replicate their DNA, we consider what kind of structures must be replicated. Structures containing genetic material can be called *genetic elements*. The **genome** is the total complement of genes in a cell or virus. Although the main genetic element is the *chromosome*, other genetic elements are found and play important roles in gene function in both prokaryotes

and eukaryotes (Table 6.1). Two key properties of genetic elements are (1) their ability to self-replicate, and (2) their genetic coding properties.

The Chromosome

In Section 3.10 we discussed the fact that a typical prokaryote has a single chromosome containing all (or most) of the genes found inside the cell. Eukaryotes have

TABLE 6.1	Kinds of genetic elements
Element	**Description**
Prokaryote	
Chromosome	Extremely long, usually circular, double-stranded DNA molecule
Plasmid	Typically a relatively short, usually circular, double-stranded DNA molecule which is extrachromosomal
Viral genome	Single- or double-stranded DNA or RNA molecule
Transposable element	Double-stranded DNA molecule always found within another DNA molecule
Eukaryote	
Chromosome	Extremely long, linear, double-stranded DNA molecule
Plasmid	Typically a relatively short circular or linear double-stranded DNA molecule which is extrachromosomal
Mitochondrion or chloroplast	Intermediate-length DNA molecules, usually circular
Viral genome	Single- or double-stranded DNA or RNA molecules
Transposable element	Double-stranded DNA molecule always found within another DNA molecule

multiple chromosomes as a part of their genome. Also, the typical prokaryotic chromosome is a circular DNA molecule, whereas the DNA in all known eukaryotic chromosomes is linear. In Table 6.2 the number, size, and configuration of chromosomes in a few known microorganisms, both prokaryotic and eukaryotic, are given. Note that the chromosome of the bacterium *Borrelia burgdorferi*, the causative agent of Lyme disease (⊙ Section 24.4) is linear. The ends of this chromosome may have hairpin repeats like those shown in Figure 6.9. Although very uncommon, linear chromosomes are now known to exist in several other Bacteria. The bacterium *Streptomyces lividans* also has a linear chromosome, but it has proteins covalently bound to its ends. We will discuss the importance of such proteins in Section 6.6.

The few examples of prokaryotes given in Table 6.2 are not random choices. They include Archaea as well as Bacteria, and examples of among the smallest and largest known prokaryotic chromosomes. We shall discuss the apparent fact that some Bacteria such as *Rhodobacter sphaeroides* has two chromosomes later.

Only a few examples of eukaryotic microorganisms are also given in Table 6.2, but all their chromosomes have linear DNA and multiple chromosomes even in the haploid state. Both these conditions are the rule in eukaryotic organisms, whatever their size. In yeast (and probably many other eukaryotic microorganisms), the length of the DNA in a single chromosome is actually shorter than that of a linearized prokaryotic chromosome. For instance, the total amount of DNA per yeast cell is only three times that in *Escherichia coli*, but yeast has 16 chromosomes, and so the average yeast DNA molecule is much shorter than the *E. coli* chromosome. In higher organisms, however, the length of the DNA molecule in a single chromosome is many times greater than that in the prokaryotic chromosome if it were opened and linearized.

TABLE 6.2   Sizes, shapes, and numbers of chromosomes in microorganisms

Organism	Comments	Size (base pairs)[a]	Chromosome Number	Chromosome Geometry
Bacteria				
Mycoplasma genitalium	Smallest known cellular genome	580,070	1	◯
Borrelia burgdorferi	Causes Lyme disease (⊙ Chapter 24)	910,725[b]	1	⋈⋈⋈⋈
Haemophilus influenzae	Gram-negative, can cause disease (⊙ Chapter 23)	1,830,137[c]	1	◯
Rhodobacter sphaeroides	Gram-negative, phototrophic	4.0×10^6	2	◯◯
Bacillus subtilis	Gram-positive, genetic model	4,214,810	1	◯
Escherichia coli K12	Gram-negative, genetic model	4,639,221[d]	1	◯
Myxococcus xanthus	Complex developmental cycle (⊙ Chapter 13)	9.5×10^6	1	◯
Archaea				
Methanococcus jannaschii	Methanogen, which grows at high temperature (⊙ Chapters 5 and 19)	1,664,976	1	◯
Thermococcus celer	Grows at high temperature (⊙ Chapters 5 and 14)	1.9×10^6	1	◯
Haloferax mediterranei	Grows in high salt (⊙ Chapters 5 and 14)	2.9×10^6	1	◯
Sulfolobus acidocaldarius	Grows at high temperature and high acidity (⊙ Chapters 5 and 14)	3.0×10^6	1	◯
Eukarya[e]				
Giardia lamblia	Flagellated protozoan that causes acute gastroenteritis (⊙ Chapters 17 and 24)	1.2×10^7	4	⋈⋈⋈⋈
Saccharomyces cerevisiae	Yeast, widely used in science and industry (⊙ Chapters 9 and 17)	12,057,500[f]	16	⋈⋈⋈⋈
Dictyostelium discoideum	Cellular slime mold, developmental model (⊙ Chapter 17)	5.0×10^7	7	⋈⋈⋈⋈
Tetrahymena thermophila	Ciliated protozoan (⊙ Chapter 17)	2.1×10^8	5	⋈⋈⋈⋈

a In the case of the Eukarya the genome sizes and chromosome number are for the haploid form.

b This is for the linear chromosome. The genome of this organism also contains at least 17 circular and linear plasmids, which themselves have a combined size of more than 533,000 base pairs.

c *Haemophilus influenzae* Rd was the first cellular organism to have its genome entirely sequenced.

d The reported sequence does not contain that of the F-plasmid (⊙ Section 9.8) nor that of bacteriophage lambda (⊙ Section 8.12), both of which would be present in a typical K-12 strain.

e All the organisms listed are single-celled.

f *Saccharomyces cerevisiae* was the first eukaryote to have its genome completely sequenced. The number given here does not include the mitochondrial genome and all the copies of some repeated sequences.

The haploid human genome has only a few more chromosomes than yeast but has over 200 times more DNA. If all the DNA of the chromosomes of a human cell were stretched end to end, the length would be close to 2 m.

The linear DNA molecule in a eukaryotic chromosome has a special DNA sequence called a *telomere* at each end and a *centromere* somewhere between the telomeres. Centromeres are important for partitioning the chromosomes during cell division. As we shall see, telomeres play an important role in the replication of these molecules (see Section 6.6). The number of chromosomes is constant within a species but varies widely among species.

In addition to *nucleosomes* (see Section 6.2) the chromosomal organization in eukaryotes involves two features not generally found in prokaryotes:

1. *Interrupted Genes.* Many eukaryotic protein-encoding genes are interrupted or split by noncoding DNA sequences inserted between the sequences that actually code for a single polypeptide (see Figure 6.2). These noncoding intervening sequences are called **introns,** and the coding sequences are called **exons.** The number of introns per gene is variable and ranges from none to more than 50. During transcription, both introns and exons are copied, and the intron sequences are subsequently cut out and removed when the messenger RNA is processed into its final form.

2. *Repetitive sequences.* Eukaryotes generally contain much more DNA per genome than is needed to encode all the proteins required for cell function. Eukaryotic DNA can be divided into several classes. **Single-copy DNA** contains the coding sequences for the main proteins of the cell (and the associated intron DNA). **Moderately repetitive DNA,** found in a few to relatively large numbers of copies, codes for some major macromolecules of the cell: histones, immunoglobulins (involved in immune mechanisms, as discussed in Chapter 20), ribosomal RNA, and transfer RNA. **Highly repetitive (satellite) DNA** is found in a very large number of copies. In humans, about 20–30% of the DNA is found in repetitive sequences, and one 300-base-pair sequence is repeated approximately 300,000 times. The function of most highly repetitive DNA is unknown.

However, as is true of so much of biology, except for the central dogma there seem to be few categorical statements. Introns have also been found in the protein-encoding genes of prokaryotes, but they are much less common and are removed from the transcript by a different process. Repetitive DNA sequences are also present in prokaryotes. These include low-copy-number repeated sequences, such as genes that encode rRNA, and even a few short but more highly repetitive sequences. For instance, there is a 38-base-pair sequence found in *Escherichia coli* K-12 that is repeated 581 times, making up over 0.5% of the genome. However, the two generalizations are still correct: Eukaryotic genomes are characterized by having interrupted genes and repetitive sequences.

Nonchromosomal Genetic Elements: Viruses and Plasmids

A number of genetic elements that are *not* cellular chromosomes have been recognized. Some nonchromosomal genetic elements that we will discuss briefly here include viruses, plasmids, the genomes of mitochondria and chloroplasts, and transposable elements.

Viruses contain genomes, either DNA or RNA, that control their own replication and transfer from cell to cell. The viral genome is also referred to as a chromosome. However, it contains genes essential to the virus, not the host cell, and is therefore clearly distinct from the cellular chromosomes. Both linear and circular viral chromosomes are known. Viruses are of special interest because they are often (but not always) responsible for disease states. We discuss viruses in Chapter 8, and virus diseases in Chapters 23 and 24.

Plasmids are typically small genetic elements that exist and replicate separately from the chromosome. The great majority of plasmids are double-stranded DNA, and although most plasmids are circular, some are linear. Plasmids differ from viruses in two ways: (1) they do not cause cellular damage (generally they are beneficial), and (2) plasmids do not have extracellular forms, whereas viruses do. Although plasmids have been recognized in only a few eukaryotes, they have been found in most prokaryotic species. We discuss plasmids in Chapter 9. Some plasmids find wide use in gene manipulation and genetic engineering, as outlined in Chapter 10.

Many prokaryotes seem to contain one or more plasmids in addition to their chromosome. Some plasmids contain genes whose protein products can confer important properties on the host cell, such as resistance to antibiotics. Many plasmids are rather small (a few kilobase pairs), but some are quite large (several megabases). None are as large as the chromosome, however. From this information, you might think that in prokaryotes a chromosome is simply defined as the largest genetic element in the cell. However, the definition of what constitutes a chromosome in Bacteria is coming to mean a genetic element that contains genes whose products are involved in essential metabolic steps *under all growth conditions.* Such genes are sometimes referred to as *housekeeping genes.* For instance, a gene encoding DNA gyrase is always required by a cell, whereas a gene that enables a bacterium to be resistant to an antibiotic is required only under certain conditions (the presence of the antibiotic). It is difficult to conclusively demonstrate that any prokaryote has more than one chromosome by this definition. Such proof requires evidence that each "chromosome" contains single-copy genes that are essential. However, even using such a stringent definition, there are several Bacteria that clearly seem to have more than one chromosome, including *Rhodobacter sphaeroides* (see Table 6.2) and members of the genus *Brucella*. This may also be true of the spirochete *Borrelia burgdorferi* (see Table 6.2), which has a complex genome containing a large, linear chromosome and several circular and linear plasmids.

Nonchromosomal Genetic Elements: Organelles and Transposable Elements

Mitochondria and **chloroplasts** contain nonchromosomal genetic elements and are found in eukaryotes. As we discussed in Section 3.16, the mitochondrion is the site of respiratory enzymes and plays a major role in energy generation in most eukaryotes. The chloroplast is a green, chlorophyll-containing structure that is the site of phototrophic ATP formation. From a genetic viewpoint, mitochondria and chloroplasts can be viewed as independently replicating genetic elements. However, these organelles are much more complex than plasmids and viruses because they contain not only DNA but also a complete machinery for protein synthesis, including 70S ribosomes, transfer RNA, and all the other components necessary for translation and formation of functional proteins. In addition, despite the fact that they contain many genes and a complete translation system, their existence is not independent of the cellular chromosomes because most proteins in them are coded not by organelle DNA but by the cell's chromosomal DNA. We discuss the genetics of mitochondria in Section 9.14.

Transposable elements are pieces of DNA having the ability to move from one site on a chromosome to another. Transposable elements are found in prokaryotes and eukaryotes and play important roles in genetic variation. There are three types of transposable elements: insertion sequences, transposons, and some special viruses. *Insertion sequences* are the simplest type and carry no genetic information other than that required for them to move into new locations. *Transposons* are larger and contain other genes. We discuss both of these types in more detail in Chapter 9. In Chapter 8 we discuss a virus, Mu, that is also a transposable element. The unique feature of transposable elements is that *they all replicate as part of some other molecule of DNA*. In spite of this, some of these elements are clearly "self-replicating" in the sense that they con-

trol their own replication and as such fit our definition of a genetic element. Others might simply be considered "jumping genes."

✓ 6.4 Concept Check

In addition to the chromosomes, a number of other genetic elements exist in cells. Plasmids are DNA molecules that exist separately from the chromosome of the cell. Mitochondria and chloroplasts contain their own DNA genomes. Viruses contain a genome, either DNA or RNA, that controls their own replication. Transposable elements exist as a part of other genetic elements.

✓ What is a *genome*?
✓ What genetic material is found in all cellular chromosomes?
✓ What is the difference between the number of chromosomes in prokaryotes and eukaryotes?

6.5

Restriction and Modification of DNA

Organisms are occasionally faced with the problem of coping with foreign DNA, generally derived from viruses, that may derange cellular metabolism or initiate processes leading to cell death. In a unicellular microorganism this problem must be dealt with without the aid of some of the complex host defense mechanisms available to higher organisms (⟳ Chapters 19 and 20). However, many prokaryotes have an extremely effective mechanism of dealing with foreign DNA, enzymatic destruction.

The enzymes involved in the destruction of foreign DNA are called **restriction endonucleases** and are remarkably specific in their action, an essential property if destruction of cellular DNA is to be avoided. Most restriction endonucleases, typically referred to as *restriction enyzmes*, combine with DNA only at sites with specific sequences of bases. Clearly such enzymes could not exist unless the DNA of the host organism were protected from attack. In order to protect itself from its own restriction enzymes, a cell has enzymes that chemically **modify** the specific sequences on its own DNA so that these sequences are not attacked. Therefore, restriction enzymes are one-half of a **restriction-modification system**.

Restriction Enzymes

There are several different kinds of restriction enzymes. The most common seem not only to recognize specific sequences of bases but also to make double-stranded breaks *within* these sequences. Many of these sequences exhibit twofold symmetry around a given point. Figure 6.13*a*

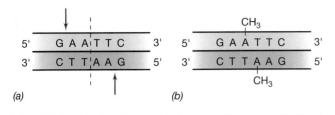

FIGURE 6.13 Restriction and modification of DNA. (a) The sequence of DNA recognized by the restriction endonuclease *Eco*RI. The red arrows indicate the bonds cleaved by the enzyme. The dashed line indicates the axis of symmetry of the sequence. (b) The same sequence after modification by the *Eco*RI methylase. The methyl groups added by this enzyme are shown.

shows the sequence that is recognized and cleaved by one restriction endonuclease from *Escherichia coli,* called *Eco*RI.

The cleavage sites are indicated by arrows, and the axis of symmetry by a dashed line. Note that the two strands have the same sequence if one is read from the left and the other from the right (or, in terms of polynucleotide strands, if both are read $5' \rightarrow 3'$ or both read $3' \rightarrow 5'$). Such a structure is called a **palindrome.** (A palindrome is a sequence of characters that reads the same when read from either right or left—for instance, *Sex at noon taxes* and *Able was I ere I saw Elba.* The term *palindrome* is derived from the Greek meaning "to run back again.")

Many restriction enzymes are composed of two identical polypeptide subunits, and each subunit recognizes and cuts the sequence on one of the strands. Since the sequences recognized by restriction enzymes are relatively short, and frequently palindromic, such enzymes always make *double-stranded* breaks, and such double-stranded breaks are not subject to correction by repair enzymes. This ensures that an invading nucleic acid will be destroyed. Interestingly, restriction enzymes that recognize *non*-palindromic sequences cut near the sequence, but not within it. Even so, they make double-strand breaks.

Almost all known restriction enzymes are from prokaryotes. The recognition sequences and cutting sites for a few restriction enzymes are given in Table 6.3. Note that the recognition sites for the enzymes listed are 4, 6, and 8 base pairs. In a "random" DNA molecule, one would expect any 4-base-pair sequence to occur approximately once every 256 base pairs based on the probability of $\frac{1}{4} \times \frac{1}{4} \times \frac{1}{4} \times \frac{1}{4}$ (assuming each base pair is equally probable in the DNA). Therefore, an enzyme that recognizes a 4-base-pair restriction site would cut a large DNA molecule into many specific fragments. A specific 6-base-pair sequence should appear every 4096 base pairs in random DNA, and an 8-base-pair sequence should appear only about once every megabase pair. The *Escherichia coli* chromosome, which is about 4.6 megabase pairs, is cut 21 times by the enzyme *Not*I, which recognizes an 8-base-pair sequence indicating that in this chromosome the *Not*I recognition sequence is used somewhat more often than one might have predicted.

Well over 2000 restriction enzymes with over 200 different specificities are now known, and more are being sought. The reason for this is not just that they are very interesting enzymes. They are also of great importance in DNA research. Note that the enzyme *Eco*RI can cut *any* double-stranded DNA that has its recognition sequence and cuts only at that sequence. This enables scientists to cut large DNA molecules into smaller fragments. Such fragments with defined termini, created as a result of the action of specific restriction enzymes, are amenable to determination of nucleotide sequences, thus permitting the working out of the complete sequence of DNA molecules (see the box).

Another use of certain restriction enzymes is that they permit the conversion of DNA molecules into fragments that can be joined by DNA ligase (see Section 6.6). This enables laboratory researchers to clone DNA, as will be discussed in Chapter 10.

Restriction enzymes are such important tools in modern molecular genetic research that they have be-

TABLE 6.3	**Recognition sequences of a few restriction endonucleases**	
Organism	**Enzyme designation**	**Recognition sequence**[a]
Bacillus subtilis	*Bsu*RI	GG↓ČC
Brevibacterium albidum	*Bal*I	TGG↓ČCA
Escherichia coli	*Eco*RI	G↓AÅTTC
Haemophilus haemolyticus	*Hha*I	GČG↓C
Haemophilus influenzae	*Hind*II	GTPy↓PuAČ
Haemophilus influenzae	*Hind*III	A↓AGCTT
Nocardia otitidis-caviarum	*Not*I	GC↓GGČCGC
Thermus aquaticus	*Taq*I	T↓CGÅ

a Arrows indicate the sites of enzymatic attack. Asterisks indicate the site of methylation (modification). G, guanine; C, cytosine; A, adenine; T, thymine; Pu, any purine; Py, any pyrimidine. Only the $5' \rightarrow 3'$ sequence is shown.

come widely available commercially. A number of companies purify and market restriction enzymes with a variety of specificities. If the DNA sequence of a particular region of a molecule is known, a research worker can generally obtain a restriction enzyme that can cut in this region.

Modification: Protection from Restriction

An integral part of the cell's restriction–modification system is the modifying enzyme, which chemically **modifies** the specific sequences on its *own* DNA so these sequences are not attacked by the cell's own restriction enzymes. Such modification generally involves *methylation* of specific bases within the recognition sequence so the restriction nuclease can no longer act. Thus, for each restriction enzyme there must also be a modification enzyme, the two enzymes being closely associated. For example, the sequence recognized by the *Eco*RI restriction enzyme (see Figure 6.13*a*) can be modified by methylation of the two most interior adenines (Figure 6.13*b*). The enzyme which performs this modification is called *Eco*RI methylase. If even a single strand is modified, the sequence is no longer a substrate for the restriction enzyme *Eco*RI.

Different Restriction Enzyme Systems in One Host

If an incoming viral genome is modified by the host, then the restriction enzyme will ignore it. The virus will replicate and release progeny (all with their DNA modified) into the environment. If these progeny viruses infect a new cell of the same type, none of them will be susceptible to the restriction enzyme. At least, none will if all members of the host species have the same restriction–modification system. However, they do not. For instance, different *strains* of *Escherichia coli* have different restriction–modification systems. Therefore, a virus propagated on one of them will be restricted if it infects another. In addition, restriction systems have been discovered where the restriction enzyme recognizes only DNA that has a modification of a type only seen in other cells. These enzymes then specifically restrict DNA previously modified in a "foreign" host. In a typical strain of *E. coli* there are at least four different restriction systems operating. One is a "classical" restriction–modification system that restricts incoming unmodified DNA, and the other three restrict DNA with different foreign modifications. Some strains of *E. coli* have even more restriction–modification systems operating. In Chapter 8 we will see how selective pressures operating on a virus and its host may lead to the accumulation of a variety of such defense mechanisms.

Restriction Enzyme Analysis of DNA

As noted, a DNA molecule can be cut at a specific location by a given restriction enzyme. Because the base sequences recognized by most restriction enzymes are four to six nucleotides long, there will generally be only a limited number of such sequences in a piece of DNA. After cleaving the DNA, the fragments can be separated by agarose gel electrophoresis, as shown in the box. The distance migrated by any band of DNA in such a gel can be determined by calibrating the electrophoresis system with DNA molecules of known size. By judicious use of several restriction enzymes of different specificities, and by use of overlapping fragments, it is possible to construct a **restriction enzyme map** in which the positions cut by each of the several restriction enzymes can be designated.

Several procedures are now available for determining the base sequences of DNA molecules. In fact, automated machines are available for sequencing DNA. Details are presented in the box. By successively determining the sequences of small overlapping fragments of DNA, it is possible to determine the sequences of very large pieces of DNA. The sequences are now known for thousands of genes, as well as for the complete genome of many viruses (∞ Chapter 8), several prokaryotes, and the eukaryotic microorganism *Saccharomyces cerevisiae* (∞ Chapter 9).

✓ 6.5 Concept Check

Restriction enzymes are cellular enzymes that recognize specific short base sequences in DNA and make two single-stranded breaks at locations within the recognition sites. A restriction enzyme does not affect the cell that produces it because its own DNA is methylated at the recognition site by a modification enzyme specific for that site.

✓ Of what use are restriction enzymes to organisms in nature?
✓ Of what use are restriction enzymes in the laboratory?

6.6

DNA Replication

The problem of DNA replication can be simply put: The nucleotide base sequence residing in each long molecule of the DNA double helix must be precisely duplicated to form a copy of the original molecule. The cell has solved this seemingly complex problem in an elegant fashion: by means of *complementary base pairing*. As we have discussed (see Figure 6.4), adenine pairs specifically with thymine and guanine pairs with cytosine. If the DNA double helix is opened up, a new strand can be synthesized as the complement of each of the parental strands. As shown in Figure 6.14, replication

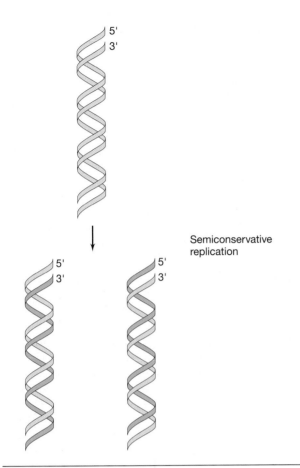

FIGURE 6.14 DNA replication is a semiconservative process. In both prokaryotes and eukaryotes the process is always semiconservative. Note that the new double helices each contain one new and one old strand.

is **semiconservative,** the two resulting double helices consisting of one progeny and one parental strand.

Templates and Primers

The DNA molecule that is copied to form a complement is called a *template*. A template is a preformed pattern that is copied. The *new* DNA molecule is not covalently connected to the *old* DNA molecule.

The chemistry of DNA, the nature of its precursors, and the activities of the enzymes involved in replication place some important restrictions on the manner in which this new strand is synthesized. The precursor of each new nucleotide in the chain is a nucleoside 5'-*tri*phosphate, of which the two terminal phosphates are removed and

the internal phosphate is attached covalently to deoxyribose of the growing chain (see Figure 6.3). The addition of the nucleotide to the growing chain requires the presence of a free hydroxyl group, and such a free hydroxyl group is available only at the 3'-end of the molecule. This chemical restriction leads to an important law that is at the basis of many facets of DNA replication: *DNA replication always proceeds from the 5' end to the 3'-hydroxyl end, the 5'-phosphate of the incoming nucleotide being attached to the 3'-hydroxyl of the previously added nucleotide.*

The enzymes that catalyze addition of the nucleotides are called **DNA polymerases.** *All* DNA polymerases synthesize new DNA in the 5' → 3' direction. *However, no known DNA polymerase can begin a new chain. All these enzymes can only add a nucleotide onto a preexisting 3'-OH group.* Therefore, for a *new* chain to be started, there must be a **primer,** a site at which the DNA polymerase can attach the first nucleotide. In most cases this primer is a short stretch of *RNA.*

When the double helix is opened up at the beginning of replication, an RNA-polymerizing enzyme acts first, resulting in formation of this RNA primer. A specific RNA-polymerizing enzyme, called *primase,* participates in primer synthesis by laying down a short stretch of RNA. At the growing end of this RNA primer is a 3'-OH group to which DNA polymerase can add the first deoxyribonucleotide. Therefore, continued extension of the molecule occurs as *DNA* rather than RNA. Thus, the newly synthesized molecule has a structure like that shown in Figure 6.15. The primer must eventually be removed, as we shall see.

To understand the complete replication of a double-stranded DNA molecule, it is easiest to choose an actual example and see how it is replicated. Most of the information on the mechanism of DNA replication has been obtained from the bacterium *Escherichia coli*, and the following discussion deals primarily with this organism.

Initiation of DNA Synthesis

As is the case for most prokaryotes, the chromosome of *Escherichia coli* is a circular DNA molecule. Also like most Bacteria, there is a single location on this chromosome where DNA synthesis is initiated, the so-called **origin of replication.** The origin of replication consists of a specific sequence of about 300 bases that is recognized by specific initiation proteins. At the ori-

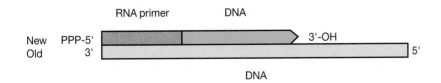

FIGURE 6.15 Structure of the RNA–DNA combination that results at the initiation of DNA synthesis.

gin of replication, the DNA double helix is opened up
and the initiation of DNA replication occurs on the two
single strands. As replication proceeds, the site of replication, called the **replication fork,** appears to move
down the DNA.

Replication is frequently bidirectional from the origin of replication, as shown in Figure 6.16, and therefore there are *two* replication forks replicating in
opposite directions. In circular DNA, bidirectional replication leads to the formation of characteristic structures
called **theta structures** (Figure 6.16). Most large DNA
molecules, whether from prokaryotes or eukaryotes,
have bidirectional replication from fixed origins. A single eukaryotic chromosome has many origins. This is
not simply because the DNA is longer because, as we
have seen, this is not always the case (see Section 6.4).
It may reflect the fact that DNA polymerases from eukaryotes do not replicate as fast as the prokaryotic enzymes. DNA replication is carefully regulated, and the
site where this regulation takes place is the origin.

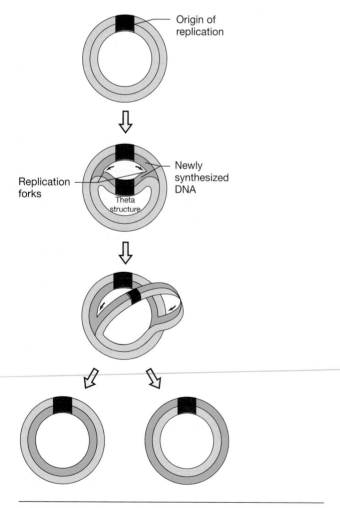

FIGURE 6.16   In circular DNA, bidirectional replication
from an origin leads to the formation of replication intermediates resembling the Greek letter theta (θ).

✓ 6.6a Concept Check

Both strands of the DNA helix serve as templates for the synthesis of two new strands. The two progeny double helices
each contain one parental strand and one new strand. The
new strands are elongated by always adding on to the 3'-end.
DNA polymerases cannot start new strands. Therefore, new
strands must start with a primer, which is usually RNA.

✓ Which end of a DNA strand being synthesized is growing?
✓ Why is a primer involved in DNA replication?

Leading and Lagging Strands

There are three different DNA polymerases in *Escherichia coli,* called DNA polymerases I, II, and III. It is
DNA polymerase III that is the primary enzyme of replication at the replicating forks. However, several other
enzymes are also involved. The details of events at the
replication fork are illustrated in Figure 6.17. At the
replication fork, the DNA double helix is unwound and
a small single-stranded region is formed by the action
of specific proteins called *helicases*. Helicases are ATP-
dependent enzymes that hydrolyze ATP as they move
down the helix in advance of the replicating fork. The
single-stranded region generated is complexed with a
special protein, the *single-strand binding protein,* which
stabilizes the single-stranded DNA, preventing the formation of intrastrand hydrogen bonds.

Figure 6.17 reveals an important difference between
replication of the two strands, which arises from the fact
that DNA replication always proceeds from 5'-phos-

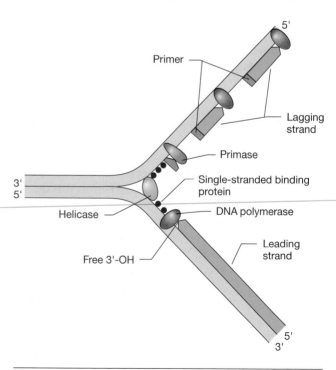

FIGURE 6.17   Events at the DNA replication fork.

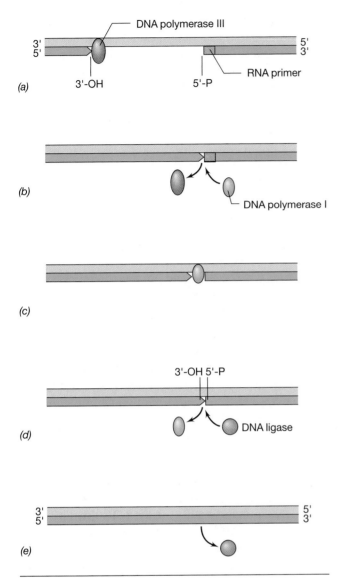

(a) DNA polymerase III   3'-OH   5'-P   RNA primer

(b) DNA polymerase I

(c)

(d) 3'-OH 5'-P   DNA ligase

(e)

FIGURE 6.18 Sealing two fragments on the lagging strand. (a) DNA polymerase III is synthesizing DNA in the 5' → 3' direction toward the RNA primer of a previously synthesized fragment on the lagging strand. (b) On reaching the fragment, DNA polymerase I replaces III. (c) DNA polymerase I continues synthesizing DNA while removing the RNA primer from the previous fragment. (d) DNA ligase replaces DNA polymerase I after the primer has been removed. (e) DNA ligase seals the two fragments together.

phate to 3'-hydroxyl (always adding a *new* nucleotide to the 3'-OH of the growing chain). On the strand growing from the 5'-phosphate to the 3'-hydroxyl, called the **leading strand,** DNA synthesis can occur *continuously* because there is always a free 3'-OH at the replication fork to which a new nucleotide can be added. But on the opposite strand, called the **lagging strand,** DNA synthesis must occur *discontinuously* (because there is no 3'-OH at the replication fork to which a new nucleotide can attach). Where is the 3'-OH on this strand? At the *opposite* end, *away* from the replication fork. Therefore, on the lagging strand, a small (11-base) RNA primer must be synthesized by primase to provide free 3'-OH groups. After synthesizing the primer, primase is replaced by the enzyme DNA polymerase III. Then deoxyribonucleotides are added until DNA polymerase III reaches the previously synthesized DNA.

At this point, DNA polymerase III stops. The next enzyme that is involved, *DNA polymerase I,* has more than one activity. It can clearly synthesize DNA. However, at the same time it is adding nucleotides on to the 3'-OH, it has an *exonuclease* activity that removes the RNA primer from in front of it (Figure 6.18). When the primer has been removed and replaced with DNA, DNA polymerase I is then released. The last phosphodiester bond is made by an enzyme called *DNA ligase.* (This enzyme can seal any nicks made in DNAs that have a 5'-phosphate and 3'-OH and along with DNA polymerase I is also involved in DNA repair.)

Each short stretch of DNA made by DNA polymerase III on the lagging strand is called an *Okazaki fragment* and is about 1000 bases long. Each of these must be primed individually. By contrast, the leading strand is primed only once, at the origin. Because of bidirectionality there are two replicating forks going in opposite directions (see Figure 6.16), which means that at the origin there are two leading strands and two lagging strands started (Figure 6.19).

As can be seen from the preceding discussion, the *Escherichia coli* enzyme called DNA polymerase II does not seem to be involved in chromosomal replication. Interestingly, however, this enzyme is closely related to the major chromosomal DNA polymerase activities of both the Archaea and the Eukarya.

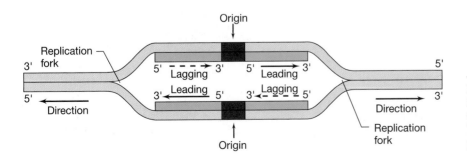

FIGURE 6.19 At an origin of replication that directs bidirectional replication, two replication forks must start. Therefore, two leading strands must be primed, one in each direction.

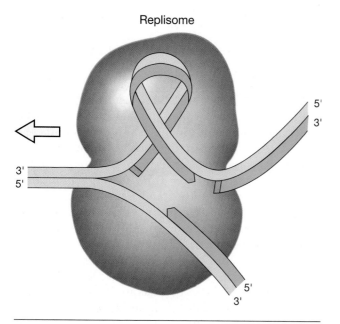

Replisome

FIGURE 6.20   Movement of the replisome, a complex of proteins and enzymes involved in DNA replication, on the double helix. All the reactions shown in Figure 6.17 are taking place, but the looping out of the lagging strand allows the complex to move forward smoothly at the replication fork.

While DNA synthesis is continuing at the replication fork, changes in the coiling of the DNA are occurring, modified by unwinding enzymes and topoisomerases (see Section 6.3). Unwinding is obviously an essential feature of DNA replication, and because supercoiled DNA is under strain, it unwinds more easily than DNA that is not supercoiled. Thus, by regulating the degree of supercoiling, topoisomerases regulate the process of replication (and also transcription, as discussed later).

Figure 6.17 shows the differences in replication of the leading and the lagging strands, and the various enzymes involved. It would appear from such a simplified drawing that each replication fork must consist of DNA polymerase moving smoothly along, synthesizing the leading strand, and one or more polymerases jumping about synthesizing the lagging strand. Actually the two strands are being synthesized by a duplex of DNA polymerases. This is made possible by a "looping" of the lagging strand as shown in Figure 6.20. The *replisome* is a complex containing helicases, primase, two DNA polymerase III molecules, and other associated proteins. Recent evidence indicates that the DNA polymerase complex is also fixed near the midpoint of the bacterial cell and serves as a *replication factory*, pulling the DNA template through it as replication occurs. Therefore, it is the DNA and not the polymerase which moves during replication.

Fidelity of DNA Replication: Proofreading

Errors in DNA replication introduce mutations. Mutation rates in living organisms are remarkably low, between 10^{-8} and 10^{-11} errors per base pair inserted. Part of the reason for this accuracy is that DNA polymerase actually gets *two* chances to incorporate the correct base at a given site. The first chance occurs when complementary bases are inserted by base-pairing rules, A with T and G with C, using the template strand as the pattern. The second chance occurs because of a second enzymatic activity, referred to as **proofreading,** associated with DNA polymerase III (Figure 6.21). In addition to inserting nucleotides in the replicating strand, DNA polymerase also contains a $3' \rightarrow 5'$ *exonuclease* activity that can remove a misinserted nucleotide and allow it to

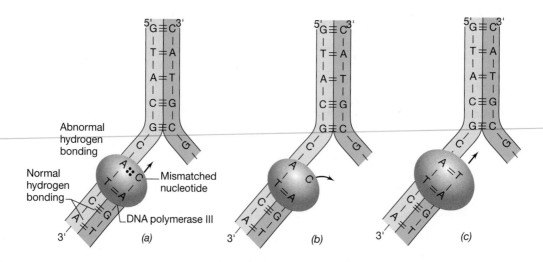

FIGURE 6.21   Proofreading by the $3' \rightarrow 5'$ exonuclease activity of DNA polymerase III. (a) A mismatch in base pairing at the terminal base pair causes the polymerase to pause briefly. This is a signal for the proofreading activity (b) to excise the mismatched nucleotide, after which the correct base is incorporated (c) by polymerase activity.

be replaced with the correct nucleotide. Proofreading activity is summoned if an incorrect base has been inserted because misinsertion creates unstable base pairing. This proofreading activity gives the polymerase activity a second chance to insert the correct base (Figure 6.21). (Note that the proofreading exonuclease activity is the *opposite* of the $5' \rightarrow 3'$ exonuclease activity of DNA polymerase I used to remove the primer from "in front" of the polymerase.)

Exonuclease proofreading occurs in prokaryotes, eukaryotes, and viral DNA replication systems. In addition to exonucleolytic proofreading capabilities, prokaryotes and eukaryotes contain *endonucleolytic* enzymes capable of removing a misinserted nucleotide long after DNA polymerase has passed the point of the error (⌀ Sections 9.1 and 9.2). The combination of endonucleolytic and exonucleolytic (proofreading) activities ensures nearly error-free replication of the extremely long DNA sequences that make up genomic DNA.

We have mentioned a number of enzymes and other proteins that combine or act on DNA. A summary of some of these enzymes is given in Table 6.4.

✓ 6.6b Concept Check

DNA synthesis begins at a unique location called the origin of replication. The double helix is unwound by helicase and is stabilized by single-stranded binding protein. Extension of the DNA occurs continuously on the leading strand but discontinuously on the lagging strand. Most errors in base pairing are corrected by proofreading functions associated with the action of DNA polymerase.

✓ What is the *leading strand*?
✓ What is the *lagging strand*?

Replicating Linear Genetic Elements

We used a circular DNA molecule in discussing the steps in DNA replication. Circular DNA molecules are common; most prokaryotic chromosomes are circular, as are most plasmids and some viruses. Almost all the steps in replication are identical whether the chromosome is linear or circular. However, there is one problem with replication of linear genetic elements that circular ones do not have, and that problem is at the extreme 5'-end of each strand. To understand the problem, refer back to Figure 6.15. Imagine that the left end of the DNA in this diagram is actually one end of a linear chromosome. Even if the RNA primer is very short and there is a special enzyme to remove it, no DNA polymerase can replace it with DNA since *all* DNA polymerases require a primer. Therefore, if nothing is done, the DNA molecule will become shorter each time it is replicated. Genetic elements that are linear have clearly solved this problem!

In fact, there are many solutions to this problem. Some viruses having linear chromosomes actually circularize themselves by their sticky ends, as shown in Figure 6.18. Some other viruses have direct repeats at each end of their chromosomes. A recombination process (a joining together of different DNA molecules) uses the repeats to join several partially replicated DNA molecules together into a very large molecule from which perfect copies are cut by endonucleases (⌀ Section 8.11). Several types of viruses and many linear plasmids solve the problem of replicating linear DNA by using not an *RNA* primer but rather a *protein* primer. Although all DNA polymerases must add each nucleotide to a free —OH group, some DNA polymerases can add the first base onto an —OH group found on specific proteins that bind to the ends of these linear

TABLE 6.4	Enzymes affecting DNA	
Enzyme	**Action**	**Function in the cell**
Restriction endonuclease	Cuts DNA at specific base sequences	Destroys foreign DNA
DNA ligase	Links DNA molecules	Completes replication process
DNA polymerase I	Attaches nucleotides to the growing DNA molecule, removes RNA primers	Fills gaps in DNA, primarily for DNA repair, and removes primers
DNA polymerase III	Attaches nucleotides to the growing DNA molecule, proofreads each inserted nucleotide	Replicates DNA
DNA gyrase (topoisomerase II)	Increases the twisting pattern of DNA, promoting supercoiling	Maintains compact structure of DNA
DNA helicase	Binds to DNA near replicating fork	Promotes DNA strand separation
DNA methylase	Places methyl groups on DNA bases, thus inhibiting restriction endonuclease action	Modifies cellular DNA so it is not affected by its own restriction endonuclease
Primase	Makes short RNA chains using a DNA template	Needed to make primer to be used by DNA polymerase
Topoisomerase I	Relaxes supercoiled DNA	Helps maintain the proper level of supercoiling

chromosomes (Figure 6.22). These proteins are encoded by the plasmid or virus, and they function to recognize the ends of the chromosomes. These protein primers are not removed, so these particular types of plasmids and viruses have proteins covalently attached to the 5′-ends of their DNA. This may also be the means by which some linear chromosomes of Bacteria, such as those of *Streptomyces lividans*, are replicated.

None of these methods of replicating linear DNA are used to complete the ends of eukaryotic chromosomes (telomeres). Telomeres of eukaryotic chromosomes contain repetitive DNA: a short sequence (often six base pairs) tandemly repeated from 20 to several hundred times (Figure 6.23*a*). The sequences from different eukaryotes are closely related, and one strand always has several guanines. This guanine-rich sequence can be added onto the 3′-end of a DNA molecule by an interesting enzyme called **telomerase** (see Figure 6.23). Telomerases add onto the 3′-ends of linear DNA. *They do not need a DNA template because they contain a small RNA template as a cofactor.* These enzymes can work repetitively to make a long extension. Once this extension is long enough, the other strand can be primed with an RNA primer in the normal

fashion. The telomeres do not need to be a precise number of repeats long, just long enough to ensure that no genetic information becomes lost during DNA replication.

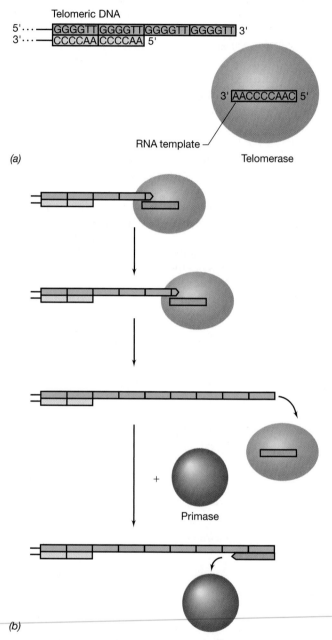

(a)

(b)

FIGURE 6.23   Model for the action of the telomerase at one end of a eukaryotic chromosome. (a) A diagram of the sequence of the end of the DNA in a telomere, with four of the guanine-rich repeats, and the enzyme telomerase, which contains a short RNA template. (b) Steps in elongation of the guanine-rich strand catalyzed by telomerase. After telomerase finishes, the lagging strand can be primed with an RNA primer by primase. The next step (not shown) would be completion of the lagging strand by DNA polymerase and ligase.

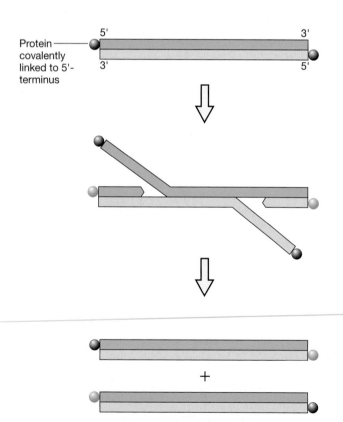

FIGURE 6.22   Replication of linear DNA using protein primers. The new strands of DNA are primed by proteins that stay covalently attached to the 5′-ends.

✓ 6.6c Concept Check

The ends of linear genetic elements present a problem to the replication machinery that circular genetic elements do not. Some prokaryotic linear elements solve this problem using a protein primer. Eukaryotes solve the problem using a special enzyme called telomerase to extend one strand of the DNA.

✓ What is a *protein primer?*
✓ What is *telomerase?*

We have given only a few of the basics of DNA replication; in particular, we have presented very little about the *regulation* of this replication. We have not discussed termination of replication of the circular molecule at all. Although all the details are not known, it is quite clear that the two replicating forks do not just smash into each other like runaway trains. There are specific DNA sequences and specific proteins involved in slowing down the replication forks and allowing replication to be completed. When the replication of the circular molecule is complete, the two circular molecules are linked together, much like the links of a chain. These can be unlinked by a topoisomerase.

One final problem remains which we also shall not discuss in any detail, the partition of the replicated double helices into the daughter cells. In Chapter 3 we outlined how cell wall synthesis is coupled with cell division (∞ Section 3.9). Obviously it is critical that, after DNA replication, the DNA is partitioned so that each daughter cell has a copy of the chromosome.

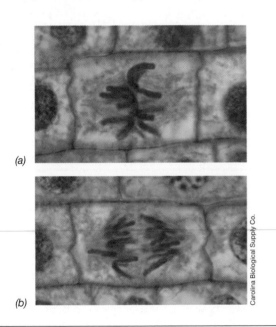

(a)

(b)

Carolina Biological Supply Co.

FIGURE 6.24 Mitosis, as seen in the light microscope. These are onion root tip cells that have been stained to reveal nucleic acid and chromosomes. (a) Metaphase. Chromosomes are paired in the center of the cell. (b) Anaphase. Chromosomes are separating.

This process has been investigated for a much longer time in higher eukaryotes because much of the cellular apparatus involved is visible in the light microscope. In eukaryotic cells, the nucleus divides following a doubling of chromosome number, a process called **mitosis,** yielding two cells, each with a full complement of chromosomes. It is during mitosis that the eukaryotic chromosomes are most compacted (see Section 6.2) and most easily visible (Figure 6.24). Small protein tubes called *microtubules* play important roles in the mitotic process. Microtubules and other protein assemblies attach to specific sequences on the chromosomes and function to form the spindle apparatus, which is the actual structure that moves the chromosomes to the two poles of the dividing cell (see Figure 6.24). There is now some evidence that in prokaryotic cells there may be proteins which attach to specific sequences of the DNA and help pull the chromosomes into the daughter cells.

6.7

Transcription: The Basic Process in Bacteria

Ribonucleic acid (RNA) plays a number of important roles in the expression of genetic information in the cell. Three major types of RNA have been recognized: **messenger RNA (mRNA), transfer RNA (tRNA),** and **ribosomal RNA (rRNA).** These are all products of *transcription* of the information in an organism's DNA. There are three key differences between the chemistry of RNA and that of DNA: (1) RNA has the sugar *ribose* instead of *deoxyribose;* (2) RNA has the base *uracil* instead of the base *thymine;* and (3) except in certain viruses, RNA is not double-stranded. A change from *deoxyribose* to *ribose* affects some of the chemical properties of a nucleic acid, and enzymes that affect DNA in general have no effect on RNA, and vice versa. The change from *thymine* to *uracil* does not affect base pairing, as the two nucleotide bases pair with adenine equally well.

It should be emphasized that RNA acts at two levels, genetic and functional. At the *genetic* level, RNA can carry the genetic information from DNA (mRNA) (or in the case of RNA viruses, play a direct genetic function). At the *functional* level, RNA acts as a macromolecule in its own right, serving a functional and structural role in ribosomes (rRNA) or an amino acid transfer role in protein synthesis (tRNA). Some RNA even has catalytic (enzymatic) activity. In this section we focus our discussion on how RNA is made.

Overview of Transcription

The transcription of genetic information from DNA to RNA is carried out through the action of the enzyme **RNA polymerase,** which catalyzes the formation of

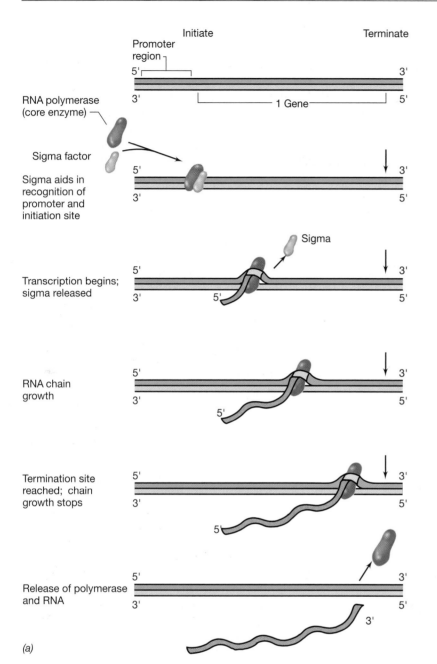

Initiate

Terminate

Promoter region

RNA polymerase (core enzyme)

Sigma factor

Sigma aids in recognition of promoter and initiation site

Sigma

Transcription begins; sigma released

RNA chain growth

Termination site reached; chain growth stops

Release of polymerase and RNA

(a)

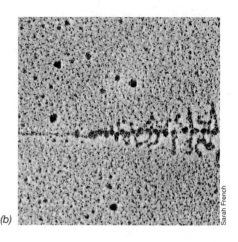

(b)

FIGURE 6.25   Transcription. (a) Steps in messenger RNA synthesis. The initiation and termination sites are specific nucleotide sequences on the DNA. RNA polymerase moves down the DNA chain, causing temporary opening of the double helix and transcription of one of the DNA strands. When a termination site is reached, chain growth stops, and the mRNA and polymerase are released. (b) Electron micrograph of transcription occurring along a gene on the *Escherichia coli* chromosome. The region of active transcription represents about two kilobase pairs of DNA. Transcription proceeds from left to right.

phosphodiester bonds between ribonucleotides. RNA polymerase requires the presence of DNA, which acts as a template. The precursors of RNA are the ribonucleoside triphosphates ATP, GTP, UTP, and CTP. The chemistry of RNA synthesis is much like the chemistry of DNA synthesis (see Figure 6.3). During elongation of an RNA chain, the nucleotides are added to the 3'-OH of the ribose of the preceding nucleotide, which are polymerized with the release of the two high energy phosphate bonds.

Thus, in RNA synthesis (as in DNA synthesis), the overall direction of chain growth is from the 5'-end to the 3'-end, and the *template* strand is antiparallel. Unlike DNA polymerase, however, *RNA polymerase can start chains* (the initial nucleotide in an RNA chain then retains all three phosphates). The first base in the RNA is almost always a purine, either adenine or guanine.

In most cases, the DNA template for RNA polymerase is a double-stranded DNA molecule, but only *one* of the two strands is transcribed for any given gene. The enzyme RNA polymerase differs markedly among Bacteria, Archaea, and Eukarya. The following discussion deals only with RNA polymerase from Bacteria, which has the simplest structure (and about which the most is known). In the next section we will discuss the types of RNA polymerases in different organisms.

All RNA polymerases from Bacteria studied are complex enzymes with closely related subunit structures. The enzyme from *Escherichia coli* has four different types of protein subunits, designated β, β′, α, and σ (sigma), with α appearing in two copies. The subunits interact to form the active enzyme, but the sigma factor is not as tightly bound as the others and easily dissociates, leading to the formation of what is called the *core enzyme* ($\alpha_2\beta\beta'$). The core enzyme alone can catalyze the formation of RNA, and the role of sigma is in *recognition* of the appropriate site on the DNA for the initiation of RNA synthesis. The process of RNA synthesis involving RNA polymerase and sigma is illustrated in Figure 6.25.

RNA polymerase is a large protein and forms contacts with the DNA over many bases simultaneously. As noted (see Section 6.2), proteins can interact specifically with DNA because parts of the base pairs are exposed in the major groove. In order to *start* an RNA chain correctly, RNA polymerase must first recognize the proper region on the DNA. These particular sites on the DNA where RNA polymerase binds are called **promoters.** Note that only *one* strand of the DNA double helix is transcribed at a time. Which strand is transcribed is determined by the orientation of the promoter sequence. RNA polymerase travels away from the promoter region, synthesizing RNA as it moves.

Once the RNA polymerase has bound, the process of transcription can proceed. In this process, the DNA double helix at the promoter is *opened up* by the RNA polymerase (Figure 6.25). As the polymerase moves, it causes the DNA to unwind in short segments, transcription of these segments occurs, and the DNA double helix closes up again. As a result of this transient unwinding, the bases of the template strand are *exposed* and then can be copied into the RNA complement. Thus, the promoter *points* the RNA polymerase in one or the other direction. When a region of DNA has two nearby promoters pointing in opposite directions, then transcription from one of the promoters occurs in one direction (on one of the strands) and transcription from the other occurs in the opposite direction (on the other strand).

Once a small portion of RNA has been formed, the sigma factor dissociates; most of the elongation is therefore carried out by the core enzyme alone (Figure 6.25). Thus, sigma is involved in the formation of only the initial RNA polymerase–DNA complex. As the newly synthesized RNA dissociates from the DNA, the opened DNA closes into the original double helix. Transcription stops at specific regions called **transcription terminators.**

Therefore, unlike replication, which involves copying an entire genome, transcription usually involves much smaller units of DNA, often a single gene. This allows the cell to transcribe different genes at very different frequencies. As we shall see in Chapter 7, regulation of the transcription of specific genes can be a very efficient mechanism of controlling gene expression.

Promoters

As we have noted, the promoter plays a key role in the initiation of RNA synthesis. Promoters are specific DNA sequences where RNA polymerase enzymes attach. The sequences of a large number of promoters from a variety of organisms have been determined. Figure 6.26 shows the sequence of a few promoters from *Escherichia coli*. It is the sigma factor, as part of the RNA polymerase, that is primarily involved in recognition of these promoters.

A single organism can have several different sigma factors, and these can recognize different promoter sequences. All the sequences in Figure 6.26 are recognized by the same sigma factor, the major sigma factor in *E. coli*. If you examine the sequences, you will see that they are not identical. However, two sequences within the promoter region *are highly conserved* between promoters, and it is these that are recognized by sigma. Both sequences precede (are *upstream* of) the site where transcription starts. One is a region 10 bases before the start of transcription, the −10 region (called the *Pribnow box*). Notice that although each promoter is slightly different, many bases are the same. When comparing the −10 regions of all the promoters recognized by this sigma to determine which base occurs most often at each position, one arrives at the *consensus sequence* TATAAT. In our example, each promoter has from three to five matches for these bases. The second region of conserved sequence is about 35 bases from the start of transcription. The consensus sequence in the −35 region is TTGACA. Once again, most of the sequences are not *exactly* the same as the consensus sequence.

Note that in Figure 6.26 the sequence of only one strand is given. By convention among geneticists the strand shown is the one oriented with its 5′-end upstream (therefore, it is *not* the strand used as the template by RNA polymerase). Showing only the sequence of one strand is simply "shorthand" to save the space of writing the other strand. It is essential, though, to re-

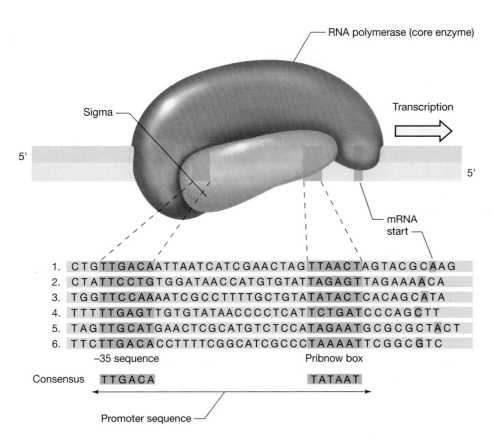

1. CTG TTGACA ATTAATCATCGAACTAG TTAACT AGTACGCAAG
2. CTA TTCCTGTGGATAACCATGTGTAT TAGAGT TAGAAAACA
3. TGG TTCCAAAATCGCCTTTTGCTGTATATACT CACAGCATA
4. TTT TTGAGT TGTGTATAACCCCTCAT TCTGATCCCAGCTT
5. TAG TTGCAT GAACTCGCATGTCTCCAT AGAAT GCGCGCTACT
6. TTC TTGACA CCTTTTCGGCATCGCCCTAAAAT TCGGCGTC

      −35 sequence   Pribnow box

Consensus   TTGACA   TATAAT

                 Promoter sequence

FIGURE 6.26 The interaction of RNA polymerase with the promoter. Shown below the diagram are six different promoter sequences identified in *Escherichia coli*. The contacts of the RNA polymerase with the −35 sequence and the Pribnow box are shown. Transcription begins at a unique base just downstream from the Pribnow box. Below the actual sequences at the −35 and Pribnow box regions are consensus sequences derived from comparing many promoters.

member that promoters are double-stranded, as is the region to be transcribed.

Other sigma factors in other organisms are sometimes much more specific; very little leeway is allowed in the critical bases that are recognized. In *E. coli*, promoters that are most like the consensus are usually more effective in binding RNA polymerase. The more effective promoters are called *strong promoters* and are of considerable value in genetic engineering, as will be discussed in Chapter 10.

Transcription Terminators

As important as initiation of transcription is *termination* of transcription. **Termination** of RNA synthesis occurs at specific base sequences on the DNA. A common termination sequence on the DNA is one containing an inverted repeat with a central nonrepeating segment (see Section 6.2 and Figure 6.7 for an explanation of inverted repeats). When such a DNA sequence is transcribed, the RNA can form a stem-loop structure by intrastrand base pairing (Figure 6.27). When such stem-loop structures *in the RNA* are followed by runs of uridines, they are effective transcription terminators. Other termination sites are regions where a GC-rich sequence is followed by an AT-rich sequence. Such kinds of structures lead to termination without addition of any extra factors and are sometimes termed *intrinsic terminators*.

Other types of terminator sequences have been discovered that require protein factors in addition to RNA polymerase in order to function. In *Escherichia coli* one type of transcription terminator requires a protein called *Rho*. Rho does not bind to RNA polymerase or to DNA but binds tightly to RNA and moves down the chain toward the RNA polymerase–DNA complex. Once RNA polymerase has paused at a *Rho-dependent termination site*, Rho can then cause the RNA and polymerase to leave the DNA, thus terminating transcription. Other proteins involved in transcription termination are, like Rho, RNA-binding proteins. In all cases the sequences involved in termination operate at the level of RNA. However, remember that RNA is transcribed from DNA, and so transcription termination is ultimately determined by *specific nucleotide sequences on the DNA*.

✓ 6.7 Concept Check

The three major types of RNA are messenger RNA (mRNA), transfer RNA (tRNA), and ribosomal RNA (rRNA). The transcription of RNA from the DNA involves the enzyme RNA polymerase, which adds bases onto 3′-ends of growing chains. However, unlike DNA polymerase, RNA polymerase can start a chain. RNA polymerase recognizes a specific start site on the DNA called the promoter. RNA synthesis stops at a transcription terminator.

✓ What is a *promoter*?
✓ What is a *transcription terminator*?

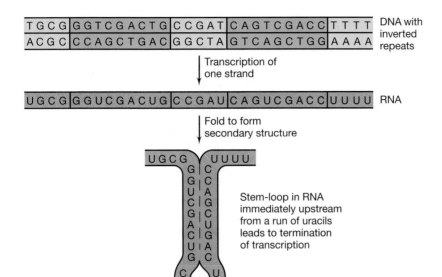

FIGURE 6.27 Inverted repeats in transcribed DNA lead to formation of a stem-loop structure in the RNA, which can result in termination of transcription.

6.8

Transcription: Other Patterns and Inhibition

In Section 6.7 we described how the RNA polymerase of Bacteria locates and transcribes a gene. In this section we will look briefly at the process in other organisms and at inhibition of RNA polymerases. We will also see that prokaryotes (Bacteria and Archaea) can sometimes have more than one gene in the same transcriptional unit.

RNA Polymerases in Eukarya and Archaea

Our overview of transcription dealt with fundamental principles of this process in all organisms, but to this point our discussion of the details has concerned only Bacteria. We mentioned that the RNA polymerase from Bacteria was the simplest. What about these enzymes from other types of organisms?

Eukaryotic organisms have three different types of RNA polymerase in their nuclei, each responsible for the synthesis of a particular type of RNA. All three enzymes in eukaryotes are relatively complex, with many different subunits. **RNA polymerase I** *synthesizes most types of rRNA;* **RNA polymerase II** *synthesizes all the mRNA,* and **RNA polymerase III** *synthesizes tRNA (and one type of rRNA).* The reason for this specificity is that each type of RNA polymerase recognizes only those promoters that occur with the particular class of gene. In Bacteria the promoter for a gene encoding a protein could well be identical to a promoter for a gene encoding a tRNA. That does not happen in eukaryotes. As is the case in Bacteria, the great majority of genes in eukaryotes encode proteins,

and therefore the great majority of genes are transcribed by RNA polymerase II.

These RNA polymerases need to interact specifically with promoter sequences; however, this interaction seems to be considerably more complex than the case in Bacteria, and many other proteins are involved. Figure 6.28 shows a representation of an RNA polymerase II interacting with a typical promoter. This particular promoter contains a *TATA box,* a conserved sequence resembling in some respects the Pribnow box in Bacteria (see Figure 6.26), and an initiator element near the transcription start site. One or both of these sequences are usually present in a promoter recognized by polymerase II, the eukaryotic polymerase most like that from Bacteria. A variety of other sequences and other proteins are typically involved in initiating transcription from these promoters. In genes transcribed by RNA

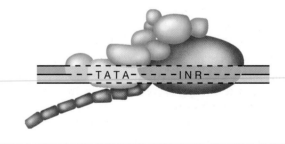

FIGURE 6.28 The interaction of eukaryotic RNA polymerase II with a promoter. The polymerase itself (shown in brown) is positioned at the initiator element (INR) of the promoter. A TATA box binding protein (shown in yellow) is shown bound at the TATA box. The polymerase has a repetitive sequence at one end (shown as a tail-like structure) that can be phosphorylated. The other proteins shown in blue are a few of the very large number of accessory factors required for initiation.

FIGURE 6.29 The sequence elements of a typical promoter from the Archaea.

polymerase II there seems to be a connection between transcription termination and a processing event that takes place at the 3′-end of the mRNA (see Section 6.9).

Like the Bacteria, the Archaea have a single RNA polymerase, but it is more closely related to the RNA polymerase II of eukaryotes than it is to the polymerase of Bacteria (∞ Figure 12.16). The promoter sequences from a variety of archaeans have been compared, and most contain a conserved AT-rich sequence of 6–8 base pairs about 18–27 base pairs upstream of the transcription start site (Figure 6.29). Once again this sequence is similar to that of the Pribnow box found in bacterial promoters (Figure 6.26), but is more similar in sequence and is at about the same position as the TATA box element found in the promoters recognized by eukaryotic RNA polymerase II. In fact, the archaeal sequence is also called a TATA box. Less is known about the transcription termination signals in the Archaea, but in some genes it seems clear that inverted repeats followed by an AT-rich sequence are involved, sequences very similar to those found in many bacterial transcription terminators. However, such sequences are not found in other archaeal genes. One other type of possible transcription terminator contains no inverted repeats, but rather the nucleotide sequence contains repeated stretches with runs of T's.

Specific Inhibitors of RNA Polymerase Action

A number of antibiotics and synthetic chemicals have been shown to specifically inhibit RNA synthesis. For example, a group of antibiotics called *rifamycins* inhibits by attacking the β subunit of the RNA polymerase enzyme. Rifamycin has marked specificity for Bacteria but also inhibits RNA synthesis in chloroplasts and mitochondria of some eukaryotes. Rifamycin has been an especially useful tool in studying nucleic acid synthesis in virus-infected cells. A group of antibiotics called *streptovaricins* is related to rifamycins in structure and function. *Streptolydigin* is an antibiotic that also inhibits transcription by binding to the β subunit of RNA polymerase, but it binds to a different site than rifamycin. Another chemical, *amanitin*, inhibits RNA synthesis in eukaryotes without affecting prokaryotes. Amanitin specifically inhibits RNA polymerase II, the polymerase that synthesizes mRNA.

Actinomycin inhibits RNA synthesis by combining with DNA and blocking elongation. Actinomycin binds most strongly to DNA at guanine-cytosine base pairs, fitting into the major groove on the double strand where RNA is synthesized.

Messenger RNA and Operons

The RNA carrying the information that is translated into a protein is called messenger RNA (mRNA). Most mRNA, in both prokaryotes and eukaryotes, is unstable and is degraded by cellular nucleases. This is in contrast to rRNA and tRNA, which are sometimes referred to as stable RNA. In prokaryotes, a single mRNA molecule often codes for more than one protein (see Figure 6.2). In *prokaryotic* (both Bacteria and Archaea) genetic elements, genes coding for related enzymes are often clustered together. In these situations the RNA polymerase proceeds down the chain and transcribes the whole series of genes into a single long mRNA molecule. An mRNA coding for such a group of genes is called a **polycistronic mRNA** (∞ Section 9.5). Subsequently, when this polycistronic mRNA participates in protein synthesis (see Section 6.12), several polypeptides coded by a single mRNA can be synthesized at one time.

We will discuss regulation of mRNA synthesis in Chapter 7, but introduce here the concept of the operon. An **operon** is a complete unit of gene expression, often involving genes coding for several polypeptides on a polycistronic mRNA or genes coding for ribosomal RNA. In some cases, the transcription of the mRNA for an operon is under the control of a specific region of the DNA, the **operator,** which is adjacent to the coding region of the first gene in the operon. As we shall see in Chapter 7, the operator functions by being able to bind certain regulatory proteins.

For the most part, polycistronic mRNA does not exist in eukaryotes. However, this is because of differences in *translation,* not in transcription. We will discuss these differences later in this chapter after we deal with additional steps often required to convert a transcript from a eukaryotic protein-encoding gene into usable mRNA.

✓ 6.8 Concept Check

Messenger RNA carries the genetic information for the synthesis of specific proteins. Transcription of several genes into a single mRNA molecule may occur in prokaryotes, and so the mRNA may contain the information for more than one polypeptide. Such genes that are transcribed together from a single promoter constitute an operon.

✓ What is *messenger RNA?*
✓ What is an *operon?*

6.9

RNA Processing and Ribozymes

As we discussed (see Section 6.7) transcription can produce several types of RNA: messenger RNA, transfer RNA, and ribosomal RNA. In prokaryotes, a transcript of a protein-encoding gene is typically the actual mRNA and is used directly to make protein. However, transcripts of other types of genes generally need to be *processed* to reach final form. In eukaryotes, all transcripts must be processed before being used. The conversion of a *precursor* RNA to a *mature* RNA is called **RNA processing.** In prokaryotes and eukaryotes, tRNAs and rRNAs are made initially as long precursor molecules, which are then cut to make the final mature RNAs. In addition, many of the bases in tRNA are modified after transcription (see Section 6.11).

In eukaryotes, and much less commonly in prokaryotes, mRNA is also the result of processing a pre-mRNA. As discussed in Section 6.4, the genes of eukaryotes are often interrupted, with noncoding intervening sequences, *introns,* separating the coding regions, *exons.* The *primary transcript* from such a gene must be extensively processed to remove the noncoding regions before the translation process can be initiated. Only a few introns have been discovered in protein-encoding genes in prokaryotes and in certain bacteriophage. The processing step by which introns are removed and exons are joined is called **splicing.**

However, although introns are found in several types of genes in both prokaryotes and eukaryotes, the splicing machinery that removes introns from eukaryotic pre-mRNA is unique. The process involves a complex containing several different ribonucleoproteins (each contains both a small RNA and several proteins) called a **spliceosome.** The spliceosome is a highly complex structure capable of removing introns and joining adjacent exons to form a mature mRNA. Figure 6.30 is a diagram of the two-step reaction by which an intron found in eukaryotic pre-mRNA is removed. Note that there are some conserved bases at the splice junctions and that the intron is removed as a lariat structure. These removed introns are degraded by the cell. In higher eukaryotes there are often many introns in a single gene, and so it is clearly important not only that they be removed but they be removed in the correct order. Some introns (particularly those found in tRNA genes and in genes of the mitochondria or chloroplasts) are removed by a different process involving just proteins. Several introns, including all of those that are found in Bacteria and bacteriophage, are *self-splicing* (ribozymes; see next subsection).

Two other unique steps occur in the processing of eukaryotic mRNA. Both steps take place in the nucleus

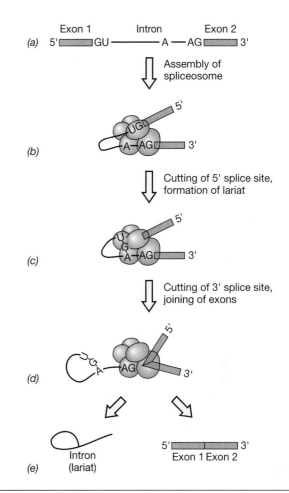

FIGURE 6.30   Removal of an intron from the transcript of a eukaryotic protein-encoding gene. (a) The pre-mRNA with a single intron. The sequence GU is conserved at the 5' splice site and AG at the 3' splice site. There is also an interior A which serves as a branch point. (b) Several small ribonucleoprotein particles (shown in brown) assemble on the RNA to form a spliceosome. Each of these particles contains distinct small RNA molecules that are involved in the splicing mechanism. (c) The 5' splice site has been cut with the simultaneous formation of a branch point. (d) The 3' splice site has been cut, while the two exons were joined. Note that overall two phosphodiester bonds were broken but two others were formed. (e) The final products are the joined exons, the mRNA, and the released intron.

before transport of the mature mRNA into the cytoplasm. The first step is called **capping** and actually occurs before transcription is complete. Capping consists of adding a methylated guanine nucleotide at the 5'-phosphate end, called the *cap* (Figure 6.31). Also, a number of the bases in mRNA are methylated after transcription, and the 2'-hydroxyl group of the ribose is occasionally methylated.

The remaining processing step consists of trimming the 3'-end of the pre-mRNA and adding a *poly-A tail.* This step is called **tailing** and, as we discussed (see Section 6.8), may occur in conjunction with termination.

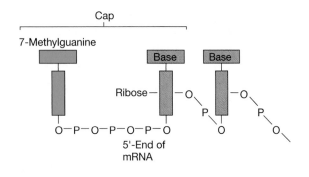

FIGURE 6.31   Structure of the cap added to the 5'-end of eukaryotic mRNA.

All three steps leading to the formation of eukaryotic mRNA are shown in Figure 6.32.

We thus see that the synthesis of a mature, functional RNA is a complex and dynamic process that involves considerably more than the simple transcription of a DNA template.

Ribozymes

We emphasized in Chapter 4 the role of proteins as biochemical catalysts. Many important cellular processes involve ribonucleoproteins, complexes including both

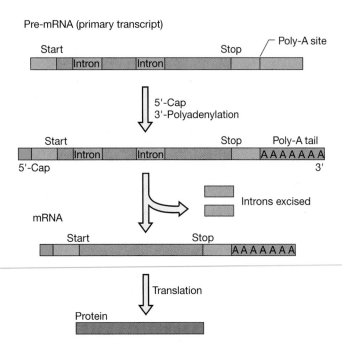

FIGURE 6.32   An overview of the processing of the pre-mRNA into mature mRNA in eukaryotes. The processing steps including adding a cap at the 5'-end, removing the introns, and clipping of the 3'-end of the transcript while adding a poly-A tail. All these steps are carried out in the nucleus. The location of the start and stop codons to be used during translation are also indicated.

RNA and protein. The role of RNA in such complexes was *assumed* to be structural (a place for the proteins to bind) or involved in base pairing with other nucleic acids. As we have seen, there is a short RNA molecule in the enzyme *telomerase* that functions as a template. However, it has now been shown that certain types of RNA can act as *enzymes* as well. Catalytic RNAs, referred to as **ribozymes,** are involved in a number of important cellular reactions. RNA enzymes work like protein enzymes in that an "active site" exists that binds the substrate and catalyzes formation of a product. Ribozymes have been discovered in both prokaryotes and eukaryotes, and in organelles, and others have now been synthesized in laboratories. Studies show that some very short RNAs, as few as 19 bases, can function as ribozymes.

Most ribozymes are **self-splicing introns.** They are *RNA-splicing enzymes* that remove themselves from an RNA molecule while joining adjacent exons together. In one well-studied case of splicing in a ribosomal RNA in *Tetrahymena* (a protozoan), a 413-nucleotide *intron* acts as a ribozyme and splices itself out of a longer precursor rRNA, joining two adjacent exons to form the final rRNA (Figure 6.33). The intron ribozyme acts as a sequence-specific endoribonuclease and, once removed from the precursor RNA, circularizes with the further removal of a short oligonucleotide fragment (Figure 6.33). This particular type of self-splicing intron (a *group I* self-splicing intron) is widespread in nature and is the only type known in Bacteria and bacteriophage. Absolute proof that these ribozymal transformations occur in the absence of specific protein has come from experiments in which the gene for the entire precursor rRNA from *Tetrahymena* has been transferred to *Escherichia coli* where the segment can be transcribed. This transcribed segment carries out the splicing reaction in the complete absence of *Tetrahymena* proteins. There is evidence that proteins play some role in the splicing reaction of members of *group II self-splicing* introns. In this type of intron the splicing reaction itself is quite like that used by the spliceosome (see Figure 6.30), but no accessory RNAs are required.

Self-splicing introns differ from most protein enzymes in that they normally can act only once. However, there is another ribozyme, RNase P, that can act repeatedly on many different substrate molecules because it does not digest itself in the reaction. RNase P is a ribonucleoprotein, but the small RNA (377 nucleotides in *E. coli*) is the catalytic component, not the protein. As is the case for proteins with enzymatic activity, all ribozymes must be folded into the proper structure for activity. In some cases, this structure might be supplied by the secondary structure of the RNA itself. In others, like RNase P, specific proteins may help keep the RNA in the active conformation. RNase P functions in the cell to modify primary transcripts coding for transfer RNAs (see Section 6.11).

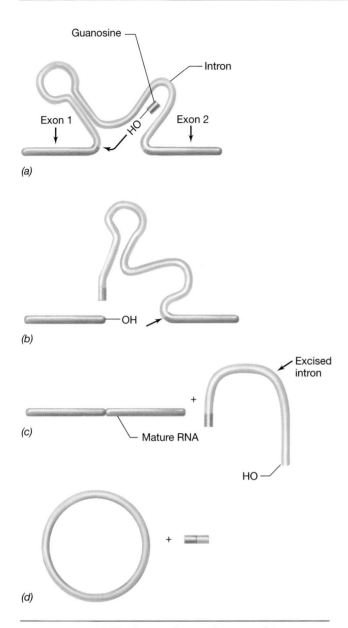

(a)

(b)

(c)

Mature RNA

Excised intron

HO

(d)

FIGURE 6.33 Self-splicing ribozymal intron of the protozoan *Tetrahymena*. There is considerable secondary structure in such molecules, which is critical for the splicing reaction. (a) A ribosomal RNA precursor contains a 413-nucleotide intron. (b) Following the addition of the nucleoside guanosine, the intron splices itself out and joins the two exons. (c) The intron is spliced out. (d) The intron circularizes with the loss of a 15-nucleotide fragment.

The discovery of ribozymes has caused a reevaluation of other cellular processes that involve RNA. For instance, it is now clear that ribosomal RNA plays an active role in protein synthesis, apparently catalyzing the formation of the peptide bonds that link amino acids together in a protein (see Section 6.12). But clearly most enzymes are protein. Why do ribozymes exist? It has been proposed that they are the vestigial remains of a simpler form of life, "RNA life,"

which may have predated the era of proteins as the cell's major catalysts. We discuss this concept in more detail in Chapter 12.

✓ 6.9 Concept Check

RNA molecules are often modified after transcription, an operation called RNA processing. All tRNAs and rRNAs are the result of processing of a longer precursor. The processing of eukaryotic pre-mRNAs is unique and can involve three distinct processing steps: splicing, capping, and tailing. Introns found in some other transcripts are self-splicing, and the RNA itself catalyzes the reaction. RNA molecules with catalytic activity are called ribozymes and play an important role in the cell.

✓ What is *splicing*?
✓ What is a *ribozyme*?

6.10

The Genetic Code

The first two steps in information transfer, *replication* and *transcription*, involve synthesis of nucleic acids using nucleic acid templates. The last step, *translation*, involves a nucleic acid template but in this case the final product is a protein. Proteins are made up of amino acid residues and not bases, so one can imagine that information transfer in translation is considerably more complicated than base pairing. In the next sections we shall deal with the mechanism of translation. In this section we discuss the correspondence between the nucleic acid template and the amino acid sequence of the protein product: the **genetic code.**

As we mentioned in Section 6.1, a triplet of three bases encodes a specific amino acid. Such a triplet is called a **codon.** It is conventional to present the genetic code as mRNA rather than as DNA because it is with mRNA that the translation process occurs. The 64 possible codons of mRNA are presented in Table 6.5. Note that in addition to the codons specifying the various amino acids, there are also special codons for starting (AUG) and for stopping (UAA, UAG, UGA) translation.

Perhaps the most interesting feature of the genetic code is that most amino acids are encoded by several different but related base triplets. This means that in most cases there is no one-to-one correspondence between the amino acid and the codon—knowing the amino acid at a given location does not mean that the codon at that location is automatically known.[†] The property of a code in which there is no one-to-one correspondence between

---

[†]The reverse is true, however. Knowing the DNA codon, one can specify the amino acid in the protein (assuming the proper reading frame is known). This permits the determination of amino acid sequences from DNA base sequences.

TABLE 6.5 The genetic code as expressed by triplet base sequences of mRNA[a]

Codon	Amino acid	Codon	Amino acid	Codon	Amino acid	Codon	Amino acid
UUU	Phenylalanine	UCU	Serine	UAU	Tyrosine	UGU	Cysteine
UUC	Phenylalanine	UCC	Serine	UAC	Tyrosine	UGC	Cysteine
UUA	Leucine	UCA	Serine	UAA	None (stop signal)	UGA	None (stop signal)
UUG	Leucine	UCG	Serine	UAG	None (stop signal)	UGG	Tryptophan
CUU	Leucine	CCU	Proline	CAU	Histidine	CGU	Arginine
CUC	Leucine	CCC	Proline	CAC	Histidine	CGC	Arginine
CUA	Leucine	CCA	Proline	CAA	Glutamine	CGA	Arginine
CUG	Leucine	CCG	Proline	CAG	Glutamine	CGG	Arginine
AUU	Isoleucine	ACU	Threonine	AAU	Asparagine	AGU	Serine
AUC	Isoleucine	ACC	Threonine	AAC	Asparagine	AGC	Serine
AUA	Isoleucine	ACA	Threonine	AAA	Lysine	AGA	Arginine
AUG (start)[b]	Methionine	ACG	Threonine	AAG	Lysine	AGG	Arginine
GUU	Valine	GCU	Alanine	GAU	Aspartic acid	GGU	Glycine
GUC	Valine	GCC	Alanine	GAC	Aspartic acid	GGC	Glycine
GUA	Valine	GCA	Alanine	GAA	Glutamic acid	GGA	Glycine
GUG	Valine	GCG	Alanine	GAG	Glutamic acid	GGG	Glycine

a The boxes of codons are colored according to the scheme: ▢ ionizable: acidic, ■ ionizable: basic, ▨ nonionizable polar, and ▢ nonpolar (🔗 Figure 2.12). The nucleotide on the left is at the 5′-end of the triplet.

b AUG encodes *N*-formylmethionine at the beginning of mRNAs of Bacteria.

word and code is called **degeneracy** (a term derived from the technical field of cryptography).

As we shall see (Section 6.12), in the cell a codon is *read* by base-pairing with a tRNA at a sequence of three bases called an **anticodon.** If the base-pairing involved was always the standard pairing of A with U and G with C, then one would expect that there must be at least one specific tRNA for each codon and, therefore, for some amino acids there must be several tRNAs. For instance there are six different tRNAs in *Escherichia coli* that carry the amino acid leucine. However, it is also true that some tRNAs can read more than one codon. For instance, there is only one tRNA in *E. coli* that carries the amino acid lysine and it can read either AAA or AAG (see Table 6.5). This is possible because in some cases, tRNA molecules form *standard* base pairs at only the first two positions of the codon, tolerating unusual base pairs at the third position. This apparent mismatch phenomenon, called **wobble,** is illustrated in Figure 6.34. The pairing between G and U is allowed at the wobble position.

Start and Stop Codons

As seen in Table 6.5, a few triplets do not correspond to any amino acid. These triplets (UAA, UAG, UGA) are the **nonsense,** or **stop, codons,** and they signal the termination of translation of the gene coding for a specific protein (see Section 6.12).

What is the mechanism by which the proper *starting point* for translation is found? The message is read by reading the **start codon, AUG,** which at the beginning of the

message, codes for the amino acid *N*-formylmethionine (or methionine in Eukarya and Archaea). The importance of having a well-defined starting point is readily understood if we consider that with a triplet code it is absolutely essential that translation begin at the correct location because if it does not, the whole **reading frame** will be shifted and an entirely different protein (or no protein at all) will be formed. By convention the "correct" reading frame, that is the one that can be translated to give the protein that is encoded by the gene, is called the 0 frame. As can be seen in Figure 6.35 the other two reading frames (−1 and +1) do not encode the same amino acid sequence. Therefore it is imperative that the ribosome find

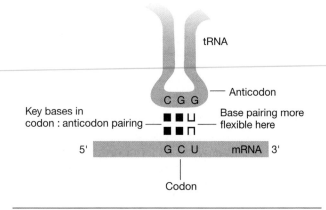

FIGURE 6.34 The wobble concept: base pairing is more flexible for the third base of the codon. Only a portion of the tRNA is shown (see Figure 6.36).

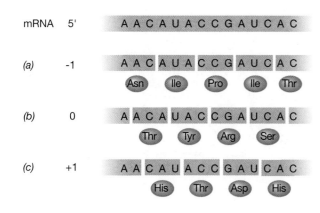

FIGURE 6.35   Possible reading frames in an mRNA. An interior sequence of an mRNA is shown. The "correct" (or 0) reading frame is determined by the start codon of the mRNA. (a) The amino acids that would be encoded by this region of the mRNA if the ribosome were in the −1 reading frame. (b) The amino acids that would be encoded if the ribosome were in the correct reading frame. (c) The amino acids that would be encoded if the ribosome were in the + 1 reading frame.

the correct start codon and that once it has done so it move down the mRNA exactly three bases at a time.

As we will discuss in Section 6.12, ribosomes from Bacteria recognize a specific AUG as a start codon with the aid of an upstream sequence on the mRNA called the Shine–Dalgarno sequence. This extra help at initiation explains why a few messages from Bacteria actually use other codons, such as GUG, for a start codon. However, even these unusual start codons specify N-formylmethionine.

Open Reading Frames

Today the genomes of several organisms have been sequenced and many more are being sequenced (including the human genome). This sequencing information would be useless unless the scientists can relatively easily determine the location of protein encoding genes. How can this be determined?

One common method is to examine each strand of the DNA sequence for **open reading frames.** Remember that mRNA is transcribed from DNA so that if one knows the sequence of DNA, one also knows the sequence of RNA which could be transcribed from it. If this RNA can be translated it must contain an open reading frame: a start codon (typically AUG) followed by some number of codons and then a stop codon in the same frame as the start codon. Such a base sequence is called an **open reading frame (ORF).** A computer can be programmed to scan long base sequences in DNA databases to look for open reading frames. The search for ORFs is very useful in genetic engineering (∞ Chapter 10) when one has isolated and sequenced an unknown piece of DNA and

is not certain whether or not it encodes protein. The computer search for ORFs permits the researcher to locate putative genes that were previously unsuspected.

✓ 6.10 Concept Check

The genetic code is expressed in terms of RNA, and a single amino acid may be encoded by several different but related codons. In addition to the nonsense, or stop, codons, there is also a specific start codon that signals the location where the translation process should begin.

✓ What is meant by a *degenerate code?*
✓ What is an *open reading frame?*

6.11

Transfer RNA

As we saw in Section 6.10, it is the anticodon on the tRNA that "reads" (base pairs) with the codon. However, a tRNA is much more than simply an anticodon (Figure 6.36). The tRNA not only has a specificity for the codon on the mRNA, it also has specificity for the appropriate amino acid. The transfer RNA and its specific amino acid are brought together by means of specific enzymes that ensure that a particular tRNA receives its correct amino acid. These enzymes, called **amino acid activating enzymes** or **aminoacyl-tRNA synthetases,** have the important function of recognizing *both* the amino acid *and* the specific tRNA for that amino acid.

Structure of tRNA

The detailed structure of tRNA is now well understood. There are about 60 different specific tRNAs in bacterial cells and 100–110 in mammalian cells. Transfer RNA molecules are short, single-stranded molecules with lengths (among different tRNAs) of 73–93 nucleotides. When compared, it has been found that certain bases and secondary structures are constant for all tRNAs, and there are other parts that are variable. Transfer RNA molecules also contain some purine and pyrimidine bases differing slightly from the normal bases found in RNA in that they are chemically modified, often methylated. These modifications are added to the bases after transcription. Some of these unusual bases are pseudouridine, inosine, dihydrouridine, ribothymidine, methyl guanosine, dimethyl guanosine, and methyl inosine. Base modifications as well as other types of processing (see Section 6.9) are necessary to make a functional tRNA from the transcript of a tRNA-encoding gene. Although the molecular structure of tRNA is single-stranded, there are extensive double-stranded regions within the molecule as a result of internal base pairing when the molecule folds back on itself.

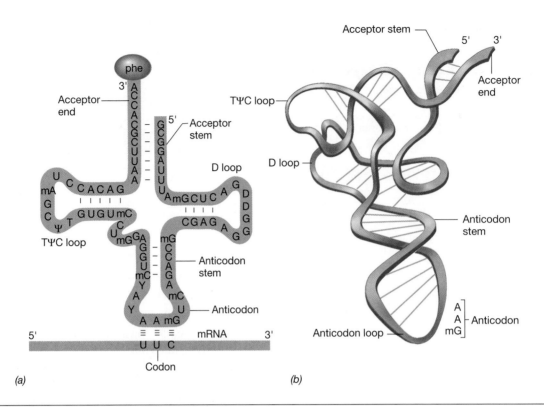

FIGURE 6.36   Structure of a transfer RNA, yeast phenylalanine tRNA. (a) The conventional cloverleaf structure. The amino acid is attached to the ribose of the terminal A at the acceptor end. A, adenine; C, cytosine; U, uracil; G, guanine; ψ, pseudouracil; D, dihydrouracil; m, methyl; Y, a modified purine. (b) In actuality the molecule folds so that the D loop and TψC loops are close together and associate by hydrophobic interactions.

The structure of tRNA is generally drawn in cloverleaf fashion, as in Figure 6.36a. Some regions of secondary structures are given names having to do either with the bases most often found there (the TψC loop and the D loop) or with specific functions (anticodon loop and acceptor end). The three-dimensional structure of a tRNA is more clearly shown in Figure 6.36b. Note that bases that appear widely separated in the cloverleaf model are actually close together when viewed in three dimensions. This means that some of the bases in the "loops" are actually paired with bases in other loops.

One of the variable parts of the tRNA molecule contains the **anticodon,** the site recognizing the codon on the mRNA. The anticodon is found in the *anticodon loop,* shown in Figure 6.36. There are just *three* nucleotides in the anticodon loop that are specifically involved in the recognition process and that base-pair with the codon (see Section 6.10). Other portions of the tRNA interact with the ribosome (both rRNA and protein), other protein factors, and the activating enzyme. At the 3'-end of all tRNAs, three nucleotides are unpaired. The sequence of these three nucleotides is always the same, cytosine-cytosine-adenine (CCA), and it is to the ribose

sugar of the terminal A that the amino acid is covalently attached via an ester linkage. From this acceptor portion of the tRNA, the amino acid is transferred to the growing polypeptide chain on the ribosome by a mechanism that will be described in the next section.

Recognition, Activation, and Charging of tRNA

Recognition of the correct tRNA by an aminoacyl-tRNA synthetase involves specific contacts between key regions of the nucleic acid and particular amino acids of its respective synthetase (Figure 6.37). As might be expected because of the unique sequence in this region, the *anticodon* of the tRNA is important in recognition by the synthetase. However, other contact sites between the tRNA and the synthetase are also important. Studies of tRNA binding to aminoacyl-tRNA synthetases in which specific bases in the tRNA have been changed by genetic mutation have shown that only a small number of key nucleotides in a tRNA besides the anticodon region are involved in recognition; these other key recognition nucleotides are often part of the acceptor stem of the tRNA molecule (see Figure 6.36). In a few cases, recognition of a tRNA by its cognate synthetase is to-

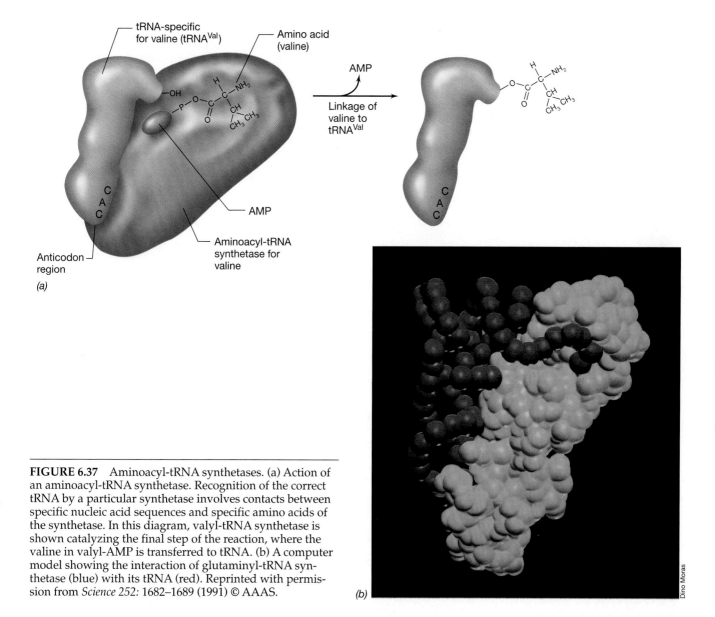

FIGURE 6.37 Aminoacyl-tRNA synthetases. (a) Action of an aminoacyl-tRNA synthetase. Recognition of the correct tRNA by a particular synthetase involves contacts between specific nucleic acid sequences and specific amino acids of the synthetase. In this diagram, valyl-tRNA synthetase is shown catalyzing the final step of the reaction, where the valine in valyl-AMP is transferred to tRNA. (b) A computer model showing the interaction of glutaminyl-tRNA synthetase (blue) with its tRNA (red). Reprinted with permission from *Science 252:* 1682–1689 (1991) © AAAS.

tally independent of the anticodon region. It should be emphasized at this point that the fidelity of this recognition process is crucial, for if the wrong amino acid is attached to the tRNA, it may be inserted in the improper place in the polypeptide, leading to the synthesis of a faulty protein.

The specific chemical reaction between amino acid and tRNA catalyzed by the aminoacyl-tRNA synthetase first involves *activation* of the amino acid by reaction with ATP:

$$\text{Amino acid} + \text{ATP} \rightleftharpoons \text{aminoacyl-AMP} + \text{P–P}$$

The aminoacyl-AMP intermediate formed normally remains bound to the enzyme until collision with the ap-

propriate tRNA molecule, and, as shown in Figure 6.37a the activated amino acid is then transferred to the tRNA to form a *charged* tRNA:

$$\text{Aminoacyl-AMP} + \text{tRNA} \rightleftharpoons \text{aminoacyl-tRNA} + \text{AMP}$$

The pyrophosphate (P–P) formed in the first reaction is split by a pyrophosphatase, forming two molecules of inorganic phosphate. Since ATP is used and AMP is formed, a total of *two* high energy phosphate bonds are required for the activation of an amino acid and charging a tRNA. Once activation and charging have occurred, the aminoacyl-tRNA (AA-tRNA) leaves the synthetase and is brought to the ribosome by a protein factor. The mechanism of protein synthesis is discussed in the next section.

✓ 6.11 Concept Check

One or more transfer RNAs exist for each amino acid found in protein. Enzymes called aminoacyl-tRNA synthetases function to attach an amino acid to a tRNA. Once the correct amino acid is attached to its tRNA, further specificity resides only in the codon-anticodon interaction.

✓ What is the function of a tRNA?
✓ What is the function of an aminoacyl-tRNA synthetase?

6.12

Translation: The Process of Protein Synthesis

It is the amino acid *sequence* that determines the structure (and ultimately the function) of the final active protein. The key objective of protein synthesis is thus placement of the proper amino acid at the proper place in the polypeptide chain. This is the role of the protein-synthesizing machinery of the cell.

Steps in Protein Synthesis

Ribosomes are the site of protein synthesis. Each ribosome is constructed of two subunits. In prokaryotes, the ribosome subunits are of 30S (Svedberg units) and 50S, yielding intact 70S ribosomes.* Each subunit is itself a ribonucleoprotein complex made up of specific ribosomal RNAs and ribosomal proteins. The 30S subunit contains 16S rRNA and about 21 proteins, while the 50S subunit contains 5S and 23S rRNA and about 34 proteins (Table 6.6 and Figure 6.38*a*). In *Escherichia coli*, there are 53 different ribosomal proteins, most present at one copy per ribosome. The actual synthesis of a protein involves a complex cycle in which the various ribosomal components play specific roles.

Although a continuous process, protein synthesis can be thought of as occurring in a number of discrete steps: **initiation, elongation, termination-release,** and **polypeptide folding.** The first two steps are outlined in Figure 6.38*b*. In addition to mRNA, tRNA, and ribosomes, the process involves a number of proteins designated initiation, elongation, and termination factors; guanosine triphosphate provides energy for the process.

---

*The numbers 30S, 50S, and 70S refer to Svedberg units, which are units of sedimentation coefficients of ribosome subunits or intact ribosomes when subjected to centrifugal force in an ultracentrifuge.

TABLE 6.6	Ribosome structure[a]	
Property	**Prokaryote**	**Eukaryote**
Overall size	70S	80S
Small subunit	30S	40S
Number of proteins	~21	~30
RNA size (number of bases)	16S (1500)	18S (2300)
Large subunit	50S	60S
Number of proteins	~34	~50
RNA size (number of bases)	23S (2900) 5S (120)	28S (4200) 5.8S (160) 5S (120)

[a] Ribosomes of mitochondria and chloroplasts of eukaryotes are similar to prokaryotic ribosomes (◯◯ Section 3.16).

Initiation of Protein Synthesis

In prokaryotes initiation always begins with a free 30S ribosome subunit, and an **initiation complex** forms consisting of a 30S ribosome subunit, mRNA, formylmethionine tRNA, and initiation factors. Guanosine triphosphate is required for this step. To this initiation complex a 50S ribosome subunit is added to make the active 70S ribosome. At the end of the translational process, the released ribosome separates again into 30S and 50S subunits. Just preceding the initiation codon on the mRNA is a sequence of from three to nine nucleotides (the so-called **Shine–Dalgarno sequence)** that is involved in the binding of the mRNA to the ribosome. This *ribosome binding site* at the 5'-end of the mRNA is complementary to the 3'-end of the 16S RNA of the ribosome, and it is thought that base pairing ensures effective formation of the ribosome–mRNA complex.

The presence of the Shine–Dalgarno site on the mRNA and its specific interaction with 16S rRNA allow prokaryotic ribosomes to use *polycistronic* mRNA because the ribosome can find each initiation site within a message (see Section 6.8). Eukaryotic ribosomes typically recognize an mRNA by its 5'-cap and initiate only at the first possible initiation codon. Therefore, they normally cannot translate polycistronic mRNA.

Initiation always begins with a special initiator aminoacyl-tRNA binding to the **start codon,** AUG. In Bacteria this is **formylmethionine** tRNA. Subsequently, the formyl group at the N-terminal end of the polypeptide is removed; the terminal amino acid of the completed protein is hence methionine. Because the Shine–Dalgarno sites (and other possible interactions between the rRNA and the mRNA) are also involved in directing the ribosome to a start site, prokaryotic mes-

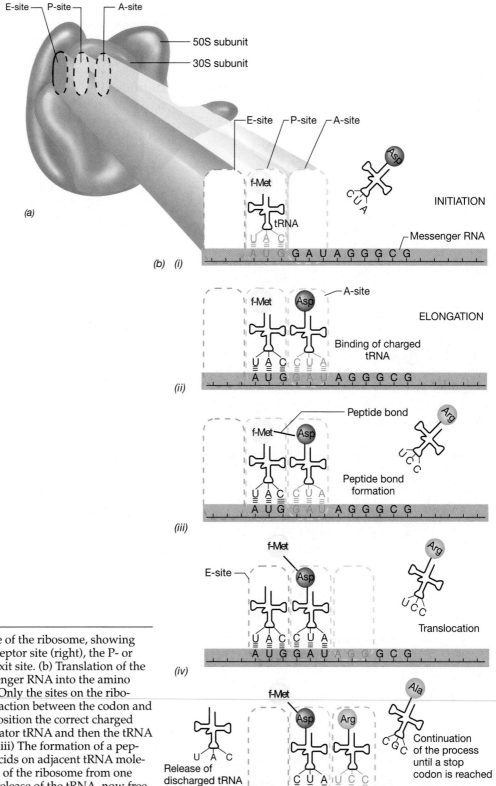

FIGURE 6.38 (a) Structure of the ribosome, showing the position of the A- or acceptor site (right), the P- or peptide site, and the E- or exit site. (b) Translation of the information from the messenger RNA into the amino acid sequence of a protein. Only the sites on the ribosome are shown. (i, ii) Interaction between the codon and anticodon brings into the position the correct charged tRNA—in this case the initiator tRNA and then the tRNA for the second amino acid. (iii) The formation of a peptide bond between amino acids on adjacent tRNA molecules. (iv) The translocation of the ribosome from one codon to the next with the release of the tRNA, now free of an amino acid, from the E-site. (v) The next charged tRNA binds to the A-site.

sages sometimes use a start codon other than AUG. The most common alternative start codon is GUG. However, when used in this unusual way, GUG calls for formylmethionine initiator tRNA (and not valine). In Eukarya and Archaea, initiation begins with methionine instead of formylmethionine. (Although all proteins are *initiated* with a methionine, this amino acid is often removed by a specific protease after translation.)

Elongation and Termination

The mRNA is threaded through the ribosome primarily bound to the 30S subunit. The ribosome contains other sites where the tRNAs interact. Two of these sites are located primarily on the 50S subunit, and they are termed the P-site and the A-site (see Figure 6.38*b*). The A-site, the **acceptor** site, is the site where the new AA-tRNA first attaches. The P-site, the **peptide** site, is the site where the growing peptide is held by a tRNA. During peptide bond formation, the peptide moves to the tRNA at the A-site as a new peptide bond is formed. Several soluble (nonribosomal) elongation factors are required for **elongation,** as well as additional molecules of GTP (to simplify Figure 6.38*b*, the elongation factors are omitted and only a portion of the ribosome is shown). The tRNA that holds the peptide must now be moved (translocated) from the A-site to the P-site, thus opening up the A-site for another AA-tRNA.

Translocation requires a specific elongation factor and one molecule of GTP per each tRNA translocated. At each translocation step the message is advanced three nucleotides, exposing a new codon at the ribosome A-site. It had been thought that translocation caused the empty tRNA to be released from the ribosome. However, it now appears that translocation pushes this empty tRNA to a third site, called the E-site. It is from this **exit** site that the tRNA is actually released from the ribosome. The precision of the translocation step is critical to the accuracy of protein synthesis. The ribosome must move exactly three bases (one codon) at each step. The ribosome is a dynamic structure: both mRNA and tRNA move through the ribosome as it carries out each of the stages of protein synthesis. Thus, the three sites we have discussed are not simply static locations on the surface but are *moving parts* of a complex biomolecular machine.

When several ribosomes are simultaneously translating a single message, the complex is called a **polysome** (Figure 6.39). Polysomes increase the speed and efficiency of mRNA translation, and because each ribosome acts independently of the others, each ribosome in a polysome complex makes a complete polypeptide (Figure 6.39). Note in Figure 6.39 how ribosomes closest to the 5′-end (the beginning) of the mRNA molecule have short polypeptides attached to them because only a few codons have been read, while ribosomes closest to the 3′-end of the message have nearly finished polypeptides.

The **termination** of protein synthesis occurs when a codon is reached that does not specify an AA-tRNA. There are three codons of this type, the **stop,** or **nonsense, codons** (see Section 6.10); they serve as the stopping points for protein synthesis. No tRNA binds to a stop codon, but instead proteins called *release factors* read the chain-terminating signal and serve to cleave the attached polypeptide from the terminal tRNA. Following this, the ribosome dissociates, and the subunits are then free to form new initiation complexes.

Role of Ribosomal RNA in Protein Synthesis

Ribosomes are composed of a series of proteins and ribosomal RNAs (see Table 6.6). Two decades ago it was assumed that ribosomal proteins were probably the functional components of the ribosome, whereas the role of the ribosomal RNAs was largely structural, that

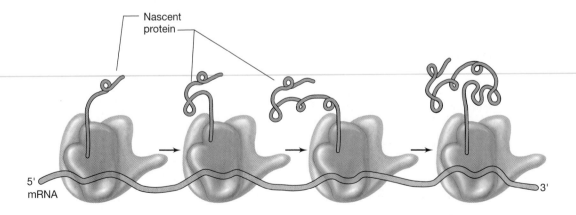

FIGURE 6.39   Translation by several ribosomes on a single messenger RNA (polysome). Note how the ribosomes nearest the 5′-end of the message are at an earlier stage in the translation process.

is, serving as a support for ribosomal proteins. However, it is now clear that rRNA plays a critical *functional* role in all stages of protein synthesis, from initiation to termination. The role of the many proteins present in the ribosome, although less clear, may be as facilitators of RNA function by stabilizing or positioning the key functional sequences in the various ribosomal RNAs.

In prokaryotes, it is clear that 16S rRNA is involved in initiation through base pairing between the ribosome binding sequence (the Shine–Dalgarno sequence) upstream of the start codon on the mRNA and a complementary sequence in 16S rRNA. There is strong evidence that mRNA and rRNA interactions also occur during elongation.

The 23S rRNA seems to play a role in translocation, and the elongation factors are known to interact with 23S rRNA. The 16S rRNA is also involved in termination, possibly interacting with the mRNA or through interactions with the release factors.

Strong evidence also exists for a role for rRNA in ribosome subunit association, as well as for tRNA positioning in the decoding (A- or P-, see Figure 6.38*b*) sites on the ribosome and even in catalyzing peptide bond formation. Charged tRNAs that enter the ribosome recognize the correct codon by codon:anticodon base pairing, but they are also physically attached to the ribosome by interactions of the anticodon stem-loop of the tRNA with specific locations within 16S rRNA. The acceptor end of the tRNA interacts with the 23S rRNA. The peptidyl transferase reaction (the actual formation of peptide bonds that occurs on the 50S subunit of the ribosome) is associated with 23S rRNA. It seems likely that the reaction itself is catalyzed by ribozymal activity (see Section 6.9).

Ribosomal RNA thus plays a major role in translation. Ribosomal function is clearly dependent on the major RNA species present.

Folding and Secreting Proteins

It was long thought that all proteins folded spontaneously into their active form while they were being synthesized (Figure 6.39). However, we now know this is not the case. Many proteins require the assistance of other proteins called **molecular chaperones** for proper folding or for assembly into larger complexes. The chaperones themselves do not become part of the assembled proteins. These proteins seem to be both extremely widespread, and their sequences highly conserved. One type of molecular chaperone, called *chaperonins,* are probably present in all living organisms. The activity of a chaperonin is shown in Figure 6.40. The unfolded or improperly folded protein enters the molecular chaperone where it is folded and then released. Energy for

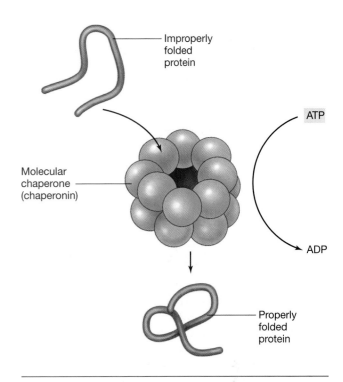

FIGURE 6.40 The action of a molecular chaperone. An improperly folded protein is taken into the barrel-like structure of the chaperonin. ATP is used to supply the energy required to fold the protein properly.

the folding comes from ATP. Other cellular chaperones are involved in carrying the unfolded protein to the chaperonin. In addition to folding newly synthesized proteins, chaperones also can refold proteins that have partially denatured in the cell. Such protein denaturation can occur because the organism has temporarily experienced high temperatures in its environment.

Many proteins are used *outside* the cell and must somehow get from the site of synthesis on ribosomes through the cytoplasmic membrane. In prokaryotes, periplasmic enzymes and extracellular enzymes are secreted or *secretory* proteins. In eukaryotes there are a large number of membrane-enclosed organelles, for example, mitochondria, into which proteins must be transported, as well as proteins that must be secreted outside the cell.

How is it possible for a cell to selectively transfer some proteins across a membrane while leaving most proteins in place in the cytoplasm? Most proteins that must be transported through membranes are synthesized with an extra N-terminal peptide sequence, about 15–20 amino acids long, which is called the **signal sequence.** In this signal sequence, hydrophobic amino acids predominate and permit the enzyme to be threaded through the hydrophobic lipid membrane. In many cases, the ribosomes that synthesize secretory proteins are bound directly to the cytoplasmic membrane, so the protein is formed and passes through the membrane simultaneously. Typically a membrane protein complex, with a pore, is involved. Once the protein has been secreted, the signal sequence is removed by a peptidase enzyme, an example of the process of **posttranslational modification.**

The study of protein secretion has important practical implications for genetic engineering (∞ Chapter 10). If bacteria are genetically engineered to serve as agents for the production of foreign proteins, it is desirable to manipulate the signal sequence in order to arrange for the desired protein to be excreted so it can be readily isolated and purified.

Effect of Antibiotics on Protein Synthesis

A large number of antibiotics inhibit protein synthesis by interacting with the ribosome. These interactions are quite specific, and many have been shown to involve rRNA. Several of these antibiotics are medically useful, and several are also effective research tools because they are specific for different steps in protein synthesis. For instance, *streptomycin* inhibits initiation, whereas *puromycin*, *chloramphenicol*, *cycloheximide*, and *tetracycline* inhibit elongation. Even when two antibiotics inhibit the same overall step in protein synthesis, the mechanisms of the inhibition can be quite different. Puromycin binds to the A-site on the ribosome, and the growing peptide is transferred to it instead of to an AA-tRNA. The puromycin–peptide is then released from the ribosome, halting elongation. Chloramphenicol inhibits elongation by blocking formation of the peptide bond.

Many antibiotics specifically inhibit ribosomes of organisms from only one or two of the phylogenetic domains. Of the antibiotics just listed, chloramphenicol and streptomycin are specific for the ribosomes of Bacteria and cycloheximide for ribosomes of Eukarya. Since mitochondria and chloroplasts have ribosomes of the prokaryotic type, it is of interest that antibiotics inhibiting protein synthesis in Bacteria also generally inhibit protein synthesis in mitochondria and chloroplasts. This is one piece of evidence supporting the hypothesis that mitochondria and chloroplasts were originally derived by intracellular infection of a eukaryotic cell by prokaryotes (∞ Sections 3.16 and 12.3). The fact that an antibiotic inhibits eukaryotic ribosomes as well as those from Bacteria does not necessarily mean it cannot be used in medicine. Although tetracycline also inhibits eukaryotic ribosomes, it apparently does not do so at the concentrations that result from administering it during antibacterial therapy.

One interesting inhibitor of protein synthesis is diphtheria toxin, an agent involved in the pathogenesis of the disease diphtheria. Diphtheria toxin inactivates an elongation factor (∞ Section 19.8) and hence is a powerful inhibitor of protein synthesis in eukaryotes and Archaea. The overall pattern of antibiotic sensitivity of the Archaea is unlike that of either the Bacteria or the Eukarya (∞ Section 15.8).

✓ 6.12 Concept Check

The ribosome plays a key role in the translation process, bringing together mRNA and amino acid–charged tRNAs. There are three sites on the ribosome: the acceptor site, where the charged tRNA first combines; the peptide site, where the growing polypeptide chain is held; and an exit site. During each step of amino acid addition, the message advances three nucleotides (one codon) and the tRNA moves from the acceptor to the peptide site. Termination of protein synthesis occurs when a stop codon, which does not code for any amino acid, is reached. Many antibiotics affect the ribosome.

✓ What are the components of a ribosome?
✓ What functional roles does rRNA play in protein synthesis?

6.13

Translation: Alternatives and Errors

In the previous three sections we have the discussed how the information in a typical protein-encoding gene is translated into a protein. In this section we will mention a few alternatives, including alternative gene structure and alternative genetic codes. We will also discuss errors in protein synthesis.

Overlapping Genes

In Figure 6.35 we showed a segment of DNA within a gene encoding a protein and the three possible reading frames. The underlying assumption is that in any such segment, there is only one correct frame. In the vast majority of cases this assumption is true: The nucleotide sequence specifying one product is separate and distinct from the sequence specifying another product. However, there are exceptions. The first evidence of this came from an analysis of the small bacterial virus φX174. It became clear that the genome of this virus was too small to encode all the proteins known to be involved in its reproduction. This is possible because in certain regions the same DNA codes for more than one product. This is made possible by reading the same nucleotide sequence in two different reading frames, beginning at different sites. A number of interesting patterns of overlapping genes are now known, most of which occur in viruses (∞ Sections 8.8 and 8.9). The use of overlapping genes seems to be a strategy to make maximal use of the small genome.

A FOCUS ON . . . Selenocysteine: The Twenty-first Amino Acid

The genetic code has codons for 20 amino acids that are assembled into proteins during translation. However, many proteins contain other amino acids. In fact, there are well over 100 different amino acids found in at least a few proteins. Until recently, it was thought that these "extra" amino acids were made by modifying one of the standard amino acids *after* it was incorporated into protein, a process called *posttranslational modification*. However, it is now clear that one of these extra amino acids is put into protein by the translational machinery itself. This one exception is *selenocysteine*.

Selenocysteine has the same structure as cysteine, but it has a selenium atom rather than a sulfur atom. It was known for some time that a few proteins contain this unusual amino acid. For example, *Escherichia coli* makes two different formate dehydrogenase enzymes and both contain a single selenocysteine residue. When the gene encoding one of these enzymes was sequenced, it was found that the codon that corresponded to the selenocysteine was a UGA. UGA is normally an efficient stop codon in *E. coli*, but it has now been demonstrated that it can be translated directly as selenocysteine in certain mRNA molecules, not only in *E. coli* but also in other prokaryotes and in eukaryotes, including humans. Therefore, selenocysteine is the twenty-first amino acid known to be encoded by the genetic code.

How can a codon sometimes be a stop codon and sometimes a sense codon *in the same chromosome?* The answer apparently lies in the *context* of the codon, the sequence of the bases surrounding the UGA codon and in their secondary structure. In certain contexts, the translational machinery interprets UGA as "selenocysteine." In all other contexts, UGA means "stop translation." Selenocysteine has its own tRNA (as do all the standard amino acids) and also has a special protein factor that brings only this tRNA to the ribosome.

Selenocysteine is even more readily oxidized than cysteine. Therefore, enzymes that contain this amino acid must be protected from oxygen. It has been proposed that UGA might once have been a normal sense codon, calling only for selenocysteine, but that the increase in oxygen in our environment following the evolution of photosynthesis (∞ Chapter 12) selected for proteins that contain cysteine (whose codons are UGU and UGC). This allowed the coding assignment of UGA to be altered except in a few special cases. ■

$$^-OOC-\underset{\underset{H}{|}}{\overset{\overset{NH_3^+}{|}}{C}}-CH_2-SH$$

Cysteine

$$^-OOC-\underset{\underset{H}{|}}{\overset{\overset{NH_3^+}{|}}{C}}-CH_2-SeH$$

Selenocysteine

Other Genetic Codes

When the genetic code was cracked during the 1960s, all the prokaryotes and eukaryotes examined were found to use the same code. When the mRNA for mammalian hemoglobin was given to the *Escherichia coli* protein-synthesizing machinery (such as ribosomes and tRNA), mammalian hemoglobin was synthesized. Therefore, the genetic code appeared to be a **universal code** in that the exact same code was used by all living systems. Once techniques became available to sequence DNA easily, genes from many organisms began to be sequenced. However, comparison of these DNA sequences to the amino acid sequence of the proteins they encode has led to a few unexpected surprises. One of these, the discovery of introns in eukaryotic genes, was very surprising but did not change our ideas about the genetic code. However, it has also been discovered that some organelles and some cells use genetic codes that are slight variations of the "universal" genetic code (see the box, Selenocysteine: The Twenty-first Amino Acid).

The original findings of these alternative codes were in the genomes of mitochondria. So far as is known, only the mitochondria of plants use the universal code without change. The other organelles in plants, the chloroplasts, also use this standard code. The mitochondria of all other eukaryotes use codes with one or a few slight differences. A few of these variations are shown in Table 6.7. Note that there is *not* simply a mitochondrial code, although there are a few common themes, such as the general use of UGA as a tryptophan codon. It is also clear that these alternate codes are very closely related to the universal code and are almost certainly derived from it evolutionarily. Several examples of cells are now known whose chromosomes also use slightly different codes, and a few examples of these are also given in Table 6.7. Note that all these alternative chromosomal genetic codes have different assignments

TABLE 6.7   Variations in the genetic code[a]

Codon	Universal code	Other codes in cellular chromosomes			Other mitochondrial codes		
		Mycoplasma	*Paramecium*	*Euplotes*	Yeast	Protozoa	Mammals
UGA	Stop	Tryptophan	Stop	Cysteine	Tryptophan	Tryptophan	Tryptophan
UAA/UAG	Stop	Stop	Glutamine	Stop	Stop	Stop	Stop
AUA	Isoleucine	Isoleucine	Isoleucine	Isoleucine	Methionine	Methionine	Methionine
CUA	Leucine	Leucine	Leucine	Leucine	Threonine	Leucine	Leucine
AGA/AGG	Arginine	Arginine	Arginine	Arginine	Arginine	Arginine	Stop

a The universal genetic code is used in the chromosomes of most cells, chloroplasts, plant mitochondria, and their viruses and plasmids. A few organisms use slightly different codes in their chromosomes (in the nucleus). The examples of these other nuclear codes are from *Mycoplasma* (Bacteria) and two different ciliated protozoa (Eukarya). All nonplant mitochondria use variations of the universal code, whereas plant mitochondria use the universal code. The examples here are only a few of the different types known.

for what are normally stop codons. These organisms simply have fewer nonsense codons because one or two are now read as sense.

If every codon has an assignment, you might imagine that it is very difficult to change the genetic code in an organism. For instance, the change of AUA from an isoleucine codon to a methionine codon means that every protein that once had an isoleucine encoded by AUA now has a methionine at this position. Such a protein may not function normally. This may not be a severe problem if *codon usage* is not random. After the genetic code had been worked out by biochemists and before any genes had actually been sequenced, it was assumed that the degenerate codons for an amino acid would be used at an equal frequency. This is another assumption that DNA sequencing has shown to be incorrect! Codon usage is highly biased, and this bias changes from organism to organism. In *Escherichia coli*, for instance, only about 1 out of 20 isoleucine residues is encoded by an AUA, the other 19 being encoded by AUU and AUC. It is thought that one of the steps that can lead to codon reassignment is that the codon becomes rarely used in a genome. This is easier to achieve in mitochondria because they have very small genomes (∞ Section 3.16 and Figure 3.42).

Mistranslation

Another problem in the translation of the genetic code is that errors sometimes occur. This means that a codon may be "read" improperly and the wrong amino acid inserted. Amino acids whose codons differ by only a single base, for example, phenylalanine (UUU) and leucine (UUA), are most likely to be mistranslated. In rare instances leucine may be added to the growing polypeptide instead of phenylalanine even when the codon is UUU. In the normal cell these rare errors occur in only a small number of all the protein molecules and hence have no detrimental effect. For example, experimental measurements of mistranslation have shown that only 1 in 10^3–10^4 codons is misread. However, certain antibiotics that act on ribosomes, such as streptomycin and neomycin, increase translation errors to such an extent that many protein molecules in the cell are abnormal and the cell can no longer function properly.

Other types of translational errors can also occur, such as a ribosome shifting into the wrong reading frame or erroneously reading a stop codon as a sense codon. Amazingly, certain genetic elements seem to have evolved to take advantage of such translational "errors" to make essential proteins (∞ Section 8.22).

✓ 6.13 Concept Check

A few genes overlap with each other, but most genes do not. Likewise, the great majority of organisms use the "universal" genetic code, but a few organisms use slightly different genetic codes.

✓ What is the "universal" genetic code?
✓ What kind of genome is most likely to contain overlapping genes?

In this chapter we have seen that biological information flows from DNA, the cell's genetic material, to RNA, and finally to protein. The mechanisms of the three main steps in information transfer, replication, transcription, and translation are quite similar in prokaryotes and eukaryotes. The information in a gene can be accurately replicated so that when a cell divides the progeny has the same genetic information. The information in the gene can also be transcribed into RNA, and for genes that encode proteins, translated into proteins. Although some RNA has catalytic activity, it is the proteins that are the components of the cell that carry out the thousands of reactions making up the cell's metabolism. We now turn our attention to how cells regulate whether a particular gene or set of genes is expressed.

REVIEW QUESTIONS

1. Describe the *central dogma* of molecular biology.

2. Genes were discovered before their chemical nature was known. Define a gene without mentioning its chemical nature. Of what is a gene composed?

3. Inverted repeats can give rise to stem-loops. Show this by giving the sequence of double-stranded DNA containing an inverted repeat and show how the transcript from this region can form a stem-loop.

4. Is the sequence 5'-GCACGGCACG-3' referred to as an inverted repeat? Explain your answer.

5. DNA molecules that are AT-rich separate into two strands more easily when the temperature is raised than do DNA molecules that are GC-rich. Write an explanation for this observation based on the properties of AT and GC base pairing.

6. What are restriction enzymes? What is the prime function of a restriction enzyme in the cell that produces it (that is, why do cells have restriction enzymes)? How is it that the restriction enzyme in a cell does not cause degradation of that cell's DNA?

7. Nucleic acid hybridization is at the basis of many modern genetic techniques. Write a short explanation for each of the following statements:

   a. The strength of the DNA hybrid is greater if two long than if two short DNA molecules are involved.

   b. Even if the base sequences are not *exactly* complementary, hybridization can still occur between two relatively long DNA molecules, but this is less likely to occur if the molecules are short.

8. A structure commonly seen in circular DNA during replication is called a *theta structure*. Draw a diagram of the replication process and show how a theta structure could arise.

9. Why are errors in DNA replication so rare? What additional enzyme activity (other than polymerization) is associated with DNA polymerase III and how does it serve to reduce errors?

10. What are ribozymes and what types of biochemical reactions are they generally associated with?

11. There are three processing steps in producing most eukaryotic mRNA but not prokaryotic mRNA. Write a short description of each of these three steps.

12. What are aminoacyl-tRNA synthetases and what types of reactions do they carry out? Approximately how many different types of these enzymes are present in the cell? How does a synthetase recognize its correct substrates?

13. Do genes for tRNAs have promoters? Do they have start codons? Explain.

14. The start and stop sites for mRNA synthesis (on the DNA) are different from the start and stop sites for protein synthesis (on the mRNA). Explain.

APPLICATION QUESTIONS

1. The genome of the bacterium *Neisseria gonorrhoeae* consists of a single double-stranded DNA molecule that contains 2220 kilobase pairs. Calculate the length of this DNA molecule in centimeters. If 85% of this DNA molecule is made up of the open reading frames of genes encoding proteins and the average protein is 300 amino acids long, how many protein-encoding genes does *Neisseria* have? What kind of information do you think might be present in the other 15% of the DNA?

2. Circular DNA molecules, such as those of most bacterial chromosomes, circumvent one problem encountered in replication of linear DNA. What is this problem? Also, having a circular chromosome results in a new problem: The two daughter chromosomes are interlocked after replication is completed. Is there any type of enzyme discussed in this chapter that might help separate these molecules so they can be partitioned?

3. Many bacterial mRNA molecules are polycistronic, each mRNA coding for more than one protein. Imagine an mRNA that codes for two proteins with an intervening noncoding region between the two coding regions. From your understanding of how the transla-

tion process works, explain why the end result would be two separate proteins rather than one mixed (hybrid) protein.

4. What would be the result (in terms of protein synthesis) if RNA polymerase initiated *transcription* one base upstream of its normal starting point? Why? What would be the result (in terms of protein synthesis) if *translation* began one base downstream of its normal starting point? Why?

5. In the bacterium *Salmonella typhimurium* glutamyl-phosphate reductase and glutamate kinase, two of the enzymes involved in the synthesis of proline, are apparently translated from a single polycistronic message. Draw a diagram of the region of the chromosome containing the two genes encoding these enzymes. Show the correct relative position(s) of that portion of the DNA that contains or encodes all of the following: promoter(s), Shine–Dalgarno sequence(s), start and stop codons, and transcription terminator(s). Do you believe these genes probably contain introns? Why or why not?

6. If the genes you diagrammed in answering Question 5 were actually two genes encoding two different tRNAs, how would your diagram be different?

The induction of the enzyme β-galactosidase by its inducer, lactose, as shown here, led Jacques Monod and his associates to begin studying the regulation of lactose metabolism in the 1940s. Throughout the 1950s, Monod, François Jacob, and colleagues performed physiological and genetic experiments on lactose metabolism in *Escherichia coli* that led to important breakthroughs in our understanding of gene expression and regulation. These studies led to the discovery of regulatory genes, such as the gene that encodes the *lac* repressor, and from there to the operon model of gene expression, which was proposed in 1960. The significance of this model was widely appreciated and for their contributions, Jacob and Monod (along with Andre Lwoff) won the 1965 Nobel Prize for Medicine.

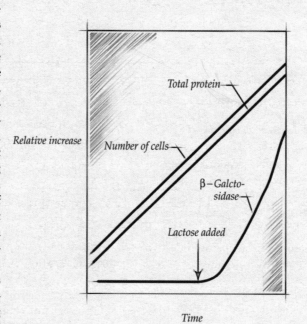

CHAPTER 7

Regulation of Gene Expression

7.1   Regulation of Enzyme Activity   213
7.2   Regulation of Transcription: Induction and Repression   217
7.3   Regulation of Transcription: Positive Control   220
7.4   DNA Binding Proteins   221
7.5   Attenuation   224
7.6   Global Control   226
7.7   Signal Transduction and Two-Component Regulatory Systems   230
7.8   Contrasts in Gene Expression between Prokaryotes and Eukaryotes   233

In the previous chapter we saw how the information stored as a sequence of bases in a region of DNA (a gene) can be transcribed into RNA and then translated to yield a specific protein. Most proteins are enzymes (∞ Section 4.5) and carry out the reactions responsible for the cell's metabolism: the reactions that allow it to process nutrients, to build new cellular material, to grow, and to divide. Hundreds of different enzymatic reactions occur simultaneously during a single cycle of cell growth. Cells and organisms also respond to changes in their environment, and many organisms have complicated developmental pathways. In order to efficiently orchestrate the numerous chemical reactions in a cell, to make maximal use of available resources, and to carry out developmental processes, cells need to *control* the level of expression of the genetic information.

Some enzymes may be needed in about the same amounts under all growth conditions and are called *constitutive enzymes*. Constitutive enzymes are generally key cellular enzymes required for growth under all nutritional conditions and are thus synthesized continuously in the growing cell. However, far more common is the situation in which a particular reaction needs to occur frequently under some conditions but not under others. Additionally, under many growth conditions a cell might not need to carry out a particular reaction at all. For instance, enzymes required for the breakdown of the sugar lactose are useful to the cell only if lactose is present in its environment. Most microorganisms have the genetic information to encode many more different kinds of proteins than are actually present in the cell under any particular condition (∞ Section 1.3). Thus, the need to regulate biochemical reactions in response to changing growth conditions, or as part of a developmental process, is clear. How does this type of regulation occur?

There are two major modes of regulation in the cell. One controls the *activity* of preexisting enzyme and one controls the *amount* (or even the complete presence or absence) of an enzyme (Figure 7.1). Regulation of the activity of an enzyme obviously happens *after* the protein has been synthesized (that is, posttranslationally). By contrast, regulation of the amount of enzyme synthesized can occur at the level of transcription (how much messenger RNA [mRNA] is made) or at the level of translation (whether or not the mRNA is translated to make the protein). Regulation of the synthesis of an enzyme is a coarser level of control than regulating activity. However, working together, these mechanisms can result in an efficient regulation of cell metabolism so energy need not be wasted carrying out unnecessary reactions.

Control systems that vary the level of *expression* of particular genes are the main subject of this chapter. However, the actual number of different regulatory mechanisms is vast, and most genes seem to be regulated by more than one. We begin by briefly discussing the processes involved in regulating the *activity* of preformed enzymes before considering how the synthesis of enzymes is controlled.

7.1

Regulation of Enzyme Activity

There are a large number of mechanisms of posttranslational regulation. In some cases an enzyme is synthesized as part of a larger inactive precursor protein, and the enzyme must be activated by removing a portion of the precursor protein (see the box, Protein Processing). Another mechanism is to reduce the level of activity by actually degrading the enzyme molecules. However,

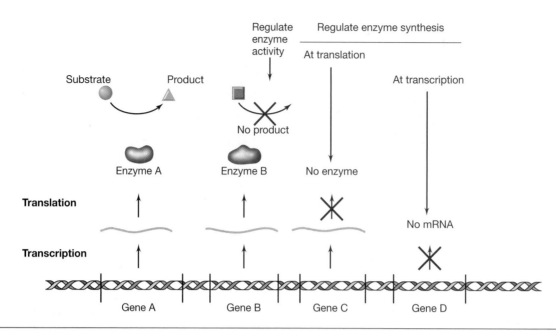

FIGURE 7.1   An overview of the mechanisms that can be used in regulation. The product of gene A is enzyme A, which is synthesized constitutively and carries out its reaction. Enzyme B is also synthesized constitutively but its activity can be inhibited. The synthesis of the product of gene C can be prevented by control at the level of translation. The synthesis of the product of gene D can be prevented by control at the level of transcription.

we discuss here a reversible and temporary form of regulation involving less drastic changes to the enzyme molecule.

Feedback Inhibition

Enzymes can often be specifically inhibited by one or more compounds. Specific enzyme inhibitors can sometimes be used in medicine or technology to control cell

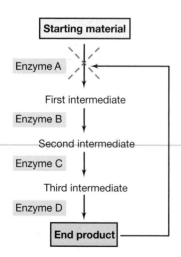

FIGURE 7.2   Feedback inhibition of enzyme activity. The activity of the first enzyme of the pathway is inhibited by the end product, thus controlling production of end product.

growth, but some are used by the cells themselves. A major mechanism for the control of enzymatic activity involves the phenomenon of **feedback inhibition.** Feedback inhibition is seen primarily in the regulation of entire biosynthetic pathways, such as the pathway involved in the synthesis of an amino acid or purine. As we have seen, such pathways involve many enzymatic steps, and the final product, the amino acid or nucleotide, is many steps removed from the starting substrate (∞ Section 4.15). Yet, this final product is able to feed back to the first step in the pathway and regulate its own biosynthesis. How?

In feedback inhibition the amino acid or other end product of the biosynthetic pathway inhibits the activity of the *first* enzyme in this pathway. Thus, as the end product builds up in the cell, its further synthesis is inhibited. If the end product is used up, however, synthesis can resume (Figure 7.2).

How is it possible for the end product to inhibit the activity of an enzyme that acts on a substrate quite unrelated to it? This occurs because of a property of the inhibited enzyme known as **allostery.** An allosteric enzyme has two important binding sites, the *active* site, where the substrate binds, and the *allosteric* site, where the inhibitor (sometimes called an "effector") binds reversibly. When an inhibitor binds, generally noncovalently, at the allosteric site, the conformation of the enzyme molecule changes so that the substrate no longer binds efficiently at the active site (Figure 7.3).

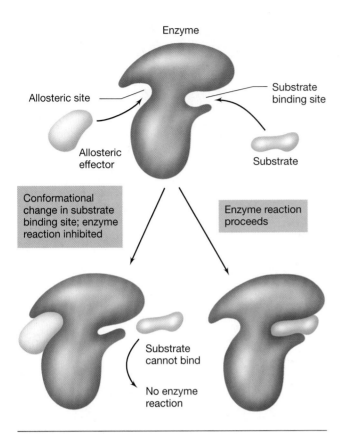

FIGURE 7.3 Mechanism of enzyme inhibition by an allosteric effector. When the effector combines with the allosteric site, the conformation of the enzyme is altered so that the substrate can no longer bind.

When the concentration of the inhibitor falls, equilibrium favors dissociation of the inhibitor from the allosteric site, returning the active site to its catalytic shape. Allosteric enzymes are very common in both anabolic and catabolic pathways and are especially important in branched pathways. For example, the amino

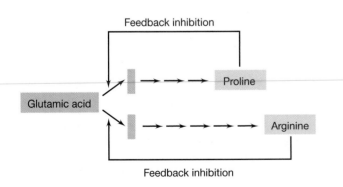

FIGURE 7.4 Feedback inhibition (solid arrows) in a branched biosynthetic pathway. A key intermediate in each pathway is shown in pink. Compare with Figure 11.14.

acids proline and arginine are both synthesized from glutamic acid. Figure 7.4 shows that these two amino acids can control the first enzyme unique to their own synthesis without affecting the other so that a surplus of proline, for example, does not cause the organism to be starved for arginine.

In addition, some biosynthetic pathways are regulated by the use of **isozymes** (short for isofunctional enzymes: *iso* means "same" or "constant"). These enzymes catalyze the same reaction but are subject to different regulatory control. An example is synthesis of the aromatic amino acids (Figure 7.5; ∞ Figure 4.25). Three different isozymes catalyze the first reaction in this pathway, and each enzyme is regulated independently by each of the three different end product amino acids. Unlike the earlier examples of feedback inhibition where inhibitors completely stopped an enzyme activity, in this case the total amount of the initial enzyme activity is diminished in a stepwise fashion and falls to zero only when all three products are present in excess.

The mechanism of feedback inhibition is of more than just academic interest. We will see in Chapter 11 how an understanding of the biochemistry of feedback inhibition has allowed industrial microbiologists to isolate mutants that have lost the ability to feedback inhibit the production of specific amino acids. These mutants are then used for the large scale commercial production of amino acids as a food supplement (∞ Section 11.7 and Figure 11.14).

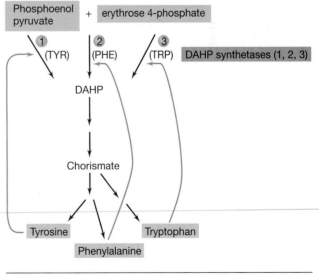

FIGURE 7.5 The common pathway leading to the synthesis of the aromatic amino acids contains three isozymes of DAHP synthetase. (DAHP is 3-deoxy-D-*arabino*-heptulosonate 7-phosphate.) Each of these enzymes is specifically feedback-inhibited by one of the aromatic amino acids. Note how an excess of all three amino acids is required to completely shut off the synthesis of DAHP.

A FOCUS ON...   Protein Processing

In Chapter 6 we discussed the processing of RNA (∞ Section 6.9). We mentioned that *all* tRNAs and rRNAs are transcribed as part of a larger RNA molecule and are then cut to size. Processing also occurs in proteins. As we also mentioned in Chapter 6, all proteins are synthesized beginning with a methionine, a formylated methionine in Bacteria. Therefore, one might expect that any protein isolated from the cell would have a methionine at its amino terminal. However, this is not the case, most proteins have the initiating methionine removed by a specific enzyme. (In Bacteria the formyl group of the methionine is removed first, and it is removed from all proteins.)

We also discussed the fact that proteins that will be secreted are synthesized with a leader sequence that is removed as part of transport (∞ Section 6.12). In some cases the final active product is the result of extensive processing of the original product of translation. This is often the case for the peptide hormones made by higher organisms. A reasonably simple example is the hormone insulin. Insulin is important in regulating carbohydrate metabolism in mammals and is used to treat diabetes. Insulin is derived

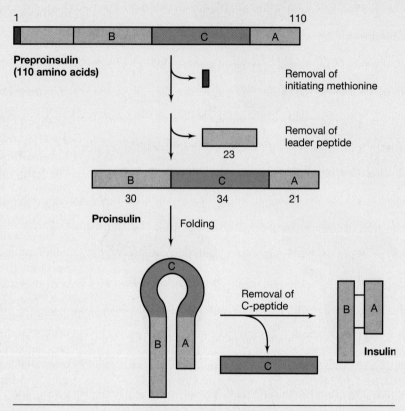

FIG. 1   Processing of preproinsulin. The initiating methionine is actually cleaved from the chain before the entire preproinsulin molecule is synthesized. After the leader peptide is removed, the proinsulin molecule folds, and the C-peptide is then removed leaving the A and B chains on insulin. These chains are held together by disulfide bridges (∞ Section 2.8).

Covalent Modification

Several examples are known in bacteria in which an enzyme is regulated by being *covalently modified,* usually by addition or deletion of some small organic molecule. As in the case of allosteric proteins, the covalent binding of the modifying group changes the conformation of the protein and can dramatically affect its catalytic activity. Removal of the modifying group returns the enzyme to an active state. Many examples of covalent modification regulatory systems are known, but the best characterized are enzymes whose activity is affected by attachment of the nucleotide adenosine monophosphate (AMP) or adenosine diphosphate (ADP), by attachment of inorganic phosphate, or by methylation.

These examples are representative of the major regulatory patterns observed; there are other, rather ele-

gant examples of regulation now known. The phenomenon of enzyme regulation may have evolved because efficient control of the rate of enzyme *activity* enables an organism to quickly adapt to changing environments.

✓ 7.1 Concept Check

Metabolic reactions can be regulated through control of the activities of the enzymes that catalyze these reactions. An important type of regulation of enzyme activity is feedback inhibition, in which the final product of a biosynthetic pathway feeds back and inhibits the first enzyme unique to that pathway. Covalent modification is a regulatory mechanism for temporarily inactivating a specific enzyme.

✓ What is *feedback inhibition?*
✓ What is an *allosteric enzyme?*

from a 110 amino acid residue protein called preproinsulin, as is shown in Fig. 1. The "pre" leader peptide is removed during secretion leaving proinsulin, which is further cleaved to yield the final active insulin molecule which has an A-chain and a B-chain.

Some genes are translated to yield a "polyprotein," that is, a long polypeptide chain that is processed to give several different active products. We will see examples of this in our discussion of poliovirus (⚬⚬ Section 8.15) and retroviruses (⚬⚬ Section 8.22).

When introns are removed during the processing of RNA, the flanking sequences, the exons, are ligated together to form a single molecule (⚬⚬ Section 6.9). Recently a number of instances have been uncovered where the "extra" information in the gene is removed at the level of the *protein* and not at the level of RNA. The process has been termed "protein splicing" and the interior peptide removed is called an **intein.** This process has been shown to occur in specific proteins from Archaea, Bacteria, and Eukaryotes. Fig. 2 shows the process in the product of the *gyrA* gene of the bacterium *Mycobacterium leprae*. Note that the flanking sequences, referred to as *exteins*, are ligated together to form a single protein, in this case a subunit of DNA gyrase. Interestingly, inteins are self-splicing and the intein protein is itself a specific protease (an enzyme that cleaves proteins). Remember that many introns are also self-splicing (⚬⚬ Section 6.9).

Many other "processing" steps are possible with proteins just as they are with RNA. We mentioned that tRNA contains many modified bases and that these bases are modified after transcription (⚬⚬ Section 6.11). Similarly there are many more than 20 (or 21, see box in Chapter 6, Selenocysteine: The Twenty-First Amino Acid) known in proteins that are the result of posttranslational covalent modifications. Covalent modifications are sometimes reversible and can be important means by which cells regulate enzyme activity (see Section 7.1). ■

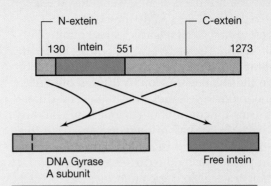

FIG. 2 Protein splicing. The protein synthesized from the *gyrA* mRNA in *Mycobacterium leprae* is 1273 amino acid residues in length (10 times the length of preproinsulin, shown in Fig. 1). The N-extein is the amino terminal extein and the C-extein is the carboxy terminal extein. Residues 131 to 550 make up an intein, which removes itself in a self-splicing reaction that generates the free intein and the DNA Gyrase A subunit.

7.2

Regulation of Transcription: Induction and Repression

Transcription is the first step in biological information flow where it is relatively simple to increase the expression of one gene relative to another. If one gene is transcribed more frequently than another, there will be a greater abundance of its mRNA in the cell and therefore a greater amount of its protein product. Even constitutive proteins can be required in different amounts. For instance, cells typically need relatively little *DNA primase* (⚬⚬ Section 6.6) but require high levels of *single-stranded binding protein* (⚬⚬ Section 6.6), even though both proteins are necessary for synthesizing DNA. From what we have discussed about the nature of promoters (⚬⚬ Section 6.7), you might predict that the gene encoding single-stranded binding protein has a promoter with a sequence similar to the *consensus sequence* and that the promoter for the primase gene does not; such an arrangement would allow the proper amounts of each of these proteins.

However, in the remainder of this chapter we will not discuss the relative strength of different promoters or the efficiency of translation initiation of a particular mRNA (⚬⚬ Section 6.12) but how the level of transcription of an individual gene can be changed in a regulated manner.

In Section 7.1 we considered how cells regulate enzyme *activity*. We now begin a discussion on how cells regulate enzyme *synthesis*. The regulation of activity is typically very rapid (seconds or less), while the regulation of enzyme synthesis is a relatively slow process (a few to several minutes). If a new enzyme needs to be synthesized, it will take some time before that enzyme is present in the cell in sufficient amounts to affect metabolism. Alternatively, if synthesis of an enzyme is stopped, it may be a considerable amount of time before the existing enzyme is diluted out sufficiently to no longer affect metabolism. Several different mechanisms for controlling enzyme synthesis are known in bacteria, and all of them are greatly influenced by the *environment* in which the organism is growing, in particular by the presence or absence of specific small molecules. These molecules can interact with specific proteins to control transcription or, more rarely, translation. We begin our discussion by describing repression and induction, simple forms of regulation that govern gene expression at the level of *transcription*.

Enzyme Repression

Often the enzymes catalyzing the synthesis of a specific product are not synthesized if this product is present in the medium. For example, the enzymes involved in formation of the amino acid arginine are synthesized only when arginine is *not* present in the culture medium; external arginine *represses* the synthesis of these enzymes. As can be seen in Figure 7.6, if arginine is added to a culture growing exponentially in a medium devoid

of arginine, growth continues at the previous rate, but the formation of the enzymes involved in arginine synthesis stops. Note that this is a *specific* effect, as the syntheses of all other enzymes in the cell continue at the same rates as previously.

Enzyme repression is a very widespread phenomenon in bacteria—it occurs as a means of controlling the synthesis of a wide variety of enzymes involved in the biosynthesis of amino acids, purines, and pyrimidines. In almost all cases it is the final product of a particular biosynthetic pathway that represses the enzymes of this pathway. In these cases repression is quite specific, and the process usually has no effect on the synthesis of enzymes other than those involved in the specific biosynthetic pathway. The value to the organism of enzyme repression is obvious because it effectively ensures that the organism does not waste energy synthesizing unneeded enzymes.

Enzyme Induction

A phenomenon complementary to repression is *enzyme induction*, the synthesis of an enzyme only when its substrate is present. Figure 7.7 shows this process in the case of the enzyme β-galactosidase, which is involved in the use of the sugar lactose. If lactose is absent from the medium the enzyme is not synthesized, but synthesis begins almost immediately after lactose is added. Enzymes involved in the catabolism of carbon and energy sources are often inducible. Again, one can see the

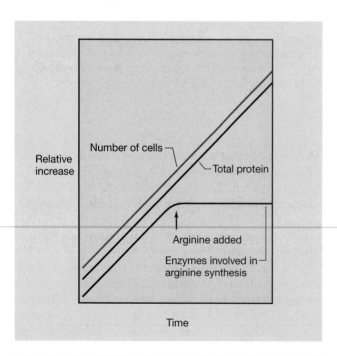

FIGURE 7.6   Repression of enzymes involved in arginine synthesis by addition of arginine to the medium. Note that the rate of total protein synthesis remains unchanged.

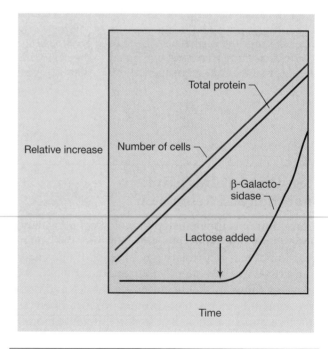

FIGURE 7.7   Induction of the enzyme β-galactosidase on the addition of lactose to the medium. Note that the rate of total protein synthesis remains unchanged.

value to the organism of such a mechanism, as it provides a means whereby the organism does not synthesize an enzyme until it is needed.

The substance that initiates enzyme induction is called an **inducer,** and a substance that represses production is called a **corepressor;** these substances, which are always small molecules, are often collectively called **effectors.** Not all inducers and corepressors are substrates or end products of the enzymes involved. For example, *analogs* of these substances may induce or repress even though they are not substrates of the enzyme. Isopropylthiogalactoside (IPTG), for instance, is an inducer of β-galactosidase even though it cannot be hydrolyzed by the enzyme. In nature, however, inducers and corepressors are probably normal cell metabolites.

Mechanism of Induction and Repression

Enzyme repression or induction acts at the level of transcription; enzyme synthesis is controlled by initiating or terminating mRNA production for a particular enzyme or group of enzymes. How can inducers and corepressors affect transcription in such a specific manner? They do this indirectly by combining with specific regulatory proteins which then in turn affect mRNA synthesis. In the case of a repressible enzyme, the corepressor (for example, arginine) combines with a specific **repressor protein,** the arginine repressor, that is present in the cell (Figure 7.8). The repressor protein is an allosteric protein (see Sections 7.1 and 7.4), its conformation being altered when the corepressor combines

with it. This altered repressor protein can then combine with a specific region of the DNA near the promoter of the gene, the **operator region.** This region gave its name to the **operon,** which as we have discussed (⌾ Section 6.8) is a cluster of genes whose expression is under the control of a single operator. All the genes in an operon are transcribed as a single unit yielding a single mRNA. The operator is adjacent to the promoter where synthesis of mRNA is initiated. If the repressor binds to the operator, the synthesis of mRNA is blocked and the protein or proteins specified by this mRNA cannot be synthesized. If the mRNA is polycistronic, *all* the proteins encoded by this mRNA will be repressed.

Enzyme induction can also be controlled by a *repressor*. In this case, the situation previously described is reversed. The specific repressor protein is active in the *absence* of the inducer, completely blocking the synthesis of mRNA, but when the inducer is added, it combines with the repressor protein and inactivates it. Inhibition of mRNA synthesis being overcome, the enzyme or enzymes can be made (Figure 7.9). All systems involving repressors have the same underlying mechanism, *inhibition* of the synthesis of mRNA by the action of specific repressor proteins that are themselves under the control of specific small-molecule inducers and repressors. Because the repressor's role is inhibitory, regulation involving repressors is often referred to as **negative control.**

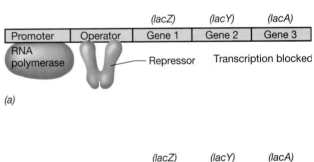

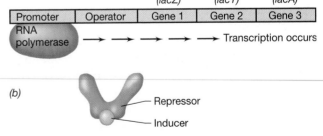

FIGURE 7.9   The process of enzyme induction using a repressor. (a) A repressor protein binds to the operator region and blocks the action of RNA polymerase. (b) An inducer molecule binds to the repressor and inactivates it. Transcription by RNA polymerase occurs and an mRNA for that operon is formed. In the case of the *lac* operon the repressor would be the *lac* repressor, and the inducer would be allolactose.

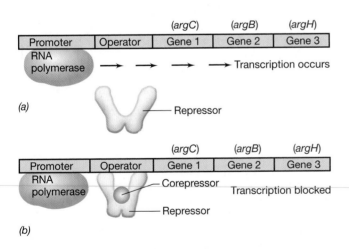

FIGURE 7.8   The process of enzyme repression. (a) Transcription of the operon occurs because the repressor is unable to bind to the operator. (b) After a corepressor (small molecule) binds to the repressor, the repressor binds to the operator and blocks transcription; mRNA and the proteins it encodes are not made. In the case of the *argCBH* operon the repressor would be the arginine repressor and the corepressor would be the amino acid arginine.

✓ 7.2 Concept Check

The amount of an enzyme in the cell can be controlled by increasing (induction) or decreasing (repression) the amount of mRNA that encodes the enzyme. This transcriptional regulation involves regulatory proteins that bind to DNA and to small molecules called effectors. For one type of transcriptional regulation, the regulatory protein is called a repressor and it functions by inhibiting mRNA synthesis.

✓ How does a repressor inhibit the synthesis of a specific mRNA?

✓ What is an operon?

7.3

Regulation of Transcription: Positive Control

Repression constitutes a kind of regulation called **negative control**. The controlling element—the repressor protein—brings about the *repression* of mRNA synthesis. Even though the repressor has a negative role, a system using a repressor can control enzyme induction, as we saw with β-galactosidase. However, another type of control has also been recognized that is called **positive control**. In positive control, a regulator protein *promotes* the binding of RNA polymerase, thus acting to *increase* mRNA synthesis. We will now consider a system that involves positive regulation, the regulation of maltose catabolism in *Escherichia coli*.

The Maltose Regulon

The enzymes for the utilization of the sugar maltose in *Escherichia coli* are synthesized only after the addition of maltose to the medium. The pattern of induction of these enzymes follows that shown for β-galactosidase in Figure 7.7, but in this case it is maltose, not lactose, that is the inducer. The synthesis of the enzymes for maltose utilization is controlled at the level of transcription, but by an **activator protein,** not by a repressor. The *maltose activator protein* cannot bind to the DNA unless the protein first binds maltose, the effector. When the activator protein does bind to DNA, it allows RNA polymerase to begin transcription (Figure 7.10). Activators, like repressors, recognize specific sequences on the DNA. The sequence that serves as the binding site of the activator is not called an operator but an *activator binding site*. Nonetheless, the genes controlled by this activator binding site *are* called an operon.

In negative control, the repressor binds to the operator and blocks transcription. How does an activator protein work? Positively controlled promoters have nucleotide sequences that are not close matches

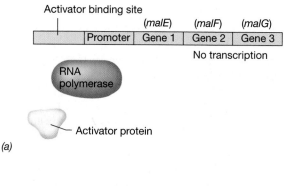

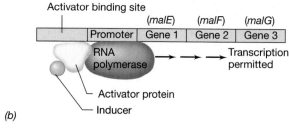

FIGURE 7.10   Positive control of enzyme induction. (a) In the absence of an inducer, neither the activator protein nor the RNA polymerase can bind to the DNA. (b) An inducer molecule binds to the activator protein, which in turn binds to the activator binding site. This allows RNA polymerase to bind to the promoter and begin transcription. In the case of the *malEFG* operon, the activator protein would be the maltose activator protein and the inducer would be the sugar maltose.

to the consensus sequence (∞ Figure 6.26). Even with the correct sigma factor, the RNA polymerase has difficulty recognizing these promoters. The activator protein, when bound to DNA, helps the RNA polymerase either recognize the promoter or begin transcription. The activator protein may cause a change in the structure of the DNA, perhaps by bending it (Figure 7.11), allowing the RNA polymerase to make the correct contacts with the DNA. The activator protein may also interact directly with the RNA polymerase. This can happen either when the activator binding site is close to the promoter (Figure 7.12a) or when it is several hundred base pairs away from the promoter (Figure 7.12b).

The genes needed for maltose utilization are spread out in several operons, each of which has an activator binding site to which the maltose activator protein can bind. Therefore, the maltose activator protein actually controls more than one operon. When more than one operon is under the primary control of the same regulatory protein, these operons are collectively known as a **regulon**. Therefore, the enzymes for maltose utilization are encoded by the *maltose regulon*. Regulons are also known for operons under negative control. The arginine biosynthetic enzymes (see Section 7.2) are en-

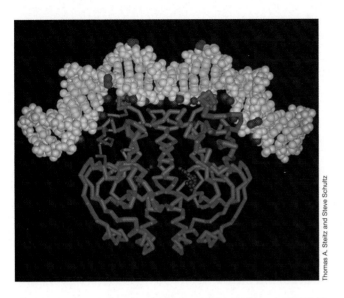

FIGURE 7.11 Computer model of the interaction of a positive regulatory protein with DNA. This figure shows the cyclic AMP binding protein, a regulatory protein involved in the control of several operons (see Section 7.6). The α-carbon backbone of this protein is shown in blue and purple. The protein is shown binding to a DNA double helix, which is shown in yellow and light blue. Note that binding of this protein to DNA has caused the DNA to be bent by almost 90°. Reprinted with permission from *Science* 253:1001–1007 (1991), © AAAS.

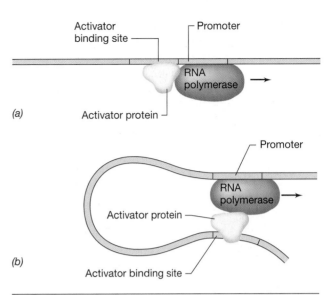

FIGURE 7.12 Some activator proteins interact with RNA polymerase. (a) The activator binding site is near the promoter. (b) The activator binding site is several hundred base pairs from the promoter. In this case, the DNA must be looped to allow the activator and the RNA polymerase to contact.

coded by the *arginine regulon* whose operons are all under the control of the arginine repressor protein. One of these operons was shown in Figure 7.8.

Many genes in *Escherichia coli* have promoters under positive control, and many have promoters under negative control. However, there are other types of regulation known. In addition, many genes (perhaps most genes) either have a promoter with multiple types of control or have more than one promoter, each with its own control system! However, negative control, positive control, and several other types of transcriptional control involve the specific binding of proteins to DNA. In the next section we discuss DNA binding proteins.

✓ 7.3 Concept Check

Positive regulators of transcription are called activator proteins. They bind to activator binding sites on the DNA and stimulate transcription by RNA polymerase. Activator protein activity, like repressor protein activity, is modified by effectors. For positive control of enzyme induction, the effector promotes the binding of the activator protein and thus stimulates mRNA synthesis.

✓ Compare and contrast the activities of an activator protein and a repressor protein.
✓ Distinguish between an *operon* and a *regulon*.

7.4

DNA Binding Proteins

Small molecules are often involved directly in the regulation of protein activity. For instance, in the example given in Figure 7.4, the amino acids proline and arginine bind directly to the enzyme and inhibit it. The situation with regard to regulating enzyme *synthesis* is quite different. Although small molecules are often involved in regulating transcription, they rarely do so directly. Instead they typically influence the binding of certain proteins, called *regulatory proteins*, to specific sites on the DNA, and it is these proteins that actually regulate transcription. In this section we shall discuss a few general properties of proteins that bind to DNA.

Interaction of Proteins with Nucleic Acids

Protein–nucleic acid interactions are central to replication, transcription, and translation, as well as to the regulation of these processes. Two general kinds of protein–nucleic acid interactions are noted: nonspecific and specific, depending on whether the protein attaches *anywhere* along the nucleic acid or whether the interaction is sequence-specific. As an example of proteins that do *not* interact in a sequence-specific fashion, we mention the **histones,** proteins that are extremely important in the structure of the eukaryotic chromosome (◯◯ Section 6.2), although less significant in prokaryotes. Histones are relatively small proteins that have a high proportion of positively charged amino acids (arginine, lysine, histidine). DNA, as we have noted, is a polynucleotide and has a high proportion of negatively charged phosphate groups, making it a negatively charged molecule. These phosphate groups are on the outside of the DNA double helix. Histones, because of their positive charge, combine strongly and relatively nonspecifically with the negatively charged DNA. In the eukaryotic cell there is generally enough histone so that all the phosphate groups of the DNA are covered. Association of histones with DNA leads to the formation of nucleosomes, the unit particles of the eukaryotic chromosome (◯◯ Section 6.2). Even these relatively nonspecific interactions can affect gene expression. If the DNA is covered with histones, other proteins such as RNA polymerase will not be able to bind and transcription cannot take place. DNA replication removes histones and allows other proteins to bind, and apparently certain other proteins can disrupt the condensed structure of DNA. However, loss of histones need not automatically lead to transcription but may simply leave the gene capable of being activated by other factors.

There are also a number of proteins that interact with DNA in a *sequence-specific* manner. These interactions occur by association of the amino acid side chains of the proteins with the bases as well as with the phosphate and sugar molecules of the DNA. The major groove in DNA, because of its size, is an important site of protein binding. In Figure 6.4 several of the atoms of the base pairs found in the major groove and known to interact with proteins are identified. In order to achieve *specificity* in such interactions, the protein must interact simultaneously with more than one nucleic acid base, frequently several. We have already described a structure in DNA called an *inverted repeat* (◯◯ Figure 6.7). Such inverted repeats are frequently the locations at which protein molecules combine specifically with DNA (Figure 7.13). Note that this interaction does not involve the formation of cruciform structures in the DNA. Proteins that interact specifically with DNA are frequently *dimers,* composed of two identical polypeptide chains. On each polypeptide chain is a region, called a *domain,* that interacts specifically with a region of DNA in the major groove. A consideration of this type of interaction provides an explanation for the fact that such proteins interact with inverted repeats: in this way, *each* of the polypeptides of the protein dimer combines with each of the DNA strands (Figure 7.13). Be-

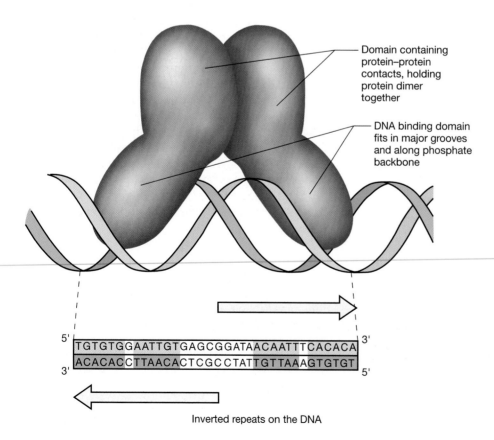

Domain containing protein–protein contacts, holding protein dimer together

DNA binding domain fits in major grooves and along phosphate backbone

5'
TGTGTGGAATTGTGAGCGGATAACAATTTCACACA 3'
3' ACACACCTTAACACTCGCCTATTGTTAAAGTGTGT
5'

Inverted repeats on the DNA

FIGURE 7.13 DNA binding proteins. Many such proteins are dimers that combine specifically with *two sites* on the DNA. The specific DNA sequences that interact with the protein are *inverted repeats*. The nucleotide sequence of the operator gene of the lactose operon is shown and the inverted repeats, which are sites at which the *lac* repressor makes contact with the DNA, are shown in shaded boxes.

cause the protein recognizes *contact points* associated with specific base pairs, its binding is sequence-specific.

Structure of DNA Binding Proteins

Studies of the structure of several DNA binding proteins from both prokaryotes and eukaryotes have revealed a few types of common protein substructures that are apparently critical for proper binding of many of these pro-

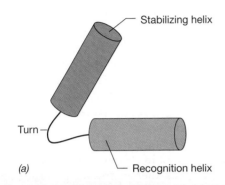

Stabilizing helix

Turn

(a)

Recognition helix

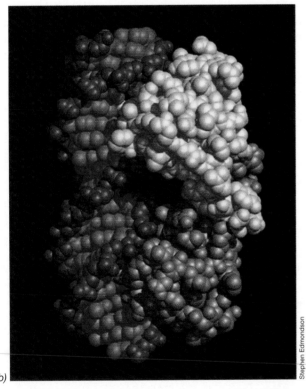

(b)

Stephen Edmondson

FIGURE 7.14   The helix-turn-helix structure of some DNA binding proteins. (a) A simple model of the helix-turn-helix elements. (b) A computer model of the bacteriophage lambda repressor, a typical helix-turn-helix protein, bound to its operator gene. One subunit of the dimeric repressor is shown in dark brown and the other in dark yellow. Each subunit contains a helix-turn-helix structure. The coordinates used to generate this image were downloaded from the Protein Data Base, Brookhaven, NY.

teins to DNA. One of these is termed the *helix-turn-helix motif* (Figure 7.14). The helix-turn-helix consists of a stretch of amino acids that form an α-helix secondary structure (the so-called recognition helix), which is joined to a short stretch of three amino acids, the first of which is usually a glycine that functions to "turn" the protein (Figure 7.14*a*). The other end of the "turn" is connected to a second helix, which stabilizes the first by interacting hydrophobically with it. Recognition of specific DNA sequences occurs by a combination of noncovalent interactions including hydrogen bonds and van der Waals contacts (Section 2.1) between the protein and base pairs on the DNA. Many different DNA binding proteins from Bacteria show the helix-turn-helix structure, including many repressor proteins such as the bacteriophage lambda repressor (Figure 7.14*b*) and the *lac* and *trp* repressors of *Escherichia coli* (see Section 7.2).

Two other types of protein substructures are commonly found in DNA binding proteins. One of these, the *zinc finger*, is frequently found in eukaryotic regulatory proteins that bind to DNA. The zinc finger is a substructure of protein that, as its name implies, binds a zinc ion (Figure 7.15*a*). It seems most likely that part of the "finger" of amino acids that is created forms an α-helix, and this interacts with the DNA in the major groove. There are typically at least two such fingers on the protein involved in binding. The other protein substructure commonly found in DNA binding proteins is the *leucine zipper*. This substructure is formed by the side chains on leucine residues spaced every seven amino acids, and it somewhat resembles a zipper. Unlike the helix-turn-helix and the zinc finger, the leucine zipper does not seem to interact with DNA itself but serves to hold two other α-helices in the correct position to bind DNA (Figure 7.15*b*).

Once a protein combines at a specific site on the DNA, a number of outcomes can occur. In some cases, the protein is an enzyme that carries out some specific action on the DNA, such as RNA polymerase, which makes RNA using DNA as the template. However, in other cases the protein that binds can *block* transcription (negative regulation, see Section 7.2) or can *activate* it (positive regulation, see Section 7.3).

✓ 7.4 Concept Check

Certain proteins can bind to DNA because of specific interactions between certain regions of the proteins and specific regions of the DNA molecule. In some cases the interactions are not sequence-specific, but in other cases they are. Proteins that bind to nucleic acid may be enzymes that use nucleic acid as substrates, or they may be regulatory proteins that affect how genes function.

✓ Why are some interactions specific to certain DNA sequences?
✓ What is a *protein domain?*

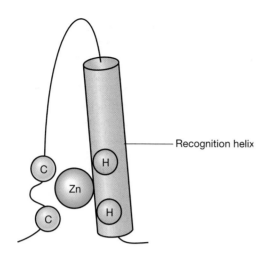

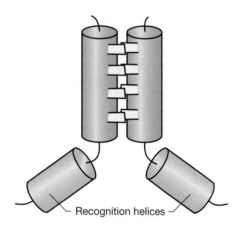

(a)

(b)

FIGURE 7.15 Simple models of protein substructures found in eukaryotic DNA binding proteins. α-Helices are represented by cylinders. Recognition helices are the domains involved in DNA binding. (a) The zinc finger structure. The amino acids holding the Zn²⁺ ion always include at least two cysteine residues (C) with the other residues being histidine (H). (b) The leucine zipper structure. The leucine residues (shown in yellow) are always spaced exactly every seven amino acids. The interaction of the leucine side chains helps hold the two helices together.

7.5

Attenuation

Some control systems to do not involve regulatory proteins that bind to the DNA. The regulatory process called **attenuation** is such a system. The word *attenuation* means "to lessen in amount." Previously we described regulating transcription at *initiation;* that is, repressors block the synthesis of RNA whereas activator proteins encourage synthesis. In transcription at-

tenuation the control occurs *after* initiation of RNA synthesis but before its completion. That is, the number of *completed* transcripts from a gene or an operon is reduced, even though the number of initiated transcripts is not. Most of the first examples of attenuation involved regulating genes controlling the biosynthesis of certain amino acids in gram-negative Bacteria. The first such system to be described was the *tryptophan operon* in *Escherichia coli,* and we focus on it here.

Attenuation and the Tryptophan Operon

The tryptophan operon contains structural genes for five proteins of the tryptophan biosynthetic pathway, plus the promoter and regulatory sequences at the beginning of the operon (Figure 7.16). Like many operons, the tryptophan operon has more than one type of regulation. One type is repression, and one of the regulatory sequences is an operator to which the tryptophan repressor can bind. In addition to promoter (P) and operator (O) regions, there is a sequence called the **leader sequence,** which codes for a polypeptide that contains tandem tryptophan codons near its terminus and functions as an **attenuator** (Figure 7.16). If tryptophan is plentiful in the cell, the leader peptide will be synthe-

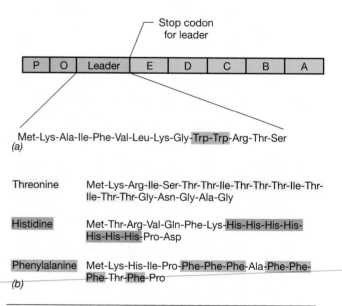

(a)

(b)

FIGURE 7.16 Structure of the tryptophan operon and of tryptophan and other leader peptides in *Escherichia coli.* (a) Arrangement of the tryptophan operon. Note that the leader encodes a short peptide containing two tryptophan residues near its terminus. The promoter is labeled P and the operator is labeled O. The genes labeled E through A encode the enzymes involved in tryptophan biosynthesis. (b) Amino acid sequence of leader peptides synthesized in some other amino acid biosynthetic operons. Because isoleucine is made from threonine, it is an important constituent of the threonine leader peptide.

sized. On the other hand, if tryptophan is in short supply, the tryptophan-rich leader peptide will *not* be synthesized. The striking fact is that synthesis of the leader peptide results in *termination* of transcription of the tryptophan structural genes, whereas if synthesis of the leader peptide is blocked by tryptophan deficiency, transcription of the tryptophan structural gene can occur.

How does *translation* of the leader peptide regulate *transcription* of the tryptophan genes downstream? This can be explained by considering that these two processes in prokaryotic cells are occurring virtually simultaneously (Figure 7.17). Thus, while *transcription* of downstream DNA sequences is still proceeding, *translation* of sequences already transcribed has begun. Apparently, as the mRNA is released from the DNA, the ribosome binds to it and translation begins. Attenuation occurs (RNA polymerase stops transcription) because a portion of the newly formed mRNA folds into a double-stranded loop that signals cessation of RNA polymerase action (Figure 6.27). The stem-loop structures formed by mRNA are brought about because two stretches of nucleotide bases near each other are complementary and can thus base-pair. If tryptophan is plentiful, the ribosome will translate the leader sequence until it comes to the stop codon. The remainder of the leader RNA can then assume a stem-loop, a *transcription pause site*, which is followed by a uracil-rich sequence that actually causes termination. However, if tryptophan is in short supply, the ribosome pauses at a tryptophan codon; the presence of the stalled ribosome at this position allows an alternative stem-loop to form (sites 2 and 3 in Figure 7.17). This stem-loop is *not* a termination signal, and it effectively prevents the terminator (sites 3 and 4 in Figure 7.17) from forming. RNA polymerase then moves past the nonfolded termination site and begins transcription of the tryptophan structural genes. Thus, we see that in attenuation there is a highly integrated system in which transcription and translation interact, with the rate of transcription being influenced by the rate of translation.

In the tryptophan biosynthetic pathway, two distinct mechanisms for the regulation of transcription exist, repression and attenuation. Repression is a mechanism that has large effects on the rate of enzyme synthesis, whereas attenuation brings about a finer control. Working together, these two mechanisms precisely regulate the synthesis of tryptophan biosynthetic enzymes, and hence the biosynthesis of tryptophan. Attenuation has also been shown to occur in *Escherichia coli* in the biosynthetic pathways for histidine, threonine-isoleucine, phenylalanine, and several other amino acids and essential metabolites as well. As shown in Figure 7.16*b*, the leader peptide for each amino acid biosynthetic operon is rich in that particular amino acid. The *his* operon is dramatic in this regard because its leader

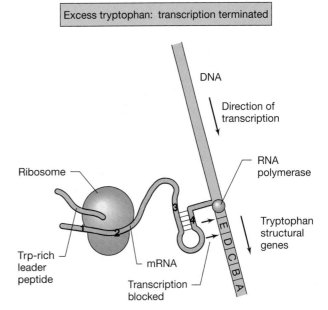

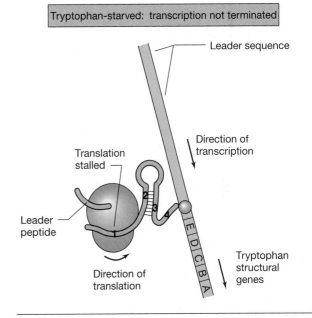

FIGURE 7.17   Control of transcription of tryptophan operon structural genes by attenuation in *Escherichia coli*. The leader peptide is coded by regions 1 and 2 of the mRNA. Two regions of the growing mRNA chain are able to form double-stranded loops, shown as 2:3 and 3:4. Under conditions of excess tryptophan, the ribosome translates the complete leader peptide, and so region 2 cannot pair with region 3. Regions 3 and 4 then pair to form a loop that blocks RNA polymerase. If translation is stalled because of tryptophan starvation, loop formation via 2:3 pairing occurs, loop 3:4 does not form, and transcription proceeds past the leader sequence.

contains *seven* histidines in a row near the end of the peptide (Figure 7.16*b*).

Other Attenuation Mechanisms

Gram-positive Bacteria, such as *Bacillus,* also use attenuation to regulate certain amino acid biosynthetic operons. As in gram-negative Bacteria, the mechanisms also involve attenuation of transcription and alternative secondary structures, which in one configuration lead to termination. However, the mechanisms are translation-independent, and rather than a translating ribosome, an RNA binding protein is involved. This protein binds as a result of interacting with an *effector,* in some cases a tRNA, but in other cases the amino acid itself. In the *Bacillus subtilis* tryptophan operon the protein is called the *trp attenuation protein,* and in the presence of the amino acid tryptophan binds to the leader and favors transcription termination. If tryptophan is limiting, the protein does not bind and transcription proceeds.

Many cases are now known where attenuation involves genes unrelated to amino acid biosynthesis, and the mechanisms obviously do not involve measuring the amount of amino acid. In *Escherichia coli* some of the operons involved in pyrimidine biosynthesis are regulated by attenuation, and the same is true for the pyrimidine biosynthetic genes of *Bacillus.* However, the mechanisms are quite different from each other, although each monitors the level of pyrimidine nucleotides in the cell. In *E. coli* a translated mRNA leader is involved, but in *Bacillus* there is no coupling of transcription and translation. Instead, an RNA binding protein controls the alternative structures of the mRNA.

Finally, a type of regulation called *translational attenuation* is also known. In these cases the translation of the leader peptide prevents the *translation* of the next gene on the polycistronic mRNA. The mechanism apparently involves accessibility of the Shine–Dalgarno sequence of the regulated gene (∞ Section 6.12). Translational attenuation is known to regulate expression of several antibiotic resistance genes in gram-positive Bacteria.

✓ 7.5 Concept Check

Attenuation is a mechanism whereby gene expression (typically at the level of transcription) is controlled *after* initiation of RNA synthesis. Most attenuation mechanisms involve a coupling of transcription and translation and can therefore occur only in prokaryotes.

✓ Why can control systems involving coupled transcription and translation occur only in *prokaryotes?*
✓ Explain how the formation of one stem-loop in the RNA can block the formation of another.

7.6
Global Control

Often an organism needs to regulate many different genes simultaneously in response to a change in its environment. For instance, when the bacterium *Escherichia coli* is starved for phosphate, over 80 different genes are transcribed in response, bringing about the synthesis of new proteins. These proteins play roles in adapting the bacterium to a phosphate-deficient environment. There are several such sets of genes in *E. coli* whose products are required to respond to particular conditions (Table 7.1). Because these control mechanisms operate on a wide cellular basis, they are referred to as *global control systems* and may include one or more regulons. Sometimes the term *modulon* is used to describe a group of genes that can respond to a common regulatory protein even though they may also be members of different regulons (and therefore have at least one other type of control).

TABLE 7.1	A few of the global control systems known in *Escherichia coli*[a]		
System	**Signal**	**Primary activity of regulatory protein**	**Number of genes regulated**
Aerobic respiration	Presence of O_2	Repressor (ArcA)	50+
Anaerobic respiration	Lack of O_2	Activator (FNR)	70
Catabolite repression	Cyclic AMP concentration	Activator (CAP)	300+
Heat shock	Temperature	Alternative sigma (σ^{32})	36
Nitrogen utilization	NH_3 limitation	Activator (NR_1)/alternative sigma (σ^{54})	12+
Oxidative stress	Oxidizing agent	Activator (OxyR)	30+
SOS response	Damaged DNA	Repressor (LexA)	20+

a For many of the global control systems, regulation is complex. A single regulatory protein can play more than one role. For instance, the regulatory protein for aerobic respiration is a repressor for many promoters but an activator for others, whereas the regulatory protein for anaerobic respiration is an activator protein for many promoters but a repressor for others. Regulation can also be indirect or require more than one regulatory protein. Some of the regulatory proteins involved are members of two-component systems (see Section 7.7). Many genes are regulated by more than one global system. (For a discussion of the SOS response, see Section 9.3.)

The term *stimulon* is sometimes used to describe a group of genes that all respond to the same environmental signal. Such a group of genes can be very large and the regulatory pathways can be very complex.

In addition to allowing an organism to respond to a signal by activating a network of genes, global regulation can be used to prevent some genes from responding unnecessarily. For instance, Sections 7.2 and 7.3 covered how the enzymes for lactose or maltose utilization can be induced by adding either lactose or maltose to the growth medium. However, it would be wasteful to induce these enzymes if the cells were already growing on a carbon source that they could use more efficiently. In fact, one of the global regulatory networks, **catabolite repression,** prevents this problem.

Catabolite Repression

In catabolite repression the syntheses of a variety of unrelated enzymes, primarily catabolic, are inhibited when cells are grown in a medium that contains a preferred energy source such as glucose. Catabolite repression has been called the **glucose effect** because glucose was the first substance shown to initiate it, although in some organisms carbon sources other than glucose cause this form of enzyme repression. Catabolite repression assures that the organism uses the more readily catabolizables carbon and energy source, such as glucose, first.

One consequence of catabolite repression is that it can lead to so-called **diauxic growth** if the two energy sources are present in the medium at the same time and if the enzyme needed for utilization of one of the energy sources is subject to catabolite repression. In diauxic growth, the organism grows first on one energy source, and there is then a temporary cessation before growth is resumed on the other energy source. This phenomenon is illustrated in Figure 7.18 for growth on a mixture of glucose and lactose. The enzyme β-galactosidase, which is responsible for utilization of lactose, is inducible, but its synthesis is also subject to catabolite repression. Thus, as long as glucose is present in the medium, β-galactosidase is not synthesized; the organism grows only on the glucose and leaves the lactose untouched. When the glucose is exhausted, catabolite repression is abolished. After a lag, β-galactosidase is synthesized and growth on lactose can occur. Notice that Figure 7.18 shows that the cells grow more rapidly on glucose. Thus, catabolite repression ensures that the cells use the *best* carbon source first.

How does catabolite repression work? Catabolite repression involves control of transcription by an activator protein (see Section 7.3). In the case of catabolite-repressible enzymes, binding of RNA polymerase to DNA occurs only if another protein, called **catabolite activator protein (CAP),** has bound first. An allosteric

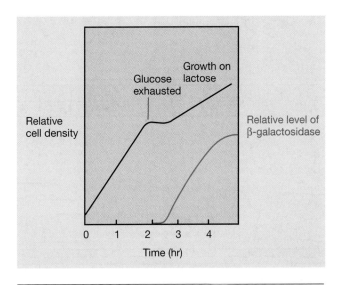

FIGURE 7.18 Diauxic growth on a mixture of glucose and lactose. Glucose represses the synthesis of β-galactosidase. After glucose is exhausted, a lag occurs until β-galactosidase is synthesized, and then growth can resume on lactose.

protein, CAP binds to DNA only if it has first bound a small molecule called *cyclic adenosine monophosphate* or **cyclic AMP** (see Figure 7.11). For this reason some scientists refer to this protein as the *cAMP-receptor protein (CRP)*. Cyclic AMP (Figure 7.19) has been shown to be a key element in a variety of control systems, not only in bacteria but in higher organisms also. Cyclic AMP is synthesized from ATP by an enzyme called *adenylate cyclase*. Glucose inhibits the synthesis of cyclic AMP and stimulates its transport out of the cell. When glucose is transported into the cell, the cyclic AMP level in the cell is lowered, and binding of RNA polymerase to the promoter does not occur. Thus, catabolite repression is really a result of a deficiency of cyclic AMP and can be overcome by adding this compound to the medium.

Although this may sound like a simple positive regulatory system (as in Figure 7.10), each of the operons that CAP controls is *also* under control of a specific regulatory protein. Therefore, catabolite repression modulates several unrelated regulatory systems and thus is an

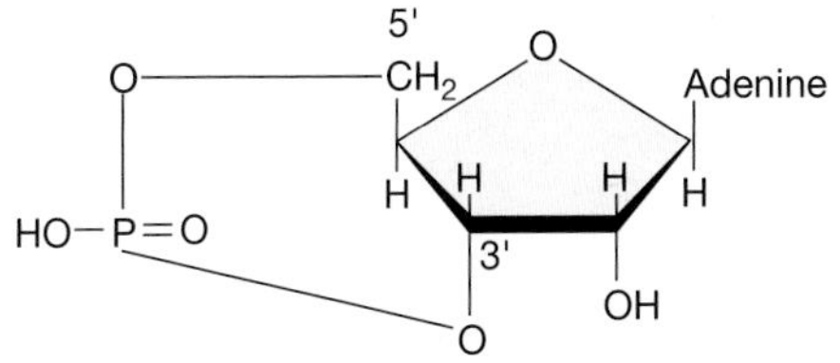

FIGURE 7.19 Cyclic adenosine monophosphate (cyclic AMP, cAMP) is produced from ATP by the enzyme adenylate cyclase.

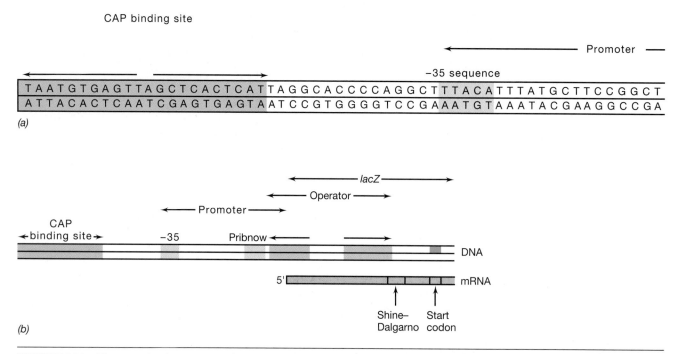

FIGURE 7.20   The genetic elements involved in regulation of the lactose operon. The first gene in this operon, *lacZ*, encodes the enzyme β-galactosidase, which breaks down lactose (see Figure 7.9). The operon contains two other genes that are also involved in lactose metabolism. (a) Part a shows the nucleotide sequence of the control region of this operon. Notice that the two halves of the operator (where the repressor would bind) are almost perfect inverted repeats. There are also inverted repeats in the CAP binding site although these are less perfect. Also shown are the transcriptional start site and the −35 sequence and the Pribnow box, which are part of the promoter (∞ Figure 6.26). In addition, the location of the base pairs encoding the Shine–Dalgarno sequence and the start codon are also given. These two sequences would function on the mRNA (∞ Section 6.10). (b) Part b shows a diagram of this region that also includes the beginning (5' end) of the mRNA that would be formed.

example of global control. As long as glucose is present, catabolite repression prevents expression of all other catabolic operons under this global controlling element. The complete regulatory region of the *lactose operon* is shown in Figure 7.20. For transcription to occur, two requirements must be met: (1) the level of cyclic AMP must be high enough so that the CAP protein binds to the CAP binding site, and (2) there must be an inducer such as lactose present so that the lactose repressor does not block transcription by binding to the operator.

Cyclic AMP has a number of regulatory roles in eukaryotes that do not involve catabolite repression and is also an extracellular signal for the aggregation process in certain cellular slime molds (∞ Section 17.5).

Quorum Sensing

Global control systems allow an organism to respond effectively to signals in its environment. One interesting "signal" is the presence of other organisms of the same species. It has been discovered that certain Bacteria have regulatory pathways that are controlled in response to the density of cells within their own population. This type of control is called **quorum sensing.**

Each bacterium that has this type of regulation has an enzyme that synthesizes a specific acylated homoserine lactone (AHL). This molecule is diffusible to the outside of the cell. Therefore, it can only reach high concentrations in the cell if there are a number of cells nearby each making the same AHL. These AHL molecules are the inducer which combines with an activator protein. Quorum sensing was first discovered as the form of regulation of bioluminescence in a number of bacteria (∞ Section 13.11 and Figure 13.29). Figure 7.21 shows colonies of the Bacterium *Vibrio fischeri* which are glowing because of the production of bacterial luciferase. The *lux* operons, which encode the proteins involved in bioluminescence, are under control of the activator protein LuxR and are induced when the concentration of N-3-oxohexanoyl homoserine lactone becomes high enough. This AHL is synthesized by the protein encoded by the *luxI* gene. Other genes are also controlled by this system. In some bacteria, for instance, *Pseudomonas aeruginosa*, it is clear that quorum sensing is a truly global response leading to the expression of a large number of different genes when the population density becomes sufficiently high.

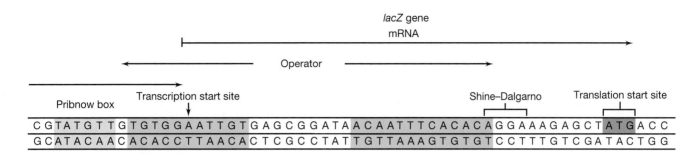

FIGURE 7.20 (continued)

Other Global Control Networks

Genes belonging to global control systems do not all use a simple combination of repressors or activators to achieve regulation. Several involve *alternative sigma factors*, including some shown in Table 7.1, and in these cases, regulation is brought about by changing either the amount or the activity of these factors. Most genes in *Escherichia coli* require the sigma factor referred to as σ^{70} (the superscript 70 indicates the size of this protein, 70 kilodaltons) for transcription and have promoters like those shown in Figure 6.26. The genes that are induced by an increase in temperature (heat shock) have promoters with a quite different sequence, and RNA polymerase requires a different sigma factor (σ^{32}) to recognize them. It is the *amount* of this alternative sigma factor in the cell that regulates the *heat shock response*, and the amount of σ^{32} itself is controlled not by transcription but by the stability of the factor, its rate of translation, and its activity.

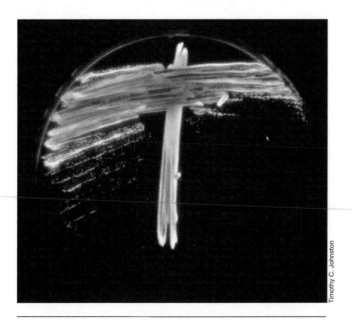

FIGURE 7.21 Bioluminescent bacteria, which are producing the enzyme luciferase. Cells of the bacterium *Vibrio fischeri* were spread on nutrient agar in a petri dish and allowed to grow overnight. The photograph was taken using only the light generated by the bacteria.

Most proteins are very stable; once made, they continue to perform their functions and are passed along at cell division. However, some proteins are unstable. They are recognized by enzymes in the cell called **proteases** and are rapidly degraded. In *E. coli*, σ^{32} is normally degraded within a minute or two after it is synthesized. However, when cells experience heat shock, this degradation process is inhibited. This means there will be more σ^{32}, and therefore it can direct more RNA polymerase to more heat shock promoters. The *translational control* of σ^{32} involves sequences on its mRNA, but the mechanism is unclear.

Global regulatory systems must have a common way to transmit a signal from the environment to the gene(s). We have seen how this is accomplished with catabolite repression, but in the case of the heat shock response, how does a bacterium know what the temperature is? This mechanism seems to involve the *heat shock proteins,* including a protein called DnaK. DnaK is essential for the normal growth of *E. coli* at any temperature, but the amount that is synthesized is increased by heat shock (recall that genes under the control of a global control system are also usually regulated in other ways). The protein DnaK is a **chaperonin,** one of a group of proteins called *molecular chaperones* (∞ Section 6.12). DnaK helps other proteins fold properly and is also involved in the degradation of σ^{32}. Possibly when the temperature is increased, the activity of DnaK is directed more toward folding proteins and is less available for the pathway leading to degradation of σ^{32}. (DnaK is also involved in inhibiting the *activity* of σ^{32}.) Certainly an increase in temperature could influence formation of the correct secondary and tertiary structures of proteins or even cause them to denature slightly (∞ Section 2.8). This would result in an increased level of σ^{32}, and the genes encoding the heat shock proteins would be transcribed. However, since the amount of DnaK increases as part of the heat shock response, it eventually builds up and σ^{32} is degraded again, bringing the cell back to its normal state.

Many genes belong to more than one global control system and therefore can have several overlapping regulatory systems. In some global control systems, more than one regulatory protein might be involved in regulating a single operon. An example is control of the *lac* operon by both the lactose repressor and the catabolite activator protein. However, in this case each regulatory protein essentially operates independently of the other. In the next section we discuss "two-component" regulatory systems, regulatory mechanisms in which at least two proteins are involved in generating a response to a single signal and which seem to be widespread in nature.

✓ 7.6 Concept Check

Cells often need to regulate many genes in response to a single environmental signal. Such cellwide regulatory responses are termed global control. In many cases the genes involved may also be under control of other regulatory circuits. Catabolite repression is an example of global control, and it serves to help cells make the most efficient use of carbon sources.

✓ Explain how catabolite repression can involve an activator protein.

✓ Why might it not be efficient to have all the genes responding to an environmental signal be in a single operon?

7.7

Signal Transduction and Two-Component Regulatory Systems

Bacteria regulate cell metabolism in response to a wide variety of environmental fluctuations, including temperature changes, changes in pH and oxygen availability, changes in the availability of nutrients, and even changes in the number of cells present. Therefore, there must be mechanisms by which bacteria receive signals from the environment and transmit them to the specific target to be regulated. We have seen in preceding sections that some signals can be small molecules that enter the cell (often by specific uptake mechanisms) and act as *effectors*. For instance, in the case of the maltose regulon (see Section 7.3) the sugar maltose binds to the maltose activator protein, causing the protein to bind to specific DNA sequences and activate transcription. However, in many cases the external signal is not transmitted directly to the regulatory protein. Instead, a signal is first detected by a sensor and then transmitted in a changed form to the rest of the regulatory machinery, a process called **signal transduction.**

Sensor Kinases and Response Regulators

Many of the regulatory systems by which cells sense and then respond to environmental signals are called **two-component systems.** Such systems are characterized by having two different proteins: (1) a specific **sen-sor protein** located in the cell membrane, and (2) a partner **response regulator protein.** The *sensor protein* has **kinase** activity and is often referred to as a *sensor kinase.* A *kinase* is an enzyme that phosphorylates compounds. Sensor kinases detect a signal from the environment on their outer surface and in response phosphorylate themselves (autophosphorylation) at a specific histidine residue on their cytoplasmic surface (see Figure 7.22). This phosphoryl group is then transmitted to another protein inside the cell, the *response regulator.* The response regulator is typically a DNA binding protein that regulates transcription. In Figure 7.22 the phosphorylated response regulator is acting as a repressor protein, while the unphosphorylated response regulator does not bind to DNA.

The mechanism used by the response regulator to control transcription depends on the system being described.

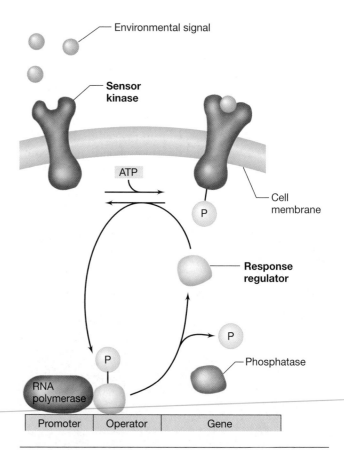

FIGURE 7.22 The control of gene expression by a two-component system. The main components of the system include a *sensor kinase* in the cell membrane that phosphorylates itself in response to an environmental signal. The phosphoryl group is then transferred to the other main component, a *response regulator.* In the system diagrammed in this figure the phosphorylated response regulator serves as a repressor. There must also be a phosphatase in the system to cycle the response regulator.

In *Escherichia coli* the osmolarity of the environment controls which of two proteins, OmpC or OmpF, is synthesized as part of the outer membrane. The response regulator of this system is OmpR. When OmpR is phosphorylated, it acts as an *activator* of transcription of the *ompC* gene and a *repressor* of transcription of the *ompF* gene.

In order to complete a regulatory circuit, there must be a way to terminate the signal. Typically, this involves a *phosphatase,* an enzyme that can remove the phosphoryl group from the response regulator protein. In some cases this reaction is carried out by the sensor kinase itself, while in other systems there is a third protein that carries out this reaction. Therefore, there are "two-component" systems with three components! Actually some systems have even more components, as the signal may be processed through several steps. However, in all cases two-component systems have a sensor kinase and a response regulator.

Two-component systems are now known to regulate a large number of genes in many different bacteria. A few examples include phosphate assimilation in *Escherichia coli,* nitrogen fixation in *Klebsiella* and *Rhizobium,* and sporulation in *Bacillus* (which has a very complex regulatory system). In *E. coli* alone it is estimated that at least 50 different two-component systems operate, and a few of these are listed in Table 7.2. Many of these systems involve global regulation (see also Table 7.1). For instance, there are over 31 genes known to be under the control of PhoB, the response regulator of the *pho* regulon. In some cases more regulatory elements are involved. As an example, in the *Ntr system,* which regulates nitrogen assimilation, the response regulator is an activator protein, NR_I (Nitrogen Regulator I), which works by allowing transcription from promoters recognized by RNA polymerase using σ^{54}, another alternative sigma factor. While the sensor kinase, NR_{II} (Nitrogen Regulator II), fills a dual role, both as the protein kinase and the phosphatase, its activity is in turn regulated by the state of phosphorylation of another protein, P_{II}. Some systems are quite complex. The Nar regulatory system involves two different sensor proteins and two different response regulators, and in addition, all the genes regulated by this system are also under control of the anaerobically active transcriptional regulatory protein FNR (see Table 7.1). Two-component systems closely related to those in bacteria have also been found in lower eukaryotes, such as the yeast *Saccharomyces cerevisiae.* Higher eukaryotes also use phosphorylation as a mechanism of signal transduction in order to respond to environmental changes.

Not all response regulators regulate transcription. We have previously discussed the fact that bacteria can move toward or away from particular chemicals, a process referred to as *chemotaxis* (∞ Section 3.12). We noted that bacteria are too small to actually sense *spatial* gradients of a chemical, but rather they respond to *temporal* gradients. That is, they can sense the change in concentration of a chemical outside the cell *over time.* Bacteria use a two-component system to sense the temporal changes in chemical concentration and regulate flagellar motion.

Mechanism of Chemotaxis

The mechanism of chemotaxis is quite complex and involves a variety of different proteins. A number of *sensory proteins* are in the cell membrane, and these sense the presence of attractants and repellants. These proteins allow the cell to sense whether, over time, the concentration of the substance increases or decreases as the cell moves. The cell thus responds to the *change* in concentration rather than the *absolute* concentration of the chemical stimulus. The sensory proteins are called *methyl-accepting chemotaxis proteins* (**MCPs**), or *receptor-transducer proteins,* or simply **transducers.** In *Escherichia coli,* four different MCPs have been identified, and each is a transmembrane protein (Figure 7.23). Each MCP can

TABLE 7.2	Some two-component regulatory systems from *Escherichia coli* that regulate transcription			
System	**Environmental signal**	**Sensor kinase**	**Response regulator**	**Activity of response regulator**[a]
Arc system	O_2	ArcB	ArcA	Repressor/Activator
Nitrate and nitrite anaerobic regulation (Nar)	Nitrate and nitrite	NarX and NarQ	NarL NarP	Activator/Repressor Activator/Repressor
Nitrogen utilization (Ntr)	NH_4^+	NR_{II}, the product of *glnL*	NR_I, the product of *glnG*	Activates RNA polymerase at promoters requiring σ^{54}.
Pho regulon	Inorganic phosphate	PhoR	PhoB	Activator
Porin regulation	Osmotic pressure	EnvZ	OmpR	Activator/Repressor

a Note that several of the response regulator proteins act as both activators and repressors depending on the genes being regulated. Although ArcA can function as either an activator or a repressor, it functions as a repressor on most operons that it regulates.

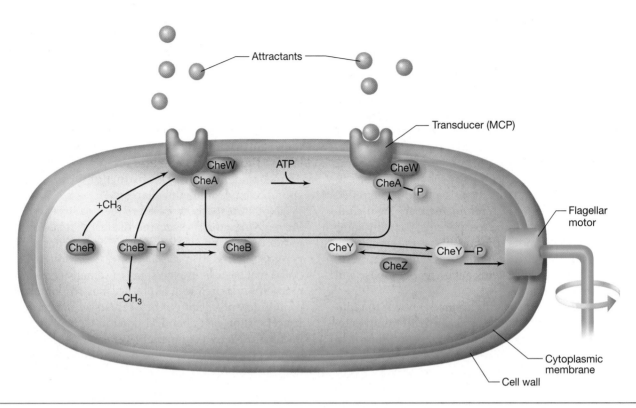

FIGURE 7.23   Interactions of transducers, chemotaxis (Che) proteins, and the flagellar motor in bacterial chemotaxis. The transducer (MCP) forms a complex with the *sensor kinase* CheA and the coupling protein CheW. This combination results in a signal-regulated autophosphorylation of CheA to CheA-P. CheA-P can then phosphorylate the *response regulators* CheB and CheY. Phosphorylated CheY (CheY-P) interacts directly with the flagellar motor switch. CheZ dephosphorylates CheY-P. CheR continually adds methyl groups to the transducer. CheB-P (but not CheB) removes them. The degree of methylation of the transducers controls their ability to respond to attractants and repellants and leads to adaptation. The structure of the flagellar motor was shown in Figure 3.48.

sense a variety of compounds. For example, the *Tar* transducer of *E. coli* can sense the attractants aspartate and maltose as well as repellants such as the heavy metals cobalt and nickel.

MCPs bind attractants or repellants directly, or in some cases indirectly, through interactions with periplasmic binding proteins. Binding of an attractant or repellant sets in play a series of interactions with cytoplasmic proteins that eventually affects flagellar rotation. If rotation of the flagellum is *counterclockwise,* the cell will continue to move in a run. If the flagellum rotates *clockwise,* however, the cell will tumble (⬀ Section 3.12).

The current model for flagellar control shows that the transducers are in contact with the cytoplasmic proteins CheW and CheA (Figure 7.23). CheA is a *sensor kinase.* When a transducer has bound a chemical, it changes conformation and (with CheW) causes a change in the autophosphorylation of CheA (forming CheA-P). *Attractants decrease* the rate of autophosphorylation, whereas *repellants increase* this rate. Phosphorylated CheA (CheA-P) then phosphorylates CheY

(forming CheY-P), a *response regulator.* CheY-P interacts with the flagellar motor to induce clockwise flagellar rotation and tumbling (the motor switch itself consists of proteins encoded by *fla* genes).

CheA-P can also phosphorylate CheB, another response regulator, but this is a much slower reaction than the phosphorylation of CheY. We shall discuss the activity of CheB-P later. Thus, CheY is the central protein in the system because it serves as the *response regulator* for chemotaxis, governing the direction of rotation of the flagellum. When CheY is phosphorylated, the flagellar motor switches from a counterclockwise to a clockwise rotation, causing the cell to tumble. If unphosphorylated, CheY cannot bind, the flagellar motor continues counterclockwise rotation, and the cell undergoes a run. Another protein, CheZ, dephosphorylates CheY, returning it to a form that allows runs instead of tumbles. Because repellants increase the level of CheY-P, they lead to tumbling, whereas attractants lead to a lower level of CheY-P and smooth swimming.

Note that the system described can *signal* that a chemical has been bound and regulate flagellar rotation but seems to be unable to note a change with the passage of time. There is a second component to chemotaxis, and this is **adaptation.**

As their name implies, MCPs can be methylated. There is a cytoplasmic protein, CheR, that continually adds methyl groups to the MCPs at a slow rate using S-adenosylmethionine as a methyl donor. The phosphorylated form of the response regulator CheB is a demethylase that removes methyl groups from the MCPs. The level of methylation of the MCPs affects their conformation and controls adaptation to a sensory signal. It allows resetting of the signaling state of the receptor even though the concentration of the chemical remains unchanged.

If the level of an attractant remains high, the level of phosphorylation of CheA (and, therefore, of CheY and CheB) will remain low, the cell will swim smoothly, and the level of methylation of the MCPs will increase (because CheB-P is not present to demethylate). However, the MCPs no longer respond to the attractant when they are fully methylated. Therefore, even though the level of attractant might remain high, the level of CheA-P (and CheB-P) increases and the cell begins to tumble. However, now the MCPs can be demethylated by CheB-P, and when this happens, the receptors can once again respond to attractants. The situation is the opposite with regard to repellants (fully methylated MCPs respond best to repellants).

The control of chemotaxis is obviously quite complicated and involves a number of regulatory switches. Unlike the case in many other two-component systems, in chemotaxis the signal transduction system regulates the activity of the gene products, not their synthesis. Signal transduction is an important regulatory mechanism in both prokaryotes and eukaryotes.

✓ 7.7 Concept Check

Signal transduction systems transmit environmental signals to the cell. In prokaryotes signal transduction typically involves two-component systems, which include a sensor protein located in the membrane and a cytoplasmic response regulator protein. The sensor protein is a kinase, and the activity of the response regulator depends on its state of phosphorylation. Most two-component systems regulate transcription, but that regulating bacterial chemotaxis operates at the level of protein activity.

✓ What are *kinases* and what is their role in two-component regulatory systems?

✓ Could a response regulator be an activator or a repressor?

7.8

Contrasts in Gene Expression between Prokaryotes and Eukaryotes

We have discussed only a few of the mechanisms by which cells can control the activity of a protein and also only a few major mechanisms that can regulate the synthesis of a protein. Most of the mechanisms we considered for regulating synthesis operate at the level of transcription, and all involve regulatory proteins. Interestingly, regulatory RNA also exists (see the box, Antisense Nucleic Acid).

Although many of the major regulatory patterns are shared between prokaryotes and eukaryotes, there are many differences. Because of the lack of compartmentation in prokaryotes, the processes of transcription and translation are *coupled*. Also, the messenger RNA of prokaryotes is frequently polycistronic, with more than one protein being translated from the same message.

In eukaryotes, on the other hand, transcription and translation take place in separate compartments in the cell, and the integration of these processes seen in prokaryotes is lacking. How about induction and repression in eukaryotes? Although many eukaryotes do exhibit repression, there is no good evidence for the kind of negative control so commonly found in prokaryotes. However, positive control mechanisms are common in eukaryotes. If operons exist in eukaryotes, they involve the control of only single enzymes, rather than the multienzyme control systems so commonly seen in prokaryotes, and there is no evidence for polycistronic RNA molecules in eukaryotes except in a few viruses. These are translated inefficiently. Post-translational protein modification is quite common in eukaryotes. Eukaryotes also have regulation involving splicing of mRNA, regulation that does not exist in prokaryotes.

It is the regulation of genes that is the basis for *development* of multicellular eukaryotic organisms, which begin life as single cells and develop into complex organisms with many very different and highly specialized cell types. This process of differentiation requires that specific sets of genes become active at precisely the correct time during the development of the organism. The complex regulatory pathways involved in development are currently under intense investigation. Understanding and controlling these pathways would allow for tremendous advances in medicine.

A FOCUS ON... Antisense Nucleic Acid

Regulation of the synthesis of proteins often involves transcriptional control. Less often, genes are controlled at the level of translation. Most control networks, whether they are transcriptional or translational, use regulatory proteins. However, it is now clear that in some cases it is a regulatory *RNA*, not a regulatory *protein*, that is involved.

One type of regulatory RNA, called **antisense RNA,** is known to be used in the regulation of several different bacterial genes. Antisense RNA acts by forming base pairs with a complementary, or sense, strand of RNA (Fig. 1). When the sense RNA is mRNA, the resulting double-stranded structure can prevent translation. Antisense RNA can be synthesized from the same gene as the sense RNA by having a second promoter oriented in the direction opposite that of the first or by having a second gene with the promoter at the other end. Antisense RNA does not have to be used only to regu-

late the synthesis of a protein. In some plasmids, it controls the initiation of DNA synthesis.

Antisense nucleic acids can be specifically designed and synthesized by scientists in the laboratory and delivered directly to cells. These short (15–25 nucleotides) synthetic chains (oligonucleotides) are usually made of DNA rather than RNA. Their sequence can be made to allow them to bind to a specific mRNA and prevent translation (or allow the molecule to be recognized by nucleases). Antisense nucleic acid can also bind to the DNA in the nucleus and prevent transcription. The latter is possible because some DNA can form a *triple* helix! The "extra" strand (the oligonucleotide) forms specific interactions with those parts of the bases that are in the major groove of a normal double helix to give **triple DNA** (Fig. 2). (Not all DNA sequences can form triple helices, at least not without the aid of special enzymes.)

Synthetic antisense oligonucleotides can be designed to be extremely specific, whether they bind to a message or to the regulatory region of a gene. The specificity arises because a sequence of only 20 bases should occur no more often than once in 10^{12} bases of "random" DNA. Therefore, it is unlikely that the antisense RNA would bind to anything other than its known target in any cell. This specificity might allow antisense nucleic acids to become an important new class of antibiotic. Antisense nucleic acids could be designed to be used against specific viruses or to regulate particular genes in either disease-causing (pathogenic) organisms or human tumor cells. Several antisense therapeutic compounds are in clinical trials, and one antisense drug has been approved to treat a viral infection of the eye. The possible utility of these molecules is just one example of how understanding the structure of a gene may have very practical applications (∞ Chapter 10). ■

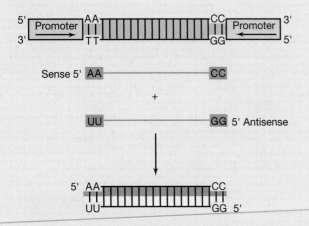

FIG. 1   A gene with promoters at either end. If the RNA made by the promoter on the left is the *sense* RNA, then the RNA made by the promoter on the right is the *antisense* RNA. If both RNA molecules are made, they will form a duplex. Only a relatively short region of overlap is necessary for strong base pairing. Usually the antisense RNA is shorter than the sense RNA, and therefore the second promoter is actually within the gene.

FIG. 2   A triple helix. The "extra" strand is shown in red and is in the major groove of a double helix.

REVIEW QUESTIONS

1. If an enzyme can be effectively inhibited by feedback inhibition, why would cells also have mechanisms to regulate its synthesis?

2. Compare the processing of proinsulin and the product of the *gyrA* gene of *Mycobacterium leprae* (see box, Protein Processing). In both cases an "interior" peptide is removed. However, only the latter case is considered protein splicing. Explain.

3. Describe why a protein that binds to a specific sequence of double-stranded DNA is unlikely to bind to the same sequence if the DNA is single-stranded.

4. Describe the regulation of two different operons, one having an effector that is an *inducer* and the other having an effector that is a *corepressor*.

5. The maltose regulon is inducible and is regulated by an activator protein. The lactose operon is inducible but is regulated by a repressor protein. Explain how induction can be brought about by either positive control (activator protein) or negative control (repressor protein).

6. In most cases operators are very close to the promoters they control, while activator binding sites can be some distance away. Explain why this should be so.

7. Describe how transcriptional attenuation works. What is actually being "attenuated"? Why hasn't the type of attenuation that controls several different amino acid biosynthetic pathways in *Escherichia coli* also been found in eukaryotes?

8. Describe the mechanism by which catabolic activator protein (CAP), the regulatory protein for catabolite repression, functions using the lactose operon as an example. For this operon the CAP protein is not a repressor. Describe the regulatory region of a gene for which the CAP protein *is* a repressor. (*Hint:* Think about your answer to Question 6.)

9. What are the two components that give the name to signal transduction regulation in prokaryotes? What is the function of each of the components?

10. One of the members of a two-component system is typically located in the cell membrane. What reason can you think of why this might be so?

11. Many genes are under multiple control systems. In the lactose operon, there is lactose-specific regulation and regulation by a global control system. Describe how each of the controls on the lactose operon actually functions. Why do you think both systems are necessary?

APPLICATION QUESTIONS

1. The amino acids isoleucine and valine share a common pathway for most steps in their biosynthesis. In *Escherichia coli* the first common step can be subject to feedback inhibition by valine but not by isoleucine. In most strains, though, the addition of valine does not cause isoleucine deprivation. However, in other strains it does (that is, adding valine causes isoleucine starvation and the cells stop growing). What explanation can you give for the difference between the normal "valine-resistant" strains and those whose growth is sensitive to valine?

2. What would happen to regulation from a promoter under negative control if the region where the regulatory protein binds were deleted? What if the promoter were under positive control?

3. Promoters from *Escherichia coli* under positive control are not close matches to the DNA consensus sequence for *E. coli* (∞ Section 6.7). Why?

4. Interestingly, the attenuation control of some of the pyrimidine biosynthetic pathway genes in *Escherichia coli* actually involves coupled transcription and translation. Can you describe a mechanism whereby the cell could somehow make use of translation to help it measure the level of pyrimidine nucleotides?

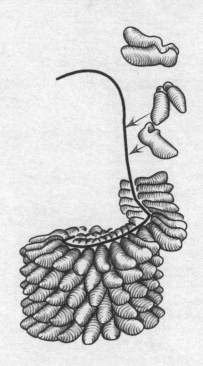

Tobacco Mosaic Virus (TMV), whose structure is shown here, was the first virus discovered. Work began on it in 1886 and the Russian microbiologist Ivanowsky showed that it passed through filters that could retain all known bacteria. The Dutch microbiologist Martinus Beijerinck proposed in 1899 that the tobacco mosaic agent was a new type of microorganism that we now call a virus. TMV was also the first virus purified and crystallized (by Wendell Stanley in 1935), and later the first large biological structure whose subunits could be shown to spontaneously reassemble. Interestingly, TMV is also a virus whose genetic material is RNA and gave early evidence of the incredible variety of strategies that viruses use to replicate themselves.

CHAPTER 8 Viruses

8.1 General Properties of Viruses 237
8.2 Nature of the Virion 238
8.3 The Virus Host 242
8.4 Quantification of Viruses 242
8.5 General Features of Virus Reproduction 244
8.6 Steps in Virus Multiplication 246
8.7 Overview of Bacterial Viruses 250
8.8 RNA Bacteriophages 251
8.9 Single-Stranded Icosahedral DNA Bacteriophages 252
8.10 Single-Stranded Filamentous DNA Bacteriophages 254
8.11 Double-Stranded DNA Bacteriophages: Lytic Viruses 255

8.12 Temperate Bacteriophages: Lysogeny and Lambda 259
8.13 A Transposable Phage: Bacteriophage Mu 265
8.14 Overview of Animal Viruses 267
8.15 Positive-Strand RNA Animal Viruses 270
8.16 Negative-Strand RNA Viruses 272
8.17 Double-Stranded RNA Viruses: Reoviruses 275
8.18 Replication of DNA Viruses of Animals 275
8.19 Herpesviruses 277
8.20 Pox Viruses 279
8.21 Adenoviruses 280
8.22 Retroviruses 281
8.23 Viroids and Prions 285

Bacteriophage   a virus that infects prokaryotic cells

Lysogen   a bacterium containing a prophage

Minus (negative)-strand nucleic acid   an RNA or DNA strand that has the opposite sense of (is complementary to) the mRNA of a virus

Oncogene   a gene whose expression causes formation of a tumor

Plaque   a zone of lysis or cell inhibition caused by virus infection of a lawn of sensitive cells

Plus (positive)-strand nucleic acid   an RNA or DNA strand that has the same sense as the mRNA of a virus

Prion   an infectious agent whose extracellular form may contain no nucleic acid

Provirus (prophage)   the genome of a temperate virus when it is replicating with, and usually integrated into, the host chromosome

Retrovirus   a virus whose RNA genome has a DNA intermediate as part of its replication cycle

Reverse transcription   the process of copying information found in RNA into DNA

Temperate virus   a virus whose genome is able to replicate along with that of its host and not cause cell death in a state called lysogeny

Transformation   a process by which a normal cell becomes a cancer cell (but see alternative usage in Chapter 9)

Virion   the complete virus particle; the nucleic acid surrounded by a protein coat and in some cases other material

Virulent virus   a virus that lyses or kills the host cell after infection; a nontemperate virus

Virus   a genetic element containing either RNA or DNA that replicates in cells but is characterized by having an extracellular state

Viruses are genetic elements that can replicate independently of a cell's chromosomes but not independently of cells themselves (∞ Section 6.4). In order to multiply, viruses must enlist a cell in which they can replicate. Such a cell is called a *host*. Viruses are characterized by also having an extracellular state.

Viruses are not the only type of genetic element that takes advantage of the metabolic machinery encoded by the cell's own chromosomes (∞ Section 6.4). Like these other elements, viruses can confer important new properties on their host cell. These properties will be inherited when the host cell divides if each new cell also inherits the viral genome. These changes are often not harmful and may even be beneficial. However, viruses, unlike genetic elements such as plasmids (∞ Sections 6.4 and 9.8), have an extracellular form that enables them to be easily transmitted from one host to another. This extracellular form has enabled some viruses to replicate themselves in a host in a way that is destructive to the host cell. This destructive replication accounts for the fact that some viruses are agents of disease. In many cases, whether a virus causes disease or hereditary change depends on the host cell and on the environmental conditions.

In this chapter we shall discuss some of the ways in which viruses can redirect the metabolism of the host cell in order to replicate. This chapter is divided into three parts. The first part deals with basic concepts of virus structure and function. The second part deals with the nature and manner of multiplication of the bacterial viruses (bacteriophages). The third part deals with important groups of animal viruses. In both the second and third parts we shall describe some of the basic molecular biology of virus multiplication. These discussions expand on the concepts of macromolecular synthesis and gene regulation we covered in Chapters 6 and 7.

Scientists have studied and continue to study viruses for what they can tell us about the genetics and biochemistry of cellular metabolism and, in the case of some viruses, the development of disease. However, as we shall see in Chapters 9 and 10, viruses are also important tools for the microbial geneticist and the genetic engineer.

8.1

General Properties of Viruses

Viruses have both an extracellular and an intracellular state. In the **extracellular** state, a virus is a minute particle containing nucleic acid surrounded by protein and occasionally containing other macromolecular components. In this extracellular state, the **virus particle,** also called the **virion,** is metabolically inert and does not carry out respiratory or biosynthetic functions. The virion is the structure by which the **virus genome** is carried from the cell in which it has been produced to another cell where the viral nucleic acid can be introduced. Once in the new cell, the **intracellular state** is initiated. In the intracellular state, **virus replication** occurs: New copies of the virus genome are produced, and the components that make up the virus coat are synthesized. When a virus genome is introduced into a host cell and reproduces, the process is called **infection.** A cell that a virus can infect and in which it can replicate is called a **host.** Viral genomes are very limited in size, and they encode primarily those functions that they cannot adapt from their hosts. Therefore, during replication inside a cell, there is a heavy dependence on host cell structural and metabolic components. The virus redirects preexisting host machinery and metabolic functions necessary for virus replication and the assembly of new virions. (Therefore, for most viruses virions can also be found inside the cell.)

As we have seen (∞ Section 6.1), all cells have double-stranded deoxyribonucleic acid (DNA) as their genetic material. In contrast, viruses can have either DNA or ribonucleic acid (RNA) as their genetic material, and it can be either single-stranded or double-stranded. Viruses are sometimes divided into two types based on whether they have DNA or RNA as their genetic material, and *all* viruses contain one or the other in the virion. However, there is a third group of viruses that use *both* DNA and RNA as their genetic material but at different stages of their reproductive cycle (Figure 8.1). The latter include the retroviruses, which contain an RNA genome in the virion but replicate through a DNA intermediate, and the human hepatitis B virus, which contains DNA in the virion but has an RNA intermediate in replication. These classes can be further subdivided according to whether the nucleic acid in the virion is single- or double-stranded (Figure 8.1). In spite of the diversity of genome structure, viruses obey the *central dogma* of molecular biology (∞ Section 6.1): All genetic information flows from nucleic acid to protein. In addition, all viruses use the cell's translational machinery, and so no matter what the genome structure of the virus, messenger RNA (mRNA) must be generated that can be translated on the host's ribosomes.

Viruses can also be classified on the basis of the hosts they infect. Thus, we have animal viruses, plant viruses, and bacterial viruses. Bacterial viruses, sometimes called *bacteriophages* (or *phage* for short, from the Greek *phagein* meaning "to eat"), have been studied primarily as convenient model systems for research on the molecular biology and genetics of virus reproduction.

Many of the basic concepts of virology were first worked out with bacterial viruses and subsequently applied to viruses of higher organisms. Because of their frequent medical importance, *animal viruses* have been extensively studied. The two groups of animal viruses most studied are those infecting insects and those infecting warm-blooded animals. *Plant viruses* are important in agriculture but have been less studied than animal viruses. In this chapter, we discuss the structure, replication, and genetics of viruses infecting bacteria and warm-blooded animals.

✓ 8.1 Concept Check

A virion is the extracellular form of a virus and contains either an RNA or a DNA genome. The virus genome is introduced into a new host cell by infection. The virus redirects the host metabolism in order to replicate.

✓ How does a *virus* differ from a *plasmid*?
✓ How does a *virion* differ from a *cell*?

8.2

Nature of the Virion

Virus particles (virions) vary widely in size and shape. Viruses are smaller than cells, ranging in size from 0.02 to 0.3 μm. A common unit of measure for viruses is the *nanometer*, which is 1000 times smaller than 1 μm and 1 million times smaller than 1 mm. Smallpox virus, one of the largest viruses, is about 200 nm in diameter (a bit smaller than the size of the smallest bacteria); poliovirus, one of the smallest, is only 28 nm in diameter (about the size of a ribosome).

As we have stated, some viruses contain RNA, others DNA, and the nucleic acid can be either double- or single-stranded, depending on the virus. Viral *genomes* are also smaller than those of cells. Most bacterial genomes are between 1000 and 5000 kilobase pairs of DNA, with the smallest known being 580 kilobase pairs. (Interestingly, the bacteria with the smallest genomes

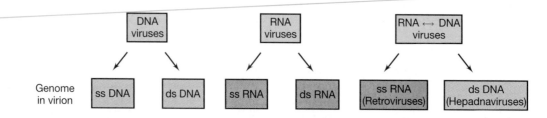

FIGURE 8.1   Viral genomes. The genomes of viruses can be composed of either DNA or RNA, and some use both as their genomic material at different stages in their life cycle. However, only one type of nucleic acid is found in the virion of any particular type of virus. This can be single-stranded (ss), double-stranded (ds), or in the case of the hepadnaviruses, partially double-stranded.

are, like viruses, parasites that replicate in other cells; ∞ Sections 13.12 and 13.26.) However, one of the largest known viral genomes, that of vaccinia, is only 190 kilobase pairs. Some viruses have genomes so small they contain less than five genes. The sizes of the genomes of a few representative types of viruses are given in Table 8.1. As can be seen in the table, the genome of some viruses, such as reovirus, is *segmented* into more than one molecule.

The structures of virions (virus particles) are quite diverse, varying widely in size, shape, and chemical composition. The nucleic acid of the virion is always located within the particle, surrounded by a protein coat called the *capsid*. The terms *coat, shell,* and *capsid* are often used interchangeably to refer to this outer layer. The protein coat is always formed of a number of individual protein molecules, called *structural subunits,* which are arranged in a precise and highly repetitive pattern around the nucleic acid (Figure 8.2). The small genome size of most viruses restricts the number of different viral proteins. A few viruses have only a single kind of protein in their capsid, but most viruses have several chemically distinct kinds of structural subunits that are themselves associated in specific ways to form larger assemblies called *capsomers*. It is the morphological unit that can be seen with the electron microscope.

The information for proper aggregation of the structural subunits into capsomers is contained within the structure of the proteins themselves, and the overall process of assembly is thus called **self-assembly.** For many viruses, this self-assembly process is assisted by *molecular chaperones,* proteins that assist in folding and assembly but that themselves are not a part of the final structure (∞ Section 6.12). A single virion generally has a large number of morphological units.

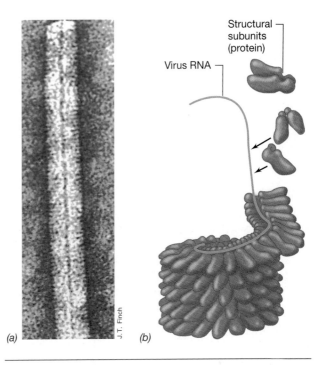

FIGURE 8.2 An example of the arrangement of virus nucleic acid and protein coat in a simple virus, tobacco mosaic virus. (a) Electron micrograph at high resolution of a portion of the virus particle. (b) Assembly of the tobacco mosaic virion. The RNA assumes a helical configuration surrounded by the protein capsid. The center of the particle is hollow.

The complete complex of nucleic acid and protein, packaged in the virus particle, is called the virus **nucleocapsid.** Although the virus structure just described is frequently the total structure of a virus particle, a number of viruses have more complex structures. These viruses are *enveloped* viruses in

TABLE 8.1	Some types of viral genomes[a]				
		Viral genome			
Virus	**Host**	**Type of nucleic acid in virion**	**Structure**	**Number of molecules**	**Size**
H-1 parvovirus	Animals	Single-stranded DNA	Linear	1	5,176 bases
φX174	Bacteria	Single-stranded DNA	Circular	1	5,386 bases
Simian virus 40 (SV40)	Animals	Double-stranded DNA	Circular	1	5,243 base pairs
Poliovirus	Animals	Single-stranded RNA	Linear	1	7,433 bases
Cauliflower mosaic virus	Plants	Double-stranded DNA	Circular	1	8,025 base pairs
Cowpea mosaic virus	Plants	Single-stranded RNA	Linear	2 different	9,370 bases (total)
Reovirus type 3	Animals	Double-stranded RNA	Linear	10 different	23,549 base pairs (total)
Bacteriophage λ	Bacteria	Double-stranded DNA	Linear	1	48,514 base pairs
Herpes simplex virus type I	Animals	Double-stranded DNA	Linear	1	152,260 base pairs

a The sizes of the viral genomes chosen for this table are known accurately because they have been sequenced. However, this accuracy can be misleading because only a particular strain or isolate of a virus was sequenced. Therefore, the sequence and exact number of bases for other isolates may be slightly different. No attempt has been made to choose the largest and smallest viruses known, but rather to give a fairly representative sampling of the sizes and structures of the genomes of viruses containing both single- and double-stranded RNA and DNA.

which the nucleocapsid is enclosed in a membrane (Figure 8.3). (Viruses without membranes are sometimes called *naked* viruses.) *Virus membranes* are generally lipid bilayer membranes (∞ Section 3.5), but associated with these membranes are often *virus-specific* proteins. Inside the virion are often one or more virus-specific *enzymes.* Such enzymes usually play a role during the infection and replication process, as we will discuss later in this chapter.

Virus Symmetry

The nucleocapsids of viruses are constructed in highly symmetric ways. Symmetry refers to the way in which the protein morphological units are arranged in the virus shell. When a symmetric structure is rotated around an axis, the same form is seen again after a certain number of degrees of rotation. Two kinds of symmetry are recognized in viruses, which correspond to the two primary shapes, rod and spherical. Rod-shaped viruses have helical symmetry, and spherical viruses have icosahedral symmetry. In all cases, the characteristic structure of the virus is determined by the structure of the protein subunits of which it is constructed.

A typical virus with **helical symmetry** is the tobacco mosaic virus (TMV) illustrated in Figure 8.2. It is an RNA virus in which the 2130 identical protein subunits are arranged in a helix. The overall dimensions of the TMV virion are 18 × 300 nm. The lengths of helical viruses are determined by the length of the nucleic acid, but the width of the helical virus particle is determined by the size and packaging of the protein subunits.

An **icosahedron** is a symmetric structure roughly spherical in shape that has 20 faces. Icosahedral symmetry is the most efficient arrangement for subunits in a closed shell because it uses the smallest number of units to build a shell. The simplest arrangement of morphological units is 3 per face, for a total of 60 units per virus particle. The 3 units at each face can be either identical or different. Most viruses have more nucleic acid than can be packed into a shell made of just 60 morphological units. The next possible structure that permits close packing contains 180 units, and many viruses have shells with this configuration. Other known configurations involve 240 units and 420 units.

Figure 8.4a shows a model of an icosahedron and Figure 8.4b shows an electron micrograph of a typical icosahedral virus.

Enveloped Viruses

Many viruses have complex membranous structures surrounding the nucleocapsid (Figure 8.5a). Enveloped viruses are common in the animal world (for example,

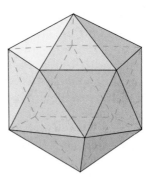

(a)

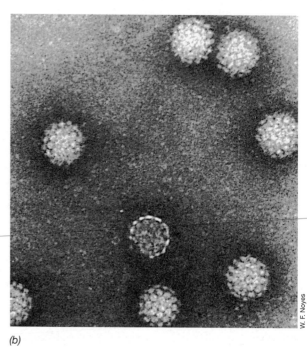

(b)

FIGURE 8.4 Icosahedral symmetry. (a) A model of an icosahedron. (b) Electron micrograph of human wart virus, a virus with icosahedral symmetry. The individual particles are about 55 nm in diameter.

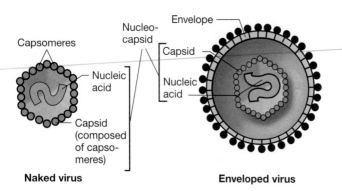

FIGURE 8.3 Comparison of naked and enveloped virus, two basic types of virus particles.

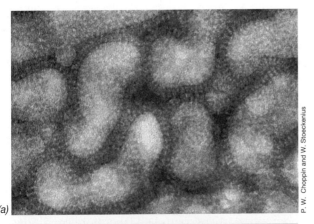

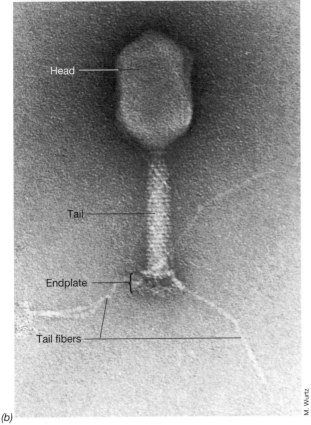

(a)

Head

Tail

Endplate

Tail fibers

(b)

P. W. Choppin and W. Stoeckenius

M. Wurtz

FIGURE 8.5 Electron micrographs of animal and bacterial viruses. (a) Influenza virus, an enveloped virus. The individual particles are about 80 nm in diameter. (b) Bacterial virus (bacteriophage) T4 of *Escherichia coli*. Note the complex structure. The tail components are involved in attachment of the virion to the host and infection of the nucleic acid. The head is about 85 nm in diameter.

influenza virus), but some enveloped bacterial viruses are also known. The virus envelope consists of a lipid bilayer with proteins, usually glycoproteins, embedded in it. The lipids of the membrane are derived from the membranes of the host cell but the proteins are encoded by the virus. The symmetry of enveloped viruses is

expressed not in terms of the virion as a whole but in terms of the nucleocapsid present inside the virus membrane.

What is the function of the membrane in a virus particle? We will discuss this in detail later but note that because of its location in the virion, the membrane is the structural component of the virus particle that interacts first with the cell. The specificity of virus infection, and some aspects of virus penetration, are controlled in part by characteristics of virus membranes.

Complex Viruses

Some virions are even more complex, being composed of several separate parts with separate shapes and symmetries. The most complicated viruses in terms of structure are some of the bacterial viruses, which possess not only icosahedral heads but also helical tails (Figure 8.5b). In some bacterial viruses, such as the T4 virus of *Escherichia coli*, the tail itself has a complex structure. For instance, T4 has almost 20 different proteins in the tail, and the T4 head has several more proteins. In such complex viruses, assembly is also complex. For instance, in T4 the complete tail is formed as a subassembly, and then the tail is added to the DNA-containing head. Finally, tail fibers formed from another protein are added to make the mature, infectious virus particle (see the discussion of T4 assembly in Section 8.11).

Enzymes in Virions

We have stated that virions do not carry out metabolic processes. Outside a host cell, a virion is metabolically inert. However, some virions do contain enzymes that play roles in the infection process. For instance, many viruses contain their own nucleic acid polymerases that transcribe the viral nucleic acid into messenger RNA once the infection process has begun. The retroviruses are RNA viruses that replicate inside the cell as DNA intermediates. These viruses possess an enzyme, an RNA-dependent DNA polymerase called *reverse transcriptase*, that transcribes the information in the incoming RNA into a DNA intermediate. A number of viruses contain enzymes that aid in entering cells or in release of the virus from the host cells in the final stages of the infection process. One group of such enzymes, called *neuraminidases*, breaks down glycosidic bonds of glycoproteins and glycolipids of the connective tissue of animal cells, thus aiding in the liberation of the virus. Virions infecting some bacteria possess an enzyme, *lysozyme* (∞ Section 3.7), that makes a small hole in the bacterial cell wall that allows the viral nucleic acid to enter. Lysozyme is produced in large amounts in the later stages of infection, causing lysis of the host cell and release of the virions. We will discuss some of these enzymes in more detail later.

✓ 8.2 Concept Check

In the virion of the naked virus, only nucleic acid (DNA or RNA) and protein are present, with the nucleic acid on the inside; the whole unit is called the nucleocapsid. Enveloped viruses have one or more lipoprotein layers surrounding the nucleocapsid. The nucleocapsid is arranged in a symmetric fashion, with a precise number and arrangement of structural subunits surrounding the virus nucleic acid. Although viruses are metabolically inert, in some viruses, one or more enzymes are present within the virion. Such enzymes play a role in the initial stages of the infection process.

✓ What is the difference between a *naked* virus and an *enveloped* virus?

✓ What kinds of enzymes can be found within the virions of specific viruses?

8.3

The Virus Host

Because viruses, like plasmids, replicate only inside living cells, research on viruses requires use of appropriate hosts. For the study of bacterial viruses, pure cultures are used either in liquid or on semisolid (agar) media. Because many bacteria are so easy to culture, it is quite easy to study bacterial viruses, and this is why such detailed knowledge of bacterial virus multiplication is available.

With animal viruses, the initial host may be a whole animal that is susceptible to the virus, but for research purposes it is desirable to have a more manageable host. Many animal viruses can be cultivated in *tissue* or *cell cultures*, and the use of such cultures has enormously facilitated research on animal viruses.

Cell Cultures

A cell culture is obtained by promoting growth of cells taken from an organ of the experimental animal. Cell cultures are generally obtained by aseptically removing pieces of the tissue in question, dissociating the cells by treatment with an enzyme that breaks apart the intercellular cement, and spreading the resulting suspension out on the bottom of a flat surface, such as a bottle or a Petri dish. The cells generally produce glycoprotein-like materials that permit them to adhere to glass surfaces. The thin layer of cells adhering to the glass or plastic dish, called a *monolayer*, is then overlaid with a suitable culture medium and the culture incubated. The culture media used for cell cultures are generally quite complex, employing a number of amino acids and vitamins, salts, glucose, and a bicarbonate buffer system. To obtain best growth, addition of a small amount of blood serum is usually necessary, and several antibiotics are generally added to prevent bacterial contamination.

Some cell cultures prepared in this way grow indefinitely and can be established as *permanent cell lines*. Such cell cultures are most convenient for virus research because cell material is continuously available for research purposes. In other cases, indefinite growth does not occur, but the culture may remain alive for a number of days. Such cultures, called *primary cell cultures*, may still be useful for virus research, although new cultures will have to be prepared from fresh sources from time to time.

In some cases, cell culture monolayers cannot be obtained, but whole organs, or pieces of organs, can be cultured. Such **organ cultures** may still be useful in virus research because they permit growth of viruses under more-or-less controlled laboratory conditions.

8.4

Quantification of Viruses

In order to obtain any significant understanding of the nature of viruses and virus replication, it is necessary to be able to *quantify* the number of virus particles. Virions are almost always too small to be seen under the light microscope. Although they can be observed under the electron microscope, the preparation of samples for observation can be cumbersome for routine study. In general, viruses are quantified by measuring their effects on the host cells that they infect. It is common to speak of a *virus infectious unit*, which is the smallest unit that causes a detectable effect when placed with a susceptible host. By determining the number of infectious units per volume of fluid, a measure of virus quantity can be obtained. We discuss here several approaches to assessment of the virus infectious unit.

Plaque Assay

When a virion initiates an infection on a layer or lawn of host cells growing spread out on a flat surface, a zone of *lysis* or a zone of *growth inhibition* may occur that results in a clear area in the lawn of growing host cells. This clearing is called a **plaque,** and it is assumed that each plaque has originated from replication events that began with one virion.

Plaques are essentially "windows" in the lawn of confluent cell growth. With bacterial viruses, plaques may be obtained when virus particles are mixed into a thin layer of host bacteria that is spread out as an agar overlay on the surface of an agar medium (Figure 8.6a). During incubation of the culture, the bacteria grow and form a turbid layer that is visible to the naked eye. However, wherever a successful viral infection has been initiated, lysis of the cells occurs, resulting in the formation of a clear zone called a *plaque* (Figure 8.6b).

The plaque procedure also permits the isolation of pure virus strains because if a plaque has arisen from a

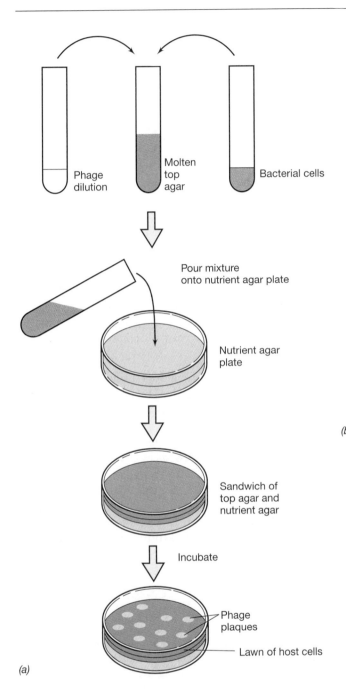

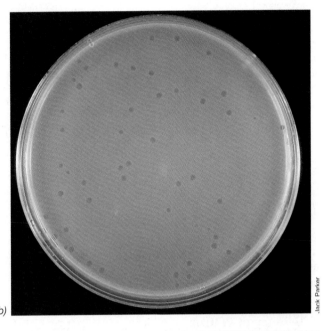

(b)

FIGURE 8.6 Quantification of bacterial virus by plaque assay using the agar overlay technique. (a) A dilution of a suspension containing the virus material is mixed in a small amount of melted agar with the sensitive host bacteria, and the mixture poured on the surface of a nutrient agar plate. The host bacteria, which have been spread uniformly throughout the top agar layer, begin to grow, and after overnight incubation form a *lawn* of confluent growth. Each virus particle that attaches to a cell and reproduces may cause cell lysis, and the virus particles released can spread to adjacent cells in the agar, infect them, be reproduced, and again lead to lysis and release. The size of the plaque formed depends on the virus, the host, and conditions of culture. (b) Photograph of a plate showing plaques formed by bacteriophage on a lawn of sensitive bacteria. The plaques shown are about 1–2 mm in diameter.

single virion, all the virions in this plaque are probably genetically identical. Some of the virions from this plaque can be picked and inoculated into a fresh bacterial culture to establish a pure virus line. The development of the plaque assay technique was as important for the advance of virology as Koch's development of solid media (∞ Section 1.8) for bacteriology.

Plaques may be obtained for animal viruses by using animal cell culture systems as hosts. A monolayer of cultured animal cells is prepared on a plate or flat bottle, and the virus suspension overlaid. Plaques are revealed by zones of destruction of the animal cells (Figure 8.7).

In some cases, the virus may not actually destroy the cells but may cause changes in morphology or growth rate that can be recognized. For instance, tumor viruses may not destroy cells but may cause the cells to grow faster than uninfected cells, a phenomenon called *transformation*. As we have noted, the general arrangement of cells in a tissue culture is a monolayer. This is because growth generally ceases when the cells, as a result of growth, come in contact with each other (a phenomenon known as *contact inhibition*). Transformed cells have altered growth requirements and continue to grow, piling up to form a small *focus of growth* (called a *focus*

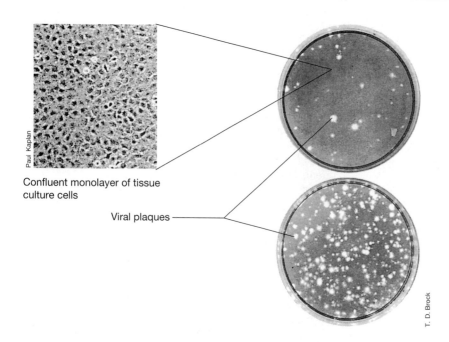

Paul Kaplan

Confluent monolayer of tissue
culture cells

Viral plaques ———

T. D. Brock

FIGURE 8.7   Cell cultures in monolayers within Petri plates. Note the presence of plaques where virus-induced cell lysis has occurred. Also shown is a photomicrograph of a cell culture.

of infection when the transformation has been brought about by virus infection). By counting foci of infection, a quantitative measure of virus may be obtained.

Efficiency of Plating

One important concept in quantitative virology involves the idea of *efficiency of plating*. Counts made by plaque assay are always lower than counts made with the electron microscope. The efficiency with which virions infect host cells is rarely 100% and may often be considerably less. This does not mean that virions that have not caused infection are inactive, although this is sometimes the case. It may merely mean that under the conditions used, successful infection with these particles has not occurred. Although with bacterial viruses, efficiency of plating is often higher than 50%, with many animal viruses it may be very low, 0.1 or 1%. Why virus particles vary in infectivity is not well understood. In some cases it is possible that the conditions used for quantification are not optimal. Because the electron microscope is not routinely used to count virions, it is sometimes difficult to assess the actual efficiency of plating, but the concept is important in both research and medical practice. Because the efficiency of plating is rarely close to 100%, when the plaque method is used to quantify virus, it is accurate to express the concentration (called the *titer*) of the virus suspension not as the absolute number of virion units but as the number of *plaque-forming units*.

Animal Infectivity Methods

Some viruses do not cause recognizable effects in cell cultures but cause death in the whole animal. In such cases, quantification can be done only by some sort of titration in infected animals. The general procedure is to carry out a serial dilution of the unknown sample, generally at 10-fold dilutions, and to inject samples of each dilution into numbers of sensitive animals. After a suitable incubation period, the fraction of dead and live animals at each dilution is tabulated and an *end point dilution* is calculated. This is the dilution at which, for example, *half* of the injected animals die. Although such serial dilution methods are much more cumbersome and much less accurate than cell culture methods, they may be essential for the study of certain types of viruses.

✓ 8.4 Concept Check

Although it requires only a single virion to initiate an infectious cycle, not all virus particles are equally infectious. One of the most accurate ways of measuring virus infectivity is by the plaque assay. Plaques are clear zones that develop on layers or lawns of host cells, each plaque due to infection by a single virus particle. The virus plaque is analogous to the bacterial colony.

✓ Give a definition of *efficiency of plating*.
✓ What is a *plaque-forming unit?*

8.5

General Features of Virus Reproduction

The basic problem of virus replication can be simply put: The virus must induce a living host cell to synthesize all the essential components needed to make more virus particles. These components must then be assembled into the proper structure, and the new virions must escape from the cell and infect other cells. The various phases of this replication process in a bacteriophage can be categorized in seven steps (Figure 8.8).

1. **Attachment** (adsorption) of the virion to a susceptible host cell.
2. **Penetration** (injection) of the virion or its nucleic acid into the cell.

3. **Early steps in replication** during which the host cell biosynthetic machinery is altered as a prelude to virus nucleic acid synthesis. Virus-specific enzymes are typically made.
4. **Replication** of the virus nucleic acid.
5. **Synthesis of proteins used as structural subunits** of the virus coat.
6. **Assembly** of structural subunits (and membrane components in enveloped viruses) and **packaging** of nucleic acid into new virus particles.
7. **Release** of mature virions from the cell.

These stages in virus replication are recognized when virus particles infect cells in culture and are illustrated in Figure 8.9, which exhibits what is called a **one-step growth curve.** In the first few minutes after infection the virus is said to undergo an *eclipse.* The virus nucleic acid has become separated from its protein coat, and so even if the infected cell had broken open, the virion no longer exists as an infectious entity. Although virus nucleic acid may be infectious, the infectivity of virus nucleic acid is many times lower than that of

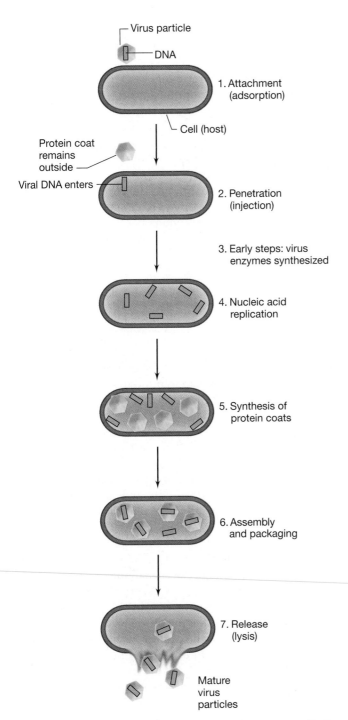

FIGURE 8.8 The replication cycle of a bacterial virus. The general stages of virus replication are indicated.

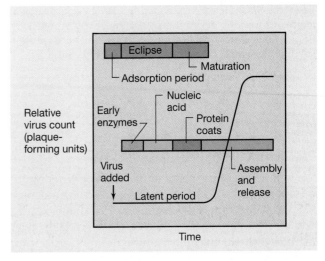

FIGURE 8.9 The one-step growth curve of virus replication. This graph displays the results of a single round of viral multiplication in a population of cells. Following adsorption, the infectivity of the virus particles disappears, a phenomenon called *eclipse.* This is due to the uncoating of the virus particles. During the *latent period,* replication of viral nucleic acid and protein occurs. The *maturation period* follows, when virus nucleic acid and protein are assembled into mature virus particles. At this time, if the cells are broken up, active virus can be detected. Finally, *release* occurs, either with or without cell lysis. The timing of the one-step growth cycle varies with the virus and host. With many bacterial viruses, the whole cycle may be complete in 20–60 min, whereas with animal viruses 8–40 hr is usually required for a complete cycle. Compare this general picture and color scheme with specific replication events shown for bacteriophage T4 in Figure 8.23.

whole virions because the machinery for bringing the virus genome into the cell is lacking. Also, outside the virion the nucleic acid is no longer protected from deleterious activities of the environment as it was when it was inside the protein coat.

The eclipse occurs during the early stages of virus replication. Maturation begins as the newly synthesized nucleic acid molecules become packaged inside protein coats. During the *maturation* phase, the titer of active virions inside the cell rises dramatically. The period of time when no infectious virions are present extracellularly is called the *latent period*. At the end of maturation, *release* of mature virions occurs, either as a result of cell *lysis* or because of some budding or excretion process. The number of virions released, called the *burst size,* varies with the particular virus and the particular host cell and can range from a few to a few thousand. The timing of this overall virus replication cycle varies from 20–60 min in many bacterial viruses to 8–40 hr in most animal viruses. We now consider each of the steps of the virus multiplication cycle in more detail.

✓ 8.5 Concept Check

The virus life cycle can be divided into seven stages: attachment (adsorption), penetration (injection), early protein synthesis, nucleic acid replication, synthesis of virus protein subunits, assembly of mature virions, and virus release.

✓ What is the *latent* period?

8.6

Steps in Virus Multiplication

We will now discuss some of the steps of virus multiplication in more detail. As we have noted, the outcome of a virus infection is the synthesis of viral nucleic acid and viral protein coats. In effect, the virus takes over the biosynthetic machinery of the host and uses it for its own synthesis. A few enzymes needed for virus replication may be present in the virion and may be introduced into the cell during the infection process, but the host supplies everything else: energy-generating system, ribosomes, amino acid activating enzymes, transfer RNA (with a few exceptions), and all soluble factors. The virus genome codes for all new proteins. Therefore, the steps in virus multiplication include mechanisms for transporting the viral genome into the cell and ensuring that it is expressed in such a way that new virions will be produced.

Attachment

There is a high specificity in the interaction between virus and host. The most common basis for host speci-

ficity involves the attachment process. The virus particle itself has one or more proteins on the outside that interact with specific cell surface components called *receptors.* The receptors on the cell surface are normal surface components of the host, such as proteins, polysaccharides, glycoproteins, and lipoprotein–polysaccharide complexes, to which the virion attaches. In the absence of the receptor site, the virus cannot adsorb, and hence cannot infect. If the receptor site is altered, the host may become resistant to virus infection. However, mutants of the virus can also arise that are able to adsorb to resistant hosts.

In general, virus receptors carry out normal functions in the cell. For example, in Bacteria some phage receptors are pili or flagella, others are cell envelope components, and others are transport binding proteins. The receptor for influenza virus is a glycoprotein found on red blood cells and on cells of the mucous membrane of susceptible animals, whereas the receptor site of poliovirus is a cell surface lipoprotein. However, many animal and plant viruses do not have specific attachment sites at all, and the virus enters passively as a result of phagocytosis or some other endocytotic process.

Penetration

The means by which the virus penetrates into the cell depends on the nature of the host cell, especially on its surface structures. Cells with cell walls, such as bacteria, are infected in a different manner from animal cells, which lack a cell wall. The most complicated penetration mechanisms have been found in viruses that infect bacteria. The bacteriophage T4, which infects *Escherichia coli,* can be used as an example.

The structure of the bacterial virus T4 was shown in Figure 8.5*b*. The virion has a **head,** within which the viral DNA is folded, and a long, fairly complex **tail,** at the end of which is a series of tail fibers. During the attachment process, the virions first attach to cells by means of the tail fibers (Figure 8.10). The ends of the fibers interact specifically with core polysaccharides that are part of the outer layer of the gram-negative cell wall (⟲ Section 3.8). These tail fibers then retract, and the core of the tail makes contact with the cell envelope of the bacterium. The action of a lysozyme-like enzyme results in the formation of a small hole. The tail sheath contracts, and the DNA of the virus passes into the cell through a hole in the tip of the tail, the majority of the coat protein remaining outside.

With animal cells, the *whole virion* penetrates the cell, being carried inside by endocytosis (phagocytosis or pinocytosis), an active cellular process, or fusion. We describe some of these processes in detail later in this chapter.

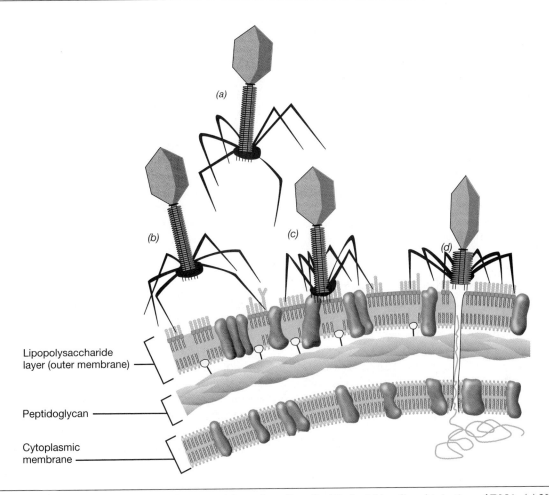

FIGURE 8.10   Attachment of T4 bacteriophage particle to the cell wall of *Escherichia coli* and injection of DNA: (a) Unattached particle. (b) Attachment to the wall by the long tail fibers interacting with core polysaccharide. (c) Contact of cell wall by the tail pins. (d) Contraction of the tail sheath and injection of the DNA. For a detailed description of the gram-negative cell wall see Section 3.8.

✓ 8.6a Concept Check

The attachment of a virion to a host cell is a highly specific process, involving the interaction of receptors on the host with proteins on the surface of the virus particle. Only after attachment has occurred can the virus or its nucleic acid penetrate the host cell.

Virus Restriction and Modification by the Host

We have already seen that one form of host resistance to virus arises when there is no receptor site on the cell surface to which the virus can attach. Another and more specific kind of host resistance occurs in prokaryotes and involves destruction of the double-stranded DNA genome of a virus after it has been injected. This destruction is brought about by host enzymes that cleave the viral DNA at one or several places, thus preventing its replication. This phenomenon is called *restriction* and is part of a general host mechanism to prevent the invasion of foreign nucleic acid. We have already dis-

cussed *restriction enzymes* and their action in some detail (∞ Section 6.5) and noted that their cellular role is in defending against foreign DNA. Restriction enzymes are highly specific, attacking only certain sequences (generally four or six base pairs). The host protects its own DNA from the action of restriction enzymes by *modifying* its DNA at the sites where the restriction enzymes act. Modification of host DNA is brought about by methylation of purine or pyrimidine bases (in such a way that their base-pairing properties are not altered).

Some viruses can overcome host restriction mechanisms by modifications of their nucleic acids so that they are no longer subject to enzymatic attack. Two kinds of chemical modifications of viral DNA have been recognized, glucosylation and methylation. For instance, the T-even bacteriophages (T2, T4, and T6) have their DNA glucosylated to varying degrees, and the glucosylation prevents or greatly reduces endonuclease attack. Many other viral nucleic acids have been found to be modified by methylation, but glucosylation has been found only in the T-even bacteriophages. It should be emphasized that modification of viral nucleic acid occurs after replication has occurred and that the modified bases are not copied directly. Other viruses, such as the bacteriophages T3 and T7, avoid restriction by encoding proteins that inhibit the host restriction systems. Some hosts have multiple restriction and methylation systems that help in preventing infection by viruses that can circumvent only one of them.

However, not all restriction systems recognize unmodified DNA. Host restriction systems are also known that restrict only *modified* DNA! Clearly the host containing this enzyme does *not* contain the modification enzyme. However, this host is protected from infection by a virus that was modified during reproduction in its previous host strain.

Hosts also contain other DNA methylases. Some of these methylases may be involved in DNA repair or in gene regulation, but others may offer protection to *host* DNA. This is because *some viruses themselves encode restriction systems.*

As we discussed in Chapter 6, a knowledge of modification and restriction systems is of considerable practical utility in studying DNA chemistry. We discuss the use of restriction enzymes in genetic engineering in Chapter 10.

✓ 8.6b Concept Check

The virus nucleic acid is foreign to the host, and the restriction–modification system of the host, which recognizes and destroys foreign DNA, is one means of defense against virus infection.

Production of Viral Nucleic Acid and Proteins

New copies of the viral genome must be replicated, and virus-specific proteins must be synthesized in order for virus multiplication to occur. Typically the production of at least some viral proteins begins very early after the viral genome has been taken up by the cell. In order for this to happen, virus-specific messenger RNA must first be made. Exactly how the virus brings about new mRNA synthesis depends on the type of virus and on the structure of its genome. Although many viruses have a double-stranded DNA genome, a great many do not, and these other types of genomes include not only single-stranded DNA but both single- and double-stranded RNA. Furthermore, we have mentioned that some viruses have one type of nucleic acid in the virion but use another as a replicative intermediate. All these "unusual" genomes present problems in understanding virus multiplication because they involve information transfers, such as RNA to RNA and RNA to DNA, that host enzymes do not perform.

The essential features of producing mRNA from double-stranded DNA were discussed in Chapter 6, and it was shown that mRNA represents a complementary copy made by RNA polymerase of one of the two strands of the DNA double helix. Which copy is read into mRNA depends on the location of the appropriate promoter, because the promoter points the direction that the RNA polymerase will follow. In cells uninfected with virus, all mRNA is made on the cell's DNA template, but when viruses are present, the situation is different.

Figure 8.11 shows the nomenclature used to describe different types of nucleic acid based on its information content. Because the nucleic acid sequence of the mRNA can be translated directly into protein, it is by convention considered to be of the *plus* (+) configuration. The sequence of the viral genome nucleic acid is then indicated by a *plus* if it is the same as the mRNA and a *minus* if it is of opposite sense.

These configurations can also be used to describe the virus. For instance, a virus whose genome is single-stranded DNA of the plus configuration is called a *positive-strand DNA virus*, while a virus with an RNA genome of the minus configuration is called a *negative-strand RNA virus*. Figure 8.12 shows how mRNA can be made from different types of genomes. As seen in Figure 8.12, if the virus has double-stranded DNA (ds DNA), then mRNA synthesis can proceed directly as in uninfected cells. However, if the virus has a single-stranded DNA (ss DNA), then it is first converted to ds DNA and the latter serves as the template for mRNA synthesis by the RNA polymerase of the cell.

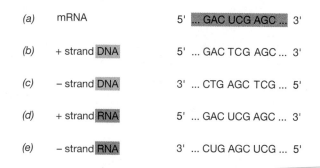

(a)	mRNA	5' ...GAC UCG AGC... 3'
(b)	+ strand DNA	5' ... GAC TCG AGC ... 3'
(c)	− strand DNA	3' ... CTG AGC TCG ... 5'
(d)	+ strand RNA	5' ... GAC UCG AGC ... 3'
(e)	− strand RNA	3' ... CUG AGC UCG ... 5'

FIGURE 8.11 Comparison of sequences of nucleic acids with different configurations. The strands of nucleic acid in viral genomes are called *plus* (+) if they are in the same configuration as mRNA and *minus* (−) if they are in a configuration opposite that of mRNA. (a) The direction and sequence of an mRNA. (b) The sequence of a + strand of a viral DNA genome that could encode that mRNA (using a − strand DNA intermediate). (c) A − strand of a viral DNA genome that could encode the mRNA. (d, e) The + and − strands of RNA genomes that could encode the mRNA. Note that a single-stranded + strand RNA genome has the same sequence and orientation as the mRNA. The 5'-end of each nucleic acid strand is highlighted in red.

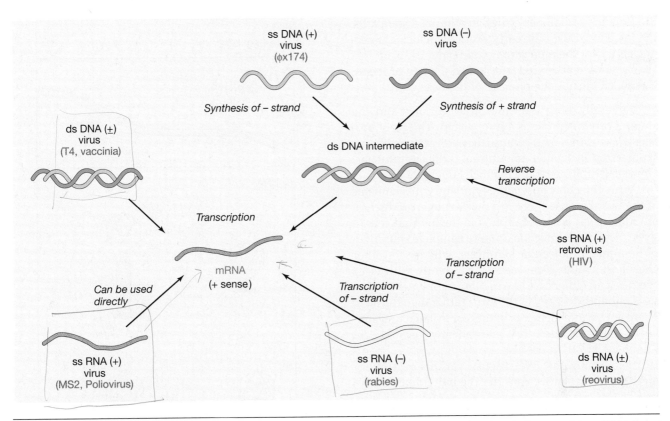

FIGURE 8.12 Formation of mRNA after infection of cells by viruses of different types. The chemical sense of the mRNA is considered as plus (+). The senses of the various virus nucleic acids are indicated as + if the same as mRNA, as − if opposite, or as ± if double-stranded. Examples are indicated next to the virus nucleic acid. Although examples of viruses containing ss DNA of the − sense are known, none are discussed in the text.

For RNA viruses, a virus-specific RNA-dependent RNA polymerase is needed because the cell RNA polymerase is DNA-dependent (requires a DNA template) and generally does not copy RNA. Several different types of viruses contain RNA as their genome, and each requires a different strategy for producing mRNA. The simplest case is the positive-strand RNA viruses in which the single incoming viral RNA strand is the *plus* strand and hence serves directly as mRNA. In addition to the other required proteins, this mRNA encodes the virus-specific RNA polymerase. This polymerase first makes complementary *minus* strands and then uses them as templates to make more plus strands. For negative-strand RNA viruses (whose virion contains only the minus strand) or double-stranded RNA viruses, the situation is more complicated. In neither case can the incoming RNA serve as mRNA, and therefore mRNA must be synthesized first. However, as mentioned earlier, cells do not typically have an RNA polymerase capable of this. To circumvent this problem, these viruses contain some of this enzyme in their virions, and it is injected into the cell along with the genomic RNA. Therefore, in these cases, the complementary plus strand is synthesized by this RNA-dependent RNA polymerase and used as message.

Retroviruses (causal agents of certain kinds of cancers and acquired immunodeficiency syndrome, AIDS) are RNA viruses that replicate through a DNA intermediate. The process of copying the information found in RNA into DNA is called **reverse transcription,** and thus these viruses require an enzyme called **reverse transcriptase.** (Telomerase is a type of reverse transcriptase; ⌒ Section 6.6.) In spite of the fact that the incoming RNA of retroviruses is the plus strand, it is not used as message, and therefore these viruses must carry reverse transcriptase in their virions. After infection, the virion ss RNA is copied to a double-stranded DNA (through an ss DNA intermediate), and the ds DNA then serves as the template for mRNA synthesis (thus, ss RNA → ss DNA → ds DNA). Retrovirus replication is of unusual complexity and is discussed in Section 8.22.

Viral Proteins

Once viral mRNA is made, viral proteins (for example, enzymes and structural subunits) can be synthesized. The proteins synthesized as a result of virus infection can be grouped into two broad categories:

1. Proteins (usually enzymes) synthesized soon after infection, called the **early proteins,** which are necessary for the replication of virus nucleic acid
2. Proteins synthesized later, called the **late proteins,** which include the proteins of the virus coat

Generally, both the time of appearance and the amount of virus proteins are regulated. The early proteins are enzymes that, because they act catalytically, are synthesized in smaller amounts, and the late proteins, often structural, are made in much larger amounts.

Virus infection upsets the regulatory mechanisms of the host because there is a marked overproduction of viral nucleic acid and protein in the infected cell. In some cases, virus infection causes a complete shutdown of host macromolecular synthesis, whereas in other cases host synthesis proceeds concurrently with virus synthesis. In either case, the regulation of virus synthesis is under the control of the virus rather than the host. There are several elements of this control that are similar to the host regulatory mechanisms discussed in Chapter 7, but there are also some uniquely viral regulatory mechanisms. We discuss various regulatory mechanisms when we consider the individual viruses later in this chapter.

✓ 8.6c Concept Check

Before replication of viral nucleic acid can occur, new virus proteins are often needed. These are encoded by messenger RNA molecules made from the virus genome. In the case of some RNA viruses, the viral RNA itself acts as mRNA. In other cases, the virus genome serves as a template for the formation of viral mRNA and certain essential enzymes are contained in the virion.

✓ Give one reason why viruses can infect only specific hosts.
✓ What is a *positive-strand RNA virus?*

8.7

Overview of Bacterial Viruses

Various kinds of bacterial viruses are illustrated in Figure 8.13. Most of the bacterial viruses that have been studied in detail infect Bacteria of the enteric group, such as *Escherichia coli* and *Salmonella typhimurium.* However, viruses are known that infect a variety of prokaryotes, both Bacteria and Archaea. A few bacterial viruses have lipid envelopes but most do not. However, many bacterial viruses are structurally complex, with head and complex tail structures. As we illustrated in Figure 8.10, the tail is involved in the injection of the nucleic acid into the cell.

We now discuss some of the bacterial viruses for which molecular details of the multiplication process

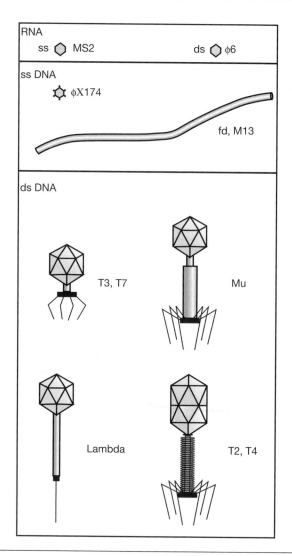

FIGURE 8.13   Schematic representations of the main types of bacterial viruses. Those discussed in detail are M13, φX174, MS2, T4, lambda, T7, and Mu. Sizes are to approximate scale.

are known. Although these bacterial viruses were first studied as *model systems* for understanding general features of virus multiplication, some of them now serve as convenient tools for *genetic engineering* (discussed in Chapter 10). Thus, the information on bacterial viruses is not only valuable as background for the discussion of animal viruses but also is essential for the material presented in the next two chapters on microbial genetics and genetic engineering.

It should be clear that a great diversity of viruses exists. It should therefore not be surprising that there is also a great diversity in the manner in which virus multiplication occurs. In the present chapter, we are able to present only some of the major types of virus replication patterns and must omit some of the interesting exceptional cases.

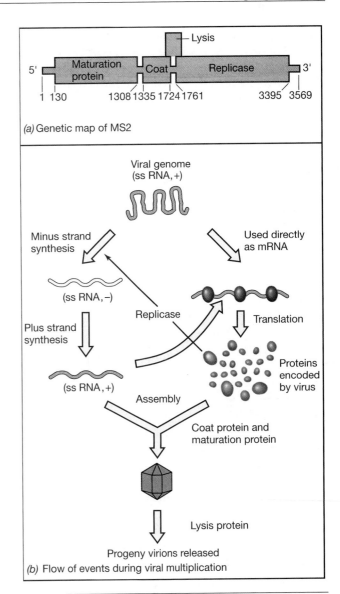

(a) Genetic map of MS2

(b) Flow of events during viral multiplication

FIGURE 8.15   (a) Genetic map of the RNA bacteriophage MS2. (b) Flow of events during multiplication. The numbers in (a) refer to the nucleotide positions on the RNA.

8.8

RNA Bacteriophages

Many viruses have RNA genomes. The best-known bacterial RNA viruses have single-stranded RNA of the plus configuration. Interestingly, the bacterial RNA viruses known in the enteric bacteria group infect only bacterial cells that contain a type of plasmid, called a *conjugative plasmid* (∞ Section 9.9), which allows the bacterial cell to function as a *donor* (sometimes referred to as a "male") in a certain type of genetic exchange (the interesting concept of male and female bacteria is discussed in Chapter 9). This restriction to male bacterial cells arises because these viruses infect bacteria by attaching to *pili* (Figure 8.14), which are encoded by the plasmid. Since such pili are absent from cells that do not carry the plasmid, these RNA viruses are unable to attach to such cells and hence do not initiate infection in these so-called female cells.

The bacterial RNA viruses are all quite small, about 26 nm in size, and they are all icosahedral, with 180 copies of coat protein per virus particle. The complete nucleotide sequences of several RNA phage genomes are known. The genome of the RNA phage MS2, which infects *Escherichia coli,* is 3569 nucleotides long. The RNA strand in the virion acts directly as mRNA on entry into the cell (see Figure 8.12).

The genetic map of MS2 is shown in Figure 8.15*a*, and the flow of events of MS2 multiplication is shown in Figure 8.15*b*. The small genome encodes only four proteins. These are the **maturation protein** (present in the mature virus particle as a single copy), **coat protein, lysis protein** (involved in the lysis process that results in release of mature virus particles), and a subunit of **RNA replicase,** the enzyme that brings about replication of the viral RNA. Interestingly, the RNA replicase is a composite protein, composed partly of the virus-encoded polypeptide and partly of host polypeptides. The host proteins involved in the formation of active viral replicase are part of the cell's normal translational machinery. Thus, the virus appears to employ host proteins that normally have entirely distinct functions and use them to make an active viral replicase.

As noted, the viral RNA is of the plus sense and can thus be translated immediately. After RNA replicase is

synthesized, it in turn can synthesize RNA of minus sense using the infecting RNA as template (see Figure 8.12). After minus RNA has been synthesized, more plus RNA is made using this minus RNA as template. The newly made plus RNA strands then serve as messengers for continued virus protein synthesis. The gene for the maturation protein is at the 5′-end of the RNA. Translation of the gene coding for the maturation protein (needed in only one copy per virus particle), occurs only from the nascent form of the plus-strand RNA as the replication process occurs. In this way, the amount of maturation protein needed is limited. The virus RNA is folded into a complex form with extensive secondary structure. Of the four AUG start sites, the most accessible to the translation

FIGURE 8.14   Electron micrograph of the pilus of a "male" bacterial cell of *Escherichia coli* showing virions of a small RNA phage attached to the pilus.

process is that for the coat protein, and translation begins there very early. The replicase mRNA is also translated early. As coat protein molecules increase in number in the cell, they combine with the RNA around the AUG start site for the replicase protein, effectively turning off synthesis of replicase. The major virus protein synthesized is coat protein, which is needed in the highest amounts.

Another interesting feature of bacteriophage MS2 is that the fourth virus protein, the *lysis* protein, is encoded by a gene that *overlaps* with both the coat protein gene and the replicase gene (see genetic map in Figure 8.15*a*). The phenomenon of **overlapping genes** is quite common in very small genomes (⚭ Section 6.13). It should be noted that although the use of overlapping genes makes possible more efficient use of genetic information, it seriously complicates the evolution process because a mutation in a region of gene overlap may affect two genes simultaneously. The start codon of this lysis gene is not easily accessible to ribosomes because of the secondary structure found in the RNA. When the ribosome terminates synthesis of the coat protein gene, the secondary structure in this region is disrupted, and sometimes this disruption allows a ribosome to begin reading the lysis gene. By restricting the efficiency of translation in this way, premature lysis of the cell is probably avoided. Only after sufficient copies of coat protein are available for the assembly of mature virus particles does lysis commence.

Ultimately, phage assembly takes place, and release of virions from the cell occurs as a result of cell lysis. The features of replication of these simple RNA viruses are themselves fairly simple. The viral RNA itself functions as an mRNA, and regulation occurs primarily by way of controlling access of ribosomes to the appropriate start sites on the viral RNA.

✓ 8.8 Concept Check

A variety of RNA viruses that infect bacteria are known. The small RNA genome of these bacterial viruses acts directly as mRNA and encodes only a few proteins.

✓ Are these viruses considered positive-strand viruses? Explain.
✓ Describe what is meant by *overlapping genes*.

8.9
Single-Stranded Icosahedral DNA Bacteriophages

A number of small bacterial viruses have genomes consisting of single-stranded DNA in circular configuration. These viruses are very small, about 25 nm in diameter, and the principal building block of the pro-

tein coat is a single protein present in 60 copies (the minimum number of protein subunits possible in an icosahedral virus), to which are attached at the vertices of the icosahedron several other proteins that make up spikelike structures (see Figure 8.13). These small DNA viruses possess only a limited amount of genetic information in their genomes, and the host cell DNA replication machinery is used in the replication of virus DNA.

The Genome of Phage φX174

The most extensively studied virus of this group is the phage designated φX174, which infects *Escherichia coli*. Its genome consists of a circular single-stranded molecule of 5386 nucleotide residues. The DNA of φX174 was the first DNA to be completely sequenced, a remarkable achievement when it was accomplished by Frederick Sanger and colleagues in 1977. Now, DNA sequencing is a routine procedure (⚭ Working with Nucleic Acids: The Tools, Chapter 6). φX174 is also of special interest because it was the first genetic element shown to have *overlapping genes*. In very small viruses such as φX174 there is insufficient DNA to code for all virus-specific proteins unless parts of certain virus nucleotide sequences are read more than once in different reading frames (Figure 8.16).

As seen in the genetic map of φX174, the sequences of genes D and E overlap each other, gene E being contained completely *within* gene D. In addition, the termination codon of gene D overlaps the initiation codon of gene J. Several other instances of gene overlap occur in the φX174 genome (Figure 8.16). Additionally, a small protein, called A*-protein, is formed by *reinitiation of translation* (not transcription) within the mRNA of gene A, with A protein being read and terminated from the same mRNA reading frame as A-protein but starting at a different in-frame start codon.

DNA Replication by the Rolling Circle Mechanism

The replication process of such a circular single-stranded DNA molecule is of considerable general interest because cellular DNA always replicates in the double-stranded configuration (⚭ Section 6.6). The DNA strand in single-stranded DNA phages is of the plus sense. On infection, this DNA strand becomes separated from the protein coat, and entrance into the cell is accompanied by the conversion of this single-stranded DNA to a double-stranded form called **replicative form** (RF) DNA (Figure 8.16*b*). Cell-coded proteins involved in the conversion of viral DNA to RF DNA consist of the enzymes *primase, DNA polymerase, ligase,* and *gyrase*. No virus-encoded proteins are involved in this process.

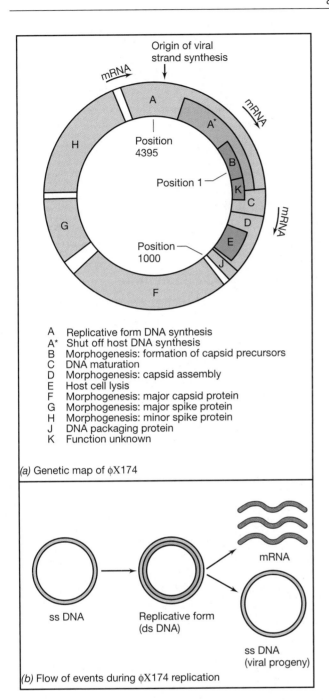

A Replicative form DNA synthesis
A* Shut off host DNA synthesis
B Morphogenesis: formation of capsid precursors
C DNA maturation
D Morphogenesis: capsid assembly
E Host cell lysis
F Morphogenesis: major capsid protein
G Morphogenesis: major spike protein
H Morphogenesis: minor spike protein
J DNA packaging protein
K Function unknown

(a) Genetic map of φX174

(b) Flow of events during φX174 replication

FIGURE 8.16 Bacteriophage φX174, a single-stranded DNA phage. (a) Genetic map. Note the regions of gene overlap (A/B, K/B, K/C, K/A, A/C, and D/E). Intergenic regions are not colored. Protein A* is formed using only part of the coding sequence of gene A by reinitiation of translation (see text). (b) Flow of events in φX174 multiplication. The production of progeny ss DNA from replicative form ds DNA involves rolling circle replication and is shown in more detail in Figure 8.17.

In cells, replication of the lagging strand involves the formation of short *RNA primers* by action of an enzyme called *primase* (⚭ Section 6.4). Such RNA primers are made at intervals on the lagging strand and are then removed and replaced with DNA by DNA polymerase (⚭ Figure 6.18). Unlike the situation with cellular DNA, however, replication of φX174 DNA begins with a single-stranded closed circle. To begin replication of this DNA, primase brings about the synthesis of a short RNA primer at one or more specific initiation sites. DNA is then synthesized by DNA polymerase III, and the primer is removed and replaced by DNA using DNA polymerase I, exactly as in the case of a lagging strand. This results in the formation of the complete, circular, double-stranded RF DNA.

Once the RF is formed, DNA replication occurs by conventional semiconservative replication, involving theta-form intermediates (⚭ Figure 6.16) and resulting in the formation of new RF DNA molecules. However, the formation of single-stranded viral genomes involves a different type of replication mechanism called **rolling circle replication** (Figure 8.17). The rolling circle arises because one strand is nicked and the 3'-end of this nick is used to prime synthesis of a new strand. Continued rotation of the circle leads to the synthesis of a linear, single-stranded structure. Note that synthesis is asymmetric because only one of the strands is serving as template. In φX174, synthesis begins when the protein encoded by gene A, called *gene A protein,* cleaves the plus strand of the RF. When the growing viral strand reaches unit length (5386 residues for φX174), gene A protein cleaves and then ligates the two ends of the newly synthesized single strand to give a circular single-stranded DNA.

Many other viruses and some plasmids also use rolling circle replication. We will see that this type of replication can also be used to synthesize double-stranded DNA.

Transcription and Translation for φX174

Viral mRNA synthesis is directed by the RF DNA. Synthesis of mRNA begins at several major promoters and terminates at a number of sites (see map, Figure 8.16). The polycistronic mRNA molecules (⚭ Section 6.8) are then translated into the various phase proteins. As we have noted, several proteins are made from mRNA transcripts formed from different reading frames from the same DNA sequences (overlapping genes). One can truly be impressed by the efficiency with which such a small genome as that of φX174 can have multiple uses.

Ultimately, assembly of mature virus particles occurs. Release of virions from the cell takes place as a result of cell lysis, which involves the participation of gene E protein.

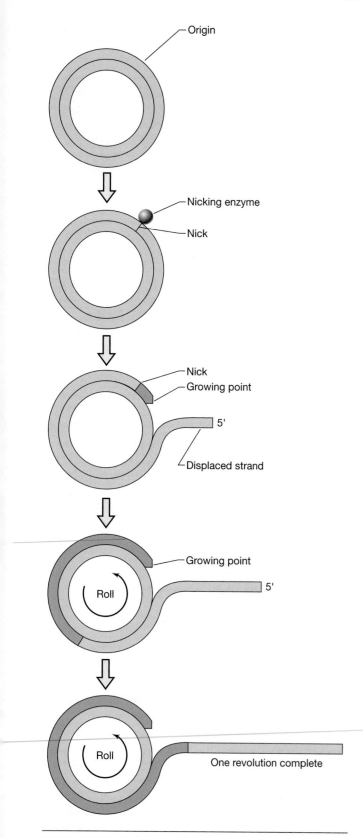

FIGURE 8.17   Rolling circle replication. Replication begins at the origin by nicking one strand of DNA (in φX174, the nicked strand is the *plus* strand, and the gene A protein makes the nick). After one new progeny strand has been synthesized (one revolution of the circle), the gene A protein cleaves the new strand and ligates its two ends.

✓ 8.9 Concept Check

The single-stranded DNA genome of the virus φX174 is so small that only the presence of overlapping genes allows it to encode all its essential functions. This virus provided the first example of overlapping genes. The production of progeny viral DNA involves a rolling circle mechanism.

- ✓ If the nucleic acid genome of φX174 is in the plus configuration, why can't it be used directly as mRNA?
- ✓ How does the replicative form of the viral nucleic acid differ from the form found in the virion?

8.10

Single-Stranded Filamentous DNA Bacteriophages

Quite distinct from φX174 are the filamentous DNA phages, which have helical rather than icosahedral symmetry. The most studied member of this group is phage M13, which infects *Escherichia coli,* but related phages include f1 and fd. As with the small RNA bacteriophages, these filamentous DNA phages infect only male cells, entering after attachment to the male-specific pilus (∞ Section 9.9). Even though these phages are linear (filamentous) in shape they possess *circular* single-stranded DNA. The DNA is not self-complementary, however, so the two adjacent halves of the molecule that run up and down the virus particle form loops at the ends but exhibit very little if any base pairing. Phage M13 has found extensive use as a cloning vector and DNA sequencing vehicle in genetic engineering (∞ Section 10.4). The virion of M13 is only 6 nm in diameter but is 860 nm long. These filamentous DNA phages have the additional interesting property of being released from the cell *without* killing the host cell. Thus, a cell infected with phage M13 can continue to grow, all the while releasing virus particles. Virus infection causes a slowing of cell growth, but otherwise a cell is able to coexist with its virus. Plaques are thus seen only as areas of reduced cell growth in the bacterial lawn.

Many aspects of DNA replication in filamentous phages are similar to that of φX174. The property of release without cell killing occurs by a budding process in which the virus particle is always released from the cell with the end containing the A-protein first (Figure 8.18). There is no accumulation of intracellular virus particles; the assembly of mature virions occurs on the inner surface of the cytoplasmic membrane, and virus assembly is coupled with the budding process.

Several features of these phages make them useful as cloning and DNA sequencing vehicles. First, they have single-stranded DNA, which means that se-

LEARNING FROM THE PAST . . .   The Phage Group

Historically, the T-even phages provided the study material for early research of the "Phage Group," a group of research workers from various universities and research institutions who spent their summers working together at the Cold Spring Harbor Laboratory on Long Island. The key members of the Phage Group, Max Delbrück, Salvador Luria, and A. D. Hershey, subsequently shared the Nobel Prize for their pioneering work. Among important concepts first uncovered from research on the T-even phages: only the nucleic acid of the virus entered the cell during infection (a discovery that provided key support for the hypothesis that DNA is the genetic material); the existence of genetic recombination in viruses; the phenomenon of restriction and modification (which led to the discovery of the restriction enzymes so important for genetic engineering); the presence in viruses of unique virus-encoded gene functions; the distinction between early and late viral functions. The first ideas of how viruses cause killing of host cells were also developed from research on the T-even phages. Selecting the T-even phages as a model system and concentrating work on them were greatly responsible for the remarkable success of the Phage Group. ■

quencing can be carried out by the Sanger dideoxynucleotide method (∞ Working with Nucleic Acids: The Tools, Chapter 6). Second, as long as infected cells are kept in the growing state, they can be maintained indefinitely with cloned DNA so a continuous source of the cloned DNA is available. Third, there is an intergenic space that does not code for protein and can be replaced by variable amounts of foreign DNA.

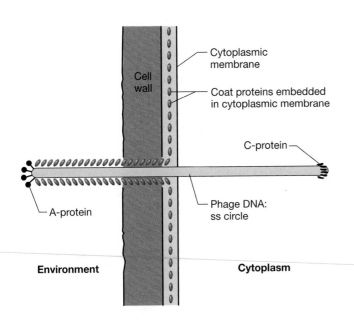

FIGURE 8.18   Illustration of the manner in which the virion of a filamentous single-stranded phage (such as M13 or fd) leaves an infected cell without lysis. The A-protein passes first through the membrane at a site on the membrane where coat protein molecules have first become embedded. The intracellular circular DNA is coated with dimers of another phage protein, which is displaced by coat protein as the DNA passes through the intact cytoplasmic membrane.

8.11

Double-Stranded DNA Bacteriophages: Lytic Viruses

The first viruses to be studied in any detail were a number of bacteriophages with linear, double-stranded DNA genomes that infect *Escherichia coli* and a number of related bacteria. A group of scientists began studying these viruses as model systems and used them to establish many of the fundamental principles of molecular biology and genetics (see the box, The Phage Group). These phages were given designations of T1, T2, and so on, up to T7. In this section we shall briefly discuss both T4 and T7, beginning with T7, the simpler of the two. Both are typical lytic viruses in that their replication leads to the destruction of the cell.

Replication of Bacteriophage T7

Bacteriophage T7 and its close relative T3 are relatively small DNA viruses that infect *Escherichia coli*. (Some strains of *Shigella* and *Pasteurella* are also hosts for phage T7.) The virion has an icosahedral head and a very small tail (see Figure 8.13).

The T7 genome is a linear double-stranded DNA molecule of 39,936 base pairs. About 92% of the DNA of T7 encodes proteins. Gene overlap also occurs in the T7 genome, as do other translational strategies such as internal translational reinitiation and internal frame-shifts within certain genes, all apparently to maximize genetic economy.

The genetic map of T7 is shown in Figure 8.19. The order of the genes on the T7 chromosome influences the regulation of virus multiplication. When the virion attaches to the bacterial cell, the DNA is injected in a linear fashion, with the genes at the "left end" of the

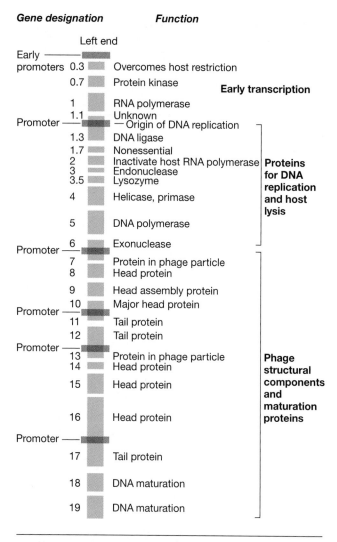

FIGURE 8.19 Genetic map of phage T7, showing gene numbers, approximate sizes, and functions of the gene products. Transcription from the early promoters involves host RNA polymerase. Transcription from all other promoters involves T7 RNA polymerase. The genes are designated by numbers.

genetic map always entering the cell first. Several genes at the left end of the DNA are transcribed immediately by the cellular RNA polymerase, using three closely spaced promoters. One of these early proteins inhibits the host restriction system. Note that this protein is synthesized before the entire T7 genome enters the cell. Another one of these early proteins is a viral RNA polymerase, called T7 RNA polymerase. Two other early mRNA molecules code for proteins that stop the action of host RNA polymerase, thus turning off the transcription of the early genes as well as the transcription of host genes. Thus, the host RNA polymerase is used just to transcribe the first few genes and to make the mRNA that codes for the phage-specific RNA polymerase and a few other proteins. The phage-specific RNA polymerase is then involved in the major tran-

scription processes of the phage. This T7 RNA polymerase uses only phage-specific promoters that are distributed along the left-center and center portions of the genome (see Figure 8.19). The T7 RNA polymerase is very specific for these promoters (whose sequence is unrelated to typical *Escherichia coli* promoters) and is also an extremely efficient enzyme. (Genetic engineers have taken advantage of this to fashion genes that can be highly expressed; ∞ Section 10.7.)

DNA replication in T7 begins at a single origin of replication (shown in Figure 8.19) and proceeds *bidirectionally* from this origin (Figure 8.20). Replicating molecules of T7 DNA can be recognized under the electron microscope by their characteristic structures. Because the origin of replication is near the left end, Y-shaped molecules are frequently seen, and earlier in replication, bubble-shaped molecules appear (Figure 8.20). Several virus-encoded proteins are involved in T7 DNA replication, unlike the situation we described for φX174 (see Section 8.9).

A structural feature of the T7 DNA that is important in DNA replication is that there is a *direct terminal repeat* of 160 base pairs at the ends of the molecule. In order to replicate DNA near the 5'-terminus, RNA primer molecules have to be removed before replication is complete. There is thus an unreplicated portion of the T7 DNA at the 5'-terminus of each strand (see lower part of Figure 8.20*a*). As discussed in Section 6.5, genetic elements with linear DNA genomes have a variety of strategies for solving this problem in DNA replication. The strategy employed by T7 involves the repeated sequence at its ends. The opposite single 3'-strands on two separate DNA molecules, being complementary, can pair with these 5'-strands, forming a DNA molecule twice as long as the original T7 DNA (Figure 8.20*b*). The unreplicated portions of this end-to-end bimolecular structure are then completed through the action of DNA polymerase and DNA ligase, resulting in a *linear bimolecule* called a *concatamer*. Continued replication and recombination can lead to concatamers of considerable length, but ultimately a phage-encoded endonuclease cuts each concatamer at a specific site, resulting in the formation of virus-sized linear molecules with terminal repeats (Figure 8.20*c*). We thus see that T7 has a more complex replication scheme than that seen for the other bacterial viruses discussed earlier.

Bacteriophage T4 and Replication of the T4 Genome

One of the most extensively studied groups of DNA viruses is the group called the *T-even phages*, which include the phages T2, T4, and T6. These phages are very closely related and among the largest and most complicated in terms of both structure and manner of replication. We will discuss primarily bacteriophage T4, the

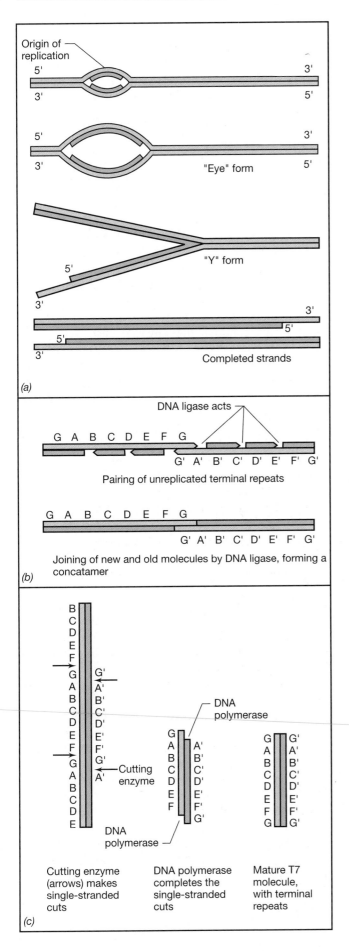

(a)

Origin of replication

"Eye" form

"Y" form

Completed strands

(b)

DNA ligase acts

G A B C D E F G
G' A' B' C' D' E' F' G'

Pairing of unreplicated terminal repeats

G A B C D E F G
G' A' B' C' D' E' F' G'

Joining of new and old molecules by DNA ligase, forming a concatamer

(c)

DNA polymerase

Cutting enzyme

DNA polymerase

Cutting enzyme (arrows) makes single-stranded cuts

DNA polymerase completes the single-stranded cuts

Mature T7 molecule, with terminal repeats

FIGURE 8.20 Replication of the linear, double-stranded DNA genome of bacteriophage T7. (a) Bidirectional replication of DNA giving rise to intermediate "eye" and "Y" forms. (b) Formation of concatamers by joining DNA molecules at the unreplicated terminal ends. The designation of the genes is arbitrary. (c) Production of mature viral DNA molecules from T7 concatamers by action of cutting enzyme, an endonuclease. Left: The enzyme makes single-stranded cuts of specific sequences (arrows); center: DNA polymerase completes the single-stranded ends; right: the mature T7 molecule with terminal repeats.

representative of this group for which the most information is available.

The virion of phage T4 is structurally complex (see Figure 8.5b). It consists of an icosahedral head that is elongated by the addition of one or two extra bands of proteins, the overall dimensions of the head being 85 × 110 nm. To this head is attached a complex tail consisting of a helical tube (25 × 110 nm) to which are connected a sheath, a connecting "neck" with "collar," and a complex endplate, to which are attached long, jointed tail fibers (see Figure 8.5b). All together, the virus particle has over 25 distinct types of proteins.

The genome of T4, like that of T7, is a double-stranded, linear DNA molecule, but it is quite large, approximately 1.7×10^5 base pairs. The genome encodes over 135 different proteins, and although no known virus encodes its own translational apparatus, T4 does encode several different tRNAs. And like T7, the T4 genome has direct terminal repeats, but they are 3000 to 6000 base pairs. The DNA of T4 has a total length about 650 times longer than the dimension of the head. This means that the DNA must be highly folded and packed very tightly within the head.

Interestingly, the DNA found in T4 is chemically distinct from cell DNA, having a unique base, *5-hydroxymethylcytosine,* instead of *cytosine* (Figure 8.21). Such a modification makes the DNA resistant to *many* host restriction enzymes. However, the hydroxyl groups of the 5-hydroxymethylcytosine are modified by addition of glucosyl residues. This *glucosylated* DNA is resistant to *virtually all* restriction endonucleases of the host. Thus, this virus-specific DNA modification plays

Site of glucosylation ——— HOH₂C

FIGURE 8.21 The unique base in the DNA of the T-even bacteriophages, 5-hydroxymethylcytosine. The site of glucosylation is shown.

an important role in the ability of the virus to attack a host cell.

The genetic map of T4 is generally represented as a circle, even though the DNA itself is linear. This "genetic circularity" arises because the DNA of the phage exhibits a phenomenon called *circular permutation*. This arises because in different T4 phage particles, the sequence of bases at each end differs (although for a given molecule the same base sequence is repeated at both ends). This structure, a consequence of the way the T4 DNA replicates, results in an appearance of genetic circularity even though the DNA itself is linear.

The process of DNA replication in T4 is similar to that in T7, but in T4 the cutting enzyme that forms virus-sized fragments from concatamers does not recognize specific locations on the long molecule but rather cuts off head-full packages of DNA irrespective of the sequence (Figure 8.22). Because of this and the "extra" room in the head, each virus DNA mole-

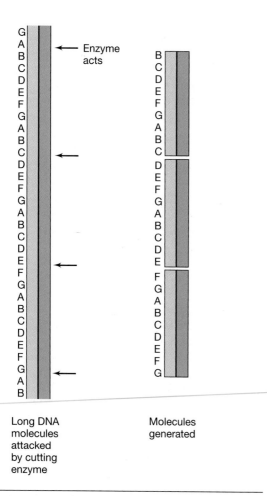

Long DNA molecules attacked by cutting enzyme

Molecules generated

FIGURE 8.22 Generation in T4 of virus length DNA molecules with permuted sequences by a cutting enzyme, which cuts off constant lengths of DNA irrespective of the sequence. Left: arrows, sites of enzyme attack; right: molecules generated.

cule not only contains repetitious ends but the nucleotide sequences at the ends of different molecules are different as well. Each molecule contains slightly more than one complete copy of the entire genome (Figure 8.22). As shown, the cutting process results in the formation of DNA molecules with permuted sequences at the ends. We have already briefly described the infection cycle of T4 in a susceptible cell (see Section 8.6 and Figures 8.8–8.10) and have noted that before DNA synthesis occurs, transcription has begun on the phage genome.

Transcription, Translation, and Regulation in Phage T4

In bacteriophage T4, the details of regulation of replication are more complex than those of T7. Phage T4 is much larger than T7 and has many more genes and phage functions. And, as previously mentioned, the DNA of T4 contains the unusual base 5-hydroxymethylcytosine (Figure 8.21), and some of the OH groups of this base are glucosylated. Thus, enzymes for the synthesis of this unusual base and for its glucosylation must be formed after phage infection, as well as formation of an enzyme that breaks down the normal DNA precursor deoxycytidine triphosphate. In addition, T4 encodes a number of enzymes that have functions similar to those of host enzymes in DNA replication but are formed in larger amounts, thus permitting faster synthesis of T4-specific DNA. In all, T4 encodes over 20 new proteins that are synthesized early after infection.

Overall, the T4 genes can be divided into three groups, one encoding early proteins, one middle proteins, and the other, late proteins (Figure 8.23). The **early** and **middle proteins** are the enzymes involved in DNA replication and transcription. The **late proteins** are the head and tail proteins and the enzymes involved in liberating the mature phage particles from the cell.

Unlike T7, T4 does not encode a new phage-specific RNA polymerase. The control of T4 mRNA synthesis involves the production of proteins that sequentially modify the specificity of the host RNA polymerase so that it recognizes different phage promoters. The early promoters are read directly by the host RNA polymerase and involve the function of host *sigma* factor. Phage-specific proteins synthesized from the early genes carry out covalent modifications on the host RNA polymerase α subunits (∞ Section 6.7), and a few phage-encoded proteins also bind to the polymerase. These modifications change the specificity of the polymerase so it now recognizes T4 middle promoters. One of the T4 early proteins, called MotA, apparently recognizes a particular DNA sequence in these promoters. Transcription from the late promoters requires a new T4-encoded sigma factor. Interestingly, it also requires

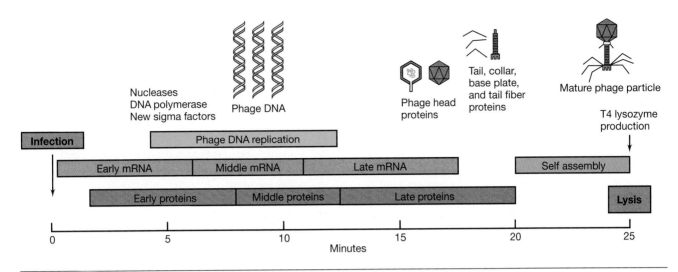

FIGURE 8.23 Time course of events in phage T4 infection. Following injection of DNA, early and middle mRNA is produced that codes for nucleases, DNA polymerase, new phage-specific sigma factors, and various other proteins involved in DNA replication. Late mRNA codes for structural proteins of the phage virion and for T4 lysozyme, needed to lyse the cell and release new phage particles.

T4 DNA synthesis. Sequential modification of host cell RNA polymerase is used to regulate gene expression by many bacteriophages.

In the case of phage T4, the entire lytic cycle takes about 25 min (Figure 8.23). Assembly of heads and tails occurs independently, DNA is packaged into the assembled head, and the tail and tail fibers are added later. Exit of the virus from the cell occurs as a result of cell lysis (Figure 8.23). The phage codes for a lytic enzyme, the *T4 lysozyme,* which attacks the peptidoglycan of the host cell.

✓ 8.11 Concept Check

The phages T4 and T7 each take over the cell's metabolic machinery, leading to the production of new virions and host cell lysis. Regulation of transcription in T7 involves the production of a virus-specific RNA polymerase, whereas in T4 the host polymerase is modified. The T4 genome is quite large and encodes many enzymes. Both viruses take advantage of repeats at the ends of their genomes to complete DNA replication.

✓ Describe the strategy employed by T4 and by T7 to ensure that late in infection only virus genes are transcribed.

✓ Describe the strategies used by T4 and by T7 to avoid the host's restriction enzymes.

8.12

Temperate Bacteriophages: Lysogeny and Lambda

Most of the bacterial viruses described previously are called **virulent** viruses, because they usually kill (lyse) the cells they infect. However, many other bacterial viruses, although also able to kill cells, frequently have more subtle effects. Such viruses are called **temperate.** These viruses can enter into a state called **lysogeny,** where most virus genes are not expressed, and the virus genome is replicated in synchrony with the host chromosome.

Thus, the phage genome is duplicated along with the host material at the time of cell division, being passed from one generation of bacteria to the next. Under certain conditions these bacteria, called **lysogens,** can spontaneously produce virions of the temperate virus. Lysogeny is probably of ecological importance because most bacteria isolated from nature are lysogens for one or more bacteriophages. Lysogeny is not limited to bacteriophages. Many animal viruses set up similar relationships with their hosts.

Overview of the Life Cycles of a Temperate Phage

Our discussions of viral reproduction so far in this chapter should make it clear that it is not the presence, or even the replication, of viral DNA that leads to the production of new virions and host cell death. Rather it is *expression* of the viral genome that is deleterious. One could imagine that host cells can harbor virus genomes without harm if the expression of the viral genes can be controlled. This is the situation found in lysogens. However, once this control has been lost, the virus enters the lytic pathway, it produces new virions, and then the host cell lyses. In a culture of lysogens at any one time, only a small fraction of the cells, 0.1–0.0001%, produce virus and lyse, while the majority of the cells neither produce virus nor lyse. Although only rarely do cells of a lysogenic strain actually produce virus, every cell has the potential for virus production. Lysogeny can thus be considered a genetic trait of a bacterial strain.

An overall view of the life cycle of a temperate bacteriophage is shown in Figure 8.24. The temperate virus does not exist in its mature, infectious state inside the cell but rather in a latent form called the **provirus** or **prophage** state. In the example shown in Figure 8.24, the prophage is integrated into the bacterial chromosome. The prophage replicates along with the host cell as long as the genes controlling its lytic pathway are not expressed. Typically this control is maintained by a phage-encoded repressor protein (indicating that at

least this gene is being expressed). The virus repressor protein not only controls the lytic genes on the prophage but also prevents the expression of any incoming genomes of the same virus. This results in the lysogens having **immunity** to infection by the same type of virus.

However, if this repressor is inactivated, or if its synthesis is prevented, the prophage is induced (center, Figure 8.24). This induction results in the production of new virions and the lysis of the host cell. In some cases (as we shall see later), induction can be brought about by environmental conditions. If the virus loses the ability to leave the host genome (because of muta-

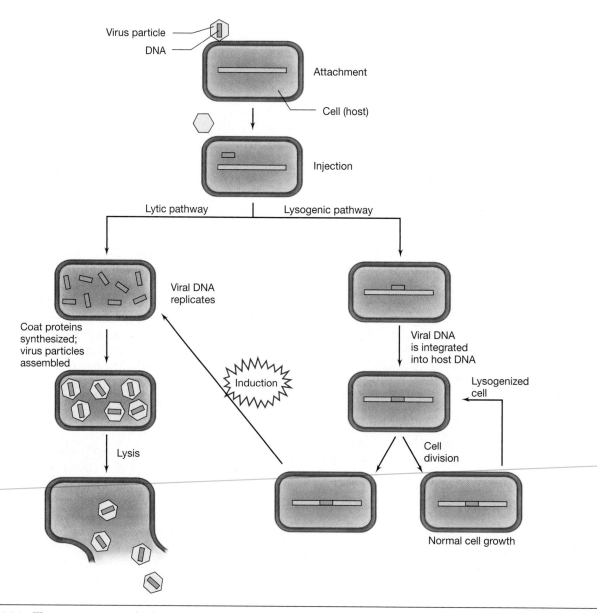

FIGURE 8.24 The consequences of infection by a temperate bacteriophage. The alternatives on infection are integration of the virus DNA into the host DNA (lysogenization) or replication and release of mature virus (lysis). The lysogenic cell can also be induced to produce mature virus and lyse.

tion), it becomes a cryptic virus. Interestingly, from sequence studies it has been shown that many bacterial chromosomes contain stretches of DNA that were clearly once part of a viral genome.

It is sometimes possible to eliminate the temperate virus (to "cure" the host) by heavy irradiation or treatment with other agents that damage DNA. Among the few survivors may be some cells that have been cured. Presumably the treatment causes the prophage to excise from the host chromosome and be lost during subsequent cell growth. Such a cured strain is no longer immune to the virus and can serve as a suitable host for study of virus replication.

Note that Figure 8.24 shows that infection of a normal cell by a temperate virus can lead to either the lytic pathway or the lysogenic pathway. In this section we shall discuss what factors favor one or the other of pathways during infections by the bacteriophage lambda.

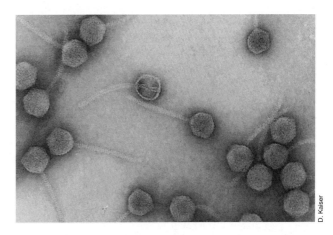

FIGURE 8.25 Electron micrograph by negative staining of bacteriophage lambda particles. The head of each particle is about 65 nm in diameter.

The Bacteriophage Lambda

One of the best-studied temperate phages is lambda, which infects *Escherichia coli,* and our knowledge of the molecular mechanisms involved in lysogenization and lytic processes in this phage is very advanced. Morphologically, lambda particles look like those of many other bacteriophages (Figure 8.25). The virus particle has an icosahedral head 64 nm in diameter and a tail 150 nm long that has helical symmetry. Attached to the tail is a single 23-nm-long fiber.

The genome of lambda consists of a linear double-stranded DNA molecule, but at the 5′-terminus of each of the single strands is a single-stranded tail 12 nucleotides long. These single-stranded ends are complementary (the ends of the DNA are said to be *cohesive*). Thus, when the two ends of the DNA are free in the host cell, they associate and the genome forms a double-stranded circle. In the circular form the DNA contains 48,502 base pairs. Figure 8.26 is a representation of the genetic map of lambda after circularization. We discuss the organization and expression of the genes later.

Lambda Infection and the Lytic Pathway

The lambda virion attaches to a specific protein in the cell wall of *Escherichia coli* (a protein involved in the uptake of the sugar maltose) and injects its DNA. The DNA circularizes almost immediately and, if the cell is not a lambda lysogen (and therefore not immune), expression of the phage genome begins. The first steps in gene expression are the same whether the final result is lysis or lysogeny.

Production of RNA using host RNA polymerase begins at a few promoters, two of which, called P_L (pro-

moter left) and P_R (promoter right), are on either side of the regulatory region shown in Figure 8.26. These yield short transcripts that are translated to give the products of the *N* and the *cro* genes. Both the proteins encoded by these genes are involved in regulatory events. The Cro protein (the product of the *cro* gene) participates in the selection between the lytic and lysogenic pathways, and we shall discuss its function later. The N protein is an *antiterminator* protein that allows the RNA polymerase to transcribe past specific terminators (marked on Figure 8.26), making the transcripts from P_L and P_R longer. These longer transcripts can be translated to yield more proteins, including the products of the *O, P, cII,* and *cIII* genes. The antiterminator is not completely effective at the terminator before the *Q* gene, and so only a small amount of the Q protein is made.

The O and P proteins are involved in the initiation of replication of lambda DNA. This early DNA synthesis is bidirectional from a single origin at a site close to the O gene (Figure 8.26) and gives rise to typical theta-like intermediates (⌖ Section 6.6). At this stage it is still possible for lambda to switch to a lysogenic cycle. However, let us consider the situation that would result if this switch were *not* made. It should be remembered that a typical infection by a temperate virus is lytic.

The Q protein is also an antiterminator protein. If its concentration becomes high enough it will allow the transcript from a nearby promoter to synthesize the transcript labeled R2 in Figure 8.26. This transcript is translated to yield the late proteins, all the necessary structural proteins to construct a virion, and the proteins necessary for cell lysis. At the same time the Q protein has built up to this level, the Cro protein (see earlier discussion) has also reached levels where it can block transcription from both P_L and P_R by binding to both

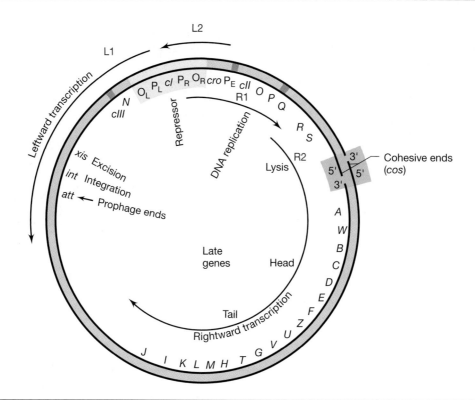

FIGURE 8.26 Genetic and molecular map of lambda. The genes are designated by letters; *att*, attachment site for phage to host chromosome. Genes of special interest: *cI*, repressor protein; O_R, operator right; P_R, promoter right; O_L, operator left; P_L, promoter left; *cro*, gene for second repressor; N, positive regulator counteracting rho-dependent termination. J through U are genes that encode tail proteins. Genes Z through A encode head proteins. The regulatory region of lambda (shown in yellow) is positioned at the top of this circular map. It is also known as the immunity region and contains the *cI* gene. The site created when the cohesive ends of the lambda genome join is called *cos* (shown in blue). Early transcription in lambda is primarily leftward (counterclockwise) from P_L and rightward (clockwise) from P_R. The main leftward transcript is labeled L1, and the main early rightward transcript is labeled R1. The three transcription terminators affected by the N-protein are shown as gray boxes on the DNA. The late rightward transcript, which encodes head and tail proteins and proteins for lytic function, is labeled R2 and begins at a promoter near the Q gene. The transcript labeled L2 is the positively regulated transcript from P_E that encodes the repressor protein.

O_L (operator left) and O_R (operator right). Therefore, Cro operates as a repressor protein.

The mechanism of Cro protein action at O_R is diagrammed in more detail in Figure 8.27. Note that there are three similar but nonidentical sites at this operator where the Cro protein can bind. It does so first at site 3, and then site 2, and only when those two sites are filled, at site 1. Note that only when bound to site 1 does it block P_R. Note also that once P_L and P_R are blocked, no more cII or cIII proteins can be synthesized. These proteins are needed to enter the lysogenic pathway (see later), and so when Cro is made in high amounts, lambda is irrevocably committed to the lytic pathway.

The shutoff of these promoters also results in a change of lambda DNA replication. At this stage when the late proteins are being synthesized (and the O and P proteins are not), long, linear concatamers are syn-

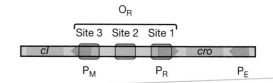

FIGURE 8.27 Both regulatory proteins Cro and the lambda repressor bind to operator right (O_R) on the lambda genome to carry out regulatory functions. The Cro protein (product of the *cro* gene) binds to the three sites in the order site 3, then site 2, and then site 1. The lambda repressor binds to these sites in the opposite order. The promoter P_R is transcribed immediately on phage entry into the cell. Rightward transcription from this promoter is necessary to produce Cro protein and other downstream genes (see Figure 8.26 for a complete map of the lambda genome). Leftward transcription from either of the promoters P_E or P_M is necessary to synthesize the lambda repressor (product of the *cI* gene). Both these promoters require activation in order to function.

thesized by **rolling circle replication.** In this mechanism, replication proceeds in only one direction and can result in very long chains of replicated DNA. Unlike the replication of φX174 DNA (see Figure 8.17), however, rolling circle replication of lambda involves synthesis of both strands (Figure 8.28). This mechanism is efficient in permitting extensive, rapid, relatively uncontrolled DNA replication; thus, it is of value in the later stages of the phage replication cycle when large amounts of DNA are needed to form mature virions. The long concatamers formed are then cut into virus-sized lengths by a DNA-cutting enzyme. In the case of lambda, the cutting enzyme makes staggered breaks at specific sites on the two strands, 12 nucleotides apart, which provide the cohesive ends involved in the cyclization process. These DNA molecules are packaged in phage heads, and then the tails and other proteins are added. The cell is then lysed by the action of phage-encoded proteins.

The lytic pathway of lambda is not much different from those of the other viruses we have discussed earlier. However, the host cell's metabolism is not irreversibly subverted early in the process, ensuring that lysogenization can take place if events favor it.

Lysis or Lysogenization?

We have seen that phage genes are controlled in such a way that viral proteins and nucleic acids are made in appropriate amounts and at appropriate times. For many viruses the patterns of expression always proceed in the same programmed manner. However, lambda and other temperate viruses have a *genetic switch* that controls whether the lytic pathway or the lysogenic pathway is followed. So far the steps we have outlined for lambda are those for the lytic pathway. We now consider how the genetic switch can be thrown to lead to lysogeny.

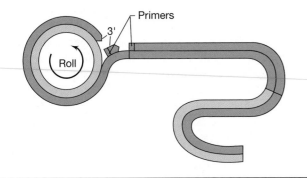

FIGURE 8.28 A late stage in the rolling circle replication of lambda. Both strands of DNA are being copied at the replicating fork, and two copies of the genome have already been synthesized. Note that this synthesis is *asymmetric* because one of the parental strands continues to serve as a template and the other is used only once.

In order to establish lysogeny, two events must happen: The production of all late proteins must be prevented, and a copy of the lambda genome must be integrated into the host chromosome. In order to prevent synthesis of the late proteins, the product of the *cI* gene must be produced. This protein is the **lambda repressor.** If it is synthesized *it will repress the synthesis of all other lambda-encoded proteins.* It is needed to establish lysogeny and to maintain the lysogenic state. The *cI* gene is located between P_L and P_R (see Figure 8.26), but these promoters are oriented in such a way that neither transcribes the *cI* gene. The promoter that can produce mRNA from the *cI* gene during infection is called P_E (promoter establishment) and is located on the map slightly to the right of the *cro* gene but facing the direction opposite that of P_R. Therefore, transcription is in the direction opposite that promoted by P_R (Figures 8.26 and 8.27). Unlike the other promoters we have previously mentioned, P_E must be *activated.* Once it is, lambda repressor protein is synthesized and the lysogenic pathway is followed.

The product of the *cII* gene is an *activator protein* (∞ Section 7.3) that activates promoter P_E (and another promoter required for the production of integrase) (see later). Although the cII protein is made early after infection, it is typically unstable in *Escherichia coli* because it is degraded by a host protease (an enzyme that degrades proteins). If the cII protein is degraded, then there is no possibility that the lysogenic pathway can be chosen. However, this protein can be stabilized by the phage-encoded cIII protein if there is no excess of host protease (or if there is an excess of cIII protein). If the cII protein is stabilized, then it will activate P_E and lambda repressor protein will be made. In a way, this rather complicated process monitors conditions in the host.

Lambda repressor binds to O_L and O_R, as does the Cro protein, but it binds to the sites within these operators in the order opposite that of Cro (see Figure 8.27). That is, it first binds to site 1, turning off P_R (and P_L by a similar mechanism). When this happens, the synthesis of all other lambda proteins is stopped. Without protein Q the late proteins cannot be expressed and lambda cannot enter the lytic pathway.

However, without the cII protein P_E no longer functions. Therefore, if the lysogenic state is to be maintained, there must be another way to transcribe the *cI* gene. Note that in Figure 8.27 there is yet another promoter shown, P_M (promoter maintenance). This promoter is facing toward the *cI* gene (in the same direction as P_E). It is *activated* when lambda repressor binds to site 1 and is repressed only when lambda repressor is bound to all three sites. Therefore, the lambda repressor is both a *repressor* and an *activator* when it binds to site 1, repressing P_R and activating P_M. This type of regulation continues to occur even after lysogenization. Only the lambda repressor is made after lambda is integrated as a prophage.

Integration

Integration of lambda DNA occurs at a unique site on the *Escherichia coli* chromosome and is required for lysogeny. Integration occurs by insertion of the virus DNA into the host genome (thus effectively lengthening the host genome by the length of the virus DNA). As illustrated in Figure 8.29, on injection, the cohesive ends of the linear lambda molecule find each other and form

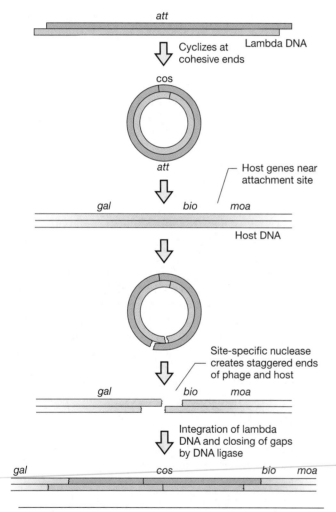

FIGURE 8.29 Integration of lambda DNA into the host. See the genetic map, Figure 8.26, for details of the gene order. Integration always occurs at a specific site on the host DNA, involving a specific attachment site (*att*) on the phage. Some of the host genes near the attachment site are given: *gal* operon, galactose utilization; *bio* operon, biotin biosynthesis; *moa* operon, molybdenum cofactor biosyntheses. A site-specific enzyme (integrase) is involved, and specific pairing of the complementary ends results in integration of phage DNA.

a circle, and it is this circular DNA that becomes integrated into the host genome (the site created when these ends join is called *cos*). To establish lysogeny, genes *cI* and *int* (encoding *integrase*) must be expressed as we discussed. The integration process requires integrase, the product of the *int* gene, which is a site-specific topoisomerase catalyzing recombination of the phage and bacterial attachment sites (labeled *att* in Figures 8.26 and 8.29). The *int* gene has a promoter that, like P_E, is activated by the cII protein.

During cell growth, the lambda repression system prevents expression of the integrated lambda gene except for the gene *cI*, which codes for the lambda repressor. During host DNA replication, the integrated lambda DNA is replicated along with the rest of the host genome and transmitted to progeny cells. When release from repression occurs (see later), the lambda productive cycle occurs. In order to be excised from the chromosome, *excisionase* (the product of the *xis* gene) and the *int* gene product are required.

Lytic Growth of Lambda After Induction

Agents that induce lambda lysogens (cells containing lambda as a prophage) to produce phage are agents that damage DNA. These include ultraviolet irradiation, X-rays, and DNA-damaging chemicals such as the nitrogen mustards. These agents interfere with the function of the lambda repressor. When DNA damage occurs a host defense mechanism called the SOS response (∞ Section 9.3) is brought into play. An array of 10–20 bacterial genes is turned on, some of which help the bacterium survive radiation. However, one result of DNA damage is that a bacterial protein called RecA (normally involved in genetic recombination) is turned into a special kind of protease that participates in the destruction of the lambda repressor. With the lambda repressor destroyed, the inhibition of expression of lambda lytic genes is abolished. We should note that the protease activity of RecA, brought about by DNA damage, normally plays an important role in the cell's response to DNA-damaging agents by participating in the breakdown of a host protein, LexA, which represses a set of host genes involved in DNA repair (∞ Section 9.3). Induction of bacteriophage lambda is thus an indirect consequence of the SOS response.

Once the lambda repressor has been inactivated, the control exerted by this repressor is abolished and new transcriptional events can be initiated. These inevitably lead to lysis because even if the lambda repressor is made, it is inactivated.

Other Strategies Used by Temperate Viruses

Lambda provides one of the best-studied examples of how a "decision" is made at the molecular level. It is also widely used in genetic engineering as a vector for carrying recombinant DNA. One feature of lambda that makes it of special use for cloning is that there is a long region of DNA, between genes *J* and *att* (Figure 8.26), that does not seem to have any essential functions for replication and can be replaced with foreign DNA. We describe the use of lambda as a cloning vector in Chapter 10. We also describe other uses of lambda as a genetic tool in Chapter 9. Other types of temperate viruses are known in bacteria, and some have also been widely studied. The virus P1 (which we will also mention in Chapter 9) is a temperate virus that maintains itself in the lysogenic state not as an integrated prophage but replicates as a circular DNA molecule in the cytoplasm, resembling a plasmid. The very interesting temperate phage named Mu is a transposable element. We discuss Mu in the next section.

✓ 8.12 Concept Check

Temperate viruses do not always cause the death of the cells they infect. The infected cell sometimes survives because the virus genome becomes a prophage (and replicates with the host chromosome), and the lytic genes of the prophage are kept under the control of a virus-encoded repressor. Sometimes this regulatory system is circumvented and prophage induction occurs, resulting in virus multiplication and lysis of the host cell. Host cells carrying temperate viruses are called lysogens.

✓ What are the two pathways available to a temperate virus?
✓ Describe how a single protein like the lambda repressor can act both as an activator and a repressor.

8.13

A Transposable Phage: Bacteriophage Mu

One of the more interesting bacteriophages is the one called **Mu**. This virus is temperate, like lambda, but has the unusual property of replicating as a transposable element (∞ Section 6.4 and Chapter 9). This phage is called Mu because it is a *mutator* phage, inducing mutations in a host genome into which it becomes integrated. This mutagenic property of Mu arises because the genome of the virus can become inserted into the middle of host genes, causing these genes to become inactive (and hence the host that has become infected with Mu behaves as a mutant). Mu is a useful phage because it can be used to generate a wide variety of bacterial mutants very easily. Also, as we will discuss in Chapter 10, Mu can be used in genetic engineering.

Transposable elements are pieces of DNA that have the ability to move from one site to another as discrete elements. They are found in both prokaryotes and eu-

karyotes and play important roles in genetic variation (see Section 9.10 for a detailed discussion of transposition). There are three types of transposable elements: *insertion elements, transposons,* and viruses like Mu (∞ Section 6.4). Mu is a very large transposable element, carrying a number of Mu genes involved in Mu multiplication.

Structure and Genetic Map of Mu

Structurally, bacteriophage Mu is a large double-stranded DNA virus with an icosahedral head, a helical tail, and six tail fibers (Figure 8.30). The genetic map of Mu is shown in Figure 8.31*a*. It can be seen that the bulk of the genetic information is involved in the synthesis of the head and tail proteins, but that important genes at each end are involved in replication and immunity. The DNA molecule found within the virion is approximately 39 kilobase pairs long, but only 37.2 kilobase pairs make up the actual Mu genome. This is because both ends of this DNA molecule contain host DNA. At the left end of the Mu DNA are 50–150 base pairs of host DNA, and at the right end are 1–2 kilobase pairs of host DNA.

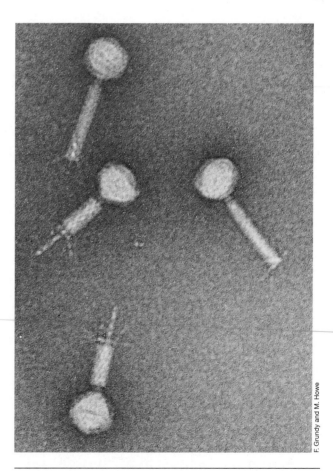

FIGURE 8.30 Electron micrograph of virions of bacteriophage Mu, the mutator phage.

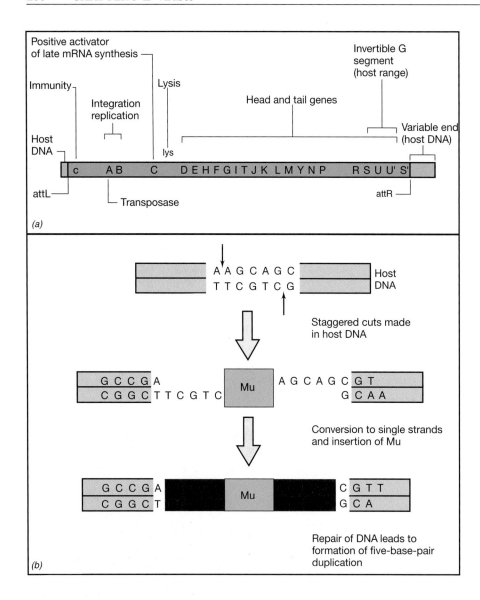

FIGURE 8.31 Bacteriophage Mu. (a) Genetic map of Mu. See text for details. (Confusingly, there are two G's, the *G* gene and the invertible G segment. These are different G's. There is also a lowercase *c* gene, which encodes a repressor, and an uppercase *C* gene, which encodes an activator protein.) (b) Integration of Mu into the host DNA, showing the generation of a five-base-pair duplication of host DNA.

These host DNA sequences are not unique and represent DNA adjacent to the location where Mu was inserted into the genome of its previous host. When a Mu phage particle is formed, a length of DNA containing the Mu genome just large enough to fill the phage head is cut out of the host, beginning at the left end. The DNA is rolled in until the head is full, but the place at the right end where the DNA is cut varies from one phage particle to another. For that reason, as shown on the genetic map, there is a variable sequence of host DNA at the right-hand end of the phage (right of the *attR* site) that represents the *host* DNA that has become packaged into the phage head. Each virion arising from a single infected cell will have a different amount of host DNA, and the host DNA base sequence in each virion from the same cell will be different.

As shown on the genetic map (Figure 8.31), a specific segment of the Mu genome called G (distinct from the G gene) is *invertible*, being present either in the ori-

entation designated G⁺ or in the inverted orientation G⁻. The orientation of this segment determines the kind of tail fibers that are made for the phage. Since adsorption to the host cell is controlled by the specificity of the tail fibers, the host range of Mu is determined by which orientation of this invertible segment is present in the phage. If the G segment is in the orientation designated G⁺, then the phage particle will infect *Escherichia coli* strain K12. If the G segment is in the G⁻ orientation, then the phage particle will infect *E. coli* strain C or several other species of enteric bacteria. The two tail fiber proteins are encoded on opposite strands within this small G segment. Left of the G segment is a promoter that directs transcription into the G segment. In the orientation G⁺, the promoter directing transcription of S and U is active, whereas in the orientation G⁻, a different promoter directs transcription of genes S' and U' on the opposite strand. Regulation involving rearrangement of DNA sequences is known in other viruses as well as both prokaryotic and eukaryotic cells.

Replication of Mu

On infection of a host cell by Mu, the DNA is injected and is protected from host restriction by a modification system in which about 15% of the adenine residues are acetoamidated. In contrast with lambda, integration of Mu DNA into the host genome is essential for both lytic and lysogenic growth. Integration requires the activity of the gene A product, which is a transposase enzyme. At the site where the Mu DNA becomes integrated, a five-base-pair duplication of the host DNA arises at the target site. As shown in Figure 8.31*b*, this host DNA duplication arises because staggered cuts are made in the host DNA at the point where Mu is inserted, and the resulting single-stranded segments are converted to the double-stranded form as part of the integration process. Duplication of short stretches of host DNA is typical of transposable element insertion (∞ Section 9.10).

Lytic growth of Mu can occur either on initial infection, if the Mu repressor (the product of the *c* gene) is not formed, or by induction of a lysogen. In either case, replication of Mu DNA involves repeated transposition of Mu to multiple sites on the host genome. Initially, transcription of only the early genes of Mu occurs, but after gene C protein, a positive activator of late RNA synthesis, is expressed, the synthesis of the Mu head and tail proteins occurs. Eventually, expression of the lytic function occurs and mature phage particles are released.

8.14

Overview of Animal Viruses

We have discussed in a general way the nature of animal viruses in the first part of this chapter. Now we examine in some detail the structure and molecular biology of a number of important animal viruses.

Viruses will be discussed that illustrate different ways of replicating, and both RNA and DNA viruses will be covered. One important group of animal viruses, those called the *retroviruses,* have both an RNA and a DNA phase of replication. Retroviruses are especially interesting not only because of their unusual mode of replication but also because they cause such important diseases as certain *cancers* and *acquired immunodeficiency syndrome* (AIDS).

Before beginning our discussion of the manner of replication of animal viruses, we should remind ourselves of the important differences that exist between eukaryotic and prokaryotic cells. Because virus replica-

tion makes use of the biosynthetic machinery of the host, these differences in cellular organization and function imply differences in the way the viruses themselves replicate.

Differences between Prokaryotes and Eukaryotes That Affect Virus Multiplication

Prokaryotes do not show compartmentation of the biosynthetic processes. The genome of a bacterium relates directly to the cytoplasm of the cell. Transcription into mRNA can lead directly to translation since the processes of transcription and translation are not carried out in separate compartments (∞ Figure 6.2*a*). Animal cells, being eukaryotic, show compartmentation. DNA replication and transcription of the genome into mRNA occur in the nucleus, whereas translation occurs in the cytoplasm (∞ Figure 6.2*b*). This compartmentation has an impact on where animal viruses replicate. For instance, a DNA virus that uses host polymerases must replicate in the nucleus. Therefore, we can expect differences in replication strategies between viruses that multiply in the nucleus and those that multiply in the cytoplasm.

Furthermore, the transcripts from eukaryotic genes must be processed and transported to the cytoplasm before they can be used as mRNA (∞ Section 6.9). This processing usually involves **splicing** out *introns* as well as adding a **poly-A tail** to the 3′-end and a methylated guanosine triphosphate, called the **cap,** to the 5′-end. The cap is required for binding of the mRNA to the ribosome. The difference between the way ribosomes recognize mRNA allows prokaryotes to use polycistronic mRNA, whereas eukaryotes use monocistronic mRNA (∞ Sections 6.8 and 6.12). All the protein-synthesizing machinery of the eukaryotic cell—the ribosomes, tRNA molecules, and accessory components—is in the cytoplasm, and the mature mRNA associates with the protein-synthesizing apparatus once it leaves the nucleus.

We might also note another important difference between animal and bacterial cells. Bacterial cells have rigid cell walls containing peptidoglycan and associated substances (∞ Section 3.7). Animal cells, on the other hand, lack cell walls. This difference is important for the way in which the virus genome enters and exits the cell. In prokaryotes, the protein coat of the virus remains on the outside of the cell and only the nucleic acid enters. In animal viruses, on the other hand, uptake of the virus often occurs by endocytosis (pinocytosis or phagocytosis), processes that are characteristic of animal cells, and so the whole virus particle enters the cell (Figure 8.32).The separation of animal virus genomes from their protein coats then occurs inside the cell.

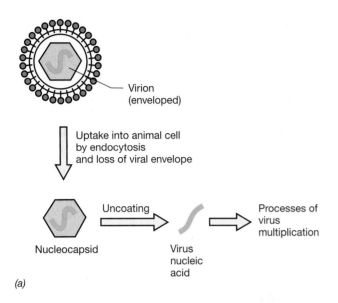

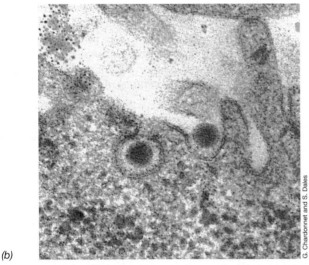

FIGURE 8.32   Uptake of an enveloped virion by an animal cell. (a) The process by which the viral nucleocapsid is separated from its envelope. (b) Electron micrograph of adenovirus virions entering a cell. Each particle is about 70 nm in diameter.

Classification of Animal Viruses

Various types of animal viruses are illustrated in Figure 8.33. Note that the major criteria used in classifying animal viruses are the type of nucleic acid, the presence or absence of an envelope, and, for certain families, the manner of replication. Most of the animal viruses that have been studied in any detail are those that have been amenable to cultivation in cell cultures. Animal viruses are known with either single- or double-stranded DNA or RNA. Some animal viruses are enveloped, others are naked. Size varies greatly, from those large enough to be just visible in the light microscope to those so tiny that they are hard to see even in the electron microscope. In

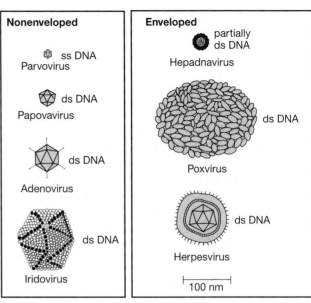

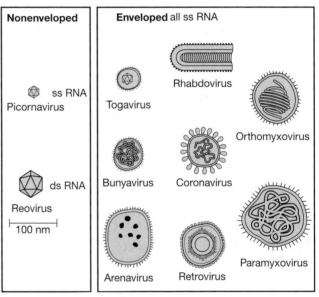

FIGURE 8.33   The shapes and relative sizes of vertebrate viruses of the major taxonomic groups. The hepadnavirus genome has one complete DNA strand and part of its complement.

the following sections, we will discuss characteristics and manner of multiplication of some of the most important and best-studied animal viruses.

Consequences of Virus Infection in Animal Cells

Viruses can have varied effects on cells. **Lytic infection** results in the destruction of the host cell (Figure 8.34). However, there are several other possible effects fol-

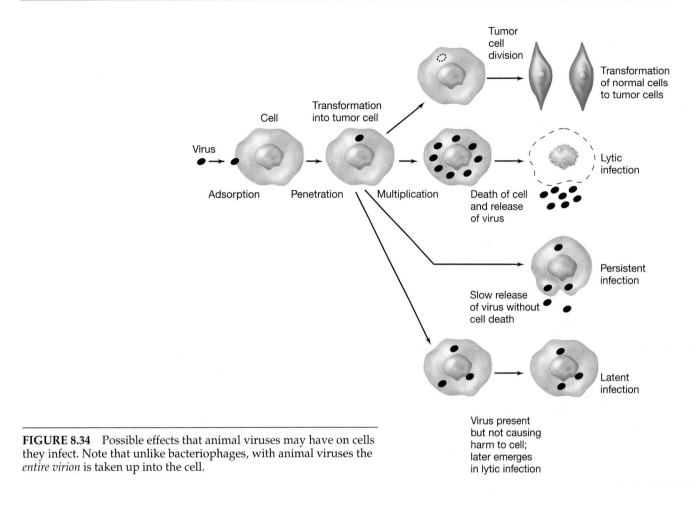

FIGURE 8.34   Possible effects that animal viruses may have on cells they infect. Note that unlike bacteriophages, with animal viruses the *entire virion* is taken up into the cell.

lowing viral infection of animal cells. In the case of enveloped viruses, release of virions, which occurs by a kind of budding process, may be slow, and the host cell may not be lysed. The cell may remain alive and continue to produce virus over a long period of time. Such infections are referred to as **persistent infections** (Figure 8.34). Viruses may also cause **latent infection** of a host. In a latent infection, there is a delay between infection by the virus and the appearance of symptoms. Fever blisters (cold sores), caused by the herpes simplex virus (see Section 8.19), result from a latent viral infection; the symptoms reappear sporadically as the virus emerges from latency. The latent stage in viral infection of an animal cell is generally not due to integration of the viral genome into the genome of the animal cell, as is the case with latent infections by temperate bacteriophages.

Viruses and Cancer

A number of animal viruses participate in the events that change a cell from a normal one to a cancer or tumor cell (Figure 8.34 and Table 8.2). **Cancer** is a cellular phenomenon of uncontrolled growth. Most cells in a mature animal, although alive, do not divide extensively, apparently because of the presence of growth-inhibiting factors that prevent them from initiating cell division. As noted in Section 8.4, infection by certain types of animal viruses leads to a process called **transformation** during which growth becomes uncontrolled. One of the key differences between normal cells and cancer cells is that the latter have different requirements for growth factors. Rapidly growing cells pile up into accumulations that are visible in culture as **foci of infection.** Because cancerous cells in the animal body have fewer growth requirements, they grow profusely, leading to the formation of large masses of cells, called *tumors.* The term **neoplasm** is often used in the medical literature to describe malignant tumors.

Not all tumors are seriously harmful. The body is able to wall off some tumors so that they do not spread; such noninvasive tumors are said to be **benign.** Other tumors, called **malignant,** invade the body and destroy normal body tissues and organs. In advanced stages of cancer, malignant tumors may develop the ability to spread to other parts of the body and initiate new tumors, a process called **metastasis.**

TABLE 8.2 Some human cancers where viruses play a role

Cancer	Virus	Family	Genome in virion
Adult T-cell leukemia	Human T-cell leukemia virus (type I)	Retrovirus	RNA
Burkitt's lymphoma	Epstein–Barr virus	Herpes	DNA
Nasopharyngeal carcinoma	Epstein–Barr virus	Herpes	DNA
Hepatocellular carcinoma (liver cancer)	Hepatitis B virus	Hepadna	DNA
Skin and cervical cancers	Papilloma virus	Papova	DNA

How does a normal cell become cancerous? The development of cancer is clearly a multistep process. There seem to be many different causes of cancer, including mutations arising either spontaneously or as the result of exposure to certain chemicals, called *carcinogens* (⚯ Section 9.4), or by physical stimuli, such as ultraviolet radiation or X-rays. Certain viruses also bring about the genetic change that results in initiation of tumor formation.

Events that cause cancer lead to a loss of the cell's normal control of its growth. The growth and division of normal cells is regulated by at least two types of genes. The first type, called *proto-oncogenes*, promote growth, but they are controlled by the second type, the growth-restraining *tumor suppressor genes*. Changes in either or both types can lead to uncontrolled cell growth and therefore to cancer.

The *initiation* event can be the activation of a proto-oncogene into an **oncogene** (a gene that causes a tumor), or the inactivation of a tumor suppressor gene. Once initiation has occurred, the potentially cancerous cell may remain dormant, but under certain conditions, generally involving some environmental alteration, it may become converted to a tumor cell, a process called **promotion.** Once a cell has been promoted to the cancerous condition, continued cell division can result in the formation of a tumor.

Although the ability of viruses to cause tumors in animals has been proved for many years, the relationship of viruses to cancer in humans has, in most cases, been uncertain. It is difficult to prove the viral origin of a human cancer because of the difficulties of carrying out the necessary experimentation. However, it is well established that certain kinds of human tumors are strongly associated with infection by specific viruses. A summary of some of the human cancers with definite viral origins is given in Table 8.2. In addition, some viral infections can lead indirectly to an increased risk of cancer, apparently by weakening the immune system's ability to detect and destroy transformed cells. This might be why infection with the retrovirus that causes AIDS increases the risk for developing certain cancers.

✓ 8.14 Concept Check

Multiplication of animal viruses differs in significant ways from the multiplication of bacterial viruses, because of differences in compartmentation of macromolecule synthesis in eukaryotes as opposed to prokaryotes. Not all infections of animal host cells result in cell lysis or death. In some cases, latent infection occurs, the virus remaining infectious but dormant inside the host and appearing spontaneously at a later time. Some animal viruses cause transformation of host cells to the cancerous state.

✓ Which macromolecules are synthesized in the nucleus and which are synthesized in the cytoplasm of eukaryotic cells?

✓ Contrast the mechanisms by which animal viruses enter cells with those used by bacterial viruses.

8.15

Positive-Strand RNA Animal Viruses

Just as there are several known positive-strand RNA bacteriophages, there are many different kinds of animal viruses with a positive-strand RNA genome. (The great majority of *plant* viruses are positive-strand RNA viruses. It has been postulated that these small genomes facilitate transfer from cell to cell within the plant.)

An important group of positive-strand RNA animal viruses is the *picornavirus family*, which contains such important viruses infecting humans as the *polioviruses,* the *rhinoviruses* that cause the common cold, and the *hepatitis A virus.* The first animal virus discovered, foot-and-mouth disease virus, is also a picornavirus. These viruses are called *picornaviruses* because they are very small viruses (30 nm in diameter) (*pico* means "small") and contain single-stranded RNA. The virus particle has a simple icosahedral structure with 60 morphological units per virion, each unit consisting of four distinct proteins (Figure 8.35*a*).

In poliovirus, the RNA is a linear molecule about 7500 bases in length (see Table 8.1). At the 5'-terminus of the viral RNA is a protein, called the *VPg protein,* that is attached covalently to the RNA. At the 3'-terminus of the RNA is a poly-A tail.

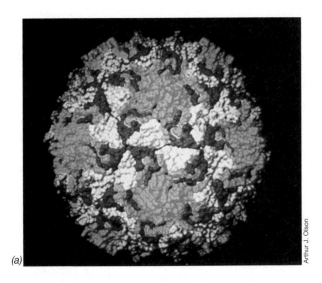

Synthesis of new plus strands

Synthesis of minus strand

Encodes structural proteins

Encodes nonstructural proteins

Vpg

Poly A

Translation

Polyprotein

Active gene products

Proteases cleave large protein

Structural coat proteins

Proteases

RNA polymerase

(b)

FIGURE 8.35 (a) The poliovirus particle. This is a computer model based on electron diffraction analysis of virus crystals. The various structural proteins are shown in distinct colors. (b) The reproduction of poliovirus. The single-stranded RNA of the virus is translated directly as a messenger RNA, with the production of one large protein molecule. This protein is cleaved, leading to production of the active viral proteins, including the structural coat proteins and the RNA polymerase that brings about replication of the poliovirus RNA. The assembly of intact poliovirus from coat protein molecules and RNA then follows.

An overview of the manner of multiplication of poliovirus is illustrated in Figure 8.35*b*. The RNA of the virus acts directly as a messenger RNA. Interestingly, this is so even though the RNA is not capped (∞ Section 6.9). The 5'-end of poliovirus RNA has a long sequence that can fold into several stem-loops (∞ Section 6.7). Somehow these permit binding of the eukaryotic ribosome. The virus RNA is monocistronic but codes for all the proteins of the virus in a single *polyprotein* that is later cleaved into the individual proteins. The whole replication process occurs in the cytoplasm.

At initial infection, the virus particle attaches to a specific receptor on the surface of a sensitive cell and enters the cell. Once inside the cell, the virus particle is uncoated, and the free RNA associates with ribosomes. The viral RNA is then translated from a single start codon into the large polyprotein mentioned earlier. This giant protein (about 2200 amino acid residues) then undergoes self-cleavage into about 20 smaller proteins (including cleavage intermediates), among which are the four *structural*

proteins of the virus particle, the RNA-linked *VPg protein*, an *RNA polymerase* responsible for synthesis of minus-strand RNA, and at least one *virus-encoded protease*, which carries out the cleavage process. This cleavage process, called *posttranslational cleavage*, occurs in a wide variety of animal viruses as well as in normal cell metabolism in animal cells (∞ box, Protein Processing, Chapter 7).

Replication of Poliovirus

Replication of viral RNA begins within a short time after infection and is catalyzed by the RNA-dependent RNA polymerase (replicase) made in the process described in the previous paragraph. This replicase transcribes the viral RNA, of plus complementarity, into an RNA molecule of minus complementarity. This minus strand then serves as a template for repeated transcription of progeny plus strands. Some of the progeny plus strands may again be transcribed into minus strands, and as many as 1000 minus strands may subsequently be present in the cell. From these minus strands, as many as a million plus strands may ultimately be formed. Both the plus and the minus strands become covalently linked to the tiny VPg protein (only 22 amino acids long), which may serve as a primer for transcription.

Once virus multiplication begins, host RNA and protein syntheses are inhibited. Host protein synthesis is inhibited as a result of destruction of a host protein, the cap-binding protein required for translation of capped mRNAs (∞ Section 6.9).

At one time, polio was a major infectious disease of humans, but the development of an effective vaccine (∞ Section 20.16) has brought the disease almost completely under control. The World Health Organization has a vaccination program intended to eradicate the disease by the year 2000.

✓ 8.15 Concept Check

In small RNA viruses such as polio, the viral RNA acts directly as a single messenger RNA, causing the production of a long polyprotein that is broken down by enzymes into the numerous small proteins necessary for nucleic acid multiplication and virus assembly.

- ✓ What is a *cap* and what is its normal function?
- ✓ How can poliovirus RNA be synthesized in the cytoplasm while host RNA must be synthesized in the nucleus?

8.16

Negative-Strand RNA Viruses

In a number of RNA viruses of animals, the RNA does not serve directly as a messenger but is transcribed into a complement that functions as the mRNA. As discussed in Section 8.6, it is conventional to express the configuration of the mRNA as *plus,* so if the viral genomic RNA is of opposite complementarity, it is called *minus* (see Figure 8.11). This group of viruses is then called minus-strand or *negative-strand RNA viruses.* We discuss here two important negative-strand viruses: rhabdoviruses, including rabies virus, and orthomyxoviruses, including influenza virus. The Ebola virus, a human pathogen responsible for an emerging infectious disease (∞ Section 22.10) is also a negative-strand RNA virus.

Rhabdoviruses

One of the most important human pathogens that is a negative-stranded RNA virus is the rabies virus, which causes the important disease rabies in animals and humans (∞ Section 24.1). Rabies virus is called a *rhabdovirus,* from *rhabdo* meaning "rod," which refers to the shape of the virus particle. Another rhabdovirus that

has been extensively studied is vesicular stomatitis virus (VSV) (Figure 8.36), a virus that causes the disease *vesicular stomatitis* in cattle, pigs, horses, and sometimes humans. Many rhabdoviruses, such as potato yellow dwarf virus, infect both insects and plants and can cause important agricultural problems.

The rhabdoviruses are enveloped viruses, with an extensive and rather complex lipid envelope surrounding the nucleocapsid (Figure 8.36). In animal rhabdoviruses the virus particle is bullet-shaped, about 70 nm in diameter and 175 nm long. The nucleocapsid is helically symmetric and makes up only a small part of the virus particle weight (about 2–3% of the virion is RNA, in contrast to the 30% or more RNA content in nonenveloped RNA viruses). The virion contains several enzymes that are essential for the infection process. One of these is an *RNA-dependent RNA polymerase.* As discussed in Section 8.6, the presence of RNA polymerase is essential because the genome of these negative-strand viruses cannot act as messengers directly, but must first be transcribed into the plus complement, and host enzymes that transcribe RNA into RNA do not exist.

The RNA of the rhabdoviruses is transcribed inside the cytoplasm of the cell into two distinct kinds of RNA (Figure 8.37). The first type of RNA synthesis results in a series of messenger RNAs made from the various genes of the virus (VSV has 5). The second is a plus-strand RNA that is a *copy* of the complete viral genome (the VSV genome is 11,162 nucleotides long). These long plus-strand RNAs then serve as templates for synthesis of the *negative-strand* RNA molecules, which will serve as genomes of progeny virions. Each mRNA is mono-

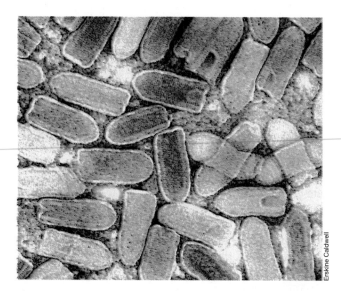

FIGURE 8.36 Electron micrograph of a rhabdovirus (vesicular stomatitis virus). A particle is about 65 nm in diameter.

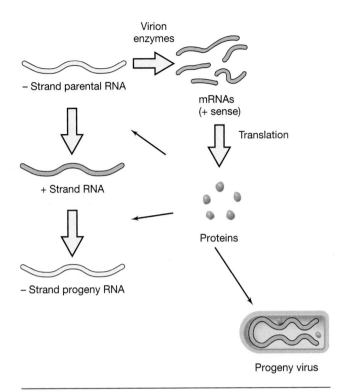

Virion
enzymes

– Strand parental RNA

mRNAs
(+ sense)

Translation

+ Strand RNA

Proteins

– Strand progeny RNA

Progeny virus

FIGURE 8.37   Flow of events during multiplication of a negative-strand RNA virus.

cistronic, coding for a single protein. Once the mRNA for the virus RNA polymerase is made in this primary transcription process, synthesis of the virus RNA polymerase can begin, leading to the formation of many *plus*-strand RNA molecules, both messengers and full-length genomic (viral) RNA templates.

Translation of viral mRNAs leads to the synthesis of viral coat proteins. *Assembly* of an enveloped virus is considerably more complex than assembly of a simple virus particle. Two kinds of coat proteins are formed, *nucleocapsid proteins* and *envelope proteins*. The nucleocapsid is formed first by association of the nucleocapsid protein molecules around the viral RNA. These nucleocapsid protein molecules are synthesized on ribosome complexes in the cytoplasm.

The envelope proteins that possess hydrophobic amino acid leader sequences at their amino-terminal ends (∞ Section 6.12) are synthesized on ribosome complexes that are themselves associated with membranes. As these proteins are synthesized, sugar residues are added, leading to the formation of *glycoproteins*. Such glycoproteins, characteristic of membrane-associated proteins, are transported to the cytoplasmic membrane (and the leader sequences are removed), where they replace host membrane proteins. Nucleocapsids then migrate to the areas on the cytoplasmic membrane where these virus-specific glycoproteins exist, recognizing the virus glycoproteins with

great specificity. The nucleocapsids then become aligned with the glycoproteins and bud through them, becoming coated by the glycoproteins in the process. The final result is an enveloped virion with a nucleocapsid center and a surrounding membrane whose lipid is derived from the host cell but whose membrane proteins are encoded by the virus. The budding process itself does not cause detectable damage to the cell, which may continue to release virions in this way for a considerable period of time. (Host damage does occur but is brought about by other unknown factors.)

Influenza and Other Orthomyxoviruses

Another group of negative-strand viruses of great importance is the group called the *orthomyxoviruses*, which contains the important human virus *influenza*. The term *myxo* refers to the fact that these viruses interact with the *mucus* or *slime* of cell surfaces. In the case of influenza virus, this mucus is at the mucous membrane of the respiratory tract, as these viruses are transmitted primarily by the respiratory route (∞ Section 23.4). The term *ortho* has been added to the influenza virus group to distinguish this group from another group of negative-strand viruses, the *paramyxovirus group.* The paramyxoviruses, which include such important human viruses as those causing mumps and measles, are actually similar in molecular biology to rhabdoviruses. The orthomyxoviruses have been extensively studied over many years, beginning with early work during and after the 1918 pandemic of influenza that caused the deaths of millions of people worldwide (∞ Section 23.4).

The orthomyxoviruses are enveloped viruses in which the viral RNA is present in the virion in a number of separate pieces. The genome of the orthomyxoviruses is thus said to be a **segmented genome.** In the case of influenza A virus, the genome is segmented into *eight* linear single-stranded molecules ranging in size from 890 to 2341 nucleotides. The influenza virus nucleocapsid is of helical symmetry, about 6–9 nm in diameter and about 60 nm long. This nucleocapsid is embedded in an envelope that has a number of virus-specific proteins as well as lipid derived from the host (Figure 8.38).

Because of the way influenza virus buds as it leaves the cell, the virus has no defined shape and is said to be *polymorphic* (Figure 8.38*a*). There are spikes on the outside of the envelope that interact with the host cell surface. One spike is called a *hemagglutinin* because it causes agglutination of red blood cells. (Agglutination is a process by which cells are caused to clump when they are mixed with an antibody or other protein or polysaccharide molecule that combines specifically with a substance on the cell surface, as described in Sections 20.14 and 21.5.) If the cells undergoing agglutination are red blood cells, then the process is called *hemagglutination.*

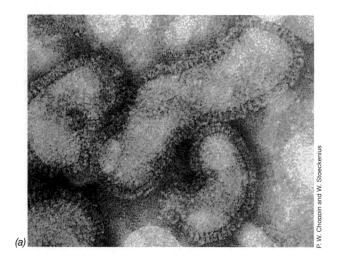

(a)

P. W. Choppin and W. Stoeckenius

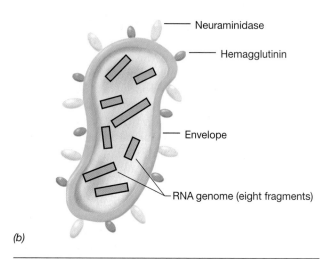

(b)

Neuraminidase

Hemagglutinin

Envelope

RNA genome (eight fragments)

FIGURE 8.38   Influenza virion structure. (a) Electron micrograph. (b) Diagram, showing some of the components.

(*Hema* is the combining form referring to *blood*.) The red blood cell is not the type of host cell the virus normally infects but contains on its surface the same type of membrane component, chemically characterized as *sialic acid*, that the mucous membrane cells of the respiratory tract contain. Thus, the red blood cell is merely a convenient cell type for measurement of agglutination activity. An important feature of the influenza virus hemagglutinin is that antibody directed against this hemagglutinin *prevents* the virus from infecting a cell. Thus, antibody directed against the hemagglutinin *neutralizes* the virus, and this is the mechanism by which immunity to influenza is brought about during the immunization process (⚬⚬ Sections 20.16 and 23.4).

A second type of spike on the virus surface is an enzyme called *neuraminidase* (see Figure 8.38). Neuraminidase breaks down the sialic acid component of the cytoplasmic membrane, which is a derivative of neuraminic acid. Neuraminidase appears to function primarily in the virus assembly process, destroying host membrane sialic acid that would otherwise block assembly or become incorporated into the mature virus particle.

In addition to the neuraminidase, the virus possesses two other enzymes, *RNA-dependent RNA polymerase*, which is involved in the conversion of a negative to a positive strand (as already discussed for the rhabdoviruses), and an *RNA endonuclease,* which cuts a primer from capped mRNA precursors.

The virus particle enters via the process of *endocytosis*. Once inside the cytoplasm, the nucleocapsid becomes separated from the envelope and migrates to the nucleus. Replication of the viral nucleic acid then occurs in the nucleus. Uncoating results in activation of the virus RNA polymerase. The mRNA molecules are then transcribed in the nucleus from the virus RNA, using oligonucleotide primers cut from the 5'-ends of newly synthesized capped cellular mRNAs. Thus, the viral mRNAs have 5'-caps. The poly-A tails of the viral mRNAs are added, and the virus mRNA molecules move to the cytoplasm.

Although influenza virus RNA replicates in the nucleus, influenza virus proteins are synthesized in the cytoplasm. *Ten* virus proteins are encoded by the *eight* segments of the virus genome. The mRNAs transcribed from six segments each encode a single protein, whereas the other two segments each encode two proteins. This is not done by using true polycistronic mRNA as in prokaryotes because eukaryotic ribosomes typically recognize only the AUG codon closest to the 5'-end of the mRNA as a start codon (⚬⚬ Section 6.12). Therefore, they can make only one protein from a given RNA. The original full-length mRNAs transcribed from these two segments are each translated to give one protein. In each case, an additional protein is translated from these messages after they have been processed by the host's RNA splicing machinery. Like overlapping genes, this is another example of how RNA viruses make maximum use of their small genome size.

Some of these proteins are involved in virus RNA replication, and others are structural proteins of the virion. The overall strategy of virus RNA synthesis resembles that of the rhabdoviruses, with primary transcription resulting in the formation of *plus*-strand templates for the formation of progeny *minus*-strand molecules. Details of assembly are still uncertain. The formation of the complete enveloped virus particle occurs by a budding-out process, as was described for the rhabdoviruses.

The segmented genome of the influenza virus has some important practical consequences. Influenza virus and other viruses of this family exhibit a phenomenon called **antigenic shift** in which pieces of the RNA genome from two genetically distinct strains that have infected the same cell become reassorted. This results in a change in the surface antigens (membrane proteins) of the virus, making the virus resistant to antibody that has been formed as a result of an immunization process (⚬⚬ Section 23.4). This antigenic shift makes it possible for the newly formed virus to infect hosts that the parent could not have infected. Antigenic shift is thought to bring about major epidemics of influenza.

✓ 8.16 Concept Check

In a number of RNA viruses, called negative-strand viruses, the virus RNA does not act directly as messenger but is copied into mRNA by an RNA-dependent RNA polymerase present in the virion. An important negative-strand virus is influenza virus.

✓ Why is it essential that negative-strand viruses carry an enzyme in their virions?
✓ What is a *segmented genome?*

8.17

Double-Stranded RNA Viruses: Reoviruses

The reoviruses, an important family of animal viruses, have a genome consisting of *double-stranded RNA*. The name *reovirus* is an acronym, derived from the terms *re*spiratory, *e*nteric, and *o*rphan. The term *orphan* was applied because the first viruses of this group to be isolated from humans were not associated with any specific disease syndrome. However, *rotavirus,* a member of the reovirus family, is probably the most common cause of diarrhea in infants from 6 to 24 months of age. Rotaviruses are also known to cause diarrhea in young animals.

As noted, the RNA of the reoviruses is double-stranded. This is the only group of animal viruses with double-stranded RNA; all other RNA virus groups have single-stranded RNA (Figure 8.33). The reovirus particle consists of a nonenveloped nucleocapsid 60–80 nm in diameter, with a *double* shell of icosahedral symmetry (Figure 8.39). Predictably, these double-stranded RNA viruses contain within the virion the virus-encoded enzymes necessary to synthesize RNA.

The genome of reoviruses consists of 10–12 molecules of double-stranded RNA. Double-stranded RNA is more difficult to unwind than double-stranded DNA, and there are no known RNA helicases (➷ Section 6.6). Replication of the RNA occurs by an asymmetric method. First, one strand is used as the template, and then the single-stranded product serves as a template to form a double helix. This difficulty in unwinding and the susceptibility of single-stranded RNA to cleavage within the cell greatly limits the size of a double-stranded molecule. Segmenting the genome into several molecules seems to be an adaptation to circumvent these problems.

Replication of reovirus RNA occurs exclusively in the *cytoplasm* of the host. The double-stranded RNA is inactive as mRNA, and the first step in reovirus replication is *transcription,* using the minus strand as a template to make mRNA. Generally, each molecule of RNA in the genome codes for a single protein, although in a few cases the protein formed is cleaved to give the final

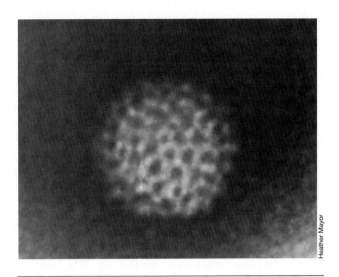

FIGURE 8.39   A reovirus particle. The diameter of the particle is about 70 nm.

product. However, one of the mRNAs produced actually encodes two proteins, and the RNA does <u>not</u> have to be modified to do so. Apparently a ribosome sometimes "misses" the start codon for the first gene in this message and travels on to the start codon of the second gene. Therefore, there are exceptions to the generalization that eukaryotic ribosomes initiate at the first AUG codon in mRNA.

Replication of the reovirus seems to occur within an intracellular equivalent of the viral core, called the *subviral particle,* which remains intact in the cell. Each of the 10 capped, single-stranded plus RNAs is assembled into this double-stranded RNA-synthesizing body. The capped single-stranded plus RNAs act as templates for the synthesis of the progeny minus genomic RNAs, yielding progeny double-stranded viral RNAs. The progeny double-stranded RNAs are further encapsidated, and when enough viral capsid proteins are present, mature virions are assembled.

In the initial infection process, the virion binds to a cellular protein. Once attachment has occurred, the virus enters the cell and is transported into lysosomes. Within the lysosome the outer shell of the virus particle is modified by removal of two proteins and cleavage of another by lysosomal enzymes. This uncoating process activates the viral RNA-dependent RNA polymerase and hence initiates the virus replication process.

8.18

Replication of DNA Viruses of Animals

Most animal viruses with DNA genomes contain double-stranded DNA (one group, the parvoviruses, contains single-stranded DNA). Among these DNA viruses of animals, the four major families are the papovaviruses, the herpesviruses, the pox viruses, and the adenoviruses. Of these, all replicate in the nucleus except for the pox viruses, which have the unique character (for DNA viruses) of replicating in the cytoplasm. In this and the following sections, we discuss the replication of each of these families briefly.

Papovaviruses: SV40

Some viruses of the papovavirus group induce tumors in animals. One of these DNA tumor viruses was first isolated from monkeys, and it was thus called *simian virus 40* or *SV40*. It was one of the first genetic elements to be studied by genetic engineering techniques and has been extensively used as a *vector* for moving genes into eukaryotic cells. (∞ Section 10.4).

The SV40 virion is a simple, nonenveloped particle 45 nm in diameter with an icosahedral head containing 72 protein subunits. There are no enzymes in the virion. In addition to the capsid proteins, however, there are host-derived *histone proteins* found complexed with the viral DNA. We have mentioned histone proteins during our discussion of chromosome structure (∞ Section 6.2) and have noted that histones play a role in neutralizing the negative charge originating from the phosphates of DNA and aid in packing of the DNA into more compact configurations.

The genome of SV40 consists of one molecule of double-stranded DNA of 5243 base pairs. The DNA is circular (Figure 8.40) and exists in a supercoiled configuration within the virion. The complete base sequence of SV40 has been determined, and the genetic map is known in some detail (Figure 8.41).

The nucleic acid is synthesized in the nucleus, but the proteins are synthesized in the cytoplasm. Final assembly of the virus particle occurs in the nucleus. The replication of these viruses can be divided into two distinct stages, *early* and *late*. During the early stage the *early region* of the viral DNA is transcribed (Figure 8.41). A single RNA molecule, the primary transcript, is made by cellular RNA polymerase, but it is processed into *two species of mRNA*, a large one and a small one. The DNA of SV40 has *introns* that are excised out of the primary RNA transcript. In the cytoplasm, viral mRNA is translated with the formation of two proteins. One of these proteins, the T-antigen, binds to the site on the parental DNA that is the *origin of replication*.

The viral DNA of SV40 is too small to code for its own DNA polymerases; host DNA polymerases are used. Replication occurs in a bidirectional fashion (so-called *theta* replication; ∞ Figure 6.16) from a single replication origin. The process involves the same events that have already been described for host cell DNA replication (∞ Section 6.6): RNA primer synthesis, formation of discontinuous DNA fragments on the lagging

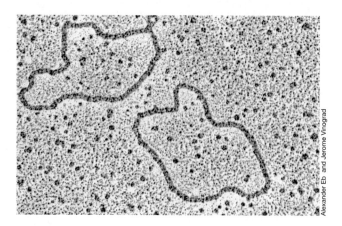

FIGURE 8.40   Electron micrograph of circular DNA from a tumor virus. The contour length of each circle is about 1.5 μm.

strand, gap filling, ligase action, and supercoiling the DNA through the action of gyrase.

Late SV40 mRNA molecules are synthesized using the strand complementary to that used for early mRNA synthesis (see Figure 8.41). Transcription begins at a promoter near the origin of replication. This late RNA is then processed by splicing and polyadenylation to give multiple forms of mRNA corresponding to the three coat proteins. These genes overlap; part of the nucleotide sequence contains information for all three pro-

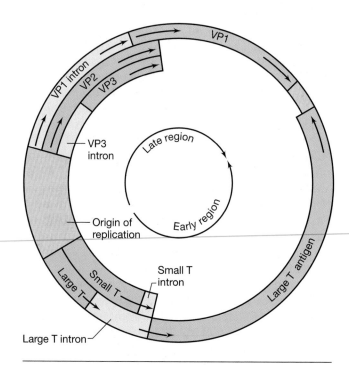

FIGURE 8.41   Genetic map of the papovavirus SV40. VP1, VP2, and VP3 are the genes coding for the three proteins that make up the coat of SV40. The arrows show the direction of transcription.

teins. These late mRNA molecules are transported to the cytoplasm and translated into the viral coat proteins. These proteins are then transported back into the nucleus where virion assembly takes place.

When a virus of the papovavirus group infects a host cell, one of two modes of replication can occur, depending on the type of host cell. In some types of host cells, known as *permissive* cells, virus infection results in the formation of new virions and the lysis of the host cell. In other types of host cells, known as *nonpermissive*, efficient multiplication does not occur, but the virus DNA becomes integrated into some of the host cells, thereby creating new, genetically altered cells. Such cells frequently show loss of growth inhibition and are called *transformed* or tumor cells.

In nonpermissive hosts, transformation can take place if the early proteins can be expressed, but the viral DNA cannot be replicated independently. Instead, in the transformed cell, the viral DNA becomes stably integrated into the DNA of the host cell (Figure 8.42). Integration can occur at many sites in the cellular and viral genome. In this integrated form, two viral proteins are made that are essential for the maintenance of a stably integrated viral DNA, but no viral structural proteins are synthesized. Some transformed cells can be converted to cells capable of producing virus, a process that probably involves excision of the viral genome from the host genome.

A study of the manner of replication of SV40 virus has given some important insights into the manner by which viruses bring about the cancerous state in host cells. However, we note that DNA viruses of other groups can also cause the cancerous transformation. Also, the important family of viruses, the *retroviruses*, are also cancer viruses but have a completely different mode of replication (see Section 8.22).

✓ **8.18 Concept Check**

Most double-stranded DNA animal viruses replicate in the nucleus. Some viruses causing cancer and other conditions in animal cells are double-stranded DNA viruses. In a permissive host, the virus may cause death and lysis, but in a nonpermissive host, it may cause transformation of the infected host cell to a tumor cell.

✓ Why wouldn't a virus like SV40 be expected to carry enzymes in the virion?
✓ How can one transcript yield more than one mRNA?

8.19

Herpesviruses

The herpesviruses are a large group of double-stranded DNA viruses that cause a wide variety of diseases in humans and animals, including fever blisters (cold

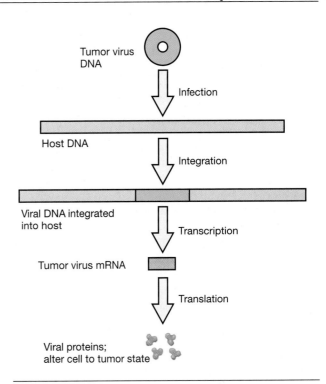

FIGURE 8.42 General scheme of molecular events involved in cell transformation by a DNA tumor virus such as SV40. All or portions of viral DNA are incorporated into host cell DNA. The viral genes that encode transforming information are transcribed and processed to viral mRNA molecules, which are transported to the cytoplasm. Here they are translated to form transforming proteins or T-antigens that code for functions that can convert host cells into cancer cells.

sores), venereal herpes, chickenpox, shingles, and infectious mononucleosis; a number of these diseases are discussed in Chapter 23. One of the interesting features of some herpesviruses is their ability to remain *latent* in the body for long periods of time, becoming active only under conditions of stress. Both *herpes simplex*, the virus that causes *fever blisters*, and *varicella-zoster virus*, the cause of *chickenpox* and *shingles*, are able to remain latent in the neurons of the sensory ganglia, from which they are able to emerge to cause infections of the skin.

An important group of herpesviruses are tumorigenic, causing clincial forms of cancer. One herpesvirus that is tumorigenic is the *Epstein–Barr virus*, which causes *Burkitt's lymphoma*, a common tumor among children in Central Africa and New Guinea. Burkitt's lymphoma was among the first human cancers to be linked to virus infection (see Table 8.2).

Molecular Biology of Herpesviruses

The *herpesvirus particle* is structurally complex, consisting of four distinct morphologic units. In herpes simplex type I, an enveloped virus about 150 nm in diameter (Figure 8.43a), the center of the virus is an electron-dense *core* consisting of double-stranded DNA. Surrounding this core is the nucleocapsid, of icosahedral symmetry, which consists of 162 capsomeres, each of which is composed of a number of distinct proteins. Outside the nucleocapsid is an amorphous layer that is called the *tegument*, a fibrous structure unique to the herpesviruses. Surrounding the tegument is an *envelope* whose outer surface contains many small *spikes*. A large number of separate proteins are present within the virion, but not all of them have been characterized.

The genome of herpes simplex type I virus consists of one large linear double-stranded DNA molecule of 152,260 base pairs (about 30 times larger than the SV40 genome). As a further indication of the complexity of herpes simplex, the DNA sequence of this virus indicates that it codes for at least 84 separate proteins.

Infection occurs by attachment of virus particles to specific cell receptors, and, following fusion of the cytoplasmic membrane with the virus envelope, the nucleocapsid is released into the cell. The nucleocapsids are transported to the nucleus, where viral DNA is uncoated. Components of the virus particle inhibit macromolecular synthesis by the host.

There are *three classes of messenger RNAs:* immediate early (also called alpha, α), which codes for five regulatory proteins; delayed early (also called beta, β), which codes for DNA replication proteins, including thymi-

dine kinase and DNA polymerase; and late (also called gamma, γ), which codes for the proteins of the virus particle (Figure 8.43b). During the *immediate early stage*, about one-third of the viral genome is transcribed by a host cell RNA polymerase. Early mRNA codes for certain positive-acting regulatory proteins that appear to stimulate the synthesis of the delayed early proteins. The second stage, *delayed early*, occurs only after the early proteins have been made. During this stage, about 40% of the viral genome is transcribed. Among the 10 proteins characterized from the delayed early stage are a *DNA polymerase*, enzymes involved in synthesis of deoxyribonucleotides, and a *DNA binding protein*. These enzymes are all involved in the process of viral DNA replication.

Herpes viral DNA synthesis itself takes place in the nucleus. After infection, the herpes genome apparently circularizes (remarkably like bacteriophage lambda) (see Section 8.12) and replicates by a rolling circle mechanism (see Figure 8.28). However, three origins seem to be involved. Long concatamers are formed that become processed to virus-length DNA during the assembly process itself (in a manner similar to that described for DNA bacteriophages) (see Sections 8.11 and 8.12).

Viral nucleocapsids are assembled in the nucleus, and acquisition of the virus envelope occurs via a budding process through the *inner membrane of the nucleus*. Mature virions are subsequently released through the

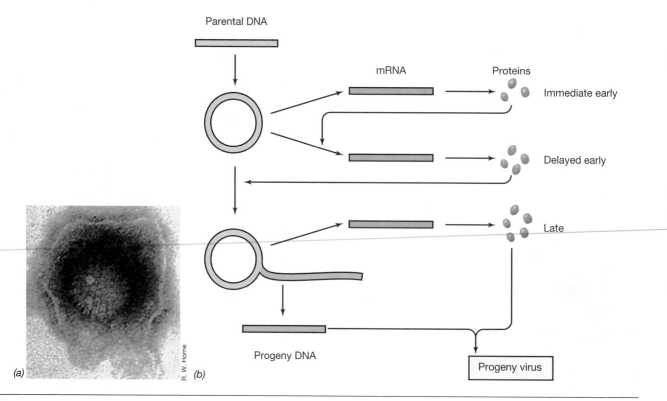

Parental DNA

mRNA   Proteins

Immediate early

Delayed early

Late

Progeny DNA

Progeny virus

(a)   R. W. Horne   (b)

FIGURE 8.43   Herpesvirus. (a) Electron micrograph of a herpesvirus particle. The diameter of the particle is about 150 nm. (b) Flow of events in multiplication of herpes simplex virus.

endoplasmic reticulum to the outside of the cell. Thus, the assembly of this enveloped DNA virus differs markedly from that of the enveloped RNA viruses, which were assembled on the *cytoplasmic membrane* instead of the nuclear membrane.

8.20

Pox Viruses

These are the most complex and largest animal viruses known (Figure 8.44) and have some characteristics that approach those of primitive cells. However, the pox viruses, like all viruses, are not able to metabolize and thus depend on the host for the complete machinery of protein synthesis. These viruses are also unique in that they are DNA viruses that replicate in the *cytoplasm*. Thus, a host cell infected with a pox virus exhibits *DNA synthesis outside the nucleus,* something that otherwise occurs only in intracellular organelles such as mitochondria.

General Properties of Pox Viruses

Pox viruses have been important medically as well as historically. *Smallpox* was the first virus to be studied in any detail and was the first virus for which a vaccine was developed (described by Edward Jenner in 1798). By diligent application of this vaccine on a worldwide basis, the disease smallpox has been *eradicated,* the first infectious disease to be eliminated in this fashion. Other pox viruses of importance are *cowpox* and *rabbit myxomatosis virus,* an important infectious agent of rabbits and one that was intentionally used in an attempt to control the Australian rabbit population (⬅ Section 22.6). Some pox viruses also cause tumors, but these tumors are generally benign.

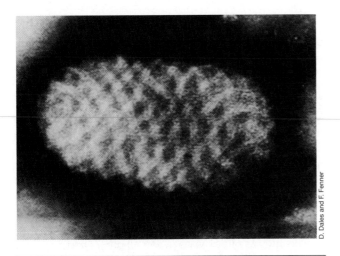

FIGURE 8.44 Electron micrograph of a negatively stained vaccinia virus virion. The virion is approximately 400 nm (0.4 μm) long.

The pox viruses are very large, so large that they can actually be seen under the light microscope. Most research has been done with *vaccinia virus,* the source of smallpox vaccine. The vaccinia virion is a brick-shaped structure about $400 \times 240 \times 200$ nm in size. The virion is covered on its outer surface with tubules or filaments arranged in a membranelike pattern, although the virus does not have a lipid membrane because the outer envelope consists of protein. Within the virion there are two lateral bodies of unknown composition and a core, the *nucleocapsid,* which contains DNA bounded by a layer of protein subunits.

The pox virus genome consists of double-stranded DNA. The vaccinia virus genome has about 185 kilobase pairs and contains between 150 and 200 genes. The pox DNA is interesting because the two strands of the double helix are cross-linked at the ends as a result of the formation of phosphodiester bonds between adjacent strands (as in the hairpin structure shown in Figure 6.12).

Replication of Pox Viruses

Vaccinia virions are taken up into cells via a phagocytic process from which the cores are liberated into the cytoplasm. Interestingly, uncoating of the virus genome requires the action of a new protein that is synthesized after infection. This protein is encoded by viral DNA, and the gene specifying this protein is transcribed by an RNA polymerase present *within* the virus particle. In addition to this uncoating gene, a number of other viral genes are transcribed. The primary transcripts are turned into mRNAs by capping and polyadenylation while they are still inside the virus core.

Once the vaccinia DNA is fully uncoated, the formation of *inclusion bodies* within the cytoplasm begins. Within these inclusion bodies, transcription, replication, and encapsidation into progeny virus particles occur. Each infecting virion initiates its own inclusion body, so the number of inclusions depends on the multiplicity of infection. Progeny DNA molecules form a pool from which individual molecules are incorporated into virions. Mature virions accumulate in the cytoplasm. There seems to be no specific release mechanism, and most virions are released only when the infected cell disintegrates.

Pox Viruses and Recombinant Vaccines

Vaccinia virus has been used as a host for genetically altered proteins of other viruses, permitting the construction of genetically engineered vaccines. As we will see in Chapter 20, a vaccine is a substance capable of eliciting an immune response in an animal and serves to protect the animal from future infection with the same agent. Vaccinia virus causes no serious health effects in humans but is highly immunogenic. Molecular cloning methods have been used to express key viral proteins of influenza virus, rabies virus, herpes simplex type I virus, and hepatitis B virus in vaccinia virus virions, and then the latter used to develop a vaccine (⌖ Section 10.13).

A similar vaccine delivery system using adenovirus (see next section) as a vehicle has been developed because, like vaccinia virus, adenoviruses are of little health consequence to humans.

8.21

Adenoviruses

The adenoviruses are a major family of icosahedral DNA-containing viruses that have unique molecular biological properties. The term *adeno* is derived from the Latin for "gland" and refers to the fact that these viruses were first isolated from the tonsils and adenoid glands of humans. Adenoviruses cause generally mild respiratory infections in humans, and a number of such viruses are isolated from apparently healthy individuals.

The genomes of the adenoviruses consist of linear double-stranded DNA of about 36 kilobase pairs. Attached in covalent linkage to the 5'-terminus of the DNA is a protein component essential for infectivity of the DNA. The DNA has inverted terminal repeats of 100–1800 base pairs (this varies with the virus strain). The DNA of the adenoviruses is six to seven times the size of the DNA of the papovavirus SV40.

Replication of the viral DNA occurs in the nucleus (Figure 8.45). After the virus particle has been transported to the nucleus, the core is released and converted to a viral DNA–histone complex. *Early transcription* is carried out by an RNA polymerase of the host, and a number of primary transcripts are made. The transcripts are spliced, capped, and polyadenylated, giving several different mRNAs.

The early proteins are involved in regulation of DNA replication; the later proteins are the virus coat protein. Viral DNA replication uses a virus-encoded protein as a primer and another virus-encoded protein as DNA polymerase. For the replication of a linear double-stranded DNA molecule such as that of adenovirus, initiation of replication can begin at either end or at both ends simultaneously (⌖ Figure 6.22). In the case of the adenoviruses, replication begins at *either* end, the two strands being replicated asynchronously. The products of a round of replication are double- and single-stranded molecules. The latter then cyclizes by means of the inverted terminal repeats, and a new complementary strand is synthesized beginning from the 5'-end, the products being another double-stranded molecule (Figure 8.45). This mechanism of replication is interesting because it does not involve the formation of discontin-

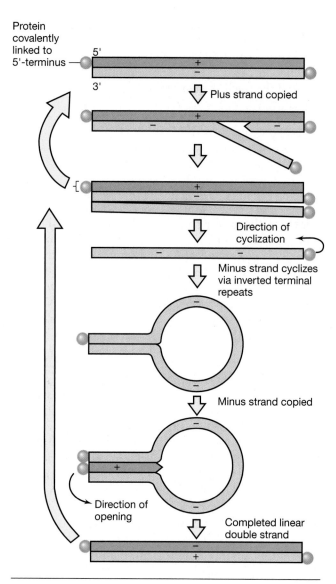

FIGURE 8.45   Replication of adenovirus DNA. See text for details.

uous fragments of DNA on the lagging strand, as occurs in conventional DNA replication (⌖ Section 6.6).

✓ 8.21 Concept Check

Most double-stranded DNA animal viruses replicate in the nucleus, although their replication strategies can be quite different. However, the pox viruses replicate in the cytoplasm using enzymes carried in the virion. Some of these viruses are now being used in genetic engineering experiments.

✓ Except in the case of pox viruses, in what cellular location is the genome of double-stranded DNA viruses replicated? Where does transcription occur?

✓ The mRNAs of pox viruses, and all other animal viruses, are translated in what cellular location?

8.22

Retroviruses

We now discuss one of the most interesting and complex families of animal viruses, the **retroviruses.** The term *retro* means "backward," and the name of this class of virus is derived from the fact that these viruses appear to have a backward mode of nucleic acid replication. The retroviruses are RNA viruses, but they *replicate by means of a DNA intermediate* using the enzyme *reverse transcriptase.* The retroviruses are interesting for a number of other reasons. First, they were the first viruses shown to cause *cancer* and have been studied most extensively for their carcinogenic characteristics. Second, one retrovirus, the one causing *acquired immunodeficiency syndrome (AIDS)* has been known only since the early 1980s but has become a major public health problem. Third, the retrovirus genome can become specifically integrated into the host genome by way of the DNA intermediate, and this integration process is being studied as a means of introducing *foreign* genes into a host, a process called *gene therapy.* Finally, the enzyme **reverse transcriptase** has become a major tool in genetic engineering.

As we will see, the retroviruses have some properties like those of RNA viruses and some like those of DNA viruses. They resemble to a considerable extent movable genetic elements and are sometimes considered to be *escaped cellular transposable elements.* In this respect, the retroviruses resemble bacterial viruses such as Mu (see Section 8.13). We should note that the use of reverse transcriptase is not restricted to the retroviruses because hepatitis B virus (a human virus) and cauliflower mosaic virus (a plant virus) also use reverse transcription in their replication processes. But in contrast to the retroviruses, these latter viruses encapsidate the DNA genome rather than the RNA genome as retroviruses do (see the box, Very Small Viruses: On The Fringe). Some transposable elements of eukaryotes, called *retrotransposons,* also encode and use reverse transcriptase as part of their replication cycle. In addition, reverse transcriptases capable of producing small multicopy DNA (ms DNA) with an RNA template have been discovered in myxobacteria and *Escherichia coli.* The reverse transcriptase in bacteria is encoded as part of a short genetic element called a *retron.* Although many copies of the ms DNA are made (which contain their RNA template covalently attached), their function is unknown.

The retroviruses are enveloped viruses (Figure 8.46*a*). There are a number of proteins in the virus coat and typically seven internal proteins, four of which

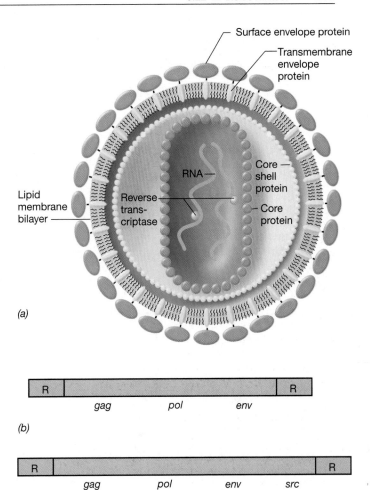

(a)

(b)

(c)

FIGURE 8.46   Retrovirus structure and function. (a) Structure of a retrovirus. (b) Genetic map of a typical retrovirus genome. (c) Genetic map of Rous sarcoma virus, a retrovirus that causes malignant tumors in birds. Each end of the genomic RNA contains direct repeats (R), and this RNA also has a 5'-cap and a 3'-poly-A tail. See text for more details.

are structural and three enzymatic. The enzymatic activities found in the virus particle are *reverse transcriptase, DNA endonuclease (integrase),* and a *protease.* The virion also contains specific cellular tRNA molecules used in *replication* (see later in this section).

Features of Retroviral Genomes and Replication

The genome of the retrovirus is unique. It consists of *two* identical single-stranded RNA molecules of plus complementarity, each 8.5–9.5 kilobases in length. The 5'-terminus of the RNA is capped and the 3'-terminus is polyadenylated, so the RNA is capable of acting di-

A FOCUS ON . . .   Very Small Viruses: On the Fringe

The life cycles of viruses contain a variety of unexpected genome structures and information transfers. Many of these seem exemplified by the hepadnaviruses, such as human hepatitis B virus. The genomes of these fascinating viruses are among the smallest known for a virus, but the virus life cycle is very complex. Like the retroviruses, these viruses use reverse transcriptase in their life cycle. In the case of hepadnaviruses the genome in the virion is DNA, but this DNA is replicated using an *RNA* intermediate, the opposite of what occurs in retroviruses.

The genomic DNA of hepadnaviruses is only *partially* double-stranded. One strand is incomplete, and both strands have breaks or gaps, but they are held in a circular form by hydrogen bonding. On entering the cytoplasm a viral polymerase carried in the virion completes the replication of this molecule. (This polymerase is quite a versatile protein; it contains DNA polymerase activity and reverse transcriptase activity and is the protein primer for synthesis of one of the DNA strands.) The figure showing the genetic map indicates the incredible compactness of the genome. Despite the small size (an average of 3200 base pairs), the genome encodes several proteins. Not only do genes overlap, but every base is part of a codon for at least one protein!

Replication of this genome involves transcription by host RNA polymerase (in the nucleus), yielding a transcript with terminal repeats. (The repeats occur because the polymerase proceeds slightly more than once around the circular molecule.) The viral polymerase then copies this into DNA, very much like the replication of retroviruses, but in this case the DNA becomes packaged into new virus particles.

Remarkably, the human hepatitis B virus is a "helper virus" for the *delta agent*. The delta agent is a "subvirus" and requires hepatitis B virus to supply the proteins necessary for its coat. Therefore, the delta agent is the parasite of a parasite! The delta agent has a circular, negative-strand RNA genome of 1679 bases. Like the viroids (see Section 8.23), it seems to be able to base-pair into a rodlike structure that can then be transcribed by the host RNA polymerase. However, unlike the situation with viroids, the transcript of the delta agent encodes at least one protein, an RNA binding protein carried by the delta agent virion. Incredibly, the RNA is also a ribozyme (∞ Section 6.9) that can cleave itself. This self-cleavage may be involved in mRNA formation.

Very small "viruses" can be very interesting indeed. The hepadnaviruses and the delta agent employ several different strategies for maximizing the genetic information carried in their very small genomes. ■

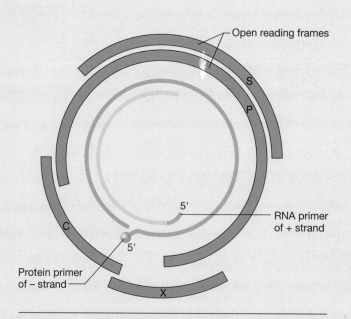

The partially double-stranded genome of human hepatitis B virus is shown in green. Note that the positive strand is not complete. The sizes of the open reading frames C, P, S, and X are also shown. Notice that all these genes overlap and that they cover every base in the genome.

rectly as mRNA but is *not* used as such. A genetic map of a typical retrovirus is shown in Figure 8.46*b*. Although there are differences between the genetic maps of different types of retroviruses, all contain the following regions and in the same order: *gag,* encoding internal structural proteins; *pol,* encoding reverse transcriptase and integrase; and *env,* encoding envelope proteins. Some, such as Rous sarcoma virus, carry a fourth gene downstream from *env* that is involved in cellular transformation and cancer (Figure 8.46*c*). The terminal repeats shown on the map play an essential role in the replication process (see later).

The overall process of replication of a retrovirus can be summarized in the following steps (Figure 8.47):

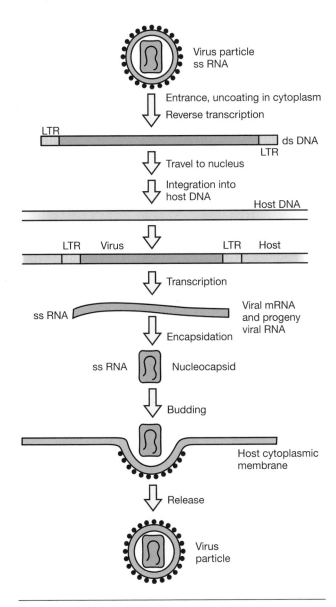

FIGURE 8.47   Replication process of a retrovirus.

1. **Entrance** into the cell.
2. **Reverse transcription** of *one* of the two RNA genomes into a single-stranded DNA that is subsequently converted to a linear double-stranded DNA by reverse transcriptase.
3. **Integration** of the DNA copy into the host genome.
4. **Transcription** of the viral DNA, leading to the formation of viral mRNAs and progeny viral RNA.
5. **Encapsidation** of the viral RNA into nucleocapsids in the cytoplasm.
6. **Budding** of enveloped virions at the cytoplasmic membrane and release from the cell.

We now discuss some aspects of the retrovirus multiplication process in detail. As we have noted, the first step after the entry of the RNA genome into the cell is reverse transcription: conversion of RNA into a DNA copy using the enzyme reverse transcriptase present in the virion. The DNA formed is a linear double-stranded molecule and is synthesized in the cytoplasm. An outline of the reverse transcription of viral RNA into DNA is given in Figure 8.48.

The enzyme reverse transcriptase is essentially a DNA polymerase, but it actually shows *three* enzymatic activities: (1) synthesis of DNA with an RNA template (reverse transcription), (2) synthesis of DNA with a DNA template, and (3) ribonuclease H activity (an activity that degrades the RNA strand of an RNA:DNA hybrid). Like all DNA polymerases, reverse transcriptase needs a primer for DNA synthesis. The primer for retrovirus reverse transcription is a specific *cellular transfer RNA (tRNA)*. The type of tRNA used as primer depends on the virus and is brought into the viron from the previous host cell. In the case of Rous sarcoma virus, the tRNA used is the *tryptophan* tRNA.

Using the tRNA primer, the 100 or so nucleotides at the 5′-terminus of the RNA are reverse-transcribed into DNA. Once transcription reaches the 5′-end of the RNA, the transcription process stops. In order to copy the remaining RNA, which is the bulk of the RNA of the virus, a different mechanism comes into play. First, terminally redundant RNA sequences at the 5′-end of the molecule are removed by the action of another enzymatic activity of reverse transcriptase, *ribonuclease H*. This leads to the formation of a small, single-stranded DNA that is complementary to the RNA segment at the *other end* of the viral RNA. The small, single-stranded piece of DNA then hybridizes with the other end of the viral RNA molecule, where copying of the viral RNA sequences continues. As summarized in Figure 8.48, continued action of reverse transcriptase and ribonuclease H leads to the formation of a double-stranded DNA molecule with long terminal repeats (LTRs) at each end. These LTRs contain strong promoters of transcription and are involved in the integration process. The *integration* of the viral DNA into the host genome is analogous to the integration of Mu (see Section 8.13) or a bacterial transposon into a bacterial genome. Integration can occur anywhere in the cellular DNA, and once integrated, the element, now called a *provirus*, is a stable genetic element. As a provirus its genetic information may be expressed, or it may remain in a latent state and not be expressed.

If the promoters in the right LTR are activated, the integrated proviral DNA is transcribed by a cellular RNA polymerase into transcripts that are capped and polyadenylated. These RNA transcripts either may be encapsidated into virus particles or may be processed and translated into virus proteins. Some virus proteins are made initially as a large primary *gag* protein, which

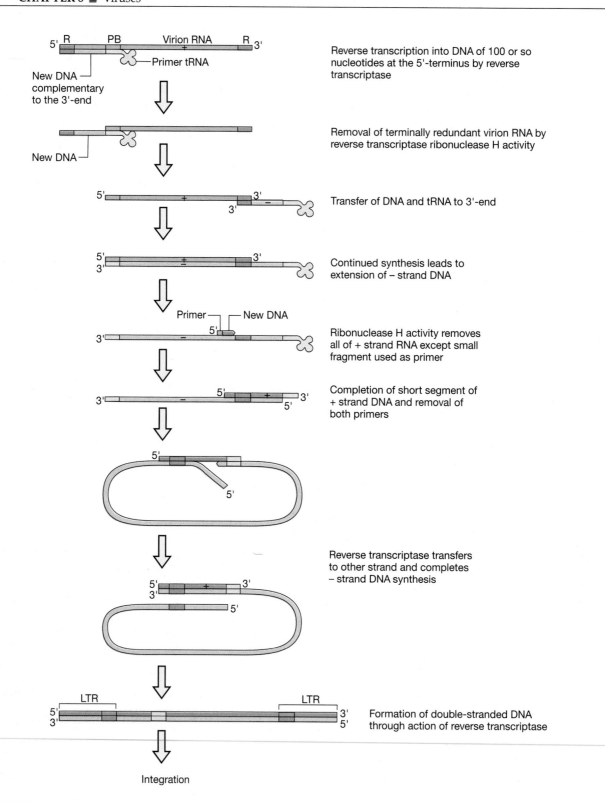

FIGURE 8.48   Overall steps in the formation of double-stranded DNA from retrovirus single-stranded RNA. The sequences labeled R on the RNA are direct repeats found at either end. The sequence labeled PB is where the primer (tRNA) binds. Note that the process of DNA synthesis has yielded longer direct repeats on the DNA than were originally on the RNA. These are called *long terminal repeats (LTRs)*.

is split by proteolytic action into the capsid proteins. Occasionally, read-through past the *gag* region (involving either inserting an amino acid in response to a stop codon or a shift in reading frame by the ribosome) leads to the translation of *pol,* the reverse transcriptase gene. Other proteins are synthesized from spliced transcripts.

When the virus proteins have accumulated in sufficient amounts, assembly of the nucleocapsid can occur. This encapsidation process leads to the formation of nucleocapsids, which move to the cytoplasmic membrane for final assembly into the enveloped virus particles.

Not all retroviruses cause cancer, but tumorigenic retroviruses are quite common. Some tumorigenic retroviruses known cause *sarcomas* or *acute leukemia;* they possess high oncogenic potential. Infection with one of these viruses can cause cellular transformation, leading to the formation of a tumor. Why are these viruses tumorigenic? It appears that they possess a transforming gene, or viral *oncogene,* that encodes a protein that brings about cellular transformation (see Section 8.14). This gene, known in Rous sarcoma virus as the *src* gene (*src* for *sarcoma*) (see Figure 8.46c), encodes a phosphoprotein that possesses protein kinase activity. Protein kinases bring about the phosphorylation of proteins, and protein phosphorylation is one mechanism for regulating the activity of proteins (∞ Section 7.7).

Transforming genes analogous to *src* have also been detected in human cancer cells. Interestingly, similar genes have also been detected in *normal* (that is, noncancerous) cells. These cell sequences are the proto-oncogenes (see Section 8.14) and have been found not only in mammalian cells but also in the cells of insects and yeast, suggesting that these sequences are of fundamental importance in the regulation of cell growth. Retroviruses are able to incorporate such normal sequences, which become altered and are abnormally expressed. Retroviruses are thus the agents by which such genes are transferred from cell to cell.

Genetic engineering with retroviruses occurs naturally by the incorporation of oncogene or proto-oncogene sequences. It is also possible to use modified retroviruses to incorporate foreign genes into cells. This is possible because substitution of such foreign genes for essential virus genes can lead to the production of virus particles carrying the foreign gene. Such particles are capable of being integrated into the host genome but are incapable of replicating or causing cancer.

As previously mentioned, one notable retrovirus is HIV, the virus causing AIDS. This virus infects a specific cell type in the human, a kind of T lymphocyte (T-helper cell) that is vital for proper functioning of the immune system. In later chapters we discuss the med-

ical and immunological aspects of AIDS (∞ Sections 22.4 and 23.7).

Because viruses are not cells but depend on cells for their replication, viral diseases pose serious medical problems; it is frequently difficult to prevent antiviral drugs from doing some damage to host cells. Despite this, certain chemotherapeutic strategies have been devised for use against viral pathogens, including retroviruses. We discuss these in some detail later along with the chemotherapy of other viral diseases (∞ Section 18.10).

✓ 8.22 **Concept Check**

Retroviruses are enveloped viruses that have complex life cycles since they are RNA viruses that replicate by means of a DNA intermediate. Important retroviruses cause cancer and acquired immunodeficiency syndrome. The retrovirus virion contains an enzyme, reverse transcriptase, that copies the information from its RNA genome into DNA. The DNA becomes integrated into the host chromosome in the manner of a temperate virus. The retrovirus DNA can be transcribed to yield mRNA (and new genomic RNA) or may remain in a latent state.

✓ What role does reverse transcriptase play in the infection cycle of a retrovirus?
✓ How does the life cycle of a temperate bacteriophage differ from that of a retrovirus?

8.23

Viroids and Prions

So far in this chapter, we have discussed some representative viral groups. Recall that our definition of a virus was a genetic element that subverts the normal cellular process for its own replication and that has an extracellular form. There are a few known entities whose properties are at variance with this definition and which most scientists would not actually consider to be viruses (see the box, When is a Virus Not a Virus). However, they seem closely related to viruses and are not considered plasmids. Two of the most important of these are *viroids* and *prions.*

Viroids are small, circular, single-stranded RNA molecules that are the smallest known pathogens (ranging from the coconut cadang-cadang viroid, which is 246 nucleotides in size, to citrus exocortis viroid, which has 375 nucleotides). Viroids cause a number of very important crop diseases. The extracellular form of the viroid is naked RNA—there is no capsid of any kind. Even more interestingly, *the RNA molecule contains no protein-encoding genes,* and therefore the viroid is totally dependent on host function for its replication. Although the viroid RNA is a single-stranded circle, there is such considerable secondary structure possible that it

A FOCUS ON . . .   When is a Virus Not a Virus—Definitions

This chapter presents summaries of the replication strategies of a few of the many types of known viruses. Our emphasis has been on viruses that infect bacteria and on animal viruses. The bacteriophages represent model genetic systems, and they infect prokaryotes, organisms that are the major focus of this text. Animal viruses are discussed in some detail because several of them cause important human diseases, and some illustrate interesting genome types.

There are many viruses that we did not discuss. These include the plant viruses, many of which cause plant diseases that have considerable impact on human affairs. We also have not discussed fungal "viruses." Partly, this choice has to do with space limitations but there is also a further problem: These "viruses" do *not* have an extracellular stage in their life cycle and, therefore, do not strictly fit the definition of virus that we are using.

All fungal "viruses" are transmitted by cell-to-cell fusion. This method of transmission may be because fungi have very thick cell walls or because fungal cell fusion is a common event in nature. These genetic elements are packaged into virion-like structures (sometimes referred to as "viruslike" particles) during their replication cycle, but most replicate benignly within the cells that carry them. In the yeast *Saccharomyces cerevisiae* there are known to be both double-stranded RNA elements and retroviral-like elements.

These genetic elements are mentioned here to draw attention to the difficulty of formulating simple definitions. In addition they help exemplify the many strategies used by genetic elements to ensure their replication. ■

---

resembles a short double-stranded molecule with closed ends (Figure 8.49). The viroid molecule seems to be replicated in the host cell nucleus, and its structure, which somewhat mimics DNA, apparently allows it to be replicated by the host RNA polymerase.

Viroids are sometimes considered "escaped" introns and, like self-splicing introns (∞ Section 6.9), appear to be remnants of an RNA world (see the discussion of the "RNA World" in Section 12.2).

Prions represent the other extreme from viroids. They have a distinct extracellular form, but the extracellular form seems to be *entirely protein*. It apparently does not contain any nucleic acid, or if it does, the molecule is not long enough to encode the single kind of protein of which the prion is composed. However, the prion protein particle is infectious, and various prions are known to cause a variety of diseases in animals, such as scrapie in sheep, bovine spongiform encephalopathy

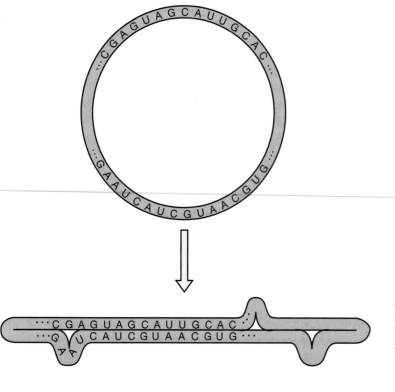

FIGURE 8.49   Structure of viroids, showing how single-stranded circular RNA can form a seemingly double-stranded structure by intrastrand base-pairing.

in cattle (BSE or "mad cow disease"), and kuru and Creutzfeldt–Jakob disease (CJD) in humans. In 1996, information became available from England that indicated that the prion that causes BSE in cattle might infect humans, resulting in a variant of CJD. Subsequent evidence seems to confirm that the BSE prion did "jump" the species barrier; however, the transmission seems to be very inefficient.

In addition to serious disease, prion infection results in production of more copies of the prion protein. Unless prions violate the central pattern of genetic information flow discussed in Chapter 6, this protein *must* be encoded by nucleic acid. Indeed, it has been discovered that the host cell contains a gene on one of its chromosomes that encodes a protein very similar to the prion protein. The host protein is normally produced and is found mostly in neurons. Apparently, the incoming prion modifies this host protein either during or after synthesis. The modification involves an alternative pattern of folding and causes the protein to lose its normal function, to become partially resistant to proteases, and to become insoluble. Therefore, prions do not simply subvert host enzymes but somehow cause a normal host gene to produce more copies of the pathogenic protein itself.

In 1997 the American scientist Stanley B. Pruisner won the Nobel Prize for his pioneering work with these diseases and with the prion proteins. Interestingly, there is now a model system for prion diseases that is much easier to study than following the disease in mammals. This model system is the yeast *Saccharomyces cerevisiae*.

It has now been shown that two characteristics of yeast seem to be transmitted by an "infectious" prion-like protein. One of these involves the PSI⁺ phenotype. PSI⁻ (normal) strains of yeast synthesize a protein called sup35, which is involved with accurate protein synthesis. In PSI⁺ strains this protein does not function correctly, even though the gene is normal. In PSI⁺ strains the protein is misfolded. This misfolding is caused by previously misfolded protein (hence the infectious nature of the process) and a molecular chaperone. This situation seems to be the same as that found in the prion diseases of mammals.

Viroids and prions do more than stretch our definition of a virus. They also demonstrate both the unexpected ways that genetic elements can replicate and the unexpected ways they can subvert the host cells. They are also of interest to us because they cause disease.

✓ 8.23 Concept Check

Viroids are circular single-stranded RNA molecules that encode no proteins and are completely dependent on host-encoded enzymes. They are the smallest known pathogens. Unlike viruses, their extracellular form is the same as their intracellular form, and they have no protein coat. Prions have an extracellular form that does contain protein, but it does not contain the nucleic acid that encodes the protein. The gene that encodes the prion protein is found in the host cell, and the prion somehow modifies this protein product.

REVIEW QUESTIONS

1. Define the term *host* as it relates to viruses.

2. Define *virus*. What are the minimal features needed to fit your definition?

3. Figure 8.40 shows circular DNA that has a contour length of 1.5 μm. How many base pairs are there in this DNA molecule (∞ Section 6.2)?

4. Under some conditions, it is possible to obtain nucleic acid-free protein coats (*capsids*) of certain viruses. Under the electron microscope these capsids look very similar to complete virions. What does this fact tell you about the role of the virus nucleic acid in the virus assembly process? Would you expect such virions to be infectious? Why?

5. Write a paragraph describing the events that occur on an agar plate containing a bacterial lawn when a single bacteriophage particle causes the formation of a *bacteriophage plaque*.

6. Describe how a *restriction endonuclease* might play a role in resistance to bacteriophage infection. Why could a restriction endonuclease play such a role whereas a generalized DNase could not?

7. One can divide the replication process of a virus into seven steps. What events are happening in each of these steps?

8. Specifically, why are both the life cycle and the virion of a positive-strand RNA virus likely to be simpler than those of a negative-strand RNA virus?

9. Is rolling circle DNA replication *bidirectional* (∞ Section 6.6)?

10. A bacterium that is missing the outer membrane protein responsible for maltose uptake is *resistant* to lambda infection. A lambda lysogen is *immune* to lambda infection. Describe the functional difference between resistance and immunity. Explain how these conditions are brought about in the examples given.

11. Some RNA bacteriophages are said to be *"male-specific."* Explain.

12. Typically, transfer RNA is used in translation. However, it also plays a role in the replication of retroviral nucleic acid. Explain this role.

13. What is unique about reovirus genomes, and what special problems does this introduce for nucleic acid replication?

14. Put together a diagram describing mRNA synthesis and nucleic acid replication for each of the following virus types: RNA tumor virus, reovirus, poliovirus, T4 phage, ϕX174 phage. For each diagram, be sure to indicate the complementarity of both the virus nucleic acid and its mRNA.

15. Many of the viruses we discussed have *early* genes and *late* genes. What is meant by these two classifications? What types of proteins tend to be encoded by early genes? What type of proteins by late genes? For any three viruses we discussed, describe how expression of the late genes is controlled.

APPLICATION QUESTIONS

1. Can you imagine any advantage for a virus having a metabolically inert extracellular stage rather than having one that is metabolically active?

2. What causes the viral plaques that appear on a bacterial lawn to stop growing larger?

3. Not all proteins are made from the RNA genome of bacteriophage MS2 in the same amounts. Can you explain why? One of the proteins functions very much like a repressor (∞ Section 7.2), but it functions at the translational level. Which protein is it and how does it function?

4. One characteristic of *temperate bacteriophages* is that they cause turbid rather than clear plaques on bacterial lawns. Can you think why this might be? (Remember the process by which a bacterial plaque develops.)

5. There are three lambda genes that, when rendered nonfunctional, turn lambda from a temperate to a virulent virus. What are these three genes and how do they normally function?

6. Figure 8.27 shows two promoters, P_R and P_E, on either side of the *cro* gene. Both transcribe through the *cro* gene, but only the transcript from P^R can be translated to yield the Cro protein. Explain.

7. The lambda cIII protein will be produced inside an infected cell in relatively high amounts if the cell is simultaneously infected by more than one lambda virion (because there will be multiple copies of the lambda chromosome being simultaneously expressed). High amounts of the cIII protein lead to lysogeny (and you should be able to explain how this is accomplished). From the point of view of the virus the "decision" to switch to the lysogenic pathway under these conditions is a good one. Explain.

8. The mechanism of replication of both strands of DNA in some viruses, such as adenoviruses, is continuous. Show how this can be without violating the "rule" learned in Chapter 6 that all DNA synthesis occurs in the overall direction of $5' \rightarrow 3'$.

9. Knowledge of the type of RNA carried in the genome of retroviruses would lead to a prediction that the virion would not carry any enzymes. Explain why one could make this prediction and why in this case it is wrong.

10. The promoters for mRNA encoding early proteins in viruses sometimes have a much different sequence than the promoters for mRNA encoding late proteins in the same virus. Explain why this might be true. (*Hint:* What type of RNA polymerase must recognize the "early" promoters?)

11. *Chemotherapeutic agents* are lacking for most virus diseases. From what you know about the stages of virus multiplication, give an explanation of why you think that may be so.

12. There are several genetic elements in yeast that have most of the characteristics of viruses, including the formation of viruslike particles containing the genome of the element. However, these genetic elements do not have an extracellular stage in their life cycle. Rather, they are transferred by cell fusion. See if you can write a simple definition of the word "virus" that would include these elements but still not include elements like plasmids. (Please note that plasmids can also be transferred from cell-to-cell, ∞ Section 9.9.)

Late in the 1940s it became apparent that bacteria had genetic systems that could be investigated and explored. The bacterium that came to be most widely used for genetic analysis was the common intestinal bacterium *Escherichia coli*, where genes were first mapped using conjugation and subsequently also using transduction. However, *E. coli* was not simply studied for its own sake, it also proved extremely useful as a model organism. In the 1970s and 1980s the techniques of DNA manipulation were developed and used for a more detailed and rapid exploration of the genome of *E. coli* and other microorganisms. The chromosome of *E. coli*, depicted here, as well as that of several other bacteria, have now been completely sequenced, and these accomplishments have spawned the era of comparative genomics.

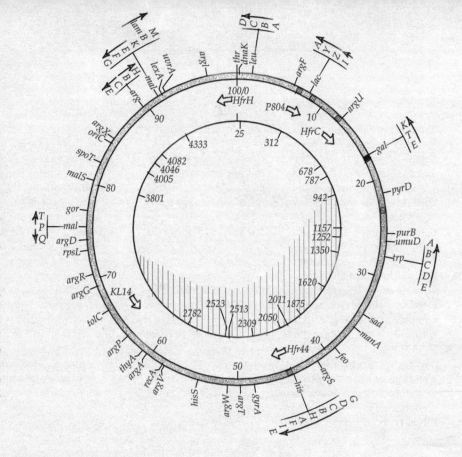

CHAPTER 9 **Microbial Genetics**

9.1 Mutations and Mutants 291
9.2 Molecular Basis of Mutation 294
9.3 Mutagens 297
9.4 Mutagenesis and Carcinogenesis: The Ames Test 300
9.5 Genetic Recombination 302
9.6 Genetic Transformation 306
9.7 Transduction 311
9.8 Plasmids 314
9.9 Conjugation and Chromosome Mobilization 319
9.10 Transposons and Insertion Sequences 324
9.11 The *Escherichia coli* Chromosome 329
9.12 Comparative Prokaryotic Genomics 333
9.13 Genetics in Eukaryotic Microorganisms 336
9.14 Yeast Genetics 337

WORKING GLOSSARY

Auxotroph an organism that has developed a nutritional requirement through mutation

Conjugation transfer of genes from one prokaryotic cell to another by a mechanism involving cell-to-cell contact and a plasmid

Diploid a eukaryotic cell or organism containing two sets of chromosomes

Electroporation the use of an electric pulse to induce cells to take up free DNA

Gametes in eukaryotic organisms, the haploid germ cells resulting from meiosis

Genetic map the arrangement of genes on a chromosome

Genome the total complement of genes of a cell or a virus

Genotype the precise genetic makeup of an organism

Haploid a cell or organism that has only one set of chromosomes

Mutagens agents that cause mutation

Mutant an organism whose genome carries a mutation

Mutation an inheritable change in the base sequence of the genome of an organism

Phenotype the observable characteristics of an organism

Plasmid an extrachromosomal genetic element that has no extracellular form

Point mutation a mutation that involves one or only a very few base pairs

Recombination the process by which parts or all of the DNA molecules from two separate sources are exchanged or brought together into a single unit

Screening any of a number of procedures that permit the sorting of organisms by phenotype or genotype, but allow the growth of those possible

Selection placing organisms under conditions where the growth of those with a particular genotype will be favored

Transduction transfer of host genes from one cell to another by a virus

Transformation transfer of bacterial genes involving free DNA (but see alternative usage in ∞ Chapter 8)

Transposable element a genetic element that has the ability to move (transpose) from one site on a chromosome to another

Transposon a type of transposable element that carries genes in addition to those involved in transposition

Now that we have introduced the main features of the molecular biology of cells and viruses, we can turn to a discussion of specific aspects of microbial genetics. In this chapter we present the basic principles of microbial genetics and then in the next chapter show how these principles apply to research in genetic engineering.

First we will discuss mutation and genetic recombination, and then we will explain how genetic material can be transferred from one organism to another. The transfer of genetic material from one organism to another can occur in a number of different ways in a microorganism, and if it is accompanied by genetic recombination, it can lead to profound changes in the organism.

Mutation is an inherited change in the base sequence of the nucleic acid comprising the genome of an organism. **Genetic recombination** is the process by which genetic elements contained in two separate genomes are brought together in one unit. Through this mechanism, new combinations of genes can arise even in the absence of mutation. Since the genetic elements brought together may enable the organism to carry out some new function, genetic recombination can result in adaptation to changing environments. Whereas mutation usually brings about only a very small amount of genetic change in a cell, genetic recombination usually involves much larger changes. Entire genes, sets of genes, or even whole chromosomes, are transferred between organisms.

The offspring of eukaryotic organisms that reproduce sexually receive a set of chromosomes from each of their parents. As a result, offspring are not exactly like either parent; they are *hybrids* and contain a combination of the traits exhibited by each parent. Prokaryotes do not reproduce sexually but there are mechanisms of genetic exchange in prokaryotes that, although considerably different from those involved in eukaryotic sexual reproduction, allow for both gene transfer and recombination.

Importance

There are a number of reasons why the study of microbial genetics is important:

1. Gene function is at the basis of cell function, and basic research in microbial genetics is necessary to understand how microorganisms function.
2. Microorganisms provide relatively simple systems for studying genetic phenomena and are thus useful tools in attempts to decipher the mechanisms underlying the genetics of all organisms.
3. Microorganisms are used for the isolation and duplication of specific genes from other organisms, a technique called **molecular cloning** (∞ Chapter 10). In molecular cloning, genes are manipulated and placed in a microorganism where they can be induced to increase in number.

4. Microorganisms produce many substances of value in industry, such as antibiotics, and genetic manipulations can be used to increase yields and improve manufacturing processes. Also, genes of higher organisms that specify the production of particular substances, such as human insulin, can be transferred by molecular cloning into microorganisms and the latter used for the production of these useful substances. The use of genetically modified microorganisms in large-scale industrial processes is an important part of the field of **biotechnology** (∞ Chapter 10).

5. Many diseases are caused by microorganisms, and genetic traits underlie these harmful activities. By understanding the genetics of disease-causing microorganisms, whether cellular or viruses, we can more readily control them and prevent their growth in the body.

6. Some of the types of genetic transfer that occur in prokaryotes, particularly conjugation (see Section 9.8), also play important roles in the spread of genes that confer properties such as resistance to antibiotics. Understanding such processes can help us to determine how genes can be transferred from one organism to another, even from one species to another.

To detect genetic exchange between two organisms, it is necessary to employ *genetic markers* whose transfer can be detected. Genetically altered strains are used for this purpose, the alteration(s) being due to one or more mutations in the DNA of the organism. We begin this chapter on microbial genetics with a consideration of the molecular mechanism of mutation and the properties of mutant microorganisms as a prelude to our discussion of genetic exchange.

9.1

Mutations and Mutants

As previously mentioned, a *mutation* is a heritable change in the base sequence of the nucleic acid genome of an organism. In all cells this nucleic acid is double-stranded DNA (∞ Section 6.1). A strain carrying such a change is called a **mutant.** A mutant by definition differs from its parental strain in **genotype,** a precise description of the genes an organism has. But in addition, the observable properties of the mutant, its **phenotype,** may also be altered relative to the parental strain. It is common to refer to a strain isolated from nature as a *wild-type* strain. Mutant derivatives can be obtained either from wild-type strains or from a strain derived from the wild type, for example, another mutant.

Depending on the mutation, a mutant may or may not show an altered phenotype from its parent. By convention in microbial genetics, the *genotype* of an organism is designated by three lowercase letters followed by a capital letter (all in italics) indicating the particular gene involved. For example, the *hisC* gene of *Escherichia coli* codes for a protein that could be called the HisC protein. In this case this protein (an enzyme in the biosynthetic pathway of histidine) is usually referred to by the name histidinol-phosphate aminotransferase, which describes its enzymatic activity. However, some proteins, such as the RecA protein (∞ Sections 8.12, 9.3, and 9.5), do not have other names, because enzyme functions can be difficult to describe in a few words. Mutations in the *hisC* gene would be designated as *hisC1*, *hisC2*, and so on, the numbers referring to the order of isolation of the mutant strains.

The *phenotype* of an organism is usually designated by a capital letter followed by two lowercase letters, with either a plus or minus superscript to indicate the presence or absence of that property. For example, a His$^+$ strain of *E. coli* is capable of making its own histidine whereas a His$^-$ strain is not. Mutations in the *hisC* gene may lead to a His$^-$ phenotype if they eliminate the function of the gene product.

Isolation of Mutants

Virtually any characteristic of an organism can be changed by mutation. However, some mutations are *selectable*, while others are *nonselectable*, even though they may lead to a very clear change in the phenotype of an organism. An example of such a nonselectable mutation is that of loss of color in a pigmented organism (Figure 9.1a). Such colonies usually have neither an advantage nor a disadvantage over the pigmented parent colonies when grown on agar plates (there may be a selective advantage for pigmented organisms in nature, however). This means that the only way we can detect such mutations is to examine large numbers of colonies and look for the "different" ones. Note that the apparent color of a colony need not be because of pigment production. Mutants colonies of *Halobacterium* lacking gas vesicles (∞ Section 3.14) transmit light differently than wild-type colonies and appear strikingly different on plates (Figure 9.1c).

Nonselectable mutants must be found by **screening** a large population of organisms, and the mutant phenotype may not be as easy to recognize as the difference between pigmented and nonpigmented colonies. A *selectable* mutation, on the other hand, confers on the mutant an advantage under certain environmental conditions, so the progeny of the mutant cell are able to outgrow and replace the parent. An example of a selectable mutation is drug resistance: An antibiot-

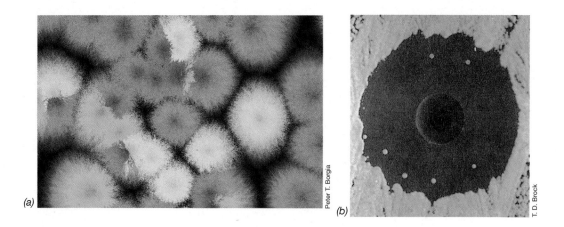

FIGURE 9.1   Observation of several kinds of mutants. (a) Pigmented mutants and nonpigmented mutants of the fungus *Aspergillus nidulans.* The wild type has a green pigment. The white or colorless mutants make no pigment, whereas the yellow mutants cannot convert the pigment they do make to the normal color. (b) Development of antibiotic-resistant mutants within the inhibition zone of an antibiotic assay disc. (c) Colonies of mutants of a species of *Halobacterium,* a member of the Archaea. The colonies that appear white are the wild type. The orange/brown colonies are mutants that are missing gas vesicles (∞ Section 3.14). Sectored colonies are the result of the mutagenic activities of transposable elements (see Section 9.10).

ic-resistant mutant can grow in the presence of antibiotic concentrations that inhibit or kill the parent (Figure 9.1*b*). However, the antibiotic-sensitive phenotype cannot be directly selected for by eliminating the antibiotic from the medium. It is relatively easy to detect and isolate selectable mutants by choosing the appropriate environmental conditions. Therefore, **selection** is an extremely powerful genetic tool, allowing the isolation of a single mutant from a population containing millions or even billions of parental organisms.

Selection of mutant or recombinant microorganisms is not just of laboratory interest. Every time antibiotics are used to kill pathogenic organisms there is a strong selection for antibiotic resistance.

Although screening is always more tedious than selection, for certain types of mutations methods are available for screening large numbers of colonies. For instance, nutritional mutants can be detected by the technique of **replica plating** (Figure 9.2*a*). With the use of sterile velveteen cloth or filter paper, an imprint of colonies from a master plate is made onto an agar plate lacking the nutrient. The colonies of the parental type will grow normally, whereas those of the mutant will not. Thus, the inability of a colony to grow on the replica plate (Figure 9.2*b*) is a signal that it is a mutant. The colony on the master plate corresponding to the vacant spot on the replica plate (Figure 9.2*b*) can then be picked, purified, and characterized. A nutritional mutant that has a requirement for a growth factor is often called an **auxotroph,** and the wild-type parent from which the auxotroph was derived is called a **prototroph.** For instance, mutants of *Escherichia coli* with a His⁻ phenotype are said to be *histidine auxotrophs.* Although of great utility, replica

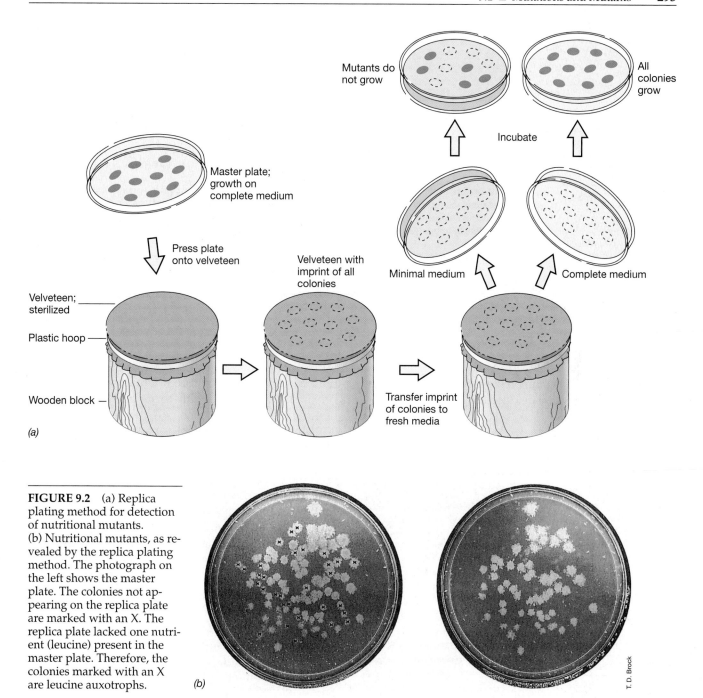

FIGURE 9.2 (a) Replica plating method for detection of nutritional mutants. (b) Nutritional mutants, as revealed by the replica plating method. The photograph on the left shows the master plate. The colonies not appearing on the replica plate are marked with an X. The replica plate lacked one nutrient (leucine) present in the master plate. Therefore, the colonies marked with an X are leucine auxotrophs. (b)

plating is a screening process, and it can be laborious to isolate mutants by screening.

An ingenious method widely used to isolate mutants that require amino acids or other growth factors is the **penicillin-selection method.** Ordinarily, mutants that require growth factors are at a disadvantage in competition with the parent cells, and so there is no direct way of isolating them. However, penicillin kills only *growing* cells, and if penicillin is added to a population growing in a medium lacking the growth factor required by the desired mutant, the parent cells will be killed, whereas the nongrowing mutant cells will be unaffected. Thus, after preliminary incubation in the absence of growth factor in a penicillin-containing medium, the population is washed free of penicillin and transferred to plates containing the growth factor. Among the colonies that grow up (including some wild-type cells that have escaped penicillin killing) should be some growth factor mutants. Penicillin selection is a kind of *negative selection;* the selection is not for the mutant but against the parental type.

TABLE 9.1	Kinds of mutants	
Description	Nature of change	Detection of mutant
Auxotroph	Loss of enzyme in biosynthetic pathway	Inability to grow on medium lacking the nutrient
Cold-sensitive	Alteration of an essential protein so it is inactivated at low temperature	Inability to grow at a low temperature (for example, 20°C) that normally supports growth
Drug-resistant	Alteration of permeability to drug or drug target or detoxification of drug	Growth on medium containing a growth-inhibitory concentration of the drug
Noncapsulated	Loss or modification of surface capsule	Small, rough colonies instead of larger, smooth colonies
Nonmotile	Loss of flagella; nonfunctional flagella	Compact colonies instead of flat, spreading colonies
Pigmentless	Loss of enzyme in biosynthetic pathway leading to loss of one or more pigments	Presence of different color or lack of color
Rough colony	Loss or change in lipopolysaccharide outer layer	Granular, irregular colonies instead of smooth, glistening colonies
Sugar fermentation	Loss of enzyme in degradative pathway	Lack of color change on agar containing sugar and a pH indicator
Temperature-sensitive	Alteration of an essential protein so it is more heat-sensitive	Inability to grow at a temperature normally supporting growth (for example, 40°C) but still growing at a lower temperature (for example, 30°C)
Virus-resistant	Loss of virus receptor	Growth in presence of large amounts of virus

Some of the most common kinds of mutants and the means by which they are detected are listed in Table 9.1.

✓ 9.1 Concept Check

Mutation, a heritable change in DNA, can lead to a change in phenotype. Selectable mutations are those that give the mutant a growth advantage under certain environmental conditions and are especially useful in genetics research.

✓ Distinguish between *mutation* and *mutant*.
✓ Distinguish between *screening* and *selection*.

9.2

Molecular Basis of Mutation

As previously mentioned, mutations arise in cells because of changes in the *base sequence* of an organism's genetic material. In many cases, mutations lead to phenotypic changes in the organism; these changes are mostly harmful or neutral although beneficial changes do occur occasionally.

Mutation can be either spontaneous or induced. **Spontaneous mutations** can occur as a result of the action of natural radiation (cosmic rays, and so on) which alters the structure of bases in the DNA. Spontaneous mutations can also occur during replication, as a result of errors in the pairing of bases, leading to changes in the replicated DNA. In fact, such errors occur at a frequency of about 10^{-7}–10^{-11} per base pair during a single round of replication. (A typical gene has about 1000 base pairs; therefore, the frequency of these errors in a gene would be 10^{-4} to 10^{-8} per generation.) Thus, in a

normal, fully grown culture of organisms having approximately 10^8 cells/ml, there are probably a number of different mutants in each milliliter of culture.

Mutations involving one (or a very few) base pairs are sometimes referred to as **point mutations.** Point mutations can result in *base-pair substitutions* in the DNA or in the insertion or deletion of a base pair (called *microinsertions* and *microdeletions*). As is the case with all mutations, the phenotypic change that comes about because of a point mutation depends on exactly where the mutation took place in the gene, what the nucleotide change was, and what product the gene normally encodes.

Base-Pair Substitutions

If a point mutation occurs within the coding region of a gene that encodes a protein, any change in the phenotype of the cell is almost certainly the result of a change in the amino acid *sequence* of the protein being produced. Figure 9.3 shows a number of base-pair substitutions that can occur in a short region of DNA within a gene that encodes a protein. The error in the DNA is transcribed into mRNA, and this erroneous mRNA in turn is used as a template and translated into protein. (Because only one strand of the DNA is used as template for the mRNA, an AT base pair does not have the same meaning as a TA base pair.) The triplet code that directs the insertion of an amino acid via a transfer RNA will thus be incorrect. What are the consequences of base-pair substitutions?

In interpreting the results of mutation, we must first recall that the genetic code is degenerate (∞ Section 6.10). Because of this degeneracy, not all mutations in

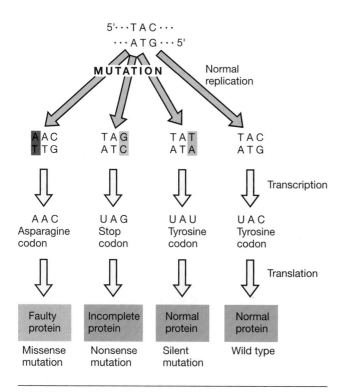

FIGURE 9.3 Possible effects of base-pair substitution in a gene encoding a protein: three different protein products from changes in the DNA for a single codon.

protein-encoding genes result in changes in protein. This is illustrated in Figure 9.3, which shows several possible results when the DNA that encodes a single *tyrosine codon* undergoes mutation. As seen, a change in the RNA from UAC to UAU would have no apparent effect because UAU is also a tyrosine codon. Mutations that give rise to such changes are called **silent mutations.** Note that silent mutations in coding regions almost always occur in the *third base* of the codon (arginine and leucine can also have silent mutations in the first position). As seen in Table 6.5, changes in the first or second base of the triplet much more often lead to significant changes in the protein. For instance, a single base change from UAC to AAC (Figure 9.3) would

result in a change in the protein from tyrosine to asparagine. This is referred to as a **missense mutation** because the chemical "sense" (sequence of amino acids) in the ensuing polypeptide has changed. If the change occurred at a critical point in the polypeptide chain, the protein could be inactive, or of reduced activity. However, not all mutations that cause amino acid substitution necessarily lead to nonfunctional proteins. The outcome depends on where in the polypeptide chain the substitution has occurred and on how it affects the folding and the catalytic activity of the protein. A missense mutation can lead to an enzyme that is temperature-sensitive, and this type of mutation is termed a **temperature-sensitive mutation.** For instance, temperature-sensitive mutants of bacteria are known that function normally at 30°C but cannot grow at 40°C, although the wild type grows well at both temperatures. Such mutations are also referred to as **conditionally lethal** because the bacteria cannot grow under one condition but can under another. Temperature-sensitive phenotypes often occur because the mutant protein can maintain its correct conformation at the low temperature but becomes partially unfolded (denatured) at the high temperature.

Another possible result of a base pair substitution is the formation of a *stop codon,* which would result in premature termination of translation, leading to an incomplete protein that would almost certainly not be functional (Figure 9.3). Mutations of this type are called **nonsense mutations** because the change is from a codon for an amino acid (sense codon) to a stop codon (nonsense codon). Unless the nonsense mutation occurs very near the end of the gene, the incomplete protein would be completely inactive.

Frameshift Mutations

Because the genetic code is read from one end in consecutive blocks of three bases, any deletion or insertion of a base pair results in a **reading frame shift,** and the translation of the gene is completely upset (Figure 9.4). Partial restoration of gene function can often be ac-

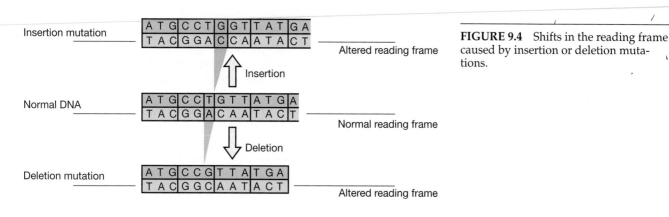

FIGURE 9.4 Shifts in the reading frame caused by insertion or deletion mutations.

complished by insertion of another base pair near the one deleted (one kind of suppressor mutation; see later in this section). After correction, depending on the exact amino acids coded by the still faulty region and the region of the protein involved, the protein formed may have some biological activity or even be completely normal.

It is important to remember that microinsertions or microdeletions are frameshift mutations only if they occur in the part of a protein-encoding gene that includes the reading frame. A single base-pair insertion in the promoter of a gene could lead to a dramatic change in the ability of the gene to function (because of increased or decreased expression or a change in regulation), but it would not be a frameshift mutation. (Similarly, base-pair substitutions that are not within the reading frame are not missense or nonsense mutations.)

Back Mutations or Reversions

Point mutations are reversible. A *revertant* is operationally defined as a strain in which the wild-type phenotype that was lost in the mutant is restored. Revertants can be of two types. In *same-site revertants,* the mutation that restores activity occurs at the same site at which the original mutation occurred. (If the back mutation is not only at the same site but also leads to the wild-type sequence, it is called a *true revertant.*) In *second-site revertants,* the mutation occurs at a different site in the DNA.

Second-site mutations may cause restoration of a wild-type phenotype because of several types of **suppressor mutations** that restore the original phenotype. Suppressor mutations are new mutations that compensate for the effect of the original mutation. Several types of suppressor mutations are known: (1) a mutation somewhere else in the same gene can restore enzyme function, such as in a frameshift mutation; (2) a mutation in another gene may restore the wild-type phenotype; and (3) the mutation may result in the production of another enzyme that can replace the mutant one by introducing a metabolic pathway different from that used by the mutant enzyme. In this last type no production of the original enzyme occurs although it does in the other types.

Mutations Involving Many Base Pairs

Deletions are mutations in which a region of the DNA has been eliminated. As we have discussed, microdeletions, the removal of one or a very few base pairs (Figure 9.4), are often frameshift mutations and can inactivate a gene. However, deletions can also involve the loss of hundreds or thousands of base pairs. Deletion of a large segment of the DNA results in complete loss of function of any gene that may be involved. Some deletions are so large that they involve several genes (if any of the genes are essential, the mutation will be lethal). Such deletions cannot be restored through further mutations but only through genetic recombination. Indeed, one way in which large deletions are distinguished from point mutations is that the latter are reversible through further mutations, whereas the former are not.

Insertions occur when new bases are added to the DNA. As we have discussed, insertions, like deletions, can involve only a single base or many bases. Microinsertions result from replication errors as do deletions, but larger insertions arise as a result of mistakes that occur during genetic recombination. Insertions inactivate the gene in which they occur. Many insertion mutations are due to the insertion of specific identifiable DNA sequences 700–1400 base pairs in length called **insertion sequences,** a type of transposable element (Section 6.4). The behavior of such insertion sequences is discussed in detail in Section 9.9. Even large insertion mutations can revert by further mutation that is, by a deletion that removes the insertion.

Other types of large-scale mutations exist that also seem to involve rearrangements brought about by mistakes in recombination. These include **translocations,** in which a large section of chromosomal DNA is moved to a new location (and in eukaryotes often to a different chromosome), and **inversions,** in which the orientation of a particular segment of DNA is reversed with respect to the surrounding DNA.

Rates of Mutation

There are wide variations in the rates at which various kinds of mutations occur. Some types of mutations occur so very rarely that they are almost impossible to detect, whereas others occur so frequently that they often present difficulties for an experimenter trying to maintain a genetically stable stock culture.

Spontaneous mutations in a gene occur at frequencies of about 10^{-6} per generation (see earlier discussion). This means that there is 1 chance in 1,000,000 that a mutation will arise at some location in a given gene during one cell cycle. Transposition events may occur more frequently, at about 10^{-4}. On the other hand, the occurrence of a nonsense mutation is less frequent, 10^{-6}–10^{-8}, because only a few codons can mutate to nonsense codons. Unless the mutant is selectable, the experimental detection of events of such rarity is difficult and much of the skill of the microbial geneticist is applied to increasing the efficiency of mutation detection. As we will see in the next section, it is possible to significantly increase the rate of mutation by the use of mutagenic treatments.

Mutations in Ribonucleic Acid (RNA) Genomes

Whereas all cells have *DNA* as their genetic material, some viruses have *RNA* genomes (for examples, ∞ Sections 8.8 and 8.15–8.17). These genomes can also mutate. Importantly, the mutation rate in RNA genomes is about *1000-fold higher* than in DNA genomes. At least some RNA polymerases have *proofreading* activities like those of DNA polymerases (∞ Section 6.6). However, while there are many repair systems for DNA that can correct many changes before they become fixed as mutations (see Section 9.3), there seem to be no comparable RNA repair mechanisms. This very high rate of mutation in RNA viruses is not merely of academic interest. The RNA genomes of viruses that cause disease can mutate very rapidly, presenting a constantly changing and evolving population of viruses.

✓ 9.2 Concept Check

Mutations, which can be either spontaneous or induced, arise because of changes in the base sequence of the nucleic acid of an organism's genome. A point mutation, which is due to a change in a single base pair, can lead to a single amino acid change in a protein or to no change at all, depending on the particular codon involved. In a nonsense mutation, the codon becomes a stop codon and an incomplete protein is made. Deletions and insertions cause more dramatic changes in the DNA, including frameshift mutations, and often result in complete loss of gene function.

✓ What does it mean to say that point mutations can spontaneously revert?

✓ Do missense mutations occur in genes encoding transfer RNAs (tRNAs)?

9.3

Mutagens

While the spontaneous rate of mutation is very low, there are a variety of chemical, physical, or biological agents that can increase the mutation rate, and are therefore said to *induce* mutations. These agents are referred to as **mutagens.** We discuss some of the major categories of mutagens and their actions here.

Chemical Mutagens

An overview of some of the major chemical mutagens and their modes of action is given in Table 9.2. Several classes of chemical mutagens exist. One variety of chemical mutagens are the **base analogs,** which resemble DNA purine and pyrimidine bases in structure yet show faulty pairing properties (Figure 9.5). When one of these base analogs is incorporated into DNA, replication may occur normally most of the time, but occasional copying errors occur, resulting in incorporation of the wrong base into the copied strand. During subsequent segregation of this strand, the mutation is revealed.

TABLE 9.2	Chemical and physical mutagens and their modes of action

Agent	Action	Result
Base analogs		
5-Bromouracil	Incorporated like T; occasional faulty pairing with G	AT pair → GC pair Occasionally GC → AT
2-Aminopurine	Incorporated like A; faulty pairing with C	AT → GC Occasionally GC → AT
Chemicals reacting with DNA		
Nitrous acid (HNO_2)	Deaminates A and C	AT → GC and GC → AT
Hydroxylamine (NH_2OH)	Reacts with C	GC → AT
Alkylating agents		
Monofunctional (for example, ethyl methane sulfonate)	Put methyl on G; faulty pairing with T	GC → AT
Bifunctional (for example, nitrogen mustards, mitomycin, nitrosoguanidine)	Cross-link DNA strands; faulty region excised by DNase	Both point mutations and deletions
Intercalative dyes		
Acridines, ethidium bromide	Insert between two base pairs	Microinsertions and microdeletions
Radiation		
Ultraviolet	Pyrimidine dimer formation	Repair may lead to error or deletion
Ionizing radiation (for example, X-rays)	Free-radical attack on DNA, breaking chain	Repair may lead to error or deletion

Analog	Substitutes for	Mutation observed
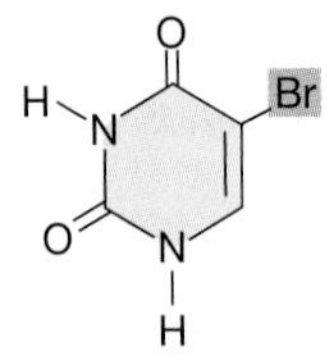 (a) 5-Bromouracil	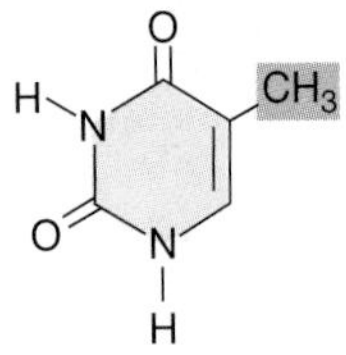 Thymine	5-Bromouracil can pair with guanine, causing AT to GC substitution
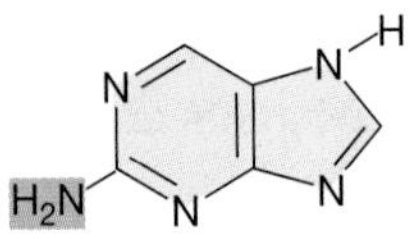 (b) 2-Aminopurine	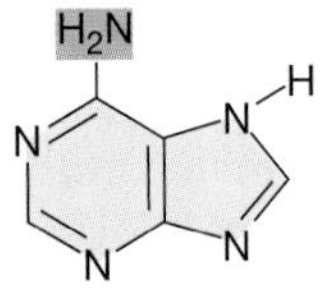 Adenine	2-Aminopurine can pair with cytosine, causing AT to GC substitution

FIGURE 9.5 Structure of two common nucleotide base analogs used to induce mutations and the normal nucleic acid bases they substitute for.

Some other chemical mutagens react directly on DNA, causing chemical changes in one base or another, which results in faulty pairing or other changes (Table 9.2). *Alkylating agents* such as nitrosoguanidine, for example, are powerful mutagens and generally induce mutations at higher frequency than base analogs. Such chemicals differ in their action from the base analogs in that the chemicals reacting on DNA are able to introduce direct changes even in nonreplicating DNA, whereas the base analogs act only after incorporation during replication. Both base analogs and alkylating agents tend to induce base-pair substitutions (see Section 9.2).

One interesting group of chemicals, the acridines, are planar molecules that act as *intercalating agents.* These mutagens become inserted between two DNA base pairs, thereby pushing them apart. During replication this abnormal conformation can lead to microinsertions or microdeletions in acridine-treated DNA. Thus acridines can induce frameshift mutations (see Section 9.2). Ethidium bromide, which is often used to detect DNA in electrophoresis (∞ see box, Working with Nucleic Acids: The Tools, Chapter 6), is also an intercalating agent that acts as a mutagen.

Radiation

Several forms of radiation are highly mutagenic. We can divide mutagenic radiation into two main categories, ionizing and nonionizing (electromagnetic) (Figure 9.6). Although both kinds of radiation are used in microbial genetics, *nonionizing* radiations find the widest use and will be discussed first.

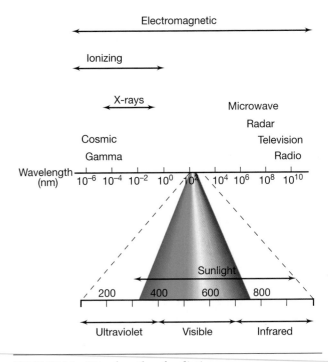

FIGURE 9.6 Wavelengths of radiation.

The purine and pyrimidine bases of the nucleic acids absorb ultraviolet (UV) radiation strongly, and the absorption maximum for DNA and RNA is at 260 nm (∞ Working with Nucleic Acids: The Tools, Chapter 6). Proteins also absorb UV but have a peak at 280 nm due to absorption of the aromatic amino acids (tryptophan, phenylalanine, tyrosine). It is now well established that killing of cells by UV radiation is due primarily to its action on DNA, and so UV radiation at 260 nm is most effective as a lethal agent. Although several effects are known, one well-established effect is the induction in DNA of **pyrimidine dimers,** a state in which two adjacent pyrimidine bases (cytosine or thymine) become covalently joined so that during replication of the DNA the probability of DNA polymerase inserting an incorrect nucleotide at this position is greatly increased.

The type of UV radiation source most frequently used for mutagenesis is the germicidal lamp, which emits large amounts of UV radiation in the 260-nm region. A dose of UV radiation is used that brings about 90–95% killing of the cell population (∞ Section 18.2), and mutants are then looked for among the survivors. If much higher doses of radiation are used, the number of viable cells will be too low, whereas if lower doses are used, insufficient damage to the DNA will be induced. UV radiation is a very useful tool in isolating mutants of microbial cultures.

Ionizing Radiation

Ionizing radiation is a more powerful form of radiation and includes short-wavelength rays such as X-rays, cosmic rays, and gamma rays (Figure 9.6). These radiations cause water and other substances to ionize, and mutagenic effects are brought about indirectly through this ionization. Among the potent chemical species formed by ionizing radiation are chemical free radicals, of which the most important is the hydroxyl radical, $OH \cdot$. Free radicals react with and inactivate macromolecules in the cell, of which the most important is DNA. DNA is probably no more sensitive to ionizing radiation than other macromolecules, but because DNA is the genetic material, changes induced in it can have a permanent effect. At low doses of ionizing radiation, only a few hits on DNA occur, but at higher doses multiple hits occur, leading to the death of the cell. In contrast to UV radiation, ionizing radiation penetrates readily through glass and other materials. Because of this, ionizing radiation is used frequently to induce mutations in animals and plants (where its penetrating power makes it possible to reach the gamete-producing cells of these organisms readily; see Section 9.13), but because ionizing radiation is more dangerous to use and is less readily available, it finds less use with microorganisms (where penetration with UV is not a problem).

Mutations That Arise from DNA Repair

Recall that a mutation is an *inheritable* change in the genetic material. Therefore, if an error in DNA synthesis can be corrected before the cell divides, there will be no mutation. Furthermore, some DNA damage clearly cannot be replicated and therefore cannot itself be a mutation. For instance, if a DNA molecule contains pyrimidine dimers, it will not be replicated to give two DNA molecules, each containing pyrimidine dimers. If such DNA damage cannot be repaired, cells often die. Most cells have a variety of different DNA repair processes to correct mistakes or repair damage. Many of these DNA repair systems do not make mistakes. However, some processes seem to be *error-prone*, and it is the repair process itself that introduces the mutation. Many kinds of mutations arise as a result of faulty repair of damage induced in DNA by some of the various agents just discussed.

Often DNA damage itself can *induce* DNA repair systems. A complex cellular mechanism, called the **SOS regulatory system,** is activated as a result of some types of DNA damage, initiating a number of DNA repair processes. However, in the SOS system, some DNA repair occurs in the absence of template instruction that is, without base-pairing, which results in the creation of many errors, hence many mutations.

In the SOS regulatory system, DNA damage serves as a distress signal to the cell, resulting in the coordinate derepression (induction) of a number of cellular functions involved in DNA repair. The SOS system is normally repressed by a protein called LexA. However, this repressor protein is inactivated by RecA, a protease that is activated as a result of DNA damage (Figure 9.7). Since one of the DNA repair mechanisms of the SOS system is inherently error-prone, many mutations arise. Thus, through the SOS regulatory system, DNA damage by various agents such as chemicals and radiation leads to mutagenesis.

Once the DNA damage has been repaired, the SOS system is switched off and further mutagenesis ceases. In addition to its effect on cellular mutagenesis, the SOS regulatory system plays a central role in the regulation of temperate virus replication (∞ Section 8.12).

It should be emphasized that not all DNA repair occurs in the absence of template instruction. Cells generally have many DNA repair systems that require template instruction and lead to proper DNA repair. These systems apparently work most of the time but are not sufficient to repair the large amounts of damage done by some of the agents previously mentioned.

Biological Mutagens

Mutations can be introduced without the use of chemical or physical agents through the process of *transposon mutagenesis*. We discuss the details of transposon mutagenesis later in this chapter (see Section 9.10) and note here only that if insertion of a transposable element occurs *within* a gene, loss of gene function generally results (see Figure 9.1c). Because transposable elements can enter the chromosome at various locations, trans-

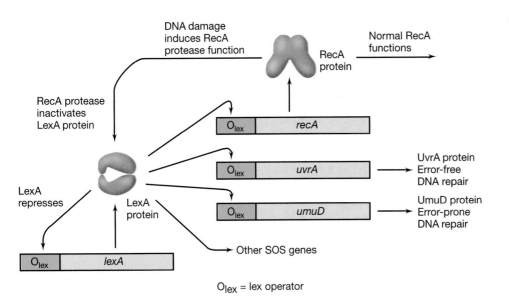

FIGURE 9.7   Mechanism of the SOS response. DNA damage results in conversion of RecA protein into a protease that cleaves LexA protein. LexA protein normally represses the activities of the *recA* gene and the DNA repair genes *uvrA* and *umuD*. Note, however, that repression is not complete. Some RecA protein is produced even in the presence of LexA protein. With LexA inactivated, these genes become active. As a protease, the RecA protein also cleaves the lambda repressor protein (∞ Section 8.12).

posons are widely used by microbial geneticists as mutagenic agents.

Site-Directed Mutagenesis

So far, the mutations that we have been discussing have been randomly directed at the genome of the microbial cell. Recombinant DNA technology and the use of synthetic DNA make it possible to induce *specific* mutations in *specific* genes. The procedures for carrying out mutagenesis of specific sites in the genome are called *site-directed mutagenesis* and will be discussed in detail in the next chapter (∞ Section 10.11). Here we briefly describe the overall principle of site-directed mutagenesis.

When the DNA containing a specific gene has been isolated and its sequence determined, it is possible to construct a modified form of this gene in which a specific base or series of bases has been changed. This modified DNA can then be inserted into a recipient cell and *mutants* selected (or screened) where the modified gene has replaced the normal gene. These mutants will differ from the wild type by the desired change at a specific site. Site-directed mutagenesis has many uses in microbial genetics and molecular biology and has been especially useful for structure–function studies of enzymes and other proteins (∞ Section 10.11).

✓ 9.3 Concept Check

Mutagens are chemical, physical, or biological agents that increase the mutation rate. Mutagens can alter DNA in many different ways. However, alterations in DNA are not mutations unless they can be inherited. Some DNA damage can lead to cell death if not repaired.

✓ How do mutagens work?
✓ Differentiate between a mutation and DNA damage.

9.4

Mutagenesis and Carcinogenesis: The Ames Test

A practical use of mutant bacterial strains has been developed to identify potentially hazardous chemicals in the environment. Because the sensitivity with which selectable mutants can be detected in large populations of bacteria is very high, bacteria can be used as screening agents for the potential mutagenicity of chemicals. This is relevant because it has been found that many mutagenic chemicals are also carcinogenic, capable of causing cancer in animals or humans.

The variety of chemicals, both natural and artificial, that the human population comes into contact with through agricultural and industrial exposure is enormous. There is considerable need for simple tests to ascertain the safety of such compounds. There is good evidence that a large proportion of human cancers have environmental causes, most likely through the agency of various chemicals, making the detection of chemical carcinogens urgent. It does not necessarily follow that because a compound is mutagenic it is also carcinogenic. The correlation, however, is quite high, and the knowledge that a compound is mutagenic in a bacterial system serves as a warning of possible danger. Similarly, the fact that a compound is not mutagenic in a bacterial system does not mean that it is not carcinogenic, because the bacterial system cannot detect all compounds active in higher animals. The development of bacterial tests for carcinogenic screening was carried out primarily by a group at the University of California in Berkeley under the direction of Bruce Ames, and the mutagenicity test for carcinogens is sometimes called the **Ames test** (Figure 9.8).

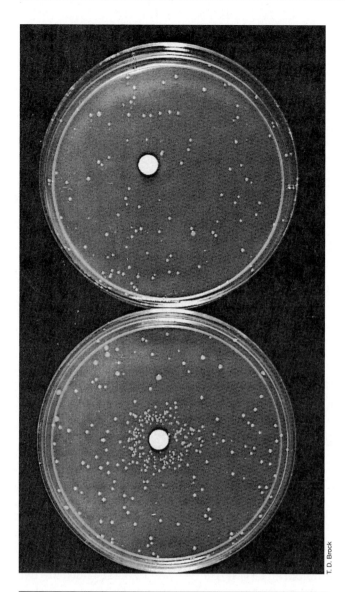

FIGURE 9.8 The Ames test is used to evaluate the mutagenicity of a chemical. Both plates were inoculated with a culture of a histidine-requiring mutant of *Salmonella typhimurium*. The medium does not contain histidine, so only cells that revert back to wild type can grow. Spontaneous revertants appear on both plates, but the chemical on the filter paper disc in the test plate (bottom) has caused an increase in the mutation rate, as shown by the large number of colonies surrounding the disc. Revertants are not seen very close to the disc because the concentration of the mutagen is so high there that it is lethal.

Protocol for an Ames Test

The standard way to test chemicals for mutagenesis has been to determine if the rate of *back* mutation (reversion) in strains of bacteria that are auxotrophic for some nutrient is increased by the suspected mutagen. It is important that the original mutation be a point mutation so reversion can occur. When cells of such an auxotrophic strain are spread on a medium lacking the re-

quired nutrient (for example, an amino acid or vitamin), no growth occurs, and even very large populations of cells can be spread on the plate without formation of visible colonies. However, if back mutants are present, those cells will be able to form colonies. Thus, if 10^8 cells are spread on the surface of a single plate, even as few as 10–20 back mutants (revertants) can be detected by the 10–20 colonies they form. If the back mutation rate has been increased by a chemical mutagen, the number of revertant colonies will also increase. Histidine auxotrophs of *Salmonella typhimurium* (Figure 9.8) and tryptophan auxotrophs of *Escherichia coli* have been the major tools of the Ames test, but a test has also been designed in which the induction of a phage lambda lysogen is used as an assay of DNA damage.

Although the simple testing of chemicals for mutagenesis in bacteria has been carried out for a long time, two elements have been introduced in the Ames test to make it much more powerful. The first of these is the use of strains of bacteria that almost exclusively use error-prone pathways to repair DNA damage (see Section 9.3). The second important element in the Ames test is the use of liver enzyme preparations to convert the chemicals into their active mutagenic (and carcinogenic) forms. It has been well established that many potent carcinogens are not directly carcinogenic or mutagenic but undergo chemical changes in the human body that convert them into active substances. These changes take place primarily in the liver, where enzymes (mixed-function oxygenases) normally involved in detoxification cause formation of epoxides or other activated forms of the compounds, which are then highly reactive with DNA.

In the Ames test, a preparation of enzymes from rat liver is first used to activate the compound. Then the activated complex is taken up on a filter-paper disk, which is placed in the center of a plate on which the proper bacterial strain has been overlaid. After overnight incubation, the mutagenicity of the compound can be detected by looking for a halo of back mutations in the area around the paper disk (Figure 9.8). It is always necessary to carry out this test with several different concentrations of the compound and with appropriate positive and negative controls because compounds vary in their mutagenic activity and are lethal at higher levels. A wide variety of chemicals have been subjected to the Ames test, and it has become one of the most useful prescreens for determining the potential carcinogenicity of a compound.

✓ 9.4 Concept Check

The Ames test employs a sensitive bacterial assay system for detecting chemical mutagens in the environment.

- ✓ Why does the Ames test measure the rate of *back* mutation rather than the rate of *forward* mutation?
- ✓ Of what significance is the detection of mutagens to the prevention of cancer?

Genetic Recombination

Genetic recombination involves the physical exchange of genetic material between genetic elements. In this section we focus on **general** or **homologous recombination,** which results in genetic exchange between *homologous* DNA sequences from two different sources (although such recombination can occur between homologous sequences on the same chromosome). Homologous DNA sequences have the same or nearly the same sequence; therefore, base pairing can occur over an extended length of the two DNA molecules. This type of recombination is involved in the process referred to as "crossing over" in classical genetics.

Homologous recombination is extremely important to all organisms. However, it is also very complex. Even in the bacterium *Escherichia coli* there are at least 25 genes involved. In addition, homologous recombination seems to be of such importance that there are several redundant pathways. Therefore, if one pathway is inhibited or nonfunctional, another may be able to supply necessary functions.

Molecular Events in Homologous Recombination

At the molecular level, recombination has been studied mostly in prokaryotes and viruses. In Bacteria, general recombination involves the participation of a specific protein called the RecA protein, which is specified by the *recA* gene. The RecA protein has been shown to be essential in nearly every homologous recombination pathway. RecA-like proteins have been identified in all prokaryotes examined, including the Archaea, as well as in yeast and in the higher Eukarya.

An overall molecular mechanism of general recombination is shown in Figure 9.9. The process begins with a *nick* (usually generated by a nuclease) in one of the DNA molecules. This nicked strand must be displaced from the other strand by proteins having helicase activity (⚭ Section 6.6). In some pathways specialized enzymes, such as the RecBCD enzyme of *E. coli,* have both nuclease and helicase activities. Single-stranded binding protein (⚭ Section 6.6) then binds to the resulting single-stranded segment. Next, the RecA protein binds to the single-stranded fragment, forming a complex that facilitates annealing with a complementary sequence in the adjacent duplex, simultaneously displacing the resident strand (Figure 9.9). This process is often referred to as *strand invasion*. As noted, recombination involves the *pairing* of DNA molecules over long stretches. Following pairing, *exchange* of homologous DNA molecules can occur, leading to the formation of recombination intermediates containing extensive **heteroduplex** regions, where each strand has originated from a different chromosome. This process also involves DNA polymerase and ligase. These regions may be extended by the RecA protein. Finally, the linked molecules are *resolved* by nucleases and DNA ligase to form two recombinant molecules.

Note that this mechanism for the formation of recombinant DNA structures is a completely natural mechanism that occurs extensively within the cell. Whether or not it leads to the formation of new genotypes depends on whether the two molecules undergoing recombination differ genetically in regions outside the region of recombination. Within limits, general recombination can be thought of as occurring at random sites throughout the genome. Thus, the probability of recombination occurring between two genes is proportional to their distance. This fact is useful for genetic mapping. By recombinational analysis it is possible to map the position of genes on chromosomes because the *farther* two genes are apart, the more likely they are to show recombination.

Until the advent of molecular techniques such as restriction enzyme mapping (⚭ Section 6.5) and DNA sequencing (⚭ Working with Nucleic Acids: The Tools, Chapter 6), recombinational analysis was the only method available for ordering genes on chromosomes and determining the distance between them. It is interesting to note that the order and relative distances as measured by mapping using recombination have been largely confirmed using modern molecular techniques.

For new genotypes to arise as a result of general recombination, it is essential that the two homologous sequences be genetically distinct. This is the case in a diploid eukaryotic cell (see Section 9.13), which has two sets of chromosomes, one from each parent. The two distinct molecules are brought together as the result of sexual reproduction, a process that occurs as part of the regular life cycles of most eukaryotic organisms (see Section 9.13). In prokaryotes, genetically distinct but homologous DNA molecules are brought together in different ways, but the process of genetic recombination is no less important. Recombination can also be critical in the life cycle of some viruses. In Chapter 8 we showed that certain bacteriophages, such as T7 and T4 (⚭ Section 8.11), require homologous recombination as a step during DNA replication.

In prokaryotes genetic recombination is observed because fragments of homologous DNA from a donor chromosome are transferred to a recipient cell by one of three processes: (1) **transformation,** which involves donor DNA free in the environment (see Section 9.6), (2) **transduction,** in which the donor DNA transfer is mediated by a virus (see Section 9.7), and (3) **conjugation,** in which the transfer involves cell-to-cell contact

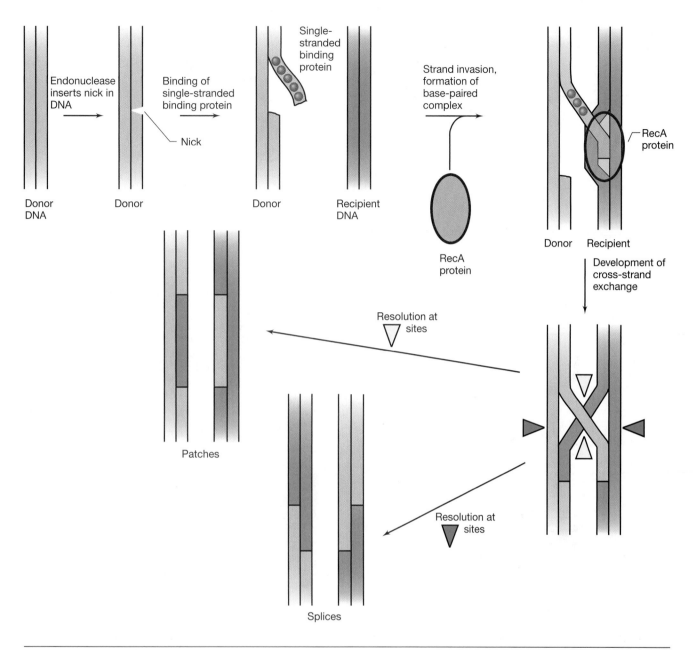

FIGURE 9.9 A simplified version of one molecular mechanism of genetic recombination. Homologous DNA molecules pair and exchange DNA segments. The mechanism involves breakage and reunion of paired segments. Two of the proteins involved, a single-stranded binding protein and the RecA protein, are shown. The diagram is not to scale: Pairing can occur over hundreds or thousands of bases. Note that there are two possible outcomes, depending on which strands are cut during the resolution process. In one outcome the recombinant molecules have patches, whereas in the other the two parental molecules appear to have been cut and then spliced together.

and a *conjugative plasmid* in the donor cell (see Section 9.9). These processes are contrasted in Figure 9.10 and will be discussed in detail later in this chapter.

It is *after* the transfer, when the DNA fragment from the host is in the recipient cell, that homologous recombination may occur. Because in prokaryotes only a chromosomal fragment is transferred, if recombination does not occur, the fragment will be lost because it cannot replicate independently. Therefore, it is important to

remember that in prokaryotes transfer is just the first step in obtaining recombinant organisms.

Detection of Recombination

In order to detect physical exchange of DNA segments, the cells resulting from recombination must be phenotypically different from the parents. In crosses involving microorganisms, one must usually use as recipients

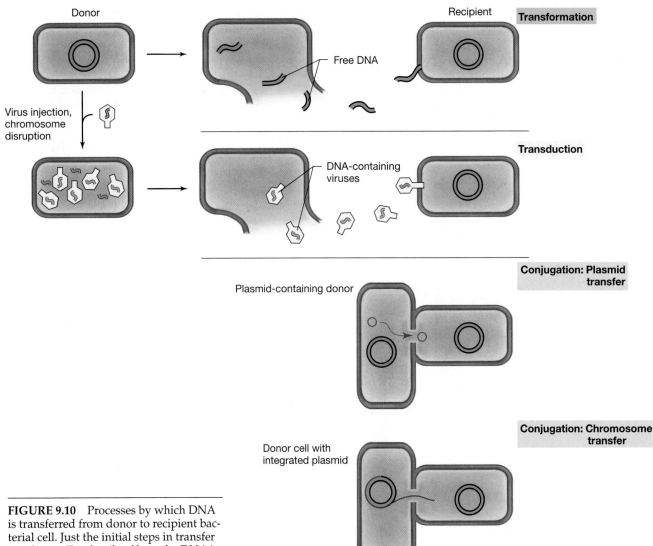

FIGURE 9.10 Processes by which DNA is transferred from donor to recipient bacterial cell. Just the initial steps in transfer are shown. For details of how the DNA is integrated into the recipient, see text.

strains that lack some selectable characteristic that the recombinants will possess. For instance, the recipient may not be able to grow on a particular medium, and genetic recombinants are selected that can. Various kinds of selectable and nonselectable markers (such as drug resistance, nutritional requirements, and so on) were discussed in Section 9.1. The exceedingly great sensitivity of the selection process is shown by the fact that 10^8 or more bacterial cells can be spread on a single plate and, if proper selective conditions are used, no parental colonies will appear, whereas even a few recombinants can form colonies (Figure 9.11). The only requirement is that the *reverse* mutation rate for the selected characteristic be low, because revertants will also form colonies. This problem can often be overcome by using double mutants because it is very unlikely that two back mutations will occur in the same cell. Much of

the skill of the bacterial geneticist is exhibited in the choice of proper mutants and selective media for efficient detection of genetic recombination. Because selection is so powerful and because crosses can be made using billions of individual cells, recombinational analysis is a very important tool to the microbial geneticist.

Complementation

When two mutant strains are genetically crossed (mated), homologous recombination can yield a wild-type recombinant unless both of the mutations include changes in exactly the same base pairs. Therefore, if two different Trp$^-$ *Escherichia coli* (strains that require the amino acid tryptophan in the medium) are crossed and Trp$^+$ recombinants are obtained, it is clear that the mutations in the two strains did not include the same base pairs. Howev-

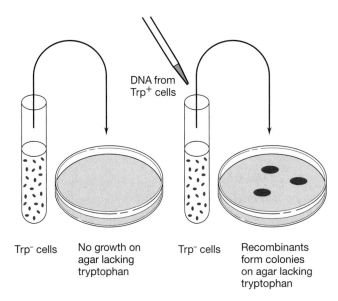

FIGURE 9.11 How a selective medium can be used to detect rare genetic recombinants among a large population of nonrecombinants. On the selective medium only the rare recombinants form colonies. Procedures such as this, which offer high resolution for genetic analyses, can ordinarily be used only with microorganisms.

er, this experiment cannot detect whether the mutations were in the same gene. This can be determined by a type of experiment called a **complementation test.**

Complementation was first used in *diploid* eukaryotic organisms. In diploid organisms the cell has *pairs* of chromosomes, one member from each parent (see Section 9.13). When the two mutations are present on separate members of a pair (that is, one mutation from each parent), they are said to be in **trans** configuration. On the other hand, if the two mutations are on the *same* chromosome (both from the same parent), then they are said to be in **cis** configuration. True diploidy does not exist in prokaryotes. For complementation tests a diploid state is achieved for a region of the chromosome by transfer of a chromosomal fragment from a donor (see earlier discussion). However, the principle of the test and the nomenclature (*cis* and *trans*) is the same.

If one of the tryptophan genes has been inactivated in a particular strain by a mutation, leading to the Trp⁻ phenotype, then putting a copy of the wild-type gene from another strain into the same cell should restore the wild-type phenotype of the cell, that is, the resulting partially diploid cell should be Trp⁺. (This will be true unless the mutation is *dominant.*) One could then say that the wild-type gene *complements* the mutation. Even though in prokaryotes only a chromosomal fragment is transferred, this fragment will typically be large enough to contain many genes. Remember that the genes for the biosynthesis of tryptophan form an operon (⚬⚬ Figure 7.16), and therefore one can easily transfer the entire

operon from the donor cell. Note then that even if the operon in the donor also has a mutation, it will complement the mutated gene in the donor *if the two mutations are in different genes.* The two mutations are then said to *complement* each other. This is shown diagrammatically in Figure 9.12. Mutations can complement

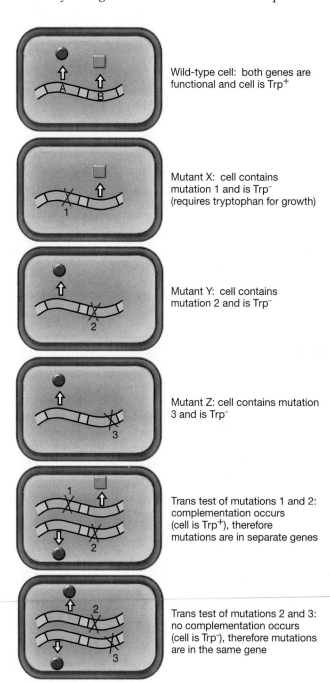

FIGURE 9.12 Complementation analysis. Mutations 1, 2, and 3 each lead to the same phenotype, a requirement for tryptophan. Complementation analysis indicates that mutations 2 and 3 are in one gene and that mutation 1 is in another. The protein products of both genes (A and B) must be required to synthesize tryptophan.

because the genes encode proteins that can diffuse through the cytoplasm of the cell. (One cannot complement mutations in regulatory sites such as promoters because these function at the DNA level.) If each homologous DNA molecule contributes a different required gene, then the cell will have all the enzymes it requires to synthesize tryptophan. Notice that complementation *does not* involve recombination. To do the test, the mutations must be in *trans*. (If one molecule has both mutations, the other is wild type and should be sufficient itself to confer the wild-type phenotype unless one of the mutations is *dominant*. Therefore, having the mutations in cis serves as a control.)

This type of complementation test, called a *cis–trans test*, is used to define whether two mutations are in the same genetic (functional) unit. The genetic unit defined by the cis–trans test is sometimes called a **cistron** (a term essentially equivalent to a gene). As noted, two mutations in the *same* cistron *cannot* complement each other, and so when complementation is found to exist, this implies that the two mutations lie in *different* cistrons (that is, different genes). The term *cistron* is now rarely used except when describing whether an mRNA has the genetic information from one gene (monocistronic mRNA) or from more than one gene (polycistronic mRNA) (⌐⊃ Section 6.8).

✓ 9.5 Concept Check

Homologous recombination arises when closely related sequences from two genetically distinct elements are combined together in the same element. Recombination is an important evolutionary process, and cells have specific mechanisms for ensuring that recombination takes place. In eukaryotes, genetic recombination occurs as a consequence of the sexual cycle. Mechanisms of recombination also occur in prokaryotes but involve DNA transfer during the processes of transformation, transduction, and conjugation.

✓ What protein, found in all prokaryotes, facilitates the pairing required for homologous recombination?
✓ Complementation tests do not involve recombination. Explain.

9.6

Genetic Transformation

As we have noted, genetic transformation is a process by which free DNA is incorporated into a recipient cell and brings about genetic change. The discovery of genetic transformation in bacteria was one of the outstanding events in biology, as it led to experiments demonstrating that DNA is the genetic material (see the box, Origins of Bacterial Genetics). This discovery became the keystone of molecular biology and modern genetics.

A number of prokaryotes have been found to be naturally transformable, including certain species of both gram-negative and gram-positive Bacteria and some species of Archaea. However, even within transformable genera, only certain strains or species are transformable. Since the DNA of prokaryotes is present in the cell as a large single molecule, when the cell is gently lysed, the DNA pours out (Figure 9.13). Because of its extreme length (1700 μm in *Bacillus subtilis*), the DNA molecule breaks easily; even after gentle extraction the *B. subtilis* chromosome of 4.2 megabase pairs is converted to fragments of about 15 kilobase pairs. Because the DNA that corresponds to an average gene is about 1000 nucleotides, each of the fragments of purified DNA has about 15 genes. A single cell usually incorporates only one or a few DNA fragments so only a small proportion of the genes of one cell can be transferred to another by a single transformation event.

Competence

A cell that is able to take up a molecule of DNA and be transformed is said to be **competent.** Only certain strains are competent; the ability seems to be an inherited property of the organism. Competence in most naturally transformable bacteria is regulated, and special proteins play a role in the uptake and processing of DNA. These competence-specific proteins may include a membrane-associated DNA binding protein, a cell

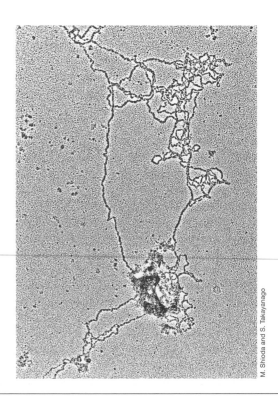

FIGURE 9.13 The prokaryotic chromosome, as shown in the electron microscope. The circular chromosome is from the hyperthermophile *Sulfolobus,* a member of the Archaea (⌐⊃ Section 14.8). See also Figure 9.18.

wall autolysin, and various nucleases. One pathway to natural competence in *Bacillus subtilis* is part of a quorum-sensing system (a regulatory system that responds to cell number, ∞ Section 7.6) regulated by a two-component regulatory system (∞ Section 7.7). Cells produce and excrete a small peptide during growth and become competent in high concentrations of this peptide, through the action of ComP, a sensory protein, and ComA, a response regulator protein. ComA is an activator protein (∞ Section 7.3) that regulates a number of genes affecting transformation. In *Bacillus*, about 20% of the cells become competent and stay that way for several hours. However, in *Streptococcus*, 100% of the cells can become competent, but only for a few minutes during the growth cycle.

Uptake of DNA

Bacteria differ in the form in which DNA is taken up. In *Haemophilus*, which is gram-negative, for example, only double-stranded DNA is taken up into the cell despite the fact that only single-stranded segments actually become incorporated into the genome by recombination. In the gram-positive Bacteria *Streptococcus* and *Bacillus*, by contrast, only a single DNA strand is taken up, while the complementary strand is simultaneously degraded. However, in all cases, double-stranded DNA binds more effectively to the cells.

During the transformation process, competent bacteria first bind DNA reversibly; soon, however, the binding becomes irreversible. Competent cells bind much more DNA than do noncompetent cells—as much as 1000 times more. As we noted earlier, the sizes of the transforming fragments are much smaller than that of the whole genome, and this DNA is further degraded during the uptake process. In *Streptococcus pneumoniae* each cell can bind only about 10 molecules of double-stranded DNA of 15–20 kilobase pairs each. However, as they are taken up, they are converted to single-stranded pieces of about 8 kilobases. The DNA fragments in the mixture compete with each other for uptake, and if excess DNA that does not contain the genetic marker is added, a decrease in the number of transformants occurs. In preparations of transforming DNA, only about 1 out of 100–200 DNA fragments contains the marker being studied. Thus, at high concentrations of DNA, the competition between DNA molecules results in saturation of the system so even under the best conditions it is impossible to transform all the cells in a population for a given genetic marker. The maximum frequency of transformation that has so far been obtained is about 20% of the population; actually the values usually obtained are between 0.1 and 1.0%. The minimum concentration of DNA yielding detectable transformants is about 0.00001 µg/ml (1×10^{-5} µg/ml), which is so low that it is undetectable chemically.

Interestingly, in *Haemophilus influenzae* there is a requirement that the DNA fragment have a particular 11-base-pair sequence for irreversible binding and uptake to occur. This sequence is found at an unexpectedly high frequency in the *Haemophilus* chromosome. Evidence such as this, and the fact that at least certain bacteria become competent in their natural environment, suggest that transformation is not a laboratory artifact but plays an important role in gene transfer in nature.

Integration of Transforming DNA

Transforming DNA is bound at the cell surface by a DNA binding protein, after which either the entire double-stranded fragment is taken up or a nuclease degrades one strand and the other is taken up (Figure 9.14). After uptake, the DNA associates with a competence-specific protein that remains attached to the DNA, presumably preventing it from nuclease attack, until it reaches the chromosome where RecA protein takes over. The DNA is then integrated into the genome of the recipient by recombinational processes (Figure 9.14; see also Figure 9.9). During replication of this heteroduplex DNA, one parental and one recombinant DNA molecule are formed. On segregation at cell division, the latter is present in the transformed cell, which is now genetically altered as compared to the parental type. The preceding discussion pertains to only small pieces of *linear* DNA. Many naturally transformable Bacteria are transformed only poorly by plasmid DNA because the plasmid must remain double-stranded and circular in order to replicate.

Transfection

Bacteria can be transformed with DNA extracted from a *bacterial virus* rather than from another bacterium, a process known as **transfection.** If the DNA is from a lytic bacteriophage, transfection leads to normal virus production and can be measured by the standard phage plaque assay (∞ Section 8.4). Transfection has become a useful tool in studying the mechanism of transformation and recombination because the small size of phage genomes allows for the isolation of a nearly homogeneous population of DNA molecules. By contrast, in conventional transformation, the transforming DNA is generally a random assortment of chromosomal DNA of various lengths, and this tends to complicate experiments designed to study the mechanism of transformation.

Artificially Induced Competence

High efficiency natural transformation is found only in a few bacteria; *Acinetobacter, Azotobacter, Bacillus, Streptococcus, Haemophilus, Neisseria,* and *Thermus,* for example, are easily transformed. Many prokaryotes are transformed only poorly or not at all under natural conditions. Deter-

Although genetic recombination in eukaryotes had been known for a long time, the discovery of genetic recombination in bacteria following transformation, transduction, and conjugation has been a relatively recent event. Of the three processes, the discovery of transformation was the most significant as it provided the first direct evidence that DNA is the genetic material. The first evidence of bacterial transformation was obtained by the British scientist Fred Griffith in the late 1920s. Griffith was working with *Streptococcus pneumoniae* (pneumococcus), a bacterium that owes its ability to invade the body in part to the presence of a polysaccharide capsule. Mutants can be isolated that lack this capsule and are thus unable to cause infection; such mutants are called R strains because their colonies appear rough on agar, in contrast to the smooth appearance of capsulated strains. A mouse infected with only a few cells of a smooth (S) strain succumbs in a day or two to pneumococcus infection, whereas even large numbers of R cells do not cause death when injected. Griffith showed that if heat-killed S cells were injected along with living R cells, a fatal infection ensued and the bacteria isolated from the dead mouse were S types. A number of different polysaccharide capsules were known in different pneumococcus S strains, and it was possible to do this experiment with heat-killed S cells from a type different from that from which the R strain was derived. Since the isolated living S cells always had the capsule type of the heat-killed S cells, the R cells had been transformed into a new type, and the process had all the properties of a genetic event. The molecular explanation for the transformation of pneumococcus types was provided by Oswald T. Avery and his associates at Rockefeller Institute in New York in a series of studies carried out during the 1930s, culminating in the now classic paper by Avery, C. M. MacLeod, and M. McCarty in 1944. Avery and his coworkers showed that under certain conditions the transformation process could be carried out in the test tube rather than the mouse and that a cell-free extract of heat-killed cells could induce transformation. In a long series of painstaking biochemical experiments, the active fraction of cell-free extracts was purified and was shown to consist of DNA. The transforming activity of purified DNA preparations was very high, and only very small amounts of material were necessary. Subsequently, others at Rockefeller showed that transformation could occur in pneumococcus not only for capsular characteristics but for other genetic characteristics of the organism as well, such as antibiotic resistance and sugar fermentation.

In 1953, James Watson and Francis Crick announced their model of the structure of DNA, providing a theoretical framework for how DNA could serve as the genetic material. Thus, two types of studies, the bacteriological and biochemical ones of Avery and the physical-chemical ones of Watson and Crick, solidified the concept of DNA as the genetic material. In the subsequent years, this work has led to the whole field of molecular genetics.

Although bacterial transformation resulted from an essentially accidental discovery, bacterial conjugation was initially shown to occur by Joshua Lederberg and E. L. Tatum in 1946 in experiments carefully designed to determine if a sexual process might occur in bacteria. Because it appeared that the process, if present, would be quite rare (no microscopic evidence for bacterial mating had ever been seen, although such evidence can easily be obtained in eukaryotes), Lederberg developed a method involving the use of nutritional mutants of *Escherichia coli*. Fortunately, these mutants had been isolated in strain K-12, one of the few wild-type strains now known to contain the F plasmid. The principle was to mix two strains, one requiring biotin and methionine, the other requiring threonine and leucine, and plate the mixture on a minimal medium lacking all four growth factors. Neither parental type could grow on this medium, but any recombinants could, and when about 10^8 cells were plated, a small but significant number of colonies was obtained. Strains with two separate nutritional requirements were employed because it would be unlikely that spontaneous back mutation of both genes would occur in a sin-

a For further reading: Brock, T. D. 1990. *The Emergence of Bacterial Genetics*. Cold Spring Harbor Laboratory Press, Cold Spring Harbor, NY.

mination of how to induce competence in such bacteria may involve considerable empirical study, with variation in culture medium, temperature, and other factors. However, to transfer DNA into cells for genetic engineering (∞ Chapter 10), it was necessary to find a way to make *Escherichia coli*, a gram-negative organism, competent. It has been found that when *E. coli* is treated with high concentrations of calcium ions and then stored in the cold, it becomes transformable at low efficiency. *Escherichia coli* treated in this manner takes up double-stranded DNA, and therefore transformation by plasmid DNA is relatively efficient. Why the calcium treatment works is not known, but this procedure also works with some other gram-negative Bacteria. However, methods such as this

O.T. Avery

E. L. Tatum (left) and J. Lederberg in 1947

gle cell. Thus, the only explanation for the phenomenon was some sort of genetic recombination. To show that the process required cell-to-cell contact, and hence could not be a type of transformation, it was shown that when culture filtrates or extracts were separated by a sintered glass disc, permeable to macromolecules but not to cells, recombination did not occur. Although initially conjugation appeared to be a very rare event, by the early 1950s a strain of *E. coli* had been isolated by the Italian scientist L. L. Cavalli-Sforza, while he was working in Lederberg's laboratory, that showed a high frequency of recombination. The British physician William Hayes, who independently isolated an Hfr strain, then showed that genetic transfer during mating between Hfr and F$^-$ was a one-way event, with the Hfr serving as donor. The interrupted mating experiment and the demonstration of the circular genetic map of *E. coli* were then carried out by Elie Wollman and Francois Jacob, working with Jacques Monod at the Pasteur Institute in Paris. The distinction between Hfr and F$^+$ was made by Lederberg, who also showed that F$^+$ behaved in an infectious manner. Lederberg coined the term *plasmid* in the 1950s to describe such apparently extrachromosomal genetic elements, although it did not find wide usage until the 1970s when infectious drug resistance became a major medical problem.

Bacterial transduction was discovered by the American scientist Norton Zinder when he was a graduate student working with Lederberg at the University of Wisconsin on genetic recombination in *Salmonella typhimurium*. The original motivation for this work was to show that conjugation occurred in an organism other than *E. coli*, and the techniques involved isolation of mutants and quantification of recombination by observing colony growth on minimal medium. However, although evidence of recombination was obtained, it could be shown that cell-to-cell contact was *not* required. Although this suggested a type of transformation, the process was not affected by DNase, and the gene transfer agent behaved like a bacteriophage. The gene transfer agent could be purified by the same procedures used to purify virus particles, and transduction occurred only with recipient cells that had receptor sites for the virus in question. Further, transducing activity could be eliminated by treatment of a lysate with substances able to adsorb the virus, such as sensitive cells or antibodies. Thus, in all cases, transducing activity and virus activity behaved in similar ways. Zinder and Lederberg coined the word *transduction* to refer to any genetic recombinational process that was only fragmentary and did not involve cell-to-cell contact, intending in this way to encompass processes involving either free DNA (transformation) or phages, but subsequently the word *transduction* has been applied only to virus-mediated genetic transfer. ■

for artificial induction of competence are being supplanted by a new method termed *electroporation*.

DNA Transfer by Electroporation

Small pores are produced in the membranes of cells exposed to pulsed electric fields. When DNA molecules are present outside the cells during the electric pulse, they can then enter the cells through these pores. This process is called **electroporation**. Electroporation requires a sophisticated power supply because the pulses must be carefully controlled and last for only milliseconds. This technique has been used to transport DNA into a large number of different species of prokaryotes, both Archaea and Bacte-

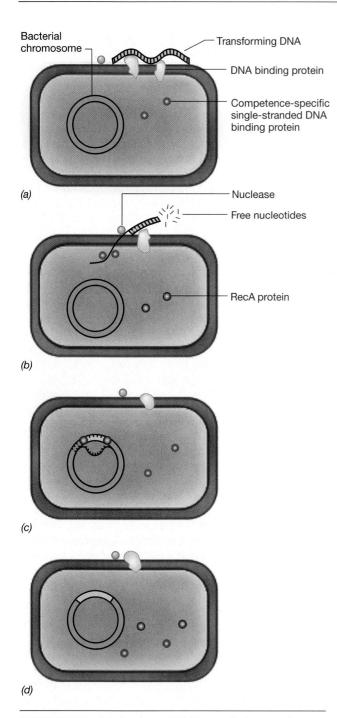

(a)

(b)

(c)

(d)

Labels: Bacterial chromosome; Transforming DNA; DNA binding protein; Competence-specific single-stranded DNA binding protein; Nuclease; Free nucleotides; RecA protein

FIGURE 9.14 Mechanism of DNA transfer by transformation in a gram-positive bacterium. (a) Binding of free DNA by a membrane-bound DNA binding protein. (b) Passage of one of the two strands into the cell while nuclease activity degrades the other strand. (c) The single strand in the cell is bound by specific proteins, and recombination with homologous regions of the bacterial chromosome mediated by RecA protein occurs. (d) Transformed cell.

ria. Additionally, electroporation allows an experimenter to transfer a plasmid directly from one cell to another if both are present during electroporation. Therefore, electroporation allows small molecules of DNA to come out

of cells as well as to go in! This type of "transformation" eliminates the steps required to isolate the plasmid from the first strain before introducing it into the second.

Transformation (Transfection) of Eukaryotic Cells

Eukaryotic microorganisms and animal and plant cells can take up DNA in a process that resembles bacterial transformation. Because the word *transformation* in mammalian cells is used to describe the conversion of cells to the malignant (tumorous, cancerous) state (⚭ Section 8.14), the introduction of DNA into mammalian cells has been called *transfection* (a term with another meaning in bacterial systems; see earlier discussion).

Transfection of cultured animal cells was originally accomplished by precipitating DNA in such a way that the cells would take it up by phagocytosis (⚭ Section 20.3) because they do not have cell walls. In yeast, which is an important organism for genetic engineering, transfection at low efficiencies can be mediated by various methods of inducing artificial competence. However, as in the case of prokaryotes, electroporation is becoming widely applied to all types of eukaryotic cells and can be used whether or not the cell wall is removed.

In addition to electroporation, a high velocity microprojectile "gun" has been developed for incorporating DNA into cells. The original **particle gun** operates somewhat like a conventional shotgun. A small steel cylinder containing a gunpowder charge is used to fire nucleic acid–coated particles at the target cells (Figure 9.15). The particles bombard the cell, piercing cell walls and membranes without actually killing the cells. The nucleic acid entering the cells can then recombine with host DNA. The particle gun has been used successfully to transfect yeast, algae, a variety of plant cells, and even mitochondria and chloroplasts. The particle gun is very useful because, unlike electroporation, it can be used on intact tissue such as plant seeds.

✓ 9.6 Concept Check

Certain prokaryotes exhibit competence, a state in which cells are able to take up free DNA released by other bacteria. This process is called transformation. Relatively few species of prokaryotes can be naturally transformed. However, certain laboratory procedures have been developed that make it possible to introduce DNA into completely unrelated organisms, even eukaryotes. Electroporation involves modification of the cytoplasmic membrane by treatment with an electric field to facilitate DNA uptake.

- ✓ The donor bacterial cell in a transformation is probably dead. Explain.
- ✓ Even in naturally transformable cells competency is usually inducible. What does this mean?

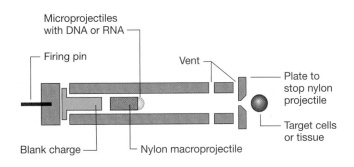

Microprojectiles with DNA or RNA

Firing pin

Vent

Plate to stop nylon projectile

Target cells or tissue

Blank charge

Nylon macroprojectile

FIGURE 9.15 Nucleic acid gun for transfection of eukaryotic cells. The inner workings of the gun show how nucleic acids attached to metal pellets are projected at target cells.

9.7

Transduction

In transduction, DNA is transferred from cell to cell through the agency of viruses. Genetic transfer of host genes by viruses can occur in two ways. In the first, called **generalized transduction,** host DNA derived from virtually any portion of the host genome becomes a part of the DNA of the mature virus particle in place of the virus genome. The second, called **specialized transduction,** occurs only in some temperate viruses; DNA from a specific region of the host chromosome is integrated directly into the virus genome—usually replacing some of the virus genes. The transducing virus particle in both specialized and generalized transduction is usually *defective* as a virus because bacterial genes have replaced some necessary viral genes.

In generalized transduction, if the donor genes do not undergo homologous recombination with the recipient bacterial chromosome, they will be lost. They cannot replicate independently and are not part of a viral genome. In specialized transduction homologous recombination may also occur. However, since the donor bacterial DNA is now actually a part of a temperate phage genome, there are two other possibilities: (1) the DNA may be integrated into the host chromosome during lysogenization, and (2) the DNA may be replicated in the recipient as part of a lytic infection.

Transduction has been found to occur in a variety of prokaryotes, including certain species of the Bacteria: *Desulfovibrio, Escherichia, Pseudomonas, Rhodococcus, Rhodobacter, Salmonella, Staphylococcus,* and *Xanthobacter,* as well as the archaean *Methanobacterium thermoautotrophicum.* Not all phages can transduce, and not all bacteria are transducible; but the phenomenon is sufficiently widespread for us to assume that it plays an important role in genetic transfer in nature.

Generalized Transduction

In generalized transduction, virtually any genetic marker can be transferred from donor to recipient. Generalized transduction was first discovered and extensively studied in the bacterium *Salmonella typhimurium* with phage P22 and has also been studied with phage P1 in *Escherichia coli.* An example of how *transducing particles* may be formed is given in Figure 9.16. When the population of sensitive bacteria is infected with a phage, the events of the phage lytic cycle may be initiated. During a lytic infection, the enzymes responsible for packaging viral DNA into the bacteriophage sometimes accidentally package host DNA. The resulting particle is called a *transducing particle.* On lysis of the cell, these particles are released along with normal virions, and so the lysate contains a mixture of normal virions and transducing particles. Because transducing particles cannot lead to a normal viral infection (they contain no viral DNA), they are said to be *defective.* When this lysate is used to infect a population of recipient cells, most of the cells become infected with normal virus. However, a small proportion of the population receives transducing particles that inject the DNA they received from the previous host bacterium. While this DNA cannot replicate, it can undergo genetic recombination with the DNA of the new host. Because only a small proportion of the particles in the lysate are of the defective transducing type and each of these contains only a small fragment of donor DNA, the probability of a transducing particle containing a particular gene is quite low, and usually only about 1 cell in 10^6–10^8 is transduced for a given marker.

Phages that form transducing particles can be either temperate or virulent, the main requirements being that they have a DNA packaging mechanism that permits accidental recognition of host DNA and that packaging occurs before the host genome is completely degraded. The detection of transduction is most certain when the multiplicity of phage to host is low, so a host cell is infected with only a single phage particle; with multiple infection, the cell will likely be killed by the normal virions in the lysate.

Specialized Transduction

Generalized transduction allows the transfer of DNA from one bacterium to another at a low frequency. However, specialized transduction can allow extremely effi-

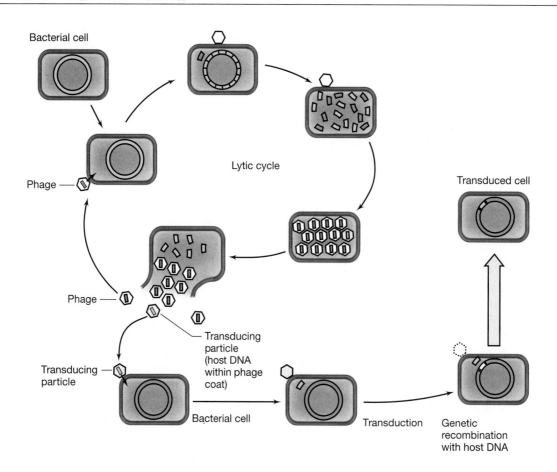

Bacterial cell

Lytic cycle

Phage

Phage

Transducing particle

Transducing particle (host DNA within phage coat)

Bacterial cell

Transduced cell

Transduction

Genetic recombination with host DNA

FIGURE 9.16   Generalized transduction: one possible mechanism by which virus (phage) particles containing host DNA can be formed.

cient transfer while also allowing a small region of a bacterial chromosome to be replicated independently of the rest. The example we shall use to discuss specialized transduction was the first to be discovered and involves transduction of the galactose genes by the temperate phage lambda of *Escherichia coli.*

As we discussed (Section 8.12), when a cell is lysogenized by lambda, the phage genome becomes integrated into the host DNA at a specific site. The region in which lambda integrates is immediately adjacent to the cluster of host genes that control the enzymes involved in galactose utilization (Figure 8.29), and the DNA of lambda is inserted into the host DNA at that site. From then on, viral DNA replication is under host control. On induction (for example, by ultraviolet radiation), the viral DNA separates from the host DNA by a process that is the reverse of integration (Figure 9.17). Ordinarily when the lysogenic cell is induced, the lambda DNA is excised as a unit. Under rare conditions, however, the phage genome is excised incorrectly. Some of the adjacent bacterial genes (for example the galactose operon) are excised along with phage DNA. At the same time, some phage genes are left behind. One type of altered phage particle, called **lambda dgal** or λ*dgal* (*dgal* means "defective, galactose"), is defective because of the phage genes lost and does not make mature phage. However, a **helper phage** can provide those functions missing in the defective particles. This "helper" is identical to the original lambda. Thus, the culture lysate obtained contains a few λ*dgal* particles mixed in with a large number of normal lambda virions.

When a galactose-negative bacterial culture is infected at high multiplicity with such a lysate and *gal*⁺ transductants are selected, many are double lysogens, carrying both lambda and λ*dgal*. (Note then that the bacterium has become a diploid for the *gal* region. Specialized transducing phage can be used in performing complementation tests in bacteria; see Section 9.5.) When such a double lysogen is induced, a lysate is produced containing about equal numbers of lambda and λ*dgal*. Such a lysate can transduce at high efficiency, although only for a restricted group of *gal* genes.

If the phage is to be viable, there is a maximum limit to the amount of phage DNA that can be replaced with host DNA, because sufficient phage DNA must be retained in order to provide the information for production of the phage protein coat and for other phage

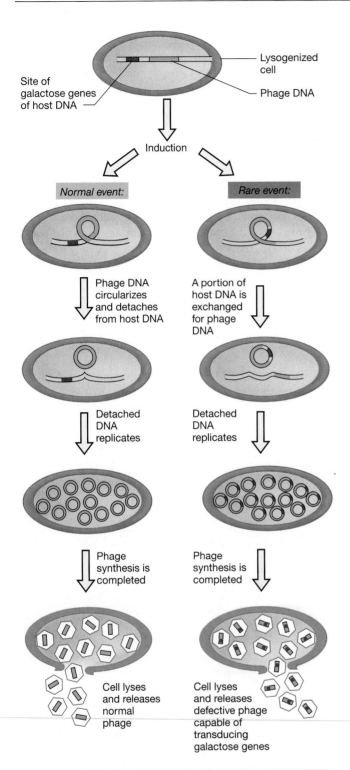

FIGURE 9.17 Normal lytic events and the production of particles transducing the galactose genes in an *Escherichia coli* cell containing a lambda prophage.

proteins needed for lysis and lysogenization. However, if a helper phage is used together with the defective phage in a mixed infection, then even less information is needed in the defective phage for transduction. Only the *att* (attachment) region, the *cos* site (cohesive ends, for packaging), and the replication origin of the lambda genome are needed for production of a transducing particle, provided a helper is used (see the genetic map of lambda ∞ Section 8.12).

One important distinction between specialized and generalized transduction is in how the transducing lysate can be formed. In specialized transduction this *must* occur by induction of a lysogen, whereas in generalized transduction it can occur either in this way or by infection of a nonlysogen by the phage, with subsequent phage replication and cell lysis.

Although we have discussed specialized transduction only in the lambda-*gal* system, phage lambda and its relative, φ80, have been widely used to form specialized transducing phages covering many specific regions of the *E. coli* genome. In addition, lambda specialized transducing phage can be constructed by the techniques of genetic engineering to contain genes from any organism (∞ Section 10.3).

Specialized transduction in Bacteria is similar to the situation we discussed with retroviruses (∞ Section 8.22) where certain host genes can be incorporated into the retroviral genome and be passed along (often in modified form) to subsequent hosts. These have been detected in retroviruses because the genes involved are *oncogenes*, involved in the development of cancer.

Phage Conversion

This is a phenomenon analogous in some ways to specialized transduction. When a normal temperate phage (that is, a nondefective one) lysogenizes a cell and its DNA is converted to the prophage state, the lysogen is immune to further infection by the same type of phage. This acquisition of immunity can be considered a change in phenotype. In certain cases other phenotypic alterations can be detected in the lysogenized cell, which seem to be unrelated to the phage immunity system. Such a change, which is brought about through lysogenization by a normal temperate phage, is called **phage conversion.**

Two cases of conversion have been especially well studied. One involves a change in structure of a polysaccharide on the cell surface of *Salmonella anatum* on lysogenization with phage ε[15]. The second involves the conversion of nontoxin-producing strains of *Corynebacterium diphtheriae* (the bacterium that causes the disease diphtheria) to toxin-producing (pathogenic) strains on lysogenization with phage β (∞ Section 23.2). In these situations the information for production of these new materials is apparently an integral part of the phage genome and hence is automatically and exclusively transferred on infection by the phage and lysogenization.

Lysogeny probably carries a strong selective value for the host cell because it confers resistance to infection by viruses of the same type. Phage conversion seems also to be of considerable evolutionary significance as it results in efficient genetic alteration of host cells. Many bacteria isolated from nature are lysogens. It seems reasonable to conclude, therefore, that lysogeny is the normal state of affairs and may often be essential for survival of the host in nature.

✓ 9.7 Concept Check

Transduction involves transfer of host genes from one bacterium to another by viruses. In generalized transduction, defective virus particles randomly incorporate fragments of the cell's chromosomal DNA; virtually any gene of the donor can be transferred, but the efficiency is low. In specialized transduction, the DNA of a temperate virus excises incorrectly and brings adjacent host genes along with it; only genes close to the integration point of the virus are transduced, but the efficiency may be high.

- ✓ What is the important difference between generalized transduction and transformation?
- ✓ In specialized transduction the donor DNA can replicate inside the recipient cell without homologous recombination taking place, but this is not true in generalized transduction. Explain.

9.8

Plasmids

Before we discuss the third method of genetic transfer, **conjugation,** we must discuss another kind of genetic element called the **plasmid.** Plasmids are genetic elements that replicate independently of the host chromosome (∞ Section 6.4).

Unlike viruses, plasmids do not have an extracellular form and exist inside cells simply as nucleic acid. However, distinguishing between viruses and plasmids can sometimes present difficulties. As we have discussed, the prophage form of some temperate viruses, such as bacteriophage P1, replicates independently of the host chromosome in a fashion analogous to plasmid replication (∞ Section 8.12).

We noted earlier (∞ Section 6.4) that it is possible to differentiate between a plasmid and a host chromosome in prokaryotes because plasmids do not carry genes required by the host under all conditions. This can be difficult to prove, and therefore it may sometimes be difficult to distinguish between chromosomes and very large plasmids in prokaryotes.

In spite of these few difficulties, literally thousands of different types of plasmids are known. Indeed, over 300 different naturally occurring plasmids have been isolated from strains of *Escherichia coli* alone. In this section we shall discuss the properties of a few of them.

Physical Nature of Plasmids

Almost all of the known plasmids are double-stranded DNA. Most plasmids are circular, but many linear plasmids are also known. Naturally occurring plasmids vary in size from approximately 1 to more than 1000 kilobase pairs. The typical plasmid is a circular double-stranded DNA molecule less than $\frac{1}{20}$ the size of the chromosome (Figure 9.18). Most of the plasmid DNA isolated from cells is in the supercoiled configuration, which is the most compact form within the cell (∞ Figure 6.10).

Isolation of plasmid DNA can generally be readily accomplished by making use of certain physical properties of supercoiled DNA molecules, which can be separated from other types of DNA using an ultracentrifuge. Although chromosomes are also supercoiled inside the cell, isolation of chromosomal DNA almost always leads to breakage of the strands and consequent loss of supercoiling. Plasmid DNA molecules of different sizes can also be readily separated by electrophoresis on agarose gels (∞ Working with Nucleic Acids: The Tools, Chapter 6). Plasmid DNA can be observed under the electron microscope (Figure 9.18).

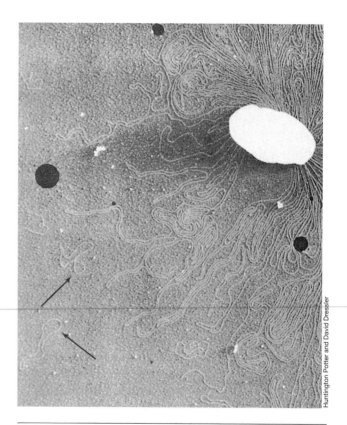

FIGURE 9.18 The bacterial chromosome and bacterial plasmids, as shown in the electron microscope. The plasmids (arrows) are the circular structures, much smaller than the main chromosomal DNA. The cell (large, white structure) was broken gently so the DNA would remain intact.

Replication of Plasmids

The enzymes involved in plasmid replication are normal cell enzymes, so the genes carried by the plasmid itself that control its replication are concerned primarily with control of the timing of the initiation process and with apportionment of the replicated plasmids between daughter cells. Also, different plasmids are present in cells in a particular number of plasmid molecules per cell; this is called the *copy number.* Some plasmids are present in the cell in only 1–3 copies, whereas others may be present in over 100 copies. Copy number is controlled by genes on the plasmid and by interactions between the host and the plasmid.

Most plasmids in gram-negative Bacteria replicate in a manner similar to that already described for the chromosome (∞ Section 6.6). This involves initiation of replication at an origin and bidirectional replication around the circle, giving a *theta* intermediate. However, some plasmids have *unidirectional* replication. Because of the small size of plasmid DNA relative to the chromosome, the whole replication process occurs very quickly, perhaps in $\frac{1}{10}$ or less of the total time of the cell division cycle.

Most plasmids of gram-positive Bacteria replicate by a rolling circle mechanism similar to that used by the phage φX174 (∞ Section 8.9 and Figure 8.17). This mechanism gives rise to a single-stranded intermediate, and thus these plasmids are sometimes referred to as *single-stranded DNA plasmids.* Most of the linear plasmids now known replicate using a mechanism involving a protein bound to the 5′-end of each strand that is used in priming DNA synthesis (∞ Figures 6.22 and 8.45).

Some individual bacterial cells may also contain several different types of plasmids, for example, *Borrelia burgdorferi* contains 17 different circular and linear plasmids. The ability of two different plasmids to both replicate in the same cell is also controlled by plasmid genes (called *inc*) involved in controlling DNA replication. When a plasmid is transferred into a cell that already carries another plasmid, a common observation is that the second plasmid may not be maintained and is lost during subsequent cell replication. The two plasmids are said to be **incompatible.** A number of incompatibility (Inc) groups have been recognized, the plasmids of one incompatibility group excluding each other but being able to coexist with plasmids from other groups. Plasmids of one incompatibility group share a common mechanism of regulating their replication and are thus *related* to one another. Therefore, although a bacterial cell may contain different kinds of plasmids, they are not closely related because they must be compatible.

Some plasmids also have the ability to become integrated into the chromosome, and under such conditions their replication comes under control of the chromosome. This situation is remarkably like that found for several viruses whose genomes can become incorporated into the host genome (for example, ∞ Sections 8.12–8.14 and 8.22). Plasmids having the ability to integrate into host chromosomes are called *episomes.* Plasmids can sometimes be eliminated from host cells by various treatments. This process, termed **curing,** apparently results from inhibition of plasmid replication without parallel inhibition of chromosome replication, and as a result of cell division the plasmid is diluted out. Curing may occur spontaneously, but it is greatly increased by use of acridine dyes, which become inserted into DNA, or other treatments that seem to interfere more with plasmid replication than with chromosome replication. Electroporation may also be used to cure a cell of plasmids (see Section 9.6).

We can exemplify many of these characteristics by a very well-characterized plasmid called the *F plasmid.* The F plasmid is a circular DNA molecule of 100 kilobase pairs. Cells containing it can be easily cured with acridine orange. Figure 9.19 shows a genetic map of the F plasmid. One region of the plasmid contains genes involved in regulating DNA replication (such as incom-

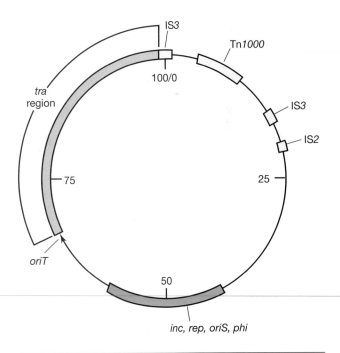

FIGURE 9.19   Genetic map of the F (fertility) plasmid of *Escherichia coli.* The numbers on the interior show the size of the plasmid in kilobase pairs. The locations of key F genes are shown: *tra,* transfer functions; *oriT,* origin of transfer; *oriS,* origin of replication; *inc,* incompatability group; *rep,* replication functions. The regions shown in yellow on F are transposable elements where integration into identical elements on the bacterial chromosome can occur and lead to the formation of different Hfr strains (see Section 9.9).

patibility, *inc*, and origin of replication, *oriS*). It also contains a number of transposable elements (see Section 9.10) involved in its ability to function as an episome. Last, it has a large region of DNA, the *tra* region, containing genes that permit it to be *tra*nsferred from one cell to another.

Cell-to-Cell Transfer of Plasmids

Because one of the defining characteristics of a plasmid is the lack of a distinct extracellular form, one can imagine that plasmids are confined almost exclusively to transferring only to daughter cells during cell division. Additionally, some prokaryotic cells can take up free DNA from the environment (see Section 9.6), so it is possible that lysis of the host, however it may happen, brings the plasmid in contact with a new host. However, this process occurs naturally in only a few bacterial species and is unlikely to account for much cell-to-cell plasmid transfer. The main mechanism of cell-to-cell transfer is **conjugation,** and it is a function encoded by some plasmids themselves. Conjugation is a replicative process, and both cells end up with copies of the plasmid (Figure 9.20).

Plasmids that govern their own transfer by cell-to-cell contact are called **conjugative,** but not all plasmids are conjugative. Transmissability by conjugation is controlled by a set of genes within the plasmid called the *tra* region. The presence of a *tra* region in a plasmid can have another important consequence if the plasmid becomes integrated into the chromosome. In that case, the plasmid can *mobilize* the transfer of chromosomal DNA from one cell to another. Strains of bacteria that transfer large amounts of chromosomal DNA during conjugation are called *Hfr* (high frequency of recombination), and the use of conjugation to transfer host genes is discussed in the next section.

Actually conjugation was discovered because the F plasmid can mobilize chromosomal genes. Therefore, we shall discuss the mechanism of DNA transfer in the next section.

Some conjugative plasmids from *Pseudomonas* have a broad host range, that is, they are transferrable to a wide variety of other gram-negative Bacteria. Some conjugative plasmids can transfer genetic information between distantly related organisms. Conjugative plasmids have been shown to transfer between gram-negative and gram-positive Bacteria, between Bacteria and plant cells, and between Bacteria and fungi. Even if the plasmid cannot replicate in the new host, the transfer of the DNA itself could have important evolutionary consequences, as well as being involved in pathogenic processes, if it can recombine into the genome of the new host.

Types of Plasmids and Their Biological Significance

Clearly all plasmids must carry genes that ensure their own replication. However, for many plasmids we know very little about what other genes they carry. These are called *cryptic plasmids* and were discovered by physical means (such as examining a cell extract using gel electrophoresis.) As we have seen, some plasmids also carry genes necessary for conjugation, and they can sometimes be detected biologically, either by the transfer functions themselves or by sensitivities to certain viruses (∞ Section 8.8). Although plasmids do not carry genes that are essential to the host under all conditions, the presence of plasmids in a cell can have a profound influence on the cell's phenotype. In some cases, plasmids encode properties that we think of as fundamental to the bacterium in question, for example, the ability of *Rhizobium* to interact with plants (∞ Section 16.23). Some pseudomonad plasmids have been shown to transfer the genetic information for biochemical pathways for the degradation of unusual organic compounds such as camphor, octane, and naphthalene.

However, plasmids can carry a very wide variety of genes. Indeed, the limitation is only that the genes they carry do not interfere either with their own replication or with the survival of the host. Because plasmids can be large and may carry many different genes, it is not always a simple matter to classify a plasmid into a simple phenotypic category. As we shall see, a single

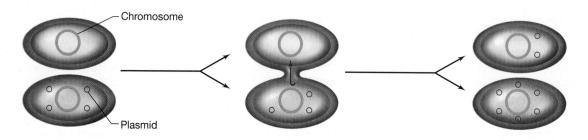

FIGURE 9.20   Plasmid transfer from cell to cell during conjugation.

TABLE 9.3	Some phenotypes conferred by plasmids in prokaryotes
Phenotype class[a]	**Organisms**[b]
Antibiotic production	*Streptomyces*
Conjugation	*Escherichia, Pseudomonas, Rhizobium, Staphylococcus,*
	Streptococcus, Sulfolobus, Vibrio
Physiological functions	
Degradation of octane, camphor, naphthalene	*Pseudomonas*
Degradation of herbicides	*Alcaligenes*
Formation of acetone and butanol (∞ Section 11.11)	*Clostridium*
Lactose, sucrose or urea utilization and nitrogen fixation	Enteric bacteria
Nodulation and symbiotic nitrogen fixation	*Rhizobium*
(∞ Section 16.16)	
Pigment production	*Erwinia, Staphylococcus*
Resistance	
Antibiotic resistance (∞ Section 18.12)	*Campylobacter,* Enteric bacteria, *Neisseria, Staphylococcus*
Resistance to cadmium, cobalt, mercury, nickel,	*Acidocella, Alcaligenes, Listeria, Pseudomonas,*
and/or zinc (∞ Section 16.20)	*Staphylococcus*
Bacteriocin resistance (and production)	*Bacillus,* Enteric bacteria, *Lactococcus, Propionibacterium*
Virulence	
Host cell invasion	*Salmonella, Shigella, Yersinia*
Coagulase, hemolysin, enterotoxin (∞ Sections 19.7–19.9)	*Staphylococcus*
Enterotoxin, K antigen (∞ Sections 13.10 and 19.9)	*Escherichia*
Tumorigenicity in plants (∞ Section 16.24)	*Agrobacterium*

a Only a few of the many phenotypes known to be associated with plasmids are given.
b Only a few well-characterized examples are given. All of the organisms given in the list are Bacteria except for *Sulfolobus*, which is a member of the Archaea.

plasmid may confer many different phenotypes on its host cell. A few of the phenotypes conferred on prokaryotes by plasmids are given in Table 9.3. It is likely that virtually all prokaryotic groups possess plasmids. In the remainder of this section we will discuss a few of the many phenotypes plasmids may confer on cells.

Resistance Plasmids

Among the most widespread and well-studied groups of plasmids are the *resistance plasmids* (*R plasmids*), which confer resistance to antibiotics and various other inhibitors of growth. R plasmids were first discovered in Japan in strains of enteric bacteria that had acquired resistance to a number of antibiotics (multiple resistance) and have since been found in other parts of the world. The emergence of bacteria resistant to several antibiotics is of considerable medical significance and was correlated with the increasing use of antibiotics for the treatment of infectious diseases. Soon after these resistant strains were isolated it was shown that they could transfer resistance to sensitive strains via cell-to-cell contact. The infectious nature of the conjugative R plasmids permits rapid spread of the characteristic through populations.

A variety of antibiotic resistance genes can be carried by an R plasmid. In general, these genes encode proteins that either inactivate the antibiotic or affect its uptake into the cell. Plasmid R100, for example, is an 89.3-kilobase-pair plasmid (Figure 9.21) that carries re-

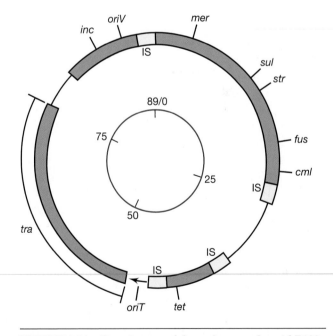

FIGURE 9.21 Genetic map of the resistance plasmid R100. The inner circle shows the size of the plasmid in kilobase pairs. The outer circle shows the location of major antibiotic resistance genes and other key functions: *inc*, incompatibility genes; *oriV*, origin of replication site; *oriT*, origin of conjugative transfer; *mer*, mercuric ion resistance; *sul*, sulfonamide resistance; *tet*, tetracycline resistance; *tra*, transfer functions. The locations of insertion sequences (IS) are also shown.

sistance genes for sulfonamides, streptomycin and spectinomycin, fusidic acid, chloramphenicol, and tetracycline. R100 also carries several genes conferring resistance to mercury (∞ Section 16.20). R100 can transfer itself between enteric bacteria of the genera *Escherichia, Klebsiella, Proteus, Salmonella,* and *Shigella* but does not transfer to the nonenteric bacterium *Pseudomonas.* R plasmids with genes for resistance to most antibiotics are known. Many drug-resistant elements on R plasmids, such as those on R100, are transposable elements (see Section 9.10) and this, plus the fact many of these plasmids are conjugative, make these plasmids a threat to traditional antibiotic therapies.

Toxins and Other Virulence Characteristics

We will discuss in Chapter 19 the physiological and genetic characteristics of microorganisms that enable them to colonize hosts and set up infections, which can lead to harm. In the present context, we merely note the two major characteristics involved in virulence: (1) the ability of microorganisms to attach to and colonize specific sites in the host; and (2) the formation of substances (toxins, enzymes, and other molecules) that cause damage to the host. It has now been well established that in several pathogenic bacteria each of these virulence characteristics is carried on plasmids. For example, enteropathogenic strains of *Escherichia coli* are characterized by an ability to colonize the small intestine and to produce a toxin that causes symptoms of diarrhea. Colonization requires the presence of a cell surface protein called the colonization factor antigen (CFA), encoded by a plasmid, which confers on the cells the ability to attach to epithelial cells of the intestine. At least two toxins in enteropathogenic *E. coli* are known to be coded for by a plasmid: the *hemolysin,* which lyses red blood cells, and the *enterotoxin,* which induces extensive secretion of water and salts into the bowel. It is the enterotoxin that is responsible for the induction of diarrhea, as will be discussed in Chapter 19. Pathogenesis in the genus *Yersinia,* the causative agent of plague (∞ Section 24.6), requires the expression and secretion of *Yersinia* outer proteins (Yops) and the V antigen. These proteins have functions that allow the bacterium to subvert host defense functions. The genes that encode and regulate these proteins are carried by a virulence plasmid.

In *Staphylococcus aureus,* some virulence-conferring properties are known to be plasmid-linked; *S. aureus* is noteworthy for the variety of enzymes and other extracellular proteins it produces that are involved in its virulence, and the production of *coagulase, hemolysin,* and *enterotoxin* is thought to be plasmid-linked. In addition, the yellow pigment in *S. aureus,* which is probably involved in its ability to resist the destructive action of singlet oxygen in the phagocyte (∞ Section 20.3), is plasmid-linked. *S. aureus* is also a notorious hospital-borne (nosocomial) pathogen, and multiple antibiotic resistance, encoded by plasmids, is common in this species (∞ Section 22.7).

Virulence factors from a variety of bacteria are known to be encoded on plasmids, and others are known to be encoded by other types of *mobile genetic elements:* transposons and bacteriophages. Several examples are now known where the genes producing a particular type of infection are present on different genetic elements within the same cell. For instance, the genes encoding the virulence determinants of the shiga-toxin-producing *E. coli* are located on the chromosome, a bacteriophage, as well as on a virulence plasmid.

Bacteriocins

Many bacteria produce agents that inhibit or kill closely related species or even different strains of the same species; these agents are called **bacteriocins** to distinguish them from the antibiotics, which have a wider spectrum of activity. Bacteriocins are ribosomally synthesized peptides (although several require extensive posttranslational modification for activity). The structural gene for the bacteriocin and the genes encoding proteins involved with processing and transporting the bacteriocin (and for conferring immunity to its action) are often carried by a plasmid or a transposon. Bacteriocins are named in accordance with the species of organism that produces them. Thus, in *Escherichia coli* we have *colicins,* coded by Col plasmids, *Bacillus subtilis* produces *subtilisin,* and so on.

The Col plasmids of *Escherichia coli* encode various colicins. Colicins released from a producing cell bind to specific receptors on the surface of susceptible cells. The receptors for colicins are generally entities whose normal function is to transport some substance, frequently a growth factor or micronutrient, through the outer membrane (the lipopolysaccharide layer) of the cell. Colicins kill cells by disrupting some critical cell function. Many colicins form channels in the cell membrane that allow potassium ions and protons to leak out, leading to a loss of the cell's energy-forming ability. However, colicin E2 is a DNA endonuclease that can cleave cellular DNA, and colicin E3 is a nuclease that cuts at a specific site in 16S rRNA and inactivates ribosomes. Col plasmids can be either conjugative or nonconjugative.

The bacteriocins or bacteriocin-like agents of gram-positive bacteria are quite different from the colicins but are also often encoded by plasmids, and some even have commercial value. For instance, the lactic acid bacteria produce the bacteriocin Nisin A, which strongly inhibits the growth of a wide range of gram-positive bacteria and is used as a preservative in the food industry.

Engineered Plasmids

The techniques of genetic engineering, discussed in the next chapter, have made possible the construction in the laboratory of a limitless number of new, artificial plasmids. Incorporation into artificial plasmids of genes from a wide variety of sources has made possible the transfer of genetic material across any species barrier. It is even possible to synthesize completely new genes and introduce them into plasmids. Such artificial plasmids are useful tools in understanding plasmid structure and function, as well as for the more practical aims of genetic engineering discussed in the next chapter.

We now turn our attention to the details of conjugation and show how certain plasmids can act to mobilize the bacterial chromosome, allowing transfer of chromosomal genes from donor to recipient.

✓ 9.8 Concept Check

The genetic information that plasmids carry is not essential for cell function under all conditions but may confer selective growth advantage under certain conditions. Examples include antibiotic resistance, enzymes for degradation of unusual organic compounds, and special metabolic pathways. Closely related plasmids cannot both replicate in the same cell. Some plasmids can transfer from cell to cell by a mechanism called conjugation.

- ✓ Are two incompatible plasmids likely to be very similar or very different?
- ✓ Conjugative plasmids tend to be large. Why?

9.9

Conjugation and Chromosome Mobilization

Bacterial conjugation (mating) is a process of genetic transfer that involves cell-to-cell contact. As we discussed previously (see Section 9.8), conjugation is a plasmid-encoded mechanism. A conjugative plasmid uses this mechanism to transfer a copy of itself to a new host. However, sometimes other genetic elements can be *mobilized* during conjugation. These other genetic elements can be other plasmids, or the host chromosome. Indeed, conjugation was discovered because the F plasmid of *Escherichia coli* (see Figure 9.19) can mobilize the host chromosome. Mechanisms of conjugative transfer may differ depending on the plasmid involved, but most plasmids in gram-negative Bacteria seem to employ a mechanism similar to that used by the F plasmid.

Conjugation involves a *donor* cell, which contains a particular type of conjugative plasmid, and a *recipient* cell, which does not. Because conjugation was discovered by performing genetic crosses, the donors were called *males* and the recipients were called *females*. The genes that control conjugation are contained in the *tra* region of the plasmid (see Section 9.8). Many genes in the *tra* region have to do with the synthesis of a surface structure, the **sex pilus** (Figure 9.22). Only donor cells have these pili, explaining why RNA bacteriophages that attach to them are *"male-specific"* (∞ Section 8.8). Different conjugative plasmids may have slightly different *tra* regions, and in some cases the pili are also different and can be distinguished immunologically. The F plasmid and its relatives encode *F pili*.

Pili allow specific pairing to take place between the donor cell and the recipient cell. All conjugation in gram-negative Bacteria is thought to depend on cell pairing brought about by pili. The pili make specific contact with a receptor on the recipient and then retract, pulling the two cells together. The contacts between the donor and recipient cells then become stabilized, probably from fusion of the outer membranes, and the DNA is then transferred from one cell to another.

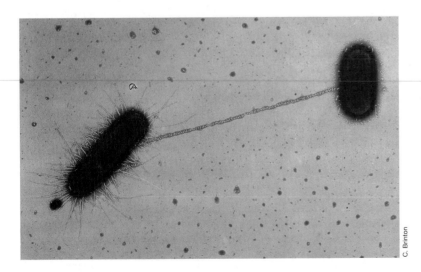

FIGURE 9.22 Direct contact between two conjugating bacteria is first made via a pilus. The cells are then drawn together for the actual transfer of DNA. Note the F-specific bacteriophages on the pilus (∞ Section 8.8).

C. Brinton

Mechanism of DNA Transfer During Conjugation

DNA synthesis is necessary for DNA transfer to occur, and the evidence suggests that one of the DNA strands is derived from the donor cell and the other is newly synthesized in the recipient during the transfer process. A mechanism of DNA synthesis in certain bacteriophages, called **rolling circle replication,** was presented in Figures 8.17 and 8.28. This model best explains DNA transfer during conjugation, and a possible mechanism for this process is outlined in Figure 9.23. The whole series of events is triggered by cell-to-cell contact, at which time one strand of the plasmid DNA circle is nicked and one parental strand is transferred. The nicking enzyme required to initiate the process, TraI, is encoded by the *tra* operon of the F plasmid. This protein also has helicase activity and is thus also involved in unwinding the strand to be transferred. As this transfer occurs, DNA synthesis by the rolling circle mechanism

replaces the transferred strand in the donor. A complementary DNA strand is also made in the recipient. The model accounts for the fact that if the DNA of the donor is labeled, some labeled DNA is transferred to the recipient but only a *single* labeled strand is transferred. Therefore, at the end of the process, both donor and recipient possess completely formed plasmids.

The high efficiency of the plasmid DNA transfer process is shown by the fact that under appropriate conditions virtually every recipient cell that pairs acquires a plasmid. When the plasmid genes can be expressed in the recipient, the recipient itself becomes a donor and can transfer the plasmid to other recipients. In this fashion, conjugative plasmids can spread rapidly between populations, behaving like infectious agents. The infectious nature of this phenomenon is of major ecological significance because a few plasmid-positive cells introduced into an appropriate population of recipients can, if they contain genes that confer a selective advantage, convert the whole recipient population into a plasmid-

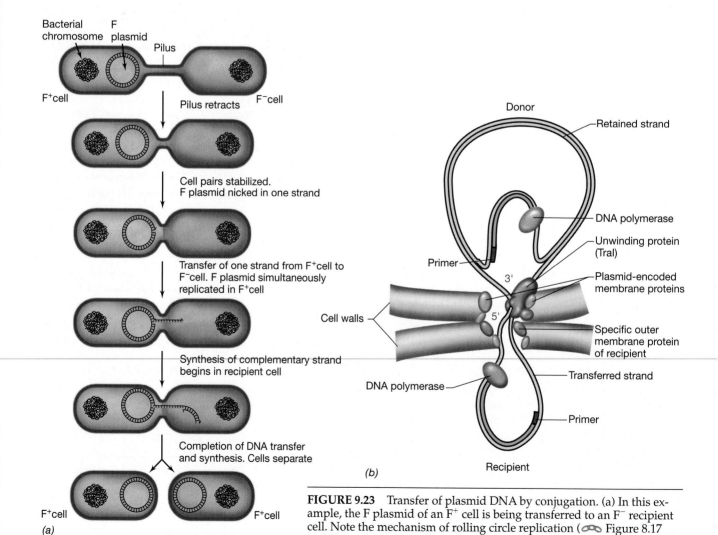

FIGURE 9.23   Transfer of plasmid DNA by conjugation. (a) In this example, the F plasmid of an F⁺ cell is being transferred to an F⁻ recipient cell. Note the mechanism of rolling circle replication (⬤⬤ Figure 8.17 and Figure 8.28). (b) Details of the replication and transfer process.

bearing population in a short period of time. The widespread occurrence of drug resistance carried by conjugative plasmids (see Section 9.8) has led to some serious problems for the chemotherapy of infectious disease (∞ Section 11.13). However, as mentioned above, plasmids can also be lost from a cell by a process called *curing*. This could happen spontaneously in natural populations when there is no selection pressure to maintain the plasmid. For example, plasmids conferring antibiotic resistance can be lost without affecting the cell's viability if there are no antibiotics in the cell's environment.

The Formation of Hfr Strains and Chromosome Mobilization

The F plasmid of *Escherichia coli* (see Section 9.8) is not only conjugative but also can mobilize the chromosome so it can be transferred during cell-to-cell contact. The F plasmid is an episome, a plasmid that can integrate into the host chromosome (see Section 9.8). When the F plasmid is integrated into the chromosome, conjugation can lead to transfer of large regions of the host chromosome and genetic recombination between donor and recipient can then be very extensive.

Cells possessing an unintegrated F plasmid are called **F⁺**, and strains that can act as recipients for F⁺ (or Hfr, see later in this section) are called **F⁻**. F⁻ cells lack the F plasmid; in general, cells that contain a plasmid are very poor recipients for the same or closely related plasmids. Bacterial strains that possess a chromosome-integrated F plasmid and show such extensive genetic recombination are called **Hfr** (for high frequency of recombination). Conjugation leads to transfer of the host chromosome in Hfr strains because the plasmid is part of the chromosome. (Plasmid integration is a very simple mechanism of mobilizing other genetic elements.)

We thus see that the presence of the F plasmid results in three distinct alterations in the properties of a cell: (1) ability to synthesize the F pilus, (2) mobilization of DNA for transfer to another cell, and (3) alteration of surface receptors so the cell is no longer able to behave as a recipient in conjugation.

The integration of the F plasmid into the host chromosome can occur at several specific sites, called IS (for *insertion sequences*). These sites are regions of homology between chromosome and F plasmid DNA (see Section 9.10 for a discussion of insertion sequences). Figure 9.24 shows the integration of an F plasmid, which involves insertion in the chromosome at such a site. In the particular Hfr shown, the integration site is between the chromosomal genes *pro* and *lac*. Once integrated, the plasmid no longer controls its own replication, but the *tra* operon still functions normally and the strain synthesizes pili. When a recipient is encountered, conjugation is triggered

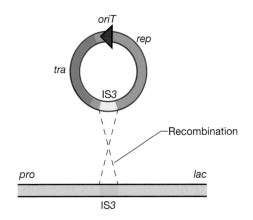

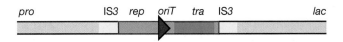

FIGURE 9.24   Integration of an F plasmid into the chromosome with the formation of an Hfr. The insertion of the F plasmid occurs at a variety of specific sites where IS elements are located, the one here being an IS3 located between the chromosomal genes *pro* and *lac̄*. Some of the genes on the F plasmid are shown. The arrow indicates the origin of transfer, *oriT*, with the arrow as the leading end. Thus, in this Hfr *pro* would be the first chromosomal gene to be transferred and *lac* would be among the last.

just as in an F⁺ cell, and DNA transfer is initiated at *oriT*. However, since the plasmid is now part of the chromosome, after part of the plasmid DNA is transferred, chromosomal genes begin to be transferred (Figure 9.25).

Thus the chromosome is said to be *mobilized*. As we mentioned, integration of a plasmid is a simple mechanism of mobilization. Other mechanisms of mobilization are known, involving not only the host chromosome but other plasmids. One such mechanism is for the mobilizable plasmid to contain an *oriT* sequence like that found on the F plasmid. Such a plasmid will be transferred by conjugation if it is in a cell with a conjugative plasmid that can supply all the other necessary genes.

Since a number of distinct insertion sites are present, a number of distinct Hfr strains are possible. A given Hfr strain always donates genes in the same order, beginning with the same position, but Hfr strains of independent origin transfer genes in different sequences. During cell division, the DNA of the Hfr replicates normally, but at the time of pairing with an F⁻ cell, a DNA strand from the Hfr is transferred to the F⁻ cell and replication occurs by the rolling circle process. After transfer, the Hfr strain still remains Hfr because it has retained a copy of the transferred genetic material.

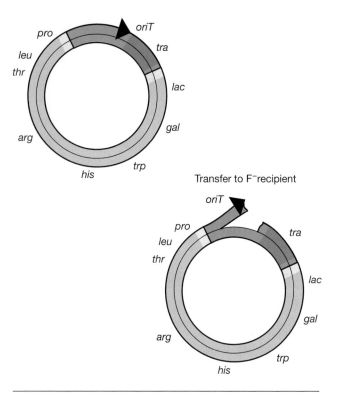

FIGURE 9.25   Breakage of the Hfr chromosome at the origin of transfer and the beginning of DNA transfer to the recipient. Replication occurs during transfer (see Figure 9.23). Please note that the figure is not drawn to scale. The inserted F plasmid is less than 3% of the size of the *Escherichia coli* chromosome.

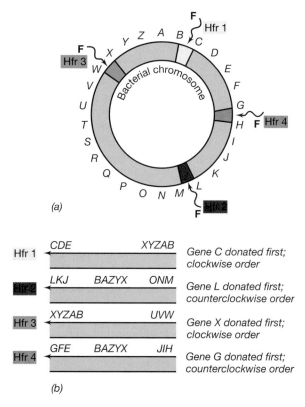

FIGURE 9.26   Manner of formation of different Hfr strains, which donate genes in different orders and from different origins. The bacterial chromosome is a circle (a) that can open at various insertion sequences, at which F plasmids become inserted. The gene orders are shown in part (b).

Usually, because of breakage of the DNA strand during transfer, only a *part* of the donor chromosome is transferred. Since only part of the chromosome is transferred, it cannot replicate in the recipient cell. Therefore, donor genes normally cannot be detected unless recombination between the incoming fragment and the recipient chromosome takes place.

Although Hfr strains transmit chromosomal genes at high frequency, they usually do not convert F⁻ cells to F⁺ or Hfr because the entire F plasmid is only rarely transferred. On the other hand, F⁺ cells efficiently convert F⁻ to F⁺ because the entire F plasmid is transferred.

At some insertion sites, the F plasmid is integrated with the origin in one direction, whereas at other sites the origin is in the opposite direction. The direction in which the F plasmid is inserted determines which of the chromosomal genes will be transferred into the recipient first. The manner in which a variety of Hfr strains can arise is illustrated in Figure 9.26. By use of various Hfr strains, it has been possible in *Escherichia coli* to determine the arrangement and orientation of a large number of chromosomal genes, as will be described in Section 9.11.

Use of Hfr Strains and the Phenomenon of Interrupted Mating

As is the case for any type of system of bacterial gene transfer, one **selects** recombinants from conjugation. However, unlike the situation in transformation and transduction, during conjugation both the donor and recipient cells are viable, so it is necessary to chose selection media where the desired recombinants can grow but where neither of the parental strains can form colonies. Typically a recipient is used that is resistant to an antibiotic but is auxotrophic for some substance, and a donor that is sensitive to the antibiotic but is prototrophic for the same substance.

For instance, in the experiment shown in Figure 9.27, an Hfr donor that is sensitive to streptomycin (Strˢ) and contains wild-type genes encoding enzymes needed for synthesis of the amino acids threonine and leucine (Thr⁺ and Leu⁺) and for utilization of the energy source lactose (Lac⁺) is mated with a recipient cell that is mutant for these genes but is resistant to streptomycin (Strʳ). The selective medium is a minimal medium containing streptomycin so that only recombinant cells can grow. The composition of each selective medium is varied de-

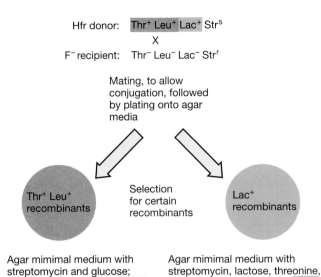

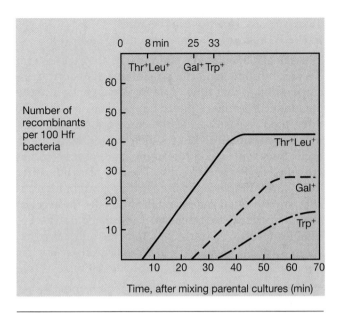

FIGURE 9.27 Laboratory procedure for the detection of genetic conjugation. Thr, threonine; Leu, leucine; Lac, lactose; Str, streptomycin. Note that each medium selects for specific classes of recombinants. The controls for the experiment are to plate samples of the donor and the recipient before they are mixed. Neither should be able to grow on the selective media used.

FIGURE 9.28 Rate of formation of recombinants containing different genes after mixing Hfr and F⁻ bacteria by the process known as interrupted mating. The location of the genes along the Hfr chromosome is shown at the upper left. Note that the genes closest to the origin (0 min) are the first ones detected in the recombinants. The experiment is done by mixing Hfr and F⁻ cells under conditions in which essentially all Hfr cells find mates. The F⁻ recipient was streptomycin-resistant but auxotrophic for the markers being scored. The Hfr donor was streptomycin-sensitive. At various times, samples of the mixture are shaken violently to separate the mating pairs and plated on a selective medium in which only the recombinants can grow and form colonies.

pending on which genotypic characteristics are desired in the recombinant, as shown in Figure 9.27. The frequency of the process is measured by counting the colonies grown on the selective medium.

If one understands the mechanism of conjugation and the fact that an Hfr strain is a cell with an F plasmid integrated into its chromosome, then it is easy to see that the transfer of the chromosomal genes will always be in the same order from a fixed site in a given Hfr strain. However, conjugation was discovered as a mechanism of transferring chromosomal genes long before it was known that a plasmid was involved or how the transfer might take place. Note that in higher organisms, such as humans, mating involves simultaneous transfer of a total set of chromosomes, not just chromosomal fragments.

The nature of the oriented, fragmentary transfer of host genes during conjugation was first shown by a procedure called **interrupted mating.** The mating pairs are rather weakly joined and can be separated by agitation in a mixer or blender. If mixtures of Hfr and F⁻ cells are agitated at various times after mixing and the genetic recombinants scored, it is found that the longer the time between pairing and agitation, the greater the number of genes of the Hfr that will appear in the F⁻ recombinant. In addition, gene transfer always occurs in a specific order in a specific Hfr. As shown in Figure 9.28,

genes present closer to the origin enter the F⁻ first and are always present in a higher percentage of the recombinants than genes that enter late. In addition to showing that gene transfer from donor to recipient is a sequential process, experiments of this kind provide a method of determining the order of the genes on the bacterial DNA (genetic mapping). The arrangement of gene loci on the chromosome is called a **genetic map** (see Section 9.11).

As in transformation and transduction, genetic recombination between Hfr genes and F⁻ genes involves homologous recombination in the recipient cell. This has been shown by the isolation of mutants of F⁻ strains that are unable to form recombinants when mated with Hfr. These mutants are Rec⁻ (recombination minus) and are deficient in the RecA protein because of a mutation in the *recA* gene (see Section 9.5). It is important to remember that recombination is not the same as DNA transfer. Both F plasmids and F' plasmids are transferred normally to a Rec⁻ cell even though recombination does not take place after transfer.

Transfer of Chromosomal Genes to the F Plasmid

Occasionally integrated F plasmids may be excised from the chromosome, and the possibility exists for the incorporation at that time of *chromosomal* genes into the liberated F plasmid. This can happen because both the integrated F plasmid and the chromosome contain a number of identical IS at which recombination can occur. Such F plasmids containing chromosomal genes are called *F′ (F prime) plasmids*. These F′ plasmids differ from normal F plasmids in that they contain identifiable chromosomal genes, and they transfer these genes at high frequency to recipients. F′-mediated transfer resembles specialized transduction in that only a restricted group of chromosomal genes can be transferred. It is often with F′ plasmids that complementation tests (see Section 9.5) are done in *Escherichia coli.*

Other Conjugation Systems

Although we have discussed conjugation almost exclusively as it occurs in *Escherichia coli*, conjugative plasmids have been found in many other gram-negative Bacteria. Indeed, conjugative plasmids of the incompatibility group (see Section 9.8) IncP can be maintained in practically all gram-negative species, and DNA transfer between species and genera can occur. Conjugative plasmids are also known in gram-positive Bacteria (for example, in *Streptococcus, Enterococcus*, and *Staphylococcus*). As a rule, the mechanism of conjugation is very similar in the other gram-negative Bacteria to what we have discussed for the F plasmid, whereas conjugation involving gram-positive Bacteria can be quite different.

Some conjugative plasmids mobilize other genetic elements, as we have discussed for the F plasmid, but this is not always the case. There are also elements called *conjugative transposons* that can transfer themselves from the chromosome of a donor host to that of a recipient and can also mobilize other genetic elements. Conjugative transposons have mostly been found in gram-positive Bacteria. They have a very wide host range and can be involved in gene transfer between different bacteria of different genera. The mechanism of conjugation used by these elements is not completely known, but it does seem to involve circular, plasmidlike intermediates. Gene transfer involving cell-to-cell contact has also been described in *Haloferax*, a member of the Archaea, but the mechanism of transfer seems to be much different from that in any other type of prokaryotic conjugation.

✓ 9.9 Concept Check

Conjugation is a mechanism of DNA transfer in prokaryotes that requires cell-to-cell contact. Conjugation is controlled by genes carried by certain plasmids (such as the F plasmid) and typically involves transfer of the plasmid from a donor cell to a recipient cell. However, other genetic elements, including the donor cell chromosome, can sometimes be mobilized and also transferred. Transfer of the host chromosome is rarely complete but can be used to map the order of the genes on the chromosome.

- ✓ How are donor and recipient cells brought into contact with each other?
- ✓ In conjugation involving the F plasmid of *Escherichia coli*, how is the host chromosome mobilized?

9.10

Transposons and Insertion Sequences

The order of genes on a bacterial chromosome can be determined by the methods of gene transfer we have considered, just as genes in eukaryotic chromosomes can be mapped by mating experiments. However, the exact arrangement of the genes along a chromosome is not necessarily permanently fixed; some genes are capable of moving under certain conditions. The process by which a gene moves from one place to another in the genome is called **transposition** and is an important process in evolution and in genetic analysis.

We should emphasize that transposition typically is a *rare* event, occurring at frequencies of 10^{-5}–10^{-7} per generation. Thus, the genes of living organisms are relatively stable.

In addition, not all genes are capable of transposition. Rather, transposition of genes is linked to the presence of special genetic elements called *transposable elements.*

Transposition was originally discovered in corn (maize) and then later in Bacteria owing to the extremely sensitive types of genetic analysis available in these organisms. It has been shown by using DNA hybridization and sequencing techniques (∞ Working with Nucleic Acids: The Tools, Chapter 6) that DNA sequences with the properties of transposable elements are widespread in nature. However, in organisms with poorly characterized genetic systems, it can be difficult to detect transposition (because it is a rare event) and actually prove that a sequence is a transposable element.

Transposable Elements

As we discussed earlier (∞ Section 6.4), there are three types of transposable elements in Bacteria: *insertion sequences, transposons*, and some special viruses (such as Mu) (∞ Section 8.13). A brief summary of different types of transposable elements found in both prokaryotes and eukaryotes is given in Table 9.4. In this section we shall confine our discussion primarily to insertion sequences and transposons found in Bacteria. Both these

TABLE 9.4	Transposable elements	
Prokaryote	**Eukaryote**	
Insertion sequence: IS	Yeast: *sigma*	
Transposon: Tn	Yeast: Ty	
	Fruit fly: copia, P	
	Maize: Ac	
Virus: Mu	Retrovirus: Rous sarcoma, human immunodeficiency virus (HIV)	

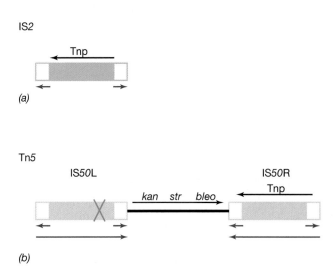

elements have two important features in common: They both carry genes encoding a **transposase,** the enzyme necessary for transposition, and both have short *inverted terminal repeats* at the ends of their DNA (remember that these "ends" are continuous with whatever DNA the element is inserted into). These repeats range in length from fewer than 20 base pairs in simple IS elements to more than 1000 base pairs in some transposons; each IS has a specific number of base pairs in its terminal repeats. Such inverted terminal repeats are involved in the transposition process (see later). Figure 9.29 shows a genetic map of an insertion element named IS2 and of a transposon named Tn5.

In prokaryotes, insertion sequences are the simplest type and carry no genetic information other than that required for them to move to new locations. Insertion sequences are short segments of DNA, about 1000 nucleotides long, that can become integrated at specific sites on the genome. Insertion sequences, abbreviated IS, are found in both chromosomal and plasmid DNA, as well as in certain bacteriophages. Several hundred distinct IS elements have been characterized, and most are designated by a number identifying its type: IS1, IS2, IS3, and so on. Because so many IS elements have been discovered, some receive designations related to the organism in which they were first identified. For example, IS Mt1 is from *Mycobacterium tuberculosis.* IS elements are scattered about the chromosome, and strains vary in the number and frequency of these elements. For instance, one strain of *Escherichia coli* has five copies of IS2 and five copies of IS3. The F plasmid also carries these insertion sequences (see Figure 9.19), and it is homologous recombination between identical sequences on the plasmid and the chromosome (*not* transposition) that allows the F plasmid to integrate into the bacterial chromosome and mobilize it (see Section 9.9). Some of the Archaea have large numbers of IS elements in their chromosomes.

Transposons are larger than insertion sequences and carry other genes, some of them conferring important properties on the organism carrying them. These often include drug resistance markers and other easily selectable

FIGURE 9.29 Maps of the transposable elements IS2 and Tn5. The red arrows underneath each map indicate the inverted repeats. The arrows above the maps show the direction of transcription of any genes on the elements. *Tnp* is the gene encoding the transposase. The transposase genes of these two elements are not closely related. (a) IS2 is an insertion sequence of 1327 base pairs with inverted repeats of 41 base pairs at its ends. (b) Tn5 is a composite transposon of 5.7 kilobase pairs with the insertion sequences IS50L and IS50R at its left and right ends, respectively. IS50L is not capable of independent transposition because there is a *nonsense mutation* (see Section 9.2) marked by a blue cross in its transposase gene. Otherwise, the two IS50 elements are very nearly identical. Note that these two IS50 elements are inverted with respect to each other. The genes *kan, str,* and *bleo,* confer resistance to the antibiotics kanamycin (and neomycin), streptomycin, and bleomycin. Interestingly, streptomycin resistance is not expressed in *Escherichia coli.*

genes. In addition, as we have mentioned (see Section 9.9), there are *conjugative transposons.* These transposons have genes allowing them not only to move from one location on a bacterial genome to another but also to transfer themselves from one bacterium to another.

Some transposons are actually composite structures containing a gene or group of genes lying between two identical insertion sequences. The existence of such *composite transposons* indicates that new transposons probably continue to arise in cells that have insertion sequences.

The Mechanism of Transposition

As mentioned previously, the inverted repeats found at the ends of transposable elements are essential for transposition. The other essential component is an enzyme called *transposase,* which recognizes these repeats. This enzyme is usually encoded by the transposable element, although some very simple IS elements use an enzyme encoded by another genetic element. The transposase apparently recognizes, cuts, and eventually ligates the DNA during transposition (Figure 9.30).

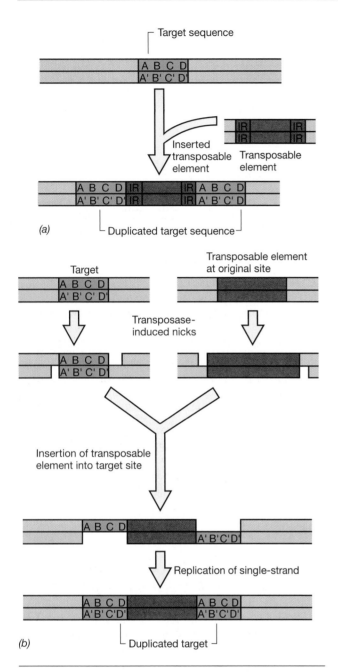

FIGURE 9.30 The transposition process. (a) Insertion of a transposable element generates a duplication of the target sequence. Note the presence of inverted repeats (IRs) at the ends of the transposable element. (b) A schematic diagram indicating how target sequences can be duplicated. For simplicity, the IRs are not marked.

When a transposable element becomes inserted into another DNA (the target DNA), a short sequence in the target DNA at the site of integration is duplicated. This target DNA sequence was not present in the transposon, but the transposable element has brought about a duplication of this DNA by the insertion process (Figure 9.30a). The duplication of the target sequence appar-

ently arises because single-stranded breaks are generated by the transposase (Figure 9.30b). The transposon is then attached to the single-stranded ends that have been generated, and repair of the single-strand portions results in the duplication.

Certain transposable elements prefer certain sequences as target sites, but others, including the bacteriophage Mu, can insert themselves almost randomly (for a representation of Mu insertion, see Figure 8.31).

Two mechanisms of transposition are known, called *conservative* and *replicative*. In conservative transposition, such as can occur in the transposon Tn5, the transposable element is excised from one location in the chromosome and becomes reinserted at a second location. The copy number of a conservative transposon therefore remains at one. By contrast, replicative transposons, such as bacteriophage Mu (∞ Section 8.13), are duplicated, and a new copy is inserted at another location. Thus, after the transposition event is completed, *one* copy of the transposing element *remains* at the original site and *another copy* is found at the new site. During this whole transposition process, the source transposon *remains* at its original site; at no time does the source transposon become free in the cell.

Although many of the molecular details of transposition are uncertain, and different transposable elements appear to have different mechanisms, one model for replicative transposition is illustrated in Figure 9.31. As seen, single-strand cuts are made at the ends of the transposon (at the sites of the inverted repeats), and staggered single-strand cuts are made at the target site. The transposon is now joined to the target site via the single-stranded ends, leading to the formation of a composite structure called a *cointegrate*. Replication repair then fills in the single-strand gaps in the target site. This process results in the formation of *direct repeats* in the target site at the ends of the transposon (in addition to the inverted repeats of the transposon). The final event is *resolution* of the cointegrate structure, leading to release of the original transposon and the presence of a new copy of the transposon at the target site. Now that the transposon is present at the new target site, it can also serve as another source of transposition.

It should be emphasized that transposition is essentially a *recombination event*, but one that does not occur between homologous sequences or use the general recombination system of the cell. It involves the special protein *transposase* rather than the RecA protein that is involved in general recombination. Because this recombination involves a *specific* base sequence, it is called *site-specific recombination* (in contrast to homologous recombination discussed earlier in this chapter).

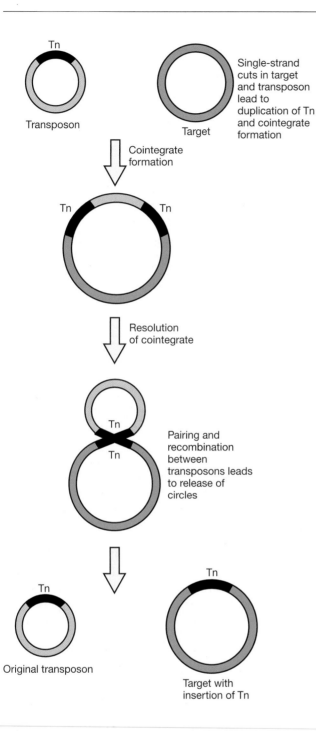

FIGURE 9.31 Replicative transposition. After the formation of single-strand cuts, a cointegrate structure arises by association of the two molecules. After recombination, resolution of the cointegrate structure leads to the release of the original transposon and duplication of the transposon in the target molecule. See Figure 9.30 for an explanation of how the duplication process can occur.

Mutagenesis with Transposable Elements

If the insertion site for a transposable element is *within* a gene, insertion of the transposon will result in loss of linear continuity of the gene, leading to mutation (Figure 9.32). Transposons thus provide a facile means of creating mutants throughout the chromosome. The most convenient element for **transposon mutagenesis** is one containing an antibiotic resistance gene. Clones containing the transposon can then be selected by the isolation of antibiotic-resistant colonies. If the antibiotic-resistant clones are selected on rich medium on which all auxotrophs can grow, they can be subsequently screened on minimal medium supplemented with various growth factors to determine if a growth factor is required.

Transposons are also useful for incorporating an auxotrophic gene marker into a wild-type organism. Normally, auxotrophic recombinants cannot be isolated by positive selection, but if the auxotrophic marker to be introduced contains a transposable element with an antibiotic resistance marker, then one can select for antibiotic-resistant clones, a positive selection procedure, and automatically obtain clones that have incorporated the auxotrophic marker.

Two transposons widely used for mutagenesis are Tn*5* (see Figure 9.29), which confers neomycin and kanamycin resistance, and Tn*10,* which contains a marker for tetracycline resistance.

The bacteriophage Mu (∞ Section 8.13) is also widely used as a *biological mutagen.* Because Mu integrates at a wide variety of host sites, it can be used to induce mutations at many locations. Also, Mu can be used to carry into the cell genes that have been derived from other host cells,

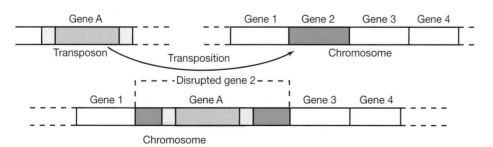

FIGURE 9.32 Transposon mutagenesis. The transposon moves into the middle of gene 2. Gene 2 is now disrupted by the transposon and is inactivated. Gene A of the transposon will be expressed in both locations.

a form of *in vivo* genetic engineering. In addition, modified Mu phages have been made artificially in which some of the lytic functions of Mu have been deleted. These phages, called Mini-Mu, are deleted for significant portions of Mu but have the ends of the phage in normal orientation. Mini-Mu phages are usually defective, unable to form plaques, and their presence must be ascertained by the presence of other genes they carry. One set of Mini-Mu phages containing the β-galactosidase gene of the host (called Mu*d-lac, d* for "defective") can be detected in the integrated state if the *lac* gene is oriented properly in relation to a host promoter. Under these conditions, the host cell forms the enzyme β-galactosidase, which can be detected in colonies by using a special color indicator. β-Galactosidase–positive colonies from a β-galactosidase–negative host are thus an indication that Mud-*lac* infection has occurred.

Integrons

The fact that many transposons contain genes conferring both antibiotic resistance and the ability to transpose to conjugative plasmids gives them the ability to pose serious problems when using traditional antibiotic therapies in the treatment of bacterial infections. Some transposons also contain other elements that make them even more formidable, **integrons.** Integrons are genetic elements that can capture and express genes from other sources. Integrons contain a gene that encodes an integrase. The integrase is an enzyme that catalyzes another type of *site-specific recombination*. Recall that the lambda genome becomes integrated into the *Escherichia coli* genome at a specific site by the action of an integrase encoded by lambda (∞ Section 8.12). The integron also contains a specific DNA sequence where the integrase can function to integrate *gene cassettes* (which are located near similar sequences on other elements) and a promoter that can then express the newly integrated gene cassette.

Integrons have been found in a number of Bacteria, for example *Acinetobacter, Citrobacter, Escherichia, Klebsiella, Pseudomonas,* and *Vibrio,* often in clinical isolates. They can occur in transposons, for example Tn7, on plasmids, or on the bacterial chromosome. Some isolates have as many as five different gene cassettes. Over 40 different antibiotic resistance genes have been identified as being located on such cassettes, as have some genes associated with virulence. Figure 9.33 shows the structure of two integrons from *Pseudomonas aeruginosa.* The clinical significance of integrons and the gene cassettes they capture seems obvious. What is much less obvious is the origin of the gene cassettes themselves. These are not simply random genes that can be captured, but genes bounded by specific DNA sequences that are recognized by the integrase and genes that are apparently not expressed until they become part of an integron and can be transcribed from the promoter on the integron.

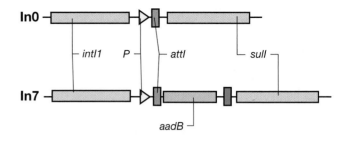

FIGURE 9.33 Structure of two naturally occurring integrons from *Pseudomonas.* The integron In0 has the basic set of genes: *intI1,* encodes integrase; *attI,* the site where site-specific integration can occur; P, a promoter; and *sulI,* a gene conferring sulfonamide resistance that contains its own promoter. The integron In7 contains all of these genes, but in addition, a gene cassette has been integrated. All cassettes contain a site (blue square) for site-specific recombination. This cassette contains *aadB,* which confers resistance to certain aminoglycoside antibiotics.

Invertible DNA and the Phenomenon of Phase Variation

Another type of *site-specific recombination* that has been recognized in some bacteria involves the **inversion** of a segment of DNA from one orientation to the other. When the segment is oriented in one direction, a particular gene is expressed, whereas when it is oriented in the opposite direction, a different gene is expressed. This "flip-flop" mechanism is catalyzed by an enzyme called a *DNA invertase* and provides an interesting example of the regulation of gene activity.

The best-studied case of gene inversion is that called *phase variation,* which has been well studied in bacteria of the genus *Salmonella.* These enteric bacteria are motile by means of peritrichous flagella. As we have noted (∞ Section 3.11), bacterial flagella are composed of a single type of protein. As a result of phase variation, the flagellar protein can be of one of two separate types. Each *Salmonella* cell has two genes, H1 and H2, coding for the two different flagellar proteins, but only one of the two genes is expressed at any one time. Thus, an individual bacterial cell makes either H1-type flagella or H2-type flagella.

The invertible element involved in expression of the flagellar proteins is a 970-base-pair segment, which contains the gene encoding the Hin invertase and also an outward facing promoter (Figure 9.34). When the invertible segment is in one orientation, the H2 gene is transcribed, but in addition, another gene is transcribed that codes for a protein that represses transcription of gene H1. Thus, when H2 is expressed, H1 is turned off. On the other hand, when the invertible segment is in the opposite orientation, the genes for H2 and the H1 repressor are no longer expressed, so H1 can now be transcribed and expressed.

Invertible segments are also known in other genetic systems. We described an invertible segment involved in

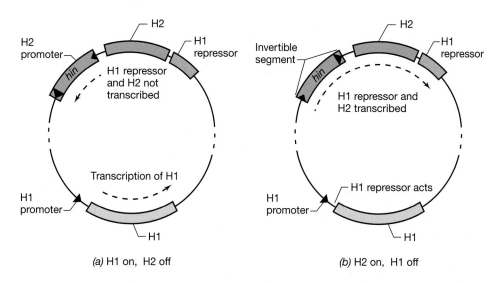

FIGURE 9.34 Site-specific inversion, the mechanism by which phase variation in *Salmonella* flagella is brought about. The dotted lines show the location and direction of transcription. (a) When the invertible segment is in one orientation, the H2 promoter points away from the H1 repressor and H2 gene; H1 is expressed. (b) In the opposite orientation, H1 repressor is made, turning off H1. At the same time, H2 is expressed. The result is a "flip-flop" between two alternate states. The gene encoding the Hin invertase is also on the invertible segment but is expressed in both orientations.

host range of bacteriophage Mu (◌◠ Section 8.13). Regulation by rearrangement also occurs in eukaryotes: We consider the regulation of mating type in yeast in Section 9.14 and we examine the complex genetic rearrangements involved in the production of antibodies (defense against infection) later (◌◠ Section 20.11 and the box, Molecular Biology of the Immune Response, in Chapter 20).

✓ 9.10 Concept Check

Transposable elements are genetic elements that can move from one location on a chromosome to another by a process called transposition, a type of site-specific recombination. Transposition can be either replicative or conservative. Two types of transposable elements are insertion sequences and transposons. Both have inverted repeats at their ends, and both encode a transposase, an enzyme involved in transposition. Transposons also carry other genes, often those encoding antibiotic resistance. Transposable elements can be used as biological mutagens.

- ✓ What features do insertion sequences and transposons have in common?
- ✓ What is a composite transposon?

9.11

The *Escherichia coli* Chromosome

The three mechanisms of genetic exchange described in this chapter, transformation, transduction, and conjugation, can be used to map the locations of various genes (actually mutations in genes) on the chromosome. In *Escherichia coli* genes were mapped to a particular region of the chromosome using conjugation. By using Hfr strains with origins at different sites, it is possible to map the whole bacterial gene complement. A circular reference map for *Escherichia coli* strain K-12 is shown in Figure 9.35. The map distances are given in minutes of

transfer, with 100 min for the whole chromosome and with "zero time" arbitrarily set as that at which the first genetic transfer (the *threonine* operon) can be detected using the original Hfr strain.

Conjugation does not permit ordering genes that are closely linked to each other. Therefore, generalized transduction was used for more fine structure mapping of the *E. coli* chromosome. Bacteriophage P1 can carry fragments of DNA equivalent to about 2 min on the map and has proved very useful for mapping genes. The genetic techniques that were used were dictated by the efficiency with which genetic transfer occurs in the organism to be studied. Transformation is a very inefficient process in *E. coli*, but is more efficient in other organisms and has proved an effective tool for mapping genes in *Bacillus*.

Although all of these "classic" techniques are still useful for the characterization and construction of strains, they have largely been supplanted in constructing genetic maps by molecular cloning (◌◠ Chapter 10), restriction mapping (◌◠ Section 6.5), and hybridization and DNA sequencing (◌◠ Working With Nucleic Acids: The Tools, Chapter 6), which have revolutionized the study of the genome organization of prokaryotes (and of eukaryotes). The map shown in Figure 9.35 shows the location of a few restriction enzyme recognition sites, and the size is given both in minutes (the original units determined by conjugation studies) and in kilobase pairs. The *E. coli* chromosome, like that of many other prokaryotes, has been completely sequenced. Indeed, sequencing of the relatively "small" bacterial chromosomes that contain "only" 3 or 4 megabases is now done by automated sequencing of random, or "shotgun," clones (◌◠ Section 10.1). Because of the enormous amount of genetic information sequencing makes available, this information can be most effectively accessed through computer databases (see Table 9.5).

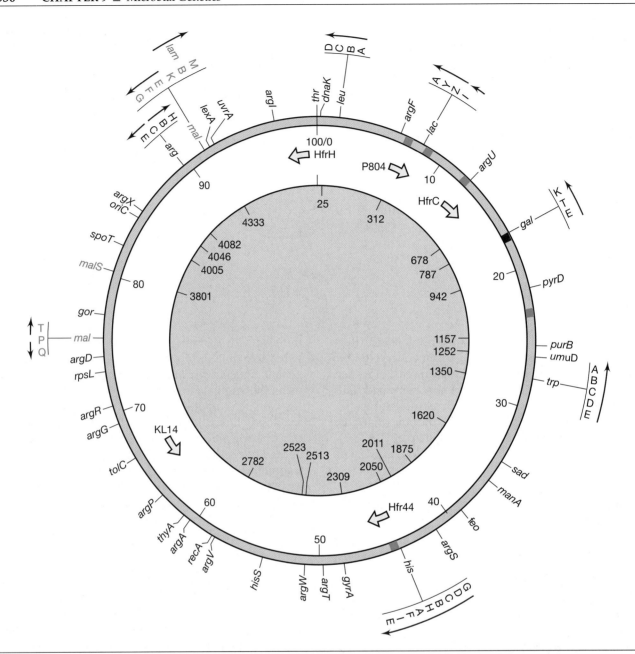

FIGURE 9.35   Circular linkage map of the chromosome of *Escherichia coli* K-12. On the outer edge of the map, the locations of a few of the mapped genes are indicated. A few operons are also shown, along with the direction in which they are transcribed. Along the inner edge of the map, the numbers from 0 to 100 refer to map position in minutes. The origin of DNA replication is marked *oriC* (84.3 min), and replication proceeds bidirectionally from this point (∞ Section 6.6 and Figures 6.16–6.19). The inner circle shows the locations, in kilobase pairs, of the sequences recognized by the restriction enzyme *Not*I. Note that 0 min and 0 kilobase pairs are both, by convention, at the *thr* locus. The origins and directions of transfer of a few Hfr strains are also shown (arrows). The positions where five copies of the transposable element IS3 have been located in a particular strain are shown in blue. This element is also found in two copies on the F plasmid and is involved in Hfr formation. The position of the site where the bacteriophage lambda prophage integrates is shown in red. If the prophage were present, it would add an extra 48.5 kilobase pairs (slightly over 1 min) to the map. The genes of the maltose regulon (∞ Section 7.6), which includes several operons, are shown in green. Although most genes in this regulon have an abbreviation beginning with *mal*, note that one of the genes is *lamB*. This gene encodes a membrane protein involved in maltose uptake by the cell, but the protein is also the receptor for bacteriophage lambda. The gene *rpsL* (73 min) encodes a ribosomal protein. The gene was once called *str* because mutations in this gene lead to streptomycin resistance.

TABLE 9.5	Prokaryotic chromosomes[a]		
Organism	Size (base pairs)[b]	ORFs[c]	Comments
Mycoplasma genitalium	580,070	470	Bacterium, smallest known cellular genome (🔗 Section 13.20)
Mycoplasma pneumoniae	816,394	677	Bacterium, causes pneumonia (🔗 Section 13.20)
Borrelia burgdorferi	910,725	853	Bacterium, spirochete, has linear chromosome,[d] causes Lyme disease (🔗 Sections 13.30 and 24.4)
Chlamydia trachomatis	1,042,519	894	Bacterium, obligate intracellular parasite, common human pathogen (🔗 Sections 13.26 and 23.6)
Rickettsia prowazekii	1,111,523	834	Bacterium, obligate intracellular parasite, causes epidemic typhus (🔗 Sections 13.12 and 24.3)
Treponema pallidum	1,138,006	1,041	Bacterium, spirochete, causes syphilis (🔗 Sections 13.30 and 23.6)
Aquifex aeolicus	1,551,335	1,512	Bacterium, hyperthermophile (🔗 Section 13.34)
Methanococcus jannaschii	1,664,976	1,738	Archaeon, methanogen (🔗 Section 14.3)
Helicobacter pylori	1,667,867	1,590	Bacterium, causes peptic ulcers (🔗 Section 24.11)
Pyrococcus horikoshii	1,738,505	2,061	Archaeon, hyperthermophile (🔗 Section 14.9)
Methanobacterium thermoautotrophicum	1,751,377	1,855	Archaeon, methanogen (🔗 Section 14.3)
Haemophilus influenzae	1,830,137	1,743	Bacterium, can cause disease (🔗 Section 23.1)
Chlorobium tepidum	2,100,000	1,900	Bacterium, anoxygenic phototrophic green bacterium (🔗 Section 13.29)
Archaeoglobus fulgidus	2,178,400	2,436	Archaeon, hyperthermophile (🔗 Section 14.6)
Synechocystis sp.	3,573,470	3,168	Bacterium, cyanobacterium (🔗 Section 13.24)
Bacillus subtilis	4,214,810	4,100	Bacterium, gram-positive genetic model (🔗 Section 13.19)
Mycobacterium tuberculosis	4,411,529	3,924	Bacterium, causes tuberculosis (🔗 Sections 13.22 and 23.3)
Escherichia coli	4,639,221	4,288	Bacterium, gram-negative genetic model (🔗 Section 13.10)

a One sometimes finds the word "genome" used to refer to a prokaryotic chromosome. While in some cases this may actually be correct, remember that the genome actually includes *all* the genes found in an organism, even those of resident plasmids and viruses. In some cases these elements may not be found, and in some cases they may be very small or the genes carried by them may be of no importance to the organism's overall metabolism. However, this need not always be the case. For example, *Borrelia burgdorferi* contains 17 different plasmids consisting of 533 kilobase pairs (🔗 Table 6.2).

b Information on these and other genomes can be found in the TIGR Microbial Database (www.tigr.org/tdb/mdb/mdb.html), a Web Site maintained by The Institute for Genomic Research (TIGR), Rockville, MD, a not-for-profit research institute. The data for *Chlorobium tepidum* are estimates.

c Open Reading Frames (🔗 Section 6.10). The purpose of reporting ORFs is to predict the total number of proteins that an organism might encode. Therefore, constraints are placed on what is reported as an ORF. Of course, genes encoding known proteins are included, as are all ORF's that could encode a protein greater than 100 amino acid residues. Smaller ORFs are typically not included unless they show similarity to a gene from another organism or unless the codon usage is typical of the organism being studied.

d All other chromosomes in this list are circular.

Escherichia coli as a Model Prokaryote

Many factors have favored the use of *Escherichia coli* as the workhorse for studies of biochemistry, genetics, and bacterial physiology. As we have seen in Chapter 8, even *E. coli* viruses have served as model systems of study. Therefore, although the chromosome of *E. coli* was not among the first prokaryotic chromosomes sequenced, this organism remains the best-known microorganism. Indeed, one can argue it is the best-known cellular organism of any kind.* In addition to its important role as a model organism, *E. coli* continues to be the organism of choice for both research and applications of genetic engineering (🔗 Chapter 10).

The strain of *E. coli* whose chromosome was originally sequenced, MG1655, is a derivative of *E. coli* K-12,

*Detailed information on the genetics and metabolism of *E. coli* can be found in the two-volume book entitled Escherichia coli *and* Salmonella: *Cellular and Molecular Biology,* by F. C. Neidhardt et al., (see Web site for this chapter) or on the World Wide Web, for example, see EcoCyc: Encyclopedia of *E. coli* Genes and Metabolism prepared by P. D. Karp and M. D. Riley (ecocyc.PangeaSystems.com/ecocyc/ecocyc.html).

the traditional strain used for genetic studies. The "wild-type" *E. coli* K-12 is a lysogen of bacteriophage lambda (🔗 Section 8.12) and also contains the F-plasmid. However, the strain sequenced had been "cured" of lambda (by radiation) and of the F-plasmid (by acridine treatment, see Section 9.8). The chromosome of this strain contains 4,639,221 base pairs. Analysis of the genomic sequence showed there to be 4,288 possibly functional *open reading frames* (🔗 Section 6.10), which account for about 88% of the genome. Approximately 1% of the genome is comprised of genes coding for tRNAs and rRNAs, and about 0.5% is comprised of noncoding, repetitive sequences (🔗 Section 6.4). The remaining 10% contain all the regulatory sequences: promoters, operators, origin and terminus of DNA replication, and so forth.

Arrangement and Expression of Genes on the *E. coli* Chromosome

Early mapping experiments and studies on the regulation of the genes that control the enzymes of a single biochemical pathway had shown that these genes were often clustered. On the genetic map in Figure 9.35 a few such clusters are shown. Notice, for instance, the *gal* gene cluster at 18 min., the *trp* gene cluster at about 28 min., and the *his* cluster at 44 min. Each of these clusters is part of an operon and is transcribed as a single *polycistronic* mRNA (∞ Section 6.8). However, genes for other biochemical pathways are not clustered, for example, the genes involved in arginine biosynthesis (*arg* genes) are scattered around the chromosome, as part of the *arg* regulon (∞ Section 7.3). Because of the early discovery of multigene operons, and their usefulness as models for the study of gene regulation, for example the *lac* operon (∞ Sections 7.2 and 7.6), one often gets the impression that such operons are the rule in prokaryotes. However, sequence analysis of the *E. coli* chromosome has shown that of the 2,584 predicted or known transcriptional units, over 70% have only a *single* gene. Only about 6% of the operons have polycistronic mRNAs encoding four or more genes.

The transcription of some operons proceeds in one direction along the chromosome, whereas with other operons transcription is in the opposite direction. The direction of transcription of a few multigene operons is shown by arrows in Figure 9.35. Because transcription always occurs in a $5' \rightarrow 3'$ direction (∞ Section 6.7), this implies that transcription off either DNA strand can occur, and evidence indicates that there are about equal numbers of operons on both strands. It had previously been known that many genes that are highly expressed in *E. coli* are oriented so that they are transcribed in the same direction that the DNA replication fork moves through them. (The two replication forks start at the origin, *oriC*, at about 84 min. on the map shown in Figure 9.35, and move bidirectionally [∞ Section 6.6] toward the terminus, which is located at approximately 34 min.

on the map.) Genomic sequencing has confirmed this, demonstrating that all seven of the rRNA operons and 53 of the 86 tRNA genes are transcribed in the direction of replication.

Almost 2,000 *E. coli* proteins, or genes encoding proteins, had been identified before the chromosome had been completely sequenced. We now know that as many as 4,288 different proteins can theoretically be encoded by this organism, although approximately 38% of them are of unknown function and/or are simply hypothetical. As was expected the "average" *E. coli* protein contains slightly more than 300 amino acid residues, but many proteins are smaller and many are much larger. The largest gene should encode a protein of 2,383 amino acid residues whose function is unknown, but the gene is similar to those involved in pathogenesis in related organisms. Although *E. coli* has very few duplicated genes, many of the genes that encode proteins seem clearly to have arisen by gene duplication during evolutionary history. There are some large *gene families*, groups of genes with related sequence and with products that have related functions. For example, there is a family of 70 genes whose products are all membrane-transport proteins. Table 9.6 shows an approximate division of the protein encoding genes into categories based on function.

Lists of genes and gene functions can be very misleading. Representations such as the ones in Table 9.6 can mislead us into thinking that the relative level of activities in the cell are divided along the same lines. Remember that a list of genes contained in a genome is simply a list of proteins that the organism <u>can</u> make under <u>some</u> conditions. Such a list of course does not distinguish between a gene that is expressed at a very low level (2 or 3 copies of the protein in a cell) and one expressed at a very high level (tens of thousands of copies per cell). In addition, although almost 40% of the

TABLE 9.6	Gene function in bacterial genomes		
	Percentage of genes on chromosome[a] in that category		
Functional categories	*Escherichia coli*	*Haemophilus influenzae*	*Mycoplasma genitalium*
Metabolism	21.0	19.0	14.6
Structural	5.5	4.7	3.6
Transport	10.0	7.0	7.3
Regulation	8.5	6.6	6.0
Translation	4.5	8.0	21.6
Transcription	1.3	1.5	2.6
Replication	2.7	4.9	6.8
Other, known	8.5	5.2	5.8
Unknown	38.1	43.0	32.0

[a] For the size of the chromosome of each of these species and the number of open reading frames that each contains, see Table 9.5.

sequenced genes of *E. coli* appear to encode proteins of unknown functions, it is estimated that we have already identified the function of at least 80% of the genes that are required by *E. coli* to live under normal conditions.

Other Features of the *E. coli* Chromosome

Even though the strain of *E. coli* sequenced had been cured of lambda and the F-plasmid, the chromosome contained many other genetic elements that replicate as part of it. There are copies of 10 different IS elements, including seven copies of IS2 and five copies of IS3 (see Section 9.10). Both of these elements are also found on the F-plasmid and both are involved in the formation of Hfr strains (see Section 9.9 and Figure 9.24). There are also several different cryptic, defective, prophages (three of which are related to lambda) in the *E. coli* chromosome and several genes that are clearly parts of other prophages now almost completely lost through deletion. Moreover, *E. coli* has likely obtained a considerable amount of its genome by *horizontal gene transfer*, that is, genes that originated in other organisms. It has been estimated that at least 18% of the *E. coli* K-12 genome has originated from such transfers. Large scale changes in the genome can still occur by such mechanisms. Strains of *E. coli* are known that contain genes involved in virulence located on large, unstable regions of the chromosome called *pathogenicity islands*, which can be acquired by horizontal transfer. Interestingly, horizontal gene transfer does not necessarily result in an ever-larger genome size. Many of the genes acquired in this way provide no selective advantage and so are lost by deletion.

Analysis of the complete sequence of a genome can yield an incredible amount of information: seemingly limited only by the questions that the investigator wishes to ask. Since the molecular genetics of *E. coli* have been seriously explored for several decades, and the genome has now been sequenced, does this mean that further analysis will be done mostly by computer and that traditional genetic and biochemical studies of *E. coli* are over? Far from it! Although computer analysis of a sequence can yield much information, in order to understand the *function* of genes, and particularly of regulatory sequences, it is often necessary to isolate mutants, map their mutations, and use biochemical and physiological analyses to determine their effects on the organism.

9.12

Comparative Prokaryotic Genomics

As has been discussed (∞ Section 6.4), the word *genome* refers to the total complement of an organism's genes. In the mid-1980's the word **genomics** was introduced to describe the discipline of mapping, sequencing, and an-

alyzing genomes. The first aim of genomics is to totally map (and/or sequence) the entire genome of an organism. Although sequencing of microbial genomes has become fairly routine, it is still a considerable undertaking and there have been important scientific and societal reasons for undertaking the sequencing of the genomes of the species that has so far been accomplished.

One can get some feeling for this by looking at the organisms listed in Table 9.5. These are the first 18 prokaryotes whose chromosomes have been completely sequenced and the sequence reported. Note that eight of these organisms are pathogens, causing widespread and severe illness. The three hyperthermophiles on the list may have important uses in biotechnology, since the enzymes in these organisms will be heat stable (∞ Section 10.9). The needs of the biotechnology industry and medicine are an important consideration for generating the interest, and the funding, to pursue genomic sequencing. However, the list also includes organisms like *Bacillus subtilis* and *Escherichia coli*, which remain widely studied model systems (see Section 9.11).

Sizes of Prokaryotic Genomes

Analyzing genomic sequences allows us to answer some fundamental biological questions. For example, analysis of the small genomes of *Mycoplasma genitalium* and *Mycoplasma pneumoniae* shows that all 470 open reading frames (ORFs) found in *M. genitalium* are also present in *M. pneumoniae*. By seeking similar genes in other organisms, some scientists have speculated that slightly over 250 specific protein-encoding genes may be sufficient to maintain a cellular existence. Perhaps someday we will find, or construct, an organism with such a minimal genome! (Construction of such an organism is not an insuperable challenge with the genetic tools discussed in this chapter and in Chapter 10.) Note that *Methanococcus jannaschii* is an *autotroph* (∞ Sections 14.3 and 15.19) and contains only 1,738 ORFs. This is enough, however, to enable it to be not only free living, but also to synthesize all of its organic cellular components from CO_2 (∞ Section 14.3).

One might think that *E. coli* represents a kind of extreme, since its 4.6-Mb chromosome is the largest prokaryotic chromosome yet sequenced (Table 9.5). However, this probably represents the fact that very few chromosomes have as yet been entirely sequenced, and sequencing strategies are simpler when one is looking at relatively small genomes. Certainly there are prokaryotes with larger chromosomes than that of *E. coli*, for example, the *Myxococcus xanthus* genome is almost twice as large (∞ Table 6.2), and most *Streptomyces* species have chromosomes of approximately 8 Mb.

Genomic Analysis

The development and use of computer programs to analyze, store, and access DNA and protein sequences is an area called *bioinformatics.* This area has developed tremendously along with the explosion of sequence data available. Without bioinformatics it would not be possible to analyze and compare genomes from prokaryotes. One benefit of being able to compare genomes using this type of analysis is that it helps us to determine a gene's function and which genes are likely to be essential to a particular organism. One might imagine, for instance, that organisms like *Treponema pallidum,* which are obligately parasitic (♋ Section 23.6), might have relatively few genes required for amino acid biosynthesis since all the amino acids they need can be supplied by their hosts. In fact this is the case, as *T. pallidum* has no genes involved in amino acid biosynthesis, although it encodes several proteases that can convert peptides taken up from the host into free amino acids. In contrast, *E. coli* has 131 genes involved in amino acid biosynthesis and metabolism, and *B. subtilis,* a soil bacterium, has over 200. Some information on the division of genes and activities in prokaryotes is given in Table 9.6 and similar information is available for all the organisms whose genomes have been sequenced.

Two organisms listed in Table 9.5 are obligate intracellular pathogens, *Chlamydia trachomatis* (♋ Section 13.26) and *Rickettsia prowazekii* (♋ Section 13.12). Their genomes are also missing many genes whose products would be involved in biosynthesizing substances that could be supplied by a host. Interestingly, several genes found in the *C. trachomatis* chromosome seem to be the result of horizontal gene transfer from a *eukaryotic* source. This is the opposite of the situation where genes in the eukaryotic nucleus seem to have been transferred there from the ancestor of the mitochondrion, an organelle found in eukaryotic cells derived evolutionarily by endosymbiosis of a bacterium (♋ Section 12.4). *Rickettsia* are related to the bacteria that are believed to be the ancestor of the mitochondrion. The genome of *R. prowazekii* seems to be undergoing a reduction similar to that which resulted in the very small mitochondrial genomes. Many of the functions retained by the rickettsial genome are involved in energy production, a function associated with mitochondria. Interestingly 24% of the *R. prowazekii* genome is noncoding, the highest noncoding fraction found in any prokaryotic genome yet examined. It seems likely that these sequences are the remnants of functional genes that have not yet been eliminated from the genome.

Although there are differences from organism to organism, in most cases the number of genes that have actually been identified is about 60% or less of the number of ORFs shown. (An ORF whose product does not resemble any known protein is sometimes referred to as an *URF,* an *unidentified reading frame.*) This does not mean that these unidentified ORFs are not actual protein-encoding genes. Rather it reflects the fact that there is still much we don't know about prokaryotic genomes. Searching a genomic database to discover new genes or new functions is called "mining" and can lead to unusual and exciting discoveries (see the box, Genomic Mining). The number of identified genes will continue to increase as more and more becomes known about a particular cell's metabolism. In this regard, it is interesting that in the very well-studied *E. coli,* functions have been assigned to about 2700 of its almost 4300 genes. Many of the genes involved in macromolecular syntheses and central metabolism essential for growth of *E. coli* have been identified. Therefore, as the function of more URFs are identified, it is likely that most of them will be nonessential, and the percent of *E. coli* genes involved in key functions in macromolecular synthesis or central metabolism will decrease (Table 9.6).

Genomic analysis can also give us new insights into the ecology of an organism. For instance, *Helicobacter pylori* encodes proteins which contain twice the amount of the basic amino acids arginine and lysine (♋ Section 2.6) than typical proteins from other prokaryotes. Presumably this helps cells of *H. pylori* survive in the acid environment of the stomach. Also, many of the genomes of pathogens, for example, *Borrelia burgdorferi, Mycobacterium tuberculosis,* and *Treponema pallidum,* that have been sequenced show the presence of (unrelated) gene families that encode proteins that might be involved in antigenic variation and hence protect the organism from the host's immune system (*e.g.,* syphilis, ♋ Section 23.6). In *B. burgdorferi* these genes are located on plasmids. Information derived from genomics can be critical in designing new strategies to protect humans and other animals from these pathogens. The ability to compare thousands of genes has resulted in considerable insight into the metabolism of the organisms listed in Table 9.5. However, at this stage there are many questions that remain unanswered. For example, there are no obvious "rules" for protein folding leading to thermostability that can be derived from comparisons of the genomes of the three hyperthermophiles that have been sequenced.

Evolution and Gene Families

As we mentioned, the first priority of genomics is to determine the number and function of genes in an organism. However, there is more to genomics than counting genes and helping us understand how an individual organism copes with its environment. Comparative genomics also allows us to understand the evolutionary relationships between different organisms. Recon-

A FOCUS ON . . . Genomic Mining

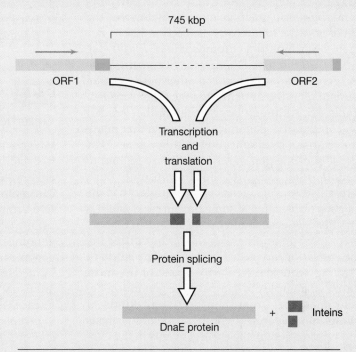

FIG. 1 The split *dnaE* gene of *Synechocystis*. The two ORFs are transcribed in opposite directions, as shown by the orange arrows. The protein products of these ORFs each contain part of an intein (note the protein product of ORF2 has been drawn in the standard amino terminal to carboxyl terminal orientation).

Searching through a database containing the complete sequence of an organism's genome to find a new gene is referred to as "mining." What kinds of things are mined? Often scientists are looking for genes that they expect to be present. But many other times they are surprised. And the DNA polymerase of the cyanobacterium *Synechocystis* is a good case in point.

All Bacteria contain a DNA polymerase similar to DNA polymerase III of *Escherichia coli*, which is used as the primary polymerase in DNA replication (⚮ Section 6.6). This is a complex enzyme containing many different subunits, and the subunit that catalyzes the actual polymerase reaction is DnaE, a product of the *dnaE* gene. Because DNA polymerases are highly conserved proteins and of prime importance to all cells, one would predict that inspection of the genome of *any* member of the Bacteria would yield a gene with sequence similarity (an *orthologous* gene) to the *dnaE* gene of *E. coli*. Interestingly, however, the original search of the genome of *Synechocystis* (see Table 9.5), failed to identify such a gene. Instead, what was found were two open reading frames (ORFs), which if combined, would form a gene with high similarity to *dnaE*. But these two ORFs were over 700,000 base pairs apart on the *Synechocystis* chromosome and located on opposite DNA strands, indicating that they were transcribed in opposite directions. Could these two ORFs be part of *dnaE*?

Careful sequence analysis of the two ORFs revealed that they each also encoded complementary halves of an **intein**, a self-splicing protein (⚮ box, Protein Processing, Chapter 7). The two *Synechocystis* ORFs were then cloned,

and it has now been demonstrated that they can be transcribed and translated and that this split intein can catalyze a splicing reaction between the two halves of DnaE to form a complete DnaE (see Fig. 1) and, presumably, a functional DNA polymerase. It is likely that this reaction also occurs in the *Synechocystis* cell.

Why does *Synechocystis* arrange its *dnaE* gene in this split fashion? This can only be guessed at this point, but one hypothesis involves regulation. Since there are only a few molecules of DNA polymerase III present in even a fast-growing cell like *E. coli*, it is likely that there are *very few* copies present in a much slower-growing cell like *Syne-*

chocystis. Thus, perhaps the unusual encoding of the *Synechocystis dnaE* gene in some way helps this organism regulate the level of this crucial enzyme. Of course it is also possible that this split gene/intein system for *dnaE*/DnaE is simply fortuitous and will be found in many other prokaryotes as well; only more complete genome sequences will confirm this possibility. But either way, it is difficult to imagine how this mechanism would have been discovered without being able to scan and compare entire genome sequences, as can be done routinely today. Mining genomes is thus likely to yield lots of surprises, and may also lead to discoveries that have practical applications as well. ■

structing evolutionary relationships helps to distinguish primitive from derived biological characteristics. This knowledge will help us better understand early life forms and, hopefully, help us eventually answer the question of how life first arose (Section 12.1).

When describing the chromosome of *E. coli* (see Section 9.11), we discussed the fact that it contains gene families, genes that are related to other genes within the organism. This is true of most of the organisms listed in Table 9.5. Although large gene families are not the rule, comparative genomics has shown that many genes have arisen by duplication of other genes. For instance, 47% of the genes in *B. subtilis* are related to one or more other genes on the chromosome. Such genes are called *paralogs,* genes whose similarity is the result of gene duplication at some time in the evolution of an organism. (Genes found in one organism that are similar to genes in another organism, but differ because of speciation, are called *orthologs.*) The study of genes having sequence similarity is one of the most important and complex in comparative genomics. Because chromosomes from many different kinds of prokaryotes have been sequenced, such comparisons can yield information on the earliest events in the evolution of cells. One interesting fact is that, as expected, the genes in the Archaea that are involved in DNA replication, transcription, and translation, are more similar to those of Eukarya than to those of Bacteria. However, unexpectedly, many of the rest of the genes in Archaea are more similar to those of Bacteria, than to those of Eukarya. (Of course, in the aggregate, Archaeal genes are most similar to those in other Archaeal genomes.) These results lend further support to the phylogenetic picture of life deduced originally by comparative ribosomal RNA sequence analysis (Section 12.7) and suggest that many genes in all organisms have common evolutionary roots.

These evolutionary considerations also have a practical component. The ability to identify genes that are clearly prokaryotic (since the genomes of eukaryotes are also being sequenced, see Section 9.13) or, in particular, genes characteristic of pathogenic bacteria, could lead to the design of very powerful and extremely specific therapeutic agents for use in clinical medicine.

✓ 9.11 and 9.12 Concept Check

Conjugation, transformation, and transduction have been important methods used to map genes on the chromosomes of Bacteria. Although still important tools, they have largely been supplanted by restriction enzyme analysis, cloning, and DNA sequencing, for study of bacterial genomes. With the availability of sequences of entire prokaryotic chromosomes, genomics is becoming an important discipline for analyzing the wealth of data. The study of genomes is yielding information on how organisms cope with their environments, what genes are essential, and how organisms are related.

✓ What is a gene family?
✓ Compare an ORF with a URF.

9.13

Genetics in Eukaryotic Microorganisms

We now turn from a discussion of genetic mechanisms in Bacteria to a consideration of analogous processes in eukaryotes, using yeast as a model system. Eukaryotes can mate during sexual reproduction, and therefore DNA transfer and recombination differ in many ways from that in prokaryotes. Unlike prokaryotes, whose genomes are single DNA molecules, eukaryotic genomes are segmented into a number of chromosomes located in the nucleus. While prokaryotic chromosomes are usually circular, eukaryotic chromosomes contain linear DNA molecules.

Typically, eukaryotic cells can be of two types, *haploid* or *diploid,* depending on the chromosome number. Diploids have two *sets* of chromosomes, so each chromosome is part of a pair. In the *haploid* phase the number of chromosomes per cell is n, and in the *diploid* phase, $2n$ (Figure 9.36). Thus, in the yeast *Saccharomyces cerevisiae* a haploid cell contains 16 chromosomes and the diploid 32. In humans, haploid and diploid cells contain 23 and 46 chromosomes, respectively. Multicellular plants and animals are usually diploid, with the haploid phase present in the germ cells (sperm and eggs, also called *gametes*), whose life spans are transitory. Eukaryotic microorganisms can be either haploid or diploid. Growing cells of *S. cerevisiae* can be either haploid or diploid (Figure 9.36). In most eukaryotic microorganisms, the diploid phase is transitory.

In diploid cells, two copies of each gene are present, one on each of the two homologous chromosomes. The term *allele* is used to refer to two alternate forms of the *same gene* present on the two homologous chromosomes. If the allele on one chromosome has a mutation preventing the normal product from being expressed, the allele on the other chromosome can continue to be expressed, and so the effect of the mutation may not be evident. Thus, the expression of one allele may be masked by the other. The gene that is expressed is said to be *dominant* to the other allele, which is said to be *recessive.* Diploidy presents difficulties in genetic research because isolation of mutants is much easier in a haploid cell, where only one form of the gene is present and, therefore, the effect of a mutation can be directly determined.

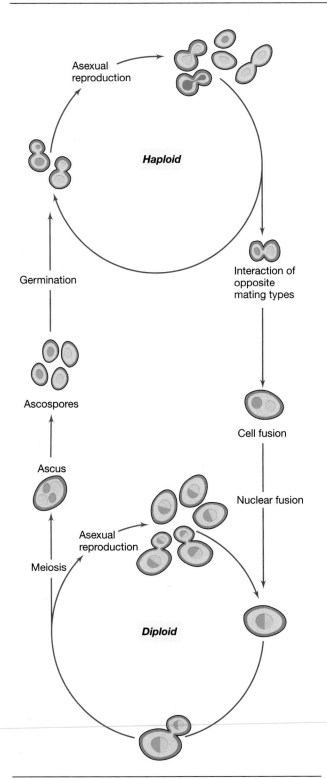

FIGURE 9.36 Life cycle of a typical yeast, *Saccharomyces cerevisiae*. A haploid cell of *S. cerevisiae* contains 16 chromosomes.

Meiosis

Mitosis is the process following DNA replication in which chromosomes condense, divide, and are sorted into two identical sets, one for each daughter cell. By contrast, **meiosis** is the process by which the change from the diploid to the haploid state is brought about. Meiosis involves two divisions. During the first meiotic division, the two sets of homologous chromosomes are segregated into two separate cells, and so the number of chromosomes is reduced from $2n$ to n, yielding haploid cells that are the precursors of the germ cells. The second meiotic division is similar to a mitotic division but involves n chromosomes in each of the precursor cells. The products of meiosis are four haploid gametes. For many microorganisms, the gametes look identical, so the distinction between egg and sperm or female and male cannot be made. Different mating types with identical appearances often exist.

If both members of a chromosome pair are genetically identical for a particular gene, the cell is said to be *homozygous* for that gene. Under such conditions, the haploid cells formed by meiosis are also identical for that gene. However, if the two chromosomes differ with respect to a particular gene, the cell is said to be *heterozygous*. Under such conditions, the four haploid gametes are not identical for that gene.

Once gametes are formed, **mating** of gametes of different type can occur, leading to the formation of diploids. The first diploid cell formed as a result of mating of haploid gametes is called a *zygote*. The diploid cell formed can then become the forerunner of a new population of genetically identical diploid organisms or, as is the case with most eukaryotic microorganisms, meiosis can generate haploid cells that can divide vegetatively.

✓ 9.13 Concept Check

Eukaryotic organisms can mate and exchange DNA during sexual reproduction. Haploid germ cells formed by meiosis can fuse to form a diploid zygote. Mitosis ensures appropriate segregation of the chromosomes during asexual cell division in eukaryotes.

✓ If the haploid number of chromosomes of an organism is 10, what is the diploid number?

✓ What does it mean to say that an allele is *dominant*?

9.14

Yeast Genetics

More is known about the molecular genetics of the yeast *Saccharomyces cerevisiae*, the common baker's and brewer's yeast (∞ box, Chapter 4), than about almost any other eukaryote. This is because yeast is easily grown in the laboratory and is an extremely favorable organism for genetic studies. This yeast is not only a useful model system for studies on eukaryotic genetics but is an important organism of commerce, and therefore studies of its genetics can be expected to have practical significance (∞ Section 11.12). Besides *S. cerevisiae*, other yeast species such as *Schizosaccharomyces pombe* and *Hansenula wingei* are also used in yeast genetics research.

A yeast grows as a single cell, and each haploid yeast cell is capable of serving as a gamete. Because yeast is unicellular and can be grown as a haploid, isolation of mutants is straightforward, and a large variety of mutant types are known. By mating mutants, genetic analyses of yeast can be carried out.

The life cycle of a typical yeast was shown in Figure 9.36. Many yeasts have two separate *mating types*, which can be considered analogous to male and female. However, the two mating types of a yeast are alike in structure and can be differentiated only by allowing them to mate. On mating of opposite types, a diploid cell is formed. In many yeasts, this diploid cell is capable of growing vegetatively, leading to the formation of a population of genetically identical, albeit diploid, cells. Under certain conditions, diploid cells of such a population can undergo meiosis and form haploid gametes. Two distinct types of gametes are formed, of opposite mating type. From a single diploid cell, a structure containing four such gametes is formed, two of each mating type. The cell in which the gametes are formed is called an *ascus*, and the cells within the ascus are called *ascospores*.

One important advantage of yeast is that genetic analysis is fairly straightforward. After mating and ascospore formation have occurred, the experimenter can dissect the four ascospores from the ascus, use each ascospore as the forerunner of a separate culture, and analyze the haploid cultures so obtained. In this way, it is possible to study recombination and *map* genes in yeast. Another advantage of yeast is that it can be transformed using exogenous DNA, and plasmids are available for use in genetic engineering.

Haploid cells of *Saccharomyces cerevisiae* have 16 chromosomes, and extensive genetic maps have been prepared for these chromosomes. These chromosomes range in size from 245 to 2200 kilobase pairs. The DNA sequence of all these chromosomes has now been determined, making this yeast the first eukaryote whose genome has been completely sequenced. Such sequence information aids genetic studies in many ways in addition to giving insight into genome evolution. The total haploid genome size of *Saccharomyces cerevisiae* is 12 *mega*base (10^6) pairs. This is only about three times that of *Escherichia coli* (see Table 9.5). However, it contains 6,034 ORFs, less than 50% more than the number found in *E. coli*.

While it is true that it is typically easier to isolate and examine mutations in haploids, there are times when it is particularly convenient to have an organism that is diploid, but from which a haploid phase can be easily derived and studied. Yeast is such an organism. Consider mutations that completely disable a gene that is essential to an organism. If such a mutation occurred in a haploid it would be invariably lethal. However, using the techniques of site-specific mutagenesis (∞ Section 10.11) one can isolate mutations in any specific gene in a diploid and then, by attempting to convert the strain to a haploid strain, determine whether the gene is essential. There is an organized effort by yeast geneticists to do this for every one of the *S. cerevisiae* ORFs.

Mating Type Genetics of Yeast

As we have noted, many yeasts have two mating types, which are indistinguishable except by their behavior in mating. The two mating types of *Saccharomyces cerevisiae* are designated α and a. Cells of type α mate only with cells of type *a*, and whether a cell is α or *a* is itself determined genetically. Some haploid strains of *S. cerevisiae* remain α or *a*, while other strains are periodically able to *switch* their mating type from one to the other. One consequence of this switching is that a pure haploid culture will produce diploids.

The switch in mating type from α to *a* and back to α has at its basis the behavior of a mobile genetic element, somewhat reminiscent of transposition or phase variation in *Salmonella* (see Section 9.10). The phenomenon in yeast is illustrated in Figure 9.37. There is a single active genetic locus, called the MAT (for *mating type*) locus, at which either gene α or gene *a* can be inserted. At this active locus, the MAT promoter controls the transcription of whichever mating type gene is present. Thus, if gene α is at that locus, then the cell is mating type α, whereas if gene *a* is at that locus, the cell is mat-

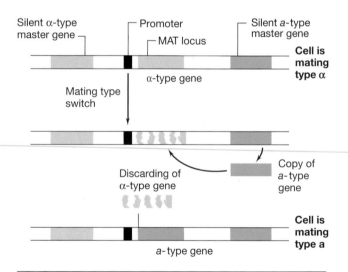

FIGURE 9.37 The cassette mechanism involved in the switch in yeast from mating type α to *a* and back again. Whichever "cassette" is inserted at the active locus (reading head) determines the mating type. The process shown is reversible, so type *a* can revert to type α.

ing type *a*. Somewhere else on the yeast genome are copies of both genes, α and *a*, which are not expressed. These *silent copies* serve as the source of the gene that is inserted when the switch occurs. When the switch occurs, the appropriate gene, α or *a*, is copied from its silent site and then inserted into the MAT location, *replacing* the gene already present. Thus, the old gene is excised out of the locus and discarded, and the new gene is inserted. This mechanism has been called the *cassette mechanism* because each gene can be considered analogous to a tape cassette inserted into a "reading head," the place on the chromosome where active transcription takes place.

The α and *a* genes are regulatory genes. Among other genes, they regulate the production of the *peptide hormones* called α factor and *a* factor, which are excreted by yeast cells undergoing mating. These hormones bind to cells of opposite mating type and bring about changes in the cell surfaces of these cells so cells of opposite mating type associate and fuse. It seems that α cells have receptors on their surfaces only for *a* factor, whereas *a* cells have receptors only for α factor. Once two cells of opposite mating type have associated, a complex series of events is initiated that leads to fusion of these two cells as well as their nuclei, resulting in the formation of a diploid zygote (see Figure 9.38).

Yeast Plasmids

We have defined plasmids as genetic elements that replicate independently of chromosomal DNA but have no extracellular form (∞ Sections 6.4 and 9.8). Yeast has a number of elements that could be considered plasmids by this strict definition. There are *retrotransposons* that accumulate viruslike particles inside the cell but are never released. In addition yeast also contains double-stranded RNA "viruses" whose particles accumulate inside the cell and are infectious only if yeast cells fuse with each other (which happens in mating). Both these types of elements are much more like viruses than typical plasmids, which are generally circular DNA molecules (with no protective particle). Howev-

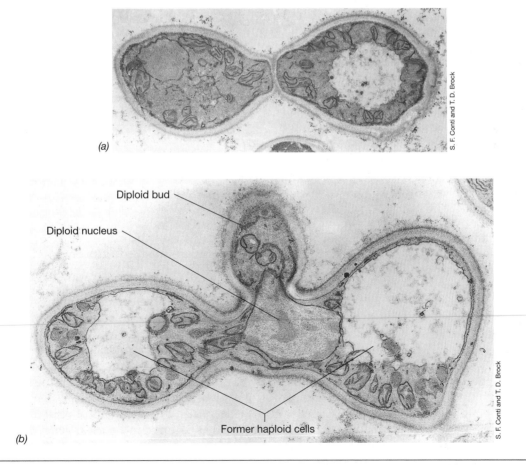

(a)

Diploid bud

Diploid nucleus

Former haploid cells

(b)

FIGURE 9.38 Electron micrographs of the mating process in a yeast, *Hansenula wingei*. (a) Two cells have fused at the point of contact and have sent out protuberances toward each other. (b) Late stage of mating. The nuclei of the two cells have fused, and the diploid bud has formed at a right angle to the conjugation tube. This bud eventually separates and becomes the forerunner of a diploid cell line.

er, most strains of yeast also contain such a typical plasmid, the so-called 2-μm circle. The 2-μm circle is a 6318-base-pair circular DNA molecule that replicates to high copy number (~70) in cells of *Saccharomyces cerevisiae*. The plasmid contains four protein-encoding genes, all involved in plasmid maintenance. It confers no phenotype on the cell.

The 2-μm circle is found within the nucleus packaged into nucleosomes (∞ Section 6.3). Segregation of the plasmid from mother to daughter cell during mitosis is random, but if the number of 2-μm circles drops to a low value, an increased rate of replication can bring the copy number back up. There is no evidence that this plasmid ever becomes integrated into the nuclear chromosomal DNA.

Although the 2-μm plasmid is a useful model for studying DNA replication in yeast, its greatest value appears to be as a vector for cloning foreign genes into yeast. The entire plasmid can be used as a vector, but, more often, DNA fragments from it are used in the construction of other yeast cloning vectors. Genes of interest can be inserted into these vectors and transformed into yeast cells with appropriate treatment (see Section 9.6). The value of vectors such as this for genetic engineering was discussed in Chapter 10.

Mitochondrial Inheritance in Yeast

One of the most interesting genetic analyses using yeast is that involving *mitochondrial inheritance*. It has been possible to isolate mutant yeast strains in which mutations have occurred in the mitochondrion rather than in the nucleus, and by genetic analysis the inheritance of the mitochondria themselves can be studied. Because inheritance of genetic characteristics via mitochondria occurs outside the nucleus, and outside the process of mitosis and meiosis, it is a form of **cytoplasmic inheritance** (sometimes called *non-Mendelian inheritance* to indicate that it does not follow Mendel's laws.)

An example of the manner in which mitochondrial characteristics are inherited in yeast is shown in Figure 9.39. As seen, when two yeast cells of opposite mating type that also differ in mitochondrial characteristics are mated, the outcome depends on which of the two types of mitochondria replicates most rapidly during subsequent cell divisions. In one class of mitochondrial mutants, called *petite*, large deletions in the mitochondrial DNA have led to abolition of all mitochondrial protein synthesis. Such mitochondria are nonfunctional, and yeast cells containing such mitochondria are unable to carry out respiration, although they are still able to grow anaerobically by fermentation processes. The term *petite* comes from the fact that these mu-

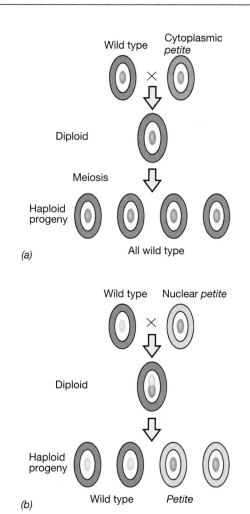

FIGURE 9.39 Mitochondrial inheritance. Outcome of crosses between petite and wild-type yeast: (a) cytoplasmic petite and (b) nuclear petite. The brown colored cells are those that can respire while petites can only generate energy by fermentation.

tants produce small colonies on agar both aerobically and anaerobically; wild-type cells form larger colonies aerobically. If a petite yeast cell is mated with a wild-type yeast cell, the diploid zygote will contain both types of mitochondria, but because in most cases the mutant mitochondria do not replicate as effectively as the wild-type mitochondria, they lose out in the competition during subsequent cell divisions. Ultimately, the cell lines derived from each of the four ascospores will all be wild type (Figure 9.39a). On the other hand, there are also petite mutants of yeast in which the mutation has occurred in a nuclear gene (since most of the proteins of the mitochondria are encoded by nuclear genes rather than mitochondrial genes). Crosses involving nuclear petites show conventional Mendelian inheritance (Figure 9.39*b*); the diploid is able to respire so the nuclear petite mutations are recessive.

Proteins encoded by yeast mitochondrial DNA include cytochrome *b*, cytochrome *c* oxidase, ATPase, and one ribosome protein. In addition, a number of tRNA molecules are encoded by yeast mitochondrial DNA, as are the ribosomal RNA molecules. Like most mitochondria, those of yeast encode relatively few of the required mitochondrial proteins. Interestingly, most of the yeast mitochondrial DNA does not encode proteins; this noncoding DNA has a very high AT content and its function is unknown.

✓ 9.14 Concept Check

The yeast *Saccharomyces cerevisiae* is a eukaryotic microorganism of great commercial value that is also widely used in genetic studies. It is the first eukaryote to have its genome completely sequenced. There are two mating types in yeast, and cells can convert from one to the other. In addition to the nuclear and mitochondrial chromosomes, the cell also has extrachromosomal elements.

✓ How does mating occur in yeast?
✓ Why do you suppose eukaryotic plasmids replicate in the *nucleus* rather than in the cytoplasm?

REVIEW QUESTIONS

1. Write a one-sentence definition of the term *genotype*. Do the same for the term *phenotype*. Does the phenotype of an organism automatically change when a change in genotype occurs? Why or why not? Can phenotype change without a change in genotype? In both cases, give some examples to support your answer.

2. Explain why an *Escherichia coli* strain that is His$^-$ is an auxotroph and one that is Lac$^-$ is not. (*Hint:* Think about what *E. coli* does with histidine and lactose.)

3. What are silent mutations and why do they occur? From your knowledge of the genetic code, why do you think most silent mutations affect the *third* position of the codon?

4. Microinsertions occur in promoters but are not frameshift mutations. Define the terms *microinsertion*, *frame shift*, *mutation*, and *promoter* (∞ Section 6.7). Explain how the statement can be true.

5. Explain how it is possible for a frameshift mutation early in a gene to be corrected by another frameshift mutation farther along the gene.

6. Give an example of one biological, one chemical, and one physical mutagen and describe the mechanism by which each causes a mutation.

7. What is site-specific mutagenesis? How can this procedure target specific genes for mutagenesis?

8. How does homologous recombination differ from site-specific recombination?

9. Why is it difficult in a single experiment using transformation to transfer a large number of genes to a cell?

10. From what you know about cell wall structure of Bacteria (∞ Sections 3.7–3.9), explain the problem a DNA molecule would encounter if the transformation process were to occur.

11. Explain why in generalized transduction one always refers to a transducing *particle* but in specialized transduction one refers to a transducing *virus* (or transducing phage).

12. Explain how it is possible to use the *interrupted mating procedure* to determine the relative order of genes on a bacterial chromosome.

13. Strains of *Escherichia coli* can be Hfr, F$^+$, or F$^-$. What are the differences between these strains and how would they behave in a mating experiment?

14. Explain why the insertion of a transposon leads to mutation.

15. The most useful transposons for isolating a variety of bacterial mutants are transposons containing antibiotic-resistance genes. Why are such transposons so useful for this purpose?

16. Compare and contrast the "flip-flop" process for *Salmonella* phase variation and the "cassette" mechanism for yeast mating type switching.

17. Differentiate clearly between the words *genome* and *chromosome* as they might be used referring to a prokaryote. In what cases would the two words have the same meaning?

18. Recombination in prokaryotes always involves chromosome fragments, whereas in eukaryotes it involves whole chromosomes. Explain.

19. Describe how certain functions in the yeast cell can be transmitted in non-Mendelian fashion.

APPLICATION QUESTIONS

1. Mutations within a gene encoding a protein can sometimes be suppressed by a mutation in another gene. One type of *suppressor mutation* involves a change in tRNA. Draw a diagram with coding sequences and amino acid sequences indicating how this occurs.

2. A constitutive mutant is a strain that continuously makes a protein that in the wild type is inducible. Describe two ways in which a change in a DNA molecule could lead to the development of a constitutive mutant. How could these two types of constitutive mutants be distinguished genetically?

3. In Chapter 6 we saw that it was critical for the ribosome to translocate with great accuracy in order to maintain the proper reading frame. However, sometimes ribosomes make frameshift errors. Compare the impact on the cell of a ribosome periodically making a frameshift error in the mRNA from a particular gene with the impact of a frameshift mutation in the same gene.

4. Although a large number of mutagenic chemicals are known, no chemical is known to induce mutations in a single gene (gene-specific mutagenesis). From what you know about mutagens, explain why it is unlikely that a gene-specific chemical mutagen will be found.

5. Describe the principle behind the Ames test. How is the test run in practice? From your knowledge of how mutants are isolated, why is the back mutation procedure used in the Ames test preferable to a forward mutation procedure?

6. What is the net result of general recombination? Why is it that the farther two genes are apart on a chromosome, the more likely they are to show recombination?

7. Defective specialized transducing phage can be replicated in a cell where a helper phage is replicating. However, a generalized transducing particle cannot replicate even if the cell is also infected with a wild-type phage. Explain how a helper phage "helps" the transducing phage and why this is not possible for the generalized transducing particle.

8. Some retroviruses (∞ Section 8.22) seem to be capable of acting as transducing viruses. Use your knowledge of the life cycle of these viruses to explain whether you think they would more likely participate in generalized or specialized transduction.

9. Transposable elements cause mutations when the element is inserted within the gene. These elements disrupt the continuity of a gene. One could also say that introns (∞ Section 6.9) disrupt the continuity of a gene yet the gene is still functional. Explain why the presence of an intron in a gene does not inactivate that gene but why insertion of a transposable element does.

10. Insertion sequences transpose (a type of site-specific recombination) in cells that have a defective *recA* gene. However, the formation of an Hfr strain from an F$^+$ strain cannot take place in a cell with a defective *recA* gene even though this is an event involving insertion sequences and recombination. Explain how Hfr formation takes place and why the RecA protein is essential.

11. Yeast can exist without mitochondria, but humans cannot. Explain.

Genetic engineering was born in the 1970s and the first vectors used to carry foreign DNA were naturally occurring plasmids and viruses. However, soon the vectors themselves were engineered and a new generation of vectors became available that contained sequences and genes that greatly enhanced the efficiency of cloning and recognizing clones. The plasmid pBR322, shown here, was one of these second generation plasmids, and one that proved to have very wide utility for cloning foreign DNA into *Escherichia coli*. Through the years, new and improved vectors have been developed while at the same time genetic engineering has evolved from being simply an improved way to map and analyze genes to become the very foundation of the ever-expanding biotechnology industry.

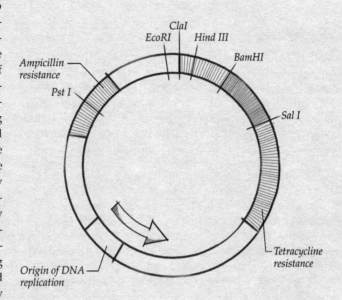

CHAPTER 10 — Genetic Engineering and Biotechnology

10.1 Molecular Cloning   345
10.2 Plasmids as Cloning Vectors   345
10.3 Bacteriophage Lambda as a Cloning Vector   347
10.4 Other Vectors   349
10.5 Hosts for Cloning Vectors   352
10.6 Finding the Right Clone   353
10.7 Expression Vectors   356
10.8 Synthetic DNA   359
10.9 Amplifying DNA: The Polymerase Chain Reaction   360
10.10 Cloning and Expression of Mammalian Genes in Bacteria   362
10.11 *In Vitro* and Site-Directed Mutagenesis   366
10.12 Practical Applications of Genetic Engineering   368
10.13 Production of Mammalian Products and Vaccines by Genetically Engineered Microorganisms   369
10.14 Genetic Engineering in Plant Agriculture   374
10.15 Genetic Engineering in Animal and Human Genetics   378
10.16 Genetic Engineering as a Microbial Research Tool   380
10.17 Summary of Principles at the Basis of Genetic Engineering   381

WORKING GLOSSARY

Biotechnology use of living organisms to carry out defined chemical processes for industrial application

Cloning vector genetic element into which genes can be recombined and replicated

DNA fingerprinting use of the techniques of genetic engineering to determine the origin of DNA in a sample of tissue

DNA library a collection of cloned DNA fragments, which in total contains cloned genes representing the entire genome of an organism; also called a gene library

Expression vector a cloning vector that contains the necessary regulatory sequences to allow transcription and translation of cloned genes

Gene disruption use of genetic techniques to inactivate a gene by inserting within it a DNA fragment containing an easily selectable marker. The inserted fragment is called a *cassette,* and the process of insertion, *cassette mutagenesis.*

Gene therapy treatment of a disease caused by a dysfunctional gene by introduction of a normally functioning copy of the gene

Molecular cloning isolation and incorporation of a fragment of DNA into a vector where it can be replicated

Nucleic acid probe a strand of nucleic acid that can be labeled and used to hybridize to a complementary molecule from a mixture of other nucleic acids

Polymerase chain reaction (PCR) a method used to amplify a specific DNA sequence *in vitro* by repeated cycles of synthesis using specific primers and DNA polymerase

Reverse translation the mental process of using a codon table and the amino acid sequence of a protein to obtain a possible sequence of the mRNA or the gene that encoded the protein

Shuttle vector a cloning vector that can replicate in two or more dissimilar hosts

Site-directed mutagenesis a technique whereby a gene with a specific mutation can be constructed *in vitro*

Synthetic DNA a DNA molecule made by a chemical process in a laboratory

T-DNA the segment of the *Agrobacterium* Ti plasmid that is transferred to plant cells

Ti plasmid a plasmid in *Agrobacterium* species capable of transferring genes from bacteria to plants

Transgenic organisms plants or animals that stably pass on cloned DNA that has been inserted into them

---

The concepts of molecular genetics, described in the four preceding chapters, have made possible the development of sophisticated procedures for the isolation, manipulation, and expression of genetic material, a field called **genetic engineering.** Genetic engineering has applications in both basic and applied research. In basic research, genetic engineering techniques are used to study the mechanisms of gene replication and expression in prokaryotes, eukaryotes, and their viruses. Some of the most important basic discoveries of molecular genetics have been made using genetic engineering techniques. For applied research, genetic engineering permits the development of microbial cultures capable of producing valuable products such as human insulin, human growth hormone, interferon, vaccines, and industrial enzymes. Genetic engineering for commercial application, sometimes called **biotechnology,** seems to have limitless potential.

Underlying genetic engineering is the isolation, purification, and replication of specific DNA fragments, a process called **molecular cloning.** Having large amounts of pure DNA allows localization, characterization, and manipulation of genes and their products. With cloned DNA we can determine the *nucleotide sequence* of a gene, from which we can derive, through the genetic code, the *amino acid sequence* of its protein product. The cloned DNA itself can be used as a *probe* to determine the structure of more complex DNA molecules like the human genome. By taking genes from organisms that are difficult or dangerous to work with and moving them into well-characterized, safe microorganisms, valuable biological substances can be produced cheaply and in quantities that were unthinkable until the advent of genetic engineering. By changing the sequence of a cloned gene in a predetermined way, genetic engineers can literally design new, possibly useful biological products that are simply unavailable from natural sources.

In this chapter we discuss the tools and processes involved in creating and purifying desired genes by techniques using *in vitro* recombination, or **recombinant DNA.** Molecular cloning involves creating recombinant DNA and introducing it into a host cell where it will be replicated. We then describe how genetic engineering is used to produce large quantities of desired gene products and how the products themselves can be altered by **site-directed mutagenesis.** We present examples of practical results derived from genetic engineering and conclude the chapter with a review of the principles that underlie biotechnology.

10.1

Molecular Cloning

Molecular cloning is at the base of most genetic-engineering procedures. The purpose of molecular cloning (also called *gene cloning*) is to isolate large quantities of specific genes in pure form. While it might be theoretically possible to physically isolate pure DNA fragments with single genes from a restriction enzyme digest of chromosomal DNA (∞ Section 6.5), a little reflection will demonstrate the impracticality of such an approach. Consider that even for a genetically simple organism like *Escherichia coli*, a specific gene represents 1–2 kilobases (kb) out of a genome of over 4600 kb. An average *E. coli* gene is thus less than 0.05% of the total DNA in the cell. In humans the problem is even worse because the coding regions of average genes are not much bigger than in *E. coli* but the genome is 1000 times larger! In contrast, the DNA of bacteriophage lambda is only 50 kb, and the DNA of some plasmids is less than 5 kb. In these genetic elements, a single average gene constitutes 2–40% of the DNA.

Thus, the basic strategy of molecular cloning is to move the desired gene from a large, complex genome to a small, simple one. Fortunately, our knowledge of DNA chemistry and enzymology allows us to break and join DNA molecules *in vitro*. This process is known as *in vitro recombination*. Restriction enzymes, DNA ligase (∞ Sections 6.5 and 6.6), and synthetic DNA (see Section 10.8) are important tools used for *in vitro* recombination.

Molecular cloning can be divided into several steps:

1. Isolation and fragmentation of the source DNA. This can be total genomic DNA from an organism of interest, DNA synthesized from an RNA template by reverse transcriptase (∞ Section 8.22), DNA synthesized by the polymerase chain reaction (see Section 10.9), or even totally synthetic DNA made *in vitro*. If genomic DNA is the source, it is generally cut with restriction enzymes to give a mixture of fragments.
2. Joining the DNA fragments to a **cloning vector** with DNA ligase. The small, independently replicating genetic elements used to replicate genes are known as cloning vectors, and most are derived from plasmids or viruses. Cloning vectors are generally designed to allow recombination of foreign DNA at a restriction site that cuts the vector in a way that does not affect its replication. If the source DNA and the vector are cut with the same restriction enzyme, joining can be mediated by annealing of the single-stranded regions called "sticky ends" (∞ Section 6.5 and Figure 6.8). "Blunt ends" generated by different restriction enzymes can also be joined, and different sticky ends

or blunt ends can be joined by the use of synthetic DNA **linkers** or **adapters.** The properties of cloning vectors are discussed in Sections 10.2–10.4.
3. Introduction and maintenance in a **host** organism. The recombinant DNA molecule made in a test tube is introduced into a host organism, for example, by DNA transformation (∞ Section 9.6) where it can replicate. Transfer of the DNA into the host usually yields a mixture of clones. Some cells contain the desired cloned gene, whereas other cells contain other clones generated by joining the source DNA to the vector. Such a mixture is known as a **DNA library** or a **gene library** because many different clones can be purified from the mixture, each containing different cloned DNA segments from the source organism. Making a gene library by cloning random fragments of a genome is called **shotgun cloning.**
4. Detection and purification of the desired clone. Often one of the most difficult tasks is finding the right clone in a mixture that may contain thousands of others. Techniques for finding the right clone will be discussed in Section 10.6.
5. Production of large numbers of cells or bacteriophage containing the desired clone for isolation and study of the cloned DNA.

✓ 10.1 Concept Check

The isolation of large quantities of a specific gene by molecular cloning is usually done using a plasmid or virus as the cloning vector. Restriction enzymes and DNA ligase are used in an *in vitro* recombination procedure to produce the hybrid DNA molecule. Once introduced into a suitable host, the target DNA can be produced in large amounts under the control of the cloning vector.

✓ What is the purpose of molecular cloning?
✓ What are the roles of a cloning vector, restriction enzymes, and DNA ligase in molecular cloning?

10.2

Plasmids as Cloning Vectors

Plasmids replicate independently of the host chromosome (∞ Section 9.8). In addition to carrying genes required for their own replication, most plasmids are natural vectors because they often carry other genes that confer important properties on their hosts (∞ Section 9.8). Such genes can be acquired by recombination within the host. In genetic engineering, geneticists add genes to a plasmid in a test tube.

Plasmids have very useful properties as cloning vectors. These properties include (1) small size, which makes the DNA easy to isolate and manipulate; (2) circular DNA, which makes the DNA more stable during chemical isolation; (3) independent origin of replication, so plasmid replication in the cell proceeds independently from direct chromosomal control; (4) multiple copy number, so they can be present in the cell in several or numerous copies, making amplification of the DNA possible; (5) the presence of selectable markers such as antibiotic resistance genes, making detection and selection of plasmid-containing clones easier.

Although in the natural environment conjugative plasmids are generally transferred by cell-to-cell contact, plasmid cloning vectors generally have been modified to prevent their transfer conjugatively in order to achieve biological containment. However, transfer in the laboratory can be brought about by transformation or electroporation (∞ Section 9.6). Depending on the host–plasmid system, replication of the plasmid may be under tight cellular control, in which case only a few copies are made, or under relaxed cellular control, in which case a large number of copies are made. Achievement of high copy number is often important in gene cloning, and by proper selection of the host–plasmid system and manipulation of cellular macromolecule synthesis, plasmid copy numbers of several thousand per cell can be obtained.

An example of a suitable cloning plasmid is pBR322, which replicates in *Escherichia coli* (Figure 10.1). Plasmid pBR322 has a number of characteristics that make it suitable as a cloning vehicle:

1. It is relatively small, only 4361 base pairs.
2. It is stably maintained in its host (*Escherichia coli*) in relatively high copy number, 20–30 copies per cell.

3. It can be amplified to a very high number (1000–3000 copies per cell, about 40% of the genome!) by inhibition of protein synthesis by the addition of chloramphenicol.
4. It is easy to isolate in the supercoiled form using a variety of simple techniques (∞ box, Working with Nucleic Acids: The Tools, Chapter 6).
5. A reasonable amount of foreign DNA can be inserted, although inserts of more than 10 kilobases lead to plasmid instability.
6. The complete base sequence of this plasmid is known, making it possible to locate sites where restriction enzymes can act.
7. There are *single* cleavage sites for various restriction enzymes such as *Pst*I, *Sal*I, *Eco*RI, *Hind*III, and *Bam*HI. It is important that only a single recognition site for at least one restriction enzyme is available so treatment with that enzyme opens the plasmid to a full-length linear molecule but does not cut it into pieces.
8. It has a gene conferring ampicillin resistance on the host and another conferring tetracycline resistance. These permit ready selection of hosts containing the plasmid. The sites recognized by some of the restriction enzymes are within one or the other of these resistance genes, facilitating the identification of plasmids carrying cloned DNA (see later).
9. It can be placed into cells easily by transformation.

The use of plasmid pBR322 in gene cloning is shown in Figure 10.2. As seen, the *Bam*HI site is within the gene for tetracycline resistance and the *Pst*I site is within the gene for ampicillin resistance. If foreign DNA is inserted into one of these sites, the antibiotic resistance conferred by the gene containing this site is lost, a phenomenon called **insertional inactivation.** Insertional inactivation is used to detect the presence of foreign DNA within the plasmid. Thus, when pBR322 is digested with *Bam*HI and linked with foreign DNA, and transformed bacterial clones then isolated, those clones that are both ampicillin-resistant and tetracycline-resistant *lack* the foreign DNA (the plasmid incorporated into these cells represents vector DNA that had recyclized without picking up foreign DNA). However, those cells still *resistant* to ampicillin but *sensitive* to tetracycline *contain* the plasmid with inserted foreign DNA. Since ampicillin resistance and tetracycline resistance can be determined independently on agar plates, isolation of bacteria containing the desired clones and elimination of cells not containing the plasmid can readily be accomplished.

The first plasmid cloning vectors used were naturally occurring plasmids. The plasmid pBR322 represents a later generation of cloning vectors, themselves constructed *in vitro*. The tetracycline resistance gene of pBR322 came from one plasmid, the replication origin from another, and the ampicillin resistance gene from the transposon Tn3 (∞ Section 9.10). There are now

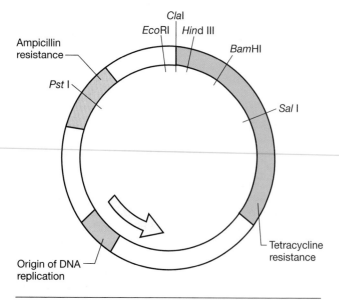

FIGURE 10.1   The structure of plasmid pBR322, a typical cloning vector, showing the essential features. The arrow indicates the direction of DNA replication from the origin.

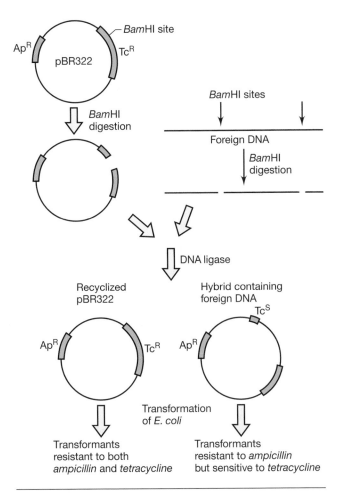

FIGURE 10.2 The use of plasmid pBR322 as a cloning vector, showing how insertion of foreign DNA causes inactivation of the tetracycline resistance gene, permitting easy identification of transformants containing the cloned DNA fragment.

newer generations of plasmid vectors that have been engineered to have even more useful features and are even simpler to use. These new features almost always include a **polylinker** or **multiple cloning site,** a short segment of DNA with many different restriction sites, each unique to the vector. This polylinker is usually contained in the coding region of a gene where insertional inactivation is very easy to monitor. Such features are also found in bacteriophage vectors, and a specific example is discussed in Section 10.3. Sometimes insertional in activation can be used as a positive selection. For example, there is a protocol that allows selection of bacteria that have lost tetracycline resistance. In some of the newly developed vectors, the gene carrying the polylinker normally produces a protein that is lethal to the host cell. Therefore, only cells containing a plasmid in which this gene has been *inactivated* can grow.

Cloning in plasmids such as pBR322 is a versatile and fairly general procedure of wide use in genetic engineering, particularly when the fragment to be cloned

is fairly small. Plasmids are often the best cloning vectors if *expression* of the cloned gene is desired (see Section 10.7). Plasmid vectors based on the 2-μm circle (∞ Section 9.14) are also available for cloning in the yeast *Saccharomyces cerevisiae*.

✓ 10.2 Concept Check

Plasmids are useful cloning vectors because they are easy to isolate and purify and are able to multiply to high copy numbers in bacterial cells. Antibiotic resistance genes of the plasmid are used to identify bacterial cells containing the plasmid.

✓ Explain why it is best to use a restriction enzyme that cuts the plasmid vector only once.
✓ What is *insertional inactivation*?

10.3

Bacteriophage Lambda as a Cloning Vector

We have discussed the fact that during specialized transduction (∞ Section 9.7) some host genes become incorporated into a bacteriophage genome. One phage that is used as a specialized transducing phage is bacteriophage lambda (∞ Section 8.12). During specialized transduction, lambda acts as a vector but the recombination occurs in the cell, not in a test tube. Lambda can also be used as a cloning vector for *in vitro* recombination. It is a particularly useful cloning vector because its molecular genetics is well known, it can hold larger amounts of DNA than most plasmids, and DNA can be efficiently packaged into phage particles *in vitro*. These can be used to infect suitable host cells, and infection is much more efficient than transformation (transfection). Lambda has a complex genetic map (∞ Figure 8.26) and a large number of genes. However, the central third of the lambda genome, between genes *J* and *N*, is not essential and can be replaced with foreign DNA.

Modified Lambda Phages

Wild-type lambda is not suitable as a cloning vector because it has too many restriction enzyme sites. To avoid this difficulty, modified lambda phages have been constructed that can be used in cloning. In one set of modified lambda phages, the so-called Charon phages, unwanted restriction enzyme sites have been removed by point mutation, deletion, or substitution. In variants that have only a single restriction site, a foreign piece of DNA can be *inserted*, whereas in variants with two sites, the foreign DNA can *replace* a specific segment of the lambda DNA. The latter variants, called **replacement vectors,** are especially useful in cloning large DNA fragments.

Figure 10.3 shows some of the essential features of a wild-type lambda and two of the Charon vectors. Whereas wild-type lambda contains five *Eco*RI sites, Charon 4A

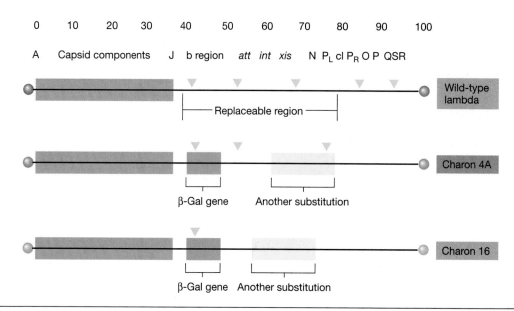

FIGURE 10.3   Molecular cloning with lambda. Abbreviated genetic map of bacteriophage lambda showing the cohesive ends as circles (◯◯ Figure 8.26). Charon 4A and 16 are both derivatives of lambda, which have various substitutions and deletions in the nonessential region. One of the substitutions in each case is a gene (β-Gal) that codes for the enzyme β-galactosidase, which permits detection of clones containing this phage. Whereas the wild-type lambda genome is 48.5 kilobase pairs, that for Charon 4A is 45.4 and that for Charon 16 is 41.7 kilobase pairs. The arrows (▼) shown above the maps of each phage indicate the sites recognized by the restriction enzyme *Eco*RI.

contains three and Charon 16 only one. Charon 4A is used as a replacement vector; the two small interior fragments are cut out and discarded during cloning. With Charon 16, the DNA to be cloned is inserted at the single *Eco*RI site. Both Charon 4A and 16 contain deletions (not shown in the figure) that not only remove some sites found in the wild-type lambda but also make the genome smaller. This allows the cloning of larger DNA fragments.

Both vectors also contain substitution mutations, which are shown in Figure 10.3. One of the substitutions is the gene for β-galactosidase. When the vectors replicate on a lactose-negative (Lac⁻) strain of *Escherichia coli*, β-galactosidase is synthesized from the phage gene and the presence of lactose-positive (Lac⁺) plaques can be detected by using a color indicator agar (see Section 10.4). If a foreign gene is inserted *into* the β-galactosidase gene, the Lac⁺ character is lost. Such Lac⁻ plaques can be readily detected as colorless plaques among a background of colored plaques.

Steps in Cloning with Lambda

Cloning with lambda replacement vectors involves the following steps (Figure 10.4):

1. Isolation of the vector DNA from phage particles and digestion with the appropriate restriction enzyme.
2. Connection of the two lambda fragments to fragments of foreign DNA using DNA ligase. Conditions are chosen so molecules are formed of a length suitable for packaging into phage particles.

3. Packaging of the DNA by adding cell extracts containing the head and tail proteins and allowing the formation of viable phage particles.
4. Infection of *E. coli* and isolation of phage clones by picking plaques on a host strain.
5. Checking recombinant phage for the presence of the desired foreign DNA sequence using nucleic acid hybridization procedures or observation of genetic properties.

Selection of recombinants is less of a problem with lambda replacement vectors (such as Charon 4A) than with plasmids because (1) the efficiency of transfer of recombinant DNA into the cell by lambda is very high, and (2) lambda fragments that have not received new DNA are too small to be incorporated into phage particles.

Although lambda is a useful cloning vector, there are limits on how much DNA can be inserted. Viability of phage particles is low if the DNA is longer than 105% of normal lambda DNA, and some lambda genes cannot be discarded and still maintain the vector's ability to replicate. Therefore, really large DNA fragments (greater than 20 kb) cannot be efficiently cloned.

Cosmids

A related type of vector that employs specific lambda genes is called a **cosmid.** Cosmids are plasmid vectors containing foreign DNA plus only the *cos* (cohesive end) site from the lambda genome. These *cos* sites are re-

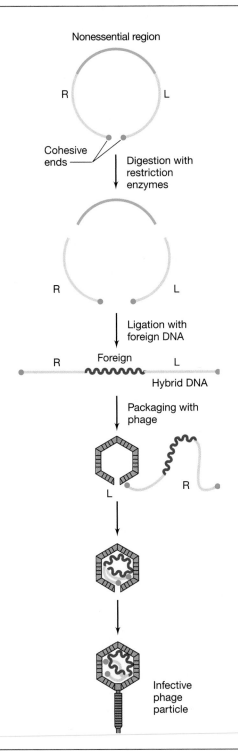

Nonessential region

R   L

Cohesive ends

Digestion with restriction enzymes

R   L

Ligation with foreign DNA

R   Foreign   L

Hybrid DNA

Packaging with phage

L   R

L

Infective phage particle

FIGURE 10.4   The use of bacteriophage lambda as a cloning vector. (See text for details.)

quired for packaging DNA into lambda virions. Cosmids are constructed from plasmids containing cloned DNA by ligating the lambda *cos* region to the plasmid DNA. The modified plasmid can then be packaged into lambda virions *in vitro* as described previously, and the phage particles used to transduce *Escherichia coli*. Cosmid construction avoids the necessity of having to trans-

form *E. coli*, which at best is an inefficient process (◯◯◯ Section 9.6).

One major advantage of cosmids is that they can be used to clone large fragments of DNA. Therefore, fewer clones are needed to obtain representation of the whole genetic element. This has been especially useful in the cloning of genes from eukaryotic chromosomes, where large amounts of DNA are involved. Another advantage of cosmids is that the DNA can be stored in phage particles instead of in plasmids. Phage particles are much more stable than plasmids, so the recombinant DNA can be kept for long periods of time (gene banking).

✓ 10.3 Concept Check

Bacteriophages such as lambda have been modified to make useful cloning vectors. Larger amounts of foreign DNA can be cloned with lambda than with many plasmids. In addition, the recombinant DNA can be packaged *in vitro* for efficient transfer to a host cell. Plasmid vectors containing the lambda *cos* sites are called cosmids, and they can carry a large fragment of foreign DNA.

✓ Why is the ability to package recombinant DNA in a test tube useful?
✓ What is a *replacement vector?*

10.4

Other Vectors

A large number of other vectors have been developed that are useful for various purposes in recombinant DNA technology. It is understandable that, in the early phases of the development of recombinant DNA techniques, the vectors used were those that had been in most widespread use in other types of genetic research. However, it does not necessarily follow that what is useful in basic research automatically provides the best system for the development of a cloning vector for specific applications beyond those needed for simply cloning a DNA fragment. In this section, we discuss briefly a few of the other kinds of vectors.

One important class of vectors, **expression vectors,** are used when one desires to obtain synthesis of the protein coded for by the foreign gene cloned into the vector. We discuss expression vectors in Section 10.7. Related to expression vectors are **secretion vectors,** in which the protein product is not only expressed but also secreted (excreted) from the cell. In these vectors the gene is cloned so the protein expressed carries a *signal sequence* (◯◯◯ Section 6.12).

For a number of reasons it is useful to be able to move DNA between completely unrelated organisms. To move DNA between unrelated organisms, a **shuttle vector** is used. A shuttle vector is one that can replicate in two different organisms. Like most specialized vectors, shuttle vectors have themselves been constructed using recombinant DNA techniques. Shuttle vectors have been developed that replicate in both *Escherichia coli* and *Bacillus subtilis, E. coli* and yeast, and *E. coli* and mammalian cells, as well as in many other pairs of organisms.

Vectors Derived from Bacteriophage M13

M13 is a filamentous phage containing single-stranded DNA and replicates without killing its host (Section 8.10). Mature particles of M13 are released from host cells by a budding process, and it is possible to obtain infected cultures that can provide continuous sources of phage DNA. An important feature of M13 is its single-stranded DNA. In the Sanger procedure for DNA sequencing (box, Working with Nucleic Acids: The Tools, Chapter 6), single-stranded DNA is needed and DNA cloned into M13 thus provides a ready source of this single-stranded DNA (although single-stranded DNA can also be generated by denaturation). Also, single-stranded DNA is very useful as a probe for detecting other nucleic acid sequences in transfer procedures such as the Southern blot, and M13 permits ready production of such single-stranded DNA probes.

However, in order to use M13 for cloning, a double-stranded form must be available because restriction enzymes work only on double-stranded DNA. Double-stranded M13 DNA can be obtained from infected cells because M13 replicates in the host as a double-stranded *replicative form* (Section 8.10).

Most of the genome of wild-type M13 contains genetic information essential for virus replication. However, there is a small region called the intergenic sequence that can be used as a cloning site. Variable lengths of foreign DNA, up to about 5 kilobase pairs (kbp), can be cloned without affecting phage viability—as the genome gets larger, the virion gets longer. M13mp18 is a derivative of M13 in which the intergenic region has been modified to facilitate cloning. A map of this vector is shown in Figure 10.5*a*.

One modification is the insertion of a functional fragment of *lac*Z, the *Escherichia coli* gene that encodes the enzyme β-galactosidase. Therefore, cells infected with M13mp18 can be easily detected by their color on indicator plates (Figure 10.5*b*). This *lac*Z gene has itself been modified to contain a 54-base-pair DNA fragment called a *polylinker*. The polylinker contains several restriction sites unique in M13 and can therefore be used for cloning. The polylinker is inserted into the beginning of the coding portion of the *lac*Z gene. This small insertion is in-frame, and the extra 18 amino acids do not affect the activity of the enzyme encoded by the gene. However, insertion of additional DNA into the polylinker during cloning inactivates the gene. Phages that contain additional DNA inserts give rise to color-

less plaques, and it is therefore very simple to identify clones (Figure 10.5*b* and *c*). Similar constructs are used in lambda cloning vectors and plasmid cloning vectors to allow identification of cells containing cloned DNA.

How then are M13 vectors used in cloning? The replicative double-stranded DNA is isolated from the infected host and treated with a restriction enzyme. The foreign DNA, also double-stranded, is treated with the same restriction enzyme. On ligation, double-stranded M13 molecules are obtained that contain the foreign DNA. When these molecules are introduced into the cell by transformation, they replicate and in time produce mature bacteriophage particles containing single-stranded DNA molecules. Only one strand of DNA is packaged into mature phage. Which of the two foreign strands the mature phage contains depends on the *orientation* in which the strand was inserted. Because foreign DNA can be inserted (in separate phage molecules) in either orientation, *both* strands of the foreign DNA can be cloned.

The single-stranded M13 DNA containing the foreign DNA can then be used in DNA sequencing. Since the base sequence where the foreign DNA is inserted is known (based on the specificity of the restriction enzyme used), it is possible to use a synthetic oligonucleotide complementary to this region as a primer and hence determine the sequence of the whole DNA downstream from this point. In this way, M13 derivatives have proved extremely useful in sequencing foreign DNA, even rather long molecules. M13 vectors have been used as a tool in the sequencing of many viral and prokaryotic genomes.

Vectors have also been constructed that are hybrids between a filamentous phage, like M13, and a plasmid. Such vectors are called **phagemids.** They contain both phage and plasmid origins of replication. Normally replication is from the plasmid origin, but when a cell containing a phagemid is infected with a wild-type phage, the phage origin on the vector is used to synthesize single-stranded DNA from the vector (and whatever cloned gene it might carry). This single-stranded DNA is packaged into virions and can easily be isolated and used for sequencing. Usually, phagemids can stably carry a larger fragment of cloned DNA than a typical M13-derived vector.

Yeast Artificial Chromosomes and the Human Genome Project

As mentioned in Sections 10.2 and 10.3, plasmid vectors usually contain less DNA than lambda vectors, and lambda vectors less than cosmid vectors. When making *gene libraries* of bacteria or simple eukaryotes, lambda or cosmid vectors can be used and the entire library will contain at most a few thousand different clones. The size of a gene library (the number of clones) re-

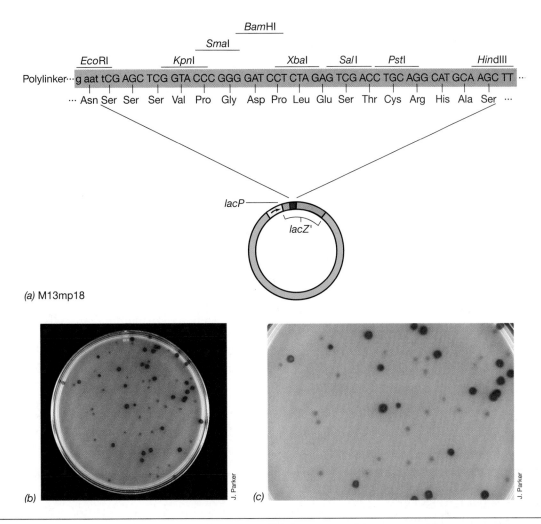

FIGURE 10.5   (a) A partial map of M13mp18, a derivative of M13 constructed for use as a cloning vector. The vector contains the *lac* promoter and a gene, *lacZ'*, which encodes a functional part of β-galactosidase. At the beginning of this gene is a polylinker that contains several restriction sites but maintains the proper reading frame. The amino acids encoded by the polylinker are shown. Most DNA fragments cloned into the polylinker disrupt the *lacZ'* gene and abolish β-galactosidase activity. (b) A plate with plaques formed by M13mp18 and by clones made using this vector on a lawn of sensitive bacteria plated on a medium containing the chemical 5-bromo-4-chloro-3-indolyl-β-D-galactopyranoside, called X-gal. When β-galactosidase hydrolyzes X-gal, it releases a relatively insoluble blue dye. Many plaques on this plate are blue, indicating the presence of vector without cloned DNA. However, many of the plaques are colorless, indicating that foreign DNA has been inserted into the vector and the *lacZ'* gene has been disrupted. (c) An enlargement of a portion of this plate.

quired to contain the complete genome of an organism depends on both the size of the genome and the size of the insert that can be placed in the vector. If the average size of the insert is 20 kb, then it will take many more clones to make a library of a typical mammalian genome (3×10^9 base pairs) than of a bacterial genome (4×10^6 base pairs). Currently, there is a concerted effort in the international genetics community to map, clone, and sequence the entire human haploid genome—the *Human Genome Project*. This is clearly a much larger undertaking than sequencing prokaryotic chromosomes (∞ Sections 9.11 and 9.12). It is useful in such a project to have a cloning vector that can hold very large segments of DNA so the size of the initial gene library can

be limited. Such vectors have been developed and are called **yeast artificial chromosomes** (YACs).

These vectors have been designed to replicate in yeast like normal chromosomes, but they have sites where DNA can be inserted. To function like normal eukaryotic chromosomes, YACs must have an *origin of DNA replication, telomeres* at the ends of the chromosome (∞ Section 6.6), and a *centromere* (the section of the chromosome required for segregation during mitosis). They must also contain a cloning site and a gene that can be used for selection after transformation into the host. Figure 10.6 shows a diagram of a YAC vector into which foreign DNA has been cloned. YAC vectors are themselves only about 10 kilobase pairs, but they can

FIGURE 10.6 Diagram of a yeast artificial chromosome (YAC) containing foreign DNA. The foreign DNA was cloned into the YAC vector at a *Not*I restriction site. The telomeres at the end of the YAC are labeled TEL and the centromere CEN. The origin of replication is labeled ARS (for autonomous replication sequence). For this vector, the gene used for selection is called URA3. The host into which the clone is transformed has a mutation in that gene so that it normally requires uracil for growth (Ura⁻). Host cells containing this YAC become Ura⁺. The diagram is not drawn to scale; the inserted DNA would normally be 200–800 kilobase pairs long and the vector about 10 kilobase pairs.

have 200–800 kilobase pairs of cloned DNA inserted. After identifying a particular gene or region in the cloned DNA on a YAC, this gene can be *subcloned* into a plasmid or bacteriophage vector for more detailed analysis.

Other Eukaryotic Vectors

In addition to the YAC vectors described previously, we have mentioned that plasmid vectors are available for use in the yeast *Saccharomyces cerevisiae* (see Section 10.2). The development of vectors for use in eukaryotes, like the same process in prokaryotes, depends on the scientific questions geneticists want to ask or the practical goals of the genetic engineer. When cloning in yeast, YACs are used to clone very large fragments of DNA, and the plasmid vectors are used for much smaller fragments. Although yeast is an extremely useful organism both for genetic studies (∞ Section 9.14) and commercial applications (∞ Sections 11.12 and 11.13), it is often important to use other eukaryotes as hosts for cloned DNA. Many cloning vectors have been developed for many different eukaryotes, including plants (see Section 10.14).

Most vectors used in the higher eukaryotes are virus vectors. The DNA virus SV40 (∞ Section 8.18), a virus causing tumors in primates, has been developed as a cloning vector for human tissue culture lines. SV40 virus has double-stranded *circular* DNA, and its entire nucleotide sequence is known. Derivatives of SV40 that do not induce tumors have been developed for cloning mammalian genes, and also for the expression of these genes. SV40 and other mammalian cloning vectors are proving very useful for understanding the mechanisms involved in gene expression in these complex organisms.

There are mammalian vectors that utilize *adenovirus* (∞ Section 8.21) and *vaccinia virus* (∞ Section 8.20). Vaccinia virus vectors have been used in the development of new vaccines (see Section 10.13). A variety of eukaryotic expression vectors have also been developed and are essentially of two kinds. One type is designed to produce a particular protein for commercial purposes. Vectors derived from *baculovirus*, a DNA virus that replicates in insect cells, can be used to make large quantities of the products of cloned genes. Other expression vectors are being developed so a cloned gene can be stably maintained and expressed in an organism or tissue, often as an approach to *gene therapy* (see Section 10.15).

The *retroviruses* (∞ Section 8.22) can be used to introduce genes into mammalian cells because these viruses replicate through a DNA form that becomes integrated into the host chromosome.

Another approach has been the development of human artificial chromosomes (HACs). Unlike YACs, whose original purpose was to carry very large fragments of DNA, HACs are seen as vectors that should be stably maintained in human cells and carry genes that can be expressed in a normal fashion. However, the ability of HACs to carry large amounts of cloned DNA can be useful because some normal human genes and their regulatory regions are very large—over a million base pairs in many cases—because of the presence of very large introns.

10.5

Hosts for Cloning Vectors

The ideal characteristics of a host for cloned genes are rapid growth, capability of growth in an inexpensive culture medium, not harmful or pathogenic, ability to take up DNA, and stability in culture. The host must have the appropriate enzymes to allow replication of the vector. The most useful hosts for cloning are microorganisms that grow well and for which we have both sufficient genetic information and the tools for genetic manipulation. These include the Bacteria *Escherichia coli* and *Bacillus subtilis* and the yeast *Saccharomyces cerevisiae*. However, many basic scientific questions can be answered only if the cloned DNA can be returned to the species of organism from which it originated. This is particularly true in studies involving gene regulation. Finally, if recombinant DNA itself is to be used therapeutically to treat human disease, the host must be a human being.

Prokaryotic Hosts

Although most molecular cloning has been done in *Escherichia coli,* some have perceived disadvantages in using this host. *Escherichia coli* presents dangers for large-scale production of products derived from cloned DNA because it is found in the human intestinal tract and wild-type strains are potentially pathogenic. Also, even nonpathogenic strains produce endotoxins that can contaminate products, an especially bad situation with pharmaceutical injectables. Finally, *E. coli* retains extracellular proteins in the periplasmic space, making isolation and purification potentially difficult. However, modified *E. coli* strains have been developed for which most of these problems have been eliminated. Because of the extensive knowledge of its genetics and biochemistry, *E. coli* remains the organism of choice for most cloning studies.

The gram-positive organism *Bacillus subtilis* can also be used as a host. *Bacillus subtilis* is not potentially pathogenic, does not produce endotoxin, and secretes proteins into the medium. Although the technology for cloning in *B. subtilis* is not nearly as well developed as that for *Escherichia coli,* plasmids and phages suitable for cloning have been developed and transformation is a well-developed procedure in *B. subtilis.* Disadvantages of using *B. subtilis* as a cloning host exist, however. Plasmid instability remains a problem: it can be difficult to maintain plasmid replication over many culture transfers. Also, foreign DNA is not well maintained in *B. subtilis* cells and so the cloned DNA is often unexpectedly lost. Adapting a bacterium for use as a host for cloning experiments is not always simple.

Often organisms used as hosts for cloning must have specific genotypes to be effective. For instance, if the vector carries the gene for β-galactosidase, then the host must have a mutation disabling this gene. Because M13 infects only bacteria with F pili (∞ Sections 8.10 and 9.8), hosts used with M13-derived vectors contain the F plasmid. These types of considerations, and others such as the ability to select for transformants, must be taken into account whether the host is prokaryotic or eukaryotic.

Eukaryotic Hosts

Cloning in *eukaryotic microorganisms* has some important uses, especially in understanding the details of gene regulation in eukaryotic systems. The yeast *Saccharomyces cerevisiae* is the best known genetically (∞ Section 9.14) and is being extensively used as a cloning host. Plasmid vectors, as well as YACs, have been developed for yeast, and genetic transformation is simple. The ability to clone appropriate genetic material in yeast has advanced our understanding of the complex transcription and translation systems of eukaryotes and provides a strong foundation for basic research.

For many purposes, gene cloning in *mammalian cells* is desirable. Mammalian cell culture systems can be handled in some ways like microbial cultures and find wide use in research on human genetics, cancer, infectious disease, and physiology. In addition, DNA can be introduced into mammalian cells by transfection or electroporation (∞ Section 9.6).

One important advantage of eukaryotic cells as hosts for cloning vectors is that they already possess the complex RNA and posttranslational processing systems involved in the production of gene products in higher organisms, and so these systems do not have to be engineered into the vector as they need to be when production of the desired product is to be carried out in a prokaryote (posttranslational processing, in particular, can create some molecular cloning problems; see Section 10.10).

A disadvantage of mammalian cells as hosts is that they are expensive and difficult to produce under large-scale conditions and expression levels of cloned genes are often low. Insect cell lines are simpler to grow, and as we have mentioned, vectors have been developed from an insect DNA virus, the baculovirus.

✓ 10.5 Concept Check

Like naturally occurring plasmids and viruses, vectors with cloned DNA must be placed in a compatible host in order to replicate. The selection of the host also depends on the nature of the studies to be performed with the cloned DNA. Fast-growing Bacteria like *Escherichia coli* are often used as hosts, but some studies require the use of eukaryotic organisms like the yeast *Saccharomyces cerevisiae* or cultured mammalian cells.

10.6

Finding the Right Clone

A crucial step in recombinant DNA technology is finding the right clone among the mixture of clones created by the recombinant DNA procedure. The foreign DNA used in the cloning procedure typically contains a large number of genes, only one or a few of which may be the genes of interest. Sections 10.3 and 10.4 discussed how one can select for hosts containing a plasmid vector by selecting for a vector marker, such as antibiotic resistance, so only these cells form colonies. For host cells containing a viral vector, one simply looks for plaques. We also discussed how these colonies or plaques can be screened for vectors that contain foreign DNA inserts by looking for the inactivation of a vector gene (see Figures 10.2 and 10.5). However, one is then left with the biggest challenge: *finding the clone* that has the gene of interest. Procedures must be available for examining colonies of bacteria or plaques of infected cells growing on agar plates and detecting those few that contain the gene of interest. It is the purpose of the present section to discuss possible approaches to finding the right clone. We consider first the situation in which the gene is *expressed* (that is, the protein is synthesized) in the cloning host. Then we discuss the situation, rather common, in which the gene is not expressed and we must look for the DNA itself.

If the Foreign Gene Is Expressed in the Cloning Host

If the foreign gene is expressed (that is, the protein product is synthesized) in the cloning host, then procedures can be used that look for the presence of this protein in recombinant colonies. The cloning host itself must *not* produce the protein being studied. If we are looking for clones that express the gene, then we are looking for the rare colonies in which this protein is present. If the protein is one that the cloning host normally produces, then the host used must be defective, that is, mutant, for the gene of interest. Then, when the foreign gene is incorporated, the expression of this foreign gene can be detected by complementation (∞ Section 9.5). If the function is required, a complementing clone can be *selected,* greatly facilitating the process. Clearly, if the host already expressed a protein with the same activity, there will be a large background of this activity against which the protein produced via the foreign gene cannot be detected. If the protein is not normally produced in host bacteria, then the host may be naturally defective. In some cases these novel activities are expressed and can be detected. A striking example is the cloning of luciferase genes from various types of bioluminescent beetles into *Escherichia coli;* in the dark, the clones glow in various colors depending on the type of luciferase present (Figure 10.7).

Antibody as a Method of Detecting the Protein

If the protein does not have a readily detectable function, then a different approach is needed. It involves the use of an antibody as a reagent that is specific for the protein of interest. We will discuss antibodies and immunology in Chapter 20. For our present purposes, we note that an antibody is a serum protein produced by a mammalian system that combines in a highly specific way with another protein, the *antigen.* In the present case, the protein of interest is the antigen, and this protein is used to produce an antibody in an experimental animal. Since the antibody combines specifically with the antigen, when the antigen is present in one or more colonies on the plate, then the locations of these colonies can be determined by observing the binding of the antibody. Because only a small amount of the protein (anti-

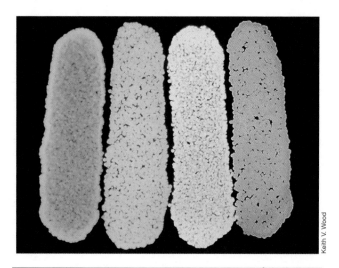

FIGURE 10.7 Bioluminescence from colony streaks of four strains of *Escherichia coli*, each containing cloned luciferase genes from a different species of click beetle. Each luciferase emits light of a different color. Reprinted with permission from *Science* 244:700–702 (1988), © AAAS.

gen) is present in the colonies, only a small amount of antibody is bound, and so a highly sensitive procedure for detecting bound antibody must be available. In practice, this is done using a system involving a radioactive agent, an agent that emits light, or an agent with a specific enzyme attached to it. The radioactivity or light can be detected by autoradiography using X-ray film. The enzymes used typically convert a colorless substrate into a colored one whose absorbance can be measured very sensitively. Such extremely sensitive techniques for detecting antigens are discussed in more detail later (∞ Section 21.8).

Note that this method of detection involves *screening,* not selection, and so thousands of clones must be examined. These can be colonies containing plasmids or plaques containing viruses that produce the cloned product. The whole procedure, utilizing plasmids and radioactive detection, is outlined in Figure 10.8a. As seen, the replica plating procedure (∞ Figure 9.2) is used to make a duplicate of the master plate, but the duplication is done onto a membrane filter and all the manipulations are done with this filter. After the duplicate colonies have grown up, they are lysed to release the protein (antigen) of interest. (Screening for expression in phage vectors eliminates this step because the bacteria are already lysed.) The antibody is then added, and the antibody–antigen reaction allowed to proceed. Unbound antibody is then washed off, and a radioactive agent is then added that is specific for the antibody. A piece of X-ray film is placed over the filter and exposed. If a radioactive colony is present, a spot on the X-ray film will be observed after the X-ray film is developed. The location of this spot on

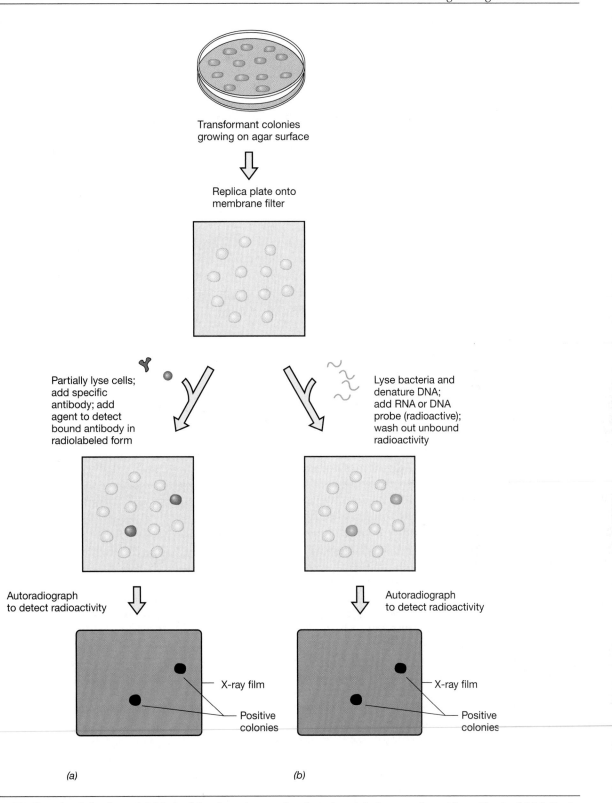

FIGURE 10.8   Finding the right clone. (a) Method for detecting production of protein by use of specific antibody. (b) Method for detection of recombinant clones by colony hybridization with a radioactive nucleic acid probe.

the film corresponds to a location on the master plate where a colony is present that produces the protein. This colony can then be picked from the master plate and cultured.

One limitation of this procedure is that an antibody must be available that is *specific* for the protein in question. As we will see in Chapter 20, antibody can be readily produced by injecting the protein (antigen) into an animal, but the protein injected must be pure; otherwise more than one antibody will be formed. Thus, one must have previously purified the protein.

Nucleic Acid Probes: Searching for the Gene Itself

Suppose that the gene is not expressed in the cloning host or that no assay or antibody is available for the gene product. How does one detect its presence in colonies? The most general way is to use a **nucleic acid probe** containing a key part of the base sequence of the gene of interest. As we have discussed (⚬⚬ Section 6.2), nucleic acid hybridization can be used as a specific means of detecting polynucleotides with specific sequences. Either DNA or RNA can be employed as a probe. The general procedure is to label the nucleic acid probe, often with radioactive phosphate, but nonisotopic techniques are increasingly being used, and allow a single-stranded probe to hybridize with single-stranded nucleic acid derived from the cloned DNA. Because of specific complementary base pairing, two single-stranded polynucleotides will hybridize only if they are fairly complementary. By using appropriate hybridization conditions, it is possible to obtain binding of the radioactive probe only to the nucleic acid of interest.

The way in which a nucleic acid probe can be used to detect the presence of recombinant DNA in colonies is shown in Figure 10.8*b*. The procedure, **colony hybridization,** again makes use of replica plating to produce a duplicate of the master plate on a membrane filter. (The same procedure can be carried out with virus vectors by blotting the plaques onto a membrane.) The cells on the filter are lysed in place to release their nucleic acid and to convert the DNA into a single-stranded form and fix it to the filter. This filter is then treated with a radioactive nucleic acid probe (either RNA or DNA) to allow hybridization, and after removal of unbound radioactive nucleic acid, the filter is subjected to autoradiography. After development, the X-ray film is examined for spots. These correspond to locations on the membrane where the radioactive probe hybridized the DNA from a particular colony. Colonies corresponding to these spots are then picked and studied further. A modification of this procedure, avoiding the use of a radioactive probe, has been developed for clinical microbiology (⚬⚬ Section 21.10).

✓ 10.6 Concept Check

Special procedures are needed for detecting the foreign gene in the cloning host. If the gene is expressed, the presence of the foreign protein itself, as detected either by its activity or by reaction with specific antibodies, is evidence that the gene is present. However, if the gene is not expressed, then its presence can be detected by use of a nucleic acid probe.

- ✓ Does use of nucleic acid probes depend on gene expression? Explain.
- ✓ Why is it necessary to lyse cells containing plasmids in order to detect the product of the cloned gene?

10.7

Expression Vectors

For practical applications it is essential that systems be available in which the cloned genes can be *expressed*. Organisms have complex regulatory systems (⚬⚬ Chapter 7) and one could expect that many cloned genes will not be expressed in a foreign host. One of the major goals of genetic engineering is the development of vectors in which *high levels* of gene expression can occur. An **expression vector** is a vector that can be used not only to clone the desired gene but also contains the necessary regulatory sequences so that expression of the gene can be subjected to experimental manipulation.

In this section we will emphasize prokaryotic expression vectors. However, expression vectors are also used in eukaryotes. The baculovirus and retrovirus vectors we mentioned (see Section 10.4) are expression vectors, as are the yeast plasmid vectors. Many of the characteristics of eukaryotic and prokaryotic expression vectors are broadly the same (and many expression vectors are also shuttle vectors). However, the details are different because there are differences in gene structure and some aspects of gene expression between prokaryotes and eukaryotes (⚬⚬ Section 6.1). In addition, the exact characteristics of the vector may be related to whether it is intended to be used in the scientific laboratory or for commercial production.

Many factors influence the level of expression of a gene, and a vector must be constructed in which all these factors are under control. In addition, a host must be used in which the expression vector is most effective. In this section we discuss the key requirements of a good expression system.

Copy Number and Replication of Expression Vector

In most cases, more product is made if many copies of the gene are present. Therefore, if very high levels of expression are desired, high copy number vectors, for example, small plasmids such as pBR322, are valuable. The replication of some vectors can be controlled in such a way that the copy number can be increased late in the growth phase of a culture so that massive numbers of plasmid can be present. Alternatively, for research purposes (or therapeutic purposes with mammalian vectors) it is sometimes desirable to have only a single copy of the cloned gene in the cell. For these cases, *integrating vectors* have been developed so the gene can recombine into the host chromosome.

Transcriptional Regulation of the Cloned Gene

One of the most important elements in the expression vector is a system that allows transcription of the cloned gene. Typically it is also important that transcription be very tightly controlled. For very high levels of expression, it is essential to produce high levels of mRNA. The promoter region is the site at which binding of RNA polymerase first occurs (∞ Section 6.7). For Bacteria, the DNA region around 10 and 35 nucleotides before the start of transcription (called the −10 and −35 regions) (∞ Figure 6.26) is especially important in the promoter. A cloned gene's native promoter may work very poorly or not at all in the new host. Promoters from eukaryotes and some other prokaryotes function poorly or not at all in *Escherichia coli*. Even some *E. coli* promoters function at low levels in *E. coli* because their sequences are not close to the consensus sequence (∞ Section 6.7). For this reason the expression vector must contain a promoter that will function efficiently in the host and one that is correctly positioned so that it can permit the transcription of the cloned gene. Promoters from *E. coli* that have been used in the construction of expression vectors include *lac* (the *lac* operon promoter), *trp* (the *trp* operon promoter), *tac* (a synthetic hybrid of the −35 region of the *trp* promoter and the −10 region of the *lac* promoter), and lambda P_L (the leftward lambda promoter) (∞ Section 8.12). Note that each of these promoters can be regulated (∞ Sections 7.2, 7.5, and 8.12).

In almost all cases it is important to be able to regulate the expression of the cloned gene. That is, although one typically wants to produce very high levels of mRNA (and have it translated), it is usually undesirable to design a vector that permits the gene to be transcribed to high levels at all times. Indeed, some proteins that are of commercial value are toxic to the host, and in the early stages of growth of the culture it may be important that the gene not be transcribed at all. The ideal situation is to be able to grow the culture containing the expression vector until a large population of cells is obtained, each containing a large copy number of the vector, and then turn on expression in all copies simultaneously by manipulation of a regulatory switch. Therefore, transcription needs to be tightly regulated.

We discussed regulatory controls of gene expression in Chapter 7. Recall the major importance of the repressor–operator system in regulating gene transcription (∞ Section 7.2). A strong repressor can completely block the synthesis of the proteins under its control by binding to the operator region. Repressor function can be turned off at the chosen time by adding an inducer, allowing transcription of the genes controlled by the operator.

For the repressor–operator system to work as a regulatory switch for the production of a foreign protein, the expression vector must contain the operator controlled by the repressor to which the cloned gene is fused. This permits proper arrangement of the sequence of genetic elements: promoter–operator–ribosome binding site–structural gene, so efficient transcription and translation can occur. In most cases the operator and promoter correspond to each other (for instance, the *lac* operator is used with the *lac* promoter), but this is not always the case. A vector could easily be constructed to contain a *trp* promoter under the control of a *lac* operator.

For vectors using the *lac* operator, the promoter is switched on by inducers such as lactose or related β-galactosides (∞ Section 7.2). For vectors using the *trp* operator, induction can be brought about by adding a tryptophan analog (such as β-indolacrylic acid) that brings about an apparent tryptophan deficiency. Phasing of cell growth and protein synthesis can thus be achieved by allowing growth to proceed in the absence of inducer until a suitable cell density is achieved, and then adding inducer to bring about synthesis of the desired proteins.

Vectors using bacteriophage lambda promoter P_L (and the corresponding operator, O_L) are controlled by having the lambda repressor protein in the cell (∞ Section 8.12). Typically the lambda repressor is encoded by a mutant gene (carried by the vector or by a prophage in the host) and is temperature-sensitive. By raising the temperature of the culture to the proper value (usually 8–10°C higher than the growth temperature), the lambda repressor is inactivated and transcription from P_L begins.

This discussion should also make it clear that the host cell must contain the repressor protein and, therefore, the gene containing the repressor. Because there may be an unusually high number of genes to repress (because of the high copy number of the plasmid), this gene may also be carried by the vector.

In some cases the control system used may not be a normal part of the host at all. An excellent example of this is the use of the bacteriophage T7 promoter and RNA polymerase as a regulatory system in an expression vector. When T7 infects *Escherichia coli*, it codes for its own RNA polymerase, which recognizes only T7 promoters, thus effectively shutting down host transcription (∞ Section 8.11). In expression vectors it is possible to place expression of cloned genes under control of a T7 promoter. However, when this is done, it is necessary to engineer into the plasmid the gene for T7 RNA polymerase as well. The latter is placed under control of an easily regulated promoter such as that of lambda or *lac*. Expression of the cloned gene(s) occurs shortly after T7 RNA polymerase transcription has been switched on. Because it recognizes only T7 promoters, T7 RNA polymerase transcribes only the cloned genes; all other host genes remain untranscribed. Because this system is so powerful and specific, induction of the T7 promoter/RNA polymerase system will cause the host to stop growing.

In addition to a strong and regulable promoter, most expression vectors contain an effective transcription terminator. This prevents transcription of the entire vector, which could interfere with vector stability. In addition, the secondary structure of some transcription terminators seems to increase message stability (∞ Section 6.7). It is important to not only make large quantities of mRNA from the cloned gene, but also to attempt to ensure that this mRNA be as stable as possible.

Translation of the Cloned Gene

Expression vectors must also be designed to ensure that the mRNA produced can be efficiently translated. In order to synthesize protein from an mRNA, it is essential that the ribosomes bind at the correct site and begin reading in the correct frame. In prokaryotes this is accomplished by having a ribosome binding site (Shine–Dalgarno sequence) (∞ Section 6.12) and a nearby start codon on the mRNA. Bacterial ribosome binding sites are not found in eukaryotic genes, and it is thus essential that the bacterial region be present in the cloned gene if high levels of gene expression are to be obtained. Part of the requirement for proper ribosome binding is the necessity for a proper distance between the ribosome binding site and the translation initiation codon. If these sites are too close or too far apart, the gene will be translated at low efficiency. In some cases even the initiation codon for the gene to be cloned is part of the expression vector. Interestingly, translation is sometimes more efficient if the cloned gene is the second gene in a bicistronic mRNA, a phenomenon called *translational coupling*. The upstream gene often encodes a small peptide, which is rapidly degraded.

Often, adjustments have to be made to ensure high efficiency translation *after* the gene has been cloned. For instance, it is important that the Shine–Dalgarno site mentioned above not be involved in a region of secondary structure, and this cannot always be predicted until the sequence of the cloned gene is known. There are also sometimes difficulties having to do with *codon usage*.

There is more than one codon for most of the 20 amino acids (∞ Table 6.5), and some codons are used more frequently than others: this preference may differ considerably between different organisms. Codon usage is partly a function of the concentration of the appropriate tRNA in the cell. Therefore, a gene whose codon usage pattern is considerably different from that of its new host may be translated inefficiently. Insertion of the appropriate codons would be difficult because it would have to be changed at all locations in the gene. However, this can be done if necessary by using synthetic DNA and site-directed mutagenesis (∞ Sections 9.3 and 10.11) to create a gene more amenable to the codon usage patterns of the host.

Finally, if the cloned gene contains introns (∞ Section 6.9), the correct protein product will not be made if the host is a prokaryote. This problem can also be corrected by site-specific mutagenesis or synthetic DNA, but there are other methods for creating an intron-free gene (see Section 10.10).

Protein Folding and Stability

Some proteins are susceptible to degradation by intracellular proteases and may be destroyed before they can be isolated. Some eukaryotic proteins are toxic to the prokaryotic host, and the host for the cloning vector may be killed before a sufficient amount of the product is synthesized. Further engineering of either the host or the vector may be necessary to eliminate these problems.

Sometimes when foreign proteins are massively overproduced they form inclusion bodies inside the host. Although inclusion bodies are relatively easy to purify because of their size, the protein found in these bodies can be very difficult to solubilize. It seems that in many cases these bodies form because the protein is incorrectly folded. One potential solution to this problem is to use a host that overproduces molecular chaperones that aid in folding (∞ Section 6.12). Interestingly, this problem, and some others, can sometimes be solved if the protein from the cloned gene be made as a fusion product with a protein encoded by the vector. This not only stabilizes the protein but might simplify purification if the portion encoded by the vector is a protein for which rapid, simple, inexpensive purification techniques are known. Several special fusion vectors are now available. The "cloned protein" is released from the fusion protein after purification by special proteases. Figure 10.9 shows an example of a fusion vector that is also an expression vector.

In some cases the desired protein can also be removed from the fusion protein by chemical means. Fusion systems can also be used for purposes other than achieving increased protein stability. One advantage of making a fusion protein is that the bacterial portion can contain the bacterial sequence coding for the *signal peptide* that enables transport of the protein across the cytoplasmic membrane (∞ Section 6.12), making possible the development of a bacterial system that not only synthesizes the mammalian protein but also actually excretes it.

This discussion of the requirements of an expression vector is by no means exhaustive, but it should give

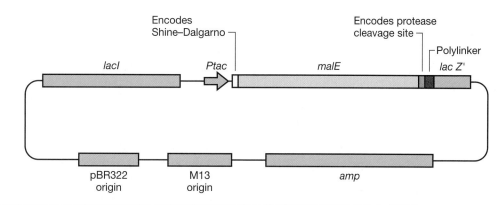

FIGURE 10.9   An expression vector for fusions. This vector was developed by the New England Biolabs Company. The gene to be cloned is inserted at the polylinker site (see also Figure 10.5) so it is in frame with the *malE* gene, which encodes the maltose binding protein. This insertion inactivates the *lacZ'* gene (see Figure 10.5). The fused gene is under control of the hybrid *tac* promoter (*Ptac*). The plasmid also contains the *lacI* gene, which encodes the *lac* repressor. Therefore, an inducer must be added to the cells in order to turn on the *tac* promoter. The fusion protein is easily purified by methods involving the affinity of the protein for maltose. Once purified, the two portions of the fusion protein can be separated by a very specific protease (factor Xa). The plasmid contains a gene conferring ampicillin resistance on its host. In addition to the plasmid origin of replication, there is a bacteriophage M13 origin. Therefore, this is a phagemid and can be propagated either as a plasmid or as a phage.

an idea of the challenges that must be met. Even with the best-designed vector, some genes are poorly expressed in a particular cell. In some cases these problems can be rectified by using a mutant host. For instance, some "foreign mRNAs" are degraded very rapidly in wild-type *Escherichia coli* but not in particular mutant strains. Using such expression systems, one can produce very high levels of foreign proteins in *Escherichia coli*. In many cases the desired protein exceeds 200,000 molecules per cell and can make up as much as 40% of the protein molecules in a cell.

✓ 10.7 Concept Check

Not all cloned genes are expressed at high efficiency in foreign hosts. Expression vectors are special cloning vectors containing various elements necessary for obtaining high levels of gene expression. Regulatory switches are also useful in expression vectors because they can be used by the investigator to turn on expression at the most favorable time in the growth cycle.

✓ List five requirements for an efficient prokaryotic expression vector.

✓ Discuss the differences in gene structure that may prevent a mammalian gene from being directly expressed in a prokaryote.

10.8

Synthetic DNA

Techniques are available for the synthesis of short fragments of DNA of specified base sequence. **Synthetic**

DNA is widely used in molecular genetics, especially in genetic engineering but also in basic research. The procedures for synthesis of DNA can be completely automated so an oligonucleotide of 30–35 bases can be easily made in a few hours and oligonucleotides of well over 100 bases in length can be made if necessary. For the synthesis of longer polynucleotides, the oligonucleotide fragments can be joined enzymatically using DNA ligase.

DNA is synthesized in a *solid-phase procedure* in which the first nucleotide in the chain is fastened to an insoluble porous support (such as silica gel with particles about 50 μm in size). The overall procedure, the chemical details of which need not concern us here, is shown in Figure 10.10. Several chemical steps are needed for the addition of each nucleotide. After each step is completed, the reaction mixtures are flushed out of the solid support and the series of reactions repeated for the addition of the next nucleotide. Once the desired length is achieved, the oligonucleotide is removed from the solid-phase support and purified to eliminate by-products and contaminants.

Synthetic DNA molecules are widely used for various purposes in genetic engineering, for instance, as **probes** to detect, via nucleic acid hybridization, specific DNA sequences (Figure 10.8*b*). We will describe later (see Section 10.11) how synthetic DNA is used in a procedure called **site-directed mutagenesis** to create mutations at specific locations on the genome. Finally, synthetic DNA is employed extensively as a source of DNA primers for the polymerase chain reaction (see next section).

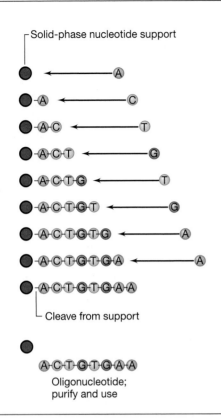

Solid-phase nucleotide support

Cleave from support

Oligonucleotide;
purify and use

FIGURE 10.10 Solid-phase procedure for synthesis of a DNA fragment of defined sequence. Chemical synthesis proceeds by adding one nucleotide at a time to the growing chain.

10.9

Amplifying DNA: The Polymerase Chain Reaction

Conventional molecular cloning methods can be considered *in vivo* DNA-amplifying tools. However, the development of synthetic DNA has spawned a new method for the rapid amplification of DNA *in vitro*, the **polymerase chain reaction (PCR)**. The polymerase chain reaction can multiply DNA molecules by up to a billionfold in the test tube, yielding large amounts of specific genes for cloning, sequencing, or mutagenesis purposes. PCR makes use of the enzyme *DNA polymerase*, which copies DNA molecules (⚬⚬ Section 6.6).

The PCR technique requires that the nucleotide sequence of a portion of the desired gene be known. This is necessary because short oligonucleotide *primers* complementary to sequences in the gene or genes of interest must be available for PCR to work. The steps in PCR amplification of DNA are as follows. (1) Two oligonucleotide primers flanking the target DNA (Figure 10.11*b*) are made on an oligonucleotide synthesizer and added in great excess to heat-denatured target DNA (Figure 10.11*a*). (2) As the mixture cools, the excess of primers relative to the target DNA ensures that most target strands anneal to a primer and not to each other (Figure 10.11*b*). (3) DNA polymerase then extends the primers using the target strands as template (Figure 10.11*c*). (4) After an appropriate incubation period, the mixture is heated again to separate the strands. The mixture is then cooled to allow the primers to hybridize with complementary regions of newly synthesized DNA, and the whole process is repeated (Figure 10.11*e*).

Thus, each PCR "cycle" involves the following: (1) heat denaturation of double-stranded target DNA, (2) cooling to allow annealing of specific primers to target DNA, and (3) primer extension by the action of DNA polymerase (Figure 10.11). Note in Figure 10.11 how the extension products of one primer can serve as a template for the other primer in the next cycle. The beauty of the PCR technique lies in the fact that each cycle literally *doubles* the content of the original target DNA. In practice, 20–30 cycles are usually run, yielding a 10^6- to 10^9-fold increase in the target sequence (Figure 10.11*f*).

PCR at High Temperature

The original PCR technique employed *Escherichia coli* DNA polymerase, but because of the high temperatures needed to denature the double-stranded copies of DNA being made, the polymerase itself also was denatured and had to be replenished every cycle. This tended to limit the number of cycles that could be run and was very expensive. This problem was solved by employing a thermostable DNA polymerase isolated from the thermophilic bacterium *Thermus aquaticus*. DNA polymerase from *T. aquaticus*, known as *Taq polymerase*, is stable to 95°C and thus is unaffected by the denaturation step employed in the PCR reaction. The use of *Taq* DNA polymerase also increased the *specificity* of the PCR reaction because the DNA is copied at 72°C rather than 37°C. At high temperatures, nonspecific hybridization of primers to nontarget DNA rarely occurs, thus making the product of *Taq* PCR more homogeneous than that obtained using the *E. coli* enzyme.

One problem with the *Taq* polymerase is that it has no proofreading function (⚬⚬ Section 6.6) and consequently makes more mistakes than the *Escherichia coli* enzyme. DNA polymerase from the hyperthermophilic archaean *Pyrococcus furiosus* (growth temperature optimum, 100°C) (⚬⚬ Sections 5.9 and 14.9), called *Pfu* polymerase (or "vent polymerase"), is also widely used and is even more thermally stable than *Taq* polymerase. *Pfu* polymerase has proofreading activity, making it a particularly good enzyme when accuracy is crucial.

Because a number of highly repetitive steps are involved in the PCR technique, machines have been developed that can be programmed to run through heating and cooling cycles automatically. Because each cycle requires only about 5 min, the automated procedure allows for large amplifications in only a few hours (by contrast, such amplification by *in vivo* cloning methods would take several days). To supply the demand for thermostable DNA polymerase in the growing PCR and DNA sequencing markets, the genes for these enzymes have been cloned into *E. coli* and produced in large quantities; the cost of the PCR method is now just a fraction of what it was when the technique was first introduced.

Applications of PCR

PCR is a powerful tool because it is very simple to perform, extremely specific, and extremely efficient. Remember that during each round of amplification the amount of product *doubles,* leading to an exponential increase. This means not only that one can get large amounts of amplified DNA in a few hours, but that only a few molecules of target DNA need to be present in the sample to start the reaction. The reaction is so specific that, with primers of 15 nucleotides and high annealing temperatures, there will be extremely low levels of "false priming" and therefore the product formed should be relatively homogeneous.

Although automated thermocycling machines are rather expensive, the technique has found a wide range of applications because what can be done with PCR (and analysis of the amplified product) can sometimes accomplish in a few hours what would have previously taken an entire laboratory several days or even weeks to do. Therefore, PCR can be extremely cost efficient and is now used in a very wide number of practical applications.

PCR is extremely valuable for obtaining DNA for cloning and sequencing because the gene or genes of interest can easily be amplified if flanking sequences are known. Because the primers used do not have to be perfectly complementary, PCR is routinely used in comparative or evolutionary studies to amplify genes from a variety of sources where the DNA has already been cloned (and sequenced) from one organism. In these cases the primers are made to regions of the gene thought to be conserved throughout a wide variety of organisms. Because of the sensitivity of PCR, it has been used to amplify and clone DNA from sources such as

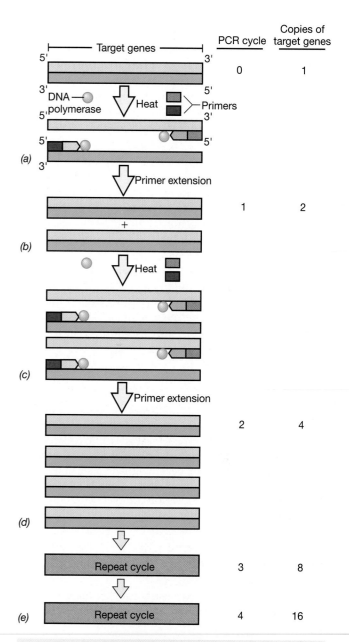

FIGURE 10.11   The polymerase chain reaction (PCR) for amplifying specific DNA sequences. (a) Target DNA is heated to separate the strands, and a large excess of two oligonucleotide primers, one complementary to the target strand and one to the complementary strand, is added along with DNA polymerase. (b) Following primer annealing, primer extension yields a copy of the original double-stranded DNA. (c) Further heating, primer annealing, and primer extension yields a second double-stranded DNA. (d) The second double-stranded DNA. (e) Two additional PCR cycles yield 8 and 16 copies, respectively, of the original DNA sequence. (f) Effect of running 20 PCR cycles on a DNA preparation originally containing 10 copies of a target gene. Note that the graph is semilogarithmic.

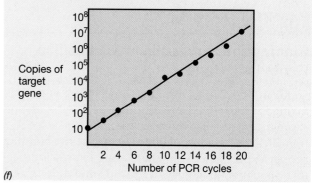

mummified human remains and even samples of extinct plants and animals.

PCR can also be used to amplify very small quantities of DNA present in a sample. Using appropriate primers, one can find and identify a single bacterial cell in a sample even if large numbers of other species are present. Please note that when amplifying DNA from a bacterium it is not necessary that the organism be grown in the laboratory. Therefore, PCR is one of the tools most responsible for revealing to scientists the enormous diversity of the microbial world and making it clear that microorganisms that have been cultured make up but a tiny fraction of all microorganisms in the environment (∞ Sections 12.6 and 16.6). Because of this ability to amplify and analyze DNA without having to grow the microorganism, PCR has also taken over a very important role in diagnostic microbiology (∞ Section 21.10). PCR has also been used in conjunction with DNA *fingerprinting,* a powerful technique that permits identification of individuals, or relationships between individuals, from small samples of their DNA (see the box, DNA Fingerprinting). PCR can also be coupled with reverse transcription in a process called RT-PCR. This can be very useful when one wants to make large amounts of DNA using RNA as a template.

✓ 10.9 Concept Check

The polymerase chain reaction, a procedure for amplifying DNA *in vitro,* makes use of a heat-stable DNA polymerase from thermophilic prokaryotes. Heat is used to denature the DNA into two single-stranded molecules, each of which is copied by the polymerase. Beginning with a small oligonucleotide that serves as a primer for the target DNA to be amplified, the polymerase copies the complete DNA to which the primer associates. After a single copy cycle, the newly formed double strands are separated by heat again, and a new round of copying permitted. At each thermal cycle, the amount of target DNA doubles.

✓ Why does PCR require primers?
✓ Why is a primer needed at each "end" of the DNA to be amplified?

10.10

Cloning and Expression of Mammalian Genes in Bacteria

So far in this chapter we have described a variety of methods for cloning DNA, finding specific clones, and expressing genes. However, as we mentioned (see Section 10.4), there are sometimes obstacles to obtaining proper expression even when using expression vectors. In mammalian genes there are almost always in-

trons (∞ Section 6.4). Indeed, some genes have over 50 introns consisting of tens of thousands of base pairs. Removing introns from genes using the techniques we have described would be difficult. In addition, mammalian genomes are vast, the human haploid genome having 3 billion base pairs. Screening DNA libraries from such a genome for a specific gene would be at best extremely laborious. In some cases very little is known about the gene, making screening almost impossible. In other cases the geneticist may wish to find genes with specific patterns of regulation; for instance, it could be valuable to clone the genes whose products are expressed only in the brain. There are a number of approaches that can be used to circumvent all these problems, and we shall discuss some of them in this section. Not only skill but also intuition and good luck play big roles in a successful outcome. A summary of approaches and procedures is given in Figure 10.12.

Reaching the Gene via Messenger RNA

One approach to isolating a gene is to get to it through its mRNA. A major advantage of using mRNA is that the noncoding information present in the DNA (introns) has been removed (∞ Section 6.9). The isolated mRNA is used to make complementary DNA (cDNA) by means of reverse transcription (∞ Section 8.22). It is likely that a tissue expressing the gene contains large amounts of the desired mRNA, although except in rare cases this certainly is not the only mRNA produced. In a fortunate situation, where a single mRNA dominates a tissue type, extraction of mRNA from that tissue provides a useful starting point for gene cloning.

In a typical mammalian cell, about 80–85% of the RNA is ribosomal, 10–15% is transfer RNA and other low-molecular-weight RNAs, and 1–5% is messenger RNA. Although low in abundance, the mRNA in a eukaryote is identifiable because of the poly-A tails found at the 3'-end (∞ Section 6.9). In maturing red blood cells, for instance, where virtually the only protein made is the globin portion of hemoglobin, from 50 to 90% of the poly-A-containing cytoplasmic RNA consists of globin mRNA. By passing a poly-A-rich RNA extract over a chromatographic column containing poly-T fragments (linked to a cellulose support), most of the mRNA of the cell can be separated from the other cellular RNA by the specific pairing of A and T bases. Elution of the RNA from the column then gives a preparation greatly enriched in mRNA.

Once the RNA message has been isolated, it is necessary to convert the information to DNA. This is accomplished by use of the enzyme *reverse transcriptase,* (∞ Section 8.22). This remarkable enzyme, an essential component of retrovirus replication, copies infor-

TECHNIQUES & APPLICATIONS... DNA fingerprinting

The techniques used in molecular genetics and genetic engineering are useful not only in determining the evolutionary relationships among organisms (∞ Chapter 15) but also in determining relationships among individuals and in establishing whether a tissue sample, even a very small one, came from a particular individual. These latter techniques are called *DNA fingerprinting*. DNA fingerprinting is made possible both by the technology that allows precise detection and amplification of very small amounts of DNA and by the fact that higher organisms contain repetitive DNA sequences that can exist in different numbers and patterns in the genome.

As discussed (∞ Section 6.4), the genomes of higher eukaryotes contain a very large amount of repetitive DNA. Some of these repeats exist in families of related sequences scattered around the genome, and members of these families have been cloned and sequenced. In order to be useful for identification purposes, a DNA sequence must have a reasonable chance of differing among different groups in a population of organisms. One family of these repetitive sequences was found to vary not simply as to sequence but also as to how many repeats of an individual sequence occurred at a single site on a particular chromosome. These sequences are called *variable number of tandem repeats* (VNTR), and several have been identified.

The use of VNTRs in DNA fingerprinting is illustrated in in Fig. 1. It shows two different alleles (alternative states of the same gene) of a eukaryotic chromosome that differ only in how many copies of the repeated sequence are present. Since the VNTR DNA has been sequenced, it is known which restriction enzyme sites are *not* found in a particular VNTR. Digestion with such an enzyme then releases the complete VNTR intact. When the DNA from two chromosomes with a different number of repeats at this

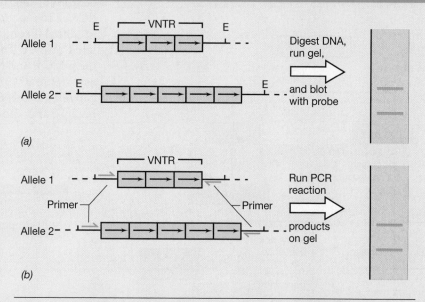

FIG. 1 DNA fingerprinting. (a) Two different alleles of a region of a single chromosome. The alleles differ only in the number of repeats in the VNTR. DNA from cells containing these chromosomes can be cut with the restriction enzyme *Eco*RI (which does not cut within the VNTR) and the fragments separated on an agarose gel. The fragments containing the VNTR are then identified after Southern blotting by hybridization with a probe specific to the VNTR. The figure shows only the result from individuals whose two chromosomes have different alleles at this site. If an individual had the same allele on both chromosomes, only a single band would be observed. (b) The same alleles, but with primers that could be used to amplify the VNTR segments by PCR. The products of the PCR reaction can be loaded directly onto the gel without restriction digestion.

particular locus is digested, the restriction fragments containing this DNA differ in size. (Such a difference is called a *restriction fragment length polymorphism*, or RFLP.) This DNA can be separated by gel electrophoresis, and the VNTR-containing fragments detected by Southern blotting (∞ box, Working with Nucleic Acids: The Tools, Chapter 6) using a probe made from the cloned VNTR sequence. Even if multiple alleles are present in a population (rather than just two as in the example), differences at a single site are insufficient to identify an individual. Many individuals in a population would be expected to have the same pattern at this site. However, it is possible to probe the digested DNA simultaneously for several different VNTR markers. With the use of these methods, the probability of identifying a particular individual

by comparing two different DNA samples is extremely high.

Notice that the polymerase chain reaction (PCR) does not need to be used in DNA fingerprinting. However, the use of PCR is essential when the amount of DNA in the sample is very small—such as that found in the cells on the root of a single hair. The use of PCR in DNA fingerprinting is also shown in the figure. To use PCR, it is necessary to know the sequences surrounding a particular site that contains a VNTR so primers can be synthesized. However, since PCR amplifies only the DNA between the primers, one does not have to cut the DNA with restriction enzymes before running it on a gel. With enough cycles of amplification, it is also sometimes possible to detect the PCR-generated bands by simply staining the gels rather than using a hybridization probe. ■

mation from RNA into DNA (Figure 10.13). As we noted, this enzyme requires a primer in order for it to begin working (in retrovirus infection the primer is a tRNA). In the present procedure, an oligo-dT primer is used that is complementary to the poly-A tail of the isolated mRNA. The oligo-dT primer is hybridized with the mRNA, and then reverse transcriptase is allowed to act (Figure 10.13). As seen, the newly synthesized DNA copy has a hairpin loop at its end, which is synthesized because after the enzyme completes copying the mRNA

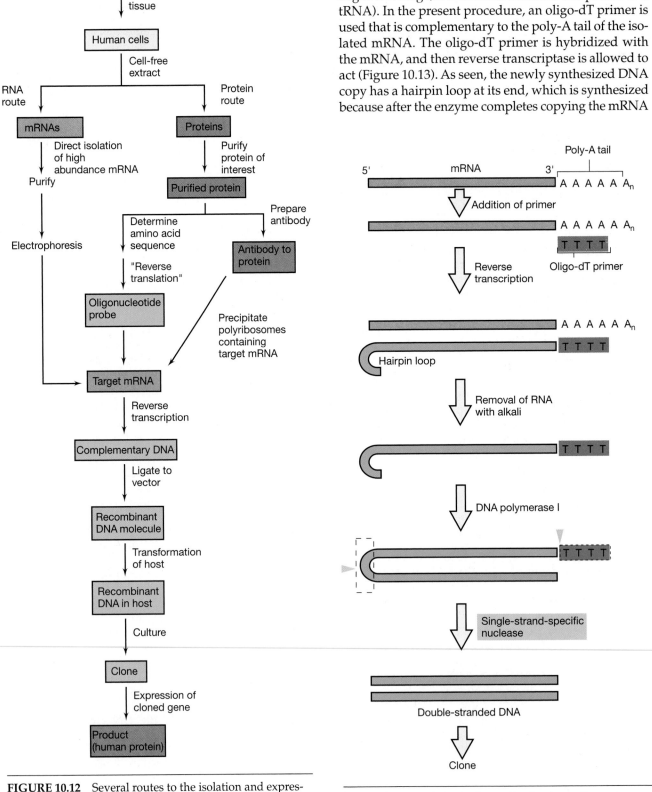

FIGURE 10.12   Several routes to the isolation and expression of mammalian genes in prokaryotes. The colors indicate whether a step involves DNA (green), RNA (orange), or protein (brown).

FIGURE 10.13   Steps in the synthesis of complementary DNA (cDNA) from an isolated mRNA using the retroviral enzyme reverse transcriptase.

it starts to copy the newly synthesized DNA. This hairpin loop, which is probably an artifact of the test tube reaction, provides a convenient primer for synthesis of the second DNA strand. The resultant double-stranded DNA, with the hairpin loop intact, is then cleaved by a single-strand-specific nuclease to produce the desired double-stranded DNA, one strand of which is complementary to the mRNA. This double-stranded DNA (the gene of interest) can then be inserted into a plasmid or other vector for cloning. The detection of specific clones makes use of the procedures discussed in Section 10.6. As mentioned in Section 10.9, one can also use RT-PCR to synthesize large amounts of cDNA without having to clone it.

The cDNA should encode the protein of interest and, therefore, can be considered its "gene." Unlike the "natural" gene on the mammalian chromosome, this one contains no introns. Although there is a start codon, there are no promoters because these are not transcribed and, therefore, their sequence won't be in the mRNA (∞ Section 6.7). The requirements for achieving high levels of expression with genes made in this manner are simply those discussed in the section on expression vectors (see Section 10.7).

One can also make *cDNA libraries* from different tissues when seeking genes whose expression is specific to those tissues. This can be extremely useful because the actual gene (chromosomal DNA) is found in almost all cells but the mRNA is found only in cells actively producing the protein. For instance, the gene encoding the

hormone insulin is found throughout the body, but insulin mRNA is found only in certain cells in the pancreas. Therefore, a library made from pancreatic cells is enriched for cDNA corresponding to the insulin gene.

Reaching the Gene via the Protein

If for some reason, mRNA for the gene of interest cannot be obtained, other approaches are possible. Two such approaches, reverse translation and polyribosome precipitation, will be discussed here.

The most widely used method of cloning low abundance mRNA is to make a **synthetic DNA** that is complementary to part of the mRNA and then use this DNA as a *probe* in a Northern blot procedure (∞ box, Working with Nucleic Acids: The Tools, Chapter 6) to pull out the mRNA of interest by hybridization. This requires that a partial or complete sequence of the protein be known. Then, from a consideration of the genetic code, the nucleotide sequence of a section of the DNA is deduced, and this piece of DNA is synthesized.

The procedure by which the nucleotide sequence is deduced from the protein sequence is called **reverse translation.** (Note that *reverse translation* is not a cellular process but a mental exercise of the genetic engineer.) The procedure of reverse translation is illustrated in Figure 10.14. From the genetic code, the nucleotide sequence of a section of the DNA is deduced, and this piece of DNA is synthesized. Unfortunately, degeneracy of the genetic code (∞ Section 6.10) somewhat

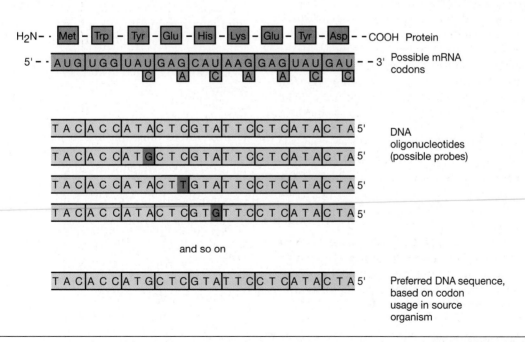

FIGURE 10.14 Reverse translation: deducing the best sequence of an oligonucleotide probe from the amino acid sequence of the protein. Because of degeneracy, many probes are possible. If codon usage by the same organism is known, then a preferred sequence can be selected. It is not essential that complete accuracy be achieved because a small amount of mismatch can be tolerated.

complicates the problem. Most amino acids are encoded by more than one codon, and codon usage varies from organism to organism. The best section of DNA to synthesize is one that corresponds to a part of the protein rich in amino acids specified by only a single codon (methionine, AUG; tryptophan, UGG) or by two codons (for example, phenylalanine, UUU, UUC; tyrosine, UAU, UAC; histidine, CAU, CAC) because this increases the chances that the synthesized DNA will be complementary or nearly complementary to the mRNA of interest. If the complete amino acid sequence of the protein is not known, then the sequence used is generally one at the amino terminus of the protein because it is at the amino terminus that sequencing of the protein begins.

Polyribosome precipitation involves separation from the tissue of *ribosome complexes* that are in the process of synthesizing the desired protein, using an antibody specific for this protein. As we describe in Chapter 20, antibodies are proteins made in response to foreign proteins able to combine with and specifically precipitate them. How is an antibody directed against a *protein* used to detect an *mRNA*?

In the ribosome complex, polypeptides of the protein of interest are still complexed with the protein-synthesizing machinery (which contains, among other things, the sought-for mRNA). When the antibody precipitates the protein in the ribosome complex, the mRNA also is precipitated. After isolation, the mRNA can be used to prepare complementary DNA as described in this section.

Synthesis of the Complete Gene

If the protein is small enough or is of sufficient economic interest to justify a major effort, the complete gene can be synthesized chemically (see Figure 10.10); this requires knowledge of the complete amino acid sequence of the protein. Chemical synthesis not only permits the acquisition of genes that cannot be obtained otherwise but also allows synthesis of *modified genes* that may make new proteins of utility. Techniques for the synthesis of DNA molecules are now well developed, and it is possible to synthesize genes coding for proteins 100–200 amino acid residues in length (300–600 nucleotides). The synthetic approach was used for production of the human hormone insulin in bacteria, as will be discussed in Section 10.13. We will discuss the use of synthetic DNA in mutagenesis in Section 10.11.

With the use of all these techniques, a large number of different human proteins have been expressed at high yield under the control of bacterial regulatory systems, including human growth hormone, insulin, virus antigens, interferon, and somatostatin (see Section 10.13).

✓ 10.10 Concept Check

To detect and isolate a gene from the large amount of DNA in the mammalian cell, it is often necessary to use special procedures. One approach is to isolate the mRNA from the mam-

malian cell and make a complementary DNA by use of the enzyme reverse transcriptase. Another approach is to synthesize a nucleic acid probe for the gene in question and use this probe to detect the cloned sequence in bacterial colonies. Under some conditions, the complete mammalian gene can be synthesized chemically and this synthetic gene cloned in bacteria.

- ✓ Draw a diagram of how cDNA is made from mRNA.
- ✓ Explain why *reverse translation* is something cells can't do (∞ Section 6.1).

10.11

In Vitro and Site-Directed Mutagenesis

Recombinant DNA technology has opened up a whole new field of mutagenesis. Whereas conventional mutagens (∞ Section 9.3) act at random, by use of synthetic DNA and recombinant DNA techniques it is possible to introduce mutations at *precisely determined sites* on genes. This process is called **site-directed mutagenesis.** Proteins made from strains carrying such mutations can be expected to have properties different from those of the wild-type proteins, properties that may, in some cases, be predicted from a knowledge of protein structure.

Site-Directed Mutagenesis

The first approaches to site-directed mutagenesis included treating transducing bacteriophage with chemical mutagens and then infecting bacteria and screening for mutants. Such a strategy had the effect of increasing the mutation rate in a limited region of a genome. New techniques are now available that allow the geneticist much greater specificity; a specific base pair in a gene can be changed to another base pair. Some of the techniques are simple and very powerful.

The basic procedure is to synthesize a short oligodeoxyribonucleotide containing the desired base change and to allow this to pair with a single-stranded DNA containing the gene of interest. Pairing is complete except for the short region of mismatch. Then the short single-stranded fragment of the synthetic oligonucleotide is extended using DNA polymerase, thus copying the rest of the gene. The double-stranded molecule obtained is inserted into a cloning host by transformation, and mutants selected by a procedure already described (∞ Section 9.6). The mutant obtained is then used in production of the modified (mutant) protein.

The whole procedure of *site-directed mutagenesis* is illustrated in Figure 10.15. Several modifications of this technique have been developed to increase the ratio of mutants recovered. As seen, it is convenient to begin with the gene of interest cloned into a single-stranded DNA vector. A widely used vector for site-directed

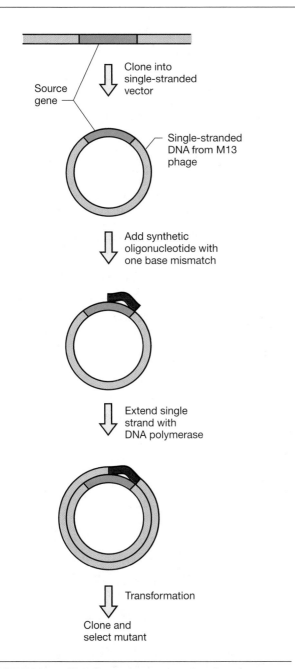

FIGURE 10.15   Site-directed mutagenesis using short synthetic oligodeoxyribonucleotide fragments.

mutagenesis is bacteriophage M13, which, as we have seen (⚭ Sections 8.10 and 10.4), has some properties that are useful in recombinant DNA technology. The target DNA is cloned into M13, from which single-stranded DNA can be purified with ease. Because cells infected with this phage remain alive, a ready source of DNA is available.

Total Synthesis of Mutant Genes

Synthetic DNA technology can be used to synthesize a complete gene with a genetic change inserted at the desired location. Although a complete gene of 1000 or so nucleotides would be difficult to synthesize in one piece, it is possible to synthesize smaller portions of the gene and then link these together to produce the final gene. By addition of appropriate sites at the ends of the gene, this gene can then be linked into a vector and cloned. The possibilities of this approach seem virtually limitless, although because of the expense involved, it is essential that one have a carefully considered rationale for the particular sequence. Typically, however, it is more advantageous to synthesize only a small part of a gene and use this to replace the same part of the wild-type gene, a process called *cassette mutagenesis.*

Cassette Mutagenesis and Gene Disruption

Because of the large number of restriction enzymes commercially available and therefore the large number of different DNA sequences that can be cut, it is usually possible to find several different restriction sites in the gene of interest. If sites for the appropriate enzyme are not found in the gene, or at the precise location required, they can also be added by site-directed mutagenesis (see Figure 10.15). If restriction sites are close together, the intervening DNA fragment can be cut out and replaced by a synthetic DNA fragment in which one or more of the bases have been changed. These synthetic fragments are called *cassettes* (or cartridges), and the process is known as **cassette mutagenesis.**

Insertion mutations can also be generated by simply inserting a cassette at a single site. When using cassettes to replace sections of genes, the cassettes are typically the same size as wild-type DNA fragments. However, the cassette used for making insertion mutations can be almost any size and can even be an entire gene. In fact, cassettes that encode proteins that confer a particular antibiotic resistance on the host are commonly used. This type of cassette mutagenesis is used in a process called **gene disruption.** The process of gene disruption is illustrated in Figure 10.16. In this case, a fragment carrying a gene conferring kanamycin resistance, the *kan* cassette, is inserted at a restriction site in a cloned gene. The vector carrying this mutant gene is then linearized by being cut with a different restriction enzyme, and the linear DNA is transformed into the host with kanamycin resistance selected. The linearized plasmid cannot replicate, and so resistant cells likely arise by homologous recombination (⚭ Section 9.5) between the mutated gene on the plasmid and the wild-type gene on the chromosome.

The cells have not only gained kanamycin resistance but have also lost the function of the gene in which the *kan* cassette was inserted. Therefore, these

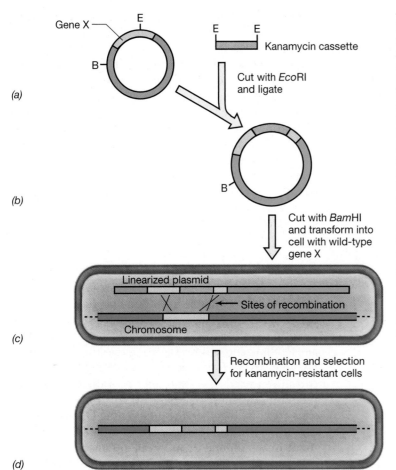

(a)

(b)

(c)

(d)

FIGURE 10.16 Gene disruption using cassette mutagenesis. (a) A plasmid containing a cloned wild-type copy of gene X is cut with the restriction enzyme *Eco*RI and mixed with a DNA fragment (the kanamycin cassette) that contains a gene capable of conferring kanamycin resistance on a cell and that has been obtained using the same restriction enzyme. The cut plasmid and the cassette are ligated. (b) The product of the ligation is a plasmid that now contains the kanamycin cassette as an insertion mutation within gene X. This new plasmid is now cut with a further restriction enzyme, *Bam*HI, and transformed into a cell containing a wild-type gene X. (c) The transformed cell contains the linearized plasmid with a disrupted gene X and its own chromosome with a wild-type copy of the gene. In some cells, homologous recombination occurs between the wild-type and mutant forms of gene X. (d) Cells that can grow in the presence of kanamycin must have the kanamycin cassette recombined into their chromosome because the linearized plasmid cannot replicate. These cells now have only a single disrupted copy of gene X. This disruption typically abolishes all gene X function.

mutations are also called "knockout" mutations. This process is similar to searching for insertion mutations made by transposons (Section 9.10), but in this case the geneticist chooses exactly which gene will receive the mutation. Note that gene disruption in haploid organisms yields viable cells only if the disrupted gene is unessential. Gene disruption experiments are now used as one mechanism to find out whether a gene is essential. Methods of obtaining gene disruption have been developed for higher organisms, including mice.

✓ **10.11 Concept Check**

Synthetic DNA molecules of desired sequence can be made *in vitro* and used to construct a mutated gene directly or to change specific base pairs within a gene via site-directed mutagenesis. Genes can also be disrupted by inserting DNA fragments, called cassettes, into them. The inserted cassette eliminates the function of the wild-type gene while conferring a new, and usually selectable, phenotype on the cell.

✓ Why might site-directed mutagenesis be useful?
✓ What are *knockout mutations*?

Practical Applications of Genetic Engineering

We note here some of the main practical applications of recombinant DNA technology and then describe some of the applications in more detail in Sections 10.13–10.15.

Genetic engineering has many commercial and practical applications. A few main areas of interest for commercial development are as follows.

1. **Microbial fermentations.** A number of important products are made industrially using microorganisms, of which the antibiotics are the most significant (Section 11.5). Genetic engineering procedures can be used to manipulate the antibiotic-producing organism in order to obtain increased yields or produce modified antibiotics.
2. **Virus vaccines.** A vaccine is a material that can induce immunity to an infectious agent (Section 20.16). Frequently, killed virus preparations are used as vaccines, and there is always a potential danger to the patient if the virus has not been com-

pletely inactivated. Because the active ingredient in the killed virus vaccine is the protein coat, it is desirable to produce the protein coat separately from the rest of the virus particle. By genetic engineering, viral coat protein genes can be cloned and expressed in bacteria, in nonpathogenic viruses, or in plants, making possible the development of safe, convenient vaccines (see Sections 10.13 and 10.14).

3. **Mammalian proteins.** A number of mammalian proteins are of great medical and commercial interest. Some of these are discussed in Section 10.13. In the case of human proteins, commercial production by direct isolation from tissues or fluids is at best complicated and expensive, and in most cases simply not feasible. By cloning the gene for a human protein in an appropriate microorganism or a culturable cell, its commercial production is possible.

4. **Transgenic plants and animals.** In addition to the production of valuable *products* by microbial means, genetic engineering promises the advent of genetically altered whole plants and animals. Some of these developments are discussed in Sections 10.14 and 10.15. Such organisms, referred to as *transgenic,* hold great promise for boosting agricultural productivity, altering the nutritional quality of meats and vegetables, and producing certain proteins not readily produced by genetically engineered microorganisms. By introducing cloned DNA into the fertilized eggs of animals, or directly into plant cells grown in tissue culture, it is now possible to grow genetically altered higher organisms.

5. **Environmental biotechnology.** Because of the enormous metabolic diversity of Bacteria (to be discussed in Chapter 16), a large gene pool exists in bacteria from natural habitats. In some cases these genes code for proteins that degrade environmental pollutants. Genes for the biodegradation of many toxic wastes and wastewater pollutants have been shown to exist in natural isolates of bacteria (◐◐ Sections 16.21 and 16.22). Genetic engineering is beginning to tap these resources for the purpose of environmental cleanup. In many cases the gene donors are bacterial strains isolated from contaminated waste sites. Some examples include genes for the biodegradation of chlorinated pesticides, like 2,4,5-trichlorophenoxyacetic acid (2,4,5-T), chlorobenzenes and related chlorophenolics, naphthalene, toluene, anilines, and various hydrocarbons (◐◐ Section 16.22). The desired genes are isolated from species of *Pseudomonas, Alcaligenes,* and a few other Bacteria and then cloned into plasmids. Some plasmids have been constructed containing genes from the biodegradation of several different toxic chemicals.

6. **Gene regulation and gene therapy.** The first use of genetic engineering in the biotechnology industry

was primarily for producing useful gene products more easily or for creating transgenic organisms. As we discuss in the following sections, this approach is yielding great benefits. However, much current research in this area involves the creation of new products ("designer drugs") and controlling the expression of specific genes. Much research is currently being directed toward design of antisense RNA (◐◐ box, Antisense Nucleic Acid, Chapter 7) and ribozymes (◐◐ Section 6.9), which regulate the expression of specific genes. (Much research is also directed toward delivering these "drugs" to, and getting them inside, the correct cells and tissues.) An additional area of research that holds great promise is *gene therapy,* the use of genetic engineering to treat genetic disease.

Despite the exciting promise of genetic engineering in biotechnology, getting a product to market is an enormous undertaking. Other than the obvious problems of correctly cloning and expressing the gene of interest in a microorganism, generally a bacterium or a yeast, and purifying the desired product, related matters such as clinical trials and governmental approval must be considered. Any microbially synthesized product intended for human use must pass extensive clinical trials. For example, human insulin produced microbially by recombinant DNA technology (see Section 10.13) had to pass strict clinical trials with human volunteers despite the fact that microbially produced insulin was shown to be identical to the protein made in humans. If all goes well in clinical trials, federal approval is usually obtained, in the United States by action of the Food and Drug Administration (FDA), but this can be a time-consuming process. In addition, unexpected problems often arise experimentally; for instance, the vector may be unstable, gene expression may be transient, or gene regulation problems may exist. Problems can also arise during the process of "scaling up" from laboratory to commercial production (◐◐ Sections 11.3 and 11.4). (This is one reason why biotechnologists should have a good knowledge of basic microbiology!) Finally, some products simply aren't effective when used in an actual commercial or clinical application. In spite of these difficulties, biotechnology is transforming many aspects of agriculture, medicine, and commerce. We now consider a few specific applications.

10.13

Production of Mammalian Products and Vaccines by Genetically Engineered Microorganisms

One of the first practical applications of genetic engineering was the use of easily grown bacteria to produce useful proteins whose genes were from organisms that are more difficult or expensive to grow. Although the special DNA polymerases used in the polymerase chain reaction were originally isolated from thermophilic bacteria (see Section 10.9), they are now produced in *Escherichia coli* from cloned genes. Most restriction enzymes are also produced in *E. coli* from cloned genes. Note that *E. coli* is not necessarily easier to grow than some of the bacteria that normally synthesize a particular restriction enzyme. However, cloning allows manipulation of expression levels (see Section 10.7) and always using the same host is more efficient, since a company can employ similar or identical culture conditions when making different products. Similarly, many proteins used industrially are now produced from cloned genes, and in some cases, the protein itself has been altered by using site-specific mutagenesis (see Section 10.11) to change the cloned gene.

Many proteins and peptides from mammalian cells have high pharmaceutical value. However, these proteins are usually present in very small amounts in normal tissue, and it is therefore extremely costly to purify them. In addition, even if the protein can be produced in cell cultures (⚙ Section 8.3), this is a much more expensive technique than growing microbial cultures. Therefore, another early effort of the biotechnology industry was to use genetic engineering to produce these proteins in microorganisms.

For many early applications, such as the production of insulin, it was known that the product would have great commercial value because of its established therapeutic value in treating a reasonably well-understood disease (in this case diabetes). However, such success is not always guaranteed, even when the protein can be produced and purified. Often this is because the disease process is complex and not well understood or the product has unexpected side effects. Nonetheless, by the late-1990s, biotechnology companies had hundreds of products in clinical trials. There are several different classes of therapeutic protein products that have been produced. These include hormones, interferons, growth factors, and vaccines. Some examples of these and other products are shown in Table 10.1.

Production of Insulin

One of the earliest and most dramatic commercial successes was the production of the hormone **insulin.** Many hormones are peptides or small proteins. These molecules are extremely important in controlling mammalian metabolism and have important therapeutic uses. Insulin is a protein produced in the pancreas that is vital for the regulation of carbohydrate metabolism in the body. Diabetes, a disease characterized by insulin deficiency, afflicts millions of people. The standard treatment for diabetes is periodic injections or oral administration of insulin, and because insulins of most

TABLE 10.1	Some therapeutic products made by recombinant DNA techniques[a]
Product	**Function**
Blood proteins	
Erythropoietin	Treats certain types of anemia
Factors VII, VIII, IX	Promote clotting
Tissue plasminogen activator	Dissolves clots
Urokinase	Blood clotting
Human hormones	
Epidermal growth factor	Wound healing
Follicle stimulating hormone	Treatment of reproductive disorders
Insulin	Treatment of diabetes
Nerve growth factor	Possible treatment of degenerative neurological disorders and stroke
Relaxin	Facilitates childbirth
Somatotropin (growth hormone)	Treatment of some types of growth failure and short stature
Immune modulators	
α-Interferon	Antiviral, antitumor, agent
β-Interferon	Treatment of multiple sclerosis
Colony stimulating factor	Treatment of infections and cancer
Lysozyme	Anti-inflammatory
Tumor necrosis factor	Antitumor agent, potential treatment of arthritis
Vaccines	
Cytomegalovirus	Prevention of infection
Hepatitis B	Prevention of serum hepatitis
Measles	Prevention of measles
Rabies	Prevention of rabies

[a] Although research is currently progressing in all these areas, not all of the listed products are yet on the market.

mammals are similar in structure, it is possible to treat human diabetes by use of insulin isolated commercially from beef or pork pancreas. However, nonhuman insulin is not as effective as *human insulin,* and the isolation process is expensive and complex. Cloning of a human insulin "gene" in bacteria has hence been carried out.

Producing hormones such as insulin in genetically engineered microorganisms is not simply a matter of cloning a gene (or even a cDNA) (see Section 10.10) in an expression vector. This is because many of these hormones are only small fragments of the polypeptides encoded by the gene. Insulin in its active form consists of two polypeptides (A and B) connected by disulfide bridges (Figure 10.17a). These two polypeptides are coded by separate parts of a single insulin gene. The insulin gene codes for *preproinsulin,* a longer polypeptide containing a signal sequence (involved in excretion of the protein) (⚙ Section 6.12), the A and B polypep-

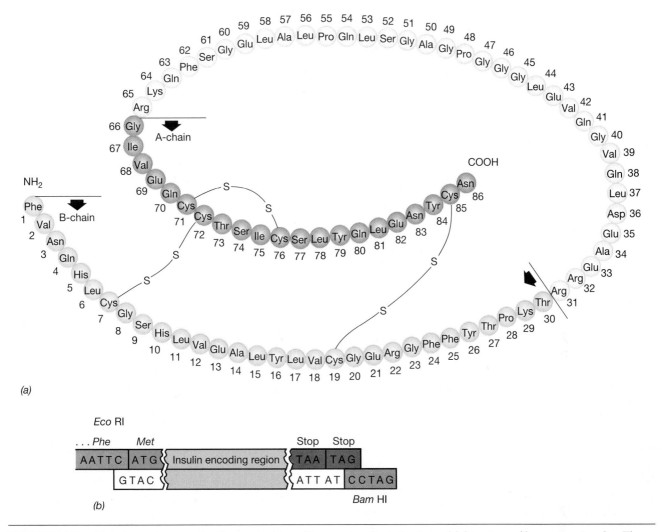

(a)

(b)

FIGURE 10.17 Genetic engineering for the production of human insulin in bacteria. (a) Structure of human proinsulin. The peptide shown in yellow must be removed from between the A and B chains in order to make insulin. (b) Chemical synthesis of the insulin gene and suitable linkers, permitting cloning and expression. The synthesized fragments were linked via restriction sites *Eco*RI and *Bam*HI in a plasmid vector in such a way that the insulin chains are formed as a fusion protein (see Section 10.7) with a portion of a gene found on the vector (note that the *Eco*RI site is part of this coding region). The methionine coding sequence was inserted to permit chemical cleavage of the A and B chains from the fused protein made in the bacteria because the reagent cyanogen bromide specifically cleaves at methionine residues and insulin does not contain methionine. Two stop codons were incorporated at the downstream end of the coding sequence.

tides of the active insulin molecule, and a connecting polypeptide that is absent from mature insulin. *Proinsulin* is formed from preproinsulin, and the conversion of proinsulin to insulin involves enzymatic cleavage of the connecting polypeptide from the A and B chains (∞ box, Protein Processing in Chapter 7).

Two approaches have been used to obtain production of human insulin in bacteria: (1) production of proinsulin and conversion to insulin by chemical cleavage, and (2) production of the A and B chains in two separate bacterial cultures, and joining of the two chains chemically to produce insulin. Because the insulin protein is fairly small, it was more convenient with either approach to synthesize the proper DNA sequence chem-

ically (see Section 10.8) rather than attempt to isolate the insulin gene from human tissue. There are 63 bases encoding the A chain and 90 bases encoding the B chain. In proinsulin there are an additional 105 bases for the peptide connecting the A and B chains (Figure 10.17b). When the polynucleotides were synthesized, suitable restriction enzyme sites were placed at each end so the polynucleotides could be ligated into a plasmid vector. To obtain effective expression, the synthesized genes were inserted downstream from a suitable *Escherichia coli* promoter but in a manner such that the insulin fragment was synthesized as part of a *fusion protein* (see Section 10.7). An important advantage of making the fusion protein is that the fusion product is much more stable in *E. coli* than insulin itself. Finally, a nucleotide triplet coding for methionine was placed at the junction joining the insulin gene to the upstream part of the fusion gene. The reason for this is that the chemical reagent *cyanogen bromide* specifically cleaves polypeptide chains at methionine residues, permitting recovery of the insulin product once the fused protein has been isolated from the bacteria. Insulin itself does not contain methionine and hence is unaffected by cyanogen bromide treatment.

When the proinsulin route is used, the proinsulin isolated from the bacteria via cyanogen bromide treatment is converted to insulin by disulfide bond formation, followed by enzymatic removal of the connecting peptide of proinsulin. Proinsulin naturally folds so the cysteine residues are opposite each other (Figure 10.17a), and chemical treatment then causes the formation of disulfide cross-links. Once this has been accomplished, the connecting peptide can be removed by treatment with the proteases trypsin and carboxypeptidase B, which have no effect on insulin itself.

When insulin is produced by way of the separate A and B peptides, each of the fusion proteins is isolated from a separate bacterial culture and the chains released by cyanogen bromide cleavage. The cleaved chains are then connected by use of chemical treatment that results in disulfide bond formation.

The final product, biosynthetic human insulin, is identical in all respects to insulin purified from the human pancreas and is being marketed commercially. Microbially produced human insulin is less expensive to make and just as effective as porcine or bovine insulin, the major source of insulin for diabetics before the advent of biotechnology.

Recombinant Vaccines

Vaccines are suspensions of killed or modified pathogenic microorganisms or specific fractions isolated from the microorganisms that when injected into an animal produce immunity to a particular disease. Often the substance that elicits the immune response is a surface protein, for instance, a virus coat protein. Genetic engineering can be applied in many different ways to the production of vaccines, including those against viral disease and others against bacterial disease. The importance of the recombinant vaccines is the fact that they can replace the killed or inactivated pathogenic organisms used as vaccines.

Genetic engineering has proved successful in the development of some *subunit vaccines*. These vaccines contain only a specific *subunit* of a protein from the pathogenic organism (usually a coat protein), and recombinant DNA techniques are used to produce these subunits in microorganisms. The highly immunogenic coat proteins are purified and used in high dosage to elicit a rapid and high level of immunity with no possibility of transmitting infection. The steps for viral gene

cloning are those outlined in the previous sections: fragmentation of viral DNA by restriction enzymes; cloning viral coat protein genes into a suitable vector; providing for proper promoters, reading frame, and ribosome binding sites; and reinsertion and expression of the viral genes in a microorganism. Unfortunately, when *Escherichia coli* is used as the cloning host, the vaccines are often poorly immunogenic and fail to protect animals from subsequent infection with the virus. The problem involves the fact that many key viral coat proteins are posttranslationally modified, generally by the addition of sugar residues (glycosylation), when the virus replicates in its normal host. However, the recombinant proteins produced by *E. coli* or other Bacteria are unglycosylated, and apparently glycosylation is necessary for the proteins to be immunologically active. Therefore, a eukaryotic host is used.

The first recombinant subunit vaccine approved for use in humans was made using yeast. The gene encoding a surface protein from hepatitis B virus was cloned and expressed in yeast. The protein was produced and formed aggregates very similar to those found in patients infected with the virus. These aggregates were purified and used to vaccinate people against hepatitis B virus. Subunit vaccines against a large variety of viruses and pathogenic organisms are also being developed employing genetic engineering. In addition to using yeast, insect cells and even cultured mammalian cells are now being used as hosts to prepare recombinant vaccines. To obtain the correct pattern of glycosylation or other modifications of the protein, it is often important to use a host that is closely related to humans. However, vaccines can also be produced in plants (see Section 10.14).

Genetic engineering can also be used to develop *recombinant live vaccines*. One way is to isolate deletion mutants of a virus lacking genes that cause disease but which is still infective and still elicits an immune response. Another method is to add genes to a virus so it can confer immunity against viral disease. In this latter category is a live recombinant virus vaccine that offers protection in poultry against both fowlpox (a disease that reduces weight gain and egg production) and Newcastle disease (a viral disease that is often lethal). The fowlpox virus, a typical pox virus (∞ Section 8.20), was first modified to delete genes that cause disease (but not those that elicit immunity). Then immunity-inducing genes from the Newcastle virus were added. This resulted in a *polyvalent vaccine*, in this case a single virus that can confer immunity to two different important diseases.

One vector used to prepare live recombinant vaccines is *vaccinia virus* (∞ Section 8.20). Cloning in vaccinia virus is done using an *Escherichia coli* plasmid containing a fragment of the vaccinia virus thymidine kinase gene (Figure 10.18a). An appropriate foreign DNA is inserted into this plasmid, and the recombinant plasmid is transformed into a host cell whose own thymidine kinase gene is inactive, but that has previously been infected with wild-type vaccinia virus (Figure 10.18*b*). If homologous recombination occurs between plasmid DNA and vaccinia genomic DNA (Figure 10.18*c*), re-

combinant virions can be obtained containing an *inactivated* thymidine kinase gene. An *active* thymidine kinase leads to growth inhibition by the compound 5-bromo-deoxyuridine. Therefore, recombinant vaccinia virions can be selected by allowing viral replication to occur in the presence of this inhibitor (Figure 10.18*d*). Although such recombinant viruses no longer express thymidine kinase, they can still infect human cells and they do express the foreign genes that have been cloned into them. Indeed, some recombinant vaccinia viruses can carry genes from four different viruses!

Vaccinia virus itself is generally not pathogenic for humans (vaccinia virus was originally used as a vaccine against the related virus smallpox). However, vaccinia virus is not completely benign (it causes severe complications in some people), and therefore more research must be done before such vaccines can be used in humans.

Genetically engineered vaccines are likely to become increasingly common because (1) they are safer than normal attenuated or killed vaccines, (2) they are more reproducible because their genetic makeup can be carefully monitored, and (3) they can be administered in high doses without fear of side effects.

Vaccines produced using genetic engineering can also usually be made much faster than those produced by more traditional methods. Recombinant vaccines made with cloned influenza virus hemagglutinin genes (∞ Section 8.16) can be made in just 2 or 3 months rather than 6–9 months. This could be of real advantage during an epidemic caused by a new strain of influenza virus (∞ Section 23.4). Recombinant vaccines are also sometimes much less expensive than those produced in other ways, and this in itself can allow for new uses. Bait carrying a recombinant rabies vaccine has been distributed over large tracts of land in Europe and has triggered a dramatic decline in the incidence of rabies among wild foxes. Such a method of vaccination would previously have been too expensive.

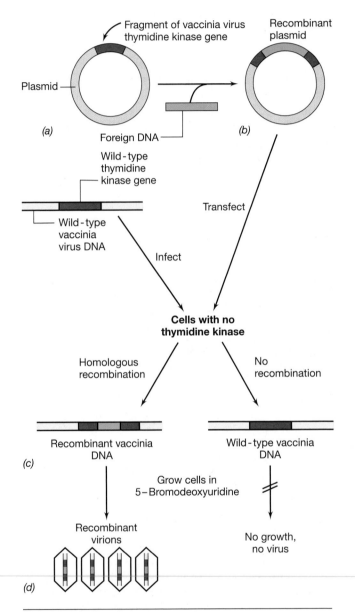

FIGURE 10.18   Production of recombinant vaccinia virus. (a) Foreign DNA is cloned into a plasmid containing a small piece of the vaccinia virus thymidine kinase gene. (b) Recombinant plasmid formed. (c) The recombinant plasmid is then used to transfect host cells already infected with wild-type vaccinia. If recombination occurs, recombinant vaccinia DNA can be produced. (d) The cells are then placed in the presence of 5-bromodeoxyuridine, a compound that is toxic to cells having an active thymidine kinase. Only recombinant virions develop under these conditions. If the recombinant vaccinia virions contain genes for other viral coat proteins, these may be expressed.

DNA Vaccines

One potentially very exciting development is the possible use of DNA itself as a vaccine. In some cases it appears that if a particular gene is delivered to an animal in such a way that it is taken up by the cells, the protein will be produced and the animal will develop immunity. A number of such DNA vaccines, or "gene vaccines," are undergoing clinical trials and many more are being developed. DNA vaccines would be both safe and inexpensive. Unlike viral vaccines, they would also escape surveillance by the host immune system.

Although genetic engineering is a powerful tool for producing a variety of novel vaccines, there are often formidable obstacles to overcome. Many laboratories have been working to develop vaccines against the virus that causes acquired immunodeficiency syndrome (AIDS) (human immunodeficiency virus, or HIV). Because of the serious consequences associated with AIDS, a safe, effective AIDS vaccine would find a huge worldwide market and thus the stakes in this area are high. Even so, no effective vaccine has yet been produced.

Other Proteins and Other Products

Table 10.1 lists a few of the mammalian proteins and products that are being produced by recombinant DNA techniques. These include a number of hormones in addition to insulin and also proteins involved in blood clotting and other blood processes. For example, *tissue plasminogen activator* (TPA) is a protein found in the blood that acts to scavenge and dissolve old blood clots in the final stages of the healing process. The clinical usefulness of TPA is primarily in heart patients or anyone suffering from poor circulation because of excessive clotting tendencies. TPA can be administered following cardiac bypass, transplant, or other open heart surgeries to prevent the development of pulmonary embolisms, which are often life-threatening. Heart disease is a leading cause of death in many developed countries, so microbially produced TPA promises to be in high demand.

In contrast to TPA, blood clotting factors VII, VIII, and IX are critically important for the *formation* of blood clots. Hemophiliacs suffer from a deficiency of one or more clotting factors and can be readily treated with the microbially produced product. Recombinant clotting factors take on added significance when one considers that hemophiliacs have in the past been treated with concentrated clotting factor extracts from pooled human blood, some of which was contaminated with the AIDS virus: this had put hemophiliacs at high risk for contracting AIDS.

A number of proteins have roles as anticancer agents or immune modulators. *Interferons* are a series of proteins made by animal cells in response to viral infection (Section 20.8) or immune activation in the case of one type of interferon.

Some mammalian proteins made via recombinant DNA technology do not fit in the categories listed in Table 10.1. For instance, human DNase I is being produced and used to treat the build-up of DNA-containing mucus in patients with cystic fibrosis. Even monoclonal antibodies are now produced in microorganisms by genetic engineering (we discuss monoclonal antibodies in some detail in Section 20.13).

Certainly not all the proteins produced using recombinant DNA techniques have therapeutic uses. Many commercial enzymes (Section 11.9) are produced in this way as well. In addition, even hormones can have other uses. *Bovine somatotropin* produced using these techniques is used to increase milk production in cattle in the United States. Often the "benefits" of genetic engineering can be quite unexpected. *Rennin*, which is used to make cheese, is an animal product. In Great Britain a "vegetarian cheese" containing a recombinant protein produced in a microorganism is being marketed and is finding wide acceptance. The first products made by genetic engineering were primarily the protein products of cloned genes, and there is still a great deal to be done with this approach. Added applications come from being able to use site-directed mutagenesis on the cloned gene so that new products with new attributes can be generated. It must also be remembered that molecules such as antibiotics are synthesized in cells in biochemical pathways using a series of enzymes (proteins). These enzymes can be modified so new antibiotics can be developed.

✓ 10.13 Concept Check

The first human protein made commercially using engineered bacteria was human insulin, but numerous other hormones and other human proteins are now being produced. Many proteins found in humans that were formerly extremely expensive to produce because they were found in human tissues in only small amounts can now be made in very large amounts from the cloned gene in a suitable expression system. In addition to useful pharmaceuticals such as anticancer agents and immune modulators, even vaccines can be produced using genetic engineering.

- ✓ Why is it sometimes important to produce proteins used for therapeutic purposes in a host closely related to humans?
- ✓ Explain why recombinant vaccines might be safer than some vaccines produced by traditional methods.

10.14

Genetic Engineering in Plant Agriculture

In the past, genetic improvement of plants has generally been a slow and difficult task, but recombinant DNA technology is leading to revolutionary changes. It is possible to use plant tissue culture techniques to select clones of plant cells that have been genetically altered and then, with proper treatments, induce these cell cultures to make whole plants that can be propagated vegetatively or by seeds. Recombinant DNA technology enters into this approach because one can transform plant cells with free DNA by either electroporation or particle gun methods (Section 9.6) or insert foreign DNA into plant cells directly via the bacterium *Agrobacterium tumefaciens*. This organism causes the plant disease *crown gall* by transferring specific genes to the plant

(∞ Section 16.24), and plant genetic engineers have used this natural transformation system as a vehicle for the introduction of foreign DNA into plants.

Vectors for Cloning in Plants

The gram-negative plant pathogen *Agrobacterium tumefaciens* contains a large plasmid called the **Ti plasmid,** which is responsible for its virulence. The plasmid contains genes that mobilize DNA for a transfer to the plant (for details of the disease process and genetic events, ∞ Section 16.24). The segment of the Ti plasmid DNA that is actually transferred to the plant is called **T-DNA.** The sequences at the ends of the T-DNA are essential for transfer, and the DNA to be transferred must be between these ends. One type of vector that has been constructed and is used for the actual transfer of genes to plants is called a *binary vector.* The word *binary* means "consisting of two parts," and a binary vector must be used in conjunction with another plasmid in order for the cloned gene to be transferred to the plant. The vector contains the two ends of the T-DNA on either side of the site used for cloning and an antibiotic resistance marker that can be used in plants. It also contains an origin of replication so that it can replicate in both *A. tumefaciens* and *Escherichia coli* (the latter serves as a host for cloning work), and another antibiotic resistance marker that is expressed in bacteria (Figure 10.19). The DNA to be cloned is inserted into the vector, which is then transformed into *E. coli*. It is then transferred to *A. tumefaciens* (usually by conjugation).

Final transfer to the plant depends on this bacterium also containing a Ti plasmid because the vector does not contain the genes required for T-DNA transfer. The cloned DNA and the kanamycin resistance marker of the vector can be mobilized by the Ti plasmid and transferred into a plant cell (Figure 10.19); note here that the Ti plasmid used for this purpose has been genetically altered in such a way as to prevent the onset of disease in the plant). Following recombination with a host chromosome, the foreign DNA can be expressed to confer new properties on the plant. Many genes are not expressed efficiently in plants unless they are cloned into an expression vector containing a plant promoter. Some of the promoters that have been used in constructing plant expression vectors include some normally found in T-DNA and a promoter from cauliflower mosaic virus, a plant DNA virus (∞ Table 8.1).

With the use of *Agrobacterium tumefaciens,* a number of transgenic plants have been produced. Most successes have come with herbaceous plants (dicots) such as tomato, potato, tobacco, soybean, alfalfa, and cotton, but *A. tumefaciens* has also been used to produce transgenic woody dicots such as walnut and apple. Transgenic crop plants from the grass family (monocots) are difficult to

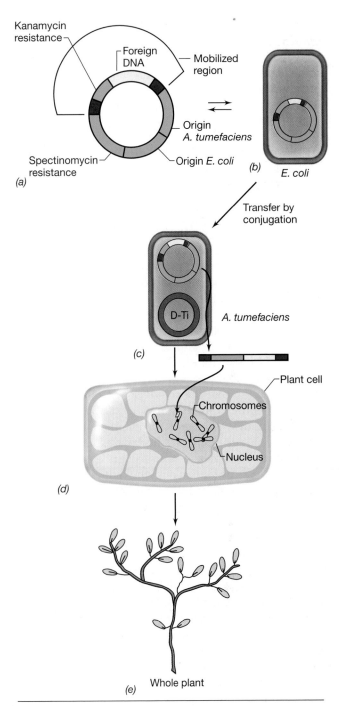

FIGURE 10.19 Production of transgenic plants using *Agrobacterium tumefaciens*. (a) Generalized plant transfection vector containing ends of T-DNA (in red), foreign DNA (in yellow), origin of replication elements for both *E. coli* and *A. tumefaciens,* and spectinomycin and kanamycin resistance markers. The kanamycin resistance marker can be selected for in plants. (b) The vector can be put into *E. coli* for cloning purposes and then transferred to *A. tumefaciens* by conjugation. (c) The resident Ti plasmid used for transferring the vector to the plant (D-Ti) is itself genetically engineered to remove key pathogenesis genes. (d) However, D-Ti can mobilize the T-DNA region of the vector for transfer to plant cells grown in tissue culture. From the recombinant cell, whole plants can be regenerated.

generate using *A. tumefaciens*, but other methods of introducing DNA, such as electroporation or a particle gun (Section 9.6), seem to work well here. In addition, some plant vectors have been developed by modifying plant viruses, but these have not yet proved as useful as the vectors derived from the Ti plasmid.

Applications in Plant Biotechnology

Major areas targeted for genetic improvement in plants include herbicide, insect, and microbial disease resistance, as well as improved product quality. Some products are already available, and genetically modified plants undergoing large-scale field trials include virus-resistant squash, herbicide-tolerant cotton, and soybeans and oilseed rape with modified oil. More than 1000 different field trials on more than 30 different plant species have been carried out in the last decade. Herbicide resistance can be obtained by genetically engineering the crop plant to no longer respond to the toxic chemical. Many herbicides act by inhibiting a key plant enzyme or protein necessary for growth. For example, the herbicide *glyphosate* kills plants by inhibiting the activity of an enzyme necessary for making aromatic amino acids. Such an herbicide kills both weeds and crop plants and thus must be used as a "preemergence herbicide," that is, before the crop plants emerge from the ground. However, some bacteria contain an enzyme that is naturally resistant to glyphosate. A gene encoding a resistant enzyme from *Agrobacterium* has been cloned, modified for expression in plants, and transferred into crop plants. When sprayed with glyphosate, plants containing the bacterial gene grow as well as unsprayed control plants (Figure 10.20). Soybeans expressing glyphosate resistance have been developed by Monsanto Company. In 1998, an estimated 25 million acres in the U.S. were planted with herbicide-resistant soybeans, and herbicide-resistant corn and cotton are also being grown.

Novel means of insect resistance have also been genetically introduced into plants. One already in wide use has been the toxic protein genes of *Bacillus thuringiensis*. This organism produces a crystalline protein (Section 13.19), called *Bt-toxin*, that is toxic to moth and butterfly larvae, and certain strains of *B. thuringiensis* produce additional proteins toxic to beetle and fly larvae and mosquitoes. Biotechnologists are using several different approaches to enhance the use of Bt-toxin for pest control in plants.

One approach is to develop a single Bt-toxin that is effective against many different insects. This can be done because the protein has separate domains for its specificity and its toxic function. The toxic domain is highly conserved in all the various Bt-toxins. Genetic

FIGURE 10.20   The photograph shows a portion of a field of soybeans that has been treated with *Roundup*, a glyphosate-based herbicide manufactured by Monsanto. The plants on the right are normal soybeans; those on the left have been genetically engineered to express glyphosate resistance.

engineers can make a gene that encodes a Bt-toxin carrying one toxic domain and several different specificity domains. In Section 10.10, we dealt with how one might have to modify a gene from a higher eukaryote before it could be efficiently expressed in a bacteria. One sometimes also has to modify bacterial genes before being expressed in a eukaryote. For instance, bacterial genes may contain sequences that lead to accidental splicing of the mRNA produced or the codon usage may have the wrong bias. Because of all the modification necessary, Bt-toxin genes now in use are typically completely synthetic.

An effective approach to achieve gene expression and stability is to transfer the gene directly into the plant genome. For example, a natural Bt-toxin gene has been cloned into a plasmid vector under control of a chloroplast rRNA promoter and transferred into tobacco plant chloroplasts by particle bombardment (Section 9.6). With this methodology plants were obtained that expressed this protein at levels that were extremely toxic to insect larvae from a number of species (Figure 10.21). Insect-resistant cotton and potatoes containing Bt-toxin are being widely grown in the U.S.

However, there have been reports of insects that have acquired resistance to Bt-toxin. Resistance to insecticides and herbicides is a common problem in agriculture, and the fact that a product has been produced by genetic engineering does not give it any magical properties. This emphasizes that many approaches must be used for pest control, and Bt-toxin is only one of many being developed by biotechnologists.

Genetic engineering has also been used to protect plants from virus infection. For example, it has been

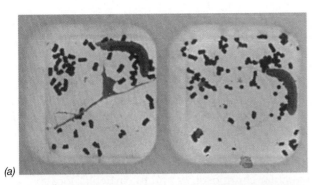

FIGURE 10.21 Panel (a) shows the results of two different assays to determine the effect of beet armyworm larvae on tobacco leaves from normal plants. Panel (b) shows the results of similar assays but using tobacco leaves taken from transgenic plants that express Bt-toxin in their chloroplasts.

discovered that transgenic plants that express the coat protein gene of a virus become resistant to infection by that virus. Although the mechanism of resistance is unknown, the presence of viral coat protein in plant cells apparently interferes with the uncoating of viral particles containing that coat protein, and this interrupts the virus replication cycle.

Not all genetic engineering is directed toward making plants disease-resistant. Genetic engineering can be used in a variety of ways for developing mutant strains of plants with desired characteristics such as delayed spoilage. In addition, transgenic plants can be genetically engineered to produce commercial or pharmaceutical products, as has been done with microorganisms (see Section 10.13) and animals (see Section 10.15). Crop plants such as tobacco and tomatoes have been engineered to produce a number of different products, such as the human protein interferon. Transgenic crop plants can be used to produce human antibodies efficiently and inexpensively (such plant-made antibodies are sometimes called "plantibodies"). These antibodies have potential as anticancer or antivirus drugs, and some are undergoing clinical trials.

Plant hosts can be useful in producing these types of products because plants typically modify proteins correctly and because crop plants can be efficiently grown and harvested.

Crop plants are also being developed for the production of vaccines. For instance, a recombinant tobacco mosaic virus has been engineered whose coat contains antigens of *Plasmodium vivax*, the organism that causes malaria (∞ Section 24.5). This recombinant virus could be used to develop a malaria vaccine that can be produced in very large amounts at very low cost by harvesting tobacco grown in fields. Another very interesting approach is to produce a vaccine in an edible plant product. Such *edible vaccines* are now under development that could immunize against diseases caused by enteric bacteria, including cholera and diarrhea (∞ Section 24.9).

Many plants have a natural ability to concentrate heavy metals and can be used to remove these contaminants from soil. Genetic engineering can be employed to enhance this ability. When the bacterial gene encoding mercuric reductase was modified and expressed in plants, the plants could grow at otherwise toxic levels of mercury. This enzyme reduces mercuric salts to elemental mercury, which is slowly released into the air. Such plants could be an important tool in removing contamination from soil. However, they would do so by releasing mercury into the air, and this may not be acceptable. This situation illustrates the careful consideration that must be given to the development of any technological solution to a problem. Good science does not necessarily translate directly into commercial viability.

It was estimated that almost 70 million acres (28 million hectares) of genetically modified crops were planted worldwide in 1998. Although public acceptance of these crops and their products seems high in the U.S., that is not always the case elsewhere. In Europe there is considerable public concern and several countries have considered bans on growing genetically engineered crops and legislation to require all genetically modified foods to be labelled. Therefore, even good science and potential commercial viability does not always ensure public acceptance.

✓ 10.14 Concept Check

Genetic engineering is being employed to make plants resistant to disease, to improve product quality, and to use crop plants as a source of recombinant proteins and even vaccines. One commonly used cloning vector for plants is the Ti plasmid of the bacterium *Agrobacterium tumefaciens*. This plasmid can transfer DNA into plant cells.

✓ Transfer of DNA by the Ti plasmid most resembles what form of bacterial gene transfer (∞ Chapter 9)?

✓ What commercial reasons could there be for producing vaccines or human proteins in a plant?

10.15

Genetic Engineering in Animal and Human Genetics

This section covers only a few highlights of some of the huge number of uses of genetic engineering in animal and human genetics. Some of these applications have to do with producing products, but more have to do with understanding gene function in mammals and in curing or treating genetic disease.

Transgenic Animals

With the use of recombinant DNA technology and microinjection techniques to deliver cloned genes to fertilized eggs, several foreign genes have been expressed in both laboratory research animals and in species important in commercial animal industries. Such animals are called **transgenic animals** and have become increasingly important in basic biomedical research for studying gene regulation and developmental biology. However, many applied aspects of transgenic animals are also of interest.

One approach is to improve the productivity or disease resistance of the animal, as in the case of transgenic plants used in agriculture. However, transgenic animals are also being used to produce proteins of pharmaceutical value—a process some scientists have called "pharming." Transgenic animals may be useful for producing human proteins that require specific posttranslational modifications for activity, such as certain blood clotting enzymes; many proteins of this type are not produced in an active form by microorganisms or by plants. Also, some proteins have been engineered to be secreted into the animals' milk, which can be readily collected and processed. These proteins include α-1-antitrypsin (used to treat lung disease) produced in sheep and tissue plasminogen activator (used to dissolve blood clots) produced in goats. The production of transgenic animals for research and commercial purposes seems likely to continue as an important area in biotechnology.

Human Genetics

Conventional genetics, involving genetic crosses or mutagenesis, cannot be done with humans. Therefore, in spite of the obvious interest, our understanding of human genetics had lagged considerably behind our understanding of the genetics of many other organisms. The advent of genetic engineering has quite simply revolutionized studies of the human genome. A detailed discussion of human genetics is beyond the scope of this book, but a few general remarks on the utility of recombinant DNA technology can be made. We have already mentioned the use of PCR for DNA fingerprinting (see the box), and considerable efforts are being made to clone and sequence the entire human genome (see Section 10.4).

However, there are some applications of genetic engineering of the human genome that are directed at treating human disease. A vast number of genetic diseases are known, but except in rare cases, little was known until recently about their molecular bases. By use of recombinant DNA technology, coupled with conventional genetic studies (following family inheritance, and so on), it is possible to localize particular defects to particular chromosomes and to particular locations on chromosomes. With the use of recombinant DNA technology, it is possible to clone the region containing the genetic defect and then to make comparisons between the base sequence in the normal gene and the defective gene. From such studies, even in the absence of knowledge of the enzyme defect, it has been possible to obtain information about the genetic change. Many genes, including those for Huntington's disease, cystic fibrosis, and Duchenne's muscular dystrophy, have been localized with these techniques and the mutation(s) in the defective genes identified.

Genetic engineering is employed to attempt to provide treatment for some of these diseases using **gene therapy.** In gene therapy, a nonfunctional or dysfunctional gene is augmented or replaced by a functional gene. Not all gene therapy is designed toward treating genetic disease; a considerable effort has been directed toward protocols for treating cancer. Major obstacles to this approach exist in trying to target the correct cells for gene therapy and in successfully transfecting cell lines that will perpetuate the genetic alteration.

The first genetic disease for which an approved gene therapy technique was used is a severe combined immune deficiency caused by the absence of adenosine deaminase (ADA), an enzyme involved in purine metabolism, in bone marrow cells. The procedure involved using a retrovirus as a vector to insert a good copy of the ADA gene into T lymphocytes (cells that are part of the immune system; ∞ Section 20.4) removed from the patient and then placing these "corrected" cells back in the body (the retrovirus also carries a marker gene, resistance to neomycin, so cells carrying the inserted retrovirus can be selected and identified). Since T lymphocytes have a limited life span, it is necessary to repeat the therapy every month or two. Attempts are being made to insert the gene into the stem cells of the bone marrow (which continue to divide) and effect a true cure for the disease.

Several other gene therapy treatments, some using other virus vectors, are currently being tested. Since the first gene therapy experiment with ADA in 1990, there have been no striking practical breakthroughs. Though gene therapy has tremendous practical potential, most applications still remain a distant prospect. Some of the current difficulties are related to the vectors being used. Although transduction using retroviral vectors gives

A FOCUS ON ...   DNA Chips

While the advent of recombinant DNA technology has revolutionized genetics, it is not the only revolution that has been occurring in science. Another involves silica-chip technology. Sometimes advances in different fields can complement each other to provide solutions to important scientific problems that had previously seemed unapproachable. An example of this is the development of the **DNA chip,** silica-chip technology that has been adapted for use in genomic analysis.

In the early 1990s it was discovered that the same process used to produce computer chips (photolithography) could be adapted to produce silica chips, each of which contained thousands of different DNA fragments. Currently, hundreds of thousands of different DNA strands of known sequence can be synthesized in an array on a 1–2 cm silica chip. DNA chips have been manufactured so that multiple sequences representing each of the 6200 protein-encoding genes from the yeast Saccharomyces cerevisiae is represented in a known location. These DNA chips can then be hybridized with DNA or RNA probes that have been tagged with a fluorescent dye or other labels. The tagged DNA or RNA binds only to the probe that is complementary to its sequence (∞ Section 6.2; box, Working with Nucleic Acids: the Tools, Chapter 6). This chip is then "read" with a laser and analyzed by a computer. Distinct patterns of hybridization will be observed, depending upon which complementary DNA or RNA sequences are present in the sample (Fig. 1).

We previously discussed hybridization using a single probe to find a corresponding gene (∞ Section 6.2). With the DNA chip one can

FIG. 1 DNA chips and their use to assay gene expression. The photo shows one fourth of the entire genome of the yeast, Saccharomyces cerevisiae affixed to a silica chip. Each gene is present in several copies and has been probed with fluorescently labelled mRNA obtained from yeast cells grown under a specific condition. The background of the chip is blue. Locations where the RNA has hybridized to the DNA is indicated by a gradation of colors up to maximum hybridization, which shows as white. Because the location of different genes on the chip is known, once the chip is scanned it will reveal which specific genes were expressed.

use very complex probes. For instance, one can ask which of the over 6,000 Saccharomyces genes are being expressed under a certain condition by using the entire population of mRNA as a probe. Of course, one can also compare expression of different genes under different conditions. The ability to relatively easily analyze the simultaneous expression of thousands of genes both qualitatively and quantitatively has tremendous potential for helping us to understand the extreme complexity of metabolism and regulation in organisms as "simple" as Escherichia coli with its 4000 genes, or as complex as ourselves, with our 75,000 different genes.

DNA chips can also be used to identify organisms. One can use fragmented genomic DNA from a particular organism as a probe and differentiate between closely related strains by differences in their hybridization patterns. This allows for very rapid identification of pathogenic viruses or bacteria, and for the detection of specific strains or mutants of these organisms. This technology is also useful for identifying and quantifying specific organisms in microbial communities in the environment, a common goal of the microbial ecologist (∞ Chapter 16). ■

stable integration of the gene, the site of insertion is un-predictable, the amount of cloned DNA is limited, and expression of the cloned gene is often transient. The vectors also have limited infectivity and are rapidly inactivated in the host. Many nonretroviral vectors, such as the adenovirus vector (∞ Section 8.21), have similar problems. These vectors are being further modified and others are being developed to attempt to overcome these problems (see Section 10.7). Promising vectors include human artificial chromosomes and highly modified versions of the HIV virus (∞ Section 8.22).

It is important to note that in the protocols being tested, the defective gene is not replaced. Rather, the retrovirus (and the good copy of the gene) simply integrates somewhere in the human genome of these cells. Actual gene replacements in germ line cells (cells that give rise to gametes) can be accomplished with some mammals, although the techniques of isolating individual animals with these changes cannot readily be applied to humans. However, attempts to change the germ cells of an individual would raise many ethical and societal questions beyond simple questions of experimental protocols.

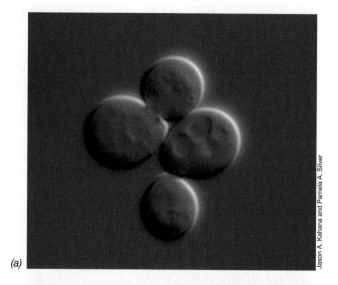

(a)

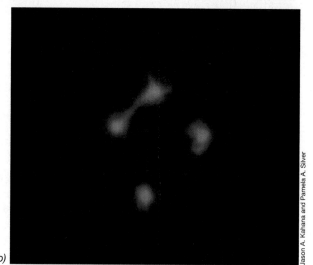

(b)

10.16

Genetic Engineering as a Microbial Research Tool

Up until now we have mostly presented the principles of genetic engineering in the light of practical goals. However, genetic engineering technology finds many uses in basic research on microorganisms. Molecular cloning and the engineering of new microbial strains provide some of the best ways to understand basic microbial processes such as structure–function relationships, cell growth, enzyme regulation, and microbial ecology. A novel microbial strain can be created that differs from the wild type in a defined manner, and the underlying process can then be studied. In this way, one can observe the importance of a particular gene product to a basic microbial process under precisely controlled conditions. Molecular cloning, restriction mapping, and sequencing also allow geneticists to quickly map and

FIGURE 10.22   The green fluorescent protein (GFP) can be used as a tag for protein localization *in vivo*. The gene encoding Pho2, a DNA binding protein from the yeast *Saccharomyces cerevisiae*, was fused to the gene encoding GFP. The recombinant gene was transformed into budding yeast cells, which could then express the fluorescent fusion protein that has localized to the nucleus. (a) Cells expressing Pho2-GFP were observed by differential interference contrast microscopy and (b) by epifluorescence microscopy (∞ Sections 3.1 and 3.2). (c) This panel is an overlay of the images in a and b.

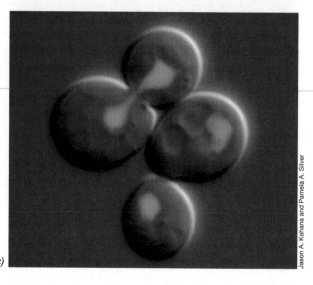

(c)

study the genomes of newly identified organisms. The development of *DNA chips* (see the box) allows scentists to explore the expression of an entire genome simply and rapidly.

We have discussed how cloned genes can be subjected to site-specific mutagenesis and how gene disruptions or knockout mutations can be introduced into a gene. These mutated genes can then be introduced into the microbial chromosome so the mutant organism can be studied. Genetic engineering also allows the researcher to "tag" a gene so that it is easier to study (Figure 10.22). For instance, if the gene product is difficult to assay or has no known function, it is often very difficult to know under what conditions the cell makes this protein. One can also use this technique to tag the organism itself so that it can be traced in the environment.

In these cases, a gene or its promoter can be fused to a *reporter gene*. Reporter genes are simple to assay, and the gene encoding β-galactosidase (whose activity can be detected on indicator plates and also assayed easily by biochemical means) is commonly used as a reporter (see Figure 10.5). However, there are other choices. For instance, *Escherichia coli* has been engineered in such a way that it produces the luciferase enzyme of the bacterium *Photobacterium* or that of several different luminescent beetles (see Figure 10.7). Because of the presence of this enzyme, the engineered *E. coli* becomes luminescent, and its colonies therefore glow in the dark. One can then detect colonies of the engineered *E. coli* on agar plates by their luminescence against a large background of other colonies. Production and detection of luciferase usually involves more than one gene and accessory factors. Recently a gene encoding a fluorescent protein, called the *green fluorescent protein* (GFP), has been isolated from the jellyfish *Aequorea victoria* and cloned. GFP needs no accessory factors and is already being used as a reporter or tag in a wide variety of organisms (Figure 10.22).

10.17

Summary of Principles at the Basis of Genetic Engineering

We have presented the fundamentals of genetic engineering and have shown how the approaches used have been derived from an understanding of basic concepts of molecular genetics and the "classic" techniques of microbial genetics. We now summarize the principles of genetic engineering by relating current knowledge back to the basic information presented in Chapters 6–9.

The following developments were essential for the development of genetic engineering; their interrelationships are diagrammed in Figure 10.23.

1. **DNA chemistry:** development of procedures for isolation, sequencing, and synthesis of DNA.
2. **DNA enzymology:** discovery of restriction endonucleases, DNA ligases, and DNA polymerases.
3. **DNA replication:** understanding how DNA replication occurs and the importance of DNA vectors capable of independent replication.
4. **Plasmids and conjugation:** discovery of plasmids, determination of the mechanisms by which plasmids replicate, and how some can transfer from cell to cell by conjugation.
5. **Temperate bacteriophage:** understanding how replication and/or integration is controlled in temperate bacteriophages and how specialized transducing phages are formed.
6. **Transformation:** discovery of methods for getting free DNA into cells.
7. **RNA chemistry and enzymology:** understanding how to work with messenger RNA, how eukaryotic mRNA is constructed, and the importance of RNA processing in the formation of mature eukaryotic mRNA.
8. **Reverse transcription:** discovery of the enzyme *reverse transcriptase* in retroviruses and its development as a means for transcribing information from mRNA back into DNA.
9. **Regulation:** understanding the factors involved in the regulation of transcription, including the discovery of promoter sites and operon control.
10. **Translation:** understanding the steps involved in translation, the importance of ribosome binding sites on mRNA, the role of the initiation codon, and the importance of a proper reading frame.
11. **Protein chemistry:** development of methods for isolation, purification, assay, and sequencing of proteins.
12. **Protein excretion and posttranslational modification:** understanding how proteins are built with signal sequences that are removed during or after excretion. Discovery of other kinds of posttranslational modification of proteins.
13. **The genetic code:** elucidation of the genetic code and the determination that it was the same in almost all organisms but that certain codons were less frequently used in some organisms than in others.

✓ 10.15–10.17 Concept Check

Genetic engineering is providing vast opportunities for improving processes of economic value and in treating disease. Medicine, agriculture, and industry are all finding important uses of cloning and gene expression technology. In addition, genetic engineering has revolutionized genetic studies in higher organisms and is a powerful tool in basic research on microbial genetics.

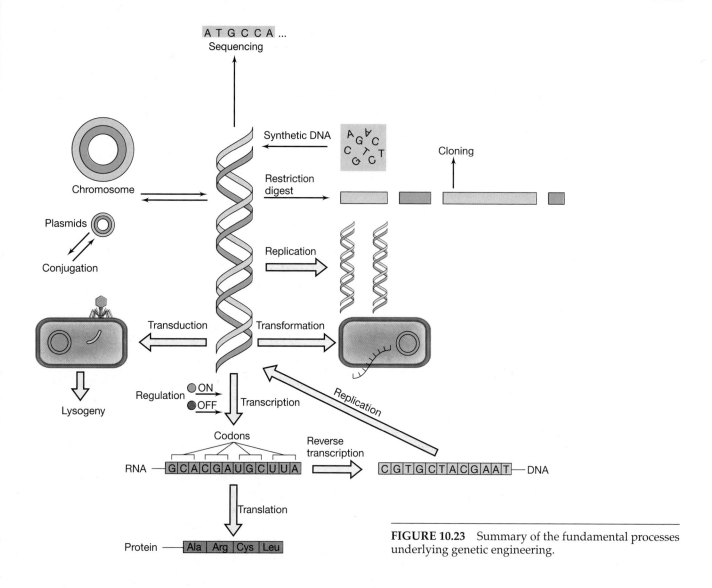

FIGURE 10.23 Summary of the fundamental processes underlying genetic engineering.

REVIEW QUESTIONS

1. What are the essential features of a cloning vector? What are the characteristics of plasmids that make them especially useful in cloning? Why aren't all plasmids equally useful for cloning?

2. Most of the lambda derivatives used as cloning vectors are virulent. Based on your knowledge of lambda genetics (⇔ Section 8.12) and the structure of the lambda Charon vectors, discuss why these viruses are no longer temperate. Why might it be useful to also have a temperate bacteriophage as a cloning vector?

3. The bacteriophage lambda can be an extremely useful cloning vector. However, it cannot be used in *Saccharomyces cerevisiae*. Explain.

4. If *insertional inactivation* is used to detect the presence of an introduced plasmid in a bacterial cell, why is it

desirable to have *two* antibiotic resistance markers in the plasmid?

5. Explain why the use of a regulatory switch is desirable for the large-scale production of a protein.

6. How has bacteriophage T7 been used in expressing foreign genes in *Escherichia coli* and what desirable features does this regulatory system possess?

7. The plasmid shown in Figure 10.9 is a cloning vector, an expression vector, a fusion vector, and a phagemid. Define each of these terms and explain how this single plasmid can have all these properties.

8. Describe the basic principles of gene amplification using the polymerase chain reaction (PCR). How have thermophilic bacteria simplified the use of PCR? How many PCR cycles need to be run to get 1000 copies of

a specific target sequence (starting from one molecule of double-stranded DNA)?

9. What are the similarities and the differences between *cassette mutagenesis* and *transposon mutagenesis* (👓 Section 9.10)?

10. What is a subunit vaccine and why are subunit vaccines considered a safer way of conferring immunity to viral pathogens than attenuated virus vaccines?

11. What is the Ti plasmid and how has it been of use in genetic engineering?

12. How do transgenic plants and animals differ from plants and animals modified by conventional breeding techniques?

13. What advantages might there be in using a transgenic plant rather than a transgenic animal to produce a protein?

APPLICATION QUESTIONS

1. Suppose you are given the task of constructing a plasmid suitable for molecular cloning in an organism of industrial interest. List the characteristics such a plasmid should have. List the steps you would use to develop such a plasmid.

2. When employing the plasmid vector pUC18, DNA fragments made using the restriction enzyme *Bam*HI inactivates the gene encoding β-galactosidase. When employing pBR322, cloning the same fragments inactivates the gene conferring tetracycline resistance. Both vectors contain a gene for ampicillin resistance that is used to select for bacterial transformants after making the recombinant molecules. Explain why it is much more efficient to use pUC18 rather than pBR322 as a cloning vector. (*Hint:* The increased efficiency has to do with finding the cells that contain vectors with cloned DNA.)

3. As mentioned in Section 10.2, some vectors have been developed so that insertional inactivation of the gene containing the polylinker can be a positive selection itself. However, simply selecting for the lack of this gene activity in transformed cells would not ensure that the cells contained any plasmid at all. Describe then what other genes must be carried by the plasmid vector and how the selection for plasmids containing foreign DNA would actually be carried out.

4. Suppose you have just determined the DNA base sequence for an especially strong promoter in *Escherichia coli* and you are interested in incorporating this sequence into an expression vector. Describe the steps you would use. What precautions would be necessary to be sure that this promoter actually works as expected in its new location?

5. Making cDNA libraries from cells undergoing a particular regulatory response can be a powerful way of detecting the genes involved. Even though it is as simple to isolate mRNA from prokaryotes as it is from eukaryotes, this technique isn't used with prokaryotes. Explain. (*Hint:* Think about how cDNA is made.)

6. Unless modified, one would expect that no more than 50% of the clones obtained using the site-directed mutagenesis technique shown in Figure 10.15 to be mutants. Why?

7. You have just discovered a protein in mice that may be an effective cure for cancer but it is present only in extremely small amounts. Describe the steps you would use to obtain production of this protein in *Escherichia coli*. If the protein produced in *E. coli* proved to be only marginally effective in clinical trials, what might you do to find a protein that is more effective?

The use of microorganisms by industry is an old practice and includes, as shown here, the production of microbial cells themselves, products from cells, and the use of microbial cells to catalyze particular reactions. From the beginnings of the field of industrial microbiology in the citric acid and penicillin fermentations of over 60 years ago, the field has grown to include the microbial production of a wide variety of different drugs, chemicals, and foods. Production of these compounds is usually a combination of science and art and often occurs on a huge scale in vessels of hundreds of thousands of liters. Effective operation of such production facilities requires the combined talents of highly skilled scientists and engineers who understand the problems that accompany the growth of microorganisms in large volumes.

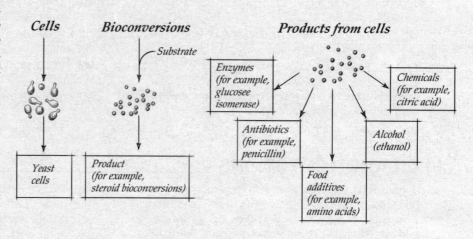

CHAPTER 11 Industrial Microbiology/Biocatalysis

11.1 Industrial Microorganisms and Products 385
11.2 Growth and Product Formation in Biocatalyses 387
11.3 Characteristics of Large-Scale Fermentations 389
11.4 Fermentation Scale-up 391
11.5 Antibiotics: Isolation and Characterization 392
11.6 Industrial Production of Penicillins and Tetracyclines 396
11.7 Vitamins and Amino Acids 399
11.8 Microbial Bioconversion 401
11.9 Enzymes 402
11.10 Vinegar 405
11.11 Citric Acid and Other Organic Compounds 406
11.12 Yeast as an Agent of Fermentation and as Food 407
11.13 Alcohol and Alcoholic Beverages 409
11.14 Mushrooms as a Food Source 413
11.15 Sewage and Wastewater Microbiology 416

Industrial microbiology is the discipline that uses microorganisms, usually grown on a large scale, to produce valuable commercial products or carry out important chemical transformations. Industrial microbiology originated with alcoholic fermentation processes, such as those for making beer and wine. Subsequently, microbial processes were developed for the production of pharmaceutical agents (such as antibiotics), food additives (such as amino acids), enzymes, and chemicals such as butanol and citric acid. All these industrial microbiological processes are enhancements of metabolic reactions that microorganisms were already capable of carrying out, with the goal in most cases of simply overproducing the product of interest. But now, in addition to traditional industrial microbiology, a new era has unfolded—that of *microbial biotechnology*. In biotechnology (∞ Chapter 10), methods for gene manipulation have given rise to new microbial products, most of which are not naturally produced by microorganisms. Microbial production arises from the fact that the microorganisms have been genetically engineered to make the desired product.

The whole field of industrial microbiology revolves around the use of microorganisms as microbial catalysts for specific chemical reactions. In fact, the term *bio-catalysis* has been used to describe the various reactions carried out by microorganisms in industrial microbiology. In this chapter we discuss a number of industrial biocatalytic reactions along with some of the problems large-scale microbial culture entails and the solutions to these problems worked out through the years by industrial microbiologists. We begin with an overview of industrial organisms and products.

11.1

Industrial Microorganisms and Products

The major organisms used in biocatalytic processes are fungi (yeasts and molds) and certain prokaryotes, in particular members of the genus *Streptomyces*. To a great extent, industrial microorganisms are metabolic specialists capable of being manipulated in large-scale culture to produce one or more products in high yield. In order to achieve this high metabolic specialization, strains of industrial microorganisms are often genetically altered by mutation or recombination, with the focus usually remaining on the *yield* of the particular product that a given strain can produce.

The ultimate source of all strains of microorganisms used in biocatalytic processes is the environment. However, actual industrial strains are often far removed from the "wild-type" condition that existed when the strain was first isolated. Once valuable industrial microorganisms have been developed, they are conserved by both microbiology laboratories in industry as well as in large national microbial culture collections (Table 11.1); in the latter, cultures of microorganisms are supplied for educational, research, and industrial purposes. When a new biocatalytic process is patented, a strain capable of carrying out the process must be deposited into one of these collections. However, for several reasons, proprietary rights chief among them, the strains deposited are *not* the actual high-yielding production strain(s) but instead a strain or strains that carry out the process at lower yield.

Properties of a Useful Industrial Microorganism

A microorganism suitable for industrial use must produce the substance of interest, of course, but there is much more to it than that. The organism must be genetically stable and must be capable of growth and product formation in large-scale culture. Moreover, the organism should preferably produce spores or some other reproductive cell form so that it can be easily inoculated into large fermentors.

An important characteristic is that the industrial organism grow rapidly and produce the desired product in a relatively short period of time. The organism must also be able to grow in a relatively inexpensive liquid culture medium obtainable in bulk quantities. Many industrial microbiological processes use waste carbon from other industries as major or supplemental ingredients for large-scale culture media. These include *corn steep liquor* (a product of the corn wet milling industry that is rich in nitrogen and growth factors), *whey* (a waste liquid of the dairy industry containing lactose and minerals), and other industrial waste materials having high organic carbon contents. In addition, an industrial microorganism should not be harmful to humans or economically important animals or plants. Because of the large population size in the industrial fermentor and the virtual impossibility of avoiding contamination of the environment outside the fermentor, a pathogen would present potentially disastrous problems.

Finally, an industrial microorganism should be amenable to genetic manipulation. In industrial microbiology, increased yields have been obtained genetically primarily by means of mutation and selection. It is also desirable for the industrial organism to be capable of genetic recombination, either by a sexual or by some sort of parasexual process. Genetic recombination permits the incorporation in a single genome of genetic traits from more than one organism. However, many industrial strains have been greatly improved by mutation and selection without use of any genetic recombination.

Examples of Industrial Products

Microbial products of industrial interest are of several major types (Figure 11.1). These include the microbial cells themselves, for example, yeast cultivated for food, baking, or brewing (see Figure 11.21), and substances produced by cells. Examples of the latter include enzymes such as glucose isomerase, pharmacologically active agents such as antibiotics, steroids, and alkaloids, specialty chemicals and food additives such as the currently popular aspartame food and drink sweetener, and commodity chemicals, such as ethanol. A summary of some important industrial products, many of which will be discussed in more detail later, is given in Figure 11.1.

TABLE 11.1	Culture collections that supply cultures of industrial microorganisms[a]	
Abbreviation	**Name**	**Location**
ATCC	American Type Culture Collection	Manassas, VA, United States
CBS	Centraalbureau voor Schimmelculturen	Baarn, The Netherlands
CCM	Czechoslovak Collection of Microorganisms	J. E. Purkyne University, Brno, Czech Republic
CDDA	Canadian Department of Agriculture	Ottawa, Canada
CMI	Commonwealth Mycological Institute	Kew, United Kingdom
DSM	Deutsche Sammlung von Mikroorganismen und Zellkulturen GmbH	Braunschweig, Germany
IAM	Institute of Applied Microbiology	University of Tokyo, Japan
NCIB	National Collection of Industrial Bacteria	Aberdeen, Scotland
NCTC	National Collection of Type Cultures	London, England
NRRL	Northern Regional Research Laboratory	Peoria, IL, United States
PCC	Pasteur Culture Collection	Paris, France

a Listed here are just a few of the general culture collections. Many universities and research laboratories maintain collections of specific microbial groups.

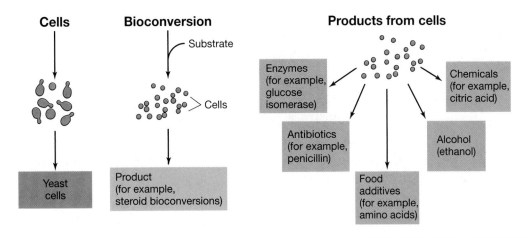

FIGURE 11.1   Products of industrial microbiology/biocatalysis. The products may be the cells themselves or products made from cells. In the case of bioconversion, cells are used to chemically convert a specific substance from one form to another.

✓ 11.1 Concept Check

An industrial microorganism must produce the product of interest in high yield, grow rapidly on inexpensive culture media available in bulk quantities, be amenable to genetic manipulation, and, if possible, be nonpathogenic. Industrial products are many and include both cells and substances made by cells.

✓ Why should industrial microorganisms be genetically manipulable?

✓ List three important products of industrial biocatalysis.

11.2

Growth and Product Formation in Biocatalyses

In Section 5.3 we discussed the microbial growth process and described the various stages: lag, exponential, and stationary. Here we describe microbial growth and product formation in an industrial context, posing the question: "When in the growth cycle is the industrially useful metabolite produced?"

Primary and Secondary Metabolites

There are two basic types of microbial metabolites: primary and secondary. A *primary metabolite* is one that is formed during the primary growth phase of the microorganism, whereas a *secondary metabolite* is one that is formed near the end of the growth phase, frequently at, near, or in the stationary phase of growth. The contrast between a primary metabolite and a secondary metabolite is illustrated in Figure 11.2.

A typical microbial process in which the product is formed during the primary growth phase is *alcohol (ethanol) fermentation.** Ethanol is a product of anaerobic metabolism of yeast and certain bacteria (⚭ Section 4.9) and is formed as part of energy metabolism. Because growth can occur only if energy production can occur, ethanol formation takes place in parallel with growth. A typical alcohol fermentation, showing the formation of microbial cells, ethanol, and sugar utilization, is illustrated in Figure 11.2*a*.

In contrast to ethanol production by yeast, in some biocatalytic processes the desired product is not produced during the primary growth phase but instead in the *stationary* phase. Metabolites produced during the stationary phase are called **secondary metabolites** and are some of the most common and important metabolites of industrial interest (Figure 11.2*b*). The following characteristics of secondary metabolites have been recognized:

1. Each secondary metabolite is formed by only a relatively few organisms.
2. Secondary metabolites are not essential for growth and reproduction.
3. The formation of secondary metabolites is extremely dependent on growth conditions, especially on the composition of the medium. Repression of secondary metabolite formation frequently occurs.
4. Secondary metabolites are often produced as a group of closely related structures. For instance, a single strain of a species of *Streptomyces* has been found to produce over 30 related but different anthracycline antibiotics.
5. It is often possible to get dramatic *overproduction* of secondary metabolites, whereas primary metabolites,

*In industrial microbiology, the term *fermentation* refers to *any* large-scale microbial process, whether or not it is biochemically a fermentation. In fact, most industrial fermentations are aerobic. The *tank* in which the industrial fermentation is carried out is called a fermen*tor*; the microorganism involved is the fermen*ter*.

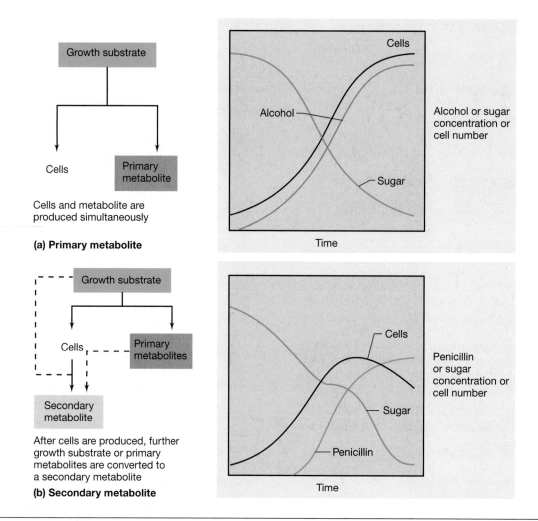

FIGURE 11.2 Contrast between production of primary and secondary metabolites. (a) Alcohol formation by yeast—an example of a primary metabolite. (b) Penicillin production by the mold *Penicillium chrysogenum*—an example of a secondary metabolite.

linked as they are to primary metabolism, usually cannot be overproduced in such a dramatic manner.

6. In secondary metabolism, the product in question may be derived not from the primary growth substrate but instead from primary metabolites. Thus, the secondary metabolite can be produced from several intermediate products that accumulate, either in the culture medium or in the cells, during primary metabolism (Figure 11.2*b*).

Relationship between Primary and Secondary Metabolism

Most secondary metabolites are complex organic molecules that require a large number of specific enzymatic reactions for synthesis. For instance, it is known that at least 72 separate enzymatic steps are involved in synthesis of the antibiotic *tetracycline* (see Section 11.6) and over 25 steps in the synthesis of *erythromycin*, none of which are reactions occurring during primary metabolism. How-

ever, the metabolic pathways of these secondary metabolites do arise out of primary metabolism, because the starting materials for secondary metabolism come from the major biosynthetic pathways. This is summarized in Figure 11.3, which shows the interrelationship of the main primary metabolic pathway for aromatic amino acid synthesis with the secondary metabolic pathways for a variety of antibiotics. As can be seen, many structurally complex secondary metabolites originate from structurally quite similar precursor molecules (Figure 11.3).

✓ 11.2 Concept Check

Primary and secondary metabolites are produced during active cell growth or near the onset of stationary phase, respectively. Many economically valuable microbial products are secondary metabolites.

✓ Is penicillin a primary or a secondary metabolite? Why?
✓ What type of metabolite, primary or secondary, can be more easily overproduced? Why?

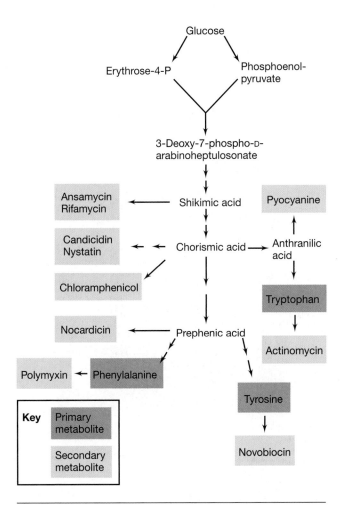

FIGURE 11.3 Relationship of the primary metabolic pathway for the synthesis of aromatic amino acids (⬡⬡ Section 4.15) and formation of a variety of secondary metabolites containing aromatic rings. Note that this is a composite scheme of processes occurring in a variety of microorganisms: no one organism produces all these secondary metabolites.

11.3

Characteristics of Large-Scale Fermentations

The vessel in which the industrial process is carried out is called a *fermentor*. Fermentors can vary in size from the small 5- to 10-liter laboratory scale (Figure 11.4*a*) to the enormous 500,000-liter industrial scale. The size of the fermentor used depends on the process and how it is operated. Processes operated in batch mode require larger fermentors than processes operated continuously or semicontinuously. A summary of fermentor sizes for some common microbial fermentations is given in Table 11.2.

Industrial fermentors can be divided into two major classes, those for *anaerobic* processes and those for *aerobic* processes. Anaerobic fermentors require little special equipment except for removal of heat generated during the fermentation, whereas aerobic fermentors require much more elaborate equipment to ensure that mixing and adequate aeration are achieved. Because most industrial fermentations are aerobic, the present discussion will be confined to aerobic fermentors.

Construction of an Aerobic Fermentor

Large-scale industrial fermentors are almost always constructed of stainless steel. Such a fermentor is essentially a large cylinder, closed at the top and bottom, into which various pipes and valves have been fitted (Figure 11.4*b*). Because sterilization of the culture medium and removal of heat are vital for successful operation, the fermentor generally has an external *cooling jacket* through which steam or cooling water can be run. For very large fermentors, insufficient heat transfer occurs through the jacket, and so *internal coils* must be provided through which either steam or cooling water can be piped.

An important part of the fermentor is the aeration system. With large-scale equipment, transfer of oxygen from the gas to the liquid is critical and elaborate precautions must be taken to ensure proper aeration. Oxygen is poorly soluble in water, and in a fermentor with a high microbial population density, there is a tremendous oxygen demand by the culture. Two separate installations are used to ensure adequate aeration: an aeration device, called a *sparger;* and a stirring device, called an *impeller* (Figure 11.4*b*). The sparger is a device, often a series of holes in a metal ring or a nozzle, through which filter-sterilized air can be passed into the fermentor under high pressure. The air enters the fermentor as a series of tiny bubbles from which the oxygen passes by diffusion into the liquid. A good sparging system should ensure that the bubbles are of very small size so diffusion of oxygen from the bubble into the liquid can occur readily.

In small fermentors use of a sparger alone may be sufficient to ensure adequate aeration, but in industrial-size fermentors, *stirring* of the fermentor with an impeller is essential (Figure 11.4*c*). Stirring accomplishes two things: it mixes the gas bubbles through the liquid; and it mixes the organism through the liquid, thus ensuring uniform access of the microbial cells to the nutrients.

The shaft that drives the impeller is attached to the motor by way of a drive shaft that must penetrate into the fermentor from outside. Because of the need to maintain sterility, it is vital that the seal connecting the drive shaft to the motor be arranged so that contaminants can-

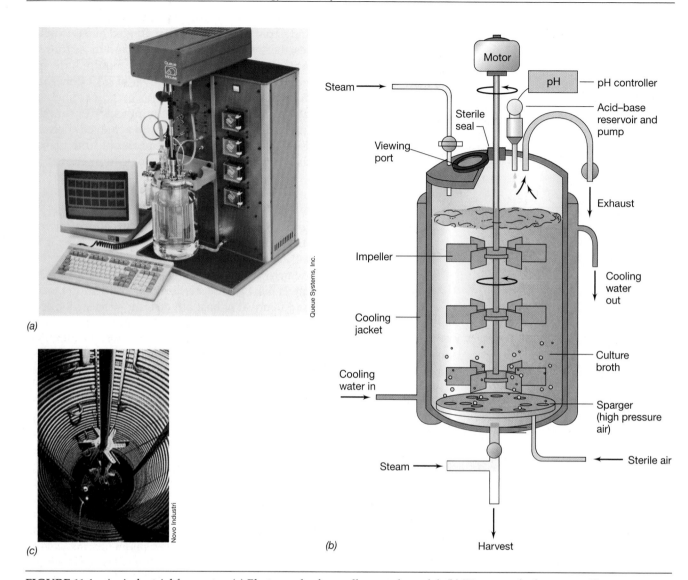

FIGURE 11.4 An industrial fermentor. (a) Photograph of a small research model. (b) Diagram of a fermentor, illustrating construction and facilities for aeration and process control. (c) Photograph of the inside of a large fermentor, showing the impeller and internal heating and cooling coils.

TABLE 11.2	Fermentor sizes for various industrial processes
Size of fermentor (liters)	**Product**
1–20,000	Diagnostic enzymes, substances for molecular biology
40–80,000	Some enzymes, antibiotics
100–150,000	Penicillin, aminoglycoside antibiotics, proteases, amylases, steroid transformations, amino acids
200,000–500,000	Amino acids (glutamic acid)

not pass through (see Figure 11.4b). A typical large-scale fermentor installation is shown in Figure 11.5.

Process Control and Monitoring

Any microbial process must be monitored to ensure that it is proceeding properly, but it is especially important that industrial fermentors be monitored carefully because there is such a major expense involved. In most cases, it is necessary not only to measure growth and product formation but also to *control* the process by altering environmental parameters as the process proceeds. Environmental factors that are frequently

FIGURE 11.5 (a) A large industrial fermentation plant. Only the tops of the fermentors, which can be several stories high, are visible. (b) Computer control room for a large fermentation plant.

monitored include temperature, oxygen concentration, pH, cell mass, and product concentration.

Computers play important roles in fermentor process control (Figure 11.5*b*). During the growth and product formation process in a large-scale fermentation it is essential, if the fermentation is to be operated properly, that data be obtained on the process *as it actually takes place.* For instance, it may be desirable to change one of the environmental parameters as the fermentation progresses, or to feed a nutrient at a rate that exactly balances growth. The computer can be used to process the data on-line and then respond as directed by a program in regard to when and how much nutrient to add. In this way, nutrient is added when needed and not before, thus avoiding potential diversion of nutrient from the desired product into unwanted products.

Computers can also be used to *model* fermentation processes. Using a mathematical model, one can test the effect of various parameters on growth and product yield quickly and interactively and then make modifications in the parameters to see how they affect the process. In this way, many variations in the fermentation can be studied inexpensively at the computer terminal, rather than expensively at the pilot plant stage or in the industrial plant (see next section).

✓ 11.3 Concept Check

Large-scale industrial fermentations present several engineering problems. In aerobic processes, ensuring adequate oxygen availability in a large industrial tank is a difficult problem, requiring installations for stirring and aeration. The microbial process must be continuously monitored in order to ensure satisfactory yields of the desired product.

✓ What types of devices are used to ensure proper aeration in a large-scale fermentation?
✓ What parameters in an industrial fermentation need to be monitored and what adjustments need to be made by computerized control?

11.4

Fermentation Scale-up

One of the most important and complicated aspects of industrial microbiology is the transfer of a process from small-scale laboratory equipment to large-scale commercial equipment, a procedure called **scale-up.** An understanding of the problems of scale-up is extremely important because rarely does a biocatalytic process behave the same way in large-scale fermentors as in small-scale laboratory equipment (Figure 11.6).

Why does a microbial process differ between large-scale and small-scale equipment? Mixing and aeration are much easier to accomplish in the small laboratory flask than in the large industrial fermentor. Oxygen transfer especially is much more difficult to obtain in a large fermentor, and because most industrial fermentations are aerobic, effective oxygen transfer is essential. With the rich culture media used in industrial processes, a high biomass is obtained, leading to a high oxygen demand. If aeration is reduced, even for a short period, the culture may experience partial anoxic conditions, with serious consequences in terms of product yield. Scale-up of an industrial process is the task of the *biochemical engineer,* one who is familiar with gas transfer, fluid dynamics, mixing, and thermodynamics.

(a)

(b)

FIGURE 11.6   (a) A bank of small research fermentors, used in process development. (b) A large bank of outdoor industrial-scale fermentors (240 m^3) used in commercial production of amino acids in Japan. Because of the great difference in their size, the same microbial fermentation would probably operate quite differently in the two different types of fermentors.

The Scale-Up Process

In transferring an industrial process from the laboratory to the commercial fermentor, several stages can be envisioned. (1) Experiments in the *laboratory flask,* which are generally the first indication that a process of commercial interest is possible. (2) The *laboratory fermentor,* a small-scale fermentor, generally of glass and generally 1- to 10-liter size, in which the first efforts at scale-up are made (Figure 11.6*a*). In the laboratory fermentor, it is possible to test variations in medium, temperature, pH, and so on, inexpensively because little cost is involved for either equipment or culture medium. (3) The *pilot plant stage,* usually carried out in equipment 300–3000 liters in size. Here, the conditions more closely approach the commercial scale; however, cost is not yet a major factor. In the pilot plant fermentor, careful instrumentation and computer control are desirable so the conditions most similar to those in the laboratory fermentor can be obtained. (4) The *commercial fermentor* itself, generally 10,000–500,000 liters (Figures 11.5*a* and 11.6*b*).

It is generally found in scale-up studies on aerobic fermentations that the oxygen transfer rate in the fermentor, usually a critical factor, is best kept *constant* as the size of the fermentor is increased. Thus, if an oxygen transfer rate of 200 millimoles (mmol) O_2 liter^{-1}hr^{-1} is required to obtain optimal yield in a small fermentor, then stirring and aeration in the large fermentor should be adjusted to ensure this

same rate. This requires more rapid stirring as well as a higher pressure of the inlet air as the culture becomes more dense. In summary, then, scale-up of a biocatalytic process can be enormously complex and requires knowledge not only of the biology of the producer organism, but also of the physics of fermentor design and operation.

We now consider the industrial production of microbial products, beginning with the antibiotics. Antibiotic production is a huge industry worldwide and one where many important principles of large-scale microbial cultures were first developed.

✓ 11.4 Concept Check

Scale-up is the process of gradually converting a useful industrial fermentation from laboratory scale to production scale. Aeration is a particularly critical aspect to monitor during scale-up studies.

✓ What are the differences in size among a typical laboratory fermentor, a pilot plant fermentor, and a commercial fermentor?

11.5

Antibiotics: Isolation and Characterization

Of the microbial products manufactured commercially, probably the most important are the antibiotics. As we will discuss in Chapter 18, antibiotics are chemical substances produced by microorganisms that kill or inhibit the growth of other microorganisms. The development of antibiotics as agents for treatment of infectious disease has probably had more impact on the practice of medicine than any other single development.

Antibiotics are typical secondary metabolites. Commercially useful antibiotics are produced primarily by filamentous fungi and by Bacteria of the actinomycete group. A listing of the most important antibiotics produced by large-scale industrial fermentation is given in Table 11.3. Frequently, a number of chemically related antibiotics exist, and so *families* of antibiotics are known. Hence, antibiotics can be classified according to their chemical structure, as will be discussed in Section 18.7.

Search for New Antibiotics

Over 8000 antibiotic substances are known, and several hundred antibiotics are discovered yearly. Are there more antibiotics waiting to be discovered? Almost certainly yes, because most of the microorganisms that have been examined for their ability to produce antibiotics are members of a few genera, such as *Streptomyces, Penicillium,* and *Bacillus.* Many antibiotic researchers believe that a vast number of new antibiotics will be discovered if other groups of microorganisms are examined. It also seems likely that genetic engineering techniques will permit the artificial construction of new antibiotics and that computer modeling will eventually replace traditional screening methods for new drugs.

To some extent this is already occurring, as new more powerful computer methods are being used for *drug design.* The approach here is to use a computer to model interactions between a drug's target (for example, a specific protein) and a modified version of a known drug. The computer can then be used to screen various hypothetical (not yet existing) drugs and actually predict their efficacy. Particularly promising drugs identified in this way can then be synthesized by chemical or biological modifications of existing drugs and tested for clinical effectiveness.

However, the traditional way in which new antibiotics are discovered is by *screening.* In the screening approach, a large number of isolates of possible antibiotic-producing microorganisms are obtained from nature in pure culture (Figure 11.7*a*), and these isolates are then tested for antibiotic production by seeing whether they produce any diffusible materials that are inhibitory to the growth of test bacteria. The test bacteria used are selected from a variety of bacterial types but are chosen to be representative of or related to bacterial pathogens. The classical procedure for testing new microbial isolates for antibiotic production is the cross-streak method, first used by Fleming in his pioneering studies on penicillin (∞ the box, Microbiology and "Magic Bullets" in Chapter 18 and Figure 11.7*b*). Those isolates that show evidence of antibiotic production are then studied further to determine if the antibiotics they produce are new. In most screening programs, most of the isolates obtained produce *known* antibiotics, so the industrial microbiologist must quickly identify producers of known antibiotics so that time and resources are not wasted in studying them. Once an organism producing a *new* antibiotic is discovered, the antibiotic is produced in sufficient amounts for structural analyses and then tested for toxicity and therapeutic activity in infected animals. Unfortunately, most new antibiotics *fail* these animal tests, but a few pass them successfully. Ultimately, only a very few of these new antibiotics prove to be medically useful and are produced commercially.

TABLE 11.3	Some antibiotics produced commercially
Antibiotic	**Producing microorganism**[a]
Bacitracin	*Bacillus licheniformis* (EFB)
Cephalosporin	*Cephalosporium* sp (F)
Chloramphenicol	Chemical synthesis (formerly produced microbially by *Streptomyces venezuelae*) (A)
Cycloheximide	*Streptomyces griseus* (A)
Cycloserine	*Streptomyces orchidaceus* (A)
Erythromycin	*Streptomyces erythreus* (A)
Griseofulvin	*Penicillium griseofulvin* (F)
Kanamycin	*Streptomyces kanamyceticus* (A)
Lincomycin	*Streptomyces lincolnensis* (A)
Neomycin	*Streptomyces fradiae* (A)
Nystatin	*Streptomyces noursei* (A)
Penicillin	*Penicillium chrysogenum* (F)
Polymyxin B	*Bacillus polymyxa* (EFB)
Streptomycin	*Streptomyces griseus* (A)
Tetracycline	*Streptomyces rimosus* (A)

a EFB, endospore-forming bacterium; F, fungus; A, actinomycete.

Steps toward Commercial Production

An antibiotic that is to be produced commercially must first be produced successfully in large-scale industrial fermentors. We discussed the general problem of scale-up earlier in this chapter. One of the most important tasks thereafter is the development of efficient purification methods. Because of the relatively small amounts of antibiotic present in the fermentation liquid, elaborate methods for extraction and purification of the antibiotic are necessary (Figure 11.8). If the antibiotic is soluble in an organic solvent, it may be relatively simple to pu-

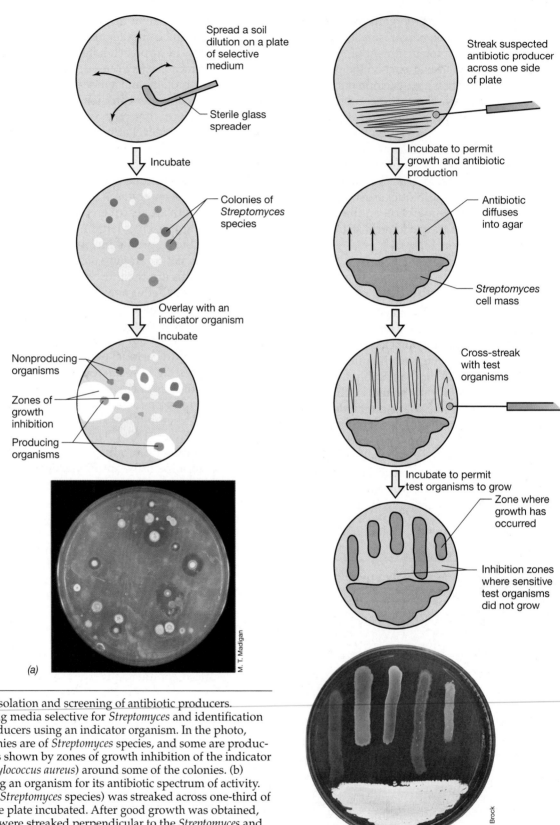

FIGURE 11.7 Isolation and screening of antibiotic producers. (a) Isolation using media selective for *Streptomyces* and identification of antibiotic producers using an indicator organism. In the photo, most of the colonies are of *Streptomyces* species, and some are producing antibiotics as shown by zones of growth inhibition of the indicator organism (*Staphylococcus aureus*) around some of the colonies. (b) Method of testing an organism for its antibiotic spectrum of activity. The producer (a *Streptomyces* species) was streaked across one-third of the plate, and the plate incubated. After good growth was obtained, the test bacteria were streaked perpendicular to the *Streptomyces* and the plate was further incubated. The failure of several organisms to grow near the mass growth of *Streptomyces* indicates that the *Streptomyces* produced an antibiotic active against these bacteria. Test organisms (left to right): *Escherichia coli, Bacillus subtilis, Staphylococcus aureus, Klebsiella pneumoniae, Mycobacterium smegmatis.*

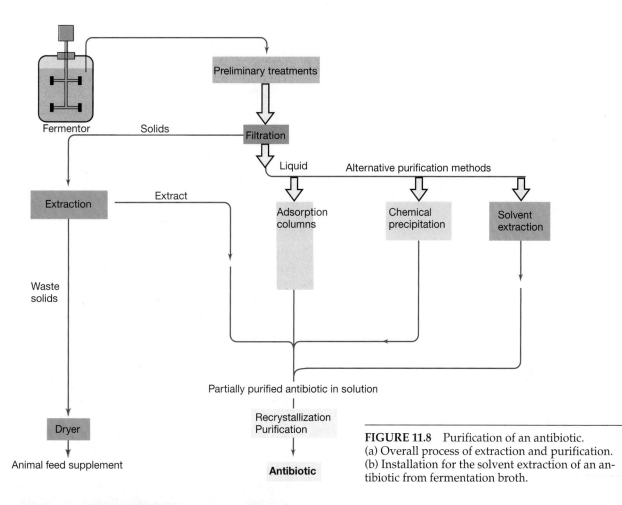

FIGURE 11.8 Purification of an antibiotic.
(a) Overall process of extraction and purification.
(b) Installation for the solvent extraction of an antibiotic from fermentation broth.

rify the antibiotic by extracting it into a small volume of the solvent, thus concentrating the antibiotic. If the antibiotic is not solvent-soluble, then it must be removed from the fermentation liquid by adsorption, ion exchange, or chemical precipitation. In all cases, the goal is to eventually obtain a crystalline product of high purity, although some antibiotics do not crystallize readily and are difficult to purify. A related problem is that cultures often produce other end products, including other antibiotics, and it is essential to end up with a product consisting of only a single chemical compound.

Rarely do antibiotic-producing strains just isolated from nature produce the desired antibiotic at sufficiently high concentration that commercial production can begin immediately. One of the main tasks of the industrial microbiologist is thus to isolate new *high-yielding strains.* The industrial microbiologist has made significant contributions to the antibiotic industry by developing high-yielding processes. Strain selection involves mutagenesis of the initial culture, plating of mutant types, and testing of these mutants for antibiotic production. However, genetic engineering has greatly improved the procedures for seeking high-yielding strains. The technique of *gene amplification* makes it possible to place additional copies of genes of interest into a cell by means of a vector such as a plasmid. Alterations in regulatory processes also may permit increased yields. However, one difficulty with using genetic procedures for increasing antibiotic yield is that the biosynthetic pathways for the synthesis of most antibiotics involve large numbers of steps with many genes, and in many cases it is not clear which genes should be altered or increased in number to increase yields. Thus, it is critical that the rate-limiting step in a given biochemical pathway be identified by basic research.

✓ 11.5 Concept Check

The industrial production of antibiotics begins with screening for antibiotic producers. Once new producers are identified, purification and chemical analyses of the antimicrobial agent are made. If the new antibiotic is biologically active *in vivo*, the industrial microbiologist may genetically modify the producing strain to increase yields to levels acceptable for commercial development.

- ✓ What is the natural habitat of most antibiotic-producing microorganisms?
- ✓ What is meant by the word *screening* in the context of finding new antibiotics?

11.6

Industrial Production of Penicillins and Tetracyclines

Once an antibiotic has been structurally characterized, has been proven medically effective in tests on experimental animals and sufficiently nontoxic, and, finally, has passed clinical trials (this sequence of events can take several years in actual practice), it is ready to be produced commercially and marketed. For antibiotics like penicillin and tetracycline, these hurdles were passed long ago; today, literally tons of these antibiotics are produced for medical and veterinary use. We focus here on the industrial production of these two antibiotics as examples of antibiotic production in general.

β-Lactam Antibiotics: Penicillin and Its Relatives

The penicillins are a class of antibiotics characterized by the β-*lactam ring* and are produced by a variety of molds (eukaryotes) of the genera *Penicillium* and *Aspergillis,* and by certain prokaryotes (Figure 11.9). Among the clinically useful penicillins several different forms are known, and these derivatives may be the result of both biocatalytic reactions and later chemical modification by the organic chemist to produce a penicillin with specific clinical properties.

The basic structure of all penicillins is *6-aminopenicillanic acid* (6-APA), which consists of a thiazolidine ring with a condensed β-lactam ring (Figures 11.9 and 11.10). The 6-APA carries a variable side chain in position 6. If the penicillin fermentation is carried out without addition of side-chain precursors, the **natural penicillins,** such as benzylpenicillin (penicillin G), are produced (Figures 11.9 and 11.10). The fermentation can be better controlled by adding to the broth a *side-chain precursor* so only one desired penicillin is produced. The product formed under these conditions is referred to as a **biosynthetic penicillin** (Figure 11.10). However, in order to produce the most useful penicillins, those with activity against *gram-negative Bacteria*, a combined fermentation and chemical approach is used that leads to the production of **semisynthetic penicillins** (Figure 11.10). In this case, microbially produced benzylpenicillin is split either chemically or enzymatically to yield 6-APA and the latter chemically modified by the addition of a side chain (Figure 11.10). Semisynthetic penicillins have many significant clinical advantages in terms of their spectrum of activity and the fact that many of them, for example, ampicillin, can be taken orally and thus do not require injection. For these reasons, semisynthetic penicillins make up the bulk of the penicillin market today.

Production Methods for β-Lactam Antibiotics

Penicillin G is produced using a submerged fermentation process in 40,000- to 200,000-liter fermentors. Penicillin production is a highly aerobic process, and efficient aeration is necessary. Penicillin is a typical secondary metabolite. During the growth phase, very little penicillin is produced, but once the carbon source has been nearly exhausted, the penicillin production phase begins (Figure 11.11). By feeding with various culture medium components, the production phase can be extended for several days (Figure 11.11).

A major ingredient of most penicillin production media is **corn steep liquor.** This substance contains the nitrogen source as well as other growth factors. The carbon source is generally *lactose* (Figure 11.11). The side chain of penicillin G is the phenylacyl moiety, and yields of penicillin G are markedly increased if phenylacetic acid (Figure 11.9) is fed as a precursor. Penicillin G is excreted into the medium, and after the cells are removed by filtration, the pH of the medium is lowered and the antibiotic extracted from the filtered broth with amyl or butyl acetate. After concentration into the solvent, the antibiotic is back-extracted into an alkaline aqueous medium, concentrated further, and crystallized. Highly purified penicillin can be readily obtained in this way.

Other β-Lactam Antibiotics

A variety of other β-lactam antibiotics are known. *Cephalosporins* are β-lactam antibiotics containing a dihydrothiazine instead of a thiazolidine ring system

Basic structures	Antibiotics	Most important producing species
Penam 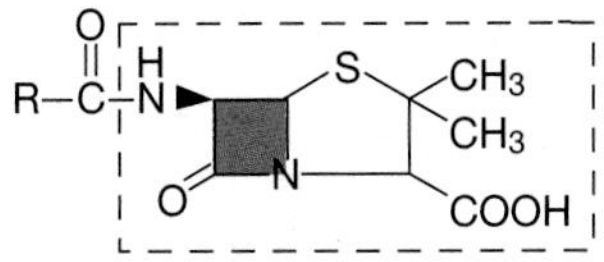	Penicillins (dashed lines outline 6-aminopenicillanic acid) If R 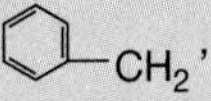 = ⬡—CH₂ , penicillin G	Penicillium chrysogenum Aspergillus nidulans Cephalosporium acremonium Streptomyces clavuligerus
Ceph-3-em 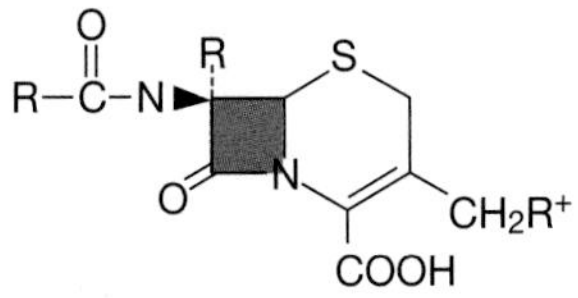	Cephalosporins	Cephalosporium acremonium Nocardia lactamdurans Streptomyces clavuligerus
Clavam 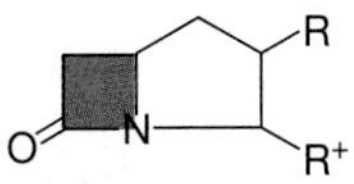	Clavulanic acids	Streptomyces clavuligerus
Carbapenem 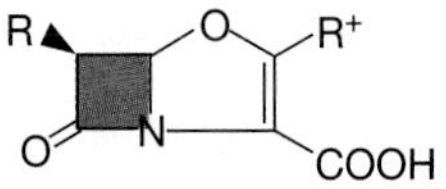	Thienamycins Olivanic acids Epithienamycins	Streptomyces cattleya Streptomyces olivaceus Streptomyces flavogriseus
Monolactam 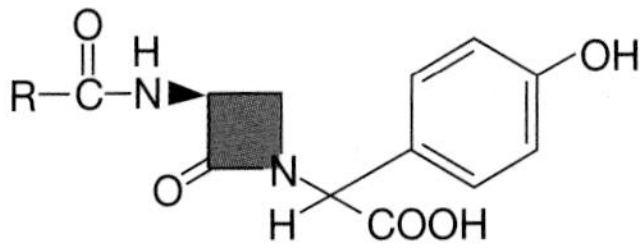	Nocardicins	Nocardia uniformis subsp. tsuyamanesis
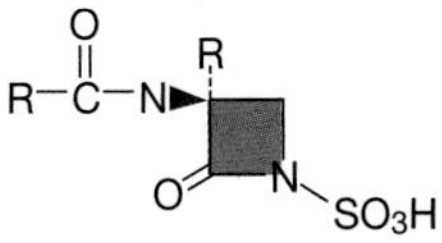	Monobactams	Gluconobacter sp. Chromobacterium violaceum Agrobacterium radiobacter Pseudomonas acidophila Pseudomonas mesoacidophila Flexibacter sp. Acetobacter sp.

FIGURE 11.9 The basic structures of the naturally occurring β-lactam antibiotics and the major producing organisms. The positions where chemical substitutions can occur are indicated by R. The β-lactam ring is shown in red. To obtain high yields of penicillin G, phenylacetate is added during the production phase. Phenylacetate is the structure shown in the top middle panel with a carboxylic acid group (COO^-) bonded to the methylene (CH_2) group.

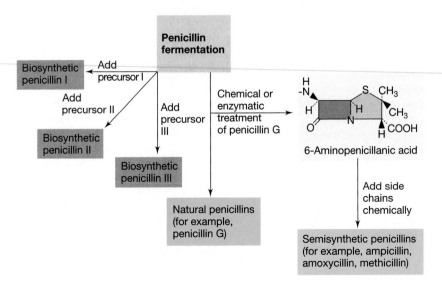

FIGURE 11.10 Industrial production of penicillins. The normal fermentation leads to the natural penicillins. If specific precursors are added during the fermentation, various biosynthetic penicillins are formed. Semisynthetic penicillins are produced by chemically adding a specific side chain to the 6-aminopenicillanic acid nucleus.

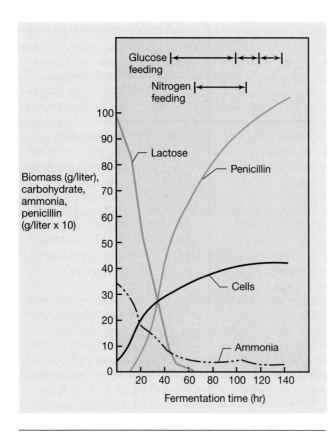

FIGURE 11.11   Kinetics of the penicillin fermentation with *Penicillium chrysogenum*. Note how the production of penicillin occurs as cells are entering stationary phase and most of the carbon and nitrogen is exhausted. Nutrient "feedings" keep penicillin production high.

(Figure 11.9). Cephalosporins were first discovered as products of the fungus *Cephalosporium acremonium*, but a number of other fungi as well as some prokaryotes also produce antibiotics with this ring system. In addition, a number of semisynthetic cephalosporins are produced. Cephalosporins are valued clinically not only because of their low toxicity but also because they are broad-spectrum antibiotics (⌁ Section 18.8).

Intensive screening for new β-lactam antibiotics has led to the development of compounds whose structures are different from those of both the penicillins and the cephalosporins. Included in this category are *nocardicin, clavulanic acid,* and *thienamycin* (Figure 11.9). Clavulanic acid is of particular interest because, although it is not very effective as an antibiotic by itself, it inhibits the activity of β-lactamases. These enzymes are produced by certain bacteria and function to destroy β-lactam antibiotics, rendering them ineffective in treating disease (⌁ Section 18.12). Thus, when used in combination with β-lactamase-sensitive penicillins and cephalosporins, clavulanic acid causes a distinct increase in the activity of these antibiotics.

Production of Tetracyclines

The biosynthesis of a tetracycline involves a large number of enzymatic steps. In the case of chlortetracycline (Figure 11.12), as many as 72 intermediate products may be involved, most of which are known in only a general way. Studies on the genetics of *Streptomyces aureofaciens,* the producer of chlortetracycline, have shown that over 300 genes are involved! With such a large number of genes, regulation of biosynthesis of this antibiotic is obviously quite complex. However, a few regulatory signals are known and production schemes are well worked out. For example, repression of chlortetracycline synthesis by both glucose and phosphate is known to occur. Phosphate repression is especially significant, and so the medium used in commercial production contains low phosphate concentrations. A production scheme for chlortetracycline is shown in Figure 11.12. Note that as in penicillin production, corn steep liquor is used in the large-scale production of chlortetracycline but that the use of glucose is avoided; chlortetracycline production is typically run

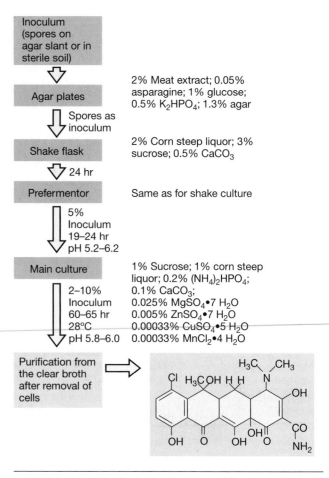

FIGURE 11.12   Production scheme for chlortetracycline with *Streptomyces aureofaciens*. The structure of chlortetracycline is shown on the bottom right.

with sucrose as the carbon source. Glucose is avoided because it causes catabolite repression (∞ Section 7.6) of antibiotic production.

✓ 11.6 Concept Check

Major antibiotics of clinical significance include the β-lactam antibiotics penicillin and cephalosporin, and the tetracyclines. All of these antibiotics are typical secondary metabolites, and their industrial production is well worked out despite the fact that the biochemistry and genetics of their biosynthesis are only partially understood.

✓ What chemical structure is common to all penicillins?
✓ In terms of penicillin production, what is meant by the term *semisynthetic? Biosynthetic?*
✓ Why is clavulanic acid of importance to clinical medicine?

11.7

Vitamins and Amino Acids

Vitamins and amino acids are growth factors that are often used pharmaceutically or are added to foods. Several important vitamins and amino acids are produced commercially by biocatalytic processes.

Vitamins

Vitamins are used as supplements for human food and animal feeds. Production of vitamins is second only to that of antibiotics in terms of total sales of pharmaceuticals—nearly $1 billion per year. Most vitamins are made commercially by chemical synthesis. However, a few are too complicated to be synthesized inexpensively but can be made by biocatalysis. Vitamin B_{12} and riboflavin are the most important of this class of vitamins.

Vitamin B_{12} (Figure 11.13a) is synthesized in nature exclusively by microorganisms. As a coenzyme, vitamin B_{12} plays an important role in animal biochemistry in various intramolecular rearrangements in which a hydrogen atom on one carbon atom and a substituent on the adjacent carbon atom exchange places. In humans, a major deficiency of vitamin B_{12} leads to a severe condition called *pernicious anemia*, characterized by low production of red blood cells and nervous system disorders. The requirements of animals for vitamin B_{12} are satisfied by food intake or by absorption of the vitamin produced in the gut of the animal by intestinal microorganisms. Plants do not produce or use vitamin B_{12}.

For industrial production of vitamin B_{12}, microbial strains are employed that have been specifically selected for their high yields of the vitamin. Members of the bacterial genus *Propionibacterium* give yields of the vitamin ranging from 19 to 23 mg/l in a two-stage process, and another bacterium, *Pseudomonas denitrificans*, pro-

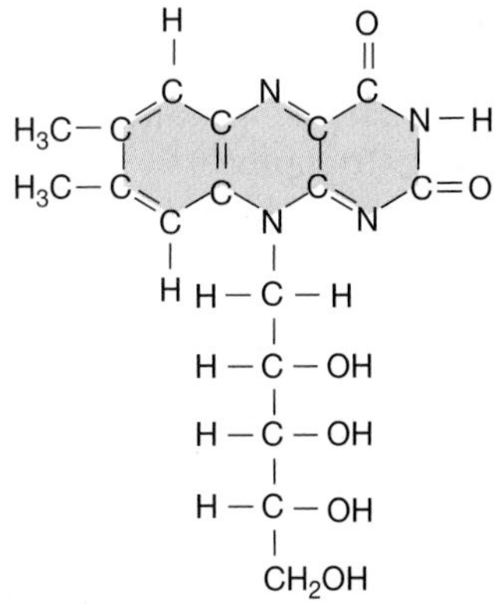

FIGURE 11.13 Vitamins produced by microorganisms on an industrial scale. (a) Vitamin B_{12}. Shown is the structure of cobalamin; note the central cobalt atom. The coenzyme form of vitamin B_{12} contains a deoxyadenosyl group attached to Co above the plane of the ring. (b) Riboflavin (vitamin B_2) (∞ Section 4.10 and Figure 4.13).

duces 60 mg/l in a one-stage process that uses sugarbeet molasses as the carbon source. Vitamin B_{12} contains cobalt as an essential part of its structure (Figure 11.13a), and yields of the vitamin are greatly increased by addition of cobalt to the culture medium.

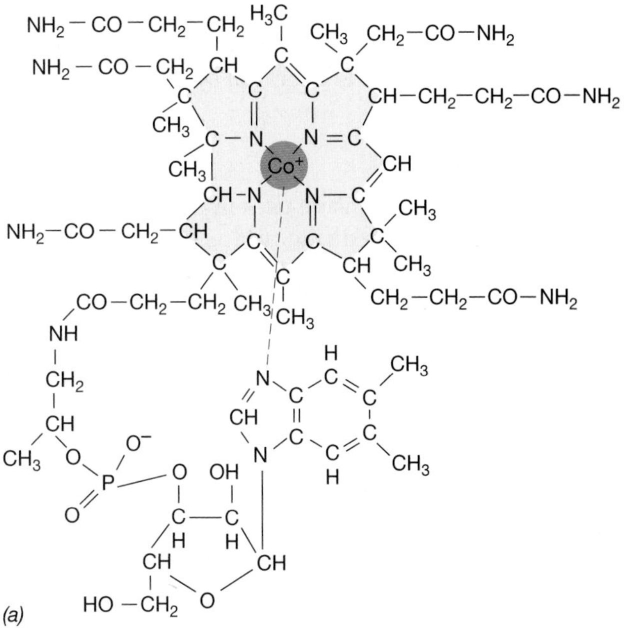

Riboflavin (Figure 11.13b) is the parent compound of the flavins, FAD and FMN, coenzymes that play important roles in enzymes involved in oxidation–reduction reactions in virtually all organisms (∞ Section 4.10). Riboflavin is synthesized by many microorganisms, including bacteria, yeasts, and fungi. The fungus *Ashbya gossypii* produces a huge amount of this vitamin (up to 7 g/l) and is therefore used for most of the microbial production processes. In spite of this good yield, there is great economic competition between this microbiological process and chemical synthesis.

Amino Acids

Amino acids have extensive uses in the food industry, as feed additives, in medicine, and as starting materials in the chemical industry (Table 11.4). The most important commercial amino acid is **glutamic acid,** which is used as a flavor enhancer [monosodium glutamate (MSG)]. Two other important amino acids, **aspartic acid** and **phenylalanine,** are the ingredients of the artificial sweetener **aspartame,** an important constituent of diet soft drinks and other foods sold as sugar-free products. **Lysine,** an essential amino acid for humans and certain farm animals, is commercially produced by the bacterium *Brevibacterium flavum* for use as a food additive. Although most of the amino acids can be made chemically, chemical synthesis results in the formation of optically inactive D, L mixtures. If the biochemically important L form (∞ Section 2.6) is desired, then a biocatalytic method of manufacturing is needed (see Figure 11.6*b*). We focus here on the production of lysine as an example of the industrial process.

We discussed amino acid biosynthesis in Chapter 4 (∞ Section 4.15), and the regulation of enzymatic pathways by both feedback inhibition and repression in Chapter 7 (∞ Sections 7.1 and 7.2). Because amino acids are used by microorganisms as building blocks of proteins, strict cellular regulation of their production generally occurs. However, for the industrial production of an amino acid, ways to circumvent these regulatory mechanisms are necessary in order to obtain an *overproducing strain*

capable of producing the amino acid economically. Such overproducing strains have been developed, and we discuss here a typical example, *Brevibacterium flavum,* used in the production of the amino acid *lysine* (Figure 11.14).

The production of lysine in *Brevibacterium flavum* is biochemically controlled at the level of the enzyme aspartokinase, in that excess lysine feedback inhibits activity of this enzyme (Figure 11.14*a*) (the general phenomenon of feedback inhibition was described in Section 7.1). However, overproduction of lysine can be obtained by isolating mutants of *B. flavum* in which aspartokinase is no longer subject to feedback inhibition. This is done by isolating mutants resistant to the lysine analog *S*-aminoethylcysteine (AEC), which binds to the allosteric site of aspartokinase and shuts down activity of the enzyme (Figure 11.14*b*). AEC-resistant mutants, which are easily obtained by positive selection, produce a mutant form of aspartokinase with an allosteric site that no longer recognizes AEC *or* lysine, and thus feedback inhibition by lysine is greatly reduced. Such mutants of *B. flavum* can produce over 60 g of lysine per liter in industrial scale reactors (see Figure 11.6*b*).

Another important factor in the commercial production of amino acids is *excretion*. In general, organisms do not excrete metabolites they have gone to great trouble to make. However, excretion of amino acids can be accomplished in various ways, in particular by altering cell permeability. For example, in *Corynebacterium glutamicum,* the commercial producer of glutamic acid, cell permeability is enhanced by a specific nutritional deprivation that causes the organism to produce a weakened cell membrane, thus allowing excretion of excess glutamic acid.

TABLE 11.4	Amino acids used in the food industry[a]			
Amino acid[b]	**Annual production worldwide (metric tons)**	**Uses**	**Purpose**	
L-Glutamate (monosodium glutamate, MSG)	370,000	Various foods	Flavor enhancer; meat tenderizer	
L-Aspartate and alanine	5,000	Fruit juices	"Round off" taste	
Glycine	6,000	Sweetened foods	Improve flavor; starting point for organic syntheses	
L-Cysteine	700	Bread	Improves quality	
		Fruit juices	Antioxidant	
L-Tryptophan + L-Histidine	400	Various foods, dried milk	Antioxidant, prevents rancidity; nutritive additive	
Aspartame (made from L-phenylalanine + L-aspartic acid)	7,000	Soft drinks	Low-calorie sweetener	
L-Lysine	70,000	Bread (Japan), feed additives	Nutritive additive	
DL-Methionine	70,000	Soy products, feed additives	Nutritive additive	

a Data from Glazer, A. N., and H. Mikaido. 1995. *Microbial Biotechnology.* W. H. Freeman, New York.

b The structures of these amino acids are shown in Figure 2.12.

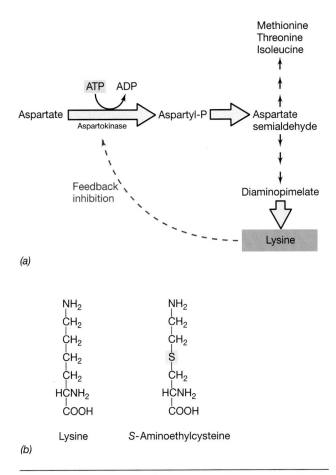

(a)

(b)

FIGURE 11.14 Industrial production of lysine using *Brevibacterium flavum*. (a) Biochemical pathway leading from aspartate to lysine; note that lysine can feedback-inhibit (∞ Section 7.1) activity of the enzyme aspartokinase, leading to cessation of lysine production. (b) Structure of lysine and the lysine analog *S*-aminoethylcysteine (AEC). AEC normally inhibits growth, but AEC-resistant mutants of *B. flavum* have an altered allosteric site on their aspartokinase and grow and overproduce lysine because feedback inhibition no longer occurs.

✓ 11.7 Concept Check

Vitamins produced microbially include vitamin B_{12} and riboflavin, whereas the most important amino acids produced commercially are glutamic acid, aspartic acid, phenylalanine, and lysine. High yields of amino acids are obtained by modifying regulatory signals that control synthesis of the particular amino acid such that overproduction occurs.

✓ What amino acid is commercially produced in the greatest amounts?

✓ How can overcoming feedback inhibition improve the yield of an amino acid?

11.8

Microbial Bioconversion

One of the most far-reaching discoveries in industrial microbiology was the understanding that microorganisms can be used to biocatalyze specific chemical reactions beyond the capabilities of organic chemistry. The use of microorganisms for this purpose is called **bioconversion** or **biotransformation** and involves growth of the organism in large fermentors, followed by the addition at an appropriate time of the chemical to be converted. Following a further incubation period during which the chemical is acted on by the organism, the fermentation broth is extracted and the desired product purified. Although in principle bioconversion may be used for a wide variety of processes, its major practical use has been in the production of certain steroid hormones (Figure 11.15).

We discussed the role of sterols in eukaryotic membranes in Section 3.5. Steroids, which are derivatives of sterols, are important hormones in animals that regulate various metabolic processes. Some steroids are also used as drugs in human medicine. Members of one group, the *adrenal cortical steroids*, reduce inflammation and hence are effective in controlling the symptoms of arthritis and allergy. Members of another group, the estrogens and an-

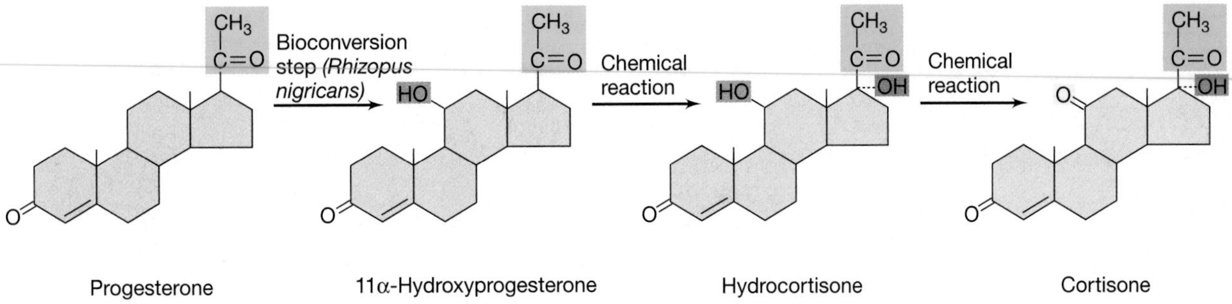

FIGURE 11.15 Cortisone production using a microorganism. The first reaction is a typical microbial bioconversion, the formation of 11α-hydroxyprogesterone from progesterone. This highly specific oxidation, carried out by the fungus *Rhizopus nigricans*, bypasses a difficult chemical synthesis. All the other steps, from progesterone to the steroid hormone cortisone, are performed chemically.

drogenic steroids, are involved in human fertility, and some of them can be used in the control of fertility. Steroids can be obtained by complete chemical synthesis, but this is a complicated and expensive process. Certain key steps in chemical synthesis can be carried out more efficiently by microorganisms, and commercial production of steroids usually has at least one microbial step.

Cortisone and Hydrocortisone

In the production of hydrocortisone and cortisone, steroids used to reduce swelling and itching from minor skin irritations, the fungus *Rhizopus nigricans* carries out a key stereospecific hydroxylation of a cortisone precursor (Figure 11.15). Most steroid bioconversions involve hydroxylations of this type, and a variety of different fungi are used industrially to carry out one or another specific hydroxylation. Steroid production is currently a big business, as worldwide sales of the four major steroids, hydrocortisone, cortisone, prednisone and prednisolone, amount to over 800 tons/year.

✓ 11.8 Concept Check

Microbial bioconversion employs microorganisms to biocatalyze a specific step or steps in an otherwise strictly chemical synthesis.

- ✓ Give an example of a microbial bioconversion. Why is this bioconversion necessary?

11.9

Enzymes

Each organism produces a large variety of enzymes, most of which are made in only small amounts and are involved in cellular processes. However, certain enzymes are produced in much larger amounts by some organisms, and instead of being held within the cell, they are excreted into the medium. *Extracellular enzymes* are usually capable of digesting insoluble polymers such as cellulose, protein, and starch, the products of digestion then being transported into the cell where they are used as nutrients for growth. Some of these extracellular enzymes are used in the food, dairy, pharmaceutical, and textile industries and are produced in large amounts by microbial synthesis (Table 11.5). They are especially useful because they often act on single chemical functional groups, they easily distinguish between similar functional groups on a single molecule, and in many cases, they catalyze reactions in a stereospecific manner producing only one of two possible enantiomers (for example, a D-sugar or an L-amino acid; ⟳ Section 2.6).

Proteases and Amylases

Enzymes are produced commercially from both fungi and bacteria. The microbial enzymes produced in the largest amounts on an industrial basis are the bacterial

TABLE 11.5	Microbial enzymes and their applications		
Enzyme	**Source**	**Application**	**Industry**
Amylase (starch-digesting)	Fungi	Bread	Baking
	Bacteria	Starch coatings	Paper
	Fungi	Syrup and glucose manufacture	Food
	Bacteria	Cold-swelling laundry starch	Starch
	Fungi	Digestive aid	Pharmaceutical
	Bacteria	Removal of coatings (desizing)	Textile
	Bacteria	Removal of stains; detergents	Laundry
Protease (protein-digesting)	Fungi	Bread	Baking
	Bacteria	Spot removal	Dry cleaning
	Bacteria	Meat tenderizing	Meat
	Bacteria	Wound cleansing	Medicine
	Bacteria	Desizing	Textile
	Bacteria	Household detergent	Laundry
Invertase (sucrose-digesting)	Yeast	Soft-center candies	Candy
Glucose oxidase	Fungi	Glucose removal, oxygen removal	Food
		Test paper for diabetes	Pharmaceutical
Glucose isomerase	Bacteria	High fructose corn syrup	Soft drink
Pectinase	Fungi	Pressing, clarification	Wine, fruit juice
Rennin	Fungi	Coagulation of milk	Cheese
Cellulase	Bacteria	Fabric softening, brightening; detergent	Laundry
Lipase	Fungi	Breaks down fat	Dairy, laundry
Lactase	Fungi	Breaks down lactose to glucose and galactose	Dairy, health foods
DNA polymerase	Bacteria	DNA replication in polymerase chain	Biological research;
	Archaea	reaction (PCR) technique (⟳ Section 10.9)	forensics

proteases, used as additives in laundry detergents. Most laundry detergents today contain enzymes, chiefly proteases but also amylases, lipases, reductases, and others. Many of these enzymes are isolated from alkaliphilic bacteria (∞ Section 5.10), mainly species of *Bacillus* like *Bacillus licheniformis* (Table 11.5). These enzymes, which have pH optima between 9 and 10, remain active at the alkaline pH of laundry detergent solutions.

Other important enzymes manufactured commercially are amylases and glucoamylases, which are used in the production of glucose from starch. The glucose so produced can then be converted by glucose isomerase to produce fructose (which is sweeter than either glucose or sucrose), resulting in the final production of a high fructose sweetener from corn, wheat, or potato starch. The use of this process in the food industry has been increasing, especially in the production of soft drinks.

Three reactions, each catalyzed by a separate microbial enzyme, operate in sequence in the conversion of cornstarch into the final product called **high fructose corn syrup.**

1. The enzyme **α-amylase** brings about the initial attack on the starch polysaccharide, shortening the chain and reducing the viscosity of the polymer. This is called the *thinning reaction.*
2. The enzyme **glucoamylase** produces glucose monomers from the shortened polysaccharides, a process called *saccharification.*
3. The enzyme **glucose isomerase** brings about the final conversion of glucose to fructose, a process called *isomerization.*

All three enzymes are produced industrially by microbial fermentation. The end product of this series of reactions is a syrup containing about equal amounts of glucose and fructose, which can be added directly to soft drinks and other food products, thereby greatly increasing their sweetness.

Extremozymes: Enzymes from Prokaryotes That Inhabit Extreme Environments

In Chapter 5 we considered aspects of microbial growth at high temperature and discovered that some prokaryotes, called *hyperthermophiles,* grow optimally at very high temperatures including, in some cases, above the boiling point of water. Hyperthermophiles are able to grow at such high temperatures because they produce heat-stable macromolecules (∞ Section 5.9) including enzymes, like some of those listed in Table 11.5 but which function at very high temperatures (Figure 11.16b). The term **extremozyme** has been coined to refer to enzymes that function at extremely high temperature (or enzymes that function optimally under any environmental extreme; for example, in the cold, in very

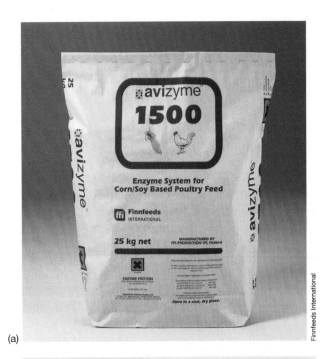

(a)

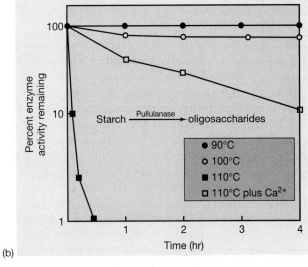

(b)

FIGURE 11.16 Extremozymes, enzymes that function under environmentally extreme conditions. (a) Acid-tolerant enzymes. An enzyme mixture used as a feed supplement for poultry (hence the trade name *avizyme*™). The enzymes function in the stomach of the bird to digest fibrous materials in the feed, thereby improving the nutritional value of the feed and promoting more rapid growth of the bird. (b) Thermostable enzymes. Thermostability of the enzyme pullulanase from *Pyrococcus woesei*, a hyperthermophile whose growth temperature optimum is 100°C (∞ Section 14.9). Calcium improves the heat stability of this enzyme and although not shown, significant activity could still be measured at 120°C with Ca^{2+}. Data from Rudiger et al., 1995. *Appl. Environ. Microbiol.* 61:567–575.

high salt, or at very acid (Figure 11.16*a*) or alkaline pH), and the organisms that produce them **extremophiles,** to indicate that they are organisms that grow best under conditions unsuitable for most microorganisms.

Because many industrial processes operate best at high temperatures, extremozymes from hyperthermophiles are becoming increasingly attractive as biocatalysts for the industrial applications shown in Table 11.5 and also for many research applications that require enzymes. Besides the *Taq* and *Pfu* DNA polymerases for use in the polymerase chain reaction (PCR) described in Sections 5.9 and 10.9, extremely thermostable proteases, amylases, cellulases, pullulanases (Figure 11.16), and xylanases have been isolated and characterized from various hyperthermophiles. The pullulanase from the hyperthermophile *Thermococcus litoralis*, which is related to the organism *Pyrococcus woesei* (Figure 11.16), is catalytically most active at a temperature of 118°C and ferredoxin (an iron–sulfur protein involved in electron transfer reactions; ∞ Section 4.10) from *Pyrococcus furiosus* is active at similar temperatures and does not denature until 140°C! Such temperature-resistant biocatalysts as well as extremozymes that are cold-active (from psychrophiles), active in the presence of high salt (from halophiles), or active at high or low pH (from alkaliphiles and acidophiles, respectively) will undoubtedly find more industrial applications in the coming years in situations that call for biocatalysis under extreme conditions. Indeed, the great specificity of enzymes and their ability to distinguish between chiral isomers make those that also function at environmental extremes particularly important to the chemical industry.

In addition to naturally occurring extremozymes, molecular biological techniques can be used to make extremozymes even better. Following cloning of the gene(s) encoding an extremozyme, random mutations can be introduced to them and the mutated gene(s) inserted into cells of a suitable host like *Escherichia coli*. Following colony formation the colonies can be screened for production of a modified extremozyme showing the desired properties. Although such a strategy usually yields extremozymes with decreased activity or that are totally inactive, occasional "positive hits" yield an improved version of the protein. This procedure of modifying existing enzymes by mutation and selection is widespread in biotechnology today and is called *directed evolution.* Using directed evolution it is possible to make heat-stable proteins even more heat stable or to improve other properties of the protein, such as its solvent stability or acid or alkali tolerance. Thus, using a naturally occurring extremozyme as a starting point, it is possible to generate customized enzymes for any particular industrial application.

Immobilized Enzymes

For some biocatalytic processes, it is frequently desirable to convert soluble enzymes into some sort of immobilized state. Immobilization not only makes it easier to carry out the enzymatic reaction under large-scale conditions but also helps stabilize the enzyme to denaturation. There are three basic approaches to enzyme immobilization (Figure 11.17):

1. **Bonding** of the enzyme to a carrier. The bonding can be through adsorption, ionic bonding, or covalent bonding. Carriers used include modified celluloses, activated carbon, clay minerals, aluminum oxide, and glass beads (Figure 11.17).
2. **Cross-linkage (polymerization)** of enzyme molecules. Linkage of enzyme molecules with each other is usually done by chemical reaction with a cross-linking agent such as glutaraldehyde. Cross-linking of enzymes involves the chemical reaction of amino groups of the enzyme protein with glutaraldehyde. If the reaction is carried out properly, the enzyme molecules can be linked in such a way that most enzymatic activity is maintained.
3. **Enzyme inclusion,** which involves incorporation of the enzyme into a *semipermeable membrane.* Enzymes can be enclosed in microcapsules, gels, semipermeable polymer membranes, or fibrous polymers such as cellulose acetate (Figure 11.17).

Each of these methods has advantages and disadvantages, and the procedure used depends on the enzyme and on the particular industrial application.

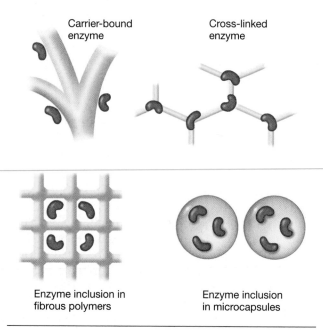

Carrier-bound enzyme

Cross-linked enzyme

Enzyme inclusion in fibrous polymers

Enzyme inclusion in microcapsules

FIGURE 11.17 Procedures for the immobilization of enzymes. In all cases, enzyme molecules are shown in red.

✓ 11.9 Concept Check

Microorganisms are ideal for the large-scale production of enzymes. Many enzymes are used in the laundry industry to remove stains from clothing, and thermostable and alkali-stable enzymes have many advantages in these markets. The production of high fructose corn syrup involves the participation of three microbial enzymes, of which glucose isomerase (which converts glucose to fructose) is the key. When an enzyme is used in a large-scale process, it may be desirable to immobilize it by bonding it to an inert substrate.

- ✓ How are enzymes of use in the laundry industry?
- ✓ How is the enzyme *glucose isomerase* used in the soft-drink industry?
- ✓ What is an *extremozyme*?

11.10

Vinegar

Vinegar is the product resulting from the conversion of ethyl alcohol to acetic acid by **acetic acid bacteria,** members of the genera *Acetobacter* and *Gluconobacter*. Vinegar can be produced from any alcoholic substance, although the usual starting material is wine or alcoholic apple juice (cider). Vinegar can also be produced from a mixture of pure alcohol in water, in which case it is called *distilled vinegar,* the term *distilled* referring to the alcohol from which the product is made rather than the vinegar itself. Vinegar is used as a flavoring ingredient in salads and other foods, and because of its acidity, it is also used in pickling. Meats and vegetables properly pickled in vinegar can be stored unrefrigerated for years.

The acetic acid bacteria are an interesting group of bacteria (∽ Section 13.7). These are strictly aerobic bacteria that differ from most other aerobes in that they do not oxidize their organic electron donors completely to CO_2 and water (Figure 11.18). Thus, when provided with

ethyl alcohol as electron donor, they oxidize it to only acetic acid, which accumulates in the medium. Acetic acid bacteria are quite acid-tolerant and are not killed by the acidity that they produce. There is a high oxygen demand during growth, and the main problem in the production of vinegar is to ensure sufficient aeration of the medium.

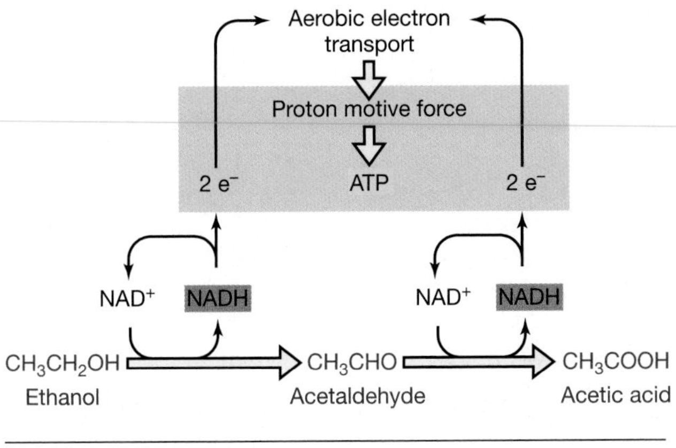

Vinegar Production

There are three different processes for the production of vinegar. The **open-vat** or **Orleans method** was the original process and is still used in France where it was developed. Wine is placed in shallow vats with considerable exposure to the air, and the acetic acid bacteria develop as a slimy layer on the top of the liquid. This process is not very efficient because the only place that the bacteria come in contact with both the air and the substrate is at the surface. The second process is the **trickle method,** in which the contact between the bacteria, air, and substrate is increased by trickling the alcoholic liquid over beechwood twigs or wood shavings packed loosely in a vat or column while a stream of air enters at the bottom and passes upward. The bacteria grow on the surface of the wood shavings and thus are maximally exposed both to air and liquid. The vat is called a *vinegar generator* (Figure 11.19), and the whole process is operated in a continuous fashion. The life of the wood shavings in a vinegar generator is long, from 5 to 30 years, depending on the kind of alcoholic liquid used in the process.

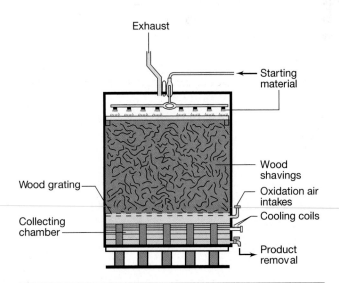

FIGURE 11.19 Diagram of a vinegar generator. The alcoholic juice is allowed to trickle through the wood shavings, and air is passed up through the shavings from the bottom. Acetic acid bacteria develop on the wood shavings and convert alcohol to acetic acid. The acetic acid solution accumulates in the collecting chamber and is removed periodically. The process can be run semicontinuously.

FIGURE 11.18 Oxidation of ethanol to acetic acid, the key process in the production of vinegar.

The third vinegar process is the **bubble method.** This is basically a submerged fermentation process such as was already described for antibiotic production. Efficient aeration is even more important with vinegar than with antibiotics, and special highly efficient aeration systems have been devised. The process is operated in a continuous fashion: alcoholic liquid is added at a rate just sufficient to balance removal of vinegar. The efficiency of the process is high, and 90–98% of the alcohol is converted to acid. One disadvantage of the bubble method is that the product must undergo more filtering to remove the bacteria, whereas in the open-vat and trickle methods the product is virtually free of bacteria because the cells are bound in the slimy layer in the former and adhere to the wood chips in the latter.

Although acetic acid can be easily made chemically from alcohol, the microbial product, *vinegar*, is a distinctive material, the flavor being due in part to other substances present in the starting material and produced in the fermentation. For this reason, the microbial process has not been supplanted by a chemical process.

✓ 11.10 Concept Check

The active ingredient in vinegar is acetic acid, which is produced by an acetic acid bacterium oxidizing an alcohol-containing fruit juice. Adequate aeration is the most important consideration in ensuring a successful vinegar process.

✓ Why is O_2 necessary in vinegar production?
✓ Why does vinegar produced by the trickle method have a more distinctive taste than vinegar produced by the bubble method?

11.11

Citric Acid and Other Organic Compounds

Many organic chemicals are produced by microorganisms in sufficient yields that they can be manufactured commercially by fermentation. *Citric acid*, used widely in foods and beverages, *itaconic acid*, used in the manufacture of acrylic resins, and *gluconic acid*, used in the form of calcium gluconate to treat calcium deficiencies in humans and industrially as a washing and softening agent, are produced by fungi. *Sorbose*, which is produced when *Acetobacter* oxidizes sorbitol, is used in the manufacture of *ascorbic acid*, vitamin C. (In fact, this sorbitol–sorbose reaction is the only biological step in the otherwise entirely nonbiological chemical synthesis of ascorbic acid.) *Gibberellin*, a plant growth hormone used to stimulate growth of plants, is produced by a fungus. *Dihydroxyacetone*, produced by allowing *Acetobacter* to oxidize glycerol, is used as a sunscreen. *Dextran*, a gum employed as a blood plasma extender and as a biochemical reagent, and *lactic acid*, used in the food industry to acidify foods

and beverages, are produced by lactic acid bacteria. *Acetone* and *butanol* can be produced in fermentations by *Clostridium acetobutylicum* but are now prepared mainly from petroleum by strictly chemical synthesis.

Citric Acid

Citric acid is produced microbiologically by a fermentation using the mold *Aspergillus niger*. Although citric acid is normally considered in connection with the citric acid cycle (∞ Section 4.12), in certain organisms such as *A. niger*, excretion of large amounts of citric acid can be obtained. The fermentation is carried out aerobically in large fermentors, and a key requirement for high citric acid yield is that the medium be *iron-deficient* because citric acid is overproduced by the fungus as a chelator to scavenge iron (Figure 11.20a). Therefore, the medium used for citric acid production is treated to remove most of the iron, and the fermentors themselves are made of stainless steel to prevent leaching of iron from the fermentor walls at the low pH values generated by citric acid accumulation.

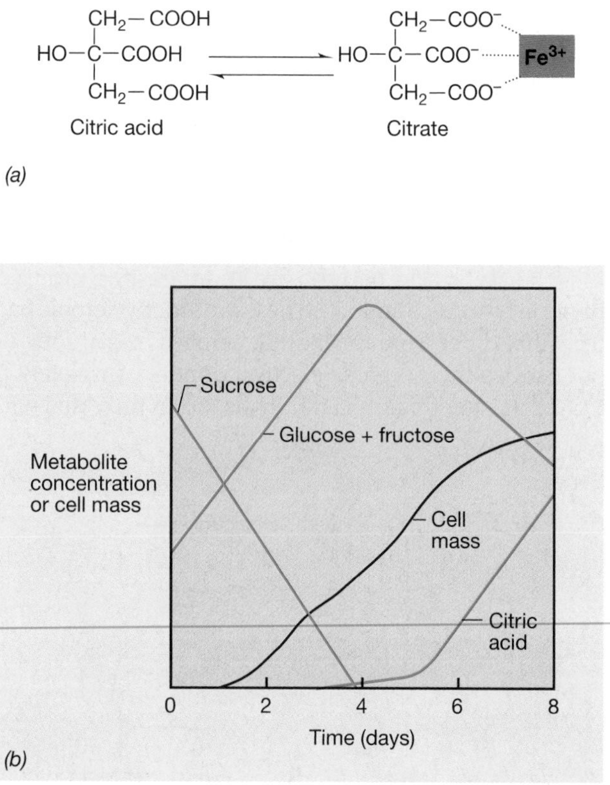

(a)

(b)

FIGURE 11.20 Citric acid fermentation. (a) Structure of citric acid. Note how the ionized form, citrate, contains three carboxylic acid groups, which can chelate ferric iron (Fe^{3+}). (b) Kinetics of citric acid fermentation. Sucrose is degraded by the enzyme *sucrase* to yield glucose plus fructose. See text for further details.

The media used for citric acid production have been highly perfected over the many years that the commercial process has been under way. A variety of starting materials can be used as carbohydrate sources: starch from potatoes, starch hydrolysates, glucose syrup from saccharified starch, sucrose (Figure 11.20b), sugarcane syrup, sugarcane molasses, and sugar beet molasses. If starch is used, amylases (see Table 11.5) formed by the producing fungus or added to the fermentation broth hydrolyze the starch to sugars. The sugars are catabolized through the glycolytic pathway (⟨∞⟩ Section 4.9) and enter the citric acid cycle where citrate production occurs.

Although both surface and submerged processes for citric acid production have been developed, most citric acid today is produced by submerged processes in large fermentors. Because *Aspergillus niger* is a strict aerobe, it is crucial to this fermentation to make sure that the culture stays properly aerated. Citric acid is produced in this way as a typical secondary metabolite. During the growth phase, sucrose is broken down into glucose and fructose, and by the time stationary phase is reached, large amounts of these hexoses remain and are converted to citric acid to counter iron starvation (Figure 11.20).

Historically, the development of a submerged process for citric acid was of great importance because it was the first *aerobic* industrial fermentation. The technology for manufacturing aerobic fermentors was perfected with the citric acid process. This technology was then applied to penicillin and the other important antibiotic fermentations. Thus, we owe some of our current success with large-scale production of antibiotics to the pioneering work done on citric acid fermentation.

✓ 11.11 Concept Check

A number of organic chemicals are produced commercially by use of microorganisms, of which the most important economically is citric acid, produced by certain fungi.

- ✓ Why is citric acid produced by *Aspergillus niger* considered a secondary metabolite (see Figure 11.20b)?
- ✓ What is the relationship between iron and citric acid production by *A. niger*?

11.12

Yeast as an Agent of Fermentation and as Food

Yeasts are the most important and the most extensively used microorganisms in industry. They are cultured for the cells themselves, for cell components, and for the end products they produce during the alcoholic fermentation (Table 11.6). Yeast cells are also used in the manufacture of bread and also as sources of food, vitamins, and other growth factors. Large-scale fermentation by yeast is re-

TABLE 11.6	Industrial uses of yeast and yeast products[a]

Production of yeast cells
  Baker's yeast, for bread making
  Dried food yeast, for food supplements
  Dried feed yeast, for animal feeds
Yeast products
  Yeast extract, for microbial culture media
  B vitamins, vitamin D
  Enzymes for food industry: invertase, galactosidase
  Biochemicals for research: ATP, NAD[+], RNA
Fermentation products from yeast
  Ethanol, for industrial alcohol and as a gasoline extender
  Glycerol
Beverage alcohol
  Beer, Wine
Distilled beverages
  Whiskey, Brandy, Vodka, Rum

a ⟨∞⟩ box, The Products of Yeast Fermentation, Chapter 4.

sponsible for the production of alcohol for industrial purposes, but yeast is better known for its role in the manufacture of alcoholic beverages: beer, wine, and liquors (⟨∞⟩ box, The Products of Yeast Fermentation, Chapter 4).

Production of yeast cells and production of alcohol by yeast are two quite different processes industrially, in that the first process requires the presence of oxygen for maximum production of cell material and hence is an *aerobic* process, whereas the alcoholic fermentation is *anaerobic* and takes place only in the absence of oxygen. However, the same or similar species of yeasts are used in virtually all industrial processes. The yeast *Saccharomyces cerevisiae* was derived from wild yeast used in ancient times for the manufacture of wine and beer. The yeasts currently used are descendants of early *S. cerevisiae*. However, because they have been cultivated in laboratories for such a long time, there has been ample opportunity for selection of strains according to particular desirable properties.

Yeast Cell Production

Bakers use yeast as a leavening agent in the rising of the dough prior to baking. A secondary contribution of yeast to bread is its flavor. In the leavening process, the yeast is mixed with the moist dough in the presence of a small amount of sugar. The yeast converts the sugar to alcohol and CO_2, and the gaseous CO_2 expands, causing the dough to rise. When the bread is baked, the heat drives off the CO_2 and the alcohol and holes are left within the bread mass, thus giving bread its characteristic light texture.

Yeast for baking or nutritional purposes is cultured in large aerated fermentors in a medium containing molasses as a major ingredient. Molasses, a by-product of sugar refining from beets or cane, still contains large amounts of sugar that serve as the source of carbon and energy. Molasses also contains minerals, vitamins, and amino acids used by the yeast. To make a complete medium for yeast growth, phosphoric acid (a phosphorus source) and ammonium sulfate (a source of nitrogen and sulfur) are added.

Fermentation vessels for yeast production range from 40,000 to 200,000 liters. Beginning with the pure stock culture, several intermediate stages are needed to scale up the inoculum to a size sufficient to inoculate the final stage (Figure 11.21a). It is undesirable to add all the molasses to the fermentor at once because this results in a sugar excess and the yeast ferments some of this surplus sugar to alcohol plus CO_2 rather than turning it into yeast cells. Therefore, only a small amount of the molasses is added initially, and then as the yeast grows and consumes this sugar, more is added.

At the end of the growth period, the yeast cells are recovered from the broth by centrifugation. The cells are usually washed by dilution with water and recentrifuged until they are light in color. Baker's yeast is marketed in two ways, either as compressed cakes or as a dry powder. *Compressed yeast* cakes (Figure 11.21b)

are made by mixing the centrifuged yeast with emulsifying agents, starch, and other additives that give it a suitable consistency and reasonable shelf life, and the product is then formed into cubes or blocks of various sizes for domestic or commercial use. A compressed yeast cake contains about 70% moisture and thus must be stored in the refrigerator so its activity is maintained. Yeast marketed in the dry state for baking is usually called *active dry yeast* (Figure 11.21b). The washed yeast is mixed with additives and dried under vacuum until its moisture is reduced to about 8%. It is then packed in airtight containers such as fiber drums, cartons, or multiwall bags, sometimes under a nitrogen atmosphere to promote long shelf life. Active dry yeast does not exhibit as great a leavening action as compressed fresh yeast but has a much longer shelf life. *Nutritional yeast,* marketed as a food supplement (Figure 11.21b), is heat-killed and usually dried. Yeast cells are rich in B vitamins and in protein, except for sulfur-containing amino acids. Yeast is added to wheat or corn flour to increase the nutritional value of these foods and is also sold in pelleted form as a health food (Figure 11.21b).

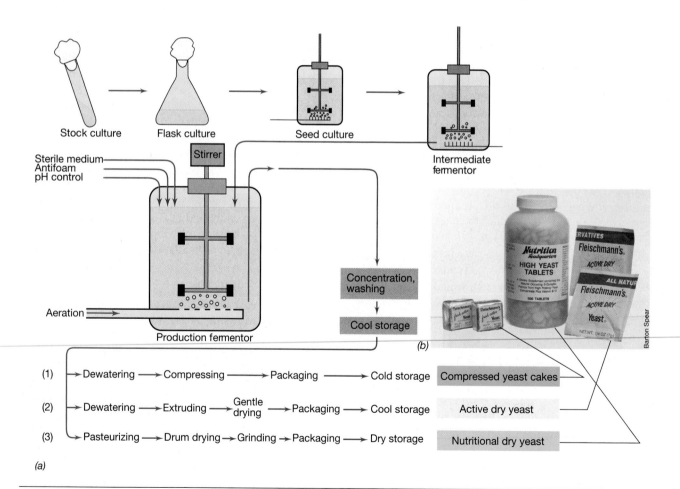

FIGURE 11.21   Industrial production of yeast cells. (a) Stages in production. (b) Photograph of common yeast products: yeast cakes; packages of active dry yeast; bottle of nutritional yeast.

Yeast as Single-Cell Protein

In recent years, there has been considerable interest in the expanded production of microorganisms, in particular cells of yeast, as food, especially in parts of the world where conventional sources of food are in short supply. Perhaps the most important potential use of microorganisms is not as a complete diet for humans but as a *protein supplement*. It is usually protein that is in shortest supply in food, and it is in the production of protein that microorganisms are perhaps the most successful. In many cases, microbial cells contain greater than 50% protein, and in at least some species this is complete protein; that is, it contains sufficient amounts of all the amino acids essential to humans. The protein produced by microorganisms as food has been called *single-cell protein* to distinguish it from the protein produced by multicellular animals and plants. The only organism presently used as a source of single-cell protein is yeast but algae, bacteria, and fungi have also been considered and in some cases have been marketed as supplements for animal feeds and as health foods (especially algae and cyanobacteria).

✓ 11.12 Concept Check

Yeast cells are grown for use in the baking and food industry. Commercial yeast is produced in large-scale aerated fermentors using molasses as the main carbon and energy source.

- ✓ Write a balanced chemical reaction that accounts for the action of yeast in bread making.
- ✓ Why is it important when growing yeast *for cells* to maintain oxic conditions in the fermentor?
- ✓ What is single-cell protein?

11.13

Alcohol and Alcoholic Beverages

The use of yeast in the production of alcoholic beverages is an ancient process. Most fruit juices undergo a natural fermentation caused by wild yeasts that are present on the fruit. From these natural fermentations, yeasts have been selected for more controlled production, and today alcoholic beverage production is a large industry worldwide. The most important alcoholic beverages are *wine*, produced by the fermentation of fruit juice; *beer*, or *ale*, produced by the fermentation of malted grains; and *distilled beverages*, produced by concentrating alcohol from a fermentation by distillation. The biochemistry of alcohol fermentation by yeast was discussed in Section 4.9.

Wine

Wine is a product of the alcoholic fermentation by yeast of fruit juices or other materials that are high in sugar. Most wine is made from grapes, and thus wine manu-

facture occurs in parts of the world where grapes can be most economically grown. The greatest wine-producing countries, in order of decreasing volume of production, are Italy, France, Spain, Algeria, Argentina, Portugal, and the United States (Figure 11.22). There are a great number of different wines, and their quality and character vary considerably. *Dry wines* are wines in which the sugars of the juice are practically all fermented, whereas in *sweet wines*, some of the sugar is left or additional sugar is added after the fermentation. A

(a)

(b)

(c)

FIGURE 11.22 Commercial wine making in California (USA). (a) Equipment for transporting grapes to the winery for crushing. (b) Large tanks where the main wine fermentation takes place. (c) Barrels where the aging process takes place.

fortified wine is one to which brandy or some other alcoholic spirit is added after the fermentation; sherry and port are the best-known fortified wines. A *sparkling wine*, such as champagne, is one in which considerable carbon dioxide is present, arising from a final fermentation by the yeast directly in the sealed bottle.

The yeasts involved in wine fermentation are of two types: the so-called wild yeasts, which are present on the grapes as they are taken from the field and are transferred to the juice, and the cultivated wine yeast, *Saccharomyces ellipsoideus,* which is added to the juice to begin the fermentation. One important distinction between wild yeasts and the cultivated wine yeast is their alcohol tolerance. Most wild yeasts can tolerate only about 4% alcohol, and when the alcohol concentration reaches this point, the fermentation stops. The best wine yeasts can tolerate up to 14% alcohol before they stop growing, although above about 10% alcohol growth can be very slow. In unfortified wine, the final alcoholic content reached is determined partly by the alcohol tolerance of the yeast and partly by the amount of sugar present in the juice. The alcohol content of unfortified wines ranges from as low as 6 to as high as 14%. Fortified wines such as sherry have an alcohol content as high as 20%, but this is achieved by adding distilled spirits such as brandy. In addition to the lower alcohol content produced, wild yeasts do not produce some of the flavor components considered desirable in the final product, and hence the presence and growth of wild yeasts during fermentation is unwanted.

Wine Production

The production of wine begins in the early fall with the harvesting of grapes. The grapes are crushed by machine, and the juice, called *must,* is squeezed out. Depending on the grapes used and on how the must is prepared, either white or red wine may be produced (Figure 11.23). A white wine is made either from white grapes or from the juice of red grapes from which the skins, containing the red coloring matter, have been removed. In the making of red wine, the *pomace* (skins, seeds, and pieces of stem) is left in during the fermentation. In addition to the color difference, red wine has a stronger flavor than white because of the presence of larger amounts of chemicals called *tannins,* which are extracted into the juice from the grape skins during the fermentation.

It is the practice in many wineries to kill the wild yeasts present in the must by adding sulfur dioxide (listed on the bottle as "sulfites") at a level of about 100 parts per million (ppm). *S. ellipsoideus* is resistant to this concentration of sulfur dioxide and is added as a starter culture from a pure culture grown on sterilized or pasteurized grape juice. The fermentation may be carried out in vats of various sizes, from 50-gallon (gal) casks to 55,000-gal tanks made of oak, cement, stone, or glass-

lined metal (see Figure 11.22*b*). The fermentor must be constructed so that the large amount of carbon dioxide produced during the fermentation can escape but air cannot enter and this is accomplished by fitting the vessel with a special one-way valve.

With a red wine, after 3–5 days of fermentation, sufficient tannin and color have been extracted from the pomace and the wine is drawn off for further fermentation in a new tank, usually for another week or two. The next step is called *racking;* the wine is separated from the sediment (called *lees*), which contains yeast cells and precipitate, and then stored at lower temperature for aging, flavor development, and further clarification. The final clarification may be hastened by the addition of materials called fining agents, such as casein, tannin, or bentonite clay, or the wine may be filtered through diatomaceous earth, asbestos, or membrane filters. The wine is then bottled and either stored for further aging or sold. Red wine is usually aged for several years or more (Figure 11.22*c*), but white wine is usually sold without much aging. During the aging process, complex chemical changes occur, including reduction of bitter components, resulting in improvement in flavor and odor, or *bouquet.*

Brewing

The manufacture of alcoholic beverages made from malted grains is called *brewing.* Typical malt beverages include beer, ale, porter, and stout. *Malt* is prepared from germinated barley seeds, and it contains natural enzymes that digest the starch of grains and convert it to sugar. Since brewing yeasts are unable to digest starch, the malting process is essential for the preparation of a fermentable material from cereal grains.

The fermentable liquid from which beer and ale are made is prepared by a process called *mashing.* The grain of the mash may consist only of malt, or other grains such as corn, rice, or wheat may be added. The mixture of ingredients in the mash is cooked and allowed to steep in a large mash tub at warm temperatures. During the heating period, enzymes from the malt cause digestion of the starches and liberate sugars, which are fermented by the yeast. Proteins and amino acids are also liberated into the liquid, as are other nutrient ingredients necessary for the growth of yeast.

After cooking, the aqueous extract, called *wort,* is separated by filtration from the husks and other grain residues of the mash. *Hops,* an herb derived from the female flowers of the hops plant, is added to the wort at this stage. Hops is a flavoring ingredient, but it also has antimicrobial properties, which probably help to prevent contamination in the subsequent fermentation. The wort is then boiled for several hours, usually in large copper kettles (Figure 11.24*a,b*), during which time desired ingredients are extracted from the hops, proteins present in the wort that are undesirable from the point of view of

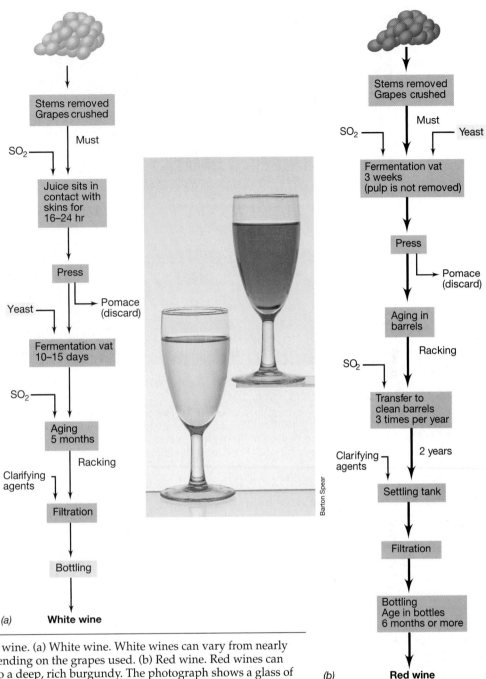

FIGURE 11.23 Production of wine. (a) White wine. White wines can vary from nearly colorless to straw-colored depending on the grapes used. (b) Red wine. Red wines can vary in color from a faint red to a deep, rich burgundy. The photograph shows a glass of Chenin Blanc, a typical white wine (left), and a glass of a light red wine (right).

beer stability are coagulated and removed, and the wort is sterilized. Heating is accomplished either by passing steam through a jacketed kettle or by direct heating of the kettle from below by fire. Then the wort is filtered again, cooled, and transferred to the fermentation vessel.

Brewery yeast strains are of two major types: top-fermenting and bottom-fermenting. The main distinction between the two is that **top-fermenting yeasts** remain uniformly distributed in the fermenting wort and are carried to the top by the CO_2 gas generated during the fermentation, whereas **bottom yeasts** settle to the bottom. Top yeasts are used in the brewing of ales,

and bottom yeasts are used to make lager beers. Bottom yeasts are usually given the species designation *Saccharomyces carlsbergensis*, and top yeasts are called *Saccharomyces cerevisiae*. Fermentation by top yeasts usually occurs at higher temperatures (14–23°C) than that by bottom yeasts (6–12°C) and is accomplished in a shorter period of time (5–7 days for top fermentation versus 8–14 days for bottom fermentation). After completion of lager beer fermentation by bottom yeasts, the beer is pumped off into large tanks where it is stored at a cold temperature (about −1°C) for several weeks (Figure 11.24c). Following the lagering process, the beer is

FIGURE 11.24 Brewing beer in a commercial brewery. (a, b) The copper brew kettle is where the wort is mixed with hops and then boiled. From the brew kettle the liquid is passed to large fermentation tanks where yeast ferments glucose to ethanol plus CO_2. (c) The fermented liquid is then stored for several weeks at low temperature in lagering tanks where settling of particulate matter including yeast cells occurs. (d) The beer is then filtered and placed in storage tanks from which it is packaged into kegs, bottles, or cans.

filtered and placed in storage tanks (Figure 11.24*d*) from which packaging occurs. Top-fermented ale is stored at a higher temperature (4–8°C), which assists in development of the characteristic ale flavor.

For more details on the brewing process, refer to the box, Home Brew.

Distilled Alcoholic Beverages

Distilled alcoholic beverages are made by heating a fermented liquid at a high temperature that volatilizes most of the alcohol. The alcohol is then condensed and collected, a process called *distilling*. A product much higher in alcohol content can be obtained by this process than is possible by direct fermentation. Virtually any alcoholic liquid can be distilled, and each yields a characteristic distilled beverage. The distillation of malt brews yields *whiskey*, distilled wine yields *brandy*, distillation of fermented molasses yields *rum*, distillation of fermented grain or potatoes yields *vodka*, and distillation of grain and juniper berries yields *gin* (Figure 11.25).

The distillate contains not only alcohol but also other volatile products arising either from the yeast fermentation or from the mash itself. Some of these other products are desirable flavor ingredients, whereas others are undesirable substances called *fusel oils*. To elim-

inate the latter, the distilled product is almost always aged, usually in wood barrels. During the aging process, fusel oils are removed and desirable new flavor ingredients develop. The fresh distillate is usually colorless, whereas the aged product is often brown or yellow (Figure 11.25). The character of the final product is partly determined by the manner and length of aging (aging times of 10 years or more are not uncommon for some distilled spirits), and the whole process of manufacturing distilled alcoholic beverages is highly complex. To a great extent, the process is carried out by traditional methods that have been found to yield a particular product, rather than by scientifically proven methods.

Commodity Ethanol

Production of ethanol as a commodity chemical is a major biocatalytic process, and today over one billion gallons (3.8 billion liters) of alcohol are produced yearly in the United States, primarily from the fermentation of cornstarch. This ethanol is used as an industrial solvent and also for the production of *gasohol*, a lead-free fuel containing 10% ethanol in gasoline. The combustion of gasohol produces lower amounts of carbon monoxide and nitrogen oxides than pure gasoline; thus gasohol is marketed as a cleaner burning fuel, and its use is encouraged in major cities where automobile pollution is extensive.

FIGURE 11.25   Typical distilled spirits. These alcoholic beverages contain not only alcohol but also distinctive volatile flavoring agents obtained from the fermented substrate. Aging in wood casks yields the distinctive amber or yellow color of certain distilled spirits. Top row (left to right): gin, vodka. Bottom row (left to right): dark rum, brandy, whiskey.

If automobile engines are modified to burn it, they can be run on pure ethanol, and this has been done in certain countries such as Brazil where sugarcane (as a fermentable substrate) is plentiful but oil is scarce.

Various yeasts have been used in commodity ethanol production, including species of *Saccharomyces*, *Kluyveromyces*, and *Candida*, but most ethanol in the United States is produced by *Saccharomyces* in the reaction

$$C_6H_{12}O_6 \rightarrow 2\ C_2H_5OH + 2\ CO_2$$

(this occurs by glycolysis) (∞ Section 4.9).

If 100% of the glucose were fermented to products, alcohol would compose 51% by weight of the original glucose. Although for technical reasons this level of alcohol production has not been achieved, commodity ethanol production has been refined to the point where ethanol distilled from the fermentation broth is obtained at 90–95% of the theoretical yield.

✓ 11.13 Concept Check

Alcoholic beverages are produced by yeast from the fermentation of sugar to ethyl alcohol and CO_2. Wine is produced from grape juice, beer from malted grain, and distilled spirits from the distillation of fermented solutions. Commodity alcohol is used as a gasoline additive and industrial solvent.

✓ How do wines differ from beer in terms of the amount of alcohol present?
✓ What are the major differences between a beer and an ale?

11.14

Mushrooms as a Food Source

Several kinds of *fungi* are sources of human food, of which the most important are the mushrooms. Mushrooms are a group of filamentous fungi that form large, complicated but edible structures called **fruiting bodies** (Figure 11.26). The fruiting body is commonly called the *mushroom* and is

FIGURE 11.26   Mushroom life cycle, showing how the fruiting body develops from underground hyphae.

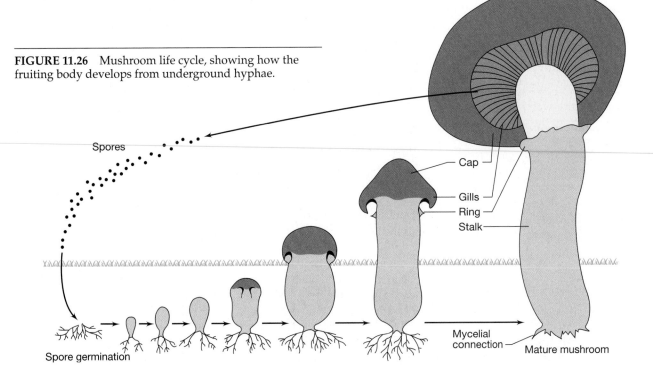

Spores

Cap

Gills
Ring
Stalk

Spore germination

Mycelial connection

Mature mushroom

The amateur brewer can make many kinds of beer, from English bitters and India pale ale to German bock and Russian Imperial stout. The necessary equipment and supplies can be purchased from a local beer and winemakers shop (the Home Wine and Beer Trade Association, 604 N. Miller Road, Valrico, FL 33594, can supply the address of a nearby shop).

The brewing process can be divided into three basic stages: making the wort, carrying out the fermentation, and bottling and aging. The character of the brew depends on many factors: the proportion of malt, sugar, hops, and grain; the kind of yeast; the temperature and duration of the fermentation; and how the aging process is carried out. The instructions provided here are for a simple and relatively foolproof beer (so-called single-stage fermentation).

The fermentor itself consists of a 20-liter (5-gal) glass jar or carboy that can be fitted with a tightly fitting closure. In order to have a good quality beer, it is essential that *everything* be sterilized that comes into contact with the wort. This includes the fermentor, tubing, stirring spoon, and bottles. The best procedure is to use a sterilizing rinse consisting of 50–60 ml of liquid bleach in 20 liters of water. Soak the items for 15 min and then rinse lightly with hot water or air-dry.

1. **Making the wort.** In commercial brewing, the wort is made by extracting fermentable sugars and yeast nutrients from malt, sugar, and hops. The process is complex and relatively difficult to carry out satisfactorily. Many home brewers make their own wort from malt, but a reasonably satisfactory beer can be made with hop-flavored malt extract purchased ready-made. Malt extracts come in a variety of flavors and colors, and the kind of beer depends on the type of malt extract used. A simple recipe for making the wort uses 5–6 lb (2.25–2.75 kg) of hop-flavored malt extract and 20 liters of water. The malt extract and 6 liters of water are brought to a boil for 15 min in an enamel or stainless steel container (aluminum heating kettles must be avoided because of the inhibiting action of metals leached from aluminum containers) (see photo *a*). The hot wort is then poured into 14–15 liters of clean, cold water that has already been added to the fermentor. After the temperature has dropped below 30°C the yeast can be added to initiate the fermentation.

2. **Carrying out the fermentation.** The process by which yeast is added to the wort is called *pitching*. Brewer's yeast can be purchased as active dry yeast from the home brew supplier. Add two packs of fresh beer yeast to the cooled wort and cover the fermentor with a rubber stopper into which a plastic hose has been inserted. The hose is directed into a bucket containing water. During the initial 2–3 days of the fermentation, large amounts of CO_2 will be given off, which will exit through the hose. The water trap is to prevent wild yeasts or bacteria from the air from getting back into the fermentor. After about 3 days, the activity will diminish as the fermentable sugars are used up. At this time, the rubber stopper and hose are replaced with an inexpensive fermentation lock. The fermentation lock (see photo *b*), which can be purchased at the home brew store, prevents contamination while permitting the small amount of gas still being produced to escape. Allow the beer to ferment for 7–10 days at 10–15°C or higher.

3. **Bottling and aging.** The fermentation should be allowed to proceed for the full 7–10 days, even if the vigorous fermentation action ceases earlier. By this time, most of the yeast should have settled to the bottom of the fermentor. Carefully siphon the beer off the yeast layer, allowing it to run into sanitized glass beer bottles. Take care that the yeast at the bottom of the fermentor is not stirred up and leave the yeast-rich liquid at the bottom. The bottles used should accept standard crown caps, and new, clean caps should be used (see photo *c*). Before capping, add $\frac{3}{4}$ teaspoon of corn syrup to each bottle. Be certain not to add more than $\frac{3}{4}$ teaspoon of syrup because if excess sugar is added, the buildup of carbon dioxide in the bottles may cause them to burst. Once the bottles are capped, turn each one upside down once to mix the sugar syrup and then allow the beer to age upright at room temperature for at least 7–10 days. If another large container is available, a better way of adding the sugar is to siphon the beer into this second container, add the proper amount of sugar for the whole brew, dissolve, and then siphon into the bottles. After this aging period, the beer may be stored at a cooler temperature.

a Primary reference source: Burch, B. 1992. *Brewing Quality Beers—The Home Brewer's Essential Guidebook*, 2nd edition. Joby Books, Fulton, CA.

All homemade beer has a natural yeast sediment in the bottom of the bottles. The beer will improve if it is allowed to age for several weeks. Aging tends to make beer smoother. Using the same basic production equipment, several different types of beer can be made, each with its own distinctive taste and character (the reference listed here contains a number of beer recipes). Dark beers, which generally contain more alcohol than lighter beers, require more malt for their production and are usually brewed from a combination of different malts such as ones obtained from darker varieties of grain or ones that have been roasted to carmelize the sugars and yield a darker color. A typical American style light lager (see photo *d*, left) contains about 3.5% alcohol (by volume) whereas a Munich style dark (see photo *d*, right) contains 4.25% alcohol and bock beers contain about 5% alcohol.

The trend toward "individuality" in beer can be attested to not only by the growing number of home brewers but also by the fact that major brewers in the United States are feeling more and more competition from new, usually very small, breweries called *microbreweries*. Although total production by a microbrewery may pale by comparison to that of a major brewer, the products themselves often have their own distinctive character and local appeal. Part of these differences probably has to do with the smaller scale on which the brewing takes place but also undoubtedly has to do with the use of different sources of ingredients, yeast strains, and brewing times. ■

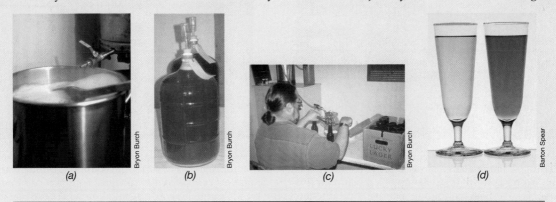

(a) Boiling the wort in a stainless steel kettle. (b) Fermentation—notice the fermentation lock. (c) Bottling and capping. (d) Typical light (left) and dark (right) beers.

formed through the association of a large number of individual hyphae to form a mycelium (Figure 11.26).

Life Cycle and Commercial Growth of Mushrooms

During most of its existence, the mushroom fungus lives as a simple mycelium, growing in soil, leaf litter, or decaying logs. However, when environmental conditions are favorable, usually following periods of wet and cool weather, the fruiting body develops, beginning first as a small button-shaped structure underground and then expanding into the full-grown fruiting body that we see above ground (Figure 11.26). Sexual spores, called **basidiospores,** are formed, borne on the underside of the fruiting body on flat plates called **gills,** which are attached to the cap of the mushroom (Figure 11.26).

The mushroom commercially grown in most parts of the world is *Agaricus bisporus,* and it is generally cultivated on mushroom farms. The organism is grown in special beds, usually in buildings where temperature and humidity are carefully controlled (Figure 11.27*a*). Beds are prepared by mixing soil with a material very rich in organic matter, such as horse manure, and the beds are then inoculated with mushroom *spawn.* The spawn is actually a pure culture of the mushroom fungus that has been grown in large bottles on an organic-rich medium. In the bed, the mycelium grows and spreads through the substrate, and after several weeks it is ready for the next step, the induction of mushroom formation. This is accomplished by adding to the surface of the bed a layer of soil called *casing soil.* The appearance of mushrooms on the surface of the bed is called a *flush* (Figure 11.27*b*), and when flushing occurs, the mushrooms must be collected immediately while still fresh. After collection they are packaged and kept cool until brought to market.

Another widely cultured mushroom is **shiitake,** *Lentinus edulus.* The most widely cultivated mushroom in the Far East, shiitake is now finding expanding demand in North America. Shiitake is a cellulose-digest-

FIGURE 11.27 Commercial mushroom production. (a) An installation for *Agaricus bisporus*, the common commercial mushroom of the Western world. (b) Close-up of a mushroom flush. (c) Shiitake, *Lentinus edulus*, the most common commercial mushroom of the Far East but finding increasing production in the West. A large Japanese installation where the mushroom is cultivated on hardwood logs. (d) Close-up of the shiitake mushroom.

ing fungus that grows well on hardwood trees and is cultivated on small logs (Figure 11.27c). The logs are soaked in water to hydrate them and then inoculated by inserting plugs of spawn into small holes drilled in them. The fungus grows through the log, and after about a year forms a flush of fruiting bodies (see Figure 11.27d). The shiitake mushroom is generally considered to have a superior taste to *A. bisporus* and because of this commands a substantially greater price.

✓ 11.14 Concept Check

The most important food produced from a microorganism is the mushroom, which is produced not for its protein but for its flavor.

✓ Why are mushrooms considered microorganisms?
✓ What is a *mushroom flush*?

11.15

Sewage and Wastewater Microbiology

Although production of a valuable commercial product is not the goal of sewage and wastewater treatment, the process itself is clearly a large-scale use of microor-

ganisms and can be considered a type of bioconversion: wastewaters enter a treatment plant and, following microbial treatment, water suitable for release to rivers and streams or to drinking water purification facilities is produced.

Wastewaters are liquid effluents derived from domestic sewage or industrial sources, which for reasons of public health and for recreational, economic, and aesthetic considerations, cannot be disposed of merely by discarding them untreated into lakes or streams. Wastewaters contain both inorganic and organic components, and microorganisms play a particularly important role in removing organic compounds. However, proper wastewater treatment also results in the elimination of pathogenic microorganisms, thus preventing these organisms from getting into rivers or other water supply sources.

About 15,000 wastewater treatment facilities exist in the United States. The vast majority of them are fairly small, treating 1 million gal (3.8 million liters) or less of wastewater per day. However, collectively, these plants treat nearly 40 *billion* gal of wastewater every day. Wastewater plants are usually constructed to handle both domestic and industrial wastes. Domestic wastewaters are made up of sewage, "gray water" (the water resulting from washing, bathing, and cooking), and wastewater from food processing. Industrial waste-waters include those from the petrochemical, pesticide, food, plastics, and pharmaceutical industries, and from metallurgical industries, such as electroplating.

Many industrial wastes contain toxic substances and must be pretreated before they can be released for wastewater treatment. Pretreatment is generally a mechanical process in which debris that could clog equipment in the wastewater treatment plant is removed. However, certain wastewaters are pretreated biologically to remove highly poisonous substances such as cyanide and high levels of heavy metals. These substances can be converted to less toxic forms by pretreating them with specific microorganisms capable of oxidizing, precipitating, or volatilizing the toxic components.

We now consider the workings of a typical waste-water treatment facility using a domestic sewage treatment facility as an example of the processes involved.

Levels of Sewage Treatment

Sewage treatment is a multistep process employing both physical and biological treatment steps (Figure 11.28). **Primary treatment** of sewage consists only of physical separations. Sewage entering the treatment plant is passed through a series of grates and screens that remove

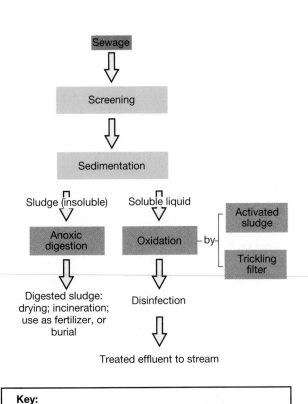

FIGURE 11.28   An overview of sewage treatment processes.

large objects, and then the effluent is left to settle for a number of hours to allow suspended solids to sediment.

Because of the high nutrient loads that remain in sewage effluent following primary treatment, municipalities that treat sewage no further than the primary stage suffer from extremely polluted water when the sewage is dumped into adjacent waterways. This is why the majority of sewage plants employ **secondary treatment** processes to reduce the organic load of the sewage to acceptable levels before releasing it to natural waterways. Secondary treatment is intimately tied to microbiological processes as described in the following sections.

Tertiary treatment is the most complete method of treating sewage but has not been widely adopted because it is so expensive. Tertiary treatment is a physicochemical process employing precipitation, filtration, and chlorination to sharply reduce the levels of inorganic nutrients, especially phosphate and nitrate, from the final effluent. Wastewater receiving proper tertiary treatment is so free of nutrients that it is unable to support extensive microbial growth.

Anoxic Secondary Treatment Processes

Anoxic sewage treatment involves a complex series of digestive and fermentative reactions carried out by a host of different bacterial species. The efficiency of a treatment process is expressed in terms of the percentage decrease in the biochemical oxygen demand (BOD), a measure of the amount of dissolved oxygen consumed by microorganisms for the oxidation of organic and inorganic matter; the higher the level of oxidizable organic and inorganic materials in the wastewater, the higher the BOD. A well-operated wastewater treatment plant can remove 95% or greater of the initial BOD.

Anoxic decomposition is usually employed in the treatment of materials that have much insoluble organic matter, such as fiber and cellulose, or of concentrated industrial wastes. The degradation process itself is carried out in large enclosed tanks called **sludge digestors** or **bioreactors** (Figure 11.29a and b), and requires the collective activities of many different types of microorganisms; the reactions are summarized in Figure 11.29c. The macromolecular components must first be digested by polysaccharidases, proteases, and lipases into soluble components. The latter are fermented to a mixture of fatty acids, H_2, and CO_2, and the fatty acids further fermented to acetate, CO_2 plus H_2. These are all substrates for methanogenic bacteria (Sections 14.3 and 15.19), which are capable of carrying out the reactions $CH_3COOH \rightarrow CH_4 + CO_2$ and $4 H_2 + CO_2 \rightarrow CH_4 + 2 H_2O$ (Figure 11.29c). Thus, major products of anoxic sewage treatment are CH_4 (methane) and CO_2. The methane is collected and either burned off or used as fuel to drive electric generators used to heat and power the treatment plant.

(a)

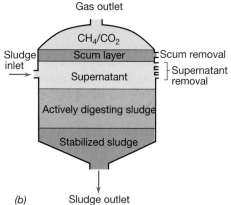

Gas outlet

CH$_4$/CO$_2$

Scum layer — Scum removal

Sludge
inlet → Supernatant — } Supernatant
removal

Actively digesting sludge

Stabilized sludge

(b)   Sludge outlet

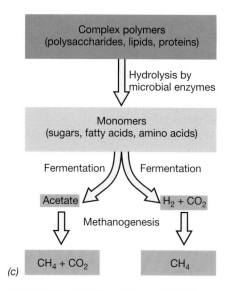

Complex polymers
(polysaccharides, lipids, proteins)

↓ Hydrolysis by
microbial enzymes

Monomers
(sugars, fatty acids, amino acids)

Fermentation   Fermentation

Acetate   H$_2$ + CO$_2$

Methanogenesis

CH$_4$ + CO$_2$ CH$_4$

(c)

FIGURE 11.29   (a) Anoxic sludge digestor. Only the top of the tank is shown; the remainder is underground. (b) Inner workings of a sludge digestor. (c) Major microbial processes occurring during sludge digestion. Note how methane (CH$_4$) is the major product of anaerobic biodegradation (◯◯ Section 16.14).

(a)

(b)

FIGURE 11.30   Aerobic sewage treatment processes. (a) Trickling filter. (b–d) Activated sludge process. (b) Aeration tank of an activated sludge installation in a metropolitan sewage treatment plant. (c) Inner workings of an activated sludge installation. (d) A small-scale activated sludge operation used to process dairy waste in Brazil.

Sewage from primary
treatment
↓

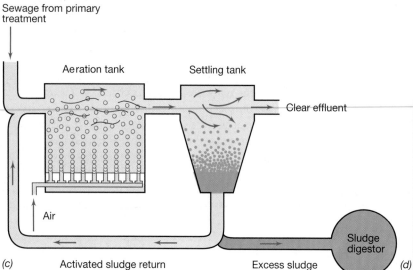

Aeration tank   Settling tank

Clear effluent

Air

(c)   Activated sludge return   Excess sludge

Sludge
digestor

(d)

Aerobic Secondary Treatment Process

Several kinds of aerobic decomposition processes are used in sewage treatment, but the trickling filter and activated sludge methods are the most common. A **trickling filter** (Figure 11.30a) is a bed of crushed rocks, about 2 m thick, on top of which the wastewater is sprayed. The liquid slowly passes through the bed, the organic matter adsorbs to the rocks, and microbial growth takes place. The complete mineralization of organic matter to carbon dioxide, ammonia, nitrate, sulfate, and phosphate occurs.

The most common aerobic treatment system is the **activated sludge** process. Here, the wastewater to be treated is mixed and aerated in a large tank (Figure 11.30b–d). Slime-forming bacteria, including the organism *Zoogloea ramigera* among others, grow and form flocs (Figure 11.31), and these flocs form the substratum to which protozoa and small animals attach. Occasionally, filamentous bacteria and fungi are also present. The basic process of oxidation is similar to that in a trick-

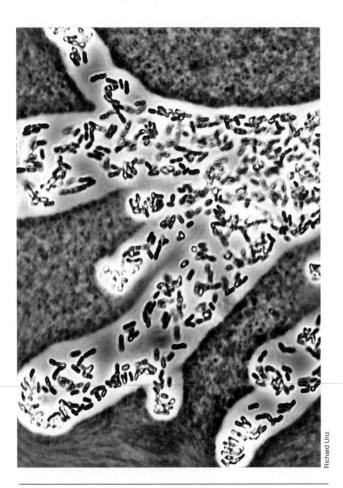

FIGURE 11.31 Photomicrograph of a floc formed by the bacterium *Zoogloea ramigera*, a characteristic organism in the activated-sludge process. Note the large number of small, rod-shaped cells of *Z. ramigera* surrounded by a polysaccharide slime layer and the characteristic fingerlike projections of the floc. Negative stain using india ink.

ling filter. The effluent containing the flocs is pumped into a holding tank or clarifier where the flocs settle. Some of the floc material is then returned to the aerator to serve as inoculum, and the rest is sent to the sludge digestor. The residence time in an activated sludge tank is generally 5–10 hr, too short for complete oxidation of organic matter. The main process occurring during this short time is *adsorption of soluble organic matter* to the floc and incorporation of some of the soluble material into microbial cell material. The BOD of the liquid is thus considerably reduced (by up to 95%) by this process, with most of the BOD now contained in the settled flocs. The main process of BOD reduction thus occurs in the anoxic sludge digestor to which the flocs are transferred and degraded.

Water Purification

Wastewaters treated as previously described are generally of such quality that they can be discharged into rivers and streams. However, such water is not *potable,* that is, suitable for drinking. Potable water requires further treatment to remove potentially pathogenic microorganisms, decrease turbidity, eliminate taste and odor, and reduce nuisance chemicals such as iron and manganese.

A typical drinking water treatment installation for a large city is shown in Figure 11.32. Water is first pumped from the source, in this case a river, to **sedimentation basins** where sand, gravel, and other particles contributing to turbidity settle out. From here the water travels to a **coagulation basin** where chemicals containing aluminum and iron are added to form a floc that traps microorganisms, absorbs organic matter and sediment, and removes them from the water. After coagulation, the clarified water is filtered to remove the remaining suspended particles and microorganisms. This is usually done by passing the water through thick layers of sand, which, when combined with previous purification steps, removes greater than 99% of the bacteria present in the original untreated water.

Chlorination is the most common method of ensuring microbiological safety in a water supply. In sufficient doses it kills microorganisms within 30 min (certain pathogenic protozoa such as *Cryptosporidium* are not easily killed by chlorine treatment and thus can be severe waterborne pathogens, ∞ Section 24.9). In addition to killing cells, chlorine reacts with organic compounds, oxidizing and effectively neutralizing them. Therefore, since most taste- and odor-producing compounds are organic in nature, chlorine treatment also improves water taste and smell. Chlorine is added to water either from a concentrated solution of sodium or calcium hypochlorite or as a gas from pressurized tanks. The latter method is used most commonly in large water treatment plants (Figure 11.32), as it is most amenable to automatic control.

FIGURE 11.32 Water purification plant. Aerial view of a water treatment plant in Louisville, Kentucky. The arrows indicate direction of flow of water through the plant.

When chlorine reacts with organic materials, it is consumed. Therefore, if a water supply is high in organic materials, sufficient chlorine must be added so there is a residual amount left to react with the microorganisms after all reactions with organic materials have occurred. The water plant operator performs chlorine analyses on the treated water to determine the residual level of chlorine. A chlorine residual of 0.2–0.6 μg/ml is an average level suitable for most water supplies. After chlorine treatment, the now-potable water is pumped to storage tanks from which it flows by gravity to the consumer.

✓ 11.15 Concept Check

Sewage treatment processes are industrial-scale microbial culture systems in which the organic materials of the sewage are converted to CO_2, CH_4, and inorganic nutrients. Two kinds of secondary treatment processes are used: anoxic, in which organic materials are converted principally to methane and carbon dioxide; and aerobic, in which organic materials are converted to microbial cells and carbon dioxide. Properly treated wastewater must be further purified before it is suitable for drinking.

✓ What is biological oxygen demand (BOD)? Why is its reduction necessary in sewage treatment?

✓ Why is chlorine added to water used for drinking purposes? What does chlorine do?

REVIEW QUESTIONS

1. In what ways do industrial microorganisms differ from conventional microorganisms? In what ways are they similar?

2. Describe some of the techniques that can be used to improve strains of industrial microorganisms.

3. List three major types of industrial products that can be obtained with microorganisms and give two examples of each.

4. Give an example of a *commodity chemical* produced by a microorganism and describe briefly the process by which this chemical is manufactured.

5. Compare and contrast *primary* and *secondary* metabolites and give an example of each. List at least two molecular explanations for why some metabolites are secondary rather than primary.

6. How does an industrial fermentor differ from a laboratory culture vessel? How does a fermentor differ from a fermenter?

7. Discuss the problems of scale-up from the viewpoints of *aeration*, *sterilization*, and *process control*. Why is sterility so much more important in an industrial fermentor than in a laboratory fermentor?

8. List three examples of *antibiotics* that are important industrially. For each of these antibiotics, list the producing organisms, the general chemical structure, and the mode of action.

9. Why are the β-lactam antibiotics so important medically? Compare and contrast the production of *natural*, *biosynthetic*, and *semisynthetic* β-lactam antibiotics.

10. Addition of what metal to the fermentation medium can markedly improve production of vitamin B_{12}?

11. What unusual characteristics must an organism have if it is to overproduce and excrete an amino acid such as *lysine*?

12. Define *microbial bioconversion* and give an example. Explain why the chemical reactions involved in microbial bioconversions are preferably carried out microbially rather than chemically.

13. List three different kinds of enzymes that are produced commercially. For each enzyme, list the organism used in commercial production, the action of the enzyme, and how the enzyme is used in commerce.

14. Describe the stages involved in the production of *high fructose syrup* and explain the role of an enzyme in each step. How is high fructose syrup used in the food industry?

15. Why is it desirable to *immobilize* enzymes? Give examples of two different immobilization procedures and describe how each is carried out.

16. What are extremozymes? What industrial uses do they have?

17. Give two reasons why stainless steel fermentors are used in the industrial production of citric acid.

18. Why are yeasts of such great industrial importance?

19. In what way is the manufacture of *beer* similar to the manufacture of *wine*? In what ways do these two processes differ? How does the production of *distilled alcoholic beverages* differ from that of beer and wine?

20. What part of the mushroom is actually consumed as food? What is contained within this structure?

21. Why is sewage that has received only primary treatment both a serious health hazard and a source of environmental pollution?

22. Compare and contrast aerobic and anoxic means of *secondary* sewage treatment. What types of compounds are best degraded by each method? What are the major products of the treatment process in each case?

APPLICATION QUESTIONS

1. You have just isolated a strain of the yeast *Saccharomyces cerevisiae* that grows over twice as fast as any known strain of *S. cerevisiae* and may thus be useful in ethanol production. However, the yeast ferments glucose to a mixture of ethanol and other products and the yield of ethanol is only about 50% that of industrial yeast strains. Recalling what you have learned in this chapter about industrial microorganisms and also your knowledge of microbial physiology (Chapter 4), genetics (Chapter 9), and genetic engineering (Chapter 10), outline a plan for converting your new yeast strain to a strain useful for the production of commodity ethanol.

2. As a researcher in a pharmaceutical company you are assigned the task of finding and developing an antibiotic effective against a new bacterial pathogen. Outline a plan for this process, starting from isolation of the low-yield producing organism to high-yield industrial production of the new antibiotic.

3. A partially consumed bottle of an "organic" (containing no preservatives) red wine is recapped and stored under refrigeration for 2 months. On tasting the wine again, you notice a distinct bitter taste, making the wine undrinkable. Using information presented in this chapter and in Section 13.7, describe (a) what microbe-mediated process occurred in the wine, and (b) a very easy way in which this process could have been prevented.

4. You wish to produce high yields of the amino acid phenylalanine for use in production of the sweetener *aspartame*. The overproducing organism you wish to use is not subject to feedback inhibition by phenylalanine but is subject to typical repression of phenylalanine biosynthesis enzymes by excess phenylalanine. Applying the principles of enzyme regulation studied in Chapter 7 and microbial genetics in Chapter 9, describe two classes of mutants you could isolate that would overcome this problem and detail the genetic lesions each would have.

Unlike their counterparts who study plants or animals, microbiologists have long suffered from the lack of an evolutionary framework within their science. Thus microbiologists used what they could, primarily comparative phenotypic analyses, in attempts to paint a picture of microbial evolution. At the present time, powerful molecular tools for the objective determination of evolutionary relationships exist, thus eliminating the need for guessing. Comparative ribosomal RNA sequencing in particular yields a broad picture of the evolution of cells, and has revealed, as shown here, that cellular life falls within three major domains: Bacteria, Archaea, and Eukarya. Combined with phenotypic determinations, microbiologists are now equipped to deduce both the taxonomy (classification and nomenclature) and the evolutionary history (phylogeny) of the organisms they work with.

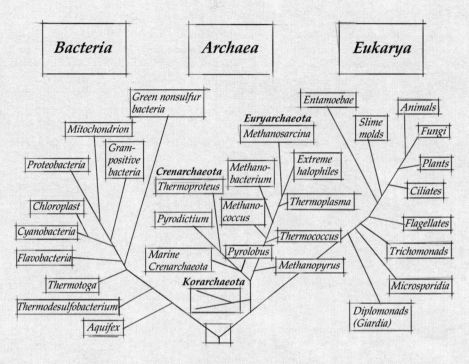

CHAPTER 12 Microbial Evolution and Systematics

12.1 Evolution of Earth and Earliest Life Forms 424
12.2 Primitive Organisms: Molecular Coding and Energy Generation 427
12.3 Eukaryotes and Organelles 430
12.4 Evolutionary Chronometers 432
12.5 Ribosomal RNA Sequences and Cellular Evolution 434
12.6 Signature Sequences, Phylogenetic Probes, and Molecular Microbial Ecology 436
12.7 Microbial Phylogeny as Revealed by Ribosomal RNA Sequencing 439
12.8 Characteristics of the Primary Domains 442
12.9 Conventional and Molecular Taxonomy 444
12.10 The Species Concept, Nomenclature, and Bergey's Manual 449

WORKING GLOSSARY

Archaea a group of phylogenetically related prokaryotes distinct from Bacteria

Bacteria a group of phylogenetically related prokaryotes distinct from Archaea

Domain in a taxonomic sense, the highest level of biological classification (see alternative usage in Chapter 20)

Endosymbiosis a theory stating that the mitochondrion and chloroplast were originally free-living Bacteria that established stable residence in primitive eukaryotic cells, eventually yielding the modern eukaryotic cell

Eukarya all eukaryotic cells: algae, protozoa, fungi, slime molds, plant and animal cells

Evolutionary distance in phylogenetic trees, the sum of the physical distance on a tree separating organisms; this distance is inversely proportional to evolutionary relatedness

Family in biological classification, an intermediate level of taxonomic hierarchy. Contains several genera, each of which consists of one or more species

FISH fluorescent *in-situ* hybridization

GC base ratio in DNA (or RNA) from any organism, the percentage of the total nucleic acid that consists of guanine and cytosine bases

Genus a collection of different species, each sharing one or more (usually several) major properties

Lateral (horizontal) gene transfer the exchange of genes between and among cells in a microbial community

Organelle unit membrane-enclosed structures of bacterial size and specialized in metabolic function found inside eukaryotic cells

Phylogenetic probe an oligonucleotide, sometimes made fluorescent by attachment of a dye, comple-mentary in sequence to some ribosomal RNA signature sequence

Phylogeny the evolutionary history of organisms

Proteobacteria a large group of phylogenetically related gram-negative Bacteria

Ribotyping a means of identifying microorganisms from analysis of DNA fragments generated from restriction enzyme digestion of genes encoding their 16S rRNA

RNA life a life-form lacking DNA and protein that may have existed on early Earth and in which RNA served both a genetic coding and a catalytic function

Signature sequence short oligonucleotides of defined sequence in 16S or 18S rRNA characteristic of specific organisms or a group of phylogenetically related organisms

16S rRNA a large polynucleotide (~1500 bases) that functions as part of the small subunit of the ribosome of prokaryotes and from whose sequence evolutionary information can be obtained; eukaryotic counterpart, 18S rRNA

Species in microbiology, a collection of strains that all share the same major properties but differ in one or more significant properties from other collections of strains; two prokaryotic species generally show differences in 16S rRNA sequence of 3% or more

Stromatolites laminated microbial mats, typically built from layers of filamentous and other microorganisms, which can become fossilized

Taxonomy the science of identification, classification, and nomenclature of organisms

Universal tree a phylogenetic tree that shows the position of representatives of all domains of living organisms

---

An integrating theme in microbiology today is the enormous diversity of microorganisms on Earth. In Chapters 13, 14, and 17 we will survey this diversity and in Chapters 15 and 16 we will revisit this diversity in terms of metabolism and ecology, respectively. In this chapter we discuss the origin of living systems and will learn how the evolutionary relationships of present-day microorganisms can be determined experimentally. We also discuss the properties of the early Earth upon which life arose and consider some key aspects of microbial taxonomy.

What is life? We touched on this issue in Chapter 1 when we considered the hallmarks of a cell (∞ Section 1.2). Life can be considered an imperfect replicating system that evolves by natural selection. It is imperfect, of course, because evolution is an ongoing process. Once self-replicating entities arose on Earth, mutation and genetic recombination coupled with natural selection came into play to evolve new types of microor-ganisms increasingly better suited to particular ecological niches. And as geochemical processes generated new microbial habitats on Earth, microbial life continued to evolve to eventually colonize every conceivable habitat compatible with life. The net result of this interaction of genetics and environment is the great microbial diversity we see today.

Why is it important to understand how life originated and to reveal the full extent of microbial diversity? One major reason is to answer the question "are we alone?"; that is, does life exist outside the boundaries of Earth's biosphere? With humans now in a position to send spacecraft to other celestial bodies, we need to understand the geochemical conditions that are compatible with living systems on Earth, how self-replicating systems originated and how they evolved, and the metabolic diversity of extant microbial life. Armed with such knowledge, scientists would have a good chance of recognizing living systems on planets such as Mars or on Jupiter's moon Europa, both places where the conditions for life either exist today or have likely existed in the past.

With this in mind, let us start our study of microbial evolution at the beginning—Earth before life was present—and see where evolution has taken us today.

12.1

Evolution of Earth and Earliest Life Forms

Origin of Earth

Earth is about 4.6 billion years old as determined by geochemical dating measurements. Our solar system is thought to have been formed when a large, very hot star exploded, generating a new star (our sun) and the other components of our galaxy. Although no rocks dating to this period have yet been discovered on Earth, presumably because they have been weathered by rain, rocks dating back to nearly 4 *billion* years ago have been found in several locations on Earth. The oldest rocks discovered thus far are those of the Itsaq gneiss complex in southwestern Greenland, which date to about 3.86 billion years ago. These rocks are of three types: sedimentary, volcanic, and carbonate. The sedimentary rocks are of particular evolutionary interest because from our understanding of how modern sedimentary rocks are formed, the presence of sedimentary rocks 3.86 billion years old strongly suggests that *liquid water* in the form of oceans or large lakes was present at that time. The presence of liquid water in turn implies that conditions on Earth at that time were likely to be compatible with life as we know it. Other rocks of ancient origin include the Warrawoona series, Towers Formation, and Pilbara supergroups in Western Australia and the Swaziland supergroup in southern Africa; all of these rock formations are about 3.5 billion years old.

Evidence for Microbial Life on Early Earth

Fossil evidence for microbial life in the oldest known rocks is scant and rests on the fossilized remains of carbonaceous materials and the isotopically "light" carbon abundant in these rocks (we discuss the use of isotopic analyses of carbon and sulfur as an indication of living processes in ⚭ Section 16.8). Some ancient rocks contain vaguely recognizable microfossils that appear bacterial shaped, usually as simple rods or cocci. However, in somewhat younger rocks, clear fossil evidence of microbial life exists and many such examples have been discovered in ancient stromatolites. *Stromatolites* are fossilized microbial mats consisting of layers of filamentous prokaryotes and trapped sediment (Figure 12.1*a* and *b*); we discuss some characteristics of microbial mats in Section 16.7. What kind of organisms were these ancient stromatolitic bacteria? Although a firm answer is not yet at hand, by comparing ancient stromatolites with modern stromatolites growing in shallow marine basins (Figure 12.1*c–e*) and in hot springs (Figure 12.1*f*; ⚭ Figure 16.16*a*), scientists conclude it is likely that ancient stromatolites were formed by filamentous phototrophic bacteria.

Figure 12.2 shows photomicrographs of thin sections of ancient rocks containing cell-like structures remarkably similar to modern filamentous bacteria and green algae (⚭ Section 17.6). In the oldest stromatolites these organisms were likely anoxygenic (nonoxygen-evolving) phototrophic bacteria (⚭ Sections 13.1, 13.19, 13.29, and 13.32) rather than the O_2-evolving cyanobacteria (⚭ Sections 13.24 and 13.25) that dominate stromatolites growing today. Nevertheless, the conclusion is inescapable that prokaryotic microorganisms had evolved an impressive morphological diversity early on in the history of life on Earth.

Conditions on Early Earth

The atmosphere of early Earth was devoid of significant amounts of O_2 and hence constituted a *reducing* environment. Besides H_2O, a variety of gases were present, the most abundant being CH_4, CO_2, N_2, and NH_3. In addition, trace amounts of CO and H_2 existed, as well as considerable amounts of sulfide, as a mixture of H_2S and FeS. It is also likely that a considerable amount of hydrogen cyanide, HCN, was produced on early Earth when NH_3 and CH_4 reacted chemically to yield HCN. Geochemical estimates of the temperature of early Earth also suggest that it was a much hotter planet than it is today. For the first half billion or so years of its existence it is likely that the surface of Earth exceeded 100°C and was bombarded by meteorites; thus free water probably did not exist on early Earth but accumulated only later as Earth cooled. How fast Earth cooled is unknown, but it has been hypothesized that self-replicating entities first appeared at a time when Earth was much hotter than it is now. Thus, early life-forms must have been quite heat-tolerant and would have resembled in this respect the hyperthermophilic prokaryotes that inhabit thermal environments today (⚭ Section 5.9 and Chapter 14). We will consider evolutionary evidence that supports this hypothesis in Section 12.6.

Origin of Life

It is now well established that the synthesis of biologically important molecules can occur if reducing atmospheres containing the aforementioned gases are subjected to intense energy sources. Of the energy sources available on primitive Earth, the most impor-

FIGURE 12.1 Ancient and modern stromatolites. (a) The oldest known stromatolite, found in a rock about 3.5 billion years old, from the Warrawoona Group in Western Australia. Shown is a vertical section through a laminated, hemispheroidal structure, which has been preserved in the rock. Scale, 10 cm. (b) Stromatolites of conical shape from 1.6-billion-year-old dolomite rock of the McArthur basin of the Northern Territory of Australia. (c) Modern stromatolites in a warm marine bay, Shark Bay, Western Australia. (d) Another view of large modern stromatolites from Shark Bay. Note the resemblance to the ancient stromatolites shown in (b). (e) Underwater photograph of modern stromatolites growing in Shark bay. The diver indicates the scale. Shown are large columns formed by a complex community of diatoms, cyanobacteria, and green algae, to which are attached various macroscopic algae. (f) Modern stromatolites composed of thermophilic cyanobacteria growing in a thermal pool in Yellowstone National Park. Each structure is about 2 cm high.

tant was probably ultraviolet (UV) radiation from the sun, but lightning discharges, radioactivity, heat from meteoritic impacts, and thermal energy from volcanic activity were also available. If gaseous mixtures resembling those thought to be present on primitive Earth are irradiated with UV or subjected to electric discharges in the laboratory, a wide variety of biochemically important molecules can be made, such as sugars, amino acids, purines, pyrimidines, various nucleotides, thioesters, and fatty acids. It has also been shown that under prebiological conditions some of these biochemical building blocks can *polymerize*, leading to the formation of polypeptides, polynucleotides, and other important macromolecules. We can therefore imagine that on primitive Earth a mixture of organic compounds eventually accumulated, and in the absence of living organisms these compounds would have been stable (that is, not consumed)

and could have persisted for countless years. Thus, with time, there could have been an extensive accumulation of organic materials, setting the stage for biological evolution.

But how could *macromolecules* have originated from monomeric constituents spontaneously in an aqueous environment? From a chemical standpoint, nucleic acids and proteins are polymerized via *dehydration* reactions; thus, it is difficult to conceive of how, in the absence of enzymes, macromolecules could have arisen spontaneously in an aqueous setting. A possibility is that relatively anhydrous *exposed surfaces* such as clays, pyrite, or basaltic glasses functioned as supports for prebiotic polymerization reactions. Such surfaces would have provided a stable, relatively dry environment for the synthesis and accumulation of macromolecules into organic films from which primitive self-replicating structures could have emerged. Pyrite (FeS_2) and Mont-

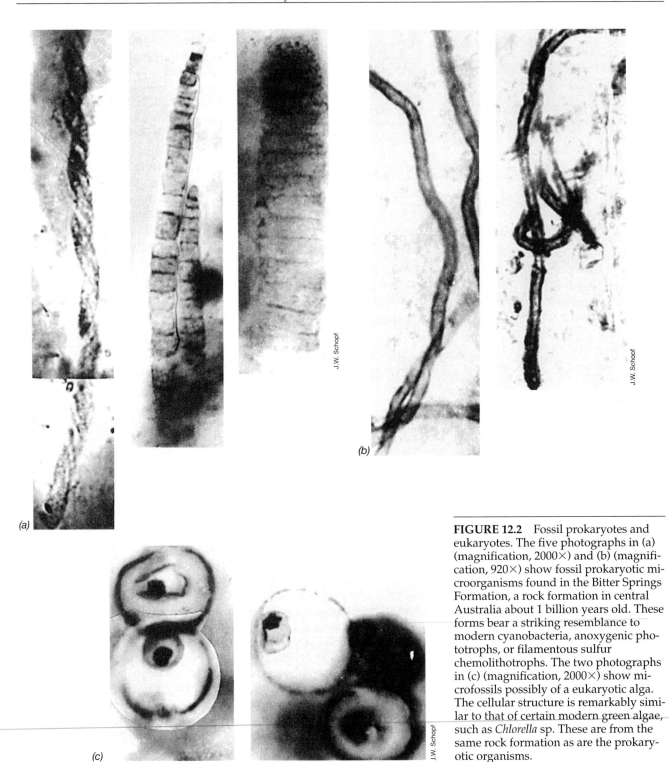

(a)

(b)

(c)

J.W. Schopf

J.W. Schopf

J.W. Schopf

FIGURE 12.2 Fossil prokaryotes and eukaryotes. The five photographs in (a) (magnification, 2000×) and (b) (magnification, 920×) show fossil prokaryotic microorganisms found in the Bitter Springs Formation, a rock formation in central Australia about 1 billion years old. These forms bear a striking resemblance to modern cyanobacteria, anoxygenic phototrophs, or filamentous sulfur chemolithotrophs. The two photographs in (c) (magnification, 2000×) show microfossils possibly of a eukaryotic alga. The cellular structure is remarkably similar to that of certain modern green algae, such as *Chlorella* sp. These are from the same rock formation as are the prokaryotic organisms.

morillonite clay have been suggested as possibilities here because of the crucial role the former may have played in early energy-generating systems (see next section) and the known ability of the latter to selectively absorb ribonucleic acid monomers and form RNA oligomers from them.

✓ 12.1 Concept Check

Earth is thought to be 4.6 billion years old; the first evidence of microbial life emerges in rocks about 3.86 billion years old. Early Earth was anoxic and much hotter than at present. The first biochemical compounds were made by abiotic syntheses that set the stage for the origin of life.

✓ How old is Earth? How old are the oldest known micro-fossils?

✓ How did the atmosphere of early Earth compare with that of Earth today?

✓ How were early biochemicals formed?

12.2

Primitive Organisms: Molecular Coding and Energy Generation

What was the first self-replicating organism like? This is an impossible question to answer at present, but based on what we know about microbial life forms today, we can predict that even the simplest self-replicating entities needed a means of obtaining energy and some form of hereditary mechanism in order to make copies of itself. But would these processes have required a cellular structure? It is tempting to extrapolate backward from the present and postulate that early organisms were much like modern cells, but contained only a very few genes (made of DNA), and had very limited transcriptional and translational abilities. But even a structure like this would have been relatively complex, much more complex than what the first self-replicating entities might have been. What would these creatures have been like? Following the discovery that certain types of ribonucleic acid (RNA) are catalytic (Section 6.7), many scientists now believe that the earliest life forms probably lacked DNA altogether; contained just a very few, if any, proteins; and consisted primarily of RNA (Figure 12.3)—simply put, this was the age of *RNA life*, where RNA functioned in both catalysis and genetic coding.

RNA Life

In the early RNA world, RNA molecules would have functioned simply to replicate themselves and would likely have carried out only the minimal number of catalytic reactions necessary for this purpose (Figure 12.3). Studies of various ribozymes have shown that several different reactions can be catalyzed, including the synthesis of nucleotides from a sugar and a nitrogenous base and other classes of nucleic acid chemistry. Thus, an era of RNA life may well have predated cellular life. These RNA life-forms may have evolved into the first *cellular* life-forms when self-replicating RNAs became enclosed within lipoprotein vesicles (Figure 12.3). These cell-like structures probably arose through the spontaneous aggregation of lipid and protein molecules to form membranous structures within which were trapped RNAs and other biomolecules. This step may have occurred countless times on the early Earth to no

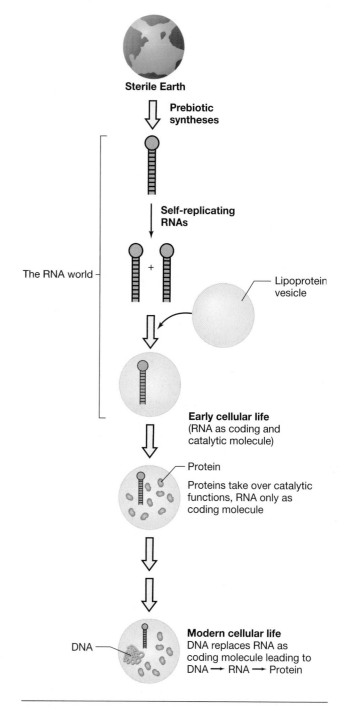

FIGURE 12.3 Possible scenario for the evolution of cellular life-forms from RNA life-forms. Self-replicating RNAs could have become cellular entities by becoming stably integrated into lipoprotein vesicles. With time, proteins replaced the catalytic functions of RNA and DNA replaced the coding functions of RNA.

avail, but eventually the proper constituents and set of circumstances came together and a primitive cellular organism arose. Although still lacking DNA and protein, this cellular life-form would otherwise have resembled a modern cell, and as such life-forms became more widespread, natural selection led to their further evolutionary development.

How would RNA life-forms have accomplished all the reactions we associate with living systems? As previously mentioned, studies of RNA chemistry have shown that many different RNAs have ribozyme activity and collectively can catalyze a wide variety of biochemical reactions, even including peptide-bond formation. Although the *specificity* of RNAs in these chemical reactions is not particularly good, their catalytic diversity is very high, such that there is no reason to believe that the suite of biochemical reactions required to maintain early self-replicating entities could not have been catalyzed by molecules of RNA alone. However, as living organisms became biochemically more complex, an evolutionary push to *proteins* as the major biocatalysts probably occurred. It is likely that proteins appeared gradually in cells, perhaps at first complexed with RNA, and as evolution selected for more and more precise biochemical catalysts, RNA was eventually replaced by protein as the major cellular enzymes (Figure 12.3).

The Modern Cell: DNA → RNA → Protein

The establishment of DNA as the genome of the cell may have resulted from the need to store genetic information in a more stable form. By storing all the genetic information in one place in the cell and processing only what was needed under a specific set of conditions (that is, *regulating gene expression*), cells would have saved energy, which would have increased their competitive fitness. However, an additional reason for the origin of DNA as the master genetic blueprint may have been the highly error-prone nature of enzymes that copy RNA. For unknown reasons, enzymes that copy RNA (RNA polymerases, ⌀⌀ Section 6.6) are inherently less precise than DNA polymerases. Thus, maintaining RNA as the cell's genetic material and relying on inherently error-prone copying systems for its replication would likely have been incompatible with increasing cellular complexity and the precise genetic demands this would entail. Evolution would therefore have favored transfer of vital coding functions to a form of nucleic acid that replicated with high fidelity, like DNA, and this would have eventually eliminated the less precise RNA life-forms (Figure 12.3).

Somewhere in the early stages of microbial evolution, the three-part system—DNA, RNA, and protein—became fixed in cellular life-forms as the best solution to biological information processing. That this system was an evolutionary success can be attested to by the fact that, as far as is known, all cells today contain all three types of these macromolecules. Thus, although modern life-forms employ DNA and proteins as essential parts of cellular function, early life-forms may well have accomplished all of this with only RNA.

Metabolism in Primitive Organisms

Recall that life is a highly ordered process. To render inherently disordered molecules into a complex biological machine, energy was necessary. How then would the energy demands of primitive self-replicating entities have been met? Until the evolution of cyanobacteria molecular oxygen was unavailable in any significant quantities on Earth (see Figure 12.5). Thus, only energy-generating mechanisms that could occur under *anoxic* conditions could be exploited to fuel the energy needs of early organisms. As we will see in Chapter 15, this restriction eliminates little from consideration, in that a variety of chemoorganotrophic and chemolithotrophic energy-generating mechanisms, as well as anoxygenic photosynthesis (⌀⌀ Chapter 15) occur anoxically. However, a simple chemical reaction involving ferrous iron (which was known to have been abundant on early Earth), may have been one of the first reactions by which primitive organisms conserved energy.

The reaction:

$$FeS + H_2S \rightarrow FeS_2 + H_2 \qquad \Delta G_0' = -42 \text{ kJ/reaction}$$

proceeds exergonically with the release of energy. This reaction also yields H_2, and it has been hypothesized that this H_2 could have been used by primitive cells to form a proton-motive force across a membrane from which a primitive ATPase could have recovered chemically useful energy as ATP (Figure 12.4). With H_2 as

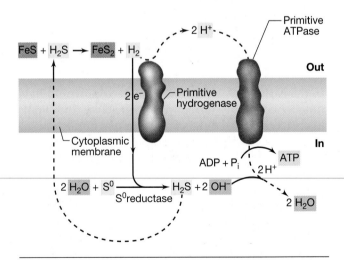

FIGURE 12.4   A hypothetical energy-generating scheme for primitive cells. Formation of pyrite leads to H_2 production and S^0 reduction, which fuels a primitive ATPase. Note how H_2S plays only a catalytic role; the net substrates would be FeS and S^0. Also note how few different proteins would be required. The $\Delta G^{0'}$ of the reaction $FeS + H_2S \rightarrow FeS_2 + H_2 = -42$ kJ.

electron donor an electron acceptor would also have been required, and this could have been elemental sulfur, S^0. As shown in Figure 12.4, this simple coupled reaction would have required few enzymes and would have been a limitless means of energy conservation as long as FeS was available. Interestingly, many hyperthermophilic Archaea (which we will see later in this chapter are the closest extant relatives of Earth's earliest organisms), can carry out the very reaction described in Figure 12.4 (∞ the box, Microbial Life Deep Underground, Chapter 16, for another example). Many other forms of anoxic metabolism could also have occurred on early Earth, including fermentations and various forms of anaerobic respiration, but these would likely have required more catalysts and biochemical complexity than that outlined in Figure 12.4.

Carbon to make cell material for primitive organisms could have come from various sources including organic carbon from prebiotic syntheses or even from CO_2, a gas that was abundant on early Earth. The latter would have required that autotrophy, the process where CO_2 is converted into all organic components in the cell (∞ Sections 15.7 and 15.8), had evolved, and although this was originally thought to be unlikely because of the great genetic demands this was thought to entail, support for an "early autotrophy" hypothesis has come from microbial genome-sequencing projects (∞ Section 9.12), where it has been shown that organisms like *Aquifex*, a hyperthermophilic autotroph that branches close to the root of the evolutionary tree of life (see Figure 12.13), contain very small genomes.

Regardless of the mechanism(s) employed for assimilating carbon, a milestone in Earth's history occurred with the evolution of oxygenic photosynthesis in the cyanobacteria. These organisms probably first appeared on Earth some three billion years ago, or even earlier, but the O_2 that they produced did not accumulate in the atmosphere because of all of the reducing substances (like FeS) still present that reacted spontaneously with O_2 to produce H_2O. However, it is virtually certain that cyanobacteria evolved from anoxygenic phototrophs through the development of a photosystem that could use H_2O as an electron donor for photosynthetic CO_2 reduction, thus releasing O_2 as a byproduct (CO_2 + 2 $H_2O \rightarrow CH_2O + H_2O + O_2$). We discuss the biochemistry of photosynthesis in Chapter 15.

The evolution of oxygenic photosynthesis had enormous consequences for the environment of Earth because as O_2 accumulated, the atmosphere gradually changed from an anoxic to an oxic one (Figure 12.5).

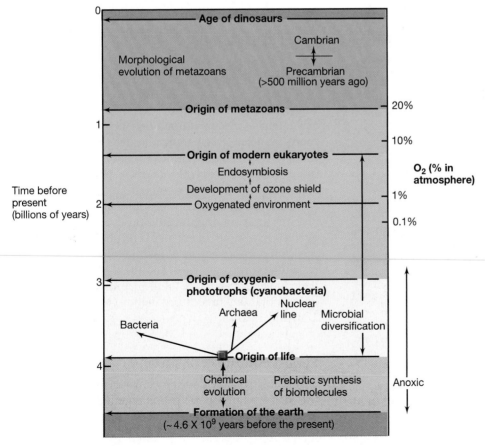

FIGURE 12.5 Major landmarks in biological evolution. The positions of the stages on the time scale are approximate. Note how the oxygenation of the atmosphere due to cyanobacterial metabolism was a gradual process, occurring over a period of about 1.5 billion years. Also note that for the bulk of Earth's history, only microbial life-forms existed.

With O_2 now available as an electron acceptor, aerobic organisms could evolve. These organisms were able to obtain more energy from the oxidation of organic compounds than anaerobes could (⟳ Chapter 15), allowing higher population densities to develop and increasing the chances for the evolution of new types of organisms and metabolic schemes. There is good evidence from the fossil record that at about the time that Earth's atmosphere became highly oxidizing there was an enormous burst in the rate of evolution, leading to the appearance of eukaryotic microorganisms with organelles and from them to the rapid diversification of metazoa (multicellular organisms) and eventually to higher animals and plants (see Figure 12.13 and Section 12.7).

The Ozone Shield

Another major consequence of the appearance of O_2 was the formation of ozone (O_3), a substance that provides a barrier preventing the intense ultraviolet radiation of the sun from reaching Earth. When O_2 is subject to short-wavelength ultraviolet radiation, it is converted to O_3, which strongly absorbs wavelengths up to 300 nm. Until an ozone shield developed, evolution could have continued only in environments protected from direct radiation from the sun, such as under rocks or in the oceans. After the photosynthetic production of O_2 and development of an ozone shield, organisms could have ranged generally over the surface of Earth, permitting evolution of a greater diversity of living organisms. A summary of the steps that could have occurred in biological evolution is shown in Figure 12.5.

✓ 12.2 Concept Check

The first organisms must have employed a simple strategy to obtain energy. Primitive metabolism was anaerobic and likely chemolithotrophic, exploiting the abundant sources of FeS and H_2S present. Fermentations and anaerobic respiration probably appeared later along with anoxygenic photosynthesis followed by oxygenic photosynthesis. The latter led to development of an oxic environment and to great bursts of biological evolution.

- ✓ What evidence supports the concept of a period of "RNA life," and why would RNA life-forms not have survived to the present?
- ✓ How could energy have been obtained from FeS + H_2S by early cells?
- ✓ What is *autotrophy*?
- ✓ Give at least two reasons why it is unlikely that oxygenic photosynthesis was the first energy-generating mechanism on Earth.

12.3

Eukaryotes and Organelles

From comparative nucleic acid sequencing studies we now know that the three main lines of cellular descent, *Bacteria, Archaea,* and *Eukarya,* were established relatively early in cellular evolution (see Section 12.7). However, the eukaryotic cell we know today undoubtedly differs structurally from primitive eukaryotic cells. Primitive eukaryotic cells consisted of structurally quite simple cells, resembling modern-day prokaryotic cells in lacking mitochondria, chloroplasts, and a membrane-enclosed nucleus. Modern eukaryotic cells, with their distinctive cellular organelles, were the result of endosymbiotic events (see later) that took place some two billion or so years after divergence of the nuclear line of descent from the universal ancestor (Figure 12.6; see also Figure 12.13 and Section 12.7).

Origin of the Nucleus

It seems likely that the eukaryotic nucleus and mitotic apparatus arose as a necessity for ensuring the replication and orderly partitioning of DNA once the genome size had increased to the point where replication as a single molecule (as in most prokaryotes) was no longer feasible. Origin of the nucleus made it possible for eukaryotic cells to manage the huge genomes needed to encode the genetic blueprint of multicellular organisms and also made possible the recombination of genomes through sexual reproduction (⟳ Sections 3.16, 9.13, and 9.14).

The widespread occurrence of plasmids in bacteria suggests that even in prokaryotes there are evolutionary advantages for segregation of genetic information into more than one DNA molecule. It is thus possible to imagine how separate chromosomes might have arisen in a primitive eukaryotic cell from plasmidlike structures and have become segregated within the cell into a membrane-enclosed nucleus. Probably spindle fibers and the mitotic apparatus would also have had to evolve at the same time. There is no obvious reason why this primitive eukaryote would have needed other typical eukaryotic organelles, and these could have arisen later. Indeed, even today there are protozoans and fungi known that lack major organelles (⟳ Sections 12.7, 17.1, and 17.2).

Endosymbiosis

Strong microscopic and molecular evidence exists that the modern eukaryotic cell is a genetic chimera, and evolved in steps through incorporation into cells of

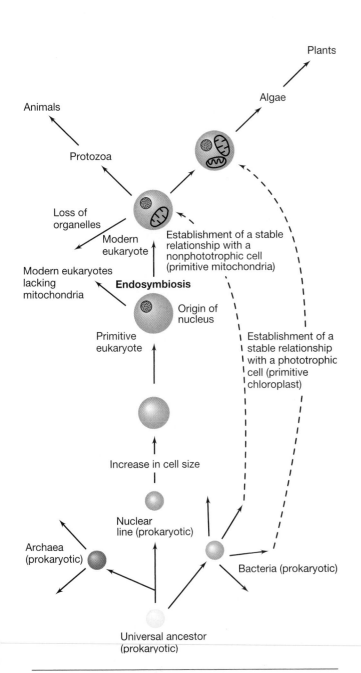

Plants

Algae

Animals

Protozoa

Loss of
organelles

Establishment of a stable
relationship with a
nonphototrophic cell
(primitive mitochondria)

Modern
eukaryote

Endosymbiosis

Modern eukaryotes
lacking
mitochondria

Origin of
nucleus

Primitive
eukaryote

Establishment of a
stable relationship
with a phototrophic
cell (primitive
chloroplast)

Increase in cell size

Nuclear
line (prokaryotic)

Archaea
(prokaryotic)

Bacteria (prokaryotic)

Universal ancestor
(prokaryotic)

FIGURE 12.6 Origin of modern eukaryotes by endosymbiotic events. Note how organelles originated from Bacteria rather than Archaea. Endosymbiosis was unlikely to have been a one-time event and probably occurred in various types of cells of the nuclear line of descent. Note, however, how some primitive eukaryotes either never underwent endosymbiotic events or permanently lost their symbionts, but otherwise maintained the basic properties of eukaryotic cells. Extant examples of such eukaryotes, all microbial, are known today (∞ Chapter 17).

chemoorganotrophic and phototrophic symbionts. This theory, referred to as the **endosymbiotic theory** of eukaryotic evolution, has through the years gathered increasing experimental support (∞ Sections 3.16, 9.12, 17.1 and 17.2). The theory postulates that an aerobic bacterium established residency within the cytoplasm of a primitive eukaryote and supplied the larger cell with energy in exchange for a stable, protected environment and a ready supply of nutrients (Figure 12.6). This aerobic bacterium would represent the forerunner of the present mitochondrion. In similar fashion, the endosymbiotic uptake of an oxygenic phototrophic prokaryote would have made the primitive eukaryote photosynthetic and no longer dependent on organic compounds for energy production. The phototrophic endosymbiont would then be considered the forerunner of the chloroplast (Figure 12.6). Following the acquisition of prokaryotic endosymbionts, eukaryotic cells underwent an explosion in biological diversity. The period from about 1.5 billion years ago to the present saw the rise of the metazoa and diversification of this highly successful group, culminating in the structurally complex higher plants and animals (see Figure 12.13 and Section 12.7).

Some eukaryotic cells either never established endosymbiotic relationships with Bacteria or did, but for some reason(s) lost or disposed of their endosymbionts, since eukaryotic cells are known today that contain a membrane-enclosed nucleus but lack organelles (Figure 12.6; ∞ Chapter 17). Thus, energy-generating endosymbionts may not have been advantageous to all types of eukaryotic cells, and in some cases the organelles might even have been a selective disadvantage to survival in particular environments. Interestingly, however, most modern eukaryotic microorganisms that lack organelles reside on early branches of the evolutionary tree of life (see Section 12.7 and Figure 12.13), suggesting that they may be relics from the past that never established endosymbiotic relationships with Bacteria.

Studies on the nucleic acids and ribosomes of the organelles of eukaryotic cells have built an impressive case for the theory of endosymbiosis. Briefly, this includes the fact that the chloroplast and mitochondrion contain ribosomes of the prokaryotic type (and whose function is inhibited by antibiotics that inhibit ribosome function in Bacteria); contain small amounts of DNA arranged in a covalently closed, circular fashion typical of prokaryotes (∞ Section 3.16); and finally, show ribosomal RNA (rRNA) sequences (see Section 12.5) characteristic of certain Bacteria.

To evolve to the modern eukaryotic stage, primitive eukaryotic cells had to abandon several advantageous features of prokaryotic cells such as a haploid genome (and the concomitant ability to adapt rapidly to new environments) and structural simplicity. Also, the greater complexity of the eukaryotic cell meant that it would face difficulty in adapting to life in extreme environments, such as thermal areas, where the prokaryote is today preeminent. For these reasons the evolution of the eukaryote did not toll the death bell for the prokaryotes. All three cell lines, Bacteria, Archaea, and Eukarya, continued to evolve, carving out their ecological niches and serving as the forerunners of the various species we know today.

Biological Evolution and Geological Time Scales

In terms of geological time, the period from the origin of the first metazoa to the present represents only one-sixth of the total time that life has existed on Earth. Put another way, five-sixths of Earth's history was restricted to *microbial* life, the bulk of this period to *prokaryotes only* (see Figure 12.5). However, because metazoa have left a considerable and highly diverse fossil record, our understanding of biological evolution from the time of the metazoa is much greater than our knowledge of evolutionary relationships among prokaryotes. Fortunately, this has changed with the advent of molecular methods for discerning evolutionary relationships. The objective of most of the remainder of this chapter is to present in some detail the modern approaches to an understanding of microbial evolution based on molecular biology and genetics. From these methods a totally new evolutionary picture of life on Earth has emerged.

✓ 12.3 Concept Check

The eukaryotic nucleus and mitotic apparatus probably arose as a necessity for ensuring the orderly partitioning of DNA in large-genome organisms. Mitochondria and chloroplasts, the principal organelles of eukaryotes, arose from prokaryotes that became symbiotic within eukaryotic cells. This process has been called endosymbiosis.

- ✓ Why did the eukaryotic nucleus evolve?
- ✓ What is the endosymbiotic theory and what evidence supports it?

12.4

Evolutionary Chronometers

It is now clear that certain cellular macromolecules are evolutionary chronometers—actual measures of evolutionary change. From studies on the sequence of monomers in certain informational macromolecules, it has been shown that the *evolutionary distance* between two organisms can be measured by differences in the nucleotide or amino acid sequence of *homologous* macromolecules isolated from them. This is so because the

number of sequence differences in a molecule is proportional to the number of stable mutational changes fixed in the DNA encoding that molecule in the two organisms as they diverged from a common ancestor. As different mutations become fixed in different populations, biological evolution occurs, and biodiversity is the end result.

Choosing the Right Chronometer

In order to determine true evolutionary relationships between organisms, it is essential that the correct molecules be chosen for sequencing studies. This is important for several reasons. First, the molecule should be *universally distributed* across the group chosen for study. Second, it must be *functionally homologous* in each organism; phylogenetic comparisons must start with molecules of *identical* function. One cannot compare the amino acid sequences of enzymes that carry out different reactions or the nucleotide sequences of nucleic acids of different functions, because functionally unrelated molecules would not be expected to show sequence similarities. Third, it is crucial in sequence comparisons to be able to properly *align* the two molecules in order to identify regions of sequence homology and sequence heterogeneity. Finally, the sequence of the molecule chosen should change at a rate commensurate with the evolutionary distance measured. And in fact the broader the phylogenetic distance being measured, the *slower* must be the rate at which the sequence changes. A molecule that has undergone too many sequence changes is essentially useless as a tool for determining evolutionary relationships because regions of common sequence would eventually be lost.

Many molecules have been proposed as molecular chronometers and comparative sequencing studies have been done on them to generate phylogenetic trees. These molecules include various cytochromes, iron–sulfur proteins like the ferredoxins, and genes for several other proteins and the ribosomal RNAs. However, some of these molecules have turned out to give useful evolutionary information while others have not. In particular, the genes encoding ribosomal RNAs, key components in the translational system (∞ Sections 6.11–6.13), ATPase, the membrane-enclosed enzyme complex that can both synthesize as well as hydrolyze ATP (∞ Section 4.11), and RecA, the protein required for genetic recombination (∞ Section 9.5), have provided highly insightful phylogenetic information about microorganisms. All of these molecules were probably essential for even rather primitive cells, and thus sequence variation in the genes encoding them allows us to probe deeply into the evolutionary record. We focus our discussion here on the most widely used of these chronometers, ribosomal RNA (Figure 12.7).

FIGURE 12.7   Ribosomal RNA. (a) Electron micrograph of 70S ribosomes from the bacterium *Escherichia coli*. (b) Parts of the ribosome; 5S, 16S, and 23S refer to different forms of RNA in the small subunit of the ribosome. (c) Primary and secondary structure of 16S ribosomal RNA (rRNA). This is the 16S rRNA from *Escherichia coli* (Bacteria); 16S rRNA from Archaea has general similarities in secondary structure (folding) but numerous differences in primary structure (sequence). The counterpart to 16S rRNA in eukaryotes is 18S rRNA present in cytoplasmic ribosomes. Using oligonucleotide primers specific for conserved rRNA sequences in organisms of one domain or the other (see Table 12.1), it is possible using PCR techniques to specifically amplify small subunit RNAs of cells of a particular domain present in a mixture with cells of other domains (see Figure 12.8). This has allowed for estimates of the diversity of particular domains of life directly in the habitat by "community sampling" techniques (see Figures 12.11 and 12.12).

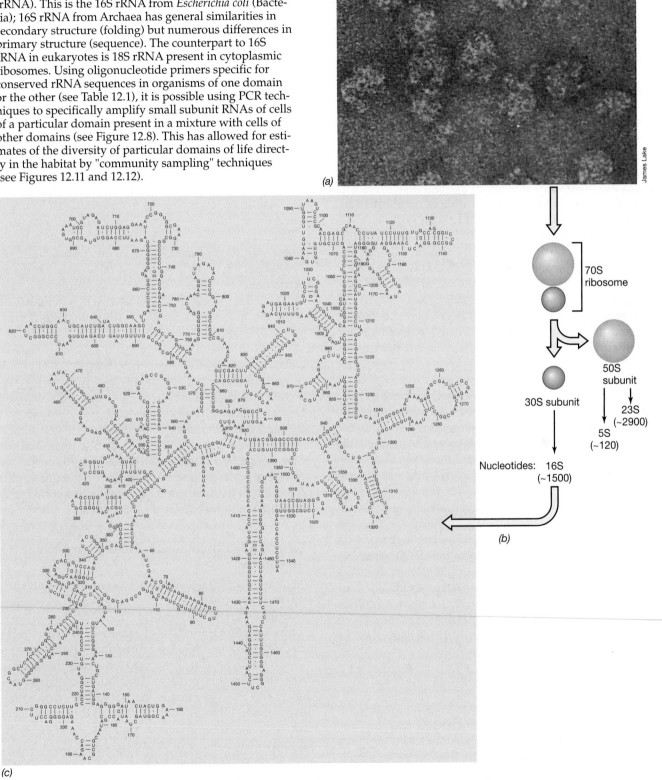

Ribosomal RNAs as Evolutionary Chronometers

Because of the likely antiquity of the protein-synthesizing machinery and for several other reasons, ribosomal RNAs are excellent molecules for discerning evolutionary relationships among living organisms. Ribosomal RNAs are ancient molecules, functionally constant, universally distributed, and moderately well conserved in sequence across broad phylogenetic distances. Also, because the *number* of different possible sequences of large molecules such as ribosomal RNAs is so large, similarity in two sequences always indicates *some* phylogenetic relationship. However, it is the *degree* of similarity in ribosomal RNA sequences between two organisms that indicates their relative evolutionary relatedness. From comparative sequence analyses, molecular genealogies can be constructed leading to phylogenetic trees that show the true evolutionary position of organisms relative to one another (see Figure 12.13).

Recall the structure of the ribosome (Figure 12.7). There are three ribosomal RNA molecules, which in prokaryotes have sizes of 5S, 16S, and 23S. The large bacterial rRNAs, 16S (Figure 12.7c) and 23S rRNA (approximately 1500 and 2900 nucleotides, respectively) contain several regions of highly conserved sequence useful for obtaining proper sequence alignments, yet contain sufficient sequence variability in other regions of the molecule to serve as excellent phylogenetic chronometers.

The 5S rRNA has also been used for phylogenetic measurements, but its small size (∼120 nucleotides) limits the information obtainable from this molecule. Because 16S RNA is more experimentally manageable than 23S RNA, it has been used extensively to develop the phylogeny of both prokaryotes and eukaryotes (using the 18S rRNA counterpart of prokaryotic 16S rRNA). Because 16S rRNA originates from the *small* (30S) *subunit* of the ribosome (Figure 12.7b), the acronym *SSU* (for *s*mall *s*ubunit) sequencing is synonymous with 16S or 18S sequencing. The database of rRNA sequences in the Ribosomal Database Project (RDP) now numbers over 10,000 and can be accessed on the World Wide Web (http://www.cme.msu.edu/RDP/). Use of 16S rRNA as a phylogenetic tool was pioneered in the early 1970s by Carl Woese at the University of Illinois, and the method is now widely used.

✓ 12.4 Concept Check

Comparisons of sequences of ribosomal RNA can be used to determine the evolutionary relationships between organisms. Phylogenetic trees based on ribosomal RNA have now been prepared for all the major prokaryotic and eukaryotic groups.

✓ Why is ribosomal RNA a good evolutionary chronometer?
✓ What is the RDP?

12.5

Ribosomal RNA Sequences and Cellular Evolution

The methods for obtaining ribosomal RNA sequences and generating phylogenetic trees are now quite routine and involve a combination of molecular biology and computer analyses. Newly generated sequences are compared with sequences in the RDP and/or with sequences obtained from other databases such as GenBank (USA) or EMBL (Germany). Then, using a treeing program, a phylogenetic tree is generated that summarizes the evolutionary information inherent in the sequences.

Sequence Methodology

We begin with the assumption that we are working with a pure culture of a microorganism although, as we will see, pure cultures are not necessary to do comparative SSU rRNA sequencing. Like life itself, the methods for ribosomal RNA sequencing have evolved through the years, and although there are several ways to obtain the actual sequence, most scientists today employ the polymerase chain reaction (PCR, ∞ Section 10.9) to amplify the genes encoding 16S ribosomal RNA from genomic DNA and then sequence the PCR product by standard dideoxy DNA sequencing (that is, *Sanger sequencing,* ∞ Nucleic acids box, Chapter 6) methods (Figure 12.8). This procedure is rapid and specific, and using synthetically produced oligonucleotide primers (available commercially at low cost) complementary to conserved sequences in SSU ribosomal RNAs as PCR primers, a tiny amount of DNA from a microbial culture can yield a huge amount of PCR product for sequencing reactions (Figure 12.8). Once the sequencing is done, either manually or by automated sequencers, the data are ready for computer analyses.

Generating Phylogenetic Trees from RNA Sequences

Several different algorithms for sequence analysis and phylogenetic tree formation are available for comparative ribosomal RNA sequencing. However, regardless of which program is used, raw sequence data must first be *aligned* using a sequence editor that aligns new sequences with previously aligned sequences. The aligned sequences are then imported into the actual treeing program and the comparative analyses done. Two widely used treeing programs are *distance* and *parsimony*. Using distance methods, sequences are aligned and an **evolutionary distance** (E_D) calculated by having the computer count every position in the data set in which there is a *difference* (Figure 12.9). From this a distance matrix can be constructed that shows the E_D between any two

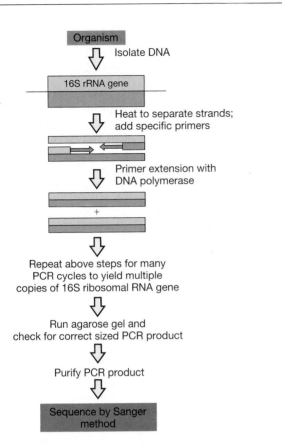

FIGURE 12.8   Ribosomal RNA sequencing of a pure culture of a microorganism using the polymerase chain reaction (PCR). The 16S rRNA gene is amplified and then sequenced by the Sanger method (⚭ Nucleic acids box, Chapter 6). The primers added are complementary to conserved sequences in one of the domains of 16S rRNA (see Figure 12.7c). A cloning step (see Figure 12.12 for an example of this) may also be used in this procedure to clone the DNA encoding the 16S rRNA following its PCR amplification.

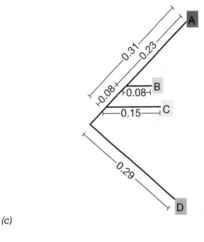

FIGURE 12.9   Preparing a phylogenetic distance-matrix tree from 16S ribosomal RNA sequences. For illustrative purposes, only short sequences are shown, but it should be considered that the sequences in (a) are representative of the entire 16S rRNA. The evolutionary distance (E_D) in (b) is calculated as the percentage of *nonidentical* sequences between the RNAs of any two organisms. The corrected E_D is a statistical correction necessary to account for either back mutations to the original genotype or additional forward mutations at the same site that could have occurred. The tree (c) is ultimately generated by computer analysis of the data to give the best fit. The total length of the branches separating any two organisms is proportional to the calculated evolutionary distance between them. In actual analyses a statistical process called *bootstrapping* is often used whereby the computer generates hundreds of versions of the tree to confirm that the final tree is indeed the best fit to the data.

sequences in the set. Once this is done, a correction is factored into the E_D that takes into account the possibility that multiple changes have occurred at any given site (Figure 12.9). For example, there is a low, but statistically significant, probability that the base that exists at any given site in two sequences could be the result of two mutational events, one that originally changed the sequence and one that returned it to the same base as before. These possibilities can be estimated and the correction factor does this. Finally, a phylogenetic tree is generated in which the length of the lines in the tree are proportional to evolutionary distances (Figure 12.9).

Parsimony, another popular method of phylogenetic analysis, creates evolutionary trees based on the assumption that only the *minimal amount* of sequence change necessary to diverge two lineages from a common ancestor actually occurred during their evolutionary divergence. This method still requires summing the number of sequence differences in a particular data set, but the algorithm handles the analysis in a somewhat different way than do distance programs. Nevertheless, the appearance of phylogenetic trees based on parsimony is similar to that of distance trees, although the branching order within a parsimonious tree can and often does differ from that in a distance tree generated from the exact same data set. Thus, no single phylogenetic tree should be considered the "final word" on the evolutionary relationships of the organisms in question but instead should be thought of as simply a "close approximation" to the true phylogeny of the group.

✓ 12.5 Concept Check

Comparative ribosomal RNA sequencing is now a routine procedure involving the amplification of the gene encoding 16S rRNA, sequencing it, and analyzing the sequence in reference to other sequences.

✓ What is an *evolutionary distance* (E_D)?
✓ How is the *polymerase chain reaction (PCR)* of use in comparative ribosomal RNA sequencing?

12.6

Signature Sequences, Phylogenetic Probes, and Molecular Microbial Ecology

In the next section we consider the "big picture" of microbial phylogeny. However, before we see how SSU ribosomal RNA sequencing has revolutionized our picture of cellular evolution, we discuss here some methods and applications of the technology itself. These include signature sequences and the design and use of ribosomal RNA probes in microbial ecology and diagnostic medicine.

Signature Sequences, Phylogenetic Probes, and FISH

Computer analyses of ribosomal RNA sequences have revealed so-called **signature sequences,** short oligonucleotides unique to certain groups of organisms. For example, signature sequences specific for each of the three domains of cellular life are known (Table 12.1).More restrictive signatures defining a specific group within a domain or, in some cases, a particular genus or even a single species are also known. The exclusivity of signature sequences are useful for several purposes; for example, signatures can be used for quickly placing newly isolated or misclassified organisms into their correct phylogenetic group. But perhaps the most extensive use of signature sequences is as a genetic blueprint for the construction of specific nucleic acid probes called *phylogenetic probes.*

Recall that we defined a nucleic acid probe as a strand of nucleic acid that can be labeled and used to hybridize to a complementary molecule from a mixture of nucleic acids (∞ Section 10.6). Probes can be general or specific. For example, universal ribosomal RNA probes are known that will bind to conserved sequences in the ribosomes of all organisms, regardless of domain (Figure 12.10*a,b*). Getting more specific, probes can be designed that will react only with cells of Bacteria because of unique signatures in their ribosomal RNA (Table 12.1). Likewise, Archaeal-specific and Eukaryl-specific (Figure 12.10*c*) probes have been synthesized that react only with the ribosomal RNA of Archaea or Eukarya, respectively. The binding of probes to cellular ribosomes can be seen microscopically when a fluorescent dye is attached (Figure 12.10). By treating cells with the appropriate reagents they become permeable and allow penetration of the oligonucleotide probe/dye mixture. Following hybridization of the probe directly

TABLE 12.1	Signature sequences from 16S or 18S rRNA defining the three domains of life				
			Occurrence among[c]		
Oligonucleotide signatures[a]	Approximate position[b]		Archaea	Bacteria	Eukarya
CACYYG	315		0	>95	0
CYAAYUNYG	510		0	>95	0
AAACUCAAA	910		3	100	0
AAACUUAAAG	910		100	0	100
NUUAAUUCG	960		0	>95	0
YUYAAUUG	960		100	<1	100
CAACCYYCR	1110		0	>95	0
UUCCCG	1380		0	>95	0
UCCCUG	1380		>95	0	100
CUCCUUG	1390		>95	0	0
UACACACCG	1400		0	>99	100
CACACACCG	1400		100	0	0

a Y, Any pyrimidine; R, any purine; N, any purine or pyrimidine.
b Refer to Figure 12.7*c* for numbering scheme of 16S rRNA.
c Occurrence refers to percentage of organisms examined in any domain that contain that sequence.

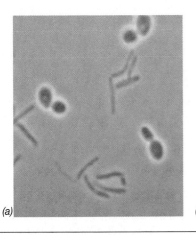
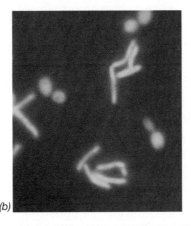

FIGURE 12.10 Fluorescently labeled ribosomal RNA probes. (a) Phase contrast photomicrograph of *Bacillus megaterium* (member of Bacteria) and the yeast *Saccharomyces cerevisiae* (Eukarya) (no probes present). (b) Same field, cells stained with a yellow-green universal rRNA probe. (c) Same field, cells stained with a red eukaryal probe (only cells of *S. cerevisiae* react). Cells of *B. megaterium* are about 1.5 μm in diameter and cells of *S. cerevisiae* are about 6 μm in diameter.

to the ribosomal RNA in the cell's ribosomes, the cells become uniformly fluorescent and can be observed under a fluorescent microscope (Figure 12.10, see also Figures 12.11 and 12.14). This technique, because it can be applied directly to cells in culture or in a natural environment, has been nicknamed *FISH,* for *fluorescent in-situ hybridization.*

Fluorescent phylogenetic probes and FISH technology are in widespread use in microbial ecology and clinical diagnostics. In ecology, FISH allows for the microscopic identification and tracking of organisms through an environment, while in clinical medicine, this method has been used for the rapid identification of specific pathogens from patient specimens (note that such a method does *not* require that an organism first be grown in laboratory culture, thus accelerating the identification process by several hours to several days). The use of multiple phylogenetic probes on a single sample, each complementary to a specific SSU signature and tagged with a different-colored fluorescing dye, has given microbial ecologists a new microscopic tool for the phylogenetic analysis of microorganisms in a microbial community (Figure 12.11). Such methods can be used to ask specific questions about the microbial composition of complex mixtures and the role a specific organism or related group of organisms plays in particular ecological processes in nature (Figure 12.11). We revisit the use of phylogenetic probes in Chapter 16.

FIGURE 12.11 Use of phylogenetic stains to make nitrifying bacteria visible in a granule of activated sewage sludge. Bottom, phase photomicrograph. Top, color photomicrograph of the same field after applying phylogenetic stains. The probe that fluoresces red is specific for a signature sequence (see Table 12.1) in the 16S rRNA of ammonia-oxidizing bacteria, while the probe that fluoresces green is specific for a sequence in nitrite-oxidizing bacteria. Both ammonia-oxidizing and nitrite-oxidizing bacteria are phylogenetically closely-related members of the Bacteria (∞ Sections 13.2 and 15.13) and carry out an interdependent series of chemolithotrophic reactions (∞ Section 16.16).

Molecular Community Analysis

Besides the use of fluorescent probes (Figures 12.10, 12.11), ribosomal RNA sequencing technology has also been used for the direct phylogenetic analysis of microbial communities. Imagine the scope of the problem here. As we will see in Chapter 16, a microbial community is a mixture of several different microorganisms that interact to effect various chemical and physical transformations on an environment. How can such a complex community be analyzed in phylogenetic terms? One way, of course, is by isolating different or-

ganisms and determining their phylogenetic status. But what if we fail to culture every representative, or even *most* representatives, of the community? Can we still determine the phylogenetic components of the ecosystem? The answer to this is yes, because if DNA extracted from the entire population in a microbial community is PCR amplified using oligonucleotide primers specific for Bacteria, Archaea, or Eukarya (or any combination of primers specific for any particular group that the experimenter wishes to study), a mixture of PCR products characteristic of the different ribosomal RNAs of the community will be obtained. Although this mixture cannot be sequenced directly, if it is first sorted out through a molecular cloning step (⟳ Section 10.1) each clone, which now contains a single SSU ribosomal RNA gene, can be PCR amplified and sequenced; provided that enough clones are sequenced, the resulting collection of sequences will contain at least one from each phylogenetically distinct component of the microbial community (Figure 12.12a).

Application of these methods to actual microbial communities has yielded interesting and rather surprising results. In virtually all cases, phylogenetic community analyses of the type described here have NOT shown the communities to be composed of organisms that have been previously cultured in other studies, but just the opposite; most natural microbial communities contain a host of phylogenetically distinct organisms whose ribosomal RNA sequences do not match any of those in the databases (Figure 12.12b). In fact, in many cases where these methods have been modified to allow for both *qualitative* (that is, "who's out there?") as well as *quantitative* (that is, "how many of each type are out there?") analyses, the result has been that the most numerically predominant members of the microbial community are ones that have so far defied laboratory culture! This problem suggests that our knowledge of microbial diversity is still in its infancy, and we revisit this problem in Chapter 16.

✓ 12.6 Concept Check

Signature sequences, short oligonucleotides found within a ribosomal RNA molecule, can be highly diagnostic of a particular organism or group of related organisms. Signature sequences can be used to generate specific phylogenetic probes for use in either direct microscopic observation of a microbial habitat or for isolating ribosomal RNA genes from organisms in a microbial community for the purpose of phylogenetically characterizing the habitat.

- ✓ Using the sequence data in Table 12.1, design a specific phylogenetic probe that would allow you to distinguish a cell of Archaea from a cell of Bacteria.
- ✓ Describe why the results of phylogenetic community analyses of microbial habitats have been rather surprising.
- ✓ How can oligonucleotide probes be made visible under the microscope?

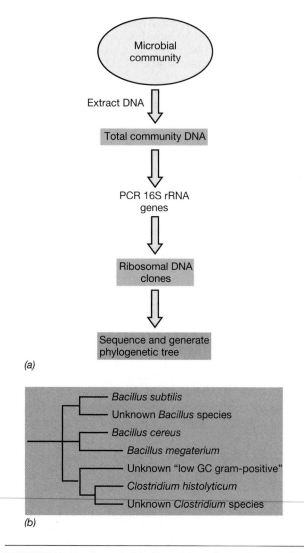

(a)

(b)

FIGURE 12.12 Steps in biodiversity analysis of a microbial community using phylogenetic probes. (a) Ribosomal DNA clones are obtained by PCR methods (see Figure 12.8). (b) Hypothetical example of a phylogenetic tree that might be generated if primers used were specific for the "low GC gram-positive" lineage of the phylogenetic domain Bacteria, which includes species of *Bacillus* and *Clostridium* (⟳ Section 13.19). Note that many of the sequences do not match any previously sequenced species and therefore represent previously unrecognized (unknown) species.

12.7

Microbial Phylogeny as Revealed by Ribosomal RNA Sequencing

We now consider what SSU ribosomal RNA sequencing has taught us about microbial evolution. Molecular sequencing has revealed a previously unsuspected phylogeny of living organisms, a phylogeny quite different from previous ones based primarily on phenotypic relationships. Biologists previously grouped the living world into five kingdoms, only one of which was prokaryotic, based on structural similarities between organisms. Molecular phylogeny, on the other hand, has revealed that the five kingdoms do not represent five major evolutionary lines. Instead, cellular life on Earth has evolved along *three* major lineages, two of which have remained exclusively microbial and are composed only of prokaryotic cells. The third line constitutes the eukaryotic lineage (Figure 12.13). The two prokaryotic lines are the *Bacteria* and the *Archaea*. The eukaryotic line is called the *Eukarya* (Figure 12.13). These terms define the three **domains** of life, the domain being the highest of biological taxons. Thus plants, animals, fungi, and protists are all kingdoms within the domain Eukarya.

The Universal Tree of Life

The universal phylogenetic tree (Figure 12.13) is the road map of life. It describes the evolutionary history of all organisms relative to all other organisms and clearly shows the three primary groupings of organisms, the phylogenetic domains. The *root* of the universal tree represents a point in evolutionary history when all extant life on Earth shared a common ancestor, the so called *Universal Ancestor*. What kind of creature was this Universal Ancestor?

Microbial genome sequencing projects (👓 Sections 9.11 and 9.12) have yielded clues about the nature of the Universal Ancestor. Complete genomic sequences have revealed the amazing array of genes present in microorganisms and have confirmed the concept of the Archaea—the latter contain hundreds of protein-encoding genes that have no counterparts in Bacteria or Eukarya. But equally important has been the finding from the genome projects that many genes are *shared* among species of Bacteria, Archaea, and Eukarya. How can these seemingly conflicting findings be reconciled?

One explanation of these results is to hypothesize that early in the history of life, before the primary lines of descent had formed, *lateral (horizontal) gene transfer* was extensive; genes encoding proteins that conferred exceptional fitness (for example, genes for core cellular functions like transcription and translation) were par-

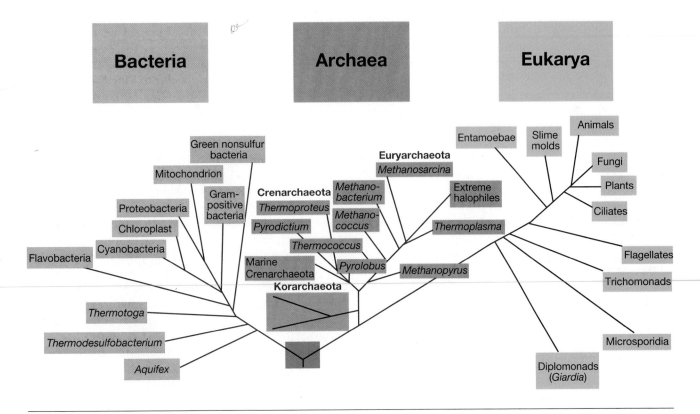

FIGURE 12.13 Universal phylogenetic tree as determined from comparative ribosomal RNA sequencing. The data support the separation of three domains, two of which (Bacteria and Archaea) contain only prokaryotic representatives. The location highlighted in red is the hypothetical root of the tree, which represents the position of the universal ancestor of all cells. See text for discussion of the Korarchaeota.

ticularly promiscuously transferred among a population of primitive organisms derived from a common ancestral cell. If this hypothesis is true, it would explain why, as genome analyses have shown, *all cells,* regardless of domain, share many key genes in common. But what about the genetic *differences* observed from complete genome sequences? Continuing with the lateral gene transfer concept, it is further hypothesized that with time, partial barriers to unrestricted gene flow evolved, perhaps from the selective colonization of habitats on the basis of the metabolic potential of different organisms to exploit them or as the result of structural barriers that in some way prevented free genetic exchange; as a result, the previously genetically promiscuous population slowly began to sort out into the primary lines of evolutionary descent (Figure 12.13). As each lineage continued to evolve, certain unique biological traits became genetically fixed within each group such that what we see today, after some 3.5 billion years of microbial evolution, are three domains of cellular life that on the one hand share many of life's key secrets while on the other, display a distinctive evolutionary history of their own.

Note how on the universal phylogenetic tree the Archaea branch off at a point closest to the root (Figure 12.13). This leads to the conclusion that Archaea remain, even today, the least evolved of organisms in the three domains, whereas the Eukarya, now the farthest away from the universal ancestor, are the most evolved (Figure 12.13). This suggests that the eukaryotic lineage evolved some important characteristic or set of characteristics that allowed for rapid, albeit late, evolutionary diversity. Perhaps this evolutionary event was something as simple as a larger cell size; this would have facilitated the large amounts of DNA needed for the development of metazoan organisms and produced a cell sufficiently large to accept prokaryotic endosymbionts. Placement of the Archaea nearest the universal ancestor is also supported by the fact that many Archaea inhabit extreme environments, such as those of high temperature, low pH, high salinity, and so on (∽ Chapter 14); these environmental extremes, especially that of high temperature, reflect the environmental conditions under which life originated (see Section 12.1). Thus, members of the Archaea may well be evolutionary relics of Earth's earliest life forms.

Finally, a point that deserves emphasis here is that *none* of the organisms living today are *primitive.* All extant life forms are *modern* organisms, well adapted to, and successful in, their ecological niches. Certain of these organisms may indeed be phenotypically similar to primitive organisms and may represent stems of the evolutionary tree that have changed little for millions if not billions of years (∽ Chapter 14); in this respect they are evolutionarily *related* to primitive organisms, but they are not themselves primitive.

Organelles

The universal tree (Figure 12.13) also tells us about other evolutionary events. For example, it is clear that mitochondria and chloroplasts arose from endosymbiotic Bacteria that established stable relationships, perhaps several different times, with cells from the nuclear line of descent (see also Figure 12.6). Molecular sequencing suggests that mitochondria arose from a small group of organisms that includes the modern prokaryotes *Agrobacterium, Rhizobium,* and the rickettsias. The latter organisms belong to a larger group of Bacteria called the purple bacteria (organisms such as *Rhodopseudomonas* and *Rhodobacter*). (This whole group is called the *Proteobacteria.*) It is interesting to note that the mitochondrion, itself an intracellular symbiont, is specifically related to organisms (*Agrobacterium, Rhizobium,* and the rickettsias) that are capable of an intracellular existence (∽ Sections 13.12, 16.24, and 16.25). The other eukaryotic endosymbiont, the chloroplast, originated from Bacteria as well. The universal tree also shows that the chloroplast and cyanobacteria *shared* a common ancestor. Thus, SSU ribosomal RNA sequencing supports the endosymbiotic theory (Figure 12.6) and has independently shown that the evolutionary roots of the mitochondrion and the chloroplast are within the domain Bacteria.

Bacteria

Among cultured organisms 14 major bacterial divisions (referred to as *kingdoms*) have been discovered and several key ones are shown in the universal tree in Figure 12.13. However, the *real number* of kingdoms of Bacteria is likely much higher than this, perhaps 50 or more, based on molecular sequencing analyses of natural microbial communities (see Figure 12.12). This once again reminds us that the bacteria we now have in laboratory culture represent only the "tip of the iceberg," so to speak, of total bacterial diversity.

Some of the lineages in the domain Bacteria are ones previously distinguished by some phenotypic property like morphology or physiology; the spirochetes and the cyanobacteria, respectively, are good examples of this. But most of the kingdoms consist of species that, although specifically related from a phylogenetic standpoint, lack strong phenotypic cohesiveness. The Proteobacteria are a good example here, because the mixture of physiologies present within the group nearly runs the gamut of known microbial physiology (∽ Chapter 15).

The branch of the Bacteria tree that emerges closest to the point of divergence from the Archaea/Eukarya contains members of the genus *Aquifex* (Figure 12.13), hyperthermophilic, chemolithotrophic prokaryotes that oxidize H_2 as a means of energy conservation (∽ Sections 13.34 and 15.10). Curiously, this is consistent with the hypothesis that early Earth was very hot and that chemolithotrophy/autotrophy may have been the lifestyle of early organisms (see Section 12.2). The complete genome sequence of a species of *Aquifex, A. aeolicus,* is known, and interestingly the biology of this organism is directed by a genome (1.55 megabases) only *one-third* the size of that of *E. coli* (⊸ Sections 9.11 and 9.12). Thus, the early position of *Aquifex* on the phylogenetic tree of Bacteria is exactly where we would expect it to be if *Aquifex* is a modern relic of early life forms. We consider the properties of major groups of Bacteria in Chapter 13.

Archaea

From a phylogenetic perspective, the domain Archaea consists of three kingdoms, the Crenarchaeota, Euryarchaeota, and Korarchaeota (Figure 12.13). Branching very close to the root of the universal tree are hyperthermophilic members of the Crenarchaeota like *Thermoproteus, Pyrolobus,* and *Pyrodictium* (Figure 12.13). They are followed by two groups of Euryarchaeota, the methane-producing (methanogenic) prokaryotes *Methanococcus–Methanobacterium* and *Methanosarcina* and the extreme halophiles *Halobacterium* and *Halococcus; Thermoplasma,* an acidophilic, thermophilic cell wall–less bacterium is loosely related to this latter group (Figure 12.13). Another methanogen branches very close to the base of the archaeal tree. This is *Methanopyrus,* a methanogen capable of growth in temperatures up to 110°C! Note that this hyperthermophilic organism is also a chemolithotroph, growing at the expense of H_2 oxidation (⊸ Sections 12.2 and 14.5). *Methanopyrus* is another good example of the type of organism that could have inhabited early Earth.

There are some branches on the Crenarchaeota lineage (Figure 12.13) that are known from community sampling of ribosomal genes from the environment (see Section 12.6) but have not been obtained from actual cultured isolates. Interestingly, however, these sequences come from environmental DNA sampling in the open oceans, including Antarctic marine waters where temperatures are far colder than in the hot-spring or deep sea hydrothermal–vent habitats of known Crenarchaeota, and also from soil and lake water. That crenarchaeotal ribosomal RNA sequences can be readily obtained from these environments tells us that cold-adapted relatives of thermophilic Crenarchaeota exist and that they may be ecologically important in many different environments. But until we know more about the physiology of organisms themselves, we can only guess at what they are doing there. We discuss cold adapted Crenarchaeota in more detail in Section 14.7.

Like cold-adapted Crenarchaeota, an entire kingdom of Archaea, the Korarchaeota, have been identified from community sampling of 16S rRNA, in this case from a specific hot spring (Obsidian Pool) in Yellowstone National Park (Wyoming, USA) (Figure 12.13).

Although pure cultures of a korarchaeotan are not yet available, stable mixed cultures containing representatives of this group (identifiable through phylogenetic staining, see Figure 12.14) can be grown in the laboratory. From the media and incubation conditions supporting these cultures we can deduce that Korarchaeota are hyperthermophiles and may have metabolic properties similar to those of hyperthermophilic Crenarchaeota (⊸ Section 14.12). We consider all kingdoms of Archaea in more detail in Chapter 14.

Eukarya

Phylogenetic trees of species in the domain Eukarya are generated from sequence data of 18S rRNA, the functional equivalent of 16S rRNA in eukaryotic cytoplasmic ribosomes. Inspection of the Eukarya tree suggests that evolution in this lineage was not a continuous process but instead occurred in major epochs. Early eukaryotes were probably similar to present-day microsporidia and diplomonads. These organisms are all obligate parasites that live in association with representatives of various groups of eukaryotes, from microorganisms to humans [for example, the pathogen *Giardia* (⊸ Section 24.9) is a member of the diplomonad group]. And interestingly, although they contain a membrane-enclosed nucleus, microsporidia and diplomonads lack mitochondria;

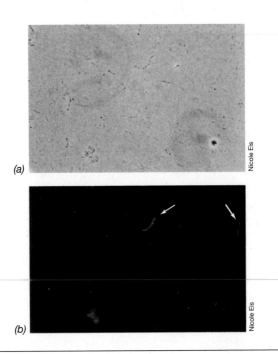

FIGURE 12.14 Identifying Korarchaeota with phylogenetic stains. (a) Phase-contrast micrograph of an enrichment culture containing korarchaeotan cells. (b) Fluorescence photomicrograph of the same field as in (a) but treated with a phylogenetic stain designed from a signature sequence (see Section 12.6) from the 16S rRNA of the Korarchaeota. The two red cells in this field are thus korarchaeotans.

in this connection they resemble the type of cell that might have first accepted stable endosymbionts (see Figure 12.6 and ∞ Sections 17.1 and 17.2).

Later branches on the Eukarya tree go from microbial forms to multicellular plants and animals (the metazoa), culminating with the largest and structurally most complex of the Eukarya, the plants and animals (Figure 12.13). When the fossil record is compared with the phylogenetic tree of Eukarya, rapid evolutionary radiation can be dated to about 1.5 billion years ago. Geochemical evidence suggests that this is the period in Earth's history in which significant oxygen levels had accumulated in the atmosphere (Figure 12.5). It is thus likely that the onset of oxic conditions and subsequent development of an ozone shield (which would have greatly expanded the number of surface habitats available for colonization) also triggered the rapid diversification of Eukarya. We consider the major groups of microbial Eukarya and their basic biology in Chapter 17.

✓ 12.7 Concept Check

The major conclusion from phylogenies derived from comparisons of ribosomal RNA sequences is that life on Earth evolved along three major lines, called domains, all derived from a common ancestor. Two of the lines, Bacteria and Archaea, remained prokaryotic, whereas the third line, Eukarya, evolved into the modern eukaryotic cell.

- ✓ Which domains contain only prokaryotic cells?
- ✓ How does the universal tree support the idea that early Earth was very hot?
- ✓ How does the universal tree support the hypothesis presented in Figure 12.6?

12.8

Characteristics of the Primary Domains

Although the primary domains—Bacteria, Archaea, and Eukarya—are defined on the basis of comparative ribosomal RNA sequencing, other studies have shown that each domain is characterized by a number of *phenotypic* properties. Some of these characteristics are unique to one domain, whereas others are found in two out of the three domains. We present here an overview of major phenotypic traits of phylogenetic value.

Cell Walls

Virtually all Bacteria have cell walls containing *peptidoglycan* (∞ Section 3.7). The only known exceptions are members of the *Planctomyces–Pirella* group (∞ Section 13.27), whose cells are composed of protein, and the *Chlamydia–Mycoplasma* groups, which lack cell walls altogether (∞ Sections 13.20 and 13.26). Peptidoglycan

can thus be considered a "signature" molecule for species of Bacteria (when assaying for peptidoglycan, it is *muramic acid* that is actually detected because this is what is unique to peptidoglycan).

Eukarya and Archaea lack peptidoglycan. In eukaryotes, if cell walls are present, they are usually made of cellulose or chitin (∞ Chapter 17). In Archaea, as previously discussed (∞ Section 3.7), various cell wall types exist, from the peptidoglycan analog pseudopeptidoglycan to walls made of polysaccharide, protein, or glycoprotein. Thus, great diversity is observed in the chemistry of microbial cell walls, but it is the presence or absence of peptidoglycan that is most useful phylogenetically.

Lipids

The chemical nature of membrane lipids is perhaps the most useful of all nongenetic criteria for differentiating Archaea from Bacteria. Bacteria and Eukarya synthesize membrane lipids with a backbone consisting of fatty acids hooked in *ester* linkage to a molecule of glycerol (∞ Figure 12.15). Although the nature of the fatty acid can be highly variable, the key point is that the chemical linkage to glycerol is an **ester link.** By contrast, archaeal lipids consist of **ether-linked** molecules (Figure 12.15). In ester-linked lipids, the fatty acids are usually straight-chain (linear) molecules, whereas in Archaea, long-chain, branched hydrocarbons, either of the phytanyl or biphytanyl type, are present in place of fatty acids and are bonded by ether linkage to glycerol molecules (Figure 12.15).

In addition to the differences in chemical *linkage* between the hydrocarbon and alcohol portions of archaeal and bacterial and eukaryotic lipids, the chirality of the glycerol moiety differs in the lipid of representatives of the three domains. The central carbon atom of the glycerol

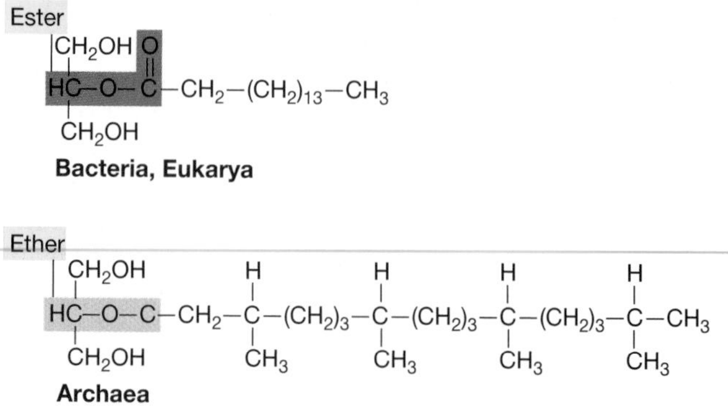

FIGURE 12.15 Lipids in Bacteria, Eukarya, and Archaea. In Bacteria and Eukarya, lipids contain fatty acids (palmitic acid is shown) bonded by *ester* linkages to glycerol. In Archaea, the side chains are branched hydrocarbon (phytanyl, C_{20}, is shown) bonded by *ether* linkages to glycerol. Phytanyl is synthesized from isoprene (∞ Figure 3.20c).

molecule is stereoisomerically of the R form in Bacteria and Eukarya and of the L form in Archaea (see Section 2.6 for more discussion about stereoisomerism and life).

RNA Polymerase

Transcription is carried out by DNA-dependent RNA polymerases in all organisms; DNA is the template, and RNA is the product (∞ Section 6.7). Cells of Bacteria contain a single type of RNA polymerase of rather simple quaternary structure. This is the classic RNA polymerase containing *four* polypeptides, α, β, β', and σ, combined in a ratio of 2:1:1:1, respectively, in the active polymerase (Figure 12.16) (∞ also Section 6.7).

Archaeal RNA polymerases are structurally more complex than those of Bacteria. The RNA polymerases of methanogens and halophiles contain *eight* polypeptides, five large ones and three smaller ones (Figure 12.16). Hyperthermophilic Archaea contain an even more complex RNA polymerase consisting of at least *10* distinct polypeptides (Figure 12.16). The major RNA polymerase of eukaryotes (there are three, only one of which makes messenger RNA) contains 10–12 polypeptides, and the relative sizes of the peptides coincide most closely with those from the hyperthermophilic Archaea (Figure 12.16). Two other RNA polymerases are known in eukaryotes, and each specializes in transcribing certain regions of the genome, one for rRNA and one for transfer RNA (tRNA). In terms of phylogenetic signatures, the $(\alpha_2\beta\beta'\sigma)$ polymerase is highly diagnostic of Bacteria whereas the remaining polymerases are too complex to be phylogenetically definitive.

Features of Protein Synthesis

Because of differences in ribosomal RNA sequences and several protein synthesis factors, it is not surprising that certain aspects of the protein-synthesizing machinery differ in representatives of the three domains. Although ribosomes of Archaea and Bacteria are the same size (70S, as compared with the 80S ribosomes in the cytoplasm of eukaryotes), several steps in archaeal protein synthesis more strongly resemble those in eukaryotes than in Bacteria. Recall that translation always begins at a unique codon, the so-called *start codon*. In Bacteria this start codon (AUG) calls for the incorporation of an initiator tRNA containing a modified methionine residue, *formyl*methionine (∞ Section 6.12). By contrast, in eukaryotes and in Archaea, the initiator tRNA carries an *unmodified* methionine.

The exotoxin produced by *Corynebacterium diphtheriae* is a potent inhibitor of eukaryotic protein synthesis because it ADP-ribosylates (adds ADP to) an elongation factor required to translocate the ribosome along

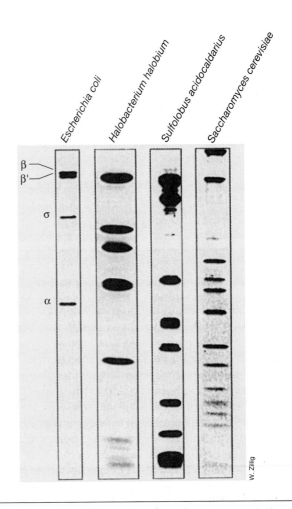

FIGURE 12.16   RNA polymerases from representatives of the three domains: *Escherichia coli* (Bacteria), *Halobacterium halobium* (Euryarchaeota, Archaea), *Sulfolobus acidocaldarius* (Crenarchaeota, Archaea), and *Saccharomyces cerevisiae* (Eukarya). The purified components of the RNA polymerases have been separated by electrophoresis on a polyacrylamide gel. The largest polypeptide subunits are on the top, and the smallest subunits are on the bottom. Only members of the Bacteria contain the simple (four-polypeptide) RNA polymerase.

the mRNA; the modified elongation factor is inactive (∞ Section 19.8). Diphtheria toxin also inhibits protein synthesis in Archaea but does not affect this process in Bacteria.

Most antibiotics that specifically affect protein synthesis in Bacteria do not affect archaeal or eukaryal protein synthesis. The sensitivity of representatives of the three domains to various protein synthesis inhibitors is shown in Table 12.2 (∞ also Section 6.12), where various antibiotics are grouped according to their modes of action in blocking protein synthesis in various domains of organisms.

TABLE 12.2   Sensitivity of representatives of the three domains to varius protein synthesis inhibitors[a]

Antibiotics	Mode of action	Archaea		Bacteria	Eukarya
		Euryarchaeota *Methanobacterium thermoautotrophicum*	Crenarchaeota *Sulfolobus acidocaldarius*	*Escherichia coli*	*Saccharomyces cerevisiae*
Fusidic acid, sparsomycin	Inhibits elongation steps	+	−	+	+
Anisomycin, narciclasine	Inhibits peptidyl transfer	+	−	−	+
Cycloheximide	Blocks initiation	−	−	−	+
Erythromycin, streptomycin, chloramphenicol	Increases error frequencies and other effects	−	−	+	−
Virginiamycin, pulvomycin	Inhibits elongation steps	+	−	+	−
Neomycin, puromycin	Causes premature termination	+	+	+	+
Rifamycin	Inhibits β subunit of RNA polymerase	−	−	+	−

a A + indicates that protein synthesis (and growth) is inhibited.

Functional similarities in the translational machinery of Archaea and Eukarya have been revealed from measurements of *in vitro* protein synthesis by *hybrid ribosomes* made up of ribosomal components from organisms of different domains. Specifically, hybrid ribosomes composed of the large subunit (50S) of ribosomes from *Sulfolobus* (Archaea) and the small subunit (40S) of ribosomes from yeast (Eukarya) have been shown to be functional *in vitro* and to actually produce polypeptides. However, if large ribosomal subunits from *Escherichia coli* (Bacteria) pair with small ribosomal subunits from yeast, no protein-synthesizing activity occurs. These results suggest that ribosomal proteins from cells of Archaea and Eukarya are more similar to each other than they are to ribosomal proteins from Bacteria and further support the position of domains relative to one another in the universal tree (Figure 12.13).

Other Features Defining the Domains

From microbiological research a number of other features, physiological and otherwise, have been discovered that can be used to differentiate organisms at the domain level; a summary of these is listed in Table 12.3. When examining this table it should be understood that not all features are universally present in a domain. For example, photosynthesis is characteristic of only some representatives of the Bacteria and the Eukarya (and in the latter only because of photosynthetic endosymbiotic Bacteria). By contrast, other distinguishing features, such as the presence of peptidoglycan in the cell walls of Bacteria, may be universal or nearly so.

✓ 12.8 Concept Check

Although the three domains of living organisms were originally defined by ribosomal RNA sequencing, subsequent studies have shown that they differ in many other ways, too. Particularly, the Bacteria and Archaea differ extensively in cell wall and lipid chemistry and in features of transcription and protein synthesis.

✓ How do the lipids from Bacteria and Eukarya differ from those of the Archaea?
✓ Organisms from which domain produce RNA polymerases that most closely resemble those of the Eukarya? What phylogenetic evidence supports this?
✓ What do the experiments with hybrid ribosomes suggest?

12.9

Conventional and Molecular Taxonomy

Taxonomy is the science of classification and consists of two major subdisciplines, *identification* and *nomenclature*. It is important to distinguish between bacterial *taxonomy* and the main topic of this chapter up to this point, bacterial *phylogeny*, for the terms really mean different things. Bacterial taxonomy has traditionally relied on *phenotypic* analyses—what does an organism look like, what does it do, what enzymes does it contain—as the basis of classification. By contrast, because bacteria are so small and contain relatively few structural clues to their evolutionary roots, the phylogeny of prokaryotes has only emerged from the *genotypic* analyses discussed in the previous sections. Nevertheless, phenotypic analyses have traditionally played an important role in bacterial identification and classification, especially in applied situations where identification may be an end in itself, for example, in clinical diagnostic microbiology. In this section we discuss conventional bacterial taxonomy along with some molecular methods that have been found useful in the classification of bacteria. These methods are often used in combination with ribosomal RNA sequencing to char-

TABLE 12.3	Summary of major differentiating features among Bacteria, Archaea, and Eukarya[a]		
Characteristic	Bacteria	Archaea	Eukarya
Prokaryotic cell structure	Yes	Yes	No
DNA present in covalently closed and circular form	Yes	Yes	No
Histone proteins present	No	Yes	Yes
Membrane-enclosed nucleus	Absent	Absent	Present
Cell wall	Muramic acid present	Muramic acid absent	Muramic acid absent
Membrane lipids	Ester-linked	Ether-linked	Ester-linked
Ribosomes	70S	70S	80S
Initiator tRNA	Formylmethionine	Methionine	Methionine
Introns in most genes	No	No	Yes
Operons	Yes	Yes	No
Capping and poly-A tailing of mRNA	No	No	Yes
Plasmids	Yes	Yes	Rare
Ribosome sensitivity to diphtheria toxin	No	Yes	Yes
RNA polymerases (see Figure 12.16)	One (4 subunits)	Several (8–12 subunits each)	Three (12–14 subunits each)
Transcription factors required (⟳ Section 6.8)	No	Yes	Yes
Promoter structure (⟳ Sections 6.7 and 6.8)	−10 and −35 sequences (Pribnow box)	TATA box	TATA box
Sensitivity to chloramphenicol, streptomycin, and kanamycin	Yes	No	No
Methanogenesis	No	Yes	No
Reduction of S^0 to H_2S or Fe^{3+} to Fe^{2+}	Yes	Yes	No
Nitrification	Yes	No	No
Denitrification	Yes	Yes	No
Nitrogen fixation	Yes	Yes	No
Chlorophyll-based photosynthesis	Yes	No	Yes (in chloroplasts)
Chemolithotrophy (Fe, S, H_2)	Yes	Yes	No
Gas vesicles	Yes	Yes	No
Synthesis of carbon storage granules composed of poly-β-hydroxyalkanoates	Yes	Yes	No
Growth above 80°C	Yes	Yes	No

[a] Note that for many features only particular representatives within a domain show the property

acterize newly isolated bacteria. The terms *chemotaxonomy* or *molecular taxonomy* are used to describe this collection of taxonomic methods, since in many cases the methods involve the chemical analysis of some cell component such as a macromolecule.

Conventional Taxonomy and GC Ratios

In conventional bacterial taxonomy, several characteristics are measured, and these traits are then used to group the organisms. Characteristics of taxonomic value that are widely used include morphology; Gram reaction; nutritional classification (phototroph, chemoorganotroph, chemolithotroph); cell wall chemistry; lipid chemistry; presence of cell inclusions and storage products; capsule chemistry, pigments, nutritional requirements; ability to use various carbon, nitrogen, and sulfur sources; fermentation products; gaseous needs; temperature and pH requirements (and tolerances); antibi-

otic sensitivity; pathogenicity; symbiotic relationships; immunological characteristics; and habitat.

To identify an organism in this way a series of criteria is used that proceeds from general to specific. An example of this is shown in Figure 12.17 (see also Table 12.4). Using a dichotomous key method and testing for several individual characteristics, the list of possible organisms is narrowed until a positive identification is made (Figure 12.17).

The determination of the guanine plus cytosine content of an organism's DNA, usually referred to as the *GC ratio*, is frequently a part of conventional bacterial taxonomy. DNA base ratios are expressed as:

$$\frac{G + C}{A + T + G + C} \times 100\%$$

GC ratios vary over a wide range, with values as low as 20% and high as nearly 80% known among prokaryotes, a somewhat broader range than for eukaryotes

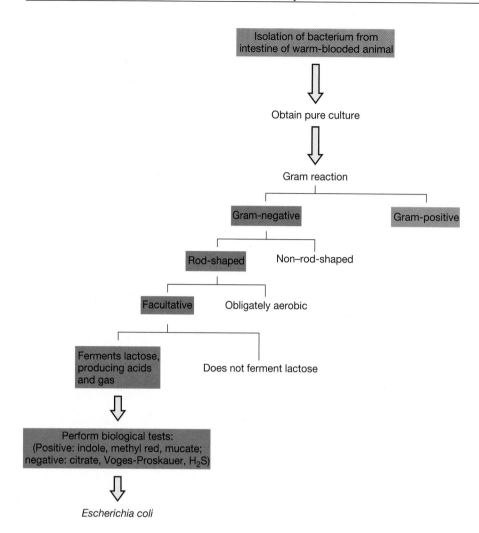

Isolation of bacterium from
intestine of warm-blooded animal

Obtain pure culture

Gram reaction

Gram-negative   Gram-positive

Rod-shaped   Non–rod-shaped

Facultative   Obligately aerobic

Ferments lactose,
producing acids
and gas   Does not ferment lactose

Perform biological tests:
(Positive: indole, methyl red, mucate;
negative: citrate, Voges-Proskauer, H$_2$S)

Escherichia coli

FIGURE 12.17 Example of methods that would be used for identification of a newly isolated enteric bacterium, using conventional microbiological methods (the example given shows the procedures that would be used for identifying *Escherichia coli*). Note that most of the analyses here require that the organisms be grown in pure culture and that solely phenotypic criteria be used in the identification. For a description of biochemical tests, ⬡⬡ Chapter 21.

(Figure 12.18). Base compositions of DNA have been determined for a wide variety of organisms, and several generalizations can be made. (1) Organisms with highly similar phenotypes often, but not always, possess similar DNA base ratios. (2) If two organisms thought to be closely related by phenotypic criteria are found to have widely different DNA base ratios, closer examination invariably indicates that the organisms are not as closely related as originally thought. (3) Two organisms can have identical GC ratios and yet be quite unrelated (both taxonomically and phylogenetically) because a variety of base *sequences* is possible with DNA of a given base composition. Thus, although occasionally quite helpful in taxonomic studies, GC ratios are usually of minimal value in the overall taxonomic characterization of an organism.

Molecular Taxonomy: DNA:DNA Hybridization

A GC base ratio describes the percentage of each nucleotide present in an organism's DNA but gives absolutely no information on the *sequence* of those nucleotides. Sequences are critical, of course, because if

two organisms have many of the same nucleotide sequences in their DNAs, they likely contain many highly similar (if not identical) genes. Such DNAs would thus be expected to *hybridize* to one another in proportion to the similarities in their gene sequences. Genomic hybridization measures the degree of sequence

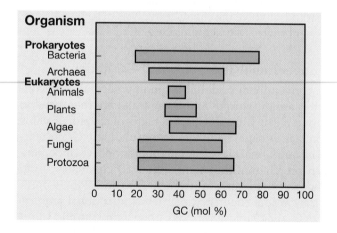

FIGURE 12.18 Ranges of DNA base composition of various organisms.

similarity and is thus useful for differentiating very closely related organisms.

We discussed the theory and methodology of nucleic acid hybridization in Chapter 6 (∞ the box, Working with Nucleic Acids: The Tools). In an actual hybridization experiment, DNA isolated from one organism is made radioactive with ^{32}P or ^{3}H, sheared to a relatively small size, heated to denature, and mixed with an excess of unlabeled DNA prepared in the same way from a second organism (Figure 12.19). The DNA mixture is then cooled to allow it to reanneal and double-stranded duplex DNA is separated from any remaining unhybridized DNA. Following this, the amount of radioactivity in the hybridized DNA is determined and compared with the control, which is taken as 100% (Figure 12.19). In addition to radioactivity, a variety of nonradioactive DNA labels have been introduced in recent years and these have the advantage that the hybridization experiment generates no radioactive wastes.

There is no fixed convention as to how much hybridization between two DNAs is necessary to assign two organisms to the same taxonomic rank. However,

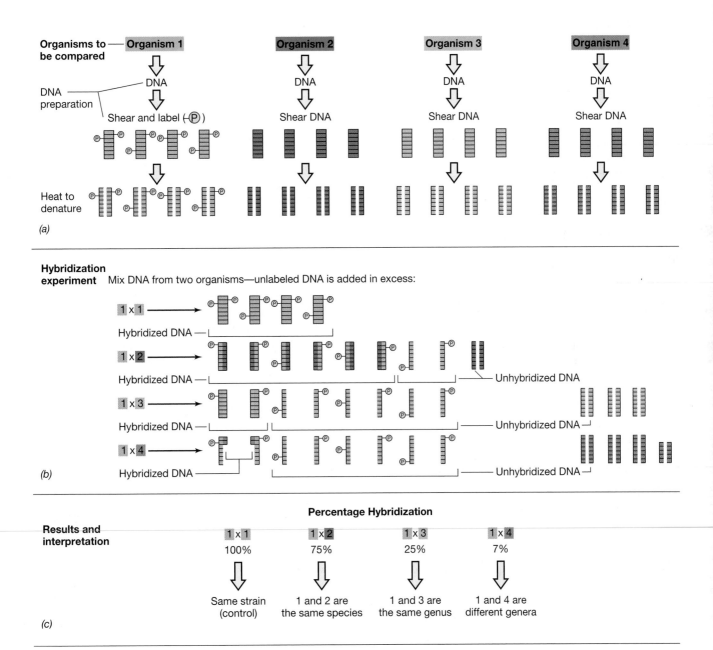

FIGURE 12.19 Genomic hybridization as a taxonomic tool. (a) DNA is isolated from test organisms. One of the DNAs is labeled (shown here as radioactive phosphate in the DNA of Organism 1). (b) Actual hybridization experiment. All combinations are tried and excess unlabeled DNA is added in each experiment to prevent labeled DNA from reannealing with itself. Following hybridization, hybridized DNA is separated from unhybridized DNA before measuring radioactivity in the hybridized DNA only. (c) Results. Radioactivity in the control is taken as the 100% hybridization value.

hybridization values of above 70% have traditionally been taken as evidence for two isolates being of the same *species* and values above 20–30% are required to seriously argue that two organisms should reside in same *genus* (see Section 12.10 for working definitions of a bacterial genus and species); hybridization of DNAs from totally unrelated organisms occurs at only background levels, usually less than 10% (Figure 12.19).

DNA:DNA hybridization is a sensitive method for revealing subtle differences in the genetic makeup of two organisms and is therefore a useful technique for differentiating only very closely related organisms. Indeed, a common application of genomic hybridization in taxonomic studies is to test two organisms that are suspected for some reason to be different species even though 16S sequencing and phenotypic analyses have failed to reveal significant differences between them.

Ribotyping

Ribotyping is a highly specific technique for bacterial identification that employs some of the methods previously discussed for ribosomal RNA-based phylogenetic characterizations (see Section 12.5). However, unlike comparative *sequencing* methods, ribotyping does not involve sequencing, but instead exploits the unique DNA restriction pattern that is generated when DNA from a particular organism is subject to restriction enzyme digestion and the resulting fragments separated on an agarose gel and then probed with a ribosomal RNA probe (Figure 12.20). Because differences in ribosomal RNA sequences between two organisms translate into the presence or absence of specific restriction enzyme cut sites, the restriction pattern of a particular bacterial species is unique (Figure 12.20); in fact, the

method is so specific that it has been given the nickname "molecular fingerprinting."

In practice, ribotyping begins when either bulk DNA or DNA encoding 16S rRNA and related genes within the ribosomal RNA operon is PCR-amplified, treated with one or more restriction enzymes, separated by electrophoresis and then probed (∞ Chapters 6 and 9 for discussion of these methods); the pattern generated from the fragments of DNA on the gel is then digitized and a computer used to make comparisons of this pattern with patterns from reference organisms in a database (Figure 12.20). Ribotyping is both a *rapid* (since it bypasses the actual sequencing, sequence alignment, and analysis requirements of ribosomal RNA sequencing methods) and *specific* method of bacterial identification and for these reasons has found many applications in clinical diagnostics and for the microbial analyses of food, water, and beverages.

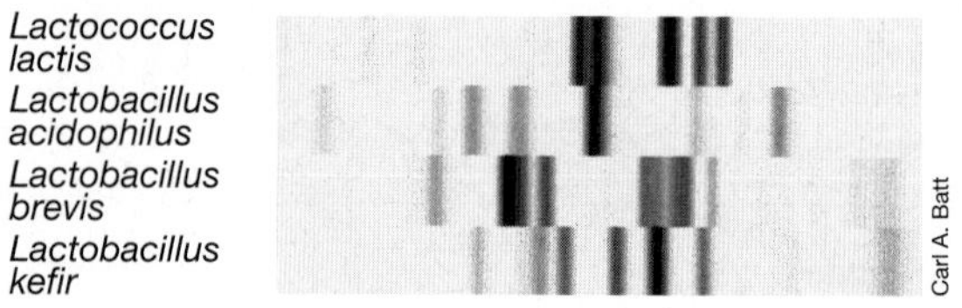

Fatty Acid Analyses: FAME

Another popular method of bacterial identification is through lipid analyses; specifically, through characterization of the types and proportions of fatty acids present in membrane and outer membrane (gram-negative bacteria, ∞ Section 3.8) lipids of cells. Because the chemistry of fatty acids can be so variable, including differences in chain length, the presence or absence of unsaturated groups and rings or branched chains, and hydroxy groups (Figure 12.21*a*), the fatty acid profile of a particular bacterium can often be useful diagnostically.

For actual analyses, fatty acids, extracted from bulk lipids of a bacterial culture grown under standardized conditions, are chemically modified to form their corresponding methyl esters; these now volatile derivatives are then identified by gas chromatography. A computerized list of the types and amounts of fatty acids from the unknown bacterium are then compared with a database containing the fatty acid profiles of thousands of reference bacteria grown under the same conditions, and the best matches to that of the unknown selected (Figure 12.21*b*).

This technique has been nicknamed *FAME*, for *f*atty *a*cid *m*ethyl *e*ster, and is in widespread use in clinical, public health, and food and water inspection laboratories where the identification of pathogens or other bacterial hazards needs to be done on a routine basis. FAME analyses require rigid standardization, since fatty acid profiles in a single organism can vary as a function of temperature and growth medium. Thus, a limitation to FAME analysis is that one must be able to grow the unknown organism on a specific medium and at a specific temperature in order to compare its fatty acid profile with those of organisms from the database that have been grown in the same way. Obviously, this is impos-

Lactococcus lactis
Lactobacillus acidophilus
Lactobacillus brevis
Lactobacillus kefir

Carl A. Batt

FIGURE 12.20 Ribotyping. Ribotype results from four different lactic acid bacteria. The pattern of DNA fragments generated from restriction enzyme digestion of DNA taken from a colony of each bacterium and then probed with 16S rRNA genes is unique to a species or even to strains within a species. The patterns generated on a gel with known organisms are digitized and stored in a database for comparisons in identifying environmental or clinical isolates. Variations in both *position* and *intensity* of the bands are important in identification.

Classes of Fatty Acids in Bacteria

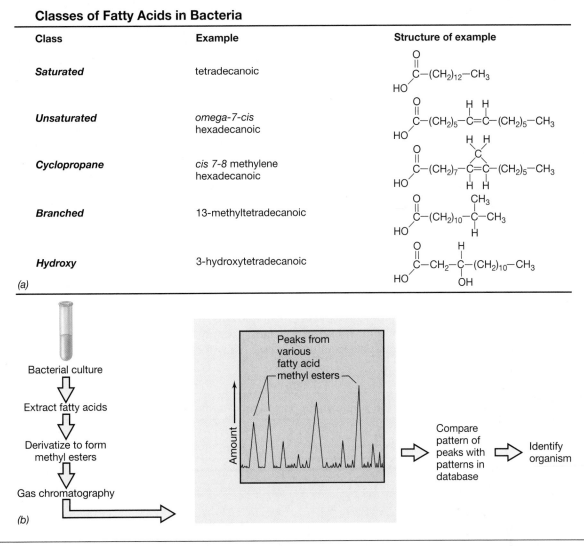

Class	Example	Structure of example
Saturated	tetradecanoic	
Unsaturated	*omega-7-cis* hexadecanoic	
Cyclopropane	*cis 7-8* methylene hexadecanoic	
Branched	13-methyltetradecanoic	
Hydroxy	3-hydroxytetradecanoic	

(a)

(b)

Bacterial culture → Extract fatty acids → Derivatize to form methyl esters → Gas chromatography

Peaks from various fatty acid methyl esters

Amount

Compare pattern of peaks with patterns in database → Identify organism

FIGURE 12.21 Fatty acid methyl ester (FAME) analysis in bacterial identification. (a) Classes of fatty acids in Bacteria. Only a single example is given of each class, but in actuality, more than 200 different fatty acids have been discovered from bacterial sources. A methyl ester contains a methyl group (CH_3) in place of the proton on the carboxylic acid group (COOH) of the fatty acid. (b) Procedure. Each peak from the gas chromatograph is due to one particular fatty acid methyl ester and the peak height is proportional to the amount.

sible for all bacteria because organisms differ in their nutrient and incubation requirements (∞ Chapters 4 and 5). Nevertheless, for the applications in which FAME analysis is currently being used, the major organisms of importance can usually be grown and quickly and reliably identified.

12.10

The Species Concept, Nomenclature, and *Bergey's Manual*

In the world of plants and animals a species is considered a population of individuals that can interbreed under natural conditions, produce fertile offspring, and that is reproductively isolated from other populations. Unfortunately, such a definition does not hold for prokaryotes; because prokaryotes are haploid and reproduce by asexual means, concepts like "the production of fertile offspring" have no meaning. Nevertheless, microbiologists traditionally refer to "species" of bacteria by formal name and regularly give new isolates of bacteria genus and species names. So what *is* a bacterial species? Although the concept of a prokaryotic species is, like prokaryotes themselves, an evolving entity, microbiologists today have turned to molecular biological methods, especially comparative ribosomal RNA sequencing and genomic hybridization, to delineate species based on genetic grounds.

Bacterial Species and Higher Taxa

It has been proposed that a prokaryote whose 16S ribosomal RNA sequence differs by more than 3% from that of all other organisms (that is, the sequence is less than 97% identical to any other sequence), should be considered a new species. Support for this proposal emerges from the observation that DNA from two bacteria whose 16S rRNA sequences are less than 97% identical usually hybridize to less than 70% (see Figure 12.19), a minimal value considered evidence for two organisms being of the same species (see Section 12.9). Although there are some exceptions to this system, the 97% dissimilarity value, along with phenotypic analyses, is now widely used to differentiate prokaryotic species.

A new species is usually defined from the characterization of several strains or clones. Use of the word *clone* in this sense can be taken to mean a population of genetically identical cells derived from a single cell. The species concept is important in microbiology because it gives the collected strains formal taxonomic identity. Groups of species are then collected into genera (sin-

gular, **genus**). What constitutes a genus by molecular criteria is more a matter of judgment than for species, but 16S sequence differences of greater than 5–7% (93–95% identity) have been taken as support for a new genus. Groups of genera are collected into **families,** families into **orders,** orders into **classes,** and so on up to the highest level taxon, the **domain** (Table 12.4). However, the family is typically the highest level taxon used routinely in taxonomic studies of prokaryotes.

In the identification of an unknown organism, it is essential that an organism satisfy all the taxonomic criteria of ranks *above* its species designation. Thus, in the example given in Table 12.4 where we show the taxonomic hierarchy for the bacterium *Chromatium warmingii*, all species of the genus *Chromatium* must be rod-shaped purple sulfur gram-negative Bacteria; if one of these criteria is not met, the organism is not considered a species of *Chromatium*. In other words, as one descends the taxonomic hierarchy from the level of domain to that of species, the criteria used to distinguish two organisms become less general and more specific (Table 12.4).

TABLE 12.4	Taxonomic hierarchy for the purple sulfur bacterium *Chromatium warmingii*		
Taxonomic division	Name	Properties	Confirmed by
Domain[a]	Bacteria	Prokaryotic cells; ribosomal RNA sequences typical of Proteobacteria (see Figure 12.13)	Microscopy; 16S ribosomal RNA sequencing; presence of unique biomarkers, for example, peptidoglycan
Class	Zymobacteria	Gram-negative bacteria	Gram-staining, microscopy
Order	Chromatiales	Phototrophic purple bacteria	Characterizing pigments (∞ Figures 15.4 and 15.9)
Family	Chromatiaceae	Purple sulfur bacteria	Ability to oxidize H_2S and store S^0 within cells; observe culture microscopically for presence of S^0 (see photo)
Genus	*Chromatium*	Rod-shaped purple sulfur bacteria	Microscopy (see photo)
Species	*warmingii*	Cells 3.5–4.0 μm × 5–11 μm; store sulfur mainly in poles of cell (see photo)	Measure cells in microscope using a micrometer; look for position of S^0 globules in cells (see photo)
Photograph of cells of *Chromatium warmingii*:			

Sulfur (S^0) globules

Norbert Pfennig

a In the new edition of *Bergey's Manual of Systematic Bacteriology,* the first volume of which appears in 2000, the term *Kingdom* is used instead of *Domain.*

Nomenclature and Formal Taxonomic Standing

Following the **binomial system** of nomenclature, all bacteria are given genus and species names. The names are either Latin or Greek derivations of some descriptive property appropriate for the species and are set in print in *italics*. For example, several species of the genus *Bacillus* have been described, including *Bacillus (B.) subtilis*, *B. cereus*, *B. stearothermophilus*, and *B. acidocaldarius*. These species names mean "slender," "waxen," "heat-loving," and "acid-thermal," respectively, and in each case refer to key morphological, physiological, or ecological traits characteristic of each organism.

When a new organism is isolated and thought to be unique, a decision must be made as to whether it is sufficiently different from other species to be described as a new species, or perhaps even sufficiently different from all described genera to warrant description as a new genus (in which case a species is automatically created). In order to achieve formal taxonomic standing as a new genus or species, a description of the isolate and the proposed name is published and a pure culture of the organism is deposited in an approved culture collection, usually the American Type Culture Collection (ATCC) or the Deutsche Sammlung von Mikroorganismen und Zellkulturen (DSMZ, German Collection for Microorganisms). (See Section 11.1 for a discussion of microbial culture collections.) The deposited strain serves as the *type* strain of the new species or genus and species and remains as the standard by which other strains thought to be the same can be compared.

Culture collections preserve the deposited culture, generally by freezing or freeze-drying. This practice differs from the botanical or zoological approach. These disciplines employ preserved (dead) specimens (either dried herbarium material or chemically fixed animal specimens) as the basis for comparison with proposed new species. Microbiologists rely on a *living type strain*, and this approach allows for more detailed and reproducible comparisons, especially at the molecular level.

If the description of a new organism is published in a journal *other than* the *International Journal of Systematic Bacteriology* (*IJSB*), the official publication of record for the taxonomy and classification of prokaryotes and yeasts, a copy of the published paper must be submitted to this journal and the name validated before it is formally accepted as a new microbiological taxon. Periodically, the *IJSB* publishes an approved list of bacterial names, which formalizes newly proposed names and paves the way for their inclusion in *Bergey's Manual of Systematic Bacteriology*, a major taxonomic treatment of prokaryotes (discussed next).

Bergey's Manual and *The Prokaryotes*

Although no source is recognized as *official* in the field of microbial taxonomy, *Bergey's Manual of Systematic Bacteriology* is widely used in this regard. *Bergey's Manual* is a compendium of standard and molecular information on all recognized species of prokaryotes and contains a number of dichotomous keys and other systematic information useful for identification purposes. The latest edition of *Bergey's Manual*, published in five volumes to appear over the next several years beginning in 2000, has incorporated many of the concepts that have emerged from ribosomal RNA sequencing studies and blends this with a wealth of conventional taxonomic information. A second major reference source is a multi-volume treatise called *The Prokaryotes*. This work, with more than 4,100 pages in its second edition (1992), is now being updated and will be available in an online edition that will be revised periodically to reflect the rapid pace at which new information is being generated on prokaryotic taxonomy and phylogeny. In combination, *Bergey's Manual* and *The Prokaryotes* offer microbiologists the foundations as well as the details of microbial taxonomy and phylogeny as we know it today, and are typically the first sources the practicing taxonomist seeks out when beginning the characterization of a newly isolated prokaryote.

✓ 12.10 Concept Check

Taxonomy is the science of identification and naming of microorganisms, based primarily on phenotypic properties. The species concept applies to prokaryotes just as it does to eukaryotes, and a similar taxonomic hierarchy exists, with the domain as the highest level taxon. Bacterial taxonomy requires testing numerous properties of a given isolate and yields a positive identification in most cases if a sufficient number of properties have been examined. *Bergey's Manual of Systematic Bacteriology* is a major taxonomic compilation of Bacteria and Archaea.

✓ How does the science of *taxonomy* differ from that of *phylogeny*?

✓ What is a *bacterial species*?

✓ Define the terms *ribotyping* and *FAME analysis*.

REVIEW QUESTIONS

1. What is the evidence that prokaryotic life forms were present on Earth billions of years before eukaryotic life forms?

2. What major features would primitive organisms have had to have in order to replicate copies of themselves. Why?

3. Why was the evolution of cyanobacteria of such importance to the further evolution of life on Earth?

4. What properties of RNA could have made possible an era of RNA life? If RNA life forms ever existed, why is there no trace of them today?

5. What might have been the advantages of abandoning the era of RNA life for cellular life based on DNA, RNA, and protein?

6. Why are macromolecules like nucleic acids or proteins excellent phylogenetic markers, whereas polysaccharides and lipids are not? (You may want to review the material in Chapter 2 before answering this question.)

7. Why are ribosomal RNAs better molecules for phylogenetic studies than proteins like ferredoxin, cytochromes, or specific enzymes?

8. What are signature sequences and of what phylogenetic value are they? How are signature sequences discerned?

9. What is FISH technology? Give an example of how it would be used.

10. Describe the methods involved in obtaining 16S rRNA sequences. How has the polymerase chain reaction benefited molecular phylogeny?

11. What major evolutionary finding has emerged from the study of ribosomal RNA sequences? How did this modify the classic view of evolution? How has this discovery changed our thinking on the origin of eukaryotic organisms?

12. What major lesson has microbiology learned from RNA sequencing concerning the use of phenotypic criteria in establishing evolutionary relationships?

13. What major physiological and biochemical properties do Archaea share with Eukarya? With Bacteria?

14. Examine the following bacterial name: *Pseudomonas aeruginosa*. What part of this name is the *species* name. What is the other name? In reference to taxonomic hierarchy, which of the two names might have several other names listed *under* it?

15. What major phenotypic properties are used to group organisms in classical bacterial taxonomy? Which, if any, of these properties have phylogenetic predictive value?

16. Why aren't GC base ratios useful for making phylogenetic determinations? In what situations are GC base ratios of use in taxonomic studies?

17. How does ribotyping differ from 16S sequencing as an identification tool?

18. What is measured in FAME analyses?

APPLICATION QUESTIONS

1. Why is it highly unlikely that life could originate today as it did billions of years ago?

2. Imagine that you are debating someone who is arguing against the theory of endosymbiosis. List five forms of evidence you would use to convince your opponent that endosymbiosis did occur. (You may wish to review Section 3.16 before writing your answer.)

3. On the basis of the following sequences, calculate an evolutionary distance between these three organisms and predict which two of the three are most closely related.

   Organism 1: AGGUACGUUA
   Organism 2: UGCCACGGUU
   Organism 3: AGGUACGGUA

Draw a phylogenetic tree that shows the approximate evolutionary relationships of these three organisms.

4. Determine the GC ratio of the following stretch of DNA:

   TAAGCCTGCAAGCTTAGCTA
   ATTCGGACGTTCGAATCGAT

5. What reference resource in your library would you check for information on the taxonomy and phylogeny of prokaryotes? Compare the tables of contents of these sources. Which has the greater emphasis on classification and nomenclature?

The Bacteria constitute a huge group of prokaryotic microorganisms that shared a distant common ancestor. In terms of pure cultures, at least fourteen major subdivisions (phyla) of Bacteria can be recognized, as shown in the tree here, but from studies of natural habitats using the tools of molecular microbial ecology, many more phyla are known to exist. The vast majority of known prokaryotes are Bacteria, including all those of medical relevance and most that are currently known to be of environmental significance.

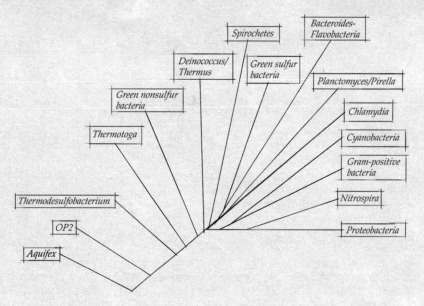

CHAPTER 13 Prokaryotic Diversity: Bacteria

Kingdom I: Proteobacteria

13.1 Purple Phototrophic Bacteria   455
13.2 The Nitrifying Bacteria   461
13.3 Sulfur- and Iron-Oxidizing Bacteria   462
13.4 Hydrogen-Oxidizing Bacteria   466
13.5 Methanotrophs and Methylotrophs   467
13.6 *Pseudomonas* and the Pseudomonads   470
13.7 Acetic Acid Bacteria   474
13.8 Free-Living Aerobic Nitrogen-Fixing Bacteria   475
13.9 *Neisseria, Chromobacterium,* and Relatives   476
13.10 Enteric Bacteria   477
13.11 *Vibrio* and *Photobacterium*   482
13.12 Rickettsias   484
13.13 Spirilla   485
13.14 Sheathed Proteobacteria: *Sphaerotilus* and *Leptothrix*   489
13.15 Budding and Prosthecate/Stalked Bacteria   491
13.16 Gliding Myxobacteria   495
13.17 Sulfate- and Sulfur-Reducing Proteobacteria   498

Kingdom II: Gram-Positive Bacteria

13.18 Nonsporulating, Low GC, Gram-Positive Bacteria   502
13.19 Endospore-Forming, Low GC, Gram-Positive Bacteria   507
13.20 Cell Wall–Less, Low GC, Gram-Positive Bacteria: The Mycoplasmas   513
13.21 High GC, Gram-Positive Bacteria: Coryneform and Propionic Acid Bacteria   515
13.22 High GC, Gram-Positive Bacteria: *Mycobacterium*   517
13.23 Filamentous, High GC, Gram-Positive Bacteria: The Actinomycetes   519

Kingdom III: Cyanobacteria, Prochlorophytes, and Chloroplasts

13.24 Cyanobacteria   524
13.25 Prochlorophytes and Chloroplasts   527

Kingdom IV: Chlamydia

13.26 The Chlamydia   529

Kingdom V: *Planctomyces/Pirella*

13.27 *Planctomyces:* A Phylogenetically Unique Stalked Bacterium   532

Kingdom VI: *Bacteroides/Flavobacteria*

13.28 *Bacteroides, Flavobacterium,* and *Cytophaga*   532

Kingdom VII: Green Sulfur Bacteria

13.29 *Chlorobium* and Other Green Sulfur Bacteria   533

Kingdom VIII: Tightly Coiled Bacteria: The Spirochetes

13.30 Spirochetes   537

Kingdom IX: Deinococci

13.31 *Deinococcus/Thermus*   540

Kingdom X: The Green Nonsulfur Bacteria

13.32 *Chloroflexus* and *Heliothrix*   542

Kingdoms XI, XII, and XIII: Hyperthermophiles

13.33 *Thermotoga* and *Thermodesulfobacterium*   543
13.34 *Aquifex* and Relatives   544

WORKING GLOSSARY

Acid fastness a property of *Mycobacterium* species in which cells stained with the dye basic fuchsin resist decolorization with acidic alcohol

Carboxysomes polyhedral cellular inclusions of crystalline ribulose bisphosphate carboxylase (RubisCO), the key enzyme of the Calvin cycle

Chemolithotrophs organisms able to oxidize inorganic compounds as energy sources

Chlorosomes cigar-shaped structures bounded by a nonunit membrane and containing the light-harvesting bacteriochlorophyll (c, d, or e) in green bacteria and *Chloroflexus*

Consortia two or more-membered association of prokaryotes, usually living in an intimate symbiotic fashion

Cyanobacteria prokaryotic oxygenic phototrophs that contain chlorophyll a and phycobilins but not chlorophyll b

Enteric bacteria a large group of gram-negative rod-shaped Bacteria characterized by a facultatively aerobic metabolism

Green bacteria anoxygenic phototrophs containing chlorosomes and bacteriochlorophyll c, c_s, d, or e as light-harvesting chlorophyll

Heliobacteria anoxygenic phototrophs containing bacteriochlorophyll g

Hyperthermophile an organism with a growth temperature optimum of greater than 80°C

Heterocyst a differentiated cyanobacterial cell that carries out nitrogen fixation but not oxygenic photosynthesis

Heterofermentative in reference to lactic acid bacteria, capable of making more than one fermentation product

Homofermentative in reference to lactic acid bacteria, producing only lactic acid as a fermentation product

Methanotroph an organism capable of oxidizing methane (CH_4)

Methylotroph an organism capable of oxidizing organic compounds that do not contain carbon–carbon bonds; if able to oxidize CH_4, also a methanotroph

Nitrifying bacteria chemolithotrophs capable of carrying out the transformation $NH_3 \rightarrow NO_2^-$ or $NO_2^- \rightarrow NO_3^-$

Purple nonsulfur bacteria a group of phototrophic prokaryotes containing bacteriochlorophylls a or b that grow best as photoheterotrophs and have a relatively low tolerance for H_2S

Prochlorophyte a prokaryotic oxygenic phototroph that contains chlorophylls a and b but lacks phycobilins

Prostheca an extrusion of cytoplasm often forming a distinct appendage, bounded by the cell wall

Proteobacteria a major lineage of Bacteria that contains a large number of gram-negative rods and cocci

Pseudomonad member of the genus *Pseudomonas*, a large group of gram-negative, obligately respiratory (never fermentative) Bacteria

Purple sulfur bacteria a group of phototrophic prokaryotes containing bacteriochlorophylls a or b and characterized by the ability to oxidize H_2S and store elemental sulfur inside the cells (or in the genus *Ectothiorhodospira*, outside the cell)

Spirochete a slender, tightly coiled gram-negative prokaryote characterized by possession of axial filaments used for motility

Stickland reaction fermentation of an amino acid pair in which one amino acid serves as an electron donor and a second serves as an electron acceptor

Sulfate-reducing bacteria a large group of anaerobic Bacteria that respire anaerobically with SO_4^{2-} as electron acceptor, producing H_2S

In the preceding chapter we stressed the evolutionary relationships among microorganisms. In this and the next two chapters we expand on these concepts with a discussion of the properties of major microbial groups. With the thousands of different species of microorganisms known, we will obviously not be able to consider them all. So, using phylogenetic trees as the focus of our discussion, we will examine particularly well-known species and ones in which much phenotypic information is available. For more detailed information on prokaryotic diversity the student should refer to *Bergey's Manual of Systematic Bacteriology* and *The Prokaryotes*.

Fourteen major lineages (kingdoms) of Bacteria have been defined on the basis of comparative 16S ribosomal RNA sequencing of laboratory cultures and a number of other kingdoms identified from community sampling of microbial habitats (∞ Section 12.6); a summary of these findings is shown in Figure 13.1. The most phylogenetically ancient kingdom of Bacteria contains the genus *Aquifex* and relatives, all of which are hyperthermophilic H_2 chemolithotrophs. Other nearby lineages defined by organisms like *Thermodesulfobacterium*, *Thermotoga*, and the green nonsulfur bacteria (*Chloroflexus* group), also contain thermophilic species.

Continuing through the domain Bacteria past the green nonsulfur bacteria, we see the deinococci and relatives, which include the highly radiation-resistant organism *Deinococcus* (see Section 13.31); the morphologically unique spirochetes (see Section 13.30); the phototrophic green sulfur bacteria (see Section 13.29); the chemoorganotrophic *Bacteroides/Flavobacterium* group (see Section 13.28); the budding bacteria *Planctomyces–Pirella*, which lack peptidoglycan in their cell walls (see Section 13.27); the *Chlamydia*, a group of obligately intracellular parasites

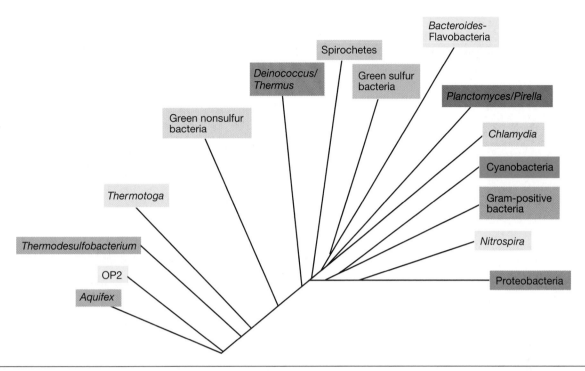

FIGURE 13.1 Detailed phylogenetic tree of the major lineages (kingdoms) of Bacteria based on 16S ribosomal RNA sequence comparisons. The relative positions of branches on this detailed tree differ slightly (for statistical reasons) from those shown in the universal tree (◯◯◯ Figure 12.13), but the branch lengths on the tree remain proportional to the corrected evolutionary distances calculated between any two groups. OP2 is an environmental sequence (◯◯◯ Section 12.6) not yet represented by a cultured bacterium.

that cause disease in humans and other animals (see Section 13.26); and the genus *Nitrospira*, a morphologically unique chemolithotroph (see Section 13.2).

The remaining kingdoms of cultured Bacteria constitute the major part of our discussion in this chapter. They include the gram-positive bacteria, the cyanobacteria, and the Proteobacteria. Each of these is a large group containing many genera and are Bacteria about which much phenotypic information is known. The gram-positive bacteria can be further separated into two subgroups, called *low GC* and *high GC*, the "GC" designation referring to the fact that the species tend to have DNA GC base ratios (◯◯◯ Section 12.9) either well below or well above 50%, respectively. The gram-positive bacteria are a large group of chemoorganotrophic bacteria and are discussed in detail in Sections 13.18–13.23. The cyanobacteria are oxygenic phototrophic prokaryotes with evolutionary roots near those of the gram-positive bacteria; these organisms are covered in Section 13.24. The final lineage on the Bacteria tree is the *Proteobacteria* (Figure 13.1). This group is the largest and most physiologically diverse of all Bacteria (see Sections 13.1–13.17). The Proteobacteria contains five clusters of species, designated by the Greek letters *alpha*, *beta*, *gamma*, *delta*, and *epsilon* (see Table 13.1). Physiologically, Proteobacteria can be either phototrophic, chemolithotrophic, or chemoorganotrophic; we will see in Chapter 15 the great diversity of energy-generating mechanisms characteristic of representatives of this group.

With this introduction to the phylogeny of the domain Bacteria, let us proceed to a description of these kingdoms. We begin with the largest group of known Bacteria: The Proteobacteria.

Kingdom I: Proteobacteria

Table 13.1 lists some of the key genera of Proteobacteria. As a group these organisms are all gram-negative, show extreme metabolic diversity, and represent the majority of known gram-negative bacteria of medical, industrial, and agricultural significance. We begin our discussion with phototrophic Proteobacteria—the purple bacteria.

13.1

Purple Phototrophic Bacteria

*Key Genera**

Chromatium
Ectothiorhodospira
Rhodobacter
Rhodospirillum

---

*In an attempt to make prokaryotic diversity more manageable for the beginning student, in this chapter and Chapters 14 and 17 each numbered head will be accompanied by a small table, as shown here, that will list some of the best-studied genera that are members of the group to be discussed. It should be understood that this list is not meant to be inclusive but only representative.

TABLE 13.1	Major genera of Proteobacteria[a]

Subdivision	Genera	
Alpha	*Acetobacter*	*Paracoccus*
	Agrobacterium	*Pseudomonas* (some species)
	Alcaligenes	*Rhodospirillum*
	Aquaspirillum	*Rhodopseudomonas*
	Azospirillum	*Rhodobacter*
	Beijerinckia	*Rhodomicrobium*
	Bradyrhizobium	*Rhodovulum*
	Brucella	*Rhodopila*
	Caulobacter	*Rhizobium*
	Ehrlichia	*Rickettsia*
	Gluconobacter	*Sphingomonas*
	Hyphomicrobium	*Thiobacillus* (some species)
		Zymomonas
Beta	*Bordetella*	*Pseudomonas* (some species)
	Chromobacterium	*Ralstonia*
	Gallionella	*Rhodocyclus*
	Leptothrix	*Sphaerotilus*
	Neisseria	*Spirillum*
	Nitrosomonas	*Thiobacillus* (some species)
	Oxalobacter	*Zoogloea*
Gamma	*Acinetobacter* (some species)	*Oceanospirillum*
	Azotobacter	*Photobacterium*
	Chromatium	*Methylococcus*
	Escherichia	*Methylobacter*
	Ectothiorhodospira	*Thiobacillus* (some species)
	Francisella	*Thiomicrospira*
	Halomonas	*Thiospirillum* and other purple
	Halorhodospira	sulfur bacteria
	Legionella	*Salmonella* and other enteric
	Leucothrix	bacteria
	Methylomonas	*Vibrio*
Delta	*Acinetobacter* (some species)	*Francisella*
	Aeromonas	*Halomonas*
	Bdellovibrio	*Moraxella*
	Desulfuromonas	*Myxococcus* and other
	Desulfovibrio and	myxobacteria
	other sulfate-	*Pelobacter*
	reducing bacteria	*Syntrophobacter*
	Erwinia	*Xanthomonas*
Epsilon	*Campylobacter*	*Thiovulum*
	Helicobacter	*Wolinella*
	Thiomicrospira	

a This table is not meant to be inclusive but only lists well-described genera of Proteobacteria. For a complete list of Proteobacteria and genera of other lineages of Bacteria, see Appendix 2.

The purple phototrophic bacteria carry out *anoxygenic* photosynthesis; unlike the cyanobacteria (see Section 13.24) no O_2 is evolved. In fact, in purple bacteria, some of which are facultative aerobes and able to grow by respiration in the dark, O_2 actually *inhibits* photosynthesis because it represses photopigment synthesis.

Purple bacteria contain chlorophyll pigments called *bacteriochlorophylls* and additionally contain any of a variety of *carotenoid* pigments. Together, these pigments give purple bacteria their spectacular colors, usually purple, red, or brown (Figure 13.2). We will examine the structure of these pigments and learn how they ac-

tually function in light-mediated energy generation (a process called *photophosphorylation*) in Chapter 15. The purple bacteria are a morphologically diverse group and the taxonomy of these organisms has been established along phylogenetic and physiological lines.

Purple bacteria synthesize intracytoplasmic photosynthetic membrane systems into which their pigments are inserted. These membranes can be of various morphologies (Figure 13.3) but in all cases originate from invaginations of the cytoplasmic membrane. These internal membranes allow purple bacteria to increase their specific pigment content and to thus better utilize the

FIGURE 13.2 Photograph of mass cultures of phototrophic bacteria showing the color of species with various carotenoid pigments. The blue culture is a carotenoid-less mutant derivative of *Rhodospirillum rubrum* showing how bacteriochlorophyll *a* is actually *blue* in color. The bottle on the far right (*Rhodobacter sphaeroides* strain G) lacks one of the carotenoids of the wild type and thus is more green in color.

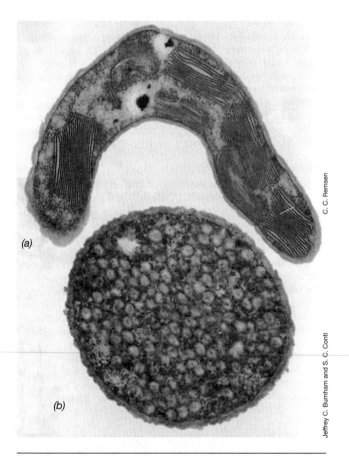

FIGURE 13.3 Membrane systems of phototrophic purple bacteria as revealed by the electron microscope. (a) Purple phototrophic bacterium, *Ectothiorhodospira mobilis*, showing the photosynthetic membranes in flat sheets (lamellae). (b) *Allochromatium vinosum*, another purple phototrophic bacterium, showing the membranes as individual vesicles.

available light; when cells are grown at high light intensities internal membranes are few and pigment contents low while at low light intensities, the cells are packed with membranes and photopigments.

Purple Sulfur Bacteria

Purple bacteria that utilize hydrogen sulfide (H_2S) as an electron donor for CO_2 reduction in photosynthesis are known as *purple sulfur* bacteria (Table 13.2). The sulfide is oxidized to elemental sulfur (S^0) that is stored in globules that appear microscopically to be inside the cells (Figure 13.4). But in actuality, the sulfur is stored in the periplasm of these gram-negative bacteria where it can be further oxidized to sulfate (SO_4^{2-}) to fuel the need for reducing power. The biochemistry of sulfide oxidation to sulfate is complex, but interestingly, turns out to be essentially the reverse of the steps used by another group of Proteobacteria, the sulfate-reducing bacteria, in their physiological process of reducing sulfate to sulfide (∞ Sections 13.17, 15.17, and 16.17). Many purple sulfur bacteria can also use other reduced sulfur compounds as photosynthetic electron donors, thiosulfate ($S_2O_3^{2-}$) being a key one used commonly to grow laboratory cultures of these organisms. All purple sulfur bacteria discovered thus far group with the gamma Proteobacteria.

Purple sulfur bacteria are generally found in illuminated anoxic zones of lakes and other aquatic habitats where H_2S accumulates, and also in "sulfur springs," where geochemically or biologically produced H_2S can trigger the formation of blooms of purple sulfur bacteria (Figure 13.5). The most favorable lakes for development of purple sulfur bacteria are *meromictic* (permanently stratified) lakes. Meromictic lakes stratify because they have denser (usually saline) water in the bottom and less dense (usually freshwater) nearer the surface. If sufficient sulfate is present to support sulfate reduction, the sulfide, produced in the sediments, diffuses upward into the anoxic bottom waters and here purple sulfur bacteria can form massive blooms, usually in association with green sulfur phototrophic bacteria (Figure 13.5c).

The genera *Ectothiorhodospira* and *Halorhodospira* are of special interest because, unlike other purple sulfur bacteria, these organisms oxidize H_2S and produce S^0 *outside* of the cell (Figure 13.6) but also because some species are extremely halophilic (salt-loving) and are among the most halophilic of all known prokaryotes. These organisms are typically found in marine environments, saline lakes, soda lakes, and salterns.

Purple Nonsulfur Bacteria

These bacteria have been called "nonsulfur" because it was originally thought that they were unable to use sulfide as an electron donor for the reduction of CO_2 to cell

TABLE 13.2	Genera and characteristics of purple sulfur bacteria[a]		
Characteristics	**Genus**	**Number of species**	**DNA (mol % GC)**
Sulfur deposited externally:			
Spirilla, polar flagella	*Ectothiorhodospira*	9	62–67
Spirilla, extreme halophiles	*Halorhodospira*	3	50–69
Sulfur deposited internally:			
Do not contain gas vesicles			
Ovals or rods, polar flagella	*Chromatium;* *Allochromatium;* *Halochromatium;* *Rhabdochromatium;* *Thermochromatium;* *Isochromatium;* *Marichromatium*	23	48–70
Spheres, diplococci, tetrads, nonmotile; cells 1.2–3 μm in diameter	*Thiocapsa*	5	63–70
Spheres or ovals, polar flagella; cells 2.5–3 μm in diameter	*Thiocystis*	4	61–68
Spheres, 1.5–2.5 μm in diameter	*Thiohalocapsa*	1	66
Spheres, 1–2 μm in diameter	*Thiorhodococcus*	1	67
Spheres, 1.2–1.5 μm in diameter	*Thiococcus*	1	69
Large spirilla, polar flagella	*Thiospirillum*	1	45
Small spirilla	*Thiorhodovibrio*	1	61–62
Contain gas vesicles			
Irregular spheres, ovals, nonmotile	*Amoebobacter*	4	63–65
Irregular spheres forming platelets of 4–16 cells	*Thiolamprovulum*	1	
Rods	*Lamprobacter*	1	64
Spheres, ovals, polar flagella	*Lamprocystis*	1	64
Rods, nonmotile; forming irregular network	*Thiodictyon*	2	65–66
Spheres, nonmotile; forming flat sheets of tetrads	*Thiopedia*	1	62–64

a From a phylogenetic standpoint, all are members of the gamma subdivision of the Proteobacteria.

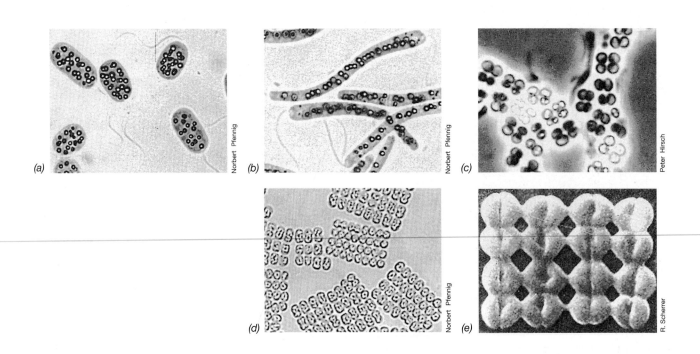

(a) Norbert Pfennig (b) Norbert Pfennig (c) Peter Hirsch (d) Norbert Pfennig (e) R. Scherrer

FIGURE 13.4 Bright-field photomicrographs of purple sulfur bacteria. (a) *Chromatium okenii;* cells are about 5 μm wide. Note the globules of elemental sulfur inside the cells. (b) *Thiospirillum jenense,* a very large, polarly flagellated spiral; cells are about 30 μm long. Note the sulfur globules. (c) *Thiocapsa;* cells are about 2 μm wide. (d) *Thiopedia rosea;* cells are about 1.5 μm wide. (e) Scanning electron micrograph of a sheet of 16 cells of *Thiopedia rosea* showing the major division planes.

(a)

T. D. Brock

(b)

Jörg Overmann

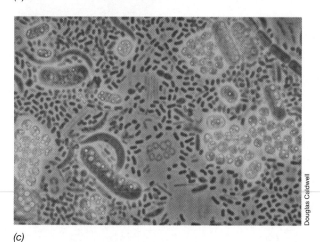

(c)

Douglas Caldwell

FIGURE 13.5 Blooms of purple sulfur bacteria. (a) *Thiopedia roseopersicinia*, in a sulfide spring in Madison, Wisconsin. The bacteria grow near the bottom of the spring pool and float to the top (by virtue of their gas vesicles) when disturbed. The green color is from cells of the eukaryotic alga *Spirogyra*. (b) Sample of water from 7 m in Lake Mahoney, British Columbia. The major organism is *Amoebobacter purpureus*. (c) Phase-contrast photomicrograph of layers of purple sulfur bacteria from a small stratified lake in Michigan. The purple sulfur bacteria include *Chromatium* species and *Thiocystis*.

bically in darkness by respiration. Under the latter conditions, the electron donor can be an organic compound or in some species even an inorganic compound, such as H_2. However, it is the great ability of this group to practice *photoheterotrophy* (where light is the energy source and an organic compound is the carbon source, ⚭ Section 15.5), that likely accounts for their competitive success in nature. Purple nonsulfur bacteria are typically nutritionally diverse in this regard, using fatty, organic, or amino acids; sugars; alcohols; and even aromatic compounds like benzoate as carbon sources. Most can also grow photoautotrophically with (CO_2 + H_2) or (CO_2 + low levels of H_2S).

Enrichment and isolation of purple nonsulfur bacteria is very easy using a mineral salts medium supplemented with an organic or fatty acid as carbon source. Such media, inoculated with a mud, lake water, or sewage sample and incubated anoxically in the light, invariably select for purple nonsulfur bacteria. Enrichment cultures can be made even more selective by omitting fixed nitrogen sources (for example, NH_4^+) or organic nitrogen sources (for example, yeast extract or peptone) from the medium and supplying a gaseous headspace of N_2; virtually all purple nonsulfur bacte-

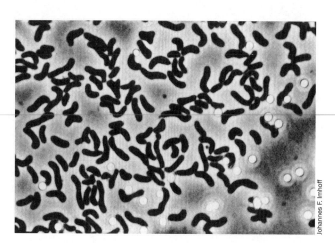

Johannes F. Imhoff

FIGURE 13.6 Phase photomicrograph of cells of *Ectothiorhodospira mobilis*. A single cell is about 0.8 μm in diameter. Notice the sulfur globules deposited outside of the cells.

material. However, sulfide *can* be used by most species, although the levels of sulfide utilized well by purple *sulfur* bacteria are toxic to most purple *nonsulfur* bacteria. Some purple nonsulfur bacteria can also grow anaerobically in the dark using fermentative or anaerobic respiratory metabolism and most can grow aero-

TABLE 13.3	Genera and characteristics of purple nonsulfur bacteria			
Characteristics	**Genus**	**Number of species**	**16S rRNA group**[a]	**DNA (mol % GC)**
Spirilla, polarly flagellated	*Rhodospirillum;*	15	Alpha	62–68
	Phaeospirillum;			
	Rhodovibrio;			
	Rhodothalassium;;			
	Roseospira;			
	Rhodospira	1	Alpha	66
Rods, polarly flagellated; divide by budding	*Rhodopseudomonas;*	15	Alpha	64–72
	Rhodoplanes;	2	Alpha	66–69
	Rhodobium	2	Alpha	61–65
Rods; divide by binary fission	*Rhodobacter*	8	Alpha	62–71
Ovoid to rod-shaped cells	*Rhodovulum*	4	Alpha	64–68
Ovals, peritrichously flagellated; growth by budding and hypha formation	*Rhodomicrobium*	1	Alpha	61–63
Large spheres, acidophilic (pH 5 optimum)	*Rhodopila*	1	Alpha	66
Ring-shaped or spirilla	*Rhodocyclus*	3	Beta	64–66
Curved rods	*Rubrivivax*	1	Beta	70–72
Curved rods	*Rhodoferax*	2	Beta	59–60

a All are members of the Proteobacteria (see Figure 13.1 and Table 13.1).

ria can fix N_2 (⌒ Section 15.29) and will thrive under such conditions, usually outcompeting other organisms.

The morphological diversity of purple nonsulfur bacteria is typical of that of purple sulfur bacteria (Table 13.3 and Figure 13.7), and it is clearly a heterogeneous group in this regard. All purple nonsulfur bacteria isolated thus far are either alpha or beta Proteobacteria.

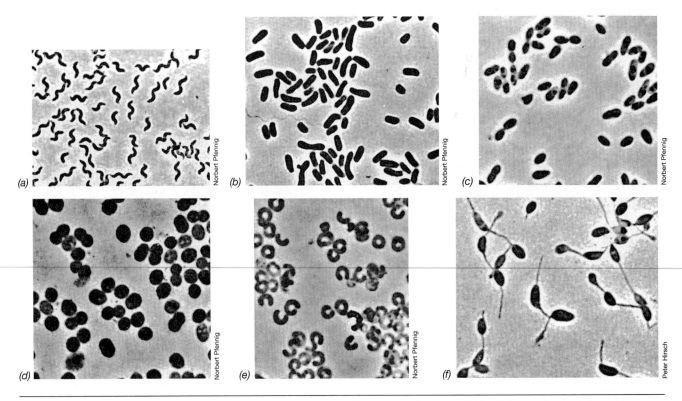

FIGURE 13.7 Representatives of several genera of purple nonsulfur bacteria (see also Table 13.3). (a) *Phaeospirillum fulvum;* cells are about 3 μm long. (b) *Rhodopseudomonas acidophila;* cells are about 4 μm long. (c) *Rhodobacter sphaeroides;* cells are about 1.5 μm wide. (d) *Rhodopila globiformis;* cells are about 1.6 μm wide. (e) *Rhodocyclus purpureus;* cells are about 0.7 μm in diameter. (f) *Rhodomicrobium vannielii;* cells are about 1.2 μm wide.

✓ 13.1 Concept Check

Purple bacteria are anoxygenic phototrophs that grow phototrophically, obtaining carbon from $CO_2 + H_2S$ (purple sulfur bacteria) or organic sources (purple nonsulfur bacteria). Some purple nonsulfur bacteria are highly physiologically diverse and, collectively, the photoautotrophic activities of purple bacteria can be of great ecological significance. The purple bacteria reside in the alpha, beta, or gamma subdivisions of the Proteobacteria.

✓ What is meant by the term *anoxygenic?*
✓ Give a major reason why photosynthesis in purple nonsulfur bacteria does not occur under aerobic conditions.
✓ Can purple bacteria grow in the absence of light?

13.2

The Nitrifying Bacteria

Key Genera

Nitrosomonas
Nitrobacter

We will discuss in Chapter 15 the conceptual basis of chemolithotrophy. Various chemolithotrophic bacteria are known, but they are all physiologically united by their ability to utilize *inorganic* electron donors as energy sources. Most chemolithotrophs are also capable of autotrophic growth and in this way share a major physiological trait with phototrophic bacteria and cyanobacteria. The best-studied chemolithotrophs are those capable of oxidizing reduced sulfur and nitrogen compounds, and the hydrogen-oxidizing bacteria, and we focus on these groups here and in the next two sections.

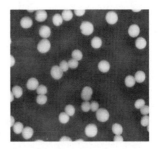 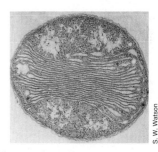

S. W. Watson

FIGURE 13.8 Phase contrast photomicrograph (left) and electron micrograph (right) of the nitrosifying bacterium *Nitrosococcus oceani.* A single cell is about 2 μm in diameter.

Bacteria able to grow chemolithotrophically at the expense of reduced inorganic nitrogen compounds are called **nitrifying bacteria.** Several genera are recognized on the basis of morphology and the particular steps in the oxidation sequences that they carry out (Table 13.4). No chemolithotroph is known that will carry out the complete oxidation of ammonia to nitrate; thus, **nitrification** of ammonia in nature results from the sequential action of two separate groups of organisms, the **ammonia-oxidizing bacteria,** the **nitrosifyers** (Figure 13.8), and the **nitrite-oxidizing bacteria,** the true **nitrifying** (nitrate-producing) bacteria (Figure 13.9). Nitrosifying bacteria typically have genus names beginning in "Nitroso," while true nitrifyers usually begin with "Nitro"; *Nitrosomonas* and *Nitrobacter* are major genera of nitrifying bacteria (Table 13.4). Historically, the nitrifying bacteria were the first organisms to be shown to grow chemolithotrophically; Winogradsky showed that they were able to produce organic matter and cell mass when

TABLE 13.4 Characteristics of the nitrifying bacteria[a]

Characteristics	Genus	DNA (mol % GC)	Habitats
Oxidize ammonia:			
Gram-negative short to long rods, motile (polar flagella) or nonmotile; peripheral membrane systems	*Nitrosomonas*	45–53	Soil, sewage, freshwater, marine
Large cocci, motile; vesicular or peripheral membranes	*Nitrosococcus*	49–50	Freshwater, marine
Spirals, motile (peritrichous flagella); no obvious membrane system	*Nitrosospira*	54	Soil
Pleomorphic, lobular, compartmented cells; motile (peritrichous flagella)	*Nitrosolobus*	54	Soil
Slender, curved rods	*Nitrosovibrio*	54	Soil
Oxidize nitrite:			
Short rods, reproduce by budding, occasionally motile (single subterminal flagellum); membrane system arranged as a polar cap	*Nitrobacter*	59–62	Soil, freshwater, marine
Long, slender rods, nonmotile; no obvious membrane system	*Nitrospina*	58	Marine
Large cocci, motile (one or two subterminal flagella); membrane system randomly arranged in tubes	*Nitrococcus*	61	Marine
Helical to vibrioid-shaped cells, nonmotile; no internal membranes	*Nitrospira*	50	Marine

a Phylogenetically, all nitrifying bacteria thus far examined are either α or γ Proteobacteria, except for *Nitrospira,* which constitutes its own phylogenetic lineage (Figure 13.1).

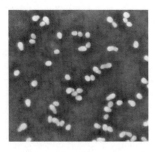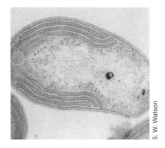

S. W. Watson

FIGURE 13.9   Phase contrast photomicrograph (left) and electron micrograph (right) of the nitrifying bacterium *Nitrobacter winogradskyi*. A cell is about 0.7 μm in diameter.

provided with CO_2 as sole carbon source (⚬⚬ box, Winogradsky's Legacy, Chapter 15).

Many species of nitrifying bacteria have remarkably complex internal membrane systems, in many respects similar to the internal membranes found in their phylogenetic close relatives, the purple anoxyphototrophs (see Section 13.1) and the methane-oxidizing (methanotrophic) bacteria (see Section 13.5). The membranes are the location of a key enzyme in NH_3 oxidation, *ammonia monooxygenase*, which oxidizes NH_3 to hydroxylamine; the latter is further oxidized to NO_2^- by the nitrosifying bacteria (Figure 13.10) and we discuss the biochemistry of this process in more detail in Section 15.14. The NO_2^- generated in this reaction is oxidized to NO_3^- by the nitrifying bacteria (Table 13.4 and Figure 13.10).

Ecology, Isolation, and Culture

The nitrifying bacteria are widespread in soil and water. They are present in highest numbers in habitats where considerable amounts of ammonia are present, such as sites where extensive protein decomposition occurs (ammonification) and in sewage treatment facilities (⚬⚬ Section 11.15). Nitrifying bacteria develop especially well in lakes and streams that receive inputs of sewage or other wastewaters because these are frequently high in ammonia.

Nitrosifying bacteria

1. $NH_3 + O_2 + 2\ e^- + 2\ H^+ \longrightarrow NH_2OH + H_2O$
2. $NH_2OH + H_2O + \frac{1}{2} O_2 \longrightarrow NO_2^- + 2\ H_2O + H^+$

Sum: $NH_3 + 1\frac{1}{2} O_2 + 2\ e^- + H^+ \longrightarrow NO_2^- + 2H_2O$

$\Delta G^{0'} = -287\ \text{kJ/reaction}$

Nitrifying bacteria

$NO_2^- + \frac{1}{2} O_2 \longrightarrow NO_3^-$

$\Delta G^{0'} = -74.1\ \text{kJ/reaction}$

FIGURE 13.10   Reactions involved in the oxidation of inorganic nitrogen compounds by chemolithotrophic nitrifying bacteria (⚬⚬ also Figures 15.32 and 15.33).

Enrichment cultures of nitrifying bacteria are readily established by using mineral salts media containing ammonia or nitrite as electron donor and bicarbonate (HCO_3^-) as sole carbon source. Because of the inefficiency of growth of these organisms (⚬⚬ Section 15.13), visible turbidity may not develop even after extensive nitrification has occurred, and so an easy means of monitoring growth is to assay for the production of nitrite (with ammonia as electron donor) or the disappearance of nitrite or production of nitrate (with nitrite as electron donor). Many nitrifying bacteria, especially the ammonia oxidizers, are inhibited by the traces of organic material present in most agar preparations. Thus, culture of these organisms on solid media requires the use of extensively washed, high-purity agar or a completely inorganic solidifying agent like *silica gel*.

Most of the nitrifying bacteria are obligate chemolithotrophs. Species of *Nitrobacter* are an exception and are able to grow chemoorganotrophically on acetate or pyruvate as sole carbon and energy source. However, although the group is somewhat heterogeneous morphologically, they are fairly tightly related phylogenetically and are either alpha or beta Proteobacteria, with the exception of the genus *Nitrospira*. The latter organism, along with a physiologically unrelated chemoorganotrophic bacterium called *Holophaga* and a number of organisms known only from community sampling (⚬⚬ Section 12.6), form their own kingdom of Bacteria (see Figure 13.1); *Nitrospira* also differs from other nitrifying bacteria by lacking the internal membranes (Figures 13.8 and 13.9) otherwise widespread among the group.

13.3

Sulfur- and Iron-Oxidizing Bacteria

Key Genera

Thiobacillus
Beggiatoa

The ability to grow chemolithotrophically on reduced sulfur compounds is a property of a diverse group of Proteobacteria (Table 13.5). Two broad ecological classes of sulfur-oxidizing bacteria can be discerned, those living at neutral pH and those living at acid pH. Some of the acidophiles also have the ability to grow chemolithotrophically using ferrous iron (Fe^{2+}) as electron donor. We discuss the biogeochemistry of acidophilic sulfur- and iron-oxidizing bacteria in Sections 16.17–16.19 and the biochemistry of these processes in Sections 15.11 and 15.12.

Thiobacillus

The genus *Thiobacillus* contains several gram-negative, rod-shaped bacteria, indistinguishable morphologically from most other gram-negative rods (Figure 13.11), and are the

TABLE 13.5	Physiological characteristics of sulfur-oxidizing chemolithotrophic prokaryotes			
Genus and/or species	Inorganic electron donor	Range of pH for growth	Proteobacterial group	DNA (mol % GC)
Thiobacillus species growing poorly if at all in organic media:				
T. thioparus	H_2S, sulfides, S^0, $S_2O_3^{2-}$	6–8	Alpha, beta, or	61–66
T. denitrificans[b]	H_2S, S^0, $S_2O_3^{2-}$	6–8	gamma	63–68
T. neapolitanus	S^0, $S_2O_3^{2-}$	6–8		52–56
T. thiooxidans	S^0	2–4		51–53
T. ferrooxidans	S^0, metal sulfides, Fe^{2+}	2–4		55–65
Thiobacillus species growing well in organic media:				
T. novellus	$S_2O_3^{2-}$	6–8		66–68
T. intermedius	$S_2O_3^{2-}$	3–7		64
Filamentous sulfur chemolithotrophs:				
Beggiatoa	H_2S, $S_2O_3^{2-}$	6–8	Gamma	37–51
Thiothrix	H_2S	6–8	Gamma	52
Thioploca[a]	H_2S, S^0	—	Gamma	—
Other genera:				
Thiomicrospira[b]	$S_2O_3^{2-}$, H_2S	6–8	Gamma	36–44
Thiosphaera[c]	H_2S, $S_2O_3^{2-}$, H_2	6–8	Alpha	66
Thermothrix[a]	H_2S, $S_2O_3^{2-}$, SO_3^-	6.5–7.5	—	—
Thiovulum	H_2S, S^0	6–8	Epsilon	—

a Facultative aerobes; use NO_3^- as electron acceptor anaerobically.

b One of its species is capable of using NO_3^- anaerobically.

c *Thiosphaera pantotropha* has the exact same 16S rRNA sequence as *Paracoccus denitrificans*.

best studied of the sulfur chemolithotrophs. Phylogenetically, species of *Thiobacillus* scatter among the Proteobacteria, with different species residing in the alpha, beta, and gamma subdivisions. The sulfur compounds most commonly used as electron donors in chemolithotrophic metabolism of *Thiobacillus* species are H_2S, S^0, and $S_2O_3^{2-}$, and the energy-yielding reactions involved are as follows:

$$H_2S + 2O_2 \rightarrow SO_4^{2-} + 2H^+$$

$$\Delta G^{0'} = -798 \text{ kJ/reaction}$$

$$S^0 + H_2O + 1\tfrac{1}{2}O_2 \rightarrow SO_4^{2-} + 2H^+$$

$$\Delta G^{0'} = -587 \text{ kJ/reaction}$$

$$S_2O_3^{2-} + H_2O + 2O_2 \rightarrow 2\,SO_4^{2-} + 2H^+$$

$$\Delta G^{0'} = -818 \text{ kJ/reaction}$$

It is obvious that large amounts of energy are released in these reactions and we will see in Section 15.11 that some of this energy can be trapped as ATP from electron transport reactions leading to a proton motive force. Moreover, these reactions generate large amounts of sulfuric acid and thus several *Thiobacillus* species are acidophilic. One acidophilic species, *T. ferrooxidans*, can also grow chemolithotrophically by the oxidation of ferrous iron and is one of the main biological agents in nature for the oxidation of this metal (∞ Section 16.18). Iron pyrite (FeS_2) is a major source of Fe^{2+} (as well as sulfide) and the oxidation of FeS_2, especially in mining operations, can be both beneficial, because leaching of the ore releases the Fe from the sulfide mineral, and ecologically disastrous, from acidification of the environment and the release of other heavy metals associated with the pyrite (∞ Sections 16.17–16.19).

Culture

Some species of *Thiobacillus* are obligate chemolithotrophs, locked into a lifestyle of using inorganic instead of organic compounds as electron donors. When growing in this

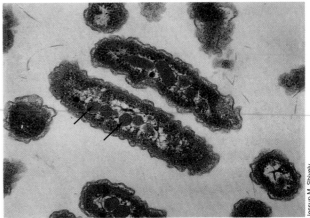

FIGURE 13.11 Transmission electron micrograph of cells of the chemolithotrophic sulfur oxidizer *Thiobacillus neapolitanus*. A single cell is about 0.4 μm in diameter. Note the polyhedral bodies (carboxysomes) distributed throughout the cell (arrows).

fashion they are also autotrophs, converting CO_2 into cell material by reactions of the Calvin cycle (∞ Section 15.7). Other *Thiobacillus* species are *facultative chemolithotrophs,* facultative in the sense that they can grow chemolithotrophically (and thus, also as autotrophs) or chemoorganotrophically (Table 13.5). In addition, however, there are organisms like *Beggiatoa* (a sulfur chemolithotroph of significant historical importance, ∞ box, Winogradsky's Legacy, Chapter 15), most species of which can obtain energy from the oxidation of inorganic sulfur compounds but lack enzymes of the Calvin cycle and thus require organic compounds as carbon source. Such a nutritional lifestyle is called *mixotrophy* (∞ Figure 15.1).

Enrichment cultures of *Thiobacillus* are fairly easy to obtain if a mineral salts medium lacking inorganic compounds but containing large amounts of $S_2O_3^{2-}$ is inoculated with soil or water, common habitats for these bacteria. Such cultures initially become highly turbid from the oxidation of $S_2O_3^{2-}$ to refractive globules of elemental sulfur (S^0), and then the turbidity decreases and the pH drops as the S^0 is oxidized to sulfate (SO_4^{2-}). Pure cultures can usually be obtained from primary enrichments but sulfur and iron chemolithotrophs, indeed virtually *all* chemolithotrophs, can be notoriously difficult to obtain in axenic culture because of the tendency of these organisms to excrete organic compounds that feed contaminating organisms.

Beggiatoa

Organisms of this genus are filamentous, gliding sulfur-oxidizing bacteria, usually quite large in diameter and long, consisting of many short cells attached end to end (Figure 13.12); filaments then flex and twist so that many filaments may become intertwined to form a complex tuft. *Beggiatoa* is found in nature primarily in habitats rich in H_2S, such as sulfur springs (Figure 13.12b), decaying seaweed beds, mud layers of lakes, and waters polluted with sewage, and in these habitats the filaments of *Beggiatoa* are usually filled with sulfur granules (Figure 13.12a). *Beggiatoa* are also common inhabitants of hydrothermal vents (∞ Section 16.12). It was with *Beggiatoa* that Winogradsky first demonstrated that a living organism could oxidize H_2S to S^0 and then to SO_4^{2-}, leading him to formulate the concept of chemolithotrophy (∞ box, Winogradsky's Legacy, Chapter 15). Although a few strains of *Beggiatoa* are truly chemolithotrophic autotrophs, most grow best mixotrophically with reduced sulfur compounds as electron donors and organic compounds as carbon sources.

An interesting habitat of *Beggiatoa* is the rhizosphere of plants (rice, cattails, and other swamp plants) living in flooded, and hence anoxic, soils. Such plants pump oxygen down into their roots so a sharply defined oxic/anoxic boundary develops between the root and the soil. *Beggiatoa* (and probably other sulfur bacteria) develops at

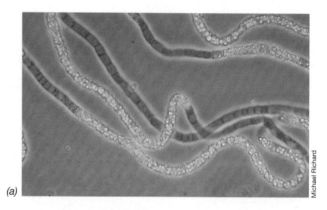

(a)

(b)

FIGURE 13.12 Filamentous sulfur-oxidizing bacteria. (a) Phase contrast photomicrograph of a *Beggiatoa* species isolated from a sewage treatment plant. Note the abundant elemental sulfur granules in some of the cells. (b) Sulfur-oxidizing bacteria in a small stream. The filamentous cells twist together to form thick streamers, and the white color is due to the abundant elemental sulfur content of the cells.

this boundary, and it has been suggested that *Beggiatoa* plays a beneficial role for the plant by oxidizing (and thus detoxifying) the H_2S. The growth of *Beggiatoa* is greatly stimulated by the addition to culture media of the enzyme *catalase* (which converts hydrogen peroxide into water and oxygen), and because plant roots contain catalase, it has also been suggested that the plant promotes the growth of *Beggiatoa* in its rhizosphere via catalase production, thus leading to the development of a loose mutualistic relationship between the plant and the bacterium.

Beggiatoa and other filamentous bacteria like *Sphaerotilus* (see Section 13.14) can cause major settling problems in sewage treatment facilities and in industrial waste lagoons such as from canning, paper pulping, brewing, and milling. These problems are generally referred to as *bulking* and occur when filamentous bacteria overgrow the normal flora of the waste system, producing a loose detrital floc instead of the normal and more easily settling tight floc containing organisms like *Zoogloea* (∞ Section 11.15). If bulking occurs, the wastewater remains improperly treated because the ef-

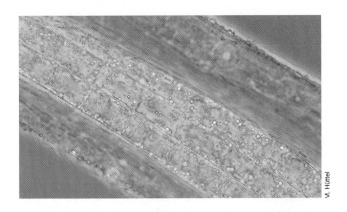

FIGURE 13.13 Cells of a large marine *Thioploca* species. Cells contain sulfur granules and are about 40–50 μm wide.

fluent discharged is still high in organic matter; in sewage treatment, for example, bulking occurs when *Beggiatoa* or other filamentous bacteria replace *Zoogloea* in the activated sludge process (∞ Section 11.15).

Thioploca and *Thiothrix*

Other filamentous sulfur-oxidizing bacteria include *Thioploca* and *Thiothrix*. *Thioploca* is a very large, filamentous sulfur-oxidizing chemolithotroph that forms cell bundles surrounded by a common sheath (Figure 13.13). Thick mats of a marine *Thioploca* species have been found on the ocean floor off the coast of Chile and Peru. Studies on the ecology of these organisms have shown that they carry out the anoxic oxidation of H_2S coupled to the reduction of nitrate (NO_3^-) presumably to N_2 (denitrification) (∞ Sections 15.16 and 16.16). Interestingly, it has been shown that cells of *Thioploca* can accumulate huge amounts of nitrate intracellularly and that this nitrate can then support extended periods of anaerobic respiration with H_2S as electron donor. It is postulated that these marine *Thioploca* mats fix substantial amounts of CO_2 and also play a major role in sulfur and nitrogen cycling. Similar mats consisting primarily of *Beggiatoa* are found near hydrothermal vents (∞ Section 16.12), but the connection with nitrate respiration in these cases is not as well established.

Thiothrix is a filamentous sulfur-oxidizing organism in which the filaments group together at their ends by way of a holdfast to form rosettes, as does the chemoorganotrophic counterpart of *Thiothrix*, *Leucothrix* (Figure 13.14). Physiologically, *Thiothrix* is a mixotroph, and in this and most other respects resembles *Beggiatoa*.

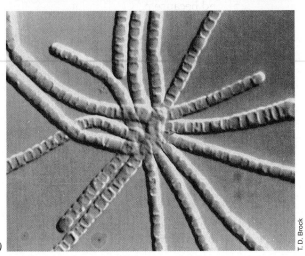

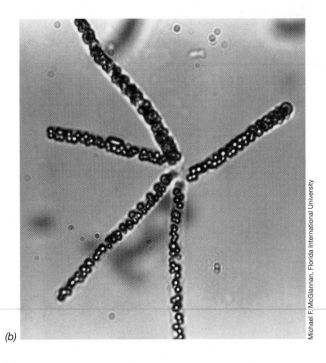

FIGURE 13.14 *Thiothrix and Leucothrix.* (a) A sulfide-containing artesian spring in Florida (USA). The outside of the spring is coated with a mat of *Thiothrix*. (b) Phase contrast photomicrograph of a rosette of cells of *Thiothrix* isolated from the spring and grown in pure culture. Note the internal sulfur globules produced from the oxidation of sulfide. (c) Nomarski interference contrast photomicrograph of a rosette of cells of *Leucothrix*. Cells of both organisms are about 2 μm in diameter.

Hydrogen-Oxidizing Bacteria

Key Genera

Ralstonia
Alcaligenes

A wide variety of bacteria are capable of growing with H_2 as sole electron donor and O_2 as electron acceptor using the "knallgas" reaction, the reduction of O_2 with H_2:

$$2\,H_2 + O_2 \rightarrow 2\,H_2O$$

Many, but not all, of these organisms can also grow autotrophically (using reactions of the Calvin cycle to incorporate CO_2) and are grouped together here as the chemolithotrophic *hydrogen-oxidizing bacteria*. Both gram-positive and gram-negative hydrogen bacteria are known, with the best-studied representatives classified in the genera *Ralstonia* (Figure 13.15), *Pseudomonas, Paracoccus,* and *Alcaligenes* (Table 13.6). All hydrogen-oxidizing bacteria contain one or more *hydrogenase* enzymes that function to bind H_2 and use it either to produce ATP or for reducing power for autotrophic growth.

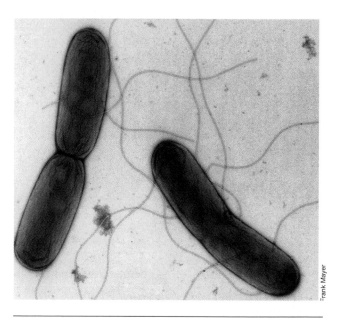

FIGURE 13.15   Transmission electron micrograph of negatively stained cells of the hydrogen-oxidizing chemolithotroph *Ralstonia eutrophus*. A cell is about 0.6 μm in diameter and contains several flagella.

TABLE 13.6	Differential characteristics of species of hydrogen-oxidizing bacteria[a]				
Genus and/or species	**Denitri-fication**	**Growth on fructose**	**Motility**	**DNA (mol % GC)**	**Other characteristics**
Gram-negative					
Acidovorax facilus	−	+	+	64	Membrane-bound hydrogenase
Ralstonia eutrophus	+	+	+	66	Membrane-bound and cytoplasmic hydrogenases
Alcaligenes xylosoxidans	−	+	+	−	Membrane-bound and cytoplasmic hydrogenases
Aquaspirillum autotrophicum	−	−	+	61	Only membrane-bound hydrogenase present
Pseudomonas carboxydovorans	−	−	+	60	Only membrane-bound hydrogenase present; also oxidizes CO
Hydrogenophaga flava	−	+	+	67	Colonies are bright yellow
Seliberia carboxydohydrogena	−	?	+	58	Also oxidizes CO
Paracoccus denitrificans	+	+	−	66	Only membrane-bound hydrogenase present; strong denitrifier
Aquifex pyrophilus	+	−	+	65	Hyperthermophile, grows microaerophilically or anaerobically (with NO_3^-), obligate chemolithotroph; also uses S^0 or $S_2O_3^{2-}$ as electron donor
Hydrogenobacter thermophilus	−	−	−	37–46	As for *Aquifex*, but obligate aerobe (microaerophile)
Gram-positive					
Bacillus schlegelii	−	−	+	66	Produces endospores; thermophile; also uses CO or $S_2O_3^{2-}$ as electron donor
Arthrobacter sp.	−	+	−	70	Only membrane-bound hydrogenase present
Mycobacterium gordonae	−	?	−	−	Acid-fast; colonies yellow to orange

a Phylogenetically, most H_2-oxidizing bacteria are α or β Proteobacteria (see Table 13.1), except for the gram-positive organisms; *Aquifex* and *Hydrogenobacter* represent one of the major lineages of Bacteria (see Figure 13.1).

Almost all hydrogen bacteria are *facultative chemolithotrophs,* meaning that they can also grow chemoorganotrophically with organic compounds as energy sources. This is a major distinction between hydrogen chemolithotrophs and many sulfur chemolithotrophs or nitrifying bacteria; most representatives of the latter two groups are *obligate chemolithotrophs*—growth does not occur in the absence of the inorganic energy source. By contrast, hydrogen chemolithotrophs can switch between chemolithotrophic and chemoorganotrophic modes of metabolism and presumably do so in nature as nutritional conditions warrant.

Physiology and Ecology of Hydrogen Bacteria

Most hydrogen bacteria grow best under microaerobic conditions when growing chemolithotrophically on H_2 because hydrogenases are oxygen-sensitive enzymes. Typically, oxygen levels of about 5–10% support best growth. *Nickel* must be present in the medium for chemolithotrophic growth of hydrogen bacteria because virtually all hydrogenases contain Ni^{2+} as a metal cofactor. A few hydrogen bacteria also fix molecular N_2, and when growing on N_2, the organisms are quite oxygen-sensitive because the enzyme nitrogenase needed for the reduction of molecular nitrogen (∞ Section 15.29) is an oxygen-sensitive enzyme.

Hydrogen-oxidizing bacteria can be enriched if a small amount of mineral salts medium containing trace metals (especially Ni^{2+} and Fe^{2+}) is inoculated with soil or water and incubated in a large, sealed flask containing a head space of 5% O_2, 10% CO_2, and 85% H_2. When the liquid becomes turbid, plates of the same medium are streaked and incubated in a glass jar containing the same gas mixture (one must exercise care in mixing gas phases, however—mixtures of O_2 and H_2 are potentially explosive).

Some hydrogen bacteria can grow on carbon monoxide, CO, as energy source, with electrons from the oxidation of CO to CO_2 entering the electron transport chain to drive ATP synthesis. CO-oxidizing bacteria, which are called *carboxydotrophic* bacteria, grow autotrophically using Calvin cycle reactions to fix the CO_2 generated from the oxidation of CO. CO is oxidized to CO_2 by the enzyme *carbon monoxide dehydrogenase,* which is a molybdenum-containing enzyme. The molybdenum in CO dehydrogenase is bound to a small cofactor consisting of a multiringed structure called a *pterin,* similar to the situation in the enzyme nitrate reductase (∞ Section 15.16).

CO consumption by carboxydotrophic bacteria is a very significant ecological process. Although much CO is generated from various human and other sources, CO levels in air have not risen significantly over many years. Microbial CO consumption is probably the reason why. Because the most significant releases of CO (primarily from automobile exhaust, incomplete combustion of fossil fuels, and the catabolism of lignin) occur in oxic environments, carboxydotrophic bacteria in the upper layers of soil probably represent the most significant sink for CO in nature. Some of the best-studied carboxydobacteria include *Pseudomonas carboxydovorans, Bacillus schlegelii,* and *Alcaligenes carboxydus* (Table 13.6). At least one carboxydobacterium can grow on CO anaerobically with nitrate as electron acceptor, but this does not seem to be a widespread property of the group. Like the hydrogen bacteria, virtually all isolates of carboxydotrophic bacteria grow chemoorganotrophically on organic substrates as well as on CO.

✓ 13.2–13.4 Concept Check

Chemolithotrophs are prokaryotes that can oxidize inorganic electron donors and in many cases use CO_2 as sole carbon source.

✓ Compare and contrast the nitrifying bacteria with the sulfur, iron, and hydrogen bacteria in terms of inorganic electron donors used, carbon sources, E_0' of electron donors (∞ Chapter 15), and habitats.

✓ What major pathway is present for assimilation of CO_2 in many chemolithotrophs?

13.5

Methanotrophs and Methylotrophs

Key Genera

Methylomonas
Methylobacter

Methane, CH_4, is found extensively in nature. It is produced in anoxic environments by methanogenic archaea (∞ Sections 14.3 and 15.19) and is a major gas of anoxic muds, marshes (∞ Figure 14.5), anoxic zones of lakes, the rumen, and the mammalian intestinal tract. Methane is the major constituent of natural gas and is also present in many coal formations. It is a relatively stable molecule; but a variety of bacteria, the **methanotrophs,** oxidize it readily, utilizing methane and a few other one-carbon compounds as electron donors for energy generation and as sole sources of carbon. These bacteria are all aerobes and are widespread in nature in soil and water. They exhibit diverse morphologies and are related both phylogenetically and in their ability to oxidize methane.

C_1 Metabolism

In addition to methane, a number of other one-carbon compounds are known to be utilized by microorganisms. A list of these compounds is given in Table 13.7.

TABLE 13.7	Substrates used by methylotrophic bacteria[a]
Substrates used for growth	**Substrates oxidized but not used for growth (cometabolism)**
Methane, CH_4	Ammonium, NH_4^+
Methanol, CH_3OH	Ethylene, $H_2C=CH_2$
Methylamine, CH_3NH_2	Chloromethane, CH_3Cl
Dimethylamine, $(CH_3)_2NH$	Bromomethane, CH_3Br
Trimethylamine, $(CH_3)_3N$	Higher hydrocarbons (ethane, propane)
Tetramethylammonium, $(CH_3)_4N^+$	
Trimethylamine N-oxide, $(CH_3)_3NO$	
Trimethylsulfonium, $(CH_3)_3S^+$	
Formate, $HCOO^-$	
Formamide, $HCONH_2$	
Carbon monoxide, CO	
Dimethyl ether, $(CH_3)_2O$	
Dimethyl carbonate, $CH_3OCOOCH_3$	
Dimethyl sulfoxide, $(CH_3)_2SO$	
Dimethylsulfide, $(CH_3)_2S$	

a A single isolate does not use all of the above, but at least one methylotrophic bacterium has been reported to oxidize each of the listed compounds.

From a biochemical viewpoint, these compounds share a key characteristic: *they contain no carbon–carbon bonds*. Thus, all carbon–carbon bonds of the cell must be synthesized de novo. Organisms that can grow using only one-carbon organic compounds are generally called **methylotrophs.** Many, but not all, methylotrophs are also methanotrophs. From the viewpoint of carbon assimilation, methanotrophs and methylotrophs have something in common with autotrophs (◯◯ Chapter 15), which also use a carbon compound lacking a carbon–carbon bond, CO_2. The two groups differ, however, in that the methylotrophs utilize a carbon compound *more reduced* than CO_2.

It is important to distinguish between methylotrophs and methanotrophs. A wide variety of bacteria are known that can grow on methanol, methylamine, or formate, but not methane, and these bacteria are members of various genera of chemoorganotrophs, for example, *Hyphomicrobium, Pseudomonas, Bacillus,* and *Vibrio.* In contrast, methanotrophs are unique in that they can grow not only on some of the more oxidized one-carbon compounds but also on methane. The methane-oxidizing bacteria possess a specific enzyme system, *methane monooxygenase,* for the introduction of an oxygen atom into the methane molecule, leading to the formation of methanol (◯◯ Section 15.14). The requirement for O_2 as a reactant in the initial oxygenation of methane thus explains why all methanotrophs are obligate aerobes. All methanotrophs also appear to be *obligate* C_1 utilizers, unable to utilize compounds containing carbon–carbon bonds. By contrast, many nonmethanotrophic methylotrophs are able to utilize organic acids, ethanol, and sugars.

Methane-oxidizing bacteria are unique among prokaryotes in possessing relatively large amounts of

sterols. As we noted in Section 3.5, sterols are found in eukaryotes as a functional part of the membrane system but are absent from most prokaryotes. In methanotrophs, sterols may be an essential part of the complex internal membrane system (see later) involved in methane oxidation.

Classification

An overview of the classification of methanotrophs is given in Table 13.8. These bacteria were initially distinguished on the basis of morphology and formation of resting stages, but it was then found that they could be divided into two major groups based on their internal cell structure and carbon assimilation pathway. *Type I* organisms assimilate one-carbon compounds via a unique pathway, the **ribulose monophosphate cycle,** whereas *Type II* organisms assimilate C_1 intermediates via the **serine pathway.** We discuss the biochemical details of these pathways in Section 15.14.

Both groups of methanotrophs contain extensive internal membrane systems, which appear to be related to their methane-oxidizing ability. Type I methanotrophs are characterized by internal membranes arranged as bundles of disc-shaped vesicles distributed throughout the cell (Figure 13.16b), whereas Type II species possess paired membranes running along the periphery of the cell (Figure 13.16a). Type I methanotrophs are also characterized by a lack of a complete citric acid cycle (the enzyme α-ketoglutarate dehydrogenase is absent), whereas Type II organisms possess a complete cycle. Absence of a complete citric acid cycle greatly diminishes the ability of an organism to grow chemoorganotrophically. If these reactions cannot be run as a cycle, NADH cannot

TABLE 13.8   Some characteristics of methanotrophic bacteria

Organism	Morphology	16S rRNA group[a]	Resting stage	Internal membranes[b]	Citric acid cycle[c]	Carbon assimilation pathway[d]	N₂ fixation	DNA (mol % GC)
Methylomonas	Rod	Gamma	Cystlike body	I	Incomplete	Ribulose monophosphate	No	50–54
Methylomicrobium	Rod	Gamma	None	I	Incomplete	Ribulose monophosphate	No	49–60
Methylobacter	Coccus to ellipsoid	Gamma	Cystlike body	I	Incomplete	Ribulose monophosphate	No	50–54
Methylococcus	Coccus	Gamma	Cystlike body	I	Incomplete	Ribulose monophosphate	Yes	62–64
Methylosinus	Rod or vibrioid	Alpha	Exospore	II	Complete	Serine	Yes	63
Methylocystis	Rod	Alpha	Exospore	II	Complete	Serine	Yes	63

a All are Proteobacteria.

b Internal membranes: Type I, bundles of disc-shaped vesicles distributed throughout the organism; Type II, paired membranes running along the periphery of the cell. See Figure 13.16.

c Organisms with an incomplete citric acid cycle lack the enzyme α-ketoglutarate dehydrogenase and thus cannot oxidize acetate to CO_2.

d See Figures 15.35 and 15.36. Unlike other methylotrophs, *Methylococcus* species contain Calvin cycle enzymes.

be generated from reactions of the cycle, thus preventing growth at the expense of organic compounds metabolized through the citric acid cycle.

Ecology and Isolation

Methanotrophs are widespread in aquatic and terrestrial environments, being found wherever stable sources of methane are present. Methane produced in the anoxic regions of lakes rises through the water column, and methanotrophs are often concentrated in a narrow band at the thermocline, where methane from the anoxic zone meets oxygen from the oxic zone. Methane-oxidizing bacteria therefore play an important role in the carbon cycle, converting methane derived from anoxic decomposition back into cell material (and CO_2).

For the initial enrichment of methanotrophs all that is needed is a mineral salts medium over which an atmosphere of 80% methane and 20% air is maintained. Once good growth is obtained, purification is carried out by streaking on mineral salts agar plates, which are incubated in a jar with the methane–air mixture. Colonies appearing on the plates are of two types, common chemoorganotrophs growing on traces of organic matter in the medium, which appear in 1–2 days, and methanotrophs, which appear after about a week. The colonies of many methanotrophs are pink in color from the production of various carotenoid pigments, and this feature can help in their isolation.

Methanotrophs are able to oxidize ammonia, although they cannot grow chemolithotrophically using ammonia as sole electron donor. In addition to methane

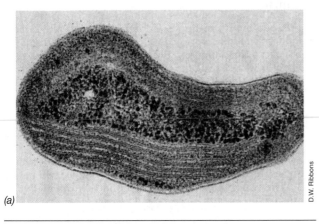

(a)

D.W. Ribbons

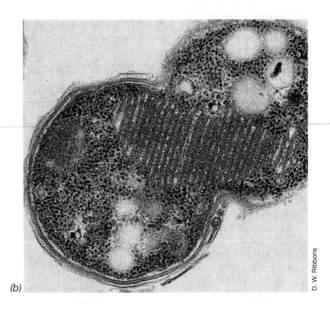

(b)

D. W. Ribbons

FIGURE 13.16   Electron micrographs of methanotrophs. (a) A *Methylosinus* species, illustrating a Type II membrane system. Cells are about 0.6 μm in diameter. (b) *Methylococcus capsulatus*, illustrating a Type I membrane system. Cells are about 1 μm in diameter.

oxidation, methane monooxygenase also functions to oxidize ammonia, and a competitive interaction between the two substrates exists. For this reason, ammonia is generally toxic to methanotrophs, and the preferred nitrogen source is nitrate. It has been speculated that methanotrophic bacteria could have evolved from the nitrosifying bacteria via genetic changes causing the conversion of ammonia oxidase to methane oxidase. The fact that both groups of bacteria have elaborate internal membrane systems (see Section 13.2) supports such a theory. In addition, however, methanotrophic bacteria contain some of the same genes and make some of the same proteins as methanogenic (methane-producing) prokaryotes, which phylogenetically are Archaea (⌀⌀ Section 14.3). We will see how the contrasting processes of methanogenesis and methanotrophy are related in this regard in Sections 15.14 and 15.19.

Methanotrophic Symbionts of Animals

A symbiotic association between methanotrophic bacteria and marine mussels and certain types of marine sponges is known to occur. Mussels live in the vicinity of hydrocarbon seeps where methane is released in substantial

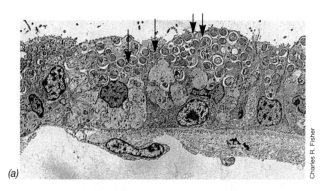

(a)

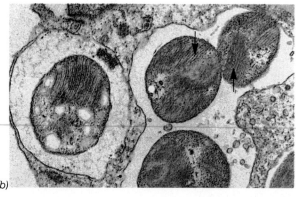

(b)

Charles R. Fisher

FIGURE 13.17 Methanotrophic symbionts of marine mussels. (a) Electron micrograph of a thin section at low magnification of gill tissue of a marine mussel living near hydrocarbon seeps in the Gulf of Mexico. Note the symbiotic methanotrophs (arrows) in the tissues. (b) High magnification view of gill tissue showing Type I methanotrophs. Note membrane bundles (arrows). The methanotrophs are about 1 μm in diameter.

amounts. Intact mussels as well as isolated mussel gill tissue consume methane at high rates in the presence of O_2. In the gill tissue of the mussel, coccoid-shaped bacteria are present in high numbers (Figure 13.17*a*). The bacterial symbionts contain stacks of intracytoplasmic membranes (Figure 13.17*b*) typical of Type I methanotrophs. The symbionts are found in vacuoles within animal cells near the gill surface, which probably ensures an effective gaseous exchange with seawater. Presumably methane assimilated by the methanotrophs is distributed throughout the animals by the excretion of carbon compounds by the methanotrophs. The methanotrophic symbiosis is therefore conceptually similar to the symbiosis established between sulfide-oxidizing chemolithotrophs and hydrothermal vent tube worms and giant clams discussed in Section 16.12. Animal–bacteria symbioses, such as the methanotrophic mussel/sponge symbiosis and the sulfide-oxidizing vent animal symbioses, thus show that prokaryotic cells can occasionally constitute the basis of a one-step food chain.

✓ 13.5 Concept Check

Methylotrophs are prokaryotes able to grow on carbon compounds that lack carbon–carbon bonds. Some methylotrophs are also methanotrophs, able to grow on CH_4. Two classes of methanotrophs are known, each having a number of structural and biochemical properties in common. Methanotrophs reside in water and soil and can also exist as symbionts of marine shellfish.

- ✓ What is the difference between a *methanotroph* and a *methylotroph*?
- ✓ What features differentiate Type I from Type II methanotrophs?
- ✓ What types of animals harbor methanotrophic symbionts and where does the methane the symbionts need come from?

13.6

Pseudomonas and the Pseudomonads

Key Genera

Pseudomonas
Burkholderia
Zymomonas
Xanthomonas

All the genera in this group are straight or slightly curved chemoorganotrophic and aerobic rods with *polar* flagella (Figure 13.18). The important genera are *Pseudomonas*, *Commamonas*, *Ralstonia*, and *Burkholderia*, discussed in some detail here. Other genera include *Xanthomonas*, primarily a plant pathogen that is responsible for a number of necrotic plant lesions and that is characterized by its yellow-colored pigments; *Zoogloea*, characterized by its formation of an extracellular fibrillar polymer, which

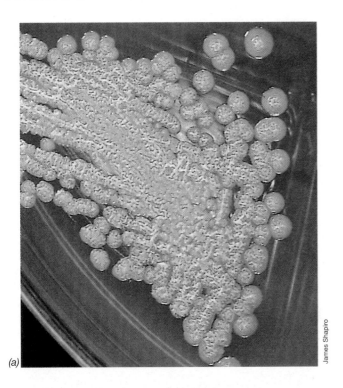

(a)

(b)

James Shapiro

Arthur Kelman

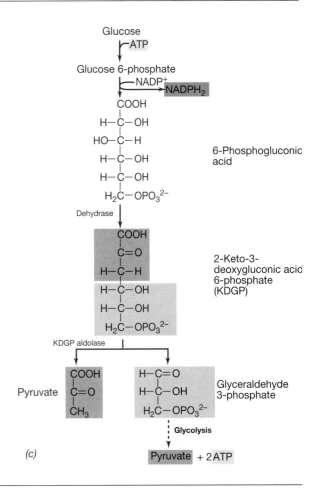

(c)

FIGURE 13.18 Typical pseudomonad colony and cell morphology and a biochemical pathway common in pseudomonads. (a) Photograph of colonies of *Burkholderia cepacia* on an agar plate. (b) Shadow-cast preparation of a *Pseudomonas* species. The cell measures about 1 μm in diameter. (c) The Entner–Doudoroff pathway, the major means of glucose catabolism in pseudomonads.

TABLE 13.9   Characteristics of pseudomonads

General characteristics:
Straight or curved rods but not vibrioid; size 0.5–1.0 μm by 1.5–4.0 μm; no spores; gram-negative; polar flagella: single or multiple; no sheaths, appendages, or buds; respiratory metabolism, never fermentative, although may produce small amounts of acid from glucose aerobically; use low-molecular-weight organic compounds, not polymers; some are chemolithotrophic, using H_2 or CO as sole electron donor; some can use nitrate as electron acceptor anaerobically; some can use arginine as energy source anaerobically

Minimal characteristics for identification:
Gram-negative, straight or slightly curved; no spores; motile (always); polar flagella (flagellar stain); oxidative-fermentative medium with glucose: tube open, acid produced; tube sealed, acid not produced; gas not produced from glucose (distinguishes them easily from enteric bacteria and *Aeromonas*); oxidase, almost always positive (enterics are oxidase-negative); catalase always positive; photosynthetic pigments absent (distinguishes them from purple nonsulfur bacteria); indole-negative; methyl red-negative; Voges–Proskauer-negative (for discussion of many of these biochemical tests, see Section 21.2)

---

causes the cells to aggregate into distinctive flocs (this organism is a dominant component of activated sewage sludge) (∞ Section 11.15), and *Gluconobacter*, characterized by its incomplete oxidation of sugars or alcohols to acids, such as the oxidation of glucose to gluconic acid or ethanol to acetic acid (this organism is discussed briefly with the other acetic acid bacteria in Section 13.7). Phylogenetically, the various genera of pseudomonads scatter within the Proteobacteria (see Table 13.1).

Characteristics

The distinguishing characteristics of the pseudomonad group are given in Table 13.9. Also listed in this table are the minimal characteristics needed to identify an organism as a pseudomonad. Key identifying characteristics are the absence of gas formation from glucose, and the positive oxidase test, both of which help to distinguish pseudomonads from enteric bacteria (see Section 13.10).

The species of the genus *Pseudomonas* are defined on the basis of phylogeny and various physiological characteristics, as outlined in Tables 13.10 and 13.11. Pseudomonads have very simple nutritional requirements and grow chemoorganotrophically at neutral pH and at temperatures in the mesophilic range. One of the striking properties of many species of pseudomonads is the wide variety of organic compounds used as carbon and energy sources. Some species utilize over *100* different compounds, and only a few species utilize fewer than 20. As an example of this versatility, a single strain of *Pseudomonas cepacia* (now *Burkholderia cepacia*) can use many different sugars, fatty acids, dicarboxylic acids, tricarboxylic acids, alcohols, polyalcohols, glycols, aromatic compounds, amino acids, and amines, plus miscellaneous organic compounds not fitting into any of

the preceding categories. On the other hand, pseudomonads generally lack the hydrolytic enzymes necessary to break down polymers into their component monomers. Nutritionally versatile pseudomonads typically contain numerous inducible operons (∞ Section 7.2) because the catabolism of unusual organic substrates often requires the activity of several different enzymes. The pseudomonads are ecologically important organisms in soil and water and are probably responsible for the degradation of many soluble compounds derived from the breakdown of plant and animal materials in oxic habitats.

Many pseudomonads, as well as a variety of other gram-negative Bacteria, metabolize glucose via the Entner–Doudoroff pathway (Figure 13.18). Two key enzymes of the Entner–Doudoroff pathway are *6-phos-*

TABLE 13.10 Characteristics of subgroups and species of the genera *Pseudomonas, Commamonas, Ralstonia,* and *Burkholderia*

Group	16s rRNA group[a]	Characteristics	DNA (mol % GC)
Fluorescent subgroup	Gamma	**Most produce water-soluble, yellow-green fluorescent pigments; do not form poly-β-hydroxybutyrate; single DNA homology group**	
Pseudomonas aeruginosa		Pyocyanin production, growth at up to 43°C, single polar flagellum, capable of denitrification	67
Pseudomonas fluorescens		Does not produce pyocyanin or grow at 43°C; tuft of polar flagella	59–61
Pseudomonas putida		Similar to *P. fluorescens* but does not liquefy gelatin and does grow on benzylamine	60–63
Pseudomonas syringae		Lacks arginine dihydrolase, oxidase-negative, pathogenic to plants	58–60
Pseudomonas stutzeri		Soil saprophyte; strong denitrifyer and nonfluorescent	62
Acidovorans subgroup	Beta	**Nonpigmented, form poly-β-hydroxybutyrate, tuft of polar flagella, do not use carbohydrates; single DNA homology group**	
Commamonas acidovorans		Uses muconic acid as sole carbon source and electron donor	67
Commamonas testosteroni		Uses testosterone as sole carbon source	62
Pseudomallei-cepacia subgroup	Beta	**No fluorescent pigments, tuft of polar flagella, forms poly-β-hydroxybutyrate; single DNA homology group**	**62**
Burkholderia cepacia		Extreme nutritional versatility; some strains pathogenic to plants	67
Burkholderia pseudomallei		Causes melioidosis in animals; nutritionally versatile	69
Burkholderia mallei		Causes glanders in animals; nonmotile; nutritionally restricted	69
Diminuta-vesicularis subgroup	Alpha	**Single flagellum of very short wavelength, require vitamins (pantothenate, biotin, B_{12})**	
Pseudomonas diminuta		Nonpigmented, does not use sugars	66–67
Pseudomonas vesicularis		Carotenoid pigment, uses sugars	66
Ralstonia **subgroup**	Beta		
Ralstonia solanacearum		Plant pathogen	66–68
Ralstonia saccharophila		Grows chemolithotrophically with H_2, digests starch	69
Pseudomonas maltophilia		Requires methionine, does not use NO_3^- as N source, oxidase-negative	67

a All pseudomonads are members of the Proteobacteria (see Table 13.1).

TABLE 13.11 Pathogenic pseudomonads

Species	Relationship to disease
Animal pathogens	
P. aeruginosa	Opportunistic pathogen, especially in hospitals; in patients with metabolic, hematologic, and malignant diseases; hospital-acquired (nosocomial) infections from catheterizations, tracheostomies, lumbar punctures, and intravenous infusions; in patients given prolonged treatment with immunosuppressive agents, corticosteroids, antibiotics and radiation; may contaminate surgical wounds, abscesses, burns, ear infections, lungs of patients treated with antibiotics; primarily a soil organism
P. fluorescens	Rarely pathogenic, as does not grow well at 37°C; may grow in and contaminate blood and blood products under refrigeration
P. maltophilia	A ubiquitous, free-living organism that is a common nosocomial pathogen
B. cepacia	Causes onion bulb rot; has also been isolated from humans and from environmental sources of medical importance
B. pseudomallei	Causes melioidosis, a disease endemic in animals and humans in Southeast Asia
B. mallei	Causes glanders, a disease of horses that is occasionally transmitted to humans
P. stutzeri	Often isolated from humans and environmental sources; may live saprophytically in the body
Plant pathogens	
R. solanacearum	Causes wilts of many cultivated plants (for example, potato, tomato, tobacco, peanut)
P. syringae	Attacks foliage, causing chlorosis and necrotic lesions on leaves; rarely found free in soil
P. marginalis	Causes soft rot of various plants; active pectinolytic species
Xanthomonas	Causes necrotic lesions on foliage, stems, fruits; also causes wilts and tissue rots; rarely found free in soil

phogluconate dehydrase and *2-keto-3-deoxyglucosephosphate aldolase* (Figure 13.18c). A survey for the presence of these enzymes in a variety of bacteria has shown that they are absent from gram-positive Bacteria (except for a few *Nocardia* isolates) but are generally present in bacteria of the genera *Pseudomonas, Rhizobium, Agrobacterium, Zymomonas,* and several other gram-negative bacteria.

Pathogenic Pseudomonads

A number of pseudomonads are pathogenic (Table 13.11). Among the fluorescent pseudomonads, the species *Pseudomonas aeruginosa* is frequently associated with infections of the urinary and respiratory tracts in humans. *Pseudomonas aeruginosa* infections are also common in patients receiving treatment for severe burns or other traumatic skin damage, and in people suffering from cystic fibrosis. *Pseudomonas aeruginosa* is not an obligate parasite, however, but appears to be primarily an opportunist, initiating infections in individuals whose resistance is low. In addition to urinary infections, it can also cause systemic infections, usually in individuals who have experienced extensive skin damage. The organism is naturally resistant to many of the widely used antibiotics, so chemotherapy is often difficult. Resistance is frequently due to a *resistance transfer plasmid (R plasmid)* (∞ Sections 9.8 and 18.12), which is a plasmid carrying genes coding for detoxification of various antibiotics. *Pseudomonas aeruginosa* is commonly found in the hospital environment and can easily infect patients receiving treatment for other illnesses (see Section 22.7 for a discussion of hospital-acquired infections). Polymyxin, an antibiotic not ordinarily

used in human therapy because of its toxicity, is effective against *P. aeruginosa* and can be used with caution.

Certain species of *Pseudomonas, Ralstonia,* and *Burkholderia* and the genus *Xanthomonas* are well-known plant pathogens (phytopathogens) (see Table 13.11). In many cases these organisms are so highly adapted to the plant environment that they can rarely be isolated from other habitats, including soil. Phytopathogens frequently inhabit nonhost plants (where disease symptoms are not apparent) and from there become transmitted to host plants and initiate infection. Disease symptoms vary considerably depending on the particular phytopathogen and host plant and are generally due to the release by the bacterium of plant toxins, lytic enzymes, plant growth factors, and other substances that destroy or distort plant tissue. In many cases the disease symptoms are highly diagnostic of the type of pseudomonad phytopathogen; thus, *Pseudomonas syringae* is frequently isolated from leaves showing chlorotic (yellowing) lesions, whereas *P. marginalis* is a typical "soft-rot" pathogen, infecting stems and shoots but rarely leaves.

Zymomonas

The genus *Zymomonas* consists of large, gram-negative rods that carry out a vigorous fermentation of sugars to ethanol. Although strictly fermentative, *Zymomonas* shows phylogenetic affiliation with the pseudomonads and contains Entner–Doudoroff pathway (Figure 13.18c) enzymes. *Zymomonas* is a common organism involved in alcoholic fermentation of various plant saps, and in many tropical areas of South and Central America,

Africa, and Asia, it occupies a position in the fermented beverage industry similar to that of *Saccharomyces cerevisiae* (yeast) in North America and Europe. *Zymomonas* is involved in the alcoholic fermentation of agave in Mexico, to form the drink, *pulque,* and palm sap in many tropical areas. It also carries out an alcoholic fermentation of sugarcane juice and honey. Although *Zymomonas* is rarely the sole organism involved in these alcoholic fermentations, it is often the dominant organism and is probably responsible for the production of most of the ethanol (the desired product) in these beverages. *Zymomonas* is also responsible for spoilage of fruit juices such as apple cider and perry and is also a constituent of the bacterial flora of spoiled beer.

Zymomonas is distinguished from *Pseudomonas* by its fermentative metabolism, microaerophilic to anaerobic nature, oxidase negativity, and other molecular taxonomic characteristics. It also resembles the acetic acid bacteria (see Section 13.7) and it is often found in nature associated with these organisms. This is of interest because, like yeast, *Zymomonas* ferments glucose to ethanol, whereas the acetic acid bacteria oxidize ethanol to acetic acid. Thus, the acetic acid bacteria probably depend on the activity of yeast and *Zymomonas* for the production of their growth substrate, ethanol. Unlike yeast, however, which ferments glucose to ethanol via the Embden–Meyerhof–Parnas (glycolytic) pathway, *Zymomonas* employs the Entner–Doudoroff pathway.

Zymomonas is of interest to the ethanol industry because it shows higher rates of glucose uptake and ethanol production and gives a higher yield of ethanol than many yeasts. *Zymomonas* is also rather tolerant of high ethanol concentrations (up to 10%) but is not quite as tolerant as some of the best yeast strains, which can grow to 12–15% ethanol. However, the fact that *Zymomonas* is a *gram-negative* bacterium and can thus be readily manipulated genetically makes it an attractive candidate for use by ethanol production industries.

13.7

Acetic Acid Bacteria

Key Genera

Acetobacter
Gluconobacter

As originally defined, the *acetic acid bacteria* comprised a group of gram-negative, aerobic, motile rods that carried out *incomplete* oxidation of alcohols and sugars, leading to the accumulation of organic acids as end products. With *ethanol* as a substrate, *acetic acid* is produced; hence the derivation of the common name for these bacteria. Another property is the relatively high tolerance to acidic conditions, most strains being able

to grow well at pH values lower than 5. This acid tolerance should of course be essential for an organism producing large amounts of acid. The acetic acid bacteria are a heterogeneous assemblage, comprising both peritrichously and polarly flagellated organisms. The *polarly* flagellated organisms are classified in the genus *Gluconobacter,* while the peritrichously flagellated species are grouped into the genus *Acetobacter.* All known acetic acid bacteria group phylogenetically with the alpha Proteobacteria (see Table 13.1).

In addition to flagellation, *Acetobacter* differs from *Gluconobacter* in being able to further oxidize the acetic acid it forms to CO_2. This difference in ability to oxidize acetic acid is related to the presence of a complete citric acid cycle. *Gluconobacter,* which *lacks* a complete citric acid cycle, is unable to oxidize acetic acid, whereas *Acetobacter,* which has all enzymes of the cycle, can oxidize it.

Ecology and Industrial Uses

The acetic acid bacteria are frequently found in association with alcoholic juices. Acetic acid bacteria can often be isolated from an alcoholic fruit juice such as hard cider or wine, or from beer. Colonies of acetic acid bacteria can be recognized on $CaCO_3$–agar plates containing ethanol, since the acetic acid produced causes a dissolution and clearing of the insoluble $CaCO_3$ (Figure 13.19). Cultures of acetic acid bacteria are used in the commercial production of vinegar (∞ Section 11.10).

In addition to ethanol, these organisms carry out an incomplete oxidation of such organic compounds as higher alcohols and sugars. For instance, glucose is oxidized only to gluconic acid, galactose to galactonic acid,

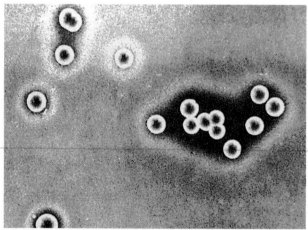

FIGURE 13.19 Photograph of colonies of *Acetobacter aceti* on calcium carbonate agar containing ethanol as energy source. Note the clearing around the colonies due to the dissolution of calcium carbonate by the acetic acid produced by the bacteria.

arabinose to arabonic acid, and so on. This property of "underoxidation" is exploited in the manufacture of ascorbic acid (vitamin C). Ascorbic acid can be formed from sorbose, but sorbose is difficult to synthesize chemically. It is, however, conveniently obtainable from acetic acid bacteria, which oxidize sorbitol (a readily available sugar alcohol) only to sorbose, a process called *bioconversion* (⌘ Section 11.8).

Another interesting property of some acetic acid bacteria is their ability to synthesize *cellulose*. The cellulose formed does not differ significantly from that of plant cellulose, with the exception that it is pure and not mixed in with other polymers like hemicelluloses, pectin, or lignins, and is formed as a matrix outside the wall where the bacteria become embedded in the tangled mass of cellulose microfibrils. When these species of acetic acid bacteria grow in an unshaken vessel, they form a surface pellicle of cellulose in which the bacteria develop. Since these bacteria are obligate aerobes, the ability to form such a pellicle may be a means by which the organisms are assured of remaining at the surface of the liquid where oxygen is readily available.

13.8

Free-Living Aerobic Nitrogen-Fixing Bacteria

Key Genera

Azotobacter
Azomonas

A variety of organisms that inhabit primarily the soil are capable of fixing N_2 *aerobically* (Table 13.12). The genus *Azotobacter* comprises large, gram-negative, obligately aerobic rods capable of fixing N_2 nonsymbiotically (Figure 13.20). The first species of this genus was discovered by the Dutch microbiologist M. W. Beijerinck early in the twentieth century, using an aerobic enrichment culture technique with a medium containing N_2 (air) but devoid of a combined nitrogen source (⌘ box, Rise of General Microbiology, Chapter 16). Most

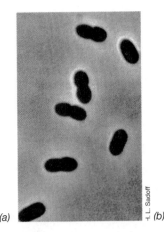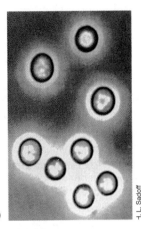

(a)   (b)   ⊣. L. Sadoff

FIGURE 13.20 *Azotobacter vinelandii:* (a) vegetative cells and (b) cysts by phase contrast microscopy. A cell measures about 2 μm in diameter, and a cyst about 3 μm. Compare with Figure 1.21*a*.

free-living nitrogen-fixing bacteria are alpha or gamma Proteobacteria.

Taxonomy

The major free-living nitrogen-fixing bacteria that have been studied include *Azotobacter, Azospirillum,* and *Beijerinckia. Azotobacter* cells are large, many isolates being almost the size of yeasts, with diameters of 2–4 μm or more. Pleomorphism is common, and a variety of cell shapes and sizes have been described. Some strains are motile by peritrichous flagella. On carbohydrate-containing media, extensive capsules or slime layers are produced by free-living N_2-fixing bacteria (Figure 13.21). *Azotobacter* is able to grow on a wide variety of carbohydrates, alcohols, and organic acids. The metabolism of carbon compounds is strictly oxidative, and acids or other fermentation products are rarely produced. All members fix nitrogen, but growth also occurs on simple forms of combined nitrogen: ammonia, urea, and nitrate.

Azotobacter can form resting structures called *cysts.* Like bacterial endospores, *Azotobacter* cysts (Figure 13.20*b*) show negligible endogenous respiration and are resistant to desiccation, mechanical disintegration, and ultraviolet and ionizing radiation. In contrast to

TABLE 13.12	Genera of free-living aerobic nitrogen-fixing bacteria[a]		
Genus	**Number of species**	**Characteristics**	**DNA (mol % GC)**
Azotobacter	9	Large rod; produces cysts; primarily found in neutral to alkaline soils	63–67
Azomonas	3	Large rod; no cysts; primarily aquatic	52–59
Azospirillum	4	Microaerophilic rod; associates with plants	69–71
Beijerinckia	4	Pear-shaped rod with large lipid bodies at each end; produces extensive slime; inhabits acidic soils	54–59
Derxia	1	Rods; form coarse, wrinkled colonies	69–73

a All species examined are members of the Proteobacteria, primarily the alpha and gamma subdivisions (see Table 13.1).

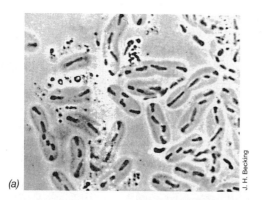

(a)

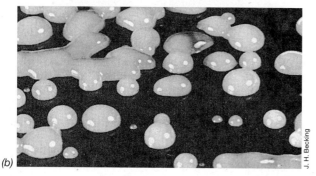

(b)

FIGURE 13.21 Examples of slime production by free-living N₂-fixing bacteria. (a) Cells of *Derxia gummosa* encased in slime. Cells are about 1–1.2 μm wide. (b) Colonies of *Beijerinckia* species growing on a carbohydrate-containing medium. Note the raised, glistening appearance of the colonies due to abundant capsular slime.

endospores, however, they are *not* especially heat-resistant, and they are not completely dormant because they rapidly oxidize exogenous energy sources.

Despite the fact that *Azotobacter* is an obligate aerobe, its nitrogenase, the enzyme that catalyzes biological N₂ fixation (⚮ Section 15.29), is O₂-sensitive. It is thought that the high respiratory rate characteristic of *Azotobacter* has something to do with protection of nitrogenase from O₂. The intracellular O₂ concentration is thus kept low enough by respiration that inactivation of nitrogenase does not occur.

The remaining genera of free-living N₂ fixers include *Azomonas*, a genus of large, rod-shaped bacteria that resemble *Azotobacter* except that they do not produce cysts and are primarily aquatic, *Beijerinckia* and *Derxia* (Figure 13.22), two genera that grow well in acidic soils, and *Azospirillum*, a spirillum-shaped nitrogen-fixing bacterium that forms nonspecific symbiotic associations with various plants, in particular, corn.

Azotobacter and Alternative Nitrogenases

The species *Azotobacter chroococcum* was the first nitrogen-fixing bacterium shown capable of growth on N₂ in the absence of molybdenum. We study the important process of biological N₂ fixation in Section 15.29. There

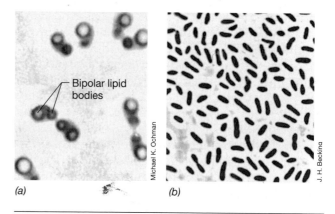

(a)   (b)

FIGURE 13.22 Phase-contrast photomicrographs of two genera of acid-tolerant, free-living N₂-fixing bacteria. (a) *Beijerinckia indica*. The cells are roughly pear-shaped, about 0.8 μm in diameter, and contain a large globule of poly-β-hydroxybutyrate at each end. (b) *Derxia gummosa*.

we will learn of the central importance of the metals molybdenum (Mo) and iron (Fe) to the enzyme nitrogenase. In *A. chroococcum*, it was first shown that in place of Mo nitrogenase, two "alternative nitrogenases" containing either vanadium (V) plus Fe or Fe only can be made under certain growth conditions; subsequent investigations of other nitrogen-fixing bacteria have shown that these "backup" nitrogenase systems, which function only when Mo is absent, are widely distributed among nitrogen-fixing bacteria.

✓ 13.6–13.8 Concept Check

Pseudomonads include many gram-negative chemoorganotrophic aerobic rods; many N₂-fixing species are phylogenetically closely related and can reduce N₂ to NH₃ in the process of nitrogen fixation. The acetic acid bacteria are also phylogenetically related to pseudomonads and are characterized by an ability to oxidize ethanol to acetate aerobically.

✓ Compare and contrast the pseudomonads, *Azotobacter*, and the acetic acid bacteria in terms of O₂ requirements, electron donors, pathogenicity, and habitats.

✓ Compare and contrast the organism *Acetobacter* with the organism *Acetobacterium* (see Section 15.18) in as many ways as you can think of.

13.9

Neisseria, Chromobacterium, and Relatives

Key Genera

Neisseria
Chromobacterium

This group of beta and gamma Proteobacteria comprises a diverse collection of organisms, related phylogenetically as well as by Gram stain, morphology, lack of motility, and aerobic metabolism. The genera *Neisse-*

TABLE 13.13	Characteristics of the genera of gram-negative cocci			
Characteristics	Genus	Number of species	Proteobacterial group	DNA (mol % GC)
I. Oxidase-positive, penicillin-sensitive:				
Cocci; complex nutrition, utilize carbohydrates, obligate aerobes	Neisseria	24	Beta	49–55
	Moraxella	8	Gamma	—
Rods or cocci; generally no growth-factor requirements, generally do not utilize carbohydrates; do not contain flagella, but some species exhibit "twitching" motility; many are commensals or pathogens of animals	Branhamella	10	Beta	40–47
	Kingella	2	Beta	47–55
II. Oxidase-negative, penicillin-resistant: some strains can utilize a restricted range of sugars, and some exhibit "twitching" motility; saprophytes in soil, water, and sewage	Acinetobacter	7	Gamma	38–47

ria, Moraxella, Kingella, and Acinetobacter are distinguished as outlined in Table 13.13.

In the genus Neisseria, the cells are always cocci (Figure 23.23), whereas cells of the other genera are rod-shaped, becoming coccoid only in the stationary phase of growth. This has led to designation of these organisms as **coccobacilli.** Organisms of the genera Neisseria, Kingella, and Moraxella are commonly isolated from animals, and some of them are pathogenic, whereas organisms of the genus Acinetobacter are common soil and water organisms, although they are occasionally found as parasites of animals and have been implicated in

nosocomial infections. Some strains of Moraxella and Acinetobacter possess the interesting property of twitching motility, exhibited as brief translocative movements or "jumps" covering distances of about 1–5μm. We discuss the clinical microbiology of Neisseria gonorrhoeae in Section 21.1 and the pathogenesis of gonorrhea itself in Section 23.6.

Chromobacterium is a close phylogenetic relative of Neisseria but is rod-shaped in morphology, resembling the pseudomonads or enteric bacteria. The best-known Chromobacterium species is C. violaceum, a purple-pigmented organism (Figure 13.23a) found in soil and water and occasionally in pus-forming infections of humans and other animals. Chromobacterium violaceum and a few other chromobacteria produce the purple pigment violacein (Figure 13.23b), a water-insoluble pigment that has antibiotic-like properties and is produced only in media containing the amino acid tryptophan. Chromobacterium is a facultative aerobe, growing fermentatively on sugars and aerobically on a variety of carbon sources.

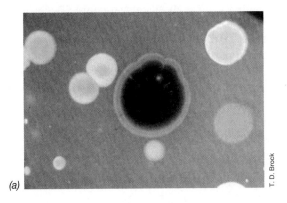
(a)

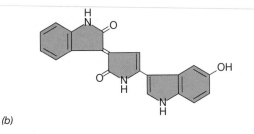
(b)

FIGURE 13.23 *Chromobacterium.* (a) A large colony of *Chromobacterium violaceum* growing among other colonies on an agar plate. (b) Structure of the pigment violacein, produced by *C. violaceum.*

13.10

Enteric Bacteria

Key Genera

Escherichia
Salmonella
Proteus
Enterobacter

The **enteric bacteria** comprise a relatively homogeneous phylogenetic group within the gamma Proteobacteria and are characterized phenotypically as follows: gram-negative, nonsporulating rods, nonmotile or motile by *peritrichous* flagella (Figure 13.24), facultative aerobes, oxidase-*negative* with relatively simple nutritional requirements, fermenting sugars to a variety of end products. The phenotypic characteristics used to separate

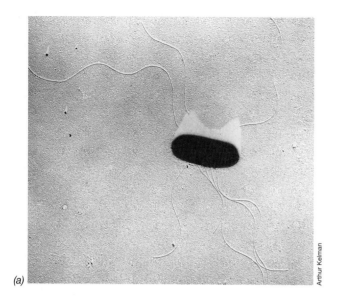

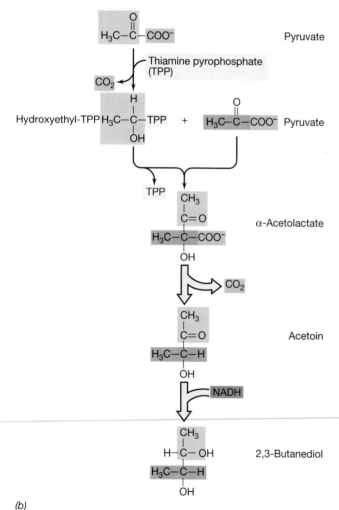

(b)

FIGURE 13.24 (a) Electron micrograph of a shadow-cast preparation of cells of the butanediol-producing enteric bacterium *Erwinia carotovora*. Cells are about 0.8 μm wide. Note the peritrichously arranged flagella. (b) Biochemical pathway for formation of butanediol from two molecules of pyruvate by butanediol fermenters.

the enteric bacteria from other bacteria of similar morphology and physiology are given in Table 13.14.

Among the enteric bacteria are many strains pathogenic to humans, animals, or plants as well as other strains of industrial importance. Undoubtedly more is known about *Escherichia coli* than about any other bacterial species (∞ Section 9.11).

Because of the medical importance of the enteric bacteria, an extremely large number of isolates have been studied and characterized, and a fair number of distinct genera have been defined. Despite the fact that there is marked genetic relatedness among many of the enteric bacteria, as shown by DNA homologies and genetic recombination, separate genera are maintained, largely for practical reasons. Since these organisms are frequently cultured from diseased states (∞ Sections 19.4, and 24.9–24.11), some means of identifying them is necessary, and thus phenotypic characteristics have traditionally been very important in distinguishing between enteric bacteria.

Fermentation Patterns in Enteric Bacteria

One of the key taxonomic characteristics separating the various genera of enteric bacteria is the type and proportion of *fermentation products* produced by anaerobic fermentation of glucose. Two broad patterns are recognized, the **mixed-acid fermentation** and the **2,3-butanediol fermentation** (Figure 13.25). In mixed-acid fermentation, *three* acids are formed in significant amounts—acetic, lactic, and succinic; ethanol, CO_2, and H_2 are also formed, but *not* butanediol. In butanediol fermentation, *smaller* amounts of acids are formed, and butanediol, ethanol, CO_2, and H_2 are the main products. As a result of a mixed-acid fermentation *equal* amounts

TABLE 13.14	Defining characteristics of the enteric bacteria

General characteristics:
Gram-negative straight rods; motile by peritrichous flagella, or nonmotile; nonsporulating; facultative aerobes, producing acid from glucose; sodium neither required nor stimulatory; catalase-positive; oxidase-negative; usually reduce nitrate to nitrite (not to N_2); 16S rRNA of gamma Proteobacteria (see Table 13.1)

Key tests to distinguish enteric bacteria from other bacteria of similar morphology:
Oxidase test, enterics always negative—separates enterics from oxidase-positive bacteria of genera *Pseudomonas, Aeromonas, Vibrio, Alcaligenes, Achromobacter, Flavobacterium, Cardiobacterium*, which may have similar morphology; nitrate reduced only to nitrite, (assay for nitrite after growth)—distinguishes enteric bacteria from bacteria that reduce nitrate to N_2 (gas formation detected), such as *Pseudomonas* and many other oxidase-positive bacteria; ability to ferment glucose—distinguishes enterics from obligately aerobic bacteria

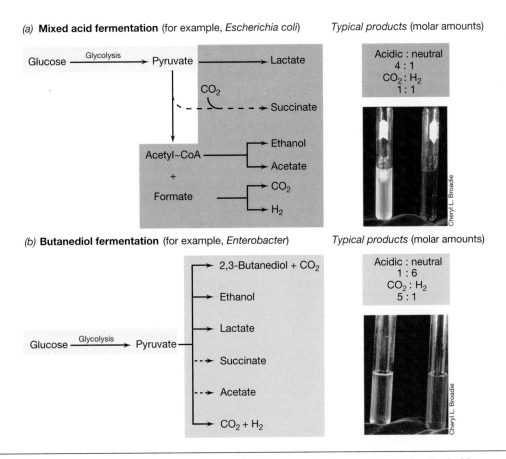

(a) **Mixed acid fermentation** (for example, *Escherichia coli*) *Typical products* (molar amounts)

(b) **Butanediol fermentation** (for example, *Enterobacter*) *Typical products* (molar amounts)

FIGURE 13.25 Distinction between mixed acid and butanediol fermentation in enteric bacteria. The bold arrows indicate reactions leading to major products. Dashed arrows indicate minor products. The upper photo shows the production of acid (yellow color) and gas (in inverted tube) in a culture of *E. coli* (purple tube was uninoculated). The bottom photo shows the pink-red color in the Voges–Proskauer (VP) test, which indicates butanediol production, following growth of *Enterobacter aerogenes*. The left (yellow) tube was uninoculated.

of CO_2 and H_2 are produced, whereas with a butanediol fermentation considerably *more* CO_2 than H_2 is produced. This is because mixed-acid fermenters produce CO_2 only from formic acid by means of the enzyme system *formic hydrogen lyase:*

$$HCOOH \rightarrow H_2 + CO_2$$

and this reaction results in equal amounts of CO_2 and H_2. The butanediol fermenters also produce CO_2 and H_2 from formic acid, but they produce two additional molecules of CO_2 during the formation of each butanediol (Figure 13.24*b*).

A variety of diagnostic tests and differential media are used to separate the various genera in the two broad groups of enteric bacteria, and these are described in detail in Tables 21.2 and 21.3. On the basis of these and other tests, genera can be defined as outlined in Tables 13.15 and 13.16.

Because enteric bacteria are genetically very closely related, their positive identification often presents considerable difficulty. In clinical laboratories, identification is frequently based on computer analysis of a large number of diagnostic tests carried out using miniaturized rapid diagnostic media kits and immunological and nucleic acid probes (Figure 21.1), with consideration being given for variable reactions of exceptional strains. Thus, the separation of genera given in Tables 13.15 and 13.16 must only be considered as approximate, for it is always possible to isolate a strain that does not possess one or another characteristic normally considered positive for the genus as a whole. With these limitations in mind, an even more simplified separation of the key genera is found in Figure 13.26. This key permits a quick decision on the likely *genus* in which to place a new isolate. We now consider some of the properties of key genera.

Escherichia, Salmonella, and *Shigella*

Members of the genus *Escherichia* are almost universal inhabitants of the intestinal tract of humans and warm-blooded animals, although they are by no means the dominant organisms in these habitats. *Escherichia* may play a nutritional role in the intestinal tract by synthesizing vitamins, particularly vitamin K. As a facultative

TABLE 13.15	Key diagnostic reactions used to separate the various genera of enteric bacteria[a]

Genus	H$_2$S (TSI)	Urease	VP[b]	Indole	Motility	Gas from glucose[b]	β-Galac-tosidase
Escherichia	–	–	–	+	+ or –	+	+
Enterobacter	–	–	+	–	+	+	+
Shigella	–	–	–	+ or –	–	–	+ or –
Edwardsiella	+	–	–	+	+	+	–
Salmonella	+	–	–	–	+	+	+ or –
Klebsiella	–	+	+ or –	–	–	+	+
Arizona	+	–	–	–	+	+	+
Citrobacter	+ or –	–	–	–	+	+	+
Proteus	+ or –	+	–	+ or –	+	+ or –	–
Providencia	–	–	–	+	+	–	–
Yersinia	–	+	–	–	+[c]	–	+
Hafnia	–	–	+	–	+	+	+ or –

Genus	KCN	Citrate	Mucate utilization	Phenyl-methyl red	Tartrate utilization	Alanine deaminase	DNA (mol % GC)
Escherichia	–	–	+	+	+	–	48–52
Enterobacter	+	+	+	–	–	–	52–60
Shigella	–	–	–	+	–	–	50
Edwardsiella	–	–	–	+ or –	–	–	53–59
Salmonella	–	+ or –	+ or –	+	+ or –	–	50–53
Klebsiella	+	+	+	–	+ or –	–	53–58
Arizona	–	+	+ or –	+	–	–	50
Citrobacter	+ or –	+	+	+	+	–	50–52
Proteus	+	+ or –	–	+	+	+	38–41
Providencia	+	+	–	+	+	+	39–42
Yersinia	–	–	–	+	–	–	46–50
Hafnia	+	+	–	+	–	–	48–49

a See Table 21.3 for the procedures for these diagnostic reactions.

b See Figure 13.25 for a photo of this reaction.

c Motile when grown at room temperature; nonmotile at 37°C.

aerobe, this organism probably also helps consume oxygen, thus rendering the large intestine anoxic. Wild-type *Escherichia* strains rarely show any growth-factor requirements and are able to grow on a wide variety of carbon and energy sources such as sugars, amino acids, organic acids, and so on. Some strains of *Escherichia* are pathogenic. The latter have been implicated in diarrhea in infants, occasionally occurring in epidemic proportions in children's nurseries or obstetric wards, and *Escherichia* may also cause urinary tract infections in older persons or in those whose resistance has been lowered by surgical treatment or by exposure to ionizing radiation. Enteropathogenic strains of *E. coli* are becoming more frequently implicated in dysentery-like infections and generalized fevers (⚬⚬ Sections 19.6, 19.9 and 24.9). As noted there, these strains form *K antigen*, permitting attachment and colonization of the small intestine, and *enterotoxin*, responsible for the symptoms of diarrhea.

Salmonella and *Escherichia* are quite closely related; the two genera have about half of their DNA sequences

TABLE 13.16	Key diagnostic reactions used to separate the various genera of 2,3-butanediol producers[a]

Genus	Ornithine decarboxyl-ase	Gelatin hydrolysis	Temperature optimum (°C)	Pigmentation	Motility	Lactose	DNase	Sorbitol	DNA (mol % GC)
Klebsiella	–	–	37–40	None	–	+	–	+	53–58
Enterobacter	+	Slow	37–40	Yellow (or none)	+	+	–	+	52–60
Serratia	+	+	37–40	Red (or none)	+	–	+	–	52–60
Erwinia	–	+ or –	27–30	Yellow (or none)	+	+ or –	–	+	50–58
Hafnia	+	–	35	None	+	–	–	–	48–49

a See Table 21.3 for a description of these diagnostic tests.

Diagnostic test	Go to number
1 MR+; VP – (mixed-acid fermenters)	2
MR –; VP + (butanediol producers)	7
2 Urease +	*Proteus*
Urease –	3
3 H₂S (TSI) +	4
H₂S (TSI) –	6
4 KCN +	*Citrobacter*
KCN –	5
5 Indole +; citrate –	*Edwardsiella*
Indole –; citrate +	*Salmonella*
6 Gas from glucose	*Escherichia*
No gas from glucose	*Shigella*
7 Nonmotile; ornithine –	*Klebsiella*
Motile; ornithine +	8
8 Gelatin+; DNAse +	*Serratia* (red pigment)
Gelatin slow; DNAse –	*Enterobacter*

Key
■ Mixed-acid fermenters
■ Butanediol producers

FIGURE 13.26 A simplified key to the main genera of enteric bacteria. Only the most common genera are given. See text for precautions in the use of this key. Diagnostic tests for use with this figure are given in Table 21.3. Other characteristics of the genera are given in Tables 13.33 and 13.34. Color coding is as in Figure 13.25.

in common. However, in contrast to most *Escherichia*, members of the genus *Salmonella* are usually pathogenic, either to humans or to other warm-blooded animals. In humans the most common diseases caused by salmonellas are *typhoid fever* and *gastroenteritis* (⚬ Sections 24.9 and 24.12). The salmonellas are characterized immunologically on the basis of three cell surface antigens, the O, or cell wall (somatic) antigen; the H, or flagellar, antigen; and the Vi (outer polysaccharide layer) antigen, found primarily in strains of *Salmonella* causing typhoid fever. The O antigens are part of the lipopolysaccharides that comprise the outer layer of the outer membrane of these organisms (⚬ Section 19.10). We discussed the chemical structure of lipopolysaccharides in Section 3.8 (⚬ Figures 3.36 and 3.37). The genus *Salmonella* contains over 1000 distinct serotypes having different antigenic specificities in their O antigens. Additional antigenic subdivisions are based on the antigenic specificities of the flagellar (H) antigens. There is little or no correlation between the antigenic type of a *Salmonella* and the disease symptoms elicited, but immunological typing permits tracing a single strain involved in an epidemic.

The shigellas are also genetically very closely related to *Escherichia*. Tests for DNA homology show strains of *Shigella* having 70 to nearly 100% homology with *Es-*

cherichia coli. In contrast to *Escherichia*, however, *Shigella* is commonly pathogenic to humans, causing a rather severe gastroenteritis usually called *bacillary dysentery*. *Shigella dysenteriae* is transmitted by food and waterborne routes and is capable of invading intestinal epithelial cells (⚬ Section 24.12). Once established, it produces both an endotoxin and a neurotoxin that exhibits enterotoxic effects.

Proteus

The genus *Proteus* is characterized by rapid motility (Figure 13.27) and by production of the enzyme *urease*. By DNA homology it shows only a distant relationship to *Escherichia coli*. *Proteus* is a frequent cause of urinary tract infections in humans, and probably benefits in this regard from its ready ability to degrade urea. Because of

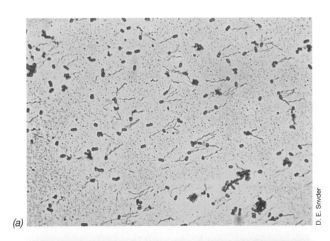
(a)

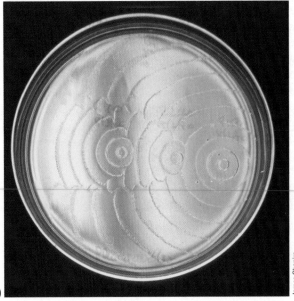
(b)

FIGURE 13.27 Swarming in *Proteus*. (a) Cells of *Proteus mirabilis* stained with a flagella stain. (b) Photo of a swarming colony of *Proteus vulgaris*. Note the concentric rings.

the rapid motility of *Proteus* cells, colonies growing on agar plates often exhibit a characteristic **swarming** phenomenon (Figure 13.27*b*). Cells at the edge of the growing colony are more rapidly motile than those in the center of the colony; the former move a short distance away from the colony in a mass and then undergo a reduction in motility, settle down, and divide, forming a new crop of motile cells that again swarm. As a result, the mature colony appears as a series of concentric rings, with higher concentrations of cells alternating with lower concentrations (Figure 13.27*b*).

Butanediol Fermenters: *Enterobacter, Klebsiella, and Serratia*

The butanediol fermenters are genetically more closely related to each other than to the mixed-acid fermenters, a finding that is in agreement with the observed physiological differences. Their DNA base composition is higher, 53–58% GC, and a current classification of this group is outlined in Table 13.16.

Enterobacter aerogenes is a common species in water and sewage as well as the intestinal tract of warm-blooded animals, and is an occasional pathogen in urinary tract infections. One species of *Klebsiella, K. pneumoniae*, occasionally causes pneumonia in humans, but klebsiellas are most commonly found in soil and water. Most *Klebsiella* strains also fix N_2 (∞ Section 15.29), a property not found among other enteric bacteria. The genus *Serratia* forms a series of red pyrrole-containing pigments called **prodigiosins** (Figure 13.28). Prodigiosin is produced in stationary phase as a secondary metabolite (∞ Section 11.2), and is of interest because it contains the pyrrole ring also found in the pigments involved in energy transfer: porphyrins, chlorophylls, and phycobilins (∞ Sections 15.3 and 15.4). However, there is no evidence that prodigiosin plays any role in energy transfer, and its exact function is unknown. Species of *Serratia* can be isolated from water and soil as well as from the gut of various insects and vertebrates and occasionally from the intestines of humans.

13.11

Vibrio and *Photobacterium*

Key Genera

Vibrio
Photobacterium

The *Vibrio* group contains gram-negative, facultatively aerobic rods and curved rods that possess a fermentative metabolism. Most of the members of the *Vibrio* group are polarly flagellated, although some are peritrichously flagellated. One key difference between the

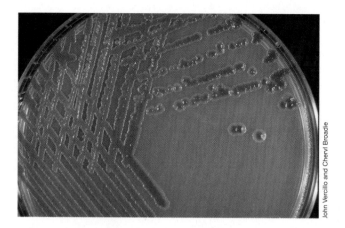

FIGURE 13.28   Colonies of *Serratia marcesens*. The orange-red pigmentation is due to the pyrrole-containing pigment prodigiosin.

Vibrio group and enteric bacteria is that members of the former are oxidase-*positive* (∞ Table 21.3), whereas members of the latter are oxidase-*negative*. Although *Pseudomonas* is also polarly flagellated and oxidase-positive, it is *not* fermentative and hence can be separated from the vibrios by simple sugar fermentation tests. The best known genera of the *Vibrio* group include *Vibrio* and *Photobacterium*.

Most vibrios and related bacteria are aquatic, found either in freshwater or marine habitats, although one important organism, *Vibrio cholerae*, is pathogenic for humans. *Vibrio cholerae* is the specific cause of the disease *cholera* in humans (∞ Sections 19.9 and 24.9); the organism does not normally infect other hosts. Cholera is one of the most common infectious human diseases in underdeveloped countries and one that has had a long history. The organism is transmitted almost exclusively via water, and studies on its distribution in the nineteenth century played a major role in demonstrating the importance of water purification in urban areas (∞ box, Snow on Cholera, Chapter 22). We discuss the pathogenesis of *V. cholerae* in Section 19.9.

Vibrio parahemolyticus is a marine organism. It is a major cause of gastroenteritis in Japan (where raw fish is widely consumed) and has also been implicated in outbreaks of gastroenteritis in other parts of the world, including the United States. The organism can be frequently isolated from seawater or from shellfish and crustaceans, and its primary habitat is probably marine animals, with human infection being a secondary development (∞ Section 24.9)

Photobacterium and Bioluminescence

A number of gram-negative, polarly flagellated rods possess the interesting property of emitting light (luminescence). Most of these bacteria have been classified in the genus *Photobacterium*, but a few *Vibrio* isolates are also luminescent (Figure 13.29). Most **bioluminescent bacteria** are marine, usually found associated with fish. Some fish possess a special organ in which bioluminescent bacteria grow (Figure 13.29*c–f*). Other bioluminescent marine bacteria live saprophytically on dead fish and occasionally form visible colonies on the fish surface. (To see bioluminescence readily, one should observe the material in a completely dark room after the eyes have become adapted to the dark, Figure 13.29*a,b*.)

Although *Photobacterium* isolates are facultative aerobes, they are bioluminescent only when O_2 is present. Several components are needed for bioluminescence: the enzyme **luciferase** and a long-chain aliphatic aldehyde (for example, *dodecanal*); flavin mononucleotide (FMN) and O_2 are also involved. The primary electron donor is NADH, and the electrons pass through luciferase. The reaction can be expressed as

Luciferase

$FMNH_2 + O_2 + RCHO \rightarrow$

$FMN + RCOOH + H_2O + Light$

The light-generating system constitutes a bypass route for shunting electrons from $FMNH_2$ to O_2, without involving other electron carriers such as quinones and cytochromes.

Regulation of Bioluminescence

The enzyme luciferase shows a unique kind of regulatory synthesis called **autoinduction.** The luminous bacteria produce a specific substance, the *autoinducer,* which accumulates in the culture medium during growth, and when the amount of this substance reaches a critical level, induction of luciferase occurs. The autoinducer in *Vibrio fischeri* has been identified as *N*-β-ketocaproylhomoserine lactone. Thus, cultures of luminous bacteria at low cell density are not luminous but only become luminous when growth reaches a sufficiently high density that the autoinducer can accumulate and function, a mechanism referred to as **quorum sensing** for the density-dependent nature of the phenomenon (∞ Section 7.6).

Because of autoinduction, it is obvious that free-living luminescent bacteria in seawater are not luminous because the autoinducer cannot accumulate, and lumi-

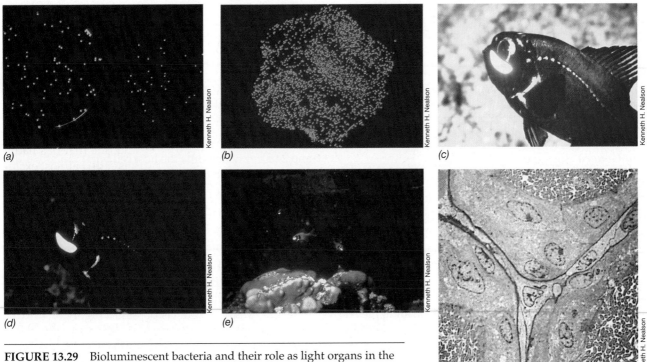

FIGURE 13.29 Bioluminescent bacteria and their role as light organs in the flashlight fish. (a) Two plates of luminous bacteria photographed by their own light. Note the different colors. Left, *Vibrio fischeri* strain MJ-1, blue light, and right, *V. fischeri* strain Y-1, green light. (b) Colonies of *Photobacterium phosphoreum* photographed by their own light. (c) The flashlight fish *Photoblepharon palpebratus;* the bright area is the light organ containing bioluminescent bacteria. (d) Same fish photographed by its own light. (e) Underwater photograph taken at night of *P. palpebratus* in coral reefs in the Gulf of Eilat. (f) Electron micrograph of a thin section through the light-emitting organ of *P. palpebratus,* showing the dense array of bioluminescent bacteria.

nescence develops only when conditions are favorable for the development of high population densities (Figure 13.29). Although it is not clear why luminescence is density-dependent in free-living bacteria, in symbiotic strains of luminescent bacteria (see Figure 13.29), the rationale for density-dependent luminescence is clear: luminescence develops only when sufficiently high population densities are reached in the light organ of the fish to allow a visible flash of light. The genetics of quorum sensing using bioluminescence as a model experimental system has been actively explored and it now appears that this type of sensing system is not limited to bioluminescent bacteria but instead is a general regulatory feature of a number of different bacteria for phenomena in which a minimum cell density is required (∞ Section 7.6).

✓ 13.10–13.11 Concept Check

The enteric bacteria are a large group of facultative aerobic rods of great medical and molecular biological significance. *Vibrio* and *Photobacterium* species are marine and many species are bioluminescent.

- ✓ How is *Escherichia coli* distinguished from *Enterobacter aerogenes* based on physiology?
- ✓ Describe two major properties of *Proteus* species that distinguish them from other enteric bacteria.
- ✓ What is necessary for an organism like *Photobacterium* to give off visible light?

13.12

Rickettsias

Key Genera

Rickettsia
Coxiella

The rickettsias are small, gram-negative, coccoid or rod-shaped Proteobacteria in the size range of 0.3–0.7 μm wide and 1–2 μm long. They are, with one exception, *obligate intracellular parasites* (Figure 13.30a) and have not yet been cultivated in the absence of host cells. Rickettsias are the causative agents of such human diseases as typhus fever, Rocky Mountain spotted fever, and Q fever (∞ Section 24.3). Electron micrographs of thin sections of rickettsias show cells with a normal bacterial morphology (Figure 13.30b); both cell wall and cell membrane are visible. The cell wall contains muramic acid and diaminopimelic acid. Both RNA and DNA are present, and the rickettsias divide by normal binary fission, with doubling times of about 8 hr. The penetration of a host cell by a rickettsial cell is an active process, requiring both host and parasite to be alive and metabolically active. Once inside the host phagocytic cell, the bacteria multiply primarily in the cytoplasm and continue replicating until

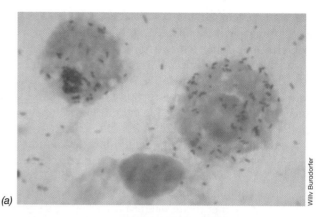

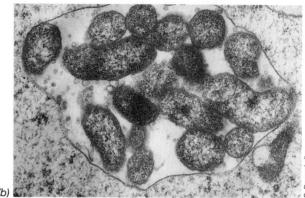

FIGURE 13.30 Rickettsias growing within host cells. (a) *Rickettsia rickettsii* in tunica vaginalis cells of the vole, *Microtus pennsylvanicus*. Cells are about 0.3 μm in diameter. (b) Electron micrograph of cells of *Rickettsiella popilliae* within a blood cell of its host, the beetle *Melolontha melolontha*. Notice that the bacteria are growing within a vacuole within the host cell.

the host cell is loaded with parasites (see Figures 13.30 and 24.3), at which time the host cell bursts and liberates the bacteria into the surrounding fluid. Several genera of rickettsias are known and the properties of four key genera are shown in Table 13.17.

Metabolism and Pathogenesis

Much attention has been directed to the metabolic activities and biochemical pathways of rickettsias in an attempt to explain why they are obligate intracellular parasites. Many rickettsias possess a highly distinctive energy metabolism, being able to oxidize only glutamate or glutamine and being unable to oxidize glucose or organic acids. However, *Coxiella burnetii* is able to utilize both glucose and pyruvate as electron donors. Rickettsias possess a respiratory chain complete with cytochromes and are able to carry out electron transport phosphorylation, using NADH as electron donor. They are also able to synthesize at least some of the small molecules needed for macromolecular synthesis and growth, and they obtain the rest of their nutrients from the host cell.

TABLE 13.17	Characteristics of rickettsias[a]					
Species	Rickettsial group	Alternate host	Cellular location	DNA (mol % GC)	Proteobacterial group	DNA hybridization to *R. rickettsii* DNA (%)[a]
Rickettsia						
R. rickettsii	Spotted fever	Tick	Cytoplasm and nucleus	32–33	Alpha	100
R. prowazekii[b]	Typhus	Louse	Cytoplasm	29–30		53
R. typhi	Typhus	Flea	Cytoplasm	29–30		36
Rochalimaea						
R. quintana	Trench fever	Louse	Epicellular	39	Alpha	30
R. vinsonii	—	Vole	Epicellular	39		30
Coxiella						
C. burnetii	Q fever	Tick	Vacuoles	43	Gamma	—
Ehrlichia	Ehrlichiosis (humans); Potomac fever (horses)	Tick or domestic animals	Mononuclear leukocytes	—	Alpha	—

a For discussion of DNA:DNA hybridization, see Section 12.9.

b The genome of this organism has been sequenced and shows several similarities to the mitochondrial genome.

Rickettsias do not survive long outside their hosts, and this may explain why they must be transmitted from animal to animal by arthropod vectors. When the arthropod obtains a blood meal from an infected vertebrate, rickettsias present in the blood are inoculated directly into the arthropod, where they penetrate to the epithelial cells of the gastrointestinal tract, multiply, and appear later in the feces. When the arthropod feeds on an uninfected individual, it then transmits the rickettsias either directly with its mouthparts or by contaminating the bite with its feces. However, the causal agent of Q fever, *Coxiella burnetii* (∞ Section 24.3), can also be transmitted to the respiratory system by aerosols. *Coxiella burnetii* is the most resistant of the rickettsias to physical damage, probably because it produces a resistant, sporelike form, and this explains its ability to survive in air. *Rochalimaea* is an atypical rickettsia because it can be grown in culture and is thus not an obligate intracellular parasite. In addition, when growing in tissue culture, cells of *Rochalimaea* grow on the *outside surface* of the eukaryotic host cells rather than within the cytoplasm or the nucleus. *Rochalimaea quintana* is the causative agent of *trench fever*, a disease that decimated troops in World War I. Species of the genus *Ehrlichia* cause disease in humans and other animals, two of which, *ehrlichiosis* in humans and *Potomac fever* in horses, can be quite severe and debilitating (∞ Section 24.3).

✓ 13.12 Concept Check

The Rickettsias are obligate intracellular parasites, many of which cause disease. Rickettsias are deficient in many metabolic functions and obtain key metabolites from their hosts.

✓ Name a disease caused by a *Rickettsia* species.
✓ What is meant by the phrase "obligate intracellular parasite?"

13.13

Spirilla

Key Genera

Spirillum
Bdellovibrio
Campylobacter

The **spirilla** are gram-negative, motile, spiral-shaped proteobacteria that show a wide variety of physiological attributes. Some of the key taxonomic criteria used are cell shape, size, kind of polar flagellation (single or multiple), relation to oxygen (obligately aerobic, microaerophilic, facultative), relationship to plants (as symbionts or plant pathogens) or animals (as pathogens), fermentative ability, and certain other physiological characteristics (nitrogen-fixing ability, halophilic nature, thermophilic nature). The genera to be covered here are given in Table 13.18.

Spirillum, Aquaspirillum, Oceanospirillum, and *Azospirillum*

These are helically curved rods, which are motile by means of polar flagella (usually tufts at both poles) (see Figure 13.31). The number of turns in the helix may vary from less than one complete turn (in which case the organism looks like a vibrio) to many turns. Spirilla with many turns can superficially resemble spirochetes (see Section 13.30) but differ both phylogenetically and in that they do not have an outer sheath and endoflagella but instead contain typical bacterial flagella.

Some spirilla are very large bacteria and were seen by early microscopists. It is likely that Antoni van Leeuwenhoek first observed a *Spirillum* species in the 1670s (∞ Section 1.8), and the genus was first creat-

TABLE 13.18	Characteristics of the genera of spiral-shaped bacteria[a]		
Genus	**Proteobacterial group**	**Characteristics**	**DNA (mol % GC)**
Spirillum	Beta	Cell diameter 1.7 μm; microaerophilic; freshwater	36–38
Aquaspirillum	Alpha or beta	Cell diameter 0.2–1.5 μm; aerobic; freshwater	49–66
Magnetospirillum	Alpha	Vibrio to spirillum-shaped; cell diameter about 0.3 μm; contains magnetosomes; microaerophilic	65
Oceanospirillum	Gamma	Cell diameter 0.3–1.2 μm; aerobic; marine (require 3% NaCl)	42–51
Azospirillum	Alpha	Cell diameter 1 μm; microaerophilic; soil and rhizosphere; fixes N_2	68–70
Herbaspirillum	—	Cell diameter 0.6–0.7 μm; microaerophilic; soil and rhizosphere; fixes N_2	66–67
Campylobacter	Epsilon	Cell diameter 0.2–0.8 μm; microaerophilic to anaerobic; pathogenic or commensal in humans and animals; single polar flagellum	30–38
Helicobacter	Epsilon	Cell diameter 0.5–1 μm; tuft of polar flagella; associated with pyloric ulcers in humans	—
Bdellovibrio	Delta	Cell diameter 0.25–0.4 μm; aerobic; predatory on other bacteria; single polar sheathed flagellum	33–52
Ancyclobacter	Bacteroides–Flavobacterium group	Cell diameter 0.5 μm; curved rods forming rings; nonmotile; aerobic; sometimes gas-vesiculate	66–69

a All are gram-negative and respiratory but never fermentative.

ed by the protozoologist Ehrenberg in 1832. The organism seen by these workers is now called *Spirillum volutans* and is a rather large bacterium (Figure 13.31*a* and *b*). A phototrophic organism resembling *S. volutans* is *Thiospirillum* (see Figure 13.4*b*). *Spirillum volutans* is microaerophilic, requiring O_2, but is inhibited by O_2 at normal levels. Another prominent characteristic of *S. volutans* is the formation of granules (volutin granules) consisting of polyphosphate (see Figure 13.31*b* and Section 3.13).

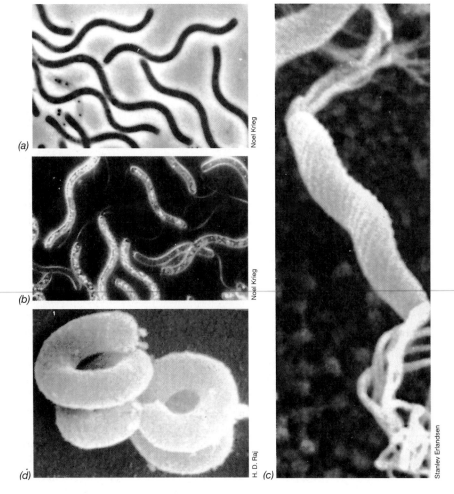

(a)

(b)

(c)

(d)

Noel Krieg

Noel Krieg

H. D. Raj

Stanley Erlandsen

FIGURE 13.31 (a) Photomicrograph by phase contrast of *Spirillum volutans*, a large spirillum. Cells are about 1.6 by 20–50 μm. (b) *Spirillum volutans*, by dark-field microscopy, showing flagellar bundles and volutin (polyphosphate) granules. (c) Scanning electron micrograph of an intestinal spirillum. Note the polar flagellar tufts and the spiral structure of the cell surface. (d) Scanning electron micrograph of cells of *Ancyclobacter linguale*. Cells are about 0.5 μm in diameter.

Azospirillum lipoferum is a nitrogen-fixing organism, which was originally described and named *Spirillum lipoferum* by Beijerinck in 1922. It has become of considerable interest in recent years because this bacterium has been found to enter into a symbiotic relationship with tropical grasses and grain crops (∞ Section 16.25). The small-diameter spirilla (which are not microaerobic) have been separated into two genera, *Aquaspirillum* and *Oceanospirillum*, the former for freshwater forms and the latter for those living in seawater and requiring NaCl for growth (Table 13.18). Numerous species of *Aquaspirillum* and *Oceanospirillum* have been described, the various species being separated on physiological grounds. These organisms undoubtedly play an important role in the recycling of organic matter in aquatic environments.

Magnetotactic Spirilla

Highly motile microaerophilic magnetic spirilla have been isolated from freshwater habitats. These organisms demonstrate a dramatic directed movement in a magnetic field referred to as **magnetotaxis.** In an artificial magnetic field magnetic spirilla quickly orient their long axis along the north–south magnetic moment of the field. Within the cells, chains of 5–40 magnetic particles consisting of magnetite (Fe_3O_4) and greigite (Fe_3S_4) called **magnetosomes** (∞ Section 3.13 and Figure 3.58) are present (Figure 13.32), and these function as internal magnets that orient the cells along a specific magnetic field. Magnetic bacteria can have one of two magnetic polarities depending on the orientation of magnetosomes within the cell. Cells in the Northern Hemisphere have the north-seeking pole of their magnetosomes forward with respect to their flagella and thus move in a northward direction. Cells in the Southern Hemisphere have the opposite polarity and move southward. Although the ecological role of bacterial magnets is unclear, it has been suggested that the ability to orient in a magnetic field is of selective advantage in maintaining these microaerophilic organisms in zones of low O_2 concentration near the oxic/anoxic interface. The spirillum *Magnetospirillum magnetotacticum* (Figure 13.32 and Table 13.18) is a major organism in this group.

Bdellovibrio

These small vibroid organisms have the unusual property of preying on other bacteria, using as nutrients the cytoplasmic constituents of their hosts. These bacterial predators are small, highly motile cells that stick to the surfaces of their prey cells. Because of the latter property, they have been given the name *Bdellovibrio* (*bdello*- is a combining form meaning "leech"). Other predatory bacteria have been isolated and given such names as *Vampirococcus*. However, *Bdellovibrio* has a unique mode

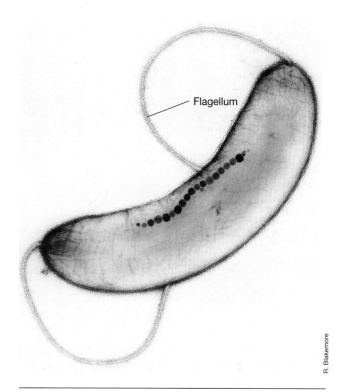

FIGURE 13.32 Negatively stained electron micrograph of a magnetotactic spirillum, *Magnetospirillum magnetotacticum.* A cell measures 0.3×2 μm. This bacterium contains particles of Fe_3O_4 (magnetite) called magnetosomes (∞ Figure 3.58) arranged in a chain; the particles align the cell along geomagnetic lines. The organism was isolated from a water treatment plant in Durham, New Hampshire.

of attack and develops intraperiplasmically. After attachment of a *Bdellovibrio* cell to its prey, the predator penetrates through the prey wall and replicates in the space between the prey wall and membrane (the periplasmic space), eventually forming a spherical structure called a **bdelloplast.** The stages of attachment and penetration are shown in electron micrographs in Figure 13.33 and diagrammatically in Figure 13.34. A wide variety of gram-negative Bacteria can be attacked by a single *Bdellovibrio* species; gram-positive cells are not attacked.

As originally isolated, *Bdellovibrio* cells grow only on living prey, but it is possible to isolate prey-independent mutants that grow on complex organic media such as yeast extract–peptone. These strains, like the wild-type strain, are unable to utilize sugars as electron donors but are proteolytic and can oxidize the amino acids liberated by protein digestion. Prey-dependent revertants can be reisolated from the prey-independent mutants by introducing a host strain.

Bdellovibrio is an obligate aerobe, obtaining its energy from the oxidation of amino acids and acetate. In addition, *Bdellovibrio* assimilates nucleotides, fatty acids, and even whole proteins in some cases directly from its host without first breaking them down. Thus it is clear

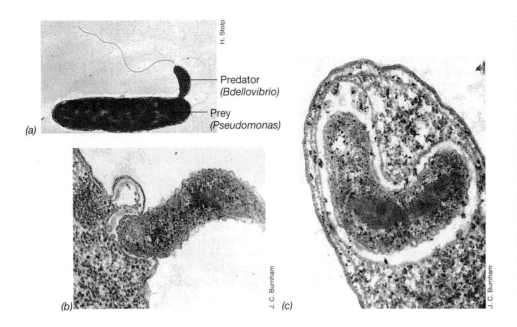

FIGURE 13.33 Stages of attachment and penetration of a prey cell by *Bdellovibrio*. A *Bdellovibrio* cell measures about 0.3 μm in diameter. (a) Electron micrograph of a shadowed whole-cell preparation showing *Bdellovibrio bacteriovorus* attacking *Pseudomonas*. (b, c) Electron micrographs of thin sections of *Bdellovibrio* attacking *Escherichia coli*; (b) early penetration; (c) complete penetration. The *Bdellovibrio* cell is enclosed in a membranous infolding of the prey cell (the bdelloplast) and replicates in the periplasmic space between the wall and the membrane.

that the predatory mode of existence has involved the development in *Bdellovibrio* of interesting and unusual biochemical processes.

Phylogenetically, bdellovibrios fall into the delta group of Proteobacteria and contain a genome about half the size of that of *E. coli*. Taxonomically, three species of *Bdellovibrio* are recognized and supported by genomic DNA:DNA hybridization. In addition to being predators themselves, bdellovibrios, are subject to attack by bacteriophages. Nicknamed *bdellophages*, most phages that plaque on lawns of *Bdellovibrio* cells are strictly lytic, single-stranded DNA phages.

Members of the genus *Bdellovibrio* are widespread in soil and water, including the marine environment. Their detection and isolation require methods reminiscent of those used in the study of bacterial viruses (∞ Section 8.4). Prey bacteria are spread on the surface of an agar plate to form a lawn and the surface is

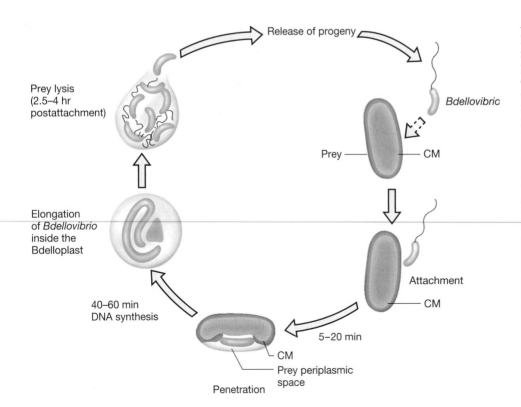

FIGURE 13.34 Developmental cycle of the bacterial predator *Bdellovibrio bacteriovorus*. Following primary contact between a highly motile *Bdellovibrio* cell and a gram-negative bacterium, attachment and penetration into the prey periplasmic space occurs. Once inside, *Bdellovibrio* elongates and within 4 hours progeny cells are released. The number of progeny cells released varies with the size of the prey bacterium. For example, 5–6 bdellovibrios are released from each infected *Escherichia coli* cell, and 20–30 for *Spirillum* sp. CM, Prey cytoplasmic membrane.

inoculated with a small amount of soil suspension that has been filtered through a membrane filter; the latter retains most bacteria but allows the small *Bdellovibrio* cells to pass. On incubation of the agar plate, plaques analogous to those produced by bacteriophages are formed at locations where *Bdellovibrio* cells are growing. However, unlike phage plaques, which continue to enlarge only as long as the bacterial host is growing, *Bdellovibrio* plaques continue to enlarge even after the prey has stopped growing, resulting in large plaques on the agar surface. Pure cultures of *Bdellovibrio* can then be isolated from these plaques. *Bdellovibrio* cultures have been obtained from a wide variety of soils and are thus common members of the soil population.

Ancyclobacter

Members of the genus *Ancyclobacter* are ring-shaped, nonmotile, extremely nutritionally diverse chemoorganotrophic bacteria (Figure 13.31*d*). They resemble very tightly coiled vibrios and are widely distributed in aquatic environments. A phototrophic counterpart to *Ancyclobacter* exists in the purple nonsulfur bacterium *Rhodocyclus purpureus* (see Figure 13.7*e*).

Campylobacter and *Helicobacter*

These two genera are key representatives of the *epsilon* subdivision of the Proteobacteria. They are gram-negative, motile spirilla, most species of which are pathogenic to humans or other animals. *Campylobacter* and *Helicobacter* are both microaerophilic and are cultured from clinical specimens in media incubated at low (3–15%) O_2 and high (3–10%) CO_2. *Campylobacter* species cause acute enteritis leading to (usually) bloody diarrhea, and pathogenesis is due to several factors including an enterotoxin that is related to cholera toxin (∞ Section 19.9). *Helicobacter pylori* is closely related to *Campylobacter* species and causes both acute and chronic gastritis, leading to the formation of *peptic ulcers*. We discuss the diseases caused by *Campylobacter* and *Helicobacter*, including modes of transmission of the organisms and clinical symptoms, in more detail in Section 24.12.

✓ 13.13 Concept Check

Spirilla are spiral-shaped, chemoorganotrophic prokaryotes widespread in the aquatic environment. The genera *Helicobacter* and *Campylobacter* are pathogenic spirilla. Spirilla are distributed among the alpha, beta, gamma, delta, and epsilon subdivisions of Proteobacteria.

- ✓ What is a *volutin granule*?
- ✓ What is unique about the spirilla *Bdellovibrio* and *Magnetospirillum*?

Sheathed Proteobacteria: *Sphaerotilus* and *Leptothrix*

Key Genera

Sphaerotilus
Leptothrix

Sheathed bacteria are filamentous beta Proteobacteria (Table 13.1) with a unique life cycle involving formation of flagellated swarmer cells within a long tube or sheath. Under certain (generally unfavorable) conditions, the swarmer cells move out and become dispersed to new environments, leaving behind the empty sheath. Under favorable conditions, vegetative growth occurs within the filament, leading to the formation of long, cell-packed sheaths. Sheathed bacteria are common in freshwater habitats that are rich in organic matter, such as polluted streams, and trickling filters and activated sludge digestors in sewage treatment plants (∞ Section 11.15), being found primarily in flowing waters. In habitats where reduced iron or manganese compounds are present, the sheaths may become coated with ferric hydroxide or manganese oxide (see, for example, Figure 15.29). Iron precipitation is probably due to chemical reactions, but some sheathed bacteria have the biochemical ability to oxidize manganous ions to manganese oxide. Two genera are the major organisms here: *Sphaerotilus*, in which manganese oxidation does not occur, and *Leptothrix*, whose members do oxidize Mn^{2+}.

Sphaerotilus

The *Sphaerotilus* filament is composed of a chain of rod-shaped cells with rounded ends enclosed in a closely fitting sheath. This thin, transparent sheath is difficult to see when it is filled with cells, but when it is partially empty, the sheath can easily be seen by phase contrast microscopy (Figure 13.35*a*) or by staining. The cells within the sheath divide by binary fission (Figure 13.35*b*), and the new cells pushed out at the end synthesize new sheath material. Thus, the sheath is always formed at the *tips* of the filaments. Individual cells are 1–2 µm wide and 3–8 µm long and stain gram-negatively. Eventually cells are liberated from the sheaths, probably when the nutrient supply is low. These free cells are actively motile, the flagella being arranged lophotrichously (in a bundle at one pole) (Figure 13.35*c*). Probably the flagella are synthesized before the cells leave the sheath and, if so, may even aid in their liberation. It is thought that the swarmer cells then migrate, attach to some solid surface, and begin to grow, each swarmer being the forerunner of a new filament. The sheath, which is devoid of mu-

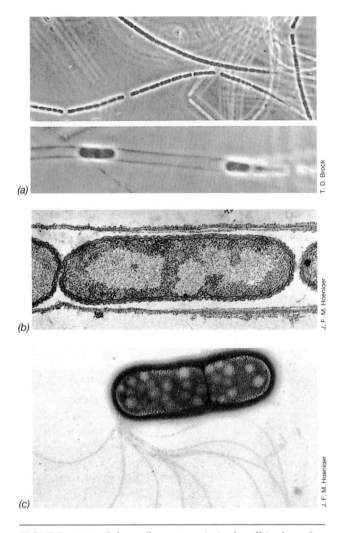

(a)

T. D. Brock

(b)

J. F. M. Hoeniger

(c)

J. F. M. Hoeniger

FIGURE 13.35 *Sphaerotilus natans.* A single cell is about 2 μm wide. (a) Phase contrast photomicrographs of material collected from a polluted stream. Active growth stage (above) and swarmer cells leaving the sheath. (b) Electron micrograph of a thin section through a filament. (c) Electron micrograph of a negatively stained swarmer cell. Notice the polar flagellar tuft.

ramic acid or other components of the peptidoglycan cell wall, is a protein–polysaccharide–lipid complex, possibly analogous to the capsules formed by many gram-negative Bacteria but differing in that it forms a linear structure.

  Sphaerotilus cultures are nutritionally versatile, able to use a wide variety of simple organic compounds as carbon and energy sources, with inorganic nitrogen sources. Befitting its habitat in flowing waters, *Sphaerotilus* is an obligate aerobe. *Sphaerotilus* blooms often occur in the fall of the year in streams and brooks when leaf litter causes a temporary increase in the organic content of the water. Its filaments are the main component of a microbial complex that

sanitary engineers call "sewage fungus," which is a filamentous slime found on the rocks in streams receiving sewage pollution. In activated sludge works in sewage treatment plants (∞ Section 11.15), *Sphaerotilus* growth, like that of *Beggiatoa* (see Section 13.3), is often responsible for a detrimental condition called "bulking." The tangled masses of *Sphaerotilus* filaments so increase the bulk of the sludge that it does not settle properly, thus presenting difficulties in sludge clarification.

Leptothrix

The ability of *Sphaerotilus* and *Leptothrix* to precipitate iron oxides on their sheaths is well established and such iron-encrusted sheaths are frequently seen in iron-rich waters (Figure 13.36). Iron precipitation usually occurs when iron, chelated to organic materials such as humic or tannic acids, is metabolized; the iron gets precipitated on the sheath while the organic constituents may get taken up and used as a carbon or energy source.

  Besides Fe^{2+} oxidation, *Leptothrix* is capable of oxidizing Mn^{2+} to Mn^{4+}. The reaction that occurs:

$$Mn^{2+} 0.5\ O_2 + H_2O \rightarrow MnO_2 + 2\ H^+ \qquad \Delta G^{0'} = -68\ kJ$$

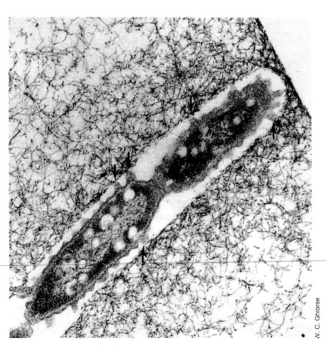

N. C. Ghiorse

FIGURE 13.36 Transmission electron micrograph of a thin section of *Leptothrix* sp. in a sample from a ferromanganese film in a swamp in Ithaca, New York. A single cell measures about 0.9 μm in diameter. Note the protuberances of the cell envelope that contact the sheath (arrows).

		Proteobacterial	
TABLE 13.19 Characteristics of stalked, appendaged (prosthecate), and budding bacteria			
Characteristics	Genus	group	DNA (mol % GC)
Stalked bacteria:			
Stalk an extension of the cytoplasm and involved in cell division	*Caulobacter*	Alpha	62–67
Stalked, fusiform-shaped cells	*Prosthecobacter*	Alpha	54–60
Stalked, but stalk an excretory product not containing cytoplasm:			
Stalk depositing iron, cell vibrioid	*Gallionella*	Beta	55
Laterally excreted gelatinous stalk not depositing iron	*Nevskia*	—	60
Appendaged (prosthecate) bacteria:			
Single or double prosthecae	*Asticcacaulis*	—	55–61
Multiple prosthecae			
Short prosthecae, multiply by fission, some with gas vesicles	*Prosthecomicrobium*	Alpha	64–70
Flat, star-shaped cells, some with gas vesicles	*Stella*	Alpha	69–74
Long prosthecae, multiply by budding, some with gas vesicles	*Ancalomicrobium*	Alpha	70–71
Budding bacteria:			
Phototrophic, produce hyphae	*Rhodomicrobium*	Alpha	61–63
Phototrophic, budding without hyphae	*Rhodopseudomonas*	Alpha	64–72
Chemoorganotrophic, rod-shaped cells	*Blastobacter*	Alpha	59–66
Chemoorganotrophic, buds on tips of slender hyphae			
Single hyphae from parent cell	*Hyphomicrobium*	Alpha	59–65
Multiple hyphae from parent cell	*Pedomicrobium*	Alpha	62–67

is exergonic, and there is evidence that Mn^{2+} oxidation is coupled to energy-yielding reactions. The gene encoding the manganese-oxidizing protein of *Leptothrix* has been isolated and from biochemical studies it has been shown that this protein resides in the sheath, not inside the cells. Exactly why *Leptothrix* oxidizes Mn^{2+} is not entirely clear, but it has been hypothesized that this occurs either for energetic benefits or possibly because the Mn^{4+} produced reacts with humic and fulvic acids, thereby releasing organic nutrients for growth.

13.15

Budding and Prosthecate/Stalked Bacteria

Key Genera

Hyphomicrobium
Caulobacter

This large and rather heterogeneous group of Proteobacteria contains organisms that form various kinds of cytoplasmic extrusions: *stalks, hyphae,* or *appendages* (Table 13.19). Extrusions of these kinds, which are smaller in diameter than the mature cell, contain cytoplasm, and are bounded by the cell wall, are called **prosthecae** (singular, **prostheca**) (Figure 13.37). Of considerable interest in this group of bacteria is that cell division often occurs as a result of unequal cell growth. In contrast to cell division in the typical bacterium,

which occurs by *binary fission* and results in the formation of two equivalent cells (Figure 13.38), cell division in the stalked and budding bacteria involves the formation of a new daughter cell with the mother cell retaining its identity after the cell division process is completed (Figure 13.38). The genera from Table 13.19 that show this unequal cell division process are indicated in Figure 13.38.

A fundamental difference between these bacteria and conventional bacteria is not the formation of buds or stalks but the formation of a new cell wall from a *single point* (polar growth) rather than throughout the whole cell (intercalary growth). Several genera not normally considered to be budding bacteria show polar growth without differentiation of cell size (Figure 13.38). An important consequence of **polar growth** is that internal structures, such as membrane complexes, are not involved in the cell division process, thus permitting the formation of more complex internal structures than in cells undergoing intercalary growth. And many budding bacteria, particularly phototrophic species, contain extensive internal membrane systems.

Budding Bacteria: *Hyphomicrobium*

The two best-studied budding bacteria are *Hyphomicrobium,* which is chemoorganotrophic, and *Rhodomicrobium,* which is phototrophic; both organisms are phylogenetically closely related and release buds from the ends of long, thin hyphae. The hypha is a direct

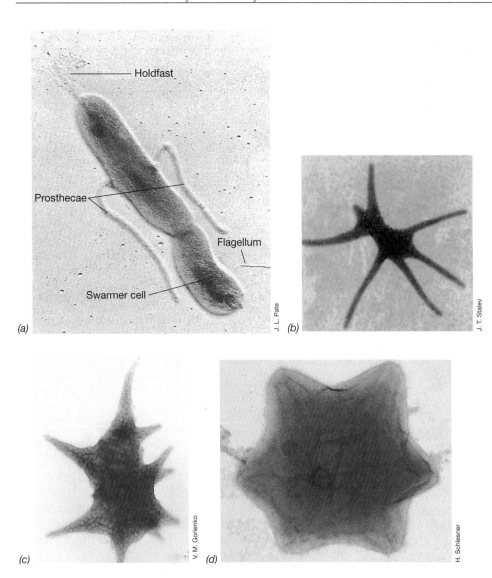

Holdfast

Prosthecae

Flagellum

Swarmer cell

(a)

J. L. Pate

(b)

J. T. Staley

(c)

V. M. Gorlenko

(d)

H. Schlesner

FIGURE 13.37 Prosthecate bacteria. (a) Electron micrograph of a shadow-cast preparation of *Asticcacaulis biprosthecum,* illustrating the location and arrangement of the prosthecae. Cells are about 0.6 μm wide. Note also the holdfast material and the swarmer cell in the process of differentiation. (b) Electron micrograph of a negatively stained preparation of a cell of the prosthecate bacterium *Ancalomicrobium adetum.* The appendages are cellular (prosthecae) because they are bounded by the cell wall and contain cytoplasm and are about 0.2 μm in diameter. (c) Electron micrograph of a whole cell of a prosthecate phototrophic green bacterium, *Ancalochloris perfilievii;* cells are about 0.7 μm in diameter. The structures within the cell are gas vesicles. (d) Electron micrograph of the star-shaped bacterium *Stella.* Cells are about 0.8 μm in diameter.

cellular extension of the mother cell (Figures 13.39 and 13.40), containing cell wall, cytoplasmic membrane, ribosomes, and occasionally DNA. The process of reproduction in *Hyphomicrobium* is illustrated in Figure 13.39. The mother cell, which is often attached by its base to a solid substrate, forms a thin outgrowth that lengthens to become a hypha, and at the end of the hypha a bud forms. This bud enlarges, forms a flagellum, breaks loose from the mother cell, and swims away. Later, the daughter cell loses its flagellum and after a period of maturation forms a hypha and buds. Further buds can also form at the hyphal tip of the mother cell leading to complex arrays of cells connected by hyphae (Figure 13.41). In some cases a bud begins to form directly from the mother cell without the intervening formation of a hypha, whereas in other cases a single cell forms hyphae from each end (Figure 13.41).

Nucleoid replication events during the budding cycle are of interest (Figure 13.39). The DNA located in the mother cell replicates, and then once the bud has formed, a copy of the circular chromosome is moved down the length of the hypha and into the bud. A cross-septum then forms, separating the still developing bud from the hypha and mother cell.

Hyphomicrobium is a methylotrophic bacterium (see Section 13.5). Preferred carbon sources are *one-carbon* compounds such as methanol, methylamine, formaldehyde, and formate. Growth on acetate, ethanol, or higher aliphatic compounds is usually slow, and growth is poor on sugars or most amino acids. Urea, amides, ammonia, nitrite, and nitrate can be utilized as nitrogen sources; vitamins are not required. *Hyphomicrobium* is widespread in freshwater, marine, and terrestrial habitats. Initial enrichment cultures can be prepared using a mineral–salts medium lacking organic carbon and nitrogen, to which a sample of natural material is added. After several weeks of incubation, the surface film that develops is streaked out on agar medium containing

Products of cell division are equal:

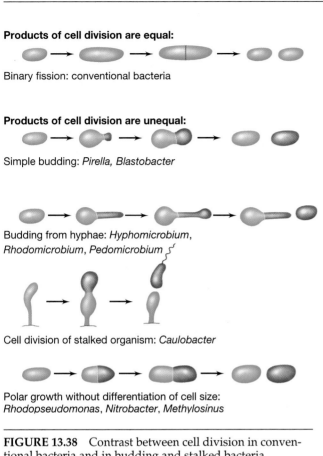

Binary fission: conventional bacteria

Products of cell division are unequal:

Simple budding: *Pirella, Blastobacter*

Budding from hyphae: *Hyphomicrobium,*
Rhodomicrobium, Pedomicrobium

Cell division of stalked organism: *Caulobacter*

Polar growth without differentiation of cell size:
Rhodopseudomonas, Nitrobacter, Methylosinus

FIGURE 13.38 Contrast between cell division in conventional bacteria and in budding and stalked bacteria.

methylamine or methanol as a sole carbon source. Colonies are then checked microscopically for the characteristic *Hyphomicrobium* cellular morphology. A fairly specific enrichment procedure for *Hyphomicrobium* uses methanol as electron donor with nitrate as electron acceptor under anoxic conditions. Virtually the only denitrifying organism using methanol is *Hyphomicrobium,* and so this procedure selects this organism out of a wide variety of environments.

Prosthecate and Stalked Bacteria

Prosthecate and stalked (Figures 13.37 and 13.42) bacteria are appendaged chemoorganotrophic aerobes that attach to particulate matter, plant material, or other microorganisms in aquatic habitats. Although a major function of these appendages is undoubtedly attachment, they also increase significantly the surface-to-volume ratio of the cells. Recall from Chapter 3 that the high surface-to-volume ratio of small cells like bacteria confers an increased ability on them to take up nutrients and expel wastes. The unusual morphology of appendaged bacteria (Figure 13.37) carries this theme to an extreme and may be an evolutionary adaptation to life in the oligotrophic (nutrient-poor) waters where these organisms are most commonly found. Selective isola-

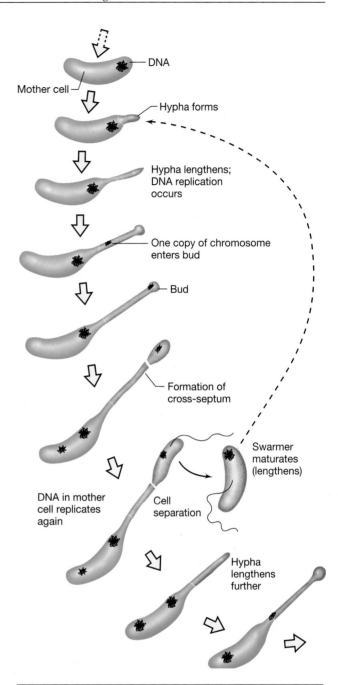

FIGURE 13.39 Stages in the *Hyphomicrobium* cell cycle. The single chromosome of *Hyphomicrobium* is circular.

tion of prosthecate/stalked bacteria can be achieved by mixing an inoculum with a very dilute nutrient solution, for example 0.01% peptone, and leaving it sit undisturbed. Within a few days a surface film often develops containing prosthecae and stalked bacteria and isolation can be achieved by streaking on agar plates.

Caulobacter and *Gallionella*

Two commonly observed stalked bacteria are *Caulobacter* (Figure 13.42) and *Gallionella* (see Figure

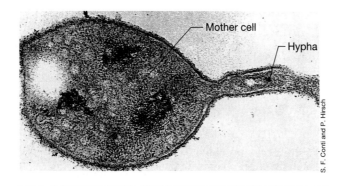

FIGURE 13.40   Electron micrograph of a thin section of a single *Hyphomicrobium* cell. The hypha is about 0.2 μm wide.

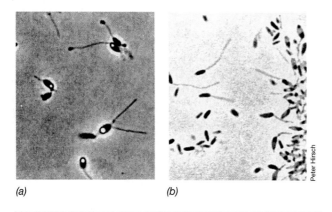

(a)   (b)

FIGURE 13.41   Photomicrographs of cells of *Hyphomicrobium*. Cells are about 0.7 μm wide. By phase contrast, showing typical fields. Notice the long hyphae and the occasional budding cell.

13.44). The former is a chemoorganotroph that produces a cytoplasm-filled stalk, that is, a prostheca, while the latter is a chemolithotrophic iron-oxidizing bacterium whose stalk is composed of ferric hydroxide. *Caulobacter* cells are often seen on surfaces in aquatic environments with the stalks of several cells attached to form *rosettes* (Figure 13.42*a*). At the end

of the stalk is a structure called a *holdfast* by which the stalk is attached to a surface.

The *Caulobacter* cell division cycle (Figure 13.43) is of special interest because it involves a process of *unequal binary fission*, and much molecular biology has been done on this organism to understand cell-division events. Cell division of a stalked cell occurs by elongation of the cell followed by fission, a single fla-

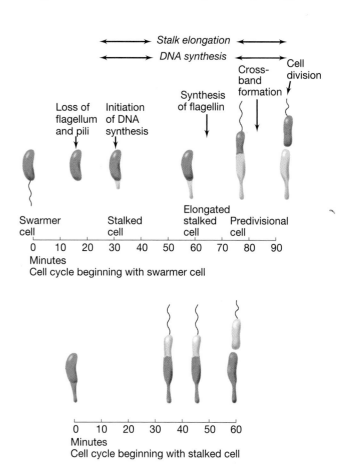

FIGURE 13.43   Stages in the *Caulobacter* cell cycle.

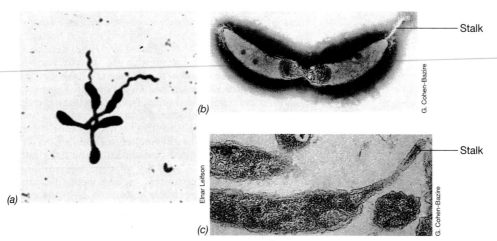

FIGURE 13.42
(a) A *Caulobacter* rosette. A single cell is about 0.5 μm wide. The five cells are attached by their stalks (prosthecae). Two of the cells have divided, and the daughter cells have formed flagella. (b, c) Electron micrographs of *Caulobacter* cells. (b) Negatively stained preparation of a cell in division. (c) A thin section. Notice that cytoplasmic constituents are present in the stalk region.

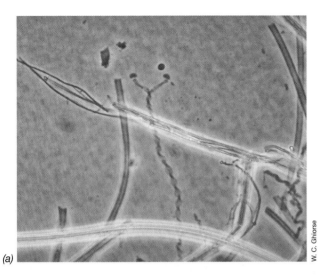

(a)

W. C. Ghiorse

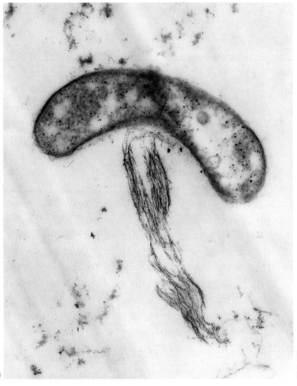

(b)

W. C. Ghiorse

FIGURE 13.44 *Gallionella ferruginea.* (a) Photomicrograph of cells from an iron seep near Ithaca, New York. Notice the twisted stalk leading to the two bean-shaped cells. (b) Transmission electron micrograph of a thin section of a cell. Cells are about 0.6 μm wide. Note the twisted stalk of ferric hydroxide emanating from the center of the cell in both photos.

gellum forming at the pole opposite the stalk. The flagellated cell so formed, called a *swarmer*, separates from the nonflagellated mother cell, swims around, and attaches to a new surface, forming a new stalk at the flagellated pole; the flagellum is then lost (Figure 13.43). Stalk formation is a necessary precursor of cell

division and is coordinated with DNA synthesis (Figure 13.43). The time span for the division of the stalked cell is thus shorter than the time span for division of the swarmer cell, owing to the requirement that a swarmer (flagellated) cell must synthesize a stalk before it divides (Figure 13.43). The cell division cycle in *Caulobacter* is thus more complex than simple binary fission because the stalked and swarmer cells have polar differentiation and the cells themselves are structurally different.

Another interesting stalked organism is *Gallionella*, which forms a twisted stalk containing ferric hydroxide (Figure 13.44). However, the stalk of *Gallionella* is not an integral part of the cell but is *excreted* from the cell surface. It contains an organic matrix on which the ferric hydroxide accumulates. *Gallionella* is frequently found in the waters draining bogs, iron springs, and other habitats where ferrous iron (Fe^{2+}) is present, usually in association with sheathed bacteria such as *Sphaerotilus*. *Gallionella* is an autotrophic chemolithotroph containing enzymes of the Calvin cycle (Section 15.7) by which CO_2 is incorporated into cell material with Fe^{2+} as electron donor.

✓ 13.14–13.15 Concept Check

Sheathed bacteria are filamentous Proteobacteria in which individual cells form chains within an outer layer called the sheath. Budding and prosthecate bacteria are appendaged cells that form stalks or prosthecae used for attachment or nutrient absorption and are primarily aquatic.

✓ Physiologically, what is unique about the sheathed bacterium *Leptothrix*?
✓ How does *budding* division differ from *binary* fission?
✓ What advantage might a prosthecate organism have in a very nutrient-poor environment?

13.16

Gliding Myxobacteria

Key Genera

Myxococcus
Stigmatella

A variety of prokaryotes exhibit a form of motility called *gliding*. These organisms, usually long rods or filaments, lack flagella but are able to move when in contact with surfaces. One group of gliding bacteria, the fruiting myxobacteria, possess the interesting property of forming multicellular structures called **fruiting bodies** and show complex developmental lifecycles involving intercellular communication. Phylogenetically, the gliding myxobacteria are members of the *delta* subdivision of Proteobacteria (Table 13.20).

TABLE 13.20	Classification of the fruiting myxobacteria[a]	
Characteristics	**Genus**	**DNA (mol % GC)**
Vegetative cells tapered		
Spherical or oval myxospores, fruiting bodies usually soft and slimy without well-defined sporangia or stalks	*Myxococcus*	68–71
Tough, cartilaginous, ridged fruiting bodies	*Corallococcus*	—
Rod-shaped myxospores:		
Myxospores not contained in sporangia, fruiting bodies without stalks	*Archangium*	67–68
Myxospores embedded in slime envelope:		
Fruiting bodies without stalks	*Cystobacter*	68
Stalked fruiting bodies, single sporangia	*Melittangium*	—
Stalked fruiting bodies, multiple sporangia	*Stigmatella*	68–69
Fruiting bodies are dark-brown clusters consisting of tiny spherical or disclike sporangia with an outer wall	*Angiococcus*	—
Vegetative cells not tapered (blunt, rounded ends); myxospores resemble vegetative cells; sporangia always produced:		
Fruiting bodies without stalks; myxospores rod-shaped	*Polyangium*	69
Fruiting bodies without stalks; myxospores oval; highly cellulolytic	*Sorangium*	—
Fruiting bodies without stalks; myxospores coccoid	*Nannocystis*	70–72
Large, solitary yellow fruiting bodies with netlike surface	*Haploangium*	—
Stalked fruiting bodies	*Chondromyces*	69–70

a Phylogenetically, those species examined fall into the delta subdivision of the Proteobacteria (see Table 13.1).

The fruiting myxobacteria exhibit the most complex behavioral patterns and life cycles of all known prokaryotic organisms. In keeping with this complexity, the chromosome of some myxobacteria is very large. *Myxococcus xanthus*, for example, has a single circular chromosome of 9500 kilobase pairs, *twice* as large as that of *Escherichia coli* (∞ Section 9.12). Indeed, this is two-thirds the size of the entire yeast genome, which is contained on 16 chromosomes (∞ Section 9.14). The vegetative cells of the fruiting myxobacteria are simple, nonflagellated, gram-negative rods (Figure 13.45*a*) that glide across surfaces and obtain their nutrients primarily by causing the lysis of other bacteria. Under appropriate conditions, a swarm of vegetative cells aggregate and construct *fruiting bodies*, within which some of the cells become converted to resting structures called **myxospores** (Figure 13.45*b*).

The fruiting bodies of the myxobacteria vary from simple globular masses of myxospores in loose slime to complex forms with a fruiting-body wall and a stalk (Figure 13.46). The fruiting bodies are often strikingly colored (Figures 13.46 and 13.47). They can often be seen with a hand lens or dissecting microscope on moist pieces of decaying wood or plant material. Fruiting bodies of myxobacteria often develop on dung pellets (for example, rabbit pellets) after they have been incubated for a few days in a moist chamber. Another way of isolating fruiting myxobacteria is to prepare Petri plates of water agar (1.5% agar in distilled water with no added nutrients) on to which is

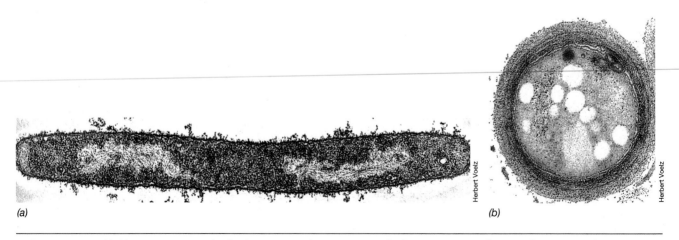

(a)   Herbert Voelz   *(b)*   Herbert Voelz

FIGURE 13.45 (a) Electron micrograph of a thin section of a vegetative cell of *Myxococcus xanthus*. A cell measures about 0.75 μm wide. (b) Myxospore of *M. xanthus*, showing the multilayered outer wall. Myxospores measure about 2 μm in diameter.

(a)

Hans Reichenbach

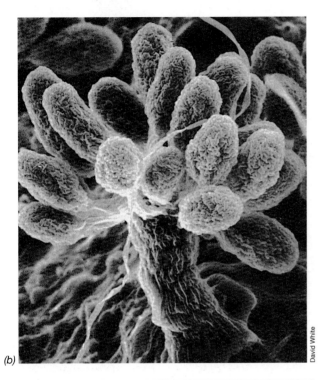

(b)

David White

FIGURE 13.46 *Stigmatella aurantiaca.* (a) Color photo of a single fruiting body. The structure is about 150 μm high. (b) Scanning electron micrograph of a fruiting body growing on a piece of wood. Note the individual cells visible in each fruiting structure.

spread a heavy suspension of virtually any bacterium. In the center of the plate a small amount of soil, decaying bark, or other natural material is placed. Myxobacteria in the inoculum lyse the bacterial cells and use their liberated products as nutrients; as they grow, they swarm out across the plate from the inoculum site. After several days to a week, the plates are examined under a dissecting microscope for myxobacterial swarms or fruiting bodies, and pure cultures are obtained by transfer of cells from the fruiting bodies or from the edge of the swarm to organic media. Many myxobacteria can be grown in the laboratory on media containing peptone or casein hydrolysate, which provides organic nutrients in the form of amino acids or small peptides. The organisms are typical aerobes with a complete citric acid cycle and cytochrome system.

Life Cycle of a Fruiting Myxobacterium

The life cycle of a typical fruiting myxobacterium is shown in Figure 13.48. A vegetative cell usually excretes slime, and as it moves across a solid surface it leaves a slime trail behind (Figure 13.49a). This trail is preferentially used by other cells in the swarm so that often a characteristic radiating pattern is soon created, with cells migrating along slime trails (Figure 13.49b). The fruiting body ultimately formed (Figures 13.46 and 13.47) is a complex structure produced by the differentiation of cells in the stalk region and in the myxospore-bearing head.

Fruiting-body formation does not occur so long as adequate nutrients for vegetative growth are present, but upon nutrient exhaustion, the vegetative swarms begin to fruit. Cells aggregate, likely through a chemotactic response, with the cells migrating toward each other and forming mounds or heaps (Figure 13.50), and a single fruiting body may have 10^9 or more cells. As the cell mounds become higher, the differentiation of the fruiting body into stalk and head begins (Figure 13.50b and c). Figure 13.50d clearly illustrates the differentiation of the fruiting body into stalk and head. The stalk is composed of slime, within which a few cells may be trapped. The majority of the cells accumulate in the fruiting-body head and undergo differentiation into *myxospores* (Figures 13.45–13.47). And, in some genera, the myxospores are enclosed in large walled structures called **cysts.** Compared to the vegetative cell, the myxospore is more resistant to drying, sonic vibration, UV radiation, and heat, but the degree of heat resistance is much less than that of the bacterial endospore. It seems likely that the main function of encysted myxospores is to enable the organism to survive desiccation during dispersal or during drying of the habitat. Upon dissemination to a suitable habitat or restoration of adequate growth conditions, the myxospore eventually germinates by a localized rupture of the capsule, with the growth and emergence of a typical vegetative rod.

Myxobacteria are usually colored by carotenoid pigments (see Figures 13.46a and 13.47), and the main pigments are carotenoid glycosides. Pigment formation is promoted by light, and at least one function of the pigment is photoprotection. Since in nature the myxobacteria usually form fruiting bodies in the light, the presence of these photoprotective pigments is understandable. In the genus *Stigmatella* (Figure 13.46), light greatly stimulates fruiting-body formation, and catalyzes production of a lipid pheromone called 2,5,8-trimethyl-8-hydroxy-nonane-4-one that initiates the aggregation step. The fruiting myxobacteria are currently classified primarily on morphological grounds using characteristics of the vegetative cells, the myxospores, and fruiting-body structure (Table 13.20).

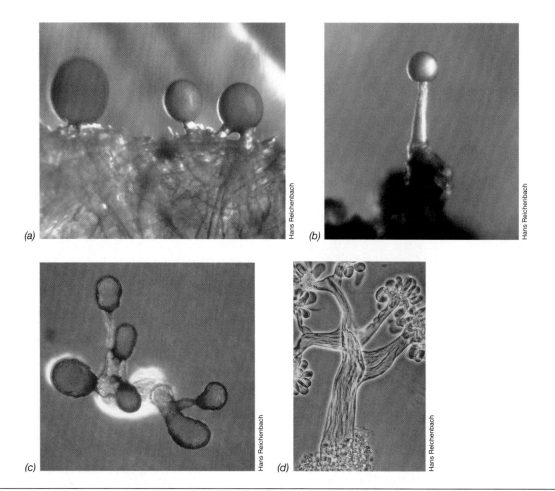

FIGURE 13.47   Fruiting bodies of four species of fruiting myxobacteria. (a) *Myxococcus fulvus* (125 μm high). (b) *Myxococcus stipitatas* (170 μm high). (c) *Mellitangium erectum* (50 μm high). (d) *Chondromyces crocatus* (560 μm high).

✓ 13.16 Concept Check

The fruiting myxobacteria are rod-shaped, gliding bacteria that aggregate to form complex masses of cells called *fruiting bodies*. Myxobacteria are chemoorganotrophic soil bacteria that live by consuming dead organic matter or other bacterial cells.

✓ What triggers fruiting body formation in myxobacteria?
✓ What is a *myxospore* and how does it compare with an *endospore*?
✓ To what specific phylogenetic group do the myxobacteria belong?

13.17

Sulfate- and Sulfur-Reducing Proteobacteria

Key Genera

Desulfovibrio
Desulfobacter
Desulfuromonas

Sulfate (SO_4^{2-}) and sulfur (S^0) function as electron acceptors under anoxic conditions by a large group of delta Proteobacteria that utilize organic compounds or H_2 as electron donors; hydrogen sulfide (H_2S) is the product of both SO_4^{2-} and S^0 reduction. Over twenty genera of these organisms, collectively known as the *dissimilative sulfate-* or *sulfur-reducing bacteria*, are known, and some of the key ones are shown in Table 13.21. The genera in group I, such as *Desulfovibrio* (Figure 13.51a, page 501), *Desulfomonas, Desulfotomaculum,* and *Desulfobulbus* (Figure 13.51c), utilize lactate, pyruvate, ethanol, or certain fatty acids as electron donors, reducing sulfate to hydrogen sulfide. The genera in group II, such as *Desulfobacter* (Figure 13.51d), *Desulfococcus, Desulfosarcina* (Figure 13.51e), and *Desulfonema* (Figure 13.51b), specialize in the oxidation of fatty acids, particularly *acetate*, reducing sulfate to sulfide. The sulfate-reducing bacteria are for the most part obligate anaerobes, and strict anoxic techniques must be used in their cultivation.

Sulfate-reducing bacteria are widespread in aquatic and terrestrial environments that become anoxic as a result of microbial decomposition processes. The best-studied genus is *Desulfovibrio* (Figure 13.51a), which is common in aquatic habitats or waterlogged soils containing abundant organic material and sufficient levels

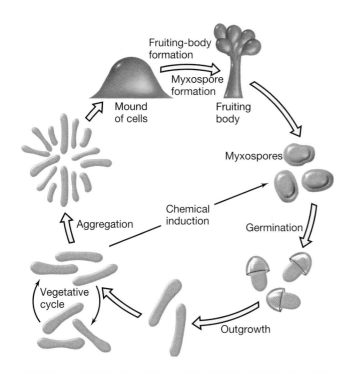

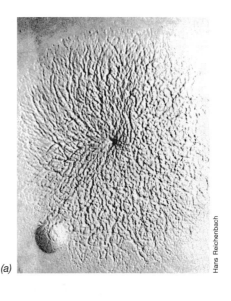

FIGURE 13.48 Life cycle of *Myxococcus xanthus*. Aggregation serves to assemble vegetative cells for fruiting-body formation. Vegetative cells undergo morphogenesis to resting cells called myxospores. The latter germinate under favorable nutritional and physical conditions to yield vegetative cells. Vegetative cells can be converted directly to myxospores without fruiting-body formation by certain chemical inducers, notably high concentrations of glycerol. See the photograph of *Myxococcus* fruiting bodies in Figure 13.47.

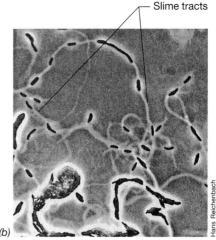

of sulfate. *Desulfotomaculum,* phylogenetically a member of the gram-positive Bacteria, consists of endospore-forming rods found primarily in soil. Growth and reduction of sulfate by *Desulfotomaculum* in certain canned foods leads to a type of spoilage called *sulfide stinker.* The remaining genera of sulfate reducers are indigenous to anoxic freshwater or marine environments, and can occasionally be isolated from the mammalian intestine.

FIGURE 13.49 (a) Photomicrograph of a swarming colony (9-mm diameter) of *Myxococcus xanthus* on agar. (b) Single cells of *Myxococcus fulvus* from an actively gliding culture, showing the characteristic slime tracks on the agar.

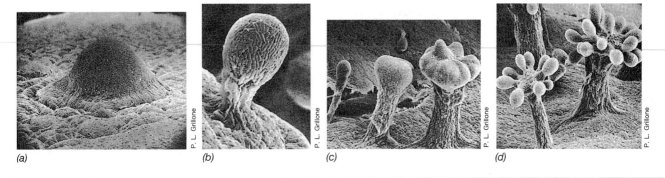

FIGURE 13.50 Scanning electron micrographs of fruiting-body formation in *Chondromyces crocatus*. (a) Early stage, showing aggregation and mound formation. (b) Initial stage of stalk formation. Slime formation in the head has not yet begun, and so the cells of which the head is composed are still visible. (c) Three stages in head formation. Note that the diameter of the stalk also increases. (d) Mature fruiting bodies. The entire fruiting structure is about 600 μm in height (see Figure 13.47*d*).

TABLE 13.21	Characteristics of sulfate- and sulfur-reducing bacteria[a]	
Genus	**Characteristics**	**DNA (mol % GC)**
Group I sulfate reducers: Nonacetate oxidizers		
Desulfovibrio	Polarly flagellated, curved rods, no spores; gram-negative; contain desulfoviridin; twelve species, one thermophilic	46–61
Desulfomicrobium	Motile rods, no spores; gram-negative; desulfoviridin absent; three species	52–57
Desulfobotulus	Vibrios; gram-negative; motile; desulfoviridin absent; one species	53
Desulfofustis	Motile rods, specializes in the degradation of glycolate and glyoxalate	56
Desulfotomaculum	Straight or curved rods; motile by peritrichous or polar flagellation; gram-negative; desulfoviridin absent; produce endospores; four species, one thermophilic; one species capable of utilizing acetate as energy source	37–46
Desulfomonile	Rod; capable of reductive dechlorination of 3-chlorobenzoate to benzoate (∞ Section 16.22)	49
Desulfobacula	Oval to coccoid cells, marine; can oxidize various aromatic compounds including the aromatic hydrocarbon toluene, to CO_2; one species	42
Archaeoglobus	Archaeon; hyperthermophile, temperature optimum, 83°C; contains some unique coenzymes of methanogenic bacteria, makes small amount of methane during growth; H_2, formate, glucose, lactate, and pyruvate are electron donors, SO_4^{2-}, $S_2O_3^{2-}$, or SO_3^{2-}, electron acceptors; two species (∞ Section 14.6)	41–46
Desulfobulbus	Ovoid or lemon-shaped cells; no spores; gram-negative; desulfoviridin absent; if motile, by single polar flagellum; utilizes propionate as electron donor with acetate + CO_2 as products; three species	59–60
Desulforhopalus	Curved rods, gas vacuolate, psychrophile; uses propionate, lactate, or alcohols as electron donor	48
Thermodesulfobacterium	Small, gram-negative rods; desulfoviridin present; thermophilic, optimum growth at 70°C; a member of the Bacteria but contains ether-linked lipids (see Section 13.33)	34
Group II sulfate reducers: Acetate oxidizers		
Desulfobacter	Rods; no spores, gram-negative; desulfoviridin absent; if motile, by single polar flagellum; utilizes only acetate as electron donor and oxidizes it to CO_2 via the citric acid cycle; four species	45–46
Desulfobacterium	Rods, some with gas vesicles, marine; capable of autotrophic growth via the acetyl-CoA pathway; three species	41–59
Desulfococcus	Spherical cells; nonmotile; gram-negative; desulfoviridin present, no spores; utilizes C_1 to C_{14} fatty acids as electron donor with complete oxidation to CO_2; capable of autotrophic growth via the acetyl-CoA pathway; two species	57
Desulfonema	Large, filamentous gliding bacteria; gram-positive, no spores; desulfoviridin present or absent; utilizes C_2 to C_{12} fatty acids as electron donor with complete oxidation to CO_2; capable of autotrophic growth via the acetyl-CoA pathway (H_2 as electron donor); two species	35–42
Desulfosarcina	Cells in packets (sarcina arrangement); gram-negative; no spores; desulfoviridin absent; utilizes C_2 to C_{14} fatty acids as electron donor with complete oxidation to CO_2; capable of autotrophic growth via the acetyl-CoA pathway (H_2 as electron donor); one species	51
Desulfoarculus	Vibrios; gram-negative; motile; desulfoviridin absent; utilizes only C_1 to C_{18} fatty acids as electron donor	66
Desulfacinum	Cocci to oval-shaped cells; gram-negative; utilizes C_1 to C_{18} fatty acids, very nutritionally diverse, capable of autotrophic growth; thermophile	64
Desulforhabdus	Rods; no spores; gram-negative; nonmotile; utilizes fatty acids with complete oxidation to CO_2	52
Thermodesulforhabdus	Gram-negative motile rods; thermophilic; uses fatty acids up to C_{18}	51
Dissimilatory sulfur reducers		
Desulfuromonas	Straight rods, single lateral flagellum; no spores; gram-negative; does not reduce sulfate; acetate, succinate, ethanol, or propanol used as electron donor; obligate anaerobe; four species, at least one of which is capable of the reductive dechlorination of trichloroethylene	50–63
Desulfurella	Motile short rods; gram-negative; requires acetate; thermophilic	31
Sulfurospirillum	Small vibrios, reduces S^0 with H_2 or formate as electron donors	—
Campylobacter	Curved, vibrio-shaped rods; polar flagella; gram-negative; no spores; unable to reduce sulfate but can reduce sulfur, sulfite, thiosulfate, nitrate, or fumarate anaerobically with acetate or a variety of other carbon or electron donor sources; facultative aerobe	40–42

[a] Phylogenetically, most sulfate- and sulfur-reducing bacteria are delta Proteobacteria.

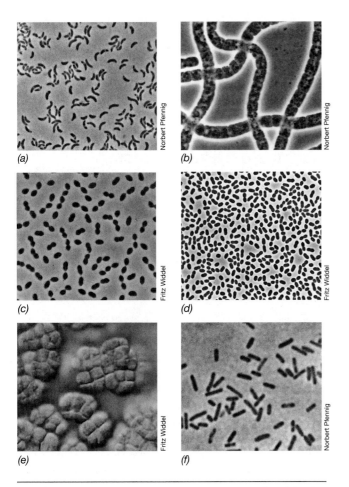

(a) Norbert Pfennig
(b) Norbert Pfennig
(c) Fritz Widdel
(d) Fritz Widdel
(e) Fritz Widdel
(f) Norbert Pfennig

FIGURE 13.51 Phase contrast photomicrographs of (a–e) representative sulfate-reducing and (f) sulfur-reducing bacteria. (a) *Desulfovibrio desulfuricans;* cell diameter about 0.7 μm. (b) *Desulfonema limicola;* cell diameter 3 μm. (c) *Desulfobulbus propionicus;* cell diameter about 1.2 μm. (d) *Desulfobacter postgatei;* cell diameter about 1.5 μm. (e) *Desulfosarcina variabilis* (interference contrast microscopy); cell diameter about 1.25 μm. (f) *Desulfuromonas acetoxidans;* cell diameter about 0.6 μm.

Sulfur Reduction

The dissimilative sulfur-reducers can reduce elemental sulfur to sulfide but are unable to reduce sulfate to sulfide. Members of the genus *Desulfuromonas* (Figure 13.51*f*) can grow anaerobically by coupling the *oxidation* of substrates such as acetate or ethanol to the *reduction* of elemental sulfur to hydrogen sulfide. However, the ability to reduce elemental sulfur, as well as other sulfur compounds such as thiosulfate, sulfite, or dimethyl sulfoxide (DMSO), is a widespread property of a variety of chemoorganotrophic, generally facultatively aerobic bacteria (for example, *Proteus, Campylobacter, Pseudomonas,* and *Salmonella*). *Desulfuromonas* differs from the latter in that it is an obligate anaerobe and utilizes *only* sulfur as an electron acceptor (see Table 13.21). Dissimilative sulfur-reducing bacteria like *Desulfuromonas* reside in many

of the same habitats as sulfate-reducing bacteria and often form associations with bacteria that oxidize H_2S to S^0, like green sulfur bacteria (see Section 13.29). The sulfur produced is then reduced back to H_2S during metabolism of the sulfur-reducer to complete a very abbreviated anaerobic sulfur cycle.

Physiology of Sulfate-Reducing Bacteria

The range of electron donors used by sulfate-reducing bacteria is broad. H_2, lactate, and pyruvate are almost universally used, and many group I species utilize malate, sulfonates, and certain primary alcohols (for example, methanol, ethanol, propanol, and butanol). Some strains of *Desulfotomaculum* utilize glucose, but this is rather rare among sulfate reducers in general. Group I sulfate reducers oxidize their energy source to the level of acetate and excrete this fatty acid as an end product. Group II organisms differ from those in group I by their ability to oxidize fatty acids, lactate, succinate, and even benzoate in some cases, all the way to CO_2. *Desulfosarcina, Desulfonema, Desulfococcus, Desulfobacterium, Desulfotomaculum,* and certain species of *Desulfovibrio,* are unique among sulfate-reducers in their ability to grow chemolithotrophically and autotrophically with H_2 as electron donor, sulfate as electron acceptor, and CO_2 as sole carbon source.

In addition to using sulfate as an electron acceptor, many sulfate-reducing bacteria can grow using *nitrate* (NO_3^-) as an electron acceptor, reducing NO_3^- to NH_3, or sulfonates, such as isethionate ($HO-CH_2-CH_2-SO_3^-$), or elemental sulfur (S^0), both of which are reduced to H_2S, or can use certain organic compounds for energy generation by fermentative pathways in the complete absence of terminal electron acceptors. The most common fermentable compound is *pyruvate,* which is converted via the phosphoroclastic reaction to acetate, CO_2, and H_2 (⊂⊃ Figure 15.53). Moreover, although thought for a long time to be *obligate* anaerobes, certain isolates of sulfate-reducing bacteria, primarily ones isolated from microbial mats where they coexist with O_2-producing cyanobacteria, are quite oxygen tolerant and actually respire with O_2 as electron acceptor. However, aerobic respiration does *not* support growth and is probably a means of removing O_2 when it is present in an environment otherwise suitable for growth of sulfate-reducing bacteria.

Isolation

The enrichment of *Desulfovibrio* is relatively easy on an anoxic lactate–sulfate medium to which ferrous iron is added. A reducing agent such as thioglycolate or ascorbate is also added to achieve a lower E_0'. When sulfate-reducing bacteria grow, the sulfide formed from sulfate reduction combines with the ferrous iron to form black, insoluble ferrous sulfide. This blackening not only indicates sulfate reduction, but the iron also ties up and detoxifies the sulfide, making possible growth to higher cell yields. After some growth has occurred as evidenced by blackening of the medium, purification is accomplished by streaking onto a tube coated on the inside surface with a thin layer of agar (called roll tubes) or on Petri plates in an anoxic glove box. Alternatively, agar shake tubes can be used for purification purposes. In the *shake tube method* a small amount of liquid from the original enrichment is added to a tube of molten agar growth medium, mixed thoroughly, and sequentially diluted through a series of molten agar tubes (∽ Section 16.4 and Figure 16.6*b*). On solidification, individual cells distributed throughout the agar form black colonies that can be removed aseptically, and the whole process is repeated until pure cultures are obtained.

✓ 13.17 Concept Check

Sulfate- and sulfur-reducing bacteria are a large group of delta Proteobacteria unified by their physiological process of reducing either SO_4^{2-} or S^0 to H_2S under anaerobic conditions. Two physiological subgroups of sulfate-reducing bacteria are known: group I, which is incapable of oxidizing acetate to CO_2, and group II, which is capable of doing so.

✓ What organic substrate would you use to enrich and isolate a group I sulfate reducer from nature?

✓ For sulfate-reducing bacteria capable of chemolithotrophic and autotrophic growth: (1) What is the electron donor, (2) what is the electron acceptor, and (3) what is the source of cell carbon?

✓ Physiologically, how does *Desulfuromonas* differ from *Desulfovibrio*?

Kingdom II: Gram-Positive Bacteria

13.18

Nonsporulating, Low GC, Gram-Positive Bacteria

Key Genera

Staphylococcus
Micrococcus
Streptococcus
Lactobacillus

As mentioned in the introduction to this chapter, gram-positive Bacteria fall into two major phylogenetic subdivisions, "low GC" and "high GC." We consider in this section the genera *Staphylococcus*, *Sarcina*, and *Micrococcus* because they are morphologically quite similar even though *Micrococcus* is actually a member of the "High GC" group, along with the lactic acid Bacteria, classical nonsporulating gram-positive rods and cocci. An overview of the properties of these organisms is given in Table 13.22.

TABLE 13.22	Distinguishing features of gram-positive cocci[a]					
Genus	**Motility**	**Arrangement of cells**	**Growth by fermentation**	**DNA (mol % GC)**	**Phylogenetic group[a]**	**Other characteristics**
Micrococcus	–	Clusters, tetrads	–	66–73	High GC	Strict aerobe
Staphylococcus	–	Clusters, pairs	+	30–39	Low GC	Only genus of this group to contain teichoic acid in cell wall
Stomatococcus	–	Clusters, pairs	+	56–60	–	Only genus of this group containing a capsule
Planococcus	+	Pairs, tetrads	–	39–52	Low GC	Primarily marine
Sarcina	–	Cuboidal packets of eight or more cells	+	28–31	Low GC	Extremely acid-tolerant; cellulose in cell wall
Ruminococcus	+	Pairs, chains	+	39–46	Low GC	Obligate anaerobe; inhabits rumen, cecum, and large intestine of many animals
Peptococcus	–	Clusters, pairs	+	50–51	Low GC	Obligate anaerobe; ferments peptone but not sugars
Peptostrepto-coccus	–	Clumps, short chains	+	28–37	Low GC	Obligate anaerobe; ferments peptone; common member of human normal flora, skin, intestine, vagina; also isolated from vaginal and purulent discharges

a All are members of the gram-positive Bacteria (see Figure 13.1).

Staphylococcus and *Micrococcus*

Staphylococcus (Figure 13.52) and *Micrococcus* are both aerobic organisms with a typical respiratory metabolism. They are catalase-positive, and this test permits their distinction from *Streptococcus* and some other genera of gram-positive cocci. Gram-positive cocci are relatively resistant to reduced water potential and tolerate drying and high salt fairly well. Their ability to grow in media with high salt provides a selective means for isolation. For example, if an appropriate inoculum is spread on an agar plate with a rich medium containing 7.5% NaCl and the plate incubated aerobically, gram-positive cocci often form the predominant colonies. Often, these organisms are pigmented, and this provides an additional aid in selecting gram-positive cocci. The two genera *Micrococcus* and *Staphylococcus* can easily be separated based on the oxidation–fermentation (O/F) (∞ Table 21.3) test. *Micrococcus* is an obligate aerobe and produces acid from glucose only aerobically, whereas *Staphylococcus* is a facultative aerobe and produces acid from glucose both aerobically and anaerobically.

Staphylococci are common parasites of humans and animals and occasionally cause serious infections. In humans, two major species are recognized, *Staphylococcus epidermidis*, a nonpigmented, nonpathogenic organism usually found on the skin or mucous membranes, and *Staphylococcus aureus*, a yellow pigmented species that is most commonly associated with pathological conditions, including boils, pimples, pneumonia, osteomyelitis, meningitis, and arthritis. We discuss the pathogenesis of *S. aureus* in Chapter 19 and staphylococcal diseases in Chapters 23 and 24. *Micrococcus* species can also be isolated from skin but are much more common on inanimate objects, dust particles, and in soil.

Sarcina

The genus *Sarcina* contains species of bacteria that divide in three perpendicular planes to yield packets of eight cells or more (Figure 13.53a). *Sarcina* are obligate anaerobes and are extremely acid-tolerant, being able to ferment sugars and grow down to pH 2. Cells of one species, *Sarcina ventriculi*, contain a thick fibrous layer of cellulose surrounding the cell wall (Figure 13.53b). The cellulose layers of adjacent cells become attached, and this functions as a cementing material to hold together packets of *S. ventriculi* cells. *Sarcina* species can be isolated from soil, mud, feces, and stomach contents. Because of its extreme acid tolerance, *S. ventriculi* is one of the only bacteria that can actually grow in the stomach of humans and other monogastric animals. Rapid growth of *S. ventriculi* is observed in the stomach of humans suffering from certain gastrointestinal pathological conditions (such as pyloric ulcerations) that retard the flow of food to the intestine.

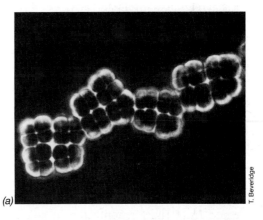
(a)

(b)

FIGURE 13.53   (a) Phase contrast photomicrograph of cells of a typical gram-positive coccus *Sarcina* sp. A single cell is about 2 μm in diameter. (b) Electron micrograph of a thin section.

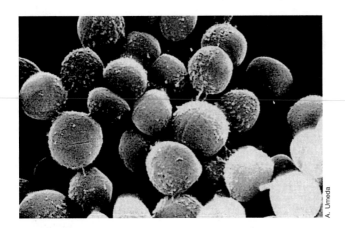

FIGURE 13.52   Scanning electron micrograph of typical *Staphylococcus*, showing the irregular arrangement of the cell clusters. Individual cells are about 0.8 μm in diameter.

TABLE 13.23	Differentiation of the principal genera of lactic acid bacteria[a]		
Genus	**Cell form and arrangement**	**Fermentation**	**DNA (mol % GC)**
Streptococcus	Cocci in chains	Homofermentative	34–46
Leuconostoc	Cocci in chains	Heterofermentative	38–41
Pediococcus	Cocci in tetrads	Homofermentative	34–42
Lactobacillus	(1) Rods, usually in chains	Homofermentative	32–53
	(2) Rods, usually in chains	Heterofermentative	34–53
Enterococcus	Cocci in chains	Homofermentative	38–40
Lactococcus	Cocci in chains	Homofermentative	38–41

a Phylogenetically, all organisms are members of the low GC subdivision of the gram-positive Bacteria.

Lactic Acid Bacteria and Lactic Acid Fermentations

The lactic acid bacteria are gram-positive rods and cocci that produce lactic acid as a major or sole fermentation product. Members of this group lack porphyrins and cytochromes, do not carry out electron transport phosphorylation, and hence obtain energy only by *substrate-level phosphorylation*. All lactic acid bacteria grow anaerobically. Unlike many anaerobes, however, most lactic acid bacteria are not sensitive to O_2 and can grow in its presence as well as in its absence; thus they are **aerotolerant anaerobes.** Most lactic acid bacteria obtain energy only from the metabolism of sugars and hence are usually restricted to habitats in which sugars are present. They usually have only limited biosynthetic ability, and their complex nutritional requirements include needs for amino acids, vitamins, purines, and pyrimidines (⚬⚬ Table 4.4).

One important difference between subgroups of the lactic acid bacteria lies in the nature of the products formed from the fermentation of sugars. One group, called **homofermentative,** produces a single fermentation product, *lactic acid,* whereas the other group, called **heterofermentative,** produces other products, mainly *ethanol* and CO_2 as well as lactate (Table 13.23). Abbreviated pathways for the fermentation of glucose by a homo- and a heterofermentative organism are shown in Figure 13.54. The differences observed in the fermentation patterns are determined by the presence or absence of the enzyme **aldolase,** the key enzyme in *glycolysis* (⚬⚬ Figure 4.12). The heterofermenters, lacking aldolase, cannot break down fructose bisphosphate to triose phosphate. Instead, they oxidize glucose 6-phosphate to 6-phosphogluconate and then decarboxylate this to pentose phosphate, which is broken down to triose phosphate and acetylphosphate by means of the enzyme **phosphoketolase** (Figure 13.54).

In heterofermenters, triose phosphate is converted ultimately to lactic acid with the production of 1 mol of ATP, while the acetylphosphate accepts electrons from the NADH generated during the production of pentose phosphate and is thereby converted to ethanol *without* yielding ATP. Because of this, heterofermenters produce only *1 mol* of ATP from glucose instead of the 2 mol produced by homofermenters. Because the heterofermenters decarboxylate 6-phosphogluconate, they produce CO_2 as a fermentation product, whereas the homofermenters produce little or no CO_2; therefore one simple way of detecting a heterofermenter is to observe for production of CO_2 in laboratory cultures.

The various genera of lactic acid bacteria have been defined on the basis of cell morphology, DNA base composition, and type of fermentative metabolism, as is shown in Table 13.23. Members of the genera *Streptococcus, Enterococcus, Lactococcus, Leuconostoc,* and *Pediococcus* have fairly similar DNA base ratio compositions; in addition, there is very little variation from strain to strain. The genus *Lactobacillus,* on the other hand, has members with widely diverse DNA compositions and hence does not constitute a homogeneous group.

Streptococcus and Other Cocci

The genus *Streptococcus* (Figure 13.55) contains a wide variety of homofermentative species with quite distinct habitats, whose activities are of considerable practical importance to humans. Some members are pathogenic to humans and animals (⚬⚬ Section 23.2). As producers of lactic acid, certain streptococci play important roles in the production of buttermilk, silage, and other fermented products (⚬⚬ Section 24.10), and certain species play a major role in dental caries formation (⚬⚬ Section 19.3). To distinguish generally nonpathogenic streptococci from human pathogenic species, three genera are recognized. The genus *Lactococcus* contains those streptococci of dairy significance, whereas the genus *Enterococcus* includes streptococci that are primarily of fecal origin.

Organisms in the genus *Streptococcus* have been divided into two groups of related species on the basis of

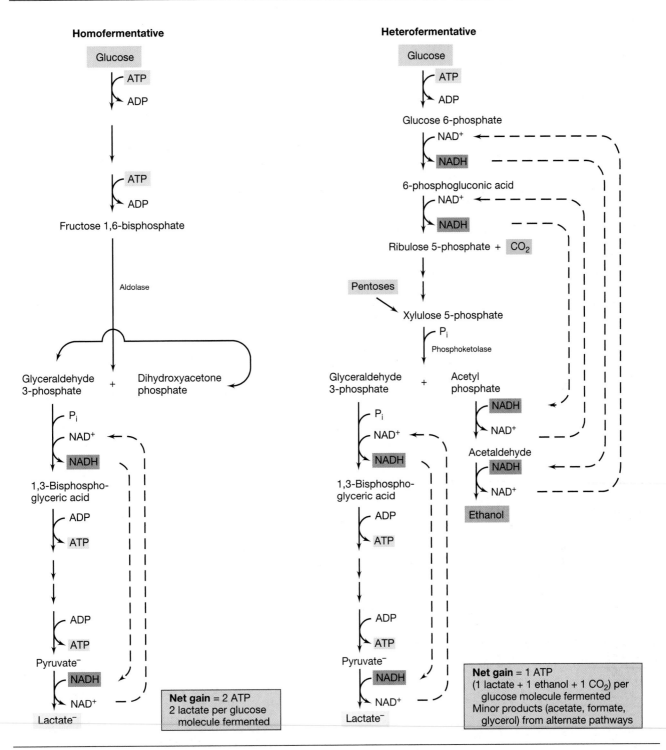

FIGURE 13.54 The fermentation of glucose in homofermentative and heterofermentative lactic acid bacteria. Note that no ATP is made in reactions leading to ethanol formation.

characteristics enumerated in Table 13.24. Hemolysis on blood agar is of considerable importance in the subdivision of the genus. Colonies of those strains producing streptolysin O or S are surrounded by a large zone of complete red blood cell hemolysis, a condition called **β hemolysis** (∞ Figure 19.16*a*). On the other hand, many streptococci and the lactococci and enterococci do not produce hemolysins but instead cause the formation of a greenish or brownish zone around colonies on blood agar, which is due not to true hemolysis but to discoloration and loss of potassium from the red cells. This type of reaction has classically been referred to as

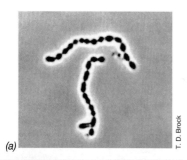

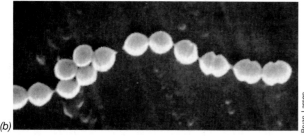

FIGURE 13.55 Phase contrast (a) and scanning electron (b) micrographs of *Streptococcus* species. (a) *Streptococcus lactis.* (b) *Streptococcus* sp. Cells in both cases are 0.5–1 μm in diameter.

α hemolysis. Streptococci and related cocci are also divided into *immunological* groups based on the presence of specific carbohydrate antigens. These antigenic groups (or **Lancefield groups** as they are commonly known, named for Rebecca Lancefield, a pioneer in *Streptococcus* taxonomy), are designated by letters; A through O are currently recognized. Those β-hemolytic streptococci found in human beings usually contain the group A antigen, while enterococci contain the group D antigen. Group B streptococci are usually found in association with animals, are a cause of mastitis in cows, and have also been implicated in human

infections. Lactococci are of antigen group N and are not pathogenic.

Placed in the genus *Leuconostoc* are heterofermentative cocci. Strains of *Leuconostoc* also produce the flavoring ingredients diacetyl and acetoin by breakdown of citrate and have been used as starter cultures in dairy fermentations. Some strains of *Leuconostoc* produce large amounts of dextran polysaccharides (α-1,6-glucan) when cultured on sucrose (⊙ Figure 15.60), and some of these have found medical use as plasma extenders in blood transfusions. Other strains of *Leuconostoc* produce other polysaccharide polymers such as fructose polymers called *levans*.

Lactobacillus

Lactobacilli are typically rod-shaped, varying from long and slender to short, bent rods (Figure 13.56). Most species are homofermentative, but some are heterofermentative. The genus has been divided into three major subgroups, and over 70 species are recognized (Table 13.25).

Lactobacilli are often found in dairy products, and some strains are used in the preparation of fermented milk products. For instance, *Lactobacillus delbrueckii* (Figure 13.56c) is used in the preparation of yogurt, *L. acidophilus* (Figure 13.56a) in the production of acidophilus milk, and other species in the production of sauerkraut, silage, and pickles (⊙ Section 24.10). The lactobacilli are usually more resistant to acidic conditions than are the other lactic acid bacteria, being able to grow well at pH values as low as 4. Because of this, they can be selectively isolated from natural materials by use of an acidic rich carbohydrate-containing medium such as tomato juice–peptone agar. The acid resistance of the lactobacilli

---

TABLE 13.24 Differential characteristics of streptococci, lactococci, and enterococci

Group	Antigenic (Lancefield) groups	Representative species	Type of hemolysis on blood agar	Good growth at 10°C	Good growth at 45°C	Survive 60°C for 30 min	Growth in Milk with 0.1% methylene blue	Growth in Broth with 40% bile	Habitat
Streptococci									
Pyogenes subgroup	A,B,C,F,G	*Streptococcus pyogenes*	Lysis (β)	−	−	−	−	−	Respiratory tract, systemic
Viridans subgroup	−	*Streptococcus mutans*	Greening (α)	−	+	−	−	−	Mouth, intestine
Enterococci	D	*Enterococcus faecalis*	Lysis (β), greening (α), or none	+	+	+	+	+	Intestine, vagina, plants
Lactococci	N	*Lactococcus lactis*	None	+	−	+	+	+	Plants, dairy products

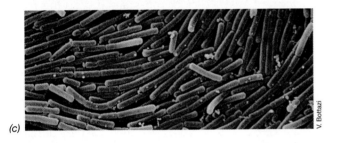

(a)

(b)

(c)

Otto Kandler

Otto Kandler

V. Bottazi

FIGURE 13.56 Phase contrast and electron micrographs of *Lactobacillus* species. (a) *Lactobacillus acidophilus.* Cells are about 0.75 μm wide. (b) *Lactobacillus brevis,* transmission electron micrograph. Cells measure about 0.8×2 μm. (c) *Lactobacillus delbrueckii,* scanning electron micrograph. Cells are about 0.7 μm in diameter.

enables them to continue growing during natural lactic fermentations when the pH value has dropped too low for other lactic acid bacteria to grow, and the lactobacilli are therefore responsible for the final stages of most lactic acid fermentations. The lactobacilli are rarely if ever pathogenic.

13.19

Endospore-Forming, Low GC, Gram-Positive Bacteria

Key Genera

Bacillus
Clostridium
Sporosarcina
Heliobacterium

Several genera of endospore-forming bacteria have been recognized (Table 13.26), distinguished on the basis of cell morphology, shape, and cellular position of the endospore (Figure 13.57), relationship to O_2, and energy metabolism; all endospore-formers show phylogenetic affiliation to the "low GC" gram-positive Bacteria. The two genera most frequently studied are *Bacillus*, species of which are aerobic or facultatively aerobic, and *Clostridium*, which contains the strictly anaerobic, fermentative species. One group of endospore-formers, the *heliobacteria*, are actually phototrophic (the word *helio* comes from the Greek word for *sun*). The structure and heat resistance of the bacterial endospore along with the process of endospore formation itself was discussed in Section 3.15.

Although considerable genetic heterogeneity exists among endospore-formers—for example, the GC ratios of *Bacillus* species alone vary over a range of 40%—all endospore-forming bacteria are *ecologically* related because they are found in nature primarily in soil. Even those species that are pathogenic to humans or other animals are primarily saprophytic soil organisms, and infect hosts only incidentally. Indeed, the ability to produce endospores should be advantageous for a soil microorganism because soil can be a highly variable environment in terms of nutrient levels, temperature, and water activity. Thus a heat- and desiccation-resistant structure capable of remaining dormant for long periods (perhaps even millions of

TABLE 13.25	Characteristics of subgroups in the genus *Lactobacillus*		
Characteristics		**Representative species**	**DNA (mol % GC)**
Homofermentative:			
Lactic acid the major product (>85% from glucose)			
No gas from glucose; aldolase present			
(1) Grow at 45°C but not at 15°C, long rods; glycerol teichoic acid		*L. delbrueckii*	49–51
		L. acidophilus	34–37
(2) Grow at 15°C, variable growth at 45°C; short rods and coryneforms; ribitol and glycerol teichoic acids; can produce more oxidized fermentation products if O_2 is present		*L. casei, L. plantarum*	45–46
		L. curvatus	42–44
Heterofermentative:			
Produce about 50% lactic acid from glucose; produce CO_2 and ethanol; aldolase absent; phosphoketolase present; long and short rods; glycerol teichoic acid		*L. fermentum*	52–54
		L. brevis, L. buchneri	44–47
		L. kefiri	41–42

TABLE 13.26	Genera of endospore-forming bacteria[a]	
Characteristics	**Genus**	**DNA (mol % GC)**
Rods		
Aerobic or facultative, catalase produced	*Bacillus*	32–69
	Paenibacillus	40–54
Microaerophilic, no catalase; homofermentative lactic acid producer	*Sporolactobacillus*	46–47
Anaerobic:		
Sulfate-reducing	*Desulfotomaculum*	38–50
Does not reduce sulfate, fermentative	*Clostridium* (see Figure 13.57)	21–54
Thermophilic, temperature optimum 65–70°C, fermentative	*Thermoanaerobacter*	31–39
Gram-negative; can grow as homoacetogen on $H_2 + CO_2$	*Sporomusa*	41–49
Halophile, isolated from the Dead Sea	*Sporohalobacter*	31
Produces up to five spores per cell; fixes N_2	*Anaerobacter*	29
Acidophile, pH optimum 3	*Alicyclobacillus*	52–60
Alkaliphile, pH optimum 9	*Amphibacillus*	36–38
Phototrophic	*Heliobacterium, Heliophilum*	50–58
Syntrophic, degrades fatty acids but only in coculture with a H_2-utilizing bacterium	*Syntrophospora*	37
Reductively dechlorinates chlorophenols	*Desulfitobacterium*	46
Cocci (usually arranged in tetrads or packets), aerobic	*Sporosarcina* (see Figure 13.61)	40–41

a Phylogenetically, all organisms are members of the low GC subdivision of the gram-positive Bacteria.

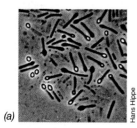 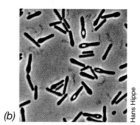 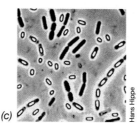

FIGURE 13.57 Phase contrast photomicrographs of various *Clostridium* species, showing the different locations of the endospore. (a) *Clostridium cadaveris*, terminal spores. Cells are about 0.9 μm wide. (b) *Clostridium sporogenes*, subterminal spores. Cells are about 1 μm wide. (c) *Clostridium bifermentans*, central spores. Cells are about 1.2 μm wide.

years, ∞ Section 3.15) should offer considerable survival value in nature.

Spore-formers can be selectively isolated from soil, food, dust, and other materials by exposing the sample to 80°C for 10 min (pasteurization), a treatment that effectively kills vegetative cells while the spores present remain viable. Streaking heat-treated samples on plates of the appropriate media and incubating either aerobically or anaerobically readily yield species of *Bacillus* or *Clostridium*, respectively.

Bacillus

Species of *Bacillus* usually grow well on defined media containing any of a number of carbon sources. Many bacilli produce extracellular hydrolytic enzymes that break down complex polymers such as polysaccharides, nucleic acids, and lipids, permitting the organisms to use these products as carbon sources and electron donors. Many bacilli produce antibiotics, of which bacitracin, polymyxin, tyrocidin, gramicidin, and circulin are examples. In most cases, antibiotic production is related to the sporulation process, the antibiotic being released when the culture enters the stationary phase of

growth and after it is committed to sporulation. An outline of the subdivision of the genus *Bacillus* is given in Table 13.27.

Several *Bacillus* species, most notably *B. popilliae* and *B. thuringiensis*, produce insect larvicides. *Bacillus popilliae* causes a fatal condition called *milky disease* in Japanese beetle larvae and larvae of closely related beetles of the family Scarabaeidae. *Bacillus thuringiensis* causes a fatal disease of larvae of many different groups of insects, although individual strains are specific as to host affected. Strains exist that are specific for lepidopterans, such as the silkworm, the cabbage worm, the tent caterpillar, and the gypsy moth. Some strains kill dipterans such as mosquitoes and black flies. Others kill coleopterans such as Colorado potato beetles. Strains of *B. thuringiensis* have also been discovered that are toxic to Japanese beetles. Spore preparations derived from *B. thuringiensis* and *B. popilliae* are commercially available as biological insecticides.

The disease caused by *Bacillus popilliae* is a septicemia, whereas the disease caused by *B. thuringiensis* is essentially an intoxication. Both of these insect pathogens form a crystalline protein during sporula-

TABLE 13.27	Characteristics of representative species of the genus *Bacillus*		
Characteristics	Species	Spore position	DNA (mol % GC)
I. Spores oval or cylindrical, facultative aerobes, casein and starch hydrolyzed; sporangia not swollen, spore wall thin			
Thermophiles and acidophiles	*B. coagulans*	Central or terminal	47
	B. acidocaldarius	Terminal	60
Mesophiles	*B. licheniformis*	Central	46
	B. cereus	Central	35
	B. anthracis	Central	33
	B. megaterium	Central	37
	B. subtilis	Central	43
Insect pathogen	*B. thuringiensis*	Central	34
Sporangia distinctly swollen, spore wall thick			
Thermophile	*B. stearothermophilus*	Terminal	52
Mesophiles	*B. polymyxa*	Terminal	44
	B. macerans	Terminal	52
	B. circulans	Central or terminal	35
Insect pathogens	*B. larvae*	Central or terminal	—
	B. popilliae	Central	41
II. Spores spherical, obligate aerobes, casein and starch not hydrolyzed			
Sporangia swollen	*B. sphaericus*	Terminal	37
Sporangia not swollen	*B. pasteurii*	Terminal	38

tion called the *parasporal body*, which is deposited within the sporangium but outside the spore proper (Figure 13.58). In the case of *B. thuringiensis*, the crystal (parasporal body) protein is a protoxin that is converted to a toxin by proteolytic cleavage in the larval gut. The toxin binds to intestinal epithelial cells and induces pore formation that causes leakage of the host cells followed by lysis.

Genes encoding crystal proteins from several *B. thuringiensis* strains have been isolated. The genes for the *B. thuringiensis* crystal protein (known commercially as "Bt-toxin") have been introduced into plants to render the plants "naturally" resistant to insects. This strategy has been shown to be effective in controlled situations, and a variety of genetically altered Bt-toxins are being developed by genetic engineering in attempts to increase toxicity and reduce resistance (∞ Section 10.14).

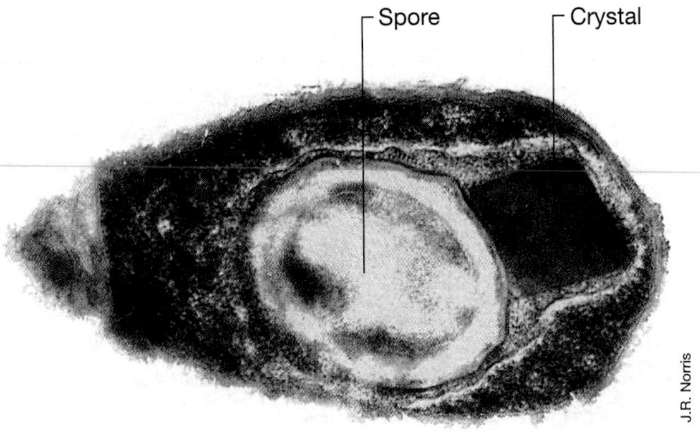

Clostridium

The clostridia lack a cytochrome system and a mechanism for electron transport phosphorylation; hence, unlike *Bacillus* species, they obtain ATP *only* by substrate-level phosphorylation. A wide variety of anaerobic energy-yielding mechanisms are known in the clostridia (fermentative diversity will be discussed in Section 15.22); indeed, the separation of the genus into subgroups is based primarily on these properties and on the nature of the fermentable substrate used (Table 13.28).

A number of clostridia ferment sugars, producing as a major end product *butyric acid*. Some of these also produce *acetone* and *butanol*, and at one time acetone-butanol fermentation by clostridia was of great industrial importance as it was the main commercial source of these products. Some clostridia of the acetone-butanol type fix N_2; the most vigorous N_2 fixer is *Clostridium pasteurianum*, which probably is responsible for most anaerobic nitrogen fixation in the soil. One group of clostridia ferments cellulose with the formation of acids and alcohols, and these are likely the major organisms decomposing cellulose anaerobically in soil.

FIGURE 13.58 Formation of the toxic parasporal crystal in the insect pathogen *Bacillus thuringiensis*. Electron micrograph of a thin section.

TABLE 13.28	Characteristics of some groups of the genus *Clostridium*		
Key characteristics	**Other characteristics**	**Species**	**DNA (mol % GC)**
I. Ferment carbohydrates			
Ferment cellulose	Fermentation products: acetate, lactate, succinate, ethanol, CO_2, H_2	*C. cellobioparum* *C. thermocellum*	28 38–39
Ferment sugars, starch, and pectin	Fermentation products: acetone, butanol, ethanol, isopropanol, butyrate, acetate, propionate, succinate, CO_2, H_2; some fix N_2	*C. butyricum* *C. acetobutylicum* *C. pasteurianum* *C. perfringens* *C. thermosulfurogenes*	27–28 28–29 26–28 24–27 33
Ferment sugars primarily to acetic acid	Total synthesis of acetate from CO_2; cytochromes present in some species	*C. aceticum* *C. thermoaceticum* *C. formicoaceticum*	33 54 34
Ferments only pentoses or methylpentoses	Ring-shaped cells form left-handed, helical chains; fermentation products: acetate, propionate, *n*-propanol, CO_2, H_2	*C. methylpentosum*	46
II. Ferment amino acids	Fermentation products: acetate, other fatty acids, NH_3, CO_2, sometimes H_2; some also ferment sugars to butyrate and acetate; may produce exotoxins	*C. sporogenes* *C. tetani* *C. botulinum* *C. tetanomorphum*	26 25–26 26–28 25–28
	Ferments three-carbon amino acids (for example, alanine) to propionate, acetate, and CO_2	*C. propionicum*	35
III. Ferments carbohydrates or amino acids	Fermentation products from glucose: acetate, formate, small amounts of isobutyrate and isovalerate	*C. bifermentans*	27
IV. Purine fermenters	Ferments uric acid and other purines, forming acetate, CO_2, NH_3	*C. acidurici*	27–30
V. Ethanol fermentation to fatty acids	Produces butyrate, caproate, and H_2; requires acetate as electron acceptor; does not use sugars, amino acids, or purines	*C. kluyveri*	30

The biochemical steps in the formation of butyric acid and butanol from sugars are well understood (Figure 13.59). Glucose is converted to pyruvate via the Embden–Meyerhof pathway, and pyruvate is split to acetyl-CoA, CO_2, and hydrogen (reduced ferredoxin) by the phosphoroclastic reaction (⚬⚬ Section 15.21 and Figure 15.53). Acetyl-CoA is then reduced to fermentation products using the NADH derived from glycolytic reactions. The proportions of the various products are influenced by the duration and the conditions of the fermentation. During the early stages, butyric and acetic acids are the predominant products, but as the pH of the medium drops, synthesis of acids ceases and the neutral products acetone and butanol begin to accumulate. However, if the pH of the medium is kept neutral with $CaCO_3$, very little of the neutral products are formed and the fermentation products consist of about three parts butyric and one part acetic acid. This makes good physiological sense because, unlike neutral product formation, acid production allows for additional ATP synthesis (Figure 13.59).

Another group of clostridia obtain their energy by fermenting *amino acids*. Some species ferment individual amino acids while others ferment only amino acid *pairs*. In this situation one functions as the electron *donor*

and is *oxidized*, whereas the other acts as the electron *acceptor* and is *reduced*. The type of coupled decomposition is known as the **Stickland reaction.** For instance, *Clostridium sporogenes* catabolizes a mixture of glycine and alanine, as outlined in Figure 13.60. Various amino acids that can function as either electron donors or acceptors in Stickland reactions are also listed in Figure 13.60. The products of Stickland oxidation are always NH_3, CO_2, and a carboxylic acid with one *less* carbon atom than the amino acid that is oxidized (Figure 13.60).

The amino acids that can be fermented singly are alanine, cysteine, glutamate, glycine, histidine, serine, or threonine. The products are generally acetate, butyrate, CO_2, and H_2. Many of the products of amino acid fermentation by clostridia are foul-smelling substances, and the odor that results from putrefaction is a result mainly of clostridial action. In addition to butyric acid, other odoriferous compounds produced are isobutyric acid, isovaleric acid, caproic acid, hydrogen sulfide, methylmercaptan (from sulfur amino acids), cadaverine (from lysine), putrescine (from ornithine), and ammonia.

The main habitat of clostridia is the soil, where they live primarily in anoxic "pockets," made anoxic by facultative organisms metabolizing various organic com-

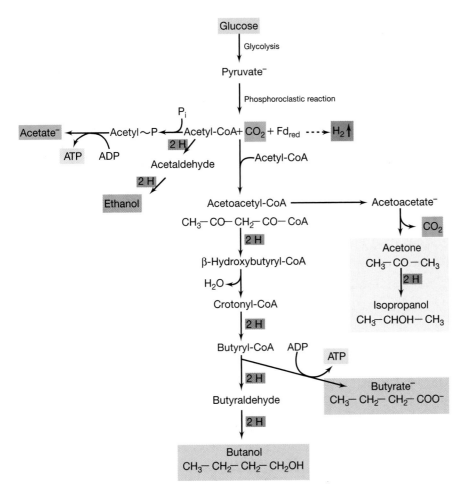

FIGURE 13.59 Pathway of formation of fermentation products from the butyric acid group of clostridia. The designation "2H" represents two electrons from one molecule of NADH.

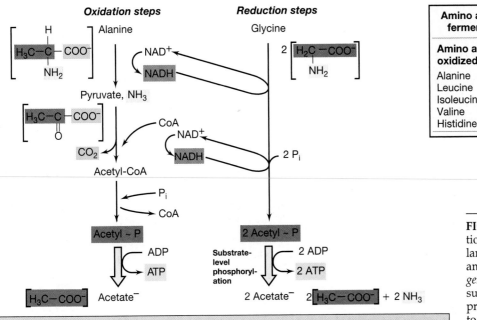

FIGURE 13.60 Coupled oxidation–reduction reaction (Stickland reaction) between alanine and glycine in *Clostridium sporogenes*. The structures of the key substrates, intermediates, and products are shown (in brackets) to allow the chemistry of the reaction to be followed.

pounds present. In addition, a number of clostridia have adapted to the anoxic environment of the mammalian intestinal tract. Also, as is discussed in Section 19.8, several clostridia are capable of causing disease in humans under specialized conditions. Botulism is caused by *Clostridium botulinum,* tetanus by *C. tetani,* and gas gangrene by *C. perfringens* and a number of other clostridia, both sugar and amino acid fermenters. These pathogenic clostridia seem in no way unusual metabolically but are distinct in that they produce specific toxins or, in the case of those causing gas gangrene, a group of toxins (∞ Section 19.8 and Table 19.4). *C. perfringens* and related species can also cause gastroenteritis in humans and domestic animals (∞ Section 24.11), and botulism occurs in sheep and ducks and a variety of other animals. An unsolved ecological problem is what role these toxins play in the natural habitat of the organism.

Sporosarcina

The genus *Sporosarcina* is unique among endospore formers because cells are *cocci* instead of rods. *Sporosarcina* consists of strictly aerobic spherical to oval cells that divide in two or three perpendicular planes to form tetrads or packets of eight or more cells (Figure 13.61). Two species of *Sporosarcina* are known, *S. ureae* and *S. halophila. Sporosarcina ureae* (Figure 13.61) can easily be enriched from soil by plating dilutions of a pasteurized soil sample on nutrient agar supplemented with 8% urea and incubating in air. Most soil bacteria are strongly inhibited by as little as 2% urea. However, *S. ureae* actively decomposes urea to CO_2 and NH_3 and in so doing can dramatically raise the pH of unbuffered media (*S. ureae* is remarkably alkaline-tolerant and grows in media up to pH 10–11). *Sporosarcina ureae* is common in soils, and studies on its distribution suggest that numbers of *S. ureae* are greatest in soils that receive inputs of urine (a source of urea), such as soils in which animals periodically urinate. Since many soil organisms are quite urea-sensitive, these results suggest that *S. ureae* is ecologically important as a major urea degrader in nature.

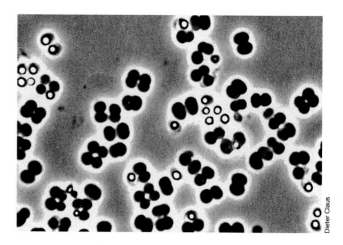

FIGURE 13.61 Phase contrast photomicrograph of cells of *Sporosarcina ureae.* A single cell is about 2 μm wide. Note bright refractile endospores. Most cell packets contain 8 cells.

Heliobacteria

Heliobacteria are *phototrophic,* low GC, gram-positive bacteria. They are anoxygenic phototrophs and produce a unique structural form of bacteriochlorophyll (∞ Section 15.3). The group contains three genera: *Heliobacterium, Heliophilum,* and *Heliobacillus.* All known heliobacteria are rod-shaped, either short or long rods, frequently with pointed ends (Figure 13.62). *Heliophilum* is especially interesting because its rod-shaped cells form into bundles (Figure 13.62b) that are motile as a unit. Heliobacteria are strict anaerobes, but in addition to anaerobic phototrophic growth, can grow chemotrophically in darkness by pyruvate fermentation (as can many clostridia, close relatives of the heliobacteria). Like the endospores of *Bacillus* or *Clostridium* species, the endospores of heliobacteria (Figure 13.62c) contain elevated Ca^{2+} levels and the signature molecule of the endospore, *dipicolinic acid* (∞ Section 3.15). Heliobacteria reside in soil, especially tropical paddy field soils, where their strong N_2-fixation activities may benefit rice productivity.

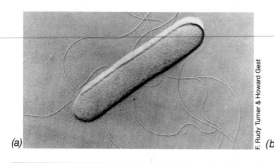 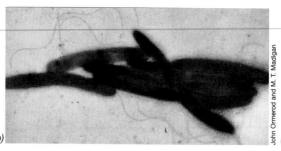 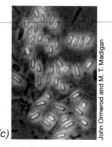

FIGURE 13.62 Cells and endospores of heliobacteria. (a) Electron micrograph of *Heliobacillus mobilis,* a peritrichously flagellated species. (b) *Heliophilum fasciatum,* bundles as observed by electron microscopy. (c) Phase micrograph of spores from *Heliobacterium gestii.*

The "Low GC," gram-positive bacteria are a large phylogenetic group that contains rods and cocci, sporulating and nonsporulating species. The production of endospores are a hallmark of the key genera *Bacillus* and *Clostridium*. Gram-positive bacteria are major agents for the degradation of organic matter in soil and a few species are pathogenic.

✓ What are the major features that differentiate *Staphylococcus* from *Bacillus*?

✓ How could you distinguish between a *heterofermentative* and a *homofermentative* lactic acid bacterium?

✓ What is the major physiological distinction between *Bacillus* and *Clostridium* species?

✓ Among endospore-producing genera, what is unique about the heliobacteria?

13.20

Cell Wall–Less, Low GC, Gram-Positive Bacteria: The Mycoplasmas

Key Genera

Mycoplasma
Spiroplasma

The mycoplasmas are organisms without cell walls that do not revert to walled organisms. They are probably the smallest organisms capable of autonomous growth and are of special evolutionary interest because of their extremely simple cell structure and small genomes. And, although not staining gram-positively since they lack cell walls, the mycoplasmas are clearly phylogenetically related to low GC, gram-positive bacteria.

Properties of Mycoplasmas

The absence of cell walls in the mycoplasmas observed by electron microscopy has been confirmed by chemical analysis, the latter showing that the key components of peptidoglycan, muramic acid and diaminopimelic acid, are missing. In Chapter 3 we discussed protoplasts and showed how these structures can be formed when cell wall-digesting enzymes act on cells that are in an osmotically protected medium, and that when the osmotic stabilizer is removed, protoplasts take up water, swell, and burst (⟳ Figure 3.34). Mycoplasmas resemble protoplasts in their lack of a cell wall, but they are more resistant to osmotic lysis and are able to survive conditions under which protoplasts lyse. This ability to resist osmotic lysis is at least partially determined by the fact that sterols are present in the mycoplasma cytoplasmic membrane, making it more stable than that of other prokaryotes. Indeed, some mycoplasmas require sterols in their growth media and this sterol requirement is a basis for separating the mycoplasmas into two groups (Table 13.29).

TABLE 13.29	Major characteristics of mycoplasmas				
Genus	Number of recognized species	Properties	DNA (mol % GC)	Genome size (kilobase pairs)	Presence of lipoglycans
Require sterols					
Mycoplasma	110	Many pathogenic; require sterols; facultative aerobes (see Figure 13.63)	23–41	600–1350	+
Anaeroplasma	4	May or may not require sterols; obligate anaerobes; degrade starch, producing acetic, lactic, and formic acids plus ethanol and CO_2; inhibited by thallium acetate; found in the bovine and ovine rumen	29–33	1500–1600	+
Spiroplasma	33	Spiral to corkscrew-shaped cells; associated with various phytopathogenic (plant disease) conditions (see Figure 13.65)	25–30	940–2200	–
Ureaplasma	6	Coccoid cells; occasional clusters and short chains; growth optimal at pH 6; strong urease reaction; associated with certain urinary tract infections in humans; inhibited by thallium acetate	27–30	750	–
Entomoplasma	5	Facultative aerobe; associated with insects and plants	27–29	790–1140	?
Do not require sterols					
Acholeplasma	16	Facultative aerobes	27–36	1500	+
Asteroleplasma	1	Obligate anaerobe; isolated from the bovine or ovine rumen	40	1500	+
Mesoplasma	12	Phylogenetically and ecologically related to *Entomoplasma*	27–30	870–1100	?

In addition to sterols, certain mycoplasmas contain compounds called **lipoglycans** (see Table 13.29). Lipoglycans are long-chain heteropolysaccharides covalently linked to membrane lipids and embedded in the cytoplasmic membrane of many mycoplasmas. Lipoglycans resemble the lipopolysaccharides (LPSs) of gram-negative Bacteria (Section 3.8) except that they lack the lipid A backbone and the phosphate typical of bacterial LPSs. Lipoglycans also function to help stabilize the membrane and have also been identified as facilitating attachment of mycoplasmas to cell surface receptors of animal cells. Like LPSs, lipoglycans stimulate antibody production when injected into experimental animals.

Growth of Mycoplasmas

Mycoplasma cells are small and they are highly pleomorphic, a consequence of their lack of a cell wall and hence rigidity. A single culture may exhibit small coccoid elements, larger, swollen forms, and filamentous forms of variable lengths, often highly branched (Figure 13.63).

The small coccoid elements (0.2–0.3 μm in size) are the smallest mycoplasma units capable of independent growth. Cellular elements of diameters close to 0.1 μm are occasionally seen in mycoplasma cultures, but these are not capable of growth. Even so, the minimum reproductive unit of 0.2–0.3 μm probably represents the smallest *free-living* cell. Additionally, the genome size of mycoplasmas is also smaller than that of most prokaryotes, between 500 and 1100 kilobase pairs of DNA, which is comparable to that of the obligately parasitic chlamydia and rickettsia (see Sections 13.26 and 13.12) and about one-fifth to one-fourth that of *Escherichia coli*. The genome of at least one *Mycoplasma* species, *M. genitalium* contains 580 kilobase pairs and has been completely sequenced (Section 9.12). This very small genome is thought to be near to the lower limit for the amount of DNA any cell must have to carry out life processes.

The mode of growth of mycoplasmas differs in liquid and agar cultures. On agar, there is a tendency for the organisms to grow so that they become embedded in the medium, and colonies of mycoplasmas on agar thus show a characteristic "fried-egg" appearance from a dense central core, which penetrates downward into the agar, surrounded by a circular spreading area that is lighter in color (Figure 13.64). As would be expected, growth of mycoplasmas is not inhibited by penicillin, cycloserine, or other antibiotics that inhibit cell wall synthesis, but the organisms are as sensitive as other Bacteria to antibiotics that act on targets other than the cell wall.

Culture media for the growth of mycoplasmas have usually been quite complex. For many species growth is poor or absent even in complex yeast extract–peptone–beef heart infusion media unless fresh serum or ascitic fluid is added; the latter provide unsaturated fatty acids and sterols. Some mycoplasmas can be cultivated on relatively simple media, however, and defined media have been developed for some strains. Most mycoplasmas use carbohydrates as energy sources and require a range of vitamins, amino acids, purines, and pyrimidines as growth factors. The energy metabolism of mycoplasmas is variable but not unique, as some species are strictly respiratory while others are facultative or even obligate anaerobes (Table 13.29).

Spiroplasma

The genus *Spiroplasma* consists of helical or spiral-shaped cells (Figure 13.65). Although they lack a cell wall and flagella, they are motile by means of a rotary (screw) motion or a slow undulation. Intracellular fibrils that are thought to play a role in motility have been demonstrated. The organism has been isolated from ticks, the hemolymph (Figure 13.65) and gut of insects, vascular plant fluids and insects that feed on fluids, and the surfaces of flowers and other plant parts. *Spiroplasma citri* has been isolated from the leaves of citrus plants, where it causes a disease called

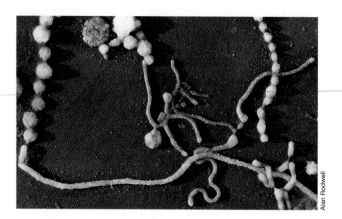

FIGURE 13.63 Electron micrograph of a metal-shadowed preparation of *Mycoplasma mycoides*. Note the coccoid and hyphalike elements. The average diameter of cells in chains is about 0.5 μm.

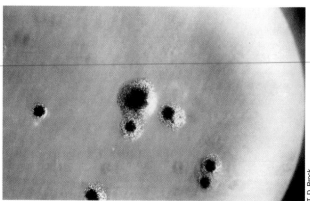

FIGURE 13.64 Typical "fried egg" appearance of mycoplasma colonies on agar. The colonies are about 0.5 mm in diameter.

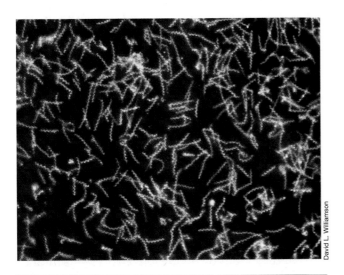

FIGURE 13.65 Dark-field micrograph of the "sex ratio" spiroplasma removed from the hemolymph of the fly *Drosophila pseudoobscura*. Female flies infected with the sex ratio spiroplasma bear only female progeny. Individual spiroplasma cells are about 0.15 μm in diameter.

citrus stubborn disease and from corn plants suffering from *corn stunt disease*. A number of other mycoplasma-like bodies have been detected in diseased plants by electron microscopy, which indicates that a large group of plant-associated mycoplasmas may exist. Four species of *Spiroplasma* are recognized that cause a variety of animal diseases such as *honeybee spiroplasmosis, suckling mouse cataract disease,* and *lethargy disease* of the beetle *Melolontha*.

✓ **13.20 Concept Check**

The mycoplasma group are organisms that lack cell walls and contain a very small genome. Many species require sterols to strengthen their membranes, and several are pathogenic for humans and other animals.

✓ Why do mycoplasmas need to have stronger cell membranes than other prokaryotes?
✓ Where do the mycoplasmas group phylogenetically?
✓ Compare and contrast the genus *Mycoplasma* with the genus *Acholeplasma* in terms of growth requirements, genome size, and metabolism.

13.21

High GC, Gram-Positive Bacteria: Coryneform and Propionic Acid Bacteria

Key Genera

Corynebacterium
Arthrobacter
Propionibacterium

High GC, gram-positive Bacteria are typically rod-shaped to filamentous primarily aerobic prokaryotes that are

common inhabitants of the soil and of plant materials. For the most part they are harmless commensals, species of *Mycobacterium* (for example *Mycobacterium tuberculosis*) being notable exceptions, and some are of great economic value in either the production of antibiotics or certain fermented dairy products. We begin with the rod-shaped representatives.

Corynebacteria

The coryneform bacteria are gram-positive, aerobic, nonmotile, rod-shaped organisms with the characteristic of forming irregular-shaped, club-shaped, or V-shaped cell arrangements during normal growth. V-shaped cell groups arise as a result of a snapping movement that occurs just after cell division (called postfission snapping movement or, simply, *snapping division*) (Figure 13.66). **Snapping division** has been shown to occur in one species because the cell wall consists of two layers; only the inner layer participates in cross-wall formation, and so after the cross-wall is formed, the two daughter cells remain attached by the outer layer of the cell wall. Localized rupture of this outer layer on one side results in a bending of the two cells away from the ruptured side (Figure 13.67) and thus development of V-shaped forms.

The main genera of coryneform bacteria are *Corynebacterium* and *Arthrobacter*. The genus *Corynebacterium* consists of an extremely diverse group of bacteria, including animal and plant pathogens as well as saprophytes. Some species, such as *C. diphtheriae*, are pathogenic (diphtheria, ⟳ Section 23.2). The genus *Arthrobacter,* consisting primarily of soil organisms, is distinguished from *Corynebacterium* on the basis of a developmental cycle in *Arthrobacter* involving conversion from rod to sphere and back to rod again (Figure 13.68). However, some corynebacteria are pleomorphic and form coccoid elements during growth, and so the distinction between the two genera on the basis of life cycle is not absolute. The *Corynebacterium* cell frequently has a swollen end, so it has a club-shaped appearance (hence the name

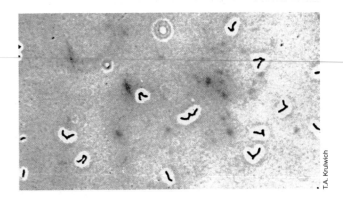

FIGURE 13.66 Photomicrograph of characteristic V-shaped cell groups in *Arthrobacter crystallopoietes*, resulting from snapping division. Cells are about 0.9 μm in diameter.

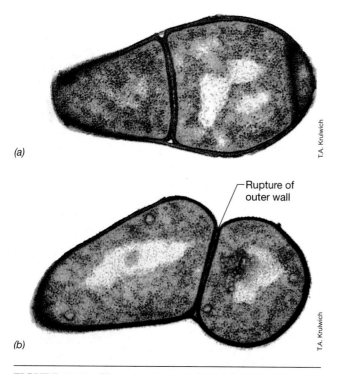

(a)

Rupture of
outer wall

(b)

T.A. Krulwich

FIGURE 13.67 Electron micrograph of cell division in *Arthrobacter crystallopoietes,* illustrating how snapping division and V-shaped cell groups arise. (a) Before rupture of the outer cell wall layer. (b) After rupture of the outer layer on one side. Cells are 0.9–1 μm in diameter.

of the genus: *koryne* is the Greek word for "club"), whereas *Arthrobacter* is less commonly club-shaped.

Organisms of the genus *Arthrobacter* are among the most common of all soil bacteria. They are remarkably resistant to desiccation and starvation, despite the fact that they do not form spores or other resting cells. Arthrobacters are a heterogeneous group that have considerable nutritional versatility, and strains have been isolated that decompose herbicides, caffeine, nicotine, phenols, and other unusual organic compounds.

Propionic Acid Bacteria

The propionic acid bacteria (genus *Propionibacterium*) were first discovered as inhabitants of Swiss (Emmentaler) cheese, where their fermentative production of CO_2 produces the characteristic holes; the presence of propionic acid is at least partly responsible for the unique flavor of the cheese. Although this acid is produced by some other bacteria, its production in large amounts by the propionic acid bacteria is a distinguishing characteristic of the genus. The bacteria in this group are gram-positive anaerobes that ferment lactic acid, carbohydrates, and polyhydroxy alcohols, producing primarily propionic acid, acetic acid, and CO_2. Their nutritional requirements are complex, and they usually grow rather slowly.

The enzymatic reactions leading from glucose to propionic acid are shown in Figure 13.69. The initial catabolism of glucose to pyruvate follows the Embden–Meyerhof pathway as in the lactic acid bacteria, but the NADH formed is oxidized as one part of a cycle in which *propionic acid* is formed. Pyruvate accepts a carboxyl group from methylmalonyl-CoA by a transcarboxylase reaction, leading to the formation of oxalacetate and propionyl-CoA. The latter substance reacts with succinate in a step catalyzed by a CoA transferase, producing succinyl-CoA and propionate. The succinyl-CoA is then isomerized to methylmalonyl-CoA, and the cycle is complete (Figure 13.69). Oxidation of NADH occurs in the steps between oxalacetate and succinate, and the oxidation–reduction balance is restored.

Propionibacterium also ferments lactate with the production of propionate, acetate, and CO_2. The anaerobic fermentation of lactic acid to propionate is of interest because lactic acid itself is an end product of fermentation for many bacteria (see Section 13.18). The propionic acid bacteria are thus able to obtain energy anaerobically from a fermentation product that other bacteria have produced; this metabolic strategy has been called a *secondary fermentation.*

It is the fermentation of lactate to propionate that is important in Swiss cheese manufacture. The starter culture consists of a mixture of homofermentative streptococci and lactobacilli, plus propionic acid bacteria. The initial fermentation of lactose to lactic acid during formation of the curd is carried out by the homofermentative organisms. After the curd (protein and fat) has been drained, the propionic acid bacteria develop rapidly and the eyes (holes) characteristic of Swiss cheese are formed by the accumulation of CO_2, the gas diffusing through the curd and gathering at weak points. In the fermentation, lactate is oxidized to pyruvate from which it is converted to propionate as shown in Figure 13.69.

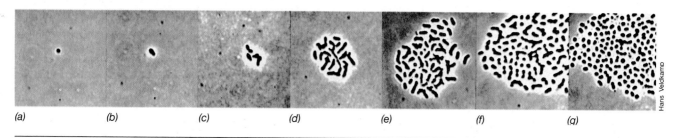

(a)   *(b)*   *(c)*   *(d)*   *(e)*   *(f)*   *(g)*

Hans Veldkamp

FIGURE 13.68 Stages in the life cycle of *Arthrobacter globiformis* as observed in slide culture: (a) Single coccoid element; (b–e) conversion to rod and growth of a microcolony consisting predominantly of rods; (f–g) conversion of rods to coccoid forms. Cells are about 0.9 μm in diameter.

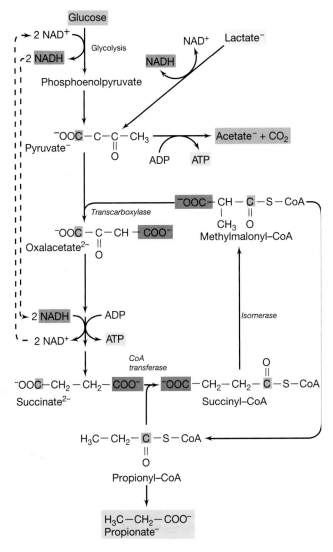

Stoichiometry from lactate:

3 Lactate⁻ ⟶ 2 Propionate⁻ + 1 Acetate⁻ + 1 CO_2 + 3–5 ATP

Propionate is also formed in the fermentation of succinate by the bacterium *Propionigenium*. This organism is phylogenetically and ecologically unrelated to *Propionibacterium* but energetic aspects of its fermentation are of considerable interest. We discuss the mechanism of the *Propionigenium* fermentation in Section 15.22.

13.22

High GC, Gram-Positive Bacteria: *Mycobacterium*

Key Genus

Mycobacterium

The genus *Mycobacterium* consists of rod-shaped organisms, which at some stage of their growth cycle possess the distinctive staining property called acid-

FIGURE 13.69　The formation of propionic acid by *Propionibacterium*. Either lactate, produced by the fermentative activities of other bacteria, or glucose can be fermented in the propionate fermentation. ATP synthesis is associated with electron transport reactions occurring during the formation of succinate and by substrate level phosphorylation in the production of acetate.

fastness. This property is due to the presence on the surface of the mycobacterial cell of unique lipid components called **mycolic acids** and is found only in the genus *Mycobacterium*. First discovered by Robert Koch during his pioneering investigations on tuberculosis (∞ Section 1.8), this unique staining property permitted the identification of the organism in tuberculous lesions; it has subsequently proved to be of great taxonomic use in defining the genus *Mycobacterium*.

Acid-Fastness (Ziehl–Neelsen Stain)

A mixture of the dye basic fuchsin and phenol is used in this staining procedure, the stain being driven into the cells by slow heating of the smear on the microscope slide to the steaming point for 2–3 min. The role of the phenol is to enhance penetration of the fuchsin into the lipids. After washing in distilled water, the preparation is decolorized with acid–alcohol; after another wash, a final counterstain of methylene blue is used. Acid-fast organisms in the final preparation appear *red*, whereas the background and non-acid-fast organisms appear *blue*.

As noted, the key component necessary for acid-fastness is a unique lipid fraction of mycobacterial cells called *mycolic acid*. Mycolic acid is actually a group of complex branched-chain hydroxy lipids with the general structure shown in Figure 13.70a; in the acid-fast stain the

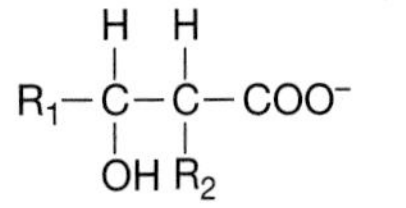

(a) Mycolic acid; R_1 and R_2 are long-chain aliphatic hydrocarbons

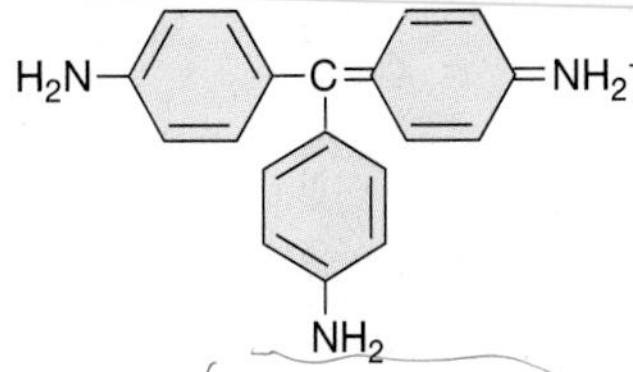

(b) Basic fuchsin

FIGURE 13.70　Structure of (a) mycolic acid and (b) basic fuchsin, the dye used in the acid-fast stain. The fuchsin dye probably combines with the mycolic acid via ionic bonds between COO⁻ and NH_2^+.

TABLE 13.30   Some characteristics of representative mycobacteria

Species	Growth in 5% NaCl	Nitrate reduction	Growth at 45°C	Human pathogen	Pigmentation
Slow-growing species					
Mycobacterium tuberculosis	−	+	−	+	None
Mycobacterium avium	−	−	−	+	Old colonies pigmented (see Figure 13.71c)
Mycobacterium bovis	−	−	+	+	None
Mycobacterium kansasii	−	+	−	+	Photochromogenic
Fast-growing species					
Mycobacterium smegmatis	+	+	+	−	None
Mycobacterium phlei	+	+	+	−	Pigmented
Mycobacterium chelonae	+	−	−	+	None
Mycobacterium parafortuitum	+	+	−	−	Photochromogenic

carboxylic acid group of the mycolic acid reacts with the fuchsin dye (Figure 13.70b). The mycolic acid is covalently bound to the peptidoglycan of the mycobacterial wall, and this complex leaves the cell surface with a waxy, hydrophobic consistency. Mycobacteria are not readily stained by the Gram method because of the high surface lipid content, but if the lipoidal portion of the cell is removed with alkaline ethanol, the intact cell remaining is non-acid-fast but instead is gram-positive.

Characteristics of Mycobacteria

Mycobacteria are rather pleomorphic and may undergo branching or filamentous growth. However, in contrast to those of the actinomycetes, filaments of the mycobacteria become fragmented into rods or coccoid elements on slight disturbance; a true mycelium is not formed. In general, mycobacteria can be separated into two major groups, *slow growers* and *fast growers* (Table 13.30). *Mycobacterium tuberculosis* is a typical slow grower, and visible colonies are produced from dilute inoculum only

after days to weeks of incubation. (The reason Koch was successful in first isolating *M. tuberculosis* was that he waited long enough after inoculating media; ∞ Section 1.8.) When growing on solid media, mycobacteria generally form tight, compact, often wrinkled colonies, the organisms piling up in a mass rather than spreading out over the surface of the agar (Figure 13.71a). This formation is probably due to the high lipid content and hydrophobic nature of the cell surface.

For the most part, mycobacteria have relatively simple nutritional requirements. Growth often occurs in simple mineral salts medium with ammonium as nitrogen source and glycerol or acetate as sole carbon source and electron donor incubated in air. Growth of *Mycobacterium tuberculosis* is stimulated by lipids and fatty acids, and egg yolk (a good source of lipids) is often added to culture media to achieve more luxuriant growth. A glycerol–whole egg medium (Lowenstein–Jensen medium) is often used in primary isolation of *M. tuberculosis* from pathological materials. Perhaps because of the high lipid content of its cell walls, *M.*

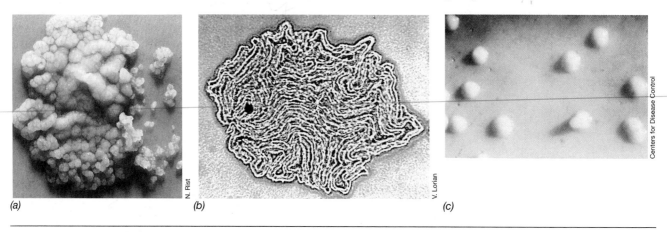

(a)   (b)   (c)

FIGURE 13.71   Characteristic colony morphology of mycobacteria. (a) *Mycobacterium tuberculosis,* showing the compact, wrinkled appearance of the colony. The colony is about 7 mm in diameter. (b) A colony of virulent *M. tuberculosis* at an early stage, showing the characteristic cordlike growth. Individual cells are about 0.5 μm in diameter. (See also the historic drawings of *M. tuberculosis* cells made by Robert Koch, in Figure 1.20). (c) Colonies of *Mycobacterium avium* from a strain of this organism isolated as an opportunistic pathogen from an AIDS patient.

tuberculosis is able to resist such chemical agents as alkali and phenol for considerable periods of time, and this property is used in the selective isolation of the organism from patient sputum and other materials that are grossly contaminated. The sputum is first treated with 1 N NaOH for 30 min and then neutralized and streaked onto an isolation medium.

A characteristic of many mycobacteria is their ability to form yellow carotenoid pigments (Figure 13.71c). Based on pigmentation, the mycobacteria can be classified into three groups: nonpigmented (including *Mycobacterium tuberculosis, M. bovis*); forming pigment only when cultured in the light, a property called *photochromogenesis* (including *M. kansasii, M. marinum*); and forming pigment even when cultured in the dark, a property called *scotochromogenesis* (including *M. gordonae, M. paraffinicum*). Photoinduction of carotenoid formation involves short-wavelength (blue) light and occurs only in the presence of O_2. The evidence indicates that the critical event in photoinduction is a light-catalyzed oxidation event, and it appears that one of the early enzymes in carotenoid biosynthesis is photoinduced. As with other carotenoid-containing bacteria, it has been suggested that carotenoids protect mycobacteria against oxidative damage involving singlet oxygen (∞ Section 5.12).

The virulence of *Mycobacterium tuberculosis* cultures has been correlated with the formation of long, cord-like structures (Figure 13.71b) on agar or in liquid medium, due to side-to-side aggregation and intertwining of long chains of bacteria. Growth in cords reflects the presence on the cell surface of a characteristic lipid, the **cord factor**, which is a glycolipid (Figure 13.72). The pathogenesis of the disease tuberculosis is discussed in detail in Section 23.3.

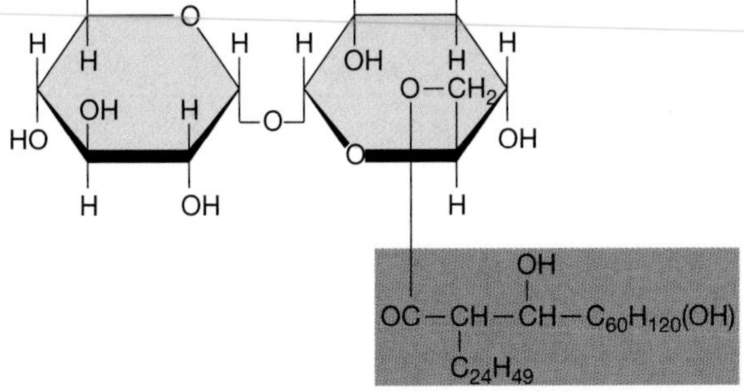

✓ **13.21–13.22 Concept Check**

"High GC," gram-positive bacteria include such organisms as *Corynebacterium, Arthrobacter, Propionibacterium* and *Mycobacterium.* They are mainly harmless soil saprophytes but *M. tuberculosis* is the causative agent of the disease tuberculosis. *M. tuberculosis* cells have a lipid-rich, waxy outer surface layer that requires special staining procedures (the acid-fast stain) in order to observe the cells microscopically.

✓ What is snapping division and what organism practices it?
✓ What organism is involved in the ripening of Swiss cheese and what chemical compound does it make that helps to flavor the cheese?
✓ What is mycolic acid, what organism produces it, and what properties does this substance confer on cells that make it?

13.23

Filamentous, High GC, Gram-Positive Bacteria: The Actinomycetes

Key Genera

Streptomyces
Actinomyces

The actinomycetes are a large group of filamentous, gram-positive Bacteria that form branching filaments. As a result of successful growth and branching, a ramifying network of filaments is formed, called a *mycelium* (Figure 13.73). Although it is of bacterial dimensions, the mycelium is in some ways analogous to the mycelium formed by the filamentous fungi. Most actinomycetes form spores; the manner of spore formation varies and is used in separating subgroups, as outlined in Table 13.31. The DNA base compositions of most members of the actinomycetes fall within the range of 63–78% GC, and organisms at the

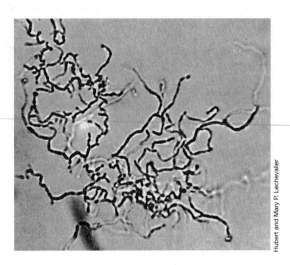

FIGURE 13.72 Structure of cord factor, a mycobacterial glycolipid: 6,6′-dimycolyltrehalose. The two identical longchain dialcohol groups are shown in purple.

FIGURE 13.73 A young colony of an actinomycete of the genus *Nocardia*, showing typical filamentous cellular structure (mycelium). Each filament is about 0.8–1 μm in diameter.

TABLE 13.31 Actinomycetes and related genera (all gram-positive)[a]	
Major groups	**DNA (mol % GC)**
Coryneform group of bacteria: rods, often club-shaped, morphologically variable; not acid-fast or filamentous; snapping cell division	
Corynebacterium: irregularly staining segments, sometimes granules; club-shaped swelling frequent; animal and plant pathogens, also soil saprophytes	51–65
Arthrobacter: coccus–rod morphogenesis; soil organisms	59–70
Cellulomonas: coryneform morphology; cellulose digested; facultative aerobe	71–73
Kurthia: rods with rounded ends occurring in chains; coccoid later	36–38
Brevibacterium: coccus–rod morphogenesis; cheese, skin	60–67
Propionic acid bacteria: anaerobic to aerotolerant; rods or filaments, branching	
Propionibacterium: nonmotile; anaerobic to aerotolerant; produce propionic acid and acetic acid; dairy products (Swiss cheese); skin, may be pathogenic	53–68
Eubacterium: obligate anaerobes; produce mixture of organic acids, including butyric, acetic, formic, and lactic; intestine, infections of soft tissue, soil; may be pathogenic; probably the predominant member of the intestinal flora	26–48
Obligate anaerobes	
Bifidobacterium: smooth microcolony, no filaments; coryneform cells common; found in intestinal tract of breast-fed infants	55–67
Acetobacterium: homoacetogen; sediments and sewage	39–43
Butyrivibrio: curved rods; rumen	36–42
Thermoanaerobacter: rods, thermophilic, found in hot springs	37–39
Actinomycetes: filamentous, often branching; highly diverse	
Group I. Actinomycetes: not acid–alcohol-fast; facultatively aerobic; mycelium not formed; branching filaments may be produced; rod, coccoid, or coryneform cells	
Actinomyces: anaerobic to facultatively aerobic; filamentous microcolony, but filaments transitory and fragment into coryneform cells; may be pathogenic for humans or animals; found in oral cavity	57–69
Other genera: *Arachnia, Bacterionema, Rothia, Agromyces*	
Group II. Mycobacteria: acid-fast, filaments transitory	
Mycobacterium: pathogens, saprophytes; obligate aerobes; lipid content of cells and cell walls high; waxes, mycolic acids; simple nutrition; growth slow; tuberculosis, leprosy, granulomas, avian tuberculosis; also soil organisms; hydrocarbon oxidizers	62–70
Group III. Nitrogen-fixing actinomycetes: nitrogen-fixing symbionts of plants; true mycelium produced	
Frankia: forms nodules of two types on various plant roots; probably microaerophilic; grows slowly; fixes N_2	67–72
Group IV. Actinoplanes: true mycelium produced; spores formed, borne inside sporangia	
Actinoplanes, Streptosporangium	69–71
Group V. Dermatophilus group: mycelial filaments divide transversely, and in at least two longitudinal planes,to form masses of motile, coccoid elements; aerial mycelium absent; occasionally responsible for epidermal infections	
Dermatophilus, Geodermatophilus	56–75
Group VI. Nocardias: mycelial filaments commonly fragment to form coccoid or elongate elements; aerial spores occasionally produced; sometimes acid-fast; lipid content of cells and cell wall very high	
Nocardia: common soil organisms; obligate aerobes; many hydrocarbon utilizers	61–72
Rhodococcus: soil saprophytes, also common in gut of various insects; utilize hydrocarbons	59–69
Group VII. Streptomycetes: mycelium remains intact, abundant aerial mycelium and long spore chains	
Streptomyces: Nearly 500 recognized species, many produce antibiotics	69–75
Other genera (differentiated morphologically): *Streptoverticillium, Sporichthya, Microcellobosporia, Kitasatoa, Chainia*	67–73
Group VIII. Micromonosporas group: mycelium remains intact; spores formed singly, in pairs, or short chains; several thermophilic; saprophytes found in soil, rotting plant debris; one species produces endospores	
Micromonospora, Microbispora, Themobispora, Thermoactinomyces, Thermomonospora	54–79

a Phylogenetically, all species (except for *Acetobacterium, Butyrivibrio,* and *Thermoanaerobacter*) fall into the high GC subdivision of the gram-positive Bacteria.

upper end of this range have the highest GC percentage of any bacteria known. Phylogenetically, the filamentous actinomycetes form a coherent group; thus, the mycelial spore-forming habit is of both phylogenetic as well as taxonomic importance. In the present discussion we concentrate on the genus *Streptomyces.*

Streptomyces

Streptomyces is a genus represented by a large number of species and varieties. Over 500 species of *Streptomyces* are recognized by *Bergey's Manual,* although GC base ratios cluster tightly between 69 and 73 mol %. *Streptomyces* filaments are usually 0.5–1.0 μm in diameter, are of indefinite length, and often lack cross-walls in the vegetative phase. Growth occurs at the tips of the filaments and is often accompanied by branching so that the vegetative phase consists of a complex, tightly woven matrix, resulting in a compact, convoluted mycelium and subsequent colony. As the colony ages, characteristic aerial filaments called *sporophores* are formed, which project above the surface of the colony and give rise to spores (Figure 13.74). *Streptomyces* spores, called **conidia,** are not related in any way to the endospores of *Bacillus* and *Clostridium* because the streptomycete spores are produced simply by the formation of cross-walls in the multinucleate sporophores followed by separation of the individual cells directly into spores (Figure 13.75). Differences in the shape and arrangement of aerial filaments and spore-bearing structures of various species are among the fundamental features used in classifying the *Streptomyces* groups (Figure 13.76). The conidia and sporophores are often pigmented and contribute a characteristic color to the mature colony (Figure 13.77*a*). The dusty appearance of the mature colony, its compact nature, and its color make detection of *Streptomyces* colonies on agar plates relatively easy (Figure 13.77*b*).

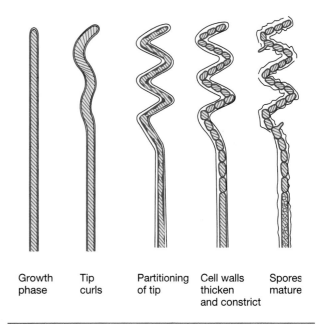

Growth phase	Tip curls	Partitioning of tip	Cell walls thicken and constrict	Spores mature

FIGURE 13.75   Diagram of stages in the conversion of a streptomycete's aerial hypha (sporophores) into spores (conidia).

Ecology and Isolation of *Streptomyces*

Although a few streptomycetes can be found in aquatic habitats, they are primarily *soil* organisms. In fact, the characteristic earthy odor of soil is caused by the production of a series of streptomycete metabolites called *geosmins*. These substances are sesquiterpenoid compounds, unsaturated ring compounds of carbon, oxygen, and hydrogen. A common geosmin is trans-1,10-di-methyl-trans-9-decalol. Geosmins are also produced by some cyanobacteria (see Section 13.24).

Alkaline and neutral soils are more favorable for the development of *Streptomyces* than are acid soils. Higher numbers of *Streptomyces* are usually found in well-drained soils (such as sandy loams, or soils covering limestone), and there is some evidence to suggest that *Streptomyces* require a lower water potential for growth than many other soil bacteria. Isolation of *Streptomyces* from soil is relatively easy: a suspension of soil in sterile water is diluted and spread on selective agar medium, and the plates are incubated at 25°C (∞ Figure 11.7*a*). Media often selective for *Streptomyces* contain the usual assortment of inorganic salts to which starch, asparagine, or calcium malate is added as a carbon source and undigested casein or potassium nitrate as a nitrogen source. After incubation for 5–7 days in air, the plates are examined for the presence of the characteristic *Streptomyces* colonies (Figure 13.77*a*), and spores of interesting colonies can be streaked to isolate pure cultures.

Nutritionally, the streptomycetes are quite versatile. Growth-factor requirements are rare, and a wide variety of carbon sources, such as sugars, alco-

(a)

Peter Hirsch

(b)

Hubert and Mary P. Lechevalier

FIGURE 13.74   Photomicrographs of several spore-bearing structures of actinomycetes. (a) *Streptomyces,* a monoverticillate type. (b) *Streptomyces,* a spiral type. Filaments are about 0.8 μm wide in both cases.

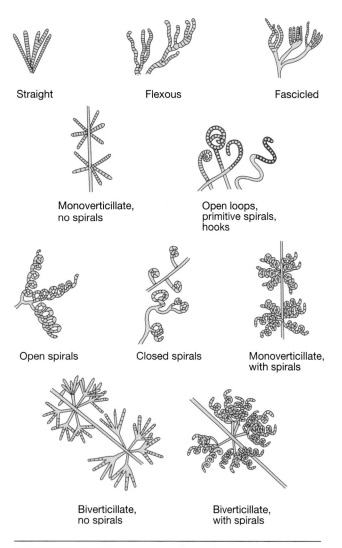

Straight

Flexous

Fascicled

Monoverticillate, no spirals

Open loops, primitive spirals, hooks

Open spirals

Closed spirals

Monoverticillate, with spirals

Biverticillate, no spirals

Biverticillate, with spirals

FIGURE 13.76 Various types of sporebearing structures in the streptomycetes.

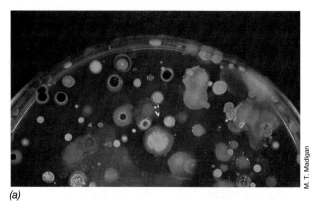

(a)

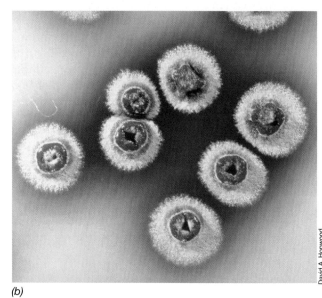

(b)

FIGURE 13.77 *Streptomyces.* (a) Colonies of *Streptomyces* and other soil bacteria derived from spreading a soil dilution on a casein–starch agar plate. The *Streptomyces* colonies are of various colors (several black *Streptomyces* colonies are in the foreground) but can easily be identified by their opaque, rough, nonspreading morphology. (b) Close up photo of colonies of *Streptomyces coelicolor.*

TABLE 13.32	Some common antibiotics synthesized by species of *Streptomyces*		
Chemical class	**Common name**	**Produced by**	**Active against**[a]
Aminoglycosides	Streptomycin	*S. griseus*	Most gram-negative Bacteria
	Spectinomycin	*Streptomyces* spp.	*M. tuberculosis*, penicillinase-producing *N. gonorrhoeae*
	Neomycin	*S. fradiae*	Broad spectrum, usually used in topical applications because of toxicity
Tetracyclines	Tetracycline	*S. aureofaciens*	Broad spectrum, gram-positive and gram-negative Bacteria, rickettsias and chlamydias, *Mycoplasma*
	Chlortetracycline	*S. aureofaciens*	As for tetracycline
Macrolides	Erythromycin	*S. erythreus*	Most gram-positive Bacteria, frequently used in place of penicillin, *Legionella*
	Clindamycin	*S. lincolnensis*	Effective against obligate anaerobes, especially *Bacteroides fragilis*
Polyenes	Nystatin	*S. noursei*	Fungi, especially *Candida* infections
	Amphocetin B	*S. nodosus*	Fungi
None	Chloramphenicol	*S. venezuelae*	Broad spectrum; drug of choice for typhoid fever

a Most antibiotics are effective against several different Bacteria. The entries in this column refer to the common clinical application of a given antibiotic. The structures and mode of action of many of these antibiotics are discussed in Sections 18.7–18.9.

hols, organic acids, amino acids, and some aromatic compounds, can be utilized. Most isolates produce extracellular hydrolytic enzymes that permit utilization of polysaccharides (starch, cellulose, hemicellulose), proteins, and fats, and some strains can use hydrocarbons, lignin, tannin, or even rubber. *Streptomyces* can often be obtained by spreading a soil dilution on an alkaline agar medium containing polymers such as casein and starch (Figure 13.77*a*). A single isolate may be able to break down over 50 distinct carbon sources. Streptomycetes are strict aerobes whose growth in liquid culture is usually markedly stimulated by forced aeration. Sporulation usually does not take place in liquid culture but only when the organism is growing on the surface of agar or another solid substrate; it can occur, however, when organisms form a pellicle on the surface of an unshaken liquid culture.

Antibiotics of *Streptomyces*

Perhaps the most striking property of the streptomycetes is the extent to which they produce **antibiotics** (Table 13.32). Evidence for antibiotic production is often seen on the agar plates used in the initial isolation of *Streptomyces:* adjacent colonies of other bacteria show zones of inhibition (Figure 13.78*a*; see also Figure 11.7*a*). In some studies, close to 50% of all *Streptomyces* isolated have proved to be antibiotic producers. Because of the great economic and medical importance of many streptomycete antibiotics, an enormous amount of work has been done on these producers. Over 500 distinct antibiotic substances have been shown to be produced by streptomycetes, and a large number of these have been identified chemically (Figure 13.78*b*). Some organisms produce more than one antibiotic, and often the several kinds produced by one organism are not even chemically related. The same antibiotic may be formed by different species found in widely scattered parts of the world. And, although an antibiotic-producing organism is resistant to its own antibiotics, it usually remains sensitive to antibiotics produced by other streptomycetes.

More than 50 streptomycete antibiotics have found practical application in human and veterinary medicine, agriculture, and industry. Some of the more common antibiotics of *Streptomyces* origin are listed in Table 13.32. They are grouped into classes based on the chemical structure of the parent molecule. The search for new streptomycete antibiotics continues because many infectious diseases are still not adequately controlled by existing antibiotics. Also, the development of antibiotic-resistant strains requires the continual discovery of new agents. Ironically, however, despite the extensive work on antibiotic-producing streptomycetes and the fact that the an-

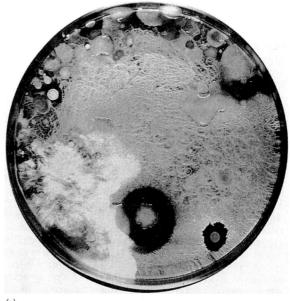

(a)

(b)

FIGURE 13.78   Antibiotics from *Streptomyces*. (a) Antibiotic action of soil microorganisms on a crowded plate. The smaller colonies surrounded by inhibition zones are streptomycetes; the larger, spreading colonies are *Bacillus* species. (b) The red-colored antibiotic undecylprodigiosin is being excreted by colonies of *Streptomyces coelicolor.*

tibiotic industry is a multibillion dollar enterprise, the ecology of *Streptomyces* remains poorly understood. The ecological rationale for why antibiotics are produced is not clear. However, one hypothesis for why *Streptomyces* species produce antibiotics is that antibiotic production, which is linked to sporulation (a process itself triggered by nutrient depletion), might be a mechanism to inhibit the growth of other organisms competing with differentiating *Streptomyces* cells for limiting nutrients. This would allow the *Streptomyces* to complete the sporulation process, thereby forming a dormant structure with increased chances of survival.

✓ 13.23 Concept Check

The Streptomycetes are a large group of filamentous, gram-positive bacteria that form spores at the end of aerial filaments. Many clinically useful antibiotics like tetracycline and neomycin have come from *Streptomyces* species.

✓ How do spores and the process of sporulation in a *Streptomyces* species differ from that in a *Bacillus* species?

✓ What energy class of organism is a *Streptomyces* and from what types of compounds do these organisms obtain their energy?

✓ Why might antibiotic production be of advantage to Streptomycetes?

Kingdom III: Cyanobacteria, Prochlorophytes, and Chloroplasts

13.24

Cyanobacteria

Key Genera

Synechococcus
Oscillatoria
Nostoc

Cyanobacteria comprise a large and morphologically heterogeneous group of phototrophic Bacteria. Cyanobacteria differ in fundamental ways from purple and green anoxyphototrophs, most notably in the fact that they are *oxygenic* phototrophs. Cyanobacteria represent one of the major kingdoms of Bacteria and show a distant relationship to gram-positive Bacteria (see Figure 13.1). The evolutionary significance of cyanobacteria was discussed in Section 12.2, and it is likely that these organisms were the first oxygen-evolving phototrophic organisms on Earth and were responsible for the conversion of the atmosphere of the Earth from anoxic to oxic.

Structure and Classification of Cyanobacteria

The morphological diversity of the cyanobacteria is considerable (Figure 13.79). Both unicellular and filamentous forms are known, and considerable variation within these morphological types occurs. Cyanobacteria can be divided into five morphological groups: unicellular dividing by binary fission (see Figure 13.79*a*); unicellular dividing by multiple fission (colonial) (see Figure 13.79*b*); filamentous containing differentiated cells called heterocysts that function in nitrogen fixation (see Figures 13.79*d* and 13.81); filamentous non-heterocystous forms (see Figure 13.79*c*); and branching filamentous types (see Figure 13.79*e*). Table 13.33 lists the major genera currently recognized in each group. Cyanobacterial cells range in size from those of typical bacteria (0.5–1 μm in diameter) to cells as large as 60 μm in diameter (in the species *Oscillatoria princeps*).

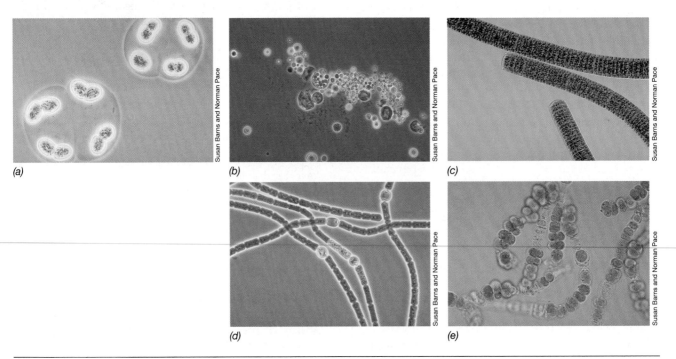

(a)   (b)   (c)

(d)   (e)

FIGURE 13.79   Morphological diversity among the cyanobacteria: the five major morphological types of cyanobacteria. (a) Unicellular, *Gloeothece*, phase contrast; a single cell measures 5–6 μm in diameter; (b) colonial, *Dermocarpa*, phase contrast; (c) filamentous, *Oscillatoria,* bright-field; a single cell measures about 15 μm wide; (d) filamentous heterocystous, *Anabaena*, phase contrast; a single cell measures about 5 μm wide; (e) filamentous branching, *Fischerella*, bright-field.

TABLE 13.33	Genera and grouping of cyanobacteria	
Group	**Genera**	**DNA (mol % GC)**
Group I—Unicellular: single cells or cell aggregates	*Gloeothece* (Figure 13.79*a*), *Gloeobacter*, *Synechococcus*, *Cyanothece*, *Gloeocapsa*, *Synechocystis*, *Chamaesiphon*	35–71
Group II—Pleurocapsalean: reproduce by formation of small spherical cells called baeocytes produced through multiple fission	*Dermocarpa* (Figure 13.79*b*), *Xenococcus*, *Dermocarpella*, *Pleurocapsa*, *Myxosarcina*, *Chroococcidiopsis*	40–46
Group III—Oscillatorian: filamentous cells that divide by binary fission in a single plane	*Oscillatoria* (Figure 13.79*c*), *Spirulina*, *Arthrospira*, *Lyngbya*, *Microcoleus*, *Pseudanabaena*	40–67
Group IV—Nostocalean: filamentous cells that produce heterocysts	*Anabaena* (Figure 13.79*d*), *Nostoc*, *Calothrix*, *Nodularia*, *Cylinodrosperum*, *Scytonema*	38–46
Group V—Branching: cells divide to form branches	*Fischerella* (Figure 13.79*e*), *Stigonema*, *Chlorogloeopsis*, *Hapalosiphon*	42–46

The fine structure of the cell wall of some cyanobacteria is similar to that of gram-negative Bacteria, and peptidoglycan is present in the walls (Figure 13.80). Many cyanobacteria produce extensive mucilaginous envelopes, or sheaths, that bind groups of cells or filaments together (see, for example, Figure 13.79*a*). The photosynthetic lamellar membrane system is often complex and multilayered (⌀ Figure 15.10*b*), although in some of the simpler cyanobacteria the lamellae are regularly arranged in concentric circles around the periphery of the cytoplasm (Figure 13.80). Cyanobacteria have only one form of chlorophyll, chlorophyll *a*, and all of them also have characteristic biliprotein pigments, **phycobilins** (⌀ Figure 15.10*a*), which function as accessory pigments in photosynthesis. One class of phycobilins, *phycocyanins,* are blue, and together with the green chlorophyll *a* are responsible for the blue-green color of the bacteria. However, some cyanobacteria produce *phycoerythrin,* a red phycobilin, and species possessing this pigment are red or brown in color.

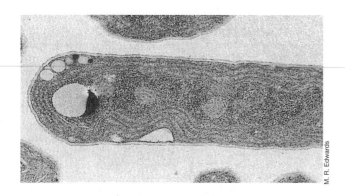

FIGURE 13.80 Electron micrograph of a thin section of the cyanobacterium *Synechococcus lividus.* A cell is about 5 μm in diameter.

Structural Variations: Gas Vesicles and Heterocysts

Among the cytoplasmic structures seen in many cyanobacteria are **gas vesicles** (⌀ Section 3.14), which are especially common in species that live in open waters (planktonic species). Their function is to regulate cell buoyancy such that it can remain in a position in the water column where light intensity is optimal for photosynthesis. Some filamentous cyanobacteria form **heterocysts,** which are rounded, seemingly empty cells, usually distributed regularly along a filament or at one end of a filament (Figure 13.81*a*). Heterocysts arise from differentiation of vegetative cells and are the sole sites of *nitrogen fixation* (the reduction of N_2 to NH_3, ⌀ Section 15.29) in heterocystous cyanobacteria. In *Anabaena,* a well-studied heterocystous cyanobacterium, complex gene rearrangements occur within the heterocyst to yield a contiguous cluster of *nif* genes that can be expressed as a unit.

Heterocysts have intercellular connections with adjacent vegetative cells, and there is mutual exchange of materials between these cells, with products of photosynthesis moving from vegetative cells to heterocysts and products of nitrogen fixation moving from heterocysts to vegetative cells (Figure 13.81*b*). Heterocysts are low in phycobilin pigments and *lack* photosystem II, the oxygen-evolving photosystem that generates reducing power from H_2O (⌀ Section 15.6). Without photosystem II they are unable to fix CO_2 and thus lack the necessary electron donor to reduce N_2 to NH_3; fixed carbon imported to the heterocyst from an adjacent vegetative cell solves this problem (Figure 13.81*b*).

Heterocysts are surrounded by a thickened cell wall containing large amounts of glycolipid, which serves to slow the diffusion of O_2 into the cell. Because of the oxygen lability of the enzyme nitrogenase (⌀ Section 15.29),

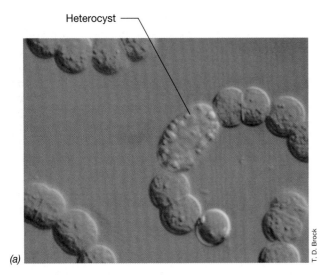

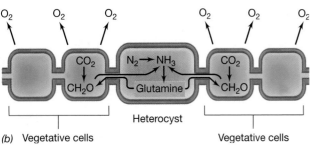

FIGURE 13.81   Heterocysts. (a) Heterocysts in the cyanobacterium *Anabaena* sp. Heterocysts are the sole site of nitrogen fixation in heterocystous cyanobacteria. (b) Model for the operation of a heterocyst. The heterocyst lacks oxygen-producing ability (Photosystem II, ∞ Section 15.6) and obtains the needed reductant for nitrogen fixation from organic matter produced by adjacent vegetative cells. Glutamine is the form of fixed nitrogen transported from heterocysts to vegetative cells.

it seems likely that the heterocyst, by maintaining an anoxic environment, stabilizes the nitrogen-fixing system in organisms that are not only aerobic but also oxygen-producing. Indeed, some nonheterocystous filamentous cyanobacteria produce nitrogenase and fix nitrogen in normal vegetative cells if they are grown anaerobically by vigorous bubbling with N_2 to remove O_2.

Cyanophycin and Other Structures

A structure called **cyanophycin** can be seen in electron micrographs of many cyanobacteria. This structure is a copolymer of aspartic acid and arginine:

$$\text{Asp} - \text{Asp} - \text{Asp} - \text{Asp} - \text{Asp} -$$
$$\quad | \qquad | \qquad | \qquad | \qquad |$$
$$\text{Arg} \quad \text{Arg} \quad \text{Arg} \quad \text{Arg} \quad \text{Arg}$$

and can constitute up to 10% of the cell mass. Cyanophycin is a nitrogen storage product in many cyanobacteria, and when nitrogen in the environment

becomes deficient, this polymer is broken down and used. Cyanophycin is also an energy reserve in cyanobacteria. Arginine, derived from cyanophycin, can be hydrolyzed to yield ornithine, with the production of ATP through the action of the enzyme *arginine dihydrolase* with carbamyl phosphate (∞ Section 15.21) occurring as an intermediate:

$$\text{Arginine} + \text{ADP} + \text{P}_i + \text{H}_2\text{O} \rightarrow$$
$$\text{Ornithine} + 2\,\text{NH}_3 + \text{CO}_2 + \text{ATP}$$

Arginine dihydrolase is present in many cyanobacteria and may function as a source of ATP for maintenance purposes during dark periods.

Many cyanobacteria exhibit gliding motility; true rotating flagella have never been found. Gliding occurs only when the cell or filament is in contact with a solid surface or with another cell or filament. In some cyanobacteria gliding is not a simple translational movement but is accompanied by rotations, reversals, and flexings of filaments. Most gliding species exhibit directional movement toward light (phototaxis) and chemotaxis (∞ Section 3.12) may occur as well.

Among the filamentous cyanobacteria, fragmentation of the filaments often occurs by formation of **hormogonia** (Figure 13.82*a* and *b*), which break away from the filaments and glide off. In some species resting spores or **akinetes** (Figure 13.82*c*) are formed, which protect the organism during periods of darkness, drying, or freezing. These are cells with thickened outer walls; they germinate through the breakdown of the outer wall and outgrowth of a new vegetative filament. However, even the vegetative cells of many cyanobacteria are relatively resistant to drying or low temperatures.

Physiology of Cyanobacteria

The nutrition of cyanobacteria is simple. Vitamins are not required, and nitrate or ammonia is used as nitrogen source. Nitrogen-fixing species are common. Most species tested are obligate phototrophs, being unable to grow in the dark on organic compounds. However, some cyanobacteria can assimilate simple organic compounds such as glucose and acetate if light is present (photoassimilation). Some cyanobacteria, mainly filamentous species, can actually grow in the dark on glucose or other sugars, using the organic material as both carbon and energy source.

Several metabolic products of cyanobacteria are of considerable practical importance. Many cyanobacteria produce potent neurotoxins, and during water blooms when massive accumulations of cyanobacteria may develop, animals ingesting such water may succumb rapidly. Fortunately, the massive accumulations needed to cause death do not occur extensively, although subclinical manifestations of cyanobacterial water blooms may be a com-

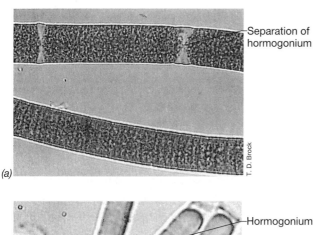

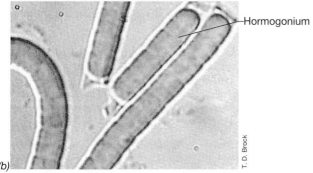

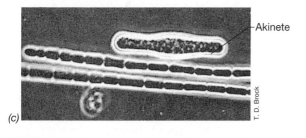

FIGURE 13.82 Structural differentiation in filamentous cyanobacteria. (a) Initial stage of hormogonium formation in *Oscillatoria*. Notice the empty spaces where the hormogonium is separating from the filament. (b) Hormogonium of a smaller *Oscillatoria* species. Notice that the cells at both ends are rounded. Nomarski interference contrast microscopy. (c) Akinete (resting spore) of *Anabaena* sp. by phase contrast.

mon, but unobserved, occurrence. Many cyanobacteria are also responsible for the production of earthy odors and flavors in fresh waters, and if such waters are used as drinking water sources, aesthetic problems may arise. The compound produced is *geosmin* (trans-1,10-dimethyl-trans-9-decalol). This substance is also produced by many actinomycetes (see the discussion in Section 13.23) and is responsible for the distinctive "earthy" odor of soil.

Ecology and Phylogeny of Cyanobacteria

Cyanobacteria are widely distributed in nature in terrestrial, freshwater, and marine habitats. In general they are more tolerant to environmental extremes than are algae and are often the dominant or sole oxygenic pho-

totrophic organisms in hot springs (∞ Table 5.1), saline lakes, and other extreme environments. Many members are found on the surfaces of rocks or soil and occasionally even within rocks themselves (∞ Figure 17.21). In desert soils subject to intense sunlight, cyanobacteria often form extensive crusts over the surface, remaining dormant during most of the year and growing during the brief winter and spring rains. In shallow marine bays, where relatively warm seawater temperatures exist, cyanobacterial mats of considerable thickness may form. Freshwater lakes, especially those that are fairly rich in nutrients, may develop blooms of cyanobacteria (∞ Figure 16.26b). A few cyanobacteria are symbionts of liverworts, ferns, and cycads; a number are found as the phototrophic component of lichens. In the case of the water fern *Azolla* (∞ Section 16.25), it has been shown that the cyanobacterial endophyte (a species of *Anabaena*) fixes nitrogen that becomes available to the plant.

Base compositions of DNA of a variety of cyanobacteria have been determined. Those of the unicellular forms vary from 35 to 71% GC, a range so wide as to suggest that this group contains many members with little genetic relationship to each other. On the other hand, the values for the heterocyst formers vary much less, from 38 to 46% GC. Phylogenetically, cyanobacteria group along morphological lines in most cases. Filamentous heterocystous and nonheterocystous species form distinct groups, as do the branching forms. However, unicellular cyanobacteria are phylogenetically highly diverse, with different representatives showing phylogenetic relationships to different morphological groups.

13.25

Prochlorophytes and Chloroplasts

Key Genera

Prochloron
Prochlorothrix

Prochlorophytes are oxygenic phototrophs that contain chlorophyll *a* and *b* but do *not* contain phycobilins. Prochlorophytes therefore resemble both cyanobacteria (because they are prokaryotic and produce chlorophyll *a*) and the green plant/alga chloroplast (because they contain chlorophyll *b* instead of phycobilins). Phylogenetically, prochlorophytes show a specific relationship to cyanobacteria, in particular the group that includes *Synechococcus* (see Figure 13.80), a unicellular cyanobacterium.

Prochloron

Prochloron was the first prochlorophyte discovered. It is found in nature as a symbiont of marine invertebrates (didemnid ascidians), and all studies of the organism

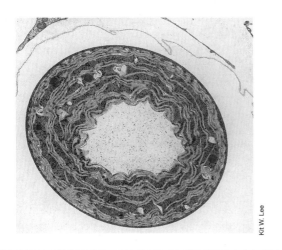

FIGURE 13.83 Electron micrograph of the prochlorophyte *Prochloron*. Note the extensive intracytoplasmic membranes (thylakoids). Cells are about 10 μm in diameter.

to date have relied on material collected from natural samples. Cells of *Prochloron* expressed from the cavities of didemnid tissue are roughly spherical in morphology (Figure 13.83), 8–10 μm in diameter. Electron micrographs of thin sections (Figure 13.83) show that *Prochloron* has an extensive thylakoid membrane system similar to that observed in the chloroplast (∞ Figure 3.72). Further evidence that *Prochloron* is phylogenetically a member of the Bacteria is the presence of muramic acid in the cell walls, indicating that peptidoglycan is present (∞ Section 3.7).

The ratio of chlorophyll *a*:chlorophyll *b* in cells of *Prochloron* is about 4–7:1, somewhat higher than the value of 1–2:1 typical of green algae. The carotenoids of *Prochloron* are similar to those of cyanobacteria, predominantly β-carotene and zeaxanthin. The GC base ratio of different samples of *Prochloron* isolated from different ascidians varies from 31 to 41%, indicating a fair bit of genetic heterogeneity. Thus, different species of *Prochloron* probably exist, but confirmation of this must await laboratory culture and study of pure strains.

Other Prochlorophytes

Prochlorothrix is a filamentous prochlorophyte (Figure 13.84) that can be grown in pure culture. Like *Prochloron*, *Prochlorothrix* contains chlorophylls *a* and *b* and lacks phycobilins. The ratio of chlorophyll *a* to chlorophyll *b* is somewhat higher in *Prochlorothrix* (about 8:1 to 9:1) than in *Prochloron*, and the thylakoid membranes are less well developed than in *Prochloron* (compare Figures 13.83 and 13.84*b*).

A novel type of prochlorophyte has been found in the euphotic zone of the open oceans. These phototrophs are extremely small cocci, measuring less than 1 μm in diameter, and have been given the genus name *Prochlorococcus*. Like those of other prochlorophytes, cells of *Prochlorococcus* contain chlorophyll *b*. However, *Prochlorococcus* lacks true chlorophyll *a* and produces instead a modified form of chlorophyll *a* called *divinyl chlorophyll a*. Cells of *Prochlorococcus* also contain α- (instead of β-) carotene, a pigment previously unknown in prokaryotes. Interestingly, the chlorophyll *a/b* ratio of *Prochlorococcus* is near 1, similar to that of the chloroplasts of marine green algae. Because their numbers in the oceans are relatively large (10^4–10^5 cells/ml), prochlorophytes like *Prochlorococcus* probably have considerable ecological significance as primary producers in open ocean waters. A variety of other prochlorophytes have been isolated including one containing chlorophyll *d* as its major pigment; the latter is present in a variety of algae (eukaryotic cells; ∞ Section 17.6).

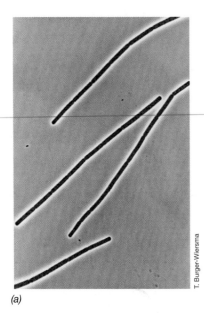

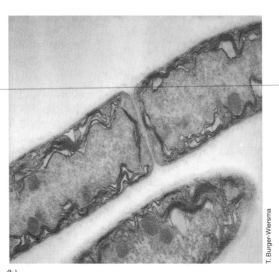

FIGURE 13.84 Phase and electron micrographs of the filamentous prochlorophyte *Prochlorothrix*. (a) Phase contrast. (b) Electron micrograph of thin section showing arrangement of membranes. The diameter of cells is about 1–2 μm. *Prochlorothrix* was first discovered as one of the dominant phototrophs in several shallow Dutch lakes but is now known to be widespread in freshwater lakes.

(a)

(b)

Prochlorophytes, Chloroplasts, and Evolution

Based on our discussion of endosymbiosis (∞ Sections 12.3, 17.1 and 17.2), the evolutionary significance of prochlorophytes should be apparent. Until the discovery of prochlorophytes, it was always assumed that the chloroplast originated from endosymbiotic association of *cyanobacteria* with a primitive eukaryotic cell. However, this hypothesis has never been scientifically satisfying for at least one major reason: how did the green plant chloroplast evolve the pigment complement it has today if it originated from a *cyanobacterial* endosymbiont that contained *phycobilins* instead of chlorophyll *b*? The hypothesis that *prochlorophytes* instead of cyanobacteria were the ancestors of the green plant chloroplast eliminates this major point of contention. However, phylogenetic analyses do not show *Prochloron, Prochlorococcus,* or *Prochlorothrix* to be the immediate ancestor of the green plant chloroplast. Instead, prochlorophytes, cyanobacteria, and the plant chloroplast all *shared* a common ancestor. Thus, although prochlorophytes arose from the same evolutionary roots as the chloroplast and are phenotypically more like chloroplasts than are cyanobacteria, prochlorophytes are not the direct "missing link" in the evolution of oxygenic phototrophs to yield the green plant chloroplast.

✓ 13.24–13.25 Concept Check

Cyanobacteria and prochlorophytes are oxygenic phototrophic prokaryotes. Prochlorophytes differ most clearly from cyanobacteria in that prochlorophytes contain chlorophyll *b* and lack phycobilins. Oxygen in Earth's atmosphere is thought to have originated from cyanobacterial photosynthesis.

✓ Describe at least three ways in which cyanobacteria differ from purple bacteria.
✓ What is a *heterocyst* and what is its function?
✓ How are cyanobacteria, prochlorophytes, and the chloroplasts of corn plants similar and how do they differ?
✓ Of what ecological significance is *Prochlorococcus?*

Kingdom IV: Chlamydia

13.26

The Chlamydia

Key Genus

Chlamydia

Organisms of the genus *Chlamydia* are obligately parasitic bacteria with little metabolic capacity that form a distinct kingdom level lineage of bacteria (see Figure 13.1). Three species of *Chlamydia* are recognized (Table 13.34): *C. psittaci,* the causative agent of the disease *psittacosis; C. trachomatis,* the causative agent of *trachoma* and a variety of other human diseases; and *C. pneumoniae,* the cause of a variety of respiratory syndromes (Table 13.34). **Psittacosis** is an epidemic disease of birds that is occasionally transmitted to humans and causes pneumonia-like symptoms. **Trachoma** is a debilitating disease of the eye char-

TABLE 13.34	Differential characteristics of species of the genus *Chlamydia*		
Characteristic	*C. trachomatis*	*C. psittaci*	*C. pneumoniae*
Hosts	Humans	Birds, mammals, occasionally humans	Humans
Usual site of infection	Mucous membrane	Multiple sites	Respiratory mucosa
Human-to-human transmission	Common	Rare	Probable
Mol % GC	42–45	39–43	40
Percent homology to *C. trachomatis* DNA by DNA:DNA hybridization[a]	100	10	10
DNA, kilobase pairs/genome (*Escherichia coli* = 4600)	1000	550	∽1000
Human diseases	Trachoma, otitis media, nongonococcal urethritis (males), urethral inflammation (females), lymphogranuloma venereum, cervicitis	Psittacosis	Respiratory syndromes
Domesticated animal diseases	—	Avian chlamydiosis (parrots, parakeets), pneumonia, synovial tissue arthritis, or conjunctivitis (kittens, lambs, calves, piglets, foals)	—

a For discussion of DNA:DNA hybridization, see Section 12.9.

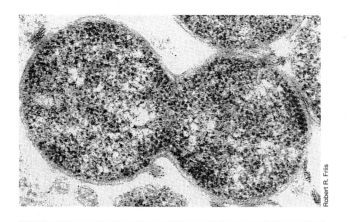

Robert R. Friis

FIGURE 13.85 Electron micrograph of a thin section of a dividing cell (reticulate body, see Figure 13.86) of *Chlamydia psittaci,* a member of the psittacosis group, within a mouse tissue culture cell. A single chlamydial cell is about 1 μm in diameter.

acterized by vascularization and scarring of the cornea. Trachoma is the leading cause of blindness in humans. Other strains of *C. trachomatis* infect the genitourinary tract, and chlamydial infections are one of the leading sexually transmitted diseases today (∞ Section 23.6). A comparison of the properties of *C. psittaci, C. trachomatis,* and *C. pneumoniae* and the diseases they cause is shown in Table 13.34.

Molecular and Metabolic Properties

Besides being disease entities, the chlamydias are intriguing because of the biological, evolutionary, and metabolic problems they pose. Biochemical studies show that the chlamydias have gram-negative-type cell walls and they have both DNA and RNA, that is, they are cells, not viruses. Electron microscopy of thin sections of in-

fected cells shows forms that clearly are undergoing binary fission (Figure 13.85). The biosynthetic capacities of the chlamydias are much more limited than even the rickettsias, the other group of obligate intracellular parasites known among the Bacteria (see Section 13.12). Indeed, for some time it was thought that chlamydias were "energy parasites," obtaining not only biosynthetic intermediates from their hosts, as do the rickettsias, but also ATP. However, this hypothesis has been questioned following the sequencing of the genome of *C. trachomatis* (∞ Section 9.12). The approximately 1 mB chromosome of *C. trachomatis* contains easily recognizable genes for ATP synthesis, and even contains a complement of genes encoding peptidoglycan biosynthetic functions, indicating that this organism may well contain peptidoglycan even though chemical analyses for this cell-wall polymer have been negative. Nevertheless, the chlamydias still probably have the simplest biochemical capacities of all known cellular organisms, and a summary of this is shown in Table 13.35.

Other interesting features of the *C. trachomatis* chromosome include the fact that it lacks a gene encoding the protein FtsZ, a key protein involved in septum formation during cell division (∞ Section 3.9) and previously thought to be indispensable for growth of all prokaryotes, both Archaea and Bacteria. Moreover, genes are present that have a distinct "eukaryotic look" to them, suggesting the *C. trachomatis* has picked up some host genes that may encode functions that assist it in its pathogenic lifestyle (Table 13.34; ∞ Sections 19.7 and 23.6).

Life Cycle of *Chlamydia*

The life cycle of a typical member of the genus *Chlamydia* is shown in Figure 13.86. Two cellular types are seen in a typical life cycle: a small, dense cell, called

TABLE 13.35	Comparison of obligate intracellular parasites: rickettsias, chlamydias, and viruses		
Property	**Rickettsias**	**Chlamydias**	**Viruses**
Structural			
Nucleic acid	RNA and DNA	RNA and DNA	Either RNA or DNA (single- or double-stranded), never both
Ribosomes	Present	Present	Absent
Cell wall	Peptidoglycan present	Peptidoglycan present[a]	No wall
Structural integrity during multiplication	Maintained	Maintained	Lost
Metabolic capacities			
Macromolecular synthesis	Carried out	Carried out	Only with use of host machinery
ATP-generating system	Present	Present (?)[a]	Absent
Capable of oxidizing glutamate	Yes	No	No
Sensitivity to antibacterial antibiotics	Sensitive	Sensitive (except for penicillin)	Resistant

a The genome of one chlamydial species, *C. trachomatis,* has been entirely sequenced (∞ Section 9.12), and genes for peptidoglycan synthesis and ATP synthesis are present.

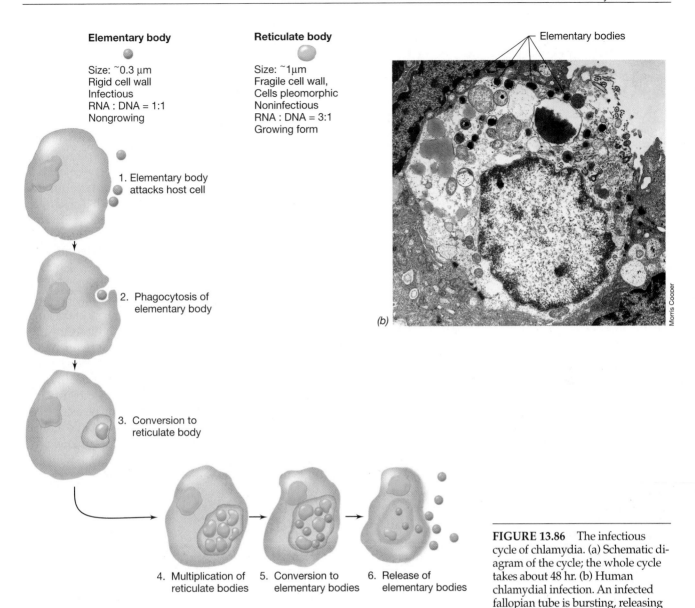

Elementary body

Size: ~0.3 μm
Rigid cell wall
Infectious
RNA : DNA = 1:1
Nongrowing

Reticulate body

Size: ~1μm
Fragile cell wall,
Cells pleomorphic
Noninfectious
RNA : DNA = 3:1
Growing form

Elementary bodies

1. Elementary body attacks host cell

2. Phagocytosis of elementary body

3. Conversion to reticulate body

4. Multiplication of reticulate bodies

5. Conversion to elementary bodies

6. Release of elementary bodies

(a)

(b)

Morris Cooper

FIGURE 13.86 The infectious cycle of chlamydia. (a) Schematic diagram of the cycle; the whole cycle takes about 48 hr. (b) Human chlamydial infection. An infected fallopian tube is bursting, releasing mature elementary bodies.

an **elementary body,** which is relatively resistant to drying and is the means of *dispersal* of the agent, and a larger, less dense cell, called a **reticulate body,** which divides by binary fission and is the *vegetative* form. Elementary bodies are nonmultiplying cells specialized for transmission, whereas reticulate bodies are noninfectious forms that specialize in intracellular multiplication. Unlike the rickettsias (see Section 13.12), the chlamydias are not transmitted by arthropods but are primarily *airborne* invaders of the respiratory system—hence the significance of resistance to drying of elementary bodies. When a virus infects a cell, it loses its structural integrity and liberates nucleic acid. When an elementary body enters a cell, however, although it changes form, it remains a structural unit and

enlarges and begins to undergo binary fission. A reticulate body is seen in Figure 13.85. After a number of divisions, the vegetative cells are converted to elementary bodies that are released when the host cell disintegrates and can then infect other cells. Generation times of 2–3 hr have been measured for reticulate bodies, which are considerably faster than those found for the rickettsias.

In sum, then, the chlamydias appear to have evolved an efficient and effective survival strategy including parasitizing the resources of the host (Table 13.35) and the production of resistant cell forms for transmission. It is thus not surprising that chlamydias have been associated with so many different disease syndromes (Table 13.34; ∞ Section 23.6).

✓ 13.26 Concept Check

Chlamydia are extremely small parasitic bacteria that cause a variety of human diseases. *Chlamydia* contain a very small genome and are deficient in many metabolic functions.

✓ Using the data of Table 13.35 as a guide, how can Chlamydias be differentiated from rickettsias? From viruses?

✓ What is the difference between an *elementary* body and a *reticulate* body?

✓ What surprises have emerged from the chlamydial genome?

Kingdom V:
Planctomyces/Pirella

13.27

Planctomyces: A Phylogenetically Unique Stalked Bacterium

Key Genera

Planctomyces
Pirella

This Kingdom contains a number of morphologically unique bacteria including the genera *Planctomyces*, *Pirella*, *Gemmata*, and *Isosphaera*. The best studied of these has been *Planctomyces* (Figure 13.87). In Section 13.15 we considered stalked bacteria such as *Caulobacter*. *Planctomyces* is also a stalked bacterium; however, unlike *Caulobacter*, the stalk of *Planctomyces* is made of protein and does not contain a cell wall or cytoplasm (compare Figure 13.87 with Figure 13.42). The *Planctomyces* stalk presumably functions in attachment but it is a much narrower and finer structure than the prosthecal stalk of *Caulobacter*.

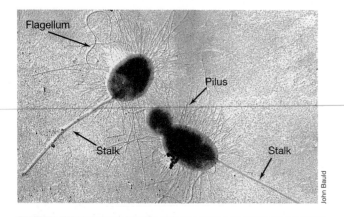

FIGURE 13.87 An electron micrograph of a metal-shadowed preparation of *Planctomyces maris*. A single cell is about 1–1.5 μm long. Note the fibrillar nature of the stalk. Pili are also abundant. Note also the flagella (curly appendages) on each cell and the bud that is developing from the nonstalked pole of one cell.

Other Features of the Group

Planctomyces and relatives are also of interest because they lack peptidoglycan and their cell walls are of an S-layer type (⚬⚬ Section 3.13) made of protein containing large amounts of cysteine (as cystine) and proline. As woud be expected of organisms lacking peptidoglycan, these organisms are resistant to antibiotics like penicillin and cephalosporin, drugs that disrupt peptidoglycan synthesis. Like *Caulobacter* (Figure 13.42), *Planctomyces* is also a budding bacterium and shows a type of "life cycle," wherein motile swarmer cells attach to a surface, grow a stalk from the attachment point, and generate a new cell from the opposite pole by budding. This daughter cell grows a flagellum, breaks away from the attached mother cell, and begins the cycle anew. Physiologically, *Planctomyces* species are typical facultatively aerobic chemoorganotrophs, growing either by fermentation or respiration of sugars. The habitat of members of the kingdom *Planctomyces* is primarily aquatic, both freshwater and marine, and the genus *Isosphaera* is a filamentous, gliding hot spring bacterium. Like for *Caulobacter* (see Section 13.15) isolation of *Planctomyces* and relatives requires dilute media and, since all known members of this group lack peptidoglycan, enrichments can be made even more selective by the addition of penicillin. Little more is known about this interesting group, but its unique phylogenetic position (see Figure 13.1) suggests that further work will reveal other novel characteristics.

Kingdom VI:
*Bacteroides/*Flavobacteria

This kingdom of Bacteria contains a mixture of physiological types from obligate aerobes to obligate anaerobes, unified by a common phylogenetic thread. The organisms inhabit many different types of environments and we focus here on the three main genera in the group.

13.28

Bacteroides, Flavobacterium, and *Cytophaga*

Key Genera

Bacteroides
Cytophaga

The genus *Bacteroides* contains obligately anaerobic, nonsporing species that are saccharolytic, fermenting sugars to primarily acetate and succinate as fermentation products. *Bacteroides* are normally commensals, found in the intestinal tract of humans and other animals (⚬⚬ Sections 16.15 and 19.4). In fact, *Bacteroides* species are thought to be the numerically dominant bacteria in the human large intestine, where measurements have shown 10^{10}–10^{11} cells per gram of human feces. However, species of *Bacteroides* can also be pathogens and are the most important anaerobic bacteria associated with human infections. Species of *Bacteroides* are also unusual among Bacteria in that they are one of the few groups of organisms to synthesize *sphingolipids,* a heterogeneous collection of lipids characterized by the long-chain amino alcohol *sphingosine* in place of glycerol (Figure 13.88). Sphingolipids such as sphingomyelin, cerebrosides, and gangliosides are common in mammalian tissues, especially in the brain and other nervous tissues.

In contrast to *Bacteroides, Flavobacterium* species are primarily found in aquatic habitats, both freshwater and marine, as well as in foods and food-processing plants. Colonies of *Flavobacterium* are frequently yellow-pigmented and physiologically these organisms are rather restricted, using glucose as carbon and energy source but very few other carbon compounds. Flavobacteria are rarely pathogenic; however, one species, *F. meningosepticum,* has been associated with cases of infant meningitis.

Cytophaga and Related Genera

Organisms of the genus *Cytophaga* are long, slender rods, often with pointed ends, that move by gliding (Figure 13.89*a,b*). The related genus, *Sporocytophaga,* is similar to *Cytophaga* in morphology and physiology but the cells form resting spherical structures called *microcysts* (Figure 13.89*c,d*) similar to those produce by some fruiting myxobacteria (see Section 13.16). They are widespread in the soil and water, often being present in great abundance. Many cytophagas digest polysaccharides like cellulose, agar (Figure 13.89*a*), or chitin. The cellulose decomposers can be easily isolated by placing small crumbs of soil on pieces of cellulose filter paper laid on the surface of mineral salts agar. The bacteria attach to and digest the cellulose fibers, forming spreading colonies that are usually yellow or orange in color (Figure 13.89*c*). The cytophagas do not produce soluble, extracellular, cellulose-digesting enzymes (cellulases); their cellulases remain attached to the cell envelope, which probably accounts for the fact that the cells must adhere to cellulose fibrils in order to digest them. In pure culture, *Cytophaga* can be cultured on agar containing embedded cellulose fibers, the presence of the organism being indicated by the clearing that occurs as the cellulose is digested (Figure 13.89*c*; see also Figure 15.58).

Species of *Cytophaga* and *Sporocytophaga* are obligately aerobic and probably account for much of the cellulose digestion by prokaryotes in oxic environments in nature. A number of *Cytophaga* species are also fish pathogens and can cause serious problems in the cultivated fish business. Two of the most important diseases are *columnaris disease,* caused by *C. columnaris,* and *cold-water disease,* caused by *C. psychrophila.* Both diseases preferentially affect stressed fish, such as those living in waters receiving pollutant discharges or living in high density confinement situations such as fish hatcheries and aquaculture operations. Infected fish show tissue destruction, frequently around the gills, and this may stem from the fact that *Cytophaga* species isolated from infected fish are frequently strongly proteolytic.

✓ 13.27–13.28 Concept Check

The *Planctomyces* group contains stalked, budding bacteria while the *Bacteroides* group contains a variety of gram-negative bacteria associated with humans, food, and the soil.

✓ What is unique about the cell wall of *Planctomyces*?
✓ How does the stalk of *Planctomyces* differ from the stalk of *Caulobacter*?
✓ Where might you find large numbers of *Bacteroides* cells in nature?
✓ Describe a method for isolating *Cytophaga* species from nature.

(a)

(b)

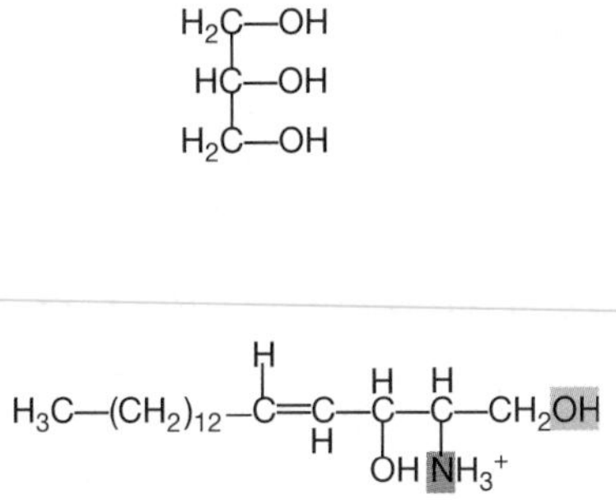

FIGURE 13.88 Comparison of (a) glycerol with (b) sphingosine. In sphingolipids, characteristic of *Bacteroides* species, sphingosine is the esterifying alcohol; a fatty acid is bonded by peptide linkage through the N atom (shown in red) and the terminal — OH group (shown in green) can contain any of a number of compounds including phosphotidyl choline (sphingomyelin) or various sugars (cerebrosides and gangliosides).

Kingdom VII: Green Sulfur Bacteria

13.29

Chlorobium and Other Green Sulfur Bacteria

Key Genera

Chlorobium
Prosthechochloris
"*Chlorochromatium*"

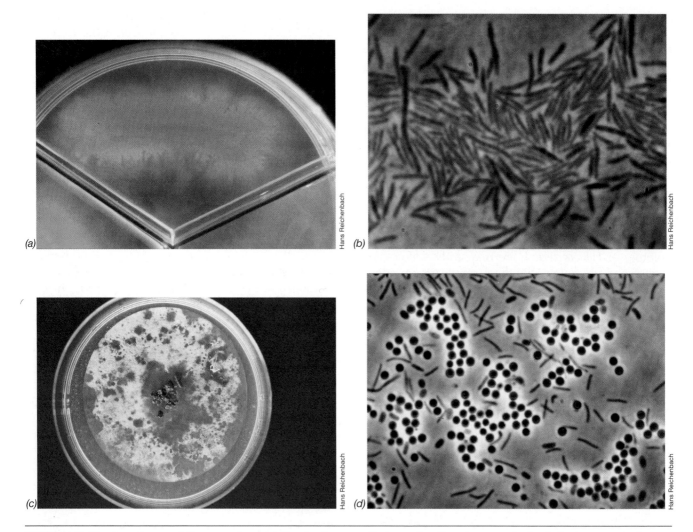

FIGURE 13.89   *Cytophaga* and *Sporocytophaga*. (a) Streak of an agarolytic marine *Cytophaga* species hydrolyzing agar in the Petri dish. (b) Phase contrast photomicrograph of cells of *C. hutchinsonii* grown on cellulose filter paper (cells are about 0.5 μm in diameter). (c) Colonies of *Sporocytophaga* growing on cellulose. Note the clearing zones where the cellulose has been degraded. (d) Phase contrast photomicrograph of the rod-shaped cells and spherical microcysts of *Sporocytophaga myxococcoides* (cells are about 0.5 μm and microcysts about 1.5 μm in diameter).

Green sulfur bacteria are a phylogenetically distinct group of nonmotile anoxygenic phototrophic bacteria that contain only obligately anaerobic and phototrophic species among cultured isolates. The group is morphologically diverse and includes short to long rods (Table 13.36 and Figure 13.90). Like purple sulfur bacteria they utilize H_2S as an electron donor, oxidizing it first to S^0 and then to SO_4^{2-}. But unlike purple sulfur bacteria, the sulfur produced by green sulfur bacteria resides *outside* the cell (∞ Figure 15.17*b*). Most species can also assimilate a few organic compounds in the light (that is, *photoheterotrophy*, ∞ Section 15.5). Strict autotrophy, however, is supported not by the reactions of the Calvin cycle as in purple bacteria, but instead by a reversal of steps in the citric acid cycle (∞ Section 15.8, reverse citric acid cycle), a unique means of autotrophy among phototrophic organisms.

The bacteriochlorophylls found in green sulfur bacteria include bacteriochlorophyll *a* and either bac-

teriochlorophylls *c*, *d*, or *e*. The latter pigments function only in light-harvesting reactions (∞ Section 15.3) and are located in a unique structure called the **chlorosome** (Figure 13.91). Chlorosomes are oblong bacteriochlorophyll-rich bodies bounded by a thin, nonunit membrane and lie attached to the cytoplasmic membrane in the periphery of the cell (Figure 13.91). Studies of energy transfer in green sulfur bacteria (∞ Section 15.3) have shown that light energy absorbed by bacteriochlorophylls *c*, *d*, or *e* in the chlorosome gets funneled to bacteriochlorophyll *a*, which resides in the cytoplasmic membrane, and it is here where photosynthetic energy conversion and ATP synthesis actually occur (∞ Figure 15.7).

Like purple sulfur bacteria (see Section 13.1), green sulfur bacteria live in anoxic aquatic environments, especially where H_2S is abundant (in general, green sulfur bacteria are more tolerant of sulfide than are purple

TABLE 13.36	Genera and characteristics of phototrophic green sulfur bacteria		
Characteristics	Genus	Number of species	DNA (mol % GC)
No gas vesicles:			
Straight or curved rods, nonmotile (see Figure 13.90a)	*Chlorobium*	7	49–58
Spheres and ovals, nonmotile, forming prosthecae (appendages)	*Prosthecochloris*	2	50–56
Contain gas vesicles:			
Branching nonmotile rods, in loose irregular network (see Figure 13.90b)	*Pelodictyon*	4	48–58
Spheres with prosthecae	*Ancalochloris*	1	—
Rods, gliding	*Chloroherpeton*	1	45–48
Filamentous, gliding, large diameter (2–2.5 μm)	*Chloronema*	1	—

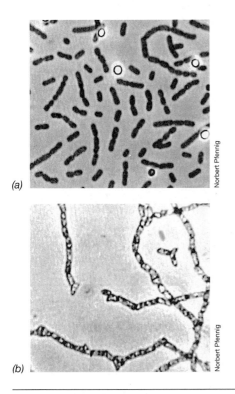

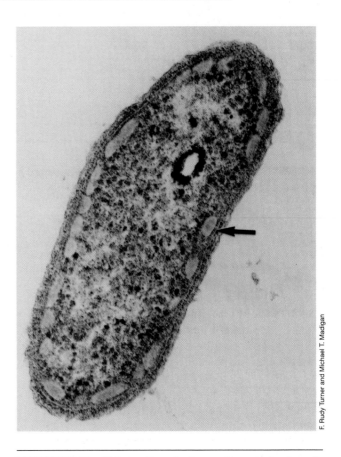

FIGURE 13.90 Phototrophic green sulfur bacteria. (a) *Chlorobium limicola*; cells are about 0.8 μm wide. Note the sulfur granules deposited *extra*cellularly. (b) *Pelodictyon clathratiforme*, a bacterium forming a three-dimensional network; cells are about 0.8 μm wide.

FIGURE 13.91 Thin section electron micrograph of a cell of the green sulfur bacterium *Chlorobium tepidum*. Note chlorosomes (arrow) in the cell periphery. A cell is about 0.7 μm wide.

bacteria). Because the chlorosome is such an efficient light-harvesting structure, little light is required to support the photosynthetic activities of green sulfur bacteria and they are thus typically found at the greatest depths in lakes of any phototrophic organisms. One species of the genus *Chlorobium*, *C. tepidum* (Figure 13.91), is thermophilic and forms dense microbial mats in high sulfide hot springs (Figure 13.92). *C. tepidum* is also notable because its genome (2 megabases) has been completely sequenced (☞ Section 9.12), the first

genome from an anoxyphototroph to have been done so (☞ Section 15.8).

Green Sulfur Bacteria Consortia

Certain green sulfur bacteria can form a tight two-membered association with a chemoorganotrophic bacterium in which each organism benefits the other; such

Richard W. Castenholz

FIGURE 13.92   A microbial mat composed almost entirely of the thermophilic green sulfur bacterium *Chlorobium tepidum* in a warm high sulfide spring in the Rotorua area of the North Island of New Zealand. The black color is iron sulfide (FeS).

associations are called **consortia.** The phototrophic component, called an *epibiont,* appears physically attached to the nonphototrophic component (Figure 13.93), although the mechanism of attachment is not clear. The name *"Chlorochromatium aggregatum"* has been used to describe one such consortium, although the name has no basis in formal prokaryotic taxonomy because it refers to two organisms rather than one.

The *"C. aggregatum"* consortium is green in color because the epibionts are green sulfur bacteria that contain bacteriochlorophyll *c* or *d* and green colored carotenoids and surround a central nonphototrophic cell. A structurally similar form referred to as

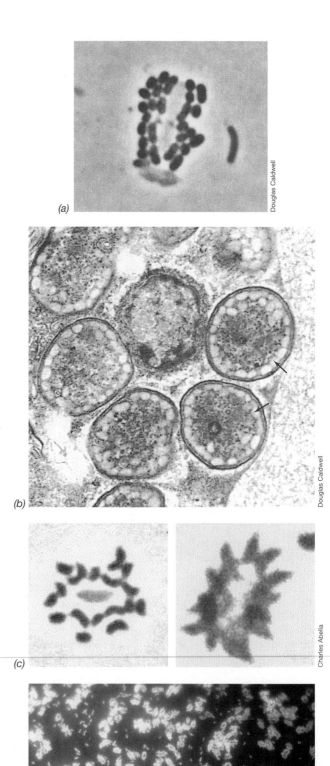

FIGURE 13.93   Green sulfur bacteria consortia. (a) Phase contrast micrograph and (b) transmission electron micrograph of the green bacterial consortium *"Chlorochromatium aggregatum."* In (a) the nonphototrophic central organism is much lighter in color than the pigmented phototrophic bacteria. Note the chlorosomes (arrows) in (b). The entire consortium is about 3×6 μm. (c) Half-moon-shaped epibionts in a *Pelochromatium* consortium from a stratified Wisconsin lake. The central cell in both photos is about 2 μm long. (d) Phylogenetic staining of *"Chlorochromatium aggregatum."* The yellow stain contained a nucleic acid probe specific for green sulfur bacteria. Note that only the epibionts are stained.

"*Pelochromatium roseum*" is brown in color because here the epibionts contain bacteriochlorophyll *e* and brown-colored carotenoids (⚭ Section 15.4 and Figure 15.9 for a discussion of the carotenoids of green sulfur bacteria). Yet other consortia are known in which the epibiont cells are half-moon shaped (Figure 13.93*c*); thus a variety of such consortia probably exist in nature.

Some green bacterial consortia have been grown in laboratory culture. On average, the "*C. aggregatum*" consortium (Figure 13.93) contains about 12 epibionts per central cell while the "*P. roseum*" consortium contains about 20. Solid evidence that the epibionts are indeed green sulfur bacteria comes from the fact that chlorosomes are visible in thin sections of the epibionts (Figure 13.93*b*) and from molecular evidence using phylogenetic probes (⚭ Section 12.6); treatment of the consortium with a fluorescent oligonucleotide probe specific for green sulfur bacterial 16S rRNA causes the epibionts, but not the central cell, to fluoresce (Figure 13.93*d*). Study of laboratory cultures have also shown that the central cell and the epibiont divide in synchrony, suggesting that the two components have some means of intercommunicating.

Although the rationale for why and the mechanism for how these associations form is not yet clear, combined laboratory and field studies suggest that the epibionts in these consortia are adapted to a very narrow regimen of light intensities and sulfide concentrations and that the role of the large central cell may be as a vehicle for these otherwise nonmotile anoxyphototrophs to position themselves in a water column where conditions for photosynthesis are optimal.

✓ 13.29 Concept Check

Green sulfur bacteria are obligately anaerobic anoxyphototrophs that produce unique structures called *chlorosomes*. These organisms can grow at very low light intensities and oxidize H_2S to S^0 and SO_4^{2-}.

✓ What pigments are found in the chlorosome?
✓ What is unique about autotrophy in *Chlorobium* (⚭ Section 15.8)?
✓ What two pieces of evidence support the idea that the epibionts of green bacterial consortia are truly green sulfur bacteria?

Kingdom VIII: Tightly Coiled Bacteria: The Spirochetes

13.30

Spirochetes

Key Genera

Spirochaeta
Treponema
Cristispira
Leptospira
Borrelia

Spirochetes are gram-negative, motile, tightly coiled Bacteria, typically slender and flexous in shape (Figure 13.94). These morphologically unique prokaryotes form a major phylogenetic lineage of Bacteria (see Figure 13.1). Spirochetes are widespread in aquatic environments and in animals, and some of them cause diseases including the important human sexually transmitted disease syphilis (⚭ Section 23.6).

The spirochete cell is made up of a "protoplasmic cylinder," consisting of the regions enclosed by the cell wall and membrane (Figure 13.95). Motility is conferred by a single to many flagella that emerge from each pole (Figure 13.95). However, unlike typical bacteria flagella (⚭ Section 3.11) spirochete flagella fold back from each pole upon the protoplasmic cylinder and remain located in the periplasm of the cell; thus they have been called *endoflagella*. Both the endoflagella and the protoplasmic cylinder are surrounded by a multilayered but flexible membrane called the *outer sheath* (Figure 13.95).

Motility of Spirochetes

Each spirochete endoflagellum is anchored at one end and extends about two-thirds of the length of the cell. Endoflagella rotate rigidly, as do typical bacterial flagella (⚭ Section 3.11). However, because the protoplasmic cylinder is also rigid whereas the outer sheath is flexible, when both endoflagella rotate in the same direction, the protoplasmic cylinder rotates in the opposite direction, placing torsion on the cell as illus-

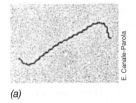
(a)

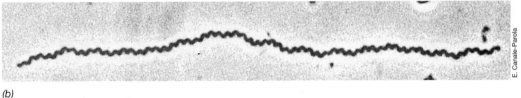
(b)

FIGURE 13.94 Two spirochetes at the same magnification, showing the wide size range in the group. (a) *Spirochaeta stenostrepta*, by phase contrast microscopy. A single cell is 0.25 μm in diameter. (b) *Spirochaeta plicatilis*. A single cell is 0.75 μm in diameter and can be up to 250 μm (0.25 mm) in length.

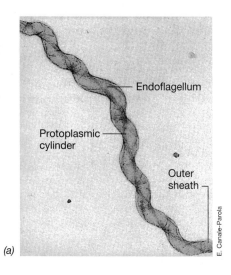

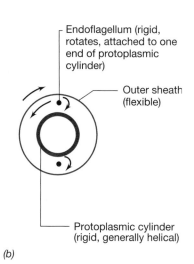

Endoflagellum (rigid, rotates, attached to one end of protoplasmic cylinder)

Outer sheath (flexible)

Protoplasmic cylinder (rigid, generally helical)

(a)   *(b)*

FIGURE 13.95   (a) Electron micrograph of a negatively stained preparation of *Spirochaeta zuelzerae,* showing the position of the endoflagellum. A single cell is about 0.3 μm in diameter. (b) Cross section of a spirochete cell, showing the arrangement of the protoplasmic cylinder, endoflagella, and external sheath and the manner in which the rotation of the rigid endoflagellum can generate rotation of the protoplasmic cylinder and (in the opposite direction) rotation of the external sheath. If the sheath is free, the cell will rotate about its longitudinal axis and move along it. If the sheath is in contact with a solid surface, the cell will creep forward. See text for details.

trated in Figure 13.95*b*. In liquid this causes the spirochete cell to move by flexing or lashing motions due to torque exerted at the ends of the protoplasmic cylinder by the rotating endoflagella (Figure 13.95*b*). Thus, despite the fact that the endoflagella of spirochetes do not extend away from the cell but instead reside inside the cell's outer membrane, the rotating action of these structures provide motility, albeit of more an irregular and jerky form than for smooth swimming, just as do the flagella of other prokaryotes.

Classification

Spirochetes are classified into eight genera primarily on the basis of habitat, pathogenicity, ribosomal RNA sequences, and morphological and physiological characteristics. Table 13.37 lists the major genera and their characteristics.

Spirochaeta and *Cristispira*

The genus *Spirochaeta* includes free-living, anaerobic, and facultatively aerobic spirochetes. These organisms are common in aquatic environments, such as the water and mud of rivers, ponds, lakes, and oceans. One species of the genus *Spirochaeta* is *S. plicatilis* (Figure 13.94*b*), a fairly large spirochete found in freshwater and marine H_2S-containing habitats and is probably anaerobic. The endoflagella of *S. plicatilis* are arranged in a bundle that winds around the coiled protoplasmic cylinder. From 18 to 20 endoflagella are inserted at each pole of this spirochete. Another species, *Spirochaeta stenostrepta,* has been cultured and is shown in Figure 13.94*a*. It is an obligate anaerobe commonly found in H_2S-rich, black muds. It ferments sugars via the glycolytic pathway to ethanol, acetate, lactate, CO_2, and H_2. The species *Spirochaeta aurantia* is an orange-pigmented facultative aerobe, ferment-

ing sugars via the glycolytic pathway under anaerobic conditions and oxidizing sugars aerobically mainly to CO_2 and acetate.

A metabolically unusual spirochete has been isolated from the hindgut of the termite. This environment is highly cellulolytic, as termites live mostly on wood and wood products, and much H_2 and CO_2 is produced from the fermentation of glucose released from the cellulose. A spirochete isolated from the termite hindgut can convert this H_2 plus CO_2 to acetate, that is, it is a *homoacetogen* (∞ Section 15.18 for a discussion of homoacetogenesis), and this is the first instance of this form of energy metabolism being found in a phylogenetic group of bacteria outside of the clostridia and relatives (see Section 13.19).

The genus *Cristispira* (Figure 13.96) contains organisms with a unique distribution, being found in nature primarily in the *crystalline style* of certain molluscs, such as clams and oysters. The crystalline style is a flexible, semisolid rod seated in a sac and rotated against a hard surface of the digestive tract, thereby mixing with and grinding the small particles of food. Being large spirochetes, the cristispiras can readily be seen microscopically within the style as they rapidly rotate forward and backward in corkscrew fashion. *Cristispira* may occur in both freshwater and marine molluscs, but not all species of molluscs possess them. Unfortunately, *Cristispira* has not been cultured, and so the physiological reason for its restriction to this unique habitat is not known.

Treponema

Anaerobic, host-associated spirochetes that are commensals or parasites of humans and animals are placed in the genus *Treponema*. *Treponema pallidum,* the causal agent of syphilis (∞ Section 23.6), is the best-known species of *Treponema*. It differs in morphology from other spirochetes; the cell is not helical but has a flat wave

TABLE 13.37   Genera of spirochetes and their characteristics

Genus	Dimensions (μm)	Number of species recognized	General characteristics	Number of endoflagella	DNA (mol % GC)	Habitat	Diseases
Cristispira	30–150 × 0.5–3.0	1	3–10 complete coils; bundle of endoflagella visible by phase contrast microscopy	>100	—	Digestive tract of molluscs; has not been cultured	None known
Spirochaeta	5–250 × 0.2–0.75	14	Anaerobic or facultatively aerobic; tightly or loosely coiled	2–40	50–65	Aquatic, free-living, freshwater and marine	None known
Treponema	5–15 × 0.1–0.4	18	Microaerophilic or anaerobic; flattened coil amplitude up to 0.5 μm	2–15	25–53	Commensal or parasitic in humans, other animals	Syphilis, yaws, swine dysentery, pinta
Borrelia	8–30 × 0.2–0.5	30	Microaerophilic; 5–7 coils of approx. 1 μm amplitude	Unknown	46	Humans and other mammals, arthropods	Relapsing fever, Lyme disease, ovine and bovine borreliosis
Leptospira	6–20 × 0.1	12	Aerobic, tightly coiled, with bent or hooked ends; requires long-chain fatty acids	2	33–43	Free-living or parasitic in humans, other mammals	Leptospirosis
Leptonema	6–20 × 0.1	1	Aerobic; does not require long-chain fatty acids	Unknown	54	Free-living	None known
Brachyspira	7–10 × 0.35–0.45	5	Anaerobe	8–28	25–27	Intestine of warm-blooded animals	Causes diarrhea in chickens and swine
Brevinema	4–5 × 0.2–0.3	1	Microaerophile by 16S rRNA sequencing, forms deep branch in spirochete lineage (see Figure 13.1)	4	34	Blood and tissue of mice and hamsters	Infectious for laboratory mice

form. The *T. pallidum* cell is remarkably thin, measuring approximately 0.2 μm in diameter. Living cells are clearly visible in the dark-field microscope or after staining with fluorescent antibody; dark-field microscopy has long been used to examine exudates from suspect-

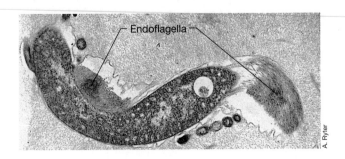

FIGURE 13.96 Electron micrograph of a thin section of *Cristispira*, a very large spirochete. A cell measures about 2 μm in diameter. Notice the numerous endoflagella.

ed syphilitic lesions (⊙⊙ Figure 23.24). In nature *T. pallidum* is restricted to humans, although artificial infections have been established in rabbits and monkeys. Although never grown in laboratory culture, it has been established from animal studies that virulent *T. pallidum* cells (purified from infected rabbits) contain a cytochrome system and are in fact microaerophiles. Such cells have also yielded sufficient DNA for the complete genome of *T. pallidum* (1.14 megabases) to be sequenced (⊙⊙ Section 9.12).

Other species of the genus *Treponema* are common commensal organisms in the oral cavity of humans and can generally be seen in material scraped from between the teeth and from the narrow space between the gums and the teeth. Three species, *Treponema denticola*, *T. macrodentium*, and *T. oralis*, have been described, differing in morphology and physiological characteristics. *Treponema denticola* ferments amino acids such as cysteine and serine, forming acetate as the major fermentation acid as well as CO_2, NH_3, and H_2S. Spirochetes are also found in the rumen. *Treponema saccharophilum* (Figure 13.97) is a large, *pectinolytic* spirochete found in the bovine rumen. *Treponema saccharophilum* is an obligate anaerobe that ferments pectin, starch, inulin, and other plant polysaccharides. This and other spirochetes may play an important role in the conversion of plant polysaccharides to volatile fatty acids, usable as energy sources by the ruminant (∞ Section 16.15).

Although the genus *Treponema* is phylogenetically a unit, the true relationship between *T. pallidum* and other *Treponema* species may be a distant one because the GC base ratio of *T. pallidum* is about 53% whereas other species of this genus cluster between 38 and 40% or 25 and 26%.

Leptospira and *Leptonema*

The genera *Leptospira* and *Leptonema* contain strictly *aerobic* spirochetes that use long-chain fatty acids (for example, oleic acid) as electron donor and carbon sources. With few exceptions, these are the only substrates utilized by leptospiras for growth. The leptospira cell is thin, finely coiled, and usually bent at each end into a semicircular hook. At present, several species are recognized in this group, some free-living and many parasitic. Two major species are *Leptospira interrogans* (parasitic) and *L. biflexa* (free-living). Strains of *L. interrogans* are parasitic for humans and animals. Rodents are the natural hosts of most leptospiras, although dogs and pigs are also important carriers of certain strains. In humans the most common leptospiral syndrome is *leptospirosis;* in this disorder the organism localizes in the kidney and can cause renal failure and death.

Leptospiras ordinarily enter the body through the mucous membranes or through breaks in the skin. After a transient multiplication in various parts of the body the organism localizes in the kidney and liver, causing nephritis and jaundice. The organism then passes out of the body in the urine and infection of another individual

is most commonly by contact with infected urine. Therapy with penicillin, streptomycin, or the tetracyclines is possible but may require extended courses to eliminate the organism from the kidney. Domestic animals such as dogs are vaccinated against **leptospirosis** with a killed virulent strain in the combined distemper–leptospira–hepatitis vaccine. In humans, prevention is effected primarily by elimination of the disease from animals.

Borrelia

The majority of species in the genus *Borrelia* are animal or human pathogens. *Borrelia recurrentis* is the causative agent of **relapsing fever** in humans and is transmitted via an insect vector, usually by the human body louse. Relapsing fever is characterized by a high fever and generalized muscular pain, which lasts for 3–7 days followed by a recovery period of 7–9 days. Left untreated, the fever returns in two to three more cycles (hence the name relapsing fever) and causes death in up to 40% of those infected. Fortunately, the organism is quite sensitive to tetracycline, and if the disease is correctly diagnosed, treatment is straightforward. Other borrelia are of veterinary importance, causing diseases in cattle, sheep, horses, and birds. In most of these diseases the organism is transmitted by ticks. *Borrelia burgdorferi* is the causative agent of the tick-borne disease called *Lyme disease,* which infects humans and other animals. Lyme disease is discussed in Section 24.4. *Borrelia burgdorferi* is also of interest because it is as yet one of the only known prokaryotes with a *linear* (as opposed to a *circular*) chromosome (∞ Section 6.4). This rather small genome of *B. burgdorferi* (1.44 megabases) has been completely sequenced (∞ Section 9.12).

✓ 13.30 Concept Check

Spirochetes are tightly coiled, motile, helical prokaryotes that contain both free-living as well as pathogenic species.

- ✓ How do the endoflagella of spirochetes compare with the flagella of *Escherichia coli*?
- ✓ Name two diseases of humans caused by spirochetes.
- ✓ What is the habitat of the spirochete *Cristispira*?

Kingdom IX: Deinococci

13.31

Deinococcus/Thermus

Key Genera

Deinococcus
Thermus

This kingdom of Bacteria contains only three genera, the best studied being *Deinococcus* and *Thermus*. The latter genus contains thermophilic chemoorgan-

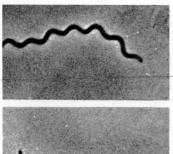
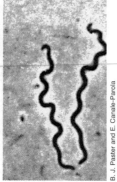

FIGURE 13.97 Phase contrast photomicrographs of *Treponema saccharophilum*, a large pectinolytic spirochete from the bovine rumen. A cell measures about 0.4 m in diameter. Left, regularly coiled cells; right, irregularly coiled cells.

otrophic bacteria including *Thermus aquaticus,* the organism from which *Taq* DNA polymerase is obtained. Because it is so heat stable, this enzyme is the major one used in the polymerase chain reaction (PCR) technique for amplifying DNA, as was discussed in Section 10.9. *Thermus* species stain gram-negative and contain a rare form of peptidoglycan in which ornithine is present in place of diaminopimelic acid in the muramic acid cross bridges (∞ Section 3.7). Interestingly, *Deinococcus* also contains ornithine. A number of species of *Thermus* have been described and all grow aerobically by catabolism of sugars, amino and organic acids, or various complex mixtures. We focus the rest of the discussion here on *Deinococcus.*

The genus *Deinococcus* contains four species of gram-positive cocci; *D. radiodurans* is the best-studied species. The *D. radiodurans* cell wall is structurally complex and consists of several layers including an outer membrane (Figure 13.98), normally present only in gram-negative bacteria (∞ Section 3.8). However, unlike the outer

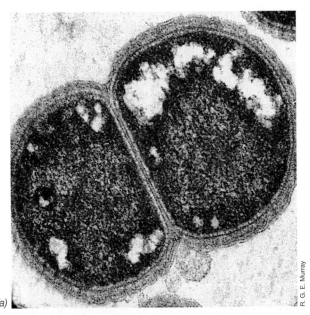

(a)

Outer membrane ⌐　Peptidoglycan ⌐　⌐ Cytoplasmic membrane

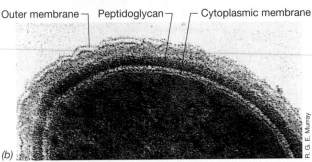

(b)

FIGURE 13.98　The radiation-resistant coccus *Deinococcus radiodurans.* An individual cell is about 2.5 μm in diameter. (a) Transmission electron micrograph of *D. radiodurans.* Note the outer membrane layer. (b) High magnification micrograph of wall layer.

membrane of gram-negative bacteria like *E. coli,* the outer membrane of *D. radiodurans* lacks lipid A. Physiologically, *D. radiodurans* is an aerobic chemoorganotroph, usually grown on complex media.

Most deinococci are red or pink in color because of the variety of carotenoids found in these organisms, and many strains are highly resistant to ultraviolet radiation and to desiccation. Resistance to radiation can be used to advantage in isolating deinococci. These remarkable organisms can be isolated from soil, ground meat, dust, and filtered air following exposure of the sample to intense ultraviolet (or even gamma) radiation and plating on a rich medium containing tryptone and yeast extract. Because many strains of *Deinococcus radiodurans* are even more resistant to radiation than bacterial endospores, treatment of a sample with strong doses of radiation effectively sterilizes the sample of organisms other than *D. radiodurans,* making isolation of deinococci relatively straightforward. For example, *D. radiodurans* cells can survive exposure to up to 30,000 Gy of ionizing radiation (1 Gy = 100 rad), sufficient to literally shatter the organism's chromosome into hundreds of fragments (by contrast, a human can be killed by exposure to less than 5 Gy). A powerful DNA repair machinery exists in *Deinococcus* cells that is able to repair the organism's chromosome, even from a fragmented state. Consistent with the fact that *D. radiodurans* is an extremely radiation-resistant bacterium, strains of this organism have been isolated from near atomic reactors and other potentially lethal radiation sources.

In addition to impressive radiation resistance, *Deinococcus radiodurans* is resistant to the mutagenic effects of many other mutagenic agents. Studies of the mutability of *D. radiodurans* have shown it to be highly efficient in repairing damaged DNA. Several different DNA repair enzymes exist in *D. radiodurans* for repairing breaks in single- or double-stranded DNA and for excising and repairing thymine dimers formed by the action of ultraviolet light. The only chemical mutagens that seem to work on *D. radiodurans* are agents like nitrosoguanidine, which tend to induce *deletions* in DNA; deletions are apparently not repaired as efficiently as point mutations in this organism and mutants of *D. radiodurans* can be isolated in this way.

Kingdom X: The Green Nonsulfur Bacteria

Key Genera

Chloroflexus
Thermomicrobium

This kingdom of Bacteria is phylogenetically distinct and contains just a few genera, the best known being the anoxyphototroph *Chloroflexus;* all cultured rep-

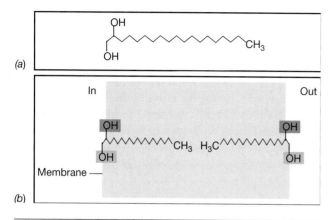

(a)

(b)

FIGURE 13.99 The unusual lipids of *Thermomicrobium.*
(a) Membrane lipids from *T. roseum* contain long chain diols like the one shown here (13-methyl-1,2 nonadecanediol). Note that unlike the lipids of other Bacteria or of Archaea (∞ Section 3.5), neither ester nor ether-linked side chains are present. (b) To form a bilayer membrane, dialcohol molecules presumably oppose each other at the methyl groups with the OH groups being the inner and outer hydrophilic surfaces. Small amounts of the diols have fatty acids esterified to the secondary —OH group (shown in red) while the primary —OH group (shown in green) can bond a hydrophilic molecule like phosphate.

resentatives are thermophilic. *Thermomicrobium* is a chemotrophic member of this group and is a strictly aerobic, gram-negative rod, growing optimally in complex media at 75°C. Besides its phylogenetic novelty, *Thermomicrobium* is also of interest because of its membrane lipids. Recall that the lipids of Bacteria and Eukarya contain fatty acids esterified to glycerol (∞ Sections 2.4 and 3.5). By contrast, the lipids of *Thermomicrobium* contain 1,2-dialcohols instead of glycerol, and have neither ester *nor* ether linkages (Figure 13.99).

13.32

Chloroflexus and *Heliothrix*

Chloroflexus and most other green nonsulfur bacteria are filamentous prokaryotes that form thick microbial mats in neutral to alkaline hot springs (Figure 13.100; see also Figure 16.16a). *Chloroflexus*-like organisms have also been found in nonthermal marine microbial mats. Although an anoxyphototroph, *Chloroflexus* is a "hybrid" phototroph in the sense that its photosynthetic mechanism shows features characteristic of both purple bacteria and green sulfur bacteria. Like the latter, *Chloroflexus* contains bacteriochlorophyll *c* and chlorosomes (see Figure 13.91 for an electron micrograph of chlorosomes). However, the bacteriochlorophyll *a* located in the cytoplasmic membrane of cells of *Chloroflexus* is arranged to form a photosynthetic reaction

center structurally identical to those of purple bacteria (by contrast, the reaction center of green sulfur bacteria is structurally quite different, ∞ Figure 15.18). It has thus been proposed that modern in *Chloroflexus* may be a vestige of a very early form of phototroph (see also later concerning autotrophy and phylogeny of this organism) that perhaps first evolved a photosynthetic reaction center and then received chlorosome-specific genes by lateral transfer.

Physiologically, *Chloroflexus* resembles purple nonsulfur bacteria in that photoautotrophy can be supported by (H_2S + CO_2) or (H_2 + CO_2). However, in *Chloroflexus,* phototrophic growth is best with organic compounds as carbon sources (photoheterotrophy). *Chloroflexus* also grows well in the dark as a chemoorganotroph by aerobic respiration. Interestingly, and this should be considered in light of the evolutionary position of *Chloroflexus* as the most phylogenetically ancient of anoxyphototrophs (see Figure 13.1), autotrophy in *Chloroflexus* is based on a CO_2 incorporation pathway, the Hydroxypropionate cycle, unique to this organism (∞ Figure 15.24b). We consider the biochemistry of this novel autotrophic pathway in Section 15.8.

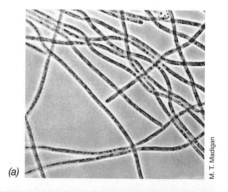

(a)

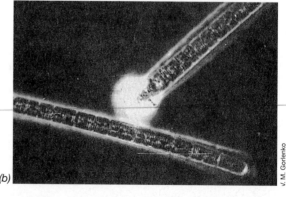

(b)

FIGURE 13.100 Green nonsulfur bacteria. (a) Phase photomicrograph of the thermophilic phototroph, *Chloroflexus aurantiacus.* Cells are about 1 μm in diameter. (b) Phase micrograph of the large phototroph *Oscillochloris.* Cells are about 5 μm wide. The brightly contrasting material is a holdfast, used for attachment.

In addition to *Chloroflexus*, other phototrophic green nonsulfur bacteria include the thermophile *Heliothrix* and the large-celled mesophile *Oscillochloris* (Figure 13.100). *Heliothrix* is of interest because it is phylogenetically and phenotypically quite similar to *Chloroflexus* except that it lacks the bacteriochlorophyll *c* and chlorosomes of this organism.

✓ 13.31 and 13.32 Concept Check

Deinococcus and *Chloroflexus* are each key genera in separate major lineages of Bacteria. *Deinococcus radiodurans* is the most radiation resistant of all known organisms and *Chloroflexus* is an anoxyphototroph that shows photosynthetic properties characteristic of both purple bacteria and green bacteria.

- ✓ How does *Deinococcus radiodurans* prevent being killed by high levels of radiation?
- ✓ In what ways does *Chloroflexus* resemble an organism like *Chlorobium*? An organism like *Rhodobacter*?
- ✓ What is unique about the organism *Thermomicrobium*?

Kingdoms XI, XII, and XIII: Hyperthermophiles

Key Genera

Thermotoga
Thermodesulfobacterium
Aquifex

The final three kingdoms of Bacteria we consider cluster deep in the phylogenetic tree of Bacteria, near to the hypothetical "root" (see Figures 12.13 and 13.1). Each kingdom consists of one or two major genera and a key physiological feature of most of them is *hyperthermophily*, that is, optimal growth at temperatures *above* 80°C (∞ Sections 5.7 and 5.9).

13.33

Thermotoga and *Thermodesulfobacterium*

Thermotoga is a rod-shaped hyperthermophile capable of growth to 90°C (optimum, 80°C). Cells of *Thermotoga* contain a sheath-like envelope (the "toga," see Figure 13.101*a*), stain gram-negatively, and are nonsporing. *Thermotoga* is an anaerobic, fermentative chemoorganotroph, catabolizing sugars and polymers such as starch and producing lactate, acetate, CO_2, and H_2 as fermentation products. Species of *Thermotoga* have been isolated from terrestrial hot springs as well as marine hydrothermal vents.

Thermodesulfobacterium (Figure 13.102) is a thermophilic, sulfate-reducing bacterium, positioned on the

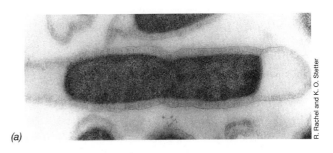

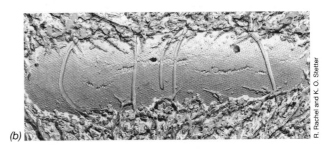

(a)

(b)

R. Rachel and K. O. Stetter

FIGURE 13.101   Hyperthermophilic Bacteria. (a) *Thermotoga maritima*—temperature optimum, 80°C. Note the outer covering on the cell (the "toga"). (b) *Aquifex pyrophilus*—temperature optimum, 85°C. Cells of *Thermotoga* (thin section) measure 0.6×3.5 μm; cells of *Aquifex* (freeze-fracture micrograph) measure 0.5×2.5 μm. Both *Thermotoga* and *Aquifex* form their own phylogenetic lineages on the tree of Bacteria (see Figure 13.1).

phylogenetic tree between *Thermotoga* and *Aquifex* (Figure 13.1). Although not a true hyperthermophile as its growth temperature optimum is only 70°C, *Thermodesulfobacterium* is the most thermophilic of all known sulfate-reducing Bacteria (the archaeon sulfate-reducer *Archaeoglobus* is a true hyperthermophile, ∞ Section 14.6). Like other "group I" sulfate reducers (see Section 13.17), *Thermodesulfobacterium* cannot utilize acetate as an electron donor in its energy metabolism and instead uses compounds like lactate, pyruvate, and ethanol, reducing SO_4^{2-} to H_2S.

An unusual biochemical feature of species of *Thermodesulfobacterium* is the presence of *ether-linked lipids*. Recall that the latter are a hallmark of the Archaea and that a poly-isoprenoid C_{20} hydrocarbon (phytanyl) replaces fatty acids as the side chains in archaeal lipids (∞ Sections 3.5 and 12.8). A very interesting situation exists in *Thermodesulfobacterium* ether-linked lipids because the glycerol side chains are not phytanyl groups but instead a unique C_{17} hydrocarbon along with some fatty acids (Figure 13.102*b*). Thus we see in *Thermodesulfobacterium* both a deep phylogenetic lineage (Figure 13.1) and a lipid profile that combines features of both the Archaea and the Bacteria (Figure 13.102*b*). However, the bacterium *Ammonifex*, a thermophilic member of the low GC, gram-positive Bacteria that

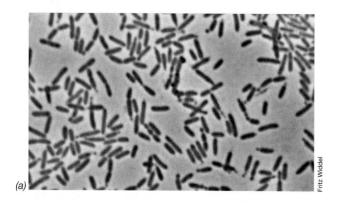

(a)

Fritz Widdel

Ether linkage

H_2C—O

CH_3

HC—O

CH_3

H_2C—R — Hydrophilic residue

(b)

FIGURE 13.102 *Thermodesulfobacterium.* (a) Phase photomicrograph of cells of *T. mobile.* (b) Structure of one of the lipids of *T. mobile.* Note that although these are ether-linked, the two hydrophobic side chains are *not* phytanyl units as in Archaea (⚬⚬ Section 3.5). The designation "R" is for a hydrophilic residue, such as a phosphate group.

grows anaerobically by H_2 oxidation coupled to the reduction of NO_3^- to NH_3, also contains lipids like those of *Thermodesulfobacterium;* thus, ether-linked lipids may be more common among Bacteria than previously thought.

13.34

Aquifex and Relatives

The genus *Aquifex* (Figure 13.101*b*) is an obligately chemolithotrophic and autotrophic hyperthermophile and is the most thermophilic of all known Bacteria. Various *Aquifex* species utilize H_2, S^0, or $S_2O_3^{2-}$ as electron donors and O_2 or NO_3^- as electron acceptors and grow up to 95°C (optimum, 85°C). Only very low O_2 concentrations are tolerated by *Aquifex*, but it remains (along with a few archaeans, ⚬⚬ Sections 14.4 and 14.8) one of the few *aerobic* hyperthermophiles known. Nutritional studies of *Aquifex* species have shown them totally unable to grow chemoorganotrophically on organic compounds including complex mixtures like yeast or meat extract.

Autotrophy in *Aquifex* is supported by the reverse citric acid cycle, a series of reactions previously found only in green sulfur bacteria (⚬⚬ Section 13.29) within the domain Bacteria. The complete genome sequence of *Aquifex aeolicus* has been determined (⚬⚬ Section 9.12) and its entirely chemolithotrophic/autotrophic lifestyle is supported by an amazingly small genome of only 1.55 megabases (one-third the size of the *E. coli* genome). This fact was previously discussed in Section 12.2 as regards the physiological properties of early life forms. The discovery that so many hyperthermophiles, both Archaea and Bacteria like *Aquifex*, are H_2 chemolithotrophs, coupled with the finding that they branch as very early lineages on their respective phylogenetic trees (⚬⚬ Figures 12.13, 13.1 and 14.1), suggests that H_2 was a key electron donor for living organisms under early Earth conditions. A simple model for the utilization of H_2 to generate a proton motive force and hence usable energy as ATP was shown in Figure 12.4.

✓ 13.33–13.34 Concept Check

Thermotoga, Thermodesulfobacterium, and *Aquifex* grow at high temperature and each spearhead a major lineage of *Bacteria. Aquifex* is an H_2-oxidizing chemolithotroph while *Thermotoga* and *Thermodesulfobacterium* are both anaerobic chemoorganotrophs.

✓ Compare the catabolic metabolism of *Thermotoga* and *Thermodesulfobacterium.*

✓ What is unusual about the lipids of *Thermodesulfobacterium?*

✓ From a genomic perspective, why is the fact that *Aquifex* is able to grow on a diet of $H_2 + CO_2 + O_2$, seem surprising?

Although the concept of a separate domain of prokaryotes called the Archaea is only some 20 years old, the phylogenetic tree of Archaea, shown here, has bushed out dramatically since that time, with many new isolations. A major unanswered question concerning the Archaea is why so many species are "extremophiles," thriving in habitats at high temperature, high salt, or high or low pH. However, even this picture is beginning to change as molecular ecology studies are showing that, like the Bacteria, "Archaea are everywhere," including relatives of extremophiles that inhabit the open oceans, soil, lakes, and ponds. Perhaps the future will also reveal archaeans that cause human disease. Nevertheless, study of extremophilic archaeans, especially the hyperthermophiles like *Methanococcus jannaschii* shown here, remains a major research focus because of the secrets these organisms hold for an understanding of early life forms on Earth.

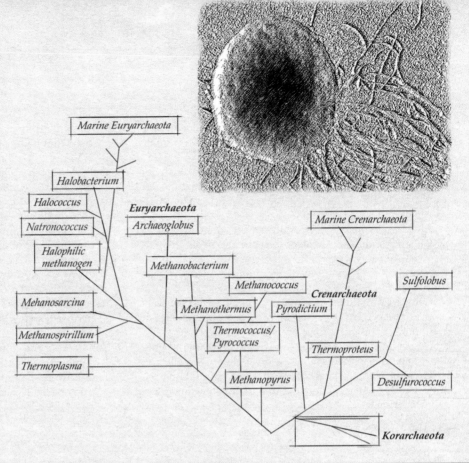

CHAPTER **14**   **Prokaryotic Diversity: The Archaea**

14.1   Phylogenetic Overview of the Archaea   546

Kingdom Euryarchaeota

14.2   Extremely Halophilic Archaea   548
14.3   Methane-Producing Archaea: Methanogens   553
14.4   Thermoplasmatales: *Thermoplasma* and *Picrophilus*   556
14.5   Hyperthermophilic Euryarchaeota: Thermococcales and *Methanopyrus*   558
14.6   Hyperthermophilic Euryarchaeota: The Archaeoglobales   560

Kingdom Crenarchaeota

14.7   Habitats and Energy Metabolism of Crenarchaeotes   561
14.8   Hyperthermophiles from Terrestrial Volcanic Habitats: Sulfolobales and Thermoproteales   564
14.9   Hyperthermophiles from Submarine Volcanic Habitats: Desulfurococcales   566

Evolution and Life at High Temperatures

14.10   Heat Stability of Biomolecules   568
14.11   Hyperthermophilic Archaea and Microbial Evolution   571
14.12   Korarchaeota, Hydrogen, and Microbial Life on Other Planets   571

WORKING GLOSSARY

Acetotrophic   acetate-consuming. When used to describe a methanogen, an organism capable of splitting acetate into CH_4 and CO_2

Acetyl-CoA (Ljungdahl-Wood) pathway   a pathway of autotrophic CO_2 fixation widespread in obligate anaerobes including methanogens, homoacetogens, and sulfate-reducing bacteria

Archaeon   a member of the phylogenetic domain Archaea

Bacteriorhodopsin   a membrane protein containing retinal, a pigment produced by certain extreme halophiles and capable of light-mediated proton motive force formation

Compatible solutes   organic or inorganic substances accumulated in the cytoplasm of halophilic organisms in order to maintain ionic pressure

Crenarchaeota   a kingdom of Archaea that contains both hyperthermophilic and cold-dwelling prokaryotes

Euryarchaeota   a kingdom of Archaea that contains primarily methanogens, the extreme halophiles, and *Thermoplasma*

Extreme halophile   an organism whose growth is dependent on large concentrations (generally >10%) of NaCl

Halorhodopsin   a light-driven chloride pump that accumulates Cl^- within the cytoplasm

Hyperthermophile   a prokaryote with a growth temperature optimum of 80°C or greater

Korarchaeota   a kingdom of hyperthermophilic Archaea that branches close to the archaeal root

Methanogen   a methane-producing prokaryote

Phytanyl   a branched-chain hydrocarbon containing 20 carbon atoms and commonly found in the lipids of Archaea

Reverse DNA gyrase   a protein universally present in hyperthermophiles that introduces positive supercoils into circular DNA

Solfatara   a hot, sulfur-rich, generally acidic environment commonly inhabited by hyperthermophilic Archaea

Thermosome   a type of heat-shock chaperonin that refolds partially heat-denatured proteins in hyperthermophiles

We now consider the domain Archaea. In Chapter 12 we emphasized the profound phylogenetic as well as phenotypic differences between Bacteria and Archaea. Here we examine the organisms themselves. As Archaea, these organisms all lack peptidoglycan in their cell walls, contain ether-linked lipids, have complex RNA polymerases, and show the other major properties of this domain as was described in Table 12.3. But as we will see, despite these unifying properties, Archaea show enormous phenotypic diversity, which we will discuss here. As in Chapter 13 on Bacteria, we begin with a phylogenetic overview to show the evolutionary relationships within the domain Archaea.

14.1

Phylogenetic Overview of the Archaea

A detailed phylogenetic tree of Archaea is shown in Figure 14.1. The tree bifurcates into two major sublineages or kingdoms, called the **Crenarchaeota** and the **Euryarchaeota.** A third kingdom, the **Korarchaeota**, branches off close to the root (Figure 14.1). Among cultured representatives, the Crenarchaeota contain mostly hyperthermophilic species including those able to grow at the highest temperatures of all known organisms. Many hyperthermophiles are chemolithotrophic autotrophs; and because their habitats are devoid of photosynthetic life, these organisms are thus the only primary producers in these harsh environments.

Hyperthermophilic crenarchaeotes tend to cluster closely together and occupy rather short branches on the 16S ribosomal RNA-based tree of life (∞ Figures 12.13 and 14.1). This suggests that these organisms have "slow evolutionary clocks" and have evolved the least away from the hypothetical universal ancestor of all life. Such organisms should thus be good models for study of early life on Earth, and we return to this theme at the end of this chapter. By contrast, cold-dwelling relatives of hyperthermophilic crenarchaeotes have been identified by community sampling (∞ Section 12.6) of ocean waters, and from a phylogenetic perspective these are more rapidly evolving and thus occupy longer branches of the tree (Figure 14.1). We consider the Crenarchaeota in more detail in Sections 14.7–14.9.

Euryarchaeota comprise a physiologically diverse group of Archaea, many of which, like crenarchaeotes, inhabit extreme environments of one sort or the other. Here we see methanogenic Archaea—prokaryotes whose metabolism is linked to the production of methane (CH_4)—related to several genera of extremely halophilic prokaryotes, the "halobacteria" (Figure 14.1). As a study in physiological contrasts, methanogens are obligate anaerobes, arguably *the* most strict of all anaerobes, while extreme halophiles are for the most part obligate aerobes. Other groups of euryarchaeotes include the hyperthermophiles *Thermococcus* and *Pyrococcus* and the methanogen *Methanopyrus*, that branch near to the root (Figure 14.1), and the cell wall–less prokaryote *Thermoplasma*, an organism similar in many respects to the

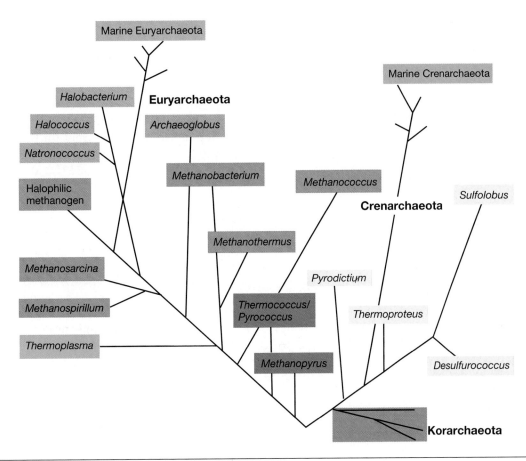

FIGURE 14.1 Detailed phylogenetic tree of the Archaea based on 16S ribosomal RNA sequence comparisons. See note in Figure 13.1 for comparisons of detailed tree with the universal tree. The marine Euryarchaeota and marine Crenarchaeota are thus far known only from community sampling (∞ Section 12.6).

mycoplasmas (∞ Section 13.20). Finally, there exists a large group of as yet uncultured marine euryarchaeotes that are positioned at the ends of long branches up near the top of the phylogenetic tree (Figure 14.1). We consider Euryarchaeota in more detail in Sections 14.2–14.6.

The Korarchaeota (Figure 14.1) were originally discovered by gene sampling (∞ Section 12.6) of an unusual Yellowstone hot spring, but now they exist in laboratory culture (see Section 14.12 and Figure 14.22). The Korarchaeota branch on the archaeal tree close to the root, and for this reason their biological properties may reveal interesting features of ancient organisms (see Section 14.12).

With this overview to the phylogeny of the Archaea, let us proceed to a brief description of the metabolic features of Archaea and then to a description of the major properties of cultured representatives.

Energy Metabolism and Metabolic Pathways in Archaea

Several Archaea are chemoorganotrophic and thus use organic compounds as energy sources for growth. Ca-

tabolism of glucose in Archaea proceeds via slight modifications of the Entner–Doudoroff (E–D) pathway (∞ Section 13.10 for a description of this pathway in Bacteria) or glycolytic pathway (∞ Section 4.9). Oxidation of acetate to CO_2 in Archaea proceeds through the citric acid cycle (∞ Section 4.12) or some slight variation of this reaction series, or by the acetyl-CoA (Ljungdahl–Wood) pathway (∞ Section 15.18). Little is known concerning biosynthesis of amino acids and other macromolecular precursors in Archaea, but presumably key monomers are produced from the central biosynthetic intermediates discussed previously for Bacteria (∞ Section 4.15). Electron transport chains including cytochromes of the a, b, and c types exist in some Archaea. Employing these and other electron carriers, chemoorganotrophic metabolism in most Archaea probably proceeds by introduction of electrons from organic electron donors into an electron transport chain, leading to the reduction of O_2, S^0, or some other electron acceptor, with concurrent establishment of a proton motive force that drives adenosine triphosphate (ATP) synthesis through membrane-bound ATPases (∞ Section 4.11 for a description of ATPase function).

Chemolithotrophy is also well established in the Archaea, with H_2 being a common electron donor.

Autotrophy is widespread among Archaea and occurs by several different means. In methanogens, and presumably in most chemolithotrophic hyperthermophiles, CO_2 is incorporated via the acetyl-CoA pathway or some modification thereof (∞ Section 15.18). In other hyperthermophiles CO_2 fixation occurs via the reverse citric acid cycle, a reaction series that functions as the autotrophic pathway in the green sulfur bacteria (∞ Sections 13.29 and 15.8), or via the Calvin cycle, the most widespread autotrophic pathway in Bacteria and eukaryotes (∞ Section 15.7). In this connection, genes encoding functional and very thermostable RubisCO enzymes (RubisCO catalyzes the first step in the Calvin cycle, ∞ Section 15.7) have been characterized from the methanogen *Methanococcus jannaschii* and from a *Pyrococcus* species, both hyperthermophiles.

We thus see that the major catabolic and anabolic sequences in the Archaea are familiar ones from our study of these processes in various Bacteria. Methanogenesis is a major exception and will be covered in detail in Chapter 15. With this as background we now begin our study of the basic biology of the major groups of Archaea.

Kingdom Euryarchaeota

14.2

Extremely Halophilic Archaea

Key Genera

Halobacterium
Haloferax
Natronobacterium

Extremely halophilic Archaea are a diverse group of prokaryotes that inhabit highly saline environments, such as solar salt evaporation ponds and natural salt lakes, or artificial saline habitats, such as the surfaces of heavily salted foods like certain fish and meats. Such habitats are often called *hypersaline*. The term *extreme* halophile is used to indicate not only that these organisms are halophilic, but also that their requirement for salt is *very high*, in some cases near saturation. An extreme halophile is an organism that requires at least 1.5 M (about 9%) NaCl for growth; most species require 2–4 M NaCl (12–23%) for optimal growth. Virtually all extreme halophiles can grow at 5.5 M NaCl (32%, the limit of saturation for NaCl), although some species grow only very slowly at this salinity.

Hypersaline Environments

Hypersaline habitats are common throughout the world, but *extremely* hypersaline habitats are rather rare. Most such environments are in hot, dry areas of the world. Salt lakes can vary considerably in ionic composition. The predominant ions in a hypersaline lake depend to a major extent on the surrounding topography, geology, and general climatic conditions. Great Salt Lake in Utah (USA) (Figure 14.2a), for example, is essentially concentrated seawater because the relative proportions of the various ions are those of seawater, although the overall concentration of ions is much higher. Sodium is the predominant cation in Great Salt Lake, whereas chloride is the predominant anion; significant levels of sulfate are also present at a slightly alkaline pH (Table 14.1). By contrast, another hypersaline basin, the Dead Sea, is relatively low in sodium but contains high levels of magnesium because of the abundance of magnesium minerals in the surrounding rocks (Table 14.1). The water chemistry of soda lakes resembles that of hypersaline lakes such as Great Salt Lake, but because high levels of *carbonate* minerals are present in the surrounding rocks, the pH of soda lakes is quite high; pH values of 10–12 are not uncommon in these environments (Table 14.1 and Figure 14.2c).

Despite what may seem like rather harsh conditions, salt lakes can be highly productive ecosystems. Archaea are not the only microorganisms found. The eukaryotic alga *Dunaliella* is the major, if not sole, oxygenic phototroph in most salt lakes. In highly alkaline soda lakes where *Dunaliella* is absent, anoxygenic phototrophic purple bacteria of the genus *Ectothiorhodospira* and *Halorhodospira* (∞ Section 13.1) predominate. Organic matter originating from primary production by oxygenic or anoxygenic phototrophs then sets the stage for development of the extremely halophilic Archaea, all of which are chemoorganotrophic. In addition, a few extremely halophilic anaerobic chemoorganotrophic Bacteria, such as *Haloanaerobium* and *Halobacteroides*, thrive in such environments.

Marine salterns are also habitats for extremely halophilic prokaryotes. Marine salterns are small enclosed basins filled with seawater that are left to evaporate, yielding NaCl and other salts of commercial value (Figure 14.2b). As salterns approach the minimum salinity limits for extreme halophiles, the waters turn a reddish purple color due to the massive growth (called a *bloom*) of halophilic Archaea (the red coloration apparent in Figures 14.2b and c comes from carotenoids and other pigments discussed later). Extreme halophiles have also been found in high salt foods such as certain sausages, marine fish, and salt pork.

(a)   (b)

(c)

FIGURE 14.2   Hypersaline habitats. (a) Great Salt Lake, Utah, a hypersaline lake in which the ratio of ions is similar to that in seawater but in which absolute concentrations of ions are about 10 times that of seawater. The green color is primarily from cells of the halophilic green alga, *Dunaliella salina*. (b) Aerial view near San Francisco Bay, California, of a series of seawater evaporating ponds where solar salt is prepared. The red-purple color is predominantly due to bacterioruberins and bacteriorhodopsin in cells of *Halobacterium*. (c) A soda lake, Lake Magadi, Kenyan section of African Rift Valley. The pink color is from haloalkaliphilic Archaea that thrive in this hypersaline and highly alkaline (pH 11) lake.

TABLE 14.1	Ionic composition of some highly saline environments			
	Concentration (g/l)			
Ion	Great Salt Lake[a]	Dead Sea	Typical soda lake[b]	Seawater (for comparison)
Na^+	105	40.1	142	10.6
K^+	6.7	7.7	2.3	0.38
Mg^{2+}	11	44	< 0.1	1.27
Ca^{2+}	0.3	17.2	< 0.1	0.40
Cl^-	181	225	155	18.9
Br^-	0.2	5.3	—	0.065
SO_4^{2-}	27	0.5	23	2.65
HCO_3^- or CO_3^{2-}	0.7	0.2	67	0.14
pH	7.7	6.1	11	8.1

a See Figure 14.2*a*

b See Figure 14.2*c*

Taxonomy and Physiology of Extremely Halophilic Archaea

Table 14.2 lists the currently recognized species of extremely halophilic Archaea. 16S ribosomal ribonucleic acid (rRNA) sequencing and other criteria have defined ten genera of extreme halophiles (Table 14.2). The extremely halophilic Archaea are frequently referred to collectively as "halobacteria," because the genus *Halobacterium* was the first in this group to be described and is still the best-studied representative of the group. *Natronobacterium, Natronosomonas,* and *Natronococcus* differ from other extreme halophiles in being extremely *alkaliphilic* as well as halophilic. As befits their soda

TABLE 14.2	Taxonomy of some extremely halophilic Archaea		
Genus	**Morphology**	**DNA (mol % GC)**	**Habitat and comments**
Halobacterium	Rods		
H. salinarum		66–71	Isolated from salted fish, hides, hypersaline lakes; related organisms, probably all the same species are *H. halobium* and *H. cutirubrum*
Halorubrum	Rods		
H. sodomense		68	Dead Sea; requires high Mg^{2+}
H. saccharovorum		71	Salterns; uses sugars
H. lacusprofundi	Pleomorphic rods	65–66	From an Antarctic lake; grows at 4°C
H. trapanicum		64	From salt works, Trapani, Italy
H. vacuolatum		63	Alkaliphilic; contains gas vesicles
Halobaculum	Rods		
H. gomorrense		70	Dead Sea; requires high Mg^{2+}, optimal NaCl about 1.5 M
Haloferax	Flattened disc or		
H. volcanii	cup-shaped	63–66	Dead Sea; requires high Mg^{2+}
H. mediterranei		60–62	Salterns; uses starch; contains gas vesicles
H. gibbonsii		62	Marine salterns in Spain
H. denitrificans		64	Baja California saltern; capable of dissimilative nitrate reduction
Haloarcula	Irregular discs,		
H. vallismortis	triangles, rectangles	65	Death Valley, CA
H. hispanica		63	Marine salterns in Spain
Halococcus	Cocci		
H. morrhuae		60–66	Salted fish
H. saccharolyticus		59	Salterns; uses sugars
Natronobacterium	Rods		
N. gregoryi		65	Isolated from highly saline soda lakes; optimum pH for growth, 9.5
Natrialba	Rods		
N. magadii		63	Alkaliphile from Lake Magadi, Kenya (see Figure 14.2c)
Natronosomonas	Rods		
N. pharaonis		64	
Natronococcus	Cocci		
N. occultus		64	Highly saline soda lakes
N. amylolyticus		63	Lake Magadi; pigmented orange-red; uses starch

lake habitat (see Table 14.1 and Figure 14.2c), growth of natronobacteria is optimal at very low Mg^{2+} concentrations and high pH (9–11). Natronobacteria also contain unusual diether lipids not found in other extreme halophiles and cluster tightly as a phylogenetic group. The cell wall of natronobacteria is also unique, being composed of a polymer of the amino acid glutamine from which oligosaccharides containing N-acetylglucosamine, glucose, and other sugars are linked via the amide group of glutamine.

All halophilic Archaea stain gram *negatively*, reproduce by binary fission, and do not form resting stages or spores. Most halobacteria are nonmotile, but a few strains are weakly motile by lophotrichous flagella. The genomic organization of *Halobacterium* and *Halococcus* is highly unusual in that large plasmids containing up to 25–30% of the total cellular DNA are frequently present and the GC base ratio of these plasmids (57–60% GC) is significantly different from that of chromosomal DNA (66–68% GC). Plasmids from extreme halophiles are among the largest naturally occurring plasmids known.

All extremely halophilic Archaea are chemoorganotrophs, and most species are obligate *aerobes*. Most halobacteria use amino acids or organic acids as energy sources and require a number of growth factors (mainly vitamins) for optimal growth. A few *Halobacterium* species oxidize carbohydrates, but this ability is relatively rare. Electron transport chains containing cytochromes *a*, *b*, and *c* are present in *Halobacterium*, and energy is conserved during aerobic growth via a proton motive force arising from membrane-mediated chemiosmotic events. Some strains of *Halobacterium* have been shown to grow anaerobically. Anoxic growth at the expense of sugar fermentation and by anaerobic respiration (∞ Section 15.16) linked to the reduction of nitrate or fumarate has been demonstrated in certain strains.

Water Balance in Extreme Halophiles

Extremely halophilic Archaea require large amounts of sodium for growth. In the case of *Halobacterium*, where detailed salinity studies have been performed, the requirement for Na^+ cannot be satisfied by replacement with another ion, even with the chemically related ion K^+. We learned in Section 5.11 that certain microorganisms can withstand the osmotic forces that accompany life in a high solute environment by accumulating or synthesizing organic compounds *intracellularly;* the latter compounds are referred to as **compatible solutes.** These compounds counteract the tendency of the cell to become dehydrated under conditions of high osmotic strength by placing the cell in positive water balance with its surroundings (Section 5.11). Ironically, however, although *Halobacterium* thrives only in an osmotically stressful environment, it produces no *organic* compatible solute. Instead, cells of *Halobacterium* pump large amounts of K^+ from the environment into the cytoplasm such that the concentration of K^+ *inside* the cell is considerably greater than the concentration of Na^+ *outside* the cell; thus, *Halobacterium* employs an *inorganic ion* as its compatible solute and in this way remains in positive water balance (Table 14.3).

The cell wall of a *Halobacterium* cell is composed of glycoprotein and stabilized by sodium ions. In electron micrographs of thin sections of *Halobacterium* (Figure 14.3), the organism appears similar to gram-negative Bacteria (compare Figure 14.3 with Figure 3.29), but Na^+ binds to the outer surface of the *Halobacterium* wall and is absolutely essential for maintaining cellular integrity; when insufficient Na^+ is present, the cell wall breaks apart and the cell lyses. The glycoprotein of the *Halobacterium* cell wall has an exceptionally high content of the *acidic* (negatively charged) amino acids aspartate and glutamate. The negative charges contributed by the carboxyl groups of these amino acids are shielded by Na^+; when Na^+ is diluted away, the negatively charged parts of the proteins actively repel each other, leading to cell lysis.

Cytoplasmic proteins of *Halobacterium* are also highly acidic, but it is K^+, not Na^+, that is required for activity. This, of course, is not surprising since K^+ is the predominate internal cation in cells of *Halobacterium* (Table 14.3). Besides a high acidic amino acid composition, halobacterial cytoplasmic proteins typically contain very low levels of *hydrophobic* amino acids. This phenomenon probably represents an evolutionary adaptation to the highly ionic cytoplasm of *Halobacterium;* in such an envi-

ronment highly polar proteins tend to remain in solution, whereas nonpolar molecules would tend to cluster and perhaps lose activity. The ribosomes of *Halobacterium* also require high K^+ levels for stability (ribosomes of nonhalophiles have no K^+ requirement). It thus appears that the extremely halophilic Archaea are highly adapted, both internally and externally, to life in a highly ionic environment. Cellular components exposed to the external environment require high Na^+ for stability, whereas internal components require high K^+. With the exception of a few extremely halophilic members of the Bacteria that also use K^+ as a compatible solute, in no other group of prokaryotes do we find this unique requirement for specific cations in such high amounts.

Bacteriorhodopsin and Light-Mediated ATP Synthesis

Certain species of extremely halophilic Archaea are capable of a *light-mediated* synthesis of ATP that does not involve chlorophyll pigments (thus it is *not* photosynthesis). We have seen the highly pigmented nature of extremely halophilic Archaea in Figure 14.2. Pigmentation is due to red- and orange-colored carotenoids, primarily C_{50} carotenoids called *bacterioruberins,* and also to inducible pigments involved in energy metabolism. Although lacking chlorophylls or bacteriochlorophylls, under conditions of low aeration, *Halobacterium salinarum* and certain other extreme halophiles synthesize and insert a protein called **bacteriorhodopsin** into their membranes. Bacteriorhodopsin was named because of its structural and functional similarity to the visual pig-

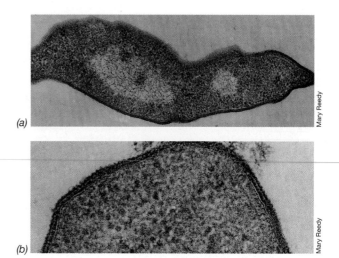

FIGURE 14.3 Electron micrographs of thin sections of the extreme halophile *Halobacterium salinarum*. A cell is about 0.8 μm in diameter. (a) Longitudinal section. (b) High magnification electron micrograph showing the regular structure of the cell wall.

TABLE 14.3	Concentration of ions in cells of *Halobacterium salinarum*	
Ion	Concentration in medium (*M*)	Concentration in cells (*M*)
Na^+	3.3	0.8
K^+	0.05	5.3
Mg^{2+}	0.13	0.12
Cl^-	3.3	3.3

ment of the eye, *rhodopsin*. Conjugated to bacteriorhodopsin is a molecule of *retinal*, a carotenoid-like molecule that can absorb light and catalyze formation of a proton motive force (Figure 14.4). Because of its retinal content, bacteriorhodopsin is purple, and cells of *Halobacterium* switched from growth under conditions of high aeration to oxygen-limiting conditions (a trigger of bacteriorhodopsin synthesis) gradually change from an orange or red color to a more reddish-purple because of the insertion of bacteriorhodopsin into the cytoplasmic membrane. Isolated purple membranes of *Halobacterium* contain about 25% lipid and 75% protein, and the bacteriorhodopsin is inserted at random on the surface of the cytoplasmic membrane; apparently bacteriorhodopsin is inserted into preexisting cytoplasmic membranes during the switch to low O_2.

Bacteriorhodopsin absorbs light strongly in the green region of the spectrum at about 570 nm. The retinal chromophore of bacteriorhodopsin, which normally exists in an *all-trans* configuration, becomes excited and converted to the *cis* form following the absorption of light (Figure 14.4). This transformation results in the transfer of protons to the *outside* surface of the membrane. The retinal molecule then relaxes and returns to its more stable all-trans isomer in the dark following the uptake of a proton from the cytoplasm, thus completing the cycle (Figure 14.4). As protons accumulate on the outer surface of the membrane, the proton motive force (∞ Section 4.11) increases until the membrane is sufficiently "charged" to drive ATP synthesis through action of a proton translocating ATPase (Figure 14.4).

Light-mediated ATP production in *Halobacterium salinarum* has been shown to support slow growth of this organism anaerobically under nutritional conditions in which other energy-generating reactions do not occur, and light has been shown to maintain the viability of anoxic cultures of *Halobacterium* incubated in the absence of organic energy sources. The light-stimulated proton pump of *H. salinarum* also functions to pump Na^+ out of the cell by action of a Na^+/H^+ antiport system (∞ Section 3.6) and to drive the uptake of a variety of nutrients, including the K^+ needed for osmotic balance. The uptake of amino acids by *H. salinarum* has been shown to be indirectly driven by light because the transport of amino acids occurs with Na^+ uptake by an amino acid–Na^+ symporter (∞ Section 3.6). Continued uptake depends on the removal of Na^+ via the (light-driven) Na^+/H^+ antiporter.

A separate light-driven pump called **halorhodopsin** is present in halobacteria to pump Cl^- into the cell as an anion for K^+. Like bacteriorhodopsin, halorhodopsin also contains retinal, and Cl^- binds to this retinal (instead of H^+ as in bacteriorhodopsin, see Figure 14.4) and is transported across the membrane from outside to inside.

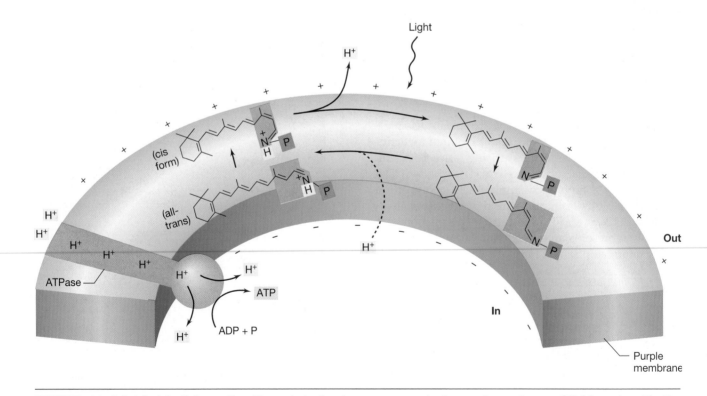

FIGURE 14.4 Model of the light-mediated bacteriorhodopsin proton pump in the purple membrane of *Halobacterium*. The P stands for the protein (bacteriorhodopsin) to which the chromophore retinal is attached. "Out" and "In" designate opposite sides of the cytoplasmic membrane.

14.3

Methane-Producing Archaea: Methanogens

Key Genera

Methanobacterium
Methanococcus
Methanosarcina

A large number of Euryarchaeota produce methane (CH_4) as an integral part of their energy metabolism.

FIGURE 14.5 The "Volta Experiment" as demonstrated in the Microbial Diversity Summer Course in Woods Hole, Massachusetts. A large inverted funnel was placed over freshwater sediments in which anaerobic decomposition was occurring in Cedar Swamp, Woods Hole. After water displaced the air in the funnel, the funnel was capped and the sediments were mixed with a stick, allowing trapped bubbles of methane to collect in the inverted funnel. Immediately after uncapping the funnel, a flame was held near the funnel port, thus igniting the methane. This experiment was performed over 200 years ago by the Italian physicist Alessandro Volta, which led him to describe methane as "combustible air."

Such organisms are called *methanogens* and the process of methane formation *methanogenesis.* Methane was first discovered as a type of "combustible air" by the Italian physicist Alessandro Volta, who collected gas from marsh sediments and showed that it was flammable (the "Volta experiment" can be easily reproduced if methane, trapped in freshwater sediments, is collected and *carefully* ignited) (Figure 14.5). In later chapters we will study the biochemically unique and amazingly complex process of methane formation (∞ Section 15.19) and then learn how methanogenesis is the terminal step in the biodegradation of organic matter in many anoxic habitats in nature such as the swamp shown in Figure 14.5 (∞ Section 16.14). The major sources of biogenic methane are listed in Table 14.4.

Diversity and Physiology of Methanogens

A wide variety of morphological types of methanogens have been described (Figure 14.6 and Table 14.5). Their taxonomy is based on both phenotypic as well as phylogenetic (comparative 16S rRNA sequencing) analyses (Table 14.5), with several orders being recognized (in taxonomy, an order contains groups of related families, ∞ Section 12.10). Structurally, methanogens are prokaryotic cells that show a diversity of cell wall chemistries. The latter include the pseudomurein (∞ Section 3.7 and Figure 3.35b) walls of *Methanobacterium* species and relatives (Figure 14.7a), the methanochondroitin (so named because of its structural resemblance to chondroitin sulfate, the connective tissue polymer of vertebrate animals) walls of *Methanosarcina* and relatives (Figure 14.7b), and the protein or glycoprotein walls of *Methanococcus* (see Figure 14.8a) and *Methanoplanus* species, respectively.

TABLE 14.4	Habitats of Methanogens

I. Anoxic sediments: marsh, swamp (see Figure 14.5), and lake sediments, paddy fields, moist landfills (∞ Section 16.14)
II. Animal digestive tracts:
  (a) Rumen of ruminant animals such as cattle, sheep, elk, deer, and camels (∞ Section 16.15)
  (b) Cecum of cecal animals such as horses and rabbits (∞ Section 16.15)
  (c) Large intestine of monogastric animals such as humans, swine, and dogs (∞ Section 19.4)
  (d) Hindgut of cellulolytic insects (for example, termites)
III. Geothermal sources of H_2 + CO_2: hydrothermal vents (∞ Section 16.12)
IV. Artificial biodegradation facilities: sewage sludge digestors (∞ Section 11.14)
V. Endosymbionts of various anaerobic protozoa (∞ Figure 16.39)

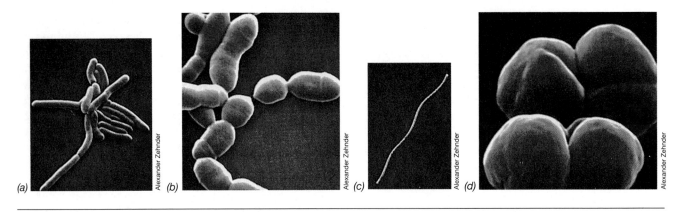

FIGURE 14.6   Scanning electron micrographs of cells of methanogenic Archaea, showing the considerable morphological diversity. (a) *Methanobrevibacter ruminantium.* A cell is about 0.7 μm in diameter. (b) *Methanobacterium* strain AZ. A cell is about 1 μm in diameter. (c) *Methanospirillum hungatii.* A cell is about 0.4 μm in diameter. (d) *Methanosarcina barkeri.* A cell is about 1.7 μm wide.

Physiologically, methanogens are anaerobes, and strict anoxic techniques are necessary to culture them. Depending on the species, laboratory cultures of methanogens can be established in a mineral salts medium under an atmosphere of H_2 plus CO_2 (in the ratio of 4:1, see the reaction shown below), or in complex media. Most known methanogens are mesophilic although "extremophilic" species growing optimally at very high (see Figures 14.8 and 14.13) or very low temperature or at very high salt concentration, have also been described.

Substrates for Methanogenesis

At least 10 substrates have been shown to be converted to methane by pure cultures of methanogens

Genus	Morphology	Number of species	Substrates for methanogenesis	DNA (mol % GC)
TABLE 14.5	**Characteristics of methanogenic Archaea**			
Methanobacteriales				
Methanobacterium	Long rods	19	$H_2 + CO_2$, formate	29–61
Methanobrevibacter	Short rods	7	$H_2 + CO_2$, formate	27–31
Methanosphaera	Cocci	2	Methanol + H_2 (both needed)	26
Methanothermus	Rods	2	$H_2 + CO_2$; can also reduce S^0; hyperthermophile	33
Methanococcales				
Methanococcus	Irregular cocci	11	$H_2 + CO_2$, pyruvate + CO_2, formate	29–34
Methanomicrobiales				
Methanomicrobium	Short rods	2	$H_2 + CO_2$, formate	45–49
Methanogenium	Irregular cocci	11	$H_2 + CO_2$, formate	51–61
Methanospirillum	Spirilla	1	$H_2 + CO_2$, formate	46–50
Methanoplanus	Plate-shaped cells—occurring as thin plates with sharp edges	3	$H_2 + CO_2$, formate	38–47
Methanocorpusculum	Irregular cocci	5	$H_2 + CO_2$, formate, alcohols	48–52
Methanoculleus	Irregular cocci	6	$H_2 + CO_2$, alcohols, formate	54–62
Methanosarcinales				
Methanosarcina	Large irregular cocci in packets	8	$H_2 + CO_2$, methanol, methylamines, acetate	41–43
Methanolobus	Irregular cocci in aggregates	5	Methanol, methylamines	38–42
Methanohalobium	Irregular cocci	1	Methanol, methylamines; halophilic	44
Methanococcoides	Irregular cocci	2	Methanol, methylamines	42
Methanohalophilus	Irregular cocci	3	Methanol, methylamines, methyl sulfides; halophile	41
Methanothrix (Methanosaeta)	Long rods to filaments	4	Acetate	52–61
Methanopyrales				
Methanopyrus	Rods in chains	1	$H_2 + CO_2$; hyperthermophile, growth at 110°C	60

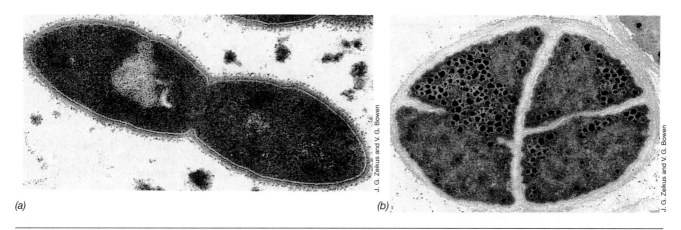

FIGURE 14.7 Transmission electron micrographs of thin sections of methanogenic Archaea. (a) *Methanobrevibacter ruminantium*. A cell is 0.7 μm in diameter. (b) *Methanosarcina barkeri*, showing the thick cell wall and the manner of cell segmentation and cross-wall formation. A cell is 1.7 μm in diameter.

(Table 14.6). Interestingly, these substrates do *not* include such common compounds as glucose and organic or fatty acids (other than acetate). However, this is not to say that a compound like glucose can never be converted to methane. We will discuss in Chapter 16 how, in cooperative reactions involving methanogens and other anaerobic bacteria, virtually any organic compound can be converted to methane and CO_2 (∞ Section 16.14).

Three classes of compounds make up the list of 10 methanogenic substrates shown in Table 14.6; these include *CO₂-type substrates*, *methyl substrates*, and *acetate*. The first class includes the important substrate CO_2, which is reduced to methane using H_2 as electron donor:

$$CO_2 + 4H_2 \rightarrow CH_4 + 2H_2O \ \Delta G^{0\prime} = -131 \text{ kJ}$$

Other substrates here include formate (which is simply $CO_2 + H_2$ in combined form) and CO, carbon monoxide.

The second class of methanogenic substrates are methyl group substances (Table 14.6). Using CH_3OH as a model methyl substrate here, the formation of CH_4 can occur in either of two ways. First, CH_3OH can be reduced using an external electron donor such as H_2:

$$CH_3OH + H_2 \rightarrow CH_4 + H_2O \ \Delta G^{0\prime} = -113 \text{ kJ}$$

Alternatively, in the absence of H_2, some CH_3OH can be oxidized to CO_2 in order to generate the electrons needed to reduce other molecules of CH_3OH to CH_4:

$$4CH_3OH \rightarrow 3CH_4 + CO_2 + 2H_2O \ \Delta G^{0\prime} = -319 \text{ kJ}$$

The final methanogenic process is the cleavage of acetate to CO_2 plus CH_4, called the *acetotrophic* reaction:

$$CH_3COO^- + H_2O \rightarrow CH_4 + HCO_3^- \ \Delta G^{0\prime} = -31 \text{ kJ}$$

Only a very few methanogens are acetotrophic (Table 14.5), yet experimental measurements of methane formation in methanogenic habitats such as sewage sludge have shown that about two-thirds of the methane formed there originates from acetate and one-third from $H_2 + CO_2$; thus, although lacking in terms of known diversity, acetotrophic methanogens are apparently very ecologically significant in nature.

As can be seen by inspection of each of the above reactions, they are all exergonic and can thus be used to synthesize ATP. We reserve until Chapter 15 the biochemical details of methanogenesis and only note here that the process is coupled to proton motive force formation and ATP synthesis by normal chemiosmotic (∞ Section 4.11) mechanisms. Concerning carbon for cellular biosynthesis, when growing on $CO_2 + H_2$, a methanogen is an autotroph, that is, CO_2 is the precursor for all cellular components. When methylated compounds or acetate are methanogenic substrates, these compounds are also incorporated into organic cell components, usually with the fixation of some CO_2. We discuss the carbon and energy metabolism of methanogens in more detail in Section 15.19.

TABLE 14.6	Substrates converted to methane by various methanogenic Archaea

I. CO₂-type substrates
Carbon dioxide, CO_2 (with electrons derived from H_2, certain alcohols, or pyruvate)
Formate, $HCOO^-$
Carbon monoxide, CO

II. Methyl substrates
Methanol, CH_3OH
Methylamine, $CH_3NH_3^+$
Dimethylamine, $(CH_3)_2NH_2^+$
Trimethylamine, $(CH_3)_3NH^+$
Methylmercaptan, CH_3SH
Dimethylsulfide, $(CH_3)_2S$

III. Acetotrophic substrate
Acetate, CH_3COO^-

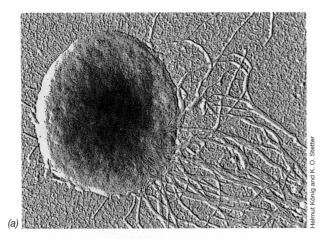

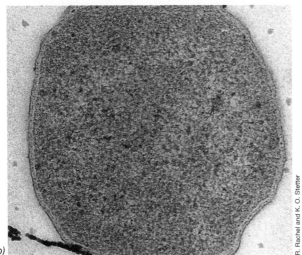

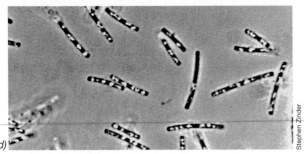

FIGURE 14.8 Hyperthermophilic and thermophilic methanogens. (a) *Methanococcus jannaschii* (temperature optimum, 85°C), shadowed preparation electron micrograph. A cell is about 1 μm in diameter. (b) *Methanococcus igneus* (temperature optimum, 88°C), thin section. A cell is about 1 μm in diameter. (c) *Methanothermus fervidus* (temperature optimum, 88°C), thin-sectioned electron micrograph. A cell is about 0.4 μm in diameter. (d) *Methanothrix (Methanosaeta) thermophila* (temperature optimum, 60°C), phase contrast micrograph. A cell is about 1 μm in diameter. The refractile bodies inside the cells are gas vesicles.

Methanococcus jannaschii as a Model Methanogen/Archaeon

As was discussed in Section 9.12, the complete genome sequence of the hyperthermophilic methanogen *Methanococcus jannaschii* (Figure 14.8*a*) has been determined. The 1.66-megabase circular genome of *M. jannaschii* contains about 1700 genes, and from sequence analyses genes encoding enzymes of methanogenesis and several other key cell functions have been identified. Interestingly, the majority of *M. jannaschii* genes encoding functions like central metabolic pathways and cell division are similar to those in Bacteria, while most of the genes encoding core molecular processes like transcription and translation more closely resemble those of eukaryotes. These findings support the evolutionary tree of life that shows the domain Archaea positioned *between* the domains Bacteria and Eukarya (∞ Section 12.7 and Figure 12.13). However, sequence analyses also showed that over 50% of *M. jannaschii* genes have no counterparts in known genes from Bacteria *or* Eukarya, suggesting that there may be many new cellular functions encoded in archaeal DNA that have yet to be discovered.

✓ 14.2–14.3 Concept Check

Extremely halophilic Archaea require high levels of NaCl for growth and store large amounts of K⁺ intracellularly as a compatible solute. Some species can generate ATP from light using a simple proton pump involving bacteriorhodopsin. Methanogenic Archaea are strictly anaerobic prokaryotes whose metabolism is tied to the production of methane (CH_4).

- ✓ How can organisms like *Halobacterium* survive in high-salt solutions whereas an organism like *Escherichia coli* cannot?
- ✓ How does bacteriorhodopsin allow for the synthesis of ATP?
- ✓ What are the major substrates for methanogenesis?

14.4

Thermoplasmatales: *Thermoplasma* and *Picrophilus*

Key Genus

Thermoplasma

A phylogenetically distinct line of Archaea contains two thermophilic and extremely acidophilic prokaryotes, *Thermoplasma* and *Picrophilus* (Figure 14.1). These organisms are among the most acidophilic of all known prokaryotes and, in the case of *Picrophilus*, even capable of growth below pH 0! They also form their own *order* of prokaryotes within the Euryarchaeota, the Thermoplasmatales. We begin with a description of the mycoplasma-like organism, *Thermoplasma*.

Thermoplasma: A Cell Wall–less Archaeon

Thermoplasma (Figure 14.9) is a chemoorganotroph and grows optimally at 55°C and pH 2 in complex media; two species have been described, *T. acidophilum* and *T. volcanium*. Species of *Thermoplasma* are facultative aerobes, growing either aerobically or anaerobically by sulfur respiration. From a morphological point of view *Thermoplasma* resembles the mycoplasmas (∞ Section 13.20) in that it lacks a cell wall. Most strains

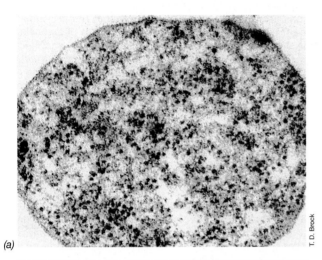

(a)

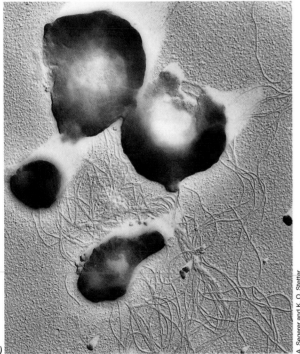

(b)

FIGURE 14.9 *Thermoplasma* species. (a) *Thermoplasma acidophilum*, an acidophilic, thermophilic mycoplasma-like archaeon. Electron micrograph of thin section. The diameter of cells is highly variable from 0.2 to 5 μm. The cell shown is about 1 μm in diameter. (b) Shadowed preparation of cells of *Thermoplasma volcanium* isolated from hot springs. Cells are 1–2 μm in diameter. Notice abundant flagella.

of *Thermoplasma* have been obtained from self-heating coal refuse piles (Figure 14.10). Coal refuse contains coal fragments, pyrite, and other organic materials extracted from coal, and when dumped into piles in coal-mining operations, tends to self-heat by spontaneous combustion (Figure 14.10). This sets the stage for growth of *Thermoplasma*, which apparently metabolizes organic compounds leached from the hot coal refuse. A second species of *Thermoplasma*, *T. volcanium* (Figure 14.9b), has been isolated from solfatara fields throughout the world. It is genetically distinct from *T. acidophilum* but resembles the latter species in many phenotypic properties. *Thermoplasma volcanium* is also highly motile, and cells show multiple flagella (Figure 14.9b).

To survive the osmotic stresses of life without a cell wall and to withstand the dual environmental extremes of low pH and high temperature, *Thermoplasma* has evolved a cell membrane of chemically unique structure. The membrane contains a lipopolysaccharide-like material (referred to as *lipoglycan* in mycoplasmas) (∞ Section 13.20) consisting of a *tetraether* lipid with mannose and glucose units (Figure 14.11). This molecule constitutes a major fraction of the total lipid composition of *Thermoplasma*. The membrane also contains glycoproteins but not sterols. Together these and other molecules render the *Thermoplasma* membrane stable to hot acid conditions.

The genome of *Thermoplasma* is of interest. Like other mycoplasmas (∞ Sections 9.12 and 13.20), *Thermoplasma* contains an extremely small genome, and has a GC content of 46% and is surrounded by a highly basic DNA binding protein that organizes the DNA into globular particles resembling the nucleosomes of eukaryotic cells (∞ Section 6.3 discusses the arrangement of DNA in eukaryotes). This protein strongly resembles the basic histone proteins of eukaryotic cells, and comparisons of amino acid sequences between the *Thermoplasma* protein and eukaryotic nuclear histones show significant sequence homology. Similar DNA-binding histonelike proteins have been found in several hyperthermophilic Euryarchaeota (see Section 14.10).

Picrophilus

A phylogenetic relative of *Thermoplasma* is the archaeon *Picrophilus*. Although *Thermoplasma* is an extreme acidophile, *Picrophilus* is even more so, growing optimally at pH 0.7 and capable of growth to as low as pH −0.06! *Picrophilus* differs from *Thermoplasma* in other ways as well, including the fact that it contains a cell wall (an S-layer; ∞ Section 3.13) and has a much lower DNA GC base ratio. Two species of *Picrophilus* have been isolated from acidic Japanese solfataras and like *Thermoplasma*, both grow heterotro-

FIGURE 14.10   A typical self-heating coal refuse pile, habitat of *Thermoplasma*. (a) Spontaneous heat production can ignite nearby vegetation. (b) Photo of a large hot refuse pile.

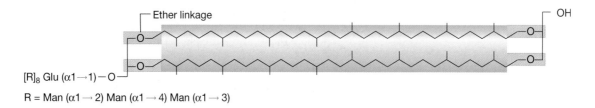

Ether linkage   OH

[R]₈ Glu (α1→1)—O

R = Man (α1 → 2) Man (α1 → 4) Man (α1 → 3)

FIGURE 14.11   Structure of the tetraether lipoglycan of *Thermoplasma acidophilum*. Glu, Glucose; Man, mannose. Note the ether linkages (shown in green) and compare with Figure 3.21*b*.

phically on complex media. The physiology of *Picrophilus* is clearly of interest as a model for acid tolerance, and in this regard, studies of the cytoplasmic membrane of this organism suggest an unusual arrangement of lipids that form an extremely acid impermeable membrane structure at optimal pH values. By contrast, at only moderately acidic pH such as pH 4, the membranes of cells of *Picrophilus* quickly become leaky and literally disintegrate, clearly indicating that this organism has evolved to survive only in highly acidic habitats.

✓ 14.4 Concept Check

Thermoplasma and *Picrophilus* are extremely acidophilic thermophilic archaeans that form their own phylogenetic family of prokaryotes inhabiting coal refuse piles and highly acidic solfataras. Cells of *Thermoplasma* lack a cell wall and thus resemble the mycoplasmas in this regard.

✓ In what ways are *Thermoplasma* and *Picrophilus* similar? In what ways do they differ?

✓ How does *Thermoplasma* strengthen its cell membrane to survive life in the absence of a cell wall?

14.5

Hyperthermophilic Euryarchaeota: Thermococcales and *Methanopyrus*

Key Genera

Thermococcus
Pyrococcus
Methanopyrus

A few euryarchaeotes thrive in thermal environments and some are extremely thermophilic; those with growth temperature optima over 80°C are called *hyperthermophiles*. We consider here three hyperthermophilic euryarchaeotes that branch on the archael tree (Figure 14.1) very near to the root. Two of these organisms, *Thermococcus* and *Pyrococcus*, show phenotypic properties very similar to those of hyperthermophilic crenarchaeotes to be discussed in Sections 14.7–14.9 and form a distinct order: the Thermococcales (see Table 14.9). The other, *Methanopyrus*, is a methanogen and closely resembles the methanogens (see Section 14.4 and Table 14.5) in its basic physiology but is unusual in its extreme requirement for heat and phylogenetic position (Figure 14.1).

Thermococcus and *Pyrococcus*

Thermococcus is a spherical hyperthermophilic euryarchaeote indigenous to anoxic thermal waters in various locations throughout the world. The spherical cells contain a tuft of polar flagella and are thus highly motile (Figure 14.12*a*). *Thermococcus* is an obligately anaerobic chemoorganotroph that grows on proteins and other complex organic mixtures (including some sugars) with S^0 as electron acceptor at temperatures from 70–95°C.

An organism morphologically similar to *Thermococcus* is *Pyrococcus* (Figure 14.12*b*). *Pyrococcus* (the Latin derivation literally means "fireball") differs from *Thermococcus* primarily by its higher temperature requirements; *Pyrococcus* grows between 70 and 106°C with an optimum of 100°C. However, metabolically *Thermococcus* and *Pyrococcus* are quite similar: proteins, starch, or maltose are oxidized as electron donors and S^0 is the terminal acceptor and is reduced to H_2S. More properties of *Thermococcus* and *Pyrococcus* are described in Tables 14.8 and 14.9.

Methanopyrus

Methanopyrus is a rod-shaped hyperthermophilic methanogen (Figure 14.13) that has been isolated from sediments near submarine hydrothermal vents and from the walls of "black smoker" hydrothermal vent chimneys (∞ Section 16.12 for a discussion of thermal vents). *Methanopyrus* occupies a unique phylogenetic position on the archaeal tree (Figure 14.1); it is one of the most ancient (least derived) of all known hyperthermophilic Archaea and shares phenotypic properties with both the hyperthermophiles

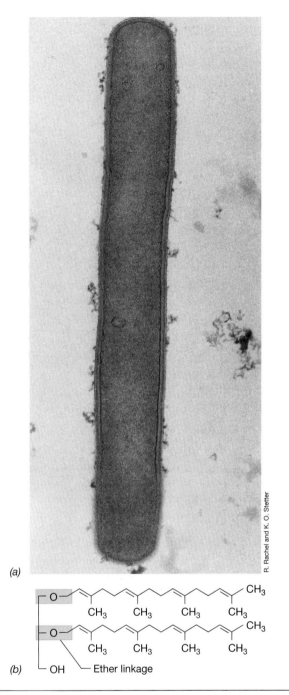

(a)

(b)

└ O ─ ... ─ CH₃
     CH₃ CH₃ CH₃ CH₃
└ O ─ ... ─ CH₃
     CH₃ CH₃ CH₃ CH₃
└ OH   └ Ether linkage

FIGURE 14.13 *Methanopyrus*. (a) Electron micrograph of a cell of *Methanopyrus kandleri*, the most thermophilic of all known methanogens (upper temperature limit, 110°C). A cell measures 0.5 × 2–14 μm. (b) Structure of the novel lipid of *M. kandleri*. This is the normal ether-linked lipid of the Archaea (∞ Section 3.5) with the exception that the side chains are an *unsaturated* form of phytanyl called *geranylgeraniol*.

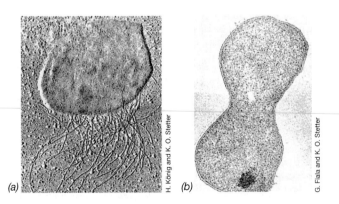

(a)   (b)

FIGURE 14.12 Spherical hyperthermophilic Archaea from submarine volcanic areas. (a) *Thermococcus celer*. Electron micrograph of shadowed cells (note tuft of flagella). (b) Dividing cell of *Pyrococcus furiosus*. Electron micrograph of thin section. Cells of both organisms are about 0.8 μm in diameter.

(growth temperature maximum, 110°C) and the methanogens (it carries out the reaction of $4 H_2 + CO_2 \rightarrow CH_4 + 2 H_2O$).

Methanopyrus produces methane only from $H_2 + CO_2$ and grows rapidly (generation time less than 1 hr) at its temperature optimum of 100°C. Unlike mesophilic methanogens, however, cells of *Methanopyrus* have large amounts of the glycolytic derivative, cyclic 2,3-diphosphoglycerate, dissolved in the cytoplasm. This compound, present at more than 1 *M* concentration in *Methanopyrus*, is thought to function as a thermostabilizing agent to prevent denaturation of enzymes and DNA inside the cell (see Section 14.9). In addition, *Methanopyrus* contains a type of membrane lipid found in no other known organism; this ether lipid is an unsaturated form of the otherwise saturated dibiphytanyl tetraethers found in other hyperthermophiles (Figure 14.13*b*; ∞ Section 3.5 and Figure 3.21).

The discovery of *Methanopyrus* may explain the origin of hydrocarbon-like materials in hot oceanic sediments previously thought to be too hot to support biogenic methanogenesis. In addition, at the depth at which *Methanopyrus* was found, approximately 2000 m, water remains liquid at temperatures up to 350°C, suggesting that other hyperthermophilic methanogens may exist capable of growth at 110°C or at perhaps even higher temperatures.

14.6

Hyperthermophilic Euryarchaeota: The Archaeoglobales

Key Genera

Archaeoglobus
Ferroglobus

We will see later that a number of hyperthermophilic Crenarchaeota carry out anaerobic respirations in which S^0 is used as an electron acceptor, being reduced to H_2S (see Table 14.8); curiously, none of these organisms are *sulfate-reducers*, that is, capable of reducing SO_4^{2-} to H_2S. However, one hyperthermophilic euryarchaeote, *Archaeoglobus*, is a true sulfate reducer and forms a phylogenetically distinct lineage within the Euryarchaeota.

Archaeoglobus and Its Genome

Archaeoglobus was isolated from hot marine sediments near hydrothermal vents and couples the oxidation of H_2, lactate, pyruvate, glucose, or complex organic compounds to the reduction of sulfate to sulfide. Cells of *Archaeoglobus* are irregular spheres (Figure 14.14*a*) and cultures grow optimally at 83°C. However, one of the most remarkable features of *Archaeoglobus* is its relationship to methanogens. We will learn in Chapter 15 about the unique biochemistry associated with methanogenesis. Briefly, this involves the use of novel coenzymes like Factor-420, coenzyme M, and many others; thus far, these novel biomolecules have only

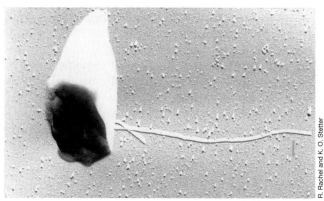

FIGURE 14.14   Archaeoglobales. (a) Shadowed preparation electron micrograph of the sulfate-reducing hyperthermophile *Archaeoglobus lithotrophicus*. The cell measures 1 × 2 μm. (b) Freeze-etched electron micrograph of *Ferroglobus placidus*, a ferrous iron–oxidizing, nitrate-reducing hyperthermophile. Cells measure about 0.8 μm in diameter.

been found in cells of methanogens (∞ Section 15.19). Surprisingly, however, *Archaeoglobus* contains traces of these coenzymes and cultures of this organism actually produce small amounts of methane during growth. Thus, *Archaeoglobus*, which also shows a rather close phylogenetic relationship to methanogens (Figure 14.1), may represent a metabolically transitional type of organism among the Archaea, one that bridged the energy-generating processes of H_2S production and methanogenesis.

A clearer picture of methanogenesis by *Archaeoglobus* has emerged from the complete genome sequence of this organism. The sequence has revealed that although *Archaeoglobus* can make methane, it lacks genes for a key enzyme of methanogenesis found in methanogens, *methyl-CoM reductase* (∞ Section 15.19). Thus, the small amounts of methane that are made by *Archaeoglobus* do not come from the activity of this enzyme. The genome of *Archaeoglobus* contains 2.1 megabases (about 2400 genes) and, interestingly, shares a number of genes with *Methanococcus* (see Section 14.3). However, the *Archaeoglobus* genome also contains a number of unique genes. This suggests that archaeal gene diversity is quite extensive and that it will likely mimic the findings of the genome projects of various Bacteria: Each new organism sequenced seems to have at least some genes never before seen in other organisms (⌒ Section 9.12).

Ferroglobus

Ferroglobus (Figure 14.14*b*) is related to *Archaeoglobus* but is not a sulfate-reducing bacterium. Instead, *Ferroglobus* is an iron-oxidizing chemolithotrophic autotroph, conserving energy from the oxidation of Fe^{2+} to Fe^{3+} coupled to the reduction of NO_3^- to NO_2^- plus NO (see Table 14.8). *Ferroglobus* can also use H_2 or H_2S as electron donors in its energy metabolism. *Ferroglobus* was isolated from a shallow marine hydrothermal vent and grows optimally at 85°C. *Ferroglobus* is interesting for several reasons, but especially so because of its ability to produce Fe^{3+} from Fe^{2+} under anoxic conditions. The mechanism for the formation of Fe^{3+} in ancient rocks, previously assumed to have been from the oxidation of Fe^{2+} by O_2 produced by cyanobacteria (⌒ Section 12.1), is now being questioned with the discovery that anoxic routes to the production of Fe^{3+} exist from the activities of organisms like *Ferroglobus* and certain anoxygenic phototrophic bacteria (⌒ Section 15.12).

✓ 14.5–14.6 Concept Check

Hyperthermophilic euryarchaeota include the Thermococcales, Archaeoglobales, and *Methanopyrus*. All organisms in this group have growth temperature optima above 80°C.

✓ How does the metabolism of Thermococcales differ from that of *Methanopyrus*?
✓ How does the metabolism of *Archaeoglobus* differ from that of the Thermococcales?
✓ How does the metabolism of *Ferroglobus* differ from that of *Archaeoglobus*?

Kingdom Crenarchaeota

Crenarchaeotes are phylogenetically distinct from Euryarchaeotes and contain representatives that live at both ends of nature's temperature extremes: boiling water and freezing water. Most cultured crenarchaeotes are hyperthermophiles (growth temperature optima above 80°C) and some actually have optima *above* the boiling point of water. We start with an overview of the habitats and energy metabolism of crenarchaeotes and then describe the properties of some key genera.

14.7

Habitats and Energy Metabolism of Crenarchaeotes

A summary of the habitats of Crenarchaeota is shown in Table 14.7. They include very hot and very cold environments. Most hyperthermophilic Archaea have been isolated from geothermally heated soils or waters containing elemental sulfur and sulfides (Figure 14.15) and most species metabolize sulfur in one way or another. In terrestrial environments, sulfur-rich springs, boiling mud, and soils may have temperatures up to 100°C and are mildly to extremely acidic owing to production of sulfuric acid (H_2SO_4) from the biological oxidation of H_2S and S^0 (⌒ Sections 15.11 and 16.19). Such hot,

TABLE 14.7	Habitats of Crenarchaeota		
Characteristic	Thermal Area*[a]*		Nonthermal Area*[b]*
	Terrestrial	Marine	
Locations	Solfataras (hot springs, fumaroles, mudpots, steam-heated soils); geothermal power plants; deep in Earth's crust	Submarine hot springs, hot sediments and vents ("black smokers"), deep oil reservoirs	Planktonic in oceans worldwide; near-shore and deep Antarctic waters; sea ice; symbionts of marine sponges
Temperature	Surface to 100°C; subsurface, above 100°C	Up to 400°C (smokers)	−2 to +4°C
Salinity/pH	Usually less than 1% NaCl; pH 0.5–9	Moderate, about 3% NaCl; pH 5–9	3–8% NaCl; pH 7–9
Gases and other nutrients	CO_2, CO, CH_4, H_2, H_2S, S^0, $S_2O_3^{2-}$, SO_4^{2-}, NH_4^+, N_2	Same as for terrestrial	CO_2, N_2, O_2; chemolithotrophic substrates such as NH_4^+

a See Figures 14.10, 14.15, and ⌒ 16.32–16.35
b See Figure 14.16

(a)

T. D. Brock

(b)

T. D. Brock

(c)

T. D. Brock

(d)

T. D. Brock

FIGURE 14.15 Habitats of hyperthermophilic Archaea. (a) A typical solfatara in Yellowstone National Park. Steam rich in hydrogen sulfide rises to the surface of the earth. Because of the heat and acidity, only prokaryotes are present. (b) Sulfur-rich hot spring, a habitat containing dense populations of *Sulfolobus*. The acidity in solfataras and sulfur springs comes from the oxidation of H_2S and S^0 to H_2SO_4 (sulfuric acid) by *Sulfolobus* and related prokaryotes. (c) A typical boiling spring of neutral pH in Yellowstone Park; Imperial Geyser. (d) An acidic iron-rich geothermal spring, another *Sulfolobus* habitat. Here the oxidation of Fe^{2+} to Fe^{3+} causes acidic conditions [$Fe^{3+} + 3H_2O \rightarrow Fe(OH)_3 + 3H^+$].

sulfur-rich environments are called *solfataras*, and are found throughout the world (Figure 14.15); extensive solfataras are present in Italy, Iceland, New Zealand, and Yellowstone National Park in Wyoming (USA). Depending on the surrounding geology, solfataras can be slightly alkaline to mildly acidic, pH 5–8, or extremely acidic, with pH values below 1 not uncommon. Hyperthermophiles have been obtained from both types of environments, but the majority of these organisms inhabit neutral or mildly acidic habitats. In addition to these natural habitats, hyperthermophilic Archaea also

thrive within artificial thermal habitats, in particular the boiling outflows of geothermal power plants.

Hyperthermophilic Crenarchaeota also thrive in undersea hot springs called *hydrothermal vents*. We discuss the geology of these habitats in Section 16.12 but for our purposes here it is only necessary to note that submarine waters can be and often are much hotter than surface waters because the water is under extreme pressure. Indeed, all hyperthermophiles with growth temperature optima above 100°C have come from submarine sources. The latter include both shallow (2–10 meters) vents such as those off the coast of Vulcano, Italy, to deep (2500–4000 meters) vents near ocean spreading centers (∞ Section 16.12). The latter are the hottest habitats so far known to yield living organisms.

Cold-Dwelling Crenarchaeotes

By contrast with the hyperthermophiles, cold-dwelling crenarchaeotes (Table 14.7) have been identified from community sampling of ribosomal RNA genes from many nonthermal environments. Using fluorescent phylogenetic probes (∞ Section 12.6) crenarchaeotes have been found in marine waters worldwide (Figure 14.16b). In stark contrast to the hyperthermophiles, marine crenarchaeotes thrive even in frigid waters, such as those of the Antarctic (Figure 14.16a). These organisms are planktonic (suspended freely or attached to suspended particles in the water column) and occur in significant numbers ($\sim 10^4$/ml) for waters that are both nutrient poor and very cold. Although their physiology is still a mystery, lipid analyses of marine crenarchaeotes filtered from seawater have shown that they contain ether-linked lipids, the hallmark of the Archaea (∞ Section 3.5), and that, in addition, these lipids are of the dibiphytanyl tetraether type (∞ Sections 3.5 and 14.10) typical of hyperthermophilic Archaea, their phylogenetic close relatives. Interestingly, such lipids have previously been found *only* in hyperthermophiles where it was thought that they were one of the molecular secrets to survival at very high temperature. The discovery of cold-dwelling Archaea with similar lipids obviously leaves this hypothesis open to refinement.

Energy Metabolism

With a few exceptions, hyperthermophilic Crenarchaeota are obligate anaerobes. Their energy-yielding metabolism can be either chemoorganotrophic or chemolithotrophic (or both, for example in *Sulfolobus*) and involves a wide diversity of different electron donors and acceptors (Table 14.8). Fermentation is rare and most bioenergetic strategies are either aerobic or anaerobic respirations (Table 14.8). Although not yet examined in a major way, it is likely that energy conservation during these respiratory processes occurs by the same general mechanism widespread in Bacteria: electron transfer within the cytoplasmic membrane leading to the formation of a proton motive force from which ATP is made by way of proton-translocating ATPases (∞ Section 4.11).

Many hyperthermophilic crenarchaeotes can grow chemolithotrophically under anoxic conditions with H_2 as electron donor and S^0 or NO_3^- as electron acceptor; a few can oxidize H_2 aerobically (Table 14.8). H_2 respi-

(a)

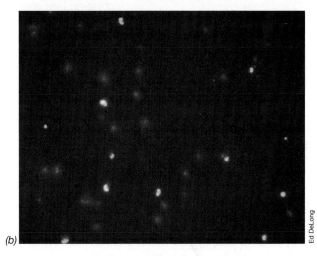
(b)

FIGURE 14.16 Cold-dwelling Crenarchaeota. (a) Photo of the Antarctic peninsula taken from shipboard. The icy waters shown here are typical habitats for cold-dwelling crenarchaeotes. (b) Fluorescence photomicrograph of seawater treated with a phylogenetic stain composed of a green fluorescing dye attached to an oligonucleotide complementary to a signature sequence in the 16S rRNA of species of Crenarchaeota (∞ Section 12.6 for a description of the methods used here). Blue cells are stained with DAPI, a stain that stains all cells (∞ Section 16.5 and Figure 16.7). Cells stained green are cold-dwelling Crenarchaeota.

ration with ferric iron (Fe^{3+}) as electron acceptor also occurs in several hyperthermophiles (Table 14.8). Other chemolithotrophic lifestyles include the oxidation of S^0 and Fe^{2+} aerobically or Fe^{2+} anaerobically with NO_3^- as acceptor (Table 14.8). Only one sulfate-reducing organism has been isolated (the euryarchaeote *Archaeoglobus*), which seems odd given the great diversity of these organisms among Bacteria (∞ Section 13.17), but others probably exist and have thus far just eluded isolation and culture. The only bioenergetic option apparently ruled out is photosynthesis. The most thermophilic phototrophic organism known can grow up to about 70°C—too cold for most hyperthermophiles!

TABLE 14.8	Energy-yielding reactions of hyperthermophilic Archaea	
Nutritional class	**Energy-yielding reaction**	**Example**
Chemoorganotrophic	Organic compound $+ S^0 \rightarrow H_2S + CO_2$	*Thermoproteus, Thermococcus, Desulfurococcus, Thermofilum, Pyrococcus*
	Organic compound $+ SO_4^{2-} \rightarrow H_2S + CO_2$	*Archaeoglobus*
	Organic compound $+ O_2 \rightarrow H_2O + CO_2$	*Sulfolobus*
	Organic compound $\rightarrow CO_2 + H_2 +$ fatty acids	*Staphylothermus, Pyrodictium*
	Pyruvate $\rightarrow CO_2 + H_2 +$ acetate	*Pyrococcus*
Chemolithotrophic	$H_2 + S^0 \rightarrow H_2S$	*Acidianus, Pyrodictium, Thermoproteus, Stygiolobus*
	$H_2 + NO_3^- \rightarrow NO_2^- + H_2O$ (NO_2^- reduced to N_2 by some species)	*Pyrobaculum, Aquifex[a]*
	$4\,H_2 + NO_3^- + H^+ \rightarrow NH_4^+ + 2\,H_2O + OH^-$	*Pyrolobus*
	$H_2 + 2\,Fe^{3+} \rightarrow 2\,Fe^{2+} + 2\,H^+$	*Pyrobaculum, Pyrodictium, Archaeoglobus*
	$2\,H_2 + O_2 \rightarrow 2\,H_2O$	*Acidianus, Sulfolobus, Pyrobaculum, Aquifex[a]*
	$2\,S^0 + 3\,O_2 + 2\,H_2O \rightarrow 2\,H_2SO_4$	*Sulfolobus, Acidianus*
	$2\,FeS_2 + 7\,O_2 + 2\,H_2O \rightarrow 2\,FeSO_4 + 2\,H_2SO_4$	*Sulfolobus, Acidianus, Metallosphaera*
	$2\,FeCO_3 + NO_3^- + 6\,H_2O \rightarrow 2\,Fe(OH)_3 + NO_2^- + 2\,HCO_3^- + 2\,H^+ + H_2O$	*Ferroglobus*
	$4\,H_2 + SO_4^{2-} + 2\,H^+ \rightarrow 4\,H_2O + H_2S$	*Archaeoglobus*
	$4\,H_2 + CO_2 \rightarrow CH_4 + 2\,H_2O$	*Methanopyrus, Methanococcus, Methanothermus*

a Member of the Bacteria.

Armed with a basic understanding of the homes and the energy metabolism of hyperthermophilic crenarchaeotes, let us now look at representative organisms.

14.8

Hyperthermophiles from Terrestrial Volcanic Habitats: Sulfolobales and Thermoproteales

Key Genera

Sulfolobus
Acidianus
Thermoproteus

Volcanic habitats can have temperatures as high as 100°C and are thus suitable for hyperthermophilic Archaea. Two phylogenetically related organisms isolated from these environments include *Sulfolobus* and *Acidianus*. These genera form the heart of an order called the Sulfolobales (Table 14.9).

Sulfolobales

Sulfolobus grows in sulfur-rich hot acid springs (Figure 14.15*b*) at temperatures up to 90°C and at pH values of 1–5*. *Sulfolobus* (Figure 14.17*a*) is an aerobic chemolitho-

troph that oxidizes H_2S or S^0 to H_2SO_4 and fixes CO_2 as sole carbon source. *Sulfolobus* can also grow chemoorganotrophically. Cells of *Sulfolobus* are more or less spherical but form distinct lobes (Figure 14.17*a*). Cells adhere tightly to sulfur crystals where they can be seen microscopically using fluorescent dyes (∞ Figure 15.26*b*). Besides the aerobic respiration of sulfur or organic compounds, *Sulfolobus* can also oxidize Fe^{2+} to Fe^{3+}, and this has been applied in the high temperature leaching of iron and copper ores (∞ Section 16.19).

A facultative aerobe resembling *Sulfolobus* also lives in acidic solfataric springs. This organism, named *Acidianus* (Figure 14.17*b*), differs from *Sulfolobus* most clearly by its ability to grow anaerobically. Remarkably, *Acidianus* is able to use S^0 in both its aerobic and anaerobic metabolism. Under aerobic conditions the organism uses S^0 as an electron *donor*, oxidizing S^0 to H_2SO_4. Anaerobically, *Acidianus* uses S^0 as an electron *acceptor* (with H_2 as electron donor) forming H_2S as the reduced product. Thus, the metabolic fate of S^0 in cultures of *Acidianus* depends on the presence or absence of O_2.

Like *Sulfolobus, Acidianus* is roughly spherical in shape but is not as lobed (Figure 14.17*b*). It grows at temperatures from 65°C up to a maximum of 95°C, with an optimum of about 90°C. Another property shared by *Sulfolobus* and *Acidianus* is an unusually low GC base ratio. The DNA of *Sulfolobus* is about 37% GC, whereas that of *Acidianus* is even lower, about 31%; many other hyperthermophiles have DNA of low GC content as well (see Table 14.9). These low GC base ratios are intriguing when one considers the hyperthermophilic nature of these organisms; how do they prevent their DNA

---

*Historical note: *Sulfolobus* was first discovered by Thomas Brock and colleagues in 1970 and formally described in 1972. The discovery of *Sulfolobus*, along with the previously isolated *Thermus aquaticus* (source of the extremely thermostable *Taq* DNA polymerase, ∞ Section 10.9), is generally credited with launching the field of hyperthermophilic microbiology. Thomas Brock was the senior author of the first seven editions of this book. In the 1980s to the present, Karl Stetter and colleagues in Germany have greatly expanded the field of hyperthermophilic microbiology with the discovery of many new genera and species.

TABLE 14.9   Properties of Hyperthermophilic Crenarchaeota

Group/Genus[a]	Morphology	Relationship to O_2[b]	DNA (mol% GC)	Temperature (°C)			Optimum pH
				Minimum	Optimum	Maximum	
Sulfolobales							
Sulfolobus	Lobed coccus	Ae	37	55	75	87	2–3
Acidianus	Coccus	Fac	31	60	88	95	2
Metallosphaera	Coccus	Ae	45	50	75	80	2
Stygiolobus	Lobed coccus	An	38	57	80	89	3
Thermoproteales							
Thermoproteus	Rod	An	56	60	88	96	6
Thermophilum	Rod	An	57	70	88	95	5.5
Pyrobaculum	Rod	Fac	46	74	100	102	6
Desulfurococcales							
Desulfurococcus	Coccus	An	51	70	85	95	6
Staphylothermus	Cocci in clusters	An	35	65	92	98	6–7
Pyrodictium	Disc-shaped with filaments	An	62	82	105	110	6
Pyrolobus	Lobed coccus	Fac	53	90	106	113	5.5
Thermodiscus	Disc-shaped	An	49	75	90	98	5.5
Igneococcus	Irregular cocci	An	—	65	90	103	5
Hyperthermus	Irregular coccus	An	56	75	102	108	7
Archaeoglobales							
Archaeoglobus	Cocus	An	46	64	83	95	7
Ferroglobus	Irregular coccus	An	43	65	85	95	7
Thermococcales[c]							
Thermococcus	Coccus	An	38–57	70	88	98	6–7
Pyrococcus	Coccus	An	38	70	100	106	6–8

a The group names ending in "ales" are order names; an order contains within it a number of families, each comprised of one or more genera (⬤◯⬤ Section 12.10).

b Ae, aerobe; An, anaerobe; Fac, facultative.

c Phylogenetically, genera in this order of hyperthermophiles are members of the Euryarchaeota (see Section 14.5).

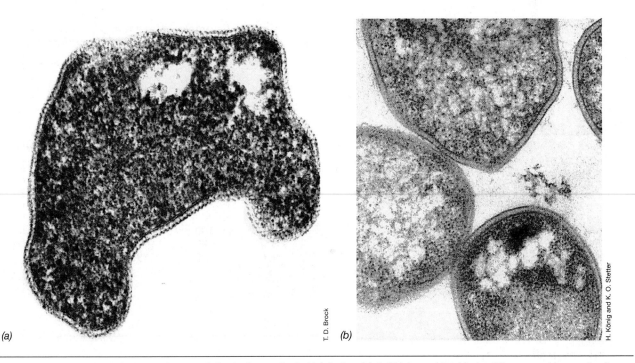

(a)   (b)

T. D. Brock   H. König and K. O. Stetter

FIGURE 14.17   Acidophilic hyperthermophilic Archaea, the Sulfolobales. (a) *Sulfolobus acidocaldarius*. Electron micrograph of a thin section. (b) *Acidianus infernus*. Electron micrograph of a thin section. Cells of both organisms vary from 0.8 to 2 μm in diameter.

from melting? In the test tube, DNA of 30–40% GC content would melt almost instantly at 90°C. Obviously hyperthermophiles have evolved protective mechanisms to prevent DNA melting *in vivo*, and we discuss these in Section 14.10.

Thermoproteales

The genera *Thermoproteus* and *Thermofilum* consist of *rod-shaped* cells that inhabit neutral or slightly acidic hot springs. Cells of *Thermoproteus* are stiff rods about 0.5 μm in diameter and are highly variable in length, ranging from short cells of 1–2 μm up to filaments 70–80 μm long (Figure 14.18*a*). Filaments of *Thermofilum* are thinner, some 0.17–0.35 μm wide with filament lengths ranging up to 100 μm (Figure 14.18*b*). Both *Thermoproteus* and *Thermofilum* are strict anaerobes that carry out a S⁰-based anaerobic respiration. Unlike most hyperthermophiles, the oxygen sensitivity of *Thermoproteus* and *Thermofilum* is extreme, comparable to that of the methanogens (see Section 14.3); thus, strict precautions must be taken in their culture. Most *Thermoproteus* isolates can grow chemolithotrophically on H_2 or chemoorganotrophically on complex carbon substrates such as yeast extract, small peptides, starch, glucose, ethanol, malate, fumarate, or formate (see Table 14.8). *Thermofilum* can be grown in either mixed culture with *Thermoproteus* or in pure culture only by the addition of a highly polar lipid fraction isolated from cells of *Thermoproteus;* in nature the two organisms probably coexist in close association. *Thermoproteus* and *Thermofilum* have similar GC base ratios (56–58% GC) but are phylogenetically distinct by nucleic acid hybridization analyses.

14.9

Hyperthermophiles from Submarine Volcanic Habitats: Desulfurococcales

Key Genera

Pyrodictium
Pyrolobus
Staphylothermus

We now turn our attention to the microbiology of *submarine* volcanic habitats, homes to the most thermophilic of all known Archaea. These habitats include both shallow water thermal springs and deep-sea hydrothermal vents. We discuss the geology of these fascinating microbial habitats in Section 16.12. The organisms to be described here make up an order-level group of Archaea called the Desulfurococcales (Table 14.9).

Pyrodictium and *Pyrolobus*

Pyrodictium, Pyrolobus, and *Pyrobaculum* are interesting examples of prokaryotes whose growth temperature optimum lies at or even above 100°C; the optimum for *Pyrodictium* is 105°C and for *Pyrolobus* is 106°C. Cells of *Pyrodictium* are irregularly disc-shaped and grow in culture as a mycelium-like layer attached to crystals of elemental sulfur; the cell mass consists of a network of fibers to which individual cells are attached (Figure 14.19*a,b*). The fibers are hollow and consist of protein arranged in a fashion similar to that of bacterial flagella (∞ Section 3.11). The filaments do not function in motility but instead as organs of attachment (see Figure 14.21), and the cell walls of *Pyrodictium* are composed of glycoprotein. Physiologically, *Pyrodictium* is a strict

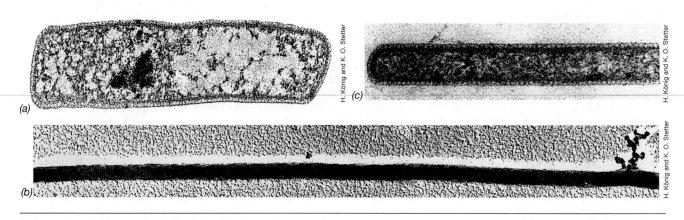

FIGURE 14.18   Rod-shaped hyperthermophilic Archaea: Thermoproteales. (a) *Thermoproteus neutrophilus.* Electron micrograph of a thin section. A cell is about 0.5 μm in diameter. (b) *Thermofilum librum.* A cell is only about 0.25 μm in diameter. Electron micrograph of shadowed cells. (c) *Thermofilum librum.* Electron micrograph of a thin section.

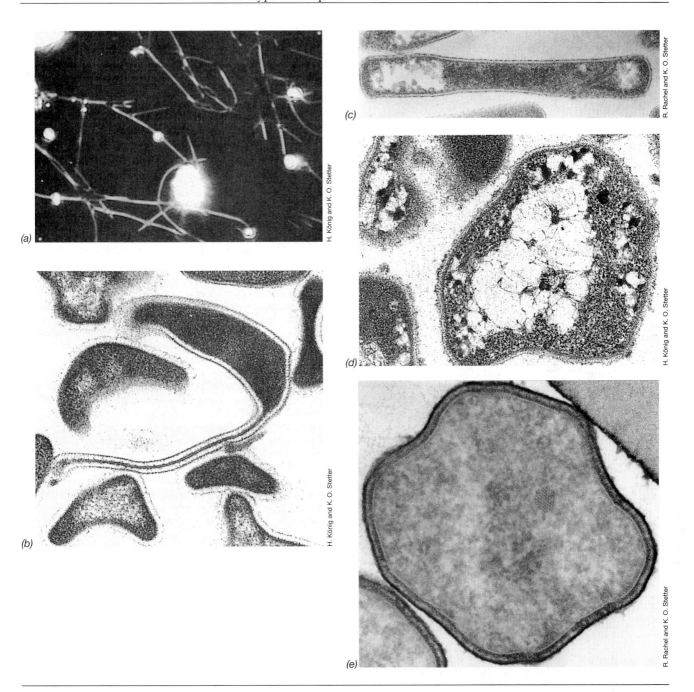

FIGURE 14.19 Hyperthermophiles from submarine volcanic habitats: Desulfurococcales. (a) *Pyrodictium occultum* (growth temperature optimum, 105°C), dark-field micrograph. (b) Thin section electron micrograph of *P. occultum*. Cells are highly variable in diameter from 0.3 to 2.5 μm. (c) Thin section of *Pyrobaculum aerophilum* (growth temperature optimum, 100°C); a cell measures 0.5 × 3.5 μm. (d) Thin section of a cell of *Desulfurococcus saccharovorans;* a cell is 0.7 μm in diameter. (e) Thin section of a cell of *Pyrolobus fumarii,* the most thermophilic of all known prokaryotes (growth temperature optimum, 106°C); a single cell is about 1.4 μm in diameter.

anaerobe that grows chemolithotrophically on H_2 with S^0 as electron acceptor or chemoorganotrophically on complex mixtures of organic compounds (see Table 14.8).

Pyrolobus fumarii (Figure 14.19e) currently holds the record for the most thermophilic of all known organisms—its growth temperature maximum is 113°

(Table 14.9) (notably, *P. fumarii* is unable to grow even at 90°C because it is too cold!). *Pyrolobus* lives in the walls of "black smoker" hydrothermal vent chimneys (⌒ Section 16.12 and Figure 16.33) where, because of its autotrophic abilities, it is possibly a major source of primary productivity in this otherwise inorganic environment. *Pyrolobus* cells are coccoid-shaped (Figure 14.19*e*) and the cell wall is composed of protein. The organism is an obligate H_2 chemolithotroph, growing at the expense of H_2 oxidation coupled to the reduction of NO_3^- (to NH_4^+), $S_2O_3^{2-}$ (to H_2S), or very low concentrations of O_2 (to H_2O). Besides its extremely thermophilic nature, *Pyrolobus* is also resistant to temperatures substantially above its growth temperature maximum. For example, cultures of *P. fumarii* survive autoclaving (121°C) for 1 hour, a condition that even bacterial endospores (Section 3.15) cannot withstand. Considering the hydrothermal vent habitat of this organism, where periodic shifts in the temperature may be a regular occurrence, high heat resistance would be of great survival value.

Pyrobaculum (Figure 14.19*d*) is a physiologically unique hyperthermophilic archaeon. Some species are capable of both aerobic respiration or anaerobic respiration with NO_3^-, Fe^{3+}, or S^0 as electron acceptors and H_2 as electron donor (that is, chemolithotrophic and autotrophic growth), while other species grow anaerobically on organic electron donors, reducing S^0 to H_2S. The growth temperature optimum of *Pyrobaculum* is right at the boiling point of water at sea level (100°C), and species of this organism have been isolated from both terrestrial hot springs as well as from hydrothermal vents.

Other notable members of this group of hyperthermophiles include *Desulfurococcus*, the genus for which the order is named (Figure 14.19*d*). *Desulfurococcus* is a strictly anaerobic S^0-reducing bacterium like *Pyrodictium* but differs from this organism in that it is much less thermophilic, growing optimally at about 85°C.

Staphylothermus

A morphologically unusual member of the order Desulfurococcales is the genus *Staphylothermus*. Cells of *Staphylothermus* are spherical and about 1 μm in diameter and form aggregates of up to 100 cells, resembling those of *Staphylococcus* (compare Figure 14.20 with Figure 3.4*b*). *Staphylothermus* is a chemoorganotroph that grows optimally at 92°C; energy is obtained from the fermentation of peptides, producing fatty acids like acetate and isovalerate as fermentation products (Table 14.8). Isolates of *Staphylothermus* have been obtained from both shallow marine vents, as well as black smoker hydrothermal vents, indicating that this organism is widely distributed in submarine thermal areas and thus may be an ecologically significant consumer of proteinaceous materials in these environments.

✓ 14.7–14.9 Concept Check

Hyperthermophilic Crenarchaeota inhabit the hottest habitats currently known to support life. Cold-dwelling phylogenetic relatives of these organisms are also known. A variety of different morphological types of Crenarchaeota are known and several different metabolic strategies are used to support growth.

✓ What are the major differences between the organisms *Sulfolobus* and *Pyrolobus*?

✓ What is unusual about the metabolic properties of *Acidianus* regarding elemental sulfur (S^0)?

✓ What two lines of support link the cold-dwelling Archaea to the Crenarchaeota?

✓ What energy class of organisms are those that use H_2 as electron donor? What Crenarchaeota use H_2, and what do they use as electron acceptors?

Evolution and Life at High Temperatures

In the case of the hyperthermophilic Archaea, we see growth at temperatures far higher than those supporting the growth of other prokaryotes. What is the nature of this extreme heat tolerance, and what are the temperature limits beyond which life is impossible? Moreover, what can hyperthermophilic prokaryotes tell us about Earth's early history? We examine these issues here.

14.10

Heat Stability of Biomolecules

Protein and DNA stability in hyperthermophiles is critical to surviving high temperature, and much research has been done on this problem in an attempt to identify basic principles. Thermostable proteins from hyperthermophiles are not particularly unusual from the

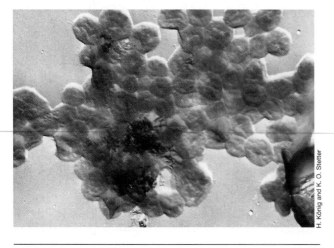

FIGURE 14.20 The hyperthermophile *Staphylothermus marinus* (growth temperature optimum, 92°C). Electron micrograph of shadowed cells. A single cell is about 1 μm in diameter.

standpoint of their amino acid composition. In fact, enzymes from hyperthermophiles often contain the same major structural features, for example the amino acid sequence in and around the active site, as their heat labile counterparts from mesophilic bacteria. Thermostable proteins do tend to have highly hydrophobic cores, which probably increases internal "sticking," and in general have more "salt bridges" (ionic interactions between amino acids) on their surfaces. But in the final analysis, it is the *folding* of the protein that will ultimately affect its heat resistance; thus, subtle changes in amino acid sequence are apparently sufficient to render heat stable an otherwise heat-labile protein.

Like all cells, hyperthermophiles also produce special proteins called *chaperonins* (heat-shock proteins, ∞ Section 6.12) that function to refold partially denatured proteins. And in hyperthermophiles, the amount of chaperonin per cell can increase dramatically near the growth temperature limit. For example, in cells of *Pyrodictium* (Figure 14.21; see also Figure 14.19) grown at 108°C (only 2°C from the upper limit and 3°C above the optimum), 80% of the cell's cytoplasmic protein is devoted to a chaperonin called the *thermosome*. This protein functions to keep the cell's other proteins functional and can actually increase the cell's overall heat resistance; when the thermosome is present, cells of *Pyrodictium* can survive 1 hour in the autoclave (121°C).

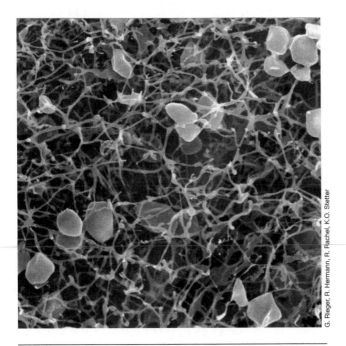

FIGURE 14.21 *Pyrodictium abyssi,* scanning electron micrograph. *Pyrodictium* has been studied as a model of macromolecular stability at high temperatures (see Sections 14.9–14.12). Cells are enmeshed in a sticky glycoprotein matrix that binds them together.

DNA Stability

How does DNA stay intact at high temperatures and keep from melting apart into single strands? A variety of mechanisms may be involved here. The cytoplasm of hyperthermophilic methanogens contains large amounts of potassium cyclic 2,3-diphosphoglycerate. This solute, especially the K^+, prevents chemical damage, like depurination, that can occur to DNA at high temperatures. However, not all hyperthermophiles produce this substance, so other DNA stabilizing mechanisms have been sought. In this regard it has been found that all hyperthermophiles produce a unique form of DNA topoisomerase called *reverse DNA gyrase* (∞ Section 6.3). Reverse gyrase introduces *positive* supercoils into DNA (in contrast to the *negative* supercoils introduced by "normal" DNA gyrase, ∞ Section 6.3), and positive supercoiling has been shown to greatly stabilize DNA to heat denaturation.

However, in addition to salts and DNA gyrase, there are other proteins in hyperthermophiles that function to maintain the integrity of duplex DNA. In this connection a small heat-stable DNA binding protein called *Sac7d* has been found in cells of *Sulfolobus*; this protein binds to the minor groove of DNA in a nonspecific manner and increases its melting temperature by some 40°C. Protein Sac7d sharply kinks the DNA and thus may also be involved in gene regulation (∞ bent DNA, Section 7.4). Although Sac7d-like proteins are limited to species of crenarchaeotes, euryarchaeotes also contain DNA binding proteins. But unlike Sac7d, these are highly basic proteins that show strong amino acid sequence homology with the core histones of the Eukarya (∞ Section 6.3). These *archaeal histones* have been particularly well studied from the hyperthermophilic methanogen *Methanothermus fervidus* where they wind and compact DNA into nucleosome-like structures (Figure 14.22) (∞ Section 6.3) that maintain DNA in a double stranded form at very high temperatures.

DNA binding proteins like Sac7d and archaeal histones may thus play key roles in stabilizing DNA to heat denaturation. And they may also explain why the transcriptional apparatus of Archaea more closely resembles that of Eukarya than that of Bacteria. Recall that the RNA polymerase of Bacteria is a rather simple protein compared to the structurally complex RNA polymerases of Archaea and Eukarya (∞ Sections 6.7, 6.8 and 12.8 and Figure 12.16). The latter, along with the requirements of Archaea and Eukarya for transcription initiation factors and the TATA-binding protein (∞ Section 6.8 and Figure 6.28), may be a consequence of the fact that both of these groups contain DNA binding/compacting proteins that make accessing DNA for transcription and regulatory events a more complex process than in Bacteria.

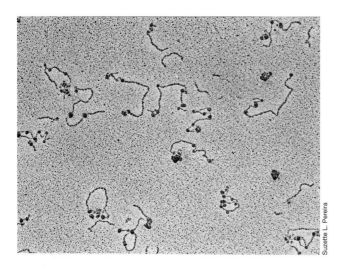

Suzette L. Pereira

FIGURE 14.22   Archaeal histones and nucleosomes. Electron micrograph of linearized plasmid DNA wrapped around copies of archaeal histone Hmf (from the hyperthermophilic methanogen *Methanothermus fervidus,* see Figure 14.8*c*) to form the roughly spherical, darkly stained nucleosome structures. Compare this micrograph with an artist's depiction of the histones and nucleosomes of Eukarya shown in Figure 6.11.

Lipid Stability

What about cellular lipids? How do hyperthermophiles prevent their membranes from peeling apart at high temperatures? Virtually all hyperthermophiles contain lipids constructed upon the dibiphytanyl tetraether model (∞ Section 3.5 for the structure of such lipids). Dibiphytanyl tetraether lipids should be naturally heat resistant because the covalent bond between phytanyl units forms a lipid *monolayer* cytoplasmic membrane instead of the normal lipid *bilayer* (∞ Section 3.5 and Figure 3.21); such a structure would be immune to the tendency for heat to pull apart a lipid bilayer constructed of fatty acids.

Stability of Monomers

Besides the stability of *macromolecules,* a second consideration is important in governing the upper temperature limits for growth of hyperthermophilic bacteria: the thermal lability of *monomers.* At temperatures above 100°C, a number of biologically significant monomers show some degree of heat lability. And, no matter how stable the macromolecules are, it should be apparent that life is impossible if the basic building blocks themselves are unstable. The thermal stability of small molecules rather than macromolecules may therefore dictate the upper temperature for

life, and even at temperatures as low as 120°C, some important small molecules are destroyed at significant rates. For example, two key molecules in energy metabolism, ATP and NAD$^+$, hydrolyze quite rapidly at these temperatures; their half-life is less than 30 minutes at 120°C and shortens dramatically at temperatures above this. With this in mind, what is the upper temperature for life?

The Limits to Microbial Existence

Because life depends on liquid water, microbial habitats in excess of 100°C are limited to environments under *pressure,* such as the sea floor. Indeed, all hyperthermophilic Archaea capable of growth over 100°C seem to be restricted to these superheated environments (see Table 14.9). Another possible habitat for such hyperthermophiles is the deep subsurface of Earth where heat from Earth's interior, coupled with immense pressure, creates conditions suitable for growth above the boiling point. The isolation of hyperthermophiles from oil-drilling waters and from deep subsurface rock samples is suggestive evidence that these organisms may actually inhabit locations deep within Earth. But even if they do, how high a temperature could they tolerate?

An answer to this question has been sought from the study of deep-sea hydrothermal vent black smokers as models for superheated natural environments. Smokers emit hydrothermal fluid at 250–350°C and form upright metallic structures called *chimneys* from metal sulfides in the fluid (∞ Figure 16.32). Although hyperthermophiles have been isolated from smoker chimney walls (which show a gradient in temperature from 250°C inside to 2°C outside) (see Sections 14.5 and 14.9 and see Figure 16.33), studies of the 250°C water itself show it to be sterile; this is consistent with laboratory measurements that have clearly shown that important biomolecules would be instantly destroyed at such a high temperature. Laboratory experiments on the heat stability of biomolecules suggest that living processes could be maintained at temperatures as high as 140–150°C, but that above this temperature organisms would probably not be able to overcome the heat lability of the essential biomolecules of life. For example, if organisms exist capable of growth at temperatures even as "low" as 150°C, their energy economy would almost certainly have to be based on something other than ATP, the energy currency of all other forms of life, because ATP is highly unstable at this temperature. Thus, if "super" hyperthermophiles capable of growth at or above 150°C are ever discovered, they will likely bring with them some fundamental new principles in biology and biochemistry.

14.11

Hyperthermophilic Archaea and Microbial Evolution

An interesting question that has surrounded the Archaea since they were first recognized as a distinct phylogenetic domain is why so many of them inhabit extreme environments. Indeed, although molecular probing of nonextreme environments (like soil and water) indicate that Archaea are all around us (∞ Section 12.6), a common theme running through cultured representatives of the Archaea is *adaptation to environmental extremes*. As we have seen in this chapter, various species growing above the boiling point or at extremes of pH or salt are known. Is this a mere coincidence or a reflection of evolutionary history?

Extreme environments of various types likely existed on early Earth as they do today, and it is within such environments that life may first have flourished. However, because at the time that cellular life first evolved it is almost certain that Earth was far hotter than it is today (∞ Section 12.1), the hyperthermophiles in particular stand out as the best extant relatives of Earth's earliest life-forms. But is there any way to experimentally support such a hypothesis? A hint lies in the phylogenetic tree (Figure 14.1).

Hyperthermophiles and Their Slow Evolutionary Clocks

Molecular sequencing suggests that many Archaea have evolved at slower rates than most Bacteria or the Eukarya. This is especially true of the hyperthermophilic Archaea, both euryarchaeotes and crenarchaeotes (see Figure 14.1). Much the same can be said about hyperthermophilic Bacteria such as *Thermotoga* and *Aquifex* (∞ Figure 13.1). This conclusion emerges from the relative lengths of the lines leading to these lineages in computer-generated evolutionary trees; lines to the hyperthermophiles are typically rather short and branch near the root (∞ Figure 13.1; see also Figure 14.1). This stands in contrast to the marine Archaea, for example (see Figures 14.1 and 14.16), that are clearly more phylogenetically derived than the hyperthermophiles.

It is not known why heat-loving Archaea have such slow "evolutionary clocks," but one hypothesis is that this is a natural consequence of their inhabiting such extreme environments. Organisms living at very high temperatures may be under unusually strong evolutionary pressure to maintain those genes that specify phenotypic characteristics critical to life there. Unlike organisms inhabiting soil or water, perhaps there is very little latitude for genes from hyperthermophiles to be significantly changed if the organism is to maintain it-

self successfully under such harsh conditions. If this hypothesis is correct, then hyperthermophilic Archaea are indeed living relics of the earliest life forms, and their continued study should reveal important principles of early Earth biology. Indeed, many of the phenotypic properties of modern hyperthermophiles (Tables 14.8 and 14.9) agree well with the predicted phenotype of primitive organisms (∞ Section 12.2).

14.12

Korarchaeota, Hydrogen, and Microbial Life on Other Planets

If hyperthermophiles were really the earliest life-forms, representatives of the kingdom **Korarchaeota** may hold important clues to an understanding of life on early Earth. Although cultures (Figure 14.23) of representatives of this group have now been obtained, little is yet known about them except that they are obviously hyperthermophiles (they grow at 85°C). Originally detected not from cultured organisms but from 16S rRNA genes cloned and sequenced (∞ Section 12.6) from an iron- and sulfur-rich Yellowstone hot spring, Obsidian Pool, the availability of cultures of Korarchaeota should allow for laboratory characterization of important properties of these phylogenetically unique Archaea. Indeed, because Korarchaeota branch very near the root of the archaeal tree (Figure 14.1), the exciting possibility exists that representatives of this kingdom may display novel biological properties of relevance to our understanding of ancient organisms.

Hydrogen as an Early Energy Source and Life on Mars?

As we leave this chapter on the Archaea, it is interesting to note how often H_2 *metabolism* enters into the metabolic picture; the ability of so many Archaea to grow anaerobically with H_2 as an electron donor and one or more of the electron acceptors S^0, NO_3^-, and Fe^{3+} (see Table 14.8) is probably more than just a coincidence. Instead, this is likely a physiological relic of ancient metabolic schemes (∞ Figure 12.4 gives an example of one) that first evolved in primitive organisms because of the ready availability of these inorganic compounds in their environment. However, the diversity of H_2-oxidizing hyperthermophilic Archaea (and Bacteria, ∞ Sections 13.33 and 13.34) today, and the fact that so many mesophilic Bacteria (distant relatives of the hyperthermophilic H_2-oxidizers; ∞ Figure 13.1) also use H_2 (∞ Sections 13.4 and 15.10), clearly indicates that the evolution of H_2 oxidation was a metabolic success story. Indeed, the discovery of H_2 utilization by various physiological groups of prokaryotes living deep in the Earth where life may

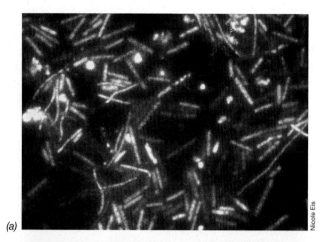

(a)

(b)

FIGURE 14.23   Korarchaeota. (a) Phase contrast photomicrograph of a stable enrichment culture containing cells of Korarchaeota from a Yellowstone hot spring and grown in the laboratory at 85°C. Various Archaea are present in this enrichment with the korarchaeotans being less than 1% of the total. However, specific staining has shown the korarchaeotan cells to be slightly curved rods 5-10 μm long. (b) Positive identification of a korarchaeotan using a red phylogenetic stain (⚬⚬ Section 12.6 and Figure 12.14).

first have evolved (⚬⚬ Sections 12.2 and 16.9; and the Chapter 16 box, Microbial Life Deep Underground), is another indication that this form of energy metabolism not only has deep evolutionary roots, but is widespread in the microbial world. Could this mean that H_2-supported microbial ecosystems even extend beyond the confines of Earth to adjacent planets, such as Mars?

Although the surface of Mars is extremely cold, deep within the mantle of Mars it may be quite hot, as is Earth. If so, hyperthermophilic organisms may exist there, thriving on the simple nutrients of hydrogen, sulfur, and iron that support the growth of so many organisms we already know of. With planned trips to Mars part of the United States space agenda in the early twenty-first century, the hypothesis that life exists anywhere in the universe that conditions permit can be carefully tested. Perhaps H_2-oxidizing prokaryotes will be discovered on Mars. If so, the science of microbiology will surely enter yet another "golden age" and a nearly inexhaustible number of interesting research problems will emerge for future microbiologists to deal with.

✓ 14.10–14.12 Concept Check

Although hyperthermophiles live at very high temperatures, in some cases above the boiling point of water, it is likely that there are temperature limits beyond which no living organism can survive. The study of modern day hyperthermophilic prokaryotes may yield clues to early life on Earth and other planets, such as Mars.

✓ How do hyperthermophiles keep important macromolecules such as proteins and DNA from being destroyed by high heat?

✓ List at least two reasons why an upper temperature limit to life undoubtedly exists. What is the likely upper temperature limit for life?

✓ What evidence suggests that extant hyperthermophiles most closely resemble ancient organisms?

✓ What form of energy metabolism was likely key to energy economies of ancient organisms?

Photosynthesis is just one of many different energy-conserving mechanisms that have evolved in prokaryotes. During photosynthetic energy conversion, as depicted here, light energy excites chlorophyll pigments producing electrons of very electronegative potentials. These electrons then travel through an electron transport chain, eventually returning to the chlorophyll molecule. During this cyclical process a proton motive force is generated that can be tapped by membrane proteins called ATPases to yield ATP. Many groups of prokaryotes are phototrophic, while others tap the energy available in chemical compounds, both organic and inorganic. The almost endless stream of energy-yielding strategies known in prokaryotes, the highlights of which make up this chapter, has clearly shaped microbial diversity and is the result of the billions of years prokaryotes have had to test the limits of thermodynamics in terms of energy conservation.

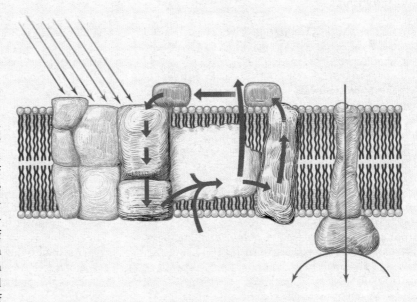

CHAPTER 15 Metabolic Diversity

15.1 Energy Conservation and Carbon Metabolism 574
15.2 Photosynthesis 575
15.3 The Role of Chlorophyll and Bacteriochlorophyll in Photosynthesis 576
15.4 Carotenoids and Phycobilins 580
15.5 Anoxygenic Photosynthesis 582
15.6 Oxygenic Photosynthesis 587
15.7 Autotrophic CO_2 Fixation: The Calvin Cycle 590
15.8 Autotrophic CO_2 Fixation: Reverse Citric Acid Cycle and the Hydroxypropionate Cycle 592
15.9 Chemolithotrophy: Energy from the Oxidation of Inorganic Electron Donors 592
15.10 Hydrogen Oxidation 594
15.11 Oxidation of Reduced Sulfur Compounds 595
15.12 Iron Oxidation 598
15.13 Nitrification 601
15.14 Methanotrophy and Methylotrophy 603
15.15 Anaerobic Respiration 605

15.16 Nitrate Reduction and the Denitrification Process 606
15.17 Sulfate Reduction 608
15.18 Acetogenesis 611
15.19 Methanogenesis 613
15.20 Ferric Iron, Manganese, Chlorate, and Organic Electron Acceptors 617
15.21 Fermentations: Energetic and Redox Considerations 620
15.22 Fermentative Diversity 622
15.23 Syntrophy 624
15.24 Hexose, Pentose, and Polysaccharide Utilization 626
15.25 Organic Acid Metabolism 629
15.26 Lipids as Microbial Nutrients 630
15.27 Molecular Oxygen (O_2) as a Reactant in Biochemical Processes 631
15.28 Hydrocarbon Transformations 632
15.29 Nitrogen Fixation 634

WORKING GLOSSARY

Anaerobic respiration respiration in which some substance such as SO_4^{2-} or NO_3^- serves as terminal electron acceptor instead of O_2

Anoxic oxygen-free

Anoxygenic photosynthesis photosynthesis in which O_2 is not produced

Bacteriochlorophyll the chlorophyll pigment of anoxygenic phototrophs

Calvin cycle the biochemical route of CO_2 fixation in many autotrophic organisms

Carotenoids hydrophobic accessory pigments present along with chlorophyll in photosynthetic membranes

Chemolithotroph a microorganism capable of oxidizing inorganic compounds as energy sources

Chlorophyll the light-sensitive, Mg-containing porphyrin of photosynthetic organisms that initiates the process of photophosphorylation

Denitrification anaerobic respiration in which NO_3^- is reduced to gaseous nitrogen compounds, primarily N_2

Disproportionation splitting of a compound into two new compounds, one more oxidized and one more reduced than the original compound

Fermentation anaerobic catabolism of an organic compound in which the compound serves as both an electron donor and an electron acceptor and in which ATP is produced by substrate-level phosphorylation

Homoacetogenesis (acetogenesis) energy metabolism involving the production of acetate from either H_2 plus CO_2 or from organic compounds

Hydrogenase an enzyme, widely distributed in anaerobic microorganisms, capable of taking up or evolving H_2

Methanogenesis the biological production of methane (CH_4)

Methylotrophy energy metabolism in which methyl groups or methane are oxidized as electron donors

Mixotrophic a nutritional state in which an inorganic compound serves as energy source (electron donor) and organic compounds serve as carbon source

Monooxygenase an enzyme that catalyzes the incorporation of one atom of O_2 into a substrate while the other atom is reduced to H_2O

Nitrification the microbial conversion of NH_3 to NO_3^-

Nitrogenase an enzyme capable of reducing N_2 to NH_3 in the process of nitrogen fixation

Nitrogen fixation the biological reduction of N_2 to NH_3 by nitrogenase

Oxygenic photosynthesis photosynthesis carried out by cyanobacteria and green plants in which O_2 is evolved

Photophosphorylation the production of ATP in photosynthesis

Phototroph an organism capable of using light as an energy source

Reaction center a photosynthetic complex containing chlorophyll (or bacteriochlorophyll) and several other components within which occurs the initial electron transfer reactions of photosynthetic electron flow

Recalcitrant resistant to microbial attack

Reversed electron transport the energy-dependent movement of electrons against the thermodynamic gradient to form a strong reductant from a weaker electron donor

Syntrophy a process whereby two or more microorganisms cooperate to degrade a substance neither can degrade alone

---

W e have seen in the preceding two chapters a large diversity of different prokaryotes—differing in morphology and physiological and evolutionary properties, as well as the habitats they occupy. We consider here in more detail the metabolic diversity of key groups of prokaryotes, with special consideration of the biochemical processes that drive this diversity.

From an ecological perspective, the metabolic processes that we are about to consider are some of the most significant on Earth. The organisms that carry them out play crucial roles in maintaining the web of life and are of great importance in agriculture and other aspects of human affairs, as we will discuss in Chapter 16. We begin with a refresher on energetics and carbon metabolism.

15.1

Energy Conservation and Carbon Metabolism

Collectively, microorganisms, especially the prokaryotes, show an impressive diversity of metabolic processes. A major focus of this chapter will be on reactions that allow for *energy conservation*. That is, as a chemical reaction proceeds, energy is released that an organism can conserve, usually as ATP. This ATP is produced by reactions leading to *substrate-level phosphorylation*, characteristic of fermentations, or by *oxidative phosphorylation*, driven by electron transport and the proton motive force (Chapter 4).

We will see that energy conservation can occur from reactions that take place under both oxic and anoxic conditions. Although oxygen is required for aerobic respiration, we will encounter a wide diversity of energy-conserving reactions that occur under anoxic conditions. For example, fermentations and anaerobic respirations all occur in the absence of O_2, and even a major form of photosynthesis, anoxygenic photosynthesis, is an anaerobic process. Indeed, the great diversity of anaerobic forms of metabolism we see in prokaryotes undoubtedly is a major reason these organisms have colonized every conceivable habitat on Earth. Thus, metabolic diversity should be looked at as the result of the evolution of microorganisms to exploit what nature has given them to work with.

Figure 15.1 reviews the basic types of energy metabolism introduced in Chapter 4 and lists some of the terms that have been used to describe organisms based on their carbon and energy metabolism. As seen here, energy conservation can be the result of either chemical reactions or photochemical reactions. Many microorganisms use *organic* chemicals as electron donors in energy metabolism; we call these organisms **chemoorganotrophs.** Those that use *inorganic* chemicals are called **chemolithotrophs.** Organisms that can use *light* are called **phototrophs** and include many microorganisms, as well as green plants.

Phototrophs and chemolithotrophs can frequently use carbon dioxide (CO_2) as the sole carbon source, and we use the term **autotroph** to describe such a lifestyle. Organisms that use organic carbon are called **heterotrophs.** However, it is frequently necessary to specify *exactly* how an organism is growing with respect to its source of energy and carbon, and when we do, we need to combine terms that refer to energy with those that refer to carbon. With phototrophs, for example, many phototrophic bacteria use light as their energy source, but organic carbon as their carbon source; such a lifestyle is called **photoheterotrophy** (compare this with photoautotrophy in Figure 15.1). Chemolithotrophs that grow using an inorganic electron donor and an organic carbon source are called **mixotrophs** (Figure 15.1). **Chemoorganotrophs,** of course, use organic carbon for both energy and carbon purposes.

With these terms to guide us, let us begin our survey of metabolic diversity with a consideration of photosynthesis, the use of light as an energy source to drive the photoautotrophic incorporation of CO_2 or photoheterotrophy.

15.2

Photosynthesis

One of the most important biological processes on Earth is **photosynthesis,** the conversion of light energy to chemical energy. As we have just said, organisms that can carry out photosynthesis are called *phototrophs.* Most phototrophic organisms are also *autotrophs,* capable of growing with CO_2 as sole carbon source. Energy from light is thus used in the reduction of CO_2 to organic compounds. The ability to photosynthesize depends on the presence of light-sensitive pigments, the *chlorophylls,* found in plants, algae, and some bacteria. Light reaches phototrophic organisms in distinct units of energy called *quanta.* Absorption of light quanta by chlorophyll pigments begins the process of photosynthetic energy conversion.

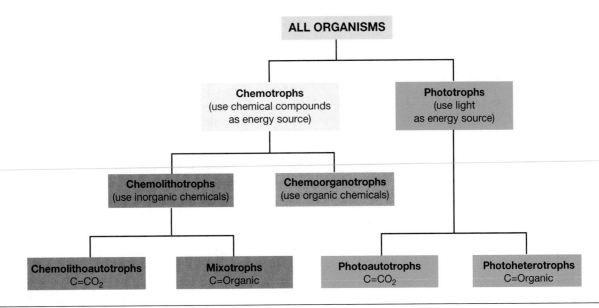

FIGURE 15.1 Classification of organisms in terms of energy and carbon sources.

Light and Dark Reactions

The growth of a photoautotroph can be characterized by two distinct sets of reactions: the **light reactions,** in which light energy is conserved as chemical energy, and the **dark reactions,** in which this chemical energy is used to reduce CO_2 to organic compounds. For autotrophic growth, energy is supplied in the form of adenosine triphosphate (ATP), while electrons for the reduction of CO_2 come from NADH or NADPH.* The latter is produced by the reduction of NAD^+ or $NADP^+$ by electrons originating from various electron donors to be discussed later.

The light reactions thus bring about the conservation of some of the energy in light to a form of chemical energy, ATP, that can be used by cells. To drive autotrophic reactions some phototrophic bacteria obtain reducing power from electron donors in their environment, typically reduced sulfur sources (H_2S, S^0, $S_2O_3^{2-}$) or H_2. By contrast, green plants, algae, and cyanobacteria use H_2O, an inherently poor electron donor (∞ Figure 4.7), as a source of reducing power to reduce $NADP^+$ to NADPH, producing molecular oxygen, O_2, as a by-product (Figure 15.2). Because O_2 is produced, photosynthesis in these organisms is called **oxygenic.** In other phototrophic bacteria no O_2 is pro-

duced, and thus the process is called **anoxygenic** photosynthesis. We will see later that although the production of NADH from substances like H_2S by anoxygenic phototrophs may or may not be directly driven by light, the oxidation of H_2O to O_2 by oxygenic phototrophs is (Figure 15.2); thus these organisms require light for *both* reducing power synthesis and energy conservation.

The "dark reactions" of photosynthesis are so named because unlike the light reactions, where light is absolutely essential to drive the photochemical reactions, the dark reactions do not require light, *per se*, but instead, the *products* of the light-driven reactions, ATP and NADH.

✓ 15.1 and 15.2 Concept Checks

Depending on the electron donors and acceptors used and the nature of the carbon source used, organisms can be classified into several groups. The major differentiating feature of energy metabolism is the use of light (phototrophs) versus chemicals (chemotrophs). Photosynthesis can be considered in terms of the light reactions, where ATP is generated, and the dark reactions, where ATP is consumed in the fixation of CO_2.

✓ How does a *chemoorganotroph* differ from a *chemolithotroph*?
✓ How are *photoautotrophs* and *photoheterotrophs* similar and how do they differ?
✓ What is the fundamental difference between an *oxygenic* and an *anoxygenic* phototroph?

15.3

The Role of Chlorophyll and Bacteriochlorophyll in Photosynthesis

Photosynthesis occurs only in organisms that possess some type of **chlorophyll.** Chlorophyll is a porphyrin, as are the cytochromes (∞ Section 4.10), but unlike the cytochromes, chlorophyll contains a *magnesium* atom instead of an iron atom at the center of the porphyrin ring. Chlorophyll also contains specific substituents bonded to the porphyrin ring, as well as a hydrophobic alcohol molecule. Because of this alcohol side chain, chlorophyll associates with lipid and hydrophobic proteins of photosynthetic membranes.

The structure of chlorophyll *a*, the principal chlorophyll of higher plants, most algae, and the cyanobacteria, is shown in Figure 15.3. Chlorophyll *a* is green in color because it *absorbs* red and blue light preferentially and *transmits* green light. We discussed the electromagnetic spectrum in Section 9.3 (∞ Figure 9.6). The spectral properties of any pigment can best be expressed by its *absorption spectrum*, which indicates the degree to which the pigment absorbs light of different wavelengths. The absorption spectrum of cells containing chlorophyll *a* shows strong absorp-

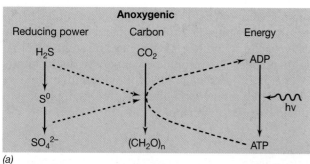

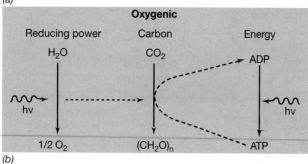

(a)

(b)

FIGURE 15.2 Energy and reducing power synthesis in (a) anoxygenic versus (b) oxygenic phototrophs. Although both types of phototrophs obtain their energy from light (hv), in oxygenic phototrophs light also drives the oxidation of water to oxygen.

_____
*In oxygenic phototrophs the reduced substance is NADPH, while in anoxygenic phototrophs it is NADH.

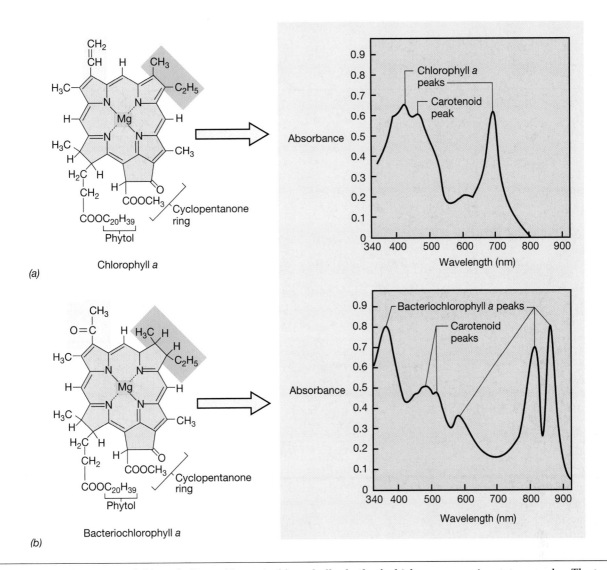

FIGURE 15.3 Structures of chlorophyll *a* and bacteriochlorophyll *a*, both of which are magnesium tetrapyrroles. The two molecules are identical except for those portions contrasted in yellow and green. The central Mg atom is shown in blue. The absorption spectra alongside each molecule are: (a) cells of the green alga *Chlamydomonas*. The peaks at 680 and 430 nm are due to chlorophyll *a*; the peak at 480 nm is due to carotenoids; (b) cells of the phototrophic purple bacterium *Rhodopseudomonas palustris*. Peaks at 870, 800, 590, and 360 nm are due to bacteriochlorophyll *a* while peaks at 525 and 475 nm are due to carotenoids.

tion of red light (maximum absorption at a wavelength of 680 nm) and blue light (maximum at 430 nm) (Figure 15.3a).

There are a number of chemically different chlorophylls that are distinguished by their different absorption spectra. Chlorophyll *b*, for instance, absorbs maximally at 660 nm rather than at 680 nm. Many plants have more than one chlorophyll, but the most common are chlorophylls *a* and *b*. Among prokaryotes, cyanobacteria have chlorophyll *a* but anoxygenic phototrophs, such as the purple and green bacteria, can have any of a number of bacteriochlorophylls (Figures 15.3 and 15.4). Bacteriochlorophyll *a* (Figure 15.3b), present in most purple bacteria (∞ Section 13.1), absorbs

maximally between 800–925 nm, depending on the species of phototroph. Other bacteriochlorophylls, whose distribution runs along phylogenetic lines, absorb in other regions of the visible and infrared spectrum (Figure 15.4).

Why do organisms have several kinds of chlorophylls absorbing light at different wavelengths? One reason appears to be to make it possible to use more of the energy of the electromagnetic spectrum. Only light energy that is *absorbed* can be used to make energy, and thus by having different pigments, two unrelated organisms can coexist in a habitat, each using wavelengths of light that the other is not using. Thus, pigment diversity has ecological significance.

Pigment	R$_1$	R$_2$	R$_3$	R$_4$	R$_5$	R$_6$	R$_7$	Infrared absorption maxima (nm) In vivo	Extract (methanol)
Bacterio-chlorophyll a (purple bacteria)	$-\overset{\displaystyle O}{\underset{\displaystyle \parallel}{C}}-CH_3$	$-CH_3{}^b$	$-CH_2-CH_3$	$-CH_3$	$-\overset{\displaystyle O}{\underset{\displaystyle \parallel}{C}}-O-CH_3$	P/Gg^a	$-H$	805 830–890	771
Bacterio-chlorophyll b (purple bacteria)	$-\overset{\displaystyle O}{\underset{\displaystyle \parallel}{C}}-CH_3$	$-CH_3{}^c$	$=\underset{\displaystyle H}{C}-CH_3$	$-CH_3$	$-\overset{\displaystyle O}{\underset{\displaystyle \parallel}{C}}-O-CH_3$	P	$-H$	835–850 1020–1040	794
Bacterio-chlorophyll c (green sulfur bacteria)	$-\overset{\displaystyle H}{\underset{\displaystyle OH}{C}}-CH_3$	$-CH_3$	$-C_2H_5$ $-C_3H_7{}^d$ $-C_4H_9$	$-C_2H_5$ $-CH_3$	$-H$	F	$-CH_3$	745–755	660–669
Bacterio-chlorophyll c_s (green nonsulfur bacteria)	$-\overset{\displaystyle H}{\underset{\displaystyle OH}{C}}-CH_3$	$-CH_3$	$-C_2H_5$	$-CH_3$	$-H$	S	$-CH_3$	740	667
Bacterio-chlorophyll d (green sulfur bacteria)	$-\overset{\displaystyle H}{\underset{\displaystyle OH}{C}}-CH_3$	$-CH_3$	$-C_2H_5$ $-C_3H_7$ $-C_4H_9$	$-C_2H_5$ $-CH_3$	$-H$	F	$-H$	705–740	654
Bacterio-chlorophyll e (green sulfur bacteria)	$-\overset{\displaystyle H}{\underset{\displaystyle OH}{C}}-CH_3$	$-\overset{\displaystyle }{\underset{\displaystyle O}{\overset{\parallel}{C}}}-H$	$-C_2H_5$ $-C_3H_7$ $-C_4H_9$	$-C_2H_5$ $-CH_3$	$-H$	F	$-CH_3$	719–726	646
Bacterio-chlorophyll g (heliobacteria)	$-\overset{\displaystyle H}{C}=CH_2$	$-CH_3{}^b$	$-C_2H_5$	$-CH_3$	$-\overset{\displaystyle O}{\underset{\displaystyle \parallel}{C}}-O-CH_3$	F	$-H$	670, 788	765

^{a}P, Phytyl ester (C$_{20}$H$_{39}$O—); F, farnesyl ester (C$_{15}$H$_{25}$O—); Gg, geranylgeraniol ester (C$_{10}$H$_{17}$O—); S, stearyl alcohol (C$_{18}$H$_{37}$O—).
^{b}No double bond between C$_3$ and C$_4$; additional H atoms are in positions C$_3$ and C$_4$.
^{c}No double bond between C$_3$ and C$_4$; an additional H atom is in position C$_3$.
^{d}Bacteriochlorophylls c, d and e consist of isomeric mixtures with the different substituents on R$_3$ as shown.

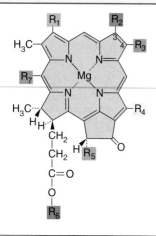

FIGURE 15.4 Structure of all known bacteriochlorophylls. The different substituents present in the positions R$_1$ to R$_7$ are given in the accompanying table.

Photosynthetic Membranes and Reaction Center versus Antenna Pigments

Just where are the chlorophyll pigments located inside the cell? These pigments, and all the other components of the light-gathering apparatus, are associated with special membrane systems, the **photosynthetic membranes.** The location of the photosynthetic membranes within the cell differs between prokaryotic and eukaryotic microorganisms. In eukaryotes, photosynthesis is associated with special intracellular organelles, the **chloroplasts** (∞ Section 3.16). The chlorophyll pigments are attached to sheetlike (lamellar) membrane structures of the chloroplast (Figure 15.5b). These photosynthetic membrane systems are called **thylakoids;** stacks of thylakoids are called *grana* (Figure 15.5b). The

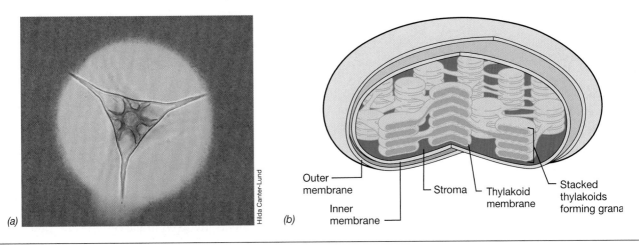

FIGURE 15.5 The chloroplast. (a) Photomicrograph of an algal cell showing chloroplasts. (b) Details of chloroplast structure, showing how the convolutions of the thylakoid membranes define an inner space called the stroma and form membrane stacks called grana.

thylakoids are so arranged that the chloroplast is divided into two regions, the matrix space that surrounds the thylakoids and the inner space within the thylakoid array (Figure 15.5b). This arrangement makes possible the development of a light-driven proton motive force that can be used to synthesize ATP, as will be described in Section 15.5.

In prokaryotes, chloroplasts are not present and photosynthetic pigments are integrated into internal membrane systems that arise from (1) invagination of the cytoplasmic membrane (purple bacteria) (see for example, Figures 13.3a and b; see also Figure 15.12), (2) the cytoplasmic membrane itself (heliobacteria) (Figure 13.62), (3) in both the cytoplasmic membrane and specialized non-unit membrane-enclosed structures called *chlorosomes* (green bacteria) (see Figure 15.7), or (4) in thylakoid membranes in cyanobacteria.

Within a photosynthetic unit membrane chlorophyll or bacteriochlorophyll molecules are associated with proteins to form complexes consisting of anywhere from 50–300 molecules (Figure 15.6). Only a very small number of these pigment molecules participate directly in the conversion of light energy to ATP—these are the **reaction center** chlorophylls or bacteriochlorophylls (Figure 15.6). These are surrounded by the more numerous **light-harvesting** or **antenna** chlorophylls. The antenna pigments function to harvest light and funnel the energy of light to the reaction center. At the low light intensities that often prevail in nature, this arrangement of pigment molecules allows for the capture and utilization of photons that would otherwise be insufficient to drive reaction center photochemistry by themselves.

The ultimate in low-light efficiency is found in the **chlorosome** of green sulfur bacteria and *Chloroflexus* (Figure 15.7). This structure functions as a giant antenna system, but unlike that described earlier, the bacteriochlorophyll molecules in the chlorosome are not

associated with proteins and instead function much like a solid state circuit, absorbing extremely low light intensities and transferring the energy to bacteriochlorophyll *a* in the reaction center located in the cytoplasmic membrane (Figure 15.7). This arrangement is highly ef-

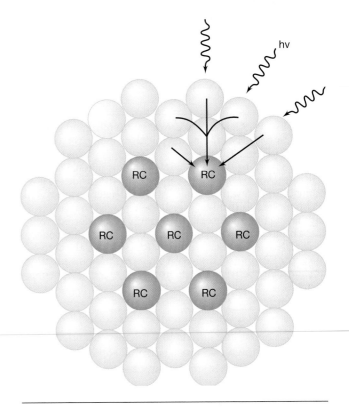

FIGURE 15.6 Model for the arrangement of light-harvesting chlorophylls/bacteriochlorophylls versus reaction centers within a photosynthetic membrane. Light energy, absorbed by light-harvesting molecules (light green), is transferred to the reaction centers (dark green, RC) where photosynthetic electron transport reactions begin. All pigment molecules are held in place within the membrane by specific pigment-binding proteins. Compare this figure to Figures 15.13 and 15.16.

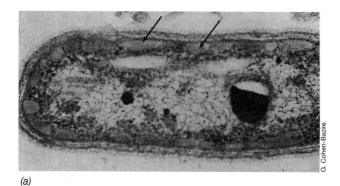

(a)

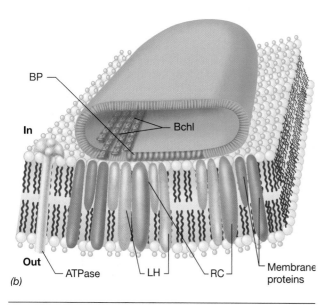

(b)

FIGURE 15.7 The chlorosome of green sulfur and green nonsulfur bacteria. (a) Electron micrograph of a cell of the green sulfur bacterium *Pelodictyon clathratiforme*. Note the chlorosomes (arrows). (b) Model of chlorosome structure. The chlorosome (green) lies appressed to the inside surface of the cytoplasmic membrane. Antenna bacteriochlorophyll (Bchl) molecules (Bchls *c, d,* or *e*) are arranged in tubelike arrays inside the chlorosome, and energy is transferred from these bacteriochlorophylls through light-harvesting Bchl *a* molecules (LH) to reaction center (RC) Bchl *a* in the cytoplasmic membrane (blue). Base plate (BP) proteins function as connectors between the chlorosome and the cytoplasmic membrane.

ficient for absorbing light at low intensities, and, indeed, it has been shown that green sulfur bacteria can grow at the lowest light intensities of any known phototrophs.

✓ **15.3 Concept Check**

The central pigment of photosynthesis is chlorophyll (or bacteriochlorophyll). Chlorophylls are located in photosynthetic membranes where the light reactions of photosynthesis are carried out. Most chlorophyll molecules are antenna molecules and function only to harvest light energy and transfer it on to reaction center chlorophylls.

✓ Why is it necessary for chlorophyll pigments to be located in membranes?
✓ What is the difference between *antenna* and *reaction center* chlorophyll molecules? Which are more abundant in the cell and why?
✓ What pigments are found within the *chlorosome*?

15.4

Carotenoids and Phycobilins

Although chlorophyll or bacteriochlorophyll is obligatory for photosynthesis, phototrophic organisms have other pigments involved in the capture and processing of light energy. These include the **carotenoids** and the **phycobilins.** They function as *accessory* pigments and primarily play a photoprotective role (carotenoids) or function in light-harvesting (phycobilins). We consider each of these groups of pigments now.

Carotenoids

The most widespread accessory pigments are the **carotenoids,** which are always found in phototrophic organisms. Carotenoids are hydrophobic pigments firmly embedded in the membrane; the structure of a typical carotenoid is shown in Figure 15.8. Carotenoids have long hydrocarbon chains with alternating C—C and C=C bonds, an arrangement called a *conjugated* double-bond system. As a rule, carotenoids are yellow, red, brown, or green (∞ Figure 13.2) and absorb light in the blue region of the spectrum (see Figure 15.3). The types and structures of carotenoids of various phototrophs have been well studied, and the major carotenoids of anoxygenic phototrophs are shown in Figure 15.9. These pigments are responsible for the brilliant colors of red, purple, pink, green, yellow, or brown that are observed in different species of anoxyphototrophs (∞ Figures 13.2 and 13.5).

Carotenoids are closely associated with chlorophyll or bacteriochlorophyll in the photosynthetic membrane but do not function directly in photophosphorylation reactions. They can, however, transfer energy to the reaction center, and this transferred energy may be used

FIGURE 15.8 Structure of β-carotene, a typical carotenoid. The conjugated double-bond system is highlighted in orange.

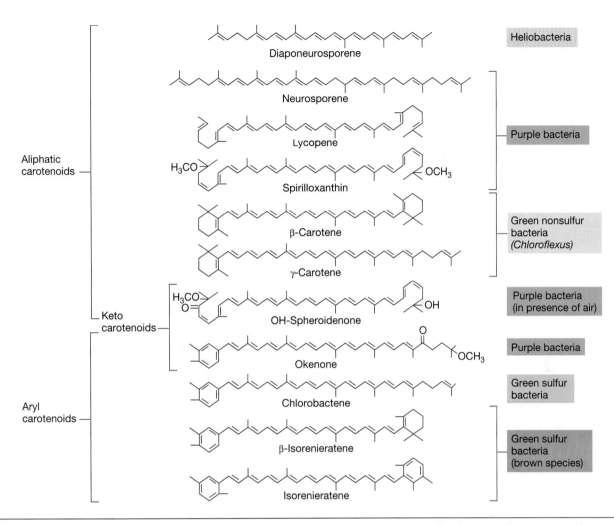

FIGURE 15.9 Structures of some common carotenoids found in anoxygenic phototrophs. Compare the structure of β-carotene portrayed in Figure 15.8 with how it is drawn here. For simplicity, in the structures shown here, methyl (CH$_3$) groups are designated by their bonds only.

in photophosphorylation in the same way as light energy captured directly by chlorophyll. Another function of the carotenoids is as photoprotective agents. Bright light can often be harmful to cells in that it causes photooxidation reactions that can lead to the production of toxic oxygen species such as singlet oxygen ($^1O_2$) (◌◌◌ Section 5.12) and destruction of the photosynthetic apparatus itself. Carotenoids quench toxic oxygen species and absorb much of this harmful light. Because phototrophic organisms must by their very nature live in the light, the photoprotective role of carotenoids is thus an obvious advantage.

Phycobilins and Phycobilisomes

Cyanobacteria and red algal chloroplasts contain **phycobiliproteins,** which are the main light-harvesting pigments of these organisms. Phycobiliproteins are red or blue and consist of open-chain tetrapyrroles coupled to proteins (Figure 15.10a). The red pigment, called *phycoerythrin,* absorbs light most strongly at wavelengths around 550 nm, whereas the blue pigment, *phycocyanin* (Figure 15.10a), absorbs most strongly at 620 (Figure 15.11). A third pigment, called *allophycocyanin* absorbs at about 650 nm.

Phycobiliproteins occur as high-molecular-weight aggregates, called **phycobilisomes,** attached to the photosynthetic membranes (Figure 15.10b). Phycobilisomes are constructed in such a way that the allophycocyanin molecules make physical contact with the photosynthetic membrane and are surrounded by molecules of phycocyanin and phycoerythrin. The latter pigments absorb shorter (higher energy) wavelengths of light and transfer the energy to allophycocyanin, which is closely linked to the reaction center chlorophyll and transfers energy to this site. The phycobilisome thus yields very efficient energy transfer from the biliprotein complex to chlorophyll *a,* which al-

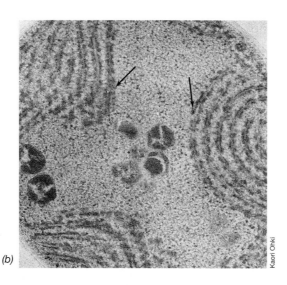

(a)   Phycocyanin   (b)

FIGURE 15.10   Phycobilins and phycobilisomes. (a) A typical phycobilin. This compound is an open-chain tetrapyrrole derived biosynthetically from a closed porphyrin ring by loss of one carbon atom as carbon monoxide. The structure shown is the prosthetic group of phycocyanin, a proteinaceous pigment found in cyanobacteria and red algae. (b) Electron micrograph of a thin section of the cyanobacterium *Synechocystis* sp. Note the darkly staining ball-like phycobilisomes (arrows) attached to the lamellar membranes.

lows for growth of cyanobacteria at fairly low light intensities. Indeed, phycobilisome content *increases* in cells of cyanobacteria as light intensity *decreases*, such that phycobilisome-rich cells are those grown at the *lowest* light intensities.

The light-gathering function of accessory pigments like carotenoids and phycobilins is an obvious advantage for the organism. Light from the sun is distributed over the whole visible range, yet chlorophylls absorb well in only part of this spectrum. By having accessory pigments, the organism is able to capture more of the available light (Figures 15.3 and 15.11).

✓ 15.4 Concept Check

Accessory pigments such as carotenoids and phycobilins can absorb light and transfer the energy to reaction center chlorophyll, thus broadening the wavelengths of light usable in photosynthesis. Carotenoids also play an important photoprotective role in preventing photooxidative damage to the cell.

✓ What are the functions of carotenoids in cells?
✓ How does the structure of a phycobilin compare with that of a chlorophyll?

15.5

Anoxygenic Photosynthesis

The process of light-mediated ATP synthesis in all phototrophic organisms involves electron transport through a sequence of electron carriers. These electron carriers are arranged in the photosynthetic membrane in series from those with electronegative to those with more electropositive reduction potentials. We now consider the structure of the photosynthetic apparatus in anoxygenic phototrophs and the details of photosynthetic electron flow in purple bacteria, where much is known concerning the molecular events of photosynthesis.

Structure of the Purple Bacterial Photosynthetic Apparatus

The photosynthetic apparatus of purple phototrophic bacteria is contained in intracytoplasmic membrane systems of various morphologies. Membrane vesicles (chromatophores) (Figure 15.12) are a commonly observed membrane type. The photosynthetic apparatus consists of four membrane-bound pigment–protein complexes plus an ATPase complex that drives

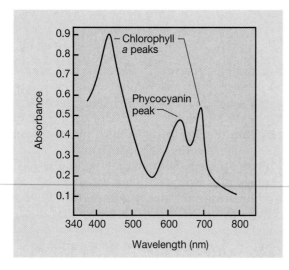

FIGURE 15.11   The absorption spectrum of a cyanobacterium that has a phycobiliprotein (phycocyanin) as an accessory pigment. Note how the presence of phycocyanin broadens the wavelengths of usable light energy (between 600 and 700 nm). Compare with Figure 15.3.

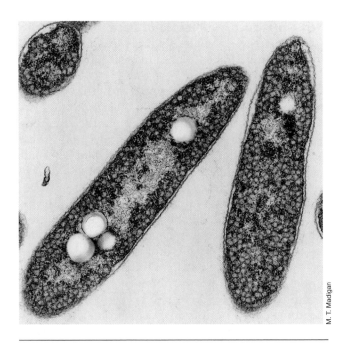

FIGURE 15.12 Chromatophores. Section through a cell of the phototrophic purple bacterium *Rhodobacter capsulatus* containing an abundance of vesicular photosynthetic membranes. The vesicles arise by invagination of the cytoplasmic membrane. The clear areas are regions in the cell in which the reserve polymer, poly-β-hydroxybutyrate (Section 3.13), was stored.

ATP synthesis at the expense of a proton motive force (Section 4.11). Three of the four complexes specific to photosynthesis are the *reaction center, light-harvesting I*, and *light-harvesting II* components. The fourth complex of the photosynthetic apparatus, the *cytochrome bc_1 complex*, is common to both respiratory and photosynthetic electron flow. We discussed the structure and function of the cytochrome bc_1 complex in Section 4.11.

The purple bacterial photosynthetic reaction center has been crystallized and its structure determined to atomic resolution by X-ray diffraction (Figure 15.13a). Reaction centers of purple bacteria contain three polypeptides, designated the L, M, and H subunits. These proteins are firmly embedded in the photosynthetic membrane and traverse the membrane several times (Figure 15.13b). The L, M, and H polypeptides bind the reaction center photochemical complex, which consists of two molecules of bacteriochlorophyll *a* called the *special pair*, two additional bacteriochlorophyll *a* molecules whose function is unknown, two molecules of *bacteriopheophytin* (bacteriochlorophyll *a* minus its magnesium atom), two molecules of quinone, and two molecules of a carotenoid pigment. All components of the reaction center are integrated in such a way that they can interact in very fast electron transfer reactions that, as we will see, ultimately result in ATP production.

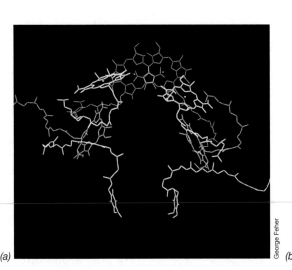

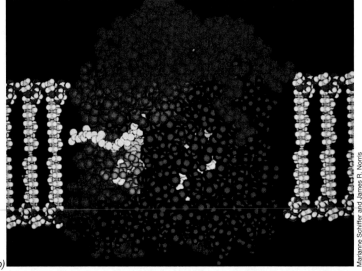

FIGURE 15.13 Structure of the reaction center of purple phototrophic bacteria. (a) Arrangement of components in the reaction center. The "special pair" of bacteriochlorophyll molecules are overlapping and shown in red, and molecules of quinone are in dark yellow and point downward in the figure. The accessory bacteriochlorophylls are in lighter yellow near the special pair, and the bacteriopheophytin molecules are shown in blue. (b) Molecular model of the protein structure of the reaction center. The pigments discussed in (a) are bound to membranes by three reaction center proteins called protein H (blue), protein M (red), and protein L (green). The reaction center pigment–protein complex is integrated into the lipid bilayer.

Photosynthetic Electron Flow in Purple Bacteria

It should be recalled that the photosynthetic reaction center is surrounded by light-harvesting antenna bacteriochlorophyll a molecules that function to funnel light energy to the reaction center (see Figure 15.13). Light energy is transferred from the antenna to the reaction center in packets called *excitons*, mobile electronic states that migrate through the antenna pigments to the reaction center at high efficiency. Photosynthesis begins when exciton energy strikes the special pair of bacteriochlorophyll a molecules (Figure 15.13a). The absorption of energy excites the special pair, converting it to a strong electron donor with a very low E_0'. Once this strong donor has been produced, the remaining steps in photosynthetic electron flow function to conserve the energy released when electrons are transported through a membrane from carriers of low E_0' to those of high E_0' (Figure 15.14).

Before excitation, the bacterial reaction center, which is referred to as *P870*, has an E_0' of about +0.5 V; after excitation it has a potential of about −1.0 V (Figure 15.14). The excited electron within P870 pro-

ceeds to reduce a molecule of bacteriopheophytin within the reaction center (Figures 15.13b and 15.14). This transition takes place incredibly fast, taking about three-trillionths of a second (3×10^{-12} sec) to occur. Once reduced, bacteriopheophytin a reduces several intermediate quinone molecules, with the electron eventually reducing a quinone in the "quinone pool" within the membrane. This transition is also very fast, taking less than one-billionth of a second (Figures 15.14 and 15.15). Relative to what has happened in the reaction center, further electron transport reactions occur rather slowly, on the order of microseconds to milliseconds. From the quinone, electrons are transported in the membrane through a series of iron–sulfur proteins and cytochromes (Figures 15.14 and 15.15), eventually returning to the reaction center. Key electron transport proteins include cytochrome bc_1 and cytochrome c_2 (Figure 15.14). Cytochrome c_2 is a periplasmic cytochrome and serves as an electron shuttle between the membrane-bound bc_1 complex and the reaction center (∞ Section 4.11 and Figures 15.14 and 15.15).

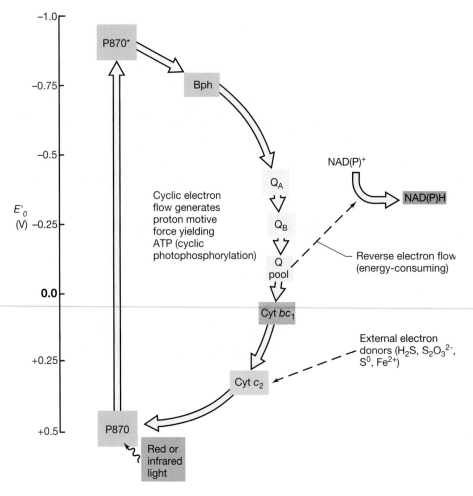

FIGURE 15.14 General scheme of electron flow in anoxygenic photosynthesis in a purple bacterium. Only a single light reaction occurs. Note how light energy converts a weak electron donor, P870, into a very strong electron donor, P870*, and that following this event, the remaining steps in photosynthetic electron flow are much the same as that of respiratory electron flow. RC, Reaction center; Bchl, bacteriochlorophyll; Bph, bacteriopheophytin; Q_A, Q_B, intermediate quinones; Q pool, quinone pool in membrane; Cyt, cytochrome.

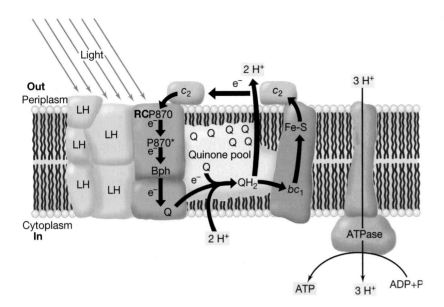

FIGURE 15.15 Arrangement of protein complexes in the photosynthetic membrane of a purple phototrophic bacterium. The light-generated proton gradient is used in the synthesis of ATP by the ATP synthase (ATPase). LH, Light-harvesting bacteriochlorophyll complexes; RC, reaction center; Bchl, bacteriochlorophyll; Bph, bacteriopheophytin; Q, quinone; FeS, iron–sulfur protein; bc_1, cytochrome bc_1 complex; c_2, cytochrome c_2. For description of the functioning of the cytochrome bc_1 complex and the Q cycle, the details of which are not shown here, see Figure 4.18.

Photophosphorylation

Synthesis of ATP during photosynthetic electron flow occurs as a result of the formation of a *proton motive force* generated by proton extrusion during electron transport and the activity of ATPases in coupling the dissipation of the proton motive force to ATP formation (∞ Section 4.11). The reaction series is completed when cytochrome c_2 donates an electron to the special pair bacteriochlorophylls (Figure 15.14), returning these molecules to their original ground state potential (E_0' = +0.5 V). The reaction center is then capable of absorbing new energy and repeating the process. This method of making ATP is called **cyclic photophosphorylation** because electrons are repeatedly moved around a closed circle. Cyclic photophosphorylation resembles respiration in that electron flow through the membrane establishes a proton motive force. However, unlike respiration, in cyclic photophosphorylation *there is no net input or consumption of electrons*; electrons simply travel a closed route.

The spatial relationships of the electron transport components in the bacterial photosynthetic membrane are illustrated in Figure 15.15. Note that as in respiratory electron flow (∞ Section 4.10), the cytochrome bc_1 complex interacts with the quinone pool during photosynthetic electron flow to allow functioning of the Q cycle (∞ Figure 4.18), a major means of establishing the proton motive force used to drive ATP synthesis (Figure 15.15).

Genetics of Bacterial Photosynthesis

Purple phototrophic bacteria are gram-negative prokaryotes (∞ Section 13.1), and certain species are readily amenable to genetic manipulation. Species of the genus *Rhodobacter*, especially *R. capsulatus* and *R. sphaeroides*, have been the main subjects of genetic research in bacterial photosynthesis. In *R. capsulatus*, most genes involved in photosynthesis are clustered in several operons that span a 45- to 50-kb region of the chromosome that has been called the **photosynthetic gene cluster** (Figure 15.16). Genes in the photosynthetic gene cluster encode proteins involved in bacteriochlorophyll biosynthesis (*bch* genes) and in carotenoid biosynthesis

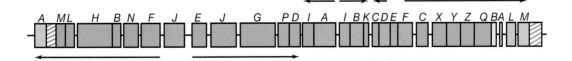

FIGURE 15.16 Map of the photosynthetic gene cluster of the purple phototrophic bacterium, *Rhodobacter capsulatus*. Genes are arranged in superoperons where transcripts of pigment biosynthesis operons extend through to include transcription of polypeptides of the photosynthetic complexes. The *bch* genes, which encode bacteriochlorophyll synthesis proteins, are shown in green, while *crt* genes, which encode proteins that synthesize carotenoids, are shown in red. Genes encoding reaction center polypeptides (*puh* and *puf* genes) are shown in blue, and genes encoding light-harvesting I polypeptides (B870 complex) (*puf* genes) are shown in yellow. Genes shown by diagonal lines are of unknown function. Not all genes have been given letter designations. Arrows indicate direction of transcription.

(*crt* genes) and polypeptides that bind pigment molecules in the reaction center and light-harvesting complexes (*puf* and *puh* genes) (Figure 15.16).

As can be imagined, synthesis of bacteriochlorophyll, carotenoids, and pigment binding proteins in phototrophic bacteria must be a highly coordinated process. When new photosynthetic complexes are synthesized, the correct proportions of each of the components of the complex must be available within the cell for final assembly. Biochemical and genetic analyses of photosynthesis in *Rhodobacter capsulatus* have shown that coordinate expression of photosynthetic components does indeed occur because the operons are arranged to form **superoperons.** Instead of terminating at the end of one operon, transcripts of pigment biosynthesis operons (*bch* and *crt*) (Figure 15.16) extend through the promoters and structural genes encoding the polypeptides of the photosynthetic complexes, yielding large transcripts encoding many proteins. Photosynthesis superoperons thus allow for transcription of many functionally related genes whose products interact and form the photosynthetic complexes that eventually integrate into the membrane. The master regulatory signal governing transcription of the photosynthetic gene cluster in these organisms is O_2. Molecular oxygen represses pigment synthesis such that photosynthesis in anoxygenic phototrophs occurs only under *anoxic* conditions.

Genetic analysis of photosynthesis in purple bacteria has been greatly assisted by the fact that *Rhodobacter* species are bioenergetically highly diverse; in addition to photosynthesis, these organisms can grow in darkness by respiration in the presence or absence of oxygen. Thus, mutants unable to photosynthesize are easily obtained and have been used in genetic exchange experiments (∞ Chapter 9) to characterize the number, organization, and expression of photosynthesis genes.

Autotrophy in Purple Bacteria: Electron Donors and Reverse Electron Flow

If a purple bacterium is going to grow autotrophically, the formation of ATP is not enough. Reducing power (NADH or NADPH) must also be made so that CO_2 can be reduced to the level of cell material. As previously mentioned, for purple sulfur bacteria this is usually H_2S, although S^0, $S_2O_3^{2-}$ and even Fe^{2+} can be used by various species. When H_2S is the electron donor in purple sulfur bacteria, globules of S^0 are stored inside the cells (Figure 15.17a). How do these substances reduce NAD^+ to NADH?

Reduced subtances like **hydrogen sulfide** (H_2S) or **thiosulfate** ($S_2O_3^{2-}$) are oxidized by cytochromes of the *c* type and electrons from them eventually end up in the "quinone pool" of the photosynthetic membrane (see Figure 15.14). However, the E_0' (about 0 volts) of quinone is insufficiently negative to reduce NAD^+ ($-.32$ volts) directly, such that electrons from the quinone pool must be forced *backward*, against the thermodynamic gradient, to reduce NAD^+ to NADH (Figure 15.14). This energy-requiring process is called *reversed electron flow* and is driven by the energy inherent in the proton motive force. Reverse electron flow is also the mechanism

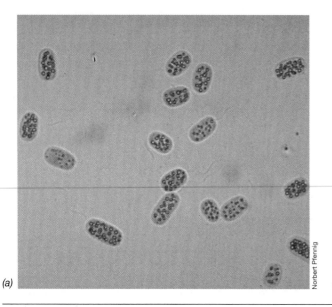

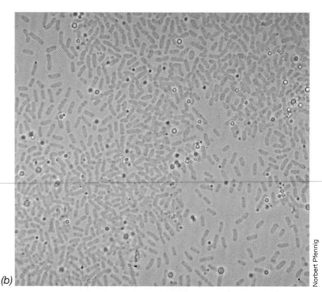

(a)   (b)

FIGURE 15.17   Photomicrographs of phototrophic bacteria taken by bright-field microscopy. (a) Purple bacterium: *Chromatium okenii*. Notice the sulfur granules deposited *inside* the cell. (b) Green bacterium: *Chlorobium limicola*. The refractile bodies are sulfur granules deposited *outside* the cell. In both cases the sulfur granules arise from the oxidation of H_2S to obtain reducing power.

by which chemolithotrophs make their reducing power, oftentimes from electron donors of extremely positive E_0' (see Sections 15.9–15.13).

Comparative Photosynthetic Electron Flow and Reducing Power Synthesis in Other Anoxyphototrophs

Our discussion of photosynthetic electron flow has thus far focused on purple bacteria. Although similar membrane components drive photophosphorylation in other anoxyphototrophs, there are differences in certain photochemical reactions that impact on reducing power biosynthesis. Figure 15.18 compares the light reactions of purple and green bacteria and the heliobacteria. Note that in the latter two groups the excited state of the reaction center bacteriochlorophylls resides at a significantly more negative E_0' and that actual *chlorophyll a* (green bacteria) or a structurally modified form of chlorophyll a called *hydroxychlorophyll a* (heliobacteria) are present in the reaction center. The significance of this is that unlike purple bacteria, where the first stable acceptor molecule (quinone) has an E_0' of about 0 volts, the acceptors of green bacteria and heliobacteria (FeS proteins) have a much more electronegative E_0', sufficiently negative in fact to reduce NAD^+ to NADH directly (via a small protein called *ferredoxin*), without the necessity for reverse electron flow (Figure 15.18). Thus, like oxygenic phototrophs (to be discussed next), in green and heliobacteria both ATP *and* reducing power

are direct products of the light reactions. When sulfide is the electron donor for reducing power synthesis in green bacteria, globules of S^0 are produced as in purple bacteria, but the globules remain *outside*, rather than *inside* the cells (Figure 15.17b).

✓ 15.5 Concept Check

A complex series of electron transport reactions occur in the photosynthetic reaction center of anoxygenic phototrophs, resulting in the formation of a proton motive force and the synthesis of ATP. The reducing power for CO_2 fixation comes from reductants present in the environment and usually requires reverse electron transport.

- ✓ How does photophosphorylation compare with electron transport phosphorylation in respiration?
- ✓ What is the value of having photosynthesis genes arranged in *superoperons*?
- ✓ What is *reverse electron flow* and why is it necessary?

15.6

Oxygenic Photosynthesis

In contrast to anoxyphototrophs, electron flow in oxygenic phototrophs involves two distinct, but interconnected, photochemical reactions. Oxygenic phototrophs use light to generate both ATP *and* NADPH, the electrons for the latter arising from the splitting of *water* into oxygen and electrons (see Figure 15.2). The two systems

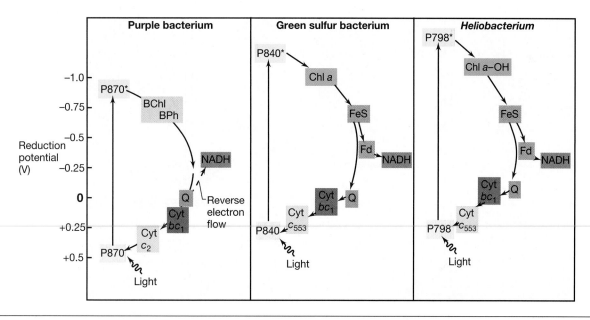

FIGURE 15.18 A comparison of electron flow in purple bacteria, green sulfur bacteria, and heliobacteria. Note how reverse electron flow in purple bacteria is necessitated by the fact that the primary acceptor (quinone, Q) is more positive in potential than the NAD^+/NADH couple. In green and heliobacteria NADH production is light-driven. Bchl, Bacteriochlorophyll; BPh, bacteriopheophytin. P870 and P840 are reaction centers of purple and green bacteria, respectively, and consist of Bchl a. The reaction center of heliobacteria (P798) contains Bchl g. The reaction center of *Chloroflexus* is similar to that of purple bacteria. Note the presence of forms of chlorophyll a in the reaction centers of green bacteria and heliobacteria.

of light reactions are called *photosystem I* and *photosystem II*, each photosystem having a spectrally distinct form of reaction center chlorophyll *a*. Photosystem I chlorophyll, called P700, absorbs light at long wavelengths (far red light), whereas photosystem II chlorophyll, called P680, absorbs at shorter wavelengths (near red light). Like anoxygenic photosynthesis, oxygenic photochemical reactions occur in membranes. In eukaryotic cells, these membranes are found in the *chloroplast*, whereas in cyanobacteria, photosynthetic membranes are arranged in stacks within the cytoplasm. In both groups of phototrophs the membranes are arranged in a similar way and the two forms of chlorophyll *a* are attached to specific proteins in the membrane and interact as shown in Figure 15.19.

Electron Flow in Oxygenic Photosynthesis

The path of electron flow in oxygenic phototrophs roughly resembles the letter Z turned on its side, and scientists studying oxygenic photosynthesis have come to refer to the electron flow of oxygenic phototrophs as the "Z" scheme. We should first note that the reduction potential of the P680 chlorophyll *a* molecule in photosystem II is very electropositive, slightly more positive than that of the O_2/H_2O couple. This facilitates the first step in oxygenic electron flow, the *splitting* of water into oxygen and hydrogen atoms (Figure 15.19), a thermodynamically unfavorable reaction. An electron from water is donated to the oxidized P680 molecule following the absorption of a quantum of light near 680 nm.

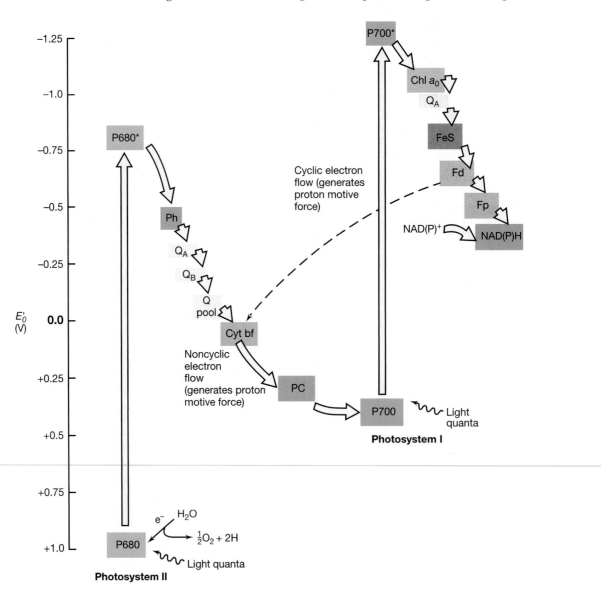

FIGURE 15.19   Electron flow in oxygenic (green plant) photosynthesis, the "Z" scheme. Two photosystems (PS) are involved, PS I and PS II. Ph, Pheophytin; Q, quinone; Chl, chlorophyll *a*; Cyt, cytochrome; PC, plastocyanin; FeS, nonheme iron–sulfur protein; Fd, ferredoxin; Fp, flavoprotein; P680 and P700 are the reaction center chlorophylls of PS II and PS I, respectively. Compare with Figure 15.14.

Light energy converts P680 into a moderately strong reductant, capable of reducing an intermediary molecule of E_0' about -0.5 V. The nature of this molecule is uncertain, but it is likely a pheophytin *a* molecule (chlorophyll *a* without the magnesium atom). From here the electron travels through several membrane carriers including quinones, cytochromes, and a copper-containing protein called **plastocyanin**; the latter donates electrons to photosystem I P700. The electron is accepted by the reaction center chlorophyll of photosystem I, P700, which has previously absorbed light quanta and begun the steps that lead to the reduction of $NADP^+$ (Figure 15.18). These involve electron transfer through several carriers of increasing E_0', terminating with the reduction of $NADP^+$ (Figure 15.19).

ATP Synthesis in Oxygenic Photosynthesis

Besides the net synthesis of reducing power (that is, NADPH), other important events take place while electrons flow in the membrane from one photosystem to another. During transfer of an electron from the acceptor in photosystem II to the reaction center chlorophyll molecule in photosystem I, electron transport occurs in a thermodynamically favorable (negative-to-positive) direction. This generates a proton motive force from which ATP can be produced. This type of ATP generation has been called *noncyclic* photophosphorylation because electrons whose transport results in ATP formation do not cycle back to reduce the oxidized P680: They are ultimately used in the reduction of $NADP^+$. When sufficient reducing power is present, ATP can also be produced in oxygenic phototrophs by *cyclic* photophosphorylation involving only photosystem I (Figure 15.18). This occurs when electrons travel from ferredoxin to the cytochrome *bf* complex from which electron transport returns the electron to P700. This flow creates a membrane potential and synthesis of additional ATP (see dashed line in Figure 15.19).

Anoxygenic Photosynthesis in Oxygenic Phototrophs

Photosystems I and II normally function together in the oxygenic process. However, under certain conditions many algae and some cyanobacteria are able to carry out cyclic photophosphorylation using *only* photosystem I, obtaining reducing power for CO_2 reduction from sources other than water. This is, in effect, photosynthesizing *anoxygenically,* as purple and green bacteria do.

A number of cyanobacteria can use H_2S as an electron donor for anoxygenic photosynthesis. When H_2S is used, it is oxidized to elemental sulfur (S^0), and sulfur granules similar to those produced by green sulfur bacteria (see Figure 15.17*b*) are deposited outside the

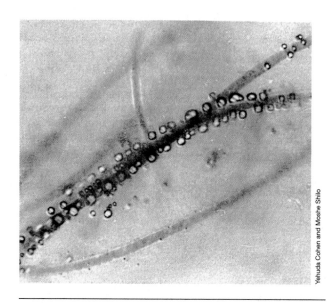

FIGURE 15.20   Cells of the filamentous cyanobacterium *Oscillatoria limnetica* grown anaerobically on sulfide as photosynthetic electron donor. Note the globules of elemental sulfur, the oxidation product of sulfide, formed *outside* the cells.

cells; an example of this is shown with the cyanobacterium *Oscillatoria limnetica* in Figure 15.20. This filamentous cyanobacterium lives in sulfide-rich saline ponds where it carries out anoxygenic photosynthesis along with photosynthetic green and purple bacteria and produces sulfur as an oxidation product of sulfide (Figure 15.20). In cultures of *O. limnetica*, electron flow from photosystem II is strongly inhibited by H_2S, thus necessitating anoxygenic photosynthesis if the organism is to survive.

From an evolutionary point of view, the existence of cyclic photophosphorylation indicates a close relationship between oxygenic and anoxygenic photosynthesis. Although organisms such as *Oscillatoria limnetica* that carry out oxygenic photosynthesis have acquired photosystem II, and hence the ability to split H_2O, they still retain the ability under certain conditions to use photosystem I alone.

✓ 15.6 Concept Check

In oxygenic photosynthesis, water is used as the source of electrons and oxygen is produced. Electron transport in oxygenic photosynthesis follows the Z scheme in which two separate light reactions are involved, photosystems I and II. Photosystem I resembles the system in anoxygenic photosynthesis. Photosystem II is responsible for splitting H_2O to yield $\frac{1}{2}O_2 + 2\,e^- + 2\,H^+$.

✓ Why is the term *noncyclic* electron flow used in reference to oxygenic photosynthesis?

✓ What is a major difference between the two reaction center chlorophyll molecules in photosystems I and II?

15.7

Autotrophic CO_2 Fixation: The Calvin Cycle

Several biochemical mechanisms are known for the fixation of CO_2 into cell material. In this section we consider the most widespread of these systems, the Calvin cycle, named for its discoverer, Melvin Calvin. The Calvin cycle requires NAD(P)H and ATP and two key enzymes, *ribulose bisphosphate carboxylase* and *phosphoribulokinase*. The remainder of the cycle is driven by a series of enzymes that are present in many organisms, both autotrophs and heterotrophs.

RubisCO and the Formation of PGA

The first step in CO_2 reduction in the Calvin Cycle is catalyzed by the enzyme ribulose bisphosphate carboxylase, or *RubisCO* for short. RubisCO is widely distributed, being present in purple bacteria, cyanobacteria, algae, and green plants, most chemolithotrophic Bacteria, and even in some Archaea, including structurally unique forms in hyperthermophilic Archaea. RubisCO catalyzes the for-

mation of two molecules of *3-phosphoglyceric acid* (PGA) from ribulose bisphosphate and CO_2 as shown in Figure 15.21. The PGA is then phosphorylated and reduced to a key intermediate of glycolysis, *glyceraldehyde 3-phosphate* (∞ Section 4.9). From here, glucose can be formed by reversal of the early steps in glycolysis. But thus far we have incorporated only one molecule of CO_2 and have consumed one molecule of ribulose bisphosphate. How do we make a full glucose molecule, and how do we regenerate the acceptor, ribulose bisphosphate?

We now consider reactions of the Calvin cycle based on the incorporation of 6 molecules of CO_2. For RubisCO to incorporate 6 molecules of CO_2, 6 ribulose bisphosphate molecules are required as acceptor molecules (Figure 15.22). This yields 12 molecules of 3-phosphoglyceric acid (a total of 36 carbon atoms). These 12 molecules serve as carbon skeletons to form 6 *new* molecules of ribulose bisphosphate (a total of 30 carbon atoms), and 1 molecule of hexose for cell biosynthesis. A complex series of rearrangements involving C_3, C_4, C_5, C_6, and C_7 intermediates finally yields the 6 molecules of ribulose 5-phosphate from which the 6 ribulose bisphosphates are generated. The final step in this regeneration is the

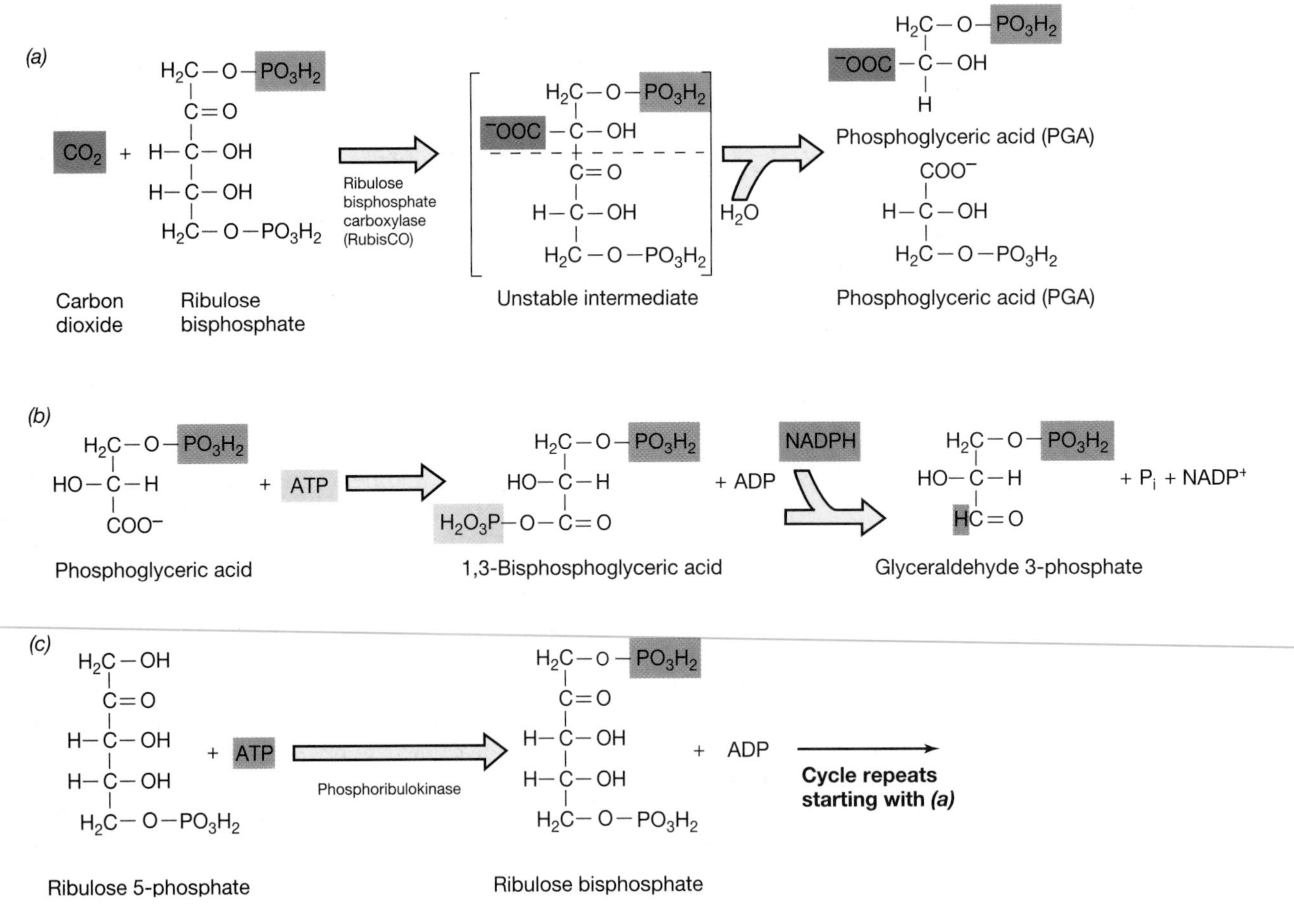

FIGURE 15.21   Key enzyme reactions of the Calvin cycle. (a) Reaction of the enzyme ribulose bisphosphate carboxylase. (b) Steps in the conversion of 3-phosphoglyceric acid (PGA) to glyceraldehyde 3-phosphate. Note that both ATP and NADPH are required. (c) Conversion of ribulose 5-phosphate to ribulose bisphosphate by the enzyme phosphoribulokinase.

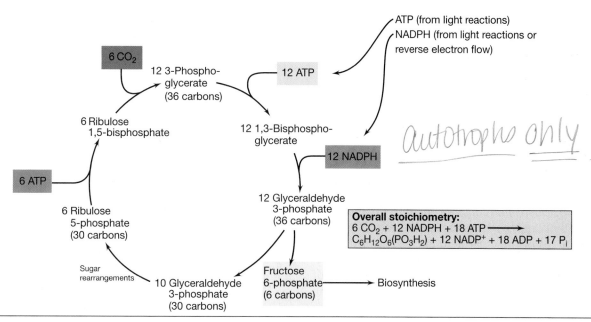

FIGURE 15.22   The Calvin cycle. For each *six* molecules of CO$_2$ incorporated, *one* fructose 6-phosphate is produced. Use the color coding here to follow the biochemical reactions occurring in Figure 15.21.

phosphorylation of ribulose 5-phosphate with ATP by the enzyme *phosphoribulokinase* (Figure 15.21c). This enzyme is another enzyme unique to the Calvin cycle.

Let us consider now the *overall* stoichiometry for conversion of 6 molecules of CO$_2$ into 1 molecule of fructose 6-phosphate (Figure 15.22). Twelve molecules each of ATP and NADPH are required for the reduction of 12 molecules of phosphoglyceric acid (PGA) to glyceraldehyde phosphate, and 6 ATP molecules are required for conversion of ribulose phosphate to ribulose bisphosphate. Thus, *12 NADPH and 18 ATP are required to synthesize 1 hexose molecule from 6 molecules of CO$_2$.* Hexose molecules can be converted to *storage polymers* such as glycogen, starch, or poly-β-hydroxyalkanoates (∞ Section 3.13) during periods when ATP and NADPH are abundant and then can be used later to build new cell material.

Carboxysomes

Several autotrophic prokaryotes that use the Calvin cycle for CO$_2$ fixation produce polyhedral cell inclusions called *carboxysomes*. The inclusions are about 100 nm in diameter; are surrounded by a thin, nonunit membrane; and consist of a tightly packed crystalline array of molecules of RubisCO (Figure 15.23). It is thought that carboxysomes are a mechanism to increase the amount of RubisCO in the cell to allow for more rapid CO$_2$ fixation without affecting the osmolarity of the cytoplasm (osmotic pressure is not affected because the carboxysome is insoluble). Carboxysomes have been found in obligately chemolithotrophic sulfur-oxidizing bacteria, the nitrifying bacteria, and cyanobacteria and prochlorophytes (∞ Sections 13.2, 13.3, 13.24, and 13.25, respectively), but not

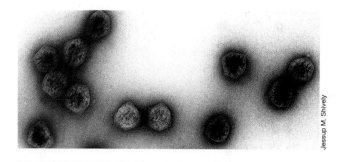

FIGURE 15.23   Carboxysomes purified from the chemolithotrophic sulfur oxidizer *Thiobacillus neapolitanus*. The structures are about 100 nm in diameter.

in facultative autotrophs (organisms that can grow either as autotrophs or as heterotrophs) like purple anoxyphototrophs, despite the fact that when these organisms grow as photoautotrophs, they use the Calvin cycle to fix CO$_2$. Thus, the carboxysome may be an evolutionary adaptation to life under strictly autotrophic conditions.

✓ 15.7 Concept Check

The fixation of CO$_2$ by most phototrophic and other autotrophic organisms occurs via the Calvin cycle, in which the enzyme ribulose bisphosphate carboxylase (RubisCO) plays a key role. The Calvin cycle is an energy-demanding process in which CO$_2$ is converted to cell material.

✓ What reaction does the enzyme *ribulose bisphosphate carboxylase* carry out?
✓ Why is reducing power needed for autotrophic growth?
✓ What is a *carboxysome*?

15.8

Autotrophic CO₂ Fixation: Reverse Citric Acid Cycle and the Hydroxypropionate Cycle

Alternative mechanisms of autotrophic CO_2 fixation are present in green sulfur bacteria and green nonsulfur bacteria. In the green sulfur bacterium *Chlorobium* (see Figures 15.7a and 15.7b) CO_2 fixation occurs by a reversal of steps in the citric acid cycle (∞ Figure 4.20), a pathway referred to as the *reverse citric acid cycle* (Figure 15.24a). *Chlorobium* contains two ferredoxin-linked enzymes that catalyze the reductive fixation of CO_2 into intermediates of the citric acid cycle. The two ferredoxin-linked reactions involve the carboxylation of succinyl-CoA to α-ketoglutarate and the carboxylation of acetyl-CoA to pyruvate (Figure 15.24a). Most of the other reactions of the reverse citric acid cycle are catalyzed by enzymes working in reverse of the normal oxidative direction of the cycle. One exception is *citrate lyase*, an ATP-dependent enzyme that cleaves citrate into acetyl-CoA and oxaloacetate in green sulfur bacteria. In the oxidative direction of the cycle, citrate is produced from these same precursors by the enzyme *citrate synthase* (∞ Figure 4.20).

The reverse citric acid cycle as a mechanism of autotrophy has also been found in certain nonphototrophic hyperthermophiles, including the Archaea *Sulfolobus* and *Thermoproteus* (∞ Section 14.8) and *Aquifex*, a very early branching autotroph on the phylogenetic tree of Bacteria (∞ Figure 13.1 and Section 13.34). This finding, of course, leaves open the possibility that this pathway is more widespread among autotrophic prokaryotes than previously thought. Interestingly, however, analysis of the completely sequenced genome of the green sulfur bacterium *Chlorobium tepidum* (∞ Section 9.12) has revealed RubisCO-like genes in this organism. If these indeed encode a functional RubisCO, the mechanism(s) of autotrophy in green bacteria and other organisms thought to use the reverse citric acid cycle may need to be reevaluated.

The green nonsulfur phototroph *Chloroflexus* grows autotrophically with either H_2 or H_2S as electron donors. However, neither the Calvin cycle nor the reverse citric acid cycle operates in this organism. Instead, two molecules of CO_2 are reduced to glyoxylate by a unique autotrophic pathway, *the hydroxypropionate pathway* (Figure 15.24b). This pathway leads to the synthesis of hydroxypropionate as a key intermediate. Thus far, the hydroxypropionate pathway has been confirmed only in *Chloroflexus*, and this is of evolutionary interest considering the fact that this organism is the earliest branching anoxyphototroph on the tree of Bacteria (∞ Figure 13.1). This suggests that the hydroxy-

propionate pathway may have been the first attempt at autotrophy in anoxygenic phototrophs and, because it is widely believed that these organisms evolved long before the cyanobacteria, perhaps it was the first autotrophic pathway in *any* phototrophic organism. However, many of the enzymes of the hydroxypropionate pathway have also been found in cells of the hyperthermophilic and nonphototrophic archaeon *Acidianus*, suggesting that, like the reverse citric acid cycle, this pathway may be more widespread than previously thought.

✓ 15.8 Concept Check

The reverse citric acid cycle and the hydroxypropionate cycle are pathways of CO_2 fixation found in green sulfur and green nonsulfur bacteria, respectively.

✓ Including the route of CO_2 fixation, discuss at least three ways that you could distinguish a purple sulfur bacterium from a green sulfur bacterium.
✓ Including the route of CO_2 fixation, what similarities and differences exist between green *sulfur* and green *nonsulfur* bacteria?

15.9

Chemolithotrophy: Energy from the Oxidation of Inorganic Electron Donors

Organisms that obtain energy from the oxidation of *inorganic* compounds are called **chemolithotrophs.** Most chemolithotrophic bacteria are also able to obtain all their carbon from CO_2, and they are therefore also autotrophs. As we have noted, two components are needed for growth on CO_2 as sole carbon source: energy in the form of ATP and reducing power. In chemolithotrophs, ATP generation is in principle similar to that in chemoorganotrophs, except that the electron donor is *inorganic* rather than organic. Thus, ATP synthesis is coupled to *oxidation* of the electron donor. Reducing power in chemolithotrophs is obtained either directly from the inorganic compound, if it has a sufficiently low reduction potential, or by *reverse electron transport reactions*, as discussed for phototrophic purple bacteria in Section 15.5.

There are many sources of inorganic electron donors for chemolithotrophs. These include geological, biological, and anthropogenic sources. Volcanic activity is a major source of reduced sulfur compounds, as is biological sulfate reduction. Agricultural and mining operations add inorganic electron donors to the environment, as does the burning of fossil fuels and the input of industrial wastes. The ecological success (∞ Chapter 16) and metabolic diversity (this chapter) of chemolithotrophs indicates

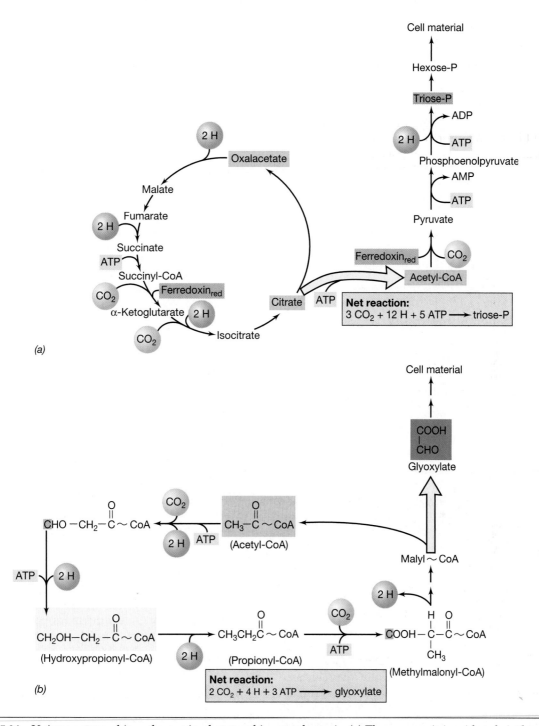

FIGURE 15.24 Unique autotrophic pathways in phototrophic green bacteria. (a) The reverse citric acid cycle is the mechanism of CO_2 fixation in the green sulfur bacterium *Chlorobium* (see also Figure 15.17*b*; ∞ Section 13.29). Ferredoxin$_{red}$ indicates carboxylation reactions requiring reduced ferredoxin (2 H each). Starting from oxalacetate, each turn of the cycle results in three molecules of CO_2 being incorporated and pyruvate as the product. The cleavage of citrate regenerates the C_4 acceptor oxalacetate and produces acetyl-CoA for biosynthesis. The conversion of pyruvate to phosphoenolpyruvate consumes two ~P equivalents. (b) The hydroxypropionate pathway is the means of autotrophy in the green nonsulfur bacterium *Chloroflexus* (∞ Section 13.32). Acetyl-CoA is carboxylated twice to yield methylmalonyl-CoA. This intermediate is rearranged to yield acetyl-CoA and glyoxylate. The latter is converted to cell material probably through a serine or glycine intermediate.

TABLE 15.1	Energy yields from the oxidation of various inorganic electron donors[a]					
Electron donor	Reaction	Type of chemolithotroph	E_0' of couple (V)	$\Delta G^{0\prime}$ (kJ/reaction)	Number of electrons	$\Delta G^{0\prime}$ (kJ/2 e$^-$)
Hydrogen	$H_2 + \frac{1}{2}O_2 \rightarrow H_2O$	Hydrogen bacteria	-0.42	-237.2	2	-237.2
Sulfide	$HS^- + H^+ + \frac{1}{2}O_2 \rightarrow S^0 + H_2O$	Sulfur bacteria	-0.27	-209.4	2	-209.4
Sulfur	$S^0 + 1\frac{1}{2}O_2 + H_2O \rightarrow SO_4^{2-} + 2\,H^+$	Sulfur bacteria	-0.25	-587.1	6	-195.7
Ammonium	$NH_4^+ + 1\frac{1}{2}O_2 \rightarrow NO_2^- + 2\,H^+ + H_2O$	Nitrifying bacteria	0^b	-274.7	6	-137.4
Nitrite	$NO_2^- + \frac{1}{2}O_2 \rightarrow NO_3^-$	Nitrifying bacteria	$+0.43$	-74.1	2	-75.8
Ferrous iron	$Fe^{2+} + H^+ + \frac{1}{4}O_2 \rightarrow Fe^{3+} + \frac{1}{2}H_2O$	Iron bacteria	$+0.77$	-32.9	1	-65.8

a Data calculated from values in Appendix 1; values for Fe^{2+} are for pH 2, and others are for pH 7. At pH 7 the Fe^{3+}/Fe^{2+} couple is about $+0.2$ V.
b E_0' of the NH_3/NH_2OH couple.

that sources and supplies of inorganic electron donors in nature are abundant.

Energetics of Chemolithotrophy

A review of reduction potentials listed in Table A1.2 reveals that a number of inorganic compounds can provide sufficient energy for ATP synthesis when O_2 is used as electron acceptor. Recall from Chapter 4 that the further apart in terms of E_0' two half reactions are, the greater the amount of energy released. For instance, the difference in reduction potential between the H^+/H_2 couple and the $\frac{1}{2}O_2/H_2O$ couple is -1.23 V, which is equivalent to a free-energy yield of -237 kJ/mol (see Appendix 1 for calculations). On the other hand, the potential difference between the H^+/H_2 couple and the NO_3^-/NO_2^- couple is less, -0.84 V, equivalent to a free-energy yield of -163 kJ/mol. This is still quite sufficient for the production of ATP (the high energy phosphate bond of ATP has a free energy of about -31.8 kJ/mol, see Table 15.5). However, a similar calculation will show that there is insufficient energy available from the oxidation of H_2S using CO_2 as electron acceptor.

From such energy calculations, it is possible to predict the kinds of chemolithotrophs that might be found in nature. Since organisms obey the laws of thermodynamics, only reactions that are thermodynamically favorable are potential energy-yielding reactions. Table 15.1 summarizes energy yields for some reactions known to be carried out by chemolithotrophic microorganisms. We discuss some of these processes in the rest of this chapter, and the organisms involved are considered in more detail in Chapters 13 and 14. We also examine ecological aspects of chemolithotrophy in Chapter 16.

✓ 15.9 Concept Check

A number of specialized prokaryotes, called chemolithotrophs, are able to oxidize inorganic chemicals as their sole sources of energy and reducing power. Most chemolithotrophs are also able to grow autotrophically.

✓ For what two purposes is a given inorganic compound used by a chemolithotroph?
✓ Why does the oxidation of H_2 yield more energy with O_2 as electron acceptor than with SO_4^{2-} as electron acceptor?

15.10

Hydrogen Oxidation

Hydrogen, H_2, is a common product of microbial metabolism, and a number of chemolithotrophs are able to use it as an electron donor in energy metabolism. A wide variety of anaerobic H_2-oxidizing Bacteria and Archaea are known, differing in the electron acceptor they use (for example, nitrate, sulfate, ferric iron, and others), and these organisms are discussed later in this chapter (see Sections 15.15–15.20). Here we consider only the *aerobic* H_2 oxidizing bacteria.

Energetics of H_2 Oxidation

Generation of ATP during H_2 oxidation comes from the oxidation of H_2 by O_2 leading to formation of a proton motive force. The overall reaction:

$$H_2 + \tfrac{1}{2}O_2 \rightarrow H_2O \qquad \Delta G^{0\prime} = -237 \text{ kJ}$$

is highly exergonic and can support the synthesis of at least one ATP. The reaction is catalyzed by the enzyme **hydrogenase,** the electrons from H_2 initially being transferred to a quinone acceptor. From here electrons pass through a series of cytochromes to eventually reduce O_2 to water (Figure 15.25). Some hydrogen bacteria have two hydrogenases, one soluble and one membrane-bound. In this case, the membrane-bound enzyme is involved in energetics while the soluble hydrogenase takes up H_2 and reduces NAD^+ to NADH directly; the reduction potential of H_2 (-0.42 V) is so low that reverse electron flow as a means of making reducing power is unnecessary (Figure 15.25). The organism *Ralstonia eutrophus* has been a model exper-

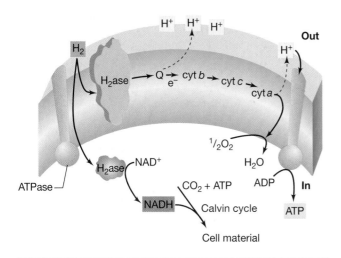

FIGURE 15.25 Bioenergetics and function of the two hydrogenases of aerobic H_2 bacteria. In *Ralstonia eutrophus*, where two hydrogenases are present, the membrane-bound hydrogenase is involved in energetics while the cytoplasmic hydrogenase makes NADH for the Calvin cycle. Note how the membrane-bound hydrogenase begins the flow of electrons leading to formation of a proton motive force. Some H_2 bacteria have only the membrane-bound hydrogenase, and in these organisms reducing power synthesis occurs from reverse electron flow. H_2ase, hydrogenase; cyt, cytochrome; Q, quinone.

imental system for studying aerobic H_2 oxidation, and we discussed some of the properties of this organism in Section 13.4.

Autotrophy in H_2 Bacteria

Although most hydrogen bacteria can also grow as chemoorganotrophs, when growing chemolithotrophically these organisms fix CO_2 by the Calvin cycle (see Section 15.7). The stoichiometry observed here is:

$$6 H_2 + 2 O_2 + CO_2 \rightarrow (CH_2O) + 5 H_2O$$

where (CH_2O) represents cell material. However, when readily used organic compounds such as glucose are present, synthesis of Calvin cycle and hydrogenase enzymes in typical aerobic H_2-oxidizing bacteria is repressed. Thus in nature, where H_2 levels in oxic environments are transient and low at best, it is likely that aerobic hydrogen bacteria must closely regulate their catabolic enzymes and shift between chemoorganotrophy and chemolithotrophy often, depending on levels of useable organic compounds and H_2 in their habitats. Moreover, because many aerobic H_2 bacteria grow best microaerobically, it is likely that these organisms are most successful as H_2 chemolithotrophs in oxic/anoxic interfaces where H_2 from fermentative metabolism would be in greater and more continuous supply than in highly oxic habitats.

Oxidation of Reduced Sulfur Compounds

Many reduced sulfur compounds can be used as electron donors by a variety of colorless sulfur bacteria [called "colorless" to distinguish them from the bacteriochlorophyll-containing (pigmented) green and purple sulfur bacteria discussed earlier in this chapter, see Figure 15.17]. Indeed, the whole concept of chemolithotrophy emerged from studies of the sulfur bacteria as the great Russian microbiologist Winogradsky first proposed the idea of chemolithotrophy from studies of these organisms (see the box, Winogradsky's Legacy).

The most common sulfur compounds used as electron donors are hydrogen sulfide (H_2S), elemental sulfur (S^0), and thiosulfate ($S_2O_3^{2-}$). The final product of sulfur oxidation in most cases is sulfate (SO_4^{2-}), and the total number of electrons involved between H_2S (oxidation state, -2) and sulfate (oxidation state, $+6$) is eight (see Table 15.3 for a summary of sulfur oxidation states). Less energy is available when one of the intermediate sulfur oxidation states is used:

$$H_2S + 2 O_2 \rightarrow SO_4^{2-} + 2 H^+$$
$$-798.2 \text{ kJ/reaction}$$

$$HS^- + \tfrac{1}{2}O_2 + H^+ \rightarrow S^0 + H_2O$$
$$-209.4 \text{ kJ/reaction}$$

$$S^0 + H_2O + 1\tfrac{1}{2}O_2 \rightarrow SO_4^{2-} + 2 H^+$$
$$-587.1 \text{ kJ/reaction}$$

$$S_2O_3^{2-} + H_2O + 2 O_2 \rightarrow 2 SO_4^{2-} + 2 H^+$$
$$-818.3 \text{ kJ/reaction}$$
$$(-409.1 \text{ kJ/S atom oxidized})$$

The oxidation of the most reduced sulfur compound, H_2S, occurs in stages, and the first oxidation step results in the formation of elemental sulfur, S^0. Some H_2S-oxidizing bacteria deposit the elemental sulfur formed inside the cell (Figure 15.26a). The sulfur deposited as a result of the initial oxidation is an energy reserve, and when the supply of H_2S has been depleted, additional energy can be obtained from the oxidation of sulfur to sulfate.

When elemental sulfur is provided externally as an electron donor, the organism must grow attached to the sulfur particle because of the extreme insolubility of elemental sulfur (Figure 15.26b). By adhering to the particle, the organism can efficiently obtain the atoms of sulfur needed. This is thought to occur through the action of membrane or periplasmic proteins that solubilize the sulfur, probably by reduction of S^0 to HS^-, from which it is transported into the cell and enters chemolithotrophic metabolism.

Note that in the sulfur oxidation reactions shown here one of the products is H^+. Production of protons results in a lowering of the pH, and one result of the oxi-

LEARNING FROM THE PAST: Winogradsky's Legacy

The discovery of autotrophy in chemolithotrophic bacteria was of major significance in the advance of our understanding of cell physiology because it showed that CO_2 could be converted to organic carbon without the intervention of chlorophyll. Previously, it had been thought that only green plants converted CO_2 to organic form. The idea of chemolithotrophic autotrophy was first developed by the great Russian microbiologist Sergei Winogradsky. Winogradsky studied sulfur bacteria because certain colorless sulfur bacteria (*Beggiatoa, Thiothrix*) are very large (see Fig. 1) and hence are easy to investigate even in the absence of pure cultures. Springs with waters rich in H_2S are fairly common around the world, and Winogradsky studied several such springs in the Bernese Oberland district of Switzerland. In the outflow channels of sulfur springs, vast populations of *Beggiatoa* and *Thiothrix* develop, and suitable material for microscopic and physiological studies could be obtained by merely lifting up the white filamentous

masses. Pure cultures were not needed for many studies. As Winogradsky noted, "This study of demipurity would be poor in an ordinary culture but is sufficient for a culture under the microscope, since it is possible to observe the development from day to day, almost from hour to hour, and see easily the presence of contaminants."

Winogradsky first showed that the colorless sulfur bacteria were present only in water containing H_2S. As the water flowed away from the source, the H_2S gradually dissipated, and sulfur bacteria were no longer present. This suggested to him that their development was dependent on the presence of H_2S. Winogradsky then showed that by starving *Beggiatoa* filaments for a while, they lost their sulfur granules; he found, however, that the granules were rapidly restored if a small amount of H_2S was added (Fig. 1). He thus concluded that H_2S was being oxidized to elemental sulfur. But what happened to the sulfur granules when the filaments were starved of H_2S? Winogradsky

showed by some clever microchemical tests that when the sulfur granules disappeared, sulfate appeared in the medium. Thus, he formulated the idea that *Beggiatoa* (and by inference other colorless sulfur bacteria) oxidize H_2S to elemental sulfur and subsequently to sulfate. Because they seemed to require H_2S for development in the springs, he postulated that this oxidation was the principal source of energy for these organisms.

Studies on *Beggiatoa* provided the first evidence that an organism could oxidize an *inorganic* substance as a possible energy source, and this was the origin of the concept of chemolithotrophy. However, the sulfur bacteria proved difficult to work with, primarily because there are a number of spontaneous chemical changes in sulfur compounds that can also occur and confuse the study. Winogradsky thus turned to a study of the nitrifying bacteria, and it was with this group that he clearly was able to show that autotrophic fixation of CO_2 was coupled to the oxidation of an inorganic

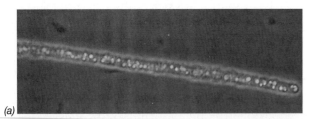

(a)

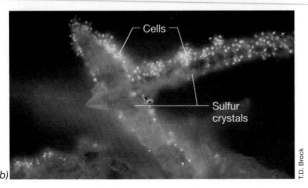

Cells

Sulfur crystals

(b)

T.D. Brock

dation of reduced sulfur compounds is the acidification of the medium. The acid formed by the sulfur bacteria is *sulfuric acid*, H_2SO_4, and sulfur bacteria are often able to bring about a marked reduction in the pH of the medium.

Biochemistry and Energetics of Sulfur Oxidation

The biochemical steps in the oxidation of various sulfur compounds are summarized in Figure 15.27. Start-

FIGURE 15.26 Sulfur bacteria. (a) Deposition of internal sulfur granules by *Beggiatoa*. (b) Attachment of the sulfur-oxidizing archaean *Sulfolobus acidocaldarius* to a crystal of elemental sulfur. Visualized by fluorescence microscopy after staining the cells with the dye acridine orange. The sulfur crystal does not fluoresce.

compound in the complete absence of light and chlorophyll. The process of nitrification had been known before Winogradsky's work from studies on the fate of sewage when added to soil. Two French soil scientists, T. Schloesing and A. Mutz, had shown soil columns to which ammonia (as NH_4^+) was allowed to trickle through yielded nitrate (NO_3^-) and that this process was biological since it could be readily arrested by saturating the soil column with chloroform vapors, a sterilant. Winogradsky picked up from here and proceeded to isolate nitrifying bacteria using completely mineral media in which CO_2 was the sole carbon source and ammonia was the sole electron donor. Because ammonia is chemically stable, it was easy to show that the oxidation of ammonia to nitrite, and subsequently to nitrate, was a strictly bacterial process. In fact, Winogradsky further showed that nitrification was a *two-step* process, with one group of organisms converting NH_4^+ to NO_2^- and a second, NO_2^- to NO_3^- (see Section 15.13). As no organic materials were present in the medium, it was also possible to show that organic matter (the bacterial cell material) was formed only from CO_2. When the ammonia or nitrite was left out of the medium, no growth occurred. Careful chemical analyses showed that the amount of organic matter formed by the bacteria was proportional to the amount of ammonia or nitrite they oxidized. Winogradsky concluded, "This [process] is contradictory to that fundamental doctrine of physiology which states that a complete synthesis of organic matter cannot take place in nature except through chlorophyll-containing plants by the action of light." His basic conclusion has been confirmed by a large number of subsequent studies. At least in one way, however, autotrophy in most chemolithotrophs and phototrophs is similar in that in both processes the pathway of CO_2 fixation follows the same biochemical steps (the Calvin cycle) involving the enzyme ribulose bisphosphate carboxylase (see Section 15.7). ■

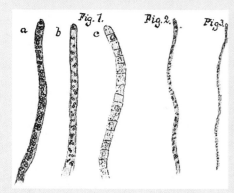

FIG. 1 Drawings made by Winogradsky of *Beggiatoa* and translation (from the French) of the legend accompanying these figures. "Fig. 1. The tip of a filament of *Beggiatoa alba:* (a) in sulfurous [sulfide-containing] water, (b) after 24 hr in water nearly depleted in H_2S, (c) after 48 hr in water without H_2S [note depletion of sulfur globules with time]. Fig. 2. The tip of a filament of *Beggiatoa media.* Fig. 3. The tip of a filament of *Beggiatoa minima.*" From Winogradsky, S. 1949. *Microbiologie du Sol.* Masson, Paris.

ing with sulfide, sulfite (SO_3^{2-}) is produced, a six electron oxidation. If S^0 is the starting substrate, sulfite is also produced, although the S^0 must first be reduced to sulfide (Figure 15.27a). There are two ways in which SO_3^{2-} can be oxidized to SO_4^{2-}. The most widespread system is that employing the enzyme *sulfite oxidase.* Sulfite oxidase transfers electrons from SO_3^{2-} directly to cytochrome *c,* and ATP is made from this during electron transport and proton motive force formation (Figure 15.27b). In addition to sulfite oxidase, a few sulfur chemolithotrophs oxidize SO_3^{2-} to SO_4^{2-} via a reversal of the activity of *adenosine phosphosulfate (APS) reductase,* an enzyme critical to the metabolism of sulfate-reducing bacteria (compare Figures 15.27a and 15.40). This reaction, run in the direction of SO_4^{2-} production by sulfur chemolithotrophs, produces one high-energy phosphate bond when AMP is converted to ADP (Figure 15.27a).

When thiosulfate is the electron donor for sulfur chemolithotrophs, it is split into S^0 and SO_3^{2-}, both of which are eventually oxidized to SO_4^{2-}.

All the electrons from reduced sulfur compounds eventually reach the electron transport system as shown in Figure 15.27b. Depending on the E_0' of the couple, electrons enter at either the flavoprotein (E_0' = ~ −0.2) or cytochrome *c* (E_0' = +0.3) level and are transported to O_2, generating a proton motive force that leads to ATP synthesis by ATPase. Electrons for autotrophic CO_2 fixation come from reverse electron flow (see Section 15.5) eventually yielding NADH, and CO_2 is actually fixed via the Calvin cycle (Figure 15.27b). Although the sulfur chemolithotrophs are primarily an aerobic group (∞ Section 13.3), some species can grow anaerobically using nitrate as an electron acceptor; *Thiobacillus denitrificans* is a classic example of this lifestyle.

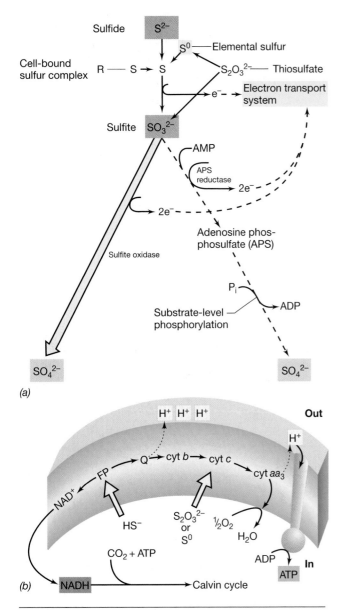

FIGURE 15.27   Oxidation of reduced sulfur compounds by sulfur chemolithotrophs. (a) Steps in the oxidation of different compounds. The sulfite oxidase pathway accounts for the majority of sulfite oxidized. (b) Electrons from sulfur compounds feed into the electron transport chain to drive a proton motive force; electrons from thiosulfate and elemental sulfur enter at the level of cytochrome c. NADH must be made by reactions of reverse electron flow since the electron donors have a more electropositive E_0' than does $NAD^+/NADH$. Cyt, cytochrome; FP, flavoprotein; Q, quinone.

✓ **15.10–15.11   Concept Checks**

Hydrogen (H_2) and reduced sulfur compounds such as H_2S and S^0 are excellent electron donors for chemolithotrophs. These compounds can be oxidized by the hydrogen bacteria or the sulfur bacteria, respectively, thereby generating a proton motive force and ATP synthesis. These chemolithotrophs are also autotrophs and fix CO_2 by the Calvin cycle.

✓   What special enzyme is needed for growth on H_2?
✓   How many electrons are available from the oxidation of H_2S if S^0 is the final product? If SO_4^{2-} is the final product?

15.12

Iron Oxidation

The aerobic oxidation of iron from the ferrous (Fe^{2+}) to the ferric (Fe^{3+}) state is an energy-yielding reaction for a few bacteria. Only a small amount of energy is available from this oxidation (see Table 15.1), and for this reason the iron bacteria must oxidize large amounts of iron in order to grow. Ferric iron forms very insoluble ferric hydroxide [$Fe(OH)_3$] precipitates in water (Figure 15.28*a*). This is in part because at neutral pH ferrous iron rapidly oxidizes nonbiologically to the ferric state and is thus stable for long periods only under *anoxic* conditions. At acid pH, however, ferrous iron is stable under oxic conditions. This explains why most iron-oxidizing bacteria are obligately acidophilic.

The best-known iron-oxidizing bacterium is *Thiobacillus ferrooxidans,* which is able to grow autotrophically using either ferrous iron (Figure 15.28*b*) or reduced sulfur compounds as electron donors. This organism is very common in acid-polluted environments such as coal-mining dumps (Figure 15.28*a*), and we will discuss its role in acid-mine pollution and mineral oxidation in Sections 16.18 and 16.19. Another iron-oxidizing prokaryote is the archaeon *Sulfolobus,* which lives in hot, acid springs at temperatures up to the boiling point of water (see Figure 15.26*b*).

Despite what was just said concerning the stability of Fe^{2+} at acidic pH, there are a number of iron-oxidizing bacteria that thrive in *neutral* pH environments, but these are usually situations where ferrous iron is moving from anoxic to oxic conditions. At interfaces between these zones iron bacteria can oxidize Fe^{2+} as it comes from an anoxic source before the Fe^{2+} oxidizes spontaneously. *Gallionella ferruginea* and *Sphaerotilus natans* are examples of organisms that live at these interfaces and are generally seen mixed in with the characteristic deposits they form (Figure 15.29; ∞ also Figure 13.44). In cells of *Gallionella,* enzymes of the Calvin cycle have been found, and thus this organism is a true iron-oxidizing chemolithoautotroph. We discuss the taxonomy of these interesting organisms in ∞ Sections 13.14 and 13.15.

Energy from Ferrous Iron Oxidation

The bioenergetics of iron oxidation by *Thiobacillus ferrooxidans* is of biochemical interest because of the very electropositive reduction potential of the Fe^{3+}/Fe^{2+} couple (+0.77 V at pH 2). The respiratory chain of *T. fer-*

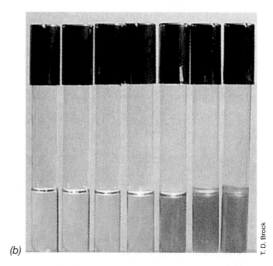

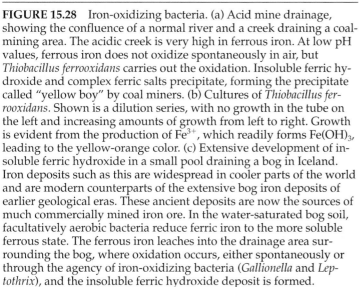

FIGURE 15.28 Iron-oxidizing bacteria. (a) Acid mine drainage, showing the confluence of a normal river and a creek draining a coal-mining area. The acidic creek is very high in ferrous iron. At low pH values, ferrous iron does not oxidize spontaneously in air, but *Thiobacillus ferrooxidans* carries out the oxidation. Insoluble ferric hydroxide and complex ferric salts precipitate, forming the precipitate called "yellow boy" by coal miners. (b) Cultures of *Thiobacillus ferrooxidans*. Shown is a dilution series, with no growth in the tube on the left and increasing amounts of growth from left to right. Growth is evident from the production of Fe^{3+}, which readily forms $Fe(OH)_3$, leading to the yellow-orange color. (c) Extensive development of insoluble ferric hydroxide in a small pool draining a bog in Iceland. Iron deposits such as this are widespread in cooler parts of the world and are modern counterparts of the extensive bog iron deposits of earlier geological eras. These ancient deposits are now the sources of much commercially mined iron ore. In the water-saturated bog soil, facultatively aerobic bacteria reduce ferric iron to the more soluble ferrous state. The ferrous iron leaches into the drainage area surrounding the bog, where oxidation occurs, either spontaneously or through the agency of iron-oxidizing bacteria (*Gallionella* and *Leptothrix*), and the insoluble ferric hydroxide deposit is formed.

rooxidans contains cytochromes of the *c* and a_1 types and a periplasmic copper-containing protein called *rusticyanin* (Figure 15.30). Because the reduction potential of the Fe^{3+}/Fe^{2+} couple is so high, the route of electron transport to oxygen ($\frac{1}{2}O_2/H_2O$, $E_0' = +0.82$ V) is obviously going to be very short. Ferrous iron oxidation begins in the periplasm where rusticyanin oxidizes Fe^{2+} to Fe^{3+}, a one electron transition. This protein then reduces cytochrome *c*, and this subsequently reduces cytochrome *a*. The latter interacts directly with O_2 to form

H_2O (Figure 15.30). ATP is then synthesized from proton-translocating ATPases in the membrane and ATP yields are relatively low because of the high potential of the electron donor.

Because of the large gradient of protons across the *T. ferrooxidans* membrane (the periplasm is pH 1–2 while the cytoplasm is pH 5.5–6), protons entering the cytoplasm via the ATPase must be consumed in order to maintain the internal pH within acceptable limits (Figure 15.30). The protons are consumed during the pro-

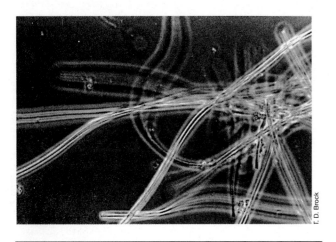

FIGURE 15.29 Phase contrast photomicrograph of empty iron-encrusted sheaths of *Sphaerotilus* collected from seepage at the edge of a small swamp.

duction of H_2O, but this reaction also requires electrons; these come from Fe^{2+} as follows:

$$2 \, Fe^{2+} + \tfrac{1}{2}O_2 + 2 \, H^+ \rightarrow 2 \, Fe^{3+} + H_2O.$$

Thus, as long as *T. ferrooxidans* has Fe^{2+} available, ATP synthesis can occur at the expense of the natural pro-

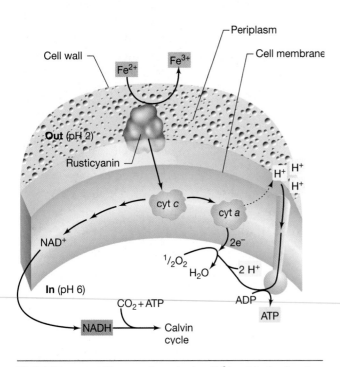

FIGURE 15.30 Electron flow during Fe^{2+} oxidation by the acidophile *Thiobacillus ferrooxidans*. The periplasmic copper-containing protein rusticyanin is the immediate acceptor of electrons from Fe^{2+}. From here, electrons travel a short electron transport chain resulting in the reduction of O_2 to H_2O. Reducing power to drive the Calvin cycle comes from reactions of reverse electron flow. Note the steep pH gradient (4–5 units) across the membrane.

ton motive force that exists across the cytoplasmic membrane (Figure 15.30).

Autotrophy in *T. ferrooxidans* is driven by the Calvin cycle, and because of the high potential of the electron donor, Fe^{2+}, much energy is consumed driving reverse electron flow reactions to obtain the reducing power necessary to drive CO_2 fixation. Thus, a relatively poor energetic picture coupled with large energetic demands means that *T. ferrooxidans* must oxidize large amounts of Fe^{2+} in order to produce even a very small amount of cell material. Thus, in environments where acidophilic Fe^{2+}-oxidizing bacteria are living, their presence is signaled not by the formation of much cell material but by the presence of large amounts of ferric iron (Figures 15.28; ∞ also Figure 16.50). We consider the important ecological processes connected with the iron-oxidizing bacteria in Sections 16.18 and 16.19.

Ferrous Iron Oxidation by Anoxygenic Phototrophs

Ferrous iron can be oxidized under *anoxic* conditions by certain anoxygenic phototrophic bacteria (Figure 15.31). The ferrous iron is used in this case not as an electron donor in energy metabolism, but as an electron donor for CO_2 reduction. At neutral pH, the Fe^{3+}/Fe^{2+} couple is much less electropositive than at pH 2, about $+0.2$ V, and thus electrons from Fe^{2+} can reduce cytochrome *c* in the photosystem of purple bacteria (see Section 15.5 for a discussion of anoxygenic photosynthesis). Anoxically, Fe^{2+} is stable at neutral pH where these organisms flourish.

In culture, iron-oxidizing anoxygenic phototrophs are supplied with Fe^{2+} in the form of $FeCO_3$, and the following reaction is observed:

$$FeCO_3 + 10 \, H_2O$$
$$\rightarrow 4 \, Fe(OH)_3 + (CH_2O) + 3 \, HCO_3^- + 3 \, H^+$$

with (CH_2O) representing new cell material (Figure 15.31a). The organisms involved, which are species of purple bacteria (Figure 15.31b), can also use FeS; under these conditions both Fe^{2+} and S^{2-} are oxidized as electron donors. Certain phototrophic green sulfur bacteria (genus *Chlorobium*, ∞ Section 13.29) can also use Fe^{2+} as a photosynthetic electron donor. Moreover, various chemotrophic denitrifying bacteria have been isolated that can couple the oxidation of Fe^{2+} to the reduction of NO_3^- to N_2 and grow under anoxic conditions. However, like the aerobic iron bacteria, in these organisms iron is an electron donor for *both* energy and reducing power needs.

The discovery of Fe^{2+}-oxidizing phototrophs has important implications for both understanding the evolution of photosynthesis and explaining the large de-

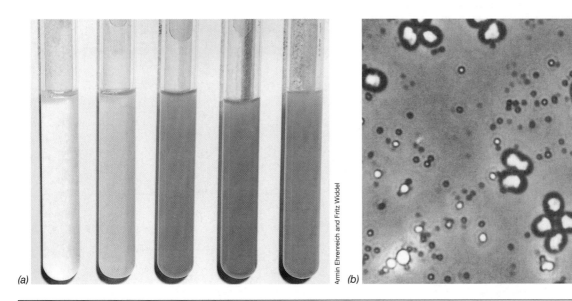

FIGURE 15.31   Ferrous iron oxidation by anoxygenic phototrophic bacteria. (a) Fe^{2+} oxidation in anoxic tube cultures. Left to right: Sterile medium, inoculated medium, growth to increasing cell densities. The brown-red color is due mainly to $Fe(OH)_3$ precipitate. (b) Phase contrast photomicrograph of an iron-oxidizing purple bacterium. The bright refractile areas within cells are gas vesicles (⚪ Section 3.14). The granules outside the cells are iron precipitates. This organism is phylogenetically related to the purple sulfur bacterium *Chromatium* (⚪ Section 13.1).

positions of ferric iron found in ancient sediments. Such ferric iron was previously thought to have been formed from the oxidation of Fe^{2+} by O_2 produced by oxygenic phototrophs (⚪ Section 12.1). However, because of the age of these sediments, it is more likely that the ferric iron was formed by anoxygenic phototrophs oxidizing Fe^{2+} in anoxic environments.

✓ 15.12 Concept Check

The iron bacteria are chemolithotrophs able to use ferrous iron (Fe^{2+}) as sole energy source. Most iron bacteria grow only at acid pH and are often associated with acid pollution from mineral and coal mining. Some phototrophic purple bacteria can oxidize Fe^{2+} to Fe^{3+} anoxically.

- ✓ Why does most iron oxidation in nature occur under *acidic* conditions?
- ✓ Why is only a very small amount of energy available from the oxidation of Fe^{2+} to Fe^{3+}?
- ✓ How can Fe^{2+} be oxidized anoxically?
- ✓ What is the function of *rusticyanin* and where is it found in the cell?

15.13

Nitrification

The most common *inorganic nitrogen compounds* used as electron donors are ammonia (NH_3) and nitrite (NO_2^-), which are oxidized aerobically by the chemolithotrophic **nitrifying bacteria** (⚪ Section 13.2). The nitrifying bacteria are widely distributed in soil and water. One group of organisms, the *nitrosofyers* (*Nitrosomonas* is one genus), oxidizes ammonia to nitrite, and another group (*Nitrobacter*) oxidizes nitrite to nitrate; the complete oxidation of ammonia to nitrate, an eight-electron transfer, is thus carried out by members of these two groups of organisms acting in sequence (see the Winogradsky box).

Bioenergetics and Enzymology of Nitrification

The electrons from nitrogen compounds enter an electron transport chain, and electron flow establishes a membrane potential and proton motive force linked to ATP synthesis. However, because of the reduction potential of their electron donors, nitrifying bacteria are faced with bioenergetic problems similar to those of the sulfur chemolithotrophs. The E_0' of the NH_2OH/NH_3 couple, the first step in the oxidation of NH_3, is about 0 V. The E_0' of the NO_3^-/NO_2^- couple is very high, about +0.43 V. These relatively high reduction potentials mean that nitrifying bacteria must donate electrons to their electron transport chains at rather late steps in the overall process. This effectively limits the amount of ATP that can be produced from each pair of electrons introduced.

Several key enzymes are involved in oxidizing reduced nitrogen compounds. In ammonia-oxidizing bacteria, NH_3 is oxidized by **ammonia monooxygenase** (see Section 15.27 for a discussion of monooxygenase enzymes) that produces NH_2OH and H_2O (Figure 15.32). *Hydroxylamine oxidoreductase* then oxidizes NH_2OH to NO_2^-, removing *four* electrons in the process. Ammonia

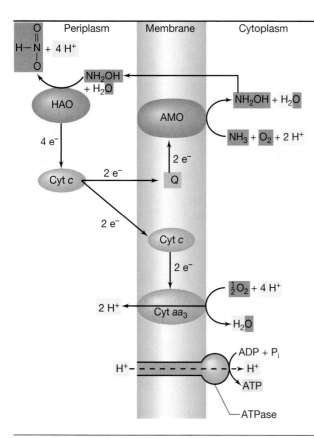

FIGURE 15.32 Oxidation of ammonia and electron flow in ammonia-oxidizing bacteria. The reactants and the products of this reaction series are highlighted. The cytochrome c (cyt c) in the periplasm is a different form of cyt c than that in the membrane. AMO, Ammonia monooxygenase; HAO, hydroxylamine oxidoreductase; Q, ubiquinone.

monooxygenase is an integral membrane protein, whereas hydroxylamine oxidoreductase is periplasmic (Figure 15.32). In the reaction carried out by ammonia monooxygenase,

$$NH_3 + O_2 + 2H^+ + 2e^- \rightarrow NH_2OH + H_2O$$

there is a need for two exogenously supplied electrons plus two protons to reduce one atom of dioxygen to water. These electrons originate from the oxidation of hydroxylamine and are supplied to ammonia monooxygenase from hydroxylamine oxidoreductase via cytochrome c and ubiquinone (Figure 15.32). Thus, for every four electrons generated from the oxidation of NH_3 to NO_2^-, only two actually reach the terminal oxidase (cytochromes aa_3, Figure 15.32).

Nitrite-oxidizing bacteria employ the enzyme *nitrite oxidase* to oxidize nitrite to nitrate, with electrons traveling a very short electron transport chain (because of the high potential of the NO_3^-/NO_2^- couple) to the terminal oxidase (Figure 15.33). Cytochromes of the a and c types are present in the electron transport chain of

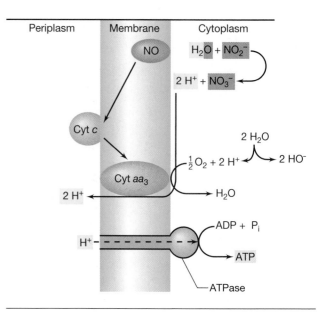

FIGURE 15.33 Oxidation of nitrite to nitrate by nitrifying bacteria. The reactants and products of this reaction series are highlighted. NO, Nitrite oxidase.

nitrite oxidizers, and generation of a proton motive force (which ultimately drives ATP synthesis) occurs through the action of cytochromes aa_3 (Figure 15.33). And like the situation with iron oxidation (see Section 15.12), only small amounts of energy are available. Thus, growth yields of nitrifying bacteria are relatively low.

Anoxic Ammonia Oxidation: Anammox

Although classical nitrifying bacteria such as *Nitrosomonas* and *Nitrobacter* are strictly aerobic bacteria, at least when growing as chemolithotrophs on ammonia or nitrite, ammonia can also be oxidized under *anoxic* conditions. Uncharacterized organisms from anoxic wastewater or sludge supplemented with nitrate to stimulate denitrification ($NO_3^- \rightarrow N_2$) (see Section 15.16) have been shown to oxidize ammonia to N_2. This reaction is nitrate-dependent and shows the following stoichiometry:

$$5NH_4^+ + 3NO_3^- \rightarrow 4N_2 + 9H_2O + 2H^+$$
$$(\Delta G^{0\prime} = -1483 \text{ kJ/reaction})$$

This reaction, referred to as *anammox* (for *anoxic ammonia oxidation*) is highly exergonic and is presumably linked to energy conservation in the microorganisms responsible. The microbiology of anammox is not yet understood, but the process is not thought to be due to anoxic metabolic activities of known nitrifying bacteria. Interestingly, however, discovery of anammox has disproven the long-held assumption that ammonia is stable under anoxic conditions, being oxidized only by the aerobic nitrifying bacteria.

Carbon Metabolism in Nitrifying Bacteria

Like sulfur- and iron-oxidizing chemolithotrophs, aerobic nitrifying bacteria employ the Calvin cycle for CO_2 fixation, and the ATP and reducing power requirements of this process place additional burdens on an already low-yielding energy-generating system (NADH to drive the Calvin cycle is formed by reverse electron flow). The energetic constraints are particularly severe for nitrite oxidizers, and it is perhaps for this reason that most nitrite oxidizers can also grow chemoorganotrophically on glucose and certain other organic substrates (∞ Section 13.2).

✓ 15.13 Concept Check

Ammonia (NH_3) and nitrite (NO_2^-) can be used as electron donors by the nitrifying bacteria. The ammonia-oxidizing bacteria produce nitrite, which is then oxidized by the nitrite-oxidizing bacteria to nitrate (NO_3^-). Anoxic NH_3 oxidation is coupled to a denitrification process.

- ✓ What is the inorganic electron donor for *Nitrosomonas*? For *Nitrobacter*?
- ✓ What are the substrates for the enzyme *ammonia monooxygenase*?
- ✓ What do nitrifying bacteria use as a carbon source?
- ✓ What is the *anammox* reaction?

15.14

Methanotrophy and Methylotrophy

In Section 13.5 we considered the unique situation, relative to carbon metabolism, of methanotrophs and methylotrophs. These organisms, although not autotrophs, use C_1 compounds for energy metabolism and biosynthesis. We focus here on the details of these two major processes using methanotrophy, the utilization of methane (CH_4) as both electron donor and carbon source, as the prime example.

Biochemistry of Methane Oxidation

The individual steps in methane oxidation to CO_2 can be summarized as:

$$CH_4 \rightarrow CH_3OH \rightarrow \mathbf{CH_2O} \rightarrow HCOO^- \rightarrow CO_2$$

Interestingly, methanotrophs, those methylotrophs that can use CH_4, assimilate either all or one-half of their carbon (depending on the organism) at the level of formaldehyde (shown in bold above), and we will see that this affects a major energy savings compared with autotrophs, where carbon is assimilated from the more oxidized CO_2. But here our focus is on energy conservation.

The initial step in the oxidation of methane involves an enzyme called **methane monooxygenase.** As we will discuss in Section 15.27, oxygenase enzymes catalyze the incorporation of oxygen from O_2 into carbon compounds (and some nitrogen compounds, see Section 15.13), and seem to be widely involved in the metabolism of hydrocarbons. Monooxygenases incorporate one atom of O_2 into the substrate while the second atom is reduced to H_2O. In the methanotroph *Methylosinus,* where the process has been most thoroughly studied, the electrons needed for the oxidation of CH_4 to CH_3OH come from cytochrome *c* (Figure 15.34). Because of this requirement for reducing power in the first step, no ATP synthesis occurs during the oxidation of methane to methanol, and this lack of ATP synthesis is consistent with the fact that growth yields (grams of cells produced per mole of substrate consumed) of methanotrophs are the same whether methane or methanol is used as the growth substrate. The other oxidation steps from CH_3OH to CO_2 supply electrons to the electron transport chain, and, from these, ATP is made from the resulting electron transport and proton motive force generated (Figure 15.34).

C_1 Assimilation into Cell Material

As was noted in Section 13.5, two classes of methanotrophs are known, *Type I* and *Type II*. Members of each

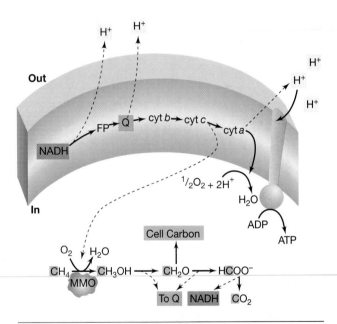

FIGURE 15.34 Oxidation of methane by methanotrophic bacteria. Methane (CH_4) is converted to methanol (CH_3OH) by the enzyme *methane monooxygenase.* The electrons needed to drive this first step come from cytochrome *c*, and no energy is conserved in this reaction. A proton motive force is established from electron flow in the membrane, and this fuels ATPase. Note how carbon for biosynthesis comes primarily from formaldehyde (CH_2O). MMO, methane monooxygenase; cyt, cytochrome; Q, quinone.

class share a number of properties in common, and each also shows a distinct mechanism for C_1 assimilation; these are the **ribulose monophosphate pathway** (Type I) and the **serine pathway** (Type II).

The serine pathway is outlined in Figure 15.35. In this pathway, a two-carbon unit, acetyl-CoA, is synthesized from one molecule of formaldehyde (produced from the oxidation of CH_4, see Figure 15.34) and one molecule of CO_2. The pathway requires the introduction of reducing power and energy in the form of two molecules each of NADH and ATP for each acetyl-CoA synthesized. The serine pathway employs a number of enzymes of the citric acid cycle and at least one enzyme, *serine transhydroxymethylase,* unique to the pathway (Figure 15.35).

The ribulose monophosphate pathway, present in Type I methanotrophs, is outlined in Figure 15.36. It is more efficient than the serine pathway because *all* of the carbon atoms for cell material are derived from

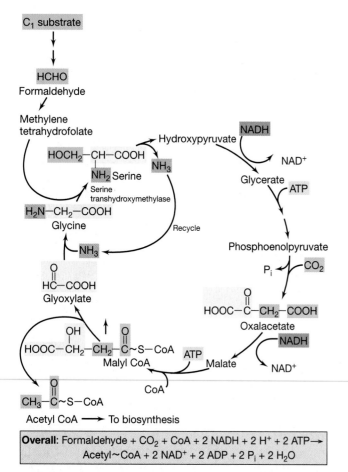

Overall: Formaldehyde + CO_2 + CoA + 2 NADH + 2 H^+ + 2 ATP→
Acetyl~CoA + 2 NAD^+ + 2 ADP + 2 P_i + 2 H_2O

FIGURE 15.35 The serine pathway for the assimilation of C_1 units into cell material by Type II methylotrophic bacteria. The product of the pathway, acetyl-CoA, is used as the starting point for making new cell material. The key enzyme of the pathway is serine transhydroxymethylase.

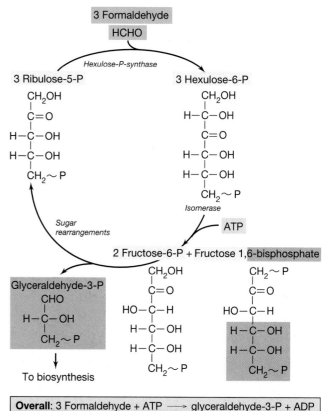

Overall: 3 Formaldehyde + ATP ⟶ glyceraldehyde-3-P + ADP

FIGURE 15.36 The ribulose monophosphate pathway for assimilation of one-carbon compounds, as found in Type I methylotrophic bacteria. The complete name of the hexulose sugar is D-erythro-L-glycero-3-hexulose 6-phosphate. Three formaldehydes are needed to carry the cycle to completion, with the net result being one molecule of glyceraldehyde-3-P. Regeneration of ribulose 5-phosphate and formation of phosphoglyceraldehyde occur via a series of pentose phosphate reactions.

formaldehyde. And, since formaldehyde is at the same oxidation level as cell material, no reducing power is needed. The ribulose monophosphate pathway requires the introduction of energy in the form of one molecule of ATP for each molecule of glyceraldehyde-3-phosphate synthesized (Figure 15.36). Consistent with the lower energy requirements of the ribulose monophosphate pathway the cell yield (grams of cells produced per mole of CH_4 oxidized) of Type I methanotrophs is higher than for Type IIs.

The enzymes *hexulosephosphate synthase,* that condenses one molecule of formaldehyde with one molecule of ribulose-5-phosphate, and *hexulose-6-P isomerase* (Figure 15.36) are unique to this pathway, while the remaining enzymes of the cycle are involved in sugar rearrangements in many different organisms. While perhaps only a coincidence, it should be noted that the substrate for the initial reaction in this pathway, ribulose-5-P, is almost identical to the C_1 acceptor of the Calvin cycle (ribulose 1,5-bisphosphate), suggesting that these two cycles may share evolutionary roots. The ribulose monophosphate pathway is also known as the *Quayle cycle* for its discoverer, the British biochemist, John Quayle.

✓ 15.14 Concept Check

Methanotrophy is the use of CH_4 as a carbon and energy source. The enzyme *methane monooxygenase* is a key enzyme in the catabolism of methane. C_1 units get assimilated into cell material at the level of formaldehyde in methanotrophs by either the ribulose monophosphate pathway or the serine pathway.

- ✓ What are the energy and reducing power requirements for the ribulose monophosphate pathway? For the serine pathway? Why do they differ?
- ✓ Why does the oxidation of CH_4 to CH_3OH require reducing power?
- ✓ Which pathway, the Calvin cycle or the ribulose monophosphate pathway, requires the greater energy input? Why? Why is it thought that one of these pathways may have evolved from the other?

15.15

Anaerobic Respiration

We examined in some detail in Chapter 4 the process of *aerobic* respiration. As we noted, molecular oxygen (O_2) functions as a terminal electron acceptor, accepting electrons from electron carriers by way of an electron transport chain. However, we noted in Chapter 4 that a variety of other electron acceptors can be used instead of O_2, in which case the process is called *anaerobic* respiration. We now consider some of these processes.

The bacteria carrying out anaerobic respiration generally possess electron transport systems containing cytochromes, quinones, iron–sulfur proteins and other typical electron transport proteins. Their respiratory systems are thus analogous to those of conventional aerobes. In some cases, such as with the denitrifying bacteria, the anaerobic respiration process competes in the same organism with an aerobic one. In such cases, if O_2 is present, aerobic respiration is usually favored, and when O_2 is depleted from the environment, the alternate electron acceptor is reduced. Other organisms carrying out anaerobic respiration are *obligate* anaerobes and are unable to use O_2.

Alternative Electron Acceptors and the Electron Tower

The energy released from the oxidation of an electron donor using O_2 as electron acceptor (⌒ Figure 4.7)

is higher than if the same compound is oxidized with an alternate electron acceptor. These energy differences are apparent if the reduction potentials of each acceptor are examined (Figure 15.37). Because the O_2/H_2O couple is the most electropositive, more energy is available when O_2 is used than when another electron acceptor is used. Other electron acceptors that are near O_2 are Fe^{3+}, NO_3^-, and NO_2^-. Farther up the scale are S^0, CO_2, and SO_4^{2-}. A summary of the most common types of anaerobic respiration is given in Figure 15.37.

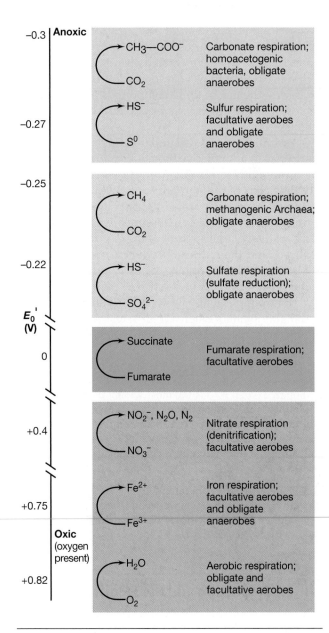

FIGURE 15.37 Examples of anaerobic respirations. The couples are arranged in order from those that are most electronegative E_0' (top) to those that are most electropositive E_0' (bottom).

Assimilative and Dissimilative Metabolism

Inorganic compounds such as NO_3^-, SO_4^{2-}, and CO_2 are reduced by many organisms as sources of cellular nitrogen, sulfur, and carbon, respectively. The end products of such reductions are primarily amino groups ($—NH_2$), sulfhydryl groups ($—SH$), and organic carbon compounds, respectively. We briefly examined the *nutrition* of microorganisms in Sections 4.2 and 4.3 and noted that all organisms need sources of N, S, and C for growth. When an inorganic compound such as NO_3^-, SO_4^{2-}, or CO_2 is reduced for use as a nutrient source, it is said to be *assimilated*, and the reduction process is called *assimilative* metabolism. We emphasize here that assimilative metabolism of NO_3^-, SO_4^{2-}, and CO_2 is quite different from the use of these compounds as electron acceptors for *energy* metabolism in anaerobic respiration. To distinguish these two kinds of reduction processes, the use of these compounds as electron acceptors in energy metabolism is called *dissimilative* metabolism.

Assimilative and dissimilative metabolism differ markedly. In assimilative metabolism, only enough of the compound (NO_3^-, SO_4^{2-}, or CO_2) is reduced to satisfy the needs of the nutrient for growth. The reduced atoms are eventually converted to cell material in the form of macromolecules. In dissimilative metabolism, a comparatively large amount of the electron acceptor is reduced, and the reduced product is *excreted* into the environment. Many organisms carry out assimilative metabolism of compounds such as NO_3^-, SO_4^{2-}, and CO_2 (for example, many Bacteria, Archaea, fungi, algae, and higher plants), whereas only a restricted variety of organisms carry out dissimilative metabolism.

✓ 15.15 Concept Check

Although oxygen (O_2) is the most widely used electron acceptor in energy-yielding metabolism, a number of other compounds can be used as electron acceptors. This process of anaerobic respiration is less energy-efficient but makes it possible for respiration to occur in environments where oxygen is absent.

✓ What is anaerobic respiration?
✓ With H_2 as electron donor, why might the reduction of NO_3^- be a more favorable reaction than the reduction of S^0?

15.16

Nitrate Reduction and the Denitrification Process

Inorganic nitrogen compounds are some of the most common electron acceptors in anaerobic respiration. A summary of the various inorganic nitrogen species with their oxidation states is given in Table 15.2. The most wide-

TABLE 15.2	Oxidation states of key nitrogen compounds
Compound	**Oxidation state**
Organic N (R — NH$_2$)	−3
Ammonia (NH$_3$)	−3
Nitrogen gas (N$_2$)	0
Nitrous oxide (N$_2$O)	+1 (average per N)
Nitrogen oxide (NO)	+2
Nitrite (NO$_2^-$)	+3
Nitrogen dioxide (NO$_2$)	+4
Nitrate (NO$_3^-$)	+5

spread inorganic nitrogen species in nature are ammonia and nitrate, both of which are formed in the atmosphere by inorganic chemical processes and nitrogen gas, N_2, also an atmospheric gas, which is the most stable form of nitrogen in nature. We discuss *nitrogen fixation*, the use of N_2 as a biosynthetic nitrogen source, later in this chapter.

One of the most common alternative electron acceptors is nitrate, NO_3^-, which is reduced to N_2O, NO, and N_2. Because these products of nitrate reduction are all gaseous, they can easily be lost from the environment, and because of this the process is called **denitrification** (Figure 15.38). The process is the main means by which gaseous N_2 is formed biologically, and because N_2 is much less readily available to organisms than nitrate as a source of nitrogen, for agricultural purposes at least, denitrification is a detrimental process. For sewage treatment (∞ Section 11.15), however, denitrification is beneficial because it converts NO_3^- to N_2, effectively decreasing the amount of available nitrogen in the sewage treatment effluent that can stimulate algal growth.

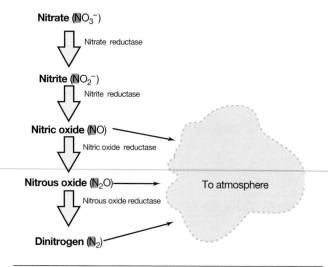

FIGURE 15.38 Steps in the dissimilative reduction of nitrate. Some organisms, for example *Escherichia coli*, can carry out only the first step. All enzymes involved are derepressed by anoxic conditions. Also, some prokaryotes are known that can reduce NO_3^- to NH_4^+ in dissimilative metabolism.

Biochemistry of Dissimilative Nitrate Reduction

The enzyme involved in the first step of dissimilative nitrate reduction, *nitrate reductase,* is a molybdenum-containing membrane-bound enzyme whose synthesis is repressed by molecular oxygen. All subsequent enzymes of the pathway (Figure 15.39) are coordinately regulated and thus also repressed by O_2, but in addition to anoxic conditions, nitrate must also be present before these enzymes are fully expressed.

The first product of nitrate reduction is nitrite (NO_2^-), and the enzyme *nitrite reductase* reduces it to nitric oxide (NO) (Figure 15.39c). Some organisms can reduce NO_2^- to ammonia (NH_3) in a dissimilative process, but it is the production of gaseous products, *denitrification,* that is of the greatest global significance because it consumes a fixed form of nitrogen (NO_3^-) readily available to plants and produces gaseous nitrogen compounds, some of which are of environmental significance (N_2O can be converted to NO by sunlight that reacts with ozone in the upper atmosphere to form ni-

trite, which returns to Earth as acid rain). The remaining biochemical steps in denitrification are shown in Figure 15.39c.

The biochemistry of dissimilative nitrate reduction has been studied in detail in several organisms including *Escherichia coli,* where NO_3^- is reduced only to NO_2^-, and *Paracoccus denitrificans* and *Pseudomonas stutzeri,* where true denitrification occurs (Figure 15.39). The *E. coli* nitrate reductase accepts electrons from a *b*-type cytochrome, and a comparison of the electron transport chains in aerobic versus anaerobic respiration in *E. coli* is shown in Figure 15.39a and b. As seen here, because of the reduction potential of the NO_3^-/NO_2^- couple (+0.43 V), only two proton translocating steps occur during nitrate reduction while three can occur in aerobic respiration ($\frac{1}{2}O_2/H_2O$, +0.82 V). In *P. denitrificans* and *P. stutzeri* nitrogen oxides are formed from nitrite by a series of enzymes including *nitrite reductase, nitric oxide reductase,* and *nitrous oxide reductase* as summarized in Figure 15.39c. During electron transport a proton motive force is established and ATP produced by ATPase in the usual fashion. Additional ATP is avail-

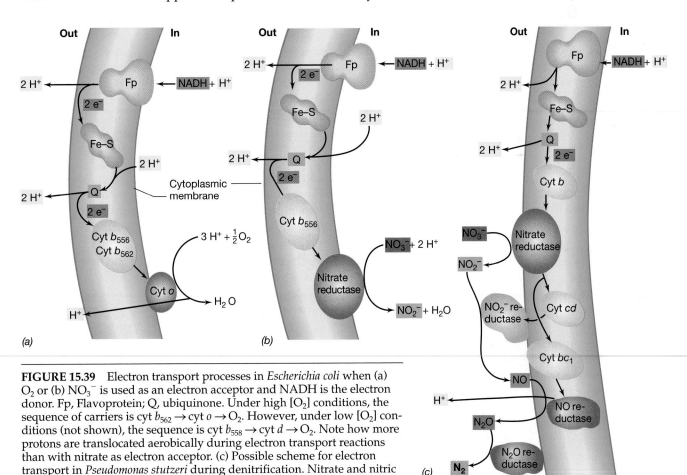

FIGURE 15.39 Electron transport processes in *Escherichia coli* when (a) O_2 or (b) NO_3^- is used as an electron acceptor and NADH is the electron donor. Fp, Flavoprotein; Q, ubiquinone. Under high [O_2] conditions, the sequence of carriers is cyt $b_{562} \rightarrow$ cyt $o \rightarrow O_2$. However, under low [O_2] conditions (not shown), the sequence is cyt $b_{558} \rightarrow$ cyt $d \rightarrow O_2$. Note how more protons are translocated aerobically during electron transport reactions than with nitrate as electron acceptor. (c) Possible scheme for electron transport in *Pseudomonas stutzeri* during denitrification. Nitrate and nitric oxide reductases are located in the cytoplasmic membrane whereas nitrite and nitrous oxide reductases are periplasmic. The immediate electron donors to the various reductases, with the exception of nitrate reductase, have not been definitively identified.

able when NO_3^- is reduced to N_2 because the NO reductase is linked to proton extrusion (Figure 15.39c).

Other Properties of Denitrifying Prokaryotes

Most denitrifying prokaryotes are phylogenetically members of the Proteobacteria (Sections 13.1–13.17) and are facultative aerobes; aerobic respiration occurs when air is present, even if nitrate is also present in the medium. Many denitrifying bacteria will also reduce other electron acceptors anaerobically such as ferric iron (Fe^{3+}) and certain organic electron acceptors (see Section 15.20). In addition, many denitrifying bacteria can grow by fermentation. Thus, one can conclude that the denitrifying bacteria are quite metabolically diverse in items of alternative energy-generating mechanisms.

✓ 15.16 Concept Check

Nitrate is a commonly used electron acceptor in anaerobic respiration. Its use involves participation of the enzyme nitrate reductase, a molybdenum-containing enzyme capable of reducing nitrate to nitrite. Many bacteria that use nitrate in anaerobic respiration eventually produce nitrogen gas (N_2), a process called *denitrification*.

- ✓ Why is more energy released when aerobic respiration occurs instead of denitrification?
- ✓ Where is the dissimilative nitrate reductase found in the cell? What metal(s) does it contain?
- ✓ Why would an organism like *Pseudomonas stutzeri* get more energy during NO_3^- respiration than would *Escherichia coli*?

15.17

Sulfate Reduction

Several inorganic sulfur compounds are important electron acceptors in anaerobic respiration. A summary of the oxidation states of the key sulfur compounds is given in Table 15.3. Sulfate, the most oxidized form of sulfur, is one of the major anions in seawater and is used by the *sulfate-reducing bacteria*, a group that is widely distributed in nature. The end product of sulfate reduction is H_2S, an important natural product that participates in many biogeochemical processes (Section 16.17). Again, as with nitrogen, it is important to distinguish between assimilative and dissimilative sulfate reduction. Many organisms, including higher plants, algae, fungi, and most prokaryotes, use sulfate as a sulfur source for biosynthesis. But the ability to use sulfate as an *electron acceptor* for energy-generating processes involves a large-scale reduction of SO_4^{2-} and is restricted to the sulfate-reducing bacteria. In assimilative sulfate reduction, the H_2S formed is immediately converted into organic sulfur in the form of amino acids,

TABLE 15.3	Sulfur compounds and electron donors for sulfate reduction
Compound	**Oxidation state**
Oxidation states of key sulfur compounds	
Organic S (R — SH)	−2
Sulfide (H_2S)	−2
Elemental sulfur (S^0)	0
Thiosulfate ($S_2O_3^{2-}$)	+2 (average per S)
Sulfur dioxide (SO_2)	+4
Sulfite (SO_3^{2-})	+4
Sulfate (SO_4^{2-})	+6
Some electron donors used for sulfate reduction	
H_2	Acetate
Lactate	Propionate
Pyruvate	Butyrate
Ethanol and other alcohols	Long-chain fatty acids
Fumarate	Benzoate
Malate	Indole
Choline	Hexadecane

and so on, but in dissimilative sulfate reduction, the H_2S is excreted.

As the reduction potential in Table A1.2 and Figure 15.37 shows, sulfate is a much less favorable electron acceptor than either O_2 or NO_3^-. However, sufficient energy to make ATP is available when an electron donor that yields NADH or FADH is used. Because of the less favorable energetics, growth yields are lower for an organism growing on SO_4^{2-} than for one growing on O_2 or NO_3^-. A list of some of the electron donors used by sulfate-reducing bacteria is given in Table 15.3. The first three compounds listed, H_2, lactate, and pyruvate, are used by a wide variety of sulfate-reducing bacteria; the others have more restricted use. However, a large variety of morphological and physiological types of sulfate-reducing bacteria are known (Section 13.17).

Biochemistry and Energetics of Sulfate Reduction

The reduction of SO_4^{2-} to H_2S, an eight-electron reduction (Table 15.3), proceeds through a number of intermediate stages. The sulfate ion is stable and cannot be reduced without first being activated. Sulfate is activated by means of ATP. The enzyme *ATP sulfurylase* catalyzes the attachment of the sulfate ion to a phosphate of ATP, leading to the formation of **adenosine phosphosulfate (APS)** as shown in Figure 15.40. In dissimilative sulfate reduction, the sulfate moiety of APS is reduced directly to sulfite (SO_3^{2-}) by the enzyme *APS reductase* with the release of AMP. In assimilative reduction, another P is added to APS to form **phosphoadenosine phosphosulfate (PAPS)** (Figure 15.40b), and only then is the sulfate moiety reduced. In both cases, the first product of sulfate reduction is *sulfite*, SO_3^{2-}.

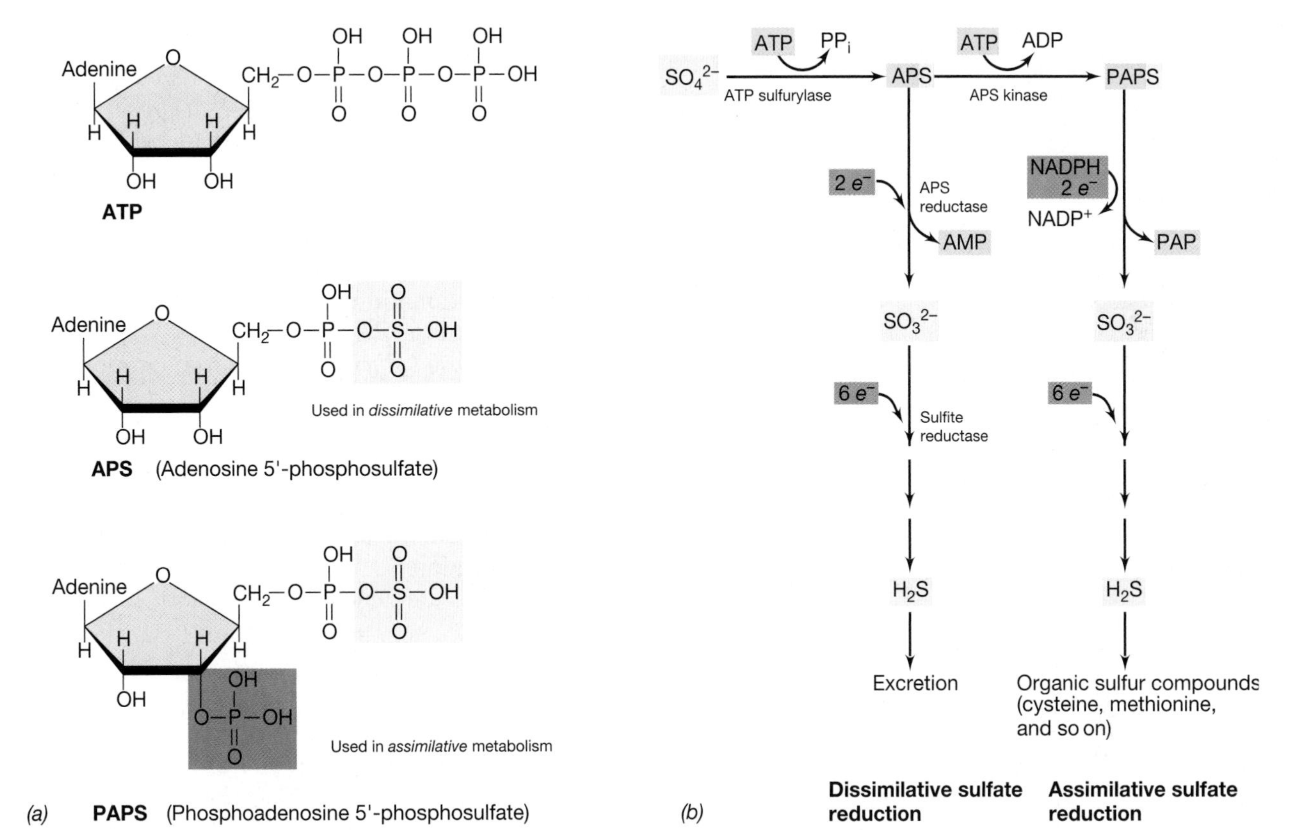

FIGURE 15.40 Biochemistry of sulfate reduction. (a) Two forms of *active sulfate*, adenosine 5'-phosphosulfate (APS) and phosphoadenosine 5'-phosphosulfate (PAPS). (b) Schemes of assimilative and dissimilative sulfate reduction.

Once SO_3^{2-} is formed, sulfide is formed by the enzyme *sulfite reductase* (Figure 15.40b).

In the process of dissimilative sulfate reduction, electron transport reactions occur leading to proton motive force formation, and this drives ATP synthesis by ATPase. A major electron carrier is cytochrome c_3, a periplasmic low potential cytochrome (Figure 15.41). Cytochrome c_3 accepts electrons from a periplasmically located hydrogenase (see below) and transfers these electrons to a membrane-associated protein complex called Hmc that carries them across the cytoplasmic membrane, thus making them available to APS reductase and sulfate reductase, which are cytoplasmic enzymes (Figure 15.41).

The enzyme hydrogenase appears to play a central role in sulfate reduction whether an organism like *Desulfovibrio desulfuricans* is growing on H_2, *per se*, or on an organic compound, like lactate. Evidence suggests that lactate is converted through pyruvate to acetate (the latter is mainly excreted because *D. desulfuricans* is a nonacetate-oxidizing sulfate reducer, (⟶ Section 13.17), with the production of H_2. The H_2 produced crosses the cytoplasmic membrane and is oxidized by the periplasmic hydrogenase to initiate a

proton motive force (Figure 15.41). Growth yields of sulfate-reducing bacteria suggest that one ATP is produced for each SO_4^{2-} reduced to HS^-. With H_2 the reaction is:

$$4 H_2 + SO_4^{2-} + H^+ \rightarrow HS^- + 4 H_2O$$

When lactate or pyruvate is the electron donor, not only is ATP produced from the proton motive force, but additional ATP is produced during the oxidation of pyruvate to acetate plus CO_2 via acetyl-CoA and acetyl-phosphate (see Section 15.21 for further discussion of this).

Acetate Use and Autotrophy

Many sulfate reducers can completely oxidize acetate to CO_2 as an electron donor for sulfate reduction (⟶ Section 13.17):

$$CH_3COO^- + SO_4^{2-} + 3 H^+ \rightarrow 2 CO_2 + H_2S + 2 H_2O \quad \Delta G^{0\prime} = -57.5 \text{ kJ}$$

Although the energetics of this process are not as well understood as H_2 or lactate metabolism by *D. desulfuri-*

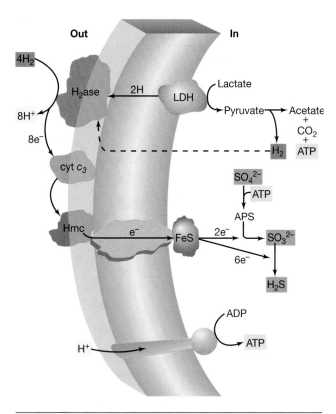

FIGURE 15.41 Electron transport and energy conservation in sulfate-reducing bacteria. In addition to external hydrogen (H_2), H_2 originating from the catabolism of organic compounds like lactate and pyruvate can fuel hydrogenase. The enzymes hydrogenase, cytochrome c_3, and a cytochrome complex (Hmc) are periplasmic proteins. A separate protein functions to shuttle electrons across the cytoplasmic membrane to a cytoplasmic iron–sulfur protein that supplies electrons to APS reductase (forming SO_3^{2-}) and sulfite reductase (forming H_2S).

cans, the mechanism for acetate oxidation is well understood. With few exceptions acetate is oxidized to CO_2 via the acetyl-CoA pathway, a series of reversible reactions used by a wide variety of anaerobes for acetate synthesis or acetate oxidation (see Section 15.18). This pathway employs the key enzyme *carbon monoxide dehydrogenase* and was first discovered in homoacetogenic bacteria that make acetate from $H_2 + CO_2$ as a mechanism of energy conservation (see Section 15.18). A few sulfate-reducing bacteria can also grow autotrophically in an anoxic mineral salts medium containing H_2 (as electron donor), SO_4^{2-} (as electron acceptor), and CO_2 (as carbon source). When growing under these conditions, autotrophic sulfate reducers use the acetyl-CoA pathway as a means of producing cell material. The acetate-oxidizing sulfate-reducing bacterium *Desulfobacter* lacks acetyl-CoA pathway enzymes and oxidizes acetate through the citric acid cycle, but this seems to be the exception rather than the rule.

Sulfur Disproportionation

Certain sulfate-reducing bacteria are capable of a unique form of energy metabolism called *disproportionation*, using sulfur compounds of intermediate oxidation state. The term disproportionation refers to the splitting of a compound into two new compounds, one of which is *more oxidized* and one of which is *more reduced* than the original substrate. In the present discussion, we describe the disproportionation of thiosulfate ($S_2O_3^{2-}$), sulfite (SO_3^{2-}), and sulfur (S^0).

Desulfovibrio sulfodismutans can disproportionate sulfur compounds as follows:

$$S_2O_3^{2-} + H_2O \rightarrow SO_4^{2-} + H_2S$$
$$\Delta G^{0\prime} = -21.9 \text{ kJ/reaction}$$

Note that one sulfur atom of $S_2O_3^{2-}$ becomes more oxidized (forming SO_4^{2-}) and the other more reduced (forming H_2S). Another disproportionation involves sulfite:

$$4 \, SO_3^{2-} + 2 \, H^+ \rightarrow 3 \, SO_4^{2-} + H_2S$$
$$\Delta G^{0\prime} = -235.6 \text{ kJ/reaction}$$

In these reactions electrons from either $S_2O_3^{2-}$ or SO_3^{2-} enter an electron transport chain and eventually reduce other molecules of $S_2O_3^{2-}$ or SO_3^{2-}, respectively, to H_2S.

Elemental sulfur (S^0) can also be disproportionated, in this case, to H_2S and SO_4^{2-}:

$$4 \, S^0 + 4 \, H_2O \rightarrow 3 \, H_2S + SO_4^{2-} + 2 \, H^+$$
$$\Delta G^{0\prime} = +40.8 \text{ kJ}$$

As written, however, this is an energetically unfavorable reaction. Nevertheless, if the H_2S that is formed is oxidized back to S^0 by chemical reaction with a metal like Mn^{4+} as

$$H_2S + MnO_2 \rightarrow S^0 + Mn^{2+} + 2 \, OH^-$$
$$\Delta G^{0\prime} = -140.87 \text{ kJ}$$

the sum of these two reactions:

$$3 \, S^0 + 4 \, H_2O + MnO_2 \rightarrow$$
$$2 \, H_2S + SO_4^{2-} + 2 \, H^+ + Mn^{2+} + 2 \, OH^-$$
$$\Delta G^{0\prime} = -100.6 \text{ kJ}$$

yields sufficient energy to support growth of sulfur-disproportionating bacteria. Thus, unlike sulfite- or thiosulfate-disproportionating bacteria, elemental sulfur-disproportionating bacteria require an electron acceptor such as Mn^{4+} to drive the energetics of their reaction.

It is likely that proton motive force formation is central to the bioenergetics of these energy-yielding reactions. In addition, however, in the case of SO_3^{2-} oxidation, substrate-level phosphorylation may increase the ATP yield as SO_3^{2-} is oxidized to SO_4^{2-} via reversal of the APS reductase system (see Figure 15.40).

The sulfate-reducing bacteria reduce sulfate to hydrogen sulfide. The reduction of sulfate requires first its activation by a reaction with ATP to form the compound adenosine phosphosulfate (APS). Electron donors for sulfate reduction include H_2 and organic compounds. Disproportionation of sulfur compounds is an additional energy-yielding strategy for certain members of this group.

✓ Define the following: S^0, SO_4^{2-}, SO_3^{2-}, $S_2O_3^{2-}$, H_2S.
✓ How is sulfate converted to sulfite?
✓ Why is H_2 of importance to sulfate-reducing bacteria?
✓ Give an example of disproportionation.

15.18

Acetogenesis

Carbon dioxide, CO_2 is common in nature and usually abundant in anoxic habitats because it is a major product of energy metabolism of chemoorganotrophs. Two major groups of strictly anaerobic prokaryotes can use CO_2 as an electron acceptor in energy metabolism, *homoacetogens* and *methanogens*. Hydrogen (H_2) is a major electron donor for both of these organisms, and an overview of the processes of methanogenesis and acetogenesis is shown in Figure 15.42. Both processes result in the generation of ion gradient, either of H^+ or Na^+, which drives ATPases in the membrane, while acetogenesis also involves energy conservation via substrate-level phosphorylation. In this section we focus an acetogenesis and in the next section, methanogenesis.

Organisms and Pathway

Homoacetogens can carry out the reaction:

$$4 H_2 + H^+ + 2 HCO_3^- \rightarrow CH_3COO^- + 4 H_2O$$

In addition to H_2, electron donors for acetogenesis include a variety of C_1 compounds, sugars, organic and amino acids, alcohols, and certain nitrogen bases, depending on the organism. Many homoacetogens can also reduce NO_3^- and $S_2O_3^{2-}$; however, CO_2 reduction is probably the major reaction of ecological significance.

The major unifying thread among homoacetogens is the pathway of CO_2 reduction. Homoacetogens convert CO_2 to acetate by the *acetyl-CoA pathway* (see later), and in many homoacetogens autotrophic growth via this pathway also occurs. The acetyl-CoA pathway is also known as the *Ljungdahl-Wood pathway* in honor of its discoverers, Lars Ljungdahl and Harland Wood. A list of the major organisms that produce acetate or oxidize acetate via the acetyl-CoA pathway is given in Table 15.4. Organisms such as *Acetobacterium woodii* and *Clostridium aceticum* can grow either chemoorganotrophically by fermentation of sugars (reaction 1) or chemolithotrophically through the reduction of CO_2 to acetate with H_2 (reaction 2) as electron donor; in either case the major product is acetate:

(1) $C_6H_{12}O_6 \rightarrow 3 CH_3COO^- + 3 H^+$
(2) $2 HCO_3^- + 4 H_2 + H^+ \rightarrow CH_3COO^- + 4 H_2O$

Homoacetogens ferment glucose via the glycolytic pathway converting glucose to two molecules of pyruvate and two molecules of NADH (the equivalent of 4 H). From this point, two molecules of acetate are produced as follows:

(3) $2 \text{ pyruvate}^- \rightarrow 2 \text{ acetate}^- + 2 CO_2 + 4 H$

The third acetate of the homoacetate fermentation comes from the reduction of the two molecules of CO_2

TABLE 15.4	Organisms employing the acetyl-CoA pathway of CO_2 fixation

I. Acetate synthesis the result of energy metabolism
    Acetoanaerobium noterae
    Acetobacterium woodii
    Acetobacterium wieringae
    Acetogenium kivui
    Acetitomaculum ruminis
    Clostridium aceticum
    Clostridium thermoaceticum
    Clostridium formicoaceticum
    Desulfotomaculum orientis
    Sporomusa paucivorans
    Eubacterium limosum (also produces butyrate)
    Treponema sp. strains ZAS-1 and ZAS-2
      (termite gut spirochetes)
II. Acetate synthesis in autotrophic metabolism
    Autotrophic homoacetogenic bacteria
    Autotrophic methanogens
    Autotrophic sulfate-reducing bacteria
III. Acetate oxidation in energy metabolism
    Reaction: Acetate $+ 2 H_2O \rightarrow 2 CO_2 + 8 H$
      Group II sulfate reducers (other than *Desulfobacter*)
    Reaction: Acetate $\rightarrow CO_2 + CH_4$
      Acetotrophic methanogens (*Methanosarcina*,
      Methanothrix)

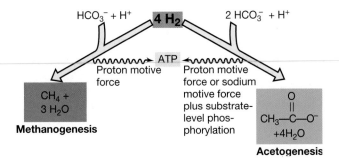

FIGURE 15.42 The contrasting processes of methanogenesis and acetogenesis. The free energy released in each reaction as drawn is: (a) methanogenesis, -136 kJ, and (b) acetogenesis, -105 kJ.

generated in reaction (3), using the four electrons generated from glycolysis *plus* the four electrons produced from the oxidation of two pyruvates to two acetates [reaction (3)]. Starting from pyruvate, then, the overall production of acetate can be written as

$$2 \text{ pyruvate}^- + 4 \text{ H} \rightarrow 3 \text{ acetate}^- + \text{H}^+$$

Most homoacetogenic bacteria that produce and excrete acetate in energy metabolism are gram-positive, and many are classified in the genus *Clostridium*. A few other gram-positive and many different gram-negative bacteria use the acetyl-CoA pathway for autotrophic purposes, reducing CO_2 to acetate for cell carbon. The acetyl-CoA pathway functions in autotrophic growth for certain sulfate-reducing bacteria (∞ Sections 13.17 and 16.17), and is also used by the methanogens, most of which can grow autotrophically on $H_2 + CO_2$ (∞ Sections 14.3, 15.19, and 16.14). By contrast, certain bacteria employ the reactions of the acetyl-CoA pathway primarily in the *reverse* direction as a means of oxidizing acetate to CO_2. These include acetotrophic methanogens (∞ Section 14.3) and sulfate-reducing bacteria (∞ Section 13.17).

Reactions of the Acetyl-CoA Pathway

Unlike other autotrophic pathways such as the Calvin cycle (see Section 15.7) or the reverse citric acid cycle (see Section 15.8), the acetyl-CoA pathway of CO_2 fixation is *not* a cycle. Instead it involves the reduction of CO_2 via two linear pathways—one molecule of CO_2 is reduced to the methyl group of acetate, and the other molecule of CO_2 is reduced to the carbonyl group—followed by their assembly at the end to form acetyl-CoA (Figure 15.43). A key enzyme of the acetyl-CoA pathway is *carbon monoxide (CO) dehydrogenase*. CO dehydrogenase is a complex enzyme that contains the metals Ni, Zn, and Fe as metal cofactors. CO dehydrogenase catalyzes the following reaction:

$$CO_2 + H_2 \rightarrow CO + H_2O$$

and the CO that is produced ends up in the *carbonyl* ($-COO^-$) position of acetate (Figure 15.43). The methyl group of acetate originates from the reduction of CO_2 by a series of reactions involving the coenzyme *tetrahydrofolate* (Figure 15.43). The methyl group that is formed is then transferred from tetrahydrofolate to an enzyme containing vitamin B_{12} as cofactor (Figure 15.43). In the final step of the pathway, the CH_3 group is combined with CO in CO dehydrogenase to form acetate. Interestingly, the reaction mechanism here involves the CH_3 group, which is attached to an atom of nickel in the enzyme, combining with CO, which is bound to an atom of Fe in the enzyme, along with coenzyme A to form the final product, acetyl-

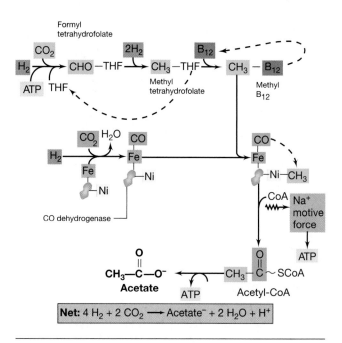

FIGURE 15.43   Reactions of the acetyl-CoA pathway. THF, Tetrahydrofolate; B_{12}, vitamin B_{12} in an enzyme-bound intermediate. CO is bound to an Fe atom in CO dehydrogenase, and the CH_3 group to a nickel atom in an organic nickel compound in CO dehydrogenase. Note how formation of acetate powers a Na^+ pump that is used to drive ATP synthesis and that ATP synthesis also occurs in the conversion of acetyl-CoA to acetate.

CoA. The significance of this resides in the fact that this reaction mechanism was the first alkyl nickel reaction to be discovered in biochemistry.

Because homoacetogens can grow at the expense of reactions of the acetyl-CoA pathway, this reaction sequence must be an overall energy-conserving one (Figure 15.43). One potential site of ATP synthesis is during the conversion of acetyl-CoA to acetate plus ATP (via acetyl-P) (see Section 15.21). However, additional energy-conserving steps occur because a Na^+ gradient (sodium motive force, analogous to a proton motive force but involving Na^+ instead of H^+) is established across the cytoplasmic membrane during acetogenesis. This energized state of the membrane allows for energy conservation via the action of a Na^+-powered ATPase. A similar situation occurs in the succinate fermenter *Propionigenium*, and we discuss the concept of energy conservation from Na^+ gradients when we discuss the metabolism of this bacterium in Section 15.22.

✓ 15.18 Concept Check

Homoacetogens are anaerobes that reduce CO_2 to acetate, usually with H_2 as electron donor. The mechanism of acetate formation is the acetyl-CoA pathway, a series of reactions widely distributed in obligate anaerobes as either a mechanism of autotrophy or for acetate catabolism.

✓ Draw the structure of acetate and identify the carbonyl group and the methyl group. What key enzyme of the acetyl-CoA pathway produces the *carbonyl* group of acetate?

✓ How do homoacetogens make ATP from the synthesis of acetate?

✓ If catabolism of fructose via glycolysis produces only *two* molecules of acetate, how can an organism like *Clostridium aceticum* ferment fructose by this pathway and produce *three* molecules of acetate?

15.19

Methanogenesis

The biological production of methane is carried out by a group of strictly anaerobic Archaea called the *methanogens*. We considered the basic properties and taxonomy of the methanogens in Section 14.3; here we focus on the biochemistry and bioenergetics of methanogenesis. Research on methanogenesis has revealed that the biological production of methane occurs through a unique series of reactions involving novel coenzymes and amazing complexity. We begin our discussion with a consideration of the coenzymes, as they are central to the reactions we will consider, and on the production of methane starting from $H_2 + CO_2$.

C_1 Carriers in Methanogenesis

The unique coenzymes in methanogenesis can be divided into two classes, those involved in carrying the C_1 unit from the initial substrate, CO_2, to the final product, CH_4, and those that function in redox reactions to supply the electrons necessary for the reduction of CO_2 to CH_4 (Figure 15.44 and see Figure 15.46).

The coenzyme **methanofuran** is involved in the first step of methanogenesis. Methanofuran contains the five-membered furan ring and an amino nitrogen atom that binds CO_2 (Figure 15.44a). **Methanopterin** (Figure 15.44b) is a methanogenic coenzyme that resembles the vitamin folic acid (∞ Figure 18.17) and is the C_1 carrier in the intermediate steps of CO_2 reduction to CH_4. **Coenzyme M (CoM)** (Figure 15.44c) is a small molecule that is involved in the terminal step of methanogenesis, the conversion of a methyl group (CH_3) to CH_4. Although not a C_1 carrier, the nickel-containing tetrapyrrole **coenzyme F_{430}** (Figure 15.44d) is also involved in the terminal step of methanogenesis as part of the methyl reductase enzyme complex (see later).

Redox Coenzymes

The coenzymes F_{420} and **7-mercaptoheptanoyl-threonine phosphate**, or **coenzyme B (CoB)** are electron *donors* in methanogenesis. Coenzyme F_{420} (Figure 15.44c)

is a flavin derivative, structurally resembling the common flavin coenzyme FMN (∞ Figure 4.13). F_{420} also plays a role in methanogenesis as the electron donor in several steps of CO_2 reduction (see Figure 15.46). The oxidized form of F_{420} absorbs light at 420 nm and fluoresces blue-green (Figure 15.45), and such fluorescence is a useful tool for microscopic identification of an organism as a methanogen. HS-HTP is involved in the terminal step of methanogenesis catalyzed by the **methyl reductase enzyme complex.** As shown in Figure 15.44f, the structure of CoB is rather simple and the coenzyme structurally resembles the vitamin pantothenic acid (part of acetyl-CoA) (∞ Figure 4.21). With this overview of the coenzymes of methanogenesis, let us now look at the reactions involved in the reduction of CO_2 to CH_4.

Biochemistry of CO_2 Reduction to CH_4

The reduction of CO_2 to CH_4 is generally H_2 dependent, but formate, carbon monoxide, and even certain organic compounds such as alcohols can supply the electrons for CO_2 reduction. For example, 2-propanol can be oxidized to acetone, yielding electrons for methanogenesis in some species. But in general, the production of CH_4 from CO_2 is driven by molecular hydrogen (H_2).

The steps in CO_2 reduction, shown in Figure 15.46, are summarized as follows:

1. CO_2 is activated by a methanofuran-containing enzyme and subsequently reduced to the formyl level.
2. The formyl group is transferred from methanofuran to an enzyme containing methanopterin (MP in Figure 15.46) and subsequently dehydrated and reduced in two separate steps to the methylene and methyl levels.
3. The methyl group is transferred from methanopterin to an enzyme containing CoM.
4. Methyl-CoM is reduced to methane by the methyl reductase system in which F_{430} and CoB are intimately involved. Coenzyme F_{430} removes the CH_3 group from CH_3-CoM, forming a Ni^{2+} — CH_3 complex. This is reduced by electrons from CoB, generating CH_4 and a disulfide complex of CoM and CoB (CoM-S — S-CoB). Free CoM and CoB are regenerated by reduction of this complex with H_2, and as we will see, it is this reaction that allows for energy conservation in methanogenesis.

Methanogenesis from Methyl Compounds and Acetate

We learned in Section 14.3 that in addition to $H_2 + CO_2$, methane can be formed from a variety of methylated compounds as well. Methyl compounds such as methanol are catabolized by donating methyl groups

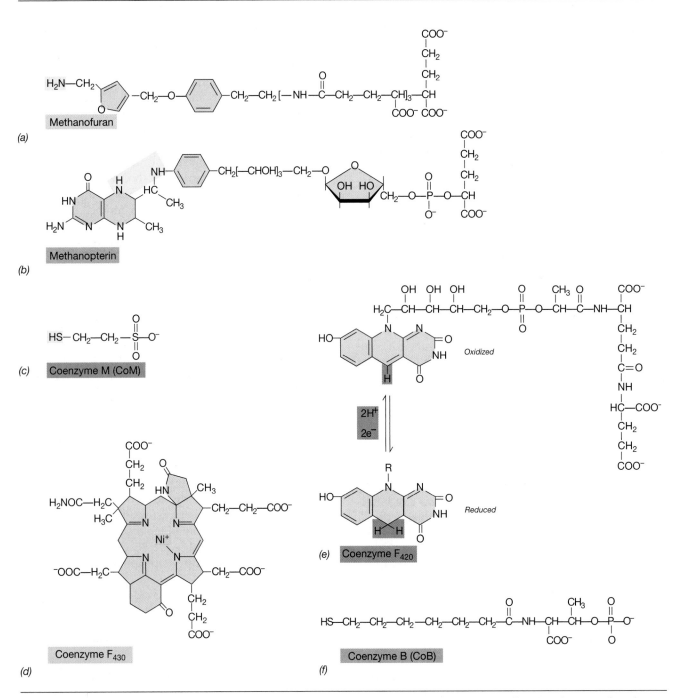

(a) Methanofuran

(b) Methanopterin

(c) Coenzyme M (CoM)

(d) Coenzyme F$_{430}$

(e) Coenzyme F$_{420}$

(f) Coenzyme B (CoB)

FIGURE 15.44 Coenzymes unique to methanogenic Archaea. The atoms shaded in brown or yellow are the sites of oxidation–reduction reactions (F$_{420}$—brown) or the position to which the C$_1$ moiety is attached during the reduction of CO$_2$ to CH$_4$ (methanofuran, methanopterin, and coenzyme M—yellow). The colors used to highlight a particular coenzyme itself (CoB is orange, for example) are used throughout in Figures 15.46–15.48 and can be used to follow the reactions in each figure.

to a corrinoid protein to form CH$_3$-corrinoid (Figure 15.47). Corrinoids are the parent structures of such compounds as vitamin B$_{12}$ and contain a porphyrin-like corrin ring with a central cobalt atom (∞ Figure 11.13a). The CH$_3$-corrinoid complex donates the methyl group to CoM to give CH$_3$-CoM from which methane is formed in the same way as in the terminal step discussed earlier (compare Figures 15.46 and 15.47a). If re-

ducing power (such as H$_2$) is not available to drive the terminal step, some of the methanol must be oxidized to CO$_2$ to yield electrons, and this occurs by reversal of steps in methanogenesis (Figure 15.47a).

When *acetate* is the substrate for methanogenesis it is first activated to acetyl-CoA, which can interact with carbon monoxide dehydrogenase of the acetyl-CoA pathway (see Section 15.18). Then, the methyl group of acetate

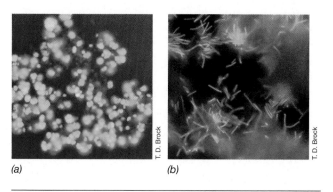

(a)　　　　　　*(b)*

FIGURE 15.45 (a) Autofluorescence in cells of the methanogen *Methanosarcina barkeri* due to the presence of the unique electron carrier F_{420}. A single cell is about 1.7 μm in diameter. The organisms were made visible with blue light in a fluorescence microscope (∞ Section 3.1 and Figure 3.6). (b) F_{420} fluorescence in cells of the methanogen *Methanobacterium formicicum*. A single cell is about 0.6 μm in diameter.

is transferred to the corrinoid enzyme to yield CH_3-corrinoid, and from there it goes through the CoM-mediated terminal step of methanogenesis (Figure 15.47*b*).

Autotrophy

Autotrophy in methanogens occurs via the reactions of the acetyl-CoA pathway discussed in Section 15.18, and as we have seen, parts of this pathway are already integrated into the catabolism of methanol and acetate

(Figure 15.47). However, methanogens lack the tetrahydrofolate-driven series of reactions of the acetyl-CoA pathway that lead to the production of a methyl group (see Figure 15.43). These, of course, are unnecessary, since methanogens either derive methyl groups directly from their electron donors (Figure 15.47) or make methyl groups during methanogenesis from $H_2 + CO_2$ (Figure 15.46); thus methyl groups are present in abundance in the cell to begin with. However, the carbonyl group of the acetate produced during autotrophic growth of methanogens is derived from CO dehydrogenase, and the terminal step in acetate synthesis occurs as described for homoacetogens (see Section 15.18 and Figure 15.43).

Energy Conservation in Methanogenesis

Under standard conditions the free energy change in the reduction of CO_2 to CH_4 with H_2 is -131 kJ/mol. This is sufficient for the synthesis of at least one ATP. We have previously discussed how energy conservation in methanogenesis is linked to the terminal step, the *methyl reductase* step (Figure 15.46). The interaction of CoB with CH_3 — CoM in this terminal step forms CH_4 and a heterodisulfide, CoM-S — S-CoB. The latter complex is then reduced with electrons supplied from F_{420} to yield CoM-SH and CoB-SH (Figure 15.46). This reduction, carried out by the enzyme *heterodisulfide reductase*, is exergonic and is associated with the extrusion of protons across the membrane, creating a proton motive force

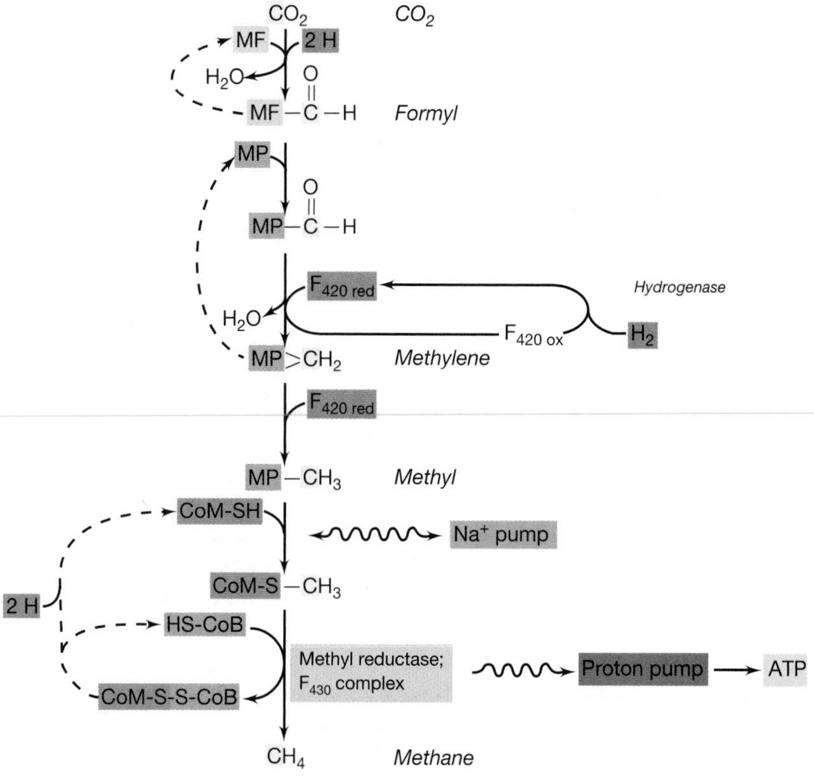

FIGURE 15.46 Pathway of methanogenesis from CO_2. MF, Methanofuran; MP, methanopterin; CoM, coenzyme M; F_{420red}, reduced coenzyme F_{420}; F_{430}, coenzyme F_{430}; CoB, 7-mercaptoheptanoylthreonine phosphate. The carbon atom reduced is shown in yellow, and the source of electrons is highlighted in brown. See Figure 15.44 for the structures of the coenzymes and the text for discussion of the reversible Na^+ pump. The immediate electron donor in the first step of methanogenesis is unknown. Electrons for CO_2 reduction generally come from H_2 but in certain methanogens, a few organic compounds can be oxidized to yield electrons for CO_2 reduction.

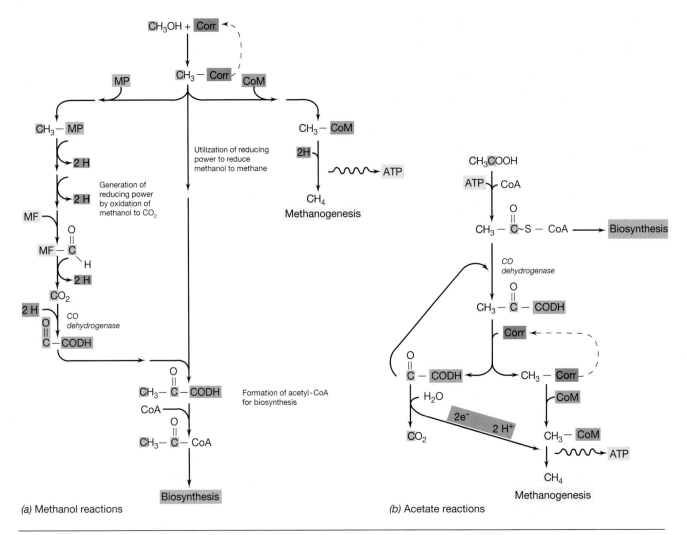

FIGURE 15.47   Utilization of reactions of the acetyl-CoA pathway during growth on methanol (a) or acetate (b) by methanogenic Archaea. For growth on methanol, most methanol carbon is converted to CH_4, while a smaller amount is converted to either CO_2 or, via formation of acetyl-CoA, is assimilated into cell material. Abbreviations and color coding are as in Figures 15.44 and 15.46.

(Figure 15.48). Dissipation of the proton gradient by a proton-translocating ATPase (⚭ Section 4.11 discusses ATPases) drives ATP synthesis during methanogenesis in the same way that this process occurs in other forms of respiration. However, electron flow to the heterodisulfide reductase involves a unique membrane-integrated electron carrier, a phenazine compound called *methanophenazine* (Figure 15.48). In the electron transport process in the terminal step, methanophenazine is alternately reduced (by F_{420}) and then oxidized (by a b type cytochrome), which is the ultimate donor of electrons to the heterodisulfide reductase (Figure 15.48).

Methanogenesis from methyl compounds is also linked to the heterodisulfide reductase proton pump, but an additional factor is involved. As previously mentioned, in the absence of H_2 methanogenesis from compounds like CH_3OH requires that some of the CH_3OH be oxidized to CO_2 to generate the electrons needed for methyl reduc-

tion to methane. This requires an energy input and occurs at the expense of a Na^+ motive force (a Na^+ energized membrane potential). The energy inherent in this potential is derived from the conversion of CH_3-MP to CH_3-CoM during methanogenesis; the reverse reaction consumes energy. Further oxidative steps in the conversion of methyl groups of CO_2 proceed by reversal of the enzymatic steps leading to CH_4 formation from CO_2 (see Figure 15.46) and show no further energy requirement. However, energy from the Na^+ motive force is also required to drive the carboxylation of methanofuran in methanogenesis (see Figure 15.46). This reaction is sufficiently endergonic that energy is needed to drive it, and this energy comes from the Na^+ gradient. Thus, in methanogens we see two types of ion pumps: a typical proton pump used to drive adenosine triphosphate (ATP) synthesis and a reversible Na^+ pump that can function to drive methyl group oxidation and methanofuran carboxylation.

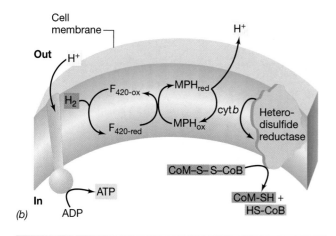

(a)

(b)

FIGURE 15.48 Energy conservation in methanogenesis. (a) Structure of methanophenazine (MPH in part b), an electron carrier in the electron transport chain leading to ATP synthesis. The central ring of the molecule can be alternately reduced and oxidized. (b) Steps in electron transport. Electrons originating from H_2 reduce F_{420} and then methanophenazine. The latter, through a cytochrome of the *b* type, reduces heterodisulfide reductase with the extrusion of protons to the outside of the membrane. In the final step, heterodisulfide reductase reduces Co-M-S — S-CoB to HS-CoM and HS-CoB. Refer to Figure 15.44 for the structures of CoM and CoB.

✓ 15.19 Concept Check

Methanogenesis is the biological production of CH_4 from either CO_2 reduction with H_2 or from methylated compounds. A variety of unique coenzymes are involved in methanogenesis and the process is a strictly anaerobic one. Energy conservation in methanogenesis involves both proton and sodium ion gradients.

- ✓ What coenzymes function as C_1 carriers in methanogenesis? As redox agents?
- ✓ Why are the steps in CH_3OH catabolism during methanogenesis different if H_2 is present than when it is not present?
- ✓ Concerning autotrophy, why are only some but not all of the enzymes of the acetyl-CoA present in methanogens?
- ✓ How is a proton motive force produced in methanogenesis?

15.20

Ferric Iron, Manganese, Chlorate, and Organic Electron Acceptors

In addition to the electron acceptors for anaerobic respiration discussed thus far, ferric iron (Fe^{3+}), manganic ion (Mn^{4+}), chlorate (ClO_3^-), and various oganic

compounds are important electron acceptors for bacteria in nature (Figure 15.49). A wide diversity of bacteria are able to reduce these acceptors, especially Fe^{3+}, and many are able to reduce other acceptors as well, such as NO_3^- and S^0 (see Sections 15.16 and 15.17).

Ferric Iron Reduction

Ferric iron is an electron acceptor for energy metabolism in a wide variety of both chemoorganotrophic and chemolithotrophic bacteria, and because Fe^{3+} is abundant in nature, its reduction is a major form of anaerobic respiration. The reduction potential of the Fe^{3+}/Fe^{2+} couple is very electropositive ($E_0' = +0.77$ V at pH 2, $E_0' = +0.2$ V at pH 7), and because of this, Fe^{3+} reduction can be coupled to the oxidation of several organic and inorganic electron donors. Various organic compounds, including aromatic compounds, can be oxidized anaerobically by ferric iron reducers with electrons traveling through electron transport chains that terminate in a ferric iron reductase system. Such electron flow establishes a proton motive force that can be used to generate ATP. Much research on the energetics of ferric iron reduction has been done in the gram-negative bacterium *Shewanella putrefaciens*, in which Fe^{3+}-dependent anaerobic growth occurs with various organic electron donors. Other important Fe^{3+} reducers include *Geobacter*, *Geospirillum*, and *Geovibrio*.

Geobacter metallireducens has been a model for study of the physiology of Fe^{3+} reduction. This organism can oxidize acetate with Fe^{3+} as an acceptor as follows:

$$Acetate^- + 8 Fe^{3+} + 4 H_2O \rightarrow$$
$$2 HCO_3^- + 8 Fe^{2+} + 9 H^+ \quad \Delta G^{0'} = -233 \text{ kJ}$$

Geobacter can also use H_2 or other organic electron donors including the aromatic hydrocarbon toluene (see Figure 15.66d for the structure of toluene); this may be of environmental significance because toluene often contaminates ferric-rich aquifers, and it has been suggested that organisms like *Geobacter* may be natural cleanup agents in such environments.

Ferric iron is one of the most common metals in soils and rocks, and its reduction leads to the production of ferrous iron, a more soluble form of iron. Bacterial iron reduction can thus lead to solubilization of iron, an important geochemical process. As was illustrated in Figure 15.28c, one type of iron deposit, called *bog iron*, is formed as a direct result of the activities of ferric iron-reducing microorganisms.

Reduction of Manganese (Mn^{4+}) and Other Inorganic Substances

The metal manganese has a number of oxidation states, of which Mn^{4+} and Mn^{2+} are the most stable and biologically relevant. Anoxic reduction of Mn^{4+} to Mn^{2+} is carried out by a variety of microorganisms, mostly

Acceptor	Reaction	E'_0 of couple (V)	Product
Chlorate	$ClO_3^- \xrightarrow[6\ H^+]{6\ e^-} Cl^- + 3\ H_2O$	+1.03	Chloride
Manganic ion	$Mn^{4+} \xrightarrow{2\ e^-} Mn^{2+}$	+0.798	Manganous ion
Ferric ion	$Fe^{3+} \xrightarrow{e^-} Fe^{2+}$	+0.77*	Ferrous ion
Selenate	$^-O-\overset{\overset{O}{\|}}{\underset{\underset{O}{\|}}{Se}}-O^- \xrightarrow[2\ H^+]{2\ e^-} Se{=}O + H_2O$ (product with O^-)	+0.475	Selenite
Dimethyl sulfoxide (DMSO)	$H_3C-\overset{\underset{\|}{\downarrow O}}{S}-CH_3 \xrightarrow[2\ H^+]{2\ e^-} (CH_3)_2S + H_2O$	+0.16	Dimethyl sulfide (DMS)
Arsenate	$^-O-\overset{\overset{O^-}{\|}}{\underset{\underset{O^-}{\|}}{As}}{=}O \xrightarrow[2\ H^+]{2\ e^-} As-O^- + H_2O$	+0.139	Arsenite
Trimethylamine-N-oxide (TMAO)	$H_3C-\overset{\overset{CH_3}{\|}}{\underset{\underset{O}{\downarrow}}{N}}-CH_3 \xrightarrow[2\ H^+]{2\ e^-} (CH_3)_3N + H_2O$	+0.13	Trimethylamine (TMA)
Fumarate	$^-O-\overset{O}{\overset{\|}{C}}-\overset{H}{\underset{\|}{C}}{=}C-\overset{O}{\overset{\|}{C}}-O^- \xrightarrow[2\ H^+]{2\ e^-} ^-O-\overset{O}{\overset{\|}{C}}-CH_2-CH_2-\overset{O}{\overset{\|}{C}}-O^-$	+0.03	Succinate

*The E_0' of the Fe^{3+}/Fe^{2+} couple is very pH dependent. The value listed is for pH 2–3; at pH 7, the E_0' is about +0.2 V.

FIGURE 15.49　Some alternative electron acceptors for anaerobic respirations. At neutral pH, where most ferric iron reducers are active, the Fe^{3+}/Fe^{2+} couple is about 0.2 V.

chemoorganotrophs. In *Shewanella putrefaciens* and a few other bacteria, anoxic growth on acetate and several other nonfermentable carbon sources occurs with Mn^{4+} as electron acceptor. The reduction potential of the Mn^{4+}/Mn^{2+} couple is extremely high (Figure 15.49); thus, several compounds should be able to donate electrons to Mn^{4+} reduction. This is also the case for chlorate, because its reduction potential is actually *more positive* than that of the O_2/H_2O couple (Figure 15.49). Several chlorate-reducing bacteria have been isolated, and most of them are facultative and thus also capable of aerobic growth.

Other inorganic substances can function as electron acceptors for anaerobic respiration. These include selenium and arsenic compounds (Figure 15.49). Although usually not present in large amounts in natural systems, arsenic and selenium compounds are occasional pollutants and can support anoxic growth of various bacteria. The reduction of SeO_4^{2-} to SeO_3^{2-} and eventually to Se^0 (metallic selenium) is an important method of selenium removal from water and has been used as a means of cleaning up (bioremediation) (⚭ Section 16.22) of selenium-contaminated soils. Most bacteria capable of selenate or arsenate reduction can also use several other electron acceptors, such as Fe^{3+}, Mn^{4+}, and organic compounds, and in most cases show a facultatively aerobic form of metabolism.

In the case of arsenic, the sulfate-reducing bacterium *Desulfotomaculum* can reduce arsenate (AsO_4^{3-}) to arsenite (AsO_3^{2-}), along with SO_4^{2-} (to HS^-), and in the process causes a mineral complex of arsenic and sulfide, As_2S_3, to precipitate spontaneously (Figure 15.50). The mineral is formed both intracellularly as well as extracellularly, and its formation is an example of *biomineralization*, the formation of a mineral by bacterial action. In this case As_2S_3 formation (Figure 15.50) also functions as a means of detoxifying what would otherwise be a toxic compound (arsenic in this case), and such activities may have practical applications for the microbial cleanup of toxic wastes.

Organic Electron Acceptors

Several organic compounds can participate as external electron acceptors in anaerobic respirations. Of those listed in Figure 15.49, the compound that has been most extensively studied is **fumarate,** which is reduced to **succinate.** An examination of the *citric acid cycle* (⚭ Figure 4.20) will indicate that fumarate and succinate are important intermediates. The role of fumarate as an electron acceptor for anaerobic respiration derives from the fact that the fumarate–succinate couple has a re-

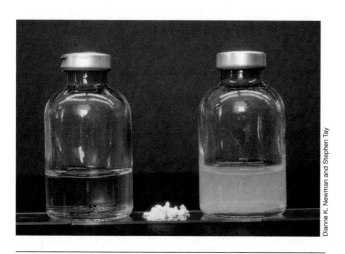

FIGURE 15.50 Production of the mineral arsenic trisulfide (As₂S₃) during arsenate reduction by the sulfate-reducing bacterium *Desulfotomaculum auripigmentum*. Left, appearance of culture bottle after inoculation. Right, following growth for two weeks. Center, synthetic sample of As₂S₃. Mineral production by microorganisms is called *biomineralization*.

duction potential near 0 V (see Figure 15.37), which allows coupling of fumarate reduction to NADH or H₂ oxidation. The energy yield is sufficient for the synthesis of one ATP. Bacteria able to use fumarate as an electron acceptor include *Wolinella succinogenes* (which can grow on H₂ as electron donor using fumarate as electron acceptor), *Desulfovibrio gigas* (a sulfate-reducing bacterium that can also grow under non–sulfate-reducing conditions), some clostridia, *Escherichia coli*, and many other bacteria.

The compound **trimethylamine oxide** shown in Figure 15.49 is an interesting organic electron acceptor. Trimethylamine oxide (TMAO) is an important osmotic solute in marine fish, where it functions in these animals as a means of excreting excess nitrogen, but a variety of bacteria are able to reduce TMAO to trimethylamine (TMA). TMA has a strong odor and flavor, and some of the odor that occurs in spoiled marine fish is due to TMA produced by bacterial action. A variety of facultatively aerobic bacteria are able to use TMAO as an alternate electron acceptor. In addition, several phototrophic purple bacteria (∞ Section 13.1) are able to use TMAO as an electron acceptor for anaerobic metabolism in darkness. A compound analogous to TMAO is **dimethyl sulfoxide** (DMSO), which is reduced by a variety of bacteria to dimethyl sulfide (DMS). DMSO is a common natural product and is found in both marine and freshwater environments. DMS has a strong, pungent odor, and bacterial reduction of DMSO to DMS is signaled by the presence of the characteristic odor of DMS. A variety of bacteria, including *Campylobacter*, *Escherichia*, and many purple bacteria, are able to use

DMSO as an electron acceptor in energy generation (∞ Section 16.17 discusses DMSO metabolism).

The reduction potentials of the TMAO/TMA and DMSO/DMS couples are similar, near +0.15 V, which means that any electron transport chain that ends with TMAO or DMSO reduction must be rather brief. Like for fumarate reduction, in most instances of TMAO and DMSO reduction, cytochromes of the *b* type (with reduction potentials near 0 V) have been identified as terminal oxidases.

Humic Substances as Electron Acceptors

Humic substances are complex mixtures of high-molecular-weight organic materials that are present in most terrestrial and aquatic environments. Humics are derived from the refractory remains of plants, animals, and microorganisms and are generally resistant to microbial metabolism. Although an exact E_0' cannot be measured for natural humic substances because they are undefined complex mixtures, model low-molecular-weight humic compounds, such as fulvic acid, have an E_0' near +0.5 V, indicating that humic substances in general should be good electron acceptors (see Figure 15.49). Indeed, humic substances have been shown to be used as electron acceptors for the anaerobic oxidation of organic compounds and H₂, yielding energy to support growth. For example, the bacterium *Geobacter metallireducens*, discussed previously in connection with Fe^{3+} reduction, grows anaerobically with acetate or H₂ as electron donors and humics as electron acceptors.

The use of humics as electron acceptors points to a biogeochemical role for these substances that was previously unsuspected. This is especially significant because it is known that reduced humics can react chemically with metals such as Fe^{3+} and various organic compounds, mobilizing these substances for transport through waterlogged soils or into groundwater. And, because many of the organisms shown to reduce humics are capable of metabolizing toxic substances such as toluene, humics may facilitate the removal of these substances from groundwater and soils contaminated with toxic wastes.

✓ 15.20 Concept Check

Besides inorganic nitrogen and sulfur compounds or CO₂, a variety of other substances, both organic and inorganic, can function as electron acceptors for anaerobic respiration. These include in particular Fe^{3+}, Mn^{4+}, and fumarate.

- ✓ With H₂ as electron donor why is reduction of Fe^{3+} a much more favorable reaction than reduction of fumarate?
- ✓ Give an example of *biomineralization*.
- ✓ Where in nature might TMAO-degrading bacteria be abundant? Why?

15.21

Fermentations: Energetic and Redox Considerations

Because oxygen is not highly soluble (9.6 mg/l distilled water in equilibrium with air at 25°C), many environments easily become anoxic. In such environments, decomposition of organic materials occurs anaerobically. If adequate supplies of electron acceptors like SO_4^{2-}, NO_3^-, Fe^{3+}, and the others considered in previous sections are not available in such anoxic environments, much of the carbon will be catabolized by fermentation. (CO_2 as an electron acceptor is an exception. It is rarely limiting in anoxic habitats but its conversion to methane requires H_2, and the latter is itself a product of fermentative reactions). We discussed the overall process of fermentation in Sections 4.8 and 4.9 and showed that it was an internally balanced oxidation–reduction process in which carbon from the same external organic compound was partially oxidized and partially reduced (Figure 15.51).

There are two problems an organism faces if it is to catabolize organic compounds in energy-yielding metabolism: (1) conserving some of the energy released as ATP, and (2) disposing of electrons removed from the electron donor. In fermentation, ATP synthesis generally occurs by way of *substrate-level phosphorylation*, a mechanism by which high energy phosphate bonds from organic intermediates of the fermentation are transferred to ADP (◐◐ Section 4.8). The second problem, that of redox balance, is solved by production and excretion by the organism of *fermentation products* generated from the original substrate (Figure 15.51). We now consider these basic principles of fermentation in more detail and highlight the enormous diversity of microbial fermentations known.

High Energy Compounds and Substrate-Level Phosphorylation

Energy conservation from substrate-level phosphorylation can occur in many different ways. However, central to the mechanism of ATP synthesis is the production of one or another *high energy compound*. These are generally organic compounds containing a phosphate group or a coenzyme-A molecule, the hydrolysis of which is highly exergonic. A list of the major high energy intermediates is given in Table 15.5. This list is not complete, but it includes most recognized high energy intermediates known to be formed during biochemical processes. Because most of the compounds listed in Table 15.5 can couple directly to ATP synthesis (−31.8 kJ/mol), if an organism can form one or another of these compounds during fermentative metabolism, it can make ATP. Substrate-level phosphorylation is a more direct way of making ATP than via a proton motive force (◐◐ Figure 4.11) but requires that the energy source couple directly to a high energy intermediate.

Pathways for the anaerobic breakdown of various fermentable substances to high energy intermediates are summarized in Figure 15.52. It should be noted that this figure is organized by the high energy compounds listed in Table 15.5 and that either one of these compounds, or a related derivative, is generated in each case and leads to ATP synthesis. Thus, Figure 15.52 and Table 15.5 should be examined together.

Energy Yields of Fermentative Organisms

How much ATP can be produced by a fermentative organism? As we have seen, glucose fermenters produce 2–3 ATPs per glucose fermented in glycolysis

TABLE 15.5	Energy-rich compounds involved in substrate-level phosphorylation[a]
Compound	**Free energy of hydrolysis, $\Delta G^{0\prime}$ (kJ/mol)[b]**
Acetyl-CoA	−35.7
Propionyl-CoA	−35.6
Butyryl-CoA	−35.6
Succinyl-CoA	−35.1
Acetylphosphate	−44.8
Butyrylphosphate	−44.8
1,3-Bisphosphoglycerate	−51.9
Carbamyl phosphate	−39.3
Phosphoenolpyruvate	−51.6
Adenosine-phosphosulfate (APS)	−88
N^{10}-formyltetrahydrofolate	−23.4
Energy of hydrolysis of ATP (ATP → ADP + P_i)	−31.8

a Data from Thauer, R. K., K. Jungermann, and K. Decker. 1977. Energy conservation in chemotrophic anaerobic bacteria. *Bacteriol. Rev. 41:* 100–180.

b The $\Delta G^{0\prime}$ values shown here are for "standard conditions," which are not necessarily those of cells. Including heat loss, the energy costs of making an ATP are more like 60 kJ than 32 kJ, and the energy of hydrolysis of the high energy compounds shown here are thus likely higher. But for simplicity and comparative purposes, the values in this table will be taken as the actual energy released per reaction.

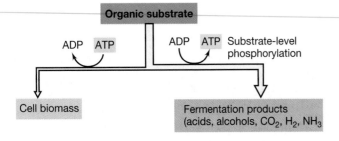

FIGURE 15.51 Overall process of fermentation. Note how in a typical fermentation, most of the carbon is excreted as a partially reduced end product of energy metabolism and only a small amount is used in biosynthesis.

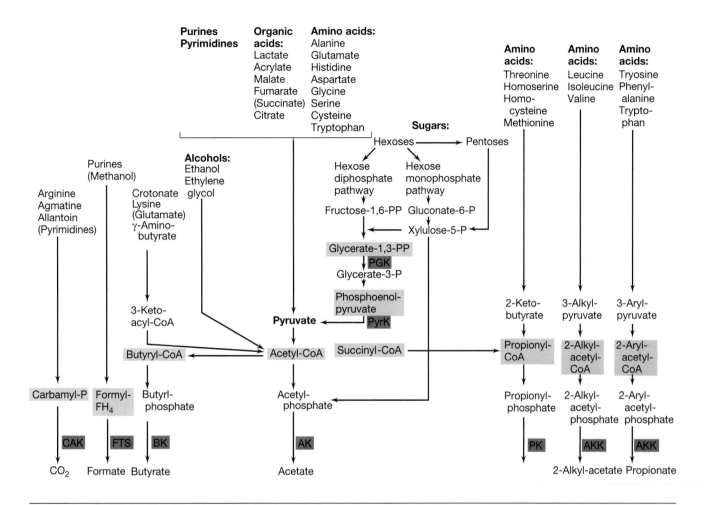

FIGURE 15.52 Major routes of the anaerobic breakdown of various fermentable substances. The sites of substrate-level phosphorylation are shown by the abbreviations of the enzymes involved. CAK, carbamyl phosphate kinase; FTS, formyltetrahydrofolate synthetase; AK, acetate kinase; PK, propionate kinase; BK, butyrate kinase; AKK, alkyl (aryl) acetate kinase; PGK, phosphoglycerate kinase; PyrK, pyruvate kinase. High energy CoA derivatives and other key high energy compounds are highlighted in blue while enzymes are in red. Refer back to Table 15.5.

(∞ Figure 4.12). This is about the maximum amount of ATP produced by fermentation; many other substrates provide less energy. The potential energy released from a particular fermentation can be calculated from the balanced reaction and from the free energy values given in Appendix 1. For instance, the fermentation of glucose to ethanol and CO_2 has a theoretical energy yield of -235 kJ/mol, enough to produce about 7 ATPs. However, only 2 ATPs are actually produced, which implies that the organism operates at considerably less than 100% efficiency, some energy being lost as heat.

Oxidation–Reduction Balance

In any fermentation reaction, there must be a *balance* between oxidation and reduction. The total number of electrons in the products on the right side of the equation must balance the number in the substrates on the left side of the equation. When fermentations are stud-

ied experimentally in the laboratory, it is conventional to calculate a *fermentation balance* to make certain that no products are missed. The fermentation balance can also be calculated theoretically from the oxidation states of the substrates and products (see Appendix 1 for the procedure for calculating oxidation states).

In a number of fermentations, electron balance is maintained by the production of molecular hydrogen, H_2. In H_2 production, protons (H^+) derived from water serve as electron acceptors. Production of H_2 is generally associated with the presence in the organism of an iron–sulfur protein called *ferredoxin,* a very electronegative electron carrier. The transfer of electrons from ferredoxin to H^+ is catalyzed by the enzyme **hydrogenase,** as illustrated in Figure 15.53. We have already discussed the enzyme hydrogenase earlier in this chapter in reference to the *utilization* of hydrogen by sulfate-reducing bacteria and the aerobic hydrogen bacteria. In the present case, hydrogenase is involved in the *production* of hydrogen.

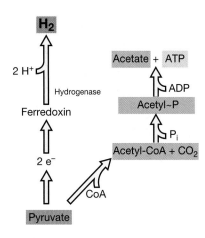

FIGURE 15.53   Production of molecular hydrogen from pyruvate. Note how production of acetate leads to ATP synthesis from hydrolysis of the high energy organic intermediate, acetyl phosphate (see Table 15.5).

The energetics of hydrogen production are actually somewhat unfavorable, and so most fermentative organisms produce only a relatively small amount of hydrogen along with other fermentation products. Hydrogen production thus functions primarily to maintain redox balance. If hydrogen production is prevented, for instance, then the oxidation–reduction balance of the other fermentation products will be shifted toward *more reduced* products. Thus, many fermentative organisms that make H_2 produce both ethanol and acetate. Since ethanol is more reduced than acetate, its formation is favored when hydrogen production is inhibited.

Numerous anaerobic bacteria produce *acetate* as one of the products of fermentation. The production of acetate or certain other fatty acids (see Table 15.5) is energetically advantageous because it allows the organism to make ATP by substrate-level phosphorylation. The key intermediate generated in acetate production is acetyl-CoA (see Table 15.5), a high energy intermediate. Acetyl-CoA can be converted to acetyl phosphate (also listed in Table 15.5), and the high energy phosphate group of acetyl phosphate subsequently transferred to adenosine diphosphate (ADP) by acetate kinase, yielding ATP. One of the main substrates that is converted to acetyl-CoA is pyruvate, a major product of glycolysis. The conversion of pyruvate to acetyl-CoA is an oxidation reaction (Figure 15.53) and the excess electrons generated must be used either to make a more reduced end product or in the production of H_2 as discussed previously.

✓ 15.21 Concept Check

In the absence of an external electron acceptor, organic compounds can be catabolized only by fermentation. Only certain compounds are fermentable, and a requirement for most fermentations is that an energy-rich organic intermediate be formed that can yield ATP by substrate-level phosphorylation. Redox balance must also be achieved in fermentations, and H_2 production is one means of disposing of excess electrons.

✓ What is *substrate-level phosphorylation*?
✓ Why is acetate formation in fermentation energetically beneficial?

15.22

Fermentative Diversity

Fermentations are classified either in terms of the substrate fermented or in terms of the fermentation products formed. Many of the specific fermentation reactions of bacteria were discussed when the individual groups were examined in Chapters 13 and 14. Here we present an overview of common fermentations.

Diversity of Fermentations

Table 15.6 summarizes some of the main types of fermentations as classified on the basis of *products formed*. Note some of the broad categories, such as alcohol, lactic acid, propionic acid, mixed acid, butyric acid, and homoacetic acid. A number of fermentations are classified on the basis of the *substrate fermented* rather than the fermentation product. For instance, many of the spore-forming anaerobic bacteria (genus *Clostridium*) ferment *amino acids* with the production of acetate, ammonia, and H_2 (∞ Figure 13.60). Other *Clostridium* species, such as *C. acidi-urici* and *C. purinolyticum*, ferment *purines* such as xanthine or adenine with the formation of acetate, formate, CO_2, and ammonia. Still other anaerobes ferment *aromatic* compounds. For example, the bacterium *Pelobacter acidigallici* ferments the aromatic compound *phloroglucinol* (1,3,5-benzenetriol, $C_6H_6O_3$) via the following overall pathway:

$$\text{Phloroglucinol (C}_6\text{H}_6\text{O}_3) + 3\,H_2O \rightarrow 3\text{ acetate}^- + 3\,H^+$$
$$\Delta G^{0\prime} = -142.5 \text{ kJ/reaction}$$

Many unusual fermentations are carried out by only a very restricted group of anaerobes and, in some cases, by only a single known bacterium. Some examples are listed in Table 15.7. Many of these bacteria can be considered metabolic specialists, having evolved biochemical capabilities to catabolize a substrate or substrates not catabolized by other bacteria. However, as for the substances listed in Table 15.6, successful fermentation of these more unusual substrates requires that the organism be able to produce a high energy intermediate, usually a coenzyme-A derivative of the type listed in Table 15.5, during the fermentation in order to conserve some of the energy released as ATP.

TABLE 15.6 Examples of common bacterial fermentations and some of the organisms carrying them out

Type	Overall reaction[a]	Organisms
Alcoholic fermentation	Hexoses $\rightarrow$ 2 Ethanol + 2 CO_2	Yeast *Zymomonas*
Homolactic fermentation	Hexose $\rightarrow$ 2 Lactate$^-$ + 2 H$^+$	*Streptococcus* Some *Lactobacillus*
Heterolactic fermentation	Hexose $\rightarrow$ Lactate$^-$ + Ethanol + CO_2 + H$^+$	*Leuconostoc* Some *Lactobacillus*
Propionic acid	Lactate$^-$ $\rightarrow$ Propionate$^-$ + Acetate$^-$ + CO_2	*Propionibacterium* *Clostridium propionicum*
Mixed acid	Hexoses $\rightarrow$ Ethanol + 2,3-Butanediol + Succinate^{2-} + Lactate $^-$ + Acetate$^-$ + Formate$^-$ + H_2 + CO_2	Enteric bacteria *Escherichia* *Salmonella* *Shigella* *Klebsiella* *Enterobacter*
Butyric acid	Hexoses $\rightarrow$ Butyrate$^-$ + Acetate$^-$ + H_2 + CO_2	*Clostridium butyricum*
Butanol	Hexoses $\rightarrow$ Butanol + Acetate$^-$ + Acetone + Ethanol + H_2 + CO_2	*Clostridium acetobutylicum*
Caproate	Ethanol + Acetate$^-$ + CO_2 $\rightarrow$ Caproate$^-$ + Butyrate$^-$ + H_2	*Clostridium kluyveri*
Homoacetogenic	Fructose $\rightarrow$ 3 Acetate$^-$ + 3 H$^+$ 4 H_2 + 2 CO_2 + H$^+$ $\rightarrow$ Acetate$^-$ + 2 H_2O	*Clostridium aceticum* *Acetobacterium*
Methanogenic	Acetate$^-$ + H_2O $\rightarrow$ CH_4 + HCO_3^-	*Methanothrix* *Methanosarcina*

a Reactions are intended as an overview of the process and are not necessarily balanced.

Fermentations Without Substrate-Level Phosphorylation: Decarboxylations of Organic Acids

With certain substrates there is insufficient energy released to couple to the synthesis of ATP directly by substrate-level phosphorylation, yet these compounds still support fermentative growth of an organism. In these cases, catabolism of the substrate is linked to ion pumps that establish a proton or sodium gradient across the membrane. Examples of this include the fermentations by *Propionigenium modestum* and *Oxalobacter formigenes;* both of these organisms couple the fermentation of dicarboxylic acids to membrane-bound energy-linked ion pumps. *Propionigenium modestum* carries out the following reaction:

TABLE 15.7 Some unusual bacterial fermentations

Type	Overall balanced reaction	Organisms
Acetylene	2 C_2H_2 + 3 H_2O $\rightarrow$ ethanol + acetate$^-$ + H$^+$	*Pelobacter acetylenicus*
Glycerol	4 Glycerol + 2 HCO_3^- $\rightarrow$ 7 acetate$^-$ + 5 H$^+$ + 4 H_2O	*Acetobacterium* spp.
Resorcinol (an aromatic compound)	2 $C_6H_4(OH)_2$ + 6 H_2O $\rightarrow$ 4 acetate$^-$ + butyrate$^-$ + 5 H$^+$	*Clostridium* spp.
Cinnamate (an aromatic compound)	2 $C_9H_7O_2^-$ + 2 H_2O $\rightarrow$ $C_9H_9O_2$ + benzoate$^-$ + acetate$^-$ + H$^+$	*Acetivibrio multivorans*
Phloroglucinol (an aromatic compound)	$C_6H_6O_3$ + 3 H_2O $\rightarrow$ 3 acetate$^-$ + 3 H$^+$	*Pelobacter massiliensis* *Pelobacter acidigallici*
Putrescine	10 $C_4H_{12}N_2$ + 26 H_2O $\rightarrow$ 6 acetate$^-$ + 7 butyrate$^-$ + 20 NH_4^+ + 16 H_2 + 13 H$^+$	Unclassified gram-positive nonsporing anaerobes
Citrate	Citrate^{3-} + 2 H_2O $\rightarrow$ formate$^-$ + 2 acetate$^-$ + HCO_3^- + H$^+$	*Bacteroides* sp.
Aconitate	Aconitate^{3-} + H$^+$ + 2 H_2O $\rightarrow$ 2 CO_2 + 2 acetate$^-$ + H_2	*Acidaminococcus fermentans*
Glyoxylate	4 Glyoxylate$^-$ + 3 H$^+$ + 3 H_2O $\rightarrow$ 6 CO_2 + 5 H_2 + glycolate$^-$	Unclassified gram-negative bacterium
Succinate	Succinate^{2-} + H_2O $\rightarrow$ propionate$^-$ + HCO_3^-	*Propionigenium modestum*
Oxalate	Oxalate^{2-} + H_2O $\rightarrow$ formate$^-$ + HCO_3^-	*Oxalobacter formigenes*
Malonate	Malonate^{2-} + H_2O $\rightarrow$ acetate$^-$ + HCO_3^-	*Malonomonas rubra* *Sporomusa malonica*

Succinate^{2-} + H$_2$O → propionate$^-$ + HCO$_3^-$

$$\Delta G^{0\prime} = -20.5 \text{ kJ/reaction}$$

This overall reaction yields insufficient free energy to couple to ATP synthesis directly by substrate-level phosphorylation, but nevertheless it serves as the sole energy-yielding reaction for growth of the organism. This is possible because the decarboxylation of succinate (via methylmalonyl-CoA and its membrane-bound decarboxylase) by *Propionigenium modestum* is coupled to the export of Na$^+$ across the cytoplasmic membrane (Figure 15.54*a*). A Na$^+$-translocating ATPase in the membrane of *P. modestum* employs this Na$^+$ gradient to drive ATP synthesis (Figure 15.54*a*).

Oxalobacter formigenes carries out the fermentation of oxalate:

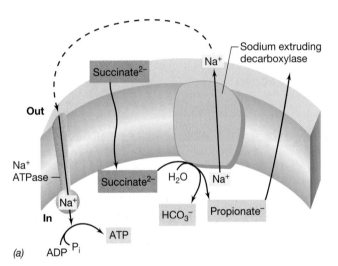

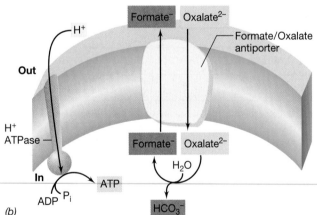

Oxalate^{2-} + H$_2$O → formate$^-$ + HCO$_3^-$

$$\Delta G^{0\prime} = -26.7 \text{ kJ/reaction}$$

At neutral pH, oxalate exists in the ionized form as oxalate^{2-}, and its decarboxylation to formate$^-$ consumes one proton. The subsequent export of formate from the cell then builds a proton motive force that can be coupled to ATP synthesis by a proton-translocating ATPase in the membrane (Figure 15.54*b*).

The interesting and unique aspect of the metabolism of both *Propionigenium modestum* and *Oxalobacter formigenes* is the fact that ATP synthesis occurs without substrate-level phosphorylation *or* electron transport occurring; however, chemiosmotic ATP formation still occurs as a result of a Na$^+$/H$^+$ pump linked to decarboxylation of organic acids. The lesson to be learned from these fermentations is the following: any chemical reaction that yields less than the 31.8 kJ required to make one ATP (see Table 15.5) or that appears unable to couple to a substrate-level phosphorylation, cannot be automatically ruled out as a potential growth-supporting reaction for a bacterium. If the reaction can be coupled to an ion gradient, ATP production (and subsequently growth) remains a possibility. However, because the influx of approximately 3 H$^+$ (or 3 Na$^+$) is required to drive ATP formation by a membrane-associated ATPase, a reaction must yield at least the energy required to pump a single H$^+$ or Na$^+$ ion to the outside of the cell membrane to be theoretically capable of supporting growth.

✓ 15.22 Concept Check

A wide variety of fermentations are known, and in many cases the product of one organism's fermentation is fermented by a second organism. Some fermentations employ ion gradients (H$^+$ or Na$^+$) as the basis of their energetics.

✓ What is unusual about the fermentation of succinate and oxalate?

✓ A common product of the fermentations shown in Table 15.7 is a fatty acid like acetate. Why is this important energetically?

15.23

Syntrophy

There are many examples in microbiology of *syntrophy*, a situation where two different organisms can together degrade some substance—and conserve energy doing it—that neither could degrade separately. We will see in Section 16.14 how syntrophy is extremely important in anoxic catabolism leading to the production of CH$_4$. Here we consider the microbiology and energetic aspects of syntrophy.

FIGURE 15.54 The unique fermentations of succinate and oxalate. (a) Succinate fermentation by *Propionigenium modestum*. A sodium-translocating ATPase produces ATP; sodium export is linked to succinate decarboxylation. (b) Oxalate fermentation by *Oxalobacter formigenes*. Oxalate import and formate export by a formate–oxalate antiporter consume protons. ATP synthesis is linked to a proton-driven ATPase. All substrates and products of a given reaction are shown in contrasting colors.

Hydrogen Consumption in Syntrophic Reactions

In most cases the nature of a syntrophic reaction involves hydrogen gas (H_2) being produced by one partner in the syntrophic relationship and getting consumed by the other. Thus, syntrophy has also been called *interspecies H_2 transfer*. The H_2 consumer can be any of a number of organisms we have already considered, sulfate-reducing bacteria, homoacetogens, and methanogens. Consider the case of syntrophy involving ethanol fermentation to acetate and eventual production of methane (Figure 15.55). As seen, the ethanol fermenter carries out a reaction that has an unfavorable (that is, positive) standard free energy change. However, the H_2 produced by the ethanol fermenter is a valuable electron donor for methanogenesis by a methanogen (Figure 15.55). And when the two reactions are summed, the overall reaction is exergonic (Figure 15.55) and supports growth of both partners in the syntrophic mixture. Another good example of syntrophy is the oxidation of butyrate to acetate plus H_2 by the fatty acid–oxidizing syntroph *Syntrophomonas*:

Butyrate$^-$ + 2 H_2O → 2 acetate$^-$ + H^+ + 2 H_2
$$\Delta G^{0\prime} = +48.2 \text{ kJ}$$

The free energy change of this reaction is highly unfavorable and in pure culture *Syntrophomonas* will not grow on butyrate. But if the H_2 produced in the reaction is immediately consumed by a partner organism (like a methanogen), *Syntrophomonas* grows luxuriantly in coculture with the H_2 consumer. How can chemical reactions whose free energy changes are positive support growth of an organism?

Energetics of Syntrophy

If growth of syntrophic organisms occurs only when H_2 is removed by a partner organism, the removal itself must obviously affect the energetics of the reaction. How does this happen? A brief review of the principles of free energy given in Appendix 1 indicates that the *actual* concentration of reactants and products in a reaction can drastically change the energetics. This is because $\Delta G^{0\prime}$ is calculated on the basis of *standard* conditions—one molar concentrations of products and reactants—whereas the related term ΔG, is calculated on the basis of *actual* concentrations of products and reactants that are present. In anoxic habitats, the products of fatty acid oxidation, especially H_2 (because it is such a powerful electron donor for anaerobic respirations), are immediately consumed, and this keeps H_2 concentrations extremely low, usually below 10^{-4} atm (⊸ Table 16.3). Thus, if the concentration of H_2 is very low, and ΔG is used to calculate the energetics of the reaction with this in mind (see Appendix 1 for how to calculate ΔG), the free energy change associated with the oxidation of

Ethanol fermentation

2 CH_3CH_2OH + 2 H_2O ⟶ 4 H_2 + 2 CH_3COO^- + 2 H^+
Ethanol   Acetate   + 19.4 kJ/reaction

Methanogenesis

4 H_2 + CO_2 ⟶ CH_4 + 2 H_2O
Methane   − 130.7 kJ/reaction

Syntrophic, coupled reaction

2 CH_3CH_2OH + CO_2 ⟶ CH_4 + 2 CH_3COO^- + 2 H^+
                                    − 111.3 kJ/reaction

FIGURE 15.55   Fermentation of ethanol to methane and acetate by syntrophic association of an ethanol-oxidizing bacterium and a H_2-consuming partner bacterium, in this case, a methanogen. Note how although the oxidation of ethanol to acetate plus H_2 is energetically unfavorable, the reaction becomes favorable when coupled to H_2 consumption by the methanogens. The two organisms thus share the energy released in the coupled reaction.

ethanol or fatty acids to acetate plus H_2 becomes exergonic, indicating that energy is released. For example, if the concentration of H_2 is kept extremely low by the activities of a H_2-consuming partner organism, the oxidation of butyrate by *Syntrophomonas* yields about −18 kJ.

The mechanism of ATP production by syntrophs probably involves both substrate-level and oxidative phosphorylations, depending on how the organisms are grown. Substrate-level phosphorylation can occur during the conversion of acetyl-CoA (generated by β-oxidation of ethanol or the fatty acid) to acetate (Figure 15.56a) and may be the *only* way ATP is produced under syntrophic conditions. However, *Syntrophomonas* can carry out an anaerobic respiration because it can be grown in pure culture on certain unsaturated fatty acids. Crotonate, for example, supports growth of *Syntrophomonas*, with some of the crotonate being oxidized to acetate and some being reduced to butyrate (Figure 15.56b). Crotonate reduction by *Syntrophomonas* is coupled to formation of a proton motive force and ATP synthesis as in other anaerobic respirations employing organic electron acceptors, such as fumarate reduction to succinate (see Section 15.20). Whether this occurs during syntrophic growth is unclear. However, regardless of how ATP is made in the syntrophic situation, there are additional energetic burdens for *Syntrophomonas* in this growth mode because the organism must somehow produce H_2 from more electropositive electron donors such as $FADH_2$ and $NADH_2$ (Figure 15.56a); this suggests that part of the ATP produced may be needed to drive reverse electron flow reactions (see Section 15.5) to yield H_2. Clearly, syntrophic alcohol and fatty-acid oxidizing bacteria are, from an energetic standpoint, living on the edge of existence.

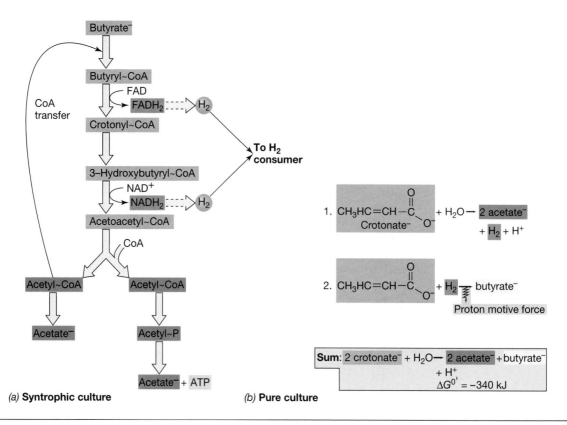

FIGURE 15.56   Energetics of growth of *Syntrophomonas wolfei* in syntrophic culture (a) and in pure culture (b). In syntrophic culture growth is dependent on the presence of a H_2-consuming organism such as a methanogen. H_2 production probably involves proton motive force–driven reverse electron flow because the E_0' of FAD/FADH$_2$ or NAD$^+$/NADH is more electropositive than that of 2 H$^+$/H$_2$ (⟳ Figure 4.7). In pure culture, energy production is linked to crotonate reduction to butyrate.

Other Aspects of Syntrophy

Syntrophic bacteria have evolved effective systems for using the highly reduced fermentation products of primary fermenters and for cooperating with other organisms to supply them with a necessary substrate. We will see how important syntrophic relationships are to the anaerobic catabolism of organic material in Section 16.14. By contrast, for *aerobic* organisms, syntrophic relationships are much less important than for anaerobic organisms. Aerobic organisms can frequently degrade even very complex molecules completely to CO_2 plus H_2O without an intimate dependence on partner bacteria, probably because with O_2 as electron acceptor the energetics of the reaction is much more favorable than for fermentative catabolism of the same substrate. Thus, syntrophic relationships are generally characteristic of anoxic transformations in which the energy available is only very small and the organisms highly specialized for exploiting energetically marginal reactions.

✓ 15.23 Concept Check

Syntrophy involves two organisms combining to degrade some compound that neither can degrade alone. This process usually involves H_2 produced by one organism being consumed by the partner. H_2 consumption can affect the energetics of the reaction carried out by the H_2 producer, allowing it to make ATP where it otherwise could not.

✓ Give an example of syntrophism. Why can it be said that both organisms benefit in this example?
✓ How is ATP made during syntrophic growth?
✓ Can a syntroph be grown in pure culture? Give an example of this and describe how the organism makes ATP under these conditions.

15.24

Hexose, Pentose, and Polysaccharide Utilization

We complete our discussion of chemoorganotrophic metabolism with consideration of a few special aspects of the catabolism of organic compounds, especially the use of polymeric substances that must first be hydrolyzed to monomeric units before energy-generating mechanisms can be employed. We begin with the microbial degradation of polysaccharides.

TABLE 15.8 Naturally occurring polysaccharides yielding hexose and pentose sugars[a]

Substance	Composition	Sources	Catabolic enzymes
Cellulose	Glucose polymer (β-1,4-)	Plants (leaves, stems)	Cellulases (β,1-4-glucanases)
Starch	Glucose polymer (α-1,4-)	Plants (leaves, seeds)	Amylase
Glycogen	Glucose polymer (α-1,4- and α-1,6-)	Animals (muscle)	Amylase, phosphorylase
Laminarin	Glucose polymer (β-1,3-)	Marine algae (Phaeophyta)	β-1,3-Glucanase (laminarinase)
Paramylon	Glucose polymer (β-1,3-)	Algae (Euglenophyta and Xanthophyta)	β-1,3-Glucanase
Agar	Galactose and galacturonic acid polymer	Marine algae (Rhodophyta)	Agarase
Chitin	N-Acetylglucosamine polymer (β-1,4-)	Fungi (cell walls) Insects (exoskeletons)	Chitinase
Pectin	Galacturonic acid polymer (from galactose)	Plants (leaves, seeds)	Pectinase (polygalacturonase)
Dextran	Glucose polymer	Capsules or slime layers of bacteria	Dextranase
Xylan	Heteropolymer of xylose and other sugars (β-1,4- and α-1,2 or α-1,3 side groups)	Plants	Xylanases
Sucrose	Glucose–fructose disaccharide	Plants (fruits, vegetables)	Invertase
Lactose	Glucose–galactose disaccharide	Milk	β-Galactosidase

a Each of these is subject to degradation by microorganisms.

Hexose and Polysaccharide Utilization

Sugars with six carbon atoms, called **hexoses,** are the most important electron donors for many chemoorganotrophs and are also important structural components of microbial cell walls, capsules, slimes, and storage products. The most common hexose sources in nature are listed in Table 15.8, from which it can be seen that most are polysaccharides, although a few are disaccharides. *Cellulose* and *starch* are two of the most important natural polysaccharides.

Although both starch and cellulose are composed of glucose units, they are connected differently (Table 15.8; ∞ Figure 2.6), and this profoundly affects their properties. Cellulose is much more insoluble than starch and is usually less rapidly digested. Cellulose forms long fibrils, and organisms that digest cellulose are often found closely associated with them (Figure 15.57). Many fungi are able to digest cellulose, and these are mainly responsible for decomposition of plant materials on the forest floor. Among bacteria, however, cellulose digestion is restricted to relatively few groups, of which the gliding bacteria such as *Sporocytophaga* and *Cytophaga* (Figures 15.57 and 15.58; see also Figure 13.89), clostridia, and actinomycetes are among the most common. Anoxic digestion of cellulose is carried out by a few *Clostridium* species, which are common in lake sediments, animal intestinal tracts, and systems for anaerobic sewage digestion. Cellulose digestion is also a major process in the rumen of ruminant animals where *Fibrobacter* and *Ruminococcus* species actively degrade cellulose (∞ Section 16.15).

Starch is digestible by many fungi and bacteria; this is illustrated for a laboratory culture in Figure 15.59. Starch-digesting enzymes, called *amylases*, are of considerable practical utility in many industrial situations where starch must be digested, such as the textile, laundry, paper, and food industries, and fungi and bacteria are the commercial sources of these enzymes (∞ Section 11.9).

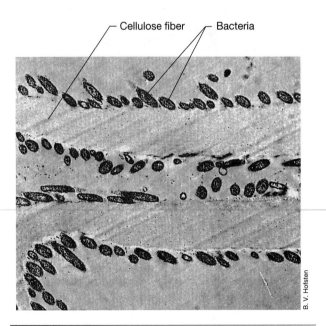

FIGURE 15.57 Transmission electron micrograph showing attachment of cellulose-digesting bacteria, *Sporocytophaga myxococcoides,* to cellulose fibers. Cells are about 0.5 μm in diameter.

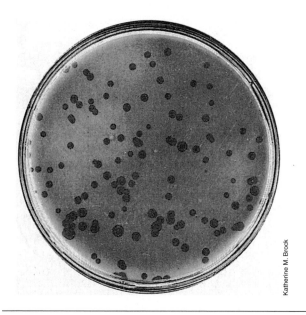

FIGURE 15.58 *Cytophaga hutchinsonii* colonies on a cellulose–agar plate. Clear areas are where cellulose has been digested.

All the polysaccharides occurring extracellularly and used as substrates are broken down to monomeric units by hydrolysis. In contrast, the polysaccharides formed within cells as storage products are broken down not by hydrolysis but by **phosphorolysis.** This

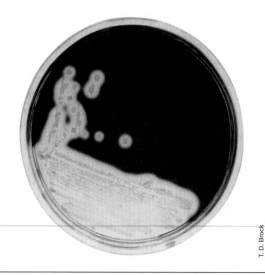

FIGURE 15.59 Demonstration of hydrolysis of starch by colonies of *Bacillus subtilis.* After incubation, the plate was flooded with Lugol's iodine solution. Where starch hydrolysis occurred, the characteristic purple color of the starch–iodine complex is absent. Hydrolysis of starch occurs at some distance from the bacterial colonies because of the production of extracellular amylase, which diffuses into the surrounding medium.

process, involving the addition of *inorganic* phosphate, results in the formation of hexose phosphate rather than the free hexose and may be summarized as follows for the degradation of starch, an α-1,4 polymer of glucose:

$$(C_6H_{12}O_6)_n + P_i \rightarrow (C_6H_{12}O_6)_{n-1} + \text{glucose 1-phosphate}$$

Because glucose 1-phosphate can be easily converted to glucose 6-phosphate, a key intermediate in glycolysis (⊙⊙ Figure 4.12), and no ATP is required to form it, phosphorolysis represents a net energy savings to the cell.

Disaccharides

Many microorganisms can use *disaccharides* for growth (Table 15.8). *Lactose* utilization by microorganisms is of considerable economic importance because milk-souring organisms produce lactic acid from lactose. *Sucrose,* the common disaccharide of higher plants, is usually first hydrolyzed to its component monosaccharides (glucose and fructose) by the enzyme *invertase,* and the monomers are then metabolized by normal pathways. *Cellobiose,* β-1,4-diglucose and a major product of cellulose digestion, is also readily degraded by a variety of bacteria that cannot degrade the cellulose polymer itself.

The microbial polysaccharide *dextran* is synthesized by some bacteria using the enzyme *dextransucrase* and sucrose as starting material:

$$n \text{ sucrose} \rightarrow (\text{glucose})_n + n \text{ fructose}$$
$$\text{dextran}$$

Dextran is formed in this way by the bacterium *Leuconostoc mesenteroides* and a few others, and the polymer formed accumulates around the cells as a massive slime or capsule (Figure 15.60). Because su-

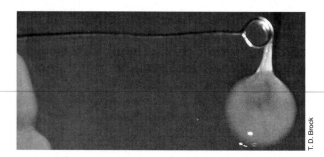

FIGURE 15.60 Slimy colony formed by the dextran-producing bacterium, *Leuconostoc mesenteroides,* growing on a sucrose-containing medium. When the same organism is grown on glucose, the colonies are small and not slimy because dextran synthesis requires sucrose (⊙⊙ Section 19.3 for further discussion).

crose is required for dextran formation, no dextran is formed when the bacterium is cultured on a medium with glucose or fructose. In nature, when cells containing dextran or other polysaccharide capsules die, these materials once again become available for attack by fermentative or other chemoorganotrophic microorganisms.

✓ 15.24 Concept Check

Polysaccharides are abundant in nature and can be broken down, usually by phosphorolysis, into hexose or pentose monomers and used as energy sources. Starch and cellulose are common polysaccharides.

✓ What is *phosphorolysis*?
✓ What disaccharides are common in nature?

15.25

Organic Acid Metabolism

A variety of organic acids can be used by microorganisms as carbon sources and electron donors. The acids of the citric acid cycle, such as *citrate, malate, fumarate,* and *succinate,* are common natural products formed by plants and are also fermentation products of microorganisms. Because the citric acid cycle has major *biosynthetic* (∞ Section 4.15) as well as *energetic* (∞ Section 4.12) functions, the complete cycle or major portions of it are nearly universal in microorganisms. Thus, it is not surprising that many microorganisms are able to use these acids as electron donors and carbon sources. Aerobic utilization of four-, five-, and six-carbon acids can be accomplished by means of enzymes of the citric acid cycle, with ATP formation by oxidative phosphorylation.

Anaerobic utilization of organic acids usually involves conversion to pyruvate followed by formation of acetate via acetyl phosphate with consequent ATP production by substrate-level phosphorylation (see Section 15.21).

Glyoxylate Cycle

Utilization of two- or three-carbon acids as carbon sources cannot occur by means of the citric acid cycle alone. This cycle can continue to operate only if the acceptor molecule, the four-carbon acid *oxalacetate*, is regenerated at each turn of the cycle; any removal of carbon compounds for biosynthetic reactions would prevent completion of the cycle. When acetate is used, the oxalacetate needed to continue the cycle is produced through the **glyoxylate cycle** (Figure 15.61), so called because glyoxylate is a key intermediate. This cycle is com-

posed of most of the citric acid cycle reactions plus two additional enzymes: *isocitrate lyase,* which splits isocitrate to succinate and glyoxylate, and *malate synthase,* which converts glyoxylate and acetyl-CoA to malate.

Biosynthesis through the glyoxylate cycle occurs as follows. The splitting of isocitrate into succinate and glyoxylate allows the succinate molecule (or another citric acid cycle intermediate derived from it) to be drawn off for biosynthesis because glyoxylate combines with acetyl-CoA to yield malate. Malate can be converted to oxalacetate to maintain the cyclic nature of the citric acid cycle despite the fact that a C_4 intermediate (succinate) has been drawn off. The succinate molecule can be used directly in the production of porphyrins,

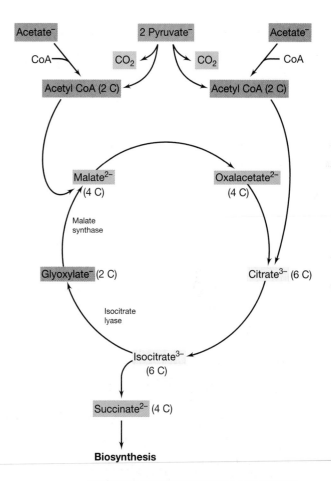

FIGURE 15.61   The glyoxylate cycle, leading to the synthesis of oxalacetate from acetate. Two unique enzymes, isocitrate lyase and malate synthase, operate with a majority of the citric acid cycle reactions. In addition to growth on pyruvate, the glyoxylate cycle can also operate during growth on acetate. All compounds containing the same number of carbons are shown in a single color.

be oxidized to oxalacetate and serve as a carbon skeleton for C_4 amino acids, or be converted (via oxalacetate and phosphoenolpyruvate) to glucose.

Pyruvate and C_3 Utilization

Three-carbon compounds such as pyruvate or compounds that can be converted to pyruvate (for example, lactate or carbohydrates) also cannot be used as energy sources through the citric acid cycle alone. Because some of the citric acid cycle intermediates are used for biosynthesis, the oxalacetate needed to keep the cycle going is synthesized from pyruvate or phosphoenolpyruvate by the addition of a carbon atom from CO_2. In some organisms this step is catalyzed by the enzyme *pyruvate carboxylase*:

$$\text{Pyruvate} + \text{ATP} + CO_2 \rightarrow \text{oxalacetate} + \text{ADP} + P_i$$

whereas in others it is catalyzed by *phosphoenolpyruvate carboxylase*:

$$\text{Phosphoenolpyruvate} + CO_2 \rightarrow \text{oxalacetate} + P_i$$

These reactions replace oxalacetate that is lost when intermediates of the citric acid cycle are removed for use in biosynthesis, and the cycle can continue to function.

✓ 15.25 Concept Check

Organic acids are frequently metabolized through the citric acid cycle or through the glyoxylate cycle. Isocitrate lyase and malate synthase are the key enzymes of the glyoxylate cycle.

✓ Why is the glyoxylate cycle necessary for growth on acetate but not on succinate?

15.26

Lipids as Microbial Nutrients

Lipids are abundant in nature. The cytoplasmic membranes of all cells contain lipids, and many microorganisms as well as macroorganisms produce lipid storage materials. These substances are all biodegradable and can be excellent substrates for microbial energy-yielding metabolism.

Fat and Phospholipid Hydrolysis

Fats are esters of glycerol and fatty acids (∞ Section 2.4). Microorganisms use fats only after hydrolysis of the ester bond, and extracellular enzymes called **lipases** are responsible for the reaction (Figure 15.62). The end result is formation of glycerol and free

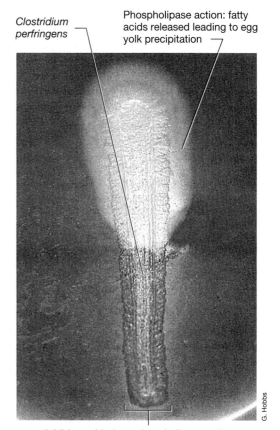

Clostridium perfringens

Phospholipase action: fatty acids released leading to egg yolk precipitation

Inhibitor added: no phospholipase action, thus no precipitation of egg yolk.

FIGURE 15.62 Action of phospholipase around streak of *Clostridium perfringens* growing on an agar medium containing egg yolk. On half of the plate, an inhibitor of phospholipase was added, preventing action of the enzyme.

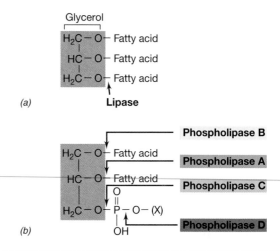

FIGURE 15.63 (a) Action of lipase on a fat. (b) Phospholipase action on phospholipid. The sites of action of the four distinct phospholipases A, B, C, and D are shown. X refers to a number of small organic molecules that may be at this position in different phospholipids. Compare this diagram to the more complete figure of a phospholipid in Figure 2.7.

fatty acids (Figures 15.62 and 15.63). Lipases are not highly specific and attack fats containing fatty acids of various chain lengths. Phospholipids are hydrolyzed by specific enzymes called *phospholipases*, given different letter designations depending on which ester bond they cleave (Figure 15.63). Phospholipases A and B cleave fatty acid esters and thus resemble the lipases described earlier, but phospholipases C and D cleave phosphate ester linkages and hence are quite different types of enzymes. The result of lipase action is the release of free fatty acids and glycerol, and all these substances can be attacked both anaerobically as well as aerobically by various chemoorganotrophic microorganisms.

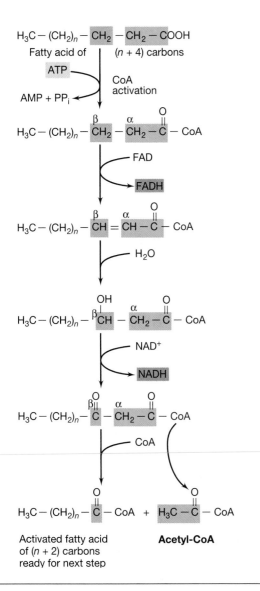

FIGURE 15.64 Mechanism of beta oxidation of a fatty acid, which leads to successive formation of two-carbon fragments of acetyl-CoA.

Fatty Acid Oxidation

Fatty acids are oxidized by a process called *beta oxidation*, in which two carbons of the fatty acid are split off at a time (Figure 15.64). In eukaryotes the enzymes are in the mitochondria, whereas in prokaryotes they are cytoplasmic. The fatty acid is first activated with coenzyme A; oxidation results in the release of *acetyl-CoA* and the formation of a fatty acid shorter by two carbons (Figure 15.64). The process of beta oxidation is then repeated, and another acetyl-CoA molecule is released. Two separate dehydrogenation reactions occur. In the first, electrons are transferred to flavin-adenine dinucleotide (FAD), whereas in the second they are transferred to NAD^+. Most fatty acids have an even number of carbon atoms, and complete oxidation yields only acetyl-CoA. The acetyl-CoA formed is then oxidized by way of the citric acid cycle or is converted to hexose and other cell constituents via the glyoxylate cycle. Fatty acids are good electron donors. For example, the anaerobic oxidation of the 16-carbon fatty acid palmitic acid results in the net synthesis of 129 ATP molecules from electron transport phosphorylation from electrons generated during the formation of acetyl-CoA from beta oxidations and from oxidation of the acetyl-CoA units themselves through the citric acid cycle (⚬⚬ Section 4.12).

✓ 15.26 Concept Check

Fats are metabolized by hydrolysis by lipases or phospholipases to free fatty acids. The latter are oxidized by beta oxidation to acetyl-CoA units, which are subsequently oxidized to CO_2 by the citric acid cycle.

✓ What are the functions of phospholipases?
✓ What is meant by the term *β-oxidation*?

15.27

Molecular Oxygen (O₂) as a Reactant in Biochemical Processes

We have discussed the role of O_2 as an *electron acceptor* in energy-generating reactions (⚬⚬ Section 4.10). Although this is by far the most important role of O_2 in cellular metabolism, O_2 plays an interesting and important role as a *direct reactant* in certain types of anabolic and catabolic processes.

Oxygenases are enzymes that catalyze the incorporation of oxygen from O_2 into organic compounds. There are two kinds of oxygenases: *dioxygenases*, which catalyze the incorporation of *both* atoms of O_2 into the molecule; and *monooxygenases*, which catalyze the transfer of *only one* of the two oxygen atoms in O_2 to an organic compound as a hydroxyl (OH) group, with the second atom of O_2 being reduced to water,

H_2O. Because monooxygenases catalyze the formation of hydroxyl groups (OH) in organic compounds, they are sometimes called *hydroxylases*. In most monooxygenases, the electron donor is NADH or NADPH, although the direct coupling to O_2 is through a flavin that is reduced by the NADH or NADPH donor. In the case of ammonia monooxygenase discussed previously (see Section 15.13), the electron donor is cytochrome *c*.

There are several types of reactions in living organisms that require O_2 as a reactant. One of the best examples is the involvement of O_2 in sterol biosynthesis. The formation of the fused sterol ring system (∞ Figure 3.19) requires the participation of molecular oxygen. Such a reaction can obviously not take place under anoxic conditions so organisms that grow anaerobically must either dispense with this reaction or obtain the required substance (sterol) preformed from their environment. The requirement of O_2 as a reactant in biosynthesis is of considerable evolutionary significance, as molecular O_2 was originally absent from the atmosphere of Earth when life evolved and became available only after the evolution of cyanobacteria, the first phototrophic organisms to produce O_2 (∞ Chapter 12). The role of O_2 in hydrocarbon utilization is discussed next.

✓ 15.27 Concept Check

In addition to its role as an electron acceptor, oxygen (O_2) is also a chemical reactant in certain biochemical processes. Enzymes called oxygenases introduce O_2 into a biochemical compound.

✓ How do *monooxygenases* differ in function from *dioxygenases*?

15.28

Hydrocarbon Transformations

Hydrocarbons are organic compounds containing only carbon and hydrogen and are highly insoluble in water. Low-molecular-weight hydrocarbons are gases, whereas those of higher molecular weight are liquids or solids at room temperature. Some hydrocarbons are aliphatic compounds, a class of carbon compounds in which the carbon atoms are joined in open chains. There is a tremendous variation among aliphatic hydrocarbons in chain length, degree of branching, and number of double bonds. Another important group of hydrocarbons contains the aromatic ring and can be viewed as derivatives of benzene.

Aliphatic Hydrocarbons

Only relatively few kinds of microorganisms (for example, *Nocardia*, *Pseudomonas*, *Mycobacterium*, and certain yeasts and molds) can use hydrocarbons for growth. For the most part, utilization of saturated aliphatic hydrocarbons is an *aerobic* process: in the absence of O_2, saturated hydrocarbons are virtually unaffected by microorganisms (a novel sulfate-reducing bacterium is an exception) (∞ Section 16.14).

The initial oxidation step of saturated aliphatic hydrocarbons involves molecular oxygen (O_2) as a reactant, and one of the atoms of the oxygen molecule is incorporated into the oxidized hydrocarbon. This reaction is carried out by a monooxygenase (see Section 15.27), and a typical reaction sequence is that shown in Figure 15.65. The end product of the reaction sequence is acetyl-CoA. However, the initial oxidation is not at the terminal carbon in all cases. Oxidation may sometimes occur at the second carbon, and then quite different subsequent reactions occur. *Unsaturated* aliphatic hydrocarbons containing a terminal double bond are not refractory to anoxic decomposition and can be oxidized by certain sulfate-reducing and other anaerobic bacteria.

Aromatic Hydrocarbons

Many aromatic hydrocarbons can be used as electron donors aerobically by microorganisms, of which bacte-

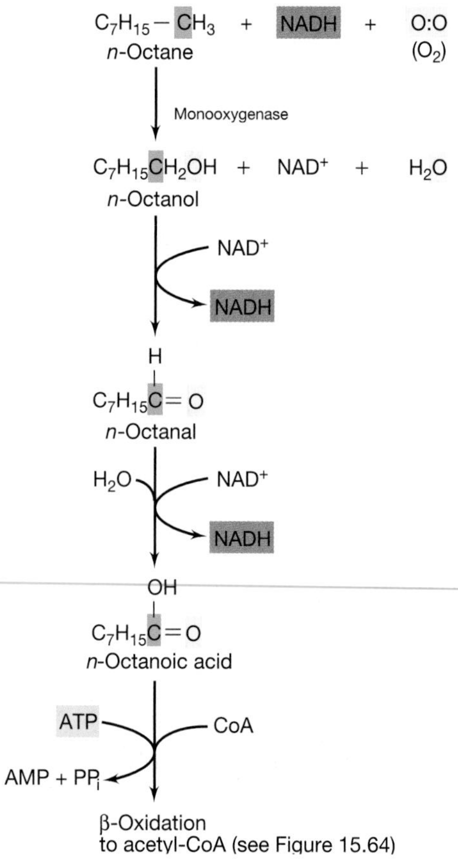

FIGURE 15.65 Steps in oxidation of an aliphatic hydrocarbon, the first of which is catalyzed by a monooxygenase.

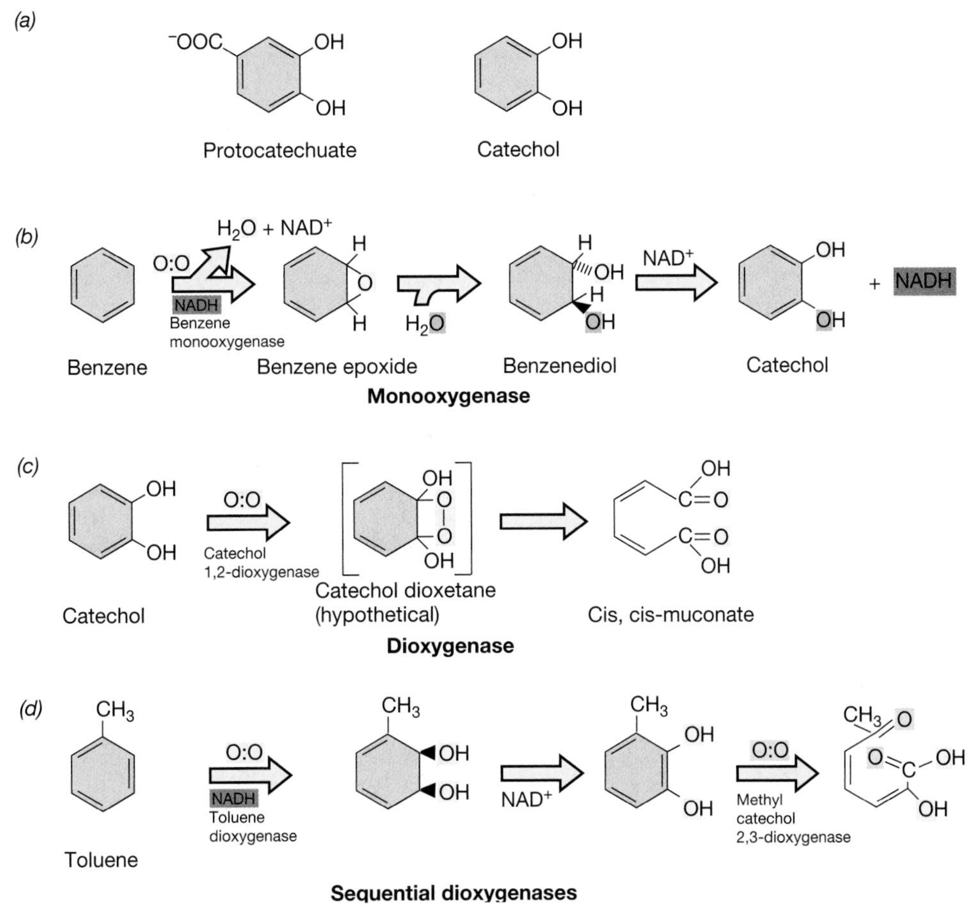

FIGURE 15.66 Roles of oxygenases in catabolism of aromatic compounds. (a) Protocatechuate and catechol, two common oxidation products of aromatic compounds. (b) Hydroxylation of benzene to catechol by a monooxygenase in which NADH is an electron donor. (c) Cleavage of catechol to *cis,cis*-muconate by a dioxygenase. Reactant oxygen atoms are shown in color in both reactions to demonstrate the different mechanisms. (d) The activity of toluene dioxygenase and methyl catechol 2,3-dioxygenase in the degradation of toluene.

ria of the genus *Pseudomonas* have been the best studied. It has been demonstrated that the metabolism of these compounds, some of which are quite large molecules, frequently has as its initial stage the formation of either *protocatechuate* or *catechol*, or a structurally related compound, as shown in Figure 15.66a.

These single-ring compounds are referred to as *starting substrates* because oxidative catabolism proceeds only after the large aromatic molecules have been converted to these more simple forms. Protocatechuate and catechol may then be further degraded to compounds that can enter the citric acid cycle: succinate, acetyl-CoA, pyruvate. Several steps in the catabolism of aromatic hydrocarbons usually require oxygenases. Figures 15.66b, c, and d show three different oxygenase-catalyzed reactions, one using a monooxygenase and two using a dioxygenase.

Aromatic compounds can also be degraded anaerobically and quite readily if they already contain an atom of oxygen. Anoxic mixed cultures have been shown to degrade benzoate and other substituted phenolic compounds, yielding CH_4 and CO_2 as final products. Substituted phenolic compounds are also degraded by certain denitrifying, phototrophic, ferric iron-reducing, and sulfate-reducing bacteria. The anoxic catabolism of aromatic compounds proceeds by *reductive* rather than oxidative ring cleavage (Figure 15.67). This involves *ring reduction* followed by *ring cleavage* to yield a straight-chain fatty acid or dicarboxylic acid. These intermediates can generally be converted to acetyl-CoA and used for both biosynthetic and energy-yielding purposes. Benzoate and benzoate derivatives are common natural products and are readily degraded anaerobically.

Evidence for the anoxic degradation of benzene and toluene, aromatic compounds lacking an oxygen atom (see Figure 15.66 for structures), has also been obtained. Catabolism of benzene occurs by anaerobic microbial consortia leading to methanogenesis, but toluene oxidation to CO_2 can occur in pure culture. In the latter instance, growth on toluene is supported by anaerobic respiration coupled to the reduction of ferric iron or

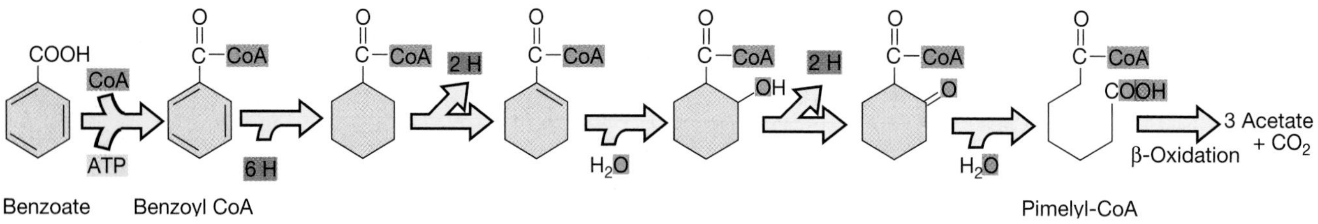

FIGURE 15.67 Anoxic degradation of benzoate by reductive-ring cleavage. Note that all intermediates of the pathway are bound to coenzyme A. The acetate produced is further catabolized in the citric acid cycle.

nitrate. Although the biochemical steps in anaerobic toluene degradation have not been completely worked out, there is good evidence that toluene is eventually converted to the benzoate derivative benzoyl-CoA and then presumably further catabolized via ring reduction as shown in Figure 15.67.

✓ 15.28 Concept Check

Many microorganisms can degrade aliphatic and aromatic hydrocarbons. Catabolism aerobically involves the action of oxygenase enzymes. Anaerobic aromatic degradation proceeds by reductive rather than oxidative pathways.

- ✓ Draw the chemical structure for benzene. Do the same for benzoate. Are these compounds aliphatic or aromatic?
- ✓ What fundamental difference exists in the *anaerobic* degradation of an aromatic compound compared with its *aerobic* metabolism. Give an example of this.

15.29

Nitrogen Fixation

The utilization of nitrogen gas (N_2) as a source of nitrogen is called *nitrogen fixation* and is a property of only certain prokaryotes. An abbreviated list of nitrogen-fixing organisms is given in Table 15.9, from which it can be seen that a variety of prokaryotes, both anaerobic and aerobic, fix nitrogen. In addition, there are some bacteria, called *symbiotic*, that fix nitrogen only in association with certain plants. As far as is currently known, no eukaryotic organisms fix nitrogen. Symbiotic nitrogen fixation will be discussed in Section 16.25.

Nitrogenase

In the fixation process, N_2 is *reduced* to ammonium and the ammonium converted to organic form. The reduction process is catalyzed by the enzyme complex **nitrogenase,** which consists of two separate proteins called *dinitrogenase* and *dinitrogenase reductase*. Both components contain iron, and dinitrogenase contains molybdenum as well. The iron and molybdenum in dinitrogenase are

TABLE 15.9	Some nitrogen-fixing organisms	
Free-living aerobes		
Chemo-organotrophs	**Phototrophs**	**Chemo-lithotrophs**
Bacteria: *Azotobacter* spp. *Azomonas* *Klebsiella*[a] *Beijerinckia* *Bacillus polymyxa* *Mycobacterium flavum* *Azospirillum lipoferum* *Citrobacter freundii* *Acetobacter diazotrophicus* *Methylomonas* *Methylococcus*	Cyanobacteria (various, but not all)	*Alcaligenes* *Thiobacillus* (some species) *Streptomyces* *thermoau-* *totrophicus*
Free-living anaerobes		
Chemo-organotrophs	**Phototrophs**	**Chemo-lithotrophs**
Bacteria: *Clostridium* spp. *Desulfovibrio* *Desulfotomaculum*	Bacteria: *Chromatium* *Thiocapsa* *Chlorobium* *Rhodospirillum* *Rhodopseudomonas* *Rhodomicrobium* *Rhodopila* *Rhodobacter* *Heliobacterium* *Heliobacillus* *Heliophilum*	Archaea: *Methanosarcina* *Methanococcus*
Symbiotic		
Leguminous plants		**Nonleguminous plants**
Soybeans, peas, clover, locust, and so on, in association with a bacterium of the genus *Rhizobium, Bradyrhizobium,* *Sinorhizobium,* or *Azorhizobium*		*Alnus, Myrica, Ceanothus,* *Comptonia, Casuarina;* in association with actinomycetes of the genus *Frankia*

a N_2 fixation occurs only under anoxic conditions.

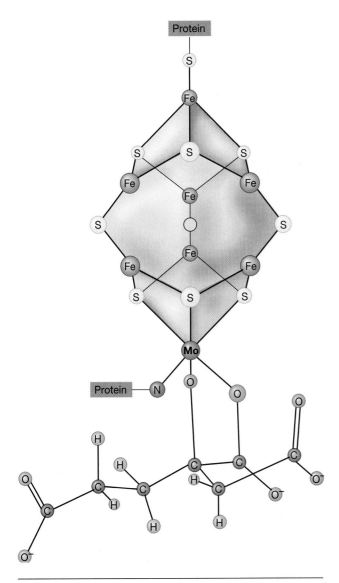

FIGURE 15.68 Structure of FeMo-co, the iron–molybdenum cofactor from nitrogenase. On the top is the Fe_7S_8 cube that binds to molybdenum along with oxygen atoms from homocitrate (bottom, all oxygen atoms shown in green) and N and S atoms from dinitrogenase. Two molecules of FeMo-co are present per molecule of nitrogenase.

contained in a cofactor known as *FeMo-co* (Figure 15.68), and the actual reduction of N_2 occurs on this iron–molybdenum center. The composition of FeMo-co is $MoFe_7S_9$ homocitrate (Figure 15.68), and FeMo-co is present in two copies per molecule of nitrogenase.

Some nitrogen-fixing bacteria can synthesize nonmolybdenum nitrogenases under certain growth conditions, and these so-called *alternative nitrogenases* do not contain molybdenum but instead contain either vanadium (and iron) or iron only. Cofactors similar to FeMo-co are present in both alternative nitrogenases as well: FeVa-co in the vanadium nitrogenase and an iron–sulfur cluster resembling FeMo-co and FeVa-co

but lacking both Mo and Va, in the iron nitrogenase. Alternative nitrogenases are not synthesized when sufficient molybdenum is present, as the molybdenum nitrogenase is generally the main nitrogenase in the cell. Alternative nitrogenases presumably serve as a backup mechanism to ensure that N_2 fixation can still occur when molybdenum is limiting in the habitat (∞ Section 13.8). A structurally and functionally novel molybdenum nitrogenase has been discovered in the thermophilic streptomycete, *Streptomyces thermoautotrophicus*; its properties are compared with those of classic Mo nitrogenases in the box, The Power of Metabolic Diversity.

Owing to the stability of the $N \equiv N$ triple bond (which has a dissociation energy of 940 kJ compared with 493 kJ for the double bond in O_2), N_2 is extremely inert and its activation is a very energy-demanding process. Six electrons must be transferred to reduce N_2 to 2 NH_3, and several intermediate steps might be visualized; it is thought that the three successive reduction steps occur directly on nitrogenase with no free intermediates accumulating (Figure 15.69a). Nitrogen fixation is highly reductive in nature, and the process is inhibited by oxygen because both dinitrogenase, and especially dinitrogenase reductase, are rapidly and irreversibly inactivated by O_2 (even when isolated from *aerobic* nitrogen fixers). In aerobic bacteria, N_2 fixation occurs in the presence of O_2 in whole cells but not in purified enzyme preparations, and nitrogenase in such organisms is protected from O_2 inactivation either by rapid removal of O_2 by respiration, the production of O_2-retarding slime layers, or by compartmentalization of nitrogenase in a special type of cell (the heterocyst) (∞ Section 13.24). In addition, although N_2 fixation does not occur in oxic cell extracts, in aerobic nitrogen fixers like *Azotobacter*, nitrogenase is protected from oxygen inactivation by complexing with a specific protein; this has been referred to as *conformational protection*.

Electron Flow in Nitrogen Fixation

The sequence of electron transfer in nitrogenase is as follows: electron donor → dinitrogenase reductase → dinitrogenase → N_2. The electrons for nitrogen reduction are transferred to dinitrogenase reductase from ferredoxin or flavodoxin, low potential iron–sulfur proteins. In *Clostridium pasteurianum*, ferredoxin is the electron donor and is reduced by phosphoroclastic splitting of pyruvate to acetyl-CoA + CO_2. In addition to reduced ferredoxin, ATP is required for N_2 fixation. In each cycle of electron transfer, dinitrogenase reductase is reduced by ferredoxin/flavodoxin and binds two molecules of ATP. ATP binding alters the conformation of dinitrogenase reductase and lowers its reduction potential, allowing it to interact with dinitrogenase. Upon

A FOCUS ON: The Power of Metabolic Diversity: A Novel Nitrogenase

Nitrogenases have been characterized from a wide variety of prokaryotes, including some Archaea, and all of them show significant sequence homology at both the gene and polypeptide level: All of them, that is, until the nitrogenase from the streptomycete *Streptomyces thermoautotrophicus* was characterized.*

S. thermoautotrophicus is a thermophilic (optimum temperature 65°C) gram-positive filamentous prokaryote that occurs naturally in burning compost and charcoal piles (Fig. 1). The organism is an aerobic chemolithotrophic H_2 bacterium

FIG. 1 Two burning charcoal piles in the Bavarian forest, Germany, containing cells of the N_2-fixing bacterium, *Streptomyces thermoautotrophicus*. The scientist shown is doing temperature measurements at various points in the piles. The piles emit gases of CO_2, CO, CH_4, and C_2H_2, and vary in temperature with depth. The surface to about 15 cm into the piles have temperatures of less than 100°C but deeper into the piles, temperatures can be over 300°C.

*Ribbe, M., D. Gadkari, and O. Meyer. 1997. N_2 fixation by *Streptomyces thermoautotrophicus* involves a molybdenum-dinitrogenase and a manganese-superoxide oxidoreductase that couple N_2 reduction to the oxidation of superoxide produced from O_2 by a molybdenum CO dehydrogenase. *J. Biol. Chem.* 272: 26627–26633.

that can also use carbon monoxide (CO) as an electron donor. Although *S. thermoautotrophicus* has been known to be a nitrogen fixer for some time, some unusual properties of its nitrogen fixation system (like the fact that ammonia did not repress nitrogenase synthesis and that the enzyme did not reduce acetylene) prompted a more detailed examination of its nitrogenase. What was found represents a totally new paradigm for N_2 fixation.

The *S. thermoautotrophicus* nitrogenase contains Mo, but unlike classical Mo nitrogenase, it is completely *insensitive* to O_2. The dinitrogenase component of the *S. thermoautotrophicus* nitrogenase, called *St1*, contains three different polypeptides that show some structural similarity to dinitrogenase polypeptides from other nitrogen-fixers, but the dinitrogenase reductase component, called *St2*, shows no similarity to other dinitrogenase reductases. However, St2 shows *very high* sequence similarity to manganese-containing superoxide dismutases. In fact St2 *is* a superoxide dismutase! Recall from Chapter 5 (◌◌ Section 5.12) that superoxide dismutases function in the cell to consume superoxide (O_2^-), forming O_2 in the process and thus preventing oxidative damage to cell components. But what does superoxide have to do with nitrogen fixation?

It has been shown that St2 supplies electrons to St1. The source of the electrons is O_2^-, and the O_2^- is

formed from the reduction of O_2 by a CO dehydrogenase (Fig. 2). Thus, in analogy to the pyruvate → flavodoxin → dinitrogenase reductase → dinitrogenase sequence in classical nitrogen-fixation (see Figure 15.69), in *S. thermoautotrophicus* the sequence is $CO \rightarrow O_2^- \rightarrow St2 \rightarrow St1$. And, astonishingly, instead of O_2 *inhibiting* nitrogenase (as it does in every nitrogenase that has ever been examined), in *S. thermoautotrophicus* O_2 is actually *required* in the reaction mechanism of the enzyme!

Clearly, *S. thermoautotrophicus* nitrogenase is a structurally and functionally unique nitrogen-fixing system. How widespread such a system is and whether its primary function in the cell is actually to fix N_2, is as yet unknown. However, the *S. thermoautotrophicus* nitrogenase system is a good example of the power of metabolic diversity in prokaryotes: Even well studied systems in which conformity prevails can occasionally yield big surprises and totally new concepts. Discovery of the *S. thermoautotrophicus* nitrogenase system has also renewed hope for the eventual genetic engineering of a nitrogenase system into agricultural crops like corn. The fact that this nitrogenase is not oxygen labile and its energy requirements are much lower than those of classical nitrogenases (measurements show the *S. thermoautotrophicus* system to use only 25–50% of the ATP of classic Mo nitrogenases) could make the dream of nitrogen-fixing row crops a reality someday. ■

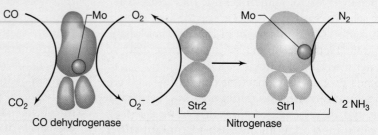

Nitrogen fixation in *Streptomyces thermoautotrophicus*

FIG. 2   Reactions of nitrogen fixation in *S. thermoautotrophicus*.

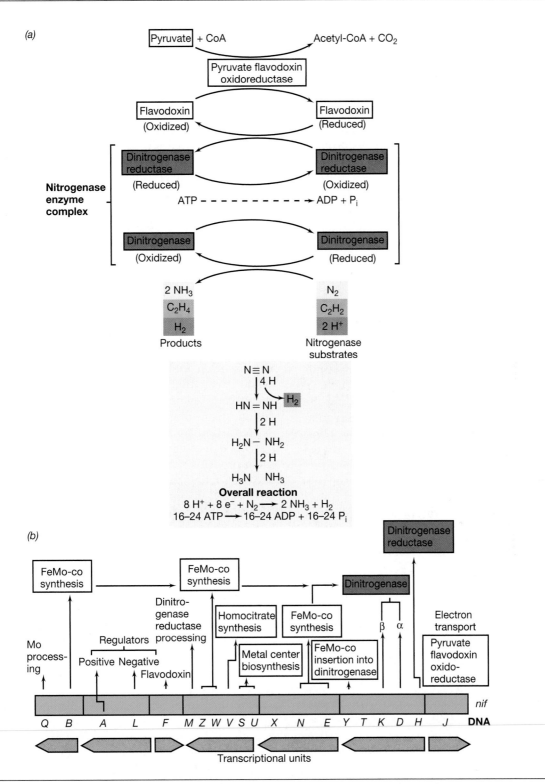

FIGURE 15.69 The nitrogenase system. (a) Steps in nitrogen fixation: reduction of N_2 to 2 NH_3. Electrons are supplied from dinitrogenase reductase to dinitrogenase one at a time, and each electron supplied is associated with the hydrolysis of two ATPs. (b) The genetic structure of the *nif* regulon in *Klebsiella pneumoniae*, the best-studied nitrogen-fixing organism. The functions of some of the genes are uncertain. The mRNA transcripts (transcriptional units) are shown below the genes; arrows indicate the direction of transcription.

electron transfer to dinitrogenase, the ATP is hydrolyzed and dinitrogenase reductase dissociates from dinitrogenase and begins another cycle of reduction and ATP binding (Figure 15.69). When appropriately reduced, dinitrogenase then reduces N_2 to NH_3, with the actual reduction occurring at the FeMo-co center. Although only *six* electrons are necessary to reduce N_2 to 2 NH_3, *eight* electrons are actually consumed in the process, *two* electrons being lost as hydrogen (H_2), for each mole of N_2 reduced (Figure 15.69a). The reason for this apparent waste is not known, but evidence is strong that H_2 evolution is an intimate part of the reaction mechanism of nitrogenase.

Genetics and Regulation of Nitrogen Fixation

The genes for dinitrogenase and dinitrogenase reductase in *Klebsiella pneumoniae*, a well-studied N_2 fixer, are part of a complex regulon (a large network of operons) called the *nif regulon* (Figure 15.69b); the *K. pneumoniae nif* regulon spans 24 kb of DNA and contains 20 genes arranged in several transcriptional units (Figure 15.69b). In addition to nitrogenase structural genes, the genes for FeMo-co, genes controlling the electron transport proteins, and a number of regulatory genes are also present in the *nif* regulon. Dinitrogenase is a complex protein made up of two subunits, α (product of the *nifD* gene) and β (product of the *nifK* gene), each of which is present in two copies. Dinitrogenase reductase is a protein dimer consisting of two identical subunits, the product of *nifH*. FeMo-co is synthesized through the participation of several genes, including *nifN, V, Z, W, E,* and *B,* as well as *Q,* which encodes a product involved in molybdenum processing. The *nifA* gene encodes a positive regulatory protein that serves to activate transcription of other *nif* genes.

Nitrogenase is subject to strict regulatory controls. Nitrogen fixation is blocked by O_2 and by fixed nitrogen, including NH_3, NO_3^-, and certain amino acids (but see the box). A major part of this regulation is at the level of transcription. The various transcriptional units of the *nif* regulon are shown in Figure 15.69b. While transcription of the *nif* structural genes is *activated* by the NifA protein (positive regulation), NifL is a negative regulator of *nif* gene expression and contains a molecule of FAD (recall that FAD is a redox coenzyme for flavoproteins (∞ Section 4.10) that is critical for O_2-sensing by the protein. In the presence of sufficient O_2, NifL functions to shut down transcription of *nif* genes in order to prevent synthesis of the oxygen labile nitrogenase.

The ammonia produced by nitrogenase does not repress enzyme synthesis because as soon as it is made, it is incorporated into organic form and used in biosynthesis. But when ammonia is in excess (as in environments high in ammonia), nitrogenase synthesis is quickly repressed. This prevents waste of ATP by not making a product already present in ample amounts. In certain nitrogen-fixing bacteria, nitrogenase *activity* is also regulated by ammonia, a phenomenon called the ammonia "switch-off" effect. In this case, excess ammonia causes a covalent modification of dinitrogenase reductase, which results in a loss of enzyme activity. When ammonia again becomes limiting, this modified protein is converted back to the active form and N_2 fixation resumes. Ammonia switch-off is thus a rapid and reversible method of controlling ATP consumption by nitrogenase.

Nitrogenase has been purified from a large number of nitrogen-fixing organisms and in all cases has been shown to be a two-protein complex (but see the box). It is of considerable evolutionary interest that among molybdenum nitrogenases, dinitrogenase from one organism usually functions with dinitrogenase reductase from another organism. This can be interpreted to mean that the structures of the nitrogenase components have not changed markedly during evolution, suggesting that the molecular requirements for N_2 reduction are fairly specific. Indeed, studies of the "HDK" gene cluster have confirmed this, in that molecular probes containing cloned *nifHDK* genes from one nitrogen-fixing bacterium hybridize to DNA from virtually all N_2 fixers, but not to DNA from non–N_2-fixing bacteria.

Assaying Nitrogenase: Acetylene Reduction

Nitrogenase is not entirely specific for N_2, because it also reduces cyanide (CN^-), acetylene ($HC \equiv CH$), and several other triply bonded compounds (for an exception, see the box). The reduction of acetylene by nitrogenase is only a two-electron process, and *ethylene* ($H_2C = CH_2$) is produced. The reduction of acetylene probably serves no useful purpose to the cell, but it does provide the experimenter with a simple and rapid way of measuring the activity of nitrogen-fixing systems because it is fairly easy to measure the reduction of acetylene to ethylene by gas chromatography (Figure 15.70). This technique is now widely used to detect nitrogen fixation in unknown systems. Previously, it was not easy to prove that an organism fixed N_2; indeed, many claims for nitrogen fixation in microorganisms were shown to be erroneous. The growth of an organism in a medium to which no nitrogen compounds have been added does not mean that the organism is fixing nitrogen from the air, because traces of fixed nitrogen compounds often occur as contami-

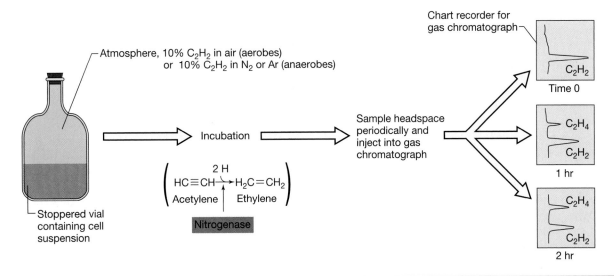

FIGURE 15.70   The acetylene reduction assay for nitrogenase activity. The results show no C_2H_4 when the experiment begins (time 0), but increasing production of C_2H_4 as the assay proceeds. Note how as C_2H_4 is produced, C_2H_2 is consumed. If the vial contained an enzyme extract, conditions would be anoxic, even if the nitrogenase were from an aerobic bacterium.

nants in the ingredients of culture media or enter the media in gaseous form or as dust particles.

Definitive proof of N_2 fixation is obtained using an isotope of nitrogen, ^{15}N, as a tracer. (^{15}N is not a radioisotope but a stable isotope. It is detected with a mass spectrometer.) The gas phase of a culture is enriched with ^{15}N, and after incubation, the cells and medium are digested, the ammonia produced being distilled off and assayed for its ^{15}N content. If there has been a significant production of ^{15}N-labeled NH_3, it is proof of nitrogen fixation. However, the acetylene reduction method is a more rapid and sensitive, albeit indirect, way of measuring nitrogen fixation. The sample, which may be soil, water, a culture, or a cell extract, is incubated with acetylene, and the gas phase of the reaction mixture is later analyzed by gas chromatography for production of ethylene (Figure 15.70). This method is far simpler and faster than other methods and can easily be adapted for field use in ecological studies of N_2-fixing bacteria directly in their habitats.

✓ 15.29 Concept Check

Nitrogen fixation, the reduction of N_2 to NH_3, involves a complex enzyme system called nitrogenase, which consists of dinitrogenase and dinitrogenase reductase. Most nitrogenases contain molybdenum or vanadium and iron as metal cofactors, and the process of nitrogen fixation is highly energy-demanding. Nitrogenase and associated regulatory proteins are encoded by the *nif* regulon. Certain artificial substrates that are structurally similar to N_2, such as acetylene and cyanide, are also reduced by nitrogenase.

✓ Write a balanced equation for the reaction carried out by the enzyme *nitrogenase*.
✓ What is *FeMo-co*?
✓ What metals are necessary for nitrogen fixation?
✓ What chemical and physical factors affect the function of nitrogenase? How does the *Streptomyces thermoautotrophicus* nitrogenase system differ in this regard?
✓ How is C_2H_2 useful for studies of nitrogen fixation?

REVIEW QUESTIONS

1. In what nutritional class would you place an organism that uses *glucose* as sole carbon and energy source? *Elemental sulfur* as an energy source? How would we refer to the latter organism if it grew with CO_2 as sole carbon source? What is the energy source for *phototrophic* organisms?

2. What is the role of light in the photosynthetic process of green and purple bacteria? Of cyanobacteria? Compare and contrast the photosynthesis process in these two groups of prokaryotes.

3. What are the functions of light-harvesting and reaction center chlorophylls? Why would a mutant incapable of making light-harvesting chlorophylls (such mutants can be readily isolated in the laboratory) probably not be a successful competitor in nature?

4. Where are the photosynthetic pigments located in a purple bacterium? A cyanobacterium? A green alga? Considering the function of chlorophyll pigments, why can't they be located elsewhere in the cell, for example, in the cytoplasm or in the cell wall?

5. How does light result in ATP production in an *anoxygenic phototroph*? In what ways are photosynthetic and respiratory electron flow similar? In what ways do they differ?

6. How is reducing power made for autotrophic growth in a purple bacterium? In a cyanobacterium?

7. How does the reduction potential of chlorophyll *a* in photosystem I and photosystem II differ? Why must the reduction potential of photosystem II chlorophyll be so highly electropositive?

8. What is the major function of carotenoids and phycobilins in phototrophic microorganisms?

9. What two enzymes are unique to organisms that carry out the Calvin cycle? What reactions do these enzymes carry out? What would be the consequences if a mutant arose that lacked either of these enzymes?

10. For conversion of 6 molecules of CO_2 into 1 fructose molecule, 18 molecules of ATP are required. Where in the Calvin cycle reactions are these ATPs consumed?

11. What organisms employ the hydroxypropionate or reverse citric acid cycles as autotrophic pathways?

12. Compare and contrast the utilization of H_2S by a purple phototrophic bacterium and by a colorless sulfur bacterium like *Beggiatoa*. What role does H_2S play in the metabolism of each organism?

13. Discuss why the growth yield (grams of cells per mole of substrate) of *Thiobacillus ferrooxidans* is considerably greater when the organism is growing aerobically on elemental sulfur than on ferrous iron as electron donor (assume the organism is growing autotrophically in both cases).

14. What is a mixotroph? Why does an organism capable of mixotrophy frequently grow better mixotrophically than chemolithotrophically or even chemoorganotrophically?

15. How do Type I and Type II methanotrophs differ in their carbon assimilation patterns? How does a *methanotroph* differ from a *methanogen*?

16. In *Escherichia coli* synthesis of the enzyme *nitrate reductase* is repressed by oxygen. On the basis of bioenergetic arguments, why do you think this repression phenomenon might have evolved?

17. Discuss at least three major differences between *assimilative* and *dissimilative* nitrate reduction.

18. Compare and contrast homoacetogens with methanogens in terms of (1) substrates and products of their energy metabolism, (2) ability to use organic compounds as electron donors in energy metabolism, (3) mechanism of autotrophy, and (4) ability to grow by aerobic respiration (you may want to review material in Section 14.3 before answering this question).

19. Define the term *substrate-level phosphorylation*. How does it differ from oxidative phosphorylation? Assuming an organism is facultative, what basic nutritional conditions dictate whether the organism obtains energy from substrate-level rather than oxidative phosphorylation?

20. Although many different compounds are theoretically fermentable, in order to support a fermentative process, most organic compounds must be eventually converted to one of a relatively small group of molecules. What are these molecules and why must they be produced?

21. To a culture of *Escherichia coli* growing fermentatively you add 1 g/l of $NaNO_3$. Would you expect the growth *yield* of the culture to increase or decrease? Why?

22. How can fermentations occur in the absence of substrate-level phosphorylation?

23. Why have hydrocarbons accumulated in large reservoirs on Earth despite the fact that they are readily degradable microbiologically under certain conditions?

24. How are xenobiotic compounds defined? Give an example of a compound you think qualifies as a xenobiotic.

25. Compare and contrast the conversion of cellulose and intracellular starch to glucose units. What enzymes are involved and which process is the more energy-efficient?

26. How do *monooxygenases* differ from *dioxygenases* in the reactions they catalyze?

27. Write out the reaction catalyzed by the enzyme *nitrogenase*. How many electrons are required in this reaction? How many are actually used? Explain.

28. What metals are found in nitrogenase? Are all nitrogenases oxygen-sensitive?

29. How does the *Streptomyces thermoautotrophicus* nitrogenase differ from that of *Azotobacter*?

APPLICATION QUESTIONS

1. Compare and contrast the absorption spectrum of chlorophyll *a* and bacteriochlorophyll *a*. What wavelengths are preferentially absorbed by each pigment and how do the absorption properties of these molecules compare with the regions of the spectrum visible to our eye? Why are most plants green?

2. The growth rate of the phototrophic purple bacterium *Rhodobacter* is about twice as fast when the organism is grown phototrophically in a medium containing malate as carbon source as when it is grown with CO_2 as carbon source (with H_2 as electron donor). Discuss the reasons why this is true and list the nutritional class we would place *Rhodobacter* in when growing under each of the two different conditions.

3. Discuss the nature of the evidence obtained from studies on the photosynthetic process of certain cyanobacteria that supports the hypothesis that these organisms evolved from anoxygenic phototrophs.

4. Although physiologically distinct, chemolithotrophs and chemoorganotrophs share a number of features with respect to the production of ATP. Discuss these common features along with reasons why the growth yield (grams of cells per mole of substrate) of a chemoorganotroph respiring glucose is so much higher than for a chemolithotroph respiring sulfur.

5. Why is the following statement, if taken literally, incorrect? "Anaerobic respiration is simply a process where an alternative electron acceptor is substituted for O_2 in the respiratory process."

6. When methane is made from CO_2 (plus H_2) or from methanol (in the absence of H_2), various steps in the pathway shown in Figure 15.46 are used. Compare and contrast methanogenesis from these two substrates and discuss why they must be metabolized in *opposite* directions?

7. Although dextran is a glucose polymer, glucose cannot be used to make dextran. Explain. How is dextran synthesis important in oral hygiene (∞ Section 19.3)?

8. *Pseudomonas fluorescens* can grow on benzoate aerobically, whereas the phototrophic bacterium *Rhodopseudomonas palustris* can grow on benzoate anaerobically. Compare and contrast the metabolism of benzoate by these two species, focusing on the following considerations: requirement for oxygenases, initial reactions leading to the opening of the ring, and product(s) formed that can feed into central metabolic pathways.

Laboratory cultures of microorganisms are rather artificial situations—in nature, resources and growth conditions are not always optimal. In fact in typical microbial habitats resources and conditions are constantly changing because of the metabolic activities of both microorganisms and the plants and animals with which they coexist. Oxygen is an excellent example. As shown here in the oxygen profile of a small soil particle (the numbers represent the percent O_2 present) the concentration of O_2, a vital nutrient for some microorganisms while a poison to others, can show dramatic spatial variation across even a very tiny distance. The "microenvironments" make possible the simultaneous activities of a great diversity of organisms, and a major goal of the microbial ecologist is to assess microbial activities in nature in order to understand the role that microorganisms play in the global ecosystem.

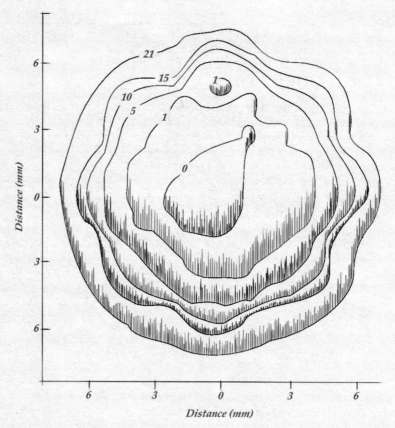

Distance (mm)

CHAPTER 16 Microbial Ecology

16.1   Populations, Guilds, and Communities   643
16.2   Microorganisms in Nature   645
16.3   Methods in Microbial Ecology   648
16.4   Enrichment and Isolation Methods   648
16.5   Viability and Quantification Using Staining Techniques   653
16.6   Genetic Stains, Community Analysis, and Optical Tweezers   655
16.7   Microbial Activity Measurements: Radioisotopes and Microelectrodes   658
16.8   Microbial Activity Measurements: Stable Isotopes   660
16.9   Terrestrial Environments   662
16.10 Aquatic Habitats   665
16.11 Deep-Sea Microbiology   667
16.12 Hydrothermal Vents   670
16.13 The Carbon Cycle   676
16.14 Ecology of Syntrophy and Methanogenesis   677
16.15 The Rumen Microbial Ecosystem   681
16.16 The Nitrogen Cycle   685
16.17 The Sulfur Cycle   686
16.18 The Iron Cycle   689
16.19 Microbial Leaching of Ores   691
16.20 Mercury and Heavy Metal Transformations   694
16.21 Petroleum Biodegradation   696
16.22 Biodegradation of Xenobiotics   698
16.23 Plant–Microorganism Interactions: Lichens and Mycorrhizae   704
16.24 *Agrobacterium* and Crown Gall Disease   706
16.25 Root Nodule Bacteria and Symbiosis with Legumes   709

Acid mine drainage acidic water containing H_2SO_4 derived from the microbial oxidation of iron sulfide minerals

Anoxic an oxygen-free environment, usually also highly reducing (low E_0')

Bacteroid morphologically misshapen *Rhizobium* cells inside a leguminous plant root nodule; can fix N_2

Barophilic an organism that grows best when placed under a pressure greater than 1 atm

Barotolerant an organism that can grow under elevated pressures but that grows best at atmospheric pressure

Biofilm colonies of microbial cells encased in slime and attached to a surface

Biogeochemistry study of biologically mediated chemical transformations

Black smoker an extremely hot (250–350°C) deep-sea hot spring emitting both hot water and various minerals

Cometabolism metabolism of a compound in the presence of a second organic compound, which is used as the primary energy source

Ecosystem a community of organisms and their natural environment

Enrichment culture a means of obtaining cultures of microorganisms from a natural environment by using highly selective culture methods

FISH fluorescent *in situ* hybridization

Guild a population of metabolically related microorganisms

Hydrothermal vent a deep-sea warm or hot spring

Infection thread in the formation of root nodules, a cellulosic tube through which *Rhizobium* cells can travel to reach and infect root cells

In situ in the environment

Interspecies hydrogen transfer the production and subsequent consumption of H_2 by different groups of microorganisms that interact closely during anaerobic catabolism

Leaching solubilization and removal of metals from an ore by microbial attack

Lichen a fungus and an alga (or cyanobacterium) living in symbiotic association

Microbial plastics biodegradable polymeric materials obtained from microorganisms that have properties similar to those of synthetic plastics

Microenvironment the immediate environmental surroundings of a microbial cell or group of cells

Mycorrhiza a symbiotic association between a fungus and the roots of a plant

Optical tweezers a laser microscope able to trap single cells and remove them from a cell mixture

Oxic an oxygen-containing environment frequently possessing a high E_0'

Primary producer an organism that uses light to synthesize new organic material from CO_2

Pyrite a common iron-containing ore, FeS_2

Reductive dechlorination removal of Cl as Cl^- from an organic compound by reducing the carbon atom from C—Cl to C—H

Rhizosphere the region immediately adjacent to plant roots

Root nodule a tumorlike growth on plant roots that contains symbiotic nitrogen-fixing bacteria

Rumen the first vessel in the multichambered stomach of ruminant animals in which cellulose digestion occurs

Ti plasmid a conjugative plasmid present in the bacterium *Agrobacterium tumefaciens* that can transfer genes into plants

Xenobiotic a totally synthetic product not naturally occurring in nature

---

Up to this point we have mainly considered microorganisms as laboratory entities. In this chapter we examine microorganisms in soil, water, and other environments and discuss how they can change the chemical and physical properties of their environments. The term *environment* refers to everything surrounding a living organism: the chemical, physical, and biological factors and forces that act on a living organism. From an ecological perspective, microorganisms are part of organismal communities called *ecosystems*. Each organism in an ecosystem interacts with its surroundings and in some cases greatly modifies the characteristics of the ecosystem in the process. This is particularly true of microorganisms, where significant chemical changes can occur because of their metabolic activities.

16.1

Populations, Guilds, and Communities

In a microbial ecosystem individual cells grow to form *populations* (Figure 16.1a). Metabolically related populations constitute groupings called *guilds*, and sets of guilds conducting complementary physiological processes interact to form microbial *communities* (Figure 16.1b). Microbial communities then interact with communities of macroorganisms to define the entire ecosystem.

Energy enters ecosystems in the form of sunlight, organic carbon, or reduced inorganic substances. Light is used by phototrophic organisms (Sections 15.2–15.6) to synthesize new organic matter (Figure 16.1b). The latter contains not only carbon but also nitrogen, sulfur, phos-

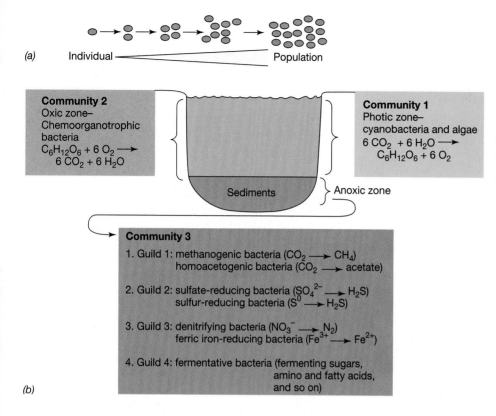

(a) Individual ——————— Population

Community 2
Oxic zone–
Chemoorganotrophic
bacteria
$C_6H_{12}O_6 + 6 O_2 \longrightarrow$
$6 CO_2 + 6 H_2O$

Community 1
Photic zone–
cyanobacteria and algae
$6 CO_2 + 6 H_2O \longrightarrow$
$C_6H_{12}O_6 + 6 O_2$

Sediments

Anoxic zone

Community 3
1. Guild 1: methanogenic bacteria ($CO_2 \longrightarrow CH_4$)
   homoacetogenic bacteria ($CO_2 \longrightarrow$ acetate)

2. Guild 2: sulfate-reducing bacteria ($SO_4^{2-} \longrightarrow H_2S$)
   sulfur-reducing bacteria ($S^0 \longrightarrow H_2S$)

3. Guild 3: denitrifying bacteria ($NO_3^- \longrightarrow N_2$)
   ferric iron-reducing bacteria ($Fe^{3+} \longrightarrow Fe^{2+}$)

4. Guild 4: fermentative bacteria (fermenting sugars,
   amino and fatty acids,
   and so on)

(b)

FIGURE 16.1   Populations, guilds, and communities—an example of microbial community structure in a lake ecosystem. (a) Microbial communities consist of populations of cells of various species that arise from cell division. (b) A lake ecosystem. For simplicity, only three microbial communities are depicted: phototrophic, aerobic chemoorganotrophic, and anaerobic. In the anaerobe community, examples of guild structure are given.

phorus, iron, and a host of other elements. This newly synthesized organic material, along with organic matter that enters the ecosystem from the outside (*allochthonous* organic matter) and reduced inorganic substances, drives the metabolic activities of chemoorganotrophic and chemolithotrophic organisms, whose metabolic diversity was discussed in the previous chapter. For the key elements of living systems, a *biogeochemical cycle* can be defined in which the element undergoes changes in oxidation state as it moves through the ecosystem (Figure 16.1*b*). Microorganisms are intimately involved in biogeochemical cycling and in many instances are the only biological agents capable of regenerating forms of the elements usable by other organisms, particularly plants. We will discuss many biogeochemical cycles in this chapter.

Microbial Ecology

The science of microbial ecology has two broad objectives: (1) to understand the *biodiversity* of microorganisms in nature and how different guilds interact in microbial communities, and (2) to measure the *activities* of microorganisms in nature and monitor their effects on ecosystems. Unfortunately, despite everything we discussed in Chapter 15 about metabolic diversity, we know relatively little about the diversity of *organisms* that carry out these reactions. In many cases only one or a few organisms able to carry out a particular metabolic transformation are known from the study of pure cultures; indeed, most microbiologists agree that there are many microorganisms left to be discovered and this is

a major goal of microbial ecology. However, measurements of microbial *activities* in nature have allowed us to assess the metabolic functioning of microbial communities even if we know little about the actual organisms that compose them. The occurrence of a specific biochemical transformation is good evidence that a corresponding microbial guild is present and metabolically active, and such information is often useful in designing procedures for attempting to isolate the responsible organisms. And with new molecular methods available (see Sections 16.5–16.8), it is becoming easier to weave biodiversity and metabolic activity objectives into a single study.

In the beginning of this chapter we discuss modern methods of assessing and tracking microbial communities in their natural habitats and consider how the activities of microorganisms in nature play important roles in biogeochemical cycling. We also examine a number of specialized microbial habitats where much is understood about the role of microorganisms in the ecosystem. We begin our treatment of microbial ecology with a consideration of microbial habitats because the habitat is where the microorganism actually lives and carries out its important functions.

✓ 16.1 Concept Check

Ecology is the study of organisms in their natural environments. Microbial communities consist of various guilds of metabolically related organisms. Microorganisms play major roles in energy transformations and biogeochemical processes. Knowledge of both microbial diversity and microbial activity is necessary in order to fully characterize a natural microbial community.

✓ How does a microbial *guild* differ from a microbial *community*?
✓ What is a *biogeochemical cycle*?

16.2

Microorganisms in Nature

The natural habitats of microorganisms are exceedingly diverse. Any habitat suitable for the growth of higher organisms can also support growth of microorganisms. But in addition, there are many habitats where, because of some physical or chemical extreme, higher organisms are absent yet microorganisms exist and occasionally even flourish. Microorganisms inhabit the surfaces of higher organisms and in some cases actually live *within* plants and animals. Microorganisms frequently reach large numbers in such habitats and may benefit the plant or animal in a nutritionally significant way. On the other hand, as we will see in Chapters 19–24, some microorganisms are pathogenic and bring harm to the host. We focus now on the microbial habitat from the standpoint of the microorganism and emphasize the heterogeneous and rapidly changing nature of typical microbial habitats.

The Microorganism and the Microenvironment

As in laboratory culture, the growth of microorganisms in nature depends on the *resources* (nutrients) available and on the growth *conditions*. Differences in the type and quantity of different resources and the physicochemical conditions (temperature, pH, water availability, light, oxygen) (∞ Chapter 5) of a habitat define the *niche* for each particular microorganism. Ecological theory states that for every organism there exists at least one niche, the *prime* niche, in which that organism is most successful. The organism may also inhabit other niches, but in these it is less ecologically successful than in its prime niche. Countless microbial niches exist on Earth and are in part responsible for the great metabolic diversity (∞ Chapter 15) and biodiversity (∞ Chapters 13, 14, and 17) of microorganisms we see today.

Because microorganisms are so small, their habitats are also small. A microbiologist must therefore learn to "think small" when considering the microorganism in its environment. For example, for a typical 3-μm rod-shaped bacterium, a distance of 3 mm in its habitat is the same that a human experiences over a distance of 2 km! And across that 3-mm distance chemical and physical gradients might exist that could greatly affect the organism. Thus, we must be more precise in our characterization of a microorganism's habitat, and microbial ecologists use the term *microenvironment* to describe where a microorganism actually lives and

metabolizes within its habitat. In a 3-mm particle of soil, for example, several different microenvironments could exist, differing chemically and physically in many ways. This can be visualized by considering the distribution of an important microbial nutrient like oxygen in a soil particle. Using microelectrodes (a technique to be described in Section 16.7) it is possible to measure oxygen concentrations throughout small soil particles. As shown in data from an actual experiment in Figure 16.2, soil particles are not homogeneous in terms of their oxygen content. The outer zones of a small soil particle may be fully oxic, whereas the center, only a very short distance away, can remain completely anoxic (Figure 16.2). This finding shows that different niches can exist across a very small spatial dimension and explains how various physiological types of microorganisms could coexist in such a soil particle. Anaerobic organisms could be active near the center of the particle shown in Figure 16.2, microaerophiles could be active further out, and obligately aerobic organisms could metabolize in the outer 2–3 mm of the particle; facultatively aerobic bacteria could be distributed throughout the particle.

Physicochemical conditions in the microenvironment can change rapidly in terms of both time and space. Because the oxygen concentrations shown in Figure 16.2 represent only "instantaneous" measurements, oxygen measurements taken following a period of microbial respiration or after an increase in soil water con-

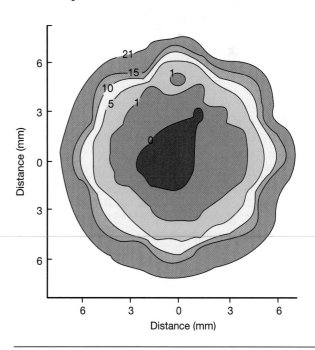

FIGURE 16.2 Contour map of O_2 concentrations in a soil particle. The axes show the dimensions of the particle. The numbers on the contours are O_2 concentrations (in percent; air is 21% O_2). In terms of oxygen relationships for microorganisms, each zone can be considered a different microenvironment.

tent could show a drastically different gradient of oxygen across the microenvironment. It can thus be said that microenvironments are *heterogeneous* and that conditions in a given microenvironment can change rapidly. Thus, because of microenvironments, high microbial diversity is possible in a relatively small physical area.

Surfaces and Biofilms

Surfaces are often of considerable importance as microbial habitats because nutrients can adsorb to them; in the microenvironment of a surface, nutrient levels may be much higher than they are in the bulk solution. This phenomenon can greatly affect the rate of microbial metabolism. On surfaces microbial numbers and activity are usually much greater than in free water because of adsorption effects. Microscope slides can serve as experimental surfaces on which organisms can attach and grow. When a slide is immersed in a microbial habitat, left for a period of time, and then retrieved and examined by microscopy, the importance of the surface to microbial development is apparent (Figure 16.3*a*). Microcolonies readily develop on such surfaces much as they do on natural surfaces in nature. Microscopic examination of immersed microscope slides can actually be used as a technique to measure growth rates of attached organisms in nature.

A surface may itself also be a nutrient, such as a particle of organic matter, where attached microorganisms catabolize organic or inorganic nutrients directly from the surface of the particle. Dead plant material, for example, is rapidly colonized by microorganisms in soil, and simple staining techniques can detect microbial populations attached to the solid surface (Figure 16.3*b*).

Studies of microbial colonization of surfaces have shown that microorganisms often grow on surfaces enclosed in **biofilms.** These are encased microcolonies of bacterial cells attached to a surface by way of adhesive polysaccharides excreted by the cells (Figure 16.4). Biofilms trap nutrients for growth of the enclosed microbial population and help prevent detachment of cells on surfaces in flowing systems. Biofilms usually contain many layers and direct microscopic examination of the microbial components in each layer can be done using scanning laser confocal microscopy (Figure 16.4*b*) (∞ Section 3.2).

Biofilms have significant implications in human medicine and commerce. In the human body, bacterial cells within a biofilm are made unavailable for attack by the immune system. This fact thus complicates the use of introduced artificial surfaces such as medical implants, which can serve as sites of development of biofilms containing pathogenic microorganisms. Biofilms are also important in oral hygiene; dental plaque, a typical biofilm, contains acid-producing bacteria responsible for dental caries (∞ Section 19.3). In industrial situations, biofilms can slow the flow of water or oil through pipelines, ac-

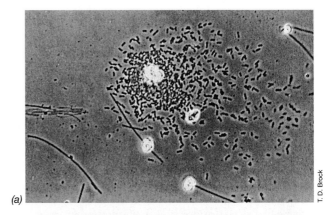

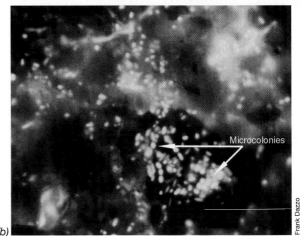

FIGURE 16.3   Microorganisms on surfaces. (a) Bacterial microcolonies developing on a microscope slide immersed in a small river. The bright particles are mineral matter. The short, rod-shaped cells are about 3 μm long. (b) Fluorescence photomicrograph of a natural microbial community colonizing plant roots in soil. Note microcolony development. The preparation has been stained with acridine orange.

celerate the corrosion of the pipes themselves, and initiate degradation of submerged objects such as offshore oil rigs, boats, and shoreline installations.

Nutrient Levels and Growth Rates

Nutrients (or in ecological terms, *resources*) often enter an ecosystem in intermittent fashion. A large pulse of nutrients—for example, an input of leaf litter or the carcass of a dead fish or animal—may be followed by a period of severe nutrient deprivation. Microorganisms in nature are thus often faced with a "feast-or-famine" type of existence, and many microorganisms have evolved biochemical systems for production of storage polymers as reserve materials. Reserve polymers store excess nutrients present under favorable growth conditions for use during periods of nutrient deprivation. Examples of reserve materials are poly-β-hydroxybutyrate and other alkanoates, polysaccharides, polyphosphate, and so on (∞ Section 3.13).

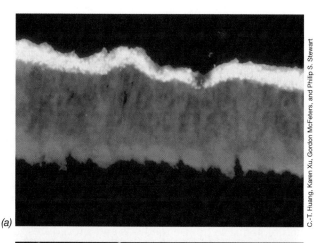

(a)

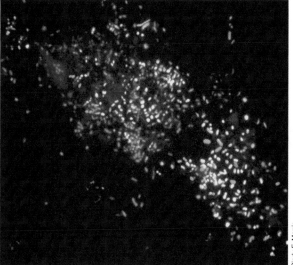

(b)

C.-T. Huang, Karen Xu, Gordon McFeters, and Philip S. Stewart

Cindy E. Morris

FIGURE 16.4   Microbial biofilms. (a) An experimental biofilm made up of cells of *Pseudomonas aeruginosa*. The yellow-green layer (about 15 μm in depth) contains cells and is stained by an enzyme activity stain for the enzyme alkaline phosphatase (b) Confocal laser scanning microscopy (∞ Section 3.2) of a natural biofilm that developed on a leaf surface. The color of the cells indicates their depth in the biofilm: red, cells on the surface; green, 9 μm depth; blue, 18 μm depth.

Extended periods of exponential growth of microorganisms in nature are rare. Growth more often occurs in spurts, linked closely to the availability of nutrients. Because physicochemical conditions in nature are rarely optimal all at the same time, growth rates of microorganisms in nature are generally well below the maximum growth rates recorded in the laboratory. For instance, the generation time of *Escherichia coli* in the intestinal tract is about 12 hr (two doublings per day), whereas in pure culture it grows much faster, with a minimum generation time of 20 min under the best of conditions. Estimates of the growth rate of certain soil bacteria have shown that they grow in nature at less than 1% of the maximal growth rate measured in the labora-

tory. On average, these slow growth rates are a reflection of the fact that (1) nutrients (resources) are frequently in low supply, (2) the distribution of nutrients throughout the microbial habitat is not uniform, and (3) except for rare instances, microorganisms do not grow in pure culture in natural environments and thus must deal with the competitive effects of other microorganisms, a situation not encountered in growth in pure culture.

Microbial Competition and Cooperation

Competition among microorganisms for available resources may be intense. In the simplest case, the outcome of a competitive interaction depends on rates of nutrient uptake, inherent metabolic rates, and, ultimately, growth rates. However, in some cases of microbial competition, one organism may inhibit growth or metabolism of other organisms. This may occur by excretion of a specific inhibitor such as an antibiotic or because of the physiological activities of an organism producing a toxic product, like acid from the fermentation of sugars.

Instead of competing for the same nutrient, some microorganisms work together to carry out a particular transformation that neither organism can carry out alone. These types of microbial interactions, called *syntrophy* (∞ Section 15.23), are crucial to the competitive success of certain anaerobic bacteria, as will be described in Section 16.14. Syntrophic relations generally require that the two or more organisms involved in the process share the same microenvironment because the product of the metabolism of one organism must be easily accessible to the second. Metabolic cooperation is also seen in the activities of groups of organisms that carry out *complementary* metabolisms. For example, in Chapter 15 we discussed metabolic transformations that involved two distinct groups of organisms, such as those of the *nitrosifying* and the *nitrifying* bacteria, which combine to oxidize NH_3 to NO_3^- although neither group is capable of doing this alone (∞ Section 15.13); such organisms typically live in tight association (see Figure 16.11). Or consider the activities of sulfate-reducing and sulfide-oxidizing bacteria; in this example the product of one organism (H_2S from the reduction of SO_4^{2-}) is the substrate for the other ($H_2S + \frac{1}{2}O_2 \rightarrow S^0 + H_2O$) (∞ Sections 15.11 and 15.17). Such types of cooperative interactions are common in microbial habitats.

✓ 16.2 Concept Check

Microorganisms are very small, and their natural environments are likewise small. The microenvironment is the place in which the microorganism actually lives. Microorganisms in nature often live a feast-or-famine existence such that only the best-adapted species survive in a given niche. Cooperation among microorganisms is also important in many microbial interrelationships.

✓ What aspects define the niche of a particular microorganism?
✓ Why can many different physiological groups of organisms live in a single habitat?
✓ What is a *biofilm?*

16.3

Methods in Microbial Ecology

As previously mentioned, microbial ecology focuses on two major issues: (1) *biodiversity*, including the isolation, identification, and quantification of microorganisms in various habitats, and (2) *microbial activity*, that is, what are microorganisms *doing* in their habitats. Although both types of studies are important, studies designed to measure microbial activities *in situ* have the advantage of allowing the microbial ecologist to make perturbations in a sample of the environment and measure what effects such chemical or physical changes have on microbial activity. However, biodiversity studies have a role in microbial ecology as well and go hand in hand with activity measurements to generate a more complete picture of the microbial ecology of a given habitat.

We begin here with a consideration of methods for assessing biodiversity through enrichment and isolation, and then consider nonculture methods of identifi-

cation and enumeration based on fluorescent probes. We go from there to a consideration of microbial activity and describe the key methods that have been developed for measuring microbial activities directly in natural environments.

16.4

Enrichment and Isolation Methods

Rarely does a natural environment contain only a single type of microorganism. In most cases, a microbial community exists, and it is particularly challenging for the microbiologist to devise methods and procedures that permit the isolation and culture of organisms of interest. The most common approach to this goal is the **enrichment culture technique.** In this method, a medium and a set of incubation conditions are used that are *selective* for the desired organism and are counterselective for the undesired organisms. We recall here the concept of *resources* and *conditions* of the microbial niche (see Section 16.1): the enrichment culture strategy is to duplicate as closely as possible resources and conditions in the niche and then probe for potential inhabitants of such a niche. An overview of some successful enrichment culture procedures is given in Table 16.1.

TABLE 16.1   Enrichment culture methods for prokaryotes[a]

Light-phototrophic bacteria: main C source, CO_2

Aerobic incubation	Organisms enriched	Inoculum
N_2 as nitrogen source	Cyanobacteria	Pond or lake water; sulfide-rich muds; stagnant water; raw sewage; moist, decomposing leaf litter; moist soil exposed to light; pasteurized soil (heliobacteria)
NO_3^- as nitrogen source, 55°C	Thermophilic cyanobacteria	Hot spring microbial mat
Anaerobic incubation		
H_2 or organic acids; N_2 as sole nitrogen source	Purple nonsulfur bacteria, heliobacteria	
H_2S as electron donor	Purple and green sulfur bacteria	

Dark-chemolithotrophic bacteria: main C source, CO_2 (medium must lack organic C)

Aerobic incubation Electron donor	Electron acceptor	Organisms enriched	Inoculum
NH_4^+	O_2	Nitrosifying bacteria (*Nitrosomonas*)	Soil, mud; sewage effluent
NO_2^-	O_2	Nitrifying bacteria (*Nitrobacter*)	
H_2	O_2	Hydrogen bacteria (various genera)	
H_2S, S^0, $S_2O_3^{2-}$	O_2	*Thiobacillus* spp.	
Fe^{2+}, low pH	O_2	*Thiobacillus ferrooxidans*	
Anaerobic incubation			**Inoculum**
S^0, $S_2O_3^{2-}$	NO_3^-	*Thiobacillus denitrificans*	Mud, lake sediments, soil
H_2	NO_3^- + yeast extract	*Paracoccus denitrificans*	

TABLE 16.1	Enrichment culture methods for prokaryotes (*continued*)

Dark-chemoorganotrophic bacteria and methanogens: main C source, organic compounds

Aerobic incubation: respiration

Electron donor and nitrogen source	Electron acceptor	Typical organisms enriched	Inoculum
Lactate + NH_4^+	O_2	*Pseudomonas fluorescens*	Soil, mud; lake sediments; decaying vegetation; pasteurize inoculum (80°C for 15 min) for all *Bacillus* enrichments
Benzoate + NH_4^+	O_2	*Pseudomonas fluorescens*	
Starch + NH_4^+	O_2	*Bacillus polymyxa*, other *Bacillus* spp.	
Ethanol (4%) + 1% yeast extract, pH 6.0	O_2	*Acetobacter, Gluconobacter*	
Urea (5%) + 1% yeast extract	O_2	*Sporosarcina ureae*	
Hydrocarbons (e.g., mineral oil) + NH_4^+	O_2	*Mycobacterium, Nocardia, Pseudomonas* (see Figure 16.57)	
Cellulose + NH_4^+	O_2	*Cytophaga, Sporocytophaga* (see Figure 13.89)	
Mannitol or benzoate, N_2 as N source	O_2	*Azotobacter*	

Anaerobic incubation: anaerobic respiration

Main ingredients	Electron acceptor	Organisms enriched	Inoculum
Organic acids	KNO_3 (0.2%)	*Pseudomonas* (denitrifying species)	Soil, mud; lake sediments
Yeast extract	KNO_3 (1%)	*Bacillus* (denitrifying species)	
Organic acids	Na_2SO_4	*Desulfovibrio, Desulfotomaculum*	
Acetate, propionate, butyrate	Na_2SO_4	Fatty-acid oxidizing sulfate reducers	As above; or sewage digestor sludge; rumen contents
Acetate, ethanol	S^0	*Desulfuromonas*	
Acetate	Fe^{3+}	*Geobacter, Geospirillum*	
Acetate	ClO_3^-	Various chlorate-reducing bacteria	
H_2	Na_2CO_3	Methanogens (chemolithotrophic species only), homoacetogens	Mud, sediments, sewage sludge
CH_3OH	Na_2CO_3	*Methanosarcina barkeri*	
CH_3NH_2	KNO_3	*Hyphomicrobium*	

Anaerobic incubation: fermentation

Electron donor and nitrogen source	Electron acceptor	Organisms enriched	Inoculum
Glutamate or histidine	No exogenous electron acceptors added	*Clostridium tetanomorphum*	Mud, lake sediments; rotting plant material; dairy products (lactic and propionic acid bacteria); rumen or intestinal contents (enteric bacteria); sewage sludge; soil
Starch + NH_4^+	None	*Clostridium* spp.	
Starch, N_2 as N source	None	*Clostridium pasteurianum*	
Lactate + yeast extract	None	*Veillonella* spp.	
Glucose or lactose + NH_4^+	None	*Escherichia, Enterobacter*, other fermentative organisms	
Glucose + yeast extract (pH 5)	None	Lactic acid bacteria (*Lactobacillus*)	
Lactate + yeast extract	None	Propionic acid bacteria	
Succinate + NaCl	None	*Propionigenium*	
Oxalate	None	*Oxalobacter*	

a All media must contain an assortment of mineral salts including N, P, S, Mg^{2+}, Mn^{2+}, Fe^{2+}, Ca^{2+}, and other trace elements (⚬⚬ Sections 4.2 and 4.3). This table is meant as an overview of enrichment methods and does not speak to the effect incubation temperature might have in isolating thermophilic (high temperature), hyperthermophilic (very high temperature), and psychrophilic (low temperature) species, or the effect that extremes of pH or salinity might have, assuming an appropriate inoculum was available.

The field of general (primarily nonmedical) microbiology owes a great debt to the Delft School of Microbiology, which was initiated by Martinus Beijerinck and continued by A. J. Kluyver and C. B. van Niel. Beijerinck, one of the world's greatest microbiologists (among other things, he was the first to characterize viruses), first devised the enrichment culture technique and used this technique in the isolation and characterization of a wide variety of bacteria (Section 1.9). Subsequently, Kluyver and van Niel used the enrichment culture technique to isolate phototrophic and chemolithotrophic bacteria and to show the fascinating physiological diversity among the bacteria. Another important figure from the Dutch school was L. G. M. Baas-Becking, who carried out the first calculations of the energetics of phototrophic and chemolithotrophic bacteria and emphasized the importance of these organisms in geochemical processes. Van Niel subsequently came to the United States and was respon-

sible for the training of a number of general bacteriologists who carried on the Delft tradition, including the late R. Y. Stanier and M. Doudoroff, and Robert E. Hungate, now retired from the University of California at Davis. In addition, a number of scientists visited van Niel's laboratory at Pacific Grove, California, in the 1950s and 1960s and learned his methods and enrichment approaches. In recent years, the "Delft tradition" has been carried on in Germany by Bernhard Schink, Karl Stetter, and Fritz Widdel, and by many others in Europe, the countries of the former Soviet Union, and the United States.

An excellent example of the application of the enrichment culture technique can be found in the isolation of *Azotobacter,* a nitrogen-fixing bacterium discovered by Beijerinck in 1901 (Fig. 1; Sections 1.9, 13.8, and 15.29). Beijerinck was interested in knowing whether any aerobic bacteria capable of fixing nitrogen existed in the soil. The only other nitrogen-fixing organism known at that

time, *Clostridium pasteurianum,* was an anaerobe discovered several years earlier by Sergei Winogradsky. To look for an aerobic nitrogen fixer, Beijerinck added a small amount of soil to an Erlenmeyer flask that contained a thin layer of mineral salts medium devoid of ammonia, nitrate, or any other form of fixed nitrogen and that contained mannitol as carbon source. Within 3 days a thin film developed on the surface of the liquid, and the liquid became quite turbid. Beijerinck observed large, rod-shaped cells that appeared quite distinct from the spore-containing rods of *C. pasteurianum.* Beijerinck streaked agar plates containing phosphate and mannitol with the turbid liquid, incubated aerobically, and within 48 hr obtained large, slimy colonies typical of *Azotobacter* (see Fig. 1). Pure cultures were obtained by picking and restreaking well-isolated colonies a number of times. Beijerinck assumed his new organism was using N_2 from air as its source of cell nitrogen and later proved

Successful enrichment requires that an appropriate *inoculum* source be used. Thus, we begin an enrichment culture protocol by going to the appropriate habitat (see Table 16.1) and sampling to obtain an enrichment inoculum. Enrichment cultures are frequently established by placing the inoculum directly into a highly selective medium; many common prokaryotes can be isolated this way. For example, by pasteurizing the inoculum to kill vegetative cells, endospore-forming organisms can be isolated on various enrichment media. The nitrogen-fixing bacterium *Azotobacter* can be easily isolated aerobically in a mannitol mineral salts medium lacking fixed forms of nitrogen (see the box, Rise of General Microbiology), and so on. Literally hundreds of different enrichment approaches have been tried in the past, and some of the more dependable ones are listed in Table 16.1. Both the culture medium *and* the general incubation conditions are listed because in devising enrichment culture protocols, one must optimize both resources and conditions in order to have the best chance at obtaining the organism of interest.

The Winogradsky Column

For isolation of purple and green phototrophic bacteria and other anaerobes, the **Winogradsky column** has traditionally been used. Named for the famous Russian microbiologist Sergei Winogradsky (the box, Winogradsky's Legacy, Chapter 15 and Section 1.9), the column was devised by him in the 1880s to study soil microorganisms. The column is a miniature anaerobic ecosystem that can be an excellent long-term source of all types of prokaryotes involved in nutrient cycling.

A Winogradsky column can be prepared by filling a large glass cylinder about one-third full with organic-rich, preferably sulfide-containing, mud (Figure 16.5). Carbon substrates are first mixed into the mud; organic additions that have been successfully used in the past include hay, shredded newsprint, sawdust, shredded leaves or roots, ground meat, hard boiled eggs, and even dead animals! The mud is also supplemented with $CaCO_3$ and $CaSO_4$ as a buffer and as a source of sulfate, respectively. The mud is packed tightly in the container, care being taken to avoid entrapping air (Figure 16.5).

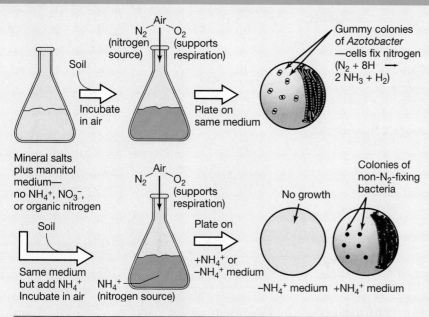

FIG. 1 The isolation of *Azotobacter*

presence of molecular oxygen. Beijerinck noted that if he placed too much liquid in his flasks or employed a more readily fermentable organic substrate, such as glucose or sucrose, in place of mannitol, his enrichment frequently turned anoxic and favored the growth of *Clostridium* rather than *Azotobacter*. The addition of ammonia or nitrate to the original enrichment never resulted in the isolation of *Azotobacter* but only in a variety of non–nitrogen-fixing bacteria instead (see the figure). Hence Beijerinck showed that *Azotobacter* has a strong selective advantage over other soil bacteria under a specific set of nutritional conditions and in the process defined the key physiological properties of *Azotobacter*: aerobiosis and nitrogen fixation. Beijerinck also demonstrated that the composition of the growth medium as well as the incubation conditions employed were of paramount importance in the proper development of an enrichment culture. ■

this by showing total nitrogen increases in pure cultures of *Azotobacter* grown in the absence of fixed nitrogen.

The selective pressure of the enrichment approach used here should be evident. By omitting from his medium any source of combined nitrogen and incubating aerobically, Beijerinck placed constraints on the microbial population. Any organism that developed had to be able to both fix its own nitrogen and tolerate the

The mud is then covered with lake, pond, or ditch water, and the top of the cylinder covered with aluminum foil. The cylinder is placed in a north window so as to receive adequate (but not excessive) sunlight and left to develop for a period of weeks (in the Southern Hemisphere use a south window).

In a typical Winogradsky column a mixture of many different types of organisms develops. Algae and cyanobacteria appear quickly in the upper portions of the water column and by producing O_2 help to keep this zone oxic. Fermentative decomposition processes in the mud quickly lead to the production of organic acids, alcohols, and H_2, suitable substrates for sulfate-reducing bacteria. As a result of the production of sulfide, purple and green patches appear on the outer layers of the mud exposed to light, the purple patches, consisting of purple sulfur bacteria, frequently developing in the upper layers, and the green patches, consisting of green sulfur bacteria, in the lower layers nearest the source of sulfide (this occurs because of differences in sulfide tolerance between green and purple bacteria). At the mud–water interface the water is frequently quite tur-

bid and may be colored as a result of the growth of purple sulfur and purple nonsulfur bacteria (Figure 16.5). Sampling of the column for phototrophic bacteria is performed by inserting a long, thin pipet into the column and removing some colored mud or water. These can be used to inoculate enrichment media (Table 16.1).

Winogradsky columns have been used to enrich for a variety of prokaryotes, both aerobes and anaerobes. The great advantage of a column, besides the ready availability of inocula for different enrichment cultures, is that it can be spiked with a particular compound whose degradation one wishes to study and then allowed to select from the inoculum for an organism or organisms that can degrade it. In addition, because the Winogradsky column more nearly resembles the natural environment than do culture media, a variety of each physiological type of microorganism is more frequently observed in Winogradsky columns than in enrichments where a liquid culture medium is inoculated directly with a natural sample. In the latter approach, rapidly growing species frequently rise to the forefront, leaving the slower growing species behind.

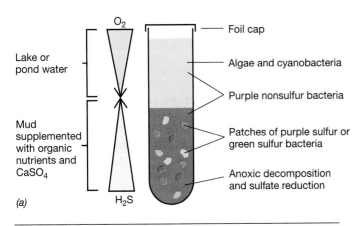

FIGURE 16.5 The Winogradsky column. (a) Schematic view of a typical column. The column is placed to receive subdued sunlight. Chemoorganotrophic bacteria grow throughout the column, aerobes and microaerophiles in the upper regions, anaerobes in the zones containing H₂S. Anoxic decomposition leading to sulfate reduction creates the gradient of H₂S. Green and purple sulfur bacteria stratify according to their tolerance for H₂S. (b) Photo of Winogradsky columns that have remained anoxic up to the top where blooms of three different phototrophic bacteria have occurred in the mud and up into the water column. Left to right: *Thiospirillum jenense, Chromatium okenii,* and *Chlorobium limicola.*

From Enrichments to Pure Cultures

The objective of an enrichment culture study is usually to obtain a pure culture. Pure cultures can be obtained in many ways, but the most frequently employed means are the streak plate, the agar shake, and liquid dilution methods. For organisms that grow well on agar plates, the **streak plate** is the method of choice (Figure 16.6). By repeated picking and restreaking of a well-isolated colony, one usually obtains a pure culture that can then be transferred to a liquid medium. With proper incubation it is possible to purify both aerobes and anaerobes on agar plates via the streak plate method.

The **agar shake tube method** (Figure 16.6*b*) involves the dilution of a mixed culture in tubes of molten agar, resulting in colonies embedded *in* the agar rather than on the surface of a plate. The shake tube method has been found useful for purifying particular types of microorganisms (for example, phototrophic sulfur bacteria and sulfate-reducing bacteria). Purification can be obtained by successively diluting a cell suspension in tubes of liquid medium until a dilution is reached in which well-isolated colonies are obtained (Figure 16.6*b*). By repeating this procedure using the highest dilution showing growth as inoculum for a new set of dilutions, it is possible in most cases to eventually obtain pure cultures.

A similar process can be done without using agar by serially diluting an inoculum in a liquid medium until

the final tube or tubes in the series show no growth. Using 10-fold serial dilutions, for example, the last tube showing growth should have originated from 10 or fewer cells. By repeating this process of *dilution to extinction* several times, pure cultures can be obtained. This method has also been used for the enumeration of microorganisms in foods, wastewater, and other samples where total cell numbers need to be assessed routinely. In these cases, the procedure is known as the *most probable number* (or *MPN*) technique.

Regardless of the method(s) used to purify a culture, once a putative pure culture has been obtained, it is essential to check its purity. This is usually done through a combination of microscopy and checking for growth in a variety of culture media that favors growth of contaminants but not the organism of interest.

Enrichment Bias

Extincting dilution methods have clearly shown that in most cases there is a bias, and usually a very severe bias, to the outcome of standard liquid enrichment cultures. In standard enrichments the most fit organism for the particular set of nutritional and incubation conditions chosen overgrows other members of the community and quickly becomes the dominant member in the liquid enrichment. Although this is exactly the goal of the enrichment culture method, to obtain an organism that is best suited to a given set of culture and incubation conditions, comparisons of liquid enrichment versus

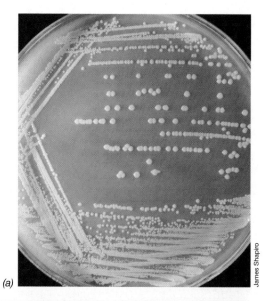

(a)

James Shapiro

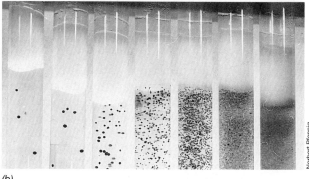

(b)

Norbert Pfennig

FIGURE 16.6   Pure culture methods (a) Organisms that form distinct colonies on plates such as those shown here are usually very easy to purify. For oxygen-tolerant anaerobes, plate isolations also work well if the plates are put under anoxic conditions immediately after streaking. (b) Colonies of oxygen-sensitive anaerobes using the agar shake technique. A dilution series was established from right to left, eventually yielding well-isolated colonies. The tubes are sealed with a sterile mixture of paraffin and mineral oil to maintain anaerobiosis.

extincting dilution methods have shown that in many cases the organism obtained in liquid enrichments is *NOT* the dominant member of the microbial community. In fact, the enriched organism often turns out to be a quantitatively insignificant one that for unknown reasons dominates the more abundant (and likely more ecologically significant) members of the microbial community, when selected for by standard liquid enrichment. Extincting dilution is one approach to overcoming this limitation—if an inoculum is sufficiently diluted, the quantitatively insignificant "weeds" will be eliminated and other members of the community can be isolated. Thus, classical enrichment methods, though powerful, may not necessarily yield an accurate picture

of the species composition of microbial communities, making dilutions often necessary for this purpose.

✓ 16.3–16.4 Concept Check

A variety of methods are used in microbial ecology to assess the biodiversity and/or activities of microorganisms in nature. The enrichment culture technique is a means of selecting microorganisms from natural samples that are capable of carrying out specific reactions. The Winogradsky column is a type of enrichment culture that is suitable for obtaining phototrophic bacteria and other anaerobes. Once a vigorous enrichment culture has been obtained, a pure culture can usually be obtained by use of conventional bacteriological procedures.

- ✓ What things are important to the success of an enrichment culture?
- ✓ How does the agar shake tube method differ from conventional streaking to obtain colonies on plates?
- ✓ What is *enrichment bias?*

16.5

Viability and Quantification Using Staining Techniques

A number of staining methods have been devised in microbiology, and some are suitable for quantifying microorganisms in natural samples. Although these methods say little about the physiological or phylogenetic composition of the populations, they are widely used and are fairly reliable for obtaining quantitative information about the *total numbers* of cells in a particular habitat.

Fluorescent Staining: Acridine Orange and DAPI

Two fluorescent dyes have been widely used to stain microorganisms in opaque habitats, acridine orange and DAPI. Acridine orange (3,6-bis [dimethylamino]-acridium chloride) binds to nucleic acids; using epifluorescence microscopy, a technique where stained cells are observed with near-ultraviolet light, acridine orange-stained cells will fluoresce either green or orange (see Figure 16.3b). A second staining method, similar in principle to acridine orange, is the use of the stain DAPI (4′,6-diamido-2-phenylindole); cells stained with DAPI fluoresce bright blue and are very easy to see and enumerate (Figure 16.7). Acridine orange and DAPI staining are widely used for the enumeration of microorganisms in environmental, food, and clinical samples.

Depending on the sample, nonspecific background staining can be a problem with either of these methods, but for many samples, soil as well as aquatic, acridine orange or DAPI staining can give a reasonable estima-

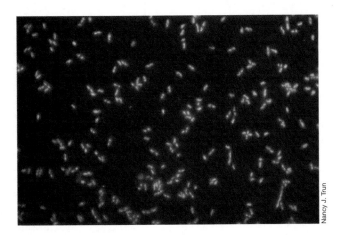

FIGURE 16.7   Cells of *Escherichia coli* made fluorescent by staining with DAPI. This staining technique is frequently used to make total microbial counts.

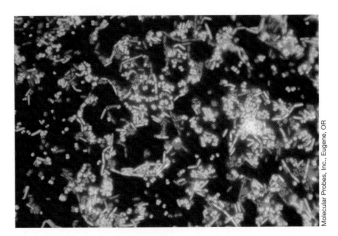

FIGURE 16.8   Viability staining. Live (green) and dead (red) cells of *Micrococcus luteus* (cocci) and *Bacillus cereus* (rods) stained by the **LIVE/DEAD Bac Light**™ **Bacterial Viability Stain.**

tion of the cell numbers present. For aquatic samples cells can be stained on the surface of a filter after filtering a given volume of liquid. These simple staining techniques have on the one hand the advantage that they are nonspecific—they stain *all* microorganisms in a sample—but on the other hand have the disadvantage that they do not differentiate between living and dead cells, nor do they allow for tracking *specific* organisms in an environment.

Viability Staining

Methods have been developed for differentiating live from dead bacterial cells using a fluorescent staining method. This gives information on not only the number of organisms present in a sample but also on their viability. The basis of the differentiation is whether or not the cytoplasmic membrane of the cells is intact. Two dyes, green and red, are added to a sample; the green fluorescing dye penetrates all cells, viable or not, while the red dye, which contains propidium iodide, penetrates only those cells whose cell membrane is no longer intact and that are therefore dead (Figure 16.8). Although useful for research situations using laboratory cultures of bacteria, the live/dead staining method is not suitable for use in the microscopic examination of natural habitats because of problems with nonspecific staining of background materials.

Fluorescent Antibodies

Specificity can be added to fluorescent-staining techniques by using *fluorescent antibodies.* We discuss the theory of fluorescent antibodies in Section 21.7. The great specificity of antibodies prepared against surface constituents of a particular organism can be exploited as a

means of identifying or tracking an organism in a complex habitat containing a mixture of many organisms, such as soil (Figure 16.9) or a clinical sample. The method requires the preparation of specific antibodies against the organism of interest, and this preparation is often a time-consuming and laborious procedure. However, at least for clinically relevant microorganisms, highly specific antibodies can be purchased commercially and are widely used in clinical microbiology laboratories (Section 21.7). Alternatively, equally

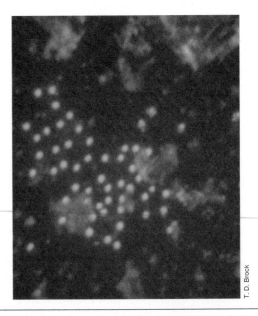

FIGURE 16.9   Visualization of bacterial microcolonies (bacterial cells appear as greenish-yellow dots) of the hyperthermophilic archaeon, *Sulfolobus acidocaldarius* on the surface of solfatara soil particles by use of the fluorescent antibody technique. Cells are about 1 μm in diameter.

specific and more easily deployed molecular biological methods for identification exist using phylogenetic and other genetic probes, and we discuss these options next.

16.6

Genetic Stains, Community Analysis, and Optical Tweezers

A very powerful and specific approach to the identification and quantification of microorganisms in nature is the use of nucleic acid probes. Recall that a nucleic acid probe is a DNA or RNA oligonucleotide complementary to a sequence in a target gene in a particular organism or organisms. As was discussed in Section 12.6, such oligonucleotides can be made to fluoresce when certain dyes are attached to them, and this now fluorescent probe can be used to identify organisms containing the complementary sequence. When this technique is applied to a natural sample it is referred to as *fluorescent in situ hybridization* (or *FISH*, for short), and three different applications of this method are described here.

Phylogenetic Staining Using FISH

Phylogenetic stains are fluorescing oligonucleotides complementary in base sequence to signature sequences in 16S (prokaryotes) or 18S (eukaryotes) ribosomal RNA (⌀ Section 12.6). Because signature sequences have been identified for individual species as well as for entire domains of organisms, the degree of specificity of the phylogenetic stain can be easily controlled (⌀ Section 12.6). For example, two different gram-negative bacteria, both members of a single lineage of the domain Bacteria, can be easily distinguished using appropriate phylogenetic stains (Figure 16.10). Using phylogenetic stains one can quickly identify and track a particular organism or group of related organisms (de-

pending on the specificity of the phylogenetic probe) in a natural sample, and thus this technology is in widespread use in microbial ecology laboratories today.

Using dual or multiple phylogenetic probes on a single sample, each probe designed to react with a particular organism or group and containing its own fluorescent dye, one can use FISH to phylogenetically characterize an entire habitat (Figure 16.11a). And, using FISH with a confocal laser microscope, a method for observing through layers of an opaque environment (⌀ Section 3.2), one can see the physical interactions of various phylogenetic groups (Figure 16.11b). Phylogenetic FISH can also be used to ask whether a particular organism is present in a particular environment, and this application has gained wide appeal in clinical diagnostics and the food industry. In this regard, a simple microscopic evaluation of a clinical sample or food product is all that is necessary to confirm a medical diagnosis or assess the microbiological safety of a food product.

We learned in Chapter 12 about phylogenetic microbial community analysis using nucleic acid probes. Recall that to do this, specific probes are used to clone ribosomal RNA genes from organisms in an environment; sequencing of the captured genes then reveals the phylogenetic diversity of the environment (⌀ Section 12.6 and Figure 12.12). From these studies the staggering conclusion has emerged that the organisms we now have in culture represent but a small fraction, probably less than 1%, of all the species of prokaryotes that exist in nature. However, using FISH technology it is possible to design specific probes to react with these as yet uncultured microorganisms and from this learn what they look like (that is, their morphology), how abundant they are in a given environment, and what other organisms they associate with in nature; such information is often helpful for devising enrichment culture methods for their laboratory culture. Alternatively, identification of the desired organisms by FISH is also use-

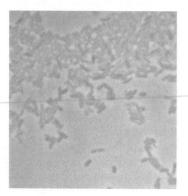

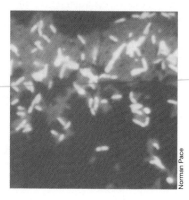

Norman Pace

FIGURE 16.10   Nucleic acid probe methods for identifying microorganisms. Differentiation of closely related gram-negative Bacteria. Left, phase micrograph of mixture of *Proteus vulgaris* and a related bacterium isolated from an insect. Center, same field stained with a 16S rRNA probe specific for all members of the Bacteria. Right, same field stained with a probe specific for the bacterium from the insect. The cells in both cases are about 0.6 μm in diameter.

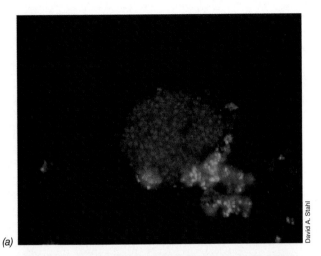

(a)

David A. Stahl

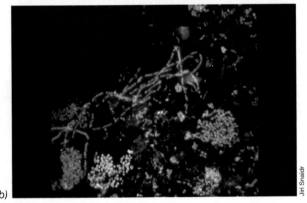

(b)

Jiri Snaidr

FIGURE 16.11 Use of multiple FISH probes on sewage samples to assess their microbial diversity (a) Nitrifying bacteria in sewage sludge. Red, ammonia-oxidizing ($NH_3 \rightarrow NO_2^-$) bacteria; green, nitrite-oxidizing ($NO_2^- \rightarrow NO_3^-$) bacteria. Note the close proximity of the two metabolic types to allow for cross-feeding of NO_2^-. (b) Confocal laser scanning micrograph of a sewage sludge sample. The sample was treated with three phylogenetic probes each containing a different fluorescent dye (green, red, or purple) and each targeting a different group of Proteobacteria (one of several major lineages in the domain Bacteria, ⚭ Chapter 13). Green-, red-, or purple-stained cells reacted with only a single probe while blue- and yellow-stained cells reacted with two probes to give the different colors. The genetic diversity of this sample is obvious from the many colors, but keep in mind that the probes were only designed to react with a relatively restricted group of Proteobacteria. No Archaea (⚭ Chapter 14) or other groups of Bacteria would have shown up under these conditions. Thus the actual genetic diversity in the sample is undoubtedly far greater.

ful for direct mechanical isolation of a single cell from a mixture (such as an enrichment culture) using optical tweezers (see later in this section).

Chromosomal Painting and *In Situ* Reverse Transcription

Two other FISH technologies are **chromosomal painting** and *in situ* **reverse transcription**. The former technique

is used to identify *specific genes* in the organism of interest while the latter focuses on identifying *specific mRNAs* (that is, *expressed genes*). For chromosomal painting, an oligonucleotide probe, a single gene, set of genes, or even an entire genome digested to yield 50–200 bp probes are fluorescently labeled. The probes are used to ask which organism(s) in an environment contain the gene or genes present in the probe (Figure 16.12*a*). For example, chromosomal painting could be used to identify bacteria that contain highly conserved genes, such as those that encode nitrogenase (⚭ Section 15.29), components of the photosynthetic reaction center (⚭ Section 15.5), or specific autotrophic pathways (⚭ Sections 15.7 and 15.8), and this information then used to enumerate nitrogen-fixing bacteria, phototrophic bacteria, or autotrophic bacteria, respectively, in a natural sample.

In situ reverse transcriptase (ISRT) FISH goes one step beyond chromosomal painting and asks not only which organisms *contain* a specific gene or genes, but which organisms are actually *expressing* the genes in nature at a given time. The method involves the use of a probe that hybridizes with a specific *RNA* previously transcribed by cells in a natural sample. Once the probe reacts, reverse transcription (⚭ Section 8.22) is carried out on the sample to produce a complementary DNA (cDNA), and then this cDNA is amplified by the polymerase chain reaction (PCR); finally, this amplified DNA reacts with the fluorescent probe (Figure 16.12*b*) (⚭ RT-PCR, Section 10.9). The advantage of ISRT FISH technology over chromosomal painting is that it allows the experimenter to explore what factors control bacterial *gene expression* in nature and to observe how experimental perturbations in a sample affect the transcription of genes in specific populations of cells.

Optical Tweezers

What if the specificity of FISH technology could be used to assist in the actual isolation of microorganisms? That is, once a particular organism was identified by FISH staining we could "reach out" and capture it? Such technology exists in the form of the *optical tweezers*.

Optical tweezers consist of an inverted light microscope equipped with a strongly focused infrared laser and a micromanipulation device (Figure 16.13). Using a specific fluorescent probe (such as a phylogenetic probe) to identify the desired organism in a mixture, a single cell can be optically trapped by the laser beam and physically moved away from contaminating organisms. If the sample was originally present in a capillary tube, the single cell can be isolated by breaking the tube at a point between the desired cell and the rest of the population and then flushing the optically trapped cell into a small tube of fresh medium (Figure 16.13). The trapping properties of the optical tweezers arise from the fact that the laser beam creates radi-

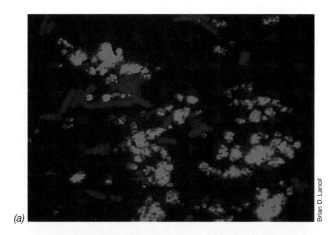

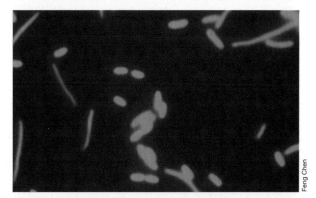

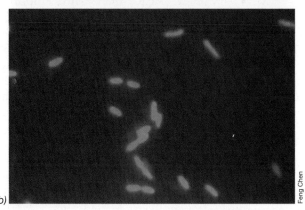

(a)

(b)

Brian D. Lanoil

Feng Chen

Feng Chen

FIGURE 16.12 Fluorescent techniques employing bacterial chromosomal painting and *in situ* reverse transcription (ISRT). (a) Chromosomal painting. Mixture of cells of *Escherichia coli* (red) and *Oceanospirillum linum* (blue). Each fluorescent dye was linked to DNA oligonucleotides formed from genomic DNA from the respective organism. (b) ISRT. The two photos show a mixed culture of the bacteria *Microbulbifer hydrolyticus* (short, fat rods) and *Sagittula stellata* (long, thin rods). Reverse transcription (⚮ Section 8.22) of a specific mRNA produced only by *M. hydrolyticus* led to production of a cDNA that was subsequently amplified by PCR and hybridized by a fluorescently labeled probe. Top photo, all cells stained with DAPI, a nonspecific stain (see Figure 16.7); bottom photo, cells stained with ISRT probe. ISRT data are useful for understanding which organism(s) in a microbial community are carring out which metabolic activities.

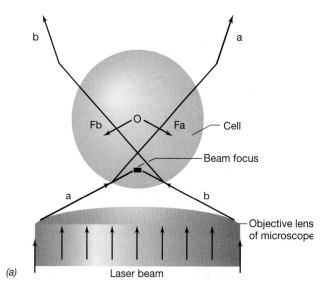

(a)

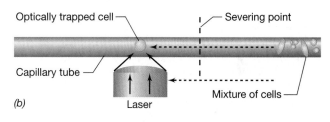

(b)

FIGURE 16.13 The optical tweezers for the isolation of bacteria. (a) Principle of the optical tweezers. A strongly focused laser beam on an object as small as a bacterial cell creates downward radiation forces (F_a, F_b) on the cell that allows the cell to be dragged in any direction as long as the beam force remains on it. (b) Isolation. The laser beam can lock onto a single cell present in a mixture in a capillary tube and drag the optically trapped cell (in the example here, the desired cell is dragged from right to left) away from the other cells. Once the desired cell is far enough away from the other cells, the capillary is severed, the laser turned off, and the cell flushed into a tube of sterile medium.

ation pressure forces that push down on small objects like bacterial cells and hold them in place; when the laser beam is moved (for example, up the capillary shown in Figure 16.13*b*), the trapped bacterial cell moves along with it.

For microbiological purposes optical tweezers offer a means for obtaining in pure culture either slow-growing bacteria that would be overgrown in typical enrichment cultures, or organisms that constitute a minority of the total population and could be missed even using dilution-based enrichment methods (see Section 16.4). Besides obtaining pure cultures of bacteria, optical tweezers have been used for many other applications as well, such as aligning egg and sperm cells for *in vitro* fertilization, laser-induced fusion of plant or mammalian cells, sorting of chromosomes, and even for the micromanipulation of individual DNA and protein molecules. The beauty of the optical tweezers, besides its ability to specifically trap a cell, is its nondestructive nature; trapped cells or molecules are not damaged and can be used in subsequent manipulations.

✓ 16.5–16.6 Concept Check

Powerful new methods based on the use of fluorescent dyes are available for identifying and tracking specific microorganisms in the environment. These include not only nonspecific stains like acridine orange and DAPI, but also highly specific stains based on the specificity of nucleic acid sequences. The latter include a variety of fluorescent *in situ* hybridization (FISH) staining procedures such as phylogenetic stains, chromosomal painting, and *in situ* reverse transcription. Besides quantification and tracking, FISH methods also allow for direct microscopic observation of the physical interactions of organisms of known phylogenetic (and in many cases also physiological) type.

✓ Why might the microbiologist for a food processor use DAPI for estimating the shelf life of a food product but a phylogenetic stain for detecting *Salmonella* in a food product?

✓ What could you conclude if by chromosomal painting you detected a specific gene in an organism from nature but by *in situ* reverse transcription of the same organism, you got no reaction?

✓ Why might the optical tweezers be of value to a microbiologist wanting to study a very slow-growing bacterium present in a natural sample?

16.7

Microbial Activity Measurements: Radioisotopes and Microelectrodes

Although the staining procedures just described are highly specific and useful for microbial identification (and in the case of the optical tweezers, for actual isolation), they do not tell us much about what the organisms in an environment are actually *doing*. To know this we must measure *microbial activities*. Two major methods have been used here, radioisotope or chemical measurements and microelectrodes.

Activity measurements are generally estimates of microbial guild processes, where the collective activities of several metabolically related populations are assessed by measuring a single transformation. Important features of these methods include *specificity* and *sensitivity*. To make sense of *in situ* activity data, the process under investigation must be one that can be specifically measured without interference from similar or competing processes and the technique used sensitive enough to detect whether the process has occurred, preferably over only very short incubation periods. We begin with a consideration of radioisotopic methods, a highly sensitive and usually highly specific measure of microbial activity.

Radioisotopes

In many situations direct chemical measurements of microbial transformations using colorimetric chemical methods or various types of analytical instrumentation is sufficient and even desirable (because no radiation is involved) for assessing microbial transformations in an environment. But when very high sensitivity is required, turnover rates need to be determined, or the fate of portions of a particular molecule need to be followed, radioisotopes are very useful and occasionally, essential. For instance, if photoautotrophy is to be measured, the light-dependent uptake of $^{14}CO_2$ into microbial cells can be measured; if sulfate reduction is of interest, the rate of conversion of $^{35}SO_4^{2-}$ to $H_2^{35}S$ can be assessed (Figure 16.14). Methanogenesis in natural environments can be studied by measuring the conversion of $^{14}CO_2$ to $^{14}CH_4$ in the presence of a suitable reductant such as H_2, or by the conversion of methyl compounds such as ^{14}C-labeled methanol or acetate (Figure 16.14*c*) to $^{14}CH_4$.

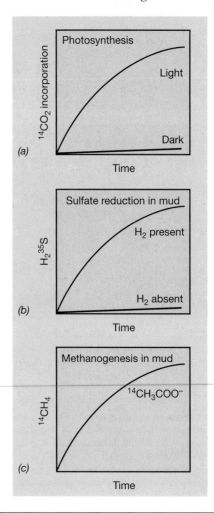

FIGURE 16.14 Use of radioisotopes to measure microbial activity in nature. (a) Photosynthesis measured in natural seawater with $^{14}CO_2$. (b) Sulfate reduction in mud measured with $^{35}SO_4^{2-}$. (c) Methanogenesis measured in mud with acetate.

Chemoorganotrophic activity can be measured by following the incorporation of ^{14}C-organic compounds into bacterial biomass or the release of $^{14}CO_2$ by the microbial population. For example, the uptake of ^{14}C-glucose or ^{14}C-amino acids by chemoorganotrophs in lake water can be studied by filtering the bacterial cells from suspension following an incubation period with the isotope. Alternatively, one might measure the extent of $^{14}CO_2$ production from the added radioisotope. In all these studies if actual *rates* of a given chemical process are to be determined, it is necessary to know the amount of *nonradioactive* as well as radioactive substrate available to the population in order for a *specific activity* of the substrate to be determined; from this, total rates of conversion can be calculated.

Isotope methods are widely used to evaluate the activity of microorganisms in nature. However, because there is always the possibility that some transformation of a labeled compound might be due to a strictly chemical, rather than a microbial, process it is essential when using isotopes to employ proper controls. The key control necessary is the *killed cell control*. It is absolutely essential to show that the transformation being measured in nature is prevented by microbicidal agents or heat treatments known to block microbial action or kill the organisms. Formalin at a final concentration of 4% is frequently used as a chemical sterilant in microbial ecology studies.

Microelectrodes

We discussed earlier the fact that the environments of microorganisms are very small; that is, they are *microenvironments*. Microbial ecologists have used small glass electrodes, referred to as **microelectrodes,** to study the activity of microorganisms in their microenvironments (we discussed the use of O_2 microelectrodes in Section 16.2) (see Figure 16.2). Although limited to measurements of chemical species that can be detected with an electrode, several types of electrodes have been used in field studies. Currently, the most popular microelectrodes are those that measure pH, oxygen, N_2O, CO_2, H_2, or H_2S.

As the name implies, microelectrodes are very small, the tips of the electrode ranging in diameter from 2 to 100 μm; O_2 microelectrodes 2–3 μm in diameter can be routinely made (Figure 16.15*a*). The electrodes are carefully inserted into the habitat using a micromanipulator, a device that allows for precise movement of the electrode through distances of a millimeter or less (Figure 16.15*b*). Microelectrodes have been used extensively in the study of chemical transformations and photosynthesis in *microbial mats*. The latter are layered microbial communities usually containing cyanobacteria in the uppermost layer, anoxygenic phototrophic bacteria in subsequent layers (until the mat becomes light-limited), and chemoorganotrophic, especially sulfate-reducing, bacteria in the lower

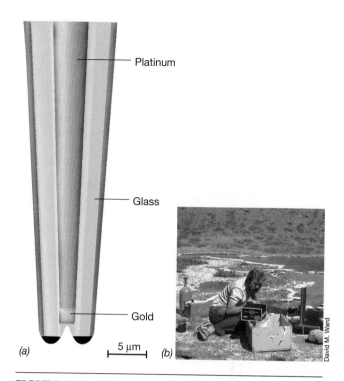

FIGURE 16.15   Microelectrodes. (a) Schematic drawing of an oxygen microelectrode. Note the scale of the electrode. (b) Photo of microelectrodes being used in a microbial mat (see Figure 16.16).

layers (Figure 16.16). Microbial mats are found in a variety of environments, especially in hot springs (Figure 16.16*a*) and shallow marine basins and are dynamic systems where photosynthesis occurring in the upper layers of the mat is balanced by decomposition from below.

With the use of O_2 microelectrodes, oxygen concentrations in microbial mats (Figure 16.16*b*) or soil particles (Figure 16.2) can be sensitively measured over extremely fine intervals. With the use of a micromanipulator, electrodes can be immersed through a microbial habitat and readings taken every 0.05–0.1 mm (50–100 μm) (Figures 16.15*b* and 16.16). With a bank of microelectrodes, each sensitive to a different chemical, simultaneous measurements of a number of microbial transformations can be made at one time (Figure 16.16*b*). Oxygen and sulfide measurements are often taken together because gradients of both chemical species form in many microbial environments as a result of photosynthesis and sulfate reduction, respectively (Figure 16.16*b*). Note also in Fig. 16.16*b* how the H_2S concentration diminishes near the oxic zone. This is a result of H_2S oxidation by phototrophic bacteria and chemolithotrophic sulfur bacteria. Thus, the concentration of H_2S at any point in the mat is a function of both H_2S production and H_2S consumption, complementary metabolisms (see Section 16.2), and an H_2S measurement is the net result of both activities at a given location in the mat.

(a)

David M. Ward

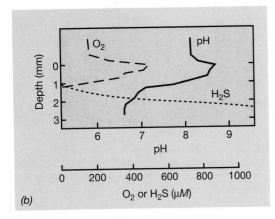

(b)

FIGURE 16.16 Microbial mats and the use of microelectrodes to study them. (a) Photograph of a core taken through the kind of hot spring microbial mat used in the experiment shown in part (b). Upper layer (dark green) contains cyanobacteria, beneath which are several layers of anoxygenic phototrophic bacteria (orange and yellow). The whole thickness of the mat is about 2 cm. (b) Oxygen, sulfide, and pH microprofiles in a hot spring microbial mat. Note the millimeter scale on the ordinate.

✓ 16.7 Concept Check

The activity of microorganisms in natural samples can be assessed very sensitively using radioisotopes or microelectrodes. In most cases these measurements are of the net activity of a microbial guild or guilds rather than of a population of a single species.

✓ Why are radioisotopes so useful in measuring microbial activities?

✓ What is a *microelectrode*?

16.8

Microbial Activity Measurements: Stable Isotopes

For many elements different isotopes exist, differing in the number of neutrons present. Certain isotopes are unstable and break down (as a result of radioactive

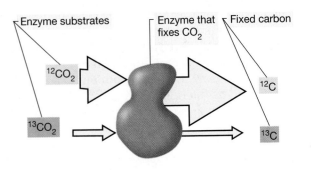

FIGURE 16.17 Mechanism of isotopic fractionation using carbon as an example. Although the ratio of natural abundance of $^{12}CO_2$ to $^{13}CO_2$ is about 19:1, enzymes that fix CO_2 preferentially fix the lighter isotope (^{12}C). This results in fixed carbon being enriched in ^{12}C and depleted in ^{13}C relative to the starting substrate. The degree of ^{13}C depletion is calculated as an isotopic fractionation (see legend to Figure 16.18 for calculation).

decay). Others, known as *stable isotopes*, can be used to study various microbial transformations.

The two chemical elements that have proven most useful for stable isotope studies in microbial ecology are carbon and sulfur. Carbon exists in nature primarily as ^{12}C. However, a small amount of carbon is found as ^{13}C. Likewise, sulfur exists primarily as ^{32}S, although some sulfur is found as ^{34}S. All these are stable isotopes. The reason stable isotope measurements are useful in microbial ecology is that most biochemical reactions involving carbon or sulfur tend to favor the *lighter* isotope. That is, the heavier isotope is discriminated against when the elements are acted on biochemically (Figure 16.17). Thus, when CO_2 is fed to a phototrophic organism, the cellular carbon becomes enriched in ^{12}C whereas the CO_2 remaining in the medium becomes enriched in ^{13}C. Likewise, sulfide produced from the bacterial reduction of sulfate is much "lighter" than sulfide of strictly geochemical origin. This phenomenon is known as *isotopic fractionation*.

Use of Isotopic Fractionation in Microbial Ecology

How can differences in isotopic composition be used to assess microbial activities? The isotopic composition of a sample contains a record of its past biological activity (Figures 16.18 and 16.19). In the case of carbon, it is easy to see that plant material and petroleum (which is derived from plant material) have similar isotopic compositions; carbon from both sources is isotopically lighter than the standard because it was fixed by a pathway that discriminated against $^{13}CO_2$ (Figure 16.18). Methane of biological origin can be extremely light, indicating that CO_2-reducing methanogens dramatically discriminate between isotopic forms of CO_2 (⟳ Section 15.19 discusses the biochemistry of methanogene-

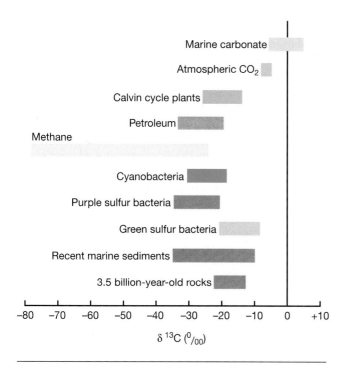

FIGURE 16.18   Carbon isotopic compositions of various substances. The values are given in parts per thousand (‰) and were calculated using the formula

$$\frac{(^{13}C/^{12}C \text{ sample}) - (^{13}C/^{12}C \text{ standard})}{(^{13}C/^{12}C \text{ standard})} \times 1000$$

The standard is a belemnite sample from the PeeDee rock formation.

sis). Marine carbonates, on the other hand, are clearly of geological origin because they are isotopically much "heavier" than biogenic carbon (Figure 16.18). This large carbon reservoir has obviously not yet passed through phototrophic organisms. Because of the differences in the proportion of ^{12}C and ^{13}C in carbon of biological and geological origin, the $^{13}C/^{12}C$ isotopic ratio of various geological strata has been used to detect the onset of

living (in this case, autotrophic) processes in ancient rocks. Interestingly, organic carbon in rocks as old as 3.5 billion years shows some isotopic lightness (Figure 16.18), suggesting that autotrophy might have evolved very early in the diversification of living organisms.

The key role of sulfate-reducing bacteria in the formation of sulfur deposits is known from studies on the fractionation of $^{34}S/^{32}S$ (Figure 16.19). As compared to a meteoritic sulfur standard, sedimentary sulfide is highly enriched in ^{32}S. Nonbiogenic sulfide (as, for instance, in igneous rocks from volcanic deposits) does not show this bias toward the lighter isotope (Figure 16.19). Also, biological oxidation of sulfide to sulfur or sulfate, either aerobically or anaerobically, shows a preference for the lighter isotope, although the fractionation is not nearly as great as that occurring during sulfate reduction.

Isotope fractionation studies thus have several uses in microbial ecology. Sulfur isotopes have been used to distinguish between biogenic and abiogenic ores (iron sulfides) and elemental sulfur deposits. Knowledge of the fractionations observed in sulfate reduction and subsequent sulfide oxidation by phototrophic or chemolithotrophic bacteria allows monitoring of important links in the sulfur cycle. As previously mentioned, carbon isotopic analyses have been used to distinguish biogenic from abiogenic organic matter, and oxygen analyses (using $^{18}O/^{16}O$ measurements of rocks of various ages) have been used to trace Earth's transition from an anoxic to an oxic environment (Earth's molecular oxygen originated from oxygenic photosynthesis by cyanobacteria) (∞ Section 12.1).

Stable isotope analyses have also been used as evidence for the lack of living processes on the moon. For example, the data in Figure 16.19 show that the sulfide isotopic composition of lunar rocks closely approximates that of meteoritic sulfide and not that typical of biogenic sulfide. Stable isotope analyses can also be used to track microbial and higher organism food chains. Because the isotopic composition of a given animal should approximate the isotopic composition of

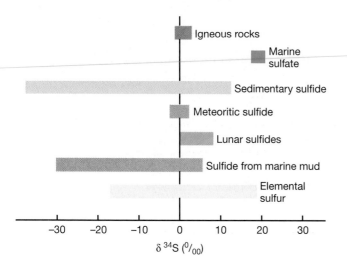

FIGURE 16.19   Summary of the isotope geochemistry of sulfur, indicating the range of values for ^{34}S and ^{32}S in various sulfur-containing substances. The values are given in parts per thousand (‰) and were calculated using the formula

$$\frac{(^{34}S/^{32}S \text{ sample}) - (^{34}S/^{32}S \text{ standard})}{(^{34}S/^{32}S \text{ standard})} \times 1000$$

The standard is an iron sulfide mineral from the Canyon Diablo meteorite. Note that sulfide and sulfur of biogenic origin tend to be depleted in ^{34}S (enriched in ^{32}S).

its major food source, it has been possible to trace the flow of carbon through an ecosystem in ways previously not possible.

✓ 16.8 Concept Check

Isotopic fractionation can yield information on the biological origin of various substances. Fractionation is a result of the activity of certain enzymes that can discriminate, usually against the heavier form of an element, when binding their substrates.

- ✓ How can the $^{12}C/^{13}C$ composition of a substance tell us anything about its possible biological origin?
- ✓ Why are lunar sulfides isotopically heavy?

16.9

Terrestrial Environments

In the consideration of terrestrial environments, our attention inevitably turns to *soil* and *plants* because it is within the soil and on or near plants that many of the key processes occur that influence the functioning of the ecosystem. The process of soil development involves complex interactions among the parent material (rock, sand, glacial drift, and so on), topography, climate, and living organisms. Soils can be divided into two broad groups—**mineral soils** and **organic soils**—depending on whether they derive initially from the weathering of rock and other inorganic material or from sedimentation in bogs and marshes, respectively. Our discussion will concentrate on mineral soils, the predominant soil in most areas.

Soil Formation

Soils form as a result of combined physical, chemical, and biological processes. An examination of almost any exposed rock reveals the presence of algae, lichens, or mosses. These organisms are able to remain dormant on the dry rock and then grow when moisture is present. They are phototrophic and produce organic matter, which supports the growth of chemoorganotrophic bacteria and fungi. The numbers of chemoorganotrophs increase directly with the degree of plant cover of the rocks. Carbon dioxide produced during respiration by chemoorganotrophs is converted to carbonic acid ($CO_2 + H_2O \rightleftharpoons H_2CO_3$), which is an important agent in the dissolution of rocks, especially those composed of limestone. Many chemoorganotrophs also excrete organic acids, which further promote the dissolution of rock into smaller particles. Freezing and thawing and other physical processes lead to the development of cracks in the rocks. In these crevices, a raw soil forms in which pioneering higher plants can develop. The plant roots penetrate farther into crevices and increase the frag-

mentation of the rock, and their excretions promote the development of a **rhizosphere** (the soil that surrounds plant roots) microflora. When the plants die, their remains are added to the soil and become nutrients for an even more extensive microbial development. Minerals are further rendered soluble, and as water percolates, it carries some of these chemical substances deeper. As weathering proceeds, the soil increases in depth, thus permitting the development of larger plants and trees. Soil animals become established and play an important role in keeping the upper layers of the soil mixed and aerated. Eventually the movement of materials downward results in the formation of layers, and a typical *soil profile* becomes evident (Figure 16.20). The

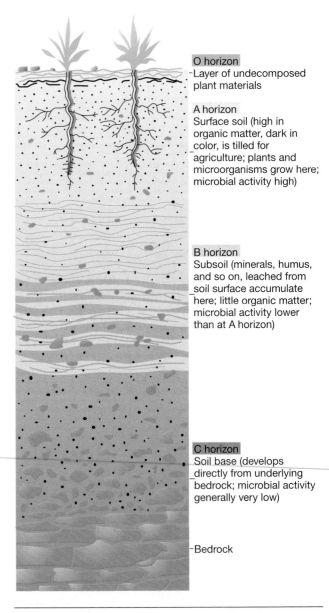

O horizon
Layer of undecomposed plant materials

A horizon
Surface soil (high in organic matter, dark in color, is tilled for agriculture; plants and microorganisms grow here; microbial activity high)

B horizon
Subsoil (minerals, humus, and so on, leached from soil surface accumulate here; little organic matter; microbial activity lower than at A horizon)

C horizon
Soil base (develops directly from underlying bedrock; microbial activity generally very low)

Bedrock

FIGURE 16.20   Profile of a mature soil. The soil horizons are soil zones as defined by soil scientists.

rate of development of a typical soil profile depends on climatic and other factors, but it is usually very slow, taking hundreds of years.

Soil as a Microbial Habitat

The most extensive microbial growth takes place on the *surfaces* of soil particles, usually within the rhizosphere (see Figures 16.2, 16.3*b*, and 16.9). As pointed out in Section 16.2, even a small soil aggregate can have many differing microenvironments (compare Figures 16.2 and 16.21), and thus several different types of microorganisms may be present. To examine soil particles directly for microorganisms, fluorescence microscopes are often used, the organisms in the soil being stained with a dye that fluoresces (see Figure 16.3*b*). To observe a *specific* microorganism in a soil particle, **fluorescent-antibody staining** (see Figure 16.9) or phylogenetic stains (see Section 16.6) can be used. Microorganisms can also be seen on such opaque surfaces as soil by means of the **scanning electron microscope** (Figure 16.22). The scanning electron microscope gives excellent information on the morphology of soil bacteria and can also be used to enumerate cells on soil particle surfaces.

One of the major factors affecting microbial activity in soil is the availability of *water,* and we have previously discussed the importance of water availability to microbial growth (∞ Section 5.11). Water is a highly variable component of soil, its presence depending on soil composition, rainfall, drainage, and plant cover. Water is held in the soil in two ways, by adsorption onto surfaces or as free water existing in thin sheets or films between soil particles. The water present in soils has a variety of materials dissolved in it, the whole mixture being referred to as the *soil solution.* In well-drained soils, air penetrates readily and oxygen concentrations can be high. In waterlogged soils, however, the only oxygen present is that dissolved in the water, and this is soon consumed by microorganisms. Such soils quickly become anoxic, showing profound changes in their

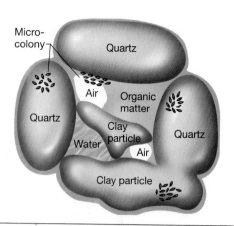

FIGURE 16.21   A soil aggregate composed of mineral and organic components, showing the localization of soil microorganisms. Very few microorganisms are found free in the soil solution; most of them occur as microcolonies attached to the soil particles.

biological properties. We discussed oxygen relationships in soil in Section 16.2 and Figure 16.2.

The *nutrient status* (resources) of a soil is the other major factor affecting microbial activity. The greatest microbial activity is in the organic-rich surface layers, especially in and around the rhizosphere. The numbers and activity of soil microorganisms depend to a great extent on the balance of nutrients present. In some soils carbon is not the limiting nutrient, but instead the availability of *inorganic* nutrients such as phosphorus and nitrogen limit microbial productivity.

Deep Subsurface Microbiology

Interest in the chemistry of groundwater and the potential leaching of pollutants and their transfer in groundwater aquifers had led to studies of the role microorganisms play in the *deep subsurface* terrestrial environment. The deep soil subsurface, which can extend for several hundred meters below the soil surface, is not a biological wasteland. A variety of microorganisms,

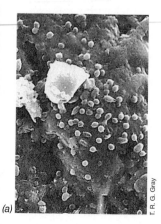
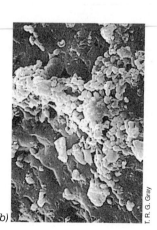
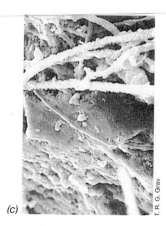

(a)   *(b)*   *(c)*

FIGURE 16.22   Visualization of microorganisms on the surface of soil particles by use of the scanning electron microscope. (a) Rod-shaped bacteria. (b) Actinomycete spores. The cells in (a) and the spores in (b) are about 1–2 μm wide. (c) Fungus hyphae. The fungal hyphae are about 4 μm wide.

A FOCUS ON . . .   Microbial Life Deep Underground

Microbiologists studying the deep terrestrial subsurface have found viable prokaryotes at depths of several thousand meters below the surface, existing in different physical and chemical environments. How are these microorganisms making a living? Initial findings indicated that these buried microorganisms were very slow growing chemoorganotrophic bacteria surviving by the slow catabolism of organic carbon deposited within the sediments. However, studies on the microbial ecology of deep basalt aquifers[a] have shown that sluggish chemoorganotrophs are not the only prokaryotes that live deep underground.

Basalts are iron-rich volcanic rocks that are essentially devoid of organic matter. In certain basalts up to 1500 m deep from the Columbia River Basin (Washington, USA), large numbers of anaerobic, *chemolithotrophic* bacteria have been found (Fig. 1*a*), including sulfate-reducing bacteria, methanogens, and homoacetogens.[a] These organisms were shown to be metabolically active by carbon-stable isotope analyses of methane (CH_4) present in the rocks and surrounding groundwaters; measurements showed a strong enrichment in the

lighter isotope of carbon (^{12}C), typical of biological methanogenesis (see Section 16.8).

A common metabolic link among these anaerobes is a thirst for H_2, an excellent electron donor for their various energy-yielding metabolisms [∞ Sections 15.17–15.19 and Fig. 1*b* below]. Hydrogen is a common product of the anoxic decomposition of organic matter (∞ Sections 15.21–15.23). But if basalts contain very little organic material, where does the H_2 come from to support metabolism of the H_2 consumers? Interestingly, H_2 in the Columbia basalts apparently originates from the *strictly chemical* interaction of water with iron minerals in the rocks [see proposed reaction in Fig. 1*b*]. Such reactions are known from inorganic chemistry, and in laboratory studies in which crushed Columbia basalt was mixed with sterile water under anoxic conditions, rapid H_2 evolution occurred.[a] H_2 was also detected *in situ* in groundwater percolating through the basalts.

From analyses of these experimental results it was hypothesized that H_2 formed in deep underground basalts is indeed the electron donor that supports growth of the substantial populations of anaerobic prokaryotes found there.

If this is true, these organisms would be living a strictly *geochemical* existence because their electron acceptor (CO_2 in the case of methanogens and homoacetogens and SO_4^{2-} in the case of sulfate-reducing bacteria) and electron donor, H_2, are all derived from inorganic materials. These buried chemolithotrophs would also be novel because of their total independence of photosynthetic primary production, which generates the molecular oxygen and/or organic matter required by virtually all surface-dwelling organisms. However, as sometimes happens in research, other findings have questioned this interpretation. Although there is not much organic matter in these basalts, there is some, and thus fermentation could be the source of some (or even most) of the H_2 used by the various subsurface H_2 chemolithotrophs.[b]

How widespread and ecologically significant H_2-based microbial ecosystems deep underground are, regardless of the source of H_2, obviously awaits further research. But it is clear that metabolically active H_2-consuming microorganisms inhabit the deep subsurface, and this in turn suggests that they may be important biogeochemical agents in these environments. ■

[a]Stevens, T. O., and J. P. McKinley. 1995. Lithoautotrophic microbial ecosystems in deep basalt aquifers. *Science 270*:450–454.

[b]Anderson, R. T., F. H. Chapelle, and D. R. Lovely. 1998. Evidence against hydrogen-based microbial ecosystems in basalt aquifers. *Science 281*:976–977.

(a)

Todd O. Stevens

FIG. 1   (a) Laser confocal photomicrograph of a microbial film attached to the surface of basalt chips. Green is reflected light from the basalt surface while the red color is from Nile-red stained bacterial cells. The cells in the film were grown on H_2 from basalt as depicted in the bottom reaction of Fig. 1*b*. (b) Key metabolic reactions of anaerobic prokaryotes growing in anoxic deep basalt aquifers.

Methanogenesis: $4 H_2 + CO_2 \longrightarrow CH_4 + 2 H_2O$

Acetogenesis: $4 H_2 + 2 HCO_3^- + H^+ \longrightarrow CH_3COO^- + 4 H_2O$

Sulfate reduction: $4 H_2 + SO_4^{2-} + H^+ \longrightarrow HS^- + 4 H_2O$

Proposed inorganic
H_2 production: $FeO + H_2O \longrightarrow H_2 + FeO_{3/2}$

(b)

primarily bacteria, are present in most deep underground soils. In samples collected aseptically from bore holes drilled down to 300 m, a diverse array of bacteria have been found including anaerobes such as sulfate-reducing bacteria, methanogens, and homoacetogens, and various aerobes and facultative aerobes. Microorganisms in the deep subsurface presumably have access to nutrients because groundwater flows through their habitats, but activity measurements suggest that metabolic rates of these bacteria are rather low in their natural habitats (however, see the box, Microbial Life Deep Underground). Compared to microorganisms in the upper layers of soil, the biogeochemical significance of deep subsurface microorganisms may thus be minimal. However, there is evidence that the metabolic activities of these buried microorganisms may over very long periods be responsible for some mineralization of organic compounds and release of products into the groundwater. The potential for *in situ* bioremediation (see Section 16.22) of toxic substances leached from soil into groundwater (for example, benzenes and agricultural chemicals) by deep subsurface microorganisms is of particular current interest.

✓ 16.9 Concept Check

The soil is a complex habitat with numerous microenvironments and niches. Microorganisms are present in the soil primarily attached to soil particles. The most important factor influencing microbial activity in surface soil is the availability of water, whereas in deep soil (the subsurface environment) nutrient availability plays a major role.

✓ What is the difference between *mineral* soils and *organic* soils?

✓ What factors govern the extent and type of microbial activity in soils?

16.10

Aquatic Habitats

Typical aquatic environments are the oceans, estuaries, salt marshes, lakes, ponds, rivers, and springs. Aquatic environments differ considerably in chemical and physical properties, and it is not surprising that their microbial species compositions also differ. The predominant phototrophic organisms in most aquatic environments are microorganisms; in oxic areas cyanobacteria and algae prevail, and in anoxic areas anoxygenic phototrophic bacteria are preponderant. Algae floating or suspended freely in the water are called **phytoplankton;** those attached to the bottom or sides are called **benthic algae.** Because these phototrophic organisms use energy from light in the initial production of organic matter, they are called **primary producers.** In the final analysis, the biological activity of

an aquatic ecosystem is dependent on the rate of primary production by the phototrophic organisms.

Marine Environments

In the oceans primary productivity is rather low because inorganic nutrients, especially nitrogen and iron, typically limit phytoplankton growth. This is especially true of iron, where experiments to seed open ocean waters with iron have been shown to trigger large increases in phototrophic CO_2 fixation. A consequence of the relatively low primary productivity in the oceans is relatively low heterotrophic microbial activities. This in turn ensures that most ocean waters remain highly oxic, even waters at great depth.

Much of the primary productivity that does occur in the open oceans is due to the photosynthetic activities of prochlorophytes, tiny oxygenic prokaryotes that contain chlorophylls *a* and *b*; the organism *Prochlorococcus* is particularly important in this regard (Section 13.25). In tropical and subtropical oceans the planktonic filamentous marine cyanobacterium *Trichodesmium* (Figure 16.23) is also widespread. Cells of *Trichodesmium* form tufts of filaments that make up a large fraction of the biomass suspended in these waters. *Trichodesmium* is a nitrogen-fixing cyanobacterium, and the production of fixed nitrogen by this organism is thought to be a major link in the nitrogen cycle in marine environments.

Inshore ocean areas are typically more nutritionally fertile and therefore support more dense populations of phytoplankton (Figure 16.24). This in turn supports higher standing crops of bacteria and aquatic animals, such as fish and shellfish. Marine bays and inlets receiving high levels of nutrients from sewage or industrial wastes can have very high phytoplankton and bacterial populations. If the pollution is severe enough, shallow marine waters can become anoxic from the con-

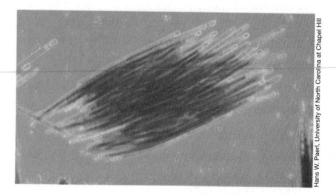

FIGURE 16.23 Photomicrograph of cells of the marine nitrogen-fixing cyanobacterium, *Trichodesmium*. Although not heterocystous (Section 13.24), this organism forms tufts of cells that actively fix N_2 in tropical marine waters. The cells shown here are a natural aggregate collected from the Gulf Stream off of North Carolina.

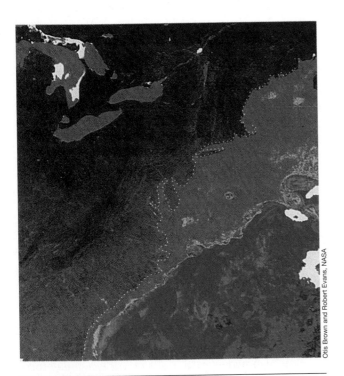

Otis Brown and Robert Evans, NASA

FIGURE 16.24 Distribution of chlorophyll in the western North Atlantic Ocean as recorded by satellite. The east coast of the United States from mid-Florida to northern Maine is shown in dotted outline. Near the center of the photograph is Chesapeake Bay; the Great Lakes are at the upper left. Areas rich in phytoplankton are shown in red (>1 mg chlorophyll/m^3); blue and purple areas have lower chlorophyll concentrations (<0.01 mg/m^3). Note the high primary productivity of coastal areas and the Great Lakes.

sumption of O_2 by heterotrophic bacteria and toxic to marine life from the production of H_2S by sulfate-reducing bacteria that can quickly develop in anoxic seawater (⚮⚮ Sections 13.17 and 16.17).

Oxygen Relationships in Lakes and Rivers

We discussed oxygen requirements and anaerobiosis in Section 5.12, the production of oxygen via photosynthesis in Section 15.6, and oxygen in microenvironments in Section 16.2. Although oxygen is one of the most plentiful gases in the atmosphere (∽21% of air), it has limited solubility in water, and in a large water mass its exchange with the atmosphere is slow. Significant photosynthetic production of oxygen occurs only in the *surface layers* of a lake or ocean, where light is available. Organic matter that is not consumed in these surface layers sinks to the depths and is decomposed by facultative microorganisms, using oxygen dissolved in the water. In lakes, once the oxygen is consumed, the deep layers become anoxic; here strictly aerobic organisms such as higher plants and animals cannot grow, and the bottom layers have a species composition restricted to anaerobic bacteria and a few kinds of microaerophilic

animals. In addition, there is a conversion from a respiratory to a fermentative metabolism, with important consequences for the carbon cycle and other nutrient cycles (see Sections 16.9–16.18 for a consideration of this).

Whether or not a body of water becomes depleted of oxygen depends on several factors. If organic matter is sparse, as it is in unproductive lakes or in the open ocean, there may be insufficient substrate available for chemoorganotrophs to consume all the oxygen. Also important is how rapidly the water from the depths exchanges with surface water. Where strong currents or turbulence occurs, the water mass may be well mixed, and consequently oxygen may be transferred to the deeper layers. However, in many lakes in temperate climates, the water mass becomes *stratified* during the summer, with the warmer and less dense surface layers, called the *epilimnion*, separated from the colder and denser bottom layers (the *hypolimnion*) (Figure 16.25). After stratification sets in, usually in early summer, the bottom layers become anoxic (Figure 16.25). In the late fall and early winter, the surface waters become colder and thus more dense than the bottom layers, and the water "turns over," leading to a reaeration of the bottom. Most lakes in temperate climates show this annual cycle in which the bottom layers of water pass from oxic to anoxic and back to oxic.

Rivers

The oxygen relations in a river are of particular interest, especially in regions where rivers receive much organic matter in the form of sewage and industrial pollution. Even though a river may be well mixed because of rapid water flow and turbulence, large amounts of added organic matter can lead to a marked oxygen deficit. This is illustrated in Figure 16.26. As the water moves away from a sewage outfall, organic matter is gradually consumed and the oxygen content returns to normal. Oxygen depletion in a body of water is undesirable because aquatic animals require O_2 and die under even very temporary anoxic conditions. Further, conversion to anoxia results in the production by anaerobic bacteria of odoriferous compounds (for example, amines, H_2S, mercaptans, fatty acids), some of which are also toxic to higher organisms.

Biochemical Oxygen Demand

Sanitary engineers term the oxygen-consuming property of a body of water its **biochemical oxygen demand** (BOD). The BOD is determined by taking a sample of water, aerating it well, placing it in a sealed bottle, incubating for a standard period of time (usually 5 days at 20°C), and determining the residual oxygen in the water at the end of incubation. A BOD determination thus gives a measure of the amount of organic material in the water that could be oxidized

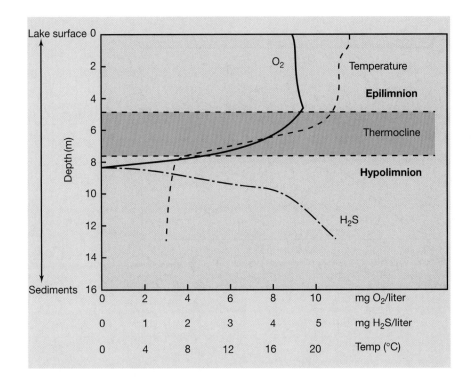

FIGURE 16.25 Development of anoxic conditions in the depths of a temperate climate lake as a result of summer stratification. The colder bottom waters are more dense and contain H_2S from bacterial sulfate reduction. The zone of rapid temperature change is referred to as the thermocline. Typically, as surface waters cool in the fall and early winter they reach the temperature and density of hypolimnetic waters and sink, displacing bottom waters and effecting "lake turnover".

by microorganisms. As a river recovers from contamination with an organic pollutant, the drop in BOD is accompanied by a corresponding increase in dissolved oxygen (Figure 16.26).

We thus see that in a water body, the oxygen and carbon cycles are interrelated, with the concentrations of organic carbon and oxygen in general being inversely related. This is particularly evident in *anoxic* environments, which are frequently rich in organic carbon (see Sections 16.13–16.15).

✓ 16.10 Concept Check

In aquatic ecosystems phototrophic microorganisms are usually the main primary producers. Most of the organic matter produced is consumed by bacteria, which can lead to depletion of oxygen in the environment. BOD is a measure of the oxygen-consuming properties of a water sample.

- ✓ What is a *primary producer*?
- ✓ In a lake, where is the *epilimnion* and where is the *hypolimnion*?
- ✓ Will addition of organic matter to a water sample increase or decrease its BOD?

16.11

Deep-Sea Microbiology

The oceans cover over three-fourths of Earth's surface, and marine microbiology is consequently an important area of study. However, because open ocean waters are

relatively poor in nutrients (see Section 16.10), microbial activity in these areas is generally not extensive. Nevertheless, marine microorganisms are of interest for several reasons, including their ability to grow at low-nutrient concentrations and cold temperatures and, in microorganisms inhabiting the deep sea, to withstand enormous hydrostatic pressures. We focus in this section on microbial life in the deep sea.

Definitions and Sampling

What is the deep sea? Visible light penetrates no further than about 300 m in open ocean waters; this upper region is referred to as the **photic zone.** Beneath the photic zone, down to a depth of about 1000 m, considerable biological activity still occurs as a result of the action of animals and chemoorganotrophic microorganisms. Water at depths greater than 1000 m is, by comparison, relatively biologically inactive and has come to be known as the "deep sea." Greater than 75% of all ocean water is in the deep sea, primarily at depths between 1000 and 6000 m.

Organisms that inhabit the deep sea are faced with three major environmental extremes: low temperature, high pressure, and low nutrient levels. Below depths of about 100 m ocean water stays a constant 2–3°C. We discussed the responses of microorganisms to changes in temperature in Section 5.7. As would be expected, bacteria isolated from depths below 100 m are *psychrophilic.* Some are *extreme* psychrophiles, growing only in a narrow range near the *in situ* temperature

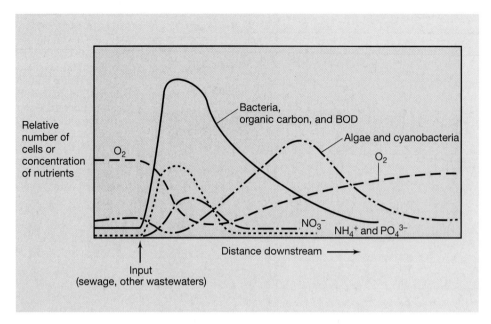

(a)

(b)

FIGURE 16.26   Effect of input of sewage or other organic-rich wastewaters into aquatic systems. (a) In a river, an increase in heterotrophic bacterial numbers and a decrease in O_2 levels occur immediately upon a spike of organic matter. If NH_4^+ is present in the input, for example, from sewage, NH_4^+ is oxidized to NO_3^- by nitrifying bacteria (Sections 13.2, 15.13, and 16.16). The rise in numbers of algae and cyanobacteria is primarily a response to inorganic nutrients, especially PO_4^{3-}. (b) A eutrophic (nutrient-rich) lake, Lake Mendota, Madison, Wisconsin, showing algae, cyanobacteria, and macrophytes (aquatic weeds) that develop in response to pollution by inorganic nutrients, much of which result from agricultural runoff into the lake watershed.

(see later). Deep-sea microorganisms must also be able to withstand the enormous hydrostatic pressures associated with great depths. Pressure increases by *1 atm* for every *10 m* depth. Thus, an organism growing at a depth of 5000 m must be able to withstand pressures of 500 atm.

Deep-sea waters and sediments can be sampled in various ways, but the most interactive is by using piloted submersibles, such as the famous *Alvin*, that operates out of the Woods Hole Oceanographic Institution (Massachusetts, USA), or for very deep dives (greater than 5000 meters), by remote controlled submersibles such as the Japanese *Kaiko*, capable of sampling even the greatest depths of the oceans (Figure 16.27).

Barotolerant and Barophilic Bacteria

Do deep-sea bacteria simply *tolerate* high pressure (that is, are they *barotolerant*) or are they actually *dependent* on pressure (that is, are they *barophilic*)? Studies of various deep-sea bacteria have shown that both patterns exist and that the distribution of barotolerant and barophilic bacteria is basically a function of depth. Organisms isolated from depths down to about 3000 m and tested for growth or metabolic activity as a function of pressure are **barotolerant;** higher metabolic rates are observed at 1 atm than at 300 atm, although growth rates at the two pressures are about the same (Figure 16.28). However, barotolerant isolates do not grow at pressures above 500 atm. By contrast, cultures derived from samples taken at greater depths, 4000–6000 m, are often **barophilic,** growing optimally at pressures of about 400 atm (Figure 16.28). Note that although barophiles grow best under pressure, they retain the ability to grow at 1 atm (Figure 16.28).

FIGURE 16.27   The sampling arm of the submersible *Kaiko* inserting a sampling tube into sediment on the sea floor of the Mariana Trench (off the Philippines, Pacific Ocean) at a depth of 10,897 meters, and collecting a sample. The tubes of sediment are then retrieved and used for enrichment and isolation of barophilic bacteria.

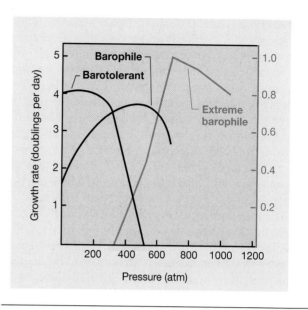

FIGURE 16.28   Growth of barotolerant, barophilic, and extremely barophilic bacteria. The extreme barophile was isolated from the Mariana Trench (see Figure 16.27). Note the much slower growth rate (at any pressure) of the extreme barophile (right ordinate) as compared to the barotolerant and barophilic bacteria (left ordinate). Note also the inability of the extreme barophile to grow at low pressures.

Samples from even deeper water (10,000 m) have yielded **extreme (obligate) barophiles.** One strain studied in detail grew fastest at a pressure of 700–800 atm and grew nearly as well at 1035 atm, the pressure it was experiencing in its natural habitat (see Figure 16.28). The unique aspect of this extremely barophilic isolate was that it not only *tolerated* pressure but actually *required* pressure for growth; the isolate would not grow at pressures of less than about 400 atm. Interestingly, this extreme barophile was not killed by decompression because it could tolerate moderate periods of decompression; however, viability was lost when the culture was left for several hours in a decompressed state.

Barotolerant and barophilic bacteria are also cold-loving, that is, psychrophilic. This property appears to be more prevalent among extremely barophilic isolates. The extreme barophile described in Figure 16.28 was found to be sensitive to temperature; the optimal growth temperature was determined to be the environmental temperature of 2°C, and temperatures above 10°C significantly reduced viability.

Molecular Effects of High Pressure

Pressure is known to affect cellular physiology and biochemistry. It has been established that increased pressure decreases the binding capacity of enzymes for their substrates. Thus, the enzymes of extreme barophiles must be folded in such a way as to minimize these pressure-related effects. Other potential pressure-sensitive targets include protein synthesis and membrane phenomena such as transport. An organism grown under high pressure has an increase in the proportion of unsaturated fatty acids in its cytoplasmic membrane. This change is presumably of adaptive significance because it makes the membrane less likely to gel at high pressures. The rather slow growth rates of extreme barophiles (see Figure 16.28) are probably due to a combination of pressure effects on cellular biochemistry and to the fact that these organisms grow only at low temperatures, where reaction rates are decreased considerably to begin with.

The use of molecular genetic tools have given new insight into the physiology of barophilism. In barophiles capable of growth up to 500–600 atm, it has been shown that growth at high pressure is accompanied by changes in the protein composition of the cell wall outer membrane. In one barophile studied in detail, a specific outer membrane protein called the OmpH protein (*outer membrane protein* H) is synthesized in cells grown at high pressures but not in cells grown at 1 atm pressure. The OmpH protein is a type of *porin*. Porins are structural proteins that form channels for the diffusion of organic molecules through the outer membrane and into the periplasm (∞ Section 3.8 and Figure 3.37). Presumably, the porin present in cells of the barophile grown at low pressures cannot function properly at high pressure, and thus a new type of porin molecule must be synthesized.

The *ompH* gene (which encodes the OmpH protein) from this barophile has been cloned into *Escherichia coli*. Studies of gene expression of the *ompH* gene in this experimental system have shown that pressure affects its transcription; messenger RNA encoding *ompH* was found to be expressed in cells of *E. coli* grown at 200 atm but not in cells grown at 1 atm (*E. coli* can grow slowly at high pressure). Further, sequencing of the *ompH* gene has shown that the OmpH protein is related but clearly distinct from the porin produced at 1 atm.

Studies of the *ompH* system thus show that pressure can affect gene expression in barophilic bacteria. How this happens is unclear but could involve the activities of pressure-sensitive repressor proteins or pressure-dependent gene activators. However, it appears that relatively few proteins are controlled by pressure in barophiles because many proteins seem to be the same in cells grown at both high and low pressure; cell wall and related structural proteins and transport proteins seem to be the major variable components. Thus, pressure acts selectively to turn on or off the transcription of specific genes encoding proteins needed for growth at high pressure. Thus, control of gene expression in barophilic bacteria is presumably affected by environmental and nutritional factors in the same manner as it is in nonbarophilic bacteria (∞ Chapter 7).

✓ 16.11 Concept Check

The deep sea is a cold, dark habitat where high hydrostatic pressure and low nutrient availability prevails. Barophiles grow best under pressure, and extreme barophiles, obtained from the greatest depths, require high pressures for growth.

✓ What is a *barophilic* bacterium?
✓ What molecular adaptations occur in barophiles that allow them to grow optimally under pressure?

16.12

Hydrothermal Vents

The conception of the deep sea as a remote, low temperature, high pressure environment capable of supporting only the slow growth of barotolerant and barophilic bacteria is generally correct, but there are some amazing exceptions. A number of dense, thriving *animal* communities supported by the activities of microorganisms exist clustered about thermal springs in deep waters throughout the world. Geologically, these springs are associated with *ocean floor spreading centers* (*rifts*), regions where hot basalt and magma near the sea floor cause the floor to slowly drift apart. Seawater seeping into these cracked regions mixes with hot minerals and is emitted from the springs (Figure 16.29); because of their unique properties these underwater hot springs have come to be known as **hydrothermal vents.**

Two major types of vents have been found. *Warm vents* emit hydrothermal fluid at temperatures of 6–23°C (into seawater at 2°C). *Hot vents*, usually referred to as "black smokers" because the mineral-rich hot water forms a dark cloud of precipitated material on mixing with seawater, emit hydrothermal fluid at 270–380°C (Figure 16.29). The flow rates of these two vent types are also characteristic; warm vents emit fluid at 0.5–2 cm/sec, whereas hot vents have higher flow rates, about 1–2 m/sec.

Animals Living at Hydrothermal Vents

Using small pressurized submarines, it is possible to visit hydrothermal vents and study the organisms associated with them. Thriving invertebrate communities have been found, with unusual worms called *tube worms* over 2 m in length and large numbers of giant clams and mussels (Figure 16.30). Considering that other locations in the deep sea are so biologically unproductive, how do dense animal communities exist in the absence of phototrophic primary producers? What is the nature of the energy source(s)?

Chemical analyses of hydrothermal fluid show large amounts of reduced inorganic materials, including H_2S, Mn^{2+}, H_2, and CO. Some vents contain little H_2S but have high levels of NH_4^+. Organic matter is not present in the fluid emitted from any of the hydrothermal

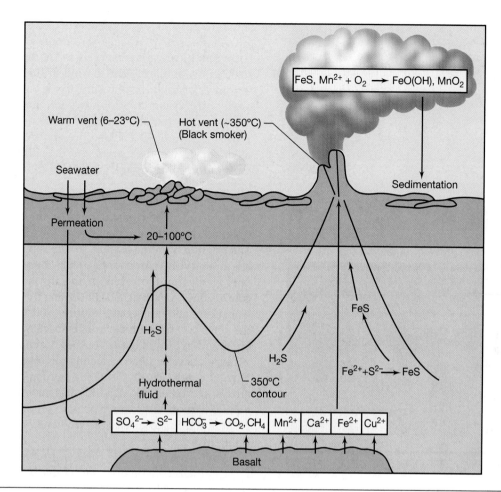

FIGURE 16.29   Schematic diagram showing the geological formations and major chemical species occurring at warm vents and black smokers. At warm vents, the hot hydrothermal fluid is cooled by cold (2–3°C) seawater permeating the sediments. In black smokers, hot hydrothermal fluid reaches the sea floor directly. Warm vents and black smokers are typically found at about 2000-meter depth. Such depths can be explored by humans in small submersibles such as *Alvin,* used by researchers at the Woods Hole Oceanographic Institution, Woods Hole, MA.

vents thus far examined. From studies of vent chemistry and associated microbial processes, it is clear that the animals are dependent on the activities of chemolithotrophic bacteria (∞ Sections 15.9–15.13), which grow at the expense of inorganic energy sources emitted from the vents. Carbon dioxide, abundant in seawater as CO_3^{2-} and HCO_3^{-}, is fixed into organic carbon by the chemolithotrophs, and the latter form the base of an extremely short food chain for hydrothermal vent animals.

Microorganisms in Hydrothermal Vents

Large numbers of sulfur-oxidizing chemolithotrophs such as *Thiobacillus, Thiomicrospira, Thiothrix,* and *Beggiatoa* (∞ Sections 13.3 and 15.11) are present in and around the vents. Samples collected from near the vents have yielded cultures of these organisms, and *in situ* experiments have shown fixation of CO_2 and oxidation of H_2S and $S_2O_3^{2-}$ by natural populations of these bacteria.

Some vents contain nitrifying bacteria, hydrogen-oxidizing bacteria, iron- and manganese-oxidizing bacteria, and methylotrophic bacteria, the latter presumably growing on the methane and carbon monoxide (CO) emitted from the vents (∞ Chapters 13–15 discusses most of these physiological groups). Table 16.2 summarizes the electron donors and electron acceptors for chemolithotrophs suspected of playing a role in hydrothermal vent ecology. However, there is no evidence that the animals at the vents actually *eat* these chemolithotrophic bacteria. Instead, it is the autotrophic abilities of these organisms that feed the animals.

Nutrition of Animals Living Near Hydrothermal Vents

Various chemolithotrophs have been found to live in symbiotic association with animals of the thermal vents. For example, the 2-m-long tube worms (see Figure 16.30) lack a mouth, gut, or anus but contain a modified gas-

FIGURE 16.30 Invertebrates from habitats near deep-sea thermal vents. (a) Tube worms (family *Pogonophora*), showing the sheath (white) and plume (red) of the worm bodies. (b) Close-up photograph showing worm plume. (c) Mussel bed in vicinity of warm vent. Note yellow deposition of elemental sulfur in and around mussels.

trointestinal tract consisting primarily of spongy tissue called the **trophosome.** Making up about 50% of the weight of the worm, trophosome tissue is loaded with sulfur granules, and microscopy of trophosome tissue

shows large numbers of prokaryotic cells (Figure 16.31), an average of 3.7×10^9 cells/g of trophosome tissue. The large spherical cells observed in the trophosome are structurally similar to the marine sulfur-oxidizing bacterium *Thiovulum*. Trophosome tissue (containing symbiotic sulfur bacteria) also shows activity of the enzyme *rhodanese*, an enzyme capable of disproportionating $S_2O_3^{2-}$ to S^0 and SO_3^{2-}. Also present are RubisCO and other enzymes of the *Calvin cycle*, the pathway by which most autotrophic organisms fix CO_2 into cellular material (∞ Section 15.7).

The chemolithotrophic bacteria supply the worm with its nourishment, the animal living off the excretory products and dead cells of its chemolithotrophic symbionts. The bright red plume (Figure 16.30b) is rich in blood vessels and serves as a trap for O_2 and H_2S (see later) for transport to chemolithotrophs in the trophosome. Similar conclusions can be reached concerning the nutrition of the giant clams and mussels (see Figure 16.30c) present around the vents because sulfur-oxidizing bacterial communities are found in the gill tissues of these animals as well. Use of nucleic acid sequencing techniques (∞ Working with Nucleic Acids: The Tools, Chapter 6) and phylogenetic analyses (∞ Sections 12.5–12.7) has shown that each vent animal harbors only one major species of bacterial symbiont and that the species of symbiont varies among the different animal types.

Further study of the tube worms has shown that these animals contain unusual soluble hemoglobins that bind H_2S as well as O_2 and transport both substrates to the trophosome where they are released to the bacterial symbiont; trapping and transporting sulfide are necessary to prevent the H_2S from poisoning the animal. The CO_2 content of tube worm blood is also high, some 20–30 mM, and presumably this is released in the trophosome as a carbon source for the symbionts. In addition, stable isotope analyses (see Section 16.8) of the elemental sulfur found within the bacterial symbionts have shown the $^{34}S/^{32}S$ isotope composition to be the same as the sulfide emitted from the vent. This ratio is distinctly different from that of seawater sulfate and serves as additional proof that geothermal sulfide is entering the worm.

A link between animal nutrition and other physiological groups of chemolithotrophs (for example, H_2 oxidizers and nitrifying bacteria) has also been suggested. Methanotrophic symbionts have been shown to play a nutritional role for animals living in symbiotic association with giant clams near natural gas seeps at relatively shallow depths in the Gulf of Mexico. Although not truly autotrophs (CH_4 is an organic compound), these symbionts support growth of the animal, in this case by the oxidation of CH_4 as an energy source.

Other chemolithotrophs, iron, H_2, and manganese oxidizers, for example, are probably not animal sym-

TABLE 16.2	Chemolithotrophic prokaryotes of potential significance to hydrothermal vent primary productivity[a]		
Chemolithotroph		**Electron donor**	**Electron acceptor**
Sulfur-oxidizing bacteria		HS^-, S^0, $S_2O_3^{2-}$	O_2, NO_3^-
Nitrifying bacteria		NH_4^+, NO_2^-	O_2
Sulfate-reducing bacteria		H_2	S^0, SO_4^{2-}
Methanogenic Archaea		H_2	CO_2
Hydrogen-oxidizing bacteria		H_2	O_2, NO_3^-
Iron and manganese-oxidizing bacteria		Fe^{2+}, Mn^{2+}	O_2
Methylotrophic bacteria		CH_4, CO	O_2

a ⬯⬯ Sections 15.9–15.19 for a discussion of chemolithotrophs.

bionts but instead exist as free-living bacteria growing at the expense of reduced substances emitted from the vents. Nevertheless, these chemolithotrophs probably contribute to overall primary productivity in the vent ecosystem.

Superheated Water: Black Smokers and Sea Mounts

The great depths of the deep sea create huge hydrostatic pressures that affect the physical properties of water. At a depth of 2600 m, water does not boil until it reaches a temperature of about 450°C. At certain vent sites superheated (but not boiling) hydrothermal fluid is emitted at temperatures of 270–380°C (see Figure 16.29) and could theoretically be a habitat for hyperthermophilic bacteria (⬯⬯ Sections 5.9 and Chapter 14). The

hydrothermal fluid emitted from black smokers contains abundant metal sulfides, especially iron sulfides, and cools quickly as it emerges into cold seawater. The precipitated metal sulfides form a tower referred to as a "chimney" about the source (Figure 16.32). Although it is very doubtful that prokaryotes actually live in the superheated hydrothermal fluid, good evidence exists that thermophilic or hyperthermophilic bacteria of various types live in the seawater or hydrothermal fluid *gradient* that forms as the hot water blends with cold ocean water. For example, the walls of smoker chimneys are teeming with hyperthermophilic prokaryotes such as *Methanopyrus*, a methanogenic archaeon that oxidizes H_2 and is of great evolutionary interest (⬯⬯ Chapter 14). Using FISH technology (see Section 16.6), the presence of both Bacteria and Archaea in smoker chimney walls can be readily detected (Figure 16.33).

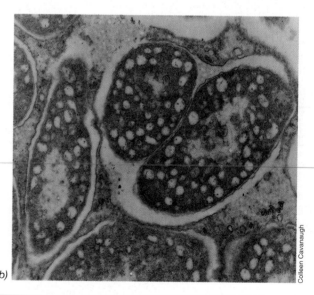

(a)   (b)

Colleen Cavanaugh

FIGURE 16.31   Chemolithotrophic sulfur-oxidizing bacteria associated with the trophosome tissue of tube worms from hydrothermal vents. (a) Scanning electron microscopy of trophosome tissue showing spherical chemolithotrophic sulfur-oxidizing bacteria. Cells are 3–5 μm in diameter. Reprinted with permission from *Science* 213:340–342 (1981), © AAAS. (b) Transmission electron micrograph of bacteria in sectioned trophosome tissue. The cells are frequently enclosed in pairs by an outer membrane of unknown origin.

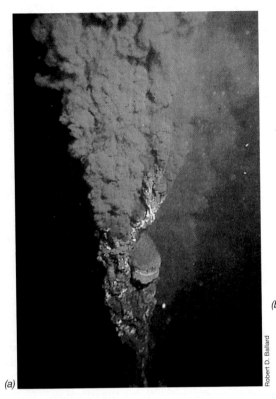

(a)

Robert D. Ballard

(b)

Dudley Foster, Woods Hole Oceanographic Institution

FIGURE 16.32   Black smokers emitting sulfide- and mineral-rich water at temperatures of 350°C. (a) The chimney is quite large, about 1 m in length. (b) The chimney is much smaller. Note the scientific equipment near the smoker in (b), indicating the relatively small size of the chimney.

In addition to black smokers, active *sea mounts* are known. These underwater volcanoes, located primarily in the Pacific Ocean on the Pacific tectonic plate, spew out ferric minerals and very hot water. Although not as extensively studied as black smoker chimneys, good evidence exists that hyperthermophilic prokaryotes exist in and about these sea mounts and are released into ocean water during eruptions.

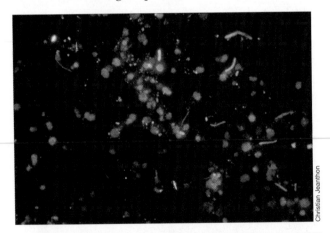

Christian Jeanthon

FIGURE 16.33   Phylogenetic staining of black smoker chimney material from a smoker at the Snake-Pit vent field of the Mid-Atlantic Ridge (3500 meters deep). A green fluorescing dye was conjugated to a probe that reacts with the 16S rRNA of all Bacteria and a red dye to a 16S rRNA probe for Archaea (see Section 16.6). Cell numbers tend to be highest in the outer regions of the chimney wall and lowest in the inner region. The hydrothermal fluid going through the center of the chimney was 300°C.

Metazoans and Vent Chimneys

Surprisingly, hydrothermal vent chimneys are also colonized by metazoans, in particular by the small colonial tube worm *Alvinella* (Figure 16.34*a*). *Alvinella*, also known as the *Pompeii worm*, grows on the outer surfaces of chimneys and, interestingly, is able to tolerate extremely hot water emitted from the vent. Measurements taken in their natural habitat have shown that temperatures inside the worm can get as high as 80°C, making the Pompeii worm the most thermotolerant of all known animals. The surface of the Pompeii worm is coated with symbiotic filamentous bacteria, possibly chemolithotrophs (Figure 16.34*b*). Although it is not known whether these epibionts nourish the worm like the symbionts that live in the large tube worms and mussels (Figure 16.30), the filamentous bacteria are clearly a food source for the Pompeii worm; scientists have observed the worms extending out of their tubes and feeding directly on the epibionts in the surrounding worm colony.

The mechanism of heat resistance in the Pompeii worm is of obvious interest. But the very existence of such a heat tolerant animal proves that eukaryotic cell structures can withstand far higher temperatures than previously thought. This in turn suggests that more thermophilic eukaryotic microorganisms may exist (the most thermophilic of all known eukaryotes, a fungus, grows only to 62°C); if such organisms are eventually discovered, their evolutionary history will be of great interest and would likely help in understanding the evolutionary progression from Archaea to Eukarya (👁 Section 12.7).

(a)

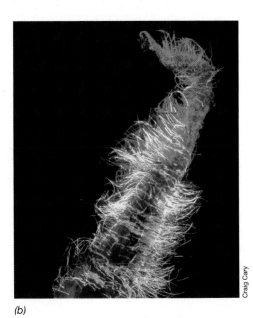

(b)

FIGURE 16.34 The black smoker chimney worm *Alvinella*. (a) A single specimen of *A. pompejana*; the worm is about 6 cm long. (b) Fluorescent photomicrograph of filamentous bacteria growing on the surface of *A. pompejana*. The cells were stained with two different phylogenetic probes, yellow–green, and red, each of which reacts with 16S rRNA from a different group of Proteobacteria (a major division of Bacteria, ⟳ Chapter 13). The diameter of the filaments is about 5 μm.

In situ Estimates of the Upper Temperature for Microbial Life

In efforts to define the upper temperature limits for life, attempts have been made to detect bacterial growth in the outflows of black smokers at various temperatures. By fitting a vent with a titanium "cap" (Figure 16.35) from which glass microscope slides and other surfaces for microbial colonization can be suspended in the

emerging hot water, evidence for colonization and growth of prokaryotes has been obtained at temperatures above 125°C to about 140°C. Similar surfaces exposed to outflows at 200°C or higher showed no microbial attachment or growth. Although hyperthermophiles from water 125°C or higher have not yet been cultured, the formation of microcolonies on artificial surfaces at these temperatures is similar to those observed at lower temperatures (see Figure 16.3*a*) and is good evidence that microbial growth is occurring here. In addition, evidence for biological sulfate reduction in hot marine sediments at temperatures to 130°C has been obtained. Collectively, these results suggest that the upper temperature limit for microbial life is likely to be higher than that of the most thermophilic of all known prokaryotes, *Pyrolobus fumarii*, which grows to 113°C, but is probably under 150°C. We discuss some of the reasons for this upper temperature limit in Section 14.10.

✓ 16.12 Concept Check

Hydrothermal vents are deep-sea hot springs where thermal (volcanic) activity generates fluids containing large amounts of inorganic energy sources that can be used by chemolithotrophic bacteria. The chemolithotrophic bacteria fix CO_2 autotrophically into organic carbon, some of which is then used by the deep-sea animals. The deep-sea hydrothermal vents are thus habitats where the primary producers are chemolithotrophic rather than phototrophic.

✓ How does a *warm hydrothermal vent* differ from a *black smoker,* both chemically and physically?
✓ How do giant tube worms receive their nutrition?
✓ What evidence is there that living organisms can grow at temperatures >100°C?

FIGURE 16.35 Methodology for studying extremely hyperthermophilic prokaryotes in black smokers. The titanium cap placed over the vent serves as a support for microscope slides submerged to regions of the vent of known temperature. With the vent cap system, evidence for organisms growing at temperatures above 125°C has been obtained. The vent cap shown covers a 160°C hydrothermal vent located at a depth of 2000 m in the Guayamas Basin (Gulf of California).

16.13

The Carbon Cycle

On a global basis, carbon is cycled through all Earth's major carbon reservoirs: the atmosphere, the land, the oceans and other aquatic environments, sediments and rocks, and biomass (Figure 16.36). As we will see, the carbon and oxygen cycles are intimately intertwined, since CO_2 fixation by oxygenic phototrophs releases O_2 and much organic matter is oxidized back to CO_2 by aerobic respiration (Figure 16.36).

The largest carbon reservoir is present in the sediments and rocks of Earth's crust, but the turnover time is so long that flux out of this compartment is relatively insignificant on a human scale. From the viewpoint of living organisms, a large amount of organic carbon is found in land plants. This represents the carbon of forests and grasslands and constitutes the major site of photosynthetic CO_2 fixation. However, more carbon is present in dead organic material, called *humus*, than in living organisms. **Humus** is a complex mixture of organic materials. It is derived partly from the protoplasmic constituents of soil microorganisms that have resisted decomposition and partly from resistant plant material. Some humic substances are fairly stable, with a global turnover time of about 40 years, although certain other humic components decompose much more rapidly than this; for example, in Section 15.20 we saw how some humic substances can be electron acceptors for anaerobic respiration.

The most rapid means of global transfer of carbon is via the CO_2 of the atmosphere. Carbon dioxide is removed from the atmosphere primarily by photosynthesis of land plants and is returned to the atmosphere by respiration of animals and chemoorganotrophic microorganisms. An analysis of the various processes suggests that the single most important contribution of CO_2 to the atmosphere is via microbial decomposition of dead organic material, including humus.

Importance of Photosynthesis in the Carbon Cycle

The only major ways in which new organic carbon is synthesized on Earth are via photosynthesis and chemosynthesis (CO_2 fixation by chemolithotrophs); most organic carbon likely comes from photosynthesis. Phototrophic organisms are therefore at the basis of the carbon cycle (Figure 16.36). Phototrophic organisms are found in nature almost exclusively in habitats where light is available. Thus, the deep sea and other permanently dark habitats are devoid of phototrophs. Oxygenic phototrophic organisms can be divided into two major groups: higher plants and microorganisms. Higher plants are the dominant phototrophic organisms of terrestrial environments, whereas phototrophic microorganisms are the most abundant photosynthesizers of aquatic environments.

The redox cycle for carbon is shown in Figure 16.37. We begin with photosynthesis. The overall equation for oxygenic photosynthesis is

$$CO_2 + H_2O \xrightarrow{\text{light}} (CH_2O) + O_2$$

where (CH_2O) represents organic matter at the oxidation state of cell material such as polysaccharides (the main form in which photosynthesized organic matter is stored in the cell). Phototrophic organisms also carry out respiration, both in the light and the dark. The overall equation for respiration is the reverse of the preceding equation:

$$(CH_2O) + O_2 \xrightarrow{\text{light or dark}} CO_2 + H_2O$$

where (CH_2O) again represents storage polysaccharides. If an organism is to grow (that is, increase in cell number or mass) phototrophically, then the rate of photosynthesis must exceed the rate of respiration. If this occurs, then some of the carbon fixed from CO_2 into polysaccharide can become the starting material for

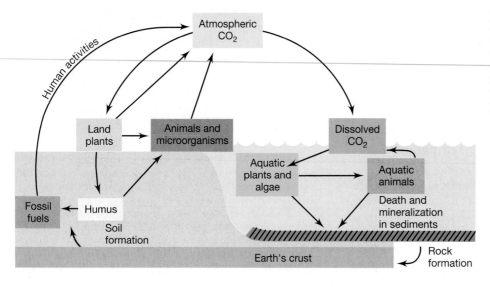

FIGURE 16.36 The carbon cycle. The carbon and oxygen cycles are closely connected, as oxygenic photosynthesis both removes CO_2 and produces O_2 while respiratory processes both produce CO_2 and remove O_2.

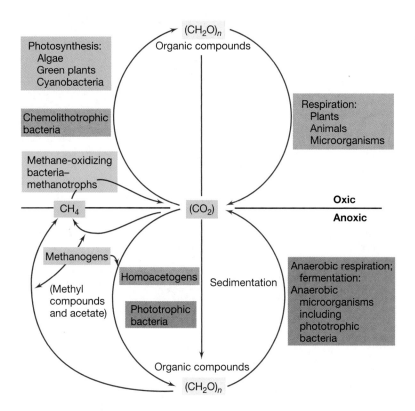

FIGURE 16.37 Redox cycle for carbon; note in particular the contrasts between autotrophic ($CO_2 \rightarrow$ organic compounds) and heterotrophic processes. Photosynthesis in oxic habitats is mainly oxygenic, whereas in anoxic environments it is mainly anoxygenic from the activities of purple and green bacteria. Under anoxic conditions, besides homoacetogens and methanogens, certain sulfate-reducing bacteria are also autotrophic.

biosynthesis. The whole carbon cycle is built on a net positive balance of the rate of photosynthesis over the rate of respiration.

Decomposition

Photosynthetically fixed carbon is eventually degraded by various organisms and two major oxidation states of carbon result: methane (CH_4) and carbon dioxide (CO_2) (Figure 16.37). These two gaseous products are formed from the activities of methanogens (CH_4) or from various chemoorganotrophs via fermentation, anaerobic respiration, or aerobic respiration (CO_2). In anoxic habitats CH_4 is produced from both the reduction of CO_2 with H_2 and from certain organic compounds like acetate. However, virtually *any* organic compound can eventually be converted to CH_4 from the combined activities of syntrophic bacteria and methanogens: H_2 generated from the fermentative degradation of organic compounds gets consumed by methanogens (∞ Section 15.23 and see next section). Methane produced in anoxic habitats is highly insoluble and thus is easily transported to oxic environments where it is oxidized to CO_2 by methanotrophs (Figure 16.37). Hence, all organic carbon eventually returns to CO_2 from which autotrophic metabolism once again begins the carbon cycle.

The balance between the oxidative and reductive portions of the carbon cycle is critical; the products of metabolism of some organisms are the substrates for others. Thus the cycle needs to keep in balance if it is to

continue as it has for many billions of years. Any significant changes in levels of gaseous forms of carbon may have serious global consequences (as we are already experiencing in the form of global warming from the increasing CO_2 levels in the atmosphere caused by deforestation and the burning of fossil fuels). In terms of decomposition, CO_2 release by microbial activities far exceeds that of eukaryotes, and this is especially true of anoxic environments, which we consider next.

✓ 16.13 Concept Check

The oxygen and carbon cycles are highly interrelated through the complementary activities of autotrophic and heterotrophic organisms. Microbial decomposition is the single largest source of CO_2 released to the atmosphere.

✓ How is new organic matter made?
✓ In what ways are oxygenic photosynthesis and respiration related?

16.14

Ecology of Syntrophy and Methanogenesis

Methane from biological methanogenesis is of great importance to carbon flow in many anoxic habitats. Methanogenesis is carried out by a group of Archaea, the methanogens, which are strict anaerobes. We discussed the biochemistry of methanogenesis in Section 15.19 and methanogens themselves in Section 14.3. Most methanogens use CO_2 as a terminal electron acceptor in anaerobic respiration, reducing it to methane with H_2. Only a very few other substrates, acetate being chief among them, can be directly converted to methane by methanogens. Thus, for the conversion of most organic compounds to CH_4, methanogens must team up with partner organisms that can supply them with their needed substrates. This is the job of the syntrophs.

We discussed the concept of syntrophy, in which two or more organisms cooperate in the degradation of some compound, in Section 15.23 and focused on the energetics behind the process. Here we consider the *ecological interactions* of syntrophic bacteria with other organisms and their significance for the whole anoxic carbon cycle. High-molecular-weight substances such as polysaccharides, proteins, and fats, are converted to CH_4 and CO_2 by the cooperative interaction of several physiological groups of prokaryotes. For the breakdown of a typical polysaccharide such as cellulose, for example (Figure 16.38 and Table 16.3), the process begins with *cellulolytic bacteria*, that cleave the high-molecular-weight cellulose molecule into cellobiose (glucose — glucose) and into free glucose. Glucose is then fermented by *primary fermenters* to a variety of fermentation products, with acetate, propionate, butyrate, succinate, alcohols, H_2, and CO_2 being the major products observed. Any H_2 produced in primary fermentative processes is immediately consumed by *H_2 consumers*, such as methanogens, homoacetogens, or sulfate-reducing bacteria (in environments containing sufficient levels of sulfate). In addition, acetate can be converted to methane by certain methanogens. But this leaves a large amount of carbon in the form of fatty acids and alcohols. The catabolism of these compounds occurs by way of the syntrophs.

Role of the Syntrophs

Key organisms in the conversion of complex organic materials to methane are *secondary fermenters*, especially the H_2-producing fatty acid-oxidizing syntrophic bacteria. For example, *Syntrophomonas wolfei* oxidizes C_4 to C_8 fatty acids yielding acetate, CO_2 (if the fatty acid contained an odd number of carbon atoms), and H_2 (Table 16.3). Other species of *Syntrophomonas* use fatty acids up to C_{18}, including some unsaturated fatty acids. *Syntrophobacter wolinii* specializes in propionate oxidation and generates acetate, CO_2, and H_2, while *Syntrophus gentiane* degrades the aromatic compound benzoate to acetate, H_2, plus CO_2 (Table 16.3). These reactions support luxuriant growth of the syntrophs in coculture with a H_2-consuming partner but not in pure culture. Why should this be so?

As was explained in Section 15.23, H_2 consumption by the partner organism is crucial for the growth of fatty

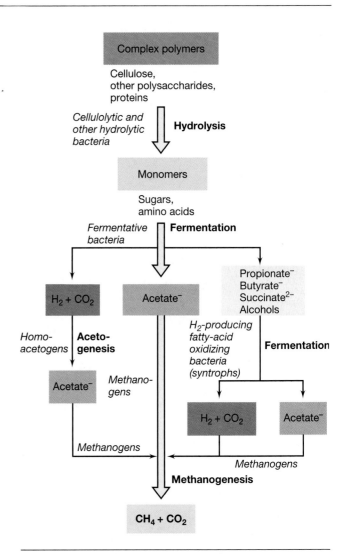

FIGURE 16.38 Overall process of anoxic decomposition, showing the manner in which various groups of fermentative anaerobes cooperate in the conversion of complex organic materials ultimately to methane (CH_4) and CO_2.

acid–oxidizing syntrophic bacteria. When the reactions in Table 16.3 are written with all reactants at standard conditions (solutes, 1 M; gases, 1 atm), the reactions yield free-energy changes that are positive in sign. That is, the $\Delta G^{0\prime}$ of these reactions do not release free energy (Table 16.3). But as was explained in Section 15.23, H_2 consumption by partner bacteria dramatically changes the energetic picture, allowing for sufficient energy conservation in the syntroph to support growth. This can be seen in Table 16.3 where the ΔG values (free-energy change measured under *actual conditions* in the habitat) are favorable for energy conservation if H_2 concentrations are kept very low by the activities of the partner organism.

Thus through the combined action of primary fermenters, syntrophs, and their H_2-consuming partners, virtually any organic compound can be degraded in anoxic habitats. The final products are CO_2 and CH_4.

TABLE 16.3	Major reactions occurring in the anoxic conversion of organic compounds to methane[a]		
		Free-energy change (kJ/reaction)	
Reaction type	Reaction	$\Delta G^{0\prime b}$	ΔG^c
Fermentation of glucose to acetate, H_2, and CO_2	Glucose + 4 $H_2O \rightarrow$ 2 acetate$^-$ + 2 HCO_3^- + 4 H^+ + 4 H_2	−207	−319
Fermentation of glucose to butyrate, CO_2, and H_2	Glucose + 2 $H_2O \rightarrow$ butyrate$^-$ + 2 HCO_3^- + 2 H_2 + 3 H^+	−135	−284
Fermentation of butyrate to acetate and H_2	Butyrate$^-$ + 2 $H_2O \rightarrow$ 2 acetate$^-$ + H^+ + 2 H_2	+48.2	−17.6
Fermentation of propionate to acetate, CO_2, and H_2	Propionate$^-$ + 3 $H_2O \rightarrow$ acetate$^-$ + HCO_3^- + H^+ + H_2	+76.2	−5.5
Fermentation of ethanol to acetate and H_2	2 Ethanol + 2 $H_2O \rightarrow$ 2 acetate$^-$ + 4 H_2 + 2 H^+	+19.4	−37
Fermentation of benzoate to acetate, CO_2 and H_2	Benzoate$^-$ + 6 $H_2O \rightarrow$ 3 acetate$^-$ + 2 H^+ + CO_2 + 3 H_2	+46.5	−18
Methanogenesis from H_2 + CO_2	4 H_2 + HCO_3^- + $H^+ \rightarrow CH_4$ + 3 H_2O	−136	−3.2
Methanogenesis from acetate	Acetate$^-$ + $H_2O \rightarrow CH_4$ + HCO_3^-	−31	−24.7
Acetogenesis from H_2 + CO_2	4 H_2 + 2 HCO_3^- + $H^+ \rightarrow$ acetate$^-$ + 4 H_2O	−105	−7.1

a Data adapted from Zinder, S. 1984. Microbiology of anaerobic conversion of organic wastes to methane: Recent developments. *Am. Soc. Microbiol. News* 50:294–298.

b Standard conditions: solutes, 1 M; gases, 1 atm.

c Concentrations of reactants in typical anoxic freshwater ecosystem: fatty acids, 1 mM; HCO_3^-, 20 mM; glucose, 10 μM; CH_4, 0.6 atm; H_2, 10^{-4} atm.

However, two groups of natural organic materials appear to be refractory to fermentative breakdown: *lignin* and *aliphatic saturated hydrocarbons*. Lignin is a complex aromatic polymer of phenylpropane units held together by C—C and C—O—C (ether) linkages and is a major component of wood; lignin is quite stable to anoxic degradation and hence does not decompose in anoxic habitats. Aliphatic hydrocarbons such as hexadecane ($C_{16}H_{34}$) are nonfermentable, although they can be oxidized to CO_2 by certain sulfate-reducing bacteria (which likely accounts for the abundance of H_2S in oil drilling operations), and of course are easily degraded aerobically (see Section 16.21). Aromatic hydrocarbons may be a special case since experiments in which ^{14}C-labeled aromatic compounds like benzene are added to anoxic freshwater sediments have shown the production of [14]CH_4, suggesting that some organisms, probably consortia containing syntrophs, can attack these highly refractory substances and generate from them the substrates for methanogenesis. However, these conversions are rather slow, and thus their ecological significance is unknown.

Methanogenic/Acetogenic Habitats

Despite the obligate anaerobiosis and specialized metabolism of methanogens, they are quite widespread on Earth. Although high levels of methanogenesis occur only in anoxic environments, such as swamps and marshes, or in the rumen (see Section 16.15), the process also occurs in habitats that normally might be considered oxic, such as forest and grassland soils. In such habitats methanogenesis occurs in anoxic microenvironments, for example, in the midst of soil crumbs (see Figure 16.2). An overview of the rates of methanogenesis in different kinds of habitats is given in Table 16.4. It should be noted that biogenic production of methane by the methanogenic Archaea exceeds the production rate from gas wells and other abiogenic sources. Eructation by ruminants (see Section 16.15) and CH_4 released from termites, paddy fields, and natural wetlands are the largest sources of biogenic methane.

Methanogens have also been found living as endosymbionts of certain protozoa. Several types of protozoa, including free-living aquatic amebas and flagellates found in the insect gut, have been shown to harbor methanogens. In termites, for example, methanogens are present primarily within cells of trichomonal protozoa inhabiting the termite hindgut (Figure 16.39). Methanogenic symbionts of protozoa resemble rod-shaped species of the genus *Methanobacterium* or *Methanobrevibacter* (∞ Section 14.3), but their exact relationship to other methanogens is unclear. In the termite hindgut, endosymbiotic methanogens are thought to benefit their protozoan hosts by consuming H_2 generated from glucose fermentation by cellulolytic protozoans. As shown in Table 16.4, termites can be a major source of biogenic CH_4.

A competing process to methanogenesis is *acetogenesis* (∞ Section 15.18). In some habitats, for example the rumen (see Section 16.15), homoacetogens appear to be poor competitors with methanogens and thus methanogenesis is the dominant H_2-consuming process. But in other habitats, such as the termite hindgut, where some methanogenesis occurs (Table 16.4 and Figure 16.39), acetogenesis is a quantitatively more important process. What governs the dominant anaerobic process

TABLE 16.4	Estimates of CH_4 released into the atmosphere[a]	
Source	**CH_4 emission (10^{12} g/year)**	
Biogenic		
Ruminants	80–100	
Termites	25–150[b]	
Paddy fields	70–120	
Natural wetlands	120–200	
Landfills	5–70	
Oceans and lakes	1–20	
Tundra	1–5	
Abiogenic		
Coal mining	10–35	
Natural gas flaring and venting	10–30	
Industrial and pipeline losses	15–45	
Biomass burning	10–40	
Methane hydrates	2–4	
Volcanoes	0.5	
Automobiles	0.5	
Total	349–820	
Total biogenic	302–665	81–86% of total
Total abiogenic	48–155	13–19% of total

a Data adapted from estimates of Tyler in Tyler, S. C. 1991. The global methane budget, pp. 7–58, in E. J. Rogers and W. B. Whitman (eds.), *Microbial Production and Consumption of Greenhouse Gases: Methane, Nitrogen Oxides, and Halomethanes,* American Society for Microbiology, Washington, DC.

b More recent estimates indicate that the lower value is probably the more accurate.

is not clearly understood; based on simple energetic considerations, methanogenesis from H_2 is a more favorable process than is acetogenesis (-136kJ versus -105kJ, respectively) and thus methanogens should have a competitive advantage. However, in the hindgut of termites acetogenesis remains the dominant process possibly because (1) homoacetogens are in some way able to position themselves in the termite gut nearer to the source of H_2 than can methanogens and thus the homoacetogens can consume the majority of H_2 produced from cellulose fermentation; (2) unlike methanogens, homoacetogens can grow chemoorganotrophically on glucose (from cellulose) as well as by CO_2 reduction by H_2 (⚙ Section 15.18); and (3) the nature of the food eaten by termites, that is, wood, contains a high content of lignin mixed in with cellulose, and this type of material may be degraded in such a way that favors acetogenesis.

Methanogenesis versus Sulfidogenesis

Methanogenesis and acetogenesis is more extensive in anoxic freshwater and terrestrial environments than in marine sediments. The reason for this is that sulfate-reducing bacteria, which are abundant in marine sediments where sulfate levels are high, can outcompete

(a)

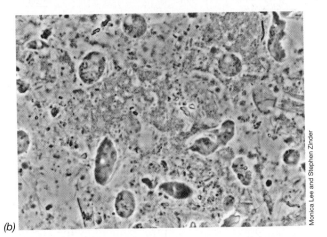
(b)

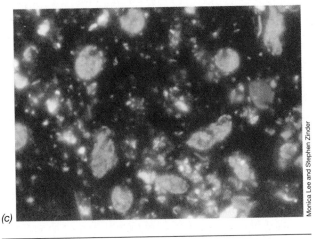
(c)

FIGURE 16.39 Termites and their carbon metabolism. (a) A common eastern (USA) subterranean termite worker larva shown beneath a hindgut extracted from a separate worker. The animal is about 0.5 cm long. Acetogenesis is the major form of anoxic carbon metabolism in these termites although methanogenesis also occurs. (b) Microorganisms from the hindgut of the termite *Zootermopsis angusticolis*. A single microscope field was photographed by two different methods. (b) Phase contrast. (c) Epifluorescence, showing color typical of methanogens due to the high content of the fluorescent coenzyme F_{420}. The methanogens are inside cells of the protozoan *Tricercomitis* sp. Plant particles fluoresce yellow. The average diameter of a protozoan cell is 15–20 μm. See Section 3.1 and Figure 3.6 for an explanation of florescence microscopy.

methanogens and homoacetogens for H_2 produced either by primary fermenters or by syntrophs. The biochemical basis for this is complex and has at least partially to do with the fact that the energetics of sulfate reduction with H_2 is significantly greater than the reduction of CO_2 with H_2 to either CH_4 or acetate (∞ Sections 15.17–15.19). In freshwater, however, where sulfidogenesis is typically very low because sulfate is limiting, methanogenesis and acetogenesis dominate. Marine sulfate-reducing bacteria also outcompete acetotrophic methanogens for acetate, thus the only methanogenesis that occurs in marine sediments is driven by methylated substrates like trimethylamine and CH_3OH, which are poorly used by sulfate reducers and homoacetogens.

✓ 16.14 Concept Check

Under anoxic conditions, organic matter is degraded principally to CH_4 and CO_2. Much CH_4 is formed from the reduction of CO_2 by H_2 supplied by H_2-producing syntrophic bacteria that depend on H_2 consumption to balance their energetics. On a global basis, biogenic CH_4 is a much larger source than abiogenic CH_4.

✓ What is it about the metabolism of butyrate by *Syntrophomonas wolfei* that makes it dependent on syntrophy?
✓ What kinds of organisms can grow in coculture with *Syntrophomonas*?
✓ What is the final product of acetogenesis? What anoxic habitat shows greater acetogenesis than methanogenesis?
✓ Why is methanogenesis from H_2 not an abundant process in ocean sediments?

16.15

The Rumen Microbial Ecosystem

Ruminants are herbivorous mammals that possess a special digestive vessel, the **rumen,** within which the digestion of cellulose and other plant polysaccharides occurs through the activity of microbial populations. Some of the most important domestic animals, the cow, sheep, and goat, are ruminants. Because the human food economy depends to a great extent on these animals, rumen microbiology is of considerable economic significance.

Rumen Anatomy and Action

The bulk of the organic matter in terrestrial plants is present in insoluble polysaccharides, of which *cellulose* is the most important. Mammals, and indeed almost all animals, lack the enzymes necessary to digest cellulose, but all mammals that subsist primarily on grasses and leafy plants can metabolize cellulose by making use of microorganisms as digestive agents. Unique features of the rumen as a site of cellulose digestion are its relatively large size (100–150 liters in a cow, 6 liters in a sheep) and its position in the alimentary tract as the organ where ingested food goes first. The high constant temperature (39°C), constant pH (6.5), and anoxic nature of the rumen are also important factors in overall rumen function. The rumen operates in a more-or-less continuous fashion and in some ways can be considered analogous to a microbial chemostat (∞ Section 5.5).

The relationship of the rumen to other parts of the ruminant digestive system is shown in Figure 16.40. The digestive processes and microbiology of the rumen are well understood, in part because it is possible to create a sampling port, called a *fistula,* into the rumen of a cow (Figure 16.40b) or a sheep and remove samples for analysis. Food enters the rumen mixed with saliva containing bicarbonate and is churned in a rotary motion during which the microbial fermentation occurs. This peristaltic action grinds the cellulose into a fine suspension, which assists in microbial attachment. The food mass then passes gradually into the reticulum where it is formed into small clumps called *cuds,* which are regurgitated into the mouth where they are chewed again. The now finely divided solids, well mixed with saliva, are swallowed again, but this time the material passes to the omasum, finally ending in the abomasum, an organ more like a true (acidic) stomach. Here chemical digestive processes begin that continue in the small and large intestine.

Microbial Fermentation in the Rumen

Food remains in the rumen about 9–12 hr. During this period cellulolytic bacteria and cellulolytic protozoa hydrolyze cellulose to the disaccharide cellobiose and to free glucose units. The released glucose then undergoes a bacterial fermentation with the production of **volatile fatty acids** (VFAs), primarily *acetic, propionic,* and *butyric,* and the gases *carbon dioxide* and *methane* (Figure 16.41). The fatty acids pass through the rumen wall into the bloodstream and are oxidized by the animal as its main source of energy. In addition to their digestive functions, rumen microorganisms synthesize amino acids and vitamins that are the main source of these essential nutrients for the animal. The rumen contains enormous numbers of prokaryotes (10^{10}–10^{11} bacteria/ml rumen fluid) plus partially digested plant materials; these proceed through the gastrointestinal tract of the animal where they undergo further digestive processes similar to those of nonruminants. Many microbial cells from the rumen are digested in the abomasum and thus are a major source of proteins and vitamins for the animal. Because microbial protein can be recovered, a ruminant is thus nutritionally superior to a nonruminant when subsisting on foods that are deficient in protein, such as grasses.

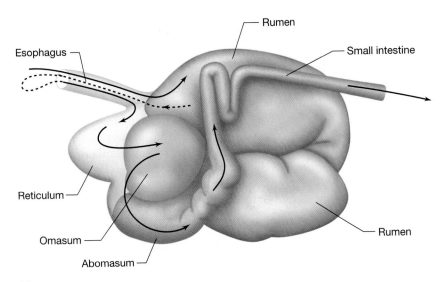

(a)

(b)

Sharisa D. Beek, Dept. Animal Science, Southern Illinois Univ.

FIGURE 16.40   The rumen. (a) Schematic diagram of the rumen and gastrointestinal system of a cow. Food travels from the esophagus to the rumen and is then regurgitated and travels to the reticulum, omasum, abomasum, and intestines, in that order. The abomasum is an acidic vessel, analogous to the stomach of monogastric animals like pigs and humans. (b) Photo of a fistulated Holstein cow. The fistula, shown unplugged, is a sampling port that allows access to the rumen. Fistulated cows and sheep have been very useful for the study of both rumen microbiology and ruminant nutrition.

Rumen Bacteria

The biochemical reactions occurring in the rumen are complex and involve the combined activities of a variety of microorganisms. Because the rumen is anoxic, anaerobic bacteria naturally dominate. Furthermore, because the conversion of cellulose to CO_2 and CH_4 involves a multistep microbial food chain, a variety of anaerobes can be expected (Table 16.5).

Several different rumen bacteria hydrolyze polymers such as cellulose to sugars and ferment the sugars to volatile fatty acids. *Fibrobacter succinogenes* and *Ruminococcus albus* are the two most abundant cellulolytic rumen anaerobes. Although both organisms produce cellulases, *Fibrobacter*, a gram-negative bacterium, contains a periplasmic cellulase (∞ Section 3.8) to break down cellulose; thus, the organism must remain attached to the cellulose fibril while digesting it. *Ruminococcus*, on the other hand, produces a cellulase that is excreted into the rumen contents where it degrades cellulose outside the bacterial cell proper. However, the

end result is the same in both cases: Free glucose is made available for fermentative anaerobes. If a ruminant is gradually switched from cellulose to a diet high in starch (grain, for instance), then starch-digesting bacteria such as *Ruminobacter amylophilus* and *Succinomonas amylolytica* develop; on a low starch diet these organisms are in a minority. If an animal is fed legume hay, which is high in pectin, then the pectin-digesting bacterium *Lachnospira multiparus* (Table 16.5) is a common member of the rumen flora.

Some of the fermentation products of the saccharolytic rumen microflora are used as energy sources by other rumen bacteria. Thus, *succinate* is converted to *propionate* and CO_2 (Figure 16.41) by *Schwartzia*, and *lactate* is fermented to *acetic* and other acids by *Selenomonas* and *Megasphaera* (Table 16.5). A number of rumen bacteria produce ethanol as a fermentation product when grown in pure culture, yet ethanol rarely accumulates in the rumen because it can be fermented to acetate + H_2. Hydrogen produced in the rumen by fermentative processes never accumulates because it is quickly used to reduce

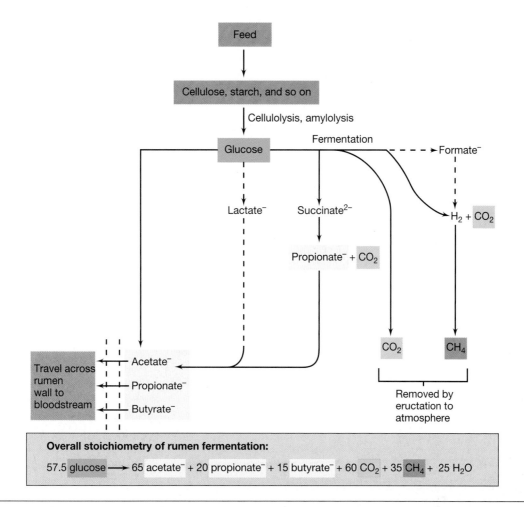

FIGURE 16.41 Biochemical reactions in the rumen. The major substrate, glucose, and end products are highlighted; dashed lines indicate minor pathways. Approximate steady state rumen levels of volatile fatty acids (VFAs) are acetate, 60 mM; propionate, 20 mM; butyrate, 10 mM.

CO_2 to CH_4 by methanogens. Another source of H_2 and CO_2 for methanogens is *formate* (Figure 16.41). Large amounts of CH_4 and CO_2 accumulate in the rumen, the average gas composition being about 65% CO_2 and 35% CH_4. These gases leave the ruminant during eructation (belching). Acetate is not converted to CH_4 in the rumen because the retention time is too short for development of acetotrophic methanogens, which typically grow slowly (∞ Section 14.3). In addition, syntrophic fatty acid-degrading bacteria (∞ Section 16.14) are not abundant in the rumen because the ruminant itself is the major sink for fatty acids (Figure 16.41).

Rumen Protozoa and Fungi

In addition to prokaryotes, the rumen has a characteristic protozoal fauna (about 10^6/ml), composed almost exclusively of ciliates (∞ Section 17.3). Many of these protozoa are obligate anaerobes, a property that is rare among eukaryotes. Although protozoa are not essential for the rumen fermentation, they definitely contribute to the overall process. Some are able to hydrolyze cellulose and starch and ferment glucose with the production of the same organic acids formed by the bacteria. Rumen protozoa also ingest rumen bacteria as food sources and are thought to play a role in controlling rumen bacterial densities.

Anaerobic fungi also inhabit the rumen and are known to play a role in ruminal digestive processes. Rumen fungi are generally species that alternate between a flagellated and a thallus form, and studies with pure cultures show that they can ferment cellulose to VFAs. *Neocalimastix,* for example, is an obligately anaerobic fungus that ferments glucose to formate, acetate, lactate, ethanol, CO_2 and H_2. Although a eukaryote, this fungus lacks mitochondria and cytochromes and thus lives an obligately fermentative existence. However, *Neocalimastix* cells do contain a redox organelle called the *hydrogenosome* that functions to evolve H_2 and has thus far only been found in evolutionarily primitive Eukarya (∞ Sections 12.3, 17.1, and 17.2). Rumen fungi play an important role in the degradation of polysac-

| TABLE 16.5 | Characteristics of some rumen prokaryotes |

Organism	Gram stain	Phylogenetic domain[a]	Shape	Motility	Fermentation products	DNA (mol % GC)
Cellulose decomposers						
Fibrobacter succinogenes[b]	Negative	B	Rod	−	Succinate, acetate, formate	45–51
Butyrivibrio fibrisolvens[b]	Negative	B	Curved rod	+	Acetate, formate, lactate, butyrate, H_2, CO_2	41
Ruminococcus albus[b]	Positive	B	Coccus	−	Acetate, formate, H_2, CO_2	43–46
Clostridium lochheadii	Positive	B	Rod (spores)	+	Acetate, formate, butyrate, H_2, CO_2	—
Starch decomposers						
Bacteroides ruminicola	Negative	B	Rod	−	Formate, acetate, succinate	40–42
Ruminobacter amylophilus	Negative	B	Rod	−	Formate, acetate, succinate	49
Selenomonas ruminantium	Negative	B	Curved rod	+	Acetate, propionate, lactate	49
Succinomonas amylolytica	Negative	B	Oval	+	Acetate, propionate, succinate	—
Streptococcus bovis	Positive	B	Coccus	−	Lactate	37–39
Lactate decomposers						
Selenomonas lactilytica	Negative	B	Curved rod	+	Acetate, succinate	50
Megasphaera elsdenii	Positive	B	Coccus	−	Acetate, propionate, butyrate, valerate, caproate, H_2, CO_2	54
Succinate decomposer						
Schwartzia succinovorans	Negative	B	Rod	+	Propionate, CO_2	46
Pectin decomposer						
Lachnospira multiparus	Positive	B	Curved rod	+	Acetate, formate, lactate, H_2, CO_2	—
Methanogens						
Methanobrevibacter ruminantium	Positive	A	Rod	−	CH_4 (from H_2 + CO_2 or formate)	31
Methanomicrobium mobile	Negative	A	Rod	+	CH_4 (from H_2 + CO_2 or formate)	49

a B, Bacteria; A, Archaea

b These species also degrade xylan, a major plant cell wall polysaccharide (⟨⟨⟩⟩ Table 15.8).

charides other than cellulose as well, including a partial degradation of lignin (the strengthening agent in the cell walls of woody plants), hemicellulose, and pectins.

Dynamics of the Rumen Ecosystem

A major feature of the rumen ecosystem is its *constancy*. Studies on various ruminant species in different parts of the world show that most animals contain the same major rumen bacterial species, with the proportions of each species varying somewhat with diet. In addition, the nature and proportions of the volatile fatty acids produced and the levels of rumen CO_2 and CH_4 are relatively constant among different ruminant species.

Occasionally, changes in the microbial composition of the rumen cause illness or even death of the animal. For example, if a cow is changed abruptly from forage to a completely grain diet, an explosive growth of *Streptococcus bovis* is observed in the rumen; the normal level of *S. bovis*, about 10^7 cells/ml (insignificant in terms of total rumen bacterial numbers), quickly expands to over 10^{10} cells/ml. This occurs because *S. bovis* grows rapidly on starch and grain contains high levels of starch, whereas grasses contain mainly cellulose. Being a lactic

acid bacterium, *S. bovis* produces large amounts of lactate from the fermentation of starch, and this acidifies the rumen (a condition called *acidosis*), killing off the normal rumen flora. Severe acidosis can cause death of the animal. To avoid acidosis, animals are switched from forage rations to grain *gradually* over a period of a few days. A slow introduction of starch selects for volatile fatty acid-producing starch degraders (Table 16.5) instead of *S. bovis*, and thus normal rumen biochemical processes are not disrupted.

Other Herbivorous Animals

The familiar ruminants are cows and sheep. However, goats, camels, buffalo, deer, reindeer, caribou, and elk are also ruminants. There is even some evidence that baleen whales have a rumenlike fermentation. Baleen whales contain a multichambered stomach consisting of a forestomach similar to the rumen. Samples of forestomach material from gray and bowhead whales show abundant volatile fatty acid production in proportions typical of the volatile fatty acids observed in the rumen of cattle or sheep. The diet of baleen whales is primarily chitinous invertebrates, small fish, and kelp. It is thought that

N-acetylglucosamine, the major monomeric unit of chitin, is the primary energy source for the forestomach microbial fermentation of baleen whales.

At least one *bird* has been shown to have a foregut fermentation that resembles that of the rumen. The *hoatzin*, a tropical bird, is one of the only obligate folivorous (leaf-eating) birds known. Probably because of its restricted diet—cellulose is its sole carbon source—the hoatzin has evolved a rumenlike forestomach to allow for cellulose digestion. In the hoatzin, a structure called the *crop* functions as the major digestive organ. The pH of the crop is near neutrality, and the organ contains high bacterial numbers and a volatile fatty acid content similar to that found in the rumen. After digestion in the crop, food travels through the esophagus and the proventriculus (an acidic organ analogous to a true stomach) to the small intestine.

Horses and rabbits are also herbivorous mammals, but they are not ruminants. Instead, these animals have only one stomach but use an organ called the **cecum,** a small digestive organ located posterior to the large intestine (just before the anus), as their cellulolytic fermentation vessel. The cecum contains a cellulolytic microflora, and digestion of cellulose occurs there. The precise microflora of the cecum is not well understood, but the species involved are not thought to be the same as those in the rumen. Nutritionally, ruminants have an advantage over horses and rabbits in that the cellulolytic microflora of the ruminant eventually passes through a true (acidic) stomach and as such is killed and is a protein source for the animal. By contrast, in horses and rabbits the cellulolytic microflora is passed out of the animal in the feces.

✓ 16.15 Concept Check

Ruminants are animals that have a special digestive organ, the rumen, that is a unique ecosystem in which anaerobic microorganisms digest insoluble feed components such as cellulose and starch. Bacteria, protozoa, and fungi of the rumen produce volatile fatty acids that are used by the ruminant. In addition to their role in the digestive process, rumen microorganisms synthesize vitamins and amino acids used by the ruminant.

✓ What physical and chemical conditions prevail in the rumen?
✓ What are *VFAs* and of what value are they to the ruminant?
✓ Why is the metabolism of *Streptococcus bovis* of special concern to ruminant nutrition?

16.16

The Nitrogen Cycle

The element nitrogen, N, a key constituent of protoplasm, exists in a number of oxidation states (∞ Table 15.2). We discussed two major processes of microbial nitrogen transformation in Chapter 15: nitrification in Section 15.13 and denitrification in Section 15.16. These and several other nitrogen transformations are summarized in the redox cycle shown in Figure 16.42.

Nitrogen Fixation

Several of the key redox reactions of nitrogen are carried out in nature almost exclusively by microorganisms, and so microbial involvement in the nitrogen cycle is of great importance (Figure 16.42). Thermodynamically, nitrogen

Key Processes and Prokaryotes in the Nitrogen Cycle	
Processes	**Example organisms**
Nitrification ($NH_4^+ \dashrightarrow NO_3^-$)	
$NH_4^+ \dashrightarrow NO_2^-$	*Nitrosomonas*
$NO_2^- \dashrightarrow NO_3^-$	*Nitrobacter*
Denitrification ($NO_3^- \dashrightarrow N_2$)	*Bacillus, Paracoccus, Pseudomonas*
N_2 Fixation ($N_2 + 8H \dashrightarrow NH_3 + H_2$)	
Free-living	
Aerobic	*Azotobacter* Cyanobacteria
Anaerobic	*Clostridium*, purple and green bacteria
Symbiotic	*Rhizobium Bradyrhizobium Frankia*
Ammonification (organic-N $\dashrightarrow NH_4^+$)	
	Many organisms can do this

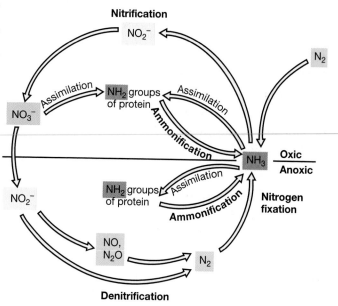

FIGURE 16.42 Redox cycle for nitrogen.

gas, N_2, is the most stable form of nitrogen, and this explains the fact that a major reservoir for nitrogen on Earth is the atmosphere. This is in contrast to carbon, for which the atmosphere is a relatively minor reservoir (CO_2, CH_4). The high energy necessary to break the $N \equiv N$ bond of molecular nitrogen (∞ Section 15.29) means that the reduction of N_2 is an energy-demanding process. Only a relatively small number of organisms are able to use N_2 in the process called **nitrogen fixation** ($N_2 + 8\,H^+ + 8\,e^- \rightarrow 2\,NH_3 + H_2$); thus, the recycling of nitrogen on Earth involves to a great extent the more easily available forms, ammonia and nitrate. However, because N_2 constitutes by far the greatest reservoir of nitrogen available to living organisms, the ability to use N_2 is of great ecological importance. In many environments, productivity is limited by the short supply of combined nitrogen compounds, putting a premium on biological nitrogen fixation. We have considered the basic process of N_2 fixation in Section 15.29 and revisit this important process in an agricultural context later in this chapter (Section 16.25) when we describe symbiotic N_2 fixation in leguminous plants.

Denitrification

We discussed the role of nitrate as an alternative electron acceptor in Section 15.16. Under most conditions, the end product of dissimilatory nitrate reduction is N_2 or N_2O, and the conversion of nitrate to gaseous nitrogen compounds is called **denitrification** (Figure 16.42). This process is the main means by which gaseous N_2 is formed biologically, and because N_2 is much less easily used by organisms than nitrate as a source of nitrogen, denitrification is a detrimental process because it removes fixed nitrogen from the environment. By contrast, denitrification can be a very beneficial process in wastewater treatment (∞ Section 11.15), where nitrate can be removed from the water, thus minimizing algal growth when the water is discharged into lakes and streams.

Ammonia Fluxes and Nitrification

Ammonia is produced during the decomposition of organic nitrogen compounds (**ammonification**) and exists at neutral pH as the ammonium ion (NH_4^+). Under anoxic conditions, ammonia is stable (but see Section 15.13), and it is in this form that nitrogen predominates in most anoxic sediments. In soils, much of the ammonia released by aerobic decomposition is rapidly recycled and converted to amino acids in plants and microorganisms. Because ammonia is volatile, some loss can occur from soils (especially highly alkaline soils) by vaporization, and major losses of ammonia to the atmosphere occur in areas of dense animal populations (for example, cattle feedlots). However, on a global basis, ammonia constitutes only about 15% of the nitrogen released to the atmos-

phere, the rest being primarily in the form of N_2 or N_2O (from denitrification).

Nitrification, the oxidation of NH_3 to NO_3^-, is a major process in nature and occurs readily in well-drained soils at neutral pH by activities of the nitrifying bacteria (∞ Sections 13.2 and 15.13) (Figure 16.42). Note that while denitrification is the process of nitrate *consumption*, nitrification is the process of nitrate *production*. If materials high in protein, such as manure or sewage, are added to soils, the rate of nitrification is increased. Although nitrate is readily assimilated by plants, it is very water-soluble and is rapidly leached from soils receiving high rainfall. Consequently, nitrification is not beneficial in agricultural practice. Ammonia, on the other hand, is cationic and consequently is strongly adsorbed to negatively charged clay minerals.

Anhydrous ammonia is used extensively as a nitrogen fertilizer, and chemicals are commonly added to the fertilizer to inhibit the nitrification process. One of the most common inhibitors of nitrification is a substituted pyridine compound called *nitrapyrin* (2-chloro-6-trichloromethylpyridine). Nitrapyrin specifically inhibits the first step in nitrification, the oxidation of NH_3 to NO_2^- (∞ Section 15.13), thus effectively inhibiting both steps in the nitrification process. The addition of nitrification inhibitors has greatly increased the efficiency of fertilization and has helped prevent pollution of waterways from nitrate leached from fertilized soils.

✓ 16.16 Concept Check

The principal form of nitrogen on Earth is N_2, which can be used as a nitrogen source only by the nitrogen-fixing bacteria. Ammonia produced by nitrogen fixation or by ammonification from organic nitrogen compounds can be assimilated into organic matter or it can be oxidized to nitrate by the nitrifying bacteria. Losses of nitrogen from the biosphere occur as a result of denitrification, in which nitrate is converted back to N_2.

- ✓ What is *nitrogen fixation* and why is it important?
- ✓ What is the process called that results in $NO_3^- \rightarrow N_2$?
- ✓ How does the compound *nitrapyrin* benefit both agriculture and the environment?

16.17

The Sulfur Cycle

Sulfur transformations are even more complex than those of nitrogen because of the variety of oxidation states of sulfur and the fact that some transformations occur at significant rates *chemically* as well as biologically. We discussed the processes of sulfate reduction and chemolithotrophic sulfur oxidation in Sections 15.11 and 15.17. The redox cycle for sulfur and the involvement of microorganisms in sulfur transformations are given in Figure 16.43. Although a number of oxidation states are pos-

Key Processes and Prokaryotes in the Sulfur Cycle

Process	Organisms
Sulfide/sulfur oxidation ($H_2S \dashrightarrow S^0 \dashrightarrow SO_4^{2-}$)	
Aerobic	Sulfur chemolithotrophs (*Thiobacillus, Beggiatoa*, many others)
Anaerobic	Purple and green phototrophic bacteria, some chemolithotrophs
Sulfate reduction (anaerobic) ($SO_4^{2-} \dashrightarrow H_2S$)	*Desulfovibrio, Desulfobacter,*
Sulfur reduction (anaerobic) ($S^0 \dashrightarrow H_2S$)	*Desulfuromonas*, many hyperthermophilic Archaea
Sulfur disproportionation ($S_2O_3^{2-} \dashrightarrow H_2S + SO_4^{2-}$)	*Desulfovibrio*, and others
Organic sulfur compound oxidation or reduction ($CH_3SH \rightarrow CO_2 + H_2S$) ($DMSO \rightarrow DMS$)	
Desulfurylation (organic–S $\dashrightarrow H_2S$)	Many organisms can do this

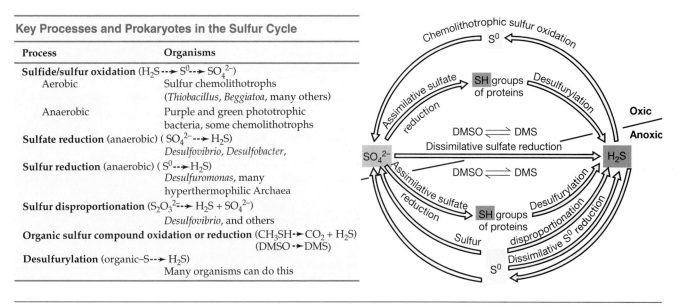

FIGURE 16.43   Redox cycle for sulfur. DMSO, dimethylsulfoxide; DMS, dimethylsulfide.

sible, only three form significant amounts of sulfur in nature, −2 (sulfhydryl, R—SH, and sulfide, HS⁻), 0 (elemental sulfur, S^0), and +6 (sulfate, SO_4^{2-}). The bulk of the sulfur of Earth is found in sediments and rocks in the form of sulfate minerals (primarily gypsum, $CaSO_4$) and sulfide minerals (primarily pyrite, FeS_2), although the oceans constitute the most significant reservoir of sulfur for the biosphere (in the form of inorganic sulfate). The global transport cycle for sulfur is given in Figure 16.44, and some of the components of this cycle are discussed later.

Hydrogen Sulfide and Sulfate Reduction

A major volatile sulfur gas is hydrogen sulfide (H_2S). This substance is produced from bacterial sulfate reduction ($SO_4^{2-} + 8 H \rightarrow H_2S + 2 H_2O + 2 OH^-$) (Figure 16.43) or is emitted from geochemical sources in sulfide springs or volcanoes. Although H_2S is volatile, the form of sulfide present in an environment is pH dependent; H_2S predominates below pH 7 while HS⁻ and S^{2-} are present above pH 7. Sulfate-reducing bacteria are widespread in nature, however in many anoxic habitats, such as freshwaters and many soils, their activities are limited by the low levels of sulfate present. And, because of the necessity of organic electron donors (or molecular hydrogen, which is a product of the fermentation of organic compounds) to drive sulfate reduction, sulfide production only occurs where significant amounts of organic material are present. In many marine sediments, the rate of sulfate reduction is carbon-limited, and the rate can be greatly increased by the addition of organic matter. This is of considerable importance for marine pollution because disposal of sewage, sewage sludge, and garbage in the sea can lead to marked increases in

organic matter in the sediments. Since HS⁻ is a toxic substance to many organisms, formation of HS⁻ by sulfate reduction is potentially detrimental. Sulfide is toxic because it combines with the iron of cytochromes and other essential iron-containing compounds in the cell. One common detoxification mechanism for sulfide in the environment is combination with iron, leading to formation of the insoluble FeS. The black color of many sediments where sulfate reduction is taking place is due to the accumulation of FeS.

Sulfide and Elemental Sulfur Oxidation/Reduction

Under oxic conditions, sulfide (HS⁻) rapidly oxidizes spontaneously at neutral pH (∞ Section 15.11). Sulfur-oxidizing bacteria are also able to catalyze the oxidation of sulfide, but because of the rapid spontaneous reaction, bacterial oxidation of sulfide occurs only in areas in which H_2S rising from anoxic areas meets O_2 descending from oxic areas. If light is available, anoxic oxidation of HS⁻ can also occur, catalyzed by the phototrophic sulfur bacteria (∞ Sections 13.1 and 15.5), but this occurs only in restricted areas, usually in lakes, where sufficient light can penetrate to anoxic zones.

Elemental sulfur, S^0, is chemically stable in most environments in the presence of oxygen but is readily oxidized by sulfur-oxidizing bacteria. Although a number of sulfur-oxidizing bacteria are known, members of the genus *Thiobacillus* (∞ Sections 13.3 and 15.11) are most commonly involved in elemental sulfur oxidation. Elemental sulfur is very insoluble, and the bacteria that oxidize it attach firmly to the sulfur crystals (∞ Figure 15.26). Oxidation of elemental sulfur results in the formation of sulfate and hydrogen ions, and sulfur oxidation

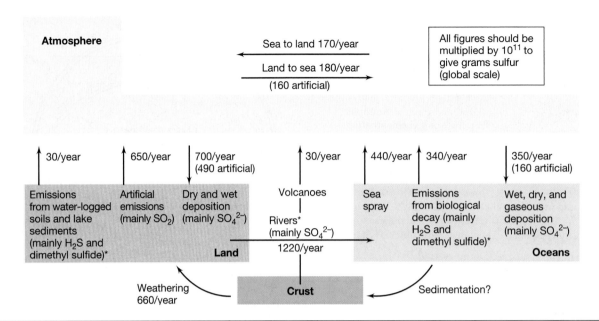

FIGURE 16.44   The global balance of sulfur. Artificial emissions are derived from human activities. An asterisk indicates a process that is partially or solely due to microbial action.

characteristically results in a *lowering* of the pH. Elemental sulfur is sometimes added to alkaline soils to effect a lowering of the pH, reliance being placed on the ubiquitous thiobacilli to carry out the acidification process.

Besides being oxidized, S^0 can also be reduced. Sulfur reduction to H_2S as a form of anaerobic respiration (∞ Sections 15.15 and 15.17) is a major ecological process, especially among hyperthermophilic Archaea (∞ Chapter 14). Although sulfate-reducing bacteria can also carry out this reaction, the bulk of S^0 reduction in nature probably occurs by the phylogenetically distinct S^0 reducers that are unable to reduce SO_4^{2-} to H_2S. However, the habitats of the S^0 reducers are generally those of the sulfate reducers, so from an ecological standpoint, the two groups coexist.

Organic Sulfur Compounds

In addition to the *inorganic* forms of sulfur whose biogeochemistry was just discussed, a vast array of *organic* sulfur compounds are also synthesized by living organisms, and these enter into biogeochemical sulfur cycling as well. Many of these foul-smelling compounds are highly volatile and can thus enter the atmosphere. The most abundant organic sulfur compound in nature is dimethyl sulfide ($H_3C-S-CH_3$). It is produced primarily in marine environments as a degradation product of dimethylsulfonium propionate, a major osmoregulatory solute in marine algae (∞ Section 5.11). Dimethylsulfonium propionate can be used as a carbon and energy source by microorganisms and is catabolized to dimethyl sulfide and acrylate; the latter compound, a derivative of the fatty acid propionate, is used to support growth.

Microbial production of dimethyl sulfide in nature is dramatic, about 45 million tons being produced annually. Dimethyl sulfide released to the atmosphere undergoes photochemical oxidation to methane sulfonate ($CH_3SO_3^-$), SO_2, and SO_4^{2-}, but dimethyl sulfide produced in anoxic habitats can be used microbiologically as a substrate for methanogenesis (yielding CH_4 and H_2S), as an electron donor for photosynthetic CO_2 fixation in phototrophic purple bacteria [yielding dimethyl sulfoxide (DMSO)], and as an electron donor in energy metabolism in certain chemoorganotrophs and chemolithotrophs (also yielding DMSO). Anaerobically, DMSO can be an electron acceptor for anaerobic respiration (∞ Section 15.20), once again yielding dimethyl sulfide. Many other organic sulfur compounds impact on the global sulfur cycle, including methanethiol (CH_3SH), dimethyl disulfide ($H_3C-S-S-CH_3$), and carbon disulfide (CS_2), but on a global basis, dimethyl sulfide production and consumption are quantitatively the most significant (Figure 16.44).

✓ 16.17 Concept Check

Bacteria play major roles in both the oxidative and reductive sides of the sulfur cycle. Sulfur- and sulfide-oxidizing bacteria *produce* sulfate, while sulfate-reducing bacteria *use* sulfate as electron acceptor in anaerobic respiration and produce hydrogen sulfide. Because sulfide is toxic and also reacts with various metals, sulfate reduction is an important biogeochemical process. Dimethyl sulfide is the major organic sulfur compound of ecological significance in nature.

✓ How many electrons are required to reduce SO_4^{2-} to H_2S?
✓ Why is acid generated from the bacterial oxidation of sulfur?
✓ What organic sulfur compound is most abundant in nature?

16.18

The Iron Cycle

Iron is one of the most abundant elements in Earth's crust. Iron exists naturally in two oxidation states, ferrous (Fe^{2+}) and ferric (Fe^{3+}). Fe^0 is a product of human activities in the smelting of ferrous or ferric iron ores to form cast iron. In nature then, iron cycles primarily between the ferrous and ferric forms, the reduction of Fe^{3+} occurring both chemically and as a form of anaerobic respiration, and the oxidation of Fe^{2+} occurring both chemically and as a form of chemolithotrophic metabolism (Figure 16.45). The only electron acceptor able to spontaneously oxidize Fe^{2+} is O_2, and we discussed in Section 15.12 how this can occur at neutral pH yielding highly insoluble ferric iron precipitates. By contrast, at low pH, Fe^{2+} is stable, and this allows the prolific growth of acidophilic iron-oxidizing bacteria.

Bacterial Iron Reduction

A number of organisms can use ferric iron as an electron acceptor. Ferric iron reduction is common in waterlogged soils, bogs, and anoxic lake sediments. Movement of iron-rich groundwater from anoxic bogs or waterlogged soils can result in the transport of large amounts of ferrous iron. Once this iron-laden water reaches oxic regions, the ferrous iron is oxidized chemically or by iron bacteria (∞ Section 15.12) and ferric compounds precipitate, leading to the formation of

brown iron deposits (∞ Figure 15.28c). The overall reaction of spontaneous ferrous iron oxidation is:

$$Fe^{2+} + \tfrac{1}{4}O_2 + 2\tfrac{1}{2}H_2O \rightarrow Fe(OH)_3 + 2H^+$$

The ferric hydroxide precipitate can interact with other nonbiological substances, such as humics (see Section 16.13), to reduce Fe^{3+} back to Fe^{2+} (Figure 16.45). Ferric iron can also form complexes with various organic constituents, thus becoming solubilized and more available to ferric iron-reducing bacteria.

Ferrous Iron and Pyrite Oxidation at Acid pH

In non-acidic habitats Fe^{2+} can be oxidized by iron bacteria such as *Gallionella* and *Leptothrix* (∞ Sections 13.14 and 13.15). This occurs primarily at interfaces between ferrous-rich anoxic ground waters and air. However, it is at low pH, where Fe^{2+} is stable, that the most extensive iron oxidation occurs, and we consider this process now. The acidophilic chemolithotroph *Thiobacillus ferrooxidans* and related acidophilic iron oxidizers can oxidize Fe^{2+} to Fe^{3+} at extremely low pH (Figure 16.46). However, because very little energy is generated in the oxidation of ferrous to ferric iron (∞ Section 15.12), these bacteria must oxidize large amounts of iron in order to grow, and consequently even a small number of cells can be responsible for precipitating a large amount of iron. This iron-oxidizing bacterium, which is a strict acidophile, is very common in acid mine drainages and in acid springs and is probably responsible for most of the ferric iron precipitated at acid pH values.

Thiobacillus ferrooxidans and *Leptospirillum ferrooxidans* live in environments in which sulfuric acid is the dominant acid and large amounts of sulfate are present. Under these conditions, ferric iron does not precipitate as the hydroxide but as a complex sulfate mineral called *jarosite* [$HFe_3(SO_4)_2(OH)_6$]. Jarosite is a yellowish or brownish precipitate and is responsible for one of the

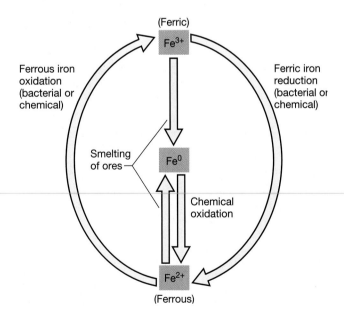

FIGURE 16.45 The redox cycle of iron. The major forms of iron in nature are Fe^{2+} and Fe^{3+}; Fe^0 is primarily a product of human activities in the smelting of iron ores. Ferrous iron oxidation occurs aerobically by the iron chemolithotrophs (or chemically at neutral pH) and anaerobically by certain anoxyphototrophic bacteria and denitrifying bacteria.

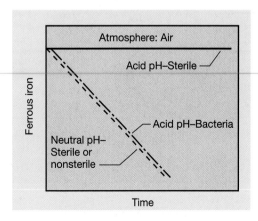

FIGURE 16.46 Oxidation of ferrous iron as a function of pH and the presence of the bacterium *Thiobacillus ferrooxidans*. Note how Fe^{2+} is stable under acidic conditions in the absence of the bacterial cells.

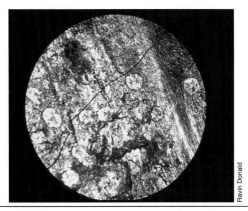

FIGURE 16.47 Pyrite in coal. Section through a piece of coal from the Black Mesa formation in northern Arizona (USA). The gold-colored spherical discs (about 1 mm in diameter) are sectioned particles of pyrite, FeS_2.

manifestations of acid mine drainage, an unsightly yellow stain called "yellow boy" by U.S. miners (◯◯ Figure 15.28a and see also Figure 16.50).

One of the most common forms of iron in nature is *pyrite*, which has the overall formula FeS_2. Pyrite is formed from the reaction of sulfur with ferrous sulfide (FeS) to form a highly insoluble crystalline structure, and pyrite is very common in bituminous coals and in many ore bodies (Figure 16.47). The bacterial oxidation of pyrite is of great significance in the development of acidic conditions in mining operations (Figure 16.48). Additionally, oxidation of pyrite by bacteria is of considerable importance in the process called *microbial leaching of ores* (see Section 16.19). The oxidation of pyrite is a combination of chemically and bacterially catalyzed reactions. Two electron acceptors are involved for this process: molecular oxygen (O_2) and ferric ions (Fe^{3+}).

When pyrite is first exposed, as in a mining operation, a slow chemical reaction with molecular oxygen occurs, as shown in Figure 16.49. This reaction, called the *initiator reaction*, leads to the oxidation of sulfide to sulfate and the development of acidic conditions under which the ferrous iron released is relatively stable in the presence of oxygen. *Thiobacillus ferrooxidans* then catalyzes the oxidation of ferrous to ferric ions. The ferric ions formed under these acidic conditions, being soluble, can readily react spontaneously with more pyrite to oxidize the pyrite to ferrous ions plus sulfate ions:

$$FeS_2 + 14\ Fe^{3+} + 8\ H_2O \rightarrow 15\ Fe^{2+} + 2\ SO_4^{2-} + 16\ H^+$$

The ferrous ions formed are again oxidized to ferric ions by the bacteria, and these ferric ions again react with more pyrite. Thus, there is a progressive, rapidly increasing rate at which pyrite is oxidized, called the *propagation cycle*, as illustrated in Figure 16.49. Under natural conditions some of the ferrous iron generated by the bacteria leaches away, being carried by ground water into surrounding streams. However, because oxygen is present in the aerated drainage, bacterial oxidation of the ferrous iron takes place in these outflows and an insoluble ferric precipitate is formed.

Bacterial oxidation of sulfide minerals is the major factor in the formation of **acid mine drainage,** a common environmental problem in coal-mining regions (Figure 16.50; ◯◯ also Figure 15.28a). Mixing of acidic mine waters with natural waters in rivers and lakes causes serious degradation in the quality of the natur-

(a)   (b)   (c)

FIGURE 16.48 Pyrite-rich microbial habitats in bituminous coal- and copper-mining environments. (a) Bituminous coal-mining operation in a strip mine. The shovel is removing the soil (overburden) to reach the coal seam. (b) The coal seam. Removal of the bituminous coal exposes the environment to air. The pyrite associated with the coal is colonized with iron-oxidizing bacteria. (c) Copper ore deposit rich in pyrite. Mining of the copper ore results in exposure of the formation to air.

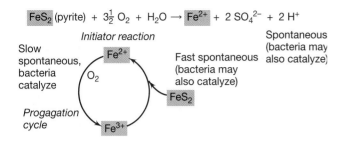

$$FeS_2 \text{ (pyrite)} + 3\tfrac{1}{2} O_2 + H_2O \rightarrow Fe^{2+} + 2 SO_4^{2-} + 2 H^+$$

FIGURE 16.49 Role of iron-oxidizing bacteria in oxidation of the mineral pyrite.

al water because both the acid and the dissolved metals are toxic to aquatic life (∞ Figures 15.28a and see also Figure 16.50). In addition, such polluted waters are unsuitable for human consumption and industrial use. The breakdown of pyrite leads ultimately to the formation of sulfuric acid and ferrous iron, and pH values can be as low as pH 2. The acid formed attacks other minerals associated with the coal and pyrite, causing breakdown of the whole rock fabric. A major rock-forming element, aluminum, is soluble only at low pH, and often several grams of Al^{3+}, which can be toxic to aquatic organisms, are present per liter of acid mine waters.

The requirement for O_2 in the oxidation of ferrous to ferric iron helps to explain how acid mine drainage

FIGURE 16.50 Acid mine drainage from a bituminous coal region. Note the yellowish-red color due to precipitated iron oxides. (∞ Figure 15.28a.)

develops. As long as the coal is unmined, oxidation of pyrite cannot occur because neither air nor the bacteria can reach it. When the coal seam is exposed (Figure 16.48), it quickly becomes contaminated with *Thiobacillus ferrooxidans*, and O_2 is introduced, making oxidation of pyrite possible. The acid formed can then leach into the surrounding streams (∞ Figure 15.28a, see also Figure 16.50).

✓ 16.18 Concept Check

Iron exists in nature primarily in two oxidation states, ferrous (Fe^{2+}) and ferric (Fe^{3+}), and bacterial and chemical transformation of these metals is of great geological and ecological importance. Bacterial ferric iron reduction occurs in anoxic environments and results in the mobilization of iron from swamps, bogs, and other iron-rich aquatic habitats. Bacterial oxidation of ferrous iron occurs on a large scale at low pH and is very common in coal-mining regions, where it results in a type of pollution called acid mine drainage.

- ✓ What form (ferrous or ferric) is iron in the mineral $Fe(OH)_3$? FeS? How is $Fe(OH)_3$ formed?
- ✓ Why does Fe^{2+} oxidation under *oxic* conditions occur mainly at *acidic* pH?

16.19

Microbial Leaching of Ores

We consider here a situation in which the process of acid production and metal solubility by acidophilic bacteria just discussed play a beneficial role in mining. Sulfide forms highly insoluble minerals with many metals, and many ores used as sources of these metals are sulfides. If the concentration of metal in the ore is low, it may not be economically feasible to concentrate the mineral by conventional chemical means. Under these conditions, **microbial leaching** is frequently practiced. Microbial leaching is especially useful for *copper* ores because copper sulfate, formed during oxidation of the copper sulfide ores, is very water-soluble. Indeed, approximately one-fourth of all copper mined worldwide is obtained from bioleaching.

We have noted that sulfide itself, HS^-, oxidizes spontaneously in air. Most metal sulfides also oxidize spontaneously, but the rate is very much slower than that of free sulfide. Bacteria such as *Thiobacillus ferrooxidans* are able to catalyze a much faster rate of oxidation of the sulfide minerals, thus aiding in solubilization of the metal. The relative rate of oxidation of a copper mineral in the presence and absence of bacteria is illustrated in Figure 16.51. The susceptibility to oxidation also varies among minerals, and those minerals that are most readily oxidized are most amenable to microbial leaching. Thus, iron and copper sulfide ores such as pyrrhotite

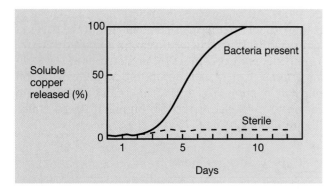

FIGURE 16.51 Effect of the bacterium *Thiobacillus ferrooxidans* on the leaching of copper from the mineral covellite. The leaching was done in a laboratory column, and the acid leach solution contained inorganic nutrients necessary for development of the bacterium. The leaching activity was monitored by assaying for soluble copper in the leach solution at the bottom of the column. The leach solution was continuously recirculated, maintaining an essentially closed system.

(FeS) and covellite (CuS) are readily leached, whereas lead and molybdenum ores are much less so.

The Leaching Process

In the microbial leaching process, low grade ore is dumped in a large pile (the leach dump) and a dilute sulfuric acid solution (pH about 2) is percolated down through the pile (Figure 16.52*a*). The liquid coming out of the bottom of the pile (Figure 16.52*b*), rich in the mineral, is collected and transported to a precipitation plant (Figure 16.52*c*) where the metal is reprecipitated and purified. The liquid is then pumped back to the top of the pile and the cycle is repeated. As needed, more acid is added to maintain the low pH.

There are several mechanisms by which the bacteria can catalyze oxidation of the sulfide minerals. To illustrate this, examples of the oxidation of two copper minerals will be used: chalcocite, Cu_2S, in which copper has a valence of +1, and covellite, CuS, in which copper has a valence of +2. As illustrated in Figure 16.53, *Thiobacillus ferrooxidans* is able to oxidize Cu^+ in chalcocite (Cu_2S) to Cu^{2+}, thus removing some of the copper in the soluble form, Cu^{2+}, and forming the mineral covellite. Note that in this reaction there is no change in the valence of sulfide, the bacteria using the reaction Cu^+ to Cu^{2+} as a source of energy. This is analogous to the oxidation by the same bacterium of Fe^{2+} to Fe^{3+}. Covellite can then be oxidized releasing sulfate and soluble Cu^{2+} (Figure 16.53).

A second mechanism, and probably the most important in most mining operations, involves *chemical* oxidation of the copper ore with *ferric* ions formed by the bacterial oxidation of ferrous ions (Figure 16.53). In almost any ore, pyrite is present, and its oxidation leads to the formation of ferric iron. Ferric iron is an excellent electron acceptor for sulfide minerals, and reaction of CuS

(a)   *(b)*   *(c)*   *(d)*

FIGURE 16.52 The leaching of low-grade copper ores using bacteria. (a) A typical leaching dump. The low-grade ore has been dumped in a large pile. Pipes distribute the acidic leach water over the surface of the pile. The acidic water slowly percolates through the pile and exits at the bottom. (b) Effluent from a copper leaching dump. The acidic water is very rich in dissolved copper. (c) Recovery of dissolved copper by passage of the copper-rich water over metallic iron in a long flume. (d) A small pile of recovered copper metal removed from the flume, ready for further purification.

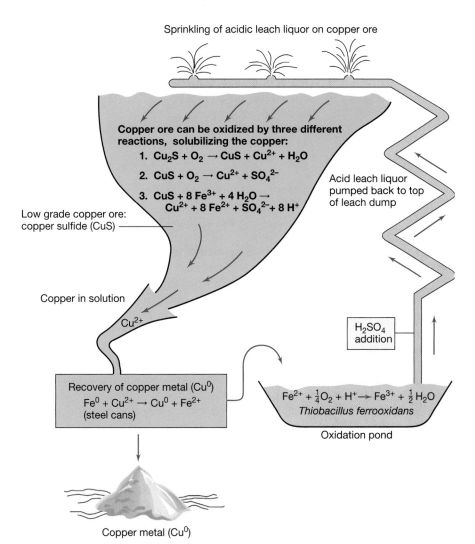

Sprinkling of acidic leach liquor on copper ore

Copper ore can be oxidized by three different reactions, solubilizing the copper:

1. $Cu_2S + O_2 \rightarrow CuS + Cu^{2+} + H_2O$
2. $CuS + O_2 \rightarrow Cu^{2+} + SO_4^{2-}$
3. $CuS + 8\,Fe^{3+} + 4\,H_2O \rightarrow Cu^{2+} + 8\,Fe^{2+} + SO_4^{2-} + 8\,H^+$

Low grade copper ore: copper sulfide (CuS)

Acid leach liquor pumped back to top of leach dump

Copper in solution

Cu^{2+}

H_2SO_4 addition

Recovery of copper metal (Cu^0)
$Fe^0 + Cu^{2+} \rightarrow Cu^0 + Fe^{2+}$
(steel cans)

$Fe^{2+} + \frac{1}{4}O_2 + H^+ \rightarrow Fe^{3+} + \frac{1}{2}H_2O$
Thiobacillus ferrooxidans

Oxidation pond

Copper metal (Cu^0)

FIGURE 16.53 Arrangement of a leaching pile and reactions involved in the microbial leaching of copper sulfide minerals to yield Cu^0 (copper metal). Reaction 1 is primarily bacterial while Reaction 2 occurs both biologically and chemically. Reaction 3 is strictly chemical, but is probably the most important reaction in copper-leaching processes. Note how it is essential for Reaction 3 to proceed that the Fe^{2+} produced (from the oxidation of sulfide in CuS to sulfate) be oxidized back to Fe^{3+} by *Thiobacillus ferrooxidans*.

with ferric iron results in solubilization of the copper and the formation of ferrous iron. In the presence of O_2, at the acid pH values involved, *Thiobacillus ferrooxidans* reoxidizes the ferrous iron back to the ferric form so it can oxidize more copper sulfide. Thus, the process is maintained by the oxidation of Fe^{2+} to Fe^{3+} by the bacterium.

Metal Recovery

Another source of iron in leaching operations is at the precipitation plant used in recovery of the soluble copper from the leaching solution (Figure 16.52c and d). Scrap iron, Fe^0, is used to recover copper from the leach liquid by the reaction shown in the lower part of Figure 16.53, and this results in the formation of Fe^{2+}. In most leaching operations, this Fe^{2+}-rich liquid remaining after the copper is removed is transferred to an oxidation pond where *Thiobacillus ferrooxidans* proliferates and forms Fe^{3+}. Acid is added to the pond to keep the pH low, thus keeping the Fe^{3+} in solution, and this ferric-rich liquid is then pumped to the top of the pile and the Fe^{3+} is available to oxidize more sulfide mineral.

Because of the huge dimensions of copper leach dumps, penetration of oxygen from air is poor, and the interior of these piles usually becomes anoxic. Although many of the reactions written in Figure 16.53 require molecular O_2, because *Thiobacillus ferrooxidans* can use Fe^{3+} as an electron *acceptor* in the absence of O_2, the oxidation of copper minerals can also proceed anaerobically; the large amounts of Fe^{3+} added to the leach solution from scrap oxidized iron drive the process forward, even under anoxic conditions. High temperature can also be a problem with leaching operations. *T. ferrooxidans* is a mesophile but temperatures inside a leach dump often rise spontaneously as a result of microbial activities. Thus thermophilic iron-oxidizing chemolithotrophs such as thermophilic *Thiobacillus* species and, at high temperatures (60–80°C), the acidophilic archaeon *Sulfolobus* (∞ Section 14.8) may be important in the leaching of ores above 40°C.

Other Leaching Processes

Microorganisms are also used in the leaching of uranium and gold-containing ores. Although *T. ferrooxidans* can oxidize U^{4+} to U^{6+} with O_2 as an electron acceptor, it is likely that the uranium leaching process depends more on the chemical oxidation of uranium by Fe^{3+}, with *T. ferrooxidans* contributing mainly through the re-oxidation of Fe^{2+} to Fe^{3+} as described for the leaching of copper ores (see Figure 16.53). The reaction observed is:

$$UO_2 + Fe_2(SO_4)_3 \rightarrow UO_2SO_4 + 2\ FeSO_4$$

$$(U^{4+}) \quad (Fe^{3+}) \quad\quad (U^{6+}) \quad\quad (Fe^{2+})$$

Unlike UO_2, the oxidized uranium mineral is soluble and can be recovered by other processes.

Gold is frequently found in nature associated with minerals containing arsenic and pyrite. In the microbial leaching of gold, *T. ferrooxidans* is used to attack and solubilize the arsenopyrite minerals and, in the process, releasing the trapped gold (Au):

$$2\ FeAsS\ [Au] + 7\ O_2 + 2\ H_2O + H_2SO_4 \rightarrow$$
$$Fe_2(SO_4)_3 + 2\ H_3AsO_4 + [Au]$$

The gold is then complexed with cyanide by traditional gold-mining methods. Unlike copper, where leaching occurs in a huge leach dump (see Figure 16.52), gold leaching usually takes place in relatively small *bioreactor tanks* (Figure 16.54); bioleaching in this fashion has been shown to release greater than 95% of the trapped gold. And, although arsenic and cyanide are toxic residues from the mining process, both are removed in the bioreactor gold-leaching process, arsenic as a ferric precipitate and cyanide (CN^-) by its microbial oxidation to CO_2 and urea in later stages of the gold recovery process. Small scale microbial gold leaching as an alternative to large scale methods is becoming more popular and is beginning to replace the more costly and environmentally damaging conventional gold-mining techniques.

✓ 16.19 Concept Check

Oxidation of copper ores by bacteria can lead to the solubilization of copper, a process called microbial leaching. Leaching is important in the recovery of copper and uranium from low grade ores. Bacterial oxidation of iron in the iron sulfide

FIGURE 16.54 Photo of the gold leaching tanks at the Ashanti Goldfields, Ghana, Africa. Within the tanks, a mixture of *Thiobacillus ferrooxidans*, *Thiobacillus thiooxidans*, and *Leptospirillum ferrooxidans* solubilizes the pyrite/arsenic mineral containing trapped gold and releases the gold.

mineral pyrite is also an important part of the microbial leaching process because the ferric iron produced is itself an oxidant of ores.

✓ How is CuS oxidized under *anoxic* conditions?
✓ Why is it important to keep the leach liquor acidic in the copper ore leaching process?
✓ From the standpoint of metal oxidation, how does the copper leach process differ from the gold leach process?

16.20

Mercury and Heavy Metal Transformations

Trace elements are elements that are present in low concentrations in rocks, soils, waters, and the atmosphere. Some trace elements (for example, cobalt, copper, zinc, nickel, molybdenum) are nutrients (∞ Section 4.2), but a number of trace elements in high concentrations are actually toxic to organisms. Of these toxic elements, several are sufficiently volatile that they exhibit significant atmospheric transport, and hence are of some environmental concern. These include mercury, lead, arsenic, cadmium, and selenium. Many of these trace elements undergo redox reactions catalyzed by microorganisms, and several are also converted to organic form via microbial action. Because of environmental concern and significant microbial involvement, we focus our discussion on the biogeochemistry of the element mercury.

Global Cycling of Mercury and Methylmercury

Although mercury is present in extremely low concentrations in most natural environments, averaging about 1 nanogram (ng)/liter, it is a widely used industrial product and is the active component of many pesticides that have been introduced into the environment. Because of its unusual ability to be concentrated in living tissues and its high toxicity, mercury is of considerable environmental importance. The mining of mercury ores and the burning of fossil fuels release about 40,000 *tons* of mercury into the environment each year; an even greater amount is released by natural geochemical processes. Other anthropogenic sources of mercury include the electronics industry, especially battery and wiring production, the chemical industry, and the burning of municipal wastes.

The major form of mercury in the atmosphere is elemental mercury (Hg^0), which is volatile and is oxidized to mercuric ion (Hg^{2+}) photochemically; most of the mercury entering aquatic environments is thus Hg^{2+} (Figure 16.55). Mercuric ion readily adsorbs to particulate matter and can be metabolized from there by microorganisms. The major microbial reaction observed is the *methylation* of mercury, yielding methylmercury, CH_3Hg^+ (Figure 16.55). Methylmercury is particularly toxic because it can be absorbed through the skin. But in addition, it is soluble and can be concentrated in the aquatic food chain, primarily in

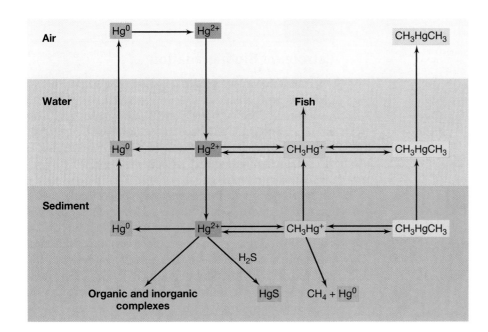

FIGURE 16.55 Biogeochemical cycling of mercury. The major reservoirs of mercury are in water and in sediments where it can be concentrated in animal tissues or precipitated out as HgS. The various forms of mercury commonly found in aquatic environments are each shown in a different color.

fish, or further methylated by microorganisms to yield the volatile compound dimethylmercury, $CH_3—Hg—CH_3$. Metabolically, methylation of mercury occurs by donation of methyl groups from $CH_3—B_{12}$.

Both methylmercury and dimethylmercury bond to proteins and tend to accumulate in animal tissues, especially muscle. Methylmercury is about 100 times more toxic than Hg^0 or Hg^{2+} and can be concentrated in fish, where it is a potent neurotoxin, eventually causing death. Methylmercury is thus a major environmental toxin, and its accumulation seems to be a particular problem in freshwater lakes where enhanced levels of methylmercury have been observed in fish caught for human consumption. Mercury can also cause liver and kidney damage in humans and other animals.

Several other mercury transformations occur on a global scale, including reactions involving sulfate-reducing bacteria ($H_2S + Hg^{2+} \rightarrow HgS$) and methanogens ($CH_3Hg^+ \rightarrow CH_4 + Hg^0$) (see Figure 16.55). The solubility of HgS is very low, so in anoxic sulfate-reducing sediments, most mercury is found as HgS. But on aeration, oxidation of HgS can occur, primarily by thiobacilli, leading to the formation of Hg^{2+} and eventually methylmercury.

Mercuric Resistance

At sufficiently high concentrations, Hg^{2+} and CH_3Hg^+ can be toxic not only to higher organisms but also to microorganisms. However several bacteria can carry out the biotransformation of toxic forms of mercury to nontoxic forms. In mercury-resistant gram-negative bacteria an NADPH-linked enzyme called *mercuric reductase* transfers two electrons to Hg^{2+}, reducing it to Hg^0. The Hg^0 produced in this reaction is volatile but is essentially nontoxic to humans and microorganisms, compared

with Hg^{2+} or CH_3Hg^+. Bacterial conversion of Hg^{2+} to Hg^0 then allows more CH_3Hg^+ to be converted to Hg^{2+}.

Mercury resistance has been intensively studied in the gram-negative bacterium *Pseudomonas aeruginosa*, where genes for mercury resistance reside on a plasmid. These genes, called *mer* genes, are arranged in an operon and are under control of the regulatory protein MerR (the product of *merR*) (Figure 16.56a). Interestingly, MerR functions as both a repressor and an activator (∞ Chapter 7). In the absence of Hg^{2+}, MerR binds to the operator region of the *mer* operon and prevents transcription of *mer TPCAD* genes (Figure 16.56a). However, when Hg^{2+} is present, it forms a complex with MerR, which then functions as an *activator* of transcription of the *mer* operon. The mercuric reductase, mentioned previously, is the product of the *merA* gene. MerD, the product of *merD*, also plays a regulatory role, whereas *merP* encodes a periplasmic Hg^{2+} binding protein (Figure 16.56b). This protein, MerP, binds Hg^{2+} and transfers it to a membrane protein MerT (the product of *merT*), which transports Hg^{2+} into the cell for reduction by mercuric reductase (Figure 16.56b). The final result is reduction of Hg^{2+} to Hg^0, which is volatile and is released from the cell (Figure 16.56b). In other organisms, additional *mer* genes have been found but the ones described here seem to be conserved in all *mer* operons examined.

Resistance to Other Heavy Metals

A variety of plasmids (∞ Section 9.8) isolated from both gram-positive and gram-negative Bacteria have been found to encode resistance to the effects of heavy metals. Certain antibiotic resistance plasmids also have genes for resistance to mercury and arsenic. Other plasmids encode only heavy-metal resistances. A large plasmid isolated from *Staphylococcus aureus* has been found

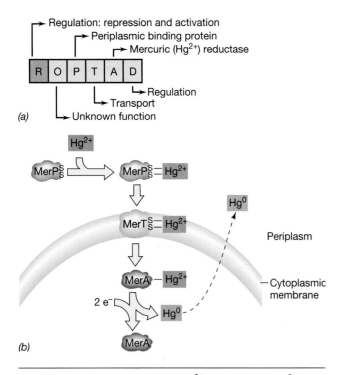

FIGURE 16.56   Mechanism of Hg^{2+} reduction to Hg^0 in *Pseudomonas aeruginosa*. (a) The *mer* operon. MerR can function as either a repressor (absence of Hg^{2+}) or transcriptional activator (presence of Hg^{2+}). (b) Transport and reduction of Hg^{2+}. Hg^{2+} is bound by cysteine residues in both proteins MerP and MerT.

to encode resistance to mercury, cadmium, arsenate, and arsenite. The mechanism of resistance to any specific metal varies. For example, arsenate and cadmium resistances are due to the action of enzymes that immediately pump out any arsenate or cadmium ions incorporated, thus preventing the metals from denaturing proteins.

Studies on nickel- and cobalt-resistant bacteria have shown that in most cases the resistance genes are plasmid-borne; resistance to both metals on a single plasmid is typical. Enrichment culture studies have shown that nickel-resistant bacteria are uncommon in soils and other environments in which this metal is absent in significant amounts. Bacteria highly resistant to nickel or other metals are most common in wastewaters of the metal processing industry or in mining operations where heavy metals are leached out along with iron or copper ores.

✓ 16.20 Concept Check

A major toxic form of mercury is methylmercury. The ability of bacteria to resist the toxicity of heavy metals is often due to the presence of specific plasmids that encode enzymes capable of detoxifying the metals.

✓ What forms of mercury are most toxic to organisms?
✓ How is mercury detoxified by bacteria?

Petroleum Biodegradation

Microbial decomposition of petroleum and petroleum products is of considerable economic and environmental importance. Because petroleum is a rich source of organic matter and the hydrocarbons within it are readily attacked aerobically by a variety of microorganisms, it is not surprising that when petroleum is brought into contact with air and moisture, it is subject to microbial attack. Under some circumstances such as in bulk storage tanks, microbial growth is not desirable. However, in other situations, such as in oil spills, microbial utilization of oil is desirable and may even be promoted by the addition of inorganic nutrients. The term *bioremediation* refers to the cleanup of oil or other pollutants by microorganisms, and in recent years the importance of bioremediation in oil spills has been amply demonstrated in several major crude oil spills in the marine environment.

Hydrocarbon Decomposition

A wide variety of bacteria, several molds and yeasts, and certain cyanobacteria and green algae have been shown to be able to oxidize hydrocarbons. Small-scale oil pollution of aquatic and terrestrial ecosystems from human as well as natural activities is very common, and hence it is not surprising that a diverse microbial community exists capable of using hydrocarbons as an electron donor. Methane, the simplest hydrocarbon, is degraded by only a specialized group of bacteria, the *methanotrophic* bacteria (⚬⚬ Sections 13.5 and 15.14), and these organisms do not oxidize higher hydrocarbons.

Hydrocarbon-oxidizing microorganisms develop rapidly on oil films and slicks. However, as we saw in Section 15.28, aliphatic hydrocarbons are not fermentable. Thus, significant aliphatic hydrocarbon oxidation occurs only in the presence of O_2. Even in oxic environments, hydrocarbon-oxidizing microorganisms are active only if other environmental conditions, such as temperature, pH, and inorganic nutrients, are adequate. Because oil is

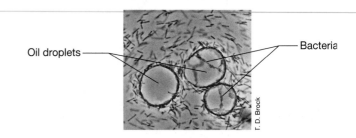

FIGURE 16.57   Hydrocarbon-oxidizing bacteria in association with oil droplets. The bacteria are concentrated in large numbers at the oil–water interface but are not within the droplet itself.

insoluble in water and is less dense, it floats to the surface and forms slicks. Hydrocarbon-oxidizing bacteria are able to attach to insoluble oil droplets and can often be seen there in large numbers (Figure 16.57). The action of these bacteria eventually leads to decomposition of the oil and dispersal of the slick.

Microorganisms participate in oil spill cleanups by oxidizing the oil to CO_2. In large spills, volatile hydrocarbon fractions evaporate quickly, leaving longer-chain aliphatic and aromatic components for cleanup crews or microorganisms to tackle. In oil spills where careful bioremediation studies have been performed, it has been shown that hydrocarbon-oxidizing bacteria increase in number 10^3–10^6 times shortly after the oil spill.

(a)

(b)

FIGURE 16.58 Environmental consequences of large oil spills in the marine environment and the effect of bioremediation. (a) A contaminated beach along the coast of Alaska containing oil from the *Exxon Valdez* spill of 1989. (b) The center rectangular plot was treated with inorganic nutrients to stimulate bioremediation of spilled oil by microorganisms, whereas areas to the left and right were untreated.

Experiments using radioisotopic hydrocarbons as tracers or O_2 uptake as a measure of heterotrophic activity have shown that under ideal conditions up to 80% of the nonvolatile components are oxidized by bacteria within a year of the spill. Certain fractions, such as branched-chain and polycyclic hydrocarbons, however, remain in the environment much longer. Spilled oil that travels to the sediments is only slowly degraded and can have a significant long-term impact on fisheries and related activities that depend on unpolluted waters for productive yields.

Large oil spills are not common, but when they do occur, such as the 11-million-gal spill from the supertanker *Exxon Valdez* that ran aground near Prince William Sound, Alaska, in March 1989, the input of oil to the environment can be ecologically devastating (Figure 16.58a) and cleanup costs staggering (estimated at $1.5 billion).

Addition of inorganic nutrients such as phosphorus and nitrogen to oil spill areas can increase bioremediation rates significantly. In the *Exxon Valdez* spill, for example, accelerated bioremediation of oil washed up on beaches was observed following spraying of the beaches with a mixture of inorganic nutrients (Figure 16.58b). For most large-scale oil spills, however, a combination of both human and microbial efforts is needed to effect a satisfactory cleanup in a reasonable amount of time.

Interfaces where oil and water meet often occur on a large scale. For example, it is virtually impossible to keep moisture from bulk-fuel storage tanks; it accumulates as a layer of water beneath the petroleum. Gasoline storage tanks (Figure 16.59) are thus potential habitats for hydrocarbon-oxidizing microorganisms, which can accumulate and grow at the oil–water interface. Gasoline generally has additional chemicals (added to aid combustion and inhibit corrosion in engines) that also inhibit the growth of microorganisms, but not all fuels have such additives.

FIGURE 16.59 Bulk fuel storage tanks, where massive microbial growth may occur at oil–water interfaces.

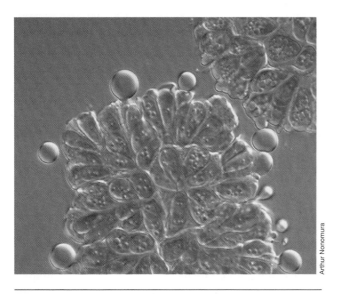

Arthur Nonomura

FIGURE 16.60   Photomicrograph by Nomarski interference contrast of cells of the green alga *Botryococcus braunii*. Note oil droplets, produced and excreted by this alga, at the margin of the cells.

Petroleum Production

Although microbial degradation of petroleum can be quite extensive, microbial *production* of hydrocarbons also occurs, particularly in certain green algae. For example, in the colonial alga *Botryococcus braunii*, growth of the alga is accompanied by the excretion of long-chain hydrocarbons (C_{30} to C_{36}) that have the consistency of oil (Figure 16.60). In *B. braunii* about 30% of the cell dry weight is petroleum, and there has been interest in using this and other oil-producing algae as renewable sources of petroleum. There is even evidence that oil in certain types of oil shale originated from green algae like *B. braunii* that grew in lake beds in ancient times.

✓ 16.21 Concept Check

Hydrocarbons are generally stable in anoxic environments but are subject to microbial attack in oxic habitats. Addition of inorganic nutrients is often important for bioremediation of spilled oil. Some algae can produce hydrocarbons.

- ✓ What is *bioremediation*?
- ✓ Why might the addition of inorganic nutrients stimulate oil degradation while the addition of glucose might not?

16.22

Biodegradation of Xenobiotics

Xenobiotics are chemically synthesized compounds that are not naturally occurring. Xenobiotics include a long list of compounds including pesticides, polychlorinated biphenyls (PCBs, used in the electric-generating and related industries), munitions, dyes, and

chlorinated solvents. Many xenobiotics are structurally related, sometimes closely structurally related, to natural compounds, and thus can be slowly degraded by enzymes that already exist to degrade these natural compounds. In other cases the compounds are structurally so different from anything organisms have previously experienced that their degradation rate in nature is extremely slow, if at all. Nevertheless, for many xenobiotic compounds microorganisms have been found that can degrade them, and we focus here on pesticides as an example of the potential of microbial degradation.

Pesticides

Some of the most widely distributed xenobiotics are the **pesticides,** which are common components of toxic wastes. Over 1000 pesticides have been marketed for chemical pest control purposes. These include primarily *herbicides, insecticides,* and *fungicides*. Pesticides are of a wide variety of chemical types, such as chlorophenoxyalkyl carboxylic acids, substituted ureas, nitrophenols, triazines, phenylcarbamates, organochlorines and organophosphates, and others (Figure 16.61). Some of these substances are suitable as carbon sources and electron donors for certain soil microorganisms, whereas others are not. If a substance can be attacked by microorganisms, it will eventually disappear from the soil. Such degradation in the soil is usually desirable because toxic accumulations of the compound are avoided. However, even closely related compounds may differ remarkably in their degradability, as is shown for the relative persistence rates of a number of herbicides in Table 16.6.

The figures in Table 16.6 are only approximate because a variety of environmental factors, such as temperature, pH, aeration, and organic matter content of the soil, influence decomposition. Some of the chlorinated insecticides are so recalcitrant that they have persisted for over 10 years. Disappearance of a pesticide from an ecosystem does not necessarily mean that it was degraded by microorganisms because pesticide loss can also occur by volatilization, leaching, or spontaneous chemical breakdown. *Bioavailability* is also a factor governing microbial attack on xenobiotic compounds. Many xenobiotics are quite hydrophobic and thus not very soluble in water, and sorption of these compounds to organic matter and clay in soils and sediments prevents access to the organism.

The organisms that are able to metabolize pesticides and herbicides are fairly diverse, including genera of both bacteria and fungi. Some pesticides can be both carbon and energy sources and are oxidized completely to CO_2. However, other compounds are much more recalcitrant and are attacked only slightly or not at all, although they may often be degraded either partially or totally provided some other organic material is pre-

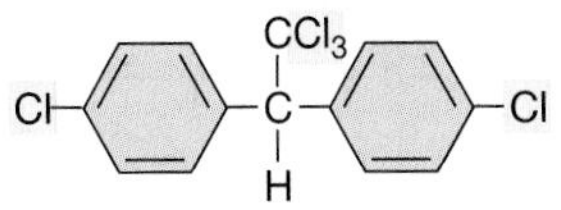

DDT; dichlorodiphenyltrichloroethane
(an organochlorine)

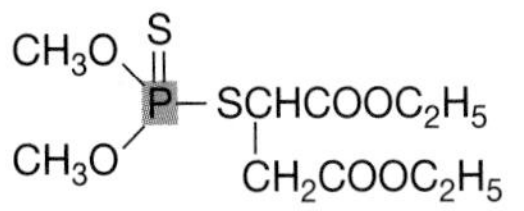

Malathion; mercaptosuccinic acid diethyl ester
(an organophosphate)

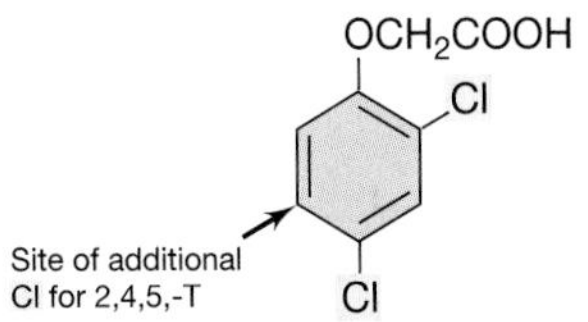

Site of additional
Cl for 2,4,5,-T

2,4-D; 2,4-dichlorophenoxy acetic acid
(a chlorophenoxy acetic acid derivative)

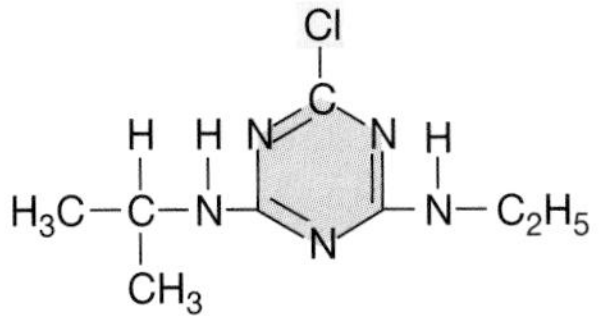

Atrazine, 2-chloro-4-ethylamino-6-isopropylaminotriazine
(a triazine derivative)

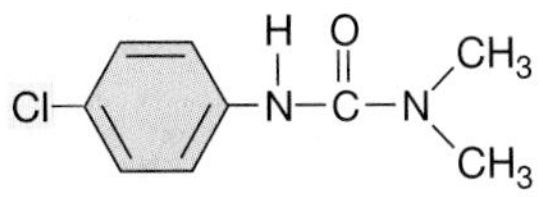

Monuron; 3-(4-chlorophenyl)-1,1-dimethylurea
(a substituted urea)

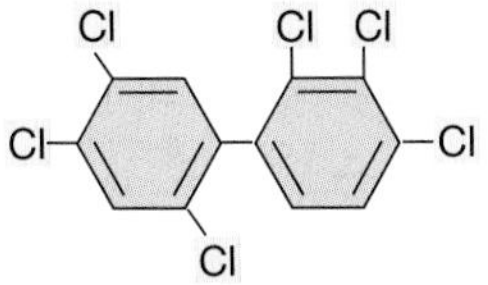

Chlorinated biphenyl (PCB);
shown is 2, 3, 4, 2', 4', 5'- Hexachlorobiphenyl

FIGURE 16.61 Some xenobiotic compounds. Although none of these compounds exist naturally, various microorganisms exist (or have been developed experimentally) that will break them down.

Substance	Time for 75–100% disappearance
Chlorinated insecticides	
DDT [1,1,1-trichloro-2,2-bis-(ρ-chlorophenyl)ethane]	4 years
Aldrin	3 years
Chlordane	5 years
Heptachlor	2 years
Lindane (hexachlorocyclohexane)	3 years
Organophosphate insecticides	
Diazinon	12 weeks
Malathion	1 week
Parathion	1 week
Herbicides	
2,4-D (2,4-dichlorophenoxyacetic acid)	4 weeks
2,4,5-T (2,4,5-trichlorophenoxyacetic acid)	20 weeks
Dalapin	8 weeks
Atrazine	40 weeks
Simazine	48 weeks
Propazine	1.5 years

TABLE 16.6 Persistence of herbicides and insecticides in soils

sent as primary energy source, a phenomenon called **cometabolism** or **cooxidation.** However, when the breakdown is only partial, the microbial degradation product of a pesticide may sometimes be even more toxic than the original compound.

Reductive Dechlorination

Significant chlorinated pesticide degradation occurs in anoxic environments. In these cases, anoxic biodegradation is linked to *reductive dechlorination* of the mole-

cule, the dechlorinated derivative being much less toxic than the original chlorinated molecule. For example, a sulfate-reducing bacterium implicated in this process, *Desulfomonile*, reduces 3-chlorobenzoate (used as a model compound for studies of chlorinated pesticide degradation) to benzoate and Cl^-:

$$\underset{\text{3-Chlorobenzoate}}{C_7H_4O_2Cl^-} + 2\,H \rightarrow \underset{\text{Benzoate}}{C_7H_5O_2^-} + HCl$$

Electrons for this reduction can come from acetate, formate, or H_2, and the reductive reaction is linked to establishment of a proton motive force that *Desulfomonile* uses to drive adenosine triphosphate (ATP) synthesis. Thus, the reductive dechlorination of 3-chlorobenzoate is a type of anaerobic respiration (∞ Section 15.15) (Table 16.7). Anoxic reductive dehalogenation can also result in methanogenesis, either from syntrophic degradation (∞ Sections 15.23 and 16.14) of the released benzoate or in the case of compounds like chloroform ($CHCl_3$), by pure cultures of methanogens.

Direct evidence for reductive dechlorination has also been obtained in the case of dichloroethylene, trichloroethylene, tetrachloroethylene (perchloroethylene), chloroform, dichloromethane, and certain brominated and fluorinated compounds. These toxic compounds, some of which (particularly trichloroethylene) are suspected of being carcinogenic, are widely used as industrial solvents and degreasing agents and are among the most frequently detected groundwater contaminants in the United States. Besides *Desulfomonile*, a variety of genera of bacteria are known to reductively dechlorinate, and some of these can use *only* chlorinated compounds as anaerobic electron

TABLE 16.7	Characteristics of major genera of bacteria capable of reductive dechlorination		
	Genus		
Property	*Dehalobacter*	*Desulfomonile*	*Desulfitobacterium*
Electron donors	H_2	H_2, formate, pyruvate, lactate, benzoate	H_2, formate, pyruvate, lactate
Electron acceptors	Trichloroethylene, tetrachloroethylene	Metachlorobenzoates, tetrachloroethylene, SO_4^{2-}, SO_3^{2-}, $S_2O_3^{2-}$	Ortho, meta or para chlorophenols, NO_3^-, fumarate, SO_3^{2-}, $S_2O_3^{2-}$, S^0
Product of reduction of tetrachloroethylene	Dichloroethylene	Dichloroethylene	Trichloroethylene
Other properties[a]	Contains cytochrome b	Contains cytochrome c_3; requires organic carbon source; can grow by fermentation of pyruvate	Can also grow by fermentation
Phylogeny[b]	Related to low GC gram-positive bacteria	Related to delta Proteobacteria	Related to low GC gram-positive bacteria

	Genus		
Property	*Dehalospirillum*	*Dehalococcoides*	*Dehalobacterium*
Electron donors	H_2, formate, pyruvate, lactate, benzoate	H_2 lactate	CH_2Cl_2 lactate, ethanol, glycerol
Electron acceptors	Trichloroethylene, tetrachloroethylene, NO_3^-, fumarate	Trichloroethylene, tetrachloroethylene	CO_2
Product of reduction of tetrachloroethylene	Dichloroethylene	Ethene	—
Other properties[a]	Contains b- and c-type cytochromes	Lacks peptidoglycan	Makes acetate and formate from CH_2Cl_2 plus CO_2. Can grow only on CH_2Cl_2 plus CO_2
Phylogeny[b]	Related to epsilon Proteobacteria	Unique lineage (kingdom?) of Bacteria	Related to low GC gram-positive bacteria

a All organisms are obligate anaerobes.
b See Chapter 13 for detailed discussion of prokaryotic phylogeny.

acceptors (Table 16.7). The product of tetrachloroethylene degradation by all of these organisms except one is dichloroethylene. The bacterium *Dehalococcoides* is unique because it removes *all* of the chlorines from polychlorinated ethylenes leaving only the gas ethene ($H_2C\!=\!CH_2$) as the final product (Table 16.7). *Dehalococcoides* is also unique because it lacks peptidoglycan in its cell wall and because phylogenetically it does not fall into any of the major lineages of Bacteria and may actually form its own kingdom within this domain.

The organism *Dehalobacterium* is also unique because it uses *only* the one-carbon compound *dichloromethane* (also known as *methylene chloride*, CH_2Cl_2) plus CO_2 in its energy metabolism (Table 16.7). *Dehalobacterium* is a strict anaerobe and converts these two substrates into the fatty acids formate and acetate as follows:

$$3\,CH_2Cl_2 + CO_2 + 4\,H_2O \rightarrow$$
$$2\,CHOOH + CH_3COOH + 6\,HCl$$

Energy conservation in *Dehalobacterium* occurs from substrate-level phosphorylation (∞ Section 15.21) during the formation of acetate and possibly also during the for-

mation of formate, since the metabolic pathway of dichloromethane degradation occurs via a tetrahydrofolate derivative, which, like acetyl phosphate, is also a high-energy compound (∞ Table 15.5). The restriction of *Dehalobacterium* to the catabolism of only dichloromethane indicates that it is an extreme nutritional specialist. Thus, this organism probably is most abundant in habitats in which methylene chloride, a widely used industrial degreasing and cleaning fluid and also a solvent in the food processing industry, is present in significant quantities, such as solvent contaminated soils, sediments, or wastewater treatment facilities.

Reductive dehalogenation of polychlorinated biphenyls (PCBs) (Figure 16.61) has been shown to occur in anoxic sediments, although pure cultures able to do this are not yet available. The removal of chlorines from PCBs leaves the more rapidly degradable biphenyl molecule, which can be readily degraded aerobically by pseudomonads and slowly catabolized anaerobically by methanogenic consortia (see Section 16.14).

Aerobic dechlorination of chlorinated organic compounds also occurs (Figure 16.62). But here, of course,

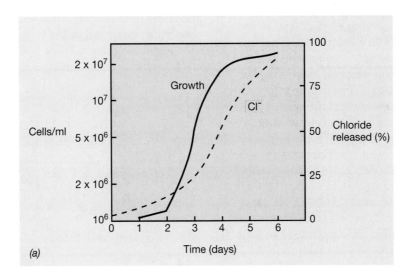

(a)

(b) 2,4,5,-T

FIGURE 16.62 Biodegradation of the herbicide 2,4,5-T. (a) Growth of *Burkholderia* (formerly *Pseudomonas*) *cepacia* on 2,4,5-T as sole source of carbon and energy. The strain was enriched from nature using a chemostat to keep the concentration of herbicide low. Growth here is aerobic on 1.5 g/l of 2,4,5-T. The release of chloride from the molecule is indicative of biodegradation. (b) Pathway of aerobic 2,4,5-T biodegradation. Note the steps in which Cl⁻ is released. The final products, succinate and acetate, are catabolized in the citric acid cycle. For the mechanism of action of a dioxygenase, see Figure 15.66c.

biochemical mechanisms involving O_2 can come into play, and it is often observed that the degradation of chlorinated aromatic compounds occurs by way of oxygenases (☞ Section 15.27). For example, in the aerobic degradation of the pesticide 2,4,5-T by pseudomonads, following dechlorination a diooxygenase enzyme breaks down the aromatic ring to generate compounds that can be metabolized by central metabolic pathways (Figure 16.62). Although aerobic breakdown of chlorinated organic xenobiotics is undoubtedly of ecological importance, reductive dechlorination is of particular environmental interest because of the rapidity with which anoxic conditions can develop in polluted microbial habitats in nature and the biochemical constraints (Figure 16.62) this puts on aerobic organisms that could otherwise degrade the compound.

Biodegradation of Xenobiotics and Microbial Evolution

The existence of organisms able to metabolize xenobiotics is of considerable evolutionary interest because these compounds are essentially new to Earth in the past 50 years or so. Observations on the rapidity with which organisms metabolizing new compounds arise can give us some idea of the rates of microbial evolution in general. In Section 18.12 we will discuss the evolution in past decades of plasmids conferring resistance to antibiotics. The evolution of pesticide-degrading bacteria seems to be a similar case. For example, enrichment cultures (see Section 16.4) yield bacteria capable of degrading 2,4,5-trichlorophenoxyacetic acid (2,4,5-T) and other recalcitrant pesticides. These organisms appear to be common *Pseudomonas* species that are now capable of growing on these pesticides as sole sources of carbon and energy (Figure 16.62). In a study of 2,4,5-T and 2,4-dichlorophenoxyacetic acid (2,4-D) biodegradation, it has been shown that portions of plasmids that code for 2,4-D biodegradation are "recruited" to form new plasmids conferring the ability to degrade 2,4,5-T. Because this can happen relatively quickly, it can be concluded that if biodegradation of a particular xenobiotic compound is possible, evolutionary events will move rather rapidly to establish microorganisms with new genetic properties to allow for breakdown of the compound. This assumes that a sufficient amount of the compound is present in the environment to maintain a selective advantage for biodegradation potential in the new population.

Biodegradation of Synthetic Polymers and the Landfill Crisis

A major area of environmental concern besides the biodegradation of toxic wastes like pesticides and other chlorinated hydrocarbons, is the disposable of *solid wastes*. Typical landfills contain large amounts of paper, food, construction and demolition debris, and plastics. Degradation rates of these materials are frequently very low because the conditions, especially the lack of moisture and probably oxygen, as well, are not suitable to rapid microbial activity. But in addition, some of the products added to landfills are inherently recalcitrant in the first place. Plastics are a typical example.

The plastics industry currently produces nearly 40 *billion* kilograms of plastic per year, approximately 40% of which is discarded in landfills. Plastics are xenobiotic polymers of various types, polyethylene, polypropylene, and polystyrene are typical examples (Figure 16.63). It is now recognized that many of these synthetic polymers are highly recalcitrant and remain essentially unaltered for decades in landfills and refuse dumps. This problem has fueled the search for *biodegradable* alternatives to the synthetic polymers now in use. Some success stories include photodegradable, starch-linked, and microbially synthesized plastics (see the box, Microbial Plastics).

Photobiodegradable plastics are polymers whose structure is altered by exposure to ultraviolet radiation (from sunlight), generating modified polymers amenable to microbial attack. Starch-based plastics incorporate starch as a linker to connect short fragments of a second biodegradable polymer. This design accelerates biodegradation because starch-digesting bac-

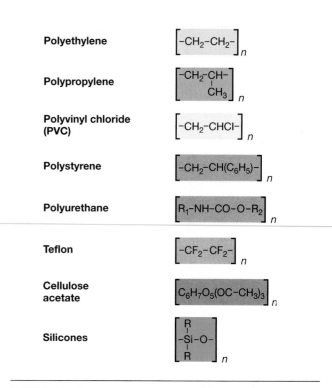

FIGURE 16.63 Synthetic polymers. Monomeric structure of a number of common synthetic polymers.

A FOCUS ON . . .   Microbial Plastics

Humans in developed countries have grown accustomed to life in a "plastic society." But unfortunately, much of this plastic is casually disposed of in the environment. How can this problem be minimized by the activities of microorganisms?

Poly-β-hydroxyalkanoates (PHAs) have been considered as synthetic plastic substitutes. PHAs are microbial storage polymers that have many of the general properties of synthetic plastics and can be synthesized by cells in various chemical forms. Depending on the length of the side chain in the monomers of the PHA polymer (a property that can be varied by modifying the growth medium or by genetically modifying the producing bacterium), PHAs of varying melting points, crystallinities, flexibilities, and tensile strengths are obtained, suitable for different plastics applications.

At a manufacturing plant in Billingham, England, the British chemical giant Imperial Chemical Industries (ICI) is producing PHA and marketing it as a packaging polymer under the trade name *Biopol*. ICI uses the bacterium *Ralstonia* (*Alcaligenes*) *eutrophus* to produce PHAs from corn syrup as feedstock. With *R. eutrophus,* ICI obtains yields of a poly-β-hydroxybutyrate/poly-β-hydroxyvalerate (PHB/PHV) copolymer (see Fig. 1) of greater than 80% of cell dry weight. The polymer is extracted from the cells and milled to a powder or pellet form for use in the production of injection-molded articles such as the shampoo bottle shown in Fig. 2. Because PHA production by ICI is still on a relatively small scale right now, about 600 tons/year, the current cost of a PHB/PHV bottle is considerably more than that of a synthetic plastic bottle. However, a major scale-up could put PHAs on a cost-competitive basis with petroleum-based plastics, especially in markets for environmentally conscious consumers. Moreover, because the metabolic diversity of bacteria is so substantial (∞ Chapter 15), a large variety of PHAs are possible; ones with unusual or particularly useful properties could boost market share, especially in niche markets where plastics of specific physical properties are often needed.

It is interesting that the field of polymer science, which has traditionally been dominated by chemical and physical scientists, has grown to include microbiologists who offer microbial solutions to the most serious problem in the synthetic polymer industry today: the ecological problem surrounding the recalcitrance of many synthetic disposable plastics. ■

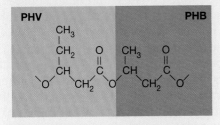

FIG. 1 PHV/PHB copolymer

FIG. 2 A brand of shampoo marketed in Europe and packaged in a bottle made of "bacterial plastic." The bottle consists of a copolymer of poly-β-hydroxybutyrate and poly-β-hydroxyvalerate (see Fig. 1). Because this material is a natural product, the bottle readily degrades both aerobically and anaerobically.

---

teria in soil attack the starch, releasing polymer fragments that are then degraded by other microorganisms. Microbially synthesized plastics include the carbon storage polymer poly-β-hydroxyalkanoate (∞ Section 3.13, and see the box). Despite their eco-friendliness, none of these biodegradable substitutes will help the landfill crisis until landfills are redesigned to make them more accessible to air and moisture. However, the use of biodegradable polymers for food and beverage containers could improve the disappearance rate of these materials when they are carelessly tossed into the environment.

✓ 16.22 Concept Check

Many chemically synthesized compounds such as insecticides, herbicides, and plastics (all called xenobiotics) are completely foreign to the environment and can persist because microorganisms capable of degrading them may not naturally occur.

✓ What chemical features are shared by the compounds shown in Figure 16.61?

✓ What is *reductive dechlorination* and in what types of environments does it occur?

✓ What advantages do *biopolymers* have over synthetic polymers?

16.23

Plant–Microorganism Interactions: Lichens and Mycorrhizae

As microbial habitats, plants are clearly vastly different from animals. Compared with warm-blooded animals, plants vary greatly in temperature, both diurnally and throughout the year, and compared with the complex circulatory system of animals, the internal communication system of the plant is only poorly developed, and so transfer of microorganisms within the plant is relatively inefficient. The aboveground parts of the plant, especially the leaves and stems, are subjected to frequent drying, and for this reason many plants have developed waxy coatings that retain moisture and keep out microorganisms. The roots, on the other hand, exist in an environment in which moisture is less variable and nutrient concentrations are higher. For this reason, the roots of plants are a main area of microbial action.

The **rhizosphere** is the region immediately outside the root; it is a zone where microbial activity is usually high. The bacterial count is almost always higher in the rhizosphere than it is in regions of the soil devoid of roots, often many times higher. This is because roots excrete significant amounts of sugars, amino acids, hormones, and vitamins, which promote such an extensive growth of bacteria and fungi that these organisms often form microcolonies on the root surface. The **phyllosphere** is the surface of the plant leaf, and under conditions of high humidity, as in wet forests in tropical and temperate zones, the microbial flora of leaves may be quite high, including fungi as well as bacteria (Figure 16.64). Many of the bacteria on leaves fix nitrogen (∞ Section 15.29, and see

Section 16.25), and nitrogen fixation presumably aids these organisms in growing with the predominantly carbohydrate nutrients provided by leaves. We proceed now to consider two highly developed associations between plants and microorganisms, lichens and mycorrhizae.

Lichens

Lichens are leafy or encrusting growths that are widespread in nature and are often found growing on bare rocks, tree trunks, house roofs, and surfaces of bare soils (Figure 16.65). The lichen plant consists of a symbiosis of two organisms, a fungus and an alga. However, little specificity resides in the relationship, as a given fungus can establish the lichen symbiosis with several different algae, and vice versa. The alga is phototrophic and is able to produce organic matter, which is then used for nutrition of the fungus. Because the fungus is unable to carry out photosynthesis, its ability to live in nature is dependent on the activity of its algal partner. Lichens are usually found on surfaces where other organisms do not grow, and their success in colonizing such environments is due to the mutual interrelationships between the alga and fungus partners.

Lichens consist of a tight association of many fungal cells within which the algal cells are embedded (Figure 16.66). The shape of the lichen is determined primarily by the fungal partner, and a wide variety of fungi are able to form lichen associations. The diversity of algal types is much smaller, and many different kinds of lichens may have the same algal component. Some lichens contain cyanobacteria, frequently N_2-fixing species, instead of algae as the phototrophic component. The algae or cyanobacteria are usually present in defined layers or clumps within the lichen structure.

The fungus clearly benefits from associating with the alga, but how does the alga benefit? The fungus provides a firm anchor within which the alga can grow protected from erosion by rain or wind. In addition, the fungus facilitates the uptake of water and absorbs from the rock or other substrate on which the lichen is living the inorganic nutrients essential for the growth of the alga. *Lichen acids,* complex organic compounds excreted by the fungus, promote the dissolution and chelation of nutrients. Another role of the fungus is to protect the phototroph from drying; most of the habitats in which lichens live are dry (rock, bare soil, roof tops) (see Figure 16.65), and fungi are in general much better able to tolerate dry conditions than are algae.

Most lichens grow extremely slowly—a 2-cm lichen observed on the surface of a rock may actually be several years old. Measurements of lichen growth vary from 1 mm or less per year to over 3 cm/year, depending on the organisms composing the symbiosis, the amount of rainfall and sunlight received, and general

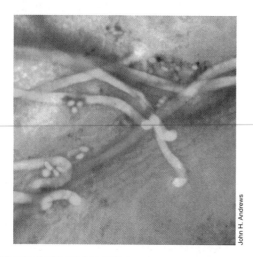

FIGURE 16.64 Fluorescent micrograph of cells of the fungus *Aureobasidium pullulans* on an apple leaf surface. Cells were stained by a phylogenetic FISH technique (see Section 16.6), and both filaments of the fungus and fungal spores show the green stain. A single fungal filament is about 7 μm wide.

John H. Andrews

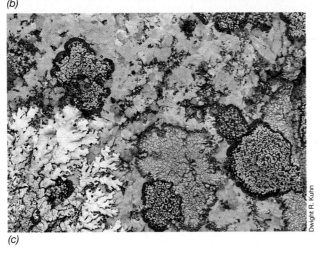

FIGURE 16.65 Lichens. (a) A lichen growing on a branch of a dead tree. (b) Lichens coating the surface of a large rock. (c) Several different lichens growing on the surface of a rock.

weather conditions. Although lichens live in nature under rather harsh conditions, they are extremely sensitive to air pollution and quickly disappear in areas experiencing heavy air pollution. One reason for this sensitivity is that they absorb and concentrate materials from rainwater and air and have no means for excreting them so lethal concentrations of compounds such as SO_2 are easily reached.

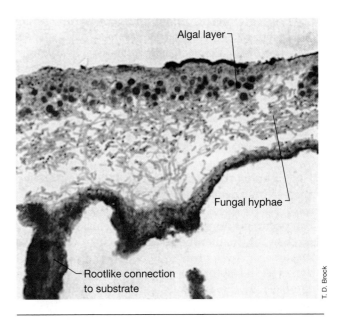

FIGURE 16.66 Photomicrograph of a cross section through a lichen.

Mycorrhizae

Mycorrhiza literally means "root fungus" and refers to the symbiotic association that exists between plant roots and fungi. Probably the roots of the majority of terrestrial plants are mycorrhizal. There are two classes of mycorrhizae: **ectomycorrhizae,** in which fungal cells form an extensive sheath around the outside of the root with only little penetration into the root tissue itself, and **endomycorrhizae,** in which the fungal mycelium is embedded within the root tissue.

Ectomycorrhizae are found mainly in forest trees, especially conifers, beeches, and oaks, and are most highly developed in boreal and temperate forests. In such forests, almost every root of every tree is mycorrhizal. The root system of a mycorrhizal tree such as pine (genus *Pinus*) is composed of both long and short roots. The short roots, which are characteristically dichotomously branched in *Pinus* (Figure 16.67a), show typical fungal colonization, and long roots are also frequently colonized. Endomycorrhizae are even more common than ectomycorrhizae. *Arbuscular mycorrhizae,* a type of endomycorrhizae, are found in the roots of over 80% of all terrestrial plant species so far examined and are thus a nearly universal plant symbiosis.

Most mycorrhizal fungi do not attack cellulose and leaf litter but instead use simple carbohydrates for growth and usually have one or more vitamin requirements; they obtain carbon from root secretions but get inorganic minerals from the soil. Mycorrhizal fungi are rarely found in nature except in association with roots and hence most can be considered obligate symbionts. Mycorrhizal fungi produce plant growth substances that induce morphological alterations in the roots, stimulating formation of the mycorrhizal state. However, despite the close rela-

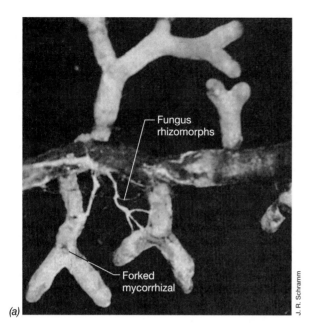

FIGURE 16.67 Mycorrhizae. (a) Typical ectomycorrhizal root of the pine, *Pinus rigida,* with rhizomorphs of the fungus *Thelophora terrestris.* (b) Seedling of *Pinus contorta* (lodgepole pine), showing extensive development of the absorptive mycelium of its fungal associate *Suillus bovinus.* This grows in fanlike formation from the ectomycorrhizal roots to capture nutrients from the soil.

tionship between fungus and root, there is little species specificity involved; a single species of pine can form a mycorrhizal association with over 40 species of fungi.

The beneficial effect on the plant of the mycorrhizal fungus is best observed in poor soils, where trees that are mycorrhizal thrive but nonmycorrhizal ones do not. For example, when trees are planted in prairie soils, which ordinarily lack a suitable fungal inoculum, trees that are artificially inoculated at the time of planting grow much more rapidly than uninoculated trees (Figure 16.68). It is well established that the mycorrhizal plant is able to absorb nutrients from its environment more efficiently than a nonmycorrhizal one, and thus the mycorrhizal plant has a competitive advantage. This improved nutrient absorption is due to the greater surface area provided by the fungal mycelium; for example, in the pine seedling shown in Figure 16.67b, the ectomycorrhizal fungal mycelium constitutes the overwhelming part of the absorptive area of the plant root system.

But in addition to simply helping plants absorb nutrients, mycorrhizae also appear to play a significant role in controlling plant diversity. Indeed, field experiments have shown that there is a positive correlation between the abundance and diversity of mycorrhizae in a soil and the extent of the plant diversity that develops in it. Thus, mycorrhizae are a prime example of a plant–microorganism symbiosis that benefits both partners—the mycorrhizal plant is better able to function physiologically and compete successfully in a species-rich plant community, while the fungus benefits from a steady supply of organic nutrients.

✓ 16.23 Concept Check

Mycorrhiza are fungi that associate with plant roots and improve their ability to absorb nutrients. Both mycorrhiza that attach outside the root and species that grow inside the root are known, but in either case their nutrition comes from excretions from the roots themselves. Mycorrhiza have a great beneficial effect on plant health and competitiveness.

✓ How do *endomycorrhiza* differ from *ectomycorrhiza?*
✓ Why are mycorrhizal associations with plants considered a type of symbiosis?

16.24

Agrobacterium and Crown Gall Disease

Some microorganisms are plant pathogens; that is, they cause plant disease. The genus *Agrobacterium* comprises organisms that cause the formation of tumorous growths on a wide variety of plants. The two species

FIGURE 16.68 Six-month-old seedlings of Monterey pine (*Pinus radiata*) growing in prairie soil: left, nonmycorrhizal; right, mycorrhizal.

FIGURE 16.69 Photograph of tumor on a tobacco plant caused by crown gall bacteria of the genus *Agrobacterium*.

most widely studied are *A. tumefaciens*, which cause *crown gall*, and *A. rhizogenes*, which causes *hairy root*.

Although plants often form benign accumulations of tissue, called a **callus**, when wounded, the growth induced by *A. tumefaciens* (Figure 16.69) is different in that the callus shows uncontrolled growth. It thus resembles tumor growth in animals, and considerable research on crown gall has been carried out with the hope that it may provide a model for how malignant growths occur in humans. Interestingly, once induced, these tumors continue to grow in the absence of *Agrobacterium* cells. Thus, once *Agrobacterium* has brought about the induction of the tumorous condition, its presence is no longer necessary.

An overview of the events in crown gall formation is given in Figure 16.70. A large plasmid called the *Ti* (*tumor induction*) *plasmid* (Figure 16.71) must be present in the *Agrobacterium* cells if they are to induce tumor formation. In *Agrobacterium rhizogenes*, a similar plasmid called the *Ri plasmid* is necessary for induction of hairy root. However, the best-studied system is that of the *A. tumefaciens*—crown gall disease—so we focus on that here. Following infection, a part of the Ti plasmid, called the *transfer DNA* (*T-DNA*), is integrated into the plant's genome. T-DNA carries the genes for tumor formation and also for the production of a number of modified amino acids called

opines. *Octopine* [N^2-(1,3-dicarboxyethyl)-*L*-arginine] and *nopaline* [N^2-(1,3-dicarboxypropyl)-*L*-arginine] are the two most common opines. Opines are produced by plant cells transformed by T-DNA and are a source of carbon and nitrogen for *Agrobacterium* cells (Figure 16.70). The Ti plasmid is also used in genetic engineering (Section 10.14 and see later).

Recognition and DNA Transfer

To initiate the tumorous state, cells of *Agrobacterium* must first attach to a wound site on the plant. The recognition of *Agrobacterium* by plant tissue involves complementary receptor molecules on the surfaces of the bacterial and plant cells. It is thought that the plant receptor molecule is a type of *pectin* (a complex polysaccharide) and that the bacterial receptor is a type of polysaccharide containing β-glucans, embedded in the cell wall lipopolysaccharide.

Studies with nontumorigenic mutants of *Agrobacterium tumefaciens* have shown that most functions necessary for attachment of the bacterium to plant are borne on the bacterial chromosome. Following attachment, rapid synthesis of cellulose microfibrils by the bacterium anchors the cells to the wound site and literally entraps the bacterial cells, forming large bacterial aggregates on the plant cell surface. This sets the stage for plasmid transfer from bacterium to plant.

The structure of the Ti plasmid is given in Figure 16.71. Note that although a number of genes are

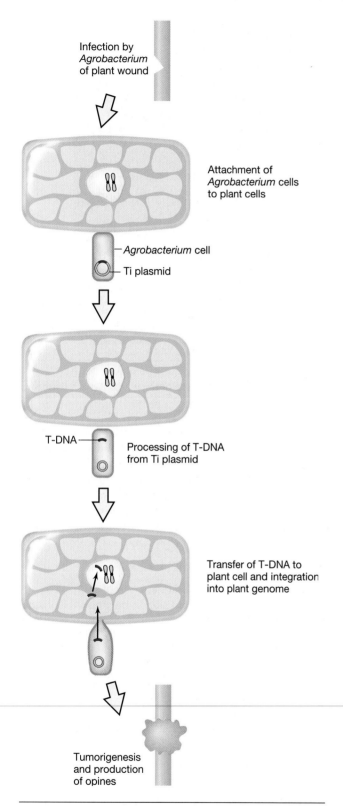

FIGURE 16.70 Overview of events of crown gall disease following infection of a susceptible plant by *Agrobacterium tumefaciens*. Note that it is only the T-DNA portion of the Ti plasmid that is transferred to the plant.

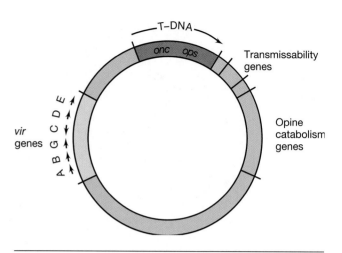

FIGURE 16.71 Structure of the Ti plasmid of *Agrobacterium tumefaciens*. T-DNA is the region actually transferred to the plant. *vir*, Virulence genes; *onc*, oncogenes (tumorigenesis genes); *ops*, opine synthesis genes. Arrows indicate the direction of transcription of each gene. The entire Ti plasmid is about 200 kb of DNA, and the T-DNA about 20 kb.

needed for infectivity, only a small portion of the Ti plasmid, a region called the *T-DNA* (Figures 16.70 and 16.71), is actually transferred to the plant. The T-DNA contains genes that encode proteins that induce tumorigenesis. The *vir* genes that reside on the Ti plasmid encode proteins that are essential for T-DNA transfer (Figure 16.71). *vir* gene expression is induced by plant signal molecules synthesized by wounded plant tissues. Some inducers that have been identified include the phenolic compounds acetosyringone, *p*-hydroxybenzoic acid, and vanillin.

The *vir* genes are the key to T-DNA transfer. The *virA* gene encodes a protein kinase that interacts with signal molecules and then phosphorylates the product of the *virG* gene (Figure 16.72). The latter becomes activated by the phosphorylation event and functions to activate other *vir* genes. The product of the *virD* gene has endonuclease activity and nicks DNA in the Ti plasmid in a region adjacent to the T-DNA (Figures 16.71 and 16.72). The product of the *virE* gene is a single-stranded DNA binding protein that binds the *single strand* of T-DNA generated from endonuclease activity and transports this small fragment of DNA into the plant cell. The *virB* gene product is located in the bacterial membrane and mediates transfer of the single strand of DNA between bacterium and plant.

T-DNA transfer occurs in a process that resembles bacterial conjugation (Figures 16.70 and 16.72). The T-DNA becomes inserted into the nuclear genome of the plant (plant cell organellar genomes are not infected) where integration can occur at any number of sites where specific inverted or direct tandem repeats are present. The tumorigenesis genes of the Ti plasmid (Figure

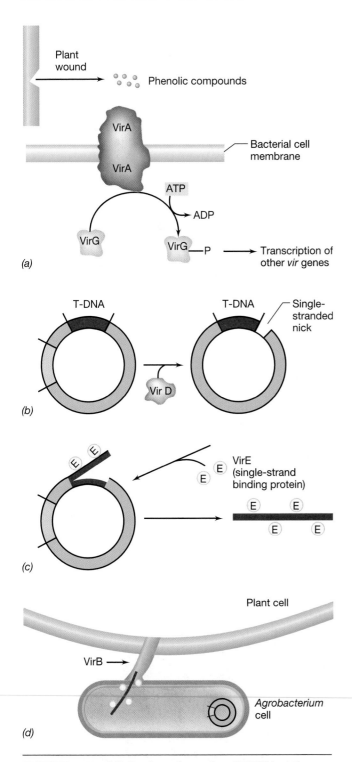

(a)

(b)

(c)

(d)

FIGURE 16.72 Mechanism of transfer of T-DNA to the plant cell by *Agrobacterium tumefaciens*. (a) Levels of VirA protein increase dramatically upon stimulation with plant phenolic inducer molecules. VirA activates VirG by phosphorylation, and VirG activates transcription of other *vir* genes. (b) VirD is an endonuclease. (c) VirE is a single-stranded binding protein. (d) VirB acts as a conjugation bridge between *Agrobacterium* and the plant cell.

16.71) encode enzymes involved in plant hormone production and at least one key enzyme of opine biosynthesis. Expression of these oncogenes leads to tumor formation. The Ri plasmid involved in hairy root disease also contains oncogenes. In this case, the oncogenes confer increased auxin responsiveness to the plant cells, which leads to overproduction of root tissue, resulting in the symptoms of the disease. The Ri plasmid also codes for several opine biosynthetic enzymes.

Genetic Engineering with the Ti Plasmid

From the standpoint of microbiology and plant pathology both crown gall and hairy root disease involve a unique type of plant–bacterial interaction in which bacterial DNA is physically transferred to plant cells. However, once the mechanism of DNA transmission was understood, it became clear that the Ti system could be used by scientists as a vector for the introduction of genetically engineered DNA into plants—that is, that Ti was a natural plant transformation system. Thus, through the years the focus of the Ti/crown gall system has shifted away from the disease process to new applications in the plant biotechnology industry.

Experience has shown that plants are fairly difficult to transform. But through the power of genetic engineering a host of modified ("disarmed") Ti plasmids are now available for the facile production of transgenic plants without transfer of disease genes. Success stories are already recorded in the areas of herbicide and insect resistance, and many other areas remain to be explored. We discuss the use of the Ti plasmid as a vector in plant biotechnology in more detail in Section 10.14.

✓ **16.24 Concept Check**

The crown gall bacterium *Agrobacterium* enters into a unique relationship with higher plants. A plasmid in the bacterium (the Ti plasmid) is able to transfer part of itself to the genome of the plant, in this way bringing about the production of the crown gall disease. The crown gall plasmid has also found extensive use in the genetic engineering of crop plants.

✓ What are *opines* and why are they produced?
✓ How do the *vir* genes differ from *T-DNA* in the Ti plasmid?
✓ How has an understanding of crown gall disease benefited the area of plant molecular biology?

16.25

Root Nodule Bacteria and Symbiosis with Legumes

One of the most interesting and important plant bacterial interactions is that between leguminous plants and certain gram-negative nitrogen-fixing bacteria. Legumes are a large group that include such economically important plants as soybeans, clover, alfalfa, beans, and peas and are defined as plants that bear seeds in pods. *Rhizobium, Bradyrhizobium, Sinorhizobium, Mesorhizobium,* and *Azorhizobium* are gram-negative motile rods. Infection of the roots of a leguminous plant with the appropriate species of one of these genera leads to the formation of **root nodules** (Figure 16.73) that are able to convert gaseous nitrogen to combined nitrogen, a process called *nitrogen fixation* (⚭ Section 15.29); *Azorhizobium* forms root nodules as well as stem nodules (see later). Nitrogen fixation by the legume–*Rhizobium* symbiosis is of considerable agricultural importance, as it leads to very significant increases in combined nitrogen in the soil. Because nitrogen deficiencies often occur in unfertilized bare soils, nodulated legumes are at a selective advantage under such conditions and can grow well in areas where other plants cannot (Figure 16.74).

Leghemoglobin and Cross-Inoculation Groups

Under normal conditions, neither legume nor *Rhizobium* alone is able to fix nitrogen; yet the interaction between the two leads to the development of nitrogen-fixing ability. In pure culture, the *Rhizobium* is able to fix N_2 alone when grown under strictly controlled *microaerophilic* conditions. Apparently *Rhizobium* needs some O_2 to generate energy for N_2 fixation, yet its nitrogenase (like those of other nitrogen-fixing organisms) (⚭ Section 15.29) is inactivated by O_2. In the nodule,

FIGURE 16.74   A field of unnodulated (left) and nodulated (right) soybean plants growing in nitrogen-poor soil.

precise O_2 levels are controlled by the O_2 binding protein **leghemoglobin.** This is a red, iron-containing protein that is always found in healthy N_2-fixing nodules (Figure 16.75). Neither plant nor *Rhizobium* alone synthesizes leghemoglobin, but formation is thought to be induced through the interaction of these two organisms. Leghemoglobin functions as an "oxygen buffer" cycling between the oxidized (Fe^{3+}) and reduced (Fe^{2+}) forms to keep free O_2 levels within the nodule at a low but constant level. The ratio of leghemoglobin-bound O_2 to free O_2 in the root nodule is on the order of 10,000:1.

About 90% of all leguminous plant species are capable of becoming nodulated. However, there is a marked specificity between species of legume and strains of *Rhizobium*. A single *Rhizobium* strain is generally able to infect certain species of legumes and not others. A group of *Rhizobium* strains able to infect a

FIGURE 16.73   Soybean root nodules. The nodules develop by infection with *Bradyrhizobium japonicum.*

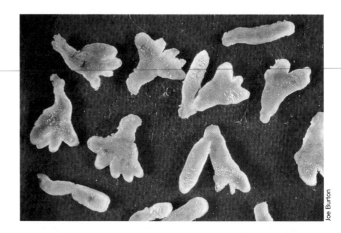

FIGURE 16.75   Sections of root nodules from the legume *Coronilla varia,* showing the reddish pigment leghemoglobin.

TABLE 16.8	Major cross-inoculation groups of leguminous plants
Host plant	**Nodulated by**
Pea	*Rhizobium leguminosarum* biovar *viciae*[a]
Bean	*Rhizobium leguminosarum* biovar *phaseoli*[a]
Bean	*Rhizobium tropici*
Lotus	*Mesorhizobium loti*
Clover	*Rhizobium leguminosarum* biovar *trifolii*[a]
Alfalfa	*Sinorhizobium meliloti*
Soybean	*Bradyrhizobium japonicum*
Soybean	*Bradyrhizobium elkanii*
Soybean	*Rhizobium fredii*
Sesbania rostrata (a tropical legume)	*Azorhizobium caulinodans*

a Several varieties (biovars) of *Rhizobium leguminosarum* exist, each capable of nodulating a different legume.

group of related legumes is called a *cross-inoculation group*. The major rhizobial cross-inoculation groups are listed in Table 16.8. Even if a *Rhizobium* strain is able to infect a certain legume, it is not always able to bring about the production of nitrogen-fixing nodules. If the strain is *ineffective*, the nodules formed will be small, greenish-white, and incapable of fixing nitrogen; if the strain is *effective*, on the other hand, the nodule will be large, reddish (Figure 16.75), and nitrogen-fixing. Effectiveness is determined by genes in the bacterium (see the discussion of *nod* genes later in this section).

Stages in Root Nodule Formation

The stages in the infection and development of root nodules are now fairly well understood (Figure 16.76). They include

1. **recognition** of the correct partner on the part of both plant and bacterium and **attachment** of the bacterium to root hairs.
2. **invasion** of the root hair by the bacterial formation of an infection thread.
3. **travel** to the main root via the infection thread.
4. formation of deformed bacterial cells, **bacteroids,** within the plant cells and development of the nitrogen-fixing state.
5. continued plant and bacterial division and formation of the mature **root nodule.**

We now explore some of these stages in nodule formation in more detail.

The roots of leguminous plants secrete a variety of organic materials that stimulate the growth of a rhizosphere microflora. This stimulation is not restricted to the rhizobia but occurs with a variety of rhizosphere bacteria. If there are rhizobia in the soil, they grow in the rhizosphere and build up to high population densities. Attachment of bacterium to plant in the legume–*Rhizobium* symbiosis is

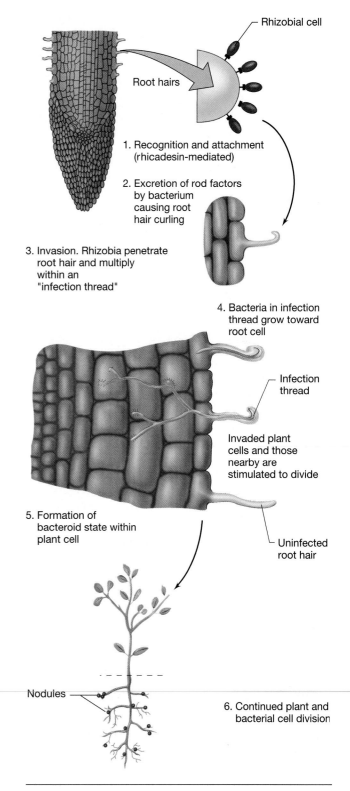

1. Recognition and attachment (rhicadesin-mediated)
2. Excretion of rod factors by bacterium causing root hair curling
3. Invasion. Rhizobia penetrate root hair and multiply within an "infection thread"
4. Bacteria in infection thread grow toward root cell
5. Formation of bacteroid state within plant cell
6. Continued plant and bacterial cell division

Rhizobial cell
Root hairs
Infection thread
Invaded plant cells and those nearby are stimulated to divide
Uninfected root hair
Nodules

FIGURE 16.76 Steps in the formation of a root nodule in a legume infected by *Rhizobium*. For a photomicrograph of an infection thread, see Figure 16.77a.

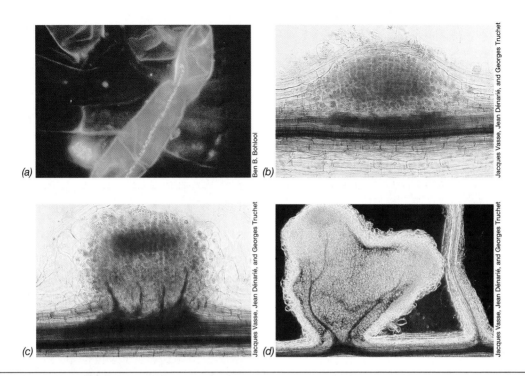

(a) Ben B. Bohlool   *(b)* Jacques Vasse, Jean Dénarié, and Georges Truchet

(c) Jacques Vasse, Jean Dénarié, and Georges Truchet   *(d)* Jacques Vasse, Jean Dénarié, and Georges Truchet

FIGURE 16.77   The infection thread and formation of root nodules. (a) An infection thread formed by cells of *Rhizobium legu-minosarum* biovar *trifolii* formed on a root hair of white clover *(Trifolium repens)*. The infection thread consists of a cellulosic tube through which bacteria move to root cells. (b–d) Nodules from alfalfa roots infected with cells of *Sinorhizobium meliloti* shown at different stages of development. Cells of both *R. leguminosarum* biovar *trifolii* and *Sinorhizobium meliloti* are about 2 μm long. Photos b–d reprinted with permission from *Nature 351*:670–673 (1991), © Macmillan Magazines Ltd.

the first step in the formation of nodules. A specific adhesion protein called *rhicadhesin* is present on the surfaces of all species of *Rhizobium* and *Bradyrhizobium*. Rhicadhesin is a calcium binding protein and may function by binding calcium complexes on the root hair surface. Other substances, such as carbohydrate-containing proteins called *lectins*, also play some role in plant–bacterium attachment. Lectins have been identified on both root hair tips and on the surface of *Rhizobium* cells, but the interaction of lectins in the binding–recognition process is thought to be less important than that of rhicadhesin.

Initial penetration of *Rhizobium* cells into the root hair is via the root hair tip. Following binding, the root hair curls as a result of the action of substances excreted by the bacterium called *Nod factors* (see later) and the bacteria enter the root hair and induce formation by the plant of a cellulosic tube, called the **infection thread,** which spreads down the root hair. Root cells adjacent to the root hairs subsequently become infected by rhizobia, and Nod factors stimulate plant cell division, eventually leading to formation of the nodule (Figures 16.77 and 16.78, see also Figures 16.73 and 16.75).

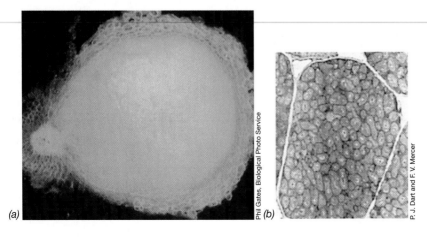

(a) Phil Gates, Biological Photo Service   *(b)* P. J. Dart and F. V. Mercer

FIGURE 16.78   (a) Cross section through a legume root nodule, as seen by fluorescence microscopy. The darkly stained region contains plant cells filled with bacteria. (b) Electron micrograph of a thin section through a single bacteria-filled cell of a subterranean clover nodule. Cells of *Rhizobium leguminosarum* biovar *trifolii* are about 2 μm long.

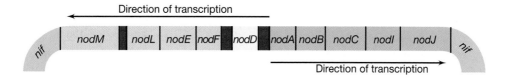

FIGURE 16.79 Organization of the *nod* gene cluster on the Sym plasmid of *Rhizobium leguminosarum* biovar *viciae*, the species that nodulates peas. The product of *nodD* controls transcription of other *nod* genes. The *nod* boxes are highlighted in red, and the arrows indicate the direction of transcription of *nod* genes. See text for further details.

The bacteria multiply rapidly within the plant cells and are transformed into swollen, misshapen, and branched forms called **bacteroids.** Bacteroids become surrounded singly or in small groups by portions of the plant cell membrane to form structures called the **symbiosome.** Only after the formation of the symbiosome does nitrogen fixation begin. (Effective nitrogen-fixing nodules can be detected by acetylene reduction) (∞ Section 15.29 and Figure 15.70). When the plant dies the nodule deteriorates, releasing bacteria into the soil. The bacteroid forms are incapable of division, but there are always a small number of dormant rod-shaped cells present. These now proliferate, using some of the products of the deteriorating nodule as nutrients, and the bacteria can initiate the infection in other roots or maintain a free-living existence in the soil.

Nodule formation: Nod genes, Nod proteins, and Nod factors

Genes directing specific steps in nodulation of a legume by a strain of *Rhizobium* have been called *nod genes* (Figure 16.79). Many *nod* genes from different *Rhizobium* species are highly conserved and are borne on large plasmids called *Sym plasmids.* In addition to *nod* genes, which direct specific nodulation events, Sym plasmids contain *specificity genes*, which restrict a strain of *Rhizobium* to a particular host plant. Indeed, cross-inoculation group specificity can be transferred across species of rhizobia by simply transferring its Sym plasmid. For example, when the Sym plasmid of *Rhizobium leguminosarum* biovar *viciae* (whose host is the pea) is transferred to *Rhizobium leguminosarum* biovar *trifolii* (whose host is clover), cells of the latter species effectively nodulate pea.

In the Sym plasmid of *Rhizobium leguminosarum* biovar *viciae*, *nod* genes are located between two clusters of genes for nitrogen fixation, the *nif* genes (in this species and in certain other *Rhizobium* species, *nif* genes are plasmid-borne). The arrangement of *nod* genes in the *R. leguminosarum* Sym plasmid is shown in Figure 16.79. Ten *nod* genes have been identified in this species. The entire *nod* region has been sequenced, and the function of many Nod proteins is known. The *nodABC* genes are involved in the production of chitinlike molecules,

called *Nod factors*, which induce root hair curling and trigger cortical plant cell division, eventually leading to formation of the nodule (Figure 16.77). Chemically, Nod factors consist of a backbone of N-acetylglucosamine to which various substituents are linked (Figure 16.80). Host specificity is determined by the precise structure of the Nod factor produced by a given species of *Rhizobium*. Thus, besides the *nodABC* genes, whose products synthesize the nod backbone, each species of *Rhizobium* contains certain unique *nod* genes responsible for introducing chemical variations on the basic Nod factor backbone. That Nod factors alone are responsible for root hair curling and initiation of the nodule has been shown in experiments in which purified Nod factors have been added to root tissue; such tissue proceeds to form a nodule, even in the absence of *Rhizobium* cells.

In *Rhizobium leguminosarum* biovar *viciae nodD* encodes a regulatory protein, NodD, which controls transcription of other *nod* genes (Figure 16.79). NodF encodes a specific acyl carrier protein used by NodA to acylate the Nod factor (Figure 16.80b), NodE and NodL are in-

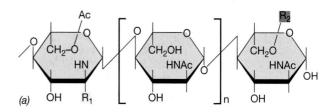

Species	R_1	R_2
Sinorhizobium meliloti	C16:2 or C16:3	SO_4^{2-}
Rhizobium leguminosarum biovar *viciae*	C18:1 or C18:4	H or Ac

FIGURE 16.80 Nod factors. (a) The general structure of the Nod factor produced by *Sinorhizobium meliloti* and *Rhizobium leguminosarum* biovar *viciae* and (b) in the table the structural differences (R_1, R_2) that define the precise Nod factor of each species. The central hexose unit can repeat up to three times. C16:2, Palmitic acid with two double bonds; C16:3, palmitic acid with three double bonds; C18:1, oleic acid with one double bond; C18:4, oleic acid with four double bonds; Ac, acetyl.

volved in host range (cross-inoculation groups), NodM is a glucosamine synthase involved in Nod factor synthesis, and NodI and NodJ are membrane proteins that function to export Nod factors from the bacterial cells.

NodD is a member of a family of regulatory proteins that function to activate transcription of other genes by bending DNA at the promoter; this enhances binding of RNA polymerase. Thus, NodD can be considered a type of *positive* regulatory element (∞ Section 7.3). It is thought that NodD binds to regions upstream of *nod* structural gene operons (regions that have been called *nod boxes*) and, after interacting with inducer molecules, bend DNA in this region, which promotes transcription. Several inducer molecules have been identified. In most cases these are plant *flavonoids*, complex organic molecules that are widespread plant products (Figure 16.81a and b). Flavonoids have many functions in plants, including growth regulation and attraction of pollinating animals. However, leguminous plants are unusual because, unlike those of other plants, their *roots* secrete large amounts of flavonoids, pre-

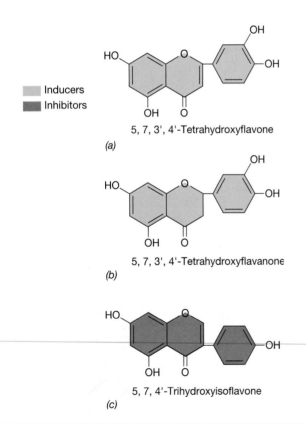

FIGURE 16.81 Structure of flavonoid molecules that are (a, b) inducers of *nod* gene expression and (c) inhibitors of *nod* gene expression in *Rhizobium leguminosarum* biovar *viciae*. Note the similarities in the structures of all three molecules. The common name of the structure shown in (a) is *luteolin*, and it is a flavone derivative. The structure in (b) is called *eriodictyol* and is chemically a flavanone. The structure in (c) is called *genistein*, and it is an isoflavone derivative.

sumably to trigger *nod* gene expression in nearby rhizobial cells in the soil. Interestingly, some flavonoids that are structurally very closely related to *nodD* inducers (Figure 16.81c) strongly *inhibit* induction of *nod* genes in certain rhizobial species, suggesting that part of the specificity observed between plant and bacterium in the *Rhizobium*–legume symbiosis could lie in the chemical nature of the flavonoids excreted by a particular plant.

Biochemistry of Nitrogen Fixation in Nodules

As discussed in Section 15.29, nitrogen fixation involves the activity of the enzyme **nitrogenase,** a large two-component protein containing iron and molybdenum. Nitrogenase in root nodules has characteristics similar to the enzyme from free-living N_2-fixing bacteria, including O_2 sensitivity and the ability to reduce acetylene as well as N_2 (∞ Figure 15.70). Nitrogenase is localized within the bacteroids themselves and is not released into the plant cytosol.

Bacteroids are totally dependent on the plant for supplying them with energy sources for N_2 fixation. The major organic compounds transported across the symbiosome membrane and into the bacteroid proper are citric acid cycle intermediates, in particular the C_4 acids *succinate, malate,* and *fumarate* (Figure 16.82). These are used as electron donors for ATP production and, following conversion to pyruvate, as the ultimate source of electrons for the reduction of N_2.

The first stable product of N_2 fixation is *ammonia,* and several lines of evidence suggest that assimilation of ammonia into organic nitrogen compounds in the root nodule is carried out primarily by the plant. Although bacteroids can assimilate some ammonia into organic form, the levels of ammonia-assimilatory enzymes in bacteroids are quite low. By contrast, the ammonia-assimilating enzyme *glutamine synthetase* (∞ Figure 4.26b) is present in high levels in the plant cell cytoplasm. Hence, ammonia transported from the bacteroid to the plant cell can be assimilated by the plant as the amino acid *glutamine*. Besides glutamine, other nitrogenous compounds, in particular other amino acid amides, such as asparagine and 4-methylene glutamine, and the ureides *allantoin* and *allantoic acid,* are synthesized by the plant and subsequently transported to plant tissues (see Figure 16.82).

Stem-Nodulating Rhizobia

Although most leguminous plants form nitrogen-fixing nodules on their *roots*, a few legume species bear nodules on their *stems*. Stem-nodulated leguminous plants are quite widespread in tropical regions where soils are often nitrogen-deficient because of leaching and intense biological activity. The major experimental system here has been the tropical legume *Sesbania*, which is nodu-

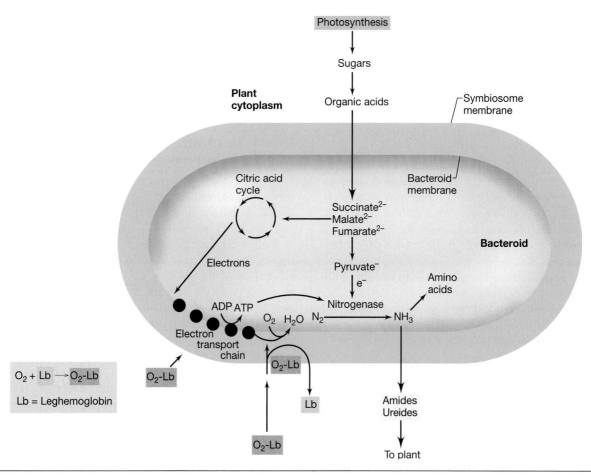

FIGURE 16.82 Schematic diagram of major metabolic reactions and nutrient exchanges occurring in the bacteroid. The symbiosome is a collection of bacteroids surrounded by a single membrane originating from the plant.

lated by the bacterium *Azorhizobium caulinodans* (Figure 16.83). Stem nodules usually form in the submerged portion of the stems or on portions of the stem just above the water level (Figure 16.83). From studies carried out thus far, the general sequence of events in formation of stem nodules strongly resembles that of root nodules. An infection thread is formed, and bacteroid formation predates the N_2-fixing state.

One curious finding is that some stem-nodulating rhizobia produce bacteriochlorophyll *a* and thus may have the potential to carry out anoxygenic photosynthesis (Section 15.5). Bacteriochlorophyll-containing rhizobia are apparently widespread in nature, particularly in association with tropical legumes, and in these species, light could drive at least part of the energy-demanding process of N_2 fixation in the bacterium.

Summary of the *Rhizobium*–Legume Symbiosis and Its Agricultural and Ecological Importance

The *Rhizobium*–legume association is a true symbiosis, because each partner clearly contributes something to the other. Fixed nitrogen benefits the plant growing in

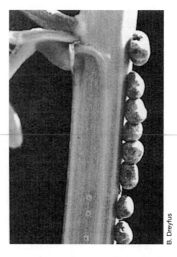

FIGURE 16.83 Stem nodules caused by stem-nodulating *Azorhizobium*. The photograph shows the stem of the tropical legume *Sesbania rostrata*. On the left side of the stem are uninoculated sites, on the right identical sites inoculated with stem-nodulating rhizobia.

nitrogen-deficient soils while the nodule is a physically protected and well-nourished environment for rhizobia. From a genetic standpoint, nitrogen fixation is an activity encoded by the bacterium while the lectins and flavonoids necessary for recognition and initiation of the association are encoded by plant genes. Some aspects of the association are genetically directed by *both* components of the symbiosis. For example, the important oxygen-binding protein in the root nodule, *leghemoglobin,* is genetically encoded in part by both the plant and the bacterium. Nowhere in microbiology do we see an example of a beneficial plant/bacterial relationship so well developed or so well understood as in the legume/root nodule symbiosis. And the agricultural benefits of the more than 120 million tons of atmospheric nitrogen that is biologically fixed to ammonia each year, are enormous. The plants listed in Table 16.8 include soybean and alfalfa, key commodities for the soy-processing industry and the feeding of domesticated animals (alfalfa hay), respectively, as well as beans and peas, major vegetables for human nutrition. Several agricultural industries revolve around leguminous crops, and the ability of legumes to grow without nitrogen fertilizer saves farmers millions of dollars in fertilizer costs yearly.

Despite the impressive gains in fixed nitrogen from biological nitrogen fixation, nitrogen fertilizer production by humans far outpaces the biological process. The use of nitrate and ammonia fertilizers is growing rapidly and is introducing new fixed nitrogen into the world's ecosystems. This in turn triggers serious ecological problems, and throws the nitrogen cycle out of balance. The major problems observed from excessive fertilization include stimulating denitrification (from nitrate fertilizers), with its production of smog-inducing nitrogen oxides and generation of acid rain (∞ Section 15.16); eutrophication of freshwaters leading to diminished water quality from algal blooms and excessive growth of macrophytes; and even reducing plant diversity in some cases where native plants adapted to nitrogen-poor soils are overtaken by exotic species that can outcompete them when nitrogen levels are high.

Nonlegume Nitrogen-Fixing Symbioses and Other Associations

Nitrogen-fixing symbioses involving microorganisms other than rhizobia occur in a variety of nonleguminous plants. Nitrogen-fixing cyanobacteria form symbioses with a variety of plants. The water fern *Azolla* contains a species of heterocystous N_2-fixing cyanobacteria called *Anabaena azollae* within small pores of its fronds (Figure 16.84). *Azolla* has been used for centuries to enrich rice paddies with fixed nitrogen. Before planting rice, the farmer allows the surface of the rice paddy to become densely covered with the fern. As the rice plants grow, they eventually crowd out the *Azolla–Anabaena* mixture, leading to death of the fern and release of fixed nitrogen, which is assimilated by the rice plants. By repeating this process each growing season, the farmer can obtain high yields of rice without the need for nitrogenous fertilizers.

The alder tree (genus *Alnus*) has nitrogen-fixing root nodules (Figure 16.85) that harbor a filamentous, streptomycete-like, nitrogen-fixing organism called *Frankia*. Although when assayed in cell extracts the nitrogenase of *Frankia* is sensitive to molecular oxygen, like intact cells of *Azotobacter* (∞ Section 13.8), intact cells of *Frankia* fix N_2 at full oxygen tensions. This is because *Frankia* protects its nitrogenase by localizing it in terminal swellings on the cells called *vesicles* (Figure 16.85*b*). The vesicles contain thick walls that retard O_2

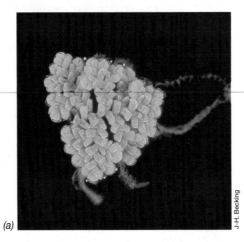

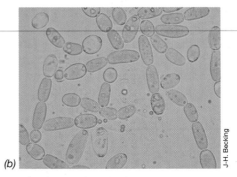

(a) (b)

FIGURE 16.84 *Azolla–Anabaena* symbiosis. (a) Intact association showing a single plant of *Azolla pinnata.* The diameter of the plant is approximately 1 cm. (b) Cyanobacterial symbiont *Anabaena azollae* as observed in crushed leaves of *A. pinnata.* Single cells of *Anabaena azollae* are about 5 μm wide. Note the oval-shaped *heterocysts* (lighter color), the site of nitrogen fixation in the cyanobacterium. For discussion of the heterocyst see Section 13.24 and Figure 13.81.

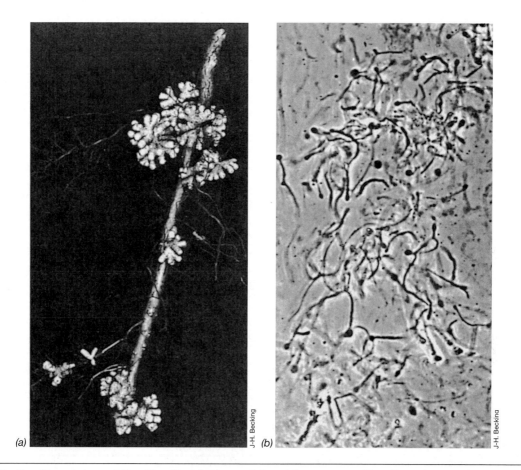

FIGURE 16.85 *Frankia* nodules and *Frankia* cells. (a) Root nodules of the common alder *Alnus glutinosa*. (b) *Frankia* culture purified from nodules of *Comptonia peregrina*. Note vesicles (spherical structures) on the tips of hyphal filaments.

diffusion, thus maintaining the O_2 tension within vesicles at levels compatible with nitrogenase activity. In this regard, *Frankia* vesicles resemble the heterocysts produced by some filamentous cyanobacteria as localized sites of N_2 fixation (see Figure 16.84b; ∾ Section 13.24 and Figure 13.81).

Alder is a characteristic pioneer tree able to colonize bare soils at nutrient-poor sites, and this is likely due to its ability to enter into a symbiotic nitrogen-fixing relationship with *Frankia*. A number of other small woody plants are nodulated by *Frankia*. Unlike the *Rhizobium*/legume situation, a single strain of *Frankia* can form nodules on several different plants, suggesting that the *Frankia*/root nodule symbiosis is less specific in this regard.

Azospirillum lipoferum is a N_2-fixing bacterium that lives in a casual association with roots of tropical grasses. It is found in the rhizosphere, where it grows on products excreted from the roots. It also has the ability to grow around the roots of cultivated grasses, such as corn (*Zea mays*), and inoculation of corn with *A. lipoferum* may lead to small increases (about 10%) in growth yield of the plant. In sugarcane, the organism *Acetobacter diazotrophicus* grows in plant vascular tissue and fixes substantial amounts of N_2, which benefits the plant.

Cultures of *A. diazotrophicus* require high sucrose concentrations, suggesting that this plant–bacterium association may be more highly developed than the *Azospirillum* association. It is likely that many casual to tight associations exist between bacteria and plants, although only in the *Rhizobium–Frankia* nodule has direct evidence for a symbiotic relationship been found.

✓ 16.25 Concept Check

One of the most widespread and important plant–microbial symbioses is that between legumes and certain nitrogen-fixing bacteria. The bacteria induce the formation of root nodules within which the nitrogen-fixing process occurs. The plant provides the organic energy source needed by the root nodule bacteria, and the bacteria provide fixed nitrogen for the growth of the plant. The root nodule bacteria play an important agricultural role because many important crop plants are legumes (alfalfa, clover, soybeans, peas, and so on).

✓ What is a *leguminous* plant?
✓ What is *leghemoglobin* and what is its function?
✓ What is a *bacteroid* and what occurs within it?
✓ What are the major similarities and differences between *Rhizobium* and *Frankia*?

REVIEW QUESTIONS

1. Explain why both obligately *anaerobic* and obligately *aerobic* bacteria can be isolated from the same soil sample.

2. What is the basis of the *enrichment culture technique*? Why is an enrichment medium usually suitable for the enrichment of only a certain group or groups of bacteria?

3. What is the principle of the Winogradsky column and what types of organisms does it serve to enrich? How might a Winogradsky column be used to study the breakdown of a xenobiotic compound?

4. Compare and contrast the use of fluorescent dyes and fluorescent antibodies for use in enumerating bacteria in natural environments. What advantages and limitations do each of these methods have?

5. Can nucleic acid probes in microbial ecology be as sensitive as culturing methods? If so, how? What advantages do nucleic acid methods have over culture methods? What disadvantages?

6. How does the technique *chromosomal painting* work? Why might it be a good method for enumerating microorganisms in a habitat capable of carrying out some specific process?

7. What are the major advantages of *radioisotopic methods* in the study of microbial ecology? What type of controls (discuss at least two) would you include in a radioisotopic experiment to show $^{14}CO_2$ incorporation by phototrophic bacteria or to show $^{35}SO_4^{2-}$ reduction by sulfate-reducing bacteria?

8. What is the difference between *barotolerant* and *barophilic* bacteria? Between these two groups and *extreme barophiles*? What properties do barotolerant, barophilic, and extremely barophilic bacteria have in common?

9. What evidence from hydrothermal vents exists to support the idea that prokaryotes are growing at extremely high temperatures? After reviewing the data of Table 5.1, why is the marine worm *Alvinella* of interest to biologists?

10. How can organisms such as *Syntrophobacter* and *Syntrophomonas* grow when their metabolism is based on thermodynamically unfavorable reactions? How are these organisms grown in laboratory culture?

11. Why is sulfate reduction the main form of anaerobic respiration in marine environments, whereas methanogenesis dominates in fresh waters? Does any methanogenesis occur in the marine environment? If so, how?

12. What is the *rumen* and how do the digestive processes operate in the ruminant digestive tract? What are the major benefits and the disadvantages of a rumen system? How does a cecal animal compare with a ruminant?

13. Why can urea or ammonia be a nitrogen source for ruminants but not for humans?

14. Compare and contrast the processes of *nitrification* and *denitrification* in terms of the organisms involved, the environmental conditions that favor each process, and the changes in nutrient availability that accompany each process.

15. What organisms are involved in cycling sulfur compounds anoxically? If sulfur chemolithotrophs had never evolved, would there be a problem in the microbial cycling of sulfur compounds? What *organic* sulfur compounds are of interest in nature?

16. Why are all iron-oxidizing chemolithotrophs obligate aerobes and why are most iron oxidizers acidophilic?

17. Explain how spontaneous chemical reactions can acidify a coal seam and how both chemical reactions and *Thiobacillus ferrooxidans* continue the production of acid thereafter.

18. How is *Thiobacillus ferrooxidans* useful in the mining of copper ores? What crucial step in the indirect oxidation of copper ores is carried out by *T. ferrooxidans*? How is copper recovered from copper solutions produced by leaching?

19. How is mercury detoxified by the *mer* system?

20. What physical and chemical conditions are necessary for the rapid microbial degradation of oil in aquatic environments? Design an experiment that would allow you to test what conditions optimized the oil oxidation process.

21. What are *xenobiotic compounds* and why might microorganisms have difficulty catabolizing them?

22. How is the organism *Desulfomonile* of benefit in solving environmental pollution problems?

23. Describe the chemical properties of a "microbial plastic" now in commercial production.

24. Compare and contrast the production of a plant tumor by *Agrobacterium tumefaciens* and a root nodule by a *Rhizobium* species. In what ways are these structures similar? In what ways are they different? Of what importance are plasmids to the development of both structures?

25. What genetic information resides on the Ti plasmid? Which gene(s) could be deleted from the Ti plasmid without affecting tumorigenesis?

26. What ecological advantages do leguminous plants have over nonlegumes?

27. What substances of plant origin are found in root nodules of legumes that are required for the rhizobial symbiont to fix nitrogen?

28. Describe the steps in the development of root nodules on a leguminous plant. What is the nature of the recognition between plant and bacterium? How does this compare with recognition in the *Agrobacterium*–plant system?

29. How does nitrogen fixed by *Rhizobium* become part of the plant's proteins?

30. Describe the *nod* operons of a typical *Rhizobium* species such as *R. leguminosarum*. Explain how *nod* genes are regulated. What are Nod factors and what do they do?

APPLICATION QUESTIONS

1. Compare and contrast a lake ecosystem with a hydrothermal vent ecosystem. How does energy enter each ecosystem? What is the basis of primary production in each ecosystem? What nutritional classes of organisms exist in each ecosystem, and how do they feed themselves?

2. How do stains like acridine orange differ from phylogenetic stains in assessing microbial diversity of a habitat?

3. Why is the enrichment for nitrifying bacteria as described in Table 16.1? Why are each of the resources and conditions necessary?

4. Design an experiment for measuring the activity of sulfur-oxidizing bacteria in soil. How would you prove that your activity measurement was due to *biological* activity?

5. You wish to identify in soil samples any organism capable of autotrophic growth using the Calvin cycle. Keeping in mind that such autotrophs contain a unique enzyme, ribulose bisphosphate carboxylase, design a procedure that could be used to identify such organisms in nature.

6. ^{14}C-Labeled cellulose is added to a vial containing a small amount of sewage sludge and sealed under anoxic conditions. A few hours later $^{14}CH_4$ appears in the vial. Discuss what has happened to yield such a result.

7. Compare and contrast the microbiological steps involved in the conversion of cellulose to methane in the rumen as compared to cellulose conversion in lake sediments. What organisms are involved in each ecosystem and why?

8. Suppose you have discovered a new animal that consumes only grass in its diet. You suspect it to be a ruminant and have available a specimen for anatomical inspection. If this animal is a ruminant, describe the position and basic components of the digestive tract you would expect to find and any key microorganisms and substances you might look for.

The eukaryotic cell is structurally and phylogenetically distinct from all prokaryotic cells, Bacteria and Archaea. As shown here, various structures can be present in the eukaryotic cell including the nucleus, mitochondrion, chloroplast, Golgi apparatus, and endoplasmic reticulum among others. The mitochondrion and chloroplast are derived from aerobic and phototrophic Bacteria, respectively, that established residency inside primitive eukaryotic cells eons ago, in a process called endosymbiosis. Telltale signs of this cozy arrangement can still be found in the DNA and ribosomes that remain in these organelles as well as the bacterial genes present in the eukaryotic nucleus that were transferred there during evolutionary conversion of the endosymbiont into an organelle. The eukaryotic cell is thus a genetic chimera, containing genetic material from more than one source.

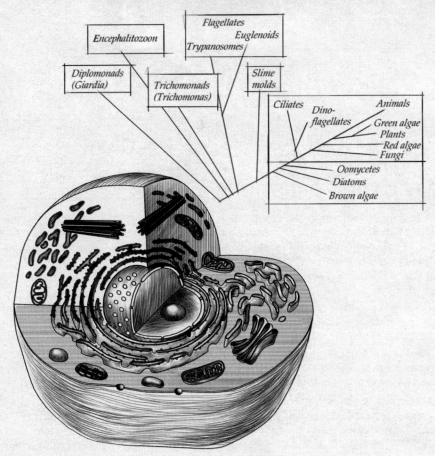

CHAPTER 17 Eukaryotic Microorganisms

17.1 Eukaryotic Cell Structure 721
17.2 Phylogenetic Overview of Eukarya 724
17.3 Protozoa 725
17.4 Fungi 729
17.5 Slime Molds 733
17.6 Algae 735

WORKING GLOSSARY

Algae   phototrophic eukaryotic microorganisms

Ameboid movement   a type of motility in which cytoplasmic streaming moves the organism forward

Chitin   a polymer of *N*-acetylglucosamine commonly found in the cell walls of algae and fungi

Chloroplast   the photosynthetic organelle of eukaryotic phototrophs

Ciliates   a group of protozoa characterized by rapid motility driven by numerous short appendages called cilia

Conidia   asexual spores of fungi

Eukarya   all eukaryotic organisms

Flagellates   a group of protozoa characterized by motility driven by the whiplike action of one or more long, thin appendages called flagella

Fungi   nonphotosynthetic eukaryotic microorganisms that contain rigid cell walls

Hydrogenosome   an organelle of endosymbiotic origin present in certain anaerobic eukaryotic microorganisms that functions to oxidize pyruvate to H_2, CO_2, and acetate, along with the production of one ATP

Mitochondrion   the respiratory organelle of eukaryotic organisms

Molds   filamentous fungi

Mushrooms   filamentous fungi that produce large, often edible structures called fruiting bodies

Phagocytosis   a mechanism for ingesting particulate food in which a portion of the cell membrane surrounds the particle and brings it into the cell

Protozoa   unicellular eukaryotic microorganisms that lack cell walls

Slime molds   nonphototrophic eukaryotic microorganisms that lack cell walls and that aggregate to form fruiting structures (cellular slime molds) or masses of protoplasm (acellular slime molds)

Sporozoa   nonmotile parasitic protozoa

Yeasts   unicellular fungi

I n this chapter we consider eukaryotic microorganisms; in particular their structure, phylogeny, and diversity. We briefly encountered the eukaryotes in Chapters 1 and 3 and here review those principles and learn a few details about this highly diverse and interesting group of microorganisms. We will see that several microbial eukaryotes occupy early branches on the phylogenetic tree of Eukarya, and thus study of their basic biology is important for both evolutionary reasons as well as for a more complete understanding of the diversity of the microbial world. In addition, however, this chapter is timely in that it precedes a series of chapters that deal with host-parasite interactions, immunology, and disease. An understanding of these important concepts requires a firmer grasp of the eukaryotic cell because eukaryotic cells will be major players in these processes. We begin this chapter with a consideration of eukaryotic cell structure.

17.1

Eukaryotic Cell Structure

Figure 17.1 shows a schematic view of a eukaryotic cell. Compared with prokaryotes, the eukaryotic cell is a structurally more complex entity. All eukaryotes contain a *membrane-enclosed nucleus,* but the complement of other intracellular structures present in a given eukaryotic cell is highly dependent on the cell type (Figure 17.1). The eukaryotic cell can be enclosed by a cell wall (plant cells, algae, fungi) or cell walls may be absent (animal cells, protozoa). Organelles like mitochondria are nearly universal among eukaryotic cells while chloroplasts are found only in photosynthetic cells; several other internal structures are also present in the typical eukaryotic cell (Figure 17.1).

The Major Organelles: Mitochondria and Chloroplasts

Mitochondria and **chloroplasts** are the size of large prokaryotes and function in energy metabolism inside the eukaryotic cell. We discussed the structure and function of these key organelles in Section 3.16. Recall that the mitochondrion is a *respiratory organelle,* oxidizing organic compounds to CO_2 primarily by way of the citric acid cycle (Section 4.12); mitochondria are present in plant and animal cells and most microbial eukaryotes. By contrast, the chloroplast is a *photosynthetic organelle,* and supplies ATP to the cell from the photosynthetic light reactions (Section 15.6); chloroplasts are present only in algae and plant cells.

Mitochondria and chloroplasts contain their own DNA and ribosomes and are phylogenetically related to Bacteria. Indeed, the molecular evidence that mitochondria and chloroplasts are derived from free-living Bacteria (Sections 3.16 and 12.3) is so strong that the theory of endosymbiosis as the origin of the modern eukaryotic cell (Section 12.3) is now accepted as fact. Because mitochondria and chloroplasts contain their own DNA, the eukaryotic cell is a *genetic chimera*— a cell that contains DNA from phylogenetically diverse sources. We will see later that some microbial eukaryotes lack mitochondria but still retain a telltale sign of previously harboring these major cellular organelles.

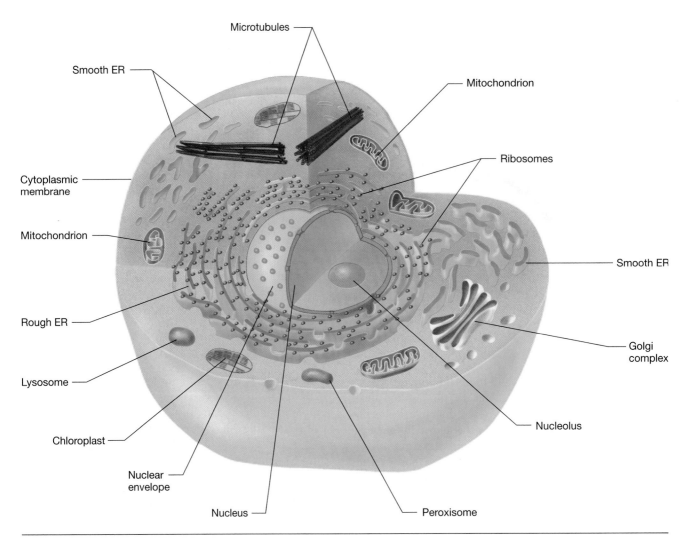

FIGURE 17.1 Schematic, cut-away view of a eukaryotic cell. Although all eukaryotic cells contain a nucleus, not all organelles or other structures shown are present in all eukaryotic cells. Chloroplasts are found only in algae and plants. The nucleolus functions in RNA synthesis and microtubules function as scaffolding to maintain cellular shape as well as in cell motility driven by cilia or flagella and chromosome/organelle movement.

Other Cytoplasmic Structures in the Eukaryotic Cell

A variety of other cytoplasmic structures are typically present in eukaryotic cells, including the endoplasmic reticulum, Golgi apparatus, lysosomes, and peroxisomes. In contrast to mitochondria and chloroplasts, these structures lack DNA and ribosomes and are not of endosymbiotic origin.

The **endoplasmic reticulum** is a network of membranes continuous with the nuclear membrane. Two types of endoplasmic reticulum (ER) are recognized: *rough*, which contains attached ribosomes, and *smooth*, which does not (Figure 17.1). Smooth ER participates in the synthesis of lipids and in some aspects of carbohydrate metabolism. Rough ER, through the activity of its ribosomes, is a major producer of glycoproteins, and

also produces new membrane material that is transported throughout the cell to enlarge the various endomembrane systems before cell division.

The **Golgi apparatus** consists of a stack of membranes (Figure 17.1) that function in concert with the ER. In the Golgi complex products of the ER are chemically modified and sorted into those destined to be secreted, for example, hormones or digestive enzymes, and those that function in other membranous structures in the cell.

Lysosomes (Figure 17.1) are membrane-enclosed sacs that contain various digestive enzymes that the cell uses to digest macromolecules like proteins, fats, and polysaccharides. The internal pH of the lysosome is about 5, two units lower than that of the cytoplasm, and the hydrolytic enzymes within the lysosome function optimally at this pH. Because these hydrolytic enzymes are nonspecific in their action and could potentially destroy key cellular macromolecules if not contained, the lysosome is a structure that allows lytic activities to be partitioned away from the cytoplasm proper. Following hydrolysis of macromolecules in the lysosome, the resulting monomers pass from the lysosome into the cytoplasm as nutrients for the cell.

The **peroxisome** is a membrane-enclosed structure (Figure 17.1) whose function is to produce hydrogen peroxide (H_2O_2) from the reduction of O_2 by various hydrogen donors including alcohols and long chain fatty acids. The H_2O_2 produced in the peroxisome is degraded to H_2O and O_2 by the enzyme catalase (∞ Section 5.12). Peroxisomes play other roles as well, such as synthesizing bile salts that aid in the absorption and digestion of fats. Peroxisomes originate in the cell by incorporating their proteins and lipids from the cytoplasm, eventually becoming a membrane-enclosed entity that can enlarge and divide in synchrony with the cell.

The Hydrogenosome

Some eukaryotic microorganisms contain a membrane-enclosed respiratory organelle distinct in both structure and function from the mitochondrion, called the **hydrogenosome** (Figure 17.2a). Although about the size of mitochondria, hydrogenosomes lack DNA and ribosomes (but see an exception later) and the extensive internal membrane systems of mitochondria (Figure 17.2a) (∞ Section 3.16). A variety of microbial eukaryotes contain hydrogenosomes but in all cases they are either obligate or aerotolerant anaerobes whose metabolism is strictly fermentative. These include a number of parasites such as the flagellate *Trichomonas vaginalis*, a sexually-transmitted pathogen (∞ Section 23.6), as well as ciliated protozoa that inhabit the anoxic rumen of ruminant animals (∞ Section 16.15) or reside in anoxic muds and sediments.

The biochemical reactions in the hydrogenosome are focused on pyruvate oxidation to yield H_2, CO_2, and acetate (Figure 17.2b). The enzyme pyruvate:ferredoxin oxidoreductase, an enzyme absent from mitochondria but widely distributed among bacteria that produce H_2 as a fermentation product, is present in the hydrogenosome, along with an iron-containing hydrogenase similar to that present in many clostridia (∞ Section 15.21 and Figure 15.53). Together, these enzymes oxidize pyruvate to the high energy intermediate acetyl-CoA (from which additional ATP is made during the formation of acetate) along with H_2 plus CO_2 (Figure 17.2b). In some anaerobic eukaryotes, H_2-consuming symbiotic prokaryotes such as methanogens reside in their cytoplasm (∞ Figure 16.39b,c) and consume the H_2 and CO_2 produced by the hydrogenosome to yield methane ($4\ H_2 + CO_2 \rightarrow CH_4 + 2\ H_2O$). However, because hydrogenosomes lack an electron transport chain and cit-

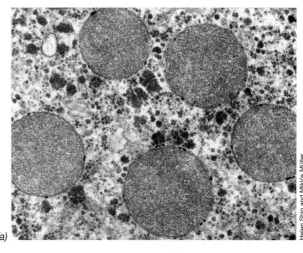

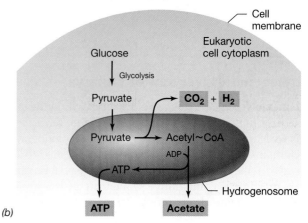

(a)

(b)

FIGURE 17.2   The hydrogenosome. (a) Electron micrograph of a thin section through a cell of the anaerobic flagellate, *Trichomonas vaginalis,* showing five hydrogenosomes. (b) Biochemistry of the hydrogenosome. Pyruvate is taken up by the hydrogenosome and H_2, CO_2, acetate, and ATP are produced. The key enzymes of the hydrogenosome are *pyruvate:ferredoxin oxidoreductase* and *hydrogenase.* Endosymbiotic methanogens are often present in the cytoplasm of hydrogenosome-containing eukaryotes, producing methane from the $H_2 + CO_2$ (∞ Figure 16.39b,c).

ric acid cycle enzymes (as well as access to an electron acceptor like O_2), they cannot oxidize the acetate produced from pyruvate like mitochondria can and thus acetate is excreted into the cytoplasm (Figure 17.2b).

What is the origin of the hydrogenosome? From molecular genetic studies it appears clear that hydrogenosomes are, like the mitochondrion and chloroplast, endosymbionts. This conclusion emerges from the fact that the nucleus of hydrogenosome-containing eukaryotes contains genes encoding specific proteins known to be of bacterial origin. Thus, like mitochondria and chloroplasts, once established by endosymbiosis, hydrogenosomes apparently also transferred genes to the nucleus. However, instead of retaining a small number of genes as in the mitochondrion, *all* genetic functions were transferred to the nucleus. It is therefore likely that hydrogenosomes originated by stable association of a pyruvate-fermenting bacterium (such as a *Clostridium* species, ⌀ Figure 15.53) with an anaerobic microbial eukaryote as a way to extract a bit more energy out of a strictly fermentative lifestyle.

Further support for an endosymbiotic origin of the hydrogenosome has emerged from the discovery that the hydrogenosomes of the ciliated protozoan *Nyctotherus ovalis*, which inhabits the hindgut of the termite (⌀ Section 16.14), *actually contain* DNA and ribosomes. This is a clear indication of the endosymbiotic origin of the hydrogenosome and has reinforced the idea that the mitochondrion and the hydrogenosome are simply different types of respiratory organelles.

17.2

Phylogenetic Overview of Eukarya

We previewed the phylogeny of Eukarya within the context of the universal phylogenetic tree in Figure 12.13. From this we learned that Eukarya form their own phylogenetic domain and as a group are more closely related to Archaea than they are to Bacteria. This phylogeny of Eukarya is deduced from comparative sequencing of 18S ribosomal RNA, obtained from cytoplasmic ribosomes (⌀ Section 12.5). But before we examine this, let us briefly consider the classical picture of phylogenetic diversity.

Five Kingdoms or Three Domains?

Eukarya have traditionally been grouped into four *kingdoms* of organisms: *plants, animals, fungi,* and the remaining microbial Eukarya, called the *protists* (or *protoctists*). Before the era of molecular phylogeny these four groups, along with all prokaryotes (that were simply lumped into one phylogenetic unit), formed the so-called *five kingdoms of living organisms.* However, molecular sequencing (⌀ Chapter 12) has shown that the five-kingdom system greatly magnified the evolutionary importance of metazoans (multicellular plants and animals). Indeed, the evolutionary picture that has emerged from comparative ribosomal RNA sequencing, that of *three domains of cellular life* (⌀ Figure 12.13), differs significantly from the five-kingdom hypothesis, in particular by showing that the bulk of evolutionary diversity resides within the microbial world. We see here that the same holds for the eukaryotes.

Figure 17.3 shows details of the Eukarya domain of the tree of life. We can see that several microbial eukaryotes occupy deep lineages on the tree, while metazoans are highly derived organisms; the latter conclusion corresponds well with the rather late appearance of metazoans in the fossil record (⌀ Section 12.1). Algae are scattered within the eukaryal tree (mainly in relatively recent lineages) while the fungi (except for the Oomycetes) form a very tight, and rather recent, phylogenetic unit (Figure 17.3).

The organisms that branch deepest in the domain Eukarya are protozoans, albeit very unusual ones.

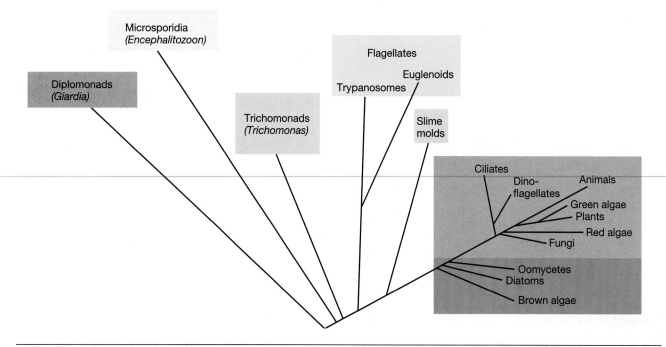

FIGURE 17.3   Phylogenetic tree of Eukarya based on comparative 18S ribosomal RNA sequences. See Figure 12.13 for comparisons of this detailed tree with the universal tree.

Diplomonads like *Giardia* (☞ Section 24.9) and *microsporidia* like *Encephalitozoon* are the most phylogenetically ancient of all known protists, with trypanosomes like *Trichomonas* not far behind (Figure 17.3). While these organisms all contain a membrane-enclosed nucleus, they lack mitochondria although many contain a hydrogenosome (see Section 17.1) (the microsporidian *Encephalitozoon* even lacks a Golgi complex and peroxisomes). Several early branching eukaryotes also contain very small genomes, smaller in some cases than that of many prokaryotes (for example, the genome of *Encephalitozoon cuniculi* is just 2.9 Mb while that of *Escherichia coli* is 4.6 Mb, ☞ Section 9.12). What does this say about their evolutionary stature?

Premitochondriate Eukaryotes?

Collectively, the characteristics of many of the early branching lineages on the tree of Eukarya fit well the predicted phenotype of premitochondriate eukaryotes, and thus it would be easy to conclude that organisms like *Giardia* are extant examples of lineages that never underwent endosymbiosis. However, this is apparently not the case because molecular methods have detected genes in the nucleus of organelle-less eukaryotes that are clearly derived from Bacteria. This leads to the inescapable conclusion that these cells once harbored endosymbionts but presumably discarded them later, leaving only a trace of their existence in the form of genes transferred from the endosymbiont to the nucleus.

Nevertheless, it seems likely that sometime before the first endosymbiotic events occurred primitive eukaryotic cells existed that never harbored endosymbionts. However, if such cells did exist, microbiologists have yet to detect any of their living ancestors. Indeed, if an organelle-less eukaryote that contains no traces of bacterially derived genes is someday found, one would predict that it would branch earlier on the phylogenetic tree than even the hydrogenosome-containing microbial eukaryotes (Figure 17.3). Such organisms would also give solid support to the hypothesis that an era of structurally primitive eukaryotes existed before the first experiments in endosymbiosis lead to the evolution of the anaerobic hydrogenosome and presumably later, the aerobic mitochondrion, as respiratory organelles for the eukaryotic cell.

With this overview of the evolutionary relationships of microbial eukaryotes in mind, let us now consider aspects of their diversity. We begin with the phylogenetically diverse protozoa.

✓ 17.1–17.2 Concept Check

The eukaryotic cell is structurally more complex than the prokaryotic cell, containing a number of membrane-enclosed structures. The key organelles—the mitochondrion, chloroplast, and hydrogenosome—are derived from endosymbiotic prokaryotes of the domain Bacteria.

- ✓ What functions occur in the *lysosome* and why is it good that these functions are compartmentalized?
- ✓ How does the mitochondrion differ from the hydrogenosome in both structure and function?
- ✓ What is structurally and genetically unusual about the microsporidian *Encephalitozoon* and why does this correlate well with its phylogeny?

17.3

Protozoa

Key Genera

Amoeba
Paramecium
Trypanosoma

Protozoa are unicellular eukaryotic microorganisms that lack cell walls (Figure 17.4). They are generally color-

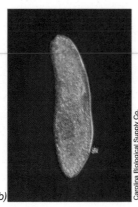
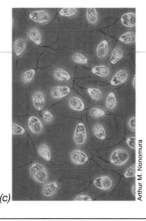
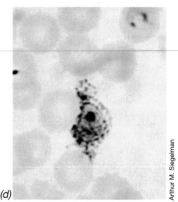

(a)   (b)   (c)   (d)

FIGURE 17.4 Typical protozoa. (a) *Amoeba*. (b) A typical ciliate, *Paramecium*. (c) A flagellate, *Dunaliella* (this flagellate contains chloroplasts and thus can also be considered an alga). (d) *Plasmodium vivax*, an apicomplexan sporozoan, growing in red blood cells.

TABLE 17.1	Characteristics of the major groups of protozoa			
Group	**Common name**	**Typical representatives**	**Habitats**	**Common diseases**
Mastigophora	Flagellates	*Trypanosoma, Giardia, Leishmania, Trichomonas*	Freshwater; parasites of animals	African sleeping sickness, giardiasis, leishmaniasis
Euglenoids[a]	Phototrophic flagellates	*Euglena*	Freshwater; some marine	None known
Sarcodina	Amebas	*Amoeba, Entamoeba*	Freshwater and marine; animal parasites	Amebic dysentery (amebiasis)
Ciliophora	Ciliates	*Balantidium, Paramecium*	Freshwater and marine; animal parasites; rumen	Dysentery
Apicomplexa	Sporozoans	*Plasmodium, Toxoplasma*	Primarily animal parasites; insects (vectors for parasitic diseases)	Malaria, toxoplasmosis

a This group is also considered with the algae (see Section 17.6 and Table 17.3).

less and motile. **Protozoa** are distinguished from prokaryotes by their eukaryotic nature and usually greater size, from algae by their lack of chlorophyll, from yeasts and other fungi by their motility and absence of a cell wall, and from the slime molds by their lack of fruiting-body formation. Also, as previously mentioned, protozoa are phylogenetically distinct from all of the aforementioned groups, existing in several lineages on the Eukarya tree (Figure 17.3).

Protozoa usually obtain food by ingesting other organisms or organic particles. Protozoa are found in a variety of freshwater and marine habitats; a large number are parasitic in other animals, including humans, and some are found growing in soil or in aerial habitats, such as on the surface of trees.

Protozoa feed by ingesting particulate or macromolecular materials. The uptake of macromolecules in solution occurs by a process called pinocytosis. Fluid droplets are sucked into a channel formed by the invagination of the cell membrane, and when portions of this channel are pinched off, the fluid becomes enclosed within a membrane-enclosed vacuole. Most protozoa are also able to ingest particulate material by **phagocytosis,** a process of surrounding a food particle with a portion of their flexible cell membrane to engulf the particle and bring it into the cell. Some protozoa can literally swallow particulate matter (such as bacterial cells) by operation of a special structure called a gullet (see Figure 17.8).

As is appropriate for organisms that "catch" their own food, most protozoa are motile. Indeed, their mechanisms of motility are key characteristics used to divide them into taxonomic groups (Table 17.1). Protozoa that move by ameboid motion are called Sarcodina; those using flagella, the Mastigophora; and those using cilia, the Ciliophora. The Apicomplexans, a fourth group, are generally nonmotile and are all parasitic for higher animals.

Mastigophora: The Flagellates

Members of this protozoal group are motile by the action of flagella (Figure 17.4c and Figure 17.5) and are phylogenetically the most ancient of protozoans (Figure 17.3). Although many flagellated protozoa are free-living organisms, a number are parasitic in, or pathogenic for, animals, including humans. The most important pathogenic Mastigophora are the *trypanosomes*. These organisms cause a number of serious diseases in humans and vertebrate animals, including the feared disease *African sleeping sickness*. In *Trypanosoma*, the genus infecting humans, the protozoa are rather small, about 20 μm in length, and are thin, crescent-shaped organisms. They have a single flagellum that originates in a basal body and folds back laterally across the cell where it is enclosed by a flap of surface membrane (Figure 17.5). Both the flagellum and the membrane participate in propelling the organism, making effective movement possible even in blood, which is rather viscous. *Trypanosoma gambiense* is the species that causes the chronic and usu-

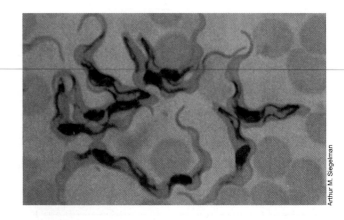

FIGURE 17.5 Photomicrograph of the flagellated protozoan *Trypanosoma gambiense*, the causative agent of African sleeping sickness, from a blood smear.

ally fatal African sleeping sickness. In humans, the parasite lives and grows primarily in the bloodstream, but in the later stages of the disease, invasion of the central nervous system occurs, causing an inflammation of the brain and spinal cord that is responsible for the characteristic neurological symptoms of the disease. The parasite is transmitted from host to host by the tsetse fly, *Glossina* sp., a bloodsucking fly found only in certain parts of Africa. The parasite proliferates in the intestinal tract of the fly and invades the insect's salivary glands and mouthparts, from which it can be transferred to a new human host following a single fly bite.

We describe *phototrophic* flagellates in Section 17.6. These are the *euglenoids*, flagellates that contain chloroplasts, which allow for photosynthetic growth. However, in darkness, cells of *Euglena* (see Figure 17.19a) can survive and grow as chemoorganotrophs, and as such these organisms become phenotypically indistinguishable from protozoa. Many euglenoids are known and they are all aquatic, inhabiting primarily fresh waters. Unlike other flagellated protozoa, the euglenoids are nonpathogenic.

Sarcodina: The Amebas

Among the sarcodines are organisms such as *Amoeba*, which are always naked in the vegetative phase (see Figure 17.4a), and the foraminifera, amebas that secrete a shell during vegetative growth (Figure 17.6). A wide variety of naked amebas are parasites of humans and other vertebrates, and their usual habitat is the oral cavity or the intestinal tract. They move in these habitats by *ameboid movement* (Figure 17.7), a mechanism previously discussed for the acellular slime molds (see Section 17.5). *Entamoeba histolytica* (∞ Figure 24.24) is a good example of a parasitic ameba. In many cases infection causes no obvious symptoms, but in some individuals it produces ulceration of the intestinal tract, which results in a diarrheal condition called *amebic dysentery*

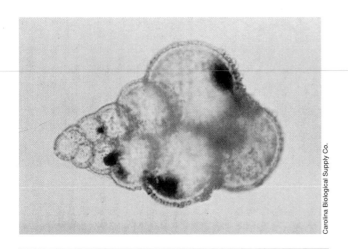

FIGURE 17.6   Shelled amebas: foraminifera.

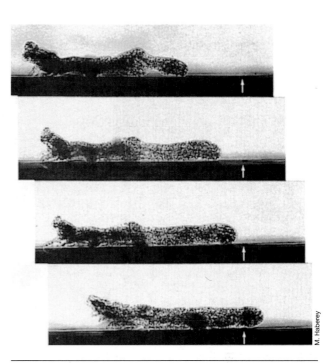

FIGURE 17.7   Side view of a moving ameba, *Amoeba proteus*, taken from a film, the time interval between frames being 2 sec. The arrows point to a fixed spot on the surface. A single cell is about 80 μm in diameter.

(amebiasis). The organism is transmitted from person to person in the cyst form by fecal contamination of water and food. We discuss the etiology and pathogenesis of amebic dysentery in Section 24.9.

Shelled sarcodines present a variety of interesting morphological forms. The best-known of the shelled forms are the *foraminifera*. Foraminifera are exclusively marine organisms, living primarily in coastal waters. The shells, called *tests*, of different species show distinctive characteristics and are often quite ornate (Figure 17.6). Tests are usually made of calcium carbonate. The cell is not firmly attached to the test, and the ameba cell may extend partway out of the shell during feeding. However, because of the weight of the test, the cell usually sinks to the bottom, and it is thought that the organisms feed on particulate deposits in the sediments, primarily bacteria and detritus. The shells of foraminifera are relatively resistant to decay and hence readily become fossilized (the White Cliffs of Dover, England, are composed to a great extent of foraminiferal shells). Because of the excellent fossil record these organisms leave, we have a better idea of their distribution through geological time than for virtually any other protozoa.

Ciliophora: The Ciliates

Ciliates are those protozoa that, in some stage of their life cycle, possess cilia (Figure 17.8). They are also

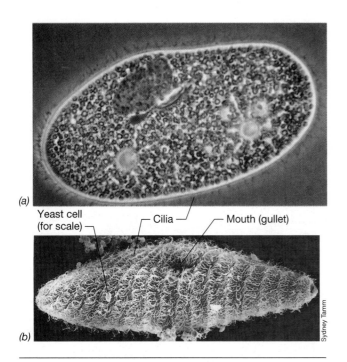

(a)

Yeast cell (for scale) — Cilia — Mouth (gullet)

(b)

Sydney Tamm

FIGURE 17.8 *Paramecium,* a ciliated protozoan. (a) Phase photomicrograph. (b) Scanning electron micrograph. Note the cilia in both micrographs. A single *Paramecium* cell is about 60 μm in diameter.

unique among protozoa in having *two kinds* of nuclei: the *micronucleus,* which is concerned only with inheritance and sexual reproduction; and the *macronucleus,* which is involved only in the production of messenger ribonucleic acid (mRNA) for various aspects of cell growth and function. Probably the best-known and most widely distributed of the ciliates are those of the genus *Paramecium* (Figure 17.8), which will be used here as an example of the group. Most ciliates obtain their food by ingesting particulate materials through a distinct oral region or mouth connected to an underlying gullet (Figure 17.8*b*). Once inside, the food particles are carried down the gullet and into the cytoplasm where they are enclosed in a food vacuole, a structure into which digestive enzymes are secreted. In addition to cilia, which function in motility, many ciliates have *trichocysts,* which are long, thin filaments of a contractile nature, anchored beneath the surface of the outer cell layer. These enable the protozoa to attach to a surface and can aid in defense by signaling the cell that it is being attacked by a predator, thus stimulating cellular defenses. In the case of predatory ciliates, such as *Didinium,* trichocysts hold onto and paralyze the prey as a prelude to ingestion.

The two kinds of nuclei in the ciliates set this group aside from all other protozoa. The micronucleus is diploid, but the macronucleus is polyploid, containing from about 40 to over 500 times as much DNA as does the micronucleus. The micronucleus plays no direct role in vegetative growth and cell division as evidenced by

the fact that strains can be developed that lack micronuclei but continue to divide normally. However, if the macronucleus is removed, the cell quickly dies. Sexual reproduction in ciliates involves conjugation between two cells of opposite mating type. Conjugation involves exchange and fusion of micronuclei with the net result being the formation of exconjugants that are hybrid for certain genes of the two strains.

Many *Paramecium* species (as well as many other protozoa) contain endosymbiotic bacteria living in the cytoplasm or the macronucleus. In some cases, evidence exists that these endosymbionts play a nutritional role for the host, synthesizing vitamins or other growth factors that would otherwise have to be obtained from the environment. In the case of protozoa existing in the termite gut, we previously discussed how endosymbiotic methanogens remove H_2 produced from pyruvate oxidation in the hydrogenosome (Figure 17.2), yielding CH_4, which is released to the atmosphere (∞ Section 16.14 and Figure 16.39*b,c*).

Although a few ciliates are parasitic for animals, this mode of existence is less extensively developed in the ciliates than it is in other groups of protozoa. The species *Balantidium coli* (Figure 17.9) is primarily a parasite of domestic animals, but occasionally it infects the intestinal tract of humans, producing intestinal dysentery symptoms similar to those caused by *Entamoeba histolytica.* In addition, there is usually a characteristic fauna of *obligately anaerobic* ciliates in the rumen, the forestomach of ruminant animals (∞ Section 16.15); these protozoa are thought to play a beneficial role in the digestive and fermentative processes that occur there.

Apicomplexa (Sporozoans)

The Apicomplexa, or *sporozoans,* comprise a large group of protozoa, all of which are obligate parasites. They are characterized by a *lack* of motile adult stages and by a nutritional mode of life in which food is generally not

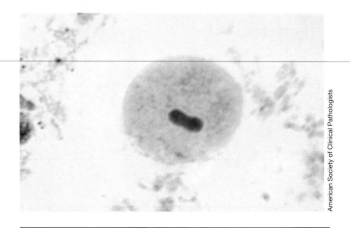

American Society of Clinical Pathologists

FIGURE 17.9 *Balantidium coli,* a ciliated protozoan that causes a dysentery-like disease in humans. The dark blue stained structure is the macronucleus.

ingested but instead is absorbed in soluble form through the outer wall, such as occurs in prokaryotes and in fungi. Although the name *sporozoa* implies the formation of spores, these organisms do not form true resting spores, like those of bacteria, algae, and fungi, but instead produce analogous structures called *sporozoites*, which are involved in transmission to a new host. Numerous kinds of vertebrates and invertebrates are hosts for Sporozoa, and in some cases an alternation of hosts takes place, with some stages of the life cycle occurring in one host and some in another. The most important members of the sporozoa are the coccidia, usually parasites of birds, and the plasmodia (malaria parasites) (Figure 17.4*d*), which infect birds and mammals, including humans. Because malaria is a major disease of humans, especially in developing countries, we devote a considerable discussion to this disease and the properties of malarial parasites in Section 24.5.

✓ 17.3 Concept Check

Protozoa are unicellular microbial Eukarya that lack cell walls and are usually motile by various means. Many protozoa are pathogenic to humans and other animals.

✓ List at least two major ways in which the protozoan *Paramecium* differs from the protozoan *Trypanosoma*.
✓ How do the Sporozoa differ from all other protozoa?
✓ Why is the alga *Euglena* also considered a protozoan?

17.4

Fungi

Key Genera

Penicillium
Aspergillis
Saccharomyces
Candida

In contrast to the protozoa, fungi contain cell walls and produce spores, among many other differences, and most described species form a relatively tight phylogenetic cluster (Figure 17.3). Three major groups of fungi are recognized: the *molds*, the *yeasts*, and the *mushrooms*.

The habitats of fungi are quite diverse. Some are aquatic, living primarily in fresh water, and a few marine fungi are also known. Most fungi, however, have terrestrial habitats, in soil or on dead plant matter, and these types often play crucial roles in the mineralization of organic carbon in nature. A large number of fungi are parasites of terrestrial plants. Indeed, fungi cause the majority of economically significant diseases of crop plants (Table 17.2). A few fungi are parasitic on animals, including humans, although in general fungi are less significant as animal pathogens than are bacteria and viruses (∞ Section 24.7 for a discussion of pathogenic fungi).

TABLE 17.2	Classification and major properties of fungi[a]					
Group	**Common name**	**Hyphae**	**Typical representatives**	**Type of sexual spore**	**Habitats**	**Common diseases**
Ascomycetes	Sac fungi	Septate	*Neurospora,* *Saccharomyces,* *Morchella* (morels)	Ascospore	Soil, decaying plant material	Dutch elm, chestnut blight, ergot, rots
Basidiomycetes	Club fungi, mushrooms	Septate	*Amanita* (poisonous mushroom), *Agaricus* (edible mushroom)	Basidiospore	Soil, decaying plant material	Black stem, wheat rust, corn smut
Zygomycetes	Bread molds	Coenocytic	*Mucor, Rhizopus* (common bread mold)	Zygospore	Soil, decaying plant material	Food spoilage; rarely involved in parasitic disease
Oomycetes	Water molds	Coenocytic	*Allomyces*	Oospore	Aquatic	Potato blight, certain fish diseases
Deuteromycetes	Fungi imperfecti	Septate	*Penicillium,* *Aspergillus,* *Candida*	None known	Soil, decaying plant material, surfaces of animal bodies	Plant wilt, infections of animals such as ringworm, athlete's foot, and other dermatomycoses, surface or systemic infections (*Candida*)

a With the exception of the Oomycetes, which are phylogenetically distinct, the other groups of fungi are closely related (see Figure 17.3).

Cell Walls and Metabolism

Fungal cell walls resemble plant cell walls architecturally, but not chemically. Although cellulose is present in the walls of certain fungi, many fungi have noncellulosic walls. **Chitin,** a polymer of the glucose derivative, N-acetylglucosamine (∞ Figure 2.5), is a common constituent of fungal cell walls. It is laid down in microfibrillar bundles like cellulose; other glucans such as mannans, galactosans, and chitosans replace chitin in some fungal cell walls. Fungal cell walls are generally 80–90% polysaccharide, with proteins, lipids, polyphosphates, and inorganic ions making up the wall-cementing matrix. An understanding of fungal cell wall chemistry is important because of the extensive biotechnological uses of fungi (∞ Chapter 11) and because the chemical nature of the fungal cell wall has been useful in classifying fungi for research and industrial purposes.

All fungi are chemoorganotrophs and generally have simple nutritional requirements. Many species can grow at environmental extremes of low pH or high temperature (up to 62°C) and this, coupled with the fact that the spores of fungi are so ubiquitous, makes these organisms common contaminants of food products, microbial culture media, and most surfaces. But it is not so much on physiological grounds that the molds and yeasts are classified but instead in their diverse array of life cycle patterns including the formation of a variety of different sexual spores.

Molds

The molds are *filamentous* fungi. They are widespread in nature and are commonly seen on stale bread, cheese, or fruit. Each filament grows mainly at the tip, by extension of the terminal cell (Figure 17.10). A single filament is called a *hypha* (plural, *hyphae*). Hyphae usually grow together across a surface and form compact tufts, collectively called a *mycelium*, which can be seen easily without a microscope. The mycelium arises because the individual hyphae form branches as they grow, and these branches intertwine, resulting in a compact mat. In most cases, the vegetative cell of a fungal hypha contains more than one nucleus—often hundreds of nuclei are present. Thus, a typical hypha is a nucleated tube containing cytoplasm (referred to as *coenocytic*). Usually there is extensive cytoplasmic movement within a hypha, generally in a direction toward the hyphal tip, and the older portions of the hypha usually become vacuolated and virtually devoid of cytoplasm. Even if a hypha has cross-walls, cytoplasmic movement is often not prevented, as there is usually a pore in the center of the septum through which nuclei and cytoplasmic particles can move.

From the fungal mycelium, other hyphal branches may reach up into the air above the surface, and on these aerial branches spores called *conidia* are formed (Figure 17.10a). Conidia are *asexual* spores (their formation does *not* involve the fusion of gametes), often highly pigmented (Figure 17.11a; ∞ Figure 9.1a) and resistant to drying, and function in the dispersal of the fungus to new habitats. When conidia form, the white color of the mycelium changes, taking on the color of the conidia, which may be black, blue-green, red, yellow, or brown. The presence of these spores gives the mycelial mat a rather dusty appearance (Figure 17.11a).

Some molds also produce *sexual* spores, formed as a result of sexual reproduction (Table 17.2). The latter occur

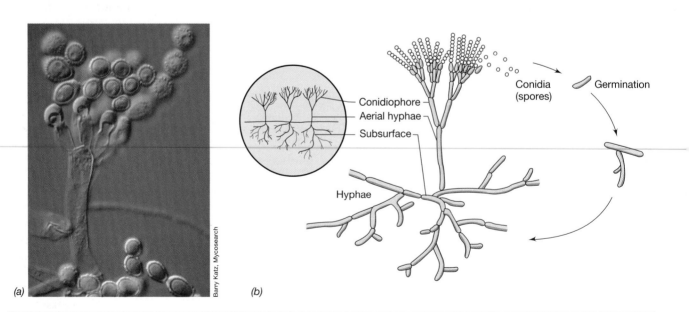

(a) Barry Katz, Mycosearch *(b)*

Conidiophore
Aerial hyphae
Subsurface
Hyphae
Conidia (spores)
Germination

FIGURE 17.10   Mold structure and growth. (a) Photomicrograph of a typical mold. Conidia are seen as the spherical structures at the ends of aerial hyphae. (b) Diagram of a mold life cycle.

Cheryl Broadie

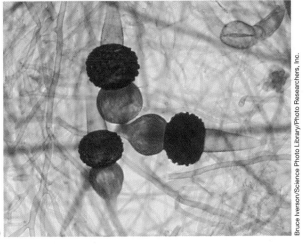

Bruce Iverson/Science Photo Library/Photo Researchers, Inc.

FIGURE 17.11 Fungi. (a) Fungal colonies growing on an agar plate. Note the appearance of the masses of mycelia and asexual spores that give the colonies a dusty, matted appearance. (b) Zygospores (sexual spores, black structures) of the common bread mold, *Rhizopus nigricans*. Zygospores result from the fusion of two nuclei and eventually release asexual spores that germinate to yield new mycelia.

from the fusion either of unicellular gametes or of specialized hyphae called *gametangia*. Alternatively, sexual spores can originate from the fusion of two haploid cells to yield a diploid cell, which then undergoes meiosis and mitosis to yield individual spores. Depending on the group to which a particular fungus belongs (see Table 17.2), different types of sexual spores are produced. Spores formed within an enclosed sac (*ascus*) are called *ascospores*, and those produced on the ends of a club-shaped structure (*basidium*) are *basidiospores* (Table 17.2). *Zygospores*, produced by zygomycetous fungi like the common bread mold *Rhizopus* (Figure 17.11*b*), are macroscopically visible structures and the result of the fusion of hyphae and genetic exchange. Eventually the zygospore matures and produces asexual spores that are dispersed by air and germinate to form new fungal mycelia.

Sexual spores of fungi are usually resistant to drying, heating, freezing, and some chemical agents. However, fungal sexual spores are not as resistant to heat as bacterial endospores (∞ Section 3.15). Either an asexual or a sexual spore of a fungus can germinate and develop into a new hypha and mycelium.

A major ecological activity of many fungi, especially members of the Basidiomycetes (see Table 17.2), is the decomposition of wood, paper, cloth, and other products derived from natural sources. Basidiomycetes that attack these products are able to utilize cellulose or lignin from the product as carbon and energy sources. Lignin is a complex polymer in which the building blocks are phenolic compounds. It is an important constituent of woody plants, and in association with cellulose it confers rigidity on them. The decomposition of lignin in nature occurs almost exclusively through the action of certain Basidiomycetes called *wood-rotting fungi*. Two types of wood rots are known: *brown rot*, in which the cellulose is attacked preferentially and the lignin left unchanged, and *white rot*, in which both cellulose and lignin are decomposed. The white rot fungi are of considerable ecological interest because they play such an important role in decomposing woody material in forests.

Mushrooms

Mushrooms are filamentous basidiomycetes that form large *fruiting bodies,* the edible part of the mushroom (Figure 17.12*a,b*). We discussed in Section 11.14 the commercial growth of mushrooms as a food source. In nature mushrooms live as saprophytes in the soil or on the trunks of trees. Mushroom basidospores (Figure 17.12*c*) are dispersed through the air and initiate mycelial growth on favorable, usually moist, substrates. From here an extensive mycelium forms following the fusion of two haploid mycelia to yield a cell containing two nuclei (a *dikaryotic* state); the latter is the beginnings of a fruiting body.

Yeasts

The yeasts are *unicellular* fungi, and most of them are classified with the Ascomycetes. Yeast cells are usually spherical, oval, or cylindrical, and cell division generally takes place by budding (Figure 17.13). In the budding process, a new cell forms as a small outgrowth of the old cell; the bud gradually enlarges and then separates (Figures 17.13 and 17.14). Although most yeasts reproduce only as single cells, under some conditions some yeasts can form filaments. And in these species, certain characteristics may be expressed only by the filamentous form. For example, the filamentous phase is essential for pathogenicity in *Candida albicans*, a yeast that can cause vaginal, oral, or lung infections and, in acquired immunodeficiency (AIDS) patients, systemic tissue damage (∞ Section 23.7).

(a)

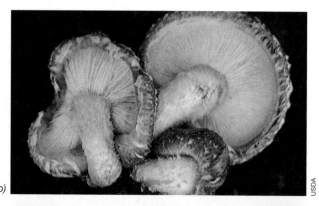

(b)

(c)

FIGURE 17.12   Mushrooms. (a) *Amanita,* a highly poisonous mushroom. (b) Gills on the underside of the mushroom contain the spore-bearing basidia. (c) Scanning electron micrograph of basidiospores released from mushroom basidia. The production of mushrooms as a food source was discussed in Section 11.14.

Yeast cells are much larger than bacterial cells and can be distinguished microscopically from bacteria by their size and by the obvious presence of internal cell structures, such as the nucleus (Figure 17.14). Some yeasts exhibit sexual reproduction by a process called *mating,* in which two yeast cells fuse. Within the fused cell, called a *zygote,* ascospores are eventually formed. We discussed the sexual cycle of a typical yeast, *Saccharomyces,* including the important property of *mating types,* in Section 9.14.

Yeasts flourish in habitats where sugars are present, such as fruits, flowers, and the bark of trees. A number of yeast species live symbiotically with animals, espe-cially insects, and a few species are pathogenetic for an-imals and humans (∞ Section 24.7). The most impor-tant commercial yeasts are the baker's and brewer's yeasts, which are members of the genus *Saccharomyces* (∞ box, The Products of Yeast Fermentation, Chapter 4). The original habitats of these yeasts were undoubt-edly fruits and fruit juices, but the commercial yeasts of today are probably quite different from wild strains because they have been greatly improved through the years by careful selection and genetic manipulation by industrial microbiologists. Baker's and brewer's yeasts are probably the best known scientifically of all fungi because they are easily manipulable eukaryotic cells, and they are thus excellent models for the study of many important problems in eukaryotic biology (∞ Section 9.14). Indeed, *S. cerevisiae* has been studied as a model eukaryote for many years and was the first

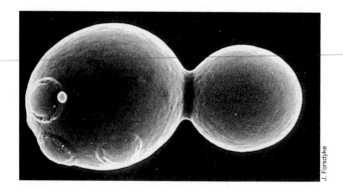

FIGURE 17.13   Scanning electron micrograph of the com-mon baker's and brewer's yeast *Saccharomyces cerevisiae.* Note the budding division and previous bud scars. A single large cell is about 8 μm in diameter.

FIGURE 17.14   Growth by budding division in *Saccha-romyces cerevisiae.* Note the pronounced nucleus. Phase-con-trast micrograph. A single cell of *S. cerevisiae* is about 8 μm in diameter.

eukaryote to have its genome completely sequenced (∞ Section 9.12).

✓ 17.4 Concept Check

Fungi include the molds and yeasts and as a group differ from protozoa by virtue of their rigid cell wall, production of spores, lack of motility, and phylogenetic position. Mushrooms are large, often edible fungi that produce fruiting bodies containing basidiospores.

- ✓ How do *molds* differ from *yeasts*?
- ✓ What is *chitin* and what is its function in fungi?
- ✓ In molds, how do *conidia* differ from *ascospores*?

17.5

Slime Molds

Key Genera

Dictyostelium
Physarum

Slime molds are microbial eukaryotes that have phenotypic similarity to both fungi and protozoa. Like fungi, slime molds undergo a life cycle and can produce spores. However, like protozoa, slime molds are motile and can move across a solid surface rather rapidly (see Figures 17.15–17.17). From a phylogenetic perspective slime molds are more ancient than fungi and some protozoa (such as the ciliates), but more derived than flagellated protozoa and their evolutionary predecessors (Figure 17.3).

The slime molds can be divided into two groups, the *cellular slime molds*, whose vegetative forms are composed of single amebalike cells, and the *acellular slime molds*, whose vegetative forms are masses of protoplasm of indefinite size and shape called *plasmodia*. Slime molds live primarily on decaying plant matter, such as leaf litter, logs,

and soil (Figure 17.15). Their food consists mainly of other microorganisms, especially bacteria, which they ingest by phagocytosis. Slime molds can maintain themselves in a vegetative state for long periods, but eventually form differentiated spore-like structures that germinate and once again generate the active amoeboid state.

Acellular Slime Molds

In the vegetative phase, acellular slime molds such as *Physarum* exist as a mass of protoplasm of indefinite size, which might be compared to a giant ameba (Figure 17.15). This structure is actively motile by *ameboid motion*, the plasmodium flowing over the surface of the substratum, engulfing food particles as it moves. Ameboid motion is the result of cytoplasmic streaming. Cytoplasm flows forward because the tip of the plasmodium is less contracted and viscous, and thus cytoplasm takes the path of least resistance. In eukaryotic cells, cytoplasmic streaming is facilitated by filaments of a protein called *actin*, which exists in a thin layer just beneath the cytoplasmic membrane of eukaryotic cells. In acellular slime molds, cytoplasmic streaming occurs in definite strands, each surrounded by the thin cytoplasmic membrane (Figure 17.15b). The strands of protoplasm do not retain their individuality indefinitely but fuse occasionally into larger masses, which then separate again into smaller strands. Slime mold plasmodia are often brilliantly colored and frequently can be seen spreading across the surface of a piece of wood that has been kept moist (Figure 17.15).

The acellular slime mold plasmodium (Figure 17.15) is genetically diploid. From this mass of protoplasm a sporangium and spores can be produced. Cells in the sporangium undergo meiotic division to form haploid spores that are released and able to remain dormant for long periods of time. Under favorable conditions, spores germinate to yield haploid swarm cells.

FIGURE 17.15   Slime molds. Plasmodia of acellular slime molds (a) growing on a decaying log and (b) the genus *Physarum*, growing on an agar surface.

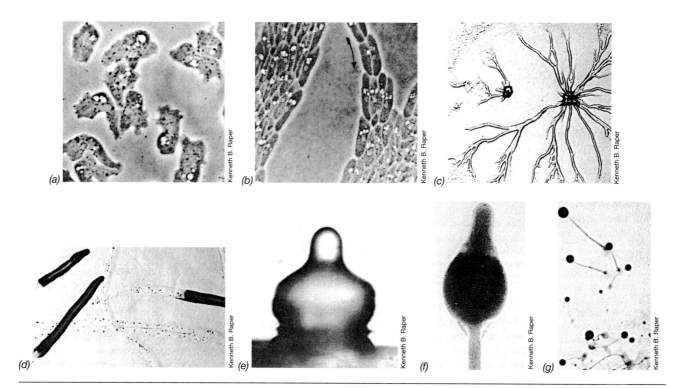

FIGURE 17.16 Photomicrographs of various stages in the life cycle of the cellular slime mold *Dictyostelium discoideum*. (a) Amebas in preaggregation stage. Note irregular shape and lack of orientation. (b) Aggregating amebas. Notice the regular shape and orientation. The cells are moving in streams in one direction. (c) Low power view of aggregating amebas. (d) Migrating pseudoplasmodia (slugs) moving on an agar surface and leaving trails of slime in their wake. (e, f) Early stage of fruiting body. (g) Mature fruiting bodies. See Figure 17.17 for sizes of these structures.

The fusion of two swarm cells then regenerates the diploid plasmodium (Figure 17.15).

Cellular Slime Molds: *Dictyostelium*

Dictyostelium discoideum, a cellular slime mold, undergoes a remarkable life cycle in which vegetative cells aggregate, migrate as a cell mass, and eventually produce fruiting bodies in which cells differentiate and form spores (Figures 17.16 and 17.17). As cells of *Dictyostelium* become starved, they aggregate and form a *pseudoplasmodium*, a structure in which the cells lose their individuality but do not fuse (Figures 17.16 and 17.17). This aggregation is triggered by the production of two compounds, *cyclic adenosine monophosphate* (cAMP) and a specific glycoprotein, both of which function as chemotactic agents (we discuss the involvement of cAMP in various regulatory systems in prokaryotes in Section 7.6). Those cells that are the first to produce these compounds serve as centers for the attraction of

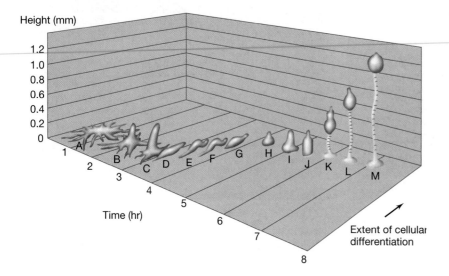

FIGURE 17.17 Stages in fruiting-body formation in the cellular slime mold *Dictyostelium discoideum*. (A–C) Aggregation of amebas; (D–G) migration of the slug formed from aggregated ameba; (H–L) culmination and formation of the fruiting body; (M) mature fruiting body.

other vegetative cells, leading to aggregating masses of cells, which come together and form a slimy migrating mass referred to as a *slug* (Figure 17.16*d*).

Fruiting-body formation begins when the slug ceases to migrate and becomes vertically oriented (Figures 17.16 and 17.17). The fruiting body then becomes differentiated into a stalk and a head; cells in the forward end of the slug become stalk cells, and those in the posterior end become spores. Cells that form stalk cells begin to secrete cellulose, which provides the rigidity of the stalk. Cells from the rear of the slug swarm up the stalk to the tip and form the head. Most of these posterior cells differentiate into spores. On maturation of the head the spores are released and dispersed. Each spore then germinates and becomes a vegetative ameba.

The cycle of fruiting-body and spore formation in *Dictyostelium* is an *asexual* process. However, *sexual* spores called *macrocysts* can also be produced. Macrocysts are formed from aggregates of amebas that become enclosed in a cellulose wall. Following the conjugation of two amebas, a single large ameba develops and proceeds to phagocytize the remaining amebas. At this point a thick cellulose wall forms around this giant ameba forming the mature macrocyst, which can remain dormant for long periods. Eventually, the diploid nucleus undergoes meiosis to form haploid nuclei that become integrated into new amebas that can once again initiate vegetative growth (Figure 17.16).

✓ 17.5 Concept Check

Acellular slime molds are masses of motile protoplasm while cellular slime molds are masses of individual cells that aggregate to form fruiting bodies from which spores are released.

✓ In what ways are the slime molds similar to *fungi* and in what way are they similar to *protozoa*?

✓ What is a *macrocyst*?

17.6

Algae

Key Genera

Chlamydomonas
Euglena
Gonyaulax

Algae are a large and diverse group of eukaryotic organisms that contain *chlorophyll* and carry out oxygenic photosynthesis (Figure 17.18). Algae should not be confused with *cyanobacteria*, which are also oxygenic pho-

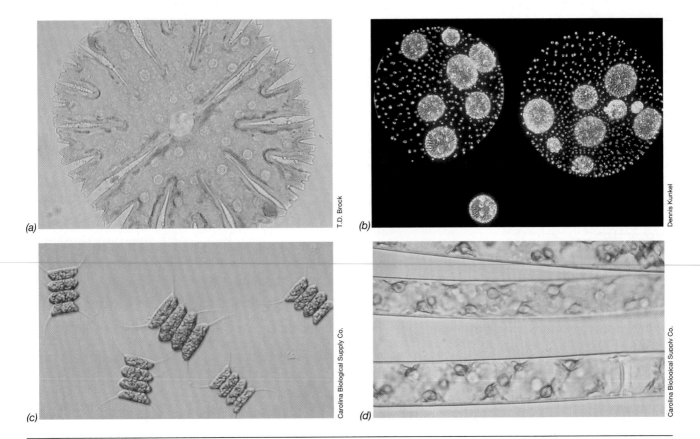

(a)   T.D. Brock

(b)   Dennis Kunkel

(c)   Carolina Biological Supply Co.

(d)   Carolina Biological Supply Co.

FIGURE 17.18 Light micrographs of representative green algae. (a) *Micrasterias*. A single cell. (b) *Volvox* colony, containing a large number of cells. (c) *Scenedesmus*. A packet of four cells. (d) *Spirogyra*. A filamentous alga. Note the green spiral-shaped chloroplasts.

totrophs but which are Bacteria and thus evolutionarily quite distinct from algae (∞ Section 13.24). Although most algae are of microscopic size and hence are clearly microorganisms, a number of forms are macroscopic, some seaweeds growing to over 30 meters in length.

Algal Diversity

Algae are either unicellular (Figure 17.18*a* and *c*) or colonial, the latter occurring as aggregates of cells (Figure 17.18*b*). When the cells are arranged end to end, the alga is said to be filamentous (Figure 17.18*d*). Among the filamentous forms, both unbranched filaments and more intricate branched filaments occur. Algae contain chlorophyll and are thus green in color. However, a few kinds of common algae are not green but appear brown or red because in addition to chlorophyll, other pigments such as xanthophylls are present that mask the green color. Algal cells contain one or more **chloroplasts,** membranous structures that house the photosynthetic pigments. Chloroplasts can often be recognized microscopically within algal cells by their distinct green color (Figure 17.18). We discussed the general structure and properties of chloroplasts in Section 3.16, and the phylogeny of the chloroplast in Sections 12.3, 13.25, and 17.2.

Examination of Figure 17.3 shows the algae to constitute a phylogenetically heterogeneous group. Although green algae (Chlorophyta, Table 17.3) and to a lesser extent red algae (Rhodophyta, Table 17.3) are quite closely related to green plants, other algal groups such as the brown algae and diatoms constitute earlier lineages (Figure 17.3). Even less derived are the euglenoids, such as the alga *Euglena* (Figure 17.19*a*). Euglenoids show a phylogenetic relationship to flagellated protozoa (Figure 17.3), and in fact, cells of *Euglena* can spontaneously lose their chloroplasts and exist as completely heterotrophic organisms (see Section 17.3). Red algae are noteworthy in that their chloroplasts contain *phycobiliproteins,* the major light-harvesting pigments of the cyanobacteria (∞ Section 15.4). It is thus not surprising that the red algal chloroplast, an endosymbiont of course (see Section 17.2), shows close phylogenetic affinities to the cyanobacteria (∞ 13.24); their host cell, by contrast, forms a distinct line of Eukarya (see Figure 17.3).

Dinoflagellates are flagellated, primarily marine algae that are close phylogenetic relatives of ciliated protozoa (Figure 17.3). Some dinoflagellates are free-

TABLE 17.3	Properties of major groups of algae						
Algal group	Common name	Morphology	Pigments	Typical representative	Carbon reserve materials	Cell wall	Major habitats
Chlorophyta	Green algae	Unicellular to leafy	Chlorophylls *a* and *b*	*Chlamydomonas*	Starch (α-1,4-glucan), sucrose	Cellulose	Freshwater, soils, a few marine
Euglenophyta[a]	Euglenoids	Unicellular, flagellated	Chlorophylls *a* and *b*	*Euglena*	Paramylon (β-1,2-glucan)	No wall present	Freshwater, a few marine
Dinoflagellata	Dinoflagellates	Unicellular, flagellated	Chlorophylls *a* and *c*, xanthophylls	*Gonyaulax*	Starch (α-1,4-glucan)	Cellulose	Mainly marine
Chrysophyta	Golden-brown algae, diatoms	Unicellular	Chlorophylls *a* and *c*	*Navicula*	Lipids	Many have two overlapping components made of silica	Freshwater, marine, soil
Phaeophyta	Brown algae	Filamentous to leafy, occasionally massive and plantlike	Chlorophylls *a* and *c*, xanthophylls	*Laminaria*	Laminarin (β-1,3-glucan), mannitol	Cellulose	Marine
Rhodophyta	Red algae	Unicellular, filamentous to leafy	Chlorophylls *a* and *d*, phycocyanin, phycoerythrin	*Polysiphonia*	Floridean starch (α-1,4- and α-1,6-glucan), fluoridoside (glycerol-galactoside)	Cellulose	Marine

a This group is also considered with the protozoa (see Section 17.3).

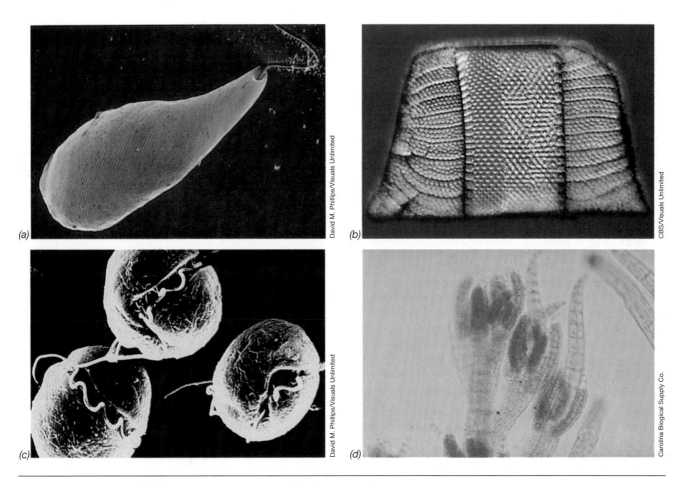

FIGURE 17.19   Scanning electron (a–c) and light (d) micrographs of algae other than Chlorophyta. (a) *Euglena*, a member of the Euglenophyta. (b) *Isthmix*, a diatom (Chrysophyta). (c) *Gonyaulax*, a dinoflagellate (Pyrrophyta)(see also Figure 17.20). (d) *Polysiphonia*. This marine red alga (Rhodophyta) grows attached to the surfaces of various marine plants.

living (Figure 17.19*c*) while others live a symbiotic existence with animals that make up coral reefs, obtaining a sheltered and protected habitat in exchange for supplying photosynthetically fixed carbon as a food source for the reef.

Dinoflagellates such as *Gonyaulax* (Figure 17.19*c*) occasionally grow in dense suspensions called *blooms* that are bright red in color from the xanthophylls present as accessory pigments in this organism. These blooms are the classical "red tides" (Figure 17.20) that occasionally develop in warm, usually polluted, coastal waters and which are associated with fish kills and poisoning in humans following consumption of contaminated water or shellfish. Toxicity occurs because of a potent neurotoxin produced by *Gonyaulax* that is sufficiently toxic to kill fish at nanogram levels.

Pigments, Energy Metabolism, and Reserve Polymers

Several characteristics are used to classify algae, including the nature of the chlorophyll(s) present, the carbon reserve polymers produced, the cell wall structure, and the type of motility. All algae contain chlorophyll *a*. Some, however, also contain other chlorophylls that differ in minor ways from chlorophyll *a*. The presence of these additional chlorophylls is characteristic of particular algal groups. The distribution of chlorophylls and other photosynthetic pigments in algae is summarized in Table 17.3.

All algae carry out oxygen-evolving (oxygenic) photosynthesis, using H_2O as an electron donor. In addition, some algae carry out photosynthesis *without* yielding oxygen, using H_2 as an electron donor. Many algae are obligate phototrophs and are thus unable to grow in darkness on organic carbon compounds. However, some species can grow chemoorganotrophically and catabolize simple sugars or organic acids in the dark. One of the organic compounds most widely used by algae is acetate, which can be used as a sole carbon and energy source by many flagellates and chlorophytes. In addition, some algae can assimilate simple organic compounds in the light (photoheterotrophy; ∽ Section 15.5) but cannot grow on them as sole energy sources.

Rita R. Colwell

FIGURE 17.20   Photograph of a red tide caused by massive growth of toxin-producing dinoflagellates. The toxin is excreted into the water and also accumulates in shellfish that feed on the dinoflagellates. The toxin is harmless to the shellfish but can poison fish and humans that eat infected shellfish.

One of the key characteristics used in the classification of algal groups is the nature of the **reserve polymer** synthesized as a result of photosynthesis. Algae of the division Chlorophyta produce starch (α-1,4-glucose) in a form very similar to that of higher plants. By contrast, algae of other groups produce a variety of reserve substances, some polymeric and some as free monomers, and the major ones are listed in Table 17.3.

Cell Walls of Algae

Algae show considerable diversity in the structure and chemistry of their cell walls. In many cases the cell wall is composed of a network of cellulose fibrils, but it is usually modified by the addition of other polysaccharides such as pectin (highly hydrated polygalacturonic acid containing small amounts of the hexose rhamnose), xylans, mannans, alginic acids, or fucinic acid. In some algae, the wall is additionally strengthened by the deposition of calcium carbonate; these forms are often called "calcareous" or "coralline" (corallike) algae. Sometimes chitin (a polymer of *N*-acetylglucosamine) is also present in the cell wall. In euglenoids (Figure 17.19a) a cell wall is absent. In diatoms (Figure 17.19b) the cell wall is composed of *silica*, to which protein and polysaccharide are added. Even after the cell dies and the organic materials have disappeared, the external structure, called the *frustule*, often remains, showing that the siliceous component is indeed responsible for the rigidity of the cell. Because of the extreme resistance to decay of these diatom frustules, they remain intact for long periods of time and constitute some of the best algal fossils ever found. From this excellent fossil record, it is known that diatoms first appeared on Earth about 200 million years ago. Diatomaceous earth, an indus-

trial filtering agent, is composed of fossilized diatom cells. This material is mined in large quantities from ancient sea beds, emphasizing the large numbers of these organisms found in prehistoric seas.

Algal cell walls contain pores some 3–5 nm wide that are large enough to pass only low-molecular-weight substances such as water, inorganic ions, gases, and other small nutrient molecules needed for metabolism and growth. Phagocytic activities are thus not possible, and in this regard, algae differ from the actively phagocytic protozoa and slime molds.

Motility and Ecology of Algae

A number of algae are motile, usually because of flagella; cilia do not occur in algae. Simple flagellate forms, such as *Euglena* (see Figure 17.19a), usually have a single polar flagellum, whereas flagellated representatives of the Chlorophyta have either two or four polar flagella. Dinoflagellates (Figure 17.19c) have two flagella of different lengths and with different points of insertion into the cell. The transverse flagellum is attached laterally, whereas the longitudinal flagellum originates from the lateral groove of the cell and extends lengthwise. In many cases, algae are nonmotile in the vegetative state and form motile gametes only during sexual reproduction. Gliding motility, widespread among filamentous cyanobacteria (∞ Section 13.24), is present in only one group of eukaryotic algae, the diatoms.

Algae abound in nature in aquatic habitats, both freshwater and marine. Algae are also found in artificial aquatic habitats like fish tanks and swimming pools, and in temporary pools of water formed from rainwater runoff. Algae are also common in moist soils while a few species thrive in dry and even extremely dry soils where the water potential (∞ Section 5.11) can be very low. Algae are often the dominant or sole phototrophic microorganisms in acidic habitats as well; below pH 4–5 cyanobacteria are absent and several algal species flourish.

Endolithic Algal Communities

Some algal species have been found growing *inside* rocks. These endolithic (*endo* means "inside") organisms exist in porous rocks, for example, those containing quartz (Figure 17.21a), and are usually found in layers near the surface. Endolithic phototrophic communities are most common in dry environments like deserts, or cold dry environments like the Antarctic (Figure 17.21a). For example, in the Antarctic Dry Valleys, where temperatures and humidity are very low, life within a rock has its advantages. Rocks are heated by the sun and water from snow melt can be absorbed and retained for relatively long periods, supplying needed moisture for growth. Moreover, water absorbed by a

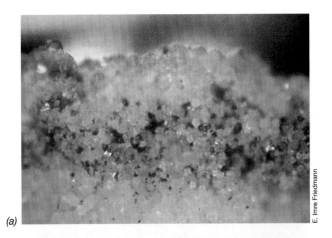

(a)

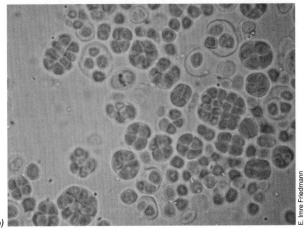

(b)

E. Imre Friedmann

FIGURE 17.21   Endolithic cyanobacteria. (a) Photograph of a limestone rock from Makhtesh Gadol, Negev Desert, Israel, showing a layer of cells of the cyanobacterium *Chroococcoidiopsis*. (b) Photomicrograph of cells of *Chroococcoidiopsis* isolated from a sandstone rock in the Negev Desert.

porous rock makes the rock more transparent, thus supplying more light to the algal layers.

A wide variety of phototrophs can form endolithic communities, including cyanobacteria (Figure 17.21*b*) and various green algae. In addition to free-living phototrophs, green algae and cyanobacteria coexist with fungi in endolithic lichen communities (∞ Section 16.23 for discussion of the lichen symbiosis). Metabolism and growth of these internal rock communities slowly weathers the rock, allowing gaps to develop where water can enter, freeze and thaw, and eventually crack the rock, producing new habitats for microbial colonization.

✓ 17.6 Concept Check

Algae are phototrophic Eukarya that contain photosynthetic pigments within a structure called the chloroplast.

✓ How do *algae* differ from *cyanobacteria?*
✓ What is unusual about red algae?
✓ What is *diatomaceous earth?*
✓ What is the primary habitat of algae?

A number of chemicals inhibit the growth of microorganisms. Some are extremely effective at inhibiting growth and a few are very useful for preventing or curing infection in the host organism. These antimicrobial agents fall into several different chemical families and come from a variety of natural sources (antibiotics) as well as from the laboratory (chemotherapeutic agents). However, because of the continuous use and overuse of these compounds, a number of resistant bacterial strains are now found, and several pathogenic microorganisms are now resistant to all available antimicrobial agents. Thus, infectious agents must be routinely checked for susceptibility to available drugs. One common method for testing susceptibility to antimicrobial agents is the minimum inhibitory concentration (MIC) test, shown here, in which the test organism is grown in the presence of sequential dilutions of the antimicrobial drug.

CHAPTER 18

Microbial Growth Control

18.1   Heat Sterilization   742
18.2   Radiation Sterilization   745
18.3   Filter Sterilization   747
18.4   Chemical Growth Control   749
18.5   Antiseptics, Disinfectants, and Sterilants   751
18.6   Synthetic Antibacterial Chemotherapeutic Agents   753
18.7   Antibiotics   758
18.8   β-Lactam Antibiotics: Penicillins and Cephalosporins   759
18.9   Antibiotics from Prokaryotes   759
18.10   Viral Control   762
18.11   Fungal Control   764
18.12   Antimicrobial Drug Resistance   765
18.13   The Search for New Antimicrobial Drugs   770

WORKING GLOSSARY

Aminoglycosides a group of antibiotics, including streptomycin, containing amino sugars linked by glycosidic bonds

Antibiotic a chemical substance produced by a microorganism that kills or inhibits the growth of another microorganism

Antibiotic resistance the acquired ability of a microorganism to grow in the presence of an antibiotic to which the microorganism is usually sensitive

Antimicrobial agent a chemical that kills or inhibits the growth of microorganisms

Antiseptic antimicrobial agents that are sufficiently nontoxic to be applied on living tissues

Autoclave a sterilizer that destroys microorganisms with temperature and steam under pressure

β-Lactam antibiotics a group of antibiotics, including penicillin, that contain the four-membered heterocyclic β-lactam ring

Bacteriocidal kills bacteria

Bacteriostatic inhibits bacterial growth

Chemotherapeutic agent an antimicrobial agent that can be used internally

Decontamination treatment that renders an object or inanimate surface safe to handle

Disinfectant an antimicrobial agent used only on inanimate objects

Disinfection the process of eliminating nearly all pathogens, but not all microorganisms, from inanimate objects or surfaces

Growth factor analog a chemical agent that is related to and blocks the uptake of a growth factor

Inhibition the reduction of microbial growth because of a decrease in the number of organisms present or alterations in the microbial environment

Interferons host-specific antiviral proteins, produced by virus-infected cells, which prevents viral infection of neighboring cells

Lysis loss of cellular integrity with release of cytoplasmic contents

Pasteurization destruction of all disease-producing microorganisms or a reduction in the number of spoilage microorganisms

Semisynthetic penicillin a natural penicillin that has been chemically altered

Sterilization the killing or removal of all living organisms and their viruses from a growth medium

Tetracycline an antibiotic characterized by a four-membered naphthacene ring

---

This chapter begins a shift in our focus on microorganisms. Up until now we have studied the basic chemistry and biology of microorganisms, the environmental factors that *promote* growth, and the wide range of organisms that contribute to the microbial community. From this chapter through the end of the book, we will be concerned with the relationships of microorganisms with humans. We start here by highlighting agents and methods used for control of microbial growth.

In general, control can be effected by limiting microbial growth, the process of **inhibition,** or by destroying the organisms with **sterilization,** the killing or removal of all viable organisms from a growth medium. Agents that destroy or kill bacteria are **bacteriocidal.** In practice, sterility is often not attainable, so in most cases we simply attempt to inhibit the rapid growth of organisms through decontamination and disinfection methods. Agents that inhibit bacterial growth are said to be **bacteriostatic.**

Why is microbial growth control necessary? We calculated that a single *Escherichia coli* cell would produce a mass of organisms much greater than the mass of Earth in less than 48 hr under optimal growth conditions (∞ Section 5.3). Of course, our studies of growth in Chapter 5 highlight the absurdity of this example, and the growth of a microorganism is fortunately limited in the long term by the exhaustion of nutrients. However, it is often desirable to stop microbial destruction of nutrients *before* they are exhausted. For example, in the food industry, food spoilage creates major economic costs (∞ Section 24.10). Therefore, considerable resources are used to control growth of microorganisms and stop the destruction of food. Another major potential "nutrient" for microorganisms are humans (and other animals). Uncontrolled microbial growth on or in human tissue causes cell destruction, a process called *infectious disease.* Microbial infectious diseases account for nearly 20 million deaths each year worldwide (∞ Table 22.1), thus control of infectious diseases is very important, both from a societal and an economic standpoint.

Microbial control measures include *decontamination, disinfection,* and *sterilization.* We continually apply decontamination methods to inhibit microbial growth. For example, the casual act of wiping a table clean after a meal removes potential microbial nutrients and contaminating microbes, preventing microbial growth. Progressing to more aggressive, directed antimicrobial measures, disinfection methods are specific procedures aimed at inhibiting microbial growth or destroying microorganisms using specialized chemical or physical agents. We routinely use disinfectant chemicals such as alcohol, for example, to cleanse and disinfect wounds. Finally, when necessary, we use rigidly controlled sterilization methods to destroy all microorganisms. Sterilization, though difficult to achieve, can completely prevent contamination and growth of microorganisms, for instance when making microbiological media or preparing surgical instruments. The goal for all of these procedures is to reduce the *microbial load*, or the number of viable microorganisms present. Here we will deal first with physical and chemical means of disinfection and sterilization to control microbial growth in the environment.

Unfortunately, standard methods used for decontamination, disinfection and sterilization are often inappropriate for control of microbial growth in human disease. However, there are a number of antimicrobial compounds that are safe and effective when used *in vivo*. We will discuss a variety of synthetic antimicrobial compounds and some naturally occurring *antibiotics*. While almost all currently used antibiotics are semisynthetic (they are chemically modified in the laboratory to increase their effectiveness), the parent antibiotic compounds have specific antimicrobial activity and are found in nature. These naturally occurring compounds are still the most productive source for discovery and development of new antimicrobial agents. Whatever their source, antimicrobial chemotherapeutic agents are extensively used for both the prevention and treatment of infectious disease. Finally, we will study some mechanisms used by microorganisms to defeat growth control efforts, with a focus on the development of antimicrobial drug resistance.

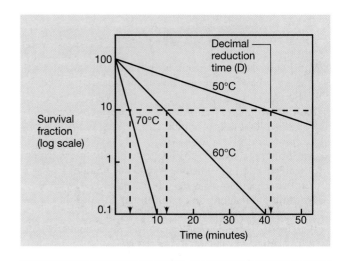

FIGURE 18.1   The effect of temperature on the viability of a mesophilic bacterium. The decimal reduction time, *D*, was obtained for the same mesophilic organism at three different temperatures. *D* is the time at which only 10% of the original population of organisms remain viable at a given temperature. For 70°C, *D* = 3 min; for 60°C, *D* = 12 min; for 50°C, *D* = 42 min.

18.1

Heat Sterilization

We now consider several physical methods of sterilization and growth inhibition including heat, filtration, and radiation. Remember that once a product is sterilized, it remains sterile indefinitely (∞ Pasteur's swan-necked flask, Figure 1.18) unless it is again contaminated by living microorganisms. Perhaps the most widespread method for controlling microbial growth is heat, which is where we begin our discussion.

Kinetics of Heat Sterilization

As the temperature rises past the maximum temperature for growth, lethal effects occur. At very high temperatures, virtually all macromolecules lose their structure and their ability to function, a process known as *denaturation*. As shown in Figure 18.1, death from

heating is an exponential (first-order) function and occurs more rapidly as the temperature is raised. The first-order relationship shown in Figure 18.1 means that the rate of death is proportional at any instant only to the concentration of organisms at that instant; the time taken for a definite fraction (for example, 90%) of the cells to be killed is independent of the initial concentration. These facts have important practical consequences. If we wish to *sterilize* a microbial population, it will take longer at lower temperatures than at higher temperatures. Thus, it is necessary to adjust the time and temperature to achieve sterilization for each specific set of conditions. The nature of the heat is also important: Moist heat has better penetrating power than dry heat and produces a faster reduction in the number of living organisms at a given temperature.

The time required for a 10-fold reduction in the population density at a given temperature, called the **decimal reduction time** or *D*, is the most useful way to characterize heat sterilization. Over the range of temperatures usually used in food preparation (*i.e.,* cooking and canning), the relationship between *D* and temperature is essentially exponential. Thus, when the logarithm of *D* is plotted against temperature, a straight line is obtained (Figure 18.2). The slope of the line provides a measure of the sensitivity of the organism to heat under the conditions employed, and the graph can be used to calculate process times to achieve sterilization, such as in canning operations (∞ Section 24.10).

Determination of decimal reduction times is a fairly lengthy procedure because it requires making a number of viable count measurements (∞ Section 5.4). An

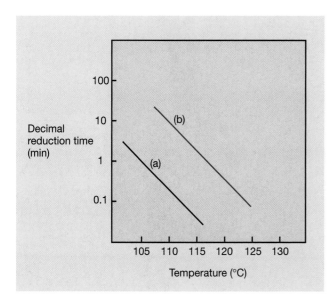

FIGURE 18.2 The relationship between the temperature and the rate of killing as indicated by the decimal reduction time for two different microorganisms. Data were obtained for decimal reduction times, *D*, at several different temperatures, as in Figure 18.1. For organism (*a*), a typical mesophile, exposure to 110°C for less than 20 sec resulted in a decimal reduction, while for organism (b), a thermophile, 10 min were required to achieve a decimal reduction.

easier way of characterizing the heat sensitivity of an organism is to determine the **thermal death time,** the *time* at which all cells are killed at a given temperature. This is done simply by heating samples of this suspension for different times, mixing the heated suspensions with culture medium, and incubating. When all cells are killed, no growth is evident in the incubated samples. Thus, the thermal death time depends on the size of the population tested because a longer time is required to kill all cells in a large population than in a small one. When the number of cells is standardized, it is possible to compare the heat sensitivities of different organisms by comparing their thermal death times, as shown graphically in Figure 18.1.

Spores and Heat Sterilization

Vegetative cells and bacterial endospores from the same organism vary considerably in heat resistance. For instance, in the autoclave, a temperature of 121°C is normally reached. Under these conditions, endospores may require 4–5 min for a decimal reduction, whereas vegetative cells may require only 0.1–0.5 min at 65°C for a decimal reduction. Thus, practically speaking, heat sterilization must involve procedures for killing endospores.

The nature of the medium in which heating takes place also influences the killing of both vegetative cells

and spores. Microbial death is more rapid at acidic pH, and acid foods such as tomatoes, fruits, and pickles are much easier to sterilize than neutral pH foods such as corn and beans. High concentrations of sugars, proteins, and fats decrease heat penetration and usually increase the resistance of organisms to heat, whereas high salt concentrations may either increase or decrease heat resistance, depending on the organism. Dry cells (and spores) are more heat-resistant than moist ones; consequently, heat sterilization of dry objects always requires higher temperatures and longer times than sterilization of moist objects.

Bacterial endospores (∞ Section 3.15) are the most heat-resistant structures known: They are able to survive heat that would rapidly kill vegetative cells of the same species. A major factor in heat resistance is the amount and state of *water* within the endospore. During endospore formation, the protoplasm is reduced to a minimum volume as a result of the accumulation of Ca^{2+}, *small acid-soluble spore proteins (SASPs),* and synthesis of dipicolinic acid, which lead to formation of a gel-like structure (∞ Section 3.15). At this stage, a thick cortex forms around the protoplast core. Contraction of the cortex results in a shrunken, dehydrated protoplast with a water content of only 10–30% of a vegetative cell. The water content of the protoplast coupled with the concentration of SASPs determines the heat resistance of the spore. If endospores have a low concentration of SASPs and high water content, they have low heat resistance; if they have a high concentration of SASPs and low water content, they have high heat resistance. Water moves freely in and out of spores, and so it is not the impermeability of the spore coat that excludes water, but the gel-like material in the spore protoplast.

The Autoclave

The **autoclave** is a sealed device that allows the entrance of steam under pressure (Figure 18.3). The use of moist heat facilitates killing of all microorganisms, including heat-resistant endospores.

Killing of the heat-resistant endospores requires heating at temperatures above boiling and the use of steam under pressure (Figure 18.3*a*). The usual procedure is to heat at 1.1 kilograms/square centimeter (kg/cm^2) [15 pounds/square inch (lb/in^2)] steam pressure, which yields a temperature of 121°C. At 121°C, the time of autoclaving to achieve sterilization is generally considered to be 10–15 min (Figure 18.3*b*). If bulky objects are being sterilized, heat transfer to the interior will be slow, and the heating time must be sufficiently long so that the *entire* object is at 121°C for 10–15 min. Extended times are also required when large volumes of liquids are being autoclaved because large volumes take longer to reach sterilization temperatures. Note that it

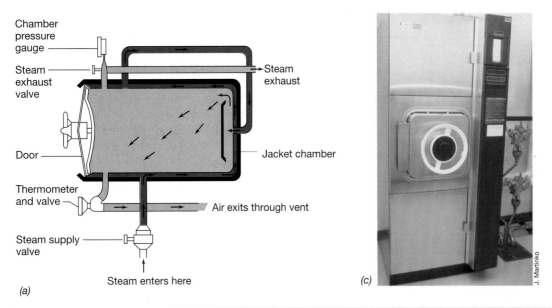

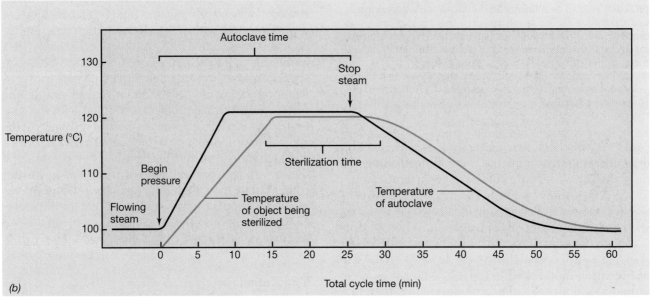

FIGURE 18.3   Use of the autoclave for sterilization. (a) The flow of steam through an autoclave. (b) A typical autoclave cycle. Shown is the sterilization of a fairly bulky object. The temperature of the object rises more slowly than the temperature of the autoclave. (c) A modern research autoclave. Note the pressure-lock door and the automatic cycle controls on the right panel. The steam inlet and exhaust fittings are on the right side of the autoclave.

is not the *pressure* of the autoclave that kills the microorganisms but the *high temperature* that can be achieved when steam is placed under pressure.

Pasteurization

Pasteurization is a process that reduces the microbial populations in milk and other heat-sensitive foods. It is named for Louis Pasteur, who first used heat for controlling the spoilage of wine (see the box, The Origin of Pasteurization). Pasteurization is not synonymous with sterilization because not all organisms are killed. Origi-

nally, pasteurization of milk was used to kill pathogenic bacteria, especially the organisms causing tuberculosis, brucellosis, Q fever, and typhoid fever, but the keeping qualities of milk were also improved following pasteurization. Although these pathogens are no longer found in food sources in the more-developed countries, the main purpose of pasteurization is still to prevent the spread of pathogens such as *Salmonella* species and *Escherichia coli* O157:H7 through common sources such as milk and juices. Pasteurization also prevents the growth of spoilage organisms, dramatically increasing the storage life of perishable liquids (Sections 24.10 and 24.11).

Pasteur's name is forever linked in the public mind with the process of pasteurization. The development of the pasteurization process has been nicely discussed by René Dubos in his book on Pasteur's life.[a] We quote from this book here:

The demonstration that microbes do not generate spontaneously encouraged the development of techniques to destroy them and to prevent or minimize subsequent contamination. Immediately these advances brought about profound technological changes in the preparation and preservation of food products and subsequently in other industrial processes as well. . . .

It was soon discovered that the introduction of microorganisms in biological products can be minimized by an intelligent and rigorous control of the technological operations, but

cannot be prevented entirely. The problem therefore was to inhibit the further development of these organisms after they had been introduced into the product. To this end, Pasteur first tried to add a variety of antiseptics, but the results were mediocre and, after much hesitation, he considered the possibility of using heat as a sterilizing agent.

Pasteur's first studies of heat as a preserving agent were carried out with wine. Pasteur had grown up in one of the best wine districts in France, and, as a connoisseur of the beverage, was much disturbed at the thought that heating might alter its flavor and bouquet. He therefore proceeded with very great caution and eventually convinced himself that heating at 55°C would not alter appreciably the bouquet of the wine. . . . These considerations led to the process of partial sterilization, which soon became known the world over under the name of "pasteurization," and which was

found applicable to wine, beer, cider, vinegar, milk, and countless other perishable beverages, foods, and organic products.

It was characteristic of Pasteur that he did not remain satisfied with formulating the theoretical basis of heat sterilization, but took an active interest in designing industrial equipment adapted to the heating of fluids in large volumes and at low cost. His treatises on vinegar, wine, and beer are illustrated with drawings and photographs of this type of equipment, and describe in detail the operations involved in the process. The word "pasteurization" is, indeed, a symbol of his scientific life; it recalls the part he played in establishing the theoretical basis of the germ theory, and the phenomenal effort that he devoted to making it useful to his fellow humans. It reminds us also of his well-known statement: "There are no such things as pure and applied science—there are only science, and the application of science." ∎

[a] René Dubos, *Pasteur and Modern Science*. 1988. Science Tech, Madison, WI.

Pasteurization of milk is usually achieved by passing the milk through a heat exchanger. Operationally, the milk is fed through tubing that is in contact with a heat source. Careful control of the milk flow rate and the size and temperature of the heat source raises the temperature of the milk to 71°C for 15 sec. The milk is then rapidly cooled. The whole process is aptly called *flash pasteurization*. Milk can also be heated in large vats to 63–66°C for 30 min. However, this *bulk pasteurization* method is less satisfactory because the milk heats and cools more slowly and is held at high temperatures for longer times. Flash pasteurization alters the flavor less, kills heat-resistant organisms more effectively, and is normally done on a continuous-flow basis. The flash pasteurization method is more adaptable to large dairy operations, and modern dairies generally employ it, often with even shorter exposure times and higher temperatures.

✓ **18.1 Concept Check**

Sterilization is the complete killing of all organisms. The most widely used method for sterilization is the application of heat. The temperature for heat sterilization is selected to eliminate the most heat-resistant organisms in the material, usually bacterial endospores. For routine sterilization, an autoclave is used; this permits application of steam heat under pressure at temperatures above the boiling point of water. Pasteurization rids the material of pathogenic microorganisms and reduces the growth of spoilage microorganisms to prolong storage life.

✓ Why is heat an effective sterilizing agent?
✓ What steps are necessary to sterilize material that may have bacterial endospores?

18.2

Radiation Sterilization

An effective way to sterilize or reduce the microbial burden in almost any substance is through the use of electromagnetic radiation. Microwaves, ultraviolet (UV) radiation, X-rays, gamma rays (γ-rays), and electrons, all forms of electromagnetic radiation (∞ Section 9.3), have the potential to control microbial growth. However, each type of radiation acts through a specific mechanism. For example, the antimicrobial effects of microwaves are due, at least in part, to thermal effects.

UV radiation, normally considered to be between 220 and 300 nm, acts by a different mechanism. UV waves have sufficient energy to cause breaks in DNA, leading to the death of the exposed organism (∞ Section 9.3). This "near-visible" light is useful for disinfecting surfaces, air, and other materials such as water that do not absorb the UV waves. For example, laboratory biological cabinets all come equipped with a "germicidal" UV light to decontaminate the surface after use (Figure 18.4). Ultraviolet radiation cannot penetrate solid, opaque, light-absorbing surfaces, and its usefulness is therefore limited to disinfection of exposed surfaces.

Ionizing Radiation

Ionizing radiation is electromagnetic radiation of sufficient energy to produce ions and other reactive molecular species from molecules with which the radiation particles collide. Ionizing radiation produces electrons, e^-, hydroxyl radicals, OH·, and hydride radicals, H·. Each of these reactive molecules is capable of degrading and altering biopolymers such as deoxyribonucleic acid (DNA) and protein. In addition, ionizing radiation can interact directly with DNA, causing breaks in the polymer. The ionization and subsequent degradation of biologically important molecules such as DNA and enzyme proteins leads to the death of irradiated cells. Several radiation sources are potentially useful for sterilization.

The unit of radiation is the *roentgen,* which is a measure of the radiation energy output from a source. The standard for biological applications such as sterilization is the *absorbed radiation dose.* The *absorbed dose* is the rad (100 erg/g), or the gray (1 Gy = 100 rad). Certain microorganisms are much more resistant to radiation than

others. Table 18.1 shows the dose of radiation necessary to reduce the numbers of some representative microorganisms or biological functions 10-fold.

In general, microorganisms are much more resistant to ionizing radiation than higher organisms. For example, the amount of energy necessary to reduce the bacterial load by 10-fold is at least 200 Gy. By contrast, the *lethal* dose of radiation for humans is considered to be 10 Gy or less! Note that the figures shown are for a one log reduction in growth of a given organism. The *D10* or *decimal reduction value* gives information similar to the decimal reduction time for heat sterilization (see Section 18.1): The relationship of the survival fraction plotted on a semilogarithmic scale versus the radiation dosage in grays is essentially linear (Figure 18.5). A standard *killing* dose for radiation sterilization can be assigned as 12 *D10* requirements for destruction of the radioresistant endospores of *Clostridium botulinum.*

Radiation Practice

Several sources of ionizing radiation are available, including X-ray machines, cathode ray tubes, and radioactive nuclides. These sources produce either X-rays or γ-rays, both of which have sufficient energy and penetrating power to efficiently inhibit microbial growth in solid and liquid media. The principal sources of commercially useful ionizing radiation are radioactive nuclides that emit γ-rays. The two most commonly used radioisotopes are ^{60}Co and ^{137}Cs, both relatively inexpensive by-products of nuclear fission.

TABLE 18.1	Radiation sensitivity of microorganisms and biological functions

Species or function	Type of microorganism	*D10*[a] (Gy)
Clostridium botulinum	Gram-positive anaerobic sporulating Bacteria	3,300
Clostridium tetani	Gram-positive anaerobic sporulating Bacteria	2,400
Bacillus subtilis	Gram-positive aerobic sporulating Bacteria	600
Salmonella typhimurium	Gram-negative Bacteria	200
Lactobacillus brevis	Gram-positive Bacteria	1,200
Deinococcus radiodurans	Gram-negative radiation-resistant Bacteria	2,200
Aspergillus niger	Mold	500
Saccharomyces cerevisiae	Yeast	500
Foot-and-mouth	Virus	13,000
Coxsackie	Virus	4,500
Enzyme inactivation	—	20,000–50,000
Insect deinfestation	—	1,000–5,000

a D10 is the amount of radiation necessary to reduce the initial population or activity level 10-fold (one logarithm). Gy = grays. 1 gray = 100 rads.

FIGURE 18.4   A biological safety cabinet, shown with an ultraviolet (UV) radiation source (mercury vapor lamp), which is used for decontamination of the inside surfaces.

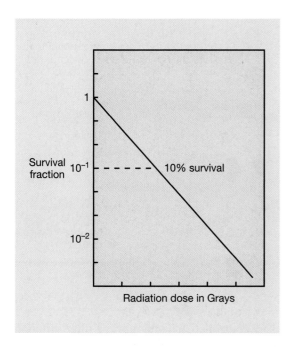

FIGURE 18.5 Relationship between the survival fraction and the radiation dose. The $D10$, or decimal reduction dose, can be interpolated from the data as shown.

Radiation is currently used for sterilization and decontamination in the medical supplies and food industries. In the United States, the Food and Drug Administration has approved the use of radiation for sterilization of such diverse items as surgical supplies, disposable labware, drugs, and even tissue grafts (Table 18.2). However, because of the costs and hazards of radiation equipment, this type of sterilization is limited to large industrial applications or very specialized facilities. In addition, food may be irradiated, although this practice is not yet widespread. However, as shown in Figure 18.6, the practice of sterilizing spices by irradiation has grown enormously in recent years, as the more dangerous alternative, ethylene oxide sterilization (see Section 18.6), has become less common. In addition to sterilization, pasteurization and insect deinfestation can be accomplished by adjusting the dose of radiation applied. Radiation is also approved by the World Health Organization as a de-

TABLE 18.2	Medical and laboratory products sterilized by radiation	
Tissue grafts	**Drugs**	**Medical and laboratory supplies**
Cartilage	Chloramphenicol	Disposable labware
Tendon	Ampicillin	Culture media
Skin	Tetracycline	Syringes
Heart valve	Atropine	Surgical equipment
	Vaccines	Sutures
	Ointments	

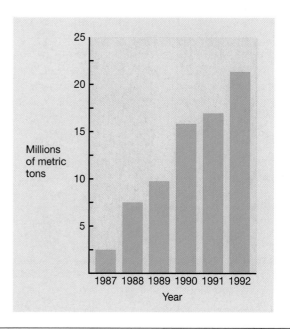

FIGURE 18.6 Commercial irradiation of spices and seasonings, worldwide.

contamination measure for use in foods particularly susceptible to microbial contamination, especially fresh meat products such as hamburger and chicken, and has recently been approved in the United States for decontamination of ground meat. The use of radiation for all these purposes is an established and accepted technology in many countries but has been slow to gain acceptance in others because of fears of possible radioactive contamination, alteration in nutritional value, production of toxic or carcinogenic products, and production of "off" tastes in irradiated food.

✓ 18.2 Concept Check

Under appropriate conditions, electromagnetic radiation effectively controls microbial growth. Ultraviolet radiation is useful for decontaminating surfaces and materials that do not absorb light, such as air and water. Ionizing radiation is necessary to penetrate solid or light-absorbing materials. Ionizing radiation is very effective for sterilization and decontamination.

✓ Define the decimal reduction dose for radiation.
✓ Why is ionizing radiation more effective than ultraviolet radiation for sterilization of food products?

18.3

Filter Sterilization

Although heat is the most common and effective way of sterilizing liquids, it cannot be used for the sterilization of heat-sensitive liquids or gases. An especially valuable technique for sterilizing such materials is filtration.

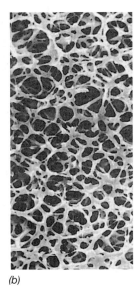

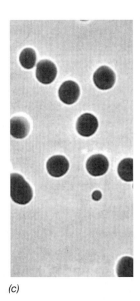

(a)   (b)   (c)

FIGURE 18.7   The structure of (a) a depth filter, (b) a conventional membrane filter, and (c) a Nucleopore filter.

A filter is a device with pores too small for the passage of microorganisms but large enough to allow the passage of the liquid or gas. The size range of particles involved in sterilization is rather broad. Some of the largest microbial cells are greater than 10 μm in diameter, but at the lower end of the size scale certain bacteria are less than 0.3 μm in diameter. Historically, selective filtration methods were also used to define and isolate infectious particles that were smaller than bacteria. These infectious particles, now known to be viruses, are very small and range from 28 nm to 200 nm in diameter (∞ Section 8.2).

Types of Filters

There are three major types of filters, which are illustrated in Figure 18.7. One of the oldest types used is the **depth filter.** A depth filter is a fibrous sheet or mat made from a random array of overlapping paper, asbestos, or glass fibers (Figure 18.7*a*). The depth filter traps particles in the tortuous paths created throughout the depth of the structure. Because they are rather porous, depth filters are often used as *prefilters* to remove larger particles from a solution so that clogging does not occur in the final filter sterilization process. They are also used for the filter sterilization of air in industrial processes (∞ Section 11.3).

The most common type of filter for sterilization in the field of microbiology is the **membrane filter** (Figure 18.7*b*). Membrane filters are composed of polymers with high tensile strength such as cellulose acetate, cellulose nitrate, or polysulfone, manufactured in such a way that they contain a large number of tiny holes. By adjusting the polymerization conditions during manufacture, the size of the holes in the membrane (and thus the size of the molecules that can pass through) can be

precisely controlled. The membrane filter differs from the depth filter in that the membrane filter functions more like a sieve, trapping many of the particles on the filter surface. The membranes are open structures with about 80–85% of the filter occupied by space. This openness provides for a relatively high fluid flow rate.

The third type of filter in common use is the **nucleation track (Nucleopore) filter.** These filters are created by treating very thin polycarbonate films (10 μm in thickness) with nuclear radiation and then etching the film with a chemical. The radiation causes localized damage in the film and the etching chemical enlarges these damaged locations into holes. The sizes of the holes can be precisely controlled by the strength of the etching solu-

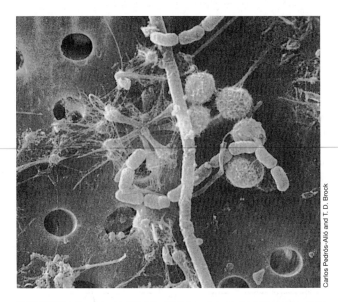

FIGURE 18.8   Scanning electron micrograph of aquatic bacteria and algae trapped on a Nucleopore filter. The pore size is 5 μm.

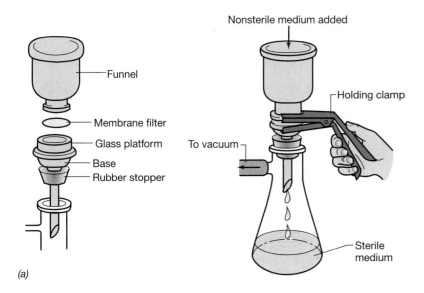

Nonsterile medium added

Funnel

Membrane filter

Glass platform

Base

Rubber stopper

Holding clamp

To vacuum

Sterile medium

(a)

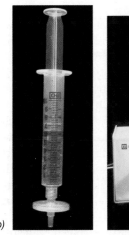

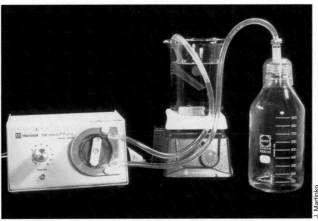

(b)

J. Martinko

FIGURE 18.9   Membrane filters. (a) Assembly of a reusable membrane filter apparatus. (b) Disposable, presterilized, and assembled membrane filter units. Left: a filter system designed for small volumes. Right: a filter system designed for larger volumes.

tion and the etching time. A typical Nucleopore filter has very uniform holes arranged almost vertically through the thin film (Figure 18.7c). Nucleopore filters are very commonly used in scanning electron microscopy of microorganisms. An organism of interest can be easily removed from a liquid by filtration, with the particles held in a uniform plane on the top of the filter where they can be readily studied in the microscope (Figure 18.8).

Membrane filters for the sterilization of a liquid are illustrated in Figure 18.9. The filter apparatus is generally sterilized separately from the filter, and the apparatus assembled aseptically at the time of filtration. The arrangement shown in Figure 18.9a is suitable for small volumes of liquid. For large-volume sterile filtration, the membrane filter material is arranged in a cartridge and placed in a stainless steel housing. Large-volume filtration of heat-sensitive fluids is very commonly done in the pharmaceutical industry.

Presterilized membrane filter assemblies for sterilization of small to medium volumes are routinely used in most laboratories (Figure 18.9b). Filtration is accomplished by using a syringe, pump, or vacuum to force the liquid through the filtration apparatus into a sterile collection vessel.

✓ 18.3 Concept Check

Filter sterilization involves the *removal* of living microorganisms from liquids. Membrane filters are widely used for sterilization of heat-sensitive liquids in the laboratory.

✓ Why are depth filters not widely used for sterilization?
✓ What advantage does the membrane filter have over the nucleation filter in sterilization?

18.4

Chemical Growth Control

The growth of microorganisms can also be controlled with chemical agents. An **antimicrobial agent** is a chemical that kills or inhibits the growth of microorganisms. Such a substance may be a synthetic chemical or a natural product. Agents that kill organisms are often called *cidal agents,* with a prefix indicating the kind of organism killed. Thus, we have **bacteriocidal, fungicidal,** and **viricidal** agents. A bacteriocidal agent kills bacteria. It may or may not kill other kinds of microorganisms. Agents that do not kill but only inhibit growth are called *static agents,* and we can speak of **bacteriostatic, fungistatic,** and **viristatic** agents.

Antimicrobial agents can vary in their **selective toxicity.** Some act in a rather nonselective manner and have similar effects on all types of cells. Others are far more selective and are more toxic to microorganisms than to animal tissues. Antimicrobial agents with selective toxicity are especially useful as *chemotherapeutic agents* in treating infectious diseases because they can be used to kill disease-causing microorganisms *in vivo* without harming the host. They will be described later in this chapter.

Effect of Antimicrobial Agents on Growth

Antimicrobial agents affect growth in a variety of ways, and a study of the action of these agents in relation to the growth curve is important in understanding their modes of action. Three distinct kinds of effects can be observed when an antimicrobial agent is added to an exponentially growing bacterial culture: bacteriostatic, bacteriocidal, and bacteriolytic. A *bacteriostatic* effect is observed when growth is inhibited, but no killing occurs (Figure 18.10*a*). Bacteriostatic agents are frequently inhibitors of protein synthesis and act by binding to ribosomes. The binding, however, is not tight, and when the concentration of the agent is lowered, the agent becomes free from the ribosome and growth is resumed. The mode of action of protein synthesis inhibitors such as the antibiotic chloramphenicol is discussed in Section 6.12. *Bacteriocidal* agents kill cells, but lysis or cell rupture does not occur (Figure 18.10*b*). Bacteriocidal agents are a class of chemical agents that generally bind tightly to their cellular targets and are not removed by dilution. *Bacteriolytic* agents induce killing by cell lysis, which is observed as a decrease in cell number or in turbidity after the agent is added (Figure 18.10*c*). Bacteriolytic agents include antibiotics that inhibit cell wall synthesis, such as penicillin (∞ Section 3.9 and see Section 18.8), as well as agents that damage the cytoplasmic membrane.

Measuring Antimicrobial Activity

Antimicrobial activity is measured by determining the smallest amount of agent needed to inhibit the growth of a test organism, a value called the **minimum inhibitory concentration (MIC).** To determine the MIC, a series of culture tubes is prepared, each tube containing medium with a different concentration of the

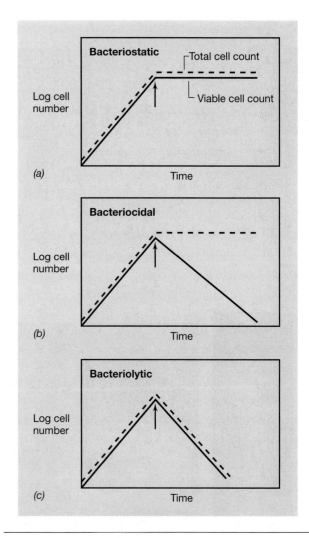

FIGURE 18.10 Three types of action of antimicrobial agents. At the time indicated by the arrow, a growth-inhibitory concentration was added to the exponentially growing culture. Note the relationships between viable and total cell counts.

agent, and all tubes of the series are inoculated. After incubation, the tubes in which growth does *not* occur (indicated by an absence of visible turbidity) are noted. The tube containing the lowest concentration of agent that completely inhibits the growth of the test organism defines the MIC (Figure 18.11). This simple and effective procedure is often called the *tube dilution technique.* The MIC is not a constant for a given agent, because it is affected by the nature of the test organism used, the inoculum size, the composition of the culture medium, the incubation time, and the conditions of incubation, such as temperature, pH, and aeration. When all conditions are rigorously standardized, it is possible to compare different antimicrobials and determine which is most effective against a given organism or to assess the activity of a single agent against a variety of organisms. This method does not distinguish

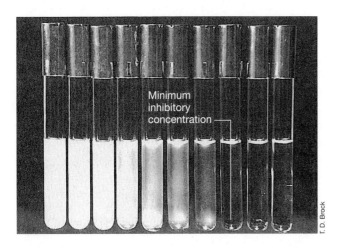

FIGURE 18.11 Antibiotic assay by tube dilution, permitting detection of the *minimum inhibitory concentration* (MIC). A series of increasing concentrations of antibiotic is prepared in the culture medium. Each tube is inoculated, and incubation is allowed to proceed. Growth (turbidity) occurs in those tubes with antibiotic concentrations below the MIC.

between a cidal and a static agent because the agent is present in the culture medium throughout the entire incubation period.

Another commonly used procedure for studying antimicrobial action is the **agar diffusion method** (Figure 18.12). A Petri plate containing an agar medium evenly inoculated with the test organism is prepared. Known amounts of the antimicrobial agent are added to filter paper discs, which are then placed on the surface of the agar. During incubation, the agent diffuses from the filter paper into the agar; the further it gets from the filter paper, the smaller the concentration of the agent. At some distance from the disc, the MIC is reached. Past this point growth occurs, but closer to the disc growth is absent. A **zone of inhibition** is thus created; the diameter of the zone is proportional to the amount of antimicrobial agent added to the disc, the solubility of the agent, the diffusion coefficient, and the overall effectiveness of the agent. This method is routinely used to test for antibiotic sensitivity in pathogens (⬿ Section 21.3).

✓ 18.4 Concept Check

Chemicals are often used to control microbial growth. Chemicals that kill organisms are called cidal agents; those that inhibit growth are called static agents. The effectiveness of an antibacterial chemical agent is assessed by determining the minimum concentration necessary to kill or inhibit bacterial growth.

✓ With regard to antibacterial agents, what is meant by *selective toxicity*?

✓ Describe how the *minimum inhibitory concentration* of a bacteriocidal agent is determined.

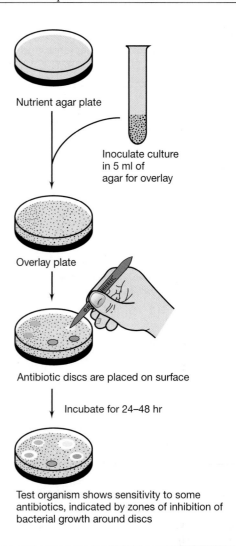

FIGURE 18.12 Agar diffusion method for assaying antibiotic activity.

18.5

Antiseptics, Disinfectants, and Sterilants

Antiseptics are chemical agents that kill or inhibit growth of microorganisms and that are sufficiently nontoxic to be applied to living tissues. Most of the compounds that fall into this category are used for handwashing or for treating surface wounds (Table 18.3). Under some circumstances, some antiseptics are also effective as disinfectants.

Disinfectants are chemicals that kill microorganisms and that are used on inanimate objects. *Sterilants* are disinfectants that, under appropriate circumstances, can kill all microbial life and can actually be used to sterilize inanimate objects and surfaces. Chemical disinfectant agents, which are frequently referred to as *germicides*, have wide use in situations where it is impractical to use heat (see Section 18.1) or radiation (see Section

TABLE 18.3	Antiseptics, disinfectants, and sterilants	
Agent	**Use**	**Mode of action**
Antiseptics		
Alcohol (60–85% ethanol or isopropanol in water)[a]	Skin	Lipid solvent and protein denaturant
Phenol-containing compounds (hexachlorophene, triclosan, chloroxylenol, chlorhexidine)	Soaps, lotions, cosmetics, body deodorants	Disrupts cell membrane
Cationic detergents, especially quaternary ammonium compounds (benzalkonium chloride)	Soaps, lotion	Interact with phospholipids of membrane
Hydrogen peroxide[a] (3% solution)	Skin	Oxidizing agent
Iodine-containing iodophor compounds in solution[a] (Betadine®)	Skin	Iodinates tyrosine residues of proteins; oxidizing agent
Organic mercury compounds[b] (mercurochrome)	Skin	Combines with -SH groups of proteins
Silver nitrate	Eyes of newborn to prevent blindness due to infection by *Neisseria gonorrhoeae*	Protein precipitant
Disinfectants and sterilants		
Alcohol (60–85% ethanol or isopropanol in water)[a]	Disinfectant and sterilant for medical instruments, laboratory surfaces	Lipid solvent and protein denaturant
Cationic detergents (quaternary ammonium compounds)	Disinfectant for medical instruments, food and dairy equipment	Interact with phospholipids
Chlorine gas	Disinfectant for purification of water supplies	Oxidizing agent
Chlorine compounds (chloramines, sodium hypochlorite, chlorine dioxide)	Disinfectant for dairy and food industry equipment, and water supplies	Oxidizing agent
Copper sulfate	Algicide in swimming pools, water supplies (disinfectant)	Protein precipitant
Ethylene oxide (gas)	Sterilant for temperature-sensitive laboratory materials such as plastics	Alkylating agent
Formaldehyde	3%–8% solution used as surface disinfectant, 37% (formalin) or vapor used as sterilant	Alkylating agent
Gluteraldehyde	2% solution used as high-level disinfectant or sterilant	Alkylating agent
Hydrogen peroxide[a]	Vapor used as sterilant	Oxidizing agent
Iodine-containing iodophor compounds in solution[a] (Wescodyne®)	Disinfectant for medical instruments, laboratory surfaces	Iodinates tyrosine residues
Mercuric dichloride[b]	Disinfectant for laboratory surfaces	Combines with -SH groups
Ozone	Disinfectant for drinking water	Strong oxidizing agent
Peracetic acid	0.2% solution used as high-level disinfectant or sterilant	Strong oxidizing agent
Phenolic compounds[b]	Disinfectant for laboratory surfaces	Protein denaturant

[a] Alcohols, hydrogen peroxide, and iodine-containing iodophor compounds can act as antiseptics, disinfectants, or even sterilants depending on concentration, length of exposure, and form of delivery.

[b] Heavy metal (mercury) compounds and phenolic compounds, especially when used at high concentrations and in high volumes, produce environmentally hazardous waste products.

TABLE 18.4	Industrial uses of disinfectants	
Industry	**Chemicals**	**Use**
Paper	Organic mercurials, phenols	To prevent microbial growth during manufacture
Leather	Heavy metals, phenols	Antimicrobial agents are present in the final product
Plastic	Cationic detergents	To prevent growth of bacteria on aqueous dispersions of plastics
Textile	Heavy metals, phenols	To prevent microbial deterioration of fabrics exposed in the environment such as awnings, tents
Wood	Phenols	To prevent deterioration of wooden structures
Metal working	Cationic detergents	To prevent growth of bacteria in aqueous cutting emulsions
Petroleum	Mercurics, phenols, cationic detergents	To prevent growth of bacteria during recovery and storage of petroleum and petroleum products
Air conditioning	Chlorine, phenols	To prevent growth of bacteria (for example, *Legionella*) in cooling towers
Electrical power	Chlorine	To prevent growth of bacteria in condensors and cooling towers
Nuclear	Chlorine	To prevent growth of radiation-resistant bacteria in nuclear reactors

18.2) for decontamination or sterilization. For example, hospitals and laboratories must be able to decontaminate floors, tables, bench tops, walls, and so on. In the food industry, floors, walls, and surfaces of equipment must often be treated with germicides to reduce the load of microorganisms. Hospitals and laboratories must also be able to sterilize heat-sensitive materials, such as thermometers, lensed instruments, polyethylene tubing, catheters, and reusable medical equipment such as respirometers. Usually, some form of *cold sterilization* is used for these purposes. Cold sterilization is performed in enclosed devices that resemble autoclaves, but employ a chemical agent such as ethylene oxide, formaldehyde, peracetic acid, or hydrogen peroxide. Finally, drinking water is commonly treated with chlorine or chlorine compounds to eliminate potentially harmful organisms (Table 18.3).

Several factors affect the efficiency of various antiseptic and disinfectant procedures. For example, many germicides are neutralized by organic materials, inhibiting their ability to kill microorganisms by effectively reducing germicide concentrations. Further, pathogens are often encased in particles or grow in large numbers as *biofilms*, covering the surfaces of tissue with several layers of microbial cells (∞ Section 16.2). As a result, penetration of a chemical agent to the viable cells may be slowed or even completely prevented. In many cases, bacterial endospores are much more resistant to germicides than are vegetative cells because of their low water content and reduced metabolism (see Section 18.1 and ∞ Section 3.15). Certain vegetative cells such as those of *Mycobacterium tuberculosis*, the causal agent of tuberculosis, are resistant to the action of germicides because of the complex nature of their cell wall (∞ Sections 13.22 and 23.3). As a result, total elimination of pathogens (sterilization) by germicidal treatment may not always occur. In practice, germicide effectiveness

can be determined only under the actual conditions of use. However, effective use of germicides ensures that the microbial load is reduced significantly, with a reasonable possibility that pathogenic organisms are eliminated. A summary of the most widely used germicides and their modes of action is given in Table 18.3.

Antiseptic and disinfectant chemicals are used in many industrial applications, where they are used to prevent microbial deterioration of a number of organic materials. In some industries, the use of antimicrobial agents is very routine and quite extensive. This frequently leads to toxic waste problems when large amounts of antimicrobial agents, such as mercury and other heavy metal compounds (paper industry), or phenols (wood preservatives), are released into the environment. Table 18.4 summarizes some of the industrial applications for disinfectants used to control microbial growth.

✓ 18.5 Concept Check

Antiseptics can be used to decontaminate living tissues. Disinfectants are chemical compounds used to decontaminate or sterilize nonliving material. These compounds are used in many commercial, health care, and industrial applications.

✓ Distinguish between an antiseptic and a disinfectant.
✓ What disinfectants are used for sterilization of water?

18.6

Synthetic Antibacterial Chemotherapeutic Agents

The preceding section dealt with chemical agents used to inhibit microbial growth *outside* the human body. Most of the chemicals mentioned were too toxic to be used in the body, although antiseptics can be used on

Antibiotic classification	Subclassification	Example	Representative structure
I. Carbohydrate-containing compounds	Pure sugars Aminoglycosides Orthosomycins N-Glycosides C-Glycosides Glycolipids	Nojirimycin Streptomycin Everninomicin Streptothricin Vancomycin Moenomycin	
II. Macrocyclic lactones	Macrolide antibiotics Polyene antibiotics Ansamycins Macrotetrolides	Erythromycin Candicidin Rifampin Tetranactin	
III. Quinones and related compounds	Tetracyclines Anthracyclines Naphthoquinones Benzoquinones	Tetracycline Adriamycin Actinorhodin Mitomycin	
IV. Amino acid and peptide analogs	Amino acid derivatives β-Lactam antibiotics Peptide antibiotics Chromopeptides Depsipeptides Chelate-forming peptides	Cycloserine Penicillin, ceftriaxone Bacitracin Actinomycin Valinomycin Bleomycin	
V. Heterocyclic compounds containing nitrogen	Nucleoside antibiotics	Polyoxins	
VI. Heterocyclic compounds containing oxygen	Polyether antibiotics	Monensin	
VII. Alicyclic derivatives	Cycloalkane derivatives Steroid antibiotics	Cycloheximide Fusidic acid	
VIII. Aromatic compounds	Benzene derivatives Condensed aromatics Aromatic ether	Chloramphenicol Griseofulvin Novobiocin	
IX. Aliphatic compounds	Compounds containing phosphorus	Fosfomycin	
X. Quinolone compound	4-Quinolone Fluoro-4-quinolones	Nalidixic acid Norfloxacin	
XI. Oxazolidinone	Cyclic lactone	2-Oxazolidinone	

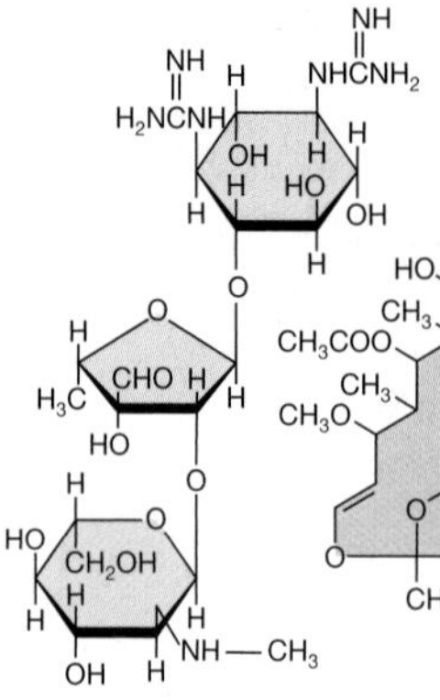

Streptomycin 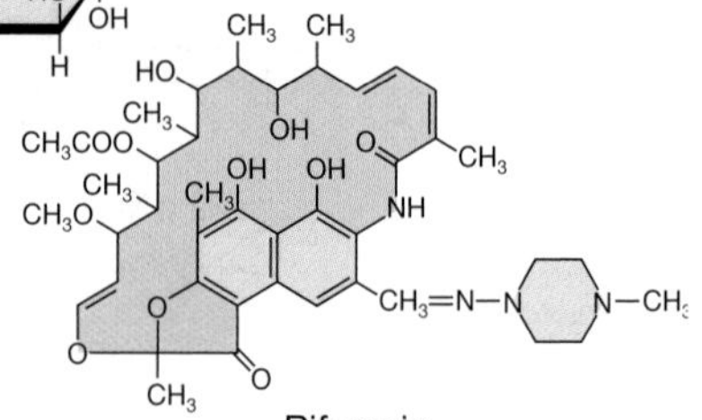 · Rifampin 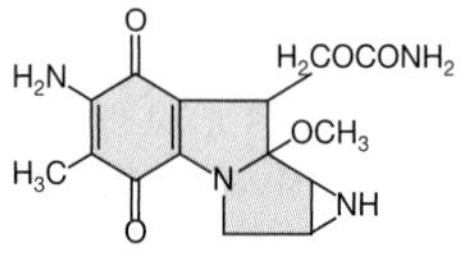 · Mitomycin C 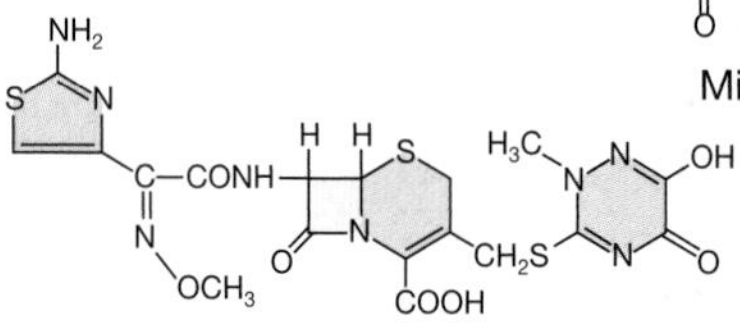 · Ceftriaxone 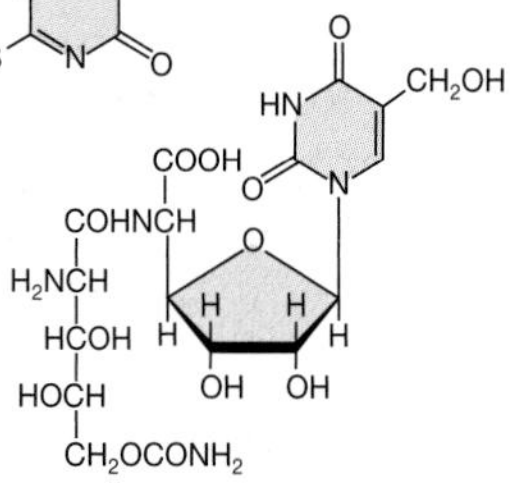 · Polyoxin B 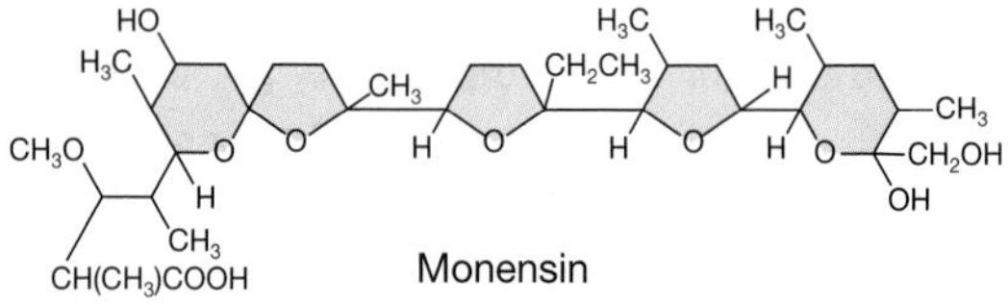 · Monensin 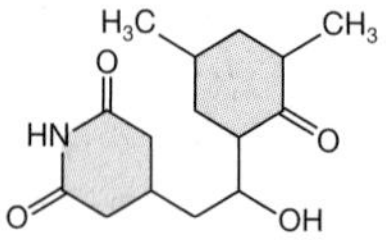 · Cycloheximide 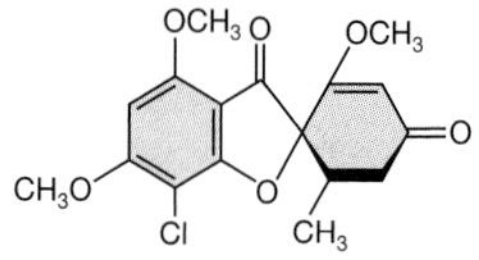 · Griseofulvin 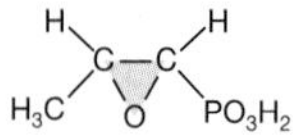 · Fosfomycin 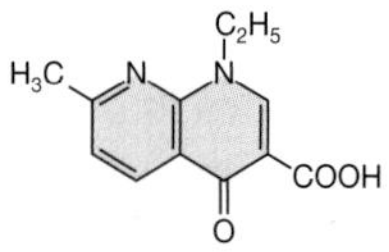 · Nalidixic acid 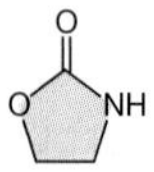 · 2-Oxazolidinone

FIGURE 18.13 Classification of antibacterial chemotherapeutic agents according to chemical structure. A representative example is shown for each group.

the skin. For control of infectious disease, chemical compounds that can be used internally are essential. Such compounds are called **chemotherapeutic agents,** and they play major roles in clinical and veterinary medicine, as well as in agriculture.

Chemotherapeutic agents are categorized based on their structure (Figure 18.13) and mode of action (Figure 18.14). Each year, more than 500 metric tons of chemotherapeutic agents of various types are manufactured and used worldwide (Figure 18.15). The key requirement of a successful chemotherapeutic agent is *selective toxicity,* the ability to inhibit bacteria or other pathogenic agents without adversely affecting the host (see the box, Microbiology and "Magic Bullets"). Each agent has a characteristic spectrum of antibacterial action (Figure 18.16). Chemotherapeutic agents fall into two general categories, the *synthetic agents* and the *antibiotics.* Here we will concentrate on the synthetic agents. We discuss antibiotics in the next three sections.

Growth Factor Analogs

In Section 4.2 we discussed growth factors and defined them as specific chemical substances *required* in the medium because the organisms cannot synthesize them. A substance that is related to a growth factor but blocks utilization of the growth factor is known as a **growth factor analog.** Growth factor analogs are usually synthetic compounds that are structurally similar to the growth factors in question, but the analogs are structurally different and cannot duplicate the function of the natural growth factor in the cell. In addition to the bacterial growth factor analogs, which we will now discuss, there are a number of growth factor analogs that are active in the treatment of viral and fungal infections, and we will discuss these in Sections 18.10 and 18.11.

Sulfa Drugs

The *sulfa drugs* were the first widely used growth factor analogs to specifically inhibit the growth of bacteria (see

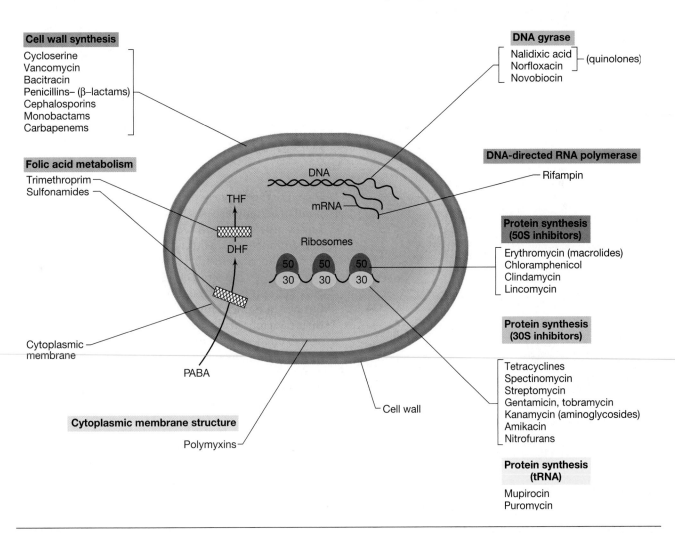

FIGURE 18.14   Mode of action of major antimicrobial chemotherapeutic agents. THF, Tetrahydrofolate; DHF, dihydrofolate; mRNA, messenger RNA; tRNA, transfer RNA.

The development of chemotherapeutic agents has had a greater impact on clinical medicine than any other discovery. Although a variety of natural chemical agents had been used earlier, the real advances in work with chemotherapeutic agents began with the German scientist Paul Ehrlich. In the early 1900s, Ehrlich developed the concept of selective toxicity. He began his work by studying the staining of microorganisms and observed that some dyes stained microorganisms but not animal tissue. He assumed that if a dye did not stain a tissue, the dye molecules were unable to combine with the cell constituents. He then reasoned that if such a dye had toxic properties, it should not affect the animal cells because it could not combine with them, but it should attack the microbial cells. In an infected animal, chemicals of this sort should behave like "magic bullets," striking the pathogen but missing the host. Ehrlich proceeded to test large numbers of chemicals for selectivity and discovered the first chemotherapeutic agents, of which Salvarsan, an arsenic-containing drug for the cure of syphilis, was the most famous (see Fig. 1).

However, no chemical agents were discovered that affected the vast majority of infectious agents until the 1930s, when Gerhard Domagk discovered the sulfa drugs. The discovery of the sulfas came

FIG. 1 Salvarsan

about through the large-scale screening of chemicals for activity in infectious diseases in experimental animals. Domagk, at the Bayer Chemical Company in Germany, tested a large variety of synthetic organic chemicals, mainly dyes, for their ability to cure streptococcal infections in mice. The first active compound was Prontosil, which was active in mice but had no activity against streptococci grown in the test tube. Domagk discovered that in the animal body, Prontosil broke down to sulfanilamide, which was the actual active agent. It was possible to embark on a program of synthesis based on the sulfanilamide structure, which yielded a large number of active drugs. D. D. Woods in England then showed that *p*-aminobenzoic acid specifically counteracted the inhibitory action of sulfanilamide, and he also showed that streptococci required *p*-aminobenzoic acid for growth. This led to the concept of the *growth factor analog*, which enabled chemists to pursue the synthesis of a wide variety of chemotherapeutic agents.

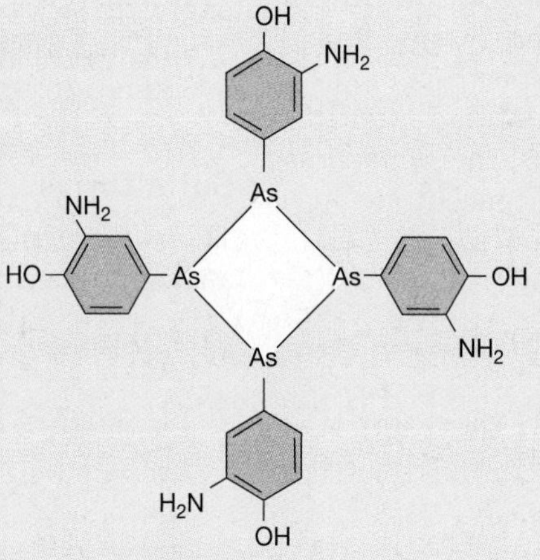

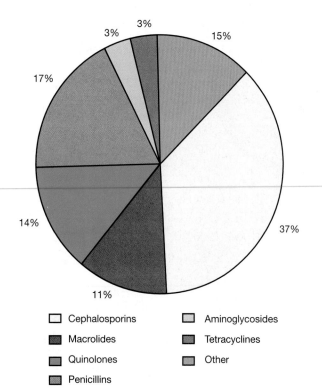

the box); they have been highly successful in the treatment of certain diseases. The simplest sulfa drug is **sulfanilamide** (Figure 18.17*a*). Sulfanilamide is an analog of *p*-aminobenzoic acid (Figure 18.17*b*), which is itself a part of the vitamin folic acid (Figure 18.17*c*). Sulfanilamide acts by blocking the synthesis of folic acid, a nucleic acid precursor. Sulfanilamide is active in Bacteria but not in higher animals because Bacteria synthesize their own folic acid, whereas higher animals obtain folic acid from their diet. Drug resistance to the clinically useful sulfanilamide derivatives, the sulfonamides, is quite common, usually because the resistant Bacteria have developed the ability to use exogenous sources of preformed folic acid (Section 18.12).

Other Growth Factor Analogs

Analogs are now known for various vitamins, amino acids, purines, pyrimidines, and other compounds.

FIGURE 18.15 Worldwide production and use of antibiotics in 1994. Each year more than 500 metric tons of chemotherapeutic agents are manufactured.

Despite the successes of the sulfa drugs, most infectious diseases were still not under chemical control. It took the discovery of the first antibiotic, penicillin, by Alexander Fleming, a Scottish physician engaged in research at St. Mary's Hospital in London, to point investigators in the right direction. Fleming's first paper on penicillin, published in 1929, begins as follows:

> While working with staphylococcus variants a number of culture plates were set aside on the laboratory bench and examined from time to time. In the examination these plates were necessarily exposed to the air and they became contaminated with various micro-organisms. It was noticed that around a large colony of contaminating mould the staphylococcus colonies became transparent and were obviously undergoing lysis. Subcultures of this mould were made and experiments conducted with a view of ascertaining something of the properties of the bacteriolytic substance which had evidently been formed in the mould culture and which had diffused into the surrounding medium.

Fleming characterized the product, and since it was produced by a fungus of the genus *Penicillium*, gave it the name *penicillin*. His work, however, did not include a process for large-scale production nor did it show that penicillin was effective in the treatment of infectious disease. This was done by a group of British scientists at Oxford University, headed by Howard Florey in 1939, motivated in part by the impending World War II and the knowledge that infectious disease was the leading cause of death among soldiers on the battlefield. Florey and his colleagues developed methods for the analysis and testing of penicillin and for its production in large quantities. They then proceeded to test penicillin against bacterial infections in humans. Penicillin was dramatically effective in controlling staphylococcal and pneumococcal infections and was also more effective for streptococcal infections than the sulfa drugs. With the effectiveness of penicillin demonstrated and the war in Europe becoming more intense, Florey brought cultures of the penicillin-producing fungus to the United States in 1941. He persuaded the U.S. government to create a large-scale research program, which led to a joint effort of the pharmaceutical industry, the U.S. Department of Agriculture at its laboratory in Peoria, Illinois, and several universities. By the end of World War II, penicillin was available in large amounts, for civilian as well as military use. As soon as the war was over, pharmaceutical companies entered into commercial production of penicillin on a competitive basis and began to look for other antibiotics. Success was quick and dramatic, and the impact on medicine has been close to phenomenal. Infant and child mortality have been greatly reduced, and many diseases that formerly had high fatality rates are now no more than medical curiosities. ∎

These analogs resemble compounds found in a variety of organisms including many eukaryotes and viruses. An important example of a clinically useful growth factor analog is *isoniazid* (Figure 23.10). Isoniazid has a very narrow spectrum of activity (Figure 18.16) and is effective only against *Mycobacterium tuberculosis*, apparently interfering with the synthesis of the mycobacterial-specific mycolic acid cell wall material. This synthetic compound, a nicotinamide (vitamin) analog, has been absolutely critical for control and treatment of tuberculosis (Figure 18.16, Section 23.3). In the examples shown in Figure 18.18, analogs have been

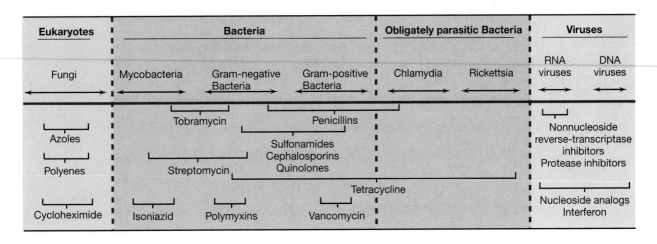

FIGURE 18.16   Antimicrobial spectrum of action for selected chemotherapeutic agents.

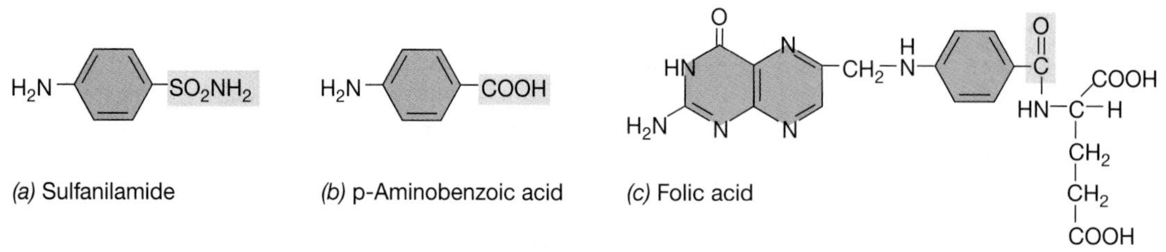

FIGURE 18.17   (a) The simplest sulfa drug, sulfanilamide. (b) Sulfanilamide is an analog of *p*-aminobenzoic acid, which itself is part of (c) the growth factor folic acid. (∞ Section 4.2 discusses growth factors.)

formed by addition of a fluorine or a bromine atom. Fluorine is a relatively small atom and does not alter the overall shape of the molecule, but it changes the chemical properties sufficiently so that the compound does not act normally in cell metabolism. Fluorouracil resembles the nucleic acid base uracil; bromouracil resembles another base, thymine. Growth factor analogs that resemble nucleic acids are widely used in the treatment of viral infections and are also used as mutagens (see Section 18.10 and ∞ Section 9.3).

Quinolones

The quinolones are not growth factor analogs, but are a class of synthetic antibacterial compounds that interact with bacterial DNA gyrase and prevent the gyrase from

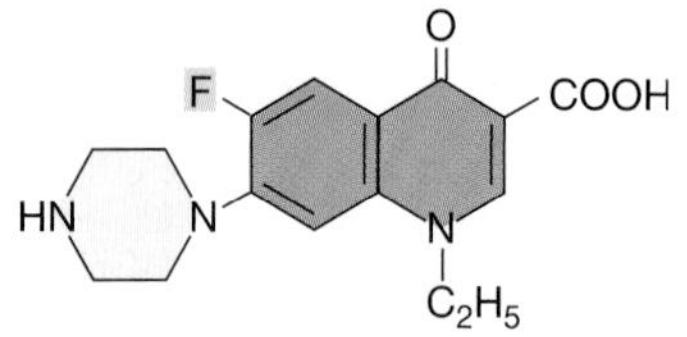

FIGURE 18.19   The structure of a fluoroquinolone, norfloxacin. Norfloxacin is a fluorinated derivative of nalidixic acid (Figure 18.13). Norfloxacin is more soluble than nalidixic acid and reaches clinically effective levels in blood and tissues.

supercoiling bacterial DNA, which is required for packaging of DNA in the bacterial cell (∞ Section 6.3). Nalidixic acid is the prototype quinolone (Figure 18.13 and Figure 18.14). Fluoroquinolone derivatives of nalidixic acid such as norfloxacin and ciprofloxacin (Figure 18.19) are routinely used to treat urinary tract infections in humans. Because DNA gyrase is found in all Bacteria, the fluoroquinolones are effective for treating both gram-positive and gram-negative bacterial infections (Figure 18.16). The fluoroquinolones are also extensively used in the beef and poultry industries for treatment of respiratory diseases.

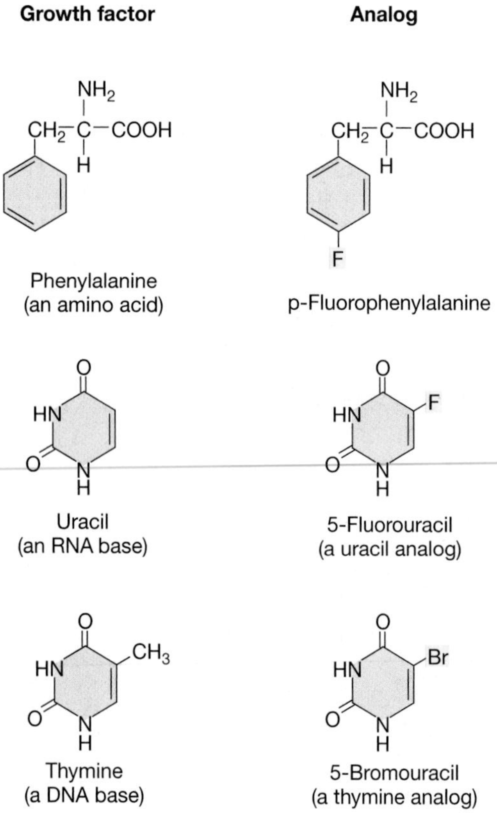

✓ 18.6 Concept Check

Synthetic chemotherapeutic agents exhibit selective toxicity for Bacteria and are safe to take internally. Growth factor analogs such as sulfanilamide are synthetic metabolic inhibitors. Quinolones inhibit the action of DNA gyrase in Bacteria.

✓ Identify at least two features that distinguish synthetic chemotherapeutic agents from antiseptics and disinfectants.
✓ What is a *competitive inhibitor*?

18.7

Antibiotics

Antibiotics are chemical substances produced by certain microorganisms that inhibit or kill other microorganisms. Antibiotics are distinguished from growth factor analogs because they are natural products (products of microbial activity) rather than synthetic chemi-

FIGURE 18.18   Growth factors and structurally similar analogs.

cals. Antibiotics constitute one of the most important classes of substances produced by large-sale microbial processes. The industrial production of antibiotics and the methods used to discover new ones were investigated in Chapter 11. In this section we present a broad overview of the antibiotics. In the next section, we discuss the structure and function of a specific group of antibiotics, the β-lactam antibiotics, which are produced by fungi. We conclude our consideration of antibacterial agents with a discussion of some important antibiotics made by prokaryotes of the genus *Streptomyces*.

Targets of Antibiotics

A very large number of antibiotics have been discovered, but less than 1% have been of practical value in medicine. However, the useful antibiotics have had a dramatic impact on the treatment of infectious diseases. Further, many antibiotics are made more effective by chemical modifications in the laboratory; these are said to be *semisynthetic antibiotics*.

The sensitivity of microorganisms to antibiotics and other chemotherapeutic agents varies (Figure 18.16). Gram-positive Bacteria are usually more sensitive to antibiotics than are gram-negative Bacteria, although some antibiotics act only on gram-negative Bacteria. An antibiotic that acts on both gram-positive and gram-negative Bacteria is called a **broad-spectrum antibiotic.** In general, a broad-spectrum antibiotic finds wider medical use than a *narrow-spectrum antibiotic*, which acts on only a single group of organisms. An antibiotic with a limited spectrum of activity may, however, be quite valuable for the control of microorganisms that fail to respond to other antibiotics. An example is vancomycin, a glycopeptide that is a bacteriocidal agent that acts against gram-positive Bacteria of the genera *Staphylococcus, Bacillus,* and *Clostridium* (Figures 18.13, 18.14, and 18.16).

Antibiotics and other chemotherapeutic agents can be grouped based on chemical structure (Figure 18.13) or on mode of action (Figure 18.14). In Bacteria, the important targets of antibiotic action are the cell wall (for example, vancomycin), the cytoplasmic membrane (polymyxins), the biosynthetic processes of protein synthesis (the macrolides and tetracyclines), and nucleic acid synthesis (rifampin).

We begin our study of the antibiotics with the β-lactam group, which includes the penicillins and related compounds of major clinical significance.

✓ 18.7 Concept Check

Antibiotics are a chemically diverse group of static or cidal compounds produced by microorganisms. They function by a variety of mechanisms to disrupt microbial metabolism. Most known antibiotics have no clinical applications.

✓ Distinguish *antibiotics* from *growth factor analogs*.
✓ What is meant by a *broad-spectrum antibiotic*?

18.8

β-Lactam Antibiotics: Penicillins and Cephalosporins

One of the most important groups of antibiotics, both historically and medically, is the β-lactam group. The β-lactam antibiotics include the penicillins, cephalosporins, and cephamycins, all medically useful antibiotics. These antibiotics all share the presence of a characteristic structural component, the β-lactam ring (Figure 18.20). Together, the penicillins and cephalosporins account for over one-half of all of the antibiotics produced and used worldwide (Figure 18.15). Penicillin is produced by fungus *Penicillium chrysogenum,* and

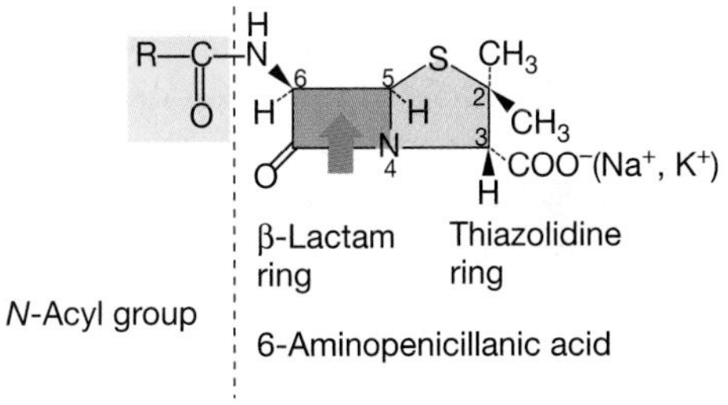

Designation	N-Acyl group
NATURAL PENICILLIN Benzylpenicillin (penicillin G) Gram-positive activity β-lactamase-sensitive	$-CH_2-CO-$
SEMISYNTHETIC PENICILLINS Methicillin acid-stable, β-lactamase-resistant	$-CO-$ (with OCH_3 groups)
Oxacillin acid-stable, β-lactamase-resistant	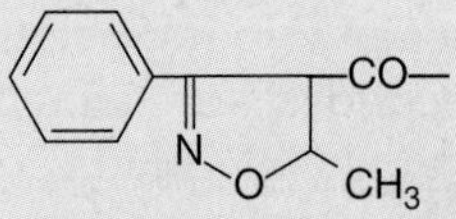 $-CO-$
Ampicillin broadened spectrum of activity (especially against gram-negative bacteria), acid-stable, β-lactamase-resistant	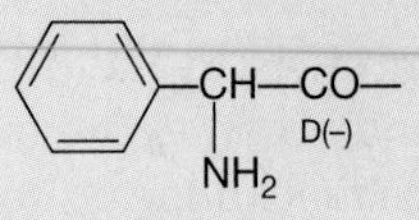 $-CH-CO-$ $D(-)$ NH_2
Carbenicillin broadened spectrum of activity (especially against *Pseudomonas aeruginosa*), acid-stable but ineffective orally, β-lactamase-sensitive	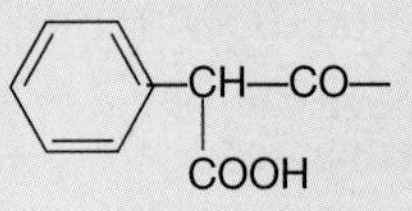 $-CH-CO-$ $COOH$

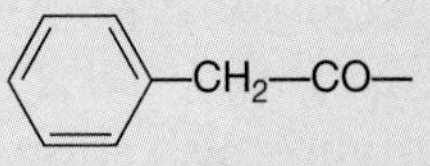

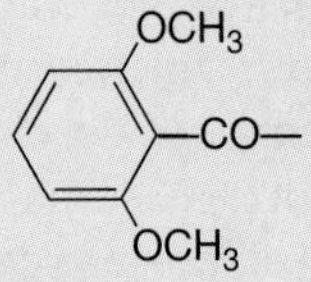

FIGURE 18.20 The structures of some important penicillins. The red arrow (top panel) is the site of action for most β-lactamases.

cephalosporin is produced by the fungus *Cephalosporium* sp. (➝ Section 11.6.)

Types of Penicillin

The first β-lactam antibiotic discovered, **penicillin G** (Figure 18.20), is active primarily against gram-positive Bacteria. Its action is restricted to gram-positive Bacteria primarily because gram-negative Bacteria are impermeable to the antibiotic. As a result of extensive research, new semisynthetic penicillins are constantly being developed and introduced, many of which are quite effective against gram-negative Bacteria.

Figure 18.20 shows the complex structures of some of these new penicillins. Modifications of the basic penicillin G structure by chemical synthesis methods, as in the semisynthetic penicillins shown, significantly change the properties of the resulting antibiotics. For example, ampicillin and carbenicillin are semisynthetic penicillins that have a broader spectrum of antibiotic activity, which includes some gram-negative Bacteria. The structural differences in the *N*-acyl groups of these semisynthetic penicillins allow them to be transported inside the gram-negative outer membrane (➝ Section 3.8) where they inhibit cell wall synthesis. Note also that penicillin G is sensitive to β-lactamase, an enzyme produced by a number of penicillin-resistant Bacteria (see Section 18.12). The semisynthetic penicillins oxacillin and methicillin are useful because they are β-lactamase resistant.

Mechanisms of Action

The β-lactam antibiotics are potent inhibitors of cell wall synthesis. As we discussed in Sections 3.7–3.9, an important feature of cell wall synthesis is the transpeptidation reaction, which results in the cross-linking of two glycan-linked peptide chains (➝ Figures 3.31 and 3.41). The enzymes that accomplish this task, the transpeptidases, are also capable of binding to penicillin or other antibiotics with the β-lactam ring. Thus, these transpeptidases are known as *penicillin binding proteins* (PBPs). The PBPs bind very tightly to penicillin and can no longer catalyze the transpeptidase reaction. The cell wall continues to be formed but is no longer cross-linked and becomes progressively weaker as the peptidoglycan backbone is laid down. In addition, the antibiotic–PBP complex stimulates the release of autolysins that digest the existing cell wall. The result is a weakened, eventually degraded cell wall. Under normal circumstances, the osmotic pressure differences inside the cell, as compared to outside, lyse the cell. By contrast, vancomycin, a glycopeptide (Figure 18.13), does not bind to PBPs but acts directly on the terminal D-alanyl-D-alanine peptide on the peptidoglycan precursors (➝ Figure 3.41), blocking the transpeptidase reaction. Because the cell wall and its synthesis mecha-

nisms are unique to Bacteria, the β-lactam antibiotics have very high specificity and are not toxic to host cells. However, because of the complex structural configurations of these antibiotics, some individuals develop serious antibody-mediated allergies to individual β-lactam compounds after repeated courses of antibiotic therapy. These anaphylactic allergic responses can be life-threatening in some cases (➝ Section 20.15).

Cephalosporins

The cephalosporins are another group of clinically important antibiotics that contain the β-lactam ring. They differ structurally from the penicillins because they have a six-member dihydrothiazine ring instead of the five-member thiazolidine ring. The cephalosporins have the same mode of action as the penicillins. That is, they bind irreversibly to the PBPs and prevent the cross-linking of peptidoglycan. The clinically important cephalosporins are semisynthetic antibiotics that generally have a broader spectrum of antibiotic activity than the penicillins and are often more resistant to the action of enzymes that destroy β-lactam rings, the β-lactamases. For example, ceftriaxone (Figure 18.13), a widely used cephalosporin, is highly resistant to the β-lactamases and has now supplanted penicillin as the drug of choice for treatment of infections due to *Neisseria gonorrhoeae* (see Section 18.12; ➝ Section 23.6) because many *N. gonorrhoeae* strains have developed β-lactamases that cleave the β-lactam rings of penicillin.

✓ 18.8 Concept Check

The β-lactam compounds are the most important clinical antibiotics. This group includes the penicillins and the cephalosporins. These antibiotics are specific for the cell wall synthesis enzymes of Bacteria and, as a group, have very low host toxicity and a very broad spectrum of activity.

✓ Draw the common structure of the β-lactam antibiotics.
✓ How do the β-lactam antibiotics function?

18.9

Antibiotics from Prokaryotes

Many antibiotics active against prokaryotes are also produced by prokaryotes. These include the aminoglycosides, the macrolides, and the tetracyclines. Many of these antibiotics have major clinical applications, and thus we discuss their general properties here.

Aminoglycoside Antibiotics

Aminoglycoside antibiotics contain amino sugars bonded by glycosidic linkage (➝ Section 2.3) to other amino sugars. A number of clinically useful antibiotics are

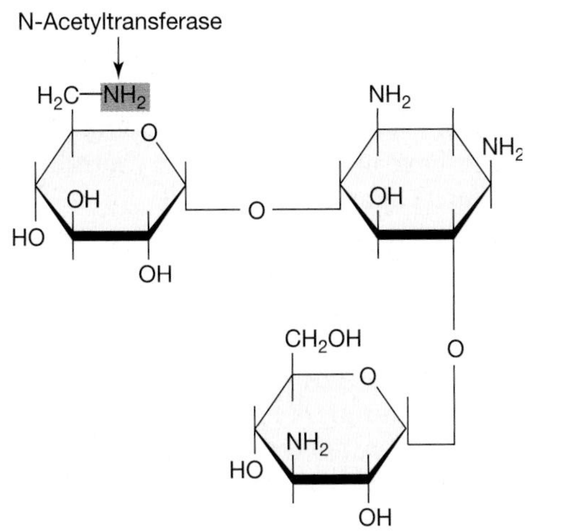

FIGURE 18.21   Structure of kanamycin, an aminoglycoside antibiotic. The amino sugars are in yellow. The site of modification by an *N*-acetyltransferase, encoded by a resistance plasmid, is indicated.

aminoglycosides, including *streptomycin* (Figure 18.13) and its relatives, *kanamycin* (Figure 18.21), *gentamicin*, and *neomycin*. The aminoglycosides act by inhibiting protein synthesis at the 30S subunit of the ribosome (Figure 18.14). The aminoglycoside antibiotics are used clinically against gram-negative Bacteria. Streptomycin has also been used extensively in the treatment of tuberculosis. Historically, the use of streptomycin for tuberculosis treatment was a major medical advance, as it was the first antibiotic capable of controlling this infectious disease. However, none of the aminoglycoside antibiotics are widely used today and together the aminoglycosides account for only about 3% of the total of all antibiotics produced and used (Figure 18.15). Streptomycin has been supplanted by several synthetic chemicals for tuberculosis treatment because streptomycin causes several serious side effects and bacterial resistance readily develops. The use of aminoglycosides for treatment of gram-negative infections has decreased since the development of the semisynthetic penicillins (see Section 18.8) and the tetracyclines (see later in this section). The aminoglycoside antibiotics are now considered reserve antibiotics used primarily when other antibiotics fail.

Macrolide Antibiotics

Macrolide antibiotics contain large lactone rings connected to sugar moieties (Figure 18.22). Variations in both the macrolide ring and the sugar moieties are known, and so a large variety of macrolide antibiotics exist. The best-known macrolide antibiotic is *erythromycin*, but other macrolides include *oleandomycin*, *spiramycin*, and *tylosin*. Together, the macrolide antibiotics account for 11% of the total world production and use (Figure 18.15).

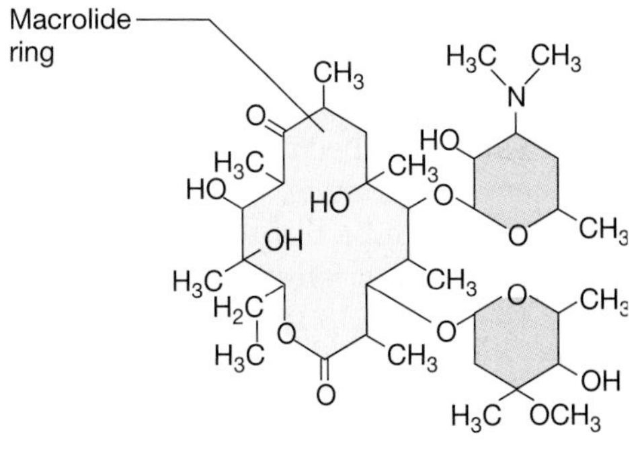

FIGURE 18.22   Structure of erythromycin, a typical macrolide antibiotic.

Erythromycin acts as a protein synthesis inhibitor at the level of the 50S subunit of the ribosome (Figure 18.14). Erythromycin is commonly used clinically in place of penicillin in patients allergic to penicillin or other β-lactam antibiotics. Erythromycin has been particularly valuable in treating cases of legionellosis (⟳ Section 24.9) because of the exquisite sensitivity of the causative agent, the bacterium *Legionella pneumophila*, to this antibiotic.

Tetracyclines

The tetracyclines are an important group of antibiotics that find widespread medical use in humans. They were some of the first so-called *broad-spectrum* antibiotics, inhibiting almost all gram-positive and gram-negative Bacteria. The basic structure of the tetracyclines consists of a naphthacene ring system (Figure 18.23). The basic naphthacene ring structure can be substituted at several positions to form new tetracycline analogs. *Chlortetracycline*, for instance, has a chlorine atom, whereas *oxytetracycline* has an additional hydroxyl (OH) group

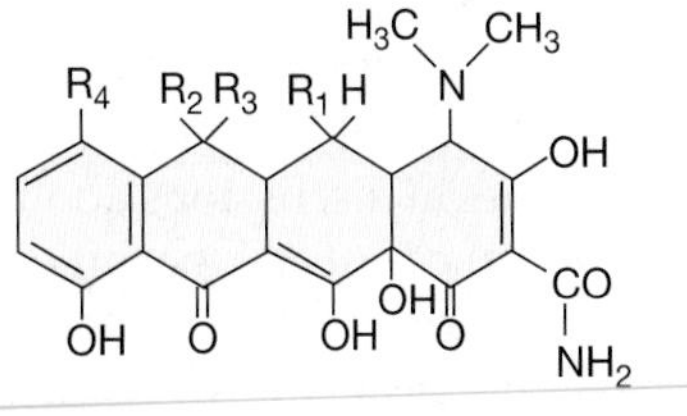

Tetracycline form	R_1	R_2	R_3	R_4
Tetracycline	H	OH	CH_3	H
7-Chlortetracycline (aureomycin)	H	OH	CH_3	Cl
5-Oxytetracycline (terramycin)	OH	OH	CH_3	H

FIGURE 18.23   Structure of tetracycline and important derivatives.

and no chlorine (Figure 18.23). All three of these antibiotics are produced microbiologically, but there are also semisynthetic tetracyclines on the market, into which other constituents have been inserted chemically into the naphthacene ring system. Like erythromycin and the aminoglycoside antibiotics, tetracycline is a protein synthesis inhibitor. It interferes with 30S ribosomal subunit function (see Figure 18.14).

The tetracyclines and the β-lactam antibiotics are the two most important groups of antibiotics in the medical field. The tetracyclines also find use in veterinary medicine, and in some countries are also used as nutritional supplements for poultry and swine. However, extensive nonmedical uses of medically important antibiotics have resulted in wide-spread antibiotic resistance and are now discouraged (see Section 18.12).

✓ 18.9 Concept Check

The aminoglycosides, macrolides, and tetracycline antibiotics are structurally complex molecules produced by prokaryotes and are active against other prokaryotes. Erythromycin and the various tetracyclines are used widely in clinical medicine.

- ✓ What are the biological sources of the aminoglycoside, tetracycline, and macrolide antibiotics?
- ✓ What is the mechanism of action for each of these classes of antibiotics?

18.10

Viral Control

In Sections 18.6–18.9, we examined a variety of control agents that are effective against Bacteria. One of the key features for every agent we discussed was selective toxicity. For example, in Section 18.8 we saw that since the β-lactam antibiotics inhibit cell wall synthesis, they are selectively toxic for Bacteria, and there is no toxicity to the animal host cell, which lacks cell walls, the target structure. We now begin discussions about the viruses. Viruses actually use the host cell machinery to perform their metabolic functions (∞ Chapter 8). Therefore, most attempts at chemical control of viruses result in toxicity for the host. However, several agents are more toxic for the virus than the host, and there are a few agents produced by the host that specifically target viruses. Here we will examine several classes of chemotherapeutic agents that have been shown to be clinically effective in controlling viral replication.

Antiviral Chemotherapeutic Agents

Because viral structures and functions are so integrated into the functions of the host cell, the therapeutic successes achieved against bacteria using extremely selec-

tive antibacterial agents have not been matched by similar achievements with antiviral agents. However, largely because of our struggle to find effective measures to control AIDS (∞ Section 23.7), some significant achievements have now been made in controlling viruses with chemical agents (Table 18.5).

The most successful and commonly used agents for antiviral chemotherapy are the nucleoside analogs (Table 18.5). The first compound to gain universal acceptance in this category was zidovudine, or azidothymidine (AZT). AZT inhibits retroviruses such as the human immunodeficiency virus (HIV), the causative agent of AIDS (∞ Section 23.7 and Figure 23.34). Azidothymidine is chemically related to thymidine but is a dideoxy derivative, lacking the 3'-hydroxyl group (thus analogous to the dideoxynucleotides used in the Sanger DNA-sequencing technique; ∞ Working with Nucleic Acids: The Tools, Chapter 6). AZT inhibits multiplication of retroviruses by blocking the synthesis of the DNA intermediate (reverse transcription) and successfully inhibits multiplication of HIV. A number of other nucleoside analogs having similar mechanisms have been developed for the treatment of HIV, as shown in Table 18.5. Nearly all nucleoside analogs work by the same mechanism, inhibiting elongation of the viral nucleic acid chain at the level of the host cell nucleic acid polymerase. Because the normal host cell function of DNA replication is targeted, these drugs almost always exhibit some level of host toxicity. Many also lose their antiviral potency with time due to the emergence of drug-resistant viruses (∞ Section 23.7). The *nucleotide* analog cidofovir works in the same way (Table 18.5).

Several other chemicals work at the level of viral polymerase. These include nevirapine, a nonnucleoside reverse transcriptase inhibitor that binds directly to reverse transcriptase and inhibits further action; phosphonoformic acid, which acts as an analog of inorganic pyrophosphate, inhibiting appropriate internucleotide linkage; and rifamycin, an antibiotic that binds and inhibits RNA polymerase.

A relatively novel class of antiviral drugs are the protease inhibitors (Table 18.5). These drugs are particularly effective for treatment of HIV. They prevent infection by binding the active site of HIV protease, inhibiting processing of viral polypeptides and virus maturation (∞ Section 23.7 and Table 23.2 and Section 18.12).

Interferon

Interferons are antiviral substances produced by many animal cells in response to infection by certain viruses. They are low-molecular-weight proteins (17,000 MW) that prevent viral multiplication in normal cells by stimulating the production of antiviral proteins. Interferons

TABLE 18.5	Antiviral chemotherapeutic compounds		
Category/drug	**Mechanism of action**	**Virus affected**
Nucleoside analogs		
Acyclovir	Viral polymerase inhibitors	Herpes viruses, *Varicella zoster*
Gancyclovir		Cytomegalovirus
Trifluridine		Herpesvirus
Valacyclovir		Herpesvirus
Vidarabine		Herpesvirus, vaccinia, hepatitis B virus
Didanosine (dideoxyinosine or ddI)	Reverse transcriptase inhibitors	HIV[a]
Lamivudine (3TC)		HIV, hepatitis B virus
Stavudine (d4T)		HIV
Zalcitabine (ddC)		HIV
Zidovudine (AZT) (∞ Figure 23.34)		HIV
Ribavirin	Blocks capping of viral RNA	Respiratory syncytial virus, influenza A and B, Lassa fever
Synthetic amines		
Amantadine	Block uncoating of virus	Influenza A
Rimantadine		Influenza A
Nucleotide analog		
Cidofovir	Viral polymerase inhibitor	Cytomegalovirus, herpesviruses
Pyrophosphate Analog		
Phosphonoformic acid	Viral polymerase inhibitor	Herpesviruses, HIV, hepatitis B virus
Nonnucleoside reverse transcriptase inhibitor		
Nevirapine	Reverse transcriptase inhibitor	HIV
RNA polymerase inhibitor		
Rifamycin	RNA polymerase inhibitor	Vaccinia, pox viruses
Protease inhibitors		
Indinavir	Protease inhibitors	HIV
Ritonavir		HIV
Saquinavir		HIV
Nelfinavir		HIV
Interferons		
Interferon α	Induces proteins that inhibit viral replication	Broad spectrum (host specific)
Interferon β		
Interferon γ		

a Human immunodeficiency virus

from virus-infected cells interact with receptors on noninfected cells, promoting the synthesis of antiviral proteins that function to prevent further virus infection. There are three molecular types, IFN-α, produced by leukocytes; IFN-β, produced by fibroblasts; and IFN-γ, produced by immune cells known as lymphocytes (∞ Section 20.8 and see Table 18.5). All three types are effective viral inhibitors. They were first discovered in the course of studies on virus interference, a phenomenon whereby infection with one virus interferes with subsequent infection with another virus, hence the name *interferon*. Interferons are formed in response to live virus, viral nucleic acids, and also to virus inactivated by radiation. Interferon is produced in larger amounts by cells infected with viruses of low virulence, but little is produced against highly virulent viruses. Apparently highly virulent viruses inhibit cell protein synthesis before any interferon can be produced. Interferon is also induced by a variety of double-stranded

RNA molecules, either natural or synthetic. Since double-stranded RNA does not exist in uninfected cells but exists as the replicative form in RNA virus-infected cells, double-stranded RNA may serve as a signal of virus infection in the animal cell and brings into action the interferon-producing system.

Interferons are not virus-specific but *host*-specific. Interferon produced by a member of one species recognizes specific receptors only on cells of the same species. Therefore, interferon produced by one type of animal (for example, chicken) in response to influenza virus inhibits multiplication of other viruses in the same species but has no effect on the multiplication of influenza virus in other animal species.

Interferons have been of interest as possible antiviral agents and possibly also as anticancer agents. However, the use of interferons as general chemotherapeutic agents has not been achieved because interferon must be present in relatively high local concentrations to stimulate the production of antiviral proteins in uninfected host cells. Thus, the clinical utility of these seemingly ideal antiviral agents depends on our ability to deliver interferon to local areas in the host. Alternatively, appropriate interferon-stimulating signals (*i.e.*, stimulation with viral nucleotides, nonvirulent viruses, or even synthetic nucleotides) given to host cells prior to viral infection might achieve the same goal.

✓ 18.10 Concept Check

Viruses use host cell metabolic machinery to replicate. However, several virus-specific enzymes and processes can be interrupted with chemotherapeutic agents to disrupt viral replication. Many clinically effective antiviral agents are nucleoside analogs and work by inhibiting viral nucleotide polymerases, but several other agents, including the protease inhibitors, interfere with viral maturation steps. Host cells also produce antiviral proteins called *interferons* that stop viral replication. However, interferons are not yet available in clinically useful forms.

✓ Why are there so few effective antiviral chemotherapeutic agents?

✓ What steps in the viral maturation process are inhibited by nucleoside analogs? By protease inhibitors? By interferons?

18.11

Fungal Control

Like the viruses, fungi pose special problems for successful chemotherapy. Since fungi are Eukarya, much of their cellular machinery is the same as in higher animals and humans, and so chemotherapeutic agents that affect metabolic pathways in fungi often affect corresponding pathways in host cells. This results in drug toxicity in higher animals. Thus, many antifungal drugs can be used only for topical (surface) applications. However, some drugs are selectively toxic for fungi. Drugs for fungal treatment are becoming increasingly important as fungal infections in immunosuppressed individuals become more prevalent (∞ Sections 23.7 and 24.7). We will look in detail at the selective toxicity of several classes of chemicals that are effective against fungi.

Ergosterol Inhibitors

Two major groups of antifungal compounds work by interacting with ergosterol or inhibiting its synthesis. In most fungi, ergosterol replaces the cholesterol component found in higher eukaryotic cell membranes (∞ Section 3.5). The first group includes the *polyenes*, a group of antibiotics produced by *Streptomyces* species. Polyenes bind to ergosterol, which disrupts membrane function, eventually causing membrane permeability and cell death (Figure 18.24). A second

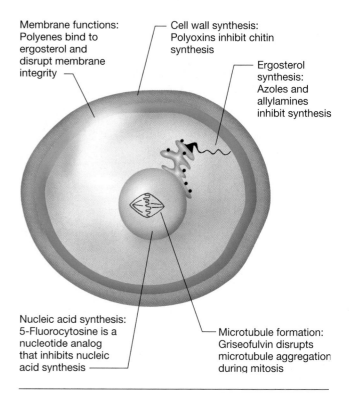

FIGURE 18.24   Sites of action of some antifungal chemotherapeutic agents. Because fungi are eukaryotic cells, antibacterial antibiotics are generally ineffective.

major group of antifungal compounds includes the *azoles* and the *allylamines*, agents that selectively inhibit ergosterol biosynthesis and therefore have broad antifungal activity. Treatment with azoles results in the inability to produce a normal membrane, leading to membrane damage and alteration of critical membrane activities such as nutrient transport. Allylamines also inhibit ergosterol biosynthesis but are useful only topically because they are not readily taken up by animal cells and tissues.

Other Antifungal Agents

A number of other antifungal drugs interfere with fungus-specific structures and functions. For example, most fungal cell walls contain *chitin*, a polymer of *N*-acetylglucosamine found only in fungi and insects (∞ Table 15.8). Several drugs such as the *polyoxins* inhibit cell wall synthesis by interfering with chitin biosynthesis. However, no cell wall targeting drugs are currently used clinically. Other drugs inhibit folate biosynthesis, interfere with DNA topology during replication, or, like *griseofulvin*, disrupt microtubule aggregation during mitosis (∞ Section 6.6). The nucleic acid analog *5-fluorocytosine* is an effective nucleic acid synthesis inhibitor in fungi. Some very effective antifungal drugs also have other biological applications. For example, *vincristine, vin-*

blastin, and *toxol* are effective antifungal agents and have known anticancer properties.

Unfortunately, the use of antifungal drugs has predictably resulted in the emergence of populations of resistant fungi and the emergence of "new" fungal pathogens. For example, *Candida* species, which are normally not pathogenic, now produce disease in individuals who have been treated with antifungal drugs. These drug-resistant *Candida* pathogens are not treatable by employing any of the currently used antifungal agents.

As the use of chemotherapeutic agents, both antibacterial and antifungal, increases, the possibilities for opportunistic fungal infections and the corresponding need for specific fungal control agents will increase.

✓ 18.11 Concept Check

Antifungal agents fall into a wide variety of chemical categories. As with viruses, selective toxicity is hard to achieve, but there are some effective chemotherapeutic agents. Treatment of fungal infection is now an important human health issue.

- ✓ Why are there very few clinically effective antifungal antibiotics?
- ✓ What factors are contributing to the apparent rise in fungal infections?

18.12

Antimicrobial Drug Resistance

We have discussed the major antibiotics and their action in a number of situations. We now come to the concept of **antimicrobial drug resistance,** the acquired ability of an organism to resist the effects of a chemotherapeutic agent to which it is normally susceptible. First, we should note that most resistance genes are acquired through a process of genetic exchange from the antibiotic producers: In order to protect themselves from the antibiotics they produce, these organisms have developed mechanisms to neutralize or destroy their own antibiotics. The existence of these mechanisms means that, under the right circumstances, resistance genes can be transferred to other organisms. First, we will look at some of the mechanisms of antibiotic resistance. Then we will explore the consequences of resistance in practical situations.

Resistance Mechanisms

Not all antibiotics act against all microorganisms. Some microorganisms are *naturally resistant* to some antibiotics. There are several reasons why microorganisms may have an inherent resistance to an antibiotic. (1) The organism may lack the structure an antibiotic inhibits. For instance, some bacteria, such as mycoplasmas, lack

a typical bacterial cell wall and are resistant to penicillins. (2) The organism may be impermeable to the antibiotic. For example, most gram-negative Bacteria are impermeable to penicillin G. (3) The organism may be able to alter the antibiotic to an inactive form. Many staphylococci contain β-lactamases that cleave the β-lactam ring of most penicillins (Figure 18.25). (4) The organism may modify the *target* of the antibiotic. (5) By genetic change, alteration may occur in a metabolic pathway that the antimicrobial agent blocks. Thus, the organism develops a resistant biochemical pathway. For example, many pathogens develop resistance to sulfonamide drugs (see Section 18.6 and Figure 18.17). Sulfon-

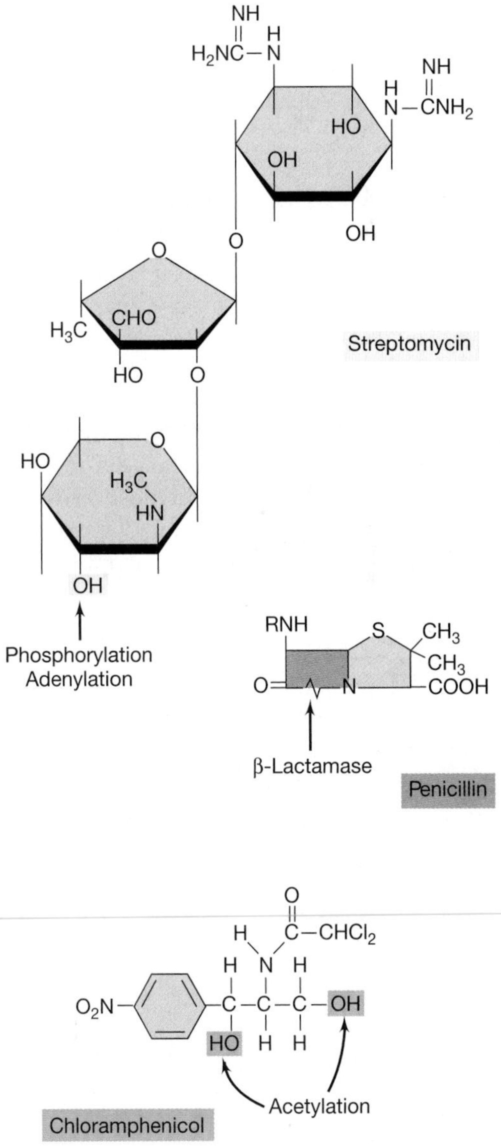

FIGURE 18.25 Sites at which antibiotics are attacked by enzymes encoded by R plasmid genes. In aminoglycoside antibiotics related to streptomycin, those with a free amino group may be inactivated by *N*-acetylation (see also Figure 18.21).

amides inhibit the production of folic acid in Bacteria, but resistant Bacteria modify their metabolism to take up preformed folic acid from the environment, avoiding the need for the pathway blocked by sulfonamides. (6) The organism may be able to pump out an antibiotic entering the cell (efflux). We give some specific examples of bacterial resistance to antibiotics in Table 18.6.

As discussed in Section 9.8, antibiotic resistance can be genetically encoded by the microorganism at either the chromosomal or the plasmid level on so-called *resistance plasmids* (*R factors*); specific types of resistance typically have a genetic basis in one location or the other (Table 18.6). Because of the development of antibiotic resistance, testing of bacteria isolated from clinical material for antibiotic sensitivity must be carried out using the minimum inhibitory concentration (MIC) method (Section 18.4 and Figures 18.11 and 18.12). Details of the antibiotic sensitivity testing of clinical isolates are described in Section 21.3.

Mechanism of Resistance Mediated by R Plasmids

In the *laboratory* antibiotic-resistant cells are often isolated from cultures that were predominantly antibiotic-sensitive. The resistance of these isolates is usually due to mutations in *chromosomal* genes. On the other hand, the majority of drug-resistant bacteria isolated from *patients* contain the drug-resistance genes on R plasmids. The mechanism of R plasmid resistance is different from that of chromosomal resistance. In most cases, antibiotic resistance mediated by chromosomal genes arises because of a modification of the *target* of antibiotic action (for example, a ribosome).

By contrast, R plasmid resistance is in most cases due to the presence in the R plasmid of genes encoding new enzymes that *inactivate* the drug (Figure 18.25) or genes that encode enzymes that either prevent uptake of the drug or actively pump it out. For instance, a number of antibiotics are known that have similar chemical structures containing aminoglycoside units. Among the aminoglycoside antibiotics are streptomycin, neomycin, kanamycin, and spectinomycin. Strains carrying R plasmids conferring resistance contain enzymes that chemically modify the antibiotics either by phosphorylation, acetylation, or adenylylation. The modified drug then lacks antibiotic activity. In the case of the penicillins, R plasmid resistance is due to the formation of penicillinase (β-lactamase), which splits the β-lactam ring, thus destroying the molecule. Chloramphenicol resistance mediated by an R plasmid arises because of the presence of an enzyme that acetylates the antibiotic. Many R plasmids can confer multiple antibiotic resistance. This is generally due to the fact that a single R plasmid may contain several different genes, each encoding a different antibiotic-inactivating enzyme.

Origin of Resistance Plasmids

Although specific evidence for the origin of multiple drug resistance R plasmids is not available, a number of lines of circumstantial evidence suggest that plasmids with R plasmid-type character existed before the antibiotic era. The widespread use of antibiotics provided selective conditions for the spread of R plasmids with one or more antibiotic resistance genes (see the box, Nonmedical Uses of Antibiotics). Indeed, a strain of *Escherichia coli* that was freeze-dried in 1946 was found to

TABLE 18.6   Mechanisms of bacterial resistance to antibiotics

Resistance mechanism	Antibiotic example	Genetic basis of resistance	Mechanism present in:
Reduced permeability	Penicillins	Chromosomal	*Pseudomonas aeruginosa* Enteric Bacteria
Inactivation of antibiotic (for example, penicillinase; modifying enzymes methylases, acetylases, and phosphorylases; and others)	Penicillins	Plasmid and chromosomal	*Staphylococcus aureus* Enteric Bacteria *Neisseria gonorrhoeae*
	Chloramphenicol	Plasmid and chromosomal	*Staphylococcus aureus* Enteric Bacteria
	Aminoglycosides	Plasmid	*Staphylococcus aureus*
Alteration of target (for example, RNA polymerase, rifamycin; ribosome, erythromycin, and streptomycin; DNA gyrase, quinolones)	Erythromycin Rifamycin Streptomycin Norfloxacin	Chromosomal	*Staphylococcus aureus* Enteric Bacteria Enteric Bacteria Enteric Bacteria *Staphylococcus aureus*
Development of resistant biochemical pathway	Sulfonamides	Chromosomal	Enteric Bacteria *Staphylococcus aureus*
Efflux	Tetracyclines Chloramphenicol	Plasmid Chromosomal	Enteric Bacteria *Staphylococcus aureus* *Bacillus subtilis*

A FOCUS ON . . .   Nonmedical Uses of Antibiotics

A major nonmedical use of antibiotics in the United States is addition to animal feed. The addition of low levels of antibiotics to animal feeds stimulates animal growth, shortening the period required to get the animal to market. For example, addition of 25 milligrams (mg) of penicillin per pound of chicken feed saves 2 billion lb (900 million kg) of feed each year because of more rapid weight gains and feeding efficiency. The antibiotics probably act by inhibiting organisms responsible for low grade infections and by reducing intestinal epithelial inflammation. Studies with germ-free animals have confirmed this idea. The growth of germ-free animals is not accelerated by antibiotic-supplemented feed. In addition, the intestinal wall of the normal animal is much thicker than that of the germ-free animal, probably because of low level inflammation caused by the normal bacterial flora. The lessening of inflammation in the gut of animals fed low levels of antibiotics probably promotes nutrient uptake and could account for the more efficient use of feed observed.

The problem with low levels of antibiotics in animal feeds is that an antibiotic-resistant microflora is selected by the constant exposure to antibiotics. The use of antibiotics in animal feed therefore expands the gene pool of antibiotic resistance in nature. Because some of the animal gut flora also inhabit the human gut, the transmission of resistant flora from animals to humans is a real possibility. Indeed, studies on antibiotic resistance in human gut flora have shown that many strains of human enteric bacteria are multiply resistant, especially among those who work in animal husbandry.

Molecular studies of resistant strains of *Salmonella* isolated from poultry have shown that the resistance resides on conjugative plasmids or transposons (⚭ Sections 9.8 and 9.10). Resistance genes are rapidly transferred between different species and even between different genera. Resistant organisms can then infect humans through contaminated meat or by contact with live animals.

Unfortunately, long-term studies on animals previously fed antibiotics and then put on antibiotic-free

rations have shown that antibiotic-resistant bacteria are not quickly lost from the gut. The resistance genes may have integrated with stable plasmids in the gut flora, and, in the absence of counterselective forces, these resistance determinants have been maintained and will remain a part of the gut flora for some time, even if supplementation of feeds with antibiotics were to stop. Nonmedical use of antibiotics has therefore reinforced, albeit painfully, a simple lesson in microbial ecology: The environment selects the best adapted species.

Although continued use of clinically useful antibiotics in animal feeds will undoubtedly widen the dissemination of resistance genes, it is not clear that halting this single practice will effectively solve the problem. Continued veterinary use of antibiotics may by itself maintain resistant animal microflora. However, in hope of reducing the spread of antibiotic resistance in Europe, most European countries have banned the use of antibiotics in animal feeds. In the United States large amounts of antibiotics continue to be used in the cattle, poultry, and swine industries. ■

---

contain a plasmid with genes conferring resistance to tetracycline and streptomycin, even though neither of these antibiotics were used clinically until several years later. Also, strains carrying R plasmid genes for resistance to semisynthetic penicillins were shown to exist before the semisynthetic penicillins had been synthesized. Of perhaps even more ecological significance, R plasmids conferring antibiotic resistance have been detected in some nonpathogenic gram-negative soil Bacteria. These resistance plasmids may confer selective advantage because major antibiotic-producing organisms (*Streptomyces*, *Penicillium*) are also normal soil organisms. Thus, it seems that R plasmids are not a recent phenomenon but existed in the natural microbial population before the antibiotic era. Later, the widespread use of antibiotics provided selective conditions for the rapid spread of these R plasmids. R plasmids are thus a predictable outcome of natural selection. They pose significant limits for the long-term use of any single antibiotic as an effective chemotherapeutic agent.

Spread of Antimicrobial Drug Resistance

Inappropriate, extensive use of antimicrobial drugs is leading to the rapid development of drug-specific resistance in disease-causing microorganisms. The discovery and clinical use of the many known antibiotics has been paralleled by the emergence of bacteria that resist their action. There are numerous examples of the overuse of antibiotics and the concomitant development of resistance. Figure 18.26*a* shows a correlation between the number of tons of antibiotics used and the percentage of bacteria resistant to each antibiotic. The resistant organisms were isolated from patients with diarrheal disease. In general, high levels of antibiotic use resulted in high levels of resistance.

There are many examples of diseases in which the drug prescribed for treatment has changed because of increased resistance of the microorganism causing the disease. A classic example is the development of resis-

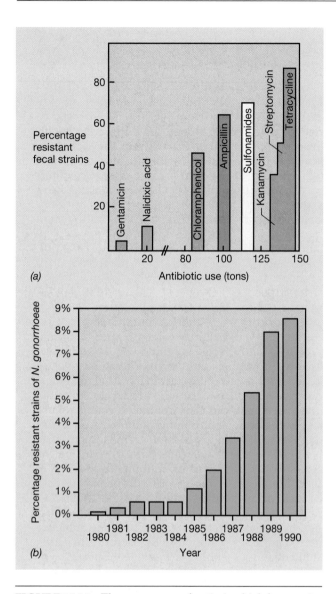

(a)

Percentage resistant fecal strains

Antibiotic use (tons)

(b)

of choice is ceftriaxone, but new treatment modalities are recommended nearly every year, simply to limit the effects of rapidly emerging resistance genes (∞ Section 23.6). Thus, resistant microorganisms are *selected* by the presence of the antibiotic in the environment.

A number of surveys worldwide suggest that antibiotics are used in clinical practice far more often than is necessary. Data indicate that antibiotic treatment is warranted in 20% of individuals who are seen for clinical infectious disease. However, antibiotics are prescribed up to 80% of the time. To add to this problem, in up to 50% of cases recommended doses or duration of treatments are not correct. This is compounded by patient noncompliance: Many patients stop taking medications, particularly antibiotics, as soon as they "feel better." For example, the emergence of isoniazid-resistant tuberculosis seems to be directly related to the fact that patients must take oral medication for up to one year, and many do not comply with the whole course of therapy (∞ Section 23.3). Thus, virulent pathogens are often subjected to sublethal doses of antibiotics for short periods of time, selecting for the emergence of resistant organisms. Largely as a result of these failures to properly use and monitor antibiotic therapy, almost all pathogenic microorganisms have developed resistance to some chemotherapeutic agents since widespread use of antimicrobial chemotherapy began in the 1950s (Figure 18.27). Penicillin and sulfa drugs, the first widely used chemotherapeutic agents, are not as widely used today because many pathogens have acquired some resistance to them. Even the organisms that are still uni-

FIGURE 18.26 The emergence of antimicrobial drug–resistant bacteria. (a) Relationship between antibiotic use and the percentage of bacteria isolated from diarrheal patients resistant to the antibiotic. Those agents that have been used in the largest amounts, as indicated by the amount produced commercially, are those for which drug-resistant strains are most frequent. (b) Percentage of reported cases of gonorrhea caused by drug-resistant strains. The actual number of reported drug-resistant cases in 1985 was 9000. This number rose to 59,000 in 1990. Greater than 95% of the reported drug-resistant cases are due to penicillinase-producing strains of *Neisseria gonorrhoeae*. Since 1990, penicillin has not been recommended for treatment of gonorrhea because of emerging drug resistance. (Source: Centers for Disease Control, Atlanta, GA).

tance to penicillin in *Neisseria gonorrhoeae*, the bacterium that causes gonorrhea (Figure 18.26b). Penicillin is no longer a useful antibiotic for treatment of gonorrhea because a large percentage of the clinical isolates produce β-lactamase and are resistant. Virtually all resistant strains have developed since 1980. The current drug

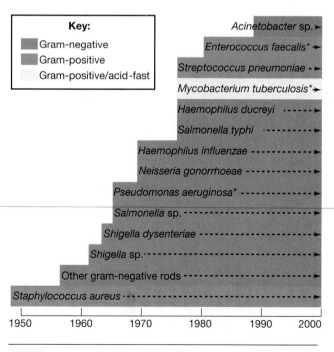

FIGURE 18.27 The appearance of antimicrobial drug resistance in some human pathogens. The * symbol indicates that some multi-drug resistant strains of these organisms are now untreatable with known antimicrobial drugs.

formly sensitive to penicillin, such as *Streptococcus pyogenes* (the bacterium that causes strep throat, scarlet fever, and rheumatic fever) (∞ Section 23.2), now need significantly more penicillin for successful treatment than a decade ago.

Other indiscriminant, nonessential uses of antibiotics may have worsened this situation. For example, antibiotics are used in agriculture both as growth-promoting substances in animal feeds and as prophylactics (to prevent the occurrence of disease rather than to treat an existing one). Several recent food poisoning outbreaks have been blamed on the use of antibiotics in animal feeds. By overloading various environments with antibiotics, rapid development of drug resistance may result.

For example, fluoroquinolones have been extensively used for only about 10 years as growth-promoting and prophylactic agents in agriculture. However, fluoroquinolone-resistant *Campylobacter jejuni* have already emerged, presumably because of the practice of treating whole flocks of poultry with fluoroquinolones to prevent the spread of respiratory diseases. New voluntary guidelines are being used by both poultry and drug producers to monitor the use of second-generation fluoroquinolones and prevent rapid emergence of resistance. Resistance can be minimized if drugs are used only for serious diseases and are given in sufficiently high doses and for sufficient lengths of time so that the microbial population level is reduced before mutants have a chance to appear. This problem can be eliminated through physician and patient education. Resistance can also be minimized by combining two unrelated chemotherapeutic agents because it is likely that a mutant strain resistant to one antibiotic will still be sensitive to the other. However, with the increasing prevalence of resistance plasmids (so-called R factors) in pathogenic bacteria (∞ Section 9.8), multiple antibiotic therapy is proving less attractive as a clinically useful strategem. These R factors are capable of transferring multiple drug resistance to all plasmid-susceptible Bacteria.

There is, however, some encouraging information concerning antibiotic resistance. Some reports suggest that if the use of a particular antibiotic is stopped, the resistance to that antibiotic will be reversed over time. At least one report suggests this phenomenon is occurring on a nationwide scale (see the box, Reversing Antibiotic Resistance). This information implies that resistance is reversible and that the efficacy of some antibiotics may be reestablished by long-term monitoring and prudent use. Finally, as we discuss below, new chemotherapeutic agents are constantly being produced using several methods for discovering and designing new drugs. These methods are capable of reviving the efficacy of well-known drugs and producing entirely new classes of chemotherapeutic agents.

TECHNIQUES & APPLICATIONS... Reversing Antibiotic Resistance

The widespread use of antibiotics has increased the number of pathogenic microorganisms that display antibiotic resistance. As a result, many antibiotics have lost effectiveness and some, including many penicillins, are no longer useful for treating certain infections. However, there are some indications that the process of selecting antibiotic-resistant organisms is reversible. In the 1980s, Hungary was extremely dependent on penicillin for treatment of ear and sinus infections, very common childhood diseases. The antibiotic was cheap, readily available, and widely used. However, the causative agent in most cases of these diseases, *Streptococcus pneumoniae*, became resistant to penicillin—in the early 1980s, 50% of the diagnosed cases were caused by penicillin-resistant pneumococci. As a result, simple infections became very difficult to treat and routine childhood illnesses became long, painful, debilitating diseases. However, the Hungarian National Institute of Public Health, through 23 national microbiology laboratories, was able to carefully monitor this trend. Eventually, physicians became convinced that penicillin was no longer useful because of their own clinical observations and the records and educational efforts of the institute. As a result, physicians began to prescribe non-β-lactam antibiotics to treat these diseases. From 1983 to 1992, the consumption of penicillin decreased by one-half. Along with the decrease in penicillin use, the levels of penicillin-resistant pneumococci surprisingly decreased from the high of 50% to 34%, presumably because there was no longer any selection for penicillin-resistant organisms. Thus, by shifting drug use away from penicillin, Hungarian physicians were able to stop and to actually reverse a trend toward complete penicillin resistance in pneumococci. If the levels of penicillin-resistant organisms continue to fall, penicillin may again become a useful chemotherapeutic agent in Hungary. This encouraging example suggests that prudent use, careful monitoring, and patient and physician education may reverse the effects of previous overuse of some antibiotics. If the same principles of education and prudent use of antibiotics are applied *before* resistance becomes a widespread problem, then the effectiveness of antibiotics can be maintained indefinitely. ■

An important side effect of the use of antimicrobial chemotherapeutic agents is the development of resistance by the targeted microorganisms. In many cases, resistance results from the selection of existing resistance genes, often through the improper and indiscriminate use of antimicrobial drugs. Many formerly useful antimicrobial agents are no longer useful because of drug resistance: A few organisms have developed resistance to all known antimicrobial drugs, prompting fears of a return to the preantibiotic era when infectious disease was untreatable.

✓ Why does antibiotic resistance occur?
✓ What practical steps should be taken to slow the development of antibiotic resistance?

18.13

The Search for New Antimicrobial Drugs

As we have already discussed, given sufficient drug exposure and time, resistance will develop to all known antimicrobial drugs. As a result, conservative, appropriate use of antibiotics is absolutely necessary to prolong the useful clinical life of these drugs. However, the long-term solution to microbial drug resistance is to develop new antimicrobial drugs. Several strategies are used to identify and produce useful analogs of existing compounds or to design or discover novel antimicrobial compounds.

New Analogs of Existing Antimicrobial Compounds

The production of new analogs of existing antimicrobial compounds is generally straightforward and often productive, largely because new compounds that are structural mimics of older ones have a predictable mechanism of action. In many cases, parameters such as solubility and affinity can be changed by introducing minor modifications to the chemical structure of a drug without altering structures critical to drug action. The new compound may actually be more potent than the parent compound, and, because resistance is based on structural recognition, the new compound may not be recognized by resistance factors. For example, Figure 18.23 shows the basic structure of tetracycline and two bioactive derivatives. Using tetracycline as a so-called *lead compound*, systematic chemical substitutions can be made at the four R group sites, generating an almost endless series of tetracycline analogs. Using this basic strategy, new β-lactam antibiotics (see Section 18.8), new tetracycline-related compounds (see Section 18.9), and new analogs of vancomycin (Figure 18.13), some up to 100 times as potent as the parent compound, are routinely synthesized and tested.

The application of computer-directed automated robotic chemistry methods to drug discovery has dramat-

ically increased our ability to rapidly generate potential new antimicrobial compounds. The automated robotic methods, referred to as *combinatorial chemistry,* employ systematic modifications of a known antimicrobial product to generate large numbers of new analogs. For instance, using automated combinatorial chemistry methods, and starting with the tetracycline lead compound, five different reagents might be used to introduce substitutions at the four different tetracycline R groups. The substituted sites would yield $5 \times 5 \times 5 \times 5$ (five derivatives at each of four sites), or 725 different tetracycline derivatives from only six different reagents, all in a few hours! These compounds are then assayed for in vitro biological activity on different microorganisms, also using automated techniques. The automated synthesis and screening processes dramatically shorten drug discovery time and increase the number of new candidate drugs by 10 times or more each year. However, toxicity and efficacy studies in both animals and humans often take several years for completion and analysis. Thus, even drugs that are rapidly created and screened *in vitro* may take several years to reach the clinic.

Computerized Drug Design

Truly novel antimicrobial compounds are much more difficult to identify than analogs of existing drugs because new antimicrobial compounds must work at unique sites in metabolism and biosynthesis, or be structurally dissimilar to existing compounds to avoid existing resistance. To find these new compounds, candidate drugs had to be isolated from natural sources and systematically screened for antimicrobial activity. However, recent advances in computer and structural graphics technology now make it possible to design a drug to interact with specific known microbial structures. Drug discovery can now begin at the computer, where new drugs can be rapidly "created" and "tested" for binding and toxicity in the computer environment at relatively low cost (∞ Section 11.5 and 23.4). One of the most dramatic recent successes in computer-directed drug design is the development of saquinavir, a protease inhibitor that is used to slow the growth of the human immunodeficiency virus (HIV) in infected individuals (Figure 18.28 and ∞ Section 23.7). HIV protease cleaves a virus-encoded precursor protein to produce the mature viral core and activate the reverse transcriptase enzyme necessary for replication (∞ Section 8.22). Saquinavir was designed by computer to fit the active site of HIV protease, based on the known three-dimensional structure of the protease-substrate complex; it is a peptide analog which displaces the HIV precursor protein, inhibiting virus maturation and slowing its growth in the human host. A number of other computer-designed protease inhibitors like saquinavir are in use as

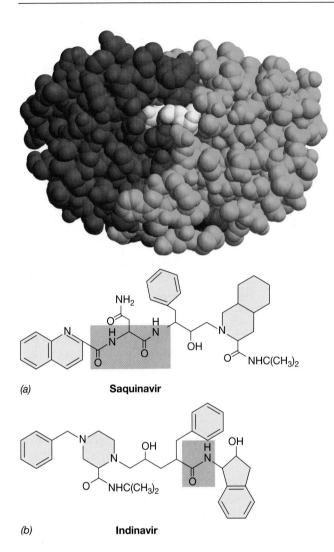

(a)   **Saquinavir**

(b)   **Indinavir**

FIGURE 18.28 Computer-generated antiviral drugs. (a) The HIV protease homodimer. Individual polypeptide chains are shown in green and blue. A peptide (yellow) is bound by the catalytic site. This protease cleaves an HIV precursor protein, a necessary step in virus maturation (∞ Section 8.22). Blocking of the protease site by the peptide shown inhibits precursor processing and HIV maturation. This structure is from the Protein Data Bank (www.rcsb.org/pdb/). (b) These anti-HIV drugs are peptide analogs which were designed by computer to block the active site of HIV protease. The areas highlighted in orange show the regions analogous to peptide bonds. Binding of these compounds by the HIV protease prevents HIV precursor processing and virus maturation. These protease inhibitors are widely used for treatment of HIV infection (see Table 18.5 and ∞ Section 23.7).

chemotherapeutic drugs for the treatment of AIDS (Figure 18.28 and Table 18.5). As this example demonstrates, computer design based on structural and biochemical modeling is a practical method for designing antimicrobial drugs, as well as being rapid and cost effective.

✓ 18.13 Concept Check

New antimicrobial compounds are constantly being developed to deal with drug-resistant organisms and enhance our ability to treat infectious diseases. Analogs of existing drugs are created by combinatorial chemistry techniques. Natural sources are still an important source of novel drugs, but computer drug design is becoming an important new tool for drug discovery.

✓ What advantages do analogs of existing drugs have for treatment of infectious disease?
✓ How can computer drug design save time and money in the search for new drugs?

REVIEW QUESTIONS

1. Why is the decimal reduction time (*D*) important in heat sterilization? How would the presence of bacterial endospores affect *D*?
2. Describe the effects of lethal irradiation at the molecular level.
3. What are the principal advantages of using membrane filters instead of depth filters?
4. Describe the procedure for obtaining the minimum inhibitory concentration (MIC) for a chemical that is bacteriocidal for *Escherichia coli.*
5. Contrast the action of disinfectants and antiseptics. Why can't disinfectants normally be used on living tissue?
6. Growth factor analogs are generally distinguished from antibiotics by a single important criterion. Explain.

7. Most antibiotics are made by only certain groups of organisms. Is this statement true? What groups of organisms make antibiotics?
8. Describe the mechanisms of action that characterizes a β-lactam antibiotic.
9. Distinguish between the mode of action of at least three of the protein synthesis-inhibiting antibiotics.
10. Why do antiviral drugs generally exhibit host toxicity?
11. Define the fungi-specific targets that allow selective toxicity of chemotherapeutic agents in fungi.
12. What is the ultimate origin of bacterial antibiotic resistance genes?
13. Starting with a parent compound, describe *combinatorial chemistry* methods used for the production of new drug analogs.

APPLICATION QUESTIONS

1. Describe in a graph the experimental results you would expect for the decimal reduction time of a very heat-sensitive organism. How would this graph be affected if the vegetative cells were heat-sensitive but the organism formed heat-resistant endospores?

2. What are some potential drawbacks to the use of radiation in food preservation? Do you think these drawbacks could be manifested as health hazards? Why or why not? How would you distinguish between radiation-damaged and radiation-contaminated food?

3. Filtration is an acceptable means of pasteurization for some liquids. Design a filtration system for pasteurization of a heat-sensitive liquid. Why might a filtration system be desirable over a heat pasteurization system?

4. Design an experiment to distinguish between a cidal and a static agent. Can you use the minimum inhibitory concentration (MIC) test in your experiments? Explain.

5. What tests would you perform to decide whether a chemical agent could be used as an antiseptic? As a disinfectant? Some chemicals serve both purposes. Describe the properties of such a chemical and give an example.

6. Although growth factor analogs may inhibit microbial metabolism, only a few of these agents are practically useful. Many potential agents, and some that are in wide use, such as azidothymidine (see Table 18.5), exhibit significant host cell toxicity. Describe a growth factor analog that is effective and has low toxicity for host cells. Why is the toxicity low for the agent you chose? Also describe a growth factor analog that is effective against an infectious disease but exhibits toxicity for host cells. Why might a toxic agent such as AZT still be used in certain situations to treat infectious diseases? What precautions would you take to limit the toxic effects of such a drug, while maximizing the antiviral activity? Explain your answer.

7. We estimated that less than 1% of all known antibiotics have any practical value for either research or clinical use. Indicate why this might be so. Do you think it is important to expand and continue searches for new antibiotics? What alternatives to antibiotic treatments are, or could be, available for the treatment of human disease?

8. Although the β-lactam antibiotics demonstrate clear selective toxicity for Bacteria, many groups of Bacteria are innately resistant to their effects. Without invoking bacterial resistance genes, indicate why gram-negative Bacteria are resistant to the effects of most, but not all, β-lactam antibiotics. Further explain why some β-lactam antibiotics are useful against these organisms.

9. What potential advantages might the aminoglycosides, macrolides, and tetracyclines have over penicillin G for chemotherapy? Explain.

10. List the features of an ideal antiviral drug, especially with regard to selective toxicity. Do such drugs exist? What factors might limit use of such a drug?

11. Like viruses, fungi present special chemotherapeutic problems. Explain the problems inherent in chemotherapy of both groups and explain whether or not you agree with the preceding statement. Give specific examples and suggest at least one group of chemotherapeutic agents that might target both types of infectious agents.

12. Explain the genetic basis of acquired resistance to β-lactam antibiotics in *Staphylococcus aureus*. Design a set of experiments to reverse resistance to the β-lactam antibiotics. Do you think this can be done? Can your experiment be applied "in the field" to promote deselection of antibiotic-resistant organisms? Explain.

13. Design a drug for the inhibition of HIV protease activity that is based on the structure in Figure 18.28. Use a mechanism *different* than competitive inhibition at the enzyme active site.

Microorganisms use a variety of mechanisms to gain access to a suitable host environment. In the illustration a number of the structural proteins involved in *Salmonella* pathogenesis are shown. These are called *virulence factors* because each protein aids in establishing infection. Among these proteins are toxins that directly damage the host cell, enzymes that confer antibiotic resistance, fimbriae and pili that provide mobility and host-cell adhesion to the pathogen, and proteins that sequester important nutrients like iron. Each of these proteins is important in helping *Salmonella* establish and maintain infection. And the presence of a large number of these virulence factors in any single organism can be an effective strategy for establishing and maintaining infection.

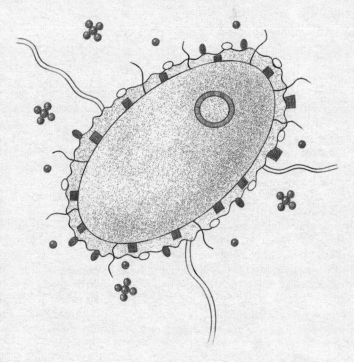

CHAPTER **19** # Host–Parasite Relationships

19.1 Microbial Interactions with Higher Organisms 775
19.2 Normal Flora of the Skin 776
19.3 Normal Flora of the Oral Cavity 777
19.4 Normal Flora of the Gastrointestinal Tract 780
19.5 Normal Flora of Other Body Regions 782
19.6 Entry of the Pathogen into the Host 784
19.7 Colonization and Growth 786
19.8 Exotoxins 788
19.9 Enterotoxins 791
19.10 Endotoxins 793
19.11 Virulence 794
19.12 Nonspecific Host Defenses 796
19.13 Inflammation and Fever 799

WORKING GLOSSARY

Adherence a property of bacteria that allows them to stick to surfaces

Attenuation decrease or loss of virulence

Capsule dense, well-defined polysaccharide or protein layer closely surrounding a cell

Colonization multiplication of a pathogen after it has gained access to host tissues

Dental caries tooth decay resulting from bacterial infection

Dental plaque bacterial cells encased in a matrix of extracellular polymers and salivary products, found on the teeth

Disease injury to the host that impairs host function

Endotoxin the lipopolysaccharide portion of the cell wall of certain gram-negative Bacteria, which acts as a toxin when solubilized

Enterotoxin protein released extracellularly by a microorganism as it grows that produces immediate damage to the small intestine of the host

Exotoxin protein released extracellularly by a microorganism as it grows that produces immediate host cell damage

Glycocalyx a loose network of polymer fibers extending outward from the cell

Host an organism that harbors a parasite

Infection growth of organisms in the host

Inflammation host response to injury or infection, characterized by redness, swelling, heat, and pain

Invasiveness pathogenicity caused by the ability of a pathogen to enter the body and spread

Leukocytes nucleated cells found in the blood (white blood cells)

Lower respiratory tract trachea, bronchi, and lungs

Normal flora microorganisms that are usually found associated with healthy body tissue

Parasite an organism that grows in or on a host

Pathogen a parasite that does harm to a host

Pathogenicity the ability of a parasite to inflict damage on the host

Slime layer a diffuse mat of polymer fibers surrounding cells that appear unattached to a single cell

Toxigenicity pathogenicity caused by toxins produced by a pathogen

Upper respiratory tract the nasopharynx, oral cavity, and throat

Virulence the degree of pathogenicity produced by a pathogen

We have completed our survey of the microbial world and have acquired a great deal of knowledge about its inhabitants. We understand microbial growth requirements and how to control growth using various physical, chemical, and biochemical means. Using this information as a starting point, we will now concentrate on the interactions of microorganisms with the animal body, a process that sometimes results in disease.

We will first outline the basic principles of microbial growth on and in the human body, emphasizing the Bacteria. Next, we will examine some of the major mechanisms used by microorganisms to damage the host. Finally, we will investigate the general resistance mechanisms used by the host to suppress or destroy microbial invaders.

The human body is constantly exposed to microorganisms at several levels. First, through normal everyday activities such as breathing, we are exposed to millions of microorganisms growing in the environment. Next, hundreds of species and billions of individual microorganisms, collectively referred to as the **normal flora,** grow on or in their human hosts and have developed intimate, beneficial, and sometimes essential relationships. Only a few of these microorganisms are potentially harmful. Organisms that live on or in another host organism, imparting no benefits to the host, are called **parasites.** In some cases, the parasite has little or no harmful effect on the host and its presence may be inapparent. However, in other cases, the parasite causes damage to the host. Such harmful organisms are called **pathogens.** The outcome of the host-parasite relationship depends on the *pathogenicity* of the parasite, that is, on the ability of the parasite to inflict damage on the host and on the *resistance* or susceptibility of the host to the parasite.

Individual pathogens differ greatly in their pathogenicity. The quantitative measure of pathogenicity is termed *virulence.* Virulence can be expressed as the cell or viral number that will elicit a pathogenic response in a host within a given time period. Neither the virulence of the pathogen nor the relative resistance of the host are constant factors. The host–parasite interaction is a dynamic, constantly changing relationship between the two parties. The virulence of the pathogen, for example, may differ under the influence of external environmental factors such as the availability of nutrients, temperature, and pH. The resistance of a complex mammalian host may also differ due to factors such as diet, age, sex, or the presence of other parasites. Finally, the host–pathogen interaction itself may influence both organisms, and this relationship may change with time: The virulence of the pathogen and the resistance of the host may change dramatically as circumstances change.

Infection refers to the *growth* of microorganisms in the host. **Disease** is damage or injury to the host that impairs host function. *Infection is not synonymous with disease* because infection does not always cause host injury. Infection simply refers to any situation in which a microorganism is established and growing in a host, whether or not the host is harmed. Thus, even the normal flora produce microbial infections, but do not normally cause disease. However, the normal flora may sometimes cause disease if host resistance is compromised, as happens in several disease states such as cancer and AIDS (∞ Section 23.7). This underscores our previous discussions concerning changes in pathogen virulence due to changes in circumstances and highlights the complexity of differentiating between normal and pathogenic microorganisms.

The ability to cause infectious disease is one of the most dramatic properties of microorganisms. Understanding the physiological and biochemical principles of infectious disease has led to therapeutic and preventive measures that have had profound influences on medicine and human affairs. We begin this chapter by considering the normal flora of the healthy human adult. By understanding the microbial ecology of the human body we will be able to appreciate the competitive forces that govern the success or failure of a potential pathogen in initiating disease. We end this chapter by discussing the innate host resistance mechanisms that limit parasitism and prevent permanent host tissue damage.

19.1

Microbial Interactions with Higher Organisms

Animal bodies provide favorable environments for the growth of many microorganisms. Animals are rich in organic nutrients and growth factors required by chemoorganotrophs, they provide relatively constant conditions of pH, osmotic pressure, and temperature. However, the animal body is not a uniform microbial environment. Each region or organ differs chemically and physically from others and thus provides a selective environment where certain microorganisms are favored over others. The skin, respiratory tract, gastrointestinal tract, and so on, each provide a wide variety of chemical and physical conditions in which different microorganisms can grow selectively. For example, the relatively dry environment of the skin favors the growth of gram-positive organisms such as *Staphylococcus aureus* (∞ Section 23.2); the highly oxygenated environment of the lungs favors the growth of the obligately aerobic *Mycobacterium tuberculosis* (∞ Section 23.3); and the anaerobic environment of the large intestine supports the growth of members of the obligately anaerobic *Clostridium* genus (∞ Section 24.11). Animals also possess a variety of defense mechanisms that

collectively prevent or inhibit microbial invasion and growth. The microorganisms that ultimately colonize the host successfully are those that have developed ways of circumventing these defense mechanisms.

Infections frequently begin at sites in the animal body called *mucous membranes.* Mucous membranes are found throughout the body including the mouth, pharynx, esophagus, and the urinary, respiratory, and gastrointestinal tracts. Mucous membranes consist of single or multiple layers of *epithelial cells,* tightly packed cells that exist in direct contact with the external environment. Mucous membranes are frequently coated with a protective layer of mucus, primarily glycoproteins, which serves to protect epithelial cells. When bacteria contact host tissues at mucous membranes, they may associate either loosely or firmly. If they associate loosely with the mucosal surface, they are usually swept away by physical processes, but they may also attach specifically to the epithelial surface as a result of specific cell–cell recognition between pathogen and host. From there, actual tissue infection may follow. When this occurs, the mucosal barrier is breached, allowing the pathogen to invade deeper tissues (Figure 19.1).

Microorganisms are almost always found in those regions of the body exposed to the environment, such as the skin, oral cavity, respiratory tract, intestinal tract, and genitourinary tract. They are not normally found in the organs, or in the blood, lymph, or nervous systems of the body; the growth of microorganisms in these usually sterile environments indicates disease.

Table 19.1 shows some of the major types of microorganisms normally found in association with body surfaces. The most visible exposed body surface, the skin (2 m^2), has a number of normal microbial inhabitants. However, mucosal surfaces have an even larger variety of associated microorganisms. This is due in part to the sheltered, moist environment of the various mucosal surfaces and also to the huge overall surface area of the mucosae (400 m^2). For example, the specialized function of a mucosal organ such as the small intestine requires a large specialized surface area for nutrient transport, and this surface also serves as a site for mi-

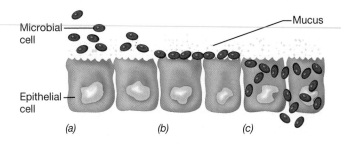

FIGURE 19.1   Bacterial interactions with mucous membranes. (a) Loose association. (b) Adhesion. (c) Invasion into submucosal cells.

TABLE 19.1	Representative genera of microorganisms in the normal flora of humans
Anatomical site	**Organism**[a]
Skin	*Staphylococcus, Corynebacterium, Acinetobacter, Pityrosporum* (yeast), *Propionibacterium, Micrococcus*
Mouth	*Streptococcus, Lactobacillus, Fusobacterium, Veillonella, Corynebacterium, Neisseria, Actinomyces*
Respiratory tract	*Streptococcus, Staphylococcus, Corynebacterium, Neisseria*
Gastrointestinal tract	*Lactobacillus, Streptococcus, Bacteroides, Bifidobacterium, Eubacterium, Peptococcus, Peptostreptococcus, Ruminococcus, Clostridium, Escherichia, Klebsiella, Proteus, Enterococcus, Staphylococcus*
Urogenital tract	*Escherichia, Klebsiella, Proteus, Neisseria, Lactobacillus* (vagina of mature females), *Corynebacterium, Staphylococcus, Candida, Provotella, Clostridium, Peptostreptococcus*

a This list is not meant to be exhaustive. New organisms are constantly being added to it, and not all of these organisms are found in every individual. Most of these organisms can contribute to disease processes under certain conditions.

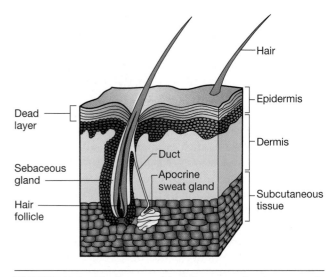

FIGURE 19.2 The human skin. Microorganisms are associated primarily with the sweat ducts and the hair follicles.

crobial growth. We shall now examine these normal microbial interactions in greater detail.

✓ 19.1 Concept Check

Microorganisms that grow in or on other organisms, imparting no benefit, are *parasites.* Infection is the process by which a parasite grows in its host. If the parasite causes harm, it is a pathogen. The ability to cause disease is influenced by the complexity of host-parasite interactions. Pathogens cause disease, but disease is limited by a variety of host defense mechanisms. Animal bodies are favorable environments for the growth of many microorganisms, including pathogens. Initial pathogen colonization is on the surface of a host, often on the mucous membranes.

- ✓ Are all *parasites* also *pathogens?* Are all *pathogens* also *parasites?*
- ✓ Distinguish between infection and disease.
- ✓ Why might one area of the body be more suitable for microbial growth than another?

19.2

Normal Flora of the Skin

An average human adult has about 2 m² of skin surface that can vary greatly in chemical composition and moisture content. Figure 19.2 shows the anatomy of the skin and regions in which bacteria may live. The skin sur-

face (epidermis) is not a favorable place for microbial growth, as it is subject to periodic drying.

Most skin microorganisms are associated directly or indirectly with the sweat glands, of which there are several kinds. The **eccrine glands** are not associated with hair follicles and are rather unevenly distributed over the body, with denser concentrations on the palms, finger pads, and soles of the feet. They are the main glands responsible for perspiration. Eccrine glands seem to be relatively devoid of microorganisms, perhaps because of the extensive flow of fluid. The **apocrine glands** are more restricted in their distribution, being confined mainly to the underarm and genital regions, the nipples, and the umbilicus. They are inactive in childhood and become fully functional only at puberty. Bacterial populations on the surface of the skin in these warm, humid places are relatively high, in contrast to the situation on the smooth, dry surface skin. *Underarm odor* develops as a result of bacterial activity on the apocrine secretions; aseptically collected apocrine secretion is odorless but develops odor on inoculation with bacteria. Each hair follicle is associated with a **sebaceous gland,** which secretes a lubricant fluid. Hair follicles provide an attractive habitat for microorganisms; a variety of aerobic and anaerobic bacteria and fungi inhabit these regions, mostly within the area just below the surface of the skin. The secretions of the skin glands are rich in microbial nutrients. Urea, amino acids, salts, lactic acid, and lipids are present in considerable amounts. The pH of human secretions is almost always acidic, the usual range being between pH 4 and 6.

The microorganisms of the normal flora of the skin are either *transient* or *resident* populations of microorganisms. The skin as an external organ is continually being inoculated with transient microorganisms, virtually all of which are unable to multiply and usually die.

Resident organisms are able to multiply, not merely survive, on the skin. The normal flora of the skin consists primarily of gram-positive Bacteria restricted to a few groups (Table 19.1). These include several species of *Staphylococcus* and a variety of both aerobic and anaerobic corynebacteria. Of the latter, *Propionibacterium acnes* is ordinarily a harmless resident but it can incite or contribute to the condition known as *acne*. Gram-negative Bacteria are almost always minor constituents of the normal flora because such intestinal organisms as *Escherichia coli* are being continually inoculated onto the surface of the skin by fecal contamination. *Acinetobacter* is an exception and is one of the few gram-negative Bacteria commonly found on skin. The lack of colonization of gram-negative Bacteria on the skin is probably due to their inability to compete with gram-positive organisms that are better adapted to the dry conditions of the skin; if the latter are eliminated by antibiotic treatment, the gram-negative Bacteria can flourish. Yeasts are uncommon on the skin surface, but the lipophilic yeast *Pityrosporum ovalis* is occasionally found on the scalp.

Although the resident microflora remains more-or-less constant, various factors can affect the nature and extent of the normal flora: (1) The weather may cause an increase in temperature and humidity, which increases the density of the skin microflora. (2) Age has an effect, and young children have a more varied microflora and carry more of the potentially pathogenic gram-negative Bacteria than adults. (3) Personal hygiene influences the resident microflora, and unclean individuals usually have higher microbial population densities on their skin. Organisms that cannot survive on the skin generally succumb from either the skin's low moisture content or low pH (due to organic acid content).

✓ 19.2 Concept Check

The skin is a dry, acidic environment that is not conducive to the growth of most microorganisms. However, moist areas, especially around sweat glands, are colonized by gram-positive Bacteria and other members of the skin normal flora.

✓ How large is the surface area of the skin?
✓ Describe the properties of microorganisms that grow well on the skin.

19.3

Normal Flora of the Oral Cavity

The oral cavity contains one of the more complex and heterogeneous microbial habitats in the body. This cavity includes the teeth and tongue. Although saliva is the most pervasive source of microbial nutrients in the oral cavity, it is not an especially good microbial culture medium. Saliva contains about 0.5% dissolved solids, about half of which are inorganic (mostly chloride, bi-

carbonate, phosphate, sodium, calcium, potassium, and trace elements); the predominant organic constituents of saliva are salivary enzymes, mucoproteins, and some serum proteins. Small amounts of carbohydrates, urea, ammonia, amino acids, and vitamins are also present. A number of antibacterial substances have been identified in saliva, of which the most important are the enzymes *lysozyme* and *lactoperoxidase*. Lysozyme is an enzyme that cleaves glycosidic linkages in peptidoglycan in the bacterial cell wall, leading to weakening of the wall and cell lysis (∞ Section 3.7). Lactoperoxidase, an enzyme present in both milk and saliva, kills bacteria by a reaction involving chloride ions and H_2O_2, in which singlet oxygen is probably generated (∞ Sections 5.12 and 20.3). The pH of saliva is controlled primarily by a bicarbonate buffering system ($H_2CO_3 \rightleftharpoons H^+ + HCO_3^-$) and varies between 5.7 and 7.0, with a mean pH near 6.7. The composition of saliva varies, and even within the same individual, variations due to physiological factors such as diet and stress are seen. Despite the activity of antibacterial substances, the presence of food particles and epithelial debris makes the oral cavity a very favorable microbial habitat.

The Teeth and Dental Plaque

The tooth consists of a mineral matrix of calcium phosphate crystals (enamel), within which the living tissue of the tooth (dentin and pulp) is present (Figure 19.3). Bacteria found in the mouth during the first year of life (when teeth are absent) are predominantly aerotolerant

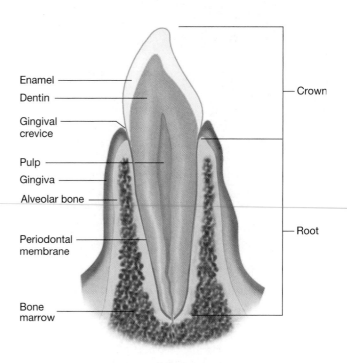

Enamel
Dentin
Gingival crevice
Pulp
Gingiva
Alveolar bone
Periodontal membrane
Bone marrow
Crown
Root

FIGURE 19.3 Section through a tooth showing the surrounding tissues that anchor the tooth in the gum.

anaerobes such as streptococci and lactobacilli, but a variety of other bacteria, including some aerobes, occur in small numbers. When the teeth appear, there is a pronounced shift in the balance of the microflora toward anaerobes, and a variety of bacteria develop that are specifically adapted for growth on surfaces and in crevices of the teeth.

The bacterial colonization of smooth tooth surfaces occurs as a result of firm attachment of single bacterial cells, followed by growth in the form of microcolonies. Beginning with a freshly cleaned tooth surface, the first event is the formation of a thin organic film several micrometers thick as a result of the attachment of acidic glycoproteins from the saliva. This film provides a firmer attachment site for the colonization and growth of bacterial microcolonies (Figure 19.4). The colonization of this glycoprotein film is highly specific, and only a few species of *Streptococcus* (primarily *S. sanguis*, *S. sobrinus*, *S. mutans*, and *S. mitis*) are involved. As a result

of extensive growth of these organisms, a thick bacterial zone, called **plaque**, is formed (Figures 19.5 and 19.6). If plaque continues to form, filamentous bacteria, usually *Fusobacterium* species, begin to grow. The filamentous bacteria are embedded in the matrix formed by the streptococci, and extend perpendicular to the tooth surface, making an ever-thicker bacterial layer. Associated with the filamentous bacteria, in addition to the streptococci, are spirochetes such as *Borrelia* species (∞ Section 13.30), gram-positive rods, and gram-negative cocci. In heavy plaque, filamentous organisms such as anaerobic *Actinomyces* species may predominate.

The anaerobic nature of the flora may seem surprising, considering that the mouth has good accessibility to oxygen. However, anoxia probably develops through the action of facultative bacteria growing aerobically on organic materials on the tooth. This plaque buildup produces a dense matrix that decreases oxygen diffusion onto the tooth surface, forming an anoxic microenvironment. Thus, the microbial populations of the dental plaque exist in a microenvironment partly of their own making and maintain themselves in the face of wide variations in the macroenvironment of the oral cavity.

Dental Caries

As dental plaque accumulates and acid products are formed, **dental caries** (tooth decay) results. Thus, tooth decay is an infectious disease caused by microorganisms. The role of the oral flora in tooth decay has now

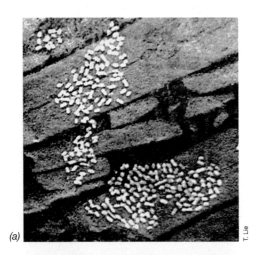

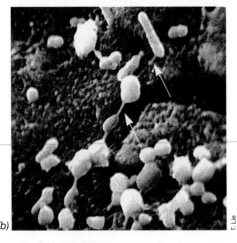

FIGURE 19.4 (a) Bacterial microcolonies growing on a model tooth surface inserted into the mouth for 6 hr. (b) Higher magnification of the preparation in (a). Note the diverse morphology of the organisms present and the slime layer (arrows) holding the organisms together.

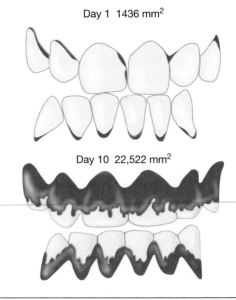

Day 1 1436 mm²

Day 10 22,522 mm²

FIGURE 19.5 Distribution of dental plaque, as revealed by use of a disclosing agent, on brushed (top) and unbrushed (bottom) teeth. The numbers give the total area of dental plaque. The red areas indicate plaque. Note that the plaque builds preferentially near the gum line, first occurring directly adjacent to the mucous membranes of the gingiva.

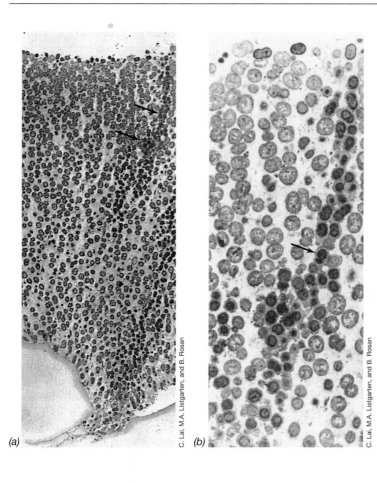

C. Lai, M.A. Listgarten, and B. Rosan

(a)

(b)

C. Lai, M.A. Listgarten, and B. Rosan

FIGURE 19.6 Electron micrographs of thin sections of dental plaque. Bottom is the base of the plaque; top is the portion exposed to the oral cavity. (a) Low power electron micrograph. Organisms are predominantly streptococci. The species *Streptococcus sobrinus* has been labeled by an antibody-microchemical technique, and these cells appear darker than the rest. They are seen as two distinct chains (arrows). The total thickness of the plaque layer shown is about 50 μm. (b) Higher power electron micrograph showing the region with *S. sobrinus* cells (dark, arrow). Note the extensive glycocalyx (see Section 19.6) surrounding the *S. sobrinus* cells.

been well established through studies on germ-free animals. The smooth surfaces of the teeth that are exposed to frequent cleaning by the tongue, cheek, saliva, or toothbrush or to the abrasive action of food mastication are relatively resistant to dental caries. The tooth surfaces in crevices, where food particles can be retained, are the sites where tooth decay usually occurs. Thus, the shape of the teeth is important. For instance, dogs are highly resistant to tooth decay because the shape of their teeth does not favor retention of food. Diets high in sugars are especially cariogenic because lactic acid bacteria ferment the sugars to lactic acid, which causes decalcification of the enamel (see Figure 19.3) of the tooth. Once breakdown of the hard tissue (tooth enamel) has begun, proteolysis of the matrix of the tooth enamel occurs through the action of proteolytic enzymes released by bacteria. Microorganisms penetrate further into the decomposing matrix, but the later stages of the process may be exceedingly slow and are often highly complex. The structure of the calcified tissue also plays an important role in the extent of dental caries. Incorporation of fluoride into the calcium phosphate crystal matrix makes the matrix more resistant to decalcification by acid. Thus, fluorides are used in drinking water and dentifrices to aid in controlling tooth decay.

Two organisms that have been implicated in dental caries are *Streptococcus sobrinus* and *Streptococcus mutans*, both lactic acid–producing bacteria. *S. sobrinus* is able to colonize smooth tooth surfaces because of its specific affinity for salivary glycoproteins (Figure 19.6), and is probably the primary organism involved in decay of smooth surfaces. *S. mutans* is found predominantly in crevices and small fissures, and its ability to attach to tooth surfaces is the result of its ability to produce a dextran polysaccharide that is strongly adhesive (Figure 19.7). *S. mutans* produces dextran only when sucrose is present, by means of the enzyme *dextransucrase*:

$$n \text{ Sucrose} \xrightarrow{\text{Dextran}} (\text{glucose})_n + n \text{ fructose}$$

Sucrose is the common table sugar ubiquitously present in the diet of most individuals in more-developed countries and is highly cariogenic because it is a substrate for dextransucrase.

Susceptibility to tooth decay varies greatly among individuals and is affected by inherent traits in the individual as well as by diet and other extraneous factors. Studies of the distribution of oral streptococci have shown a direct correlation between the presence of *S. mutans*, and to a lesser degree *S. sobrinus*, in humans

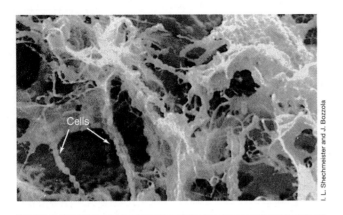

I. L. Shechmeister and J. Bozzola

FIGURE 19.7   Scanning electron micrograph of the cariogenic bacterium *Streptococcus mutans*. The sticky dextran material holds the cells together as filaments. Individual cells are about 1 μm in diameter.

and the extent of dental caries. In the United States and Western Europe, for example, 80–90% of all people have their teeth colonized by *S. mutans*, and dental caries is a nearly universal phenomenon. By contrast, dental caries do not occur in Tanzanian children, presumably because sucrose is almost completely absent in their diets, and *S. mutans* is absent from the plaque of these individuals.

Although tooth decay is an infectious disease, we tend to place it in a different category from other infectious diseases. However, microorganisms in the mouth can also cause other infections. The areas along the periodontal membrane at or below the gingival crevice (*periodontal pockets*) (Figure 19.3) can become infected with a variety of microorganisms, causing inflammation (gingivitis) and more serious tissue and bone-destroying periodontal disease. Some of the genera involved are the fusiform anaerobic bacterium *Capnocytophaga* and the aerobic bacterium *Rothia*.

✓ 19.3 Concept Check

Bacteria can grow on tooth surfaces in thick layers called plaque. The microorganisms in plaque produce adherent substances that encourage further colonization. Acid-producing organisms in plaque damage tooth surfaces, and dental caries form. A variety of microorganisms can contribute to caries and periodontal disease.

- ✓ How do anaerobic microorganisms become established in the mouth?
- ✓ Is dental caries an infectious disease? Give at least one reason for your answer.

19.4

Normal Flora of the Gastrointestinal Tract

The general anatomy of the gastrointestinal tract is shown in Figure 19.8. The human gastrointestinal tract, the site of food digestion, consists of the stomach, small intestine, and large intestine. The pH of stomach fluids is low, about pH 2. The stomach can thus be viewed as a chemical barrier against entry of foreign bacteria into the intestinal tract. The bacterial count of the stomach contents is generally very low, and so the human stomach is devoid of any significant normal flora. However, organisms such as *Helicobacter pylori* can colonize the stomach wall, resulting in ulcers (∞ Section 24.12).

The Intestinal Tract

The intestinal tract consists of the *small intestine* and the *large intestine,* each of which is further subdivided into different anatomical structures.

The **small intestine** is separated into two parts, the *duodenum* and the *ileum,* with the *jejunum* connecting them. The duodenum, adjacent to the stomach, is fairly acidic and resembles the stomach in its lack of a microbial flora. From the duodenum to the ileum the pH gradually becomes less acidic and bacterial numbers increase. In the lower ileum, bacteria are found in the intestinal cavity (the lumen), mixed with digestive material. Cell numbers of 10^5–10^7 per gram are common.

The ileum empties into the *cecum,* the first portion of the **large intestine.** The *colon* makes up the rest of the large intestine. In the colon, bacteria are present in enormous numbers, and this region can be viewed as a specialized fermentation vessel. Many bacteria live within the lumen itself, probably using as nutrients some products of the digestion of food (Table 19.1). Facultative aerobes, such as *Escherichia coli*, are present but in smaller numbers than other bacteria; total counts of facultative aerobes are generally less than 10^7 per gram of intestinal contents. The activities of facultative aerobes consume any oxygen present, making the environment of the large intestine strictly anaerobic and favorable for the profuse growth of obligate anaerobes. Many of these anaerobes are long, thin, gram-negative rods with tapering ends (called *fusiform*) and are attached end-on to small indentations in the intestinal wall (Figure 19.9). Other obligate anaerobes include species of *Clostridium* and *Bacteroides*. The total number of obligate anaerobes is enormous. Counts of 10^{10}–10^{11} cells/g of intestinal contents are not uncommon, with various species of *Bacteroides* accounting for the majority of intestinal obligate anaerobes. In addition, *Enterococcus faecalis* is almost always present in significant numbers.

The normal flora of the gastrointestinal tract varies among species. For example, in guinea pigs lactobacilli make up 80% of the intestinal flora, whereas the same organisms are only minor components of human gastrointestinal flora. The gut flora in humans can also vary qualitatively depending on the diet. Persons who consume a considerable amount of meat show higher numbers of *Bacteroides* and lower numbers of coliforms and

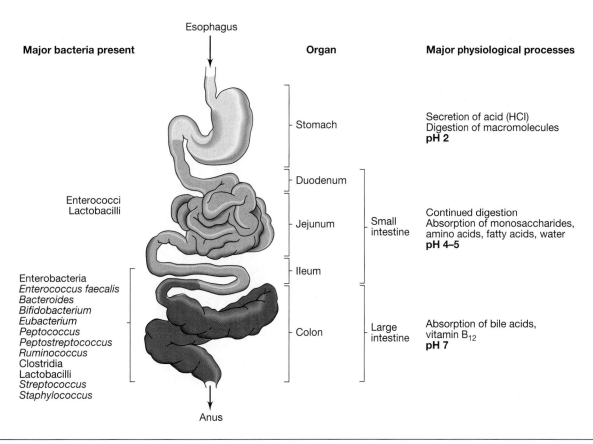

Major bacteria present

Enterococci
Lactobacilli

Enterobacteria
Enterococcus faecalis
Bacteroides
Bifidobacterium
Eubacterium
Peptococcus
Peptostreptococcus
Ruminococcus
Clostridia
Lactobacilli
Streptococcus
Staphylococcus

Esophagus

Organ

Stomach

Duodenum

Jejunum   Small intestine

Ileum

Colon   Large intestine

Anus

Major physiological processes

Secretion of acid (HCl)
Digestion of macromolecules
pH 2

Continued digestion
Absorption of monosaccharides,
amino acids, fatty acids, water
pH 4–5

Absorption of bile acids,
vitamin B_{12}
pH 7

FIGURE 19.8   The human gastrointestinal tract showing functions and the distribution of nonpathenogenic microorganisms usually found in normal individuals.

lactic acid bacteria than those on a vegetable diet. An overview of microorganisms that inhabit the gastrointestinal tract is given in Figure 19.8.

The intestinal flora has a profound influence on the animal, carrying out a wide variety of metabolic reactions (Table 19.2). Not all microorganisms carry out these reactions, and changes in the intestinal flora due to diet or disease may thus affect the animal. Of special note in Table 19.2 are the roles of the intestinal flora in steroid metabolism. Steroids are produced in the liver

and secreted into the intestine via the gallbladder as bile acids. Their role is to promote emulsification of fats in the diet so the fats can be effectively digested. Intestinal microorganisms cause a variety of transformations of these bile acids so the materials excreted in the feces are quite different from the original bile acids. Other products of microbial fermentation are the odor-producing substances listed in Table 19.2. Composition of the intestinal microflora as well as diet influences the amount of gas and the amount of odoriferous materials present.

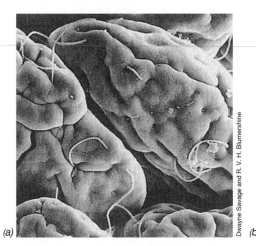

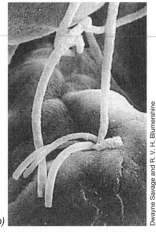

(a)   (b)

Dwayne Savage and R. V. H. Blumershine

FIGURE 19.9   Scanning electron micrographs of the microbial community on the surface of the columnar epithelium in the mouse ileum. (a) An overview at low magnification. Note the long, filamentous *fusiform* bacteria lying on the surface. (b) Higher magnification, showing several filaments attached at a single depression. Note that the attachment is at the end of the filaments only. Individual cells are 10–15 μm long.

TABLE 19.2	Biochemical/metabolic contributions of intestinal microorganisms
Vitamin synthesis	Product: thiamine, riboflavin, pyridoxine, B$_{12}$, K
Gas production	Product: CO$_2$, CH$_4$, H$_2$
Odor production	Product: H$_2$S, NH$_3$, amines, indole, (bile acids)skatole, butyric acid
Organic acid production	Product: acetic, propionic, butyric acids
Glycosidase reactions	Enzyme: β-glucuronidase, β-galactosidase, β-glucosidase, α-glucosidase, α-galactosidase
Steroid metabolism (bile acids)	Process: esterification, dehydroxylation, oxidation, reduction, inversion

During the passage of food through the gastrointestinal tract, water is withdrawn from the digested material, and it gradually becomes more concentrated and is converted to feces. Bacteria make up about one-third of the weight of fecal matter. Organisms living in the lumen of the large intestine are continuously being displaced downward by the flow of material, and if bacterial numbers are to be maintained, those bacteria that are lost must be replaced by new growth. Thus, the large intestine resembles a chemostat (∞ Section 5.5). The time needed for passage of material through the complete gastrointestinal tract is about 24 hr in humans; the growth rate of bacteria in the lumen is one to two doublings per day.

When an antibiotic is given orally, it may inhibit the growth of the normal flora as well as pathogens; continued movement of the intestinal contents then leads to loss of the preexisting bacteria and virtual sterilization of the intestinal tract. In the absence of the normal flora, opportunistic microorganisms such as antibiotic-resistant *Staphylococcus*, *Proteus*, or the yeast *Candida albicans* may become established. These organisms usually do not grow in the intestinal tract because they cannot compete with the normal flora. Occasionally, establishment of these opportunistic pathogens can lead to a harmful alteration in digestive function or even to disease. After antibiotic therapy stops, the normal flora eventually becomes reestablished, but often only after a considerable period.

Intestinal Gas

The gas produced within the intestines, called *flatus*, is the result of the action of fermentative and methanogenic microorganisms. Some foods can be metabolized by fermentative bacteria in the intestines, resulting in the production of hydrogen (H$_2$) and carbon dioxide (CO$_2$). Methanogens (∞ Section 14.3) are found in the intestines of over one-third of normal adults. The methanogens convert to H$_2$ and CO$_2$ produced by the other intestinal microorganisms to methane (CH$_4$).

Normal adults expel several hundred milliliters of gas, of which about half is N$_2$ from swallowed air, from the intestines every day.

✓ 19.4 Concept Check

The stomach is very acidic and is a barrier to most microbial growth. The intestinal tract is slightly acid to neutral and supports a diverse population of microorganisms in a variety of nutritional and environmental conditions.

✓ Why might the small intestine be more suitable for growth of facultative aerobes than the large intestine?

✓ Why might the small intestine be more suitable for growth of acid-tolerant microorganisms than the large intestine?

19.5

Normal Flora of Other Body Regions

Each individual mucous membrane supports the growth of a specialized group of microorganisms. These organisms are part of the normal local environment and are characteristic of healthy tissue. In many cases, potentially pathogenic microorganisms cannot colonize mucous membranes because of the presence of the normal resident population of microorganisms. In this section, we discuss two such mucosal environments and their resident microorganisms.

Respiratory Tract

The anatomy of the respiratory tract is shown in Figure 19.10 (∞ Figure 23.2). In the **upper respiratory tract** (nasopharynx, oral cavity, and throat) microorganisms live primarily in areas bathed with the secre-

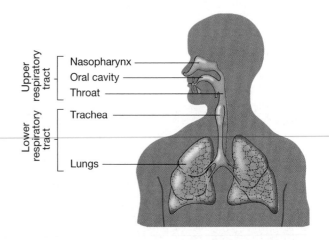

FIGURE 19.10 The respiratory tract. The upper respiratory tract is populated with a large variety and number of microorganisms, but the lower respiratory tract has relatively few microbial inhabitants, unless there is an ongoing active infection.

tions of the mucous membranes. Bacteria enter the upper respiratory tract from the air during breathing, but most of them are trapped in the nasal passages and expelled again with the nasal secretions. The resident organisms most commonly found are staphylococci, streptococci, diphtheroid bacilli, and gram-negative cocci. Potentially harmful bacteria, such as *Staphylococcus aureus*, *Streptococcus pneumoniae*, *Streptococcus pyogenes*, and *Corynebacterium diphtheriae* are often part of the normal flora of the nasopharynx of healthy individuals (Table 19.1). These individuals are *carriers* of the pathogens but do not normally acquire disease, presumably because the other resident microorganisms compete successfully for resources and limit pathogen growth. The local immune system (◌◌ Section 20.5) is particularly active at mucosal surfaces and may also inhibit the growth of pathogens.

The **lower respiratory tract** (trachea, bronchi, and lungs) is essentially sterile, in spite of the large numbers of organisms potentially able to reach this region during breathing. Dust particles, which are fairly large, settle out in the upper respiratory tract. As the air passes into the lower respiratory tract, its rate of flow decreases markedly, and organisms settle onto the walls of the passages.

The walls of the entire respiratory tract are lined with ciliated epithelium, and the cilia, beating upward, push bacteria and other particulate matter toward the upper respiratory tract where they are then expelled in the saliva and nasal secretions. Only particles smaller than about 10 μm in diameter are able to reach the lungs.

Urogenital Tract

The main anatomical features of the male and female urogenital tracts are shown in Figure 19.11*a*. In both male and female, the bladder itself is usually sterile, but the epithelial cells lining the urethra are colonized by facultatively aerobic gram-negative rods and cocci (Table 19.1). These organisms, including *Escherichia coli* and *Proteus mirabilis* and others, can occasionally become *opportunistic pathogens*. These organisms are normally present in the body or in the local environment, but they are not pathogenic under normal circumstances. Changes in the body, such as local pH changes or decreases in immune system functions (◌◌ Section 20.15), allow the organisms to multiply and become pathogenic. Such organisms frequently cause urinary tract infections, especially in women.

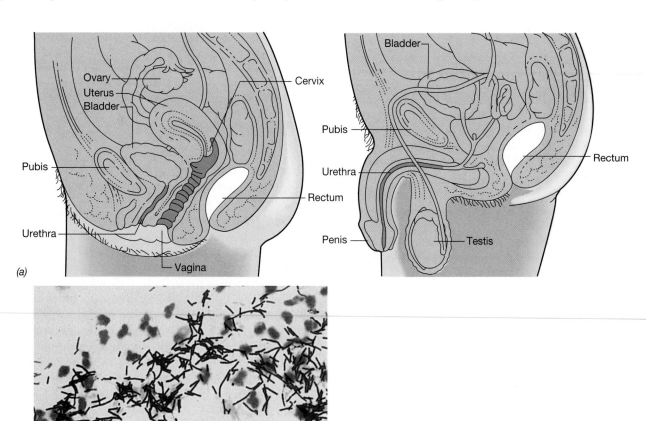

John Durham

FIGURE 19.11 (a) The genitourinary tracts of the human female and male, showing regions (color) where microorganisms often grow. (b) Gram stain of *Lactobacillus acidophilus*, the predominant organism in the vagina of women. Individual rods are 3–4 μm long.

The vagina of the adult female generally is weakly acidic and contains significant amounts of the polysaccharide glycogen. *Lactobacillus acidophilus,* which is a resident organism in the vagina, ferments glycogen to produce lactic acid and lower the pH of the vagina (Figure 19.11) (∞ Section 13.18). Other organisms—yeasts (*Torulopsis* and *Candida* species), streptococci, and *E. coli*—may also be present. Before puberty, the female vagina is alkaline and does not produce glycogen, *L. acidophilus* is absent, and the flora consists predominantly of staphylococci, streptococci, diphtheroids, and *E. coli*. After menopause, glycogen disappears, the pH rises, and the flora again resembles that found before puberty.

✓ 19.5 Concept Check

The presence of a population of normal nonpathogenic microorganisms in the respiratory and urogenital tracts is essential for normal organ function and often prevents the colonization of pathogens.

- ✓ Pathogens are sometimes found in the normal flora of the upper respiratory tract. Why do they not cause disease in some cases?
- ✓ Why is *Lactobacillus* found in the urogenital tract of normal adult women?

19.6

Entry of the Pathogen into the Host

We now consider mechanisms used by pathogens to alter host function and cause disease. **Pathogenesis,** the ability of microorganisms to initiate disease, includes *entry, colonization,* and *growth* of microorganisms in the host, resulting in changes in host function that damage the host. As we have discussed previously, many organisms grow on or in the host, and most have little or no *obvious* effect on the host. However, disease-producing microorganisms elicit host changes, and in the process they use several strategies to establish *virulence,* the relative ability of a pathogen to cause disease (Figure 19.12). We start our discussion by considering the factors responsible for entry of a pathogen into a host.

A pathogen must usually gain access to host tissues and multiply before damage can be done. In most cases, this requires that the organism penetrate the skin, mucous membranes, or intestinal epithelium, surfaces that normally act as microbial barriers. Passage through the skin into subcutaneous layers almost always occurs through wounds; in rare instances pathogens penetrate through the unbroken skin.

Specific Adherence

Most microbial infections begin on the mucous membranes of the respiratory, alimentary, or genitourinary tract. There is considerable evidence that bacteria or

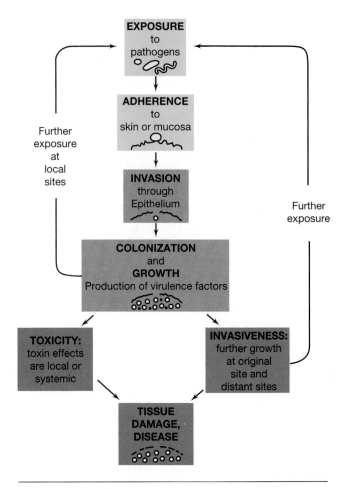

FIGURE 19.12 Microorganisms and pathogenesis. The presence of microorganisms on the host does not always lead to disease.

viruses able to initiate infection can adhere specifically to epithelial cells (Figure 19.13). The evidence for specificity is of several types. First, there is *tissue specificity.* An infecting microorganism does not adhere to all epithelial cells equally but selectively adheres to cells in the particular region of the body where it normally gains entrance. For example, *Neisseria gonorrhoeae,* the causative agent of the sexually transmitted disease gonorrhea, adheres much more strongly to urogenital epithelia than to other tissues. Second, there is *host specificity.* A bacterial strain that normally infects humans adheres more strongly to the appropriate human epithelial cells than to similar cells in another animal (for example, the rat), whereas a strain that specifically colonizes the rat adheres more firmly to rat cells than to human cells.

Many bacteria possess surface macromolecules that bind to receptors on the surface of host tissues. Some of these macromolecules are not covalently attached to the bacteria. They are usually polysaccharides synthesized and secreted by the bacteria (∞ Section 3.13). A polymer coat consisting of a dense, well-defined layer close-

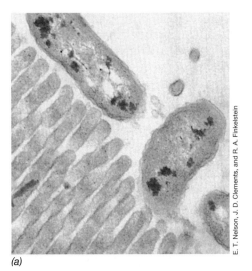

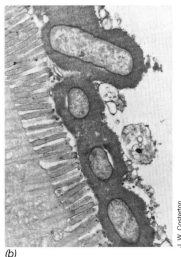

(a)   (b)

E. T. Nelson, J. D. Clements, and R. A. Finkelstein

J. W. Costerton

FIGURE 19.13 Adherence of pathogens to animal tissues. (a) Transmission electron micrograph of a thin section of *Vibrio cholerae* adhering to the brush border of rabbit villi. Note the absence of the glycocalyx. (b) Enteropathogenic *Escherichia coli* in a fatal model infection in the newborn calf. The bacterial cells are attached to the brush border of calf villi via an extensive glycocalyx. The rods are about 0.5 μm in diameter.

ly surrounding the cell is known as a **capsule** (∞ Figure 23.5). A loose network of polymer fibers extending outward from a cell is known as a **glycocalyx** (Figure 19.13), while a diffuse mass of polymer fibers, seemingly unattached to any single cell, is called a **slime layer** (see Figure 19.4*b*). These structures may be important for adherence not only to host tissues, but also between other bacteria. In addition, these layers may protect the bacteria from host defense mechanisms such as phagocytosis by macrophages and other cells (∞ Section 20.3).

Fimbriae and **pili** (∞ Section 3.13) are bacterial cell surface protein structures (Figure 19.14) that may also function in the attachment process. For instance, the pili of *Neisseria gonorrhoeae* play a key role in the attachment of this organism to urogenital epithelium, and fimbriated strains of *Escherichia coli* (Figure 19.14) are much more frequent causes of urinary tract infections than strains lacking fimbriae. Among the best-characterized fimbriae are the so-called *type I fimbriae* of enteric bacteria (*Escherichia, Klebsiella, Salmonella, Shigella*). Type I fimbriae are uniformly distributed on the surface of cells. Pili are generally longer than fimbriae, with fewer pili found on the cell surface. Both pili and fimbriae function by binding host cell glycoproteins, initiating the attachment event.

Evidence of the specific interaction between mucosal epithelium and the pathogen comes from studies of diarrhea caused by *Escherichia coli*. Most strains of *E. coli* are nonpathogenic and are part of the normal flora of the *large* intestine. A few strains (only a handful of the 160 different *E. coli* serotypes known) are *enteropathogenic*, possessing the ability to colonize the *small* intestine and initiate diarrhea. Such strains possess specific surface structures, the **colonization factor antigens (CFA)**, which are fimbrial proteins involved in specific attachment to intestinal mucosa. Thus, two kinds of *E. coli* can be recognized: pathogenic strains, which are able to adhere to the mucosal surface of the small intestine and cause disease symptoms, and "normal" *E. coli* strains, which are unable to adhere to the small intestine or produce enterotoxin (see Section 19.9). Nonpathogenic strains of *E. coli* are always present in the large intestine (cecum and colon) as part of the normal flora, and several different strains are present at the same time. On average, each strain remains in the intestine for several weeks, and so the normal *E. coli* population is constantly changing. A summary of major factors in microbial adherence is given in Table 19.3.

Invasion

A few microorganisms are pathogenic solely because of the toxins they produce. These organisms do not need to gain access to host tissues, and we will discuss them separately (see Sections 19.8 and 19.9). However, most pathogens penetrate the epithelium to initiate pathogenicity, a process called *invasion*. At the point of entry, usually at small breaks or lesions in the skin or in mu-

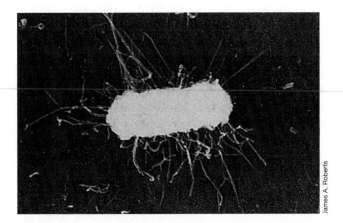

James A. Roberts

FIGURE 19.14 Shadow-cast electron micrograph of the bacterium *Escherichia coli*, showing type P fimbriae. Type P fimbriae resemble type I fimbriae but are somewhat longer. The cell shown is about 0.5 μm wide.

TABLE 19.3	Major adherence factors used to facilitate attachment of microbial pathogens to host tissues[a]
Factor	**Example**
Glycocalyx/capsule/ slime layer	Enterotoxic *Escherichia coli*—glycocalyx promotes adherence to the brush border of intestinal villi
	Streptococcus mutans—dextran glycocalyx promotes binding to tooth surfaces
Adherence proteins	*Streptococcus pyogenes*—M protein on the cell binds to receptors on respiratory mucosa
	Neisseria gonorrhoeae—Opa protein on the cell binds to receptors on urogenital epithelium
Lipoteichoic acid	*Streptococcus pyogenes*—facilitates binding to respiratory mucosal receptor (along with M protein)
Fimbriae (pili)	*Neisseria gonorrhoeae*—pili facilitate binding to urogenital epithelium
	Salmonella species—type I fimbriae facilitate binding to epithelium of small intestine
	Enterotoxic *Escherichia coli* (ETEC)— colonization factor antigens (CFAs), which are fimbrial, facilitate binding to epithelium of small intestine

[a] Most receptor sites on host tissues are glycoproteins or complex lipids such as gangliosides or globosides.

cosal surfaces, growth is often established in the submucosa. Growth may also be established on intact mucosal surfaces, especially if the normal flora is altered or eliminated, for example, by antimicrobial chemotherapy. Pathogens may then more readily colonize the tissue and begin the invasion process. Pathogen growth may also be established at sites distant from the original point of entry. Access to distant, usually interior, sites is through the blood or lymphatic circulatory system (⌀ Section 20.1).

✓ 19.6 Concept Check

Pathogens may first gain access to host tissues by adherence to specific host molecules, usually on mucosal surfaces. Invasion starts at the site of adherence and may spread throughout the host via the circulatory systems.

✓ Distinguish between *adherence* and *invasion*.
✓ Why is invasion usually necessary to establish pathogenicity?

19.7

Colonization and Growth

If a pathogen gains access to tissues, it may multiply, a process called **colonization.** Because the initial inoculum of a pathogen is rarely sufficient to cause damage,

a pathogen must *grow* within host tissues in order to produce a disease. If the pathogen is to grow, it must find appropriate nutrients and environmental conditions. Temperature, pH, and reduction potential are environmental factors that affect pathogen growth, but the availability of microbial nutrients in host tissues is most important. Although a vertebrate host might seem to be a nutritional paradise for microorganisms, not all nutrients are plentiful. Soluble nutrients such as sugars, amino acids, and organic acids may often be in short supply, and organisms able to use complex nutrient sources such as glycogen may be favored. Not all vitamins and other growth factors are necessarily in adequate supply in all tissues at all times. *Brucella abortus,* for example, grows very slowly in most tissues of infected cattle but grows very rapidly in the placenta, where it causes abortion. This is due to the elevated concentration of erythritol found in the placenta, a nutrient that enhances growth of *B. abortus* (see Table 19.6).

Trace elements may also be in short supply and can influence establishment of the pathogen. For example, considerable evidence exists for the influence of **iron** on microbial growth. Specific proteins called *transferrin* and *lactoferrin,* present in animals, bind iron tightly and transfer it through the body. These proteins have such high affinity for iron that microbial iron deficiency may be common; indeed, a soluble iron salt fed or injected into an infected animal may greatly increase the virulence of some pathogens. As we noted in Section 4.2, many bacteria produce iron-chelating compounds (**siderophores**) that help them obtain iron from the environment. Some iron chelators isolated from pathogenic bacteria are so efficient that they can remove iron from animal iron-binding proteins. For example, a siderophore called *aerobactin,* produced by certain strains of *Escherichia coli* and encoded by the Col V plasmid (⌀ Section 9.8), readily removes iron bound to transferrin.

Localization in the Body

After initial entry, the organism often remains localized and multiplies, producing a small focus of infection such as the boil, carbuncle, or pimple that commonly arises from *Staphylococcus* infections of the skin (⌀ Section 23.2). Alternatively, the organisms may pass through the lymphatic vessels and be deposited in lymph nodes. If an organism reaches the blood, it will be distributed to distant parts of the body, usually concentrating in the liver or spleen. Spread of the pathogen through the blood and lymph systems can result in a generalized (systemic) infection of the body, with the organism growing in a variety of tissues. If extensive bacterial growth in tissues occurs, some of the organisms are usually shed into the bloodstream in large numbers, a condition called **bacteremia.** Generalized infections of this type almost always start as a localized infection at a specific organ site but fortunately are rare.

Virulence Factors

A number of pathogen-produced extracellular proteins aid in the establishment and maintenance of disease. These proteins, which are mostly enzymes, are called *virulence factors*. For example, streptococci, staphylococci, pneumococci, and certain clostridia produce **hyaluronidase** (Table 19.4), an enzyme that promotes spreading of organisms in tissues by breaking down hyaluronic acid, a polysaccharide that functions in the body as a tissue cement. Production of this enzyme enables these organisms to spread from an initial focus. Streptococci and staphylococci also produce a vast array of proteases, nucleases, and lipases that serve to depolymerize host proteins, nucleic acids, and fats, respectively. Clostridia that cause gas gangrene produce

TABLE 19.4 Exotoxins and extracellular virulence factors produced by human pathogens

Organism	Disease	Toxin or factor[a]	Action
Bacillus anthracis	Anthrax	Lethal factor (LF) Edema factor (EF) Protective antigen (PA) (AB)	PA is the cell-binding B component, EF causes edema, LF causes cell death
Bacillus cereus	Food poisoning	Enterotoxin (?)	Induces fluid loss from intestinal cells
Bordetella pertussis	Whooping cough	Pertussis toxin (AB)	Blocks G protein signal transduction, kills cells
Clostridium botulinum	Botulism	Neurotoxin (AB)	Flaccid paralysis (see Figure 19.17)
Clostridium tetani	Tetanus	Neurotoxin (AB)	Spastic paralysis (see Figure 19.18)
Clostridium perfringens	Gas gangrene, food poisoning	α-Toxin (CT)	Hemolysis (lecithinase, see Figure 19.15b)
		β-Toxin (CT)	Hemolysis
		γ-Toxin (CT)	Hemolysis
		δ-Toxin (CT)	Hemolysis (cardiotoxin)
		κ-Toxin (E)	Collagenase
		λ-Toxin (E)	Protease
		Enterotoxin (CT)	Alters permeability of intestinal epithelium
Corynebacterium diphtheriae	Diphtheria	Diphtheria toxin (AB)	Inhibits protein synthesis in eukaryotes (Figure 19.16)
Escherichia coli (enteropathogenic strains only)	Gastroenteritis	Enterotoxin (AB)	Induces fluid loss from intestinal cells
Pseudomonas aeruginosa	*P. aeruginosa* infections	Exotoxin A (AB)	Inhibits protein synthesis
Salmonella spp.	Salmonellosis, typhoid fever, paratyphoid fever	Enterotoxin (AB)	Inhibits protein synthesis and lyses host cells
		Cytotoxin (CT)	Induces fluid loss from intestinal cells
Shigella dysenteriae	Bacterial dysentery	Enterotoxin (AB)	Inhibits protein synthesis
Staphylococcus aureus	Pyrogenic (pus-forming) infections (boils, and so on), respiratory infections, food poisoning, toxic shock syndrome, scalded skin syndrome	α-Toxin (CT)	Hemolysis
		Toxic shock syndrome toxin (SA)	Systemic shock
		Exfoliating toxin A and B (SA)	Peeling of skin, shock
		Leukocidin (CT)	Destroys leukocytes
		β-Toxin (CT)	Hemolysis
		γ-Toxin (CT)	Kills cells
		δ-Toxin (CT)	Hemolysis, leukolysis
		Enterotoxin A, B, C, D, and E (SA)	Induce vomiting, diarrhea, shock
		Coagulase (E)	Induces fibrin clotting
Streptococcus pyogenes	Pyrogenic infections, tonsillitis, scarlet fever	Streptolysin O (CT)	Hemolysin
		Streptolysin S (CT)	Hemolysin (see Figure 19.15a)
		Erythrogenic toxin (SA)	Causes scarlet fever rash
		Streptokinase (E)	Dissolves fibrin clots
		Hyaluronidase (E)	Dissolves hyaluronic acid in connective tissue
Vibrio cholerae	Cholera	Enterotoxin (AB)	Induces fluid loss from intestinal cells (Figure 19.19)

a AB, A-B toxin; CT, cytolytic toxin; E, enzymatic virulence factor; SA, superantigen toxin; ?, not classified.

collagenase, or κ-toxin (Table 19.4), which breaks down the collagen network supporting the tissues; the resulting dissolution of tissue is a factor in enabling these organisms to spread through the body. Fibrin clots are often formed by the host in a region of microbial invasion to wall off the organism, preventing its spread through the body. Some organisms are able to produce fibrinolytic enzymes to dissolve these clots and make further invasion possible. One such fibrinolytic substance, produced by *Streptococcus pyogenes,* is known as **streptokinase** (Table 19.4). On the other hand, some organisms produce enzymes that actually promote fibrin clotting, which causes localization of the organism rather than its spread. The best-studied fibrin-clotting enzyme is **coagulase** (Table 19.4), produced by pathogenic *Staphylococcus aureus.* Coagulase causes fibrin to be deposited on the cocci and may protect coated bacteria from attack by host cells. The fibrin matrix produced as a result of coagulase activity probably accounts for localization of many staphylococcal infections in boils and pimples (∞ Figure 23.6).

✓ 19.7 Concept Check

A pathogen must gain access to nutrients, colonize, and grow in substantial numbers in host tissue to cause disease. A number of pathogen-produced enzymes are extracellular virulence factors that increase access to host nutrients and aid in colonization and growth on the host.

✓ Why are colonization and growth necessary for the success of most pathogens?
✓ Why do bacterial enzymes attack structural components of host cells?

19.8

Exotoxins

Pathogens cause host damage in a variety of ways. Symptoms of a disease are rarely due only to the presence of large numbers of microorganisms. Although a large mass of cells could block vessels, or heart valves, or clog air passages in the lungs, in most cases pathogens produce a variety of molecules that promote pathogenesis. Among these molecules are the **toxins,** products of microorganisms that produce immediate host cell damage.

Exotoxins are proteins released extracellularly as the organism grows. These toxins may travel from a focus of infection to distant parts of the body and cause damage in regions far removed from the site of microbial growth. Table 19.4 provides a summary of the properties and actions of some of the best-known exotoxins as well as other extracellular virulence factors.

Most exotoxins fall into three categories, the *cytolytic toxins,* the *A-B toxins,* and the *superantigen toxins.* The cytolytic toxins work by enzymatically attacking cell constituents, causing lysis. The A-B toxins consist of two covalently bonded subunits, A and B. The B component generally binds to a cell surface receptor, allowing the transfer of the A subunit across the targeted cell membrane, where it functions to damage the cell. The superantigens work by stimulating large numbers of immune response cells, resulting in massive, even systemic, inflammatory reactions (∞ Section 20.15). Here we examine the mechanisms of several of the cytolytic toxins and A-B toxins.

Cytolytic Toxins

Various pathogens produce proteins that are able to act on the animal cytoplasmic membrane, causing cell lysis and hence cell death. The action of these toxins is most easily detected with red blood cells (erythrocytes), hence they are often called **hemolysins** (Table 19.4); in probably all cases, however, they also work on cells other than erythrocytes. The production of such toxins is most readily demonstrated in the laboratory by streaking the organism on a *blood agar plate.* During growth of the colonies, hemolysin is released and lyses the surrounding red blood cells, typically creating a zone of hemolysis (Figure 19.15*a*). Some hemolysins attack the phospholipid of the host cytoplasmic membrane. Because the phospholipid lecithin (phosphatidylcholine) is often used as a substrate, these enzymes are called **lecithinases** or **phospholipases** (Figure 19.15*b*). An example is the α-toxin of *Clostridium perfringens,* which is a lecithinase that dissolves membrane

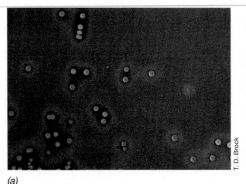
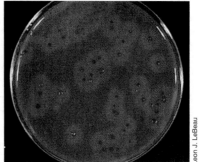

(a)   (b)

FIGURE 19.15   (a) Zones of hemolysis around colonies of *Streptococcus pyogenes* growing on a blood agar plate. (b) Action of lecithinase, a phospholipase, around colonies of *Clostridium perfringens,* growing on an agar medium containing egg yolk, a source of lecithin.

lipids, resulting in cell lysis (Table 19.4). Since the cytoplasmic membranes of all organisms, both prokaryotes and eukaryotes, contain phospholipids, phospholipases sometimes destroy bacterial as well as animal cytoplasmic membranes. Some hemolysins are not phospholipases, however. Streptolysin O, a hemolysin produced by streptococci, affects the sterols of the host cytoplasmic membrane, and its action is neutralized by addition of cholesterol or other sterols. **Leukocidins** (Table 19.4) are lytic agents capable of lysing white blood cells and may decrease host resistance (⌀ Section 20.1).

Diphtheria Toxin

The toxin produced by *Corynebacterium diphtheriae*, the causal agent of diphtheria, was the first exotoxin to be discovered. Diphtheria toxin is an important factor in the pathogenesis of diphtheria, discussed in Section 23.2. The toxin has different effects on different animal species; rats and mice are relatively resistant, but humans, rabbits, guinea pigs, and birds are susceptible, with only a single toxin molecule required to kill a cell in these species. Diphtheria toxin is an A-B toxin secreted by cells of *C. diphtheriae* as a polypeptide of 62,000 molecular weight. Fragment B promotes specific binding of the toxin to a host cell receptor (Figure 19.16). After binding, proteolytic cleavage between Fragment A and B allows entry of Fragment A (21,000 molecular weight) into the host cytoplasm. Fragment A then disrupts protein synthesis by blocking transfer of an amino

acid from a transfer ribonucleic acid (tRNA) to the growing polypeptide chain. The toxin specifically inactivates elongation factor 2 (a protein involved in growth of the polypeptide chain) by catalyzing the attachment of adenosine diphosphate (ADP) ribose from NAD^+; following ADP-ribosylation, the activity of the modified elongation factor 2 decreases dramatically and protein synthesis stops (Figure 19.16).

Diphtheria toxin is formed only by strains of *C. diphtheriae* that are lysogenized by a bacteriophage called phage β, which carries the toxin-encoding *tox* gene. Nontoxigenic, nonpathogenic strains of *C. diphtheriae* can be converted to pathogenic strains by infection with phage β (the process of *phage conversion*) (⌀ Section 9.7).

Another factor in diphtheria toxin production is the concentration of *iron* in the environment. In media containing sufficient iron for optimal growth, no toxin is produced. When the iron concentration is reduced to growth-limiting levels, toxin production occurs. The role of iron is to bind to a regulatory protein in *C. diphtheriae* (that is, act as a *negative control element*) (⌀ Section 7.2). The iron-binding protein then combines with a control region of the DNA of β phage and prevents expression of the diphtheria toxin. When iron is limiting, the regulatory protein does not act, and toxin synthesis occurs.

This strategy for toxin production is also found in other microorganisms. For example, exotoxin A of *Pseudomonas aeruginosa* (Table 19.4) functions similarly to diphtheria toxin, also modifying elongation factor 2 by ADP-ribosylation.

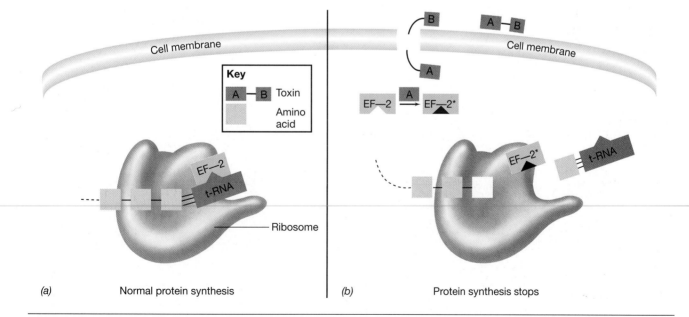

(a)   Normal protein synthesis   *(b)*   Protein synthesis stops

FIGURE 19.16   The action of diphtheria toxin from *Corynebacterium diphtheriae*. (a) Elongation factor 2 (EF-2) normally binds to the ribosome and brings an amino acid-charged t-RNA to the ribosome, causing protein elongation. (b) Diphtheria toxin binds to the cell membrane, where it is cleaved and the A peptide is internalized. The A peptide catalyzes the ADP-ribosylation of elongation factor 2 (EF-2*). The modified elongation factor no longer aids transfer of amino acids to the growing polypeptide chain, resulting in shutdown of protein synthesis and death of the cell.

Tetanus and Botulinum Toxins

These toxins are produced by two species of obligately anaerobic bacteria, *Clostridium tetani* and *Clostridium botulinum,* which are normal soil organisms that occasionally cause disease in animals (∞ Section 24.11). *C. botulinum* rarely grows directly in the body, but it does grow and produce toxin in improperly preserved foods. The fatality rate from *botulism* food poisoning can approach 100% in untreated individuals, depending largely on the amount of toxin ingested. Death from botulism is usually due to respiratory failure from muscle paralysis. However, fatalities can be significantly reduced by quick administration of an antitoxin antibody (∞ Section 20.16) and by use of an artificial respirator to prevent respiratory failure. In *infant botulism,* infection of the intestinal tract of children less than six months old from a *C. botulinum* spore–containing food product such as raw honey results in chronic infection by the toxin-producing organism. Production of the toxin then causes botulism. *C. botulinum* infection is rare in normal adults because the normal flora and the immune response prevent colonization of the intestinal tract by the pathogen. *C. tetani* grows in the body in deep wound punctures that become anoxic, and although *C. tetani* does not invade the body from the initial site of infection, the toxin can spread through the neural cells and cause the severe neurological symptoms of *tetanus* that can result in death.

Botulinum toxin is a series of seven related A-B toxins that are the most toxic substances known. One milligram of pure botulinum toxin is enough to kill more than 1 million guinea pigs. Of the seven distinct botulinum toxins known, at least two are encoded on lysogenic bacteriophages specific for *Clostridium botulinum.* The major toxin is a protein of about 150,000 molecular weight, which readily forms complexes with nontoxic botulinum proteins to give a bioactive toxin of almost 10^6 molecular weight. Toxicity occurs because the toxin binds to presynaptic membranes on the termini of the stimulatory motor neurons at the neuromuscular junction, blocking the release of acetylcholine. Because transmission of the nerve impulse to the muscle is through acetylcholine interaction with a muscle receptor, the botulinum-poisoned muscle does not receive an excitatory signal and contraction is prevented, causing a flaccid paralysis (Figure 19.17).

Tetanus toxin is a protein of molecular weight 150,000, containing linked A-B polypeptides. On contact with the central nervous system this toxin is transported through the motor neurons back to the spinal cord, where it binds specifically to ganglioside lipids at the termini of the inhibitory interneurons. The inhibitory interneurons normally work by releasing an inhibitory neurotransmitter, usually glycine, that binds to receptors on the motor neurons. Normally, the glycine signal from the inhibitory interneurons stops the release of acetylcholine by the motor neurons and inhibits muscle contraction, allowing relaxation of the

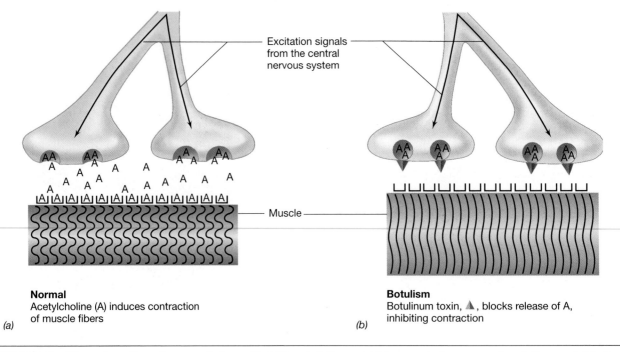

Normal
Acetylcholine (A) induces contraction of muscle fibers

(a)

Botulism
Botulinum toxin, ▲, blocks release of A, inhibiting contraction

(b)

— Excitation signals from the central nervous system

— Muscle

FIGURE 19.17   The action of botulinum toxin from *Clostridium botulinum.* (a) Upon central nervous stimulation, acetylcholine (A) is normally released from vesicles at the neural side of the motor end plate. Acetylcholine then binds to specific receptors on the muscle, inducing contraction. (b) Botulinum toxin acts at the motor end plate to prevent release of acetylcholine (A) from vesicles, resulting in a lack of stimulus to the muscle fibers, irreversible relaxation of the muscles, and flaccid paralysis.

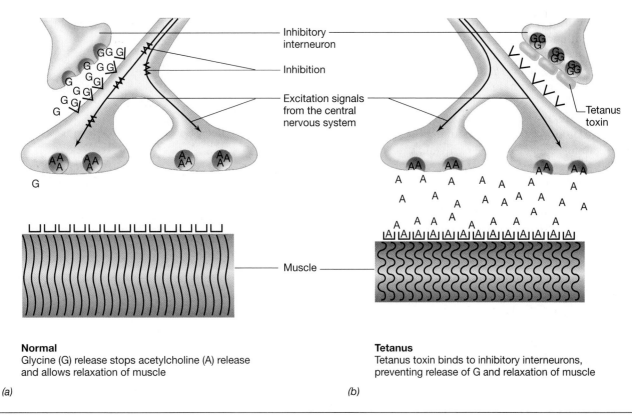

FIGURE 19.18 The action of tetanus toxin from *Clostridium tetani*. (a) Muscle relaxation is normally induced by glycine (G) release from inhibitory interneurons. Glycine acts on the motor neurons to block excitation and release of acetylcholine (A) at the motor end plate. (b) Tetanus toxin binds to the interneuron to prevent release of glycine from vesicles, resulting in a lack of inhibitory signals to the motor neurons, constant release of acetylcholine to the muscle fibers, irreversible contraction of the muscles, and spastic paralysis.

muscle fibers. However, if tetanus toxin blocks the release of glycine, the motor neurons cannot be inhibited, resulting in continual release of acetylcholine and uncontrolled contraction of the poisoned muscles (Figure 19.18). The outcome is a spastic, twitching paralysis, with affected muscles constantly contracted. If the muscles of the mouth are involved, the prolonged spasm restricts the mouth's movement, resulting in the condition known as *trismus* or *lockjaw*. If respiratory muscles are involved, death may occur due to asphyxiation. Thus, tetanus toxin and botulinum toxin both block release of neurotransmitters involved in muscle control, but the outcome is quite different, depending on the particular neurotransmitters involved.

✓ 19.8 Concept Check

The most potent biological toxins known are the exotoxins produced by microorganisms. Each exotoxin acts on specific host cells or molecules, resulting in specific impairment of a major host cell function.

✓ What key features are shared by all exotoxins? By the A-B exotoxins?

✓ Are infections necessary for the production of toxins?

19.9

Enterotoxins

Enterotoxins are exotoxins that act on the *small* intestine, generally causing massive secretion of fluid into the intestinal lumen, leading to the symptoms of diarrhea. Enterotoxins are produced by a variety of bacteria, including the food-poisoning organisms *Staphylococcus aureus*, *Clostridium perfringens*, and *Bacillus cereus*, and the intestinal pathogens *Vibrio cholerae*, *Escherichia coli*, and *Salmonella enteritidis*.

Cholera Toxin

The enterotoxin produced by *Vibrio cholerae*, the causal agent of cholera, is the best understood enterotoxin. Cholera toxin is an A-B toxin (see Section 19.8) and consists of an A component of 27,200 molecular weight, with five B subunits, each of 11,600 molecular weight (82,200 molecular weight for the entire complex). The B subunit contains the binding site by which the cholera toxin combines specifically with the ganglioside GM1 (a complex glycolipid) in the epithelial cytoplasmic membrane (Figure 19.19*a*), but the B subunit itself does not cause an alteration in membrane permeability. Rather, the toxic action is in the A chain, which activates the cellular enzyme *adenyl cyclase,* causing the conversion of adenosine triphosphate (ATP) to cyclic adenosine monophosphate (cAMP).

As we discussed in Section 7.6, cyclic AMP is a specific mediator of a variety of regulatory systems in cells. In mammals, cyclic AMP is involved in the action of a variety of hormones, as well as in synaptic transmission in the nervous system, and in inflammatory and immune reactions of tissues, including allergies. Although the A subunit of cholera toxin is responsible for activation of adenyl cyclase, A must first be activated by a cellular enzyme that requires NAD^+ and ATP. In the action of cholera enterotoxin, the increased cyclic AMP levels bring about the active secretion of chloride and bicarbonate ions from the mucosal cells into the intestinal lumen. This change in ionic balance leads to the secretion of large amounts of water into the lumen (Figure 19.19*b*). In the acute phase of cholera, the rate of water loss into the small intestine is greater than reabsorption of water by the large intestine, and so massive net fluid loss occurs. Cholera victims generally die from extreme dehydration, and the best treatment for the disease is the oral administration of electrolyte solutions containing solutes (Figure 19.19) to replace the lost fluid and ions.

At the molecular level, cholera enterotoxin has a mode of action (formation of cyclic AMP) identical to that of some normal mammalian hormones, and it has been suggested that cholera toxin may represent an ancestral hormone. Because cholera enterotoxin activates adenyl cyclase in a variety of cells and tissues, pathological manifestations of cholera toxin are related more to the specific site at which it binds, the epithelial cells of the small intestine, than to toxin activation of adenyl cyclase. Indeed, purified B subunits devoid of adenyl cyclase activity can actually *prevent* the action of cholera enterotoxin, if they are administered first, because they bind to the specific cholera receptors on the mucosal cells and block the binding of the complete toxin.

Genetic studies of cholera toxin have shown that the cholera enterotoxin is encoded by two genes, *ctxA* and *ctxB*. Expression of *ctxA* and *ctxB* is controlled by a positive regulatory element, a protein encoded by the *toxR* gene. The *toxR* gene product is a transmembrane protein that controls not only cholera toxin production but also several other important virulence factors, such as outer membrane proteins and pili required for successful colonization of *Vibrio cholerae* in the small intestine.

Other Enterotoxins

There is good evidence that the enterotoxins produced by enteropathogenic *Escherichia* and *Salmonella* have modes of action similar to that of cholera toxin, and an-

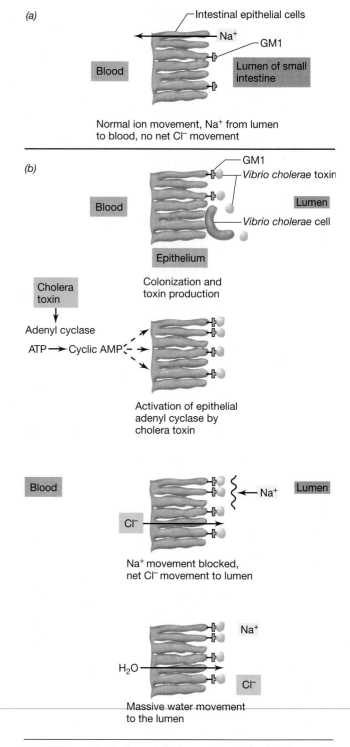

FIGURE 19.19 Action of cholera enterotoxin. (a) The normal process of ion movement in the intestine and colonization of the epithelium by *Vibrio cholerae* followed by binding of the enterotoxin by specific interaction with the GM1 ganglioside on host cells. (b) The A-B toxin acts by internalizing the toxic A component and activating adenyl cyclase, leading to disruption of normal sodium (Na^+) influx, loss of H_2O into the lumen, and diarrhea. Treatment for cholera is by ion replacement and rehydration therapy. Antibiotic treatment may shorten the course of the disease by limiting *V. cholerae* growth, but has no affect on toxin that has already been produced.

tibody against cholera enterotoxin also inactivates these other enterotoxins, suggesting a similar structure. The sequence of the cholera toxin genes *ctxA* and *B* further supports this relationship: Cholera toxin genes show greater than 75% sequence homology with the genes encoding the heat-labile enterotoxin produced by enteropathogenic *Escherichia coli*. As discussed in Section 9.8, *Escherichia* enterotoxin is controlled by a conjugative plasmid, but the enterotoxin gene of *Vibrio cholerae* is chromosomal, although transmissible by conjugation. However, enterotoxins produced by some of the food-poisoning bacteria (*Staphylococcus aureus, Clostridium perfringens, Bacillus cereus*) may be quite different in their modes of action. For example, *Staphylococcus aureus* enterotoxin is a superantigen toxin (Table 19.4) and has a completely different mechanism of action, stimulating large numbers of immune lymphocytes and causing systemic as well as intestinal inflammatory responses (∞ Section 20.15).

✓ 19.9 Concept Check

Enterotoxins are exotoxins that act specifically on the small intestine, causing changes in intestinal permeability that lead to diarrhea. Many food-poisoning and food infection microorganisms produce enterotoxins.

✓ What key features are shared by all enterotoxins? By the A-B exotoxins?

✓ Describe the action of *Vibrio cholerae* toxin on the small intestine.

19.10

Endotoxins

Gram-negative Bacteria produce lipopolysaccharides as part of the outer layer of their cell walls (∞ Figures 3.36 and 3.37), which under many conditions are toxic.

These are called **endotoxins** because they are generally cell-bound and released in large amounts only when cells lyse. In most cases, endotoxin can be equated with lipopolysaccharide toxin. Endotoxins have been studied primarily in *Escherichia, Shigella*, and especially *Salmonella*. The major differences between exotoxins and endotoxins are listed in Table 19.5.

Endotoxin Structure and Function

When injected into an animal, endotoxins cause a variety of physiological effects. *Fever* is an almost universal symptom because endotoxin stimulates host cells to release proteins called *endogenous pyrogens*, which affect the temperature-controlling center of the brain. In addition, however, the animal may develop diarrhea, experience a rapid decrease in lymphocyte, leukocyte, and platelet numbers, and enter into a generalized inflammatory state. Large doses of endotoxin can cause death, primarily through hemorrhagic shock and tissue necrosis. However, the toxicity of endotoxins is much *lower* than that of exotoxins. For instance, in the mouse the amount of endotoxin required to kill 50% of a population of test animals (the so-called LD_{50}) is 200–400 μg per mouse, whereas the LD_{50} for botulinum toxin is about 25 picograms (pg) per mouse, about 10 million times less! (A picogram is 10^{-12} g or 10^{-6} μg.)

The overall structure of lipopolysaccharide (LPS) was diagrammed in Figure 3.36. Lipopolysaccharide consists of lipid A, a core polysaccharide consisting of ketodeoxyoctonate, seven-carbon sugars (heptoses), glucose, galactose, and *N*-acetylglucosamine, and the *O-polysaccharide*, a highly variable molecule that usually contains galactose, glucose, rhamnose, and mannose and generally contains one or more unusual dideoxy sugars such as abequose, colitose, paratose, or tyvelose. The sugars of the *O*-polysaccharide are connected in four- to five-sugar sequences (often branched), which then re-

TABLE 19.5	Basic properties of exotoxins and endotoxins	
Property	**Exotoxins**	**Endotoxins**
Chemical properties	Proteins, excreted by certain gram-positive or gram-negative Bacteria; generally heat-labile	Lipopolysaccharide–lipoprotein complexes (∞ Figures 3.36 and 3.37); released on cell lysis as part of the outer membrane of gram-negative Bacteria; extremely heat-stable
Mode of action; symptoms	Specific; usually binds to specific cell receptors or structures; either cytotoxin, enterotoxin, or neurotoxin with defined specific action on cells or tissues	General; fever, diarrhea, vomiting
Toxicity	Highly toxic, often fatal	Weakly toxic, rarely fatal
Immunogenicity	Highly immunogenic; stimulate the production of neutralizing antibody (antitoxin)	Relatively poor immunogen; immune response not sufficient to neutralize toxin
Toxoid potential	Treatment of toxin with formaldehyde will destroy toxicity, but treated toxin (toxoid) remains immunogenic	None
Fever potential	Do not produce fever in host	Pyrogenic, often produce fever in host

peat to form the complete molecule (∞ Figures 3.36 and 3.37). Lipid A is composed of fatty acids connected by ester linkages to N-acetylglucosamine. Fatty acids frequently found in the lipid include β-hydroxymyristic, lauric, myristic, and palmitic acids. Studies using fractions of the lipopolysaccharide indicate that the lipid fraction is responsible for toxicity, while the polysaccharide fraction is necessary to make the complex water soluble and immunogenic (∞ Section 20.2). Animal studies indicate that both the lipid and polysaccharide fractions are necessary to obtain an *in vivo* toxic effect.

Limulus Assay for Endotoxin

Because endotoxins are fever inducers, pharmaceuticals such as antibiotics and intravenous solutions must be endotoxin-free. An endotoxin assay of very high sensitivity has been developed using lysates of amebocytes from the horseshoe crab, *Limulus polyphemus*. Although the mechanism of this assay is not understood, endo-

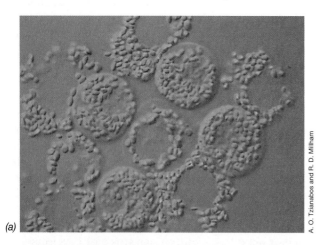

(a)

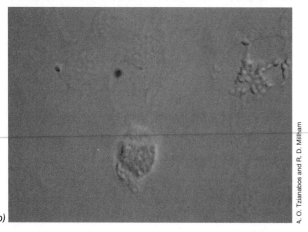

(b)

FIGURE 19.20 Photomicrographs of *Limulus* amebocytes. (a) Normal amebocytes. (b) Amebocytes following exposure to bacterial lipopolysaccharide. Treatment with lipopolysaccharide causes degranulation of the cells, and this response can be used as an assay for lipopolysaccharide content.

toxin specifically causes lysis of amebocytes (Figure 19.20). In a commercial assay, amebocyte extracts are mixed with the solution to be tested. If endotoxin is present, the amebocyte extract gels and precipitates, causing a change in turbidity. This reaction can be measured quantitatively with a spectrophotometer. A measurable reaction can be obtained with as little as 10–20 pg/ml of lipopolysaccharide. Apparently the active component of the *Limulus* extract reacts with the lipid component of lipopolysaccharide. The *Limulus* assay has been used to detect the presence of minute quantities of endotoxin in serum, cerebrospinal fluid, drinking water, and fluids used for injection.

The *Limulus* test is so sensitive that considerable care must be taken to avoid contamination of the equipment, solutions, and reagents with the gram-negative Bacteria in the laboratory and clinical environment, for example, as contaminants in distilled water. In clinical work, detection of endotoxin by the *Limulus* assay in serum or cerebrospinal fluid is presumptive evidence of gram-negative infection of these body fluids.

✓ 19.10 Concept Check

Endotoxins are lipopolysaccharides derived from gram-negative Bacteria. Released upon lysis of the Bacteria, endotoxins cause systemic toxic effects such as fever in the host. Endotoxins are generally less toxic than exotoxins.

✓ Why do gram-positive Bacteria not produce endotoxins?
✓ Why are the endotoxins generally not as potent as the exotoxins?

19.11

Virulence

Virulence is the relative ability of a parasite to cause disease. In the last five sections, we described several specific virulence factors, all of which dealt with the ability of a pathogen either to *invade* a host or to cause damage by producing *toxins*. In this section, we will deal with specific examples of particularly virulent organisms and we will apply our knowledge of virulence factors to explain the virulence of these organisms.

Invasiveness and *toxicity* are measurable features of pathogenesis, and each pathogen uses a unique blend of these properties to cause disease. For example, *Clostridium tetani* is virtually noninvasive, but produces a potent exotoxin (see Section 19.8), and this organism establishes its virulence almost exclusively with the toxin. The cells of *C. tetani* rarely leave the wound where they were first deposited; yet they are able to bring about death of the host because they produce tetanus exotoxin, which can move to distant parts of the body and initiate paralysis. On the other hand, a weakly toxigenic organism may still be able to produce disease if it is highly invasive. For example, the only known virulence factor for *Streptococcus pneumoniae* is the polysaccharide capsule that prevents the phagocytosis of pathogenic strains (see Section 19.6, ∞ Section 20.3, and ∞ Figure 23.5), defeating a major defense mechanism used by the host to prevent invasion. Encapsulated strains of *S. pneumoniae* do not produce any toxin, but are able to cause extensive host damage because they are highly invasive and are able to grow in lung tissues in enormous numbers where they initiate host responses that lead to pneumonia. These two organisms exemplify the extremes of invasiveness and toxigenicity; most pathogens fall somewhere between these extremes.

Virulence Factors of *Salmonella* Species

By contrast to *Streptococcus pneumoniae*, *Salmonella* spp. employ a *mixture* of toxins and other virulence factors to enhance their pathogenicity. First, several toxins contribute to the virulence of *Salmonella*, and at least three toxins are produced: enterotoxin (Table 19.4), endotoxin (∞ Section 3.8 and Figure 3.36), and *cytotoxin*. Cytotoxin acts by inhibiting host cell protein synthesis and allowing Ca^{2+} to escape from host cells. In addition, a number of virulence factors are involved in invasion. Adherence factors are the cell surface polysaccharide O antigen (∞ Figure 3.36) and the flagellar H antigen. Fimbriae may also enhance adherence. The capsular Vi polysaccharide inhibits complement binding and antibody-mediated killing (∞ Section 20.12). The *inv*

genes of *Salmonella* encode at least 10 different proteins involved in invasion. For example, *invH* encodes a surface adhesin, while *invC*, *invG*, *invI*, and *invJ* encode proteins involved in assembly of the *surface appendages*, protein structures used for host-cell binding.

Salmonella readily establish infections through intracellular parasitism, growing inside the cells that line the intestine and even in macrophages, a group of white blood cells that normally ingest (phagocytize) and kill bacteria (∞ Section 20.3). The toxic oxygen products of macrophages are neutralized by proteins induced by the *Salmonella oxyR* gene. Macrophage-produced antibacterial molecules called *defensins* are neutralized by the products of the *phoP* and *phoQ* genes. Thus, the *oxy* and *pho* genes allow *Salmonella* to become intracellular pathogens, since these genes neutralize host defense factors that normally limit intracellular bacterial growth. Several plasmidborne virulence factors are also involved in persistence and spread in most *Salmonella* species. Finally, *Salmonella* also produce *siderophores*, bacterial iron-chelating proteins that sequester iron (∞ Section 4.2 and see Section 19.7), a valuable growth factor. Thus, *Salmonella*, and probably most other pathogens, use several virulence factors to initiate infection. Figure 19.21 summarizes the known virulence factors of *Salmonella*.

Measuring Virulence

The virulence of a pathogen can be estimated from experimental studies of the LD_{50}. Highly virulent pathogens frequently show little difference in the num-

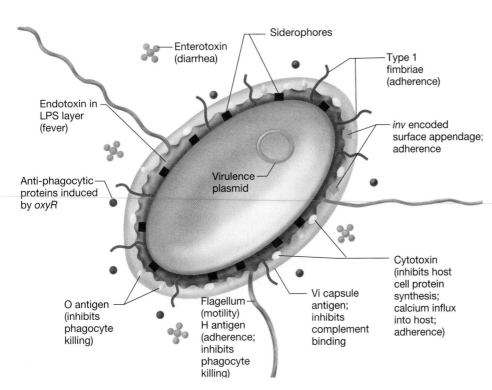

FIGURE 19.21 Virulence factors important in *Salmonella* pathogenesis. Structural elements known to be important in pathogenesis are shown. Protein products of the *pho* gene system have not been identified, but are known to neutralize the effects of the detergent-like, macrophage-produced defensins.

ber of cells required to kill 100% of the population as compared to the number required to kill 50% of the population. This is illustrated in Figure 19.22 for experimental *Streptococcus* and *Salmonella* infections in mice. Only a few cells of *Streptococcus pneumoniae* are required to establish a fatal infection in mice. Fewer than 100 cells per mouse are necessary to kill every member of a test population once the virulence of a particular strain has been established. In fact, the LD_{50} for this organism is hard to determine because so few organisms are needed to produce a lethal infection. By contrast, the LD_{50} for *Salmonella typhimurium*, also a mouse pathogen but a much less virulent one, is much higher than for *S. pneumoniae*, and the number of cells required to kill 100% of the population is more than 100 times greater than the number of cells needed to achieve the LD_{50}.

When pathogens are kept in laboratory culture and not passed through animals, their virulence is often decreased or even completely lost. Such organisms are said to be **attenuated.** Attenuation probably occurs because nonvirulent mutants may grow faster and, through successive transfers to fresh media, such mutants are selectively favored. Attenuation often occurs more readily when culture conditions are not optimal for the species. If an attenuated culture is reinoculated into an animal, virulent organisms are sometimes reisolated, but in many cases loss of virulence is permanent. Attenuated strains are often used in the production of vaccines, especially viral vaccines (∞ Section 20.16). Measles and mumps vaccines, for example, are composed of attenuated viruses.

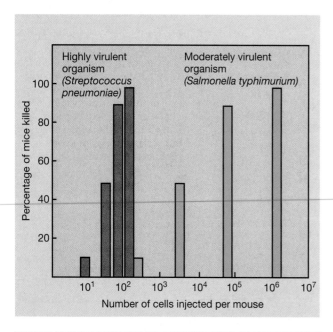

FIGURE 19.22 Comparison of differences in microbial virulence based on the number of cells of *Streptococcus pneumoniae* or *Salmonella typhimurium* required to kill a mouse population.

✓ 19.11 Concept Check

Virulence is determined by the invasiveness and toxigenicity of a pathogen. In most pathogens, a number of factors contribute to virulence. Attenuation is loss of virulence.

✓ What virulence factors are important for pathogenicity in *Streptococcus pneumoniae*? In *Salmonella* species?

✓ Explain how an organism may become attenuated. Define the role of attenuated organisms in vaccine production.

19.12

Nonspecific Host Defenses

Many of the mechanisms responsible for the suppression of pathogens are innate "resistance factors." These resistance factors can be divided into two categories: *specific* host defenses, which are directed against individual species or strains of pathogens, and *nonspecific* host defenses, directed against a variety of pathogens. In this chapter we consider the major *nonspecific* host defenses that have been identified as important in preserving the healthy state of the host. In the following chapter we consider *specific* host defenses—the immune response.

Natural Host Resistance

The ability of a particular pathogen to cause disease in different animal species is highly variable. In *rabies*, for instance, death usually occurs in all species of mammals once symptoms of the disease develop. Nevertheless, certain animal species are much more susceptible to rabies than others. Raccoons and skunks, for example, are extremely susceptible to rabies infection as compared to opossums, which rarely acquire the disease. *Anthrax* infects a variety of animals and causes disease symptoms varying from mild pustules in humans to a fatal blood poisoning in cattle. However, birds are totally resistant to anthrax. In addition, diseases of warm-blooded animals are rarely transmitted to cold-blooded species, and vice versa. Why should this be so?

Resistance to certain diseases and susceptibility to others is an innate property of a given host species and is governed by complex and interdependent factors. Differences in physiology and nutrition as well as anatomical differences are important, as is variation in tissue surface receptors, as discussed later. The net result is that even very closely related species may show completely different susceptibilities to the same pathogen.

Age, Stress, and Diet

Age is an important factor in susceptibility to infectious disease. Infectious diseases are more common in the very young and in the aged. In the infant, for example, development of an intestinal microflora occurs quite quickly, but the normal flora of a young infant is not the same as that of the adult. Before the development of an adult flora, and especially in the days immediately following birth, pathogens have a greater opportunity to become established and produce disease. Thus, diarrhea caused by enteropathogenic strains of *Escherichia coli* (∞ Section 24.12) or *Pseudomonas aeruginosa* is frequently encountered in infants under the age of 1 year. The underdeveloped state of the infant's microflora provides poor competition for the pathogenic species and disease may result. As we discussed previously, infant botulism is encountered only in very young infants because establishment of the intestinal normal flora in older children precludes intestinal infection with *Clostridium botulinum,* which causes the disease (see Section 19.8, and ∞ Section 24.11).

In individuals over the age of 65, infectious diseases are much more common than in younger adults. For example, the elderly are much more susceptible to respiratory infections, particularly influenza (∞ Section 23.4), probably because of a declining ability to make an effective immune response to respiratory pathogens. In addition, anatomical changes associated with age may also encourage infection. Enlargement of the prostate gland, a common condition in men over the age of 50, frequently leads to a decreased urine flow. This, in turn, allows pathogens to colonize the male urinary tract (Figure 19.12) more readily, leading to an increase in urinary tract infections in elderly men.

Stress can predispose a normally healthy individual to disease. In studies with rats and mice, fatigue, exertion, poor diet, dehydration, or drastic climatic changes, all sources of physiological stress, increase the incidence and severity of infectious diseases. For example, rats subjected to intense physical activity for long periods of time show a higher mortality rate from experimental *Salmonella* infections than well-rested control animals. Hormones that are produced under stress influence the immune system and may play a role in stress-mediated disease. The hormone *cortisone,* for example, is produced at much higher levels in times of stress than during normal periods, and this hormone is an effective anti-inflammatory agent. Suppression of inflammation removes one of the normal defenses against disease (see Section 19.13).

Diet plays a role in host resistance. The correlation between famine and infectious disease has been known for centuries. Diets low in protein may alter the composition of the normal flora, thus allowing opportunistic pathogens a better chance to multiply. For example, the number of *Vibrio cholerae* necessary to produce the disease cholera in an exposed individual is drastically reduced if the exposed individual is malnourished. On the other hand, the number of *V. cholerae* required to cause infection is drastically reduced when the *V. choler-*

ae is ingested in food, presumably because the food neutralizes the stomach acids that would normally destroy the pathogen (∞ Section 24.11). Overeating may be harmful as well. Studies on clostridial diseases of sheep, in particular *bloats* caused by excessive gas accumulation, indicate that constant overeating affects the composition of the normal flora, leading to massive growth of bacterial species normally present in low numbers.

In some cases, lack of a particular dietary substance may *prevent* disease by depriving a pathogen of critical nutrients. The best example here is the effect sucrose has on the development of dental caries. As explained in Section 19.3, absence of sucrose from the diet (along with good oral hygiene) virtually eliminates tooth decay. Without a dietary source of sucrose, the highly cariogenic bacteria *Streptococcus mutans* and *S. sobrinus* are unable to synthesize the gummy outer surface polysaccharide needed to keep the bacterial cells attached to the teeth.

Anatomical Defenses

The structural integrity of tissue surfaces poses a barrier to penetration by microorganisms. In the skin and mucosal tissues, potential pathogens must not only bind to tissue surfaces but also grow at these sites before traveling elsewhere in the body. Intact surfaces usually prevent colonization, but damaged surfaces (for example, abraded skin) are often readily colonized, promoting invasion. Resistance to colonization and invasion is due to the production of host defense substances and to various anatomical mechanisms that disrupt colonization. A summary of the major anatomical defenses is shown in Figure 19.23.

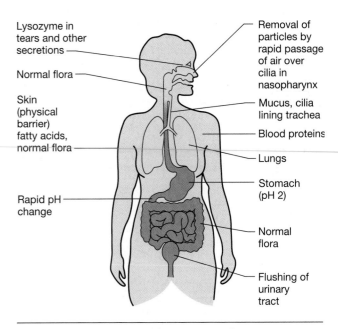

Lysozyme in tears and other secretions

Normal flora

Skin (physical barrier) fatty acids, normal flora

Rapid pH change

Removal of particles by rapid passage of air over cilia in nasopharynx

Mucus, cilia lining trachea

Blood proteins

Lungs

Stomach (pH 2)

Normal flora

Flushing of urinary tract

FIGURE 19.23   Anatomical barriers to infection.

The **skin** is an effective barrier to the penetration of microorganisms. *Sebaceous glands* in the skin (Figure 19.2) secrete fatty acids and lactic acid, which lower skin pH and inhibit colonization of pathogenic bacteria. Microorganisms inhaled through the *nose* or *mouth* are removed by the action of ciliated epithelial cells in the mucous surfaces of the nasopharynx and tracheal regions. Cilia push bacterial cells upward until they are caught in oral secretions and either are expectorated or are swallowed and killed in the stomach. Potential pathogens entering the stomach must survive the *acidity* (which is about pH 2) and then successfully compete with the increasingly abundant resident microflora present in the small intestine (which is about pH 5) and finally in the large intestine (pH 6–7). The large intestine contains bacterial numbers of about 10^{10} per gram of intestinal contents in a normal adult (see Section 19.4), making establishment of new microorganisms very difficult.

In a healthy adult, the kidney and the surface of the eye are constantly bathed with secretions containing *lysozyme* that markedly reduce microbial populations. Extracellular fluids such as blood plasma also contain bactericidal substances. For example, blood proteins called β-lysins bind and destroy microbial cells. β-Lysins act by disrupting the bacterial cytoplasmic membrane, leading to leakage of cytoplasmic constituents and cell death.

However effective these defenses may be, damage to physical barriers and changes in other nonspecific defenses can quickly lead to growth of the pathogen, invasion, and initiation of disease.

Tissue Specificity

Most pathogens must first adhere and colonize at the site of infection. If the site is not compatible with their nutritional and environmental needs, the organisms cannot multiply. Thus, if *Clostridium tetani* were ingested, it would not bring about tetanus because the pathogen is killed by the acidity of the stomach. If, on the other hand, *C. tetani* cells were introduced into a deep wound, the

organism would grow in the anoxic zones created by localized tissue death and produce tetanus toxin (see Section 19.8). By contrast, enteric bacteria such as *Salmonella* and *Shigella* do not cause wound infections but successfully colonize the intestinal tract. Table 19.6 summarizes a number of examples of tissue specificity.

The Compromised Host

The term *compromised host* refers to hosts in which one or more resistance mechanisms are inactive and in which the probability of infection is therefore increased.

Hospital patients are often compromised hosts. Many hospital procedures such as catheterization, hypodermic injection, spinal puncture, and biopsy can inadvertently introduce pathogens into the patient. Surgical procedures expose highly susceptible parts of the body to sources of contamination. The stress of surgery also diminishes the resistance of the patient to infection. Finally, in organ transplant procedures (⚭ Section 20.7), drugs are used that suppress the immune system to prevent rejection of the transplant. Immunosuppressive drugs greatly increase susceptibility to infection. Thus, many hospital patients with noninfectious primary ailments (for example, cancer and heart disease) die of microbial infection because they are compromised hosts (⚭ Section 22.7).

Compromised hosts exist even outside the hospital. Smoking, excess consumption of alcohol, intravenous drug use, lack of sleep, poor nutrition, and current infection with another agent are conditions that compromise host resistance to infection. A good example is HIV (human immunodeficiency virus). HIV causes acquired immunodeficiency syndrome (AIDS) by destroying one type of immune cell, the CD4 T lymphocytes (⚭ Section 20.4), involved in the immune response. Because of this preexisting host damage from HIV infection, AIDS patients are unable to mount effective resistance to infection; death is generally due to some infectious agent (⚭ Sections 22.4 and 23.7).

Finally, there are certain genetic conditions that may compromise the host, such as genetic diseases that elim-

TABLE 19.6	Tissue specificity as a factor in infectious disease	
Disease	**Tissue infected**	**Organism**
Diphtheria	Throat epithelium	*Corynebacterium diphtheriae*
Gonorrhea	Urogenital epithelium	*Neisseria gonorrhoeae*
Cholera	Small intestine epithelium	*Vibrio cholerae*
Pyelonephritis	Kidney medulla	*Proteus* sp.
Dental caries	Oral epithelium	*Streptococcus mutans, S. sobrinus, S. sanguis, S. mitis*
Spontaneous abortion (cattle)	Placenta	*Brucella abortus*
Acquired immunodeficiency syndrome (AIDS)	T helper lymphocytes	Human immunodeficiency virus (HIV)
Malaria	Blood (erythrocytes)	*Plasmodium* sp.

inate important parts of the immune system. Individuals with such conditions frequently die at an early age, not from the genetic condition itself but from microbial infection.

✓ 19.12 Concept Check

Nonspecific physical, anatomical, and chemical barriers prevent colonization of the host by most pathogens. Breakdown in these defenses results in a compromised host who is more susceptible to infection.

✓ How can diet and smoking influence host resistance to a pathogen?
✓ How might preexisting infection compromise an otherwise healthy host?

19.13

Inflammation and Fever

Inflammation is a general nonspecific reaction to foreign particles and other noxious stimuli such as toxins and pathogens. The characteristic inflammatory response results in redness, swelling, pain, and heat, which are localized at the site of infection (∞ Section 23.3 and Figure 20.25). The mediators of inflammation include a group of proteins called *cytokines* (∞ Section 20.8), which are produced by white blood cells or *leukocytes* (∞ Section 20.3). Leukocytes are also involved in pathogen-specific responses to noxious stimuli in the immune response, which we will discuss in Chapter 20. The most important outcome of the inflammatory response is the immediate localization of the pathogen, often by the production of a fibrin clot at the site of inflammation.

Inflammation is one of the most important and ubiquitous aspects of host defense against invading microorganisms. However, inflammation may also aid microbial pathogenesis because the inflammatory response elicited by an invading microorganism can result in considerable host damage, making nutrients available and providing access to host tissues.

Fever

The healthy human body maintains a surprisingly constant temperature. Over an average 24-hr period, body temperature varies over the narrow range of 1–1.5°C. However, individuals vary in their "normal" temperatures, and although 37°C is considered normal, the actual normal temperature in some individuals may be as low as 36°C or as high as 38°C. Also, body temperature varies with physical activity and can be as much as 2°C below normal in sleep and as much as 4°C above normal during strenuous exercise.

Fever is defined as an abnormal increase in body temperature. Although fever can be caused by noninfectious disease, most fevers are caused by infection. At least one reason why fever occurs during many infections is that certain products of pathogenic organisms are *pyrogenic* (fever-inducing). The best-studied pyrogenic agents are the endotoxins of gram-negative Bacteria (see Section 19.10). However, many organisms that do not produce endotoxins are able to cause fever on infection. These organisms release proteins known as *endogenous pyrogens* when leukocytes destroy them (∞ Section 20.3). Fevers may be beneficial to the host. Slight temperature increases benefit the host by accelerating phagocytic and antibody responses. However, strong fevers of 40°C (104°F) or greater usually benefit the pathogen because host tissues are further damaged.

Three characteristic fever patterns have been recognized in infectious disease. (1) *Continuous fever* is that condition in which the body temperature remains elevated over a whole 24-hr period, and the total range of variation in temperature is less than 1°C. Continuous fever is seen in *typhoid fever* (∞ Section 24.12) and *typhus fever* (∞ Section 24.3). (2) A *remittent fever* is one in which the body temperature is abnormal over the whole of a 24-hr period, and the daily range shows swings greater than 1°C. This occurs in some *pyogenic infections* (∞ Section 23.2) and in *tuberculosis* (∞ Section 23.3). (3) An *intermittent fever* is one in which the temperature is normal for part of the day and then rises above normal. Most infectious diseases elicit some intermittent fever, and the condition is a diagnostic characteristic of malaria (∞ Section 24.5), a protozoan infection. *Relapsing fever,* caused by various *Borrelia* species (∞ Sections 24.4 and 13.12) is an intermittent fever in which the temperature remains normal for a long period of time, followed by a new attack of fever. This is characteristic of an incomplete recovery from an infectious disease, the fever arising when the infection periodically reestablishes itself.

✓ 19.13 Concept Check

Inflammation and fever are nonspecific responses to noxious stimuli such as pathogens. These host responses can result in accelerated isolation and destruction of the pathogen or the further destruction of host tissue.

✓ Describe the chief symptoms of inflammation.
✓ Describe the three types of fever.

REVIEW QUESTIONS

1. Distinguish between a parasite and a pathogen. Distinguish between infection and disease.

2. Which organs of the human body are normally colonized with microorganisms? What do these organs have in common? Which organs are normally devoid of microorganisms? What do these organs have in common?

3. Distinguish between resident and transient microorganisms at a body site. How could you distinguish between resident and transient microorganisms experimentally?

4. Why are members of the genus *Streptococcus* instrumental in forming dental caries? Why are they more capable of causing caries than other organisms?

5. What region of the gastrointestinal tract has the highest concentrations of microorganisms? What region has the lowest concentrations? Why?

6. Describe the relationship between *Lactobacillus acidophilus* and glycogen in the vaginal tract. Why do adult females have a different local flora in the vaginal tract than juvenile females?

7. Give two examples of adherence factors important for pathogen attachment. At least one example should not be a protein.

8. Distinguish between the three modes of action for exotoxins. Give an example of each category of toxin. How does each toxin category promote disease?

9. Review the mode of action of cholera toxin. What is the appropriate therapy for this disease, and why is antibiotic treatment generally not effective?

10. Describe the structure of a typical endotoxin. How does endotoxin induce fever?

11. Give an example of a microorganism that is pathogenic almost solely because of its toxin-producing ability. Define the toxin and its mode of action. Give an example of a microorganism that is pathogenic almost solely because of its invasive characteristics. What factor or factors confer invasive qualities on this microorganism?

12. How do temperature and pH work to limit bacterial infections? Where in or on the body might you find temperature and pH values that are different from standard body conditions? What organisms might benefit from differences in temperature or pH?

13. Distinguish between a continuous fever and an intermittent fever. Which type most commonly occurs in infectious diseases?

APPLICATION QUESTIONS

1. How does mucus inhibit the growth of most microorganisms? Describe experiments to demonstrate the effects of mucus in protection against bacterial colonization.

2. The skin is an effective barrier against colonization and growth of microorganisms. However, it is colonized with a variety of different microorganisms that occasionally cause disease. Explain these seemingly contradictory statements.

3. What steps are involved in the formation of dental plaque? Describe and discuss experiments that demonstrate the buildup of plaque on toothlike surfaces. Design experiments to illustrate biological methods for removal of plaque.

4. Obligately anaerobic bacteria are very common in the large intestine, yet they are able to colonize and grow only if facultatively aerobic bacteria are also present. Explain. How could you test the validity of your answer in the laboratory?

5. Antibiotic therapy can sterilize the genitourinary tract, at least with respect to normal flora. Unfortunately such episodes are often followed by infections due to opportunistic pathogens, many of which are caused by organisms causing opportunistic infections in individuals with AIDS (∞ Section 23.7). What pathogens are involved? Why are individuals who have undergone antibiotic therapy susceptible to these pathogens?

What measures could be taken to limit these opportunistic pathogens?

6. How would you operationally and experimentally distinguish between the three major modes of toxin action for a new exotoxin?

7. Describe how enteropathogenic strains of *Escherichia coli* differ from normal strains of *E. coli*. Include a discussion of structural and ecological variables.

8. Although mutants incapable of producing exotoxins are relatively easy to isolate, mutants incapable of producing endotoxins are much harder to isolate. From what you know of the structure and function of these types of toxins, explain the differences in mutant recovery.

9. *Salmonella* species produce *at least* 10 different gene products that act to increase the virulence of the pathogen. *Streptococcus pneumoniae*, arguably a more virulent pathogen (see Figure 19.22), relies on a single virulence factor. Review the critical virulence factors for each organism and explain why a single factor is so important and effective for *S. pneumoniae*. Why haven't all human pathogens adopted *S. pneumoniae*-like virulence schemes?

10. Should fever always be treated? Give reasons for your answer based on your knowledge of the importance of the inflammatory response in limiting the spread of infection.

The immune response is designed to recognize virtually any protein. In the illustration, antibodies are shown interacting with a number of different sites on a particular protein, or antigen. There are a number of key features for a typical immune response. First, the response is specific: Antibodies interact with only a single protein antigen and do not interact with other proteins. Second, after the first exposure to antigen, subsequent exposure induces a more rapid response and also produces a higher quantity of the antigen-specific antibody. Finally, during embryological development, cells that produce antibodies that react against host proteins are eliminated. The immune response has thus evolved to tolerate self antigens while destroying foreign ones, resulting in a unique and effective pathogen-specific host defense system.

CHAPTER 20   # Concepts of Immunology

20.1 Cells and Organs of the Immune System   804
20.2 Immunogens and Antigens   806
20.3 Nonspecific Immunity: Phagocytes and Phagocytosis   808
20.4 Specific Immunity: Lymphocytes   811
20.5 Immunoglobulins (Antibodies)   812
20.6 T Cell Receptors   823
20.7 Histocompatibility Proteins   824
20.8 Cytokines   826
20.9 Cell-Mediated Immunity   829
20.10 Clonal Selection and Immune Tolerance   830
20.11 Mechanism of Immunoglobulin Formation   832
20.12 The Complement System   835
20.13 Polyclonal and Monoclonal Antibodies   837
20.14 Antigen–Antibody Reactions   840
20.15 Immune Diseases   843
20.16 Immunity to Infectious Diseases   847
20.17 Alternative Immunization Strategies   851

Antibody a soluble protein, produced by B cells, that interacts with antigen; also called immunoglobulin

Antigen a molecule capable of interacting with specific components of the immune system

Antigenic determinant that portion of an antigen that is reactive with a specific antibody or T cell receptor; also called an epitope

Antigen-presenting cell (APC) any cell that functions primarily to present antigen to a T cell

Autoantibody an antibody that reacts to self antigens

B Cell a lymphocyte that produces immunoglobulin

Cell-mediated immunity (CMI) immunity resulting from direct interaction with antigen-specific T cells

Chemokine a low-molecular-weight soluble immune response modulator protein produced by a variety of cells

Class I MHC proteins antigen-presenting molecules found on all nucleated vertebrate cells

Class II MHC proteins antigen-presenting molecules found primarily on macrophages, B cells, and dendrocytes

Clonal selection a theory that each B or T cell, when stimulated by antigen, produces copies of itself

Complement a series of proteins that react in a sequential manner with antibody–antigen complexes to amplify or potentiate their activity

Cytokine a soluble immune response modulator produced by leukocytes

Domain a region of a protein having a defined structure and function

Hapten a low-molecular-weight substance that combines with specific antibodies but that is incapable of eliciting an immune response by itself

Humoral immunity immunity resulting from direct interaction with antibodies

Hybridoma an artificially fused product of two unrelated cells that exhibits properties of both cells; used to produce monoclonal antibodies

Hypersensitivity an immune response leading to damage to host tissues, sometimes referred to as allergies

Immunity the ability of an organism to resist infection

Immunization (vaccination) inoculation of a host with inactive or weakened pathogens or pathogen products to stimulate protective immunity

Immunodeficient having a dysfunctional or completely nonfunctional immune system

Immunogen a molecule capable of eliciting an immune response

Immunoglobulin (Ig) a soluble protein, produced by B cells, that interacts with antigens; also called antibody

Immunological memory ability to rapidly produce large quantities of specific immune cells or antibodies after subsequent exposure to a previously encountered antigen

Interleukin (IL) soluble cytokine mediator secreted by leukocytes

Leukocytes nucleated cells found in the blood (white blood cells)

Lymph a fluid similar to blood but which lacks red blood cells and travels through a separate circulatory system (the lymphatic system) containing lymph nodes, which filter out particulate materials such as bacterial cells

Lymphocytes a subset of nucleated cells found in the blood that are involved in the immune response

Macrophage a type of large leukocyte that has phagocytic properties

Major histocompatibility complex (MHC) a genetic complex responsible for encoding several cell surface proteins important in antigen presentation

Monoclonal antibody an antibody that is the product of a single B cell clone

Natural killer (NK) cell a specialized lymphocyte that recognizes and destroys foreign cells or infected host cells in a nonspecific manner

Plasma the liquid portion of the blood with cells removed and clotting proteins deactivated

Polyclonal antiserum a serum containing antibodies derived from many different B cell clones, as occurs in a normal immune response

Polymorphonuclear leukocyte (PMN) a type of leukocyte exhibiting phagocytic properties, a granular cytoplasm, and a multilobed nucleus; a neutrophil

Primary antibody response antibodies made on first exposure to antigen; mostly of the class IgM

Secondary antibody response antibody made on second (or any subsequent) exposure to antigen; mostly of the class IgG

Serology the study of antigen–antibody reactions *in vitro*

Serum the liquid portion of the blood with clotting proteins and cells removed

Specificity the ability of the immune response to interact with individual antigens

T cell a lymphocyte responsible for antigen-specific cellular interactions; T cells are divided into functional subsets including T_C cytotoxic T cells and T_H helper T cells. T_H cells are further subdivided into T_H1 inflammatory cells and T_H2 helper cells, which aid B cells in antibody formation

T cell receptor (TCR) antigen-specific receptor protein on the surface of T cells

Tolerance inability to mount an immune response to specific antigens

In the previous chapter, we discussed some general mechanisms that the human body uses to prevent colonization and infection by microorganisms. In this chapter, we will examine the events that take place when physical and chemical host defense factors are not sufficient to control invading microorganisms.

We will first consider the cells and mechanisms responsible for **immunity,** the ability of an organism to resist infection. Next, we will examine *nonspecific immunity,* the general ability of certain cells to resist most pathogenic viruses, bacteria, and fungi. Nonspecific reactions with pathogens trigger the *specific immune response,* a highly sophisticated mechanism found in vertebrates for developing resistance to individual pathogens. A specific immune response can be made against a wide variety of individual molecules that are foreign to the vertebrate host. These foreign molecules, collectively known as **immunogens,** are usually macromolecular components of the pathogens, such as surface proteins, or secreted proteins, such as toxins. Foreign molecules are known as **antigens** when they are recognized by the immune system.

Antigens are passed along to antigen-specific cells, the *T lymphocytes* or *T cells,* by a process called *antigen presentation.* Certain T cells interact directly with antigen-bearing cells and destroy the cells directly (T_C, or T cytotoxic cells). Other T cells, the T_H, or T-helper cells, interact with antigen and secrete cell-activating proteins called **cytokines.** Through the cytokines, T_H cells activate a variety of different effector cells, which destroy the antigen-bearing cells. Collectively, these cells produce the immune responses known as *cell-mediated immunity.* Still other T_H cells interact with a second group of antigen-specific cells, the *B lymphocytes* or *B cells,* which then make proteins called *immunoglobulins* or *antibodies.* The antibodies, generally found as soluble proteins in serum or body secretions, react with the antigen to destroy or neutralize it. Antibody-mediated immunity is known as *humoral immunity.* The cell-mediated responses are produced by T cells that react directly to antigens or microorganisms *within host cells,* while the antibodies interact with pathogens or their products found *outside the host cells.*

The adaptive immune response has three major characteristics: specificity, memory, and tolerance (Figure 20.1). First, the **specificity** of the antigen–antibody or antigen–T cell interaction is unlike the other host resistance mechanisms we have discussed previously. Anatomical and nonspecific host responses are present to challenge virtually any invading microorganism, even those the host has never before encountered. In the specific immune response, however, each new microorganism must interact with the immune system *before* a response occurs. No specific immune response can be detected for several days after the first contact with the pathogen. However, when

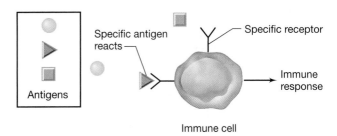

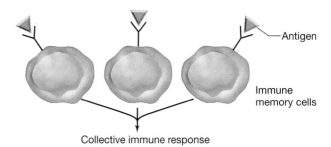

Specificity: Immune cells recognize and react with individual molecules (antigens) via direct molecular interactions.

Memory: The immune response to a specific molecule is faster and stronger upon subsequent exposure because the initial antigen exposure induced growth and division of antigen-reactive cells, resulting in multiple copies of antigen-reactive cells.

Tolerance: Immune cells are not able to react with self antigen. Self-reactive clones are destroyed during development of the immune response.

FIGURE 20.1   Key features of the adaptive immune response.

the immune response occurs, it is directed against antigens on that particular microorganism.

Second, once the immune system produces a specific type of antibody or activated T cell, challenge by further exposure to the same microorganism results in the rapid production of large amounts of the same antibody or large numbers of antigen-reactive T cells. The specific immune effectors, either T cells or antibodies, then interact with the invader and destroy it. This capacity for responding to challenge after additional exposure to a pathogen is known as **immunological memory.** Memory allows the host to resist reinfection by specific pathogens that have been previously encountered. We take advantage of this principle by employing the procedure of **immunization,** the practice of inoculating the host with inactive or weakened pathogens to artificially stimulate immunity and actively enhance specific protection against individual pathogens.

Finally, **tolerance,** the inability to make an immune response to certain antigens, occurs because macromolecules in the host are also potential antigens. Tolerance, like the immune response, is learned by antigen exposure. However, in this case, the immune system learns to *not* recognize the host antigens. Host molecules would be damaged if they were recognized by specific antibodies and activated T cells. Through tolerance, the host immune response distinguishes between foreign macromolecules (nonself and potentially dangerous) and host macromolecules (self and not dangerous) and interacts appropriately with them.

Figure 20.2 is an overview of the humoral and cell-mediated aspects of specific immunity. We will first discuss the cells and molecules of the immune system and the molecular basis for antigen recognition. Then we will explain how the individual components of the immune system interact to produce specific immunity. Next, we will look at instances where this process breaks down and the host acquires diseases resulting from inappropriate immune responses. Finally, we will examine specific instances where we can manipulate the immune response to

provide protection against a pathogen and prevent a serious infection or disease.

✓ **Concept Check**

The adaptive immune response is a concerted effort by the body to resist and destroy individual pathogens. It is characterized by specificity for an individual foreign molecule (antigen), the ability to react to that molecule more rapidly and forcefully upon subsequent exposure to that antigen (memory), and the *inability* to react with self antigens (tolerance). Cellular immunity targets pathogens and their antigens found on and in host cells, while humoral or antibody-mediated immunity is directed against extracellular pathogens and their products.

✓ Why is the immune response not normally directed against self antigens?
✓ What kinds of molecules might be recognized by the cell-mediated response? By the humoral response?

20.1

Cells and Organs of the Immune System

Specific and nonspecific immunity result from the actions of cells found circulating in the blood and lymphatic systems. In this section, we will discuss some

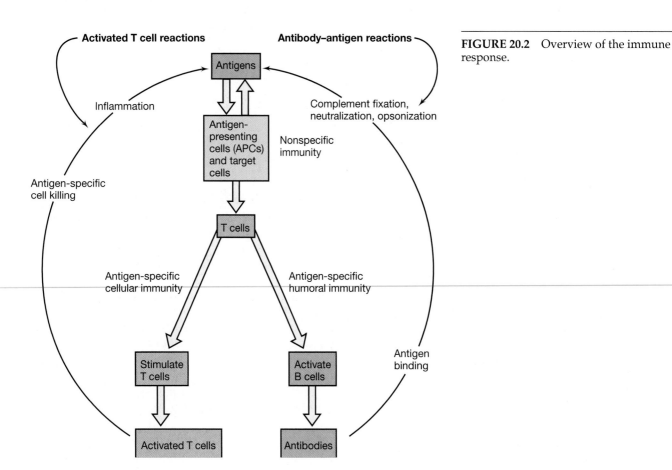

FIGURE 20.2   Overview of the immune response.

of the important cells involved in immunity and explore their origin and functions in the body. Most of the cells involved in the immune response are found in the *blood* and *lymph,* two major body fluids that directly or indirectly interact with every major organ system.

Blood and Lymph

Blood consists of cellular and noncellular components. Changes in blood properties and constituents are sensitive reflections of body changes, including disease states. Blood also contains many of the cells and molecules involved in the immune response. Because blood can be easily and safely obtained from patients, it is a valuable source of material for clinical analytical procedures, including most immune response assays. The most numerous cells in human blood are *erythrocytes* (red blood cells), which are nonnucleated cells that function to carry oxygen from the lungs to the tissues (Table 20.1). White blood cells, or *leukocytes,* include a variety of phagocytic cells such as *monocytes,* as well as cells called *lymphocytes,* which are involved in antibody production and cell-mediated immunity. Erythrocytes outnumber leukocytes by roughly a factor of 1000. Platelets are small cell-like constituents that lack a nucleus and play an important role in preventing leakage of blood from damaged blood vessels. Platelets clump together to form a temporary plug in a damaged vessel until a permanent clot forms through the action of various clotting agents, some of which are released from the platelets themselves. **Lymph** is a fluid similar to blood but lacks red blood cells.

All the blood and lymph components have a common origin. As shown in Figure 20.3, common stem cells in the bone marrow are the progenitors of all the mature cells. Stem cells differentiate to produce mature cells largely because of the influence of a group of soluble cell proteins known as **cytokines** (see Section 20.8).

When cells and platelets are removed from blood, the remaining fluid is called *plasma.* An important component of plasma is the protein fibrinogen, which undergoes a complex set of reactions during the formation of a fibrin clot. Clotting can be prevented by the addition of an anticoagulant such as potassium oxalate, potassium citrate, or heparin. Plasma is stable only when such an anticoagulant is added. If no anticoagulant is added, whole blood or plasma quickly forms a clot. The fluid left behind is called *serum.* Serum consists of proteins and the other noncellular components of plasma, except for fibrin. Since serum contains a high concentration of antibody proteins, it is widely used in immunological investigations (∞ Section 21.4).

Blood is pumped by the heart through a network of arteries and capillaries to various parts of the body and is returned through the veins (Figure 20.4). The circulatory system carries nutrients (including O_2) as well as the components of the blood involved in host resistance to infection. Figure 20.4 shows the capillary beds, a site where leukocytes may pass to and from the blood into the *lymphatic system*, a separate circulatory system through which lymph circulates.

Lymph drains from extravascular tissues into lymphatic capillaries and then into **lymph nodes** (Figure 20.4*d*) found at various locations throughout the lymph system. Lymph nodes contain high concentrations of leukocytes, arranged in such a way that they filter out microorganisms and antigens. The spleen serves an analogous function in the blood circulatory system. As a result of this filtration activity, lymph nodes may become sites of infection because organisms collected by the filtering mechanisms may proliferate if they are not destroyed. The lymph nodes and the spleen are the most important sites of most immune responses. Lymph, carrying antibodies and immune cells, eventually flows back into the circulatory system via the thoracic lymph duct.

Leukocytes are nucleated white blood cells found in the blood and the lymph. There are several distinct kinds of leukocytes (Table 20.1), but all participate in nonspecific or specific immune functions (see Sections 20.3 and 20.4). Specialized white blood cells called **macrophages** are found in abundance in the lymph nodes and carry out the lymph filtering action, as will be described later (see Section 20.3). **Lymphocytes** are specialized leukocytes involved in the specific immune response (see Section 20.4). They are concentrated in the lymph nodes and spleen and interact there with the macrophages. Lymphocytes and other leukocytes can travel throughout the body and pass freely from blood to interstitial spaces to lymph and back, a process called *extravasation* (Figure 20.4*c*). Later we will examine these leukocyte types and define how each functional leukocyte contributes to the overall immune response.

TABLE 20.1	Major cells and formed elements in normal human blood
Cell type	**Cells per milliliter**
Erythrocytes	$4.2–6.2 \times 10^9$
Leukocytes	$4.5–11 \times 10^6$
Lymphocytes	$1.0–4.8 \times 10^6$
Monocytes	Up to 8.0×10^5
Platelets	$1.5–4.0 \times 10^8$

Source: Henry, J. B. 1996. *Clinical Diagnosis and Management by Laboratory Methods,* 19th edition. W. B. Saunders, Philadelphia.

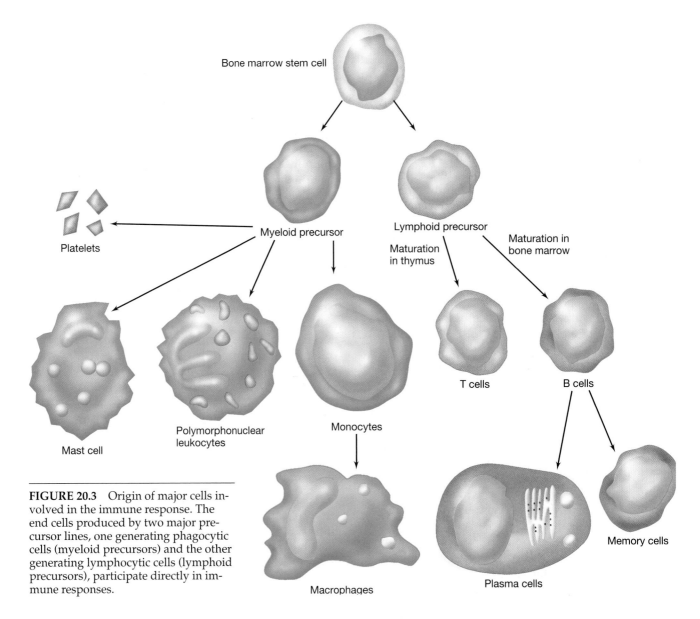

FIGURE 20.3 Origin of major cells involved in the immune response. The end cells produced by two major precursor lines, one generating phagocytic cells (myeloid precursors) and the other generating lymphocytic cells (lymphoid precursors), participate directly in immune responses.

✓ 20.1 Concept Check

All the cells involved in immunity originate from a common stem cell. The blood and lymph circulation systems include cellular and noncellular elements that are important for a functional immune system. A variety of leukocytes participate in immune responses.

✓ Where are leukocytes found in circulation? In tissues?
✓ Describe the circulation of a leukocyte from the blood to the lymph and back to the blood.

20.2

Immunogens and Antigens

Immunogens are substances that, when administered to an animal in the appropriate manner, induce an immune response. The immune response may involve an-

tibody production, the activation of specific T cells, or both. **Antigens** are substances that react with either antibodies or antigen-specific receptors known as **T cell receptors (TCRs)** that are found on T cells. Most antigens are also immunogens. However, some substances recognized by immune systems are not true immunogens. For example, **haptens** are low-molecular-weight substances that combine with specific antibody molecules but do not by themselves induce antibody formation. Haptens include such molecules as sugars, amino acids, and other low-molecular-weight organic compounds.

An enormous variety of macromolecules that are foreign to the host act as immunogens under appropriate conditions. These include all proteins and lipoproteins, many polysaccharides, some nucleic acids, and certain teichoic acids. One important requirement is that the molecules must be of fairly high molecular weight,

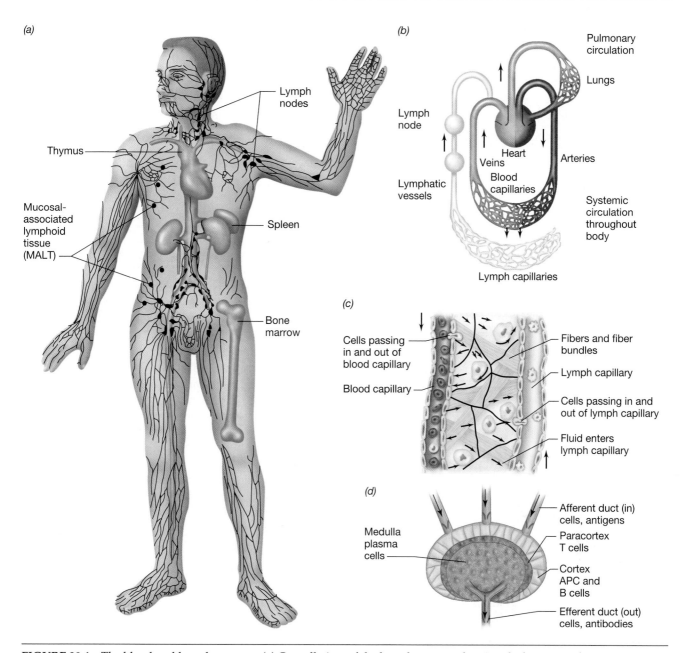

FIGURE 20.4 The blood and lymph systems. (a) Overall view of the lymph system, showing the locations of major organs. (b) Communication between the lymph and blood systems. Blood flows from the veins to the heart, then to the lungs where it becomes oxygenated, and then through the arteries to the tissues. (c) Connection between the blood and lymph systems is shown microscopically. Both blood and lymph capillaries are closed vessels, but cells and fluids can pass from one vessel to another by a process known as extravasation. (d) A lymph node. The diagram depicts major anatomical areas and the immune cells present.

usually greater than 10,000. However, the antibody or TCR does not interact with the antigenic macromolecule as a whole but only against distinct portions of the molecule that are called its **antigenic determinants** or **epitopes** (Figure 20.5). Antigenic determinants may include sugars, amino acids and other hydrocarbons. Thus, the haptens mentioned previously are actually examples of individual antigenic determinants. Antibodies are formed most readily to determinants of a polymer chain. In proteins, for example, the majority

of antibodies react with accessible surface determinants. A sequence of four to six amino acids is sufficient to define an antigenic determinant on a protein. As a result, the surface of a protein consists of a continuous array of overlapping antigenic determinants. In some cases, antibodies may even recognize epitopes that are composed of amino acids from two portions of the molecule that are distant in terms of their primary structure, but are brought together by the secondary, tertiary, or quaternary structure of the macromolecule

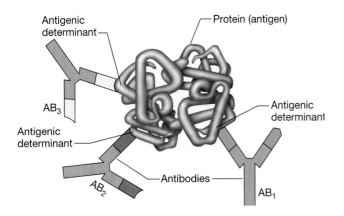

FIGURE 20.5   Antigens and antigenic determinants for antibodies. Antigens may contain several different antigenic determinants, each capable of reacting with a specific antibody. The antigenic determinant recognized by AB$_1$ is a *conformational determinant* consisting of two different parts of the same polypeptide chain. The polypeptide chain is folded to bring two distant parts of the protein together to make a single determinant. AB$_2$ and AB$_3$ recognize linear sequences on the polypeptide.

(Figure 20.5; ∞ Sections 2.7 and 2.8). These **conformational determinants (epitopes)** add to the antigenic complexity of macromolecules. The surface of a bacterial cell or virus consists of a mosaic of proteins, polysaccharides, and other macromolecules, all with individual antigenic determinants. Thus, the antigenic makeup of a typical microorganism, or even of a single protein, is extremely complex.

Antibody specificity is sensitive enough to distinguish between closely related epitopes. For instance, antibodies can distinguish between the sugars glucose and galactose, which differ only in the orientation of a hydroxyl group. However, specificity is not absolute, and an antibody may react, at least to some extent, with other epitopes. The antigen that induced the antibody is called the **homologous antigen,** and other antigens that react with the antibody are called **heterologous antigens.** The interaction between an antibody and a heterologous antigen is called a *cross-reaction.*

While antibodies generally recognize epitopes expressed on macromolecular surfaces, TCRs recognize determinants only after the macromolecules have been partially degraded. This degraded or "processed" antigen is then presented to T cells on the surface of specialized antigen-presenting cells (APCs) or target cells (see Section 20.7 and Figure 20.13). Since antigen processing and presentation normally destroy the conformational structure of an antigen, T cell epitopes consist of sequential linear portions of proteins rather than the conformational epitopes recognized by antibodies.

Immunogens are foreign molecules of sufficient size, complexity, and accessibility that are recognized by the immune system. Epitopes are the individual molecular sites at which immune molecules interact with antigens. Antibody epitopes are conformational antigen structures, whereas T cell epitopes are generally linear.

✓ Distinguish among *immunogens, antigens,* and *haptens.*
✓ How do *antibody* and *TCR* epitopes differ?

20.3

Nonspecific Immunity: Phagocytes and Phagocytosis

On rare occasions, pathogens break through the host physical and chemical defense mechanisms described in Chapter 19 (∞ Section 19.12). The pathogen can then invade host tissues and begin to colonize and infect the host (∞ Section 19.7). At this point, the immune system must become mobilized.

The starting point for immunity, whether the final effect is specific or nonspecific, cellular or humoral, is contact of a cell with the pathogen or immunogenic protein. The cell type involved in this initial contact is a *phagocyte* (literally, a cell that eats). The primary function of a phagocyte is to engulf and destroy pathogens and digest their remains. In this process, antigens are often generated and presented to antigen-specific immune cells (see Section 20.7). In this section, we will examine some of the important phagocytic cells and discuss their cytolytic and antigen-processing functions.

Phagocytes

Some of the leukocytes found in blood are phagocytes, and phagocytes are also found in various tissues and fluids of the body. Phagocytes are usually motile and move by ameboid action. Most have granular inclusions called *lysosomes*, which contain bactericidal substances such as hydrogen peroxide, lysozyme, proteases, phosphatases, nucleases, and lipases. Attracted to microorganisms by chemotactic esterases and agents such as complement proteins (see Section 20.12), phagocytes can trap a pathogen on a surface such as a blood vessel wall or a fibrin clot. After adhering to the cell, the phagocyte's cytoplasmic membrane invaginates and engulfs the foreign cell. The entire complex pinches off and eventually fuses with the lysosomes, forming a new inclusion, a *phagolysosome.* The toxic substances and enzymes inside the phagolysosome are usually capable of killing and digesting the engulfed microorganism.

One group of phagocytes, the **neutrophils,** or **polymorphonuclear leukocytes** (sometimes abbreviated

PMNs), are actively motile cells containing large numbers of lysosomes (Figure 20.6a). PMNs are short-lived cells (2–3 days) that are found predominantly in the bloodstream and bone marrow but may appear in large numbers at sites of active infection in tissues. They can move rapidly, up to 40 μm/min, and are attracted to bacteria and bacterial components, often through chemotactic proteins induced by an immune response (see Sections 20.8 and 20.12). In general, large numbers of PMNs in the blood or at a site of inflammation indicate an active infection.

Macrophages and **monocytes** are the other major groups of phagocytic cells. Macrophages are large cells capable of ingesting and destroying most pathogens and antigens as well as cooperating with lymphocytes in the production of specific immunity. Monocytes are circulating cells that differentiate to become macrophages (Figures 20.6a and 20.7). Thus, the term *macrophage* is generally used to describe phagocytes that are fixed to tissue surfaces, and the term *monocyte* describes the circulating precursor. Macrophages are up to 10 times larger than monocytes and are abundant in lymphoid tissue and spleen, whereas monocytes circulate in the blood and lymph. Macrophages are important *antigen-presenting cells* (APCs); they can present partially degraded foreign antigens to antigen-specific T cells, the first step in antibody production (see Section 20.11). This specialized feature of macrophages makes them a very important component of antigen-specific immunity, and we will examine their role as APCs in more detail in Section 20.7.

During the process of phagocytosis, phagocytes convert from aerobic respiration to anaerobic metabolism. Anaerobic glycolysis results in the formation of lactic acid and a consequent drop in pH. This lowered pH is partly responsible for the death of the ingested microbial cell because the hydrolytic lysosomal enzymes all have acid pH optima. Finally, the initial act of phagocytosis conditions the phagocyte so that it is more efficient—a cell that has recently phagocytized can take up bacteria about 10 times more effectively than a cell that has not.

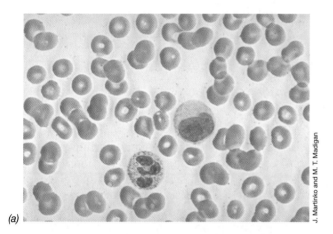

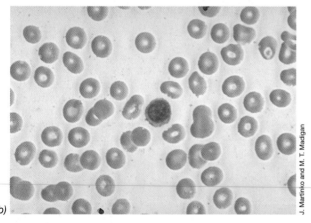

FIGURE 20.6 Major immune cell types. (a) Phagocytic cells. The nucleated cell in the lower left center is a neutrophil (PMN), characterized by a segmented nucleus (violet stain) and granulated cytoplasm. The nucleated cell to the right and slightly above the PMN is a monocyte. Phagocytes are 12–15 μm in diameter. The nonnucleated red blood cells are about 6 μm in diameter. (b) The nucleated cell is a circulating lymphocyte. The lymphocyte has almost no visible cytoplasm and is smaller than the phagocytes, about 10 μm in diameter.

Oxygen-Dependent Phagocytic Killing

As discussed in Section 5.12, various biochemical reactions can lead to the formation of toxic oxygen-containing compounds including hydrogen peroxide (H_2O_2), superoxide anions (O_2^-), hydroxyl radicals (OH·), and singlet oxygen ($^1O_2$). Phagocytic cells make use of toxic forms of oxygen to kill ingested bacterial cells. Superoxide, formed by the reduction of O_2 by NADPH oxidase, reacts at the acid pH of the phagolysosome to yield singlet oxygen and hydrogen peroxide (Figure 20.8). The phagocytic enzyme myeloperoxidase forms hypochlorous acid (HOCl) from chloride ions and H_2O_2, and the HOCl reacts with a second molecule of H_2O_2 to yield additional singlet oxygen. Activation of macrophages in some mammalian species catalyzes the production of nitric oxide (NO) by the inducible enzyme *nitric oxide synthase*. NO directly inhibits viral replication and is toxic to bacteria. It may also react with other toxic oxygen compounds, producing highly reactive and bactericidal *peroxynitrate* radicals. The combined action of these oxygen-dependent phagocyte enzymes forms sufficient levels of toxic oxygen compounds to kill ingested bacterial cells by oxidizing key cellular constituents. These reactions occur within the phagocytic cell itself, which is not damaged by the toxic oxygen products. The action of phagocytic cells in oxygen-mediated killing is summarized in Figure 20.8.

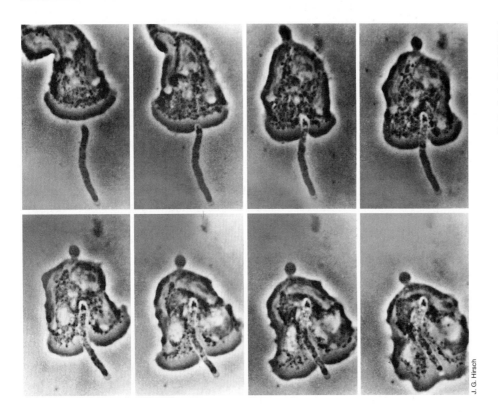

J. G. Hirsch

FIGURE 20.7 Phagocytosis: time-lapse photomicrographs of the engulfment and digestion of a chain of *Bacillus megaterium* cells by a human macrophage, observed by phase contrast microscopy. The bacterial chain is about 18–20 μm long.

Phagocyte Failure

In some cases, pathogens have developed mechanisms for neutralizing the effects of toxic phagocyte products, for killing the phagocyte, or for avoiding phagocytosis. For example, *Staphylococcus aureus* (∞ Section 23.2) produces pigmented compounds called *carotenoids*, which quench singlet oxygen and prevent killing (∞ Section 5.12). Intracellular pathogens such as *Mycobacterium leprae* (leprosy bacillus) and *Mycobacterium tuberculosis* (tuberculosis bacillus) grow and persist within phagocytic cells (∞ Section 23.3). They apparently use cell wall-associated phenolic glycolipids (∞ Section 13.22) to scavenge toxic oxygen compounds. These glycolipids are highly effective in removing hydroxyl radicals and superoxide anions, the most damaging of the toxic oxygen species produced by phagocytic cells.

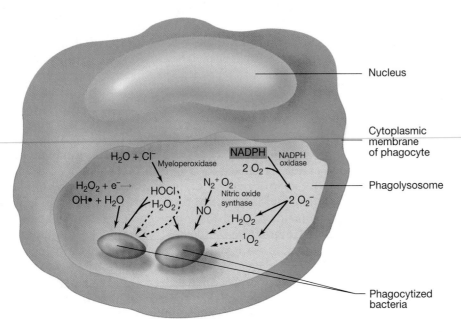

Nucleus

Cytoplasmic membrane of phagocyte

Phagolysosome

Phagocytized bacteria

FIGURE 20.8 Action of phagocyte enzymes in generating toxic oxygen species. These include hydrogen peroxide (H_2O_2), the hydroxyl radical (OH·), hypochlorous acid (HOCl), the superoxide anion (O_2^-), singlet oxygen ($^1O_2$), and nitric oxide (NO).

Other intracellular pathogens produce proteins called *leukocidins,* which destroy phagocytes. In such cases, the pathogen is not killed when ingested but instead kills the phagocyte and is then released. *Streptococcus pyogenes* and *Staphylococcus aureus* are the major leukocidin producers. Destroyed phagocytes make up much of the material of *pus;* organisms that produce leukocidins are therefore usually *pyogenic* (pus-forming) and cause localized infections resulting in boils or abscesses (∞ Sections 19.7 and 23.2).

Another important microbial defense against phagocytosis is the bacterial capsule (∞ Section 3.13). Capsulated bacteria are often highly resistant to phagocytosis, apparently because the capsule somehow prevents adherence of the phagocyte to the bacterial cell. The clearest case of the importance of a capsule that prevents phagocytosis is that of *Streptococcus pneumoniae.* If only a few cells of a capsulated strain of this species are injected into a mouse, an infection is initiated that leads to death in a few days (∞ Figure 19.22). On the other hand, even large numbers of a noncapsulated mutant strain are completely avirulent. Surface components other than capsules can also inhibit phagocytosis. For instance, pathogenic *Streptococcus pyogenes* produces a specific substance, the M-protein, on both the cell surface and fimbriae (∞ Section 23.2). M-protein apparently alters the surface properties of the bacterial cell in such a way that phagocytes cannot act.

Antibodies to capsules or other cell surface molecules often reverse the protective effect of these bacterial defense mechanisms and enhance phagocytosis, a process known as *opsonization* (see Section 20.12).

✓ 20.3 Concept Check

The phagocytes are the first cells involved in protection against invading pathogens after physical and chemical barriers have been breached. They are capable of destroying most, but not all, pathogens. In many cases, phagocytes also process and present molecules to antigen-specific leukocytes. Many pathogens have developed mechanisms for inhibiting phagocytosis and killing.

✓ Describe separately the functions of *PMNs* and *macrophages.*
✓ What oxygen-dependent mechanisms are used by phagocytes to kill pathogens?

20.4

Specific Immunity: Lymphocytes

As we have discussed, one of the functions of phagocytes is to present antigens to other cells, the antigen-specific leukocytes known as **lymphocytes.** Lymphocytes are one of the most prevalent mammalian cell types (the aver-

age adult human has about 10^{12} lymphocytes) (Figure 20.6b). Both B lymphocytes and T lymphocytes are involved in antigen-specific immune responses and are derived from stem cells in the bone marrow. The differentiation of stem cells into mature lymphocytes is determined by the organ in which precursor lymphocytes become established (Figure 20.3). B cells mature in the bone marrow in mammals, but develop in a special organ called the *Bursa of Fabricius* in birds (hence, the designation "B"). T cells mature in the thymus (thus, the designation "T"). Because of their role in the initial development and maturation of B and T cells, the bone marrow, bursa (birds only), and thymus are called *primary lymphoid organs.* After maturation, B and T cells are dispersed throughout the body via the blood and lymph (Figure 20.4) and come to reside in the lymph nodes, spleen or *mucosa-associated lymphoid tissue (MALT)* (Figure 20.4a), which are collectively known as *secondary lymphoid organs.* The spleen and lymph nodes are positioned in the blood and lymph and act as filters where they trap antigens. The MALT interacts with antigens that are found on mucosal surfaces. The B and T cells in these organs can then produce an immune response.

B Lymphocytes

B cells are responsible for antigen interaction, antibody production, and immune memory. B lymphocytes are readily distinguished from T lymphocytes by the presence of immunoglobulin molecules on their surface. These surface immunoglobulins are copies of the single type of antibody that a given B cell will produce later in its development. Surface immunoglobulins on B cells recognize antigen in native conformation, usually on the pathogen surface (see Section 20.11). B cells are concentrated in the cortex of the lymph nodes where they can contact antigens. After antigen exposure, B cells divide into memory cells or plasma cells. The memory cells are long-lived and may remain in the cortical area for years. If reexposed to the same antigen, memory cells quickly proliferate to produce more memory cells and plasma cells. The differentiated, antibody-producing plasma cells live for only several days. They are found in the medulla where the immunoglobulins can drain directly into an efferent lymph vessel (Figure 20.4d).

T Lymphocytes

The situation with T cells is more complex. All T cells have antigen-specific T cell receptors (TCRs) on their surface and interact specifically with antigen. In the case of TCRs, the antigen is always "presented" by another cell, often a macrophage. Because this antigen is always in a processed form, degraded by the lysosomal proteases, the determinant is always a short, linear peptide derived from the intact immunogen (see Sections 20.2 and 20.7).

Several functionally distinct subsets of T cells have been identified. Two major subpopulations are distinguished from each other by the presence of specific cell surface proteins called CD4 or CD8: A single mature T cell has only one of these proteins (Figure 20.9). The CD4 population is subdivided into two functional subsets called T_H1, or T helper 1 cells, and T_H2, or T helper 2 cells. T_H1 cells participate in cell-mediated immunity and are responsible for recruiting and activating nonspecific effector cells such as phagocytes. They are thus often called *T inflammatory* cells. T_H2 cells stimulate B lymphocytes to produce large amounts of antibody. In most cases, little if any antibody is made by B cells without T_H2 cooperation (see Section 20.11). The second major T cell population, CD8 cells, have only a single functional T cell set, the *T cytotoxic* (T_C) cells, also known as *CTLs* or *cytotoxic T lym-*

phocytes. The T_C cells kill antigen-bearing cells directly and specifically through interaction between a cell-surface antigen on the target cell and the antigen-specific T-cell receptor. In addition to functional differences and surface proteins, T cell subsets are differentiated from one another by their unique pattern of secreted **cytokines,** a group of molecules that influences the metabolic and functional activities of leukocytes (see Section 20.8). Table 20.2 compares functional properties, surface antigens, and cytokine production of B and T lymphocytes.

✓ 20.4 Concept Check

Lymphocytes are antigen-specific leukocytes. B lymphocytes have immunoglobulin antigen receptors on their surface, whereas T cells have antigen-specific T cell receptors. T cells are divided into a number of subsets based on their surface proteins and functional characteristics.

✓ What are the chief differences between *B cells* and *T cells*?
✓ Differentiate between T_H cells and T_C cells. Are their surface proteins different? Are their secreted cytokines different?

20.5

Immunoglobulins (Antibodies)

The next three sections of this chapter discuss the antigen-specific molecules of the immune system. We begin with immunoglobulins because we understand their structure and function in detail. They serve as the model for an antigen-specific receptor molecule. *Immunoglobulins (antibodies)* are protein molecules that are able to combine with antigenic determinants. They are found in the serum and in other body fluids such as gastric secretions and milk. Serum containing antigen-specific antibody is often called **antiserum.** Immunoglobulins (Ig's) can be separated into five major classes on the basis of their physical, chemical, and immunological properties: **IgG, IgA, IgM, IgD,** and **IgE** (Table 20.3). Immunoglobulin class IgG has been further divided into four immunologically distinct subclasses called IgG_1, IgG_2, IgG_3, and IgG_4. These subclasses are genetically, structurally, and functionally different from one another. On initial immunization, the first immunoglobulin to appear is IgM, a pentameric immunoglobulin with a molecular weight of about 970,000; IgG appears later. In most individuals about 80% of the serum immunoglobulins are IgG proteins, and these have therefore been studied extensively.

Immunoglobulin Structure

Immunoglobulin G is the most common circulating antibody, and thus we will discuss its structure in some detail. Immunoglobulin G has a molecular weight of

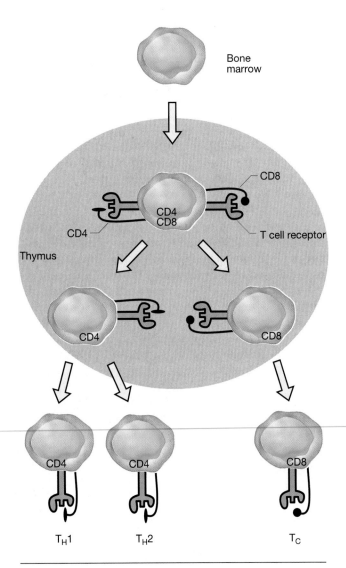

FIGURE 20.9 T cell subsets. T cell subsets develop in the thymus from bone marrow precursor cells. T_H1, T inflammatory cells; T_H2, T helper 2 cells; T_C, T cytotoxic cells.

TABLE 20.2	Comparison of B and T Lymphocytes
T Cells	**B Cells**
Origin: bone marrow	Origin: bone marrow
Maturation: thymus	Maturation: bone marrow
Long-lived: months to years	Long-lived: years (memory cells); short-lived: days (plasma cells)
Mobile	Relatively immobile (stationary)
T cell antigen receptor (TCR) on surface	Immunoglobulin antigen receptor on surface
CD4 or CD8 on surface	Complement receptors on surface (C3R)
CD3 on surface	
Restricted antigenic specificity	Restricted antigenic specificity
Proliferate on antigenic stimulation	Proliferate on antigenic stimulation into plasma cells and memory cells
T_H1 cells secrete IL-2, TNF-β, and IFN-γ cytokines, which promote cell-mediated immune functions	Synthesize immunoglobulin (antibody)
T_H2 cells secrete IL-4, IL-5, IL-6, and IL-10 cytokines, which promote B-cell activation	May serve as an antigen-presenting cell (APC)
T_C cells secrete perforin and granzyme, cytokines that are toxic for target cells	

about 150,000 and is composed of four polypeptide chains (Figure 20.10a). Both intrachain and interchain disulfide (S—S) bridges are present. The two light (short) chains of about 25,000 molecular weight are identical in amino acid sequence, as are the two heavy (longer, 50,000 molecular weight) chains. The molecule as a whole is thus symmetric. Each light chain consists of about 220 amino acids, and each heavy chain consists of about 450 amino acids.

When an IgG molecule is treated with the proteolytic enzyme *papain*, it breaks into several fragments (Figure 20.10b). Two fragments contain the complete

TABLE 20.3	Properties of human immunoglobulins				
Class/ H chain isotype[a]	Molecular weight/ formula[b]	Serum (mg/ml)	Antigen binding sites	Properties	Distribution
IgG γ	150,000 2(H + L)	13.5	2	Major circulating antibody; four subclasses: IgG_1, IgG_2, IgG_3, IgG_4; IgG_1 and IgG_3, activate complement	Extracellular fluid; blood and lymph; crosses placenta
IgM μ	970,000 (pentamer) 5[2(H + L)] + J	1.5	10	First antibody to appear after immunization; strong complement activator	Blood and lymph; monomer is B cell-surface receptor
	175,000 (monomer) 2(H + L)	0	2		
IgA α	150,000 2(H + L)	3.5	2	Important serum antibody	Secretions (saliva, colostrum, cellular and blood fluids); monomer in serum and dimer in secretions.
	385,000 (secreted dimer) 2[2(H + L)] + J + SC	0.05	4	Major secretory antibody	
IgD δ	180,000 2(H + L)	0.03	2	Minor circulating antibody	Blood and lymph; B lymphocyte surfaces
IgE ϵ	190,000 2(H + L)	0.00005	2	Involved in allergic reactions; CH4 contains mast cell binding fragment	Blood and lymph; binds to mast cell surfaces

a All immunoglobulins may have either λ or κ light chain types, but not both.

b Based on the number and arrangement of heavy (H) and light (L) chains in each functional molecule. J is a joining protein present in serum IgM and secretory IgA. SC is the secretory component found in secreted IgA. See Figure 20.10 for a diagram of each immunoglobulin.

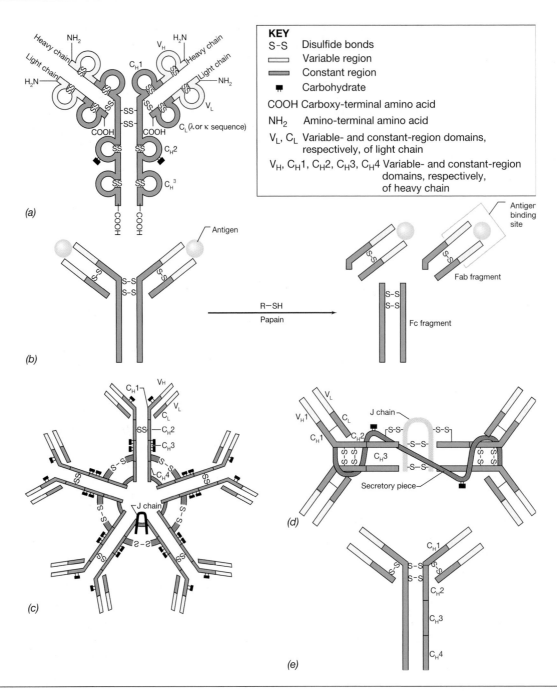

FIGURE 20.10 Structure of immunoglobulin molecules. (a) Structure of IgG showing disulfide links within (intrachain disulfide bonds) and between chains (interchain disulfide bonds). The monomer consists of two γ heavy chains and two κ or λ light chains. IgD has a similar structure, with a δ heavy chain. (b) Papain (an enzyme) digestion products of IgG. Note the production of two antigen binding fragments, Fab. (c) Structure of serum IgM, a large immunoglobulin composed of five individual molecules (a pentamer) having 10 antigen binding sites. The intrachain disulfide bonds are omitted to simplify the structure. Note the presence of a C_H4 domain in the μ heavy chain and the addition of a J (joining) chain, which covalently links two IgM monomers in the pentamer. (d) Structure of secretory IgA. Note the J chain and the secretory piece, a polypeptide that aids in transfer of IgA across mucosal cell membranes. Both additional polypeptides are covalently bound to the α heavy chain. (e) Structure of IgE. Note the presence of a C_H4 domain in the ε heavy chain.

light chain plus the amino-terminal half of the heavy chain. These portions combine with antigen and are called *Fab* fragments (*fragment of antigen binding*). The fragment containing the carboxy-terminal half of both heavy chains, called *Fc* (*fragment crystallizable*), does *not* combine with antigen. Therefore, each antibody molecule of the IgG class contains *two antigen combining sites* (and is thus *bivalent*). This bivalency is of considerable importance in understanding the manner in which some antigen–antibody reactions occur (see Section 20.14). The antigen binding site is in the amino-terminal portion of both the heavy and the light chains (Figure 20.10). Immunoglobulins also contain small amounts of complex carbohydrates attached to portions of the heavy chain; the carbohydrate is not involved in the antigen binding site.

Although the view of the immunoglobulin molecules shown in Figure 20.10 is adequate for conveying the general structure of this molecule, immunoglobulins are very complex proteins that are folded into a complex three-dimensional structure. The details of immunoglobulin structure and their unique genetic make-up are discussed further in the box, Molecular Biology of the Immune Response.

Light Chains of IgG

Each IgG light chain contains two amino acid domains, the *variable domain* and the *constant domain*. The sequences of amino acids in a major portion of the light chains of immunoglobulins of the class IgG are frequently identical, even in IgGs directed against completely different antigenic determinants. This is because the amino acid sequence in the carboxy-terminal half of the light chain constitutes one of two specific and constant sequences, referred to as the *lambda* (λ) sequence and the *kappa* (κ) sequence. One IgG molecule has either two λ chains or two κ chains but never one of each (Figure 20.10a). By contrast, light-chain *variable* domains (V_L), located in the amino-terminal half of the light chain, always differ in amino acid sequence from one IgG molecule to the next unless both molecules are produced by the same cell or clone of cells.

Heavy Chains of IgG

Each IgG heavy chain contains one variable and three constant domains. Each domain is approximately 110 amino acids long. Analogous to the situation that exists in the light chain, all immunoglobulins of the IgG class have a portion of their heavy chain (the three carboxy-terminal domains) in which the amino acid sequence is identical (C_H1, C_H2, and C_H3) (Figure 20.10a) from one IgG molecule to another. In addition, each

heavy chain has a region in the amino-terminal domain (the heavy chain variable domain, V_H) (Figure 20.10a) where considerable amino acid sequence variation occurs from one IgG to the next. The specificity of a given antibody molecule for a particular antigen lies in the unique three-dimensional structure of the antigen binding site, formed by a combination of the amino acid sequences in the variable domains of the heavy and light chains (see the box, Molecular Biology of the Immune Response).

Other Classes of Immunoglobulins

How do immunoglobulins of the other classes differ from IgG? The heavy-chain *constant* domain of a given immunoglobulin molecule defines its class and can have one of five amino acid sequences: gamma (γ), alpha (α), mu (μ), delta (δ), or epsilon (ε). These sequences constitute the carboxy-terminal three-fourths of the heavy chains of immunoglobulins of the class IgG, IgA, and IgD, respectively and four-fifths of the heavy chains of IgM and IgE (Figure 20.10). Each antibody of the IgM class, for example, contains amino acids in its heavy-chain constant domains that constitute the mu sequence. If two immunoglobulins of *different classes* react with the same antigenic determinant, then the variable domains of their heavy and light chains can be identical, but their class-determining sequences, specific to their *heavy* chains, would be different. It is not unusual in a typical immune response to observe the production of antibodies of two different classes with the same heavy chain variable domain.

The structure of **immunoglobulin M** (IgM) is shown in Figure 20.10c. It is usually found as an aggregate of five immunoglobulin molecules attached by at least one *J chain*. IgM accounts for 5–10% of the total serum immunoglobulins. Each heavy chain of IgM contains an extra constant domain (C_H4), and IgM has a very high carbohydrate content. IgM is the first class of immunoglobulin made in a typical immune response to a bacterial infection, but immunoglobulins of this class are generally of low affinity. Antigen binding strength is enhanced to some degree, however, by the high *valency* of the pentameric IgM molecule; 10 binding sites are available for interaction with antigen (Table 20.3 and Figure 20.10c). The term *avidity* is used to describe the *strength of binding* by multivalent antigen binding molecules; thus, IgM is said to be of *low* affinity but *high* avidity. IgM monomers are also found on the surface of B cells, where they bind antigen.

Immunoglobulin A (IgA), in the dimeric form (Figure 20.10d), is present in body secretions. It is the dominant antibody in all fluids bathing organs and tissues in contact with the outside world. IgA is present in saliva, tears, breast milk and colostrum, gastrointestinal secre-

Three kinds of proteins make specific contact with antigens during an immune response. These are the **immunoglobulins** (antibodies, or Igs), the **T-cell receptors** (TCRs), and products of the genes of the major histocompatibility complex (MHC), the **MHC proteins.** Each consists of two nonidentical polypeptides that associate to form a functional protein, and all are expressed on the surface of cells as antigen receptors. However, as we discuss in the text, the specific functions of these molecules are quite different. The *immunoglobulins,* anchored on B-cell surfaces, bind to surface proteins on bacteria and viruses and to bacterial products such as toxins; Igs are also produced in large quantities as soluble serum proteins. The *T-cell receptors,* found exclusively on T lymphocytes, in-

*Here we provide further information about the protein structure and molecular genetics of the antigen-binding molecules in the immune response. We discuss this information separately because these details are not necessary for understanding the *basic* features of the immune response. However, in-depth studies of these molecules provide insight into the molecular and genetic workings of eukaryotic cells. Immunoglobulins, T-cell receptors, and MHC proteins also provide elegant molecular solutions to the problem of generating antigen-receptor diversity. The molecular models that result are paradigms for evolutionary, genetic, and structural studies. Finally, as we shall see, molecular studies provide the details necessary to manipulate the immune response.

teract with antigenic peptides presented by target or antigen-presenting cells. The *MHC proteins* are responsible for bringing antigenic peptides to the surface of the target cells or antigen-presenting cells, where the peptides can interact with TCRs on T cells.

These differences in function and location of the TCRs, Igs, and MHC proteins are mirrored by differences in their structure and genetic organization. However, these proteins have several shared structural features and are also evolutionarily related. These molecules are members of a family of genes that have evolved by duplication and selection of primordial antigen-receptor genes, the *immunoglobulin gene superfamily.* The basic features of selected Ig superfamily proteins are shown in Fig. 1. Each protein has at least one region of highly conserved amino acid sequence, a so-called "constant" (C) domain about 100 amino acids long containing an intrachain disulfide bond spanning 50 to 70 amino acids. Beta-2 microglobulin (β2m), part of the class I MHC protein, consists of a single C domain. The "variable" (V) domains of TCRs, Igs, and MHC, are about the same size as the constant domains, but V domains are considerably different from one another and from the C

domains, based on their variable primary structure (thus their name) and their function. C domains provide structural support for the immune molecules, holding the V domains away from the membrane and giving the proteins their characteristic shape. C domains may also act as targets for other accessory molecules. For example, constant domains of IgG and IgM bind Complement (Section 20.12); MHC class I constant domains bind to the accessory CD8 protein on T_C cells, and homologous class II constant domains binds CD4 on T_H cells (Section 20.7). Variable domains interact with a wide variety of antigenic proteins from many sources.

Antigen-Binding Protein Structure

MHC proteins were discovered because of their role as the major target molecules for transplantation rejection. Most individuals have different MHC proteins, and tissue from a donor is immunologically rejected when transplanted to a recipient unless the MHC proteins of the donor and recipient are matched for structural and immunological identity. A single amino acid sequence difference in these MHC proteins may result in rejection of tissue transplanted be-

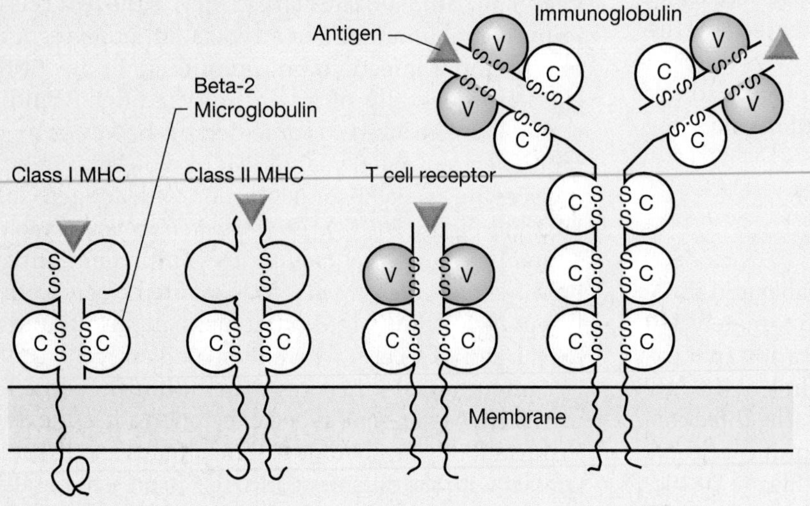

FIG. 1 Proteins encoded by the immunoglobulin gene superfamily. Constant domains (C) are regions of homologous or nearly homologous amino acid sequence and higher-order structure. The presence of the Ig-like C domains identifies the molecules as members of the Ig superfamily and strongly suggests evolutionary relatedness. The variable domains (V), however, cannot be identified as Ig superfamily genes because of their variable, nonconserved structure. Presumably, they have evolved from Ig-like domains, but because of the evolutionary pressure and selection to bind certain forms of antigen, most recognizable structural homology to the Ig-like C domain has disappeared.

tween individuals. More importantly, these sequence differences lead to changes in the ability of MHC proteins to present antigens to T cells (see Section 20.7). The MHC proteins are divided into two structural classes. Class II MHC proteins are found on the surfaces of *all* nucleated cells. Class I MHC proteins are found only on the surface of B lymphocytes, macrophages, and dendrocytes, the dedicated antigen-presenting cells (APCs). This cell-specific distribution of class I and class II proteins has important functional implications (see Section 20.7).

Class I MHC proteins consist of two polypeptides (Fig. 1 and 2*a*). The membrane-integrated alpha (α) chain of 42,000 molecular weight is encoded in the MHC gene region on human chromosome 6. The other class I polypeptide is the noncovalently associated 12,000-molecular-weight protein β2m. Class II MHC proteins consist of two noncovalently linked, membrane-integrated polypeptides, called α and β, of 33,000 and 28,000 molecular weight, respectively (Fig. 1 and 2*b*). Class II molecules are usually found in *pairs*, which enhances their ability to bind to TCRs. The three-dimensional

structures of class I and class II MHC molecules reveal a distinctive shape that suggests how these proteins interact with antigens and the cells of the immune system (Fig. 2*a* and 2*c*). The class I α chain folds to form a large groove between the α1 and α2 domains, and it is within this groove that the MHC molecule binds peptide antigens. The α1 and β1 domains of the class II protein interact to form a similar groove (Fig. 2*b*).

Immunoglobulins are found as cell-surface antigen receptors on B cells, or in soluble form in high concentrations in serum and other body fluids, where Igs function to neu-

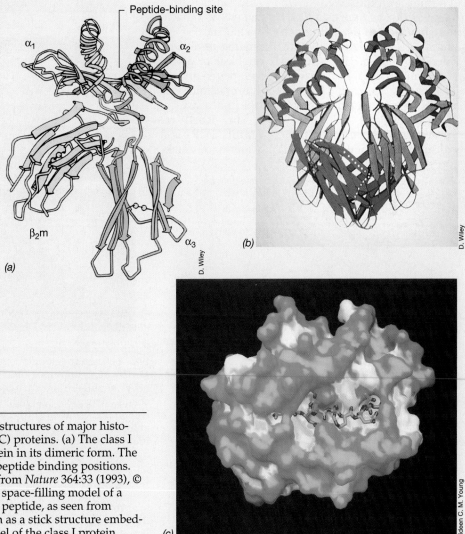

FIG. 2 Three-dimensional structures of major histocompatibility complex (MHC) proteins. (a) The class I protein. (b) The class II protein in its dimeric form. The yellow arrows indicate the peptide binding positions. Reprinted with permission from *Nature* 364:33 (1993), © Macmillan Magazines. (c) A space-filling model of a class I protein with a bound peptide, as seen from above. The peptide is shown as a stick structure embedded in the space-filling model of the class I protein.

tralize and opsonize foreign antigens (see Section 20.5). Variations in amino acid sequence occur in the *variable domains* (V domains) of different immunoglobulins (Fig. 3). This amino acid variability is especially apparent in several so-called **hypervariable regions.** These hypervariable regions, also termed *complementarity determining regions (CDRs)*, provide most of the molecular contacts with antigen. Each V domain in the light and heavy chains has three CDRs. CDR1 and CDR2 are variable between different immunoglobulins, but the CDR3 of the heavy chain has a particularly complex structure. Amino acid sequence studies have shown that the CDR3 consists of the COOH terminal portion of the V domain, followed by a short "diversity" (D) segment of 3 amino acids, and a longer joining (J) region of about 13 to 15 amino acids long. The light chain has a similar arrangement, but lacks the D segment.

An immunoglobulin three-dimensional structure is shown in Fig. 4a.

The principle of all antibody reactions lies in the *specific* combination of determinants on the antigen with the *variable* region. The combining site of an antibody molecule, formed by the association of heavy and light chains, measures about 2×3 nm and is capable of binding to a small number of amino acids, about 10–15, on a protein antigen. The recognition of antigen is ultimately governed by the immunoglobulin folding pattern of the heavy and light polypeptide chains. The result is a *unique* and *specific* antigen-binding site (Fig. 4b). Binding of antigen can block or distort the antigen and reduce or eliminate its biological activity.

The **T-cell receptor** (TCR) is a dimeric, membrane-integrated α, β dimer found only on T cells, where it functions to recognize foreign peptides embedded in major histocompatibility complex (MHC) molecules on the surface of APCs or target cells (Section 20.6 and Section 20.7). The TCRs accomplish this dual binding

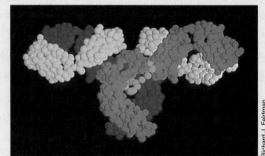

(a)

(b)

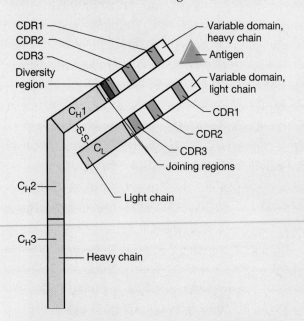

FIG. 3 The variable regions of immunoglobulin light and heavy chains. Only one half of a typical immunoglobulin molecule is shown. C_H and C_L are constant regions of heavy and light chains, respectively.

FIG. 4 Immunoglobulin structure. (a) Three-dimensional view of an Immunoglobulin G (IgG) molecule. The heavy chains are shown in red and dark blue. The light chains are in green and light blue. (b) Structure of the combining site of an antigen and immunoglobulin. The antigen (lysozyme) is in green. The variable region of the immunoglobulin heavy chain is shown in blue, and the light chain in yellow. The amino acid shown in red is a glutamine residue of lysozyme. The glutamine residue fits into a pocket on the immunoglobulin molecule, but the overall antigen–antibody recognition involves contacts made between several other amino acids on both the immunoglobulin and antigen as well. Reprinted with permission from *Science* 233:747 (1986) ©AAAS.

function using an Ig-like structure involving cooperation of the α and β chain hypervariable regions CDR1, CDR2, and CDR3. The three-dimensional structure of the TCR has been solved as part of a complex with the peptide-MHC and is shown in Fig. 5. An important feature of this interaction is that both TCR and MHC bind directly to peptide antigen. The MHC binds one face of the processed peptide, the *agretope*, while the TCR interacts with the other peptide face, known as the *epitope*. The CDR3 portion of the α and β chains of the TCR make contact with the epitope. Finally, the TCR and MHC make direct contact and interact with one another, with the CDR1 and CDR2 regions of the α and β TCR chains contacting the MHC proteins.

Genetics of the Antigen-Binding Molecules

The human **MHC** is known as the **HLA** (*h*uman *l*eukocyte *a*ntigen) complex and is located on chromosome 6 (Fig. 6). Several genes in the HLA complex encode MHC proteins and, as mentioned above, there is considerable diversity in MHC proteins in humans. The sequence differences distributed within a species are called *polymorphisms*. In humans, for example, there may be 100 different MHC polymorphic alleles for each MHC gene position (locus), but each individual has only two of these alleles (one allele is of paternal origin, and one is of maternal origin). The two variant proteins expressed by these alleles are expressed codominantly (equally). There are at least

three gene loci for MHC class I genes, HLA-A, B, and C (Fig. 6), which have a high degree of polymorphism. Likewise, there are at least three highly polymorphic loci for class II genes, HLA-DR, DP, and DQ (Fig. 6). Thus, an individual usually has *six* class I and *six* class II proteins of different sequences. These genetic variations in MHC proteins are the major barrier to successful intraspecies transplants. The genetic organization and expression of MHC molecules is very simple: One gene encodes one protein. However, the polymorphic nature of the MHC genes ensures that, within the human population, there is a large and varied pool of MHCs available for antigen presentation. With immunoglobulins, however, the situation is quite different.

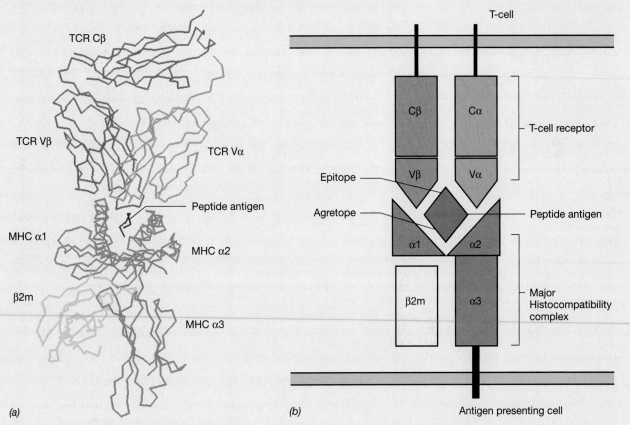

FIG. 5   The T cell receptor–peptide–MHC protein complex. (a) The three-dimensional α-carbon backbone structure, showing the orientation of TCR, an antigen peptide (red), and MHC. This structure was derived from data deposited in the Protein Data Bank (www.rcsb.org/pdb/). (b) A diagrammatic interpretation of the MHC–peptide–TCR structure.

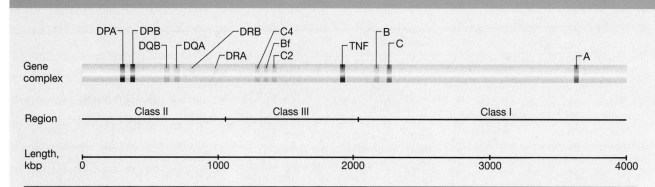

FIG. 6 Map of the HLA complex (human MHC). The gene complex, located on chromosome 6, shown in kilobase pairs (kbp), is more than 4 million bases in length. Class II genes DPA and DPB encode class II proteins DPα and DPβ; DQA and DQB encode DQα and DQβ; DRA and DRB encode DRα and DRβ. Class III genes encode several proteins associated with immune recognition functions, as well as other nonimmune-related proteins. C4 and C2 encode complement proteins C4 and C2 (see Section 20.12), while Bf encodes an alternative pathway component, factor B. The TNF gene encodes a cytokine, *tumor necrosis factor* (see Section 20.8). The class I MHC proteins HLA-B, HLA-C and HLA-A are encoded in the class I region by genes B, C, and A. Over 200 genes, many of which are involved in antigen recognition or processing, have been mapped to the HLA complex.

If one B lymphocyte produces one **immunoglobulin,** one gene should encode the light chain and one gene should encode the heavy chain. However, this is *not* the case. A single light or heavy chain is actually encoded by *several* genes that undergo a complex series of gene rearrangements (recombination followed by deletion of intervening sequences) as B cells develop. This gene rearrangement strategy turns out to be more efficient for Igs (and TCRs) than the one gene = one protein strategy used for most proteins and allows the formation of a virtually infinite number of antibodies from a very limited set of about 400 genes.

Molecular studies verified this "genes in pieces" hypothesis and also confirmed that genes for V, D, J, and C regions were separated from one another in the genome and brought together to form a mature immunoglobulin gene in a developed antigen-reactive B cell (Fig. 7). The V gene encodes CDR1 and CDR2. CDR3 is encoded by a mosaic consisting of the 3' end of the V gene, and the entire D and J genes. Finally, the class-defining constant domain of the immunoglobulin molecule is encoded by its own gene, the C gene. Thus, four differ-

ent genes, V, D, J, and C, recombine to form one functional H gene. Similarly, light chains are encoded by their own V, J, and C genes.

The genetic material for *all* antibodies exists in each lymphocyte as it develops from the stem cells of the bone marrow. As shown in Fig. 7, each B cell contains about 150 tandemly arranged light-chain V genes and 5 distinct J genes; about 200 tandem V genes, 50 D genes, and 4 J genes exist for the heavy chains. In addition, the heavy-chain constant-region genes and the light-chain constant-region genes are present. The V, D, J, and C genes are not located adjacent to one another but are separated by noncoding sequences (introns) typical of gene arrangements in eukaryotes (Section 6.9). During maturation of B lymphocytes, genetic recombination in each B cell occurs. Randomly selected V, D, and J segments are fused in each B cell by enzymes that delete all intervening DNA, resulting in the construction of an active heavy-chain gene and an active light-chain gene. The active gene (containing an intervening sequence between the VDJ gene segment and the C gene segment) is transcribed, and the resulting pri-

mary RNA transcript is spliced to yield the final messenger RNA. The RNA is then translated to make the heavy and light chains of the immunoglobulin molecule.

The final light and heavy gene in a given B cell is largely a matter of *chance rearrangement*. For example, based on the numbers of genes at the κ light-chain loci, there are 150 V × 5 J possible recombinations, or 750 different possible light chains. At the H chain loci, there are approximately 200 V × 50 D × 4 J, or 4,000 possible heavy-chain sequences. Assuming that each H chain and L chain has an equal chance to be expressed in each cell, there are 750 × 4,000 or 3,000,000 possible antibodies that can be expressed. Additionally, the DNA-joining mechanism appears to be rather imprecise and frequently varies the site of VDJ fusions by a few nucleotides. This genetic imprecision is sufficient to change an amino acid or two, and leads to even greater antibody diversity. Finally, the heavy chain D regions can be read in all three reading frames, adding considerably to the diversity generated by somatic events. However, in each B cell, only a single protein-producing rearrangement occurs for both the

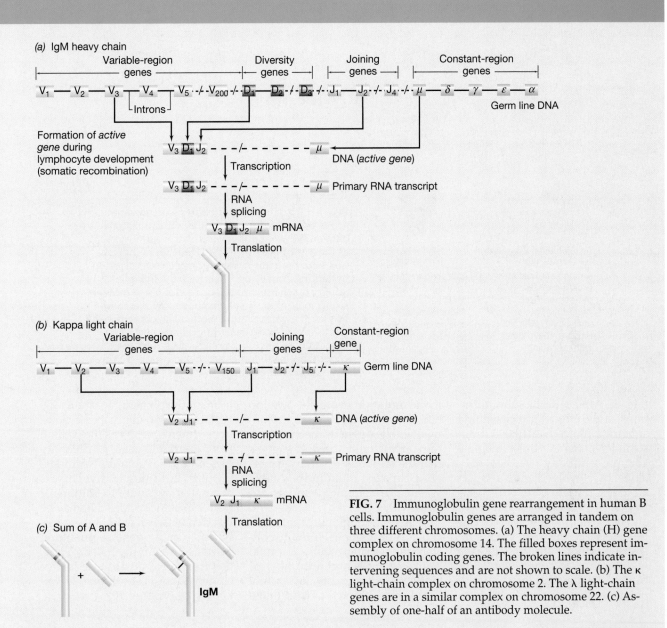

FIG. 7 Immunoglobulin gene rearrangement in human B cells. Immunoglobulin genes are arranged in tandem on three different chromosomes. (a) The heavy chain (H) gene complex on chromosome 14. The filled boxes represent immunoglobulin coding genes. The broken lines indicate intervening sequences and are not shown to scale. (b) The κ light-chain complex on chromosome 2. The λ light-chain genes are in a similar complex on chromosome 22. (c) Assembly of one-half of an antibody molecule.

heavy- and light-chain genes. This principle of *allelic exclusion* ensures that each B cell produces only a single immunoglobulin. In other words, *each B cell makes copies of a single, virtually unique, immunoglobulin.*

Additional antibody diversity is generated in B cells by mutations arising after second exposure to the immunizing antigen. Secondary challenge with antigen results in a change in the predominant antibody class produced—usually IgM

to IgG—as well as in an increase in antibody–antigen binding strength (affinity). This *affinity maturation* is one of the factors responsible for the dramatically stronger secondary response in immunity (see Section 20.11 and Figure 20.18). Compelling evidence for mutations leading to affinity maturation comes from experimental studies on *abzymes*, antibodies that function as enzymes. Abzymes perform catalytic reactions on bound substrates, resulting in covalent modification of the substrate and formation of a product. Abzyme mechanisms seem to be identical to the mechanisms involved in traditional enzyme-substrate reactions (∞ Section 4.5). Unfortunately, most abzymes have very low substrate affinity and do not efficiently convert substrate to product. However, when abzyme-producing B-cell clones are reexposed to antigen (secondary immunization), the substrate affinity of some of the abzymes increases dramatically

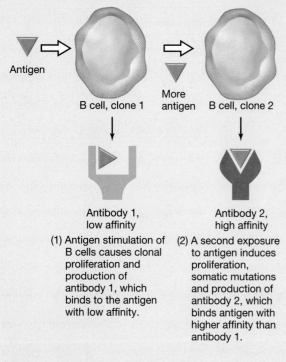

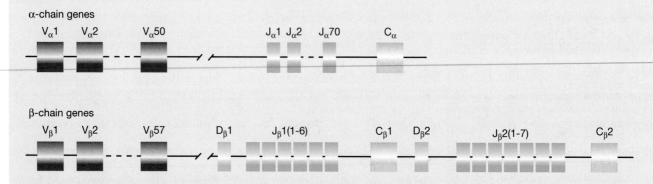

Antigen

B cell, clone 1

More antigen

B cell, clone 2

Antibody 1, low affinity

Antibody 2, high affinity

(1) Antigen stimulation of B cells causes clonal proliferation and production of antibody 1, which binds to the antigen with low affinity.

(2) A second exposure to antigen induces proliferation, somatic mutations and production of antibody 2, which binds antigen with higher affinity than antibody 1.

FIG. 8 Abzymes are antibodies with enzymatic activity. Primary immunization with a substrate antigen induces antibodies with low affinity, such as those produced by clone 1. Further immunization with the same substrate antigen selects for mutations from clone 1 that have much higher affinity for the substrate, such as for clone 2. This seems to be a universal phenomenon for all antibodies: A hypermutation mechanism continuously works on antibody genes to generate new mutations, and antigen acts as a selecting force to choose those antibodies and clones that react most strongly with the antigen.

(Fig. 8). After repeated immunizations, some cloned abzymes bind substrate more than 10,000 times stronger than the original abzyme/antibody. Studies of a clone that evolved into a better abzyme revealed several basic principles. Mutations in the heavy and light chain V region genes correlated with the increased substrate-binding affinity in the high-binding abzyme clone. Thus, from an immunological standpoint, antigen selection acts on so-matic mutation events to develop effective antibody responses. Coupled with the genetic mechanisms, the somatic mutation data suggest that the antibody diversity-generating capacity of an individual is essentially unlimited and is extremely flexible.

Like immunoglobulin genes, T-cell receptor genes contain constant and variable regions of amino acid sequences and a number of tandemly arranged genes encode the variable regions of TCRs. In the human, the α chain has about 150 variable-region genes and 70 joining segments, whereas the β chain has 57 variable genes, 2 diversity genes, and 13 joining segments (Fig. 9). Additional diversity is generated by random additions of from 1 to 6 nucleotides that may be inserted between variable, diversity, and joining gene segments in TCRs (N-region diversity). Thus, the number of possible sequence combinations is enormous, probably on the order of 10^{15}. ■

α-chain genes

$V_\alpha 1$ $V_\alpha 2$ $V_\alpha 50$ $J_\alpha 1$ $J_\alpha 2$ $J_\alpha 70$ C_α

β-chain genes

$V_\beta 1$ $V_\beta 2$ $V_\beta 57$ $D_\beta 1$ $J_\beta 1(1-6)$ $C_\beta 1$ $D_\beta 2$ $J_\beta 2(1-7)$ $C_\beta 2$

FIG. 9 Organization of the human T-cell receptor α and β chain genes. The α chain genes are located on chromosome 14 and the β chain genes are on chromosome 6. // indicates that a large segment of non α-chain or non β-chain encoding DNA is not shown.

tions, and mucus secretions of the respiratory and genitourinary tracts. The mucosal surfaces of the human body total about 400 m^2. All these mucosal surfaces are associated with the MALT (Figure 20.4) that secrete IgA. As a result, the total amount of *secretory* IgA produced by the body is higher than the amount of *serum* IgG. IgA is also present in the second highest concentration in serum as a monomer (Table 20.3). The secretory form of IgA has an altered molecular structure, consisting of a dimeric immunoglobulin attached to a protein high in carbohydrate, called the *secretory piece*, and a J chain peptide (Figure 20.10*d*). These proteins help hold the dimeric immunoglobulin molecule together and aid in the transport of IgA across membranes and into secretions.

Immunoglobulin E (IgE) is found in serum in extremely small amounts (in an average human about 1 of every 50,000 serum immunoglobulin molecules is IgE). Despite its low concentration, it is important because immediate-type hypersensitivities (allergies) (see Section 20.15) are mediated by IgE. The molecular weight of an IgE molecule is significantly higher than most other immunoglobulins (Table 20.3) because, like IgM, it contains an additional constant region (Figure 20.10*e*). This additional constant region functions to bind IgE to mast cell surfaces (see Figure 20.24), an important prerequisite for certain allergic reactions.

Immunoglobulin D (IgD) is also present in low concentrations, and its function in the overall immune response is unclear. IgD is abundant on the surfaces of B cells and plays a role along with monomeric IgM in binding antigen to B cells.

Immunoglobulins, either as serum proteins, or as B-cell antigen receptors, are capable of recognizing a vast array of foreign antigens. A single individual recognizes thousands of different antigens during their lifetime and has the genetic capacity to form more than 1 *billion* different Ig binding sites. How is this accomplished? Several hundred immunoglobulin genes are randomly recombined in somatic B cells using a novel gene shuffling mechanism to produce a virtually limitless number of functional Ig genes and proteins. The details of this mechanism are discussed in the box, Molecular Biology of the Immune Response.

✓ **20.5 Concept Check**

Immunoglobulin (antibody) proteins consist of four chains, two heavy and two light. The antigen binding site is formed by the interaction of variable regions of heavy and light chains. Each class of immunoglobulin has different structural and functional characteristics.

✓ What immunoglobulin domains are involved in antigen binding?
✓ What functional and structural characteristics differentiate Ig classes?

T Cell Receptors

As we have seen, T cells play a variety of complex roles in the overall immune response. Although T cells do not produce antibody, they interact with antigen through antigen-specific receptor molecules located on T cell surfaces called **T cell receptors (TCRs).** T cell receptors have antigen specificity, but unlike serum immunoglobulins, they are integrated into the T cell membrane. Both CD4 and CD8 lymphocytes have TCRs on their cell surface. Thus, all T cells are equipped to recognize specific antigens with their TCRs. How is this accomplished at the molecular level?

Structure of the T Cell Antigen Receptor

T cells must be structurally diverse. For example, a different TCR must be available to distinguish each different virus that infects the body. Thus, TCRs must have at least as many different antigen binding sites as antibodies. Although TCRs are not immunoglobulin molecules, they resemble immunoglobulins in many ways. TCRs and immunoglobulins are evolutionarily and structurally related and the mechanisms for generating multiple antigen-binding sites is very similar for both molecules (see the box, Molecular Biology of the Immune Response).

The TCR consists of two disulfide-linked peptides, called alpha (α) and (β). The alpha chain is about 40,000 molecular weight, and the beta chain is about 43,000 molecular weight. The α-β heterodimer is found on 85% of mature T cells. The remaining T cells have analogous polypeptide chains designated gamma (γ) and delta (δ). TCRs contain regions of highly variable amino acid sequences. These *variable* domains (which are the amino-terminal portion of each of the two polypeptide chains) combine to form the actual antigen binding site, much as a combination of the heavy- and light-chain variable regions forms the antigen binding site in immunoglobulins. Each TCR polypeptide also contains one constant domain in which the amino acid sequence is invariant from chain to chain within a corresponding type. Thus, all α chain and β chain constant-region sequences are the same. The variable and constant domains are roughly equivalent in size to the immunoglobulin domains (see Section 20.5). A comparison of a TCR with an analogous antibody molecule is shown in Figure 20.11. Note that the TCR is an integral membrane protein, with both chains spanning the membrane.

Although the TCR is structurally quite similar to immunoglobulin, there are very significant differences

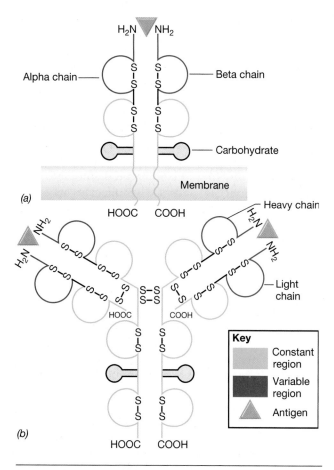

FIGURE 20.11 Structural comparison of (a) the T cell receptor with (b) an immunoglobulin. Note the presence of variable regions.

in the way that it binds antigen. As we shall discuss in detail in the following section, TCR can bind antigen effectively only if it is presented on the surface of another cell, in direct association with a self protein.

✓ **20.6 Concept Check**

T cell receptors are antigen-specific proteins found on the surface of T cells. They are structurally and evolutionarily related to immunoglobulins. T cells recognize antigen only in the context of other proteins found on cell surfaces.

✓ Why are TCRs found only on the membranes of T cells?
✓ How do TCRs differ from serum immunoglobulins in structure? How are they similar?

20.7

Histocompatibility Proteins

Antibodies recognize antigens in solution. However, although they have much in common structurally with antibodies, T cell receptors (TCRs) can only recognize an antigen that is bound to a set of *self* proteins found on

the surface of normal cells. These proteins are encoded by a genetic region, present in all vertebrates, called the *major histocompatibility complex* (MHC). MHC proteins are produced by a number of genes in this complex and are collectively called *human leukocyte antigens* or HLAs. MHC molecules were first discovered as the major target molecules for transplantation rejection; if tissues from one animal, a donor, are immunologically rejected when transplanted to another animal, a recipient, then their MHC proteins are different. We now know that MHC proteins function as antigen-presenting molecules and interact with both the antigen and the TCR. Thus, MHC proteins are a *third* set of antigen binding molecules and play an integral role in the immune response.

Structure of Human Major Histocompatibility Complex Proteins

The MHC genes encode two distinct types of proteins known as *class I* and *class II*. These class I and class II molecules are cell surface proteins and are intimately involved in immune recognition events. Class I MHC proteins are found on the surfaces of *all* nucleated cells. Class II MHC proteins are found only on the surface of B lymphocytes,

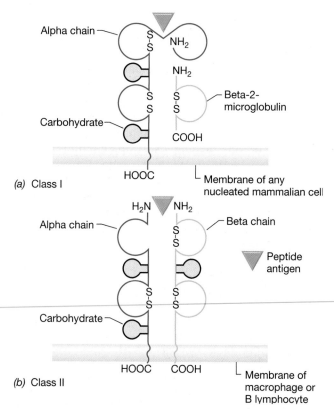

FIGURE 20.12 Structure of major histocompatibility complex (MHC) proteins. (a) Class I type. (b) Class II type. Note that class I molecules are present on the surfaces of all nucleated cells and that class II molecules are present only on certain specialized antigen-presenting cells.

macrophages, and other antigen-presenting cells (APCs). The reasons for this differential distribution will become apparent when we discuss the function of these molecules.

Class I MHC proteins consist of two polypeptides (Figure 20.12), one, an alpha (α) chain of 42,000 molecular weight, is encoded in the MHC gene region. The other class I polypeptide, a 12,000-molecular-weight protein called *β-2 microglobulin (β₂m)*, is encoded by a non-MHC gene. The MHC-encoded polypeptide is a glycoprotein firmly anchored in the cell membrane; β₂m is noncovalently linked to class I (Figure 20.12*a*). Class II MHC proteins consist of two noncovalently linked glycosylated polypeptides, called α and β, of 33,000 and 28,000 molecular weight, respectively. Like class I molecules, these polypeptides are embedded in the cytoplasmic membrane and project outward from the cell surface (Figure 20.12*b*).

MHC proteins are not structurally identical *within* a given species. Different individuals often show subtle differences in the amino acid sequence of their MHC molecules. These limited sequence variations are called *polymorphisms*. There are several hundred different MHC genes in humans and the cell surface products of these genes are the major reason why tissues transplanted from one individual or one species to another seldom match, are recognized as nonself, and are rejected. The detailed molecular structure and genetic organization of the MHC genes and proteins are presented in the box, Molecular Biology of the Immune Response.

Functions of MHC Proteins

MHC proteins serve as molecular reference points that permit T cells to identify foreign antigens. In a normal animal, T cells, through their TCRs, constantly interact with proteins or other potential antigens. They must be able to discriminate self from nonself antigens. The T cell, through its TCR, binds to MHC molecules and can then recognize foreign antigens embedded in the MHC structure; a T cell cannot recognize a foreign antigen unless it is presented in the context of an MHC protein.

How does this happen? As we discussed previously, when a foreign antigen is taken up by a host cell, the cell "processes" or degrades it. This processed antigen then becomes embedded, or bound, to the MHC protein, and the complex is passed through the cytoplasmic membrane and expressed on the surface of the cell. Two distinct antigen-processing schemes are known, one for class I antigen presentation and one for class II antigen presentation (Figure 20.13). In the class I scheme (Figure 20.13*a*), antigens that are manufactured by host degradation reactions are bound by class I proteins in the endoplasmic reticulum. The actual processed peptide is about 10 amino acids long. This method of antigen contact is very important in virus infections, where

the host cell manufactures viral proteins. Degraded viral peptides, which are nonself, then complex with class I proteins. The complex moves to the cell surface and is recognized by peptide-specific T cells through the TCRs.

In effect, the MHC molecules act as a *platform* on which the foreign antigen is bound. The viral infection of a cell leads to the embedding of viral antigens in class I molecules on the infected cell's surface (Figure 20.13*a*). T_C cells are constantly exposed to the entire cell population. These T_C cells have specificity for nonself peptides in the context of self MHC. Normal, healthy cells all express class I proteins on their surface, but the class I molecules contain self peptides, which are not recognized by the T cells. However, the T cells recognize the virus-infected cell because it exhibits the nonself viral antigen bound by the self class I MHC molecule (Figure 20.13*a*). Thus, the TCR on the surface of the T cell interacts with both antigen (nonself) and MHC molecule (self) sites. This specific interaction induces the T_C cell to produce cytotoxic proteins called *perforins* (see Section 20.8) that kill the virus-containing *target cell*.

A second antigen presentation scheme involves the class II molecule (Figure 20.13*b*). In this case, class II molecules, complete with a self peptide called *Ii*, or invariant chain, line the cell vacuoles (lysosomes) (see Section 20.3) that degrade antigens phagocytized by APCs. When the phagosome containing the foreign antigen fuses with the lysosome forming a *phagolysosome*, the antigens are digested by proteolytic enzymes along with the Ii. The foreign peptides, generally about 11 to 15 amino acids in length (slightly larger than class I-binding peptides), are then bound by the newly opened class II antigen binding site, and the whole complex is eventually expressed on the external cytoplasmic membrane where it is presented to the T_H cells. The T_H cells, through the TCR and the CD4 coreceptor, then recognize the class II MHC–foreign peptide complex on the surface of the APCs. The T_H2 cell is activated by contact with foreign antigen and secretes molecules that either stimulate antibody production by specific B cell clones or secrete a battery of inflammatory cytokines (see Section 20.4 and Section 20.11).

Finally, MHC proteins obviously do not recognize every antigenic peptide with an individual MHC molecule as do TCRs and antibodies. Because of the limited individual variation in the MHC proteins (only two per locus and only three loci each for class I and class II), a different mechanism for binding large numbers of different peptides must be used. The peptides bound by a single MHC protein share common structural patterns, or **motifs,** and each different MHC molecule can bind to a different motif. For example, one class I protein binds all peptides having tyrosine at position 2 and leucine at position 7. Thus, this single MHC protein is able to present every peptide with the amino acid sequence X-tyrosine-X-X-X-X-

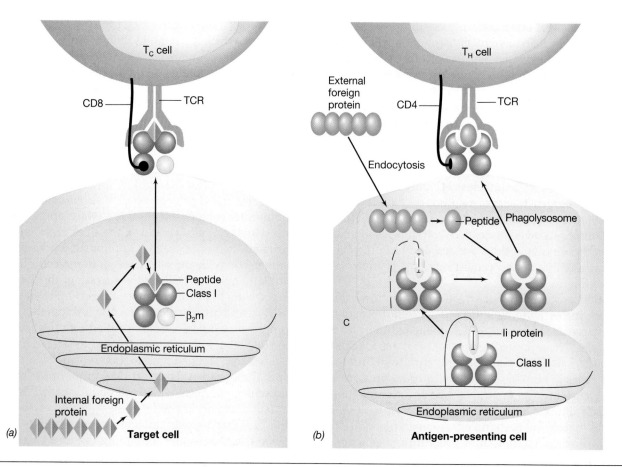

FIGURE 20.13   Antigen presentation by class I and class II major histocompatibility complex (MHC) proteins. (a) In the class I antigen presentation pathway, the class I proteins are made and assembled in the endoplasmic reticulum. Protein antigens manufactured within the cell, for instance from viruses or tumors, are degraded in the cytoplasm and transported across the endoplasmic reticulum membrane, where the peptides then bind to class I, are transported to the cell surface, and interact with T cell receptors (TCRs) on the surface of T_C cells. The CD8 coreceptor on the T_C cell also engages the class I MHC, resulting in a stronger complex. The T_C cells then release cytokines and cytotoxins, proteins that kill the target cell. Any nucleated cell can act as a target cell for class I. (b) In the class II antigen presentation pathway, class II proteins are produced in the endoplasmic reticulum and are assembled with a blocking protein, Ii (invariant chain), which prevents class II from complexing with other peptides made in the endoplasmic reticulum of antigen-presenting cells (APCs). Class II then goes to the phagolysosome where the Ii and foreign proteins, imported from outside the cell (by endocytosis), are digested. The class II protein then binds to the processed foreign peptides, and the complex is transported to the cell surface where it interacts with TCRs and the CD4 coreceptor on T_H cells. The T_H cells then release cytokines, which act on other cells to activate an immune response. The APCs include certain phagocytes and B cells.

leucine-X-X-X, where X is any amino acid. In this way, each MHC protein can present a large number of different peptides with a limited number of peptide binding sites.

✓ 20.7 **Concept Check**

Proteins encoded by the MHC are molecular reference markers that permit T cells to interact with antigen. Found on the surfaces of APCs and target cells, MHC proteins embed processed antigen and present it to T cells. This is the only way T cells can recognize nonself antigens and initiate the specific immune response.

✓ On what cells are class I and class II MHC proteins found?
✓ What are the differences between antigens presented by class I MHC and those presented by class II MHC?

✓ How can MHC proteins recognize peptides from several different pathogens?

20.8

Cytokines

Section 20.1 described how the cells of the immune system develop from a variety of leukocyte cell types (Figure 20.4). In order to accomplish individual tasks, the cells in the immune system must communicate, and one method for doing this is through a number of soluble proteins known as **cytokines.** Cytokines are a group of soluble proteins that regulate cellular functions. Specialized cytokines produced by lymphocytes are sometimes known as *lymphokines*. In general, cytokines are secreted from one cell and bind to corresponding specific receptors on a target cell. Some cytokines bind to receptors on the cell that produced them. Thus, these cytokines have *autocrine* (self-stimulation) abilities. The receptors are responsible for signal transduction (⚬ Section 7.7) and send information inside the cell to either increase or decrease metabolic activity such as protein synthesis and cell division. These signals can ultimately result in differentiation and clonal proliferation of leukocytes. Many of the cytokines are designated *interleukins* (ILs) because they are molecules that mediate interactions between leukocytes. Table 20.4 lists some of the immunologically important cytokines, their producer cells, their target cells, and their biological effects. In all, there are nearly 40 known cytokines, most of which are produced by either T_H cells or monocytes and macrophages. Cytokines are usually small proteins of less than 30,000 molecular weight, and most belong to one of four distinct families, as defined by their protein structure. As examples of cytokine function, we now explore the action of three cytokines that are essential for the specific immune response.

IL-1, IL-2, and IL-4

As we have seen, macrophages are responsible for antigen uptake, processing, and presentation (see Sections 20.3 and 20.7). In addition, they secrete a potent cytokine known as IL-1, which acts on several different cell types (Table 20.4 and Figure 20.14). IL-1 is a key component of the immune response because T_H cells are one of its main targets. Because of the proximity of T_H cells and macrophages in the lymph nodes, nearby T_H cells are the most likely to be activated. IL-1 binding by the IL-1 receptors (IL-1R) on the T_H cell acts as an activation signal. The activated T_H cell, in turn, responds by producing IL-2, which is secreted and bound by the IL-2R on the surface of the T_H cells. Thus, IL-2 is *autocrine*: IL-2 can activate the

TABLE 20.4	Properties of some major cytokines and chemokines		
Cytokine/Chemokine[a]	Producer cells	Target cells	Effect
IL-1	Monocytes, Macrophages, B cells	T_H / B	Activation / Maturation, expansion
IL-2	T cells	T cells	Induce proliferation
IL-4	T_H2	B (antigen-primed)	Activation
		B (activated)	Proliferation / Class switching / IgE synthesis
IL-5	T_H2	B	Differentiation / IgA synthesis
IL-6	T_H2	Proliferating B cells	Differentiation to plasma cells
	Monocytes	Plasma cells	Antibody secretion
	Macrophages	Myeloid stem cells	Differentiation
IL-8	Connective tissue	T / PMNs	Chemoattractant, activation
IL-10	T_H2	Macrophages	Suppress IL-1 production
IL-12	Macrophages, B cells	T_C / T_H1	Differentiation to T_C / Proliferation
IFN-α	Leukocytes	Normal cells	Antiviral
IFN-γ	T_H1, T_C, NK	Macrophages / Normal cells	Activation / Antiviral
GM-CSF	Macrophages, T_H1	Myeloid stem cells	Differentiation to granulocytes, monocytes
MCAF	Connective tissue	T / Macrophages	Chemoattractant, activation
TNF-α	Macrophages, NK	Tumor cells	Cytotoxic
TNF-β	T_H1, B	Tumor cells	Cytotoxic

a IL, Interleukin; IFN, interferon; GM-CSF, granulocyte, monocyte colony stimulating factor; TNF, tumor necrosis factor; MCAF, macrophage chemoattractant and activating factor. IL-8 and MCAF are classified as *chemokines*. All others are *cytokines*.

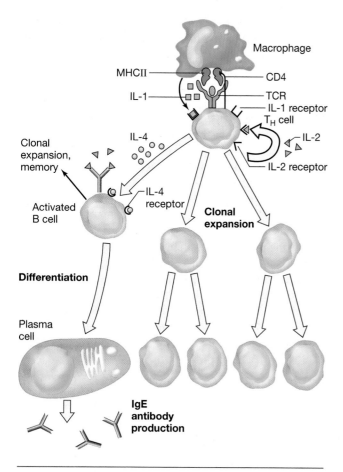

FIGURE 20.14   Some major effects and interactions of IL-1, IL-2, and IL-4 cytokines in the antibody response.

same cell that secreted it. Under the influence of IL-2, the cell divides, making clonal copies. In the process, the T_H cell also makes other cytokines, in this case IL-4, which then binds to the IL-4R on B cells. IL-4 stimulates the B cells to differentiate and produce plasma cells, which ultimately produce antibodies. Thus, these cytokines are soluble mediators and activators for macrophages, T_H lymphocytes, and B cells, and are essential components for the initiation of an immune response.

Other Cytokines

Table 20.4 shows the action of several cytokines. Many of these proteins affect cells involved in specific immunity. For example, IL-6 acts primarily on B cells. However, several cytokines affect nonimmune cells and serve as important modulators of nonspecific host responses. For example, interferon (IFN-α and IFN-γ) is produced by leukocytes and inhibits viral replication in virtually any cell in the body. Tumor necrosis factors (TNF-α and TNF-β) are capable of destroying a variety of tumors provided that the TNF-producing cells have access to the tumor. Interferons and TNFs appear to have no target cell specificity, but because they are produced by T cells, they augment the effects of immune cells.

Cytokines are important mediators of a variety of immune functions. Some cytokines have broader activities, interacting with a variety of normal cells. However, the unifying feature of these heterogeneous proteins is that they are all produced by leukocytes.

Chemokines

Chemokines are a group of proteins, usually of 8–12,000 molecular weight, that function as chemoattractants for phagocytic cells and T cells. They are produced by lymphocytes and a wide variety of other host cells in response to bacterial products, viruses, and other agents that cause damage to host cells. Chemokines attract phagocytes and T cells to the site of injury, stimulating an inflammatory response as well as an immune response. Perhaps the best-studied chemokines are *interleukin-8* (IL-8) and *macrophage chemoattractant and activating factor* (MCAF; Table 20.4). IL-8 is produced by monocytes, macrophages, fibroblasts (connective tissue cells), and keratinocytes (skin cells) in response to tissue injury or contact with pathogens. IL-8 is secreted by the affected cells and binds to the surrounding tissue, where it is a chemoattractant for T cells and PMNs (see Section 20.4), resulting in an inflammatory response (PMNs) followed by a specific immune response (T cells). Membrane-integrated chemokine receptors on the target cells act through signal transduction pathways (⊙⊙ Section 7.7) to induce activation of the target phagocytes or T cells. MCAF is produced by the same cell types and also attracts T cells. However, MCAF attracts and activates macrophages instead of PMNs, thereby attracting another source of inflammatory mediators as well as potentially organizing an immune response. Thus, chemokines produced by a wide variety of cell types are potent initiators of inflammatory reactions and also enhance opportunities for specific immune interactions.

✓ 20.8 Concept Check

Cytokines are soluble mediators, produced by leukocytes, that regulate interactions between cells. Several, such as IL-1 and IL-4, affect leukocytes and are critical components in the generation of specific immune responses. Others, such as IFN and TNF, affect a wide variety of cell types. Chemokines are also soluble mediators, but are produced by a variety of cell types in response to injury. Chemokines function as attractants for nonspecific inflammatory cells and T cells and are involved in both nonspecific and specific aspects of the host response.

✓ Compare the target cells for IL-1, IL-4, IL-12, IFN-γ, and TNF-β.

✓ How do cytokines and chemokines differ in their cell source? Their cell target?

✓ What events stimulate cytokine production by macrophages? What events stimulate chemokine production by macrophages?

20.9

Cell-Mediated Immunity

The terms **cell-mediated immunity (CMI)** and **cellular immunity** are used to describe any immune response in which antigen-specific cells of the immune system are directly involved but in which antibody production or activity is of minor importance.

Cellular immunity cannot be transferred from animal to animal by antibodies or antiserum but only by lymphocytes. The lymphocytes that function in the transfer of cellular immunity are activated T cells.

The events that occur when an animal is exposed to an antigen are summarized in Figure 20.15. We now turn our attention to the cytotoxic T cells (T_C), the natural killer cells (NK), and the inflammatory T cells (T_H1) and discuss the role each plays in cellular immunity.

Cytotoxic T Lymphocytes

Cytotoxic T cells (T_C cells) (see Section 20.4) are involved in the destruction of cells that display foreign antigens on their surfaces. As previously mentioned in connection with the major histocompatibility complex (MHC) (see Section 20.7), T_C cells recognize foreign antigens embedded in MHC class I molecules. Any cell carrying a foreign antigen, such as those introduced on cells harboring viruses, can be lysed by T_C cells (recognition of viral antigens by T_C cells was discussed in Section 20.6 in connection with the T cell receptor).

Contact between a T_C cell and the target cell is required for lysis. The contact is initiated by the TCR: Ag-MHC complex (Figure 20.13a). On contact with the target cell, granules in the T_C cell are drawn to the con-

tact site and the contents of the granules are released. The granules contain *perforin* monomers, which enter the membrane of the target cell, polymerize, and form a pore. The cellular contents leak out and the target cell dies, but the T cell remains unaffected. T_C cells lyse only the specific target cells displaying the foreign antigen; killing of other neighboring cells seldom occurs. Presumably, the concentration of perforin that escapes from the cell contact area is not sufficient to damage by-stander cells.

Natural Killer Cells

Killer cells are an additional class of lymphocytes distinct from cytotoxic T cells that play a role in destroying foreign cells. *Natural killer* (NK) cells are neither T cells nor B cells. Their numbers are not enhanced, nor do they exhibit memory after stimulation. Nevertheless, NK cells resemble T_C cells in their ability to kill foreign cells. For example, NK cells also use perforin to lyse their targets. However, NK cells differ from T_C cells in that they kill targets in the *absence* of stimulation by a specific antigen. NK cells are capable of destroying malignant and virus-infected cells *in vitro* without previous exposure or contact with the foreign antigen. The molecular target for NK cells seems to be the *lack* of appropriate MHC class I proteins. NK cells recognize normal cells and their class I proteins through a set of special class I receptors. Binding of these receptors to class I *deactivates* the lytic mechanism, but in the absence of binding, the lytic mechanism remains active and the NK cell lyses the unrecognized target. The main targets of NK cells, tumor cells and virus infected cells, often have reduced or altered class I expression.

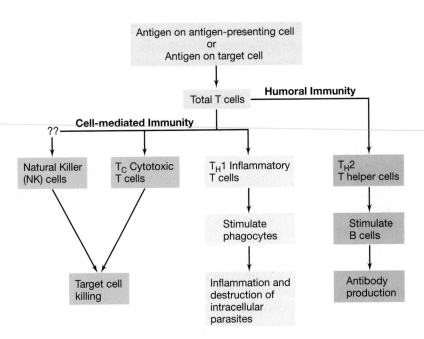

FIGURE 20.15 Overview of the role of T cells in cell-mediated immunity and humoral immunity. T cells play a central role in initiating and maintaining all antigen-specific immune responses. The cellular origin of the lymphocyte-like NK cells is unclear.

T$_H$1 Cells and Macrophage Activation

Macrophages play a central role as **antigen-presenting cells (APCs)** in both antibody-mediated and cell-mediated immunity. As illustrated in Figure 20.13*b*, macrophages bind, process, and present antigen to T cells. As phagocytic cells, however, macrophages also take up and kill certain foreign cells by themselves, and this ability can be stimulated by T$_H$1 cells. A key property of activated macrophages is that they can kill intracellular bacteria that would normally multiply. As we have noted (see Section 20.3), some bacteria survive and multiply within macrophages, whereas most bacteria taken into macrophages are killed and digested. Bacteria multiplying within macrophages include *Mycobacterium tuberculosis*, *M. leprae* (causal agents of tuberculosis and leprosy, respectively), *Listeria monocytogenes* (causal agent of listeriosis), and various *Brucella* species (causal agents of undulant fever and infectious abortion). Animals given a moderate dose of *M. tuberculosis* are able to overcome the infection and become immune because of the development of a T cell-mediated immune response. The cells involved are the T inflammatory cells, the T$_H$1 subset. They activate macrophages and other nonspecific phagocytes by secreting a number of cytokines, including IFN-γ (interferon gamma) (see Section 20.8 and Table 20.4). Surprisingly, such immunized animals also phagocytize and kill unrelated organisms such as *Listeria*, because macrophages in the immunized animal have been activated so that they more readily kill the secondary invader.

Macrophages not only kill foreign pathogens but are also involved in the destruction of foreign mammalian cells. This shows up in the development of transplantation immunity, and is a major problem in the transplantation of organs and tissues from one person to another. Macrophages also target tumor cells in some cases. Tumor cells contain some specific antigens not seen on normal cells, and tumors function like self-inflicted transplants. There is considerable evidence that tumor cells are normally recognized as foreign and are destroyed primarily by macrophages which are activated by cytokines from T$_H$1 cells.

✓ 20.9 Concept Check

Cell-mediated immunity involves a variety of T cells, which are activated by interaction with MHC–antigen complexes. Activated T cells may interact directly to kill target cells, or they may facilitate involvement of other cellular effector cells, particularly macrophages, with cytokines.

✓ What antigens do T$_C$ cells recognize?
✓ How do NK cells recognize targets?
✓ How can macrophages become involved in the response to specific antigens?

Clonal Selection and Immune Tolerance

We have now developed an overview of the immune response, especially in relation to effector molecules and mechanisms. In this section, we examine the mechanisms used to focus this very powerful, specific response on potentially dangerous nonself antigens. We will also examine the mechanisms used to avoid interactions with the equally complex but nonthreatening self antigens.

Clonal Selection

The **clonal selection theory** states that each antigen-reactive B cell or T cell has only a single type, or specificity, of antigen-specific receptor on its surface. When stimulated by interaction with a specific antigen, each cell is capable of dividing, making a copy of itself. The antigen-driven B and T cells continue to divide, whereas cells that are not antigen-stimulated do not divide; each antigen-stimulated cell divides and makes an exact copy of itself, resulting in multiple copies, or *clones* (Figure 20.16).

Because of the infinite variety of antigens available, a large number of antigen-reactive cells are available in the body, and each cell is capable of expanding into an antigen-reactive clone. However, antigen-reactive cells must avoid interactions and subsequent immune reactions with *self* antigens in the host. How does this occur? The clonal selection theory again proposes an answer. Ultimately, the immune response must develop the ability to discriminate between foreign invaders (nonself and potentially dangerous) and host (self and not dangerous) antigens by deleting or inactivating self-reactive clones. This capacity for antigen-specific immune *un*responsiveness to self antigens is known as **tolerance** and occurs in both T and B cells. We consider the normal development of tolerance now. In Section 20.16, we will discuss the failure of tolerance and the subsequent development of autoimmunity.

Immune Tolerance

The development of T cell tolerance occurs in the thymus, a walnut-sized organ that lies on top of the heart. The thymus has a central role in the maturation and development of T cells and is therefore a primary lymphoid organ (Figure 20.4). In the first T cell maturation stage, called *positive selection*, lymphocytes that will become T cells leave the bone marrow and enter the thymus from the lymphatic ducts (Figure 20.17). After entry, some of these immature T cells interact, using their newly developed T cell receptors (TCRs) to bind

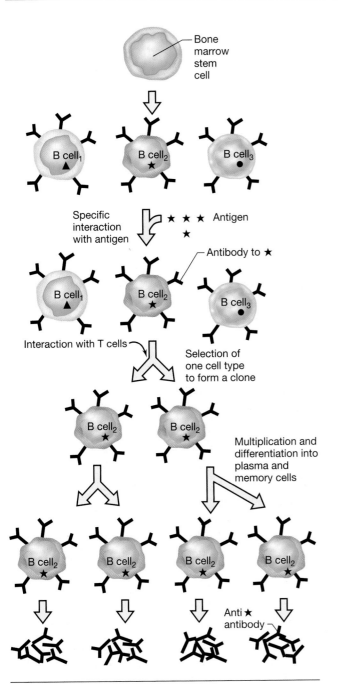

FIGURE 20.16 Clonal selection. Individual B cells, specific for a single antigen, proliferate and expand to form a clone after interaction with the specific antigen. The specific antigen acts as the selection agent, driving selection and then proliferation of the individual antigen-specific B cell. Clonal copies of the antigen-reactive cell all have the same antigen-specific surface receptor. Continued exposure to antigen results in continued expansion of the clone. An analogous situation exists for T cells.

to self major histocompatibility complex (MHC) molecules on the thymus. The T cells that do not bind MHC molecules are then programmed to die, a process called *apoptosis;* the T cells that bind thymic MHC proteins continue to mature and proliferate. Thus, positive selection *retains* T cells that *do* recognize self MHC proteins and *deletes* T cells that *do not* recognize self MHC proteins.

The second stage of T cell development is termed *negative selection.* After proliferation, positively selected T cells continue to react with MHC molecules, which are complexed with antigen (see Section 20.7). The antigens in the thymus are mostly of self origin. T cells that react with the self antigens in the thymus are potentially dangerous because they can lead to destruction of self tissue. Therefore, these *autoreactive* T cells must be eliminated. The autoreactive T cells bind very tightly to thymus tissues and cannot leave; they remain bound to the thymus and eventually die. However, T cells that are destined to interact with nonself antigens do not bind as tightly, presumably because there are no nonself antigens in the thymus. These T cells do not die, but leave the thymus and migrate to the spleen and lymph nodes where they can contact foreign antigens presented by B lymphocytes and other antigen-presenting cells (APCs) (Figure 20.17).

This mechanism for inducing tolerance is called *clonal deletion;* precursors of T cell clones that are useless or harmful are deleted during development. There is also evidence for another tolerance mechanism. In some cases, self-reactive T cell clones avoid clonal deletion but become unresponsive to self antigen through interactions in the secondary lymphoid organs. This mechanism is known as *clonal anergy* or *clonal paralysis.* However, these clones remain in circulation and, in certain individuals, are reactivated later in life. These reactivated self-reactive clones are often associated with autoimmune diseases (see Section 20.15).

The acquisition of immune tolerance in B cells is also necessary because antibodies produced by self-reactive B cells **(autoantibodies)** may damage host tissue (see Section 20.15). B cells also undergo clonal deletion; many self-reactive B cells are eliminated during development. In addition, clonal anergy also plays a role; cells that are self-reactive do not develop when exposed to antigen. Several explanations are possible for induction of B cell clonal anergy. It can result when high concentrations of self antigens are always present. These antigens constantly occupy all the reactive B cell surface receptors, making the B cell unable to react with antigen in a normal manner. B cell anergy can also result if antigen-reactive T cells are not available (because they were eliminated during T cell maturation) to help B cells in the production of antibodies, as we will discuss in the next section.

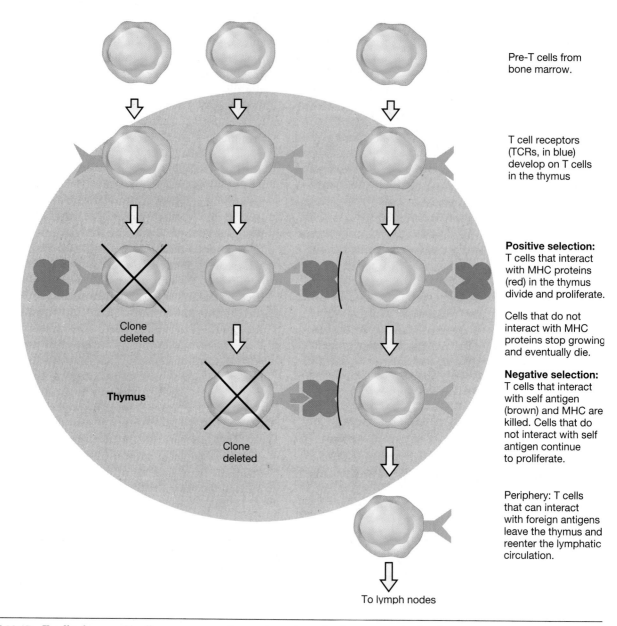

Pre-T cells from bone marrow.

T cell receptors (TCRs, in blue) develop on T cells in the thymus

Positive selection: T cells that interact with MHC proteins (red) in the thymus divide and proliferate.

Cells that do not interact with MHC proteins stop growing and eventually die.

Negative selection: T cells that interact with self antigen (brown) and MHC are killed. Cells that do not interact with self antigen continue to proliferate.

Periphery: T cells that can interact with foreign antigens leave the thymus and reenter the lymphatic circulation.

Clone deleted

Thymus

Clone deleted

To lymph nodes

FIGURE 20.17 T-cell tolerance. T cell precursors undergo a two-step selection process in the thymus. The selection ensures that mature T cells express antigen-specific T-cell receptors that react with foreign antigen in the context of self MHC proteins. T cell precursors that are not capable of reacting with any antigen are useless and are eliminated (positive selection); T cell precursors that react with self antigens are dangerous and are also eliminated (negative selection).

✓ 20.10 **Concept Check**

The sole function of the thymus is to produce T cell clones that react only with nonself antigens. Most self-reactive T cells are deleted during development and maturation in the thymus. Other self-reactive T cells as well as self-reactive B cells are anergized in the secondary lymphoid organs but occasionally become reactivated and cause autoimmune disease.

✓ Distinguish between *positive* and *negative* T cell selection.
✓ How do clonal *deletion* and clonal *anergy* differ?

20.11

Mechanism of Immunoglobulin Formation

Antibody formation is a complex process. In this section, we will examine a typical antibody response, especially with regard to the cell interactions necessary to produce effective immunity. We discuss the unique genetic mechanisms used to generate the incredible antibody diversity necessary for antigen-specific host defense in the box, Molecular Biology of the Immune Response.

The production of immunoglobulins in response to stimulation by antigen involves T cells, B cells, and antigen-presenting cells (APCs such as macrophages) and the intimate interaction of the various cell surface molecules discussed previously. As we have described, APCs act nonspecifically to ingest foreign antigens. Once the antigen is processed, the APC presents it to T_H cells. Following antigen interaction, T_H cells release IL-2 that acts in an autocrine fashion via IL-2 receptors (see Section 20.8), stimulating T_H proliferation and differentiation to T_H2 cells. The T_H2 cells, in turn, secrete cytokines such as IL-4, IL-5, IL-6, and IL-10. These cytokines stimulate B cells to begin antibody synthesis. Although the initial interaction between antigen and the APC may be nonspecific, all further steps in the process of antibody production are *highly* specific; receptor molecules on the surfaces of both T cells (T cell receptors) (see Section 20.6) and B cells (antibody molecules) (see Section 20.5) are highly specific for that antigen.

The genetic control of antibody production is also a highly complex process. An animal may be capable of making over 1 billion structurally distinct antibody molecules. Although at first this seems like an enormous demand on an organism's genetic coding potential, we saw that only a relatively small number of genes is required to encode this immense antibody diversity because of a phenomenon known as *gene rearrangement*. During development of lymphocytes in the bone marrow, gene rearrangements occur in B cells to produce two complete transcriptional units, one of which codes for the synthesis of a specific heavy chain and one for a light chain of the antibody molecule. The number of possible gene rearrangements is sufficient to account for the diversity of immunoglobulin molecules. Similar rearrangements occur during T cell development and result in the diversity of T cell receptors observed (see the box, Molecular Biology of the Immune Response). We now detail the steps in antibody production, beginning with the introduction of an antigen and ending with the production of a specific antibody that will react with that antigen.

Exposure to Antigen

Antigens may be introduced into any part of the body. After introduction, antigens are spread via the lymphatic and blood circulatory systems to neighboring secondary lymphoid organs such as lymph nodes, spleen, or MALT (Section 20.3 and Figure 20.4). For example, if antigen is injected intravenously, it travels via the blood to the spleen, where antibodies are formed. If antigen is introduced subcutaneously, intradermally, topically, or intraperitoneally, the lymphatic system drains the site and deposits antigen in adjacent lymph nodes, where an immune response is initiated. Likewise, antigen introduction to mucosal surfaces, for example, ingestion of antigen into the gut by feeding,

delivers antigen to the neighboring MALT and results in antigen-specific antibody production in the gut.

Following antigen introduction, there is a lapse of time (latent period) before specific antibody appears in the blood, followed by a gradual increase in antibody *titer* (that is, concentration) and then a slow fall. This reaction to a single antigen exposure is called the **primary antibody response** (Figure 20.18). When a second exposure to antigen is made some days or weeks later, the titer rises rapidly to a maximum of 10–100 times above the titer achieved following the first exposure. This large rise in antibody titer is referred to as the **secondary antibody response** (Figure 20.18). With time, the titer slowly drops again, but later exposures to the same antigen can bring it back up. The secondary response is the basis for the vaccination procedure known as a "booster shot" (for example, the yearly rabies shot given to domestic animals). Periodic reimmunization maintains high levels of circulating antibody specific for a certain antigen (see Section 20.16).

Cell Interactions

How do B cells, T_H2 cells, and APCs cooperate to produce antibodies? The APCs for antibody production include macrophages, dendritic cells, and, surprisingly, B cells. All APCs express class II MHC proteins on their surface (see Section 20.7). All APCs are efficient phagocytes: They ingest, process and display antigens very efficiently. *Dendritic cells* are found in large numbers in the lymph node cortex (see Figure 20.4c), where they

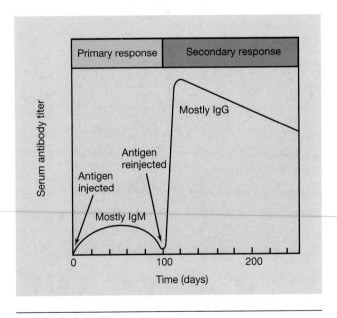

FIGURE 20.18 Primary and secondary antibody responses. The antigen injected at day 0 and day 100 must be identical to induce the secondary response. The secondary response, also called a booster response, may be over 10-fold greater than the primary response. Note the class switch from IgM production in the primary response to IgG production in the secondary response.

are very efficient non-specific phagocytes and antigen presenters. *Macrophages* are also very efficient phagocytes and non-specific antigen presenters, and macrophages also kill most ingested pathogens. Thus, dendritic cells and macrophages ingest, process and present antigen to T cells in a non-specific manner. B cells, however, are the most efficient APCs because B cells *specifically* bind antigens through the antigen-binding sites of their surface immunoglobulins (Figure 20.19).

The Ig-bound antigen is then phagocytized and digested into small peptides in the phagolysosomes. The resulting peptides bind to class II MHC and the peptide-MHC complex then moves to the cell surface, where it is presented to a T cell (see Figure 20.13*b*).

Interaction of the MHC-peptide complex with a T-cell receptor causes activation of the T_H1 cell, resulting in cytokine production (see Section 20.8). The cytokines cause activation and *proliferation* of the individual B cell,

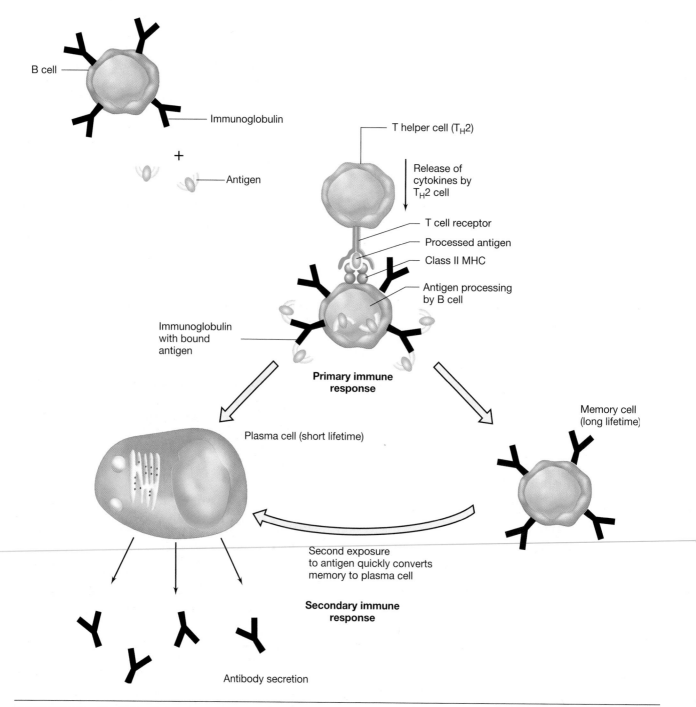

FIGURE 20.19 T cell–B cell interaction and antibody production. The B cell functions as an antigen-presenting cell (APC) and concentrates antigen with specific immunoglobulin receptors. After processing, antigen is presented to the T_H2 cell by a class II MHC molecule. The T_H2 cell in turn signals the same B cell to proliferate and form plasma cells (antibody producers) or memory cells. After subsequent antigen exposure, memory cells quickly convert to plasma cells.

resulting in an antigen-specific B-cell clone. Each activated B cell clone differentiates to form both antibody-secreting **plasma cells** and **memory cells** (Figure 20.19). Plasma cells are relatively short-lived (less than 1 week), but produce and secrete large amounts of IgM antibody (see Section 20.5) in this *primary response* to antigen.

By contrast, memory cells are resting cells that may live for several years. If reexposure to the immunizing antigen occurs, memory cells need no further T cell help; they quickly transform to plasma cells and begin producing IgG antibody (see Section 20.5). This secondary response, or *immunological memory,* results in more rapid, more abundant antibody production than the primary response, as well as the antibody *class switch* from IgM to IgG (Figure 20.18).

Certain antigens can stimulate low-level antibody production in the *absence* of previous T cell interactions (these are the so-called *T-independent antigens*). Most T-independent antigens are large polymeric molecules with repeating antigenic determinants (for example, polysaccharides). The immunoglobulins produced to T-independent antigens are usually IgM and are low affinity. B cells that respond to T-independent antigens do not have immunological memory.

The preceding discussion describes the general principles of antibody production. Each antigen (strictly speaking, each antigenic determinant) catalyzes, through the activity of APCs and T_H2 cells, the growth of a *different* B cell line, which is capable of producing antibodies that react specifically *only* with that antigen. In this fashion, the normal animal can respond to perhaps as many as 1 billion distinct antigens by developing a specific B cell clone in response to stimulation by antigen. We discuss the genetic and structural basis for this diversity in the box, Molecular Biology of the Immune Response.

✓ 20.11 Concept Check

Antibody production is initiated by antigen contact, through an APC, with an antigen-specific T_H2 cell. The T_H2 cell, in turn, signals an antigen-specific B cell to produce antibody. The antigen acts as a selective agent, inducing proliferation of clones of cells that react with that particular antigen. Clones remain active for years as memory cells and can be immediately stimulated after reexposure to antigen to produce large quantities (high titers) of antibodies.

✓ What APCs are involved in antigen presentation?
✓ How do T_H2 cells activate antigen-specific B cells?

20.12

The Complement System

Complement acts with specific antigen–antibody complexes to bring about reactions that would not otherwise occur. Complement is an important effector mechanism and, as we shall see, it enhances the effects of antibody–antigen interactions. (These reactions will be described in Table 20.6.)

Complement is composed of a number of proteins, many with enzymatic activity, that interact in a sequential, ordered fashion with bacterial cells or other foreign material, causing lysis or leakage of cellular constituents as a result of damage to the cell membrane. These proteins, found at comparable levels in the serum of all individuals, are only activated when an antibody–antigen reaction occurs. In fact, a major function of antibody is to recognize invading cells and activate the complement system for attack. Thus, a wide variety of antibodies, each specific for a single antigen, can recruit the complement system proteins and the body does not need separate enzymes to attack each invading agent.

Some reactions in which complement participates include (1) bacterial lysis, especially in gram-negative Bacteria, when specific antibody combines with antigen on bacterial cells in the presence of complement; (2) microbial killing, even in the absence of lysis; and (3) phagocytosis, which may not occur during infection if the invading microorganism possesses a capsule or other surface structure that prevents the phagocyte from acting (see Section 20.3). However, when antibody and complement react with antigen on the surface of a cell, the cell is much more likely to be phagocytized. This is because most phagocytes, including macrophages and B cells, have high affinity receptors, the C3 receptors (C3R), specific for the C3 complement protein (Table 20.2). This amplification of the normal phagocytic process by antibody and complement is called **opsonization.**

Activation of the Complement System

Complement is a system of 11 proteins, designated C1, C2, C3, and so on. Activation of complement occurs only with antibodies of the IgG and IgM classes (see Table 20.3). When such antibodies combine with their respective antigens, especially on cell surfaces, the antibodies assume new conformations in their constant-region domains, which allow them to fix (bind) the ever-present complement proteins (Figure 20.20). The complement proteins act in a cascade fashion, with activation of one component resulting in activation of the next. The key steps are (1) binding of antibody to antigen (initiation); (2) recognition of antigen–antibody complex by C1; (3) C4-C2 binding to an adjacent membrane site; (4) activation and binding of C3; (5) formation of the C5-C6-C7 complex at another membrane site; and (6) formation of the C8-C9 complex at the same site, causing cell lysis.

As seen in Figure 20.20, complement reactions at the C3 level result in chemotactic attraction of phagocytes followed by phagocytosis. Reaction of C5 also leads to leukocyte attraction. The terminal series of reactions from C5 through C9 results in cell lysis and death. Lysis results from the formation of holes in the membrane at

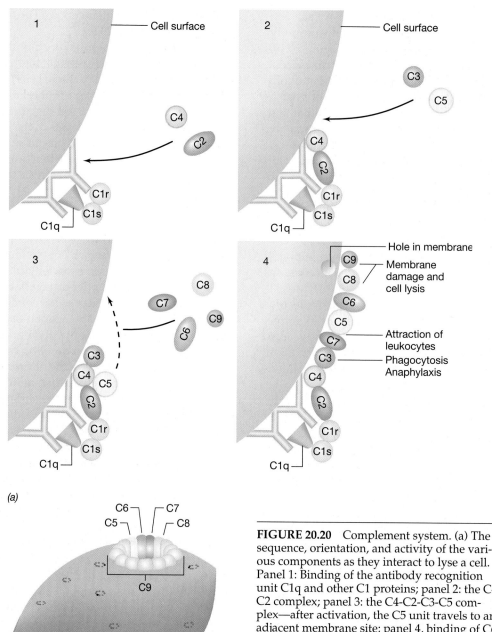

(a)

(b)

FIGURE 20.20   Complement system. (a) The sequence, orientation, and activity of the various components as they interact to lyse a cell. Panel 1: Binding of the antibody recognition unit C1q and other C1 proteins; panel 2: the C4-C2 complex; panel 3: the C4-C2-C3-C5 complex—after activation, the C5 unit travels to an adjacent membrane site; panel 4, binding of C6, C7, C8, and C9, resulting in membrane damage. (b) Three-dimensional view of the hole formed by complement components C5 through C9.

the sites of the C5–9 complexes. Cytoplasm leaks from the holes, resulting in cell death (Figure 20.21).

Complement is effective for the bactericidal and lytic actions of antibodies against many gram-negative Bacteria. Gram-positive Bacteria are not killed by specific antibody, in either the presence or absence of complement, although gram-positive Bacteria are opsonized. Death of gram-negative Bacteria involves antibodies against antigens on the surface of the cell; complement perhaps brings about an actual change in the cell surface, possibly by an enzymelike reaction, after antibodies have prepared the way. No cytocidal or lytic effect is seen when cells and complement are mixed alone, but if antibody has bound to cells first, death or lysis occurs rapidly after complement is added. The presence of antibodies is absolutely required to fix complement.

✓ 20.12 Concept Check

The complement protein cascade is a nonspecific mechanism for cell destruction and opsonization. It is triggered by specific immune antibody interactions and is a critical component of host defense.

✓ What classes of immunoglobulin fix complement?
✓ What complement components are necessary to complete cell lysis? To promote opsonization?

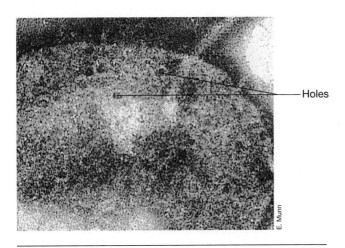

                                                                    — Holes

E. Munn

FIGURE 20.21 Electron micrograph of a negatively stained preparation of *Salmonella paratyphi*, showing holes created in the cell envelope as a result of a reaction involving cell envelope antigens, specific antibody, and complement.

TABLE 20.5	Characteristics of monoclonal and polyclonal antibody production
Polyclonal	**Monoclonal**
Contains many antibodies recognizing many determinants on an antigen	Contains a single antibody recognizing only a single determinant
Various classes of antibodies are present (IgG, IgM, and so on)	Single class of antibody produced
Can make a specific antibody using only a highly purified antigen	Can make a specific antibody using an impure antigen
Reproducibility and standardization difficult	Highly reproducible

20.13

Polyclonal and Monoclonal Antibodies

A typical immune response results in the production of a broad spectrum of immunoglobulin molecules of various classes, affinities, and specificities, all directed at the numerous determinants present on the antigen. The immunoglobulins directed toward a single determinant represent only a portion of the total antibody pool. Each immunoglobulin is produced by a single clone of B cells following antigen stimulation; an antiserum containing such a mixture of antibodies is thus referred to as a **polyclonal antiserum.** However, it is possible to produce antibodies of only a *single* specificity, derived from a single B cell clone. A variety of such monospecific antibodies, called **monoclonal antibodies,** have been generated for use in research and clinical medicine. Table 20.5 compares the properties of antibodies prepared against an antigen in the usual way—by preparing a polyclonal antiserum—with monoclonal antibodies.

Hybridomas and Monoclonal Antibodies

Each antigen-activated B cell is genetically programmed to become a clone that produces a monoclonal antibody (see the box, Molecular Biology of the Immune Response). But we consider here how to isolate and grow a single B cell clone for production of monoclonal antibodies. Normal lymphocytes cannot be grown and maintained in cell culture; they usually die after several weeks. However, B lymphocytes can be fused with myeloma (tumor) cells to form hybrid cell lines that will grow in culture and retain the ability to produce antibodies. The tumor cells grow indefinitely in culture and the cell fusion technique transfers this property to the antibody-producing B cells. This technique of B cell–myeloma cell

fusion is known as the **hybridoma technique** and is summarized in Figure 20.22.

First, a mouse is immunized with the antigen of interest and a period of weeks allowed for specific B cell clones to proliferate and begin producing antibody by the normal sequence of events. Then, spleen tissue, rich in B lymphocytes, is removed from the mouse and the B cells are fused with myeloma cells (Figure 20.22). Although true fusions represent only a small fraction of the total cell population remaining in the mixture, addition of the compounds hypoxanthine, aminopterin, and thymidine to the medium (the so-called HAT medium) strongly selects for *fused* hybrids. This occurs because unfused myeloma cells are unable to use the metabolites hypoxanthine and thymidine to bypass a metabolic block caused by aminopterin, a cell poison. By contrast, fused hybrid cells are able to use hypoxanthine and thymidine. The hybrids grow normally in HAT medium because they received the genetic information for this metabolic pathway from the normal cell fusion partner. Unfused B cells die off in a week or two because they are unable to grow in culture for more than a few cell divisions. Following fusion, the clones of interest must be identified; that is, if individual fused cells are placed in the wells of microtiter plates and allowed to grow and produce antibody, which cells produce the antibodies of interest?

A variety of assay techniques (see Sections 20.14 and 21.8) can be used to identify clones producing monoclonal antibodies. From a typical fusion, several distinct clones are isolated, each making a monoclonal antibody to a different determinant on the antigen. Once the clones of interest are identified, they can be grown in cell culture or they can be injected into mice and be perpetuated as a mouse myeloma tumor. As a mouse tumor line, hybridomas secrete large amounts of monoclonal antibody (in mice over 10 mg of pure monoclonal antibody can be obtained per milliliter of mouse peritoneal fluid). Specific hybridomas of interest can also be frozen. The hybridoma can then be thawed and

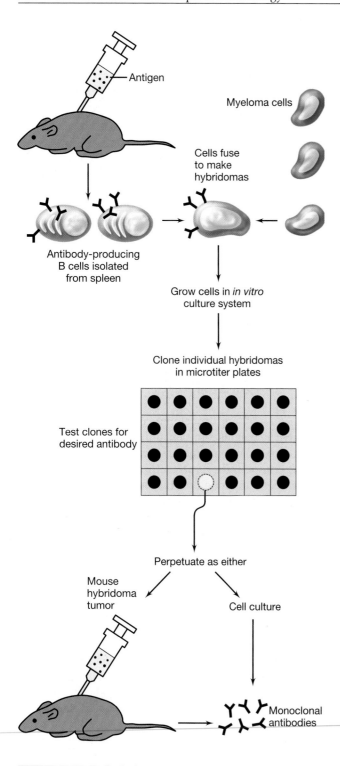

FIGURE 20.22 The hybridoma technique and production of monoclonal antibodies. The hybridoma can be indefinitely cultured or passed through animals as a tumor. The hybridoma cells can also be stored as frozen tumor cells and reconstituted when needed in tissue culture or in a suitable animal host.

injected into mice, or grown in tissue culture to produce more of the specific monoclonal antibody.

Monoclonal antibodies are extremely useful reagents for research and medical science. For research purposes, monoclonal antibodies directed against specific cell markers, for example, the CD4 and CD8 lymphocyte markers, can be used to identify and separate mixtures of T cells from one another. We will examine methods for using the discriminatory abilities of monoclonal antibodies for clinical applications in Chapter 21. Monoclonal antibodies are also useful for the immunological typing of bacteria and for the identification of cells containing foreign surface antigens (for example, a virus-infected cell). Monoclonal antibodies have also been used in genetic engineering for identifying and measuring levels of gene products not detectable by other methods.

Monoclonal Antibodies and Medicine

Monoclonal antibodies are very important reagents for clinical diagnostic and therapeutic applications. We discuss the use of monoclonal antibodies in clinical diagnostics in Section 21.4. In therapeutics, perhaps the boldest application of monoclonal antibodies may be in the detection and treatment of human cancer. Malignant cells contain a variety of surface antigens not expressed on the surfaces of normal cells. Tumor antigens include differentiation antigens expressed during fetal development but not in the normal adult, and unique, tumor-specific cell surface antigens (tumor-specific antigens were originally detected by monoclonal antibodies made against preparations derived from whole tumor cell membrane preparations). Because monoclonal antibodies prepared against cancer-related antigenic determinants are able to distinguish between normal and malignant cells, monoclonal antibodies serve as vehicles for delivering toxins directly and specifically to malignant cells. Such tumor-specific antibodies, covalently linked to toxins, are now undergoing clinical testing. The specificity of the monoclonal antibody treatment may greatly improve cancer chemotherapy by offering an alternative to the highly toxic chemical and radiation treatments used in conventional cancer chemotherapy.

Monoclonal antibodies have also been prepared that can distinguish human transplantation antigens, thus improving the specificity of the tissue matching process critical for successful transplantations. Monoclonal antibodies also show great promise for increasing the specificity of a variety of conventional clinical tests including blood typing, rheumatoid factor determination, and even pregnancy determination. For example, the pregnancy test employs monoclonal antibodies prepared against the specific hormones associated with pregnancy (see the box, Over-the-Counter Immunodiagnostic Kits). Other monoclonals

A FOCUS ON . . . Over-the-Counter Immunodiagnostic Kits

A number of manufacturers now market immunodiagnostic kits for use by the general public. For example, virtually every pharmacy now carries several brands of pregnancy tests and many also stock kits to determine the onset of ovulation in women.

These kits are all based on principles involving detection of a hormone excreted in the urine. All pregnancy tests detect the presence of human chorionic gonadotropin (HCG). When the ovum is fertilized, it produces HCG, which functions to maintain the corpus luteum and sustain pregnancy. As pregnancy continues, the new embryo produces and releases increasing amounts of HCG. The hormone, by this time being produced in massive amounts, is released into the bloodstream, removed by the kidneys, and excreted in the urine. Normal menstruation, which occurs about 10–14 days after ovulation if the ovum is not fertilized, does not take place, and at this time the level of HCG present in the urine of the pregnant woman is high enough to be detected easily by these tests. Thus, detectable HCG in the urine means that the individual is pregnant. The tests are advertised to be sensitive enough to detect pregnancy on the first day of a missed menstrual period, that is, 10–14 days *after* fertiliza-

tion, but *not* on the first day of a pregnancy.

The key to marketing these kits is their simplicity. Since most people using the tests are not trained in standard laboratory procedures, the test methods must be straightforward and the results must be easy to interpret. Typically, the patient must first provide a urine sample. In the easiest test, the urine sample is simply applied to a test strip. In others, the first step involves pouring the sample through a filter or immersing a "dipstick" in the urine, followed by exposing the filter or dipstick to a second solution. The results are interpreted as a simple color change: a positive test turns the strip, dipstick, or filter from white to pink (see figures) within 30 min or less. All tests share certain principles. At some point, the HCG in the urine is bound to an HCG-specific monoclonal antibody that is immobilized on a solid support—the dipstick, test strip, or filter. Next, a second HCG-specific monoclonal antibody, provided in a solution for the dipstick or filter test, and already soaked onto the test strip for the one-step test, reacts with the immobilized antibody–HCG complex. This second antibody is chemically linked to colloidal gold particles. If this second an-

tibody reacts with the HCG, it becomes immobilized and the whole complex appears colored.

These kinds of tests, all modifications of sol particle immunoassay (SPIA) tests, are also extensively used in doctor's offices to help diagnose diseases. For example, a number of available tests are specific for cell surface antigens on *Streptococcus pyogenes*, which causes strep throat. Formerly, the only way to positively identify a streptococcal infection was to culture and identify the organism, a process that involved skilled personnel and a laboratory. Because of the requirement for incubation of the culture, results were not available for at least 16 hr. Today, results are available in minutes, using material from a throat swab as the antigen source. Several test kits are also available for detecting serum antibodies that react with self proteins. These kits are useful in diagnosing autoimmune diseases such as rheumatoid arthritis.

These tests have several advantages. They are simple to perform, require no technical expertise, are relatively inexpensive, and are reasonably accurate. They also have a long shelf life, and no special storage conditions are required. However, there are disadvantages. They are qualitative tests, which means that they can be used only when the outcome is absolute. For example, one is either pregnant or not pregnant. However, quantitative measures of antigen levels might be desirable in some cases, such as in the tests that identify autoimmune antibodies. In addition, many manufacturers do not provide adequate positive and negative controls to ensure that the test works properly. Thus, results should always be interpreted with caution, and negative or positive results may be in error because of improper test procedures or faulty test reagents. ■

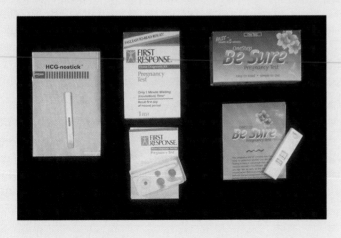

are being developed for *in vivo* diagnostics. For example, monoclonal antibodies have been used experimentally to detect exposed myosin in heart muscle; myosin is a major heart muscle protein that becomes exposed and detectable only following damage to the heart muscle. A myosin-specific monoclonal antibody would therefore be useful for diagnosing the extent of heart damage following a heart attack.

As with any clinical reagent, the greater the specificity, the more useful it is in clinical diagnostics. Monoclonal antibodies offer the physician and clinical microbiologist a reproduceable, highly specific biological reagent with unlimited applications (for another example, see the box, Urine Testing for Drug Abuse, for a description of the use of monoclonal antibodies for detecting illicit drugs).

✓ 20.13 Concept Check

Both polyclonal and monoclonal antibodies are useful for research and clinical applications. Monoclonal technology, however, makes reproducible, monospecific antibodies available for these procedures.

✓ Can a single monoclonal antibody recognize a variety of epitopes?
✓ Are polyclonal antibodies useful for diagnostic tests? Why or why not?

20.14

Antigen–Antibody Reactions

The study of antigen–antibody reactions *in vitro* is called **serology** and is especially important in clinical diagnostic microbiology. A variety of serological reactions can be observed (Table 20.6), depending on the specific properties of the antigen and antibody and on the conditions chosen for reaction. Serology has many applications, only a few of which will be discussed here. Further practical applications of these basic reactions will be discussed in detail in the next chapter.

The principle of all antigen–antibody reactions lies in the specific combination of determinants on the antigen with the *variable* region of the antibody molecule. The details of antigen binding are discussed in the box, Molecular Biology of the Immune Response. The end result of this complex process is a specific antibody molecule that combines with a specific antigenic determinant. Interaction with antibody can block or distort the antigen sufficiently to reduce or totally eliminate its biological activity.

Some typical antigen–antibody reactions are illustrated in Figure 20.23. We will discuss a number of antigen–antibody detection systems in the context of clinical immunology in Chapter 21.

Neutralization

Neutralization of microbial toxins by specific antibody can occur when toxin molecules and antibody molecules directed against the toxin combine in such a way that the active portion of the toxin is blocked (Figure 20.23a). Neutralization reactions of this type occur for a wide variety of exotoxins, including most of those listed in Table 19.4. An antiserum containing an antibody that neutralizes a toxin is referred to as an **antitoxin.** Similar neutralization reactions also occur with viruses and their specific antibodies. For instance, antibodies directed against the protein coats of viruses may prevent the adsorption of the viruses to host cells. This is one means by which a viral vaccine can act. If the vaccine stimulates antibodies that react to free virus, the virus is unable to attach to host cells and cannot initiate infection.

Precipitation

Since antibody molecules generally have two combining sites (that is, they are bivalent) (see Figure 20.10), it is possible for each site to combine with a separate antigen molecule, and if the antigen also has more than one available epitope, a **precipitate** may develop consisting

TABLE 20.6	Types of antigen–antibody reactions	
Location of antigen	**Accessory factors required**	**Reaction observed**
Soluble	None	Precipitation
On cell or inert particle	None	Agglutination
Flagellum	None	Immobilization or agglutination
On bacterial cell	Complement	Lysis
On bacterial cell	Complement	Killing
On erythrocyte	Complement	Hemolysis
Toxin	None	Neutralization
Virus	None	Neutralization
On bacterial cell	Phagocyte, complement	Phagocytosis (opsonization)

TECHNIQUES & APPLICATIONS ...   Urine Testing for Drug Abuse

Many government agencies and private employers carry out drug testing programs in an effort to control drug abuse in the workplace. If a person uses a drug, metabolites of the drug are excreted into the urine. Testing urine for the presence of such metabolites thus permits detection of drug use. Common drug tests include those for cannabinoids (found in marijuana), cocaine, phencyclidine, amphetamines, propoxyphene, benzodiazepine, opiates, steroids, and barbiturates.

Because of the minute quantities of drugs or drug metabolites that occur in the urine, extremely sensitive detection methods are needed, but these methods must also be very specific. Immunological procedures are among the most sensitive and specific methods known for testing urine. Two types of immunoassays are usually employed for drug testing, radioimmunoassay (RIA) and enzyme-linked immunosorbent assay (ELISA). For a drug immunoassay, an antibody must first be prepared against the drug or drug metabolite to be assayed. The antibody thus specifically recognizes an antigenic determinant of the drug. The assay is based on the principle of *competition* between a labeled antigen present in the drug testing system and an unlabeled antigen (the drug or drug metabolite in the urine) for binding sites on the specific antibody. In both the RIA and ELISA tests, the *greater* the amount of drug in the urine, the *less* of the labeled drug will be bound.

In RIA drug testing, known amounts of radiolabeled drug are added to a urine sample together with known amounts of an antibody specific for the drug being tested. The presence of the drug is measured by determining the radioactivity of the antibody bound to a solid substrate. As the concentration of drug in the urine goes up,

the radioactivity bound goes down. Concentrations of drug as low as 1–5 μg/ml can be detected in 1–5 hr. The RIA test is very suitable for large-scale screening programs because automatic pipetting and counting procedures can be used.

In the ELISA drug testing procedure, the antigen (drug) is covalently linked to an enzyme, commonly glucose-6-phosphate dehydrogenase, whose activity can be detected by a simple and sensitive colorimetric assay. If the enzyme, via its linked drug ligand, becomes associated with the antibody, it loses its enzymatic activity, but free enzyme can react with the substrate. Urine is mixed with a reagent consisting of antibody to the drug, the glucose-6-phosphate dehydrogenase–drug derivative, and glucose 6-phosphate. If drug is present in the urine, it competes with the enzyme–drug derivative for specific antibody, and more enzyme is then free to react with its substrate. Thus, the more drug present in the subject's urine, the more intense the color. The ELISA procedure has the advantage that the analysis time is short, and the color change can be detected simply. However, the ELISA procedure is sometimes less sensitive than the RIA procedure. The ELISA procedure is especially valuable where a small number of samples are to be analyzed at the site where the urine is collected (for instance, away from a laboratory). In some

cases, the reagents have been incorporated into paper strips, providing means for urine testing that can be used even by those lacking laboratory experience (see the photo). For large-scale testing, robotic devices have the capability of processing up to 18,000 urine samples per hour with each sample identified by a bar code identification sticker.

Either polyclonal or monoclonal antibodies can be used for urine testing. Monoclonal antibodies have the advantage of defined specificity but may not necessarily have as high an affinity for the drug as a selected polyclonal antibody. The selection of an antibody depends on the specificity and sensitivity needed, the cost of the product, and the ease of use.

One problem with the use of any of these immunological methods for urine testing is cross-reaction with other chemicals that might be present in the urine. For instance, the pain killer ibuprofen cross-reacts with marijuana metabolites in some tests, and the antihistamine drug diphenhydramine can cross-react in the ELISA test for methadone, a controlled narcotic. Because of such false-positive reactions, positive tests must be confirmed by independent tests, such as the use of gas chromatography or high performance liquid chromatography. Highly sensitive combined mass spectrometer–gas chromatographic methods are also available and are valuable for identifying exotic drugs or drugs that elicit a poor immune response.

Immunological tests for drugs in the urine have provided a new and potentially widespread technique for controlling the use of illegal drugs in the workplace. Urine drug testing is another example of the power and utility of immunology in modern society. ■

Keystone Diagnostics, Inc.

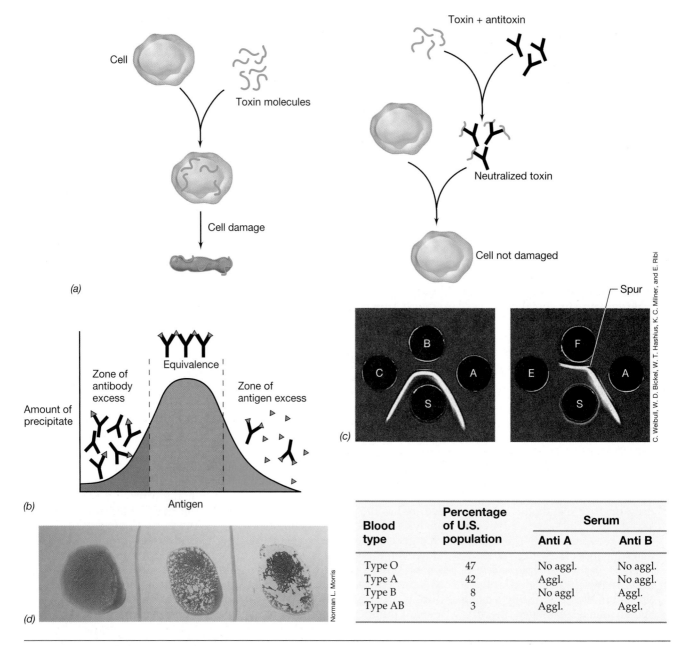

FIGURE 20.23   Some antigen–antibody reactions. (a) Neutralization of toxin by antitoxin. (b) Precipitation reaction between soluble antigen and antibody. The graph shows the extent of precipitation as a function of antigen and antibody concentration. (c) Precipitation in an agar gel, a process known as immunodiffusion. Wells in agar labeled S contained antibodies to cells of *Proteus mirabilis*. Wells labeled A, B, and C contained extracts of *P. mirabilis* cells. (Left) A line of identity is observed. (Right) Antigen E does not react and antigen F shows a line of partial identity with antigen A. (d) Agglutination. Typing of human blood in the ABO system. (Left) Typical agglutination pattern. The reaction on the left is negative and shows no agglutination. The center reaction is positive for the B blood group. The agglutinate is apparent, but somewhat diffuse. The reaction on the right is positive for the A blood group, with large, clumped agglutinates of all red blood cells. (Right) Table of expected blood-typing results.

of aggregates of antibody and antigen molecules (Figure 20.23*b*). Because they are easily observed *in vitro*, precipitation reactions are very useful serological tests, especially in the quantitative measurement of antibody concentrations. Precipitation occurs maximally only when there are optimal proportions of the two reacting

substances. The presence of either antigen or antibody in excess results in the formation of only small, soluble complexes (Figure 20.23*b*).

Precipitation reactions carried out in agar gels, referred to as *immunodiffusion*, are useful in the study of the specificity of antigen–antibody reactions. Both antigen and antibody diffuse outward from separate wells cut in the agar gel, and precipitation bands form in the region where antibody and antigen meet in optimal proportions (Figure 20.23c). The shapes of the precipitation bands are characteristic for the reacting substances, and two antigens reacting against antibodies in an antiserum can be tested for relatedness by observing the bands formed when the two antigens are placed in adjacent wells near the antiserum well. For example, if two antigens in adjacent wells are identical, they will form a single fused precipitin band. This is referred to as a *line of identity*. If, on the other hand, adjacent wells contain one antigen in common but one well contains a second antigen, a *line of partial identity* will form (Figure 20.23c). The small protruding precipitin line (representing a reaction between the antiserum and the second antigen) is referred to as a *spur*. Immunodiffusion is used as a tool in biochemical research to assess the relatedness of proteins obtained from different sources. Unfortunately, the readily visible precipitation reactions are not very sensitive. Microgram quantities of antibody are necessary to see a precipitate (∞ Table 21.5). Therefore, precipitation reactions are not widely used for clinical or diagnostic applications.

Agglutination

When antigen is not in solution but is present on the surface of a cell or other particle, an antigen–antibody reaction can lead to clumping of the particles, called **agglutination.**

Soluble antigens can be adsorbed or coupled chemically to cells or other particles such as latex beads, and they can then be detected by agglutination reactions, the cell or particle serving only as an inert carrier. This greatly increases the ability to detect the presence of antibodies against soluble antigens because agglutination is much more sensitive than precipitation.

Agglutination of red blood cells is referred to as *hemagglutination*. Red blood cells contain a variety of antigens, and people differ in the antigens present on their red blood cells. The classification of blood antigens is called *blood typing* and is a good example of an agglutination reaction (Figure 20.23d). The principle of blood typing is that antibodies, specific for particular erythrocyte surface antigens, cause red blood cells to visibly clump. The human red cell antigens of most clinical interest are called A, B, and D (also known as Rh). Humans make antibodies to some nonself blood group antigens. For example, type A humans have antibodies to group B antigens, while type B individuals have antibodies to group A antigens. Type AB individuals have neither A nor B antibodies, whereas type O individuals have anti-A and anti-B activity. Antibodies against A and B antigens are called *natural antibodies* because they

appear to be produced by most individuals in response to ubiquitous antigen sources such as enteric Bacteria.

The reason for blood typing before blood transfusion is to prevent the red blood cell destruction that would occur if blood containing a particular antigen were transfused into a person having an antibody against this antigen. If agglutination occurred, the clumps of cells could lodge in blood vessels or arteries and block the flow of blood, causing serious illness or death; if the antibodies caused hemolysis through the action of complement, anemia could result.

The presence or absence of the A and B antigens or antibodies is the basis for the ABO blood grouping assay (Figure 20.23d). A small drop of blood is placed on a microscope slide containing antiserum prepared against either the A or B antigens of human erythrocytes. These commercially available high titer antisera are obtained from human donors who have been immunized to these antigens by natural or artificial means. The agglutination reaction observed in the ABO blood grouping test is rapid and easy to interpret (Figure 20.23d); hence this simple assay is used routinely to group human blood.

The agglutination reaction is about 100-fold more sensitive than precipitation, and is used in a wide variety of clinical and diagnostic tests (∞ Table 21.6).

✓ 20.14 Concept Check

Antigen–antibody reactions all require that antibody bind to antigen. Precipitation and agglutination reactions are examples of antibody binding to soluble or insoluble macromolecular antigens, respectively.

- ✓ What are the minimum antigen and antibody requirements for a precipitation reaction? For an agglutination reaction?
- ✓ Is agglutination *more* or *less* sensitive than precipitation? Why?

20.15

Immune Diseases

Immune diseases occur when immune reactions cause host cell damage. In this section, we deal with **hypersensitivity,** the overreaction of the immune response resulting in host damage. First, we will describe antibody-mediated *immediate hypersensitivity*. Next, we will focus on hypersensitivity reactions that are involved in autoimmune diseases, a group of diseases that result from immune reactions directed against "self" antigens (see Section 20.10). These clinically important immune diseases have been classified as separate hypersensitivity syndromes (Table 20.7). Then we will look at cell-mediated *delayed-type hypersensitivity*. Finally, we will discuss a group of toxins called **super-**

TABLE 20.7	Classification of hypersensitivity diseases			
Classification	Description	Immune mechanism	Time of latency	Examples
Type I	Immediate	IgE sensitization of mast cells	Minutes	Reaction to bee venom (sting) Hay fever
Type II	Cytotoxic[a]	IgG interaction with cell surface antigen	Hours	Drug allergies (Penicillin)
Type III	Immune complex[a]	IgG interaction with soluble or circulating antigen	Hours	Systemic lupus erythematosis (SLE)
Type IV	Delayed Type	T_H1 inflammatory cells	Days	Poison ivy Tuberculin test

a Most autoimmune diseases are caused by Type II or Type III reactions.

antigens, a group of proteins produced by bacteria and viruses that cause massive stimulation of immune cells, resulting in host damage.

Immediate Hypersensitivity (Type I Hypersensitivity)

Many hypersensitivity reactions occur within minutes after exposure to antigen. This form of hypersensitivity is referred to as **immediate hypersensitivity** and is mediated by *antibodies* (Table 20.7). Immediate hypersensitivities are commonly called **allergies.** Depending on the individual and the antigen, immediate hypersensitivity reactions may be very mild, or cause extremely severe, even life-threatening reactions, a process known as *anaphylaxis.* Antigens that cause these hypersensitivities are known as *allergens.*

Up to 20% of the population suffers from immediate hypersensitivities, involving allergic (*anaphylactic*) reactions to specific allergens such as pollens, animal dander, and a variety of other agents (Table 20.8). In a typical anaphylactic reaction, an allergen elicits (on first exposure) the production of immunoglobulins of the IgE class. Instead of circulating like IgG or IgM, IgE molecules bind via a cell-binding constant domain (see Section 20.5) to specific receptors on mast cells and basophils (Figure 20.24). **Mast cells** are nonmotile connective tissue cells found adjacent to capillaries throughout the body, and **basophils** are motile white blood cells (leukocytes) that

TABLE 20.8	Common immediate-type hypersensitivity allergens

Pollen and fungal spores (hay fever)
Insect venoms (bee sting)
Certain foods
Animal dander
Mites in house dust

make up about 1% of the total leukocyte population. On subsequent exposure to antigen, the cell-bound IgE molecules bind the antigen. Thus, immediate hypersensitivity reactions are the result of secondary memory responses. This triggers the release of several soluble allergic mediators from mast cells and basophils.

The primary chemical mediators released from mast cells and basophils are **histamine** and **serotonin** (both are modified amino acids). Several other anaphylactic mediators have been characterized as small peptides. The release of histamine and serotonin causes dilation of blood vessels and contraction of smooth muscle, which initiate the typical symptoms of anaphylaxis. These symptoms include difficulty in breathing, flushed skin, copious mucus production, sneezing, and itchy, watery eyes. In general, the symptoms are relatively short-lived, but once initially sensitized by an allergen, an individual can respond repeatedly on subsequent exposure to the antigen. The magnitude of the anaphylactic reactions may vary from mild symptoms (or none), to such severe symptoms that the individual goes into **anaphylactic shock.** In humans, anaphylactic shock is characterized by severe respiratory distress, capillary dilation (causing a sharp drop in blood pressure), and flushing and itching of the skin. If severe cases of anaphylactic shock are not treated immediately with adrenalin to counter smooth muscle contraction, increase blood pressure, and promote breathing, death can occur.

Autoimmune Diseases (Type II and Type III Hypersensitivities)

We saw in Section 20.10 how T and B cells destined to react with self antigens are eliminated or anergized during the process of lymphocyte maturation. However, in some individuals, anergized T and B lymphocytes can become reactivated. This resurgence of self-reactive clones leads to *autoimmune diseases.* Most are slow, pro-

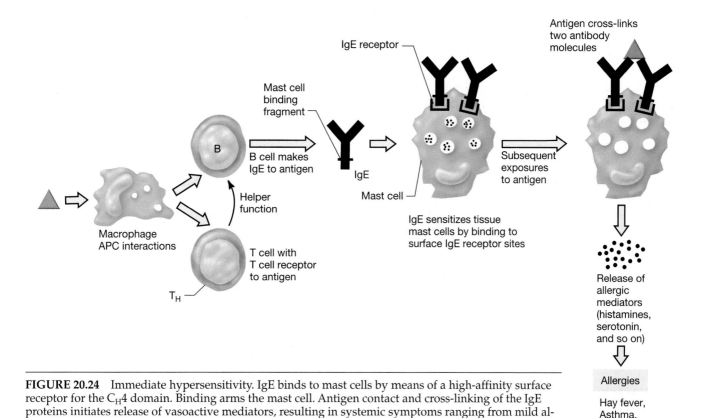

FIGURE 20.24 Immediate hypersensitivity. IgE binds to mast cells by means of a high-affinity surface receptor for the C_H4 domain. Binding arms the mast cell. Antigen contact and cross-linking of the IgE proteins initiates release of vasoactive mediators, resulting in systemic symptoms ranging from mild allergies to life-threatening anaphylaxis.

gressive disorders in which the function of some specific organ or set of tissues becomes increasingly poorer with time (Table 20.9). Depending on the specific disorder, autoimmunity may involve autoantibodies or a cellular immune response to self constituents. Certain autoimmune diseases are highly organ-specific. For example, in *Hashimoto's disease,* autoantibodies are made against thyroglobulin and other thyroid proteins. The disease affects thyroid function and is classified as a Type II disorder because antibodies interact with antigens found on the surface of thyroid cells and cause destruction of the thyroid gland. In *juvenile diabetes*

TABLE 20.9	Some autoimmune diseases of humans	
Disease	**Organ or area affected**	**Mechanism (Hypersensitivity Type)**
Juvenile diabetes (insulin-dependent diabetes mellitus)	Pancreas	Cell- mediated immunity and autoantibodies against surface and cytoplasmic antigens of islets of Langerhans (II and IV)
Myasthenia gravis	Skeletal muscle	Autoantibodies against acetylcholine receptors on skeletal muscle (II)
Goodpasture's syndrome	Kidney	Autoantibodies against basement membrane of kidney glomeruli (II)
Rheumatoid arthritis	Cartilage	Autoantibodies against self IgG antibodies, which form complexes deposited in joint tissue, leading to inflammation and cartilage destruction (III)
Hashimoto's disease (hypothyroidism)	Thyroid	Autoantibodies to thyroid surface antigens (II)
Male infertility (some cases)	Sperm cells	Autoantibodies agglutinate host sperm cells (II)
Pernicious anemia	Intrinsic factor	Autoantibodies prevent absorption of vitamin B_{12} (III)
Systemic lupus erythematosis	DNA, cardiolipin, nucleoprotein, blood clotting factors	Massive autoantibody response to various cellular constituents results in immune complex formation (III)
Addison's disease	Adrenal glands	Autoantibodies to adrenal cell antigens (II)
Allergic encephalomyelitis	Brain	Cell-mediated response against brain tissue (IV)
Multiple sclerosis	Brain	Cell-mediated and autoantibody response against central nervous system (II and IV)

(insulin-dependent diabetes mellitus), autoantibodies against the insulin-producing cells, the islets of Langerhans, are observed, but tissue destruction clearly occurs via T_H1 cells. Such antibodies may help cause the disease by destroying the cells, or they may be the result of antigens released from islet cells damaged by the cellular mechanisms. *Systemic lupus erythematosis* (SLE) involves a large-scale production of autoantibodies against many self constituents, including DNA. This disease and others like it are induced by circulating antigen–antibody complexes that may deposit in several different body tissues, such as the kidney and the spleen. Complement fixation and the resulting lytic and inflammatory responses cause local but often severe cell damage at the site of the complex deposition. SLE is a clear example of a Type III immune complex disorder. Organ-specific autoimmune diseases are sometimes more easily controlled clinically because the product of organ function, such as thyroxin in hypothyroidism or insulin in diabetes, can often be supplied in pure form from another source. More generalized syndromes such as SLE can be controlled by immunosuppressive therapy, but this approach is risky because of the increased chance of opportunistic infections.

Evidence is accumulating that heredity has an important influence on the incidence, type, and severity of autoimmune diseases. An inherited tendency to develop certain autoimmune diseases is known to exist; many autoimmune diseases correlate strongly with the presence of certain major histocompatibility complex (MHC) antigens (see Section 20.7). Studies of model autoimmune diseases in mice support such a genetic link, but the precise conditions necessary for developing autoimmunity are also dependent on other factors, including hormone levels and the presence of infectious agents such as pathogenic bacteria and viruses.

While most autoimmune diseases fall into the category of Type II or III hypersensitivities, juvenile diabetes and allergic encephalomyelitis (Table 20.9) are thought to have cell-mediated autoimmune mechanisms, which we discuss next.

Delayed-Type Hypersensitivity (Type IV Hypersensitivity)

Delayed-type hypersensitivity (DTH) is cell-mediated and involves the activities of the T_H1 inflammatory cell subset (Table 20.7). Symptoms begin to appear several hours after secondary exposure to the eliciting antigen, with a maximal response usually occurring in 24–48 hr. The eliciting antigens include certain microorganisms, a few self-antigens (Table 20.9), and a group of chemicals that covalently bind to the skin, creating new antigens. Hypersensitivity to these newly created antigens is known as **contact dermatitis** and is responsible for the allergic skin reactions to poison ivy, jewelry, cosmetics, and certain chemicals and drugs. Shortly after exposure to the agent, the skin feels itchy at the site of contact, and within several hours reddening and swelling appear, indicative of a general inflammatory response. Localized tissue destruction, often in the form of blistering, occurs as a result of the activities of immune cells.

A good example of a delayed-type hypersensitivity reaction is the development of immunity to the causal agent of tuberculosis, *Mycobacterium tuberculosis* (Figure 20.25). This cellular immune response was first discovered by Robert Koch during his classic work on tuberculosis and has been widely studied. Antigens derived from the bacterium, when injected subcutaneously into an animal previously infected with *M. tuberculosis*, elicit a characteristic skin reaction that develops fully only after a period of 24–48 hr. (In contrast, skin reactions to IgE-mediated responses as discussed previously, develop almost immediately after antigen injection.) In the region of the injected antigen, T_H1 cells become stimulated by the antigen and release cytokines such as IFN-γ, which attract and activate large numbers of macrophages. The macrophages are responsible for the ingestion and digestion of the invading antigen. The characteristic skin reaction seen at the site of injection includes induration (hardening), edema (swelling), erythema (reddening), pain, and localized heating. These are common features of an inflammatory response resulting from the release of cytokines by activated, antigen-stimulated T_H1 cells. This skin response serves as the basis for the **tuberculin test** for determining prior exposure to *M. tuberculosis* (Figure 20.25).

A number of microbial infections elicit delayed-type hypersensitivity reactions. In addition to tuberculosis,

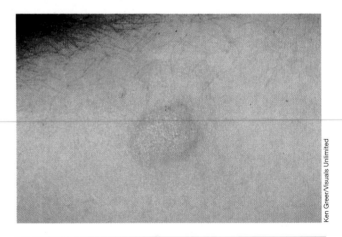

FIGURE 20.25 Cell-mediated immunity. A positive tuberculin test, typical for delayed hypersensitivity, and the result of the action of T_H1 effector cells. The raised area of inflammation is 1.5 cm in diameter.

these include leprosy, brucellosis, psittacosis (all caused by Bacteria), mumps (caused by a virus), and coccidioidomycosis, histoplasmosis, and blastomycosis (caused by fungi). In all these cellular immune reactions, characteristic skin reactions occur after injection of antigens derived from the pathogens, and these skin reactions are used to diagnose prior exposure to the pathogen.

Superantigens

In Sections 19.8 and 19.9, we discussed the effects of bacterial exotoxins and enterotoxins on the host. As we noted, not all exotoxins or even enterotoxins work by the method described for cholera toxin (Figure 19.19). In fact, a completely different mechanism of action is used by a family of exotoxins known as **superantigens.** These toxins are produced by the staphylococci and streptococci. They include *toxic shock syndrome toxin, staphylococcal enterotoxin,* and *exfoliating toxin* (Table 19.4), and all have a toxic mechanism dependent on the immune system. Viral superantigens bind to a site on the β chain of the class II MHC molecule that is outside the normal antigen-binding site, while bacterial superantigens bind to a site on the α chain of the class II MHC molecule. Both sets of superantigens also bind to the Vβ domain of the TCR, again outside of the normal antigen binding site (Figure 20.26). In some cases, this binding may stimulate greater than 10% of the normal T cells in the affected individual. A normal antigen, even a very potent one, stimulates less than 0.1% of the total T cell population. Superantigen stimulation results in simultaneous activation of all reactive T cells and an overwhelming cell-mediated response characterized by systemic inflammatory effects, sometimes resulting in generalized shock (such as toxic shock syndrome). This mechanism for stimulating the host immune system to induce host damage may be a common pathogenic strategy for the staphylococci and streptococci.

✓ 20.15 Concept Check

Allergy results when foreign antigens stimulate cellular or humoral immunity and effector cells react so vigorously that host tissue is damaged by the resulting inflammation. Autoimmunity results when the immune response reacts or cross-reacts with self antigens. Superantigens cause massive over-stimulation of the immune response and serious inflammatory consequences.

- ✓ Distinguish between *hay fever* and *poison ivy* allergies.
- ✓ Why is insulin-dependent diabetes mellitus considered an autoimmune disease?
- ✓ How can superantigen reactions promote the spread of bacterial disease?

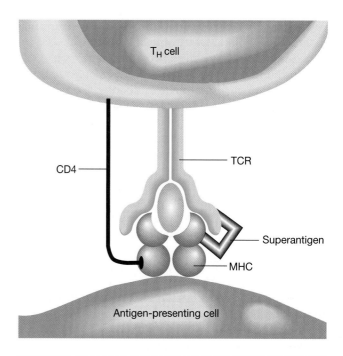

FIGURE 20.26 Bacterial superantigens act by binding to both the MHC protein and the TCR at positions outside the normal binding site. Because the superantigen binding sites are found on conserved regions of MHC and TCR proteins, the superantigen can interact with large numbers of cells, stimulating massive T cell activation, cytokine release, and even systemic inflammation.

20.16

Immunity to Infectious Diseases

The major role of the immune response in the body is to protect the animal from the consequences of infection. The importance of antibodies in disease resistance is shown dramatically in individuals with the genetic disorder **agammaglobulinemia,** in whom antibodies are not produced because their B cells are defective. Such individuals suffer from recurrent, even life-threatening, bacterial infections. The general lack of an antibody response is also observed in those suffering from acquired immunodeficiency syndrome (AIDS). However, in this case the problem is not due to defective B cells. Instead, AIDS patients suffer from a virtually total cessation of CD4 (T_H) cell activities (Section 23.7). The crucial importance of T cells in immunity is clearly evident in AIDS patients: The inability to mount an immune response causes death, usually from infectious diseases (Section 22.4).

The purposeful artificial induction of specific immunity to infectious diseases is a major contribution of microbiology to the treatment and prevention of infectious diseases. However, an animal or human may acquire immunity to a disease in several ways. (1) The individual may acquire infection and develop immunity. This is **natural active immunity** because the immunization was a natural outcome of infection and the infected individual produced the immune response. (2) The individual may be given injections of an antigen known to induce formation of antibodies, a type of immunity known as **artificial active immunity** because the individual in question produced the antibodies. This process is commonly known as **vaccination,** but is more properly termed **immunization,** since vaccination refers specifically to immunization with vaccinia virus, the agent used to immunize against infection by the smallpox virus (Section 8.20). (3) Alternatively, the individual may receive injections of an antiserum derived from another individual who has previously formed antibodies against the antigen in question. This is called **artificial passive immunity** because the individual receiving the antibodies played no active part in the antibody-producing process. (4) Finally, **natural passive immunity** also occurs. For several months after birth, newborns have maternal IgG antibodies in their blood. These antibodies, acquired across the placenta before birth, provide valuable disease protection while the immune system of the newborn is maturing. Active and passive immunity are contrasted in Table 20.10.

In *active* immunity, introduction of antigen produces fundamental changes in the host: The immune cells produce large quantities of antigen-specific immune effector molecules (Igs) and cells. A second (booster) dose of the same antigen results in a faster, higher titer secondary response (see Figure 20.18). Active immunity often remains throughout life. A *passively* immunized individual never has more antibodies than it received in the initial injection, and these antibodies gradually disappear from the body; moreover, a later exposure to the antigen does not elicit a secondary response. Active artificial immunity is usually used as a *prophylactic* measure, to protect a person against future attack by a pathogen. Passive artificial immunity is usually *therapeutic,* designed to cure a person who is presently suffering from the disease. For example, tetanus toxoid (see the following section) actively immunizes an individual against future encounters with *Clostridium tetani* exotoxin, whereas tetanus antiserum

(antitoxin) (see later) is administered to passively immunize an individual suspected of coming in contact with *C. tetani* exotoxin via growth of the organism in a penetrating wound.

Immunization

The material used in inducing artificial active immunity, the antigen or mixture of antigens, is known as a **vaccine** or an **immunogen,** and the process of generating such an immune response is **immunization.** However, to induce active immunity to toxin-caused diseases, the toxin itself is not injected. Many exotoxins can be modified chemically so they retain their antigenicity but are no longer toxic. Such a modified exotoxin is called a **toxoid.** Toxoids are usually not such efficient antigens as the original exotoxin, but they can be given safely and in high doses. When immunization against whole microorganisms is necessary, such as for endotoxin-producing organisms, the microorganism is usually killed by an agent such as formaldehyde, phenol, or heat, and the dead cells are then injected. Endotoxin-caused diseases for which vaccines are made routinely are whooping cough and typhoid fever. Formaldehyde treatment is also used to inactivate viruses in preparing some vaccines, such as the Salk polio vaccine.

Immunization with live cells or virus is usually more effective than with dead or inactivated material. Often it is possible to isolate a mutant strain of a pathogen that has lost its virulence but still retains the immunizing antigens; strains of this type are called **attenuated strains** (Section 19.11).

A summary of vaccines available for use in humans is given in Table 20.11. Most effective viral vaccines are in the attenuated form. Killed virus vaccines tend to provide only short-lived immune responses, without the desirable long-term memory response. Bacterial vaccines, on the other hand, are nearly always in killed form. The killed vaccines are adequate to induce long-term antibacterial protection in most cases. While the attenuated vaccines may be more effective (and in the case of many viruses are necessary), attenuated strains are difficult to select, standardize, and maintain. Live

TABLE 20.10	Comparison of active and passive immunity	
Active immunity	**Passive immunity**
Exposure to antigen; immunity achieved by injecting antigen	No exposure to antigen; immunity achieved by injecting antibodies or antigen-reactive T cells
Specific response made by individual achieving immunity	Specific response made in a secondary host
Immune system activated to antigen; immunological memory in effect	No immune system activation; no immunological memory
Immune response can be maintained via stimulation of memory cells (*i.e.,* booster immunization)	Immune response cannot be maintained and decays rapidly
Immune state develops over a period of weeks	Immune state develops immediately

TABLE 20.11	Available vaccines for infectious diseases in humans
Disease	**Type of vaccine used**
Bacterial diseases	
Diphtheria	Toxoid
Tetanus	Toxoid
Pertussis	Killed bacteria (*Bordetella pertussis*) *or* acellular proteins
Typhoid fever	Killed bacteria (*Salmonella typhi*)
Paratyphoid fever	Killed bacteria (*Salmonella paratyphi*)
Cholera	Killed cells or cell extract (*Vibrio cholerae*)
Plague	Killed cells or cell extract (*Yersinia pestis*)
Tuberculosis	Attenuated strain of *Mycobacterium tuberculosis* (BCG)
Meningitis	Purified polysaccharide from *Neisseria meningitidis*
Bacterial pneumonia	Purified polysaccharide from *Streptococcus pneumoniae*
Typhus fever	Killed bacteria (*Rickettsia prowazekii*)
Haemophilus influenzae meningitis	Conjugated vaccine (polysaccharide of *Haemophilus influenzae* conjugated to protein)
Lyme disease	Recombinant membrane protein of *Borrelia burgdorferi*
Viral diseases	
Yellow fever	Attenuated virus
Measles	Attenuated virus
Mumps	Attenuated virus
Rubella	Attenuated virus
Polio	Attenuated virus (Sabin) or inactivated virus (Salk)
Influenza	Inactivated virus
Rabies	Inactivated virus (human) *or* attenuated virus (dogs and other animals)
Hepatitis A	Recombinant DNA vaccine
Hepatitis B	Recombinant DNA vaccine *or* inactivated virus
Varicella (chickenpox)	Attenuated virus

vaccines also have a limited shelf life and usually require refrigeration for adequate storage.

Immunization Practices

Infants possess antibodies derived from their mothers across the placenta or, after birth, in colostrum and are relatively immune to infectious disease during the first 6 months of life. However, they are immunized for key infectious diseases as soon as possible so that their own *active* immunity can replace the maternal *passive* immunity. As discussed in Section 20.11, a single injection of antigen does not lead to a high antibody titer; after the initial innoculation, a series of "booster" injections are given to produce a high antibody titer. The recommended vaccination schedule for children in the United States from birth to adulthood is shown in Figure 20.27.

The importance of immunization in controlling infectious diseases is well established. Introduction of a specific immunization procedure into a population often dramatically reduces the incidence of the disease (Figure 20.28). The degree of immunity obtained by vaccination varies greatly, depending on the individual and on the quality and quantity of the vaccine. However, lifelong immunity is rarely achieved by means of a single injection, or even a series of injections, and the immune cells induced by immunization gradually disappear from the body. One way in which antigenic stimulation occurs even in the absence of immuniza-

tion is by nonsymptomatic or minor infections. A natural infection results in a booster response, leading to an increase in activated antibody-producing cells and to production of antibody. In the complete absence of antigenic stimulation, the immune period varies considerably with different antigens. However, active immunity to certain vaccines such as tetanus toxoid can last many years, and in some cases, a lifetime.

Immunization procedures are not only beneficial to the individual but are effective public health procedures because disease spreads poorly through a population with a large proportion of immune individuals (∞ Section 22.6).

Passive Immunity

The material used in inducing passive immunity—the serum containing antibodies—is known as a *serum*, an *antiserum*, or an *antitoxin* (the last applies to a serum containing antibodies directed against a toxin). Antisera are obtained either from large immunized animals, such as horses, or from humans who have high antibody titers. These individuals are said to be **hyperimmune.** The antiserum or antitoxin is standardized to contain a known antibody titer; a sufficient number of units of antiserum must be injected to neutralize any antigen that might be present in the body. Sometimes the immunoglobulin fraction of pooled human serum is used as a source of antibodies. It contains a wide va-

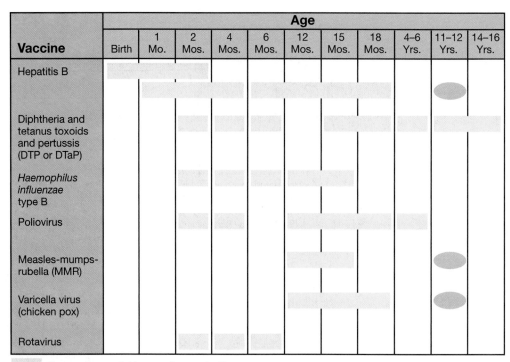

FIGURE 20.27   Recommended childhood immunizations in the United States as specified by the Centers for Disease Control and Prevention, Atlanta, GA, for 1999. For pertussis, the DTP vaccine contains a whole-cell *Bordetella pertussis* preparation while the DtaP vaccine is an acellular *Bordetella pertussis* preparation. Both vaccines are effective in preventing disease. For polio, killed poliovirus immunizations should be used at 2 and 4 months while live attenuated preparations may be used for subsequent immunizations.

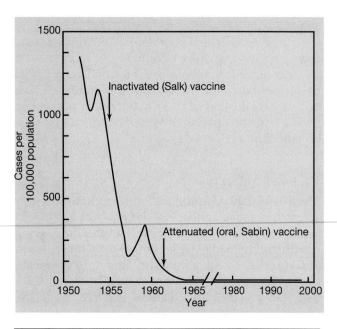

FIGURE 20.28   Cases of polio in the United States 1950–1990, showing the consequences of the polio vaccine. There have been no cases of endemic polio in the United States since 1990. Several cases have been imported from other countries or have been vaccine induced.

riety of antibodies that normal people have formed by artificial or natural exposure to various antigens. Pooled sera are used when hyperimmune antisera are not available.

The most common use of passive immunization is in the *prophylaxis* (prevention) of infectious hepatitis (resulting from hepatitis A virus) (⌖ Section 24.12). Pooled human immunoglobulin, often referred to as *gamma globulin,* contains fairly high titers of antibody against hepatitis A virus because infection with hepatitis A virus is widespread in the population. Travelers to areas where the incidence of infectious hepatitis is high, such as tropical Africa and North Africa, the Middle East, Asia, and parts of South America, may be given prophylactic doses of pooled human gamma globulin. Pooled human gamma globulin may also be of value in the *therapy* of infectious hepatitis if given early in the incubation period.

✓ 20.16 Concept Check

Immunity to infectious disease can be either passive or active, natural or artificial. Immunization, a form of artificial active immunity, is widely employed to prevent infectious diseases. Most agents used for immunization are either attenuated or inactivated microorganisms or viruses or inactivated forms of natural microbial products.

✓ Provide an example of natural passive immunity. How does natural passive immunity benefit the immunized individual?

✓ Provide an example of artificial active immunity. How does artificial active immunity benefit the immunized individual?

20.17

Alternative Immunization Strategies

Most immunization preparations currently in use are produced from whole organisms or toxoids, as described earlier in this section. However, there are several other methods for producing antigens suitable for immunization. Agents made by some of these methods are replacing more traditional immunization agents.

Synthetic and Genetically Engineered Immunizing Agents

The simplest alternate approach to vaccine development is the use of *synthetic peptides*. To make a vaccine, a peptide can be synthesized that corresponds to a known epitope on an infectious agent. For example, the structure of the protein antigen responsible for immunity to foot-and-mouth virus, an important animal pathogen, is known. A synthetic peptide of 20 amino acids constituting the antigenic portion of the protein has been made and attached to suitable carrier molecules. This synthetic vaccine evokes an excellent neutralizing antibody response to foot-and-mouth virus. However, as a general method, this approach has one major problem: The entire antigenic profile of the protein must be known to make an effective vaccine. Although this condition has been met for foot-and-mouth virus, very few pathogens have such a well-defined antigenic profile.

Sophisticated molecular biology techniques can also be used to make vaccines. As shown in Figure 10.18, genes that encode antigens from virtually any virus can be cloned into the vaccinia virus genome and expressed. The new *genetically engineered* vaccinia virus can then be used directly to induce immunity to the product of the cloned gene. Such vaccines for hepatitis B surface antigen and chickenpox are now in use. This method depends on the availability of the cloned gene that encodes the antigen and also on the ability of the vaccinia virus to express the cloned gene as an antigenic protein. The use of recombinant DNA methods to develop vaccines was discussed in Section 10.13.

DNA Immunization

A unique and surprisingly simple method for immunization has recently been discovered. Plasmids containing cDNA, which encodes foreign proteins, can be injected intramuscularly into a host animal. After several weeks, the host responds with antibody and CTL (T_C) immune responses directed to the protein encoded by the cDNA. Several questions and numerous possibilities arise from these experiments. First, how can foreign cDNA be transcribed and translated in the host? Based on current information, the cDNA must transfect local APCs, leading to the production of the antigenic protein (∞ Section 9.6) and the stimulation of the immune response. If this method proves to be widely applicable and safe, it offers considerable advantages over many conventional immunization protocols. For instance, since only a single foreign gene is injected, there is no chance of infection as there might be with an attenuated vaccine. Second, antigens and even single antigenic determinants could be used in a vaccine to target the immune response a particular cell component. For example, tumor-specific antigens could be cloned and cDNA incorporated into plasmids. A host immunized with this DNA tumor vaccine could then kill its own tumors. Although we can raise an effective immune response with synthetic peptide antigens, the ease of producing and manipulating cDNA makes the cDNA vaccine method far less expensive and more practical. Finally, this is the only method known where immunization with single bioengineered components is capable of eliciting a CTL response, in addition to the antibody response. This is presumably because the transfected gene causes expression of antigenic peptides within the host cell. These peptides are then processed and presented through the class I antigen presentation pathway, producing a class I-peptide complex that is presented at the cell surface to CD8 T_C cells (see Section 20.7).

Idiotype Immunization

Finally, a novel experimental approach to immunization has resulted from our knowledge of antibody structure. The antigen binding site of an antibody molecule consists of portions of the variable region of the antibody. The unique sequences and determinants formed by the variable region are referred to as *idiotypes*. Because they are structurally unique (recall there are about 10^9 different variable regions), idiotypes can serve as *antigens* in the same animal that produced them, yielding *anti-idiotypic* antibodies (Figure 20.29). If the variable region of an anti-idiotypic antibody is *complementary* to the idiotype, this region will conformationally resemble the original antigen itself. Anti-idiotype responses appear to be a normal part of the immune response and because they bind to individual antibody

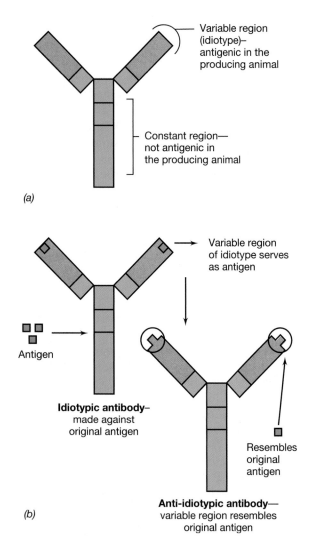

(a)

Variable region
(idiotype)–
antigenic in the
producing animal

Constant region—
not antigenic in
the producing animal

Variable region
of idiotype serves
as antigen

Antigen

Idiotypic antibody–
made against
original antigen

Resembles
original
antigen

Anti-idiotypic antibody—
variable region resembles
original antigen

(b)

FIGURE 20.29   Generation of idiotypic and anti-idiotypic antibodies during an immune response. (a) Structure of an immunoglobulin showing portions of the molecule that are immunogenic in the producing animal. (b) Generation of anti-idiotypic antibodies. Note how the *variable* region of the anti-idiotypic antibody resembles the original antigen to which the idiotypic antibody was formed.

determinants, may play a role in curbing overproduction of specific idiotypes.

Idiotypes and anti-idiotypes have potential uses as vaccines. Because the variable region of an anti-idiotypic antibody mimics the conformation of the original antigen, there is considerable interest in the use of these antibodies as *antigen-free vaccines.* The anti-idiotypic antibody cannot cause disease, but can act as a surrogate antigen to elicit a protective immune response. Vaccines of this type are developed by isolating a monoclonal antibody (a single idiotype) reactive against a specific pathogen and then using this antibody as an *antigen* to produce an anti-idiotypic antibody (Figure 20.29). The anti-idiotypic antibody, when properly purified, can then be employed as a vaccine against the pathogen. Anti-idiotype vaccines to *Escherichia coli*, rabies virus, hepatitis virus, and *Trypanosoma cruzi*, a parasite, have been effective in experimental animal models. If antigen-free vaccines can be developed for a number of human diseases, genetic engineering might also enter the picture because cloned genes coding for anti-idiotypic antibodies (obtained by standard genetic engineering methods) (⚭ Section 10.13) would allow antigen-free vaccines to be produced in microorganisms.

Antigen-free vaccines based on anti-idiotypic antibodies represent a novel and safe means of disease prophylaxis. This practical alternative has emerged from basic knowledge of the structure of antibody molecules and the nature of the immune response.

✓ 20.17 Concept Check

Advances in biotechnology and immunology have resulted in alternate immunization strategies that eliminate exposure to microorganisms and, in some cases, even to antigen!

✓ Provide at least two examples of alternate immunization strategies. What is the advantage of each alternative immunization strategy over current immunization procedures?

Rᴇᴠɪᴇᴡ Qᴜᴇsᴛɪᴏɴs

1. What are three basic features of the adaptive immune response?

2. What is the origin of the cells involved in the immune response? Where do B cells and T cells mature?

3. What substances induce immune responses? What substances do not? What properties are necessary for a substance to induce an immune response?

4. Explain how phagocytes engulf and kill microorganisms, with particular attention to oxygen-dependent mechanisms.

5. Differentiate between *T* and *B* lymphocytes. Differentiate between T cell subsets. *Hint:* Define each cell type by surface markers. Define each subset by surface markers, secreted products, and function.

6. Describe the basic protein structure of an IgG molecule. How do molecules of the other major Ig classes differ?

7. Describe the basic structure of the T cell receptor (TCR). Why do TCRs need to recognize as many epitopes as do the immunoglobulins?

8. Describe the basic structure of class I and class II major histocompatibility complex (MHC) proteins. In what structural and functional ways do they differ? How are they the same?

9. Explain how up to 10^9 immunoglobulin binding sites are generated from the approximately 400 genes encoding portions of the variable region.

10. Describe the action of IL-4 on cells in the immune system. In which cell is it produced? What cells does it activate? What is the final product of the activated cells?

11. What important protective functions are ascribed to cell-mediated immunity? How do T_C cells kill their targets?

12. How does positive and negative selection of the T cell repertoire result in tolerance? Why does tolerance sometimes break down?

13. Describe the various antigen-presenting cells (APCs) that may be involved in antibody formation. What do they have in common?

14. Describe the complement cascade. Is the order of protein interaction important? Why or why not?

15. Describe two cells that would be used to generate a hybridoma. How can you identify a fused hybrid cell?

16. How does *precipitation* differ from *agglutination?* What do these two basic reactions have in common? What are their chief limitations?

17. Define the differences between *immediate* and *delayed-type* hypersensitivity in terms of immune effectors, target tissues, antigens, and clinical outcome.

18. List the diseases for which you have been vaccinated. List the disease for which you may have naturally acquired immunity.

APPLICATION QUESTIONS

1. While *specificity* and *tolerance* are necessary qualities for an adaptive immune response, *memory* seems to be less critical, at least at first glance. Define the role of memory and explain how the production and maintenance of memory cells might benefit the host in the long term.

2. Trace the path of a stem cell that becomes a macrophage, a B lymphocyte, or a T lymphocyte. What environmental factors induce stem cells to become one of these end cells?

3. All immunogens are antigens. However, all antigens are not immunogens. Explain these statements, and illustrate your answer using molecular examples.

4. Phagocytes, dendritic cells, and B cells are involved as dedicated APCs in antigen processing and presentation. Explain the differences in the antigens that may be processed by each cell type.

5. T lymphocytes interact with a variety of cells. Describe the interaction of T_H1, T_H2, and T_C cells at the level of the cell surface receptors and cytokines involved. What do all these reactions have in common? How do they differ?

6. Antibodies of the IgA class are probably more prevalent than those of the IgG class. Explain why this is true and what advantage this provides for the host.

7. How does TCR interaction with antigen differ from that of immunoglobulin (Ig) interaction? What fundamental difference does this make in the kinds of antigens recognized by the TCR and Ig?

8. Do TCRs need a diversity level comparable to that of Igs? Why or why not? Trace the evolutionary development of each of these molecules back to a theoretical precursor molecule. What functions and structural features would a common precursor molecule have in common with both TCRs and Igs?

9. Why, in your opinion, do MHC proteins not need the level of diversity found in TCRs and Igs? What might happen to the human population if there was no significant diversity in the MHC proteins? Explain the value of diversity for TCRs and Igs.

10. Some cytokines have very general functions, such as cell killing. Why are the cells producing such cytotoxins (*i.e.*, T_C cells) not killed?

11. Cell-mediated immunity has often been implicated in tumor surveillance and defense. Explain how this might occur. What cells might be involved? What advantages, if any, might cell-mediated immune surveillance for tumors have, as compared to an antibody-mediated tumor surveillance system.

12. Why might tolerance break down, and what potential outcomes would result? *Hint:* Think about cross-reactions between pathogens and host tissue antigens.

13. In certain situations, B cells are activated in the absence of T cells. Why might T cells not be able to "help" B cells with certain antigens? *Hint:* Think about the antigen processing pathways. Are polysaccharide antigens processed in the same way as protein antigens?

14. Complement has been regarded as a critical humoral defense mechanism. Do you agree with this statement? Explain your answer. What might happen to individuals who lack complement component C3? C5?

15. Would a passively administered monoclonal antibody against a viral surface protein be protective against the disease? Could this antibody protect individuals during an epidemic outbreak of the virus? Would you expect these monoclonal antibodies to be effective over a long time period? Why or why not?

16. How can the specificity (that is, the individual antigen detected) be increased in antigen–antibody reactions? *Hint:* What do you know about the specificity of polyclonal and monoclonal antibodies?

17. Immune diseases are becoming much more prevalent, especially in people over 50 years of age. Which of these diseases are most common among the older age group, and why is this happening?

18. Many infectious diseases have no effective vaccines. Pick several of these diseases (for example, AIDS, malaria, the common cold) and explain why current vaccine strategies have not been effective. Prepare some alternate strategies for immunization against the diseases you have chosen.

$\mathcal{D}$iagnostic microbiology involves the application of laboratory methods for the identification of microorganisms. The most common and proven methods for identifying microorganisms are growth-dependent culture methods. The growth-dependent method shown first in the illustration is a miniaturized version of the traditional larger-scale agar plate/broth tube methods. However, in many cases growth of pathogenic microorganisms is either impossible or impractical because of constraints such as time. Rapid nucleic acid probe methods shown in the second part of the illustration do not require growth of the organism: If a small amount of pathogen DNA can be captured, a positive identification of the organism can be made, sometimes within minutes. These traditional and modern clinical diagnostic methods are used side by side in virtually all clinical microbiology laboratories to efficiently identify pathogenic microorganisms.

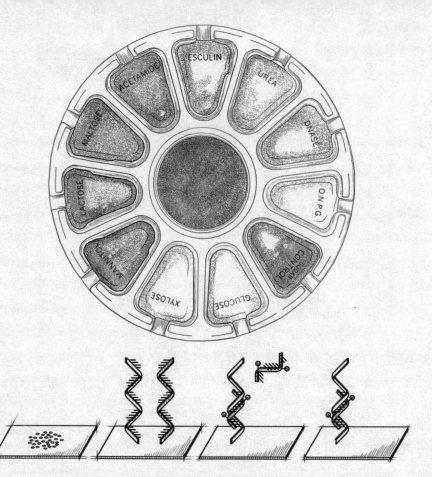

CHAPTER 21

Clinical and Diagnostic Microbiology and Immunology

21.1 Isolation of Pathogens from Clinical Specimens 855
21.2 Growth-Dependent Identification Methods 861
21.3 Testing Cultures for Antimicrobial Drug Sensitivity 865
21.4 Immunodiagnostics 867
21.5 Agglutination 870
21.6 Immunoelectron Microscopy 871
21.7 Fluorescent Antibodies 871
21.8 Enzyme-Linked Immunosorbent Assay and Radioimmunoassay 875
21.9 Immunoblot Procedures 879
21.10 Nucleic Acid Probes in Clinical Diagnostics 882
21.11 Diagnostic Virology 887
21.12 Safety in the Clinical Laboratory 888

The most important activity of the microbiologist in medicine is to isolate and identify the agents that cause infectious disease. This major area of microbiology is called **clinical** or **diagnostic microbiology.** There is increasing awareness of the importance of precise identification of the pathogen for proper treatment of infectious disease, and new methods are being continually developed. Clinical laboratories can isolate, identify, and determine the antibiotic sensitivity of most routinely encountered pathogenic bacteria within 48 hr of sampling. However, recent advances in rapid diagnostic methods make it possible to identify some pathogens in minutes and antibiotic susceptibility patterns in hours. Diagnostic tests based on immunology and molecular biology make it possible to identify many pathogens without culturing the organism. Rapid diagnostic methods are particularly important for the diagnosis of viral and protozoal infections, diseases that are typically difficult to identify because of the difficulty of culturing the causal agent.

21.1

Isolation of Pathogens from Clinical Specimens

The physician, following clinical examination of the patient, may suspect that an infectious disease is present. Samples of infected tissues or fluids are then collected for microbiological, immunological, and molecular biological analyses (Figure 21.1). Depending on the kind of infection, materials collected may include blood, urine, feces, sputum, cerebrospinal fluid, or pus. A sterile swab may be used to sample a suspected infected area (Figure 21.2). The swab is then streaked over the surface of an agar plate or placed directly in a liquid culture medium. In some cases, small pieces of living tissue may be aseptically removed (biopsy) for culture. Table 21.1 summarizes current recommendations for initial culture of organisms isolated from typical clinical specimens.

If clinically relevant organisms are to be isolated and a correct diagnosis made, care must be taken in obtaining samples of clinical specimens. The physician must ensure that the specimen is removed from the *actual site of the infection.* Recovery or detection of pathogens may not be possible if insufficient inoculum is available. The sample must also be taken under aseptic conditions so that contamination is avoided. Care must also be taken to ensure that metabolic requirements for certain organisms, such as anoxic conditions, are maintained. Once taken, the sample is analyzed as soon as possible. If a sample cannot be analyzed immediately, it is usually refrigerated to slow down deterioration. In the rest of this section, we describe some of the most common microbiological procedures used to obtain and culture microorganisms in the clinical laboratory.

Blood Cultures

Bacteremia is the presence of bacteria in the blood (∞ Section 19.7). Bacteria are normally cleared from the bloodstream rapidly. Therefore, bacteremia is uncommon in healthy individuals, and the presence of bacteria in the blood is generally indicative of systemic infection. The most common pathogens found in blood include *Pseudomonas aeruginosa*, enteric bacteria, especially *Escherichia coli* and *Klebsiella pneumoniae*, and the gram-positive cocci *Staphylococcus aureus* and *Strepto-*

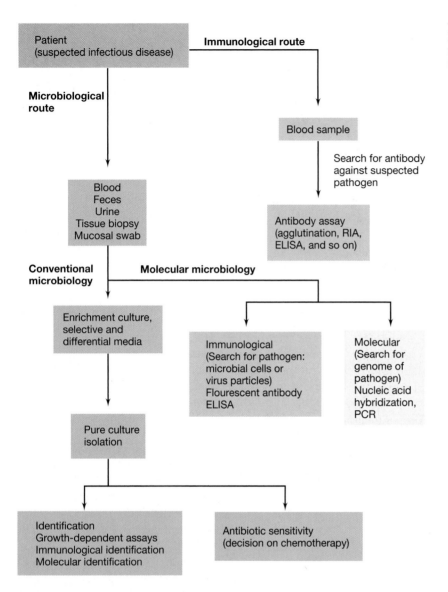

FIGURE 21.1 Clinical and diagnostic methods used for isolation and identification of infectious pathogens.

coccus pyogenes. **Septicemia** is a blood infection resulting from the growth of a virulent organism entering the blood from a focus of infection, multiplying, and traveling to various body tissues to initiate new infections. Septicemia is indicated by the presence of severe systemic symptoms, including fever and chills, followed by prostration. In many disease situations, blood cultures provide the only immediate way of isolating and identifying the causal agent, and diagnosis therefore depends on careful and proper blood culture.

The standard blood culture procedure is to remove 10 ml of blood aseptically from a vein and inject it into a

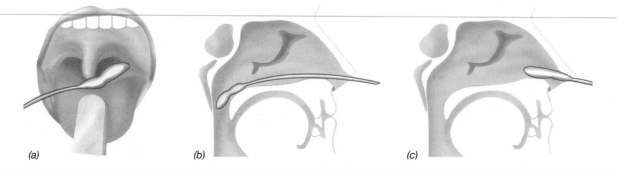

FIGURE 21.2 Methods for obtaining specimens from the upper respiratory tract. (a) Throat swab. (b) Nasopharyngeal swab passed through the nose. (c) Swabbing the inside of the nose.

TABLE 21.1 Recommended enrichment and selective media for primary isolation of pathogens[a]

Specimen	Media[b]				
	Blood agar	Enteric agar	CA	MTM	ANA
Fluids: chest, abdomen, pericardium	+	+	+	−	+
Feces: rectal swabs	+	+	+	−	−
Surgical tissue biopsies: lung, lymph nodes	+	+	−	−	+
Throat: swabs, sputum, tonsil, nasopharynx	+	+	+	−	−
Genitourinary swabs: urethra, vagina, cervix	+	+	−	+	−
Urine	+	+	−	−	−
Blood	+	−	−	−	+
Swabs: wounds, abscesses, exudates	+	+	+	−	+

a Data from Murray, P. R., E. J. Baron, M. A. Pfaller, F. C. Tenover, and R. H. Yolken. 1999. *Manual of Clinical Microbiology,* 7th edition. American Society for Microbiology, Washington, DC.

b Blood agar, 5% whole sheep blood added to trypticase soy agar; enteric agar, either eosin–methylene blue (EMB) agar or MacConkey agar; CA, chocolate (heated blood) agar; MTM, Modified Thayer–Martin agar; ANA, anaerobic agar, thioglycolate-containing blood agar or supplemented thioglycolate agar incubated anaerobically.

blood culture bottle containing an anticoagulant and an all-purpose culture medium. Two cultures are set up; one bottle is incubated aerobically and one anaerobically. Media used are all relatively rich, containing protein digests and other complex ingredients. Blood culture bottles are incubated at 35°C and examined daily for up to 7 days. Most clinically significant bacteria are recovered within this period. Some blood isolation systems employ a chemical that lyses red and white blood cells, releasing potential intracellular pathogens that might otherwise be overlooked. Microorganisms in blood cultures are commonly detected by visual inspection (turbidity), microscopic examination, and subculture. Automated blood culture systems detect growth by continuously monitoring carbon dioxide production and turbidity.

Because a certain amount of skin contamination is unavoidable during initial drawing of the blood, a contamination rate of 2–3% can be expected. Contamination may be indicated if certain organisms commonly found on the skin are isolated, such as *Staphylococcus epidermidis*, coryneform bacteria, or propionibacteria, although even these organisms can occasionally cause infection of the wall of the heart (subacute bacterial endocarditis). Thus, considerable microbiological and clinical experience is necessary when interpreting blood cultures.

Urine Cultures

Urinary tract infections are very common, and because the causal agents are often identical or similar to bacteria of the normal flora (for example, *Escherichia coli*), considerable care must be taken in the bacteriological analysis of urine. Since urine supports extensive bacterial growth under many conditions, fairly high cell numbers are often found in urinary infection. In most cases, the infection occurs as a result of an organism ascending the urethra from the outside. Occasionally, even the bladder may become infected. Urinary tract infec-

tions are the most common form of *nosocomial* (hospital-acquired) infection (∞ Section 22.7).

Significant urinary infection generally results in bacterial counts of 10^5 or more organisms per milliliter of a clean-voided midstream specimen, whereas in the absence of infection, contamination of the urine from the external genitalia (almost unavoidable to some extent) results in less than 10^3 organisms per milliliter. The most common urinary tract pathogens are members of the enteric bacteria, with *E. coli* accounting for about 90% of the cases. Other urinary tract pathogens include *Klebsiella, Enterobacter, Proteus, Pseudomonas, Staphylococcus saprophyticus,* and *Enterococcus faecalis. Neisseria gonorrhoeae,* the causal agent of gonorrhea, does not grow in the urine itself, but in the urethral epithelium, and must be diagnosed by different methods (see later).

Direct microscopic examination of urine may be used to indicate *bacturia,* the presence of abnormal numbers of bacteria in the urine. However, because nearly all urine contains some level of bacterial growth, significant bacturia is most commonly monitored by using a variety of commercially available dipstick tests. For example, one dipstick test monitors the reduction of nitrate by detecting the reduction product, nitrite. A positive test is indicated by a color change on the dipstick (Figure 21.3). Since nitrite production occurs only when significant numbers ($>10^5$ per milliliter) of enteric organisms are present, the method is a virtually instantaneous check for urinary tract infections. Other dipstick tests for urinary tract infections, often used in conjunction with nitrate reduction, detect esterase (produced by leukocytes) (∞ Section 20.3) and peroxidase (produced by a variety of bacteria) (∞ Sections 5.12 and 15.27). A positive test is then followed by a urine culture.

To culture potential urinary tract pathogens, two media are used: blood agar as a nonselective general medium and a medium selective for enteric bacteria,

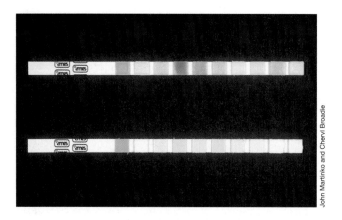

FIGURE 21.3 Urinalysis dipstick test. A control strip is shown underneath the test strip. From left to right, the strip measures abnormal levels of glucose, bilirubin, ketones, specific gravity, blood, pH, protein, urobilinogen, nitrite, and leukocytes (esterase) in a urine sample. Abnormal readings for esterase (trace positive, far right) and nitrite (strong positive, second from right) indicate bacturia. Subsequent culture of this sample indicated the presence of *Escherichia coli*.

FIGURE 21.4 An eosin–methylene blue (EMB) agar plate showing a lactose fermenter, *Escherichia coli* (left), and a nonlactose fermenter, *Pseudomonas aeruginosa* (right). Note the green metallic sheen of the *E. coli* colonies.

such as MacConkey or eosin–methylene blue agar (EMB) (see Section 21.2 and Figure 21.4). These specialized media permit the initial differentiation of lactose fermenters from nonfermenters, and the growth of gram-positive organisms such as *Staphylococcus* spp. (common skin contaminants) is inhibited. Experienced clinical microbiologists may make a tentative identification of an isolate by observing the color and morphology of colonies of the suspected pathogen

growth on various selective media as described in Table 21.2. Such an identification must be followed up with more detailed analyses, but clinical microbiologists use this information in conjunction with more de-

TABLE 21.2	Colony characteristics of frequently isolated gram-negative rods cultured on various clinically useful media[a]			
	Agar media[b]			
Organism	**EMB**	**MC**	**SS**	**BS**
Escherichia coli	Dark center with greenish metallic sheen (see Figure 21.4)	Red or pink	Red to pink	Mostly inhibited
Enterobacter	Similar to *E. coli*, but colonies are larger	Red or pink	White or beige	Mucoid colonies with silver sheen
Klebsiella	Large, mucoid, brownish	Pink	Red to pink	Mostly inhibited
Proteus	Translucent, colorless	Transparent, colorless	Black center, clear periphery	Green
Pseudomonas	Translucent, colorless to gold (see Figure 21.4)	Transparent, colorless	Mostly inhibited	No growth
Salmonella	Translucent, colorless to gold	Translucent, colorless	Opaque	Black to dark green
Shigella	Translucent, colorless to gold	Transparent, colorless	Opaque	Brown or inhibited

a Adapted from Murray, P. R., E. J. Baron, M. A. Pfaller, F. C. Tenover, and R. H. Yolken. 1999. *Manual of Clinical Microbiology,* 7th edition. American Society for Microbiology, Washington, DC.

b BS, Bismuth sulfite agar; EMB, eosin–methylene blue agar; MC, MacConkey agar; SS, *Salmonella–Shigella* agar.

tailed test results, discussed throughout the remainder of this chapter, to make a positive identification.

Finally, if no bacterial growth is obtained in spite of persistent urinary tract infection symptoms, a clinician may request direct cultures for a number of fastidious organisms, especially *Neisseria gonorrhoeae, Chlamydia trachomatis, Branhamella* spp., mycoplasma, or several anaerobic organisms (∞ Section 23.6).

Fecal Cultures

Proper collection and preservation of feces is important in the isolation of intestinal pathogens. During storage, fecal acidity increases and thus an extended delay between sampling and sample processing must be avoided. This is especially critical for the isolation of *Shigella* and *Salmonella* species, both of which are rather sensitive to acid pH. Samples, collected from feces freshly voided into a sterile plastic cup, are placed in a vial containing phosphate buffer for transport to the lab. If a patient has a bloody or pus-containing stool, this material is always sampled; such discharges generally contain a large number of the organisms of interest. In the case of suspected foodborne or waterborne infections, fecal samples should be inoculated into a variety of selective media (see Section 21.2) for the isolation of specific bacteria or characterization of intestinal parasites. Positive identifications are made by the techniques described in later sections.

Wounds and Abscesses

Infections associated with traumatic injuries such as animal or human bites, burns, cuts, or the penetration of foreign objects, must be carefully sampled in order to recover the relevant pathogen. This is because wound infections and abscesses are frequently contaminated with members of the normal flora. Swab samples of such lesions are frequently misleading. The best sampling method is to aspirate purulent (pus-containing) lesions with a sterile syringe and needle following disinfection of the skin surface with 70% ethyl or isopropyl alcohol. Internal purulent discharges are usually sampled by biopsy or from tissues removed in surgery.

A variety of pathogens can be associated with wound infections, and because some of these are anaerobes, samples must be transported from the collection site under anaerobic conditions. A common pathogen associated with purulent discharges is *Staphylococcus aureus,* but enteric bacteria, *Pseudomonas aeruginosa,* and the anaerobes from the genera *Bacteroides* and *Clostridium* are also commonly encountered. The major isolation media are blood agar, several selective media for enteric bacteria (Tables 21.1 and 21.2), and blood agar containing additional supplements and reducing agents for obligate anaerobes. Smears from such specimens should also be examined directly by microscopy.

Genital Specimens and the Laboratory Diagnosis of Gonorrhea

In males, a purulent urethral discharge is the classic symptom of the sexually transmitted disease gonorrhea (∞ Section 23.6). If no discharge is present, a suitable sample can be obtained using a sterile narrow-diameter cotton swab that is inserted into the anterior urethra, left in place a few seconds to absorb any exudate, and then removed for culture of *Neisseria gonorrhoeae,* the causative agent of gonorrhea. Alternatively, a sample of the first early morning urine of an infected individual usually contains viable cells of *N. gonorrhoeae.* In females suspected of having gonorrhea or other genital infections, samples are usually obtained by swab from the cervix and the urethra.

Gonorrhea is one of the most common infectious diseases in adults, and clinical microbiology procedures are central to its diagnosis. *N. gonorrhoeae* (referred to clinically as *gonococcus*) colonizes mucosal surfaces of the urethra, uterine cervix, anal canal, throat, and conjunctiva. The organism is quite sensitive to drying and therefore is transmitted almost exclusively by direct person-to-person contact, usually by sexual intercourse. The major goal of public health measures to control gonorrhea involves identification of asymptomatic carriers, and this requires microbiological analysis.

Because the gonococcus is a gram-negative coccus usually observed as diplococci and similar organisms are not very common in the normal flora of the urogenital tract, direct microscopy of Gram-stained material is of value. Observation of gram-negative diplococci in a urethral discharge or in a vaginal or cervical smear is presumptive evidence for gonorrhea. In acute gonorrhea, microscopy usually reveals phagocytized gram-negative diplococci in the polymorphonuclear leukocytes (∞ Section 20.3), with virtually no other organisms present (Figure 21.5*a*).

Culture methods have a higher degree of sensitivity than microscopic analysis. Most nonselective enrichment media for the isolation of *N. gonorrhoeae* contain heat-lysed blood and are often called *chocolate agar* because of the deep brown appearance. The heated blood interacts with the media components, absorbing compounds that are normally toxic for *N. gonorrhoeae.* One of several selective media used for primary isolation is Modified Thayer-Martin (MTM) agar (Figure 21.5*b*). This medium incorporates vancomycin, nystatin, trimethoprim, and colistin to suppress the growth of normal flora, but these antibiotics do not suppress either *N. gonorrhoeae* or *N. meningitidis,* which causes bacterial meningitis.

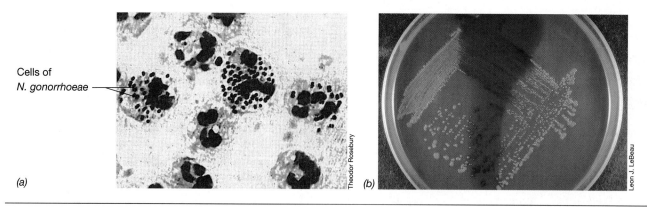

FIGURE 21.5   (a) Photomicrograph of *Neisseria gonorrhoeae* within human polymorphonuclear leukocytes from a cervical smear. Note the paired diplococci (arrows). (b) *N. gonorrhoeae* growing on Thayer–Martin agar. The plate has been stained in the middle with a reagent that turns colonies blue if cells contain cytochrome *c* (the oxidase test).

After streaking, the plates must be incubated in a humid environment in an atmosphere containing 3–7% CO_2 (CO_2 is required for growth of gonococci). The plates are examined after 24 and 48 hr, and should be immediately tested by the oxidase test because all *Neisseria* are oxidase-positive (Figure 21.5*b*) (see Section 21.2). Oxidase-positive gram-negative diplococci growing on chocolate agar or selective media can be presumed to be gonococci if the inoculum was derived from genitourinary sources, but definitive identification requires determination of carbohydrate utilization patterns or immunological or nucleic acid probe tests (see Sections 21.4–21.10).

Culture of Anaerobes

Obligately anaerobic bacteria are common causes of infection and are completely missed in clinical diagnosis unless special precautions are taken for their isolation and culture. We have discussed anaerobes in general in Section 5.12, and we noted that many anaerobes are extremely susceptible to oxygen. Therefore, specimen collection, handling, and processing require special attention if obligate anaerobic organisms are to be isolated. There are several habitats in the body (for example, the oral cavity and the intestinal tract) (Sections 19.3 and 19.4) that are generally anoxic and in which obligately anaerobic bacteria can be found as part of the normal flora. However, other parts of the body can become anoxic as a result of tissue injury or trauma, which results in reduction of blood supply to the injured site. These anaerobic sites can then be colonized by obligate anaerobes. In general, pathogenic anaerobic bacteria are part of the normal flora and are only opportunistic pathogens, although two important pathogenic anaerobes, *Clostridium tetani* (causal agent of tetanus) and *Clostridium perfringens* (causal agent of gas gangrene and one type of food poisoning), both endospore-forming Bacteria, are predominantly soil organisms.

With anaerobic culture, the microbiologist is presented not only with the usual problems of obtaining and maintaining an uncontaminated specimen but also with ensuring that the specimen not come in contact with air. Samples collected by suction or biopsy must be immediately placed in a tube containing oxygen-free gas, preferably containing a small amount of a dilute salts solution with a reducing agent such as thioglycolate and the redox indicator resazurin. This dye is colorless when reduced and becomes pink when oxidized, quickly indicating any oxygen contamination of the specimen. If a proper anaerobic transport tube is not available, the syringe itself can be used to transport the specimen, the needle being inserted into a sterile rubber stopper so that no air enters the syringe.

For anaerobic incubation, agar plates are placed in a sealed jar, which is made anoxic by either replacing the atmosphere in the jar with an oxygen-free gas mixture (a mixture of N_2 and CO_2 is frequently employed) or by adding some compound to the enclosed vessel that removes O_2 from the atmosphere. For example, as shown in Figure 21.6, H_2 is generated and, in the presence of a suitable catalyst, usually palladium, the H_2 is combined with free O_2 to form H_2O, thus removing the contaminating oxygen. Alternative means for providing anaerobic conditions include the use of culture media containing reducing agents or the use of anoxic "glove boxes." Glove boxes are large gas-impermeable bags filled with an oxygen-free gas such as nitrogen or hydrogen and fitted with an airlock for inserting and removing cultures (Figure 5.22). The advantage of an anoxic glove box is that manipulations can be done on a laboratory bench. However, because of their specialized nature, anoxic glove boxes are not employed extensively in clinical laboratories but are in widespread use in research laboratories that specialize in anaerobic microorganisms.

In general, media for anaerobes do not differ greatly from those used for aerobes, except that they are generally richer in organic constituents, and contain

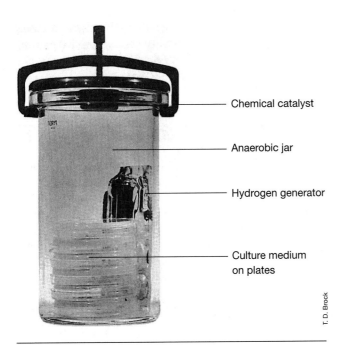

Chemical catalyst

Anaerobic jar

Hydrogen generator

Culture medium on plates

T. D. Brock

FIGURE 21.6 Sealed jar for incubating cultures under anoxic conditions.

reducing agents (usually cysteine or thioglycolate) and a redox indicator such as resazurin.

✓ 21.1 Concept Check

Proper sampling and culture of the suspected pathogen is the most reliable way to identify an organism that causes a disease. The selection of appropriate sampling and culture conditions requires knowledge of bacterial ecology, physiology, and nutrition.

- ✓ Why are urine cultures almost always positive for bacterial growth?
- ✓ Describe precautions required for successful isolation of anaerobic pathogens.

21.2

Growth-Dependent Identification Methods

If the inoculation of a primary medium results in bacterial growth, the clinical microbiologist must identify the organism or organisms present. Identification of a clinical isolate can frequently be made using a variety of growth-dependent assays. We discuss some of these methods here.

Growth on Selective and Differential Media

On the basis of growth characteristics in primary isolation media, an unknown pathogen is usually subcultured on perhaps several of the dozens of available,

diagnostically useful culture media. Many of these media are available in miniaturized kits containing a number of different media in separate wells, all of which can be inoculated at one time (Figure 21.7).

The battery of media employed are selective, differential, or both. A *selective medium* is one to which compounds have been added to selectively inhibit the growth of certain microorganisms but not others. A *differential medium* is one to which some sort of indicator, usually a dye, has been added, which allows the clinician to differentiate between various chemical reactions carried out during growth. Eosin–methylene blue (EMB) agar, for example, is a widely used selective *and* differential medium. EMB agar is used for the isolation of gram-negative enteric Bacteria. Methylene blue is present to inhibit gram-positive Bacteria; small amounts of this dye effectively inhibit the growth of most gram-positive Bacteria. Eosin is a dye that responds to changes in pH, going from colorless to black under acidic conditions. EMB agar medium contains lactose and sucrose, but not glucose, as energy sources. Lactose-fermenting (generally enteric) bacteria, such as *Escherichia coli, Klebsiella,* and *Enterobacter,* acidify the medium and the colonies appear black with a greenish sheen. Colonies of lactose nonfermenters, such as *Salmonella, Shigella,* and *Pseudomonas,* are translucent or pink (Figure 21.4).

In the battery of tests performed to help identify an organism, many different biochemical reactions can be measured. The most important tests are summarized in Table 21.3. These tests measure the presence or absence of *enzymes* involved in catabolism of the substrate or substrates in the differential medium. Fermentation of sugars is measured by incorporating pH indicator dyes that change color on acidification (Figure 21.7a). Production of hydrogen gas and/or carbon dioxide during sugar fermentation is assayed by observing gas production either in gas collection vials or in agar (Figure 21.7a and b). Hydrogen sulfide production is indicated following growth in a medium containing ferric iron. If sulfide is produced, ferric iron complexes with H_2S to form a black precipitate of iron sulfide (Figure 21.7b). Utilization of citric acid, a six-carbon acid containing three carboxylic acid groups, is accompanied by a pH increase, and a specific dye incorporated into the citric acid test medium changes color as conditions become alkaline (Figure 21.7c). Hundreds of differential tests have been developed for clinical use, but only about 20 are used routinely (Figure 21.7d).

The typical reaction patterns for large numbers of strains of various pathogens have been published, and in the modern clinical microbiology laboratory, the information is stored in a computer. The results of the differential tests on an unknown pathogen are entered, and the computer makes the best match by comparing the characteristics of the unknown with the known meta-

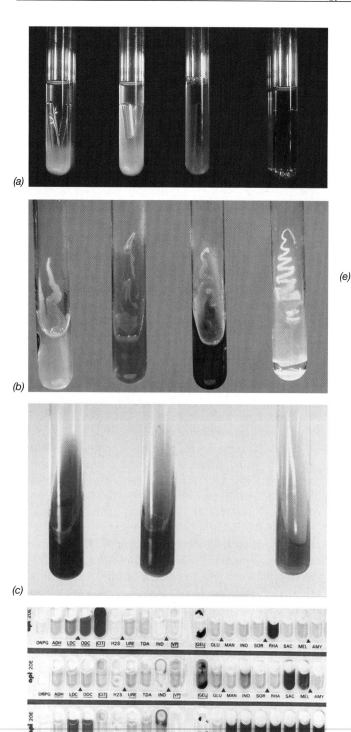

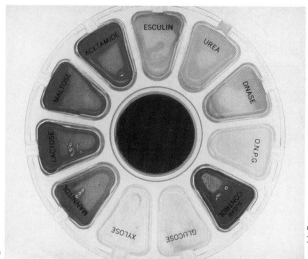

FIGURE 21.7   Growth-dependent diagnostic methods used for the identification of clinical isolates by color changes in various diagnostic media. (a) Use of a differential medium to assess sugar fermentation. Acid production is indicated by color change of the pH-indicating dye added to the liquid medium. If gas production occurs, a bubble appears in the inverted vial in each tube. From left to right: acid, acid and gas, negative, uninoculated. (b) A conventional diagnostic test for enteric bacteria in a medium called *triple sugar iron (TSI) agar.* The medium is inoculated both on the surface of the slant and by stabbing into the butt. The medium contains a small amount of glucose and a large amount of lactose and sucrose. Organisms able to ferment only the glucose cause acid formation only in the butt, whereas lactose or sucrose-fermenting organisms also cause acid formation in the top. Gas formation is indicated by the breaking up of the agar in the butt. Hydrogen sulfide formation (either from protein degradation or from reduction of thiosulfate in the medium) is indicated by a blackening due to reaction of the H_2S with ferrous iron in the medium. From left to right: fermentation of glucose only; no reaction; hydrogen sulfide formation; fermentation of glucose and another sugar. (c) Measurement of citrate utilization by *Salmonella* on Simmons citrate agar. The change in pH causes a change in the color of the indicating dye. From left to right: positive, negative, uninoculated. (d) Media kits used for the rapid identification of clinical isolates. The principle is the same as in (a), but the whole arrangement has been miniaturized so that a number of tests can be run at the same time. Four separate strips, each with a separate culture, are shown. (e) Another arrangement of a miniaturized test kit. This one defines sugar utilization in nonfermentative organisms.

TABLE 21.3	Important clinical diagnostic tests for bacteria		
Test	**Principle**	**Procedure**	**Most common use**
Carbohydrate fermentation	Acid and/or gas produced during fermentative growth with sugars or sugar alcohols	Broth medium with carbohydrate and phenol red as pH indicator; inverted tube for gas	Enteric bacteria differentiation (also several other genera or species separations with some individual sugars) (Figure 21.7)
Catalase	Enzyme decomposes hydrogen peroxide, H_2O_2	Add drop of H_2O_2 to dense culture and look for bubbles (O_2) (⚮ Figure 5.25)	*Bacillus* (+) from *Clostridium* (−); *Streptococcus* (−) from *Micrococcus–Staphylococcus* (+)
Citrate utilization	Utilization of citrate as sole carbon source, results in alkalinization of medium	Citrate medium with bromthymol blue as pH indicator. Look for intense blue color (alkaline pH)	*Klebsiella–Enterobacter* (+) from *Escherichia* (−), *Edwardsiella* (−) from *Salmonella* (+) (Figure 21.7)
Coagulase	Enzyme causes clotting of blood plasma	Mix dense liquid suspension of bacteria with plasma, incubate, and look for fibrin clot	*Staphylococcus aureus* (+) from *S. epidermidis* (−)
Decarboxylases (lysine, ornithine, arginine)	Decarboxylation of amino acid releases CO_2 and amine	Medium enriched with amino acids. Bromcresol purple pH indicator becomes purple (alkaline pH) if there is enzyme action	Aid in determining bacterial group among the enteric bacteria
β-Galactosidase (ONPG) test	Orthonitrophenyl-β-galactoside (ONPG) is an artificial substrate for the enzyme. When hydrolyzed, nitrophenol (yellow) is formed	Incubate heavy suspension of lysed culture with ONPG. Look for yellow color	*Citrobacter* and *Arizona* (+) from *Salmonella* (−). Identifying some *Shigella* and *Pseudomonas* species
Gelatin liquefaction	Many proteases hydrolyze gelatin and destroy the gel	Incubate in broth with 12% gelatin. Cool to check for gel formation. If gelatin is hydrolyzed, tube remains liquid on cooling	To aid in identification of *Serratia*, *Pseudomonas*, *Flavobacterium*, *Clostridium*
Hydrogen sulfide (H_2S) production	H_2S produced by breakdown of sulfur amino acids or reduction of thiosulfate	H_2S detected in iron-rich medium from formation of black ferrous sulfide (many variants: Kliger's iron agar and triple sugar iron agar also detect carbohydrate fermentation)	In enteric bacteria, to aid in identifying *Salmonella*, *Arizona*, *Edwardsiella*, and *Proteus* (Figure 21.7)
Indole test	Tryptophan from proteins converted to indole	Detect indole in culture medium with dimethyl-aminobenzaldehyde (red color)	To distinguish *Escherichia* (+) from *Klebsiella* (−) and *Enterobacter* (−); *Edwardsiella* (+) from *Salmonella* (−)
Methyl red test	Mixed-acid fermenters produce sufficient acid to lower pH below 4.3	Glucose-broth medium. Add methyl red indicator to a sample after incubation	To differentiate *Escherichia* (+, culture red) from *Enterobacter* and *Klebsiella* (usually −, culture yellow)
Nitrate reduction	Nitrate as alternate electron acceptor, reduced to NO_2^- or N_2	Broth with nitrate. After incubation, detect nitrite with α-naphthylamine-sulfanilic acid (red color). If negative, confirm that NO_3^- still present by adding zinc dust to reduce NO_3^- to NO_2^-. If no color after zinc, then $NO_3^- \rightarrow N_2$	To aid in identification of enteric bacteria (usually +)

TABLE 21.3	Important clinical diagnostic tests for bacteria (continued)		
Test	**Principle**	**Procedure**	**Most common use**
Oxidase test	Cytochrome c oxidizes artificial electron acceptor: tetramethyl (or dimethyl)-p-phenylenediamine	Broth or agar. Oxidase-positive colonies on agar can be detected by flooding plate with reagent and looking for blue or brown colonies	To separate *Neisseria* and *Moraxella* (+) from *Acinetobacter* (−). To separate enteric bacteria (all −) from pseudo-monads (+). To aid in identification of *Aeromonas* (+)
Oxidation–fermentation (O/F) test	Some organisms produce acid only when growing aerobically	Acid production in top part of sugar-containing culture tube; soft agar used to restrict mixing during incubation	To differentiate *Micrococcus* (acid produced aerobically only) from *Staphylococcus* (acid produced anaerobically). To characterize *Pseudomonas* (aerobic acid production) from enteric bacteria (acid produced anaerobically)
Phenylalanine deaminase test	Deamination produces phenylpyruvic acid, which is detected in a colorimetric test	Medium enriched in phenylalanine. After growth, add ferric chloride reagent and look for green color	To characterize the genus *Proteus* and the *Providencia*
Starch hydrolysis	Iodine-iodide gives blue color with starch	Grow organism on plate containing starch. Flood plate with Gram's iodine and look for clear zones around colonies	To identify typical starch hydrolyzers such as *Bacillus* spp.
Urease test	Urea (H_2N—CO—NH_2) split to $2\,NH_3 + CO_2$	Medium with 2% urea and phenol red indicator. Ammonia release raises pH, intense pink-red color	To distinguish *Klebsiella* (+) from *Escherichia* (−). To distinguish *Proteus* (+) from *Providencia* (−)
Voges–Proskauer test	Acetoin produced from sugar fermentation	Chemical test for acetoin using α-naphthol	To separate *Klebsiella* and *Enterobacter* (+) from *Escherichia* (−). To characterize members of genus *Bacillus*

bolic patterns of the species in the data bank. For many organisms, as few as three or four key tests are all that are required to make an unambiguous identification. In cases of a dubious match, however, more sophisticated identification procedures may be called for, especially if the chemotherapy regimens are different for several pathogens with similar growth characteristics.

Clinical Diagnosis

Many companies market their own versions of growth-dependent rapid identification systems (Figure 21.7). Such systems are frequently designed for use in identifying enteric bacteria because enterics are frequently implicated in routine urinary tract and intestinal infections (see Section 21.1).

Other growth-dependent rapid identification kits are available for other bacterial groups or even for single bacterial species. For example, commercial kits containing a battery of tests have been developed for *Staphylococcus aureus*, *Streptococcus pyogenes*, *Neisseria*

gonorrhoeae, *Haemophilus influenzae*, and *Mycobacterium tuberculosis*. Other kits are available for identification of the pathogenic fungi *Candida albicans* and *Cryptococcus neoformans* (∞ Section 24.7).

The decision to use a specific diagnostic test is usually made by the clinical microbiologist. This individual takes into consideration the nature of the clinical specimen, basic characteristics (especially the Gram stain) of pure cultures obtained, and previous experience with similar cases. For example, an enteric identification kit would be useless in identifying a gram-positive coccus isolated from an abscess. Instead, a *Staphylococcus aureus* or *Streptococcus pyogenes* kit would be used to make a positive identification.

✓ 21.2 Concept Check

Traditional methods for identifying pathogens depend on observing metabolic changes induced as a result of growth. These growth-dependent methods provide rapid and accurate means of diagnosing many infectious diseases.

✓ Distinguish between *selective* and *differential* identification methods. Give an example of a medium used for each purpose.

✓ What parameters would a clinical microbiologist use to prescribe a specific diagnostic test kit for identification of an infectious agent?

21.3

Testing Cultures for Antimicrobial Drug Sensitivity

In medical practice, microbial cultures are isolated from diseased patients to confirm diagnoses and to aid in decisions on therapy. Determination of the sensitivity of microbial isolates to antimicrobial agents is one of the most important tasks of the clinical microbiologist.

We discussed the principles for the measurement of antimicrobial activity in Chapter 18. The sensitivity of a culture can be most easily determined by an agar diffusion method or by using a tube dilution technique to determine the *minimum inhibitory concentration* (MIC) of an agent that is necessary to inhibit growth (∞ Section 18.4). Food and Drug Administration (FDA) regulations now control the procedures used for sensitivity testing in the United States, and similar regulations exist in other countries. A recommended agar diffusion procedure is called the *Kirby–Bauer method*, named after the workers who developed it (Figure 21.8). Culture media are inoculated by spreading a sample of culture evenly across the agar surface. Filter paper discs containing

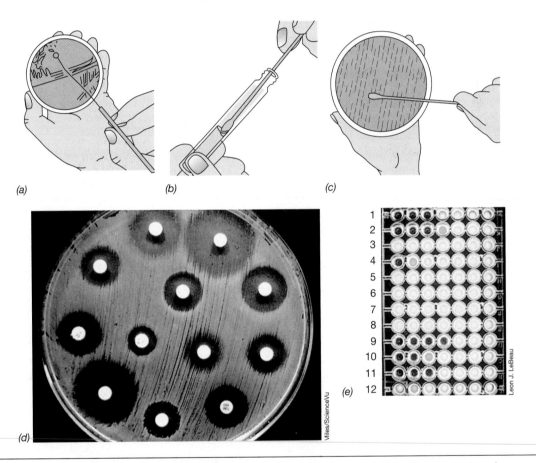

(a)   (b)   (c)

(d)   (e)

Viles/ScienceVu

Leon J. LeBeau

FIGURE 21.8   Antibiotic sensitivity testing. (a–d) The Kirby–Bauer procedure for determining the sensitivity of an organism to antibiotics. (a) A colony is picked from an agar plate. It is inoculated into a tube of liquid culture medium and allowed to grow to a specified density. (b) A swab is dipped in the liquid culture. (c) The swab is streaked evenly over a plate of sterile agar medium. (d) Discs containing known amounts of different antibiotics are placed on the plate. After incubation, inhibition zones are observed. The susceptibility of the organism is determined by reference to a chart of zone sizes (Table 21.4). (e) Antibiotic sensitivity determined by the dilution method. The organism is *Pseudomonas aeruginosa*. Each row has a different antibiotic. The use of the microtiter plate enables automation of these tests. The end point is read as the well with the lowest concentration of antibiotic that shows no evidence of bacterial growth. The highest concentration of antibiotic is in the well at the left; serial dilutions are made in the wells to the right. For example, in rows 1 and 2, the end point is the third well. In row 3, the antibiotic is ineffective at the concentrations tested, since there is bacterial growth in all the wells. In row 4, the end point is in the first well. The lowest concentration of antibiotic that completely inhibits bacterial growth defines the minimum inhibitory concentration (MIC) for that agent (∞ Section 18.4).

known concentrations of different antimicrobial agents are then placed on the plate. The concentration of each agent on the disc is specified, and after incubation, the presence and size of inhibition zones around the discs of the different agents are noted. Table 21.4 presents typical zone sizes for several antibiotics. Zones observed on the plate are measured and compared to standard data to determine if the isolate is truly sensitive to a given antibiotic.

The MIC procedure for antibiotic sensitivity testing involves an *antibiotic dilution assay*, either in culture tubes (☜ Figure 18.11) or in the wells of a microtiter plate (Figure 21.8e). A series of twofold dilutions of each antibiotic are made in the wells, and then all wells are inoculated with a standard amount of the same test organism. After incubation, the inhibition of growth by the various antibiotics is observed by measuring turbidity. Sensitivity is usually expressed as the *highest dilution* (lowest concentration) of antibiotic that completely inhibits growth. The dilution assay, because it can be performed in microtiter plates, is readily automated.

Because of the widespread occurrence of antimicrobial drug resistance (☜ Section 18.12), an antibiotic sensitivity test is essential for pathogens isolated from each patient. Data such as those in Table 21.4 are useful for choosing the best antibiotic for a specific bacterial infection. Although many potentially serious pathogens are susceptible to a number of different antibiotics, some pathogens, for example, *Pseudomonas aeruginosa*, are sensitive to very few drugs. Other pathogens, such as some encountered in hospital environments, have developed antibiotic resistance and a few are completely resistant to all known antibiotics (☜ Section 18.12). Thus, antibiotic sensitivity testing for these organisms is absolutely essential for effective chemotherapy. Using the drug sensitivity information gathered in this fashion, the clinical microbiologist generates periodic reports called *antibiograms* that indicate the sensitivity of clinically isolated organisms to the antibiotics in current use. Antibiograms are particularly valuable for tracking the emergence of antibiotic-resistant strains of pathogens in facilities such as hospitals and nursing homes.

✓ 21.3 Concept Check

Antimicrobial drugs are in wide use for the treatment of infectious diseases. Pathogens should be tested for sensitivity to individual antibiotics *before treatment* to ensure appropriate chemotherapy. This rigorous approach to antimicrobial drug treatment is usually applied only in hospital settings.

✓ Describe the Kirby–Bauer technique. What does it indicate?

✓ Why is antimicrobial drug sensitivity testing important for the clinical microbiologist, the physician, and the patient?

TABLE 21.4	**Zone sizes for some antimicrobial disc susceptibility tests**			
		Inhibition zone diameter (mm)[a]		
Antibiotic	**Amount on disc**	**Resistant**	**Intermediate**	**Sensitive**
Ampicillin[b]	10 μg	11 or less	12–13	14 or more
Ampicillin[c]	10 μg	28 or less	—	29 or more
Cephoxitin	30 μg	14 or less	15–17	18 or more
Cephalothin	30 μg	14 or less	15–17	18 or more
Chloramphenicol	30 μg	12 or less	13–17	18 or more
Clindamycin	2 μg	14 or less	15–16	17 or more
Erythromycin	15 μg	13 or less	14–17	18 or more
Gentamicin	10 μg	12 or less	13–14	15 or more
Kanamycin	30 μg	13 or less	14–17	18 or more
Methicillin[c]	5 μg	9 or less	10–13	14 or more
Neomycin	30 μg	12 or less	13–16	17 or more
Nitrofurantoin	300 μg	14 or less	15–16	17 or more
Penicillin G[d]	10 units	28 or less	—	29 or more
Penicillin G[e]	10 units	11 or less	12–21	22 or more
Polymyxin B	300 units	8 or less	9–11	12 or more
Streptomycin	10 μg	11 or less	12–14	15 or more
Tetracycline	30 μg	14 or less	15–18	19 or more
Trimethoprim-sulfamethoxazole	1.25/23.75 μg	10 or less	11–15	16 or more
Tobramycin	10 μg	12 or less	13–14	15 or more

a See Figure 21.8*d* for an illustration of a typical test.

b For gram-negative organisms and enterococci.

c For staphylococci and highly penicillin-sensitive organisms.

d For staphylococci.

e For organisms other than staphylococci. Includes some organisms, such as enterococci and some gram-negative rods, that may cause some systemic infections treatable with high doses of penicillin G.

21.4

Immunodiagnostics

In this section, we will apply the principles of immunity to the diagnosis of infectious diseases. First, we will examine the requirements for an immunodiagnostic test. Next, we will briefly review the immune response to pathogens. Finally, we will examine immunological reagents that are useful for diagnostic applications. In the following sections, we will examine specific applications of these reagents.

Specificity and Sensitivity

For any immunodiagnostic test, important features are **specificity** and **sensitivity.** *Specificity* is the ability of an antibody preparation to recognize a single antigen. A desirable level of specificity implies that the antibody is specific for a single antigen, will not cross-react with any other antigen, and therefore will not provide *false positive* results. Specificity must be defined in terms of reactions with positive and negative control antigens. Specificity for each test must be determined.

Sensitivity defines the *lowest amount* of an antigen that can be detected. The most desirable level of sensitivity implies that the antibody in the test be capable of identifying a single antigen molecule. High sensitivity prevents *false negative* reactions. The sensitivity of some common tests are shown in Table 21.5. Note that this sensitivity is shown in terms of the amount of antibody necessary to detect antigen, but the amount of antigen detected by each test system is proportional to the amount of antibody used in the system. For example, immune precipitation reactions generally detect 0.1–1 *mg* quantities of antigen, while ELISA tests are up to one million times more sensitive and detect antigen in 0.1–1 *ng* quantities.

TABLE 21.5	Sensitivity of immunodiagnostic assays	
Assay		**Sensitivity (μg antibody/ml)[a]**
Precipitin reaction		
In fluids		24–160
In gels (double immunodiffusion)		24–160
Agglutination reactions		
Direct		0.4
Passive		0.08
Radioimmunoassay (RIA)		0.0008–0.008
Enzyme-linked immunosorbent assay (ELISA)		0.0008–0.008
Immunofluorescence		8.0

a The smallest amount of antibody necessary to give a positive reaction in the presence of antigen.

Immunity to Infection: Overview and Review

The immune response was discussed in Chapter 20. A summary of the major aspects of immunity is shown in Figure 20.2. The body responds to pathogens in a three-step process. For a pathogen that the body has never before encountered, the pathogen must first be recognized, usually by a group of cells called phagocytes (Section 20.3). Fortunately, phagocytes ingest and destroy most pathogens (a process called *phagocytosis*). Phagocytosis is *nonspecific,* and the target may be any foreign substance, including the pathogens and their components.

In the second phase of immunity, the phagocytes present pathogen-derived *antigens* (proteins obtained from the destroyed pathogen) to antigen-specific immune lymphocytes known as T cells (Section 20.4). Some T cells known as T helper (T_H) cells do not act directly on the pathogen but recruit and stimulate (help) other cells. One type of T_H cell, the antigen-specific T_H1 cell, attracts and activates phagocytes such as macrophages and neutrophils, causing an inflammatory reaction and limiting the infection (Section 20.3). T_H2 cells, another T_H subset, activate other antigen-specific lymphocytes, the B cells. The B cells then respond by producing soluble, antigen-binding proteins known as *antibodies* (Sections 20.5 and 20.11). A *primary antibody response* generally occurs within 5 days, but antibodies do not reach peak quantities for several weeks. The antibody proteins are pathogen-specific and are critical components of the immune response.

The antibodies interact specifically with the antigen on target cells, but cannot kill the cells. A group of proteins, known collectively as complement (Section 20.12), may attach to antibodies bound to the pathogen and lyse all cells with attached antibody. For example, antibodies specific for cell surface proteins of *Salmonella* spp. interact only with *Salmonella:* complement causes lysis of the antibody-sensitized *Salmonella* cell, but not of an *Escherichia coli* cell that is not antibody-sensitized. Thus, the immune response is *specific* for individual antigens, by virtue of specific antibodies, but the final effect may occur by means of nonspecific mechanisms such as complement.

In many cases, antibody-mediated immunity is not an effective mechanism for controlling the spread of infection. Some infectious agents parasitize the body from *within* cells. For example, animal viruses reproduce using host cell systems and, therefore, spend a large portion of their life cycle within the host cells (Section 8.14). Likewise, bacteria such as *Mycobacterium tuberculosis,* the causative agent of tuberculosis, live in phagocytes (Sections 20.3 and 23.3). Because antibodies are geared to recognize pathogens in the blood or at mucosal cell surfaces, the infected host cells must be identified and destroyed by other means, usually involving the cell-to-cell interactions of the *cell-mediated immunity.* Fortunately, intracellular pathogens produce antigens that are in turn presented on the surface of infected target cells. T cytotoxic cells (T_C) recognize the antigen and act directly on the infected target cell by secreting cytolytic proteins called *perforins,* which destroy the infected cell (Section 20.9).

No useful specific immunity exists before exposure to antigen, but after the first antigen exposure, specific immune T and B cells are present and specific immunity may persist for years. More importantly, a second antigen stimulation of these cells generates a very rapid and very strong immune response, which peaks within several days (Section 20.11). This *secondary response* quickly targets and destroys the pathogen. Thus, the immune response has *memory.* Memory is characterized by a rapid rise in immune *titer,* or quantity. *Antibody titer* is routinely used to track infections.

Antibody Titers, Skin Tests, and the Diagnosis of Infectious Disease

In the diagnosis of an infectious disease, isolation of the pathogen is not always possible. Thus, one alternative is to measure antibody titer (quantity) for a suspected pathogen. As we discussed earlier, if an individual is infected with a suspected pathogen, the antibody titer to that pathogen should be elevated. Antibody titer can be measured by precipitation, agglutination, or any of the methods in Sections 21.7 or 21.8. The general procedure is to set up a series of dilutions of patient serum (usually twofold dilutions: 1:2, 1:4, 1:8, 1:16, 1:32, and so on) and to determine the *highest* dilution at which the antigen–antibody reaction occurs. These methods are all termed *serological tests* because they make use of patient *serum.*

A *single* measure of antibody titer does not indicate active infection. Many antibodies remain at high titer for long periods after infection; to establish that an acute illness is due to a particular pathogen, it is essential to show a *rise* in antibody titer in successive samples of serum from the same patient. Frequently, the antibody titer is low during the acute stage of the infection and rises during convalescence (Figure 21.9). Such a rise in antibody titer is the best indication that the illness is due to the suspected agent and is also useful in diagnosis of infectious diseases of a rather chronic nature, such as typhoid fever and brucellosis. In some cases, however, the mere presence of antibody may be sufficient to indicate infection. This is the case for a pathogen that is quite rare in a population, and so the presence of antibody is sufficient to indicate that the individual has experienced an infection. A relevant example here is acquired immunodeficiency syndrome (AIDS) (Section 23.7). We will discuss methods for determining HIV antibody levels in Section 21.8.

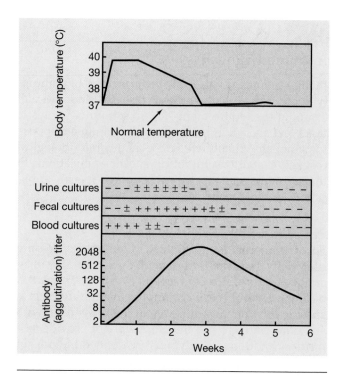

FIGURE 21.9   The course of infection in a typical untreated typhoid fever patient. Measurement of body temperature provides a measure of the course of clinical symptoms. The antibody titer was measured by determining the highest dilution (twofold series) causing agglutination of a test strain of *Salmonella typhi.* Presence of viable bacteria in blood, feces, and urine was determined from periodic cultures. Note that the pathogen clears from the blood as the antibody titer rises, and clearance from feces and urine requires a longer time. Body temperature gradually drops to normal as the antibody titer rises. The data given do not represent a single patient but are a composite of the picture seen in large numbers of patients.

Unfortunately, not all infections result in formation of systemic antibody. If a pathogen is extremely localized, there may be little induction of an immunological response and no rise in antibody titer even if the pathogen is proliferating profusely at its site of infection. A good example is the disease gonorrhea. Infection with *Neisseria gonorrhoeae,* the causative agent of gonorrhea, does not elicit a systemic immune response, and thus reinfection of a cured individual is not uncommon (see Sections 21.1 and 23.6). In other cases, the presence of antibody in the serum may have been due to a recent immunization. In fact, measurement of the rise in antibody titer following immunization is one of the best ways of determining that the immunization is effective.

Skin testing is another method for determining exposure to a pathogen. The most commonly used skin test is the **tuberculin test,** which consists of an intradermal injection of a soluble extract of *Mycobacterium tuberculosis.* A positive inflammatory reaction at the site of infection indicates current infection or previous exposure to *M. tuberculosis.* This test identifies delayed-type hypersensitivity responses and, thus, the presence of pathogen-specific T_H1 cells (Section 20.15). Skin tests are routinely used for diagnosing tuberculosis and leprosy (Section 23.3), as well as a variety of fungal diseases (Section 24.7).

Some of the most common immunodiagnostic tests for pathogens are shown in Table 21.6.

Monoclonal Antibodies for Immunodiagnostics

As we discussed in Section 20.13, the normal antibody response to an antigen is *polyclonal;* that is, many B cells are stimulated to produce antibodies to a complex antigen. The resulting antiserum consists of a mixture of different antibodies. Although this antibody population can give adequate immune protection to the host, it is usually not specific for a single epitope and is not precisely reproducible because it was produced in an indi-

vidual animal at a single time. For immunodiagnostic procedures, these types of antibodies, while they may be very potent, are extremely difficult to standardize. *Monoclonal* antibodies, on the other hand, are products of clones of single cells, and because the cell clones can be stored and later reconstituted for use, reproducibility and standardization are straightforward. Monoclonal antibody technology, therefore, has supplanted standard polyclonal techniques for many immunodiagnostic applications.

A monoclonal antibody is generally highly specific for a *single* antigenic determinant. For example, fluorescent antibodies (see Section 21.7) against *Chlamydia trachomatis* membrane proteins can be used to detect this organism in host tissues (*C. trachomatis* causes a variety of sexually transmitted diseases as well as trachoma, a serious eye disease) (Sections 13.26 and 23.6). These monoclonal antibodies react with *C. trachomatis* but they fail to react with even the closely

TABLE 21.6	Some clinical immunological procedures for identification of infectious agents	
Pathogen/disease	**Antigen**	**Procedure**[a]
HIV (AIDS)	Human immunodeficiency virus (HIV)	ELISA
Borrelia burgdorferi (Lyme disease)	Flagellin	ELISA
	Surface proteins	Immunoblot Bactericidal test (Section 24.3)
Brucella	Cell wall antigen	Agglutination
Candida albicans (yeast infections)	Soluble extract of fungal proteins	Skin test
Corynebacterium diphtheriae (diphtheria)	Toxin	Skin test (Schick test)
Influenza virus (influenza)	Influenza virus suspensions	Complement-based assay
	Nasopharynx cells containing influenza virus	Immunofluorescence
Mycobacterium leprae (leprosy)	Lepromin (soluble extract of bacterial proteins)	Skin test
Mycobacterium tuberculosis (tuberculosis)	Tuberculin (partially purified bacterial proteins, PPD)	Skin test
Neisseria meningitidis (meningitis)	Capsular polysaccharide	Passive hemagglutination (*N. meningitidis* polysaccharide adsorbed to red cells)
Pneumocystis carinii (lung infection)	*P. carinii* cells	Immunofluorescence
Rickettsial diseases (Q fever, typhus, Rocky Mountain spotted fever)	Killed rickettsial cells	Complement-based assay or cell agglutination tests ELISA
Salmonella (gastroenteritus)	O and H antigen	Agglutination (Widal test) ELISA
Streptococcus (group A) (strep throat, scarlet fever)	Streptolysin O (exotoxin) DNase (extracellular protein)	Neutralization of hemolysis Neutralization of enzyme
Treponema pallidum (syphilis)	Cardiolipin-lecithin-cholesterol	Flocculation [Venereal Disease Research Laboratory (VDRL) test]
Vibrio cholerae (cholera)	O antigen	Agglutination Bactericidal test (in presence of complement) ELISA

[a] Immunofluorescence tests use preformed antibody to detect the presence of the indicated pathogen in a patient specimen. Skin tests for *C. albicans, M. tuberculosis,* and *M. leprae* indicate T_H1-mediated delayed-type hypersensitivity. The *C. diphtheriae* Schick test detects serum antibodies with a toxin-neutralization skin test. All other tests are used to measure serum antibody levels.

related species *C. psittaci*. *C. trachomatis* is also an obligate intracellular parasite and is not easily cultured because it is dependent on the host cell to complete its life cycle. Use of the fluorescent anti–*C. trachomatis* monoclonal antibody on cervical scrapings or urethral or vaginal exudates makes positive identification of chlamydial infections almost routine.

Other monoclonal antibodies have been developed against outer membrane proteins of *Neisseria gonorrhoeae*. These probes are not only monospecific, reacting only with *N. gonorrhoeae*, but are also capable of differentiating strains of *N. gonorrhoeae*. The use of fluorescent monoclonal antibodies therefore eliminates much of the cross-reactivity problem observed when polyclonal sera are used.

✓ 21.4 Concept Check

An immune response is a natural outcome of infection. A specific immune response to a pathogen can be used as a diagnostic aid. Monoclonal antibodies are widely used for immunodiagnostic applications.

- ✓ How might a patient develop an antibody titer to an organism?
- ✓ Why does the antibody titer to an organism rise during convalescence?
- ✓ What advantages do *monoclonal* antibodies have over *polyclonal* antibodies in immunodiagnostic tests?

21.5

Agglutination

Agglutination involves the binding of a particulate antigen by antibody. Agglutination reactions were discussed in Section 20.14, with the well-known ABO blood grouping reaction serving as a prime example of direct agglutination. Many other agglutination reactions are used for the detection of antigens or antibodies. Although not as sensitive as some other immunoassays (Table 21.5), agglutination remains useful in clinical diagnostics as a simple, inexpensive, highly specific, rapid immunoassay.

Coated-Particle Agglutination

The agglutination of antigen-coated or antibody-coated latex beads by complementary antibody or antigen from a patient is a typical method of rapid diagnosis. Small (0.8-μm) latex beads coated with a specific antigen or antibody are mixed with patient serum on a microscope slide and incubated for a short period. If patient antibody complementary to the molecule bound to the bead surface is present, the milky-white latex suspension will become visibly clumped, indicating a positive agglutination reaction. Latex agglutination is also used to de-

tect bacterial surface antigens by mixing a small amount of a bacterial colony with antibody-coated latex beads. For example, a commercially available suspension of latex beads containing antibodies to protein A and clumping factor, two molecules found exclusively on the surface of *Staphylococcus aureus*, is virtually 100% accurate in identifying clinical isolates of *S. aureus*. Unlike traditional growth-dependent tests for *S. aureus*, identification of *S. aureus* by the latex bead assay takes only 30 seconds (Figure 21.10). Other latex bead agglutination assays have been developed to identify *Streptococcus pyogenes*, *Neisseria gonorrhoeae*, *Haemophilus influenzae*, *Campylobacter* spp., and the yeasts *Cryptococcus neoformans* and *Candida albicans*.

A very widely employed latex agglutination assay is that used for detecting specific serum antibodies for *rheumatoid factor*, an antibody directed against the body's own immunoglobulins and associated with the autoimmune disease *rheumatoid arthritis* (∞ Section 20.15). Latex beads coated with human immunoglobulin are mixed with whole blood or serum, and agglutination is scored versus positive and negative control sera run in parallel. Latex bead assays are simple and specific. In addition, the inexpensive nature of the assays makes them suitable for large-scale screening purposes; the widespread use of the rheumatoid test is a good example of this. Because they require no expensive equipment or particular expertise, they are in wide use in virtually all clinical settings.

Some agglutination assays use a suspension of activated charcoal as the carrier. For example, a rapid diagnostic test for detection of the virus *Herpes simplex*, frequently associated with oral fever blisters or genital sores (∞ Section 23.6), employs antibodies to *H. simplex* virus that are adsorbed to small particles of activated charcoal. Cotton swabs used to sample suspected

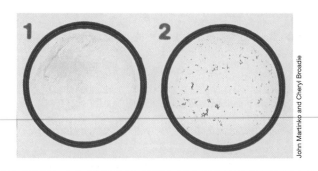

FIGURE 21.10 Latex bead agglutination test for *Staphylococcus aureus*. Panel 1 shows a negative control. Note the uniform pink color of the suspended latex beads coated with antibodies to protein A and clumping factor, two antigens found exclusively on the surface of *S. aureus* cells. Panel 2 shows the same suspension after a loopful of material from a bacterial colony was mixed into the suspension. The bright red clumps indicate a positive agglutination reaction took place and indicates that the colony is *S. aureus*.

herpes lesions are placed in a buffer solution, and samples of the solution, possibly now containing virus, are tested with the antibody-coated charcoal. A positive test is indicated by visible clumping of the charcoal into large, black aggregates. Because of the specificity of the monoclonal antibodies used, complicating cross-reactions with related pathogens are not a problem. Like latex beads, charcoal agglutination tests can be rapid and cost-effective diagnostic tools.

Coated-particle tests are *passive* agglutination reactions and are up to five times more sensitive than the direct agglutination tests we discussed in Section 20.14 (Table 21.5).

✓ 21.5 Concept Check

A number of clinically useful agglutination tests are available. These tests are rapid, relatively sensitive, and inexpensive methods for identifying a variety of pathogens.

- ✓ Distinguish between *direct* and *passive* agglutination. Which tests are more sensitive?
- ✓ What advantages do agglutination tests have over other immunoassays? What disadvantages?

21.6

Immunoelectron Microscopy

Antibodies to which heavy metals have been chemically conjugated can be used to locate antigens in cells by electron microscopy. This is possible because heavy metals scatter the electron beam of the electron microscope. This technique, called *immunoelectron microscopy*, is used primarily in research where there is a need to determine where a specific antigen (usually a protein) is localized in a particular region of the cell (Figure 21.11). Cells, following chemical fixation and other preparations necessary for observation by the electron microscope, are treated with antibodies covalently conjugated to a heavy metal, usually gold or platinum. The electron-dense metals scatter electrons, and thus the presence of bound antibody can be detected by dense black spots in photographs of the preparation.

In immunoelectron microscopy, although the cell is dead and chemically fixed, most protein antigens retain sufficient native structure and antibodies still react with little nonspecific cross-reaction. This technique has been used extensively to pinpoint the location of enzymes in cells, especially those suspected to be associated with the cytoplasmic membrane or some other internal structure (Figure 21.11).

Although immunoelectron microscopy can be used for identifying pathogens such as human immunodeficiency virus (HIV) in cells (∞ Figure 23.31), the time, expense, expertise, and specialized equipment involved

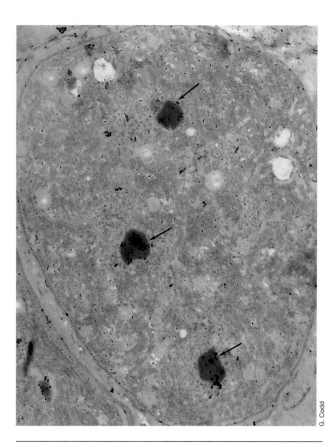

FIGURE 21.11 Immunoelectron microscopy. Antibodies made in rabbits to the enzyme ribulose-1,5-bisphosphate carboxylase from the cyanobacterium *Chlorogloeopsis fritschii* were added to thin sections of *C. fritschii* and the preparation treated with goat anti-rabbit IgG conjugated to 20-nm colloidal gold particles. The concentration of the particles around large inclusions called carboxysomes (arrows) indicate that these sites contain large amounts of the enzyme.

make it impractical for diagnostic procedures in all but the most specialized clinical research settings.

✓ 21.6 Concept Check

Immunoelectron microscopy is a research tool used for localizing antigens in cells.

- ✓ Why are heavy metal–antibody conjugates used for immunoelectron microscopy?

21.7

Fluorescent Antibodies

In this section, we will discuss the use of antibodies chemically modified with fluorescent dyes. This procedure makes it possible to detect reactions of antibodies with single cells. Virtually all well-equipped clinical laboratories and many research laboratories

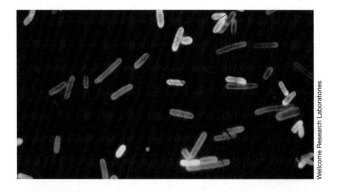

FIGURE 21.12   Fluorescent antibody reactions. Cells of *Clostridium septicum* were tested with antibody conjugated with fluorescein isothiocyanate, which fluoresces yellow-green. Cells of *Clostridium chauvei* were stained with antibody conjugated with rhodamine B, which fluoresces red.

make extensive use of fluorescent antibodies for diagnostic procedures.

Fluorescent Methods

Antibody molecules can be made fluorescent by covalently attaching them to fluorescent organic compounds such as rhodamine B, which fluoresces red, or fluorescein isothiocyanate, which fluoresces yellow-green. This does not alter the specificity of the antibody but makes it possible to detect the antibody bound to cell or tissue surface antigens by use of the fluorescence microscope (Figure 21.12). Cells with bound fluorescent antibodies emit a bright fluorescent color, usually red-orange or yellow-green, depending on the dye used. Fluorescent antibodies are used in diagnostic microbiology because they permit the identification of a microorganism directly in a patient specimen (*in situ*) and avoid the isolation and culturing of the organism (see below). The fluorescent antibody technique is also very useful in microbial ecology as a method for directly viewing microbial cells in natural environments.

Two distinct fluorescent antibody procedures, the **direct** and the **indirect** staining methods, are used. In the direct method, the antibody against the organism itself is fluorescent. In the indirect method, the presence of a nonfluorescent antibody on the surface of the cell is detected by the use of a fluorescent antibody directed against the nonfluorescent antibody (Figure 21.13). This is done by immunizing one animal species, for example, a goat, with antibodies from a second species, for example, a rabbit, and then conjugating the fluorescent dye to the goat antibodies. The fluorescent goat anti-rabbit antibodies can then be used to detect the presence of any rabbit immunoglobulin previously bound to cells.

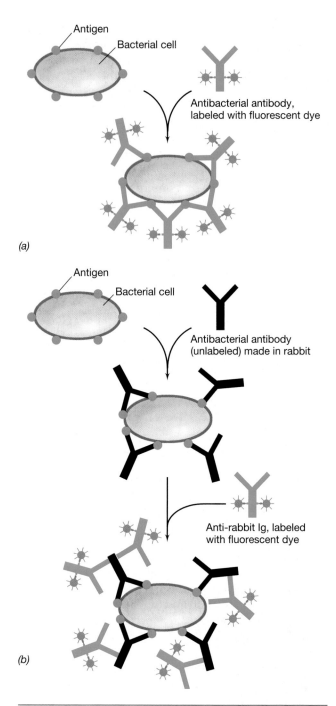

FIGURE 21.13   Fluorescent antibody methods for detection of microbial surface antigens. (a) Direct staining method. (b) Indirect staining method.

Clinical Applications

In a typical clinical test using fluorescent antibodies, a specimen containing a suspected pathogen is allowed to react with a specific fluorescent antibody and observed with a fluorescent microscope. If the pathogen contains surface antigens reactive with the antibody, the cells will fluoresce (Figure 21.14).

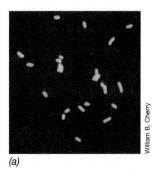
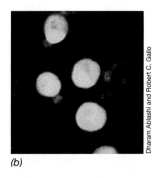

(a)   *(b)*

FIGURE 21.14   Examples of the use of fluorescent antibodies in clinical microbiology. (a) Immunofluorescent stained cells of *Legionella pneumophila*, the cause of legionellosis. The specimen was taken from biopsied lung tissue. The individual organisms are 2–5 μm in length. (b) Detection of virus-infected cells by immunofluorescence. Human B lymphotrophic virus (HBLV)-infected spleen cells were incubated with serum containing antibodies to HBLV from a patient with a lymphoproliferative disorder. Cells were then treated with fluorescein isothiocyanate-conjugated anti-human IgG antibodies. HBLV-infected cells fluoresce bright yellow. Cells in the background did not react with the patient's serum. Individual cells are about 10–15 μm in diameter.

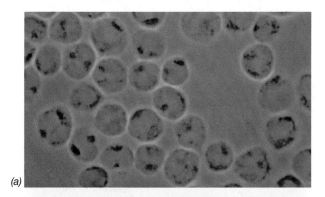

(a)

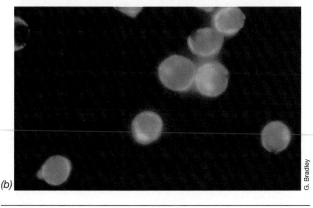

(b)

FIGURE 21.15   Use of fluorescent antibodies in noninfectious disease diagnostics. (a) Human leukemic cells, some of which are sensitive to a toxic anticancer drug and some of which are not, appear indistinguishable. (b) When the cells in (a) are treated with a fluorescent monoclonal antibody that binds specifically to a protein found only on the surface of drug-resistant cells, the latter fluoresce whereas drug-sensitive cells do not. Individual cells are about 10–12 μm in diameter.

Fluorescent antibodies can also be applied directly to infected host tissues, permitting diagnosis long before primary isolation techniques yield a suspected organism. For example, in diagnosing legionellosis (⟳ Section 24.9) a positive diagnosis can be made by staining biopsied lung tissue with fluorescent antibodies prepared against cell walls of *Legionella pneumophila*, the causative agent of legionellosis (Figure 21.14*a*). Likewise, a fluorescent antibody against the capsule of *Bacillus anthracis* can be used in the microscopic diagnosis for anthrax. Fluorescent antibody reactions can also be used in diagnosis of viral infection (Figure 21.14*b*) and in a variety of noninfectious diseases. For example, in identifying cell types expressing a particular antigen, such as malignant cells, fluorescent antibodies may be very valuable in following the course of the disease (Figure 21.15).

Fluorescent antibodies can also be used to separate mixtures of cells into relatively pure populations or to define the numbers of certain cell types in complex mixtures such as blood. Fluorescent-labeled monoclonal antibodies directed against the CD4 and CD8 surface antigens of T lymphocytes (⟳ Section 20.4) are routinely used to identify and enumerate these cells in the blood leukocyte population (Figure 21.16). For example, the definition of acquired immunodeficiency syn-

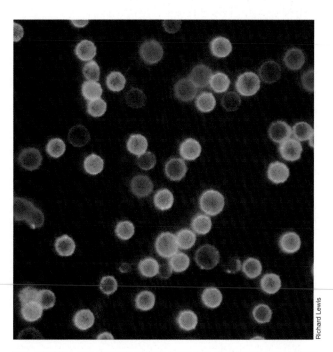

FIGURE 21.16   T lymphocytes stained with fluorescent-tagged monoclonal antibodies to specific surface markers. Yellow-green cells are cytotoxic (CD8) T cells; red-orange cells are T helper (CD4) cells. The different-colored cells can be separated from one another by a fluorescence-activated cell sorter to yield enriched populations of different cell types. Individual cells are about 10–12 μm in diameter. Reprinted with permission from *Science 239:* Cover (Feb. 12, 1988), © AAAS.

drome (AIDS) includes a reduction in CD4 cells. In addition, the CD4 T cell number changes during the progression of AIDS and is diagnostic for the disease. Thus, by defining the CD4 numbers, the clinician can identify the reduction in CD4 cells compared to normal values and, with successive assays over time, can follow the progress of the disease (∞ Section 23.7).

Fluorescing cells can be visualized, counted, and separated with an instrument called a *fluorescence spectrometer*, often referred to as a fluorescence-activated cell sorter (FACS). The FACS uses a laser beam to activate fluorescent antibody bound to cells, placing a charge on the labeled cells. Following laser exposure, an electric field is applied to the cell mixture. The fluorescing and nonfluorescing cells are then deflected to opposite ends of the electric field where each cell population is counted and deposited in a tube. The use of several antibodies, each labeled with a different fluorescent dye, can result in the simultaneous identification of several cell markers. A typical application used for identifying CD3 and CD4 surface proteins on T cells in normal and AIDS patients is shown in Figure 21.17.

FACS analysis is also useful for research applications. For example, immunologists routinely use FACS methods to separate complex mixtures of immune cells. They can then study the properties of the highly enriched cell populations.

Under appropriate conditions, fluorescent antibodies yield rapid, highly specific, useful information about a variety of clinical conditions. However, fluorescent antibody techniques are not without their pitfalls. Nonspecific staining can be a problem because of surface antigens that may *cross-react* between various bacterial species, some of which may be members of the normal flora. This is a major problem among enteric bacteria, where lipopolysaccharide antigens (∞ Section 3.8) are frequently sufficiently similar among species to cause binding or partial binding of the fluorescent probe (cross-reactions, ∞ Section 20.14). The clinical microbiologist must therefore be careful to perform controls using nonspecific sera and confirm all positive immunofluorescent findings by other immunological or microbiological tests.

✓ 21.7 Concept Check

Fluorescent antibodies are used for quick, accurate identification of pathogens and other antigenic substances in tissue samples and other complex environments. Fluorescent antibodies can be used for quantitative enumeration and sorting of a variety of cell types.

✓ Are fluorescent antibodies more sensitive for detecting antigens than normal antibodies?
✓ How are fluorescent antibodies used to identify specific cells in complex mixtures like blood?

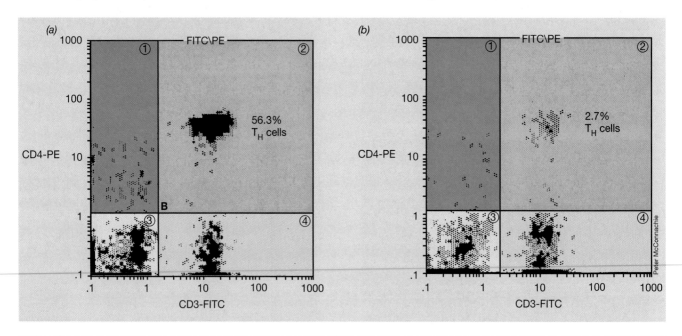

FIGURE 21.17   CD3 and CD4 cell enumeration from a healthy human (a) and from a human with acquired immunodeficiency syndrome (AIDS) (b) using a fluorescence-activated cell sorter (FACS). Each dot represents a single cell. Peripheral blood cells were simultaneously labeled with monoclonal antibody to CD4 conjugated to phycoerythrin (PE) and with monoclonal antibody to CD3 conjugated to fluorescein isothiocyanate (FITC). CD3 is found on all T cells. CD4 is found on T helper (T_H) cells only. Quadrant 3 shows cells that were stained with neither antibody. Quadrant 1 shows cells stained with only anti-CD4. Quadrant 4 shows cells stained with only anti-CD3. Quadrant 2 shows cells stained with both anti-CD3 and anti-CD4. (a) Results from a healthy human. In this case, 56.3% of the T cells were T_H cells. Thus, quadrant 2 shows a dense staining pattern. (b) Results from a patient with clinical AIDS. In this case, only 2.7% of the total T cells are T_H cells. This is indicated by the very light staining pattern in quadrant 2. [Original data from Peter McConnachie, used with permission.]

Enzyme-Linked Immunosorbent Assay and Radioimmunoassay

The specificity of antibodies is such that the limiting factor in most of the immunological reactions discussed thus far is not *specificity* but *sensitivity* (see Section 21.4 and Table 21.5). Because of their high sensitivity, radioimmunoassay (RIA) and enzyme-linked immunosorbent assay (ELISA) are two widely used immunological techniques. These methods employ radioisotopes and enzymes, respectively, to label antibody molecules used for antigen detection. Because radioactivity and the products of certain enzymatic reactions can be measured in very small amounts, the attachment of radioactive or enzyme ligands to antibody molecules serves to decrease the amount of antigen–antibody complex required to detect a reaction. This increased sensitivity has been extremely helpful in clinical diagnostics and research and has opened the door to the development of a variety of new immunological tests, previously impossible because the detection methods available were not sufficiently sensitive (∞ Urine Testing for Drug Abuse, and Over-the-Counter Immunodiagnostic Kits, Chapter 20).

ELISA

The covalent attachment of enzymes to antibody molecules creates an immunological tool possessing both high specificity and high sensitivity. The technique, called **ELISA** (for *enzyme-linked immunosorbent assay*), makes use of antibodies to which enzymes have been covalently bound such that the enzyme's catalytic properties and the antibody's specificity are unaltered. Typical linked enzymes include peroxidase, alkaline phosphatase, and β-galactosidase, all of which catalyze reactions whose products are colored and can be detected in very low amounts.

Two basic ELISA methodologies have been developed, one for detecting antigen (*direct* ELISA) and the other for detecting antibodies (*indirect* ELISA). For detecting antigens such as virus particles from a blood or fecal sample, the direct ELISA method is used. In this procedure the antigen is "trapped" between two layers of antibodies (Figure 21.18). Thus, this method is sometimes called the "sandwich ELISA." The specimen is added to the wells of a microtiter plate (see the box, The Microtiter Plate and Immunoassays) previously coated with antibodies specific for the antigen to be detected. If the antigen (virus particle) is present in the sample, it will be trapped by the antigen-binding sites on the antibodies. After washing unbound material away, a second antibody containing a conjugated enzyme is added. The second antibody is also specific for the antigen, and so it binds to any remaining exposed determinants. Following a wash, the enzyme activity of the bound material in each microtiter well is determined by adding the substrate of the enzyme. The color formed is proportional to the amount of antigen present (Figure 21.18).

A FOCUS ON... The Microtiter Plate and Immunoassays

A number of immunoassays have been developed requiring that either the antigen or the antibody be bound to a solid support. A variety of solid phase carriers such as latex beads or plastic tubes have been used, but the plastic, disposable *microtiter plate* has been the most successful solid support in modern immunoassays.

Microtiter plates are small, plastic trays containing a number of wells. The plates are made of polystyrene or polypropylene, both of which are transparent plastics. Proteins, either antigens or antibodies, adsorb to the plastic surface of the microtiter plate as a result of interactions between hydrophobic regions of the protein and the nonpolar plastic surface. Once bound, the proteins are not easily removed by washing or other manipulations, and thus bound proteins can be treated with various reagents and washed repeatedly without being removed from the plate surface. The standard microtiter plate contains 96 wells, which allows for several replicates of each sample to be run. For quantitation, several dilutions are prepared. Although only small amounts of protein can be adsorbed to each well, the sensitivity of ELISAs and radioimmunoassays are such that only small amounts of antigen or antibody are needed anyway.

Automatic machines are available for adding reagents and preparing dilutions. Depending on whether the microtiter plate assay is an ELISA or a radioimmunoassay, special machines are available to read the absorbance of a colored product (ELISA) or amount of radioactivity (radioimmunoassay) in each of the wells of the plate. Such automation allows these immunoassays to be performed both rapidly and routinely at relatively low cost. The microtiter plate has therefore helped to revolutionize research and diagnostic immunoassays. ELISAs in particular have become part of the everyday activities of the clinical laboratory. ∎

Perkin-Elmer Cetus

Procedure

1. Antibodies (Y) to virus (★) bound to wells of microtiter plate

2. Add patient sample (secretions, serum, and so on) suspected of containing virus particles or virus antigens and wash wells with buffer

3. Add antivirus antibody containing conjugated enzyme

(E┼E)

4. Wash with buffer

5. Add substrate for enzyme and measure amount of colored product (●).

Results

Color

Quantitation

Colored product produced is proportional to amount of antigen.

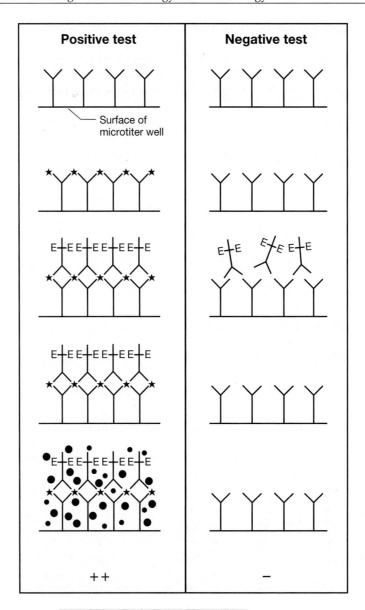

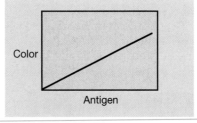

FIGURE 21.18 Detection of viruses by a direct ELISA test.

To detect *antibodies* in human serum, an indirect ELISA is employed. An indirect ELISA test is widely used to detect antibodies to human immunodeficiency virus (HIV), and we will discuss this test in detail because the principles involved are applicable to all indirect ELISA tests.

The HIV-ELISA

The causal agent of AIDS, the human immunodeficiency virus (HIV) (∞ Section 23.7), is transmitted by bodily fluids including blood. Sensitive, specific, rapid, and cost-effective screening tools are needed to test blood samples from individuals exposed to HIV and to ensure that HIV is not being inadvertently transmitted during blood transfusions or through the transfer of blood products. An ELISA test is used for the routine screening of blood for signs of exposure to HIV (and hence possible AIDS).

The HIV-ELISA test is an *indirect* ELISA designed to measure *antibodies* to HIV present in serum. Initial infection with HIV leads to the production of antibodies to several HIV antigens, in particular those of the

HIV envelope. These antibodies can be detected by the HIV-ELISA test (Figure 21.19).

To carry out an HIV-ELISA test, microtiter plates are first coated with a disrupted preparation of HIV particles; about 200 ng of disrupted HIV is required in each well. Following a brief incubation period to ensure bind-ing of the antigens to the surface of the microtiter wells, a diluted serum sample is added and the mixture incubated to allow HIV-specific antibodies to bind to HIV antigens. To detect the presence of antigen–antibody complexes, a second antibody is then added. This second antibody is an enzyme-conjugated anti-human IgG

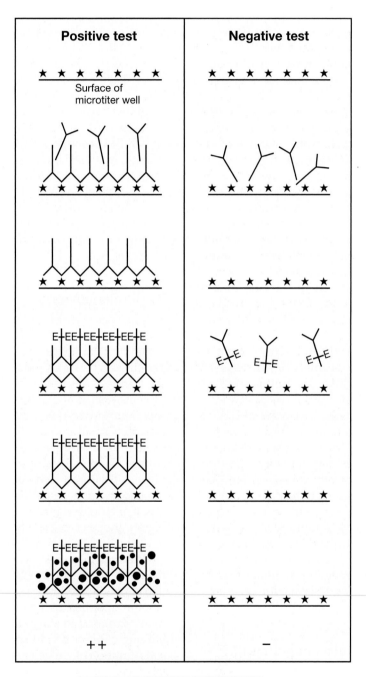

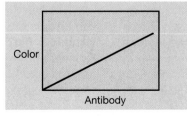

FIGURE 21.19   Indirect ELISA test for de-tecting antibodies to human immunodeficien-cy virus (HIV), the causal agent of acquired immunodeficiency syndrome (AIDS).

preparation. Following addition of the second antibody, the enzyme activity is assayed (the anti-human IgG antibodies bind to any HIV-specific IgG antibodies previously bound to the HIV antigen preparation). A color is obtained in the enzyme assay in proportion to the amount of anti-human IgG antibody bound (Figure 21.19). The binding of the second antibody is an indication that antibodies from the patient's serum recognized the HIV antigens, the patient has antibodies to HIV, and the patient has been exposed to HIV. Control sera (known to be HIV-negative) are assayed in parallel with any samples to measure the extent of background absorbance in the assay.

The HIV-ELISA test is a rapid, highly sensitive, specific method for detecting exposure to HIV. Since ELISAs in general are highly adaptable to mass screening and automation, the HIV-ELISA test is used as a standard screening method for blood. However, this test method can give erroneous results under certain circumstances.

For example, the test occasionally gives false positive results. Because a number of factors can contribute to these results, none of which are related to exposure to HIV, all positive HIV-ELISA tests *must* be confirmed by another independent test, usually the Western blot (immunoblot) test (see Section 21.9). A positive HIV Western blot test after a positive HIV-ELISA test is considered proof of HIV infection.

A final drawback to the HIV-ELISA test is the possibility of obtaining false negative results. As we learned in Section 20.11, it takes the immune system some time to develop an effective antibody response with a detectable antibody titer. In the case of HIV infection, this lag time is estimated to be 6 weeks to a year. Therefore, individuals who have been recently infected with HIV may not yet be producing detectable amounts of antibody when they are tested. Another reason for a false negative result in the HIV-ELISA test is the total destruction of the immune system seen in advanced cases of AIDS; if no immune cells are left in the body, no antibodies can be made and the ELISA test is not useful. However, at this stage of disease, a diagnosis is possible based on other information (Section 23.7) and the ELISA test is useful only as a confirmatory indicator.

Other ELISA Tests of Clinical Importance

Besides the ELISA test for HIV, literally hundreds of clinically useful ELISAs have been developed. Some of these are direct ELISAs for detecting antigens. Direct ELISAs for detecting bacterial toxins such as cholera toxin, enteropathogenic *Escherichia coli* toxin, and *Staphylococcus aureus* enterotoxin have been developed. Viruses currently detected using direct ELISA tech-

niques include rotavirus, hepatitis viruses, rubella virus, bunyavirus, measles and mumps viruses, and parainfluenza virus.

Indirect ELISAs have been developed for detecting antibodies to a variety of clinically important bacteria. Although not meant to be a complete list, ELISAs for detecting serum antibodies to *Salmonella* (gastrointestinal diseases), *Yersinia* (plague), *Brucella* (brucellosis), a variety of rickettsias (Rocky Mountain spotted fever, typhus, Q fever), *Vibrio cholerae* (cholera), *Mycobacterium tuberculosis* (tuberculosis), *Mycobacterium leprae* (leprosy), *Legionella pneumophila* (legionellosis), *Borrelia burgdorferi* (Lyme disease), and *Treponema pallidum* (syphilis) have been developed. ELISAs have also been developed for detecting antibodies to *Candida* (yeast) and antibodies to a variety of parasites, including those causing amebiasis, Chagas' disease, schistosomiasis, toxoplasmosis, and malaria.

The speed, low cost, lack of radioactive waste, and long shelf life make ELISA tests particularly attractive for many laboratories. But it is the extreme *sensitivity* of ELISAs that really make them important immunodiagnostic tools.

Radioimmunoassay

Radioimmunoassay (RIA) employs radioisotopes instead of enzymes as antibody conjugates. The isotope iodine-125 is the most commonly used detection system, because proteins can be readily iodinated without disrupting their specificity. RIA is used clinically to measure rare serum proteins such as human growth hormone, glucagon, vasopressin, testosterone, and insulin present in humans in extremely small amounts (Figure 21.20) and also in some urine tests for drug abuse (Urine Testing for Drug Abuse, Chapter 20). In most cases a *direct* RIA is employed. The direct assay is a two-step procedure. First, radioactive antigen-specific antibodies are added to a series of microtiter wells containing known concentrations of pure antigen (such as a hormone), which is first bound to the wells. The radioactivity in each of these standard wells is then measured. Next, the antigen sample from a patient is allowed to bind to another well, and radioactive antibodies are added and measured, as before. The amount of radioactivity bound by the patient sample is then compared to a standard plot generated from the binding data obtained using the pure antigen, and the concentration of antigen in the patient serum is interpolated from the standard plot (Figure 21.20).

RIA has the same sensitivity range as ELISA and can also be performed very rapidly. However, the instruments used to detect radioactivity are quite specialized and expensive. RIA generates a considerable amount of radioactive waste, and the radioactive decay time (half life) of the radioisotopes used for detection

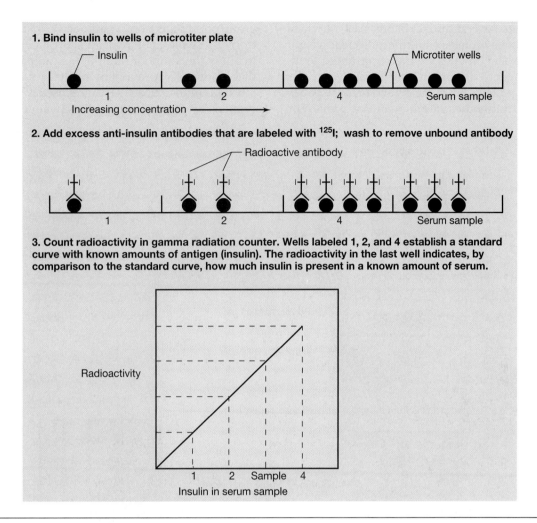

FIGURE 21.20 Radioimmunoassay (RIA). Using RIA to detect insulin levels in human serum. Following establishment of a standard curve, the insulin concentration in a serum sample can be estimated.

may limit the useful life of the test kit. As a result, RIA is often used only when ELISA is not sufficiently accurate or sensitive. For example, RIA is often more useful than ELISA for detecting serum protein levels (as described earlier) because some serum components may inhibit ELISA enzyme–substrate reactions or antibodies. Thus, for certain applications, each test system has clear advantages.

✓ **21.8 Concept Check**

ELISA and RIA methods are the most sensitive known immunoassay techniques. Both involve linking a detection system, either an enzyme or a radioactive molecule, to an antibody or antigen, enhancing sensitivity. ELISA and RIA are used for clinical and research work; tests have been designed to detect either antibody or antigen in a vast number of applications.

✓ Why are ELISA and RIA techniques more sensitive than standard immunoassays such as precipitation and agglutination?
✓ What hazards are associated with radioimmunoassays?

21.9

Immunoblot Procedures

Antibodies can be also used in clinical diagnostics to identify individual specific *proteins* associated with specific pathogens. The procedure employs three techniques discussed previously: (1) the separation of proteins on polyacrylamide gels, (2) the transfer (blotting) of proteins from gels to nitrocellulose paper (∞ Working with Nucleic Acids: The Tools, Chapter 6), and (3) identification of the proteins by specific antibodies. Protein blotting and the subsequent identification of the proteins by specific antibodies is also called the "Western" blot technique to distinguish it from the (DNA) "Southern" blot technique.

The immunoblot is a very sensitive method for detecting specific proteins in complex mixtures. In the first step of an immunoblot, a protein mixture is subjected to electrophoresis on a polyacrylamide gel. This separates the proteins into several distinct bands, each of which represents a single protein of specific molecular weight (Figure 21.21). The proteins are then transferred to nitrocellulose paper by an electrophoretic transfer process that forces the proteins out of the gel and onto the paper. At this point, antibodies raised against a protein or group of proteins from a pathogen are added to the nitrocellulose blot. Following a short incubation period to allow the antibodies to bind, a radioactive marker that binds antigen–antibody complexes is added. The most common radioactive marker used is *Staphylococcus* protein A iodinated with radioactive iodine, ^{125}I. Pro-

1. Denature proteins by boiling in detergent

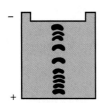

2. Subject to electrophoresis; proteins separate by molecular weight

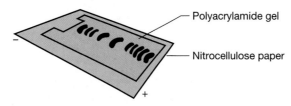

— Polyacrylamide gel

— Nitrocellulose paper

3. Blot the separated proteins from the gel to nitrocellulose paper

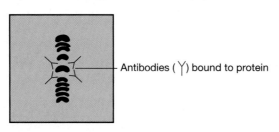

— Antibodies (Y) bound to protein

4. Treat nitrocellulose paper containing blotted proteins with antibodies; each antibody recognizes and binds to a specific protein

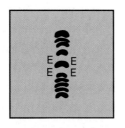

5. Add marker to bind to antigen-antibody complexes, either (left) radioactive *Staphylococcus* protein A–^{125}I, or (right) antibody containing conjugated enzyme

X-ray film

Nitrocellulose with enzyme-produced colored spot

(a)

(b)

←— GP41-45

←— P24

Victor Tsang

FIGURE 21.21   The Western blot (immunoblot) and its use in the diagnosis of human immunodeficiency virus (HIV) infection. (a) Protocol for an immunoblot. (b) Developed HIV immunoblot. The proteins P24 and GP41-45 are coat proteins of the virus and are diagnostic for HIV. Lane 1, Positive control serum (from known AIDS patients); lane 2, negative control serum (from healthy volunteer); lane 3, strong positive from patient sample; lane 4, weak positive from patient sample; lane 5, reagent blank to check for background binding.

tein A has a strong affinity for antibody and binds firmly. Once the radioactive marker has bound, its vertical position on the blot can be detected by exposing the nitrocellulose blot to X-ray film; the gamma rays emitted by the ^{125}I expose the film only in the region where the radioactive antibody has bound to antigen–antibody complexes (Figure 21.21).

For many clinical applications, immunoblots employ enzyme-linked immunosorbent assay (ELISA) technology (see Section 21.8) for detection of bound antigen–antibody complexes. Following treatment of the blotted proteins with specific antibody, the paper is washed and then treated with a second antibody, which binds to the first. For example, if antibodies from a human were used in the first step, then the second antibody could be a rabbit anti-human antibody. Covalently attached to this second antibody is an enzyme. The original antigen–antibody complexes are seen when the enzyme is assayed because the product of the enzyme reaction leaves a colored product on the nitrocellulose filter at any spot where rabbit antibodies are bound to the human antibodies. By comparing the location of the color bands on the nitrocellulose paper with the position of colored bands from control samples, a protein associated with a given pathogen can be positively identified.

The immunoblot procedure can be used to detect either antigen (*direct* evidence for pathogen presence) or antibody (*indirect* evidence for pathogen exposure). Thus, this very sensitive, extremely accurate method is analogous to the direct and indirect ELISA procedures detailed in Section 21.8. We now examine a widely used immunoblot test designed to identify exposure to HIV.

The HIV Immunoblot

Immunoblots have had a significant clinical impact on the diagnosis and confirmation of cases of AIDS. Because an immunoblot is more laborious, more time-consuming, less sensitive and more costly than the ELISA test, HIV-ELISA tests have been widely used for screening purposes. However, the HIV-ELISA test occasionally yields false positive results. Thus, an immunoblot is used to confirm positive ELISA results.

Like the HIV-ELISA, the HIV immunoblot is designed to detect the presence of *antibodies* to HIV in a serum sample. To perform the immunoblot, a purified preparation of HIV is treated with the detergent sodium dodecyl sulfate (SDS), which solubilizes HIV proteins and also renders the virus inactive. HIV proteins are then resolved by polyacrylamide gel electrophoresis. The HIV proteins are then blotted from the gel onto sheets of nitrocellulose paper (Figure 21.21). At least seven major HIV proteins are resolved by electrophoresis, and two of them, designated P24

and GP41-45, are used as specific diagnostic proteins in the AIDS immunoblot. Protein P24 is the HIV core protein, and proteins GP41-45 are HIV coat proteins (∞ Section 8.22).

Following blotting of the proteins, the nitrocellulose strips are incubated with a serum sample previously identified as HIV-positive (a positive control) by HIV-ELISA. If the sample is truly HIV-positive, antibodies against HIV proteins will be present and will bind to the HIV proteins separated on the nitrocellulose paper (Figure 21.21). To detect whether antibodies from the serum sample have bound to HIV antigens, a *detecting antibody*, anti-human IgG conjugated to the enzyme peroxidase, is added to the strips. If detecting antibody binds, the activity of the conjugated enzyme will form a brown band on the strip at the site of antibody binding after addition of substrate. The serum from the HIV-ELISA-positive patient is assayed in parallel with the positive control serum. The patient can be confirmed as HIV-positive if the position of the bands in the patient and the positive control sera are identical; negative control sera are also analyzed in parallel and must show no bands (Figure 21.21).

Although the intensity of the bands obtained in the HIV immunoblot varies somewhat from sample to sample (Figure 21.21*b*), the interpretation of an immunoblot is generally unequivocal, and thus the test is valuable for confirming HIV-ELISA positives and eliminating HIV-ELISA false positives. To make the HIV immunoblot clinically accessible, nitrocellulose strips containing inactivated HIV antigens (previously separated by electrophoresis) are available commercially. Separate strips can be incubated directly with the patient and control serum samples and subsequently treated with the detecting antibody.

The immunoblot technique is also used to confirm the specificity of tests for the Lyme disease antibody (see Table 21.6). However, because of the expense, technical requirements, lower *sensitivity* (but higher *specificity*), and time involved, immunoblot tests are not likely to supplant the rapid, low cost ELISA methods for general screening purposes.

✓ 21.9 Concept Check

Immunoblot procedures can be used to detect antibodies to specific antigens or to detect the presence of the antigens themselves. The antigens are electrophoresed, transferred (blotted) to a filter, and exposed to antibody. Immune complexes are made visible through the use of enzyme-labeled or radioactive second antibodies. Immunoblots are extremely sensitive *and* accurate, but procedures are complex and time-consuming.

✓ What advantage does the immunoblot have over immunoassays such as ELISA and RIA?

✓ Why is the immunoblot not used for general screening for HIV exposure?

21.10

Nucleic Acid Probes in Clinical Diagnostics

The emergence of molecular biology has given rise to new molecular tools that are rapidly being adapted to the field of diagnostic microbiology. *DNA diagnostics,* as this area has come to be known, is revolutionizing the whole approach to identifying and monitoring infectious diseases, genetic and malignant diseases, and other medical conditions such as coronary artery disease and diabetes. This new approach uses *genotypic* rather than *phenotypic* factors to identify specific pathogens. The power of DNA diagnostics is a consequence of two facts: (1) nucleic acids can be rapidly and sensitively measured, and (2) the *sequence* of nucleotides in a given DNA molecule is so specific that hybridization analyses can be used for reliable clinical diagnoses.

Automation is helping to make DNA analysis virtually routine in the clinical setting. Automated DNA extractors, polymerase chain reaction (PCR) machines (∞ Section 10.9), DNA sequencers, and pulse field gel electrophoresis equipment (for separating large DNA segments such as whole chromosomes) (∞ Working with Nucleic Acids: The Tools, Chapter 6) are all available for

use in the diagnosis of diseases. The **nucleic acid probe** is now a major molecular tool in clinical laboratories.

Nucleic Acid Probes

One of the most powerful analytical tools available to clinical microbiologists is *nucleic acid hybridization.* Instead of detecting a whole organism or its products (for example, antigens), hybridization detects the presence of *specific DNA sequences* associated with a specific organism. To identify a microorganism through DNA analysis, the clinical microbiologist must have available a *nucleic acid probe* to that microorganism, a *single strand* of DNA containing sequences unique to the organism. The probe may be up to several kilobases in length, but many synthetic oligonucleotides consist of 20 bases or less and are still highly specific. If a microorganism in a clinical specimen contains DNA sequences complementary to the probe, the two sequences can hybridize (following appropriate sample preparation to yield single-stranded DNA from the microorganism), forming a *double-stranded* molecule (Figure 21.22). To detect that a reaction has occurred, the probe is labeled with a *reporter molecule,* a radioisotope, an enzyme, or a fluorescent compound that can be measured in small amounts

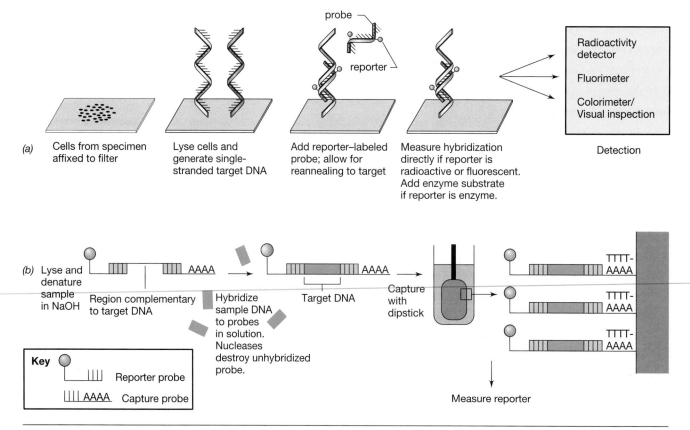

FIGURE 21.22 Nucleic acid probe methodology in clinical diagnostics. (a) Membrane filter assay. The detecting system (reporter) can be a radioisotope, a fluorescent dye, or an enzyme. (b) Dipstick assay. In the dipstick assay a dual reporter or capture probe is used. The capture probe contains a poly-dA tail that hybridizes to a poly-dT oligonucleotide affixed to the dipstick.

following hybridization. Depending on the reporter used (radioisotopes are the most sensitive), as little as 0.25 µg of DNA per sample can be detected.

Nucleic acid probes offer many advantages over clinical immunological assays. Nucleic acids are much more stable than proteins at high temperatures and at high pH, and are more resistant to organic solvents and other chemicals. This means that a clinical sample can be treated in a relatively harsh manner to destroy most interfering material, leaving behind the nucleic acid. Because of the relative chemical stability of the target nucleic acids, nucleic acid probe technology can even be used to positively identify organisms that are no longer alive. Additionally, nucleic acid probes may be even more specific than antibodies, since they can detect single base pair differences between DNA sequences. Finally, since many probes are made in the laboratory, new probes can be synthesized whenever necessary, avoiding the complexities of the biological systems necessary to produce antibodies.

Nucleic acid probes are also very sensitive. With current technology it is possible to detect less than 1 µg of nucleic acid per sample. This translates into about 10^6 bacterial cells or virus particles. Although probes used in this fashion are not as sensitive as direct culture (where as few as 1–10 cells per sample can be detected), probe methods can be useful in situations where culture of the organism is difficult or even impossible.

PCR and Nucleic Acid Probes

Perhaps the most widespread recent advance in clinical diagnostic technology has come from the application of sequence-specific probes for polymerase chain reaction (PCR) amplification of DNA or ribonucleic acid

(RNA) from specific pathogens. As we discussed in Section 10.9, PCR uses two sequence-specific oligonucleotides to amplify target DNA. A DNA amplification of a millionfold or more increases the probe sensitivity and theoretically makes this procedure capable of detecting DNA from a single bacterial cell. For example, probes for a pathogen might be used to examine DNA derived from suspected infected tissue, even in the absence of an observable, culturable pathogen. These methods are particularly useful for identifying viral and intracellular infections. The presence of the appropriate amplified gene segment (Figure 21.23) confirms the presence of the pathogen. Several of the specific organisms for which either hybridization or PCR methods are in use are listed in Table 21.7. We now consider several specific examples of nucleic acid probes and discuss some applications in more detail.

HIV and Viral Load

One of the most useful applications of the PCR method is the test for *viral load* in individuals infected with HIV. Viral load is the number of HIV RNA strands in the plasma or serum of an HIV-infected person. As we will discuss in Section 23.7, HIV is found in the serum of infected individuals at all times after infection. The amount of HIV present, or the viral load, is an indicator of the progression of the disease, with a high viral load indicating a poor prognosis and a low viral load indicating a good prognosis (Figure 21.24). Viral load is also an excellent indication of the efficacy of anti-HIV drug therapy (Figure 21.24; ∞ Section 20.7 and Section 18.10).

Quantitation of viral load is possible using a commercially available quantitative PCR assay (for example, the widely used Amplicor HIV-1 Monitor Test®, man-

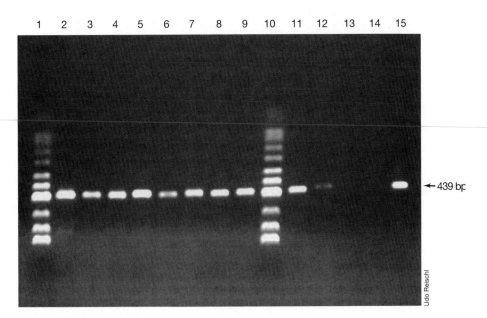

FIGURE 21.23 Polymerase chain reaction (PCR) analysis of patient sputum for *Mycobacterium tuberculosis* in the diagnosis of tuberculosis. Sputum samples from patients were used as a source of DNA. Amplification was initiated with a primer pair, which produced the indicated 439-base pair product when a pure culture of *M. tuberculosis* was used as the DNA source (lane 15). Lanes 2–9, 11 and 12 are from sputums positive for *M. tuberculosis* (lane 12 is a weak positive). Lanes 13 and 14 are from *M. tuberculosis*-negative sputum samples. Lanes 1 and 10 are molecular weight reference markers.

TABLE 21.7	Pathogens that can be identified with nucleic acid probes including PCR methods

Pathogen	Diseases
Bacteria	
Campylobacter spp.	Food infections
Chlamydia trachomatis	Venereal syndromes; trachoma
Enterococcus spp.	Nosocomial infections
Escherichia coli (enteropathogenic strains)	Gastrointestinal disease
Haemophilus influenzae	Infectious meningitis
Legionella pneumophila	Pneumonia
Listeria monocytogenes	Listeriosis
Mycobacterium avium	Tuberculosis
Mycobacterium tuberculosis	Tuberculosis
Mycoplasma hominis	Urinary tract infection; pelvic inflammatory disease
Mycoplasma pneumoniae	Pneumonia
Neisseria gonorrhoeae	Gonorrhea
Neisseria meningitidis	Neurological disease
Rickettsia spp.	Typhus, hemmorhagic fever, etc.
Salmonella spp.	Gastrointestinal disease
Shigella spp.	Gastrointestinal disease
Staphylococcus aureus	Purulent discharges (boils, blisters, pus-forming skin infections)
Streptococcus pyogenes	Scarlet, rheumatic fever; "strep throat"
Streptococcus pneumoniae	Pneumonia
Treponema pallidum	Syphilis
Fungi	
Blastomyces dermatitidis	Blastomycosis
Candida spp.	Candidiasis, thrush
Coccidioides immitis	Coccidioidomycosis
Histoplasma capsulatum	Histoplasmosis
Viruses	
Cytomegalovirus	Congenital viral infections
Epstein-Barr virus	Burkitt's lymphoma; mononucleosis
Hepatitis viruses A, B, C, D, E	Hepatitis
Herpes virus (types I and II)	Cold sores; genital herpes
Human immunodeficiency virus (HIV)	Acquired immunodeficiency syndrome (AIDS)
Human papilloma virus	Genital warts; cervical cancer
Influenza	Respiratory disease
Polyoma virus	Neurological disease
Rotavirus	Gastrointestinal disease
Protozoa	
Leishmania donovani	Leishmaniasis
Plasmodium spp.	Malaria
Pneumocystis carinii	Pneumonia
Trichomonas vaginalis	Trichomoniasis
Trypanosoma spp.	Trypanosomiasis

ufactured by Roche Molecular Systems, Inc.). The method employed is known as RT-PCR (reverse-transcription PCR) and is used to make an amplifiable DNA template from RNA viral genomes (∞ Section 10.9). First, plasma samples are treated to lyse HIV and release the RNA viral genome. The RNA, after precipitation with isopropanol, is reverse transcribed with reverse transcriptase enzyme to make a DNA copy. The DNA copy then serves as the template in a PCR reaction, with primers directed to a portion of the HIV *gag* gene (∞ Section 8.22), yielding a 155 bp *amplicon* (am-

plified product) in HIV-positive samples. Using a standard DNA template of known quantity, the HIV signal present in the sample can be compared to the DNA standard, and the amount of HIV can be quantitated. By comparing viral load over time, a relatively accurate prognosis can be made for each patient (Figure 21.24).

The most important use of this technique is to monitor antiviral chemotherapy. HIV, after drug treatment, can be suppressed to virus levels below the limits of the test (less than 500 copies per milliliter of blood), as indicated in Figure 21.25. However, the few remaining

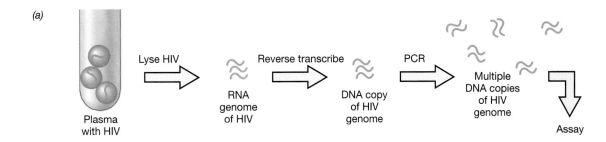

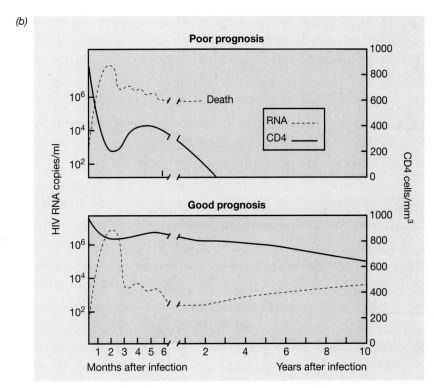

FIGURE 21.24 Monitoring of HIV load. (a) Detection of HIV via the RT-PCR test (Reverse Transcription-Polymerase Chain Reaction). The HIV copies obtained are compared quantitatively with DNA copies from a control template that is amplified in the same PCR amplification. HIV load is expressed as the number of HIV copies per milliliter of patient plasma. For a discussion of reverse transcription and the PCR technique, see Sections 8.22 and 10.9, respectively. (b) Time course for HIV infection as monitored by HIV RT-PCR. Progression of infection is estimated based on viral load at successive times after infection. CD4 T cell counts are measured in cells per cubic millimeter (see Figure 21.17). In the upper panel, a viral load of greater than 10^4 copies per milliliter correlates with below normal CD4 cell numbers (normal = 600–1500/mm³), indicating a poor prognosis and early death of the patient. In the lower panel, a viral load of less than 10^4 copies per milliliter correlates with normal CD4 cell numbers, indicating a good prognosis and extended survival of the patient. This test is used to monitor the course of HIV infection and is particularly useful for tracking the efficacy of drug treatment protocols. Data are adapted from the Centers for Disease Control and Prevention, Atlanta, GA, USA.

HIV particles may acquire mutations that result in drug resistance and resurgence of the infection, resulting once again in a high viral load. The HIV monitor system is useful for identifying the emergence of drug-resistant mutant HIV strains, necessitating a change in the chemotherapy. Several other systems based on amplification techniques are also used to monitor HIV infections, and work is progressing to routinely allow detection of HIV in the range of 0–500 copies per milliliter of blood, a necessity for supporting any claims that HIV cure is possible using drug therapies.

Other Clinical Laboratory Probes

In most clinical probe assays, colonies from plates or pieces of infected tissue are treated with strong alkali, usually NaOH, to lyse the cells and partially denature the DNA, forming single-stranded molecules (Figure 21.22). This mixture is then affixed to a filter or left in solution (for dipstick assays, see later), and the labeled probe added. Hybridization is allowed to occur at a temperature at which sequence homology between target DNA and probe DNA is necessary to form a stable duplex (the actual temperature used in a given probe assay is governed by the length and nucleic acid composition of the probe and target DNA). Following a wash to remove unhybridized probe DNA, the extent of hybridization is measured using the reporter molecule attached to the probe. Depending on how the probe was labeled, this involves measurement of radioactivity, enzyme activity, or fluorescence.

Nucleic acid probes have been marketed for the identification of several major microbial pathogens and are in widespread use for the detection of *Neisseria gon-*

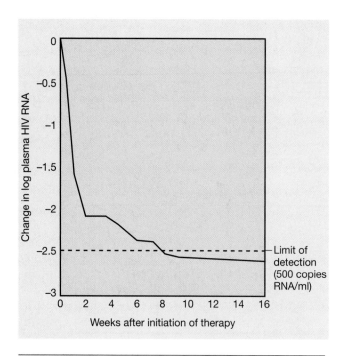

FIGURE 21.25 Rate of change in plasma HIV RNA copy numbers after initiation of antiretroviral drug therapy. The RT-PCR system was used to monitor viral load over a 16-week period. The data are shown as the change in the $\log_{10}$ copies of HIV RNA and indicates a steady, rapid decrease to levels below the detection limits of the test (500 copies/milliliter of blood). Data were provided by the Centers for Disease Control and Prevention, Atlanta, GA, USA.

orrhoeae and *Chlamydia trachomatis* (see Table 21.7 and ⌘ Section 23.6). However, in addition to their clinical usefulness, probes are finding widespread application in food industries and in food regulatory agencies. Probe detection systems can be used to monitor foods for their content of important pathogens such as *Salmonella* and *Staphylococcus*. In probe assays of food, an enrichment period is usually employed to allow low numbers of cells in the food to multiply to a sufficient number to be detectable by the probe. However, the use of PCR gene amplification techniques eliminates the need for the enrichment period.

Probes designed for use in the food industry employ probe dipsticks to remove hybridized DNA from solution. Two component probes are used here, one serving as a *reporter probe* and the other as a *capture probe* (Figure 21.22). Following hybridization of the reporter or capture probe to target DNA, the dipstick, which contains a sequence complementary to the capture probe (usually poly dT to capture poly dA on the probe) (see Figure 21.22b) is inserted into the hybridization solution, and it traps hybridized DNA for removal and measurement.

Probes for detecting certain cancer viruses are also being developed. For example, a probe is now available to detect DNA sequences unique to human papilloma viruses. These viruses sometimes cause skin and cervical cancer in humans (⌘ Section 8.14), and a specific group of papilloma viruses causes genital warts (⌘ Section 23.6). In women, an increased incidence of genital papilloma virus infection is associated with an increased risk of cervical cancer. The DNA probe developed for papilloma viruses can be used to search by hybridization for papilloma virus sequences in tissues removed during a cervical exam. Early detection and treatment of papilloma virus infections decreases the risk of cervical cancer.

The development of new nucleic acid probes is a major activity in pharmaceutical, biotechnology, and clinical diagnostic companies. Although developing probes for diseases for which no probe yet exists is a top priority, a major goal of probe research and development is to make existing probes even more specific and to continue to simplify the procedures necessary for implementing probe-based assays in the clinical setting.

Recent advances in our understanding of bacterial phylogenetics based on 16S ribosomal RNA (rRNA) sequencing (⌘ Section 12.7) now allow new and more specific nucleic acid probes to be constructed. For example, *ribotyping* is a method based on DNA probes that recognize conserved RNA operon genes (⌘ Section 12.9 and Figure 12.20). Ribotyping is essentially a Southern blot analysis in which strains are characterized for restriction fragment length polymorphisms (RFLPs; ⌘ DNA Fingerprinting, Chapter 10) of their individual ribosomal genes. Within a species, and particularly within a strain, the DNA sequences and restriction digest patterns of genes encoding ribosomal RNAs are highly conserved and are a molecular fingerprint for that organism. Since *all* organisms have ribosomal genes, this technique is universally applicable and is finding wide acceptance as a clinical and phylogenetic tool both for the identification of *species* of organisms and for differentiation between and among *strains* within a species.

✓ 21.10 Concept Check

Nucleic acid hybridization is a powerful laboratory tool used for identification of microorganisms. A nucleic acid sequence specific for the microorganism of interest must be available in order to design a probe. Perhaps the most widespread use of probe-based technology is in the application of the gene amplification (PCR) methods. Various probe-based methodologies are currently used in clinical, food, and environmental laboratories.

✓ What advantage does nucleic acid hybridization have over standard culture methods for identification of microorganisms? What disadvantages?

✓ Cite an example where information about a microorganism can be obtained with a nucleic acid probe instead of with standard growth-dependent assays.

21.11

Diagnostic Virology

Because of their unique characteristics as cellular parasites, identification and diagnosis of pathogenic viruses pose significantly different problems from the bacterial pathogens on which we have focused (∞ Chapter 8). For example, viruses cannot be directly cultured on artificial media, but they can be grown in host cells. Thus, viral growth-dependent assays are extremely complex. However, immunodiagnostic methods (see Sections 21.4–21.9) and nucleic acid hybridization methods (see Section 21.10) are widely used for viral identification. Finally, electron microscopy is often used for direct examination of viral specimens. In this section, we will concentrate on specific methods useful for viral identification, especially for the pathogenic viruses we will discuss in Chapter 23 and Chapter 24.

Laboratory cultivation of viruses from clinical materials is more difficult, time-consuming, and specialized than the cultivation of most bacterial pathogens. This is because viruses grow only in living cells. We discussed the use of cell cultures for the growth of viruses in Section 8.3, and such cultures are commonly used in diagnostic virology. A common cell line is a human diploid fibroblast culture called WI-38, which grows rapidly and reproducibly in cell culture medium. Another cell line sometimes used is HeLa, a cell culture derived initially from a human cancer. This cell line has been maintained *in vitro* for so many successive transfers that it has greatly changed its character (for instance, it is no longer diploid). In addition to these and other cell lines, which can be maintained in the laboratory indefinitely, cultures are also made from Rhesus monkey kidneys. Monkey kidney cell lines are called *primary* because they are not maintained by successive transfer in the laboratory. Primary monkey kidney cells support growth of a number of pathogenic viruses and are therefore of value in initial isolation of unknown viruses, but are routinely used only in specialized laboratories. Because of the technical expertise and expense involved (for example, laboratories that do primary culture must maintain or purchase Rhesus monkeys), most diagnostic virology laboratories are located at specialized government facilities or major clinical research institutions.

Although the best diagnostic technique for most viral infections is isolation, growth, and identification of the virus, this is not practical in most clinical settings. Instead, several immunological tests and nucleic acid probes for viruses have been developed. Most of the immunological tests are either direct enzyme-linked immunosorbent assays (ELISAs) that detect viral particles (see Section 21.8 and Figure 21.18) or fluorescent antibody methods in which antibodies made against viral antigens are used to detect cells containing viruses (see Section 21.7 and Figure 21.14*b*). Agglutination tests are also available in which antiviral antibodies conjugated to latex beads or activated charcoal (see Section 21.5) are used to test for agglutination of viral antigens released from lysed tissue samples. Nucleic acid probes and PCR assays for detection and identification of clinically significant viruses are also becoming a major diagnostic tool in clinical virology (see Section 21.10).

Electron Microscopy

In addition to the tools described previously, diagnostic virology can be done by electron microscopy. Because many viruses have distinctive morphologies (∞ Chapter 8), their presence in clinical specimens can often be detected by observing the sample under the electron microscope (Figure 21.26). In most specimens the virus particles must first be concentrated and separated from human tissues, and a variety of techniques, generally employing centrifugation and filtration, are used to obtain a sample enriched in virus particles. Although not as reliable as immunological or nucleic acid probe methods, the observation of virus particles of a specific morphology in a particular type of human tissue is presumptive evidence for a disease. Viral antibodies against particular viruses can be used to increase the sensitivity and specificity of this method; such antibodies may cause the viral particles to agglutinate, and this makes them easier to distinguish from cellular debris under the electron microscope. Viruses can also be seen

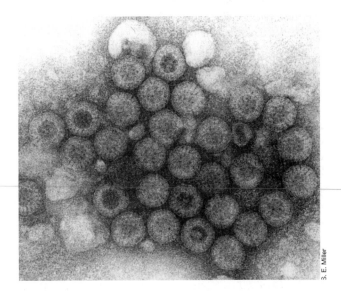

FIGURE 21.26 Electron microscopic observation of clinical specimens to detect viruses. Human rotavirus from a fecal sample. The distinct spherical nature of the virus, coupled with the source, is a highly diagnostic criterion. Each rotavirus particle is approximately 75 nm in diameter.

TABLE 21.8	Some laboratory procedures used in diagnostic virology[a]		
Condition	**Possible viral cause**	**Sample source**	**Inoculation procedure**
Upper respiratory infection	Rhinovirus Coronavirus Adenovirus	Nasopharyngeal or tracheal fluid (aspirate)	Human fibroblast culture
Pneumonia	Influenza	Nasopharyngeal fluid or swab	Human fibroblast cultures or embryonated eggs
Measles	Measles virus	Nasopharyngeal fluid or swab	Monkey kidney cells
Vesicular rash	Herpes simplex	Vesicular fluid by aspiration	Human fibroblast culture

Diarrhea	Rotavirus (infants)
	Norwalk agent (adults)
		Feces or rectal swab	Look for characteristic
			    virus particles with the
			    electron microscope
			    (Figure 21.26)

Nonbacterial meningitis	Enterovirus Mumps Herpes simplex	Spinal fluid	Human fibroblast or monkey kidney cultures

a Immunological methods and nucleic acid probe methods are also widely used in the diagnosis of viral infections (see Sections 21.4–21.11).

following treatment of specimens with antiviral antibodies containing conjugated heavy metals (see Section 21.6). With the use of negative staining techniques (Figure 21.26), results from electron microscopic analysis can be available 20 minutes after collection of the specimen.

A summary of some laboratory procedures used in diagnostic virology is given in Table 21.8. Most of these procedures are used only under special circumstances. For routine virus infections, diagnoses are made by assessing symptoms or by immunological or other indirect means. For example, testing for human immunodeficiency virus (HIV) infection involves the ELISA test and the immunoblot test. Both these methods detect the presence of *antibodies* to HIV, not HIV itself; the *viral load* test discussed in Section 21.10 is used to directly detect the presence of HIV.

✓ 21.11 Concept Check

Virus culture can be accomplished only in susceptible tissue or organs. Therefore, most diagnostic techniques for viral identification are not growth-dependent but routinely rely on immunoassays and nucleic acid hybridization techniques. Electron microscopy techniques are useful for direct observation of virus in host samples.

✓ Why must viruses be grown in tissue or organ culture and not on artificial, inert media?

✓ How can individual pathogenic viruses be identified?

21.12

Safety in the Clinical Laboratory

By their very nature, clinical laboratories are areas in which potentially dangerous biological specimens must be handled on a routine basis. Hence, a defined proto-

col for handling clinical samples must be established to avoid laboratory accidents. In the United States, every clinical and research institution that deals with human or primate tissue is required by law to have an occupational exposure control plan in place for the handling of all bloodborne pathogens. This law was specifically designed to protect workers from infection by hepatitis B virus (HBV) (Section 24.12) and human immunodeficiency virus (HIV) (Section 23.7) but effectively protects workers from infection by virtually all pathogens because of the stringent precautions.

Studies of laboratory-associated infections have indicated that most such infections do not result from known exposures or accidents but instead from routine handling of patient specimens. The two most common causes of laboratory accidents are ignorance and carelessness. Infectious aerosols, generated during processing of the specimen, are the most likely cause of laboratory infections. In attempts to minimize the exposure of clinicians to infectious agents and to thereby reduce the number of nonaccident-associated laboratory infections, well-run clinical laboratories stress the safety rules outlined here. These rules, if applied stringently, ensure the prevention of pathogen spread and meet the legal requirements of the United States law.

1. Laboratories handling hazardous materials must restrict access to laboratory and support personnel. These individuals must have knowledge of the biological risks involved in the laboratory and act accordingly.
2. Effective procedures for decontaminating infectious materials or wastes, including specimens, syringes and needles, inoculated media, bacterial cultures, tissue cultures, experimental animals, glassware, instruments, and surfaces must be in place and be practiced without compromise. A 5.25% (full

strength) chlorine bleach solution or other approved disinfectant is recommended for decontaminating spilled infectious material. All potentially infectious waste must be burned in a certified incinerator or handled by a licensed waste handler.

3. Personnel working with hazardous infectious agents or vaccines (for example, rabies, polio, or diphtheria-pertussis-tetanus vaccines) must be properly vaccinated against the agent. Persons working with human or primate tissue must be vaccinated against HBV.

4. All clinical specimens should be considered potentially infectious and handled in the appropriate manner. This is especially important for preventing laboratory-acquired hepatitis because of the relative frequency with which hepatitis viruses are present in clinical specimens.

5. All pipetting must be done with automatic pipetting devices (not by mouth), and devices such as syringes, needles, and clinical centrifuges must always be used with proper biological containment equipment.

6. Animals should be handled only by trained laboratory personnel, and anesthetics and/or tranquilizers should be used to avoid injury to both personnel and animals.

7. Laboratory personnel must wear laboratory coats or gowns, sealed shoes, rubber gloves, masks, eye protection, respiratory devices when needed, and other barrier protection as deemed appropriate by the level of exposure and the severity of the potential infection. These barrier devices must also be properly stored and decontaminated after use. Laboratory personnel must also practice good personal hygiene with respect to hand washing. Eating and drinking, applying cosmetics or lip balm, or wearing contact lenses is never permitted in the clinical laboratory.

8. Because of the special risks associated with AIDS, all clinical (human) specimens should be treated as if they contain HIV (which they might). Latex or vinyl gloves should be worn whenever handling specimens of *any* kind. Masks and/or full-face shields must be worn any time there is a possibility of generating an aerosol during specimen preparation. Needles must not be resheathed, bent, or broken; they should be placed in a labeled container designated expressly for this purpose that can be sealed and autoclaved before disposal.

These safety rules should be the norm for all clinical laboratories. Specialized clinical laboratories may have additional rules to ensure a safe work environment. For example, if laboratory personnel handle extremely hazardous airborne pathogens (such as the causative agent of tuberculosis, *Mycobacterium tuberculosis*) on a routine basis, the laboratory should be fitted with special features, such as negatively pressurized rooms, biological safety cabinets (∞ Figure 18.4), and air filters, to prevent accidental release of the pathogen from the laboratory. In the final analysis, however, it is the attitude of the personnel that makes the laboratory a safe or an unsafe place to work. Any clinical laboratory is a potentially hazardous place for untrained personnel or those unwilling to take the necessary steps to prevent laboratory-acquired infection.

✓ 21.12 Concept Check

Safety in the clinical laboratory requires effective training, planning, and care to prevent the infection of laboratory workers with pathogens. Materials such as inoculated culture media, needles, and patient specimens require specific precautions for safe handling.

✓ What are the major precautions necessary to prevent spread of a bloodborne pathogen to laboratory personnel?
✓ What are the major causes of laboratory infections?

REVIEW QUESTIONS

1. Describe the standard procedure for obtaining and culturing a blood sample for bacteria.

2. Why is the *number* of bacterial cells in urine, rather than simply the *presence* of bacteria in urine, of significance? What organism is responsible for most urinary tract infections? Why?

3. Describe the procedures used for culturing anaerobic microorganisms. Why is it important to process all clinical specimens quickly? What special procedures and precautions are necessary for the isolation and culture of anaerobes?

4. Differentiate between *selective* and *differential* media. Is eosin–methylene blue (EMB) agar selective or differential? How and why is it used in a clinical laboratory?

5. Describe the Kirby–Bauer test for antibiotic sensitivity. Why should potential pathogens from patient isolates be tested by this method?

6. Why does the antibody titer rise after infection? Why is it necessary to draw two serum samples to monitor infections? Is a high antibody titer indicative of an ongoing infection? Why or why not?

7. How are fluorescent antibodies used for the diagnosis of viral diseases? What advantages do they have over unlabeled antibodies?

8. Agglutination tests are significantly more sensitive than precipitation tests. Why is this the case?

9. Likewise, radioimmunoassay (RIA) and enzyme-linked immunosorbent assay (ELISA) tests are extremely sensitive, as compared to agglutination. Why?

10. What advantages do *monoclonal* antibodies have over *polyclonal* antibody preparations, especially with regard to standardization of antibody preparations?

11. Why is the immunoblot (Western blot) procedure used to confirm positive human immunodeficiency virus (HIV)-ELISA results?

12. What information is essential for the design of a pathogen-specific nucleotide probe? Where can one obtain such information? Is this information available for all pathogens?

13. What is a primary cell line? Why do some animal viruses grow in primary cell lines but not in cell lines such as HeLa cells?

14. How are most laboratory-associated infections contracted? What action can be taken to prevent laboratory infections?

APPLICATION QUESTIONS

1. From a blood culture, you obtain a culture positive for *Staphylococcus aureus*. Interpret and explain the results. Is it likely that the patient has a *S. aureus* bacteremia? Why or why not?

2. Describe the microscopic and cultural evidence that would support a diagnosis of gonorrhea. Why is Thayer–Martin agar a "better" medium than chocolate agar for the isolation of *Neisseria gonorrhoeae?*

3. Compare the changes in color due to pH-sensitive dyes in tests for carbohydrate fermentation and citrate utilization. Is the same dye used in both tests? Why or why not?

4. Why should it *not* be a common medical practice to treat an infectious disease with antibiotics before isolating the suspected pathogen? What further steps should be taken before antibiotic therapy is initiated? Why are these steps seldom taken outside a hospital environment?

5. What are the advantages of rapid identification systems such as agglutination tests as compared with growth-dependent clinical diagnostic procedures? Also, discuss the potential disadvantages of the rapid, non–culture-based tests.

6. Design a fluorescent antibody assay for confirming an initial diagnosis of "strep throat" (*Streptococcus pyo-*

genes is the causative agent of strep throat). Discuss all aspects of the assay, including preparation of antisera, necessary controls, and clinical interpretation.

7. Design an ELISA test for detecting hepatitis A virus in fecal samples. Likewise, design an *indirect* hepatitis A virus test for detection of exposure. Would either test require anti-human IgG antibodies? Why or why not?

8. What are the major advantages of using DNA probes in diagnostic microbiology? Discuss at least four aspects of probe technology that benefit clinical medicine. Where can you find information to design polymerase chain reaction (PCR) assay probes for the hepatitis A virus in Question 7? Remember, the probes must be sequence-specific for the virus.

9. PCR tests are generally considered very sensitive, but are often considered less specific than antibody-based tests. Why might this be the case?

10. Discuss the importance of *viral load* in assessing the treatment and progression of HIV infection.

11. As a professional in a clinical laboratory, you are assigned the task of formalizing the laboratory safety requirements to prevent infectious diseases. Explain how you would monitor and enforce the recommendations outlined in Section 21.12.

Infectious disease epidemics are characterized by a very high incidence of a particular disease in a particular geographic location. As the illustration shows for a model epidemic, the disease incidence has increased at several sites, but still has not spread over the entire map. In some cases, new epidemics result from new pathogens, but many established pathogens periodically reemerge and cause new epidemics. An example of a new epidemic due to a reemerging pathogen is the yearly influenza epidemic. Recent epidemics due to new emerging pathogens are Ebola virus and human immunodeficiency virus (HIV). In all probability, infectious disease epidemics can never be completely suppressed. Infectious disease surveillance and response capabilities must be enhanced if we are to control infectious diseases in a world with overcrowding and rapid international shipping and travel.

CHAPTER 22   # Epidemiology and Public Health Microbiology

22.1   The Science of Epidemiology   892
22.2   Epidemiological Terminology   893
22.3   Disease Reservoirs   895
22.4   Epidemiology of AIDS: An Example of How Epidemiological Research Is Done   898
22.5   Infectious Disease Transmission   900
22.6   The Host Community   902
22.7   Hospital-Acquired (Nosocomial) Infections   906
22.8   Public Health Measures for the Control of Disease   908
22.9   Global Health Considerations   911
22.10   Emerging and Reemerging Infectious Diseases   913

WORKING GLOSSARY

Acute short-term infection usually characterized by dramatic onset and rapid recovery

Carrier subclinically infected individuals who may spread a disease

Chronic long-term infection

Common-source epidemic an epidemic resulting from infection of a large number of people from a single contaminated source

Emerging infections infectious diseases whose incidence has increased in the past 20 years or whose incidence threatens to increase in the near future

Endemic disease constantly present, usually in low numbers

Epidemic the occurrence of a disease in unusually high numbers in a localized region

Epidemiology the study of the occurrence, distribution, and control of diseases

Fomites inanimate objects that, when contaminated with a viable pathogen, can transfer the pathogen to a host

Herd immunity resistance of a group to a pathogen as a result of the immunity of a large portion of the group

Host-to-host epidemic an epidemic resulting from person-to-person contact, characterized by a gradual rise and fall in numbers of cases

Incidence the number of cases of disease in a population

Morbidity incidence of illness in a population

Mortality incidence of death in a population

Nosocomial infection hospital-acquired infection

Outbreak the occurrence of a large number of cases of a disease in a short period of time

Pandemic a worldwide epidemic

Prevalence the proportion or percentage of individuals in the population having a disease

Public health the health of the population as a whole

Quarantine the practice of restricting the movement of individuals with highly contagious serious infections to prevent spread of the disease

Reemergent infections infectious diseases, thought to be under control, that produce new epidemics

Reservoir sites in which viable infectious agents remain and from which infection of individuals may occur

Surveillance observation, recognition, and reporting of diseases as they occur

Vector a living agent that transfers a pathogen (note alternative usage in Chapter 10)

Vehicle nonliving source of pathogens that infect large numbers of individuals; common vehicles are food and water

Zoonosis a disease that occurs primarily in animals but can be transmitted to humans

---

In Chapters 19 and 20 we considered the general principles of infectious disease and how the host responds to microbial infection. In Chapter 21, we discussed the methods for diagnosis of infectious diseases. In this chapter, we consider how a pathogen spreads from an infected individual to others in a population. Thus, we are dealing here with *public health*. In the next two chapters we consider the diseases themselves. The principles put forward in this chapter are vital for understanding the spread and control of infectious disease.

One measure of our success in the control of infectious disease was shown by the data presented in Figure 1.18, which compared the present causes of death in the United States with those at the beginning of the twentieth century. Many microbial diseases are no longer perceived to be the threat to public health they once were in developed countries. Hoever, we will discuss a number of infectious diseases that are emerging as important public health problems, even in developed countries. In developing countries, which include about 75% of the world's population, infectious diseases are still a major problem. Worldwide, infectious diseases account for nearly one-third of the total of 52 million annual estimated deaths. Table 22.1 shows the most prevalent causes of death. Note that, for many of these diseases (for example, measles, whooping cough, and

even tuberculosis), effective vaccines are manufactured (∞ Section 20.16) but are sometimes not available in developing countries. Clearly, infectious diseases remain an important, but approachable, public health problem throughout the world. The current acquired immunodeficiency syndrome (AIDS) epidemic, which has spread worldwide in 20 years or less, is only one example of the devastating consequences of a new infectious disease in a global theater. Eradication or even effective control of infectious diseases must involve scientific, economic, political, and educational solutions, and ultimately, global cooperation.

22.1

The Science of Epidemiology

The most visible aspect of microbial disease is the actual diseased individual. However, individuals do not live alone, and when we consider infectious diseases in populations, some new factors arise. The study of the occurrence, distribution, and control of disease in populations is the field of **epidemiology.**

A pathogen must be able to grow and reproduce to maintain its existence. For this reason, an important aspect of the epidemiology of any disease is a considera-

TABLE 22.1	Infectious diseases: The leading human killers	
Cause of death	Estimated yearly deaths[a]	Infectious agents
Acute lower respiratory infections	3,700,000	Bacteria, viruses, protozoa, fungi
Tuberculosis	2,900,000	Bacteria
Diarrheal diseases	2,500,000	Bacteria, viruses
Acquired immunodeficiency syndrome (AIDS)	2,300,000	Virus
Malaria	1,500,000–2,700,000	Protozoa
Hepatitis (all types)	1,000,000–2,000,000	Viruses
Measles	220,000	Virus
Meningitis, bacterial	200,000	Bacteria
Schistosomiasis	200,000	Parasitic worm
Pertussis (whooping cough)	100,000	Bacterium
Amebiasis	40,000–100,000	Protozoa
Hookworm	50,000–60,000	Parasitic worm
Rabies	35,000	Virus
Yellow fever	30,000	Virus
African trypanosomiasis (sleeping sickness)	20,000 or more	Protozoan

a Data represent estimates for yearly worldwide total deaths each year from each disease. There are approximately 52 million deaths per year, worldwide, from all causes. About 19 million deaths are caused by infectious disease each year, nearly all in developing countries.

Source: World Health Organization.

tion of the natural history of the pathogen. In many cases the pathogen cannot grow outside the host, and if the host dies, the pathogen will also die. Pathogens that kill the host before they are transmitted to a new host would thus become extinct. How, then, do pathogens continue to exist if they kill the host? Actually, a well-adapted parasite lives in synchrony with its host, taking only what it needs for existence and causing only a minimum of harm. However, serious host damage often occurs when new varieties of pathogens arise for which the host has not developed resistance, or when the resistance of the host changes because of the factors discussed in Chapter 19. Thus, pathogens act as selective forces in the evolution of the host, just as hosts act as selective forces in the evolution of pathogens. When equilibrium between host and pathogen exists, both host and pathogen survive in a stable relationship. However, in cases where the pathogen is not dependent on the host for survival, the pathogen can cause devastating disease without affecting its normal reservoir. Organisms in the genus *Clostridium,* for example, ubiquitous inhabitants of the soil (∞ Section 13.19), occasionally accidentally infect humans, causing such devastating diseases as tetanus and botulism (∞ Section 19.8 and Section 24.11).

The epidemiologist traces the spread of a disease to identify its origin and mode of transmission. Epidemiologic data are obtained from clinical studies, disease-reporting surveys, insurance questionnaires, and interviews with patients to define common factors that constitute a disease. The science of epidemiology has been referred to as "medical ecology" because the study of a disease in populations is really a study of a disease in its natural environment. This is in contrast to the clin-ical or laboratory study of disease, where the focus is on treating the individual patient. Knowledge of both the clinical aspects and ecological aspects of a given disease are important if public health measures to control diseases are to be effective.

✓ 22.1 Concept Check

Epidemiology is the study of disease in populations. For infectious disease, the epidemiologist develops methods for the control of infectious disease by defining the interactions of the pathogen in the host population.

✓ How does an epidemiologist differ from a microbiologist?
✓ Why do epidemiologists acquire data for infectious diseases within a population?

22.2

Epidemiological Terminology

A number of terms having specific meanings are used by the epidemiologist to describe patterns of disease. The **prevalence** of a disease in a population is defined as the proportion (or percentage) of diseased individuals in a population at any one time. The **incidence** of a disease is the *number* of diseased individuals in a population. A disease is said to be **epidemic** when it occurs in an unusually high number of individuals in a population at the same time; a **pandemic** is a widely distributed epidemic (Figure 22.1). By contrast, an **endemic** disease is one that is constantly present, usually at low incidence, in a population. In an endemic disease, the pathogen may not be highly virulent, or the majority of the individuals may be immune, re-

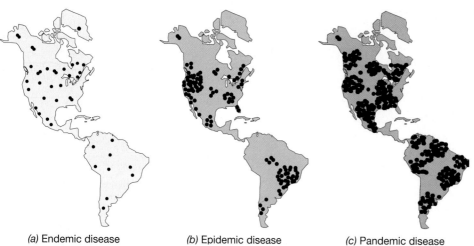

(a) Endemic disease   (b) Epidemic disease   (c) Pandemic disease

FIGURE 22.1 Classification of disease by incidence. Each dot represents several cases of a particular disease.

sulting in low disease incidence. However, as long as an endemic situation lasts, a few infected individuals remain who serve as reservoirs of infection.

Sporadic cases of a disease occur when individual cases are recorded in geographically separated areas, implying that the incidents are not related. A disease **outbreak,** on the other hand, occurs when a number of cases are observed, usually in a relatively short period of time, in an area previously experiencing only sporadic cases of the disease. Finally, diseased individuals who show no or only mild symptoms have *subclinical infections.* Subclinically infected individuals are frequently identified as **carriers** of a particular disease, because even though they themselves show few or perhaps no symptoms, they may still be actively carrying and shedding the pathogenic agent (see Section 22.3).

Mortality and Morbidity

The incidence and prevalence of disease is determined from statistics of illness and death. From these data, a picture of the public health in a population can be obtained. The population might be the total global population of humans or the population of a localized region of a country or district. Public health concerns vary with location and time; thus, assessment of public health at a given moment provides only a snapshot of the situation. By continuing to examine health statistics over many years, it is possible to assess the value of various public health policies that may influence the incidence of disease.

Mortality expresses the incidence of *death* in the population. Infectious diseases were the major causes of death in 1900 in developed countries (➔ Figure 1.15), whereas currently they are of much less significance; noninfectious diseases such as heart disease and cancer are of greater importance. However, the current

situation could rapidly change if a breakdown in public health measures were to occur and, in fact, does not mirror the worldwide situation. In developing countries, infectious diseases are still the major killers (Table 22.1 and Section 22.9).

Morbidity refers to the incidence of *disease* in populations and includes both fatal and nonfatal diseases. Clearly, morbidity statistics define the health of the population more precisely than mortality statistics because many diseases that affect health in important ways have relatively low mortality. The major causes of illness are quite different from the major causes of death. Major illnesses include acute respiratory diseases (the common cold, for instance) and acute digestive system conditions, which are generally due to infectious agents and seldom cause death.

Disease Progression

In terms of clinical symptoms, the course of a typical disease can be divided into stages:

1. *Infection:* the organism begins to grow in the host.
2. *Incubation period:* the time between infection and the appearance of disease symptoms. Some diseases, like influenza (➔ Section 23.4), have short incubation periods, measured in days; others, like AIDS, have longer ones, measured in years (➔ Section 23.7). The incubation period for a given disease is determined by inoculum size, virulence of the pathogen, resistance of the host, and distance of the site of entrance from the focus of infection (➔ Sections 19.6 and 19.7). At the end of incubation, the first symptoms, such as headache and a feeling of illness, appear.
3. *Acute period:* the disease is at its height, with overt symptoms such as fever and chills.
4. *Decline period:* disease symptoms are subsiding, the temperature falls, usually following a period of in-

tense sweating, and a feeling of well-being develops. The decline may be rapid (within 1 day), in which case it is said to occur by *crisis,* or it may be slower, extending over several days, in which case it is said to be by *lysis.*

5. *Convalescent period:* the patient regains strength and returns to normal.

During the later stages of the infection cycle, the immune mechanisms of the host become increasingly important, and in most cases complete recovery from the disease requires and results in active immunity.

✓ 22.2 Concept Check

An endemic disease is constantly present in low numbers in a specific population. In epidemics, an unusually high incidence of disease occurs in a specific population. Infectious diseases cause morbidity (illness) and may cause mortality (death). An infectious disease follows a predictable clinical pattern in the host.

- ✓ Distinguish between an endemic disease, an epidemic disease, and a pandemic disease.
- ✓ Distinguish between morbidity and mortality. How might high host morbidity be advantageous for the pathogen?

22.3

Disease Reservoirs

Reservoirs are sites in which viable infectious agents remain alive and from which infection of individuals may occur. Reservoirs may be either animate or inanimate. Table 22.2 lists some common human diseases and their reservoirs. Some pathogens are primarily saprophytic (living on dead matter) and only incidentally infect and cause disease. For example, *Clostridium tetani* (the causal agent of tetanus) normally inhabits the soil. Infection of animals by this organism is an accidental event; infection of a host is not essential for its continued existence and even if there were no susceptible hosts, *C. tetani* would still survive in nature.

However, many pathogens use other living organisms as their only reservoirs. In these cases, the reservoir is an essential component of the natural life cycle of the infectious agent. Some infections occur only in humans, and maintenance of the cycle involves person-to-person transmission. This type of pathogen cycle is common for viral and bacterial respiratory diseases, sexually transmitted diseases, staphylococcal and streptococcal infections, diphtheria, typhoid fever, and mumps. As we shall see, pathogens that live their entire life cycle dependent on a single host can be eradicated (see Section 22.8).

Zoonosis

A number of infectious diseases that occur in humans also occur in animals. A disease that occurs primarily in animals but is occasionally transmitted to humans is called a **zoonosis.** Because public health measures for animal populations are much less developed than for humans, the infection rate for many diseases is much higher in animals, and animal-to-animal transmission is the rule. However, occasionally transmission is from animal to human. It is less likely for transmission to also occur from person to person in such diseases. Thus, maintenance of the pathogen in nature depends on animal-to-animal transfer. However, control of a zoonosis in the human population in no way eliminates it as a public health problem. Control of the human disease can generally be achieved only through elimination of the disease in the animal reservoir. Considerable success has been achieved in the control of two diseases that were often transferred to humans from domestic animals, bovine tuberculosis and brucellosis. Control was achieved primarily by identifying and destroying infected animals. Pasteurization of milk was also of considerable importance in the prevention of the spread of bovine tuberculosis to humans because milk was the main vehicle of transmission.

Certain infectious diseases, particularly those caused by organisms such as protozoa, have more complex cycles, involving an obligate transfer from animal to human to animal (for example, malaria, ∞ Section 24.5). In such cases, the disease may potentially be controlled in either humans or the alternate animal host.

Carriers

A carrier is an infected individual with no obvious signs of clinical disease. Carriers are potential sources of infection for others and are critically important for the spread of disease. Carriers may be individuals in the incubation period of the disease, in which case the carrier state precedes the development of actual symptoms. These individuals are prime sources of respiratory infections because they are not yet aware of their infection and so are not taking any precautions against infecting others. Such persons are **acute carriers** because the carrier state lasts for only a short time. On the other hand, **chronic carriers** may remain infected and carry disease for extended periods of time. Chronic carriers usually appear perfectly healthy. They may be individuals who have recovered from a clinical disease, but still harbor viable pathogens, or they may be individuals with inapparent infections.

Carriers can be identified in populations by using a variety of diagnostic techniques, such as mass culture surveys (∞ Section 21.1) or mass serological (anti-

TABLE 22.2	Epidemic diseases: Agents, sources, reservoirs, and control			
Disease	**Causative agent**[a]	**Infection sources**	**Reservoirs**	**Control measures**
Common-source epidemics[b]				
Anthrax	*Bacillus anthracis* (B)	Milk or meat from infected animals	Cattle, swine, goats, sheep, horses	Destruction of infected animals
Bacillary dysentery	*Shigella dysenteriae* (B)	Fecal contamination of food and water	Humans	Detection and control of carriers; oversight of food handlers; decontamination of water supplies
Botulism	*Clostridium botulinum* (B)	Soil-contaminated food	Soil	Proper preservation of food
Brucellosis	*Brucella melitensis* (B)	Milk or meat from infected animals	Cattle, swine, goats, sheep, horses	Pasteurization of milk; control of infection in animals
Cholera	*Vibrio cholerae* (B)	Fecal contamination of food and water	Humans	Decontamination of public water sources; immunization
E. coli O157:H7 food infection	*Escherichia coli* O157:H7 (B)	Fecal contamination of food and water	Humans, cattle	Decontamination of public water sources; oversight of food handlers; pasteurization of beverages
Giardiasis	*Giardia* spp. (P)	Fecal contamination of water	Wild mammals	Decontamination of public water sources
Hepatitis	Hepatitis A, B, C, D, E (V)	Infected humans	Humans	Decontamination of contaminated fluids and fomites, immunization if available (A, B, and C only)
Legionnaire's disease	*Legionella pneumophila* (B)	Contaminated water	High-moisture environments	Decontamination of air-conditioning cooling towers, etc.
Paratyphoid	*Salmonella paratyphi* (B)	Fecal contamination of food and water	Humans	Decontamination of public water sources; oversight of food handlers; immunization
Typhoid fever	*Salmonella typhi* (B)	Fecal contamination of food and water	Humans	Decontamination of public water sources; oversight of food handlers; pasteurization of milk; immunization
Host-to-host epidemics *Respiratory diseases*				
Diphtheria	*Corynebacterium diphtheriae* (B)	Human cases and carriers; infected food and fomites	Humans	Immunization; quarantine of infected individuals
Hantavirus pulmonary syndrome	Hantavirus (V)	Inhalation of contaminated fecal material	Rodents	Control rodent population and exposure
Hemorrhagic fever	Ebola virus (V)	Infected body fluids	Unknown	Quarantine of active cases
Meningicoccal meningitis	*Neisseria meningitidis* (B)	Human cases and carriers	Humans	Exposure treated with sulfadiazine for susceptible strains
Pneumococcal pneumonia	*Streptococcus pneumoniae* (B)	Human carriers	Humans	Antibiotic treatment; isolation of cases for period of communicability
Tuberculosis	*Mycobacterium tuberculosis* (B)	Sputum from human cases; contaminated milk	Humans, cattle	Treatment with isoniazid; pasteurization of milk
Whooping cough	*Bordetella pertussis* (B)	Human cases	Humans	Immunization; case isolation

TABLE 22.2 (continued)

Disease	Causative agent[a]	Infection sources	Reservoirs	Control measures
German measles	Rubella virus (V)	Human cases	Humans	Immunization; avoid contact between infected individuals and pregnant women
Influenza	Influenza virus (V)	Human cases	Humans, animals	Immunization
Measles	Measles virus (V)	Human cases	Humans	Immunization
Sexually transmitted diseases[c]				
Acquired immuno-deficiency syndrome (AIDS)	Human immuno-deficiency virus (HIV)	Infected body fluids, especially blood and semen	Humans	Treatment with reverse transcriptase inhibitors, protease inhibitors (not curative)(◯◯ Section 23.7).
Chlamydia	*Chlamydia trachomatis* (B)	Urethral, vaginal, and anal secretions	Humans	Testing for organism during routine pelvic examinations; chemotherapy of carriers and potential contacts; case tracing and treatment
Gonorrhea	*Neisseria gonorrhoeae* (B)	Urethral and vaginal secretions	Humans	Chemotherapy of carriers and potential contacts; case tracing and treatment
Syphilis	*Treponema pallidum* (B)	Infected exudate or blood	Humans	Identification by serological tests; antibiotic treatment of seropositive individuals
Trichomoniasis	*Trichomonas vaginalis* (P)	Urethral, vaginal, prostate secretions	Humans	Chemotherapy of infected individuals and contacts
Vectorborne diseases				
Epidemic typhus	*Rickettsia prowazekii* (B)	Bite by infected louse	Humans, lice	Control louse population
Lyme disease	*Borrelia burgdorferi* (B)	Bite from infected tick	Rodents, deer, ticks	Avoid tick exposure; treat infected individuals with antibiotics
Malaria	*Plasmodium* spp. (P)	Bite from *Anopheles* mosquito	Humans, mosquito	Control mosquito population; treat infected humans with antimalarial drugs
Plague	*Yersinia pestis* (B)	Bite by flea	Wild rodents	Control rodent populations, immunization
Rocky Mountain spotted fever	*Rickettsia rickettsii* (B)	Bite by infected tick	Ticks, rabbits, mice	Avoid tick exposure; treat infected individuals with antibiotics
Direct-contact diseases				
Psittacosis	*Chlamydia psittaci* (B)	Contact with birds or bird excrement	Wild and domestic birds	Avoid contact with birds; treat infected individuals with antibiotics
Rabies	Rabies virus (V)	Bite by carnivores	Wild and domestic carnivores	Avoid animal bites; immunization of animal handlers and exposed individuals
Tularemia	*Franciscella tularensis* (B)	Contact with rabbits	Rabbits	Avoid contact with rabbits; treat infected individuals with antibiotics

a B, Bacteria; V, virus; P, protozoan.

b Some common-source diseases can also be spread from host to host.

c Sexually transmitted diseases can also be controlled by effective use of condoms and by sexual abstinence.

body) surveys (◯◯ Section 21.4). Skin tests using bacterial antigens to test for delayed hypersensitivity (and thus exposure and previous or current infection) are also widely used to define the carrier state.

Diseases in which carriers are important for the spread of infection include hepatitis (◯◯ Section 24.12), tuberculosis (◯◯ Section 23.3), and typhoid fever (see the box, The Tragic Case of Typhoid Mary). Surveys of food handlers and health care workers are sometimes made to identify and remove from public contact those individuals who are common sources of infection.

LEARNING FROM THE PAST...   The Tragic Case of Typhoid Mary

The classic example of a chronic carrier was the woman known as "Typhoid Mary," a cook in New York City and Long Island in the early part of the twentieth century. Typhoid Mary (her real name was Mary Mallon) was employed in a number of households and institutions, and as a cook she was in a central position to infect large numbers of people. Extensive epidemiological investigation of a number of typhoid outbreaks by Dr. George Soper revealed that Mary was the likely source of contamination. When her feces were examined bacteriologically, she had very high numbers of the typhoid bacterium, *Salmonella typhi*. She remained a carrier for many years, probably because her gallbladder was infected, and organisms were continuously being excreted from there into her intestine. Public health authorities offered to remove her gallbladder, but she refused the operation, and to prevent her from continuing to serve as a source of infection, she was imprisoned. After almost 3 years in prison, she was released on the pledge that she would not cook or handle food for others and that she was to report to the health department every 3 months. She promptly disappeared, changed her name, and cooked in hotels, restaurants, and sanitariums, leaving behind a wake of typhoid fever. After 5 years she was captured as a result of the investigation of an epidemic at a New York hospital. She was again arrested and imprisoned and remained in custody on North Brother Island in the East River of New York City for 23 years. She died in 1938, 32 years after epidemiologists had first discovered she was a chronic typhoid carrier. ■

✓ 22.3 Concept Check

To understand how diseases develop, the pathogen reservoir must be known. Some pathogens exist in soil, water, or animals. Other pathogens exist only in humans and are maintained solely by person-to-person contact. An understanding of disease carriers and pathogen life cycles is critical for controlling disease.

- ✓ What is a disease reservoir?
- ✓ Distinguish between *acute* and *chronic* carriers.

22.4

Epidemiology of AIDS: An Example of How Epidemiological Research Is Done

Cases of acquired immunodeficiency syndrome (AIDS), the virus-mediated infectious disease that severely cripples the body's immune system (∞ Section 23.7), were first reported in the United States in 1981. Approximately 680,000 AIDS cases have been diagnosed since that time (Figure 22.2). The very high numbers for 1992–1993 are due to a change in the definition of AIDS (∞ A Definition of AIDS, Chapter 23) but *at least 40,000 new cases of AIDS will be diagnosed every year* in the United States for the foreseeable future. *Worldwide, 30 to 40 million individuals are now infected with human immunodeficiency virus (HIV)* (∞ Section 23.7). *Over 2 million people now die each year from AIDS* (Table 22.1). A vast majority of the new infections and deaths will occur in developing countries.

Tracking an Epidemic

Initial case studies suggested an unusually high AIDS prevalence among homosexual men and intravenous drug abusers. This in turn strongly implicated a trans-

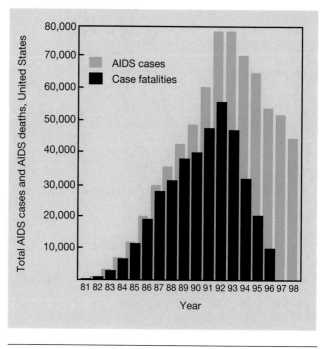

FIGURE 22.2   Total cases of acquired immunodeficiency syndrome (AIDS), by year, in the United States since 1981. The numbers of deaths in AIDS patients are also shown, by the year of the original diagnosis. Cumulatively, there have been approximately 680,000 cases of AIDS and 400,000 deaths since 1981. Data for fatalities in 1997 and 1998 are incomplete.

missible agent, presumably transferred during sexual activity or by contaminated needles. Individuals receiving blood or blood products were also at high risk: Hemophiliacs who required infusions of blood products and a small number of individuals who received blood transfusions or tissue transplants before 1982 acquired AIDS (today less than 1% of the total current

AIDS cases are attributable to these modes of transmission) (Figure 22.3). The connection between transfer of AIDS with blood or tissue further reinforced the case for an infectious transmissible agent.

Soon after the discovery of HIV, laboratory tests were developed to detect antibodies to the virus in serum (∞ Sections 21.8 and 21.9). This made possible more extensive surveys of the incidence of HIV infection in different populations and also served as a screening method to ensure that new cases of AIDS were not transmitted by blood transfusions. Such tests revealed a fourth group of individuals at high risk for AIDS— the children of mothers who are themselves at high risk for AIDS. Since the beginning of the AIDS epidemic in the United States, there have been over 8,000 cases of pediatric AIDS. In over 90% of these cases, the moth-

ers' behavior (Figure 22.3b) was the only identifiable risk factor, again supporting the idea that AIDS is an infectious disease.

The pattern illustrated in Figure 22.3 is typical of an agent transmissible by sexual activity or by blood, and the association of AIDS cases with *specific* groups was an important epidemiological finding. The identification of certain well-defined high risk groups implied that AIDS was *not* transmitted from person to person by casual contact, such as the respiratory route, or by contaminated food or water. Instead, epidemiological findings pointed clearly to body fluids, primarily blood and semen, as the major vehicles for transmission of HIV.

In the United States, the number of AIDS cases have been disproportionately high in homosexual men (Figure 22.3a), but the patterns in women and in certain

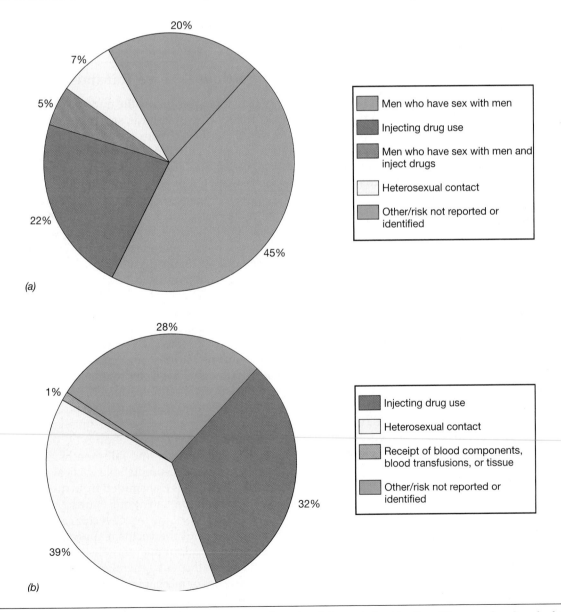

FIGURE 22.3   Distribution of AIDS cases by risk group and sex in adolescents (greater than 13 years of age) and adults, 1997. The total number of cases reported was 60,161. (a) AIDS in men, N = 47,056. (b) AIDS in women, N = 13,105.

racial and ethnic minorities indicate that homosexuality is not the predisposing factor for acquiring AIDS. For example, among women the heterosexual contact group is now the largest risk group (Figure 22.3*b*), while in African-American and Hispanic men, intravenous drug use is nearly as important as homosexual activity as a risk factor. If we consider all risk groups, **heterosexual activity is the single fastest growing category for new AIDS cases among all adults,** regardless of sex or ethnic/racial background. This confusing array of individuals who are at high risk for acquiring AIDS prompts us to look for a set of factors common to all, and virtually all individuals who are at risk for acquiring HIV share two specific behavior patterns. First, they engage in sex or drug use that involves **transfer of body fluids,** usually semen or blood. Second, individuals who acquire HIV often exchange body fluids with **multiple partners,** either through sexual activity or through needle-sharing drug activity (or both). Thus, they increase their probability of exchanging body fluids with an HIV-infected individual and, therefore, their chance of acquiring HIV infection and AIDS.

The incidence of AIDS in hemophiliac and blood transfusion recipients has been nearly eliminated in recent years (Figure 22.3). This is due not only to screening of the blood supply but also because many blood clotting factors needed by hemophiliacs can withstand a heat treatment sufficient to inactivate HIV. Pediatric AIDS cases are still a major concern. In 1997, there were 473 new cases among this group. HIV can be transmitted to the fetus by infected mothers and probably also in mother's milk. All infants born to HIV-infected mothers have maternally derived antibodies to HIV in their blood, but a clear diagnosis of HIV infection in infants must wait a year or more after birth because about 70% of infants showing maternal HIV antibodies at birth do not go on to develop HIV infection.

Epidemiological studies of AIDS in Africa, where the disease is thought to have originated, have clearly shown that transmission of AIDS is not linked to particular sexual practices, such as homosexuality, but instead to person-to-person transfer of HIV-infected fluids. In Africa heterosexual transmission of AIDS seems the norm, with about equal numbers of men and women infected. Unfortunately, reliable global statistics on AIDS are not available because of differences in reporting in various countries. Current estimates are about 10 million AIDS cases worldwide with up to 40 million infected people who do not (yet) show clinical symptoms; that is, they are carriers. The identification of high risk groups through epidemiological studies led to the development of health education campaigns to inform the public of how AIDS is transmitted and what activities constitute high risk behavior (∞ Sexual Activity and AIDS, Chapter 23). Because no cure or effec-

tive immunization for AIDS is yet available, public health education offers the most effective approach to the control of AIDS and is the major weapon for preventing the spread of infection. We discuss the pathology and therapy of AIDS in Section 23.7.

✓ 22.4 Concept Check

AIDS is one of the newest and most studied pandemics. AIDS will continue to be a major public health problem, especially in developing countries. There is no effective cure or immunization to prevent AIDS, although we now know a great deal about its pathology and spread.

✓ Describe the major risk factors for acquiring AIDS.
✓ Estimate the total number of individuals in the United States who now have AIDS and predict how many will be living with AIDS in the year 2005.

22.5

Infectious Disease Transmission

Epidemiologists follow the transmission of a disease by correlating geographical, seasonal, and age group incidence of a disease with possible modes of transmission. A disease limited to a restricted geographical location may suggest a particular vector; malaria, for example, is transmitted by a mosquito found mainly in tropical regions. A marked seasonality to a disease is often indicative of certain modes of transmission, such as in the case of chickenpox, where the number of cases jumps sharply when children enter school and come in close contact (∞ Figure 23.20).

Different pathogens have different modes of transmission, which are usually related to the habitats of the organisms in the body. For instance, respiratory pathogens are generally airborne, whereas intestinal pathogens are spread by food or water. If the pathogen is to survive, it must undergo transmission from one host to another. Even environmental factors may play a role in survival of the pathogen, and such variables as weather patterns may influence exposure to a pathogen. For example, California encephalitis, caused by a group of Bunyaviruses (∞ Section 8.14) occurs primarily during the summer and fall months, and disappears every winter, in a predictable cyclical pattern (Figure 22.4). The virus is transmitted from mosquitoes which, of course, are prevalent only during warmer months. Disappearance of the insect vector cause the disease to disappear until the vector reappears and retransmits the virus in the summer months.

Thus pathogens generally are associated with specific features or mechanisms that permit or ensure transmittal. Transmission involves three stages: (1) escape from the host, (2) travel, and (3) entry into a new host.

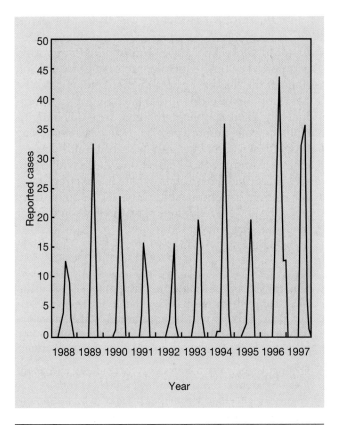

FIGURE 22.4 The incidence of California encephalitis, a viral disease transmissible through mosquitoes, in the United States, by month of onset. Note the sharp rise in cases in the summer, followed by a complete decline in winter. The disease cycle follows the yearly cycle of the mosquito vector prevalence. Data are from the Centers for Disease Control and Prevention, Atlanta, GA, USA.

We give here a brief overview of transmission mechanisms. Several of these will be discussed in detail for certain diseases in the next chapter.

Direct Host-to-Host Transmission

Host-to-host transmission occurs whenever an infected host transmits the disease to a susceptible host. Transmission by the respiratory route and by direct contact is very common. Transmission by infectious droplets is the most frequent means by which upper respiratory infections such as the common cold and influenza are propagated. However, some pathogens are so sensitive to environmental influences that they are unable to survive for significant periods of time away from the host and must be transmitted from host to host by direct contact. The best examples of pathogens transmitted in this way are those responsible for sexually transmitted diseases, such as *Treponema pallidum* (syphilis) and *Neisseria gonorrhoeae* (gonorrhea). These agents are extremely sensitive to drying and do not survive away from the body, even for a few moments. Intimate person-to-person contact, such as kissing or sexual intercourse, provides a direct means for the transmission of such pathogens. However, such transfer can occur only if the viable pathogen is present on the transmitting person at the body site that comes in direct contact with that of the recipient. Thus, pathogens causing sexually transmitted diseases live in genitalia, the mouth, or the anus because these are the sites involved in sexual contact.

Direct contact is also involved in the transmittal of skin pathogens, such as staphylococci (boils and pimples) and fungi (ringworm). However, these pathogens are relatively resistant to environmental influences such as drying, and intimate person-to-person contact is not the only means of transmission. Many respiratory pathogens are also transmitted by direct means because they are spread by droplets resulting from sneezing or coughing. However, many of these droplets do not remain airborne for long. Transmission, therefore, requires close, although not necessarily intimate, person-to-person contact.

Indirect Host-to-Host Transmission

Indirect transmission can occur by either living or inanimate means. Living agents transmitting pathogens are called **vectors;** they are generally arthropods (for example, insects, mites, ticks, or fleas) or vertebrates (for example, dogs, cats, or rodents). Arthropod vectors may not be hosts for the disease but simply carry the agent from one host to another. Large numbers of arthropods obtain nourishment by biting, and if the pathogen is present in the blood, the arthropod vector may ingest the pathogen and transmit it when biting another individual. In some cases, the pathogen actually replicates in the arthropod, which is then considered an alternate host. Such replication leads to an increase in pathogen numbers, increasing the probability that a subsequent bite will lead to infection.

Inanimate agents such as bedding, toys, books, and surgical instruments can also transmit disease. These inanimate objects are collectively referred to as **fomites.** Food and water are referred to as disease **vehicles.** Fomites can also be disease vehicles, but major epidemics originating from a single source are usually traced to food or water because these are actively consumed in large amounts by a number of individuals in a population.

Epidemics

Two major types of epidemics can be distinguished: common-source and host-to-host. These two types are contrasted in Figure 22.5. A **common-source epidemic** arises as the result of infection (or intoxication) of a large number of people from a contaminated common source,

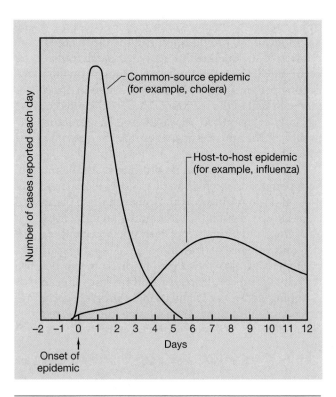

FIGURE 22.5 Origins of epidemics. The shape of the epidemic curve helps to distinguish the likely origin. In a common-source epidemic, such as from contaminated food or water, the curve is characterized by a sharp rise to a peak, with a rapid decline, which is less abrupt than the rise. Cases continue to be reported for a period approximately equal to the duration of one incubation period of the disease. In a host-to-host epidemic, the curve is characterized by a relatively slow, progressive rise, and cases continue to be reported over a period equivalent to several incubation periods of the disease.

such as food or water. Usually such contamination occurs because of a malfunction in the sanitation of a central distribution system. Foodborne and waterborne diseases are primarily *intestinal* diseases; the pathogen leaves the body in fecal material, contaminates food or water via improper sanitary procedures, and then enters the intestinal tract of the recipient during ingestion. Because foodborne and waterborne diseases are some of those that are most amenable to control by public health measures, we shall discuss them in some detail in Chapter 23 (also see the boxes, Snow on Cholera, and The Tragic Case of Typhoid Mary in this chapter). The disease incidence for a common-source outbreak is characterized by a rapid rise to a peak because a large number of individuals ingest contaminated food or water and become ill within a relatively brief period of time (Figure 22.5). The common-source illness also declines rapidly, although the decline is less rapid than the rise. Cases continue to be reported for a period of time approximately equal to the duration of one incubation period of the disease.

In a **host-to-host epidemic,** the disease incidence shows a relatively slow, progressive rise (Figure 22.5) and a gradual decline. Cases continue to be reported over a period of time equivalent to several incubation periods of the disease. A host-to-host epidemic can be initiated by the introduction of a single infected individual into a susceptible population, with this individual infecting one or more people in the population. The pathogen then replicates in susceptible individuals, reaches a communicable stage, and is transferred to other susceptible individuals, where it again replicates and becomes communicable. Table 22.2 summarizes some of the key epidemiological features of some major epidemic diseases observed today.

✓ 22.5 Concept Check

A pathogen can be transmitted directly from one host to another, or indirectly by means of another living agent called a vector. Pathogens can also be transmitted by inanimate objects (fomites) and common vehicles such as food and water. Epidemics may be of common-source or host-to-host origin.

✓ Distinguish between a *common-source* epidemic and a *host-to-host* epidemic. Cite at least one example of each.

✓ Suggest one method for halting the spread of a common-source epidemic and a host-to-host epidemic.

22.6

The Host Community

The colonization of a susceptible, unimmunized host by a parasite may first lead to an explosive infection and an epidemic. As the host population develops resistance, however, the spread of the parasite is checked, and eventually a *balance* is reached in which host and parasite are in equilibrium. In an extreme case, failure to reach equilibrium could result in death and eventual extinction of the host species. If the pathogen has no other host, then the extinction of the host could also result in extinction of the pathogen. Thus, the evolutionary success of a pathogen is best measured by its ability to establish a balanced equilibrium with the host, rather than by a pathogen's ability to destroy the host. In effect, the host and parasite are affecting each other's evolution; that is, the host and parasite are *coevolving*.

Coevolution of a Host and a Parasite

An excellent experimental example of coevolution of host and parasite occurred when a virus was intentionally introduced for purposes of population control in the wild rabbits of Australia.

Wild rabbits were introduced into Australia from Europe in 1859 and quickly spread until they were over-

LEARNING FROM THE PAST... Snow on Cholera

The importance of drinking water as a vehicle for the spread of cholera was first shown in 1855 by British physician John Snow, who at that time had no knowledge of the bacterial causation of the disease. Snow's study is one of the great classics of epidemiology and serves as a model for how a careful study can lead to clear and meaningful conclusions.

In London, the water supplies to different parts of the city were from different sources and were transmitted in different ways. In a large area south of the Thames River, across the river from Westminster Abbey and the Parliament Building, the water was supplied to houses by two competing private water companies, the Southwark and Vauxhall Company, and the Lambeth Company. It was the water of the former company that was the major vehicle for the transmission of cholera. When Snow began to suspect the water supply of the Southwark and Vauxhall Company, he made a careful survey of the residence of every cholera death in this district and determined which company supplied the water to that residence. In some parts of the area served by these two companies, each had a monopoly, but in a fairly large area the two companies competed directly, each having run independent water pipes along the various streets. Houses had the option of connecting with either supply, and the distribution of houses between the two companies was random. The clear-cut results of Snow's survey were completely convincing, even to those skeptical about the importance of polluted water in the transmission of cholera: In the first seven weeks of the epidemic, there were 315 deaths per 10,000 houses supplied by the Southwark and Vauxhall Company, and only 37 per 10,000 houses supplied by the Lambeth Company. In the rest of London, there were 59 deaths per 10,000 houses, showing that those supplied by the Lambeth Company had fewer deaths than the general population. In the districts where each company had exclusive rights, it could of course be argued that it was not the water, but some other factor (soil, air, general layout of houses, and so on), that might have been responsible for the differences in disease incidence, but in the districts where the two companies competed, all of these other factors were the same, yet the incidence was high for those supplied with Southwark and Vauxhall water and low for those supplied with Lambeth water. Snow attempted to relate these differences in disease incidence to the sources of the waters used by the two companies. Since he suspected that the excrement and evacuations from cholera patients were highly infectious, he considered that sewage contamination of the water supply might exist. In those days, sewage treatment did not exist and raw sewage was dumped directly into the Thames River. The Southwark and Vauxhall Company obtained its water supply from the Thames right in the heart of London, where sewage contamination could occur, while the Lambeth Company obtained its water from a point on the river considerably above the city, and hence was relatively free of pollution. It was this difference in source that accounted for the difference in disease incidence. In Snow's words:

As there is no difference whatever, either in the houses or the people receiving the supply of the two Water Companies, or in any of the physical conditions with which they are surrounded, it is obvious that no experiment could have been devised which would more thoroughly test the effect of water supply on the progress of cholera than this. . . . The experiment, too, was on the grandest scale. No fewer than three hundred thousand people of both sexes, of every age and occupation, and of every rank and station, from gentlefolk down to the very poor, were divided into two groups without their choice, and, in most cases, without their knowledge; one group being supplied with water containing the sewage of London, and, amongst it, whatever might have come from cholera patients, the other group having water quite free from such impurity. ■

running large parts of the continent. Myxoma virus was discovered in South American rabbits, which are a different species from the European rabbits in Australia. In South America the virus and its hosts are apparently in equilibrium, and the virus causes only a minor disease. However, this same virus is extremely virulent in the European rabbit and almost always causes a fatal infection. The virus is spread from rabbit to rabbit by mosquitoes and other biting insects, and is capable of rapid spread in areas with appropriate insect vectors.

Myxoma virus was introduced into Australian rabbits in 1950 to control the rabbit population. Within several months, the virus was well established in the population and spread over an area in Australia as large as all of Western Europe. The disease showed a marked seasonal pattern, rising to a peak in the summer when the mosquito vectors were present and declining in the winter. The epidemiology of myxoma virus was studied as a model of a virus-induced epidemic by Australian scientists. Virus was isolated from wild rabbits, and the isolated strains were characterized for virulence with laboratory rabbits. At the same time, baby rabbits were removed from their dens before infection could occur and reared in the laboratory. Then these wild rabbits were infected with standard virulent strains of myxoma virus to determine their susceptibility. The results of this large-scale model study are shown in Figure 22.6.

During the first year of the epidemic, over 95% of the infected rabbits died. However, within 6 years both the virus and the rabbit population had changed. Over this interval, rabbit mortality dropped to about 84%, and the virus isolated was of decreased virulence. In addition, changes were noted in the resistance of the rabbit. In parts of Australia where the virus was first introduced, the remaining rabbit population was subjected to selective pressure by the virus for several years. As seen in Figure 22.6, within years the resistance of the rabbits had increased dramatically. This resistance was due to innate changes in the rabbit population and not to immunological responses, for the rabbits tested had been removed from their mothers at birth and had never been in contact with the virus. Their resistance was due to some genetic change in the animal that made it less susceptible to myxoma virus.

As a result of the introduction of myxoma virus, the Australian rabbit population was initially controlled, but the genetic changes in virus and host prevented complete eradication of the rabbit from Australia. In time, most of the surviving rabbits acquired the resistance factors, and by the 1980s the rabbit population in

Australia was nearing the premyxomatosis levels, with widespread environmental destruction and pressure on native plants and animals. As a result, Australian authorities, starting in 1995, released another rabbit pathogen, the rabbit hemorrhagic disease virus (RHDV), a member of the Calciviridae, a single-stranded positive RNA virus (Section 8.15). This virus has been released at 80 specific sites via direct inoculation into rabbits, with eventual plans for up to 280 controlled releases. Because RHDV is spread largely by direct contact and kills animals usually within three days of initial infection, authorities believed the infections would be self-limiting, killing all the rabbits in a local population. Thus, they reasoned, the virus–host equilibrium could be monitored and controlled more reliably than the arthropod-borne myxomatosis virus; resistant rabbit populations should not arise and spread easily in the wild. However, although early reports indicate that RHDV was initially very effective at rapidly reducing local rabbit populations, instances of resistant rabbit populations have already been reported, suggesting that the host and pathogen are already coevolving. The myxomatosis virus was initially a very effective population control agent, but did not eliminate the rabbit hosts because of coevolutionary events. It seems that the RHDV release strategy is also working to rapidly reduce the rabbit population, but may also be destined to coevolve with the host, losing its lethality and ability to control the rabbit population.

While coevolution of host and pathogen may be the norm for diseases that rely on host-to-host transmission, for those pathogens that do not rely on host-to-host transmission, as we have already mentioned for *Clostridium* spp. (see Section 22.1), there is no selection for decreased virulence to support mutual coexistence. Vectorborne pathogens that are transmitted by the bite of arthropods or ticks are also under no evolutionary pressure to spare the human host. As long as the vector can obtain its blood meal before the host dies, the pathogen can maintain a high level of virulence, decimating the human host in the process of infection. For example, the malaria parasites *Plasmodium* spp. show antigenic variations in their coat proteins that aid in avoiding the immune response of the host. This genetic trick to avoid the host responses *increases* virulence within that specific host.

Other evidence for the phenomenon of continually increasing pathogen virulence comes from studies of diarrheal diseases in newborns (Section 24.9). In hospital situations, *Escherichia coli* can cause severe diarrheal illness and even death, and virulence seems to increase with each passage of the pathogen through a hospital patient. The *E. coli* organisms replicate in one host and are then transferred through carriers and "vectors" such as hospital attendants or soiled bedding and furniture to another patient. Even if that host dies or

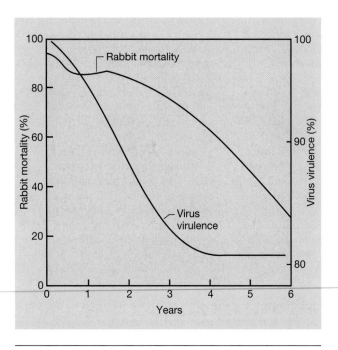

FIGURE 22.6 Changes in virulence of myxoma virus and in susceptibility of the Australian rabbit during the years after the virus was introduced into Australia in 1950. Virus virulence is given as the average mortality in standard laboratory rabbits for virus recovered from the field each year. Rabbit susceptibility was determined by removing young rabbits from their dens and infecting them with a virus strain of moderately high virulence, which killed 90–95% of normal laboratory rabbits.

cannot contact others to transfer the disease, the virulent *E. coli* strain infects others through alternate means of transmission rather than relying on the person-to-person route. Sometimes it takes extraordinary efforts such as completely washing the nursery and furniture with disinfectant and transferring staff to interrupt the cycle of these super-virulent infections.

Herd Immunity

An analysis of the immune state of a group is of great importance in understanding the role of immunity in the development of epidemics. **Herd immunity** is the resistance of a group to infection and spread of a pathogen resulting from immunity of a high proportion of the members of the group. If the proportion of immune individuals is sufficiently great, then the whole population will be protected. The fraction of resistant individuals necessary to prevent an epidemic is higher for a highly virulent agent or one with a long period of infectivity and lower for a mildly virulent agent or one with a brief period of infectivity.

The proportion of the population that must be immune to prevent infection of the rest of the population can be estimated from data on poliovirus immunization in the United States. From epidemiological studies of the incidence of polio in large populations, it appears that if a population is 70% immunized, polio will be essentially absent in the population. The immunized individuals protect the rest of the population because they cannot acquire and pass on the pathogen, thus breaking the cycle of infection (Figure 22.7). For a highly infectious disease such as influenza, the proportion of immune individuals necessary to confer herd immunity is higher, about 90–95%. A value of about 70% has also been estimated for diphtheria, but further study of several small diphtheria outbreaks has shown that in densely settled areas a much higher proportion must be immunized to prevent development of an epidemic. Apparently, in dense populations, person-to-person transmission can occur even if the agent is not highly infectious. In the case of diphtheria, an additional complication arises because immunized persons can still harbor the pathogen (inapparent infection) and thus act as chronic carriers, serving as potential sources of infection.

Cycles of Disease

The concepts of epidemics and herd immunity can also explain why certain diseases occur in *cycles*. A good example of a cyclical disease is chickenpox, which occurs in a high proportion of school children. Because the chickenpox virus (∞ Section 8.19) is transmitted by

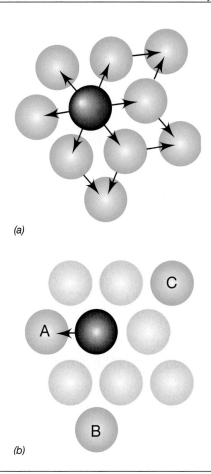

(a)

(b)

FIGURE 22.7   Herd immunity: a model for spread of infection. Immunity in some individuals protects nonimmune individuals from infection. (a) In an unprotected population, an infected individual (black) can successfully infect (arrows) all of the susceptible individuals (blue). Newly infected individuals will, in turn, transfer the disease to other susceptible individuals. (b) For the case of a moderately transmissible pathogen such as *Corynebacterium diphtheriae* (diphtheria) in a population of moderate density, the infected individual cannot transfer the disease to all susceptible individuals because resistant individuals (yellow), immune by virtue of previous exposure or immunization, break the cycle of pathogen transfer. Even if susceptible individual A acquires the disease, the other susceptible individuals, B and C, are protected. For the average pathogen, 70% resistance confers immunity to the entire population. For very easily transmitted pathogens, higher levels of protection are necessary.

the respiratory route, its infectivity is high in crowded situations such as schools. On entry into school at age 5, most children are susceptible, so that on the introduction of virus into the school, an explosive propagated epidemic results. Virtually every individual becomes infected and develops immunity, and as the immune population builds up, the epidemic dies down. Chickenpox shows an annual cycle (∞ Figure 23.20) probably because a new group of nonimmune children arrives each year; the epidemic starts again in the early fall, when children begin school and come into close contact with one another. In such a situation, a single infected child can initiate an epidemic causing disease in nearly all previously unexposed children.

✓ 22.6 Concept Check

Hosts and pathogens coevolve with time and arrive at a steady state that favors the continued survival of both. With herd immunity, a large fraction of a population is immune to a given disease, and it is difficult for the disease to spread. Disease cycles occur when a large, recurring, nonimmune population such as children entering school is exposed to a pathogen.

- ✓ Explain coevolution of host and pathogen. Cite a specific example.
- ✓ How does herd immunity prevent a nonimmune individual from acquiring a disease?

22.7

Hospital-Acquired (Nosocomial) Infections

A hospital may not only be a place where sick people get well but may also be a place where sick people get sicker. Cross-infection from patient to patient or from hospital personnel to patients presents a constant hazard. Hospital infections are called *nosocomial infections* (*nosocomium* is the Latin word for "hospital") and occur in about 5% of all patients admitted. In certain clinical services, such as intensive care units, up to 10% of the patients acquire a nosocomial infection. In all, there are about 2 million nosocomial infections each year in the United States, leading directly or indirectly to 80,000 deaths. Hospital infections are partly due to the prevalence of diseased patients but are more often due to the presence of pathogenic microorganisms that are selected for and maintained within the hospital environment. Even multiple-drug-resistant organisms are often spread from host to host as part of the normal flora. Therefore, virtually all nosocomial pathogens are normal flora in either patients or hospital staff.

The Hospital Environment

Infectious diseases are spread easily and rapidly in hospital environments for several reasons. (1) Many patients have weakened resistance to infectious disease because of their illness (compromised hosts) (∞ Section 19.12). (2) Hospitals treat patients suffering from infectious disease, and these patients may be reservoirs of highly virulent pathogens. (3) The housing of multiple patients in rooms and wards increases the chance of cross-infection. (4) Hospital personnel move from patient to patient, increasing the probability of transfer of pathogens. (5) Many hospital procedures, such as

catheterization, hypodermic injection, spinal puncture, and removal of tissue samples (biopsy) or fluids, breach the skin barrier and carry with them the risk of introducing pathogens to the patient. (6) In maternity wards of hospitals, newborn infants are unusually susceptible to certain kinds of infection because they lack well-developed defense mechanisms. (7) Surgical procedures are a major hazard because internal organs are exposed to sources of contamination and the stress of surgery often diminishes the resistance of the patient to infection (∞ Section 19.12). (8) Certain therapeutic drugs, such as those used for immunosuppression in transplant patients, increase susceptibility to infection. (9) Use of antibiotics to control infection selects for antibiotic-resistant organisms, which may not be easily controlled if they cause further infection (∞ Section 18.12). Figure 22.8 summarizes information concerning the most prevalent hospital-acquired infections.

Hospital Pathogens

Hospital pathogens preferentially infect several sites, notably the urinary tract, blood, wounds, and the respiratory tract. A relatively small number of pathogens cause the majority of nosocomial infections at these sites (Figure 22.8).

One of the most important and widespread hospital pathogens is *Staphylococcus aureus*. It is the most common cause of surgical wound infections and pneumonia and the second most common cause of blood infections. *S. aureus* is also particularly problematic in nurseries. Many strains are unusually virulent and are also resistant to common antibiotics (∞ Section 18.12), making their treatment very difficult. In addition to *S. aureus*, other *Staphylococcus* species are now the largest collective cause of hospital-acquired blood infections and are also very prevalent as the causal agents of wound infections. Most members of the genus *Staphylococcus* are found in the upper respiratory tract or on the skin where they are a part of the normal flora in many individuals, including hospital patients and personnel.

Escherichia coli is the most common cause of urinary tract infections in hospitals, but *Enterococcus* species, *Pseudomonas aeruginosa*, *Candida albicans*, and *Klebsiella pneumoniae* infections are also very common. While *Enterococcus*, *E. coli*, and *K. pneumoniae* are normally found only in the human body, *Candida* and *Pseudomonas* are good examples of *opportunistic pathogens:* They are found relatively ubiquitously in nature, and cause disease only in individuals whose defenses are somehow debilitated (∞ Section 19.12). In addition, *P. aeruginosa* from hospital isolates is often resistant to multiple antibiotics, complicating treatment. *E. coli* has the potential for antibiotic resistance, but is generally sensitive to at least some common antibiotics.

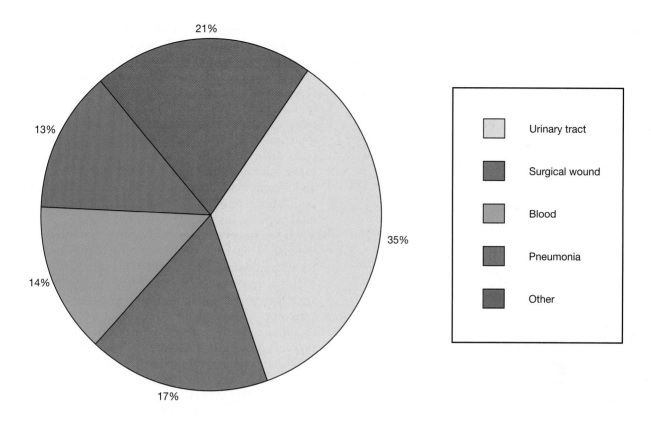

Pathogen	All sites n=101,821	Urinary tract n=35,079	Surgical wound n=17,671	Blood infection n=14,424	Pneumonia n=13,433	Other sites n=21,214
Staphylococcus aureus	13	2	20	16	19	18
Escherichia coli	12	24	8	5	4	4
Staphylococcus spp.	11	4	14	31	2	14
Enterococcus spp.	10	16	12	9	2	5
Pseudomonas aeruginosa	9	11	8	3	17	7
Enterobacter spp.	6	5	7	4	11	4
Candida albicans	5	8	3	5	5	4
Klebsiella pneumoniae	5	8	3	5	8	3
Gram–positive anaerobes	4	0	1	1	0	19
Proteus mirabilis	3	5	3	1	2	2
Streptococcus spp.	2	1	3	3	1	2
Candida spp.	2	3	1	3	1	1
Fungi	2	3	0	1	1	1
All other pathogens*	12	9	13	11	21	12

*1% or less of the total isolates from all sites

FIGURE 22.8   Percentage distribution of nosocomial pathogens, by infection site, January 1990–March 1996.

✓ 22.7 Concept Check

Many common microorganisms have the potential to be pathogens in a hospital environment. Hospital patients are unusually sensitive to infectious disease and are exposed to a variety of potential pathogens, including opportunistic pathogens, in the hospital environment. Treatment of these infections is complicated by antibiotic resistance.

- ✓ Why are hospital patients more susceptible than normal individuals to pathogens?
- ✓ Why is antibiotic resistance a major problem in hospital environments?
- ✓ What is the source of opportunistic pathogens?

22.8

Public Health Measures for the Control of Disease

An understanding of the epidemiology of an infectious disease makes it possible to develop methods for control of the disease. **Public health** refers to the health of the general population and to the activities of public health authorities in the control of disease. However, the incidence of many infectious diseases has dropped dramatically over the past 100 years, especially in developed countries, not because of public health efforts but because of general increases in the well-being of the population. Better nutrition, less crowded living quarters, and lighter work loads have probably done as much as public health measures to control diseases such as tuberculosis, primarily by reducing the risk factors related to disease (∞ Section 19.12). However, diseases such as typhoid fever, diphtheria, brucellosis, and poliomyelitis owe their low incidence to active and specific public health measures.

Overall public health depends on application of control measures to prevent the spread of infectious disease, and we discuss these here.

Controls Directed against the Reservoir

If the disease occurs primarily in *domestic animals*, then infection of humans can be prevented if the disease is eliminated from the infected animal population. Immunization procedures or destruction of infected animals may be used to eliminate the disease in animals. These procedures have been quite effective in eliminating brucellosis and bovine tuberculosis from humans. Recently, these procedures have been used to eliminate bovine spongiform encephalitis in cattle in the United Kingdom. Not incidentally, the health of the domestic animal population is also enhanced, with likely long-term economic benefits to the farmer.

When the reservoir is a *wild animal*, then eradication is much more difficult. **Rabies** is a disease that occurs in both wild and domestic animals but is transmitted to domestic animals primarily by wild animals. Thus *control* of rabies can be achieved by immunization of domestic animals. However, the majority of rabies cases are in wild rather than domestic animals, at least in the United States (∞ Figure 24.1). Therefore, *eradication* of rabies would require the immunization or destruction of all wild animal reservoirs, which includes such diverse species as raccoons, bats, skunks, and foxes. Although oral rabies immunization is practical and recommended for rabies control in limited animal populations, its efficacy is untested in large, diverse animal populations.

If the reservoir is an *insect* (such as a mosquito in the case of malaria), effective control of the disease can be accomplished by eliminating the reservoir with chemical insecticides or other lethal agents. However, such use must be balanced with environmental concerns about the use of toxic or carcinogenic chemicals—in some cases the elimination of one public health problem only creates another. For example, the insecticide dichlorodiphenyltrichloroethane (DDT) (∞ Figure 16.61) is very effective against mosquitoes and is credited with eradicating yellow fever and malaria in North America. However, its use is currently banned in the United States because of environmental concerns. DDT is still widely used in many developing countries to control mosquito-borne diseases.

When *humans* are the reservoir (for example, AIDS), then control and eradication can be much more difficult, especially if there are asymptomatic carriers. On the other hand, if a disease limited to humans has no asymptomatic phase and can be prevented through immunization or treated with chemotherapy, then the reservoir of infected individuals can be limited by treating or quarantining every case of disease and immunizing all contacts, stopping disease transmission and eventually eradicating the disease. Such a strategy was successfully employed by the World Health Organization to eradicate smallpox and is currently being used to target the eradication of polio (see below).

Controls Directed against Transmission of the Pathogen

If the organism is transmitted via food or water, then public health procedures can be instituted either to prevent contamination of these vehicles or to destroy the pathogen in the vehicle. Water purification methods (∞ Section 24.8) have been responsible for dramatic reductions in the incidence of typhoid fever, and the pasteurization of milk has helped in the control of bovine tuberculosis in humans. Food protection laws have been devised that greatly decrease the probability of transmission of a number of enteric pathogens to humans (∞ Sections 24.10 and

24.11). Transmission of respiratory pathogens is much more difficult to prevent. Attempts at chemical disinfection of air have been unsuccessful. However, air filtration is a viable method, but is limited to small enclosed areas (∞ Section 18.3). In Japan, many individuals wear face masks when they have upper respiratory infections to prevent transmission to others, but such methods, although effective, are voluntary, and would be difficult to institute as public health measures.

Immunization

Smallpox, diphtheria, tetanus, pertussis (whooping cough), and poliomyelitis have been controlled primarily by means of immunization. As we discussed in Section 22.6, 100% immunization is not necessary in order to control the disease in a population, although the percentage needed to ensure disease control varies with the virulence of the pathogen and with the condition of the population (for example, crowding).

Over the past several decades, the proportion of children immunized for diphtheria, tetanus, pertussis, measles, mumps, rubella, and polio has been decreasing, apparently because the public has become less fearful of contracting these diseases because of their very low incidence in the population. However, because none of these diseases has been eradicated from the United States (the reservoir of tetanus is the soil, and so it will never be eradicated), and with a decrease in the proportion of individuals immunized, the protection afforded by herd immunity (see Section 22.6) can be overcome, and these infectious diseases could reemerge in epidemic form. For example, Table 22.3 shows the vaccination rate for measles in selected countries in the Americas. In the United States, nearly 30,000 cases of measles were reported in 1990 (∞ Section 23.4), but renewed efforts to increase vaccination levels in preschool children have reversed this alarming trend (∞ Figure 23.19). Presumably, as the percentage of the immunized preschool population was reduced to only 70% in the United States, the benefits of herd immunity disappeared but have now been reestablished as the result of an aggressive immunization program. In countries such as Haiti, measles remains a significant cause of morbidity and mortality because of inadequate vaccination standards.

Many *adults* are inadequately immunized to a variety of infectious agents, either because they received low titer vaccines when they were children or because their immunity has gradually disappeared with age. In the United States, up to 80% of adults may lack effective immunity to important infectious diseases. When these so-called childhood diseases occur in adults, they can have severe effects. If a woman contracts rubella (a viral disease) (∞ Section 23.4) during pregnancy, the unborn child can be affected by serious developmental

TABLE 22.3 Infants immunized against measles in the Americas (1990)[a]

Country	Immunized (%)
Panama	99
Chile	98
Dominican Republic	96
Argentina	94
Cuba	94
Honduras	91
Bahamas	86
Costa Rica	85
Colombia	82
Uruguay	82
Belize	81
Nicaragua	81
Brazil	77
Paraguay	77
El Salvador	75
Jamaica	74
United States	70
Guatemala	68
Mexico	66
Peru	64
Venezuela	64
Ecuador	62
Bolivia	53
Haiti	31

a Data are for children less than 2 years of age.

and neurological disorders. Measles and polio are also much more serious diseases in adults than in children.

All adults are advised to review their immunization status, checking their medical records (if available) to ascertain dates of immunizations. *Tetanus* immunizations, for example, must be renewed at least every 10 years to provide effective immunity. Surveys of adult populations have shown that more than 10% of adults under the age of 40 and over 50% of those over 60 are not adequately immunized. *Measles* immunity in adults also needs to be reviewed. People born before 1957 probably had measles as children and are immune. Those born after 1956 may have been immunized, but the effectiveness of early vaccines was variable and effective immunity may not be present, especially if the immunization was given before 1 year of age. Reimmunization for polio is not recommended for adults unless they are traveling to countries in Africa and Asia where polio may still occur.

Immunization practices and procedures have been discussed in Section 20.16, and those for particular infections will be discussed in Chapters 23 and 24.

Quarantine

Quarantine involves restricting the movement of individuals with active infections to prevent spread of disease to other members of the population. The *time limit* *of quarantine is the longest period of communicability of the given disease.* Quarantine must be done in such a manner that the infected individual cannot contact unexposed individuals. Quarantine is not as severe a measure as strict isolation, which is used for unusually infectious diseases in hospital situations.

By international agreement, six diseases are considered quarantinable: smallpox, cholera, plague, yellow fever, typhoid fever, and relapsing fever. Although smallpox has been eliminated from the world, quarantine for the other five diseases is still mandated. Each of them is considered a highly serious, particularly communicable disease. Spread of certain other highly contagious diseases, such as Ebola hemorrhagic fever and meningitis, may also be controlled using quarantine methods (see Table 22.6 and Section 22.10).

Surveillance

Surveillance is the observation, recognition, and reporting of diseases as they occur. The diseases that are under surveillance in the United States are listed in Table 22.4. Note that several of the epidemic diseases listed in Table 22.2 are not on the surveillance list. Several diseases like influenza are, however, surveyed through regional laboratories that identify *index cases*—those cases that exhibit new syndromes, characteristics, or pathogens indicating high potential for new epidemics.

Pathogen Eradication

Disease eradication can be accomplished in some specific cases and has been successful in the case of smallpox. As we mentioned earlier in this section, smallpox was a disease with a human reservoir consisting of the individuals suffering from acute smallpox infections, and transmission of smallpox was exclusively person-to-person. Infected individuals transmitted the disease through direct contact with previously unexposed members of the population. Although smallpox, a viral disease, cannot be treated, immunization practices were very effective: Vaccination with a related viral strain conferred virtually complete immunity to lethal smallpox infection. In 1967, the World Health Organization (WHO) used this knowledge to develop a plan to eradicate smallpox. Because of successful vaccination programs throughout the developed countries, endemic smallpox was then largely confined to Africa, the Middle East, and the Indian subcontinent. After a preliminary program to vaccinate everyone in these endemic

---

TABLE 22.4	Reportable infectious diseases in the United States
Diseases caused by Bacteria	**Diseases caused by Bacteria**
Anthrax	Toxic shock syndrome
Botulism	Tuberculosis
Brucellosis	Typhoid fever
Chancroid	**Diseases caused by fungi (molds, yeast)**
Chlamydia trachomatis, genital infections	Coccidiomycosis
Cholera	Cryptosporidiosis
Diphtheria	**Diseases caused by viruses**
Escherichia coli O157:H7	Acquired immunodeficiency syndrome
Gonorrhea	(AIDS) and pediatric HIV infection
Haemophilus influenzae, invasive disease	Encephalitis
Hansen's disease (leprosy)	California serogroup
Hemolytic uremic syndrome, postdiarrheal	Eastern equine
Legionellosis	St. Louis
Lyme disease	Western equine
Meningococcal infections	Hantavirus pulmonary syndrome
Pertussis	Hepatitis A, B, C/non A, non B
Plague	Measles
Psittacosis	Mumps
Rocky Mountain spotted fever	Poliomyelitis, paralytic
Salmonellosis	Rabies, animal, human
Shigellosis	Rubella, acute and congenital syndrome
Streptococcal diseases, invasive, Group A	Yellow fever
Streptococcus pneumoniae, drug-resistant invasive disease	**Diseases caused by a protozoan**
	Malaria
Streptococcal toxic shock syndrome	**Diseases caused by a helminth**
Syphilis, acute and congenital	Trichinosis
Tetanus	

areas, every suspected smallpox outbreak was targeted by a team of WHO personnel who traveled to the outbreak site, quarantined individuals with active disease, and vaccinated all contacts and even the contacts of contacts. This aggressive policy resulted in the elimination of the active disease within a decade, and in 1980 the WHO declared the eradication of smallpox.

Polio, another viral disease with a very effective immunization program, is also targeted for eradication (polio is already eradicated from the Western hemisphere). Using much the same strategy to target polio as was used for smallpox, the WHO immunized 420 million individuals in 1996 alone, and projects eradication of the disease by the year 2000. Leprosy, another disease with a human reservoir, is also targeted for eradication, but this is because active cases can now be effectively treated with a multidrug therapy that cures the patient and also prevents spread of *Mycobacterium leprae*, the causal agent (⌒ Section 23.3). Other diseases that are being targeted for eradication are Chagas' disease (treat active cases and destroy the insect vector of this parasitic worm) and dracunculiasis (treat drinking water to prevent transmission of the Guinea worm parasite). Other diseases that are thought to be reasonable candidates for eradication include syphilis (⌒ Section 23.6) and rabies (⌒ Section 24.1).

✓ 22.8 Concept Check

Food and water purity regulations, vector control, immunization, quarantine, disease surveillance, and pathogen eradication are public health measures that play a major role in reduction of disease incidence.

- ✓ Compare public measures for controlling infectious disease caused by insect reservoirs and by human carriers.
- ✓ What public health methods can be used to halt the spread of an epidemic disease once it has begun?

22.9

Global Health Considerations

The United States is typical of countries where public health protection is highly developed. Other countries with similar characteristics include Japan, Australia, New Zealand, Israel, and the western European countries. However, only about one-quarter of the nearly 6 billion people in the world live in these developed countries. In quite another category as far as infectious disease is concerned are the developing countries, a category that includes most of the countries in Africa, Central and South America, and Asia. In these countries, infectious diseases are still major causes of death (Figure 22.9).

Infectious Disease in Developing Countries

There is a sharp contrast in the degree of importance of infectious diseases as causes of death in developing versus developed nations. In developing regions of the world, infectious diseases account for 46% of deaths, whereas infectious diseases account for about 1% of deaths in developed regions (Figure 22.8). Diseases that were leading causes of death in the United States nearly a century ago, such as tuberculosis and gastroenteritis, are still leading causes of death in developing countries today (see Figure 22.9 and ⌒ Figure 1.15). Furthermore, the majority of deaths due to infectious disease in developing regions occur among infants and children. Thus, the average age of individuals dying as a result of infectious diseases in developing versus developed countries is also dramatically different.

The distinct differences in the health status of people in different regions of the world are due in part to general nutritional deficiency in individuals in developing countries and to a lower overall standard of living. As discussed in Section 19.12, factors such as physical stress and diet play important roles in the ability of the host to ward off infection. Thus, it is not surprising that in developing countries death from infection is about 10-fold more likely. In addition, the generally lower levels of public health protection and lack of economic resources for implementing widespread immunization and food and water purity programs in developing countries make infection more likely in the first place. Statistics on disease in developed countries show that control of many diseases is possible. However, statistics on the worldwide incidence of disease show that infectious disease remains an important public health problem.

Travel to Endemic Areas

The high incidence of disease in many parts of the world is also a concern for people traveling to such areas. It is possible to be immunized against many of the diseases that are endemic in foreign countries. Some typical recommendations for immunization for those traveling abroad are shown in Table 22.5. Many foreign countries currently require immunization certificates for yellow fever, but most other immunizations are recommended only for people who are expected to be at high risk. There is also risk, in many parts of the world, for being exposed to a variety of diseases (for example, Ebola hemorrhagic fever, dengue fever, amebiasis, encephalitis, malaria, and typhus) for which there are no effective immunizations. Travelers are advised to take reasonable precautions such as avoiding insect and animal bites, drinking only

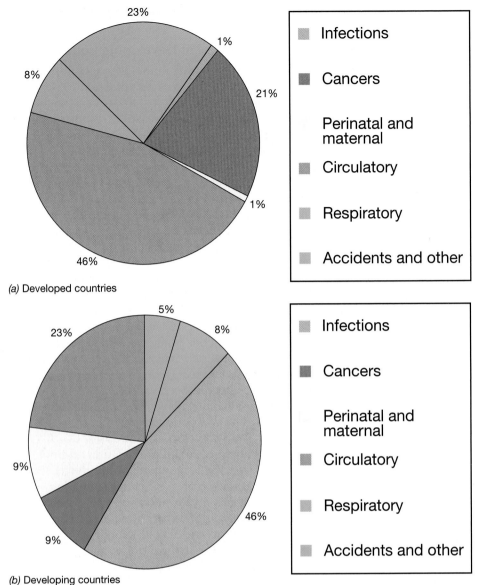

FIGURE 22.9 Leading causes of death in developed and developing countries, by classification of disease. (a) Developed countries. Of about 12 million deaths per year, only about 120,000 deaths are from infectious diseases, nearly all due to pneumonia. (b) Developing countries. Of about 40 million deaths per year, about 19 million deaths are from infectious disease. For the major diseases in this category, see Table 22.1.

(a) Developed countries

(b) Developing countries

TABLE 22.5	Immunizations required or recommended for travel to developing countries[a]	
Disease	**Destination**	**Recommendation**
Cholera	Many central African nations, India, Pakistan, South Korea, Albania, Malta, endemic areas in South America	*Immunization recommended* if entering from or continuing to endemic areas
Yellow fever	Tropical and subtropical countries, worldwide	*Immunization often required* for entry; or if entering from or continuing to endemic areas
Plague	Mostly rural mountainous and upland areas of Africa, Asia, and South America	*Immunization recommended* if direct contact with rodents is anticipated
Infectious hepatitis (A)	Specific tropical areas and many developing countries	*Immunization recommended*
Serum hepatitis (B)	Africa, Indochina, eastern and southern Europe, countries in the former Soviet Union, Central and South America	*Immunization recommended*
Typhoid fever	Many African, Asian, Central and South American countries	*Immunization recommended*

a Current Health Information for International Travelers, U.S. Department of Health and Human Services.

Vaccinations are also recommended for diphtheria, pertussis, tetanus, polio, measles, mumps, and rubella. Most U.S. citizens are already immunized against these diseases through normal immunization practices.

water that has been properly treated and eating food properly stored and prepared, and undergoing antibiotic and chemotherapeutic programs when exposure is suspected.

✓ 22.9 Concept Check

Infectious diseases account for over one-third of all worldwide mortality. Most infectious diseases occur in developing countries. Travelers to endemic disease areas should be immunized when possible and should take appropriate precautions to prevent infection.

- ✓ Contrast mortality due to infectious diseases in developing and developed countries.
- ✓ List a series of infectious diseases for which you have *not* been immunized and with which you could come into contact next year.

22.10

Emerging and Reemerging Infectious Diseases

Infectious diseases are *global* health problems, and the scope and focus of these problems are constantly changing. In this section, we will examine some recent changes in patterns of infectious disease outbreaks, the reasons for the changing patterns, and the methods used by epidemiologists to identify and deal with new threats to public health.

The worldwide distribution of diseases can change dramatically and rapidly. Alterations in the pathogen, the environment, or the host population can contribute to the rapid spread of new diseases, with potential for high morbidity and mortality among infected individuals. We refer to diseases that suddenly become prevalent as *emerging* diseases. Emerging infections are not limited to "new" diseases but also include *reemergence* of diseases thought to be controlled, especially when antibiotics become less effective and public health systems fail. Some of the most recent, dramatic examples of emerging and resurgent disease are shown in Figure 22.10 on a global scale. Some recently occurring emerging and reemerging diseases are described in Table 22.6. In addition, the epidemic diseases listed in Table 22.2 all have the potential to reemerge and cause widespread epidemics and even pandemics in newly susceptible populations.

The phenomenon of suddenly emerging diseases of epidemic proportions is not new. Some of the diseases that suddenly emerged into prominence in the past were syphilis (caused by *Treponema pallidum*) (∞ Section 23.6) and plague (caused by *Yersinia pestis*) (∞ Section 24.6). In the Middle Ages, up to one-third of all living humans were killed by the plague epidemics that swept Europe, Asia, and Africa. More recently, influenza (∞ Section 23.4) became a major public health threat in the early part of the twentieth century. In the 1980s, legionellosis (caused by *Legionella pneumophila*) (∞ Section 24.9), acquired immunodeficiency syn-

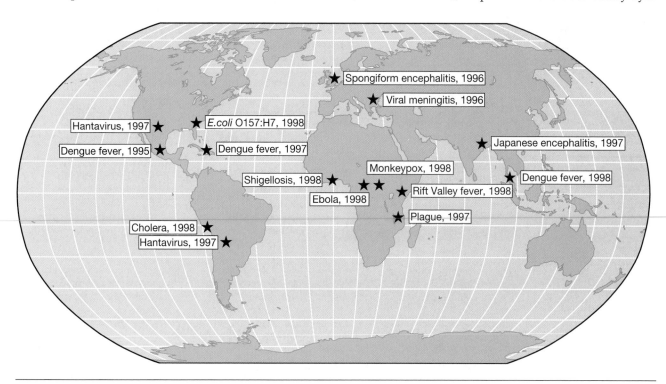

FIGURE 22.10   Recent outbreaks of emerging and reemerging infectious diseases on a global scale.

TABLE 22.6	Some emerging and reemerging infectious diseases		
Agent	**Disease and symptoms**	**Mode of transmission**	**Cause(s) of emergence**
Bacteria, Rickettsias, and Chlamydias			
Borrelia burgdorferi	Lyme disease: rash, fever, neurological and cardiac abnormalities, arthritis	Bite of infective *Ixodes* tick	Increase in deer and human populations in wooded areas
Campylobacter jejuni	Campylobacter enteritis: abdominal pain, diarrhea, fever	Ingestion of contaminated food, water, or milk; fecal-oral spread from infected person or animal	Increased recognition; consumption of undercooked poultry
Chlamydia trachomatis	Trachoma, genital infections, conjunctivitis, infant pneumonia	Sexual intercourse	Increased sexual activity; changes in sanitation
Escherichia coli O157:H7	Hemorrhagic colitis; thrombocytopenia; hemolytic uremic syndrome	Ingestion of contaminated food, especially undercooked beef and raw milk	Development of a new pathogen
Haemophilus influenzae biogroup *aegyptus*	Brazilian purpuric fever: purulent conjunctivitis, fever, vomiting	Discharges of infected persons; flies are suspected vectors	Possible increase in virulence due to mutation
Helicobacter pylori	Gastritis, peptic ulcers, possibly stomach cancer	Contaminated food or water, especially unpasteurized milk; contact with infected pets	Increased recognition
Legionella pneumophila	Legionnaires' disease: malaise, myalgia, fever, headache, respiratory illness	Air-cooling systems, water supplies	Recognition in an epidemic situation
Mycobacterium tuberculosis	Tuberculosis: cough, weight loss, lung lesions; infection can spread to other organ systems	Sputum droplets (exhaled through a cough or sneeze) of a person with active disease	Immunosuppression, immunodeficiency
Neisseria meningitidis	Bacterial meningitis	Person-to-person contact	Urbanization, breakdown/ lack of local public health surveillance
Staphylococcus aureus	Abscesses, pneumonia, endocarditis, toxic shock	Contact with the organism in a purulent lesion or on the hands	Recognition in an epidemic situation; possibly mutation
Streptococcus pyogenes	Scarlet fever, rheumatic fever, toxic shock	Direct contact with infected persons or carriers; ingestion of contaminated foods	Change in virulence of the bacteria; possibly mutation
Vibrio cholerae	Cholera: severe diarrhea, rapid dehydration	Water contaminated with the feces of infected persons; food exposed to contaminated water	Poor sanitation and hygiene; possibly introduced via bilge water from cargo ships
Viruses			
Dengue	Hemorrhagic fever	Bite of an infected mosquito (primarily *Aedes aegypti*)	Poor mosquito control; increased urbanization in tropics; increased air travel
Filoviruses (Marburg, Ebola)	Fulminant, high mortality, hemorrhagic fever	Direct contact with infected blood, organs, secretions, and semen	Unknown; in Europe and the United States, virus-infected monkeys shipped from developing countries via air
Hantaviruses	Abdominal pain, vomiting, hemorrhagic fever	Inhalation of aerosolized rodent urine and feces	Human intrusion into virus ecological niche
Hepatitis B	Nausea, vomiting, jaundice; chronic infection leads to hepatocellular carcinoma and cirrhosis	Contact with saliva, semen, blood, or vaginal fluids of an infected person; mode of transmission to children not known	Probably increased sexual activity and intravenous drug abuse; transfusion (before 1978)
Hepatitis C	Nausea, vomiting, jaundice; chronic infection leads to hepatocellular carcinoma and cirrhosis	Exposure (percutaneous) to contaminated blood or plasma; sexual transmission	Recognition through molecular virology applications; blood transfusion practices, especially in Japan
Hepatitis E	Fever, abdominal pain, jaundice	Contaminated water	Newly recognized
Human immuno-deficiency viruses: HIV-1 and HIV-2	HIV disease, including AIDS: severe immune system dysfunction, opportunistic infections	Sexual contact with or exposure to blood or tissues of an infected person; vertical transmission	Urbanization; changes in lifestyle or mores; increased intravenous drug use; international travel; medical technology (transfusions and transplants)

TABLE 22.6	(continued)		
Agent	**Disease and symptoms**	**Mode of transmission**	**Cause(s) of emergence**
Viruses (cont.)			
Human papillomavirus	Skin and mucous membrane lesions (often, warts); strongly linked to cancer of the cervix and penis	Direct contact (sexual contact or contact with contaminated surfaces)	Newly recognized; perhaps changes in sexual lifestyle
Human T-cell lymphotrophic viruses (HTLV-I and HTLV-II)	Leukemias and lymphomas	Vertical transmission through blood or breast milk; exposure to contaminated blood products; sexual transmission	Increased intravenous drug abuse; medical technology (transfusion and transplantation)
Influenza pandemic	Fever, headache, cough, pneumonia	Airborne; especially in crowded, enclosed spaces	Animal–human virus reassortment; antigenic shift
Lassa	Fever, headache, sore throat, nausea	Contact with urine or feces of infected rodents	Urbanization and conditions favoring infestation by rodents
Measles	Fever, conjunctivitis, cough, red blotchy rash	Airborne; direct contact with respiratory secretions of infected persons	Deterioration of public health infrastructure supporting immunization
Monkey pox	Rash, lymphadenopathy, pulmonary distress	Direct contact with infected primates	Travel to endemic areas, consumption and handling of infected primates
Norwalk and Norwalk-like agents	Gastroenteritis; epidemic diarrhea	Most likely fecal-oral; vehicles may include drinking and swimming water, and uncooked foods	Increased recognition
Rabies	Acute viral encephalomyelitis	Bite of a rabid animal	Introduction of infected host reservoir to new areas
Rift Valley	Febrile illness	Bite of an infective mosquito	Importation of infected mosquitoes and/or animals; development (dams, irrigation)
Rotavirus	Enteritis: diarrhea, vomiting, dehydration, and low grade fever	Primarily fecal-oral; fecal-respiratory transmission can also occur	Increased recognition
Venezuelan equine encephalitis	Encephalitis	Bite of an infective mosquito	Movement of mosquitoes and hosts (horses)
West Nile virus	Meningitis, encephalitis	*Culex pipiens* mosquito	Agricultural development, increase in mosquito breeding areas
Yellow fever	Fever, headache, muscle pain, nausea, vomiting	Bite of an infective mosquito (*Aedes aegypti*)	Lack of effective mosquito control and widespread vaccination; urbanization in tropics; increased air travel
Protozoa and Fungi			
Candida	Candidiasis: fungal infections of the gastrointestinal tract, vagina, and oral cavity	Endogenous flora; contact with secretions or excretions from infected persons	Immunosuppression; medical management (catheters); antibiotic use
Cryptococcus	Meningitis; sometimes infections of the lungs, kidneys, prostate, liver	Inhalation	Immunosuppression
Cryptosporidium	Cryptosporidiosis: infection of epithelial cells in the gastrointestinal and respiratory tracts	Fecal-oral, person to person, waterborne	Development near watershed areas; immunosuppression
Giardia lamblia	Giardiasis: infection of the upper small intestine, diarrhea, bloating	Ingestion of fecally contaminated food or water	Inadequate control in some water supply systems; immunosuppression; international travel
Microsporidia	Gastrointestinal illness, diarrhea; wasting in immunosuppressed persons	Unknown; probably ingestion of fecally contaminated food or water	Immunosuppression; recognition

TABLE 22.6 (continued)			
Agent	**Disease and symptoms**	**Mode of transmission**	**Cause(s) of emergence**
Protozoa and Fungi (cont.)			
Plasmodium	Malaria	Bite of an infective *Anopheles* mosquito	Urbanization; changing parasite biology; environmental changes; drug resistance; air travel
Pneumocystis carinii	Acute pneumonia	Unknown; possibly reactivation of latent infection	Immunosuppression
Toxoplasma gondii	Toxoplasmosis: fever, lymphadenopathy, lymphocytosis	Exposure to feces of cats carrying the protozoan; sometimes foodborne	Immunosuppression; increase in cats as pets
Other Agents			
Bovine prions	Bovine spongiform encephalitis (animal and human)	Foodborne	Consumption of contaminated beef

Biological Warfare and Bioterrorism: An Emerging Plague?

BIOHAZARD

The photo shown is a monument to the plague victims of Budapest, Hungary. Plague was responsible for decimating the population of Europe at several times in the Middle Ages (∞ Section 24.6). Ancient plague epidemics resulted from shifts in the rat populations within medieval cities and were affected by such variables as weather and crop yields. Today we have the specter of new plagues that may just as

certainly decimate the cities of the world. However, the new plagues are not influenced by natural phenomena; rather, they are influenced by the modern variables of political, economic, and military ambition. These new plagues are the agents of biological warfare—the use of biological agents to incapacitate or kill a military or civilian population in an act of war or terrorism.

Biological warfare is receiving a lot of public attention these days because biological weapons and weapons-making facilities are known or suspected to be in the hands of several rogue governments and extremist groups that

support or use terrorism. In fact, although bioweapons are potentially useful in the hands of conventional military forces, the greatest likelihood of bioweapons use is probably by terrorist groups. This is in part because any well-trained microbiologist possesses the laboratory skills necessary for the propagation of most of the organisms useful for biological warfare. Any organization, even the poorest developing country or the most extreme political group, can acquire the modest finances and technical expertise to obtain and use biological weapons. Thus, these agents of potential mass destruction have been dubbed the equivalent of a "poor man's atomic bomb."

Virtually any pathogenic bacterium or virus is potentially useful for biological warfare, and most of the candidate organisms are very easy to obtain, grow and disseminate. The most commonly mentioned candidate is *Bacillus anthracis*, the causal agent of anthrax. This is because *B. anthracis* produces endospores (∞ Section 3.15) which, when aerosolized, can be a very effective means of distributing the bacterium. Inhalation of the spores or the live bacteria re-

Jolynn and Gerard Smith

drome (AIDS) (∞ Section 23.7), and Lyme disease (∞ Section 24.4) emerged as major new diseases.

Emergence Factors

Some factors responsible for the emergence of new pathogens are (1) human population shifts (demographics) and behavior, (2) technology and industry, (3) economic development and land use, (4) international travel and commerce, (5) microbial adaptation and change, (6) breakdown of public health measures, and (7) abnormal natural occurrences that upset the usual host–pathogen balance. Finally, another situation that threatens to expose large populations to new or established diseases is the possibility of a biological warfare incident. In such a case, large numbers of individuals would be rapidly exposed to a virulent pathogen, cre-

ating a scenario of high morbidity and mortality in an unprepared population (see the box, Biological Warfare and Bioterrorism: An Emerging Plague?)

The *demographics* of human populations have changed dramatically in the last two centuries. In 1800, less than 2% of the world's population lived in urban areas. Today, nearly one-half of the world's population lives in cities. The numbers, sizes, and population densities of modern urban centers make disease transmission much easier. For example, dengue fever (Figure 22.9 and Table 22.6) is now recognized as a serious hemorrhagic disease in tropical cities, largely because of the spread of dengue virus by the mosquito *Aedes aegypti*. The disease now spreads as an epidemic in tropical urban areas. Prior to 1950, dengue fever was rare, presumably because the virus was not easily spread among a more dispersed, smaller population.

sults in pulmonary infections, which have a mortality rate of nearly 100%.

Smallpox virus (∞ Section 8.20) is also an important bioweapons candidate. Although an extremely effective smallpox vaccine exists, it has not been in regular use for more than 20 years because smallpox was eradicated worldwide in 1980. As a result, over 90% of the current worldwide population are now inadequately vaccinated and are susceptible to the disease. Could terrorist groups or even conventional military forces gain access to smallpox virus? Since the known remaining stocks of smallpox virus were scheduled to be destroyed in 1999, we can only hope that this agent is permanently off the list of potential bioweapons.

Other bioweapons candidates include *Yersinia pestis*, the organism responsible for the medieval plague pandemics (∞ Section 24.6), *Brucella abortus* (fever and bacteremia), *Francisella tularensis* ("rabbit fever"), *Salmonella* (food- and waterborne illnesses; ∞ Section 24.9 and 24.12) and *Clostridium botulinum* (botulism, ∞ Section 19.8). The more exotic candidate agents include rabies virus (∞ Section 24.1) and Ebola virus

(see Section 22.10). Most of these agents cause disease coupled with high death rates within days to weeks after exposure.

One common feature of these bioweapons candidates is that they can be spread in an aerosolized form, making dissemination and infection simple and rapid. For example, in 1962 one of the last outbreaks of smallpox in developed countries occurred in Germany. An infected German worker, after returning from Pakistan, developed smallpox and was immediately hospitalized and quarantined. Because the patient had a cough, he aerosolized the virus and infected 19 *vaccinated* individuals; at least one individual died. In another incident, anthrax spores were inadvertently released into the atmosphere from a bioweapons facility in Sverdlovsk, Russia, in 1979. Less than one gram of spores was released and everyone in the surrounding area was immunized and given prophylactic antibiotic therapy as soon as the first anthrax case was diagnosed. However, 77 individuals *outside* the facility contracted pulmonary anthrax and 66 died.

Planned bioterrorist attacks have already occurred. In 1984 in The

Dalles, Oregon (United States), an extremist religious cult inoculated salad bars with *Salmonella typhimurium* at 10 local restaurants, causing 751 cases of foodborne salmonellosis in a region that usually has less than 10 cases per year. In 1995, an extremist group released Saran nerve gas into a Tokyo subway killing several people and injuring scores of others. This group also had anthrax cultures, bacteriological media, drone airplanes, and spray tanks in its terrorist arsenal. Clearly, the knowledge and will to use bioweapons is widespread.

Proactive measures against bioweapons have already begun with efforts to update the international agreements of the 1972 Biological and Toxin Weapons Convention. However, at the practical level, little is being done to fund the large-scale production and distribution of vaccines or the development of other strategic and tactical measures to prevent and contain the effects of bioweapons. Bioterrorism is a real threat in a world of rapid international travel, widespread technical knowledge, and overnight mailings. This new plague emphasizes the problems associated with irresponsible use of our knowledge of microorganisms. ■

Human behavior, especially in large population centers, also contributes to disease spread. For example, sexual promiscuity and the use of injectable drugs, centered mainly in large urban areas, have been a major contributing factor to the spread of AIDS and hepatitis (Table 22.6; ∞ Sections 23.6 and 23.7).

Although *technological advances* and *industrial development* have had a generally positive impact on living standards worldwide, in some cases these advances have contributed to the spread of diseases. For example, one of the chief technological advances of the twentieth century has been in the health care area. However, as we noted in Section 22.7, the health care environment, especially in hospitals, has resulted in an explosive increase in nosocomial infections. For example, during the 1980s there was a threefold rise in hospital-associated bacteremias in the United States (see Section 22.7 and Figure 22.8). Antimicrobial drug resistance in microorganisms is another negative outcome of modern health care practices; vancomycin-resistant enterococci and multiple-drug-resistant *Streptococcus pneumoniae* have become important emerging diseases, affecting especially developed countries.

Transportation, bulk processing, and central distribution methods have become an important factor for quality assurance and economy in the food industry. However, these same factors can increase the potential for common-source epidemics when sanitation measures fail. For example, a single meat processing plant spread *Escherichia coli* O157:H7 (Table 22.6) to at least 500 individuals in four states in the United States. Finally, the food source, ground beef, was recalled and the epidemic was curtailed, but not before several people died (∞ Section 24.12).

Economic development and changes in land use also have potential implications for promoting disease spread. For example, Rift Valley fever, a mosquito-borne viral infection, has been on the increase since completion of the Aswan High Dam in Egypt in 1970. The dam created 2 million acres of flooded land, which dramatically increased mosquito breeding grounds at the edge of the new reservoir. The first major epidemic of Rift Valley fever occurred in Egypt in 1977 when an estimated 200,000 people became ill and 598 died. Several epidemic outbreaks have occurred since then including a major outbreak in 1998 (Figure 22.10), and the disease has become endemic near the reservoir.

Lyme disease, the most common vector-borne disease in the United States, is probably on the rise because of *changes in land use.* Reforestation and the concomitant increase in the numbers of deer (the natural host for the disease-producing *Borrelia burgdorferi*) have resulted in greater numbers of infected ticks, the arthropod vector (∞ Section 24.4). In addition, increasing numbers of people are building homes and pursuing recreational activities in and near forests, resulting in increased contact between the infected ticks and humans and, consequently, increased disease.

International travel and commerce can also affect the spread of pathogens. For example, filoviruses (Filoviridae), a group of ribonucleic acid (RNA) viruses, (∞ Section 8.16), cause fevers culminating in hemorrhagic disease in infected hosts. These diseases, because of their viral origin, are not readily treatable because there are a very limited number of antiviral chemotherapeutic agents (∞ Section 18.10). They generally have a mortality rate higher than 20%. Most outbreaks of these diseases have been restricted to equatorial central Africa, where the still-unidentified natural hosts and vectors undoubtedly live (Figure 22.10). Travel of potential hosts to or from endemic areas is usually implicated in disease transmission. For example, one of these viruses was imported into Marburg, Germany, with a shipment of African green monkeys, a species used for laboratory work. The virus quickly spread from the primate vector to some of the human handlers. Twenty-five people were initially infected, and six more developed disease as a result of contact with the human cases. Seven people died in this outbreak of what came to be known as the Marburg virus. Another shipment of laboratory monkeys brought a filovirus to the United States. At least four individuals who worked with the imported monkeys were infected with what is now called the Reston virus (named for Reston, Virginia, the site of the outbreak). The Reston virus was highly contagious and spread through the monkeys, presumably by a respiratory route. However, only four humans were infected and none developed clinical disease. Fortunately, this virus did not cause significant human disease. These two filoviruses are closely related to the Ebola virus (Table 22.6 and Figure 22.10). Periodic, but sporadic Ebola outbreaks in central Africa, often characterized by mortality rates of greater than 50%, have again underscored the existence of highly virulent human pathogens for which there is little or no immunity or therapy. These pathogens could potentially be disseminated via air travel throughout the world in a matter of days. As a worst-case scenario, a single agent that combines the highly contagious respiratory transmission route of the Reston virus and the high mortality rate of the Ebola virus could start a major pandemic that could devastate population centers worldwide in a matter of weeks.

Microbial adaptation and change also contribute to pathogen emergence. For example, nearly all RNA viruses, including influenza and human immunodeficiency virus (HIV), undergo rapid, unpredictable genetic mutations. RNA viruses lack correction mechanisms for replication steps, and so they incorporate genomic mutations at an extremely high rate compared to most DNA viruses, presenting major epidemiological problems because of their constantly changing genomes.

Bacteria also have genetic mechanisms that enhance virulence and promote emergence of new epidemics. One group of virulence-enhancing mechanisms are the mobile genetic elements: bacteriophages, plasmids, and transposons (∞ Sections 8.7, 9.8, and 9.10, respectively). Table 22.7 shows some representative virulence factors that are carried on these mobile genetic elements and contribute to pathogen emergence.

Antimicrobial drug resistance is also a major factor in bacterial pathogen resurgence (∞ Section 18.12). Drug resistance is also a factor for virus emergence. Although several drugs are effective against certain viral diseases (∞ Section 18.10), resistance to these drugs is very common, especially among the RNA viruses. For example, most strains of HIV develop resistance to azidothymidine very rapidly unless it is used in combination with other drugs (∞ Section 23.7).

A *breakdown of public health measures* is sometimes responsible for the emergence or reemergence of diseases. For instance, cholera (caused by *Vibrio cholerae*) (see Figure 22.10 and ∞ Section 24.9) can be adequately controlled, even in endemic areas, by providing proper sanitation, especially for water sources. However, contaminated municipal water supplies in Peru led to a major cholera pandemic, involving nearly 400,000 people by 1991, with almost 4000 deaths. In another case, the municipal water supply of Milwaukee, Wisconsin, was contaminated with the chlorine-resistant protozoan *Cryptosporidium* in 1993. The contamination resulted in 370,000 cases of intestinal disease, 4000 of which required hospitalization. More effective treatment procedures including enhanced filtration systems were required to rid the water supply of the pathogen.

Inadequate public vaccination programs are an important potential reason for the reemergence of some previously controlled diseases. For example, recent outbreaks of diphtheria (caused by *Corynebacterium diphtheriae*) (∞ Section 23.2) in the former Soviet Union are the result of inadequate immunization of susceptible children resulting from the breakdown of the formerly centralized public health infrastructure. Pertussis, another vaccine-preventable childhood respiratory disease (caused by *Bordetella pertussis*) (∞ Section 23.2), has also increased recently in Eastern Europe because of inadequate immunization and record keeping.

Finally, *abnormal natural occurrences* such as rapid environmental changes sometimes upset the usual host–pathogen balance. For example, hantavirus is a well-known human pathogen that occurs in many rodent populations, even in laboratory animals (∞ Section 24.2). Over the last decade, several isolated cases of hantavirus infection have occurred in laboratory animal handlers. However, a number of lethal cases of hantavirus infection were reported in 1993 in the American Southwest and were linked to exposure to wild animal droppings. Abundant rainfall and a long growing season, coupled with a mild winter, caused a tremendous increase in the number of mice in 1993. Virtually everyone who acquired the hantavirus infection had been exposed to rodents or their droppings. Thus, increased human contact with the larger-than-normal mouse population resulted in propagation and transfer of a deadly virus to a large number of human hosts, all because of abnormally mild weather conditions.

TABLE 22.7	Virulence factors encoded by bacteriophages, plasmids, and transposons	
Genetic element	**Organism**	**Virulence factors**
Bacteriophage	*Streptococcus pyogenes*	Erythrogenic toxin
	Escherichia coli	Shiga-like toxin
	Staphylococcus aureus	Enterotoxins A, D, E, staphylokinase, toxic shock syndrome toxin-1 (TSST-1)
	Clostridium botulinum	Neurotoxins C, D, E
	Corynebacterium diphtheriae	Diphtheria toxin
Plasmid	*Escherichia coli*	Enterotoxins, pili colonization factor, hemolysin, urease, serum resistance factor, adherence factors, cell invasion factors
	Bacillus anthracis	Edema factor, lethal factor, protective antigen, poly-D-glutamic acid capsule
	Yersinia pestis	Coagulase, fibrinolysin, murine toxin
Transposon	*Escherichia coli*	Heat-stable enterotoxins, aerobactin siderophores, hemolysin and pili operons
	Shigella dysenteriae	Shiga toxin
	Vibrio cholerae	Cholera toxin

Addressing Emerging Diseases

Many of the emerging diseases we have discussed are absent from the official notifiable disease list for the United States (Table 22.4). How then do public health officials define and deal with emerging diseases to prevent major epidemics? The key features for addressing emerging diseases are *recognition* of the disease and *intervention* to prevent spread of the disease.

The first step in *disease recognition* is surveillance. *Epidemic* diseases that exhibit particular *clinical syndromes* warrant intensive public health surveillance. These syndromes are (1) acute respiratory diseases, (2) encephalitis and aseptic meningitis, (3) hemorrhagic fever, (4) acute diarrhea, (5) clusterings of high fever cases, (6) unusual clusterings of any disease or deaths, and (7) resistance to common drugs or treatment. Thus, new diseases are recognized because of their epidemic incidence, clusterings, and syndromes. As the prevalence and pathology of an emerging disease are recognized, it is added to the notifiable disease list. For example, AIDS was recognized as a disease in 1981 and was added to the notifiable disease list in 1984. Lyme disease was first recognized as a separate clinical disease in the 1980s and added to the notifiable disease list in 1991. Likewise, outbreaks of gastrointestinal disease due to enteropathogenic *Escherchia coli* O157:H7 have been increasing in recent years, and the strain was added to the notifiable disease list in 1995.

Intervention to prevent spread of emerging infections must be a public health response involving a variety of methods. General strategies such as strengthening the public health system and supporting research and training are useful, but disease-specific intervention is the key to controlling individual outbreaks. Methods such as quarantine, immunization, and drug treatment must be applied to contain individual outbreaks of specific diseases. Finally, since a number of emerging diseases are propagated or transferred by nonhuman animals, we must identify the nonhuman host or vector for each disease and develop methods to intervene in the life cycle of the pathogen, ultimately interrupting transfer to humans.

In the following chapters, we will examine a number of infectious diseases, including several emerging and reemerging diseases. We will define their individual effects on the host and identify specific intervention strategies.

✓ 22.10 Concept Check

Emerging and reemerging diseases are of major global concern. Changes in host, vector, or pathogen conditions, whether natural or artificial, can result in conditions that encourage the explosive emergence of certain infectious diseases. Global surveillance and intervention programs must be maintained and enhanced to prevent major new epidemics and pandemics.

✓ What factors are important in the emergence or reemergence of potential pathogens?

✓ Indicate general and specific methods that would be useful for dealing with perceived and actual emerging infectious diseases.

REVIEW QUESTIONS

1. What are the six most common causes of mortality due to infectious diseases throughout the world?

2. Distinguish between *mortality* and *morbidity*, *prevalence* and *incidence*, and *epidemic* and *pandemic*, as these terms relate to infectious disease.

3. Explain the difference between a chronic carrier and an acute carrier of an infectious disease.

4. Identify the major risk factors for acquiring human immunodeficiency virus (HIV) infection. Based on these risk factors, what is the likely source or sources of HIV infection?

5. Give examples of host-to-host transmission of disease via direct contact. Also give examples of indirect host-to-host transmission of disease.

6. How can immunity to an infection by a large proportion of the population protect the nonimmune members of the population from acquiring a disease? Will this herd immunity work for diseases that have a common source, such as water? Why or why not?

7. Hospitals are particularly hazardous environments for the spread of infectious diseases. Review the reasons for the enhanced spread of infection in hospitals. What is the source of most nosocomial infections?

8. Many factors that can control the spread of infection are important considerations for public health personnel. Describe the major methods used to control the spread of infectious diseases.

9. Compare the role of infectious diseases on mortality in developed and developing countries.

10. Review the major reasons for the emergence of new infectious diseases. Review the major methods available for controlling the emergence of new infectious diseases.

APPLICATION QUESTIONS

1. How would an epidemiologist acquire data concerning a potential common-source epidemic? What resources are currently at the epidemiologist's disposal and what resources must be enhanced to better define serious infectious disease outbreaks?

2. If an infectious disease causes high *mortality*, then *morbidity* may be quite low. On the other hand, diseases characterized by high morbidity often have very low mortality. Explain these statements and present examples to support your explanation. Can you identify any infectious diseases that do not fit these generalizations?

3. Smallpox, a disease that was limited to humans, was eradicated. Plague, a disease with a zoonotic reservoir in rodents (Table 22.2) will never be eradicated. Explain this statement and why you agree or disagree with the possibility of eradicating plague. Devise a plan to eradicate plague in a limited environment such as a town or city. Be sure to use methods that involve the reservoir, the pathogen, and the host.

4. Acquired immunodeficiency syndrome (AIDS) transmission is considered to be person to person. How did epidemiologists determine this fact? AIDS is a candidate for a disease that can be eliminated because it is propagated by person-to-person contact and there are no known animal reservoirs. Design a program for eliminating AIDS in a developed country and in a developing country. How would these programs differ from one another? What factors would work against the success of your program, both in terms of human behavior and in terms of the AIDS disease itself? Why are the numbers of HIV-infected and AIDS patients continuing to grow, especially in developing countries?

5. Transmission of many epidemic diseases is host to host, whereas other epidemics are spread via a common source. Some epidemics can be transmitted by both routes. Explain how this might happen, using specific infectious agents (at least one bacterium and one virus) as examples.

6. What is the overall advantage of pathogen-host co-evolution in terms of species survival? Is it beneficial to the pathogen to cause high mortality in the host? Why or why not? What diseases in Table 22.2 have caused high mortality? Compare their reservoir or host to diseases with low mortality.

7. Why are diseases due to antibiotic-resistant pathogens more common in hospital environments than in the general population? Why are diseases caused by so-called "normal flora" such as *Staphylococcus* species more common in hospital environments than in the general population? What special precautions must one take when diagnosing and treating infectious diseases in a hospital setting?

8. As a public health official, you are faced with a common-source epidemic, and you believe the source is the municipal water supply. How would you use your resources to stop the epidemic? (*Hint:* do not focus on treatment of the disease unless you, as a public health officer, believe that treatment will stop disease spread.) List the steps you would take in priority order. Perform the same exercise for a host-to-host epidemic caused by a pathogen for which there are available vaccines and chemotherapeutic agents.

9. Travel to developing countries involves a certain amount of exposure to infectious diseases. What general precautions should you take before, during, and after visits to developing countries? Where can you obtain information on the infectious disease status in a specific foreign country? When you return from a foreign country, are you a disease risk to your family or your associates? Explain.

10. Although many factors may be involved in the emergence of an infectious disease, some diseases develop to the pandemic stage while others never get beyond localized epidemics. Examples of this are pandemic HIV and epidemic Ebola virus, which were both identified within the last 20 years. What factors do these viruses share that led to their emergence? What specific factors cause them to be quite different in terms of their spread, and how have these factors contributed to the worldwide spread of HIV, while limiting the spread of Ebola virus to sporadic, dramatic outbreaks?

Aerosols, such as those generated by the sneeze shown in the illustration, are an important means of person-to-person transmission for many infectious diseases. Most respiratory diseases are spread almost exclusively in this fashion. Some microorganisms, such as *Mycobacterium tuberculosis,* are very successful using this strategy—*M. tuberculosis* has infected at least one-third of the world's population in this manner. The viruses responsible for colds and influenza are so successful that virtually everyone acquires more than one of these infections every year. Another category of diseases that are transmitted from person to person are the sexually transmitted diseases. In most cases serious diseases that spread exclusively from person to person are susceptible to antimicrobial drugs or immunizations that limit both the severity and transmission of the disease.

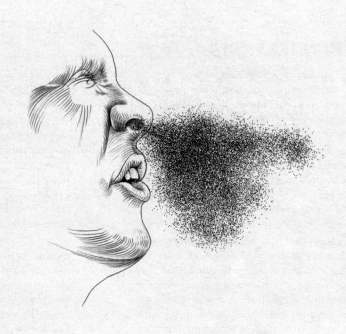

CHAPTER 23 # Person-to-Person Microbial Diseases

23.1   Airborne Transmission of Pathogens   923
23.2   Respiratory Pathogens: Bacterial   925
23.3   Mycobacterium and Tuberculosis   931
23.4   Respiratory Pathogens: Viral   934
23.5   Why Are Respiratory Infections So Common?   940
23.6   Sexually Transmitted Diseases   940
23.7   Acquired Immunodeficiency Syndrome   946

Antigenic drift minor changes in antigens due to gene mutations in influenza virus.

Antigenic shift major changes in antigens due to gene reassortment in influenza virus

Congenital syphilis syphilis contracted by an infant from its mother during birth

Nonnucleoside reverse transcriptase inhibitor (NNRTI) a nonnucleoside compound that inhibits the action of viral reverse transcriptase by binding directly to the catalytic site

Nucleoside reverse transcriptase inhibitor (NRTI) a nucleoside analog compound that inhibits the action of viral reverse transcriptase by competing with nucleosides

Protease inhibitor a compound that inhibits the action of viral protease by binding directly to the catalytic site, preventing viral protein processing

Rheumatic fever an inflammatory autoimmune disease triggered by an immune response to infection by *Streptococcus pyogenes*

Scarlet fever characteristic reddish rash resulting from an exotoxin produced by *Streptococcus pyogenes*

Sexually transmitted disease (STD) a disease that is usually transmitted by sexual contact

Toxic shock syndrome acute shock resulting from a host response to an exotoxin produced by *Staphylococcus aureus*

Tuberculin test a skin test for previous infection with *Mycobacterium tuberculosis*

Viral load a quantitative assessment of the amount of virus in a host organism

---

A t least 500,000 microbial species exist in nature, and probably many more (∞ Section 12.6), but only a few hundred species are potential human pathogens. Most microorganisms carry out essential life-supporting activities independently, and many others are closely associated with plants or animals in stable, beneficial relationships (∞ Chapter 16). However, the pathogenic species have profoundly negative effects on host organisms and, as a result, have been intensively studied in many cases. In the next two chapters, we will examine representative examples of human pathogens and study the prevention, treatment, and pathology of the diseases that they cause. For this study, we will divide the pathogens based first on their *mode of transmission*. We will then subdivide the pathogens based on the organ systems they affect, discussing each infectious disease in relation to the ecology of the pathogen. In this chapter we consider *respiratory* and *sexually transmitted* diseases because their mode of transmission is via direct *person-to-person* contact. In the next chapter we will consider several diseases that use environmental modes of transmission, either through living *vectors* or through *common sources* such as food, water, and soil.

By dividing pathogens according to their modes of transmission and the diseases they cause, we are able to connect seemingly unrelated organisms with one another. For example, influenza and streptococcal pharyngitis, which we will discuss together, are spread person-to-person via the respiratory route. They produce primary diseases with very similar symptoms, although the causal agents, one viral and one bacterial,

are markedly different. Using this approach, we hope to provide a logical framework for making connections among biologically diverse but ecologically and pathogenically related agents.

23.1

Airborne Transmission of Pathogens

Air is not a suitable medium for the growth of microorganisms; organisms found in air are derived from soil, water, plants, animals, people, or other sources. In outdoor air, soil organisms predominate. Microbial numbers indoors are considerably higher than those outdoors, and the organisms are mostly those commonly found in the human respiratory tract.

Windblown dust carries with it significant microbial populations that can travel long distances. Most microorganisms survive poorly in air, and so effective transmittal to a suitable habitat (another human) occurs only over short distances. However, certain human pathogens (*Staphylococcus, Streptococcus*) survive under dry conditions fairly well and may remain alive in dust for long periods of time. Gram-positive Bacteria are in general more resistant to drying than gram-negative Bacteria because of their thicker, more rigid cell wall. Spore-forming Bacteria are extremely resistant to drying but are not generally passed from human to human in the spore form. However, airborne *Clostridium* spp. are derived from soil, are spore formers, and are occasional pathogens, although they are seldom respiratory pathogens (∞ Section 13.19 and 24.11).

FIGURE 23.1   High speed photograph of an unstifled sneeze.

An enormous number of droplets of moisture are expelled during sneezing (Figure 23.1), and a considerable number are expelled during coughing or even merely talking. Each infectious droplet has a size of about 10 μm and contains one or two bacteria. The speed of the droplet movement is about 100 m/sec (more than 200 mi/hr) in a sneeze and about 16–48 m/sec during coughing or shouting. The number of bacteria in a single sneeze varies from 10,000 to 100,000. Because of the small size of the droplets, the moisture evaporates quickly in the air, leaving behind a nucleus of organic matter and mucus to which bacterial cells are attached.

Respiratory Infection

The average human breathes nearly 500 million liters of air in a lifetime, much of it containing microorganism-laden dust, which is a potential source of inoculum for upper respiratory infections. The speed at which air moves through the respiratory tract varies, and in the lower respiratory tract the rate is quite slow. As the air slows down, particles in it stop moving and settle. The larger particles settle first and the smaller ones later, and only particles smaller than 3 μm travel as far as the bronchioles in the lower respiratory tract (Figure 23.2). Different organisms reach different levels in the tract, thus accounting for the differences in the kinds of infections that occur in the upper and lower respiratory tracts.

✓ 23.1 Concept Check

Many respiratory pathogens are gram-positive Bacteria. Because gram-positive Bacteria are very resistant to drying, they are easily transmitted in air. Less hardy respiratory

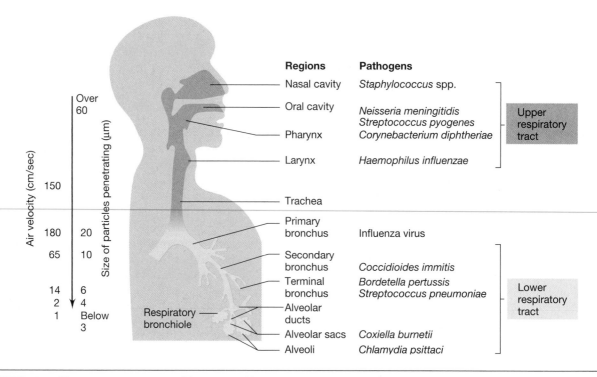

FIGURE 23.2   The respiratory system of humans and locations at which various organisms generally initiate infections.

pathogens can be transferred from person to person via respiratory aerosols generated by coughing, sneezing, talking, or breathing.

✓ What physical features of gram-positive Bacteria allow them to survive for long periods in air?
✓ Why are certain pathogens more commonly found in the upper respiratory tract? Why are certain pathogens more commonly found in the lower respiratory tract?

23.2

Respiratory Pathogens: Bacterial

A variety of bacterial pathogens affect the respiratory tract. Most inhabit only humans, and thus their mode of transmission must be person to person. A few respiratory pathogens such as *Legionella pneumophila* are found in water or soil; because they share common sources with other pathogens, they will be discussed separately in Chapter 24. As we explained in Section 23.1, many of the person-to-person respiratory pathogens are gram-positive Bacteria because the gram-positive Bacteria can survive in dry conditions outside the host for long periods of time by virtue of the thick gram-positive cell wall. Bacterial respiratory infections, while serious by themselves, often initiate secondary problems that can be life-threatening. Thus, it is critical to quickly and accurately diagnose and treat bacterial respiratory infections to limit host damage. Fortunately, most respiratory bacterial pathogens respond readily to antibiotic therapy and many can also be controlled by immunization. Nevertheless, bacterial respiratory infections are still rather common, and we begin here with a consideration of two of the most common bacterial respiratory pathogens, *Streptococcus pyogenes* and *Streptococcus pneumoniae*.

Streptococcal Diseases

Streptococci are gram-positive cocci that typically grow in elongated chains (Section 13.18). Several species of streptococci are potent human pathogens. Of particular importance are *Streptococcus pyogenes* and *S. pneumoniae*.

Streptococcus pyogenes and *S. pneumoniae* are frequently isolated from the upper respiratory tract of healthy adults. Although numbers of *S. pyogenes* and *S. pneumoniae* are usually low here, if the host's defenses are weakened or a new, highly virulent strain is introduced, acute streptococcal bacterial infections are possible. *Streptococcus pyogenes* is the cause of streptococcal pharyngitis, so-called strep throat (the pharynx is the tube that connects the oral cavity to the larynx and the esophagus) (Figure 23.2). Most isolates from clinical cases of strep throat produce a toxin that lyses red blood cells, a condition called *β-hemolysis* (Figure 19.15). Streptococcal pharyngitis is characterized by a severe

sore throat, enlarged tonsils, tonsillar exudate, tender cervical lymph nodes, a mild fever, and a general feeling of malaise (tonsilitis). *Streptococcus pyogenes* can also cause related infections of the inner ear (otitis media), the mammary glands (mastitis), and infections of the superficial layers of the skin, a condition referred to as impetigo (however, most cases of impetigo are caused by *Staphylococcus aureus*) (Figure 23.3).

About half of the clinical cases of severe sore throat turn out to be due to *Streptococcus pyogenes*, the remainder being of viral origin. An accurate, prompt diagnosis is important because if the sore throat is due to a virus, treatment with antibacterial chemotherapeutic agents ("antibiotics") will be useless, whereas if the sore throat is due to *S. pyogenes*, immediate antibacterial therapy is indicated. Penicillin and semisynthetic derivatives (Section 18.8) are still the agents of choice for treating *S. pyogenes* infections, since resistant organisms have not yet appeared. Other drugs such as erythromycin are used for treatment in individuals with penicillin allergies (Section 20.15). Rapid, complete treatment of streptococcal sore throat is important because it can occasionally lead to more serious streptococcal syndromes such as scarlet fever, rheumatic fever, acute glomerulonephritis, and toxic shock syndrome (see later).

Certain strains of *Streptococcus pyogenes* carry a lysogenic bacteriophage that encodes production of erythrogenic toxin, an exotoxin responsible for most of the

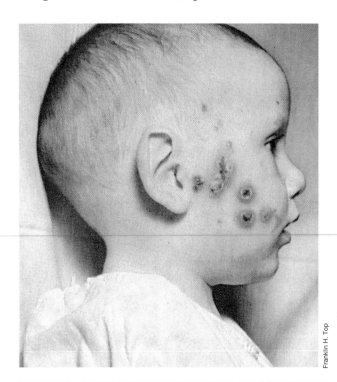

FIGURE 23.3 Typical lesions of impetigo, commonly caused by *Streptococcus pyogenes* or *Staphylococcus aureus*.

symptoms of **scarlet fever.** Erythrogenic toxin causes a pink-red rash to develop (Figure 23.4) and also acts to damage small blood vessels and initiate fever. The condition is acute and easily treated with antibacterial agents.

Occasionally, *Streptococcus pyogenes* may cause fulminant systemic infections, often marked by necrotizing fasciitis, a rapid and progressive infection of subcutaneous tissue. These infections are responsible for the dramatic, but fortunately rare, reports of "flesh-eating bacteria." In these cases, exotoxins A and B and the surface M-protein act as superantigens (∞ Section 20.15) that recruit massive numbers of T cells to the infected tissues. The T cells then secrete cytokines, which activate large numbers of effector cells, resulting in massive systemic inflammation, tissue destruction, and death in up to 30% of the cases.

Untreated or insufficiently treated cases of *Streptococcus pyogenes* infection may lead to severe *delayed sequelae,* or follow-up diseases. **Rheumatic fever,** one of these delayed sequelae, is caused by *rheumatogenic* strains of *S. pyogenes* containing cell-surface antigens that are similar to certain human cell-surface antigens. The immune response to the invading pathogen produces antibodies that cross-react with host tissues, in particular those of the heart, joints, and kidneys, resulting in tissue destruction. Rheumatic fever is a type of autoimmune disease, with antibodies reacting with self constituents (∞ Section 20.15). Damage may be

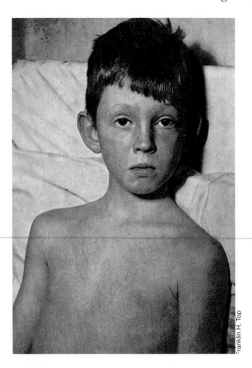

FIGURE 23.4 The typical rash of scarlet fever, resulting from the action of the erythrogenic toxin produced by *Streptococcus pyogenes.*

permanent and is often compounded by later streptococcal infections and subsequent bouts of rheumatic fever.

Another potential delayed sequela of *Streptococcus pyogenes* infection is *acute glomerulonephritis,* a painful disease of the kidney. This is an immune complex disease (∞ Section 20.15) resulting from the formation of streptococcal antigen–antibody complexes in the bloodstream. The immune complexes lodge in the *glomeruli,* or filtration membranes of the kidney, causing inflammation of the kidney (*nephritis*) accompanied by severe kidney pain. Within several days, these complexes are usually dissolved and the patient quickly returns to normal. Unfortunately, even timely antibacterial treatment does not prevent glomerulonephritis. However, only a few strains of *S. pyogenes*—so-called *nephritogenic* strains—produce this painful disease.

Fortunately, reinfection by a particular *S. pyogenes* strain is rare, but there are over 60 different strains defined by the antigenically distinct, strain-specific, cell surface M-proteins. Thus, an individual can be infected over 60 times by different *S. pyogenes* strains. Unfortunately, there are no available vaccines to prevent *S. pyogenes* infections.

Because serious host damage following a streptococcal sore throat can occur, several rapid antigen detection (RAD) systems have been developed for identification of *S. pyogenes.* Surface antigens are first extracted by enzymatic or chemical means directly from a swab of the patient's throat. Immunological methods such as latex bead agglutination, enzyme-linked immunoassay (ELISA), or fluorescent antibody staining (∞ Sections 21.5, 21.7, and 21.8) using antibodies specific for surface proteins unique to *S. pyogenes,* are employed. Specimens are taken directly from a patient throat swab and are processed and analyzed in minutes. These rapid diagnostic procedures allow the physician to immediately initiate appropriate antibiotic therapy in order to avoid complications such as rheumatic fever. However, a more accurate confirmation of infection by pathogenic streptococci is a positive *S. pyogenes* culture from the throat. In general, throat culture techniques are up to three times as sensitive as RAD tests, but take up to several days to process. Finally, the most sensitive methods for identifying recent streptococcal infections are serology tests, where patients are examined for the presence or increase (rise in titer) of antibodies to various streptococcal antigens (∞ Section 21.4). The presence of new antibodies or an increase in the quantity of an existing antibody confirms a very recent streptococcal infection.

The other major pathogenic streptococcal species, *Streptococcus pneumoniae,* causes lung infections that often develop as secondary infections to other respiratory disorders. A characteristic of *S. pneumoniae* is that cells are typically present in pairs (or short chains) and are surrounded by a large capsule (Figure 23.5). The

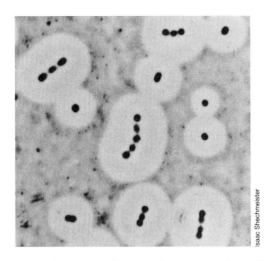

FIGURE 23.5   India ink negatively stained preparation of cells of *Streptococcus pneumoniae*. Note the extensive capsule surrounding the cells.

capsule enables the cells to resist phagocytosis; capsulated strains of *S. pneumoniae* are very invasive. Cells invade alveolar tissues (lower respiratory tract) of the lung and elicit a strong host inflammatory response. Reduced lung function can result from accumulation of phagocytic cells and fluid, and the *S. pneumoniae* cells can spread from the focus of infection as a bacteremia, sometimes resulting in bone infections, inner ear infections, and endocarditis. Pneumococcal pneumonia is a serious infection, untreated cases having a mortality rate of about 30%. Even with aggressive antimicrobial treatment, individuals hospitalized with pneumococcal pneumonia have a 5–10% fatality rate.

Laboratory diagnosis of *S. pneumoniae* involves the culture of the diagnostic gram-positive diplococci from either patient sputum or blood. There are 90 different serotypes of *S. pneumoniae*, defined by unique antigenic determinants on the capsular polysaccharides. Since antibodies to one serotype are protective, an individual cannot be reinfected by the same strain. Effective multivalent vaccines are available that confer immunity to at least two-thirds of the known strains, including virtually all of the common strains. These vaccines are recommended for the elderly, individuals with compromised immune systems, health care workers, and others at particular risk for respiratory infections.

Most strains of *S. pneumoniae* respond to penicillin therapy. However, there are penicillin-resistant strains, especially among those acquired in hospitals (∞ Section 22.7), and individual isolates must be checked for penicillin sensitivity (∞ Section 21.3). Erythromycin is the drug of choice for penicillin-resistant organisms, but cephalosporin, fluoroquinolone, ceftriaxone, cefotaxime, or vancomycin (∞ Sections 18.7–18.9) may also be used. However, strains with resistance to each of

these drugs, and strains with multiple drug resistance, have been found, underscoring the need to test each isolate individually.

✓ 23.2a Concept Check

Two respiratory diseases caused by streptococci are streptococcal sore throat and pneumococcal pneumonia. Under certain conditions, simple *Streptococcus pyogenes* infections can develop into more serious conditions such as scarlet fever and rheumatic fever. Pneumonia caused by *Streptococcus pneumoniae* is always a serious disease.

✓ Why does *Streptococcus pyogenes* infection occasionally cause rheumatic fever?
✓ What is the primary virulence factor for *Streptococcus pneumoniae*?

Staphylococcus

The genus *Staphylococcus* contains common pathogens of humans and animals, including some that are occasionally serious pathogens. Staphylococci are gram-positive cocci that divide in several planes to form irregular clumps (∞ Section 16.24 and Figure 16.82). Staphylococci are resistant to drying and are readily dispersed in dust particles through the air. In humans, two species are important: *Staphylococcus epidermidis*, a nonpigmented, nonpathogenic form usually found on the skin or mucous membranes, and *S. aureus*, a yellow-pigmented form associated with pathological conditions, including boils, pimples, and impetigo (Figures 23.3 and 23.6), pneumonia, osteomyelitis, carditis, meningitis, and arthritis. Both species often occur as part of the normal microbial flora in the upper respiratory tract (Figure 23.2).

Those strains of *Staphylococcus aureus* most frequently causing human disease produce a number of extracellular enzymes or toxins (∞ Sections 19.8 and 19.9). At least four different *hemolysins* have been recognized, a single strain often being capable of producing more than one hemolysin. The production of these is responsible for the hemolysis seen around colonies on blood agar plates. *S. aureus* is also capable of producing an *enterotoxin*, commonly associated with foodborne illness (∞ Sections 19.9, 20.15, and 24.11).

Another substance produced by *Staphylococcus aureus* is *coagulase*, an enzymelike factor that causes fibrin to coagulate and form a clot (∞ Section 19.8). The production of coagulase is generally associated with pathogenicity. Clotting induced by coagulase results in the accumulation of fibrin around the bacterial cells and makes them resistant to phagocytosis (Figure 23.6). In addition, the formation of such fibrin clots results in isolation of the infected area, making it difficult for host defense agents to come into contact with the bacteria. Most *S. aureus* strains also produce *leukocidin*, which causes the destruction of leukocytes, allowing the *S. au-*

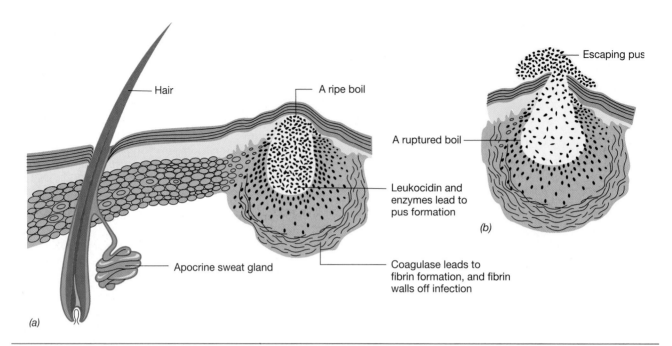

FIGURE 23.6 The structure of a boil. (a) Staphylococci initiate a localized infection of the skin and become walled off by co-agulated blood and fibrin through the action of coagulase. (b) The rupture of the boil releases pus and bacteria.

reus cells to escape phagocytosis unharmed. Production of leukocidin in skin lesions such as boils and pimples results in much cell destruction and is one of the factors responsible for pus formation (Figure 23.6). Certain strains of *S. aureus* also produce proteolytic enzymes, hyaluronidase, fibrinolysin, lipase, ribonuclease, and deoxyribonuclease.

The most common habitat of *Staphylococcus aureus* is the upper respiratory tract, especially the nose and throat, as well as the surface of the skin. Many healthy people are carriers of this organism. Most infants become infected during the first week of life from the mother or from another close human contact. In most cases, these strains do not cause disease. Serious staphylococcal infections occur only when the resistance of the host is low because of hormonal changes, debilitating illness, wounds to the skin's surface, or treatment with steroids or other anti-inflammatory drugs. Extensive use of antibiotics has resulted in the natural selection of resistant strains of *Staphylococcus aureus*. Hospital epidemics with these antibiotic-resistant staphylococci often occur in patients whose resistance to infection is lowered due to other diseases, surgical procedures, or drug therapy (∞ Section 22.7). These patients acquire staphylococci from hospital personnel, who are often normal carriers of antibiotic-resistant strains. Control of such hospital epidemics requires careful attention to the maintenance of asepsis.

Certain strains of *Staphylococcus aureus* have been implicated as the agents responsible for **toxic shock syn-**

drome (TSS), a severe result of staphylococcal infection characterized by high fever, rash, vomiting, diarrhea, and occasionally death. Toxic shock was first recognized in menstruating women and was associated with use of tampons. In addition, cases of toxic shock have been reported in both men and women from staphylococcal infections following surgery. In females, blood and mucus in the vagina become colonized by hemolytic *S. aureus* from the skin, and the presence of a tampon concentrates this material, creating ideal microbial growth conditions.

The symptoms of TSS result indirectly from an exotoxin called *toxic shock toxin*. This toxin is a superantigen (∞ Section 20.15). The exotoxin is released by the growing staphylococci, causing the massive T cell reaction and inflammatory response characteristic of superantigen reactions. Since 1981, the incidence of TSS has declined sharply, primarily because of changes in the absorbent materials used in tampons, proper use of superabsorbent tampons, and public awareness of TSS symptoms. However, there is increasing evidence that TSS may also be caused by different exotoxins and Bacteria, including *Streptococcus pyogenes* (see earlier discussion).

Staphylococcal enterotoxin A, which causes the most prevalent form of food poisoning in the United States, is also a superantigen. Presumably, after ingestion of toxin-contaminated food, the toxin stimulates T cells localized along the intestine, resulting in a massive T cell response, release of mediators, and increased permeability of the intestine. The final outcome is the severe but short-lived diarrhea and vomiting associated with this type of food poisoning (∞ Section 24.11).

Finally, appropriate antibiotic therapy for *Staphylococcus aureus* infections is a problem. Although some community-acquired infections are treatable with penicillin, nosocomial (hospital-acquired) infections due to *S. aureus* are often drug-resistant (∞ Section 22.7). Therefore, disease-producing isolates of *S. aureus* must be individually checked for antibiotic sensitivity.

✓ 23.2b Concept Check

Although staphylococci are usually harmless inhabitants of the upper respiratory tract and skin, several serious diseases can result from infection, including some caused by staphylococcal toxins that act as superantigens.

✓ What is the normal habitat of *Staphylococcus aureus*?
✓ What is the role of the superantigen TSST in toxic shock syndrome?

Diphtheria

Corynebacterium diphtheriae, the causative agent of diphtheria, is a gram-positive irregular rod (∞ Section 13.21). *Corynebacterium diphtheriae* enters the body via the respiratory route with cells lodging in the throat and tonsils. Infection is usually spread from healthy carriers or infected individuals to susceptible individuals. Prior infection or immunization (see below) provides complete resistance to infection. Although limited information is available concerning the mechanism of adherence of *C. diphtheriae* to these tissues, the organism produces a neuraminidase capable of splitting *N*-acetylneuraminic acid (a component of glycoproteins found on animal cell surfaces), and this may enhance the invasion process. The inflammatory response of throat tissues to *C. diphtheriae* infection results in formation of a characteristic lesion called a *pseudomembrane* (Figure 23.7), which consists of damaged host cells and cells of *C. diphtheriae.* As described in Section 19.8, certain strains of *C. diphtheriae* are lysogenized by bacteriophage β, and these strains produce a powerful exotoxin, the *diphtheria toxin.* Diphtheria toxin inhibits eukaryotic protein synthesis and thus kills cells (∞ Section 19.8).

The pseudomembrane that forms in diphtheria may block the passage of air, and death from diphtheria is usually due to a combination of the effects of partial suffocation and tissue destruction by exotoxin. Although diphtheria was once a major childhood disease, it is now rarely encountered because an effective vaccine is available. Worldwide, there are still more than 50,000 cases of diphtheria per year, largely because of a lack of immunization. Recent diphtheria outbreaks in South-

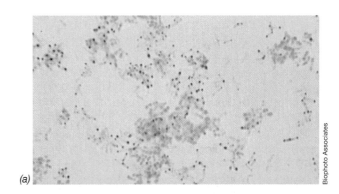

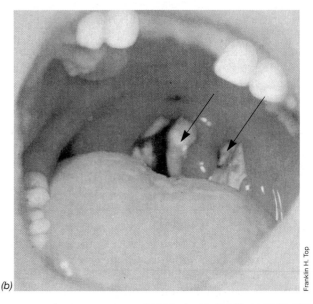

FIGURE 23.7 Diphtheria. (a) Cells of *Corynebacterium diphtheriae* stained to show metachromatic (polyphosphate) granules. (b) Pseudomembrane (arrows) in an active case of diphtheria caused by the bacterium *C. diphtheriae.*

east Asia and in Eastern Europe have been attributed to a lack of vaccination programs or a breakdown in existing vaccination programs, respectively. This vaccine is made by treating the diphtheria exotoxin with formalin to yield an immunogenic, yet nontoxic, toxoid (∞ Section 20.16 discusses toxoids). Diphtheria toxoid is part of the *DTP* (diphtheria, tetanus, pertussis) vaccine (∞ Section 20.16).

A patient diagnosed as having diphtheria by culture of *Corynebacterium diphtheriae* from a pseudomembrane in the throat is usually treated simultaneously with antibiotics and diphtheria antitoxin (an antitoxin contains neutralizing antibodies formed in another animal) (∞ Section 20.16 discusses antitoxins). Penicillin, erythromycin, or gentamicin is generally effective for diphtheria therapy. Early administration of both antibiotics and antitoxin is necessary for effective control of the disease.

✓ 23.2c Concept Check

Diphtheria is caused by the gram-positive bacterium *Corynebacterium diphtheriae*. A standard early childhood vaccine (DTP) is very effective in preventing this very serious respiratory disease.

✓ Is the pathogenesis of diphtheria due to infection?
✓ How can the spread of diphtheria be prevented?

Whooping Cough

Whooping cough (pertussis) is an acute, highly infectious respiratory disease generally observed in children under 1 year of age. Whooping cough is caused by a small, gram-negative, strictly aerobic coccobacillus, *Bordetella pertussis*. The organism attaches to cells of the upper respiratory tract by producing a specific adherence factor called *filamentous hemagglutinin antigen*, which recognizes a complementary molecule on the surface of host cells. Once attached, *B. pertussis* grows and produces pertussis exotoxin that induces synthesis of cyclic adenosine monophosphate (cyclic AMP) (∞ Section 7.6), which is at least partially responsible for the events that lead to host tissue damage. *Bordetella pertussis* also produces an endotoxin, which also may induce some of the symptoms of whooping cough. Clinically, whooping cough is characterized by a recurrent, violent cough that usually lasts up to 6 weeks. The spasmodic coughing gives the disease its name, for a whooping sound results from the patient inhaling in deep breaths to obtain sufficient air.

A vaccine consisting of killed whole cells or proteins derived from *Bordetella pertussis* is part of the routinely administered DTP vaccine. This vaccine, while normally very effective, must be given to susceptible individuals, usually children, at appropriate intervals beginning soon after birth (∞ Section 20.16). In the United States, up to 50% of children who acquire whooping cough have not been properly immunized. The threat of this very communicable disease remains high, as illustrated by epidemic outbreaks in several cities in the United States and in the former Soviet Union, presumably as a result of recent breakdowns in public health programs. In recent years, there has been an alarming upward trend in the numbers of cases of pertussis in the United States (Figure 23.8).

Because of undesirable side effects of pertussis vaccine, including local swelling and redness, fever, and occasional more serious problems such as encephalitis and convulsions, a "second generation" pertussis vaccine, containing purified cell fractions of *Bordetella pertussis* rather than whole cells, is now approved for use in the United States (∞ Figure 20.27).

Diagnosis of whooping cough can be made by fluorescent antibody staining of throat smears or by actually culturing the organism. For culture of *Bordetella pertussis*,

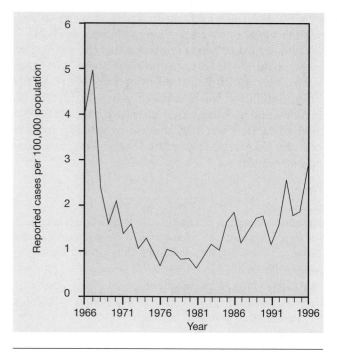

FIGURE 23.8 Pertussis (whooping cough) prevalence in the United States. Note the general upward trend starting about 1982 and the three to four year periodicity. In 1996, there were 7796 cases of pertussis, the highest incidence reported since 1967.

the "cough-plate" method is used. The patient is asked to cough directly into a blood–glycerol–potato extract agar plate (although not selective, this medium supports good recovery of *B. pertussis*). Alternatively, throat and nose swabs (if streaked immediately after sampling) can be used. β-Hemolytic colonies containing small gram-negative coccobacilli are tested for *B. pertussis* by a latex bead agglutination test or are stained with an anti-*B. pertussis* fluorescent antibody for positive identification (∞ Sections 21.5 and 21.7). Cultures of *B. pertussis* are killed by ampicillin, tetracycline, and erythromycin, although antibiotics alone do not seem to be sufficient to kill the pathogen *in vivo*. Because a patient with whooping cough remains infectious for up to 2 weeks following commencement of antibiotic therapy, the immune response may be as important, if not more so, than antibiotics, in the elimination of *B. pertussis* from the body.

✓ 23.2d Concept Check

In the United States, there has been a disturbing increase in the number of annual cases of whooping cough in the last decade. From 1975 to 1982, there was an average of less than 2000 cases per year, but that number has risen to more than 4000 cases per year at present. Up to 60% of preschool children in some cities are inadequately immunized, creating a major potential public health threat.

✓ What measures can be taken to decrease the current incidence of whooping cough in a population?

23.3

Mycobacterium and Tuberculosis

The famous German microbiologist Robert Koch isolated and described the causative agent of tuberculosis, *Mycobacterium tuberculosis,* in 1882 (see the box, Discoverers of the Main Bacterial Pathogens). At one time, tuberculosis was the single most important infectious disease of humans and accounted for one-seventh of all deaths worldwide. At present in the United States, nearly 20,000 new cases of tuberculosis are diagnosed each year. Worldwide, tuberculosis still accounts for almost 3 million deaths per year, more than 5% of all deaths, and up to one-third of the world's population have been infected with *M. tuberculosis* (∞ Table 22.1). In recent years, many of the new tuberculosis cases in the United States result at least in part from the elevated incidence of tuberculosis in acquired immunodeficiency syndrome (AIDS) patients. As many as 2000 people die every year from tuberculosis in the United States alone.

Pathology of Tuberculosis

The microbiology of *Mycobacterium tuberculosis* is discussed in Section 13.22. The interaction of the human host and *M. tuberculosis* is extremely complex, being determined in part by the virulence of the strain but also by the specific and nonspecific resistance of the host. Cell-mediated immunity plays an important role in the development of disease symptoms. It is convenient to distinguish between two kinds of human tuberculosis infections: *primary* and *postprimary* (or reinfection). Primary infection is the first infection that an individual acquires and usually results from inhalation of droplets containing viable bacteria from an individual with an active pulmonary infection. Dust particles that have become contaminated from sputum of tubercular individuals are another source of primary infection. The bacteria settle in the lungs and grow. A delayed-type hypersensitivity reaction (∞ Sections 20.9 and 20.15) results in the formation of aggregates of activated macrophages, called *tubercles,* characteristic of tuberculosis. However, the bacteria are often able to survive

---

LEARNING FROM THE PAST . . .   Discoverers of the Main Bacterial Pathogens

The history of the discovery of the microbial role in infectious disease was described in Chapter 1. Once the concept of specific microbial disease agents was clarified and the procedures for culture of microorganisms developed, it was a relatively simple procedure to isolate a large number of microbial pathogens. The decades surrounding the formulation of Koch's postulates (1884) were indeed fruitful for medical microbiology. The rapid development of this field is indicated by the accompanying table, which lists the main bacterial pathogens isolated during the "golden age of bacteriology." ■

Year	Disease	Organism	Discoverer
1873	Leprosy	*Mycobacterium leprae*	Hansen, G. A.
1877	Anthrax	*Bacillus anthracis*	Koch, R.
1878	Suppuration	*Staphylococcus*	Koch, R.
1879	Gonorrhea	*Neisseria gonorrhoeae*	Neisser, A. L. S.
1880	Typhoid fever	*Salmonella typhi*	Eberth, C. J.
1881	Suppuration	*Streptococcus*	Ogston, A.
1882	Tuberculosis	*Mycobacterium tuberculosis*	Koch, R.
1883	Cholera	*Vibrio cholerae*	Koch, R.
1883	Diphtheria	*Corynebacterium diphtheriae*	Klebs, T. A. E.
1884	Tetanus	*Clostridium tetani*	Nicolaier, A.
1885	Diarrhea	*Escherichia coli*	Escherich, T.
1886	Pneumonia	*Streptococcus pneumoniae*	Fraenkel, A.
1887	Meningitis	*Neisseria meningitidis*	Weichselbaum, A.
1888	Food poisoning	*Salmonella enteritidis*	Gaertner, A. A. H.
1892	Gas gangrene	*Clostridium perfringens*	Welch, W. H.
1894	Plague	*Yersinia pestis*	Kitasato, S., Yersin, A. J. E. (independently)
1896	Botulism	*Clostridium botulinum*	van Ermengem, E. M. P.
1898	Dysentery	*Shigella dysenteriae*	Shiga, K.
1900	Paratyphoid	*Salmonella paratyphi*	Schottmüller, H.
1903	Syphilis	*Treponema pallidum*	Schaudinn, F. R., and Hoffman, E.
1906	Whooping cough	*Bordetella pertussis*	Bordet, J., and Gengou, O.

and grow to some extent within the macrophages. In individuals with low resistance, the bacteria are not effectively controlled, and an acute pulmonary infection occurs, which can lead to the extensive destruction of lung tissue, the spread of the bacteria to other parts of the body, and death.

In most cases of tuberculosis, however, acute infection does not occur, and the infection remains localized and is usually inapparent; later it subsides. But this initial infection hypersensitizes the individual to the bacteria or their products and consequently alters the response of the individual to subsequent *M. tuberculosis* exposures. A diagnostic test, called the **tuberculin test,** can be used to measure this hypersensitivity. When *tuberculin,* a protein fraction extracted from *Mycobacterium tuberculosis,* is injected intradermally into a hypersensitive individual, it elicits a localized immune reaction within 1–3 days at the site of injection. The reaction is characterized by *induration* (hardening) and *edema* (swelling) (∞ Figure 20.25). An individual exhibiting this reaction is said to be *tuberculin-positive,* and many healthy adults give positive reactions as a result of previous inapparent infections. A positive tuberculin test does not indicate active disease but only that the individual has been exposed to the organism in the past and has generated a cell-mediated immune response.

For most individuals, this immunity is protective and life-long. However, some tuberculin-positive patients develop postprimary tuberculosis through reinfection from outside sources or as a result of reactivation of bacteria that have remained alive but dormant in lung macrophages, often for years. Factors such as aging, malnutrition, overcrowding, stress, and hormonal imbalance all play a role in predisposing individuals to reinfection by reducing effective immunity (∞ Section 19.12) and allowing reactivation of dormant infections.

When renewed pulmonary infections occur, they often progress to chronic infections that result in destruction of lung tissue, followed by partial healing and calcification at the infection site. Thus, chronic postprimary tuberculosis often results in a gradual spread of tubercular lesions in the lungs. Areas of destroyed tissue are seen by X-ray examination (Figure 23.9), but bacteria are found in the sputum only in individuals with extensive tissue destruction.

Control and Treatment

Individuals who have active cases of tuberculosis may spread the disease simply by coughing on or speaking to uninfected individuals. Because tuberculosis is so highly contagious, the United States Occupational Safety and Health Administration has stringent requirements for the protection of health care workers who are responsible for

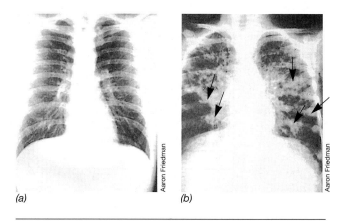

(a)   *(b)*

FIGURE 23.9 X-ray photographs. (a) Normal chest X-ray. The faint white lines are arteries and other blood vessels. The heart is visible as a white bulge in the lower right quadrant. (b) An advanced case of pulmonary tuberculosis; white patches (arrows) indicate areas of disease. These patches, or tubercles, may contain live *Mycobacterium tuberculosis.* Lung tissue and function is permanently destroyed by these lesions.

tuberculosis patient care. For example, patients with infectious tuberculosis must be hospitalized in negative-pressure rooms. In addition, health care workers who have patient contact must be provided with personally fitted face masks with high energy particulate air (HEPA) filters. These special filters prevent the passage of *Mycobacterium tuberculosis* in sputum or on dust particles.

Chemotherapy of tuberculosis has been a major factor in control of the disease. The initial success in chemotherapy occurred with the introduction of streptomycin, but the real revolution in tuberculosis treatment came with the discovery of isonicotinic acid hydrazide (isoniazid or INH) (Figure 23.10), a nicotinamide derivative virtually specific for mycobacteria. This agent is not only effective and free from toxicity but is also inexpensive and readily absorbed when given orally. Although the mode of action of INH is not completely understood, it apparently affects the synthesis of mycolic acid by *Mycobacterium* (mycolic acid is a complex lipid that complexes with the peptidoglycan of the mycobacterial cell wall) (∞ Section 13.22). INH may

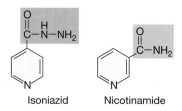

Isoniazid   Nicotinamide

FIGURE 23.10 Structure of isoniazid (isonicotinic acid hydrazide), an effective chemotherapeutic agent for tuberculosis. Note the structural similarity to nicotinamide.

act by mimicking the activity of a structurally related molecule, nicotinamide (Figure 23.10), becoming incorporated in place of nicotinamide and thus inactivating enzymes requiring this compound for activity. Treatment of mycobacteria with very small amounts of INH (as little as 5 picomoles [pmol] per 10^9 cells) results in complete inhibition of mycolic acid synthesis, and continued incubation results in a complete loss of outer membrane areas of the cell, a loss of cellular integrity, and death. Following treatment with INH, mycobacteria lose their acid–alcohol fastness, in keeping with the role of mycolic acid in this staining property (⚬ Section 13.22). However, mycobacterial resistance to INH and other drugs is increasing at an alarming rate, especially in AIDS patients (see Section 23.7).

The development of drug resistance may be encouraged by patients themselves because many patients do not complete their therapy. The typical course of INH treatment is 12 months, and patients must take daily medication throughout this time to eradicate the tubercle bacilli. Failure to complete the entire prescribed treatment may reactivate the infection, and the reactivated organisms often are resistant to the original treatment drug. In certain populations, such as in hospitals and nursing homes, patients are routinely treated with several drugs simultaneously because of the emergence and prevalence of multiple-drug–resistant tuberculosis.

Mycobacterium leprae and Hansen's Disease

Mycobacterium leprae is the causative agent of the ancient and dreaded *Hansen's disease,* or **leprosy**. *M. leprae* is the only *Mycobacterium* species that has not been grown on artificial media. The only experimental animal that has been successfully used to grow *M. leprae* and reproduce a similar disease is the armadillo. The most serious form of Hansen's disease is characterized by folded, bulblike lesions on the body, especially on the face and extremities (Figure 23.11) due to growth of *M. leprae* cells in the skin. The lesions contain up to 10^9 bacterial cells per gram of tissue. Like *M. tuberculosis, M. leprae* from the lesions stain deep red with carbol fuschin in the acid-fast staining procedure, providing a rapid, definitive demonstration of active infection (⚬ Section 13.22). This *multibacillary* or *lepromatous* form of leprosy has a very poor prognosis. In severe cases the disfiguring lesions lead to destruction of peripheral nerves and loss of motor function. Many patients exhibit less pronounced lesions from which no bacterial cells can be recovered. These individuals have the *tubercular* or *paucibacillary* form of the disease. Tubercular leprosy is characterized by a vigorous delayed-type hypersensitivity response (⚬ Section 20.15) and a good prognosis for spontaneous recovery. Hansen's disease of

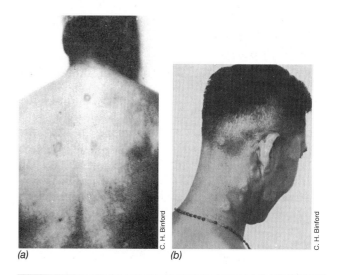

FIGURE 23.11 Leprosy lesions on the skin due to infection with *Mycobacterium leprae.*

either form, and the continuum of intermediate forms between these extremes, is treated using a multiple drug therapy (MDT) protocol, which includes some combination of dapsone (4,4′-sulfonylbisbenzeneamine), rifampin, and clofazimine. As in the case of tuberculosis, drug-resistant organisms have appeared, especially with single drugs or inadequate treatment. Extended drug therapy of up to one year with an MDT protocol is required for eradication of the organism.

The pathogenicity of *M. leprae* is probably due to a combination of delayed hypersensitivity (⚬ Section 20.15) and the invasiveness of the organism. Transmission probably involves both the direct contact and respiratory routes, and incubation time varies from several weeks to years or even decades. *M. leprae* grows within macrophages, causing an intracellular infection that can result in the enormous population of bacteria within the skin. In many areas of the world, the incidence of Hansen's disease is very low. For example, in the United States, fewer than 300 cases are diagnosed every year. Worldwide, areas of high disease incidence are concentrated in central and South America, Africa, and Southeast Asia. Hansen's disease has currently been diagnosed in 1.2 million people, with about 500,000 new cases occurring every year. However, leprosy may go unreported in as many as 12 million people.

Other Pathogenic *Mycobacterium* spp.

A common pathogen of dairy cattle, *Mycobacterium bovis,* is pathogenic for humans as well as other animals. *M. bovis* enters humans via the intestinal tract, typically from the ingestion of raw milk. After a localized intestinal infection, the organism eventually spreads to the respiratory tract and initiates the classic symptoms of tuberculosis. We do not know whether *M. bovis* is really a different organism from *M. tuberculosis* because the two organisms are nearly 100% homologous at the DNA level when examined and compared by DNA hybridization methods (⬥ Section 12.9). Pasteurization of milk and elimination of diseased cattle have essentially eradicated bovine-to-human transmission of tuberculosis.

A number of other *Mycobacterium* species are also occasional human pathogens. For example, *M. kansasii*, *M. scrofulaceum*, *M. chelonae*, and other members of the genus (⬥ Section 13.22) cause disease. Tuberculosis due to the *M. avium* complex (MAC) group of organisms is particularly prevalent in AIDS patients as compared to members of the normal population (see Section 23.7).

✓ 23.3 Concept Check

Tuberculosis is one of the most prevalent and dangerous single diseases in the world. Its incidence is on the increase in developed countries, in part because of the emergence of drug-resistant strains. The pathology of tuberculosis and leprosy is influenced by the cellular immune response.

✓ Why is *Mycobacterium tuberculosis* such a widespread respiratory pathogen?

✓ Describe factors that contribute to the incidence of drug resistance in mycobacterial infections.

23.4

Respiratory Pathogens: Viral

As we discussed (⬥ Section 18.10), viruses are less easily controlled by chemotherapeutic means than bacteria or other microorganisms because the growth of viruses is intimately tied to host cell functions. Most chemotherapeutic agents that specifically attack viruses cause at least some harm to host cells as well. Not surprisingly, therefore, the most prevalent infectious diseases, especially in developed countries, are of viral etiology (Figure 23.12). Fortunately, most viral diseases are acute, self-limiting infections that are rarely problematic in normal healthy adults. In addition, serious viral diseases such as smallpox and rabies have been effectively controlled by immunization. We begin here by describing the two most common viral infections, the common cold and influenza, and proceed to discuss measles, mumps, and chickenpox; these viral diseases are all transmitted in infectious droplets by an airborne route.

The Common Cold

The common cold is one of the most prevalent diseases of children and adults. Estimates suggest that each person averages two to five colds per year throughout his or her lifetime. The symptoms include rhinitis (inflam-

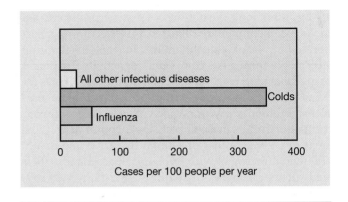

FIGURE 23.12 Viruses are the leading causes of acute infectious disease in the United States. The data are typical for recent years.

mation of the nasal region, especially the mucous membranes), nasal obstruction, watery nasal discharges, and a general feel of malaise, usually without an accompanying fever. Rhinoviruses, single-stranded ribonucleic acid (RNA) viruses of the picornavirus group (see Figure 23.13a and ⬥ Section 8.15), are the most common causes of colds. At least 115 different serotypes of rhinoviruses have been identified. Another group of single-stranded RNA viruses, the coronaviruses (Figure 23.13b), are responsible for about 15% of all colds in adults. A variety of other viruses including adenoviruses, coxsackie viruses, respiratory syncytial virus, and orthomyxoviruses, are responsible for about 10% of common colds. Colds generally induce a specific, local, neutralizing IgA response (⬥ Section 20.5). However, the number of potential infectious agents makes immunity via immunization or previous exposure very unlikely.

Aerosol transmission of the virus is probably the major means of spreading colds, although experiments with human volunteers suggest that direct contact and/or fomite contact is also an important method of transmission. For example, one effective experimental method for the prevention of rhinovirus spread is through the use of disposable tissues impregnated with antiviral disinfectant agents. Most antiviral drugs are ineffective, but a pyrazidine derivative (Figure 23.14a) has proven promising for preventing colds in volunteers. In addition, new experimental antiviral drugs are being designed based on information derived from three-dimensional structures. For example, the antirhinovirus drug WIN 52084 (Figure 23.14b) binds to the virus, changing its three-dimensional surface configuration and disrupting a cellular binding site, thus preventing infection. Interferon-α, a cytokine (⬥ Section 20.8), is also effective in preventing the onset of colds. Thus, there are several experimental possibilities for cold prevention and treatment, although none are widely accepted as effective and safe. The accepted treatment

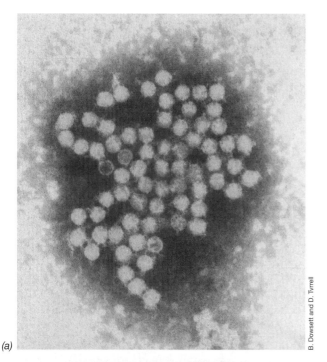

(a)

B. Dowsett and D. Tyrrell

(b)

Heather Davies and D. Tyrrell

FIGURE 23.13 Electron micrographs of some common cold viruses. (a) Human rhinovirus. (b) Human coronavirus. Each rhinovirus virion is about 30 nm in diameter. Each coronavirus virion is about 60 nm in diameter.

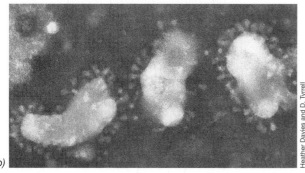

(a)

(b)

FIGURE 23.14 Experimental antirhinovirus drugs. (a) The structure of 3-methoxy-6-[4-(3-methylphenyl)]-1-piper-azinyl. (b) The structure of WIN 52084, a receptor-blocking drug.

for colds is to treat the symptoms, especially nasal discharges, with a variety of antihistamine and decongestant drugs.

Influenza

Influenza is caused by an RNA virus of the orthomyxovirus group (Section 8.16). Influenza virus is an enveloped virus, the single-stranded, helical RNA genome being surrounded by an envelope made up of protein, a lipid bilayer, and external glycoproteins (see Figure 23.15 and Figure 8.38 and Section 8.16). Human influenza virus exists in nature only in humans. It is transmitted from person to person through the air, primarily in droplets expelled during coughing and sneezing. The virus infects the mucous membranes of the upper respiratory tract and occasionally invades the lungs. Symptoms include a low grade fever for 3–7 days, chills, fatigue, headache, and general aching (see the box, Is It a Cold or Is It the Flu?) Recovery is usually spontaneous and rapid. Most of the serious consequences of influenza infection do not occur because of the viral infection but because bacterial invaders may be able to develop as secondary infections in persons whose resistance has been lowered. Especially in infants and elderly people, influenza is often followed by bacterial pneumonia; death, if it occurs, is usually due to the bacterial infection.

Influenza occurs every year in epidemics and often occurs in pandemics. Early pandemics, of which the one in 1918 is the most famous, occurred before knowledge was sufficiently advanced to make careful analysis possible, but the 1957 pandemic of the so-called Asiatic flu provided an opportunity for careful study of how a

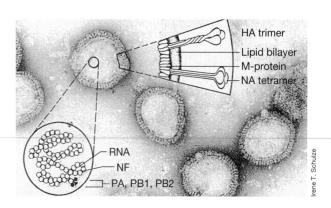

Irene T. Schulze

FIGURE 23.15 Electron micrograph of the influenza virus, showing the location of the major viral coat proteins and the nucleic acid. Each virion is about 100 nm in diameter. HA, Hemagglutinin (three copies make up the HA coat spike); NA, neuraminidase (four copies make up the NA coat spike); M, coat protein; NP, nucleoprotein; PA, PB1, PB2, other internal proteins, some of which may have enzymatic functions.

A FOCUS ON . . .   Is It a Cold or Is It the Flu?

The symptoms of a common cold and symptoms of "the flu" (influenza) often seem similar, but the two diseases are distinct and caused by quite different viruses. A typical common cold caused by a rhinovirus is associated with nasal discharges, cough, chills, and perhaps a sore throat. Influenza, caused by an orthomyxovirus, is generally associated with a different set of symptoms. Although either condition may make one feel miserable for a period, colds are usually of shorter duration and the symptoms are milder. The following can serve as a guideline for determining whether you have "caught a cold" or "caught the flu." ■

Symptoms	Common cold	Influenza
Fever	Rare	Common (39–40°C); sudden onset
Headache	Rare	Common
General malaise	Slight	Common; often quite severe; can last several weeks
Nasal discharge	Common and abundant	Less common; usually not abundant
Sore throat	Common	Much less common
Vomiting and/or diarrhea	Rare	Common

worldwide epidemic develops (Figure 23.16). The epidemic probably arose when a virulent mutant virus strain that differed from all previous strains in antigenicity appeared in the population. Since immunity to this strain was not present, the virus was able to spread rapidly throughout the world. It first appeared in the interior of China in late February 1957 and by early April had been brought to Hong Kong by refugees. It spread from Hong Kong along air and naval routes and was apparently transferred to San Diego, California, by naval ships. In May, an outbreak occurred in Newport, Rhode Island, on a naval vessel. Other outbreaks occurred in various parts of the United States. Peak incidence occurred in the last 2 weeks of October, during which time 22 million new cases developed. Afterward, there was a progressive decline.

The genetic material of influenza virus, single-stranded RNA, is arranged in a highly unusual manner. As discussed in Section 8.16, the influenza virus genome is *segmented*, with genes found on each of eight distinct fragments of its single-stranded RNA (👓 Figure 8.38 and Section 8.16). Such an arrangement allows the rapid

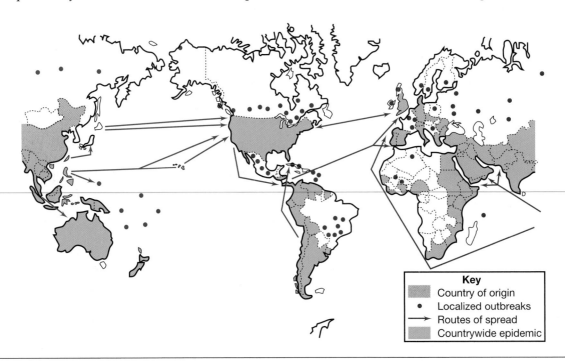

Key
- ▨ Country of origin
- • Localized outbreaks
- → Routes of spread
- ▤ Countrywide epidemic

FIGURE 23.16   Route of spread of a major influenza epidemic, the Asian flu pandemic of 1957.

and constant reassortment of genes with genes from a different strain of influenza virus because more than one strain of influenza virus can infect a cell at one time. This allows reassorted viruses to arise at frequent intervals and prevents complete control of influenza by immunization. This reassortment of genes in different strains of influenza virus usually manifests itself in the phenomenon called *antigenic shift*. Antigenic shift refers to modifications in the protein coat of the virions, especially to two proteins important in the attachment and eventual release of virus from host cells, hemagglutinin and neuraminidase, respectively (∞ Figures 8.38 and 23.15). Immunity to influenza in humans is largely dependent on the production of secretory antibody (IgA) (∞ Section 20.5), especially to antigenic determinants of the hemagglutinin and neuraminidase proteins.

Once a strain of influenza virus has passed through the population, a majority of the people are immune to that strain, and it is impossible for a strain of similar antigenic type to cause an epidemic for about 3 years. During this time, the hemagglutinin and neuraminidase antigens exhibit frequent minor antigenic variation because of genetic mutations that result in the change of one or more amino acids. This phenomenon is known as *antigenic drift* and is responsible for the recurrence of minor epidemics of influenza in a 2- to 3-year cycle. There is some evidence that in 1918 (the year of the unusually serious *pandemic*, or worldwide epidemic), the strain may have originated from a related virus that infects swine (swine flu) and that the 1957 strain may have arisen from a similar animal reservoir (perhaps a wild animal) somewhere in Asia.

Influenza epidemics can be controlled by immunization, but the choice of appropriate vaccines is complicated by the large number of existing strains and the ability of these strains to undergo antigenic drift or antigenic shift. When new strains evolve, vaccines are not immediately available, but through careful worldwide surveillance (∞ Section 22.8), samples of the major emerging strains of influenza virus are usually obtained before epidemic outbreaks occur. In the United States, inactivated viral preparations from three candidate strains are mixed to prepare a *polyvalent* vaccine that is then used for immunization prior to the next "flu season," which usually starts in late autumn and continues through winter. Influenza immunization is recommended for those individuals most likely to succumb or be exposed to serious secondary illnesses, such as the elderly (more than 65 years of age), those suffering from chronic debilitating diseases (e.g., AIDS patients, see Section 23.7), and health care workers. The duration of effective artificial immunity from the inactivated influenza vaccine is usually only a few years, and of course it is strain-specific. Therefore, annual immunization is usually necessary.

Influenza may also be controlled by use of the chemicals *amantadine* and *ramantadine*. These drugs inhibit viral replication and have been used as chemoprophylactic agents to prevent the spread of influenza to those at high risk. They are also used in treatment to shorten the course and severity of infection. The treatment of influenza with aspirin is not recommended, as there is evidence of a link between aspirin treatment of influenza and Reye's syndrome (a rare but occasionally fatal affliction involving the central nervous system) in children.

Measles

Measles (rubeola) virus causes an acute highly infectious childhood disease characterized by nasal discharges, redness of the eyes, cough, and fever. The measles virus is a paramyxovirus (∞ Section 8.16) that enters the nose and throat by airborne transmission, quickly leading to systemic viremia. As the disease progresses, fever and cough appear and rapidly intensify, and a rash appears (Figure 23.17); in most cases measles lasts a total of 7–10 days. Circulating antibodies to measles virus are measurable about 5 days after initiation of infection, and both serum antibodies and cytotoxic T lymphocytes (∞ Sections 20.5 and 20.9) combine to eliminate the virus from the system. A variety of complications may occur due to measles infection, including inner ear infection, pneumonia, and, in rare cases, measles encephalomyelitis. Encephalomyelitis can cause neurological disorders and a form of epilepsy, and has a mortality rate of nearly 20%.

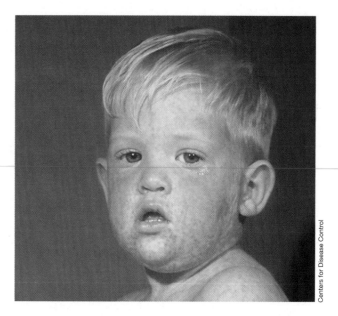

FIGURE 23.17 Typical rash associated with measles in children.

Although once a common childhood illness, measles generally occurs nowadays in rather isolated outbreaks because of widespread immunization programs begun in the mid-1960s (Figure 23.18*a*). In the United States, all public school systems require proof of measles immunization before allowing children to enroll because of the highly infectious nature of the disease. Active immunization is done with the MMR (measles, mumps, and rubella) vaccine (⌘ Table 20.11 and Figure 20.27). A childhood case of measles generally confers lifelong immunity to reinfection.

Mumps

Mumps is caused by a different paramyxovirus than that causing measles and is also highly infectious. Mumps is spread by airborne droplets, and the disease is characterized by inflammation of the salivary glands leading to swelling of the jaws and neck (Figure 23.19). The virus spreads through the bloodstream and may infect other organs including the brain, testes, and pancreas. Severe complications may include encephalitis and sterility. The host immune response produces antibodies to mumps virus surface proteins, and this generally leads to a quick recovery. An attenuated mumps vaccine is highly effective in preventing the disease (Fig-

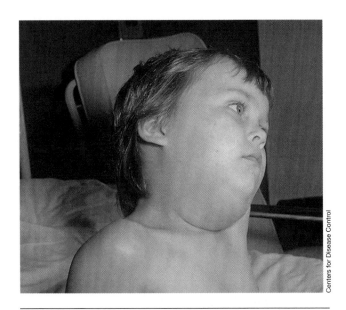

FIGURE 23.19 Typical glandular swelling associated with mumps.

ure 23.18*b*). Hence, like measles, the incidence of mumps in developed countries has been greatly reduced in the last three decades, with mumps epidemics usually restricted to those individuals who did not receive the MMR vaccine during childhood.

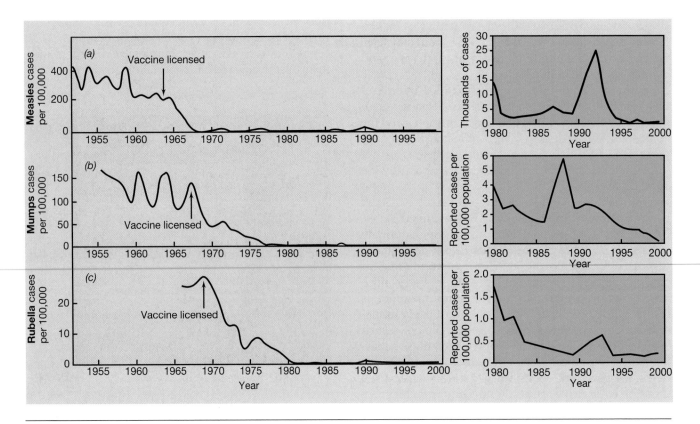

FIGURE 23.18 The effect of vaccines on the prevalence in the United States of the major childhood viral diseases now controlled by the MMR (measles, mumps, rubella) vaccine. (a) Measles. (b) Mumps. (c) Rubella. The insets show a more detailed picture of these diseases in the past two decades. Data were obtained from U.S. Centers for Disease Control, Atlanta, GA.

Rubella

Rubella (*German measles*) is caused by a single-stranded RNA virus of the togavirus group (∞ Section 8.15). The symptoms of the disease resemble those of measles but are generally milder. Rubella is less contagious than true measles, and thus a good proportion of the population has never been infected. However, during the first 3 months of pregnancy rubella virus can infect the fetus by placental transmission and cause a host of serious fetal abnormalities. Rubella can cause stillbirth, or deafness, heart and eye defects, and brain damage in live births. Thus, it is important that pregnant women not be immunized with the rubella vaccine, or contract rubella during this period. For this reason, routine childhood immunization against rubella should be practiced. An attenuated virus vaccine is administered with attenuated measles and mumps viruses in the MMR vaccine mentioned previously (see Figure 23.18c and ∞ Table 20.11 and Figure 20.27).

Chickenpox and Shingles

Chickenpox (varicella) is a common childhood disease caused by a herpes virus (∞ Section 8.19). Chickenpox is highly contagious and is transmitted by infectious droplets, especially when susceptible individuals are in close contact. The incidence of chickenpox shows a disease cycle typical of a respiratory infection (Figure 23.20).

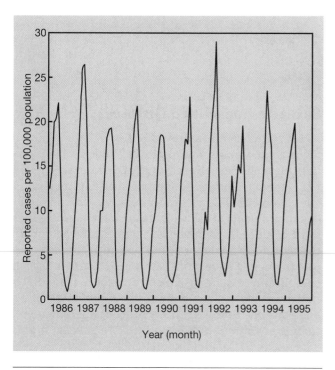

FIGURE 23.20 Reported prevalence of chickenpox (varicella) by month in the United States, 1986–1995. Note the high winter/spring seasonal incidence, typical of a disease transmitted by the respiratory route. In 1996 there were 83,511 cases reported, estimated to be about 5% of the total number of cases.

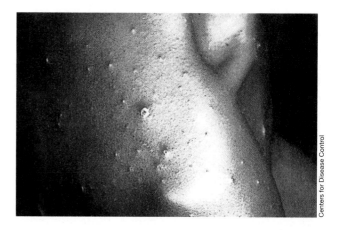

FIGURE 23.21 Mild papular rash associated with the disease chickenpox.

In schoolchildren, for example, close confinement during the winter months leads to the spread of chickenpox by airborne secretions from infected classmates and through contact with contaminated fomites. The virus enters the respiratory tract, multiplies, and is quickly disseminated via the bloodstream, resulting in a systemic papular rash that quickly heals, rarely leaving disfiguring marks (Figure 23.21). For unknown reasons, early chickenpox vaccines have been poorly immunogenic, but a successful and highly protective attenuated virus vaccine is now recommended for use in the United States (∞ Figure 20.27). Although the immunization has been in widespread use only since 1994, the incidence of varicella cases has already been reduced to about one-third of the historical annual incidence.

The chickenpox virus can remain dormant in nerve cells for years with no apparent symptoms. The virus occasionally migrates from this reservoir to the skin surface, causing a painful skin eruption referred to as **shingles** (zoster). Shingles most commonly strikes immunosuppressed individuals or the elderly. Studies with human volunteers suggest that T cells are important in destroying the virus. The prophylactic use of human hyperimmune globulin prepared against the virus is useful for preventing the onset of symptoms of shingles. Such therapy is advised only for patients where secondary infections occasionally associated with shingles, such as pneumonia or encephalitis, may be life-threatening.

✓ 23.4 Concept Check

Most respiratory infections are caused by viruses. The most prevalent is the common cold, and the most serious is influenza, which occurs in cyclical epidemics. These diseases all share a common mode of transmission, the airborne route.

✓ List the viral respiratory diseases for which an effective vaccine is *not* available.
✓ Suggest a treatment protocol for acute influenza.

23.5

Why Are Respiratory Infections So Common?

Respiratory infections have several common features. The infectious agents can be dispersed in the air in microdroplets, or they can be transmitted in dried form on dust or on the surfaces of fomites. Respiratory disease incidence often shows dramatic seasonal fluctuations. Because respiratory infections are transmitted by the normal everyday human activities of coughing, sneezing, talking, and even breathing, they are the most common person-to-person infectious diseases.

Respiratory infections are difficult to control for several reasons. First, many respiratory infections are caused by viruses; antibiotics and most other therapeutic agents are completely ineffective in controlling viral diseases. Second, most respiratory diseases are acute, showing a rapid onset of symptoms and rapid recovery. Although this rapid disease cycle may seem beneficial for the host, it creates special problems for uninfected individuals because it allows transmission of the infectious agent from individuals who appear healthy but are already infected and spreading the disease. Patients suffering from the common cold and from chickenpox, for example, can spread live virus before symptoms begin; because of this presymptomatic infectious stage, chickenpox spreads in an explosive fashion from a few infected asymptomatic individuals to previously unexposed school children every winter (Figure 23.20). Even tuberculosis, arguably the most serious infectious respiratory disease, is spread largely by individuals who have active infection but exhibit relatively benign symptoms, such as a mild cough. Third, effective immunizations are not available for a number of respiratory diseases. This can be due to unusual biological properties of the pathogen. For the influenza virus, antigenic drift and antigenic shift continually generate new influenza strains for which entire populations lack effective immunity, thus making it necessary to produce new vaccines every year. Also, unusually large numbers of genetic variants of a particular pathogen make vaccine production very difficult. For example, there are 115 serotypes of rhinovirus, the agent responsible for more than 50% of common colds, and more than 60 serotypes of *Streptococcus pyogenes*, the agent responsible for scarlet fever, rheumatic fever, and necrotizing fasciitis. A single multivalent vaccine would probably not be effective for all variants of each pathogen, and it would be prohibitively expensive and inconvenient to deliver all possible single-antigen vaccines. In addition, vaccines against some pathogens, for example against *S. pyogenes*, might actually harm the host. Immune reactions to antigens on *S. pyogenes* might

cross react with host cell antigens and destroy host cells (➤ Section 20.15). Finally, many respiratory infections are transmitted from healthy *carriers* to susceptible hosts. For example, in the case of streptococcal infections, up to 20% of the adult population carry strains of *Streptococcus pyogenes* in their nose and throat, apparently without ill effect.

Respiratory diseases are a major concern for public health officials. Because many respiratory diseases cause discomfort but are not life-threatening, the tendency for many individuals suffering a respiratory illness is to simply carry on normal activities. Unfortunately, these individuals are infecting others. Thus, respiratory pathogens present a variety of practical problems that complicate control measures. Respiratory pathogens that spread exclusively by the person-to-person route have evolved a highly successful high-morbidity (illness), low-mortality (death) strategy for maintaining themselves in nature.

✓ 23.5 Concept Check

Acute respiratory diseases spread in a predictable seasonal pattern due largely to normal human activities. Many respiratory infections are very difficult to prevent, control, or treat.

✓ Give an example of seasonal incidence and inability to control the spread of a viral or bacterial respiratory pathogen. What factors play a role in promoting spread of the pathogen?

✓ Give another example of a viral or bacterial respiratory pathogen that has been controlled. How was control achieved?

23.6

Sexually Transmitted Diseases

Various sexually transmitted diseases (STDs) are caused by bacteria, viruses, and protozoa. Table 23.1 summarizes the major sexually transmitted diseases seen today. STDs are of major medical importance worldwide.

Control of STDs presents unusually difficult public health problems for several reasons. First, nearly one-third of all cases involve teenagers. Sexual activity in this age group is common, and sexually active individuals frequently have more than one sex partner, further complicating the disease picture. Second, many STDs initially cause no symptoms, or the symptoms that develop may be confused with the symptoms of non-sexually transmitted diseases. Third, the social stigma attached to STDs inhibits some individuals from seeking prompt medical care.

On the other hand, there are many reasons to seek medical treatment of STDs. First of all, most sexually transmitted diseases are curable or at least can be controlled with medical intervention. As shown in Table 23.1, antibiotic and chemotherapeutic agents are highly effective in

TABLE 23.1	Summary of some sexually transmitted diseases and treatment guidelines	
Disease	**Causative organisms**[a]	**Recommended treatment**[b]
Gonorrhea	*Neisseria gonorrhoeae* (B)	Cefixime or ceftriaxone, *and* azithromycin or doxycycline
Syphilis	*Treponema pallidum* (B)	Benzathine penicillin G
Chlamydia trachomatis infections	*Chlamydia trachomatis* (B)	Doxycycline or azithromycin
Nongonococcal urethritis	*C. trachomatis* (B) or *Ureaplasma urealyticum* (B) or *Mycoplasma genitalium* (B) or *Trichomonas vaginalis* (P)	Azithromycin
Lymphogranuloma venereum	*C. trachomatis* (B)	Doxycycline
Chancroid	*Haemophilus ducreyi* (B)	Azithromycin
Genital herpes	Herpes simplex type 2 (V)	No known cure; symptoms can be controlled with topical application of acyclovir (see Figure 23.28).
Genital warts	Papilloma virus (certain strains)	No known cure; symptomatic warts can be removed surgically, chemically, or by cryotherapy.
Trichomoniasis	*Trichomonas vaginalis* (P)	Metronidazole
Acquired immunodeficiency syndrome (AIDS)	Human immunodeficiency virus (HIV)	No known cure; nucleotide base analogs, protease inhibitors, and nonnucleoside reverse transcriptase inhibitors are clinically useful in some treatments (see Table 23.2).
Pelvic inflammatory disease	*N. gonorrhoeae* (B) or *C. trachomatis* (B)	Cefotefan
Vulvovaginal candidiasis	*Candida albicans* (F)	Butoconazole

a B, Bacterium; V, virus; P, protozoan; F, fungus.

b Recommendations of the U.S. Department of Health and Human Services, Public Health Service. For many drugs, there are a number of acceptable alternatives.

treating *most* STDs. Second, the long-term health consequences of many STDs can be quite serious. For example, certain STDs can cause secondary health problems such as pelvic inflammatory disease (a major cause of infertility), cervical cancer, and heart and nerve damage. Pregnant women infected with a sexually transmitted pathogen can pass the agent on to the fetus, resulting in birth defects or even stillbirths. Third, diagnosis and treatment of one STD often reveals the presence of a second, inapparent infection that can still be transmitted after treatment for the first (for example, inapparent chlamydial infections are often diagnosed in individuals treated for gonorrhea).

Despite the fact that most sexually transmitted diseases can be controlled, the incidence of many of these diseases is still quite high; sexually transmitted diseases are obviously a social as well as a medical problem (Figure 23.22). We begin our discussion here with the disease gonorrhea because, despite the value of antibiotics in treating this disease, the prevalence of inapparent infections and the use of birth control pills have made gonorrhea the most widespread of the reportable sexually transmitted diseases.

Gonorrhea

Gonorrhea is one of the most widespread sexually transmitted diseases, and in spite of the availability of excellent treatment it is still very common (Figure 23.22).

The disease symptoms of gonorrhea are quite different in the male and female. In the female the symptoms are usually a mild vaginitis that is difficult to distinguish from vaginal infections caused by other organisms, and the infection may easily go unnoticed; in the male, however, the organism causes a painful infection of the urethral canal (∞ Figure 19.11). The causal agent of gonorrhea, *Neisseria gonorrhoeae*, is killed quite rapidly by drying, sunlight, and ultraviolet light. Because of this extreme sensitivity to environmental conditions, the organism can only be transmitted by intimate person-to-person contact. In addition to gonorrhea, the organism also causes eye infections in the newborn and adult. Infants born of infected mothers may acquire eye infections during birth. Therefore, prophylactic treatment of the eyes of all newborns with silver nitrate or an ointment containing penicillin is generally mandatory to control infection in infants. Complications from untreated gonorrhea include pelvic inflammatory disease and damage to heart valves and joint tissues.

We discussed the clinical microbiology of *Neisseria gonorrhoeae* in Section 21.1, and the general bacteriology of the genus *Neisseria* is described in Section 13.9. The pathogen enters the body by way of the mucous membranes of the genitourinary tract (Figure 23.23*a*), being transmitted during sexual intercourse. Treatment of the infection with penicillin has been successful in the past. However, strains of *N. gonorrhoeae* resistant to

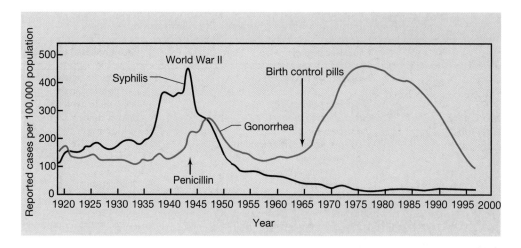

FIGURE 23.22   Reported cases of gonorrhea and syphilis (primary and secondary cases only) per 100,000 population in the United States. Note the downward trend in disease incidence after the introduction of antibiotics and the upward trend in the incidence of gonorrhea after the introduction of birth control pills.

penicillin arose in the 1970s (Figure 23.23b), and are now widespread. This resistance is due to a plasmid-encoded penicillinase. In the United States, more than 8% of all clinical isolates are now penicillinase-producing. Fortunately, the majority of penicillinase-producing strains respond to alternative antibiotic therapy, with a single dose of cefixime or ceftriaxone. Azithromycin or doxycycline is often given at the same time because they are antichlamydial agents and nearly 50% of gonorrhea patients are also infected with the harder-to-diagnose *Chlamydia trachomatis* organism (Table 23.1).

Despite the ease with which gonorrhea can be cured, the incidence of gonococcus infection remains relatively high. The reasons for this are threefold. (1) Acquired immunity does not exist; hence repeated reinfection is possible (whether this is due to lack of

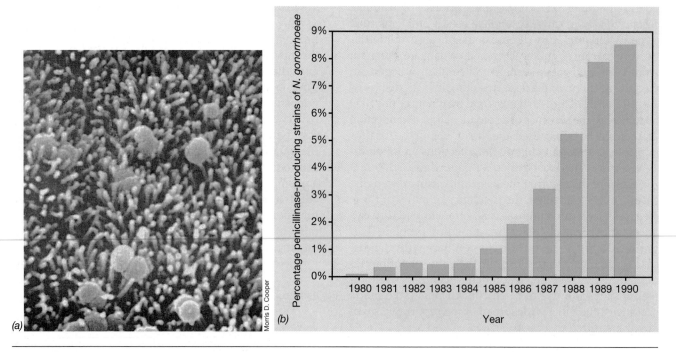

FIGURE 23.23   The causative agent of gonorrhea, *Neisseria gonorrhoeae*, and the prevalence of penicillinase production in this organism. (a) Scanning electron micrograph of the microvilli of human fallopian tube mucosa, showing how cells of *N. gonorrhoeae* attach to the surfaces of epithelial cells. Note the distinct diplococcus morphology. Cells of *N. gonorrhoeae* are about 0.8 μm in diameter. (b) Reported penicillinase-producing *N. gonorrhoeae* (PPNG) in the United States. Note the rapid rise in resistance. As a result, penicillin is no longer used for treatment of gonorrhea (see Table 23.1 for current treatment guidelines).

local immunity or to the fact that at least 16 distinct serotypes of *Neisseria gonorrhoeae* have been isolated is not understood). (2) The use of oral contraceptives alters the local mucosal environment in favor of the pathogen. Oral contraceptives induce the body to mimic pregnancy, which results, among other things, in a lack of glycogen production in the vagina and a raising of the vaginal pH. Lactic acid bacteria normally found in the adult vagina (⚙ Section 19.5) fail to develop under such circumstances, and this allows *N. gonorrhoeae* transmitted from an infected partner to colonize more easily than in an acidic vagina. (3) Symptoms in the female are so mild that the disease may be unrecognized, and a promiscuous infected female can serve as a reservoir for the infection of many males. The disease can be controlled if the sexual contacts of infected persons are quickly identified and treated, but it is often difficult to obtain this information and even more difficult to arrange treatment.

Syphilis

The sexually transmitted disease syphilis is potentially much more serious than gonorrhea, but because of differences in pathobiology, the incidence of syphilis in the United States has been much lower since the introduction of effective antibiotic therapy (Figure 23.22). Syphilis is caused by a spirochete, *Treponema pallidum*, an organism that has been very difficult to cultivate (Figure 23.24). We mentioned the clinical immunology and diagnostic methods for syphilis in Table 21.6, and the biology of the spirochetes is discussed in Section 13.30.

Syphilis exhibits variable symptoms. The organism does not pass through unbroken skin, and initial infection most probably takes place through tiny breaks in the epidermal layer. In the male, initial infection is usually on the penis; in the female it is most often in the vagina, cervix, or perineal region. In about 10% of cases, infection is extragenital, usually in the oral region. During pregnancy, the organism can be transmitted from an infected woman to the fetus; the disease acquired in this way by an infant is called **congenital syphilis**. *Treponema pallidum* multiplies at the initial site of entry, and a characteristic *primary* lesion known as a **chancre** (Figure 23.25) is formed within 2 weeks to 2 months. Dark-field microscopy of the exudate from syphilitic chancres often reveals the actively motile spirochetes (Figure 23.24*a*). In most cases the chancre heals spontaneously and the organisms disappear from the site. Some cells, however, spread from the initial site to various parts of the body, such as the mucous membranes, the eyes, joints, bones, or central nervous system, and extensive multiplication occurs. A hypersensitivity reaction to the treponeme then often takes place, revealed by the development of a generalized skin rash; this rash is the

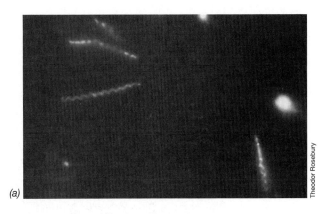

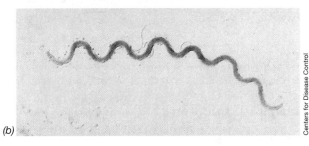

FIGURE 23.24 The spirochete of syphilis, *Treponema pallidum*. (a) Dark-field microscopy of an exudate. *Treponema pallidum* cells measure 0.15 μm wide and 10–15 μm long. (b) Shadowcast electron micrograph of a cell of *T. pallidum*. Note the endoflagella, typical of spirochetes (⚙ Section 13.30).

key symptom of the *secondary* stage of the disease. The patient's condition may now be highly infectious, but eventually the organism disappears from secondary lesions and infectiousness ceases.

The subsequent course of the disease in the absence of treatment is highly variable. About one-fourth of infected individuals appear to undergo a cure, and another one-fourth do not exhibit any further symptoms, although a demonstrable infection may persist. In about half of the patients the disease enters the *tertiary* stage, with symptoms ranging from relatively mild infections of the skin and bone to serious or fatal infections of the cardiovascular system or central nervous system. Involvement of the nervous system is the most serious phase of the illness because generalized paralysis or other severe neurological damage may result. In the tertiary stage very few organisms are present, and most of the symptoms probably result from delayed hypersensitivity reactions (⚙ Sections 20.9 and 20.15) to the spirochetes.

Penicillin is highly effective in syphilis therapy, and the primary and secondary stages of the disease can usually be controlled by a single injection of benzathine penicillin G. In tertiary syphilis, penicillin treatment must extend for longer periods of time. The incidence of primary and secondary syphilis in the United States has decreased significantly over the last two decades and is now at the lowest level since record keeping began.

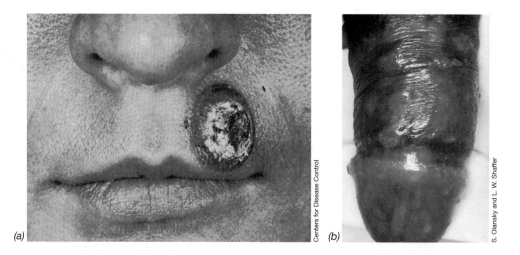

FIGURE 23.25   Primary syphilis lesions. (a) Chancre on lip. (b) Several chancres on penis.

Chlamydial Infections

Now a *reportable* disease, as are gonorrhea and syphilis (∞ Table 22.4), a host of sexually transmitted syndromes can be ascribed to infection by the obligate intracellular bacterium *Chlamydia trachomatis* (Figure 23.26 and Section 13.26). The total incidence of sexually transmitted *C. trachomatis* infections probably greatly outnumbers the incidence of gonorrhea. There may be up to 3 million new *C. trachomatis* sexually transmitted infections every year, making this organism the most prevalent cause of venereal disease. *C. trachomatis* also causes a serious disease of the eye called *trachoma* (∞ Section 13.26), but the strains of *C. trachomatis* responsible for venereal infections are distinct from those causing trachoma. Chlamydial infections may also be transmitted congenitally to the newborn from contamination in the birth canal, causing newborn conjunctivitis and pneumonia. Finally, chlamydial infections are now implicated in the development of arterial plaque and coronary artery disease.

Chlamydial nongonococcal urethritis (NGU) is one of the most frequently observed sexually transmitted diseases today. *Chlamydia trachomatis* causes urethritis in males and urethritis, cervicitis, and pelvic inflammatory disease in females. In both the male and female, inapparent chlamydial infections are common. In a small percentage of cases, NGU can lead to serious

FIGURE 23.26   Cells of *Chlamydia trachomatis* (arrows) attached to human fallopian tube tissues. (a) Cells attached to microvilli of fallopian tube. (b) Damaged fallopian tube containing a cell of *C. trachomatis* (arrow) in the lesion.

acute complications, including testicular swelling and prostate inflammation in men, and pelvic inflammatory diseases and fallopian tube damage in women; cells of *C. trachomatis* attach to microvilli of fallopian tube cells, enter, multiply, and eventually lyse the cells (Figure 23.26*b*). Untreated NGU can cause infertility.

Chlamydia NGU is relatively difficult to diagnose by traditional isolation and identification methods. To expedite diagnoses, a variety of immunological tests have been developed for identifying *Chlamydia trachomatis* from a vaginal or pelvic swab or from discharges. These clinical tests include fluorescent monoclonal antibodies and various enzyme-linked immunosorbent assay (ELISA) tests for detecting specific *C. trachomatis* antigens. If a chlamydial infection is suspected, treatment is initiated with azithromycin or doxycycline. Penicillin is ineffective against *C. trachomatis* because the organisms lack peptidoglycan, the target of penicillin (⌘ Section 13.26 and Table 23.1).

Chlamydial NGU is frequently observed as a secondary event following gonorrhea infection. If both *Neisseria gonorrhoeae* and *Chlamydia trachomatis* are transmitted to a new host in a single event, treatment of gonorrhea with cefixime or ceftriaxone is usually successful but does not eliminate the chlamydia. Although cured of gonorrhea, such patients are still infected with chlamydia and eventually experience an apparent recurrence of gonorrhea that is instead a case of NGU. Thus, current recommendations are to *also* treat gonorrhea patients with azithromycin or doxycycline to treat the potential coexisting, but usually undiagnosed, *C. trachomatis* infection.

Lymphogranuloma venereum is a sexually transmitted disease also caused by a specific strain of *Chlamydia trachomatis*. The disease, which occurs most frequently in males, consists of a swelling of the lymph nodes in and about the groin. From the infected lymph nodes, chlamy-

dial cells may travel to the rectum and cause a painful inflammation of rectal tissues called *proctitis*. Because of the potential for regional lymph node damage and the complications of proctitis, lymphogranuloma venereum is considered to be one of the most serious sexually transmitted chlamydial syndromes.

Herpes

We discussed the molecular biology of herpesviruses in Section 8.19. Herpes simplex viruses are responsible for fever blisters and cold sores but can also cause genital infections. There are two main types of herpesviruses and many serotypes of each main type. **Herpesvirus type 1** (HV1) is generally associated with cold sores and fever blisters in and around the mouth and lips (Figure 23.27*a*). The incubation period of HV1 infections is short (3–5 days), and the lesions heal without treatment in 2–3 weeks. Relapses of HV1 infections are relatively common, and it is thought that the virus is spread primarily via contact with infectious lesions. Latent herpes infections are apparently quite common, with the virus persisting in low numbers in nerve tissue. Recurrent acute herpes infections are due to a periodic triggering of virus activity by unknown causes.

Herpesvirus type 2 (HV2) infections are associated primarily with the anogenital region, where the virus causes painful blisters on the penis of males or on the cervix, vulva, or vagina of females (Figure 23.27*b*). HV2 infections are transmitted by direct sexual contact, and the disease is most easily transmitted during the active blister stage rather than during periods of inapparent (presumably latent) infection.

Genital herpes infections are presently incurable, although a limited number of drugs have been successful in controlling the infectious blister stages. The guanine analog **acyclovir** (Figure 23.28), in ointment form, is par-

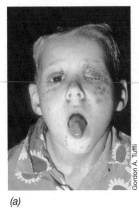
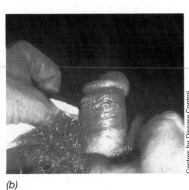
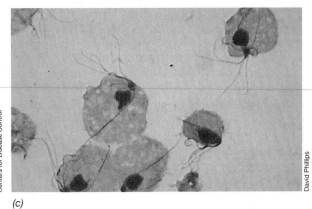

(a)   (b)   (c)

FIGURE 23.27   Nonbacterial sexually transmitted pathogens: herpesvirus and *Trichomonas*. (a) A severe case of herpes blisters on the face due to infection with herpesvirus type 1. (b) Genital herpes due to infection on the penis with herpesvirus type 2. (c) Cells of the flagellated protozoan *Trichomonas vaginalis*.

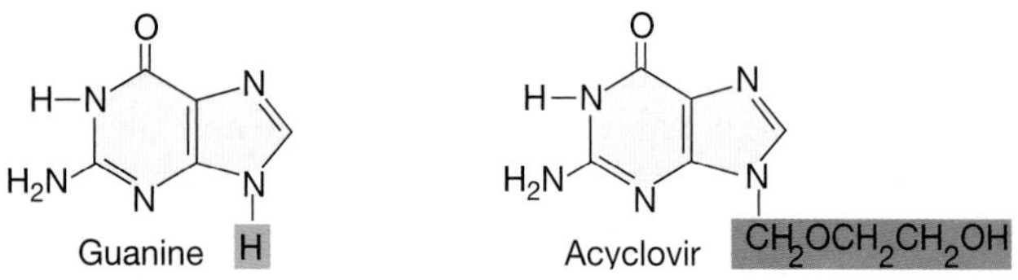

FIGURE 23.28 Structure of guanine and the guanine analog acyclovir. Acyclovir has been used therapeutically to control genital herpes (HV2) blisters.

ticularly effective in limiting the shed of active virus from blisters and promoting the healing of blistering lesions. Acyclovir specifically interferes with herpesvirus DNA polymerase, inhibiting viral DNA replication.

The long-term effects of genital herpes infections are not yet understood. Oral herpes is quite common and apparently has no harmful effects on the host beyond the oral blisters. However, epidemiological studies have shown a significant correlation between *genital* herpes infections and cervical cancer in females. In addition, herpesvirus type 2 can be transmitted to a newborn at birth by contact with herpetic lesions in the birth canal. The disease in the newborn varies from latent infections with no apparent damage to systemic disease that can result in brain damage or death. To avoid passing herpes infections to newborns, delivery by caesarean section is advised for pregnant women with genital herpes infections.

Trichomoniasis

Nongonococcal urethritis may also be caused by infections with the protozoan *Trichomonas vaginalis* (Figure 23.27c). Although many protozoa produce resting cells called *cysts*, *T. vaginalis* does not produce cysts. Thus transmission must be from person to person, generally by sexual intercourse. However, cells of *T. vaginalis* can survive a few hours outside the host, provided they do not dry out. Thus, transmission of *T. vaginalis* by contaminated toilet seats, sauna benches, and paper towels occasionally occurs. *Trichomonas vaginalis* infects the vagina in women, the prostate and seminal vesicles of men, and the urethra of both males and females.

Many cases of trichomoniasis are totally asymptomatic in males. In women trichomoniasis is characterized by a vaginal discharge, vaginitis, and painful urination. The infection is more common in females; surveys indicate that 25 to 50% of sexually active women are infected, while only about 5% of men are infected. The male partner of an infected female should be examined for *Trichomonas vaginalis* and treated if necessary because promiscuous asymptomatic males can serve as reservoirs, transmitting the infection to several females. Trichomoniasis is diagnosed by preparing and microscopically examining a wet mount of fluid

discharged from the patient for the motile protozoa. The antiprotozoal drug *metronidazole* is particularly effective in treating trichomoniasis (Table 23.1).

Ecology of Sexually Transmitted Diseases

Thus far we have seen that a variety of microorganisms can cause sexually transmitted diseases. What properties do these organisms share that limit their distribution to the human genitourinary tract and their mode of transmission to sexual activity? Unlike respiratory infections where large numbers of infectious particles may be expelled by an individual, sexually transmitted pathogens are generally *not* shed in large numbers other than during sexual activity. Consequently, transmission is limited to physical contact, generally during sexual intercourse. This is very clearly shown by the fact that spread of venereal diseases is controlled very effectively by sexual abstinence (no fluid exchange) or by the use of barriers such as condoms that stop the exchange of body fluids during sexual activity. In addition, many sexually transmitted pathogens are very sensitive to drying. Their habitat, the human genitourinary tract, is generally a moist environment. Thus, these organisms colonize a specific habitat that is found only in the genitourinary tract of the host.

✓ 23.6 Concept Check

Pathogens that cause STDs are spread only by sexual contact because of their sensitivity to environmental factors such as drying. Most bacterial and protozoan STDs are curable with chemotherapy, while viral STDs are not. Nonreportable STDs such as trichomoniasis are much more prevalent than reportable diseases.

✓ What is the current accepted treatment for syphilis? For gonorrhea? For herpesvirus? Do these treatments produce cures?

✓ What is the most common sexually transmitted disease?

23.7

Acquired Immunodeficiency Syndrome

Acquired immunodeficiency syndrome (AIDS) was recognized as a disease in 1981. More than 600,000 cases of AIDS have been reported since then in the United States alone, and more than 400,000 people have died (∞ Section 22.4 and Figure 22.2). Up to 900,000 people in the United States may now be infected with HIV. Worldwide, the outlook is even more serious, with more than 30 million people already infected with the human immunodeficiency virus (HIV), the causative agent of AIDS. HIV is divided into two major types, HIV-1 and HIV-2. HIV-1 is genetically similar, but distinct from HIV-2. HIV-2, discovered in West Africa in 1985, has reduced virulence as compared to HIV-1, but also causes an AIDS-like disease. Currently, more than 99% of global AIDS cases are due to HIV-1, and, therefore, our discussions of AIDS will center on infections with HIV-1.

The numbers of HIV-infected individuals will continue to rise dramatically unless effective treatment or prevention methods are discovered. We have already discussed the epidemiology of AIDS (∞ Section 22.4), including these grim predictions, and the clinical diagnostic methods for identifying and tracking HIV infection (∞ Sections 21.8–21.10). In this section, we will concentrate on the pathogenesis of AIDS.

AIDS was first suspected of being a disease of the immune system because a startling increase in the number of so-called opportunistic infections was observed in certain populations (∞ Section 22.4)(see the box, A Definition of AIDS). *Opportunistic infections* are defined as infections rarely observed in humans with normal immune responses (∞ Section 19.12).

The most common AIDS-associated opportunistic infections include pneumonia caused by the fungus *Pneumocystis carinii* (Figure 23.29a) protozoal infections such as cryptosporidiosis, caused by *Cryptosporidium* species (Figure 23.29b), and toxoplasmosis, caused by *Toxoplasma gondii* (Figure 23.29c), systemic yeast infections due to *Cryptococcus neoformans* (Figure 23.29d), *Candida albicans* (Figure 20.29e), and *Histoplasma capsulatum* (Figure 23.29f), viral infections due to herpes simplex (∞ Figure 8.43a) or cytomegalovirus (CMV), tuberculosis, and other mycobacterial infections (Figure 23.29g), and enteric helminthic infections due to *Strongyloides stercoralis* (Figure 23.29h). However, *Pneumocystis carinii* pneumonia is by far the most common

opportunistic disease encountered, being observed at some time in nearly two-thirds of all AIDS patients.

Besides the opportunistic infections associated with many AIDS cases, a rare form of cancer called *Kaposi's sarcoma* is also observed in many AIDS patients. Kaposi's sarcoma is a cancer of the cells lining blood vessel walls and is diagnosed by the characteristic purplish patches it leaves on the surface of the skin (Figure 23.30). This cancer, 20,000 times more prevalent in HIV-infected homosexual males than in the general population, may be caused by coinfection with a newly defined herpes virus, Kaposi's sarcoma-associated herpes virus, or HHV8.

Human Immunodeficiency Virus

The disease AIDS is caused by human immunodeficiency virus. HIV-1 is a retrovirus (∞ Section 8.22) containing 9749 nucleotides in each of its two identical single-stranded RNA genomes. Using the enzyme *reverse transcriptase*, which is present in the intact virion, HIV forms a complementary single-stranded DNA molecule using RNA as a template and converts the complementary DNA (cDNA) formed into double-stranded DNA, which can enter the host cell genome. Here we consider the natural course of HIV infection and the effects of HIV on the immune system, leading to the development of AIDS.

HIV:Cell Interactions and Infection

HIV has the ability to infect cells displaying the CD4 cell-surface protein. Although a number of cells,

A FOCUS ON . . .   A Definition of AIDS[a]

The current case definition for acquired immunodeficiency syndrome (AIDS) includes *individuals who test positive for human immunodeficiency virus (HIV) AND*

1. have a CD4 T cell number of less than 200/mm^3 (the normal count is 600–1000/mm^3) of whole blood, or a CD4 T cell/total lymphocytes percentage of less than 14%, or
2. have a CD4 T cell number of ≥200/mm^3 and any of the following conditions: fungal diseases including candidiasis, coccidioidomycosis, cryptococcosis, histoplasmosis, isosporiasis, *Pneumocystis carinii* pneumonia, cryptosporidiosis, or toxoplas-

mosis of the brain; bacterial diseases including pulmonary tuberculosis and other *Mycobacterium* spp. infections, or recurrent *Salmonella* septicemia; viral diseases including cytomegalovirus infection, HIV-related encephalopathy, HIV wasting syndrome, chronic ulcers or bronchitis due to *Herpes simplex*, or progressive multifocal leukoencephalopathy; malignant diseases such as invasive cervical cancer, Kaposi's sarcoma, Burkitt's lymphoma, primary lymphoma of the brain, or immunoblastic lymphoma; recurrent pneumonia due to any agent.

The above definition was adopted in 1992 to more precisely de-

fine AIDS cases in the United States. As a result of this new, more inclusive definition, several thousand more AIDS cases were diagnosed in 1992 and 1993, as compared with AIDS cases reported in previous years (∞ Figure 22.2). The rise in new cases was not totally attributable to the new definition: The actual number of new AIDS patients, by any definition, increased through 1993, but there has been a gradual decline since then. For the latest information on AIDS, check the AIDS website at http://www.cdcnac.org. ■

[a] Source: CDC *Morbidity and Mortality Weekly Report* Vol. 41, No. RR-17, December 18, 1992.

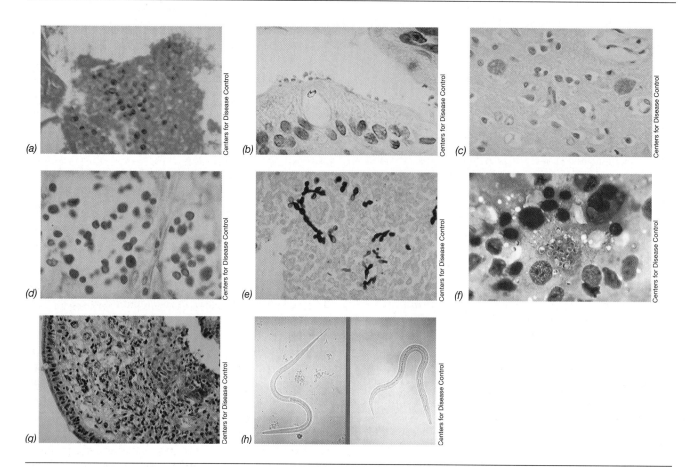

FIGURE 23.29 Opportunistic pathogens associated with cases of acquired immunodeficiency syndrome (AIDS).
a) *Pneumocystis carinii*, from patient with pulmonary pneumocystosis. (b) *Cryptosporidium* sp., from biopsy of small intestine.
(c) *Toxoplasma gondii*, from brain tissue of patient with toxoplasmosis. (d) *Cryptococcus neoformans*, from liver tissue of patient with cryptococcosis. (e) *Candida albicans*, from heart tissue of patient with systemic *Candida* infection. (f) *Histoplasma capsulatum*, from liver tissue of patient with histoplasmosis. (g) Mycobacterial infection of small bowel, acid-fast stain. (h) *Strongyloides stercoralis*, filariform larvae.

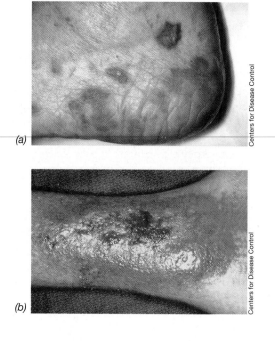

including B cells and certain brain and intestinal cells, have very low levels of CD4 on their surfaces, the two cell types most commonly infected are macrophages and T_H cells, both of which are important components of the immune system (⟳ Sections 20.3 and 20.4). Infected macrophages and T cells produce and release large numbers of HIV particles, which in turn infect other cells that display CD4 (Figure 23.31).

HIV infection normally occurs first in macrophages, an antigen-presenting cell (APC) that has a very low level of CD4 on its surface (⟳ Section 20.3) (Figure 23.32). At the cell surface, the macrophage CD4 molecule binds to the gp120 protein of HIV. The viral gp120 protein then interacts with another macrophage protein, the membrane-spanning chemokine receptor CCR5 (⟳ Section 20.8). CCR5 acts as a coreceptor for HIV and, together

FIGURE 23.30 Kaposi's sarcoma lesions as they appear on (a) the heel and lateral foot, and (b) the distal leg and ankle.

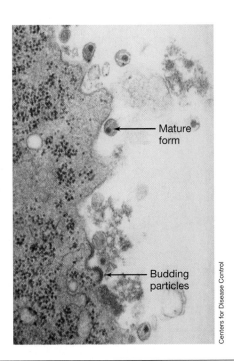

FIGURE 23.31 Transmission electron micrograph of a thin section of a lymphocyte releasing human immunodeficiency virus (HIV). Cells were from a hemophiliac patient who developed AIDS. HIV particles are 90–120 nm in diameter.

with CD4, forms the docking site where the HIV envelope fuses with the host cell membrane, allowing insertion of the viral nucleocapsid.

Evidence for the requirement for the CCR5 coreceptor comes from two sources. First, some individuals have a variant CCR5 gene that stops expression of CCR5. Individuals who are homozygous for this variant gene cannot get infected with HIV even if they engage in risky behavior because they do not express the proper HIV coreceptor. The variant coreceptor gene is found in light-skinned races at a frequency of 0.09, indicating that almost 1% of whites are homozygous for this trait and are protected from HIV infection. Second, *in vitro* experiments show that chemokines that bind to CCR5 inhibit the binding of HIV to target cells.

After HIV has infected the macrophage APCs, a different form of gp120 is made, which in turn binds to a different coreceptor, the CXCR4 chemokine receptor on T cells. HIV then enters and destroys the CD4 T helper lymphocytes, the T_H1 and T_H2 cells that are responsible for cell-mediated inflammatory responses and B-cell help, respectively (Sections 20.9 and 20.11). Thus, HIV starts as a macrophage-tropic (M-tropic) infection and progresses to a T-cell-tropic (T-tropic) infection. The net result of HIV infection is the systematic destruction of macrophages and T cells, leading to a catastrophic breakdown of immunity. Specific knowledge of the coreceptors involved in HIV infection in macrophages

and T cells may be used to design HIV-specific chemokine-receptor blocking agents to prevent the attachment of HIV to either the macrophage or the T cell, thus preventing infection.

In cases of clinical AIDS, CD4 lymphocytes are greatly reduced in number. However, unlike lytic animal viruses, HIV usually does not immediately kill and lyse its host cell. Following reverse transcription to produce DNA from the RNA genome, the viral cDNA integrates into host chromosomal DNA and exists as a provirus. The cell may show no outward sign of infection, and HIV DNA can remain in a latent state for long periods. Eventually, however, productive virus synthesis occurs and new HIV particles are produced and released from the cell. T cells producing HIV no longer divide and eventually die.

Accelerated destruction of CD4 cells occurs following the processing of HIV antigens by infected T cells. Such cells insert molecules of gp120 from HIV particles on their cell surfaces. The embedded gp120 protein on the infected cells then sticks to uninfected T cells by binding to the CD4 molecule. Eventually, numerous cells of each type fuse to produce multinucleate giant cells called *syncytia*. One HIV-infected T cell may eventually bind and fuse with up to 50 uninfected T cells. Shortly after syncytia formation occurs, the resulting cells lose immune function and die.

The end result of HIV infection is that CD4 cells progressively decline in number. This has serious health consequences. In a normal human, CD4 cells constitute about 70% of the total T cell pool; in AIDS patients, the number of CD4 cells steadily decreases, and by the time opportunistic infections set in, CD4 cells may be almost absent (Figures 21.17 and 23.33). As CD4 cells decline in number, there is a concomitant loss in the cytokines they produce. Since cytokines influence the production and maturation of other lymphocytes, this leads to the gradual reduction of uninfected T cells and eventually of all other lymphocytes, effectively shutting down the immune system in those suffering from clinical AIDS. This loss of both humoral and cellular immune function is readily apparent in the opportunistic infections observed. Systemic infections by fungi and mycobacteria (Figure 23.29) point to a loss in T_H1 cellular immunity (Section 20.9). Other opportunistic infections, such as the various viral and bacterial infections associated with AIDS, indicate the loss of humoral immunity; decline in antibody production is due to the loss of T_H2 cells necessary to stimulate antibody production by B cells (Section 20.11).

The overall picture of typical untreated AIDS progression indicates that during the clinical latency period, a very active infectious process is proceeding. First, there is an intense immune response to HIV: about 1 billion virions are destroyed each day and HIV numbers drop. However, this means that HIV is replicating at a very high rate, and this replication results in the corre-

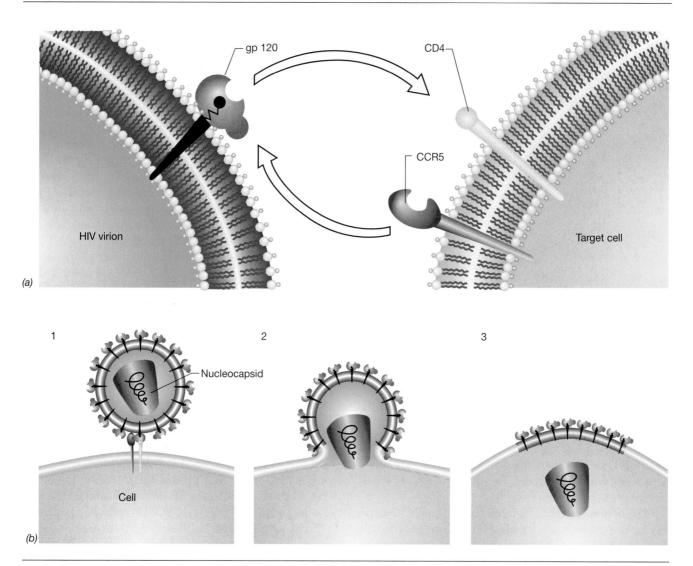

FIGURE 23.32 Infection of a CD4 target cell with the human immunodeficiency virus (HIV). (a) Interaction of HIV with a target cell through specific binding of HIV gp120 to the CD4 receptor and a coreceptor on target cells such as macrophages or T cells. The coreceptor shown is the membrane-spanning chemokine receptor CCR5 found on macrophages. A similar coreceptor, CXCR4, is found on CD4 T cells. (b) Fusion of the HIV envelope with the host cell membrane and insertion of the nucleocapsid. The actual site of virus entry is the site formed by the receptor and coreceptor. The coreceptor is necessary for viral insertion and infection because cells that do not express the coreceptor do not get infected with HIV.

sponding destruction of about 100 million CD4 T cells each day. Eventually, the immune response is simply overwhelmed, HIV levels increase, and the T cells are finally completely destroyed, crippling the immune response and allowing the emergence of opportunistic infections. The example in Figure 23.33 documents T cell destruction and the increase in HIV over a typical time course.

Diagnosis of AIDS

Diagnosis of AIDS is based on clinical and laboratory findings, with the key elements being a positive test for HIV in the blood, a depressed level of CD4 T cells in the blood, and the presence or recent history of one or more opportunistic infections (see Definition of AIDS box). The presence of HIV antibodies in the blood is the usual indicator for HIV exposure and infection (∞ Sections 21.8 and 21.9). In addition, several laboratory tests have been developed, based on the *reverse transcriptase-polymerase chain reaction (RT-PCR,* ∞ Section 16.6) that identifies HIV RNA directly and quantitatively from blood samples (∞ Sections 21.10 and 10.9). The RT-PCR estimates the number of viruses present in the blood, or the so-called *viral load.* The RT-PCR test indicates the magnitude of HIV replication and correlates with the rate of CD4 T cell destruction. The CD4 test indicates the extent of HIV-induced immune damage already suffered, a direct indicator of the magnitude of destruc-

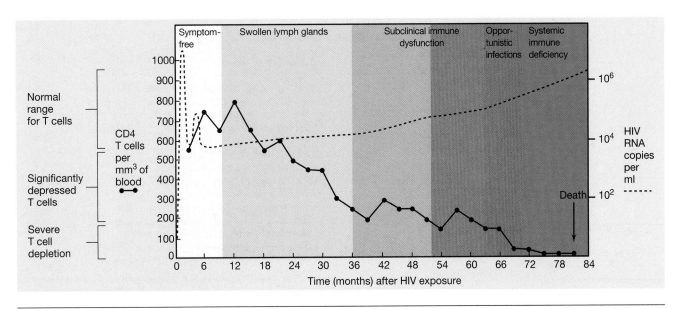

FIGURE 23.33   Decline of CD4 T lymphocytes and progress of HIV infection, based on viral numbers (viral load) in the blood of a typical untreated AIDS patient (∞ Section 21.10). During the typical progression of AIDS, there is a gradual loss in the number and functional ability of the CD4 T cells, while the viral load, measured as HIV-specific RNA per milliliter of blood, gradually increases after an initial decline.

tion of the immune system. The RT-PCR test is not routinely used to screen for HIV because it is costly and technically demanding. After initial discovery of infection, however, the RT-PCR test is used to monitor progression of AIDS and the effectiveness of chemotherapy (Figure 23.33; see below).

The prognosis of an untreated HIV-infected individual is not encouraging. Opportunistic pathogens or malignancies (Figures 23.29 and 23.30) eventually kill most AIDS patients. Long-term studies of AIDS patients indicate that the average person infected with HIV progresses through several stages of decreasing immune function, with CD4 cells dropping from a normal range of 600–1000/mm³ of blood to near zero over a period of 5–7 years (Figure 23.33). Although the *rate* of decline in immune function varies from one HIV-infected individual to another, it is rare for an HIV-positive individual to live for more than 10 years without some form of chemotherapeutic intervention.

Treatment of AIDS

No *cure* for HIV infection is known, although research is very intensive in the areas of vaccine production and chemotherapy. Several drugs have been identified as helpful in delaying symptoms of AIDS and in some cases in prolonging the life of those infected with HIV (Table 23.2).

Effective anti-HIV drugs fall into three major categories. The first of these are a group of *nucleoside analogs*

that act as reverse transcriptase inhibitors. **Reverse transcriptase** is the enzyme that converts the single-stranded RNA genetic information into complementary DNA (∞ Section 8.22). The oldest effective anti-HIV drug *azidothymidine* (*AZT*), is an effective inhibitor of HIV replication because it closely resembles the nucleoside thymidine but lacks the correct attachment site for the next base in an extending nucleotide chain. Thus, AZT and the other nucleoside analogs are *DNA chain terminators*, effectively stopping HIV replication (Figure 23.34a; Table 23.2). When these drugs are given to HIV-infected individuals, the result is a rapid decrease in viral load. However, within several weeks, drug resistant strains of HIV arise in each patient as a result of mutation and selection. Although this process seems very rapid, HIV replicates very quickly (see above) and as little as four mutations result in resistance to a given nucleoside analog.

The second category of anti-HIV drugs are the *non-nucleoside reverse transcriptase inhibitors* (*NNRTIs*) (Figure 23.34b; Table 23.2). These compounds directly inhibit the action of reverse transcriptase by interacting with the protein and altering the conformation of the catalytic site. Unfortunately, a single mutation in the reverse transcriptase gene is often sufficient to reduce the effectiveness of these drugs. The final category of anti-HIV drugs are the *protease inhibitors* (Figure 23.34c; Table 23.2). The protease inhibitors are computer-designed peptide analogs designed to bind to the active site of HIV protease, inhibiting processing of viral polypeptides (∞ Section 18.13); this inhibits virus maturation.

TABLE 23.2	Chemotherapeutic agents approved for HIV/AIDS treatment
Drug	**Mechanism of action**
Azidothymidine (AZT, ZDV, or Zidovudine)(Figure 23.34a)	*Nucleoside analog;* reverse transcriptase inhibitor; nucleotide chain synthesis terminator; increases survival time and reduces incidence of opportunistic infection in AIDS patients; toxic to bone marrow cells; may be used in combination with other drugs in multiple drug treatment protocols.
Dideoxycytidine (ddC or zalcitabine) Dideoxyinosine (ddI or didanosine) Stavudine (d4T) Lamivudine (3TC)	*Nucleoside analogs;* reverse transcriptase inhibitors; mechanism of action and effects are the same as AZT; may have less toxicity than AZT in some patients; may be used in combination with other drugs in multiple drug treatment protocols.
Efavirenz Nevirapine (Figure 23.34b) Delavirdine	*Nonnucleoside reverse transcriptase inhibitors (NNRTIs);* bind directly to reverse transcriptase and disrupts the catalytic site; do not compete with nucleosides; may be used in combination with other drugs in multiple drug treatment protocols.
Indinavir Nelfinavir Ritonavir Saquinavir (Figure 23.34c)	*Protease inhibitors;* computer-designed peptide analogs designed to bind to the active site of HIV protease, inhibiting processing of viral polypeptides and virus maturation; may be used in combination with other drugs in multiple drug treatment protocols.

However, as with the other enzyme-targeted chemotherapy strategy (reverse transcriptase; see above), a single mutation in the HIV protease gene is capable of rendering these drugs nonfunctional.

Because of the problems with drug resistance, a typical recommended protocol for treatment of an individual with established HIV infection includes at least one protease inhibitor *plus* a combination of two nucleoside analogs (Table 23.2). Multiple drug therapy (so-called "drug cocktails") reduces the possibility that a drug-resistant virus could emerge because the virus would need to develop resistance to three drugs simultaneously. This combination therapy is then monitored by the RT-PCR methods to track changes in viral load. An effective protocol reduces viral load to nondetectable levels (less than 500 copies of HIV per milliliter of blood) within several days. The therapy is continued and monitored for viral load indefinitely. If the viral load again reaches de-

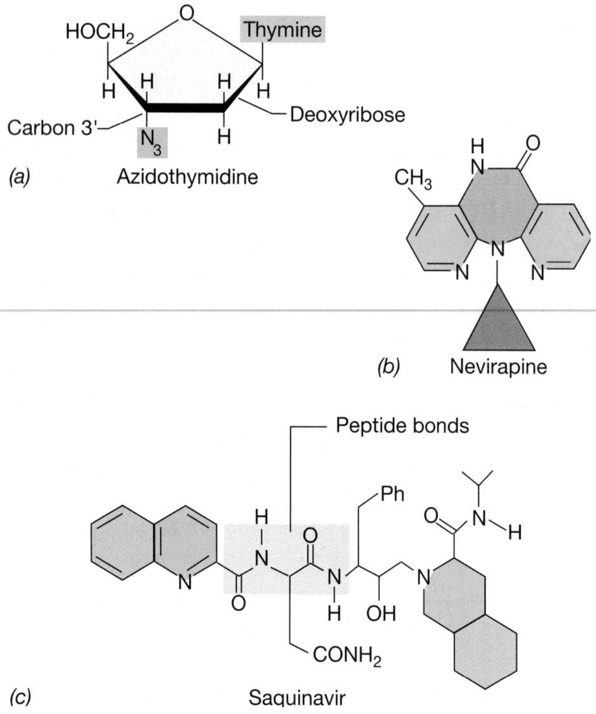

FIGURE 23.34 Examples of the three categories of HIV/AIDS chemotherapeutic drugs. (a) Azidothymidine (AZT), a nucleoside analog. The lack of a —OH group on the 3' carbon causes nucleotide chain elongation to cease when this analog is incorporated, inhibiting virus replication. (b) Nevirapine, a non-nucleoside reverse transcriptase inhibitor, binds directly to the catalytic site, also inhibiting elongation of the nucleotide chain. (c) Saquinavir, a protease inhibitor, was designed by computer modeling to fit the active site of the HIV protease (∞ Section 18.13). Saquinavir is a peptide analog: the tan highlighted area shows the region analogous to peptide bonds. Blocking the activity of HIV protease prevents the processing of HIV proteins and maturation of the virus (∞ Figure 18.28).

tectable limits, the drug cocktail is changed because an increase in viral load indicates the emergence of a drug-resistant virus population (∞ Section 21.10).

In addition to drug resistance, some of the antiviral drugs are toxic to the host. In many cases, nucleoside analogs are not well tolerated by patients, presumably because they interfere with host functions such as cell division (Table 23.2). In general, the NNRTIs and the protease inhibitors are better tolerated because they interfere with only virus-specific functions. However, drug resistance coupled with host toxicity are major problems in HIV therapy, and new chemotherapeutic agents and drug protocols are constantly being developed to avoid these pitfalls.

AIDS Immunization

The genetic variability of HIV has thus far hampered the development of an AIDS immunization protocol. One strategy is to make antibodies to the envelope protein, gp120, and use these antibodies to block CD4–gp120 interactions (Figure 23.32) and thus block infection. However, this approach has not been successful thus far because the gene encoding gp120 mutates frequently, forming antigenic variants of the protein that are not recognized by antibodies made to a different form. The most impressive results from clinical immunization trials have emerged from *subunit vaccines* (∞ Section 10.13), where genes for several HIV envelope proteins have been engineered into vaccinia virus or adenovirus particles. Using these harmless viruses as expression vectors and vehicles for delivery of HIV antigens, several subunit immunization procedures have been shown to elicit a potent humoral and cellular immune response to HIV. Clinical trials of subunit immunization procedures are under way.

Other potential immunization candidates include *killed intact HIV*. These inactivated vaccines are restricted to use in HIV-infected individuals because inactivation procedures may not kill 100% of the HIV; it would be unethical to expose uninfected individuals to even a small risk of HIV infection. Next, some laboratories are exploring the possibilities of producing *live attenuated virus* for use as an immunizing agent. This strategy is bolstered by the finding that individuals infected with HIV-2, a related virus that causes a very mild form of AIDS with a very long latent period, protects carriers from infection with HIV-1, the strain responsible for severe AIDS. However, there are serious potential risks. For example, integrated virus could cause cancer, mutations might reactivate virulence, and so on. Another approach uses *anti-idiotypic antibodies*. In this approach, antibodies to CD4 are used as an antigen. Antibodies raised to the idiotype (binding site) should resemble the molecular configuration of CD4 (∞ Section 20.17). In sufficient quantity, anti-

idiotypic antibodies could then bind HIV particles by gp120–CD4 type interactions.

In all, there are about 20 different AIDS immunization protocols undergoing clinical trials, but no immunization has yet been found to be effective. Additionally, AIDS immunization would most probably not be useful in *treating* most patients that already have the disease because these individuals already have a debilitated immune system or are dealing with large amounts of virus. Most of these individuals lack significant immune function and thus would not respond to a vaccine. Thus, despite considerable advances in our molecular understanding of HIV and in our clinical understanding of the AIDS disease process, public education about AIDS and avoidance of high risk behavior are still the major tools to combat AIDS today (see the box, Sexual Activity and AIDS).

Detection of HIV Infection

Exposure to HIV can be diagnosed by immunological means. Both radioimmunoassay (RIA) and enzyme-linked immunosorbent assay (ELISA) tests (∞ Section 21.8) have been developed for screening blood samples to detect anti-HIV antibodies. The ELISA test has proven particularly valuable for large-scale screening of donated blood to prevent transfusion-associated HIV. Statistics have shown that about 0.25% (2–3 per thousand) of all blood donated by volunteer donors in the United States tests HIV-positive in the ELISA assay. A positive HIV-ELISA test must be confirmed by a second procedure called immunoblotting (Western blotting), a technique that combines the analytical tools of protein purification and immunology (∞ Section 21.9).

A number of other rapid tests are being developed and marketed to identify individuals who are infected with HIV. One test uses a single drop of patient blood and a single reagent. The reagent is a bioengineered antibody in which one binding site is directed to a red blood cell antigen, while the other site is directed to the gp41 HIV surface antigen. In a positive test, the bifunctional antibody cross-links the red blood cells to the HIV, resulting in a visible agglutination (∞ Section 21.5). In another test, saliva is used as a source of secretory antibody to HIV. The saliva is expelled onto a cartridge containing immobilized HIV antigens. A second antibody, reactive with the bound antibody and conjugated to an enzyme, is then added. After addition of the enzyme substrate, a positive reaction shows a colored product, as in the ELISA methods (∞ Section 21.8). The rapid tests are designed to provide maximum convenience, speed (*minutes* instead of the hours or days required for ELISA or immunoblot), extended shelf life, portability, and ease of application and interpretation. However, in general, the rapid tests are not as sensitive or accurate as

A FOCUS ON . . .   Sexual Activity and AIDS

Sexual promiscuity has always been associated with sexually transmitted diseases, but the acquired immunodeficiency syndrome (AIDS) epidemic, discussed in this chapter and elsewhere in this book, has focused attention on the dangers of multiple sex partners and on the high risk associated with certain sex practices. AIDS, caused by the human immunodeficiency virus (HIV), is only one type of sexually transmitted disease. Others include gonorrhea, syphilis, herpes simplex, nonspecific urethritis (caused by *Chlamydia*), protozoal vaginitis (caused by *Trichomonas vaginalis*), fungal vaginitis (caused by *Candida albicans*), and venereal warts (caused by the human papilloma virus). Some of these sexually transmitted diseases have been associated with human society for all of recorded history. However, AIDS is unique. There are no drugs or immunizations to cure or prevent AIDS. Drugs used for treatment are costly and available only in developed countries, so 90% of the 30 million or more HIV-infected individuals do not have access to therapy. AIDS already kills more than 2 million people every year, and this number will grow as people who have harbored and spread HIV for up to 10 years develop full-blown AIDS.

Because AIDS is linked to certain sex practices, prevention means avoidance of these sex practices. The United States Surgeon General has issued a report that makes specific recommendations that individuals can follow if they wish to reduce the likelihood of AIDS infection. Among the recommendations are

1. Avoid mouth contact with penis, vagina, or rectum.
2. Avoid all sexual activities that could cause cuts or tears in the linings of the rectum, vagina, or penis.
3. Avoid sexual activities with individuals from high risk groups. These include prosti-

tutes (both male and female), promiscuous homosexual men, bisexual individuals, and intravenous drug users.
4. If a person has had sex with a member of one of the high risk groups, a blood test should be done to determine if infection with HIV has occurred. If the test is positive, then it is essential that sexual partners of an HIV-positive individual be protected by use of a condom during sexual intercourse.

It is important to emphasize that AIDS is *not* just a disease of male homosexuals. In certain cultures, AIDS is as common in women as in men. The disease is linked to promiscuous sexual activities and other activities that involve exchange of body fluids, which include not only male homosexuality but also female prostitution and intravenous drug use.

Is it possible, then, to have sex without incurring the risk of AIDS? Certain sex practices are inherently much safer than others. Safe sex practices include dry kissing (mouths closed), mutual masturbation (in the absence of breaks in the skin), and intercourse protected by a condom. Dangerous sex practices include wet kissing (mouths open), masturbation where breaks in the skin occur, oral sex (either male or female), and unprotected sexual intercourse (either vaginal or anal). The U.S. Surgeon General has recommended that if the health status of the partner is unknown, a condom be used for all sex practices in which exchange of body fluids occurs.

The AIDS epidemic has focused new attention on the condom (see photo). Condoms have always played two roles in sexual activity: disease protection and prevention of pregnancy. Although the best way to avoid AIDS is to avoid dangerous sex practices, if sexual intercourse is to be carried out with an individual whose infection status is

unknown, then a latex condom should be used. The U.S. Surgeon General strongly recommends the use of condoms for all extramarital sexual activity. In certain countries, advertising campaigns to promote the use of condoms are widespread.

Moralistic statements alone (prescriptions for monogamy, abstinence, avoidance of sexual activity outside of matrimony), will *not* control the AIDS epidemic. Epidemiological studies on all previously known sexually transmitted diseases have shown that fear of disease is not, by itself, sufficient to prevent sexual activities that put an individual at risk for a sexually transmitted disease. The sex drive in some individuals is so strong that it will suppress the fear of disease, even a disease like AIDS. Every individual must therefore take the responsibility for protecting himself or herself from this widespread and extremely dangerous infectious disease.

For more information on prevention of AIDS, see the *Surgeon General's Report on Acquired Immune Deficiency Syndrome*, U.S. Department of Health and Human Services. For more information on protection against AIDS, the Public Health Service has established a toll-free telephone number, called the PHS AIDS Hotline. The number to call is 800-342-2437. The CDC National AIDS Clearinghouse can be contacted at http://www.cdcnac.org. ■

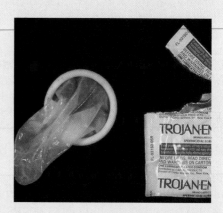

the standard HIV-ELISA and HIV-immunoblot tests (∞ Sections 21.8 and 21.9).

All tests, no matter how sensitive or accurate, can fail to detect HIV-positive individuals who have recently acquired the virus but have not yet made an antibody response, which may be six weeks to a year after exposure to HIV. In spite of this drawback, these tests ensure the general safety of the blood supply, and statistics indicate that the risk of contracting HIV through contaminated blood or blood products is now very low. Sexual promiscuity and group intravenous drug use are the major routes of HIV infection today (∞ Figure 22.3).

✓ 23.7 Concept Check

AIDS is one of the most prevalent infectious diseases in the human population. HIV first destroys a central part of the immune system, and opportunistic pathogens then kill the host. There is still no effective vaccine for HIV. However, several antiviral drugs are available that slow the progress of AIDS. The only prevention for the spread of HIV infection is through total avoidance of risky behavior such as intravenous drug use (needle sharing) and unsafe sexual practices.

✓ Review the definition of AIDS. What diagnostic features are shared by all AIDS patients?

✓ What is the current treatment for AIDS? Is it effective? What are the side effects?

REVIEW QUESTIONS

1. Why do so many different gram-positive Bacteria cause respiratory diseases?

2. What are the typical symptoms of a staphylococcal skin infection and what properties of *Staphylococcus aureus* are responsible for these symptoms?

3. Describe the process of infection by *Mycobacterium tuberculosis*. Does infection always lead to active disease? Why or why not?

4. Why are colds and influenza such common respiratory diseases? Discuss at least five reasons for their high incidence.

5. Compare and contrast the diseases measles, mumps, and rubella. Include in your discussion a description of the pathogen, major symptoms encountered, and any potentially serious consequences of these infections.

Why is it important that women be vaccinated against rubella before puberty?

6. Discuss at least three reasons why the incidence of gonorrhea rose dramatically in the mid-1960s, whereas the incidence of syphilis did not rise.

7. For the sexually transmitted diseases syphilis, gonorrhea, and trichomoniasis, describe the methods for treating each infection. In each case, is the treatment an effective cure? Why or why not? Why is trichomoniasis not treatable with antibiotics such as penicillin and tetracycline?

8. Describe how human immunodeficiency virus (HIV) effectively shuts down most aspects of both humoral immunity and cell-mediated immunity. Are there any parts of the immune system that remain functional in cases of acquired immunodeficiency syndrome (AIDS)?

APPLICATION QUESTIONS

1. Most sneezes generate an aerosol of mucous droplets containing microorganisms. Why are infections due to sneeze transmission more like to be *upper* rather than *lower* respiratory tract infections?

2. Explain why *rheumatic fever* is considered a type of *autoimmune* disease.

3. Why does the disease *tuberculosis* often lead to a permanent reduction in lung capacity, whereas most other respiratory diseases cause only temporary respiratory problems?

4. Discuss the molecular biology of *antigenic shift* and comment on the immunological consequences for the host. Why does antigenic shift prevent the production of a single universally effective vaccine for influenza control? Why do cycles of influenza occur about every 3–5 years? Next, compare antigenic *shift* to antigenic *drift*. Which mechanism is more important for the influenza virus? Which causes the greatest antigenic change? Which creates the biggest problems for vaccine developers? Why?

5. Despite the ease with which gonorrhea can be diagnosed and cured, its incidence remains very high. Why?

6. As the director of your dormitory's public health advisory group, you are charged to present information on sexually transmitted diseases (STDs), including prevention, symptoms, notification, and treatment. Outline your information program for three separate STDs. Will your program for each disease overlap? For each of the diseases, discuss the social, legal, and public health issues that must be considered for reporting an occurrence of the disease to sexual partners of the infected individual.

7. What is the current basis for treatment of AIDS? Does it work? Why or why not? Describe the various immunization strategies for AIDS. How might each immunization protocol actually induce immunity to the disease? Is passive immunity of potential use for treatment of AIDS? Why or why not? What risks may be involved?

Vectorborne diseases such as malaria and plague have played a significant role in human history. While these diseases are still important today, other vectorborne diseases such as Lyme disease, transmitted through the ticks shown in the illustration, are emerging as new and important infectious diseases. In addition to the vectorborne diseases, infectious agents that spread directly from animals, such as hantavirus hemorrhagic fever, and from common sources, particularly food and water, are major emerging and ongoing public health concerns. Because most of these diseases are spread by pathogens that have nonhuman reservoirs, the possibility of eradicating them is slight. Control efforts include public health measures such as educating the public, monitoring disease outbreaks, assuring high quality food and water, and providing adequate, effective immunizations.

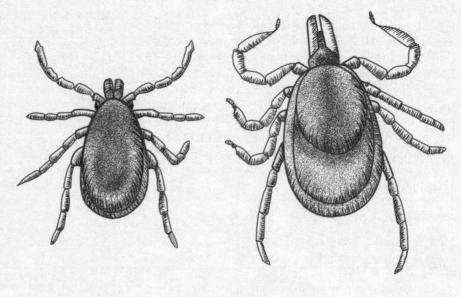

CHAPTER 24 Animal-Transmitted, Vectorborne, and Common-Source Microbial Diseases

Animal-Transmitted and Vectorborne Diseases

24.1 Animal-Transmitted Diseases: Rabies   957
24.2 Animal-Transmitted Diseases: Hantavirus Pulmonary Syndrome   959
24.3 Insect- and Tick-Transmitted Diseases: Rickettsias   960
24.4 Tick-Transmitted Diseases: Lyme Disease   963
24.5 Insect-Transmitted Diseases: Malaria   966
24.6 Insect-Transmitted Diseases: Plague   969

Common-Source Diseases

24.7 Soil Microorganisms: The Pathogenic Fungi   971
24.8 Public Health and Water Quality   974
24.9 Waterborne Diseases   976
24.10 Microbial Growth in Food   980
24.11 Foodborne Diseases: Food Poisoning   983
24.12 Foodborne Diseases: Food Infection   986

WORKING GLOSSARY

Chlorination a highly effective disinfectant procedure for drinking water using chlorine gas or other chlorine-containing compounds

Coliform a large group of gram-negative, facultative Bacteria

***Escherichia coli* O157:H7** an emerging enterotoxigenic *E. coli* spread by fecal contamination of animal or human origin to food and water

Ehrlichiosis an emerging tick-transmitted disease caused by rickettsia of the *Ehrlichia* genus

Food infection microbial infection resulting from ingestion of contaminated food

Food poisoning disease resulting from ingestion of a bacterial exotoxin found in contaminated food. Also referred to as *food intoxication.*

Food spoilage any change in a food product that makes it unacceptable to the consumer

Hantavirus pulmonary syndrome (HPS) an emerging acute viral disease characterized by respiratory pneumonia, obtained by transmission of hantavirus from rodents

Lyme disease an emerging tick-transmitted disease caused by the spirochete *Borrelia burgdorferi*

Mycoses infections caused by fungi

Rickettsias obligate intracellular parasites that cause a variety of diseases including typhus, Rocky Mountain spotted fever, and ehrlichiosis

Thallasemia a genetic trait that confers resistance to malaria but causes a reduction in the efficiency of red blood cells by altering a red blood cell enzyme

Sickle cell anemia a genetic trait that confers resistance to malaria but causes a reduction in the efficiency of red blood cells by reducing the oxygen-binding affinity of hemoglobin

Animal-Transmitted and Vectorborne Diseases

In this chapter we will first discuss several examples of animal-transmitted and vectorborne pathogens and the diseases they cause. The natural host animal for the animal-transmitted pathogens is a nonhuman vertebrate, usually a wild rodent. Populations of wild rodents act as reservoirs for these diseases, making pathogen irradication unlikely or impossible. As we shall see, when these infected rodent populations contact humans, the result is often an accidental human infection. The animal-transmitted diseases are generally spread to accidental hosts such as humans by direct contact, aerosols, or bites. In the case of vectorborne diseases, pathogens are spread to uninfected hosts from the bite of an arthropod vector that last fed on a pathogen-infected host. Again, humans are often accidental hosts in the life cycles of the pathogens, but infected humans may also be a major component of the disease reservoir, as is the case with malaria.

24.1

Animal-Transmitted Diseases: Rabies

Animals can contract a number of infectious diseases, some of which can be passed to humans. Fortunately, the use of effective immunization practices and good veterinary care prevents most infectious disease in domestic animals and limits the transfer of disease to humans. However, the situation is different in the wild (feral) animal population. Wild animals cannot be routinely immunized and do not receive veterinary care. Thus, animal diseases transmissible to humans *(zoonoses)* occur in wild animal populations, usually on a periodic, cyclic basis.

Epidemiology, Biology, and Pathology of Rabies

Most zoonoses are of little serious consequence to human health but **rabies** is a notable exception. Rabies is one of a handful of diseases that occurs primarily in animals but is spread to humans under certain conditions. The major reservoir of rabies in the United States is in wild animals, primarily raccoons, skunks, coyotes, foxes, and bats. However, a small but significant number of cases are also seen in domestic animals (Figure 24.1). Rabies is still a major disease in humans worldwide: Approximately 35,000 people die every year from this disease, primarily in developing countries where rabies is still endemic in domestic animals such as dogs (∞ Table 22.1), and about 1 million people worldwide receive rabies treatment for animal bites. Rabies is caused by a member of the Rhabdovirus family (a negative-stranded RNA virus) (∞ Section 8.16) that infects cells in the central nervous system of most warm-blooded animals, almost invariably leading to death if not treated. The virus, present in the saliva of rabid animals, enters the body through a bite wound. Rabies virus multiplies at the site of inoculation and then travels to the central nervous system. The incubation period for the onset of symptoms is highly variable, depending on the size, location, and depth of the wound and the actual number of viral particles transmitted in the bite. In dogs,

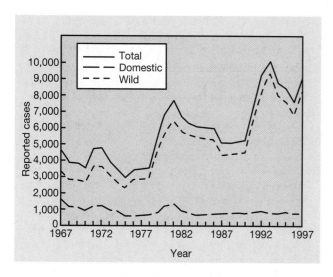

FIGURE 24.1 Rabies cases in wild and domestic animals in the United States. Since 1973, there have been a total of 49 human rabies cases, or just under 2 per year. Rabies is endemic in wild animal populations in the United States. The recent rise in reported cases following several years of decline is the result of periodic reemergence of rabies in wild animals, especially in raccoons in the Northeastern United States. The data were supplied by the Centers for Disease Control and Prevention, Atlanta, GA, USA.

the incubation period averages 10–14 days. In humans, up to nine months may pass before the onset of rabies symptoms. The virus proliferates in the brain (especially in the thalamus and hypothalamus), leading to fever, excitation, dilation of the pupils, excessive salivation, and anxiety. A fear of swallowing (hydrophobia) develops from uncontrollable spasms of the throat muscles; death eventually results from respiratory paralysis. In humans, an *untreated* rabies infection that progresses to the symptomatic stage is nearly always fatal.

Treatment, Diagnosis, and Prevention of Rabies

Because of the lethal nature of rabies, any contact with potentially rabid animals must be taken seriously. A wild animal suspected of being rabid should be cap-

tured, sacrificed, and immediately examined for evidence of rabies in order to expedite treatment of those exposed. If a domestic animal, generally a dog, cat, or ferret, bites a human, especially if the bite is unprovoked, the animal is generally held 10 days for observation for development of clinical signs of rabies. If the animal is wild, shows signs of rabies symptoms, or a determination cannot be made after 10 days, the patient will be passively immunized with rabies immune globulin (purified anti-rabies virus antibodies obtained from a hyperimmune individual) (∞ Section 20.16), injected at both the site of the bite and intramuscularly. The patient will also be vaccinated with an inactivated rabies virus preparation. A summary of guidelines for treating possible human exposure to rabies is shown in Table 24.1. Because of the very slow progression of rabies in humans, this combination therapy is nearly 100% effective, stopping the onset of the active disease. Rabies is diagnosed in the laboratory by examining tissue samples. Fluorescent antibody tests (∞ Section 21.7) that recognize rabies-infected brain or corneal tissue are used for confirming a clinical diagnosis of rabies, either in a potentially rabid animal or in postmortem examination of a human case. In addition, characteristic virus inclusion bodies in the cytoplasm of nerve cells, called *Negri bodies*, also obtained by biopsy or postmortem sampling, are taken as confirmation of rabies.

Rabies is prevented largely through effective immunization practices. A number of effective inactivated rabies vaccines are used in the United States for both human and domestic animal immunizations, and a variety of effective inactivated and attenuated virus preparations are also used worldwide. As we discussed above, most cases of human rabies are preventable through prompt immunization: Because of the long incubation period, passive and active immunization of potentially exposed individuals is sufficient to prevent disease. Therefore, prophylactic rabies immunization is not generally recommended for humans, except for individuals at high risk such as veterinarians and animal control personnel.

TABLE 24.1	Guidelines for treating possible human exposure to rabies virus

Unprovoked bite by a domestic animal

Animal suspected of rabies	*Animal not suspected of rabies*
1. Sacrifice animal and test for rabies.	1. Hold for 10 days. If no symptoms, do not treat human.
2. Begin treatment of human immediately.[a]	2. If symptoms develop, treat human immediately.[a]

Bite by wild carnivore (for example, skunk, bat, fox, raccoon, coyote)

Regard animal as rabid
1. Sacrifice animal and test for rabies.
2. Begin treatment of human immediately.[a]

Bite by wild rodent, squirrel, livestock, rabbit

Consult local or state public health officials about possible recent cases of rabies transmitted by these animals (these animals rarely transmit rabies). If no reports, do not treat human.

a All bites should be thoroughly cleansed with soap and water. Treatment is generally a combination of rabies immunoglobulin and human diploid cell rabies vaccine (five injections intramuscularly).

The rabies prevention strategy just described has been extremely successful, and an average of only about 2 cases of human rabies are reported in the United States each year, nearly always the result of a bite by a wild animal. Domestic animals are another matter. Since domestic animals often have some exposure to wild animals, all dogs and cats should be vaccinated beginning at 3 months old, and booster inoculations should be given yearly or triannually. Other domestic animals, including large farm animals, are also often immunized with rabies vaccines. However, the key to effective rabies prevention and possible eradication, at least in the United States, lies in control of the disease in the large wild animal reservoir of rabies virus (Figure 24.1). If all or even most members of the potential disease reservoir are immune, the disease can be stopped and possibly even eradicated. Currently, subunit vaccines consisting of rabies virus genes that encode rabies coat proteins expressed in vaccinia virus (∞ Section 10.13) are available. Because effective doses can be given orally, subunit vaccines can be included in food "baits" and used to immunize local populations of susceptible animals to reduce the incidence and spread of rabies. Such vaccines are a completely safe means of controlling rabies in the wild animal reservoir and, if applicable to large populations, could lead to eventual eradication of the disease.

✓ 24.1 Concept Check

Rabies occurs primarily in wild animals in the United States but can be transmitted to domestic animals or to humans. Vaccination of dogs and cats is of central importance for the control of rabies. In developing countries, rabies is often endemic in domestic animals and is still an important human disease.

✓ Why is rabies vaccine not given routinely to humans in the United States?

✓ What is the procedure for treating a patient bitten by an animal if the animal cannot be found?

✓ What advantages might an oral vaccine have over a parenteral vaccine for rabies control in animals?

24.2

Animal-Transmitted Diseases: Hantavirus Pulmonary Syndrome

In 1993, a hantavirus strain emerged in the United States as the cause of an acute respiratory syndrome now known as **hantavirus pulmonary syndrome** (HPS). The outbreak of hantavirus occurred in the "Four Corners" region of the United States (Arizona, Colorado, New Mexico, and Utah) and was eventually traced to the deer mouse (*Peromyscus maniculatis*) population in the area. The outbreak caused 32 deaths in 53 infected adults, and it underscored the potential danger of outbreaks due to diseases that are directly transmitted from a variety of different animal reservoirs, sometimes under new or unusual circumstances. Other diseases, such as Ebola hemorrhagic fever (∞ Section 22.10), *may* also be spread to humans directly from an animal reservoir, although the animal reservoir and means of transmission have not yet been identified.

Biology and Epidemiology of Hantaviruses

The genus *Hantavirus* is a member of the Bunyaviridae, a family of enveloped single-stranded RNA viruses (∞ Section 8.14 and Figure 24.2). The genus includes a number of viruses that cause either *hantavirus pul-*

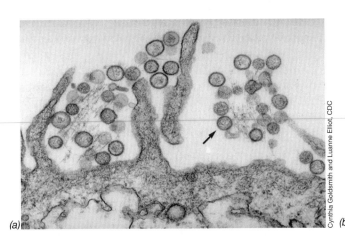
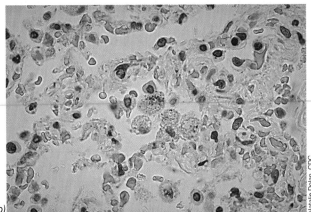

FIGURE 24.2   Hantavirus. (a) An electron micrograph of the Sin Nombre hantavirus. The arrow indicates one of several viruses. The virus is approximately 0.1 micrometers in diameter. (b) Immunostaining of Andes hantavirus antigens in alveolar macrophages. Each granular dark blue stained area indicates cellular infection of an individual macrophage (approximately 15 micrometers in diameter). Hantaviruses belong to the Bunyaviridae and, like many phylogenetically similar viruses (for example, Rift Valley virus and Ebola virus), cause hemorrhagic fevers with very high human mortality.

monary syndrome (HPS) or *hemorrhagic fever with renal syndrome (HFRS).* The viruses are primarily found in rodents, including mice and rats of several species, lemmings, and voles, and are occasionally found in other animals. The HFRS strains are more common in the Eastern hemisphere and have been implicated in a number of HFRS outbreaks in recent years. Up to 200,000 cases per year are recognized, chiefly in China, Korea, and Russia. The HPS strains are more prevalent in the Western hemisphere, and continued investigation of the ecology of these viruses is likely to uncover a number of other pathogenic strains. The viruses are found in rodents throughout the world, with strains being limited to particular geographic areas. Hantaviruses are most commonly transmitted by inhalation of virus-contaminated rodent excreta. Humans seem to be an accidental host and are infected only when they come into contact with rodents and their feces. For example, all of the individuals who acquired HPS in the Four Corners outbreak had been exposed to mice or their droppings, the result of a warm winter and an unusual increase in numbers of rodents in 1993 (⟳ Section 22.10). The *aerosol route* of infection is most common, with most aerosols being in the form of dust generated from mouse droppings or dried urine. However, there are rare reports of person-to-person transmission, as well as a few incidents where HPS or HFRS was spread by a rodent bite.

Pathology, Diagnosis, Treatment and Prevention of Hantavirus Pulmonary Syndrome

Hantavirus pulmonary syndrome is characterized by a sudden onset of fever, myalgia, thrombocytopenia (reduction in the number of blood platelets), leukocytosis (an increase in the number of circulating lymphocytes), and pulmonary capillary leakage. Death occurs in a matter of several days in greater than 40% of cases, usually due to shock and cardiac complications precipitated by pulmonary edema (leakage of fluid into the lungs, causing suffocation). These symptoms are typical of the Sin Nombre virus, which caused the Four Corners outbreak, but a variety of other symptoms may be evident, depending on the strain of virus causing the disease. For example, the Bayou strain common in rodents in the southeastern United States also causes kidney failure.

If hantavirus from candidate infections can be grown in tissue culture (⟳ Section 8.4), the strain can be identified by serological techniques including a virus plaque-reduction neutralization assay. More commonly, ELISAs (enzyme-linked immunoassays) are performed on patient blood to identify antibodies, indicating exposure and an immune response (⟳ Section 21.4); or the presence of the viral genome, indicating infection, is detected based on a PCR (polymerase

chain reaction) assay using patient tissue or blood specimens (⟳ Section 21.1).

There is no virus-specific treatment or vaccine for any of the hantaviruses. However, the disease can be prevented by avoiding contact with rodents and rodent habitat, since its major mode of transmission is through exposure to rodent feces. Reduction in exposure can be accomplished by destroying mouse habitat, restricting food supplies (*e.g.,* keeping food in sealed containers), and aggressive rodent extermination measures. The long-term prognosis for disease eradication is very poor because a considerable percentage of the rodent population in a given geographical area is generally infected with the local hantavirus strain. For example, retrospective serological testing of the deer mice in the Four Corners area in 1993 indicated that 30% of the local mouse population had the Sin Nombre hantavirus.

✓ 24.2 Concept Check

Hantaviruses occur worldwide in rodent populations and cause serious diseases such as hantavirus pulmonary syndrome (HPS) when accidentally spread to humans. In the United States, hantavirus infections were not recognized until 1993.

- ✓ Why were hantaviruses not recognized as a major public health problem until 1993 in the United States?
- ✓ Describe the spread of hantaviruses to humans. What are some effective measures for preventing infection by hantaviruses?

24.3

Insect- and Tick-Transmitted Diseases: Rickettsias

The rickettsias are small bacteria that have a strictly intracellular existence in vertebrates, usually in mammals, and are also associated at some point in their natural cycle with blood-sucking arthropods such as fleas, lice, or ticks. We discussed the biology of rickettsias in Section 13.12. Rickettsias cause a variety of diseases in humans and animals, of which the most important are typhus fever, Rocky Mountain spotted fever, and ehrlichiosis. Rickettsias take their name from Howard Ricketts, a scientist at the University of Chicago who first provided evidence for their existence and who died from infection with the rickettsia that causes typhus fever, *Rickettsia prowazekii.* Rickettsias have not been cultured in artificial media but can be cultured in laboratory animals, lice, mammalian tissue culture cells, and the yolk sac of chick embryos. In animals, growth takes place primarily in phagocytic cells (⟳ Section 20.3). Although the rickettsias have not been grown in pure culture, the 1.1 kb genome of *Rickettsia prowazekii* has been sequenced (⟳ Section 9.12). Based on homology with other genomic sequences, these intracellular parasites are closely related to human mitochondria. Like the mitochondria, the rickettsial genome has been reduced in size to encode a set of genes tailored for intracellular dependency and does not have many of the genes necessary for independent energy metabolism and structural biosynthesis. The rickettsial genome also contains a set of virulence genes closely related to the *virB* operon of the plant pathogen *Agrobacterium tumefaciens* (∞ Section 16.24 and Figures 16.71 and 16.72). This operon encodes components of virulence factors involved in DNA transfer and protein export systems. Thus, the genomic sequence provides evidence for the intracellular dependence and the virulence of these pathogens.

Rickettsias are divided into three groups, based loosely on the types of clinical disease they produce. The groups are (1) the *typhus group*, typified by *Rickettsia prowazekii*; (2) the *spotted fever group*, typified by *Rickettsia rickettsii*; and (3) the *ehrlichiosis group*, characterized by *Ehrlichia chaffeensis*. Here, we will examine one example of a pathogen from each group.

The Typhus Group: *Rickettsia prowazekii*

Typhus fever is caused by *Rickettsia prowazekii*. Epidemic typhus is transmitted from human to human by the common body or head louse. The known mammalian reservoirs for typhus are humans. Typhus can be a serious disease. During World War I, an epidemic of typhus spread throughout eastern Europe and claimed almost 3 million lives. Typhus has frequently been a problem among military troops during wartime. Because of the unsanitary, cramped conditions characteristic of wartime military infantry operations, lice spread easily among soldiers, and typhus spread in epidemic proportions. Up until World War II, typhus claimed more lives than combat on the battlefield.

Cells of *R. prowazekii* are introduced through the skin when the puncture caused by the louse bite becomes contaminated with louse feces, the major source of rickettsial cells. During an incubation period of 1–3 weeks, the organism multiplies inside cells lining the small blood vessels. Symptoms of typhus (fever, headache, and general body weakness) then begin to appear. Five to nine days later a characteristic *rash* is observed in the armpits and generally spreads throughout the body *except* for the face, palms of the hands, and soles of the feet. Complications from untreated typhus may develop involving damage to the central nervous system, lungs, kidneys, and heart. Several drugs are effective against rickettsias. Epidemic typhus has a mortality rate of 6–30%. Tetracycline and chloramphenicol are most commonly used to control *R. prowazekii*.

Rickettsia typhi, the organism that causes murine typhus, is another important pathogen in the typhus group.

The Spotted Fever Group: *Rickettsia rickettsii*

Rocky Mountain spotted fever was first recognized in the western United States about 1900 but is actually more common today in the southeastern United States. Rocky Mountain spotted fever is caused by *Rickettsia rickettsii* and is transmitted to humans by various species of ticks, most commonly the dog and wood ticks (Figure 24.3). Humans acquire the pathogen from tick fecal matter, which is injected into the body during a bite, or by rubbing infectious material into the skin by scratching. Cells of *R. rickettsii*, unlike other rickettsias, grow within the nucleus of the host cell as well as in host cell cytoplasm (Figure 24.3a and 24.3b). Following an incubation period of 3–12 days, an abrupt onset of symptoms occurs, including fever and a severe headache. Three to five days later, a rash is observed that is present on the palms of the hands and soles of the feet (Figure 24.3c) (the location of rickettsial rashes is of some diagnostic value). Gastrointestinal problems such as diarrhea and vomiting are usually observed as well, and the clinical symptoms of Rocky Mountain spotted fever may exist for over 2 weeks if the disease is left untreated. Tetracycline or chloramphenicol generally promotes a prompt recovery from Rocky Mountain spotted fever if administered early in the course of the infection.

The Ehrlichiosis Group: *Ehrlichia*

The genus *Ehrlichia* (∞ Section 13.12) is responsible for two emerging tickborne diseases in the United States, *human monocytic ehrlichiosis* (HME) and *human granulocytic ehrlichiosis* (HGE). The rickettsias that cause the diseases are *E. chaffeensis* and an organism similar or identical to *E. equi*, respectively. The onset of these clinically indistinguishable ehrlichioses is characterized by flulike symptoms including fever, headache, malaise, and frequently leukopenia (decreased number of leukocytes) and/or thrombocytopenia. Laboratory findings frequently document changes in liver function, characterized by an increase in the enzyme hepatic transaminase. In addition, peripheral blood leukocytes characteristically have visible inclusions of *Ehrlichia* cells (Figure 24.4). The symptoms, except for the inclusions, are similar to other rickettsioses, but the *Ehrlichia* genus is antigenically distinct from members of the other major rickettsial groups. Infections range from subclinical to fatal. Long-term complications for progressive untreated cases may include respiratory and renal insufficiency and serious neurological involvement. Di-

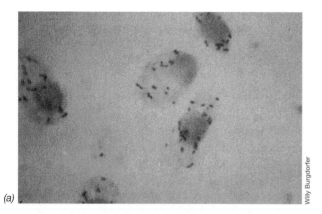

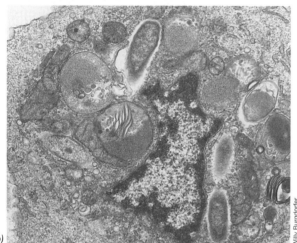

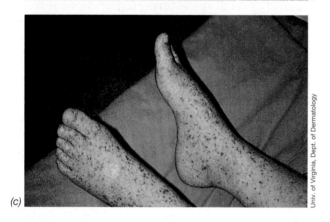

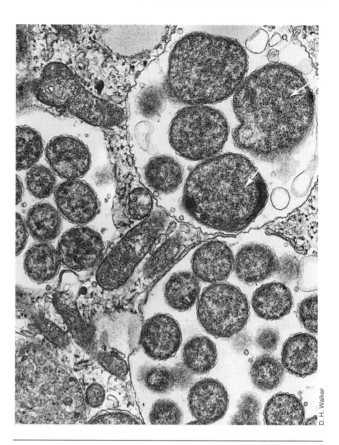

FIGURE 24.4 *Ehrlichia chaffeensis*, the causative agent of human monocytic ehrlichiosis (HME). The electron micrograph shows inclusions in a human monocyte which contain large numbers of *E. chaffeensis*. The arrows indicate two of the many bacteria in each inclusion. *E. chaffeensis* ranges from about 300–900 nm in diameter.

FIGURE 24.3 *Rickettsia rickettsii*, the causative agent of Rocky Mountain spotted fever. (a) Cells of *R. rickettsii*, growing in the cytoplasm and nucleus of tick hemocytes. Individual cells are about 0.4 μm in diameter. (b) Cells of *R. rickettsii* in a granular hemocyte of an infected wood tick, *Dermacentor andersoni*. Transmission electron micrograph. (c) Rash of the disease on the feet.

agnosis is based on an indirect fluorescence antibody assay (IFA) of patient serum (👓 Section 21.7) and also on polymerase chain reaction (PCR) tests of whole blood or serum to detect the presence of *Ehrlichia* DNA (👓 Section 21.10).

The infection is spread by the bites of infected ticks; the mammalian reservoirs includes deer and possibly rodents, in addition to the human hosts. Retrospective serological analyses in areas with relatively high incidence of tickborne disease indicate that HGE may be a more prevalent disease than Rocky Mountain spotted fever. Most infections are not properly identified because of the variable nature of the symptoms. However, since 1985 more than 500 cases in the United States have been verified by the Centers for Disease Control. Although these numbers are not yet extremely high, a surveillance program in New York and Connecticut for 1994–1997 found 398 confirmed cases. These patterns suggest that while ehrlichiosis will be reported more frequently as physicians become more familiar with the disease, the numbers of cases are rising, and this represents, like Lyme disease (see Section 24.4), an emerging tickborne disease.

As with other tickborne illnesses, outdoor activities in tick-infested habitat are the major predisposing factors in acquiring ehrlichiosis: Golfers and hikers are particularly prone to infection. Prevention of ehrlichiosis at the individual level involves reducing exposure to ticks and tick bites by avoiding tick habitat, wearing tick-proof clothing, and applying appropriate insect repellents. On a communal level, tick densities can be successfully reduced through areawide application of acaricides (chemicals specifically toxic for ticks and related arthropods) and removal of tick habitat such as leaves and brush. Doxycycline, a semisynthetic tetracycline derivative, is the antibiotic of choice for the treatment of ehrlichiosis.

Other Rickettsial Diseases

Q fever is a pneumonia-like infection caused by an obligate intracellular parasite, *Coxiella burnetii,* related to the rickettsias (⟳ Section 13.12). Although not transmitted to *humans* directly by an insect bite, the agent of Q fever is transmitted to *animals* by insect bites, and various arthropod species serve as a reservoir of infection. Domestic animals generally have inapparent infections, but may shed large quantities of *C. burnetii* cells in their urine, feces, milk, and other body fluids, which then act as sources of infection for humans. The resulting influenza-like illness may progress to include prolonged fever, headache, chills, chest pains, pneumonia, and endocarditis. Diagnosis of Q fever can be made by immunological tests designed to measure host antibodies to the organisms (an indirect enzyme-linked immunosorbent assay, ELISA, test has made the diagnosis of Q fever almost routine) (⟳ Section 21.8). *C. burnetii* infections respond quite dramatically to the antibiotic tetracycline, and therapy is usually begun quickly in any suspected human case of Q fever in order to prevent heart damage. Finally, Q fever is one of the infectious diseases that has been studied as a possible agent for biological warfare (⟳ box, Biological Warfare, Chapter 22).

Scrub typhus, or *tsutsugamushi disease,* is restricted to Asia, the Indian subcontinent, and Australia, and is caused by *Orientia tsutsugamushi.* Although the disease is similar to typhus, it is transmitted by *mites* to its normal rodent hosts.

Diagnosis and Control of Rickettsial Diseases

In the past, rickettsial infections have been difficult to diagnose because the characteristic rash associated with many rickettsial diseases may be mistaken for measles, scarlet fever, or adverse drug reactions. Clinical confirmation of rickettsial diseases has now been greatly aided by the introduction of specific immunological reagents. These include polyclonal or monoclonal antibodies that detect rickettsial surface antigens by immunofluorescence, latex bead agglutination assays, and ELISA analyses (⟳ Sections 21.5, 21.7, and

21.8) and PCR assays (⟳ 21.10). Control of most rickettsial diseases requires control of the vectors: lice, fleas, and ticks. For humans traveling in wooded or grassy areas, the use of insect repellants usually prevents tick attachment. Firmly attached ticks should be removed gently with forceps, care being taken to remove all the mouth parts. A solvent such as ethanol applied to a tick with a saturated swab usually expedites removal. Although a vaccine is available for the prevention of typhus, the few cases reported do not warrant its general administration. No vaccines are currently available for the prevention of Rocky Mountain spotted fever or ehrlichiosis.

✓ 24.3 Concept Check

Rickettsias are obligate intracellular parasitic Bacteria that are transmitted by insect or tick bite. Most rickettsial infections are readily controlled by antibiotic therapy.

✓ What are the arthropod vectors for typhus, Rocky Mountain spotted fever, and ehrlichiosis?
✓ What are the normal mammalian hosts for these same diseases?

24.4

Tick-Transmitted Diseases: Lyme Disease

Lyme disease is a rapidly emerging tickborne disease that affects humans and other animals. Lyme disease was named for Old Lyme, Connecticut, where cases were first recognized, and has rapidly become the most prevalent tickborne disease in the United States. Lyme disease is caused by a spirochete, *Borrelia burgdorferi* (Figure 24.5), which is spread primarily by the deer tick, *Ixodes dammini,* but can also be spread by the common dog (wood) tick and other species of ticks (Figure 24.6). The ticks that carry *B. burgdorferi* cells feed on the blood of birds, domesticated animals, various wild animals, and occasionally humans; deer and the white-footed field mouse are prime mammalian reservoirs of *B. burgdorferi* in the northeastern portions of the United States. However, in other parts of the country, different species of rodents and ticks are involved in the transmission of Lyme disease. Lyme disease has now also been identified in Europe and Asia, again with its life cycle in different rodent reservoirs and different species of the *Ixodes* genus of tick. The deer tick and other members of the genus are much smaller than many other types of ticks and thus are easy to overlook (Figure 24.6). Unlike the case with other tickborne diseases, a very high percentage (up to 50% in certain regions of the Northeast) of deer ticks carry *B. burgdorferi* cells. Thus, extended contact with a vector gives a high probability of disease transmission.

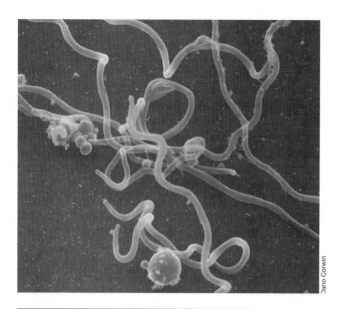

Dano Corwin

FIGURE 24.5   Electron micrograph of the Lyme spirochete, *Borrelia burgdorferi*. The diameter of a single cell is approximately 0.4 μm.

Although most cases of Lyme disease have been reported from the northeastern and upper midwestern United States, cases have been observed in nearly every state, and Lyme disease is spreading west and south. For example, in California, Lyme disease is becoming more prevalent; different species of ticks and mammals (in this case, the dusky-footed woodrat) have been implicated as nonhuman reservoirs. Figure 24.7 shows the recent spread of Lyme disease across the continental United States and the alarming and rapid rise in the total number of cases.

Pfizer Research

FIGURE 24.6   Deer ticks *(Ixodes dammini)*, the major vectors of Lyme disease. Left to right, male and female adult ticks, nymph, and larval forms. The length of an adult female is about 3 mm. All forms feed on humans and are capable of transmitting *Borrelia burgdorferi*.

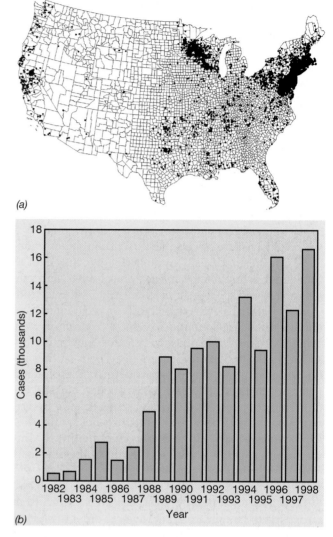

(a)

(b)

FIGURE 24.7   Lyme disease in the United States. (a) Incidence of Lyme disease in the United States in 1997. Each dot represents a single case, placed in the county of origin. Lyme disease is spreading from its original endemic focus in the northeast. (b) Incidence of Lyme disease by year in the United States. Lyme disease is reported through the National Notifiable Diseases Surveillance System of the Centers for Disease Control and Prevention.

Transmission and Pathogenesis of Lyme Disease

Cells of *Borrelia burgdorferi* are transmitted to humans while the tick is obtaining a blood meal (Figure 24.8). A systemic infection develops, leading to the main symptoms of Lyme disease, which include an acute headache, backache, chills, and fatigue. In about 75% of all cases, a large rash, known as *erythema chronicum migrans (ECM)*, is observed at the site of the tick bite (Figure 24.8). If a correct diagnosis is made at this point, Lyme disease is treatable with tetracycline or penicillin. However, if not treated properly, Lyme disease may progress to a chron-

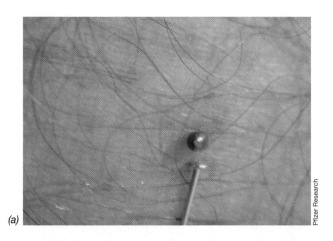

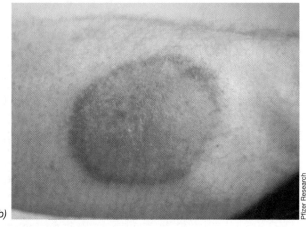

FIGURE 24.8 (a) Deer tick obtaining a blood meal from a human. (b) Characteristic circular rash associated with Lyme disease. The rash, known as *erythema chronicum migrans* (ECM), typically starts at the site of the bite and grows in a circular fashion over a period of several days. This typical example is about 5 cm in diameter.

ic stage. This chronic stage can begin weeks to months later and is characterized by arthritis in 40–60% of patients. Neurological involvement such as palsy, weakness in the limbs, and facial ticks occurs in 15–20% of patients. Cardiac damage results in about 8% of all cases. Treatment at this stage generally requires intravenous antibiotics. The drug ceftriaxone, a highly active β-lactam antibiotic, is used to treat chronic Lyme disease because it is one of the few antibiotics that can cross the blood–brain barrier and can thus attack spirochetes residing in the central nervous system. If no treatment is obtained, cells of *B. burgdorferi* infecting the central nervous system may lie dormant for long periods before eliciting a variety of additional chronic symptoms, including visual disturbances, facial paralysis, and seizures. There is also some indication that Lyme disease can cause miscarriages and stillbirths.

Because the disease is relatively new—the causative agent was first identified in 1982—little is known about the pathogenesis of Lyme disease. No toxins or other virulence factors have yet been identified. In many respects the latent symptoms of Lyme disease resemble those of syphilis, caused by a different spirochete, *Treponema pallidum*. Indeed, some of the neurological symptoms of Lyme disease resemble those of chronic syphilis (∞ Section 23.6). However, unlike syphilis, Lyme disease has not been reported to be spread by sexual intercourse or other types of human contact. Small numbers of *Borrelia burgdorferi* cells are shed in the urine of infected individuals, and there is some indication that Lyme disease can spread from domestic animal populations, particularly cattle, by infected urine.

Detection, Treatment, and Prevention of Lyme Disease

Serological tests have been developed for detection of antibodies to *Borrelia burgdorferi*. Antibodies appear about 4–6 weeks after infection and can be detected by an indirect enzyme-linked immunosorbent assay (ELISA) or a fluorescent antibody assay (∞ Sections 21.7 and 21.8). However, the most definitive serological test for Lyme disease is a Western blot (∞ Section 21.9). Unfortunately, the immune response to Lyme disease is frequently not strong (as is also true of syphilis) and antibodies against the pathogen are difficult to detect in many cases of the disease. In addition, the surface antigens of *B. burgdorferi* are not well defined, and cross reactions often occur with related organisms such as *Treponema pallidum* (the causative agent of syphilis) and *Rickettsia rickettsii* (the causative agent of Rocky Mountain spotted fever). As a result, the serology-based Gunderson Lyme test is one of the most accurate and sensitive indicators of *B. burgdorferi* exposure. Patient serum and complement (∞ Section 20.12) are incubated with *B. burgdorferi* cells. If antigen is bound by the patient serum, the *B. burgdorferi* cell membrane is damaged and the cell contents begin to escape. A dye, acridine orange, stains escaping DNA and confirms the presence of antibodies to *B. burgdorferi* in the serum, indicating that the patient has been exposed to the pathogen.

A polymerase chain reaction (PCR) assay (∞ Section 21.10) has also been developed for the detection of *Borrelia burgdorferi* from many body fluids and tissues. While they are rapid and sensitive, the PCR methods cannot differentiate between live *B. burgdorferi* in active disease and dead *B. burgdorferi* found in treated or inactive disease.

Early detection and treatment of Lyme disease generally afford complete recovery without further symptoms. In cases where a typical Lyme rash is observed surrounding a tick bite (Figure 24.8), antibiotic therapy is usually initiated without waiting for an antibody titer to develop. If no rash is apparent but Lyme disease is still suspected, a Lyme ELISA test or the Gunderson Lyme test is done after an appropriate interval.

Prevention of Lyme disease requires proper precautions to prevent tick attachment. In tick-infested areas like woods, tall grass, and brush, it is advisable to wear protective clothing such as shoes, long pants, and a long-sleeved shirt with a snug collar and cuffs. Tucking the pants into tight-fitting socks worn with boots forms an effective barrier to tick attachment. After spending time in a tick-infested environment, individuals should check themselves carefully for ticks and gently remove any attached ticks (including the head). Insect repellants containing diethyl-*m*-toluamide (DEET) are very effective if applied to both skin and clothing. Finally, an effective Lyme disease vaccine is currently available and is highly recommended for individuals who are occupationally or recreationally exposed to ticks or tick habitats. This highly protective vaccine consists of a *B. burgdorferi* cell-surface protein antigen (OspA) that is expressed in *Escherichia coli* by a bioengineered recombinant gene (⌘ Sections 10.13 and 20.16).

✓ 24.4 Concept Check

Lyme disease is now the most prevalent arthropod-borne disease in the United States. It is transmitted from several mammalian host vectors to humans via ticks. Prevention and treatment of Lyme disease are straightforward, but proper diagnosis is a major problem.

- ✓ What are the primary symptoms of Lyme disease?
- ✓ What antibiotics can be used to treat Lyme disease?

24.5

Insect-Transmitted Diseases: Malaria

Malaria is a disease caused by a protozoan, a member of the Sporozoa group. We discussed sporozoa as a group in Section 18.4. The malaria parasite is one of the most important human pathogens and has played an extremely significant role in the development and spread of human culture. Indeed, as we will see, malaria has even affected human evolution. Malaria is still a significant human disease even though there are several effective treatments available: There are over *100 million* people worldwide who now have malaria, and each year over 1 million of these will die (⌘ Section 22.1).

Ecology and Pathogenesis of Malaria

The major mammalian reservoir for malaria is humans. Four species of sporozoa infect humans, of which the most widespread is *Plasmodium vivax* and the most serious is *Plasmodium falciparum.* This parasite carries out part of its life cycle in humans, and part in the mosquito vector, which spreads the parasite from person to person. Only female mosquitoes of the genus *Anopheles* transmit malaria. *Anopholes* mosquitoes inhabit primarily warmer parts of the world and, therefore, malaria occurs predominantly in the tropics and subtropics. Malaria did not exist in the northern regions of North America prior to settlement by Europeans but was a major problem in the South, where appropriate breeding grounds for the mosquito existed. The disease is associated with swampy low-lying areas, and the name *malaria* is derived from the Italian words for "bad air."

The life cycle of the malaria parasite is complex (Figure 24.9). First, the human host is infected by plasmodial **sporozoites,** small, elongated cells produced in the mosquito, which localize in the salivary gland of the insect. The mosquito injects saliva (containing an anticoagulant) along with the sporozoites. The sporozoites travel through the bloodstream to the liver, where they may remain quiescent, or they may replicate and become enlarged in a stage known as a *schizont.* The schizonts then segment into a number of small cells called **merozoites;** these cells are liberated from the liver into the bloodstream. Some of the merozoites then infect red blood cells (erythrocytes). The cycle in erythrocytes proceeds as in the liver and usually repeats at regular intervals of 48 hr in the case of *Plasmodium vivax.* It is during this period that the defining clinical symptoms of malaria occur, characterized by a fever of up to 40°C (104°F) followed by chills. The chills occur when a new generation of *P. vivax* cells is liberated from erythrocytes. Vomiting and severe headache may accompany the fever–chill cycles, but asymptomatic periods generally alternate with periods in which the characteristic symptoms are present. Because of the loss of red blood cells, malaria generally causes anemia and some enlargement of the spleen as well.

Not all protozoal cells liberated from red blood cells are able to infect other erythrocytes; those that cannot, called *gametocytes,* are infective only for the mosquito. These gametocytes are ingested when another *Anopheles* mosquito bites the infected person and they mature within the mosquito into *gametes*. Two gametes fuse, and a zygote is formed; the zygote migrates by ameboid motility to the outer wall of the insect's intestine where it enlarges and forms a number of sporozoites. These are released and some reach the salivary gland of the mosquito, from where they can be inoculated into another person; the cycle then begins again.

Conclusive evidence for the diagnosis of malaria in humans is obtained by examining blood smears for the presence of infected erythrocytes (Figure 24.10). *Chloroquine* is the drug of choice for treating parasites *within*

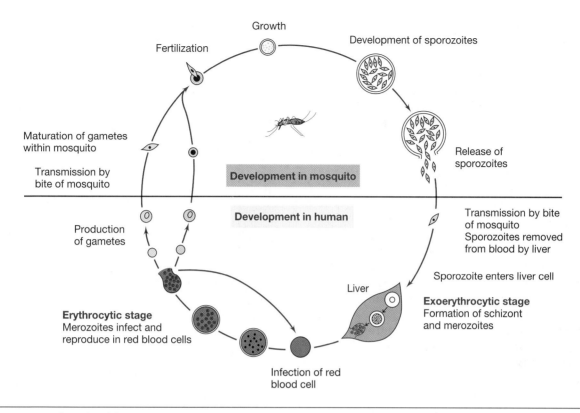

FIGURE 24.9 Life cycle of the malaria parasite, *Plasmodium vivax.*

red blood cells, but this quinine derivative does not kill stages of the malarial parasite residing *outside* erythrocytes. The related drug *primaquine* effectively eliminates the sporozoites, merozoites, and gametes *outside* the cells, and treatment of malarial patients with chloroquine and primaquine together effects a complete cure. However, because recurrences of malaria many years after a primary infection are not uncommon, small numbers of sporozoites apparently survive in the liver, protected from the effects of quinine drugs, and release merozoites months or years later to reinitiate the disease.

Plasmodium vivax has been very difficult to grow in culture, and because of its specificity for humans, experimental infection has also been difficult. For these reasons, most of the experimental work on malarial parasites has been done with species of *Plasmodium* that infect birds or rats, and it is with these species that most of the studies on the development of new drugs have been carried out. However, scientists have been able to grow *Plasmodium falciparum,* another human pathogen, in laboratory culture, and this organism is the current model for experimental malarial research.

Eradication of Malaria

Although quinine derivatives are effective in treating human cases of malaria, because of the obligatory alternation of hosts, *control* of malaria can be best effected by elimination of the *Anopheles* mosquito. Two approaches to mosquito control are possible: (1) elimination of habitat by drainage of swamps and similar breeding areas, (2) elimination of the mosquito by insecticides, and treat-

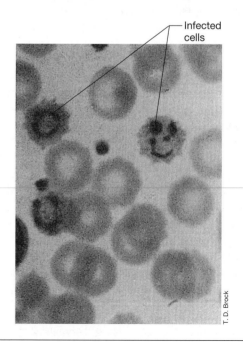

FIGURE 24.10 *Plasmodium vivax,* the causative agent of malaria, growing inside human red blood cells.

ing patients with primaquine, thereby breaking the *Plasmodium* life cycle. During the 1930s, about 33,000 miles of ditches were dug in 16 southern states in the United States, removing 544,000 acres of mosquito breeding area. Millions of gallons of oil were also spread on swamps to reduce the oxygen supply to mosquito larvae. With the discovery of the insecticide dichlorodiphenyltrichloroethane (DDT) (📷 Figure 16.61), chemical control of both larvae and adult mosquitoes was possible. During World War II, the Public Health Service organized an Office of Mosquito Control in War Areas, and because many U.S. military bases were in the southern states, this organization carried out an extensive eradication program in the United States as well as overseas. In 1946, a year in which there were 48,610 cases of malaria in the United States, Congress established a 5-year malaria eradication program. In endemic areas, the program involved drug prophylaxis and treatment regimens for individuals, along with DDT treatment of mosquito infestations. By 1953 there were only 1310 malaria cases. In 1935 there were about 4000 deaths from malaria; in 1952 there were only 25 deaths. Thus, the overall public health threat from malaria in the United States is now minimal. However, *endemic* malaria has occurred in recent years, albeit in very low numbers, as far north as New York City. Increases in malaria incidence still occur as a result of influxes of imported malaria due to soldiers or immigrants coming from malaria-endemic areas (Figure 24.11), but there is, on average, less than one annual death due to malaria in the United States.

In other parts of the world, eradication has been much slower, but the same control measures are used and are still effective. Reduction of mosquito habitat, control of mosquitoes via insecticides (DDT is no longer used due to environmental concerns, but other effective insecticides are available), and treatment of infected individuals with drugs both for cure and prophylaxis, are still the major strategies for controlling malaria. Additionally, several experimental prophylactic vaccines are now in trial stages.

Malaria and Biochemical Evolution of Humans

Malaria has undoubtedly been endemic in Africa for thousands of years. In West Africans, resistance to malaria caused by *Plasmodium falciparum* is associated with the presence in their red blood cells of an altered blood protein, hemoglobin S, which differs from normal hemoglobin A at only a single amino acid in each of the two identical subunits of the molecule. In hemoglobin S, the neutral amino acid *valine* is substituted for the acidic amino acid *glutamic acid* in hemoglobin A. Red blood cells containing hemoglobin S have reduced affinity for oxygen, and the malaria parasite, having a highly aerobic metabolism, cannot grow as well in these red blood cells as it can in normal ones. Individuals who are *heterozygous* for the hemoglobin S trait use oxygen less efficiently than individuals with normal hemoglobin A and are less able

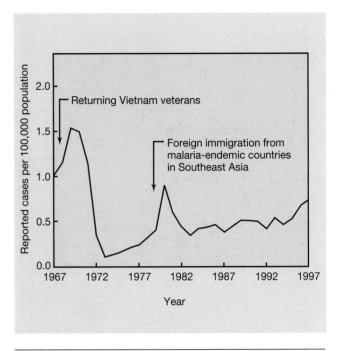

FIGURE 24.11 The prevalence of malaria in the United States. There are currently about 1000 cases of malaria per year. Nearly all cases are imported, with only a few cases from endemic sources. Prior to 1947, there were at least 48,000 cases per year, mostly from endemic sources. The data are from the Centers for Disease Control and Prevention, Atlanta, GA, USA.

to survive at high altitudes, where oxygen pressures are lower. However, in tropical lowland Africa, where oxygen tension is high, this is not a significant disadvantage. Hemoglobin S–containing red cells sometimes appear as C-shaped cells, rather than the round, biconcave red blood cells seen in normal blood (📷 Figure 20.6). As a result, this phenotype is known as the *sickle cell trait*. Individuals who are *homozygous* for the sickle cell trait cannot process oxygen well, even in the lowlands, and normally do not live beyond childhood. In West Africans, resistance to another malarial parasite, *Plasmodium vivax*, is associated with the presence of another abnormal hemoglobin, hemoglobin E. In certain Mediterranean regions where malaria is endemic, resistance to *P. falciparum* is associated with a deficiency in the red blood cells of the enzyme glucose-6-phosphate dehydrogenase (GPD). The faulty GPD leads to higher levels of oxidants, which damage parasite membranes. In many Mediterranean populations, a diverse group of genetic abnormalities affect hemoglobin production and efficiency. These are known collectively as the thalassemias, and all involve some aspect of hemoglobin synthesis, expression, or function. Like hemoglobins S and E and the GPD traits, the thalassemias are genetic mutations that confer a level of resistance to malaria infections and thus are selected in the population in spite of the fact that the carriers all have red blood cell and oxygen-processing defects.

Another case in which the malaria parasite influences biochemical evolution involves the major histocompatibility complex (MHC) and the immune system (∞ Section 20.7). As discussed previously, the MHC class I and class II proteins present antigens to T cells for initiation of an immune response. In malaria-prone equtorial West Africa, individuals are very likely to have one particular MHC class I gene and one particular set of class II genes and gene sets. These particular selected MHC genes are more common in the West African population and are virtually unknown in other human population groups. Individuals who express these genes have as much resistance to severe fatal malaria infections as those with the hemoglobin S trait. These particular MHC proteins are exceptionally good antigen-presenting molecules for certain malarial antigens and confer protective resistance to *Plasmodium* sp. infection as a result of the powerful immune response they help initiate.

Like the hemoglobin variants, the parasite acts as a strong selection agent for individual genes important for host survival: Individuals with particular MHC genes have a measurable survival advantage and are more likely to reproduce and pass the resistance-conferring genes to their progeny.

Thus, several lines of evidence involving several genes and proteins indicate that malaria has been a selective agent in human evolution. We suspect that other microbial parasites such as *Mycobacterium tuberculosis* (tuberculosis, ∞ Section 23.3) and *Yersinia pestis* (plague, see also Section 24.6) have also promoted selective changes in humans, but in no case is the evidence as clear as it is for malaria.

✓ 24.5 Concept Check

Malaria is a widespread, mosquitoborne infectious disease occurring mainly in tropical and subtropical portions of the world. It is a major cause of morbidity and mortality in developing countries and has been responsible for the evolution of several resistance genes. The disease is preventable with a combination of public health and chemotherapy measures.

✓ How can malaria be prevented?
✓ Review genetic mechanisms responsible for malaria resistance. Why are anti-malarial genes not found in all humans?

24.6

Insect-Transmitted Diseases: Plague

Pandemic occurrences of **plague** have been directly responsible for more human deaths than any other infectious disease other than malaria. Plague killed between 25% and 33% of Europe's population in individual epidemics in the Middle Ages. Plague is caused by a gram-negative, facultatively aerobic rod called *Yersinia pestis* (∞ Section 13.10). Plague is a natural disease of domestic and wild rodents, but rats are the primary disease reservoir. Humans are accidental hosts and are not critical for the maintenance of the disease. Fleas are intermediate hosts and act as vectors, spreading plague between the mammalian hosts (Figure 24.12). Most infected rats die soon after symptoms appear, but a low proportion develop a chronic infection and can serve as a source of virulent *Y. pestis*. The majority of cases of human plague in the United States occur in the southwestern states, where the disease is endemic among wild rodents (sylvatic plague).

Bubonic Plague

Plague is transmitted by the rat flea (*Xenopsylla cheopis*), which ingests *Yersinia pestis* cells by sucking blood from an infected animal. Cells multiply in the flea's intestine and can be transmitted to a healthy animal in the next bite. As the disease spreads, rat mortality becomes so great that infected fleas seek new hosts, including humans. Once in humans, cells of *Y. pestis* usually travel to the lymph nodes, where they cause the formation of swollen areas referred to as *buboes*. For this reason the disease is frequently referred to as **bubonic plague** (Figure 24.13a). The buboes become filled with *Y. pestis*, but the distinct capsule on cells of *Y. pestis* prevents them from being phagocytized (∞ Section 20.3). Secondary buboes form in peripheral lymph nodes, and

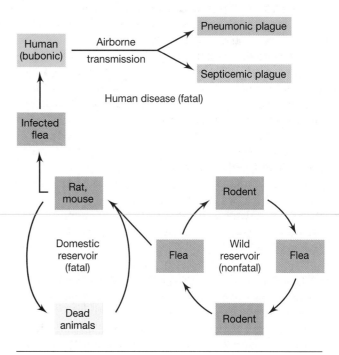

FIGURE 24.12 The ecology of plague. Plague in most animals is generally a mild infection. Plague in rats and humans is frequently fatal.

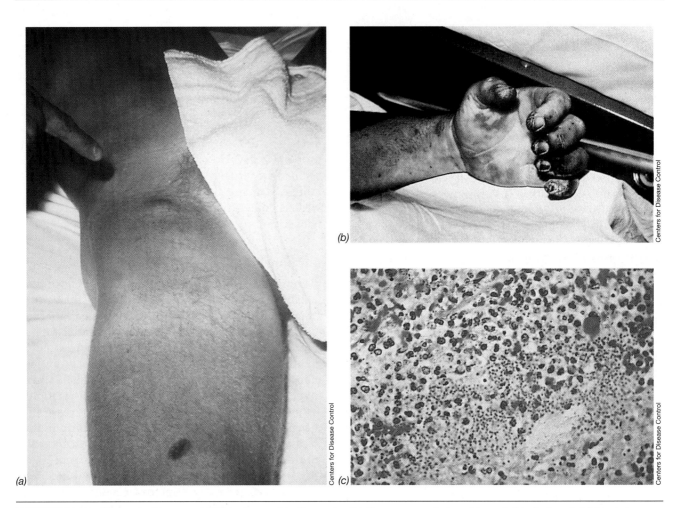

FIGURE 24.13 Plague in humans. (a) Bubo formed in the groin. (b) Gangrene and sloughing of skin in hand. (c) *Yersinia pestis*, the causative agent of plague. Cells are seen as very small blue cells from the lung tissue of a pneumonic plague victim.

cells eventually enter the bloodstream, causing a generalized septicemia. Multiple hemorrhages produce dark splotches on the skin giving plague its historical nickname, the "Black Death" (Figure 24.13b). If not treated prior to the septicemic stage, the symptoms of plague, including extreme lymph node pain, prostration, shock, and delirium, progress and usually cause death within 3–5 days.

The pathogenesis of plague is not clearly understood, but it is known that cells of *Yersinia pestis* produce a number of antigenically distinct molecules, including toxins, that undoubtedly contribute to the disease process. The V and W antigens of *Y. pestis* cell walls are protein–lipoprotein complexes that serve to prevent phagocytosis. Other envelope proteins are also present. An exotoxin called *murine toxin*, because of its extreme toxicity for mice, is produced by all virulent strains of *Y. pestis*. Murine toxin is a respiratory inhibitor that blocks mitochondrial electron transport reactions at the point of coenzyme Q (⚭ Section 4.11). Although it is not clear that murine toxin is involved in the pathogenesis of human plague (murine toxin is highly toxic

for certain animal species but not for others), it produces systemic shock, liver damage, and respiratory distress in mice. These symptoms are similar to those in human cases of plague. *Y. pestis* also produces a highly immunogenic *endotoxin* that may also play a role in the disease process.

Other Forms of Plague

Pneumonic plague occurs when cells of *Yersinia pestis* are either inhaled directly or reach the lungs during bubonic plague (Figure 24.13c). Symptoms are usually absent until the last day or two of the disease when large amounts of bloody sputum are emitted. Untreated cases rarely survive more than 2 days. Pneumonic plague, as one might expect, is a highly contagious disease and can spread rapidly via the person-to-person respiratory route if infected individuals are not immediately quarantined. **Septicemic plague** involves the rapid spread of *Y. pestis* throughout the body via the bloodstream without the formation of buboes and usually causes death before a diagnosis can be made.

Treatment and Control

Plague can be successfully treated if rapidly diagnosed. *Yersinia pestis* infection is preferentially treated with streptomycin, and alternatively with tetracycline. For individuals exposed to pneumonic plague, tetracycline is recommended for prophylaxis. If treatment is started promptly, mortality from bubonic plague can be reduced to as few as 1–5% of those infected. Pneumonic and septicemic plague can also be treated, but these forms progress so rapidly that antibiotic therapy in the latter stages of the disease is usually too late. Although potentially a devastating disease, there have been only 91 cases of plague in the United States in the last decade (Figure 24.14). Unfortunately, 55 of these patients died (a fatality rate of 60%). Worldwide, there are fewer than 1500 confirmed cases and fewer than 300 deaths per year.

Plague control is accomplished through surveillance of infected animals, vectors, and human contacts. Plague-infected animal populations are destroyed when identified. Undoubtedly, improved public health practices and the overall control of rodent populations have limited human exposure to plague. Prevention can also be accomplished by immunization with a formalin-killed vaccine, but, because the incidence of disease is low, immunization is only recommended for individuals with high exposure risk.

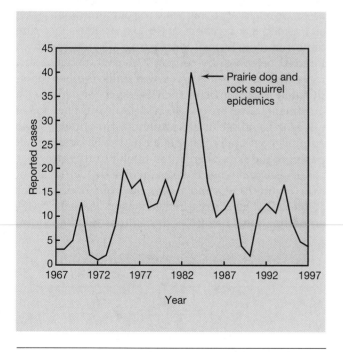

FIGURE 24.14 Plague in the United States. From 1988–1997, only 90 cases and 8 deaths were reported. The data are from the Centers for Disease Control and Prevention, Atlanta, GA, USA.

✓ 24.6 Concept Check

Plague is largely confined to individuals who come into contact with rodent populations that are endemic reservoirs for *Yersinia pestis*. A disseminated systemic infection often leads to rapid death, but localized infections are treatable with antibiotics.

✓ Distinguish among *bubonic, septicemic,* and *pneumonic* plague.
✓ What is the insect reservoir, natural host, and treatment for plague?

Common-Source Diseases

We now begin a discussion of a variety of diseases that come from *common sources:* environmental reservoirs of pathogens with which humans have constant contact. We will discuss diseases normally spread through soil, water, and food. As we shall see, organisms like *Aspergillus* spp., a mold, are present in the environment at all times; molds are common and ubiquitous members of the community of microorganisms found worldwide in soil, and a few can act as pathogens. However, certain pathogens such as *Escherichia coli* O157:H7 must be introduced by human activity into a common source such as food or water. The major challenges for protecting public health come from a variety of pathogens that can be transmitted by common sources. Finally, keep in mind that many diseases can be spread by both the person-to-person route, as well as through common sources. For example, tuberculosis in the United States, in addition to person-to-person spread, was formerly spread through dairy products originating from *Mycobacterium bovis*–infected dairy herds. The pasteurization of dairy products and eradication of tuberculosis-infected dairy cattle were public health measures aimed at eliminating a common source of infection, namely tuberculosis-infected dairy and meat products. These measures were probably more important in reducing tuberculosis in the United States than the advent and use of antibiotics. Indeed, probably the most important contributions to overall public health have been made by ensuring adequate supplies of pathogen-free food and water.

24.7

Soil Microorganisms: The Pathogenic Fungi

Many microorganisms occur in nature as free-living saprophytes. Among these are a few that occasionally act as pathogens. These include Bacteria from the genus *Clostridium* and a group of eukaryotic microorganisms collectively called *fungi.* We will discuss *Clostridium* in the context of food microbiology (see Section 24.11) because of their importance as food poisoning agents. Here we will discuss fungi that act as accidental, often opportunistic, pathogens.

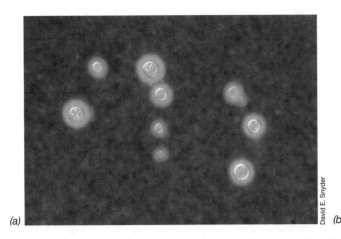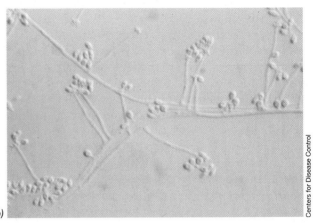

(a)   David E. Snyder

(b)   Centers for Disease Control

FIGURE 24.15 Typical forms of pathogenic fungi. (a) Yeast form of encapsulated *Cryptococcus neoformans*, stained with india ink. The cell itself ranges from 4 to 20 μm. (b) *Sporothrix schenckii*, showing the branching, or hyphae, characteristic of the mold form of fungi. The round basidiospores are about 2 μm in diameter.

Fungi grow in some form in nearly every ecological niche, but are particularly prevalent growing on the nonliving organic materials in soil. The fungi include the eukaryotic organisms we commonly refer to as *yeasts*, which normally grow as single cells (Figure 24.15*a*), and *molds,* which grow in branching chains called *hyphae* (Figure 24.15*b*). The taxonomy and biological diversity of these organisms were discussed in Section 17.4. Fortunately, most fungi are harmless to humans. Only about 50 species cause human disease, and the overall incidence of serious fungal infections is rather low, although certain superficial fungal infections are quite common.

Fungi cause disease through three major mechanisms. First, some fungi cause immune responses that can result in allergic (hypersensitivity) reactions following exposure to specific fungal antigens (∞ Section 20.15). Exposure to fungi, whether growing on the host or in the environment, may cause allergic symptoms on reexposure. For example, *Aspergillus* spp., a common saprophyte often found in nature as a leaf mold, is a potent and common allergen, often causing asthma and other hypersensitivity reactions. As we shall see later, *Aspergillus* also has other mechanisms for producing disease.

A second fungal disease-producing mechanism involves the production and action of *mycotoxins,* a large, diverse group of fungal exotoxins. The best-known examples of mycotoxins are produced by *Aspergillus flavus,* an organism that commonly grows on improperly stored food such as grain. The toxins produced by *A. flavus* are known as *aflatoxins* (Figure 24.16). Aflatoxins are highly toxic and induce tumors in some animals, especially in birds that feed on contaminated grain. Their direct role in human disease is not well defined.

The third fungal disease-producing mechanism is through infection. The growth of a fungus on or in the body is called a *mycosis* (plural, *mycoses*). Mycoses can range in severity from relatively innocuous, superficial diseases to serious, life-threatening diseases.

Mycoses

Mycoses are subdivided into three categories. The first of these is the *superficial mycoses.* These diseases involve colonization of the skin, hair, or nails and infect only the surface layers (Figure 24.17*a*). Table 24.2 lists some of the common superficial fungi. In general, these diseases are relatively benign and self-limiting. Some, such as *Trichophyton* infections of the feet (athlete's foot), are quite common. Spread is by personal contact with an infected person or by contact with contaminated surfaces such as bathtubs or shower stalls or other contaminated shared articles such as towels or bed linens. Treatment for severe cases is with topical application of miconazole nitrate or griseofulvin. Griseofulvin is also administered orally. After entering the bloodstream, it passes to the skin where it can inhibit fungal growth.

The *subcutaneous mycoses* involve deeper layers of skin (Figure 24.17*b*) and a different group of organisms (Table 24.2). One disease in this category is *sporotrichosis,* an occupational hazard of agricultural workers, min-

FIGURE 24.16 Structure of aflatoxin B1. This toxin is one of a group of related compounds produced by *Aspergillus flavus.*

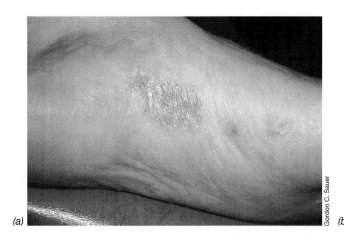

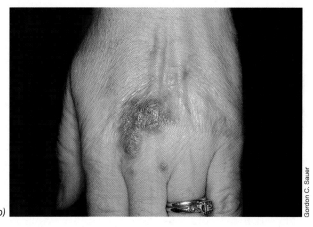

FIGURE 24.17   Fungal infections. (a) Superficial mycosis of the foot (athlete's foot) due to infection with *Trichophyton rubrum*. (b) Sporotrichosis, a subcutaneous infection due to *Sporothrix schenckii*.

ers, and other workers who come into contact with the soil. The causative organism is found as a ubiquitous saprophyte on wood and in soil. The lesions, usually initiated by fungal infection of a small wound or abrasion, resemble the ulcers and chancres seen in ulcerating diseases such as syphilis (compare Figure 24.17*b* and ∞ Figure 23.25). The causative organism, *Sporothrix schenckii*, can readily be isolated from the lesion and cultured *in vitro*. Treatment is with oral potassium iodide or oral ketoconazole.

The *systemic mycoses* involve fungal growth in internal organs of the body. These are subclassified as *primary* or *secondary* infections. A primary infection is one resulting directly from the fungal pathogen in otherwise normal, healthy individuals. A secondary infection involves infection with the pathogen only in hosts with a predisposing condition such as antibiotic therapy or immunosuppression. In the United States, the most widespread primary fungal infections are *histoplasmosis*, caused by *Histoplasma capsulatum*, and *coccidioidomycosis* (San Joaquin Valley fever), caused by *Coccidioides immitis*. Both of these organisms normally live in soil. These are both respiratory diseases in which the host becomes infected by breathing in airborne spores, which germinate and grow in the lungs. Histoplasmosis is primarily a disease of rural areas in the midwestern United States, especially in the Ohio and Mississippi river valleys. Most cases are mild and are often mistaken for more common respiratory infections. San Joaquin Valley fever is generally restricted to the desert regions of the southwestern United States. The fungus lives in desert soils, and the spores are disseminated on dry, windblown particles that are inhaled. In some areas in the southwestern United States, as many as 80% of the inhabitants may be infected, although most individuals suffer no apparent ill effects.

TABLE 24.2	Some pathogenic fungi and the diseases they cause	
Disease	**Causal organism**	**Main disease foci**
Superficial mycoses (dermatomycoses)		
Ringworm	*Microsporum*	Scalp of children
Favus	*Trichophyton*	Scalp
Athlete's foot	*Epidermophyton, Trichophyton*	Between toes, skin
Jock itch	*Trichophyton, Epidermophyton*	Genital region
Subcutaneous mycoses		
Sporotrichosis	*Sporothrix schenckii*	Arms, hands
Chromoblastomycosis	Several fungal genera	Legs, feet
Systemic mycoses		
Cryptococcosis[a]	*Cryptococcus neoformans*	Lungs, meninges
Coccidioidomycosis[a]	*Coccidioides immitis*	Lungs
Histoplasmosis[a]	*Histoplasma capsulatum*	Lungs
Blastomycosis	*Blastomyces dermatitidis*	Lungs, skin
Candidiasis[a]	*Candida albicans*	Oral cavity, intestinal tract

a Considered opportunistic pathogens and have been implicated in the pathogenesis of AIDS (∞ Section 23.7).

A number of fungal infections, including histoplasmosis and coccidioidomycosis, are especially serious and common in individuals whose immune systems have been impaired, either by acquired immunodeficiency syndrome (AIDS) (∞ Section 23.7) or by immunosuppressive drugs. These are secondary fungal diseases because normal individuals either do not get the disease or generally have a less severe form. Examples of other fungal organisms involved as secondary pathogens are given in Table 24.2. These fungi are known as *opportunistic* pathogens because of their particular ability to cause serious infections only in individuals with impaired defense mechanisms (in particular AIDS patients) (∞ Section 23.7, the box, A Definition of AIDS, and Figure 23.29).

Effective chemotherapy against systemic fungal infections is very difficult (∞ Section 18.11). Most antibiotics that inhibit fungi also harm other eukaryotic organisms, including the human host. One of the most effective antibiotics, amphotericin B, is widely used to treat systemic fungal infections of humans, but serious side effects such as kidney toxicity may occur.

Control of infections by elimination of fungal pathogens from the environment is impractical. As with many common-source pathogens, control of fungal growth is very difficult because there is a limitless reservoir: Exposure to fungi cannot be eliminated, except through efforts such as decontamination and air filtration in restricted local environments.

✓ 24.7 Concept Check

Fungal diseases are often difficult to control because of a lack of suitable fungus-specific antibiotics. Serious systemic infections due to the normally nonpathogenic *Candida albicans* and other fungi have been routinely seen in AIDS patients as opportunistic pathogens. The three classes of fungal infections (mycoses) include the superficial, the subcutaneous, and the systemic mycoses.

- ✓ Distinguish a superficial mycosis from a systemic mycosis.
- ✓ Why are mycoses so difficult to treat with antibiotics?

24.8

Public Health and Water Quality

Perhaps the most important common source of infection is water. Even water that looks clear and pure may be sufficiently contaminated with pathogenic microorganisms to be a health hazard. Some means are necessary to ensure that drinking water is safe. Unfortunately, it is not usually practical to examine water for the various pathogenic organisms that may be present.

However, it *is* practical to sample water supplies for the overall presence of microorganisms. While a few nonpathogenic microorganisms might be tolerable in a water

supply, the presence of specific *indicator organisms* signals that a given water source might be contaminated with pathogens. These indicator organisms are usually associated with the intestinal tract and their presence indicates possible fecal contamination of the water source. The most widely used indicator is the **coliform group** of organisms. This group is defined in water bacteriology as all the aerobic and facultatively aerobic, gram-negative, non–spore-forming, rod-shaped Bacteria that ferment lactose with gas formation within 48 hr at 35°C. This is an operational rather than a taxonomic definition, and the coliform group includes a variety of organisms, *mostly* of intestinal origin. In practice, the coliform organisms are almost always members of the enteric bacterial group (∞ Section 13.10). The coliform group includes the organism *Escherichia coli*, a common intestinal organism, and the organism *Klebsiella pneumoniae*, a less common intestinal inhabitant. The definition also currently includes organisms of the species *Enterobacter aerogenes* not generally associated with the intestine.

The coliform group of organisms are suitable as indicators because they are common inhabitants of the intestinal tract, both of humans and warm-blooded animals, and are present in the intestinal tract in large numbers. When excreted into water, the coliforms eventually die, but they do not die at a faster rate than the pathogenic bacteria *Salmonella* and *Shigella*, and both the coliforms and the pathogens behave similarly during water purification processes. Thus, it is likely that if coliforms are found in a water sample, the water has received fecal contamination and may be unsafe for drinking purposes. Finally, the coliform group includes organisms derived not only from humans but also from other warm-blooded animals. Because many of the pathogens (for example, *Salmonella, Leptospira*) found in warm-blooded animals also infect humans, an indicator of both human and animal pollution is desirable.

The Coliform Test

There are two types of procedures that are used for the coliform test. These are the **most-probable-number** (MPN) procedure and the **membrane filter** (MF) procedure. The MPN procedure employs liquid culture medium in test tubes, the samples of drinking water being added to the tubes of media. In the more common MF procedure, the sample of drinking water is passed through a sterile membrane filter, which removes the bacteria (∞ Section 18.3), and the filter is then placed on a culture medium for incubation. When using the membrane filter method with drinking water, at least 100 ml of water should be filtered, although in clean water systems, even larger volumes can be filtered. After filtration of a known volume of water, the filter is placed on the surface of a plate of eosin–methylene blue (EMB) culture medium, which is highly selective for coliform organisms (⚭ Section 21.2 and Figure 21.4). The coliform colonies (Figure 24.18) are counted, and from this value the number of coliforms in the original water sample can be determined. In well-regulated water supply systems, coliform tests are always negative. If the coliform tests are not uniformly negative, a breakdown in the system has occurred (such as in chlorination) or in the distribution network (pipelines).

Drinking water standards in the United States are specified under the Safe Drinking Water Act, which provides a framework for the development of drinking water standards by the Environmental Protection Agency (EPA). Current standards prescribe that when the MF technique is used, 100-ml samples must be filtered and the number of coliform bacteria shall not exceed any of the following: (1) 1 per 100 ml as the arithmetic mean of all samples examined per month; (2) 4 per 100 ml in more than one sample when less than 20 are examined per month; or (3) 4 per 100 ml in more than 5% of the samples when 20 or more are examined per month. Water utilities report their results to the EPA, and if they do not meet the prescribed standards, they must notify the public and take steps to correct the problem. Many smaller communities and even large cities sometimes fail to meet the standards.

Public Health Significance of Drinking Water Purification

Today the incidence of waterborne disease in developed countries is so low that it is difficult to appreciate the significance of treatment practices and drinking water standards. Most intestinal infection today is not due to transmission by the water route but via food (see Sections 24.11 and 24.12). It was not always so. Until the twentieth century, effective water treatment practices did not exist, and there were no bacteriological methods for evaluating the health significance of polluted drinking water (⚭ box, Snow on Cholera, Chapter 22). The first coliform counting procedures were introduced about 1905. Until then, water purification, if practiced at all, was primarily for aesthetic purposes, to remove turbidity. Actually, turbidity removal by filtration provides a significant decrease in the microbial load of water, and so filtration did play a part in providing safer drinking water. But filtration alone was of only partial value because many organisms passed through the filters. It was the discovery of chlorine as an extremely efficient water disinfectant, in about 1910, that had major impact. Chlorine is so effective and so inexpensive that its use spread widely; chlorination was of major significance in reducing the incidence of waterborne disease. However, the effectiveness of chlorination would not have been realized, and the necessary doses could not have been determined, if standard methods for assessing the coliform content of drinking water had not been developed. Thus, engineering and microbiology moved forward together.

The significance of filtration and chlorination for ensuring the safety of drinking water cannot be overemphasized. Figure 24.19 illustrates the dramatic drop in incidence of typhoid fever in a major American city after these two purification procedures were introduced. Similar results were obtained in other major cities. The

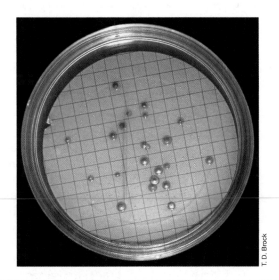

FIGURE 24.18 Coliform colonies growing on a membrane filter. A drinking water sample has been passed through the filter. The filter was placed on eosin–methylene blue (EMB) media that is both selective and differential for lactose-fermenting bacteria (coliforms) (⚭ Section 21.2). The dark color of the colonies is characteristic of coliforms. A count of the number of colonies gives a measure of the coliform count of the original water sample.

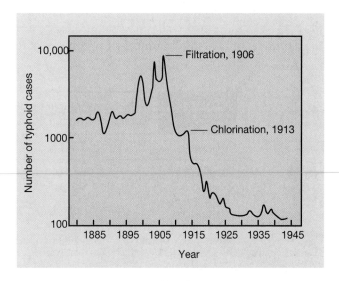

FIGURE 24.19 The dramatic effect of water purification on the incidence of waterborne disease. The graph shows the incidence of typhoid fever in Philadelphia (PA, USA) during the early part of the twentieth century. Note the dramatic reduction in incidence of the disease after the introduction of filtration and chlorination.

dramatic improvement in the health of the American people in the early decades of the twentieth century was due to a large extent to the establishment of satisfactory water purification procedures.

✓ 24.8 Concept Check

Drinking water quality is assessed by counting coliform bacteria. Strict microbiological standards make this method a reliable indicator of fecal contamination. The introduction of water purification measures is probably the single most important public health measure ever devised.

✓ Why is the coliform test an indicator of water quality?
✓ Why is filtration an important component in water purification?

24.9

Waterborne Diseases

Many human pathogens are transmitted by water used for cooking, drinking, and even swimming. In the United States, a number of different bacterial and protozoan pathogens are occasionally transmitted in drinking water (Table 24.3). Worldwide, as we shall see, viruses and other bacteria are also implicated as waterborne pathogens. Microorganisms transmitted in water generally grow in the intestines and leave the body in feces. Fecal pollution of water supplies may then occur, and if the contamination is not identified (for example, by the coliform test, see Section 24.8) or eliminated by disinfection (see Section 24.8), then a new host may consume the water and the pathogen may colonize the intestine and cause disease. Thus, waterborne diseases are *infections*. Keep in mind that water may be infectious even if it contains only a small number of organisms; the

exact numbers necessary to cause disease are functions of the virulence of the pathogen and the ability of the host to resist infection (∞ Sections 19.11 and 19.12). As a common source of pathogen dissemination, water has the most potential for the catastrophic epidemic spread of disease, as we shall see.

A few waterborne pathogens are not transmitted by the drinking water route. The best known of these is *Legionella pneumophila*, the bacterium that causes Legionellosis. We start by examining the emergence of this organism as an important waterborne pathogen.

Legionellosis (Legionnaires' Disease)

This disease, caused by the organism *Legionella pneumophilia*, derives its name from the fact that it was first recognized as a disease entity from an outbreak of pneumonia occurring during a convention of the American Legion in the summer of 1976. *Legionella* is a thin gram-negative rod (Figure 24.20) with complex nutritional requirements, including an unusually high iron requirement, and is immunologically distinct from any other pathogen associated with respiratory infections. *Legionella* can be detected by immunofluorescence techniques (∞ Section 21.7 and Figure 21.14a) and can be isolated from terrestrial and aquatic habitats as well as from patients suffering from legionellosis. *Legionella* is a common inhabitant of the cooling towers and evaporative condensers of large air conditioning systems. The pathogen apparently grows in the water and is disseminated in humidified aerosols. Human infection is via airborne droplets, but the infection is not spread person to person. Consistent with these findings is the fact that multiple outbreaks of clustered cases of legionellosis tend to peak in midsummer to late summer months when air conditioners are most extensively

TABLE 24.3	Waterborne infectious disease outbreaks associated with drinking water in the United States [a]		
Disease	**Agent**	**Outbreaks**	**Cases**
Bacteria			
Shigellosis	*Shigella* spp.	2	266
Salmonellosis	*Salmonella typhimurium*	1	625
Gastroenteritis	*Campylobacter* spp.	3	223
Cholera	*Vibrio cholerae*	1	11
Protozoa			
Giardiasis	*Giardia lamblia*	5	385
Cryptosporidiosis[b]	*Cryptosporidium parvum*	3	403,237
Unknown			
Gastroenteritis		5	495

a Compiled from data provided by the Centers for Disease Control for 1993–1994.

b A single outbreak of cryptosporidiosis caused illness in about 400,000 individuals. This was the largest single recorded outbreak of a waterborne disease in history. The outbreak was due to a failure in a large municipal water treatment facility in Milwaukee, Wisconsin (USA).

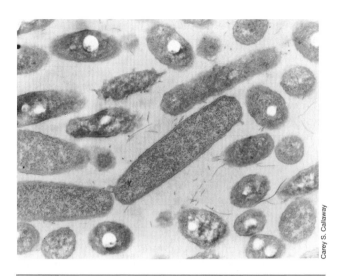

FIGURE 24.20 Transmission electron micrograph of *Legionella pneumophila*. Cells are approximately 0.6 μm in diameter.

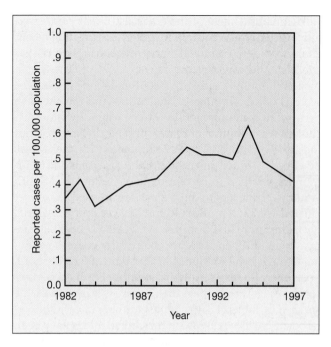

FIGURE 24.21 The annual prevalence of legionellosis in the United States. Note the general upward trend. In 1997 the total number of cases of legionellosis was 1163. The data are from the Centers for Disease Control and Prevention, Atlanta, GA, USA.

used. This is in contrast to an airborne disease such as chickenpox, which is spread from person to person and peaks in the winter months (co Figure 23.20) when people are more frequently indoors and in close contact. *Legionella has* also been found in other water systems in hot water tanks and whirlpool spas. It grows to high numbers in any warm, stagnant water at 35–45°C. Epidemiological studies now indicate that *Legionella* infections occur at various times of the year, primarily as a result of aerosols generated by such common practices as showering or bathing. Overall, the incidence of reported cases of legionellosis has been steadily increasing, but epidemiological studies suggest that upwards of 90% of cases are not diagnosed or reported properly (Figure 24.21). *Legionella* is thus an important emerging pathogen. Prevention of legionellosis centers on improving the maintenance and design of water-dependent cooling and heating systems.

Legionella infections may be totally asymptomatic and occasionally result in only mild symptoms such as headache and fever. These mild, self-limiting cases, called *Pontiac fever,* need no treatment, and resolve in 2–5 days. The majority of cases of *Legionella* pneumonia are in elderly individuals whose resistance has been previously compromised. In addition, certain serotypes of *Legionella* (eight are known) are more strongly associated with the pneumonic form of the illness than others. Prior to the onset of pneumonia, intestinal disorders are common, followed by high fever, chills, and muscle aches. These symptoms precede the dry cough and chest and abdominal pains typical of legionellosis. Death occurs in 5–15% of cases and is usually due to respiratory failure. Clinical detection of *Legionella pneumophila* is now straightforward because of the development of serological methods to measure changes in antibody titer (co Section 22.3). In addition, *Legionella* antigens can be detected in patient urine. However, the most definitive test is the detection of *Legionella* in the sputum of patients. *Legionella pneumophila* is sensitive to the antibiotics rifampicin and erythromycin, and intravenous administration of erythromycin is the treatment of choice in most cases.

Pathogenic Organisms Transmitted by Drinking Water

Probably the most important pathogenic *bacteria* transmitted by the water route are *Salmonella typhi*, the organism causing typhoid fever, and *Vibrio cholerae*, the organism causing cholera. Although the causal agent of typhoid fever may also be transmitted by contaminated food (see Section 24.11) and by direct contact from infected people (co Chapter 22, box, The Tragic Case of Typhoid Mary), the most common and serious means of transmission is the water route. Typhoid fever has been virtually eliminated in many parts of the world and is no longer a problem in developed countries, primarily as a result of the development of effective water treatment methods. However, a breakdown in water purification methods, contamination of water during floods, earthquakes, and other disasters, or cross-contamination of water pipes from leaking sewer lines occasionally results in epidemics of typhoid fever.

The causal agent of cholera is also usually transmitted by the water route. However, during the recent cholera epidemic in the Americas, consumption of raw shellfish and raw vegetables was also implicated as a means of cholera spread. Presumably, the vegetables were washed in contaminated water and the shellfish beds were contaminated by raw sewage. At one time, cholera was common in Europe and North America, but the disease has been virtually eliminated from these areas by effective water purification. The disease is still common in Asia and in certain parts of Central and South America. Travelers to these areas are advised to be vaccinated for cholera. Both *Vibrio cholerae* and *Salmonella typhi* are eliminated from sewage during proper sewage treatment and hence do not enter water courses receiving treated sewage effluent. More frequent than typhoid, but generally less serious, is salmonellosis caused by species of *Salmonella* other than *S. typhi* (see Section 24.11). As seen in Table 24.3, the largest number of cases of waterborne bacterial disease in the United States have been due to *Salmonella* spp. infections, except for a single, massive outbreak due to *Cryptosporidium parvum* (see later).

Enteric bacteria are effectively eliminated from water during water purification and so they should never be present in properly treated drinking water. Most outbreaks of waterborne disease in the United States are due to breakdowns in treatment systems or are a result of postcontamination in pipelines. This latter problem can be controlled by maintaining a detectable level of free chlorine in pipelines.

Viruses can also be transmitted in water. Quite commonly, enteroviruses such as polio virus (∞ Section 8.15), Norwalk virus (∞ Table 22.6), and hepatitis A virus (see below) are shed into the water in fecal material. The most serious of these was poliovirus, but the disease polio has been eliminated from the Western Hemisphere. Although these agents can survive in water for relatively long periods, they are easily neutralized by ordinary water purification schemes such as chlorination, and the maintenance of 0.6 parts per million (ppm) of chlorine (see Section 24.8) ensures the safety of the water supply.

Cholera

We discussed the action of cholera enterotoxin in Section 19.9. The disease cholera is caused by *Vibrio cholerae*, a gram-negative, curved rod transmitted almost exclusively via contaminated water. Cholera enterotoxin catalyzes a life-threatening diarrhea that can result in dehydration and death unless the patient is given fluid and electrolyte therapy. The disease has swept the world in seven major pandemics, the most recent of which began in South America in 1991. Since then, more than one million cases of cholera and more than 10,000

deaths have been reported, and health officials believe that the actual number of cases and deaths are much higher. Today, the disease is threatening to take hold in developed countries (for instance, cases have been reported along the Gulf coast of the United States where it is apparently endemic, although the incidence is low) and is very common in developing countries in the Americas, especially in coastal areas where sewage treatment is either not practiced or poorly performed.

Two major biotypes of *Vibrio cholerae* have been recognized, the *classic* and the *El Tor* types. Each biotype has two major serotypes. The classic strain, such as the *V. cholerae* strain first isolated by Robert Koch in 1883, was more prevalent in cholera outbreaks before 1960, whereas the El Tor strain has been more frequently observed since that time. Following ingestion of a substantial inoculum, the *V. cholerae* cells take up residence in the *small* intestine. Studies with human volunteers have shown that the acidity of the stomach is responsible for the large inoculum needed to initiate cholera. Although the ingestion of 10^8–10^9 cholera vibrios is generally required to cause cholera, human volunteers given bicarbonate to neutralize gastric acidity developed the disease when only 10^4 cells were administered. Far lower cell numbers are required to initiate infection if *V. cholerae* is ingested with food. Cholera vibrios attach firmly to small intestinal epithelium and grow and release enterotoxin. The enterotoxin causes huge fluid losses of up to 20 liters per day. If untreated, the mortality rate can be as high as 60%. Intravenous or oral liquid and electrolyte replacement therapy is the major means of treatment (∞ Section 19.9). Streptomycin or tetracycline may shorten the course of cholera, but antibiotics are of little benefit without simultaneous fluid and electrolyte replacement.

Control of cholera depends primarily on satisfactory sanitation measures, particularly in the treatment of sewage and the purification of drinking water. *Vibrio cholerae* organisms can adhere to normal flora in fresh water and can survive for long periods of time (Figure 24.22). In the last 10 years sporadic outbreaks of cholera (less than 250 total cases) have been reported in the United States, and some evidence exists that raw shellfish may be an alternate vehicle, as mentioned previously. Cholera is heavily endemic in India, Pakistan, Bangladesh, and Central and South America, with occasional epidemics emerging from time to time. Although most of these cases are due to water contaminated with human feces, cholera is also spread within households in these regions by fecally contaminated food.

Giardiasis

Giardiasis is an acute form of gastroenteritis caused by the protozoan parasite *Giardia lamblia* (Figure 24.23). *G. lamblia* is a flagellated protozoan that is transmitted to

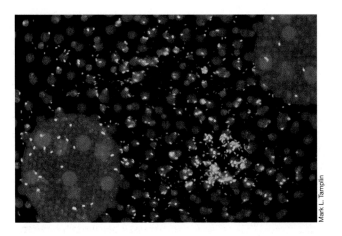

FIGURE 24.22 Cells of *Vibrio cholerae* attached to the surface of *Volvox,* a freshwater alga. The isolate was from a cholera-endemic area in Bangladesh. The green *V. cholerae* cells are stained with a monoclonal antibody to bacterial cell surface proteins. The red color is due to the fluorescence of chlorophyll *a* in algal cells (∞ Figures 3.6 and 17.18*b*).

humans primarily by contaminated water, although foodborne and even sexual transmission of giardiasis has been documented (∞ Section 17.3). The protozoal cells, called trophozoites (Figure 24.23*a*) produce a resting stage called a cyst (Figure 24.23*b*), and this is the primary form transmitted by water. Cysts germinate in the gastrointestinal tract and bring about the symptoms of giardiasis: an explosive, foul-smelling, watery diarrhea and intestinal cramps, flatulence, nausea, and malaise. The foul-smelling nature of the diarrhea and the absence of blood or mucus in the stool are diagnostically helpful in distinguishing giardiasis from diarrhea of bacterial or viral origin. The drugs quinacrine and metronidazole are useful in treating the disease.

In 1993–1994, *Giardia* was implicated in 25% of the drinking-water–borne infectious disease outbreaks in the United States (Table 24.3) and was also the factor in gastrointestinal illnesses in 4 of 14 recreational water disease

outbreaks. These diseases were acquired after accidental ingestion of water from swimming pools or lakes. *Giardia* cysts have been found in up to 97% of surface water sources in the United States. The cysts are fairly resistant to chlorine, and many outbreaks have been associated with water systems using only chlorination as a means of water purification. Water subjected to proper sedimentation, filtration, and chlorination (∞ Section 11.15) is generally free of *Giardia* cysts. Several isolated cases of giardiasis have been associated with untreated drinking water in wilderness areas. Studies of wild animals have indicated that beavers and muskrats are major carriers of *Giardia* and may transmit cells or cysts to water supplies. As a safety precaution, all water consumed from rivers and streams, for example, during a camping or hiking trip, should be filtered *and* treated with iodine or chlorine, or filtered *and* boiled. Boiling is the preferred method of rendering water microbially safe.

Laboratory diagnostic methods include the demonstration of *Giardia* cysts or trophozoites in the stool, or the demonstration of *Giardia* antigens in the stool using a direct ELISA (enzyme-linked immunosorbent assay) test (∞ Section 21.8).

Cryptosporidiosis

The largest single outbreak of a waterborne disease ever recorded took place in Milwaukee, Wisconsin, in the spring of 1993 (Table 24.3). About 400,000 people developed a diarrheal illness that was traced to the municipal water supply. Apparently, spring rains and runoff from surrounding farmland into Lake Michigan had overburdened the water supply system, leading to contamination by the protozoan *Cryptosporidium parvum.* The protozoan is a significant intestinal pathogen in dairy cattle, a likely source of this outbreak.

Cryptosporidiosis is characterized by severe diarrhea, but in normal individuals this is self-limiting and most people recover without incident within 2 weeks.

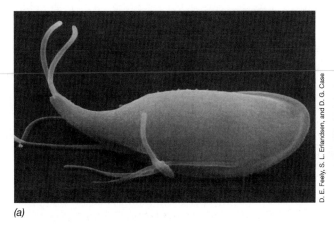
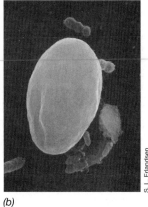

(a)   *(b)*

FIGURE 24.23 Scanning electron micrographs of the parasite *Giardia.* (a) Motile trophozoite. (b) Cyst.

However, individuals with impaired immunity such as acquired immunodeficiency syndrome (AIDS) patients or the very young or old can develop serious complications. In this case, about 4400 people required hospital care, and several died from complications of the disease, including severe dehydration.

Cryptosporidium is highly resistant to chlorine (up to 14 times as resistant as the chlorine-resistant *Giardia*), and therefore methods for removing it rely on sedimentation and filtration. The Milwaukee outbreak points out the fragility of water purification systems, the need for constant water monitoring and surveillance, and the consequences of failure in a large supply system.

As for *Giardia*, laboratory diagnostic methods for cryptosporidiosis include the demonstration of *Cryptosporidium* cysts in the stool, or the demonstration of *Cryptosporidium* antigens in the stool using a direct ELISA (enzyme-linked immunosorbent assay) test (∞ Section 21.8).

Amebiasis

A number of different amebas inhabit the tissues of humans and other vertebrates, usually in the oral cavity or intestinal tract. Some of these are pathogenic. We discussed the general properties of ameboid protozoa in Section 17.3. *Entamoeba histolytica*, the causative agent of amebiasis, is a common pathogenic protozoan transmitted to humans primarily by contaminated water and occasionally by the foodborne route (Figure 24.24). *E. histolytica* is an *anaerobic* ameba, the trophozoites lacking mitochondria. Like *Giardia*, the trophozoites of *Entamoeba* produce cysts. The cysts cause infestation, and cyst germination occurs in the intestine, where amebic cells grow both on and in intestinal mucosal cells. Many infections are asymptomatic, but continued growth may lead to ulceration of intestinal mucosa, causing diarrhea and severe intestinal cramps. Diarrhea is replaced by a condition referred to as *dysentery*, characterized by fever, the passage of intestinal exudates, blood, and mucus. If not treated, trophozoites of *E. histolytica* can migrate to the liver, lung, and brain. Growth in these tissues can cause abscesses and other tissue damage.

Amebiasis can be treated with the drugs dehydroemetine for invasive disease and diloxanide furoate for certain asymptomatic cases, such as in immune compromised individuals, but amebicidal drugs are not universally effective. Spontaneous cures do occur, implying that the host immune system plays some role in ending the infection. However, protective immunity is not guaranteed by primary infection because reinfection is not uncommon. The disease occurs at very low incidence in regions that practice adequate sewage treatment. Ineffective sewage treatment and use of untreated surface waters for drinking purposes are the usual scenarios for cases of amebiasis.

Laboratory diagnosis involves the demonstration of *Entamoeba* cysts in the stool, trophozoites in tissue, or the demonstration of *Entamoeba* antibodies in the blood using an ELISA (enzyme-linked immunosorbent assay) test (∞ Section 21.8).

✓ 24.9 Concept Check

Common-source waterborne diseases are a very significant source of morbidity, especially in developing countries. A variety of bacteria, viruses, and protozoa are involved. Recent massive outbreaks in developed countries and the newest cholera pandemic emphasize the need for maintaining high standards for water quality.

✓ Outline at least two different sources of waterborne infection other than drinking water. What organisms are generally associated with each source?
✓ Why are protozoans often associated with waterborne diseases, even in developed countries? Outline steps to reduce their impact.

24.10

Microbial Growth in Food

Microbial growth destroys vast quantities of food, causing economic problems and loss of significant nutrient sources. Even more importantly, as we will discuss in Section 24.11, consumption of microbially contaminated food can cause serious infections or poisoning (intoxication). In this section, we will look at food spoilage and growth of microorganisms, with the goal of defining how certain microorganisms colonize food. We will then examine a number of processes that inhibit or stop microbial growth in food, allowing for food storage and preservation. Here, the goal is to define methods for stopping the growth of microorganisms, including

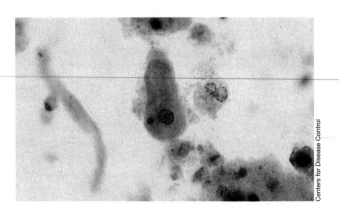

FIGURE 24.24 Trophozoites of *Entamoeba histolytica*, the causative agent of amebiasis. The small red structures are red blood cells.

pathogens, by intervening at the earliest possible time. As we shall see in Section 24.11, failure to adequately preserve food, both for the long term and the short term, may allow the growth of pathogens in the food, resulting in significant morbidity and occasional mortality, even in developed countries.

Food Spoilage

Food spoilage is defined as any change in the visual appearance, smell, or taste of a food product that makes it unacceptable to the consumer. From a health standpoint, spoiled food is not necessarily food that is unsafe to eat. However, unpalatable food will not be purchased by the average consumer. Foods are attacked by microorganisms in a variety of ways that are harmful to the quality of the food. Foods are organic and hence provide adequate nutrients for the growth of a wide variety of chemoorganotrophic bacteria. The physical and chemical characteristics of the food and how it is stored determine its degree of susceptibility to microbial attack. Foods can be classified into three major categories: (1) *highly perishable foods* such as meats, fish, poultry, eggs, milk, and most fruits and vegetables; (2) *semiperishable foods* such as potatoes, some apples, and nuts; and (3) *stable* or *nonperishable foods* such as sugar, flour, rice, and dry beans. These categories differ with regard to *moisture content*, which is related, as we saw in Section 5.11, to water activity, a_w. Stable foods have *low* water activity and can generally be stored for considerable lengths of time without deterioration. Perishable and semiperishable foods are those with *high* water activity. These foods must be stored under conditions that slow or stop microbial growth.

Fresh foods are spoiled by a variety of both bacteria and fungi, but each type of fresh food is typically attacked only by particular microorganisms (Table 24.4). This is because the chemical properties of foods vary widely, and different foods are colonized by the indigenous spoilage organisms that are best able to use the nutrients available.

For example, enteric bacteria such as *Salmonella, Shigella Escherichia,* and *Campylobacter,* all potential pathogens, are rarely implicated in fruit or vegetable *spoilage* but can spoil meat products because their habitat is the gut of warm-blooded animals: They can easily contaminate the meat when the animal is slaughtered. Likewise, lactic acid bacteria, the most common microorganisms in dairy products, are the major spoilage organisms of milk and milk products. *Pseudomonas* species inhabit both soil and the animal body and are thus widely involved in the spoilage of various fresh foods.

Microbial growth in foods follows the standard pattern for a bacterial growth curve (◠◠ Figure 5.4). The lag phase may be of variable duration in a food, depending on the contaminating organism and its previous growth history. The rate of growth during the exponential phase depends on the temperature, the available nutrients, and other conditions of growth. The time required for the population density to reach a significant level in a given food product depends on both the initial inoculum and the rate of growth during the exponential phase. However, it is only when the microbial population density reaches a substantial level that harmful spoilage effects are usually observed. Indeed, throughout much of the exponential phase of growth, population densities may be low enough that no perceptible effect can be observed, but because of the nature of exponential growth (◠◠ Figure 5.2), it is only the *last* doubling or two that leads to observable spoilage. Thus, for much of the period of microbial growth in a food, the observer is unaware of impending problems.

Food Preservation

Besides moisture, one of the most crucial factors affecting microbial growth in food is **temperature** (◠◠ Section 5.7). In general, the *lower* the storage temperature, the *less* rapid the spoilage rate, although, as we have seen, psychrotolerant microorganisms can grow well at refrigerator temperatures. Therefore, storage of perish-

TABLE 24.4	Microbial spoilage of fresh food[a]	
Food product	**Type of microorganism**	**Common spoilage organisms**
Fruits and vegetables	Bacteria	*Erwinia, Pseudomonas, Corynebacterium* (mainly vegetable pathogens; rarely spoil fruit)
	Fungi	***Aspergillus,*** *Botrytis, Geotrichium, Rhizopus, Penicillium, Cladosporium Alternaria, Phytophora,* various yeasts
Fresh meat, poultry, and seafood	Bacteria	*Acinetobacter, Aeromonas,* ***Pseudomonas,*** *Micrococcus, Achromobacter, Flavobacterium, Proteus,* ***Salmonella, Escherichia, Campylobacter,*** *Listeria*
	Fungi	*Cladosporium, Mucor, Rhizopus, Penicillium, Geotrichium,* ***Sporotrichium, Candida,*** *Torula, Rhodotorula*
Milk	Bacteria	***Streptococcus,*** *Leuconostoc, Lactococcus, Lactobacillus, Pseudomonas, Proteus*
High sugar foods	Bacteria	***Clostridium, Bacillus,*** *Flavobacterium*
	Fungi	*Saccharomyces, Torula, Penicillium*

a Several other organisms not listed here have also been isolated, but the organisms listed are the most commonly observed spoilage agents of the fresh foods indicated. Genera in bold face are also possible human pathogens.

able food products for long periods of time (greater than several days) is possible only at temperatures below freezing. Freezing alters the physical structure of many food products and therefore cannot be universally used, but it is widely and successfully used for the preservation of meats and many vegetables and fruits. Freezers providing a temperature of −20°C are most commonly used. At −20°C, storage for weeks or months is possible, but some microbial growth may occur in pockets of liquid water trapped within the frozen mass. For very long-term storage, temperatures lower than −20°C are necessary, such as −80°C (dry ice temperature), but maintenance at such low temperatures is expensive and consequently is not used for routine food storage.

Another major factor affecting microbial growth in food is **pH** or **acidity.** Foods vary widely in pH, although most are neutral or acidic. As we have seen (∞ Section 5.10), microorganisms differ in their ability to grow under acidic conditions: Most food spoilage bacteria do not grow at pH values below 5. Therefore, acid is often used in food preservation, in the process called *pickling.* Foods commonly pickled include cucumbers (sweet, sour, and dill pickles), cabbage (sauerkraut), and some meats and fruits. The food can be made acid either by addition of vinegar or by allowing acidity to develop directly in the food through microbial action, in which case the product is called a *fermented food.* The microorganisms involved in food fermentations are the lactic acid bacteria, the acetic acid bacteria, and the propionic acid bacteria. But even these bacteria cannot grow below about pH 4, so the food fermentation is self-limiting. Vinegar, frequently added to food to lower the pH, is essentially dilute acetic acid. Vinegar itself is a product of the action of acetic acid bacteria; its industrial production was discussed in Section 11.10.

Because microorganisms do not grow at low water activities, microbial growth can be controlled by lowering the water activity of the product by drying or by adding salt. Natural or artificial heat is often used, but the least damaging way of drying foods is freeze-drying (lyophilization). Milk, meats, fish, vegetables, fruits, eggs, and other economically important foods are all commonly preserved by some form of drying.

A number of foods are preserved by addition of salt or sugar to lower water activity. Foods preserved by addition of sugar are mainly fruits (jams, jellies, and preserves). Salted products are primarily meats and fish. Sausage and ham are preserved by salt, although these products vary in water activity depending on how much salt is added and how much the meat has been dried.

Canning

Canning is a process in which a food is sealed and heated so as to kill all living organisms, or at least to ensure that there will be no growth of residual organisms in the can. Canning is hence a type of heat sterilization (∞ Section 18.1). When the can is properly sealed and heated, the food should remain stable and unspoiled indefinitely, even when stored in the absence of refrigeration.

The temperature–time relationships for canning depend on the type of food, its pH, the size of the container, and the consistency or bulkiness of the food. Because heat must penetrate completely to the center of the food within the can, heating times must be longer

(a) (b) (c) (d)

FIGURE 24.25 Changes in cans as a result of microbial spoilage. (a) Normal can; note that the top of the can is indented due to negative pressure (vacuum) inside. (b) Slight swell resulting from minimal gas production. Note that the lid is slightly raised. (c) Severe swell due to extensive gas production. Note the great deformation of the can. (d) The can shown in (c) was dropped and the gas pressure resulted in a violent explosion. Note that the lid has been torn apart.

for large cans or very viscous foods. Acid foods can often be canned effectively by heating just to boiling, 100°C, whereas nonacid foods must be heated to autoclave temperatures. However, heating times long enough to guarantee absolute sterility of every can would change the food so greatly that it would likely be unpalatable and nutritionally altered. Therefore, even properly canned foods may not be sterile.

The environment inside a can is anoxic, and microbial growth in a canned food frequently is the result of fermentative organisms that produce extensive amounts of gas. This can result in pressure buildup inside the can, resulting in bulges or, in severe cases, even an explosion of the can (Figure 24.25). Because many of the anoxic bacteria that grow in canned foods are powerful toxin producers of the genus *Clostidium* (see Table 24.4 and Section 24.11), food from a visibly altered can should never be eaten.

Chemical Food Preservation

A number of chemicals are used commercially to control microbial growth in food. These are classified by the U.S. Food and Drug Administration as "generally recognized as safe" and find wide application in the food industry (Table 24.5). Many of these chemicals, like sodium propionate, have been used for many years with no evidence of human toxicity. Others, like nitrites (carcinogen precursor), ethylene or propylene oxides (mutagens; ∞ Section 9.4), or antibiotics (development of antibiotic-resistant pathogens; ∞ Section 18.12), are more controversial food supplements because of evidence that these compounds are detrimental to human health.

Because of lengthy and costly testing programs for any new chemical proposed as a food preservative today, it is unlikely that new compounds will be added to the list in Table 24.5 in the near future. An alternative to chemical preservatives, preservation by ionizing radiation, was discussed in Section 18.2 and is now approved for use in some fresh foods in the United States.

✓ 24.10 Concept Check

Food microbiology deals with methods for keeping microorganisms, including pathogens, from growing in food during processing and storage. Foods vary considerably in their sensitivity to microbial growth, depending on their nutrient content, water content, and pH. Microbial growth in foods can be controlled by heat, refrigeration, chemical agents, or irradiation.

- ✓ Are food spoilage organisms also pathogens? Give examples to support your answer.
- ✓ Outline at least four methods of food preservation. How does each method limit growth of microorganisms?

24.11

Foodborne Diseases: Food Poisoning

We now begin a discussion of foodborne illnesses. A summary of the most prevalent foodborne diseases and the microorganisms that cause them is shown in Table 24.6. These common diseases can be separated into two categories. The first category is *food poisoning*, or *food intoxication*. Food poisoning results from ingestion of foods containing preformed microbial toxins. The microorganisms that produced the toxins do not have to grow in the host and are often not alive at the time the contaminated food is consumed. The illness is due to ingestion and subsequent action of preformed bioactive toxin. We previously discussed some of these toxins, notably the exotoxins of *Clostridium* spp. (∞ Section 19.8) and the superantigen toxins of *Staphylococcus aureus* (∞ Section 20.15). Here we will discuss the basic biology, pathology, prevention, and treatment of these diseases. In the next section, we will discuss the second foodborne disease category, that of *foodborne infections*. Foodborne infections are host infections resulting from ingestion of pathogen-contaminated food.

We begin our discussion with the most common foodborne illness, staphylococcal food poisoning.

Staphylococcal Food Poisoning

The most common food poisoning is caused by the gram-positive coccus *Staphylococcus aureus*. This organism produces several *enterotoxins* (∞ Section 19.9) that are released into the surrounding medium or food; if food containing the toxin is ingested, severe reactions are observed within 1–6 hr, including nausea with vomiting and diarrhea. Seven types of *S. aureus* enterotoxin have been identified, A, B, C_1, C_2, C_3, D, and E. Entero-

TABLE 24.5	Chemical food preservatives
Chemical	**Foods**
Sodium or calcium propionate	Bread
Sodium benzoate	Carbonated beverages, fruit juices, pickles, margarine, preserves
Sorbic acid	Citrus products, cheese, pickles, salads
Sulfur dioxide, sulfites, bisulfites	Dried fruits and vegetables; wine
Formaldehyde (from food-smoking process)	Meat, fish
Ethylene and propylene oxides	Spices, dried fruits, nuts
Sodium nitrite	Smoked ham, bacon

TABLE 24.6	Major food-poisoning and food infection bacteria	
Organism	**Disease**[a]	**Foods usually involved**
Bacillus cereus	FP	Rice and other starchy foods
Campylobacter jejuni	FI	Poultry, dairy products
Clostridium botulinum	FP	Home-canned vegetables (especially beans and corn), smoked fish
Clostridium perfringens	FP	Cooked and reheated meats and meat products
Escherichia coli O157:H7	FI	Meat, especially ground meat
Listeria monocytogenes	FI	Ready-to-eat meat and dairy products
Salmonella spp.	FI	Poultry, other meats, dairy products, eggs
Staphylococcus aureus	FP	Meat dishes, desserts
Vibrio parahaemolyticus	FI	Seafood
Yersinia enterocolitica	FI	Pork, milk

a FP, food poisoning; FI, food infection

toxin A is most frequently associated with outbreaks of staphylococcal food poisoning. Enterotoxin A is a superantigen (∞ Section 20.15). Superantigens work by stimulating large numbers of T cells, which in turn release chemicals called *cytokines*, activating physiological responses such as inflammation, vomiting, and diarrhea. *S. aureus* enterotoxin A is a small single peptide of 30,000 molecular weight that is encoded by a chromosomal gene. Cloning and sequencing of this gene, the *entA* gene, and of several other *S. aureus* enterotoxin genes, show that this family of toxins is genetically related. Although the *entA* gene is chromosomally located, the B- and C-type *S. aureus* enterotoxins may be plasmid- or transposon-encoded, or alternatively encoded by a lysogenic bacteriophage. We discussed the importance of accessory genetic elements such as plasmids and bacteriophages as vectors for toxin production in Sections 19.8 and 19.9.

The kinds of foods most commonly involved in *Staphylococcus aureus* food poisoning are custard- and cream-filled baked goods, poultry, meat and meat products, gravies, egg and meat salads, puddings, and creamy salad dressings. If such foods are kept refrigerated after preparation, they remain relatively safe, as *Staphylococcus* growth is markedly reduced at low temperatures. However, foods of this type are often not refrigerated, and are frequently kept in warm kitchens or outdoors at summer picnics. Under these conditions, *Staphylococcus*, which might have entered the food from a food handler during preparation, grows and produces enterotoxin. Many of the foods involved in staphylococcal food poisoning are not cooked again before eating, but even if they are, this toxin is relatively heat-stable and may remain active. Staphylococcal food poisoning can be prevented by careful sanitation methods or by storage of the food at low temperatures to prevent bacterial growth, and by the discarding of foods stored for any period of hours above 4°C (refrigerator temperature).

Because *S. aureus* food poisoning is very common, several assays, based on the detection of either enterotoxin (ELISA detection of antigens on enterotoxin; ∞ Section 21.8) or *S. aureus* exonuclease (an enzyme that degrades DNA) are available to detect *S. aureus* in food at levels known to be dangerous for human consumption.

Clostridium perfringens Food Poisoning

Clostridium perfringens is a major cause of food poisoning in the United States. *Clostridium perfringens*, an obligate anaerobe, produces an enterotoxin (∞ Table 19.4) that elicits diarrhea and intestinal cramps, but nausea and vomiting in only one-third of those infected. The symptoms last for about 24 hr, and fatalities are rare. The disease results from the ingestion of a large dose ($>10^8$ cells) of *C. perfringens* and is most frequently associated with the consumption of tainted meat or meat products.

Clostridium perfringens is quite common in a variety of cooked and uncooked foods, especially meat, poultry, and fish, and in soil and sewage. The organism is present naturally in low numbers in the human gut but apparently does not reach large numbers in the presence of the other competing intestinal microorganisms. Large doses of *C. perfringens* are most likely to be obtained from meat dishes cooked in bulk lots (heat penetration in these situations is often slow and insufficient) and then left at 20–40°C for short periods. Spores of *C. perfringens* germinate under *anoxic* conditions and grow quickly in the meat. Upon consumption of the contaminated meat, sporulation begins in the intestine and toxin is produced, altering the permeability of the intestinal epithelium and leading to gastrointestinal symptoms. The onset of perfringens food poisoning begins about 7–15 hr after consumption of the contaminated food. Diagnosis of perfringens food poisoning is made by isolation of *C. perfringens* from the gut or, more reliably, by a direct enzyme-linked immunosorbent assay (ELISA) to detect *C. perfringens* enterotoxin in feces (◠◠ Section 21.8).

Botulism

Botulism is the most severe type of food poisoning; it is often fatal and occurs following the consumption of food containing the exotoxin produced by the anaerobic bacterium *Clostridium botulinum*. This bacterium normally lives in soil or water, but its spores may contaminate raw foods before harvest or slaughter. If the foods are properly processed so that the *C. botulinum* spores are killed, no problem arises; but if viable spores are present, they may initiate growth and even a small amount of the resultant neurotoxin can be extremely poisonous.

We discussed the nature and action of botulinum toxin in Section 19.8 (◠◠ Figure 19.17). At least seven distinct types of botulinum toxin are known, most of which are toxic to humans. The toxins are destroyed by heat (80°C for 10 min), and so properly cooked food should be harmless, even if it originally contained toxin. Most cases of botulism occur as a result of eating foods that are not cooked after processing (Figure 24.26a). For example, canned vegetables and beans are often used without cooking in making cold salads. Similarly, smoked fish and meat and most vacuum-packed sliced meats are often eaten directly, without heating. If these products contain the botulinum toxin, then ingestion of even a small amount will result in this severe and highly dangerous type of food poisoning. In the United States, the disease is fortunately quite rare and is almost always associated with improperly preserved or stored food. In Japan cases of botulism are often linked to the consumption of *sushi*, a raw fish preparation.

Infant botulism occurs when spores of *Clostridium botulinum* are ingested, often from raw honey (Figure 24.26b). If the infant's normal flora is not well developed or if the infant is undergoing antibiotic therapy, the spores may germinate and *C. botulinum* cells may grow and release toxin. Most cases of infant botulism occur between the first week of life and 2 months of age; infant botulism is rare in children older than 6 months when the intestinal flora is more developed. It has been suggested that infant botulism is one possible cause of *sudden infant death symdrome* (SIDS).

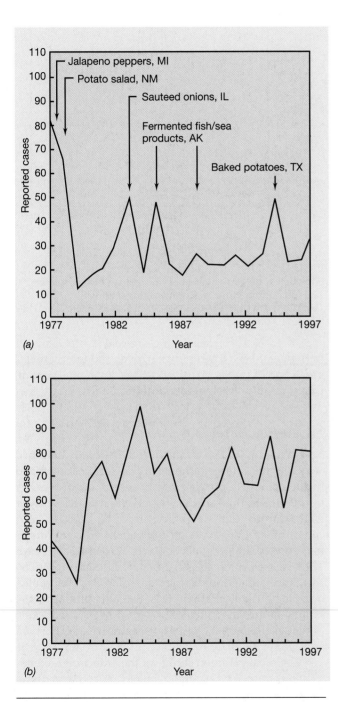

(a)

(b)

FIGURE 24.26   The incidence of botulism in the United States. (a) Foodborne botulism. In years with high numbers of cases, major outbreaks that account for the increase are indicated. (b) Infant botulism. More than half of the cases of infant botulism in the United States occur in California. Data are from the Centers for Disease Control and Prevention, Atlanta, GA, USA.

✓ 24.11 Concept Check

Many microbial diseases are foodborne. Some foodborne diseases are food poisonings in which the causal agent produces an exotoxin or enterotoxin. Even in the absence of infection, the toxin may be ingested with the food and cause the food-poisoning symptoms.

✓ What are some agents that cause food poisoning?
✓ How can you treat a food poisoning?

24.12

Foodborne Diseases: Food Infection

In addition to the passive transfer of microbial toxins, food may contain sufficient numbers of living pathogenic microorganisms to cause infection and disease in the host. *Food infection*, the growth of microorganisms in the body after ingestion of contaminated food, is a very common type of foodborne illness. In this section we present several examples of microorganisms and viruses that cause food infections (Table 24.6). Many of these agents also cause waterborne diseases (see Section 24.9).

Salmonellosis

Although sometimes called food poisoning, gastrointestinal disease due to foodborne *Salmonella* is more aptly called *Salmonella* **food infection** because symptoms arise only after the pathogen grows in the intestine (hence symptoms can begin several *days* after eating a contaminated food). The symptoms of salmonellosis include the sudden onset of headache, chills, vomiting, and diarrhea, followed by a fever that lasts a few days. Diagnosis is made by observation of clinical symptoms, history of recent food consumption, and by culture of the organism from feces. Virtually all species of *Salmonella* are pathogenic for humans: One, *S. typhi*, causes the serious human disease typhoid fever, and a small number of other species cause foodborne gastroenteritis. *Salmonella typhimurium* is the most common cause of salmonellosis in humans.

The ultimate sources of the foodborne salmonellas are humans and warm-blooded animals. The organism reaches most food by contamination from food handlers (⊂⊃ the box, the Tragic Case of Typhoid Mary, Chapter 22). For foods such as eggs or meat, the animal that produced the food may be the source of contamination. The foods most commonly implicated are meats and meat products (such as meat pies, sausage, and cured meats), poultry, eggs, and milk and milk products. *Salmonella* food infections are often traced to products made with *uncooked* eggs such as custards, cream cakes, meringues, pies, and eggnog. Properly cooked foods are safe if consumed immediately or if stored immediately at 4°C or less. However, cooked or canned foods that become contaminated by an infected food handler can support the growth of *Salmonella*, especially if the foods are held for long periods of time without refrigeration. *Salmonella* infection is more common in summer than in winter, probably because warm environmental conditions are more favorable for growth of microorganisms in foods. The reported incidence and prevalence of salmonellosis has been very steady over the last decade (Figure 24.27),

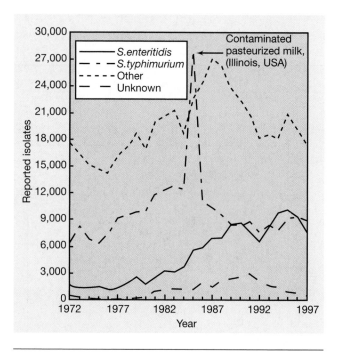

FIGURE 24.27   The incidence of salmonellosis in the United States. Cases are divided by the species of the isolate. The total number of cases is averaging about 40,000 per year. Data are from the Centers for Disease Control and Prevention, Atlanta, GA, USA.

with over 40,000 cases per year, even though there is a great deal of public health effort to educate the general public about microbial food infections.

Escherichia coli

Several strains of *Escherichia coli* have recently emerged as potent foodborne pathogens. They are characterized by their ability to produce potent enterotoxins (⊂⊃ Section 19.9) and thus are designated *enterotoxic E. coli*, or ETEC strains. One particular strain, *E. coli* O157:H7, causes at least 20,000 cases (only about 10% are probably reported) of food infection and 250 deaths each year. The disease produced is characterized by bloody diarrhea and is therefore further classified as an *enterohemorrhagic E. coli* (EHEC). It is also a leading cause of kidney failure in children. *E. coli* O157:H7 produces a potent exotoxin that causes both hemorrhagic diarrhea and kidney failure.

The most common source of this infection is contaminated uncooked or undercooked meat, particularly mass-process ground meat. For example, several major outbreaks have involved infected ground beef. In several outbreaks, regional distribution centers were involved and the infected product caused disease in several states. Another outbreak involved processed, cured, but uncooked, beef. The major source of contamination seems to be the source of the beef, and contamination probably originated from *E. coli* O157:H7 strains originating from the slaughtered beef carcasses. Recently, the United States has approved the irradiation of ground meat as an acceptable means of dealing with food infection bacteria, largely because *E. coli* O157:H7 has been implicated in these common-source foodborne epidemics. Penetrating irradiation is considered the only effective means to ensure decontamination after the grinding process because grinding may distribute the pathogens throughout the meat, not simply on the surface (∞ Section 18.2).

Diagnosis of infection by *Escherichia coli* O157:H7 involves culture from the feces and identification of the O and H antigens and toxins by serology. Subtyping of strains is also done using molecular methods such as restriction fragment length polymorphism (RFLP) and pulse field gel electrophoresis (∞ the box, DNA Fingerprinting, Chapter 10, and the box, Working with Nucleic Acids: The Tools, Chapter 6). This is now a nationally reportable infectious disease (∞ Table 22.4). The most effective way to prevent infection with *E. coli* O157:H7 is to make sure that meat is cooked thoroughly, which means that it should appear gray or brown and juices should be clear.

Since *E. coli* O157:H7 grows in the intestines, it is also a potential source of waterborne disease. There have been several cases of serious *E. coli* O157:H7 infection occurring because of fecally contaminated public swimming areas.

"Traveler's diarrhea" is another extremely common enteric infection in North Americans and Europeans traveling to developing countries. The primary causal agents are the enteropathogenic *Escherichia coli*, although *Salmonella* and *Shigella* species are sometimes implicated. Several studies have been done on groups of U.S. citizens traveling in Mexico. Such studies have shown that the infection rate in travelers is often greater than 50% and that the prime vehicles are foods, such as uncooked vegetables (for example, lettuce in salads) and water. The very high infection rate in travelers is due to contamination of local public water supplies. The local population is usually resistant to the infecting strains, undoubtedly because they have lived with the agent for a long period of time. Secretory antibodies present in the bowel may prevent successful colonization of the pathogen in local residents, but when the organism colonizes the intestine of a nonimmune person, it finds a hospitable environment. Also, stomach acidity, so often a barrier to intestinal infection, may not be able to act if the organism is consumed in small amounts of liquid (as, for instance, the melting ice of a cocktail) because small amounts of liquid induce rapid emptying of the stomach and hence pass through so quickly that stomach acidity may have no effect on an enteric pathogen present in the liquid.

Campylobacter

Campylobacter species is a gram-negative, curved rod that grows at reduced oxygen tensions, that is, as a microaerophile (∞ Section 13.13). Two major species are recognized, *Campylobacter jejuni* and *C. fetus,* and these species probably account for the majority of cases of bacterial diarrhea in children. *Campylobacter fetus* is also of economic importance because it is a major cause of sterility and spontaneous abortion in cattle and sheep. The symptoms of *Campylobacter* infection include a high fever (usually greater than 104°F or 40°C), nausea, abdominal cramps, and a watery, frequently bloody, stool. Diagnosis requires isolation of the organism from stool samples and identification by growth-dependent tests or immunological assays. Because of the frequency with which *C. jejuni* infections are observed in infants, a variety of selective media and highly specific immunological methods have been developed for positive identification of this organism.

Campylobacter is transmitted to humans via contaminated food, most frequently in poultry, pork, raw clams, and other shellfish, or by a water route in surface waters not subjected to chlorination. Poultry is a major reservoir of *C. jejuni,* and virtually all chicken and turkey carcasses contain this organism (Table 24.6). Beef, on the other hand, is rarely a vehicle for *Campylobacter.* Proper washing of uncooked poultry (and any utensils coming in contact with uncooked poultry) and thorough cooking of the meat eliminate the possiblity of *Campylobacter* infection. *Campylobacter* species also infect domestic animals such as dogs, causing a milder form of diarrhea than that observed in humans. Infant cases of *Campylobacter* infection are frequently traced to infected domestic animals, especially dogs.

Listeriosis

Listeria monocytogenes, the cause of **listeriosis,** is emerging as an important foodborne pathogen. *L. monocytogenes* is an acid-tolerant, psychrotolerant (cold-tolerant), and salt-tolerant bacterium (∞ Chapter 5). Because of its widespread distribution in soil and water, virtually no fresh food source is immune to possible *L. monocytogenes* contamination. As a result, food can become contaminated at any stage during food growth or processing and methods such as refrigeration, which ordinarily slow microbial growth, are ineffective in limiting growth of the organism. Thus, meat, dairy products, and fresh produce are often contaminated with this pathogen.

L. monocytogenes enters the body through the gastrointestinal tract after ingestion of contaminated food. Uptake of the pathogen by phagocytes results in growth and proliferation, followed by lysis of the phagocyte and spread to surrounding cells. *L. monocy-*

togenes is thus an intracellular pathogen (⟨∞ Section 19.7); immunity to *L. monocytogenes* is mainly cell-mediated via T$_H$1 cells (⟨∞ Section 20.3 and Section 20.9). Individuals having weakened cellular immunity, including the elderly, neonates, patients undergoing immunosuppressive drug treatment (e.g., steroid treatment), or who have immunosuppressive diseases such as AIDS, have increased susceptibility to listeriosis (⟨∞ Section 19.12).

Although exposure to *L. monocytogenes* is undoubtedly very common, listeriosis is actually quite rare. However, listeriosis is a serious disease because it has a very high mortality rate of about 20–30%. The disease is usually characterized by bacteremia and meningitis, but can be treated with ampicillin, erythromycin, or trimethoprim-sulfamethoxazole. *L. monocytogenes* can be identified in food by direct culture, or by a variety of molecular methods such as ribotyping (⟨∞ Section 12.9) and the polymerase chain reaction (PCR) (⟨∞ Sections 10.9 and 21.10, and Table 21.7).

Potential sources of listeriosis are ready-to-eat processed foods such as meat products and unpasteurized dairy products that are stored for long periods at refrigerator temperature (4°C). Prevention measures include recalling contaminated food and taking steps to limit *L. monocytogenes* contamination at the food processing site. Since *L. monocytogenes* is susceptible to heat and radiation, raw food and food handling equipment can be readily decontaminated. However, without sterilizing the finished food product, the risk of food contamination cannot be totally eliminated because of the widespread distribution of the pathogen.

Helicobacter pylori **and Peptic Ulcers**

Helicobacter pylori is a gram-negative spiral-shaped bacterium that is related to *Campylobacter* (⟨∞ Section 13.13). *H. pylori*, first identified in human intestinal biopsies in 1983, is associated with gastritis and peptic ulcers. Up to 80% of ulcer patients have concomitant *H. pylori* infection, and about 50% of all adults, especially in developing countries, are chronically infected. Although the mode of transmission has not been established, there is no known nonhuman reservoir of *H. pylori*. In addition, infection occurs in high incidence in certain families, and the overall incidence in the population increases with age. These factors suggest a host-to-host type of transmission (⟨∞ Sections 22.3 and 22.5). However, infections sometimes occur in epidemic clusters, suggesting that a common source such as food or water may also be involved (⟨∞ Section 22.5).

Helicobacter pylori is not invasive, but colonizes the gastric mucosal surfaces, where a host response causes inflammation. Researchers speculate that severe and untreated inflammation leads to tissue destruction and ulceration. Antibodies to *H. pylori* are usually present in infected individuals, but do not prevent colonization. Therefore, individuals who acquire *H. pylori* tend to have chronic, long-term infections.

More evidence for a causal association between *Helicobacter pylori* and ulcers comes from treatments for the disease. Long-term treatment of most ulcers with antacid preparations has never been uniformly successful, and most patients relapse within one year. However, by treating ulcers as an infectious disease, permanent cures are often possible. As a consequence, physicians now routinely use antimicrobial drugs to treat ulcers. Treatment usually consists of a combination of drugs including metranidazole, a second antibiotic such as tetracycline, and a bismuth-containing antacid preparation. The combination treatment abolishes the *H. pylori* infection and seems to cure the ulcers on a long-term basis. However, even though the current battery of circumstantial evidence points more and more to *H. pylori* as a major cause of ulcers, the causal relationship between *H. pylori* infection and ulcers, though widely accepted, has not been unequivocally established.

Hepatitis

Infectious hepatitis (hepatitis A) is a virus-mediated inflammation of the liver caused by a picornavirus (positive, single-strand RNA virus) (⟨∞ Section 8.15) (Figure 24.28). The virus is transmitted primarily through fecal

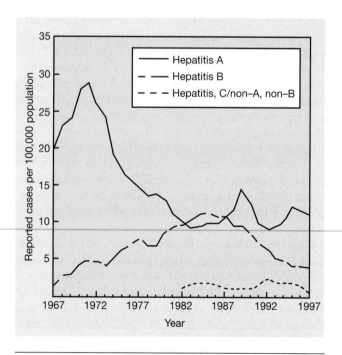

FIGURE 24.28   The prevalence of hepatitis in the United States. Cases are divided based on the strain of hepatitis virus. Data are from the Centers for Disease Control and Prevention, Atlanta, GA, USA.

contamination of water, food, or milk. **Hepatitis A** infection can be subclinical in mild cases or can lead to severe liver damage in chronic infections. The type A virus spreads from the intestine via the bloodstream to the liver and usually results in jaundice, a yellowing of the skin and eyes, and a browning of the urine as a result of stimulation of bile pigment production by infected liver cells. An immune response is initiated against the hepatitis A virus, and this eventually brings the condition under control. However, in severe cases, permanent loss of a portion of liver function can occur.

The most significant food vehicles for type A hepatitis are shellfish (oysters and clams) harvested from waters polluted with human feces. As filter feeders, shellfish living in such environments tend to concentrate the hepatitis virus. Only *raw* or undercooked shellfish are a problem because hepatitis A virus is destroyed by sufficient heating. Therefore, the best means of controlling hepatitis transmission include sound sanitary practices, especially sewage treatment, the prevention of fecal contamination of food products by infected food handlers, and avoiding the consumption of raw or undercooked shellfish. There is also an effective subunit vaccine available (Table 20.11).

Serum hepatitis (hepatitis B) is caused by a *DNA-containing* hepatitis virus transmitted primarily by infected blood or blood products (Section 21.12 and see Figure 24.28). The type B virus can also be spread by maternal transmission *in utero* and from one infected individual to another by sexual intercourse. Serum hepatitis frequently results in more severe liver damage than infectious hepatitis, leading to death in up to 10% of those infected; hepatitis A rarely causes death. Long-term infection with type B virus is also associated with increased risk of liver cancer. Those at high risk for serum hepatitis include individuals who require frequent blood transfusions, dialysis patients, health care workers, and intravenous drug abusers. As in the case of infectious hepatitis, no specific therapy for serum hepatitis is currently available. However, because of the seriousness of the infection, a vaccine against serum hepatitis is now available and is recommended or required in most immunization programs (Figure 20.27). Use of the serum hepatitis vaccine is recommended for those in the high risk groups already mentioned, as well as for health personnel who come in frequent contact with blood or blood products. A variety of immunological tests, especially ELISAs (Section 21.8), have been devised for the diagnosis of infectious and serum hepatitis by either antibody detection or direct viral detection methods.

A third type of hepatitis virus, unrelated to type A or type B and referred to as type C hepatitis virus, is also known. Although not yet reliably grown in tissue culture, the type C virus is widespread in the human population and is now the most common hepatitis virus encountered in cases of hepatitis mediated by blood transfusion (Figure 24.28). Like type B, type C hepatitis virus is spread primarily by contaminated blood and possibly also by sexual transmission. Although less likely to cause serious liver damage than type A or type B hepatitis viruses, the type C virus can cause chronic ailments such as cirrhosis. In addition, *at least* two other strains, hepatitis D and hepatitis E, are known (Table 22.6).

Altogether, the various forms of hepatitis account for over 40,000 illnesses per year (Figure 24.28). Control of these infections is based primarily on hygiene and cleanliness measures, but vaccines are also important. As mentioned above, there is now an effective vaccine in widespread use for hepatitis B and a vaccine is also available for hepatitis A (Section 20.16), but none is yet available for hepatitis C, D, or E.

Assessing Microbial Content of Foods

All fresh foods have some variable microorganisms present. The purpose of assay methods is to detect evidence of abnormal microbial growth in foods or to detect the presence of specific organisms of public health concern, such as *E. coli* O157:H7, *Salmonella*, *Staphylococcus*, and *Clostridium botulinum*. We discussed in Section 21.10 the use of nucleic acid probes for the detection of specific foodborne pathogens, and these methods are finding increasing use. For growth studies of nonliquid food products, preliminary treatment is usually required to suspend microorganisms embedded or entrapped within the food in a liquid medium. The most suitable method for treatment is high-speed blending. Examination of the food should be done as soon after sampling as possible, and if examination cannot begin within 1 hr of sampling, the food should be refrigerated. A frozen food should be thawed in its original container in a refrigerator and examined as soon as possible after thawing is complete. For *Salmonella*, several selective media are available (Section 21.2), and tests for its presence are most commonly done on animal food products, such as raw meat, poultry, eggs, and powdered milk, because *Salmonella* from production animals is the usual source of food contamination.

For detection of *Staphylococcus aureus* food poisoning, we already mentioned rapid techniques based on the immunological detection of *S. aureus* toxins or the enzymatic activity of *S. aureus* enzymes. However, these rapid tests are largely qualitative in nature, confirming only the presence or absence of *S. aureus* above the detection limits of the assay. To obtain more quantitative data, bacterial plate counts are usually required. For staphyloccal counts, a high-salt medium (either sodium chloride or lithium chloride at a final concentration of 7.5%) is used. Of the organisms present in foods, staphylococci are the only common ones tolerant of such levels of salt. Because *Staphylococcus aureus* is responsible for one of the most common types of food poisoning, staphylococcal counts are of considerable importance for determining the extent of food contamination.

✓ 24.12 Concept Check

Among the microbial foodborne diseases are a number of food infections in which microbially contaminated food is the agent by which the pathogen is transmitted to the host. Raw and undercooked foods are the most common vehicles for transmission of food infections.

✓ What are some agents that cause food infections?
✓ How can you treat a food infection?

REVIEW QUESTIONS

1. What animals are likely to carry rabies in the United States? What immunization programs are in place for the treatment of rabies? For the prevention of rabies?

2. Why has hantavirus pulmonary syndrome (HPS) emerged as a human pathogen in the United States? How can HPS be prevented?

3. What are the three major categories of organisms that cause rickettsial diseases? For typhus, Rocky Mountain spotted fever, and ehrlichiosis, identify the most common reservoir and vector.

4. Identify the most common reservoir and vector for Lyme disease in the United States. How can the spread of Lyme disease be controlled? How can Lyme disease be treated?

5. Classic malaria includes severe long-term fever followed by chills. These symptoms are related to activities of the pathogen *Plasmodium vivax* or *P. falciparum*. Describe the growth stages of this pathogen in the human host and relate them to the fever-chill pattern. Why might a person of Western European descent be more susceptible to malaria than a person of African or Southern European descent?

6. Distinguish between bubonic, septicemic, and pneumonic plague. Which is most serious? How is each acquired?

7. What is the reservoir for most fungal pathogens? How can fungal exposure be controlled? What particular problems, especially in terms of therapy, do fungi pose for the clinician?

8. Describe the steps used to carry out a coliform test on a sample of drinking water. What are the acceptable limits for coliforms in drinking water?

9. Why are *Cryptosporidium parvum* and *Giardia lamblia* now two of the most common waterborne pathogens? How can they be eliminated as sources of drinking water contamination?

10. Review the methods used for preventing or inhibiting the growth of microorganisms in food. What specific factors influence the rate of growth of food spoilage organisms?

11. Write a scenario for a typical food poisoning outbreak due to Staphylococcus. What conditions might lead to the development of the toxic food, how might the outbreak arise, and how should patients be treated?

12. Compare and contrast *Clostridium perfringens* and *C. botulinum* food poisoning. Discuss similarities and differences between the organisms involved, the role of infection in the disease, the toxins produced, and the symptoms observed.

APPLICATION QUESTIONS

1. Describe the sequence of events you would take if a child received an unprovoked bite from a stray dog. Present one scenario where you were able to catch and detain the dog and another for a dog that escaped. How would these procedures differ from a situation where the child was bitten by a dog with up-to-date rabies vaccinations?

2. Why are diseases like hantavirus pulmonary syndrome (HPS) emerging as important infectious diseases, even in developed countries? Include in your considerations a discussion of social, economic, environmental, and public health issues.

3. Discuss at least three common properties of the disease agents and review the disease process for Rocky Mountain spotted fever, typhus, and ehrlichiosis. Why is ehrlichiosis emerging as an important rickettsial disease? Compare its emergence to that of Lyme disease.

4. Discuss the causative agent, mode of transmission, symptoms, therapy, prevention, and diagnosis of Lyme disease. Predict the future history of this disease in the United States for the next decade, considering that a universally effective vaccine is now available.

5. Malaria eradication has been a goal of public health programs for at least 100 years. What factors preclude our ability to eradicate malaria? If an effective vaccine was developed, could malaria be eradicated? Compare this possibility to the possibility of eradicating plague.

6. Bubonic plague may have killed up to 30% of the entire human population during several pandemics occurring as recently as the 19th century. Why are plague pandemics no longer occurring? Speculate on the possibility that humans have evolved to resist plague. Can you supply any evidence for this hypothesis? Name some potential emerging diseases that could have the same destructive effect on the human population in today's world.

7. With regard to infections, why are fungi often secondary opportunistic pathogens in immunologically compromised individuals? Devise an environment that eliminates all exposure to fungi. Is such an environment a realistic possibility for isolation of patients with immune deficiency? Do you know of examples where this has been accomplished?

8. Why was the development of the coliform test an important step in the development of effective methods for drinking water purification? Describe a molecular biology–based test of your own design that could replace the coliform test. What organisms would you try to identify?

9. Contrast *food spoilage*, *food infection*, and *food poisoning*. Give an example of each and indicate how each could be effectively controlled.

10. Compare and contrast the toxins responsible for cholera, botulism, and staphylococcal food poisoning. What is the mechanism of action of each toxin? In each case, is *infection* with the causative organism necessary for the initiation of disease symptoms?

APPENDIX 1   Energy Calculations in Microbial Bioenergetics

The information in Appendix 1 is intended to help students calculate changes in free energy accompanying chemical reactions carried out by microorganisms. It begins with definitions of the terms required to make such calculations and proceeds to show how knowledge of redox state, atomic and charge balance, and other factors are necessary to calculate free-energy problems successfully.

Definitions

1. ΔG^0 = standard free-energy change of the reaction at 1 atm pressure and 1 M concentrations; ΔG = free-energy change under the conditions specified; $\Delta G^{0\prime}$ = free-energy change under standard conditions at pH 7.

2. Calculation of ΔG^0 for a chemical reaction from the free energy of formation, G_f^0, of products and reactants:

$$\Delta G^0 = \Sigma \, \Delta G_f^0 \, (\text{products}) - \Sigma \, \Delta G_f^0 \, (\text{reactants})$$

That is, sum the ΔG_f^0 of products, sum the ΔG_f^0 of reactants, and subtract the latter from the former.

3. For energy-yielding reactions involving H^+, converting from standard conditions (pH 0) to biochemical conditions (pH 7):

$$\Delta G^{0\prime} = \Delta G^0 + m \, \Delta G'_f(H^+)$$

where m is the net number of protons in the reaction (m is negative when more protons are consumed than formed) and $\Delta G'_f(H^+)$ is the free energy of formation of a proton at pH 7 = -39.83 kJ at 25°C.

4. Effect of concentrations on ΔG: with soluble substrates, the concentration ratios of products formed to exogenous substrates used are generally equal to or greater than 10^{-2} at the beginning of growth and equal to or less than 10^{-2} at the end of growth. From the relation between ΔG and the equilibrium constant (see item 8), it can be calculated that ΔG for the free-energy yield in practical situations differs from the free-energy yield under standard conditions by at most 11.7 kJ, a rather small amount, and so for a first approximation, standard free-energy yields can be used in most situations. However, with H_2 as a product, H_2-consuming bacteria present may keep the concentration of H_2 so low that the free-energy yield is significantly affected. Thus, in the fermentation of ethanol to acetate and H_2 ($C_2H_5OH + H_2O \rightarrow C_2H_3O_2^- + 2\,H_2 + H^+$), the $\Delta G^{0\prime}$ at 1 atm H_2 is +9.68 kJ, but at 10^{-4} atm H_2 it is -36.03 kJ. With H_2-consuming bacteria present, therefore, the ethanol fermentation becomes useful. (See also item 9.)

5. Reduction potentials: by convention, electrode equations are written in the direction, oxidant + $ne^- \rightarrow$ reductant (that is, as reductions), where n is the number of electrons transferred. The standard potential (E_0) of the hydrogen electrode, $2\,H^+ + 2\,e^- \rightarrow H_2$ is set by definition at 0.0 V at 1.0 atm pressure of H_2 gas and 1.0 M H^+, at 25°C. E_0' is the standard reduction potential at pH 7. See also Table A1.2.

6. Relation of free energy to reduction potential:

$$\Delta G^{0\prime} = -nF \, \Delta E_0'$$

where n is the number of electrons transferred, F is the Faraday constant (96.48 kJ/V), and $\Delta E_0'$ is the E_0' of the electron-accepting couple minus the E_0' of the electron-donating couple.

7. Equilibrium constant, K. For the generalized reaction $aA + bB \rightleftharpoons cC + dD$,

$$K = \frac{[C]^c \, [D]^d}{[A]^a [B]^b}$$

where A, B, C, and D represent reactants and products; a, b, c, and d represent number of molecules of each; and brackets indicate concentrations. This is true only when the chemical system is in equilibrium.

8. Relation of equilibrium constant, K, to free-energy change. At constant temperature and pressure,

$$\Delta G = \Delta G^0 + RT \ln K$$

where R is a constant (8.29 J/mol/°K) and T is the absolute temperature (in °K).

9. Two substances can react in a redox reaction even if the standard potentials are unfavorable, provided that the concentrations are appropriate.

Assume that normally the reduced form of A would donate electrons to the oxidized form of B. However, if the concentration of the reduced form of A were low and the concentration of the reduced form of B were high, it would be possible for the reduced form of B to donate electrons to the oxidized form of A. Thus, the reaction would proceed in the direction opposite that predicted from standard potentials. A practical example of this is the utilization of H^+ as an electron acceptor to produce H_2. Normally, H_2 production in fermentative bacteria is not extensive because H^+ is a poor electron acceptor; the E_0' of the $2\,H^+/H_2$ pair is -0.41 V. However, if the concentration of H_2 is kept low by continually removing it (a process done by methanogenic prokaryotes, which use $H_2 + CO_2$ to produce methane, CH_4, or by many other anaerobes capable of consuming H_2 anaerobically), the potential will be more positive and then H^+ will serve as a suitable electron acceptor.

TABLE A1.1   **Free energies of formation, $G^0{}_f$, for some substances (kJ/mol)[a]**

Carbon compound	Metal	Nonmetal	Nitrogen compound
CO, -137.34	Cu^+, $+50.28$	H_2, 0	N_2, 0
CO_2, -394.4	Cu^{2+}, $+64.94$	H^+, 0 at pH 0;	NO, $+86.57$
CH_4, -50.75	CuS, -49.02	-39.83 at pH 7	NO_2, $+51.95$
H_2CO_3, -623.16	Fe^{2+}, -78.87	(-5.69 per pH unit)	NO_2^-, -37.2
HCO_3^-, -586.85	Fe^{3+}, -4.6	O_2, 0	NO_3^-, -111.34
CO_3^{2-}, -527.90	$FeCO_3$, -673.23	OH^-, -157.3 at pH 14;	NH_3, -26.57
Acetate, -369.41	FeS_2, -150.84	-198.76 at pH 7;	NH_4^+, -79.37
Alanine, -371.54	$FeSO_4$, -829.62	-237.57 at pH 0	N_2O, $+104.18$
Aspartate, -700.4	PbS, -92.59	H_2O, -237.17	
Benzoic acid, -245.6	Mn^{2+}, -227.93	H_2O_2, -134.1	
Butyrate, -352.63	Mn^{3+}, -82.12	PO_4^{3-}, -1026.55	
Caproate, -335.96	MnO_4^{2-}, -506.57	Se^0, 0	
Citrate, -1168.34	MnO_2, -456.71	H_2Se, -77.09	
Crotonate, -277.4	$MnSO_4$, -955.32	SeO_4^{2-}, -439.95	
Cysteine, -339.8	HgS, -49.02	S^0, 0	
Ethanol, -181.75	MoS_2, -225.42	SO_3^{2-}, -486.6	
Formaldehyde, -130.54	ZnS, -198.60	SO_4^{2-}, -744.6	
Formate, -351.04		$S_2O_3^{2-}$, -513.4	
Fructose, -915.38		H_2S, -27.87	
Fumarate, -604.21		HS^-, $+12.05$	
Gluconate, -1128.3		S^{2-}, $+85.8$	
Glucose, -917.22			
Glutamate, -699.6			
Glycerate, -658.1			
Glycerol, -488.52			
Glycine, -314.96			
Glycolate, -530.95			
Guanine, $+46.99$			
Lactate, -517.81			
Lactose, -1515.24			
Malate, -845.08			
Mannitol, -942.61			
Methanol, -175.39			
Oxalate, -674.04			
Phenol, -47.6			
n-Propanol, -175.81			
Propionate, -361.08			
Pyruvate, -474.63			
Ribose, -757.3			
Succinate, -690.23			
Sucrose, -370.90			
Urea, -203.76			
Valerate, -344.34			

a Values for free energy of formation of various compounds can be found in Dean, J. A. 1973. *Lange's Handbook of Chemistry*, 11th edition. McGraw-Hill, New York; Garrels, R. M., and C. L. Christ. 1965. *Solutions, Minerals, and Equilibria*. Harper and Row, New York; Burton, K. 1957. In Krebs,

H. A., and H. L. Kornberg. Energy transformations in living matter, *Ergebnisse der Physiologie* (appendix). Springer-Verlag, Berlin; and Thauer, R. K., K. Jungermann, and H. Decker. 1977. Energy conservation in anaerobic chemotrophic bacteria. *Bacteriol. Rev.* 41:100–180.

Oxidation State or Number

1. The oxidation state of an element in an elementary substance (for example, H_2, O_2) is zero.

2. The oxidation state of the ion of an element is equal to its charge (for example, $Na^+ = +1$, $Fe^{3+} = +3$, $O^{2-} = -2$).

3. The sum of oxidation numbers of all atoms in a neutral molecule is zero. Thus, H_2O is neutral because it has two H at +1 each and one O at −2.

4. In an ion, the sum of oxidation numbers of all atoms is equal to the charge on that ion. Thus, in the OH^- ion, $O(-2) + H(+1) = -1$.

5. In compounds, the oxidation state of O is virtually always −2, and that of H is +1.

6. In simple carbon compounds, the oxidation state of C can be calculated by adding up the H and O atoms present and using the oxidation states of these elements as given in item 5, because in a neutral compound the sum of all oxi-

dation numbers must be zero. Thus, the oxidation state of carbon in methane, CH_4, is -4 (4 H at $+1$ each $= +4$); in carbon dioxide, CO_2, the oxidation state of carbon is $+ 4$ (2 O at -2 each $= -4$).

7. In organic compounds with more than one C atom, it may not be possible to assign a specific oxidation number to each C atom, but it is still useful to calculate the oxidation state of the compound as a whole. The same conventions are used. Thus, the oxidation state of carbon in glucose, $C_6H_{12}O_6$, is zero (12 H at $+1 = 12$; 6 O at $-2 = -12$) and the oxidation state of carbon in ethanol, C_2H_6O, is -2 each (6 H at $+1 = +6$; one O at -2).

8. In all oxidation–reduction reactions there is a balance between the oxidized and reduced products. To calculate an oxidation–reduction balance, the number of molecules of each product is multiplied by its oxidation state. For instance, in calculating the oxidation–reduction balance for the alcoholic fermentation, there are two molecules of ethanol at $-4 = -8$ and two molecules of CO_2 at $+4 = +8$ so the net balance is zero. When constructing model reactions, it is useful to calculate redox balances to be certain that the reaction is possible.

Calculating Free-Energy Yields for Hypothetical Reactions

Energy yields can be calculated either from free energies of formation of the reactants and products or from differences in reduction potentials of electron-donating and electron-accepting partial reactions.

Calculations from Free Energy

Free energies of formation are given in Table A1.1. The procedure to use for calculating energy yields of reactions follows.

1. *Balancing reactions*   In all cases, it is essential to ascertain that the coupled oxidation–reduction reaction is balanced. Balancing involves three things: (*a*) the total number of each kind of atom must be identical on both sides of the equation; (*b*) there must be an ionic balance so that when positive and negative ions are added up on the right side of the equation, the total ionic charge (whether positive, negative, or neutral) exactly balances the ionic charge on the left side of the equation; and (*c*) there must be an oxidation–reduction balance so that all the electrons removed from one substance must be transferred to another substance. In general, when constructing balanced reactions, one proceeds in the reverse of the three steps just listed. Usually, if steps (*c*) and (*b*) have been properly handled, step (*a*) becomes correct automatically.

2. *Examples*   (*a*) What is the balanced reaction for the oxidation of H_2S to SO_4^{2-} with O_2? First, decide how many electrons are involved in the oxidation of H_2S to SO_4^{2-}. This can be most easily calculated from the oxidation states of the compounds, using the rules given previously. Because H has an oxidation state of $+1$, the oxidation state of S in H_2S is -2. Because O has an oxidation state of -2, the oxidation state of S in SO_4^{2-} is $+6$ (because it is an ion, using the rules given in items 4 and 5 of the previous section). Thus, the oxidation of H_2S to SO_4^{2-} involves an eight-elec-

tron transfer (from -2 to $+6$). Because each O atom can accept two electrons (the oxidation state of O in O_2 is zero, but in H_2O is -2), this means that two molecules of molecular oxygen, O_2, are required to provide sufficient electron-accepting capacity. Thus, at this point, we know that the reaction requires 1 H_2S and 2 O_2 on the left side of the equation, and 1 SO_4^{2-} on the right side. To achieve an ionic balance, we must have two positive charges on the right side of the equation to balance the two negative charges of SO_4^{2-}. Thus, 2 H^+ must be added to the right side of the equation, making the overall reaction

$$H_2S + 2\,O_2 \rightarrow SO_4^{2-} + 2\,H^+$$

By inspection, it can be seen that this equation is also balanced in terms of the total number of atoms of each kind on each side of the equation.

(*b*) What is the balanced reaction for the oxidation of H_2S to SO_4^{2-} with Fe^{3+} as electron acceptor? We have just ascertained that the oxidation of H_2S to SO_4^{2-} is an eight-electron transfer. Because the reduction of Fe^{3+} to Fe^{2+} is only a one-electron transfer, 8 Fe^{3+} will be required. At this point, the reaction looks like

$$H_2S + 8\,Fe^{3+} \rightarrow 8\,Fe^{2+} + SO_4^{2-} \qquad \text{(not balanced)}$$

We note that the ionic balance is incorrect. We have 24 positive charges on the left and 14 positive charges on the right ($16+$ from Fe, $2-$ from sulfate). To equalize the charges, we add 10 H^+ on the right. Now our equation looks like

$$H_2S + 8\,Fe^{3+} \rightarrow 8\,Fe^{2+} + 10\,H^+ + SO_4^{2-}$$
$$\text{(not balanced)}$$

To provide the necessary hydrogen for the H^+ and oxygen for the sulfate, we add 4 H_2O to the left and find that the equation is now balanced:

$$H_2S + 4\,H_2O + 8\,Fe^{3+} \rightarrow 8\,Fe^{2+} + 10\,H^+ + SO_4^{2-}$$
$$\text{(balanced)}$$

In general, in microbiological reactions, ionic balance can be achieved by adding H^+ or OH^- to the left or right side of the equation, and because all reactions take place in an aqueous medium, H_2O molecules can be added where needed. Whether H^+ or OH^- is added generally depends on whether the reaction is taking place under acid or alkaline conditions.

3. *Calculation of energy yield for balanced equations from free energies of formation*   Once an equation has been balanced, the free-energy yield can be calculated by inserting the values for the free energy of formation of each reactant and product from Table A1.1 and using the formula in item 2 of the first section of this appendix.

For instance, for the equation

$$H_2S + 2\,O_2 \rightarrow SO_4^{2-} + 2\,H^+$$

$$G'_f \text{ values} \rightarrow (-27.87) + (0)\ (-744.6) + 2\,(-39.83)$$
$$\text{(assuming pH 7)}$$

$$\Delta G^{0\prime} = -796.39 \text{ kJ/rx}$$

The values on the left are summed and subtracted from the values on the left, taking care to ensure that the signs are correct. From the data in Table A1.1, a wide variety of free-energy yields for reactions of microbiological interest can be calculated.

Calculation of Free-Energy Yield from Reduction Potential

Reduction potentials of some important redox pairs are given in Table A1.2. The amount of energy that can be released from two half reactions can be calculated from the differences in reduction potentials of the two reactions and from the number of electrons transferred. The further apart the two half reactions are, and the greater the number of electrons, the more energy released. The conversion of potential difference to free energy is given by the formula $\Delta G^0 = -nF\,\Delta E_0'$, where n is the number of electrons, F is the Faraday constant (96.48 kJ/V), and $\Delta E^{0'}$ is the difference in potentials. Thus, the $2\,H^+/H_2$ couple has a potential of -0.41 V and the $\frac{1}{2}O_2/H_2O$ pair has a potential of $+0.82$V, and so the potential difference is 1.23, which (because two electrons are involved) is equivalent to a free-energy yield (ΔG^0) of -237.34 kJ. On the other hand, the potential difference between the $2\,H^+/H_2$ and the NO_3^-/NO_2^- reactions is less, 0.84 V, which is equivalent to a free-energy yield of -162.08 kJ.

Because many biochemical reactions are two-electron transfers, it is often useful to give energy yields for two-electron reactions, even if more electrons are involved. Thus, the SO_4^{2-}/H_2 redox pair involves eight electrons, and complete reduction of SO_4^{2-} with H_2 requires $4\,H_2$ (equivalent to eight electrons). From the reduction potential difference between $2\,H^+/H_2$ and SO_4^{2-}/H_2S (0.19 V), a free-energy yield of -146.64 kJ is calculated, or -36.66 kJ per two electrons. By convention, reduction potentials are given for conditions in which equal concentrations of oxidized and reduced forms are present. In actual practice, the concentrations of these two forms may be quite different. As discussed earlier in this appendix (item 9, first section), it is possible to couple half reactions even if the potential difference is unfavorable, providing the concentrations of the reacting species are appropriate.

TABLE A1.2	Microbiologically important reduction potentials[a]

Redox pair	E_0' (V)
SO_4^{2-}/HSO_3^-	-0.52
$CO_2/\text{formate}^-$	-0.43
$2\,H^+/H_2$	-0.41
$S_2O_3^{2-}/HS^- + HSO_3^-$	-0.40
Ferredoxin ox/red	-0.39
Flavodoxin ox/red[b]	-0.37
$NAD^+/NADH$	-0.32
Cytochrome c_3 ox/red	-0.29
$CO_2/\text{acetate}^-$	-0.29
S^0/HS^-	-0.27
CO_2/CH_4	-0.24
$FAD/FADH$	-0.22
SO_4^{2-}/HS^-	-0.217
Acetaldehyde/ethanol	-0.197
$\text{Pyruvate}^-/\text{lactate}^-$	-0.19
$FMN/FMNH$	-0.19
Dihydroxyacetone phosphate/glycerolphosphate	-0.19
$HSO_3^-/S_3O_6^{2-}$	-0.17
Flavodoxin ox/red[b]	-0.12
HSO_3^-/HS^-	-0.116
Menaquinone ox/red	-0.075
$APS/AMP + HSO_3^-$	-0.060
Rubredoxin ox/red	-0.057
Acrylyl-CoA/propionyl-CoA	-0.015
Glycine/acetate$^-$ + NH_4^+	-0.010
$S_4O_6^{2-}/S_2O_3^{2-}$	$+0.024$
$\text{Fumarate}^{2-}/\text{succinate}^{2-}$	$+0.033$
Cytochrome b ox/red	$+0.035$
Ubiquinone ox/red	$+0.113$
AsO_4^{3-}/AsO_3^{3-}	$+0.139$
Dimethyl sulfoxide (DMSO)/dimethylsulfide (DMS)	$+0.16$
$Fe(OH)_3 + HCO_3^-/FeCO_3$	$+0.20$
$S_3O_6^{2-}/S_2O_3^{2-} + HSO_3^-$	$+0.225$
Cytochrome c_1 ox/red	$+0.23$
NO_2^-/NO	$+0.36$
Cytochrome a_3 ox/red	$+0.385$
NO_3^-/NO_2^-	$+0.43$
SeO_4^{2-}/SeO_3^{2-}	$+0.475$
Fe^{3+}/Fe^{2+}	$+0.77$
Mn^{4+}/Mn^{2+}	$+0.798$
O_2/H_2O	$+0.82$
$ClO_3^-Cl^-$	$+1.03$
NO/N_2O	$+1.18$
N_2O/N_2	$+1.36$

[a] Data from Thauer, R. K., K. Jungermann, and K. Decker, 1977. Energy conservation in anaerobic chemotrophic bacteria. *Bacteriol. Rev.* 41:100–180.

[b] Separate potentials are given for each electron transfer in this potentially two-electron transfer.

Bergey's Manual of Systematic Bacteriology, Second Edition

VOLUME 1
The Archaea, Cyanobacteria, Phototrophs and Deeply Branching Genera

THE ARCHAEA

Kingdom*I: *Archaeota*
  Phylum I: *Crenarchaeota*
  SECTION I   Thermoprotei, sulfolobi and barophiles
    Class I: *Thermoprotei*
      Order I: *Thermoproteales*
        Family I: *Thermoproteaceae*

Pyrobaculum	3 spp.
Sulfophobococcus	1 sp.
Thermocladium	1 sp.
Thermoproteus	2 spp.

        Family II: *Thermofilaceae*

Thermofilum	1 sp.

      Order II: *'Pyrodictiales'*
        Family I: *Desulfurococcaceae*

Aeropyrum	1 sp.
Desulfurococcus	2 spp.
Igneococcus	1 sp.
Staphylothermus	1 sp.
Thermodiscus	1 sp.
Thermosphaera	1 sp.

        Family II: *'Pyrodictiaceae'*

Hyperthermus	1 sp.
Pyrodictium	3 spp.
Pyrolobus	1 sp.

    Class II: *'Sulfolobi'*
      Order I: *'Sulfolobales'*
        Family I: *Sulfolobaceae*

Acidianus	3 spp.
Metallosphaera	2 spp.
Stetteria	1 sp.
Stygiolobus	1 sp.
Sulfolobus	6 spp.
Sulfurisphaera	1 sp.
Sulfurococcus	2 spp.

  SECTION II   The Methanogens
  Phylum II: *'Euryarchaeota'*
    Class I: *'Methanobacteria'*
      Order I: *Methanobacteriales*
        Family I: *Methanobacteriaceae*

Methanobacterium	19 spp.
Methanobrevibacter	7 spp.
Methanosphaera	2 spp.
Methanothermus	2 spp.

    Class II: *'Methanococci'*
      Order I: *Methanococcales*
        Family I: *Methanococcaceae*

Methanococcus	11 spp

      Order II: *Methanomicrobiales*
        Family I: *Methanomicrobiaceae*

Methanocalculus	1 sp.
Methanocorpusculum	5 spp.
Methanoculleus	6 spp.
Methanogenium	11 spp.
Methanolacinia	1 sp.
Methanomicrobium	2 spp.
Methanoplanus	3 spp.
Methanospirillum	1 sp.
Methanofollis	2 spp.

        Family II: *Methanosarcinaceae*

Methanococcoides	2 spp.
Methanohalobium	1 sp.
Methanohalophilus	5 spp.
Methanolobus	5 spp.
Methanosarcina	8 spp.
Methanothrix	4 spp.

        Family III: *'Uncertain Placement'*

Halomethanococcus	1 sp.

  SECTION III   The Halobacteria
    Class I: *'Halobacteria'*
      Order I: *Halobacteriales*
        Family I: *Halobacteriaceae*

Haloarcula	6 spp.
Halobacterium	13 spp.
Halobaculum	1 sp.
Halococcus	4 spp.
Haloferax	4 spp.
Halogeometricum	1 sp.
Halorubrum	7 spp.
Natrialba	2 spp.
Natrinema	2 spp.
Natronobacterium	4 spp.
Natronococcus	2 spp.
Natronomonas	1 sp.
Haloterrigena	1 sp.
Natronorubrum	2 spp.

  SECTION IV   The Thermoplasms
    Class I: *'Thermoplasma'*
      Order I: *'Thermoplasmatales'*
        Family I: *'Thermoplasmataceae'*

Thermoplasma	2 spp.

        Family II: *Picrophilaceae*

Picrophilus	2 spp.

  SECTION V   The Thermococci
    Class I: *'Thermococci'*
      Order I: *'Archaeoglobales'*
        Family I: *'Archaeoglobaceae'*

Archaeoglobus	3 spp.
Ferroglobus	1 sp.

      Order II: *Thermococcales*
        Family I: *Thermococcaceae*

Pyrococcus	2 spp.
Thermococcus	12 spp.

    Class II: *'Methanopyri'*
      Order I: *'Methanopyrales'*
        Family I: *'Methanopyraceae'*

Methanopyrus	1 sp.

        Family II: *'Genera incertiae sedis'*

*The taxonomic hierarchy employed in the second edition of *Bergey's Manual* uses the term *Kingdom* instead of *Domain*. The list of genera and higher order taxa shown here are the organisms currently recognized, with the numbers besides each genus name giving the number of species recognized. Genera or higher order taxa in quotation marks or set in roman instead of italics are recognized taxa whose names have not yet been validated. Because bacterial taxonomy is a work in progress, updates to the list shown here occur as new genera and species are described and as new data support new taxonomic arrangements. The contents shown here were kindly provided by Dr. George Garrity, Editor-in-Chief, Bergey's Manual Trust, prior to publication of the second edition of *Bergey's Manual of Systematic Bacteriology*, and may therefore vary slightly from the final version actually published. However, any errors in this list are the sole responsibility of the authors of *BBOM 9/e*.

THE DEEPLY BRANCHING GENERA OF BACTERIA

Kingdom II: *Bacteria*
SECTION VI   Aquifex and relatives
  Phylum I: *'Aquificae'*
    Class I: *'Aquificae'*
      Order I: *'Aquificales'*
        Family I: *'Aquificaceae'*
          Aquifex   2 spp.
          Calderobacterium   1 sp.
          Hydrogenobacter   2 spp.
SECTION VII   Thermotogas and Geotogas
    Class I: *'Thermotogae'*
      Order I: Thermotogales
        Family I: *'Thermotogaceae'*
          Fervidobacterium   4 spp.
          Geotoga   2 spp.
          Petrotoga   2 spp.
          Thermosipho   2 spp.
          Thermotoga   5 spp.
    Class II: *'Thermodesulfobacteria'*
      Order I: *'Thermodesulfobacteriales'*
        Family I: *'Thermodesulfobacteriaceae'*
          Carboxydothermus   1 sp.
          Coprothermobacter   2 spp.
          Dictyoglomus   2 spp.
          Thermodesulfobacterium   2 spp.
SECTION VIII   The Deinococci
  Phylum II: *'Xenobacteria'*
    Class I: *'Deinococci'*
      Order I: *Deinococcales*
        Family I: *Deinococcaceae*
          Deinococcus   8 spp.
SECTION IX   Thermi and genera of uncertain affiliation
    Class I: *'Thermi'*
      Order I: *'Thermales'*
        Family I: *'Thermaceae'*
          Meiothermus   4 spp.
          Thermus   9 spp.
        Family II: 'The paraphyletic assemblage'
          Deferribacter   1 sp.
          Nitrospira   1 sp.
          Synergistes   1 sp.
          Thermodesulfovibrio   1 sp.
          'Flexistipes'   1 sp.
          'Geovibrio'   1 sp.
          'Leptospirillum'   1 sp.
          'Magnetobacterium'   1 sp.
SECTION X   Chrysiogenes
  Phylum III: *'Chrysiogenetes'*
    Class I: *'Chrysiogenetes'*
      Order I: *'Chrysiogenales'*
        Family I: *'Chrysiogenaceae'*
          Chrysiogenes   1 sp.
SECTION XI   The Chloroflexi and Herpetosiphons
  Phylum IV: *'Thermomicrobia'*
    Class I: *'Chloroflexi'*
      Order I: *'Chloroflexales'*
        Family I: *'Chloroflexaceae'*
          Chloroflexus   2 spp.
          Chloronema   1 sp.
          Heliothrix   1 sp.
          Oscillochloris   2 spp.
      Order II: *'Herpetosiphonales'*
        Family I: *'Herpetosiphonaceae'*
          Herpetosiphon   5 spp.
SECTION XII   Thermomicrobia
    Class I: *'Thermomicrobia'*
      Order I: *'Thermomicrobiales'*
        Family I: *'Thermomicrobiaceae'*
          Thermomicrobium   2 spp.
SECTION XIII   The *Cyanobacteria*
  Phylum V: *'Cyanobacteria'*

Class I: *'Prochlorophyta'*
  Order I: *'Chroococcales'*
    Family I:
      'Chamaesiphon'   2 spp.
      'Chroococcus'   1 sp.
      'Cyanobacterium'   1 sp.
      'Cyanothece'   1 sp.
      'Dactylococcopsis (Myxobaktron)' 1 sp.
      'Gloeobacter'   1 sp.
      'Gloeocapsa'   1 sp.
      'Gloeothece'   1 sp.
      'Microcytis'   0 spp.
      'Prochlorococcus'   1 sp.
      Prochloron   1 sp.
      'Synechococcus'   3 spp.
      'Synechocystis'   1 sp.
  Order II: 'Pleurocapsales'
    Family I:
      'Chroococcidiopsis'   2 spp.
      'Cyanocystis'   1 sp.
      'Dermocarpa'   1 sp.
      'Dermocarpella'   1 sp.
      'Hyella'   1 sp.
      'Myxosarcina'   1 sp.
      'Pleurocapsa'   1 sp.
      'Stanieria'   1 sp.
      'Xenococcus'   1 sp.
  Order III: 'Oscillatoriales'
    Family I:
      'Arthrospira'   2 spp.
      'Borzia'   1 sp.
      'Crinalium'   1 sp.
      'Geitlerinema'   1 sp.
      'Hormoscilla'   1 sp.
      'Isocystis'   1 sp.
      'Leptolyngbia'   1 sp.
      'Limnothrix'   1 sp.
      'Lyngbya'   1 sp.
      'Microcoleus'   2 spp.
      'Oscillatoria'   6 spp.
      'Planktothrix'   1 sp.
      Prochlorothrix   1 sp.
      'Pseudoanabaena'   1 sp.
      'Spirulina'   1 sp.
      'Starria'   1 sp.
      'Symploca'   1 sp.
      'Trichodesmium'   1 sp.
      'Tychonema'   1 sp.
  Order IV: 'Nostocales'
    Family I: 'Nostocaceae'
      'Anabaena'   1 sp.
      'Anabaenopsis'   1 sp.
      'Aphanizomenon'   1 sp.
      'Cyanospira'   1 sp.
      'Cylindrospermum'   2 spp.
      'Microchaete'   1 sp.
      'Nodularia'   1 sp.
      'Nostoc'   4 spp.
    Family II: 'Scytonemataceae'
      'Scytonema'   1 sp.
      'Tolypothrix'   2 spp.
    Family III: 'Rivulariaceae'
      'Calothrix'   1 sp.
      'Rivularia'   1 sp.
  Order V: 'Stigonematales'
    Family IV:
      'Chlorogloeopsis'   1 sp.
      'Fischerella'   1 sp.
      'Geitleria'   1 sp.
      'Hapalosiphon'   1 sp.
      'Loriella'   1 sp.
      'Mastigocladus'   1 sp.
      'Mastigocoleus'   1 sp.

'Matteia'	1 sp.
'Nostochopsis'	1 sp.
'Stigonema'	1 sp.

SECTION XIV   Chlorobia
   Phylum VI: 'Chlorobia'
      Class I: 'Chlorobia'
         Order I: Chlorobiales
            Family I: Chlorobiaceae

Ancalochloris	1 sp.
Chlorobium	7 spp.
Chloroherpeton	1 sp.
Pelodictyon	4 spp.
Prosthecochloris	1 sp.

Phototrophic bacteria that also appear elsewhere in the manual

Heliobacterium	3 spp.

VOLUME 2
The Proteobacteria

Kingdom II: Bacteria
   Phylum VII: 'Proteobacteria'
SECTION XV   The α-Proteobacteria
   Class I: 'Rhodospirilli'
      Order I: Rhodospirillales
         Family I: 'Rhodospirillaceae'

Azospirillum	7 spp.
Magnetospirillum	2 spp.
Phaeospirillum	2 spp.
Rhodocista	1 sp.
Rhodospira	1 sp.
Rhodospirillum	9 spp.
Rhodothalassium	1 spp.
Rhodovibrio	2 spp.
Roseospira	1 sp.

         Family II: Acetobacteraceae

Acetobacter	20 spp.
Acidiphilium	8 spp.
Acidocella	2 spp.
Acidomonas	1 sp.
Craurococcus	1 sp.
Frateuria	1 sp.
Gluconacetobacter	6 spp.
Gluconobacter	8 spp.
Paracraurococcus	1 sp.
Rhodopila	1 sp.
Roseococcus	1 sp.
Stella	2 spp.
Zavarzinia	1 sp.

      Order II: Rickettsiales
         Family I: Rickettsiaceae

Orientia	1 sp.
Rickettsia	22 spp.

         Family II: Ehrlichiaceae

Aegyptianella	1 sp.
Anaplasma	4 spp.
Cowdria	1 sp.
Ehrlichia	8 spp.
Eperythrozoon	5 spp.
Haemobartonella	3 spp.
Neorickettsia	1 sp.

         Family III: 'Holosporaceae'

Caedibacter	5 spp.
Holospora	4 spp.
Lyticum	2 spp.
Polynucleobacter	1 sp.
Pseudocaedibacter	3 spp.
Symbiotes	1 sp.
Tectibacter	1 sp.

      Order III: 'Rhodobacterales'
         Family I: 'Rhodobacteraceae'

Amaricoccus	4 spp.
Antarctobacter	1 sp.
Gemmobacter	1 sp.
Hirschia	1 sp.
Hyphomonas	5 spp.
Octadecabacter	2 spp.
Paracoccus	13 spp.
Rhodobacter	8 spp.
Rhodovulum	4 spp.
Roseobacter	4 spp.
Sagittula	1 sp.
Sulfitobacter	1 sp.
Ahrensia	1 sp.
Roseovarius	1 sp.
Rubrimonas	1 sp.
Ruegeria	3 spp.
Stappia	2 spp.

      Order IV: 'Sphingomonadales'
         Family I: 'Sphingomonadaceae'

Blastomonas	1 sp.
Erythrobacter	2 spp.
Erythromicrobium	1 sp.
Erythromonas	1 sp.
Porphyrobacter	2 spp.
Rhizomonas	1 sp.
Sandaracinobacter	1 sp.
Sphingomonas	19 spp.
Zymomonas	2 spp.
Rhodanobacter	1 sp.

      Order V: Caulobacterales
         Family I: Caulobacteraceae

Asticcacaulis	2 spp.
Brevundimonas	2 spp.
Caulobacter	11 spp.

      Order VI: 'Rhizobiales'
         Family I: Rhizobiaceae

Agrobacterium	10 spp.
Carbophilus	1 sp.
Chelatobacter	1 sp.
Ensifer	1 sp.
Rhizobium	20 spp.
Sinorhizobium	6 spp.

         Family II: 'Phyllobacteriaceae'

Mesorhizobium	7 spp.
Phyllobacterium	2 spp.

         Family III: Bartonellaceae

Bartonella	14 spp.

         Family IV: Brucellaceae

Brucella	6 spp.
Mycoplana	4 spp.
Ochrobactrum	2 spp.

         Family VI: 'Rhodobiaceae'

Rhodobium	2 spp.

         Family V: Methylobacteriacea

Methylorhabdus	1 sp.
Protomonas	1 sp.
Roseomonas	3 spp.

         Family VI: 'Methylocystaceae'

Methylocystis	2 spp.
Methylosinus	2 spp.

         Family VII: 'Beijerinckiaceae'

Beijerinckia	6 spp.
Chelatococcus	1 sp.
Derxia	1 sp.

         Family VIII: Hyphomicrobiaceae

Ancalomicrobium	1 sp.
Ancylobacter	1 sp.
Angulomicrobium	1 sp.
Aquabacter	1 sp.
Azorhizobium	1 sp.
Blastochloris	2 spp.
Devosia	1 sp.
Dichotomicrobium	1 sp.
Filomicrobium	1 sp.
Gemmiger	1 sp.
Hyphomicrobium	12 spp.

Labrys	1 sp.	Order IV: 'Nitrosomonadales'	
Nevskia	1 sp.	Family I: 'Nitrosomonadaceae'	
Pedomicrobium	4 spp.	*Nitrosomonas*	1 sp.
Prosthecomicrobium	4 spp.	*Nitrosospira*	3 spp.
Rhodomicrobium	1 sp.	Family II: *Spirillaceae*	
Rhodoplanes	2 spp.	*Spirillum*	1 sp.
Seliberia	1 sp.	*Thiobacillus*	21 spp.
Xanthobacter	4 spp.	Family III: *Gallionellacea*	
Family IX: 'Bradyrhizobiaceae'		*Gallionella*	1 sp.
Afipia	3 spp.	Order V: 'Methylophilales'	
Blastobacter	5 spp.	Family I: 'Methylophilaceae'	
Bosea	1 sp.	*Methylobacillus*	2 spp.
Bradyrhizobium	3 spp.	*Methylophaga*	3 spp.
Nitrobacter	1 sp.	*Methylophilus*	1 sp.
Nitrococcus	1 sp.	*Methylovorus*	1 sp.
Nitrospina	1 sp.	SECTION XVII The γ-Proteobacteria	
Oligotropha	1 sp.	Class I: 'Zymobacteria'	
Rhodopseudomonas	15 spp.	Order I: 'Chromatiales'	
SECTION XVI The β-Proteobacteria		Family I: *Chromatiaceae*	
Class I: 'Neisseriae'		*Allochromatium*	3 spp.
Order I: 'Neisseriales'		*Amoebobacter*	4 spp.
Family I: *Neisseriaceae*		*Chromatium*	13 spp.
Alysiella	1 sp.	*Halochromatium*	2 spp.
Aquaspirillum	21 spp.	*Isochromatium*	1 sp.
Catenococcus	1 sp.	*Lamprobacter*	1 sp.
Chromobacterium	2 spp.	*Lamprocystis*	1 sp.
Eikenella	1 sp.	*Marichromatium*	2 spp.
Iodobacter	1 sp.	*Nitrosococcus*	2 spp.
Kingella	4 spp.	*Rhabdochromatium*	1 sp.
Microvirgula	1 sp.	*Thermochromatium*	1 sp.
Neisseria	24 spp.	*Thiocapsa*	5 spp.
Prolinoborus	1 sp.	*Thiococcus*	1 sp.
Simonsiella	3 spp.	*Thiocystis*	4 spp.
Vogesella	1 sp.	*Thiodictyon*	2 spp.
Order II: 'Burkholderiales'		*Thiohalocapsa*	1 sp.
Family I: 'Burkholderiaceae'		*Thiolamprovum*	1 sp.
Burkholderia	20 spp.	*Thiopedia*	1 sp.
Cupriavidus	1 sp.	*Thiorhodococcus*	1 sp.
Ralstonia	3 spp.	*Thiorhodovibrio*	1 sp.
Thermothrix	2 spp.	*Thiospirillum*	1 sp.
Family II: 'Oxalobacteraceae'		Family II: *Ectothiorhodospiraceae*	
Duganella	1 sp.	*Arhodomonas*	1 sp.
Janthinobacterium	1 sp.	*Ectothiorhodospira*	9 spp.
Lautropia	1 sp.	*Halorhodospira*	3 spp.
Oxalobacter	2 spp.	Order II: 'Xanthomonadales'	
Telluria	2 spp.	Family I: 'Xanthomonadaceae'	
Family III: *Alcaligenaceae*		*Stenotrophomonas*	2 spp.
Achromobacter	4 spp.	*Xanthomonas*	24 spp.
Alcaligenes	16 spp.	*Xylella*	1 sp.
Bordetella	7 spp.	Order III: 'Cardiobacteriales'	
Pelistega	1 sp.	Family I: *Cardiobacteriaceae*	
Sutterella	1 sp.	*Cardiobacterium*	1 sp.
Taylorella	1 sp.	*Dichelobacter*	1 sp.
Family VI: *Comamonadaceae*		*Suttonella*	1 sp.
Acidovorax	7 spp.	Order IV: 'Thiotrichales'	
Brachymonas	1 sp.	Family I: 'Thiotrichaceae'	
Comamonas	3 spp.	*Achromatium*	1 sp.
Herbaspirillum	2 spp.	*Beggiatoa*	1 sp.
Hydrogenophaga	4 spp.	*Leucothrix*	1 sp.
Ideonella	1 sp.	*Macromonas*	2 spp.
Leptothrix	5 spp.	*Thiobacterium*	1 sp.
Polaromonas	1 sp.	*Thioploca*	4 spp.
Rhodoferax	1 sp.	*Thiospira*	1 sp.
Rubrivivax	1 sp.	*Thiothrix*	1 sp.
Sphaerotilus	1 sp.	*Vitreoscilla*	3 spp.
Thiomonas	4 spp.	Family II: *Piscirickettsiaceae*	
Variovorax	1 sp.	*Cycloclasticus*	1 sp.
Order III: 'Rhodocyclales'		*Hydrogenovibrio*	1 sp.
Family I: 'Rhodocyclaceae'		*Piscirickettsia*	1 sp.
Azoarcus	5 spp.	*Thiomicrospira*	4 spp.
Rhodocyclus	3 spp.	Family III: 'Francisellaceae'	
Thauera	4 spp.	*Francisella*	5 spp.
Zoogloea	1 sp.	Order V: 'Legionellales'	

Family I: *Legionellaceae*
  Legionella 44 spp.
Family II: '*Coxiellaceae*'
  Coxiella 1 sp.
  Rickettsiella 4 spp.
Order VI: 'Methylococcales'
  Family I: *Methylococcaceae*
    Methylobacter 6 spp.
    Methylobacterium 9 spp.
    Methylocaldum 3 spp.
    Methylococcus 8 spp.
    Methylomicrobium 3 spp.
    Methylomonas 4 spp.
    Methylosphaera 1 sp.
Order VII: 'Oceanospirillales'
  Family I: '*Oceanospirillaceae*'
    Balneatrix 1 sp.
    Marinomonas 2 spp.
    Marinospirillum 2 spp.
    Oceanospirillum 16 spp.
  Family II: *Halomonadaceae*
    Alcanivorax 1 sp.
    Carnimonas 1 sp.
    Chromohalobacter 1 sp.
    Deleya 8 spp.
    Halomonas 19 spp.
    Zymobacter 1 sp.
Order VIII: *Pseudomonadales*
  Family I: *Pseudomonadaceae*
    Agromonas 1 sp.
    Aminobacter 3 spp.
    Azomonas 3 spp.
    Azotobacter 9 spp.
    Cellvibrio 2 spp.
    Chryseomonas 2 spp.
    Flavimonas 1 sp.
    Lampropedia 1 sp.
    Lysobacter 5 spp.
    Mesophilobacter 1 sp.
    Morococcus 1 sp.
    Oligella 2 spp.
    Phenylobacterium 1 sp.
    Pseudomonas 117 spp.
    Rhizobacter 1 sp.
    Rugamonas 1 sp.
    Serpens 1 sp.
    Thermoleophilum 2 spp.
    Xylophilus 1 sp.
  Family II: *Moraxellaceae*
    Acinetobacter 7 spp.
    Moraxella 8 spp.
    Moraxella (Branhamella) 4 spp.
    Moraxella (Moraxella) 6 spp.
    Psychrobacter 5 spp.
Order IX: '*Alteromonadales*'
  Family I: '*Alteromonadaceae*'
    Alteromonas 21 spp.
    Colwellia 7 spp.
    Ferrimonas 1 sp.
    Marinobacter 1 sp.
    Marinobacterium 1 sp.
    Microbulbifer 1 sp.
    Pseudoalteromonas 17 spp.
    Shewanella 10 spp.
Order X: '*Vibrionales*'
  Family I: *Vibrionaceae*
    Allomonas 1 sp.
    Enhydrobacter 1 sp.
    Listonella 3 spp.
    Photobacterium 10 spp.
    Salinivibrio 1 sp.
    Vibrio 46 spp.
Order XI: '*Aeromonadales*'

Family I: *Aeromonadaceae*
  Aeromonas 23 spp.
  Ruminobacter 1 sp.
  Tolumonas 1 sp.
Order XII: '*Enterobacteriales*'
  Family I: *Enterobacteriaceae*
    Arsenophonus 1 sp.
    Buchnera 1 sp.
    Budvicia 1 sp.
    Buttiauxella 7 spp.
    Calymmatobacterium 1 sp.
    Cedecea 3 spp.
    Citrobacter 10 spp.
    Edwardsiella 4 spp.
    Enterobacter 15 spp.
    Erwinia 30 spp.
    Escherichia 6 spp.
    Ewingella 1 sp.
    Hafnia 1 sp.
    Klebsiella 11 spp.
    Kluyvera 4 spp.
    Leclercia 1 sp.
    Leminorella 2 spp.
    Moellerella 1 sp.
    Morganella 2 spp.
    Obesumbacterium 1 sp.
    Pantoea 8 spp.
    Photorhabdus 1 sp.
    Plesiomonas 1 sp.
    Pragia 1 sp.
    Proteus 7 spp.
    Providencia 6 spp.
    Rahnella 1 sp.
    Saccharobacter 1 sp.
    Salmonella 12 spp.
    Serratia 12 spp.
    Shigella 4 spp.
    Tatumella 1 sp.
    Trabulsiella 1 sp.
    Wigglesworthia 1 sp.
    Xenorhabdus 9 spp.
    Yersinia 12 spp.
    Yokenella 1 sp.
    Brenneria 6 spp.
    Pectobacterium 11 spp.
    Sodalis 1 sp.
Order XIII: '*Pasteurellales*'
  Family I: *Pasteurellaceae*
    Actinobacillus 17 spp.
    Haemophilus 20 spp.
    Lonepinella 1 sp.
    Pasteurella 23 spp.
    Mannheimia 5 spp.
SECTION XVIII The δ-Proteobacteria
  Class I: '*Predibacteria*'
    Order I: '*Desulfovibrionales*'
      Family I: '*Desulfovibrionaceae*'
        Bilophila 1 sp.
        Desulfomicrobium 4 spp.
        Desulfomonas 1 sp.
        Desulfonatronovibrio 1 sp.
        Desulfovibrio 29 spp.
        Lawsonia 1 sp.
      Family II: '*Desulfurellaceae*'
        Desulfurella 4 spp.
      Family III: '*Desulfobacteraceae*'
        Desulfobacter 6 spp.
        Desulfobacterium 7 spp.
        Desulfococcus 2 spp.
        Desulfosarcina 1 sp.
        Desulfospira 1 sp.
        Desulfocella 1 sp.
      Family IV: '*Desulfobulbaceae*'

Desulfacinum 1 sp.
Desulfobulbus 3 spp.
Desulfocapsa 1 sp.
Desulfofustis 1 sp.
Desulfomonile 1 sp.
Desulforhabdus 1 sp.
Syntrophobacter 3 spp.
Thermodesulforhabdus 1 sp.
Family V: 'Geobacteraceae'
Desulfohalobium 1 sp.
Desulfonema 2 spp.
Desulfuromonas 4 spp.
Desulfuromusa 3 spp.
Geobacter 2 spp.
Pelobacter 6 spp.
Family VI: 'Bdellovibrionaceae'
Bdellovibrio 3 spp.
Micavibrio 1 sp.
Vampirovibrio 1 sp.
Order II: *Myxococcales*
Family I: *Myxococcaceae*
Angiococcus 1 sp.
Myxococcus 8 spp.
Family II: *Archangiaceae*
Archangium 1 sp.
Family III: *Cystobacteraceae*
Cystobacter 3 spp.
Melittangium 3 spp.
Stigmatella 2 spp.
Family IV: *Polyangiaceae*
Chondromyces 5 spp.
Nannocystis 1 sp.
Polyangium 10 spp.
SECTION XIX The ε-Proteobacteria
Class I: 'Campylobacteres'
Order I: 'Campylobacterales'
Family I: *Campylobacteraceae*
Arcobacter 4 spp.
Campylobacter 28 spp.
Sulfurospirillum 2 spp.
Thiovulum 1 sp.
Family II: 'Helicobacteraceae'
Helicobacter 18 spp.
Wolinella 3 spp.

VOLUME 3
The Low G + C Gram Positives
Kingdom II: *Bacteria*
Phylum VIII: 'Firmicutes'
SECTION XX The *Clostridia* and relatives
Class I: 'Clostridia'
Order I: *Clostridiales*
Family I: *Clostridiaceae*
Anaerobacter 1 sp.
Anaerofilum 2 spp.
Caloramator 3 spp.
Clostridium 146 spp.
Ilyobacter 3 spp.
Oxobacter 1 sp.
Sarcina 2 spp.
Sporobacter 1 sp.
Thermobrachium 1 sp.
Family II: 'Peptostreptococcace'
Filifactor 1 sp.
Peptostreptococcus 18 spp.
Tissierella 3 spp.
Family III: 'Eubacteriaceae'
Eubacterium 52 spp.
Helcococcus 1 sp.
Pseudoramibacter 1 sp.
Family IV: 'Lachnospiraceae'
Acetitomaculum 1 sp.

Butyrivibrio 2 spp.
Catonella 1 sp.
Coprococcus 3 spp.
Johnsonella 1 sp.
Lachnospira 2 spp.
Pseudobutyrivibrio 1 sp.
Roseburia 1 sp.
Ruminococcus 14 spp.
Family V: *Peptococcaceae*
Acetonema 1 sp.
Acidaminococcus 1 sp.
Ammonifex 1 sp.
Desulfitobacterium 3 spp.
Desulfosporosinus 1 sp.
Desulfotomaculum 18 spp.
Dialister 1 sp.
Heliobacillus 1 sp.
Heliobacterium 3 spp.
Heliophilum 1 sp.
Megasphaera 2 spp.
Mitsuokella 2 spp.
Pectinatus 2 spp.
Peptococcus 8 spp.
Phascolarctobacterium 1 sp.
Propionispira 1 sp.
Quinella 1 sp.
Schwartzia 1 sp.
Selenomonas 11 spp.
Sporomusa 7 spp.
Succiniclasticum 1 sp.
Succinivibrio 1 sp.
Syntrophobotulus 1 sp.
Thermoterrabacterium 1 sp.
Veillonella 14 spp.
Zymophilus 2 spp.
Family VI: *Syntrophomonadaceae*
Acetogenium 1 sp.
Anaerobaculum 1 sp.
Anaerobranca 1 sp.
Caldicellulosiruptor 3 spp.
Dethiosulfovibrio 1 sp.
Moorella 3 spp.
Sporotomaculum 1 sp.
Syntrophomonas 3 spp.
Syntrophospora 1 sp.
Thermoanaerobacter 13 spp.
Thermoanaerobacterium 5 spp.
Thermoanaerobium 2 spp.
Thermohydrogenium 1 sp.
Thermosyntropha 1 sp.
Order II: *Haloanaerobiales*
Family I: *Haloanaerobiaceae*
Haloanaerobium 8 spp.
Halocella 1 sp.
Halothermothrix 1 sp.
Natroniella 1 sp.
Family II: *Halobacteroidaceae*
Acetohalobium 1 sp.
Haloanaerobacter 3 spp.
Halobacteroides 4 spp.
Orenia 1 sp.
Sporohalobacter 2 spp.
SECTION XXI The Mollicutes
Class I: *Mollicutes*
Order I: *Mycoplasmatales*
Family I: *Mycoplasmataceae*
Mycoplasma 110 spp.
Ureaplasma 6 spp.
Family II: 'Spiroplasma group'
Entomoplasma 6 spp.
Mesoplasma 12 spp.
Spiroplasma 33 spp.
Family III: *Acholeplasmataceae*
Acholeplasma 16 spp.

Anaeroplasma	4 spp.
Family IV: *'Asteroleplasmataceae'*	
Asteroleplasma	1 sp.
Family V: *'Erysipelothrichaeceae'*	
Erysipelothrix	2 spp.
Holdemania	1 sp.

SECTION XXII   The Bacilli and Lactobacilli
  Class I: *'Bacilli'*
    Order I: *Bacillales*
      Family I: *Bacillaceae*

Ammoniphilus	2 spp.
Aneurinibacillus	3 spp.
Bacillus	114 spp.
Brevibacillus	10 spp.
Exiguobacterium	2 spp.
Halobacillus	3 spp.
Oxalophagus	1 sp.
Saccharococcus	1 sp.
Virgibacillus	1 sp.

      Family II: *Planococcaceae*

Filibacter	1 sp.
Kurthia	3 spp.
Planococcus	5 spp.
Sporosarcina	2 spp.

      Family III: *Caryophanaceae*

Caryophanon	2 spp.

      Family IV: *'Sporolactobacillaceae'*

Marinococcus	3 spp.
Sporolactobacillus	6 spp.

      Family V: *'Paenibacillaceae'*

Paenibacillus	27 spp.

      Family VI: *'Alicyclobacillaceae'*

Alicyclobacillus	3 spp.
Sulfobacillus	3 spp.

      Family VII: *'Thermoactinomycetaceae'*

Thermoactinomyces	8 spp.

    Order II: *'Lactobacillales'*
      Family I: *Lactobacillaceae*

Lactobacillus	100 spp.

      Family II: *'Leuconostocaceae'*

Leuconostoc	15 spp.
Oenococcus	1 sp.
Weissella	7 spp.

      Family III: *Streptococcaceae*

Lactococcus	7 spp.
Pediococcus	8 spp.
Streptococcus	68 spp.

      Family IV: *'Enterococcaceae'*

Enterococcus	20 spp.
Melissococcus	1 sp.
Vagococcus	2 spp.

      Family V: *'Carnobacteriaceae'*

Agitococcus	1 sp.
Alloiococcus	1 sp.
Carnobacterium	6 spp.
Dolosigranulum	1 sp.
Lactosphaera	1 sp.
Desemzia	1 sp.

      Family VI: *'Aerococcaceae'*

Abiotrophia	3 spp.
Aerococcus	2 spp.
Facklamia	2 spp.
Globicatella	1 sp.
Tetragenococcus	2 spp.
Ignavigranum	1 sp.

      Family VII: *'Listeriaceae'*

Brochothrix	2 spp.
Listeria	9 spp.

      Family VIII: *'Staphylococcaceae'*

Gemella	4 spp.
Macrococcus	4 spp.
Salinicoccus	2 spp.
Staphylococcus	47 spp.

      Family IX: *'Genera incertae sedis'*

Acetoanaerobium	1 sp.
Acetobacterium	7 spp.
Amphibacillus	1 sp.
Oscillospira	1 sp.
Pasteuria	4 spp.
Syntrophococcus	1 sp.
Trichococcus	1 sp.

VOLUME 4
The High G + C Gram Positives

Kingdom II: *Bacteria*
  Phylum VIII: *'Firmicutes'*
SECTION XXIII   The Actinobacteria
  Class I: *Actinobacteria*
    Subclass I: *Acidimicrobidae*
      Order I: *Acidimicrobiales*
        Suborder I: *Acidimicrobidae*
          Family I: *Acidimicrobiaceae*

Acidimicrobium	1 sp.

    Subclass II: *Rubrobacteridae*
      Order II: *Rubrobacterales*
        Suborder I: *Rubrobacteridae*
          Family I: *Rubrobacteraceae*

Rubrobacter	2 spp.

    Subclass III: *Coriobacteridae*
      Order III: *Coriobacteriales*
        Suborder I: *Coriobacteridae*
          Family I: *Coriobacteriaceae*

Atopobium	3 spp.
Coriobacterium	1 sp.

    Subclass IV: *Sphaerobacteridae*
      Order IV: *'Sphaerobacterales'*
        Suborder I: *Sphaerobacteridae*
          Family I: *Sphaerobacteraceae*

Sphaerobacter	1 sp.

    Subclass V: *Actinobacteridae*
      Order V: *Actinomycetales*
        Suborder I: *Actinomycineae*
          Family I: *Actinomycetaceae*

Actinobaculum	2 spp.
Actinomyces	23 spp.
Arcanobacterium	4 spp.
Mobiluncus	3 spp.

        Suborder II: *Micrococcineae*
          Family II: *Micrococcaceae*

Arthrobacter	32 spp.
Bogoriella	1 sp.
Demetria	1 sp.
Kocuria	6 spp.
Leucobacter	1 sp.
Micrococcus	9 spp.
Nesterenkonia	1 sp.
Renibacterium	1 sp.
Rothia	1 sp.
Stomatococcus	1 sp.
Terracoccus	1 sp.

          Family III: *'Brevibacteraceae'*

Brevibacterium	26 spp.

          Family IV: *Cellulomonadaceae*

Cellulomonas	11 spp.
Oerskovia	2 spp.
Rarobacter	2 spp.

          Family V: *Dermabacteraceae*

Brachybacterium	7 spp.
Dermabacter	1 sp.

          Family VI: *Dermatophilaceae*

Dermacoccus	1 sp.
Dermatophilus	2 spp.
Kytococcus	1 sp.

          Family VII: *Intrasporangiaceae*

Intrasporangium	1 sp.
Janibacter	1 sp.

Sanguibacter	3 spp.
Terrabacter	1 sp.
Family VIII: *Jonesiaceae*	
Jonesia	1 sp.
Family IX: *Microbacteriaceae*	
Agrococcus	1 sp.
Agromyces	6 spp.
Aureobacterium	14 spp.
Clavibacter	11 spp.
Cryobacterium	1 sp.
Curtobacterium	8 spp.
Microbacterium	27 spp.
Rathayibacter	4 spp.
Family X: *Promicromonosporaceae*	
Promicromonospora	3 spp.
Suborder III: *Corynebacterineae*	
Family I: *Corynebacteriaceae*	
Corynebacterium	67 spp.
Family II: *Dietziaceae*	
Dietzia	2 spp.
Family III: *Gordoniaceae*	
Gordonia	9 spp.
Skermania	1 sp.
Family IV: *Mycobacteriaceae*	
Mycobacterium	85 spp.
Family V: *Nocardiaceae*	
Micropolyspora	5 spp.
Nocardia	29 spp.
Rhodococcus	25 spp.
Family VI: *Tsukamurellaceae*	
Tsukamurella	5 spp.
Suborder IV: *Micromonosporaceae*	
Family I: *Micromonosporaceae*	
Actinoplanes	23 spp.
Catellatospora	5 spp.
Catenuloplanes	6 spp.
Couchioplanes	2 spp.
Dactylosporangium	6 spp.
Micromonospora	19 spp.
Pilimelia	4 spp.
Spirilliplanes	1 sp.
Verrucosispora	1 sp.
Suborder VI: *Propionibacterineae*	
Family I: *Propionibacteriaceae*	
Luteococcus	1 sp.
Microlunatus	1 sp.
Propionibacterium	12 spp.
Propioniferax	1 sp.
Family II: *Nocardioidaceae*	
Aeromicrobium	2 spp.
Friedmanniella	1 sp.
Nocardioides	7 spp.
Suborder VII: *Pseudonocardineae*	
Family I: *Pseudonocardiaceae*	
Actinopolyspora	3 spp.
Actinosynnema	3 spp.
Amycolatopsis	11 spp.
Kibdelosporangium	4 spp.
Kutzneria	3 spp.
Lentzea	1 sp.
Micropolyspora	5 spp.
Pseudonocardia	12 spp.
Saccharomonospora	5 spp.
Saccharopolyspora	10 spp.
Saccharothrix	15 spp.
Streptoalloteichus	1 sp.
Thermobispora	1 sp.
Thermocrispum	2 spp.
Suborder VIII: *Streptomycineae*	
Family I: *Streptomycetaceae*	
Streptomyces	509 spp.
Suborder IX: *Streptosporangineae*	
Family I: *Streptosporangiaceae*	

Herbidospora	1 sp.
Microbispora	15 spp.
Micropolyspora	5 spp.
Microtetraspora	20 spp.
Nonomuria	15 spp.
Planobispora	2 spp.
Planomonospora	5 spp.
Planopolyspora	1 sp.
Planotetraspora	1 sp.
Streptosporangium	19 spp.
Family II: *Nocardiopsaceae*	
Nocardiopsis	17 spp.
Thermobifida	2 spp.
Family III: *Thermomonosporaceae*	
Actinomadura	51 spp.
Spirillospora	2 spp.
Thermomonospora	7 spp.
Suborder X: *Frankineae*	
Family I: *Frankiaceae*	
Frankia	1 sp.
Family II: *Geodermatophilaceae*	
Blastococcus	1 sp.
Geodermatophilus	1 sp.
Family III: *Microsphaeraceae*	
Microsphaera	1 sp.
Family IV: *Sporichthyaceae*	
Sporichthya	1 sp.
Family V: *Acidothermaceae*	
Acidothermus	1 sp.
Family VI: 'Incertae sedis'	
Cryptosporangium	2 spp.
Kineococcus	1 sp.
Kineosporia	5 spp.
Suborder XI: *Glycomycineae*	
Family I: *Glycomycetaceae*	
Glycomyces	3 spp.
Order VI: *Bifidobacteriales*	
Family I: *Bifidobacteriaceae*	
Bifidobacterium	33 spp.
Falcivibrio	2 spp.
Gardnerella	1 sp.
Family II: 'Unknown Affiliation'	
Actinobispora	1 sp.
Actinocorallia	1 sp.
Actinokineospora	5 spp.
Excellospora	1 sp.
Pelczaria	1 sp.
Turicella	1 sp.

VOLUME 5
The Planctomycetes, Spirochaetes, Fibrobacteres, Bacteroides and Fusobacteria

Kingdom II: *Bacteria*	
Phylum IX: 'Wall-less forms'	
SECTION XXIV The Planctomycetes and Chlamydia	
Class I: 'Planctomycetacia'	
Order I: *Planctomycetales*	
Family I: *Planctomycetaceae*	
Gemmata	1 sp.
Isosphaera	1 sp.
Pirellula	2 spp.
Planctomyces	6 spp.
Order II: *Chlamydiales*	
Family I: *Chlamydiaceae*	
Chlamydia	4 spp.
SECTION XXV The Spirochetes	
Phylum X: 'Spirochetes'	
Class I: 'Spirochaetes'	
Order I: 'Spirochaetales'	
Family I: *Spirochaetaceae*	
Borrelia	30 spp.
Brevinema	1 sp.
Clevelandina	1 sp.

Cristispira	1 sp.
Diplocalyx	1 sp.
Hollandina	1 sp.
Pillotina	1 sp.
Spirochaeta	14 spp.
Treponema	18 spp.
Family II: *Serpulinaceae*	
Brachyspira	5 spp.
Serpulina	6 spp.
Family III: *Leptospiraceae*	
Leptonema	1 sp.
Leptospira	12 spp.

SECTION XXVI   The Fibrobacters
Phylum XI: 'Fibrobacter'
Class I: 'Fibrobacteres'
Order I: 'Fibrobacterales'
Family I: 'Fibrobacteraceae'

Fibrobacter	3 spp.

Family II: 'Acidobacteriaceae'

Acidobacterium	1 sp.
Holophaga	1 sp.

SECTION XXVII   The Bacteroides
Phylum XII: 'Bacteroids'
Class I: 'Bacteroides'
Order I: 'Bacteroidales'
Family I: *Bacteroidaceae*

Acetivibrio	4 spp.
Acetofilamentum	1 sp.
Acetomicrobium	2 spp.
Acetothermus	1 sp.
Acidaminobacter	1 sp.
Anaerobiospirillum	2 spp.
Anaerorhabdus	1 sp.
Anaerovibrio	3 spp.
Bacteroides	65 spp.
Centipeda	1 sp.
Formivibrio	1 sp.
Malonomonas	1 sp.
Megamonas	1 sp.
Propionivibrio	1 sp.
Succinimonas	1 sp.
Syntrophus	2 spp.

Family II: 'Rikenellaceae'

Marinilabilia	2 spp.
Rikenella	1 sp.

Family III: 'Porphyromonadaceae'

Porphyromonas	13 spp.

Family IV: *Prevotellaceae*

Prevotella	26 spp.

SECTION XXVIII   The Flavobacteria
Phylum XIII: 'Flavobacteria'
Class I: 'Flavobacteria'
Order I: 'Flavobacteriales'
Family I: *Flavobacteriaceae*

Bergeyella	1 sp.
Capnocytophaga	7 spp.
Chryseobacterium	6 spp.
Empedobacter	1 sp.
Flavobacterium	40 spp.
Gelidibacter	1 sp.
Ornithobacterium	1 sp.

Polaribacter	4 spp.
Psychroserpens	1 sp.
Riemerella	2 spp.
Weeksella	2 spp.
Psychroflexus	2 spp.
Family II: 'Myroideaceae'	
Myroides	2 spp.
Psychromonas	1 sp.
Family III: 'Blattabacteriaceae'	
Blattabacterium	1 sp.

SECTION XXIX   The Sphingobacteria
Phylum XIV: 'Sphingobacteria'
Class I: 'Sphingobacteria'
Order I: 'Sphingobacteriales'
Family I: *Sphingobacteriaceae*

Pedobacter	4 spp.
Sphingobacterium	8 spp.

Family II: 'Saprospiraceae'

Haliscomenobacter	1 sp.
Lewinella	3 spp.
Saprospira	1 sp.

Family III: 'Flexibacteraceae'

Cyclobacterium	1 sp.
Cytophaga	23 spp.
Flectobacillus	3 spp.
Flexibacter	17 spp.
Meniscus	1 sp.
Microscilla	1 sp.
Runella	1 sp.
Spirosoma	1 sp.
Sporocytophaga	1 sp.

Family IV: 'Flammeovirgaceae'

Flammeovirga	1 sp.
Persicobacter	1 sp.
Thermonema	2 spp.

Family V: 'Crenothrichaeceae'

Chitinophaga	1 sp.
Crenothrix	1 sp.
Flexithrix	1 sp.
Rhodothermus	2 spp.
Toxothrix	1 sp.

SECTION XXX   The Fusiforms
Phylum XV: 'Fusobacteria'
Class I: 'Fusobacteria'
Order I: 'Fusobacteriales'
Family I: 'Fusobacteriaceae'

Fusobacterium	23 spp.
Leptotrichia	1 sp.
Propionigenium	2 spp.
Sebaldella	1 sp.
Streptobacillus	1 sp.

Family II: 'Genera incertae sedis'

Cetobacterium	1 sp.

SECTION XXXI   Verrucomicrobium and relatives
Phylum XVI: 'Verrucomicrobia'
Class I: *Verrucomicrobiae*
Order I: *Verrucomicrobiales*
Family I: *Verrucomicrobiaceae*

Prosthecobacter	4 spp.
Verrucomicrobium	1 sp.

Glossary

Only the major terms and concepts are included. If a term is not here, consult the index.

ABC transporter   A membrane transport system consisting of three proteins, one of which hydrolyzes ATP, one of which binds the substrate, and one of which functions as the transport channel through the membrane.

Abscess   A localized infection characterized by production of pus.

Acetotrophic   Splitting of acetate into CH_4 plus CO_2 by certain methanogens.

Acetyl-CoA (Ljungdahl–Wood) pathway   A pathway of autotrophic CO_2 fixation widespread in obligate anaerobes including methanogens, homoacetogens, and sulfate-reducing bacteria.

Acetylene reduction assay   Method of measuring activity of nitrogenase by substituting acetylene for the natural substrate of the enzyme, N_2. Acetylene is reduced to ethylene or ethane.

Acid fastness   A staining property of *Mycobacterium* species where cells stained with hot carbolfuschin do not decolorize with acid–alcohol.

Acid mine drainage   Acidic water containing H_2SO_4 derived from the microbial oxidation of iron sulfide minerals.

Acidophile   Organism that grows best at acidic pH values.

Activation energy   Energy needed to make substrate molecules more reactive; enzymes function by lowering activation energy.

Activator protein   A regulatory protein that binds to specific sites on DNA and stimulates transcription; involved in positive control.

Active immunity   An immune state achieved by self-production of antibodies. Compare with *Passive immunity*.

Active site   The portion of an enzyme that is directly involved in binding substrate(s).

Active transport   The energy-dependent process of transporting substances into or out of the cell in which the transported substances are chemically unchanged.

Acute   In reference to infections, short-term, usually characterized by dramatic onset and rapid recovery.

Adherence   A property of bacteria that allows them to stick to host surfaces.

Aerobe   An organism that grows in the presence of O_2; may be facultative, obligate, or microaerobic.

Aerosol   Suspension of particles in airborne water droplets.

Aerotolerant   Of an anaerobe, not being inhibited by O_2.

Agglutination   Reaction between antibody and particle-bound antigen resulting in clumping of the particles.

Algae   Phototrophic eukaryotic microorganisms.

Alkaliphile   An organism that grows best at high pH.

Allergy   A harmful immune reaction, usually caused by a foreign antigen in food, pollen, or chemicals; immediate-type or delayed-type hypersensitivity.

Allosteric enzyme   An enzyme that contains two combining sites, the active site (where the substrate binds) and the allosteric site (where an effector molecule binds).

Ameboid movement   A type of motility in which cytoplasmic streaming moves the organism forward.

Aminoacyl-tRNA synthetase   An enzyme that catalyzes the attachment of the correct amino acid to the correct tRNA.

Aminoglycoside   An antibiotic such as streptomycin that consists of amino sugars linked by glycosidic bonds.

Anabolism   The biochemical processes involved in the synthesis of cell constituents from simpler molecules, usually requiring energy.

Anaerobe   An organism that grows in the absence of O_2; some may even be killed by O_2.

Anaerobic respiration   Use of an electron acceptor other than O_2 in an electron transport-based oxidation and leading to a proton motive force.

Anaphylatoxins   The C3a and C5a fractions of complement that act to mimic some of the reactions of anaphylaxis.

Anaphylaxis (anaphylactic shock)   A violent allergic reaction caused by an antigen–antibody reaction.

Anoxic   Absence of oxygen. Usually used in reference to a microbial habitat.

Anoxygenic photosynthesis   Use of light energy to synthesize ATP by cyclic photophosphorylation without O_2 production.

Antibiotic   A chemical agent produced by one organism that is harmful to other organisms.

Antibiotic resistance   The acquired ability of a microorganism to grow in the presence of an antibiotic to which the microorganism is usually sensitive.

Antibody   A protein present in serum or other body fluid that combines specifically with antigen. An immunoglobulin.

Anticodon   A sequence of three bases in transfer RNA that base-pairs with a codon in messenger RNA during protein synthesis.

Antigen   A substance that interacts with a T cell receptor or an immunoglobulin.

Antigen-presenting cell (APC)   A cell that processes and presents antigen to T lymphocytes.

Antigenic determinant   The portion of an antigen that interacts with an immunoglobulin or T cell receptor. Also called an *epitope*.

Antigenic drift   In influenza virus, minor changes in viral proteins (antigens) due to gene mutation.

Antigenic shift   In influenza virus, major changes in viral proteins (antigens) due to gene reassortment.

Antimicrobial   Harmful to microorganisms by either killing or inhibiting growth.

Antimicrobial agent   A chemical that kills or inhibits the growth of microorganisms.

Antiparallel   In reference to double-stranded DNA, one strand runs $5' \rightarrow 3'$, the other $3' \rightarrow 5'$.

Antiseptic   An agent that kills or inhibits microbial growth but is not harmful to human tissue.

Antiserum   A serum containing antibodies.

Antitoxin   An antibody that specifically interacts with and neutralizes a toxin.

Archaea   A phylogenetic domain of prokaryotes consisting of the methanogens, most extreme halophiles and hyperthermophiles, and *Thermoplasma*.

Aseptic technique   Manipulation of sterile instruments or culture media in such a way as to maintain sterility.

ATP   Adenosine triphosphate, the principal energy carrier of the cell.

Attenuation   Selection of nonvirulent strains of a pathogen still capable of immunizing. Also, a mechanism for controlling gene expression. Typically transcription is terminated after initiation but before a full-length mRNA is produced.

Autoantibody   An antibody that reacts to self antigens.

Autoclave   A sterilizer that destroys microorganisms by high temperature using steam under pressure.

Autoimmunity   Immune reactions of a host against its own self antigens.

Autolysis   The lysis of a cell brought about by the activity of the cell itself.

Autoradiography   Detection of radioactivity in a sample, for example, a cell or gel, by placing it in contact with a photographic film.

Autotroph   Organism able to utilize CO_2 as a sole source of carbon.

Auxotroph   An organism that has developed a nutritional requirement through mutation. Contrast with a *Prototroph*.

B lymphocyte   A cell of the immune system that differentiates into an immunoglobulin-producing cell.

Bacteremia   The transient appearance of bacteria in the blood.

Bacteria   All prokaryotes that are not members of the domain Archaea.

Bacteriocidal   Capable of killing bacteria.

Bacteriocins   Agents produced by certain bacteria that inhibit or kill closely related species.

Bacteriophage   A virus that infects prokaryotic cells.

Bacteriorhodopsin   A protein containing retinal that is found in the membranes of certain extremely halophilic Archaea and that is involved in light-mediated ATP synthesis.

Bacteriostatic   Capable of inhibiting bacterial growth without killing.

Bacteroid   A swollen, deformed *Rhizobium* cell found in the root nodule; capable of nitrogen fixation.

Barophile   An organism that lives optimally at high hydrostatic pressure.

Barotolerant   An organism able to tolerate high hydrostatic pressure, although growing better at 1 atm.

Base composition   In reference to nucleic acids, the proportion of the total bases consisting of guanine plus cytosine or thymine plus adenine base pairs. Usually expressed as a guanine + cytosine (G + C) value for example, 60% G + C.

Batch culture   A closed-system microbial culture of fixed volume.

Beta-lactam   An antibiotic such as penicillin that contains the four-membered heterocyclic beta-lactam ring.

Biocatalysis   The use of microorganisms to synthesize a product or carry out a specific chemical transformation.

Bioconversion   In industrial microbiology, use of microorganisms to convert a substance to a chemically modified form.

Biofilm   Microbial colonies encased in an adhesive, usually polysaccharide material, and attached to a surface.

Biogeochemistry   Study of microbially mediated chemical transformations of geochemical interest, for example, nitrogen or sulfur cycling.

Bioremediation   Use of microorganisms to remove or detoxify toxic or unwanted chemicals in an environment.

Biosynthesis   The production of needed cellular constituents from other (usually simpler) molecules.

Biosynthetic penicillin   Production of a particular form of penicillin by supplying the producing organism with specific side chain precursors.

Biotechnology   The use of living organisms to carry out defined chemical processes for industrial application.

Black smoker   A thermal vent emitting very hot (250–400°C) water and minerals.

Brewing   The manufacture of alcoholic beverages such as beer from the fermentation of malted grains.

Calvin cycle   The biochemical route of CO_2 fixation in many autotrophic organisms.

Capsid   The protein coat of a virus.

Capsomere   An individual protein subunit of the virus capsid.

Capsule   Dense, well-defined polysaccharide or protein layer closely surrounding a cell.

Carboxysomes   Polyhedral cellular inclusions of crystalline ribulose bisphosphate carboxylase (RubisCO), the key enzyme of the Calvin cycle.

Carcinogen   A substance that causes the initiation of tumor formation. Frequently a mutagen.

Carrier   An individual that harbors infectious organisms but does not show symptoms of disease.

Catabolism   The biochemical processes involved in the breakdown of organic or inorganic compounds, usually leading to the production of energy.

Catabolite repression   Repression of a variety of unrelated enzymes when cells are grown in a medium containing glucose.

Catalysis   Increase in rate of a chemical reaction.

Catalyst   A substance that promotes a chemical reaction without itself being changed in the end.

CD4 cells   T helper cells. They are targets for HIV infection.

Cell   The fundamental unit of life.

Cell-mediated immunity   An immune response generated by the activities of non-antibody-producing cells such as T cells. Compare with *Humoral immunity.*

Chemiosmosis   The use of ion gradients, especially proton gradients, across membranes to generate ATP. See *Proton motive force.*

Chemokine   A low-molecular-weight soluble immune response modulator protein produced by a variety of cells.

Chemolithotroph   An organism obtaining its energy from the oxidation of inorganic compounds.

Chemoorganotroph   An organism obtaining its energy from the oxidation of organic compounds.

Chemostat   A continuous culture device controlled by the concentration of limiting nutrient and dilution rate.

Chemotaxis   Movement toward or away from a chemical.

Chemotherapeutic agent   An antimicrobial agent that can be used internally.

Chemotherapy   Treatment of infectious disease with chemicals or antibiotics.

Chlorination   A highly effective disinfectant procedure for drinking water using chlorine gas or other chlorine-containing compounds as disinfectant.

Chlorophyll and bacteriochlorophyll   Pigments of phototrophic organisms consisting of light-sensitive magnesium tetrapyrroles.

Chloroplast   The chlorophyll-containing organelle of phototrophic eukaryotes.

Chlorosomes   Cigar-shaped structures enclosed by a nonunit membrane and containing the light-harvesting bacteriochlorophyll (c, c_s, d, or e) in green sulfur bacteria and in *Chloroflexus.*

Chromogenic   Producing color; a chromogenic colony is a pigmented colony.

Chromosome   A genetic element carrying genes essential to cellular function. Prokaryotes typically have a single chromosome consisting of a circular DNA molecule. Eukaryotes typically have several chromosomes, each containing a linear DNA molecule.

Chronic   Long-term.

Cidal   Lethal or killing.

Cilium   Short, filamentous structure that beats with many others to make a cell move.

Citric acid cycle   A cyclical series of reactions resulting in the conversion of acetate to CO_2 and NADH. Also called the *Tricarboxylic acid cycle* or the *Kreb's cycle.*

Class I MHC proteins   Antigen-presenting molecules found on all nucleated vertebrate cells.

Class II MHC proteins   Antigen-presenting molecules found primarily on macrophages and B lymphocytes in vertebrates.

Clonal selection   A theory that each B or T lymphocyte, when stimulated by antigen, divides to form a clone of itself.

Clone   A population of cells all descended from a single cell; a number of copies of a DNA fragment obtained by allowing an inserted DNA fragment to be replicated by a phage or plasmid.

Cloning vectors   Genetic elements into which genes can be recombined and replicated.

Coccoid   Sphere-shaped.

Coccus   A spherical bacterium.

Codon   A sequence of three bases in messenger RNA that encodes a specific amino acid.

Coenzyme   A low-molecular-weight molecule that participates in an enzymatic reaction by accepting and donating electrons or functional groups. Examples: NAD^+, FAD.

Coliforms   Gram-negative, nonsporing, facultative rods that ferment lactose with gas formation within 48 hr at 35°C.

Colonization   Multiplication of a microorganism after it has attached to host tissues or other surfaces.

Colony   A macroscopically visible population of cells growing on solid medium, arising from a single cell.

Cometabolism   The metabolic transformation of a substance while a second substance serves as primary energy or carbon source.

Commodity chemicals   Chemicals such as ethanol that have low monetary value and thus are sold primarily in bulk.

Common-source epidemic   An epidemic resulting from infection of a large number of people from a single contaminated source.

Compatible solutes   Organic compounds that serve as cytoplasmic solutes to balance water relations for cells growing in environments of high salt or sugar.

Competence   Ability to take up DNA and become genetically transformed.

Complement   A complex of proteins in the blood serum that interacts sequentially with specific antigen–antibody complexes.

Complement fixation   The consumption of complement by an antibody–antigen reaction.

Complementary   Nucleic acid sequences that can base-pair with each other.

Complex media   Culture media whose precise chemical composition is unknown. Also called *undefined media.*

Concatamer   A DNA molecule consisting of two or more separate molecules linked end to end to form a long, linear structure.

Congenital syphilis   Syphilis contracted by an infant from its mother during birth.

Conjugation   Transfer of genes from one prokaryotic cell to another by a mechanism involving cell-to-cell contact.

Consensus sequence   A nucleic acid sequence in which the base present in a given position is that base most commonly found when many experimentally determined sequences are compared.

Consortium   A two- (or more) membered bacterial culture (or natural assemblage) in which each organism benefits from the others.

Contagious   Transmissible.

Cortex   The region inside the spore coat of an endospore, around the core.

Covalent bond   A nonionic chemical bond formed by a sharing of electrons between two atoms.

Crista   Inner membrane in a mitochondrion, site of respiration.

Culture   A particular strain or kind of organism growing in a laboratory medium.

Culture medium   An aqueous solution of various nutrients suitable for the growth of microorganisms.

Cutaneous   Relating to the skin.

Cyanobacteria   Prokaryotic oxygenic phototrophs containing chlorophyll *a* and phycobilins.

Cyst   A resting stage formed by some bacteria and protozoa in which the whole cell is surrounded by a protective layer; not the same as a spore.

Cytochrome   Iron-containing porphyrin complexed with proteins, which functions as an electron carrier in the electron transport system.

Cytokine   A soluble immune response modulator produced by cells other than lymphocytes, usually phagocytic cells.

Cytoplasm   Cellular contents inside the cytoplasmic membrane, excluding the nucleus.

Cytoplasmic membrane   The permeability barrier of the cell, separating the cytoplasm from the environment.

Decontamination   Treatment that renders an object or inanimate surface safe to handle.

Defined media   Culture media whose exact chemical composition is known. Compare with *Complex media.*

Degeneracy   In relation to the genetic code, the fact that more than one codon can code for the same amino acid.

Deletion   Removal of a portion of a gene.

Denaturation   Irreversible destruction of a macromolecule, as for example, the destruction of a protein by heat.

Denitrification   Conversion of nitrate into nitrogen gases under anoxic conditions.

Dental caries   Tooth decay resulting from bacterial infection.

Dental plaque   Bacterial cells encased in a matrix of extracellular polymers, found on the teeth.

Deoxyribonucleic acid (DNA)   A polymer of nucleotides connected via a phosphate–deoxyribose sugar backbone; the genetic material of cells and some viruses.

Desiccation   Drying.

Dideoxynucleotide   A nucleotide lacking the 3′-hydroxyl group on the deoxyribose sugar. Used in the Sanger method of DNA sequencing.

Differentiation   The modification of a cell in terms of structure and/or function occurring during the course of development.

Diploid   In eukaryotes, an organism or cell with two chromosome complements, one derived from each haploid gamete.

Disease   Injury to the host that impairs host function.

Disinfectant   An agent that kills microorganisms but may also be harmful to human tissue.

Disinfection   The process of eliminating nearly all pathogens, but not all microorganisms, from inanimate objects or surfaces.

Disproportionation   The splitting of a chemical compound into two new compounds, one more oxidized and one more reduced than the original compound.

DNA fingerprinting   Use of genetic engineering to determine the origin of DNA in a sample of tissue.

DNA library   A collection of cloned DNA fragments that

in total contain genes from the entire genome of an organism; also called a *gene library.*

DNA polymerase   An enzyme that synthesizes a new strand of DNA in the 5′ → 3′ direction using an antiparallel DNA strand as a template.

Domain   The highest level of biological classification. The three domains of biological organisms are the Bacteria, the Archaea, and the Eukarya. Also used to describe a region of a protein having a distinct function.

Doubling time   The time needed for a population to double. See also *Generation time.*

Downstream position   Refers to nucleic acid sequences on the 3′ side of a given site on the DNA or RNA molecule. Compare with *Upstream position.*

Ecology   Study of the interrelationships between organisms and their environments.

Ecosystem   A community of organisms and their natural environment.

Ehrlichiosis   An emerging tick-transmitted disease caused by rickettsia of the *Ehrlichia* genus.

Electron acceptor   A substance that accepts electrons during an oxidation–reduction reaction.

Electron donor   A compound that donates electrons in an oxidation–reduction reaction.

Electron transport phosphorylation   Synthesis of ATP involving a membrane-associated electron transport chain and the creation of a proton motive force. Also called *Oxidative phosphorylation.* See also *Chemiosmosis.*

Electrophoresis   Separation of charged molecules in an electric field.

Electroporation   The use of an electric pulse to enable cells to take up DNA.

ELISA   Enzyme-linked immunosorbent assay. An immunoassay that uses specific antibodies to detect antigens or antibodies in body fluids. The antibody-containing complexes are visualized through enzyme coupled to the antibody. Addition of substrate to the enzyme–antibody–antigen complex results in a colored product.

Emerging infection   An infectious disease that has increased in incidence in the last 20 years or threatens to increase in incidence in the future.

Endemic   A disease that is constantly present in low numbers in a population. Compare with *Epidemic.*

Endergonic reaction   A chemical reaction requiring an input of energy to proceed.

Endocytosis   A process in which a particle such as a virus is taken intact into an animal cell. Phagocytosis and pinocytosis are two kinds of endocytosis.

Endoplasmic reticulum   An extensive array of internal membranes in eukaryotes.

Endospore   A differentiated cell formed within the cells of certain gram-positive bacteria that is extremely resistant to heat as well as to other harmful agents.

Endosymbiosis   The hypothesis that mitochondria, chloroplasts, and hydrogenosomes are the descendants of ancient prokaryotic organisms from the domain Bacteria.

Endotoxin   A toxin not released from the cell; bound to the cell surface or intracellular. Compare with *Exotoxin.*

Enrichment culture   Use of selective culture media and incubation conditions to isolate microorganisms from natural samples.

Enteric   Intestinal.

Enterotoxin   A toxin affecting the intestine.

Entropy   A measure of the degree of disorder in a system; entropy always increases in a closed system.

Enzyme   A catalyst, usually composed of protein, that promotes specific reactions or groups of reactions.

Epidemic   A disease occurring in an unusually high number of individuals in a population at the same time. Compare with *Endemic*.

Epidemiology   The study of the incidence and prevalence of disease in populations.

Epitope   Antigenic determinant.

Escherichia coli **O157:H7**   An emerging enterotoxigenic strain of *E. coli* spread by fecal contamination of animal or human origin to food and water.

Eukarya   The phylogenetic domain containing all eukaryotic organisms.

Eukaryote   A cell or organism having a unit membrane-enclosed (true) nucleus and usually other organelles.

Evolution   Change in a line of descent over time leading to the production of new species or varieties within a species.

Evolutionary distance   In phylogenetic trees, the sum of the physical distance on a tree separating organisms; this distance is inversely proportional to evolutionary relatedness.

Exergonic reaction   A chemical reaction that proceeds with the liberation of energy.

Exons   The coding sequences in a split gene. Contrast with *Introns*, the intervening noncoding regions.

Exotoxin   A toxin released extracellularly. Compare with *Endotoxin*.

Exponential growth   Growth of a microorganism where the cell number doubles within a fixed time period.

Exponential phase   A period during the growth cycle of a population in which growth increases at an exponential rate.

Expression   The ability of a gene to function within a cell in such a way that the gene product is formed.

Expression vector   A cloning vector that contains the necessary regulatory sequences allowing transcription and translation of a cloned gene or genes.

Extreme halophile   An organism whose growth is dependent on large amounts (generally >10%) of NaCl.

Extremophile   An organism that grows optimally under one or more chemical or physical extremes, such as high or low temperature or pH.

Facultative   A qualifying adjective indicating that an organism is able to grow in either the presence or absence of an environmental factor (for example, "facultative aerobe").

Feedback inhibition   A decrease in the activity of the first enzyme of a biochemical pathway caused by the final product of the pathway.

Fermentation   Catabolic reactions producing ATP in which organic compounds serve as both primary electron donor and ultimate electron acceptor and ATP is produced by substrate-level phosphorylation.

Fermentation (industrial)   A large-scale microbial process.

Fermenter   An organism that carries out the process of fermentation.

Fermentor   A growth vessel, usually quite large, used to culture microorganisms for the production of some commercially valuable product.

Ferredoxin   An electron carrier of low reduction potential; small protein containing iron–sulfur clusters.

Fever   A rise of body temperature above normal.

Filamentous   In the form of very long rods, many times longer than wide.

Fimbria (plural fimbriae)   Short, filamentous structure on a bacterial cell; although flagella-like in structure, generally present in many copies and not involved in motility. Plays a role in adherence to surfaces and in the formation of pellicles. See also *Pilus*.

FISH   Fluorescent *in-situ* hybridization; a process in which a cell is made fluorescent by labeling it with a specific nucleic acid probe that contains an attached fluorescent dye.

Flagellum (plural flagella)   A thin, filamentous organ of motility in prokaryotes that functions by rotating.

Flavoprotein   A protein containing a derivative of riboflavin, which functions as electron carrier in the electron transport system.

Fluorescent   Having the ability to emit light of a certain wavelength when activated by light of another wavelength.

Fluorescent antibody   Immunoglobulin molecule that has been coupled with a fluorescent molecule so that it exhibits fluorescence.

Fomites   Inanimate objects that, when contaminated with a viable pathogen, can transfer the pathogen to a host.

Food infection   Microbial infection resulting from ingestion of contaminated food.

Food poisoning   Disease resulting from ingestion of food contaminated with a toxin produced by a microorganism.

Food spoilage   Any change in a food product that makes it unacceptable to the consumer.

Frame shift   A type of mutation. Because the genetic code is read three bases at a time, if reading begins at either the second or third base of a codon, a faulty product usually results.

Free energy   Energy available to do useful work.

Fruiting body   A macroscopic reproductive structure produced by some fungi (for example, mushrooms) and some Bacteria (for example, myxobacteria), each distinct in size, shape, and coloration.

Fungi   Nonphototrophic eukaryotic microorganisms that contain rigid cell walls.

Fusion protein   The result of translation of two or more genes joined such that they retain their correct reading frames but make a single protein.

G + C base ratio   In DNA (or RNA) from any organism, the percentage of the total nucleic acid that consists of guanine plus cytosine bases (expressed as mol % GC).

Gametes   In eukaryotes, the haploid germ cells that result from meiosis.

Gas vesicle   A gas-filled structure made of protein that confers ability on a cell to float.

Gel   An inert polymer, usually made of agarose or poly-acrylamide, used for separating macromolecules such as nucleic acids and proteins by electrophoresis.

Gene   A unit of heredity; a segment of DNA specifying a particular protein or polypeptide chain, a tRNA or an rRNA.

Gene cloning   See *Molecular cloning.*

Gene disruption   Use of genetic techniques to inactivate a gene by inserting within it a DNA fragment containing an easily selectable marker. The inserted fragment is called a *cassette,* and the process of insertion, *cassette mutagenesis.*

Gene library   A collection of cloned DNA fragments that contains all the genetic information for a particular organism.

Gene therapy   Treatment of a disease caused by a dysfunctional gene by introduction of a normally functioning copy of the gene.

Generation time   Time needed for a population to double. See also *Doubling time.*

Genetic engineering   The use of *in vitro* techniques in the isolation, manipulation, recombination, and expression of DNA.

Genetic map   The arrangement of genes on a chromosome.

Genetics   Heredity and variation of organisms.

Genome   The complete set of genes present in an organism.

Genotype   The precise genetic constitution of an organism. Compare with *Phenotype.*

Genus   A taxonomic group of related species.

Germicide   A substance that inhibits or kills microorganisms.

Glycocalyx   A term that describes polysaccharide components outside of the bacterial cell wall; usually a loose network of polymer fibers extending outward from the cell.

Glycolysis   Reactions of the Embden–Meyerhof pathway in which glucose is oxidized to pyruvate.

Glycosidic bond   A type of covalent bond that links sugar units together in a polysaccharide.

Gonococcus   *Neisseria gonorrhoeae,* the gram-negative diplococcus that causes the disease gonorrhea.

Gram-negative cell   A prokaryotic cell whose cell wall contains relatively little peptidoglycan but has an outer membrane composed of lipopolysaccharide, lipoprotein, and other complex macromolecules.

Gram-positive cell   A prokaryotic cell whose cell wall consists chiefly of peptidoglycan and lacks the outer membrane of gram-negative cells.

Growth   In microbiology, an increase in cell number.

Growth-factor analog   A chemical agent that is related to and blocks the uptake or utilization of a growth factor.

Growth rate   The rate at which growth occurs, usually expressed as the generation time.

Guild   A group of metabolically related organisms.

Habitat   The location in nature where an organism resides.

Halophile   An organism requiring salt (NaCl) for growth.

Halotolerant   Capable of growing in the presence of NaCl, but not requiring it.

Hantavirus pulmonary syndrome (HPS)   An emerging acute viral disease characterized by respiratory pneumonia, obtained by transmission of hantavirus from rodents.

Haploid   An organism or cell containing only one set of chromosomes.

Hapten   A low-molecular-weight substance not inducing antibody formation itself but still able to combine with a specific antibody.

Helix   A spiral structure in a macromolecule that contains a repeating pattern.

Hemagglutination   Agglutination of red blood cells.

Hemolysins   Bacterial toxins capable of lysing red blood cells.

Hemolysis   Lysis of red blood cells.

Herd immunity   Resistance of a group to a pathogen as a result of the immunity of a large proportion of the group to that pathogen.

Heterocyst   A differentiated cyanobacterial cell that carries out nitrogen fixation.

Heteroduplex   A double-stranded DNA in which one strand is from one source and the other strand is from another, usually related, source.

Heterofermentation   Fermentation of glucose or another sugar to a mixture of reduced products.

Heterotroph   Chemoorganotroph.

Homoacetogens   Bacteria that produce acetate as the sole product of sugar fermentation or from $H_2 + CO_2$.

Homofermentation   Fermentation of glucose or other sugar leading to a single product, lactic acid.

Homologous antigen   An antigen that reacts with the antibody it has induced.

Host   An organism capable of supporting the growth of a virus or other parasite.

Host-to-host epidemic   An epidemic resulting from host-to-host contact, characterized by a gradual rise and fall in disease incidence.

Humoral immunity   An immune response involving antibodies.

Hybridization   The natural formation or artificial construction of a duplex nucleic acid molecule by complementary base pairing between two nucleic acid strands derived from different sources.

Hybridoma   The fusion of an immortal (tumor) cell with a lymphocyte to produce an immortal lymphocyte.

Hydrogen bond   A weak chemical bond between a hydrogen atom and a second, more electronegative element, usually an oxygen or nitrogen atom.

Hydrogenosome   An organelle of endosymbiotic origin in the cytoplasm of certain anaerobic eukaryotes that functions to oxidize pyruvate to $H_2 + CO_2$ + acetate.

Hydrolysis   Breakdown of a polymer into smaller units, usually monomers, by addition of water; digestion.

Hydrophobic interactions   Attractive forces between molecules due to the close positioning of nonhydrophilic portions of the two molecules.

Hydrothermal vents   Warm or hot water-emitting springs associated with crustal spreading centers on the sea floor.

Hypersensitivity   An immune reaction, usually harmful to the animal, caused either by antigen–antibody reactions or cellular immune processes (see *Allergy*).

Hyperthermophile   A prokaryote having a growth temperature optimum of 80°C or higher.

Icosahedron   A geometrical shape occurring in many virus particles, with 20 triangular faces and 12 corners.

Immobilized enzyme   An enzyme attached to a solid support over which substrate is passed and converted to product.

Immune   Able to resist infectious disease.

Immunity   The ability of an organism to resist infection.

Immunization   Induction of specific immunity by injecting antigens, antibodies, or immune cells into an animal.

Immunoblot (Western blot)   Detection of proteins immobilized on a filter by complementary reaction with specific antibody. Compare with *Southern* and *Northern blot*.

Immunodeficiency   Having a dysfunctional or completely nonfunctional immune system.

Immunogen   An antigen that can induce the production of an immune response.

Immunoglobulin   Antibody

Immunological memory   The ability to rapidly produce large quantities of specific immune cells following reexposure to a previously encountered antigen.

In vitro   In glass, away from the living organism.

In vivo   In the body, in a living organism.

Incidence   In reference to disease transmission, the number of cases of the disease in a specific subset of the population.

Induced enzyme   An enzyme subject to induction.

Induction   The process by which an enzyme is synthesized in response to the presence of an external substance, the inducer.

Infection   Growth of an organism within the body.

Infection thread   In the formation of root nodules, a cellulosic tube through which *Rhizobium* cells travel to reach and infect root cells.

Inflammation   Characteristic reaction to foreign particles and noxious stimuli, resulting in redness, swelling, heat, and pain.

Inhibition   Prevention of growth or function.

Inoculum   Material used to initiate a microbial culture.

Insertion   A genetic phenomenon in which a piece of DNA is inserted into the middle of a gene.

Insertion sequence (IS elements)   The simplest type of transposable element. Has only genes involved in transposition.

Integration   The process by which a DNA molecule becomes incorporated into another genome.

Interferon   Host-specific protein produced by virus-infected cells, which prevents viral infection of neighboring cells.

Interleukin (IL)   Soluble cytokine mediator secreted by leukocytes.

Interspecies hydrogen transfer   The process by which organic matter is degraded by the interaction of several groups of microorganisms in which H_2 production and H_2 consumption are closely coupled.

Introns   The intervening noncoding sequences in a split gene. Contrasted with *Exons*, the coding sequences.

Invasiveness   The degree to which an organism is able to spread through the body from a focus of infection.

Ionophore   A compound that can cause the leakage of ions across membranes.

Isotopes   Different forms of the same element containing the same number of protons and electrons but differing in the number of neutrons.

Joule (J)   A unit of energy equal to 10^7 ergs; 1000 Joules equal 1 kilojoule (kJ).

Kilobase (kb)   A 1000-base fragment of nucleic acid. A *kilobase pair* is a fragment containing 1000 base pairs.

Kinase   An enzyme that adds a phosphoryl group to a compound.

Lag phase   The period after inoculation of a population before growth begins.

Latent virus   A virus present in a cell, yet not causing any detectable effect.

Lateral gene transfer   The exchange of genes between and among cells in a microbial community not involving vertical transfer from one cell to its offspring. Also called *horizontal* gene transfer.

Leaching   Removal of valuable metals from ores by microbial action.

Leukocidin   A substance able to destroy phagocytes.

Leukocyte   A white blood cell.

Lichen   A fungus and an alga (or a cyanobacterium) living in symbiotic association.

Lipid   Water-insoluble organic molecules important in structure of the cytoplasmic membrane and (in some organisms) the cell wall. See also *Phospholipid*.

Lipopolysaccharide (LPS)   Complex lipid structure containing unusual sugars and fatty acids found in many gram-negative Bacteria and constituting the chemical structure of the outer membrane.

Lophotrichous   Having a tuft of polar flagella.

Lower respiratory tract   Trachea, bronchi, and lungs.

Luminescence   Production of light.

Lymph   A clear, yellowish fluid found in the lymphatic vessels that carries various white (but not red) blood cells.

Lymphocyte   A white blood cell involved in antibody formation or cellular immune responses.

Lysin   An antibody that induces lysis.

Lysis   Rupture of a cell, resulting in a loss of cell contents.

Lysogen   A prokaryote containing a prophage. See also *Temperate virus*.

Lysosome   A cell organelle containing digestive enzymes.

Macromolecule   A large molecule (polymer) formed by the connection of a number of small molecules (monomers).

Macrophage Large, noncirculating phagocytic cells involved in both phagocytosis and the antibody production process.

Magnetosomes Small particles of Fe_3O_4 present in cells that exhibit magnetotaxis (magnetic bacteria).

Magnetotaxis Directed movement of bacterial cells by a magnetic field.

Major histocompatability complex (MHC) A cluster of genes encoding cell surface proteins important for antigen presentation to T cells in the immune response.

Malignant In reference to a tumor, an infiltrating metastasizing growth no longer under normal growth control.

Mast cells Tissue cells adjoining blood vessels throughout the body that contain granules with inflammatory mediators.

Medium (plural media) In microbiology, the nutrient solution(s) used to grow microorganisms.

Megabase (Mb) One million nucleotide bases.

Meiosis In eukaryotes, reduction division, the process by which the change from diploid to haploid occurs.

Membrane Any thin sheet or layer. See especially *Cytoplasmic membrane.*

Memory cell A differentiated B lymphocyte capable of rapid conversion to an antibody-producing plasma cell on subsequent stimulation with antigen.

Mesophile Organism living in the temperature range near that of warm-blooded animals, and usually showing a growth temperature optimum between 25 and 40°C.

Messenger RNA (mRNA) An RNA molecule transcribed from DNA that contains the genetic information necessary to encode a particular protein.

Metabolism All biochemical reactions in a cell, both anabolic and catabolic.

Methanogen A methane-producing prokaryote; member of the Archaea.

Methanogenesis The biological production of methane (CH_4).

Methanotroph An organism capable of oxidizing methane.

Methylotroph An organism capable of oxidizing organic compounds that do not contain carbon–carbon bonds; if able to oxidize CH_4, also a methanotroph.

Microaerophilic Requiring O_2 but at a level lower than atmospheric.

Microenvironment The immediate physical and chemical surroundings of a microorganism.

Micrometer One-millionth of a meter, or 10^{-6} m (abbreviated μm), the unit used for measuring microorganisms.

Microorganism A microscopic organism consisting of a single cell or cell cluster, also including the viruses.

Microtubules Tubes that are the structural entity for eukaryotic flagella, have a role in maintaining cell shape, and function as mitotic spindle fibers.

Minus (negative)-strand nucleic acid An RNA or DNA strand that has the opposite sense of (would be complementary to) the mRNA of a virus.

Mitochondrion Eukaryotic organelle responsible for the processes of respiration and electron transport phosphorylation.

Mitosis A highly ordered process by which the nucleus divides in eukaryotes.

Mixotroph An organism able to assimilate organic compounds as carbon sources while using inorganic compounds as electron donors for energy metabolism.

Molds Filamentous fungi.

Molecular chaperone A protein that helps other proteins fold or refold properly.

Molecular cloning Isolation and incorporation of a fragment of DNA into a vector where it can be replicated.

Molecule Two or more atoms chemically bonded to one another.

Monoclonal antibody An antibody produced from a single clone of cells. This antibody has uniform structure and specificity.

Monocytes Circulating white blood cells that contain many lysosomes and can differentiate into macrophages.

Monomer A building block of a polymer.

Monotrichous Having a single polar flagellum.

Morbidity Incidence of disease in a population, including both fatal and nonfatal cases.

Mortality Incidence of death in a population.

Motility The property of movement of a cell under its own power.

Mushrooms Filamentous fungi that produce large, often edible structures called fruiting bodies.

Mutagen An agent that induces mutation, such as radiation or certain chemicals.

Mutant A strain differing from its parent because of mutation.

Mutation An inheritable change in the base sequence of the genome of an organism.

Mycorrhiza A symbiotic association between a fungus and the roots of a plant.

Mycoses Infections caused by fungi.

Myeloma A malignant tumor of a plasma cell (antibody-producing cell).

Natural killer (NK) cell A specialized lymphocyte that recognizes and destroys foreign cells or infected host cells in a nonspecific manner.

Neutrophil A polymorphonuclear leukocyte.

Neutrophile An organism that grows best around pH 7.

Nitrification The microbiological conversion of ammonia to nitrate.

Nitrogen fixation Reduction of nitrogen gas to ammonia by the enzyme nitrogenase.

Nodule A tumorlike structure produced by the roots of symbiotic nitrogen-fixing plants. Contains the nitrogen-fixing microbial component of the symbiosis.

Nonnucleoside reverse transcriptase inhibitor (NNRTI) A nonnucleoside compound that inhibits the action of retroviral reverse transcriptase by binding directly to the catalytic site.

Nonpolar Possessing hydrophobic (water-repelling) characteristics and not easily dissolved in water.

Nonsense mutation A mutation that changes a sense codon into one that does not code for an amino acid.

Normal flora   Microorganisms that are usually found associated with healthy body tissue.

Northern blot   Hybridization of a single strand of nucleic acid (DNA or RNA) to RNA fragments immobilized on a filter. Compare with *Southern blot* and *Western blot*.

Nosocomial infection   Hospital-acquired infection.

Nucleic acid   A polymer of nucleotides. See *Deoxyribonucleic acid* and *Ribonucleic acid*.

Nucleic acid probe   A strand of nucleic acid that can be labeled and used to hybridize to a complementary molecule from a mixture of other nucleic acids. In clinical microbiology or microbial ecology, short oligonucleotides of unique sequences used as hybridization probes for identifying pathogens or other organisms of interest.

Nucleoid   The aggregated mass of DNA that makes up the chromosome of prokaryotic cells.

Nucleoside   A nucleotide minus phosphate.

Nucleoside reverse transcriptase inhibitor (NRTI)   A nucleoside analog compound that inhibits the action of viral reverse transcriptase by competing with nucleosides.

Nucleotide   A monomeric unit of nucleic acid, consisting of a sugar, a phosphate, and a nitrogenous base.

Nucleus   A membrane-enclosed structure in eukaryotes containing the genetic material (DNA) organized in chromosomes.

Nutrient   A substance taken by a cell from its environment and used in catabolic or anabolic reactions.

Obligate   A qualifying adjective referring to an environmental factor always required for growth (for example, "obligate anaerobe").

Oligonucleotide   A short nucleic acid molecule, either obtained from an organism or synthesized chemically.

Oligotrophic   Describing a habitat in which nutrients are in low supply.

Oncogene   A gene whose expression causes formation of a tumor.

Open reading frame (ORF)   The entire length of a DNA molecule that starts with a start codon and ends with a stop codon.

Operator   A specific region of the DNA at the initial end of a gene, where the repressor protein binds and blocks mRNA synthesis.

Operon   A cluster of genes whose expression is controlled by a single operator. Typical of prokaryotic cells.

Opsonization   Promotion of phagocytosis by a specific antibody in combination with complement.

Organelle   A membrane-enclosed structure found in eukaryotic cells.

Osmosis   Diffusion of water through a membrane from a region of low solute concentration to one of higher concentration.

Outbreak   The occurrence of a large number of cases of a disease in a short period of time.

Oxic   Containing oxygen; aerobic. Usually used in reference to a microbial habitat.

Oxidation   A process by which a compound gives up electrons (or H atoms) and becomes oxidized.

Oxidation–reduction (redox) reaction   A pair of reactions in which one compound becomes oxidized while another becomes reduced and takes up the electrons released in the oxidation reaction.

Oxidative (electron transport) phosphorylation   The non-phototrophic production of ATP at the expense of a proton motive force formed by electron transport.

Oxygenic photosynthesis   Use of light energy to synthesize ATP and NADPH by noncyclic photophosphorylation with the production of O_2 from water.

Palindrome   A nucleotide sequence on a DNA molecule in which the same sequence is found on each strand but in the opposite direction.

Pandemic   A worldwide epidemic.

Parasite   An organism able to live on and cause damage to another organism.

Passive immunity   Immunity resulting from transfer of antibodies or immune cells from an immune to a nonimmune individual.

Pasteurization   A process using mild heat (usually 80°C for 15 min) to reduce the microbial level in heat-sensitive materials.

Pathogen   An organism able to inflict damage on a host it infects.

Pathogenicity   The ability of a parasite to inflict damage on the host.

Peptide bond   A type of covalent bond joining amino acids in a polypeptide.

Peptidoglycan   The rigid layer of the cell walls of Bacteria, a thin sheet composed of N-acetylglucosamine, N-acetylmuramic acid, and a few amino acids. Also called *murein*.

Periplasmic space   The area between the cytoplasmic membrane and the outer membrane in gram-negative Bacteria.

Peritrichous flagellation   Condition of having flagella attached to many places on the cell surface.

Phage   See *Bacteriophage*.

Phagemid   A cloning vector that can replicate either as a plasmid or as a bacteriophage.

Phagocyte   A body cell able to ingest and digest foreign particles.

Phagocytosis   Ingestion of particulate material such as bacteria by protozoa and phagocytic cells of higher organisms.

Phenotype   The observable characteristics of an organism. Compare with *Genotype*.

Phosphodiester bond   A type of covalent bond linking nucleotides together in a polynucleotide.

Phospholipid   Lipids containing a substituted phosphate group and two fatty acid chains on a glycerol backbone.

Photoautotroph   An organism able to use light as its sole source of energy and CO_2 as its sole carbon source.

Photoheterotroph   An organism using light as a source of energy and organic materials as carbon source.

Photophosphorylation   Synthesis of high energy phosphate bonds as ATP, using light energy.

Photosynthesis   The use of light energy to drive the incorporation of CO_2 into cell material. See also *Anoxygenic photosynthesis* and *Oxygenic photosynthesis*.

Phototaxis   Movement toward light.

Phototroph   An organism that obtains energy from light.

Phylogeny   The ordering of species into higher taxa and the construction of evolutionary trees based on evolutionary (natural) relationships.

Phytanyl   A branched-chain hydrocarbon containing 20 carbon atoms, commonly found in the lipids of Archaea.

Pilus   A fimbria-like structure that is present on fertile cells, both Hfr and F$^+$, and is involved in DNA transfer during conjugation. Sometimes called a *sex pilus*. See also *Fimbria*.

Pinocytosis   In eukaryotes, phagocytosis of soluble molecules.

Plaque   A zone of lysis or cell inhibition caused by virus infection on a lawn of cells.

Plasma   The noncellular portion of blood.

Plasma cell   A large, differentiated, short-lived B lymphocyte specializing in abundant (but short-term) antibody production.

Plasmid   An extrachromosomal genetic element that is not essential for growth and has no extracellular form.

Platelet   A noncellular disc-shaped structure containing protoplasm found in large numbers in blood and functioning in the blood clotting process.

Plus-strand nucleic acid   An RNA or DNA strand that has the same sense as the mRNA of a virus.

Point mutation   A mutation that involves one or only a very few base pairs.

Polar   Possessing hydrophilic characteristics and generally water-soluble.

Polar flagellation   Condition of having flagella attached at one end or both ends of the cell.

Poly-β-hydroxybutyrate (PHB)   A common storage material of prokaryotic cells consisting of a polymer of β-hydroxybutyrate (PHB) or other β-alkanoic acids (PHA).

Polyclonal antiserum   A mixture of antibodies to a variety of antigens or to a variety of determinants on a single antigen.

Polymer   A large molecule formed by polymerization of monomeric units.

Polymerase chain reaction (PCR)   A method used to amplify a specific DNA sequence *in vitro* by repeated cycles of synthesis using specific primers and DNA polymerase.

Polymorphonuclear leukocyte (PMN)   Motile white blood cells containing many lysosomes and specializing in phagocytosis. Characterized by a distinct segmented nucleus. Also, a neutrophil.

Polynucleotide   A polymer of nucleotides bonded to one another by phosphodiester bonds.

Polypeptide   Several amino acids linked together by peptide bonds.

Polysaccharide   A long chain of monosaccharides (sugars) linked by glycosidic bonds.

Porins   Protein channels in the outer membrane of gram-negative Bacteria through which small to medium-sized molecules can flow.

Porters   Membrane proteins that function to transport substances into and out of the cell.

Precipitation   A reaction between antibody and soluble antigen resulting in visible antibody–antigen complexes.

Prevalence   The *proportion* of individuals in a population having a disease.

Pribnow box   The consensus sequence TATAAT located approximately 10 base pairs upstream from the transcriptional start site. A binding site for RNA polymerase.

Primary antibody response   Antibodies made on first exposure to antigen; mostly of the class IgM.

Primary metabolite   A metabolite excreted during the growth phase.

Primary producer   An organism that uses light to synthesize new organic material from CO_2.

Primary structure   In an informational macromolecule, such as a polypeptide or a nucleic acid, the *precise sequence* of monomeric units.

Primary transcript   An unprocessed RNA molecule that is the direct product of transcription.

Primer   A molecule (usually a polynucleotide) to which DNA polymerase can attach the first deoxyribonucleotide during DNA replication.

Prion   An infectious agent whose extracellular form may contain no nucleic acid.

Probe   See *Nucleic acid probe*.

Prochlorophyte   A prokaryotic oxygenic phototroph that contains chlorophylls *a* and *b* but lacks phycobilins.

Prokaryote   A cell or organism lacking a nucleus and other membrane-enclosed organelles, usually having its DNA in a single circular molecule.

Promoter   The site on DNA where the RNA polymerase binds and begins transcription.

Prophage   The state of the genome of a temperate virus when it is replicating in synchrony with that of the host, typically integrated into the host genome.

Prophylactic   Treatment, usually immunological or chemotherapeutic, designed to protect an individual from a future attack by a pathogen.

Prostheca   A cytoplasmic extrusion from a cell such as a bud, hypha, or stalk.

Prosthetic group   The tightly bound, nonprotein portion of an enzyme; not the same as a *Coenzyme*.

Protease inhibitor   A compound that inhibits the action of viral protease by binding directly to the catalytic site, preventing viral protein processing.

Protein   A polymeric molecule consisting of one or more polypeptides.

Proton motive force   An energized state of a membrane created by expulsion of protons usually occurring through action of an electron transport chain. See also *Chemiosmosis*.

Protoplasm   The complete cellular contents, cytoplasmic membrane, cytoplasm, and nucleus/nucleoid.

Protoplast   A cell from which the wall has been removed.

Prototroph   The parent from which an auxotrophic mutant has been derived. Contrast with *Auxotroph*.

Protozoa   Unicellular eukaryotic microorganisms that lack cell walls.

Provirus   See *Prophage*.

Psychrophile An organism able to grow at low temperatures and showing a growth temperature optimum of <15°C.

Psychrotolerant Able to grow at low temperature but having a growth temperature optimum of <15°C.

Public health The health of the population as a whole.

Pure culture A culture containing a single kind of microorganism.

Pyogenic Pus-forming; causing abscesses.

Pyrite A common iron ore, FeS_2.

Pyrogenic Fever-inducing.

Quarantine The limitation on the freedom of movement of an infected individual to prevent spread of a disease to other members of a population.

Quaternary structure In proteins, the number and arrangement of individual polypeptides in the final protein molecule.

Radioimmunoassay An immunological assay employing radioactive antibody or antigen for the detection of antigen or antibody binding.

Radioisotope An isotope of an element that undergoes spontaneous decay with the release of radioactive particles.

Reaction center A photosynthetic complex containing chlorophyll (or bacteriochlorophyll) and other components, within which occurs the initial electron transfer reactions of photophosphorylation.

Reading-frame shift See *Frame shift.*

Recalcitrant Resistant to microbial attack.

Recombinant DNA A DNA molecule containing DNA originating from two or more sources.

Recombination Process by which genetic elements in two separate genomes are brought together in one unit.

Redox See *Oxidation–reduction reaction.*

Reduction A process by which a compound accepts electrons to become reduced.

Reduction potential (E_0') The inherent tendency, measured in volts, of a compound to act as an electron donor or an electron acceptor.

Reductive dechlorination Removal of Cl as Cl^- from an organic compound by reducing the carbon atom from C — Cl to C — H.

Reemergent infections Infectious diseases, thought to be under control, that produce new epidemics.

Regulation Processes that control the rates of synthesis of proteins. Induction and repression are examples of regulation.

Regulon A set of operons that are all controlled by the same regulatory protein (repressor or activator).

Replacement vector A cloning vector, such as a bacteriophage, in which some of the DNA of the vector can be replaced with foreign DNA.

Replication Synthesis of DNA using DNA as a template.

Repression The process by which the synthesis of an enzyme is inhibited by the presence of an external substance, the repressor.

Repressor protein A regulatory protein that binds to specific sites on DNA and blocks transcription; involved in negative control.

Reservoir In epidemiology, the organism or environment that normally harbors a pathogen.

Respiration Catabolic reactions producing ATP in which either organic or inorganic compounds are primary electron donors and organic or inorganic compounds are ultimate electron acceptors.

Response regulator protein One of the members of a two-component system; a regulatory protein that is phosphorylated by a sensor protein (see *Sensor protein*).

Restriction endonucleases (restriction enzymes) Enzymes that recognize and cleave specific DNA sequences, generating either blunt or single-stranded (sticky) ends.

Retrovirus A virus containing single-stranded RNA as its genetic material, which produces a complementary DNA by activity of the enzyme reverse transcriptase.

Reverse electron transport The energy-dependent movement of electrons against the thermodynamic gradient to form a strong electron donor from a weaker electron donor.

Reverse transcription The process of copying information found in RNA into DNA.

Reverse translation The mental process of using a codon table and the amino acid sequence of a protein to obtain a possible sequence of the mRNA or the gene that encoded the protein.

Rheumatic fever An inflammatory autoimmune disease triggered by an immune response to infection by *Streptococcus pyogenes.*

Rhizosphere The region immediately adjacent to plant roots.

Ribonucleic acid (RNA) A polymer of nucleotides connected via a phosphate–ribose backbone; involved in protein synthesis or as genetic material of some viruses.

Ribosomal RNA (rRNA) Type of RNA found in the ribosome; some rRNAs participate actively in the process of protein synthesis.

Ribosome A cytoplasmic particle composed of ribosomal RNA and protein, which is part of the protein-synthesizing machinery of the cell.

Ribozyme An RNA molecule that can catalyze a chemical reaction.

Rickettsias Obligate intracellular parasites that cause a variety of diseases, including typhus, Rocky Mountain spotted fever, and ehrlichiosis.

RNA life A hypothetical ancient life form lacking DNA and protein, in which RNA had both a genetic coding and a catalytic function.

RNA polymerase An enzyme that synthesizes RNA in the $5' \rightarrow 3'$ direction using an antiparallel DNA strand as a template.

RNA processing The conversion of a precursor RNA to its mature form.

Root nodule A tumorlike growth on certain plant roots that contains symbiotic nitrogen-fixing bacteria.

Rumen The forestomach of ruminant animals in which cellulose digestion occurs.

S-layer A paracrystalline outer wall layer composed of protein or glycoprotein and found in many prokaryotes.

Scale-up Conversion of an industrial process from a small laboratory setup to a large commercial fermentation.

Scarlet fever Disease characterized by high fever and a reddish skin rash resulting from an exotoxin produced by cells of *Streptococcus pyogenes*.

Screening Any of a number of procedures that permits the sorting of organisms by phenotype or genotype by allowing growth of some types but not others.

Secondary antibody response Antibody made on second (subsequent) exposure to antigen; mostly of the class IgG.

Secondary metabolite A product excreted by a microorganism near the end of the growth phase or during the stationary phase.

Secondary structure The initial pattern of folding of a polypeptide or a polynucleotide, usually the result of hydrogen bonding.

Secondary treatment In sewage treatment, either the aerobic or anaerobic decomposition of sewage following the removal of nondegradable objects by primary treatment.

Secretion vector A DNA vector in which the protein product is both expressed in and secreted (excreted) from the cell.

Selection Placing organisms under conditions where the growth of those with a particular genotype will be favored.

Semiconservative replication DNA synthesis yielding new double helices, each consisting of one parental and one progeny strand.

Semisynthetic penicillin A natural penicillin that has been chemically altered.

Sensitivity The lowest amount of antigen that can be detected in an immunological assay.

Sensor protein One of the members of a two-component system; a kinase found in the cell membrane that phosphorylates itself in response to an external signal and then passes the phosphoryl group to a response regulator protein (see *Response regulator protein*).

Septicemia Infection of the bloodstream by microorganisms.

Serology The study of antigen–antibody reactions *in vitro*.

Serum Fluid portion of blood remaining after the blood cells and materials responsible for clotting are removed.

Sexually transmitted disease (STD) A disease whose usual means of transmission is by sexual contact.

Shine–Dalgarno sequence A short stretch of nucleotides on a prokaryotic mRNA molecule upstream of the translational start site that binds to ribosomal RNA and thereby brings the ribosome to the initiation codon on the mRNA.

Shuttle vector A cloning vector that can replicate in two different organisms; used for moving DNA between unrelated organisms.

Sickle-cell anemia A genetic trait that confers resistance to malaria but causes a reduction in the number and efficiency of red blood cells.

Siderophore An iron chelator that can bind iron present at very low concentrations.

Signal sequence A short stretch of amino acids found at the beginning of proteins that are typically excreted from the cell. The signal sequence is usually rich in hydrophobic amino acids, which helps transport the entire polypeptide through the membrane.

Signature sequence Short oligonucleotides of unique sequence found in 16S ribosomal RNA of a particular group of prokaryotes.

Single-cell protein Protein derived from microbial cells for use as food or a food supplement.

Site-directed mutagenesis A technique whereby a gene with a specific mutation can be constructed *in vitro*.

16S rRNA A large polynucleotide ($\sim$1500 bases) that functions as a part of the small subunit of the ribosome of prokaryotes (Bacteria and Archaea) and from whose sequence evolutionary relationships can be obtained; eukaryotic counterpart, 18S rRNA.

Slime layer A diffuse mat of polymer fibers surrounding cells that appear unattached to a single cell.

Slime molds Nonphototrophic eukaryotic microorganisms lacking cell walls, which aggregate to form fruiting structures (cellular slime molds) or simply masses of protoplasm (acellular slime molds).

Solfatara A hot, sulfur-rich, generally acidic environment, commonly inhabited by hyperthermophilic Archaea.

Southern blot Hybridization of a single strand of nucleic acid (DNA or RNA) to DNA fragments immobilized on a filter. Compare with *Northern blot* and *Western blot*.

Species Of prokaryotes, a collection of closely related (>97% 16S rRNA sequence homology and >70% genomic hybridization) strains sufficiently different from all other strains to be recognized as a distinct unit.

Specificity The ability of the immune response to interact with individual antigens.

Spheroplast A spherical, osmotically sensitive cell derived from a bacterium by loss of some but not all of the rigid wall layer. If all the rigid wall layer has been completely lost, the structure is called a *Protoplast*.

Spontaneous generation The hypothesis that living organisms can originate from nonliving matter.

Spore A general term for resistant resting structures formed by many prokaryotes and fungi.

Sporozoa Nonmotile parasitic protozoa.

Stalk An elongate structure, either cellular or excreted, that anchors a cell to a surface.

Static Inhibitory.

Stationary phase The period during the growth cycle of a population in which growth ceases.

Stereoisomers Mirror image forms of two molecules having the same molecular and structural formulas.

Sterile Free of living organisms and viruses.

Sterilization Treatment resulting in death of all living organisms and viruses in a material.

Strain A population of cells of a single species all descended from a single cell; a clone.

Stromatolites Laminated microbial mats, typically built from layers of filamentous and other microorganisms which can become fossilized.

Substrate The molecule undergoing reaction with an enzyme.

Substrate-level phosphorylation Synthesis of high energy phosphate bonds through reaction of inorganic phosphate with an activated organic substrate.

Supercoil   Highly twisted form of circular DNA.

Superoxide anion (O_2^-)   A derivative of O_2 capable of oxidative destruction of cell components.

Suppressor   A mutation that restores a wild-type phenotype without altering the original mutation, usually arising by mutation in another gene.

Surveillance   Observation, recognition, and reporting of diseases as they occur.

Symbiosis   A relationship between two organisms.

Synthetic DNA   A DNA molecule that has been made by a chemical process in a laboratory.

Syntrophy   A nutritional situation in which two or more organisms combine their metabolic capabilities to catabolize a substance not capable of being catabolized by either one alone.

Systemic   Not localized in the body; an infection disseminated widely through the body is said to be systemic.

T cell   A lymphocyte responsible for antigen-specific cellular interactions. T cells are divided into functional subsets including T_C cytotoxic T cells and T_H helper T cells. T_H cells are further subdivided into T_H1 (inflammatory) and T_H2 (helper cells) that aid B cells in antibody formation.

T cell receptor   The antigen-specific receptor on the surface of T lymphocytes.

T-DNA   The segment of the *Agrobacterium* Ti plasmid that is transferred to plant cells.

Taxis   Movement toward or away from a stimulus.

Taxonomy   The study of scientific classification and nomenclature.

Temperate virus   A virus that on infection of a host does not necessarily cause lysis but whose genome may replicate in synchrony with that of the host. See *Lysogen*.

Tertiary structure   The final folded structure of a polypeptide that has previously attained secondary structure.

Tetracycline   A member of a class of antibiotics containing the four-membered naphthacene ring.

Thallasemia   A genetic trait that confers resistance to malaria, but causes a reduction in the efficiency of red blood cells by altering a red blood cell enzyme.

Thermocline   Zone of water in a stratified lake in which temperature and oxygen concentration drop precipitously with depth.

Thermophile   An organism with a growth temperature optimum between 45 and 80°C.

Ti plasmid   A conjugative plasmid present in the bacterium *Agrobacterium tumefaciens* that can transfer genes into plants.

Titer   Measure of antibody quantity.

Tolerance   In reference to immunology, the acquisition of nonresponsiveness to a molecule normally recognized by the immune system.

Toxic shock syndrome   Acute shock resulting from host response to an exotoxin produced by *Staphylococcus aureus*.

Toxigenicity   The degree to which an organism is able to elicit toxic symptoms.

Toxin   A microbial substance able to induce host damage.

Toxoid   A toxin modified so that it is no longer toxic but is still able to induce antibody formation.

Transcription   Synthesis of an RNA molecule complementary to one of the two strands of a double-stranded DNA molecule.

Transduction   Transfer of host genes from one cell to another by a virus.

Transfection   The transformation of a prokaryotic cell by DNA or RNA from a virus. Used also to describe the process of genetic transformation in eukaryotic cells.

Transfer RNA (tRNA)   A type of RNA that carries amino acids to the ribosome during translation; contains the anticodon.

Transformation   Transfer of genetic information via free DNA. Also, a process, sometimes initiated by infection with certain viruses, whereby a normal animal cell becomes a cancer cell.

Transgenic organisms   Plants or animals that stably pass on cloned DNA that has been inserted into them.

Translation   The synthesis of protein using the genetic information in a messenger RNA as a template.

Transpeptidation   The formation of peptide bonds between the short peptides present in the cell wall polymer, peptidoglycan.

Transposable element   A genetic element that has the ability to move (transpose) from one site on a chromosome to another.

Transposon   A type of transposable element that, in addition to genes involved in transposition, carries other genes; often genes conferring selectable phenotypes such as antibiotic resistance.

Transposon mutagenesis   Insertion of a transposon into a gene; this inactivates the host gene, leading to a mutant phenotype, and also confers the phenotype associated with the transposon gene.

Tuberculin test   A test for previous infection with *Mycobacterium tuberculosis* characterized by an inflammatory cell-mediated immune response.

Two-component system   A regulatory system containing a sensor protein and a response regulator protein (see *Sensor protein* and *Response regulator protein*).

Upper respiratory tract   The nasopharynx, oral cavity, and throat.

Upstream position   Refers to nucleic acid sequences on the 5' side of a given site on a DNA or RNA molecule. Compare with *Downstream position*.

Vaccination   Inoculation of a host with inactive, killed, or weakened pathogens or pathogen products to stimulate protective immunity.

Vaccine   Material used to induce specific protective immunity to a pathogen.

Vacuole   A small space in a cell that contains fluid and is surrounded by a membrane. In contrast to a vesicle, a vacuole is not rigid.

Vector   An agent, usually an insect or other animal, able to carry pathogens from one host to another. Also, a genetic element able to incorporate DNA and cause it to be replicated in another cell.

Vehicle Nonliving source of pathogens that infect large numbers of individuals; common vehicles are food and water.

Viable Alive; able to reproduce.

Viable count Measurement of the concentration of live cells in a microbial population.

Viral load The number of viral genome copies in the tissue of an infected host, providing a quantitative assessment of the amount of virus in the host.

Virion A virus particle; the virus nucleic acid surrounded by a protein coat and in some cases other material.

Viroid A small RNA molecule with viruslike properties.

Virulence Degree of pathogenicity of a parasite.

Virus A genetic element containing either DNA or RNA that replicates in cells but is characterized by having an extracellular state.

Water activity (a_w) An expression of the relative availability of water in a substance. Pure water has an a_w of 1.000.

Western blot See *Immunoblot*.

Wild type A strain of microorganism isolated from nature. The usual or native form of a gene or organism.

Wobble In reference to reading the genetic code, the concept that nonstandard base pairing is allowed between the anticodon and the third position of the codon.

Xenobiotic A completely synthetic chemical compound not naturally occurring on Earth.

Xerophile An organism adapted to growth at very low water potentials.

Yeasts Unicellular fungi.

Zoonoses Diseases primarily of animals that are occasionally transmitted to humans.

Zygote In eukaryotes, the single diploid cell resulting from the union of two haploid gametes.

Index

A-B toxins, 788, 789, 791–792
ABC system, 65, 67
Abscesses, 859
Absorbed radiation dose, 746
Abzymes, 821–822
Acceptor site, 206
Acellular slime molds, 733–734
Acetate, 498, 609–610, 622, 737
  methanogenesis, 613–615
Acetic acid, 682
Acetic acid bacteria, 405, 474–475
Acetobacter, 405, 406
Acetobacter aceti, 474
Acetobacter diazotrophicus, 717
Acetobacterium woodii, 611
Acetogenesis, 611–613, 679–680
Acetone, 406, 509
Acetyl phosphate, structure, 116
Acetyl-CoA, 127–128, 631
  structure, 127
Acetyl-CoA pathway, 546, 547, 611–613
Acetylene, structure, 31
Acetylene reduction, 638–639
N-Acetylglucosamine, 69, 72
  structure, 35
N-Acetylmuramic acid, 69
  structure, 35, 76
N-Acetyltalosaminuronic acid, 72
Acid fastness, 517–518
Acid fastness, defined, 454
Acid mine drainage, 643, 690–691
Acid pH, ferrous iron and pyrite oxidation
  at, 689–691
Acidianus, 564
Acidianus infernus, 565
Acidic pH, 154
Acidity, food preservation and, 982
Acidophiles, 136, 155, 404
Acidosis, 684
Acinetobacter, 87, 307, 328, 477, 777
Acne, 777
ACP, *see* Acyl carrier proteins
Acquired immunodeficiency syndrome, *see*
  AIDS
Acridine dyes, 315
Acridine orange, 653–654
Actin, 733
Actinomyces, 778
Actinomycetes, 519–524
Actinomycin, 196
Activated sludge sewage treatment, 418, 419
Activation, of spores, 95
Activation energy, 103, 110
Activator protein, 213, 220, 263
Active dry yeast, 408
Active immunity, vs. passive, 848
Active site, 214
  enzymatic, 110
Acute carriers, 895
Acute glomerulonephritis, 926
Acute leukemia, 285
Acute period, 894
Acyclovir, 945–946

Acyl carrier proteins, 133
Acylated homoserine lactone, 229
Adaptation, 233
Adapters, 345
Adenine, 37–38, 167–168
  structure, 298
Adenosine diphosphoglucose, 131
Adenosine phosphosulfate, 608–609
Adenosine phosphosulfate reductase, 597
Adenosine triphosphate, *see* ATP
Adenovirus, 268, 276, 280, 352
Adenylate cyclase, 227
Adherence, defined, 774
ADP, 216
ADPG, *see* Adenosine diphosphoglucose
Adrenal cortical steroids, 401
Aequorea victoria, 381
Aeration:
  citric acid, 406–407
  vinegar, 405–406
Aerobe:
  vs. anaerobes, 158
  defined, 103, 136
Aerobic fermentor, 389–390
Aerobic respiration, 121–128, 129
  net energy storage, 128
Aerobic sewage treatment processes, 418, 419
Aerosol transmission, 922, 934, 960
Aerotaxis, 85
Aerotolerant anaerobes, 158, 504
Affinity maturation, 821
Aflatoxin B1, structure, 972
Aflatoxins, 972
AFM, *see* Atomic force microscopy
African sleeping sickness, 726–727
Agammaglobulinemia, 847
Agar, 13, 22, 23
Agar diffusion method, 751
Agar plate/broth tube method, 854
Agar shake tube method, 652, 653
Agaricus bisporus, 415, 416
Age, resistance and, 796–797
Agents, 896–897
Agglutination, 842, 843, 870–871, 887
Agretope, 819
Agricultural microbiology, 26
Agriculture, role of microorganisms, 15, 16
Agrobacterium, 440, 473, 706–709
Agrobacterium rhizogenes, 707
Agrobacterium tumefaciens, 374–376, 707–709,
  961
AHL, *see* Acylated homoserine lactone
AIDS, 267, 270, 762, 847, 873–874, 913, 917,
  946–947
  defined, 947
  detection of HIV infection, 953, 955
  diagnosis, 950–951
  epidemiology, 898–900
  HIV, 947–950
  immunization, 953
  retroviruses and, 281, 285
  sexual activity and, 954
  treatment, 951–953
  vaccines, 374

Air pollution, lichen and, 705
Airborne transmission of pathogens,
  923–925, 934, 960
Akinetes, 526
Alanine, stereoisomers, 43
D-Alanine, 43, 69, 76
L-Alanine, 43, 69
Alcaligenes, 369, 466
Alcaligenes carboxydus, 467
Alcaligenes eutrophus, 703
Alcohol:
  industrial production, 387, 409–415
  production by yeast, 407
Alder tree, 716–717
Aldolase, 118, 504
Ale, 409, 410–412, 414–415
Algae, 11, 58, 724, 735–739
  in lichen, 704–705
  magnetosomes, 88
Aliphatic hydrocarbons, 632–633
Aliphatic saturated hydrocarbons, 679
Alkaline pH, 154
Alkaliphiles, 155, 404
Alkaliphilic, 549
Alkylating agents, 298
Allantoic acid, 714
Allantoin, 714
Allele, 336
Allelic exclusion, 821
Allergens, 844
Allergies, 844
Allochromatium vinosum, 457
Allochthonous organic matter, 644
Allophycocyanin, 581
Allosteric site, 214–215
Allostery, 214
Allylamines, 764
Alnus, 716–717
Alnus glutinosa, 717
Alternative nitrogenases, 635
Alternative sigma factors, 229
Alvin, 669, 671
Alvinella, 674, 675
Alvinella pompejana, 675
Amanita, 732
Amanitin, 196
Amantadine, 937
Amebiasis, 980
Amebic dysentery, 727
Ameboid movement, 727, 733
  defined, 721
American Type Culture Collection, 451
Ames, Bruce, 300
Ames test, 300–301
Amino acid activating enzymes, 201
Amino acids, 33, 40–43, 106, 131–132, 165
  codon and, 199–200
  fermentation, 510
  industrial production, 400–401
  selenocysteine, 209
  structure, 41
D-Amino acids, 42–43
L-Amino acids, 42–43

Aminoacyl-tRNA synthetases, 201, 202–203
Aminoglycoside antibiotics, 760–761
6-Aminopenicillanic acid, 396
  structure, 397
2-Aminopurine, structure, 298
Aminotransferase reaction, 131
Ammonia, 129, 469–470, 601, 714
  amino acid derivation and, 131, 132
  in nitrogen fixation, 638
Ammonia fluxes, 686
Ammonia monooxygenase, 462, 601–602
Ammonia-oxidizing bacteria, 461
Ammonification, 686
Amoeba, 725, 727
Amoeba proteus, 727
Amoebobacter purpureus, 459
AMP, 216
Amphotericin B, 974
Ampicillin, 346, 396, 760, 930, 988
  structure, 759
Amplicon, 884
α-Amylase, 403
Amylases, 627
  industrial production, 402–403
Anabaena, 90, 524–527
Anabaena azollae, 716
Anabaena flos-aquae, 90
Anabolism, 103, 130–133
  allosteric enzymes and, 215
Anacalomicrobium adetum, 492
Anaerobes, 161
  vs. aerobes, 158
  culture of, 860–861
Anaerobic jar, 159
Anaerobic respiration, 129, 605–606
Analogs, 770
Anammox, 602
Anaphase, 191
Anaphylactic shock, 844
Anaphylaxis, 844
Anatomical defenses, 797–798
Ancalochloris perflievii, 492
Ancyclobacter, 489
Ancyclobacter aquaticus, 90
Ancyclobacter linguale, 486
Anhydride bonds, 116
Animal cells, 6
Animal feed, antibiotics in, 767, 769
Animal infectivity methods, 244
Animal pathogens, pseudomonads, 473
Animals:
  as disease reservoirs, 908
  as hosts, 775
  in hydrothermal vents, 670–671
  methanotrophic symbionts, 470
  transgenic, 369, 378
Animal-transmitted diseases:
  hantavirus pulmonary syndrome, 959–960
  rabies, 957–959
Animal viruses, 238, 267–270
  adenoviruses, 280
  double-stranded RNA, 275
  herpesviruses, 277–279
  negative-strand RNA, 272–275
  positive-strand RNA, 270–272
  pox viruses, 279–280
  replication of DNA viruses, 275–277
  retroviruses, 281–285
Anopheles, 966–968
Anoxic ammonia oxidation, 602
Anoxic decomposition, 678–680
Anoxic glove boxes, 160

Anoxic sewage treatment, 417, 418
Anoxygenic photosynthesis, 456, 576, 582–587, 589
  defined, 574
Anoxygenic phototrophs, ferrous iron oxidation, 600–601
Antenna pigments, 578–580
Anthracycline, 387
Anthrax, 20, 796, 873, 916
Antibiograms, 866
Antibiotic dilution assay, 866
Antibiotic resistance, 291–292, 327, 346, 398, 765–770, see also Drug resistance
  defined, 741
  plasmids and, 317–318
  reversing, 769
Antibiotic sensitivity testing, 865–866
Antibiotics, 196, 387, 742, 755, 758–759
  antisense RNA and, 234
  ß-lactam, 759–760
  commercial production, 368, 393–396, 508
  effect on protein synthesis, 208
  isolation and characterization, 392–396
  nonmedical uses, 767, 769
  from prokaryotes, 760–762
  as protein synthesis inhibitors, 443–444
  synthesized by Streptomyces, 522–523
  worldwide production and use, 756
Antibodies, 801, 803, 844, 867, see also Immunoglobulin
  detecting protein using, 354–356
  fluorescent, 871–874
Antibody titer, 868–869
Anticancer agents, 763–765
Anticodon, 166, 200, 202, 205
Anticodon loop, 202
Antigen-antibody reactions, 841–843
Antigen-Binding Cassette transporter, see ABC system
Antigen-binding protein structure, 816–819
Antigen exposure, 833
Antigen-free vaccines, 852
Antigenic determinant, 807
Antigenic drift, 937
Antigenic shift, 274, 937
Antigen presentation, 803
Antigen-presenting cell, 802, 809, 826, 830, 833
Antigens, 801, 803, 806–808, 867
Antigen-specific T cell receptors, 811
Anti-idiotypic antibodies, 851–852, 953
Antimicrobial activity, measuring, 750–751
Antimicrobial agent, 749–751
  defined, 741
Antimicrobial drugs, search for, 770–771
Antiparallel structure, of DNA, 168
Antiporters, 65
Antisense nucleic acid, 234
Antisense RNA, 234, 369
Antiseptics, 751–753
Antiserum, 812, 849
Antiterminator protein, 261
Antitoxin, 841, 849
Antiviral chemotherapeutic agents, 762–764
Antiviral chemotherapy monitoring, 884–885
APC, see Antigen-presenting cell
Apicomplexa, 728–729
Apocrine glands, 776
Apoptosis, 831
Appendaged bacteria, 57
Appendages, 491
Applied sciences, 2

APS, see Adenosine phosphosulfate
APS reductase, 608
Aquaspirillum, 485–487
Aquaspirillum serpens, 72
Aquatic bacteria, 532–535, 538
Aquatic environments, 665–670, 738
Aquatic microbiology, 26
Aquifex, 429, 440–441, 454, 544, 571, 592
Aquifex aeolicus, 544
Aquifex pyrophilus, 543
Arbuscular mycorrhizae, 705
Archaea:
  alkaliphiles in, 155
  chromosomes in, 180
  differentiating features, 445
  evolution of, 9–11, 440, 441
  histones, 570
  lipids in, 133
  membranes, 62
  phylogeny, 545–548
  pseudopeptidoglycan in, 72–73
  RNA polymerases, 195–196
  S-layers in, 86
  thermophiles, 153
Archaeoglobales, 560–561
Archaeoglobus, 560–561, 563
Archaeoglobus lithotrophicus, 560
Arenavirus, 268
arg regulon, 332
Arginine, 218
Arginine dihydrolase, 526
Aromatic hydrocarbons, 633–634
Arsenic, 618, 694
Arthrobacter crystallopoietes, 68, 515, 516
Arthrobacter globiformis, 516
Arthropod vectors, 485
Artificial active immunity, 848
Artificial chromosomes, 350–352
Artificial passive immunity, 848
Artificially induced competence, 307–309
Ascomycetes, 731
Ascorbic acid, 406, 475
Ascospores, 338, 731
Ascus, 338, 731
-ase, enzyme naming, 112
Aseptic technique, 14–15
Ashbya gossypii, 399–400
Asian flu pandemic, 935–936
Aspartame, 400
Aspartic acid, 400
Aspergillis, 396
Aspergillus flavus, 972
Aspergillus nidulans, mutants, 292
Aspergillus niger, 406–407
Assembly, 245
Assimilative metabolism, 606
Asticcacaulis biprosthecum, 492
ATCC, see American Type Culture Collection
Athlete's foot, 972
Atomic force microscopy, 54
ATP, 116–117, 574–575
  cell yield and, 128
  glycolysis and, 118–120
  structure, 39, 116
ATP formation, proton motive force and, 125–126
ATP sulfurylase, 608
ATP synthase, 102, 125–126
ATP synthesis, see also Fermentation; Photosynthesis; Respiration
  light-mediated, 551–552
  in oxygenic photosynthesis, 589

ATPase, 102, 125–126, 432
Atrazine, structure, 699
Attachment, 245, 246
Attenuated pathogens, 796
Attenuated strains, 848
Attenuation, 224–226
Attenuator, 224
Aureobasidium pullulans, 704
Autoantibodies, 831
Autoclave, 743–744
Autocrine ability, 827
Autoimmune diseases, 839, 844–846
Autoinducer, 483
Autoinduction, 483
Autolysins, 76
Autolysis, 76
Autophosphorylation, 232
Autoradiography, 172–174
Autoreactive T cells, 831
Autotrophic carbon dioxide fixation,
    590–592
Autotrophs, 26, 104, 130, 575, 592
Autotrophy, 429, 542, 544, 586–587, 615
    acetate use and, 609–610
    Archaea, 548
    discovery, 596–597
    in hydrogen bacteria, 595
Auxotroph, 292
Avery, Oswald T., 308, 309
Avizyme, 403
Azidothymidine, *see* AZT
Azithromycin, 942, 945
Azoles, 764
Azolla, 527
Azolla pinnata, 716
Azolla-Anabaena symbiosis, 716
Azomonas, 476
Azorhizobium, 710
Azorhizobium caulinodans, 715
Azospirillum, 475, 476, 485–487
Azospirillum lipoferum, 487, 717
Azotobacter, 307
    isolation of, 650
Azotobacter chroococcum, 25, 476
AZT, 762, 951
    structure, 952

B cells, *see* B lymphocytes
B lymphocytes, 803, 811, 867
    vs. T lymphocytes, 813
B lymphocyte-T lymphocyte interaction, 834
Bacillary dysentery, 481
Bacillus, 155, 231, 307, 468, 508–509, 759
    antibiotic production, 393
    endospores and, 91
Bacillus anthracis, 20, 873, 916
Bacillus cereus, 52, 654, 791
Bacillus licheniformis, 403
Bacillus megaterium, 93, 437, 810
Bacillus popilliae, 508
Bacillus schlegelii, 467
Bacillus subtilis, 56, 68, 93, 394, 628
    attenuation in, 226
    genome, 333–334, 336
    spore formation, 94
    transformation in, 306–307
    as vector host, 352, 353
Bacillus thuringiensis, 376, 508–509
Bacitracin, 508
Back mutation, 296, 301
Bacteremia, 786, 855
Bacteria, *see also* Gram-negative bacteria,
    Gram-positive bacteria; *individual*

species; Proteobacteria
cell surface structures and inclusions,
    85–89
cell walls, 68
chromosomes in, 180
cloning mammalian genes in, 362,
    364–366
diagnostic tests, 863–864
differentiating features, 445
enzyme repression in, 218
evolution of, 9–11, 440–441
fatty acid classes, 449
magnetic, 88–89
phylogenetic tree, 453, 455
respiratory pathogens, 925–930
rumen, 682–683
transcription in, 191–195
Bacterial behavior, 83–85
Bacterial biochemistry, 26–27
Bacterial capsule, 811
Bacterial chromosome, 7
Bacterial cytology, 26
Bacterial diseases, vaccines for, 849
Bacterial flagella, 79–80
Bacterial genetics, 27, 585–586
    origins, 308–309
Bacterial iron reduction, 689
Bacterial photosynthesis, 585–586
Bacterial physiology, 26
Bacterial species, higher taxa and, 450
Bacterial taxonomy, 26
Bacterial viruses, 250
Bacteriochlorophyll, 456, 576–580
Bacteriochlorophyll *a*, 534, 577, 715
Bacteriochlorophyll *c*, 542
Bacteriocidal agents, 750
Bacteriocins, 318
Bacteriophage lambda, 9
Bacteriophage T4, 241, 246, 247
Bacteriophages, 27, 238, 318, 919
    double-stranded DNA, 255–259
    RNA, 251–252
    single-stranded filamentous DNA,
        254–255
    single-stranded icosahedral DNA,
        252–254
    temperate, 259–265
Bacteriopheophytin, 583
Bacteriorhodopsin, 551–552
Bacterioruberins, 551
Bacteriostatic agents, 750
Bacteroides, 532–533, 780, 859
Bacteroides/Flavobacterium, 454
Bacteroids, 711, 713–715
Bactoprenol, 76
Bacturia, 857
Baculovirus, 352, 353
Baker's yeast, 408
Balantidium coli, 728
Baleen whales, 684–685
Barophilic bacteria, 669
Barotolerant bacteria, 669
Basal body, 81, 82
Basalts, 664
Base analogs, 297
Base pairs, mutations and, 296
Base-pair substitutions, 294–295
Base sequences, 165
Basic pH, 154
Basidiospores, 415, 731
Basophils, 844
Bass-Becking, L. G. M., 650
Batch culture, 139

Bayer Chemical Company, 756
Bdellophages, 488
Bdelloplast, 487
Bdellovibrio, 487–489
Bdellovibrio bacteriovorus, 488
Beer, 409, 410–412, 414–415, 474
Beggiatoa, 464–465, 596, 671
Behavior, AIDS and, 954
Beijerinck, Martinus W., 24–25, 236, 475, 487,
    650
Beijerinckia, 475, 476
Beijerinckia indica, 476
Benign tumor, 269
Bent DNA, 169
Benthic algae, 665
Benzene catabolism, 634
Benzylpenicillin, 396
Bergey's Manual of Systematic Bacteriology,
    451, 454, 520
Beta oxidation, 631
Binary fission, 137, 530
    unequal, 494–495
Binary vector, 375
Binding proteins, 75
Binomial system, 451
Bioavailability, 698
Biocatalyses, 387–389
Biochemical engineer, 391
Biochemical oxygen demand, 417, 419,
    666–667
Biochemicals, formation of, 424–425
Bioconversion, 401–402, 475
Biodegradation, 369
    of petroleum, 696–698
    of synthetic polymers, 702–703
    of xenobiotics, 698–703
Biodiversity, 648
Bioenergetics, nitrification, 601–602
Biofilms, 646–647, 752
Biofuels, 17
Biogeochemical cycle, 644
Bioinformatics, 334
Biological and Toxin Weapons Convention,
    917
Biological mutagens, 299–300, 327–328
Biological warfare, 916–917, 963
Bioluminescence, 482–484
Bioluminescent bacteria, 482–484
Biomass, 17
Biomineralization, 618, 619
Biopol, 703
Bioreactor tanks, 694
Bioreactors, 417
Bioremediation, 17, 618, 619, 665, 696–698
Biosynthesis, 103
    citric acid cycle and, 127–128
    of monomers, 130–133
Biosynthetic capacity, nutrition and,
    107–108
Biosynthetic penicillins, 396
Biotechnology, 2, 15, 17, 28, 291, 344
    environmental, 369
    plant, 376–377
Bioterrorism, 916–917
Biotin, 106, 107
Biotransformation, 401–402
Birds, fermentation in, 685
Black death, 970
Black smokers, 570, 671, 673–675
Bloats, 797
Blood, 805–807
Blood agar plate, 788
Blood clotting factors, 374

Blood cultures, 855–857
Blood typing, 843
Blooms, 548, 735, 737
BOD, *see* Biochemical oxygen demand
Bog iron, 617
Boil, 928
Bonding, enzyme to carrier, 404
Bonds, 30–33
Bordetella pertussis, 850, 919, 930
Borrelia, 540, 778, 799
Borrelia burgdorferi, 180, 182, 315, 540, 878, 918, 963–964
   genome, 334
Borrelia recurrentis, 540
Botryococcus braunii, 698
Bottom yeasts, 411
Botulinum toxins, 790–791
Botulism, 512, 790, 917, 985
Bovine somatotropin, 374
Bovine spongiform encephalopathy, 286–287
Bradyrhizobium, 710, 712
Bradyrhizobium japonicum, 710
Brandy, 412
Branhamella, 859
Brevibacterium flavum, 400–401
Brewing, 410–412, 414–415
Bright-field microscopy, 50
   increasing contrast, 51–52
Broad-spectrum antibiotics, 759, 761
5-Bromouracil, structure, 298
Brown rot, 731
Brucella, 182, 830, 878
Brucella abortus, 786, 917
Brucellosis, 878
BSE, *see* Bovine spongiform encephalopathy
Bt-toxin, 376–377, 509
Bubble method, 406
Buboes, 969–970
Bubonic plague, 969–970
Budding, 491–493, 532
   virion, 283
   in yeasts, 731–732
Buffers, 155
Bulk pasteurization, 745
Bulking, 464–465, 490
Bunyavirus, 268, 878
Buoyant density method, 172
Burkholderia, 470, 472, 473
Burkholderia cepacia, 471, 472, 701
Burkitt's lymphoma, 277
Bursa of Fabricius, 811
Burst size, 246
Butanediol, structure, 478
2,3-Butanediol fermentation, 478–480
Butanediol fermentors, 482
Butanol, 406, 509–510
Butyric acid, 509–510

C₁ metabolism, 467–468
Cadmium, 694
Calcium, 105
California encephalitis, incidence, 900–901
Callus, 707
Calvin, Melvin, 590
Calvin cycle, 548, 590–591, 597, 600, 672
Calvulanic acid, 397, 398
cAMP, structure, 227
cAMP-receptor protein, 227
Campylobacter, 489, 870, 987
Campylobacter fetus, 987
Campylobacter jejuni, 769, 987
Cancer:

gene therapy and, 378
   retroviruses and, 281
   Rous sarcoma virus and, 282
   viruses and, 267, 269–270, 886
Candida, 413, 765, 784, 878
Candida albicans, 731, 782, 864, 870, 947, 948
   as hospital pathogen, 906–907
Canning, 982–983
Cap, 197–198, 267
CAP, *see* Catabolite activator protein
Capillary technique, 84
Capnocytophaga, 780
Capping, 197
Capsid, 239
Capsomers, 239
Capsules, 86–87, 785
Capture probe, 886
Carbapenem, structure, 397
Carbenicillin, 760
   structure, 759
Carbohydrates, 34
Carbon, 32–33, 104–105
   isotope studies, 660–662
   use by primitive organisms, 429
Carbon assimilation, into cell material, 603–605
Carbon cycle, 469, 675–677
Carbon dioxide, 417, 504, 675–677
   citric acid cycle and, 126–128
   fermentation and, 120
   phototrophy and, 130
   in rumen, 682
Carbon dioxide fixation:
   acetyl-CoA pathway, 611–613
   autotrophic, 592
Carbon flow, 126–128
Carbon metabolism, 574–575
   in nitrifying bacteria, 603
Carbon monoxide, 467
Carbon monoxide dehydrogenase, 467, 612
Carbon storage polymers, 87
Carbonate, 129
Carboxydotrophic bacteria, 467
Carboxysomes, 591, 871
Carcinogenesis, 300–301
Carcinogens, 270
Cardinal temperatures, 147
Caretenoids, 169, 456, 497, 519, 580–581, 810
ß-Carotene, structure, 580
Carrier-mediated transport, 64–68
Carriers, 114–116, 894, 895, 897–898, 940
Casing soil, 415
Cassette mechanism, 338–339
Cassette mutagenesis, 367–368
Catabolism, 103, 130
   allosteric enzymes and, 215
Catabolite activator protein, 227
Catabolite repression, 227–228
Catalase, 160–161, 464
Catalysis, 110–112
Catalyst, 110
Catechol, 633
Cauliflower mosaic virus, 281
Caulobacter, 493–495
Cavalli-Sforza, L. L., 309
CCR5 receptor, 948–950
CD4 cells, 948–950
cDNA, 851
   libraries, 385
   synthesis, 384
Cecum, 685
Cefixime, 942
Cefotaxime, 927

Ceftriaxone, 760, 768, 927, 942, 965
   structure, 754
Cell cultures, viral, 242
Cell division, 493, *see also* Budding
   snapping division, 515–516
Cell growth, 136–137, see also Growth, Population growth
Cell lines, 887
Cell-mediated immunity, 803, 829–830, 867–868
Cell membrane, 4
   Thermoplasma, 557
Cell number, turbidimetric measurements, 143–145
Cellobiose, 628
Cell size, 7, 58–60
Cell structure, 6–8, 57–60
Cell wall, 6, 7, 442
   algal, 738
   antibiotics and, 760
   Archaea, 72–73
   bacterial, 68
   cyanobacterial, 525
   Deinococcus, 541
   effect of pressure, 670
   fungal, 730
   gram stain and, 75
   Halobacterium, 551
   prokaryotes vs. eukaryotes and, 100
   prokaryotic, 57, 68–73
   synthesis and division, 76–77
   viruses and, 267
Cells:
   macromolecules and, 46
   microorganisms as, 4–6
   modern, 428
   origin, 4
   surface structures, 85–89
Cellular evolution, 434–436
Cellular immunity, 829–830
Cellular life forms, 427–428
Cellular slime molds, 733, 734–735
Cellulolytic bacteria, 678
Cellulose, 731, 738
   structure, 36
   synthesis, 475
Cellulose digestion, 627–628, 678
   rumen and, 681–685
Centers for Disease Control and Prevention, 850, 962
Central dogma of molecular biology, 166
   viruses and, 238
Centromere, 181, 351
Ceph-3-em, structure, 397
Cephalosporins, 396, 398, 759–760, 927
Cephalosporium, 760
Cephalosporium acremonium, 398
Cephamycins, 759
CFA, *see* Colonization factor antigen
Chagas' disease, 911
Chancre, 943–944
Chaperonins, 207, 229, 569
Charon phages, 347–348
Chemical carcinogens, 300–301
Chemical energy, 108–109
Chemical growth control, 749–751
Chemical mutagens, 297–298
Chemical signaling, 5
Chemically defined media, 107
Chemicals, as industrial product, 387
Chemiosmosis, 123–124
Chemiosmotic theory, 124
Chemokines, 827, 828

Chemolithotrophic bacteria, 664
Chemolithotrophic prokaryotes, 673
Chemolithotrophs, 104, 461–467, 575
Chemolithotrophy, 25, 129–130, 592–594
Chemoorganotrophs, 104, 575
Chemoreceptors, 75, 83–84
Chemostat, 145–146
Chemotaxis, 83–84, 231–233
Chemotaxonomy, 445
Chemotherapeutic agents, synthetic, 753–758
Chemotrophs, 104
Chickenpox, 277, 905–906, 939, 940
    incidence, 939
Childhood immunization, 850
    effect on incidence of disease, 938
Chimney, hydrothermal vents, 570, 673
Chiral isomers, 404
Chitin, 730, 738, 764
Chlamydia, 454, 529–532
Chlamydia pneumonae, 529
Chlamydia psittaci, 529, 530
Chlamydia trachomatis, 529–530, 859, 869–870, 886, 942, 944–945
    genome, 334
Chlamydial infections, 944–945
Chlamydial nongonococcal urethritis, 944–945
Chlamydia-Mycoplasma, 442
Chlamydomonas nivalis, 149–150
Chloramphenicol, 208, 318, 961
Chlorella, 426
Chlorination, 419–420, 957
Chlorine, 752, 975–976
Chlorobium, 533–537, 592
Chlorobium limicola, 535, 586, 652
Chlorobium tepidum, 535, 536, 592
Chlorochromatium aggregatum, 536, 537
Chloroflexus, 454, 542–543, 579–580, 592–593
Chloroflexus aurantiacus, 58, 542
Chlorophyll, 576–580, 735
Chlorophyll a, 525, 527, 528, 576–577, 737
Chlorophyll b, 527, 528, 577
Chlorophyll distribution, 666
Chloroplast, 58, 98, 179, 182, 431, 529, 578–579, 588, 721–722, 735
Chloroquine, 966–967
Chlorosome, 534, 542, 579
Chlortetracycline, 761
    structure, 398
Chocolate agar, 859
Cholera, 23, 482, 878, 902, 978
Cholera toxin, 791–792
Cholesterol, structure, 62
Chondromyces crocatus, 498, 499
Christispira, 538, 539
Chromatin, 178
Chromatium, 26, 459
Chromatium buderi, 88
Chromatium okenii, 458, 586, 652
Chromatium warmingii, 450
Chromatophores, 582–583
Chromobacterium, 476–477
Chromobacterium violaceum, 477
Chromosomal copy number, 79–82
Chromosomal gene transfer, F plasmid and, 324
Chromosomal painting, 656
Chromosome, 7–8, 179–181
    artificial, 350–352
    E. coli, 329–333
    in eukaryotes, 57

in prokaryotes, 57
Chromosome mobilization, 319–324
Chromosome structure, 77–79
Chronic carriers, 895
Chroococcoidiopsis, 739
Cidal agents, 750
Cidofovir, 762
Ciliates, 721, 727–728
Ciliophora, 727–728
Ciprofloxacin, 178, 758
Circular permutation, 258
Circulin, 508
Cis configuration, 305
Cis-trans test, 305–306
Cistron, 306
Citrate lyase, 592
Citric acid:
    industrial production, 406–407
    structure, 406
Citric acid cycle, 126–128, 468–469, 474, 497, 534, 547, 618–619, 629, 633
Citrobacter, 328
Citrus stubborn disease, 515
CJD, see Creutzfeldt-Jakob disease
Clams, 672
Class I MHC proteins, 824–826
Class II MHC proteins, 824–826
Class switch, 835
Classes, 450
Classic cholera, 978
Classification, 10–11
Clavam, structure, 397
Clinical microbiology, 855
Clinical specimens, isolating pathogens, 855–861
Clofazimine, 933
Clonal anergy, 831
Clonal deletion, 831
Clonal paralysis, 831
Clonal selection, 830–831
Clone, 343, 450
    selecting, 353–356
Cloned gene:
    transcriptional regulation, 357–358
    translation, 358
Cloning:
    mammalian genes in bacteria, 362, 364–366
    molecular, 290
    single-stranded filamentous DNA
        bacteriophages and, 254–255
Cloning vectors, 345, 349–352
    cosmids, 348–349
    expression vectors, 356–359
    hosts for, 352–353
    phage lambda, 347–349
    in plants, 375–376
    plasmids as, 345–347
Clostridia, 787
Clostridium, 161, 509–512, 62, 627, 759, 775, 780, 859, 983
    endospores and, 91
Clostridium aceticum, 92, 611
Clostridium acetobutylicum, 406
Clostridium botulinum, 512, 790, 797, 917, 985
Clostridium chauvei, 872
Clostridium pascui, 94
Clostridium pasteurianum, 26, 509, 635, 650
Clostridium perfringens, 512, 630, 788, 791, 860, 984–985
Clostridium septicum, 872
Clostridium sporogenes, 510
Clostridium tetani, 512, 790, 794, 848, 860
CMI, see Cell-mediated immunity

Coagulase, 318, 788, 927
Coagulation basin, 419
Coal mining, 690–691
Coal refuse pile, 557–558
Coat protein, 251
Coated-particle agglutination, 870–871
Cobalamin, 106
Coccidia, 729
    in plants, 375–376
Coccidioides immitis, 973
Coccidioidomycosis, 973
Coccobacilli, 477
Coccus, 57
Coding, cellular, 5–6
Codon, 166, 199–200, 205, 209–210
Codon usage, 358
Coenocytic, 730
Coenzyme A, 117, 126
Coenzyme B, 613
    structure, 614
Coenzyme F$_{420}$, structure, 614
Coenzyme F$_{430}$, 613
    structure, 614
Coenzyme M, 613
    structure, 614
Coenzyme Q, 124
    structure, 123
Coenzymes, 106, 111, 115–116
    redox, 613, 614
Coevolution, host and parasite, 902–905
Cointegrate, 326
Cold sterilization, 752
Cold temperature, growth and, 148–151
Cold-water disease, 533
Colicins, 318
Coliform, 957
Coliform group, 974
Coliform test, 974–975
Collagenase, 788
Colon, 780
Colonies, bacterial, 13, 23
Colonization, 774
    of pathogens, 786–788
Colonization factor antigen, 318, 785
Colony characteristics, gram-negative rods, 858
Colony count, 142
Colony-forming units, 143
Colony hybridization, 355–356
Columbia River Basin, 664
Columnaris disease, 533
Combinatorial chemistry, 770
Cometabolism, 698
Commamonas, 470, 472
Commensals, 538
Commercial fermentor, 392
Commodity chemicals, 385
Commodity ethanol, production, 412–413
Common cold, 934–935
Common-source diseases:
    foodborne, 983–990
    pathogenic fungi, 971–974
    waterborne, 976–980
Common-source epidemic, 901–902
Communication, cellular, 5
Communities, 643–645
    microbial, 4, 11–12
Community analysis, 655–658
    microbial, 438
Compatible solutes, 157–158, 551
Competence, 306–307
    artificially induced, 307–309
Competition, 647

Complement, 867
Complement fixation, 855
Complement system, 835–837
Complementarity determining regions, 818
Complementary base pairing, 184
Complementary bases, 168
Complementary DNA, synthesis, 384
Complementary metabolisms, 647
Complementation, 304–306
Complementation test, 305
Complex lipids, 36–37
Complex medium, 103
Complex viruses, 241
Composite transposons, 325
Compound light microscope, 50
Compressed yeast, 408
Compromised host, 798–799
Comptonia peregrina, 717
Computers, role in fermentor control, 391
Concatamer, 256–258, 262–263, 278
Conditionally lethal mutations, 295
Conditions, 645
Confocal scanning laser microscopy, 55
Conformational determinants, 808
Conformational protection, 635
Congenital syphilis, 923, 943
Conidia, 521, 730
Conjugation, 302–304, 319–324
  in genetic mapping, 329
  in plasmids, 316
Conjugative plasmid, 251, 303, 767
Conjugative transposons, 324, 325
Consensus sequence, 193–194
Conservative transposition, 326
Consortia, 535–537
Constitutive enzymes, 213
Contact dermatitis, 846
Contact inhibition, 243
Contagious disease, theory of, 20–21
Contamination, of culture medium, 13–14
Continuous culture, 145–146
Continuous fever, 799
Control:
  of plague, 971
  of rickettsial disease, 963
  of epidemics, 896–897
Convalescent period, 895
Cooling jacket, fermentor, 389
Cooperation, 647
Cooxidation, 698
Copper mining, 690–693
Copy number, 315
  of expression vector, 356–357
Cord factor, 519
Core, 93–94
Core enzyme, 193
Core polysaccharide, 73
Corepressor, 219
Corn, 717
Corn steep liquor, 386, 396, 398
Corn stunt disease, 515
Coronavirus, 268, 934–935
Coronilla varia, 710
Corrinoids, 614
Cortex, 93
Cortisone, 797
  industrial production, 401–402
Corynebacteria, 515–516
Corynebacterium, 515
Corynebacterium diphtheriae, 313, 443, 515,
  783, 789, 919, 929
Corynebacterium glutamicum, 400
Cosmids, 348–349

Coughing, 924
Cough-plate method, 930
Counting chambers, 141
Covalent bond, 30
Covalent modification, 216
Cowpox, 279
Coxiella burnetii, 484, 485, 963
Crenarchaeota, 441, 546
  cold-dwelling, 563
  habitats and energy metabolism, 561–564
  hyperthermophiles, 564–568
Creutzfeldt-Jakob disease, 287
Crick, Francis, 308
Crisis, 895
Cristae, 97
Crop, 685
Crop diseases, viruses and, 286
Cross-inoculation groups, 710–711
Cross-linkage, 404
Cross-reaction, 808
Cross-streak method, 393
Crotonate reduction, 625
Crown gall disease, 374–375, 706–709
CRP, *see* cAMP-receptor protein
Cryoprotectants, 151
Cryptic growth, 140
Cryptic plasmids, 316
Cryptic virus, 261
Cryptococcus neoformans, 864, 870, 947, 948,
  972
Cryptosporidiosis, 979–980
Cryptosporidium, 419, 919, 947, 948
Cryptosporidium parvum, 978–980
Crystal violet dye, 51
CSLM, *see* Confocal scanning laser
  microscopy
Cuds, 681
Culture:
  continuous, 145–146
  enrichment, 24–25
  oxygen effects, 159–160
  for testing drug sensitivity, 865–866
  viral, 242
Culture collections, industrial use, 386
Culture media, 13–14, 106–108, 514, 518,
  521, 523, 526, 532, 533, *see also*
  Isolation media
Curing, 315, 321
Cyanobacteria, 3, 89–90, 152, 429, 524–527,
  529, 704, 738–739
Cyanogen bromide, 372
Cyanophycin, 526
Cycle AMP, 227
Cyclic adenosine monophosphate, 734
Cyclic photophosphorylation, 585
Cycloheximide, 208
  structure, 754
Cysts, 475, 497
Cytochrome bc_1, 124
Cytochrome bc_1 complex, 583
Cytochrome c, 609
Cytochromes, 121–122, 547, 550
  structure, 122
Cytokines, 799, 802, 803, 805, 812, 826–828,
  984
Cytolytic toxins, 788–789
Cytomegalovirus, 947
Cytophaga, 532–534, 627
Cytophaga columnaris, 533
Cytophaga hutchinsonii, 534, 628
Cytophaga psychrophila, 533
Cytoplasm, 6
Cytoplasmic inheritance, 340–341

Cytoplasmic membrane, 6, 7
  in acidophiles, 155
  effect of temperature, 147
  function, 63–67
  in prokaryotes, 57
  in psychrophiles, 150
  structure, 60–63
  in thermophiles, 153–154
Cytoplasmic streaming, 733
Cytoplasmic structure, prokaryotes vs.
  eukaryotes, 100
Cytosine, 37–38, 167–168
Cytotoxic T lymphocytes, 812, 829
Cytotoxin, 795

2,4-D, structure, 699
DAHP synthetase, 215
DAP, *see* Diaminopimelic acid
DAPI, 653–654
Dapsone, 933
Dark-field microscopy, 53
Dark reaction, 576
DDT, 908, 968
  structure, 699
Dead Sea, 548
Death:
  from infectious diseases, 893
  leading causes in developed and
    developing countries, 912
  leading causes in U.S., 16
Death phase, 140–141
Decarboxylations, of organic acids, 623–624
Decimal reduction time, 742–743
Decimal reduction value, 746–747
Decline period, 894–895
Decomposition, 677
Deep sea, defined, 667
Deep-sea microbiology, 667–670
Deep subsurface microbiology, 663–665
Deer tick, 963–964
DEET, 966
Defensins, 795
Defined medium, 103
Degeneracy, 199–200, 294–295, 366
Dehalobacter, 700
Dehalobacterium, 700
Dehalococcoides, 700, 701
Dehalospirillum, 700
Dehydroemetine, 980
Dehydrogenation, 115
Deinococcus, 454, 540–541
Deinococcus radiodurans, 541
Delayed sequelae, 926
Delayed-type hypersensitivity, 843, 846–847
Delbrück, Max, 255
Deletions, 296
Delft School of Microbiology, 650
Delta agent, 282
Delta subdivision, 495
Demographics, infectious disease and, 917
Denaturation, 46, 742
  nucleic acid, 173
Dendritic cells, 833–834
Dengue fever, 917
Denitrification, 606–608, 686
Density-gradient centrifigation, 171–172
Dental caries, 774, 778–780
Dental plaque, 774, 777–778, 779
Deoxyribonucleic acid, see DNA
Deoxyribose, 167
  structure, 34
Depth filter, 748
Dermacentor andersoni, 962

Dermocarpa, 524
Derxia gummosa, 476
Desulfitobacterium, 700
Desulfobacter, 498
Desulfobacter postgatei, 501
Desulfobulbus, 498
Desulfobulbus propionicus, 501
Desulfococcus, 498
Desulfomonas, 498
Desulfomonile, 699, 700
Desulfonema, 498
Desulfonema limicola, 501
Desulfosarcina, 498
Desulfosarcina variabilis, 501
Desulfotomaculum, 498, 499, 618
Desulfotomaculum auripigmentum, 619
Desulfovibrio, 311, 498–499
Desulfovibrio desulfuricans, 501, 609
Desulfovibrio gigas, 619
Desulfovibrio sulfodismutans, 610
Desulfurococcales, 566–568
Desulfurococcus saccharovorans, 567
Desulfuromonas acetoxidans, 58, 501
Detecting antibody, 881
Detecting DNA, 171
Deutsche Sammlung von Mikroorganismen und Zellkulturen, 451
Developing countries, infectious disease in, 911–913
Dextran, 406, 628
Diagnosis, 864
    AIDS, 950–951
    diphtheria, 929
    hantavirus pulmonary syndrome, 960
    of infectious disease, 868–870
    malaria, 966–967
    rabies, 958–959
    rickettsial disease, 963
    streptococcal diseases, 926–927
    tuberculosis, 932
    whooping cough, 930
Diagnostic microbiology, 855
Diagnostic tests, bacteria, 480, 863–864
Diagnostic virology, 887–888
Diaminopimelic acid, 70, 76–77
    structure, 69
Diarrhea, 793, *see also* Cholera
Diatoms, 149
Diauxic growth, 227
DIC, *see* Differential interference contrast microscopy
Dichloromethane, 701
Dictyostelium, 734–735
Dictyostelium discoideum, 734
Didinium, 728
Diet, resistance and, 797
Diethers, 62
Differential interference contrast microscopy, 54
Differential media, 861–864
Differential stains, 52
Differentiation, cellular, 5
Dihydroxyacetone, 406
Diloxanide furoate, 980
Dilutions, 142–143
Dimers, 222
Dimethyl sulfide, 501, 687–688
Dimethyl sulfoxide, 619
Dimethylsulfonium propionate, structure, 157
Dinoflagellates, 735–738
Dioxygenases, 631, 633
Diphtheria, 515, 919, 929

herd immunity and, 905
Diphtheria toxin, 208, 789, 929
Dipicolinic acid, 93, 512
Diploidy, 336
Diplomonads, 441–442
Diploptene, 62
Dipstick assay, 882
Direct ELISA, 875
Direct microscopic count, 141
Direct repeats, 326
Direct RIA, 878
Direct staining method, 872
Directed evolution, 404
Disaccharides, 628–629
Discovery, bacterial pathogens, 931
Disease, 774–775, *see also* Common-source diseases, *individual diseases*
    animal-transmitted, 957–960
    foodborne, 983–990
    insect- and tick-transmitted, 960–963
    insect-transmitted, 966–971
    microorganisms and, 2
    tick-transmitted, 963–966
    waterborne, 976–980
Disease control, public health measures, 908–911, *see also* Control, Prevention, Treatment
Disease cycles, 905–906
Disease progression, 894–895
Disease reservoirs, 895–898
Disease resistance, plants, 376–377
Disinfectants, 751–753
Disproportionation, 574
Dissimilative metabolism, 606
Dissimilative sulfate-reducing bacteria, 498
Distance, in phylogenetic tree, 434
Distilled alcoholic beverages, production of, 412
Distilled beverages, 409
Distilled spirits, fermentation and, 120
Disulfide linkages, 45
Diversity, microbial, 11
Divinyl chlorophyll *a*, 528
DMS, 687
DMSO, *see* Dimethyl sulfoxide
DNA, 33, 37–40, 165, *see also* Transcription, Translation
    amplifying, 360–362
    arrangement in prokaryotes and eukaryotes, 7–8
    discovery of, 308
    enzymes affecting, 189
    invertible, 328–329
    in modern cell, 428
    in organelles, 98
    plasmids, 314
    prokaryotic vs. eukaryotes, 100
    prokaryotic, 77–79
    restriction and modification, 182–184
    restriction enzyme analysis of, 184
    secondary structure, 169–170
    structure, 167–177
    supercoiling, 177–179
    synthetic, 359–360
    transforming, 307
    triple, 234
    viral, 238
DNA animal viruses, 268
DNA binding proteins, 221–224, 278
DNA chain terminators, 951
DNA chips, 379
DNA diagnostics, 882–886
DNA endonuclease, 281

DNA fingerprinting, 362, 363
DNA gyrase, 178
DNA immunization, 851
DNA invertase, 328
DNA library, 345
DNA ligase, 187, 345
DNA polymerase, 175, 185, 186, 252, 278, 361
DNA repair, mutations and, 299–300
DNA replication, 184–185
    initiation of synthesis, 185–186
    leading and lagging strands, 186–188
    proofreading, 188–189
    replicating linear elements, 189–191
    rolling circle mechanism, 252–254
    templates and primers, 185
DNA sequence, determining, 174–176
DNA sequencing, 329
    single-stranded filamentous DNA bacteriophages and, 254–255
DNA stability, at high temperatures, 569–570
DNA transfer:
    in conjugation, 320–321
    in crown gall disease, 707–709
    by electroporation, 309–310
DNA uptake, 307
DNA vaccines, 373–374
DNA:DNA hybridization, 446–449
Domagk, Gerhard, 756
Domain, 44, 222, 439, 442–444, 450, 724
Dominant gene, 336
Dominant mutations, 305
Donor, 251
Double helix, 165, 167–177
Double-stranded breaks, 183
Double-stranded DNA bacteriophages, 255–259
Double-stranded RNA viruses, 249, 275
Doubling time, 137
Doudoroff, M., 650
Doxycycline, 942, 945, 963
DPA, *see* Dipicolinic acid
Dracunculiasis, 911
Drinking water:
    pathogen transmitted by, 977–978
    purification, 975–976
Drug abuse testing, 840
Drug cocktails, 952
Drug design, 393
    computerized, 770–771
Drug resistance, 291–292, 765–770, *see also* Antibiotic resistance
    AIDS treatment and, 952–953
    gonorrhea and, 941–943
    pathogenic resurgence and, 919
    tuberculosis, 933
Drug sensitivity, testing cultures, 865–866
Drugs, *see* Antibiotics, Fungal control, Sulfa drugs, Viral control
DSMZ, *see* Deutsche Sammlung von Mikroorganismen und Zellkulturen
Dubos, René, 745
Dunaliella, 548, 725
Dunaliella salina, 549
Dyes, 51–52, 88, 93
Dysentery, 980

Early proteins, 250, 258
Earth, evolution of, 424–427
Ebola virus, 917, 918
Eccrine glands, 776
Eclipse, 245

ECM, *see* Erythema chronicum migrans
Ecology, 644
 acetic acid bacteria, 474–475
 algae, 738
 cyanobacteria, 527
 hydrogen bacteria, 467
 isotopic fractionation in, 660–662
 malaria, 966–967
 methanotrophs, methylotrophs, 469–470
 methods, 648
 molecular microbial, 436–438
 nitrifying bacteria, 462
 plague, 969
 sexually transmitted disease, 946
 Streptomyces, 521–523
 syntrophy and methanogenesis, 677–681
Economic development, infectious disease
  and, 918
EcoRI, 183
Ecosystem, 11–12, 643
 rumen, 681–685
Ectoine, 157
 structure, 157
Ectomycorrhizae, 705
Ectothiorhodospira, 457, 548
Ectothiorhodospira halochloris, 60
Ectothiorhodospira mobilis, 459
Edible vaccines, 377
Effectors, 219
Efficiency of plating, 244
Ehrenberg, 486
Ehrlich, Paul, 756
Ehrlichia, 485
Ehrlichia chaffeensis, 961–962
Ehrlichia equi, 961
Ehrlichiosis, 485, 957, 960
Ehrlichiosis group, 961–962
Euplotes, alternative genetic codes, 210
El Tor cholera, 978
Electrochemical potential, 123–124
Electron acceptors, 112
Electron carriers, 114–116
Electron donors, 112, 586–587
Electron flow, 584
 in nitrogen fixation, 635, 637–638
 in oxygenic photosynthesis, 588–589
Electron micrography, 56
Electron microscope, 50
Electron microscopy, 55–57
 diagnostic virology and, 887–888
Electron towers, 113–114, 605
Electron transport, 121–126
 carbon dioxide release and, 126–127
Electron transport system, 127
Electrophoresis, 170, 172
Electroporation, 309–310, 374
 cloning and, 346
Elemental sulfur, 88, 595
 oxidation, 687–688
Elementary body, 531
ELISA, *see* Enzyme-linked immunosorbent
  assay
Elongation, 206
Embden-Meyerhof pathway, 118–121, 516
EMBL, 434
Emerging infectious diseases, 892, 913–920
Enantiomers, 41
Encephalitozoon, 725
End labeling, 173
End point dilution, 244
Endemic disease, 893–894
Endemic malaria, 968
Endergonic reactions, 109–111

Endocytosis, 274
 viruses and, 267–268
Endoflagella, 537–538, 943
Endogenous pyrogens, 793, 799
Endolithic algal communities, 738–739
Endomycorrhizae, 705
Endonucleolytic enzymes, 189
Endoplasmic reticulum, 722
Endospore-forming, low GC, gram-positive
  bacteria, 507–513
Endospores, 91–95, 96, 752
 prokaryotes vs. eukaryotes, 100
 vs. vegetative cells, 94
Endosymbionts, 679
Endosymbiosis, 9, 98, 430–432, 529
Endosymbiotic bacteria, 728
Endosymbiotic theory, 431
Endotoxin, 74–75, 774, 793–795
Endotoxin complex, 73
Energetics, 108–109
 chemolithotrophy, 594
 fermentation, 620–622
 hydrogen oxidation, 594–595
 sulfate reduction, 608–609
 sulfur oxidation, 596–598
 syntrophy, 625–626
Energy:
 activation, 110
 in ecosystems, 643–644
 free, 109
 role of microorganisms, 15, 17
Energy classes, of microorganisms, 104
Energy conservation, 116–117, 123–126,
  574–575
Energy metabolism, crenarchaeotes,
  563–564
Energy-rich compounds, substrate-level
  phosphorylation and, 620
Energy source, 114
Energy storage, 128
Energy yields:
 fermentative organisms, 620–621
 oxidation, 594
Enrichment bias, 652–653
Enrichment culture, 24–25
Enrichment culture technique, 648–650
Entamoeba histolytica, 727, 908
Enteric bacteria, 477–482
Enterobacter, 857, 861
Enterobacter aerogenes, 479, 482, 974
Enterobactins, 105
Enterococcus, 324, 504, 506
 as hospital pathogen, 906–907
Enterococcus faecalis, 780, 857
Enteropathogenic *E. coli*, 785
Enterotoxin, 318, 480, 774, 791–793, 795,
  983–984
Enterotoxic *E. coli*, 986
Entner-Doudoroff pathway, 471–474,
  547
Entropy, 2
Envelope proteins, 273
Enveloped viruses, 239–241
Environment:
 acetogenic, 679–680
 aquatic, 665–670, 738
 cold, 148–149
 Crenarchaeota, 561
 effect on growth, 146–147
 methanogenic, 679–680
 methanogens, 553
 role of microorganisms, 15, 17
 terrestrial, 662–665

Environmental biotechnology, 369
Environmental Protection Agency, 975
Enzyme activity, regulation of, 213–216
Enzyme catalysis, 110–111
Enzyme inclusion, 404
Enzyme induction, 218–219
 positive control, 220–221
Enzyme-linked immunosorbent assay
  (ELISA), 840, 875–878, 887, 926, 945,
  953, 960, 963, 965
 immunoblots and, 881
Enzyme repression, 218
Enzymes, 5, 42, 43, 105, 861
 affecting DNA, 189
 catalysis and, 110–112
 gene expression and, 213
 immobilized, 404
 industrial production and, 387, 402–405
 structure, 111–112
 toxic oxygen and, 160–161
 viral, 240
 in virions, 241
Enzyme-substrate complex, 110
Enzymology, nitrification, 601–602
Eosin-methylene blue agar, 861
Eosin-methylene blue agar plate, 858
EPA, *see* Environmental Protection Agency
Epibiont, 536
Epidemic diseases, 896–897
Epidemics, 891, 901–902
 influenza, 935–936
 tracing, 898–900
Epidemiological terminology, 893–895
Epidemiology, 892–893
 AIDS, 898–900
 hantaviruses, 959–960
 rabies, 957–958
Epilimnion, 666
Episomes, 315, 321
Epithelial cells, 775
Epitopes, 807, 819
Epstein-Barr virus, 277
Epulopiscium fishelsoni, 58, 59
ER, *see* Endoplasmic reticulum
Ergosterol inhibitors, 764
Eriodictyol, structure, 714
Erwinia carotovora, 478
Erythema chronicum migrans, 964–965
Erythrocytes, 805
Erythromycin, 388, 925, 927, 929, 930, 977,
  988
 structure, 761
Escaped cellular transposable elements, 281
Escherichia, 74, 328, 479–481
Escherichia coli, 8, 52, 58, 68, 250, 394, 619,
  744, 780, 791, 797, 852, 855, 857, 861,
  878, 974, 986–987
 alternative genetic codes in, 209–210
 Ames test and, 301
 cardinal temperatures, 148
 chromosomal painting, 657
 chromosome, 289, 329–333
 DAPI stain, 654
 discovery of recombination in, 308–309
 DNA, 77–78
 DNA replication in, 185–187
 electron transport chain, 124
 enterobactin, 106, 107
 enteropathic strains, 785
 flagellar genes, 81–82
 gene regulation, 231
 genetic engineering in, 381
 genome, 333–334, 336

global control systems in, 226
as hospital pathogen, 906–907
lactose metabolism, 212
MCPs, 231–232
metabolic regulation, 220–221
as model prokaryote, 331
nitrate reduction, 607
in normal flora, 777
nutritional requirements, 107–108
periplasm, 75
phage lambda and, 261
phage T7 and, 256
pili and, 86, 251
restriction-modification system, 184
ribosomal RNA, 433
ribosomes, 204
RNA polymerase, 443–444
sigma factor, 229
toxins, 318
transcription in, 192–194
transduction in, 311
transposition in, 325
tryptophan operon in, 224–226
as vector host, 352–354
Escherichia coli O157:H7, 957, 986–987
Escherichia enterotoxin, 792
Ester, structure, 442
Ester bonds, 116
Ester link, 442
Ester linkage, 62
Estrogens, 401
ETEC strains, 986
Ethanol, 17, 118, 120, 387–388, 474, 504, 682
   commodity, 412–413
      production, 474
      vinegar production and, 405
Ethanol fermentation, 625
Ether, structure, 442
Ether linkage, 62, 63
Ether-linked lipids, 543
Ethidium bromide, 171, 172, 298
Ethylene, 638
   structure, 31
Euglena, 727, 735, 737, 738
Euglenoids, 727
Eukarya, evolutionary relationships, 9–11
Eukaryote hosts, 353
Eukaryote vectors, 352
Eukaryotes, 6–7, *see also* Algae; Fungi;
      Protozoa
   cell structure, 57–58, 720–724
   chloroplasts, 578
   chromosomes in, 180
   differentiating features, 445
   DNA, 7–8, 177–178
   evolution of, 441–442
   fossil, 426
   gene expression in, 233
   genetics, 166–167, 179, 336–337
   heat resistance, 674–675
   mRNA in, 197
   nucleus and organelles, 95–99
   organelles, 430–432
   phylogenetic overview, 724–725
   vs. prokaryotes, 99–100
   recombination in, 302
   ribosome structure, 204
   RNA polymerases, 195–196
   transformation in, 310–311
   virus multiplication and, 267–268
Euryarchaeota, 441, 546–547
   Archaeoglobales, 560–561
   extremely halophilic Archaea, 548–552

hyperthermophilic euryarchaeota,
      558–560
   methanogens, 553–556
   Thermoplasmatales, 556–558
Evolution, 5
   of Archaea, 440, 441
   of Bacteria, 440–441
   cellular, 434–436
   degradation of xenobiotics and, 702
   directed, 404
   of earth and early life forms, 424–427
   of eukaryotes, 430–432, 441–442
   gene families and, 334, 336
   at high temperature, 568–572
   hyperthermophilic Archaea and, 571
   landmarks, 429
   malaria and human, 968–969
   of organelles, 440
   prochlorophytes and, 529
Evolutionary chronometers, 432–434
Evolutionary clocks, 571
Evolutionary distance, 432, 434–435
Evolutionary history, 422
Evolutionary relationships, 8–11
Excitons, 584
Excretion, of amino acids, 400
Exergonic reactions, 109–111
Exfoliating toxin, 847
Exit site, 206
Exons, 167, 181, 197
Exonuclease, 187, 188
Exosporium, 93
Exotoxin, 774, 787–791, 793
Exponential growth, 135, 137–138
Exponential phase, 140
Expressed genes, 656
Expression vector, 349, 356–359
Extein, 217
Extincting dilution, 653
Extracellular enzymes, 402
Extracellular state, 237
Extraction, of DNA, 171
Extravasation, 805
Extreme barophiles, 669
Extreme halophiles, 157
Extremely halophilic Archaea, 548–552
Extremophiles, 404, 545
Extremozymes, 403–404
Exxon Valdez, 697

F plasmid, 315–316, 330
   chromosomal gene transfer, 324
   conjugation and, 319–324
   genetic map, 315
F prime plasmids, 324
Fab fragments, 815
FACS, *see* Fluorescence-activated cell sorter
Facultative aerobes, 158, 159
Facultative chemolithotrophs, 464, 467
FAD, *see* Flavin-adenine dinucleotide
False negative reaction, 867
False positive results, ELISA and, 878
FAME, 448–449
Families, 450
Fat hydrolysis, 630–631
Fatty acid analyses, 448–449
Fatty acid oxidation, 631
Fatty acids, 34, 36–37, 133
   volatile, 681
FDA, *see* Food and Drug Administration
Fecal cultures, 859
Feedback inhibition, 214–215
FeMo-co, structure, 635

Fermentation, 117, 118–121
   brewing, 410–412, 414–415
   citric acid, 406–407
   distilled alcoholic beverages, 412
   energetics, 620–622
   enteric bacteria, 478–482
   lactic acid, 504–507
   large-scale, 389–391
   vs. respiration, 126–127
   in rumen, 681
   vinegar, 405–406
   of wine, 410
   yeast, 120, 407–409
Fermentation balance, 621–622
Fermentation products, 118
Fermentation scale-up, 391–392
Fermentative diversity, 622–624
Fermented foods, 982
Fermentor, 389
Ferredoxin, 122, 587, 621
Ferric enterobactin, 107
Ferric iron, 129, 692–693
Ferric iron reduction, 617, 689
Ferroglobus, 561
Ferroglobus placidus, 560
Ferrous iron, 129
Ferrous iron oxidation, 600–601, 689–691
FeVa-co, 635
Fever, 793, 799
Fever blisters, 277
Fibrin, 788
Fibrin clots, 927
Fibrinogen, 805
Fibrobacter, 627
Fibrobacter succinogenes, 682
Filamentous bacteria, 57
Filter sterilization, 747–749
Filtration, 975–976
Fimbriae, 86, 785
Fischerella, 524
FISH, 436–438, 655–656
Fish, mercury in, 695
Fish diseases, 533
Fission, 137
Fistula, 681, 682
Flagella, 79–82
Flagellates, 721, 726–727
Flagellin, 80
Flagellum, 57
Flash pasteurization, 745
Flatus, 782
Flavin enzymes, 124
Flavin mononucleotide, 483
   structure, 121
Flavin-adenine dinucleotide, 121
Flavobacteria, 532–533
Flavobacterium, 532–533
Flavobacterium meningosepticum, 533
Flavonoids, 714
Flavoproteins, 121, 160
Fleas, 969
Fleming, Alexander, 757
Flesh-eating bacteria, 926
Fli proteins, 81
Floc, 419
Florey, Howard, 757
Fluorescence-activated cell sorter, 874
Fluorescence microscopy, 53
Fluorescence spectrometer, 874
Fluorescent antibodies, 654–655, 855,
      871–874
Fluorescent antibody assay, 965
Fluorescent-antibody staining, 663, 926

Fluorescent dyes, 871
Fluorescent *in situ* hybridization, see FISH
Fluorescent staining, 653–654
5-Fluorocytosine, 764
Fluoroquinolone, 178, 769, 927
Flush, 415
FMN, *see* Flavin mononucleotide
Focus of infection, 243–244, 269
Folding, 569
    protein, 207–208
Fomites, 892, 901, 934
Food:
    assessing microbial content, 989–990
    microbial growth in, 980–983
    yeast as, 407–409
Food additives, 387
Food and Drug Administration, 369, 747,
    865, 983
Food industry, 886
    amino acids used in, 400
    role of microorganisms, 15, 16–17
Food infection, 986–990
Food intoxication, *see* Food poisoning
Food poisoning, 928–929, 983–985
Food preservation, 981–982
    chemical, 983
Food spoilage, defined, 957
Food supplements, 409
Foodborne diseases:
    food infection, 986–990
    food poisoning, 983–985
Foraminifera, 727
Formate, 683
Formic hydrogen lyase, 479
Formylmethionine, 443
N-Formylmethionine, 200
Formylmethionine tRNA, 204, 206
Fortified wine, 410
Fosfomycin, structure, 754
Fossil evidence, early life, 424, 426
Fossil record, algal, 738
Fowlpox, 372
Frameshift mutations, 295–296, 298
Francisella tularensis, 917
Frankia, 716–717
Free energy, 109
Free energy of formation, 109
Free-living aerobic nitrogen-fixing bacteria,
    475–476
Freezing, 150–151
Fructose, structure, 34
Fructose 1,6-bisphosphate, 118–119
Fructose bisphospate aldolase, catalytic
    cycle, 111
Fruiting bodies, 413, 495–497, 734
Frustule, 738
Fuchsin, structure, 517
Fucinic acid, 738
Fumarate, 618, 714
Functional groups, 32–33
Fungal control, 764–765
Fungal viruses, 286
Fungi, 11, 721, 724, 729–733
    classification, 729
    industrial uses, 385
    lichen, 704–705
    mushrooms, 413, 415–416
    pathogenic, 971–974
    rumen, 683–684
Fungicidal agents, 750
Fungicides, biodegradation of, 698
Fungistatic agents, 750
Fusel oils, 412

Fusidic acid, 318
Fusiform bacteria, 780–781
Fusion protein, 371–372
Fusions, expression vector, 359
Fusobacterium, 778

G, *see* Free energy
gag protein, 282, 283
Galactose utilization, 312–313
ß-Galactosidase, 212, 218
Gallionella, 493–495, 689
Gallionella ferruginea, 495, 598
Gametangia, 731
Gametes, 336
Gamma globulin, 850
γ-rays, 746
Gas gangrene, 512
Gas vesicles, 89–91, 525–526
    prokaryotes vs. eukaryotes, 100
Gasohol, 412–413
Gastroenteritis, 481, 482, 878
Gastrointestinal tract, normal flora, 780–782
GC ratios, taxonomy and, 445–446
Gel electrophoresis, studying DNA and,
    172, 174–175
Gelatin, 22
Gemmata, 532
GenBank, 434
Gene A protein, 253
Gene amplification, 395–396
Gene cassettes, 328
Gene cloning, *see* Molecular cloning
Gene disruption, 367–368
Gene expression, 8
    in *E. coli*, 331–333
    prokaryotes vs. eukaryotes, 233
Gene families, 332
    evolution and, 334, 336
Gene library, 345, 350
Gene rearrangement, 833
Gene regulation, 369
Gene synthesis, 366–367
Gene therapy, 281, 352, 369, 378, 380
Gene vaccines, 373–374
Genes:
    cloned, 357–358
    flagellar synthesis, 81–82
    function, 165
    interrupted, 181
    overlapping, 208, 252
General microbiology, 24
General recombination, 302–306
Generalized transduction, 311
Generation, 137
Generation time, 137
    calculating, 138–139
Genetic chimera, 721
Genetic code, 166, 199–201, 209–210
Genetic elements, 179–182
    replicating linear, 189–191
Genetic engineering, 17, 344, *see also*
    Molecular cloning; Vectors
    bacterial viruses and, 250
    competence and, 308
    engineered plasmids, 319
    Mu phages and, 327–328
    origins, 343
    phage M13 and, 254
    practical applications, 368–378
    protein secretion and, 208
    retroviruses and, 281, 285
    SV40 and, 276
    temperate viruses and, 264–265

    with Ti plasmid, 709
Genetic information, macromolecules and,
    165–167
Genetic map, 323
    F plasmid, 315
    plasmid R100, 317
    S. cerevisiae, 338
Genetic mapping, *see also* Genome,
    Genomics
    E. coli, 329–333
    recombinational analysis and, 302
Genetic markers, 291
Genetic recombination, 290, 302–306
Genetic stains, 655–658
Genetic switch, 263
Genetic transformation, 302, 304, 306–311
Genetically engineered immunizing agents,
    851
Genetics, 164–165, 290–291
    antigen-binding molecules, 819–822
    of bacterial photosynthesis, 585–586
    in eukaryotes, 336–337
    genetic engineering in animal and human
        genetics, 378–380
    of nitrogen fixation, 638
    origin of bacterial genetics, 308–309
    prokaryotes and eukaryotes, 166–167
    yeast, 337–341
Genistein, structure, 714
Genital specimens, 859
Genome, 179
    Archaeoglobus, 560–561
    eukaryotic, 78
    human genome project, 350–352
    prokaryotic, 333
    Thermoplasma, 557
    viral, 238–239
Genomic hybridization, 446–449
Genomics, comparative, 333–336
Genotype, designation, 291
Gentamicin, 761, 929
Genus, 10–11, 450
    genomic hybridization and classification
        of, 448
Geobacter metallireducens, 617, 619
Geochemical existence, 664
Geological time scales, 432
Geosmin, 521, 527
Geospirillum, 617
Geovibrio, 617
Geranylgeraniol, 559
Germ theory of disease, 20–21
German measles, 938, 939
Germicides, 751
Germination, 94, 95, 96
GFP, *see* Green fluorescent protein
Giardia, 441, 725
Giardia lamblia, 978–979
Giardiasis, 978–979
Gibberellin, 406
Gills, 415
Gin, 412
Gliding motility, 526
Gliding myxobacteria, 495–498
Global balance of sulfur, 688
Global control systems, 226–230
Global cycling, mercury and
    methylmercury, 694–695
Global health considerations, 911–913
Gloeothece, 524
Glove boxes, 860
Glucoamylase, 403
Gluconeogenesis, 131

Gluconic acid, 406
Gluconobacter, 405
Glucose, 34, 547
    fermentation of, 505
    glycolysis, 118–120
    stereoisomers, 43
    structure, 34
D-Glucose, 43
L-Glucose, 43
Glucose effect, 227
Glucose isomerase, 403
Glucose metabolism, 516
Glucose-6-phosphate, structure, 116
Glucosylation, 247–248, 258
Glutamate, 131
Glutamate dehydrogenase, 131
Glutamic acid, 400
D-Glutamic acid, 69
Glutamine, 131, 714
Glutamine synthesis, 714
Glutamine synthetase, 131
Glyceraldehyde 3-phosphate, 590
Glyceraldehyde-3-phosphate
    dehydrogenase, 118
Glycerol, 62, 157
    structure, 157, 533
Glycine betaine, 157
    structure, 157
Glycocalyx, 86–87, 774, 785
Glycogen, 35–36, 87, 128, 130–131, 784
    structure, 36
Glycolipids, 36
Glycolysis, 118–121, 504
Glycolytic pathway, 407, 547
Glycoproteins, 36, 241, 273
Glycosidic bond, 35–36
Glycosylation, 372
Glyoxylate cycle, 629–630
Glyphosate, 376
Gold leaching, 693–694
Golgi apparatus, 722
Gonococcus, 855
Gonorrhea, 859–860, 901, 941–943
Gonyaulax, 735, 737
Gram stain, 52
    cell wall structure and, 75
Gramicidin, 508
Gram-negative bacteria, 52, 68, 69, 477
    complement and, 836
    conjugation in, 319
    flagella, 81
    lipopolysaccharide layer, 73–75
    plasmids in, 315, 316
Gram-positive bacteria, 52, 68, 69
    actinomycetes, 519–524
    airborne transmission and, 923
    cell wall, 71
    coryneform and propionic acid bacteria,
        515–517
    endospore-forming, low GC, gram-
        positive bacteria, 507–513
    flagella, 81
    Mycobacterium, 517–519
    mycoplasmas, 513–515
    plasmids in, 315, 316
Gram-positive bacteria, nonsporulating,
    low GC, gram-positive bacteria,
    502–507
Grana, 98, 578
Great Salt Lake, 548, 549
Green algae, 149, 736
Green fluorescent protein, 381
Green phototrophic bacteria, 650–652

Green sulfur bacteria, 533–537, 541–543,
    579–580
Green sulfur bacteria consortia, 535–537
Griffith, Fred, 308
Griseofulvin, 764, 972
    structure, 754
Group I self-splicing introns, 198
Group II self-splicing introns, 198
Group translocation, 65, 66–67
Growth, *see also* Population growth
    cellular, 136–137
    cold temperatures and, 148–151
    effect of environment, 146–147
    effect of temperature, 147–148
    high temperatures and, 151–154
    measuring, 141–145
    molecular biology and, 161–162
    nutrient levels and, 646–647
    osmotic effects, 156–158
    oxygen and, 158–162
    of pathogens, 786–788
    pH and, 154–155
Growth control, chemical, 749–751
Growth curve, 135, 139–141
Growth-dependent culture methods, 854
Growth-dependent identification methods,
    861–865
Growth factor analogs, 755–758
Growth factors, 106–107
Growth inhibition, 242
Growth rate, 137
Growth rate constant, 139
GTP, 128, 204
Guanine, 37–38, 167–168
    structure, 946
Guanosine triphosphate, *see* GTP
Guilds, 643–645
Gunderson Lyme test, 965
GvpA, 91
GvpC, 91
Gyrase, 252

H antigens, 481
H$_2$ consumers, 678
Habitat, 11–12, see also Environment
    microorganisms, 645–648
Haemophilus, 307
Haemophilus influenzae, 307, 864, 870
    gene function, 332
Hairpin, 169–170
Hairpin loop, 385
Hairy root, 707
Haloanaerobium, 548
Halobacteria, 546, *see also* Extremely
        halophilic Archaea
Halobacterium, 441, 549, 550
    mutants, 292
Halobacterium halobium, 443
Halobacterium salinarum, 551, 552
Halobacteroides, 548
Halococcus, 73, 441, 550
Haloferax, 324
Halophiles, 156–157, 404
Halorhodopsin, 552
Halorhodospira, 457, 548
Halotolerant organisms, 156–157
Hansen's disease, 933
Hansenula wingei, 338, 339
Hantavirus, 919, 959
Hantavirus pulmonary syndrome, 957,
    959–960
Haploid, 79

Haptens, 806
Hashimoto's disease, 845
Hayes, William, 309
Head, virion, 246
Heat shock protein, 229
Heat shock response, 229
Heat stability, biomolecules, 568–569
Heat sterilization, 742–745
Heavy chain, immunoglobulin, 815, 818
Heavy metal transformations, 684–696
HeLa, 887
Helical symmetry, 240
Helicase activity, 302, 320
Helicases, 186
Helicobacter, 489
Helicobacter pylori, 489, 780, 988
    genome, 334
Heliobacillus, 512
Heliobacteria, 507, 512–513
Heliobacterium gestii, 512
Heliobacterium modesticaldum, 10
Heliophilum fasciatum, 512
Heliothrix, 543
α-Helix, 44, 45, 150
Helix-turn-helix motif, 223
Helper phage, 312
Hemagglutination, 273, 843
Hemagglutinin, 273
Heme, 121
Hemolysin, 318, 788, 927
α Hemolysis, 506
ß Hemolysis, 505, 925
Hemorrhagic fever with renal syndrome,
    960
Hepadnaviruses, 268, 282
Hepatitis, 852, 878, 988–989
    carriers, 897
Hepatitis A, 850, 988–989, 978
Hepatitis A virus, 270
Hepatitis B, 888, 989
Hepatitis B virus, 238, 281, 282
Hepatitis C, 989
Herbicides, biodegradation of, 698, 699
Herbivore resistance, 376
Herd immunity, 905
Heredity, autoimmune diseases and, 846
Herpes, 945–946
Herpes simplex, 277, 870–871, 947
Herpesvirus, 268, 276–279
Herpesvirus type 1, 945
Herpesvirus type 2, 945
Hershey, A. D., 255
Hesse, Walter, 22
Heterocysts, 525–526
Heterodisulfide reductase, 615
Heteroduplex regions, 302
Heterofermentative, 504
Heterologous antigen, 808
Heterotrophs, 575
Heterozygous cell, 337
Hexose catabolism, 627–628
Hexoses, 34, 130–131
Hexulose-6-P isomerase, 604
Hexulosephosphate synthase, 604
Hfr formation, 330
Hfr strains, 321–322
    interrupted mating and, 322–323
HFRS, *see* Hemorrhagic fever with renal
        syndrome
HGE, *see* Human granulocytic ehrlichiosis
High energy compounds, 116–117
High energy phosphate bonds, 116
High fructose corn syrup, 403

High temperature, polymerase chain reaction and, 361
Highly repetitive DNA, 181
Histamine, 844
Histocompatibility proteins, 824–826
Histone proteins, 276
Histones, 178, 221–222, 569
Histoplasma capsulatum, 947, 948, 973
Histoplasmosis, 973
HIV, 285, 762, 770, 798, 876, 880, 888, 947–950
   vaccines, 374
   viral load and, 883–885
HIV-ELISA, 876–878
HIV immunoblot, 880–881
HIV protease homodimer, 771
HLA complex, 819–820
HME, *see* Human monocytic ehrlichiosis
Hoatzin, 685
Holdfast, 494
Holophaga, 462
Homoacetogen, 538, 611–613
Homofermentative, 504
Homologous antigen, 808
Homologous recombination, 302–306
Homozygous cell, 337
Honeybee spiroplasmosis, 515
Hook, 81
Hooke, Robert, 17–18
Hopanoids, 61–62
Hops, 410
Horizontal gene transfer, 333
Hormogonia, 526
Horses, cellulose digestion in, 685
Hospital-acquired infection, *see* Nosocomial infection
Host, 237, 345
   cloning vectors for, 352–353
   coevolution with parasite, 902–905
   compromised, 798–799
   entry of pathogen, 784–786
   eukaryotic, 353
   plant, 377
   prokaryotic, 353
   resistance, 796–799
   viral, 242
   viral restriction and modification by, 247–248
Host community, 902–906
Host defenses, nonspecific, 796–799
Host-parasite interaction, 774
Host specificity, 784
Host-to-host epidemic, 902
Host-to-host transmission, 901
Host toxicity, 953
Hot springs, 151–153, 527
Hot vents, 670
Housekeeping genes, 181–182
HPS, *see* Hantavirus pulmonary syndrome
Human B lymphotrophic virus, 873
Human behavior, infectious disease and, 918
Human cancers, viral role, 270
Human DNase I, 374
Human exposure, rabies, 958
Human genetics, genetic engineering and, 378, 380
Human genome project, 350–352
Human granulocytic ehrlichiosis, 961
Human immunodeficiency virus, *see* HIV
Human insulin, production of, 370–372
Human leukemic cells, 873
Human leukocyte antigens, 824

Human leukocytes antigen complex, *see* HLA complex
Human monocytic ehrlichiosis, 961
Human pathogens, exotoxins, 787
Humans:
   as disease reservoirs, 908
   major histocompatibility complex proteins, 824–826
   malaria and evolution of, 968–969
   microbial impact on, 15–17
   microbial interactions, 775–776
   normal flora, 776
   normal gastrointestinal tract flora, 780–782
   normal oral cavity flora, 777–780
   normal respiratory tract flora, 782–783
   normal skin flora, 776–777
   normal urogenital tract flora, 783–784
Humic substances, as electron acceptors, 619
Humoral immunity, 803, 829
Humus, 675
Hungarian National Institute of Public Health, 769
Hungate, Robert E., 650
Hyaluronidase, 787
Hybrid ribosomes, 444
Hybridization, 329
   nucleic acid, 173–174, 177
Hybridoma technique, 837
Hybridomas, 837–838, 841
Hydrocarbon decomposition, 696–697
Hydrocarbon transformations, 632–634
Hydrocortisone, industrial production, 402
Hydrogen, 664
   syntrophic reactions and, 625
Hydrogen bacteria, autotrophy in, 595
Hydrogen bonds, 30–34, 170
   secondary structure and, 44
Hydrogen gas, 129
Hydrogen metabolism, 571
Hydrogen oxidation, 594–595
Hydrogen-oxidizing bacteria, 466–467
Hydrogen peroxide, 160
Hydrogen production, 622
Hydrogen sulfide, 129, 586, 595
   sulfate reduction, 687
Hydrogenase, 584, 621, 723
Hydrogenosome, 97, 683, 723–724
Hydrolytic enzymes, 75
Hydrophilic proteins, 60–61
Hydrophobic interactions, 32
Hydrophobic proteins, 60–61
Hydrostatic pressures, 669–670
Hydrothermal vents, 562–563, 670–675
Hydroxamate, structure, 107
Hydroxychlorophyll *a,* 587
Hydroxyl radical, 160
Hydroxylases, 632
5-Hydroxymethylcytosine, structure, 257
Hydroxypropionate cycle, 542, 592–593
Hyperimmune, 849
Hypersaline environments, 548–549
Hypersensitivity, 843
Hyperthermophiles, 148, 151–153, 403, 441, 543–544
Hyperthermophilic Archaea, 571
Hyperthermophilic Crenarchaeotes, 564–568
Hyperthermophilic Euryarchaeota, 558–561
Hypervariable regions, 818
Hyphae, 491, 730
Hyphomicrobium, 468, 491–494

Hypolimnion, 666

Icosahedron, 240
Identification, 444
Identification methods, growth-dependent, 861–865
Idiotype immunization, 851–852
Ig, *see* Immunoglobulin
IgA, *see* Immunoglobulin A
IgD, *see* Immunoglobulin D
IgE, *see* Immunoglobulin E
IgG, *see* Immunoglobulin G
IgM, *see* Immunoglobulin M
IJSB, *see International Journal of Systematic Bacteriology*
IL, *see* Interleukin
IL-8, *see* Interleukin-8
Imaging, three-dimensional, 54–55
Immediate hypersensitivity, 843–845
Immobilized enzymes, 404
Immune diseases, 843–847
Immune response, 796, 801, 804
Immune system, cells and organs, 804–806
Immune tolerance, 830–832
Immunity, 260, 803
   active vs. passive, 848
   cell-mediated, 829–830
   herd, 905
   humoral, 803, 829
   to infectious disease, 847–851
   nonspecific, 808–811
   specific, 811–812
Immunization, 803–804, 848–849, 909
   AIDS, 953
   alternative strategies, 851–852
   childhood, 850
   influenza, 937
   for travel to developing countries, 912
Immunoblot, 174, 855, 878–881, 953
Immunodiagnostic kits, 839
Immunodiagnostics, 867–870
Immunodiffusion, 842–843
Immunoelectron microscopy, 871
Immunogen, 803, 806–808, 848
Immunoglobulin, 803, 812–813, 817–818
   structure, 812–815, 818
   vs. T cell receptors, 823–824
Immunoglobulin A, 812–823, 937
Immunoglobulin D, 812, 814, 823
Immunoglobulin E, 812, 814, 823
Immunoglobulin formation, 832–835
Immunoglobulin G, 812
   heavy chains, 815
   light chains, 815
   structure, 812–815, 818
Immunoglobulin gene superfamily, 816
Immunoglobulin M, 812, 814, 815
Immunological memory, 803, 835, 868
Immunology, 26
Impeller, 389
Imperial Chemical Industries, 703
Impetigo, 925, 927
Importins, 96
In situ, defined, 643
In situ reverse transcription, 656
in vitro mutagenesis, 366–368
in vitro recombination, 345
Incidence:
   chickenpox, 939
   defined, 893
   hepatitis in U.S., 988
   legionellosis, 977
   Lyme disease, 964

malaria in U.S., 968
plague in U.S., 971
Inclusion bodies, 279
Inclusions, 57, 85–89
Incompatibility groups, 315
Incubation period, 894
Index cases, 910
Indicator organisms, 974
Indinavir, structure, 771
Indirect ELISA, 876
Indirect staining method, 872
Inducer, 219
Induction, 217–220
lytic growth of lambda after, 264
Industrial development, infectious disease and, 918
Industrial microbiology, 2, 26, 384, 385
Industrial microorganisms, products, 385–387
Industrial products, growth and formation, 387–388
Infant botulism, 790, 985
Infection, 237, 774–775, 894
respiratory, 924
Infection cycle, 894–895
Chlamydia, 531
Infection thread, 712
Infection types, 269
Infectious disease, 741, 893
in developing countries, 911–913
emerging and reemerging, 913–920
immunity to, 847–851
reportable in U.S., 910
transmission, 900–902
Inflammation, 799
Influenza virus, 241, 273–275, 913, 935–937
genome, 936–937
Information flow, 165–166
Informational macromolecules, 165
INH, *see* Isoniazid
Inhibition, defined, 741
Inhibitors, 126
of RNA polymerase action, 196
Initiation:
of DNA synthesis, 185–186
of protein synthesis, 204–206
Initiation complex, 204
Initiator reaction, 690
Inoculum, 650
Inorganic nutrients, 663
Inosinic acid, structure, 132
Insect cell lines, 353
Insecticide resistance, 376
Insecticides, 508
biodegradation of, 698, 699
Insects, as disease reservoirs, 908
Insect-transmitted diseases:
malaria, 966–969
plague, 969–971
rickettsias, 960–963
Insertion elements, 265
Insertion mutations, 367–368
Insertion sequences, 182, 296, 321, 324–329
Insertional inactivation, 346
Insertions, 296
Insulin, 216–217
production, 369–372
Integrase, 281
Integrating vectors, 357
Integration, lambda DNA, 264
Integrons, 328
Intein, 217, 335
Intercalary growth, 491

Intercalating agents, 298
Interfaces, 697
Interferon, 374, 762–764, 828
Interferon-α, 934
Interleukin, 802, 827
Interleukin-8, 828
Intermittent fever, 799
Internal coils, fermentor, 389
International Journal of Systematic Bacteriology, 451
International travel, infectious disease and, 918
Interrupted genes, 181
Interrupted mating, 322–323
Interspecies hydrogen gas transfer, 625
Intestinal gas, 782
Intestinal tract, 780–782
Intracellular state, 237
Intrinsic terminators, 194
Introns, 166, 167, 181, 197, 286, 358, 362, 365
SV40, 276
Invasion, pathogenic, 785–786
Invasiveness, 774
Inversions, 296
Invertase, 628
Inverted repeat, 169–170, 222
Invertible DNA, phase variation and, 328–329
Ion concentration, *Halobacterium salinarum*, 551
Ionic composition, hypersaline environments, 549
Ionizing radiation, 299, 746
Iridovirus, 268
Iron, 105, 476
diphtheria toxin and, 789
influence on microbial growth, 786
Iron cycle, 689–691
Iron oxidation, 598–601
Iron-oxidizing bacteria, 462–465
Irradiation, 747
Isocitrate lyase, 629
Isolation:
antibiotics, 392–396
of mutants, 291–294
Isolation culture, anaerobes, 860–861
Isolation media, 857
gonorrhea, 859
urinary tract pathogens, 857–858
wounds, 859
Isomerization, 403
Isomers, 41
Isoniazid, 757
structure, 932
Isonicotinic acid hydrazide, 932
Isoprene, structure, 62
Isosphaera, 532
Isotopic fractionation, 660–662
Isozymes, 215
Isthmix, 737
Itaconic acid, 406
Ixodes dammini, 963–964

J chain, 823
Jacob, François, 212
Jacob, Jacques, 309
Jarosite, 689–690
Jenner, Edward, 279
Juvenile diabetes, 845–846

K antigen, 480
Kaiko, 669
Kanamycin, 327, 766
structure, 761

Kanamycin resistance, 367–368
Karposi's sarcoma, 947–948
2-Keto-3-deoxyglucosephosphate aldolase, 473
α-Ketoglutarate dehydrogenase, 469
Killed cell control, 659
Killed intact HIV, 953
Kilobase, 169
Kilobase pairs, 169
Kilocalorie, 109
Kilojoule, 109
Kinases, 230–231
Kingdoms, 440, 724
bacteria, 453, 454
Kingella, 477
Kirby-Bauer procedure, 865–866
Klebsiella, 231, 328, 857, 861
Klebsiella pneumoniae, 394, 482, 637, 638, 855, 974
as hospital pathogen, 906–907
Klett units, 143–144
Kluyver, A. J., 650
Kluyveromyces, 413
Knallgas reaction, 466
Knockout mutations, 368
Koch, Robert, 20–24, 517, 846, 931, 978
Koch's postulates, 20–21
Korarchaeota, 441, 546, 547, 571–572
Kuru, 287

Labeling, nucleic acids, 173
Laboratory culture, of microorganisms, 12–15
Laboratory fermentor, 392
Laboratory flask, 392
Laboratory safety, 888–889
lac operon, 230, 357–358
lac repressor, 222
Lachnospira multiparus, 682
ß-Lactam antibiotics. 396–398, 741, 759–760, *see also* Penicillin
Lactate, 118, 682
Lactic acid, 406
Lactic acid bacteria, 106, 504–507
Lactobacillus, 25, 106, 504, 506–507
Lactobacillus acidophilus, 507, 783–784
Lactobacillus brevis, 507
Lactobacillus delbrueckii, 506, 507
Lactococcus, 504, 506
Lactoferrin, 786
Lactoperoxidase, 777
Lactose, 212, 628
LacY permease, 66
lacZ, 350, 351
Lag phase, 135, 139
Lagging strands, 186–188
Lake ecosystem, 644
Lake turnover, 667
Lakes, 666, 668
Lambda dgal, 312
Lambda promoter, 357–358
Lambda repressor, 263
Lancefield, Rebecca, 506
Lancefield groups, 506
Land use changes, infectious disease and, 918
Landfill crisis, 702–703
Large intestine, 780
microbial interactions, 775
Late proteins, 250, 258
Latent infection, 269
Latent period, in viral replication, 245–246
Lateral gene transfer, 439–440

Latex bead agglutination, 870, 926, 963
Leaching, 690, 691–694
Lead, 694
Leader sequence, 224
Leading strands, 186–188
Lecithinases, 788
Lectins, 712
Lederberg, Joshua, 308–309
Lees, 410
Leghemoglobin, 710–711, 716
Legionella pneumophila, 761, 873, 878, 913, 976–977
Legionellosis, 761, 878, 913, 976–977
Legionnaires' disease, *see* Legionellosis
Legumes, 16, 709–717
Lentinus edulus, 415–416
Lepromatous leprosy, 933
Leprosy, 878, 911, 933
Leptonema, 540
Leptospira, 540, 974
Leptospira biflexa, 540
Leptospira interrogans, 540
Leptospirillum ferrooxidans, 689, 694
Leptospirosis, 540
Leptothrix, 490–491, 689
Lethargy disease, 515
Leucine zipper, 223–224
Leuconostoc, 106, 504, 506
Leuconostoc mesenteroides, 628
    nutritional requirements, 107–108
Leucothrix, 465
Leucothrix mucor, 53, 68
Leukocidin, 927–928
Leukocidins, 789, 811
Leukocytes, 774, 799, 805
Levans, 506
Lichen acids, 704
Lichens, 704–705, 739
Life:
    defined, 423
    origin, 424–426
Life cycle:
    Chlamydia, 530–531
    fruiting myxobacterium, 497–498, 499
Life processes, microbiology and, 2
Ligase, 252
Light chain, immunoglobulin, 815, 818
Light-harvesting chlorophylls, 579
Light-harvesting I, 583
Light-harvesting II, 583
Light microscopy, 50–53
Light reaction, 576
Lignin, 678, 731
Limulus assay, for endotoxin, 794
Limulus polyphemus, 794
Line of identity, 843
Line of partial identity, 843
Linear bimolecule, 256
Linear genetic elements, replicating, 189–191
Linkers, 345
Linnaeus, 10
Lipases, 630
Lipid A, 73–74
Lipid bilayer, 48, 60, 63
Lipid monolayer, in thermophiles, 154
Lipid monomers, 133
Lipid stability, high temperatures and, 570
Lipids, 4, 34, 36–37, 442–443
    ether-linked, 543
    as microbial nutrients, 630–631
    Thermomicrobium roseum, 542
    in thermophiles, 153–154

Lipoglycan, 514, 557–558
Lipopolysaccharide, 793–794
    structure, 73
Lipopolysaccharide layer, 73–75
Lipoteichoic acids, *see* Teichoic acids
Liquid dilution methods, 652
Lister, Joseph, 20
Listeria monocytogenes, 830, 987–988
Listeriosis, 987–988
Live attenuated virus, 953
Living system characteristics, 4–5
Living type strain, 451
Ljungdahl, Lars, 611
Ljungdahl-Wood pathway, *see* Acetyl-CoA pathway
Localization, of infection, 786
Lockjaw, 791
Lophotrichous flagellation, 79, 82
Louse, 961
Lowenstein-Jensen medium, 518
Lower respiratory tract, 783
LPS, *see* Lipopolysaccharide
Luciferase, 483
Lungs, microbial interactions, 775
Luria, Salvador, 255
Luteolin, structure, 714
Lwoff, Andre, 212
Lyme disease, 540, 878, 881, 917, 918, 963–966
    incidence in U.S., 964
Lymph, 802, 805–807
Lymph nodes, 805
Lymphatic system, 805, 807
Lymphocytes, 805, 808–809
Lymphogranuloma venereum, 945
Lymphokines, 827
Lysine, 70, 400–401
    structure, 69
Lysis, 71, 242, 246, 835–836, 895
    vs. lysogenization, 263–264
Lysis protein, 251
Lysogen, 313
Lysogenization, vs. lysis, 263–264
Lysogens, 259
Lysogeny, 259, 313–314
Lysosomes, 722–723, 808
Lysozyme, 71, 111, 241, 777, 798
Lytic infection, in animals, 268–269
Lytic pathway, 261–263
Lytic viruses, 255–259

MacLeod, C. M., 308
Macrocysts, 735
Macrolide antibiotics, 761
Macrolides, 759
Macromolecules, 4
    cells and, 46
    genetic information and, 165–167
    water and, 33–34
Macronucleus, 728
Macronutrients, 104, 105
Macrophage, 795, 805, 809, 834
Macrophage activation, 830
Macrophage chemoattractant and activating factor, 828
Mad cow disease, 287
Magnesium, 105
Magnetic bacteria, 88–89
Magnetosomes, 88–89, 487
    prokaryotes vs. eukaryotes, 100
Magnetospirillum magnetotacticum, 89, 487
Magnetotactic spirilla, 487
Magnetotaxis, 88–89, 487

Magnification, 50–51
Major groove, 168, 193
Major histocompatibility complex, 824–826, 969
Malaria, 966–969
Malate, 714
Malate synthase, 629
Malathion, structure, 699
Malignant tumor, 269
Malonate, 133
Malt, 410
Maltose regulon, 220–221
Mammalian cells, gene cloning in, 353
Mammalian genes, cloning in bacteria, 362, 364–366
Mammalian proteins, 369
Manganese oxidation, 489, 490
Manganese reduction, 617–618
Mannitol, structure, 157
Marburg virus, 918
Mariana Trench, 669
Marine environments, 665–666
Marine salterns, 548
Mars, possibility of life on, 571–572
Mashing, 410
Mass culture surveys, 895
Mass serological surveys, 895, 897
Mast cells, 844
Mastigophora, 726–727
MAT, *see* Mating type locus
Mating, 337, 732
Mating type locus, 338–339
Mating types, 732
Matrix, 97
Maturation period, in viral replication, 245–246
Maturation protein, 251
Maxam-Gilbert procedure, 174–175
Maximum temperature, 147
MCAF, *see* Macrophage chemoattractant and activating factor
McCarty, M., 308
MCPs, 231–233
Measles, 878, 937–938
    infants immunized in Americas, 909
Measuring microbial activity, 750–751
    population growth, 141–145
    radioisotopes and microelectrodes, 658–660
    stable isotopes, 660–662
    virulence, 795–796
Medical microbiology, 26
Medicine, monoclonal antibodies and, 838, 841
Megabase pairs, 169
Megasphaera, 682
Meiosis, 8, 337
Mellitangium erectum, 498
Melolontha melolontha, 484
Melting, 173
    DNA, 170, 177
Membrane filter, 748
Membrane filter assay, 882
Membrane filter procedure, 974
Membrane transport proteins, structure and function, 65
Membranes, 48
    Archaeal, 62
    chemical composition, 60–61
    prokaryotes vs. eukaryotes, 100
    viral, 240–241
Memory cells, 835
Meningitis, 533

mer genes, 695
mer TPCAS genes, 695
7-Mercatoheptanoyl-threonine phosphate, 613
Mercuric reductase, 695
Mercuric resistance, 695
Mercury, 377
Mercury transformations, 694–696
Meromictic lakes, 457
Merozoites, 966
Meso-diaminopimelic acid, 69
Mesophiles, 148
Mesorhizobium, 710
Messenger RNA, *see* mRNA
Metabolic cooperation, 647
Metabolic pathways, Archaea, 547–548
Metabolism, 4, 5, 103–104
  assimilative and dissimilative, 606
  fungal, 730
  organic acid, 629–630
  in primitive organisms, 428–430
  rickettsias, 484–485
Metachromasy, 88
Metachromatic granules, 88
Metal recovery, 693
Metals, as nutrients, 105
Metaphase, 8, 191
Metastasis, 269
Metazoans, vent chimneys and, 674–675
Methane, 17, 417, 467, 553, 559–560, 677
  anoxic conversion and, 678–680
  carbon dioxide reduction and, 613
  rumen and, 682–683
Methane monooxygenase, 468, 470, 603
Methane oxidation, 603
Methanobacterium, 553, 554, 679
Methanobacterium formicicum, 615
Methanobacterium thermoautotrophicum, 311
Methanobrevibacter, 679
Methanobrevibacter ruminantium, 554, 555
Methanochondroitin, 553
Methanococcus, 553
Methanococcus igneus, 556
Methanococcus jannaschii, 333, 548, 556
Methanococcus-Methanobacterium, 441
Methanofuran, 613
  structure, 614
Methanogenesis, 553–555, 560–561, 613–617, 625
  ecology of, 677–681
  from methyl compounds and acetate, 613–615
  vs. sulfidogenesis, 680–681
Methanogens, 546, 548, 553–556, 611–613
Methanophenazine, 616
Methanoplanus, 553
Methanopterin, 613
  structure, 614
Methanopyrus, 441, 546, 559–560, 673
Methanopyrus kandleri, 10, 559
Methanosarcina, 72–73, 553
Methanosarcina barkeri, 554, 555, 615
Methanospirillum hungatii, 554
Methanothermus fervidus, 178–179, 556, 569–570
Methanothrix thermophila, 556
Methanotrophic bacteria, 696
Methanotrophic symbionts, 470
Methanotrophs, 467–470
Methanotrophy, 603–605
Methanotropic symbionts, 672
Methicillin, 760
  structure, 759
Methionine, 200, 206, 216, 371–372, 443

Methyl-accepting chemotaxis proteins, *see* MCPs
Methylation, 216, 247
  of bases, 184
  of mercury, 694–695
Methyl-CoM reductase, 560
Methyl compounds, methanogenesis from, 613–615
Methylene blue, 51
Methylene chloride, 701
Methylmercury, 694–695
Methylococcus capsulatus, 469
Methylosinus, 469, 603
Methylotrophs, 467–470
Methylotrophy, 603–605
Methyl reductase enzyme complex, 613
Methyl reductase step, 615
Metranidazole, 946, 979, 988
MHC, *see* Major histocompatibility complex, Minimum inhibitory concentration test
MHC antigens, autoimmune diseases and, 846
MHC complex, 831
MHC proteins, 816–819
MIC, *see* Minimum inhibitory concentration
Miconazole nitrate, 972
Micrasterias, 736
Microaerophiles, 158, 159
Microbial activity, 648
Microbial adaptation, infectious disease and, 918–919
Microbial communities, 11–12
Microbial diversity, 11
Microbial ecology, 26
Microbial existence, limits to, 570
Microbial fermentation, 368
Microbial growth, populations, 137–139
Microbial life, on early earth, 424
Microbial load, 742
Microbial mats, 659
Microbial plastics, 643
Microbiologie du Sol, 26
Microbiology
  history, 17–24
  subdisciplines, 26–28
Microbreweries, 415
Microbulbifer hydrolyticus, 657
Micrococcus, 503
Micrococcus luteus, 15, 654
Microcystis, 90
Microcysts, 533
Microdeletions, 294, 296
Microelectrodes, 658–660
Microenvironment, microorganisms and, 645–646
Microinsertions, 294, 296
Micronucleus, 728
Micronutrients, 104–106
Microorganisms:
  energy classes, 104
  in hydrothermal vents, 671
  impact on humans, 15–17
  role in web of life, 3–4
  size and, 7
Microscopy:
  atomic force, 54
  bright-field, 50–52
  confocal scanning laser, 55
  dark-field, 53
  differential contrast, 54
  electron, 55–57, 887–888
  fluorescence, 53

immunoelectron, 871
  light, 50–53
  phase contrast, 52–53
  scanning electron 55–56, 749
Microsporidia, 441–442, 725
Microtiter plate, 875
Microtubules, 191
Microtus pennsylvanicus, 484
Microwaves, 745–746
Middle proteins, 258
Milky disease, 508
Mineral soils, 662
Minimum inhibitory concentration, 750, 865
Minimum inhibitory concentration test, 740
Minimum temperature, 147
Mining, 334, 335
Minor groove, 168
Minus-strand nucleic acid, 237
Missense mutations, 295
Mistranslation, 210
Mitchell, Peter, 124
Mites, 963
Mitochondria, 58, 97, 98, 179, 182, 431, 721–722
  alternative genetic codes in, 209
Mitochondrial inheritance, in yeast, 340–341
Mitomycin C, structure, 754
Mitosis, 8, 57, 191, 337
Mixed-acid fermentation, 478–479
Mixotrophs, 575
Mixotrophy, 464
Moderately repetitive DNA, 181
Modern cell, 428
Modified genes, 366
Modified Thayer-Martin agar, 859
Modulon, 226
Moisture content, food spoilage and, 981
Molds, 729, 730–731
Molecular adaptations:
  to psychrophily, 150
  to thermophily, 153–154
Molecular biology, 27
  central dogma of, 166, 238
  growth and, 161–162
  of herpesviruses, 277–279
Molecular chaperones, 207, 229, 239, 287
Molecular cloning, 290, 329, 344, 345
Molecular fingerprinting, 448
Molecular microbial ecology, 436–438
Molecular oxygen, 631–632
Molecular taxonomy, 445, 446–449
Molecular weight
  of DNA, 168–169
  of immunogens, 806–807
Molluscs, 538
Molybdenum, 476
Monensin, structure, 754
Monocistronic mRNA, 166, 267, 273
Monoclonal antibodies, 837–841
  immunodiagnostics and, 869–870
Monocytes, 805, 809
Monod, Jacques, 212
Monolactam, structure, 397
Monolayers, 62, 242
Monomer stability, high temperatures and, 570
Monomers, 33
  biosynthesis of, 130–133
Monooxygenases, 631, 633
Monosodium glutamate, 400
Monuron, structure, 699
Moraxella, 477
Morbidity, 894

Mortality, 894
Mosquitoes, 966–968
Most probable number technique, 652, 974
Mot proteins, 81
Motifs, 825–826
Motility, 79–82
  algae, 738
  prokaryotes vs. eukaryotes, 100
  protozoa, 726
  spirochetes, 537–538
mRNA, 39, 163, 165–167, 191–192, 195, 219
  formation after viral infection, 248
  identifying, 656
  operons and, 196
  reaching gene via, 362, 364–365
mRNA classes, 278
MS2, 251–252
MSG, see Monosodium glutamate
Mucosa-associated lymphoid tissue, 811
Mucous membranes, 775, 797
Multibacillary leprosy, 933
Multiple cloning site, 347
Mumps, 878, 938
Muramic acid, 442
Murein, see Peptidoglycan
Murine toxin, 970
Mushrooms, 413, 415–416, 729, 731–734
Mussels, 470, 672
Must, 410
Mutagenesis, 300–301
  cassette, 367–368
  in vitro, 366–368
  site-directed, 366–368
  transposon, 327–328
Mutagens, 297–300
Mutant genes, synthesis, 367
Mutants, 291–294
  isolation of, 291–294
  types, 294
Mutation rates, 296–297
Mutations, 188, 291–294
  molecular basis, 294–297
Mutator phage, 265–267
Mutz, A., 597
Mycelium, 519, 730
Mycobacterium, 517–519, 632
Mycobacterium avium, 518, 935
Mycobacterium bovis, 519, 933–934
Mycobacterium chelonae, 935
Mycobacterium gordonae, 519
Mycobacterium kansasii, 519, 935
Mycobacterium leprae, 217, 810, 830, 878, 933
Mycobacterium marinum, 519
Mycobacterium paraffinicum, 519
Mycobacterium scrofulaceum, 935
Mycobacterium smegmatis, 394
Mycobacterium tuberculosis, 22, 23–24, 515,
    518–519, 752, 757, 775, 810, 830, 846,
    864, 867, 878, 883, 922, 931–934
  genome, 334
  transposition in, 325
Mycolic acid, 517
Mycoplasma, alternative genetic codes, 210
Mycoplasma genitalium, 514
  gene function, 332
  genome, 333
Mycoplasma mycoides, 514
Mycoplasma pneumoniae, genome, 333
Mycoplasmas, 71–72, 513–515
Mycorrhizae, 705–706
Mycoses, 957, 972–974
Mycotoxins, 972
Myxobacteria, 495–498

Myxococcus fulvus, 498, 499
Myxococcus stipitatas, 498
Myxococcus xanthus, 496, 499
  genome, 333
Myxoma virus, coevolution with rabbits,
    902–905
Myxospores, 496, 497

NAD+, see Nicotinamide adenine
    dinucleotide
NADH, 118, 126
NADH dehydrogenases, 121
NADP+, 115
NAD-phosphate, see NADP+
Naked viruses, 240
Nalidixic acid, 178, 758
  structure, 754
Nasopharyngeal swab, 856
Natronobacterium, 549
Natronococcus, 549
Natronosomonas, 549
Natural active immunity, 848
Natural antibodies, 843
Natural killer cells, 829
Natural passive immunity, 848
Natural penicillins, 396
Negative control, 220
Negative selection, 293, 831, 832
Negative staining, 55–56
Negative-strand RNA virus, 248–249,
    272–275
Negative supercoiling, 177
Negri bodies, 958
Neisseria, 307, 476–477
Neisseria gonorrhoeae, 760, 768, 784, 785, 857,
    859, 864, 868, 870, 885–886, 901,
    941–943
Neisseria meningitidis, 859
Neocalimastix, 683
Neomycin, 210, 327, 761, 766
Neoplasm, 269
Neuraminadases, 241, 274
Neurotoxins, 526
Neutralization, 841
Neutrophiles, 155
Neutrophils, 808–809
Nevirapine, 762
  structure, 952
New England Biolabs Company, 359
Newcastle disease, 372
NGU, see Chlamydial nongonococcal
    urethritis
Niche, 645
Nick, 177, 178, 253–254, 302
Nickel, 467
Nicotinamide adenine dinucleotide, 115
  structure, 114
nif regulon, 637–638
Nitrapyrin, 686
Nitrate, 129, 501
Nitrate reductase, 607
Nitrate reduction, 606–608
Nitric oxide synthase, 809
Nitrification, 25, 461, 597, 601–603
  ammonia fluxes and, 686
Nitrifying bacteria, 461–462, 601, 656
  carbon metabolism, 603
Nitrite, 601
Nitrite-oxidizing bacteria, 461
Nitrobacter, 461, 462, 601
Nitrobacter winogradskyi, 462
Nitrogen, 104–105
  redox cycle, 685

Nitrogen compounds, oxidation states, 606
Nitrogen cycle, 685–686
Nitrogen fixation, 634–639, 685–686, 704
  cyanobacteria and, 525–527
  legumes, 709–717
Nitrogen fixing, 467, 509
Nitrogen-fixing bacteria, 105, 475–476
Nitrogenase system, 637
Nitrogenases, 636–638, 714
Nitrosifyers, 461
Nitrosococcus oceani, 461
Nitrosofyers, 601
Nitrosomonas, 461, 601
Nitrospira, 462, 454
NNRTI, see Nonnucleoside reverse
    transcriptase inhibitor
Nocardia, 473, 519, 632
Nocardicin, 397, 398
Nod boxes, 714
Nod factors, 712, 713–714
Nod genes, 713–714
Nod proteins, 713–714
Nodules, 16
Nomenclature, 10–11, 444, 451
Noncyclic photophosphorylation, 589
Nonheme iron-sulfur proteins, 122
Nonionizing radiations, 298–299
Non-Mendelian inheritance, 340–341
Nonnucleoside reverse transcriptase
    inhibitor, 923, 951
Nonpermissive cells, 277
Nonpolar, 30
Nonsense codons, 200–201, 206
Nonsense mutation rate, 296
Nonsense mutations, 295
Nonspecific immunity, 803, 808–811
Nonspecific porins, 75
Nonsporulating, low GC, gram-positive
    bacteria, 502–507
Nopaline, 707
Norfloxacin, structure, 758
Normal flora, 776
  gastrointestinal tract, 780–782
  oral cavity, 777–780
  respiratory tract, 782–783
  skin, 776–777
  urogenital tract, 783–784
Northern blot, 174, 365
Norwalk virus, 978
Nosocomial infection, 857, 892, 906–908
Novobiocin, 178
NRTI, see Nucleoside reverse transcriptase
    inhibitor
Nuclear pores, 96
Nuclear structure, prokaryotes vs.
    eukaryotes, 100
Nucleation track, 748–749
Nucleic acid hybridization, 173–174, 447,
    882
Nucleic acid monomers, 132
Nucleic acid probe, 355–356, 855, 882–887
Nucleic acid probe method, 854
Nucleic acids, 4, 33, 34, 37–40
  antisense, 234
  denaturing, 173
  hybridization, 177
  labeling, 172–173
  protein interaction, 221–223
  viral, 248–249
  working with, 171–176
Nucleocapsid proteins, 273
Nucleocapsids, 239, 278–279
Nucleoid, 4, 7, 77–78

in prokaryotes, 57
Nucleolus, 97
Nucleopore filter, 748–749
Nucleoside, 38
Nucleoside analogs, 762, 951
Nucleoside reverse transcriptase inhibitor, 923
Nucleosomes, 178, 181, 340, 569
Nucleotides, 33, 37–38, 132
Nucleus, 4, 95–97, 721, 728
    in eukaryotes, 57
    origin of, 430
Numerical aperture, 51
Nutrient levels, growth rates and, 646–647
Nutrient status, 663
Nutrients, 104
Nutrition, 104–106
    of animals living near hydrothermal vents, 671–673
Nutritional yeast, 408
Nyctotherus ovalis, 724

O antigens, 481
Obligate anaerobes, 158, 159
Obligate barophiles, 669
Obligate chemolithotrophs, 467
Obligate intracellular parasites, 484–485, 530
Obsidian Pool, 571
Occupational exposure control plan, 888–889
Occupational Safety and Health Administration, 932
Oceanospirillum, 485–487
Oceanospirillum linum, 657
Ochromonas danica, 98
Octopine, 707
Office of Mosquito Control in War Areas, 968
Oil spill cleanups, 696–697
Oil-immersion lenses, 51
Okazaki fragment, 187
Oleandomycin, 761
ompH gene, 670
Oncogene, 270, 285, 708–709
One-step growth curve, 245
Open reading frame, 201, 331, 334, 335
Operator, 196
Operator region, 219
Operon, 219
    mRNA and, 196
Opines, 707
Opportunistic infections, 947, 948
Opportunistic pathogens, 906, 974
Opsonization, 811, 835
Optical density units, 143
Optical isomers, 42
Optical tweezers, 656–658
Optimum temperature, 147
Oral cavity, normal flora, 777–780
Orders, 450
Ore, microbial leaching, 690, 691–694
ORF, see Open reading frame
Organ cultures, 242
Organ transplantation, 830, 838
Organelles, 6
    eukaryotic, 57–58, 97–99, 430–432
    evolution of, 440
    prokaryotic, 98–99
Organic acid metabolism, 629–630
Organic acids, decarboxylations, 623–624
Organic electron acceptors, 618–619
Organic soils, 662

Organic sulfur compounds, 688
Organisms, temperature classes, 147–148
Orientia tsutsugamushi, 963
Origin of replication, 185–186, 276, 351
Orleans method, for vinegar, 405
Orotic acid, structure, 132
Orthologous gene, 335
Orthologs, 336
Orthomyxovirus, 268, 273–275
Oscillatoria, 527
Oscillatoria limnetica, 589
Oscillatoria princeps, 524
Oscillochloris, 542–543
Osmophiles, 157
Osmosis, effect on growth, 156–158
Osmotaxis, 85
Outbreaks, 894, 913
    waterborne diseases, 976
Outer sheath, 537
Outgrowth, of spores, 95
Oven-vat method, 405
Overlapping genes, 208, 252
Oxacillin, 760
    structure, 759
Oxalacetate, 128, 629
Oxalate, 624
Oxalobacter formigenes, 623–624
2-Oxazolidinone, structure, 754
Oxic, 643
Oxidase-positive, 482
Oxidation:
    anoxic ammonia, 602
    fatty acid, 631
    ferrous iron and pyrite, 689–691
    hydrocarbons, 632–634
    iron, 598–601
    methane, 603
    reduced sulfur compounds, 595–598
    sulfide and elemental sulfur, 687–688
    sulfur, 595–598
Oxidation-fermentation test, 503
Oxidation-reduction balance, 620, 621–622
Oxidation-reduction reactions, 112–114
Oxidative phosphorylation, 117, 125–126, 574
Oxygen:
    effects on culture, 159–160
    growth and, 158–162
    molecular, 631–632
    toxic forms, 160–161
Oxygen profile, 642
Oxygen relationships:
    in lakes and rivers, 666
    of microorganisms, 158
Oxygenases, 631–633
Oxygen-dependent phagocytic killing, 809–810
Oxygenic photosynthesis, 576, 587–589
Oxygenic phototrophs, anoxygenic photosynthesis in, 589
Oxytetracycline, 761
Ozone shield, 430

P870, 584
Packaging, 245
Palindrome, 183
Palmitate, structure, 133
Pandemics, influenza, 935–936
Papers, key works in microbiology, 27
Papilloma viruses, 886
Papovavirus, 268, 276–277
PAPS, *see* Phosphoadenosine phosphosulfate

Paracoccus, 466
Paracoccus denitrificans, 607
    electron transport chain, 124
Paracrystalline surface layers, 86
Parainfluenza, 878
Paralogs, 336
Paramecium, 725, 728
    alternative genetic codes, 210
Paramyxovirus, 268, 937–938
Paramyxovirus group, 273
Parasites, 538, 728–729
    coevolution with host, 902–905
    fungal, 729
    obligate intracellular, 484–485, 530
Parasporal body, 509
Parsimony, 434–436
Particle gun, 310–311
Particle gun method, 374
Parvovirus, 268
Passive immunity, 849–850
    vs. active, 848
Pasteur, Louis, 17–20, 42, 744–745
Pasteur flask, 19–20
Pasteurization, 744–745, 908, 934
Pathogen eradication, 910–911
Pathogenicity islands, 333
Pathogens:
    airborne transmission, 923–925
    bacterial, 925–930
    colonization and growth, 786–788
    discovery of main bacterial, 931
    drug resistance in, 768
    entry into host, 784–786
    hospital, 906–907
    immunodiagnostic tests, 869
    isolation from clinical specimens, 855–861
    Lyme disease, 964–965
    malaria, 966–967
    microorganisms as, 15–16
    nucleic acid probe identification of, 884
    pseudomonads, 473
    respiratory, 934–939
    rickettsias, 484–485
Pathology, rabies, 957–958
Paucibacillary leprosy, 933
PCB, *see* Polychlorinated biphenyl
PCR, *see* Polymerase chain reaction
Pectin, 707, 738
Pediococcus, 504
Pelobacter acidigallici, 622
Pelochromatium, 536
Pelochromatium roseum, 537
Pelodictyon clathratiforme, 535, 580
Penam, structure, 397
Penetration, 246–247
Penicillin, 759–760, 925, 927, 929, 941, 943, 964
    discovery, 757
    industrial production, 396–399
    transpeptidation and, 76–77
Penicillin binding proteins, 760
Penicillin G, 396, 760
    structure, 759
Penicillin production, 388
Penicillinase production, 942
Penicillin-selection method, 293
Penicillium, 757
    antibiotic production and, 393
Penicillium chrysogenum, 388, 398, 759
Pentoses, 34, 130–131
Peptic ulcers, 489, 988
Peptide bond, 40, 43–44
Peptide hormones, 339

Peptide site, 206
Peptidoglycan, 34, 35, 69–70, 442
  biosynthesis of, 76
  diversity in, 70–71
Perforin, 829, 868
Peridontal pockets, 780
Periplasm, 75
Periplasmic binding protein-dependent
    transport, 67
Peritrichous flagella, 477, 482
Peritrichous flagellation, 79–80, 82
Permanent cell lines, 242
Permeability, cytoplasmic membrane and,
    63–64
Permissive cells, 277
Pernicious anemia, 399
Peromyscus maniculatis, 959
Peroxidase, 160–161
Peroxisome, 723
Peroxynitrate radicals, 809
Persistent infections, 269
Pertussis, 919
Pesticides, biodegradation of, 698–699
Petri, Richard, 22
Petri dishes, 13
Petroff-Hausser counting chamber, 141
Petroleum biodegradation, 696–698
Petroleum production, 698
PFGE, *see* Pulse field gel electrophoresis
Pfu polymerase, 361
PGA, formation of, 590–591
pH:
  food preservation and, 982
  growth and, 154–155
pH gradient, 123–124
pH scale, 154
PHA, *see* Poly-ß-hydroxyalkanoate
Phaeospirillum fulvum, 460
Phage ß, 789
Phage conversion, 313–314, 789
Phage group, 255
Phage lambda, 261
  Ames test and, 301
  as cloning vector, 347–349
  transduction and, 312–313
Phage M13, 154, 353, 367
  cloning vectors from, 350, 351
Phage Mu, 265–267, 327–328
Phage φχ174, genome, 252–253
Phage T4, 256–259
Phage T7, 255–257
Phagemids, 350
Phagocyte failure, 810–811
Phagocytes, 808–811
Phagocytosis, 726, 808–811, 835, 867
Phagolysosome, 808, 825
Pharming, 378
Phase contrast microscopy, 52–53
Phase variation, 328–329
PHB, *see* Poly-ß-hydroxybutyrate, Poly-ß-
    hydroxybutyric acid
Phenotypes:
  conferred by plasmids, 317
  designation, 291
Phenotypic traits, classifying microbes
    using, 442–444
Phenylalanine, 400
Phloroglucinol, 622
Phosphatase, 231
Phosphoadenosine phosphosulfate, 608–609
Phosphodiester, 39
Phosphoenolpyruvate, 67, 131
  structure, 116

Phosphoenolpyruvate carboxylase, 630
6-Phosphogluconate dehydrase, 473
3-Phosphoglyceric acid, *see* PGA
Phosphoketolase, 504
Phospholipases, 631, 788
Phospholipid bilayer, 60
Phospholipid hydrolysis, 630–631
Phospholipids, 37
Phosphonoformic acid, 762
Phosphoribulokinase, 590, 591
Phosphoroclastic reaction, 501, 510
Phosphorolysis, 628
Phosphorus, 105
Phosphotransferase system, 66–67
Photic zone, 667
Photobacterium, 482–484
Photobacterium phosphoreum, 483
Photobiodegradable plastics, 702–703
Photoblepharon palpebratus, 483
Photochromogenesis, 519
Photoheterotrophy, 459, 534, 542, 575
Photometer, 143
Photophosphorylation, 456, 585
Photoreceptor, 85
Photosynthesis, 573, 575–580
  algal, 737
  anoxygenic, 582–587, 589
  in carbon cycle, 675–677
  carotenoids and phycobilins, 580–582
  oxygenic, 587–589
Photosynthetic gene cluster, 585–586
Photosynthetic membranes, 578–580
Photosystem I, 588
Photosystem II, 588
Phototaxis, 84–85
Phototrophs, 104, 575
Phototrophy, 11, 129, 130
Phycobilins, 525, 581–582
Phycobiliproteins, 735
Phycobilisomes, 581–582
Phycocyanin, 525, 581–582
Phycoerythrin, 525, 581
Phyllosphere, 704
Phylogenetic probes, 436–438
Phylogenetic staining, using FISH, 655–656
Phylogenetic trees, 434–436
  Archaea, 545, 547
  Bacteria, 453, 455
  Eukarya, 724
Phylogeny, 422, 444
  cyanobacteria, 527
  microbial, 439–442
Physarum, 733
Physical mutagens, 297
Phytanyl, 546
Phytoplankton, 665
Pickling, 982
Picornavirus, 268
Picornavirus family, 270
Picrophilus, 557–558
Picrophilus oshimae, 155
Pigments, 737
Pili, 86, 251, 319, 785
Pilot plant stage, 392
Pinocytosis, 726
Pinus, 705–706
Pinus contorta, 706
Pinus radiata, 707
Pinus rigida, 706
Pirella, 532
Pitching, yeast in brewing, 414
Pityrosporum ovalis, 777
Plague, 878, 913, 969–971

Planctomyces, 532
Planctomyces maris, 532
Planctomyces-Pirella, 442, 454
Plant agriculture, genetic engineering in,
    374–377
Plant biotechnology, 376–377
Plant cells, 6
Plant diseases, 515
Plant hosts, 377
Plant nutrition, role of microorganisms, 16
Plant pathogens, pseudomonds, 473
Plant viruses, 238, 270, 286
Plant-microorganism interactions, 704–706
Plants:
  cloning vectors, 375–376
  transgenic, 369
  virus infection, 376–377
Plaque, 242–244
Plaque assay, 242–244
Plaque-forming units, 244
Plasma, 802, 805
Plasma cells, 835
Plasmid, 7, 77, 179, 181–182, 307, 314–319,
    702, 919
  antibiotic resistance, 695
  bacteriocins, 318
  as cloning vectors, 345–347
  conjugation in, 319–324
  discovery, 309
  engineered, 319
  resistance, 317–318
  vs. viruses, 237
  yeast, 339–340
Plasmid pBR322, 343, 346–347
Plasmid R100, 317–318
Plasmodia, 729, 733
Plasmodium falciparum, 966–968
Plasmodium vivax, 377, 725, 966–968
Plastics, degradation of, 702–703
Plastocyanin, 589
Plate count, 142
  sources of error, 143
Platelets, 805
Plus-strand nucleic acid, defined, 237
Pneumocystis carinii, 947, 948
Pneumonic plague, 970
Pogonophora, 672
Point mutations, 294
Polar, 30
Polar flagella, 470–471, 482
Polar flagellation, 79–80, 82
Polar growth, 491
Polio virus, 238, 270–272, 911, 978
  herd immunity and, 905
  incidence in United States, 850
Poly-A tail, 197–198, 267
Polyacrylamide gels, 879
Poly-ß-hydroxyalkanoate, 87, 703
Poly-ß-hydroxybutyrate, 49, 128
Poly-ß-hydroxybutyric acid, 87–88
Polychlorinated biphenyl, 698, 701
  structure, 699
Polycistronic mRNA, 166, 196, 204, 226, 253,
    267, 332
Polyclonal antibodies, 837–841
Polyclonal antiserum, 802, 837
Polyenes, 764
Polylinker, 347, 350
Polymer formation, energy storage and, 128
Polymerase chain reaction, 154, 360–362,
    404, 434, 541, 887, 960, 962, 963, 965,
    988
  nucleid acid probes and, 882, 883

Polymerization, 404
Polymerization reactions, 136
Polymorphic virus, 273
Polymorphisms, 819, 825
Polymorphonuclear leukocyte, 802, 808–809
Polymyxin, 473, 508, 759
Polynucleotides, 37, 38
Polyoxin B, structure, 754
Polyoxins, 764
Polypeptide, 43, 165
Polyphosphate, 87–88
Polyprotein, 271
Polyribosome precipitation, 366
Polysaccharide catabolism, 627–628
Polysaccharide layers, 86–87
Polysaccharides, 4, 34–36, 130–131
Polysiphonia, 737
Polysome, 206
Polyvalent vaccine, 372
Pomace, 410
Pompeii worm, 674
Pontiac fever, 977
Population growth, 137–139, *see also* Growth
    cycle of, 139–141
    measuring, 141–145
Populations, 643–645
    microbial, 11–12
Porin, 75, 670
Porin proteins, 74
Porphyrin ring, structure, 122
Positive control, 220–221
Positive selection, 830, 832
Positive-strand DNA virus, 248–249
Positive-strand RNA animal viruses,
    270–272
Positive supercoils, 178–179, 569
Posttranslational cleavage, 271
Posttranslational modification, 208, 209
Potassium, 105
Potato yellow dwarf virus, 272
Potomac fever, 485
Pour plate method, 142
Pox viruses, 268, 276, 279–280
Precipitation, 841–843
Precipitation reaction, 842
Pregnancy testing, monoclonal antibodies,
    and, 839
Pre-mRNA, 167
Preproinsulin, 370
Pressure, molecular effects, 669–670
Prevalence:
    botulism, 985
    defined, 893
    ehrlichiosis, 962–963
    hantavirus pulmonary syndrome, 960
    Lyme disease, 964
Prevention:
    Lyme disease, 965–966
    malaria, 967–968
    rabies, 958–959
Pribnow box, 193–194, 196
Primaquine, 967
Primary antibody response, 833, 867
Primary cell culture, 242
Primary domains, 442–444
Primary electron donor, 114–115
Primary fermenters, 678
Primary lymphoid organs, 811
Primary metabolites, 387–389
Primary producers, 665
Primary productivity, 673
Primary structure, 43–44
    DNA, 39

Primary transcript, 167, 197
Primary treatment, sewage, 417
Primase, 185, 252–253
Prime niche, 645
Primers, 185, 189, 361
Primitive organisms, 427–430
Prince William Sound, Alaska, 697
Prions, 285–287
Probe, 173, 359
Prochlorococcus, 528, 665
Prochloron, 527–528
Prochlorophytes, 527–529
Prochlorothrix, 528
Prodigiosins, 482
Product, 110
Proinsulin, 371
Prokaryote hosts, 353
Prokaryote, 6–7, *see also* Archaea, Bacteria
    antibiotics from, 760–762
    cell structure, 57
    cell wall, 68–73
    chemical composition, 33
    chemolithotrophic, 673
    chlorosomes, 579
    chromosomes, 331
    denitrifying, 608
    DNA, 7–8, 77–79, 177, 178
    E. coli as model, 331
    enrichment culture methods for, 648–649
    vs. eukaryotes, 99–100
    extremozymes and, 403–404
    fossil, 426
    gene expression in, 233
    genetic elements, 179
    genetics, 166–167
    heat resistance, 674–675
    host resistance, 247
    morphology, 57–58
    nitrogen cycle in, 685
    organelles and, 98–99
    phenotypes, plasmids, and, 317
    plasmids in, 181–182
    recombination in, 302
    restriction enzymes in, 183
    ribosome structure, 204
    rRNA, 207
    rumen, 684
    transposition in, 325
    virus multiplication and, 267–268
    Winogradsky column and, 651
The Prokaryotes, 451, 454
Prokaryotic genomics, comparative, 333–336
Proline, 157
Promoters, 193–194, 196, 262
Promotion, 270
Prontosil, 756
Proofreading, 188–189
    mutations and, 297
Propagation cycle, 690
Prophage, 260
Prophylactic, 848
Prophylaxis, 850
Propionate, 682
Propionibacterium, 399, 516
Propionibacterium acnes, 777
Propionic acid, 516–517
Propionic acid bacteria, 516–517, 520
Propionigenium, 517, 612
Propionigenium modestum, 623
Prosthecae, 491
Prosthecate bacteria, 491–495
Prosthetic groups, 111
Protease inhibitors, 762, 923, 951–953

Proteases, 229, 281
    industrial production, 402–403
Protein Data Bank, 771
Protein folding, 358–359, 569
Protein monomers, 131–132
Protein primer, 189
Protein processing, 216–217
Protein splicing, 217
Protein stabiliity, 358–359
Protein structure, 29
Protein synthesis, 204–208, 443–444
    effect of antibiotics, 208
    errors, 210
    role of ribosomal RNA, 206–207
Proteins, 33, 34, 43–44, 46
    appearance of, 428
    cellular, 4
    DNA binding, 221–224
    folding and secreting, 207–208
    genes and, 165
    higher structure, 44–45
    production of viral, 248–249
    reaching gene via, 365–366
    sensor, 230
    transport, 64
    viral, 249–250
    yeast as, 409
Proteobacteria, 440
    acetic acid bacteria, 474–475
    budding and prosthecate/stalked
        bacteria, 491–495
    enteric bacteria, 477–482
    free-living aerobic nitrogen-fixing
        bacteria, 475–476
    gliding myxobacteria, 495–498
    hydrogen-oxidizing bacteria, 466–467
    major genera, 456
    methanotrophs and methylotrophs, 467–470
    Neisseria, Chromobacterium, 476–477
    nitrifying bacteria, 461–462
    Pseudomonas and Pseudomonads, 470–474
    purple phototrophic bacteria, 455–460
    rickettsias, 484–485
    sheathed proteobacteria, 489–491
    spirilla, 485–489
    sulfate- and sulfur-reducing, 498–502
    sulfur- and iron-oxidizing bacteria,
        462–465
    Vibrio, Photobacterium, 482–484
Proteus, 481–482, 782, 857
Proteus mirabilis, 481, 842, 907
Proteus vulgaris, 481, 655
Protists, 724
Protocatechuate, 633
Protoctists, 724
Proton motive force, 63, 123–125
    in alternate bioenergetic strategies, 130
    and ATP formation, 125–126
Proto-oncogenes, 270
Protoplasmic cylinder, 537
Protoplast formation, 71–72
Protoplasts, 513
Prototroph, 292
Protozoa, 11, 721, 724–729
    alternative mitochondrial codes, 210
    rumen, 683–684
Provirus, 260, 283
Pruisner, Stanley, 287
Pseudomembrane, 929
Pseudomonads, 470–474
Pseudomonas, 311, 369, 466, 468, 470–474,
    632, 633, 702, 857, 861
    plasmids, 316

Pseudomonas aeruginosa, 13, 52, 55, 228, 328, 473, 647, 695, 696, 789, 797, 855, 859, 865–866
  as hospital pathogen, 906–907
Pseudomonas carboxydovorans, 467
Pseudomonas cepacia, 472, 701
Pseudomonas denitrificans, 399
Pseudomonas marginalis, 473
Pseudomonas stutzeri, 607
Pseudomonas syringae, 473
Pseudomurein, 553
Pseudopeptidoglycan, 72–73
Pseudoplasmodium, 734
Psittacosis, 529
Psychrophiles, 148–150, 404
Psychrophilic bactera, 667–669
Psychrotolerant microorganisms, 149–150
Pterin, 467
Public health, water quality and, 974–976
Public health measures:
  disease control, 908–911
  infectious disease and, 919
Public Health Service, 968
Pulque, 474
Pulse field gel electrophoresis, 172, 987
Pure culture, 12–13, 652
Purification:
  of antibiotics, 393, 395
  of DNA, 171
Purine bases, 37–38, 165
Purines, 106, 132
Puromycin, 208
Purple bacteria, 3, 440
  photosynthetic apparatus, 582–587
Purple nonsulfur bacteria, 457–460
Purple phototrophic bacteria, 455–460, 650–652
Purple sulfur bacteria, 457, 458
Pus, 811
Pyogenic infections, 799
Pyridoxine, 106
Pyrimidine bases, 38, 165
Pyrimidine dimers, 299
Pyrimidines, 106, 132
Pyrite, 692
Pyrite oxidation, 689–691
Pyrobaculum, 566
Pyrobaculum aerophilum, 567
Pyrococcus, 546, 548, 559
Pyrococcus furiosus, 361, 559
Pyrococcus woesei, 403, 404
Pyrodictium, 441, 566–569
Pyrodictium abyssi, 569
Pyrodictium occultum, 567
Pyrogenic organisms, 799
Pyrolobus, 441, 566–568
Pyrolobus fumarii, 147, 567, 675
Pyrrole ring, 482
  structure, 122
Pyruvate, 118–119, 501, 510, 516, 622, 630
  in citric acid cycle, 126–128
Pyruvate carboxylase, 630
Pyruvate oxidation, 723
Pyruvate:ferredoxin oxidoreductase, 723

Q cycle, 124–125
Q fever, 484, 485, 878, 963
Quantification, staining techniques and cell numbers, 653–655
Quarantine, 909–910
Quaternary structure, 45–46
Quayle, John, 605
Quayle cycle, 605

Quinacrine, 979
Quinolones, 178, 758
Quinones, 121–124, 160
Quorum sensing, 228–229, 483–484
Quorum-sensing system, 307

R plasmid, 473, 765–767, 769
Rabbit fever, 917
Rabbit hemorrhagic disease virus, 904
Rabbit myxomatosis virus, 279
Rabbits:
  cellulose digestion in, 685
  coevolution with myxoma, 902–905
Rabies, 852, 908, 911, 917, 957–959
Rabies virus, 272
Racemases, 43
Racking, 410
RAD systems, *see* Rapid antigen detection systems
Radiation, 297–299
  endospores and, 92
Radiation practice, 746–747
Radiation sterilization, 745–747
Radioimmunoassay, 840, 878–879, 953
Radioisotopes, 658–659
Ralstonia, 470, 472, 473
Ralstonia eutrophus, 466, 594–595, 703
Ramantadine, 937
Random clones, 329
Rapid antigen detection systems, 926
Rats, 969
RDP, *see* Ribosomal Database Project
Reaction center, 578–580, 583
Reading frame, 200–201
Reading frame shift, 295–296
recA gene, 323
RecA protein, 302, 303, 432
Recalcitrant, defined, 574
Receptors, viral, 246
Receptor-transducer proteins, 231–233
Recessive gene, 336
Recombinant DNA, 344
Recombinant DNA techniques, therapeutic products, 370
Recombinant live vaccines, 372
Recombinant vaccines, 372–373
  pox viruses and, 279–280
Recombination, detecting, 303–304
Recombination event, 326
Red tides, 735, 737, 738
Redox balance, 620, 621–622
Redox coenzymes, 613, 614
Redox cycle:
  carbon, 676–677
  iron, 689
  nitrogen, 685
  sulfur, 687
Redox reaction, 112–114
  in glycolysis, 118–119
Reduced sulfur compounds, oxidation of, 595–598
Reduction potentials, 112
Reduction, 112
  acetylene, 638–639
  bacterial iron, 689
  sulfide and elemental sulfur, 687–688
Reductive dechlorination, 699–702
Reemergent infections, 892
Reemerging infectious disease, 913–920
Regulation:
  enzyme activity, 213–216
  phage T4, 258–259
  transcription, 217–221

Regulatory proteins, 221
Regulon, 220–221
Relapsing fever, 540, 799
Relaxed DNA, 177
Release, of virions, 245–246
Release factors, 206
Remittent fever, 799
Rennin, 374
Reovirus, 268, 275
Replacement vectors, 347
Replica plating, 292–293
Replica plating procedure, 354
Replication, 165
  adenoviruses, 280
  of bacteriophage T7, 255–257
  of DNA animal viruses, 275–277
  of expression vector, 356–357
  of phage T4 genome, 256–258
  of plasmids, 315–316
  of poliovirus, 271–272
  of pox viruses, 279
  phage Mu, 267
  retroviruses, 281–285
  viral, 245
Replication fork, 186–188
Replicative form DNA, 252
Replicative transposition, 326–327
Replisome, 188
Reporter gene, 381
Reporter molecule, 882
Reporter probe, 886
Repression, 217–220
Repressor protein, 219
Repressor-operator system, 357
Reproduction, 4–5
  prokaryotes vs. eukaryotes, 100
  viral, 244–246
Resazurin, 159
Research tool, genetic engineering as, 380–381
Reserve polymers, 737
Reservoir, 892, 895–898
  disease controls, 908
Resistance, 774
  natural host, 796–799
Resistance factors, 765–766
Resistance plasmids, 317–318
Resistance transfer plasmid, *see* R plasmid
Resolution, 50–51
Resources, 645–647
Respiration, 117
  aerobic, 121–129
  anaerobic, 129, 605–606
  carbon cycle, 675–677
  vs. fermentation, 126–127
Respiratory infection, 924
  control of, 940
Respiratory pathogens:
  bacterial, 925–930
  viral, 934–939
Respiratory route, 901
Respiratory tract, normal flora, 782–783
Response regulator, 230–232
Response regulator protein, 230
Reston virus, 918
Restriction endonucleases, 172, 182
Restriction enzyme map, 184
Restriction enzymes, 182–184, 247, 345
Restriction fragment length polymorphism, 363, 987
Restriction mapping, 329
Restriction-modification system, 182
Reticulate body, 531

Retinal, 552
Retron, 281
Retrotransposons, 339
Retrovirus, 238, 241, 249, 267–268, 281–285, 352, 762
Reverse citric acid cycle, 544, 548, 592–593
Reverse DNA gyrase, 569
Reverse electron flow, 586–587
Reverse electron transport reactions, 592
Reverse gyrase, 178–179
Reverse transcriptase, 176, 241, 249, 281, 282, 345, 362–364, 947, 951
Reverse transcriptase-polymerase chain reaction, 950–951
Reverse transcription, 166, 249, 362
Reverse translation, 365–366
Reversions, 296
Revertant, 296
RFLP, *see* Restriction fragment length polymorphism
Rhabdovirus, 268, 272–273
RHDV, *see* Rabbit hemorrhagic disease virus
Rheumatic fever, 923, 926
Rheumatoid arthritis, 839, 870
Rheumatoid factor, 870
Rhicadhesin, 712
Rhinoviruses, 270, 934–935
Rhizobium, 231, 440, 473, 710–712
plasmids and, 316
Rhizobium leguminosarum biovar *trifolii*, 712, 713
Rhizobium leguminosarum biovar *viciae*, 713
Rhizobium trifolii, 87
Rhizopus nigricans, 401, 402, 731
Rhizosphere, 464, 662, 704
Rhodanese, 672
Rhodobacter, 311, 440
Rhodobacter blasticus, 74
Rhodobacter capsulatus, 583, 585–586
Rhodobacter sphaeroides, 180, 182, 460, 585
Rhodococcus, 311
Rhodocyclus purpureus, 460
Rhodomicrobium, 491
Rhodomicrobium vannielii, 58, 460
Rhodopila globiformis, 460
Rhodopseudomonas, 440
Rhodopseudomonas acidophila, 460
Rhodospirillum centenum, 80, 84, 85
Rhodospirillum photometricum, 80
Rhodospirillum rubrum, 58
Rhodospirillum sodomense, 56, 88
Ri plasmid, 707
RIA, *see* Radioimmunoassay
Riboflavin, 399–400
structure, 399
Ribonuclease H, 283
Ribonucleic acid, *see* RNA
Ribonucleotide reductase, 131
Ribose, 191
structure, 34
Ribosomal Database Project, 434
Ribosomal RNA, *see* rRNA
Ribosomal RNA sequences, 434–436, 439–442
Ribosomes, 6,7, 99, 166, 167, 204
in prokaryotes, 57
prokaryotes vs. eukaryotes, 100
Ribotyping, 448, 886, 988
Ribozyme, 198–199, 282, 369
Ribulose bisphosphate carboxylase, *see* RubisCO
Ribulose monophosphate cycle, 468
Ribulose monophosphate pathway, 604–605
Rice, 716

Ricketts, Howard, 960
Rickettsia prowazekii, 960–961
genome, 334
Rickettsia rickettsii, 484, 961, 965
Rickettsia typhi, 961
Rickettsias, 440, 484–485, 530, 878, 957, 960–963
Rickettsiella popilliae, 484
Rifampicin, 977
Rifampin, 759, 933
structure, 754
Rifamycin, 196, 762
Rift Valley fever, 918
Rifts, 670
Ring cleavage, 633–634
Ring reduction, 633–634
Rivers, 666, 668
RNA, 33, 37–40, 97, 165, *see also* Transcription, Translation
protein processing and, 216–217
regulatory, 234
sequencing, 175–176
translation of, 204–208
viral, 238
RNA animal viruses, 268
RNA bacteriophages, 251–252
RNA endonuclease, 274
RNA genomes, mutations in, 297
RNA life, 427–428
RNA polymerase, 191–193, 195–196, 219, 220, 225, 256, 271, 428, 443
RNA primers, 253
RNA processing, 197–199
RNA replicase, 251
RNA synthesis, 191–195
RNA-dependent RNA polymerase, 272, 274
Rochalimaea quintana, 485
Rocky Mountain spotted fever, 484, 878, 960, 961
Rod, 57
Roentgen, 746
Rolling circle mechanism, 252–254, 278, 315
Rolling circle replication, 262–263, 320
Root nodule bacteria, 709–717
Root nodules, 710–714
Rosettes, 494
Rotavirus, 275, 878
Rothia, 780
Rough endoplasmic reticulum, 722
Roundup, 376
Rous sarcoma virus, 281–283, 285
Royal Society of London, 17
rRNA, 39, 165–167, 169, 191, 195, 432–434
16S rRNA, defined, 423
RT-PCR, *see* Reverse transcriptase-polymerase chain reaction
Rubella, 878, 938, 939
RubisCO, 98, 590–591
Rum, 412
Rumen, 16, 643, 681–685, 728
Ruminobacter amylophilus, 682
Ruminococcus, 627
Ruminococcus albus, 682
Runs, 83–84
Rusticyanin, 599

Sac7d, 569–570
Saccharification, 403
Saccharomyces, 413
Saccharomyces carlsbergensis, 411
Saccharomyces cerevisiae, 7, 10, 53, 99, 120, 231, 286, 287, 407–409, 411, 437, 732
DNA, 180

DNA chips and, 379
genetics, 337–341
as model eukaryote, 336–337
RNA polymerase, 443
as vector host, 352
Saccharomyces ellipsoideus, 410
Safe Drinking Water Act, 975
Safety, laboratory, 888–889
Safranin, 51
Sagittula stellata, 657
Saliva, 777
Salmonella, 74, 479–481, 744, 767, 859, 861–862, 878, 886, 917, 974
DNA invertase, 328–329
lipopolysaccharide layer in, 73
virulence factors, 795
Salmonella anatum, 313
Salmonella enteritidis, 791
Salmonella paratyphi, 837
Salmonella pathogenesis, 773
Salmonella typhi, 86, 898, 977–978
agglutination, 868
Salmonella typhimurium, 250, 311, 796, 917, 986
Ames test and, 301
discovery of recombination in, 309
flagellar genes, 81–82
Salmonellosis, 978, 986
Salt bridges, 153, 569
Salt tolerance, 156
Salvarsan, structure, 756
Same-site revertant, 296
San Joaquin Valley fever, 973
Sandwich ELISA, 875
Sanger, Frederick, 252
Sanger dideoxynucleotide method, 175–176, 255, 434
Saquinavir, 770
structure, 771, 952
Sarcina, 503
Sarcina ventriculi, 503
Sarcodina, 727
Sarcomas, 285
SASPs, *see* Small acid-soluble spore proteins
Satellite DNA, 181
Saturation effect, 64
Scale-up process, 392
Scanning electron microscope, 663
Scanning electron microscopy, 55–56, 749
Scarlet fever, 923, 926
Scenedesmus, 736
Schink, Bernhard, 650
Schizosaccharomyces pombe, 338
Schloesing, T., 597
Schwartzia, 682
Scotochromogenesis, 519
Scotophobotaxis, 84
Scrapie, 286
Screening, 291, 354, 393
Scrub typhus, 963
Sea mounts, 673–674
Sebaceous gland, 776, 798
Secondary antibody response, 833
Secondary fermentation, 516
Secondary fermenters, 678–679
Secondary lymphoid organs, 811
Secondary metabolites, 387–388
antibiotics as, 393
Secondary response, 868
Secondary structure, 44
RNA, 39
Secondary treatment, sewage, 417
Second-site revertants, 296

Secreting proteins, 207–208
Secretion vectors, 349
Secretory piece, 823
Sedimentation basins, 419
Segmented genome, 273
Selectable mutation, 291
Selection, 292
Selective culture, *see* Enrichment culture
Selective media, 861–862, 864
Selective medium, 304, 305
Selective toxicity, 750, 755, 756
Selenium, 694
Selenocysteine, 209
Selenomonas, 682
Self-assembly, 82, 239
Self-splicing introns, 198
SEM, *see* Scanning electron microscopy
Semiconservative replication, 185
Semilogarithmic graph, 138
Semisynthetic antibiotics, 759
Semisynthetic penicillin, 396, 741
Semmelweis, Ignaz, 20
Sense codon, 209
Sensitivity, 867
Sensor kinases, 230–231
Sensor protein, 230
Septicemia, 856
Septicemic plague, 970
Septum, 137
Serial dilutions, 142–143
Serine pathway, 468, 604
Serine transhydroxymethylase, 604
Serological tests, 868
Serology, 841
Serotonin, 844
Serratia, 482
Serratia marcesens, 13
Serum, 805, 812, 849
Sesbania, 714–715
Sesbania rostrata, 715
Sewage, 668
Sewage fungus, 490
Sewage microbiology, 416–420
Sex pilus, 319
Sexual activity, AIDS and, 954
Sexual spores, 730–731
Sexually transmitted disease, 530, 901, 940–946
Shake tube method, 502
β-Sheet, 44, 45, 150
Shewanella putrefaciens, 617–618
Shigella, 74, 479–481, 859, 861, 974
Shigella dysentery, 481
Shigella flexneri, 13
Shiitake, 415–416
Shine-Dalgarno sequence, 201, 204, 207, 226, 358
Shingles, 277, 939
Shotgun clones, 329
Shotgun cloning, 345
Shuttle vector, 349–350
Sialic acid, 274
Sickle cell anemia, 957
Sickle cell trait, 968
Siderophores, 105, 786, 795
SIDS, *see* Sudden infant death syndrome
Sigma factor, 193–194, 258
Signal peptide, 358
Signal sequence, 207, 349
Signal transduction, 230–233
Signature sequences, 436–438
Silent copies, 339
Silent mutations, 295

Silica, 738
Silica gel, 462
Simmons citrate agar, 862
Single-cell protein, 409
Single-copy DNA, 181
Single-stranded binding protein, 186, 302, 303
Single-stranded DNA plasmids, 315
Single-stranded filamentous DNA bacteriophages, 254–255
Single-stranded icosahedral DNA bacteriophages, 252–254
Singlet oxygen, 160
Sinorhizobium, 710
Sinorhizobium meliloti, 712, 713
Site-directed mutagenesis, 300, 344, 359, 366–368
Site-specific recombination, 326, 328
Size:
    of microbial cells, 58–60
    of prokaryotes vs. eukaryotes, 100
    of prokaryotic genomes, 333
    of viruses, 238
Skin, 797–798
    microbial interactions, 775
    normal flora, 776–777
Skin test, 868–869
S-layer, 72–73, 86
SLE, *see* Systemic lupus erythematosis, 846
Slime, 497
Slime layer, 86–87, 774, 785
Slime molds, 733–735
Sludge digestors, 417
Slug, 735
Small acid-soluble spore proteins, 93–94, 743
Small intestine, 780
Smallpox, 279, 910–911, 917
Smallpox virus, 238
Smooth endoplasmic reticulum, 722
Snapping division, 515
Sneeze, 922, 924
Snow, John, 902
Snow algae, 149–150
Soda lakes, 548–550
Sodium, 105
Soil:
    as microbial habitat, 663
    persistence of herbicides and insecticides, 699
Soil bacteria, 507, 510, 512, 515–516, 521, 527
Soil formation, 662–663
Soil microorganisms, 971–974
Soil profile, 662
Soil solution, 663
Sol particle immunoassay tests, 839
Solfataras, 561–562
Solid-phase procedure, 359
Solid wastes, 702
Soper, George, 898
Sorbose, 406
SOS regulatory system, 299–300
SOS response, 264
Sources, 896–897
Southern blot, 174, 886
    DNA fingerprinting, 363
Southern, E. M., 174
Soybean, root nodules, 710
Sparger, 389
Sparkling wine, 410
Spawn, 415
Specialized transduction, 311–313
Species, 10–11

genomic hybridization and classification of, 448
Species concept, 449–451
Specific immunity, 803, 811–812
Specificity, 803, 867
    tissue, 798
Spectinomycin, 318, 766
Spectrophotometer, 143–144
Sphaerotilus, 464, 489–490, 495, 600
Sphaerotilus natans, 490, 598
Spheroplast, 71
Sphingolipids, 533
Sphingosine, 533
SPIA, *see* Sol particle immunoassay tests
Spikes, 278
Spiramycin, 761
Spirilla, 57, 485–489
Spirillum, 485–487
Spirillum lipoferum, 487
Spirillum volutans, 486
Spirochaeta, 538
Spirochaeta aurantia, 538
Spirochaeta plicatilis, 537, 538
Spirochaeta stenostrepta, 58, 537, 538
Spirochaeta zuelzerae, 538
Spirochetes, 57, 537–540
Spirogyra, 98, 459, 736
Spiroplasma, 514–515
Spiroplasma citri, 514
Spliceosome, 197
Splicing, 197, 267
Sponges, 470
Spontaneous generation, 18–20
Spontaneous mutation rate, 296
Spontaneous mutations, 294
Spore coats, 93
Spore protoplast, 93
Spores, 519–522, 733
    heat sterilization and, 743
    as property of industrial microorganism, 386
    sexual, 730–731
Sporocytophaga, 533, 534, 627
Sporocytophaga myxococcoides, 534, 627
Sporophores, 521
Sporosarcina, 512
Sporosarcina halophila, 512
Sporosarcina ureae, 512
Sporothrix schenckii, 972, 973
Sporotrichosis, 972–973
Sporozoans, 728–729
Sporozoites, 729, 966
Sporulation, antibiotic production and, 523
Spotted fever group, rickettsias, 961
Spread plate method, 142
Spur, 843
Stable isotopes, 660–662
Staining, 51–52
Staining techniques, 653–655
Stains, genetic, 655–658
Stalked bacteria, 491, 493–495, 532
Stalks, 491
Standard curve, in bacterial growth measurements, 143
Stanier, R. Y., 650
Stanley, Wendell, 236
Staphylococcal enterotoxin, 847
Staphylococcal enterotoxin A, 928–929
Staphylococci, 787
Staphylococcus, 157, 311, 324, 503, 759, 777, 782, 886, 927–929
Staphylococcus aureus, 52, 394, 503, 695–696, 775, 783, 788, 791, 810, 811, 855, 859,

864, 870, 878, 925, 927–928
detecting in food, 989–990
food poisoning, 983–985
as hospital pathogen, 906–907
peptidoglycan in, 70
toxins, 318
Staphylococcus aureus enterotoxin, 792
Staphylococcus epidermidis, 503, 857, 927
Staphylococcus pneumoniae, 795
Staphylococcus saprophyticus, 857
Staphylothermus, 568
Staphylothermus marinus, 568
Starch, 128, 130–131, 737
   structure, 36
Starch digestion, 627–628
Start codon, 200–201, 204, 206, 443
Starting substrates, 633
Static agents, 750
Stationary phase, 135, 140
STD, see Sexually transmitted disease
Steady state, 145
Stella, 492
Stem cells, 805–806
Stem-loop, 169–170, 225, 271
Stem-nodulating rhizobia, 714–715
Stephanodiscus, 98
Stereoisomerism, 41, 43
Sterilants, 751–753
Sterilization, 19, 741
   of culture medium, 14
   filter, 747–749
   heat, 742–745
   radiation, 745–747
Steroid metabolism, 781
Steroids, industrial production, 401–402
Sterol biosynthesis, 632
Sterols, 61–62, 468, 513
   structure, 62
Stetter, Karl, 650
Stickland reaction, 454
Stigmatella aurantiaca, 497
Stimulon, 227
Stop codon, 206, 209
   mutations and, 295
Strand invasion, 302
Streak plate, 15, 652
Strep throat, 925
Streptococcal diseases, 925–927
Streptococci, 787
Streptococcus, 106, 307, 324, 504–506, 778
Streptococcus bovis, 684
Streptococcus hemolyticus, 76
Streptococcus mutans, 779–780
Streptococcus pneumoniae, 307, 308, 769, 783, 796, 811, 925–927
Streptococcus pyogenes, 769, 783, 788, 811, 839, 855–856, 864, 870, 925–926, 928, 940
Streptococcus sobrinus, 779
Streptokinase, 788
Streptolydigin, 196
Streptomyces, 520–523, 764
   antibiotic production, 393, 394
   genome, 333
   industrial uses, 385
Streptomyces aurefaciens, 398
Streptomyces coelicolor, 522, 523
Streptomyces lividans, 180, 190
Streptomyces thermoautotrophicus, 635, 636
Streptomycin, 208, 210, 318, 761, 766, 932, 971, 978
   structure, 754
Streptovaricins, 196
Stress, resistance and, 797

Strickland reaction, 510–511
Stroma, 98
Stromatolites, 424, 425
Strong promotors, 194
Strongyloides stercoralis, 947, 948
Structural subunits, 239
Subclinical infection, 894
Subcutaneous mycoses, 972–973
Substrate-level phosphorylation, 117, 504, 574, 620
Substrates, 110
   methanogenesis, 554–555
Subtilisin, 318
Subunit vaccines, 372, 953
Subviral particle, 275
Succinate, 618, 682, 714
   fermentation of, 623–624
Succinomonas amylolytica, 682
Suckling mouse cataract disease, 515
Sucrose, 628, 779
   structure, 157
Sudden infant death syndrome, 985
Sugarcane, 717
Sugars, 34, 130–131
   fermentation of, 504
Suillus bovinus, 706
Sulfa drugs, 755–756
Sulfanilamide, 756
   structure, 758
Sulfate, 129
Sulfate-reducers, 560
Sulfate-reducing bacteria, 454
Sulfate-reducing proteobacteria, 498–502
Sulfate reduction, 608–611
   hydrogen sulfide, 687
Sulfide oxidation, 687–688
Sulfide stinker, 499
Sulfides, 561
Sulfidogenesis, vs. methanogenesis, 680–681
Sulfite oxidase, 597
Sulfite reductase, 609
Sulfolobales, 564–566
Sulfolobus, 155, 306, 562, 563, 569, 592, 598, 693
Sulfolobus acidocaldarius, 565, 596, 654
   RNA polymerase, 443
Sulfonamides, 318
Sulfur, 105, 128, 561
   isotope studies, 660–662
Sulfur compounds, organic, 688
Sulfur cycle, 686–688
Sulfur disproportionation, 610
Sulfur oxidation, 595–598
Sulfur oxidation/reduction, 687–688
Sulfur-oxidizing bacteria, 462–465, 673
Sulfur-reducing proteobacteria, 498–502
Sulfur reduction, 501
Sulfur respiration, 564, 566
Superantigen toxin, 788, 792
Superantigens, 843–844, 847, 928
Supercoiled domains, 78, 178
Supercoiling, 77–79
   DNA, 177–179, 188
Superficial mycoses, 972
Superoperons, 586
Superoxidase dismutase, 161
Superoxide anion, 160
Supressor mutation, 296
sur genes, 140
Surfaces, 646
Surgeon General's Report on Acquired Immune Deficiency Syndrome, 954
Surveillance, 892, 910, 920

Survival fraction, 747
SV40, 276–277, 352
Swarmer, 495
Swarming, 482
Swiss cheese, 516
Symbionts, 527, 672
Symbiosis, legumes, 709–717
Symbiosome, 713
Symbiotic bacteria, 634–639
Symporters, 65, 66
Syncytia, 949
Synechococcus, 527
Synechococcus lividus, 525
Synechocystis, 582
   genomic mining, 335
Synthetic agents, 755
Synthetic antibacterial chemotherapeutic agents, 753–758
Synthetic DNA, 345, 359–360, 365
Synthetic immunizing agents, 851
Synthetic peptides, 851
Synthetic polymers, biodegradation of, 702–703
Syntrophobacter wolinii, 678
Syntrophomonas, 625
Syntrophomonas wolfei, 626, 678
Syntrophus gentiane, 678
Syntrophy, 624–626, 647
   ecology of, 677–681
Syphilis, 538–539, 878, 901, 911, 913, 943–944, 965
Systemic lupus erythematosis, 846
Systemic mycoses, 973

2,4,4–T, 701–702
T cell antigen receptor, structure, 823–824
T cell receptor, 806–808, 823–824
T cells, *see* T lymphocytes
T cytotoxic cells, 803, 812, 826, 829, 868
T helper cells, 803, 812, 825, 827–828, 867
T helper 1 cells, 813, 867
   macrophage activation and, 830
T helper 2 cells, 813, 834
T inflammatory cells, 812
T lymphocyte–B lymphocyte interaction, 834
T lymphocytes, 803, 811–812, 847, 867, 873
   vs. B lymphocytes, 813
   cytotoxic, 829
   HIV infection and, 949–950
T4 infection, time course, 259
T4 lysozyme, 259
tac, 357
Tail, virion, 246
Tailing, 197–198
Tannins, 410
Taq polymerase, 154, 361
Target cell, 825–826
TATA box, 195–196
Tatum, E. L., 308–309
Taxonomy, 422
   conventional, 444–446
   extremely halophilic Archaea, 549–550
   molecular, 445–449
TC cells, *see* T cytotoxic cells
T-cell receptor, 816, 818–819, 826
TCR, *see* T cell receptor
T-DNA, 375, 707–708
Technological advances, infectious disease and, 918
Teeth, dental plaque and, 777–779
Tegument, 278
Teichoic acids, 71

Telomerase, 190, 198
Telomere, 181, 190, 351
TEM, see Transmission electron microscope, 55
Temperate bacteriophages, 259–265
Temperate viruses, 259
Temperature:
    effect of cold, 148–152
    effect of high temperatures, 151–154
    effect on DNA structure, 170, 177
    effect on growth, 147–148
    food preservation and, 981–982
    upper limits for growth, 153
Temperature–sensitive mutations, 295
Templates, 185, 192
Terminal electron donor, 115
Termination, of protein synthesis, 206
Termites, 538, 679–680
Terrestrial environments, 662–665
Tertiary structure, 45
Tertiary treatment, sewage, 417
Tests, of foraminifera, 727
Tetanus, 512, 790–791, 909
Tetracycline, 208, 318, 327, 346, 388, 759, 761–762, 930, 961, 963, 964, 971, 978, 988
    industrial production, 398–399
    structure, 761
Tetraethers, 62
Tetrahydrofolate, 612
Tetrahymena, 198–199
T-even bacteriophages, 247–248
T-even phages, 256–258
T_H cells, see T helper cells
T_H1 cells, see T helper 1 cells
T_H2 cells, see T helper 2 cells
Thallassemias, 968
Thelophora terrestris, 706
Therapeutic, 848
Thermal death time, 743
Thermal environments, 151
Thermal gradient, 153
Thermoactinomyces, 92
Thermococcales, 558–560
Thermococcus, 546, 559
Thermococcus celer, 559
Thermococcus litoralis, 404
Thermodesulfobacterium, 454, 543–544
Thermodesulfobacterium mobile, 544
Thermofilum, 566
Thermofilum librum, 566
Thermomicrobium roseum, 542
Thermophiles, 148, 151–154
Thermoplasma, 73, 155, 441, 546, 556–557
Thermoplasma acidophilum, 557, 558
Thermoplasma volcanium, 557
Thermoplasmatales, 556–558
Thermoproteales, 566
Thermoproteus, 441, 566, 592
Thermoproteus neutrophilus, 566
Thermosome, 569
Thermotoga, 454, 543–544, 571
Thermotoga maritima, 543
Thermus, 307, 540–541
Thermus aquaticus, 153, 154, 361, 541
Theta intermediate, 315
Theta replication, 276
Theta structures, 186
Thiamine, 106, 107
Thienamycin, 397, 398
Thin sectioning, 55
Thinning reaction, 403
Thiobacillus, 462–464, 671, 687

Thiobacillus denitrificans, 597
Thiobacillus ferrooxidans, 155, 463, 598–600, 689–694
Thiobacillus neapolitanus, 463, 591
Thiobacillus thiooxidans, 694
Thiocapsa, 458
Thiocapsa roseopersicina, 58
Thiocystis, 459
Thioglycolate, 159
Thioglycolate broth, 159
Thiols, 160
Thiomicrospira, 671
Thiopedia rosea, 458
Thiopedia roseopersicinia, 459
Thioploca, 465
Thiospirillum, 486
Thiospirillum jenense, 85, 458, 652
Thiosulfate, 586, 595
Thiothrix, 465, 596, 671
Thiovulum, 672
Three–dimensional imaging, 54–55
Throat swab, 856
Thylakoid membrane, 528
Thylakoids, 98, 578–579
Thymine, 37–38, 167–168
    structure, 298
Thymus, 811, 830
Ti plasmid, 375, 707–708
Tick, 956, 961
Tick-transmitted diseases, 960–966
T-independent antigens, 835
Tissue plasminogen activator, 374
Tissue specificity, 798
Titer, 855
    of virus, 244
TMA, see Trimethylamine
TMAO, see Trimethylamine oxide
TMV, see Tobacco mosaic virus
TNF, see Tumor necrosis factor
Tobacco mosaic disease, 25, 236, 240
Togavirus, 268, 939
Tolerance, 803–804, 830–832
Toluene catabolism, 634
Top-fermenting yeasts, 411
Topoisomerases, 178–179, 188, 191
Torulopsis, 784
Total cell count, 141–142
Toxic oxygen, 160–161
Toxic oxygen species, 810
Toxic shock syndrome, 928
Toxic shock syndrome toxin, 847, 928
Toxigenicity, 774
Toxins, 318, 788
Toxoid, 848
Toxol, 765
Toxoplasma gondii, 947, 948
Toxoplasmosis, 947
TPA, see Tissue plasminogen activator
tra operon, 319, 321
Tra region, 316
Trachoma, 529–530, 944
Trans configuration, 305
Transaminase reaction, 131
Transcription, 165, 167, 195–196
    in bacteria, 191–195
    phage φχ 174, 253
    phage T4, 258–259
    regulation of, 217–221
    tryptophan operon and, 225
Transcription pause site, 225
Transcription terminators, 193, 194
Transcriptional regulation, of cloned gene, 357–358

Transducers, 231–233
Transducing particles, 311
Transduction, 302, 304, 311–314
    discovery, 309
Transfection, 307, 310–311
Transfer RNA, see tRNA
Transferrin, 786
Transformation, 243, 269, 302, 304–311, see also Transfection
    cloning and, 346
    in genetic mapping, 329
Transformed cells, 277
Transforming DNA, integrating, 307
Transgenic animals, 369, 378
Transgenic plants, 369
Translation, 166, 167
    of cloned gene, 358
    errors, 210
    phage φχ174, 253
    phage T4, 258–259
    of RNA, 204–208
    tryptophan operon and, 225
Translational attenuation, 226
Translational coupling, 358
Translocation, 205, 296
Transmission:
    airborne, 923–925
    disease, 900–902
    host-to-host, 901
    leprosy, 933
    public health measures, 908–909
Transmission electron microscope, 55
Transpeptidation, 76–77
Transport proteins, 64
Transportation, infectious disease and, 918
Transposable elements, 179, 182, 265, 324–325
Transposable phage, 265–267
Transposase, 325–326
Transposition, 324–327
Transposition rate, 296
Transposon mutagenesis, 299, 327
Transposons, 182, 265, 318, 324–329, 767, 919
Travel, to endemic areas, 911–913
Traveler's diarrhea, 987
Treatment:
    AIDS, 951–953
    hantavirus pulmonary syndrome, 960
    Lyme diseae, 965–966
    malaria, 966–967
    plague, 971
    rabies, 958–959
Trehalose, structure, 157
Trench fever, 485
Treponema, 538–540
Treponema denticola, 539–540
Treponema macrodentium, 539
Treponema oralis, 539
Treponema pallidum, 334, 538–539, 878, 901, 913, 943–944, 965
    genome, 334
Treponema saccharophilum, 540
Tricercomitis, 680
Trichloroethylene, 699, 701
Trichocysts, 728
Trichodesmium, 665
Trichomonas, 725
Trichomonas vaginalis, 723, 945, 946
Trichomoniasis, 945, 946
Trichophyton, 972
Trichophyton rubrum, 973
Trickle method, 405
Trickling filter sewage treatment, 418, 419

Trifolium repens, 712
Triglycerides, 36–37
Trimethoprim-sulfamethoxazole, 988
Trimethylamine oxide, 619
Trimethylamine, 619
Triple DNA, 234
Triple helix, 234
Triple sugar iron agar, 862
Triplet oxygen, 160
Trismus, 791
tRNA, 39, 165–167, 169, 191, 195, 201–204
    recognition, activation, and charging, 202–204
    in retrovirus replication, 281–285
troph, defined. 104
Trophosome, 672, 673
trp attentuation protein, 226
TR-PCR test, 885
True revertant, 296
Trypanosoma cruzi, 852
Trypanosoma gambiense, 726
Trypanosomes, 726
Tryptophan operon, 224–226
Tsetse fly, 727
TSS, *see* Toxic shock syndrome
Tsutsugamushi disease, 963
Tube dilution technique, 750
Tube worms, 670–674
Tubercles, 931
Tubercular leprosy, 933
Tuberculin, 23
Tuberculin test, 846, 868–869, 932
Tuberculosis, 799, 846, 878, 931–934
    carriers, 897
    Koch and, 23–24
    treatment, 761
Tubulin, 8
Tumbles, 83–84
Tumor necrosis factor, 820, 828
Tumor supressor genes, 270
Tumors, 830
    natural killer cells and, 829
    pox viruses and, 279
    SV40 and, 276–277
Turbidimetric measurements, 143–145
Twitching motility, 477
Two-component regulatory systems, 230–233, 307
Tylosin, 761
Type strains, 451
Typhoid fever, 481, 799, 977
    carriers, 897
    water purification and, 975
Typhoid Mary, 898
Typhus fever, 484, 799, 878, 960
Typhus group, rickettsias, 961
Tyrocidin, 508

UDPG, *see* Uridine diphosphoglucose
Ultraviolet radiation, early life and, 425
Uncouplers, 126
Undefined media, 107
Unequal binary fission, 494–495
Unidentified reading frame, 334
Uniporters, 65
Unit membrane, 60
Universal ancestor, 4, 9, 439–440
Universal code, 209
Universal tree of life, 439–440
Upper respiratory tract, 782–783
    obtaining specimens, 856
Upper temperature limits, microbial life, 675
Uracil, 37–38, 191

Uranium leaching, 693–694
Urease, 481
URF, *see* Unidentified reading frame
Uridine diphosphoglucose, structure, 131
Uridylate, structure, 132
Urine cultures, 857–859
Urine testing, 840
Urogenital tract, normal flora, 783–784
U.S. Department of Agriculture, 757
U.S. Surgeon General, 954
UV radiation, 746

Vaccination, 848
Vaccine production, 368–374
Vaccines, 796, 848, 849
    DNA, 373–374
    effect on U.S. incidence of childhood viral disease, 938
    Pasteur and, 19
    plants for production of, 377
    recombinant, 279–280, 372–373
Vaccinia virus, 279, 352, 372–373
van der Waals forces, 31–32
van Leeuwenhoek, Antoni, 17–19, 485
van Niel, C. B., 650
Vancomycin, 759, 927
Variable domains, immunoglobulins, 818
Variable number of tandem repeats, 363
Varicella-zoster virus, 277
Vectors, 343, 892, 901
    eukaryotic, 352
Vegetative cells, vs. endospores, 94
Vehicles, 892, 901
Vent chimneys, 674–675
Vent polymerase, 361
Vesicles, 716
Vesicular stomatitis virus, 272
VFAs, *see* Volatile fatty acids
Viability, staining techniques and, 653–655
Viability staining, 654
Viable cells, 143
Viable count, 142
Vibrio, 328, 468, 482–484
Vibrio cholerae, 23, 482, 785, 791, 797, 878, 977–979
Vibrio fischeri, 228–229, 483
Vibrio parahemolyticus, 482
Vinblastin, 764–756
Vincristine, 764
Vinegar, 474
    industrial production, 405–406
Violacein, structure, 477
vir genes, 708
Viral control, 762–764, 923, 951–953
Viral diseases, vaccines, 849
Viral genomes, 238–239
Viral load, 950–953
    HIV and, 883–885
Viral nucleic acid, production of, 248–249
Viral pathogens, respiratory, 934–939
Viral proteins, 249–250
Viral structure, 6–8
Viricidal agents, 750
Virion, 238–242
Viristatic agents, 750
Viroids, 285–287
Virology, diagnostic, 887–888
Virulence, 794–796
    measuring, 795–796
Virulence characteristics, 318
Virulence factors, 773, 787–788, 919
Virulent viruses, 259
Virus genome, 237

Virus infections, major histocompatibility complex proteins and, 825
Virus infectious unit, 242
Virus membranes, 240–241
Virus particle, 237
Virus replication, 237
Virus restriction and modification by host, 247–248
Virus vaccines, 368–369
Virus vectors, 352
Virus-encoded protease, 271
Viruses, 8, 9, 27–28, 181–182, 530, *see also* Animal viruses, Plant viruses
    animal, 267–270
    bacterial, 250
    cancer and 269–270
    classification, 238
    complex, 241
    control, 867
    discovery, 236
    ELISA and, 876
    enveloped, 240–241
    general properties, 237–238
    host, 242
    multiplication, 246–250
    natural killer cells and, 829
    neutralization and, 841
    overlapping genes, 208
    pili and, 86
    plant infection, 376–377
    prokaryotes vs. eukaryotes, 267–268
    quantification of, 242–244
    reproduction, 244–246
    symmetry, 240
    transduction and, 311–314
    in water, 978
Vitamin B$_{12}$, structure, 399
Vitamin C, 406, 475
Vitamin K, 479
Vitamins, 106, 107
    industrial production, 399–400
VNTR, *see* Variable number of tandem repeats
Vodka, 412
Voges-Proskauer test, 479
Volatile fatty acids, 681
Volcanic habitats, 564–568
Volta, Alessandro, 553
Volta experiment, 553
Volutin granules, 486
Volvox, 736
VPg protein, 270–272
VSV, *see* Vesicular stomatitus virus

Warm vents, 670–671
Waste products, 103
Wastewater microbiology, 416–420
Water:
    as biological solvent, 34
    heat resistance and, 743
    macromolecules and, 33–34
    microbial activity and, 663
    purification, 975–976
Water activity, 156–157
Water balance, in extreme halophiles, 550–551
Water molecules, bonding, 31
Water purification, 419–420, 908
Water quality, public health and, 974–976
Waterborne diseases, 976–980
Watson, James, 308
Wavelength, flagellar, 80
Western blot, 174, 855, 878–880, 953, 965

Whey, 386
Whiskey, 412
White rot, 731
WHO, *see* World Health Organization
Whooping cough, 930
WI-38, 887
Widdel, Fritz, 650
Wieringa, K. T., 92
Wild yeasts, 410
Wild-type strain, 291
WIN 52084, 934
    structure, 935
Wine, 474
    fermentation and, 120
    pasteurization of, 745
    production of, 409–410
Winogradsky, Sergei, 1, 24, 25–26, 461–462, 464, 595–597, 650
Winogradsky column, 650–652
Wobble concept, 200
Woese, Carl, 434
Wolinella succinogenes, 619
Wollman, Elie, 309
Wood, D. D., 756

Wood, Harland, 611
Wood-rotting fungi, 731
Woods Hole Oceanographic Institution, 669, 671
World Health Organization, 272, 747, 908, 910
Wort, 410, 414
Wounds, 859

Xanthobacter, 311
Xanthomonas, 473
Xenobiotics:
    biodegradation of, 698–703
    degradation of and evolution, 702
Xenopsylla cheopis, 969
Xerophiles, 157
X-rays, 299, 746
Xylans, 738

Yeast, 729
    alternative mitochondrial codes, 210
    genetics, 337–341
    industrial production, 407–409
    as model eukaryotes, 336–337

Yeast artificial chromosomes, 350–352
Yeast fermentation, 120
Yellow boy, 690
Yellowstone National Park, 441, 562
Yersinia, 318, 878
Yersinia pestis, 913, 917, 969–971

Zea mays, 717
Zidovudine, 762
Ziehl-Neelsen statin, 517–518
Zinc finger, 223–224
Zinder, Norton, 309
Zone of inhibition, 751
Zoogloea, 464–465, 470
Zoogloea ramigera, 419
Zoonosis, 892, 895, 957
Zootermopsis angusticolis, 680
Zoster, 939
Zygospores, 731
Zygote, 337, 732
Zymomonas, 473–474

Phylogeny of the Living World–Archaea

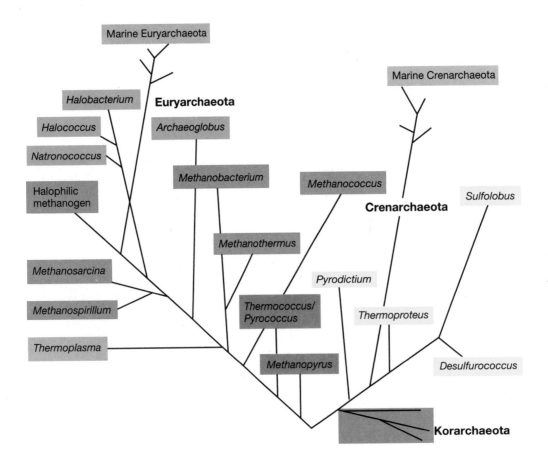

PHYLOGENETIC TREE OF ARCHAEA. This tree is derived from 16S ribosomal RNA sequences. Three major kingdoms of Archaea can be defined: the Crenarchaeota, which consists of both hyperthermophiles and cold-dwelling species; the Euryarchaeota, which contains methanogenic and extremely halophilic prokaryotes; and the Korarchaeota, which are, as far as is known, hyperthermophiles. See Sections 12.4–12.8 for further information on ribosomal RNA-based phylogenies. *Data for the tree obtained from the Ribosomal Database Project* http://www.cme.msu.edu/RDP/